MECHANICAL AND ELECTRICAL EQUIPMENT FOR BUILDINGS

MECHANICAL AND ELECTRICAL EQUIPMENT FOR BUILDINGS

Eighth Edition

Benjamin Stein
Consulting Architectural Engineer

John S. Reynolds
Professor of Architecture
University of Oregon

John Wiley & Sons, Inc.

New York Chichester Brisbane Toronto Singapore

DISCLAIMER

The information in this book has been derived and extracted from a multitude of sources including building codes, fire codes, industry codes and standards, manufacturer's literature, engineering reference works and personal professional experience. It is presented in good faith, but although the authors and the publisher have made every reasonable effort to make the information presented accurate and authoritative, they do not warrant, and assume no liability for its accuracy or completeness or its fitness for any specific purpose. It is primarily intended as a learning and teaching aid, and not as a source of information for the actual final design of building systems. It is the responsibility of users to apply their professional knowledge in the use of the information presented in this book, and to consult original sources for current detailed information as needed, in actual design situations.

Library of Congress Cataloging in Publication Data:

Stein, Benjamin

Mechanical and electrical equipment for buildings/Benjamin Stein, John S. Reynolds. — 8th ed.

p. cm.

Rev. ed. of: Mechanical and electrical equipment for buildings/ Benjamin Stein, John S. Reynolds, William J. McGuiness. 7th ed. 1986.

Includes index.

ISBN 0-471-52502-2 (alk. paper)

1. Buildings—Mechanical equipment. 2. Buildings—Electric equipment. 3. Buildings—Environmental engineering. I. Reynolds, John, 1938— . II. McGuinness, William J. Mechanical and electrical equipment for buildings. III. Title.

TH6010.S74 1992

696—dc20 91-39364

CIP

Printed in the United States of America

10 9 8 7 6 5 4 3 2 1

PREFACE

Change and Progress

Since the publication of the first edition of *Mechanical and Electrical Equipment for Buildings* in 1937, each succeeding interval of six to eight years between editions has resulted in what has essentially been, except for unchanging basics, a new book. This edition is no exception. It has been written during a period that has seen pronounced changes in the fields of architectural engineering—changes both in design philosophy and in design execution.

Traditionally, Mechanical and Electrical Equipment for Buildings has been concerned primarily with the relationship between mechanical and electrical systems and the buildings they serve. This edition also emphasizes the role that environmental control systems play in the design of buildings as well as in building performance.

Objectives of the Book

The book is both a textbook and a professional reference book. Since the book presents basic theory, preliminary design guidelines, and detailed design procedures, it is suitable as a text at both introductory and advanced levels, for courses and seminars in most aspects of environmental design and architectural engineering as taught in schools of architecture. These include heating and cooling, comfort control, water and waste, daylighting and electric lighting, electrical system design, architectural acoustics, and vertical transportation. Further, because of the large amount of design information and reference data it provides, the book is also appropriate for the practicing professional. A background of college-level mathematics and physics is assumed. Each subject is logically developed with a minimum of mathematical treatment; emphasis is placed instead on design approach and equipment applications.

The book typically serves as an multi-course textbook for students of architecture, architectural engineering, and environmental control, who then retain it for use as a reference in professional practice. The book is recognized as a particularly important reference for the national architectural licensing examination.

The Eighth Edition

This edition extends the philosophy of the seventh edition, emphasizing the themes of energy conservation and the use of renewable energy sources while keeping the reader informed of the major changes in equipment technology wrought by the microprocessor and the computer.

Because of our commitment to the conservation of the earth's resources, we present current practice for assessing on-site energy sources and their usage: daylighting, solar heating, passive cooling, solid waste,—even rainwater. These renewable energy sources can have important impacts on the building design process; such effects are discussed in stages, beginning with the most approximate and early "rules of thumb" or building organizational strategies. Because renewable resource usage usually increases first cost, we include an appendix on economic analysis to encourage life cycle cost analyses.

Along with our emphasis on the architectural design implications of systems, the approach to system design follows logical development. Throughout the book, sizing information is presented: first the approximate information, then the more detailed. Thus, for instance, for heat gain a rule-of-thumb calculation is given first;

later we present the much more detailed heat gain calculations that recognize thermal mass effects through elapsed time. Similarly, both rules-of-thumb and detailed procedures are presented for solar heating, daylighting, passive cooling, and cisterns. Further, new material on strategies for both outside (site) design as well as interior building design are presented.

Design guidelines throughout the heating/cooling material (Chapter 5) utilize the same sample building. This simplifies and clarifies the presentation of detailed design/equipment information while simultaneously emphasizing the integrative nature of design decisions involving daylighting, energy conservation in the light of new energy codes (ASHRAE/IES 90.1-1989), passive heating and cooling and equipment.

On-site resources such as daylighting, solar heating, cooling, rainwater, and even solid waste are discussed, with theory, design guidelines and examples of contemporary buildings utilizing such resources. Too, detailed information on current criteria and equipment for maintenance of satisfactory indoor air quality is presented.

The portion of the text devoted to fire safety reflects the most up-to-date design techniques in compartmentation, sprinklering, and smoke venting. Also included here are a discussion of lightning protection and an extensive discussion of fire alarm systems. This latter involves not only principles of fire and smoke detection, but also explanations of the operation and application of modern microprocessor controlled and computerized systems utilizing "smart" addressable devices and multiplexed wiring arrangements.

The section on modern electrical building design has been updated to conform to the 1990 National Electric Code while emphasizing design strategies that are economical of both energy and cost. Included is a comprehensive discussion of electric demand control per se, and as part of overall energy management systems, which are also discussed. Cognizance is taken throughout of the application of microprocessor controls as well as the pronounced impact of data processing equipment on the electrical design of commercial spaces.

The section on natural (day) lighting has been expanded to include material on the use of scale models in daylighting studies and the use of readily available computer programs. Output of these programs is given in both numerical and graphic forms for ease of analysis. The extensive section on electric lighting emphasizes energy conserving design strategies in the light of current energy codes (ASHRAE/IES 90.1) and illuminance recommendations of the Illuminating Engineering Society of North America (IESNA) and other authoritative bodies. Operating characteristics and application guidelines for all types of light sources are presented, including compact fluorescents and high efficacy HID sources. Detailed design examples are developed utilizing both manual and computerized calculation procedures. Particular emphasis is placed on the design of commercial interiors utilizing visual display terminals (VDTs). Cost analyses are included here, as in other sections of the book, to demonstrate the practical aspects of design alternatives.

The signal section which deals with systems of signalling, communication and surveillance in buildings has been thoroughly updated and presents current design practice and equipment. This is followed by a section on transportation that contains complete current material on low-energy-use solid-state controls for elevators and escalators, along with our traditional extensive coverage of elevator selection, and the design of moving stairs and ramps. New in this section is the application of computer programs in elevator selection and a discussion of variable frequency motor drives. Other new material includes a novel elevator drive system which obviates the need for a machine room, and comparative power and energy use data. The discussion of materials handling in commercial buildings has also been updated.

Noise is our society's most pervasive pollution problem. The section on architectural acoustics discusses noise control at length in addition to standard topics of acoustics design and analysis.

The expanded appendices present detailed tabular data on climate, passive solar design, acoustic design, metrication, sunshading, economic analysis, and computer-aided design. One new appendix discusses building automation systems and intelligent building design, while another presents data extracted from energy standard ASHRAE/IES 90.1 1989 required for building system design that will meet this very widely accepted energy code. Finally, an expanded index—along with a detailed contents and table list—simplify use of this large and data-filled text.

From an overall point of view, new architecturally oriented illustrations make this edition more appealing to visually oriented students. This edition considers psychological as well as physiological design influences in lighting, thermal and acoustic topics.

Benjamin Stein
John S. Reynolds

ACKNOWLEDGMENTS

A good many people have contributed to this book. We begin with those from whose work we have borrowed at length: J. Douglas Balcomb, Alexander Kira, William Lowry, and Murray Milne, as well as organizations such as ASHRAE (the American Society of Heating, Refrigerating and Air-Conditioning Engineers), the American Solar Energy Society, the International Association of Plumbing and Mechanical Officials, the National Fire Protection Association, the American National Standards Institute, the Illuminating Engineering Society of North American (IESNA), and the Carrier Corporation.

We are grateful to those who made helpful and clarifying suggestions, some of whom carefully read outlines and drafts. These include Robert B. Allen, Washington State University, Dale Brentrup, University of North Carolina, Charlotte; Virginia Cartwright, University of Oregon; and Bruce Haglund, University of Idaho. Those whose work contributed indirectly include Carol Venolia and Barbara Jo Novitski.

A number of professionals provided valuable assistance in assembling materials, and clarifying ideas and details. These include Dr. Yaakov J. Stein (computer application) and John A. Van Deusen (vertical transportation). A particularly important acknowledgement is due to Michael Cockram who prepared many of the new illustrations throughout the book. His drawings present technical information with a fine blend of clarity, informality and occasional humor. A special role is played by the students in our classes, with whom ideas are tested and rules of thumb developed. Students at the University of Oregon were especially helpful.

The heaviest and most tedious burdens fall upon supporting staff. Among these we wish particularly to note the research assistance provided by Don Appelmann and Robert Iwersen. For handling correspondence, typing the manuscript, assisting in proofreading and other production-related items, and continuous encouragement and support, we thank Lila Stein.

We are indebted to the staff of John Wiley for their diligent and highly professional work—to Stephen Kliment and Everett Smethurst for managing the project, to Diana Cisek for editorial and production; to Dean Gonzalez for his work on the hundreds of illustrations; and to Geraldine Spellissy for the design of the book.

CONTENTS

PART I ENERGY OVERVIEW

PART II THERMAL CONTROL

PART III WATER AND WASTE

PART IV FIRE PROTECTION

PART V ELECTRICITY

PART VI ILLUMINATION

CHAPTER 18 LIGHTING FUNDAMENTALS 911

CHAPTER 19 LIGHT SOURCES: THEIR CHARACTERISTICS AND APPLICATION 966

CHAPTER 21 LIGHTING APPLICATION 1140

PART VII SIGNAL EQUIPMENT

CHAPTER 22 SIGNAL SYSTEMS 1195

PART VIII TRANSPORTATION

PART IX ACOUSTICS

PART X APPENDICES

TABLES

PART I
ENERGY OVERVIEW

Often mechanical and electrical equipment for buildings is not considered until many important design decisions have been made. Often, too, such equipment is considered to have a corrective function, which permits a building to "work" in a climate it essentially ignores.

Part I discusses some topics that encourage designers to *include* both climate and comfort in their early decisions. Chapter 1 discusses the fuel and resource relationship design through demolition. C human comfort, the variety seem "comfortable," and s building design of a more b fort zone." Chapter 3 en building site as a collec sources, to be used and heating, and cooling of b

1

ENERGY SOURCES FOR BUILDINGS

Today, some buildings in the United States are labeled as energy gluttons, while others are praised for their low annual energy consumption. Our society uses more energy per capita than most other countries, even some that have "high standards of living" and colder winters than ours. Although buildings are not our only users of energy—automobiles, industry, and agriculture are some others—buildings and their energy-related equipment are the subjects of this book. One way to begin is to examine the path that architecture has followed, from simple shelters that have adapted well to their climate, to sophisticated and tightly controlled internal climates that strive to ignore the conditions outside.

1.1 Energy and Architectural History

Few books on architectural history deal with the influence of changing fuel sources on the development of building. Style—including the proportion of spaces and the elements of the facade—has often been more influential, particularly on exterior appearance. Structural innovation is more visible and more permanent than fuel sources in architecture, and therefore easier to trace back to its earliest forms. Most buildings constructed before the latter part of this century relied heavily on the sun for light and sometimes for heat, and on breezes or massive construction to temper the hot portions of the day. These energy-related distinctions were relatively subtle, particularly among the buildings in given climatic regions. Size and placement of windows were energy related; yet, the size, placement, and, particularly, the shape of the windows were more clearly attributable to prevailing customs of proportion and materials than to a building's energy supply. The amount of sun or breeze admitted through a window, or the amount of building heat lost through it, has less often been cited as an influence on window design.

The impact of energy on buildings in the past has been both considerable and visible, and it is becoming so again. Space heating in buildings beyond that provided by the sun began with the burning of available fuel (Figs. 1.1 and 1.2). From the central open fire—around which all inhabitants slept each night—building heating dispersed to individual fireplaces. This began the continuing choice between concepts of central versus dispersed climate control and is credited for hastening the separation between social classes as well (since the masters got the fireplaces, while the servants still slept around the central fire).

Building heating then proceeded by the gradual use of more imported fuels and less visible equipment, until the obvious architectural im-

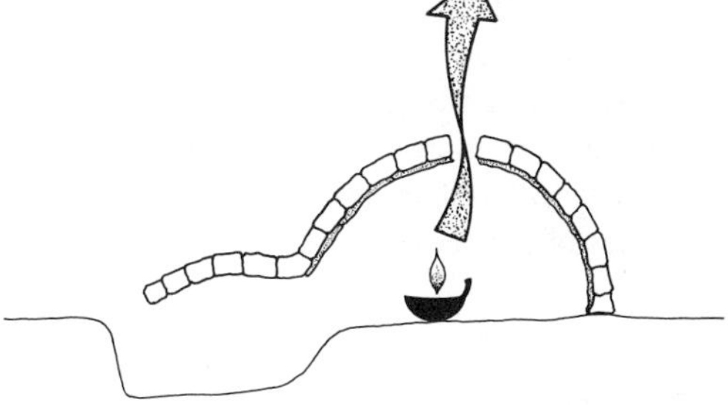

Fig. 1.1 *The portable lamp (as used in the igloo) has a slight architectural impact: a storage place is needed for fuel; the design must admit combustion air (to feed oxygen to the flame) and must allow ventilation for the gases of combustion; there is a residue of soot on interior surfaces.*

Fig. 1.2 *The indoor fire (as used in the tipi) introduces two evident architectural impacts: the* adjustable smoke flap, *to encourage venting of smoke by prevailing winds and to minimize the entrance to rain, and the* interior liner, *which forces the cold combustion air to rise along the sides of the tipi, gaining some warmth, before it moves across the occupants on its way to the fire. Again, some fuel storage is needed, and there is a residue of soot indoors.*

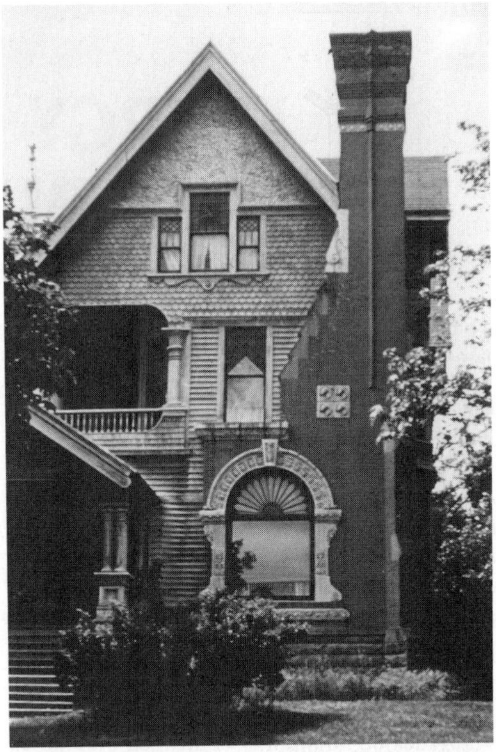

Fig. 1.3 *The fireplace and the more efficient wood stove increase the impact of heating on architecture. Their permanent location within a space both allows a massive and visible chimney and determines the placement of other furnishings within the space. As the enclosed fire burns hotter and heats more space, the needs for fuel storage and for providing combustion air intensify. (Photo by William Johnston.)*

pacts (Fig. 1.3) nearly disappeared (Figs. 1.4 and 1.5). Today, solar collection devices, such as large windows, Trombe walls, and/or flat-plate collectors (Fig. 1.6), have reintroduced the significant architectural impact of heating; their large glass areas, significant thermal mass, and the slopes of collectors combine to produce one of the most visible influences of any heating system on building form.

This is not the only sudden and recent shift in long-term trends concerning energy in buildings. A new awareness of the sources, characteristics, and limitations of energy supplies, and the results of their consumption, is resulting in new directions in building design, away from many practices of the recent past.

1.2 Trends, Recent Shifts, and Some Challenges

In the following discussion of energy in today's architecture, two assumptions are made: (1) designers can have a positive impact on society through energy and nonrenewable resource conservation, and (2) buildings that encourage their users to directly experience the natural environment will both facilitate energy conservation and enrich the user's architectural experience.

(a) A Trend from Local, Renewable Energy Sources to Imported, Nonrenewable Ones. Renewable fuel sources are those that are available indefinitely but arrive at a relatively fixed rate; the influx of solar energy varies from day to day, but on the average it continues at a steady rate. A woodlot will produce a limited amount of wood per year, but it will do so for centuries if properly managed. A popular analogy for renewable fuel sources is a fixed-but-steady monetary income, such as a salary with no raises.

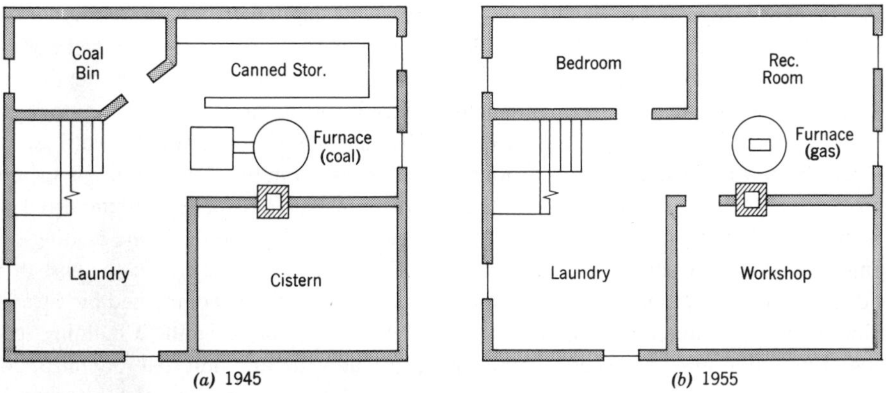

Fig. 1.4 *The furnace (or boiler) begins the trend toward less visible, more automatic heating. These plans show a basement remodeling that resulted from changing fuel sources.* (a) *A central large "machine" can be located in a basement along with its fuel storage area. A single chimney exhausts gases, and combustion air can infiltrate the basement without causing undue discomfort upstairs. An extensive ductwork system is threaded through floors and walls to simple grille openings in each room.* (b) *The use of piped-in fuel, such as natural gas, eliminates on-site fuel storage, and furnace sizes diminish with technical developments. With the electric furnace, even the chimney disappears along with the gases of combustion.*

Using the same analogy, nonrenewable fuel sources are like savings accounts that draw no interest—once spent, they are gone. The nonrenewable resources we utilize can be bought and used in large quantities all at once, which make possible many processes that are more difficult to attain with low-concentration, steady, renewable resources. Thus, the use of nonrenewable fossil and nuclear fuels in high-temperature and high-concentration processes—for example, in power-generating plants—is widespread and is likely to continue until resource depletion is more closely approached.

Plentiful, locally available fuels such as wood and solar energy are cheap and convenient up to a point. As population density increases, how-

Fig. 1.5 *Radiant heat in ceiling or floor, provided through pipes or cables, reduces the visible architectural impact of heating to a thermostat on the wall—and occasional cracks in the heating surfaces!*

Fig. 1.6 *Solar energy used for space heating brings the visible impact of large glass areas, in this case windows facing south, sloping back to express their relationship to the low winter sun. Sundown House at Sea Ranch, California, David Wright, AIA, architect.*

ever, such supplies dwindle per capita, and locally available quantities can become inadequate. As in the case of firewood in urban areas, prices can also rise. At that point, the conveniences of an established distribution network and a precisely measured heat content, as found with electricity, natural gas, and fuel oil, become attractive. The source of pollution that was formerly evident with some local fuels, such as woodsmoke, can be relocated to a distant generating plant. Eventually, of course, the problems of pollution must be faced, wherever they occur.

Even where local fuels are plentiful and "free," imported fuels have often supplanted them. Figure 1.7 shows one of thousands of residential solar water heaters formerly in use in Florida. When automatic washing machines were introduced, their greater appetite for hot water taxed the capacity of these heaters, which had been sized for laundry practices that reused hot soapy water rather than discharging it after one washing cycle. Natural gas utilities promoted the fast temperature recovery of their water heaters, and solar water heating was replaced by natural gas. In another climate, Eskimo families have turned, with government subsidies, from burning seal oil—a locally available, renewable resource—to imported and nonrenewable kerosene for home heating.

This changing pattern in the use of energy sources has been accompanied by a trend in design: away from designing a building so that its external skin and internal organization work with the surrounding climate for "natural" (passive) heating, cooling, and lighting, toward buildings much more dependent on mechanical and electrical equipment for their interior comfort.

Shifts back to renewable and local fuel sources, such as the sun in areas like Florida and California, are occurring rapidly. For example, California has banned new hookups for natural gas swimming pool heaters, creating an early

Fig. 1.7 *Solar water heaters are widely used and are available in many versions. This house in Miami, Florida, incorporates a storage tank above the collector, which is enclosed in a chimneylike form. When the lower collector is warmer than the upper storage tank, water circulates without the need for a pump. (Photo by M. Steven Baker.)*

and heavy demand for solar pool heaters. Although the shift in the United States to solar heating has only begun, it is nonetheless highly visible. Figure 1.8 shows an award-winning passive solar house in Connecticut, whose large south-facing windows admit winter sun to warm the thermally massive, earth-bermed walls. A skylight washes the north interior wall with daylight, avoiding the ''cave effect'' sometimes associated with passive solar, earth-sheltered housing.

Government subsidies have added solar heating to thousands of buildings, just as such subsidies have encouraged mining and drilling for fossil fuels and have supported uranium enrichment. Our current attitude seems tied to the world price of oil; the higher the oil price, the greater our interest in solar energy utilization.

Buildings designed today may outlast the supplies of the fuels that currently support them. This possibility provides two major challenges for designers:

Fig. 1.8 *Passive solar-heated, balanced-daylight residence of architect Donald Watson, FAIA, near New Canaan, Connecticut.* (a) *South facade in winter, with hillside providing earth sheltering.* (b) *Daylight model shows balanced light between south windows, and a skylight washing the north wall with daylight down through two stories.* (c) *First floor plan, showing elongation along east–west access, and art gallery to take advantage of subdued natural light at north wall.* (d) *Upper floor plan. (Courtesy of Donald Watson, FAIA, architect.)*

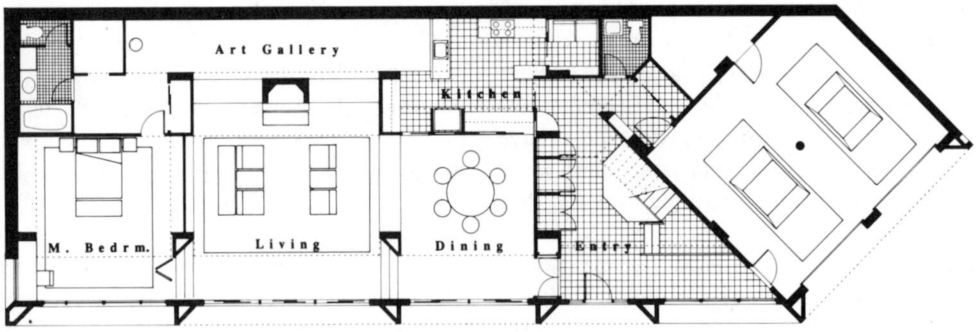

(c) First Floor Plan

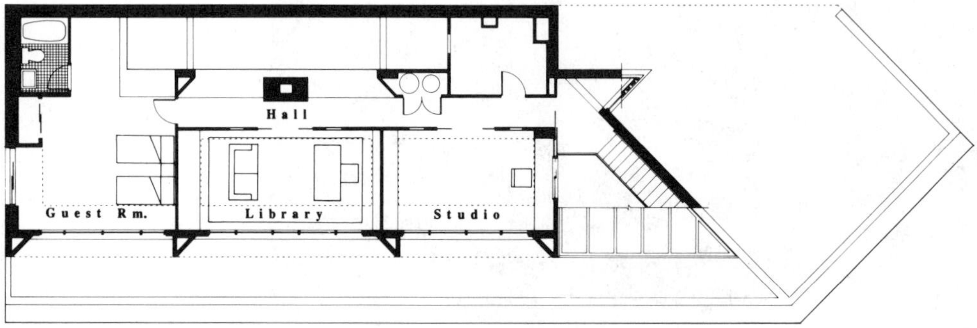

(d) Second Floor Plan

Fig. 1.8 (*continued*)

1. To design our buildings not only to save energy, but so that they can also eventually be weaned away from dependence on nonrenewable fuels.
2. To use energy wisely; to expect only a "fair share" of locally available renewable fuels, recognizing that such sources are limited even though they are continually available.

For example, with supplies stretched tight by increasing density, it is tempting to erect a larger solar collector to intercept sunlight that would otherwise be utilized by the neighboring building. The temptation grows stronger as the building is designed to rely more heavily on the sun. The concept of a "solar envelope" to protect each building's fair share is discussed in Section 3.4.

(b) The Trend from Labor-Intensive to Energy-Intensive Practices Inside Buildings. All of us know the convenience of energy-driven motors that do what we otherwise would do manually. As appliances that replace human labor become widespread, we design them into our buildings, often without provision for the now-obsolete human-powered practices they replace.

Consider the effective but time-consuming and labor-intensive practice of solar clothes drying. The use of the energy-intensive alternative, the mechanical clothes dryer, was increasing U.S. energy consumption for this purpose by about 10% per year in the 1970s. A designer who considers weather unpredictability, cultural expectations, and energy scarcity can provide an outdoor clothesline that is both visible and easily accessible from the mechanical dryer.

Another trend to energy-intensive practices is evident in the thermal control of buildings. Figure 1.9 shows the plan of a pioneer house that is dependent on fireplaces and the kitchen's wood stove for heat. Spaces containing fire-

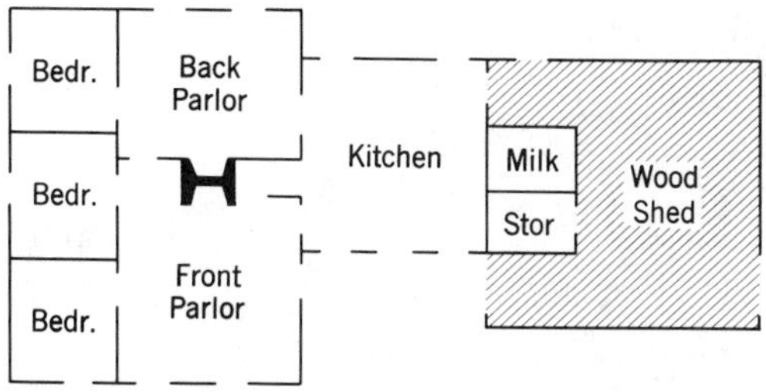

Fig. 1.9 Labor-intensive heat regulation. The house dependent on fireplaces or wood stoves also depends on someone to tend the fire. The warm area near the fire in this early Oregon farmhouse was used for social purposes; the colder spaces at the extremities served for sleeping areas and for storage of food and fuel. (Based on a plan drawn by Philip Dole.)

places could be kept warm by practices requiring periodic labor on the part of the inhabitants (adding wood to the fire or stirring it). Spaces without fireplaces are less controlled, with a temperature somewhere between that of the fireplace rooms and that of the outdoors. Today, the common approach is to provide thermal control via a thermostat, only occasionally adjusted by users. One such device often governs several rooms, keeping them all near one temperature, whether or not they are occupied.

Recent shifts to labor-intensive practices are scarce. The manually operated thermal shades or shutters of passively heated buildings are the most widespread examples; the popular wood stove is another.

Labor intensiveness is not so much an energy saver—most controls need very little power—as it is an education. A user who understands how a building gets and conserves its heat in cold weather is more likely to respond by lowering the indoor temperature and tightening up any heat leaks.

As we face a continued increase in world population and continued depletion of renewable resources, it seems obvious that labor-intensive alternatives must be made available. The designer's challenge is to make these alternatives attractive and rewarding to use.

(c) The Trend from Energy Awareness to Unconscious Energy Use. With human labor being greatly reduced and the fuel that replaces it being brought into our buildings from far away, we have slipped into an unconscious use of energy. Until 1973, with energy prices declining, the cheap, seemingly endless, and unobtrusive supply of energy encouraged increased use. Even our attitudes toward on-site energy sources, such as the sun, have evolved to exclude user awareness (Fig. 1.10).

With the apparent depletion of nonrenewable energy resources, a "scarcity ethic" has developed. This encourages increased self-reliance in the face of eventually decreasing imports, emphasizes conservation rather than increased production, and brings "user control" of on-site-energy sources back among architectural design considerations.

With the widespread availability of computer-controlled heating, cooling, and lighting systems, sophisticated thermal performance is possible without getting people involved. (See many examples in Chapters 6 and 7.) Thus, the overall trend toward user unconsciousness continues, except where visible devices such as sunshading automatically change position during the day.

This is the challenge to the designer: to engage the users in indoor environmental control without making them slaves to a building's thermal regulation.

(d) The Trend from Building–Climate Connection to Isolation. With fuels provided on demand from networks, and with devices to regulate building temperature automatically, it became ever more attractive to think of a building's interior as a self-contained climate that need bear little relationship to the climate outdoors. As urban conditions outdoors deteriorated due to air pollution and noise, this trend toward isolation strengthened. The outdoor climate is unpredictable, capable of sudden and dramatic changes; why not protect a building from such threats by making it self-contained?

With the advent of energy conservation brought on by the oil embargo in the 1970s, a shift in this trend occurred. For some designers, energy conservation was best served by increasing the isolation between climate and building, a building envelope dominated by *barriers* between inside and out. For other designers, energy conservation must rely in part on supplanting imported fuels with on-site sources, and the

Fig. 1.10 *Varying provisions for user awareness in schools. The famous early solar-heated school at Wallasey, England* (a *and* b) *depends on custodians and users to make the adjustments that alternately admit or exclude direct sun, control the extent of diffused light and outdoor air admitted, and insulate against loss of stored heat at night (see also Fig. 2.23). (Photos by Reg McDonald.) In a similar climate, a windowless school in the Pacific Northwest* (c) *excludes the unpredictable outdoors. Here the users' consciousness is directed away from energy.*

result is often a building envelope dominated by *switches* that encourage interaction with the climate when it is benign, and allow isolation when the climate is threatening. These concepts are discussed in Chapter 2.

One trend toward isolation was especially popular: when the difference between indoor and outdoor temperatures is great (hottest summer, coldest winter), reduce the flow of air between outside and inside. Building codes, revised to encourage energy conservation, sanctioned this trend toward reduced airflows.

This practice of isolating building interiors, supported by climate-making machines and regulated by automatic devices, has produced large numbers of buildings with unopenable windows and mostly recirculated air. Some of these buildings have become unhealthy places to work, due to a number of factors (see Section 2.8), one of which often is an inadequate supply of outdoor air. Thanks to the so-called *sick building syndrome,* a shift toward increased building–climate connection is under way, centered on the provision of outdoor air.

On a much larger scale, the indoor–outdoor connection has regained the attention of designers due to changes in the earth's climate. Two current developments are a growing "hole" in the upper atmosphere ozone layer in the southern hemisphere and a worldwide increase in carbon dioxide and other "greenhouse gases," which many predict will result in global warming (see Table 3.3 and Fig. 3.20).

Depletion of the upper ozone layer will result in more radiation at the earth's surface. The result will be increased skin cancer among Caucasians, and possibly depressed immune systems for all human beings. Some plants are also affected, with the threat of decreased photosynthesis and crop yield. Phytoplankton, the foundation of the ocean's food web, are particularly vulnerable. Buildings are also threatened; many synthetic materials, particularly plastics, are made brittle by increased radiation.

Global warming already seems under way, and if it continues to accelerate, will threaten our supplies of both food and water, our forests, and our fisheries. The effects on buildings will be less dramatic: they will experience increased cooling needs, decreased heating needs. This shift is expected to be more evident at latitudes farther from the equator.

There are two challenges for designers here. The first challenge is to strike the proper balance between indoors and out, to allow a building's occupants an optimum contact with the climate beyond, using isolation only when appropriate due either to building function or to climate severity. Chapters 2, 4, and 5 deal with this topic in detail. The second challenge is to design so as to minimize the production of upper-ozone-threatening gases and of greenhouse gases. This involves both energy conservation and the choice of building materials and equipment. Some guidelines are found in Chapters 3 and 6.

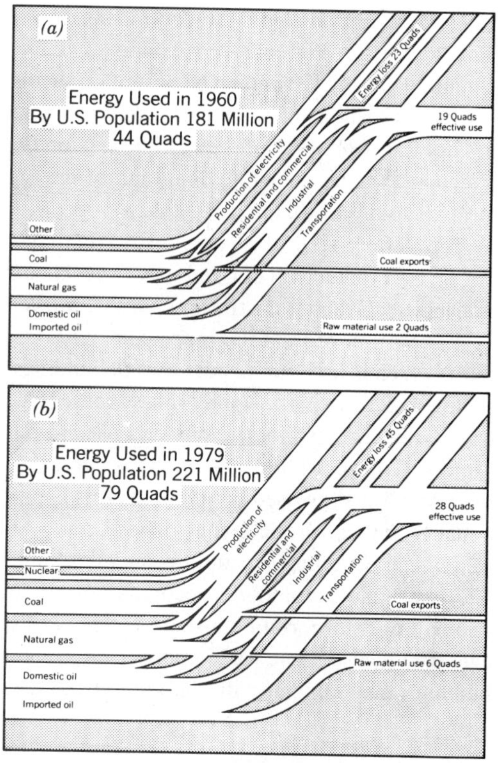

Fig. 1.11 *Flow of energy, from source to end use ("effective use"); graphs show the growing percentage that is wasted in the conversion of energy from one form (low-grade resource) to another (higher-grade resource, such as electricity). "Unmetered" fuels, such as solar and wood energy used in building heating are not included. One quad = 10^{15} Btu. [Energy used in 1989 by U.S. population (248 million), 81 quads; energy conservation's impact is evident.] [Based on* National Geographic Society *(1981).]*

1.3 Energy Sources and Uses

Before taking a detailed look at the way we use energy in buildings, it is helpful to get perspective on how energy sources are converted to end uses for all purposes. Figure 1.11 shows this energy pattern in the United States from 1960 (during the war in Vietnam) to 1979 (6 years after the start of the OPEC oil embargo). Note that solar energy and wood fuel used in space heating are *not* included; only fuel sources "measurable" (by sales) are included here. Several trends are evident; in those 19 years, we gradually required more energy per person (from 243 million to 357 million Btu per person, annually); we gradually wasted a higher percentage of our energy sources in converting them to useful work (from 52 to 57%); and we gradually imported a much higher percentage of the oil we use. The first two trends are directly traceable in large part to our increasing reliance upon electricity: this will be discussed further in Section 1.4.

A note about units of energy: 1 Btu (British thermal unit) is about the amount of energy expended in burning a standard wooden match: a quad is 1 quadrillion (10^{15}) Btu, about equal to 44 million short tons of coal, or 1 trillion cubic feet of natural gas, or 170 million barrels of crude oil, or 8 billion gallons of gasoline, or 63 short tons of dry firewood—or 1 quadrillion wooden matches. A more scientific definition of energy terms is found in Chapter 4, Table 4.1.

The "effective use" figures at the right side of Fig. 1.11 include all our end uses, such as for transportation and industrial processes, as well as that portion of the total (approximately one-third) which is used for all purposes in buildings. Figure 1.12 graphs the approximate percentage of gross-energy use for various end uses in our society. Transportation, 25% of gross usage, is linked with energy used in buildings; this is particularly evident in the layout of residential areas, which essentially require frequent automobile use. The automobile accounts for about one-half of transportation's share of energy usage. The goal of renewable energy sources for buildings may conflict with the goal of reduced automobile usage, since solar energy is a somewhat thinly spread resource. This tends to limit the density of buildings; decreasing density usually results in the increasing use of transportation, as people travel farther between places of home and work. This conflict has several possible resolutions that may include a merging of work and home sites, a substitution of communication for transportation, or a provision for transportation to be powered by renewable resources.

A summary of our U.S. energy sources, their characteristics, and their impacts on architecture and the environment is presented in Table 1.1. Several widely discussed, but still uncertain, energy sources omitted from this table include nuclear fusion, nuclear fission with breeder technology, oil shale, and solar energy from sea-thermal gradients. The uncertainties inherent in these sources include a combination of technical, political, and environmental problems (especially with the nuclear breeder and

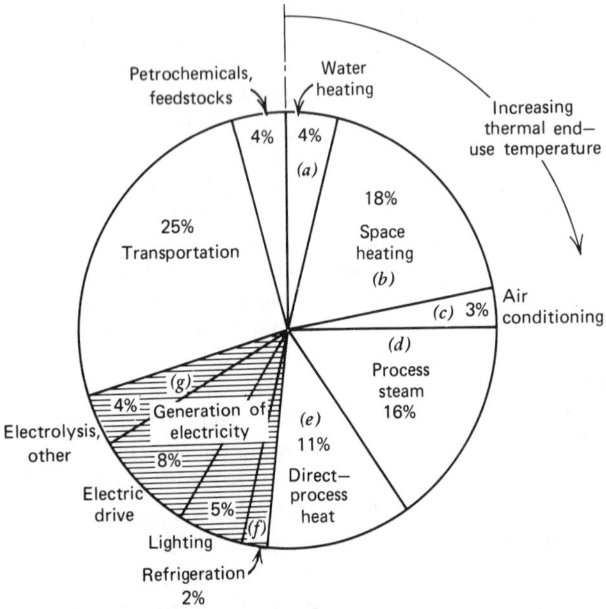

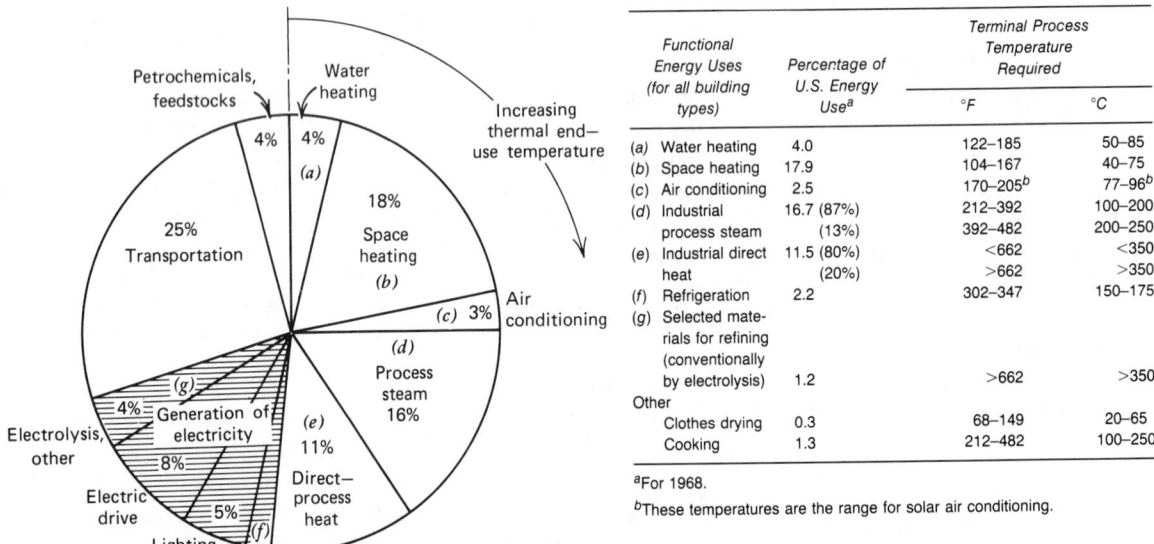

Functional Energy Uses (for all building types)	Percentage of U.S. Energy Use[a]	Terminal Process Temperature Required °F	°C
(a) Water heating	4.0	122–185	50–85
(b) Space heating	17.9	104–167	40–75
(c) Air conditioning	2.5	170–205[b]	77–96[b]
(d) Industrial	16.7 (87%)	212–392	100–200
process steam	(13%)	392–482	200–250
(e) Industrial direct	11.5 (80%)	<662	<350
heat	(20%)	>662	>350
(f) Refrigeration	2.2	302–347	150–175
(g) Selected materials for refining (conventionally by electrolysis)	1.2	>662	>350
Other			
Clothes drying	0.3	68–149	20–65
Cooking	1.3	212–482	100–250

[a]For 1968.
[b]These temperatures are the range for solar air conditioning.

***Fig. 1.12** Percentage and quality of energy use in the United States as of 1968. The categories of water and space heating, air conditioning, refrigeration, and lighting are those most influenced by architectural design. "Quality" is related to terminal process temperatures; "low-grade" uses, at lower temperatures, can be fueled by a variety of energy sources. As temperatures rise, the variety of possible fuels diminishes. At the top of the "high-grade" uses (shaded on graph), electricity is produced by the expenditure of much lower grade fuel (see also Fig. 1.19). (From E. Lazslo and J. Bierman, Eds.,* Goals in a Global Community, *published by Pergamon Press, Inc., Fairview Park, Elmsford, N.Y. Copyright © 1977 by State University of New York.)*

TABLE 1.1 **U.S. Energy Sources**

Part A. ***Nonrenewable Sources***

Nonrenewable Source	*Estimated U.S. Reserves and Resources*	*Notable Characteristics*	*Form in Which Supplied to Consumer*	*Typical Architectural Utilization as Fuel (not including electricity)*	*Architectural Impacts*	*Environmental Impacts*
Coal	U.S. Proven coal resources,[a] total 4577 Quads 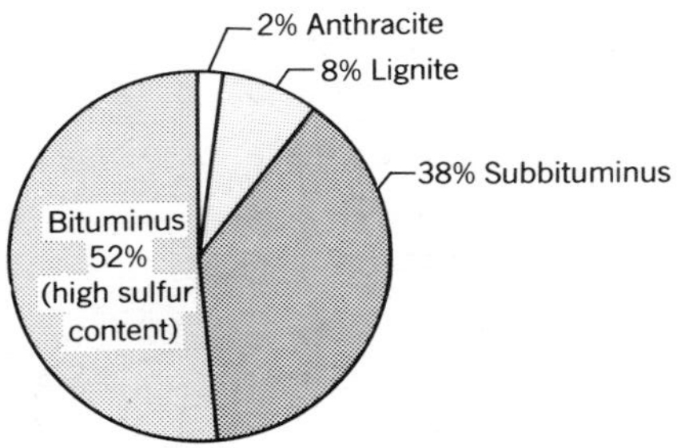 U.S. 1989 use: 19 quads[b]	• Least flammable fossil fuel to transport, but bulky (less fuel value for its weight) • Coal dust presents mining, transport, and storage problems • Coal gasification can supply cleaner fuel to electric generating plants	• Mostly as electricity • Some as coal	• Some central steam plants (for campuses and other building complexes)	• Combustion air, stack, air pollution control equipment, and cooling tower or pond • District heating possibilities for waste heat from steam plants • Large and "dusty" fuel storage and delivery areas • Ash collection and disposal	Recovery • Strip mining's long-term ecological disruption, including acid drainage to streams • Underground mine safety problems, and subsidence of surface over mines Transport • For slurry pipelines, water pollution Use • Air pollution (particularly with high-sulfur coal); acid rain • Thermal pollution Waste • Landfills for ash disposal

TABLE 1.1 **U.S. Energy Sources (*Continued*)**

Part A. ***Nonrenewable Sources*** **(*Continued*)**

Oil	U.S. Oil resources[a] (estimated recoverable in billions of barrels) 1 quad = 0.17 billion barrels 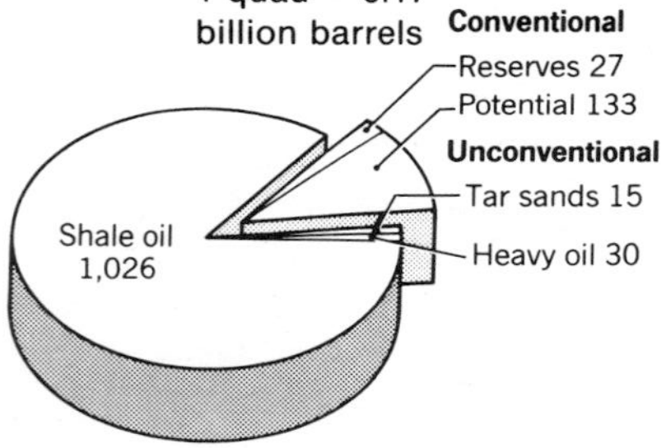 U.S. 1989 use: 34 quads	• Flammable but less explosive than natural gas, readily transported • Stored on site where used • Fuel-price manipulation by cartels	• Mostly as gasoline • Much as fuel oil • Much as electricity • Some for nonenergy usages	• Space heating • Some water heating • Some central steam plants • Some large motors	• Combustion air, flue (and waste heat recovery opportunities) • Oil storage tank and delivery area	Recovery • Offshore oil spills • Saltwater intrusion and surface subsidence near oil fields Transport • Oil spills, especially on water • Pipeline causes ecological disruption, especially during construction Refining • Air and water pollution Use • Air pollution • Thermal pollution
Natural gas	U.S. Natural gas resources[a] (in trillion cubic feet) 1 quad = 1 billion ft^3 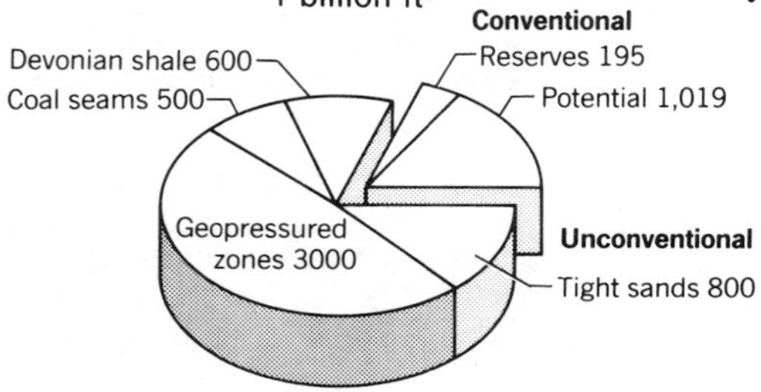 U.S. 1989 use: 19 quads	• Explosive, harder to transport and to store • High fuel content burns relatively cleanly • Distributed by piped networks (except as propane)	• Mostly as natural gas • Much as electricity • Some for nonenergy usages	• Space heating • Water heating • Cooking • (Some lighting, cooling) • Some central steam plants	• Combustion air, flue (and waste heat recovery cpportunities) • Meter accessible to utility • Care in locating gas lines, especially underground on site • (For propane, outdoor storage tank, and delivery area)	Transport • Pipeline causes ecological disruption • For liquefied natural gas, potential for a spreading icy vapor cloud, which is highly explosive • For bottled gas, hazard of explosive contents under pressure on site with buildings Use • Air pollution (though less than with oil, coal) • Thermal pollution • In fuel cells, almost no pollution

Nonrenewable Source	Estimated U.S. Reserves and Resources	Notable Characteristics	Form in Which Supplied to Consumer	Typical Architectural Utilization as Fuel (not including electricity)	Architectural Impacts	Environmental Impacts
Uranium	U.S. uranium recoverable reserves: 220 Quads[c] U.S. 1989 Use: 5.7 quads	• Light-water reactors now operating use very little of the energy contained in the enriched fuel • Fuel enrichment itself consumes large amounts of energy • An especially controversial source, involving both environmental and arms control issues • Fuel/price manipulation by cartels	• Almost all as electricity	• (Radioactive materials utilized in some medical and scientific equipment)	• (Radioactive materials require heavy shielding for delivery, storage, use, and waste pickup areas) • "Fallout shelter" designation for qualifying spaces	Recovery • Mining problems similar to coal; less bulky, but with additional problem of radioactive tailings Transport • Radiation shielding and security needs for nuclear wastes Refining • Large amount of energy consumed in fuel enrichment • Radiation shielding and security needs for waste reprocessing Use • Potential for catastrophic radiation release • Radiation shielding and security needs • Thermal pollution Waste • Long-term radiation shielding and security needs
Geothermal energy	**U.S. potential[a], estimated 18.5 quads/year**	• Renewable, from earth's core, but millions of years after heat is removed from outermost layers of rock	• Electricity • Hot water	• Direct heating of buildings	• "Hot springs" water often highly corrosive • Source usually must be very near to point of use	• Unknown impacts of large-scale removal of heat from outer crust • Subsidence of surface is possible where water is removed without recharging to the earth

TABLE 1.1 **Energy Sources (*continued*)**

Part B. ***Renewable Sources***

Renewable Source	*Estimated Maximum U.S. Arrival Rate [quads (10^{15} Btu) per year]*	*Notable Characteristics*	*Form in Which Supplied to Consumer*	*Typical Architectural Utilization as Fuel (not including electricity)*	*Architectural Impacts*	*Environmental Impacts*
Solar radiation	47,000[d] Estimated U.S. yearly capture[d] by year 2000	• Thinly spread, with large daily and seasonal variations in most of the United States		• Daylighting • Space and water heating • Some cooling	• Window and skylight areas for daylighting • Internal arrangement of thermal storage materials in passive solar buildings • Large collectors, storage volumes • Less densely built complexes for solar buildings	• Ecological disruption where large land areas are converted to solar collection • Soil depletion where organic waste is not returned as fertilizer • River ecosystem disruption by hydroelectric dams
Fuel wood	3[d]		• Wood fuel	• Some cooking (wood and gas)		
Farm waste	6[d]	• Methane gas highly explosive	• Methane gas • Gas or oil			
Photosynthesis fuel	15[d]					
Solar heating and cooling of buildings	9[e]		• Heated air or water			
Photovoltaics	7[e]	• Varies with seasonal river flows	• Electricity			
Hydropower	9[d] U.S. 1989 use: 3 quads		• Electricity			
Wind power	6[e]	• Varies considerably between sites	• Electricity			
Waste paper and plastic incineration	2[f]		• Electricity			• Air pollution • Thermal pollution
Tidal power	3[d]		• Electricity			• Ecological disruption of estuaries

[a]National Geoegraphic Society (1981).

[b]U.S. Department of Energy, *Monthly Energy Review*.

[c]From Fisher (1974). This material was based on U.S. Geologic Survey Circular 650, *Energy Resources of the United States* (1972).

[d]From Starr (1971).

[e]Jewell (1978).

[f]R. Berry and H. Makino, "Energy Thrift in Packaging and Marketing," *Technology Review*, 1974. © 1974, by the Alumni Association of the Massachusetts Institute of Technology, Cambridge, Mass.

with oil shale) and involve doubt about whether more energy would eventually be produced by some of these sources than is required to develop them.

Similarly, the known total U.S. quantities of natural gas, crude oil, coal, and uranium in the ground are greater than the portion called *recoverable* reserves; the numbers of quads of this most easily tapped portion for each of the nonrenewable energy sources are shown in Table 1.1. The remainder of our U.S. supply of these resources is called by names such as "submarginal" (difficult to utilize, expensive to extract, etc.) or "undiscovered."

Despite the significant total quads of solar radiation that the United States receives each year, we utilize relatively little for energy in buildings. The relatively high initial cost of using such a plentiful but thinly spread resource has discouraged its use, but as the submarginal, nonrenewable resources must be worked to replace the dwindling recoverable portions, the difference in cost between solar and nonrenewable energy sources will shrink.

Probably even before nationwide utilization of solar energy is achieved, the United States is expected to shift away from reliance on gas and oil to coal as its major energy source. This is evident from a comparison of Figs. 1.13 and 1.14 and Table 1.1.

In our past, we shifted from almost complete reliance on renewable resources (wood and work animals) to nonrenewable (fossil fuel) resources. More recently, we shifted from heavy reliance on coal to reliance on oil and gas. At present, electricity is the dominant form by which the energy of coal is delivered to buildings. Thus part of the prediction for electricity's rapid growth (Section 1.4) can be traced to the relatively plentiful coal reserves in the United States. The waste heat and other environmental impacts associated with the production of electricity are significant and reinforce the necessity of careful and appropriate use of this versatile energy form by the designers of buildings.

In our future, it is clear that we must once again rely primarily upon renewable energy sources; this time, however, we have many more people, accustomed to a much higher per capita energy usage. Figure 1.15 projects one scenario of a transition, by the year 2020, to renewable and more environmentally benign energy sources. This scenario assumes no overall growth in energy consumption, which may be reasonable in view of the trend since 1973. It also assumes that "low-tech" sources such as

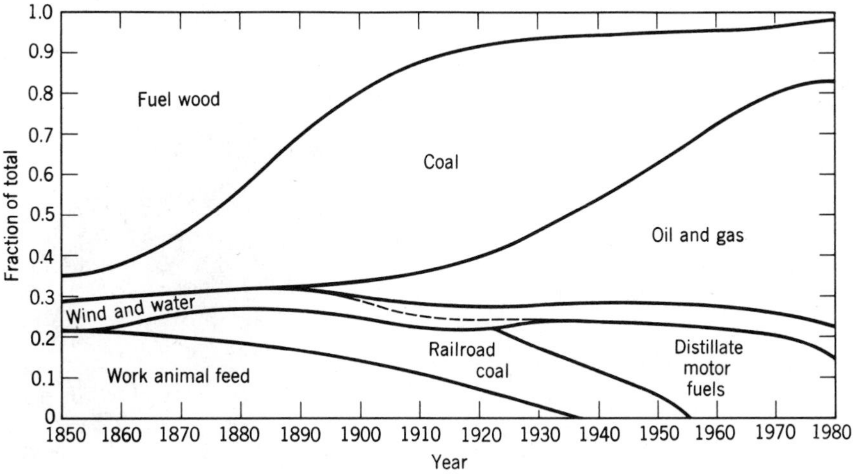

Fig. 1.13 *U.S. fuel input, 1850 to 1980, illustrates the progression from dependence on renewable fuels (wood and work animal) to fossil fuels (coal, then oil and gas). Wind and water power changed their form from "mills" to hydroelectricity between 1890 and 1940. Similarly, though not shown here, much fossil fuel is now converted to electricity before use. [From Fisher (1974) and Meyers (1983).]*

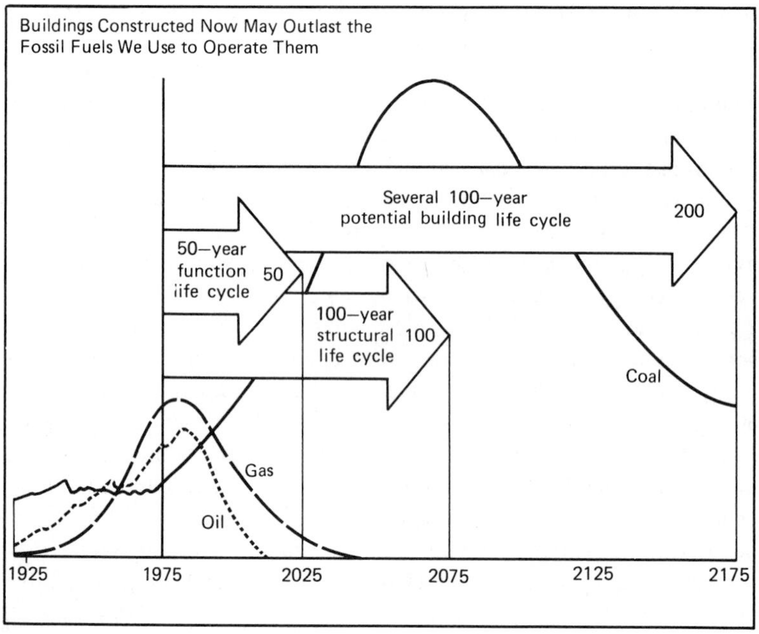

Fig. 1.14 Life cycles of buildings and of fossil-fuel reserves, showing our relative dependence on three major energy sources. In 1925 we used mostly coal, some oil, and very little natural gas. In 1975 natural gas ranked first, followed by oil, then coal. By 2025, we may be almost entirely dependent on coal. [From California, Office of the State Architect (1976).]

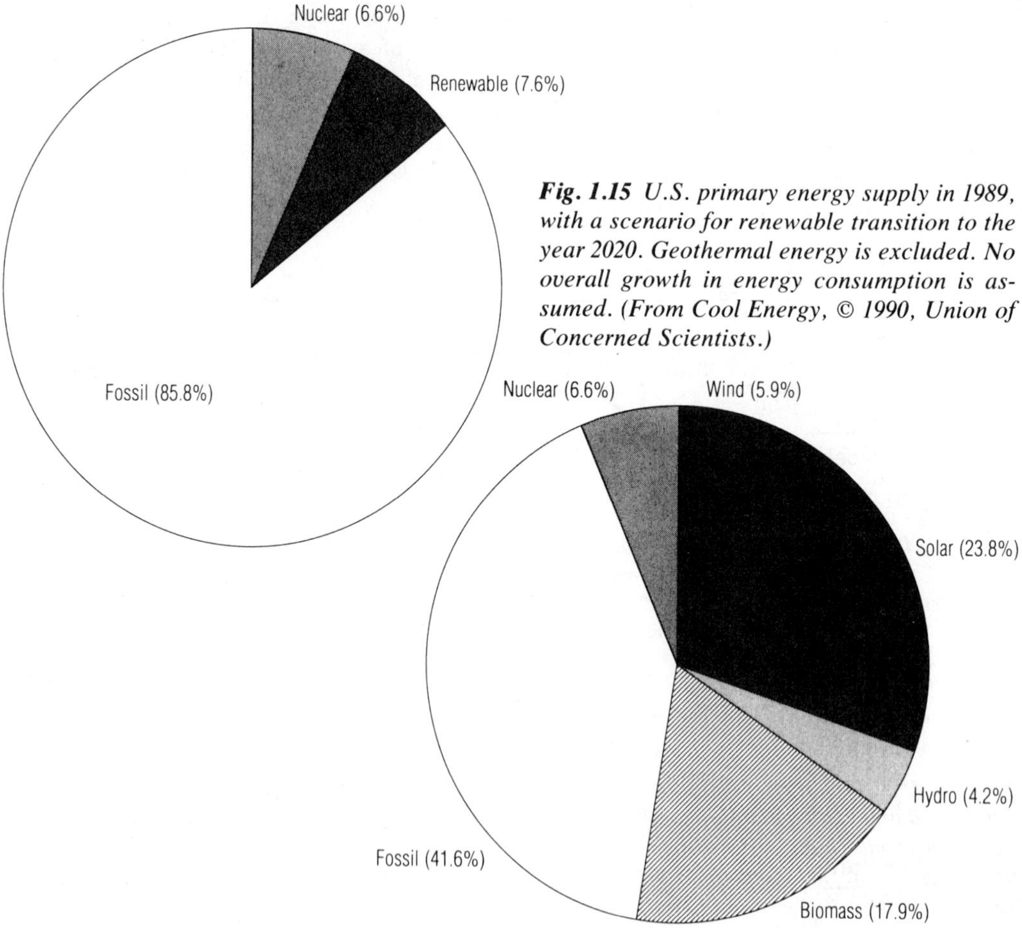

Fig. 1.15 U.S. primary energy supply in 1989, with a scenario for renewable transition to the year 2020. Geothermal energy is excluded. No overall growth in energy consumption is assumed. (From Cool Energy, © 1990, Union of Concerned Scientists.)

biomass—the combustion of wood and waste products—will develop faster than "high-tech" sources such as solar photovoltaic electricity production. Biomass has potentially serious consequences for global warming; vegetation converted to fuel must be replaced. Therefore, careful forest and agricultural management must accompany biomass energy.

One of the more interesting transitions might be from natural gas and gasoline to hydrogen. This highly combustible fuel could be provided by solar thermal (also photovoltaic) electricity, used to separate water into hydrogen and oxygen. Concentrations of solar electrical generating plants in the sunny southwest could supply a hydrogen pipeline throughout the United States. When burned with oxygen, hydrogen creates the single by-product of water.

But is solar energy adequate for our energy needs? Table 1.2 compares the earth's receipt of solar energy at the surface in a single day with some other energy phenomena.

The environmental impacts listed in Table 1.1 show only the most obvious or large-scale impacts for each energy source. Although each source has environmentally damaging consequences when used, the relatively mild impacts of some sources make them more environmentally attractive than others. The details of such consequences of energy utilization are not treated further in this book, except for a summary of air pollution in the United States and its impacts on buildings and their occupants in Chapter 3 (Fig. 3.21).

1.4 Architecture and Energy Usage

(a) The Recent Past. The "energy-rich" decades of the 1950s and 1960s saw a dramatic increase in recommended levels of heating and lighting for buildings. The evolution of higher recommended indoor temperatures is shown in Table 1.3. Lighting levels rose even more rapidly, as a look at past editions of this book (Table 1.4) reveals. (Footcandles are defined in Chapter 18.) These lighting levels have been reassessed, as shown by later scaled-down recommendations of the Federal Energy Administration. These recommended increases and their impact on energy consumption reached a peak in the years around 1970, which we could call the "pre-energy-awareness" period in recent U.S. history. One example of the extent to which increased lighting has been carried is seen in an electric utility headquarters building, renovated and enlarged in the mid-1960s (see Fig. 1.16). Among its features are:

- 300 to 350 footcandles (fc) of light in office areas.
- Up to 550 fc in display areas.

TABLE 1.2 **Daily Arrival of Solar Energy on Earth Compared to Other Energy Quantities**

Solar energy received each day	1
Melting of an average winter's snow during the spring	1/10
A monsoon circulation between ocean and continent	1/100
Use of energy by all mankind in a year	1/100
A midlatitude cyclone	1/1,000
A tropical cyclone	1/10,000
Kinetic energy of motion in earth's general circulation	1/100,000
The first H bomb	1/100,000
A squall line containing thunderstorms and perhaps tornados	1/1,000,000
A thunderstorm	1/100,000,000
The first A bomb	1/100,000,000
The daily output of Boulder Dam	1/100,000,000
A typical local rain shower	1/10,000,000,000
A tornado	1/100,000,000,000
Lighting New York City for one night	1/100,000,000,000

Source: Reprinted by permission from Lowry (1988).

TABLE 1.3 **Criteria for Thermal Comfort Since 1900**

Date	*Environmental Specifications*[a]	*Dry-Bulb Temperature (Fahrenheit) for 40% rh*
Before 1900	65–70 F DBT	65–70
Early 1900	56 F WBT	70
1914	68 F DBT 40% rh	68
1923	66–72 F DBT, 19–61% rh	66–72
1923	62–69 ET 64 ET (optimum)	68
1925	63–71 ET 66 ET (optimum)	71
1929	66–75 ET 71 ET (optimum)	77
1939	64.8–76.0 F ET 71.8 F ET (optimum)	78
1941	68 ET (optimum)	74
1938–1956	73–77 DBT, 25–60% rh	73–77
1960	77.6 F DBT, 30% rh 76.5 F DBT, 85% rh	77
1965[b]	73–77 DBT, less than 60% rh	73–77
1965[c]	77 F DBT, 70% rh 79.5 F DBT, 20% rh	78
1975[d]	72 F DBT, (30%) winter 78 F DBT, summer	
1980s	See Figs. 2.2–2.4	

[a]DBT = dry bulb temperature, WBT = wet bulb temperature, ET = effective temperature, similar to "operative temperature" defined in Section 2.2.

[b]ASHRAE Standards, Series 55.

[c]For lightly clothed subjects engaged in sedentary activity. Reprinted by permission from *Criteria for Thermal Comfort* by R. G. Nevins, Institute for Environmental Research, Kansas State University.

[d]ASHRAE Standard 90–75.

TABLE 1.4 **Trends in Recommended Minimum Lighting Levels (footcandles)**

Category	*From* Mechanical and Electrical Equipment for Buildings: *2nd Edition (1945)*	*From* Mechanical and Electrical Equipment for Buildings: *5th Edition (1971)*	*Federal Energy Administration (1976)*
Offices			
Accounting, bookkeeping	30	150	
Regular office work	20	100	50
Conference rooms	10	(Not listed)	30
Corridors, stairs	5	30 (But not less than one fifth the level of the adjacent area)	10
Schools			
Auditoriums	10	30	
Classrooms, regular deskwork	20	70	
Drafting, drawing	30–50	100	
Sewing	50–100	150	
Libraries			
Reading room	20	70	

Fig. 1.16 Office building for an electric utility, designed in the mid-1960s when energy seemed to be in plentiful supply. Extraordinarily high interior lighting levels as well as (now unused) 500-W exterior night lights in each *windowsill are remnants of a less energy-conscious era. (Photo by Stan A. Adams.)*

- Up to 600 fc in conference and demonstration rooms.
- 500-watt (W) luminaires in each of 288 window sills, for nighttime facade lighting.

During the design phase of this building, the architect calculated that construction costs would be reduced by about $1 million if lighting levels were cut to 150 fc. This savings would be evident in the lower number of luminaires and in the reduced size of cooling equipment necessary to remove the surplus heat. The utility, however, expressed its interest in leading the trend to higher lighting levels, and retained the 300-plus footcandle level. Some of the heat emitted by this lighting is captured and used when needed to heat the office building. ("Heat-of-light" systems are discussed in Chapter 7, Section 7.4.) Yet this high a lighting level produces a need for cooling, not for additional heating, in a typical office building on most winter days. The lighting thus provides surplus heat for almost the entire year and requires that still more energy be spent to remove surplus heat. This utility has since discontinued the lighting of its facade at night but continues to provide very high lighting levels for its interiors.

(b) The Near Future. Most projections of future U.S. energy consumption assume continued, though less rapid, growth in demand. These views of energy supply and demand are constantly shifting, and this year's estimate does not assure next year's performance, nor certainly the next decade's. Consider the impact of the 1973 oil embargo both on the reality of our subsequent consumption and on the revised forecasts for the future. Figure 1.17 plots the predicted impact on an "active conservation program," including improved energy efficiency for automobiles, homes, and office buildings. Compare this 1976 forecast of about 105 quads for 1988 with the actual total of 80 quads used (Fig. 1.18). Clearly, energy efficiency in buildings can have an impact on future consumption patterns.

A more detailed look at building energy efficiency is found in many parts of this book, especially Part II (Thermal Control) and Part VI (Illumination). The relationship between indoor climate and illumination is likely to be particularly influential on building energy consumption,

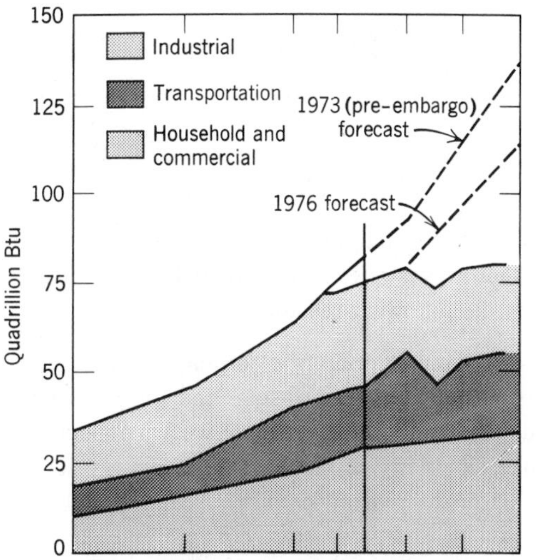

Fig. 1.17 U.S. energy consumption, comparing forecast to actual. The 1973 oil embargo lowered the 1976 forecast (data from Federal Energy Administration) but actual savings from conservation lowered consumption even more. (Actual consumption 1976–1988 from Energy Information Administration, Monthly Energy Review, *U.S. Department of Energy.)*

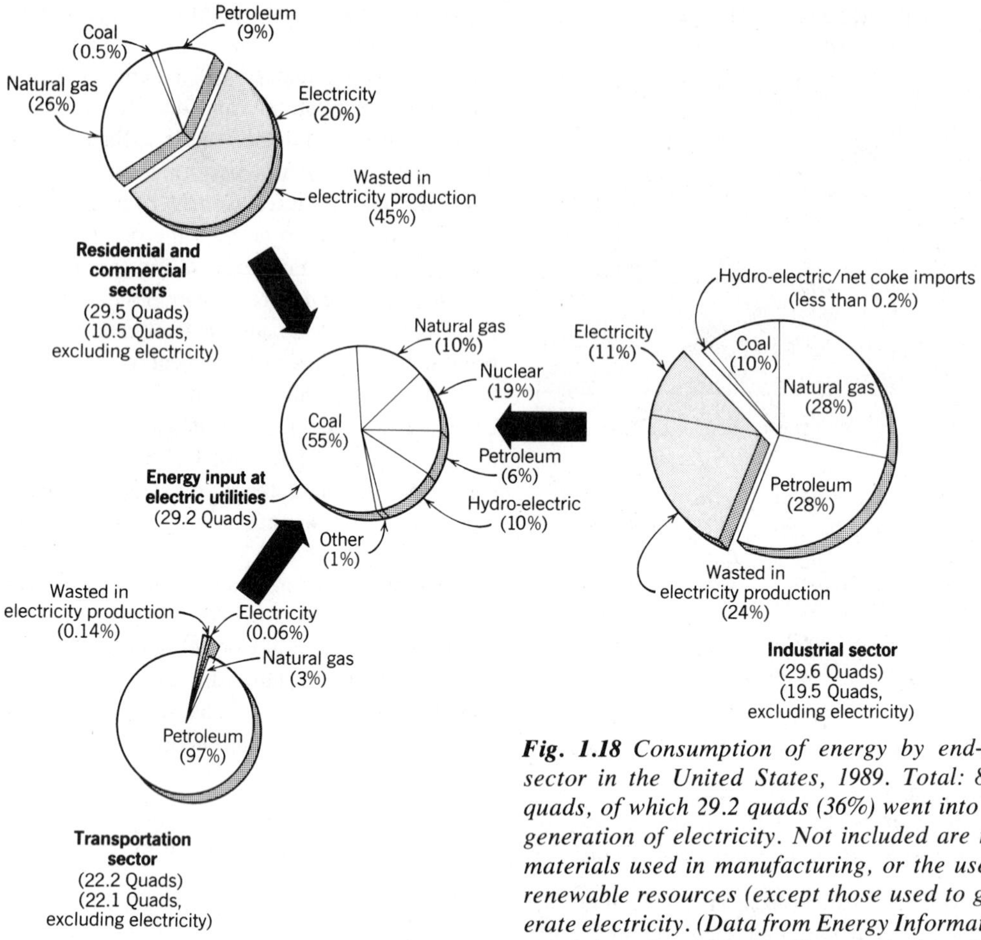

***Fig. 1.18** Consumption of energy by end-use sector in the United States, 1989. Total: 81.3 quads, of which 29.2 quads (36%) went into the generation of electricity. Not included are raw materials used in manufacturing, or the use of renewable resources (except those used to generate electricity. (Data from Energy Information Administration,* Monthly Energy Review, *U.S. Department of Energy.)*

as the office building cited above demonstrates. Electricity is one of the keys to this relationship, for the following reasons.

1. Consumption of electricity is expected to rise about twice as fast as the overall energy demand, and more and more we are expected to use electricity in place of other energy forms. Part of the reason for this is seen in Table 1.1; for some energy sources, conversion to electricity before distribution to buildings is the most convenient option.

2. Other than daylight, electricity is almost the only source of illumination in buildings; the pre-1973 annual growth rate (about 7%) of electricity consumption was boosted by the steeply increased illumination recommendations and the cooling that accompanied them.

3. Electricity is a convenient and versatile energy form; it not only serves such high-temperature and highly concentrated (or "high-grade") energy tasks as lighting and motive power, but it can also serve the low-temperature simpler (or "low-grade") tasks, such as cooking, heating water, and space heating for buildings (see again Fig. 1.12). "All-electric" buildings have become commonplace, even though they are subject to paralysis in blackouts, as is any building dependent on a single energy source.

4. Electricity generated by thermal processes delivers only about one-third the total energy that goes into its production; the other two-thirds is usually lost as waste heat at the

generating plant. See Fig. 1.18 for proportions of energy thus wasted.

Thus, the relationships among greater energy demands, electricity, lighting, and interior climate are complex. In the future, solar (photovoltaic) generation of electricity will decentralize much of the production and remove most of the environmental costs associated with electricity generation. Until then, as we increase our use of electricity, we doubly increase our use of the nonrenewable fuels that serve almost all today's electrical generating plants.

The building designer interested in slowing our energy-demand growth rate can begin with a careful utilization of electricity; low-grade energy tasks might be better served by nonelectric fuel sources. As an example, consider the furnace in a residence (Fig. 1.19). The low-grade alternatives to electric-resistance furnaces include not only fossil-fuel-burning furnaces but also renewable heat sources, such as wood and solar energy.

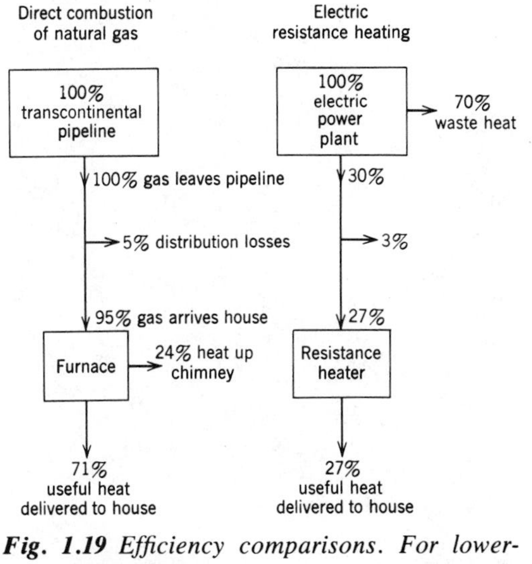

***Fig. 1.19** Efficiency comparisons. For lower-grade heating tasks, more of the energy available in fossil fuels is utilized by the furnace, which burns the fuel directly. Much more efficient furnaces are now available (see Chapter 6; see also Fig. 6.64). The electric furnace using resistance heaters is fully efficient in the home, but the higher-grade energy it receives has produced waste heat at the generating plant. [From Fisher (1974).]*

Although the low efficiency of electrical generation is obvious, the central generating plant has some advantages over many fuel-burning furnaces. Air pollution control is more efficient at a central plant, and the massive concentration of its waste heat could be coupled with certain industrial processes (cogeneration) or piped to nearby buildings for space heating or absorption cooling (see Sections 7.7 and 7.8).

The higher-grade tasks for which electricity is suitable also need careful consideration. The higher the level of electric lighting, the more cooling is needed; depending on the climate, every unit of energy expended in lighting will require another half-unit of energy for removing the surplus heat thus generated. The annual energy growth rate for the cooling of buildings is about 10%—more than three times as fast as our overall energy growth shown previously in Fig. 1.17.

One way to reduce the electricity consumed by lighting is to substitute daylight for electric lighting. This is most applicable at the perimeters of buildings; yet, even the interiors of low-rise buildings can be served with daylight via skylights for the general or overall illumination, using small individually controlled electric lights only where and when needed (Fig. 1.20).

As with many such substitutions, the designer must consider the trade-off: will more glass area to admit daylight produce greater heat loss on winter nights and undesired heat gain on summer days? (Calculations of heat gain and heat loss are presented in Part II.) In Chapter 6 techniques are discussed for protecting glass against heat loss with insulating shutters and for designing windows to minimize summer solar gain.

In summary, today's designer is aware of the overall price increases and fuel reserve decreases associated with nonrenewable fuels. The design response, however, is pulled in seemingly opposite directions: a tendency to *close in* the building to conserve energy, versus the notion of *opening* it to daylighting and passive heating and cooling opportunities. This balancing act between conservation and passive design will be a continuing theme in the next five chapters.

***Fig. 1.20** Combining general daylight with specific-task electric light.* (a) *Library designed by Alvar Aalto at Mount Angel Abbey (Oregon) utilizes a central north-facing skylight to provide daylight on two levels of the interior. The individual electric lights* (b), *each with a pull chain, can be used when reading fine print makes more light desirable.*

1.5 Energy in Building Construction

In the United States, the total energy expended in *constructing* new buildings, as a percentage of the total used for all purposes, appears relatively small (Fig. 1.21). The decisions that designers make regarding the building envelope and the structural and mechanical systems generally have greater energy consequences in building operation over many years than in the shorter construction period. Most of the energy embodied in construction is invested in the manufacture of materials and components (Fig. 1.22).

Some building types are much more "energy intensive" in the construction phase than others, as shown in Fig. 1.23. Single-family residences are relatively low in "energy intensiveness," compared to the total energy invested in them (from Fig. 1.21); such residences constitute almost one third of the total square footage of all new buildings constructed in the years for

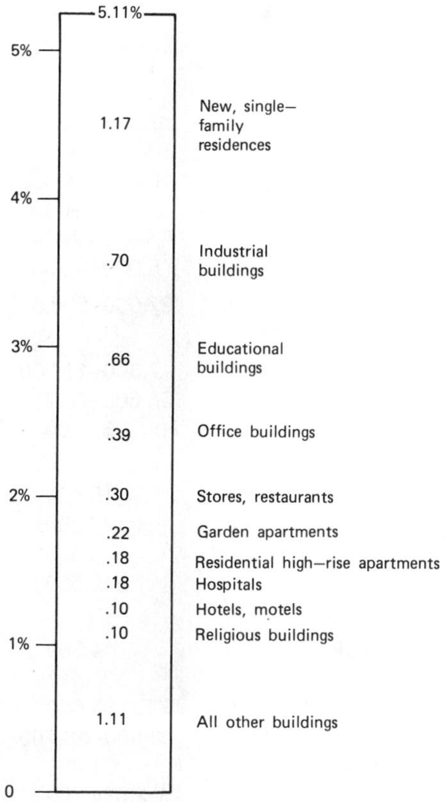

Fig. 1.21 Energy in new building construction, as percentage of U.S. total energy consumption, including energy to manufacture, transport, and erect new buildings. [From Hannon et al. (1977).]

which these statistics are compiled. The single-family residence and the closely related garden apartments (two- to four-family residential and

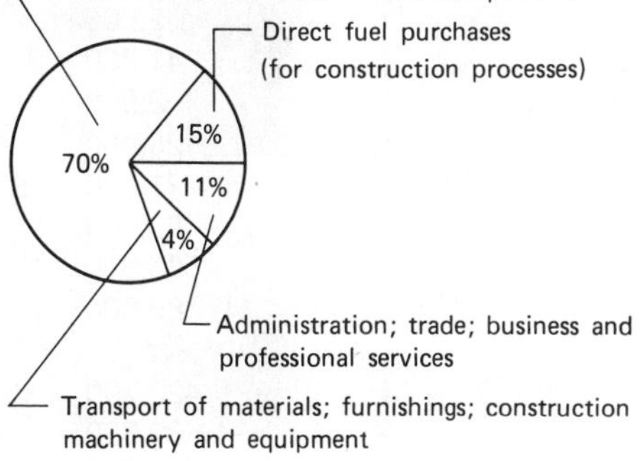

Fig. 1.22 Energy embodied in building construction goes largely into the manufacture of building materials and other components. [Based on Hannon et al. (1977).]

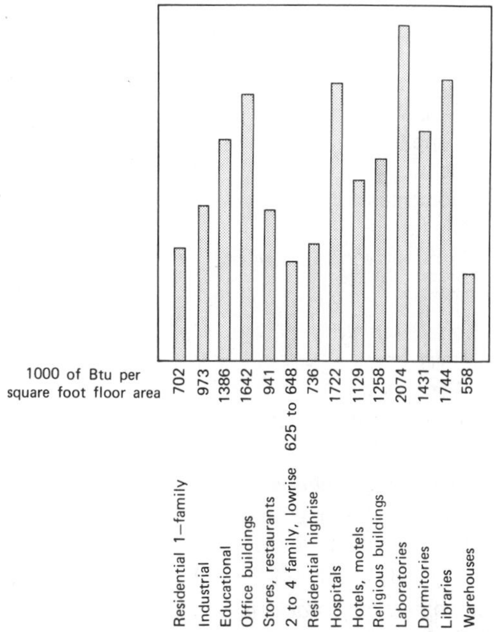

Fig. 1.23 Energy embodied per square foot of floor area, for various building types. These comparisons of 1000 Btu per unit area are for the building types shown in Fig. 1.21. [From Hannon et al. (1977).]

low-rise residential) are low in energy intensiveness primarily because they utilize so much wood in their construction, and wood is a low-energy material. Table 1.5 compares the approximate total energy embodied in selected materials, delivered to the job site. Figure 1.24 illustrates the effects of added insulation on the total energy embodied per square foot of typical residential wall construction. (For the annual energy savings, see Fig. 1.26, following.)

Until recently, building material choices were rarely influenced by the amount of energy invested in those materials. Now that such information is becoming more widely available, energy content can be included with other factors—aesthetics, durability, fire resistance, labor intensiveness, and installed cost—that a designer considers in choosing materials for a building.

A related consideration is the monetary cost of the energy-consuming equipment in a building. Table 1.6 indicates the portion of recent construction costs that are assigned to the building elements discussed in this text: plumbing,

TABLE 1.5 **Total Energy Embodiment in Selected Building Materials per Unit of Material at Job Site**

Material	*Unit*	*Total Embodied Energy (Btu/unit) at Job Site*
Wood products		
Lumber	Board foot	7,600–9,800
Shingles		7,300
Flooring		10,300–14,300
Mouldings		17,900
Glu-lam		16,700
3/8-in. plywood (softwood)	Square foot	5,000–5,800
Paints	Gallon	437,000–508,500
Asphalt roofing		
Rolls	Square foot	7,800–11,000
Shingles		25,600–29,700
Mineral-surface insulating board siding		67,500
Glass		
Flat glass: double strength	Square foot	15,430
Flat glass: tempered		72,600
Plate and float glass, 1/8 to 1/4-in. thick		48,000
Laminated plate glass, 1/4-in. and over		212,500
Stone and clay products		
Common brick	Per brick	14,300
Ceramic glazed brick		33,413
Quarry tile	Square foot	51,000
Ceramic mosaic tile and accessories, glazed		63,600–68,700
Concrete block	Per block	31,800
Ready-mix concrete	Cubic yard	2,594,300
Gypsum board-3/8 in.	Square foot	5,300
Mineral wool insulation, 4 1/2-in. thick	Square foot	8,300
Iron and steel		
Steel sheets: 22 gauge	Square foot	29,400
16 gauge		58,800
Galvanized sheets: 22 gauge		49,800
16 gauge		98,500
Steel shapes: W 12 × 65, carbon	Lineal foot	1,217,800
W 12 × 65, alloy		1,749,200
WT 6 × 27, carbon		543,344
WT 6 × 27, alloy		780,400
Reinforcing bars: #2		2,600
#8		41,800
Welded wire mesh: 2 × 4, 14/14	Pound	3,900
2 × 12, 8/8		25,400
Pipe, carbon steel: 1/2-in. diameter	Lineal foot	21,900
6-in. diameter		489,700
Stainless steel sheets: cold rolled	Pound	138,300
hot rolled		80,800
Aluminum		
1/4-in. plate	Square foot	420,700
1-in. plate		1,680,300
1/8-in. sheet		174,800
Standard shapes: 8 I 8.81	Lineal foot	811,800
6 I 5.10		469,900

Source: Hannon et al. (1977).

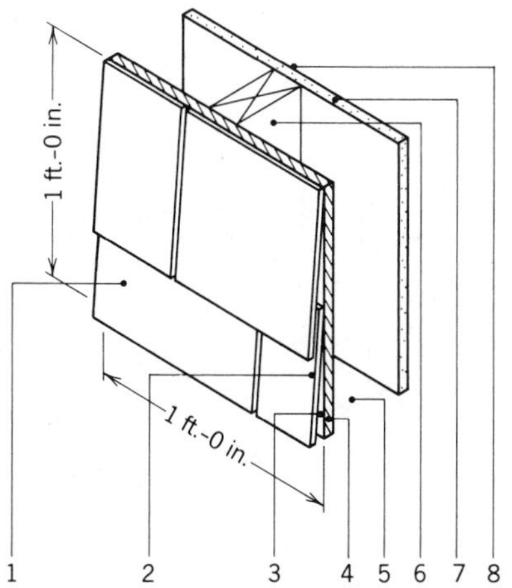

(a)

Construction	*R Value*		*Embodied Energy (Btu/ft²) in Building Section*
1. Outside surface (15-mph wind)	0.17		—
2. Wood shingles (½ in. × 8 in. lapped)	0.87		7,315
3. Building paper (asphalt)	0.15		—
4. Plywood (½ in.)	0.62		7,705
5. 4 in. Airspace	0.97		—
6. 2 in. × 4 in. at 16 in. o.c.	—	4.35	3,486
7. Gypsum wallboard (½ in.)	0.45		6,920
8. Inside surface (still air)	0.68		—
	3.91	4.35	25,426
$U = 1/R =$	0.26	$U = 0.23$ at framing	
Adjusted U (to account for framing) $=$	0.25		
Addition of Insulation			
Add 3½-in. batt insulation	11.00		Add 6,860
Deduct R value of air space	0.97		
	10.03		
Add to above R value	3.91		
	13.79		32,286
$U = 1/R =$	0.07	$U = 0.23$ at framing	
Adjusted U (to account for framing) $=$	0.085		

Fig. 1.24 *Energy embodied per square foot in typical residential wall constructions.* (a) *With the addition of insulation, frame walls experience a 29% increase in embodied energy—and a 73% decrease in heat loss rate.* (b) *Similarly, insulation added to a brick veneer or frame wall increases embodied energy by only 3%, while decreasing heat loss rate by 72%. See also Fig. 1.26. [From Hannon et al. (1977).]*

HVAC (heating, ventilating, air conditioning), and electrical equipment. The thermal and other environmental control systems for buildings are already a major influence on construction costs, as well as on energy consumption in operating the building. As the techniques and equipment

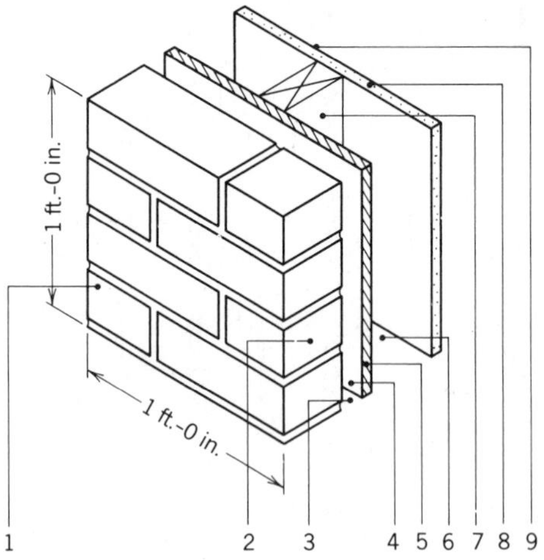

(b)

Construction	R Value		Embodied Energy (Btu/ft^2) in Building Section
1. Outside surface (15-mph wind)	0.17		—
2. Brick and masonry (4 in.)	0.44		105,004
3. 1-in. Air space	0.97		—
4. Building paper (asphalt)	0.15		—
5. Plywood (3/8 in.)	0.47		5,779
6. 4-in. Air space	0.97		—
7. 2 in. × 4 in. at 16 in. o.c.	—	4.35	3,486
8. Gypsum wallboard (3/8 in.)	0.32		5,297
9. Inside surface	0.68		—
	3.98	4.35	119,566
$U = 1/R =$	0.25	$U = 0.23$ at framing	
Adjusted U (to account for framing) =	0.24		
Addition of Insulation			
Add 3½-in. batt insulation	11.00		Add 6,860
Deduct R value of air space	0.97		
	10.03		
Add to above R value	3.98		
	14.01		126,426
$U = 1/R =$	0.07	$U = 0.23$ at framing	
Adjusted U (to account for framing) =	0.085		

Fig. 1.24 *(continued)*

for reducing energy consumption are developed through such steps as recovery of waste heat and the substitution of solar energy for fossil fuels, the percentage of mechanical and electrical system cost will probably become an even greater influence on the cost of constructing new buildings.

TABLE 1.6 **Construction Costs for Mechanical and Electrical Equipment**

	Portion of Total Construction Dollar Cost (%)		
Building Type	*Plumbing*	*Heating, Ventilating, Air Conditioning*	*Electrical*
Apartments, mid-rise (4–7 story)	7	6	7
Auditoriums	7	16	9
Banks	4	7	10
Churches	5	10	9
College			
Classrooms and administrative offices	6	12	10
Laboratories	7	14	10
Community centers	7	10	9
Department stores	4	12	12
Hospitals	9	13	12
Libraries	5	11	11
Motels	11	6	8
Offices, mid-rise (5–10 story)	4	9	8
Restaurants	8	12	11
Retail stores	5	9	10
Schools			
High schools	7	12	10
Elementary schools	7	11	10
Supermarkets	6	9	12

Source: © Robert Snow Means Co., Inc. Reproduced with permission from *Building Construction Cost Data* 1990.

1.6 Energy in Building Operation

The typical building will consume more energy in its lifetime of operation than is used in its construction. The estimated energy needs for the construction, operation, and demolition of an office building in Albany, New York, are shown in Fig. 1.25.

There are several proposals to establish an upper limit for building energy consumption in operation. In their simplest form, such regulations would specify the maximum allowable energy input per building-area unit per year: British thermal units per square foot per year (Btu/ft^2-year), or megajoules per square meter per year (MJ/m^2-yr). A more complicated set of regulations to define—but relatively easy to apply to the design of walls, roofs, and so on—would establish maximum overall thermal transfer values. This means that the designer would choose a combination of wall materials—for instance, those that did not allow more than a specified rate of heat transfer. These more detailed regulations would also include HVAC system performance (indoor temperatures, controls, ventilation rates, humidity control, zoning, pipe and duct insulation, equipment performance ratings, etc.), hot water service, electrical distribution systems, and lighting. Details of such proposals are found in Section 5.2 and Appendix K.

Better insulated walls and roofs require some energy to build, but save much more energy over their lifetimes, as shown in Fig. 1.26.

The building envelope is not the only opportunity for energy conservation. The significant energy savings now available in more efficient appliances and mechanical equipment are detailed in Chapters 6 and 7; Fig. 1.27 summarizes such opportunities within residences.

As energy used in operating buildings becomes more expensive and more difficult to obtain, its influence on the design process will become evident over a wider range of the

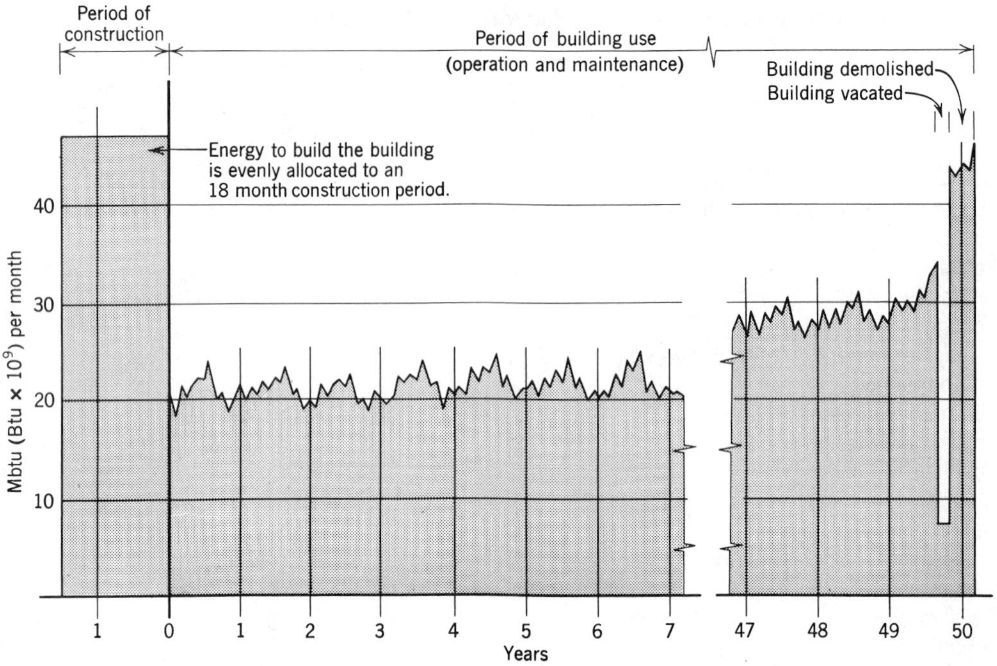

Fig. 1.25 *Energy use in constructing, operating, and demolishing an office building. This shows the approximate energy used over an estimated 50-year life span for a 650,000-ft² office building in Albany, New York. Energy in construction includes total energy embodied in the materials and is shown as constant through the construction period. The peaks and valleys of energy use in each year of operation correspond to the expected weather, peaks during the extremes of winter cold and summer heat, valleys in the milder spring and fall. As the building ages and energy leaks become more numerous, the consumption is assumed to rise gradually. (From* Architecture and Energy, *by Richard G. Stein. Copyright © 1977 by Richard G. Stein. Used by permission of Doubleday & Company, Inc.)*

Fig. 1.26 *Estimated energy for constructing and operating a residence based on a one-story, flat-roofed house with a 1500-ft² floor area, located in an area with about a 5000-degree-day heating season, near the band shown in the map* (a). *(Degree days are discussed in Chapter 5.) In areas north of the line, yearly energy consumed—and saved, by the more efficient house—would be even greater.*

	Energy Embodied	*Annual Energy Demand*	*Demand over 20 Years*
Type I	169 million Btu	109 million Btu	2180 million Btu
Type II	179 million Btu	77 million Btu	1540 million Btu
Difference between types I and II	10 million Btu more to build type II	32 million Btu less to operate type II yearly	640 million Btu less to operate type II for 20 years

The annual energy demand includes heat lost to cold air infiltrating the house. The extra energy used in building the 5½-in. insulation and the double glazing is quickly repaid—in about one-third of one heating season! [Based on Hannon et al. (1977).]

designers' choices. We can expect buildings to be placed not only to take advantage of natural energy sources, but to be so conscious of the energy they consume that significant efforts will be made to utilize formerly wasted heat. It is possible that even a single-family residence will be so thoroughly insulated and so well equipped to recover otherwise wasted heat that a furnace (or other space heating equipment) will become unnecessary. For many brightly lighted office buildings, this has already occurred.

1.7 Energy in Building Reuse or Demolition

Compared to construction and operation, relatively little energy is invested in a building's demolition (see again Fig. 1.25). Yet another look at Fig. 1.22 is a reminder that about 70% of the total energy invested in construction is embodied in a building's materials and components. If more of a building can be recycled, more energy can be recovered. At present, the

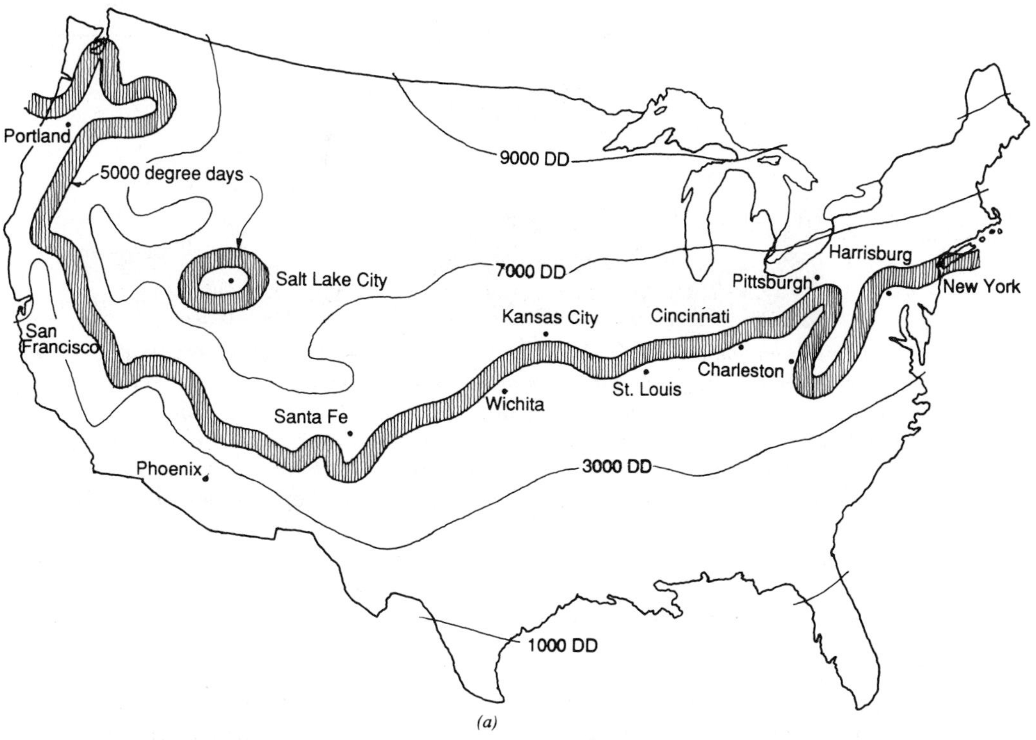

(a)

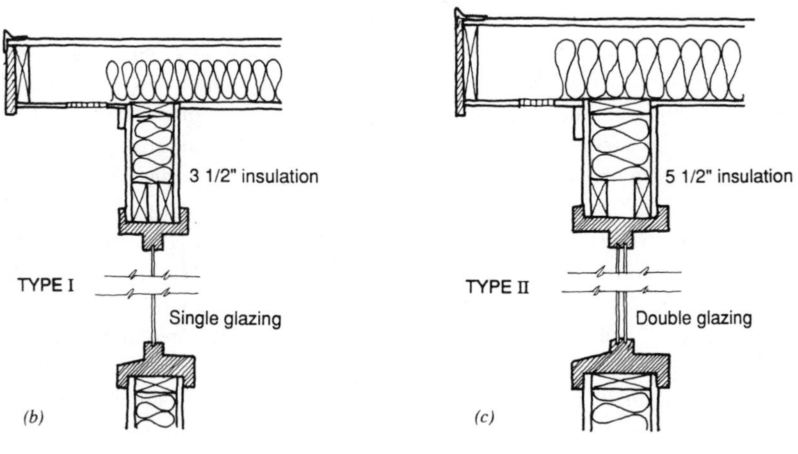

(b) (c)

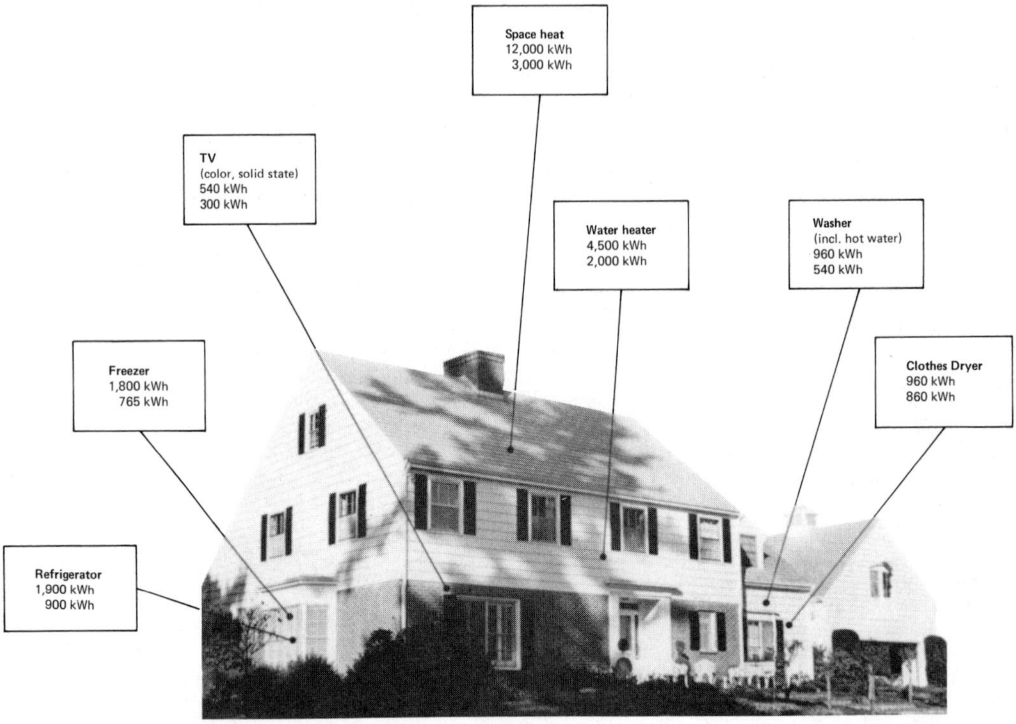

Fig. 1.27 *Appliance and equipment efficiency improvements in residences. Upper kWh figures are "before" and lower ones "after" the initiation of conservation measures. This is an example of energy conservation by "leak plugging," rather than "belt tightening"—the standard of living is not diminished. Space heat (typical for a rainy, cool Pacific northwest winter) reduction is due to difference between newer northwest energy codes, compared to those of the mid-1970s. Water heater reduction is due to the use of a heat-pump water heater (see Chapter 9). All other reductions are due to increased efficiency in appliances. When all other, more minor electricity uses are added in, the annual 1982 regional averages are 26,860 kWh; in the most efficient available home, 11,715 kWh—a reduction of 56%. (Courtesy of* Northwest Energy News, *Vol. 1, No. 4, July 1982, published by the Northwest Power Planning Council.)*

recovery of usable materials from demolition is limited. The cost of labor is high, and the cost of energy is relatively low; it is currently easier, quicker, and cheaper to reduce a building to rubble and haul it to a landfill. As landfill capacity becomes scarce, design regulations concerning recycled materials can be expected.

The purest recycling involves the reuse of a building component "as is." Some other opportunities include crushed wallboard as a replacement for lime as an agricultural land conditioner, pulverized wood as an aid to composting at sludge treatment facilities, and crushed concrete and glass for use in road construction. Building materials are now increasingly made from recycled materials; reinforcing bars from ferrous scrap metal, parking lot bumper strips, fenceposts and park benches from mixed recycled plastics, and nonstructural concrete from incinerator ash.

Designing for the recycling of buildings is a two-part balancing act. First, the designer should provide enough flexibility to prolong the useful life of the building by enabling it to easily adapt to changed usage. Flexibility can be expensive to implement physically and can result in a bland "sameness" throughout the structure. The latter characteristic is easier for the designer to change than the former. Second, the design can allow for demounting parts so that the structure can safely remain intact, while reusable materials and components are removed. Yet this can result in heavier buildings, where floor systems are not structurally integral with beams. It also discourages the integration of mechanical and structural systems, as is discussed in Chapter 6. Furthermore, a demountable building is perhaps especially subject to energy leaks, such as cracks widening around self-contained components of the facade.

Some initial design guidelines for recyclable buildings are as follows:

1. Design the structure to be separable from everything else and, of itself, to be easily disassembled. Extensive remodeling is then possible without major structural modifications, and at the end of a building's life, elements of its structure can be reused elsewhere.

2. Design for "breathing room" where possible, between a building and its neighbors, or between major spaces within a building. Some expansion is thus possible without rebuilding. This could include designing the columns and footings to support an extra floor or two, for vertical expansion.

3. Maximize the utilization of on-site (or natural) forces such as sun and wind. The less sophisticated the mechanical and electrical equipment, the less obvious will be the obsolescence of such equipment with the passing of time.

4. Use the materials and components distinctly; avoid combinations that make recycling of these elements difficult. A steel or plastic pipe embedded in a concrete slab is neither repaired nor recycled easily; some "sandwiches" or panels of building materials do not allow metals, plastics, and other products they contain to be separated for reuse at the end of the panel's life.

Although maximum savings of energy can be realized when a building component can be used "as is," even the crushing and reprocessing of some (separated) building materials will save energy when compared to their original manufacture when virgin material (see Chapter 12).

REFERENCES

Publications of the American Society of Heating, Refrigerating, and Air-Conditioning Engineers, Atlanta, Ga.:

ASHRAE Standard 55 is the series "Thermal Environmental Conditions for Human Occupancy."

ASHRAE Standard 90 is the series "Energy Conservation in New Building Design." (ASHRAE Standard 90.1–1989, for example, was published in 1989.)

Banham, R. (1969). *The Architecture of the Well-Tempered Environment,* University of Chicago Press, Chicago, and Architectural Press, London.

Berry, R., and Makino, H. (1974). "Energy Thrift in Packaging and Marketing," *Technology Review.*

California, Office of the State Architect. (1976). *Building Value, Energy Guidelines for State Buildings,* Sacramento, Calif.

Commoner, B. (1976). *The Poverty of Power: Energy and the Economic Crisis,* Knopf, New York (distributed by Random House, New York).

Federal Energy Administration (1976). *National Energy Outlook,* FEA, Washington, D.C.

Fisher, J. (1974). *The Energy Crisis in Perspective,* Wiley, New York.

Hannon, B. M., Stein, R. G., Segal, B. Z., and Serber, D. (1977). *Energy Use for Building Construction,* Center for Advanced Computation, University of Illinois, Champaign-Urbana, Ill.

Jewell, W. J. (1978). "Biomass Fuels—Past, Present, and Future," presented at the American Section Conference of the International Solar Energy Society, Denver, Colo.

Lowry, W. P. (1988). *Atmosphere Ecology for Designers and Planners,* Van Nostrand Reinhold, New York.

Lazslo, E., and Bierman, J., Editors (1977). *Goals in a Global Community,* Pergamon Press, Elmsford, N.Y.

Meyers, R. (1983). *Handbook of Energy Technology and Economics,* Wiley, New York, 1983.

National Geographic Society (1981). *Special Report; Energy,* National Geographic, Washington, D.C.

Nevins, R. G. *Criteria for Thermal Comfort,* Institute for Environmental Comfort, Kansas State University, Manhattan, Kans.

Starr, C. (1971). "Energy and Power," *Scientific American,* September.

Stein, R. G. (1977). *Architecture and Energy.* Anchor Press–Doubleday, New York.

U.S. Department of Energy, Energy Information Administration. *Monthly Energy Review.*

2 CLIMATE, COMFORT, AND DESIGN STRATEGIES

One of the earliest reasons for building was to create shelter from the climate; to enhance thermal comfort. This chapter introduces this interrelationship between bodies, buildings, and climate by discussing bodily heat flow, then thermal comfort, then the design strategies that are appropriate to various climates. The final topic is the role of building skin elements as they change the outside climate to the one inside.

2.1 The Body

Because we are alive, we are always generating bodily heat. And because the cores of our bodies need to stay within a narrow temperature range, we nearly always need to lose this internally produced heat to our environment. The rate at which we produce heat changes frequently, as does the environment's ability to accept or reject heat. To regulate our bodily heat loss, we have available three common layers between our body cores and our environment: the first skin, our own; the second skin, clothing; and the third skin, a building.

(a) Metabolism. The rate at which we generate heat (our metabolic rate) depends mostly on our level of muscular activity, partly on what we ate and drank (and when), and partly on where we are in our normal daily cycle. Our heat production is measured in metabolic or *met* units (see Table 2.1). One met is defined as 58.2 W/m^2, or 18.4 Btu/h·ft^2; it is the energy produced per unit of surface area, by a seated person at rest. (The total heat thus produced for the normal adult is about 117 W, or 400 Btu/h.) The more active we are, the more heat we produce, and our own first skin is the most important regulator of heat flow.

Of the many interactions between our skin and the rest of our body, consider these three: touch, blood, and water. The sensation of *touch* includes pressure and pain, as well as heat and cold. The sensations of heat and cold are produced by contact with surfaces or moving air, as well as by radiant heat. They are frequently the signals for shifts in bodily heat regulation, which is controlled by the thermostat in our brains, called the hypothalamus.

In response to signals from our surface, and to changes in our core temperature, the hypothalamus calls for changes in our *blood* distribution system. If we are too cold, we need to decrease our rate of heat loss, so the flow of blood toward the surface of our skin decreases. Blood carries heat; the less exposure to cool air at the surface, the less heat is lost. This decreased blood flow toward the surface is called vasoconstriction, and is triggered primarily by cold signals from our skin. In this condition, less *water* is forced to the skin surface by our sweat glands, which reduces evaporation (and accompanying heat loss).

If these cold conditions worsen, we get "goose bumps," symptoms of our skin's unsuccessful attempt to create insulation by fluffing up our body hair. Other furred creatures probably find our efforts amusing. Since the added insulation does not work, we soon increase our metabolic rate, or burn more fuel, by shivering, muscular tension, or increased muscular activity. At this point, we seek help from our second, then third skins of clothing and architecture.

The opposite occurs when we are too hot; first, blood flow toward the skin surface increases (vasodilation), triggered primarily by warm signals from our core. Then the sweat glands greatly increase their secretion of water and salt to the skin surface. This increases heat

TABLE 2.1 **Metabolic Rates for Typical Tasks**

Activity	*Metabolic Rate*[a] *(met units)*[b]	*Btu/h-ft*^{2} [c]
Resting		
Sleeping	0.7	13
Reclining	0.8	15
Seated, quiet	1.0	18
Standing, relaxed	1.2	22
Walking (on the level)		
0.89 m/s	2.0	37
1.34 m/s	2.6	48
1.79 m/s	3.8	70
Office activities		
Reading, seated	1.0	18
Writing	1.0	18
Typing	1.1	20
Filing, seated	1.2	22
Filing, standing	1.4	26
Walking about	1.7	31
Lifting/packing	2.1	39
Driving/flying		
Car	1.0–2.0	18–37
Aircraft, routine	1.2	22
Aircraft, instrument landing	1.8	33
Aircraft, combat	2.4	44
Heavy vehicle	3.2	59
Miscellaneous occupational activities		
Cooking	1.6–2.0	29–37
House cleaning	2.0–3.4	37–63
Seated, heavy limb movement	2.2	41
Machine work		
Sawing (table saw)	1.8	33
Light (electrical industry)	2.0–2.4	37–44
Heavy	4.0	74
Handling 50-kg bags	4.0	74
Pick and shovel work	4.0–4.8	74–88
Miscellaneous leisure activities		
Dancing, social	2.4–4.4	44–81
Calisthenics/exercise	3.0–4.0	55–74
Tennis, singles	3.6–4.0	66–74
Basketball	5.0–7.6	90–140
Wrestling, competitive	7.0–8.7	130–160

Source: Copyright © by the American Society of Heating, Refrigerating, and Air-Conditioning Engineers, Inc., Atlanta, Ga.; reprinted by permission from 1989 *Handbook of Fundamentals.*

[a]For whole-body average heat production in watts and Btu per hour, see Table 5.22.

[b]One met = 58.2 W/m^2 = 18.4 Btu/h-ft^2.

[c]5 ft 8 in. 154-lb male, surface area 19.6 ft^2.

loss by evaporation. As we get hotter and wetter, the roles of our second and third skins change. In a hot, humid environment, our first skin needs exposure to air to encourage heat loss, but protection from the sun's radiant heat; we need a simple second or third skin, which acts as a sunshade. In addition, thanks to mechanical equipment, our third skin can create and enclose a less humid body of air. In a hot, arid environment, our second skin may keep us from losing too much valuable water, while our third skin might help us with stored "coolth" from the often chilly night air. Both can also play the vital role of shading.

(b) Migration. An important design principle has been demonstrated here: that of *zoning*. Our bodies strive to maintain, at all costs, a nearly constant core temperature for our vital organs. This most-protected zone takes thermal precedence over the less-vital zone of our extremities, such as arms and legs; next down in priority are our fingers and toes. The further from our central body mass (fingers and toes) and the greater the surface area per unit mass (ears), the faster the temperature will drop in cold conditions. The most variable thermal zone of all is our skin surface. Similarly, buildings are frequently thermally zoned, and users (paralleling human blood flow) can retreat from—or advance to—the less-protected zones as conditions demand. On a larger scale, migration occurs from one climate to another as seasons change.

(c) Heat Flow. Once the blood and water get our surplus heat to the skin surface, we have four ways to pass it to the environment: *convection* (air molecules pass by our surface, absorbing heat), *conduction* (we touch cooler surfaces, and heat is transferred), *radiation* (when our skin surface is hotter than other surfaces "seen" but not touched, heat is radiated to these cooler surfaces); and *evaporation* (a liquid can evaporate only by removing large amounts of heat from the surface it is leaving). The amount of heat we lose by each of these four methods depends on the interaction of our metabolism, our clothing, and our environment; Fig. 2.1 illustrates the typical situation of a person at rest in a changing environment. As air and surface temperatures approach our own body temperature, we lose the options of convection, conduction, and radiation. Evaporation is essential, so access to dry, moving air is greatly appreciated. As air and surface temperatures fall, evaporation drops while convection, conduction, and particularly radiation increase. Under the "normal" conditions in Fig. 2.1 of about 70 F, the proportions of bodily heat loss per hour are:

Radiation, convection, and conduction	72%
Evaporation	
From skin surface	15%
From lungs (exhaled air)	7%
Warming of air inhaled to lungs	3%
Heat contained in feces and urine expelled	3%

(d) Clothing. Usually clothing acts as an insulating layer and is particularly effective at retarding radiation, convection, and conduction. As air and surface temperatures fall well below our own, we adjust the second skin. The insulating value of clothing is measured in Clo units, 1.0 Clo being equivalent to the typical American man's business suit in 1941, when Clo was born (see Table 2.2). The total Clo of what you're now wearing can be estimated by assuming 0.35 Clo per kilogram (0.15 Clo per pound) of your clothing. Our second skin is just as likely as our third skin to be dominated by considerations of style more than of thermal regulation; we can't always count on clothing—or buildings—to increase our comfort. A basic discussion of the third skin begins in Section 2.5.

2.2 Comfort

A positive definition of comfort is "a feeling of well-being." However, the more common experience of thermal comfort is a lack of discomfort—or being unconscious of how you are losing heat to your environment. There are three categories of factors that affect comfort: personal, measurable environmental, and psychological. Most *personal* (or physical) factors are under your control; they are the metabolism, migration, and clothing factors just discussed. *Measurable environmental* factors are the familiar tools of the designer and engineer: air temperature, surface temperature, air motion, and humidity. *Psychological* factors are also familiar designers' tools, but they are harder to measure for comfort: color, texture, sound, light, movement, and aroma. These psychological factors are often overlooked as we strive to meet the

Heat generated, Btuh	400	400	400	400	400 (curve 1)
Heat lost by:					
Radiation and convection	350	300	200	100	0 (curve 2)
Evaporation	50	100	200	300	400 (curve 3)
Total, Btuh	400	400	400	400	400 Total (curve 1)

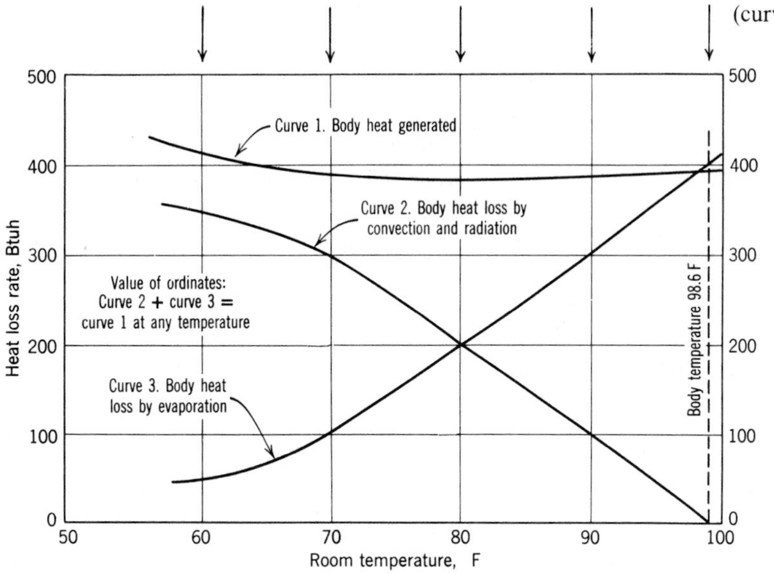

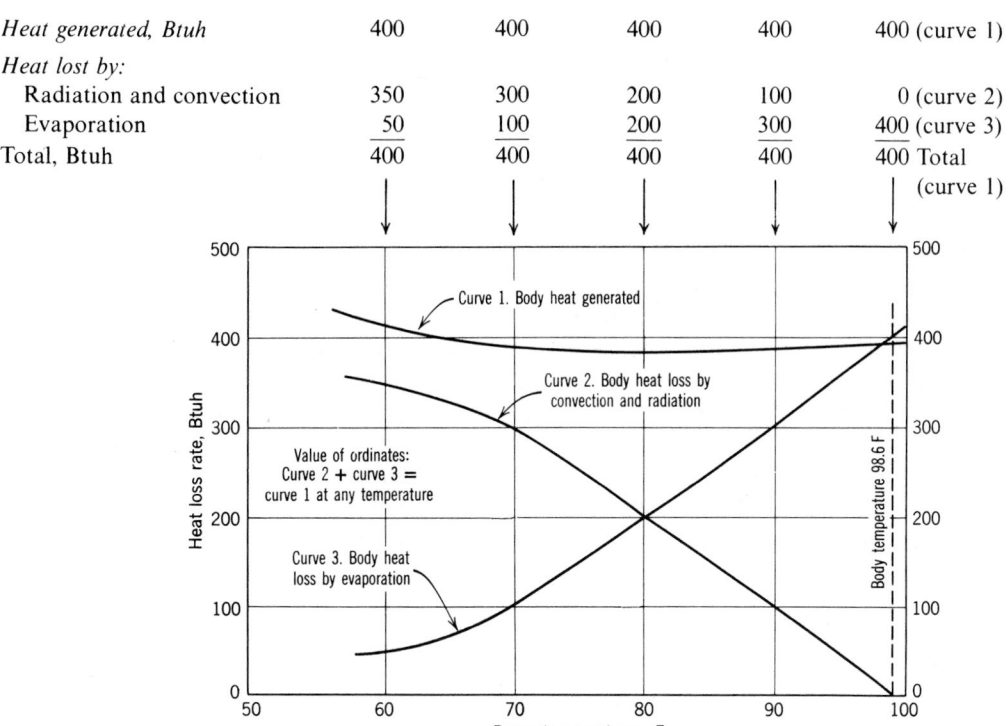

Fig. 2.1 Heat generated and lost (approximate) by a person at rest (rh fixed at 45%).

TABLE 2.2 **Typical Insulation Values for Clothing Ensembles**

Ensemble Description[a]	*Clo*[b]
Walking shorts, short-sleeved shirt	0.41
Fitted trousers, short-sleeved shirt	0.50
Fitted trousers, long-sleeved shirt	0.62
Same as above, plus suit jacket	0.96
Loose trousers, long-sleeved shirt, long-sleeved sweater, T-shirt	1.01
Loose trousers, long-sleeved shirt, long-sleeved sweater, suit jacket, long underwear bottoms, T-shirt	1.30
Sweat pants, sweat shirt	0.77
Knee-length skirt, short-sleeved shirt, panty hose (no socks), sandals	0.54
Knee-length skirt, long-sleeved shirt, full slip, panty hose (no socks)	0.67
Knee-length skirt, long-sleeved shirt, half-slip, panty hose (no socks), long-sleeved sweater	1.10
Ankle-length skirt, long-sleeved shirt, suit jacket, panty hose (no socks)	1.10
Long-sleeved coveralls, T-shirt	0.72
Overalls, long-sleeved shirt, long underwear tops and bottoms, flannel long-sleeved shirt	1.00

Source: Based on 1989 *Handbook of Fundamentals,* published by the American Society of Heating, Refrigerating, and Air-Conditioning Engineers, Inc., Atlanta, Ga.

[a]Unless otherwise noted, all ensembles include briefs or panties, shoes, and socks.

[b]One Clo = 0.88°F-ft²-h/Btu.

numerical physical criteria for comfort. However, our primary objective is to heat or cool *people;* buildings are our means to that end. Consider a courtyard in a hot climate. Its fountain *suggests* coolness in the color and texture of its water; running water provides splashing sounds and sparkles of light. Plants provide shade while their leaves sway in the slightest breeze; their blossoms yield a cool fragrance. Then consider a fireplace in a cold climate. The fire's color is intensely warm, it dances and casts a flickering light; it crackles, it yields a smoky aroma. Few textures *seem* hotter than that of glowing coals. Psychological factors can create an impression of comfort even when measurable factors indicate that mild thermal discomfort is expected. In our society, designers are often tempted to take numerical data quite literally. The measurable environmental factors have been extensively tested in laboratories—but such testing is done by *excluding* other factors. A holistic view of designing for comfort will consider numbers as a nonabsolute guide; common sense and the designers' own thermal experience play important roles as well. Lisa Heschong's *Thermal Delight in Architecture* (1979) is an excellent extension of these ideas.

Ultimately, our buildings will be expected to demonstrate success with regard to the measurable environmental factors of comfort, so it is necessary to understand how air and surface temperatures, air motion, and humidity are related to heat transfer.

Heat Transferred by:	*Is Primarily Dependent on:*
Conduction	Surface temperature
Convection	Air temperature, then air motion, then humidity
Radiation	Surface temperature
Evaporation	Humidity, air motion, air temperature

From this comparison, it is evident that humidity is of relatively low importance in cold conditions, where heat loss by conduction, convection, and radiation is dominant. But humidity is of primary importance in hot conditions, dominated by evaporative heat loss. This is further evident in comfort studies, which show that skin temperature is an important factor in cold conditions, while skin wettedness (percent covered by water) is of most importance in hot conditions.

A detailed analysis of a space's surface temperature can be performed using the *mean radiant temperature* (MRT). Such a calculation involves determining the temperatures of all surfaces and the position within the space at which to measure MRT, then determining the solid angle that is formed between the measuring position and the outer edges of each surface. These values are averaged to find the mean radiant temperature. However, a more useful term—both for impact on comfort and because it can be physically measured—is *operative temperature* (t_{op}), defined below. For details on MRT calculations, see Chapter 8 of the 1989 *Handbook of Fundamentals* published by the American Society of Heating, Refrigerating, and Air-Conditioning Engineers (ASHRAE).

The interaction between comfort and those environmental factors can be generally summarized in Fig. 2.2. The "comfort zone" represents combinations of air temperature and relative humidity that most often produce comfort for a seated North American adult in shirtsleeves (0.6 Clo), in the shade. Surface temperatures are not markedly different from air temperatures in this zone, and minimal air motion is assumed. However, at air temperatures *below* (to the left of) this comfort zone, comfort is still attainable *if* added radiant heat (increased surface temperatures) such as increased exposure to sun is provided. (Also, more activity and/or more clothing is a possible option.) Similarly, at air temperatures *above* (to the right of) this comfort zone, comfort is still attainable *if* added air motion is provided. (Less activity and clothing may also be an option.) In both cases, limits are eventually reached; but the important point is that the basic comfort zone can be extended by utilizing more sun or more wind and, in very dry climates, by adding more moisture to the air.

Designers of buildings can aim for specific comfort goals, especially where the measurable environmental factors are concerned. The following combinations of air and surface tempera-

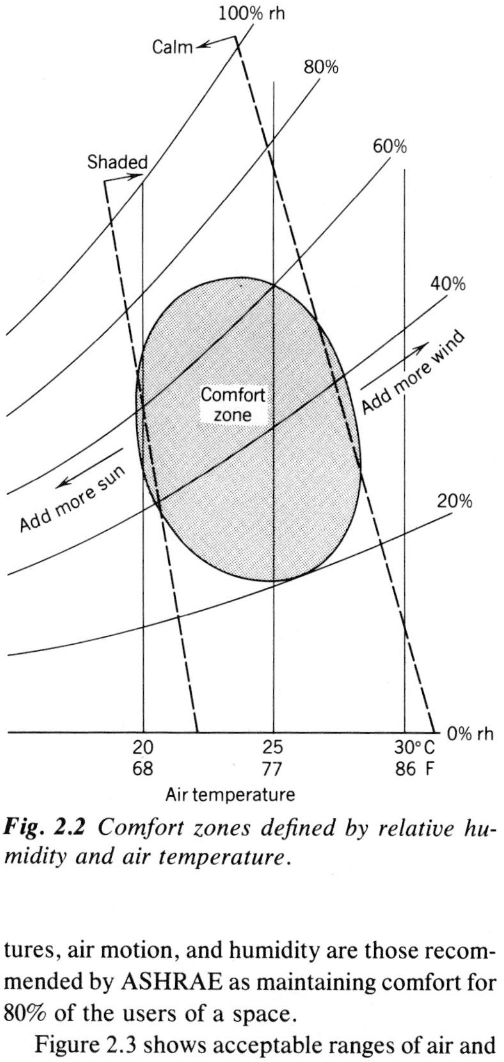

Fig. 2.2 *Comfort zones defined by relative humidity and air temperature.*

tures, air motion, and humidity are those recommended by ASHRAE as maintaining comfort for 80% of the users of a space.

Figure 2.3 shows acceptable ranges of air and surface temperatures and humidity for persons wearing typical clothing for both summer and winter. Air motion is considered later in this chapter. The terms used here are defined as follows:

Operative Temperature (t_{op}). Essentially an average of the air temperature of a space and the average of the various surface temperatures surrounding the space (or MRT). (To get a working grasp of operative temperature, get a 6-in.-diameter metal toilet float, paint it flat black, and drill a small hole in it. Then through a rubber stopper, insert an ordinary air-temperature thermometer about 5 in. long, so that its bulb is near the center of the toilet float. The thermometer will read approximate operative temperature; it will rise dramatically when passed from shade to sun; it will fall slightly in a breeze. Such a device is called a globe thermometer. Ordinary air temperature by itself is often not a good comfort indicator, *especially* in passively heated or cooled spaces where radiant temperature or air motion may be more influential than air temperature.)

Relative Humidity (rh). This familiar weather report term is approximately the ratio of the density of water vapor in air to the maximum density of water vapor that such air *could* contain, at the same temperature, if it were saturated.

The criteria for operative temperature-relative humidity combinations must be specific for combinations of activity and clothing, since such personal factors are just as important in determining comfort. Also, air motion must be included in any comfort standard. When using *any* standard, remember that conditions can change with the *age* of the users of buildings. Elderly people tend to be more sedentary, thus preferring warmer temperatures than do more active people. Another influence may be the degree of *acclimatization* that they have achieved. Very hot weather may be more bearable at the end of summer than at its beginning.

Figure 2.3*a* shows the ASHRAE comfort standard range for typical home or office activity and clothing levels, along with fixed air motion conditions. Numerically, these are as follows:

ASHRAE 55—1981:

Activity: 1.2 met, mainly sedentary

Winter: 0.9 Clo, heavy slacks, long-sleeved shirt and sweater. Air motion at maximum = 0.15 m/s (30 fpm). The lower the temperature, the less air motion is desirable. *Note:* Optimum operative temperature is 22.7°C (71 F).

Summer: 0.5 Clo, light slacks and short-sleeved shirt or blouse. Air motion is assumed at 0.25 m/s (50 fpm); however, this summer zone can be extended by increasing this air motion 0.275 m/s for each degree kelvin, up to a maximum temperature of 28°C at 0.8 m/s of air motion (or increasing this air motion 30 fpm for each degree Fahrenheit, up

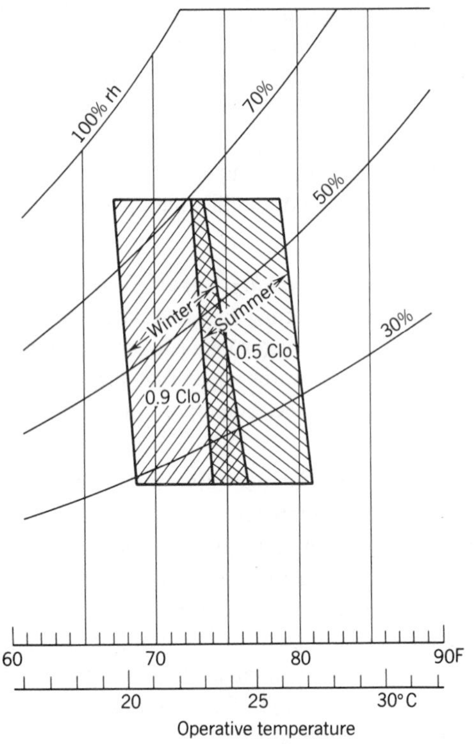

(a)

to a maximum temperature of 82.5 F at 160 fpm of air motion). At this point, loose paper, hair, and other light objects might start to be blown about (however, see Table 2.3). *Note:* Optimum summer operative temperature is 24.4°C (76 F). If minimal clothing is worm (0.05 Clo), this optimum operative temperature is 27.2°C (81 F).

The converging edges of the "comfort zones" in Fig. 2.2 show that less tolerance of hotter air temperatures is expected at higher relative humidities; it's harder to sweat successfully in a humid environment, because the moisture can't easily evaporate. The horizontal edges of these comfort zones are based on practical humidity limits as follows: *above* the upper edge, excessive indoor moisture with mold problems can be expected. *Below* the lower edge, respiratory discomfort is expected from excessively dry air; coughs and nosebleeds can result. (For those acclimatized to "extreme" conditions, such as discomfort is minimized.)

Figure 2.3*b* presents the relationship between

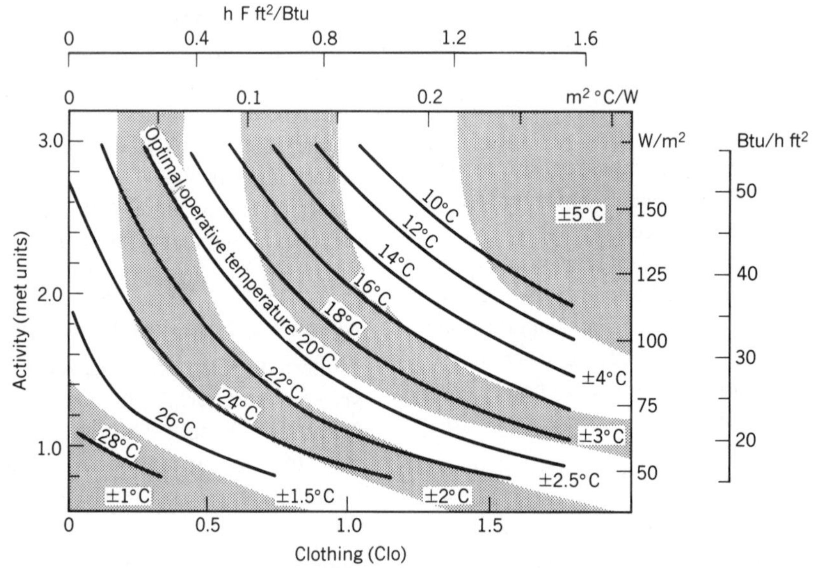

(b)

Fig. 2.3 (a). *The winter and summer comfort zones for light activity (1.2 met) that conform to ASHRAE Standard 55-1981. (Copyright © by the American Society of Heating, Refrigerating, and Air-Conditioning Engineers, Inc., Atlanta, Georgia; reprinted by permission from ASHRAE Standard 55-1981.)* (b) *Relationships between preferred (optimal) operative temperature, clothing, and activity. The shaded bands represent the degree to which departures from the optimal are tolerable; that is, nearly all people will find the departures only "slightly cool" or "slightly warm." (P. O. Fanger, Laboratory of Heating and Air Conditioning, Technical University of Denmark, Copenhagen.)*

TABLE 2.3 **Indoor Air Velocity and Comfort**

Air Velocity	*Possible "Lower Temperature" Comfort Sensation (between 80 and 90 F; larger numbers correspond to high-humidity areas)*	*Probable Impact*
Up to 50 fpm (0.25 m/s)	No change in comfort sensation	Unnoticed
50–100 fpm (0.25–0.51 m/s)	2–3 F lower (1.1–1.7°C)	Pleasant
100–200 fpm (0.51–1.02 m/s)	4–5 F lower (2.2–2.8°C)	Generally pleasant, but causing a constant awareness of air movement
200–300 fpm (1.02–1.52 m/s)	5–7 F lower (2.8–3.9°C)	From slightly drafty to annoyingly drafty
Above 300 fpm (1.52 m/s)	More than 5–7 F lower (2.8–3.9°C)	Requires corrective measures if work is to be efficient and health secured

Source: Adapted from Victor Olgyay, *Design with Climate: Bioclimatic Approach to Architectural Regionalism*, Copyright © 1963, Princeton University Press. Reprinted by permission.

clothing, activity, and comfort. It shows the expected tolerance of lower temperatures with increased clothing and activity; importantly, it also shows a greater tolerance of temperatures somewhat different from the optimum with increased clothing and activity.

A different approach to comfort standards has been proposed for buildings that take a more passive approach to heating and cooling. In these buildings, direct sun might add significantly to body warming in winter; strong air currents might be expected as a part of summer body cooling. Arens et al. (1980) demonstrated a wider range of comfort conditions, and considerably more tolerance for summer air motion, than the 1981 ASHRAE standard range. A summary of these proposals is shown in Fig. 2.4. Such graphs are called "bioclimatic" charts because of their interrelation of climate and human comfort factors. Where the users of buildings are expected to adjust to the wider temperature swings associated with passive buildings, these standards should be used. Some differences in the basic assumptions of Arens et al. from ASHRAE 1981 should be noted:

Activity: 1.3 met (ASHRAE, 1.2 met)

Winter: 0.8 Clo (ASHRAE, 0.9 Clo)

Summer: 0.4 Clo (ASHRAE, 0.5 Clo)

Note particularly that rather than "operative temperature," ordinary air temperature is charted here. Added radiant heat below the shading line can be utilized to extend the comfort zone, as can added air motion above the minimum air motion line. *Radiant heat* to be added to an extended comfort zone in Fig. 2.4 is shown in two quantities: effective radiant field (ERF) is a measure of the net radiant heat flux to the body from all surfaces at temperatures other than air temperature. However, a more convenient figure for designers may be the total solar radiation (or *insolation*) on a horizontal surface, termed I_{TH}. Such values are more readily available, and the added-radiation lines of Fig. 2.4 represents I_{TH} converted to its approximate radiant impact on a human body's surface area, when the sun is at an altitude of 45°. Arens et al. (1980) give further details on the effects of solar radiation on comfort (e.g., at other sun altitudes or for other activity and clothing combinations).

Added air motion is shown for a wide range of velocities; while Table 2.3 indicates that people easily tolerate air motion outdoors up to about 3 m/s (about 600 fpm), indoor studies have indicated 2 m/s (about 400 fpm) maximum for overhead fans. Different comfort zones are also to be expected in other cultures and climates; tolerance for heat is higher in the always-hot cultures of the tropics; cold is much less bothersome to those acclimated to Arctic conditions. Proper clothing is highly influential in climatic adaptation.

Although the designer can select a range of air and surface temperatures and relative humidity to achieve average comfort conditions in a

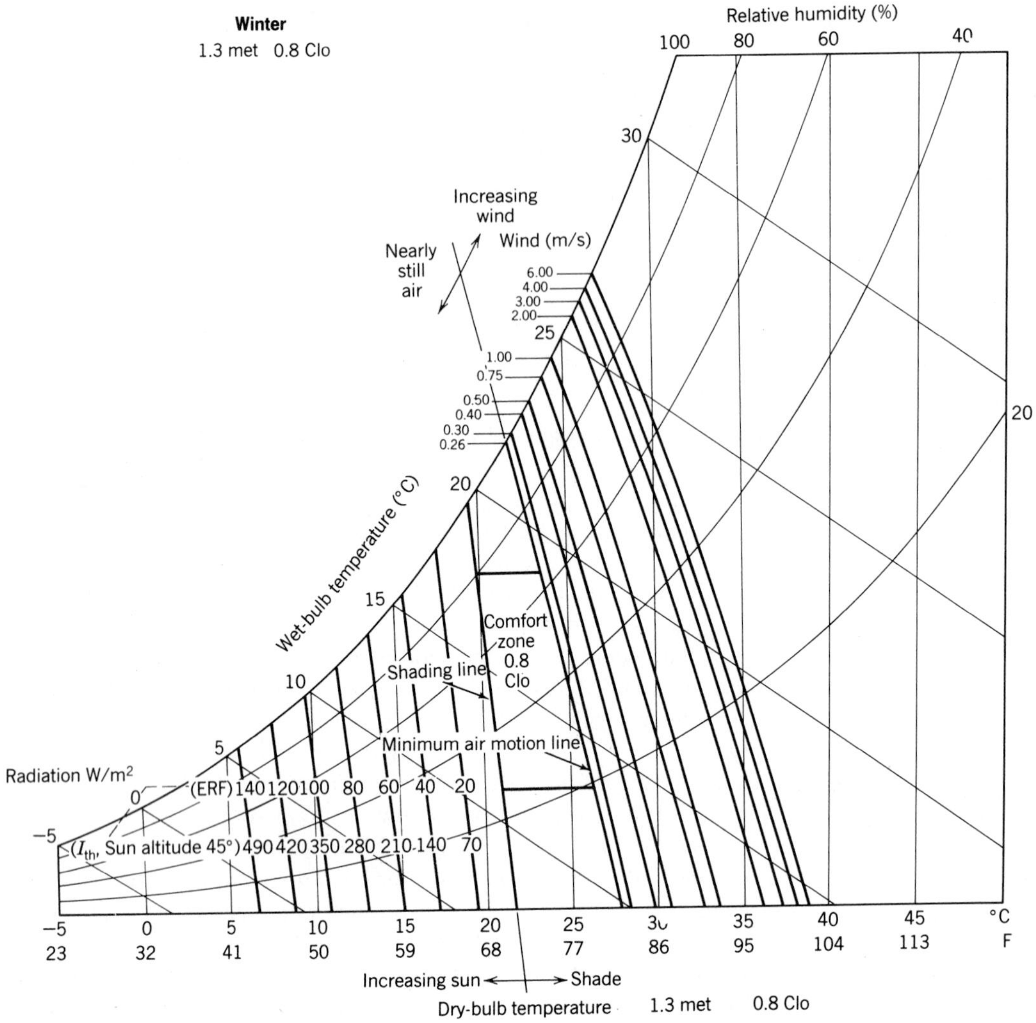

(*a*)

Fig. 2.4 *Comfort zones that encourage passively heated and cooled buildings. The winter comfort zone* (a) *and the summer* (b) *are combined for the year-round zone shown in* (c). *[Based on Arens et al. (1980).]*

space, there remains the more personal matter of comfort at each work station or location where people spend longer time periods. The location of heating, cooling, and ventilating components is therefore an important detail. A closer look at the body, air and surface temperatures, and air motion and humidity follows.

The human body is most affected by its thermal environment in its thermally sensitive places. Although we sweat and are sensitive to heat or cold over nearly all our skin surface, we are most thermally sensitive due to concentrations in these places:

Heat receptors: Fingertips, nose, elbows.

Cold receptors: Upper lip, nose, chin, chest, fingers.

So, on a hot and humid day, cool air moving across the face is a particularly strong promise of comfort, while on a cold day, a burst of heat (such as radiant heat from a light, a heater, or a cup of coffee) to the face and fingers is quickly effective.

A note here about our sense of touch: our fingertips are really most sensitive to the *rate* at which heat is being *conducted* to colder objects,

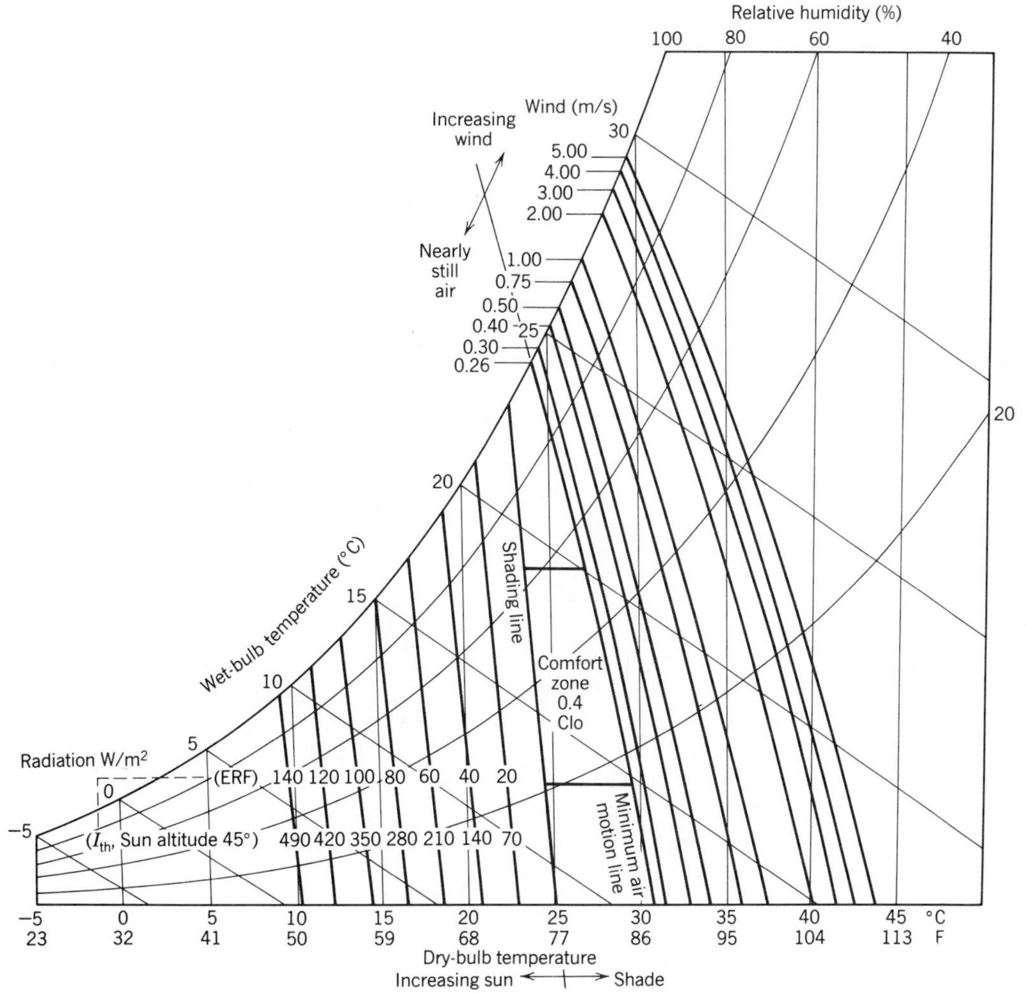

(b)

***Fig. 2.4** (continued)*

or from hotter objects. Since the temperature of our skin at fingertips in ordinary conditions is in the high 20s on the Celsius scale (high 80s Fahrenheit), our sense of touch works *against* many passively heated surfaces, which feel "cool" even though they are successfully warmer than room air temperature. These surfaces are made of materials that conduct heat rapidly, so that they can soak up solar radiation without overheating the room. This high conductivity makes them eager to accept human warmth as well, and persuades us that they are cooler than in fact they are. Everyone has walked barefoot from a rug to a tile floor; even if a surface temperature thermometer shows the same temperature for both rug and tile, our feet will signal "colder" for the tile. This same characteristic works *for* passively *cooled* surfaces; their conductivity persuades us that they are even cooler than our passive design has in fact achieved.

Since warm air is less dense than cold air, it rises; this principle is often applied in locating heating and cooling sources within spaces. Hot air outlets are placed near the floor, so that the hottest incoming air can induce cooler air at the floor to mix with it and rise slowly to warm the center of spaces. Cold air outlets are placed in the ceiling or high in walls, so that the reverse

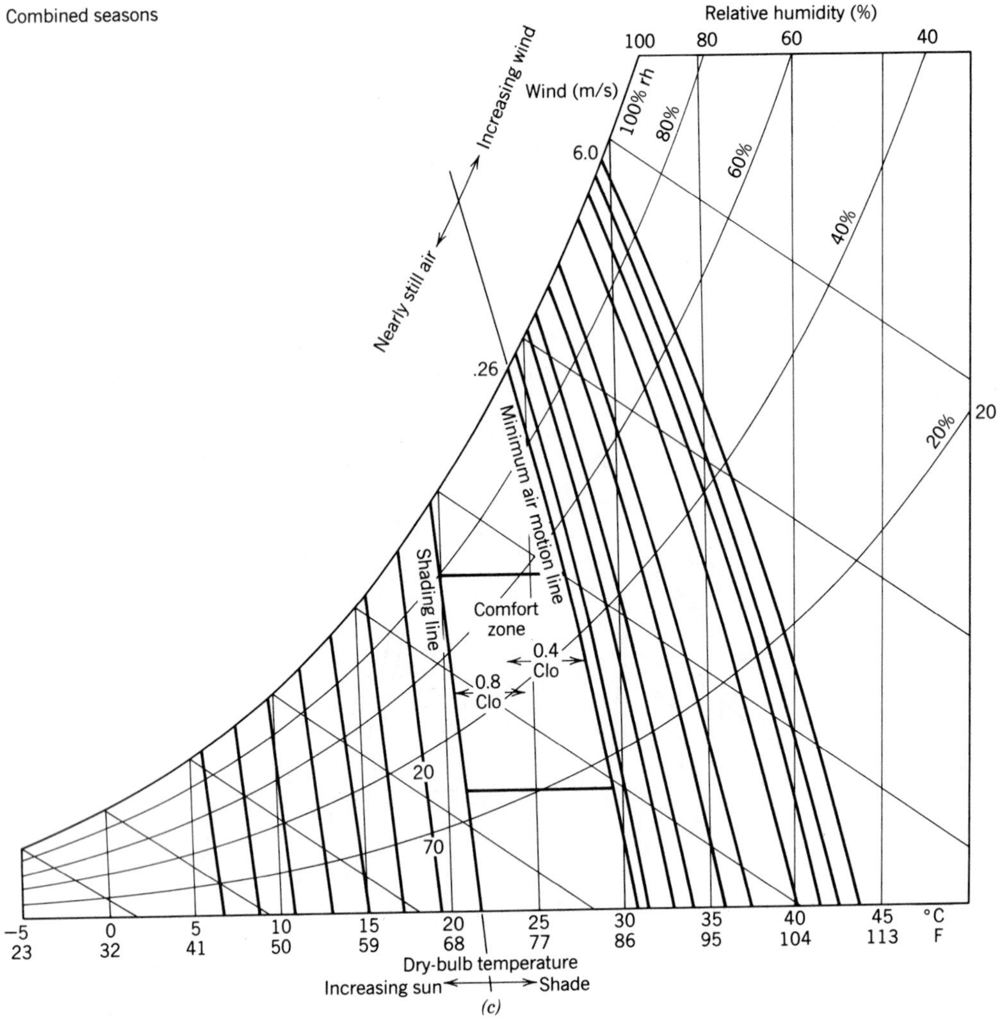

Fig. 2.4 *(continued)*

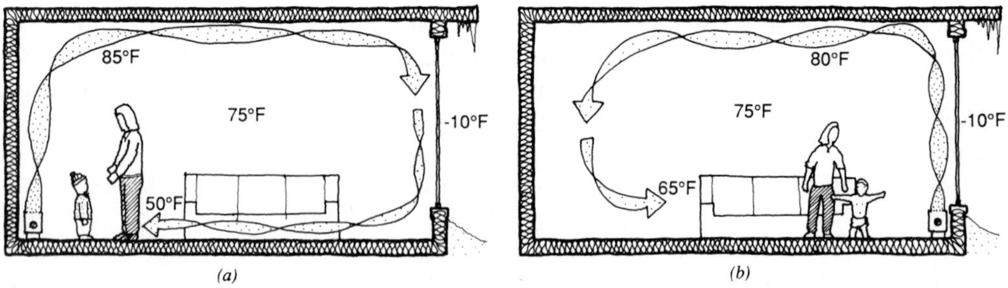

(a) (b)

Fig. 2.5 Dealing with outdoor conditions. Convection is inevitable. It can work for or against you. The temperatures are approximate. (a) *The stove is not merely in the wrong place. It accelerates the "downslip" of cold air from the glass.* (b) *The convector strip moves air up to warm the glass and provides local radiant warmth. The location of wood stoves should be carefully chosen.*

procedure can occur. Again, more detailed thermal factors enter in; Fig. 2.5 shows how heat sources near the floor need also to be near colder exterior surfaces to avoid unpleasant drafts. The overall goal is usually to provide an evenness in the thermal environment.

There is a trade-off here between thermal comfort and energy conservation. Moving, warmed air entering the space below a cold window surface raises the heat loss rate through the window both by raising the window interior surface temperature and by disturbing the insulating layer of still air against that inner surface. (Heat flow rate is explained in Chapter 4.)

Another comfort factor is *radiant asymmetry,* a difference in surface temperature between one surface and all the other surfaces in a space. Figure 2.6 shows which surfaces are better tolerated when of a different radiant temperature. These experimental results favor warmer walls (typical in passive solar buildings with thermally massive interior walls in sun) and cooler ceilings (typical of roof pond cooled spaces). They are less favorable to cooler walls (interior window surfaces in winter) and to warmer ceilings (radiant ceiling heat).

Thermal comfort is not the only comfort issue; air quality is of increasing importance and is discussed in Section 2.8.

2.3 Climates

Our most familiar names for climates describe their most severe season, as shown in Figure 2.7. This is a convenient description, but can be misleading for designers. "Cold" climates can have very hot, sometimes humid summer days; hot-arid climates can have bitterly cold winter conditions. Before designing the buildings that modify exterior conditions to provide indoor comfort, we should know when and how much of this modifying is appropriate.

We can utilize the bioclimatic chart of Fig. 2.4 to compare specific climate information with human comfort. Table 2.4 shows the kind of information available at every local climatological station maintained by the National Oceanic and Atmospheric Administration (NOAA) of the U.S. Department of Commerce. These climate summaries, often called LCDs, are available from the National Climatic Data Center (Federal Building, Asheville, NC 28801). In Fig. 2.8, this average information for Dodge City, Kansas, is graphed on the bioclimatic chart.

EXAMPLE. Compare the coldest and hottest average months for Dodge City, Kansas, to the basic comfort zone.

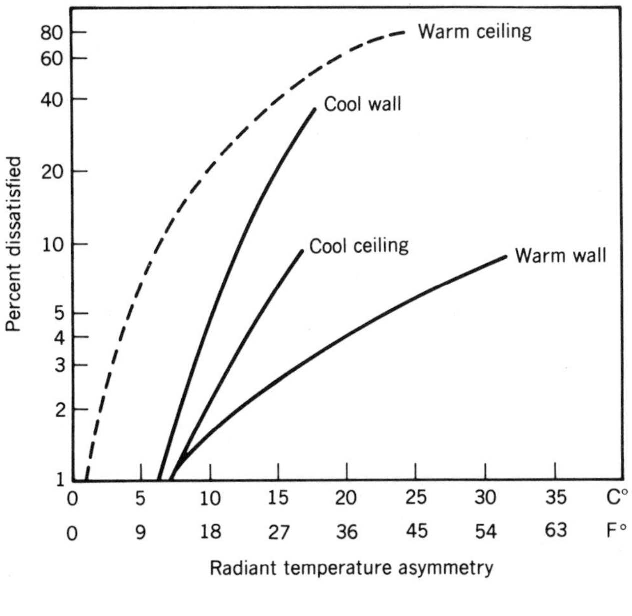

Fig. 2.6 Percentage of people expressing discomfort due to asymmetrical radiation. (Copyright © by the American Society of Heating, Refrigerating, and Air-Conditioning Engineers, Inc., Atlanta, Georgia. Reprinted by permission from the 1989 Handbook of Fundamentals.*)*

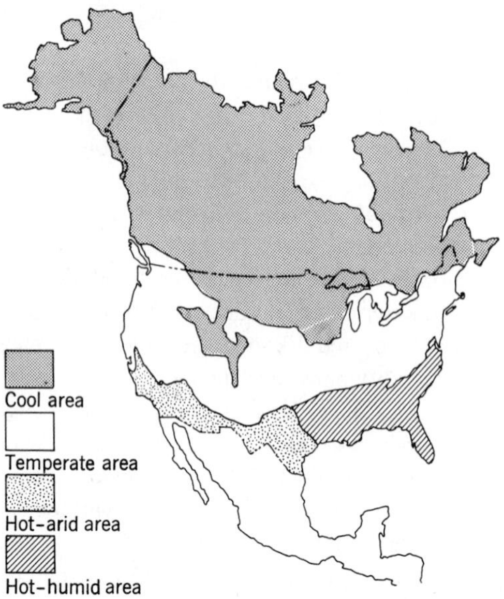

Fig. 2.7 *Regional climate zones of the North American continent. (From Victor Olgyay,* Design with Climate: Bioclimatic Approach to Architectural Regionalism, *Copyright © 1963 by Princeton University Press. Reprinted by permission.)*

SOLUTION. By inspection of Table 2.4, the coldest month is January. To chart the typical January day, find the approximate relative humidity (rh) for the coldest and warmest hours. Since rh is listed only at four times, select the probable coldest hour (usually the highest rh), which is 6 A.M.; and the warmest hour (usually lowest rh), which is noon. Then plot these combinations for January:

42.6 F, 58% rh (at 12 noon)

19.0 F, 76% rh (at 6 A.M.)

Similarly, the hottest month is July:

91.4 F, 42% rh (at 6 P.M.)

66.9 F, 70% rh (at 6 A.M.)

These plots, and the average-day line that connects them, are shown in Figure 2.8.

From the analysis in Fig. 2.8, it is evident that outdoor conditions in the coldest months are not normally within even the sun-extended comfort zone. (Remember, however, that the clothing and activity levels assumed are indoor, sedentary ones.) In the hottest month, most of the cooler half of the normal day falls within the (shaded) comfort zone, although some hours are very humid; in the hotter half of this day, comfort outdoors can be extended by adding air motion (as needed to about 2 m/s). From these averages of the "worst" months, it appears that Dodge City will need well-enclosed buildings in winter, while air motion at 2 m/s (about 4½ mph) seems achievable within buildings where the average summer wind speed is almost 13 mph. Remember that Dodge City is described as a "temperate" climate (Fig. 2.7), despite its *average* range of 19 to 91.4 F!

But what about extreme, rather than average, conditions and their impact on design? How much of the rest of the year is normally within the comfort zone? How many summer design strategies besides open windows might be ap-

TABLE 2.4 **Normal Data from Annual Summary[a] of Local Climatological Data for Dodge City, Kansas**

	Daily Temperatures (F)				*Relative Humidity (%) at Hour*				*Wind*	
	Normal		*Extreme*							
Month	*Max*	*Min*	*Max*	*Min*	*00*	*06*	*12*	*18*	*Mean Speed (mph)*	*Prevailing Direction*
December	44.6	22.2	86	−7	72	76	56	60	14.0	South
January	42.6	19.0	78	−12	72	76	58	59	13.6	South
February	47.1	23.2	85	−15	70	75	54	52	14.1	North
June	86.0	61.4	108	41	63	75	44	38	14.4	South
July	91.4	66.9	109	47	66	78	46	42	12.9	South
August	90.4	65.7	107	47	71	80	50	46	12.7	South

[a]Records as of 1979.

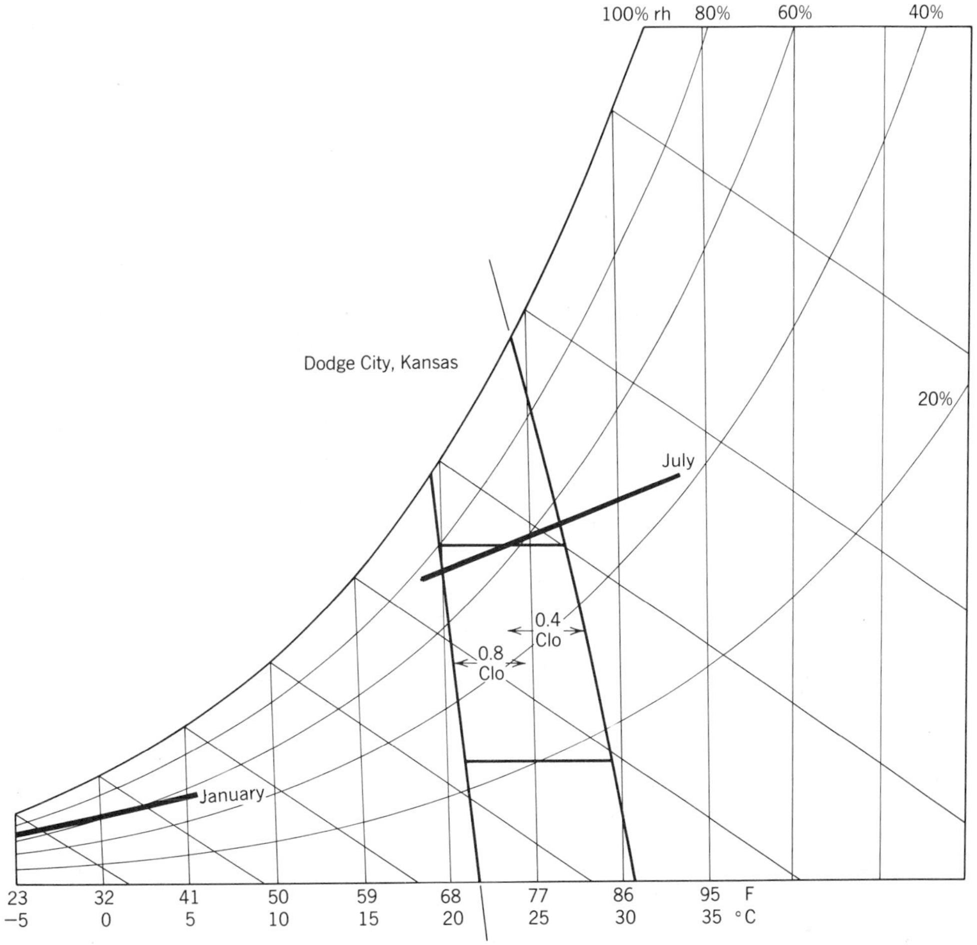

Fig. 2.8 Comfort zone, with the hottest and coldest months' daily ranges, for Dodge City, Kansas.

propriate, and how reasonable is solar heating in winter?

2.4 Design Strategies

The "building bioclimatic chart" shown in Fig. 2.9 was developed by Milne and Givoni (1979) to relate climate and design strategies. The average monthly data (as in Table 2.4) can be plotted here, thus revealing some appropriate strategies. Although the edges of each strategy zone are shown as lines, these boundaries are much more broad and vague than such lines suggest. (See Milne and Givoni for a discussion of how these boundaries were drawn.)

Before discussing these design strategies, a further elaboration on the bioclimatic chart has occurred. What we have called ordinary air temperature is now called *dry-bulb* temperature; and a new set of lines called *wet-bulb* temperatures have appeared. These terms are derived from the instrument shown in Fig. 2.10, the sling psychrometer.

Dry-Bulb Temperature (DB). The temperature of the ambient mixture of air and water vapor measured in the normal way with a simple thermometer.

Wet-Bulb Temperature (WB). The temperature shown by a thermometer with a wetted bulb rotated rapidly in the air to cause evaporation of its moisture. In dry air the moisture evaporates and draws heat out of the thermometer to produce a large *wet-bulb depression* (difference between dry- and wet-bulb temperatures). This is an index of low relative humidity. Slow evapora-

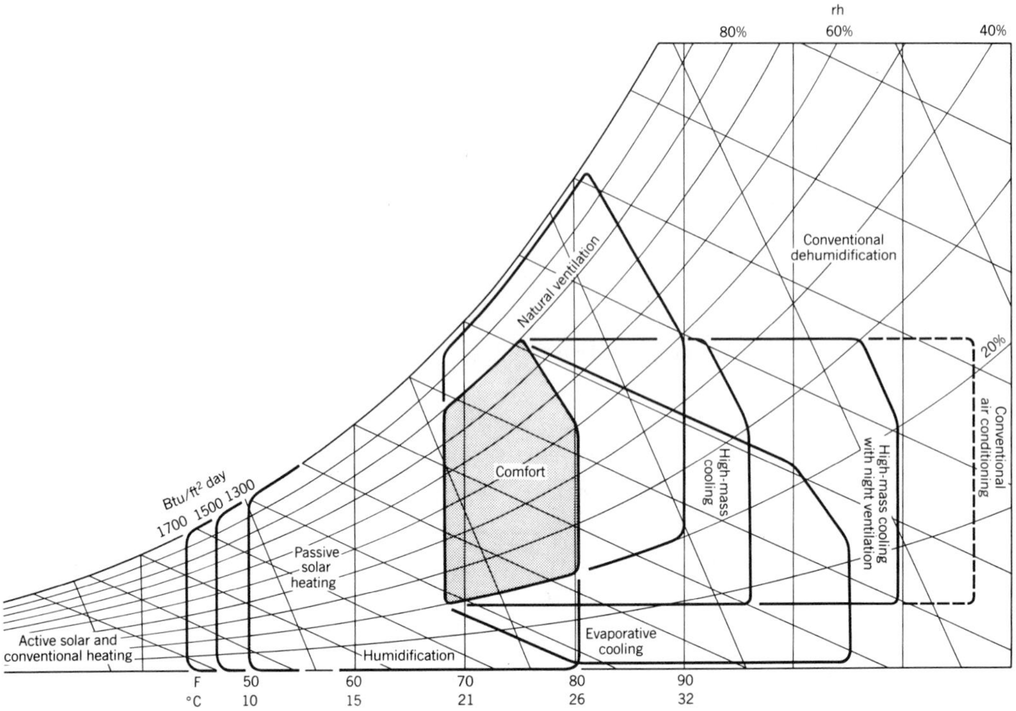

Fig. 2.9 *Design strategies by climate. Buildings usually contain sources of heat. The more heat occurs within the building, the more an* artificially warmer *climate is created. Thus, after plotting outdoor climate data on this chart, consider how* shifting *these plots would affect your choice of design strategy. The more:*

- *Solar gains allowed inside*
- *Electric lights*
- *Business machines, etc.*

the more your plots move horizontally to the right. The more:

- *People*
- *Cooking, bathing*
- *Other heat-plus-moisture sources*

the more your plots move up and to the right, following the rh curves. (Reprinted from Milne and Givoni, in Energy Conservation Through Building Design, *edited by Donald Watson, with the permission of McGraw-Hill, Inc.)*

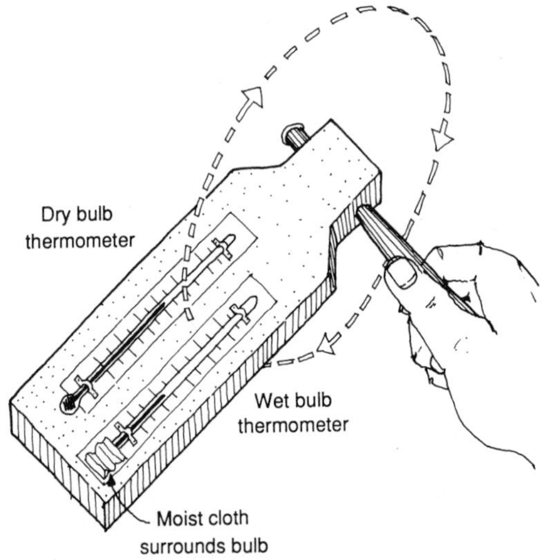

Fig. 2.10 *Sling psychrometer and its usage. Air motion encourages evaporation from the moist cloth, lowering the wet-bulb temperature while the dry-bulb temperature stays constant.*

tion when the air is already moisture laden results in a small wet-bulb depression and indicates a condition of high relative humidity.

The wet-bulb lines have many uses, which become more evident when discussing evaporative cooling; for now, they are added because climate data often are given in terms of WB rather than rh. Note that at 100% rh, DB and WB are equal. The design strategies shown in Fig. 2.9 include the following.

(a) Active Solar Heating and Conventional Heating. These are the familiar heating methods that can concentrate heat (from burning fuel, a heat storage tank, electric resistance coils, etc.) on demand. They are frequently used as backup heat to passive solar systems, as well as stand-alone systems.

(b) Passive Solar Heating. The collection and storage of solar heat is integrated within the surfaces of the building. Large (optimum) areas of south-facing glass admit winter sun, which strikes surfaces of high thermal mass (brick, quarry tile, concrete, water containers, etc.). In colder climates, thermal shades or shutters protect the windows against nighttime heat loss. The lines of required insolation shown in Fig. 2.9 indicate the point above which 100% of space heating by passive solar energy is likely; below these lines, some backup heat will probably be needed. These insolation levels may be checked for your approximate latitude in Appendix B.

Passive solar heating systems encourage large areas of south glass, with much smaller windows at all other orientations. These systems are described and compared in Table 5.15, which also describes their potential for integration with daylighting and cooling design strategies. A rule-of-thumb sizing procedure is given in Section 5.3*a*.

(c) Humidification. In this infrequently occurring zone, moisture should be added to indoor air to avoid respiratory problems and to reduce static electricity.

For the following cooling strategies, the exclusion of direct sun, by shading devices, is assumed.

(d) High-Mass Cooling. This is an especially useful strategy in warm, dry climates, where the extremes of hot days are tempered by the still-cool thermal mass of the building. Cool nights slowly drain away the heat that such mass accumulates during the day. The thermal mass can be on floors, walls, or roofs; the roof has the advantage of radiating to the cold night sky, but should, like all other thermal mass in these conditions, be protected from the sun by day. As in all these cooling strategies, daylighting is usually designed to occur without direct sun, since solar space heating is unneeded in these conditions.

Roof ponds are a form of high mass cooling for one- or two-story buildings. Because they involve only the roof, they allow for considerable design freedom in walls and fenestration. At lower latitudes with high winter sun altitudes, they can be used for passive solar heating as well. Roof ponds are described in Figs. 6.26 and 6.27. A rule-of-thumb sizing procedure is given in Section 5.3*c*.

(e) High-Mass Cooling with Night Ventilation. This hot-dry climate design strategy must use the air at comfortable nighttime temperatures to flush away the heat stored in daytime. The fewer the comfortable night hours, the more thermally massive surfaces must be provided to store the day's heat. Also, ventilation must occur more quickly and thoroughly, perhaps with fans. The building switches from a closed condition by day (to exclude sun and hot air) to an open one at night (to allow ventilation to cool the mass). *Note:* Nighttime outdoor temperatures should be *cooler* than the comfort zone, if this strategy is to be effective.

This cooling strategy is highly compatible with passive solar heating strategies that rely on large areas of thermal mass, such as *direct gain*. It is suitable for large, high buildings, particularly those that have concrete (thermally massive) structural systems. An example of this strategy is given in Fig. 5.4. A rule-of-thumb sizing procedure is given in Section 5.3*c*.

(f) Natural Ventilation Cooling. This is the most obvious strategy suggested by the comfort charts presented earlier, where higher air temperatures were offset by increased air motion. It is most appropriate in more humid, hot climates where temperatures are only slightly lower by night than by day. Buildings should be very open to breezes while simultaneously closed to direct sun. They may be thermally lightweight as well, since night air is not cool enough to remove much stored daytime heat.

Because this cooling strategy relies on rather narrow plans with large ventilation openings on either side, it is naturally compatible with daylighting. However, when ventilation openings face east and west, daylighting problems occur. Some components used in natural ventilation are shown in Figs. 6.29, 6.30, and 6.31. A rule-of-thumb sizing procedure is given in Section 5.3*c*.

(g) Evaporative Cooling. This design strategy relies on the principle that when moisture is added to air, the air *increases* in relative humidity while *decreasing* in dry-bulb air temperature. (On the bioclimatic chart, this pattern exactly follows the WB line, upward to the left.) In conditions that are more uncomfortably dry than uncomfortably hot, higher humidity is gladly exchanged for lower air temperature. However, large quantities of both water and outdoor air are needed; fan-driven evaporative coolers (see Section 6.7) are the most common way to provide this kind of cooling.

Evaporative cooling imposes few constraints on designers, as the equipment resembles conventional HVAC systems. Rather high indoor air velocities and their associated sounds are typical of these systems. Evaporative cooling equipment is shown in Fig. 6.69. A rule-of-thumb sizing procedure is given in Section 5.3*c*.

(h) Conventional Dehumidification and Air Conditioning. These are the familiar air conditioning methods, which rely on machinery and can cool on demand. Buildings utilizing such equipment should be very closed to both wind and direct sun. Thus, conventional air conditioners can be used as backup to passive systems, which imply closed buildings, such as high mass with night ventilation. Where evaporative coolers are used, conventional air conditioners can also be used to back up their cooling.

Returning to the analysis of the climate of Dodge City, the winter maximum of about 43 F, 58% rh indicates that passive solar heating will probably need a backup source almost all day in the coldest month. The summer night minimum of about 67 F, 78% rh is slightly below the comfort zone, while the daytime maximum of about 91 F, 42% rh allows several possible strategies: high thermal mass, high mass with night ventilation, and (only slightly beyond) natural ventilation. To help in making this choice, consider two more factors: temperature extremes and averages for the other hot months.

Temperature extremes are available from sources such as shown back in Table 2.4. A more useful guide, however, is the *design conditions* information provided in Appendix A. This condition shows the climate near its "worst," while avoiding the freakish temperatures found in the all-time record highs and lows. In Fig. 2.11, this design condition is plotted on the building bioclimatic chart, along with normal conditions for the other two hot months. The design condition indicates high thermal mass to be appropriate; simple natural ventilation is not sufficient (although nighttime ventilation could be useful). Normal conditions for other hot months, however, reinforce the option of simple natural ventilation, along with the options of high mass and night ventilation.

Building function is a potentially large influence on the choice of design strategy. By how much will internally generated heat produce an artificially warmer climate? A Dodge City office building that excludes daylight and utilizes 100% electric lighting may shift its indoor climate plot so dramatically that 100% passive solar heating is possible in winter, and high mass with night ventilation is strained to (or beyond) its limit to provide summer cooling. Such an approach would eliminate both simple natural ventilation and simple high thermal mass as strategies, introducing the probability of conventional air conditioning. An office building that uses only daylight (excluding direct sun except in winter) and allows for the use of outdoor conditions when appropriate would not appreciably shift its indoor plot. A more detailed treatment of the daylight–heat–cool tradeoff for various buildings and climates is given in Section 5.1.

The actual choice of design strategy will be influenced by many factors in addition to climate and function. These include the peculiarities of the site, first cost versus lifetime costs, the desired architectural characteristics of the building's envelope and its internal spaces, and simple personal preference. A closer look at climate could also help in choosing a design strategy.

Climate varies not only from day to day but *during* the day as well. Many buildings are not expected to provide comfort conditions all the

Fig. 2.11 Hot month daily ranges and design condition, for Dodge City, Kansas, superimposed on design strategy diagram (from Fig. 2.9).

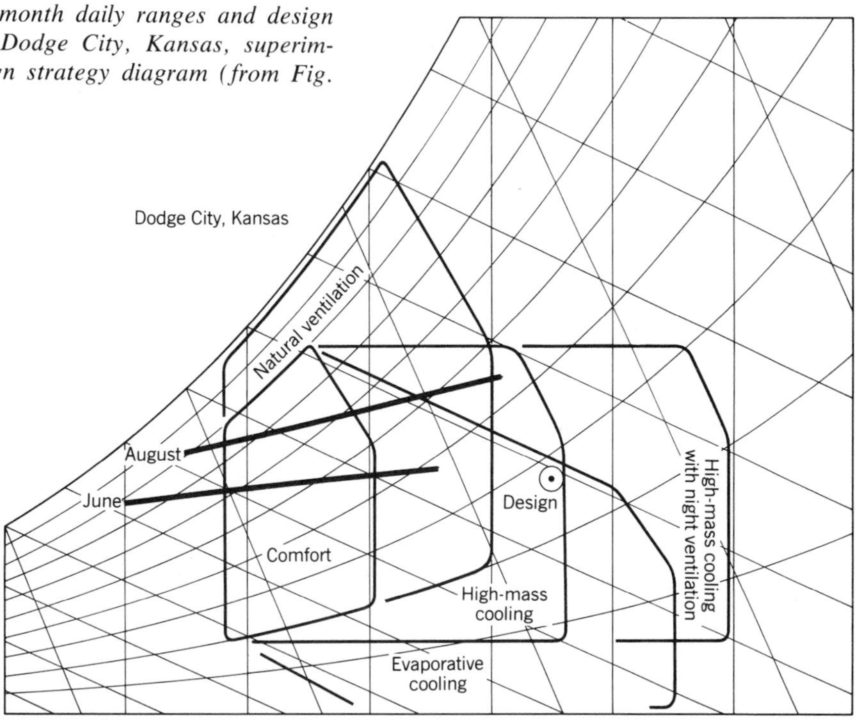

time; office buildings, for example, are used almost entirely within a 9- or 10-h period centered around noon. A view of climate as a daily pattern, linked to hours of operation, can reveal a picture quite different from the "monthly average day" that has so far been considered here.

An example of such an approach to climate analysis was devised by Olgyay (1963), shown in Fig. 2.12. The isolines correspond roughly to the comfort lines shown earlier in Fig. 2.4*a*; those labeled "100, 200, and 300" are Btu/h-ft^2 of radiation needed to remain comfortable outdoors, and correspond roughly to dry-bulb temperatures of 61, 53, and 45 F, respectively, though the higher the relative humidity, the lower the dry-bulb temperature to which these isolines correspond. The "shading line" corresponds to about 70 F at lowest rh, to about 68 F at highest rh. The darker shaded ("overheated") zone is outlined by an isoline that begins at about 82 F at lowest rh, to about 68 F at highest rh, indicating a need for air motion as well as shade to remain comfortable outdoors. Further isolines within this "overheated" zone correspond to a thermal comfort need for 300- and 700-fpm air motion; from Table 2.3 it is evident that even at 300 fpm, excessive air motion for most work situations is likely.

The timetable for New York City shows that about one-fifth of the year needs shade outdoors, with most hours of mid-July to mid-August in the overheating period; little relief is expected by night in summer. Perhaps one-third of the year is "too cold," at temperatures too low for shirtsleeve thermal comfort outdoors (needing more than 300 Btu/h-ft^2 radiation).

The timetable for Phoenix, Arizona, shows that about one-half the year needs shade outdoors, and most of that time is also in the overheated period. Temperatures are so extreme that over 700-fpm wind (unrealistic for most activities) is needed about one-eighth of the year. Like New York, there is little relief on summer nights, although the daily temperature range is much greater in arid Phoenix. This high daily range is evident throughout the year.

The timetable for Miami, Florida, shows that about four-fifths of the year needs shade outdoors, and about three-fifths is in the overheated period. Again, there is a large portion needing

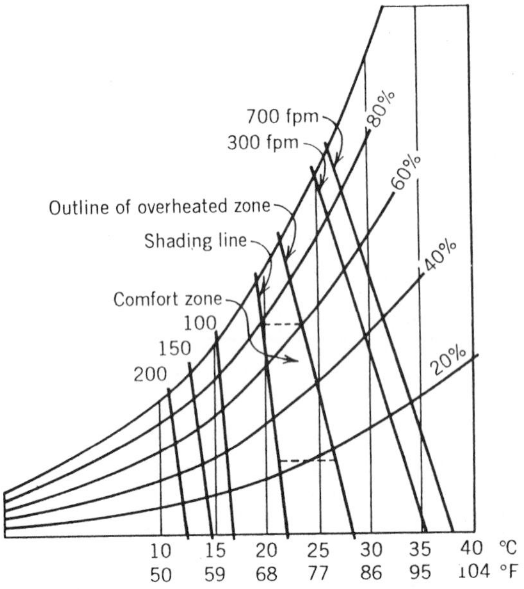

(a) The isolines on the timetables roughly correspond to these lines on the comfort zone (see Fig. 2.4)

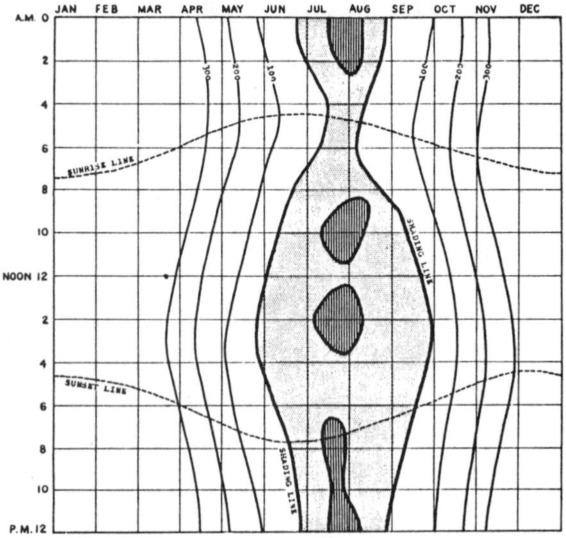

(b) Timetable for New York, NY

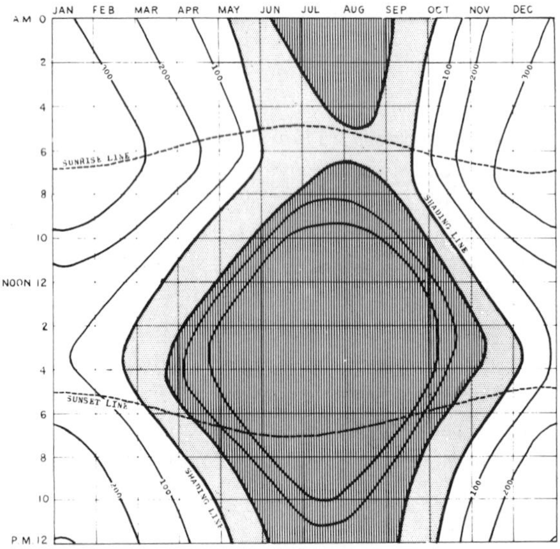

(c) Timetable for Phoenix, AZ

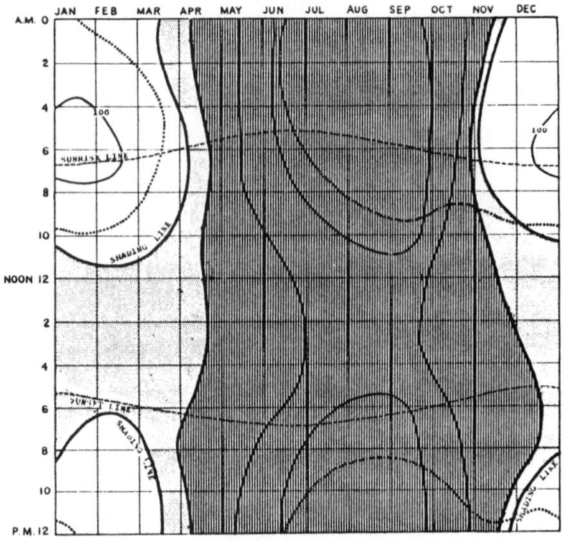

(d) Timetable for Miami, FL

Fig. 2.12 *Climatic timetables related to outdoor thermal comfort. [*(b–d) *From Victor Olgyay,* Design with Climate: Bioclimatic Approach to Architectural Regionalism, *copyright © 1963 by Princeton University Press; reprinted by permission.]*

over 700-fpm wind, though unlike Phoenix, these hours tend to occur also by night. The added dotted line indicates a need for wind to counteract high vapor pressure, independent of high temperature. Clearly, wind is important to comfort in humid Miami.

When the building schedule is overlaid on such climatic timetables, the fit between climate and building is made more clear. An example of temperature, daylight, and building schedule in timetable format is shown in Fig. 5.5. Another collection of design strategies shown for various climates is found in *Climatic Design* (Watson and Labs, 1983; see Fig. 3.5).

You now have some appropriate design strategies identified, and an idea of the extent to which your building can be closed or opened to exterior conditions. Next, the site and the skin or envelope of the building can be more closely examined. Following are some basic factors concerning envelopes; the details of envelope thermal design procedures are then found in Chapter 4. The site is the subject of the next chapter.

2.5 The Building Envelope

The envelope of a building is not merely a set of two-dimensional exterior surfaces; it is a transition space—a theater where the interaction between outdoor forces and indoor conditions can be watched. Some of these interactions include the ways in which sun and daylight are admitted or redirected to the interior, the channeling of breezes and sounds, and the deflection of rain. This transition space, which forms the envelope, is a place where people indoors experience something of what the outdoors is like at the moment, and where people outside get a glimpse of the functions within. The more suited the outdoors to comfort, the more easily indoor activity can move into this outdoor transition space. At entries, where there is a space created in the transition from one environment to another, a person will be especially aware of the difference between outdoors and indoors; an example of entry as space, not just surface, is shown in Fig. 2.13.

The envelope also has a fourth dimension; it changes with time. Daily changes were discussed in the preceding section; seasonal changes have a marked effect on the entry in Fig. 2.13, and a more subtle effect on the east-facing balconies of the apartments in Fig. 2.14. The year-round usable volume of these apartments is increased by making the balcony into a sun porch, as has been done for many apartments in Fig. 2.14. This is a direct response to the prevalence of "block wind–admit sun" conditions in Oregon's long but mild winter. Not all buildings encourage such flexibility; an unchanging envelope can be symbolic of stability and is considered appropriate for some governmental and religious monuments. Generally, the more that users are involved in decisions about how much of the outside to bring inside, the more changeable the building envelope will be. Further examples appear in Figs. 2.21 and 2.22.

2.6 Connectors, Filters, and Barriers

Called by their familiar names, the basic components of the envelope are windows, doors, walls, and roofs. On closer inspection we find that windows can include skylights, clerestories, screens, shutters, drapes, blinds, diffusing glass, and reflecting glass—an array of components that determines more exactly how the envelope does its job of making the transition between inside and outside. Norberg-Schulz (1965) suggests that a component can also be thought of by its function in the exchange of energies: as a *filter, connector, barrier,* or *switch.*

> In general, we define a *connector* as a means to establish a direct connection, a *filter* as a means to make the connection indirect (controlled), a *switch* as a regulating connector, and a *barrier* as a separating element. . . . An opaque wall thus serves as a filter to heat and cold, and as a barrier to light. Doors and windows have the character of switches, because they can stop or connect at will.

This approach to the choice of components can be illustrated by two opposite concepts of envelope design: the *open frame* and the *closed shell.* In harsh climates (or where unwanted external influences such as noise or intruding activities abound), the designer frequently conceives of a building's envelope as a closed shell and proceeds to selectively punch holes in it to make limited and special contacts with the outdoors. In the hot-humid regions (or where external conditions are very close to the desired internal ones), the envelope begins as an open structural frame, with pieces of building skin selectively added to modify only a few outdoor forces. The open-frame or the closed-shell approach to envelope design, when combined with material availability and the influence of local culture, can produce a distinct regional architecture (Fig. 2.15).

Fig. 2.13 *The envelope is more than a surface. This south-facing entry to an architect's office in Oregon becomes a microclimate that buffers the transition between the indoors and outdoors.* (a) *Three kinds of entry conditions are visible: the awning (over the windows of a restaurant), the gable roof with bare rafters (over the planting in the architect's entry), and the arcade, a second story carried out over a covered walkway, that links shops.* (b) *Detail of architect's entry, shown on March 21.* (c) *The change of seasons brings deep shade to this entry in summer and early fall. (Wilmsen & Endicott, Architects.)*

Fig. 2.14 *Envelopes change with time. This east-facing building envelope shows changes between seasons and between years.* (a) *December 21, 8:30 A.M., the sun is low and shining from the southeast; shading occurs not from balcony overhang but from vertical balcony-divider walls.* (b) *June 21, 8:30 A.M., the sun is high and nearly due east, so shading occurs only from balcony overhangs and from blinds at the railings.* (c) *An August morning, several years later, showing the conversion of balconies by many of the occupants. (Cascade Manor, Eugene, Oregon; John Graham & Associates, Architects.)*

With a wide range of energy sources, building materials, and mechanical equipment available, it is possible to design connector-dominated buildings anywhere, despite the climate. The consequences of the resulting energy consumption can be severe. By contrast, if defending against outdoor conditions becomes an overriding consideration, barrier-dominated envelopes can occur in any climate. The resulting fitness for human usage—and the potential of using natural energy sources—can be reduced. The designer's combinations of connectors, filters, and barriers (and the switches that allow these elements to respond to changing conditions) are basic to the design of building exteriors and can give them the liveliness that makes a building an attractive addition to its neighborhood. *Note:* A connector, filter, or barrier for one natural energy may change its role for another: glazing

Hot Humid Climate: To the open frame, a barrier roof of local plant materials is added to reject rain and sun. A raised floor avoids damp earth and its creatures and allows breezes to pass over and under its users.

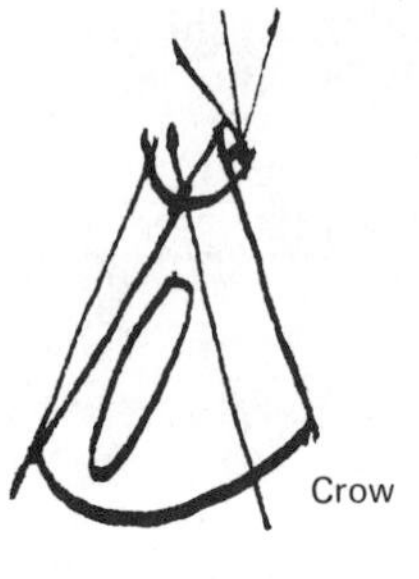

Temperate Climate: This open frame is wrapped in light–filtering animal skin, doubled near the ground. Wind and rain are rejected; protection against cold is provided by users' clothing (blankets) more than by the envelope. The switch at the crown controls smoke (see also Fig. 2.20). Portability of shelter is a cultural factor here.

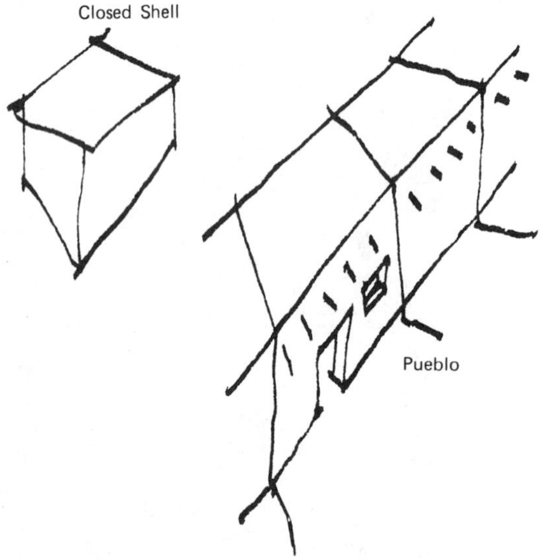

Arid Climate: The closed shell of mud block is a barrier to wind and sunlight; it filters heat by both delaying and reducing its impact on the interior. Some light and heat are admitted directly by small connectors: the door and window, typically south–facing.
By early morning, the cold interiors are abandoned in favor of rapidly warming south terraces.

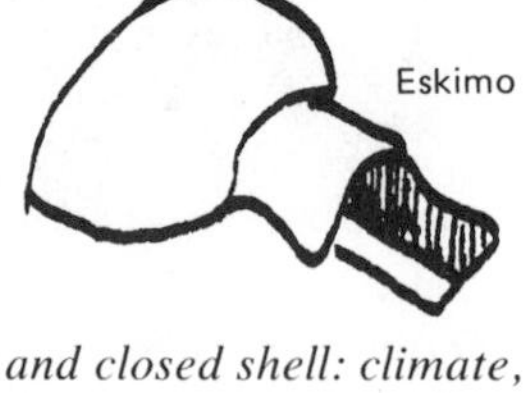

Cold Climate: The igloo's closed shell of ice is a filter to light and heat, a barrier to wind. Holes for entry and for smoke are allowed, but sparingly. Fur–bearing hides hung inside can increase thermal comfort for users.

Fig. 2.15 *Open frame and closed shell: climate, materials, and culture. The influence of climate is clear, but material availability and cultural patterns are also strong determinants in the choice of envelope design in these examples.*

Fig. 2.16 *Connectors for a corridor. An old cannery building in the mild climate of San Francisco is converted into a series of shops; a corridor is given a near-outdoors environment by the connectors at the right. Across the corridor, filters give greater internal climate control to the shops beyond. (The Cannery; Joseph Esherick & Associates, Architects. Photo by Richard A. Cooke III.)*

may be a connector to daylight but a barrier to wind.

Connectors are strong indicators that something outside is welcome inside. They are characteristic of regional architecture in milder climates, but sun connectors are dominant in solar-heated buildings anywhere. Connectors, being open to outside influences, are often one position of a switch that in other positions becomes a filter or a barrier. Or, as in Fig. 2.16, a connector is sometimes followed by a combination of filters and barriers somewhere further inside the space that forms a building's envelope.

Filters represent decisions about how much or what kind of outdoor condition is to be admitted. They are found in some form in all building envelopes and in all climates, and they include a wide variety of types. Because they admit desired amounts or qualities of light, air, and sound, they offer an opportunity for an enhanced awareness of selected outside conditions from inside the building. For example, the stained glass of a church enhances the blue of a north sky, or the warm reds and oranges of a sunset; the *texture* of the sky's cloud patterns is not admitted to the interior—only its color comes through. Often the filter is one of the positions of a switch, as in the case of the windows in a famous building by Le Corbusier (Fig. 2.17).

Barriers are more drastic in their complete severance of the outdoor–indoor relationship. They are characteristic of regional architecture in harsh climates, but are also common to spaces needing a tightly controlled environment (such as an auditorium). Barriers to rain are an almost universal building feature; barriers to wind are at least seasonally common in all climates, except hot-humid ones. Barriers to sun are more likely to be one position of a switch, unless a building is suffused with electric light or other plentiful sources of internal heat that make solar heat permanently unwelcome. In practice, cultural influences often override those of climate; barriers to sun are erected even in cold, damp environments, such as those of Canadian Pacific (Fig. 2.18).

Fig. 2.17 *Filters in various positions of a switch. Corbusier's Pavillon Suisse, 1930–1932, at the University of Paris. (Photo by Nicolai Shur.)*

Fig. 2.18 *Barriers, but not to heat. These barrier-dominated exteriors, typical of Haida villages in coastal British Columbia and southern Alaska, shed rain, wind, and light. The walls are filters, however, to heat, which passes with relative ease through the thick planks. (In winter, the narrow spaces between the houses can be enclosed, producing a more heat-conserving "row-house" configuration.) The central fires of each house are fed by plentiful wood fuel, and firelight is preferred over daylight for illuminating the elaborate masks used in Haida ceremonies. (Photo courtesy of the Milwaukee Public Museum.)*

2.7 Switches and Users' Choice

Switches are a valued component for designers who appreciate that climate and functions change, and that people are unpredictable. As much as a designer might prefer that a building have a certain identity, unchanging in its expression of a set of ideas, time will pass and changes will occur. Switches allow the designer to give control of the building to its users. If the designer has carefully considered the range of choices that the switch should provide, successful user control will be possible. Buildings whose skins are plentifully supplied with switches become continuing demonstrations that architecture is a performing art, not just sculpture.

In Section 2.4, switches were called on to alternately admit or block the sun or wind, so that a building's envelope might fully utilize the climate rather than simply exclude it. Passive solar heating systems rely on switches to throttle down the incoming sun on warmer days; especially in cold climates, such systems sometimes rely on switches to insulate windows by night (Fig. 2.19). Ventilating switches can be integral to a building's envelope in ways other than as sets of opening windows, as in Fig. 2.20.

Daylighting switches are perhaps our most common and visible ones. Awnings might block direct sun at some times, admit it at others. Opaque drapes might expose all the window to incoming daylight on dark winter days, but only part of the window on bright summer days, blocking the window entirely at night. Translucent curtains might change bright sun into a diffuse light for the interior, or be drawn back to allow strong shadows to be cast inside the room.

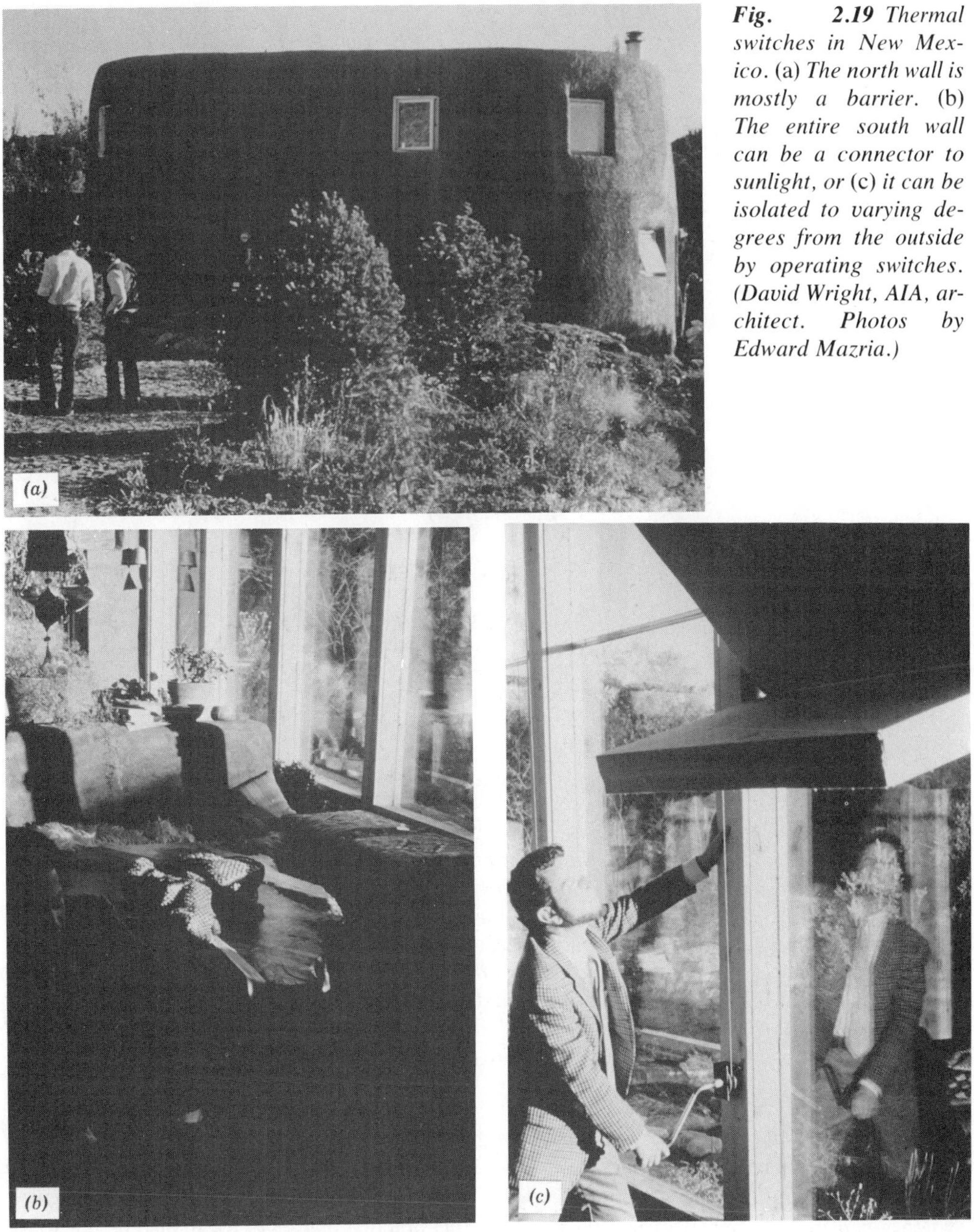

Fig. 2.19 *Thermal switches in New Mexico.* (a) *The north wall is mostly a barrier.* (b) *The entire south wall can be a connector to sunlight, or* (c) *it can be isolated to varying degrees from the outside by operating switches. (David Wright, AIA, architect. Photos by Edward Mazria.)*

Electric lighting can be turned off while daylight is plentiful.

Visually, switches are a particularly promising source of three- and four-dimensional interest on the building's exterior. For the buildings in Fig. 2.21, on which facade would daily and seasonal changes be most visually evident? If it is easier to imagine human beings working behind the windows of one building than of the other, might that suggest a more satisfactory work environment?

Switches allow change, and thus encourage interaction between users and their environments. This is usually satisfying to the users,

ENERGY OVERVIEW

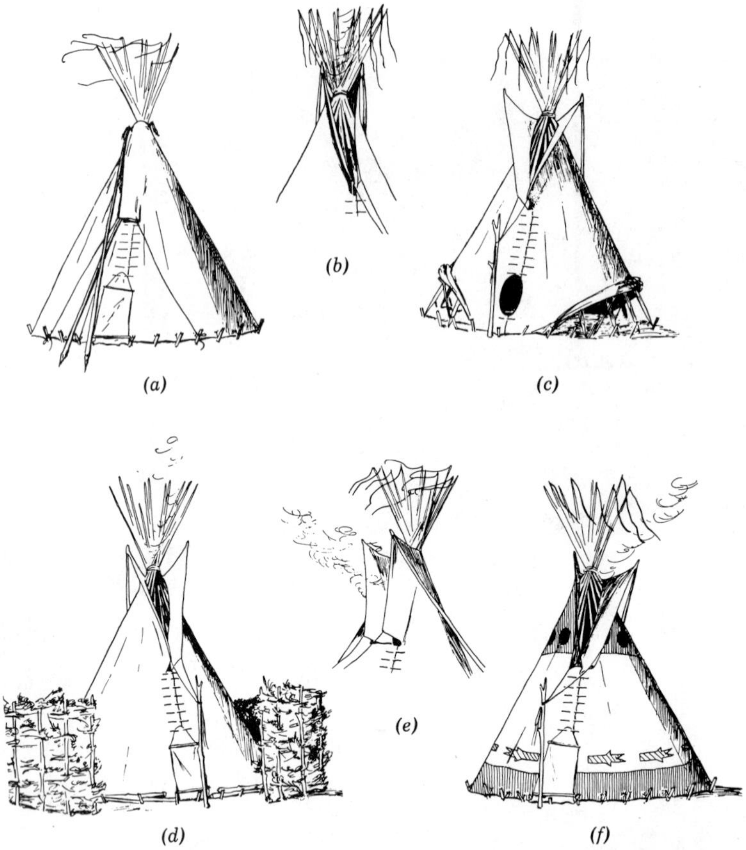

Fig. 2.20 *Thermal switches on the Great Plains. Six adjustments to the lightweight, translucent, and portable tipi are shown; the tipis in these diagrams are east facing, with their backs to the prevailing westerly winds.* (a) *In severe rainstorms, the smoke flaps can be closed.* (b) *For ordinary conditions of west wind, the smoke flaps block the wind, thus creating a suction at the opening to draw out the smoke.* (c) *For hot weather, breezes are admitted under the cover at the ground.* (d) *For extremely cold weather, a temporary windbreak can be added.* (e) *For the unusual wind or* (f) *southwest wind, the smokeflaps are manipulated to block this wind, thus encouraging smoke draw-out, as in* (b). *(From* The Indian Tipi: Its History, Construction, and Use, *by Reginald and Gladys Laubin. Copyright 1957 by the University of Oklahoma Press.)*

who are able to select the desired exposure to climate at that moment. Yet without automation, supervision, or training in their uses, switches can be detrimental to system performance. Examples are a thermal shade left to cover a passive solar collecting window on a cold, sunny morning, a vent left open during the hottest hours in a high-mass, night ventilation building, or an awning rolled up to leave a window exposed to summer sun. People in buildings often utilize switches in unexpected ways (see Fig. 2.22). Conventionally air-conditioned buildings typically do away with ventilating switches for users (operating windows), so that the system will function with a closely controlled amount of outdoor air. Thermally efficient as this practice might be, it can also be the source of widespread user dissatisfaction with air-conditioned spaces. Sealed windows greatly curtail people's contact with the sounds, smells, and breezes of the outdoors. This is frequently beneficial in urban areas, yet on beautiful days it can be very frustrating. A lack of switches contributes to a feeling of helplessness about the internal environment.

Switches frequently offer more variety than a

Fig. 2.21 *Sun control for Boston offices.* (a) *Switches both inside and outside the glass are evident, and the windows themselves are operable. Awnings are more likely to be fully extended as the windows sit higher up in the canyon of the street. (New England Merchants' Bank; Shepley Rutan & Coolidge, Architects; demolished 1966.)* (b) *Three-dimensional filters in the form of overhangs dominate the south facade (right) of this building at Boston University; the west windows (where overhangs are less effective) have internal switches, shown here in a variety of positions. (Law and Education Building; Sert, Jackson and Gourley, Architects.)* (c) *The two-dimensional filter of reflective glass sheathes all faces of this office tower, sending a beam of reflected sunlight to the neighborhood below. The switches in this case are thermostats; users are not involved in the outside–inside decisions. (John Hancock Headquarters; I.M. Pei and Partners, Architects. Photo by Stephen Tang.)*

simple "on–off" choice. The solar-heated St. George's school at Wallasey, England, was shown in Fig. 1.10; in Fig. 2.23 is a diagram of the operable windows that occur periodically on the school's south facade. When the air outside is very cold, minimal fresh air is admitted along other, less direct paths through the two-layered south glass wall; the operable window is firmly closed to infiltration. As the indoor–outdoor temperatures reach levels where more fresh air is desired, three degrees of openness are available to admit more and more outdoor air.

Passive heating and cooling techniques are especially reliant on switches, hence on the un-

Fig. 2.22 *Which is spring, which is summer?* (a) *The awnings for this office building in Eugene, Oregon, are joined by overhangs and side walls (or fins) to provide sun control for these south-facing windows.* (b) *The summer sun, at a high angle is readily blocked by the overhang; the awnings can be rolled up with less risk of direct glare into the users' eyes. (Moreland-Unruh-Smith, Architects.)*

derstanding and cooperation of the people within these spaces. These users often must base their actions of the moment on what will be needed later, by manipulating thermal switches. This practice, called "thermal sailing," is similar to actions of outdoor workers in the far north, who learn to unbutton their coats in the cold early hours of the workday *before* they begin to sweat and rebutton in the relative warmth of the late afternoon before the rapidly falling temperatures near dusk can chill the skin. Misjudgments in passive solar-heated homes can result in extraordinarily high temperatures on a sunny day, or nights without stored solar heat. For the building closely related to its climate, the design of switches is also the design of an educational process for users. The challenge is to involve, but not enslave, the users in the management of their environment. Automated controls are a partial answer to this challenge;

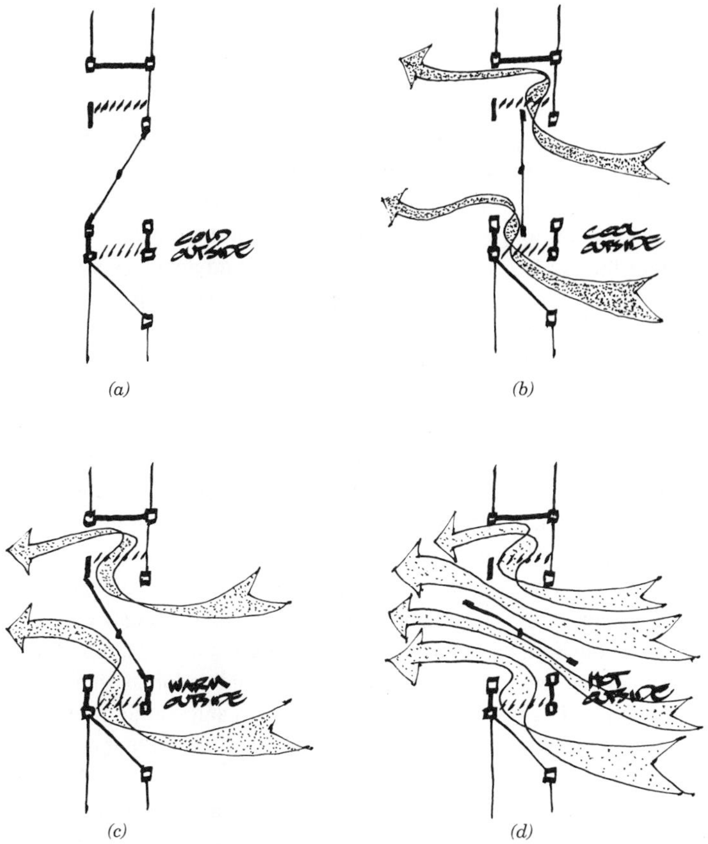

***Fig. 2.23** Positions of a classroom's operable south window at St. George's School in Wallasey, England (see also Fig. 1.10). The window pivots on a center horizontal axis.* (a) *When no air is desired through the window (very cold outside, or too hot outside to ventilate), the window is tightly closed.* (b) *Minimum fresh air enters at waist and head levels with the window in a vertical plane.* (c) *More fresh air enters as the window top tilts back. The window can be locked in this position for night ventilation with security.* (d) *Maximum ventilation is achieved with the window nearly horizontal.*

switches that are easy and fun to use are another.

2.8 The Building Interior

Returning to the subject of human comfort indoors, in addition to thermal considerations we find the factor of *interior air quality*. This is not new—the smoke flaps of the tipi in Fig. 2.20 are eloquent examples of air quality's early impact on building form—but it has assumed new significance in today's buildings. Interior air quality (IAQ) was brought into focus by the emphasis on energy conservation of the mid-1970s. As designers struggled to save energy by reducing heat flow through the building skin, it became apparent that enormous heat flows were also associated with the building's "lungs." By curtailing the introduction of outdoor air, large energy savings were possible when the indoor–outdoor temperature difference was large. By itself, this may not have caused immediate problems. However, designers were also reducing the capacity of heating and cooling systems (and their fans), since equipment oversizing often led to inefficient operation. Thus less air, outdoor or recirculated, was flowing through a building. At the same time, newer systems such as variable air volume were tending to further reduce airflow within spaces. Finally, some popular new materials were found to be *outgassing* vapors that contributed significantly to IAQ problems. This combination of conservation, systems, and materials produced a series of "sick buildings," so called because of occupant dissatisfaction with their indoor air quality. A lack as yet of generally accepted standards and guidelines complicates the designer's task of responding to this threat.

Problematic new materials include many wood products, carpeting, and paints. One of the most common pollutants from building materials is formaldehyde, a colorless water-soluble gas that is emitted most strongly when the material is new, diminishing gradually with time. It appears that formaldehyde emissions increase temporarily with increased indoor temperature, humidity, and (oddly) ventilation. (The most spectacular problem material was urea-formaldehyde foam insulation (UFFI), widely used until resulting occupant sickness forced its ban in Canada and the collapse of its market for use in buildings in the United States.) As yet, emissions data are not readily available from product manufacturers. This makes difficult the task of specifying materials for IAQ control.

People and their activities are also causes of IAQ problems. People continuously emit carbon dioxide, moisture, and bioeffluents (hair and skin flakes, viruses, and bacteria). Tobacco smoke, body odors, and odors associated with personal grooming products are common. Wood stoves and unvented gas appliances such as ranges emit harmful products of combustion. Copy machines emit irritants such as ozone and hydrocarbon compounds. Cleaning and maintenance activities produce irritating organic solvents and ammonia. Inevitably, some processes of living within a building make it less livable.

Comfort was defined (Section 2.2) primarily in thermal terms, yet the influence of visual, acoustic, and olfactory aspects is quite strong (as in the examples of fireplaces and fountains). Air quality interacts subtly with thermal comfort; particularly common are complaints about "stuffiness" that accompany too-hot or too-humid environments. Problems unrelated to physical criteria, such as unhappiness with one's work, can lower one's tolerance to air pollutants; such discontent is particularly likely at the end of a workday. Again, we are reminded that comfort is as much psychological as physiological.

Sick building syndrome is said to occur when a significant portion (for example, 20%) of the occupants complain during a 2-week period of a set of symptoms, including headaches, fatigue, nausea, eye irritation, and throat irritation, that are alleviated by leaving the building and are not known to be caused by any specific contaminants (SMACNA, 1988). These buildings illustrate two types of occupant responses: (1) air quality problems are perceived, and result in physiological stress, and less commonly (2) social problems are perceived and the resulting stress is psychological rather than physiological; this sometimes results in the transfer of complaints to the physical environment. A third, insidious occupant response: (3) air quality problems are *not* perceived but still cause physiological strains. An example of the latter category is radon gas, undetectable by human beings, but no less dangerous than odorous formaldehyde.

Diagnosis of sick building syndrome can be complex; complaints by occupants are often the first indication of a problem, and interviews with occupants can be the single most useful source of information. A simple test is to measure the concentration of carbon dioxide, often used as a first indicator of general indoor air pollution problems. More complicated equipment and testing procedures are available for gaseous contaminants such as carbon monoxide, nitrogen dioxide, formaldehyde, and radon. Other equipment is used for organic and particulate contaminants. A summary of equipment and testing procedures is found in SMACNA (1988).

The most common causes of sick building syndrome are insufficient outdoor air (in over half the cases to date), biological contaminants within the HVAC system, exhaust air reentering (or contaminated air entering) through outdoor air intakes, and contaminant migration from "dirty" to "clean" spaces within buildings.

The most common remedy is to increase the quantity of outdoor air, but effective ventilation is not just a matter of more air. In "task ventilation," for example, individual exhaust fans (as for bathrooms and kitchens) are installed to pull contaminated air from office copiers or other odor-producing equipment rooms. This method is particularly effective when barriers can be erected to contain the pollution at its source. Another common remedy consists of various types of air filters; these are discussed in detail in Section 6.6. Two distinct types of filtering systems are usually called for: one for particulates, another for treating gases. Finally, increased maintenance is often helpful, not only of the occupied building spaces, but of the heating, ventilating, and air-conditioning equipment.

REFERENCES

Publications of the American Society of Heating, Refrigerating, and Air-Conditioning Engineers, Atlanta, Ga.:

ASHRAE *Handbook of Fundamentals,* 1981, 1989.

ASHRAE Standard 55-1981, *Thermal Environmental Conditions for Human Occupancy,* 1981.

Arens, E., Gonzalez, R., Berglund, L., McNall, P., and Zeren, L. (1980). "A New Bioclimatic Chart for Passive Solar Design," in Vol. 5.2, *Proceedings of the Fifth National Passive Solar Conference,* American Section of the International Solar Energy Society, Boulder, Colo., pp. 1202–1206.

Heschong, L. (1979). *Thermal Delight in Architecture,* MIT Press, Cambridge, Mass.

Laubin, R., and Laubin, G. (1975). *The Indian Tipi: Its History, Construction and Use,* University of Oklahoma Press, Norman, Okla.

Milne, M., and Givoni, B. (1979). "Architectural Design Based on Climate," in *Energy Conservation Through Building Design,* D. Watson, Editor, McGraw-Hill, New York.

Norberg-Schulz, C. (1965). *Intentions in Architecture,* MIT Press, Cambridge, Mass.

Olgyay, Victor (1963). *Design with Climate; Bioclimatic Approach to Architectural Regionalism,* Princeton University Press, Princeton, N.J.

Sheet Metal and Air-conditioning Contractors National Association (SMACNA) (1988). *Indoor Air Quality.* SMACNA, Vienna, Virginia.

3

SITES AND RESOURCES

In the first chapter, the supply of ultimately limited nonrenewable energy was contrasted with the supply of timeless but limited-delivery renewable resources. In the second chapter, climate was examined for its suitability for indoor comfort, so that buildings might open themselves often to this free energy available on site, rather than retreating permanently within an indoor climate dependent on purchased energy. This chapter looks at microclimates; the localized climates near buildings, where inside meets outside.

A designer's early decisions in site planning will influence the later choices of both the building's mechanical and electrical equipment and its overall consumption of energy. If the site is seen as a collection of resources (sun, wind, water, plants) and also as a part of the environment we all share, the buildings we design can approach self-sufficiency of energy supply, without limiting the availability of local energy resources for neighboring buildings. The use of on-site resources not only can reduce the amount of energy needed to maintain the interior climate, it also can produce outdoor spaces that become especially pleasant to use. Such spaces can direct winter sun to a glass wall while blocking the wind, or funnel the summer breeze through shade to an open window. Site planning is greatly influenced by economic considerations, zoning regulations, and adjacent developments, all of which can interfere with the design of a site to utilize the sun and the wind. Integration of all these concerns at the site-planning stage is the first step in adapting a building to its climate.

The microclimate is influenced by the interaction of both site characteristics and climate characteristics:

Site	*Climate*
Soil type	Sun
Ground surface	Temperature
Topography (profile)	Humidity
Vegetation	Precipitation
Water presence	Air motion
View	Air quality
Human activity (heat, noise, etc.)	

This chapter looks briefly at some results of site–climate interactions.

3.1 Climates Within Climates

The climate at a particular site can be quite different from the averages that are published for the whole region. This is particularly evident where a neighboring hill blocks wind or winter sun, or an adjacent lake cools summer breezes or adds a damp chill to winter air. It may be less obvious that urban sites are under the influence of an urban subclimate, a "heat island" that differs from the conditions of surrounding countryside unaffected by urbanization. (The cities' climatological stations are often at a more rural site, such as an outlying airport.) The effects of this urban heat island are summarized in Table 3.1 and Fig. 3.1.

The most obvious reason for the city's relative warmth year-round is its concentration of *heat sources:* the air conditioners, furnaces, and internal combustion engines in cars and buildings. This "internal" urban heat production is shown in Table 3.2 and Fig. 3.2. It seems that cities in the world's temperate zones release less internal heat as people live closer together; to what extent this is due to less transportation, to less building heating, to climate, or to culture is unclear. The "average" rate of 30 kW/m^2 is greatly exceeded by that of Moscow and Manhattan and is vastly more than that of Hong Kong.

The *rain* that falls on the city and countryside

TABLE 3.1 **Average Changes in Climatic Effects Caused by Urbanization**[a]

Effect	*Comparison with Rural Environment*
Contaminants	
Condensation nuclei and particulates	10 times more
Gaseous admixtures	5 to 25 times more
Cloudiness	
Cover	5 to 10% more
Fog, winter	100% more
Fog, summer	30% more
Precipitation	
Totals	5 to 10% more
Days with less than 5 mm	10% more
Snowfall	5% less
Relative humidity	
Winter	2% less
Summer	8% less
Radiation	
Global	15 to 20% less
Ultraviolet, winter	30% less
Ultraviolet, summer	5% less
Sunshine duration	5 to 15% less
Temperature	
Annual mean	0.5 to 1.0°C more
Winter minima (average)	1 to 2°C more
Heating degree days	10% less
Wind speed	
Annual mean	20 to 30% less
Extreme gusts	10 to 20% less
Calms	5 to 20% more

Source: H. E. Landsberg, "Climates and Urban Planning," *Urban Climates,* World Meteorological Organization, 1970. WMO Technical Note No. 108, p. 372.
[a]These effects vary from city to city and from day to day.

can be an effective cooling mechanism, especially as water evaporates from wet surfaces; but the materials of streets and buildings are usually designed to shed water quickly and thoroughly, so evaporative cooling for these surfaces is minimized.

The city also changes the overall cooling action of the *wind* by channeling it into narrow streets. The geometry of high vertical walls and narrow streets also increases the summer heat in cities as the high *sun* is reflected downward and is absorbed, and then reradiated, by the often rocklike street and building surfaces. In winter, however, this geometry puts most urban surfaces at a solar disadvantage, since the low sun strikes only the upper portion of south-facing walls. This lack of access to the sky—whether for solar gain and daylight, or for radiant loss to the cold night sky—is a key element in the formation of the urban heat island, as summarized in Fig. 3.3. Sky access is discussed in more detail in Section 3.4.

A more subtle climate influence is *contaminated air;* small particles in the city's air can keep some sunlight from reaching the city; yet it can help to keep the city's heat from radiating outward. These particles also form additional nuclei for fog droplets; Table 3.1 shows that up to twice as much fog occurs in the city in winter as in the surrounding countryside. Trees and greenery are not as available in the city to act as crude filters to airborne dust.

The city thus changes its climate from that of its surroundings—to the city's decided disadvantage in summer. In winter, the city's additional warmth reduces the need for heating buildings. The typical means of providing additional winter heating (by fossil-fuel-burning furnaces and power plants) contribute to airborne particles and urban fog. A switch to solar-assisted heating would diminish this pollution—and with less air pollution, more sun would reach a solar collector.

Site and urban planning responses to these urban climate characteristics can sometimes lead in contradictory directions. For example, the provision of greenways within cities would bring softer ground surfaces cooled in summer by shading, breezes, and evaporation. However its winter impact might be an increase in fog via evaporation of retained water; or perhaps fog would be discouraged by the increased wind speeds facilitated by such greenways. The summer impact of the greenway is also not entirely positive; increased evaporation could mean higher relative humidity—but with slightly lower temperature. On balance, the greenway within a city seems to be a positive and esthetic step in ameliorating the urban climate, but the complexity of such a question deserves further research.

While considering how the sun, wind, and other climatic elements can be utilized on a site for the benefit of a building, it is important to remember the need to protect the access of others to these same resources. This is effectively illustrated in Garrett Hardin's "Tragedy

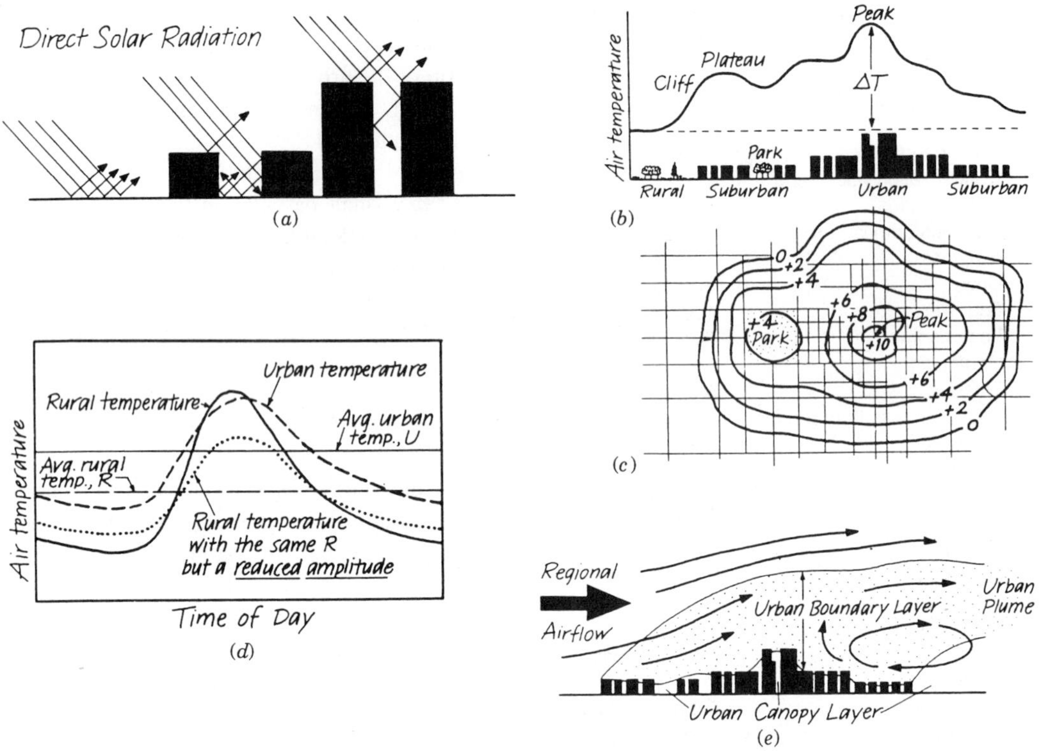

Fig. 3.1 *Urban heat island; a difference between the original countryside climate and that of the new urban area within it.* (a) *Direct solar radiation is less likely to be reflected by the city, thereby increasing solar gain to urban areas.* (b) *Idealized profile of the air temperature difference between urban and rural, at times of peak differences: calm, clear nights.* (c) *Based on* (b), *typical isotherms of equal temperature provide a "contour map" of the urban heat island.* (d) *Differences over a typical day. (The "reduced amplitude rural" is an intermediate case resulting from heat-conducting materials and a thin cloud of polluted air.)* (e) *The urban heat island can affect the countryside "downstream." [Reprinted by permission from Lowry (1988).]*

of the Commons" (1968), often reprinted since its original publication in *Science*. In his example, the commons are meadows publicly owned and shared by many herders. Each herder realizes that his personal wealth will increase as animals are added to his herd, so all herders increase their livestock. But the meadow capacity is not increased; overgrazing occurs, and as a result the commons become unable to support any animals. The following discussions of the resources of sun, air, and water found on a site are influenced both by the "private" needs of a building and the "public" patterns of these resources, which should remain accessible to all.

3.2 Analyzing the Site

One of the designer's first concerns is to recognize the resources that exist on and around a site (or in a place) and to decide how best to integrate these resources into the final design, while making the design a successful addition to the larger patterns of its surroundings. Schematic site plans are typically used as a kind of inventory; overlaid "bubble diagrams" can test possible design arrangements that can relate rooms and functions to their surroundings in the plan. Sun and wind conditions (in both summer and winter), noise sources, and water runoff patterns can be included in this schematic plan. It is

TABLE 3.2 **Heat Generated Within Cities**

City	Population[a] (millions)	Area[a] (km^2)	Population Density ($10^3/km^2$)	Energy per Capita (kW/capita)	Energy per Unit Area[a] (W/m^2)[b]
Hong Kong	4.4	92	47.8	0.06	3
Manhattan	1.7	59	28.8	5.5	159
Sheffield	0.5	49	10.2	2.0	20
West Berlin	2.3	233	9.87	2.1	21
Moscow	6.4	878	7.29	17.4	127
Sydney (center city)	0.1	24	4.17	13.7	57
Budapest	2.0	525	3.81	11.3	43
St. Louis	0.75	250	3.00	5.3	16
Cincinnati (summer)	0.54	200	2.70	9.6	26
Brussels	1.0	400	2.50	11.2	28
Los Angeles	7.0	3500	2.00	10.5	21
Chicago	3.5	1800	1.94	27.2	53
Fairbanks	0.03	19	1.57	12.1	19

Source: Reprinted by permission from Lowry (1988).

[a]Data from H. E. Landsberg, (1981), *The Urban Climate,* Academic Press, New York, 1981, as corrected by T. R. Oke.

[b]0.317 W/m^2 = 1 Btu/h-ft^2.

particularly important to identify microclimates on the site, the places that have special characteristics differing from the regional climate. Microclimates can present opportunities to utilize the expanded comfort zone discussed in Section 2.4; they can sometimes represent building sites where less energy is consumed because the winter is warmer, or the summer cooler. Or microclimates can be problem areas to be avoided for building or outdoor activity, if possible, or where special design measures need to be taken to correct their difficulties.

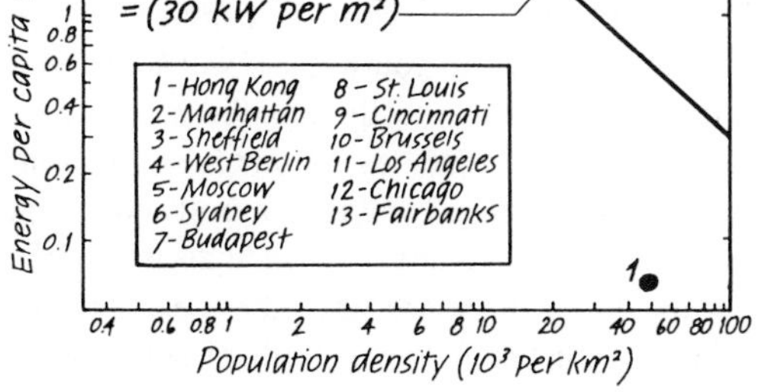

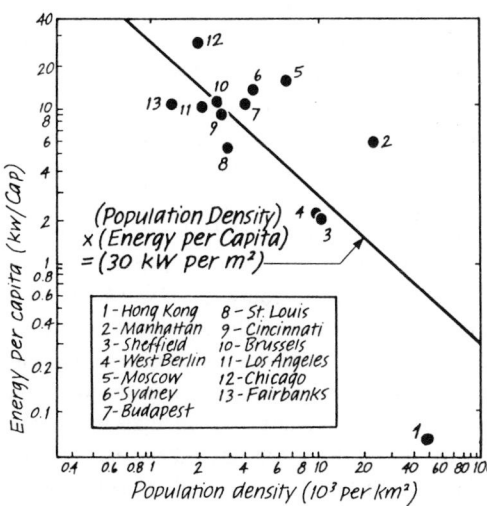

Fig. 3.2 *Heat released from cities: population density and energy usage per capita. [Reprinted by permission from Lowry (1988).]*

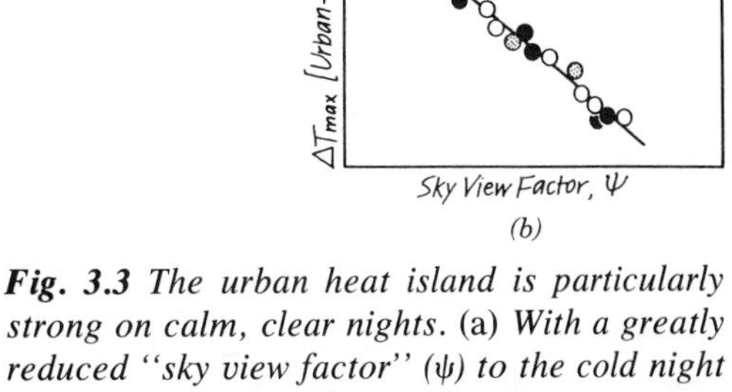

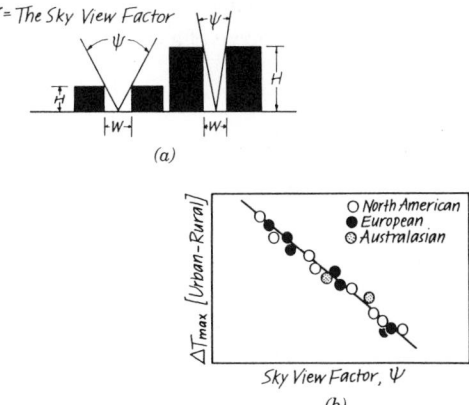

Fig. 3.3 *The urban heat island is particularly strong on calm, clear nights.* (a) *With a greatly reduced "sky view factor" (ψ) to the cold night sky, the walls and floors of urban canyons—streets—cannot lose heat as readily as can the open countryside.* (b) *The more narrow the ψ, the more pronounced is the effect of the urban heat island, ΔT, in cities throughout the world. [Reprinted by permission from Lowry (1988).]*

The site's collection of microclimates is not limited to those visible in plan, on the surface. Conditions of privacy and accessibility, of view, heat, light, air motion, sound, and water, all change with vertical distance from the surface (Fig. 3.4*a*).

To minimize energy consumption in constructing and using buildings, and to integrate them with their surroundings, the conditions best suited to various functions should approach the characteristics of the "layer" of the site in which they are located. Consequently, both horizontal and vertical site analyses are needed. A lecture hall, needing both an isolated and a closely controlled environment, is a candidate for the subsurface layer. Electrical equipment, which benefits from cool environments, also is suitable to the subsurface layer.

One building that derives its form from these layers is Boston's City Hall (Fig. 3.4*b*). Activities with the most frequent public interaction are located near the skylit, high-ceilinged lobby on surface levels, whereas special ceremonial functions are elevated to distinctive forms in the near-surface layer. The city offices with less frequent public contact occupy several floors in the sky layer, and storage and mechanical functions are in the subsurface layer. (An aesthetic equivalent of this horizontal layering is architect Louis Sullivan's concept of a facade as a "base, body, and capital.")

3.3 Site Design Strategies

In the preceding chapter, appropriate building design strategies were identified by analyzing climate data. Sites can also be organized to aid in the heating, lighting, cooling, and noise characteristics of buildings; a few of the most common site elements are analyzed by Watson and Labs (1983) for their thermal contribution in Fig. 3.5. Although housing is suggested in these graphics, they apply to any function that can utilize the extended comfort zone. Later in this chapter, several of these approaches are discussed in more detail.

One example of how the information in Fig. 3.4 can be applied (and manipulated) in the planning of a building and its site is a house designed by Frank Lloyd Wright nearly 40 years ago (Fig. 3.6). The direct gain of solar heat through its windows in winter makes this an early example of "passive solar heating" (by contrast, "active solar heating" includes collectors and a separate storage volume for solar heat). This house, known as the Solar Hemicycle, was built in 1948 near Madison, Wisconsin, which lies between the cool and temperate climate zones. Winter heating is the dominant thermal influence in this area. The house stands on a hilltop site, particularly vulnerable to winter winds. In construction, earth was scooped from in front of the south face of the house and bermed against the entire curved north wall, almost to roof level. This gives winter wind protection from northeast to northwest and provides further insulation behind the almost windowless north wall. The north wall is made of stone, which absorbs and stores the winter solar heat that comes in through the floor-to-ceiling, southeast-to-southwest-facing glass. The concrete floor also stores solar heat in winter. The impression that this house and site are sun collectors is heightened by an entrance tunnel at the northeast end of the house, which leads from the parking area through the berm wall and onto the sunny, protected south terrace and front door.

This house is longer in the east–west direction (about 1 : 3). This elongation is typical of passive solar designs, which are able to store and use the winter solar heat gain, and thus profit from having long south-facing glass walls to act as collectors. Had the large windows been well insulated at night, and the roof and walls insulated up to the present standards, this house would be a very up-to-date example of passive solar heating.

In the Solar Hemicycle house, protection from summer overheating is provided by an overhang along the south glass walls, the cool thermal mass of the north stone/berm wall combination (which receives no direct sun in summer), and the high windows in both north and south walls, which allow warmed air to rise and escape. Over the years since these early photographs were taken, the growth of some plants around the south rim of the scooped pocket has diminished the summer "heat trap" effect of this pleasant outdoor space.

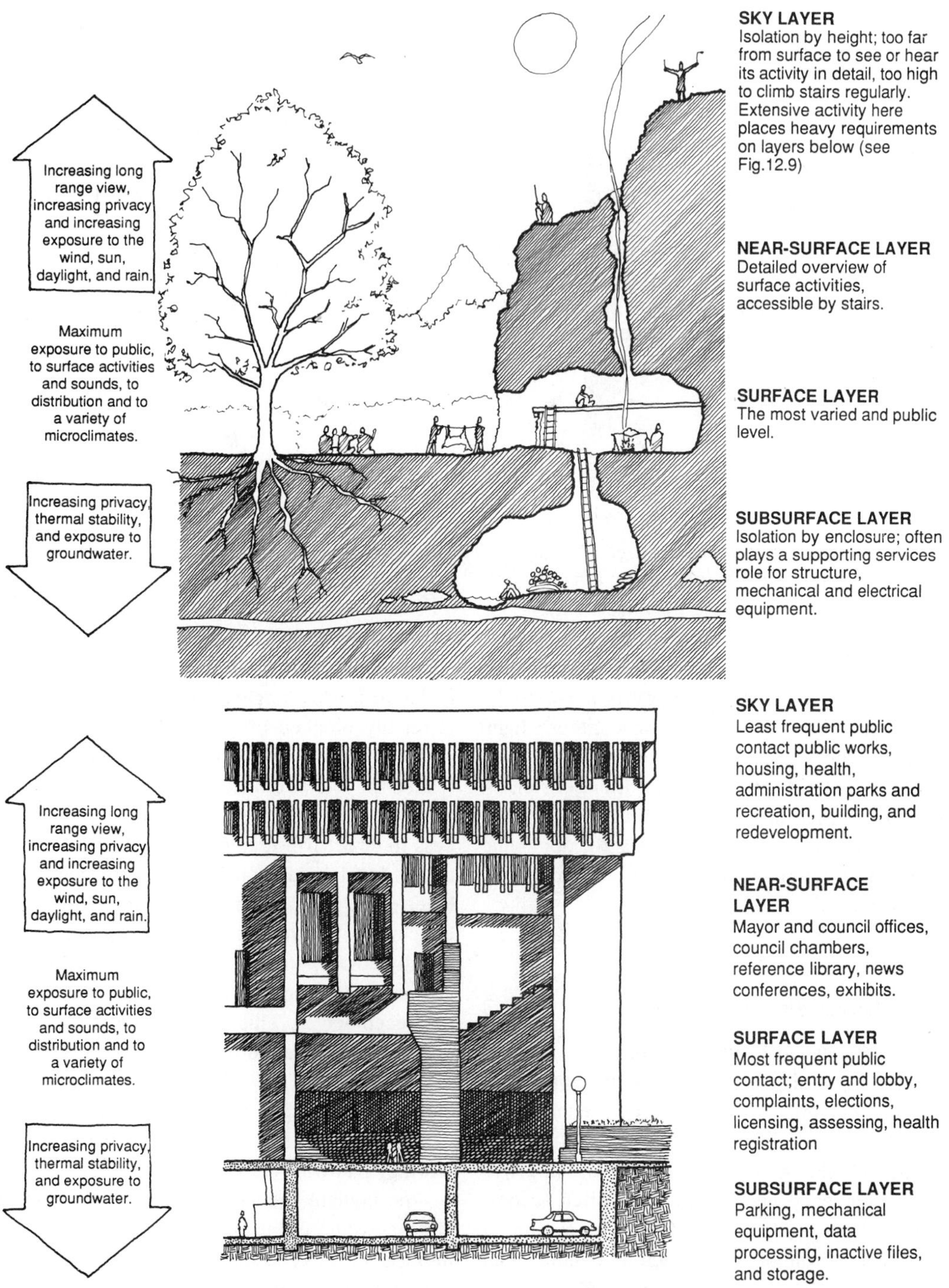
Increasing long range view, increasing privacy and increasing exposure to the wind, sun, daylight, and rain.
Maximum exposure to public, to surface activities and sounds, to distribution and to a variety of microclimates.
Increasing privacy, thermal stability, and exposure to groundwater.
SKY LAYER
Isolation by height; too far from surface to see or hear its activity in detail, too high to climb stairs regularly. Extensive activity here places heavy requirements on layers below (see Fig.12.9)
NEAR-SURFACE LAYER
Detailed overview of surface activities, accessible by stairs.
SURFACE LAYER
The most varied and public level.
SUBSURFACE LAYER
Isolation by enclosure; often plays a supporting services role for structure, mechanical and electrical equipment.
Increasing long range view, increasing privacy and increasing exposure to the wind, sun, daylight, and rain.
Maximum exposure to public, to surface activities and sounds, to distribution and to a variety of microclimates.
Increasing privacy, thermal stability, and exposure to groundwater.
SKY LAYER
Least frequent public contact public works, housing, health, administration parks and recreation, building, and redevelopment.
NEAR-SURFACE LAYER
Mayor and council offices, council chambers, reference library, news conferences, exhibits.
SURFACE LAYER
Most frequent public contact; entry and lobby, complaints, elections, licensing, assessing, health registration
SUBSURFACE LAYER
Parking, mechanical equipment, data processing, inactive files, and storage.

Fig. 3.4 (a) *Characteristics of horizontal layers for site analysis.* (b) *Layers and form: Boston City Hall, 1969. (Kallmann, McKinnell and Knowles, Architects.)*

Almost all of today's energy sources have the sun as an ancestor. The earth receives a very small percentage of the sun's daily energy output; the amount of solar energy available to each site varies both seasonally and daily. Typically, the closer the sun to a position directly over-

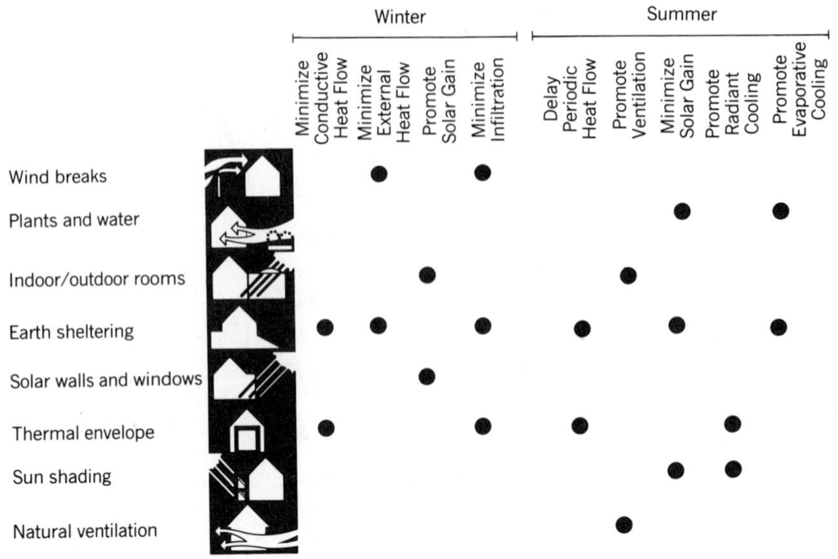

Fig. 3.5 *Generic bioclimatic design concepts. (Reprinted from* Passive Cooling *by permission of the publisher, American Solar Energy Society, Inc.)*

head, the more solar energy reaches the site. Direct sunlight's radiant energy is not its only resource; direct sun is our most intense light source, whose usage is complicated by its slow but constant change of position. Indirect sun, as on an overcast day, is a diffuse and readily utilized light source. Daylighting was the dominant method of lighting buildings until the middle of this century.

3.4 Direct Sun and Daylight

(a) Access to Light and Sun. The value of daylight (and air) to buildings has long been recognized in zoning laws, which require that minimum distances (setbacks) be maintained between a building and the property line in lower density areas. Height restrictions often accompany these setbacks, defining a maximum "buildable volume" in which a building can grow (Fig. 3.7). As buildings become taller and density increases, daylight reflecting down between them is diminished; in response, the maximum buildable volume approximates a pyramid (Figs. 3.7 and 3.8).

When *direct sun* in winter is desirable at the ground floor of each site, this pyramid changes shape and its volume decreases (Fig. 3.7*c*); this is due to the low angle of the winter sun, which is readily blocked by taller objects south of a given window. This most-restricted pyramid (called the "solar envelope") is at present rarely achieved, but various proposals to guarantee access to direct winter sun for solar collection are under development on federal, state, and local levels. The most restrictive feature of the solar envelope is the low slope of its northern face, usually corresponding to the altitude of the sun above the horizontal at about 2 hours from noon on December 21. This allows 4 hours of access to direct sun on even the shortest day.

The impact of solar zoning on site utilization can be considerable. Figure 3.9 presents further development of the solar envelope based on the work of Knowles (1981).

For daylight access, building surfaces can be almost as important as building geometry. The importance of reflected light from vertical and horizontal surfaces will be apparent in the daylighting calculations in Chapter 19; lighter colored surfaces produce more internal light, especially in crowded urban conditions where a view of the sky from windows is not common. An increase in daylight will result for all surfaces on an urban street, if the building and site surfaces

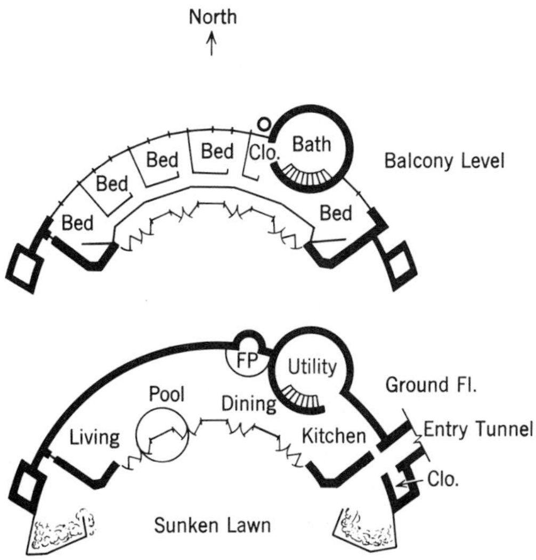

(a)

are light colors. This can conflict with the use of planting for shading and evaporative cooling, although lighter colored ground cover plants are available. While daylight reflected in a diffused way from building and ground surfaces is a potential benefit, the harsh specular reflections from mirror surfaces are often unwelcome (see Section 3.4*e*).

Fig. 3.6 *An early passive solar-heated home, Frank Lloyd Wright's Solar Hemicycle near Madison, Wisconsin (between the cool and temperate zones), designed in the early 1940s, built in 1948.* (a) *A representative floor plan.* (b) *Early view of the northeast berm wall.* (c) *Passive solar-heated interior. (Photos by Ezra Stoller. Copyright © ESTO.)*

(b)

(c)

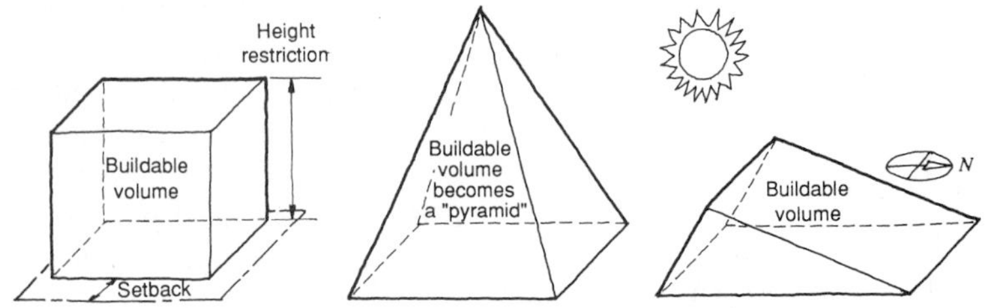

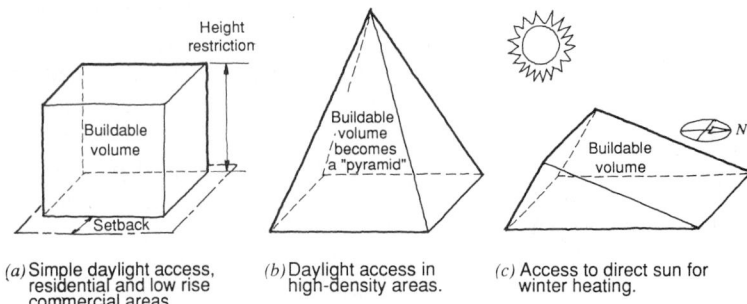

(*a*) Simple daylight access, residential and low rise commercial areas. (*b*) Daylight access in high-density areas. (*c*) Access to direct sun for winter heating.

Fig. 3.7 *Access to light. Some methods of compromising private optima (e.g., maximum rentable floor space) with public optima (e.g., daylight at street level).*

(b) Charting the Sun. The details of sun position and solar utilization are found in later chapters on solar heating (Chapters 5 and 6) and daylighting (Chapter 19). Sun position and intensity data are found in Appendices B and C. For site planning, one of the earliest frequent tasks is to determine whether direct sun reaches a building (or playground, etc.) in the winter.

Presented here are charts of the sun's position for two latitudes (Fig. 3.10). These charts are actually graphs superimposed on the eye of an observer. At the vertical center line of each chart, the observer is looking due south. The horizon (a horizontal plane at observer eye level) is the line at the bottom of the chart.

Fig. 3.8 *Some results of zoning for "pyramids" in densely built areas.*

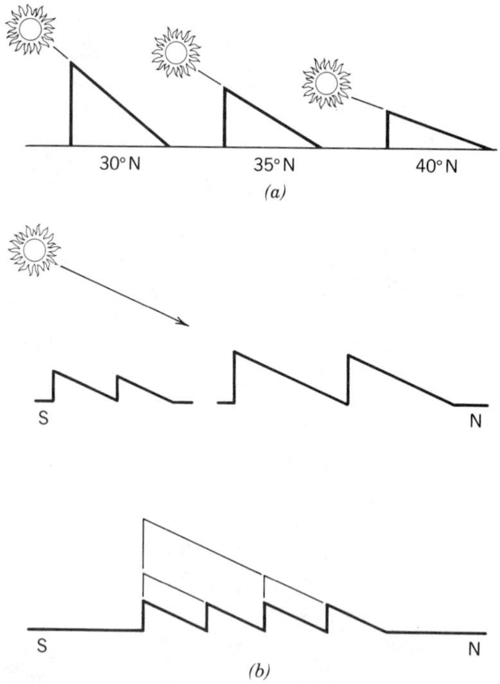

Fig. 3.9 *These solar envelopes are refinements of the solar access "pyramid" of Figs. 3.7 and 3.8.* (a) *The slope of the solar envelope changes with latitude.* (b) *The larger the site, the greater the volume of the solar envelope. (Reprinted by permission from R. Knowles,* Sun, Rhythm, Form, *© 1981, MIT Press.)*

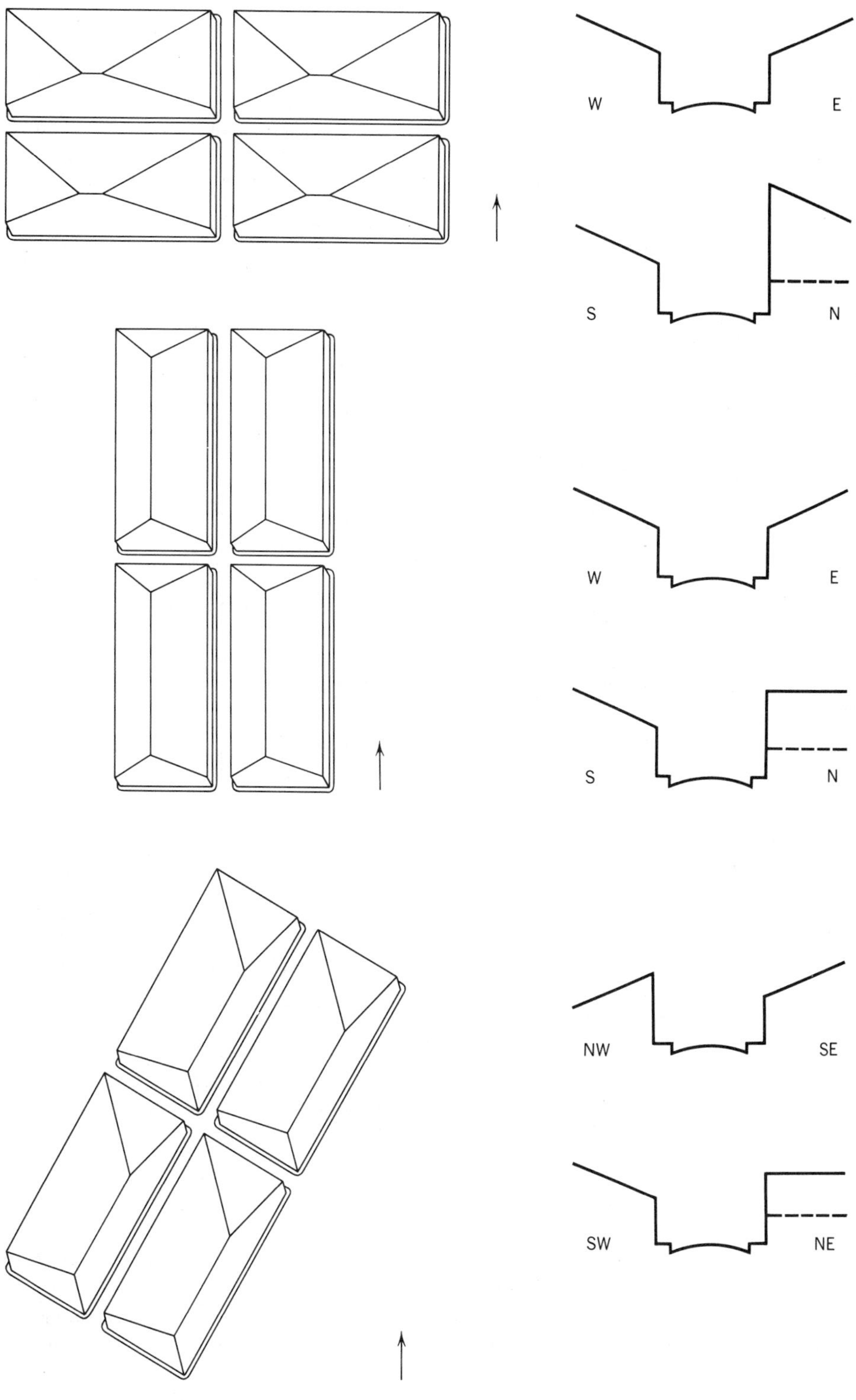

(c)

Fig. 3.9 (c) *Solar envelopes for various orientations of individual sites. (Reprinted by permission of R. Knowles,* Sun, Rhythm, Form, *© 1981, MIT Press.)*

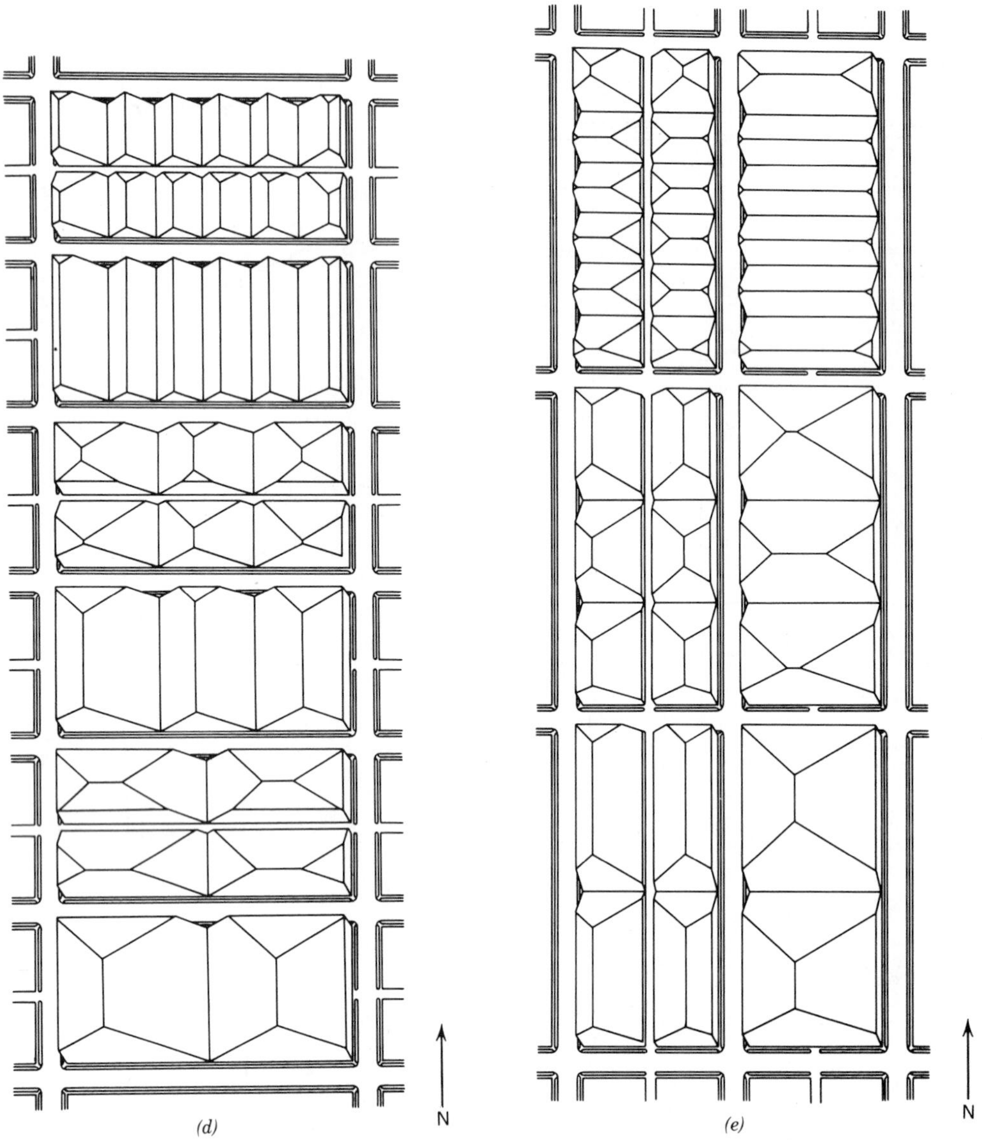

(d) (e)

Fig. 3.9 (d) *Solar envelopes for east–west elongated blocks.* (e) *Solar envelopes for north–south elongated blocks. (Reprinted by permission of R. Knowles,* Sun, Rhythm, Form, *© 1981, MIT Press.)*

(c) Skylines and Winter Sun. The actual skyline as seen from a given place can be charted to determine access to direct sun at any time of year (Fig. 3.11). Such an analysis should precede the planning of the location of a solar collector. Obstructions within the four "best" hours (10 A.M. to 2 P.M.) are particularly serious and should be minimized in collector siting.

At the same time, consideration of the neighbors' access to direct sun is appropriate. This can be checked by charting another skyline, somewhere along the northern boundary of your site. The building you are designing should be included on this skyline and modified, if necessary, to preserve solar access for your neighbor.

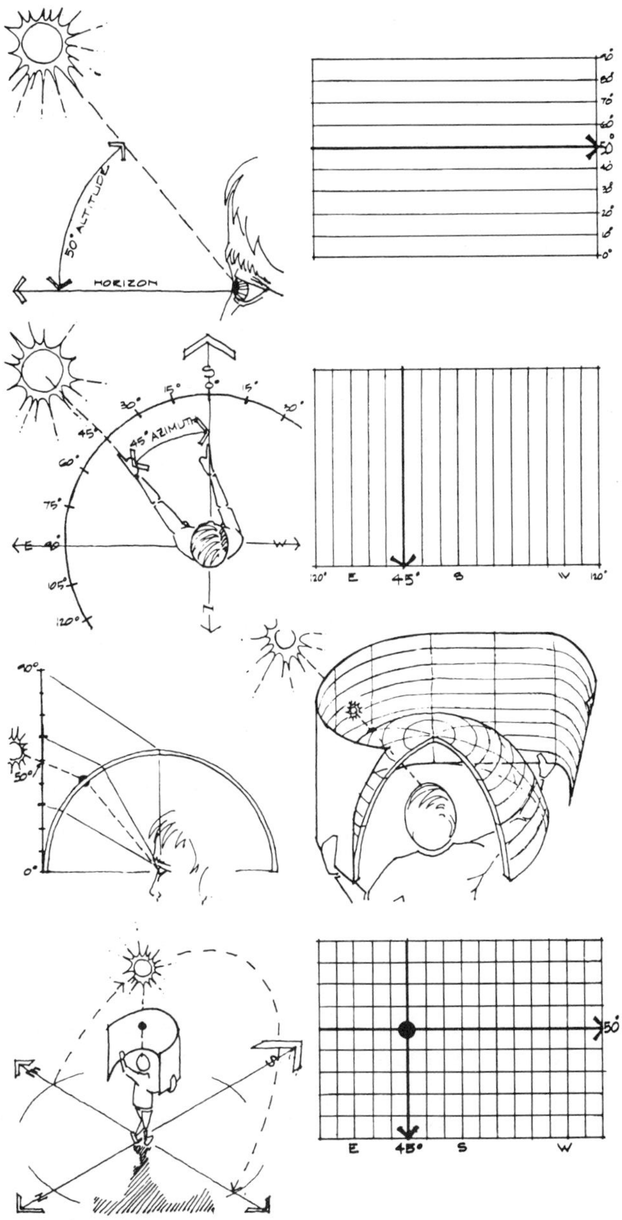

(a) Altitude

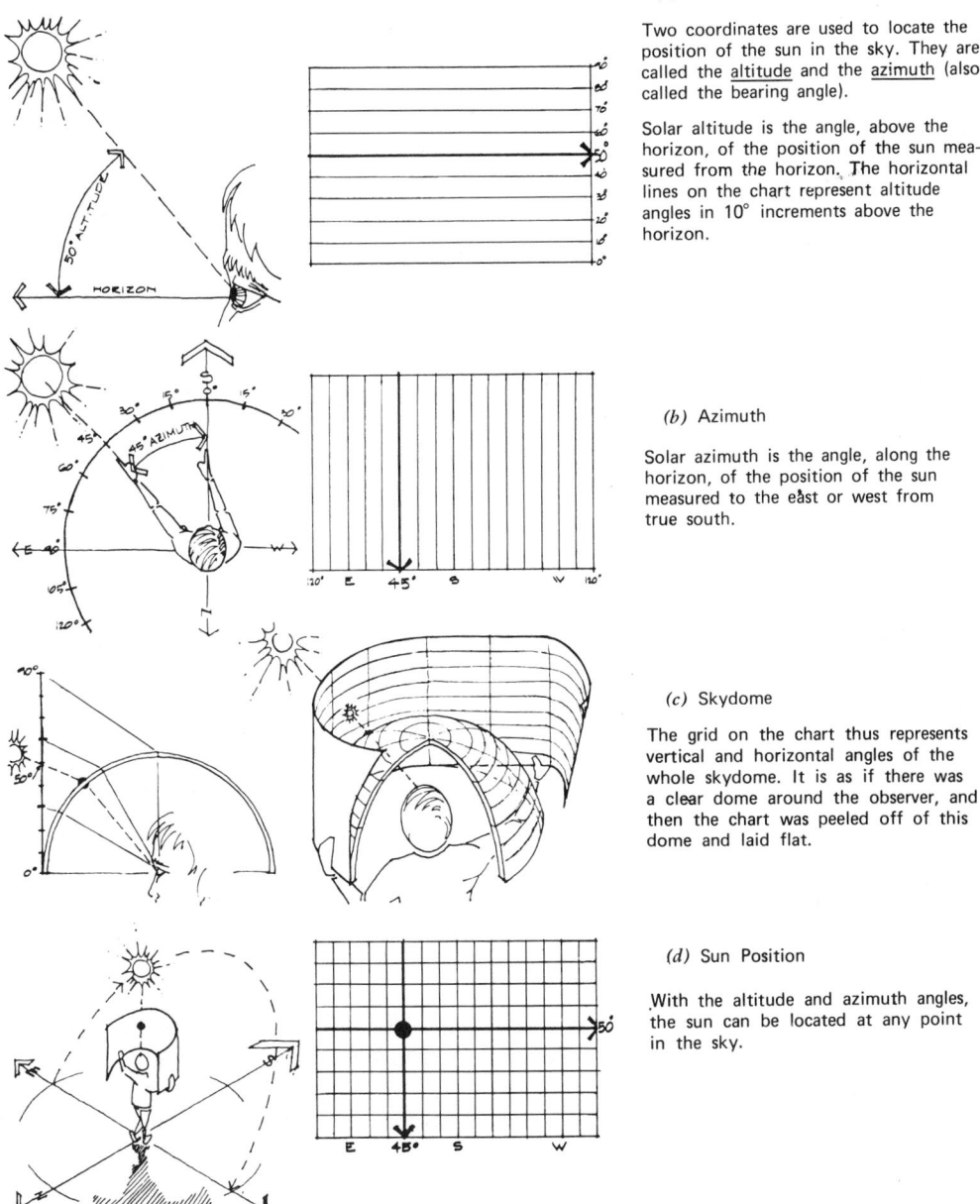

Two coordinates are used to locate the position of the sun in the sky. They are called the altitude and the azimuth (also called the bearing angle).

Solar altitude is the angle, above the horizon, of the position of the sun measured from the horizon. The horizontal lines on the chart represent altitude angles in 10° increments above the horizon.

(b) Azimuth

Solar azimuth is the angle, along the horizon, of the position of the sun measured to the east or west from true south.

(c) Skydome

The grid on the chart thus represents vertical and horizontal angles of the whole skydome. It is as if there was a clear dome around the observer, and then the chart was peeled off of this dome and laid flat.

(d) Sun Position

With the altitude and azimuth angles, the sun can be located at any point in the sky.

Fig. 3.10 *Sun charts. Illustrations* (a) *through* (g) *describe the development of the charts shown in* (h) *and* (i), *for two latitudes that cross the United States (see map, Fig. 3.12). (From Edward Mazria and David Winitsky,* Solar Guide and Calculator, *Center for Environmental Research, University of Oregon, 1976, and Edward Mazria,* The Passive Solar Energy Book, *Rodale Press, Emmaus, Pa., 1978.)*

(d) Sun and Shadows: Model Technique. The graphic techniques just discussed have a limitation: each graph applies to only *one* particular location. To study multiple locations, multiple graphs must be constructed. By contrast, a three-dimensional model used in conjunction with a sun shadow plot (Fig. 3.12) can yield the three-dimensional sun penetration pat-

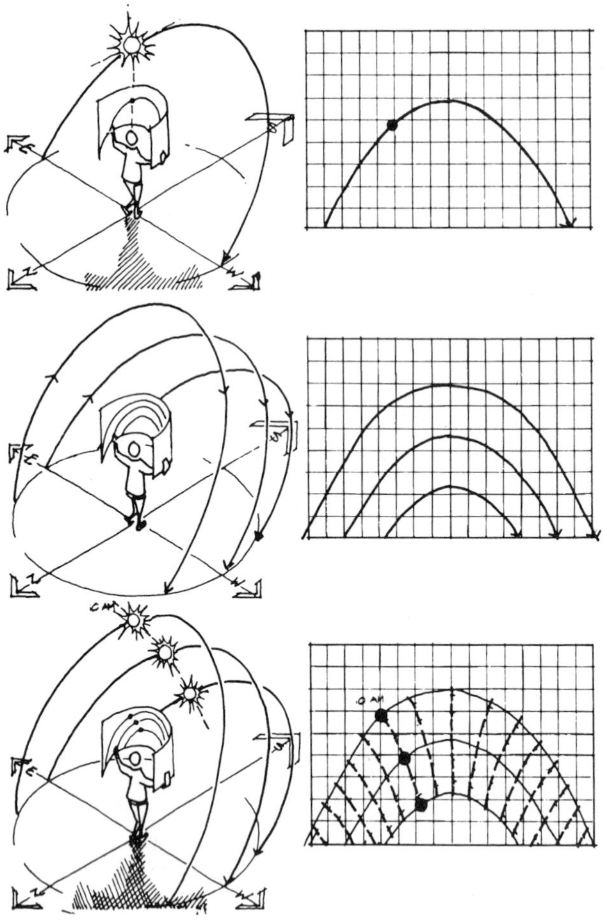

(e) Sun Path

By connecting the points of the location of the sun throughout various times of the day, the sun's path for that day can be drawn.

(f) Seasonal Paths

Thus, we can plot the sun's path for any day of the year. Those paths shown represent summer, fall/spring, and winter. The sun path is greatest during the summer when it reaches its highest altitude and rises and sets with the widest azimuth angle from true south. During winter, the sun is much lower in the sky and rises and sets with the narrowest azimuth angles from true south.

(g) Times of Day

Finally, if we connect the hour lines on each path, we get the heavy dotted line which represents the hours of the day. This completes the sun chart. NOTE: The times on the sun chart are for solar time. This may vary from standard time as much as an hour and fifteen minutes for different locations and different times of the year. This is adequate for most practical uses of the sun chart. It is important to remember to at least use standard time when using the charts.

Fig. 3.10 *(continued)*

terns as they change over time, for *any* location, indoors or out, on the model. Models are initially time consuming to build, but can save time when considering alternate locations on a site, alternate window and space combinations, alternate shading devices, and so on. Perhaps most important to the designer, models suggest three-dimensional alternative solutions, because you are testing *space,* not merely plan or section.

However, if obstructions to needed sun exist far from the site (such as nearby mountains), you might want to rely on graphs rather than to try to include such large obstructions on your model.

In Fig. 3.12, "sun peg shadow plots" are found for six latitudes. A copy of such plots, correctly attached to a model *of any scale,* will allow the designer to quickly determine exact sun penetration patterns at many times, for any date. These plots are one of the most important tools for the solar designer, both early and late in the design process.

(e) Controlling Solar Reflections. The use of highly reflective (or "mirror") glass to reduce heat gain in office buildings, and the rapid spread of solar collectors on walls and sloped roofs, has increased the frequency of annoying solar reflections from buildings (Fig. 3.13). The farther the sun's rays are from being perpendicular to any surface, the more of the sun's light is reflected, rather than absorbed, by that surface. Thus, the intensity of the reflection is greatest when the sun's rays are nearly paral-

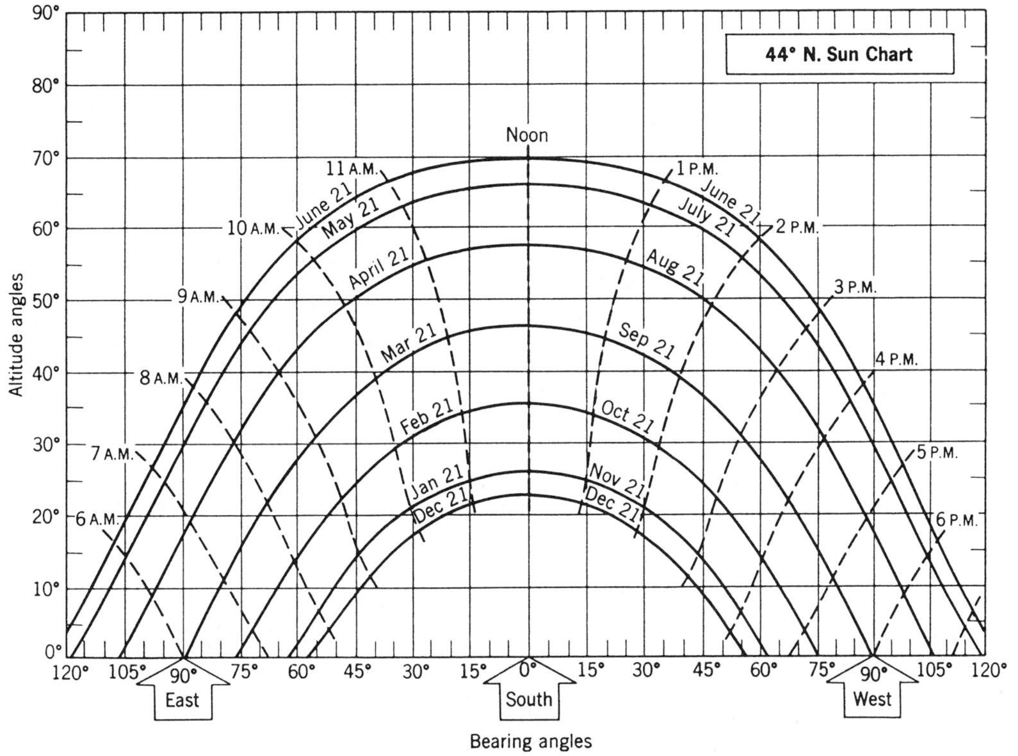

(h)

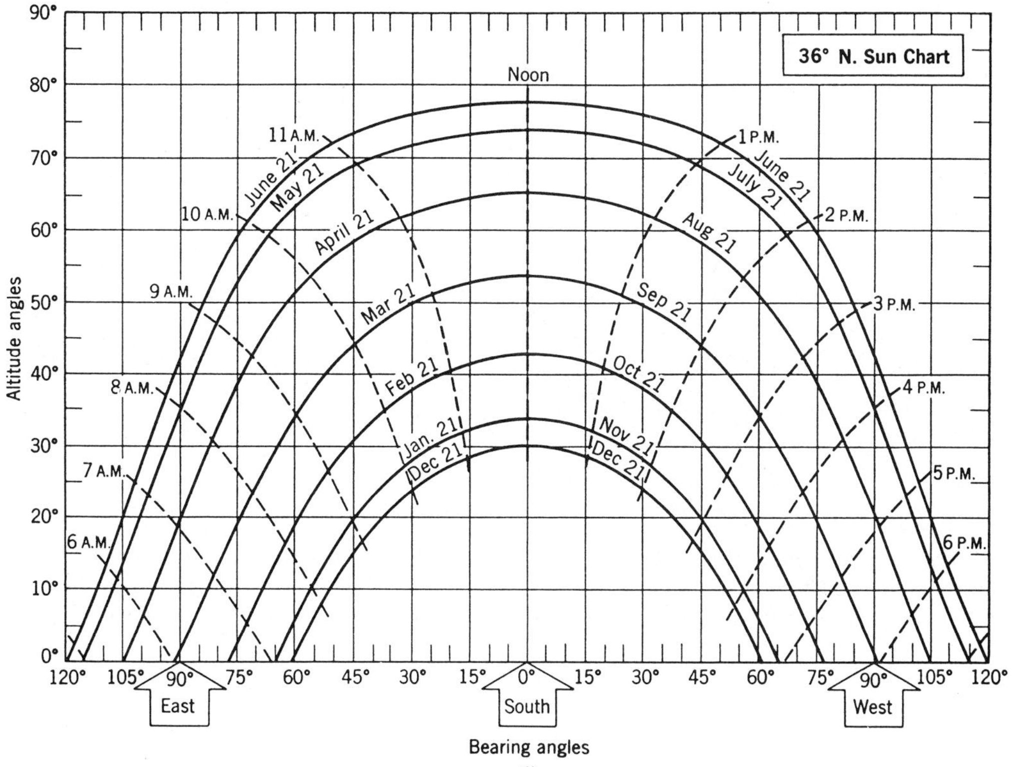

(i)

***Fig. 3.10** (continued)*

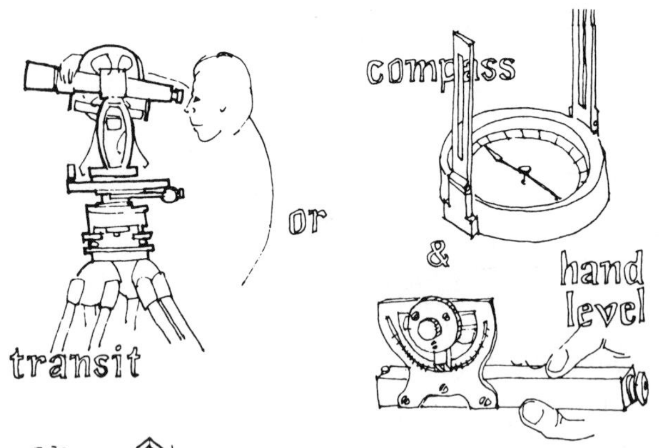

To find the amount of direct sun that a specific location receives, it is necessary to plot obstructions that block the sun as seen from that point. This plotting is called the "skyline" and it is done in a similar way in which the sun's path was plotted. If the skyline is low, that is with no tall trees, adjacent buildings, abruptly rising hills, or other obstructions, the following procedure is unnecessary.
You will need: (1) a transit or compass for the bearing readings and a hand level for altitude readings, and (2) a copy of the sun chart for your latitude.
Place yourself at the approximate location where you wish to know the solar radiation This may require a ladder if the location is on the second floor or roof of a building—to—be.

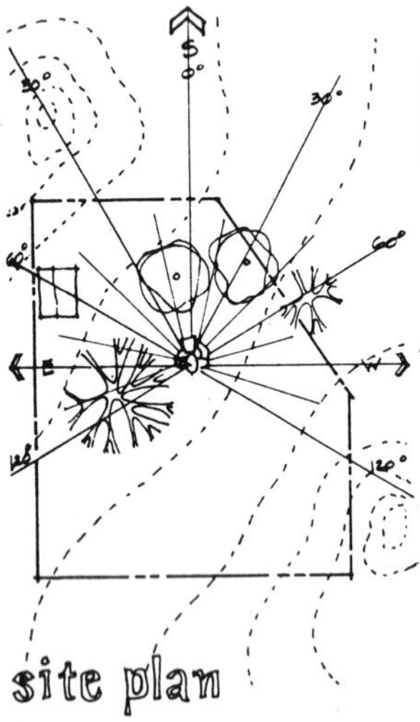

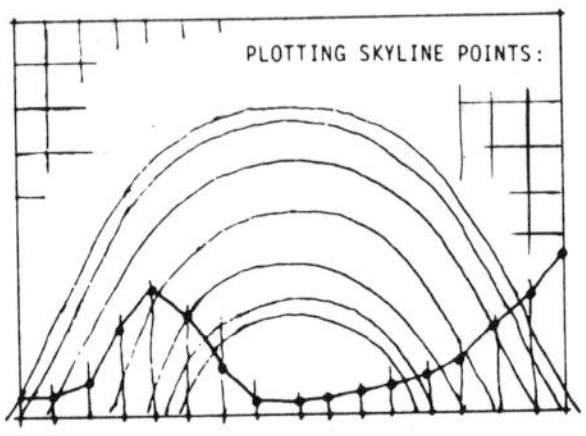

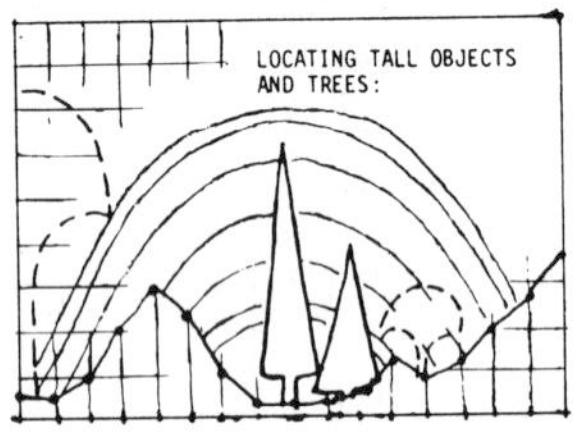

Find the altitude of the skyline as follows:

- Determine which direction (bearing) is true south.
- Aiming your level true south, determine the altitude (angle above the horizon) of the skyline. Plot this point on your chart above 0° true south.
- Similarly, find and record the altitudes of skyline for each 15° bearing, both east and west of south, to at least 120°. This is a total of 17 altitude readings. Plot these readings for the respective bearing angles on the sun chart and connect them.
- For isolated tall objects that block the sun such as tall evergreen trees, find both the bearing angle and the altitude for each object and record them at the appropriate point on the chart.

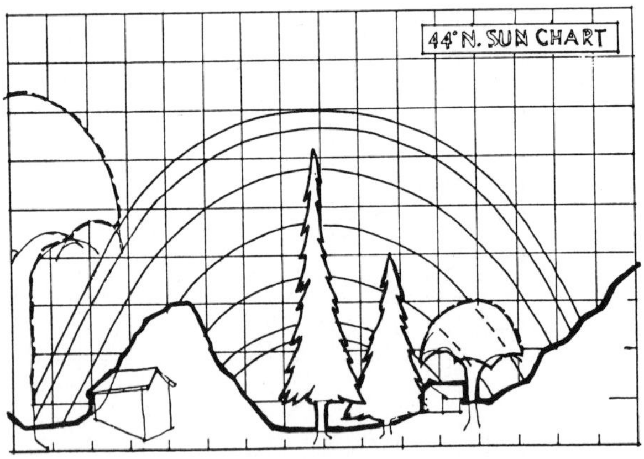

- Finally, plot the deciduous trees in the skyline with a dotted line. These are of a special nature because they will block the sun during spring through fall and let most of the sun pass through when their leaves are gone, late fall through early spring. This completes the skyline. The open areas on your completed chart are those times when the sun will reach this specific location.

Fig. 3.11 *Charting the skyline from a specific location. (From Edward Mazria and David Winitsky,* Solar Guide and Calculator, *Center for Environmental Research, University of Oregon, 1976, and Edward Mazria,* The Passive Solar Energy Book, *Rodale Press, Emmaus, Pa., 1978.)*

28° N

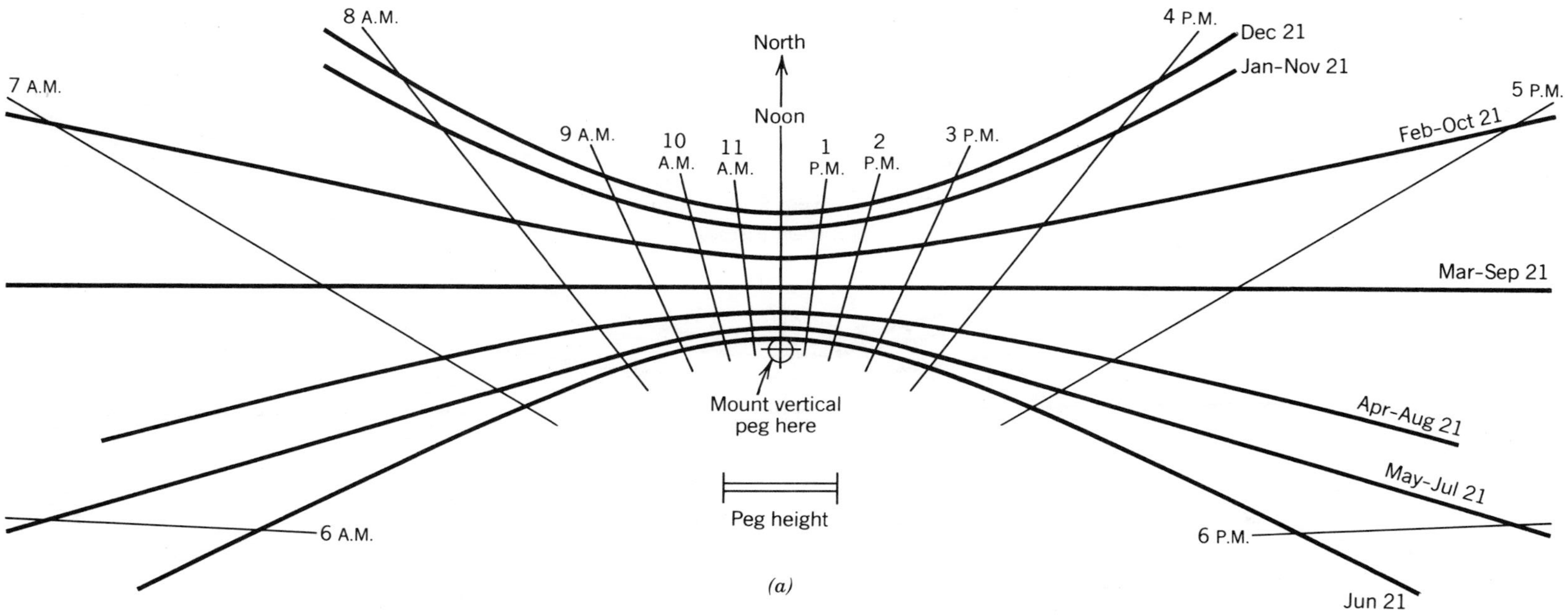

(a)

Fig. 3.12 (a–f) *Sun peg shadow charts. These charts will show the exact position of sunlight and shadow on a model of any scale, on any date, at any time of day between shortly after sunrise and shortly before sunset.*

1. *Find the chart nearest your latitude using Fig. 3.12g.*
2. *Make a copy of this chart. (Don't worry if the copier changes the chart size, because the peg height will be changed in the same proportion.)*
3. *Construct a "peg" that will stand—and remain—perfectly vertical, and whose finished height above the chart surface corresponds exactly to the "peg height" shown on your copy of the chart for your latitude.*
4. *Mount your copy of the chart on the model to be tested. It is important that the chart be perfectly horizontal over its entire surface, and that the north arrow on the chart point to true north for the model.*
5. *Mount your vertical peg at the location shown on the chart. Be sure it's vertical, and will stay vertical.*
6. *Choose a test time and date. Take your model out into direct sunlight. Then* tilt *the model until the shadow of the peg points toward the intersection of the chosen time's line and the chosen date's curve. When the end of the peg touches this intersection, your model will show the same* sun–shadow pat*terns as would occur on the time and date at the intersection you choose.*

(Reprinted by permission from Brown, Reynolds, and Ubbelohde, InsideOut: Design Procedures for Passive Environmental Technologies, *John Wiley and Sons, © 1982.)*

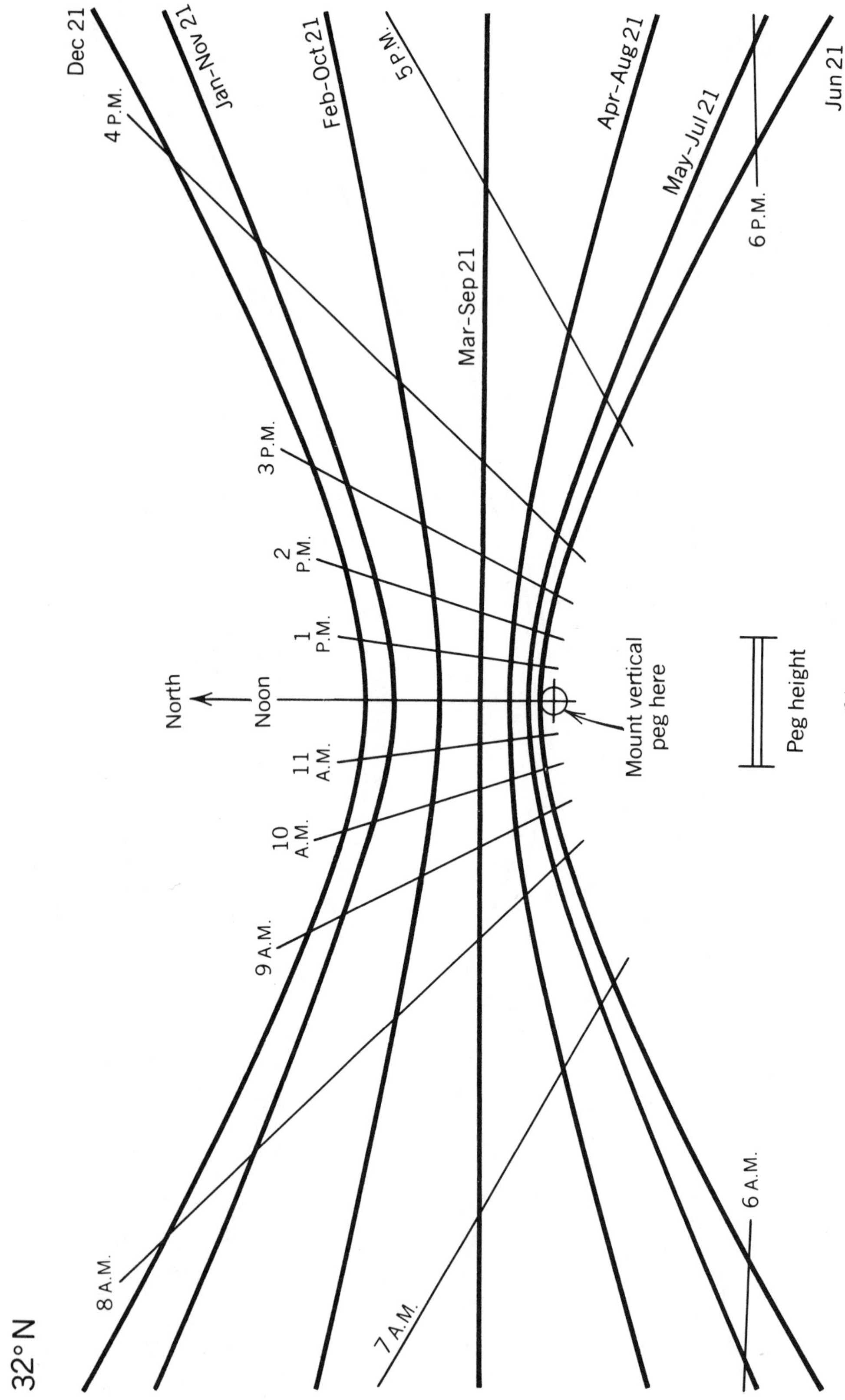
32° N
Dec 21
Jan–Nov 21
Feb–Oct 21
Mar–Sep 21
Apr–Aug 21
May–Jul 21
Jun 21
North
Noon
8 A.M.
7 A.M.
6 A.M.
9 A.M.
10 A.M.
11 A.M.
1 P.M.
2 P.M.
3 P.M.
4 P.M.
5 P.M.
6 P.M.
Mount vertical peg here
Peg height

(b)

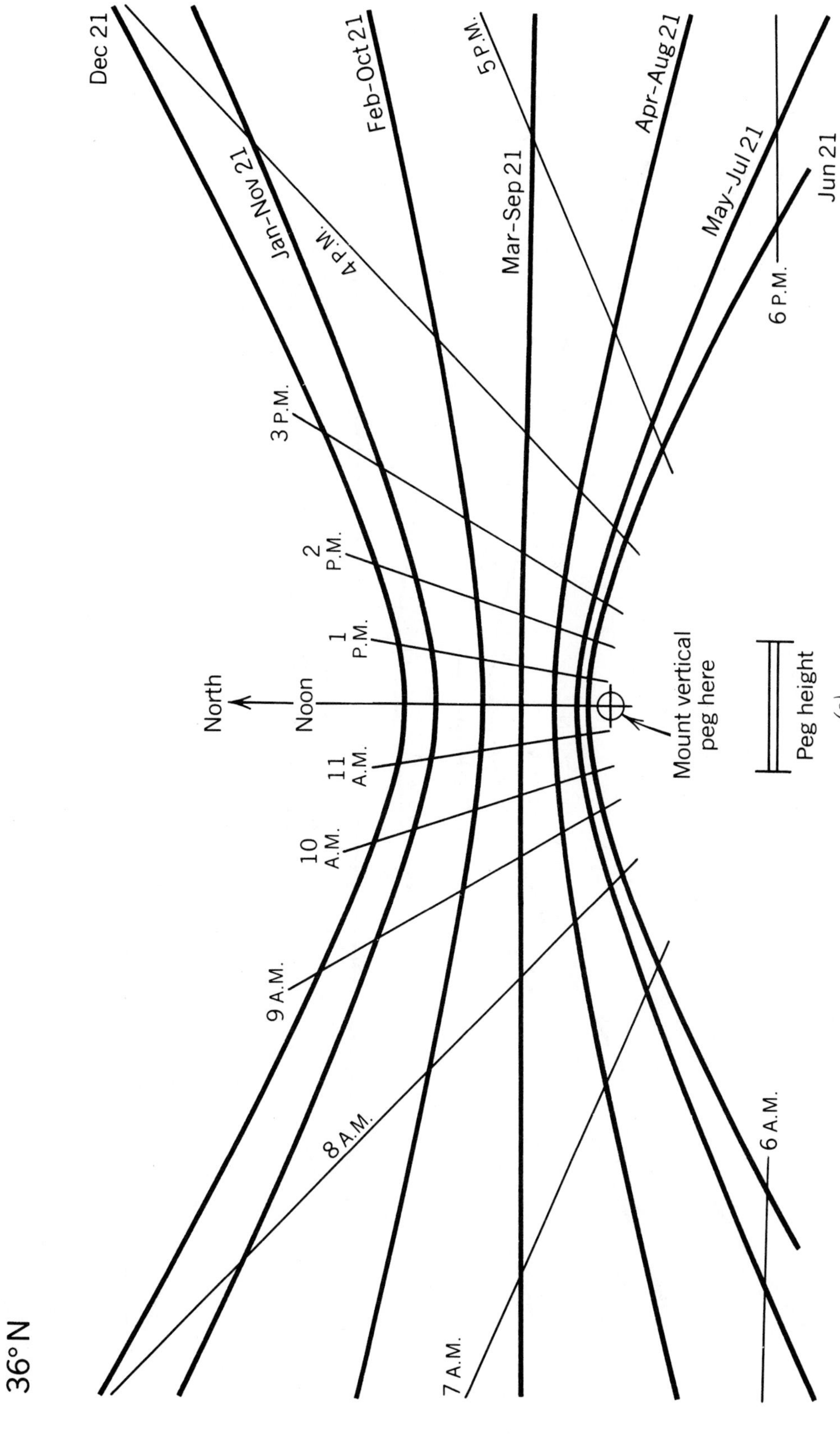

Fig. 3.12 *(continued)*

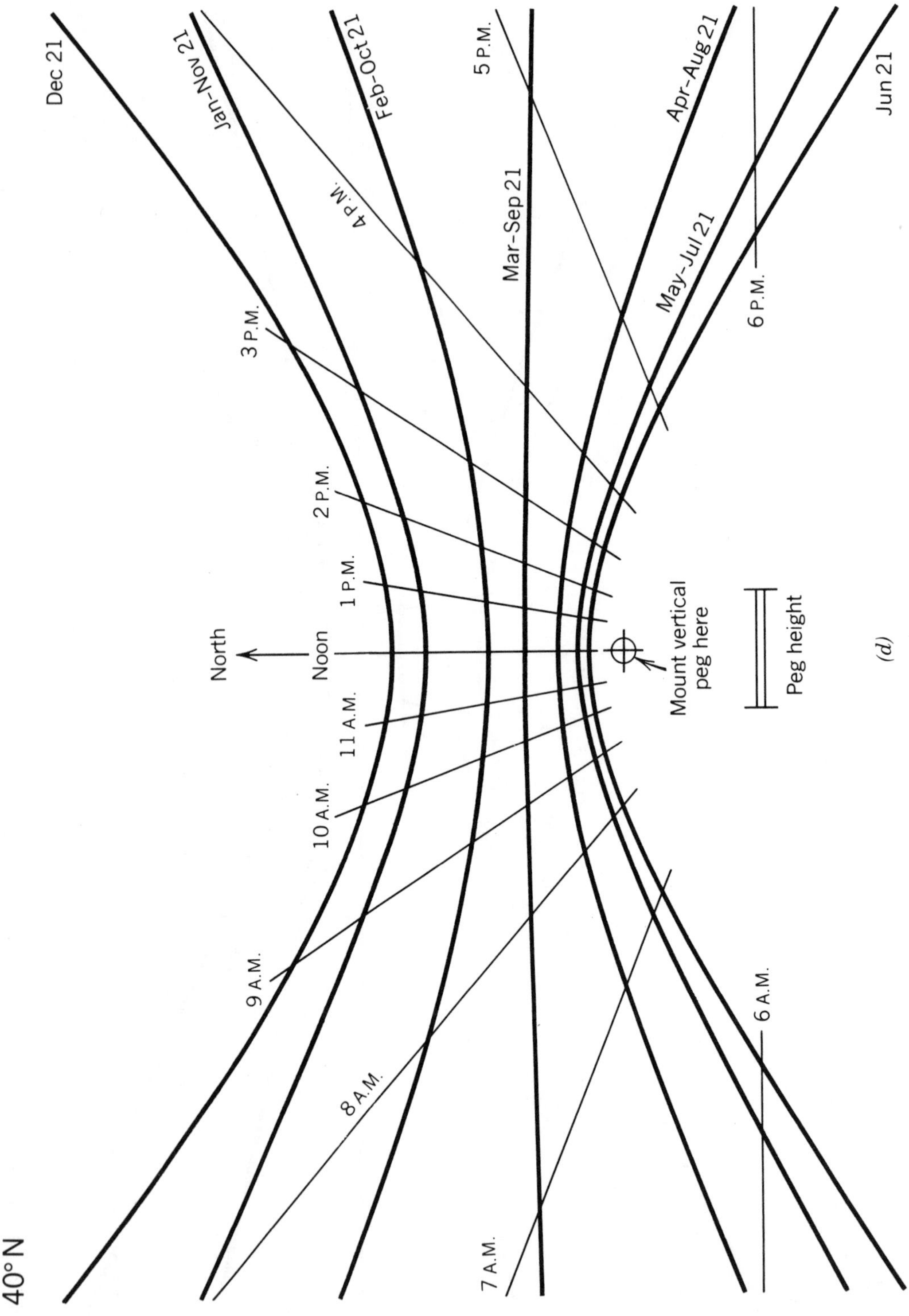
40°N
Dec 21
Jan-Nov 21
Feb-Oct 21
Mar-Sep 21
Apr-Aug 21
May-Jul 21
Jun 21
North
Noon
1 P.M.
2 P.M.
3 P.M.
4 P.M.
5 P.M.
6 P.M.
11 A.M.
10 A.M.
9 A.M.
8 A.M.
7 A.M.
6 A.M.
Mount vertical peg here
Peg height

(*d*)

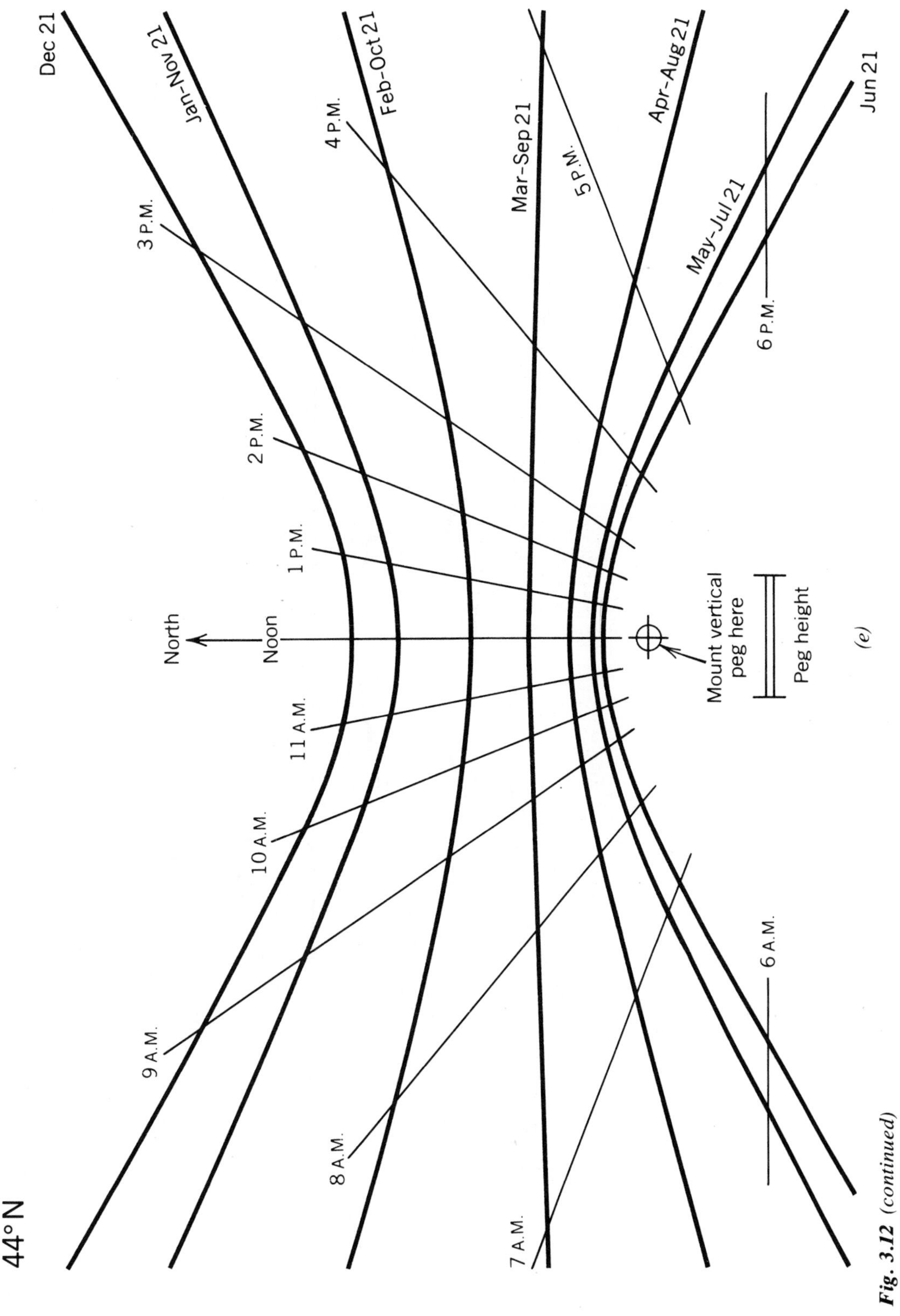

Fig. 3.12 *(continued)*

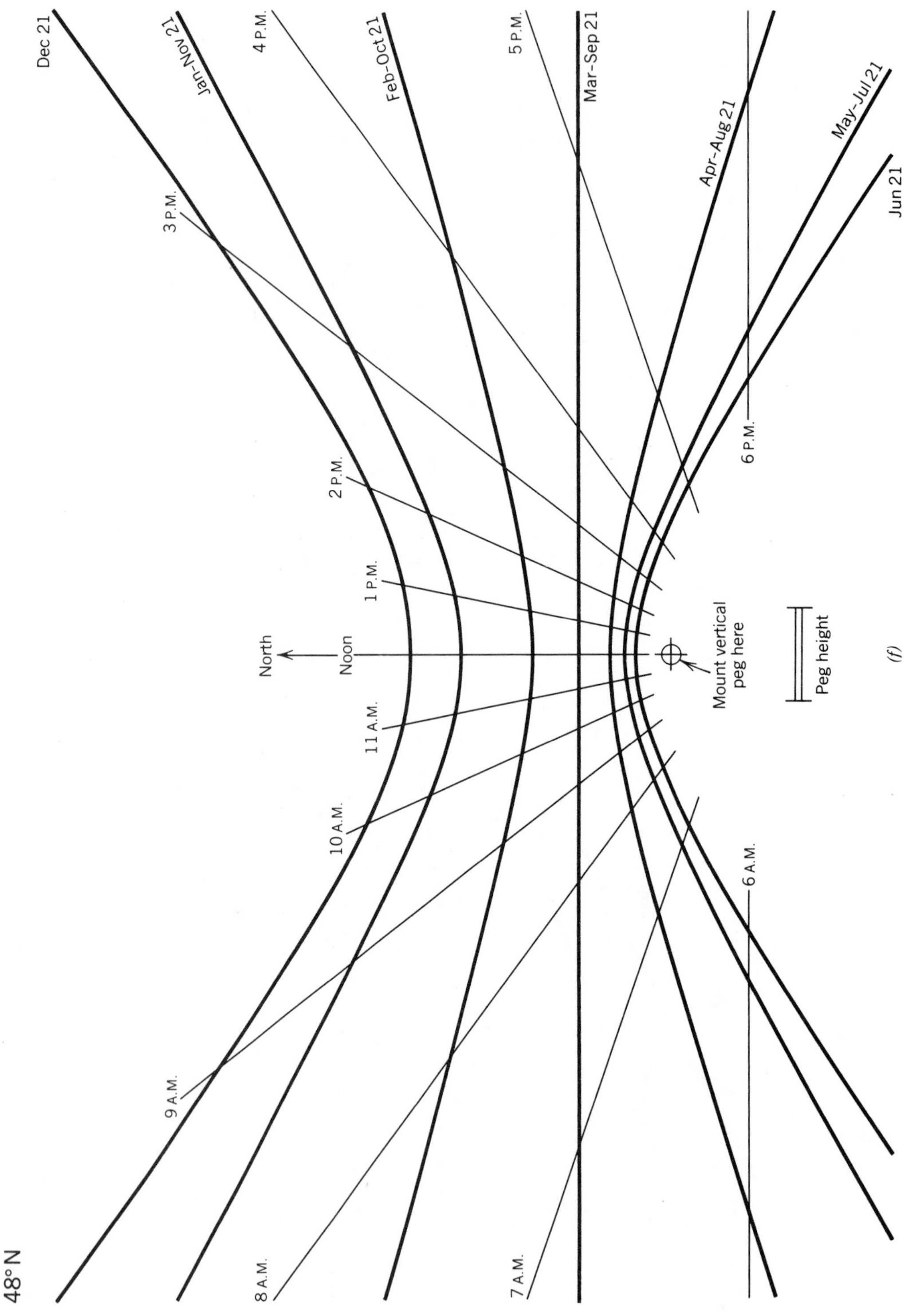

Fig. 3.12 *(continued)*

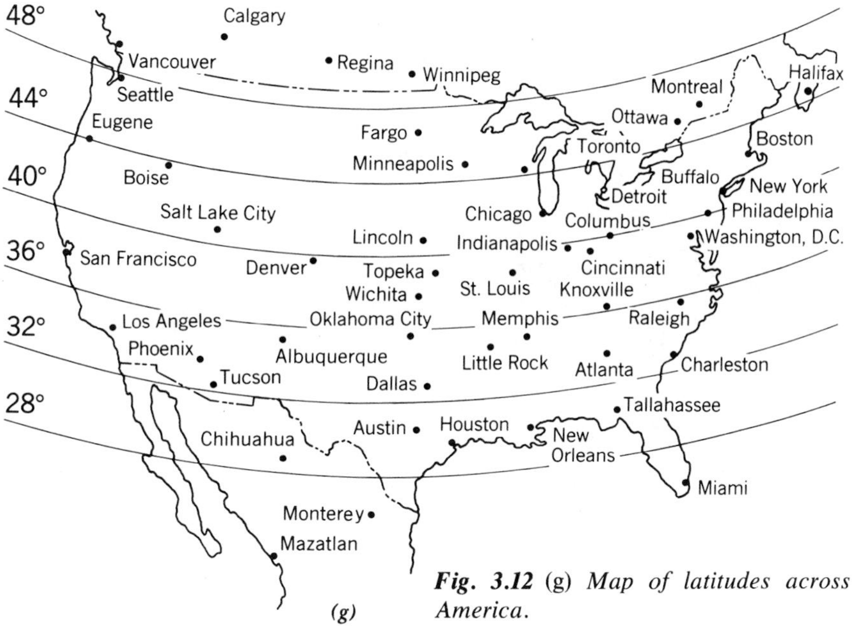

(g)

Fig. 3.12 (g) *Map of latitudes across North America.*

lel to a surface. Fortunately, this is the poorest time for solar collection through that surface; curtailed reflections need not mean curtailed collections.

The path or direction of the reflection is a "mirror image" of the sun ray's path. Figure 3.14 shows the intensity (percentage of sunlight striking a surface that is reflected) and the path (on horizontal ground) of reflections from a vertical south wall at 44°N latitude. Such an analysis can be used to minimize the most intense solar reflections.

Since these most intense reflections occur where the sun's rays are nearly parallel to a wall, they are easily blocked. For example, foliage (Fig. 3.15) can intercept reflected sunlight. In another approach, the designer can call for external projections around windows (Fig. 3.16).

Fig. 3.13 Reflections. Mirror-glass windows in a newer office building (left) in Milwaukee, Wisconsin, cast strong reflections on the north-facing wall of the older hotel next door. Although this reflected heat might be welcome in winter, the glare is intense. In summer, the older building is particularly disadvantaged.

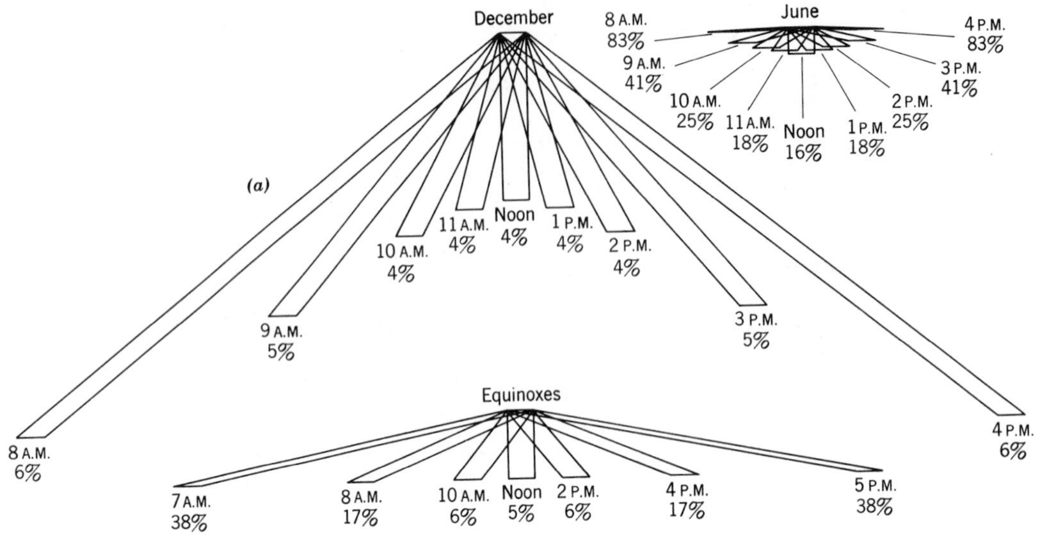

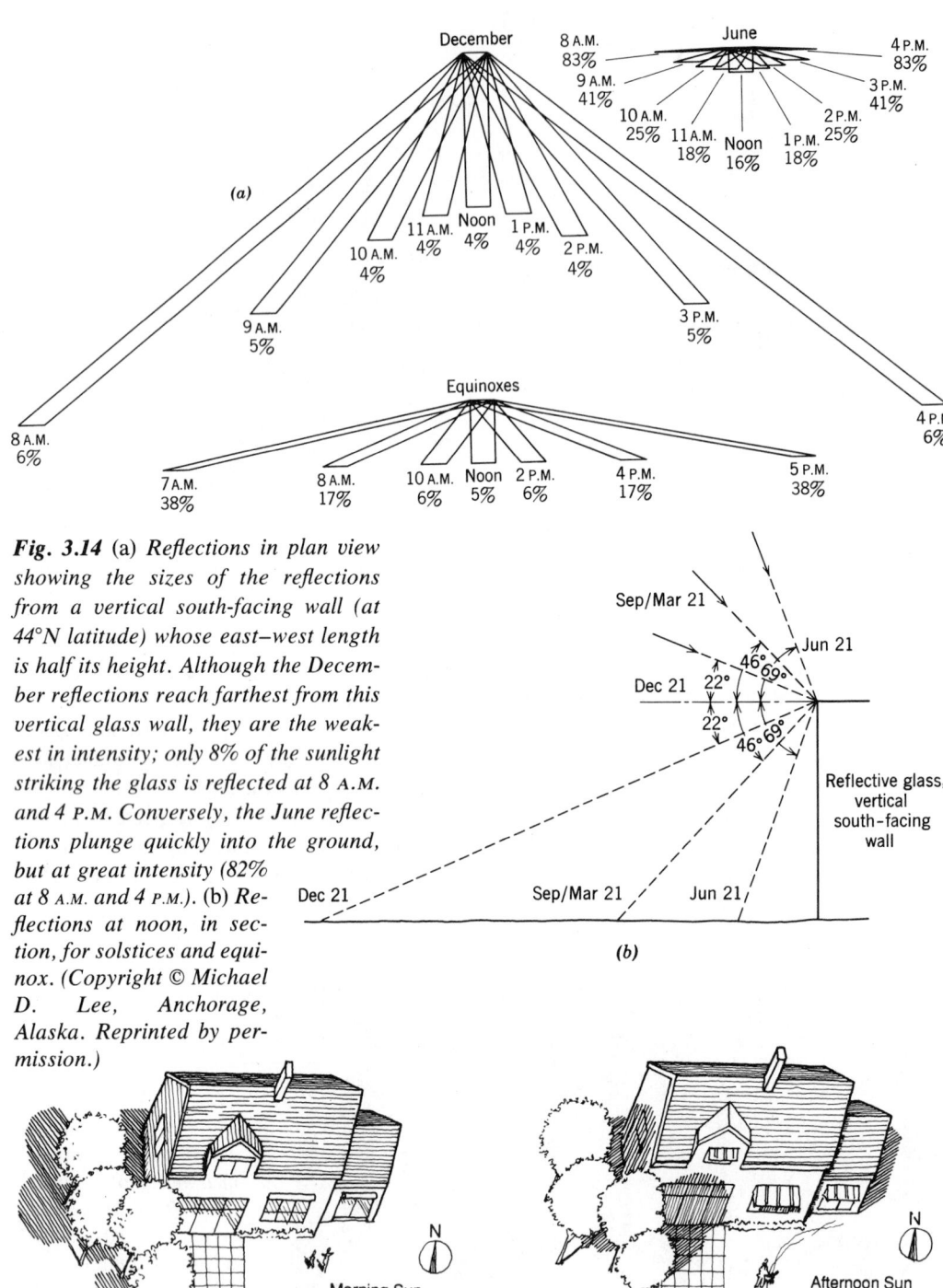

Fig. 3.14 (a) *Reflections in plan view showing the sizes of the reflections from a vertical south-facing wall (at 44°N latitude) whose east–west length is half its height. Although the December reflections reach farthest from this vertical glass wall, they are the weakest in intensity; only 8% of the sunlight striking the glass is reflected at 8* A.M. *and 4* P.M. *Conversely, the June reflections plunge quickly into the ground, but at great intensity (82% at 8* A.M. *and 4* P.M.*).* (b) *Reflections at noon, in section, for solstices and equinox. (Copyright © Michael D. Lee, Anchorage, Alaska. Reprinted by permission.)*

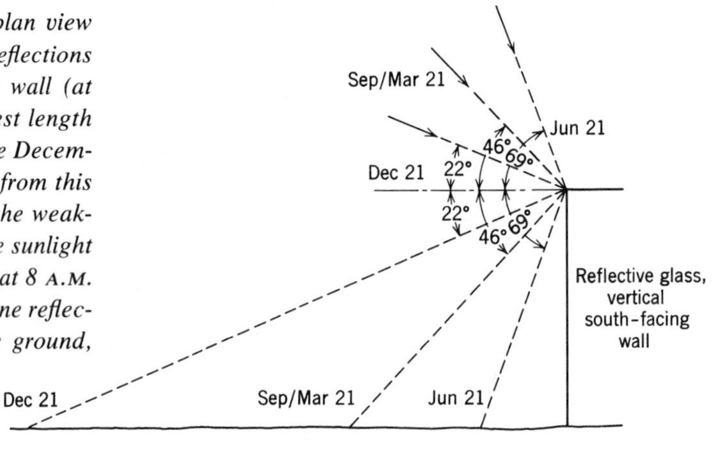

(b)

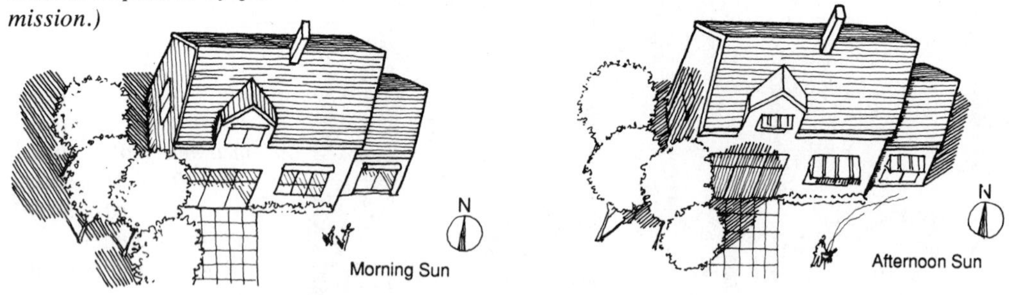

(a) *(b)*

Fig. 3.15 Selective protection from reflections. (a) *The tree standing west of this south window wall does not interfere with solar access during the best hours for solar gain (around noon), nor does it prevent early morning sun from entering the windows. The reflections of early morning are intercepted by the tree, before they can escape to annoy elsewhere.* (b) *The late afternoon sun is blocked by the tree, before either heat gain or reflections can occur.*

Fig. 3.16 *The "eggcrate" shading devices on the southwest corner of this University of Oregon building prevent reflections by blocking the sun from either side of the glass. However, different shading devices are appropriate for different orientations of facades.*

Fig. 3.17 *Apartment buildings in a series straddle the approach ramps to New York's George Washington Bridge. These buildings were the scene of a study linking noise level with reading disabilities (Cohen et al., 1973).*

3.5 Sound and Air

Sound and air are considered together because they are so difficult to separate. Many buildings that could be opened to ventilation or cooling by breezes rely instead on forced ventilation because of the noise that would accompany the breeze through an open window. Polluted air is another potential deterrent to "natural" ventilation. Almost any device that reduces sound will also reduce the velocity of the breeze, as is true of most filtering devices to remove dirt particles.

(a) Noise. The building in Fig. 3.17 is unusual both in providing an opportunity for wind ventilation—air moves freely below it as well as around and over it—and in the extraordinary intensity of noise that is carried through such air.

Two characteristics of cities (see Section 3.1) contribute to increased noise at street level: *hard surfaces* reflect rather than absorb sound, and *parallel walls* intensify noise between them rather than dissipating it. Increasing horizontal and vertical distance from urban noise sources affects the outdoor noise level on the street in different ways. See the graphs of sound level versus distance (and height) of Fig. 3.18. Although building surfaces are generally hard for durability in weathering, softer and multiplaned materials (such as plants) are desirable from a public noise reduction viewpoint. Their impact on measured noise may be slight, but visually softer surfaces reinforce our perception of acoustically softer environments (much as the sound of running water reinforces our perception of cool environments). Furthermore, the sounds made by plants as wind moves through them can help mask the unwanted noises of the street beyond. Fountains are especially useful sources of such *masking sound;* they can be kept running as long as the noise persists, and they enhance the cooling function of natural ventilation, especially in drier climates.

Where site conditions allow, barriers to street noise can be installed that cast "sound shadows" on the surface of the site (Fig. 3.19). Such barriers do little to reduce noise levels at upper windows, but surface activities can be given much lower noise levels, especially very near the barrier. Many cities now require such barriers between new housing developments and highways or railroads.

Another urban noise source is the mechanical equipment of buildings themselves. Many of the noise complaints against buildings involve air conditioning equipment (the compressive refrigeration cycle and its year-round utilization as a heat pump are described in Chapter 6). Noise is generated both by the compressor and by the great quantity of exterior air that must be rapidly pushed through outdoor coils (further heating outdoor air so that the indoors may be cooled). When densely packed buildings are forced to rely on mechanical cooling, such closeness makes the noise of the systems even more annoying. The machine's appetite for outdoor air is so enormous that attempts to surround it with noise shields can greatly hinder its efficient operation and shorten its life.

In residential neighborhoods, the greater distance between buildings might be expected to lessen these difficulties. Yet the much lower ambient (or background) noise level of residential areas is one of their more desirable characteristics, and an intruding compressor can trigger lawsuits from neighbors, who formerly enjoyed cool—and quiet—night breezes.

(b) Air Pollution. While one of the deterrents to operable windows is the dirt that they could admit to the interior, such air pollution is a threat to people and their buildings in other ways, as well.

Problems of global importance are resulting from air pollution; these are summarized in Table 3.3. The greenhouse effect (Fig. 3.20) has entered the public consciousness, with its threat of global warming. This occurs because gases

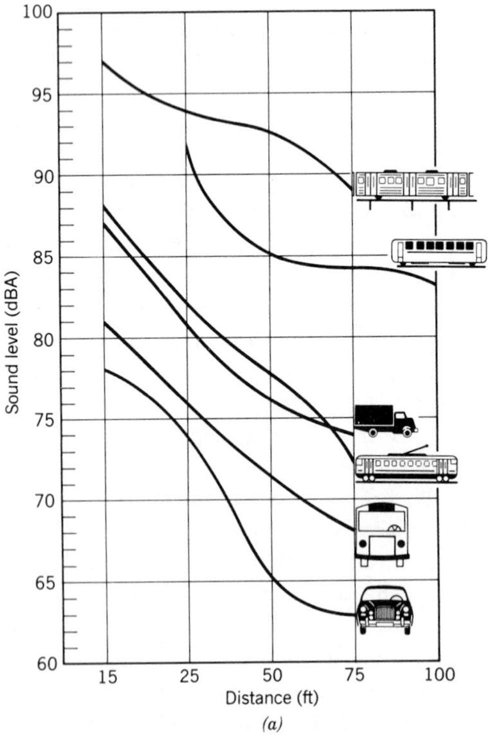

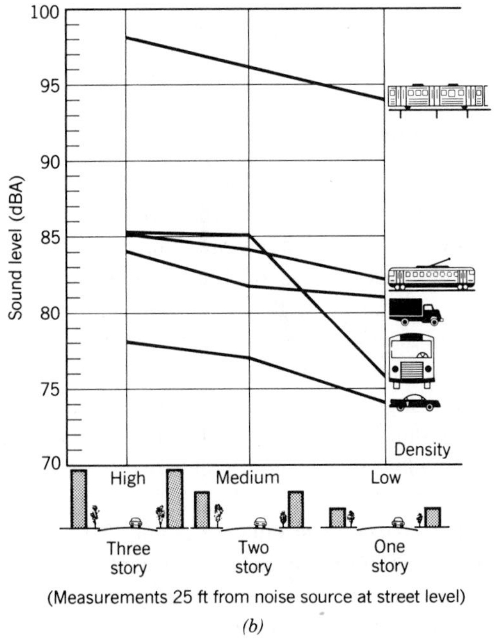

Fig. 3.18 *Predicting noise levels outdoors.* (a) *Distance as a factor in noise intensity.* (b) *Building height as a factor in noise propagation. (From Clifford R. Bragdon,* Noise Pollution: The Unquiet Crisis, *University of Pennsylvania Press, 1971. Reprinted by permission.)*

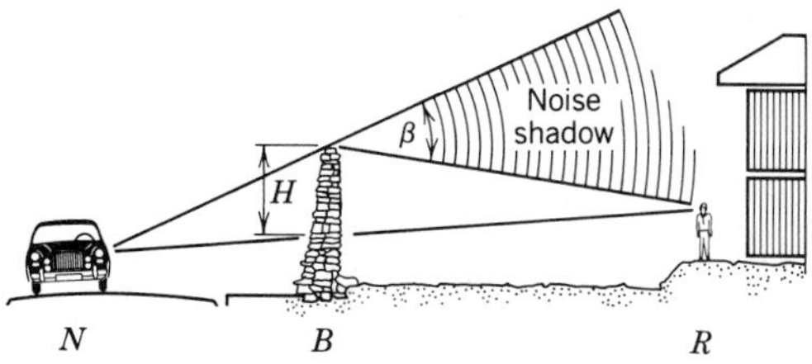

(a)

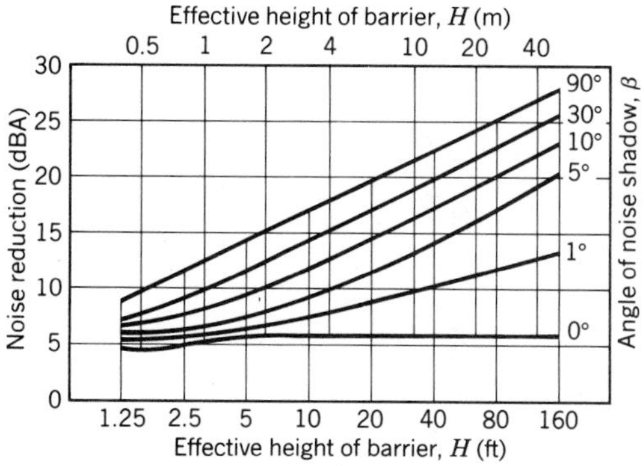

(b)

Fig. 3.19 Barriers to outdoor noise sources. (a) *To determine the approximate reduction in noise in decibels (dBA) due to an outdoor barrier, construct a section locating the noise source* (N), *the solid barrier* (B), *and the receiver's location* (R). *On this section, determine effective height* (H) *of the barrier, and the diffraction angle* (β) *with the resulting "noise shadow." Enter the graph* (b) *with* H *and* β; *where their lines intersect, determines the noise reduction in dBA (left margin). Note significant noise reduction from simply breaking the line of sight* (β = 1°). *(From Leslie Doelle,* Environmental Acoustics, *McGraw-Hill Book Company, 1972. Reprinted by permission.) For a more detailed procedure, see Fig. 27.29.* (c) *A sound barrier along a freeway near Washington, D.C. Columns fill the space between the giant slabs, creating a solid wall.*

that absorb long-wave radiation (heat) from the earth's surface are accumulating in the atmosphere. Energy production and use (especially of fossil fuels) are contributing heavily to these greenhouse gases, which include carbon dioxide, methane, nitrous oxide, ozone, and CFCs (chlorofluorocarbons).

The designers of buildings are affected in several ways. First, we can help to greatly reduce the air pollution caused by electric power plants and by fuel burning in buildings, by designing both for *greater energy conservation* and by utilizing *clean and renewable energy sources* within our buildings; Chapter 5 is concerned largely with this topic. Then we can specify materials and equipment that, through their manufacture or operation, lessen air pollution. This suggests *avoiding fuel combustion* (coal, oil, trash, wood, and natural gas, roughly in descending order of air pollution threat). It encour-

TABLE 3.3 **Air Pollution and the Atmosphere**

Part A. Effects of Various Gases on the Atmosphere[a]

Gas	Greenhouse Effect	Stratospheric Ozone Depletion	Acid Deposition	Smog	Corrosion	Decreased Visibility	Decreased Self-Cleansing of Atmosphere
Carbon monoxide (CO)							+
Carbon dioxide (CO_2)	+	+/−[b]					
Methane (CH_4)	+	+/−[c]					+/−[c]
NO_x: nitric oxide (NO) and nitrogen dioxide (NO_2)		+/−[b]	+	+		+	−
Nitrous oxide (N_2O)	+	+/−[b]					
Sulfur dioxide (SO_2)	−		+		+	+	
Chlorofluorocarbons	+	+					
Ozone (O_3)	+			+			−

Part B. Concentrations of Gases Today and Predicted for the Future

Gas	Major Anthropogenic Sources	Anthropogenic/Total Emissions per Year (Millions of Tons)	Average Residence Time in Atmosphere	Average Concentration 100 Years Ago (ppb)	Approximate Current Concentration (ppb)	Projected Concentration in Year 2030 (ppb)
Carbon monoxide (CO)	Fossil-fuel combustion, biomass burning	700/2000	Months	?, n. hem. 40–80, s. hem. (clean atmospheres)	100–200, n. hem. 40–80, s. hem. (clean atmospheres)	Probably increasing
Carbon dioxide (CO_2)	Fossil-fuel combustion, deforestation	5500/~5500	100 years	290,000	350,000	400,000–550,000
Methane (CH_4)	Rice fields, cattle, landfills, fossil-fuel production	300–400/550	10 years	900	1700	2200–2500
NO_x gases	Fossil-fuel combustion, biomass burning	20–30/30–50	Days	0.001–? (clean to industrial)	0.001–50 (clean to industrial)	0.001–50[d] (clean to industrial)
Nitrous oxide (N_2O)	Nitrogenous fertilizers, deforestation, biomass burning	6/25	170 years	285	310	330 to 350
Sulfur dioxide (SO_2)	Fossil-fuel combustion, ore smelting	100–130/150–200	Days to weeks	0.03–? (clean to industrial)	0.03–50 (clean to industrial)	0.03–50[d] (clean to industrial)
Chlorofluorocarbons	Aerosol sprays, refrigerants, foams	~1/1	60 to 100 years	0	About 3[e] (chlorine atoms)	2.4–6[e] (chlorine atoms)

Source: "The Changing Atmosphere," by Thomas E. Graedel and Paul J. Crutzen, in *Scientific American*, September 1989.

[a] +, Gas contributes to the effect; −, gas ameliorates the effect; +/−, effect of the gas varies.

[b] Effects depend on the altitude.

[c] Generally ameliorates the effect in the northern hemisphere, contributes to effect in the southern hemisphere.

[d] Concentrations over highly industrial sites may not rise much, but the number of such sites is expected to grow, particularly in developing nations.

[e] Concentrations given in chlorine atoms because the molecules generally contain more than one ozone-destroying chlorine atom.

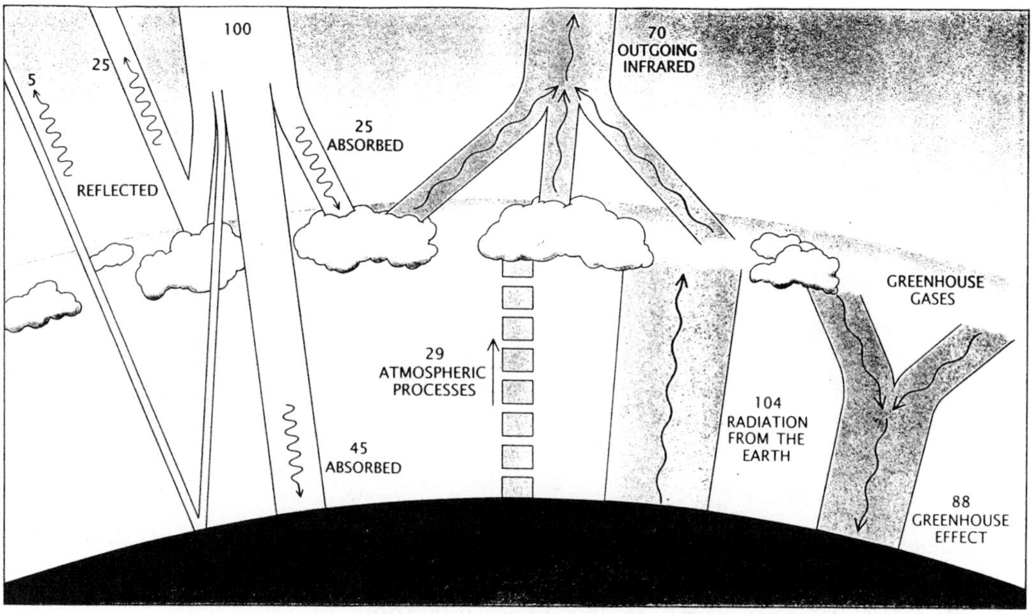

(a)

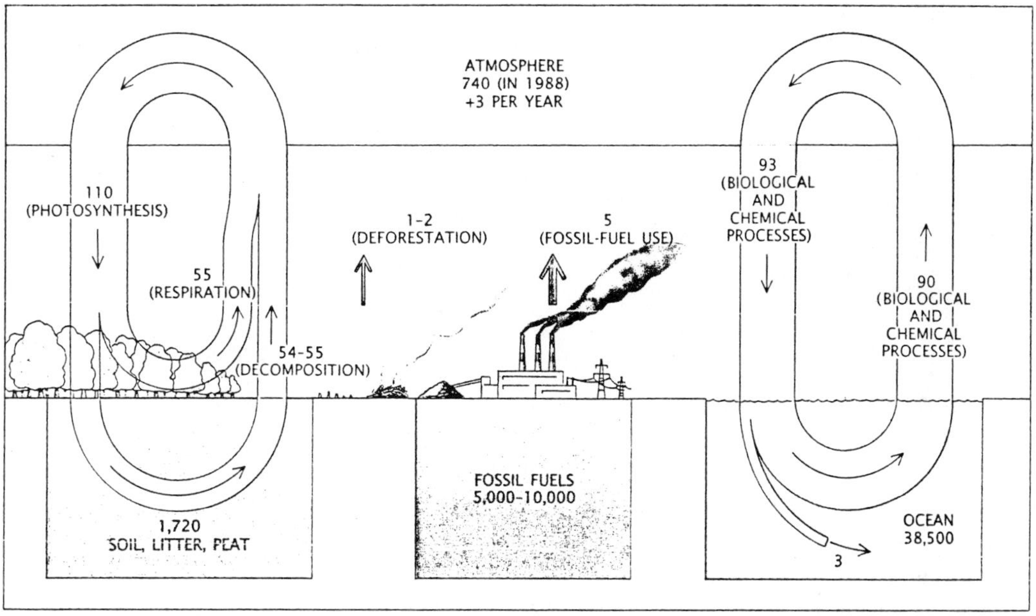

(b)

Fig. 3.20 *"Greenhouse effect," trapping heat in the earth's upper atmosphere.* (a) *Clouds and particles in the atmosphere reflect about 30% of incoming solar radiation, while they are returning about 66% of the heat that the earth would otherwise lose to outer space. Currently, the atmosphere keeps earth about 33 C° (60 F°) warmer than it would be without heat trapping.* (b) *Yearly carbon exchange (in billions of metric tons) between earth and atmosphere is roughly in balance for the natural cycles on land and in oceans; the human activities of fossil-fuel burning and deforestation are currently increasing atmospheric carbon by about 3 billion metric tons yearly. (From "The Changing Climate" by Stephen F. Schneider, in* Scientific American, *September 1989. Copyright © 1989 by Scientific American, Inc. All rights reserved.)*

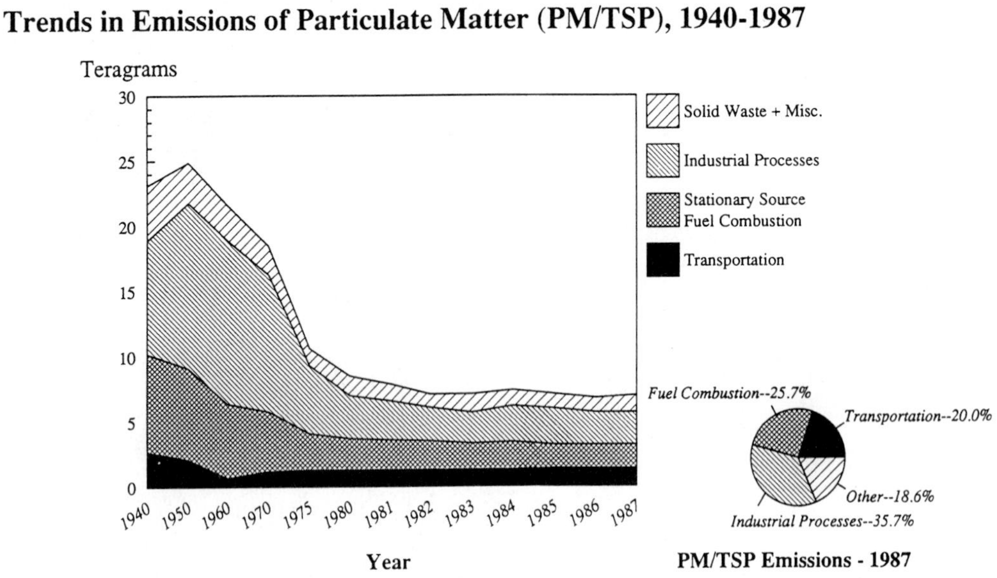

(*a*) Particulate matter

Particulate Matter Impacts: Visibly dirty air, reducing sunlight and visibility and contributing to dirty surfaces; respiratory irritation.

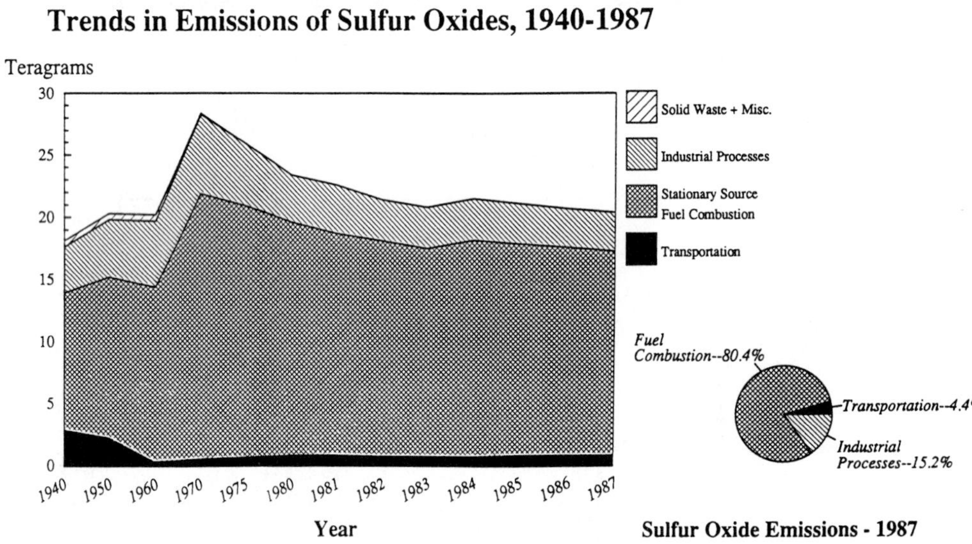

(*b*) Sulfur oxides

Sulfur Oxides Impacts: Harshly irritating to respiratory passages, contributing to diseases such as asthma, bronchitis, and emphysema. They also react with moisture to form acids that accelerate iron and steel corrosion and plant leaf droppage; "acid rain."

Fig. 3.21 *Air pollution in the United States, 1940–1987. Calculated total emissions of pollutants by source category. One teragram = 10^6 metric tons, or about 1100 short tons. [From U.S. Environmental Protection Agency (1989).]*

ages specifying that refrigeration equipment use *non-CFC refrigerants,* and that urethane foam insulation or upholstery products be made with non-CFC blowing agents. Until these practices are mandated, it remains the designer's responsibility to seek out environmentally benign products.

Figure 3.21 summarizes the impacts of air pollution in the United States and indicates hope for improvement, provided we continue to enforce the Clean Air Act. Our buildings are substantial contributors to air pollution: the fuel combustion within, the electric power plants that supply, and the incinerators and landfills

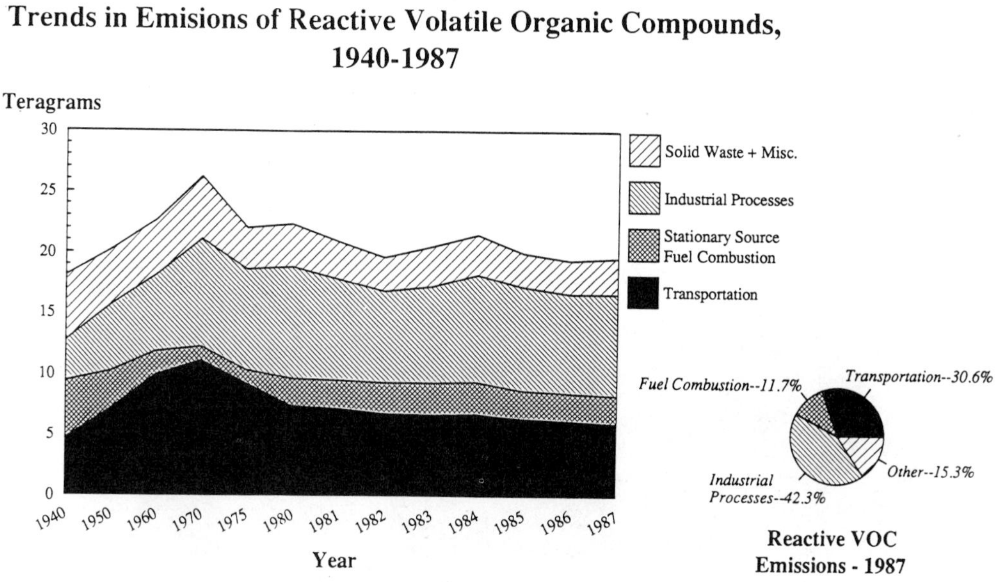

(*c*) Volatile organic compounds

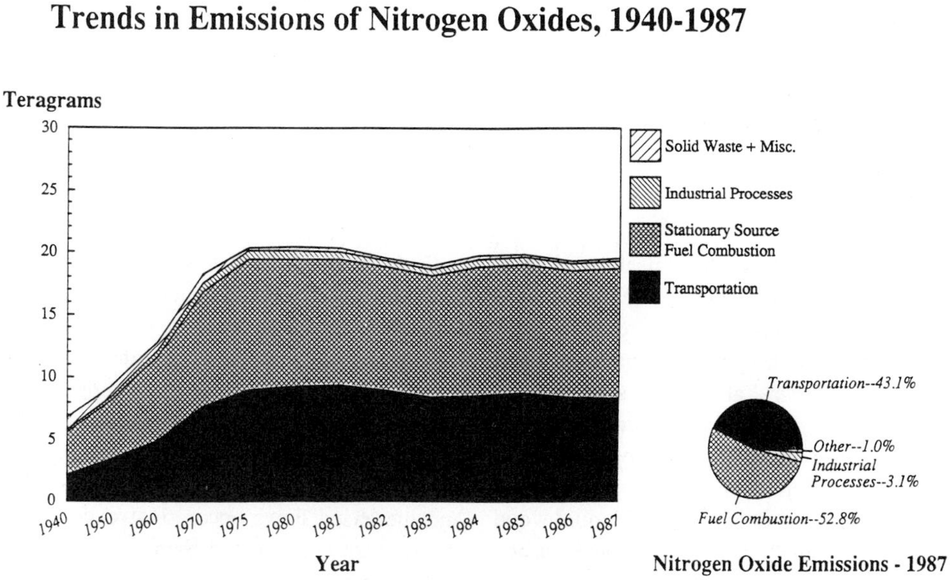

(*d*) Nitrogen oxides

Volatile Organic Compounds and Nitrogen Oxides Impacts: These combine with oxygen and sunlight to produce smog. Smog contributes to eye, nose, and throat irritations; it impairs normal lung function. Nitrogen oxides in rainfall contribute to increased acidity, which is damaging both to ecosystems and to buildings. Nitrous oxide is one of the greenhouse gases.

***Fig. 3.21** (continued)*

that receive wastes from buildings. Buildings/power plants are major contributors to the greenhouse effect and to the primary causes of acid rain (sulfur oxides) and of smog (nitrogen oxides). Indirectly, the transportation that takes people to and from buildings is another major air pollution source. We design buildings to utilize "fresh" air, whether by natural or forced ventilation; we must then also design buildings to preserve our fresh air resource. The less energy they require, the cleaner their outdoor air will be.

Particulate matter emissions from buildings/power plants are declining and were about one-fourth of the national total in 1987. These would come mostly from incinerators and wood

Trends in Emissions of Carbon Monoxide, 1940-1987

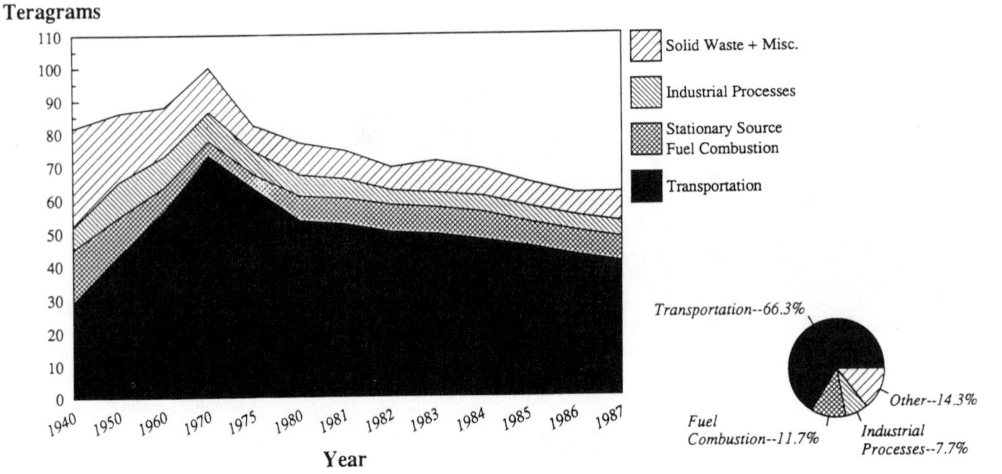

(*e*) Carbon monoxide

Carbon Monoxide Emissions - 1987

Carbon Monoxide Impacts: Tends to cause suffocation by interfering with the blood's ability to supply oxygen to the body; as a result, heart and respiratory systems are forced to work harder. Symptoms include headache, loss of vision, decreased muscular coordination, nausea, and abdominal pain.

Fig. 3.21 *(continued)*

stoves. *Sulfur oxide* levels have declined slightly nationwide, although "acid rain" remains a serious problem; there has been a decrease in sulfur dioxide in the urban centers, counterbalanced by an increase in rural areas from sources such as electric generating plants that burn fossil fuel. *Carbon monoxide* levels have changed little nationwide, but the concentration of this pollutant in urban centers has declined. Residential wood stoves are a contributor to increased carbon monoxide pollution since 1970.

The pollutants that cause smog continue to be troublesome, since levels of *nitrogen dioxide* have increased because of the increased total automobile miles traveled, and because of increasing electrical generation in fossil-fueled plants. *Volatile organic compounds* are released from buildings by combustion; the EPA cautions that its estimates from wood stoves (stationary source fuel combustion) are overstated in Fig. 3.21, perhaps by a factor of 3. *Lead* emissions have not been graphed; they are due in large part to transportation and industrial processes.

(c) Wind Control. For most buildings, wind (like sun) changes from friend to enemy with the change of seasons. Control of wind often means utilizing wind-sheltered areas in winter, while encouraging increased wind speeds for the building in summer. Outdoor spaces can benefit from this block-to-admit changeover on a daily basis. The generalized patterns of wind flow around thin windbreaks and thicker buildings (Fig. 3.22) help us understand where shelter and increased flows occur. However, they are much more complicated than they first appear and are highly influenced by objects upstream, to the sides, and downstream of the wind-directing object being analyzed. Wind tunnel tests are far more reliable than these patterns; unfortunately, such tests are expensive and still fraught with opportunity for misprediction. With these warnings in mind, the site can be analyzed graphically for seasonal wind utilization.

Wind will ultimately return to its original flow pattern after encountering an obstacle such as a windbreak or a building. Before it reaches the obstacle, it slows, builds pressure, and turns upward or sideways; it increases its speed as it passes, and reduced or negative pressure results at the sides and behind the obstacle. These pressure differences, flow patterns, and the size and

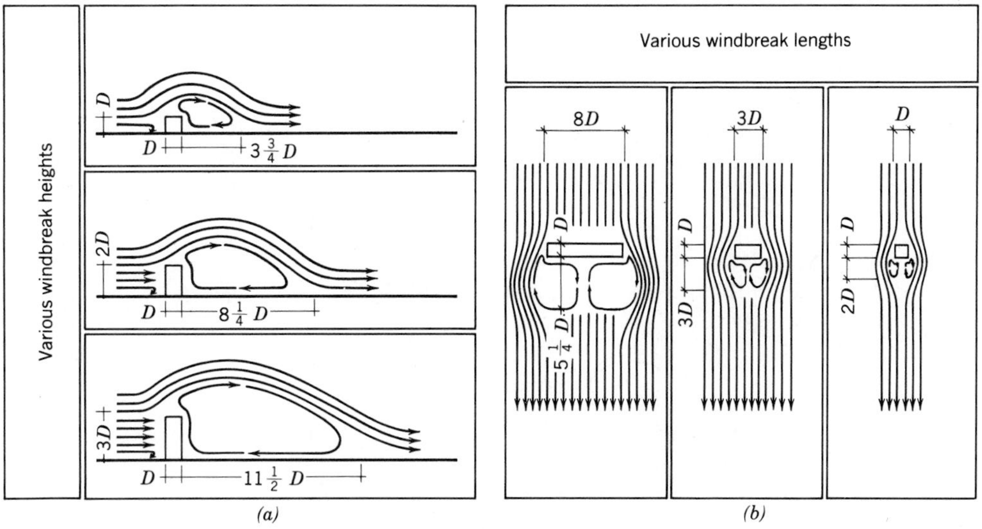

Density of Belt	Average Over First 50 Yd	Average Over First 100 Yd	Average Over First 150 Yd	Average Over First 300 Yd
	Wind Speed Reduction (%)			
Very open	18	24	25	18
Open	54	46	37	20
Medium	60	56	48	28
Dense	66	55	44	25
Very dense	66	48	37	20

(d)

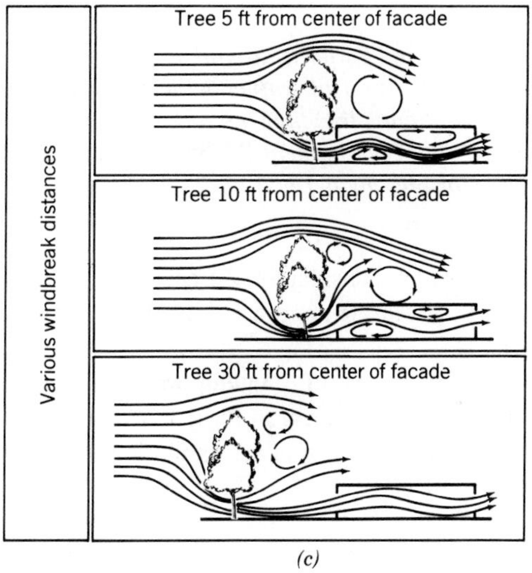

Fig. 3.22 *Approximate patterns of wind around objects.* (a) *Difference in barrier height.* (b) *Difference in barrier distance.* (c) *Difference in windbreak distance.* (d) *Reduction in wind speed due to density. (Reproduced with the permission of the American Institute of Architects, © 1981, AIA.)*

shape of wind-protected areas behind the obstacle are all utilizable for control of air motion, inside and out.

Wind around buildings is a complex matter; some rules of thumb for shelter areas were shown in Fig. 3.22, while Figs. 3.23 and 3.24 show some wind behavior to be expected in some typical building combinations. Beranek (1980) has discussed methods for charting the shelter areas in these building groups.

Wind within buildings is discussed in detail in Chapter 4, Section 4.6.

(d) Ventilation and Cooling. A distinction must be made between two common architectural introductions of outdoor air: ventilation and cooling. *Ventilation* involves the provision of "fresh air" to interiors to replenish the oxygen used by people and to help carry away their by-products of carbon dioxide and bodily odors. Ventilation is desirable all year round; recommended minimum rates for providing fresh air are found in Chapter 4 (Table 4.25). *Cooling* (with outdoor air) replaces heated indoor air with cooler outdoor air. Thus cooling by breezes is a seasonal opportunity, limited to times when the outdoor air temperature is lower than the indoor air temperature. Cooling can require far greater quantities of air than ventilation, and its influence on building siting and window size and

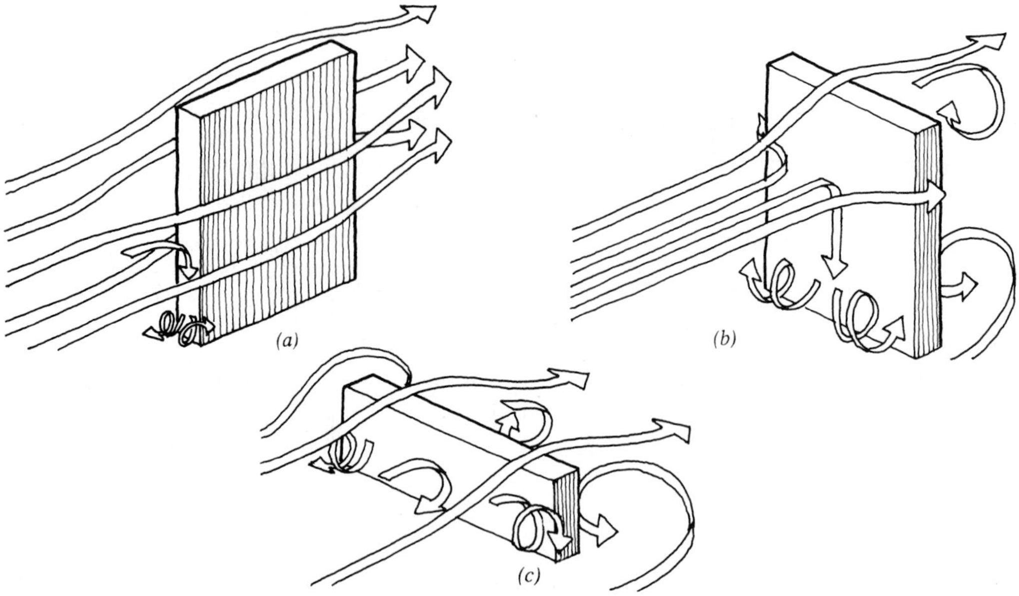

Fig. 3.23 *Wind patterns around single buildings.* (a) *Tall, slender buildings; height greater than 2.5 times width.* (b) *Tall, rather wide buildings; height between 2.5 and 0.6 times width.* (c) *Long buildings; height less than 0.6 times width. (From W. J. Beranek, "General Rules of the Determination of Wind Environment,"* Wind Engineering, *J. E. Cermak, ed., Vol. 1, © 1980, Pergamon Press Ltd., reprinted by permission.)*

placement is considerable. Figure 3.25 illustrates the differing approaches to these two usages.

Two examples of contemporary buildings that use the wind appear in Figs. 3.26 and 3.27*a*. The first served as a summer exhibition space in Montreal, Canada, and relied on prevailing winds to remove the air heated by display lighting and crowds of people. The second is a church in a quiet residential area in southern California (Fig. 3.27). It is sited to take advantage of prevailing southerly winds and has ex-

Fig. 3.24 *Wind patterns among building clusters.* (a) *Increased wind speeds can occur, especially where wind is at 45° to the face. With the bar effect, the downward spinning wind behind these buildings can reach 1.4 times the speed of the average wind.* (b) *Venturi effect: Where few obstructions occur upwind or downwind from the narrow neck of this plan illustrating the Venturi effect, wind speeds through the neck can reach 1.3 times the average, up to heights of 30 m, and 1.6 times the average at about 50 m height.* (c) *The gap effect begins to occur with perpendicular winds at buildings with more than 5 stories (15 m) height; by 7 stories, wind speeds 1.2 times the average can occur through the gaps; by 60 stories, gap wind speeds can be 1.5 times the average.* (d) *For higher buildings, increased wind speeds occur at the corners (localized within a radius, equal to the width of the building* d, *around the corner). Where height is 15 m, wind speed can reach 1.2 times the average; for heights above 35 m, wind speed can be 1.5 times the average. Where two towers approach each other, increased wind around corners and between the towers can go as high as 2.2 times the average for towers 100 m high.* (e) *Increased wind speed and turbulence within the (shaded) wake of buildings can be especially serious for towers where at heights from 16 to 30 stories, wind speeds can reach 1.4 to 2.2 times the average. (From J. Gandemer, "Wind Environment Around Buildings,"* Wind Effects on Buildings and Structures, *K. J. Eaton, ed., © 1977, Cambridge University Press. Reprinted by permission.)*

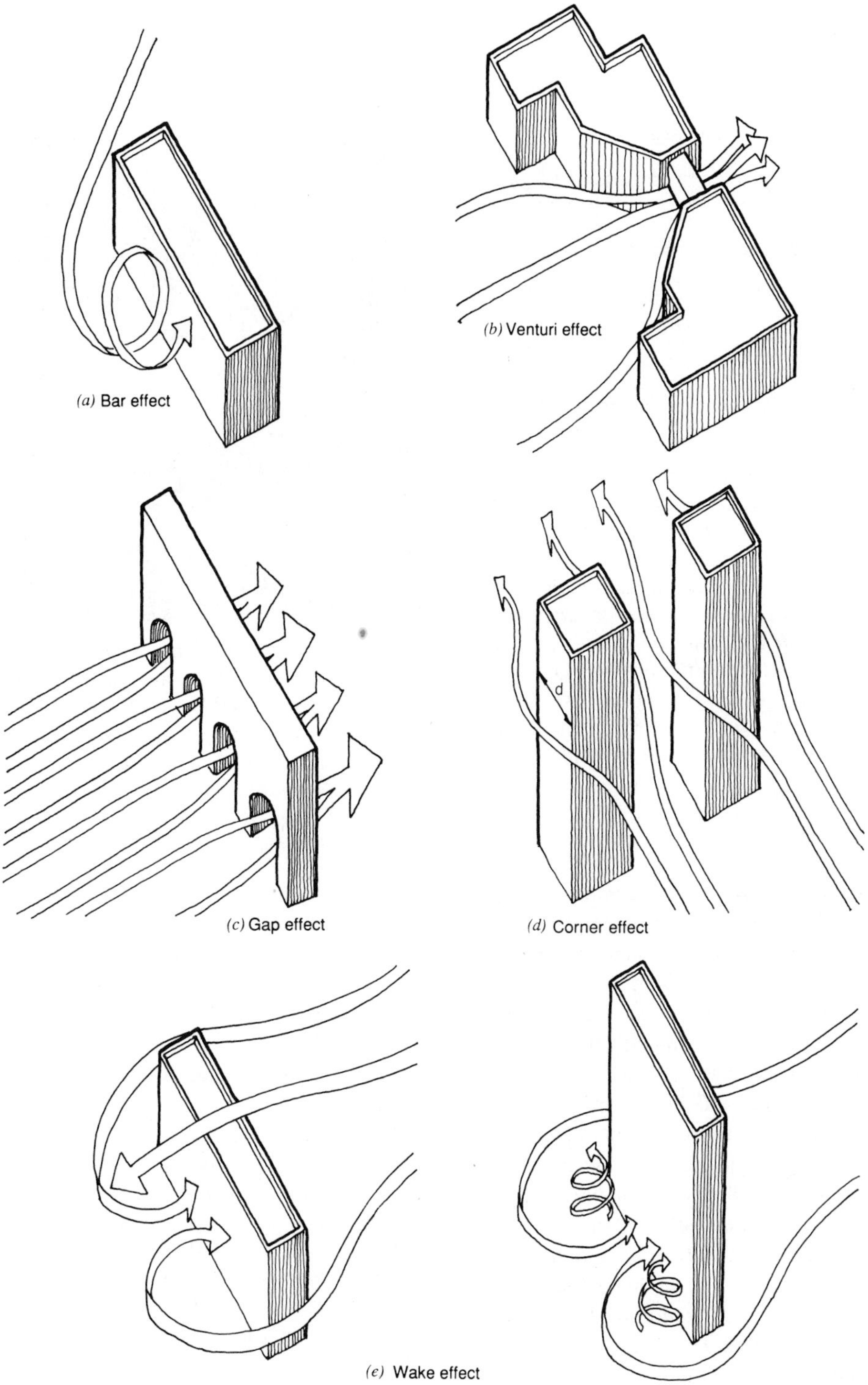
(a) Bar effect
(b) Venturi effect
(c) Gap effect
d
(d) Corner effect
(e) Wake effect

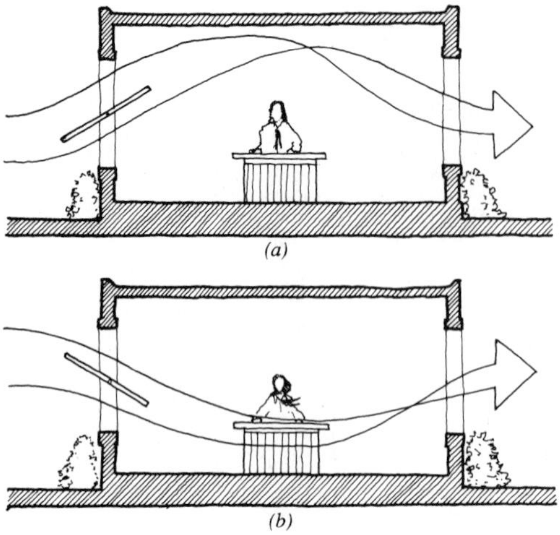

Fig. 3.25 *Ventilation, with and without cooling. The window's placement and size can influence the stream of air within a space, providing either ventilation alone or adding a cooling function to moving air.* (a) *Ventilation: the pivoting window sends the breeze toward the ceiling, where it replaces inside air. It has minimum contact with people below.* (b) *Ventilating and cooling: this position of the window directs the breeze toward the floor, where it encounters people and provides a direct cooling effect from air motion.*

tensive openings in the walls on all sides. In addition, a 15-ft^2 north-facing opening at the center peak of the sloping roof provides an outlet for rising air heated by people and/or solar gains in the church. This stack effect (discussed in Section 4.6) is augmented by negative pressure created by southerly winds outside the north-facing opening. In both Figs. 3.26 and 3.27, the negative pressure created by the prevailing wind pulls heated air out through the opening.

Another example (Fig. 3.28) is a faculty office building in England, whose windows can be manipulated to provide either a small amount of ventilation air or a more thorough scouring for cooling purposes. The building that relies on prevailing wind for cooling must be sited with attention to wind direction. Again, one building should not be erected to obstruct the next building's access to breeze. As seen previously, obstacles upstream from intake openings, or downstream near the outlets, can substantially reduce the velocity—and thereby the cooling effect—of the wind. But although wind cools people in hot weather primarily by speeding the evaporation of sweat from skin, it can itself become an irritant at high velocity (Tables 2.3 and 3.4). Manual controls of openings are a necessary part of

Fig. 3.26 *Cooling with the wind. The air inside this exhibition hall is heated by lights and crowds of people. It then rises and is sucked out of the large clerestory opening by the prevailing winds. Note that these winds do not blow* into *this high opening, but encourage outward flow of the overheated interior's air. (African Place, at Expo '67, Montreal; John Andrews, architect.)*

Fig. 3.27 *The Corpus Christi church in Bonita, California, takes advantage of the mild southern California climate by cooling with natural ventilation.* (a) *Located uphill from any obstruction to the prevailing breeze, the church is quite open on all four sides* (b). (c) *In addition, a central "crystal ventilator" (an artwork of colored glass and stainless steel) provides an opening for hot air to exit via both the stack effect and negative pressure created by prevailing winds. The ventilator can be closed during the coldest months. (Courtesy of F•A•D Architecture and Planning, Encinitas, California.)*

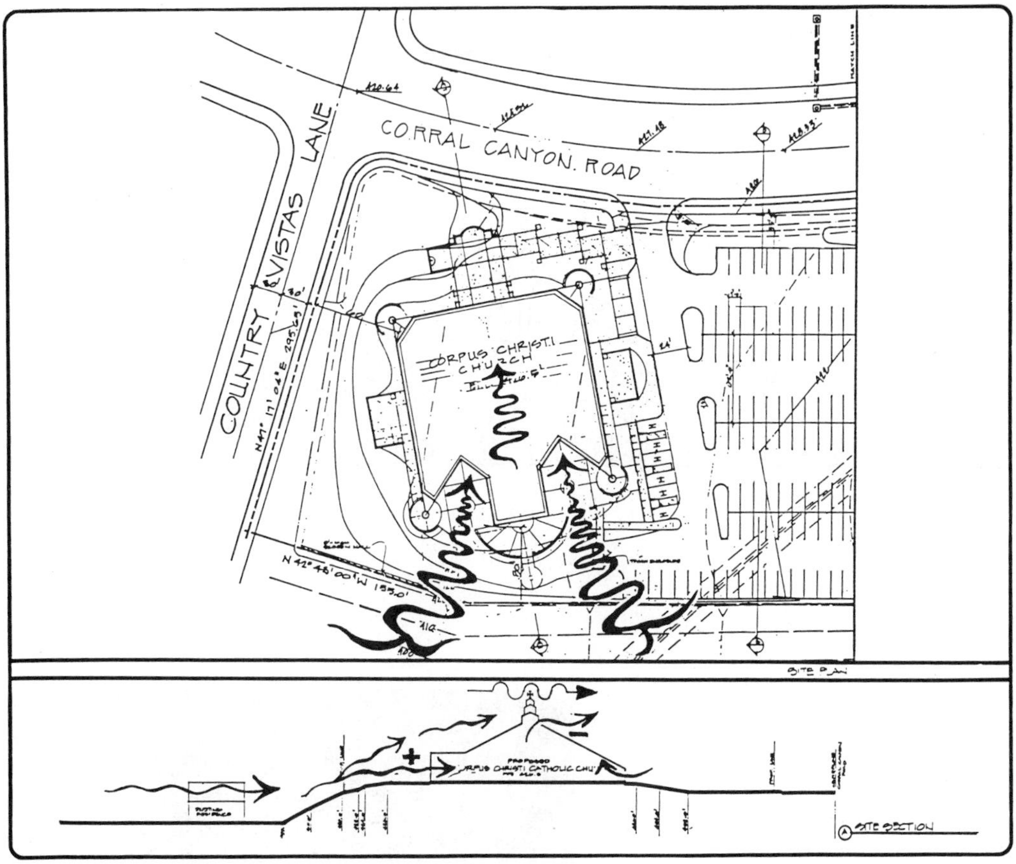
CORRAL CANYON ROAD
COUNTRY VISTAS LANE
CORPUS CHRISTI CHURCH
N 42° 46' 00" W 155.0'
SITE PLAN
PROPOSED CORPUS CHRISTI CATHOLIC CHURCH
SITE SECTION

(*a*)

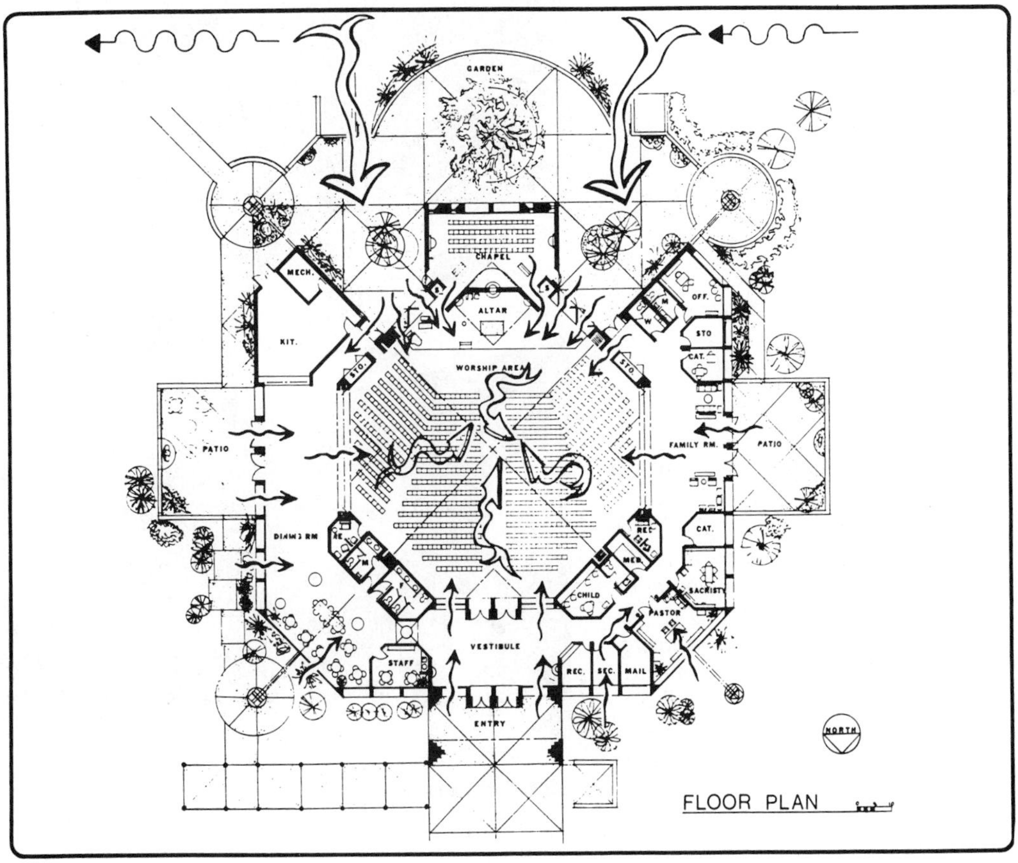
GARDEN
CHAPEL
ALTAR
MECH.
KIT.
STO
WORSHIP AREA
OFF.
CAT.
PATIO
FAMILY RM.
DINING RM
CHILD
SACRISTY
PASTOR
STAFF
VESTIBULE
REC.
SEC.
MAIL
ENTRY
NORTH
FLOOR PLAN

(*b*)

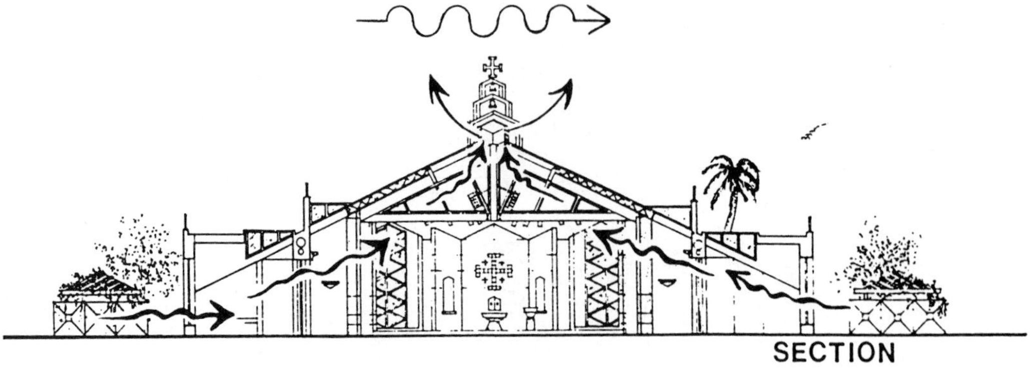

(c)

Fig. 3.27 *(continued)*

Fig. 3.28 (a) *These faculty offices and seminar rooms in England have ventilation and cooling options.* (b) *They can admit a small amount of winter ventilating air or a large amount of air to directly cool the interior. The prefabricated window unit* (c) *is hoisted into position during construction* (d). *(Leicester University's Attenborough Building; Arup Associates, Architects, Engineers and Quantity Surveyors, London.)*

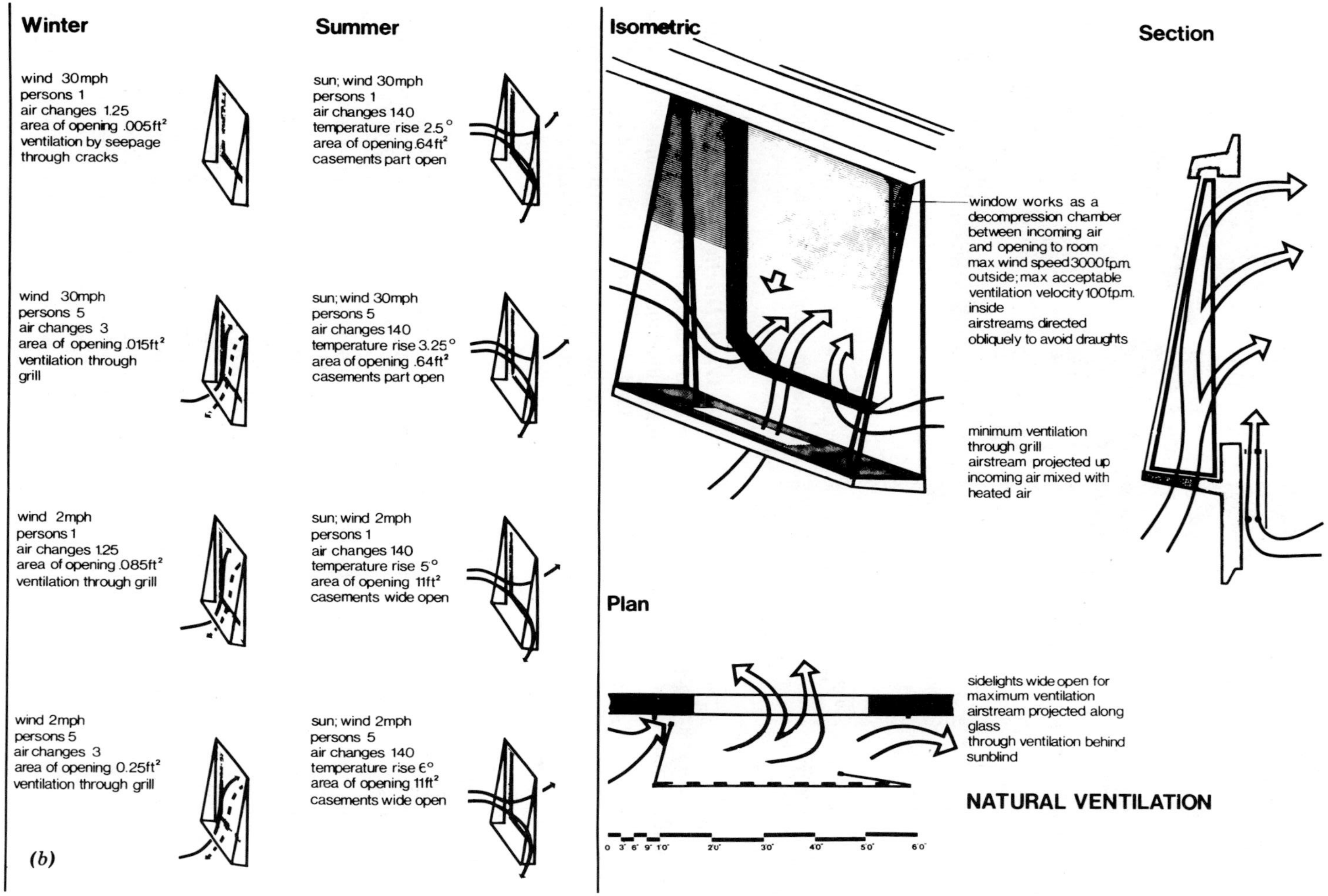
Winter
wind 30mph
persons 1
air changes 1.25
area of opening .005ft²
ventilation by seepage
through cracks
wind 30mph
persons 5
air changes 3
area of opening .015ft²
ventilation through
grill
wind 2mph
persons 1
air changes 1.25
area of opening .085ft²
ventilation through grill
wind 2mph
persons 5
air changes 3
area of opening 0.25ft²
ventilation through grill
Summer
sun; wind 30mph
persons 1
air changes 140
temperature rise 2.5°
area of opening .64ft²
casements part open
sun; wind 30mph
persons 5
air changes 140
temperature rise 3.25°
area of opening .64ft²
casements part open
sun; wind 2mph
persons 1
air changes 140
temperature rise 5°
area of opening 11ft²
casements wide open
sun; wind 2mph
persons 5
air changes 140
temperature rise 6°
area of opening 11ft²
casements wide open
Isometric
Section
window works as a
decompression chamber
between incoming air
and opening to room
max wind speed 3000 f.p.m.
outside; max acceptable
ventilation velocity 100 f.p.m.
inside
airstreams directed
obliquely to avoid draughts
minimum ventilation
through grill
airstream projected up
incoming air mixed with
heated air
Plan
sidelights wide open for
maximum ventilation
airstream projected along
glass
through ventilation behind
sunblind
NATURAL VENTILATION
0 3' 6' 9' 10' 20' 30' 40' 50' 60'

(b)

Fig. 3.28 (continued)

Fig. 3.28 (continued)

natural ventilation equipment. And finally, the proper size and placement of openings, with a ventilation path through the building, must be provided.

(e) Wind, Daylight, and Sun. When daylight and wind ventilation are desired, they combine to limit the width of buildings. This is particularly evident in multistory office buildings, where increasing urban density and reliance upon electric lighting and cooling have changed

TABLE 3.4 **Beaufort Scale (Lower Speeds Only)**

Beaufort Number	Speed 6 m (19.7 ft) Above Ground			Description of Effects Outdoors	
	m/s	*fpm*	*mph*	*On Land*	*Over Water*
0	0.3	Less than 88	Less than 1	Smoke rises; no perceptible movement	Smooth sea
1	0.6–1.7	88–264	1–3	Smoke drift shows wind direction; tree leaves barely move	Scalelike ripples
2	1.8–3.3	352–616	4–7	Wind felt on face; leaves rustle	Small wavelets
3	3.4–5.2	704–968	8–11	Leaves, twigs in constant motion; hair is disturbed; wind extends light flag	Large wavelets; occasional white foam crests
4	5.3–7.4	1056–1408	12–16	Small branches move; dust rises; hair disarranged	Small waves become longer

Source: Reprinted from *Passive Cooling* by permission of the publisher, American Solar Energy Society, Inc.

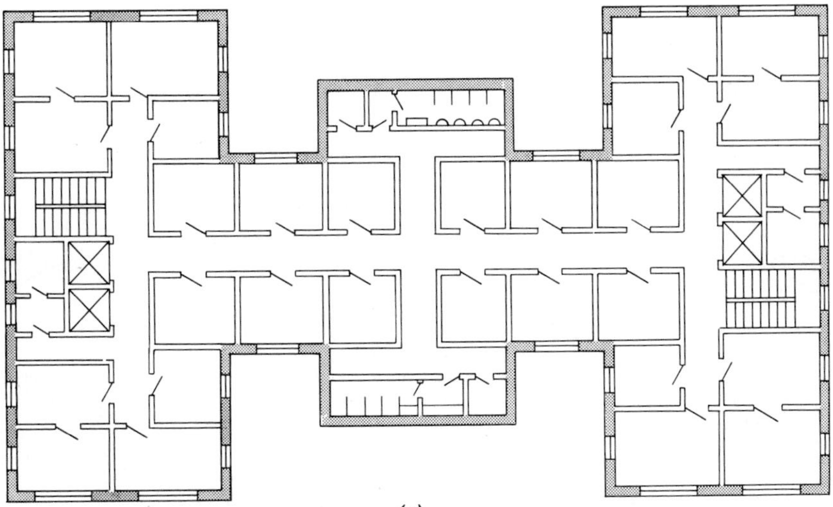

(a)

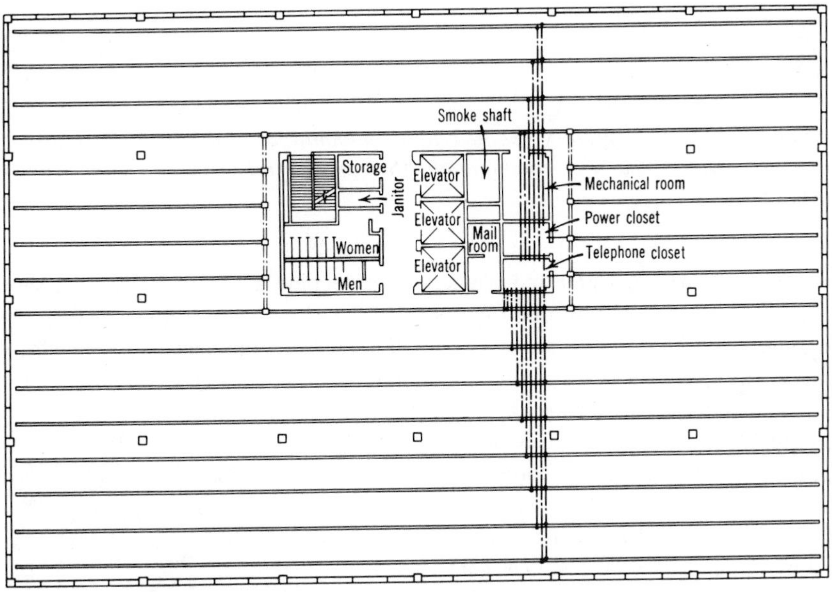

(b)

Fig. 3.29 *In contrast to building plan* (a), *which uses daylight and wind ventilation in each office, office building* (b) *receives cooler, filtered air, is less subject to the noise of the city, typically provides both constant light and temperature throughout, and provides for more rentable floor space on the site. It also allows less daylight to reach street level, requires much more electricity (though probably less heating fuel), and thus contributes more heat (and possibly noise from mechanical equipment) to the city all year round.*

the form of buildings considerably in the past few decades (Fig. 3.29). Another impact of combined wind and sun is shown in Fig. 3.30, which represents the factors a designer would weigh in determining the orientation of a clerestory window.

3.6 Rain and Groundwater

Most buildings interact with water in three forms: rainwater, groundwater, and potable water (brought to and taken from urban sites by utilities). A detailed treatment of water within

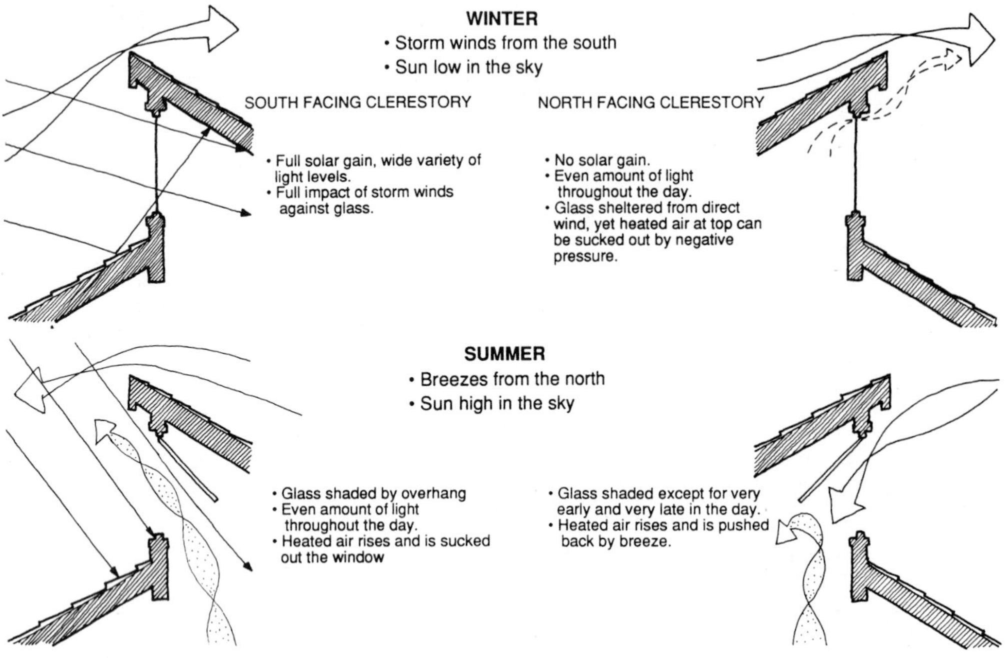

WINTER
• Storm winds from the south
• Sun low in the sky
SOUTH FACING CLERESTORY
NORTH FACING CLERESTORY
• Full solar gain, wide variety of light levels.
• Full impact of storm winds against glass.
• No solar gain.
• Even amount of light throughout the day.
• Glass sheltered from direct wind, yet heated air at top can be sucked out by negative pressure.
SUMMER
• Breezes from the north
• Sun high in the sky
• Glass shaded by overhang
• Even amount of light throughout the day.
• Heated air rises and is sucked out the window
• Glass shaded except for very early and very late in the day.
• Heated air rises and is pushed back by breeze.

Fig. 3.30 *Which way for a clerestory? Some relative advantages of north versus south orientation for a clerestory window/shed roof combination. (Seasonal wind directions shown prevail in the Pacific northwest.)*

buildings is found in Part III, Water and Waste, and a close look at the utilization of rainwater is found in Chapter 8.

Like solar energy, *rainwater* is a diffuse, intermittent, and often seasonal resource. As a source of water, it is most often collected and used where other water sources are scarce, or of poor quality. Rain also has an influence on building design: heavy rains and pitched roofs have long been found in the same locales. Overhangs may extend further beyond walls exposed to storm winds; even gutter and downspout details can become a design feature, as shown in the examples in Chapter 8. A building that reflects the combined influences of daylight, wind, and rain is shown in Fig. 3.31.

Fig. 3.31 *Rain, wind, sun, daylight, and design. This covered outdoor tennis facility at the University of Oregon, Eugene, was designed to ward off the rain-bearing south winds in winter. Direct sun is also unwelcome; instead, north skylight is admitted along with reflected light from roof surfaces. Cool outdoor temperatures are maintained, appropriate to strenuous activity. Gutters at the lower edge of each roof plane carry away rain; the courts stay dry for all but a few days of the year. (Unthank, Seder, Poticha Architects.)*

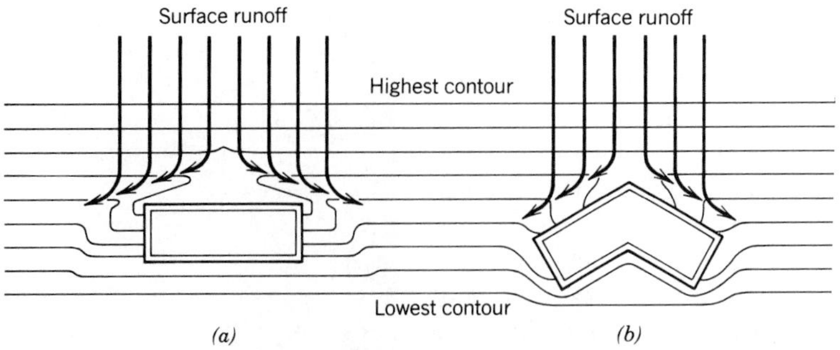

Fig. 3.32 *Rain as surface flow.* (a) *Where buildings intercept surface water, provisions for diversion are necessary. The building oriented as in* (b) *needs less elaborate provisions.*

Rainwater's impact on site design can be thought of as similar to that of wind. Once on the surface, it will flow downhill in a wide path, and buildings that obstruct this path must make provisions for diverting it. Slight, shallow ditches called swales are frequently used in such diversion; orientation of the building to the slope can affect surface water diversion, as shown in Fig. 3.32.

Surface water can be used to advantage in thermal, acoustic, and daylighting roles. Hot, dry breezes that pass over water surfaces (and especially through misty sprays above ponds) gain substantial moisture while undergoing a drop in drybulb temperature, as seen in the preceding discussion on evaporative cooling (Section 2.4). Such air can provide improved comfort in hot, dry conditions. If the water body also provides the sound of running water, it can serve to mask unwanted sounds such as freeway traffic, or conversations in adjacent offices. Surface water has a more complex role in daylighting, due to the reflection characteristics of water. The surface of water is highly reflective to light that strikes it at *low* angles of incidence; the setting sun off the west coast can throw a blinding sheet of light across the ocean surface. Conversely, sunlight near noon on summer strikes a water body at *high* angles of incidence, nearly perpendicular at southern U.S. latitudes. Water is highly absorptive of light at these angles, reflecting relatively little light. Therefore, water bodies east, south, and west of our buildings can provide increased reflected light on sunny winter days, and somewhat decreased light in summer, relative to grass surfaces. However, on heavily *overcast* days, when the sky is uniformly gray (and least daylight is available), water surfaces will not be particularly helpful. Furthermore, the reflection of sun off water tends to be in sparkling patches of always-changing light. This can be fascinating, or it can be annoying, either as glare to the eye or as distraction from a visual task. Reflected off a matte-finish ceiling, such dancing light might be welcome; directly on eyes or work surface, it easily becomes a problem. Sparkle for one viewer may be glare for another.

Generally avoided by designers where possible, *groundwater* is a threat to foundations and underground spaces. This avoidance carries over into site planning, and marshy places are usually unwelcome near buildings. Urban dryness is intensified as ponds are drained and streams are piped away. However, architects have used groundwater as a heat sink, discharging building heat to groundwater in summer and removing the groundwater heat for the building's winter benefit. The quantity of groundwater available to act thus as a heat sink varies with subsurface conditions, and little information is yet available to indicate its limits. More buildings discharging more heat will eventually raise groundwater temperatures; but how many degrees, and for how long, remains unknown in many areas. Another risk is that of pollution, through increased handling of groundwater. Many localities strictly regulate the use of groundwater heat sinks as a result.

The ability of water to conduct and store

heat, which encourages its use as a heat sink, also makes it a thermal enemy of heat storage tanks or bins in solar buildings. There is little point in collecting solar energy to store in an underground tank for later use if groundwater is allowed to rob the tank of its heat. On the other hand, groundwater helps storage tanks that are to be kept as cool as possible. Such tanks provide cool water to buildings at peak-heat hours of hot days and can greatly lessen the demand for electricity to run conventional air conditioners.

Snow has special implications; it delays runoff, provides a blanket of thermal insulation, absorbs sound, and reflects more light than almost any other naturally occurring surface. Wind patterns can deposit snow much more thickly in some places, for better or worse. Snow hampers the moving of external devices such as thermal shutters (or roof pond covers) and can create a disabling glare when it reflects the low winter sun into windows at eye level.

Fig. 3.33 *Fixed overhang sun control. This south-facing corridor in Oregon is in the open air all year round. The low winter sun fills the space, and some of its heat is stored by the tile floor. However, in cold weather the offices that it connects are disadvantaged by exposure to cold air. In summer, little of the corridor is exposed to sun. Sun control is identical in the spring and fall.*

3.7 Plants

Plants have several roles: they affect the absorptivity and emissivity of the earth surface; they are part of both the food and water cycles; they turn carbon dioxide into oxygen by day; they provide organic matter suitable for building materials; and they help us tell time both by growth and by change with the seasons. Our associations with plants are mostly pleasant ones, and they contribute to our enjoyment of the places where they grow.

Plants are also of immediate practical value to buildings because they enhance privacy, slow the winter wind, reduce the glare of strong daylight, or prevent summer sun from entering and overheating buildings. In this latter role, plants are particularly noteworthy, because they enhance a feeling of coolness when breezes rustle or sway their leaves, and especially because they respond more to *cycles of outdoor temperature* than to those of *sun position*. Unlike fixed sunscreens on buildings, plants thus provide deepest shade in the hottest weather.

To illustrate this contrast, a fixed sunscreen (here an overhang for south-facing windows) is shown in Fig. 3.33. Such sunscreens block the sun for some portion of the year, centered on June 21; that is, maximum shade is provided at the summer solstice. A typical approach to such sunscreens in the U.S. temperate zone is to shade at least half a south window in a residence (or all the window in an office with internal heat sources), from March 21 to September 21, or from equinox to equinox. Yet, March is on the average a colder month than September; March 21 is the last day of winter, while September 21 is the last day of summer. Full solar radiation is more welcome in early spring than in early fall, yet sun position is identical at these times. In contrast to fixed sunscreens, deciduous plants do most of their shading from the middle of June

Fig. 3.34 *Deciduous vines, temperature, and sun position. The sun's path through the sky is identical in late May and late July (see Fig. 3.12)* (a *and* b). *Identical lower sun paths occur in November and January* (c *and* d). *This deciduous vine responds more to the temperature of its Oregon climate than to the sun's position at 44°N latitude, which makes it particularly useful as a potential sun control device. [From Reynolds (1976).]*

(a)
(b)

Fig. 3.35 A deciduous tree and the equinoxes. The sun position (4:45 P.M. sun time) is identical on September 21 (a) *and March 21* (b); *yet this hour's average temperature (Eugene, Oregon) is 75 F in September and 53 F in March. The second-floor overhang provides identical shading, while the tree shades only in September. [From Reynolds (1976).]*

to early October, giving windows access to solar radiation through much of the spring (Figs. 3.34 and 3.35).

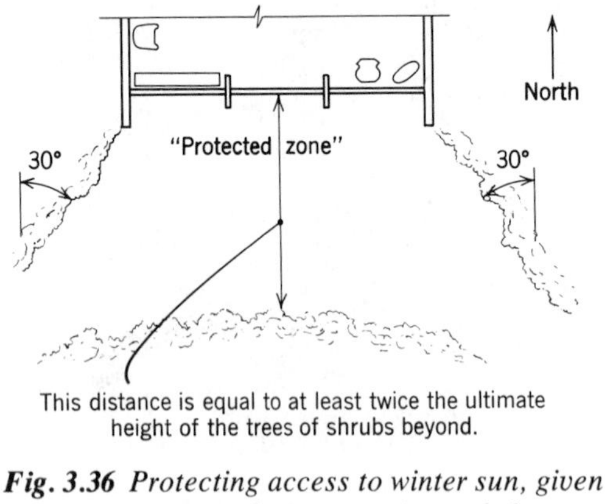
North
"Protected zone"
30°
30°
This distance is equal to at least twice the ultimate height of the trees of shrubs beyond.

Fig. 3.36 Protecting access to winter sun, given a lawn or terrace of minimum size and shape to the south of solar collecting surfaces. Deciduous plants within this "protected zone" should be avoided, unless they are very low growing or are a reliably early-defoliating species. Summer sun protection for these south windows is best provided by flexible architectural controls, such as awnings or hanging screens.

Deciduous trees have a potential solar heating disadvantage in that certain species (such as some oaks) hold onto their leaves well into the heating season, a tendency increased by fertilizing or irrigating near the tree. Others have a dense branch structure, blocking a surprisingly high percentage of solar radiation even when bare. A listing of shading density for various trees is found in Moffat and Schiler (1981). Agricultural extension services in most areas can provide information on late-defoliating trees. For solar collection, avoiding trees or large shrubs within the area shown in Fig. 3.36 is recommended.

Many large buildings have a relatively short heating season, since they have constant internal heat sources. So, the choice of trees, shrubs, and vines for the improvement of a building's microclimate can best be made after an analysis of building heating and cooling needs by month.

Clearly, a tree or vine that unfolds its leaves early in spring and holds them until late fall is advantageous to these internal-heat-dominated buildings.

REFERENCES

American Solar Energy Society (1981). *Passive Cooling,* ASES, Boulder, Colo.

Beranek, W. J. (1980). "General Rules for the Determination of Wind Environment," in *Wind Engineering,* J. E. Cermak, Editor, Pergamon Press, Elmsford, N.Y.

Bragdon, C. (1970). *Noise Pollution, The Unquiet Crisis,* University of Pennsylvania Press, Philadelphia.

Brown, G. Z., Reynolds, J. S., and Ubbelohde, M. S. (1982). *Inside/Out: Design Procedures for Passive Environmental Technologies,* Wiley, New York.

Cohen, S., Glass, D. C., and Singer, J. E. (1973). "Apartment Noise, Auditory Discrimination, and Reading Ability in Children," *Journal of Experimental Social Psychology,* Vol. 9.

Doelle, L. (1972). *Environmental Acoustics,* McGraw-Hill, New York.

Gandemer, J. (1977). "Wind Environment Around Buildings: Aerodynamic Concepts," in *Wind Effects on Buildings and Structures,* K. J. Eaton, Editor, Cambridge University Press, New York.

Gandemer, J. (1981). "The Aerodynamic Characteristics of Windbreaks, Resulting in Empirical Design Rules," *Journal of Wind Engineering and Industrial Aerodynamics,* Vol. 1, July. Elsevier Scientific Publishing Co., Amsterdam.

Knowles, R. (1981). *Sun, Rhythm, Form,* MIT Press, Cambridge, Mass.

Landsberg, H. E. (1970). "Climates and Urban Planning," in *Urban Climates,* World Meteorological Organization Technical Note 108, Geneva.

Lowry, W. P. (1988). *Atmospheric Ecology for Designers and Planners,* Van Nostrand Reinhold, New York.

Moffat, A. S., and Schiler, M. (1981). *Landscape Design That Saves Energy,* William Morrow, New York.

Olgyay, A., and Olgyay, V. (1957). *Solar Control and Shading Devices,* Princeton University Press, Princeton, N.J.

Reynolds, J. S. (1976). *Solar Energy for Pacific Northwest Buildings,* Center for Environmental Research, University of Oregon, Eugene, Oreg.

U.S. Environmental Protection Agency (1989). *National Air Pollutant Emission Estimates,* U.S. EPA, Research Triangle Park, N.C.

Watson, D., and Labs, K. (1983). *Climatic Design.* McGraw-Hill, New York.

PART II
THERMAL CONTROL

The next four chapters deal with the topics of heating and cooling. The objective continues to be the *comfort of people,* for whom we heat and cool buildings; thermal comfort criteria were discussed in Chapter 2. In Chapter 4, the basic theory of heat flow is discussed; whether we are calculating heat gain in summer or heat loss in winter, the process is based on the same principles of heat flow. Chapter 5 presents criteria for building performance, followed by detailed heat loss and heat gain procedures. Chapters 6 and 7 introduce some of the extensive array of available systems and equipment for heating and cooling.

It is important to keep our perspective during the process of calculating thermal loads for buildings. Throughout these lengthy chapters, keep the following points in mind:

1. Human comfort is the ultimate objective of our calculations and resulting architectural manipulations; this comfort is not attainable by numbers alone. Most of us have experienced situations in which psychological factors seemed to have at least as much influence on comfort as temperature, humidity, or air motion. Achieving numerical adequacy is important, but it is not by itself a guarantee of comfort.

2. These calculations are a rare opportunity to numerically evaluate a building design. When the numbers are all cranked out, sit back and take a larger view. What design strategies do they suggest? We have worked hard to get them; now let them work for us.

3. There are two specific objectives to our calculations, which yield different kinds of answers and require different amounts of work. (a) How big should a device be to furnish enough heating (or cooling) in the coldest (or hottest) conditions we are likely to encounter for our building? (b) About how much energy will be used by our building in a typical season?

For the first question, we deal with extreme "design" conditions, as discussed in Section 2.4. We calculate our hourly winter heat loss (or summer gain) under these special conditions, then adjust our buildings, probably recalculate, and finally size our equipment according to these calculations. We have, in effect, a snapshot of the building's performance at a special high-stress time—one that occurs rarely in the building's lifetime.

By contrast, the second question looks at normal conditions, as they change over a long period of time. Our calculations for design-hour performance can be used, but must be adjusted for normal and continuous-change conditions. These calculations take much longer, and are more suited to automatic computation (Appendix H) than simple design-hour answers. Once completed, we again have the opportunity to adjust our buildings. This time, adjustments are aimed more at reducing long-term energy use than at equipment sizing. We have, in effect, a series of photos that become a movie of our building's performance over time, under normal circumstances. We get another perspective on architecture as a performing art.

4. Buildings, like our bodies, exchange heat with the outside environment in two distinct ways: *through the envelope,* akin to heat exchange through human skin, and with *incoming fresh air,* akin to our heat exchange with the air we inhale. When calculating envelope (skin) losses or gains, materials, areas, and rates of heat flow through the envelope are used. When calculating fresh air (lung) losses or gains, volumes of space and rates of fresh air exchange are used. For heavily insulated buildings in cold climates, it is not unusual for "lung" losses to exceed "skin" heat losses. Skin and lung losses (or gains) must be added together to determine the total rate of a building's heat exchanges with its environment; both can be manipulated

through design to improve a building's thermal performance.

5. Buildings nearly always experience internal heat production; this is helpful in meeting cold weather demands, but harmful in hot weather. For most buildings, it will be helpful to manipulate the envelope to *minimize* internal gains (e.g., to reduce the need for electric lights by specifying larger, better-placed windows), since the reduction in internal heat is so helpful in summer, and can so often be replaced by an increased reliance on passive solar heating in winter.

6. An occupied building can be very different thermally from an unoccupied one. Many buildings (such as offices and schools) experience high internal gains and rates of fresh air for only about a third of a typical day, and only 5 days per week; this amounts to less than a quarter of the year. It is important to remember that although we are designing to provide thermal comfort while a building is being used, we cannot depart too far from comfortable interior temperatures during the night (or weekend) without straining the heating and cooling systems early in the morning on a working day. Said another way, one hour's surplus heat can be another hour's needed warmth.

4

HEAT FLOW

Most of this chapter discusses the theory of heat flow; applications are generally found in Chapter 5. Heat flow through the skin—by convection, conduction, and radiation—is contrasted to heat flow due to "breathing"—by infiltration and provisions for fresh air. Finally, there is psychrometry, which combines the properties of air and moisture.

Note: Except for the introductory Table 4.1 (terms and definitions), the tables that accompany this chapter are grouped together at the chapter's end. They begin on page 136.

4.1 Convection, Conduction, and Radiation

Whenever an object is at a temperature different from its surroundings, heat flows from the hotter to the colder. Chapter 2 discussed the human body's means of disposing of its surplus heat to the cooler environment. Buildings, like bodies, experience heat loss to, and heat gain from, the environment by *convection* (molecules of cool air absorb heat from a warm surface, rise, and carry it away), *conduction* (heat is transferred directly from molecules of hot building surfaces to the molecules of cooler solids in contact with the building, such as earth or water), and *radiation* (heat flows in electromagnetic waves from hotter surfaces to detached, distant colder ones, through any transparent medium, even empty space). Evaporation is also involved (from wet surfaces), but much less influential for most buildings than for our bodies.

This combination of heat flow by convection, conduction, and radiation is illustrated in Fig. 4.1 through some typical combinations of materials. Notice that multiple air spaces and reflective surfaces are an inexpensive but particularly useful way to slow the flow of heat. Some of the most effective insulating materials therefore combine multiple dead-air spaces and layers of reflective films. Heat flow through the various components of a building's skin involves both heat flow through solids and heat flow through films and layers of air.

4.2 Heat Flow Through Solids

Certain terms are fundamental to any discussion of heat flow; Table 4.1 presents terms of energy, terms of power (energy used over a given time span), and terms of heat flow rate.

Each individual material has a characteristic rate at which heat will flow through it. For such individual, or homogeneous, solids this rate is called its *conductivity,* designated as k; the number of British thermal units per hour (Btu/h) that flow through 1 square foot (ft^2) of material, 1 in. thick, when the temperature drop through this material is 1F° (under conditions of steady heat flow). The units are Btu-in./h ft^2 F. The SI equivalent is the rate at which watts flow through 1 square meter (m^2) of material, when the temperature drop is 1K° (equal to 1C°), so units are W/m k (or W/m°C; these terms are used interchangeably in the tables of this chapter). Conductivity is an important factor in passive heating or cooling design, which depends heavily upon the rate at which heat is conducted through a material from its surface; when we touch the solid surface, we sense the material's conductivity. Conductivity for each solid is established by tests and is published as a basic rating.

Many solids (common brick, wood siding, thermal batt insulation, gypsum board, etc.) are widely available in standard thicknesses. For such materials, it is useful to know the rate of heat flow for the standard thickness instead of the rate per inch. *Conductance,* designated as C, is the number of Btu per hour that flow through 1

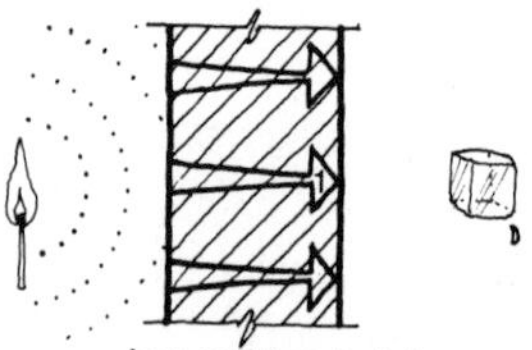

Any solid material wall, floor or roof

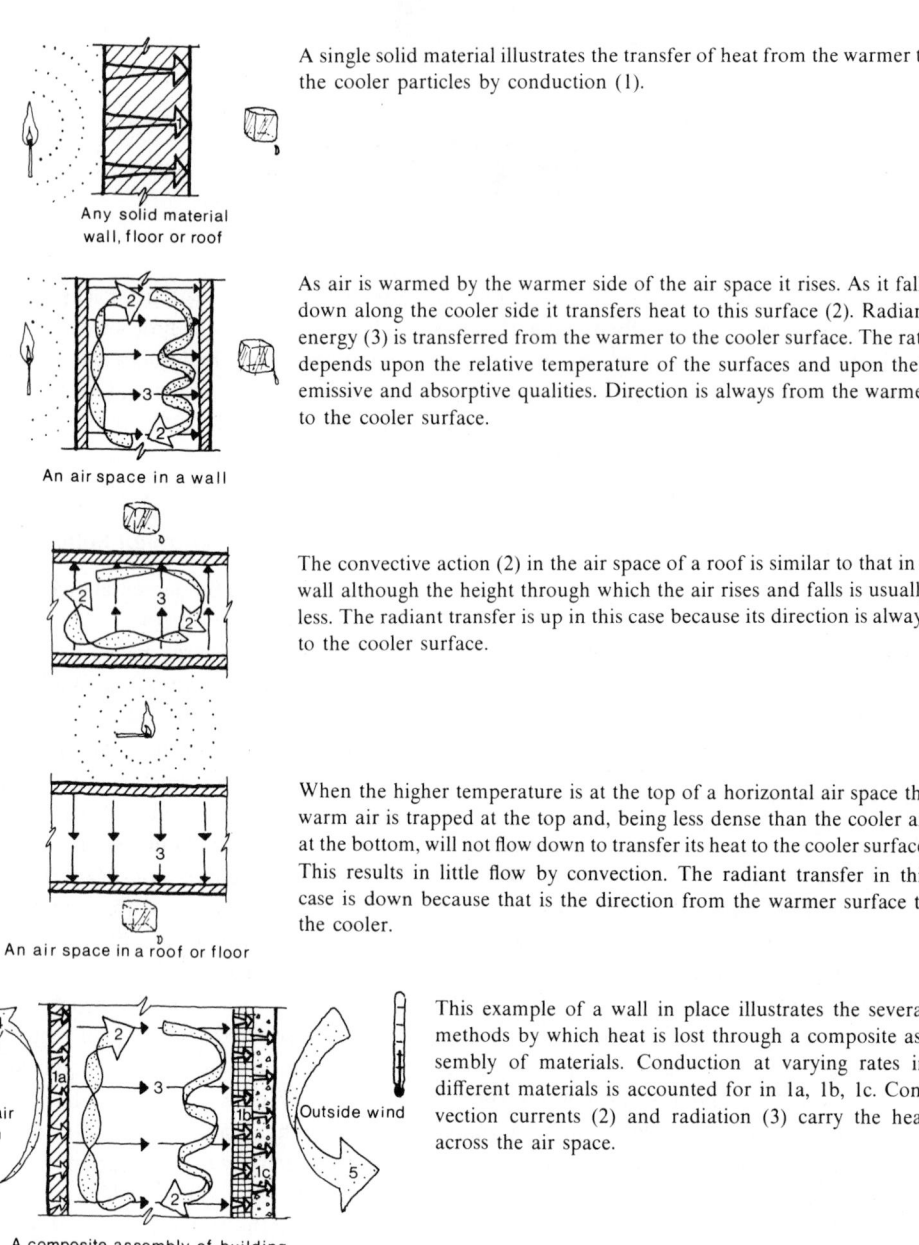

A single solid material illustrates the transfer of heat from the warmer to the cooler particles by conduction (1).

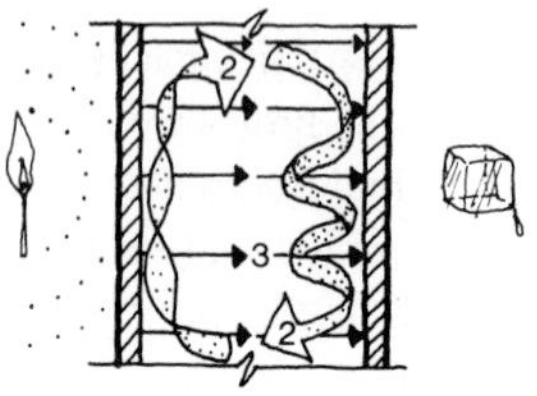

An air space in a wall

As air is warmed by the warmer side of the air space it rises. As it falls down along the cooler side it transfers heat to this surface (2). Radiant energy (3) is transferred from the warmer to the cooler surface. The rate depends upon the relative temperature of the surfaces and upon their emissive and absorptive qualities. Direction is always from the warmer to the cooler surface.

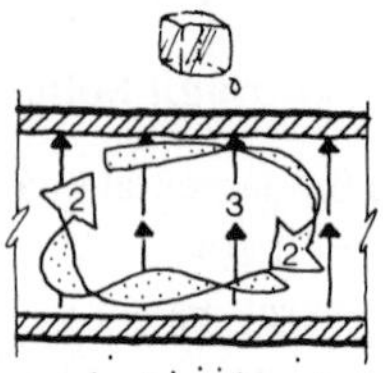

The convective action (2) in the air space of a roof is similar to that in a wall although the height through which the air rises and falls is usually less. The radiant transfer is up in this case because its direction is always to the cooler surface.

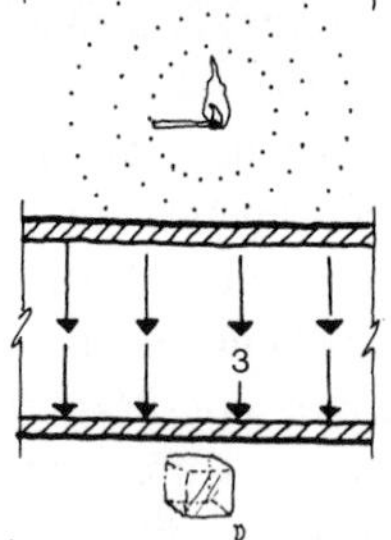

An air space in a roof or floor

When the higher temperature is at the top of a horizontal air space the warm air is trapped at the top and, being less dense than the cooler air at the bottom, will not flow down to transfer its heat to the cooler surface. This results in little flow by convection. The radiant transfer in this case is down because that is the direction from the warmer surface to the cooler.

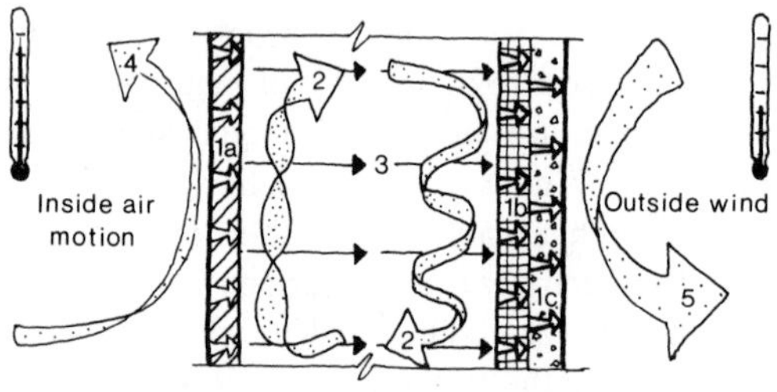

A composite assembly of building materials in place at the site

This example of a wall in place illustrates the several methods by which heat is lost through a composite assembly of materials. Conduction at varying rates in different materials is accounted for in 1a, 1b, 1c. Convection currents (2) and radiation (3) carry the heat across the air space.

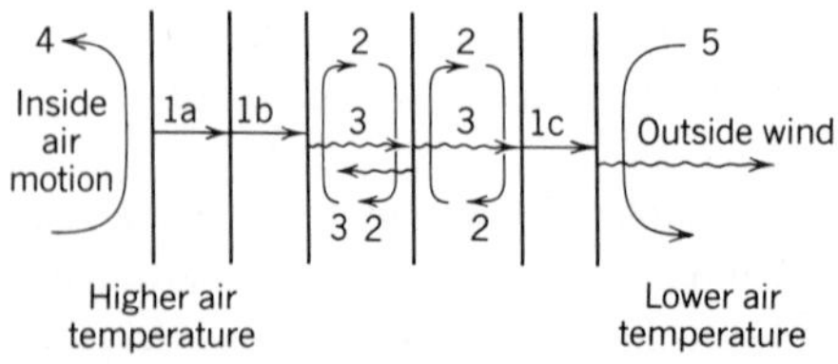

Heat is conducted from the room air by warm air currents that strike the inside wall. Heat is conducted away from the exterior surface of the wall by the action of the wind.

Fig. 4.1 *Nature of heat flow through materials, air spaces, and assembled structures. Thermal action is identified by the following numbers: 1, conduction; 2, convection; 3, radiation; 4, inside surface conductance; and 5, outside surface conductance.*

TABLE 4.1 **Heat Symbols, Terms, and Definitions**[a]

Conventional Units			SI (Système International) Units		
Symbol	*Term*	*Definition or Usage; Conversions*	*Symbol*	*Term*	*Definition or Usage; Conversions*
Part A. ***Energy***					
Btu	British thermal unit	The amount of heat required to raise 1 pound of water by 1F°. (It roughly corresponds to the energy released in burning an ordinary wooden match.) 1 Btu = 1.055 kilojoules	J	joule	Newton-meter, or a force of one newton (1 N) acting through a distance of one meter. In terms of heat, a joule is 1/4.184 of the amount of heat required to raise a gram of water by 1C°.
			kJ	kilojoule	1000 joules; 1 kJ = 0.9478 Btu
Part B. ***Power***					
Btu/h	British thermal units per hour	Commonly used to express the total heat loss or gain of a building, and to express the size of heating or cooling equipment. 1 Btu/h = 0.2929 W	W	watt	The power required to produce energy at the rate of one joule per second. Commonly used to specify size for light bulbs, furnaces, air conditioners, and other heat producers. 1 W = 3.412 Btu/h
hp	horsepower	1 hp = 746 W			

Part C. ***Rate of Heat Flow***

Conventional Units	Symbol	Definition or Usage; Conversions	SI Units
Btu/h ft² F		1 Btu/h ft² F × 5.67 = 1 W/m² K 1 W/m² K × 0.176 = 1 Btu/h ft² F These are the units of conductance and of overall heat transmission for the following terms:	W/m² K (or W/m² °C)
	C	(conductance). The rate of heat flow through a homogeneous material (or combination of materials) of a stated thickness.	
	a	(air space conductance). The rate of heat flow through an air space bounded by two surfaces.	
	h	(film or surface conductance coefficient). The rate of heat flow from a surface due to air (or other fluid) motion against the surface. (Identical to *f*, the film conductance.)	
	U	(overall coefficient of heat transmission). The overall rate of heat flow through any combination of materials, air layers, and air spaces. It is equal to the reciprocal of the sum of all resistances *R* (see below) that are involved in this combination. This is the term used directly in building envelope heat loss or gain calculations.	
Btu-in./h-ft²-F	*k*	(conductivity). The rate of heat flow through a homogeneous material, per unit of thickness.	W/m K (or W/m °C)

TABLE 4.1 **Heat Symbols, Terms, and Definitions**[a] ***(Continued)***

Conventional Units		*Definitions or Usage; Conversions*	*SI Units*
Part C. ***Rate of Heat Flow*** **(*Continued*)**			
h ft^2 F/Btu	R	(resistance). A measure of resistance to the passage of heat; the reciprocal of conductance.	m^2 K/W
	ε	(emittance). The ratio of the radiant flux from a given surface to that of a blackbody (a "perfect" emitter) at the same temperature.	
	E	(effective emittance). The combined effect of the emittance of parallel surfaces bounding an air space.	

[a]A more complete listing of conventional–SI conversion units is found in Appendix I.

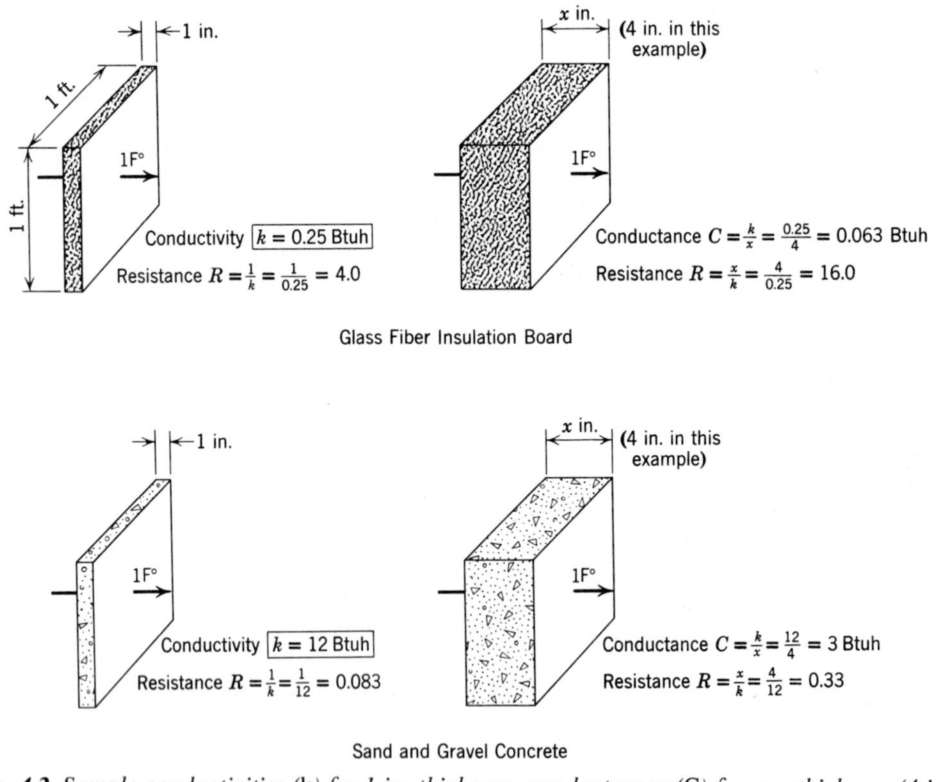

Fig. 4.2 *Sample conductivities* (k) *for 1-in. thickness, conductances* (C) *for any thickness (4 in. in this example), and resistance* (R) *for two materials. Glass fiber is a material of low conductivity; concrete is a material of high conductivity.* Note: *Standard unit of area is 1 square foot; standard unit temperature differential is 1 degree Fahrenheit (shown as 1F°).*

square foot of a given thickness of material when the temperature drop is 1F°. The units are Btu/h-ft^2-°F. The SI units of conductance are W/m^2-K.

Conductivity and conductance are compared in Table 4.1 and Fig. 4.2, which include another especially useful term, *resistance*. Resistance, designated R, tells us how effective a solid is as an insulator. The reciprocal of conductivity (or conductance), R is measured in *hours* needed for 1 Btu to flow through one inch (or a given thickness) of a solid when the temperature difference is 1F°; its units are h-ft^2-°F/Btu-in. (or h-ft^2-°F/Btu). In SI, units are m^2 K/W. Resistance is especially useful when comparing insulating materials, since the greater the R value, the more effective the insulator. Resistances and other important thermal properties are listed for many materials in Table 4.2 (page 136). For a nonhomogeneous material, such as concrete block, conductance is a better indicator of performance than resistance.

Architectural materials interact with heat flow either as *insulators* to retard the flow of heat, or as *conductors* to encourage heat flow for purposes of thermal storage. It is quite common to find both insulators and conductors in the same construction, as for example a wall whose inner layer is highly conductive and thermally massive, while its outer layer is highly insulative.

Materials used for insulation fall into three categories: (1) *inorganic,* fibrous or cellular (such as glass, rock wool, perlite, vermiculite), (2) *organic* fibrous or cellular (such as cotton, synthetic fibers, cork, foamed rubber, polystyrene), and (3) *metallic* or metallized organic *reflective* membranes (which must face an airspace to be effective).

Insulating materials are available in a wide variety of forms. (Common R values and thicknesses are listed in Table 4.8.) Most adaptable are *loose fill* (as above a ceiling on the floor of an attic), *insulating cement,* a loose material mixed with a binder and troweled onto a surface, and *formed-in-place* materials such as expandable pellets or liquids that are poured, frothed, or sprayed in place. Less adaptable and more common are the batts and blankets of *flexible* and *semirigid* insulation, with varying degrees of compressibility and adaptability.

Rigid insulation is applied in blocks, boards, or sheets, and can be preformed to nonplanar surfaces such as pipes. Exterior insulation and finish systems have become very popular, both for retrofit and for new construction. Utilizing expanded polystyrene rigid boards applied to exterior gypsum, plywood, or cementitious substrates, then covered with fabric-reinforced acrylic, this method achieves slightly more than R-4 per inch of thickness.

Reflective materials are found in sheets and rolls of either single or multiple layers, sometimes as preformed shapes with integral airspaces. When used without attachment to blanket or batt insulation, a reflective layer is called a *radiant barrier* and is especially applicable to roofs in the warmer climates nearer the equator. Radiant barriers are also useful in east- and west-facing walls in such climates; details of applications in the southern United States are found in Chandra et al. (1986).

Materials used as conductors are typically dense and diffuse heat readily. Table 4.2 lists density, conductivity (or conductance), and specific heat for many common materials. For a given material, the higher the numerical values of these three characteristics, the more successful will be the material's performance as a conductor. Some specifics on thermal mass are found in Section 5.3.

This discussion has so far concentrated on heat flow by conduction, but convection and radiation are also involved. Both the latter methods become more evident when looking at heat flow through air.

4.3 Heat Flow Through Air

Where surfaces of solids are exposed to air, heat transfer takes place both by convection and by radiation. (Evaporation also can occur, occasionally with thermally significant results, as in the case of roof ponds for passive cooling.) Convection is highly dependent on air motion, so wind outdoors must be considered. Also, since warm air rises and cold air falls, vertical surfaces that encourage this kind of airflow will exchange heat faster than the same surfaces placed horizontally, unless the direction of heat flow is *upward* through this horizontal air layer, as was evident in Fig. 4.1.

When air motion along surfaces is minimal, a surprisingly effective insulating layer of air is created. The resistance of a layer of still air along a vertical surface is numerically equal to that of ½-in. plywood, for example. When this air layer is disturbed, its resistance drops quickly; with a 15-mph (6.7 m/s) wind, resistance drops to about one quarter of the still-air value (see Table 4.3). Similar drops in resistance will occur when forced-air grilles are located immediately above or below windows.

These surface layers of air are often evaluated by their conductance, the opposite of resistance, because we often wish to *encourage* heat transfer between solids and air. For example, passive heating and cooling systems are dependent on large building surfaces as heat exchangers. Active (conventional) systems typically concentrate the heat exchange within mechanical equipment.

Surface conductances are termed h_i for interior air layers and h_o for exterior or outside layers; sometimes f_i and f_o are used instead. Like other conductances, they are expressed in Btu per hour per square foot for 1F° temperature difference (in SI units, watts per square meter for 1K°). The variations of resistances and surfaces conductances can be seen in Table 4.3.

Radiation is also highly influenced by surface characteristics; shiny materials are much less able to radiate than are common rough building materials, This characteristic is called *emittance,* the *ratio* of the radiation emitted by a given material to that emitted by a blackbody at the same temperature. The impact of shiny versus ordinary surfaces can be seen in Tables 4.3 and 4.4, which deal with air layers and airspaces; the lower the emittance, the lower the radiative heat exchange. For most materials, emittance is related to absorptance; a highly absorptive (low reflectance) material will usually have high emittance as well.

The combination of dead air spaces and reflective surfaces produces some of our most effective insulating products, especially when they are made of lightweight materials of low conductivity. Glass fiber, cellular glass, expanded styrenes (foamed plastics), and mineral fibers all exhibit the characteristics of enclosing vast numbers of dead-air spaces per unit volume. When bonded to reflective films and properly installed (facing an airspace), high resistance to heat flow is achieved. Resistances to heat flow can be compared between various building materials, including insulation products (Table 4.2) and air films and airspaces (Tables 4.3 and 4.4).

4.4 Heat Flow Through the Building Envelope

When the process of heat flow is understood, the calculations can begin. Initially, the *hourly* heat loss or heat gain is calculated, since these rates can later be used either to establish equipment sizing (design conditions) or approximate energy consumption (normal conditions).

(a) Heat Flow Due to Temperature Difference. To calculate hourly heat flow through a building's envelope, these factors are necessary:

1. The *rate* at which heat flows through the various assemblies of materials that make up the envelope.
2. The *area* of each of these assemblies.
3. The *temperature difference* between inside and outside for the hour being calculated.

The *areas* of walls, roofs, windows, and floors of various kinds are determined from preliminary architectural design drawings; final versions of these drawings should be subject to changes suggested by these calculations. The *temperature differences* for design conditions are listed in Appendix A; for energy consumption, see Section 5.5. The *rates* at which heat flows are discussed in this section.

The variety of terms used so far to express heat flow is potentially bewildering. These terms are but parts of a larger picture; what is needed is *one* final, overall expression of the steady-state rate at which heat flows through architectural skin elements (walls, roofs, floors, etc.) This is provided by the *U value,* where *U* is thermal transmittance, again expressed in the familiar terms of Btu/h-ft^2-°F (W/m^2-K). Since *U* values are both common and rather complex, considerable space is devoted here to listing them for familiar constructions of walls, floors, roofs, doors, and windows. See Tables 4.5 to 4.26 at the end of this chapter. *U* values are also used in some design criteria, presented in Section 5.2.

These U values are also readily calculated for a particular (floor, roof, wall, etc.) by finding the resistances of each of its materials, its air layers, and its internal airspaces, then adding all these resistances (obtaining ΣR) and finding the reciprocal:

$$U = \frac{1}{\Sigma R}$$

This procedure is illustrated in Fig. 4.3. Where framing interrupts the insulation, an averaged U value must be found. Examples of typical constructions are shown in Tables 4.5 through 4.7 and 4.9 to 4.12. The coefficients in these tables are expressed in Btu/h-ft^2-°F (i.e., per degree Fahrenheit difference in temperature between the air on the two sides) and are based on an outside wind velocity of 15 mph, except as noted in Table 4.12.

Table 4.8 deals specifically with insulation; the R value of some common insulation types are shown, as are corrections to insulation R values due to thermal bridging through metal studs and roof trusses. (A more extensive and complicated procedure for calculating thermal bridging due to metal framing is found in Chapters 20 and 22 of the 1989 ASHRAE *Handbook of Fundamentals*.) Averaged U values for movable insulation over windows is shown in Table 4.17. U values for doors are listed in Table 4.15.

When the building envelope is in contact with the ground instead of outdoor air, the heat flow calculation procedure changes. The ground is often at a temperature different from outdoor air, and earth is more conductive than air. Table 4.13 shows the heat flow through the edge of a concrete slab on grade, with heat flow rate in F_2 units rather than U values, expressed per unit of perimeter length. Four edge insulation conditions are shown, with values of F_2 for three climate zones; interpolation can be used to estimate F_2 for other levels of degree days.

What if more slab edge insulation is provided than is listed in Table 4.13? An approximate extrapolation method is to convert the values of F_2 to their approximate ΣR equivalent ($F_2 = 1/\Sigma R$), then use your judgment to adjust ΣR to include the added slab edge insulation.

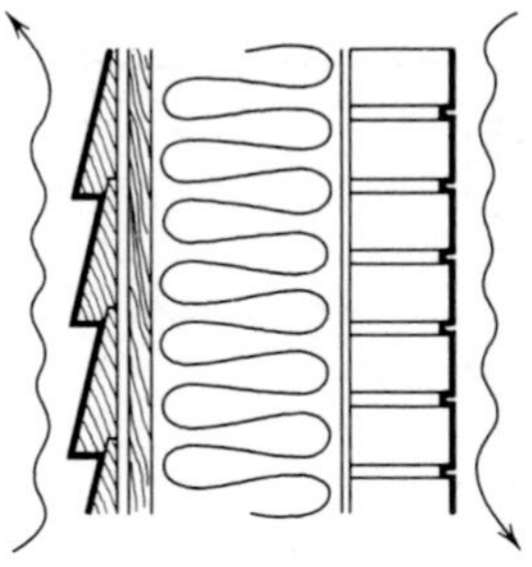

Component	R	Reference Table
Inside air layer	0.68	Table 4.3
Common brick	0.20	Table 4.2
Nominal 6-in. batt fiberglass	19.00	Table 4.2
½-in. Plywood	0.62	Table 4.2
1-in. Wood siding	0.79	Table 4.2
Outside air layer	0.17	Table 4.3
ΣR	21.46	

$$U = \frac{1}{\Sigma R} = \frac{1}{21.46} = 0.046$$

Fig. 4.3 Procedure for determining U *values (overall thermal transmittance).*

EXAMPLE 4.1. Refer to Table 4.13. For the construction type C (metal studs, stucco, and dry wall), in the zone for 5350 DD, find the approximate F_1 when the listed slab insulation is doubled; that is, insulation R = 10.8.

SOLUTION.	*Given* F_2	*Equivalent* ΣR
Uninsulated	1.20	0.83
With R = 5.4	0.53	1.89

Equivalent

added R = 1.89–0.83 = 1.06

If we assume that another identical thickness of slab edge insulation will add another equivalent R = 1.06, then the new equivalent ΣR = 1.89 + 1.06 = 2.95. Therefore, the new F_2 = $1/\Sigma$R = 0.34.

This calculation method will produce differing results for each construction type and each climate zone.

Heat flow through crawl spaces is most simply estimated by insulating the floor above, and assuming that the crawl space is at outdoor temperature. It is rarely energy conserving, especially in colder climates, to insulate the walls of the crawl space instead of the floor above. If a conditioned crawl space is desired, see the ASHRAE 1989 *Handbook of Fundamentals,* Chapter 25.

Heat flow through basement walls and floors is complicated by increasing length of path of heat flow with increased depth, shown in Fig. 4.4. A further complication is that the surface temperature of the earth varies. To obtain a "design temperature" of the earth surface, first approximate the mean winter temperature at your location. This can be done from Appendix B, Table B.15; take the average between the TA (ambient temperature) for January and for the year. Then *subtract* the value of constant amplitude, from Fig. 4.4*c*, from this mean winter temperature, to obtain the design temperature. Table 4.14 is then used to determine the heat flow rate.

EXAMPLE 4.2. Calculate the heat loss for a basement in Minneapolis, Minnesota, which is 28 ft wide by 30 ft long, sunk 6 ft below grade. An insulation of *R*-8.34 is applied to the top 2 ft of the wall below grade. An internal temperature of 70 F is to be maintained.

SOLUTION. The earth surface design temperature is estimated from Table B.15; for Minneapolis, Minnesota, the TA in January is 12 F; TA for the year is 44 F; the average of these is 28 F. From Fig. 4.4*c*, the amplitude is about 24F° at Minneapolis. The design temperature at the earth surface is therefore 28 F − 24F° = 4 F.

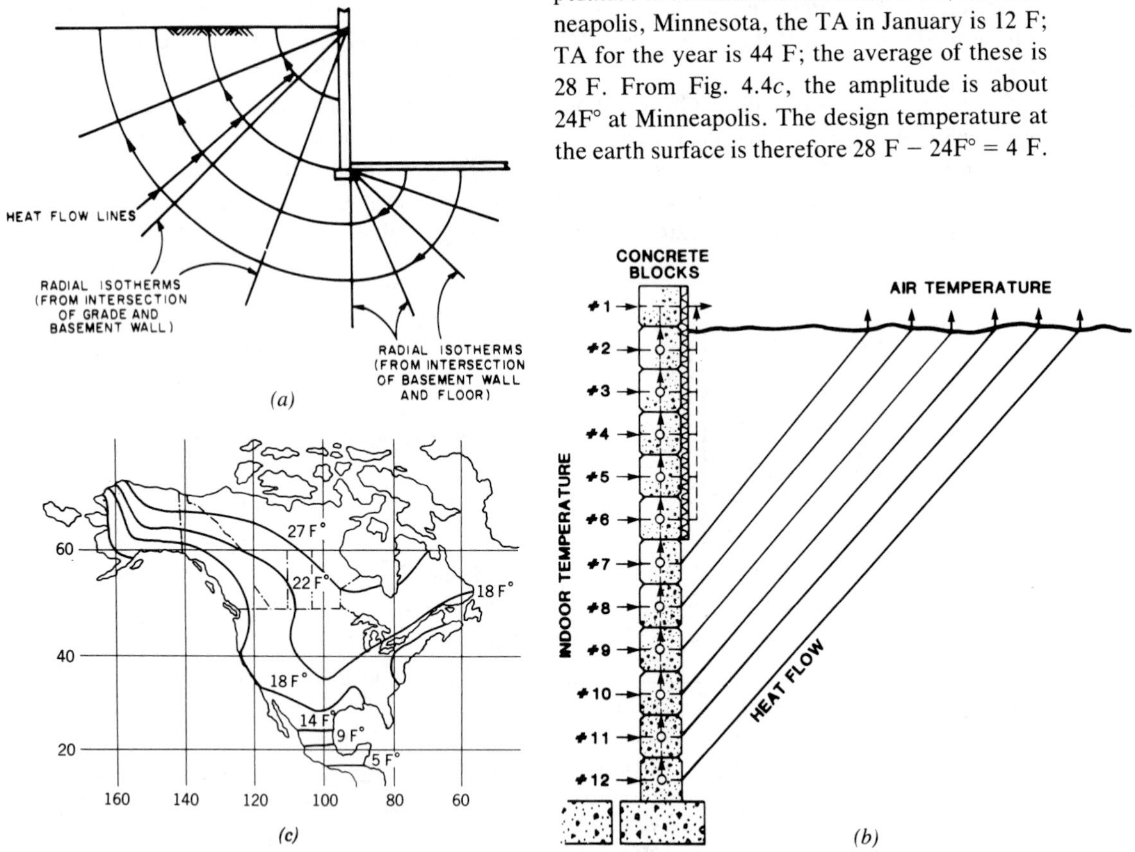

Fig. 4.4 Heat flow below grade. (a) *The paths of heat flow through basement walls and floor follow radial isotherms as shown.* (b) *Heat flow paths for a partially insulated basement wall.* (c) *Lines of constant amplitude of the ground temperature; just below the earth's surface, the yearly average ground temperature will be this much higher in midsummer and this much lower in midwinter. (Reprinted by permission from the ASHRAE 1989* Handbook of Fundamentals, *copyright © by the American Society of Heating, Refrigerating, and Air-Conditioning Engineers, Inc., Atlanta, Georgia.)*

The heat loss through the basement wall is calculated using Table 4.14, Part A;

1st ft below grade (insulated)	0.093 Btu/h-ft-°F
2nd ft below grade (insulated)	0.079
3rd ft below grade (uninsulated)	0.155
4th ft below grade (uninsulated)	0.119
5th ft below grade (uninsulated)	0.096
6th ft below grade (uninsulated)	0.079
Total per 1-ft length of wall	0.621 Btu/h-ft-°F

Basement perimeter = 2(28 ft + 30 ft) = 116 ft

Total wall heat loss = 0.621 Btu/h-ft-°F × 116 ft = 72 Btu/h-°F

The heat loss through the basement floor is calculated using Table 4.14, Part B;

Average heat loss per ft² floor
= 0.025 Btu/h-ft²-°F
(for 28 ft width, 6 ft depth);

Floor area = 28 ft × 30 ft = 840 ft²

Total floor heat loss
= 0.025 Btu/h-ft²-°F × 840 ft²
= 21 Btu/h-°F

Total heat loss for the basement, below grade:

walls 72 + floor 21 = 93 Btu/h-°F

Design temperature difference:
70 F inside − 4 F at earth surface = 66F°

Maximum rate of heat loss below grade
= 93 Btu/h-°F × 66F° = 6138 Btu/h

Calculating the flow of heat through windows is more complicated than through most other opaque exterior surfaces. Windows with more than one glazing layer have different *U* values for the center of the glass than for the area near the glass edges, and the window frame has yet another *U* value. To simplify the potentially tedious process of three *U*-value computations for one window, overall *U* values (U_0) are calculated for two typical "product types" for a wide range of glazing constructions (see Table 4.16). Product type R (residential) is a typical 3 ft by 4 ft operable window with two glazing units separated by a central mullion, while product type C (commercial) is a typical 4 ft by 6 ft one-unit operable window. An important characteristic of types R and C are their relative proportions of center-of-glass area, edge-of-glass area, and frame area; these are listed at the bottom of Table 4.16. The key to determining whether to use the R or C values of U_0 is the size of the glazing unit(s) within the window rather than the size of the overall window. For glazing units of 16 ft² or less, use type R. Also use type R when large one-unit windows have mullions or grilles within the window that form a thermal bridge from the inner to the outer layer of glass. When storm sash is added over the primary (initial) glazing, it frequently has a frame of a different *U* value. In this case, use the *U* value of the primary unit's frame in Table 4.16.

Emittance, ε, is assumed in Table 4.16 at 0.84 for uncoated glass. Two other values of ε are listed at 0.40 (for hard-coated low-emittance glass) and 0.15 (for soft-coated or sputtered low-emittance glass). Manufacturers of glass often list the emittance of their product; when a glass value of ε is other than 0.84, 0.40, or 0.15, interpolation in Table 4.16 is recommended.

The variety of glazing products and their associated *U* values, averaged for center-of-glass, edge-of-glass, and frame conditions, is a potentially bewildering amount of information. New products appear frequently; the designer has only the manufacturer's literature for thermal performance information, with a basis for determining such performance that varies with each company. Fortunately, the city of Seattle publishes at regular intervals a list of glazing test reports, using a common test procedure so that the results are comparable; currently over 700 products are listed. To obtain the latest copy of the *Complete List of Acceptable Glazing U-Value Test Reports,* contact: John Hogan, Major Products Energy Analyst, Seattle Department of Construction and Land Use, 400 Municipal Building, Seattle, WA 98104.

The *U* value must be determined for each

element (or construction type) of the building envelope; then the hourly conductive heat loss (or gain) through a building's skin can be calculated as follows:

$$q = \Sigma(U \cdot A)\ \Delta t$$

where

q is the total heat exchange conducted through building's skin (Btu/h or W).

U and A (U values and areas) are specific to each skin element.

Σ indicates that all UA for these elements are to be added together.

Δt is the temperature difference between indoors and outdoors. For *design* conditions, use outdoor temperature from Appendix A. For *average* conditions, use average temperatures (such as from Appendix B, Part 3, or local climatological data).

One important use of this quantity, q, will be shown in Section 4.6.

Another use of this overall thermal transmittance q is in determining temperatures at various points within walls, roofs, or floors. This is particularly useful information for surfaces such as windows on cold days, which have a strong influence on human comfort. It is also useful for avoiding moisture problems as discussed in the next section.

The gradual change of temperature through a wall, roof, or floor from inside to outside is known as its *thermal gradient;* it can be charted by proportioning the amount of temperature change to the amount of thermal resistance at any point. This procedure is illustrated in Fig. 4.5.

(b) Heat Gain Due to Solar Radiation. This section deals only with heat flow into the building, from solar radiation through fenestration. This quantity is complex, including variations in type of window, shading devices, and the incident angles and intensities of sunlight. Monthly average hourly values of solar gain through one layer of double-strength clear sheet glass, called *solar heat gain factors* (SHGF), are presented in Appendix B for a variety of latitudes. These basic heat gain values must be further modified by the various shading devices that are a likely part of a window. This modifica-

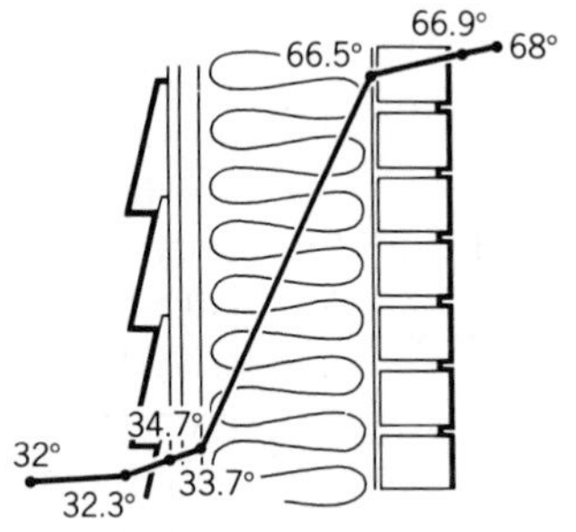

R	Component	ΣR *from Interior*	*Temperature Drop from Interior (F°)*	*Temperature at Outer Edge of Component (F)*
0.68	Inside air layer	0.68	0.68/21.46 × 36 = 1.1	66.9
0.20	Common brick	0.88	0.88/21.46 × 36 = 1.5	66.5
19.00	Nominal 6-in. insulation	19.88	19.88/21.46 × 36 = 33.3	34.7
0.62	½-in. Plywood	20.5	20.5/21.46 × 36 = 34.3	33.7
0.79	1-in. Wood siding	21.29	21.29/21.46 × 36 = 35.7	32.3
0.17	Outside air	21.46		32

Fig. 4.5 *Procedure for calculating the thermal gradient through construction. Assume that outside and inside temperatures are 32 and 68 F (0 and 20°C):* $\Delta t = 68 - 32 = 36F°$.

tion is called the *shading coefficient* (SC), a dimensionless ratio defined as

$$SC = \frac{\text{solar heat gain of fenestration}}{\text{solar heat gain of double-strength glass}}$$

and Tables 4.18 to 4.24 present SC for many window assemblies. The hourly heat gain due to solar radiation through glass, Btu/h-ft^2, is

$$q\ (\text{Btu/h}) = [\text{window area}(\text{ft}^2)] \times [\text{SHGF}(\text{Btu/h-ft}^2)] \times \text{SC}$$

Shading coefficients are based on still air against the inside window surface and a 7.5-mph wind against the outside surface. For a discussion of how an SC is changed by other air motion conditions, see Chapter 27 of the ASHRAE 1989 *Handbook of Fundamentals*.

4.5 Moisture and Infiltration

Buildings, like our bodies, exchange moisture and air with the environment, as well as exchanging heat. Although most of this moisture exchange occurs during the exchange of fresh air, some moisture is exchanged through a building's skin. This can cause problems, either in hot, humid climates or very cold ones.

In hot-humid conditions, cool inside surfaces are often encountered—for example, the ceiling directly below a roof pond used for passive cooling. As hot and humid air contacts such a surface, *condensation* can occur; the moisture vapor in the air condenses to form visible droplets of water on the ceiling. The result can be mildly annoying as water drips on the head, or serious as water stains occur, and, eventually, mold grows on surfaces. For these reasons, such large cooled surfaces are often avoided in hot-humid climates. Even in less humid but hot conditions, condensation can occur within the rock beds that store heat for passively solar heated or cooled buildings, creating problems of odor and bacterial growth.

In cold climates, cold interior surfaces also occur, especially at windows. Although the air indoors may not be particularly humid (40 to 50% rh is common), it contains enough moisture to permit condensation on cold surfaces. (For details, see the discussion on psychrometry in Section 4.7.) Again, mild annoyance or more serious damage can result. A much less visible moisture threat occurs *within* walls, ceilings, or floors. Almost all common building materials, including gypsum board, concrete, clay masonry, and wood, are easily permeated by moisture. Most surface finishes are also permeable. In cold climates, the air outside contains relatively little moisture, even though the rh may be high. By contrast, inside air contains much more moisture per unit of volume, despite its probably lower rh. (This is evident from the psychrometric chart, Fig. 4.8, later in this chapter.) The result is a flow of vapor from high vapor pressure to low vapor pressure (typically warm to cold).

The problem with such a flow occurs when the temperature within the wall (floor, etc.) drops low enough for this vapor to condense. Insulation can then become wet and thereby less effective, since water conducts heat far better than the air pockets it has filled. If wet insulation compacts, these air pockets are permanently lost. Worse yet, moisture damage can occur, such as dry rot in wood structural members.

The usual remedy for such a potential problem is to install a *vapor barrier* within the building envelope; since very low permeability is desired, these barriers are commonly of plastic film installed with as few holes as possible. Since the moister air on the warm side is the source of the problem, this barrier needs to be installed as *close to the warm side* as possible—typically, just behind interior surfaces (gypsum board, wood floors, etc.). A less desirable approach is to use vinyl wallpaper or vapor barrier paints on interior surfaces. These cannot give the around-the-corners, wrapped protection that is obtained with properly installed plastic films. This disadvantage is shared by aluminum-foil-faced insulation; it is very effective thermally, but less effective as a vapor barrier. (Also, the aluminum foil must face an air space if it is to be effective.)

A substantial benefit of plastic films is that they reduce airflow through construction. Outdoor air is always infiltrating a building, gradually replacing the indoor air. This unintentional source of fresh air becomes a problem when temperatures outside are very different from those inside, especially when strong winds force outdoor air indoors fast enough to produce noticeably cold (or hot) drafts. Some fresh air is always desirable in buildings, but so is user control of how and where it is admitted. Therefore, the moisture-tight and infiltration-tight characteristics of plastic film vapor barriers are usually beneficial. When good vapor barriers are installed, the smaller air-change values that accompany "tight" construction may be assumed in the following calculations of heat flow due to infiltration and ventilation.

4.6 Heat Flow by Ventilation

Two common terms for outdoor air that enters a building are:

Infiltration. The "accidental" influx of outdoor air due to air leakage through the building skin.

Ventilation. The deliberate, designed introduction of outdoor air.

Winter heat loss (and summer heat gain in closed, cooled buildings) occurs when "fresh" outdoor air enters the building to replace "stale" indoor air. This heat exchange, analogous to human lung heat losses or gains, must also be calculated when sizing heating or cooling equipment, or when estimating energy use per season.

It is important to provide at least a minimum amount of fresh air indoors, both for comfort and for health. Odors and a sense of staleness can be uncomfortable, and dangerous buildups of radon gas and other pollutants such as formaldehyde can be produced within buildings. These pollutants are most effectively removed with air changes through rooms. Table 4.25 lists recommended design outdoor airflow rates. See Section 6.6 for a discussion of ventilation for acceptable indoor air quality.

The calculation for the heat lost (or gained) by the introduction of outdoor air into spaces is

$$q_v = (V)(1.08)(\Delta t)$$

where

q_v = sensible heat exchange due to ventilation (Btu/h)

V = volume flow rate, in cubic feet per minute (cfm) of outdoor air introduced

Δt = temperature difference between outdoor and indoor air (F°)

1.08 = a constant derived from the density of air at 0.075 lb/ft^3 under "average" conditions, multiplied by the specific heat of air (heat required to raise 1 lb of air 1F°), which is 0.24 Btu/lb-°F, and by 60 min/h. The units of this constant are Btu-min/ft^3 °F h.

In SI units, this equation becomes

$$q_v = (V)(1200)(\Delta t)$$

where q_v is in watts, V is in liters per second, 1200 is the approximate volumetric specific heat capacity of air (J-s/m^3-°C-h), and Δt is in degrees Celsius (or K).

The next task is to determine how much air, V (in cfm or L/s), should be assumed for this calculation. Four basic approaches that represent various degrees of design control over ventilation are considered here.

(a) Natural Infiltration by Approximation. For residences and small commercial buildings without significant internal heat sources, many building codes do not require mechanical or forced ventilation. In some codes, if the openable (or operable) window area is 5% of the floor area of each room in these buildings, adequate natural ventilation is assumed to be achievable. The two methods that follow are typically used with design condition calculations; that is, to find worst-hour loss (or gain) when windows and doors are closed.

There are two approximation methods available: the *air-change* method and the *crack method;* the air-change method is very quick (but tends to overestimate); a glance at Table 4.26 is sufficient to find an assumed number of air changes per hour (ACH) for a space; then,

$$V = \frac{\text{(ACH)(room volume)}}{60 \text{ min/h}}$$

where V is in cfm.

The crack method takes longer; it assumes that data on window and door construction and wind velocities are known. This method assumes that doors and openable windows represent all the cracks by which outdoor air infiltrates a closed room under worst-hour conditions. Table 4.27 shows how many cfm per foot of crack should be assumed, *on the windward exposure(s) only,* to arrive at a total cfm. To convert to SI units, use

$$\text{cfm} \times 0.4719 = \text{L/s}$$

It is now much more common to measure buildings for infiltration or air leakage, and methods of calculating air exchange due to infiltration, based on the data from such tests, is described in Chapter 23 of the ASHRAE 1989 *Handbook of Fundamentals*. Often, designers aim at removing excess indoor heat by deliberately admitting cooler outdoor air. This procedure is usually applied in passive cooling situations utilizing an "open" building, where outdoor air "as is" is desired indoors.

(b) Cooling by Natural Ventilation. It is important to remember that natural ventilation cooling—whether by windows or by stack effect—will only work when the *outside is cooler than the inside*. More details of natural cooling calculations are found in Chapter 5. For window ventilation, the quantity of outdoor air admitted is called Q (analogous to V in the preceding paragraphs):

$$Q = C_v A v$$

where

Q = volume flow rate of air (cfm)

A = area of openable windows on inlet side or sides (ft²). *Note:* Outlet side(s) should have at least this much openable area as well.

C_v = effectiveness factor that adjusts for different wind orientations (dimensionless): 0.5 to 0.6 for winds perpendicular to window openings; 0.25 to 0.35 for wind diagonal to window openings

v = velocity of wind in *feet per minute.* feet per minute = (miles per hour)(88); feet per minute = (meters per second)(196.86)

In SI units,

$$Q = 1000 C_v A v$$

where Q is in liters per second, A is in square meters, and v is in meters per second.

A further modification to these cooling formulas is usually necessary, because wind data are usually taken at airports at a height of 10 m (about 30 ft) above open, unobstructed ground. Rarely will buildings relying on wind ventilation be so favorably situated. Figure 4.6 shows correction factors to be applied to airport wind velocities, for variations in terrain and in height above the ground. This dimensionless correction factor should then be applied to the quantity Q, usually resulting in reduced volume flow rate (cfm).

(c) Natural Ventilation Through Stack Effect. Another way of deliberately ventilating and cooling buildings is to take advantage of the lighter weight of hot air, which allows it to rise. When there is a difference in height between inlet openings low in walls (or in floors) and outlets through roofs, *and* when outdoor air is cooler than indoor air, natural ventilation will occur through the *stack effect* of warm air rising and leaving through the higher openings. For stack effect,

$$Q = CA\sqrt{\frac{h(t_i - t_o)}{t_i}}$$

where

Q = airflow (cfm)

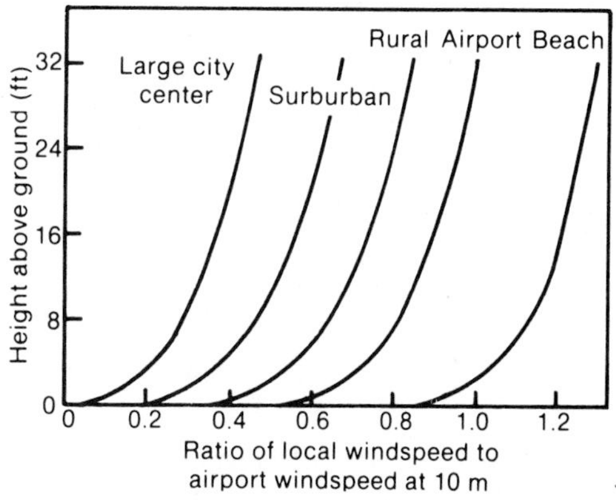

Fig. 4.6 *Windspeed variation with height for various terrains. (From* Cooling with Natural Ventilation, *Florida Solar Energy Center, published by the Solar Energy Research Institute.)*

C = constant of proportionality = 313. *Note:* This assumes a value of 65% of the maximum theoretical flow, due to limited effectiveness of actual openings. With less favorable conditions (due to indirect paths from openings to the stack, etc.), the effectiveness drops to 50%, and C = 240.

A = area of cross section through stack, or outlets (ft^2). *Note:* Inlet area must be at least equal to this amount.

h = height difference, between inlets and outlets (ft)

t_i = (higher) temperature inside, within the height h, (F)

t_o = (lower) temperature outside (F)

In SI units, Q is in cubic meters per second, C = 91.1 at 65% effectiveness (70 at 50% effectiveness), A is in square meters, h is in meters, and t_i and t_o are in kelvin.

Clearly, the taller the stack and the greater the temperature difference, the more air will flow. Increased area of openings is even more influential. Figure 4.7 presents solutions for some small stack effect combinations in a convenient graph.

The stack effect is more pronounced in tall buildings, and can result in more infiltration of outdoor air to the lower floors and an exfiltration of air from the upper floors. A detailed stack effect calculation procedure will take into account pressure differences at a variety of levels and identifies the horizontal plane at which neutral pressure—neither infiltration nor exfiltration due to stack effect—occurs. This procedure is found in Chapter 23 of the ASHRAE 1989 *Handbook of Fundamentals*.

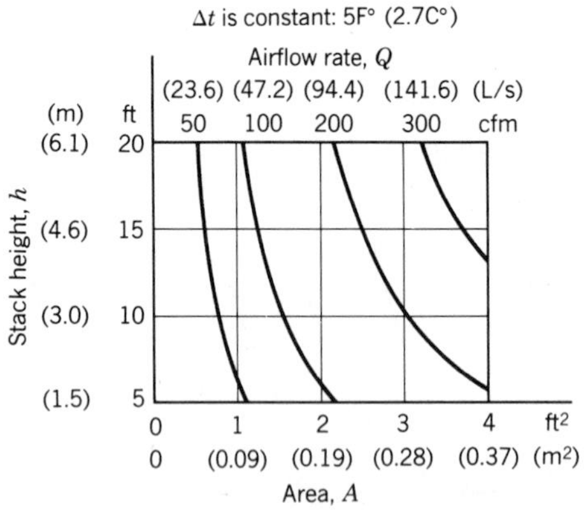

Fig. 4.7 *Approximate rate of natural ventilation due to stack effect. Indoor temperatures are near the comfort zone, and Δt is assumed to be 5F° (2.7C°). Reprinted by permission from S. V. Szokolay,* Environmental Science Handbook, *Longman Group Ltd., London (1980).*

When *both* wind and stack effect are used together, their combined flow rate effect is *less* than the simple sum of their individual flow rates:

$$Q_{ws} = \sqrt{(Q_w^2 + Q_s^2)}$$

where

Q_{ws} = flow rate of the combined effects of wind and stack ventilation

Q_w = flow rate from wind effect

Q_s = flow rate from stack effect

(d) Forced Ventilation. Fans can be used to forcibly introduce the desired amount of outdoor air directly into spaces. Fan manufacturers list their capacity in cubic feet per hour (cfh) (or in cfm). This outdoor air can be blown into (or sucked out of) spaces, or it can be mixed with air being recirculated so that the different temperature of outdoor air is less noticeable. An important energy conservation opportunity arises with forced ventilation: a heat exchanger may be used. Outgoing and incoming airstreams can be kept as separate airstreams, but heat can be transferred from one stream to the other. Thus, incoming very cold outdoor air can be given the heat, but not the pollutants, of outgoing warm indoor air; this flow of heat can be reversed in the hot, humid summer. (Our bodies do the same thing with arteries and veins, where arms and legs intersect the trunk. That is, in cold weather, cooler blood from the extremities is warmed before it enters the thermally protected core of the body.) The equipment involved in heat exchange is discussed in Sections 6.4 and 7.3.

To approximate the size of the fan, where Q is the desired flow rate (from Table 4.25):

$$Q = \text{(cfm outdoor air person)} \\ \text{(number of people)};$$

or

$$Q = (\text{cfm/ft}^2 \text{ floor area})(\text{ft}^2 \text{ floor area}).$$

In SI units, substitute liters per second for cfm, and square meters for square feet.

4.7 Psychrometry

Before going into the next chapters' details of hourly heat gain and seasonal energy use, it is necessary to better understand the interaction of heat, moisture, and air. Moisture is a relatively minor component in winter comfort and usually can be safely ignored in calculating heat losses. By contrast, moisture is highly influential on heat gain, both for comfort and in calculations.

Air, moisture, and heat interactions are rather complex; as air temperature rises, its capacity to hold moisture rises also; and warmer air becomes less dense. These combined interactions are described by *psychrometry,* the study of moist air. Fortunately, these interactions can be combined in a single chart (Fig. 4.8*a* and *b*).

The basis of the psychrometric chart has been developed in Sections 2.2, 2.3, and 2.4, so that relative humidity (rh) and dry-bulb (DB) and wet-bulb (WB) temperatures are familiar terms. They are combined in the basic chart of Fig. 4.9, where the term "saturation line," at 100% rh, is also introduced. This is also called the "dew point" (DP) because dew will form (water vapor will condense) when saturated air touches any surface at or below the air's dew point temperature. This condensation is sometimes undesirable (as within walls or on ceiling, air duct, or glass surfaces), but often desirable, as on air conditioner coils, where a reduction of the moisture content in the air is deliberate.

The psychrometric chart may be used to graph a wide variety of processes, which are summarized in Fig. 4.10. To understand these processes, we must add to the basic chart of Fig. 4.9. The first addition is *humidity ratio,* which indicates the amount of moisture by weight within a given weight of air. Air treatment processes that travel along these horizontal lines of constant humidity ratio (Fig. 4.11) are the familiar processes of simple heating (air passing through the heating coil of a furnace, or through a solar collector) and simple (sensible) cooling (air passes through the cooling coil of an air conditioner, before saturation). The humidity ratio will be used later in calculating latent heat gains from outdoor air (see Section 6.8).

The next addition shows how the *density* of air varies as its temperature and moisture content vary. These lines are those of *specific volume,* the reciprocal of density, a useful quantity in air conditioning calculations and helpful in understanding how the stack effect (Section 4.6) works. The specific volume is given in ft^3/lb (m^3/kg) of dry air. It is evident from these lines in Fig. 4.12 that a pound of hot air is larger (has more volume) than a pound of cold air. This larger volume per pound increases buoyancy; thus hot air rises while cold air sinks.

The next addition (Fig. 4.13) contains some of the most important information of all: *enthalpy,* the total amount of both *sensible* and *latent* heat in the air–moisture mixture.

Sensible heat is the kind of heat that *increases the temperature* of air; the glowing coil of an electric range is adding sensible heat to the air of the kitchen.

Latent heat is the kind of heat that is present in *increased moisture* in air; the boiling water in a pan on top of the electric range is evaporating, and as it does it increases the latent heat in the kitchen air. This moisture has the potential to condense, thereby releasing its heat of vaporization. (Note that 180 Btu is required to raise the temperature of a pound of water from 32 F to 212 F; 1061 Btu is required to then evaporate the pound of water, without raising its temperature.)

Enthalpy is the sum of the sensible and latent heat content of an air–moisture mixture, relative to the sensible plus latent heat in air at 0 F (0°C in SI units) at standard atmospheric pressure. Its units are Btu/lb of dry air (in SI units, kJ/kg of dry air).

Enthalpy lines almost parallel those of WB temperature. To avoid further complicating the visual appearance of the psychrometric chart, enthalpy lines (Fig. 4.13) are shown as solid lines beyond the saturation point only.

Perhaps the most familiar air treatment process that travels along lines of constant enthalpy is evaporative cooling, where increased mois-

Fig. 4.8 (a) *Psychrometric chart, conventional units. Conditions at sea level, standard atmospheric pressure (29.921 in. Hg). (Courtesy of the Carrier Corporation.)*

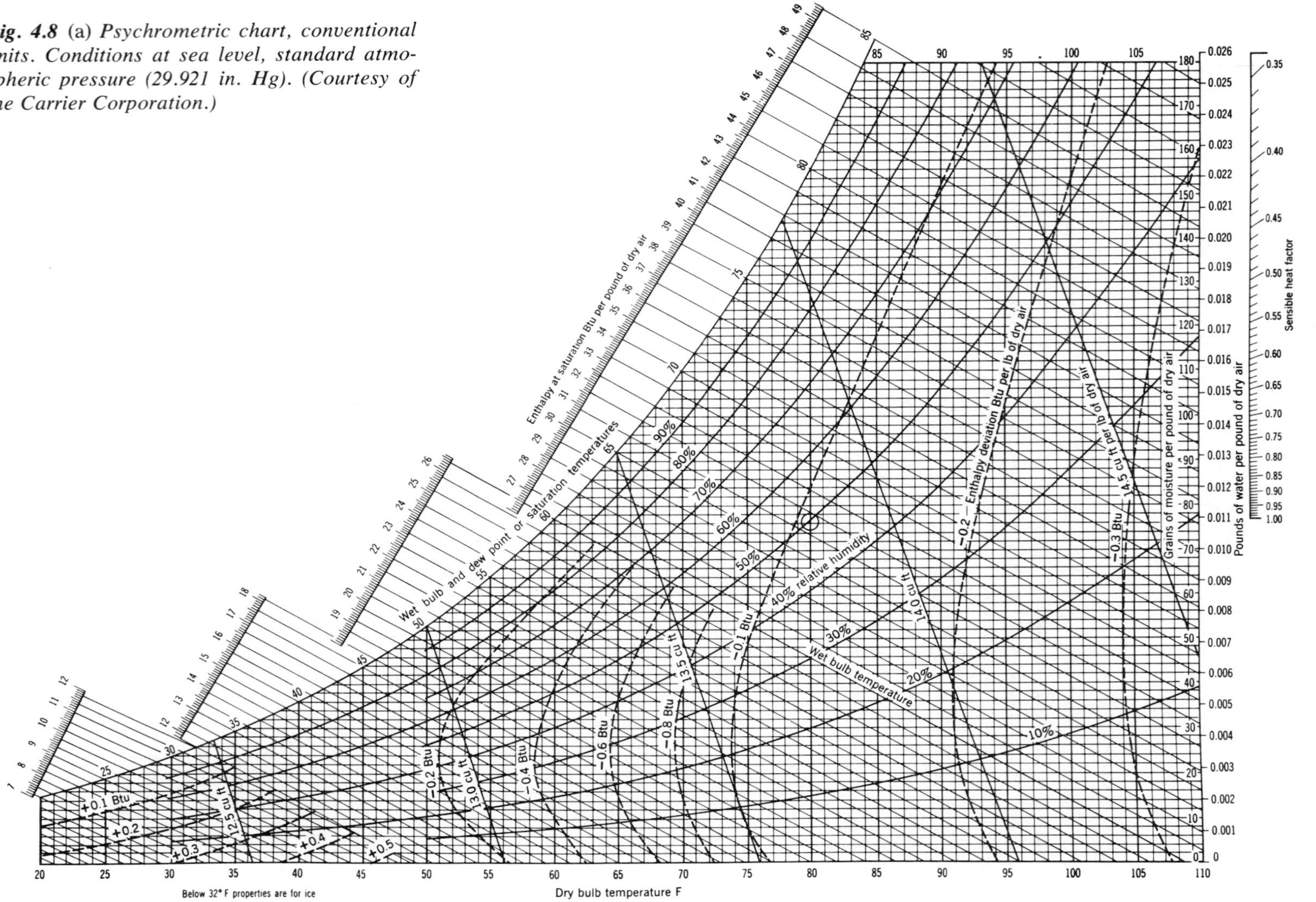

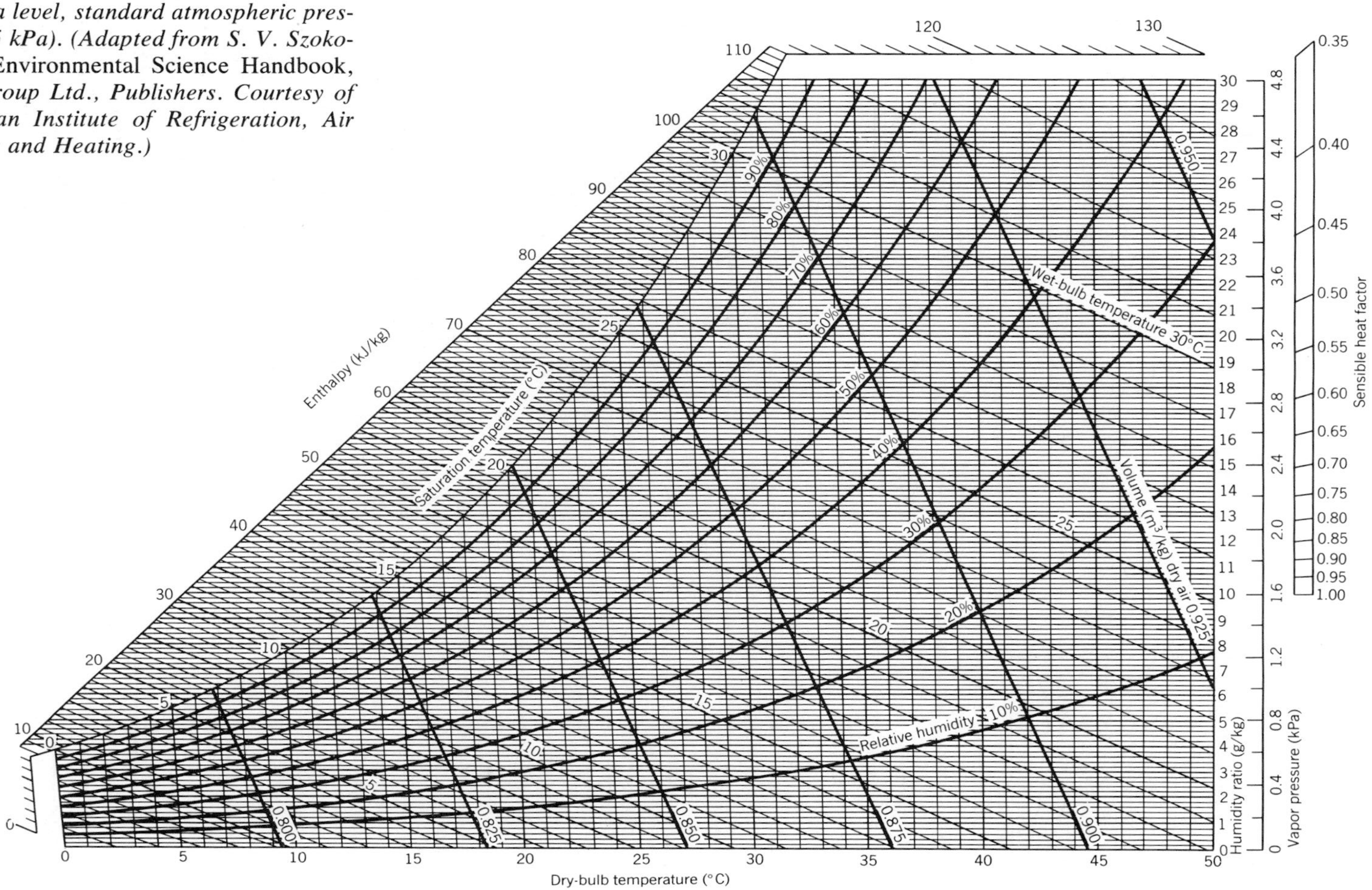

Fig. 4.8 (b) *Psychrometric chart, SI units. Conditions at sea level, standard atmospheric pressure (101.325 kPa). (Adapted from S. V. Szokolay (1980),* Environmental Science Handbook, *Longman Group Ltd., Publishers. Courtesy of the Australian Institute of Refrigeration, Air Conditioning and Heating.)*

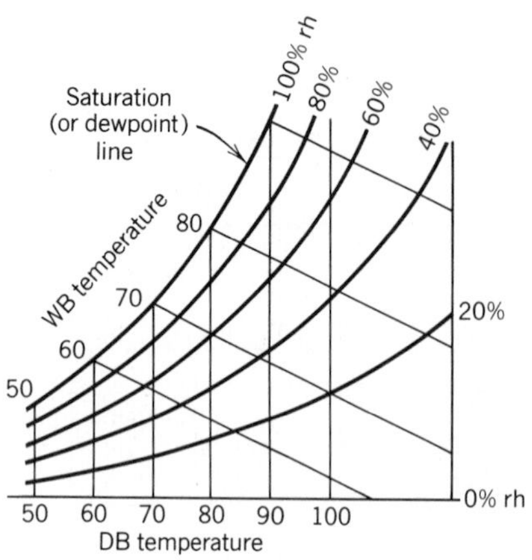

Fig. 4.9 Some basic components of the psychrometric chart.

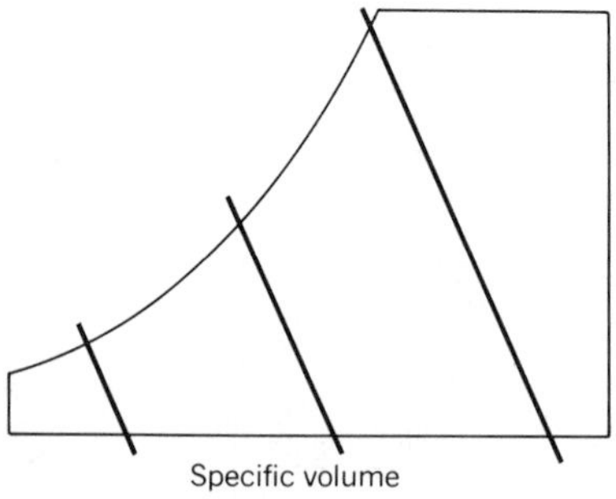

Fig. 4.12 Specific volume on the psychrometric chart: conventional units, ft^3/lb dry air; SI units, m^3/kg dry air.

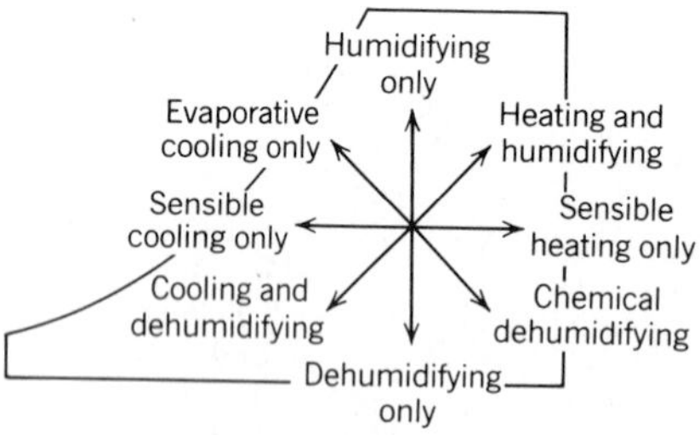

Fig. 4.10 Climatic-conditioning processes, expressed on the psychrometric chart. [Adapted from "Architectural Design Based on Climate," by M. Milne and B. Givoni, in Watson (ed.), Energy Conservation Through Building Design. *Reprinted with the permission of the publisher, McGraw-Hill, Inc.]*

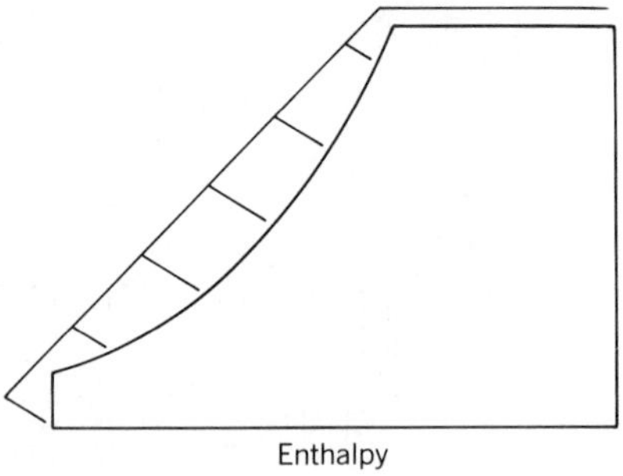

Fig. 4.13 Enthalpy on the psychrometric chart: conventional units, Btu/lb; SI units, kJ/kg.

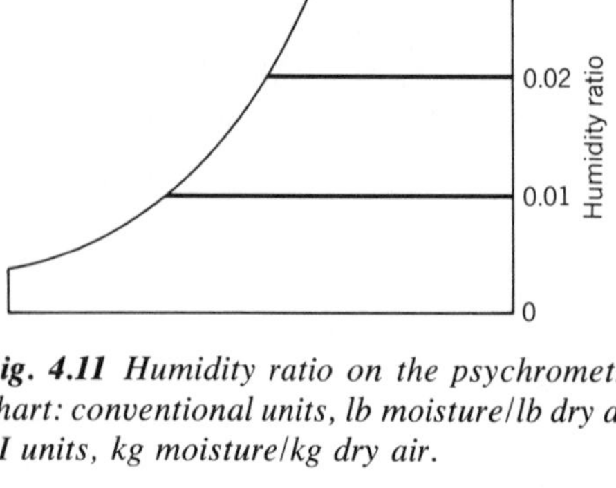

Fig. 4.11 Humidity ratio on the psychrometric chart: conventional units, lb moisture/lb dry air; SI units, kg moisture/kg dry air.

ture and lower air temperature are obtained *without changing the enthalpy* (total heat content) of the air. There is indeed a drop in sensible heat as the temperature drops, but in terms of heat content, it is matched by an increase in latent heat as the moisture content increases. From the standpoint of human comfort, very hot, very dry air is improved by becoming less hot and more humid, up to a point. Thus evaporative cooling "works" in hot dry climates even though it doesn't change the total heat (enthalpy) of the air at all. The opposite process is chemical dehumidifying, where decreased moisture content is obtained at the price of increased temperature; again, no change in total heat (enthalpy) occurs.

The work to be done by mechanical air conditioning equipment is measured by the total change in enthalpy that must occur within the air that is treated by such equipment. The psychrometric chart is used to accurately size an air conditioner. This process is shown in Section 6.8.

First, consider the problem of determining the total change in enthalpy. Assume outdoor conditions of 90 F dry bulb, 76 F wet bulb, with desired indoor conditions at 75 F DB, 50% rh (which translates to 62.7 F WB). In the simple case of cooling 100% outside air, the total heat to be removed is determined as shown in Fig. 4.14. (Once-through systems cooling 100% outdoor air are rare, largely because they require substantial energy. They do occur in hospital surgical rooms and in laboratories with many fume hoods.) How much total heat will be removed, in getting from 90 F DB, 76 F WB to 75 F DB, 62.7 F WB?

For every pound of "dry" air (it is not really dry, but the values are based on the weight of the air alone) that is cooled and dehumidified, 39.6 − 28.3 = 11.3 Btu must be extracted. Similarly, for every pound of air so treated, 0.0162 − 0.0093 = 0.0069 lb of condensed moisture must be disposed of.

Each pound of moist air contains heat that is, by custom, measured above the value of 0 F. It consists of the sensible heat of the air and the water vapor, and the latent heat of the water vapor. Using 1061 Btu as an average value of latent heat per pound of moisture, one may check the enthalpy values of Fig. 4.14.

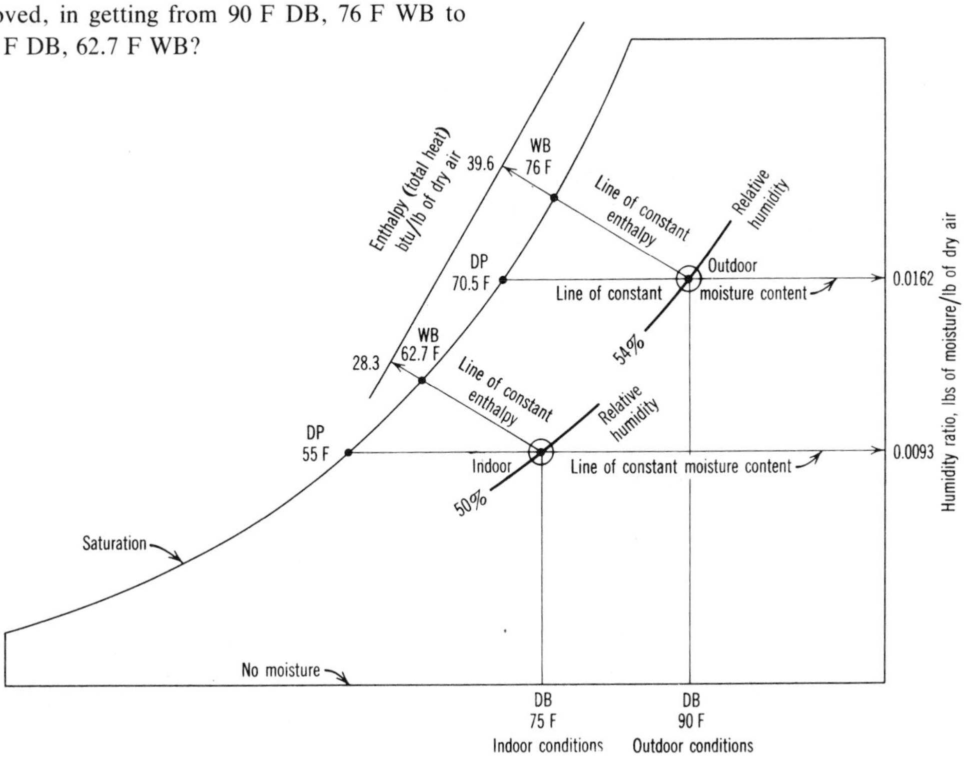

Verification of the Enthalpy Values Above

Item	*Indoor Conditions*		*Outdoor Conditions*	
Latent heat, vapor	0.0093 × 1061[a] =	9.90	0.0162 × 1061[a] =	17.25
Sensible heat, vapor	0.0093 × 0.444 × 75 =	0.31	0.0162 × 0.444 × 90 =	0.65
Sensible heat, air (1 lb)	1 × 0.241 × 75 =	18.10	1 × 0.241 × 90 =	21.70
Enthalpy (Total heat)		28.30		39.60

[a]Latent heat of vaporization, Btu per pound (approximate).

Fig. 4.14 *Use of the psychrometric chart to determine the change in enthalpy between given outdoor and indoor air conditions.*

The actual cooling process is more complex, as shown in Fig. 4.15. The conditioned air must be introduced to the space at conditions of temperature and relative humidity lower in value than those of the indoor conditions so that the entering air may "soak up" heat and moisture and leave through the return grills no worse than the design characteristics of 75 F and 50% relative humidity. So another set of conditions lower in dry bulb and humidity ratio will be established, and the values will depend on the rates at which the air is introduced and the amounts of heat and moisture to be absorbed by the air passing through. The heat and moisture are, of course, the sensible and latent heat gain of the space to be conditioned.

Within the cooling equipment, the outdoor air follows a complex path; the lines labeled 1, 2, and 3 in Fig. 4.15 trace the cooling and dehumidifying steps. Air is cooled without loss of moisture until it reaches a temperature at which it is saturated (its dew point). The cooling then continues. The air that had been losing sensible heat in step 1 continues to lose it in step 2, but additionally, moisture is condensed, which also requires the extraction of heat. When step 3 begins (it is known as a reheating process), the remaining smaller amount of moisture and the air are both heated to 65 F, 40% rh—slightly below the conditions to be maintained in the room. The changes in the heat content of the air and moisture at the various stages are measured along the enthalpy line.

The psychrometric chart can also be used to track the process of heating outdoor air in winter (Fig. 4.16). The chart shows that air at low temperatures in winter often has a humidity ratio so low that this moisture content would be unsatisfactory when the mixture was warmed to acceptable room temperature. Moisture must be added. Often this is in the form of a warm water spray. For our purpose, however, consider an adiabatic spray (no gain or loss of total heat) accomplished by first warming the outdoor air and vapor to a predetermined temperature, in this case 67 F (step 1). Then it is sprayed with water at room temperature to saturation. During this process (step 2), the added water that is evaporated mechanically draws sensible heat from the air, cooling it. The water acquires an equal amount of heat for its change of state to vapor—latent heat. Thus the value at the en-

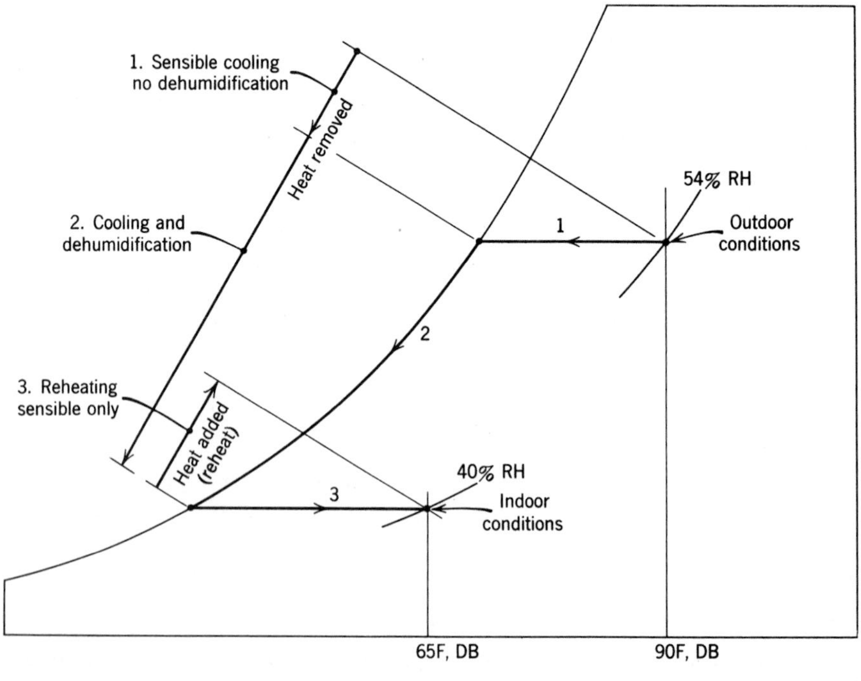

Fig. 4.15 Process of cooling and dehumidifying outdoor air, summer conditions.

thalpy scale does not change. The saturated mixture is then heated (step 3) with no addition of moisture until, at 70 F, it has a reasonably suitable relative humidity. Again the heat changes are read on the enthalpy scale. The action in Fig. 4.15 is known as cooling and dehumidifying with a reheat process. Figure 4.16 illustrates a preheat, spray, and reheat arrangement.

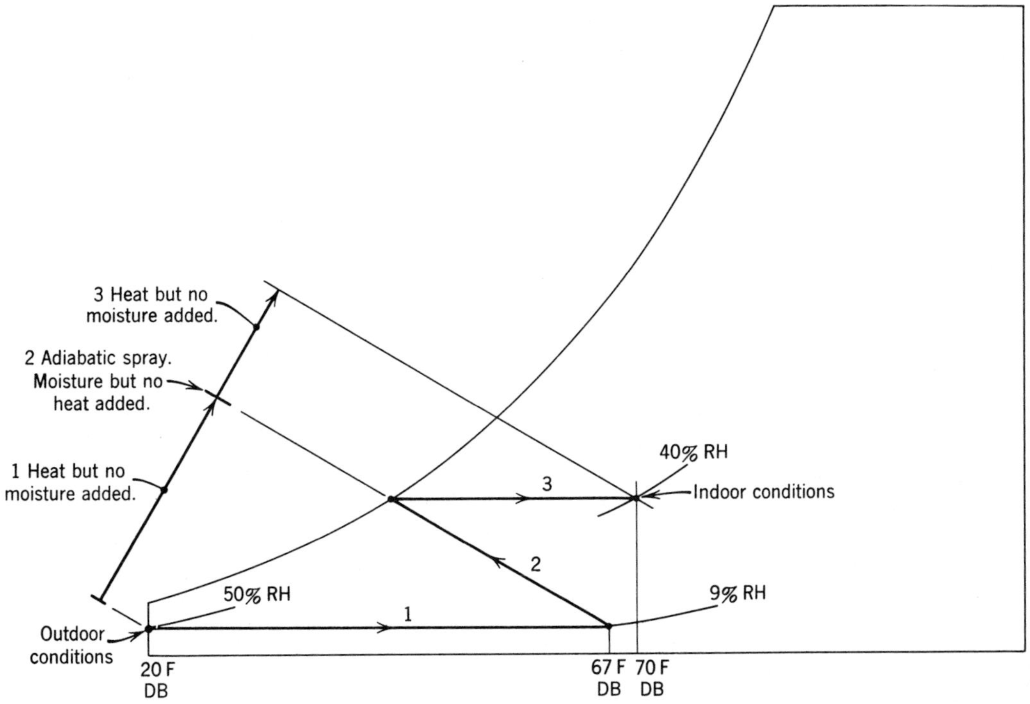

Fig. 4.16 *Process of heating and humidifying outdoor air, winter conditions.*

REFERENCES

Publications of the American Society of Heating, Refrigerating, and Air-Conditioning Engineers, Inc., Atlanta, Ga.:

ASHRAE *Cooling and Heating Load Calculation Manual.*

ASHRAE *Handbook of Fundamentals.*

ASHRAE *Standard 62-1989,* Ventilation for Acceptable Indoor Air Quality, 1989.

ASHRAE/IES *Standard 90.1-1989,* Energy Efficient Design of New Buildings Except New Low-Rise Residential Buildings, 1989.

Balcomb, J. D., Barley, D., McFarland, R., Perry, J., Jr., Wray, W., and Noll, S. (1980). *Passive Solar Design Handbook,* Vol. 2, *Passive Solar Design Analysis,* U.S. Department of Energy, Washington, D.C.

Balcomb, J. D., Jones, R. W. (ed.), Kosiewicz, C. E., Lazarus, G. S., McFarland, R. D., and Wray, W. O. (1983). *Passive Solar Design Handbook,* Vol. 3, American Solar Energy Society, Boulder, Colo.

Chandra, S., Fairey, P. W., and Houston, M. (1986). *Cooling with Ventilation,* Solar Energy Research Institute, Golden, Colo.

Johnson, T. (1981). *Solar Architecture: The Direct Gain Approach,* McGraw-Hill, New York.

Milne, M., and Givoni, B. (1979). "Architectural Design Based on Climate," in *Energy Conservation Through Building Design,* D. Watson, Editor. McGraw-Hill, New York.

Szokolay, S. V. (1980). *Environmental Science Handbook,* Longman Group, London.

TABLE 4.2 **Thermal Properties of Typical Building and Insulating Materials (design values)**[a]

NOTE: The customary units for resistance (*R*), either per inch (1/*k*) or for thickness stated (1/*C*), are given in Table 4.1. The SI units for resistance (last two columns) were calculated by taking the values from the two resistance columns under Customary Unit, multiplying by the factor 1/*k*(*r*/in.) and 1/*C*(*R*) for the appropriate conversion factor. Author's note: Actual (on-site) resistance values frequently are lower than the test-cell-determined "design" values listed in this table.

Description	Density (lb/ft³)	Conductivity, k (Btu-in./h-ft²-°F)	Conductance, C (h-ft²-°F)	Customary Unit Resistance, R[b]: Per Inch Thickness 1/k	Customary Unit Resistance, R[b]: For Thickness Listed, 1/C	Specific Heat, (Btu/lb-°F)	SI Unit Resistance, R[b]: (m-K)/W	SI Unit Resistance, R[b]: (m²-K)/W
BUILDING BOARD								
Boards, Panels, Subflooring, Sheathing								
Woodboard Panel Products								
Asbestos-cement board	120	4.0	—	0.25	—	0.24	1.73	
Asbestos-cement board . . . 0.125 in.	120	—	33.00	—	0.03			0.005
Asbestos-cement board . . . 0.25 in.	120	—	16.50	—	0.06			0.01
Gypsum or plaster board . . . 0.375 in.	50	—	3.10	—	0.32	0.26		0.06
Gypsum or plaster board . . . 0.5 in.	50	—	2.22	—	0.45			0.08
Gypsum or plaster board . . . 0.625 in.	50	—	1.78	—	0.56			0.10
Plywood (Douglas fir)	34	0.80	—	1.25	—	0.29	8.66	
Plywood (Douglas fir) . . . 0.25 in.	34	—	3.20	—	0.31			0.05
Plywood (Douglas fir) . . . 0.375 in.	34	—	2.13	—	0.47			0.08
Plywood (Douglas fir) . . . 0.5 in.	34	—	1.60	—	0.62			0.11
Plywood (Douglas fir) . . . 0.625 in.	34	—	1.29	—	0.77			0.19
Plywood or wood panels . . . 0.75 in.	34	—	1.07	—	0.93	0.29		0.16
Vegetable fiberboard								
Sheathing, regular density[c] . . . 0.5 in.	18	—	0.76	—	1.32	0.31		0.23
. . . 0.78125 in.	18	—	0.49	—	2.06			0.36
Sheathing intermediate density[c] . . . 0.5 in.	22	—	0.82	—	1.22	0.31		0.21
Nail-base sheathing[c] . . . 0.5 in.	25	—	0.88	—	1.14	0.31		0.20
Shingle backer . . . 0.375 in.	18	—	1.06	—	0.94	0.31		0.17
Shingle backer . . . 0.3125 in.	18	—	1.28	—	0.78			0.14
Sound deadening board . . . 0.5 in.	15	—	0.74	—	1.35	0.30		0.24
Tile and lay-in panels, plain or acoustic	18	0.40	—	2.50	—	0.14	17.33	
. . . 0.5 in.	18	—	0.80	—	1.25			0.22
. . . 0.75 in.	18	—	0.53	—	1.89			0.33
Laminated paperboard	30	0.50	—	2.00	—	0.33	13.86	
Homogeneous board from repulped paper	30	0.50	—	2.00	—	0.28	13.86	
Hardboard[c]								
Medium density	50	0.73	—	1.37	—	0.31	9.49	
High density, service temp. service underlay	55	0.82	—	1.22	—	0.32	8.46	

High density, std. tempered	63	1.00	—	1.00	—	0.32	6.93	
Particleboard[c]								
Low density	37	0.54	—	1.85	—	0.31	12.82	
Medium density	50	0.94	—	1.06	—	0.31	7.35	
High density	62.5	1.18	—	0.85	—	0.31	5.89	
Underlayment 0.625 in.	40	—	1.22	—	0.82	0.29		0.14
Wood subfloor 0.75 in.		—	1.06	—	0.94	0.33		0.17
BUILDING MEMBRANE								
Vapor—permeable felt	—	—	16.70	—	0.06			0.01
Vapor—seal, 2 layers of mopped 15-lb felt	—	—	8.35	—	0.12			0.02
Vapor—seal, plastic film	—	—	—	—	Negl.			
FINISH FLOORING MATERIALS								
Carpet and fibrous pad	—	—	0.48	—	2.08	0.34		0.37
Carpet and rubber pad	—	—	0.81	—	1.23	0.33		0.22
Cork tile 0.125 in.	—	—	3.60	—	0.28	0.48		0.05
Terrazzo 1 in.	—	—	12.50	—	0.08	0.19		0.01
Tile—asphalt, linoleum, vinyl, rubber	—	—	20.00	—	0.05	0.30		0.01
Vinyl asbestos						0.24		
Ceramic						0.19		
Wood, hardwood finish 0.75 in.			1.47		0.68			0.12
INSULATING MATERIALS								
Blanket and Batt[d,e]								
Mineral fiber, fibrous form processed from rock, slag, or glass								
Approx. 3–4 in.	0.3–2.0	—	0.091	—	11			1.94
Approx. 3.5 in.	0.3–2.0	—	0.077	—	13			2.29
Approx. 5.5–6.5 in.	0.3–2.0	—	0.053	—	19			3.35
Approx. 6–7.5 in.	0.3–2.0	—	0.045	—	22			3.87
Approx. 9–10 in.	0.3–2.0	—	0.033	—	30			5.28
Approx. 12–13 in.	0.3–2.0	—	0.026	—	38			6.69
Board and Slabs								
Cellular glass	8.5	0.35	—	2.86	—	0.18	19.81	
Glass fiber, organic bonded	4–9	0.25	—	4.00	—	0.23	27.72	
Expanded perlite, organic bonded	1.0	0.36	—	2.78	—	0.30	19.26	
Expanded rubber (rigid)	4.5	0.22	—	4.55	—	0.40	31.53	
Expanded polystyrene extruded								
Smooth skin surface (CFC-12 exp.)	1.8–3.5	0.20	—	5.00	—	0.29	34.65	
Expanded polystyrene, molded beads	1.0	0.26	—	—	—	—	26.3	3.8
	1.25	0.25	—	—	—	—	27.8	4.0
	1.5	0.24	—	—	—	—	29.1	4.2
	1.75	0.24	—	—	—	—	29.1	4.2
	2.0	0.23	—	—	—	—	29.8	4.3

TABLE 4.2 **Thermal Properties of Typical Building and Insulating Materials (design values)[a] (*Continued*)**

Description	Density (lb/ft³)	Conductivity, k (Btu-in./h-ft²-°F)	Conductance, C (h-ft²-°F)	Customary Unit: Resistance, R[b]: Per Inch Thickness 1/k	Customary Unit: Resistance, R[b]: For Thickness Listed, 1/C	Specific Heat, (Btu/lb-°F)	SI Unit: Resistance, R[b]: (m-K)/W	SI Unit: Resistance, R[b]: (m²-K)/W
Cellular polyurethane/polyisocyanurate[f] (CFC-11 exp.) (unfaced)	1.5	0.16–0.18	—	6.25–5.56	—	0.38	43.82–38.98	
Cellular polyisocyanurate[f] (CFC-11 exp.) (gas-permeable facers)	1.5–2.5	0.16–0.18	—	6.25–5.56	—	0.22	43.82–38.98	
Cellular polyisocyanurate[g] (CFC-11 exp.) (gas-impermeable facers)	2.0	0.14	—	7.20	—	0.22	51.75	
Cellular phenolic (closed cell) (CFC-11, CFC-113 exp.)	3.0	0.12	—	8.20	—	—	58.94	
Cellular phenolic (open cell)	1.8–2.2	0.23	—	4.40	—	—	31.62	
Mineral fiber with resin binder	15	0.29	—	3.45	—	0.17	23.91	
Mineral fiberboard, wet felted								
Core or roof insulation	16–17	0.34	—	2.94	—		20.38	
Acoustical tile	18	0.35	—	2.86	—	0.19	19.82	
Acoustical tile	21	0.37	—	2.70	—		18.71	
Mineral fiberboard, wet molded								
Acoustical tile[h]	23	0.42	—	2.38	—	0.14	16.49	
Wood or cane fiberboard								
Acoustical tile[h] 0.5 in.	—	—	0.80	—	1.25	0.31		0.22
Acoustical tile[h] 0.75 in.	—	—	0.53	—	1.89			0.33
Interior finish (plank, tile)	15	0.35	—	2.86	—	0.32	19.82	
Cement fiber slabs (shredded wood with portland cement binder	25–27	0.50–0.53	—	2.0–1.89	—	—	13.87	
Cement fiber slabs (shredded wood with magnesia oxysulfide binder)	22	0.57	—	1.75	—	0.31	12.16	
LOOSE FILL								
Cellulosic insulation (milled paper or wood pulp)	2.3–3.2	0.27–0.32	—	3.13–3.70	—	0.33	21.69–25.64	
Sawdust or shavings	8.0–15.0	0.45	—	2.22	—	0.33	15.39	
Wood fiber, softwoods	2.0–3.5	0.30	—	3.33	—	0.33	23.08	
Perlite, expanded			—	2.70	—	0.26	18.71	
	2.0–4.1	0.27–0.31	3.7–3.3					
	4.1–7.4	0.31–0.36	3.3–2.8					
	7.4–11.0	0.36–0.42	2.8–2.4					
Mineral fiber (rock, slag or glass)								
Approx. 3.5 in. (closed sidewall application)	2.0–3.5	—	—		12–14			
Approx.[e] 3.75–5 in.	0.6–2.0	—	—		11	0.17		1.94

Approx.[e] 6.5–8.75 in.	0.6–2.0	—	—		19			3.35
Approx.[e] 7.5–10 in.	0.6–2.0	—	—		22			3.87
Approx.[e] 10.25–13.75 in.	0.6–2.0	—	—		30			5.28
Vermiculite, exfoliated	7.0–8.2	0.47	—	2.13	—	3.20	14.76	
	4.0–6.0	0.44	—	2.27	—		15.73	
MASONRY MATERIALS								
Concretes								
Cement mortar	105–135	5.0–10.5	—	0.20–0.10	—	—	1.39–0.69	
Gypsum-fiber concrete 87.5% gypsum, 12.5% wood chips	51	1.66	—	0.60	—	0.21	4.16	
Lightweight aggregates including expanded shale, clay or slate; expanded slags; cinders; pumice; vermiculite; also cellular concretes	120	5.5–11.0	—	0.18–0.09	—	—	1.25–0.62	
	100	3.7–5.9	—	0.27–0.17	—	0.20	1.87–1.18	
	80	2.5–3.5	—	0.40–0.29	—	0.20	2.77–2.01	
	60	1.6–1.8	—	0.63–0.56	—	—	4.36–3.88	
	40	0.93–1.11	—	1.08–0.90	—	—	7.49–6.24	
	30	0.75–0.91	—	1.33–1.10	—	0.20	9.22–7.63	
	20	0.63–0.83	—	1.59–1.20	—	—	11.02–8.32	
Perlite, expanded	50	1.4–1.8	—	0.71–0.56	—	—	4.92–3.88	
	40	0.93		1.08			7.48	
	30	0.71		1.41			9.77	
	20	0.50		2.00		0.32	13.86	
Sand and gravel or stone aggregate (oven dried)	140	8.0–16.0	—	0.13–0.06	—	0.18–0.22	0.90–0.42	
Sand and gravel or stone aggregate (not dried)	140	10.0–20.0	—	0.10–0.05	—	0.19–0.24	0.69–0.35	
Stucco	116	5.0	—	0.20			1.39	
MASONRY UNITS								
Brick, common	80	2.2–3.2	—	0.45–0.31	—	—	3.12–2.15	
	90	2.7–3.7	—	0.37–0.27	—	—	2.56–1.87	
	100	3.3–4.3	—	0.30–0.23	—	—	2.08–1.59	
	110	3.5–5.5	—	0.29–0.18	—	—	2.01–1.25	
	120	4.4–6.4	—	0.23–0.16	—	0.19	1.59–1.11	
	130	5.4–9.0	—	0.19–0.11	—	—	1.32–0.76	
Clay tile, hollow:								
1 cell deep 3 in.	—	—	1.25	—	0.80	0.21		0.14
1 cell deep 4 in.	—	—	0.90	—	1.11			0.20
2 cells deep 6 in.	—	—	0.66	—	1.52			0.27
2 cells deep 8 in.	—	—	0.54	—	1.85			0.33
2 cells deep 10 in.	—	—	0.45	—	2.22			0.39
3 cells deep 12 in.	—	—	0.40	—	2.50			0.44
Concrete blocks[i]								
Limestone aggregate								
8 in., 36 lb, 138 lb/ft³ concrete, 2 cores	—	—	—	—	—	—		
Same with perlite-filled cores	—	—	0.48	—	2.1	—		
12 in., 55 lb, 138 lb/ft³ concrete, 2 cores	—	—	—	—	—	—		
Same with perlite-filled cores	—	—	0.27	—	3.7	—		

TABLE 4.2 **Thermal Properties of Typical Building and Insulating Materials (design values)[a] (*Continued*)**

Description	Density (lb/ft³)	Conductivity, k (Btu-in./h-ft²-°F)	Conductance, C (h-ft²-°F)	Customary Unit: Resistance, R[b]: Per Inch Thickness 1/k	Customary Unit: Resistance, R[b]: For Thickness Listed, 1/C	Specific Heat, (Btu/lb-°F)	SI Unit: Resistance, R[b]: (m-K)/W	SI Unit: Resistance, R[b]: (m²-K)/W
Normal weight aggregate (sand and gravel)								
8 in., 33–36 lb, 126–136 lb/ft³ concrete, 2 or 3 cores	—	—	0.90–1.03	—	1.11–0.97	0.22		
Same with perlite-filled cores	—	—	0.50	—	2.0	—		
Same with verm. filled cores	—	—	0.52–0.73	—	1.92–1.37	—		
12 in., 50 lb, 125 lb/ft³ concrete, 2 cores	—	—	0.81	—	1.23	0.22		
Medium weight aggregate (combinations of normal weight and lightweight aggregate)								
8 in., 26–29 lb, 97–112 lb/ft³ concrete, 2 or 3 cores	—	—	0.58–0.78	—	1.71–1.28	—		
Same with perlite-filled cores	—	—	0.27–0.44	—	3.7–2.3	—		
Same with verm. filled cores	—	—	0.30	—	3.3	—		
Same with molded EPS (beads) filled cores	—	—	0.32	—	3.2	—		
Same with molded EPS inserts in cores	—	—	0.37	—	2.7	—		
Lightweight aggregate (expanded shale, clay, slate or slag, pumice)								
6 in., 16–17 lb, 85–87 lb/ft³ concrete, 2 or 3 cores	—	—	0.52–0.61	—	1.93–1.65	—		
Same with perlite-filled cores	—	—	0.24	—	4.2	—		
Same with verm. filled cores	—	—	0.33	—	3.0	—		
8 in., 19–22 lb, 72–86 lb/ft³ concrete,	—	—	0.32–0.54	—	3.2–1.90	0.21		
Same with perlite-filled cores	—	—	0.15–0.23	—	6.8–4.4	—		
Same with verm. filled cores	—	—	0.19–0.26	—	5.3–3.9	—		
Same with molded EPS (beads) filled cores	—	—	0.21	—	4.8	—		
Same with UF foam-filled cores	—	—	0.22	—	4.5	—		
Same with molded EPS inserts in cores	—	—	0.29	—	3.5	—		
12 in., 32–36 lb, 80–90 lb/ft³ concrete, 2 or 3 cores	—	—	0.38–0.44	—	2.6–2.3	—		
Same with perlite-filled cores	—	—	0.11–0.16	—	9.2–6.3	—		
Same with verm. filled cores	—	—	0.17	—	5.8	—		
Stone, lime or sand	—	12.50	—	0.08	—	0.19	0.55	
Gypsum partition tile:								
3 × 12 × 30 in. solid	—	—	0.79	—	1.26	0.19		0.22
3 × 12 × 30 in. 4-cell	—	—	0.74	—	1.35			0.24
4 × 12 × 30 in. 3-cell	—	—	0.60	—	1.67			0.29
METALS (See ASHRAE Handbook of Fundamentals)								
PLASTERING MATERIALS								
Cement plaster, sand aggregate	116	5.0	—	0.20	—	0.20	1.39	

Sand aggregate . . . . . 0.375 in.	—	—	13.3	—	0.08	0.20		0.01
Sand aggregate . . . . . 0.75 in.	—	—	6.66	—	0.15	0.20		0.03
Gypsum plaster:								
Lightweight aggregate . . . . . 0.5 in.	45	—	3.12	—	0.32			0.06
Lightweight aggregate . . . . . 0.625 in.	45	—	2.67	—	0.39			0.07
Lightweight agg. on metal lath . . . . . 0.75 in.	—	—	2.13	—	0.47			0.08
Perlite aggregate	45	1.5	—	0.67	—	0.32	4.64	
Sand aggregate	105	5.6	—	0.18	—	0.20	1.25	
Sand aggregate . . . . . 0.5 in.	105	—	11.10	—	0.09			0.02
Sand aggregate . . . . . 0.625 in.	105	—	9.10	—	0.11			0.02
Sand aggregate on metal lath . . . . . 0.75 in.	—	—	7.70	—	0.13			0.02
Vermiculite aggregate	45	1.7	—	0.59	—		4.09	
ROOFING								
Asbestos-cement shingles	120	—	4.76	—	0.21	0.24		0.04
Asphalt roll roofing	70	—	6.50	—	0.15	0.36		0.03
Asphalt shingles	70	—	2.27	—	0.44	0.30		0.08
Built-up roofing . . . . . 0.375 in.	70	—	3.00	—	0.33	0.35		0.06
Slate . . . . . 0.5 in.	—	—	20.00	—	0.05	0.30		0.01
Wood shingles, plain and plastic film faced	—	—	1.06	—	0.94	0.31		0.17
Spray Applied								
Polyurethane foam	1.5–2.5	0.16–0.18	—	6.25–5.56	—	43.33–38.54		
Ureaformaldehyde foam	0.7–1.6	0.22–0.28	—	4.55–3.57	—	31.54–24.75		
Cellulosic fiber	3.5–6.0	0.29–0.34	—	3.45–2.94	—	23.92–20.38		
Glass fiber	3.5–4.5	0.26–0.27	—	3.85–3.70	—	26.69–25.65		
SIDING MATERIALS (ON FLAT SURFACE)								
Shingles								
Asbestos-cement	120	—	4.75	—	0.21			0.04
Wood, 16 in., 7.5 exposure	—	—	1.15	—	0.87	0.31		0.15
Wood, double, 16-in., 12-in. exposure	—	—	0.84	—	1.19	0.28		0.21
Wood, plus insul. backer board, 0.3125 in.	—	—	0.71	—	1.40	0.31		0.25
Siding								
Asbestos-cement, 0.25 in., lapped	—	—	4.76	—	0.21	0.24		0.04
Asphalt roll siding	—	—	6.50	—	0.15	0.35		0.03
Asphalt insulating siding (0.5 in. bed.)	—	—	0.69	—	1.46	0.35		0.26
Hardboard siding, 0.4375 in.	40	—	0.49	—	0.67	0.28	4.65	
Wood, drop, 1 × 8 in.	—	—	1.27	—	0.79	0.28		0.14
Wood, bevel, 0.5 × 8 in., lapped	—	—	1.23	—	0.81	0.28		0.14
Wood, bevel, 0.75 × 10 in., lapped	—	—	0.95	—	1.05	0.28		0.18
Wood, plywood, 0.375 in., lapped	—	—	1.59	—	0.59	0.29		0.10
Aluminum or steel,[i] over sheathing								
Hollow-backed	—	—	1.61	—	0.61	0.29		0.11

TABLE 4.2 **Thermal Properties of Typical Building and Insulating Materials (design values)[a] (*Continued*)**

Description	Density (lb/ft³)	Conductivity, k (Btu-in./h-ft²-°F)	Conductance, C (h-ft²-°F)	Customary Unit Resistance, R[b] Per Inch Thickness 1/k	Customary Unit Resistance, R[b] For Thickness Listed, 1/C	Specific Heat, (Btu/lb-°F)	SI Unit Resistance, R[b] (m-K)/W	SI Unit Resistance, R[b] (m²-K)/W
Insulating-board backed nominal 0.375 in.	—	—	0.55	—	1.82	0.32		0.32
Insulating-board backed nominal 0.375 in., foil backed			0.34		2.96			0.52
Architectural glass	—	—	10.00	—	0.10	0.20		0.02
WOODS (12% Moisture Content)[c,k]								
Hardwoods						0.39		
Oak	41.2–46.8	1.12–1.25	—	0.89–0.80	—		6.17–5.55	
Birch	42.6–45.4	1.16–1.22	—	0.87–0.82	—		6.03–5.68	
Maple	39.8–44.0	1.09–1.19	—	0.92–0.84	—		5.68–5.82	
Ash	38.4–41.9	1.06–1.14	—	0.94–0.88	—		6.51–6.10	
Softwoods						0.39		
Southern pine	35.6–41.2	1.00–1.12	—	1.00–0.89	—		6.93–6.17	
Douglas fir–larch	33.5–36.3	0.95–1.01	—	1.06–0.99	—		7.39–6.86	
Southern cypress	31.4–32.1	0.90–0.92	—	1.11–1.09	—		7.69–7.56	
Hem–fir, spruce–pine–fir	24.5–31.4	0.74–0.90	—	1.35–1.11	—		9.36–7.69	
West Coast woods, cedars	21.7–31.4	0.68–0.90	—	1.48–1.11	—		10.26–7.69	
California redwood	24.5–28.0	0.74–0.82	—	1.35–1.22	—		9.36–8.46	

Source: Copyright © by the American Society of Heating, Refrigerating, and Air-Conditioning Engineers, Inc., Atlanta, Ga. Reprinted by permission from *1989 Handbook of Fundamentals.*

[a] Representative values for dry materials at 75 F. They are intended as design (not specification) values for materials in normal use. Insulation materials in actual service may have thermal values which vary from design values depending on their in-situ properties such as density and moisture content. For properties of a particular product, use the value supplied by the manufacturer or by unbiased tests.

[b] Resistance values are the reciprocals of C before rounding off C to two decimal places.

[c] Forest Products Laboratory Wood Handbook, USDA Handbook 72, 1974, Tables 3 and 4.

[d] Does not include paper backing and facing, if any. Where insulation forms a boundary (reflective or otherwise) of an air space, see Tables 4.3 and 4.4 for the insulating value of air space for the appropriate effective emittance and temperature conditions of the space.

[e] Conductivity varies with fiber diameter. Insulation is produced in different densities; therefore, there is a wide variation in thickness for the same R value among manufacturers. No effort should be made to relate any specific R value to any specific thickness. Commercial thicknesses generally available range from 2 to 8.5.

[f] Values are for aged, unfaced, board stock. For change in conductivity with age of expanded urethane, see 1989 *Handbook of Fundamentals,* chapter 20.

[g] Time-aged values for board stock with gas-barrier quality (0.001 in. thickness or greater) aluminum foil facers on two major surfaces.

[h] Insulating values of acoustical tile vary, depending on density of the board and on type, size, and depth of perforations.

[i] Values for fully grouted block may be approximated using values for concrete with a similar unit weight.

[j] Values for metal siding applied over flat surfaces vary widely, depending on amount of ventilation of air space beneath the siding; whether airspace is reflective or nonreflective; and on thickness, type, and application of insulating backing board used. Values given are averages for use as design guides, and were obtained from several guarded hotbox tests (ASTM C236) or calibrated hotbox (ASTM 967) on hollow-backed types and types made using backing boards of wood fiber, foamed plastic, and glass fiber. Departures of ±50% or more from the values given may occur.

[k] L. Adams: Supporting cryogenic equipment with wood (*Chemical Engineering,* May 17, 1971). Conductivity values listed are for heat transfer across the grain.

TABLE 4.3 **Surface Conductances, *h* (Btu/h-ft²-F), and Resistances, *R*, for Air**[a]

Position of Surface	Direction of Heat Flow	Non-reflective $\varepsilon = 0.90$ h_i	R	Reflective $\varepsilon = 0.20$ h_i	R	Reflective $\varepsilon = 0.05$ h_i	R
Still air							
Horizontal.........	Upward	1.63	0.61	0.91	1.10	0.76	1.32
Sloping (45°)	Upward	1.60	0.62	0.88	1.14	0.73	1.37
Vertical...........	Horizontal	1.46	0.68	0.74	1.35	0.59	1.70
Sloping (45°)	Downward	1.32	0.76	0.60	1.67	0.45	2.22
Horizontal.........	Downward	1.08	0.92	0.37	2.70	0.22	4.55
		h_o	R	h_o	R	h_o	R
Moving air (any position)							
15-mph wind (for winter)	Any	6.00	0.17				
7.5-mph wind (for summer)	Any	4.00	0.25				

NOTE: A surface cannot take credit for both an air space resistance value and a surface resistance value. No credit for an air space value can be taken for any surface facing an air space of less than 0.5 in.

Source: Copyright © by the American Society of Heating, Refrigerating and Air Conditioning Engineers, Inc., Atlanta, GA. Reprinted by permission, from 1989 *Handbook of Fundamentals.*

[a]Conductances are for surfaces of the stated emittance facing virtual blackbody surroundings at the same temperature as ambient air. Values are based on a surface-air temperature difference of 10 F° and for surface temperature of 70 F.

TABLE 4.4 **Thermal Resistances of Plane[a] Air Spaces**

SECTION A

All resistance values expressed in $ft^2 \cdot F \cdot h/Btu$

Values apply only to air spaces of uniform thickness bounded by plane, smooth, parallel surfaces with no leakage of air to or from the space. These conditions are not normally present in standard building construction. When accurate values are required, use overall U-factors determined for your particular construction through calibrated hot box (ASTM C976) or guarded hot box (ASTM-C-236) testing. Thermal resistance values for multiple air spaces must be based on careful estimates of mean temperature differences for each air space.

Position of Air Space	Direction of Heat Flow	Air Space Mean Temp,[b] (F)	Air Space Temp Diff,[b] (deg F)	0.5-in. Air Space[d] Value of $E^{b,c}$ 0.03	0.05	0.2	0.5	0.82	0.75-in. Air Space[d] Value of $E^{b,c}$ 0.03	0.05	0.2	0.5	0.82
Horiz.	Up ↑	90	10	2.13	2.03	1.51	0.99	0.73	2.34	2.22	1.61	1.04	0.75
		50	30	1.62	1.57	1.29	0.96	0.75	1.71	1.66	1.35	0.99	0.77
		50	10	2.13	2.05	1.60	1.11	0.84	2.30	2.21	1.70	1.16	0.87
		0	20	1.73	1.70	1.45	1.12	0.91	1.83	1.79	1.52	1.16	0.93
		0	10	2.10	2.04	1.70	1.27	1.00	2.23	2.16	1.78	1.31	1.02
		−50	20	1.69	1.66	1.49	1.23	1.04	1.77	1.74	1.55	1.27	1.07
		−50	10	2.04	2.00	1.75	1.40	1.16	2.16	2.11	1.84	1.46	1.20
45° Slope	Up ↗	90	10	2.44	2.31	1.65	1.06	0.76	2.96	2.78	1.88	1.15	0.81
		50	30	2.06	1.98	1.56	1.10	0.83	1.99	1.92	1.52	1.08	0.82
		50	10	2.55	2.44	1.83	1.22	0.90	2.90	2.75	2.00	1.29	0.94
		0	20	2.20	2.14	1.76	1.30	1.02	2.13	2.07	1.72	1.28	1.00
		0	10	2.63	2.54	2.03	1.44	1.10	2.72	2.62	2.08	1.47	1.12
		−50	20	2.08	2.04	1.78	1.42	1.17	2.05	2.01	1.76	1.41	1.16
		−50	10	2.62	2.56	2.17	1.66	1.33	2.53	2.47	2.10	1.62	1.30
Vertical	Horiz. →	90	10	2.47	2.34	1.67	1.06	0.77	3.50	3.24	2.08	1.22	0.84
		50	30	2.57	2.46	1.84	1.23	0.90	2.91	2.77	2.01	1.30	0.94
		50	10	2.66	2.54	1.88	1.24	0.91	3.70	3.46	2.35	1.43	1.01
		0	20	2.82	2.72	2.14	1.50	1.13	3.14	3.02	2.32	1.58	1.18
		0	10	2.93	2.82	2.20	1.53	1.15	3.77	3.59	2.64	1.73	1.26
		−50	20	2.90	2.82	2.35	1.76	1.39	2.90	2.83	2.36	1.77	1.39
		−50	10	3.20	3.10	2.54	1.87	1.46	3.72	3.60	2.87	2.04	1.56
45° Slope	Down ↘	90	10	2.48	2.34	1.67	1.06	0.77	3.53	3.27	2.10	1.22	0.84
		50	30	2.64	2.52	1.87	1.24	0.91	3.43	3.23	2.24	1.39	0.99
		50	10	2.67	2.55	1.89	1.25	0.92	3.81	3.57	2.40	1.45	1.02
		0	20	2.91	2.80	2.19	1.52	1.15	3.75	3.57	2.63	1.72	1.26
		0	10	2.94	2.83	2.21	1.53	1.15	4.12	3.91	2.81	1.80	1.30
		−50	20	3.16	3.07	2.52	1.86	1.45	3.78	3.65	2.90	2.05	1.57
		−50	10	3.26	3.16	2.58	1.89	1.47	4.35	4.18	3.22	2.21	1.66
Horiz.	Down ↓	90	10	2.48	2.34	1.67	1.06	0.77	3.55	3.29	2.10	1.22	0.85
		50	30	2.66	2.54	1.88	1.24	0.91	3.77	3.52	2.38	1.44	1.02
		50	10	2.67	2.55	1.89	1.25	0.92	3.84	3.59	2.41	1.45	1.02
		0	20	2.94	2.83	2.20	1.53	1.15	4.18	3.96	2.83	1.81	1.30
		0	10	2.96	2.85	2.22	1.53	1.16	4.25	4.02	2.87	1.82	1.31
		−50	20	3.25	3.15	2.58	1.89	1.47	4.60	4.41	3.36	2.28	1.69
		−50	10	3.28	3.18	2.60	1.90	1.47	4.71	4.51	3.42	2.30	1.71

Position of Air Space	Direction of Heat Flow	Air Space Mean Temp,[b] (F)	Air Space Temp Diff,[b] (deg F)	1.5-in. Air Space[d] Value of $E^{b,c}$ 0.03	0.05	0.2	0.5	0.82	3.5-in. Air Space[d] Value of $E^{b,c}$ 0.03	0.05	0.2	0.5	0.82
Horiz	Up ↑	90	10	2.55	2.41	1.71	1.08	0.77	2.84	2.66	1.83	1.13	0.80
		50	30	1.87	1.81	1.45	1.04	0.80	2.09	2.01	1.58	1.10	0.84
		50	10	2.50	2.40	1.81	1.21	0.89	2.80	2.66	1.95	1.28	0.93
		0	20	2.01	1.95	1.63	1.23	0.97	2.25	2.18	1.79	1.32	1.03
		0	10	2.43	2.35	1.90	1.38	1.06	2.71	2.62	2.07	1.47	1.12
		−50	20	1.94	1.91	1.68	1.36	1.13	2.19	2.14	1.86	1.47	1.20
		−50	10	2.37	2.31	1.99	1.55	1.26	2.65	2.58	2.18	1.67	1.33
45° Slope	Up ↗	90	10	2.92	2.73	1.86	1.14	0.80	3.18	2.96	1.97	1.18	0.82
		50	30	2.14	2.06	1.61	1.12	0.84	2.26	2.17	1.67	1.15	0.86
		50	10	2.88	2.74	1.99	1.29	0.94	3.12	2.95	2.10	1.34	0.96
		0	20	2.30	2.23	1.82	1.34	1.04	2.42	2.35	1.90	1.38	1.06
		0	10	2.79	2.69	2.12	1.49	1.13	2.98	2.87	2.23	1.54	1.16
		−50	20	2.22	2.17	1.88	1.49	1.21	2.34	2.29	1.97	1.54	1.25
		−50	10	2.71	2.64	2.23	1.69	1.35	2.87	2.79	2.33	1.75	1.39
Vertical	Horiz. →	90	10	3.99	3.66	2.25	1.27	0.87	3.69	3.40	2.15	1.24	0.85
		50	30	2.58	2.46	1.84	1.23	0.90	2.67	2.55	1.89	1.25	0.91
		50	10	3.79	3.55	2.39	1.45	1.02	3.63	3.40	2.32	1.42	1.01
		0	20	2.76	2.66	2.10	1.48	1.12	2.88	2.78	2.17	1.51	1.14
		0	10	3.51	3.35	2.51	1.67	1.23	3.49	3.33	2.50	1.67	1.23
		−50	20	2.64	2.58	2.18	1.66	1.33	2.82	2.75	2.30	1.73	1.37
		−50	10	3.31	3.21	2.62	1.91	1.48	3.40	3.30	2.67	1.94	1.50
45° Slope	Down ↘	90	10	5.07	4.55	2.56	1.36	0.91	4.81	4.33	2.49	1.34	0.90
		50	30	3.58	3.36	2.31	1.42	1.00	3.51	3.30	2.28	1.40	1.00
		50	10	5.10	4.66	2.85	1.60	1.09	4.74	4.36	2.73	1.57	1.08
		0	20	3.85	3.66	2.68	1.74	1.27	3.81	3.63	2.66	1.74	1.27
		0	10	4.92	4.62	3.16	1.94	1.37	4.59	4.32	3.02	1.88	1.34
		−50	20	3.62	3.50	2.80	2.01	1.54	3.77	3.64	2.90	2.05	1.57
		−50	10	4.67	4.47	3.40	2.29	1.70	4.50	4.32	3.31	2.25	1.68
Horiz.	Down ↓	90	10	6.09	5.35	2.79	1.43	0.94	10.07	8.19	3.41	1.57	1.00
		50	30	6.27	5.63	3.18	1.70	1.14	9.60	8.17	3.86	1.88	1.22
		50	10	6.61	5.90	3.27	1.73	1.15	11.15	9.27	4.09	1.93	1.24
		0	20	7.03	6.43	3.91	2.19	1.49	10.90	9.52	4.87	2.47	1.62
		0	10	7.31	6.66	4.00	2.22	1.51	11.97	10.32	5.08	2.52	1.64
		−50	20	7.73	7.20	4.77	2.85	1.99	11.64	10.49	6.02	3.25	2.18
		−50	10	8.09	7.52	4.91	2.89	2.01	12.98	11.56	6.36	3.34	2.22

(continued)

TABLE 4.4 **Thermal Resistance of Plane[a] Air Spaces (*Continued*)**

Section B. ***Reflectivity and Emittance Values of Various Surfaces and Effective Emittances of Airspaces[f]***

Surface	Average Emittance ε	Effective Emittance E of Airspace: One Surface Emittance ε; the Other 0.90	Effective Emittance E of Airspace: Both Surfaces Emittances ε
Aluminum foil, bright	0.05	0.05	0.03
Aluminum foil, with condensate just visible (>0.7 gr/ft^2)	0.30[g]	0.29	—
Aluminum foil, with condensate clearly visible (>2.9 gr/ft^2)	0.70[g]	0.65	—
Aluminum sheet	0.12	0.12	0.06
Aluminum coated paper, polished	0.20	0.20	0.11
Steel, galvanized, bright	0.25	0.24	0.15
Aluminum paint	0.50	0.47	0.35
Building materials: wood, paper, masonry, nonmetallic paints	0.90	0.82	0.82
Regular glass	0.84	0.77	0.72

Source: Copyright © by the American Society of Heating, Refrigerating, and Air-Conditioning Engineers, Inc., Atlanta, Ga. Reprinted by permission from 1989 *Handbook of Fundamentals.*

[a] Thermal resistance values were determined from the relation $R = 1/C$, where $C = h_c + Eh_r$, h_c is the conduction–convection coefficient, Eh_r, is the radiation coefficient $\cong$ 0.00686E $[(t_m + 460)/100]^3$, and t_m is the mean temperature of the airspace. Values for h_c were determined from research data (National Bureau of Standards), such as those presented in 1954 in Housing Research Paper 32 (HRP No. 32) by the Housing and Home Finance Agency (Government Printing Office, Washington, D.C.). For interpolation from Table 4.4 to airspace thicknesses less than 0.5 in. (as in insulating window glass), assume that

$$h_c = 0.159(1 + 0.0016t_m)/l$$

where l is the thickness in inches, and h_c is assumed to represent heat transfer by conduction alone through air.

[b] Interpolation is permissible for other values of mean temperature, temperature differences, and effective emittance E. Interpolation and moderate extrapolation for airspaces greater than 3.5 in. are also permissible.

[c] Effective emittance of the space E is given by $1/E = 1/e_1 + 1/e_2 - 1$, *where* e_1 and e_2 are the emittances of the surfaces of the airspace.

[d] Credit for an airspace resistance value cannot be taken more than once and only for the boundary conditions established.

[e] Resistances of horizontal spaces with heat flow downward are substantially independent of temperature difference.

[f] These values apply in the 4- to 40-μm range of the electromagnetic spectrum.

[g] From M. R. Bassett, and H. A. Trethowen, 1984, "Effect of Condensation on Emittance of Reflective Insulation," *Journal of Thermal Insulation,* Vol. 8, October, p. 127.

TABLE 4.5 **Coefficients of Transmission, U (Btu/h ft^2 F), of Wood Frame Walls**

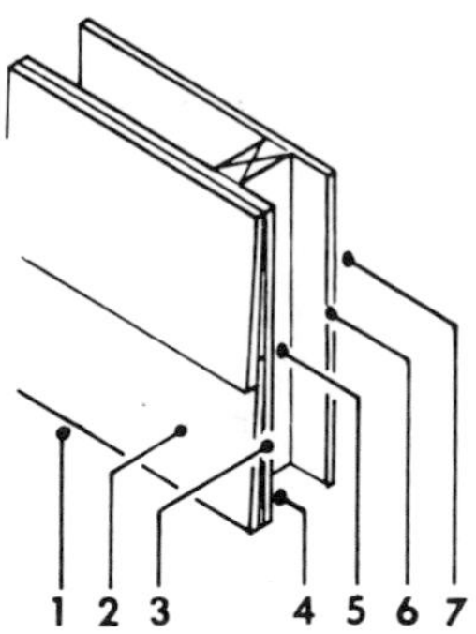

Base Case Resistance: Between Framing, R_i	Base Case Resistance: At Framing, R_s	*Construction*	Resistance[a]: New Item 4 ("Construction"): Between Framing, R_i	Resistance[a]: New Item 4 ("Construction"): At Framing, R_s
0.17	0.17	1. Outside surface (15-mph wind)	0.17	0.17
0.81	0.81	2. Siding, wood, 0.5 in. × 8 in. lapped (average)	0.81	0.81
1.32	1.32	3. Sheathing, 0.5-in. vegetable fiber board	1.32	1.32
1.01	—	4. Nonreflective air space, 3.5 in. (50 F mean; 10 F temperature difference)	11.00	—
—	4.35	5. Nominal 2-in. × 4-in. wood stud	—	4.35
0.45	0.45	6. Gypsum wallboard, 0.5 in.	0.45	0.45
0.68	0.68	7. Inside surface (still air)	0.68	0.68
4.44	7.78	*Total thermal resistance, R*	14.43	7.78

Base Case

$$U_i = \frac{1}{4.44} = 0.225$$

New Item 4

$$U_i = \frac{1}{14.43} = 0.069$$

To adjust U values for the effect of framing, see the example below. With 15% framing (typical of 2-in. × 4-in. studs at 16-in. o.c.) these adjusted U values are, respectively:

$U_{av} = 0.199$ (12% less heat loss) $U_{av} = 0.081$ (17% more heat loss)

EXAMPLE. To adjust U values for the effect of framing, average the U_i and U_s values in proportion to their areas. For the constructions above, with 15% of the wall area in framing and 85% in insulation:

$$U_i = \frac{1}{4.44} = 0.225 \qquad U_i = \frac{1}{14.43} = 0.069$$

$$U_s = \frac{1}{7.78} = 0.128 \qquad U_s = \frac{1}{7.78} = 0.128$$

$$U_{av} = 0.85(0.225) + 0.15(0.128) = 0.199 \qquad U_{av} = 0.85(0.069) + 0.15(0.128) = 0.081$$

[a]When air space replaced with 3.5-in. R-11 blanket insulation.

TABLE 4.6 **Coefficients of Transmission, *U* (Btu/h ft² F), of Solid Masonry Walls**

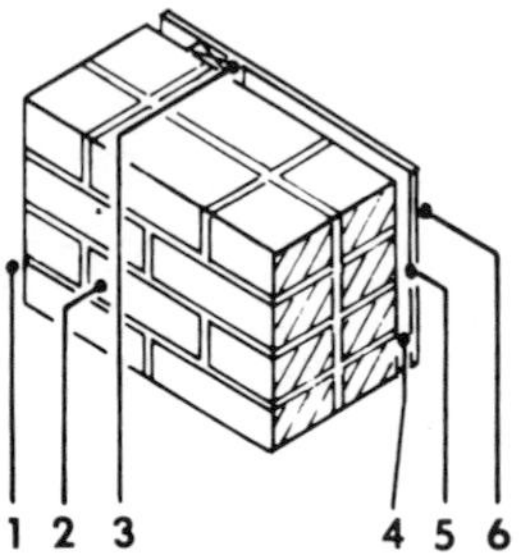

Base Case Resistance: Between Furring, R_i	Base Case Resistance: At Furring, R_s	*Construction*	*Resistance[a]: New Item 4 ("Construction")*
0.17	0.17	1. Outside surface (15-mph wind)	0.17
1.60	1.60	2. Common brick, 8 in.	1.60
—	0.94	3. Nominal 1-in. × 3-in. vertical furring	—
1.01	—	4. Nonreflective air space, 0.75 in. (50 F mean; 10 F° temperature difference)	5.00
0.45	0.45	5. Gypsum wallboard, 0.5 in.	0.45
0.68	0.68	6. Inside surface (still air)	0.68
3.91	3.84	*Total thermal resistance, R*	$R_i = 7.90 = R_s$

Base Case

$$U_i = \frac{1}{3.91} = 0.256$$

New Item 4

$$U = \frac{1}{7.90} = 0.127$$

To adjust *U* values for the effect of furring strips, see the example given in Table 4.5. With 20% furring (typical of 1-in. × 3-in. vertical furring on masonry at 16-in. o.c.):

$$U_{av} = 0.257 \qquad \text{(about 1\% more heat loss)}$$

Source: Copyright © by the American Society of Heating, Refrigerating and Air Conditioning Engineers, Inc., Atlanta, GA. Reprinted by permission from *1981 Handbook of Fundamentals.*

[a]When furring strips and air space replaced with 1-in. expanded polystyrene extruded, smooth skin surface, 2.2 lb/ft³.

TABLE 4.7 **Coefficients of Transmission, *U* (Btu/h ft^2 F), of Masonry Cavity Walls**

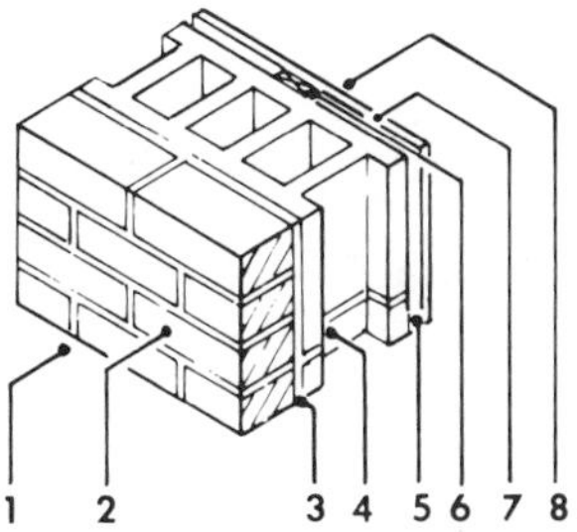

Base Case Resistance: Between Furring, R_i	Base Case Resistance: At Furring, R_s	*Construction*	*Resistance[c]: New Items 3 and 7 ("Construction")*
0.17	0.17	1. Outside surface (15-mph wind)	0.17
0.80	0.80	2. Common brick, 4 in.	0.80
1.10[a]	1.10[a]	3. Nonreflective air space, 2.5 in. (30 F mean; 10 F° temperature difference)	5.32[b]
0.71	0.71	4. Concrete block, three-oval core, stone and gravel aggregate, 4 in.	0.71
1.01	—	5. Nonreflective air space 0.75 in. (50 F mean; 10 F° temperature difference)	—
—	0.94	6. Nominal 1-in. × 3-in vertical furring	—
0.45	0.45	7. Gypsum wallboard, 0.5 in.	0.11
0.68	0.68	8. Inside surface (still air)	0.68
4.92	4.85	*Total thermal resistance, R*	$R_i = R_s = 7.79$

Base Case

$$U_i = \frac{1}{4.92} = 0.203$$

New Items 3 and 7

$$U = \frac{1}{7.79} = 0.128$$

To adjust *U* values for the effect of furring strips, see the example given in Table 4.5; with 20% furring (typical of 1-in. × 3-in. vertical furring on masonry at 16-in. o.c.):

$$U_{av} = 0.204 \quad \text{(about 1\% more heat loss)}$$

[a]Interpolated from vertical air space values in Table 4.4.

[b]Calculated from vermiculite *R* values in Table 4.2.

[c]Resistance when furring strips and gypsum wallboard replaced with 0.625-in. plaster (sand aggregate) applied directly to concrete block and 2.5-in. air space filled with vermiculite insulation, 7–8.2 lb/ft^3.

Source: Copyright © by the American Society of Heating, Refrigerating and Air Conditioning Engineers, Inc., Atlanta, GA. Reprinted by permission from *1981 Handbook of Fundamentals.*

TABLE 4.8 **Thermal Resistance (*R* Value) of Insulation**

*Part A. **Approximate Thickness (in.) of Insulation for Thermal Resistances, R (ft² F h/Btu)***

Thermal Resistance of Insulation	Batts or Blankets: Glass Fiber	Batts or Blankets: Rock Wool	Loose Fill: Glass Fiber	Loose Fill: Rock Wool	Loose Fill: Cellulosic	Boards and Slabs: Polyurethane	Boards and Slabs: Cellular Glass
R-7	2¼–2¾	2	3–4	2–3	2	1	2⅝
R-11	3½–4	3	5	4	3	1¾	4¼
R-13	3⅝	3½	6	4–5	4	2	5
R-14[a]			3½				
R-15[b]	3½						
R-19	6–6½	5¼	8–9	6–7	5	3	7¼
R-21[b]	5½						
R-22	6½	6	10	7–8	6	3½	8⅜
R-22[a]			5½				
R-29[a]			7¼				
R-30	9½–10½	9	13–14	10–11	8	4¾	11⅜
R-30[b]	8¼						
R-37[a]			9¼				
R-38	12–13	10½	17–18	13–14	10–11	6	14½
R-45[a]			11¼				

*Part B. **Corrected R Values[c] for Wall Sections with Metal Studs***

Size of Members	Gauge of Stud	Spacing of Framing (in o.c.)	Cavity Insulation R Value	Correction Factor	Corrected R-Value
2 × 4	18-16	16	*R*-11	0.50	*R*-5.5
2 × 4	18-16	24	*R*-11	0.60	*R*-6.6
2 × 6	18-16	16	*R*-19	0.40	*R*-7.6
2 × 6	18-16	24	*R*-19	0.45	*R*-8.6

*Part C. **Corrected R Values[c] for Roofs with Metal Trusses[d]***

Bridged R Value	*0*	*5*	*10*	*15*	*20*	*25*	*30*	*35*	*40*	*45*	*50*	*55*
Correction Factor	1.0	0.96	0.92	0.88	0.85	0.81	0.79	0.76	0.73	0.71	0.69	0.67
Corrected *R* Value	*R*-0	*R*-4.8	*R*-9.2	*R*-13.2	*R*-17	*R*-20.3	*R*-23.7	*R*-26.6	*R*-29.2	*R*-32	*R*-34.5	*R*-36.9

Source: Adapted from *Cooling and Heating Load Calculation Manual,* and ASHRAE/IES Standard 90.1-1989, *Energy Efficient Design of New Buildings Except Low-Rise Residential Buildings,* © 1989 by the American Society of Heating, Refrigerating, and Air-Conditioning Engineers, Inc., Atlanta, Ga.

[a] Blow-In-Blanket System, fiberglass with adhesive to prevent settling.

[b] Specially designed to fit without compression into cavity of standard construction.

[c] These correction factors utilize the parallel path calculation method, see Chapter 20 of the ASHRAE 1989 *Handbook of Fundamentals.*

[d] These correction factors are for metal trusses at 4 ft on center that penetrate the insulation, with 0.66-in.-diameter cross members at 1 ft on center.

TABLE 4.9 **Coefficients of Transmission, U (Btu/h ft^2 F), of Flat Masonry Roofs with Built-Up Roofing, With and Without Suspended Ceilings: Winter Conditions, Upward Flow**

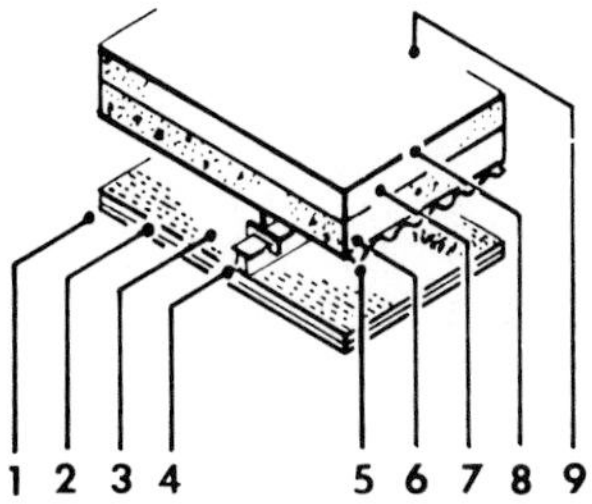

Base Case Resistance R	Construction (heat flow up)	Resistance R[c]: New Item 7 ("Construction")
0.61	1. Inside surface (still air)	0.61
0.47	2. Metal lath and lightweight aggregate plaster, 0.75 in.	0.47
0.93[a]	3. Nonreflective air space, greater than 3.5 in. (50 F mean; 10 F° temperature difference)	0.93[a]
0[b]	4. Metal ceiling suspension system with metal hanger rods	0[b]
0	5. Corrugated metal deck	0
2.22	6. Concrete slab, lightweight aggregate, 2 in. (30 lb/ft^3)	2.22
—	7. Rigid roof deck insulation (none)	4.17
0.33	8. Built-up roofing, 0.375 in.	0.33
0.17	9. Outside surface (15-mph wind)	0.17
4.73	*Total Thermal Resistance, R*	8.90

Base Case

$$U = \frac{1}{4.73} = 0.211$$

New Item 7

$$U = \frac{1}{8.90} = 0.112$$

[a]Use largest air space (3.5 in.) value shown in Table 4.4.

[b]Area of hanger rods is negligible in relation to ceiling area.

[c]When rigid roof deck insulation added, $C = 0.24$ ($R = 1/C = 4.17$).

Source: Copyright © by the American Society of Heating, Refrigerating and Air Conditioning Engineers, Inc., Atlanta, GA. Reprinted by permission from *1981 Handbook of Fundamentals*.

TABLE 4.10 **Coefficients of Transmission, U (Btu/h ft^2 F), of Metal Construction Flat Roofs and Ceilings: Winter Conditions, Upward Flow**

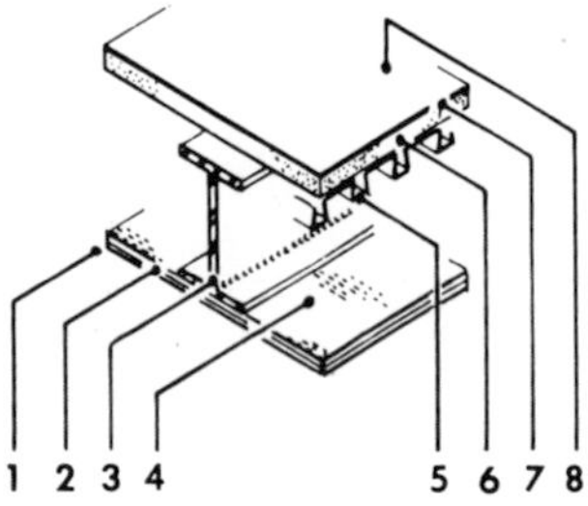

Base Case Resistance, R	*Construction (heat flow up)*	*Resistance* R[c]*: New Items 2 and 6 ("Construction")*
0.61	1. Inside surface (still air)	0.61
0.13	2. Metal lath and sand aggregate plaster, 0.75 in.	0.47
0.00[a]	3. Structural beam	0.00[a]
0.93[b]	4. Nonreflective air space (50 F mean; 10 F° temperature difference)	0.93[b]
0.00[a]	5. Metal deck	0.00[a]
4.17	6. Rigid roof deck insulation, $C = 0.24 (R = 1/C)$	2.78
0.33	7. Built-up roofing, 0.375 in.	0.33
0.17	8. Outside surface (15-mph wind)	0.17
6.34	*Total Thermal Resistance, R*	5.29

Base Case

$$U = \frac{1}{6.34} = 0.158$$

New Items 2 and 6

$$U = \frac{1}{5.29} = 0.189$$

[a]If thermal effects of structural beams and metal deck are to be considered, see *ASHRAE 1981 Handbook of Fundamentals,* Chapter 23.

[b]Use larger air space (3.5 in.) value shown in Table 4.4.

[c]When rigid roof deck insulation (C = 0.24) and sand aggregate plaster replaced with rigid roof deck insulation (C = 0.36) and lightweight aggregate plaster, 0.75 in., on metal lath.

TABLE 4.11 **Coefficients of Transmission, *U* (Btu/h ft² F), of Wood Construction Flat Roofs and Ceilings: Winter Conditions, Upward Flow**

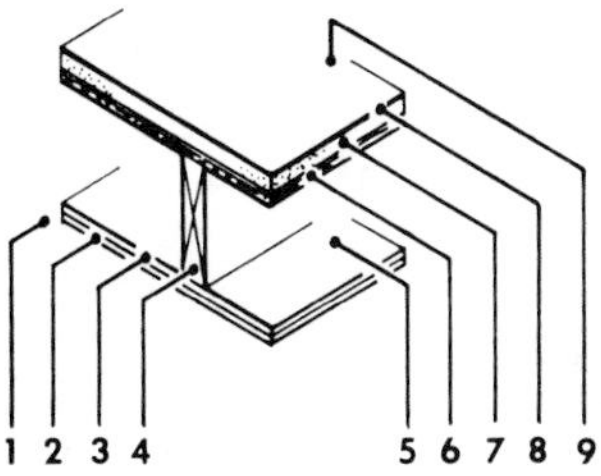

Base Case Resistance: Between Joists, R_i	Base Case Resistance: At Joists, R_s	Construction (heat flow up)	Resistance[c]: New Items 5 and 7 ("Construction"): Between Joists, R_i	Resistance[c]: New Items 5 and 7 ("Construction"): At Joists, R_s
0.61	0.61	1. Inside surface (still air)	0.61	0.61
1.25	1.25	2. Acoustical tile, fiberboard, 0.5 in.	1.25	1.25
0.45	0.45	3. Gypsum wallboard, 0.5 in.	0.45	0.45
—	9.06	4. Nominal 2-in. × 8-in. ceiling joists	—	9.06
0.93[a]	—	5. Nonreflective air space, 7.25 in. (50 F mean; 10 F° temperature difference)	1.05[b]	—
0.78	0.78	6. Plywood deck, 0.625 in.	0.78	0.78
1.39	1.39	7. Rigid roof deck insulation, C = 0.72 (*R* = 1/*C*)	19.00	—
0.33	0.33	8. Built-up roof	0.33	0.33
0.17	0.17	9. Outside surface (15-mph wind)	0.17	0.17
5.91	14.04	*Total thermal resistance, R*	23.64	12.65

Base Case

$$U_i = \frac{1}{5.91} = 0.169$$

New Items 5 and 7

$$U_i = \frac{1}{23.64} = 0.042$$

To adjust *U*-values for the effect of framing, see the example given in Table 4.5. With 10% framing (typical of 2-in. joists at 16-in. o.c.), these adjusted U_{av} values are, respectively:

$U_{av} = 0.159$ (6% less heat loss) $U_{av} = 0.046$ (10% more heat loss)

[a]Use largest air space (3.5 in.) value shown in Table 4.4.

[b]Interpolated value (0 F mean; 10 F° temperature difference).

[c]When roof deck insulation replaced and 7.25-in. air space partially filled with 6-in. R-19 blanket insulation and 1.25-in. air space.

Source: Copyright © by the American Society of Heating, Refrigerating and Air Conditioning Engineers, Inc., Atlanta, GA. Reprinted by permission from *1981 Handbook of Fundamentals.*

TABLE 4.12 **Coefficients of Transmission, U (Btu/h ft^2 F), of 45° Pitched Roofs**

Part A. ***Reflective Air Space***

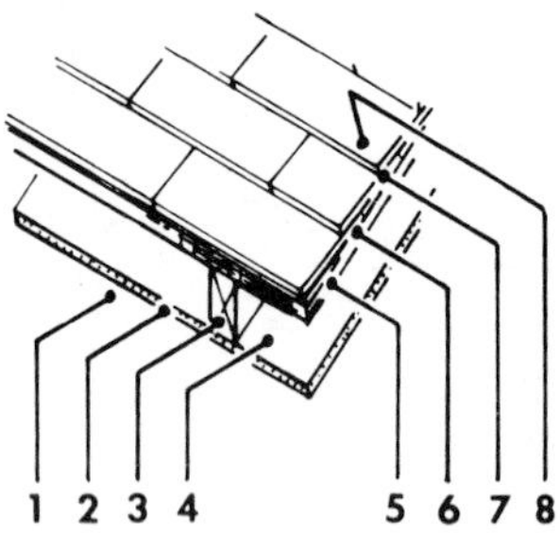

Resistance for Heat Flow Up: Winter Conditions			Resistance for Heat Flow Down: Summer Conditions	
Between Rafters, R_i	*At Rafters,* R_s	*Construction*	*Between Rafters,* R_i	*At Rafters,* R_s
0.62	0.62	1. Inside surface (still air)	0.76	0.76
0.45	0.45	2. Gypsum wallboard 0.5 in., foil backed	0.45	0.45
—	4.35	3. Nominal 2-in. × 4-in. ceiling rafter	—	4.35
2.17	—	4. 45° slope reflective air space, 3.5 in. (50 F mean, 30 F° temperature difference), E = 0.05	4.33[a]	—
0.77	0.77	5. Plywood sheathing, 0.625 in.	0.77	0.77
0.06	0.06	6. Permeable felt building membrane	0.06	0.06
0.44	0.44	7. Asphalt shingle roofing	0.44	0.44
0.17	0.17	8. Outside surface (15-mph wind)	0.25[b]	0.25[b]
4.68	6.86	*Total Thermal Resistance, R*	7.06	7.08

Heat Flow Up

$$U_i = \frac{1}{4.68} = 0.213$$

Heat Flow Down

$$U_i = \frac{1}{7.06} = 0.141$$

To adjust U values for the effect of framing, see the example given in Table 4.5. With 10% framing (typical of 2-in. rafters at 16-in. o.c.), these adjusted U_{av} values are, respectively:

$U_{av} = 0.206$ (3% less heat loss) $U_{av} = 0.141$ (unchanged)

Part B. ***Nonreflective Air Space***

Resistance for Heat Flow Up: Winter Conditions			Resistance for Heat Flow Down: Summer Conditions	
Between Rafters, R_i	*At Rafters,* R_s		*Between Rafters,* R_i	*At Rafters,* R_s
0.62	0.62	1. Inside surface (still air)	0.76	0.76
0.45	0.45	2. Gypsum wallboard, 0.5 in.	0.45	0.45
—	4.35	3. Nominal 2-in. × 4-in. ceiling rafter	—	4.35
0.96	—	4. 45° slope, nonreflective air space, 3.5 in. (50 F mean, 10 F° temperature difference)	0.90[a]	—

TABLE 4.12 **Coefficients of Transmission, U (Btu/h ft² F), of 45° Pitched Roofs (*Continued*)**

Part B. Nonreflective Air Space (Continued)

Resistance for Heat Flow Up: Winter Conditions			Resistance for Heat Flow Down: Summer Conditions	
Between Rafters, R_i	At Rafters, R_s		Between Rafters, R_i	At Rafters, R_s
0.77	0.77	5. Plywood sheathing, 0.625 in.	0.77	0.77
0.06	0.06	6. Permeable felt building membrane	0.06	0.06
0.44	0.44	7. Asphalt shingle roofing	0.44	0.44
0.17	0.17	8. Outside surface	0.25[b]	0.25[b]
3.47	6.86	*Total Thermal Resistance, R*	3.63	7.08

Heat Flow Up

$$U_i = \frac{1}{3.47} = 0.287$$

Heat Flow Down

$$U_i = \frac{1}{3.63} = 0.275$$

Adjusted for 10% framing, as above:

$U_{av} = 0.273$ (5% less heat loss)

$U_{av} = 0.262$ (5% less heat loss)

[a] Air space value of 90 F mean, 10F° temperature difference.

[b] Outside wind velocity 7.5 mph.

Source: Copyright © by the American Society of Heating, Refrigerating and Air Conditioning Engineers, Inc., Atlanta, GA. Reprinted by permission from *1981 Handbook of Fundamentals.*

THERMAL CONTROL

TABLE 4.13 **Heat Flow Coefficients of Slab Floor Construction, F_2: Units are W/K per meter of perimeter (Btu/h F per foot of perimeter)**

Construction	*Insulation*	*Degree Days (65 F base)*		
		7433	*5350*	*2950*
(a) 8-in. block wall, brick facing	Uninsulated	1.07 (0.62)	1.17 (0.68)	1.24 (0.72)
	R = 0.95 (5.4) from edge to footer	0.83 (0.48)	0.86 (0.50)	0.97 (0.56)
(b) 4-in. block wall, brick facing	Uninsulated	1.38 (0.80)	1.45 (0.84)	1.61 (0.93)
	R = 0.95 (5.4) from edge to footer	0.81 (0.47)	0.85 (0.49)	0.93 (0.54)
(c) Metal stud wall, stucco	Uninsulated	1.99 (1.15)	2.07 (1.20)	2.32 (1.34)
	R = 0.95 (5.4) from edge to footer	0.88 (0.51)	0.92 (0.53)	1.00 (0.58)
(d) Poured concrete wall, with duct near perimeter[a]	Uninsulated	3.18 (1.84)	3.67 (2.12)	4.72 (2.73)
	R = 0.95 (5.4) from edge to footer, 0.91 m (3 ft) under floor	1.11 (0.64)	1.24 (0.72)	1.56 (0.90)

[a]Weighted average temperature of the heating duct was assumed at 43°C (110 F) during the heating season [outdoor air temperature less than 18°C (65 F)].

NOTE: To use this table: $q = F_2 P \, \Delta t$

where q = heat loss through perimeter [W (Btu/h)]
F_2 = heat loss coefficients, from above
P = perimeter of exposed edge of floor slab [m (ft)]
Δt = temperature difference between indoors and outdoors air

Do not assume additional heat losses from the slab to the earth below (as would be calculated from Table 4.14). Heat gains are never assumed to occur, either through perimeter or from the earth below.

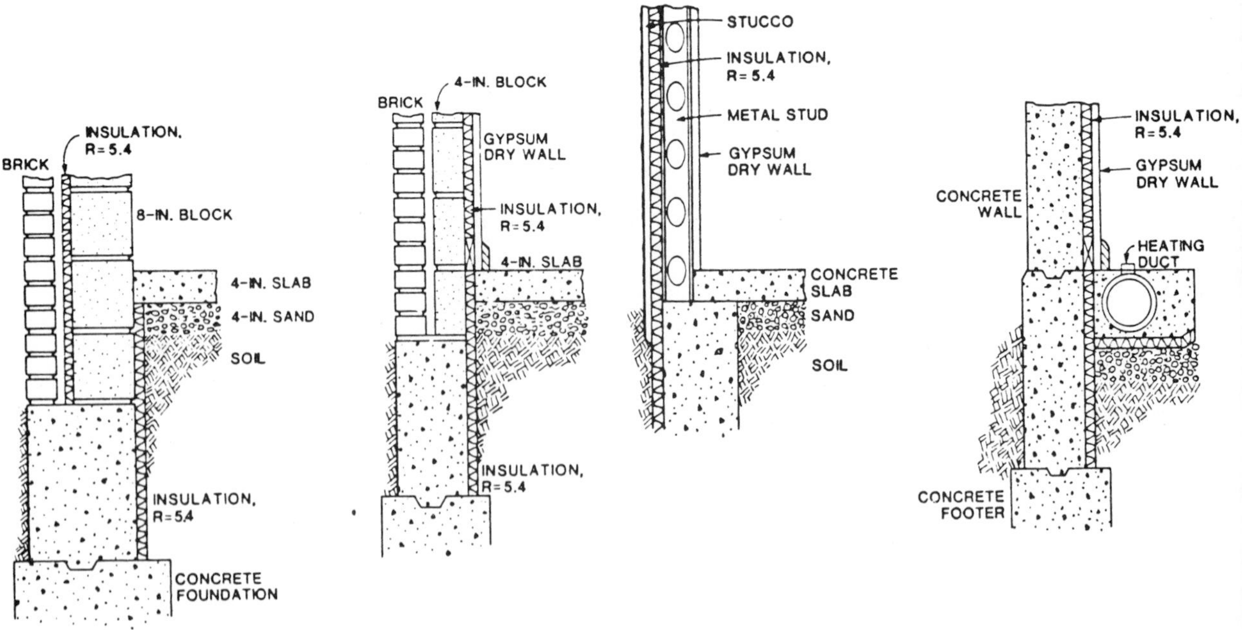

(a) 8-IN. BLOCK WALL (b) 4-IN. BLOCK WALL (c) METAL STUD WALL (d) CONCRETE WALL

TABLE 4.14 **Heat Flow Below Grade**[a]

Part A. ***Basement Walls***

Heat Loss[b] (W/m² K)				Depth		Path Length Through Soil		Heat Loss[b] (Btu/h ft² F)			
Uninsulated	R = 0.73	R = 1.47	R = 2.20	m	(ft)	m	(ft)	Uninsulated	R = 4.17	R = 8.34	R = 12.51
2.33	0.86	0.53	0.38	0–0.30	(0–1)	0.20	(0.68)	0.410	0.152	0.093	0.067
1.26	0.66	0.45	0.36	0.30–0.61	(1–2)	0.69	(2.27)	0.222	0.116	0.079	0.059
0.88	0.53	0.38	0.30	0.61–0.91	(2–3)	1.18	(3.88)	0.155	0.094	0.068	0.053
0.67	0.45	0.34	0.27	0.91–1.22	(3–4)	1.68	(5.52)	0.119	0.079	0.060	0.048
0.54	0.39	0.30	0.25	1.22–1.52	(4–5)	2.15	(7.05)	0.096	0.069	0.053	0.044
0.45	0.34	0.27	0.23	1.52–1.83	(5–6)	2.64	(8.65)	0.079	0.060	0.048	0.040
0.39	0.30	0.25	0.21	1.83–2.13	(6–7)	3.13	(10.28)	0.069	0.054	0.044	0.037

Part B. ***Basement Floors***

Heat Loss (W/m² K), Width of House in Meters				Depth of Foundation Wall Below Grade	Heat Loss (Btu/h ft² F), Width of House in Feet			
6	7.3	8.5	9.7	m (ft)	20	24	28	32
0.18	0.16	0.15	0.13	1.50 (5)	0.032	0.029	0.026	0.023
0.17	0.15	0.14	0.12	1.83 (6)	0.030	0.027	0.025	0.022
0.16	0.15	0.13	0.12	2.10 (7)	0.029	0.026	0.023	0.021

[a]Only heat losses are assumed, since the interior temperature is normally above the ground temperature.

[b] Soil conductivity assumed at 9.6 Btu-in./h-ft² F.

Source: Copyright © by the American Society of Heating, Refrigerating and Air Conditioning Engineers, Inc., Atlanta, GA. Reprinted by permission from 1989 *Handbook of Fundamentals.*

TABLE 4.15 **Coefficients of Transmission, U, of Doors (Btu/h ft^2 F)**

*Part A. **For Wood Doors**[a]*

		Winter[b]			*Summer*[c]
Door Thickness (in.)[d]	*Description*	*No Storm Door*	*Wood Storm Door*[e]	*Metal Storm Door*[f]	*No Storm Door*
1–3/8	Hollow-core flush door	0.47	0.30	0.32	0.45
1–3/8	Solid-core flush door	0.39	0.26	0.28	0.38
1–3/8	Panel door with 7/16-in. panels	0.57	0.33	0.37	0.54
1–3/4	Hollow core flush door with single glazing[g]	0.46	0.29	0.32	0.44
		0.56	0.33	0.36	0.54
1–3/4	Solid-core flush door	0.33	0.28	0.25	0.32
	With single glazing[g]	0.46	0.29	0.32	0.44
	With insulating glass[g]	0.37	0.25	0.27	0.36
1–3/4	Panel door with 7/16-in. panels[h]	0.54	0.32	0.36	0.52
	With single glazing[i]	0.67	0.36	0.41	0.63
	With insulating glass[i]	0.50	0.31	0.34	0.48
1–3/4	Panel door with 1 1/8-in. panels[h]	0.39	0.26	0.28	0.38
	With single glazing[i]	0.61	0.34	0.38	0.58
	With insulating glass[i]	0.44	0.28	0.31	0.42
2–1/4	Solid core flush door	0.27	0.20	0.21	0.26
	With single glazing[g]	0.41	0.27	0.29	0.40
	With insulating glass[g]	0.33	0.23	0.25	0.32

TABLE 4.15 **Coefficients of Transmission, *U*, of Doors (Btu/h ft² F) (*Continued*)**

*Part B. **For Steel Doors, 1¾ in. Thick***[b]

Door Thickness (in.)	Description	No Storm Door
1¾	Fiberglass or mineral wool core with steel stiffeners, no thermal break[j]	0.60
1¾	Paper honeycomb core without thermal break[j]	0.56
1¾	Solid urethane foam core without thermal break[k]	0.40
1¾	Solid fire rated mineral fiberboard core without thermal break[j]	0.38
1¾	Polystyrene core without thermal break (18 gauge commercial steel)[j]	0.35
1¾	Polystyrene core without thermal break (18 gauge commercial steel)[j]	0.35
1¾	Polyurethane core without thermal break (18 gauge commercial steel)[j]	0.29
1¾	Polyurethane core without thermal break (24 gauge commercial steel)[j]	0.29
1¾	Polyurethane core without thermal break and wood perimeter (24 gauge residential steel)[j]	0.20
1¾	Solid urethane foam core with thermal break[k]	0.19

Source: Copyright © by the American Society of Heating, Refrigerating and Air Conditioning Engineers, Inc., Atlanta, GA. Reprinted by permission from *1981* and *1989 Handbook of Fundamentals.*

[a] Values for doors are based on nominal 2′8-in. × 6′8-in. door size. Interpolation and moderate extrapolation are permitted for glazing areas and door thicknesses other than those specified.

[b] 15 mph outdoor air velocity; 0 F outdoor air; 70 F inside air temperature, natural convection.

[c] 7.5 mph outdoor air velocity; 89 F outdoor air; 75 F inside air temperature, natural convection.

[d] Nominal thickness.

[e] Values for wood storm door are approximately 50% glass area.

[f] Values for metal storm door are for any percent of glass area.

[g] 17% exposed glass area; insulating glass contains 0.25-in. air space.

[h] 55% panel area.

[i] 33% glass area; insulating glass contains 0.25-in. air space.

[j] ASTM C 236 hot box data on a nominal 3 ft by 7 ft door size with no glazing.

[k] Values for nominal 32-in × 80-in door with no glazing.

THERMAL CONTROL

TABLE 4.16 **Overall Coefficients of Heat Transmission, U_0, of Various Fenestration Products**

*Part A. **U Values for Vertical Installation[a] (Btu/h-ft²°F)***

Glazing Type[b]	Glass Only		Aluminum Frame, No Thermal Break (U_f = 1.9)		Aluminum Frame, Thermal Break (U_f = 1.0)		Wood or Vinyl Frame (U_f = 0.4)	
	Center of Glass	Edge of Glass[c]	Product Type		Product Type		Product Type	
			R	C	R	C	R	C
Single glazing								
glass	(1.11)	(n/a)	1.31	1.23	1.09	1.10	0.90	0.98
⅛-in. acrylic	(1.03)	(n/a)	1.26	1.16	1.02	1.03	0.84	0.92
Double glass								
¼-in. airspace	(0.57)	(0.66)	0.92	0.78	0.70	0.65	0.54	0.55
⅜-in. airspace	(0.52)	(0.62)	0.88	0.74	0.66	0.60	0.50	0.51
½-in. and greater airspace	(0.49)	(0.59)	0.87	0.72	0.64	0.59	0.49	0.49
Double glass, ε = 0.40 on surface 2 or 3								
¼-in. airspace	(0.50)	(0.60)	0.87	0.73	0.65	0.59	0.49	0.50
⅜-in. airspace	(0.43)	(0.55)	0.83	0.67	0.60	0.54	0.45	0.45
½-in. and greater airspace	(0.41)	(0.54)	0.81	0.65	0.58	0.52	0.43	0.42
Double glass, ε = 0.15 on surface 2 or 3								
¼-in. airspace	(0.45)	(0.56)	0.84	0.68	0.61	0.55	0.46	0.46
⅜-in. airspace	(0.36)	(0.51)	0.78	0.62	0.56	0.48	0.41	0.39
½-in. airspace	(0.34)	(0.50)	0.76	0.60	0.54	0.46	0.39	0.37
Double glass								
¼-in. argon space	(0.52)	(0.62)	0.88	0.74	0.66	0.61	0.50	0.51
⅛-in. argon space	(0.48)	(0.59)	0.86	0.71	0.63	0.57	0.48	0.48
½-in. and greater argon space	(0.46)	(0.57)	0.82	0.69	0.62	0.56	0.47	0.47
Double glass, ε = 0.40 on surface 2 or 3								
¼-in. argon space	(0.43)	(0.55)	0.83	0.67	0.60	0.54	0.45	0.45
⅜-in. argon space	(0.38)	(0.52)	0.79	0.63	0.57	0.49	0.42	0.40
½-in. and greater argon space	(0.36)	(0.51)	0.78	0.62	0.56	0.48	0.41	0.39
Double glass, ε = 0.15 on surface 2 or 3								
¼-in. argon space	(0.36)	(0.51)	0.78	0.62	0.56	0.48	0.41	0.39
⅜-in. argon space	(0.30)	(0.48)	0.74	0.57	0.51	0.43	0.37	0.34
½-in. and greater argon space	(0.28)	(0.47)	0.73	0.55	0.50	0.42	0.36	0.33
Double glazing, ⅛-in. acrylic or polycarbonate								
¼-in. airspace	(0.52)	(0.62)	0.89	0.74	0.67	0.61	0.51	0.51
⅜-in. airspace	(0.48)	(0.59)	0.86	0.71	0.64	0.57	0.48	0.48
½-in. and greater airspace	(0.46)	(0.57)	0.85	0.69	0.62	0.56	0.47	0.47
Double glazing, ¼-in. acrylic or polycarbonate								
¼-in. airspace	(0.48)	(0.59)	0.86	0.71	0.64	0.57	0.48	0.48
⅜-in. airspace	(0.44)	(0.56)	0.84	0.68	0.61	0.54	0.46	0.45
½-in. and greater airspace	(0.42)	(0.54)	0.82	0.66	0.60	0.53	0.45	0.43

TABLE 4.16 **Overall Coefficients of Heat Transmission, U_0, of Various Fenestration Products (*Continued*)**

Part A. ***U Values for Vertical Installation[a] (Btu/h-ft²°F) (Continued)***

Glazing Type[b]	*Glass Only: Center of Glass*	*Glass Only: Edge of Glass[c]*	*Aluminum Frame, No Thermal Break (U_f = 1.9) Product Type: R*	*C*	*Aluminum Frame, Thermal Break (U_f = 1.0) Product Type: R*	*C*	*Wood or Vinyl Frame (U_f = 0.4) Product Type: R*	*C*
Triple glass								
¼-in. airspace	(0.38)	(0.52)	0.79	0.64	0.57	0.50	0.42	0.41
⅜-in. airspace	(0.34)	(0.50)	0.76	0.60	0.54	0.46	0.39	0.38
½-in. and greater airspace	(0.32)	(0.49)	0.75	0.58	0.53	0.45	0.38	0.36
Triple glass, ε =0.40 on surface 2, 3, 4 or 5								
¼-in. airspaces	(0.35)	(0.50)	0.77	0.61	0.55	0.48	0.40	0.39
⅜-in. airspaces	(0.30)	(0.48)	0.74	0.57	0.52	0.44	0.37	0.35
½-in. and greater airspaces	(0.28)	(0.47)	0.72	0.55	0.50	0.41	0.36	0.33
Triple glass or double glass with polyester film suspended in between, ε = 0.40 on surface 2, 3, 4, or 5								
¼-in. airspaces	(0.33)	(0.49)	0.76	0.59	0.53	0.45	0.39	0.37
⅜-in. airspaces	(0.27)	(0.46)	0.72	0.54	0.50	0.41	0.35	0.32
½-in. and greater airspaces	(0.24)	(0.45)	0.70	0.52	0.48	0.39	0.34	0.30
Triple glass or double glass with polyester film suspended in between, ε = 0.15 on surfaces 2 or 3 and 4 or 5								
¼-in. airspaces	(0.28)	(0.47)	0.73	0.55	0.53	0.42	0.36	0.33
⅜-in. airspaces	(0.22)	(0.45)	0.69	0.51	0.47	0.37	0.32	0.29
½-in. and greater airspaces	(0.19)	(0.44)	0.67	0.48	0.45	0.35	0.31	0.26
Triple glass								
¼-in. argon spaces	(0.34)	(0.50)	0.77	0.60	0.54	0.46	0.39	0.38
⅜-in. argon spaces	(0.31)	(0.48)	0.74	0.57	0.52	0.44	0.37	0.35
½-in. and greater argon spaces	(0.29)	(0.47)	0.73	0.56	0.51	0.42	0.36	0.34
Triple glass, ε = 0.40 on surface 2, 3, 4, or 5								
¼-in. argon spaces	(0.30)	(0.48)	0.74	0.57	0.52	0.44	0.37	0.35
⅜-in. argon spaces	(0.26)	(0.46)	0.72	0.54	0.49	0.41	0.35	0.32
½-in. and greater argon spaces	(0.25)	(0.46)	0.71	0.53	0.48	0.39	0.34	0.31
Triple glass or double glass with polyester film suspended in between, ε = 0.40 on surface 2, 3, 4, or 5								
¼-in. argon spaces	(0.27)	(0.46)	0.72	0.54	0.50	0.41	0.35	0.32
⅜-in. argon spaces	(0.22)	(0.45)	0.69	0.51	0.47	0.37	0.33	0.29
½-in. and greater argon spaces	(0.20)	(0.44)	0.68	0.50	0.46	0.36	0.31	0.28
Triple glass or double glass with polyester film suspended in between, $\varepsilon \doteq$ 0.15 on surfaces 2 or 3 and 4 or 5								
¼-in. argon spaces	(0.22)	(0.45)	0.69	0.51	0.47	0.37	0.32	0.29
⅜-in argon spaces	(0.17)	(0.43)	0.66	0.47	0.44	0.34	0.30	0.25
½-in. and greater argon spaces	(0.15)	(0.43)	0.65	0.46	0.43	0.32	0.29	0.24

TABLE 4.16 **Overall Coefficients of Heat Transmission, U_0, of Various Fenestration Products (*Continued*)**

Part B. ***U-Value Conversion Table for Sloped and Horizontal Glazing for Upward Heat Flow***

Slope	*U Value (Btu/h-ft² F)*												
90° (vertical)	0.10	0.20	0.30	0.40	0.50	0.60	0.70	0.80	0.90	1.00	1.10	1.20	1.30
45°	0.14	0.25	0.36	0.47	0.57	0.68	0.79	0.90	1.00	1.11	1.22	1.33	1.44
0 (horiz.)	0.19	0.29	0.40	0.51	0.61	0.72	0.82	0.93	1.04	1.14	1.25	1.35	1.46

Source: Copyright © by the American Society of Heating, Refrigerating and Air-Conditioning Engineers, Inc., Atlanta, Ga. Reprinted by permission from 1989 *Handbook of Fundamentals.*

[a] All U values are based on standard ASHRAE winter conditions of 70 F indoor and 0 F outdoor air temperature with 15 mph outdoor air velocity and zero solar flux. The outside surface coefficient at these conditions is approximately 5.1 Btu/hr-ft²-°F, depending on the glass surface temperature. With the exception of single glazing, small changes in the interior and exterior temperatures do not significantly affect overall U values.

[b] Glazing layer surfaces are numbered from the outside to the inside. Double and triple refer to the number of glazing lites. All data are based on ⅛-in. glass unless otherwise noted. Thermal conductivities are: 0.53 Btu/h-ft-°F for glass, and 0.11 Btu/h-ft-°F for acrylic and polycarbonate.

[c] Based on aluminum spacers data. Edge-of-glass effect assumed to extend over the 2.5-in. band around perimeter of each glazing unit as seen below.

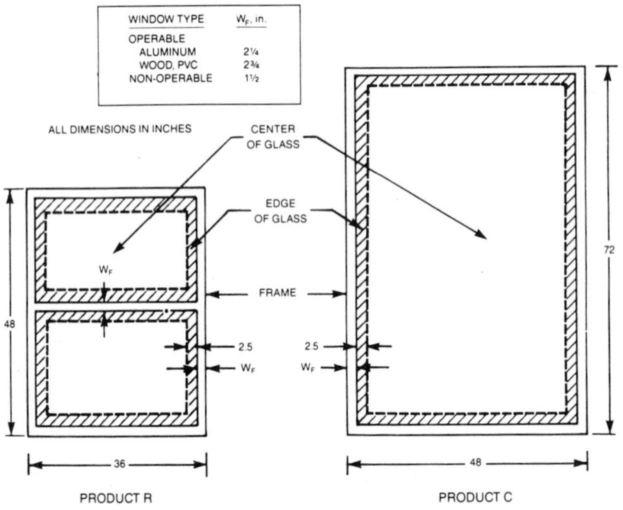

R (Residential) and C (Commercial) Category Frame Dimensions

	R	*C*
Overall window area	6 ft²	24 ft²
Aluminum: frame area	25%	15%
edge-of-glass area	27%	15%
center-of-glass area	48%	70%
Wood and vinyl: frame area	30%	18%
	26%	15%
	44%	67%

TABLE 4.17 **Coefficients of Transmission, *U*, of Windows for Passive Solar Collecting: Units are W/m²-°C (Btu/h-ft² F)**

Component	U, *Winter*[a]		
	No Night Insulation	*With Night Insulation*[b]	
		R-4	*R-9*
Direct gain[c]	3.1 (0.55)	1.9 (0.33)	1.6 (0.28)
Trombé wall, 18 in. thick[c]	1.2 (0.22)	0.9 (0.16)	0.7 (0.13)
Water wall[c]	1.9 (0.33)	1.2 (0.22)	1.1 (0.19)
Selective transmitter film, with glass spaced ½ in[d]	1.6 (0.29) (not used with night insulation)		

[a]Summer-day *U* can be developed, from the designer's assumptions about the hours for which the insulation is in place.

[b]Insulation assumed in place from 5:30 P.M. to 7:30 A.M., solar time. These *U* values are thus *average,* rather than instantaneous, values, unlike the others throughout this table.

[c]From Balcomb et al. (1980).

[d]From Johnson (1981).

TABLE 4.18 **Shading Coefficients for Glazing Without Interior or Exterior Shading**[a]

Part A. ***Single Glass***

Type of Glass	*Nominal Thickness*[b]	*Solar Trans.*[b]	*Shading Coefficient, SC*[c]
Clear	3 mm (⅛ in.)	0.86	1.00
	6 mm (¼ in.)	0.78	0.94
	10 mm (⅜ in.)	0.72	0.90
	12 mm (½ in.)	0.67	0.87
Heat absorbing	3 mm (⅛ in.)	0.64	0.83
	6 mm (¼ in.)	0.46	0.69
	10 mm (⅜ in.)	0.33	0.60
	12 mm (½ in.)	0.24	0.53

Part B. ***Insulating Glass***

Clear out, clear in	3 mm (⅛ in.)[d]	0.71[e]	0.88
Clear out, clear in	6 mm (¼ in.)	0.61	0.81
Heat absorbing[f] out, clear in	6 mm (¼ in.)	0.36	0.55

[a]Refers to factory-fabricated units with 5-, 6-, or 12-mm (3/16-, ¼-, or ½-in.) air space or to prime windows plus storm sash.

[b]Refer to manufacturer's literature for values.

[c]Based on outdoor air at 12 km/h (7.5 mph) and still air indoors.

[d]Thickness of each pane of glass, not thickness of assembled unit.

[e]Combined transmittance for assembled unit.

[f]Refers to gray-, bronze-, and green-tinted heat-absorbing float glass.

TABLE 4.18 **Shading Coefficients for Glazing Without Interior or Exterior Shading (*Continued*)**

Part C. ***Solar Optical Properties and Shading Coefficients of Transparent Plastic Sheeting***

Type of Plastic	Transmittance		SC
	Visible	Solar	
Acrylic			
Clear	0.92	0.85	0.98
Gray tint	0.16	0.27	0.52
Gray tint	0.33	0.41	0.63
Gray tint	0.45	0.55	0.74
Gray tint	0.59	0.62	0.80
Gray tint	0.76	0.74	0.89
Bronze tint	0.10	0.20	0.46
Bronze tint	0.27	0.35	0.58
Bronze tint	0.49	0.56	0.75
Bronze tint	0.61	0.62	0.80
Bronze tint	0.75	0.75	0.90
Reflective[a]	0.14	0.12	0.21
Polycarbonate			
Clear, 3-mm (1/8-in.)	0.88	0.82	0.98
Gray, 3-mm (1/8-in.)	0.50	0.57	0.74
Bronze, 3-mm (1/8-in.)	0.50	0.57	0.74

Source: Copyright © by the American Society of Heating, Refrigerating and Air Conditioning Engineers, Inc., Atlanta, GA. Reprinted by permission from the *ASHRAE Handbook of Fundamentals*, 1981.

[a]Aluminum metallized polyester film on plastic.

TABLE 4.19 **Shading Coefficients (SC)**

Part A. ***Shading Coefficients for Single Glass with Indoor Shading by Venetian Blinds or Roller Shades***

			Type of Shading				
			Venetian blinds		Roller Shade		
					Opaque		Translucent
Type of Glass	*Nominal Thickness[a] [mm (in.)]*	*Solar Trans.*	*Medium*	*Light*	*Dark*	*White*	*Light*
Clear	2.5 to 6 (3/32 to 1/4)	0.87 to 0.79	0.74	0.67 (open 45°)			
Clear	6 to 12 (1/4 to 1/2)	0.80 to 0.71					
Clear pattern	3 to 12 (1/8 to 1/2)	0.87 to 0.79	0.63	0.58 (closed)[b]	0.59	0.25	0.39
Heat-absorbing pattern	3 (1/8)	—					
Tinted	5 to 5.5 (3/16, 7/32)	0.74, 0.71	0.29	0.25 (tightly closed)[c]			
Heat-absorbing[d]	5 to 6 (3/16, 1/4)	0.46					
Heat-absorbing pattern	5 to 6 (3/16, 1/4)	—	0.57	0.53	0.45	0.30	0.36
Tinted	3 to 5.5 (1/8, 7/32)	0.59, 0.45					
Heat-absorbing or pattern	—	0.44 to 0.30	0.54	0.52	0.40	0.28	0.32
Heat-absorbing[d]	10 (3/8)	0.34					
Heat-absorbing or pattern	—	0.29 to 0.15 0.24	0.42	0.40	0.36	0.28	0.31
Reflective coated glass							
SC[e] = 0.30			0.25	0.23			
0.40			0.33	0.29			
0.50			0.42	0.38			
0.60			0.50	0.44			

THERMAL CONTROL

TABLE 4.19 **Shading Coefficients (SC) (*Continued*)**

Part B. ***Shading Coefficients for Insulating Glass[f] with Indoor Shading by Venetian Blinds or Roller Shades***

Type of Glass	Nominal Thickness, Each Light	Solar Trans.[a] Outer Pane	Solar Trans.[a] Inner Pane	Type of Shading: Venetian Blinds[g] Medium	Venetian Blinds[g] Light	Roller Shade: Opaque Dark	Roller Shade: Opaque White	Roller Shade: Translucent Light
Clear out Clear in	2.5, 3 mm (3/32, 1/8 in.)	0.87	0.87	0.62	0.58 (45° open)	0.71	0.35	0.40
Clear out Clear in	6 mm (1/4 in.)	0.80	0.80	0.63	0.58 (closed)[b]			
Heat-absorbing[d] out Clear in	6 mm (1/4 in.)	0.46	0.80	0.39	0.36	0.40	0.22	0.30
Reflective coated glass								
SC[e] = 0.20				0.19	0.18			
0.30				0.27	0.26			
0.40				0.34	0.33			

Part C. ***Shading Coefficients for Double Glazing with Between-Glass Shading***

Type of Glass	Nominal Thickness, Each Pane	Solar Trans.[a] Outer Pane	Solar Trans.[a] Inner Pane	Description of Air Space	Type of Shading: Venetian Blinds Light	Type of Shading: Venetian Blinds Medium	Type of Shading: Louvered Sun Screen
Clear out / Clear in	2.5, 3 mm (3/32, 1/8 in.)	0.87	0.87	Shade in contact with glass or shade separated from glass by airspace	0.33	0.36	0.43
Clear out / Clear in	6 mm (1/4 in.)	0.80	0.80	Shade in contact with glass; voids filled with plastic	—	—	0.49
Heat-absorbing[d] out				Shade in contact with glass or shade separated from glass by airspace	0.28	0.30	0.37
Clear in	6 mm (1/4 in.)	0.46	0.80	Shade in contact with glass; voids filled with plastic	—	—	0.41

Source: Copyright © by the American Society of Heating, Refrigerating, and Air-Conditioning Engineers, Inc., Atlanta, Ga. Reprinted by permission from 1989 *Handbook of Fundamentals.*

[a] Refer to manufacturer's literature for exact values.

[b] Use for automated operation, for solar gain reduction (as opposed to daylight use).

[c] For vertical blinds with opaque white and beige louvres, tightly closed, with glass of 0.71 to 0.80 transmittance.

[d] Refers to gray-, bronze-, and green-tinted heat-absorbing glass.

[e] SC for glass with no shading device.

[f] Refers to factory-manufactured units with 5-, 6-, or 13-mm (3/16-, 1/4-, or 1/2-in.) airspace, or to prime windows plus storm windows.

[g] For vertical blinds with opaque white or beige louvres, tightly closed, SC is approximately the same as for opaque white roller shades.

TABLE 4.20 **Shading Coefficients for Single and Insulating Glass with Draperies**

	Range of Shading Coefficients			
			Drapery Fabrics[b]	
Glazing	*Glass Trans.*	*Glass SC*[a]	*High Transmittance, Low Reflectance*[c]	*Low Transmittance, High Reflectance*[c]
Single Glass				
3-mm (1/8-in.) clear	0.86	1.00	0.87	0.37
6 mm (1/4 in.) clear	0.80	0.95	0.80	0.35
12.mm (1/2 in.) clear	0.71	0.88	0.74	0.35
6 mm (1/4 in.) heat abs.	0.46	0.67	0.57	0.33
12 mm (1/2 in.) heat abs.	0.24	0.50	0.43	0.30
Reflective coated (see	—	0.60	0.57	0.33
manufacturer's literature for	—	0.50	0.46	0.31
exact values)	—	0.40	0.36	0.26
	—	0.30	0.25	0.20
Insulating Glass, 12-mm (1/2-in.) air				
space clear out and clear in	0.64	0.83	0.66	0.35
Heat abs. out and clear in	0.37	0.55	0.49	0.32
Reflective coated (see	—	0.40	0.38	0.28
manufacturers' literature for	—	0.30	0.29	0.24
exact values)	—	0.20	0.19	0.15

Source: Based on *Handbook of Fundamentals,* published by the American Society of Heating, Refrigerating and Air Conditioning Engineers, Inc., Atlanta, GA.

[a]For glass alone, with no drapery.

[b]Drapes of 100% fullness, loose hanging.

[c]See *ASHRAE Handbook of Fundamentals,* 1989, Chapter 27, Table 29, for more detailed listings.

TABLE 4.21 **Shading Coefficients for Domed Skylights**

Dome	Light Diffuser (Translucent)	Curb (See Below) Height, [mm (in.)]	Curb (See Below) Width-to-Height Ratio	Shading Coefficient	U Factor at Center of Skylight
Clear	Yes	0 (0)	∞	0.61	2.6 (0.46)
$\tau = 0.86$[a]	$\tau = 0.58$	230 (9)	5	0.58	2.4 (0.43)
		460 (18)	2.5	0.50	2.3 (0.40)
Clear		0 (0)	∞	0.99	4.5 (0.80)
$\tau = 0.86$	None	230 (9)	5	0.88	4.3 (0.75)
		460 (18)	2.5	0.80	4.0 (0.70)
Translucent	None	0 (0)	∞	0.57	4.5 (0.80)
$\tau = 0.52$		460 (18)	2.5	0.46	4.0 (0.70)
Translucent	None	0 (0)	∞	0.34	4.5 (0.80)
$\tau = 0.27$		230 (9)	5	0.30	4.3 (0.75)
		460 (18)	2.5	0.28	4.0 (0.70)

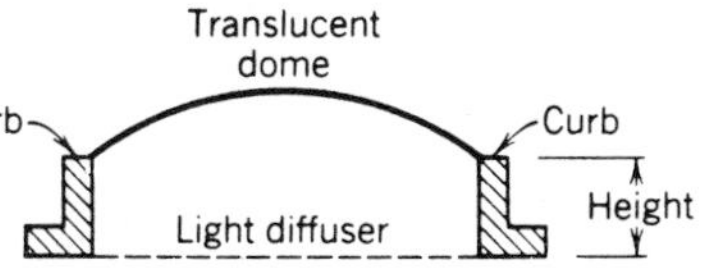

Source: Copyright © by the American Society of Heating, Refrigerating and Air Conditioning Engineers, Inc., Atlanta, GA. Reprinted by permission from the *ASHRAE Handbook of Fundamentals,* 1981 and 1989.

[a] τ = transmittance.

TABLE 4.22 **Approximate Shading Coefficients of External Shading Devices**

Awnings	
Of venetian blind type, ⅔ drawn	0.43
fully drawn	0.15
Dark or medium canvas	0.25
Shading screens	0.28–0.23
Louvres, movable	0.15–0.10
Overhang: continuous, completely shading window	0.25
Dense tree casting heavy shade	0.25

Source: Adapted from Olgyay and Olgyay (1957).

TABLE 4.23 **Shading Coefficients and *U* Values for Glass Block**

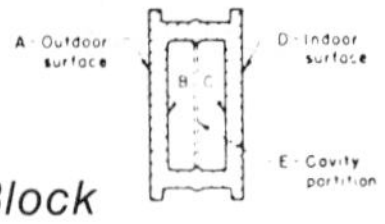

Type of Glass Block[a]	Description of Glass Block	Shading Coefficient[b] Panels[c] in the Sun	Panels[d] in the Shade (N, NW, W, SW)	U Value[e]
Type I	Glass colorless or aqua A, D: Smooth B, C: Smooth or wide ribs, or flutes horizontal or vertical, or shallow configuration E: None	0.65	0.40	0.51
Type IA	Same as type I except A: Ceramic enamel on exterior face	0.27	0.20	0.51
Type II	Same as type I except E: Glass fiber screen	0.44	0.34	0.48
Type III	Glass colorless or aqua A, D: Narrow vertical ribs or flutes. B, C: Horizontal light-diffusing prisms, or horizontal light-directing prisms E: Glass fiber screen	0.33	0.27	0.48
Type IIIA	Same as type III except E: Glass fiber screen with green ceramic spray coating or glass fiber screen and gray glass or glass fiber screen with light-selecting prisms	0.25	0.18	0.48
Type IV	Same as type I except A: Reflective oxide coating on face	0.16	0.12	0.51

Source: Copyright © American Society of Heating, Refrigerating, and Air-Conditioning Engineers, Inc., Atlanta, Ga. Reprinted by permission from 1989 *Handbook of Fundamentals.*

[a] All values are for 7.75 × 7.75 × 3.88 in. block, set in light-colored mortar. For 11.75 × 11.75 × 3.88 in. block, increase coefficients by 15%, and for 5.75 × 5.75 × 3.88 in. block, reduce coefficients by 15%.

[b] Shading coefficients are applied to heat gain factors for 1 h earlier than the load calculation time to allow for heat storage in the panel.

[c] Shading coefficients are peak load condition, but closely approximate other conditions. Smith and Pennington, *ASHRAE Journal,* Vol. 5, December 1964, p. 31, give values for other conditions.

[d] For NE, E, and SE panels in the shade, add 50% to the values listed for panels in the shade.

[e] Values are the same for all block sizes; Btu/h-ft^2 F.

TABLE 4.24 **Shading Coefficients for Louvered Sun Screens**[a]

Profile Angle, (deg)	Group 1		Group 2		Group 3		Group 4		Group 5		Group 6	
	Trans-mittance	SC	Trans-mittance	SC	Trans-mittance	SC	Trans-mittance	SC	Trans-mittance	SC	Trans-mittance	SC
10	0.23	0.35	0.25	0.33	0.40	0.51	0.48	0.59	0.15	0.27	0.26	0.45
20	0.06	0.17	0.14	0.23	0.32	0.42	0.39	0.50	0.04	0.11	0.20	0.35
30	0.04	0.15	0.12	0.21	0.21	0.31	0.28	0.38	0.03	0.10	0.13	0.26
⩾ 40	0.04	0.15	0.11	0.20	0.07	0.18	0.20	0.30	0.03	0.10	0.04	0.13

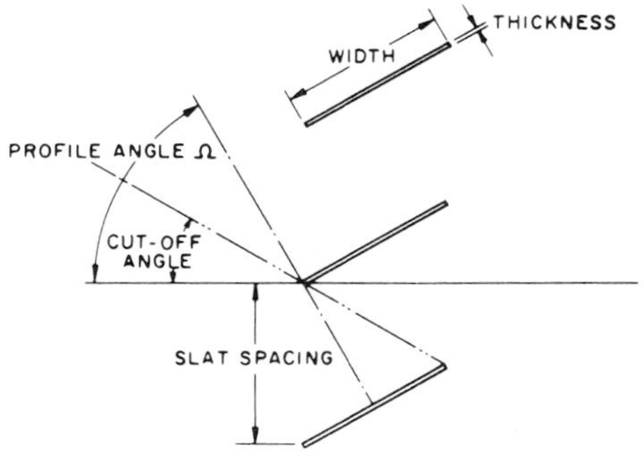

Source: Copyright © American Society of Heating, Refrigerating, and Air-Conditioning Engineers, Inc., Atlanta, Ga. Reprinted by permission from 1989 *Handbook of Fundamentals.*

[a]**Group 1.** Black, width over spacing ratio 1.15/1; 23 louvers/in.
Group 2. Light color; high reflectance, otherwise same as group 1.
Group 3. Black or dark color; w/s ratio 0.85/1; 17 louvers/in.
Group 4. Light color or unpainted aluminum; high reflectance; otherwise same as group 3.
Group 5. Same as group 1, except two lights of 0.25-in. clear glass with 0.5-in. airspace.
Group 6. Same as group 3, except two lights of 0.25-in. clear glass with 0.5-in. airspace.
U value = **0.85 Btu/h-ft²-F** for all groups when used with single glazing.

THERMAL CONTROL

TABLE 4.25 **Recommended Outdoor Air Requirements for Ventilation**[a]

Part A. ***Commercial Facilities (Offices, Stores, Shops, Hotels, Sports Facilities, etc.)***

Application	*Estimated Occupancy*[b] *(Persons per 1000 ft² or 100 m² Floor Area): Use Only When Design Occupancy Is Not Known*	*Outdoor Air Requirements* *cfm per person*	*L/s per person*	*cfm per ft²*	*L/s per m²*	*Comments*
Dry Cleaners, Laundries						Dry cleaning processes may require more air.
Commercial laundries	10	25	13			
Commercial dry cleaners	30	30	15			
Storage, pickup	30	35	18			
Coin-op laundries	20	15	8			
Coin-op dry cleaners	20	15	8			
Food and Beverage Service						
Dining rooms	70	20	10			
Cafeterias, fast food	100	20	10			
Bars, Cocktail lounges	100	30	15			Supplementary smoke removal equipment may be required.
Kitchens (cooking)	20	15	8			Makeup air for hood exhaust may require more ventilating air. The sum of the outdoor air and transfer air of acceptable quality from adjacent spaces shall be sufficient to provide an exhaust rate of not less than 1.5 cfm/ft² (7.5 L/s per m²)
Garages, Repair Service Stations						
Enclosed parking garages				1.50	7.50	Distribution among people must consider worker location and concentration of run-

Auto repair rooms				1.50	7.50	ning engines; stands where engines are run must incorporate systems for positive engine exhaust withdrawal. Contaminant sensors may be used to control ventilation.
Hotels, Motels, Resorts Dormitories						See also food and beverage services, merchandising, barber and beauty shops, garages.
Dormitory sleeping areas	20	15	8			
				cfm/room	L/s/room	Independent of room size.
Bedrooms				30	15	
Living rooms				30	15	
Bathrooms				35	18	Installed capacity for intermittent use.
Lobbies	30	15	8			
Conference rooms	50	20	10			
Assembly rooms	120	15	8			
Gambling casinos	120	30	15			Supplementary smoke removal equipment may be required.
Offices						Some office equipment may require local exhaust.
Office space	7	20	10			
Reception areas	60	15	8			Supplementary smoke removal equipment may be required
Telecommunication centers and data entry areas	60	20	10			
Conference rooms	50	20	10			
Public Spaces						
Corridors and utility rooms				0.50	0.25	
Public rest rooms cfm/wc or urinal		50	25			Mechanical exhaust with no recirculation is recommended

TABLE 4.25 **Recommended Outdoor Air Requirements for Ventilation**[a] ***(Continued)***

Part A. ***Commercial Facilities (Offices, Stores, Shops, Hotels, Sports Facilities, etc.) (Continued)***

Application	Estimated Occupancy[b] (Persons per 1000 ft^2 or 100 m^2 Floor Area): Use Only When Design Occupancy Is Not Known	Outdoor Air Requirements: cfm per person	Outdoor Air Requirements: L/s per person	Outdoor Air Requirements: cfm per ft^2	Outdoor Air Requirements: L/s per m^2	Comments
Locker and dressing rooms				0.5	2.5	
Smoking lounge	70	60	30			Normally supplied by transfer air, local mechanical exhaust. Exhaust with no recirculation recommended.
Elevators				1.00	5.00	Normally supplied by transfer air.
Retail Stores Sales Floors and Showroom Floors						
Basement and street	30			0.30	1.50	
Upper floors	20			0.20	1.00	
Storage rooms	15			0.15	0.75	
Dressing rooms				0.20	1.00	
Malls and arcades	20			0.20	1.00	
Shipping and receiving	10			0.15	0.75	
Warehouses	5			0.05	0.25	
Smoking lounge	70	60	30			Normally supplied by transfer air, local mechanical exhaust. Exhaust with no recirculation recommended.
Specialty Shops						
Barber	25	15	8			
Beauty	25	25	13			
Reducing salons	20	15	8			
Florists	8	15	8			Ventilation to optimize plant growth may dictate requirements.

Clothiers, furniture				0.30	1.50	
Hardware, drugs, fabric	8	15	8			
Supermarkets	8	15	8			
Pet shops				1.00	5.00	
Sports and Amusement						
Spectator areas	150	15	8			
Game rooms	70	25	13			When internal combustion engines are operated for maintenance of playing surfaces, increased ventilation rates may be required.
Ice arenas (playing area)				0.50	2.50	
Swimming pools (pool and deck area)				0.50	2.50	Higher values may be required for humidity control.
Playing floors (gymnasium)	30	20	10			
Ballrooms and discos	100	25	13			
Bowling alleys (seating areas)	70	25	13			
Theaters						
Ticket booths	60	20	10			Special ventilation will be needed to eliminate special stage effects (e.g., dry ice vapors, mists, etc.)
Lobbies	150	20	10			
Auditorium	150	15	8			
Stages, studios	70	15	8			
Transportation						Ventilation within vehicles may require special consideration.
Waiting rooms	100	15	8			
Platforms	100	15	8			
Vehicles	150	15	8			
Workrooms						Spaces maintained at low temperatures, −10 to +50 F (−23° to +10°C) are not covered by these requirements unless the occupancy is continuous. Ventilation from adjoining
Meat processing	10	15	8			

TABLE 4.25 **Recommended Outdoor Air Requirements for Ventilation**[a] **(*Continued*)**

Part A. ***Commercial Facilities (Offices, Stores, Shops, Hotels, Sports Facilities, etc.) (Continued)***

Application	*Estimated Occupancy*[b] *(Persons per 1000 ft^2 or 100 m^2 Floor Area): Use Only When Design Occupancy Is Not Known*	*Outdoor Air Requirements* *cfm per person*	*L/s per person*	*cfm per ft^2*	*L/s per m^2*	*Comments*
						spaces is permissible. When the occupancy is intermittent, infiltration will normally exceed the ventilation requirement.
Photo studios	10	15	8			
Darkrooms	10			0.50	2.50	
Pharmacy	20	15	8			
Bank vaults	5	15	8			
Duplication printing				0.50	2.50	Installed equipment must incorporate positive exhaust and control (as required) of undesirable contaminants (toxic or otherwise).

Part B. ***Institutional Facilities***

Application	*Estimated Max. Occupancy*[b] *(Person per 1000 ft^2 or 100 m^2*	*Outdoor Air Requirements* *cfm per person*	*L/s per person*	*cfm per ft^2*	*L/s per m^2*	*Comments*
Education						
Classroom	50	15	8			
Laboratories	30	20	10			Special contaminant control systems may be required for processes or functions

						including laboratory animal occupancy.
Training shop	30	20	10			
Music rooms	50	15	8			
Libraries	20	15	8			
Locker rooms				0.50	2.50	
Corridors				0.10	0.50	
Auditoriums	150	15	8			
Smoking lounges	70	60	30			Normally supplied by transfer air, local mechanical exhaust. Exhaust with no recirculation recommended.
Hospitals, Nursing and Convalescent Homes						
Patient rooms	10	25	13			Special requirements or codes and pressure relationships may determine minimum ventilation rates and filter efficiency.
Medical procedure	20	15	8			
Operating rooms	20	30	15			Procedures generating contaminants may require higher rates.
Recovery and ICU	20	15	8			
Autopsy rooms				0.50	2.50	Air shall not be recirculated into other spaces.
Physical therapy	20	15	8			
Correctional Facilities						
Cells	20	20	10			
Dining halls	100	15	8			
Guard stations	40	15	8			

TABLE 4.25 **Recommended Outdoor Air Requirements for Ventilation**[a] ***(Continued)***

Part C. ***Residential Facilities (Private Dwellings, Single, Multiple)***

Application	*Outdoor Air Requirements*[c]	*Comments*
Living areas	0.35 air changes per hour but not less than 15 cfm (7.5 L/s) per person	For calculating the air changes per hour, the volume of the living areas shall include all areas within the conditioned space. The ventilation is normally satisfied by infiltration and natural ventilation. Dwellings with tight enclosures may require supplemental ventilation supply for fuel-burning appliances, including fireplaces and mechanically exhausted appliances. Occupant loading shall be based on the number of bedrooms as follows: first bedroom, two persons; each additional bedroom one person. Where higher occupancies are known, they should be used.
Kitchens[d]	100 cfm (50 L/s) intermittent —or— 25 cfm (12 L/s) continuous —or— Openable windows	Installed mechanical exhaust capacity.[e]
Baths, toilets[d]	50 cfm (25 L/s) intermittent —or— 25 cfm (12 L/s) continuous —or— Openable windows	Installed mechanical exhaust capacity.[e]
Garages		
Separate for each dwelling unit	100 cfm (50 L/s) per car	Normally satisfied by infiltration or natural ventilation.
Common for several units	1.5 cfm/sq ft (7.5 L/s/m^2)	See "Enclosed Parking Garages," Part A.

Source: Reprinted by permission from ASHRAE Standard 62-1989, *Ventilation for Acceptable Indoor Air Quality,* published by the American Society of Heating, Refrigerating, and Air-Conditioning Engineers, Inc., Atlanta, Ga.

[a] This table prescribes supply rates of acceptable outdoor air required for acceptable indoor air quality. These values have been chosen to control carbon dioxide and other contaminants with an adequate margin of safety, and to account for health variations among people, varied activity levels, and a moderate amount of smoking.

[b] Net occupiable spaces.

[c] The outdoor air is assumed to be acceptable.

[d] Climatic conditions may affect choice of ventilation option chosen.

[e] The air exhausted from kitchens, bath and toilet rooms may utilize air supplied through adjacent living areas to compensate for the air exhausted. The air supplied shall be of sufficient quantity to meet the requirements of this table.

TABLE 4.26 **Estimated Overall Infiltration Rates for Small Buildings**

Part A. Construction Types

Construction Type	*Description*
Tight	New buildings where there is close supervision of workmanship and special precautions are taken to prevent infiltration. Descriptions for tight windows and doors are given in Table 4.27.
Medium	Building is constructed using conventional construction procedures. Medium-fitting windows and doors are described in Table 4.27.
Loose	Buildings constructed with poor workmanship or older buildings where joints have separated. Loose windows and doors are described in Table 4.27.

Part B. Design Infiltration Rate (ACH) for Winter; Heating; Wind Speed = 15 mph

Type of Construction	*Winter Outdoor Design Temperature (F)*									
	50	*40*	*30*	*20*	*10*	*0*	*−10*	*−20*	*−30*	*−40*
Tight	0.4	0.5	0.6	0.6	0.7	0.8	0.8	0.9	0.9	1.0
Medium	0.6	0.7	0.8	0.9	1.0	1.1	1.2	1.2	1.3	1.4
Loose	0.8	0.9	1.0	1.2	1.3	1.4	1.5	1.6	1.8	1.9

TABLE 4.26 **Estimated Overall Infiltration Rates for Small Buildings (*Continued*)**

Part C. ***Design Infiltration Rate (ACH) for Summer; Cooling; Wind Speed = 7.5 mph***

Type of Construction	Summer Outdoor Design Temperature (F)					
	85	90	95	100	105	110
Tight	0.3	0.3	0.3	0.4	0.4	0.4
Medium	0.4	0.4	0.5	0.5	0.5	0.6
Loose	0.4	0.5	0.6	0.6	0.7	0.8

Part D. ***Infiltration per Square Foot of Floor Area***

Ceiling Height (ft)	Air Changes per Hour																	
	0.3	0.4	0.5	0.6	0.7	0.8	0.9	1.0	1.1	1.2	1.3	1.4	1.5	1.6	1.7	1.8	1.9	2.0
	cfm/ft²																	
7.5	0.04	0.05	0.06	0.08	0.09	0.10	0.11	0.13	0.14	0.15	0.16	0.18	0.19	0.20	0.21	0.23	0.24	0.25
8	0.04	0.05	0.07	0.08	0.09	0.11	0.12	0.13	0.15	0.16	0.17	0.19	0.20	0.21	0.23	0.24	0.26	0.27
8.5	0.04	0.06	0.07	0.09	0.10	0.11	0.13	0.14	0.16	0.17	0.18	0.20	0.21	0.23	0.24	0.26	0.27	0.28
9	0.05	0.06	0.08	0.09	0.11	0.12	0.14	0.15	0.17	0.18	0.20	0.21	0.23	0.24	0.26	0.27	0.29	0.30
	Btu/h ft² F																	
7.5	0.04	0.05	0.07	0.08	0.09	0.11	0.12	0.14	0.15	0.16	0.18	0.20	0.20	0.22	0.23	0.24	0.26	0.27
8	0.04	0.06	0.07	0.09	0.10	0.12	0.13	0.14	0.16	0.17	0.19	0.22	0.22	0.23	0.24	0.26	0.27	0.29
8.5	0.05	0.06	0.08	0.09	0.11	0.12	0.14	0.15	0.17	0.18	0.20	0.23	0.23	0.24	0.26	0.28	0.29	0.30
9	0.05	0.06	0.08	0.10	0.11	0.13	0.15	0.16	0.18	0.19	0.21	0.24	0.24	0.26	0.28	0.29	0.31	0.32

Source: Copyright © American Society of Heating, Refrigerating and Air Conditioning Engineers, Inc., Atlanta, GA. Reprinted by permission from *Cooling and Heating Load Calculation Manual,* 1979.

TABLE 4.27 **Approximate Infiltration Through Doors and Windows of Small Buildings**

*Part A. **Converting Wind Speed to Velocity Head Factor***

NOTE: Typical design assumptions:
Winter wind V_w = 15 mph = velocity head factor of 0.105
Summer wind V_w = 7.5 mph = velocity head factor of 0.028

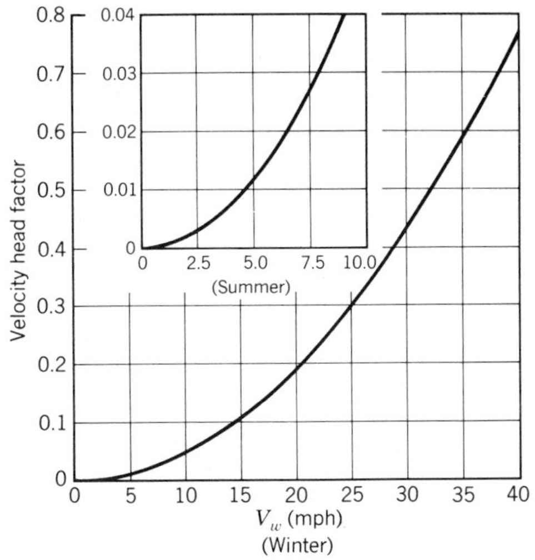

*Part B. **Infiltration Rates for Velocity Head Factors***

NOTE: Enter this graph with velocity head factor (from Part A) to find infiltration rate in cfm/ft of crack (using values of *k* found in Part C or D).

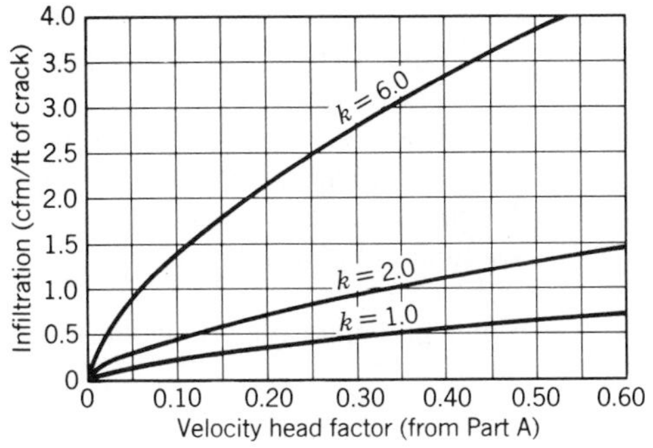

TABLE 4.27 **Approximate Infiltration Through Doors and Windows of Small Buildings (*Continued*)**

*Part C. **Classifications of Windows for Infiltration***

Window Fit	Wood Double-Hung (Locked)	Other Types
Tight, $k = 1.0$	Weather-stripped; average gap (1/64-in. crack)	Wood casement and awning windows; weather-stripped.
		Metal casement windows; weather-stripped.
Average, $k = 2.0$	Nonweather-stripped; average gap (1/64-in. crack) or weather-stripped; large gap (3/32-in. crack)	All types of vertical and horizontal sliding windows; weather-stripped. If average gap (1/64-in. crack), this could be tight-fitting window.
		Metal casement windows; non-weather-stripped. If large gap (3/32-in. crack), this could be a loose-fitting window.
Loose, $k = 6.0$	Non-weather-stripped; large gap (3/32-in. crack)	Vertical and horizontal sliding windows; non-weather-stripped.

*Part D. **Classification of Residential-type Doors for Infiltration***

Door Fit	Comments
Tight, $k = 1.0$	Very small perimeter gap and perfect fit weather-stripping—often characteristic of new doors
Average, $k = 2.0$	Small perimeter gap having stop trim fitting properly around door; weather-stripped
Loose, $k = 6.0$	Large perimeter gap having poor fitting stop trim; weather-stripped or Small perimeter gap; no weather-stripping

Source: Copyright © American Society of Heating, Refrigerating and Air Conditioning Engineers, Inc., Atlanta, GA. Reprinted by permission from *Cooling and Heating Load Calculation Manual,* 1979.

5

DESIGNING FOR HEATING AND COOLING

Chapters 1 to 4 have provided the necessary preparation for design: a perspective on national energy resources, an understanding of human comfort, an analysis of the climate resources available on site, a list of the general design strategies appropriate to various climates, and a discussion of the basics of heat transfer calculations. Hence, we now know much about the variables outside a building, about the desired conditions inside, and about individual components of a building's skin. In this chapter, all these variables are integrated into the process of designing for heating and cooling with the important related factor of daylighting. The design of mechanical support systems is introduced in Chapter 6.

Chapter 5 has two major parts: Sections 5.1 to 5.3 accompany preliminary design, where quick decisions are made based on broad guidelines. Sections 5.4 to 5.9 accompany final design, where more detail is known and more accuracy is appropriate.

5.1 Organizing the Problem

How should your building's skin respond to the sometimes conflicting needs for heating, cooling, and daylighting? Heating and cooling design strategies were related to climate in Sections 2.3 and 2.4, in which it became evident that internal heat gains can shift the appropriate design strategies for a building toward cooling, perhaps eliminating heating needs entirely. Typically, much of this internal heat is provided by electric lighting. Daylighting can replace electric lighting for most of the typical working day in most building types—*if* the building is designed to allow daylight to reach most of the interior. Daylighting is accompanied by large glass areas, which increase a building's heating needs in winter; in such buildings, however, less heat from electric lighting is available to fill those needs. Another complication is that adequate daylight requires much larger glass areas under dim winter skies than under bright summer skies. If the glass area is sized for winter daylighting, then excessive daylight—and along with it, excessive heat—will be admitted in summer. With proper controls, daylighting can reduce summer cooling loads, relative to electric lights, but it will usually increase winter heating loads. Where passive solar heating or surplus heat from another source is readily available, this trade-off is attractive. One of the earliest and most difficult questions for the designer is what relative weights to give to considerations for heating, cooling, and daylighting.

The most desirable balance among the energy needs for heating, cooling, and daylighting will vary both by building type and by climate. Table 5.1 shows the usual ranking of these three energy usages, as well as of power and process energy use and domestic hot water. The rankings are shown for six typical building types and three categories of winter heating needs. The table indicates only rank, not size; ranks 1 and 2 may be very close in size, or quite different.

The ranking of probable energy use for your building in its climate is the first step toward the design of the building's envelope and mechanical support systems. In the next step, we look at the building as a collection of thermal conditions, or zones.

The thermal zoning of a building recognizes that different envelope and support systems may be required within the building. The more carefully zoning is considered in these early design stages, the better will be the thermal performance and the lower will be the annual energy

TABLE 5.1 **Ranking the Energy Use in Buildings**

Rank 1 = most energy used in typical building of this type, relative to other uses

	Categories of Energy Usage				
Building Types	*Heating and Ventilating*	*Cooling and Ventilating*	*Lighting*	*Power and Process*[a]	*Domestic Hot Water*
Schools					
A	4	3	1	5	2
B	1	4	2	5	3
C	1	4	2	5	3
Colleges					
A	5	2	1	4	3
B	1	3	2	5	4
C	1	5	2	4	3
Office buildings					
A	3	1	2	4	5
B	1	3	2	4	5
C	1	3	2	4	5
Stores					
A	3	1	2	4	5
B	2	3	1	4	5
C	1	3	2	3	5
Religious buildings					
A	3	2	1	4	5
B	1	3	2	4	5
C	1	3	2	4	5
Hospitals					
A	4	1	2	5	3
B	1	3	4	5	2
C	1	5	3	4	2

Source: F. Dubin and G. C. Long, *Energy Conservation Standards for Building Design, Construction and Operation,* © 1978, McGraw-Hill, New York.

[a]Extensive use of computer terminals in workplaces may increase the ranking of this category.

A = climate with fewer than 2500 heating degree days.
B = climate with 2500 to 5500 degree days.
C = climate with 5500 to 9500 degree days.
(See Appendix B for degree day information.)

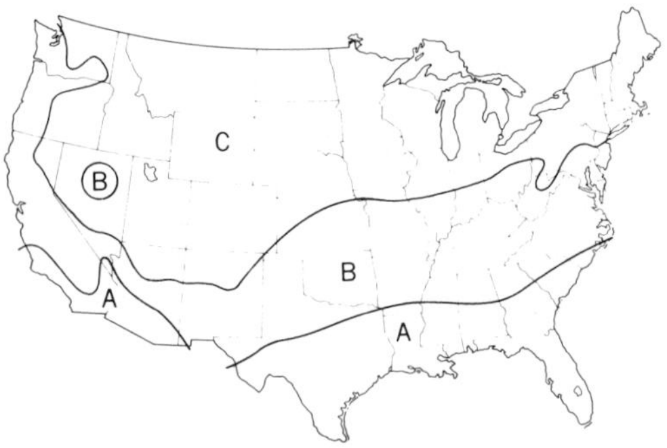

consumption. (Also, the less likely it will be that all sides of a building will have an identical appearance.)

Zoning is most often influenced by the following factors.

1. *Function.* Particularly important because of the variations in internal heat gains between functions, this factor may also influence the vertical organization of a building, as was shown in Fig. 3.1. Comfort conditions may vary considerably between functions; air temperatures can be lower for a strenuous activity than for a sedentary activity, or heat tolerance may be greater for some activities (kitchens) than for others.

2. *Schedule.* This factor, closely related to function, can influence both the envelope and the support system. An activity scheduled only between 9 A.M. and 4 P.M. can often be entirely

daylit, at a time when the outside temperatures are the warmest of the daily cycle. By contrast, activity that takes place only from 9 P.M. to 4 A.M. will be entirely dependent on electric lighting, whose heat can be used to overcome the chill of the outside temperatures during these hours in winter. (In the summer, such heat can be flushed away with the cool outside night air in many U.S. locations.) Support systems are often divided by scheduling considerations: if one activity has operating hours different from those of the remainder of the building, a separate mechanical system is often provided. This saves energy, because large equipment will not be underused to provide heating or cooling for only one zone.

3. *Orientation*. The degree of exposure to daylight, direct sun, and wind is obviously important to thermal zoning. Consider a block-square office building on a cold, sunny, and windy day. Perimeter spaces with direct sun through the windows may gain more heat than is lost, and thus need cooling. This might be done by the opening of windows, but too much cold air (especially on the windy side of a building) may make the workers near the windows uncomfortable. Perimeter spaces without direct sun may have a net heat loss, due to heat loss through glass, infiltration, and a lack of electric lights (since daylight is adequate). These spaces will need heat from a mechanical support system. Interior (no daylight) spaces are overheated by electric lights, since they cannot lose heat. These spaces will need cooling from the support system.

Zoning considerations are among the most important influences on building form and external design, along with the familiar esthetic, social, legal, economic, and technical influences that combine in a tug-of-war familiar to the building designer. Some of the most basic design concepts related to thermal zoning are building form, building envelope, and support system design.

Building form, at its simplest, can be reduced to questions of tall or short, thick or thin. Figures 5.1 and 5.2 compare these form variations to the relative importance of heating, cooling, and daylighting. It is apparent that thicker, taller buildings have more floor space away from climate influences; being electrically lit rather than daylit, they generate heat and need cooling all year. These buildings are called *internal load dominated* (ILD) (Fig. 5.1). In contrast, thinner buildings in which nearly all spaces have an exterior wall will need heating in cold weather and cooling in hot weather; electric lights by day are largely unnecessary. These buildings are called *skin load dominated* (SLD) (Fig. 5.2).

By comparing the rankings of energy use by function and climate given in Table 5.1 with the forms shown in Figs. 5.1 and 5.2, we can identify a range of design opportunities and their attendant thermal problems. The ultimate choice of building form is determined by a combination of design issues; the energy use issue can help in the selection process. Once the building form has been chosen, the functions can be distributed according to typical architectural criteria, including the thermal zoning considerations mentioned above.

In the selection of a building form, some particularly important questions accompany each energy use. Because the question of daylighting versus electric lighting frequently is so influential in determining whether heating or cooling will be the dominant concern, we begin with daylighting.

For *daylighting:*

1. What will be the relative emphasis on sidelighting (characterized by uneven distribution and glare in the visual field but little glare on horizontal surfaces) and toplighting (the reverse characteristics)?
2. What role will direct sun play in daylighting? In winter, can heat without glare be admitted?
3. How can seasonal adjustments be made in the size of daylight openings?
4. To what extent will daily changes in daylighting control be necessary?
5. How can adequate daylight be admitted in an even way, such that unwelcome dark-appearing places are avoided?

For *heating:*

1. Can the sun be used to heat spaces? If so, how will south wall design be affected?
2. How can openings in walls facing other

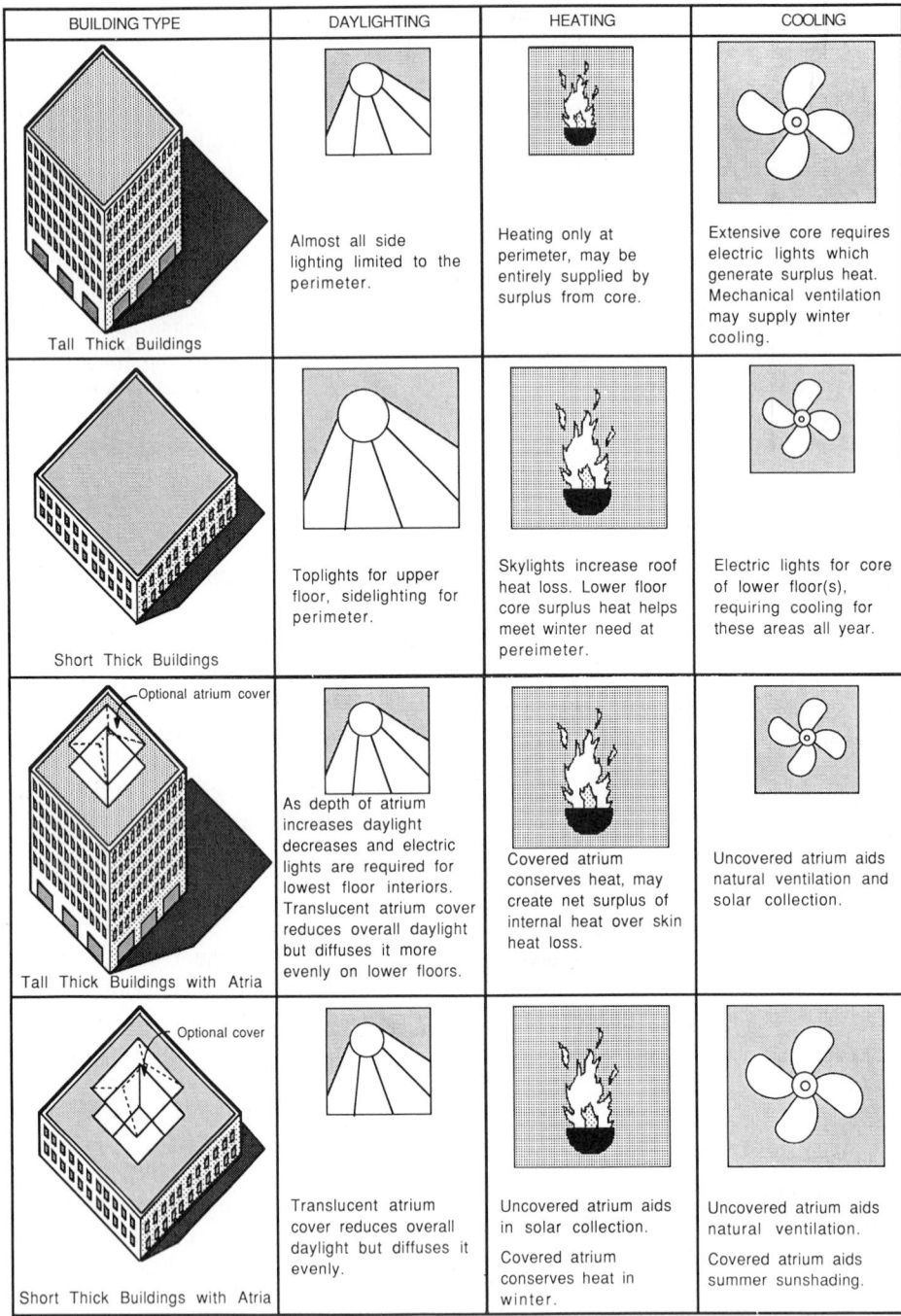

BUILDING TYPE	DAYLIGHTING	HEATING	COOLING
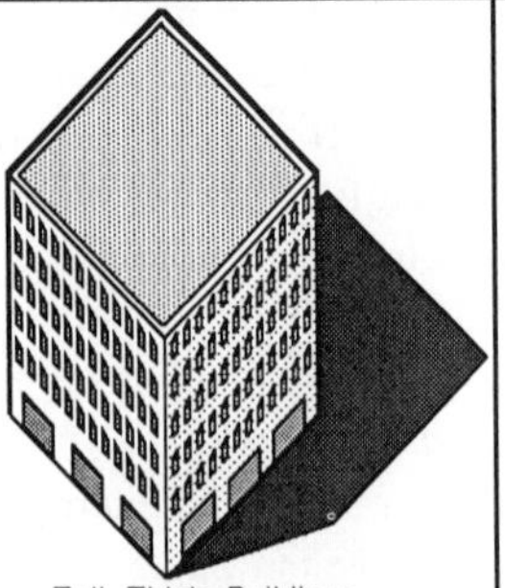 Tall Thick Buildings	 Almost all side lighting limited to the perimeter.	 Heating only at perimeter, may be entirely supplied by surplus from core.	 Extensive core requires electric lights which generate surplus heat. Mechanical ventilation may supply winter cooling.
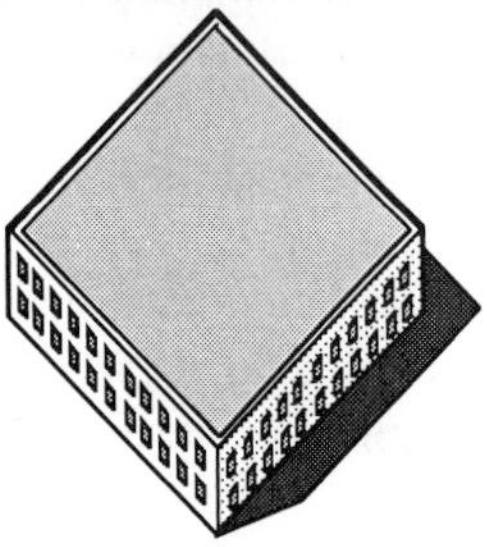 Short Thick Buildings	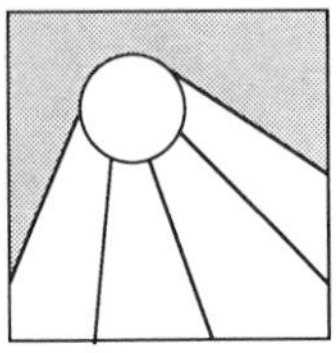 Toplights for upper floor, sidelighting for perimeter.	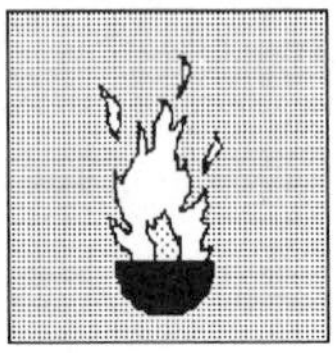 Skylights increase roof heat loss. Lower floor core surplus heat helps meet winter need at pereimeter.	 Electric lights for core of lower floor(s), requiring cooling for these areas all year.
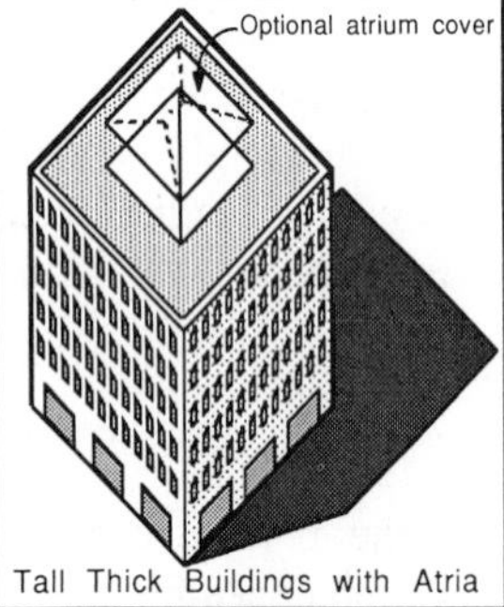 Tall Thick Buildings with Atria	 As depth of atrium increases daylight decreases and electric lights are required for lowest floor interiors. Translucent atrium cover reduces overall daylight but diffuses it more evenly on lower floors.	 Covered atrium conserves heat, may create net surplus of internal heat over skin heat loss.	 Uncovered atrium aids natural ventilation and solar collection.
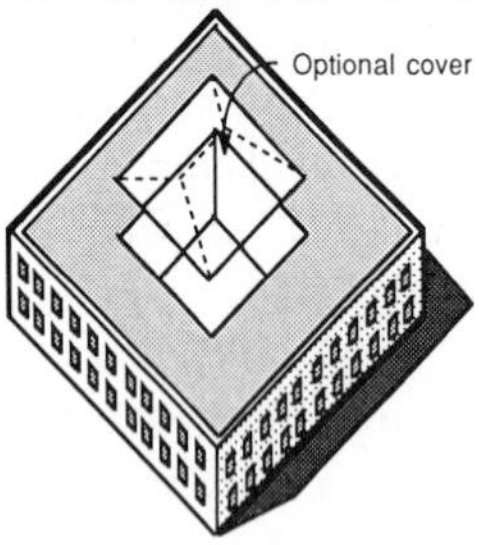 Short Thick Buildings with Atria	 Translucent atrium cover reduces overall daylight but diffuses it evenly.	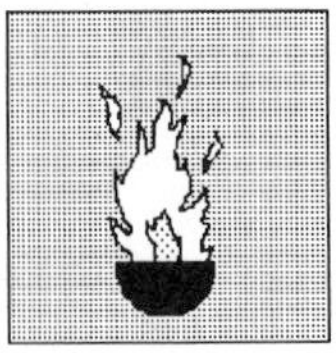 Uncovered atrium aids in solar collection. Covered atrium conserves heat in winter.	 Uncovered atrium aids natural ventilation. Covered atrium aids summer sunshading.

Fig. 5.1 Internal-load-dominated (ILD) buildings: the relative importance of cooling, daylighting, and heating varies with building form, as do opportunities for the use of on-site or natural energy sources (see also Table 5.1).

directions be kept to a minimum without daylight being shut out?

3. What role will direct sun through south glass or skylights play in daylighting?
4. How can daylight be admitted but the chilling effects of large, cold glass surfaces be minimized?
5. How can incoming fresh air be warmed

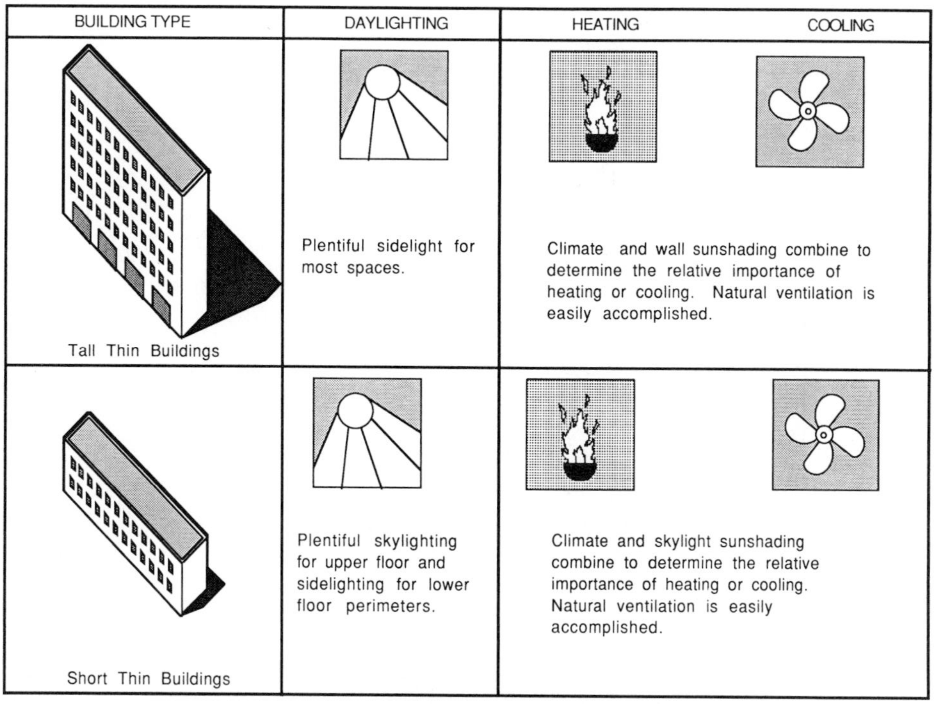

Fig. 5.2 *Skin-load-dominated (SLD) buildings: the relative importance of cooling, daylighting, and heating varies with building form, as do opportunities for the use of on-site or natural energy sources (see also Table 5.1).*

before it chills the people sitting near the fresh air opening?

6. Is there surplus heat elsewhere in the building that can be used to help warm perimeter spaces?

For *cooling:*

1. Will the strategy be to open the building to breeze or to close the building for "coolth" retention, or to use a combination of these alternatives (open by night, closed by day)?
2. How can direct sun be kept out of the building? Can east and west windows be minimized and adequate daylight still provided?
3. How can adequate daylight be admitted for winter conditions without overlighting (and thus overheating) for summer conditions?
4. When can cooling be provided by outdoor air, rather than by a refrigeration cycle?
5. Can the operation of refrigeration machinery be concentrated during the coldest (nighttime) hours?
6. How can incoming fresh air be cooled before it warms the people sitting near the fresh air opening?
7. Can the structure of the building be used to absorb heat by day, then be flushed with night air in climates with cool nights?

Building envelope, the next design step, involves relating the climate to the perimeter of the building zones in the design of the building's skin. Each skin element provides an opportunity for thermal and luminous exchange between inside and out; heating, cooling, ventilating, and daylighting devices can be mixed as needed. Figure 5.3 shows some of the most common of these devices for varying orientations. Section 5.2 gives numerical criteria for sizing these skin elements.

Support systems are considered in detail in the next chapter. At this design stage, the most important concept is that of the "distribution

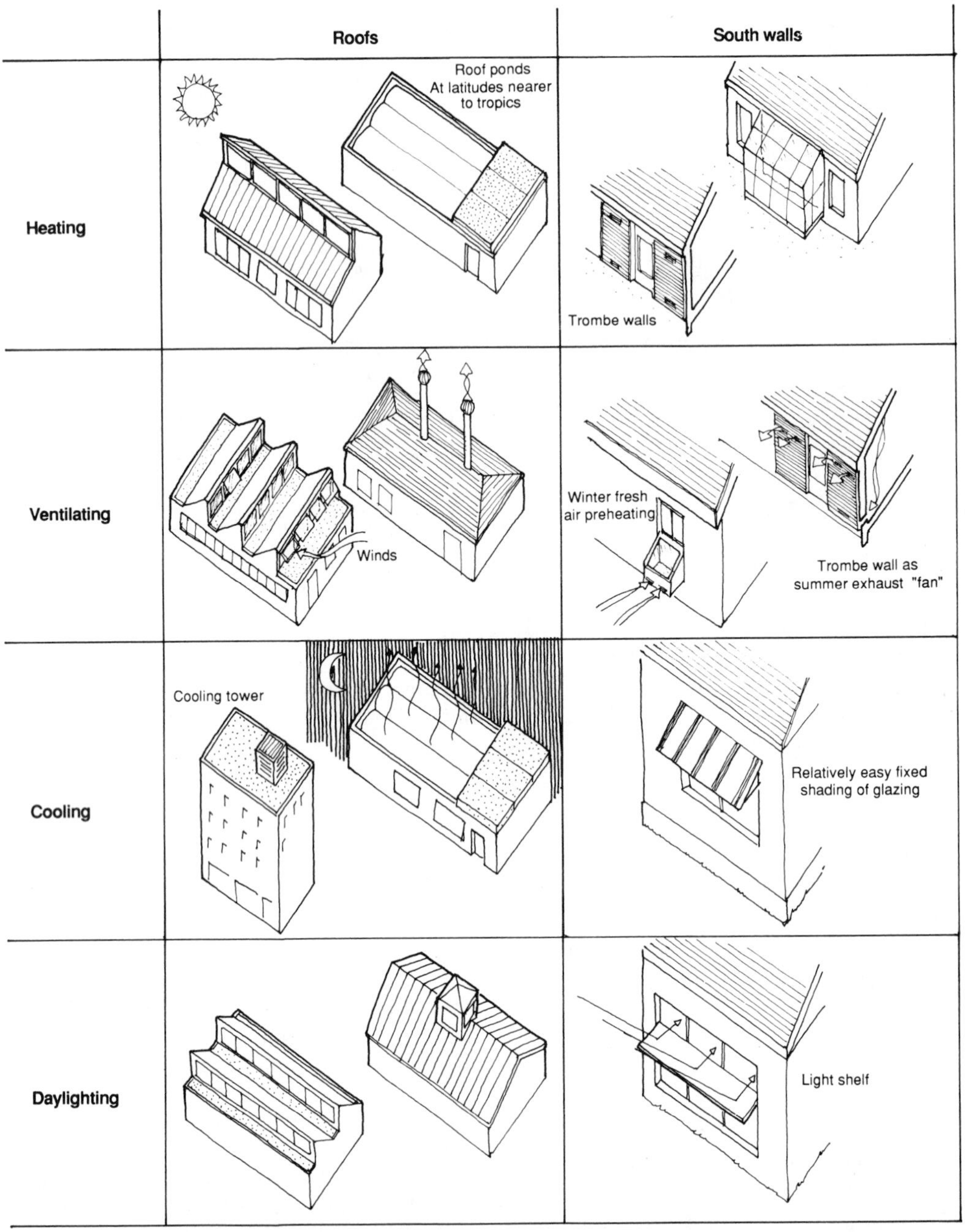

Fig. 5.3 *The components of a building's envelope can be used both to conserve energy and to admit on-site or natural energy sources. Some of the many opportunities for heating, ventilating, cooling, and daylighting are shown here.*

tree.'' Mechanical support systems often produce heating and cooling in one place, then distribute them to other building spaces according to respective needs. The distribution tree is the means for delivering heating and cooling: the ''roots'' are the machines that provide heat and cold, the ''trunk'' is the main duct or pipe from the mechanical equipment to the zone to be

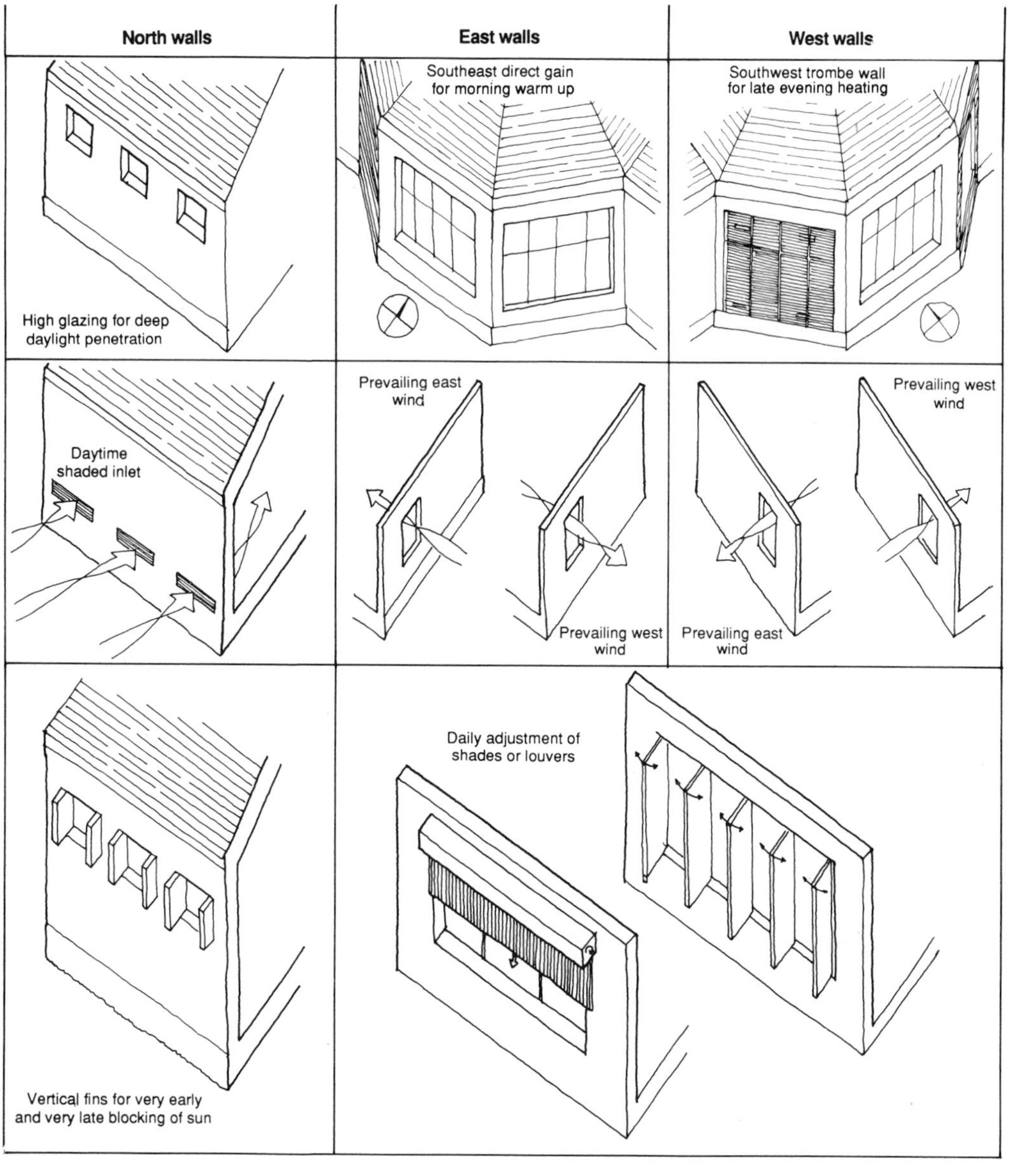

Fig. 5.3 (continued)

served, and the "branches" are the many smaller ducts or pipes that lead to individual spaces.

For now, the questions to be answered about distribution trees for buildings are How many? What kind? and Where? A building can have one giant distribution tree, several medium-sized trees, or an orchard of much smaller trees. At one extreme, a large mechanical room is the scene of all heating and cooling production; leading from this room is a very large trunk duct with perhaps hundreds of branches. At the other extreme, each zone has its own mechanical equipment (such as a rooftop heat pump), with short trunks and relatively few branches on each tree.

What kind of distribution tree? Most simply, air (ducts) or water (pipes). Air distribution trees are bulky and therefore likely to have major visual impacts unless they are concealed above ceilings or within vertical chases. Water distribution trees consume much less space (a given volume of water carries vastly more heat than does the same volume of air at the same temperature) and can be easily integrated within structural members, such as columns. Both air and water trees can be sources of noise.

Where does the distribution tree fit in? On the exterior, it can lend a three-dimensional organizational structure to a facade, as can be seen in the Blue Cross–Blue Shield office building in Boston (see Fig. 7.53). Exterior trees can take up a smaller amount of rentable floor space but require expensive cladding and are subject to heat losses and gains, which could increase energy usage. Interior trees are often combined with other continuous vertical spaces, such as elevator shafts and stairways. If the choice is an exterior distribution tree, its potential contribution to facade performance should be considered. For example, the distribution tree might act as a sunshade or as a light reflector.

To carry the tree analogy to its logical conclusion, consider the "leaves," the points of interchange between the piped or ducted heating or cooling and the spaces served. One example is a large, bulky device such as a fan-coil unit on the exterior wall, below windows. In contrast, a perforated ceiling system with thousands of small holes acting as a widely spread grille is essentially invisible.

Distribution trees are rich in form possibilities; their variations are enormous. For now, assume that the choices of how many, what kind, and where have tentatively been made and that the details and variations are to be considered later. The designer who has chosen the basic form, types of envelope components, and a support system distribution tree can now consider questions of size. The remainder of this chapter deals with such questions.

5.2 Guidelines and Criteria

After approximating the degree to which lighting, heating, and cooling are necessary for your building and climate, you can begin the detailed design of the envelope. Chapter 4 presented the procedures for determining the rate at which heat flows through the many parts of the envelope; now you must make the *choice* of which components to combine for your building.

The information in this section can be used either before preliminary design, as guidelines to window sizing and roof/wall/floor insulation, or after preliminary design, as criteria for measurement of the predicted performance of a building's envelope. Several kinds of design guidelines are available for energy-conserving construction. Presented here are some of the most widely used guidelines, ranging from very rough—but quickly formulated—criteria to very detailed criteria that require computer processing.

EXAMPLE 5.1, PART A. Many of the design guidelines in this chapter will be illustrated by applying them to a typical bay of the building in Fig. 5.4. (Example 5.1 will continue to appear in parts throughout this chapter.) This 24,000-ft^2 office building for an electric utility is located in the wet-winter, dry-summer climate of the U.S. Pacific northwest. From Table 5.1 we see that Eugene, in western Oregon, is in zone B, where an office building can expect that heating will be the greatest energy user, followed by lighting, then by cooling. Energy conservation and daylighting considerations played a major role in this building's design and were integrated with solar heating, night ventilation of mass, structure, and acoustics. Table 5.2 presents the types of design statistics that will be needed for any building to which this chapter's design guidelines will be applied.

The various applications of design guidelines to this building can be found as follows:

Example 5.1, Part B. Daylighting design, page 193.

Example 5.1, Part C. Overall rate of Btu/DD ft^2 heat loss, page 200.

Example 5.1, Part D. ASHRAE/IES Standard 90.1-1989, page 201.

Example 5.1, Part E. Approximate SSF (solar savings fraction), page 211.

Example 5.1, Part F. Approximate heat gain, page 213.

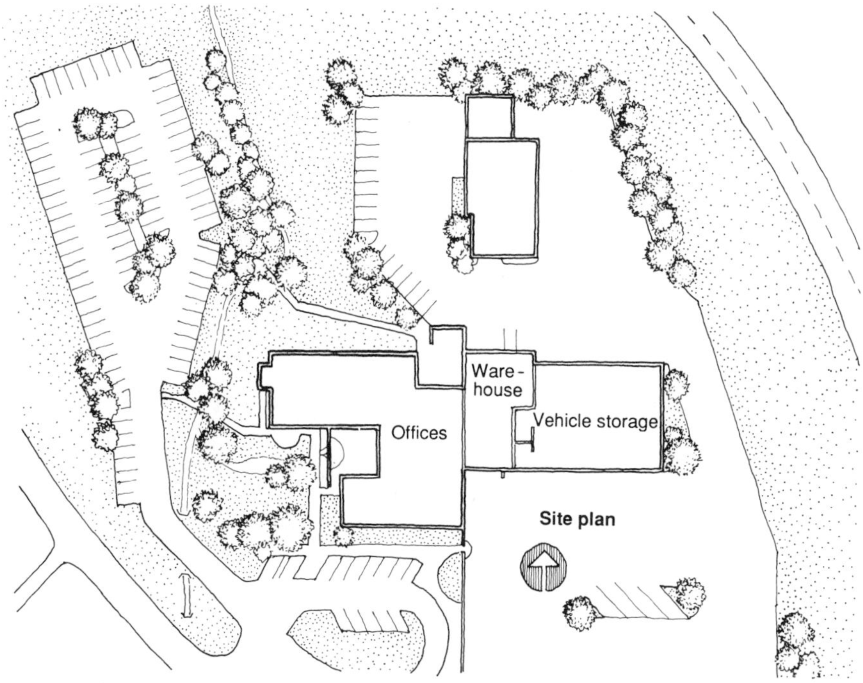

(a) Site plan

***Fig. 5.4** The office building for the Emerald People's Utility District near Eugene, Oregon, is elongated east–west for maximum window areas facing north–south* (a). *The ground floor plan* (b) *and second floor plan* (c) *and diagrammatic daylight section* (d) *show how almost all workstations are daylit. The section* (e) *shows relation between windows, trellises for shading, and exposed concrete construction. A cutaway axonometric* (f) *of a typical open-office two-story bay shows air flush through hollow cores of exposed precast concrete slabs, windows and shading, and suspended sound-absorbing baffles. Lightshelves* (g) *reduce the contrast between lots of daylight near windows, too little daylight farther in. The resulting windows are T-shaped* (h) *with more glass above lightshelves, less glass below. The deciduous vines on trellises at south facade change the building's appearance by season;* (i) *at summer solstice, the stripes will soften as vines leaf out with age;* (j) *at fall equinox the leaves will still shade, while at spring equinox the branches will be bare to allow sun to reach the windows;* (k) *at winter solstice most of the window is exposed to sun. (Drawings* c *and* g *courtesy of Virginia Cartwright.) (Courtesy of Equinox Design, Inc., and WEGroup, PC, Architects, Eugene Oregon.)*

Example 5.1, Part G. Cross ventilation, page 216.

Example 5.1, Part H. Night ventilation of thermal mass, page 220.

Example 5.1, Part I. Approximate sizes of equipment, page 228.

Example 5.1, Part J. January balance point temperature, page 232.

Example 5.1, Part K. Annual SSF, page 237.

Example 5.1, Part L. Clear January day temperature swing, page 246.

Example 5.1, Part M. Detailed night-cooling calculation, page 266.

Example 7.1. Detailed duct sizing, page 431.

(a) Daylighting. When a building is designed to rely heavily on daylighting, a prime design concern is the *daylight factor* (DF),

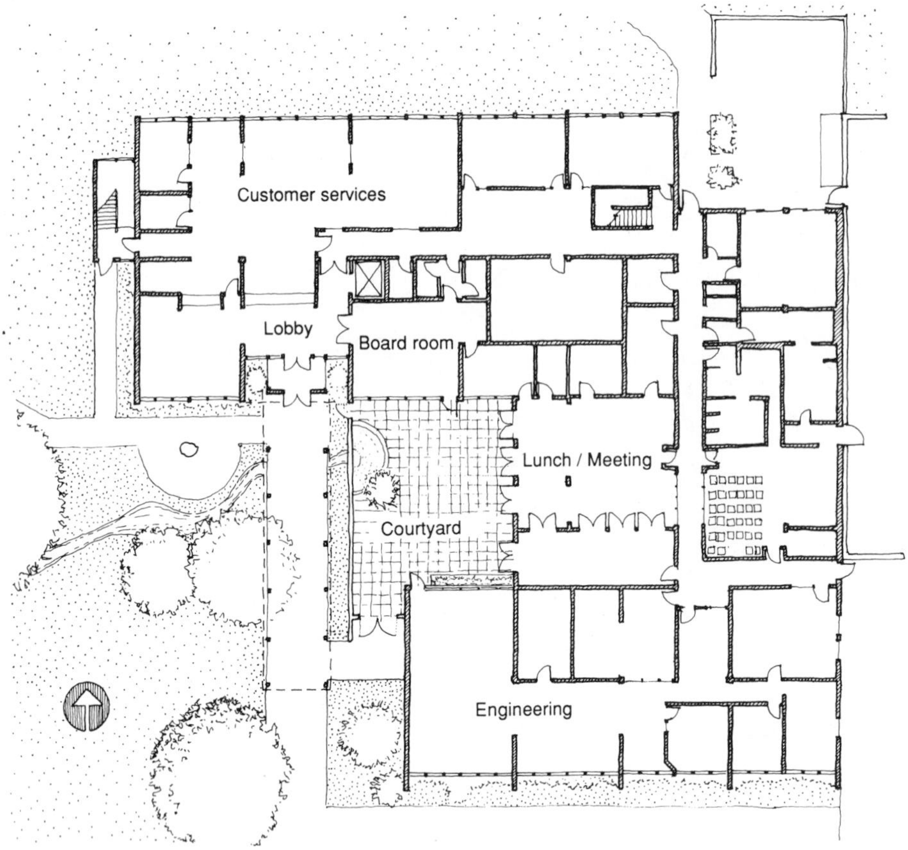

(*b*) Ground floor plan

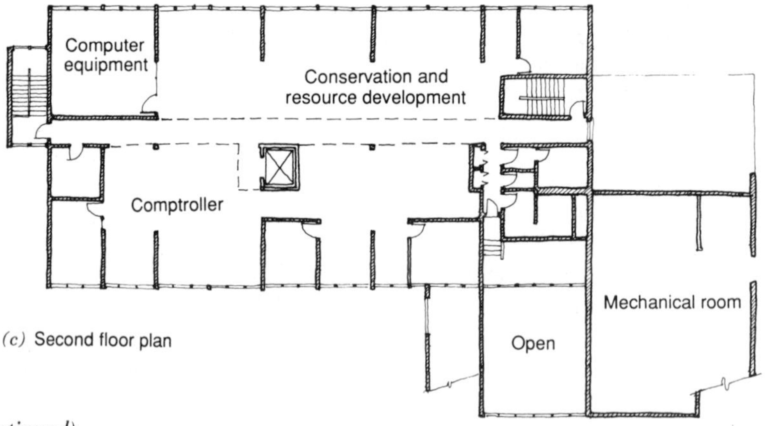

(*c*) Second floor plan

Fig. 5.4 *(continued)*

which is expressed as a percentage of the outdoor light that is available indoors.

$$DF = \frac{\text{indoor illumination from daylight}}{\text{outdoor illumination}} \times 100\%$$

Chapter 19 presents detailed procedures for determining the DF for any point within a building. For now, some simple target DFs are presented in Table 5.3. Table 5.4 gives the simplest design rules of thumb that yield these DF results. Typi-

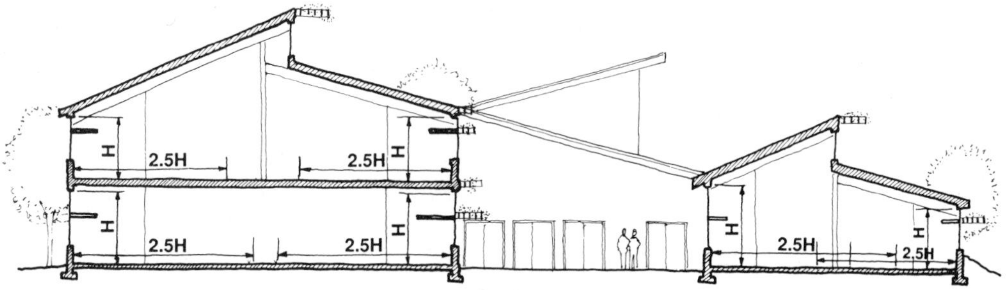

(d) Daylighting cross section (south to right)

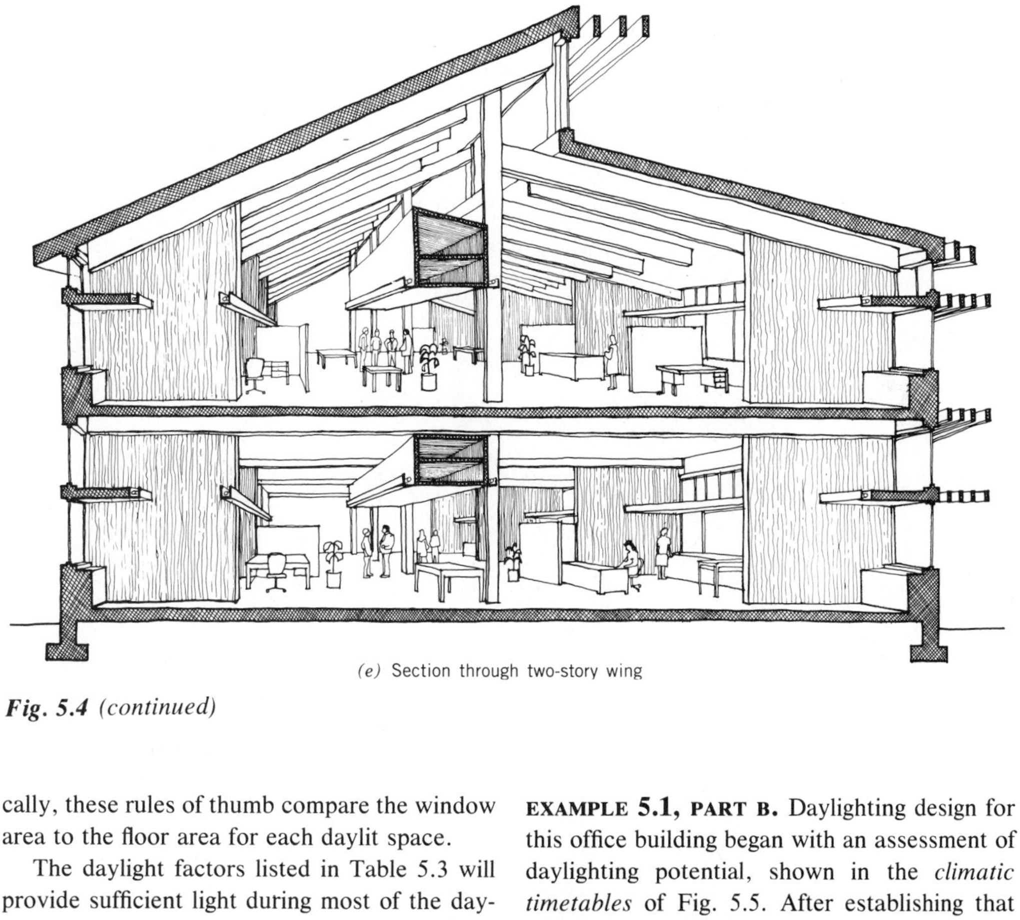

(e) Section through two-story wing

Fig. 5.4 *(continued)*

cally, these rules of thumb compare the window area to the floor area for each daylit space.

The daylight factors listed in Table 5.3 will provide sufficient light during most of the daylight hours on overcast winter days. Obviously, much more light will be available on summer days—more light than is needed, bringing heat along with it. When sizing windows and skylights, remember that controlling direct sun is necessary and that less opening area is needed in summer than in winter. Further details on sunshading and daylighting control devices can be found in Section 6.4.

EXAMPLE 5.1, PART B. Daylighting design for this office building began with an assessment of daylighting potential, shown in the *climatic timetables* of Fig. 5.5. After establishing that daylighting was available during almost all normal working hours, the building plan was then organized so that almost all windows faced either south or north (Fig. 5.4) to avoid the problems of low-altitude sun (year-round glare and summer heat gain) that accompany east- and west-facing windows. Then the building section was designed so that the height of the windows

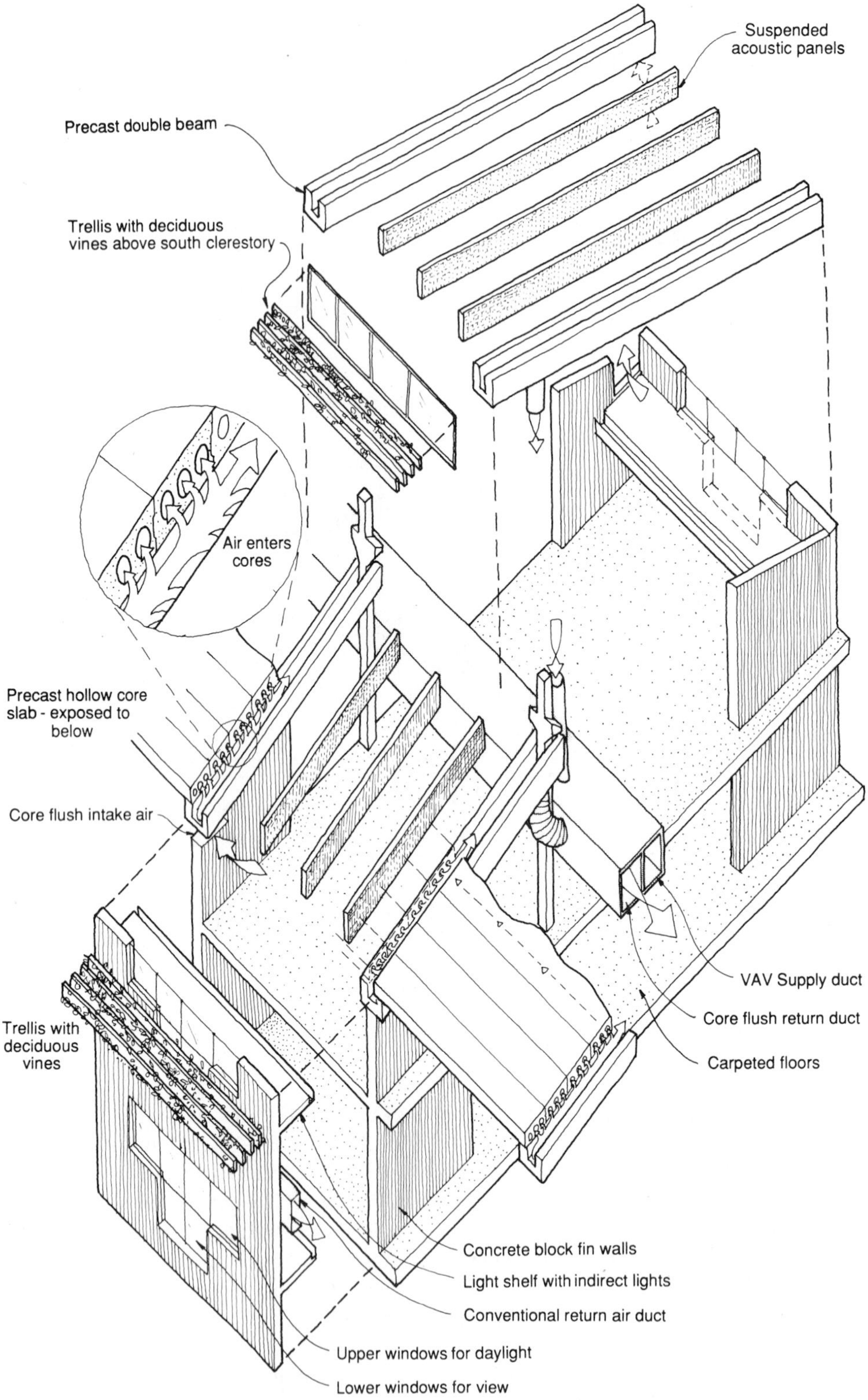

(*f*) Axonometric of a typical two-story bay

Fig. 5.4 *(continued)*

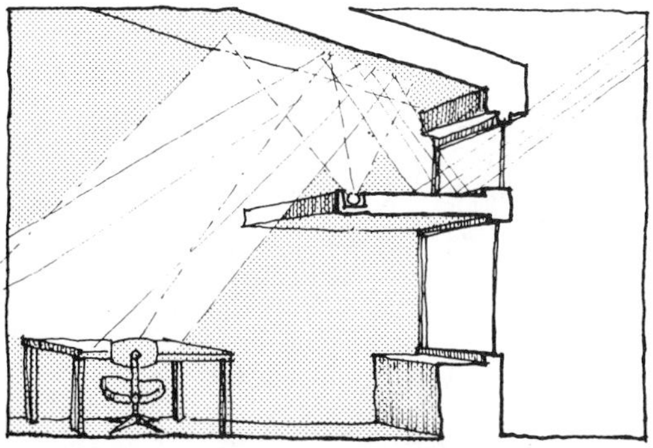

(g) Lightshelves

(h) North Façade

Fig. 5.4 (continued)

(H) was related to the depth of the floor plan served by those windows ($2.5H$), shown in Fig. 5.4*d*.

After choosing a generous target $DF_{av} = 4.0\%$ from Table 5.2, the next step was to size the windows and clerestories (as skylights), using the DF rules of thumb from Fig. 5.3. Applicable formulas for both sidelighting and for vertical monitor skylights:

$$DF_{av} = 0.2 \frac{\text{window (or skylight) area}}{\text{floor area}}$$

Applied to the entire typical bay,

$$DF_{av} = 0.2 \frac{97 \text{ ft}^2 \text{ north} + 97 \text{ ft}^2 \text{ south} + 132 \text{ ft}^2 \text{ clerestory}}{1440 \text{ ft}^2}$$

$$= 0.045, \text{ or } 4.5\%$$

slightly above the target DF_{av} of 4.0%.

Note that in this example, DF_{min} from the sidelighting will occur near the center of the building, about the point where the most light will be available from the skylight. Therefore, relatively even daylighting distribution is expected. This is helped by the use of lightshelves and the T-shaped windows, shown in Fig. 5.4*g* and *h*.

(i) South Façade at Summer Solstice

(j) South Façade at the Spring Equinox

(k) South Façade Near the Winter Solstice

Fig. 5.4 *(continued)*

TABLE 5.2 **Design Data for the Office Building in Fig. 5.4**

Component of Envelope	*U Value (Btu/h-ft²-°F)*	*Area (ft²)*	*U × A (Btu/h-°F)*
For the typical second-floor bay shown in Fig. 5.4*f*:			
Roof[a]	0.028	1512	42
Exterior opaque walls[b] (north and south)	0.084	296	25
North window[c]	0.74	97	72
South window[c]	0.74	97	(72)[d]
South clerestory[c]	0.74	132	(98)[d]
Floor	[e]	1440	
Exposed interior thermal mass (ceiling)	[f]	1512	
(bearing walls, beams, and columns)	[f]	720	
Volume of the enclosed space is 22,900 ft³.			
Occupancy is 8 persons.			
Ventilation is provided at a rate of 20 cfm per person.			
Openable window area: 16 ft² on north, 16 ft² on south.			

[a] *R*-40 over concrete precast slab; assume that metal roof thin supports are half as thermally bridging as those in Table 4.8, so corrected $R = 35$ and $U = 0.28$.

[b] 4-in. concrete block outside, 1-in. airspace, *R*-19 batt with metal studs at 24 in. and gypsum inside; corrected *R*-8.6 (see Table 4.8); approx. $\Sigma R = 11.9$ and $U = 0.084$.

[c] Clear double glazed, ⅜-in. airspace, non-thermal-break aluminum, product type C (see Table 4.16).

[d] $U \times A$ is usually not calculated for south glass area in passively solar heated buildings (see Section 5.2*c*).

[e] The carpeted floor surface, over the occupied ground floor, is not involved in heat exchange with the exterior or with heat storage on the interior.

[f] *U* value is not stated here because heat exchange with the exterior is not relevant; see later, Example 5.1, Parts E, H, K, and M.

(b) Electric Lighting. To control the electricity usage (and the building overheating that frequently results) that accompanies electric lights, a "power budget" is established. This budget sets a *maximum* of installed lighting, expressed in watts of electric lighting per square foot of floor area. Several state building codes specify lighting power budgets; check the applicable code before using the recommendations given in Appendix K, Table K.2, or Chapter 20.

TABLE 5.3 **Recommended Daylight Factors**[a]

Task	*DF*
Ordinary seeing tasks, such as reading, filing, and easy office work	1.5–2.5%
Moderately difficult tasks, such as prolonged reading, stenographic work, normal machine tool work	2.5–4.0%
Difficult, prolonged tasks, such as drafting, proofreading poor copy, fine machine work, and fine inspection	4.0–8.0%

Source: Millet and Bedrick (1980).

[a] Use the smaller DF values for southern latitudes with plentiful winter daylight.

(c) Heating: Whole-Building Criteria. The recommended maximum rates of heat loss [in Btu per degree days (DD) per square foot] shown in Table 5.5 were the basis for development of the passive solar heating rules of thumb in Section 5.3*a*. They have since proven useful as a quick check on overall envelope performance in residential or small commercial buildings. Some states have adopted similar criteria as part of their building codes; be sure to check the applicable code before using these numbers. The heat loss rates are shown for two conditions, as follows.

TABLE 5.4 **Daylight Factor (DF) Design Rules of Thumb for Overcast Sky Conditions**

For spaces with sidelighting[a,b,c]

$$DF_{av} = 0.2\left(\frac{\text{window area}}{\text{floor area}}\right)$$

$$DF_{min} = 0.1\left(\frac{\text{window area}}{\text{floor area}}\right)$$

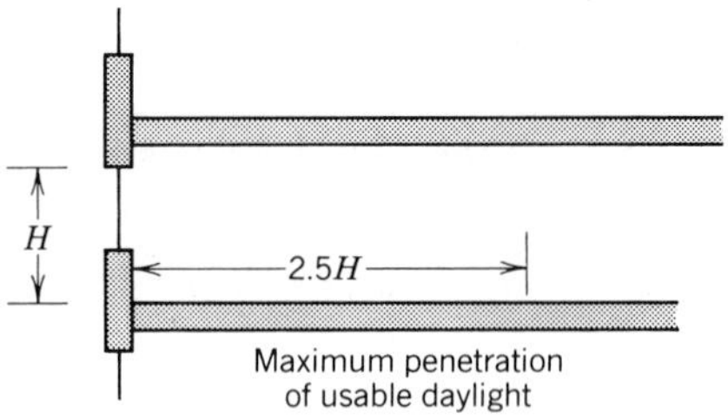

For spaces with toplighting[c,d]

Vertical monitors:

$$DF_{av} = 0.2\left(\frac{\text{skylight glazing area}}{\text{floor area}}\right)$$

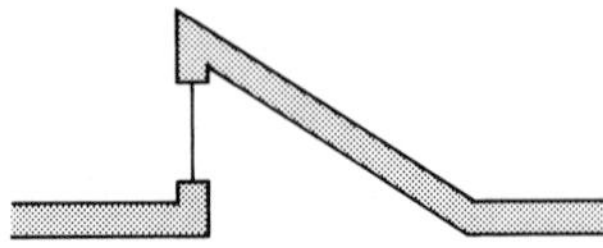

North-facing sawtooth:

$$DF_{av} = 0.33\left(\frac{\text{skylight glazing area}}{\text{floor area}}\right)$$

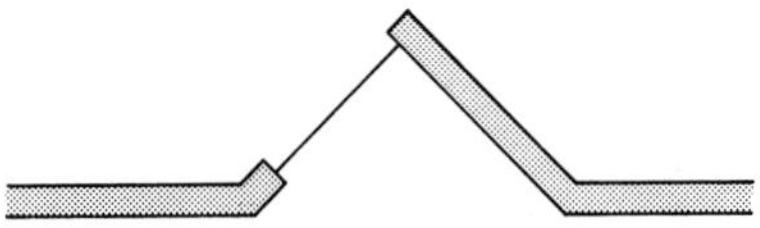

Horizontal skylights:

$$DF_{av} = 0.5\left(\frac{\text{skylight glazing area}}{\text{floor area}}\right)$$

Source: Millet and Bedrick (1980).

[a]Assumes windows in one wall of a room with relatively light-colored surfaces.

[b]Window height/room depth relationships based on the works of R. G. Hopkinson and others at the British Research Station.

[c]For sunny, clear winter sky conditions, use same formula for north, east or west glazing. For south or horizontal glazing, divide each formula's constant by 3.

[d]Assumes an even distribution of such skylights in the roof, so that an even distribution of light results in room below; thus, only average DF are listed.

1. Conventional (nonpassively solar heated) small buildings. The overall rate of Btu/DD-ft^2 is based on *total* heat loss, including all portions of the envelope *and* infiltration. To determine the whole-building heat loss rate, list for each envelope component (roof, walls, floor, windows, etc.) the U value (Chapter 4) and the total exposed area A, then simply multiply $U \times A$. (For slab floors on grade, see Chapter 4 for determining perimeter heat losses.) For the special cases of walls below grade, such as berm walls, an approximation is needed. During the coldest weather, the temperature outside such walls will nearly always be higher than the outdoor air temperature, so any procedure based on Btu/DD will *overpredict* the heat loss through these walls. As a result, designers often calculate the UA of below grade walls by using their actual U value, but using only *half* of their actual area; this lesser UA roughly compensates for the lesser ΔT through these walls.

For infiltration, determine the number of air changes per hour (ACH) under winter design conditions (Section 4.6) and multiply this infiltration (or fresh air) rate by a constant that accounts for density and specific heat:

$$\left.\begin{array}{l}\text{ACH (volume, ft}^3) \times 0.018 = \\ \textit{or} \\ \text{ACH (volume, m}^3) \times 0.33 =\end{array}\right\} UA \text{ for infiltration}$$

Add the envelope UA values to those for infiltration, multiply by 24 h/day to account for degree

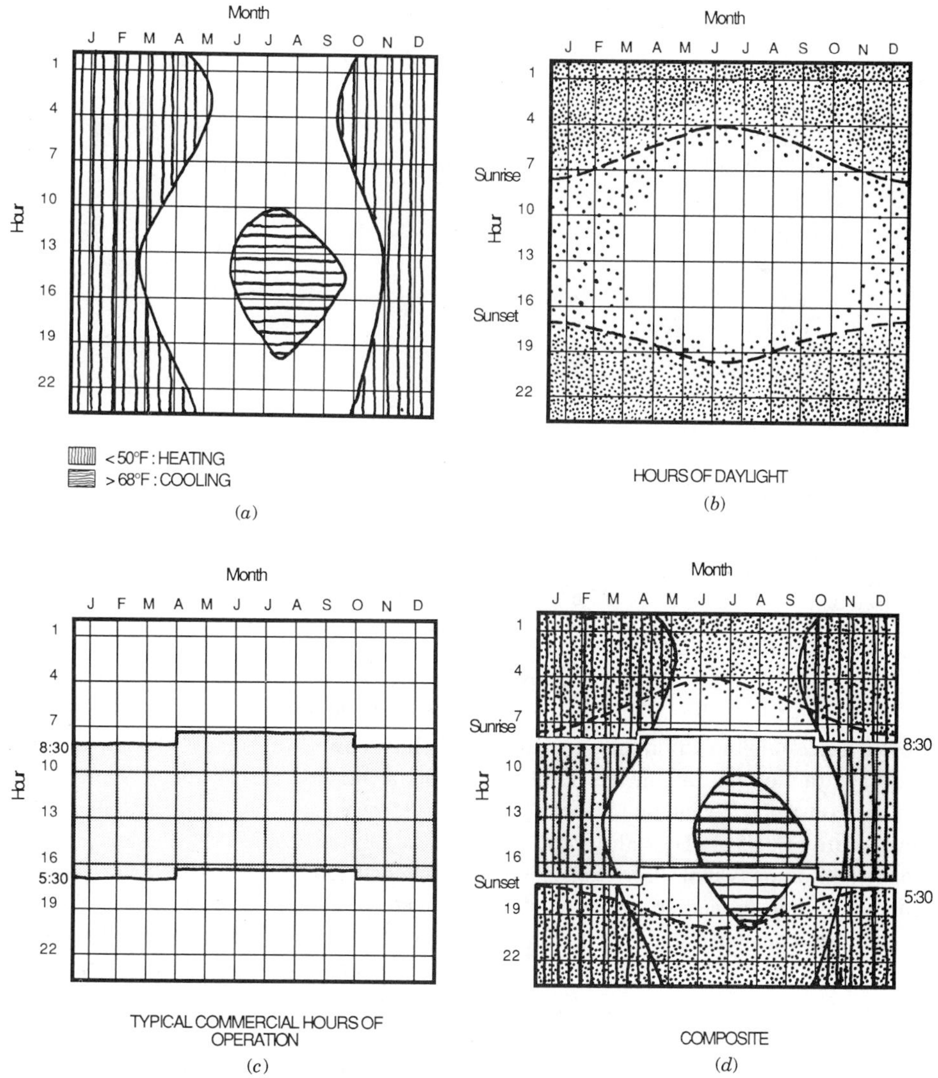

Fig. 5.5 *A series of climate "timetables" (see again Fig. 2.12) were prepared while designing the office building for the Emerald People's Utility District near Eugene, Oregon.* (a) *Average outdoor temperatures show little daily variation in winter, large daily range in summer.* (b) *Daylight peaks around noon near summer solstice.* (c) *Work schedule shows the impact of summer's daylight savings time.* (d) *A composite timetable indicates that daylight is available during almost all working hours, and that half of the hottest hours occur* after *the building closes on summer afternoons—thanks to daylight savings time. Cool nighttime temperatures make night ventilation of thermal mass an attractive cooling option. (Courtesy of Equinox Design, Inc., and WEGroup, PC, Architects, Eugene, Oregon.)*

days (DD), and divide by the building's total heated floor area:

$$\frac{(UA_{\text{envelope}} + UA_{\text{infiltration}}) \times 24 \text{ h}}{\text{total heated floor area (ft}^2)} = \text{Btu/DD-ft}^2$$

2. Passively solar heated buildings. Here the overall rate of Btu/DD-ft^2 *excludes* the solar collecting portion(s) of the envelope; otherwise, it is also based on total heat loss from all other portions of the envelope, and it includes infiltra-

TABLE 5.5 **Overall Heat Loss Criteria**

Annual Heating Degree Days (base 65 F)	Maximum Heat Loss (BTU/DD ft²) Conventional Buildings	Passively Solar Heated Buildings, Exclusive of Solar Wall
Less than 1000	9	7.6
1000–3000	8	6.6
3000–5000	7	5.6
5000–7000	6	4.6
Greater than 7000	5	3.6

Source: Balcomb et al. (1980).

tion. The equation used to determine the overall rate is

$$\frac{(UA_{\text{envelope, except south glass}} + UA_{\text{infiltration}}) \times 24 \text{ h}}{\text{total heated floor area (ft}^2)} = \text{Btu/DD-ft}^2$$

One of the biggest unknowns in this procedure is the assumed rate of infiltration. A carefully designed and constructed small building can easily achieve a rate of 0.75 ACH; with increased attention to infiltration (vapor) barrier installation, caulking of all cracks, and so on, rates of 0.33 ACH have been demonstrated.

EXAMPLE 5.1, PART C. This office building (Fig. 5.4) is passively solar heated, so the overall rate of heat loss, Btu/DD-ft², is determined by using the total $U \times A$ of the non-south-glass envelope. From Table 5.2, this is 42 + 25 + 72 = 139 Btu/h-°F.

To this must be added the effects of ventilation. From the same table, outdoor air is shown to be supplied at the rate of

$$20 \text{ cfm/person} \times 8 \text{ persons} = 160 \text{ cfm}$$

which can also be expressed as 160 cfm × 60 min/h = 9600 cfh. Comparing this hourly rate to the volume of the typical bay, we have

$$\frac{9600 \text{ ft}^3/\text{h}}{22{,}000 \text{ ft}^3} = 0.44 \text{ ACH}$$

The $U \times A$ for ventilation is therefore

$$0.44 \text{ ACH} \times 22{,}000 \text{ ft}^3 \times 0.018 = 174 \text{ Btu/h-°F}$$

and the total rate of heat loss is

$$\frac{(139 \text{ Btu/h-°F} + 174 \text{ Btu/h-°F}) \times 24 \text{ h/day}}{1440 \text{ ft}^2} = 5.2 \text{ Btu/DD-ft}^2$$

This building is located in Eugene, Oregon, which corresponds most closely to the location of Salem, Oregon as listed in Appendix B, Table B.15, where we find DD65 in Salem = 4852 per year. From Table 5.5, the maximum recommended rate of heat loss for a passively solar heated building in a 3000 to 5000-DD climate is 5.6 Btu/DD-ft². This building's heat loss rate is under the maximum, enhancing energy conservation.

(d) Heating and Cooling Criteria for Components of the Envelope: Nonresidential Buildings. The preceding whole-building, single-number criteria have several limitations; they do not satisfactorily account for differences in building size (and the accompanying differences in the likely rates of internal heat gains), nor do they distinguish between the thermal roles of roofs and walls, which can be quite different seasonally. The design guidelines for nonresidential buildings are taken from ASHRAE/IES Standard 90.1-1989, *Energy Efficient Design of New Buildings Except Low-Rise Residential Buildings,* published by the American Society of Heating, Refrigerating, and Air-Conditioning Engineers, Inc., Atlanta, GA. They are found in Appendix K, Section K.1.

(e) Heating and Cooling Criteria for Components of the Envelope: Residential Buildings. These design guidelines are taken from ASHRAE Standard 90A-1980, Energy Conservation in New Building Design, published by the American Society of Heating, Refrigerating, and Air-Conditioning Engineers, Inc., Atlanta, GA. As this book went to press, the newer BSR/ASHRAE Standard 90.2-1990, *Energy Efficient Design of New Low-Rise Residential Buildings,* was in the approval process. The newer standard should be used, once approval is final. The residential design criteria are found in Appendix K, Sections K.2 to K.4.

EXAMPLE 5.1, PART D. A typical bay of the Oregon office building shown in Fig. 5.4*f* will be examined for compliance with ASHRAE/IES Standard 90.1-1989; see Appendix K for this standard.

STEP 1. *Select ACP table from Appendix K.* The climate of Eugene, Oregon, is very similar to Salem, Oregon, so ACP Table K.18 (page 1587) will be used.

STEP 2. *Select maximum allowable percent fenestration.* Following the substeps:

(a) ILD = LPD + EPD + OLA

where LPD, lighting power density, will be considered as conforming to the maximum for office buildings from Table K.2, Part A office area 1440 ft^2 (below 2000 ft^2), = 1.9 W/ft^2; where EPD, equipment power density, is taken from Table K.2, Part B, = 0.75 W/ft^2 for offices, and where OLA, occupant load allowance, will be left at 0 W/ft^2, which allows for a built-in assumption of 0.6 W/ft^2.

$$\text{ILD} = \text{LPD} + \text{EPD} + \text{OLA}$$

$$= 1.9 + 0.75 + 0 = 2.65\ \text{W/ft}^2$$

The electric lighting actually installed in the typical bay consists of six rows of indirect fluorescent luminaires, with a total of 24 fluorescent lamps (48 in.) at 35 W each (including the ballast), and 44 fluorescent lamps (48 in.) at 46 W each (including the ballast), so the total installed lighting wattage per bay is about 2860 W.

Actual power density

$$= \frac{2860\ \text{W lighting}}{1440\ \text{ft}^2\ \text{floor area}} = 1.99\ \text{W/ft}^2$$

Therefore, the ACP table will be entered in the ILD range 1.51 to 3.00.

(b) *Select the PF (external projecting shading factor).* Trellises are used on all south-facing windows, but not on the north. These trellises are exactly as deep as the windows are high, and the bottom of the trellis is at the top of the window. Thus, for all south-facing windows, the PF is calculated at

$$\text{PF} = \frac{P_d}{H}$$

where P_d is here equal to H; therefore,

$$\text{PF} = 1.0$$

No projecting shading devices on north windows (97 ft^2), combined with PF = 1.0 on south windows (97 + 132 ft^2), indicates that an average PF for all windows of somewhat more than 0.5; therefore, continue into the ACP table in the PF range "0.5+."

(c) *Select the shading coefficient of the window (SC_x).* These windows are double glazed with clear glass, and interior white venetian blinds are provided below all lightshelves. From Table 4.19, the SC can be assumed at 0.58. Therefore, continue into the ACP table in the SC_x range 0.599 to 0.50.

(d) *Enter the vertical "fenestration" column.* Unfortunately for this daylit building, no thermal credit can be taken for daylight in this climate using this prescriptive method. Therefore, only the "base case" columns are available.

(e) *Select appropriate fenestration type.* From Table 5.2, the U value of these double-glazed, aluminum frame (no thermal break), ¼-in. glass windows with a ⅜-in. airspace, product type C (commercial), is 0.74. Therefore, enter the vertical fenestration base case column in the U_{of} range 0.81 to 0.46. Finally, the maximum allowable % fenestration can be found at the intersection of the horizontal line of ILD 1.51 to 3.00, PF 0.5+, SC_x 0.599 to 0.50, and the vertical column of base case U_{of} 0.81 to 0.46:

Maximum 35%, fenestration area to gross wall area.

Does this typical bay comply, even without counting the windowless east–west end walls?

Glazing area:

$$\text{North } 97 \text{ ft}^2 + \text{south } 97 \text{ ft}^2 + \text{clerestory } 132 \text{ ft}^2 = 326 \text{ ft}^2$$

Gross wall area:

$$326 \text{ ft}^2 \text{ glazing} + 296 \text{ ft}^2 \text{ opaque} = 622 \text{ ft}^2$$

$$\% \text{ fenestration: } 326 \text{ ft}^2/622 \text{ ft}^2 = 52\%$$

This is more than the 35% maximum, so the typical bay does *not* comply with maximum allowable percent fenestration.

The problem here is the window, with a very high U value. From Table 4.16, assume a double-glazed type C window with a wood or vinyl frame, with $\varepsilon = 0.40$ on surface 2 or 3, 3/8-in. airspace; this would yield $U = 0.45$. (Alternatively, assume a thermal break aluminum window frame, low-emittance glass of $\varepsilon = 0.15$ on either surface 2 or 3, and argon-filled 3/8-in. airspace; then $U = 0.43$.) In either case, now enter the vertical fenestration base case column in the U_{of} range 0.45 to 0.39. Then the maximum allowable percent fenestration can be found at the intersection of the horizontal line of ILD 1.51 to 3.00, PF 0.5+, SC_x 0.599 to 0.50, and the vertical column of base case U_{of} 0.45 to 0.39:

Maximum 57%, fenestration area to gross wall area.

The actual, 52%, is less than the 57% maximum, so the typical bay complies with maximum allowable % fenestration with an improved U-value window.

STEP 3. *Determine the maximum U_{ow} for the opaque wall assembly.* The north and south exterior walls of this typical bay are (from Table 5.2) concrete block exterior, metal frame interior. This exceeds "lightweight" construction, so the mass wall data columns will be used. The HC may be approximated as that of the concrete block. From Table 4.2, the HC = density of this normal-weight aggregate concrete block at 136 lb/ft^3 × thickness, 0.33 ft × specific heat, 0.22 Btu/lb-°F = 9.9 Btu/ft^2-°F. Therefore, enter the horizontal line, still within the ILD range 1.51 to 3.00, at HC ≥ 5. The insulation separates the mass from the interior, so use the vertical column INT INS. Next, select the PCT FEN (% fenestration) group; we have 52%, and the listed limits are a low of 22%, a high of 70%. A simple straight-line interpolation is appropriate; in this climate for HC ≥ 5, the U_{ow} is 0.0975. Therefore, $U_{ow} = 0.0975$. Actual $U_{ow} = 0.084$ (from Table 5.2), less than the maximum allowable, so the typical bay complies.

STEP 4. *Determine minimum R value for insulation on walls below grade, or for a slab on grade.* The typical bay uses neither, but for the lower floor of this building, assume that in this mild climate (less than 5000 HDD65), a depth of 24 in. for slab insulation is applied to the foundation stem wall. Then, for an unheated slab, the minimum insulation value required would be R-8.

STEP 5. *Note the maximum overall thermal transmittance for the roof assembly, U_{or}.* From the ACP table, this is 0.064. The actual U_{or} is 0.028 for the opaque roof (from Table 5.2), and no skylights are included; the vertical clerestories were calculated in the wall and fenestration process. So the typical bay easily complies.

STEP 6. *Note maximum overall thermal transmittance value for walls adjacent to unconditioned space.* This condition does not occur in the typical bay; if it did, the U_o would be a maximum of 0.14.

STEP 7. *Note maximum overall thermal transmittance value for the floor assembly, over outside or unconditioned space, U_{off}.* This condition does not occur in the typical bay; if it did, the U_{off} would be a maximum of 0.056.

The typical bay, with an improved window, would comply with ASHRAE/IES Standard 90.1-1989, *Energy Efficient Design of New*

Buildings Except Low-Rise Residential Buildings.

(f) Yearly Total Energy: Whole-Building Criteria. When computer assistance is available, the total yearly average consumption of energy can be estimated for the combination of heating, cooling, ventilation, lights, vertical transportation, and domestic hot water in each new building. By varying the proposed design, we can estimate the resulting changes in this total energy consumption. This approach, like the earlier single-number criteria for building heating (Table 5.5), has the advantage of allowing *many combinations* of energy conservation measures, so long as they produce acceptable final results. They also encourage the use of renewable energy, by exempting such sources from budget limitations. Obviously, the calculations that produce this result are lengthy.

One of the most widely used computer programs is the DOE 2-series, developed by the U.S. Department of Energy. More information on such programs is given in Appendix H.

(g) Indoor Air Quality. The building designer has a more elusive task when air quality is at issue, because so little can be predicted. Heat flow rates, occupancy schedules, and typical weather patterns can be combined to predict with confidence how much energy will be consumed by a building; construction types can then be altered in the design stage to yield predictably different results. As yet, designers have few tables that yield rates of outgassing for various materials at given temperatures, nor can we predict accurately the quality of outdoor air any more than we can predict its temperature and humidity.

There are some design guidelines that can help to provide for indoor air quality. They range from siting a building to avoid sources of contamination, to zoning the interior so that the cleanest air reaches the cleanest areas, to specifying the materials and the equipment systems that will yield the highest air quality.

1. *Avoid exposed earth on the interior.* In many of the buildings with high levels of radon, a cancer-causing gas, the problem has been traced to exposed earth. Radon penetrates through cracks and openings around plumbing, and below-grade spaces are particularly at risk. Penetrations of below-grade walls and floors should be both minimized and well sealed.

2. *Separate fresh air intakes from sources of contamination.* A surprisingly frequent contributor to the sick building syndrome is vehicle exhaust fumes drawn into outdoor air intakes. These intakes should be kept well away from driveways and especially loading docks, where truck motors may idle and garbage may be stored. Sometimes the fume hood exhausts from one building will be drawn into another building's intake. Less frequently, a building's conventional exhaust air openings have been located too close to its intakes, resulting in reentrainment of stale air. The mechanical equipment room is the location for both intake and exhaust; energy conservation devices such as heat exchangers benefit from close proximity of intake and exhaust. Most animals use the same "ducts" to breathe in and exhale, obviously inviting such air reentrainment. But for a building, separation of these openings is prudent design. If there is a prevailing wind direction, keep intakes upwind from exhausts.

3. *Zone for appropriate air quality.* Identify which areas of a building are likely sources of air contamination, then isolate the more sensitive areas from the contaminators. This is sometimes difficult, as in "open offices" where walls are unwelcome but copying machines are essential. In such cases, erect as much of a barrier as is possible around the offender, then "task ventilate" to remove the contaminated air immediately. Many health care and laboratory buildings have "clean" and "dirty" zones, even separate circulation pathways. Apply this analogy to air quality zoning where possible.

4. *Consider heating and cooling systems that facilitate clean air.* In general, indoor air quality will be easier to achieve if the heating and cooling systems utilize forced air motion, since some filtering is built in to the air handling equipment. However, separate air cleaning systems are increasingly common, so radiant-only heating systems with added forced-air cleaning

can yield high indoor air quality. For cooling, the economizer cycle (see Fig. 7.30) provides up to 100% outdoor air at times, and cooling by night ventilation of thermal mass provides many complete air changes, during the nightly building maintenance activities that are so fume producing.

5. *Specify interior finishes with minimal outgassing.* This will become easier as manufacturers provide the necessary information. At present, the designer must view with suspicion such ordinary items as paints, adhesives, sealants, office furniture, carpeting, and vinyl wall coverings.

5.3 Rules of Thumb for Preliminary Design

With the aid of the guidelines provided in this section, the designer can make early decisions that will lead to energy-efficient buildings that, where possible, utilize on-site energy sources. From the criteria given in the preceding section and these guidelines, a rough estimate of energy performance can often be made. Remember that the simpler the rule of thumb, the cruder the result: the largest divergence from actual performance can be expected from the easiest shortcut! Rules of thumb for daylighting and electric lighting were presented along with the criteria in Section 5.2; the similar guidelines that follow cover various approaches to heating and cooling.

(a) Passive Solar Heating. The energy conservation criteria for passively solar heated buildings were presented in Table 5.5. Given that the U_o wall criterion is difficult to meet with a high percentage of window area, the first design question becomes, what mix of passive collecting area and opaque wall insulation will meet these criteria?

Passive solar heating and energy conservation have a complex relationship. Relative to "conventional" buildings, passively solar heated buildings usually conserve purchased energy; yet, buildings that aim at very high percentages of solar heating can use more *total* energy than is used by those with smaller window areas and "superinsulated" walls, floors, and roofs. Designers interested primarily in saving purchased energy may aim at lower solar percentages and more insulation; those interested in buildings that closely relate to climate and climatic changes may aim at higher solar percentages, along with higher thermal masses and, probably, greater ranges of indoor temperature.

The solar savings fraction (SSF) is used to evaluate a building's solar performance. The solar savings is the extent to which a solar design reduces a building's auxiliary heat requirement relative to a "reference" building—one that has, instead of a solar wall, an energy-neutral wall that experiences neither solar gain nor heat loss; otherwise, the solar building and the reference building are identical. The solar savings fraction compares the auxiliary energy needed by the solar building to the auxiliary energy needed by the reference building, as illustrated in Fig. 5.6. Remember that the SSF is *not* the percentage of the solar building's heat supplied by the sun; typically, the sun provides a much *higher* fraction of a building's heat than does the SSF. Rather, the SSF is more a measure of the solar building's *conservation* advantage.

A starting point for passive solar preliminary design is Table 5.6. For your location, both a range of SSF values and a range of areas of south glass can be determined. The table shows these areas as a ratio of the total floor areas of solar-heated buildings and shows the SSF ranges for both uninsulated and night-insulated solar openings. This is shown graphically for a few locations in Fig. 5.7.

Another early design question involves the amount of thermal mass necessary to store the solar heat admitted by day. Table 5.7 details the simple relationship between SSF and weight of water or masonry. The *distribution* of the thermal mass is also important, however. In Trombe wall and water wall systems, the thermal mass usually sits in full sun for the entire day, often just inside the glazing. In direct-gain systems, this thermal mass should be within (or should enclose) the direct-gain-heated space, and the exposed surface area of the mass should be at least three times the glazing area. Masonry surfaces are less thermally effective beyond a depth of 4 to 6 in. Note that thermal storage is rela-

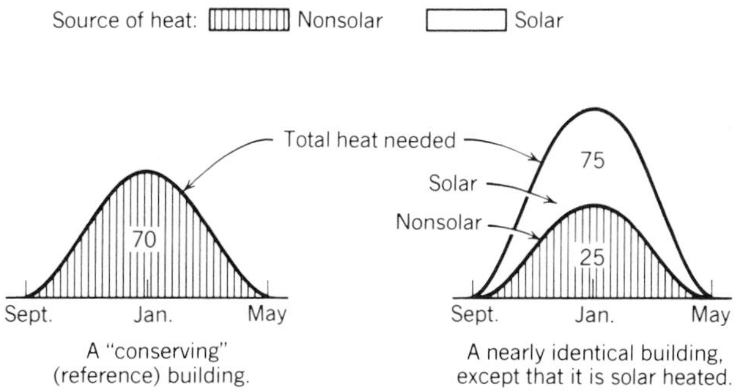

Fig. 5.6 *The solar savings fraction (SSF) compares the auxiliary heat needed by solar-heated buildings to that needed by a nonsolar but energy-conserving building that is otherwise similar, called the "reference" building. In this example, the solar building needs 25 units of auxiliary heat, whereas the reference building needs 70 units. The difference is 70 − 25 = 45, or 64% of the reference 70 units. Therefore, SSF = 64%. (Note, however, that the solar building is 75% solar heated.) (From Brown, Reynolds, and Ubbelohde* InsideOut: Design Procedures for Passive Environmental Technologies, *© 1982, John Wiley & Sons, New York. Reprinted by permission.)*

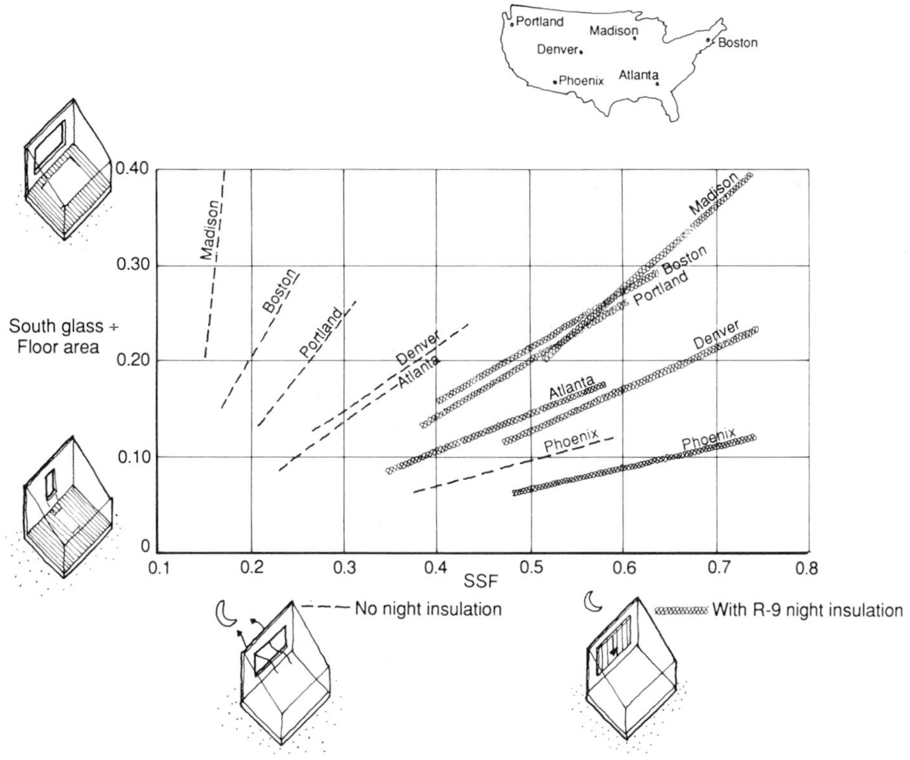

Fig. 5.7 *Rules of thumb for passive solar heating: ratio of south glass area to floor area. The information of Table 5.6 is presented graphically for six U.S. cities.*

THERMAL CONTROL

TABLE 5.6 **Rules of Thumb for Passive Solar Glazing Area**

Location	Area of Solar Glazing[a] as Ratio of Floor Area		Approximate SSF Values: No Night Insulation		Approximate SSF Values: With R9 Night Insulation[b]	
	Low	High	Low	High	Low	High
Birmingham, Alabama	0.09	0.18	22	37	34	58
Mobile, Alabama	0.06	0.12	26	44	34	60
Montgomery, Alabama	0.07	0.15	24	41	34	59
Phoenix, Arizona	0.06	0.12	37	60	48	75
Prescott, Arizona	0.10	0.20	29	48	44	72
Tucson, Arizona	0.06	0.12	35	57	45	73
Winslow, Arizona	0.12	0.24	30	47	48	74
Yuma, Arizona	0.04	0.09	43	66	51	78
Fort Smith, Arkansas	0.10	0.20	24	39	38	64
Little Rock, Arkansas	0.10	0.19	23	38	37	62
Bakersfield, California	0.08	0.15	31	50	42	67
Daggett, California	0.07	0.15	35	56	46	73
Fresno, California	0.09	0.17	29	46	41	65
Long Beach, California	0.05	0.10	35	58	44	72
Los Angeles, California	0.05	0.09	36	58	44	72
Mount Shasta, California	0.11	0.21	24	38	42	67
Needles, California	0.06	0.12	39	61	49	76
Oakland, California	0.07	0.15	35	55	46	72
Red Bluff, California	0.09	0.18	29	46	41	65
Sacramento, California	0.09	0.18	29	47	41	66
San Diego, California	0.04	0.09	37	61	46	74
San Francisco, California	0.06	0.13	34	54	45	71
Santa Maria, California	0.05	0.11	31	53	42	69
Colorado Springs, Colorado	0.12	0.24	27	42	47	74
Denver, Colorado	0.12	0.23	27	43	47	74
Eagle, Colorado	0.14	0.29	25	35	53	77
Grand Junction, Colorado	0.13	0.27	29	43	50	76
Pueblo, Colorado	0.11	0.23	29	45	48	75
Hartford, Connecticut	0.17	0.35	14	19	40	64
Wilmington, Delaware	0.15	0.29	19	30	39	63
Washington, D.C.	0.12	0.23	18	28	37	61
Apalachicola, Florida	0.05	0.10	28	47	36	61
Daytona Beach, Florida	0.04	0.08	30	51	36	63
Jacksonville, Florida	0.05	0.09	27	47	35	62
Miami, Florida	0.01	0.02	27	48	31	54
Orlando, Florida	0.03	0.06	30	52	37	63
Tallahassee, Florida	0.05	0.11	26	45	35	60
Tampa, Florida	0.03	0.06	30	52	36	63
West Palm Beach, Florida	0.01	0.03	30	51	34	59
Atlanta, Georgia	0.08	0.17	22	36	34	58
Augusta, Georgia	0.08	0.16	24	40	35	60
Macon, Georgia	0.07	0.15	25	41	35	59
Savannah, Georgia	0.06	0.13	25	43	35	60
Boise, Idaho	0.14	0.28	27	38	48	71
Lewiston, Idaho	0.15	0.29	22	29	44	65
Pocatello, Idaho	0.13	0.26	25	35	51	74
Chicago, Illinois	0.17	0.35	17	23	43	67
Moline, Illinois	0.20	0.39	17	22	46	70
Springfield, Illinois	0.15	0.30	19	28	42	67

TABLE 5.6 **Rules of Thumb for Passive Solar Glazing Area (*continued*)**

Location	Area of Solar Glazing[a] as Ratio of Floor Area		Approximate SSF Values			
			No Night Insulation		With R9 Night Insulation[b]	
	Low	High	Low	High	Low	High
Evansville, Indiana	0.14	0.27	19	29	37	61
Fort Wayne, Indiana	0.16	0.33	13	17	37	60
Indianapolis, Indiana	0.14	0.28	15	21	37	60
South Bend, Indiana	0.18	0.35	12	15	39	61
Burlington, Iowa	0.18	0.36	20	27	47	71
Des Moines, Iowa	0.21	0.43	19	25	50	75
Mason City, Iowa	0.22	0.44	18	19	56	79
Sioux City, Iowa	0.23	0.46	20	24	53	76
Dodge City, Kansas	0.12	0.23	27	42	46	73
Goodland, Kansas	0.13	0.27	26	39	47	74
Topeka, Kansas	0.14	0.28	24	35	45	71
Wichita, Kansas	0.14	0.28	26	41	45	72
Lexington, Kentucky	0.13	0.27	17	26	35	58
Louisville, Kentucky	0.13	0.27	18	27	35	59
Baton Rouge, Louisiana	0.06	0.12	26	43	34	59
Lake Charles, Louisiana	0.06	0.11	24	41	32	57
New Orleans, Louisiana	0.05	0.11	27	46	35	61
Shreveport, Louisiana	0.08	0.15	26	43	36	61
Caribou, Maine	0.25	0.50	—	NR[c] —	53	74
Portland, Maine	0.17	0.34	14	17	45	69
Baltimore, Maryland	0.14	0.27	19	30	38	62
Boston, Massachusetts	0.15	0.29	17	25	40	64
Alpena, Michigan	0.21	0.42	—	NR —	47	69
Detroit, Michigan	0.17	0.34	13	17	39	61
Flint, Michigan	0.15	0.31	11	12	40	62
Grand Rapids, Michigan	0.19	0.38	12	13	39	61
Sault Ste. Marie, Michigan	0.25	0.50	—	NR —	50	70
Traverse City, Michigan	0.18	0.36	—	NR —	42	62
Duluth, Minnesota	0.25	0.50	—	NR —	50	70
International Falls, Minnesota	0.25	0.50	—	NR —	47	66
Minneapolis–St. Paul, Minnesota	0.25	0.50	—	NR —	55	76
Rochester, Minnesota	0.24	0.49	—	NR —	54	76
Jackson, Mississippi	0.08	0.15	24	40	34	59
Meridian, Mississippi	0.08	0.15	23	39	34	58
Columbia, Missouri	0.13	0.26	20	30	41	66
Kansas City, Missouri	0.14	0.29	22	32	44	70
Saint Louis, Missouri	0.15	0.29	21	33	41	65
Springfield, Missouri	0.13	0.26	22	34	40	65
Billings, Montana	0.16	0.32	24	31	53	76
Cut Bank, Montana	0.24	0.49	22	23	62	81
Dillon, Montana	0.16	0.32	24	32	54	77
Glasgow, Montana	0.25	0.50	—	NR —	55	75
Great Falls, Montana	0.18	0.37	23	28	56	77
Helena, Montana	0.20	0.39	21	25	55	77
Lewistown, Montana	0.19	0.38	21	25	54	76
Miles City, Montana	0.23	0.47	21	23	60	80
Missoula, Montana	0.18	0.36	15	16	47	68
Grand Island, Nebraska	0.18	0.36	24	33	51	76
North Omaha, Nebraska	0.20	0.40	21	29	51	76

THERMAL CONTROL

TABLE 5.6 **Rules of Thumb for Passive Solar Glazing Area (*continued*)**

	Area of Solar Glazing[a] as Ratio of Floor Area		*Approximate SSF Values*			
			No Night Insulation		*With R9 Night Insulation[b]*	
Location	*Low*	*High*	*Low*	*High*	*Low*	*High*
North Platte, Nebraska	0.17	0.34	25	36	50	76
Scotts Bluff, Nebraska	0.16	0.31	24	36	49	74
Elko, Nevada	0.12	0.25	27	39	52	76
Ely, Nevada	0.12	0.23	27	41	50	77
Las Vegas, Nevada	0.09	0.18	35	56	48	75
Lovelock, Nevada	0.13	0.25	32	48	53	78
Reno, Nevada	0.11	0.22	31	48	49	76
Tonopah, Nevada	0.11	0.23	31	48	51	77
Winnemucca, Nevada	0.13	0.26	28	42	49	75
Concord, New Hampshire	0.17	0.34	13	15	45	68
Newark, New Jersey	0.13	0.25	19	29	39	64
Albuquerque, New Mexico	0.11	0.22	29	47	46	73
Clayton, New Mexico	0.10	0.20	28	45	45	73
Farmington, New Mexico	0.12	0.24	29	45	49	76
Los Alamos, New Mexico	0.11	0.22	25	40	44	72
Roswell, New Mexico	0.10	0.19	30	49	45	73
Truth or Consequences, New Mexico	0.09	0.17	32	51	46	73
Tucumcari, New Mexico	0.10	0.20	30	48	45	73
Zuñi, New Mexico	0.11	0.21	27	43	45	73
Albany, New York	0.21	0.41	13	15	43	66
Binghamton, New York	0.15	0.30	— NR	—	35	56
Buffalo, New York	0.19	0.37	— NR	—	36	57
Massena, New York	0.25	0.50	— NR	—	50	71
New York (Central Park), NY	0.15	0.30	16	25	36	59
Rochester, New York	0.18	0.37	— NR	—	37	58
Syracuse, New York	0.19	0.38	— NR	—	37	59
Asheville, North Carolina	0.10	0.20	21	35	36	61
Cape Hatteras, North Carolina	0.09	0.17	24	40	36	60
Charlotte, North Carolina	0.08	0.17	23	38	36	60
Greensboro, North Carolina	0.10	0.20	23	37	37	63
Raleigh–Durham, North Carolina	0.09	0.19	22	37	36	61
Bismarck, North Dakota	0.25	0.50	— NR	—	56	77
Fargo, North Dakota	0.25	0.50	— NR	—	51	72
Minot, North Dakota	0.25	0.50	— NR	—	52	72
Akron–Canton, Ohio	0.15	0.31	12	16	35	57
Cincinnati, Ohio	0.12	0.24	15	23	35	57
Cleveland, Ohio	0.15	0.31	11	14	34	55
Columbus, Ohio	0.14	0.28	13	18	35	57
Dayton, Ohio	0.14	0.28	14	20	36	59
Toledo, Ohio	0.17	0.34	13	17	38	61
Youngstown, Ohio	0.16	0.32	— NR	—	34	54
Oklahoma City, Oklahoma	0.11	0.22	25	41	41	67
Tulsa, Oklahoma	0.11	0.22	24	38	40	65
Astoria, Oregon	0.09	0.19	21	34	37	60
Burns, Oregon	0.13	0.25	23	32	47	71
Medford, Oregon	0.12	0.24	21	32	38	60
North Bend, Oregon	0.09	0.17	25	42	38	64
Pendleton, Oregon	0.14	0.27	22	30	43	64
Portland, Oregon	0.13	0.26	21	31	38	60

TABLE 5.6 **Rules of Thumb for Passive Solar Glazing Area (*continued*)**

Location	Area of Solar Glazing[a] as Ratio of Floor Area		Approximate SSF Values			
			No Night Insulation		With R9 Night Insulation[b]	
	Low	High	Low	High	Low	High
Redmond, Oregon	0.13	0.27	26	38	47	71
Salem, Oregon	0.12	0.24	21	32	37	59
Allentown, Pennsylvania	0.15	0.29	16	24	39	63
Erie, Pennsylvania	0.17	0.34	— NR	—	35	55
Harrisburg, Pennsylvania	0.13	0.26	17	26	38	62
Philadelphia, Pennsylvania	0.15	0.29	19	29	38	62
Pittsburgh, Pennsylvania	0.14	0.28	12	16	33	55
Wilkes Barre–Scranton, PA	0.16	0.32	13	18	37	60
Providence, Rhode Island	0.15	0.30	17	24	40	64
Charleston, South Carolina	0.07	0.14	25	41	34	59
Columbia, South Carolina	0.08	0.17	25	41	36	61
Greenville–Spartanburg, SC	0.08	0.17	23	38	36	60
Huron, South Dakota	0.25	0.50	— NR	—	58	79
Pierre, South Dakota	0.22	0.43	21	23	58	80
Rapid City, South Dakota	0.15	0.30	23	32	51	76
Sioux Falls, South Dakota	0.22	0.45	18	19	57	79
Chattanooga, Tennessee	0.09	0.19	19	32	33	56
Knoxville, Tennessee	0.09	0.18	20	33	33	56
Memphis, Tennessee	0.09	0.19	22	36	36	60
Nashville, Tennessee	0.10	0.21	19	30	33	55
Abilene, Texas	0.09	0.18	29	47	41	68
Amarillo, Texas	0.11	0.22	29	46	45	72
Austin, Texas	0.06	0.13	27	46	37	63
Brownsville, Texas	0.03	0.06	27	46	32	57
Corpus Christi, Texas	0.05	0.09	29	49	36	63
Dallas, Texas	0.08	0.17	27	44	38	64
Del Rio, Texas	0.06	0.12	30	50	39	66
El Paso, Texas	0.09	0.17	32	53	45	72
Fort Worth, Texas	0.09	0.17	26	44	38	64
Houston, Texas	0.06	0.11	25	43	34	59
Laredo, Texas	0.05	0.09	31	52	39	64
Lubbock, Texas	0.09	0.19	30	49	44	72
Lufkin, Texas	0.07	0.14	26	43	35	61
Midland–Odessa, Texas	0.09	0.18	32	52	44	72
Port Arthur, Texas	0.06	0.11	26	44	34	60
San Angelo, Texas	0.08	0.15	29	48	40	67
San Antonio, Texas	0.06	0.12	28	48	38	64
Sherman, Texas	0.10	0.20	25	41	38	64
Waco, Texas	0.08	0.15	27	45	38	64
Wichita Falls, Texas	0.10	0.20	27	45	41	67
Bryce Canyon, Utah	0.13	0.25	26	39	52	78
Cedar City, Utah	0.12	0.24	28	43	48	75
Salt Lake City, Utah	0.13	0.26	27	39	48	72
Burlington, Vermont	0.22	0.43	— NR	—	46	68
Norfolk, Virginia	0.09	0.19	23	38	37	62
Richmond, Virginia	0.11	0.22	21	34	37	61
Roanoke, Virginia	0.11	0.23	21	34	37	61
Olympia, Washington	0.12	0.23	20	29	38	59
Seattle–Tacoma, Washington	0.11	0.22	21	30	39	59

TABLE 5.6 **Rules of Thumb for Passive Solar Glazing Area (*continued*)**

Location	*Area of Solar Glazing[a] as Ratio of Floor Area*		*Approximate SSF Values*			
			No Night Insulation		*With R9 Night Insulation[b]*	
	Low	*High*	*Low*	*High*	*Low*	*High*
Spokane, Washington	0.20	0.39	20	24	48	68
Yakima, Washington	0.18	0.36	24	31	49	70
Charleston, West Virginia	0.13	0.25	16	24	32	54
Huntington, West Virginia	0.13	0.25	17	27	34	57
Eau Claire, Wisconsin	0.25	0.50	— NR —		53	75
Green Bay, Wisconsin	0.23	0.46	— NR —		53	75
La Crosse, Wisconsin	0.21	0.43	— NR —		52	75
Madison, Wisconsin	0.20	0.40	15	17	51	74
Milwaukee, Wisconsin	0.18	0.35	15	18	48	71
Casper, Wyoming	0.13	0.26	27	39	53	78
Cheyenne, Wyoming	0.11	0.21	25	39	47	74
Rock Springs, Wyoming	0.14	0.28	26	38	54	79
Sheridan, Wyoming	0.16	0.31	22	30	52	75
Canada						
Edmonton, Alberta	0.25	0.50	— NR —		54	72
Suffield, Alberta	0.25	0.50	28	30	67	85
Nanaimo, British Columbia	0.13	0.26	26	35	45	66
Vancouver, British Columbia	0.13	0.26	20	28	40	60
Winnipeg, Manitoba	0.25	0.50	— NR —		54	74
Dartmouth, Nova Scotia	0.14	0.28	17	24	45	70
Moosonee, Ontario	0.25	0.50	— NR —		48	67
Ottawa, Ontario	0.25	0.50	— NR —		59	80
Toronto, Ontario	0.18	0.36	17	23	44	68
Normandin, Quebec	0.25	0.50	— NR —		54	74

Source: Balcomb et al. (1980).

[a] Double-glazed, due-south-facing solar openings are assumed. Higher-percentage glazings, with correspondingly higher SSF values, can, of course, be designed.

[b] Night insulation in place from 5:30 P.M. to 7:30 A.M., solar time.

[c] NR = Not recommended.

tively unimportant at low SSF values, but as the SSF increases, so does the relative proportion of thermal mass to solar glazing.

Other types of thermal mass include phase change materials and rock beds. Thin, horizontal tiles packed with phase change materials can store great quantities of heat with the phase change from solid to liquid. This change can be formulated to occur in the low 70s F, to prevent overheating of the space. As a preliminary guide, the phase-change tile surface area equals one to three times the area of solar opening. Additional information on these materials can be found in Johnson (1981).

Rock beds are frequently used to store the excess heat that sun spaces often generate. The general guidelines cited by Mazria (1979) are

Rock bed volume, in cubic feet of rock per square foot of solar opening:

Cold climates: 3/4 to 1 1/2

Temperate climates: 1 1/2 to 3

TABLE 5.7 **Rules of Thumb for Thermal Mass**

Expected Solar Savings Fraction (%)	Recommended Effective Thermal Storage per Square Foot of Solar Collection Area			
	Water (lb)	*(ft² area)[a]*	*Masonry (lb)*	*(ft² area)[b]*
10	6	(0.1)	30	(0.7)
20	12	(0.2)	60	(1.5)
30	18	(0.3)	90	(2.2)
40	24	(0.4)	120	(2.9)
50	30	(0.5)	150	(3.7)
60	36	(0.6)	180	(4.4)
70	42	(0.7)	210	(5.1)
80	48	(0.8)	240	(5.9)
90	54	(0.9)	270	(6.6)

[a] For a water container 12 in. wide.

[b] For 4-in.-thick brick, density 123 lb/ft³.

Rock bed surface area, in contact with floor above:

Cold climates: 75 to 100% of floor area above

Temperate climates: 50 to 75% of floor area above

Yet another early design question involves orientation. How important is it that the solar opening face due south? The general recommendation is that this orientation be *within 30° of south*. In Volume 2 of *The Passive Solar Design Handbook,* the average penalties for off-south orientation were listed as follows.

5% decrease in SSF at 18° east or 30° west of true south.

10% decrease in SSF at 28° east or 40° west of true south.

20% decrease in SSF at 42° east or 54° west of true south.

See Section 5.6 for more detailed coverage of passive heating performance, including the question of the expected annual auxiliary energy needed by a passively solar-heated building.

EXAMPLE 5.1, PART E. There is no night insulation on the windows of this Oregon office building (Fig. 5.4), so the approximate SSF can be found from Tables 5.2 and 5.6 as follows:

$$\frac{\text{south glass area } 97\ \text{ft}^2 + 132\ \text{ft}^2}{\text{floor area } 1440\ \text{ft}^2} = 0.16$$

For Salem, Oregon, without night insulation, 0.12 to 0.24 ratio glass/floor yields SSF 21 to 32%, so in this case, the approximate SSF = 25%. This is consistent with the daylighting design considerations, and there is not much more south wall area that could be utilized for additional collection.

Table 5.7 shows that a thermal mass appropriate to the SSF = 25% would be about 75 lb of masonry per ft² of south glass; or about 2 ft² of mass surface per ft² of south glass. From Table 5.2, the total exposed area of thermal mass is 1512 + 720 = 2212 ft²:

$$2212\ \text{ft}^2\text{mass}/219\ \text{ft}^2\ \text{south glass} = 10\ \text{ft}^2\ \text{mass}/\text{ft}^2\ \text{glass}$$

This is much more mass area than the minimum recommended. This excess of mass will help to prevent overheating on sunny winter days and contribute to thermal stability the year around. For a more detailed analysis of this building's SSF, see Example 5.1, Part K, page 237.

EXAMPLE 5.2. A small office building in Omaha, Nebraska, is to be passively solar heated. Table 5.6 shows that an Omaha building with a south glass area equal to 20% of its floor area can expect an SSF of 51 if its solar openings are insulated at night.

Select 20% of floor area in south glass, with night insulation; SSF = 51. Thermal mass can be estimated from Table 5.7; for SSF = 50, either

30 lb of water or 150 lb of masonry (per square foot of glass) should be provided. Using brick as a thermally massive surface, 3.7 ft^2 of brick should be provided for each ft^2 of south glass.

Roof ponds provide another approach to passive solar heating, one not covered by the preceding guidelines. In general, roof ponds are used in warmer, less humid areas of the southern United States, where snow will not impede the movement of roof insulating panels and the winter sun is higher in the sky than at northern latitudes. They are frequently sized for their *summer cooling* performance; for this region of the country, a pond sized for cooling will usually be adequate to absorb the needed winter sun. As a check for the pond's heating capacity, Mazria (1979) recommends the following guidelines.

> Roof pond area = 85 to 100% of floor area for winter average outdoor temperatures of 25 to 35 F.
>
> Roof pond area = 60 to 90% of floor area for winter average outdoor temperatures of 35 to 45 F.

These figures are for roof ponds that have two layers of enclosing material between the water and the sky (i.e., are "doubled glazed") and movable night insulation.

The roof-pond-cooling rule of thumb, which usually governs, is discussed later in this section.

(b) Active Solar Heating. In contrast to passive systems, which incorporate sun collection and storage as part of a building's walls, floors, or ceilings, active solar heating uses mechanical equipment to collect and store solar energy. The most common early design questions for such systems involve the area of the solar collectors, their tilt and azimuth, and the size of the thermal storage. [Domestic hot water (DHW) solar heating information is given in Section 9.4.]

For active solar space-heating (SH) systems, the rules of thumb are more complex. Building heating needs vary by function and climate. The percentage of space heating that can economically be provided with active solar systems is another big variable. Nevertheless, a rough guide is desirable as a design starting point. For collector area,

> collector area = the smaller of the two floor area percentages listed in Table 5.6.

This should provide a portion of the annual heating load somewhere in the range of the listed SSF for night insulation passive systems (also from Table 5.6). Larger arrays of collectors can be designed, of course, but they rarely will be economically attractive.

For collector tilt and azimuth optima,

> optimum tilt = latitude plus 10 to 15°; optimum azimuth is from due south to 15° W of south

where the tilt angle is measured up from horizontal. The orientation somewhat west of south is attractive in climates with frequent morning fog. Also, because air temperatures are higher in the afternoon, collectors lose less heat then and therefore operate more efficiently.

The rules of thumb for storage size are

> 2 gal of water storage per square foot of collector, or 0.5 to 0.75 ft^3 of rock bed per square foot of collector

The large arrays of collectors necessary for space heating must be served by correspondingly large pipes or ducts. Whereas pipe size rarely influences design, the air ducts for air-type collectors can consume large amounts of space. Therefore, the following flow rates are typical.

> water flow rate of 0.25 to 0.5 gpm per square foot of collector, or airflow rate of 2 cfm per square foot of collector

Approximate pipe sizes can be determined from Chapter 9, and approximate duct sizing is shown in Section 7.4.

(c) Passive Cooling Heat Gains. The rules of thumb for passive cooling are much newer and less tested than are those for passive heating. Design guidelines for cooling are further complicated by the fact that cooling loads frequently are more related to individual building characteristics than to climate: sunshading and internal heat gains are particularly influential on cooling loads. Thus, the following rules of thumb are especially crude. It is important that the designer *first* check the match between the climate and the cooling strategy, as was done in Chapter 2 using Fig. 2.9. Second, since these rules of thumb are expressed in heat to be removed per unit of floor area, it is necessary to estimate the extent of the heat gain problem. Later in this chapter, more detailed passive cooling procedures are shown. A comparison of the rules-of-thumb with these more detailed procedures is given in Table 5.9, page 225.

Detailed calculations for heat gain are presented in Sections 5.7 and 5.8. Table 5.8 gives a quick *approximation*. Many buildings (restaurants, factories, stores selling heating appliances, etc.) have special heat sources within. For these unusually heat-loaded situations, Table 5.8 will be inadequate. As a starting point for preliminary passive-cooling sizing for typical buildings, however, it should be helpful.

Note that "open" buildings, such as those that are naturally ventilated, do not have heat gains from infiltration, because they assumedly maintain internal temperatures that are slightly *above* exterior temperatures. However, these buildings do experience heat gains through windows, walls, and roofs, due to solar impacts on these surfaces. For "closed" buildings, heat gain from infiltration or ventilation must be added, since these structures maintain internal temperatures lower than outside temperatures.

EXAMPLE 5.1, PART F. The approximate heat gain calculation for the Oregon office building in Fig. 5.4 (from Table 5.8, next page):

A. People and equipment — 5.9 Btu/h-ft²

B. Electric lighting (with a 4% DF, from Example 5.1, Part B; this procedure assumes most electric lights are off when daylight is available) — 0.5 Btu/h-ft²

C. Envelope (Eugene's design temperature 89 F)

South glass, shaded by vines

$$\frac{(97\ \text{ft}^2 + 132\ \text{ft}^2) \times 16}{1440\ \text{ft}^2} \qquad 2.54\ \text{Btu/h ft}^2$$

North glass, unshaded (Table 5.20 Part B), regular double glass, venetian blinds

$$\frac{97\ \text{ft}^2 \times 14\ \text{Btu/h-ft}^2}{1440\ \text{ft}^2} \qquad 0.94\ \text{Btu/h-ft}^2$$

Walls:

$$\frac{296\ \text{ft}^2 \times 0.084\ \text{Btu/h-ft}^2\text{-°F} \times 15}{1440\ \text{ft}^2} \qquad 0.26\ \text{Btu/h-ft}^2$$

Roof:

$$\frac{1512\ \text{ft}^2 \times 0.028\ \text{Btu/h-ft}^2 \times 35}{1440\ \text{ft}^2} \qquad \underline{1.03}\ \text{Btu/h-ft}^2$$

D. Heat gains, internal and through building skin: — 11.17 Btu/h-ft²

E. Heat gains from ventilation

$$\frac{20\ \text{cfm/person} \times 8\ \text{persons} \times 16}{1440\ \text{ft}^2} \qquad \underline{1.8}\ \text{Btu/h-ft}^2$$

Total approximate heat gains — 12.97 Btu/h-ft²

(d) Cross Ventilation. The inlet areas, expressed as percentage of total floor area, are related to wind speed and resulting heat removal in Fig. 5.8*a*. Remember that an equal (or greater) area of outlet openings must also be provided. The assumptions about wind direction and indoor–outdoor temperature differences that were used to produce these guidelines are explained in the figure.

The ΔT of 3F° used in Fig. 5.8 is deliberately kept small, to encourage "open" strategies in milder summer climates. Thus, an interior temperature of 83 F, which is comfortable if sufficient air motion, a lower-percentage relative humidity, and comfortable surface temperatures are present, would be obtainable with an outside temperature of 80 F. However, a greater ΔT is often appropriate—for example, for spring or fall cooling of office buildings, or for summer cooling of factories or kitchens where internal temperatures may remain in the low 90s. In such

TABLE 5.8 **Approximate Heat Gains**

Part A. ***Internal Heat Sources—People and Equipment***

Function	*Area per Person (ft^2)*	*Sensible Heat Gain (Btu/h ft² of Floor Area)*		
		People[a]	*Equipment*[b]	*Total*
Office	100	2.5	3.4	5.9
School: elementary	100	2.5	3.4	5.9
School: secondary, college	150	1.7	3.4	5.1
Hospital	100	2.5	Varies	2.5 plus
Clinic	50	5.0	Varies	5.0 plus
Assembly: theater[c]	15	15.3	—	15.3
Assembly: arena[c]	15	16.7	—	16.7
Restaurant	25	11.0	Varies	11.0 plus
Mercantile	50	5.0	Varies	5.0 plus
Warehouse	1000	0.4	—	0.4
Hotels, nursing homes	300	0.8	3.4	4.2
Apartments[d]	300	0.8	(see note *d*)	(see note *d*)

Part B. ***Internal Heat Sources—Lighting, Daylight and Electric***

Function	*Sensible Heat Gain (Btu/h ft² of Floor Area)*[e]		
	DF < 1	*1 < DF < 4*	*DF > 4*
Office	5.1	2.0	0.5
School: elementary	6.3–6.8	2.5–2.7	0.6–0.7
School: secondary, college	6.3–6.8	2.5–2.7	0.6–0.7
Hospital	6.8	2.7	0.7
Clinic	6.8	2.7	0.7
Assembly: theater[c]	3.8	1.5	0.4
Assembly: arena[c]	3.8	1.5	0.4
Restaurant	6.3	2.5	0.6
Mercantile	5.1–6.8	2.0–2.7	0.5–0.7
Warehouse	2.4	1.0	0.2
Hotels, nursing homes	6.8	2.7	0.7
Apartments[d]	Up to 6.8	Up to 2.7	Up to 0.7

Part C. ***Heat Gains through Envelope***[f] ***(Btu/h ft² of Floor Area)***

		Outdoor Design Temperature	
		90 F	*100 F*
I. Gains through externally shaded windows: Find $\dfrac{\text{total window area}}{\text{total floor area}}$,	then multiply by	16	21
II. Gains through opaque walls: Find $\dfrac{\text{total opaque wall area}}{\text{total floor area}} \times (U_{\text{wall}})$,	then multiply by	15	25
III. Gains through roofs: Find $\dfrac{\text{total opaque roof area}}{\text{total floor area}} \times (U_{\text{roof}})$,	then multiply by	35	45

TABLE 5.8 **Approximate Heat Gains (*continued*)**

*Part D. **Summary Gains (Btu/h ft² of Floor Area)***

I. Passive cooling systems for "open" buildings:
  Cross ventilation
  Stack ventilation
  Nighttime or "open" hours of thermal mass/night ventilation
  Total: add Parts A, B, and C gains to obtain total cooling load
II. Passive cooling system for "closed" buildings:
  Roof ponds
  Evaporative cooling
  Daytime or "closed" hours of thermal mass/night ventilation
  Total: add Parts A, B, C and Part E (below) gains, to obtain total cooling load

*Part E. **Gains from Infiltration/Ventilation of "Closed" Buildings (Btu/h ft² of Floor Area)***

	Outdoor Design Temperature	
	90 F	*100 F*
Find $\frac{\text{total window + opaque wall area}}{\text{total floor area}}$, then multiply by	1.0	1.9
OR		
Find $\frac{\text{known total cfm of outdoor air}}{\text{total floor area}}$, then multiply by	16.0	27.0

[a]Adapted from Buehrer (1978).

[b]The usual load of 1 W/ft² is assumed here. However, heavy use of computers can produce loads of up to 6 W/ft².

[c]Gains listed for these functions are only for the seating areas, not for lobbies, stage areas, kitchens, and so on.

[d]Residential internal gains often assumed at 225 Btu/h per occupant plus 1200 Btu/h total from appliances. See Section 5.7.

[e]Adapted from Northwest Power Planning Council, *Maximum Lighting Standards,* 1983.

[f]Averaged from the more specific data found in Table 5.20 (Section 5.7).

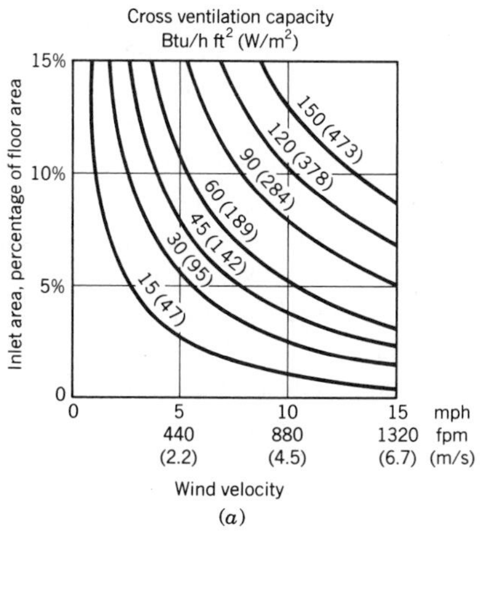

Fig. 5.8 (a) *Cross ventilation rule of thumb, for heat removed per unit floor area, and relationship of area of inlet openings and wind speed. Total inlet opening area is expressed as a percent of total floor area.* Note: *Outlet areas must* also *be at least equal to this area. The figure assumes that the internal temperature is 3F° (1.7C°) above the exterior temperature and that wind is not quite perpendicular to the inlet opening, for a wind effectiveness factor of 0.4 (see Section 4.6).*

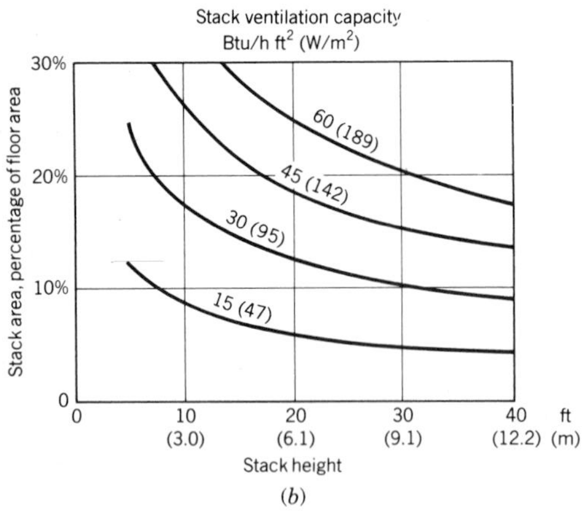

Fig. 5.8 (b) *Stack ventilation rule of thumb, for heat removed per unit floor area, and relationship area of stack and height of stack. Total stack area is expressed as a percent of total floor area.* Note: *Stack area refers to minimum area of inlets* and *of cross section through the vertical stack* and *of outlets. The figure assumes an internal temperature 3F° (1.7C°) above the exterior temperature (see also Fig. 4.7).*

cases, find the percentage of inlet area required, then multiply by the ratio

$$\frac{3\text{F}^\circ}{\text{actual } \Delta T}$$

to obtain required cross-ventilation areas for any specific temperature difference (see Example 5.1, Part H).

See Table 5.9 for a comparison of rule-of-thumb and detailed calculations for both cross and stack ventilation.

(e) Stack Ventilation. The inlet areas, expressed as percentage of total floor area, are related to stack height and the resulting heat removal in Fig. 5.8*b*. Remember that an equal (or greater) area of stack outlet openings, as well as at least an equal cross-sectional area through the vertical stack, are also required. The assumptions about indoor–outdoor temperature differences that were used to produce this guideline are explained in the figure.

Adjustment of the ΔT for this rule of thumb is similar to the procedure for cross ventilation. It requires multiplication of the required percentage of stack area by the ratio

$$\sqrt{\frac{3\text{F}^\circ}{\text{actual } \Delta T}}$$

to obtain the required stack ventilation areas for any specific temperature difference (see Example 5.3).

EXAMPLE 5.1, PART G. The office building in Fig. 5.4 is located in Eugene, Oregon, whose monthly average wind velocities may be obtained from NOAA Local Climatological Data. For July, the hottest month, the average wind velocity is about 8 mph, from a nearly-due-north average direction. From Table 5.2 the inlet area (= outlet area also, in this case) is 16 ft^2. The inlet area, as a percentage of floor area, is therefore

$$16 \text{ ft}^2/1440 \text{ ft}^2 = 1\%$$

Entering Fig. 5.8 with 8-mph wind velocity and 1% inlet area, we find that somewhat less than 15 Btu/h-ft^2 heat gain can be removed by cross ventilation. In Example 5.1, Part F, we approximated a heat gain (excluding ventilation) of about 11 Btu/h-ft^2. Therefore, under average July conditions, the window area is probably adequate to remove heat gains and maintain an indoor temperature 3F° *above* outdoor temperature.

Since Eugene's average July daily high is 83 F, this would produce an interior daily high of 86 F. From Fig. 2.4*b*, such a temperature would require somewhat more than minimum air motion in order to feel comfortable. At Eugene's summer *design* temperature of 89 F, the interior temperature would be 92 F. For these reasons this building was designed to be night ventilated, relying on thermal mass rather than upon cross ventilation for hot-day comfort.

In milder weather, these windows can achieve substantial cooling because the temperature difference, ΔT, is more than the 3 F° assumed in Fig. 5.8. For example, when the temperature indoors is 75 F, outdoors 65 F, the fully opened windows with the 8-mph wind would result in a cooling rate of

$$\frac{\text{about 15 Btu/h-ft}^2 \times 10\text{F}^\circ}{3\text{F}^\circ} = \text{about 50 Btu/h-ft}^2$$

which is considerably more than the actual gain of 11 Btu/h-ft^2. Therefore, under such conditions the windows are only partially opened.

A large building utilizing both cross and fan-assisted stack ventilation is shown in Fig. 5.9. The St. Vincent de Paul/Joan Kroc center covers a block in downtown San Diego. Built to serve the homeless of that city, it can house 450 persons per night and serve 1500 meals a day. The 110,000-ft^2 structure relies on natural ventilation for the rooms on the second and third floors. There are openable, shaded windows in each room, and bell towers at the shelter's main entrance also enhance ventilation. One faces upwind to act as an intake for sea breezes that occur 70% of the time at a speed of between 8 and 11 mph; the other faces downwind and acts as a stack effect exhaust opening, supplemented by negative pressure from the breezes. A large central courtyard provides daylight, cross ventilation, and a secure outdoor place for the residents. The building also features advanced evaporative cooling equipment for the ground-floor facilities and a peaking/backup power plant that provides space heating and hot water when it operates.

(f) Night Ventilation of Thermal Mass. This procedure is shown in Fig. 5.10, in which climate data are related to two representative types of thermally massive building. For each type, the graphs show the daily Btu per square foot of floor area that can be stored.

The climate data (maximum summer design DB temperature and mean daily range) are given in Appendix A. These data also allow calculation of the *minimum* summer design DB temperature; that is, maximum design DB temperature minus mean daily range. This minimum temperature is of interest here because the thermal mass of the building will be lowered toward (but not quite to) it during night ventilation. For high daily range climates, the lowest temperature ob-

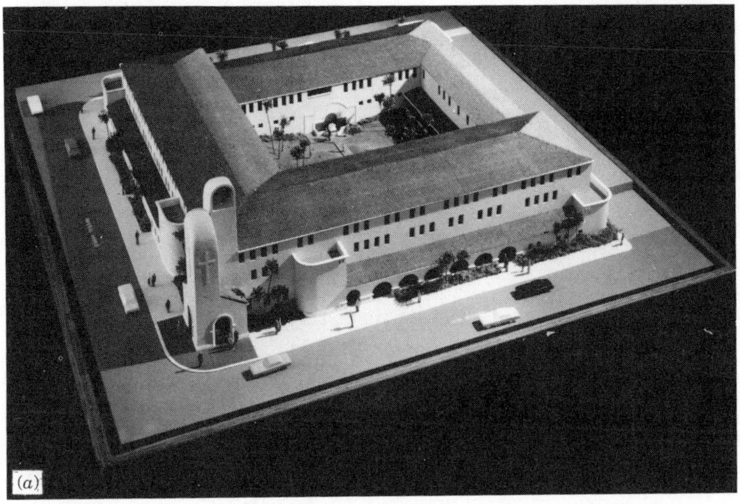

Fig. 5.9 *San Diego's facility for the homeless* (a), *the St. Vincent de Paul/Joan Kroc Center, uses fan-assisted natural ventilation to cool the top two floors.* (b) *The west elevation shows the opening at the top of the bell tower that is an intake for the sea breezes; the other tower faces downwind, thus acts as an exhaust.* (c) *A diagram of the fresh air cooling system also shows the large central courtyard that encourages daylight and cross ventilation. (Courtesy of F·A·D Architecture and Planning, Encinitas, California.)*

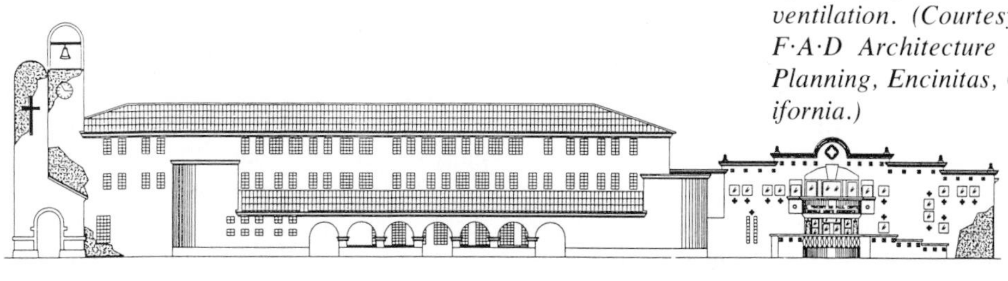

(b) *WEST ELEVATION*

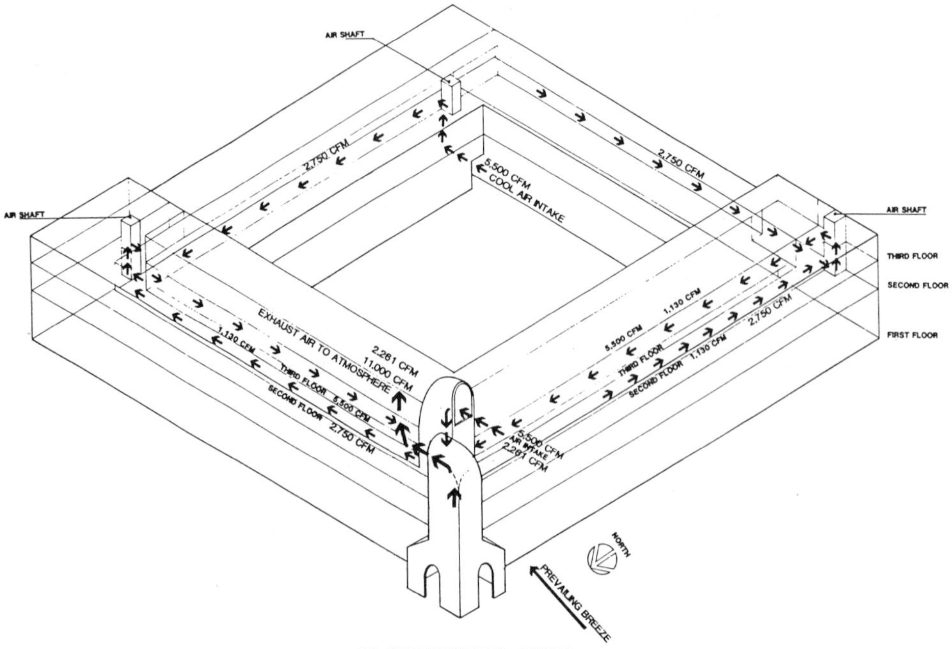

ST. VINCENT DE PAUL CENTER

ISOMETRIC (DIAGRAM OF THE VENTILATION TOWERS)

(*c*)

Fig. 5.9 *(continued)*

tained by the thermal mass will be one-fourth of the mean daily range *above* the minimum air temperature (for lower daily range climates—30F° or less—one-fifth of the mean daily range).

EXAMPLE. Sacramento, California:

highest DB = 98 F (Appendix A)

mean daily range = 36 (Appendix A)

lowest DB = 98 − 36 = 62 F

approximate lowest mass temperature: ¼ × 36 = 9; 62 + 9 = 71 F

Oklahoma City, Oklahoma:

highest DB = 97 F

mean daily range = 23

lowest DB = 97 − 23 = 74 F

approximate lowest mass temperature: ⅕ × 23 = 4.6; 74 + 4.6 = 78.6 F

From these data, it appears that Sacramento, with a comfortably low 71 F mass temperature, is a likely site for night ventilation of heat stored in the building's mass. However, Oklahoma City's 78.6 F lowest mass temperature is at the middle of the comfort zone, a climate in which this passive strategy will be harder to achieve. In either location, performance studies can best be carried out with the details of this mass cooling procedure, found in Section 5.9.

Before these rules of thumb can be utilized, the number of hours for which the building must be "closed" must be determined. During these hours, heat will be stored in the structure, up to the maximum indicated in Fig. 5.10. Typically, these buildings will go into the "closed" mode (allowing minimum ventilation only) at 6 A.M. for 100 F maximum, or 8 A.M. for 85 F maximum, remaining in the closed mode until the outdoor temperature drops below 80 F. (To approximate this hour, assume that the midpoint outdoor temperature, between daily high and daily low, occurs around 10 P.M.) Thus, the typical office building remains "closed" during the eight to nine hours of summer occupancy. All the heat generated during the "closed" mode—8 to 9 daytime hours—must be stored in the structure:

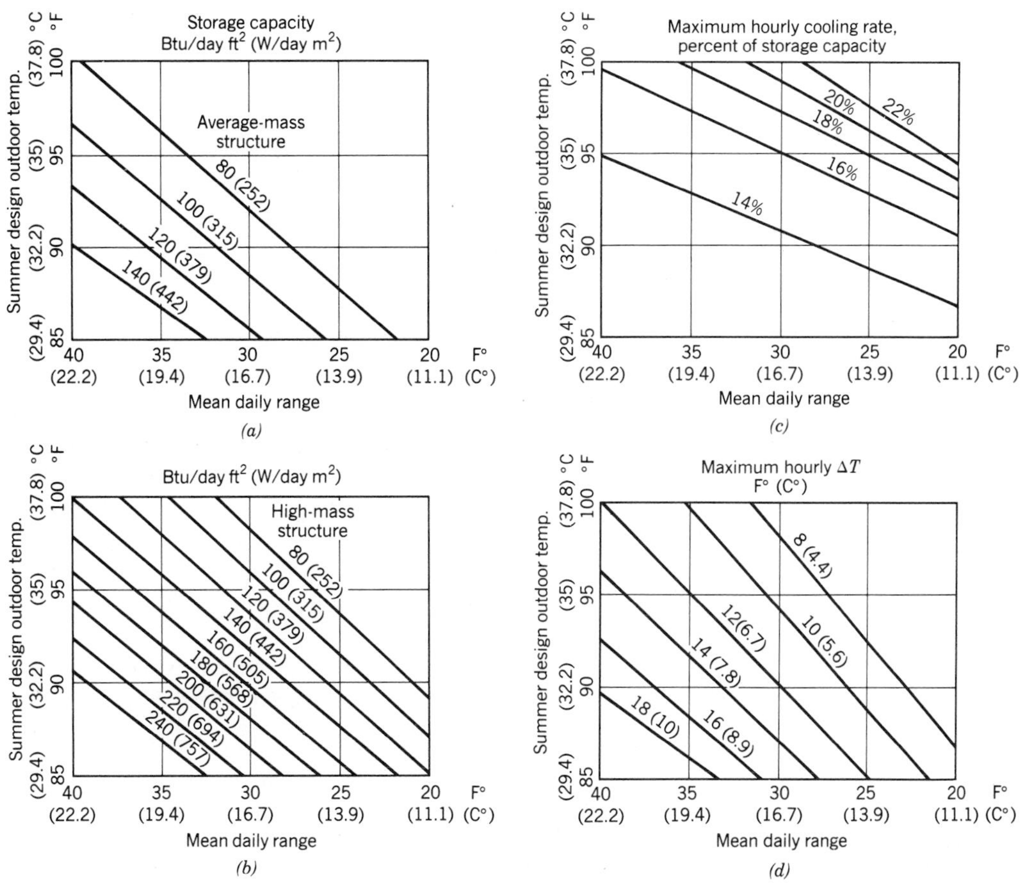

Fig. 5.10. *Night ventilation of thermal mass rule of thumb, for Btu/day removed per ft² of floor area. Parts* a *and* b *also relate daily maximum air temperature, mean daily outdoor temperature range, and degree of thermal massiveness of the building. Assumptions: "average" mass structure* (a) *has 1 ft² of surface exposed per ft² of floor area, of a 4-in. ordinary-density concrete slab; no other thermal mass is included. "High"-mass structure* (b) *has 2 ft² of surface exposed per ft² of floor area, of 3-in. ordinary-density concrete (or, alternatively, both sides of a 6-in. concrete wall or slab exposed); no other thermal mass is included. Both buildings go into "open" mode when outdoor temperature drops below 80 F (therefore, the highest mass temperature is assumed to be 80 F). Part* (c) *shows the approximate percent of the total daily stored heat that is removed during the hour of maximum cooling. Part* (d) *shows the* ΔT *between outside air and inside mass that exists during the hour of maximum cooling. Required ventilation rates can be determined from* (c) *and* (d).

heat to be stored/ft² floor area =

$$\begin{bmatrix}\text{hours occupied} \\ \text{in ``closed''} \\ \text{mode}\end{bmatrix} \times \begin{bmatrix}\text{heat gain, Btu/h ft}^2\text{ floor} \\ \text{(from Table 5.8)}\end{bmatrix}$$

Figure 5.10 shows two types of thermally massive buildings and explains the details of their construction. The "average" mass building is represented by a building with an exposed concrete floor 4 in. thick. The "high" mass building is similar to the typical passively solar heated, direct-gain building, or to a multistory building with an exposed concrete structure, in which both sides of the floor slab are available for thermal storage.

Having found the amount of heat that can be stored each day, the designer next must solve the problem of removing the heat by night. Either natural or forced ventilation can be used. The ventilation rate is determined by the "best hour" of cooling during the night: that is, the hour during which the temperature difference between inside mass and outside air is greatest, hence the most heat is removed. The "best-hour" information can also be found in Fig. 5.10*c* and *d*. The ventilation equations presented in Section 4.6 can then be used to find the volume of air needed, or the cross or stack ventilation rules of thumb can be used to size the openings.

See Table 5.9 for a comparison of rule-of-thumb and detailed calculation procedures for night ventilation of thermal mass.

EXAMPLE 5.1, PART H. The office building in Fig. 5.4 is located in Eugene, Oregon; from Appendix A, this climate has a summer design temperature of 89 F and a mean daily range of 31F°. From Table 5.2, the typical bay has a ratio of exposed thermal mass/floor area of

$$\frac{1512 \text{ ft}^2 + 720 \text{ ft}^2}{1440 \text{ ft}^2} = 1.55$$

which places this example halfway between "average" mass (ratio 1.0) and "high" mass (ratio 2.0). Entering Fig. 5.10*a*, with Eugene's climate data, we find that an average mass building can store about 100 Btu/ft^2-day, while a high mass building (Fig. 5.10*b*) can store about 170 Btu/ft^2-day. This building can therefore store about 135 Btu/ft^2-day.

Assuming a 9-h typical working day (including noon hour), at the average approximate heat gain rate of 13 Btu/h-ft^2 (Example 5.1 Part F), the daily heat gain in this building is

$$9 \text{ h} \times 13 \text{ Btu/h-ft}^2 = 117 \text{ Btu/ft}^2\text{-day}$$

This average heat gain is less than the storage capacity of 135 Btu/ft^2-day, so the building should perform satisfactorily with this method of cooling.

Although this building was designed for nighttime forced ventilation, we will check to see whether the openable windows are adequate for natural night cross ventilation. In Fig. 5.10*c*, we find that during the hour of maximum night cooling, something less than 14% of the total stored heat gains must be removed:

$$14\% \times 117 \text{ Btu/ft}^2\text{-day} = 16 \text{ Btu/h-ft}^2$$

From Fig. 5.10*d*, the maximum hourly ΔT is about 13F°.

In Example 5.1, Part G, we determine that less than 15 Btu/h-ft^2 could be removed by cross ventilation under average July conditions, at a 3F° ΔT. During this hour of maximum cooling, the indoor–outdoor ΔT is 13F°, so cross ventilation could remove

$$\frac{13\text{F}° \times 15 \text{ Btu/h-ft}^2}{3\text{F}°} = 65 \text{ Btu/h-ft}^2$$

Cross ventilation by night appears feasible for this building, since excess stored heat could be removed during this hour of maximum cooling. Unfortunately, Eugene's summer *nighttime* average wind velocity is nearly zero. Forced ventilation was therefore chosen, also in order to flush all interior mass surfaces thoroughly with cool air.

The rate of forced ventilation during the hour of maximum cooling can be estimated from Section 4.6:

$$q_v = \frac{V(1.08 \text{ Btu-min/ft}^3\text{-°F-h})(\Delta T)}{60 \text{ min/h}}$$

Expressed per ft^2 floor area in this building,

$$16 \text{ Btu/h-ft}^2 = V \text{ cfh } [0.018 \text{ Btu-min/ft}^3\text{-°F-h}]\ 13\text{F}°$$

$$V = \frac{16}{0.018 \times 13} = 68.4 \text{ cfh outdoor air/ft}^2 \text{ floor area}$$

This rate of forced outdoor air ventilation produces this air change: Average ceiling height is 15.3 ft, so

$$\frac{68.4 \text{ cfh/ft}^2 \text{ floor}}{15.3 \text{ ft}^3\text{/ft}^2 \text{ floor}} = 4.47 \text{ ACH}$$

a rate considerably greater than that required for minimum outdoor air per person by day. The duct sizes required for this rate of night flushing will be approximated in Example 5.1, Part I.

(g) Evaporative Cooling. This is not strictly a passive cooling technique, as it depends on a fan to force large quantities of outdoor air through a wet filter, lowering the air's temperature and raising its relative humidity, before delivering the air to the space to be cooled. In hot and arid climates, the energy used by the fan in evaporative systems is less than the energy needed to achieve conventional cooling based on the compressive refrigeration cycle. Although this process requires quantities of water, it does not use refrigerants that pose a threat to the earth's ozone layer. Before using this rule of thumb, be sure to check the cooling strategy chart in Fig. 2.9 to determine whether evaporative cooling is appropriate for your climate. It is unlikely that evaporative cooling will be helpful in the humid southeast United States.

The rates of evaporative cooling presented in Fig. 5.11 are based on a rather high airflow of 2.67 cfm per ft^2 of floor area. (In conventional cooling, an airflow rate closer to 1 cfm per ft^2 is more common.) With this amount of air motion, a highest indoor air temperature of 83 F is assumed; so the evaporatively cooled air, after absorbing the heat from the space, exits at 83 F. To use Fig. 5.11, first find in Appendix A, Table A.1, the summer design dry-bulb and mean coincident wet-bulb temperatures for your location. Enter the graph at these two data points, and at their intersection, find the approximate amount of heat in Btu/h-ft^2 that evaporative cooling can remove. As wet-bulb temperatures surpass 68 F (represented by dotted lines in the graph), indoor conditions will become increasingly humid, reaching almost certain discomfort indoors by wet-bulb 75 F.

A more thorough method for evaluating evaporative cooling potential is presented in Section 5.9, where various airflow rates and temperatures of supply air and exit air can be examined. See Table 5.9 for a comparison of this rule-of-thumb and the more thorough method.

EXAMPLE 5.3. Evaluate the potential for evaporative cooling for a 3000-ft^2 retail store in Denver, Colorado. Due to large electric lighting and equipment display loads, the approximate heat gain is 30 Btu/h-ft^2.

SOLUTION. From Table A.1, Denver has a summer design dry-bulb temperature of 91 F, with a coincident wet-bulb temperature of 59 F. Checking back with Fig. 2.9, these design conditions fall well within the area served by evaporative cooling. Next, enter Fig. 5.11 at the points of 91 DB and 59 WB; the intersection of these data lines shows that at such conditions, heat gains can be removed by evaporative cooling at a rate of about 37 Btu/h-ft^2 of floor area.

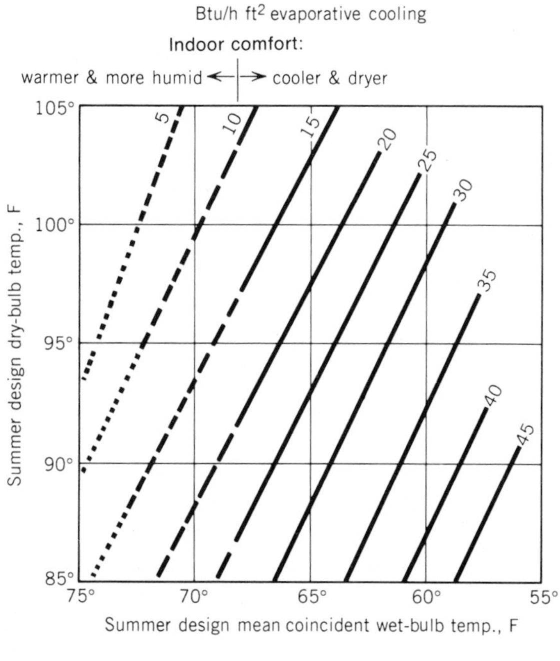

Fig. 5.11 Evaporative cooling rule of thumb, for Btu/h removed per ft^2 of floor area. A forced airflow rate of 2.67 cfm per ft^2 of floor area is assumed, as is an exiting indoor (exhaust) air temperature of 83 F. The dotted lines toward the left side of the graph represent increasingly humid indoor air, which may cause discomfort.

37 Btu/h-ft^2 capacity
> 30 Btu/h-ft^2 approximate heat gains

Therefore, evaporative coolers can meet the need. Since Fig. 5.11 assumes an air flow rate of 2.67 cf/m ft^2, the cooler's approximate size will be

$$3000 \text{ ft}^2 \times 2.67 \text{ cfm/ft}^2 = 8000 \text{ cfm}$$

This cooler size could be reduced by the ratio of cooling capacity to heat removal need:

$$8000 \text{ cfm} \times \frac{30 \text{ Btu/h-ft}^2 \text{ need}}{37 \text{ Btu/h-ft}^2 \text{ capacity}} = \text{about } 6500 \text{ cfm}$$

However, the larger capacity will allow a lower indoor air exit temperature than the 83 F assumed in Fig. 5.11, so the larger capacity is a safer approximation.

(h) Roof Ponds. Roof ponds sized for cooling frequently are nearly equal in area to the floors of the buildings they cool. Average pond depth is between 3 and 6 in. A more detailed and precise procedure for determining pond area is presented in Section 5.9, but here is the quick approximation:

pond maximum temperature: 80 F

pond minimum temperature = minimum night-time DB (= max. daytime temp − mean daily range, from Appendix A)

pond ΔT = pond maximum − pond minimum

pond's allowable daily heat stored (from building), Btu/day-ft^2 = 0.7 × pond ΔT × pond depth in feet, not to exceed 0.75 × 62.5 lb/ft^3 water (assuming 70% of the pond's heat gain from the building below and 30% through the insulated panels above the pond)

required size of pond (ft^2 pond per ft^2 floor area)

$$= \frac{\text{building total heat gain per day (Btu/day-ft}^2\text{ floor area)}}{\text{pond allowable heat stored per day (Btu/day-ft}^2\text{ floor area)}}$$

See Table 5.9 for a comparison of the rule-of-thumb and the more detailed calculations for roof ponds.

EXAMPLE 5.4, PART A. An office building in Albuquerque, New Mexico, is to be cooled by a roof pond 4 in. deep. An hourly heat gain of 15 Btu/h-ft^2 is assumed (excluding heat gain through the roof) somewhat more than the Oregon office building of Example 5.1. For Albuquerque,

maximum temperature = 94 F (Appendix A)

mean daily range = 27 (Appendix A)

minimum (night) temperature
= 94 − 27 = 67 F

Therefore,

pond ΔT = 80 maximum − 67 minimum
= 13 F°

pond storage capacity for building heat = 0.7(13)(0.33 ft)(62.5 lb/ft^3) = 188 Btu/day-ft^2

The required pond size, then, is

$$\frac{15 \text{ Btu/h-ft}^2 \times 9 \text{ h/day}}{188 \text{ Btu/day-ft}^2} = 0.72 \text{ ft}^2/\text{ft}^2 \text{ floor area}$$

Therefore, a 4-in. pond covering 72% of the one-story building's floor area will be approximately large enough to cool the building.

(i) Earth Tubes. These provide a way to cool fresh air before it enters a building. Because earth tubes need to be well underground and of rather great length in order to cool outdoor air, it is rarely economical to install enough earth tubes to completely meet a building's need for cooling. However, that component of building heat gain that is represented by cooling fresh air (Table 5.8, Part E) can sometimes be met using this strategy.

The estimate of cooling for the earth tubes described in Fig. 5.12 is based on Abrams (1986), who assumed a soil conductivity equal to that of heavy, damp earth. One of the most influential variables in earth tube performance is soil conductivity, so Fig. 5.13 (also from Abrams) compares cooling performance of 6-in. diameter earth tubes when soil conductivity changes. Note that this rule of thumb presents total Btu/h (for the entire building), rather than Btu/h-ft^2 (for a typical ft^2 floor area). See Section 5.9*d* for a more detailed calculation procedure for earth

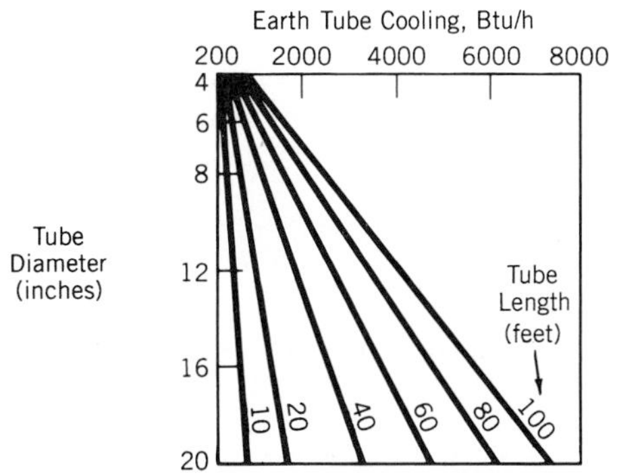

Fig. 5.12 *Cooling provided by an earth tube, in Btu/h. This tube is set in heavy, damp soil (*k = *0.75 Btu/h-°F-ft) whose temperature at the tube is 65 F. The outdoor air entering the tube is at 85 F and is moved at a flow velocity of 500 fpm. (Adapted by permission from Donald W. Abrams,* Low Energy Cooling, *© 1986, Van Nostrand Reinhold Company.)*

tubes (and earth contact tempering in general) in which tube depth, soil temperature, and soil conductivity are available. See Table 5.9 for a comparison of the rule-of-thumb with the more detailed calculation procedure.

EXAMPLE 5.5. A partially underground nature center of 4000 ft^2 in northern Pennsylvania has an hourly heat gain of 9.3 Btu/h-ft^2, of which 2.2 Btu/h-ft^2 are due to fresh air. If earth tubes are used only for cooling the fresh air, how many tubes at what length will be needed?

SOLUTION. The total cooling needed from the earth tubes is (4000 ft^2)(2.2 Btu/h ft^2) = 8800 Btu/h. From Fig. 5.12 assuming heavy damp soil, a 60-ft-long tube of 20 in. in diameter will deliver about 4500 Btu/h; two such tubes will deliver about 9000 Btu/h.

Table 5.9 compares the rules-of-thumb for passive cooling with the more detailed calculation procedures that are given in Chapter 4 and later in Chapter 5.

(j) Equipment Space Allocations. The design impact of the passive systems described above would be felt throughout a building—in a certain percentage of the floor area that must be provided in south glass, or inlets and outlets, and in the emphasis on spreading thermal mass throughout the structure. In contrast, a conventional heating and cooling system has a *concentrated* impact on a building: a central mechanical room and the distribution tree connecting it with a building are the initial design impacts.

An important early design decision is whether to integrate or separate the *heating/cooling* equipment and the *air handling* equipment (see Fig. 5.14). If integrated, one or a few central mechanical room(s) can serve many floors and each mechanical room will need area and height sufficient for both heating/cooling and air handling equipment. If separated, one (or several, in tall buildings) large space for heating/cooling equipment is typically located in the basement or the penthouse, with a smaller fan room on each floor. Each mechanical room should have both a central location relative to the area it serves and direct access to the out-

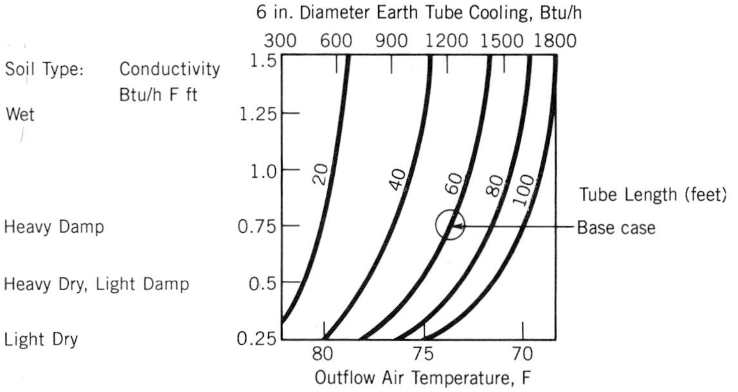

Fig. 5.13 *6-inch diameter earth tube cooling capacity as soil conductivity varies. The "base case" is a 6 in. diameter earth tube, 60 ft long, shown in Fig. 5.12 to deliver 1200 Btu/h of cooling at soil conductivity 0.75, with other conditions as specified in Fig. 5.12. Adapted by permission from Donald W. Abrams,* Low Energy Cooling, *© 1986, Van Nostrand Reinhold Company.)*

TABLE 5.9 **Comparison of Passive Cooling Rules-of-Thumb and Detailed Calculation Procedures**

Rule of Thumb	*Detailed Procedure*
Part A. ***Cross Ventilation***	
Section 5.3(d) • Assume 3F° ΔT • Assume window orientation to wind	*Section 4.6 (b)* • Use actual ΔT • Use actual window orientation to wind
Part B. ***Stack Ventilation***	
Section 5.3 (e) • Assume 3F° ΔT	*Section 4.6 (c)* • Use actual ΔT
Part C. ***Night Ventilation of Thermal Mass***	
Section 5.3 (f) • Assume ratio, mass area/floor area • Assume cooling during hour of max. ΔT • Assume max ΔT, for natural ventilation estimation	*Section 5.9 (a)* • Use actual exposed mass area • Use actual mass heat capacity • Use actual hour by hour chart of air and mass temperatures • Find total cooling and required flow rate
Part D. ***Evaporative Cooling***	
Section 5.3 (g) • Assume 2.67 cfm/ft^2 floor • Assume 83 F exhaust air	*Section 5.9 (b)* • Use actual outdoor temperature • Find actual indoor supply temperature • Find allowable ΔT as air passes through indoors • Then cfm = (Btu/h)/(1.08)(ΔT) • Size cooler(s) based on cfm
Part E. ***Roof Ponds***	
Section 5.3 (h) • Assume pond max. 80 F • Estimate pond minimum temperature • Assume 30% gain thru roof insulation • Assume pond depth between 3 in. and 6 in. • Find area of pond	*Section 5.9 (c)* • Actual outdoor temperature; resulting pond temperature • Actual heat gain thru roof insulation • Actual hours of internal heat gains • Choice of fan-forced or still air layer at ceiling • Evaporation or not, on pond upper surface • Find actual pond depth and area • Find size of backup cooling
Part F. ***Earth Tubes***	
Section 5.3 (i) • Assume 65 F soil temperature	*Section 5.9 (d)* • Actual underground temperature; assume resulting tube temperature is within 4F°

TABLE 5.9 **Comparison of Passive Cooling Rules-of-Thumb and Detailed Calculation Procedures (*continued*)**

Rule of Thumb	*Detailed Procedure*
Part F. ***Earth Tubes* (*continued*)**	
• Assume soil conductivity	• Actual soil conductivity
• Does not specify depth	• Actual depth
• Assume 85 F outdoor air	• Actual outdoor air temperature
• Assume 500 fpm velocity	• Assume 500 fpm velocity
• Choose diam., length to match cooling load	• Calculate actual cooling (assume 1.3 Btu/h F ft. length × ΔT)

side—contradictory requirements in many cases.

Central locations within the area served minimize distribution tree size; access to the outdoors facilitates the use of outdoor air as a heat source (winter) or sink (summer) and allows equipment to be installed or removed in later remodelings. Mechanical rooms serving both heating/cooling and air handling equipment need relatively high ceilings (12 ft clear is a typical minimum, 20 ft clear a typical maximum). Tables 5.10 and 5.11 present the approximate space requirements for conventional mechanical systems. (For a more detailed look at space requirements, see Table 7.2.)

The sizing graphs shown in Tables 5.10 and

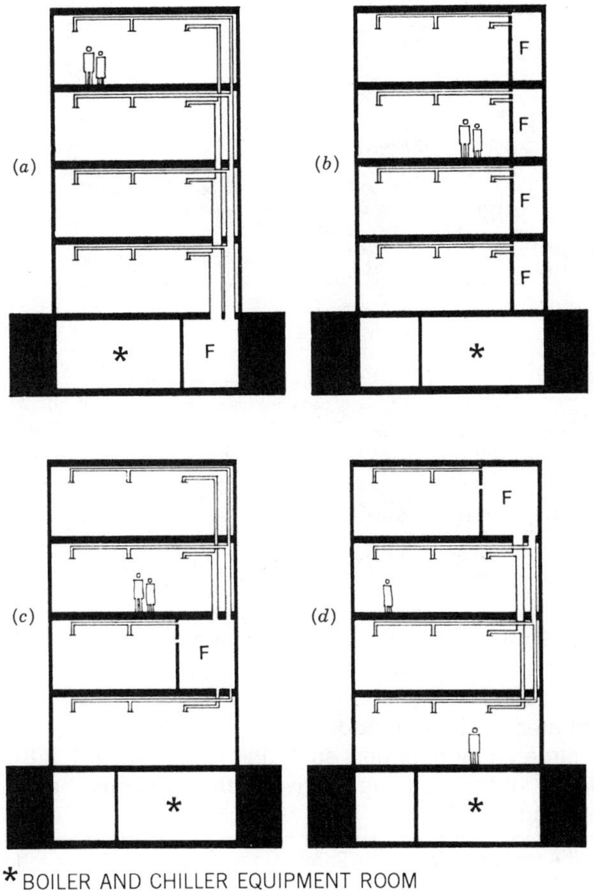

Fig. 5.14 *Fan rooms* (F) *can be either combined with, or separated from, boiler/chiller equipment rooms.* (a) *Common location for a central, combined equipment room.* (b) *Increasingly common arrangement of small fan room on each floor, equipment room in basement.* (c) *An intermediate floor may be able to give space to a central fan room, while the heavier and noisier equipment remains in the basement.* (d) *With a top floor central fan room, the equipment may be located on the roof or in a mechanical penthouse; or remain in the basement. (Adapted by permission from Edward Allen,* The Architect's Studio Companion, *© 1989 by John Wiley & Sons, New York.)*

TABLE 5.10 **Approximate Space Sizes for Major Heating and Cooling Equipment**

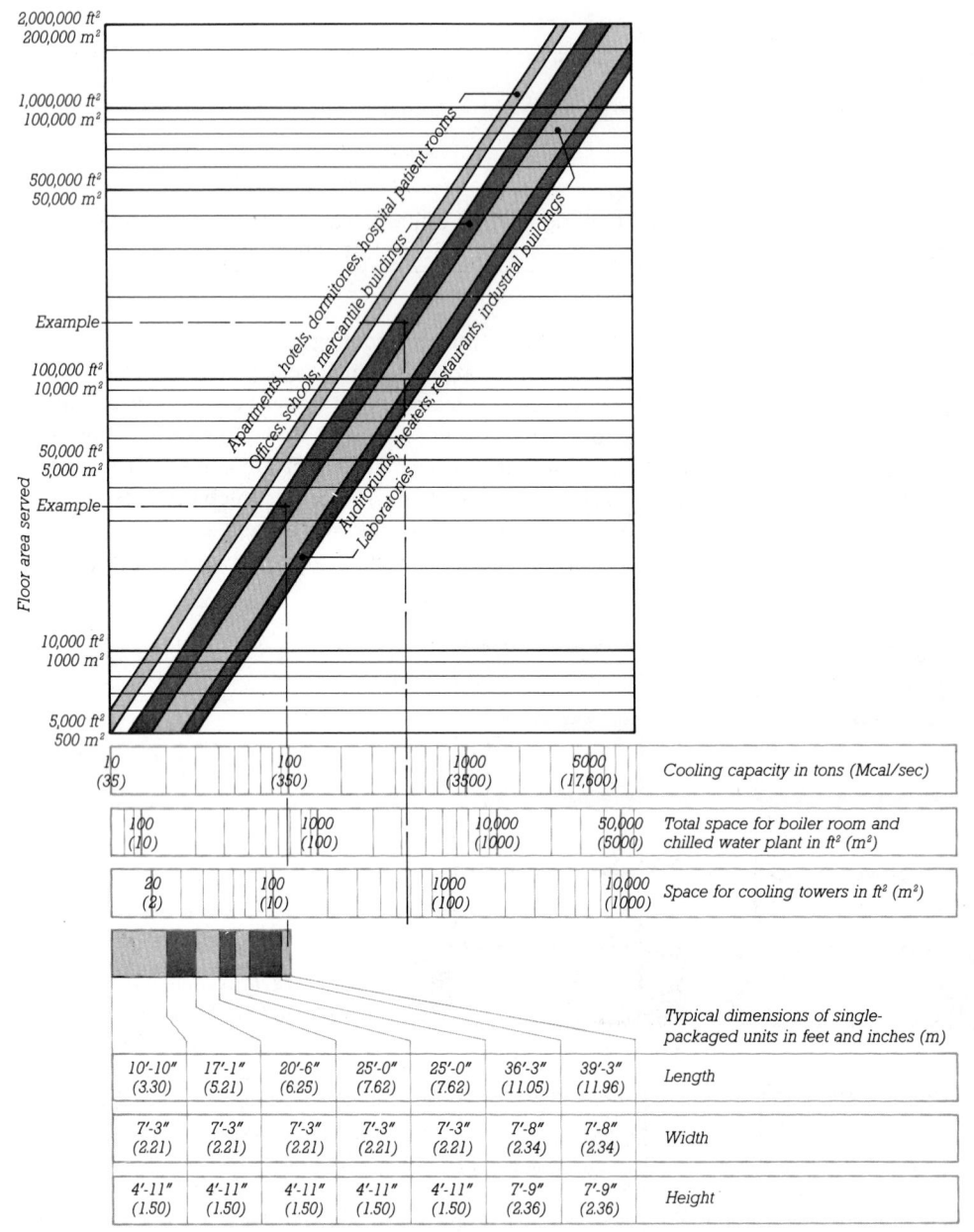

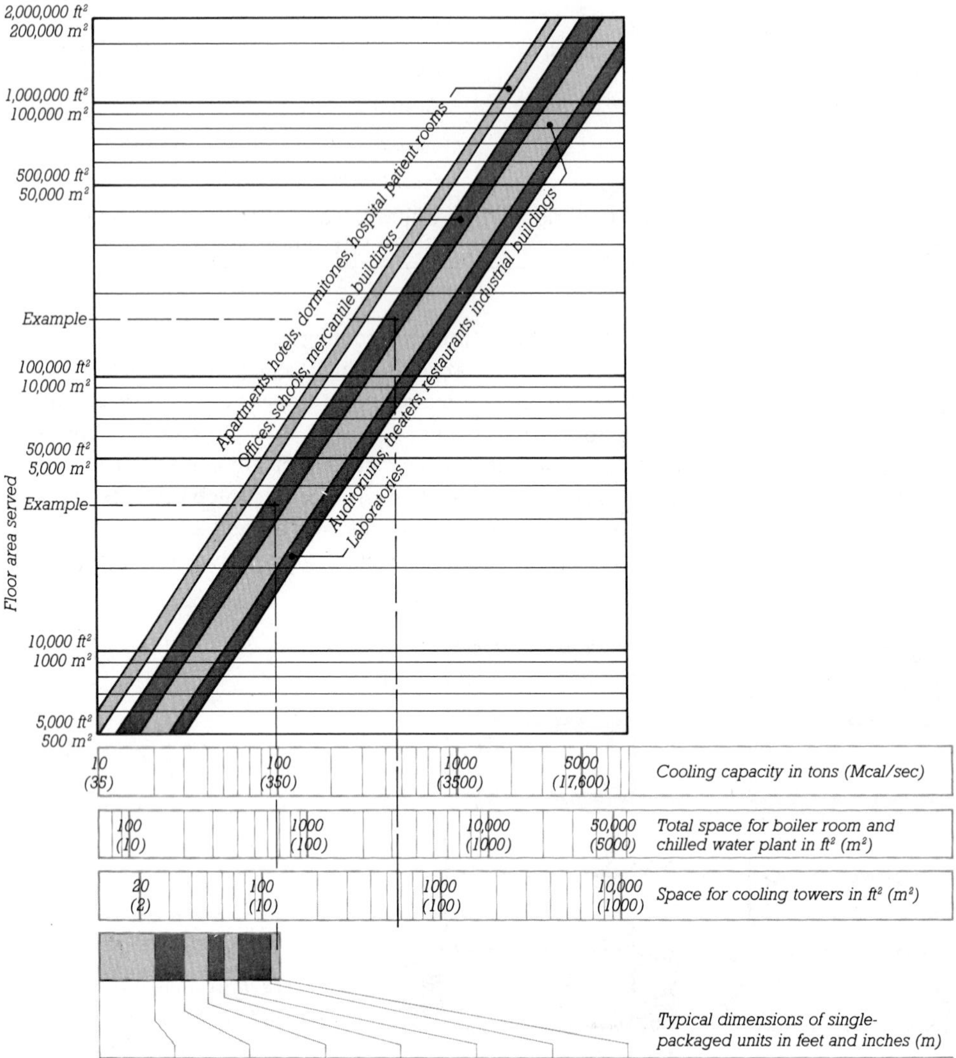

10′-10″ (3.30)	17′-1″ (5.21)	20′-6″ (6.25)	25′-0″ (7.62)	25′-0″ (7.62)	36′-3″ (11.05)	39′-3″ (11.96)	Length
7′-3″ (2.21)	7′-3″ (2.21)	7′-3″ (2.21)	7′-3″ (2.21)	7′-3″ (2.21)	7′-8″ (2.34)	7′-8″ (2.34)	Width
4′-11″ (1.50)	4′-11″ (1.50)	4′-11″ (1.50)	4′-11″ (1.50)	4′-11″ (1.50)	7′-9″ (2.36)	7′-9″ (2.36)	Height

Example: Approximate sizes for all-air system in a 150,000-ft² department store:
Cooling capacity: about 450 tons:
Total space boiler/chiller room: about 3200 ft²
Cooling towers: about 560 ft² of roof area (or other outdoor space)
Packaged unit sizes: no single unit large enough; largest such unit (39 ft 3 in. × 7 ft 8 in. × 7 ft 9 in.) serves about 33,000 ft² of this building, so five such units could be arranged on rooftop, to meet total load.

Source: Reprinted by permission from Edward Allen, *The Architect's Studio Companion,* © 1989 by John Wiley & Sons, New York.

TABLE 5.11 **Approximate Space Sizes for Air-Handling Equipment**

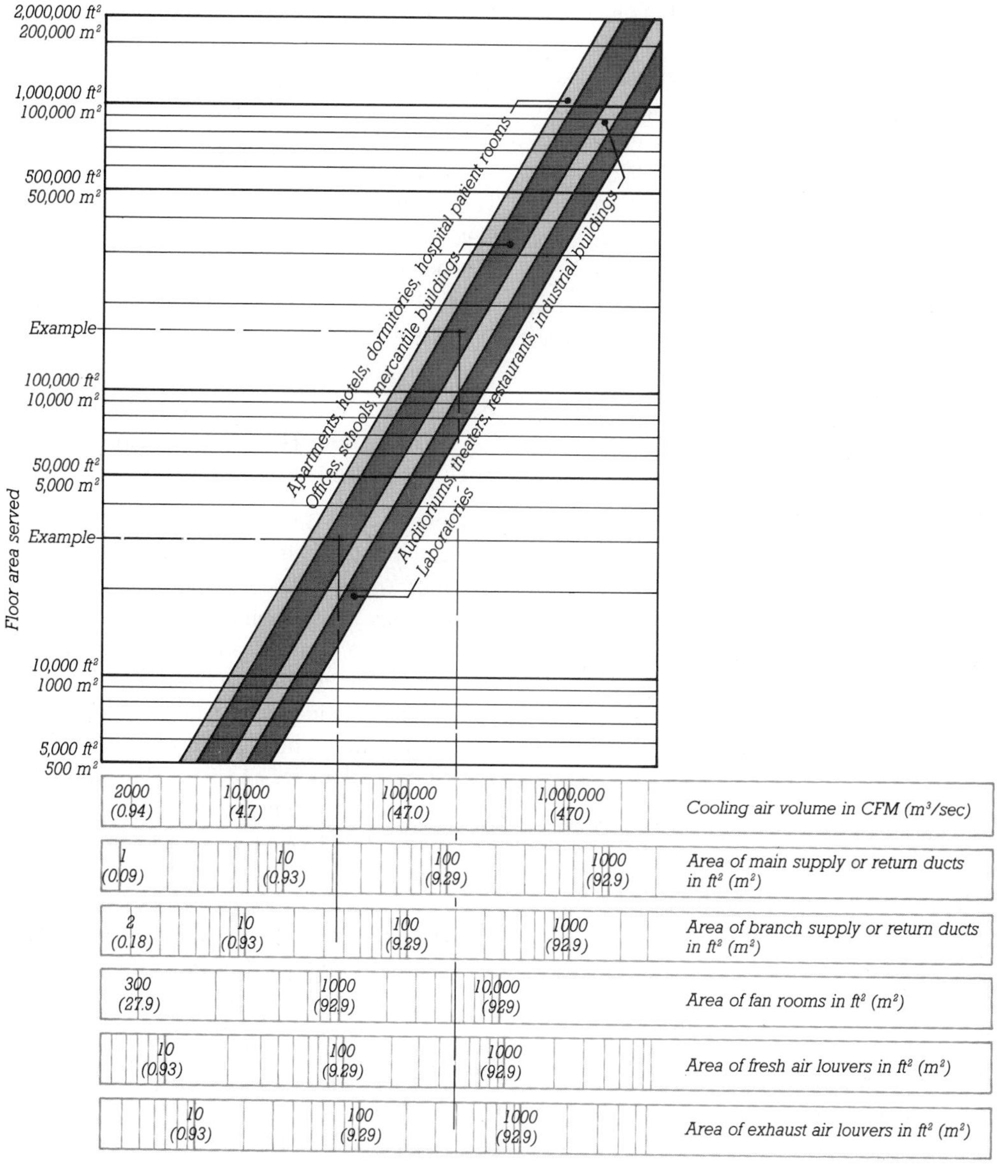

Example: Approximate sizes for all-air system in a 150,000-ft² department store:

Central fan: moves a volume of about 220,000 cfm
Main supply/return ducts: each, 120 ft² total cross-sectional area
Branch supply/return ducts: each, 300 ft² total cross-sectional area
Fan rooms: 5200 ft² total area
Fresh air louvers: 500 ft² total area
Exhaust air louvers: 400 ft² total area
If rooftop packaged units are used, each unit needs both supply and return main duct area of about 21 ft²; and both supply and return branch duct area of about 35 ft².

Source: Reprinted by permission from Edward Allen, *The Architect's Studio Companion*, © 1989 by John Wiley & Sons, New York.

5.11 are generous if buildings are designed with energy conservation in mind; this becomes evident in Example 5.1, Part I. For buildings with heavy heat gains or losses, these graphs may slightly undersize the areas needed. These graphs are intended to give a very fast approximation of areas; much more detailed procedures for sizing are presented in Section 6.3 (systems and zones), Section 7.3 (boilers, chillers, fan rooms) and Section 7.4 (air ducts).

EXAMPLE 5.1, PART I. Approximate the central equipment room needed to serve the Oregon office building (Fig. 5.4).

SOLUTION. Enter Table 5.10, spaces for heating and cooling, for an office building at 24,000 ft^2. The space for the boiler and chiller is estimated at about 500 ft^2. Enter Table 5.11, spaces for air handling, for an office building at 24,000 ft^2. The space for the fan room is estimated at about 800 ft^2. Thus a central mechanical room (boiler, chiller, and main fans) is estimated at 500 + 800 = 1300 ft^2. (The actual size of the mechanical room in this building is 1200 ft^2, a very close prediction.)

Now, approximate the largest duct sizes required. On the building's upper floor are five bays, each 1440 ft^2, for a total of 7200 ft^2. The building's mechanical room is beyond the east end of this upper floor, so all the HVAC system supply air must enter through one end of the central duct. Enter Table 5.11, spaces for air handling, for an office building at 7200 ft^2. The following ranges of sizes are indicated:

Volume of air: 7000–11,000 cfm

Area, main supply or return ducts: 4–6.5 ft^2

(The actual central supply duct maximum size in this building's upper floor is 2.6 ft^2. The table thus overestimated duct size by about 150% in this case of a well-insulated, shaded, and daylit building.)

For the thermally well-designed residence of today, a rule of thumb is

1 ton cooling/1000 ft^2 of floor area

For commercial buildings, see estimates in Table 5.10.

Yet another common unit used in the sizing of cooling equipment is the total (latent plus sensible) heat capacity of the machine. A quick *approximate* estimate of your building's *total* cooling load can be obtained from

sensible heat gain (Btu/h) × 1.3
= total (latent + sensible) Btu/h

This additional 30% of latent heat represents a *typical* office or commercial building's mixture of sensible and latent heat; it does not account for food preparation or other activities that produce great amounts of moisture.

5.4 Calculating Hourly Heat Loss and Fuel Consumption

Sections 4.1 to 4.6 showed how heat is exchanged by conductance through a building's envelope (q) and through the ventilation of a building (q_v). By combining these rates of heat exchange, we can obtain a building's total hourly heat loss in winter. (To calculate the total hourly heat gain in summer, other gains must be added; this more complicated procedure is discussed in Sections 5.7 and 5.8). The total hourly heat loss of a building can be calculated under several different assumptions reflecting different purposes.

(a) Maximum Hourly Loss: Sizing Conventional Heating Equipment. The most typical use of $q + q_v$ is to determine the maximum amount of heat per hour that heaters must provide. Two important assumptions are usually made:

1. No internal heat gains (lights, people, etc.) and no solar gains are present in the building.
2. The lowest outdoor temperature ("design condition"—see Appendix A) is occurring.

These are conservative assumptions that lead to the installation of heating equipment that is rarely used to capacity. Such equipment does provide a safety margin, for those rare times when lower than design temperatures occur or when windows are inadvertently left open or other temporary and unexpected heat leaks occur in very cold weather.

Thus, to obtain design hourly heat loss, calculate

$$q_{\text{total}} = q + q_v$$

where

$$q = (\Sigma UA)\,\Delta T$$

with ΣUA the sum, for all exposed components of the building's envelope, of $U \times A$; ΔT the design condition (from Appendix A); and

$$q_v = 1.08V\,\Delta T \qquad (1200V\,\Delta T;\ \text{SI units})$$

where V is the volume in cfm (L/s) of outdoor air introduced.

(b) Maximum Hourly Heat Loss: Sizing Auxiliary Heating for Passive Solar Buildings. The one important difference between the maximum heat loss calculations for conventional and passive solar buildings is the following assumption.

> No internal heat gains (lights, people, etc.) are present, but there is sufficient stored solar energy to at least cancel out the heat losses through the south solar collection area.

Otherwise, the procedure is the same as that for conventional buildings:

$$q_{\text{total}} = q + q_v$$

where

$$q = \Sigma UA_{\text{ns}}\,\Delta T$$

$$q_v = 0.018\text{V}\,\Delta T \qquad (\Delta T \text{ is the design condition from Appendix A})$$

$$(1200\,\Delta T;\ \text{SI units})$$

ΣUA_{ns} *excludes* the solar collector area

This can lead to occasional chilly interiors, as when several days of heavily overcast skies coincide with design-condition outdoor temperatures. In some locations, therefore, designers use the more conservative, conventional procedure for sizing the auxiliary heaters for passive solar buildings.

(c) Maximum Hourly Loss: Checking Design Criteria. The calculations that produce q and q_v are also useful in reviewing a building's design. By showing where most of the heat loss is occurring, they can quickly pinpoint opportunities for energy conservation and increased comfort. If much of the building's heat loss is occurring through the large windows in one wall, for example, consider the following options.

1. Reduce window size. (Architectural and daylighting considerations may override.)
2. Add a layer of glazing, to reduce heat loss and increase the surface temperature of this large glass area. (Cost and detailing considerations may override.)
3. Add thermal shades or shutters, to dramatically reduce heat loss and increase surface temperature. (Architectural, view, cost, and detailing considerations may override.)

If much of the building's heat loss occurs through ventilation air, consider these options.

1. Reduce infiltration, by tighter construction (or reduce mechanical ventilation toward the code-required minimum).
2. Add a heat exchanger between outgoing and incoming air.

These calculations can also be used to check your building against published criteria for thermal performance (see Section 5.2). Redesign of building envelopes to meet such criteria is fairly common in the early stages of building design.

(d) Hourly Rates of Fuel Consumption. When outdoor temperatures in winter drop below the building balance point (see Section 5.5), heating systems usually begin to operate. The hourly rate of fuel consumption depends on the hourly heat loss from the interior space. If the boiler (or furnace) is selected to run continuously at the outdoor, critical winter design temperature, then it will cycle (run intermittently) at higher outdoor temperatures. The equipment, however, *is* selected on the basis of the maximum winter demand rate and therefore relates to the calculated heat loss at the design temperature.

TABLE 5.12 **Approximate Heat Values of Fuels**

NOTE: This table includes the thermal value of electricity used on-site (but not losses in fuel energy at the electrical generating plant). Approximate seasonal efficiencies of typical burner-boiler equipment are also shown.

	Heat Value		Typical Seasonal
Fuel	*Conventional Units*	*SI Units*	*Efficiency, Percent*
Anthracite coal	14,600 Btu/lb	33,980 kJ/kg	65–75
No. 2 oil	141,000 Btu/gal	39,300 kJ/L	70–80[c]
Natural gas[a]	1,050 Btu/ft^3	39,100 kJ/m^3	70–80[c]
Propane	2,500 Btu/ft^3	93,150 kJ/m^3	70–80[c]
Electricity	3,413 Btu/kW	1 kW	95–100
Wood[b]	7,000 Btu/lb	16,290 kJ/kg	30–50

[a]Natural gas is frequently sold in "therms"; one therm = 100,000 Btu/h (SI units, 29.3 kW).

[b]At 20% moisture content.

[c]Higher efficiency furnaces are now available; see Fig. 6.53.

EXAMPLE 5.6. Calculate the rates of burning for several fuels (or the rate of using electricity) to make up the hourly heat loss, under design conditions, of a mercantile store. Its maximum hourly heat loss is 159,840 Btu/h. For fuel values, refer to the data in Table 5.12.

SOLUTION. If, for instance, coal were used, the statement would be

$$\text{lb/h} \times \text{Btu/lb heat value} \times \text{efficiency} = \text{Btu/h heat loss}$$

Transposing, we have

$$\text{lb/h} = \frac{\text{Btu/h heat loss}}{\text{Btu/lb} \times \text{efficiency}}$$

(Other efficiency statements are similar.) Applying values to this and to statements for the other fuels, we obtain the rates

$$\text{Coal:} \quad \frac{159{,}840}{14{,}600 \times 0.70} = 15.6 \text{ lb/h}$$

$$\text{Oil:} \quad \frac{159{,}840}{141{,}000 \times 0.75} = 1.51 \text{ gal/h}$$

$$\text{Gas:} \quad \frac{159{,}840}{1052 \times 0.75} = 203 \text{ cfh}$$

$$\text{Electricity:} \quad \frac{159{,}840}{3.41 \times 1.00} = 46{,}600 \text{ W (46.6 kW)}$$

The foregoing results are based on the assumption that the boiler and its piping are enclosed within the useful volume of the store, as is most usual. If they were in cold basements, or if the ducts or pipes ran through unheated space, more fuel would be used and system efficiency would decline. The rates established set the values by which the fuel-burning apparatus is selected. For instance, if oil were used, a nozzle that would inject oil at the rate of about 1½ gph should be tried. These rates are for design (extreme) conditions and are not typical of the lower average rate of operation throughout the winter.

5.5 Calculations for Heating-Season Fuel Consumption (Conventional Buildings)

The following method of estimating the fuel used for space heating in a typical season best applies to residences and small commercial buildings that are skin load dominated and not passively solar heated beyond SSF = 10. To the extent that the combination of internal and solar gains can be predicted accurately, this method yields a reasonable estimate of annual fuel consumption for any building. For passively solar heated buildings, use instead the method given in Section 5.6.

Internal and solar gains make almost any building warmer than the outdoors during the heating season. The furnace (or other space heating device) is not needed until the outdoor temperature drops to the point at which these internal and solar gains are insufficient to heat the building by themselves: that is, when the heat lost through the building's skin and infiltration matches the heat gained through solar plus internal loads. This particular outdoor temperature is called the *balance point;* it represents the beginning of the need for space heating equipment.

To estimate the annual energy needed for a building's space heating, it is necessary to know

- The building's heat loss rate (envelope and infiltration).
- The building's internal plus solar gain rate.
- The building's balance point temperature.
- The time period during which the outside temperature falls below the building's balance point temperature ("degree days").

(a) The Balance Point Temperature. This measurement can be expressed as

$$Q_i = UA(T_i - T_b)$$

or, frequently,

$$T_b = T_i - \frac{Q_i}{UA}$$

where

T_b = balance point temperature
T_i = average interior temperature over 24 h, winter
Q_i = internal gains plus solar gains (Btu/h, or W)
UA = total heat loss rate (envelope plus infiltration)(Btu/h-°F, or W/°C)

To determine the total heat loss rate UA, combine the envelope (or skin) losses ΣUA (as was done in Sections 4.4 and 5.4) and the infiltration (or ventilation) losses $1.08V$ (or, in SI units, $1200V$), where V is the volume in cfm (or L/s).

The quantity Q_i cannot be determined so straightforwardly. The internal gains can be estimated as shown in Table 5.8 or calculated more precisely from known building population, lighting, and equipment data. For residences, the following *daily total* internal gains are considered typical:

People: Two adults and two children (average times of occupancy):

23,000 to 24,500 Btu/day
(6.7 to 7.2 kWh/day)

Lights and equipment: See Table 5.13 for individual heat sources, but if actual appliances are unknown, then

53,000 Btu/day (15.5 kWh/day) for energy efficient (as of 1980) equipment

100,000 Btu/day (29.3 kWh/day) for typical (as of 1970) equipment

For more details on internal gains in residences, see the ASHRAE 1989 *Handbook of Fundamentals,* Chapter 28.

The solar gains are elusive; each month has a different average gain. For simplicity (but less accuracy), use the average January daily insolation on a vertical surface (found in Appendix B) with this approach.

$$Q_i = \frac{\text{internal gains (Btu/day)}}{24\text{ h}} + \frac{\left[\begin{array}{l}\text{January insolation} \\ \text{(Btu/ft}^2\text{ day average),} \\ \text{vertical surface}\end{array} \times \begin{array}{l}\text{area (ft}^2\text{),} \\ \text{south} \\ \text{glass}\end{array}\right]}{24\text{ h}}$$

The balance point temperature T_b can be used to do several things besides predict fuel consumption. By adding the T_b to a graph of monthly outdoor temperatures (such as in Fig. 2.8), the designer can quickly see whether heating or cooling will be the major problem for a specific building in a given climate. Also, it can be used to gain a better understanding of how zones in a building interrelate. Once the T_b is calculated for each zone, it can be determined when the entire building needs heating (outdoor temperature below any zone's T_b) or cooling (outdoor temperature above any zone's T_b), or when one zone's surplus heat can be another zone's space heating source (outdoor temperature higher than one zone's T_b, but lower than that of another). If thermal exchange between zones occurs for a major portion of the typical year, the choice of either a heating or a cooling strategy could be influenced.

TABLE 5.13 **Typical Residential Daily Internal Heat Gains from Appliances and Lighting**

Part A. ***Electric Appliances***

Heat Source	*Btu/day*	*kWh/day*
Frost-free refrigerator	16,000	4.7
Freezer	10,900–15,700	3.2–4.6
Dryer[a]	8,900–11,600	2.6–3.4
Range[a]	6,500–11,300	1.9–3.3
Television[b]	3,400–4,400	1.0–1.3
Dishwasher	2,400	0.7
Lights and miscellaneous[c]	9,500–18,800	2.8–5.5
Water heater[d]	13,700–27,400	4.0–8.0

Part B. ***Gas Appliances***

Heat Source	*Btu/day*
Water heater	13,700–27,400
Dryer[a]	12,000–19,000
Range[a]	12,000–27,000

Source: A New Prosperity: The SERI Solar Conservation Study, Brick House Publishing, Andover, Mass., 1981. Adapted by permission from the 1989 *Handbook of Fundamentals,* published by the American Society of Heating, Refrigerating, and Air-Conditioning Engineers, Inc., Atlanta, Ga.

[a]These are for appliance's consumption per day; heat gain to house is less, depending on the amount of heated exhaust air.

[b]Total daily use of TV per household.

[c]These figures are for an average 1350-ft^2 house; rate per ft^2 floor area may be extrapolated.

[d]Standby heat loss from water heater to house.

EXAMPLE 5.1, PART J. Calculate the January balance point temperature for a typical bay of the office building in Fig. 5.4. Assume that the average indoor January temperature over a 24-h period is 68 F.

SOLUTION. The internal gains are, from Example 5.1, Part F, 5.9 + 0.5 = 6.4 Btu/h-ft^2, from people, equipment, and electric lights. The average internal gains are

$$\frac{6.4 \text{ Btu/h-ft}^2 \times 9 \text{ h}}{24 \text{ h}} = 2.4 \text{ Btu/h-ft}^2$$

The January solar gain, VS, from Appendix B, Table B.15, for Salem, Oregon, is 471 Btu/ft^2-day per ft^2 of south glass:

$$\frac{471 \text{ Btu/h-ft}^2 \times (97 \text{ ft}^2 + 132 \text{ ft}^2)\text{south glass}}{24 \text{ h} \times 1440 \text{ ft}^2} = 3.1 \text{ Btu/h-ft}^2$$

Therefore,

$$Q_i = 2.4 + 3.1 = 5.5 \text{ Btu/h-ft}^2$$

UA for the typical bay (from Table 5.2 and Example 5.1, Part C) includes, for this calculation, all elements of the building envelope: (42 + 25 + 72 + 72 + 98 + 174 infiltration) = 483 Btu/h-°F, expressed per ft^2 floor area, 483/1440 = 0.34 Btu/h-°F-ft^2. Therefore,

$$T_b = 68 \text{ F} - \frac{5.5 \text{ Btu/h-ft}^2}{0.34 \text{ Btu/h-ft}^2} = 68 - 16 = 52 \text{ F}$$

(b) Degree Days. These data are published for each climate station and are calculated to various *base* temperatures (found in Appendix B). Heating degree days to a specified base temperature, such as HDD65, refer to average temperatures *below* the base temperature of 65 F. Cooling degree days to a specified base temperature, such as CDD65, refer to average temperatures *above* the base temperature of 65 F. For smaller buildings such as residences, heating degree days are in much wider usage than are cooling degree days, and many data sources simply list DD65; this notation is interchangeable with HDD65, and will be used for the remainder of this section. Until recently, degree days (DD) were always based on 65 F, because older, indifferently insulated buildings, with low internal gains, had a typical balance point of about 65 F. The combination of much higher levels of insulation and much more electric equipment has shoved the average building's balance point temperature downward; hence DD50, DD55, and DD60 are included with the traditional DD65 in Appendix B.

To derive heating DD for a particular climate and X base temperature DDX, each day's mean temperature (halfway between high and low) is subtracted from the base temperature; the result is the number of DDX for that day. If the mean temperature equals or exceeds the base temperature, no DDX are recorded. Then the DDX are totaled for an average year.

For example, assume that a day in Troy, New York, had a high of 60 F and a low of 34 F. The mean temperature was (60 − 34)/2 + 34 = 47 F.

$$65 - 47 = 18 \text{ DD65}$$

$$60 - 47 = 13 \text{ DD60}$$

$$55 - 47 = 8 \text{ DD55}$$

$$50 - 47 = 3 \text{ DD50}$$

Clearly, a building with a 65 F balance point will need more heat that day than will a building with a 50 F balance point temperature.

To convert DD conventional to DD SI, simply multiply DD conventional by 5/9:

$$\text{DDSI} = 0.56 \text{ DD conventional}$$

To obtain the "DD balance point" needed to estimate your particular building's heating needs, interpolate between the various base DDs as required. (If your balance point is below 50 F, get lower DD base figures from your local weather station. Do not extrapolate!)

(c) Yearly Space Heating Energy. To estimate the energy needed over an average year for a building E, calculate

$$E = \frac{(UA)(\text{DD balance point})(24 \text{ h})}{kV}$$

where

E is in units of fuel consumed per year (e.g., therms of gas or kWh of electricity).

UA is the total heat loss rate, envelope + infiltration (Btu/h-°F, W/°C).

DD balance point is obtained as just described.

k is a factor that includes the effects of equipment efficiency at both full and potential capacity, of energy conservation devices, of energy-efficient buildings and equipment (see Table 5.14).

V is the heating value of the fuel, from Table 5.12.

EXAMPLE 5.7. A residence in Springfield, Illinois, has a total heat loss rate UA of 544 Btu/h-°F and a balance point temperature of 55 F. It will have an induced draft, natural gas forced-air furnace, for which $k = 0.8$. For Springfield, DD55 = 3434. The average annual energy used for space heating is

$$E_{\text{therms}} = \frac{(544 \text{ Btu/h-°F})(3434 \text{ DD})(24 \text{ h/day})}{0.8 \times 100{,}000 \text{ Btu/therm}} = 560 \text{ therms}$$

TABLE 5.14 **Heating Equipment Efficiency Factors, k^a**

Space Heating Equipment	*k*
Electric resistance heaters within the heated space	1.0
Gas-fired boiler, fully condensing	0.9
Gas furnace: forced-air, spark-ignited, and induced-draft	0.8
Oil-fired boiler, energy-efficient model	0.65
Gas furnace: forced-air, atmospherically vented	0.60

Source: Based on 1981 *Handbook of Fundamentals*, published by American Society of Heating, Refrigerating, and Air-Conditioning Engineers, Inc., Atlanta, Ga.

[a]For use in predicting annual energy consumption for space heating. A detailed list of certified furnace and boiler efficiency ratings is available from the American Gas Association (1515 Wilson Blvd., Alexandria, VA 22209).

5.6 Passive Solar Heating Performance

Section 5.3 presented the rules of thumb for determining solar opening size, thermal mass, and the solar savings fraction (SSF). As a building design takes shape, more-detailed information becomes useful: Which passive system matches the architectural program? Which performs better thermally? For a given solar opening, what exactly is the resulting SSF? How much auxiliary fuel consumption per year must accompany that SSF? If a building overheats on sunny winter days, how hot will it get?

An example of a passively solar heated building is the Visitor Center for the Antelope Valley, California, Poppy Reserve, about 85 miles northeast of Los Angeles (Fig. 5.15). Set in a remote desert where winter temperatures sometimes fall below 20 F, the building is 100% passively solar heated by a combination of direct-gain and Trombe wall strategies, interior thermal shades, and thermally massive ceiling, walls, and floor. An earth-sheltered, thermally massive building was chosen for its suitability to the environmentally sensitive site, as well as for its all-year thermal advantage in a desert climate. On summer days, when air temperatures frequently exceed 100 F, a combination of underground air tubes, evaporative cooling, and stack ventilation provides a modest supply of fresh air without overheating. Continuous power ventilating of the building at night draws the cool outdoor air over the thermally massive surfaces to provide added cooling capacity for the following day. The 2100-ft^2 building also is served by an 8-kW wind generator and has its own well.

To this point, passive solar heating has been treated as a single approach; the rules of thumb for SSF distinguished only between systems with night insulation and those without. Important architectural differences, however, characterize the various passive solar heating approaches, as summarized in Table 5.15. On the basis of the wider architectural implications presented in Table 5.15 and the detailed sizing information found in Appendix C, the designer can select a passive solar heating approach with some confidence in its applicability and its yearly need for auxiliary space heating.

The following procedure is based on Balcomb et al. (1984) *Passive Solar Heating Analysis,* published by ASHRAE, and reprinted by permission. The reference offers a much wider variety of passive systems and a wider network of location listings than can be presented in this book. Along with more "sensitivity curves," to allow prediction for nonstandard passive systems, the reference also provides a much more time-consuming and detailed method for calculating the *monthly* SSF and auxiliary energy needs; the method presented here gives annual results only. Thus, Balcomb (1984) is an important, perhaps indispensable reference for the serious passive solar designer.

(a) Load Collector Ratio (LCR) Annual Performance. The method presented here, called the load collector ratio (LCR), yields the annual SSF and auxiliary energy needs for a building. This method has the following steps:

STEP 1. Choose the location and the reference passive system that most closely coincide with your building and its site. The locations listed in Appendices B and C are shown in Fig. B.1; the reference passive systems for which performance data are available (Appendix C) are summarized in Table C.1. If your system differs significantly from the closest reference system, see Section 5.6*b*, "Variations on Reference Systems."

STEP 2. Tentatively select a size for the solar openings, balancing the rule of thumb for SSF (Table 5.6) with that for daylighting (Table 5.4).

STEP 3. Calculate the UA_{ns} for the building design—one that *excludes* the solar openings but *includes* all other envelope losses, as well as the infiltration loss. Then multiply UA_{ns} by 24 h to obtain Btu/DD; this is called the building load coefficient (BLC).

$$\text{BLC} = 24 \times UA_{ns}$$

STEP 4. Check your building's overall loss rate against the criteria from Table 5.5:

$$\text{Btu/DD ft}^2 = \frac{\text{BLC}}{\text{floor area (ft}^2)}$$

Does your building envelope conserve energy

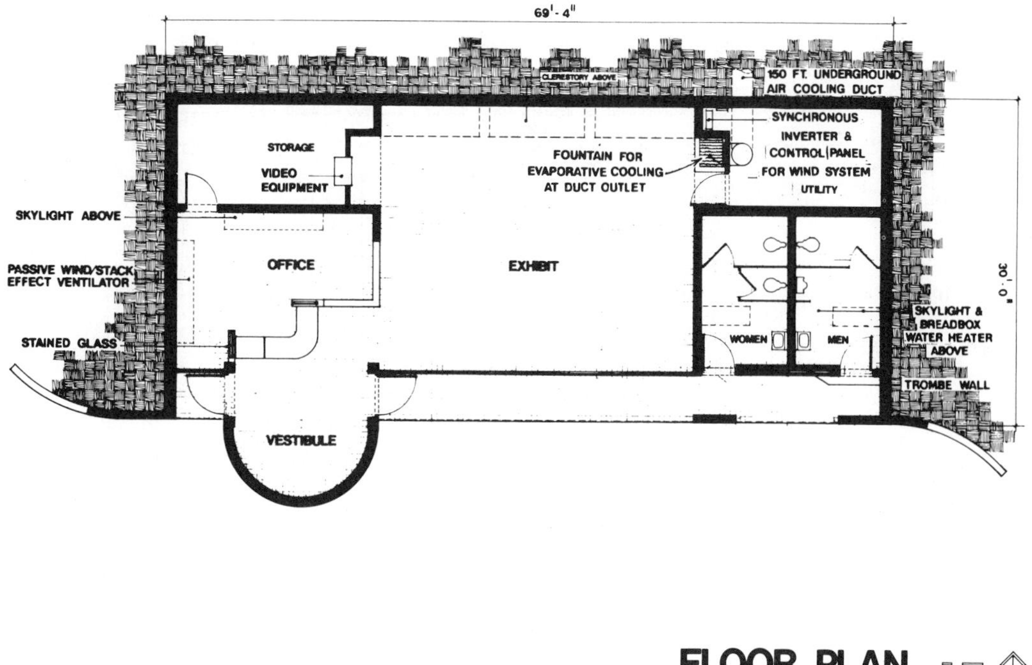

(*a*)

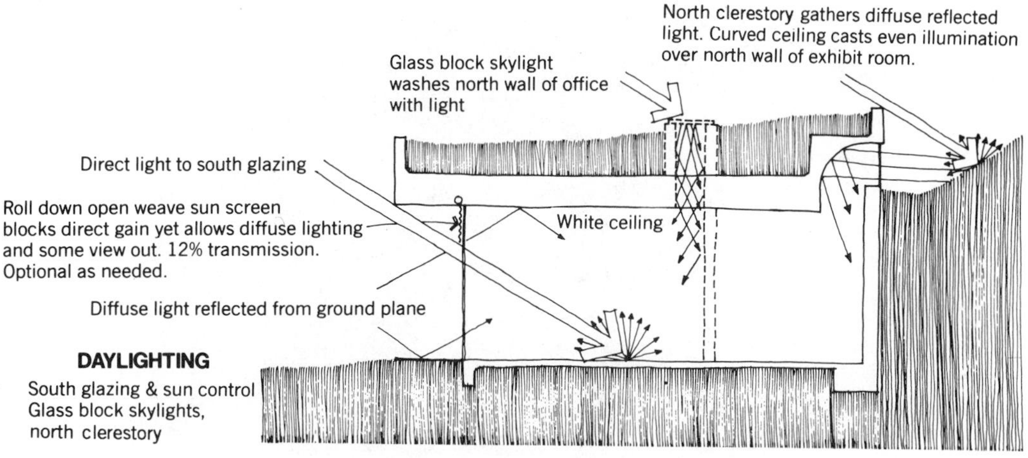

(*b*)

Fig. 5.15 *Visitor Center for the Antelope Valley California Poppy Reserve, the Colyer/Freeman Group, architects.* (a) *The plan shows earth sheltering for a somewhat elongated east–west axis, served by both direct-gain and Trombé wall heating strategies and by earth tube intake/stack exhaust for passive cooling.* (b) *Solar energy enters mostly through the south glass; the glass block skylights and north clerestory are primarily for even distribution of daylighting. At rest rooms, a Trombé wall is used. Thermal shades on the interior protect the south glass and north clerestory on winter nights; a roll-down exterior sunscreen nearly eliminates direct sun through south glass in summer. Concrete block walls, concrete roof, and floor provide thermal mass, as useful for summer cooling as for winter heating.*

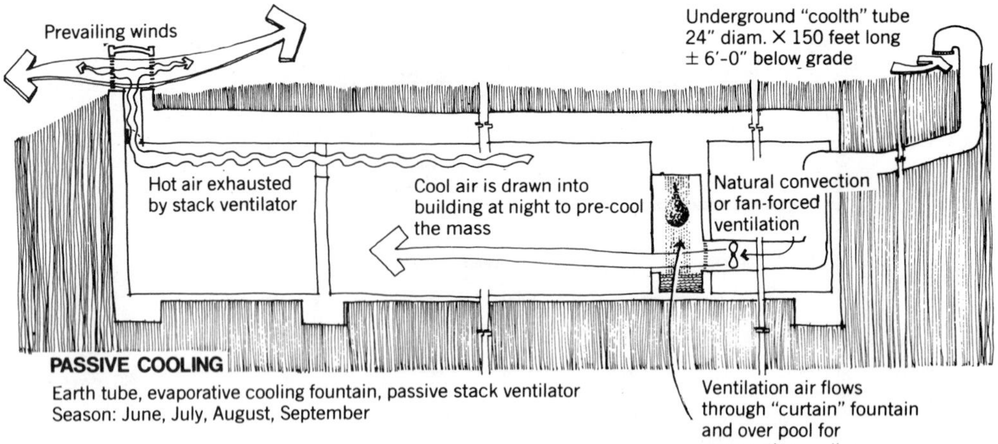

(*c*)

***Fig. 5.15** (continued)* (c) *In summer, the desert air first passes through an underground tube, then is evaporatively cooled within the building, in sight of visitors. This cooled air picks up the building's heat and is exhausted from the stack ventilator, assisted by a fan when necessary.* (d) *The most obvious feature of the Visitor Center, which is set into a hillside, is the 8-kW wind generator to supply display lights and video equipment, a fan, and a pump for the well. The cooling tube inlet is visible at the base of the tower.*

sufficiently, or do you need more insulation (or less nonsouth glass, or less infiltration)?

STEP 5. Determine the *vertical projection* of the solar opening area A_p. (For a vertical solar opening, A_p is identical to the actual opening area; for a 45°-inclined solar opening, $A_p = 0.707$ actual area.)

STEP 6. Find the load collector ratio LCR, expressed in Btu/DD-ft²:

$$\text{LCR} = \frac{\text{BLC}}{A_p}$$

STEP 7. For the reference system that most closely approaches your design, consult Table C.2 (Appendix C) at the appropriate location. By interpolation, find the annual SSF that corresponds to your passive system's listed LCR. Note also the annual degree days listed in this table. (These are DD65.)

STEP 8. Finally, determine the approximate annual auxiliary heating Q required:

$$Q = (1 - \text{SSF}) \times \text{BLC} \times \text{DD}$$

Although this quick annual results method is based on the DD65 listed in Table C.2, it is possible to adjust Q to *approximately* account for higher internal gains or better-insulated envelopes. In this adjustment, use DD balance point instead of DD65 in the equation above for Q.

STEP 9. Now compare the rule-of-thumb predicted relationship between collector size and SSF to the actual one you've just calculated. If the SSF is *smaller* than you had hoped, can you decrease BLC (improve conservation) or increase collector size, or switch to another passive system with a more favorable SSF for the same LCR? If the SSF is *larger,* will you be happy with the increased fuel savings, or will you reduce the collector size, or consider another, less-efficient passive system that has some architectural advantage over the one for which you calculated SSF? Table 5.16 shows four early and well known passive solar buildings, with system type and approximate LCR.

EXAMPLE 5.1, PART K. What is the annual SSF for the Oregon office building of Fig. 5.4?

SOLUTION. The reference passive system most closely corresponding to the direct gain in a typical bay is found in Appendix C, Table C.1, Part D: DGC1 has a high thermal storage capacity, based on an average thermal mass thickness of 4 in., and a ratio of mass to south glass area of 6 : 1. It has double glazed windows with no night insulation. [The actual building's mass to south glass area ratio is (1512 + 720)/(97 + 132) = almost 10 : 1.] The BLC of the typical bay is

$$24 \text{ h/day} \times (42 + 25 + 72 + 174 \text{ infiltration}) \text{ Btu/h-°F} = 7512 \text{ Btu/DD}$$

The LCR of the typical bay is therefore

$$\frac{7512 \text{ BLC}}{(97 + 132) \text{ collector area}} = 32.8$$

From Appendix C, Table C.2, for Salem, Oregon, with LCR = 33, system DGC1, SSF = about 27%, compared to the rule of thumb (Example 5.1, Part E) of SSF = 25%.

Note that if moderate night insulation (R-4) had been added on all windows in the typical bay, the BLC would be reduced. The north window UA (see Table 4.17) would become 97 ft² × 0.33 Btu/h-ft²-°F = 32. Thus

$$\text{BLC} = 24 \text{ h/day} \times (42 + 25 + 32 + 174 \text{ infiltration}) \text{ Btu/h-°F} = 6552 \text{ Btu/DD}$$

The passive system now closest to the actual building is DGC3, since night insulation has been added. The new LCR is

$$\frac{6552 \text{ BLC}}{97 + 132 \text{ collector}} = 28.6$$

Therefore, DGC3 in Salem, Oregon, with LCR of 28.6 yields SSF = 49%, a substantial improvement that nearly doubles the energy saved.

EXAMPLE 5.8. We will now take a more detailed look at the solar collecting area required for the passively solar heated building in Omaha, Nebraska, discussed in Example 5.2.

From the rules of thumb used in Example 5.2, we expected that 20% of the floor area in south glass with night insulation would yield a 51% SSF. More detailed characteristics of this building are shown in Fig. 5.16. From a program

TABLE 5.15 **Passive Solar Heating Systems Compared**

	Influence on Plan	*Heating Characteristics*
Direct gain (DG) 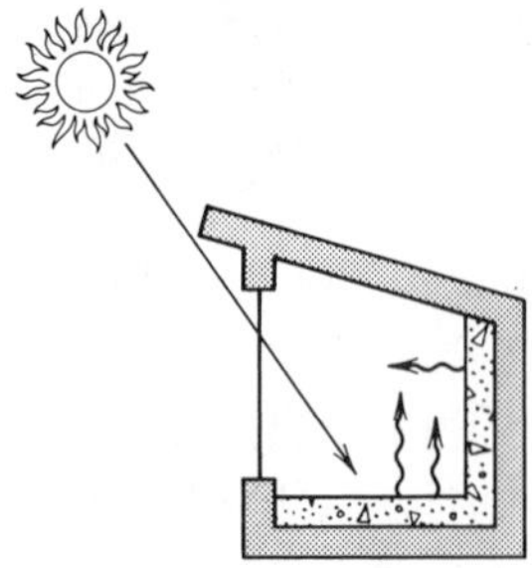	Sun can enter through south windows or skylights; open plan can allow sun and stored heat to serve entire top floor of building. Large areas of thermal mass surface should be darker colored and free of rugs, wall hangings, etc. Light-colored surfaces near glass reduce glare. Outdoor view and access to south are encouraged.	Quick to warm up in the morning, fast response to sun. Tendency to overheat at midday; large temperature swings. Much radiant loss to bare window by night; movable insulation encouraged (or triple glazing or selective film). Warmth spread throughout space along with thermal mass.
Thermal storage wall Trombe (masonry) Wall (TW) Water wall (WW) 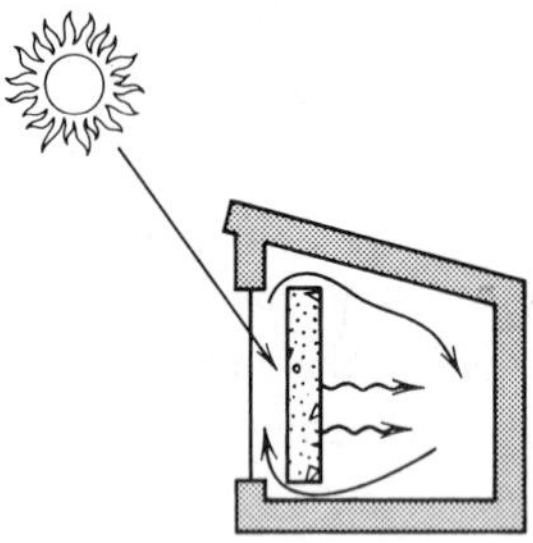	Needs to be on south wall of building. Inner wall of TW or WW should be kept clear of hangings, furniture, etc., but rest of space is unrestricted. Not much solar impact beyond about 25 ft from TW or WW. Outdoor view and access to south are discouraged.	Unvented TW, WW are slow to warm by day, slow to cool by night; small temperature swings. Most radiant heat arrives in evening; comfort is most likely near the TW or WW surface. (Vented TW behave midway between unvented TW and DG systems.)
Sunspace (SS) 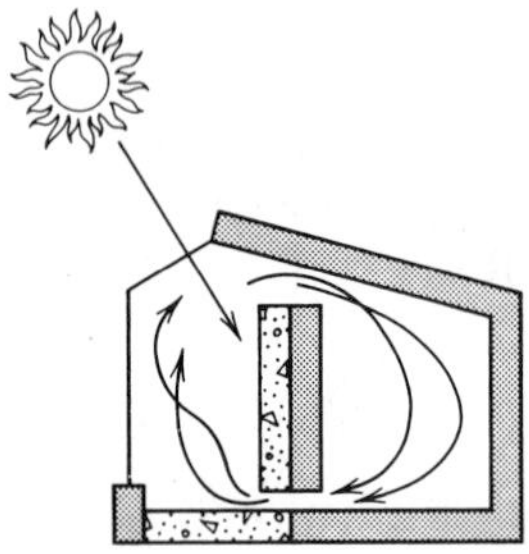	Same influences as TW and WW above; or sunspace can be insulated from building, becoming less efficient heat source for a rock bed. Floor above rock bed should be kept free of rugs. Sunspace becomes a special place with different characteristics from rest of building. Access to sunspace thereby encouraged. Access to south encouraged, view to south filtered.	Sunspace thermally like DG, but with extreme temperature swings and accentuated radiant loss by night. Movable insulation often omitted and night use curtailed. Building beyond is thermally like TW or WW, depending on its connection with sunspace. Warm floor above rock bed in months with SS surplus heat.
Roof pond 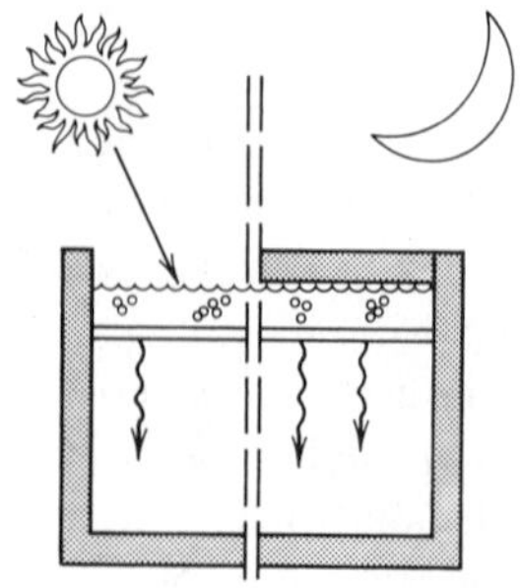	Flat or nearly flat roof is desirable. Skylight is discouraged. Plan is completely unrestricted; sidelighting and views, access to outdoors encouraged.	Low temperature swings; steady temperatures both summer and winter. Winter air stratification possible, since warmest surface is also highest surface in the space.

TABLE 5.15 **Passive Solar Heating Systems Compared (*continued*)**

	Daylighting	*Cooling*
DG	Very high DF possible, with high glare potential at lower windows. Possible conflict between light-colored surfaces for glare reduction and dark-colored surfaces for solar absorption. Summer shading greatly reduces DF. Encourages both side and top lighting.	Cross ventilation: encouraged by large windows to south. Stack ventilation: helped when clerestories are used. Night vent./thermal mass: excellent potential, much mass surface and capacity. Other closed-bldg. cooling: large south windows are big threat. Shading (and lower DF) necessary in overheating months.
TW, WW	No daylighting through TW, unless interrupted by windows (high glare potential). Diffuse light possible through translucent WW. Discourages sidelighting.	Cross ventilation: discouraged by solid TW, WW. Stack ventilation: TW or WW can itself produce stack effect, but with risk of evening overheating Night vent./thermal mass: limited interior, mass surface exposure, but lots of mass capacity. Other closed-bldg. cooling: with summer shading of TW or WW, good match with controlled cooling; thermal mass delays and reduces peak heat gains.
SS	Very high DF within sunspace; little or no daylight through common wall, except as encouraged by view and access to surface. Summer shading of sunspace reduces DF.	Cross ventilation: only to extent that common wall is penetrated for view and access. Stack ventilation: SS can become moderately effective stack. Night vent./thermal mass: both surfaces of common mass wall are available, but two spaces must be cooled. Other closed-bldg. cooling: with summer shading of SS, good match with controlled cooling (except in SS itself).
Roof pond	Discourages toplighting, encourages sidelighting, with very light color on underside of roof.	Cross ventilation: excellent potential. Stack ventilation: discouraged. Night vent./thermal mass: easily achieved with roof undersurface. Other closed-bldg. cooling: roof pond night sky cooling is often sufficient by itself, making other cooling unnecessary.

Source: Illustrations from AIA, Ramsey/Sleeper, *Architectural Graphic Standards* 7th ed., © 1981 John Wiley & Sons, Inc., New York. Reprinted by permission.

requirement of 2900 ft^2 of floor area, the 20% south glass area equals 0.20 × 2900 = 580 ft^2. For daylighting by sidelight only, DF_{av} = 0.2 (window area/floor area); if all the south glass area is available for daylighting (as with direct gain systems), DF_{av} = 0.2 × 20%, or 4%. This is adequate for office work. However, because you want to avoid dark areas near the rear walls of these spaces and to investigate Trombé walls, and so on, some north light is desirable. Choose about 3% of the floor area in north glass (say, 90 ft^2 glass); added DF_{av} = 0.2 × 3%, or 0.6%.

Wall and roof insulation, and overall percent wall area in fenestration, will be checked against ASHRAE 90.1-1989, from Appendix K. Omaha, Nebraska, data are given in ACP Table K.24. Assume these design features:

- Moderate ILD range of 1.51 to 3.00 W/ft^2.
- No external shading projections (therefore, use 0 to 0.249).
- Internal SC range of 0.709 to 0.60 (conventional venetian blinds).
- Daylight-sensing controls are utilized to turn off electric lights.
- Vinyl-frame commercial-type double-glazed windows, U range 0.45 to 0.29.
- Thermally massive exterior wálls HC ≥ 10, with exterior insulation.

With these data, Table K.24 yields these requirements:

- Maximum fenestration area = 32% wall area.
- Wall U_{ow} maximum of 0.12 at 19% fenestration, 0.11 at 67% fenestration.
- Roof insulation maximum U = 0.053.
- Minimum R value for 48 in. of slab insulation: vertical, R = 4; horizontal, R = 11.

Check the fenestration area:

$$\text{south glass } 580 \text{ ft}^2 + \text{north glass } 90 \text{ ft}^2 = 670 \text{ ft}^2$$

Check the total wall area:

$$\text{north } 1100 \text{ ft}^2 + \text{south } 1100 \text{ ft}^2 + \text{east } 600 \text{ ft}^2 + \text{west } 600 \text{ ft}^2 = 3400 \text{ ft}^2$$

Fenestration/wall area ratio = 670 ft^2/3400 ft^2 = 20%, well under the maximum allowable 32% of the total wall area.

If we proceed with the maximum allowable U values from ASHRAE 90.1-1989 and assume that tight construction (with few entries/exits per hour) allows us to assume 0.5 ACH infiltration:

STEP 1. Given the emphasis on daylight and the original assumption of night ventilation on the windows, choose system DGC3. This has double-glazed, night-insulated windows and will require six times as much thermally massive surface area (minimum 4 in. thick) as south glass area;

$$6 \times 580 \text{ ft}^2 \text{ south glass} = 3480 \text{ ft}^2 \text{ exposed thermal mass, minimum}$$

If all interior surfaces of exterior walls are exposed concrete block, we obtain

$$3400 \text{ ft}^2 \text{ wall total} - 670 \text{ ft}^2 \text{ window} = 2730 \text{ ft}^2 \text{ interior mass area}$$

which is not sufficient. So add at least 750 ft^2

TABLE 5.16 **Design Data for Some Early Passive Solar Buildings**

	Name, Function, Location	*Floor Area (ft^2)*	*Area Ratio, South Glass ÷ Floor*	*System Type (from Appendix C)*	*Approx. LCR: at ACH = 1.0*	*Approx. LCR: at ACH = 0.5*
	Dove Publications, Pecos, New Mexico, Office	2660	0.38	DGB1 and WWB4	37	25
	Warehouse	5040	0.11	DGC1		
	Karen Terry House, Santa Fe, New Mexico	850	0.45	DGC3	22	17
	Kelbaugh House, Princeton, New Jersey	1570[a]	0.51[b]	TWE2, DGB1, and SSB4	13	9
	First Village, Unit 1, Santa Fe, New Mexico	1800[a]	0.22[b]	SSE1	49	31

[a]Not including sunspace floor area.

[b]Includes sunspace south glazing, but not sunspace floor area.

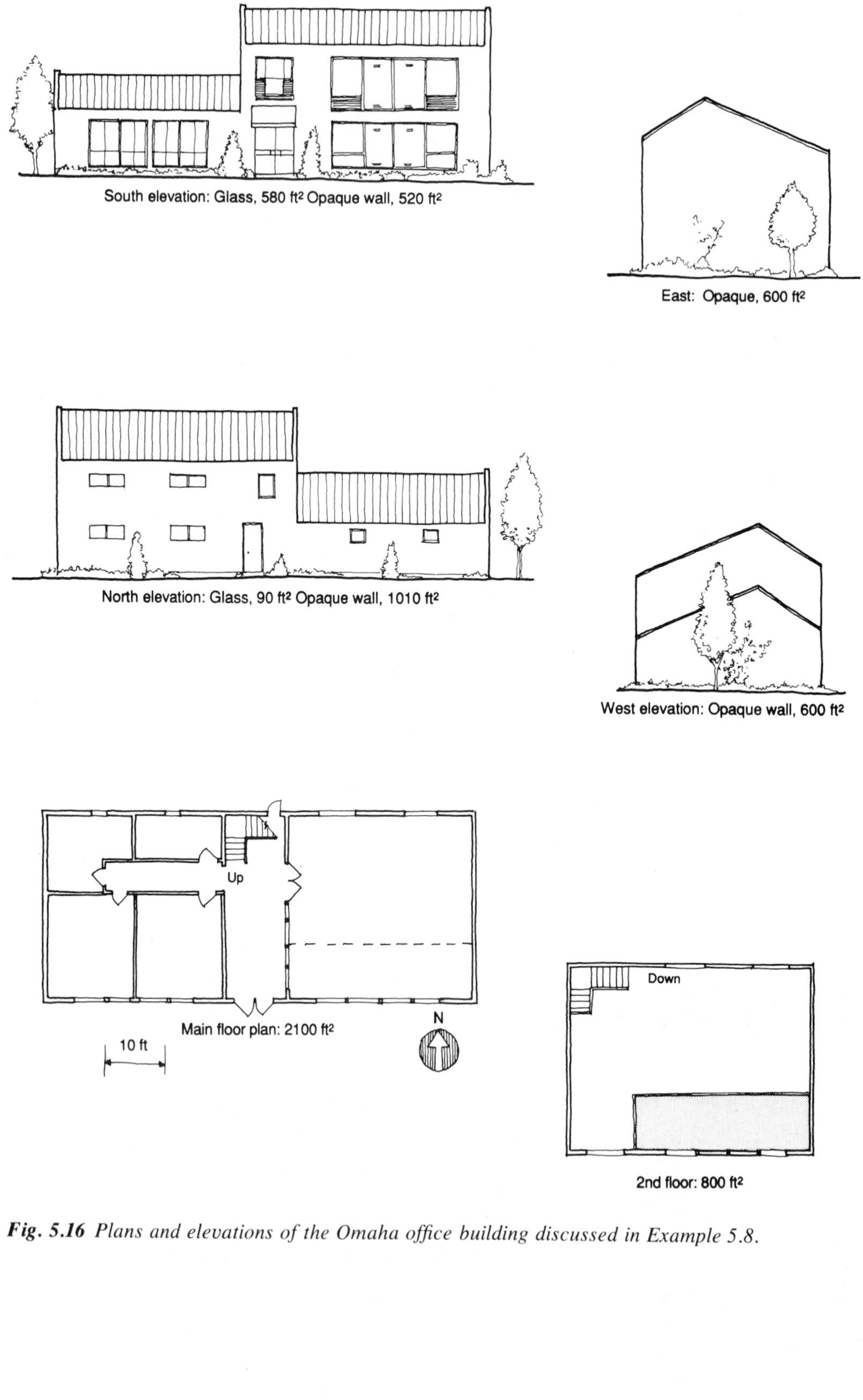

Fig. 5.16 *Plans and elevations of the Omaha office building discussed in Example 5.8.*

exposed concrete slab floor area, such as a 10-ft-wide strip just inside the south windows, for the entire length of the building.

STEP 2. The solar opening was selected at 20% floor area, or 580 ft^2.

STEP 3. Calculate UA_{ns}.

North, east, and west opaque walls: 2210 ft^2 × 0.12 = 265

South opaque wall: 520 ft^2 × 0.12 = 62

Roof: 1974 ft^2 × 0.053 = 105

Slab perimeter: 200 lin. ft × 0.49* Btu/h-°F-lin. ft = 98

North glass (assume that $U = 0.45$): 90 ft^2 × 0.45 = 41

Infiltration: 0.5 ACH × 31,000 ft^3 × 1.08/60 min = 279

Total 850 Btu/h-°F

*From Table 4.13, assuming insulated construction type (a), $R = 5.4$, and interpolating for 6601 DD65.

Then

$$BLC = 850 \text{ Btu/h-°F} \times 24 \text{ h/DD} = 20{,}400 \text{ Btu/DD}$$

STEP 4. Calculate overall heat loss rate:

$$\frac{20400 \text{ Btu/DD}}{2900 \text{ ft}^2 \text{ floor}} = 7.0 \text{ Btu/DD-ft}^2$$

This is considerably in excess of the Table 5.5 criterion of 4.6 Btu/DD-ft^2 for this 6601 DD65 climate. Unless further conservation steps are taken beyond the minimum to meet ASHRAE 90.1-1989, the annual SSF is likely to be much lower than was predicted by the rule of thumb. Promising targets: more wall insulation and more roof insulation.

STEP 5. Determine the vertical projection of the solar opening. The south windows are vertical; therefore, A_p is identical to the actual area, 580 ft^2.

STEP 6. Find the LCR:

$$\frac{20{,}400 \text{ Btu/DD}}{580 \text{ ft}^2} = 35.2$$

STEP 7. Enter Table C.2 for Omaha, Nebraska, system DGC3, and LCR = 35.2; interpolate between LCR 30 and 40. The annual SSF is found to be about SSF = 37.5%.

STEP 8. The approximate annual auxiliary heating required is

$$Q = (1 - 0.375) \times 20{,}400 \times 6601 \text{ DD} = \text{about } 84{,}160{,}000 \text{ Btu}$$

STEP 9. Compare actual SSF with predicted: SSF = 37.5% is below the 51% SSF predicted by the rule of thumb, due to inadequate conservation measures as explained in step 4. Obviously, a lower LCR yields a higher SSF; a larger collector area would lower the LCR. However, step 4 indicated that lower BLC, not higher A_p, is the better way to a lower LCR in this case.

Table C.2 can be used to compare the relative efficiencies of many passive heating systems. Using the column of LCR = 30, for instance, the best-performing systems are:

Omaha, Nebraska, LCR = 30	
WWB4	47% SSF
WWC2	45%
SSE2	44%
DGC3	42%
TWD4	42%
TWE2	42%

If any of these systems are more architecturally compatible with the building program and/or design intent, they could be used instead of DGC3. In this case DGC3 gives substantial daylighting, with little impact on the exterior appearance; the large area of thermal mass necessary was relatively easily obtained. DGC3 appears to be a reasonable choice for this building in this climate.

What if two or more passive systems are used in the same building? In that case, first do calculations for the entire building assuming that *one* system has *all* the solar area, and find the SSF. Next, do the calculations assuming that the second system has all the solar area, and find its SSF. Then average the SSF values according to the relative solar areas of the two systems.

EXAMPLE 5.9. A veterinary clinic in Buffalo, New York, is using two systems: water walls for examining rooms and direct gain for the waiting/reception area. The building's characteristics are

$$\text{balance point} = 50 \text{ F}$$

$$UA_{\text{ns}} = 356$$

A_p water wall

$$= 240 \text{ ft}^2 \text{ (reference system WWB4)}$$

A_p direct gain

$$= 150 \text{ ft}^2 \text{ (reference system DGA3)}$$

$$\text{total floor area} = 1900 \text{ ft}^2$$

predicted SSF from rule of thumb

= about 36%, with night insulation

Checking the overall heat loss criteria (Table 5.5),

$$\frac{356 \times 24}{1900 \text{ ft}^2} = 4.5$$

This is less than 4.6 (for 5000 to 7000 DD), so it is acceptable.

First, calculate as though all A_p (390 ft^2) were system WWB4:

$$\text{BLC} = 24 \times 356 = 8544 \text{ Btu/DD}$$

$$\text{LCR} = \frac{8544}{390} = 22$$

For Buffalo, LCR 20 yields SSF = 0.37, and LCR 25 yields SSF = 0.31, so SSF = 0.35, by interpolation.

Next, calculate as though all A_p were system DGA3. For LCR 22, Buffalo SSF = 0.28, by interpolation.

Now calculate the average SSF, given that water wall comprises 62% of total solar opening and direct gain comprises 38%:

$$\frac{0.35(62\%) + 0.28(38\%)}{100\%} = 0.217 + 0.106 = 0.323; \text{ SSF} = 0.32.$$

Result: The SSF is about 90% of what was predicted (0.32 as opposed to 0.36); since the glass area is already large, program requirements probably prevented further increases in it. The building already seems to conserve energy well, since it meets the heat loss criteria.

The approximate annual auxiliary heating energy required is

$$Q = (1 - 0.323) \times 8544 \text{ Btu/DD} \times 3322 \text{ DD50 (balance point)} = 19 \text{ million Btu}$$

(This is equivalent to about 5600 kWh, or about 250 therms of natural gas burned at 75% efficiency.)

(b) Variations on Reference Systems. A particularly wide set of choices faces the designer of direct-gain and sunspace systems, in which mass distribution and glass orientation can assume thousands of different combinations. The "sensitivity curves" furnished in Balcomb et al. (1984) give some guidance on how a predicted SSF might vary as an actual passive system departs from a reference system.

Sensitivity curves can serve as early general design guidelines. Looking at the curves for your location, which design changes yield dramatic results, and which make little difference? The curves may also be used to adjust the SSF you found for a reference design.

(c) Thermal Lag Through Mass Walls. The time necessary for solar heat to pass through various thermally massive materials is shown in Table 5.17. This time lag can be put to use when the time of maximum solar heat is different from the time of maximum internal heat need. A typical example is for a residence's living room in winter, where the late evening sedentary entertainment hours occur with cold temperatures outside, yet maximum warmth desired inside. A Trombé wall, for instance, made of solid grouted concrete block and 12 in. thick, will delay the arrival of maximum solar gain to the interior by almost 8 h. If maximum solar heat gain occurs at about 1 P.M. (maximum sun at noon, but highest temperatures at about 2 P.M.), then such a Trombé wall would deliver the maximum heat to the inner surface at about 9 P.M.

(d) Internal Temperatures. Two quantities are of particular interest to passive solar designers: How much higher, compared to outdoors, will the average indoor temperature be from solar heating alone? And, how widely will

TABLE 5.17 **Time Lag Through Homogeneous[a] Walls**

Material	*Thickness (in.)*	*U Value[b] (Btu/h-ft²)*	*Time Lag (h)*
Stone	8	0.67	5.5
	12	0.55	8.0
	16	0.47	10.5
	24	0.36	15.5
Solid concrete	2	0.98	1.1
	4	0.84	2.5
	6	0.74	3.8
	8	0.66	5.1
	12	0.54	7.8
	16	0.46	10.2
Common brick	4	0.60	2.3
	8	0.41	5.5
	12	0.31	8.5
	16	0.25	12.0
Face brick	4	0.77	2.4
Wood	½	0.68	0.17
	1	0.48	0.45
	2	0.30	1.3
Insulating board	½	0.42	0.08
	1	0.26	0.23
	2	0.14	0.77
	4	0.08	2.7
	6	0.05	5.0

Source: Victor Olgyay, *Design with Climate: Bioclimatic Approach to Architectural Regionalism,* Copyright © 1963 by Princeton University Press. Reprinted by permission.

[a]For composite constructions, add an estimated additional time lag to the sum of the individual materials' time lags, as follows:

Two-layer, light construction: ½ h more
Three or more layers: 1 h more
Very heavy construction: 1 h more

[b]The *U* value is based on outdoor surface conductance of 4.0 and an indoor surface conductance of 1.65 Btu/h-ft²-°F.

this internal temperature vary (swing) on a clear winter day?

The approximate temperature difference between inside and outside on a *clear* January day, called ΔT solar, can be estimated from Fig. 5.17; it varies with latitude and with the LCR. Sunspaces are not shown in the figure. Although the temperature within the sunspace cannot be easily approximated, the ΔT solar for the room beyond the sunspace can be approximated by using Fig. 5.17*a*, if these spaces are connected by vents, or Fig. 5.17*b*, if they are connected only by a solid masonry wall.

To determine the average winter indoor temperature,

1. Find the average January ambient (outdoor) temperature, *TA*, from Appendix B.
2. Find ΔT solar for the building and its site (Fig. 5.17).
3. Find the ΔT due to internal heat sources:

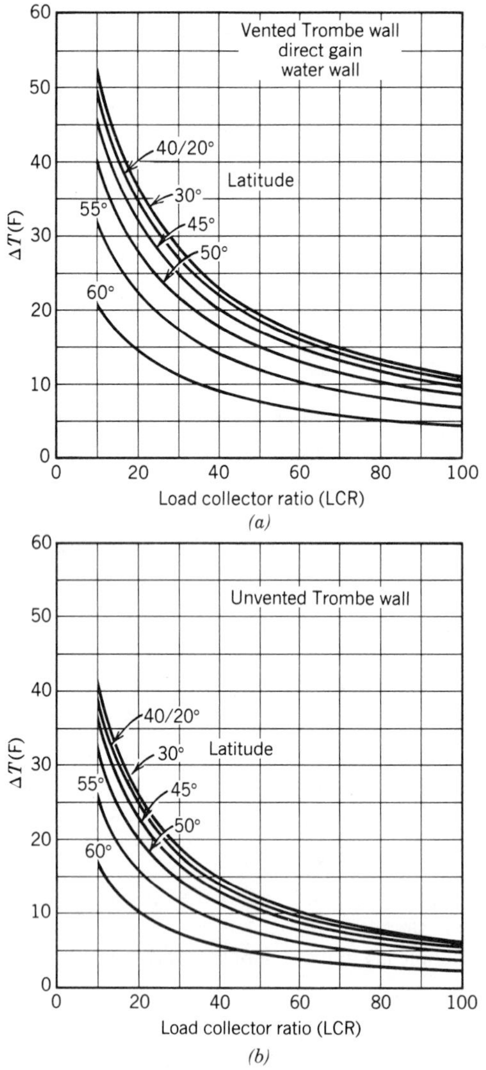

Fig. 5.17 Graphs of "ΔT solar," the temperature difference to be expected between the average inside temperature and the average outside temperature on a clear January day. The curve marked "40/20°" applies to both 40° latitude and 20° latitude. (a) *ΔT solar for direct gain, water wall, or vented Trombé wall systems.* (b) *ΔT solar for unvented Trombé wall systems. [From Balcomb et al. (1980).]*

$$\Delta T \text{ internal} = \frac{\text{total internal gains (Btu/day)}}{\text{BLC} + (UA_s \times 24)}$$

where UA_s is for the solar area only. *Note:* ΔT internal averages 5 to 7F° for residences.

4. Add the quantities from the first three steps to find the average January clear-day indoor temperature.

When internal gains are high, there is less need for ΔT solar; the building is mostly "heating itself" (becoming an ILD rather than an SLD building). If the average indoor temperature is too high, smaller solar openings should be considered unless your climate is predominantly cloudy (clear days rare) in November, December, and January.

The other important comfort question is the size of *temperature swing* due to passive solar heating. Controlled by the sun and by the actions of users, rather than by a thermostat, passive solar buildings typically experience larger daily variations (swings) in indoor temperature than do conventional buildings, especially on clear days. To estimate your building's clear-day January temperature swing, or ΔT swing, see Table 5.18. The average indoor temperature determined previously will fall in the middle of this ΔT swing.

EXAMPLE 5.1, PART L. Find the clear January day temperature swing in the typical bay of the office building in Fig. 5.4.

1. (Appendix B) Salem's average January outdoor temperature = 39 F.
2. (Fig. 5.17*a*) At 44°NL, using a direct gain system, with LCR = 33, ΔT solar = 24F°.
3. ΔT internal = (from Example 5.1, Parts F and K)

$$\frac{6.4 \text{ Btu/h-ft}^2 \times 1440 \text{ ft}^2 \times 9 \text{ h/day}}{7512 \text{ Btu/DD} + (229 \text{ ft}^2 \times 0.74 \text{ Btu/h-ft}^2\text{-°F} \times 24 \text{ h/day})} = \frac{82{,}944}{11{,}579} = 7.1\text{F°}$$

4. Average January clear day temperature indoors is

$$39 \text{ F} + 24\text{F°} + 7.1\text{F°} = 70.1 \text{ F (within the comfort zone)}$$

5. (Table 5.18) Indoor temperature swing, for direct gain with a 9 : 1 ratio of mass to south glass area, = 0.37 × ΔT solar:

$$0.37 \times 24\text{F°} = 8.88\text{F° } \Delta T \text{ swing}$$

TABLE 5.18 **Indoor Temperature Swing, ΔT Swing**

NOTE: These swings are based on a thermal storage mass capacity of 45 Btu/ft² F; ΔT solar can be found in Fig. 5.17.

Passive Solar System	ΔT *Swing*
Direct gain: $\frac{\text{mass area}}{\text{glass area}} = 1.5$	1.11 × ΔT solar
$= 3$	0.74 × ΔT solar
$= 9$	0.37 × ΔT solar
Water wall	0.39 × ΔT solar
Trombe wall, vented for 3% of wall area	0.65 × ΔT solar
Trombe wall, unvented	0.13 × ΔT solar

Source: Balcomb et al. (1980).

Therefore, the clear January day interior will vary by 8.88/2 on either side of the average temperature:

$$70.1 + 8.88/2 = 74.5 \text{ F high}$$

$$70.1 - 8.88/2 = 65.7 \text{ F low}$$

This appears to be a quite comfortable range. Since the Pacific northwest is largely overcast in winter, this daily range typically will be even less wide, but the ΔT solar will also be smaller. Thus the average January day temperature indoors will be lower.

5.7 Approximate Method for Calculating Heat Gain (Cooling Load)

Unlike the calculations for winter worst-hour heat loss, which simply assume nighttime conditions and few if any internal gains, summer worst-hour heat gain calculations are very complex. The difference in air temperature between inside and outside, which was so influential in winter, is much less important in summer. Solar gains and internal gains from lights, people, and equipment *must* be included. Summer calculations are complicated further by the fact that the hourly change in summer load can be very great. Also, in summer the thermal mass of the building becomes influential, delaying the impact of the radiant component of heat gains from all sources. The rule-of-thumb approach represented in Table 5.8 cannot respond to hourly changes or to thermally massive construction.

A simplified heat gain procedure has been developed for residential buildings; with some risk and some judgment, it can also be used for a quick *approximation* of the conditions in commercial buildings. This simplified method was devised to be used with buildings that, like residences,

1. Are occupied and air conditioned (internal temperatures closely controlled) for 24 h/day, including weekends.
2. Derive much of their gains through the building envelope and ventilation, rather than internally.
3. Are tolerant of undersized cooling equipment, with the result that unusually hot weather means noticeably higher indoor temperatures (and that interior temperatures will vary, or "swing," during a typical summer day).

Since many commercial buildings do *not* have these characteristics, this method should not be applied if very accurate estimates are desired for such buildings. But the method is rapid, if risky. Compare the results from this procedure (Example 5.10) with those from the more exact procedure for commercial buildings in Table 5.24.

Note: This procedure is based on ASHRAE tables that consistently list SI units (with conventional units in parentheses); therefore, in Sections 5.7 and 5.8 *only* we will use the order SI (conventional).

(a) Gains Through Roof and Walls. The sensible heat gains through opaque parts of a building's envelope are calculated with the equation

$$q = U \times A \times \text{DETD}$$

where the U value sometimes differs from winter and summer conditions (e.g., Table 4.12). DETD (design equivalent temperature differences) values are listed for broad categories of construction in Table 5.19; the DETD values are based on an average indoor temperature of 23.8°C (75 F) and the outdoor conditions listed. A means of correcting DETD for other temperatures is as follows:

Where the design temperature difference [outdoor design temperature, minus 23.8°C (75 F)] is not an even increment of 2.8C° (5F°), the equivalent temperature difference should be corrected 0.5C° (1F°) for each 0.5C° (1F°) difference from the tabulated values.

For rapid approximation, however, select the DETD directly from the table for the conditions nearest to your building/climate combination. The design temperature for your climate is listed in Appendix A.

(b) Gains Through Glass. The quick way to approximate these gains is

$$q = A \times \text{DCLF}$$

DCLF (design cooling load factors) values are listed in Table 5.20 and *include the U values* as well as the equivalent temperature differences. (These *DCLF* values do *not* correspond to the worst-hour gains; they were obtained by averaging the hours from 5:30 A.M. to 6:30 P.M., at both 30°N and 40°N latitudes.) Again, the DCLF values were based on an inside temperature of 23.8°C (75 F) and outside temperatures as listed. (For *rapid* approximation, ignore the corrections procedure shown in note *a*.)

Glass protected by exterior shading devices that exclude all direct sun may be considered equal to the table's values for north glass protected by awnings.

(c) Gains from Outdoor Air. In residences, outdoor air usually enters by infiltration. In many other buildings, codes require the deliberate introduction of outdoor air by mechanical ventilation. Whichever way outdoor air enters your building in summer, its sensible heat gain can be calculated by either

$$q_{\text{infiltration}} = (A_{\text{exposed}})(\text{infiltration factor})$$

where A_{exposed} is the total area of exposed wall surface, including windows and doors and the infiltration factor is found from Table 5.21: or

$$q_{\text{mechanical ventilation}} = (Q)(\text{ventilation factor})$$

where Q is the volume of outdoor air (cfm or L/s; see Section 4.6), and the ventilation factor is found from Table 5.21.

(d) Gains from People. Only the sensible gains are tabulated, since a simple overall factor for latent gains is included later. The rate of heat gain from people in various activities is shown in Table 5.22; values in the "Sensible Heat" column are generally used. [For residences, sensible heat gain per occupant is assumed at 66 W (225 Btu/h)].

$$q_{\text{people}} = (\text{number of occupants})(\text{sensible gain per occupant})$$

(e) Gains from Lights. The power supplied to electric lights (those that normally are on while cooling equipment is functioning) can be added directly to the sensible heat gain. Be sure to include ballast heat gains along with fluorescent lights; that is usually done by taking *from 1.12 to 1.2 times the total bulb wattage* of such lights (use the lower figure with energy-efficient ballasts).

(f) Gains from Equipment. In residences, a standard assumption is that 350 W (1200 Btu/h) of sensible heat gain is produced by kitchen appliances. (Other residential heat loads are assumed to be vented. See Table 5.13 for more detailed gains.) For other buildings, see the estimated range of gains given in Tables 5.23 and 5.24.

EXAMPLE **5.10.** (This example is adapted from the *ASHRAE Handbook of Fundamentals*, 1989. Reprinted by permission of the American Society of Heating, Refrigerating, and Air-Conditioning Engineers, Inc., Atlanta, Ga.)

TABLE 5.19 **Design Equivalent Temperature Differences**

29.4°C		32.2			35.0			37.7		40.5	43.3	Outdoor Design Temperature / Daily Temperature Range[a]	85 F		90			95			100		105	110
L	M	L	M	H	L	M	H	M	H	H	H		L	M	L	M	H	L	M	H	M	H	H	H
												Walls and Doors												
9.7	7.5	12.5	10.3	7.5	15.3	13.1	10.3	15.8	13.1	15.8	18.6	1. Frame and veneer-on-frame	17.6	13.6	22.6	18.6	13.6	27.6	23.6	18.6	28.6	23.6	28.6	33.6
5.7	3.5	8.5	6.3	3.5	11.3	9.1	6.3	11.8	9.1	11.8	14.6	2. Masonry walls, 203.2-mm (8-in.) block or brick	10.3	6.3	15.3	11.3	6.3	20.3	16.3	11.3	21.3	16.3	21.3	26.3
5.0	2.7	7.7	5.5	2.7	10.5	8.3	5.5	11.1	8.3	11.1	13.8	3. Partitions, frame	9.0	5.0	14.0	10.0	5.0	19.0	15.0	10.0	20.0	15.0	20.0	25.0
1.4	0.0	4.2	1.9	0.0	6.9	4.7	1.9	7.5	4.7	7.5	10.3	masonry	2.5	0	7.5	3.5	0	12.5	8.5	3.5	13.5	8.5	13.5	18.5
7.7	7.5	12.5	10.3	7.5	15.3	13.1	10.3	15.8	13.1	15.8	18.6	4. Wood doors	17.6	13.6	22.6	18.6	13.6	27.6	23.6	18.6	28.6	23.6	28.6	33.6
												Ceilings and Roofs[b]												
21.1	18.8	23.8	21.6	18.8	26.6	24.4	21.6	27.2	24.4	27.2	30.0	1. Ceilings under naturally vented attic or vented flat roof—dark	38.0	34.0	43.0	39.0	34.0	48.0	44.0	39.0	49.0	44.0	49.0	54.0
16.6	14.9	19.4	17.2	14.4	22.2	20.0	17.2	22.7	20.0	22.7	25.5	—light	30.0	26.0	35.0	31.0	26.0	40.0	36.0	31.0	41.0	36.0	41.0	46.0
21.1	18.8	23.3	21.6	18.8	26.6	24.4	21.6	27.2	24.4	27.2	30.0	2. Built-up roof, no ceiling—dark	38.0	34.0	43.0	39.0	34.0	48.0	44.0	39.0	49.0	44.0	49.0	54.0
16.6	14.4	19.4	17.2	14.4	22.2	20.0	17.2	22.7	20.0	22.7	25.5	—light	30.0	26.0	35.0	31.0	26.0	40.0	36.0	31.0	41.0	36.0	41.0	46.0
5.0	2.7	7.7	5.5	2.7	10.5	8.3	5.5	11.1	8.3	11.1	13.8	3. Ceilings under unconditioned rooms	9.0	5.0	14.0	10.0	5.0	19.0	15.0	10.0	20.0	15.0	20.0	25.0
												Floors												
5.0	2.7	7.7	5.5	2.7	10.5	8.3	5.5	11.1	8.3	11.1	13.8	1. Over unconditioned rooms	9.0	5.0	14.0	10.0	5.0	19.0	15.0	10.0	20.0	15.0	20.0	25.0
0.0	0.0	0.0	0.0	0.0	0.0	0.0	0.0	0.0	0.0	0.0	0.0	2. Over basement, enclosed crawl space or concrete slab on ground	0	0	0	0	0	0	0	0	0	0	0	0
5.0	2.7	7.7	5.5	2.7	10.5	8.3	5.5	11.1	8.3	11.1	13.8	3. Over open crawl space	9.0	5.0	14.0	10.0	5.0	19.0	15.0	10.0	20.0	15.0	20.0	25.0

Source: Copyright © by the American Society of Heating, Refrigerating and Air Conditioning Engineers, Inc., Atlanta, GA. Reprinted by permission from the *ASHRAE Handbook of Fundamentals,* 1981.

[a]Daily temperature range: L (Low) calculation value: 6.7C° (12F°) Applicable range: less than 8.3C° (15F°); M (Medium) calculation value: 11.1C° (20F°) Applicable range: 8.3 to 13.8C° (15 to 25F°); H (High) calculation value: 16.7C° (30F°) Applicable range: more than 13.8C° (25F°)

[b]For roofs in shade, 18-h average = 6.1C° (11F°) temperature differential. At 32.2°C (90F) design and medium daily range, equivalent temperature differential for light-colored roof equals 6.1 + (0.71)(21.6 − 6.1) = 17.1C° [11 + (0.71)(39 − 11) = 31F°].

TABLE 5.20 **Design Cooling Load Factors Through Glass**

Part A. **W/m²**

Outdoor Design Temp[a]	Regular Single Glass						Regular Double Glass						Heat-Absorbing Double Glass						Clear Triple Glass		
	29.4	32.2	35.0	37.7	40.5	43.3	29.4	32.2	35.0	37.7	40.5	43.3	29.4	32.2	35.0	37.7	40.5	43.3	29.4	32.2	35.0
No Awnings or Inside Shading																					
North	72.6	85.2	97.8	110.4	123.0	138.8	59.9	66.2	75.8	82.0	88.3	94.7	37.9	44.1	53.7	60.0	66.2	72.6	53.7	60.0	63.1
NE and NW	176.6	189.3	202.0	214.6	227.1	243.0	145.1	151.4	161.0	167.2	173.6	179.9	85.1	91.4	101.0	107.2	113.6	119.9	132.6	135.7	138.9
East and west	255.5	268.1	280.8	293.4	306.0	321.9	214.6	220.9	230.3	236.7	243.0	249.2	132.6	138.9	148.2	154.6	160.9	167.2	195.7	198.8	202.6
SE and SW	220.9	233.4	246.0	258.8	271.3	287.1	186.1	192.4	202.0	208.2	214.6	220.9	110.4	116.8	126.2	132.6	138.9	145.1	167.2	173.6	176.7
South	126.2	138.9	151.4	164.0	176.7	192.4	104.1	110.4	119.9	126.2	132.6	138.9	66.0	75.8	75.8	82.0	88.3	94.7	94.7	97.9	104.1
Horiz. skylight	504.8	517.4	530.0	542.7	555.2	571.0	438.6	444.9	454.3	460.7	467.0	473.2	280.8	296.6	296.6	302.9	309.1	315.6	397.6	400.7	407.0
Draperies or Venetian Blinds																					
North	47.3	60.0	72.6	85.1	97.9	113.6	37.9	44.1	53.7	60.0	66.2	72.6	28.3	34.7	44.1	50.4	56.8	63.1	34.7	37.9	44.2
NE and NW	101.0	113.6	126.2	138.9	151.4	167.2	85.1	91.4	101.0	107.2	113.6	119.9	63.1	69.4	78.9	85.1	91.4	97.9	75.8	82.0	85.1
East and west	151.4	164.0	176.7	189.3	202.0	217.7	132.6	138.9	148.2	154.6	161.0	167.2	94.7	101.0	110.4	116.8	123.0	129.3	119.9	123.0	129.3
SE and SW	126.2	138.9	151.4	164.0	176.7	192.4	110.4	116.8	126.2	132.6	138.9	145.1	75.8	82.0	91.4	97.9	104.1	110.4	101.0	104.1	107.2
South	72.5	85.1	97.9	110.4	123.0	138.9	63.1	69.4	78.9	85.1	91.4	97.9	47.3	53.7	63.1	75.8	75.8	82.0	56.8	60.0	66.2
Roller Shades Half-Drawn																					
North	56.8	69.4	82.0	92.7	102.2	123.0	47.3	53.7	63.1	69.4	75.8	82.0	31.6	37.9	47.3	53.7	60.0	66.2	41.0	44.2	47.3
NE and NW	126.2	138.9	151.4	164.0	176.7	192.4	119.9	126.2	135.7	142.0	148.2	154.6	75.8	82.0	91.4	97.9	104.1	110.4	107.2	110.4	110.4
East and west	192.5	205.0	217.7	230.3	243.0	258.8	170.3	176.7	186.1	192.4	198.8	205.0	110.4	116.8	126.2	132.5	138.9	145.1	154.6	154.6	157.8
SE and SW	164.0	176.7	189.3	202.0	214.6	230.3	145.1	151.4	161.0	167.2	173.6	179.9	94.7	101.0	110.4	123.0	123.0	129.3	129.3	132.6	135.7
South	91.4	104.1	116.8	129.3	142.0	157.8	85.1	91.4	101.0	107.2	113.6	119.9	56.8	63.1	72.6	78.9	85.1	91.4	78.9	82.0	82.0
Awnings[b]																					
North	63.1	75.8	88.3	101.0	113.6	129.3	41.0	47.3	56.8	63.1	69.4	75.8	31.6	37.9	47.3	53.7	60.0	66.2	34.7	37.9	41.0
NE and NW	66.2	78.9	91.4	104.1	116.8	132.6	44.2	50.4	60.0	66.2	72.6	78.9	34.7	41.0	50.4	56.8	63.1	69.4	37.9	41.0	44.2
East and west	69.4	82.0	94.7	107.2	119.9	135.7	44.2	50.4	60.0	66.2	72.6	78.9	37.9	44.2	53.6	60.0	66.2	72.6	37.9	41.0	44.2
SE and SW	66.2	78.9	91.4	104.1	116.8	132.6	44.2	5.04	60.0	66.2	72.6	78.9	34.7	41.0	50.4	56.8	63.1	69.4	37.9	41.0	44.2
South	66.2	75.8	88.3	101.0	113.6	129.3	41.0	47.3	56.8	63.1	69.4	75.8	34.7	41.0	50.4	56.8	63.1	69.4	34.7	37.9	41.0

Part B. **Btu/h ft²**

Outdoor Design Temp.[a]	Regular Single Glass						Regular Double Glass						Heat-Absorbing Double Glass						Clear Triple Glass		
	85	90	95	100	105	110	85	90	95	100	105	110	85	90	95	100	105	110	85	90	95
										No Awnings or Inside Shading											
North	23	27	31	35	39	44	19	21	24	26	28	30	12	14	17	19	21	23	17	19	20
NE and NW	56	60	64	68	72	77	46	48	51	53	55	57	27	29	32	34	36	38	42	43	44
East and west	81	85	89	93	97	102	68	70	73	75	77	79	42	44	47	49	51	53	62	63	64
SE and SW	70	74	78	82	86	91	59	61	64	66	68	70	35	37	40	42	44	46	53	55	56
South	40	44	48	52	56	61	33	35	38	40	42	44	19	21	24	26	28	30	30	31	33
Horiz. skylight	160	164	168	172	176	181	139	141	144	146	148	150	89	91	94	96	98	100	126	127	129
										Draperies or Venetian Blinds											
North	15	19	23	27	31	36	12	14	17	19	21	23	9	11	14	16	18	20	11	12	14
NE and NW	32	36	40	44	48	53	27	29	32	34	36	38	20	22	25	27	29	31	24	26	27
East and west	48	52	56	60	64	69	42	44	47	49	51	53	30	32	35	37	39	41	38	39	41
SE and SW	40	44	48	52	56	61	35	37	40	42	44	46	24	26	29	31	33	35	32	33	34
South	23	27	31	35	39	44	20	22	25	27	29	31	15	17	20	22	24	26	18	19	21
										Roller Shades Half-Drawn											
North	18	22	26	30	34	39	15	17	20	22	24	26	10	12	15	17	19	21	13	14	15
NE and NW	40	44	48	52	56	61	38	40	43	45	47	49	24	26	29	31	33	35	34	35	35
East and west	61	65	69	73	77	82	54	56	59	61	63	65	35	37	40	42	44	46	49	49	50
SE and SW	52	56	60	64	68	73	46	48	51	53	55	57	30	32	35	37	39	41	41	42	43
South	29	33	37	41	45	50	27	29	32	34	36	38	18	20	23	25	27	29	25	26	26
										Awnings[b]											
North	20	24	28	32	36	41	13	15	18	20	22	24	10	12	15	17	19	21	11	12	13
NE and NW	21	25	29	33	37	42	14	16	19	21	23	25	11	13	16	18	20	22	12	13	14
East and west	22	26	30	34	38	43	14	16	19	21	23	25	12	14	17	19	21	23	12	13	14
SE and SW	21	25	29	33	37	42	14	16	19	21	23	25	11	13	16	18	20	22	12	13	14
South	21	24	28	32	36	41	13	15	18	20	22	24	11	13	16	18	20	22	11	12	13

Source: Copyright © by the American Society of Heating, Refrigerating and Air Conditioning Engineers, Inc., Atlanta, GA. Reprinted by permission from the *ASHRAE Handbook of Fundamentals*, 1981.

[a]Based upon indoor design temperature of 23.8°C (75 F) and outdoor design temperatures as indicated. Interpolate to obtain factors for outdoor design temperatures other than those given.

[b]For other external shading devices which completely shade the glass at any orientation, use the values for "Awnings, north."

TABLE 5.21 **Sensible Cooling Load Factors Due to Infiltration and Ventilation**

°C: 29.4	*32.2*	*35.0*	*37.7*	*41.5*	*43.3*	*(Units)*	*Design Temperature*	*(Units) F:*	*85*	*90*	*95*	*100*	*105*	*110*
2.2	3.5	4.7	6.0	6.9	8.2	W/m^2	Infiltration, per gross exposed wall area	Btu/h ft^2	0.7	1.1	1.5	1.9	2.2	2.6
6.8	9.9	13.6	16.7	19.8	23.6	W per L/s	Mechanical ventilation	Btu/h cfm	11.0	16.0	22.0	27.0	32.0	38.0

Source: Copyright © by the American Society of Heating, Refrigerating and Air Conditioning Engineers, Inc., Atlanta, GA. Reprinted by permission from the *ASHRAE Handbook of Fundamentals,* 1981.

TABLE 5.22 **Rates of Heat Gain from Occupants of Conditioned Spaces**[a]

Degree of Activity	*Typical Application*	*Total Heat Adults, Male*		*Total Heat Adjusted*[b]		*Sensible Heat*		*Latent Heat*	
		W	*Btu/h*	*W*	*Btu/h*	*W*	*Btu/h*	*W*	*Btu/h*
Seated at rest	Theater, movie	115	400	100	350	**60**	**210**	40	140
Seated, very light work writing	Offices, hotels, apts.	140	480	120	420	**65**	**230**	55	190
Seated, eating	Restaurant[c]	150	520	170	580[c]	**75**	**255**[c]	95	325[c]
Seated, light work, typing	Offices, hotels, apts.	185	640	150	510	**75**	**255**	75	255
Standing, light work or walking slowly	Retail store, bank	235	800	185	640	**90**	**315**	95	325
Light bench work	Factory	255	880	230	780	**100**	**345**	130	435
Walking, 1.3 m/s (3 mph), light machine work	Factory	305	1040	305	1040	**100**	**345**	205	695
Bowling[d]	Bowling alley	350	1200	280	960	**100**	**345**	180	615
Moderate dancing	Dance hall	400	1360	375	1280	**120**	**405**	255	875
Heavy work, heavy machine work, lifting	Factory	470	1600	470	1600	**165**	**565**	300	1035
Heavy work, athletics	Gymnasium	585	2000	525	1800	**185**	**635**	340	1165

NOTE: All values rounded to nearest 5 watts or to nearest 10 Btu/h.

Source: Copyright © by the American Society of Heating, Refrigerating and Air Conditioning Engineers, Inc., Atlanta, GA. Reprinted by permission from the *ASHRAE Handbook of Fundamentals,* 1981.

[a]Tabulated values are based on 25.5°C (78 F) room dry-bulb temperature. For 26.6°C (80 F) room dry-bulb, the total heat remains the same, but the sensible heat value should be decreased by approximately 8% and the latent heat values increased accordingly.

[b]Adjusted total heat gain is based on normal percentage of men, women, and children for the application listed, with the postulate that the gain from an adult female is 85% of that for an adult male, and that the gain from a child is 75% of that for an adult male.

[c]Adjusted total heat value for eating in a restaurant; includes 17.6 W (60 Btu/h) for food per individual [8.8 W (30 Btu/h) sensible and 8.8 W (30 Btu/h) latent].

[d]For bowling, figure one person per alley as actually bowling and all others as sitting 117 W (400 Btu/h) or standing and walking slowly 231 W (790 Btu/h).

TABLE 5.23 **Approximate Rates of Internal Sensible Heat Gain from Equipment**

NOTE: These values do *not* include gains from electric lighting or from people.

Offices	*Btu/h ft²*	*W/m²*
General offices	3 to 4	9.5 to 12.6
Purchasing and accounting	6 to 7	19 to 22
Office with computer display units	up to 15	up to 47
Computer areas		
Digital	75 to 175	237 to 552
Analog	50 to 150	158 to 475
(For equipment with all transistors, multiply values by 0.75)		
Laboratories	15 to 70	47 to 220
Residences (see Table 5.13)		
Restaurants,[a]		
per meal served in dining area		
Sensible	(30 Btu/h)	(8.8 W)
Latent	(30 Btu/h)	(8.8 W)

Source: ASHRAE Handbook of Fundamentals (1981), Chapter 26.

[a] Note that in Table 5.22, these gains are *already included* as heat gains from people eating in restaurants.

A one-story office building (Fig. 5.18) is located in an eastern state near 40°N latitude. The adjoining buildings on the north and west are not conditioned, and their inside air temperatures are basically equal to the outdoor air temperature at any time of day.

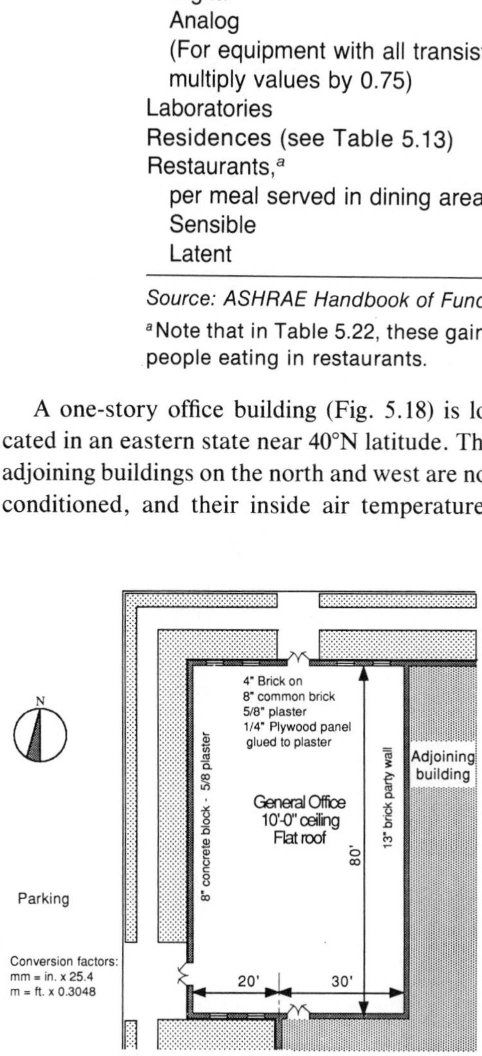

Fig. 5.18 *Plan of the office building for which heat gain calculations are shown in Examples 5.10 and 5.11. (Copyright © by the American Society of Heating, Refrigerating, and Air-Conditioning Engineers, Inc., Atlanta, GA. Reprinted by permission from the* Handbook of Fundamentals, *1989.)*

Roof construction. 114-mm (4.5-in.) flat roof deck of 50.8-mm (2-in.) gypsum slab [equivalent to four 12.7-mm (0.5-in.) layers of gypsum board], 50.8-mm (2-in.) preformed above-deck roof insulation, and two layers of mopped 6.8-kg (15-lb) felt vapor seal over insulation, no false ceiling [weight approximately 60 kg/m² (12 lb/ft²)] [summer U = 0.51 W/m²-°C (0.09 Btu/h-ft²-°F)].

South wall construction. 101.6-mm (4-in.) face brick, 203.2-mm (8-in.) common brick, 15.9-mm (0.625-in.) plaster, 6.4-mm (0.25-in.) plywood panel glued to plaster [Summer U = 1.36 W/m²-°C (0.24 Btu/h-ft²-°F)].

West wall and adjoining north party wall construction. 330-mm (13-in.) solid brick, no plaster [U = 1.42 W/m²-°C (0.25 Btu/h-ft²-°F)].

North exposed wall and east wall construction. 203.2-mm (8-in.) concrete block and 15.9-mm (0.625-in.) plaster [summer U = 2.7 W/m²-°C (0.48 Btu/h-ft² °F)].

Floor construction. 101.6-mm (4-in.) concrete on ground.

Fenestration. 0.91-m × 1.52-m (3-ft × 5-ft) window of regular plate glass with light-colored

TABLE 5.24 **Rate of Heat Gain from Miscellaneous Appliances**[a]

Miscellaneous Data	Manufacturer's Rating		Recommended Rate of Heat Gain, (W)			Miscellaneous Data	Manufacturer's Rating		Recomended Rate of Heat Gain (Btu/h)	
	Watts	Sensible	Latent	Total			Btu/h	Sensible	Latent	Total
					Electrical Appliances					
Blower type	1580	675	120	785	Hair dryer	Blower type	5,400	2,300	400	2,700
Helmet type 60 heaters @ 25 W	700	550	100	650	Hair dryer	Helmet type 60 heaters @ 25 W	2,400	1,870	330	2,200
91.44-cm normal use	1500	250	50	300	Permanent wave machine	36-in. normal use	5,000	850	150	1,000
1.27-cm diameter		28		28	Neon sign, per linear meter of tube[b]	0.5-in. diameter		30		30
0.95-cm diameter		56		56		0.375-in. diameter		60		60
	1100	190	350	540	Sterilizer, instrument		3,750	650	1200	1,850
					Magnetic card typewriter		690	350	0	350
Running	1760	1760	0	1760	Small copier	Running	6,000	6,000	0	6,000
Standby	880	880	0	880		Standby	3,000	3,000	0	3,000
Running	3515	3515	0	3515	Large copier	Running	12,000	12,000	0	12,000
Standby	1760	1760	0	1760		Standby	6,000	6,000	0	6,000
					Gas-Burning Appliances					
					Lab burners					
1.1-cm barrel	880	495	125	620	Bunsen	0.4375-in. barrel	3,000	1,680	430	2,100
3.8-cm wide	1465	820	205	1025	Fishtail	1.5-in. wide	5,000	2,800	700	3,500
2.54-cm diameter	1760	985	245	1230	Meeker	1-in. diameter	6,000	3,360	840	4,200
Mantle type	585	530	60	590	Gas light, per burner	Mantle type	2,000	1,800	200	2,000
Continuous flame	730	265	30	295	Cigar lighter	Continuous flame	2,500	900	100	1,000

Source: Copyright © by the American Society of Heating, Refrigerating and Air Conditioning Engineers, Inc., Atlanta, GA. Reprinted by permission from the *ASHRAE Handbook of Fundamentals*, 1981.

[a] For Residential appliances, see Table 5.13.

[b] Conventional (Btu/h) values are per linear foot of tube.

venetian blinds, not openable [U = 4.6 W/m²-°C (0.81 Btu/h-ft²-°F)].

Front doors. Two, 0.76 m × 2.13 m (2.5 × 7 ft).

Side doors. Two, 0.76 m × 2.13 m (2.5 ft × 7 ft).

Rear doors. Two, 0.76 m × 2.13 m (2.5 ft × 7 ft).

Door construction. Light colored 44.5-mm (1.75-in.) steel door with solid urethane core and thermal break [summer U = 1.08 W/m²-°C (0.19 Btu/h-ft²)].

U values for all doors and outside walls were calculated assuming a wind speed of 3.35 m/s (7.5 mph). For party and inside walls, still air was assumed.

Outdoor design conditions. Dry-bulb temperature, 34.4°C (94 F); daily range, 11.1C° (20F°). Wet-bulb temperature, 25°C (77 F).

Indoor design conditions. Dry-bulb temperature, 23.9°C (75 F); wet-bulb temperature, 16.9°C (62.5 F).

Occupancy. 85 office workers from 8 A.M. to 5 P.M.

Lights. 17,500 W, fluorescent, from 8 A.M. to 5 P.M.; 4000 W, tungsten, continuous.

The conditioning equipment is located in the adjoining structure to the north.

Determine the sensible, latent, and total space cooling load at design conditions.

Note: This example is compared to a much more detailed calculation procedure in Example 5.14.

Before beginning the calculations, *estimate* the heat gain using the rule of thumb procedure of Table 5.8.

People and equipment:

$$5.9 \times 2^{a} = 11.8 \text{ Btu/h-ft}^2$$

Lighting[b]: 21,500 W/4000 ft²

$$= 5.4 \text{ W/ft}^2 \times 3.41 \text{ Btu/h-W} = 18.4$$

Envelope gains:

Windows: 90 ft²/4000 ft²

$$\times 18^{c} = 0.4$$

Walls: South: 405 ft²/4000 ft²

$$\times 0.24 \times 19^{c} = 0.5$$

East, north exposed;

$$935 \text{ ft}^2/4000 \text{ ft}^2 \times 0.48 \times 19^{c} = 2.1$$

Party:

$$1065 \text{ ft}^2/4000 \text{ ft}^2 \times 0.25 \times 19^{c} = 1.3$$

Roof: 4000 ft²/4000 ft²

$$\times 0.09 \times 39^{c} = 3.5$$

Ventilation:

$$1275 \text{ cfm}/4000 \text{ ft}^2 \times 21^{c} = 6.7$$

Total 44.7 Btu/h-ft²

[a] 4000 ft²/85 people = 47 ft² per person, more than double the heat gain at a density of 100 ft²/person assumed in this approximate procedure, so increase by factor of 2.

[b] This is considerably in excess of the lighting limitations on which this approximate procedure was based. Without the detailed knowledge of installed electric lighting, the rule of thumb would have suggested, based on

$$\text{DF} = 0.2 \times \frac{90 \text{ ft}^2 \text{ windows}}{4000 \text{ ft}^2 \text{ floor}} = 0.5\% \text{ DF},$$

<1, a gain of only 5.1 Btu/h-ft²

[c] Interpolate for 94 F, between 90 and 100 F outdoor temperature.

Cooling Load Due to Heat Gain Through Building Envelope. The heat gains through roof, exposed walls, and doors shown in Table 5.25 were calculated by

$$q = U \times A \times \text{DETD}$$

for which DETD values are taken from Table 5.19. In line with the approximate (and therefore rapid) nature of this calculation method, *no corrections* were made to DETD values to adjust for either outside or inside design temperatures differing from those listed, and actual gains were rounded to nearest 10. The climate's daily temperature range is medium; the roof is dark (typical of commercial buildings in air-polluted areas). Outside design temperatures of 35°C (95 F) were used. To keep things simple, the temperature difference through party walls was made equal to that on which the table was based; in this case, 35 − 23.8 = 11.2 (95 − 75 = 20).

Cooling Load Due to Heat Gain Through Glass. These heat gains were calculated by

$$q = A \times \text{DCLF}$$

for which the DCLF values (which include U value) were taken from Table 5.20. Again, *no*

THERMAL CONTROL

TABLE 5.25 **Cooling Load Through Building Envelope (Example 5.10)**

Net Area (m^2)	U Value (W/m^2-°C)	ΔT (C°)	DETD (C°)	Reference	DCLF	Cooling Load (W)	Section	Net Area (ft^2)	U Value (Btu/h-ft^2 °F)	ΔT (F°)	DETD (F°)	Reference	DCLF	Cooling Load (Btu/h)
371.6	0.51		24.4	Table 5.19		4620	Roof	4000	0.09		44.0	Table 5.19		15,840
37.6[a]	1.36		9.1	Table 5.19		470	South wall	405[a]	0.24		16.3	Table 5.19		1,580
71.1[a]	2.7		9.1	Table 5.19		1750	East wall	765[a]	0.48		16.3	Table 5.19		5,990
15.8[a]	2.7		9.1	Table 5.19		390	North exposed wall	170[a]	0.48		16.3	Table 5.19		1,330
98.9[a]	1.41	11.2[b]				1,560	Party walls	1065[a]	0.25	20[b]				5,330
3.25	1.08		13.1	Table 5.19		50	Doors: S	35	0.19		23.6	Table 5.19		160
3.25	1.08	11.2[b]				40	N	35	0.19	20[b]				130
3.25	1.08		13.1	Table 5.19		50	E	35	0.19		23.6	Table 5.19		160
5.6	(4.59)[c]			Table 5.20	97.9	550	Windows: S	60	(0.81)[c]			Table 5.20	31	1,860
2.8	(4.59)[c]			Table 5.20	72.6	200	N	30	(0.81)[c]			Table 5.20	23	690
						Total: 9680								Total: 33,070

[a]Calculated from gross wall area, less windows and doors.

[b]Design temperature difference, inside to outside.

[c]DCLF for glass *includes* *U* value.

corrections were made, and gains were rounded to nearest 10.

Cooling Load Due to Heat Gain from Lights, People, and Equipment. The rate of heat gain from the lighting can be approximated simply by taking the energy input (including extra power required by ballasts for fluorescents). For fluorescent lights,

17,500 W × 1.2 ballast factor
= 21,000 W (71,650 Btu/h)

For incandescent lights,

4000 W (13,650 Btu/h)

The sensible gains from people can be determined from Table 5.22. For office work, assume 75 W (255 Btu/h):

85 people × 75 = 6380 W (21,680 Btu/h)

Heat gains from office equipment are ignored in this example. (See Table 5.8 for typical ranges of such gains.) The sensible gains from lights, people and equipment thus total 31,380 W (106,980 Btu/h).

Cooling Load Due to Ventilation or Infiltration. Since this is a commercial building incorporating deliberate introduction of outdoor air, infiltration will be ignored. Although the mechanical ventilation suggested in Table 4.25 is 9.3 L/s (20 cfm) per person, this example assumes 7 L/s (15 cfm) per person. Total mechanical ventilation is therefore 85 people × 7 L/s = 595 L/s (85 × 15 = 1275 cfm). Sensible heat gains (from Table 5.21):

mechanical ventilation 13.6 × 595 L/s = 8090 W

[22.0 × 1275 cfm = 28,050 Btu/h]

Latent Heat Gains. The approximate method estimates latent heat at 30% of sensible gains (in dry climates, 20%); the climate description suggests a typical northeastern U.S. climate, for which 30% is appropriate.

	W	Btu/h
Sensible gains through building envelope	9,680	33,070
Sensible gains, lights, people, (equipment)	31,380	106,980
Sensible gains, ventilation	8,090	28,050
Sensible total	49,150 W	168,100 Btu/h

Total gains due to sensible plus latent heat = 1.3 sensible = 63,900 W (218,530 Btu/h)

5.8 Detailed Hourly Heat Gain (Cooling Load) Calculations

Ordinarily, the approximate method of calculating the peak cooling load (Section 5.7) is sufficient for a designer's preliminary estimates of cooling equipment size. However, much more detailed procedures are used by engineers to actually size the equipment and to assess the peak-load impact of various design options such as shading devices.

The 1989 ASHRAE *Handbook of Fundamentals* (in its Chapter 26) presents three detailed methods for calculating heat gains and equipment cooling loads. The transfer function method (TFM) and the total equivalent temperature differential, time-averaging method (TETD/TA) each yield hourly values over a 24-h day. They account for thermal storage by building mass, which can significantly shift impact of instantaneous heat gains on the actual cooling load for the HVAC equipment, as shown in Fig. 5.19. These detailed methods are so tedious and/or complicated as to require the use of a computer. Another more direct, one-step method, cooling load temperature differential/cooling load factor (CLTD/CLF), more quickly yields a 1-h value that also accounts for the effects of thermal mass; however, considerable judgment is necessary to choose the "correct" hour for analysis. These methods are used by engineers, rarely by architects, and are not presented here. However, a summary of the results of the calculations by each method for the office building of Example 5.10 is shown in Table 5.26.

The quick and uncomplicated heat gain methods shown in Table 5.8 and in Section 5.7 served

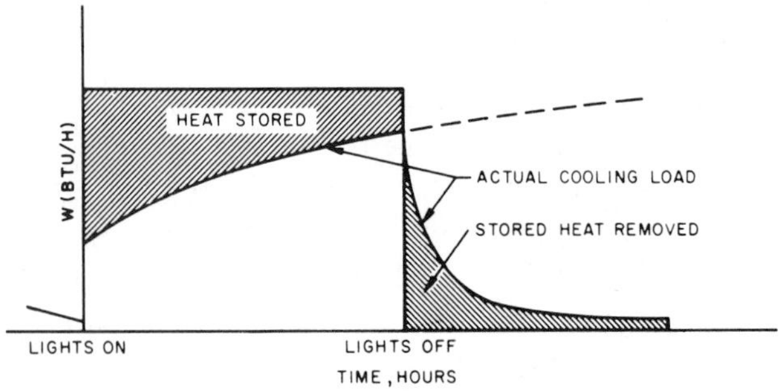

Fig. 5.19 *Thermal storage effect in the cooling load caused by electric lights. Copyright © by the American Society of Heating, Refrigerating, and Air-Conditioning Engineers, Inc., Atlanta, Ga. Reprinted by permission from* Handbook of Fundamentals, *1981.)*

the simple building in Example 5.10 rather well. As a quick approximation for estimating overall equipment size or passive cooling approach, they are adequate for smaller and more simple buildings. When utilizing these quicker methods, keep in mind that some judgment may require subdividing the calculations, or even changing some of the multipliers to reflect unusual conditions. A designer's checklist of considerations for heat gain should include:

1. Characteristics of the building's envelope: materials, sizes, external surface colors, and shapes.
2. Building location and orientation, as well as the extent of external shading of the building by trees or adjacent structures.
3. Outdoor design conditions (see Appendix A).
4. Indoor design conditions: DB, WB, and ventilation rate.
5. The schedule of lighting, occupancy, equipment, and any other processes that contribute to internal heat gain.
6. Thermal zoning requirements.

5.9 Passive Cooling Calculation Procedures

When outdoor air is 85 F or below, it is possible to cool buildings by simple ventilation, maintaining conditions indoors that are within the "comfort zone." The equations for sizing windows and stacks for either cross ventilation or stack ventilation were presented in Section 4.6.

In the many climates in which the outdoor temperature is above 80 F for a large number of

TABLE 5.26 **Comparison of Methods of Heat Gain Calculations, Example 5.10**

	Maximum Hourly Gain (Btu/h)			*24-H Average Gain (Btu/h-ft²)*		
Method	*Sensible*	*Latent*	*Total*	*Sensible*	*Latent*	*Total*
Rule of thumb (Table 5.8)	44.7 Btu/h-ft²					
Approximate[a] (Section 5.7)	168,100 (42 Btu/h-ft²)	50,430	218,530			
Transfer function method[b] (worst hour 4 P.M.)	153,765 (38.4 Btu/h-ft²)	61,168	214,933	21.9	12.5	34.4
CLTD/CLF[b] (for 4 P.M.)	156,187 (39.0 Btu/h-ft²)	63,208	219,395			
TETD/TA[b] (worst hour 4 P.M.)	174,041 (43.5 Btu/h-ft²)	61,168	235,209	21.9	12.5	34.4

[a]See Table 5.25 for calculations.

[b]From ASHRAE 1989 *Handbook of Fundamentals*, Chapter 26.

working-day hours (see Table 5.27), buildings that are closed to the hot exterior during those hours are practical. [Between 80 and 85 F, outdoor air can keep people within the comfort zone if it moves across the body fast enough (see Fig. 2.4). In areas with reliable winds, open buildings are feasible up to 85 F.] From the advent of air conditioning to recent years, the standard response of designers was to turn to mechanical cooling, utilizing the refrigeration cycle and forced air. Recently, some passive cooling alternatives have proved to be effective in climates that have clear or cool summer nights. The following procedures go beyond the quicker rules of thumb that appeared in Section 5.3 and allow the designer to better adjust a preliminary design to the program and the climate.

As an introduction to today's passively cooled buildings, consider the Bateson Building in Sacramento, California, the first in an impressive series of state office buildings that are largely passively heated and cooled. The Bateson Building (Fig. 5.20*a–c*) is a four-story structure of 285,000 ft^2 occupied by 1200 workers. Although most of the building's needed heat is internally produced by lights, people, and equipment (Fig. 5.20*d*), direct-gain passive heating is also used. Winter sun enters some office windows but comes mostly through the banks of clerestories that cover the atrium. Heat is stored in the thermally massive structure; hot air from the atrium's ceiling can be blown down to floor level through huge, bright yellow tubes, or forced into two 660-ton rock beds below the building for storage. Night insulation is not utilized because night heating is a relatively minor issue. Cooling is the major problem for such a building in Sacramento, so shading devices are plentiful. Movable exterior vertical louvres shade the south-facing clerestories; horizontal louvres shade the north-sloped glass that admits more daylight to the atrium. During the hotter months, bright orange roll-down fabric shades are automatically deployed to block direct sun in the morning on east windows and in the afternoon on west windows. From March through September, simple fixed overhangs block all south sun from offices.

Daylighting is an important design strategy: every portion of the building is within 40 ft of a window or of the daylit atrium. Each window has venetian blinds to control daylight distribution, and direct sun entering the atrium is intercepted and diffused by hanging banners before it can annoy office workers. The supplementary electric lighting still required is provided as indirect light, so that much of its heat is immediately absorbed by the exposed-concrete, thermally massive ceiling structure. Direct task lights can be used as required.

The exposed concrete structure stores the building's heat by day; on really hot days, warm indoor air can be blown into the rock beds in exchange for their stored cool air. At night, cool outdoor air is blown throughout the building by a forced-air distribution system, capable of 6 ACH, thus removing the structure's stored heat. The Bateson Building is designed to require 19,900 Btu/ft^2 per year.

(a) Night Ventilation of Thermal Mass. Before doing these detailed calculations, be sure that (1) you have checked your summer climate against the passive cooling design strategies (Fig. 2.9) and that night ventilation of thermal mass is appropriate and (2) your building and climate have been checked for approximate performance based on rules of thumb (Fig. 5.10) for this cooling strategy. Also have in mind a positive night ventilation strategy; Fig. 5.21 shows an example of a forced air system integral with a concrete joist and girder structural system. The following procedure is adapted from one developed by Karen Crowther for a workshop at the Fifth National Passive Solar Conference in Amherst, Massachusetts, 1980.

STEP 1. In column II of Table 5.28, list the hourly outdoor temperatures for the design condition; (these may be approximated from the summer DB temperature and mean daily range given in Appendix A). This will give the worst-day performance. (To get average-day performance, list average hourly temperatures, which are available from local weather service.) You need not list temperatures above 80 F.

STEP 2. Calculate the 24-h heat gain for the building, in Btu. Find the sum of all the hourly heat gains through the envelope, the minimum ventilation while "closed," and the internal gains while operating. List on line H of Table 5.28.

STEP 3. Find the total area of the thermal mass surface that is *exposed* (no rugs, etc.) both to the space to be cooled *and* to moving night air during the ventilation ("open") cycle; list on line A of Table 5.28. *Note:* The larger the mass area exposed, the better the performance. This is why direct-gain solar-heated buildings often make such good candidates for night ventilation cooling. Two additional comments on mass surface: place it where people "see" it, so that it can readily receive their radiant heat; and keep direct sun off the mass (and out of the building) during the cooling season.

STEP 4. Find the mass heat capacity for the entire space to be cooled: mass volume × density × specific heat. Table 4.2 lists both density and specific heat for most common building materials; Table 5.29 shows a quick way to get mass heat capacities for the most common thermal mass materials. Enter this total mass heat capacity on line B of Table 5.28.

STEP 5. For "supplementary" cooling due to surfaces *other* than those of the principal thermal mass, list the space's *floor area*. This step should only be taken for spaces with a significant amount of roof, wall, or floor area *in addition* to the thermal mass areas counted in step 3. *Examples:* If your space has exposed concrete ceilings, walls, and floors—all counted already as thermal mass—skip this step. If all your thermal mass is in the form of free-standing water containers, enter the entire floor area. If the entire ceiling is thermal mass, but the walls or the floors are not, enter half the floor area. If the entire ceiling is thermal mass, but the floor is not, and there are few or no walls (e.g., open office plan), enter one-third to one-fourth of the floor area.

STEP 6. Complete column III hour by hour, after determining the mass temperature (column IV for the preceding hour):

$$\text{cooling Btu/h} = \left[\begin{array}{c}\text{previous hour mass temp., F} \\ \text{(col. IV)}\end{array} - \begin{array}{c}\text{outside temp., F} \\ \text{(col. II)}\end{array}\right] \times \begin{array}{c}\text{mass surface area, ft}^2 \\ \text{(line A)}\end{array} \times \begin{array}{c}\text{surface conductance,} \\ \text{Btu/h-ft}^2\text{-°F}\end{array}$$

For the first hour, assume that the mass temperature is 80 F. The surface conductance is here assumed as 1.0 Btu/h-ft²-°F. See Table 4.3 for

TABLE 5.27 **Cooling Degree Days (CDD), to Base 18.3°C (65 F), and Cooling Hours (CH) over 26.7°C (80 F)**[a]

City, State	*CDD [°C-days (F-days)]*	*CH (h)*
Albuquerque, New Mexico	747.2 (1345)	922
Albany, New York	273.2 (492)	267
Amarillo, Texas	698.9 (1258)	951
Atlanta, Georgia	816.1 (1469)	695
Birmingham, Alabama	918.9 (1654)	1020
Bismarck, North Dakota	293.3 (528)	545
Brownsville, Texas	2139.4 (3851)	2504
Boise, Idaho	416.1 (749)	715
Boston, Massachusetts	374.4 (674)	310
Buffalo, New York	222.2 (400)	157
Burlington, Vermont	204.4 (368)	232
Charleston, South Carolina	1101.7 (1983)	1084
Cheyenne, Wyoming	171.1 (308)	299
Chicago, Illinois	396.1 (713)	377
Cincinnati, Ohio	637.2 (1147)	663

TABLE 5.27 **Cooling Degree Days (CDD), to Base 18.3°C (65 F), and Cooling Hours (CH) over 26.7°C (80 F)**[a] ***(continued)***

City, State	CDD [°C-days (F-days)]	CH (h)
Cleveland, Ohio	377.3 (670)	364
Columbia, Missouri	669.4 (1205)	765
Detroit, Michigan	381.7 (687)	372
Dodge City, Kansas	736.7 (1326)	986
El Paso, Texas	1083.9 (1951)	1576
Fort Worth, Texas	1388.9 (2500)	1742
Fresno, California	910.6 (1639)	1377
Great Falls, Montana	190.6 (343)	299
Houston, Texas	1525.0 (2745)	1593
Indianapolis, Indiana	501.1 (902)	542
Jackson, Mississippi	1311.7 (2361)	1490
Jacksonville, Florida	1517.2 (2731)	1509
Kansas City, Missouri	819.4 (1475)	1034
Lake Charles, Louisiana	1462.8 (2633)	1463
Los Angeles, California	198.3 (357)	73
Louisville, Kentucky	670.6 (1207)	696
Lubbock, Texas	876.1 (1557)	1234
Madison, Wisconsin	235.6 (424)	293
Medford, Oregon	253.3 (456)	558
Memphis, Tennessee	1040.0 (1872)	1272
Miami, Florida	2327.2 (4189)	2495
Minneapolis, Minnesota	496.7 (894)	607
Nashville, Tennessee	831.1 (1496)	893
New Orleans, Louisiana	1473.9 (2653)	1419
New York, New York	570.6 (1027)	377
Norfolk, Virginia	754.4 (1358)	634
Oklahoma City, Oklahoma	1045.6 (1882)	1280
Omaha, Nebraska	559.4 (1007)	738
Philadelphia, Pennsylvania	600.6 (1081)	563
Phoenix, Arizona	1852.2 (3334)	2710
Pittsburgh, Pennsylvania	406.7 (732)	390
Portland, Maine	162.2 (292)	181
Portland, Oregon	137.8 (248)	200
Raleigh, North Carolina	752.8 (1355)	750
Richmond, Virginia	686.7 (1236)	684
Sacramento, California	540.6 (973)	851
Salt Lake City, Utah	532.2 (958)	844
San Antonio, Texas	1588.9 (2860)	1816
San Diego, California	333.3 (600)	55
San Francisco, California	54.4 (98)	58
Saint Louis, Missouri	772.2 (1390)	927
Seattle–Tacoma, Washington	74. (134)	116
Tampa, Florida	1751.1 (3152)	1794
Tulsa, Oklahoma	955.0 (1719)	1142
Washington, D.C.	828.3 (1491)	939

Source: Copyright © by the American Society of Heating, Refrigerating and Air Conditioning Engineers, Inc., Atlanta, GA. Reprinted by permission from the *ASHRAE Handbook of Fundamentals,* 1981.

[a]From NOAA's test reference year (TRY) weather tapes.

(a)

(b)

***Fig. 5.20** The Bateson Building, an office building for Sacramento's climate. (Reprinted courtesy of the Office of the State Architect, Sacramento.) A variety of approaches to envelope and equipment distinguishes this design, developed by the staff of the California State Architect's office.* (a) *From the exterior the east facade (and its counterpart on the west) is brightened by roll-down fabric shades. These disappear daily, when direct sun no longer threatens to overheat the perimeter offices. (Sun control on the south facade is a simple trellis.)* (b) *The interior court is roofed to admit south sun in winter for passive solar heating; movable louvres can close out direct sun whenever overheating is a problem.*

other surface conductances under various conditions.

STEP 7. Complete column IV hour by hour, after calculating the cooling Btu/h (column III):

$$\text{mass temp.} = \text{previous hour mass temp. (col. IV)} - \frac{\text{cooling Btu/h (col. III)}}{\text{mass heat capacity, Btu/°F (line B)}}$$

STEP 8. Continue this hourly process using columns III and IV *until* the falling temperature of the mass equals the rising temperature of the outdoor air. At that point, continuing with plentiful ventilation will only rob the mass of its "coolth"; the building therefore switches to closed mode, with minimal ventilation.

STEP 9. Add all the hourly cooling Btu/h values (column III) to obtain the total mass cooling, Btu (line D).

STEP 10. Note the final mass temperature, from column IV. This is likely to be at least 5F° *above* the *lowest* air temperature of the night (column II). If the mass temperature is significantly higher, consider redesigning for more exposed mass surface area.

STEP 11. If supplementary cooling is appropriate (see step 5), calculate it as follows.

$$\text{supplementary cooling Btu} = \left[80\text{ F} - \text{final mass temperature (line E)} \right] \times 2.25 \times \left[\text{floor area (line C)} \right]$$

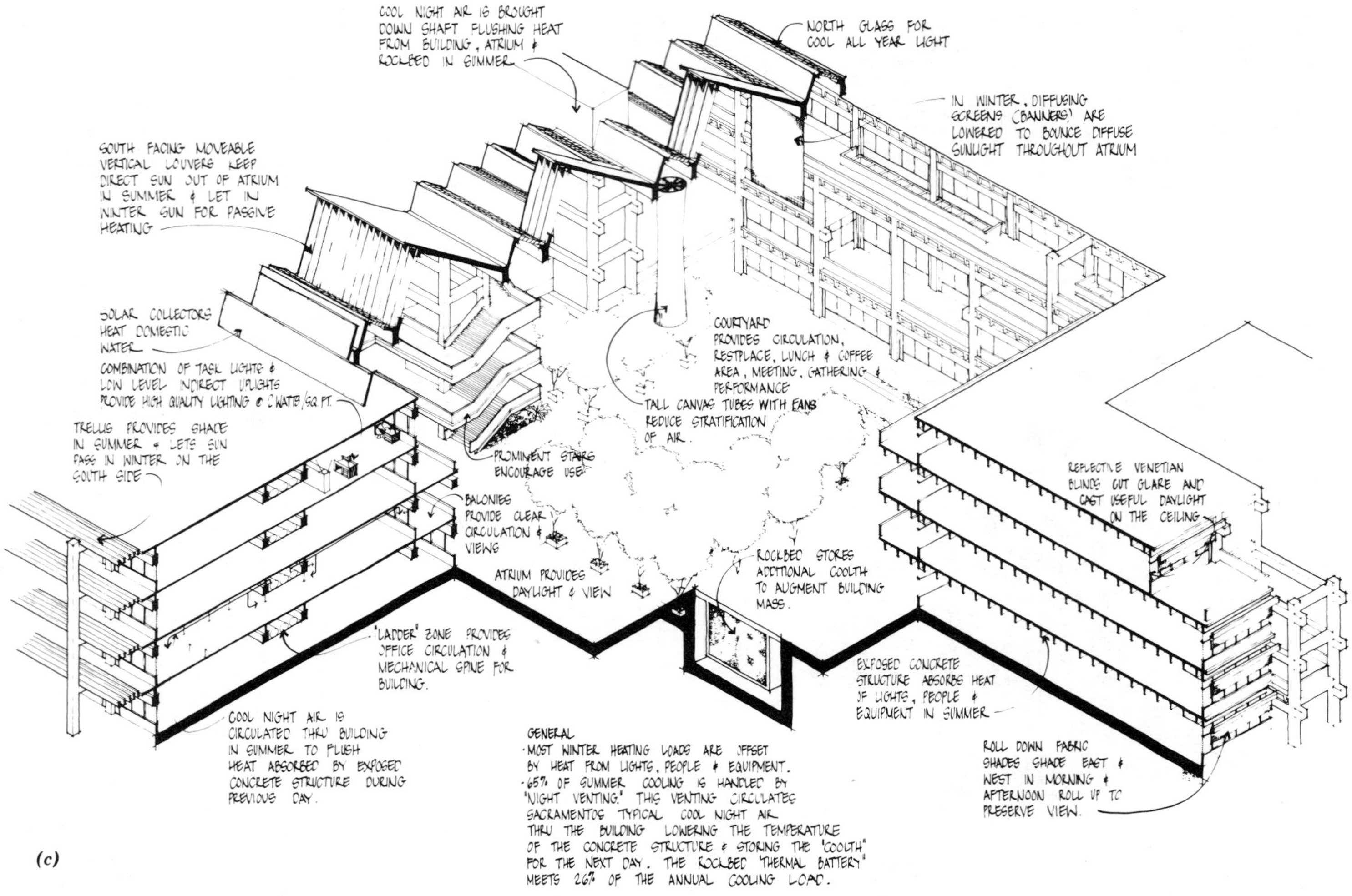

(c)

Fig. 5.20 (continued)

THERMAL CONTROL

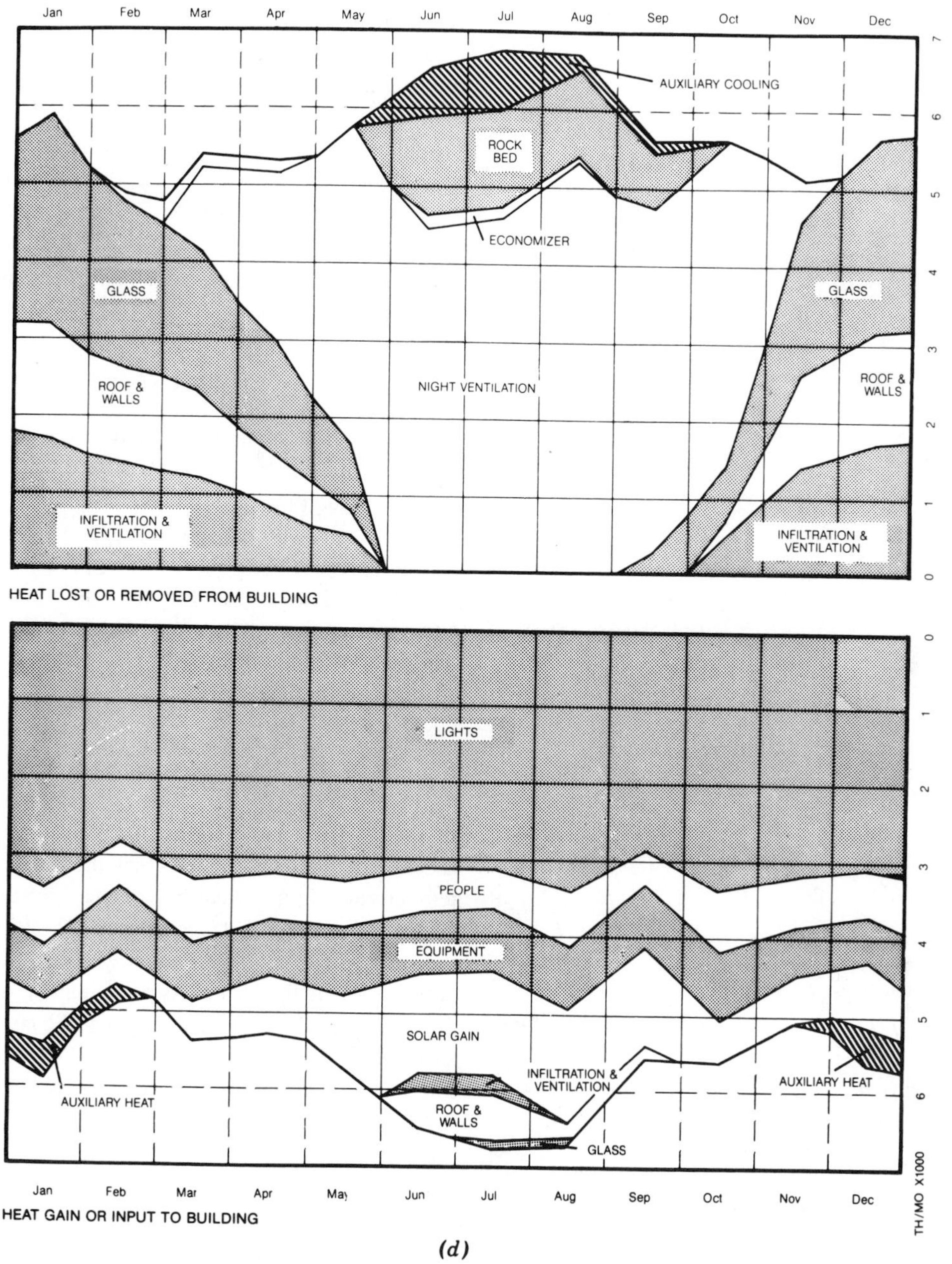

(d)

***Fig. 5.20** (continued)* (c) *The range of heating and cooling devices, including solar collectors for domestic water heating, rock bed, and building mass to store "coolth" (obtained from venting with Sacramento's cool night air), and vertical canvas tubes with fans to reduce the courtyard's air stratification.* (d) *Graphs of the average annual heat losses and gains. Note the very small portion of the total that is required of the auxiliary heating and cooling equipment. Electric lights in this daylight-emphasizing example still constitute nearly half the cooling problem in the summer—and nearly half the heating source in the winter.*

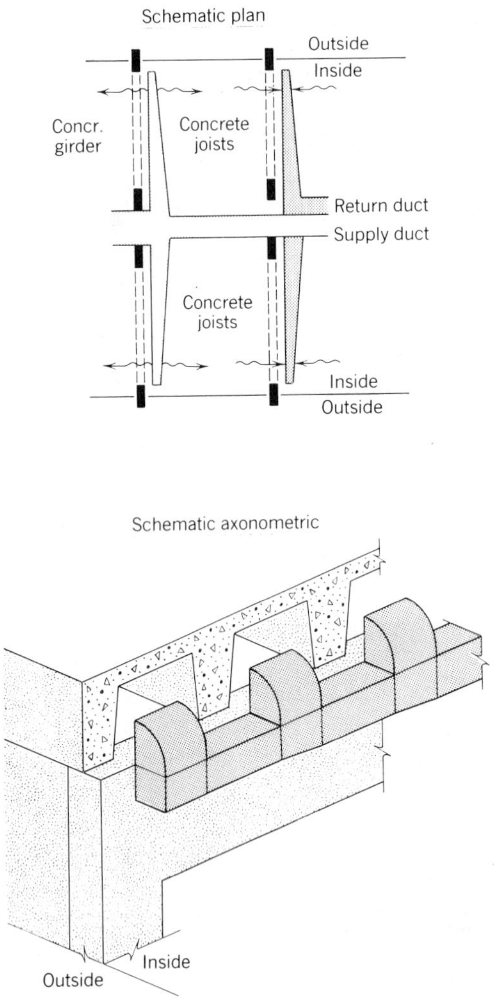

Fig. 5.21 Example of ductwork for night ventilation and the thermal mass of a concrete structural system.

The 2.25 factor assumes a modest role for the other, less thermally massive surfaces. Enter this supplementary cooling on line F.

STEP 12. Obtain total cooling by adding lines D and F; enter total on line G.

STEP 13. In step 2, you entered the 24-h heat gain for the building on line H. Compare this cooling Btu needed to the total provided (line G). If you haven't provided enough cooling, and the final mass temperature is more than 7 F° above the lowest nighttime outdoor temperature, consider redistributing the building mass over a wider surface (e.g., 3000 ft² of 4-in. slab rather than 2000 ft² of 6-in. slab) and trying again. If you do not have enough cooling, and the final mass temperature is 5 to 7 F° above the lowest nighttime outdoor temperature, consider providing both more mass and more surface area (e.g., 3000 ft² of 4-in. slab rather than 2000 ft² of 4-in. slab) and trying again.

STEP 14. Determine and enter on line I the approximate flow required for night ventilating air. Use the formula

$$\text{cfh} = \frac{\text{Btu/h}}{0.018\ \Delta T}$$

where cfh is the minimum required flow rate of night air, Btu/h is the cooling Btu for the hour of maximum cooling during the night (column III), and ΔT is the temperature difference between the final mass temperature (column IV) and the outdoor air (column II) for that same hour of maximum cooling. It is often useful to express this night ventilation flow rate in terms of air changes per hour (ACH).

$$\text{ACH} = \frac{\text{cfh required}}{\text{building volume (ft}^3)}$$

EXAMPLE 5.1, PART M. How much heat will be removed, and what is the final thermal mass temperature, in a typical bay of the Oregon office building of Fig. 5.4? (See pages 191–196.) In this procedure, assume that only the exposed surfaces of the thermal mass participate in night flush cooling. (In the actual building, the interior surfaces of the continuous cores in the precast floor slabs are also flushed with night air, yielding added cooling.)

SOLUTION.

STEP 1. The design temperature for Eugene, Oregon, is 89 F, with a mean daily range of 31 F°. This is distributed by hour (using the sine curve equation) as shown in Table 5.30, with the minimum daily temperature assumed at 4 A.M., maximum at 4 P.M.

STEP 2. The 24-h heat gain is estimated from Example 5.1, Part F. During the 9-h workday, the heat gains total 13 Btu/h-ft². During the unoccupied hours, there are negligible gains from electric lights, people and equipment, windows, or ventilation. However, the *opaque* envelope

TABLE 5.28 **Night Ventilation of Thermal Mass**

Calculation Procedure

A. Mass surface area (from step 3) ________ ft^2
B. Mass heat capacity (from step 4) ________ Btu/F
C. Floor area (supplementary cooling, step 5), and ________ ft^2
Total building volume ________ ft^3

(I) Hour	*(II)* Outside Air Temperature (F)	*(III)* Cooling (Btu)	*(IV)* Mass Temperature (F)
8 P.M.	________	________	________
9 P.M.	________	________	________
10 P.M.	________	________	________
11 P.M.	________	________	________
Midnight	________	________	________
1 A.M.	________	________	________
2 A.M.	________	________	________
3 A.M.	________	________	________
4 A.M.	________	________	________
5 A.M.	________	________	________
6 A.M.	________	________	________
7 A.M.	________	________	________
8 A.M.	________	________	________
9 A.M.	________	________	________

D. Total mass cooling ________ Btu
E. Final mass temperature ________ F
F. Supplementary cooling (see Steps 5 and 11) ________ Btu
G. Total cooling, D + F ________ Btu
H. 24-h heat gain, from step 2 ________ Btu
I. Flow rate required for night ventilation ________ cfh or ________ ACH

gains remain since this envelope stores daily heat and releases it gradually. These unoccupied gains total (from Table 5.2) 0.26 Btu/h-ft^2 from the walls, plus 1.03 Btu/h-ft^2 from the roof, totaling 1.29 Btu/h-ft^2.

Occupied:	9 h × 13 =	117	Btu/ft^2
Unoccupied:	15 h × 1.29 =	19.35	Btu/ft^2
		136.35	Btu/day-ft^2

For the entire typical bay,

TABLE 5.29 **Common Mass Heat Capacities**

Conventional Units			*S.I. Units*
ft^3	*× (Btu/ft^3 F) = Btu/F*		*m^3 × (kJ/m^3K) = kJ/K*
Volume	× (62.4)	Water	volume × (4181)
Volume	× (22.5)	Ordinary concrete	volume × (1507)
Volume	× (18.7)	Masonry, grout-filled	volume × (1253)
Volume	× (15.6)	Brick	volume × (1045)

Source: Adapted from Crowther, "Night Ventilation Cooling of Mass," in Miller (1980).

TABLE 5.30 **Oregon Office Building (Fig. 5.4) Performance: Cooling by Night Ventilation of Thermal Mass for a Typical Bay**

A. Mass surface area	2,232 ft^2
B. Mass heat capacity	16,573 Btu/°F
C. Floor area (supplementary cooling)	360 ft^2
Total building (bay) volume	22,900 ft^3

(I) Hour	(II) Outside Air Temperature (F)	(III) Cooling[a] (Btu/h)		(IV) Mass Temperature (F)
P.M.				
8	80	No heat removed		80
9	77	(80 − 77)2232 (line A)	= 6,696	$80 - \frac{6696}{16{,}573 \text{ (line B)}} = 79.6$
10	74	(79.6 − 74)2232	= 12,499	$79.6 - \frac{12{,}499}{16{,}573} = 78.8$
11	71	(78.8 − 71)2232	= 17,410	$78.8 - \frac{17{,}410}{16{,}573} = 77.7$
12	68	(77.7 − 68)2232	= 21,650	$77.7 - \frac{21{,}650}{16{,}573} = 76.4$
A.M.				
1	65	(76.4 − 65)2232	= 25,445	$76.4 - \frac{25{,}445}{16{,}573} = 74.9$
2	62	(74.9 − 62)2232	= 28,793	$74.9 - \frac{28{,}793}{16{,}573} = 73.2$
3	60	(73.2 − 60)2232	= 29,462	$73.2 - \frac{29{,}462}{16{,}573} = 71.4$
4	58	(71.4 − 58)2232	= 29,909	$71.4 - \frac{29{,}909}{16{,}573} = 69.6$
5	60	(69.6 − 60)2232	= 21,427[b]	$69.6 - \frac{21{,}427}{16{,}573} = 68.3$
6	62	(68.3 − 62)2232	= 14,062	$68.3 - \frac{14{,}062}{16{,}573} = 67.5$
7	65	(67.5 − 65)2232	= 5,580	$67.5 - \frac{5580}{16{,}573} = 67.2$
8	68	Stop flush: mass temperature is now below outdoor temperature		

D. Total mass cooling	212,933 Btu
E. Final mass temperature	67.2 F
F. Supplementary cooling	10,368 Btu
G. Total cooling (212,933 + 10,368)	223,301 Btu
H. Compare to 24-h heat gain	196,344 Btu
I. Flow rate required for night ventilation	About 5.5 ACH

[a]A surface conductance of 1.0 Btu/h-ft²-°F is assumed in this calculation.

[b]At this point, enough heat has been removed to meet the typical bay's design-day heat gain.

$$136.35 \text{ Btu/day-ft}^2 \times 1440 \text{ ft}^2 = 196{,}344 \text{ Btu/day}$$

STEP 3. The total area of exposed mass, from Table 5.2, is 1512 ft^2 for ceiling + 720 ft^2 for beams, bearing walls, and columns, for a total of 2232 ft^2. The concrete floor is covered by carpet and is therefore assumed disconnected from thermal storage for this space. The wallboard interior surfaces of the exterior walls are also ignored.

STEP 4. The mass heat capacity will be estimated assuming that all elements are solid concrete at 150 lb/ft^3 and 4 in. deep. The bearing walls are 8 in. thick, exposed on both sides, with concrete fill in the cores of all concrete blocks. The beams are 8 in. thick, but their inner faces are unexposed (they form the air duct for the night flush system). The precast ceiling slabs are 10 in. thick, but much of this thickness consists of air-filled cores; therefore, only the lower 4 in. is assumed to participate as "exposed" mass. From Table 5.29:

$$\text{Mass heat capacity} = 2232\ \text{ft}^2 \times 0.33\ \text{ft} \times 22.5\ \text{Btu/ft}^3\text{-°F} = 16{,}573\ \text{Btu/°F}$$

STEP 5. Little supplementary cooling is assumed, although the carpeted floor provides some mass, as does the wallboard surface of the exterior walls. Assuming that only one-fourth of the floor area contributes to cooling,

$$0.25 \times 1440\ \text{ft}^2 = 360\ \text{ft}^2$$

STEPS 6–10. See Table 5.30.

STEP 11. Determine supplementary cooling:

$$(80\ \text{F} - 67.2\ \text{F}) \times 2.25 \times 360\ \text{ft}^2 = 10{,}368\ \text{Btu}$$

STEP 12. See Table 5.30.

STEP 13. The building has more than adequate cooling capacity to meet this 89 F design condition; on typical summer days, with a high of only 83 F, there will be even more excess mass capacity. (A more conservative and very detailed analysis, assuming more electric lights left on, extreme high temperatures, and a less favorable coefficient of heat transfer between mass and air, led to the decision to also flush the precast cores at night.)

Because of this excess capacity, it appears advisable to begin the night flush a few hours later, when the ΔT between indoors and outdoors is greater and therefore more cooling is achieved per hour of fan operation. (Repeat the procedure above, beginning the flush at midnight with a ΔT of $80 - 68 = 12$F°, to see how similar end results can be obtained with 3 fewer hours of fan operation.) The disadvantage is that the interior will remain hotter for a longer period in the evening, which could discomfort a late worker. Alternatively, the flush could be stopped after 5 A.M., by which time the stored heat has been removed.

STEP 14. The airflow required will be checked at 4 A.M., the hour of maximum cooling:

$$\frac{29{,}909\ \text{Btu/h}}{0.018\ \text{Btu/ft}^3\text{-°F-h} \times (71.4 - 58)\text{F°}} = 124{,}000\ \text{cfh}$$

$$\text{ACH} = \frac{124{,}000\ \text{cfh}}{22{,}900\ \text{ft}^3} = 5.41$$

which is somewhat more than that predicted by the rule of thumb in Exercise 5.1, Part H, of 4.47 ACH.

(b) Evaporative Cooling. Before beginning the following calculations, be sure that:

1. You have checked your summer climate against the passive cooling design strategies (Fig. 2.9) and that the "evaporative cooling" strategy is appropriate.
2. Your building and climate have been checked for approximate performance based on the rules of thumb for evaporative cooling (Section 5.3*c*).

First find the total sensible heat gain in Btu/h that is to be removed from your building by evaporative cooling. Generally, this is calculated at your climate's summer design dry-bulb and mean coincident wet-bulb temperatures (Table A.1). The psychrometric chart is then used to plot the progress of evaporatively cooled air, as shown in Fig. 5.22. (The complete and complex psychrometric chart is found in Fig. 4.8.)

STEP 1. *Determine outdoor air conditions.* Enter the psychrometric chart at the summer design dry-bulb and mean coincident wet-bulb temperatures for your climate (point *A* in Fig. 5.22). As the air moves through the evaporative cooler, it proceeds along the constant wet-bulb line toward saturation or 100% rh (from point *A* toward point *B* in Fig. 5.22).

STEP 2. *Determine supply air temperature.* The most efficient evaporative cooler will be able to cool and humidify the outdoor air to a

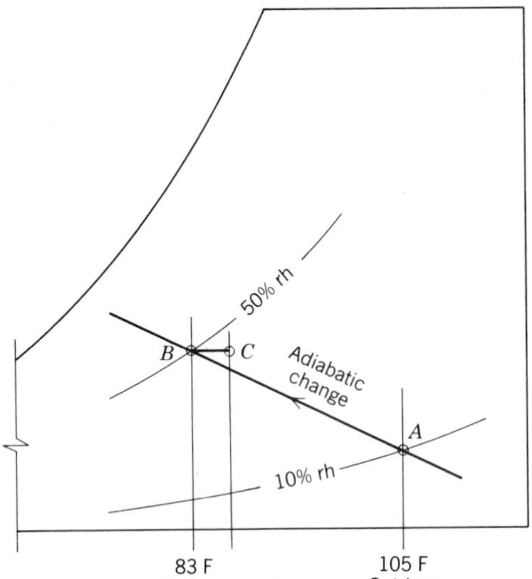

Fig. 5.22 *Once-through cycle of outdoor air through an evaporative cooler. Values are transcribed from the psychrometric chart in Fig. 4.8 (page 130). Hot-dry air can be humidified adiabatically (from point* A *to point* B*) to indoor conditions that fall within, or close to, the optimum comfort envelope shown in Fig. 2.2, without the high-energy-consuming refrigeration cycle. As the humidified air moves through the space (point* B *to point* C*) it removes some heat.*

point about three-fourths of the total distance on the chart from entering air conditions to saturation (the three-fourths point is reached at point *B* in Fig. 5.22); a more reasonable operating assumption may be two-thirds of that total distance.

STEP 3. *Determine ΔT indoors.* At these conditions, the air leaves the evaporative cooler and enters the building. It immediately begins to pick up sensible heat (moving from point *B* toward point *C* in Fig. 5.22). When the air has picked up enough heat to reach an uncomfortably warm temperature, it is exhausted to the outdoors (point *C* in Fig. 5.22).

STEP 4. *Calculate heat removed and airflow rate indoors.* The heat removed from the building by this exhaust air depends on two factors: (1) the dry-bulb temperature difference (ΔT) between the supply air and the exhaust air (point *B* to point *C* in Fig. 5.22), and (2) the airflow rate usually expressed in cubic feet per minute:

$$\text{Btu/h removed} = (\text{cfm airflow}) \times (1.08\ \text{Btu min/ft}^3\text{-°F-h}) \times (\Delta T,\ \text{F°})$$

The designer can vary the cfm airflow rate by choosing larger or smaller evaporative coolers; any point less than three-fourths of the total distance from outdoor air to saturation may also be chosen.

EXAMPLE 5.11. A 4000-ft^2 retail store near Tucson, Arizona, has been calculated to have a sensible heat gain of 100,000 Btu/h at summer design conditions (105 DB, 66 WB for this location).

SOLUTION.

STEP 1. Enter Fig. 5.22 at 105 DB, 66 WB (point *A*).

STEP 2. Measure along the constant 66-WB line to saturation, and determine that three-fourths of this distance corresponds to 76 DB, 66 WB (point *B*).

STEP 3. ΔT indoors depends on the assumed exhaust air dry-bulb temperature; assume 83 F in this example (point *C*, corresponding to 83 DB, 68 WB). Then, $\Delta T = 83 - 76 = 7$ F°.

STEP 4. Calculate heat removed and airflow rate indoors.

$$100{,}000\ \text{Btu/h removed} = (\text{cfm airflow}) \times (1.08\ \text{Btu-min/ft}^3\text{-°F-h}) \times (7\ \text{F°})$$

Rearranging the equation yields

$$\text{cfm} = \frac{100{,}000\ \text{Btu/h}}{1.08\ \text{Btu-min/ft}^3\text{-°F-h} \times 7\ \text{F°}} = 13{,}227\ \text{cfm}$$

This airflow rate can be provided by one larger or several smaller cooling units, depending on possible zones within the building. For instance, two evaporative coolers at 7000 cfm each would meet the requirements for this retail store; one might serve a smaller area zone in front with show windows and lots of solar gain, another serve a larger area interior zone with less solar gain.

Some typical evaporative coolers are shown in Section 6.7*g*.

(c) Roof Pond Cooling. Before doing the following detailed calculations be sure that (1) you have checked your summer climate against the passive cooling design strategies (Fig. 2.9) and that either the "evaporative" or the "high thermal mass" strategy is appropriate and (2) your building and climate have been checked for approximate performance based on the rules of thumb. Note, however, that the rule of thumb is based only upon summer night DB temperature. With evaporative cooling (such as a light spray onto the surface of the roof pond containers), better performance can be expected, as the following calculation procedure, based on Fleischhacker, Bentley, and Clark "A Simple Verified Methodology for Thermal Design of Roof Pond Cooled Buildings," in ASES (1982) shows. This procedure can be used to check the size and depth required for your building's roof pond. It assumes an *optimum pond depth of 4 in.*, but allows for other depths as well.

STEP 1. First, assemble the following data about your climate.

Maximum DB temperature (Appendix A).

Mean daily range (Appendix A).

Minimum DB temperature (= max DB − mean daily range).

Design Wet-Bulb (2½%) (Appendix A).

Average maximum rh for July (from local climatological data, such as those given in Table 2.4).

July average temperature (TA July, in Appendix B).

From these data, determine two further characteristics for your climate: (1) minimum WB temperature; (2) average July operating hours for residential air conditioning, or *N*.

Minimum WB temperature can be determined from the psychrometric chart shown in Fig. 4.8. Enter the chart with minimum DB and move vertically along the constant DB line until you reach the average maximum rh for July. At that intersection, refer to the diagonal WB temperature lines, from which minimum WB temperature can be determined (see Fig. 5.23).

Average July operating hours, N, can be estimated from Fig. 5.24. Enter at July TA for your climate. Your climate's *N* will fall somewhere between the maximum *N* and minimum *N* lines; as a guide to estimating *N* for your climate compare your climate's mean daily range and design wet-bulb to those of the cities shown in Fig. 5.24. The higher the mean daily range, and the lower the design wet-bulb, the lower the *N*.

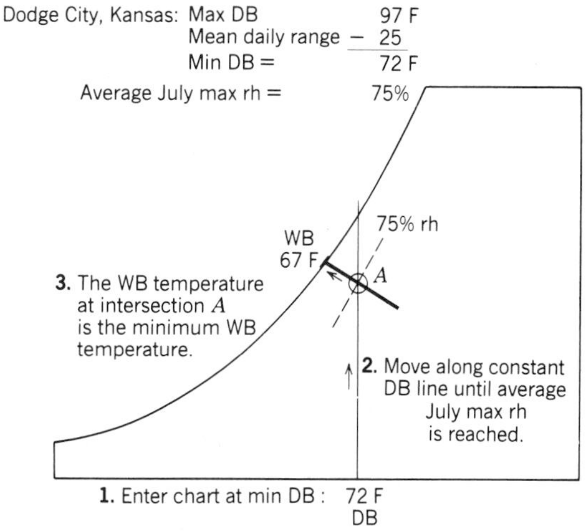

Fig. 5.23 Finding minimum WB temperature for Dodge City, Kansas.

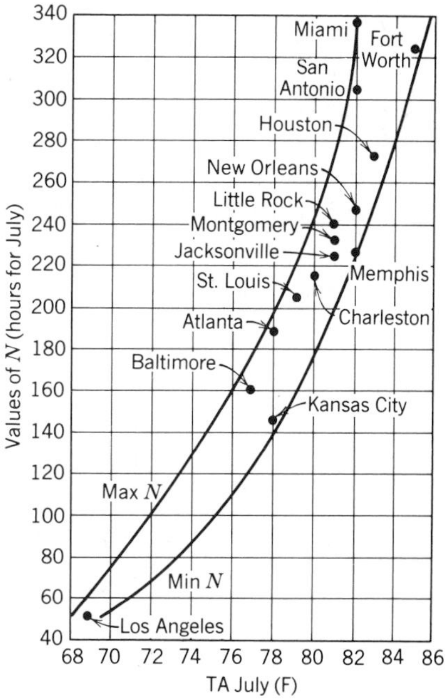

Fig. 5.24 Range of N, *average July operating hours for residential air conditioners.*

STEP 2. Calculate the peak hourly heat gain, *but exclude internal gains*. (The method given in Section 5.7 is appropriate for this roof-pond-sizing procedure.)

STEP 3. Approximate the daily total heat (excluding internal heat) to be stored in the roof pond, or Q_E, in Btu.

$$Q_E = \frac{\text{(Btu/h, peak hourly gain)}(N\text{, hours for July})}{31\text{, days in July}}$$

STEP 4. Determine the rate of the internal gain, in Btu/h, while the building is occupied.

STEP 5. Approximate the daily total internal heat to be stored in the roof pond, or Q_I, in Btu.

$$Q_I = \text{hourly internal gains, in Btu/h} \times \text{daily hours of building occupation}$$

STEP 6. Calculate the daily heat gain directly to the roof pond through its insulated covers, or Q_P, in Btu. This formula assumes $R16$ insulating panels with white upper surfaces and foil faces on the under surface.

$$Q_P = 0.4(A_c)(4 \times \text{DB max., F} - \text{DB min., F} - 200)$$

where A_c is the horizontal surface area of pond in square feet.

STEP 7. Consider whether fans will be used to stir the room air below the roof pond (to help the heat exchange between the pond and the room, as well as to aid comfort) and determine the value to be used for h, the overall heat transfer coefficient. A corrugated steel deck ceiling is assumed.

Air velocity	fpm (m/s)	0 (0)	44 (0.22)	73 (0.37)	115 (0.58)
Value of h		1.25	1.47	1.53	1.70

STEP 8. Determine the highest comfortable internal air temperature, T_{op}, from the comfort criteria provided in Fig. 2.4.

STEP 9. Calculate the maximum allowable internal air temperature, T_{imax}, which will be higher than T_{op} by a factor F that is related to your climate's characteristics.

$$T_{imax} = T_{op} + F$$

Values of *F*

	Miami	*San Antonio*	*Phoenix*
(Max. DB, coincident WB)	(90/77)	(97/73)	(107/71)
(Mean daily range)	(15)	(19)	(27)
For 0 fpm interior air motion, F =	2.0	3.0	4.0
For 115 fpm interior air motion, F =	1.0	1.5	2.0

STEP 10. Determine the allowable temperature swing of the roof pond.

(a) Calculate the maximum pond T:

$$\text{max pond } T = T_{imax} - \frac{\left(\begin{array}{c}\text{peak total hourly heat gain,}\\ \text{including internal gains, Btu/h}\end{array}\right)}{h(A_c)}$$

where T_{imax} is from step 9, h is from step 7, and A_c is the horizontal surface area of the pond in square feet.

(b) Calculate the minimum pond T, which depends on whether or not evaporative cooling will be used to help lower the pond's temperature and on several characteristics of the pond and the building.

1. For a *dry* pond surface,

$$\text{min pond } T_{dry} = \text{DB}_{min} + 1.5 \pm \text{corrections F° (if any)}$$

No Corrections Required	*Corrections Required to Minimum (Dry) Pond Temperature*								
	Pond Depth (in.)			*Night Internal Load*			*Portion Exposed to Sky*		
For a 4-in. deep pond, 2 Btu/hr ft² night internal load, 100% exposed to sky, F° corrections:	2	6	10	0	4 Btu/h ft²	8 Btu/h ft²	⅓	½	⅔
+0	−1.5	+0.7	+1.0	+0	+0.8	+2.5	+3.3	+2.1	+1.2

2. For a *wet* pond surface,

$$\text{min pond } T_{\text{wet}} = \text{DB}_{\text{min}} - \frac{\text{DB}_{\text{min}} - \text{WB}_{\text{min}}}{2}$$

(c) Find the pond temperature swing ΔT_P:

$$\Delta T_{p\text{dry}} = \text{max pond } T - \text{min pond } T_{\text{dry}}$$
$$\Delta T_{p\text{wet}} = \text{max pond } T - \text{min pond } T_{\text{wet}}$$

(Obviously, if neither minimum pond T dry nor minimum pond T wet is lower than maximum pond T, a roof pond cannot be used for cooling!)

STEP 11. Determine the required pond depth D (in inches).

$$D = \frac{(0.19)(Q_E + Q_I + Q_P)}{(\Delta T_p)(A_c)}$$

where Q_E, Q_I, and Q_P are the daily pond heat gains, from steps 3, 5, and 6; ΔT_p is from step 10; and A_c is the horizontal surface area of the pond in square feet.

Note: If D is less than 4 in., consider reducing the pond size and recalculating. If D is much more than 4 in., consider a larger pond area, more air motion, or the use of a wet pond surface to more closely approach the optimum 4-in. depth, or see optional step 12.

OPTIONAL STEP 12. Auxiliary mechanical air conditioning may offer a more economical alternative than increased pond size. The size, in "tons" of air conditioning required, is determined as follows.

(a) Desired D is 4 in., optimum.

(b) $\Delta T_p = \left(\frac{0.19}{D}\right)\frac{(Q_E + Q_I + Q_P)}{A_c}$

(c) So

$$\text{max pond } T = \text{min pond } T + \Delta T_p$$

(d) And

$$\text{tons of AC} = \frac{\text{peak total hourly heat gain (Btu/h)} - [h(A_c)](T_{i\text{max}} - \text{max pond } T)}{6000}$$

where the peak total hourly heat gain *includes* internal gains, h is from step 7, A_c is the horizontal surface area of the pond in square feet, $T_{i\text{max}}$ is from step 9, and max pond T is from step 10.

EXAMPLE 5.4, PART B. Size the roof pond for the Albuquerque office building for which we predicted that a 4-in.-deep pond equal to three-fourths of the building's floor area would be sufficient. Assume that the one about story office is 4000 ft² in area. Try a 4-in.-deep pond of 3000 ft².

SOLUTION.

STEP 1. Albuquerque,

DB_{max} is 94 F (Appendix A).

MDR is 27 F° (Appendix A).

DB min is 67 F.

July rh maximum is approximately 50% (local data).

July TA is 79 F (Appendix B).

WB_{min}, from Fig. 4.8 at the intersection of the 67 F DB line and 50% rh, is 56 F.

July operating hours N for this dry, high area should be close to the minimum for TA = 79; from Fig. 5.24, this is about 160 h.

STEP 2. Determine the peak hourly gain. Earlier in this exercise, a total gain of 15 Btu/h ft² was assumed as an average. Assume that 9 Btu/h ft² of this total represents the load from electric light, people, and equipment, and that 6 Btu/h-ft² is due to envelope and ventilation gains. (No heat gains are assumed through the roof pond.) Then

$$\text{peak hourly gains (excluding internal)} = 6 \times 4000 \text{ ft}^2 = 24{,}000 \text{ Btu/h}$$

STEP 3

$$Q_E = \frac{24{,}000 \text{ Btu/h} \times 160 \text{ h}}{31 \text{ days}} = 123{,}870 \text{ Btu}$$

STEP 4

$$\text{internal gains} = 9 \text{ Btu/h-ft}^2 \times 4000 \text{ ft}^2 = 36{,}000 \text{ Btu/ft}^2$$

STEP 5

$$Q_I = (36{,}000 \text{ Btu/h})(9 \text{ h operation}) = 324{,}000 \text{ Btu}$$

STEP 6

$$Q_p = 0.4(3000 \text{ ft}^2)(4 \times 94 \text{ F} - 67 \text{ F} - 200)$$
$$= 1200 \times 109 = 130{,}800 \text{ Btu}$$

STEP 7. Fans will be used, at 115 fpm, for added comfort as well as for heat transfer. Therefore, $h = 1.7$.

STEP 8. From Fig. 2.4, with 115 fpm (0.58 m/s) airspeed, it appears that 83 F is within the comfort zone. So $T_{op} = 83$ F.

STEP 9. $T_{imax} = T_{op} + F$. Albuquerque's conditions appear to be somewhere between those of San Antonio and those of Phoenix, so assume F to be 1.75.

$$T_{imax} = 83 + 1.75 = 84.75 \text{ F}$$

STEP 10

(a) max pond T $= T_{imax} - \dfrac{\text{total hourly gain}}{h(A_c)}$

$$= 84.75 - \frac{24{,}000 + 36{,}000}{1.7 \times 3000}$$
$$= 84.75 - 11.75 = 73 \text{ F}$$

(b) Assume a dry pond surface. Since this is a 4-in.-deep pond that is fully exposed to sky, with no night load, there is no correction factor.
min pond T_{dry} =
DB min + 1.5 = 67 + 1.5 = 68.5 F
For a wet pond,

min pond T_{wet} =

$$DB_{min} - \frac{DB_{min} - WB_{min}}{2}$$
$$= 67 - \frac{67 - 56}{2}$$
$$= 61.5 \text{ F}$$

(c) the pond temperature swing is therefore

$$\Delta T_p \text{ dry} = 73 - 68.5 = 4.5 \text{ F}^\circ$$

$$(\Delta T_p \text{ wet} = 73 - 61.5 = 11.5 \text{ F}^\circ)$$

STEP 11

$$D_{rqd} = \frac{0.19(123{,}870 + 324{,}000 + 130{,}800)}{4.5 \times 3000}$$
$$= \frac{109{,}947}{13{,}500} = 8.1 \text{ in., for a dry pond}$$

(For a wet pond, $D_{rqd} = 3.2$ in.)

It appears that a 4-in. deep pond with a wet surface of somewhat less than 3000 ft² would be sufficient for this building in this climate.

(d) Earth Contact Tempering. This cooling approach is still in a very early development stage. The most direct application of this approach would be an underground building with uninsulated concrete walls set against the soil. The problem, of course, is that winter heat loss will exceed summer loss; if solar heat can be admitted (e.g., through skylights) to counterbalance the increased winter loss, uninsulated walls sized for the desired summer loss become more attractive. (However, except in dry climates, condensation on walls may pose a seasonal or even year-round problem.)

The long-term potential of the earth as a heat sink is lessened by the fact that soils are relatively slow heat conductors. For example, we ignore heat loss through concrete slabs to the earth below, calculating instead the heat losses through the slab's exposed perimeter. A factor accounting for lower long-term heat flow rate is included in the following rule of thumb.

To estimate the cooling potential of uninsulated walls against the earth, proceed as follows.

1. Determine average underground earth temperature (about equal to well water temperature) from the map in Fig. 5.25.
2. Decide whether the soil outside your underground wall is dry, average, or wet during the cooling period.
3. Adjust the average earth temperature by comparing it to the average amplitude of surface temperature, using the map in Fig. 4.4, page 122. These amplitudes, or seasonal variations in earth temperature, diminish to near zero at about these depths:

Dry soil	14 ft
Average soil	18 ft
Wet soil	22 ft

Thus, the *summer* ground temperature at the surface will be equal to the average ground temperature (Fig. 5.25) *plus* the amplitude (Fig. 4.4); but at the depths listed above, the ground temperature re-

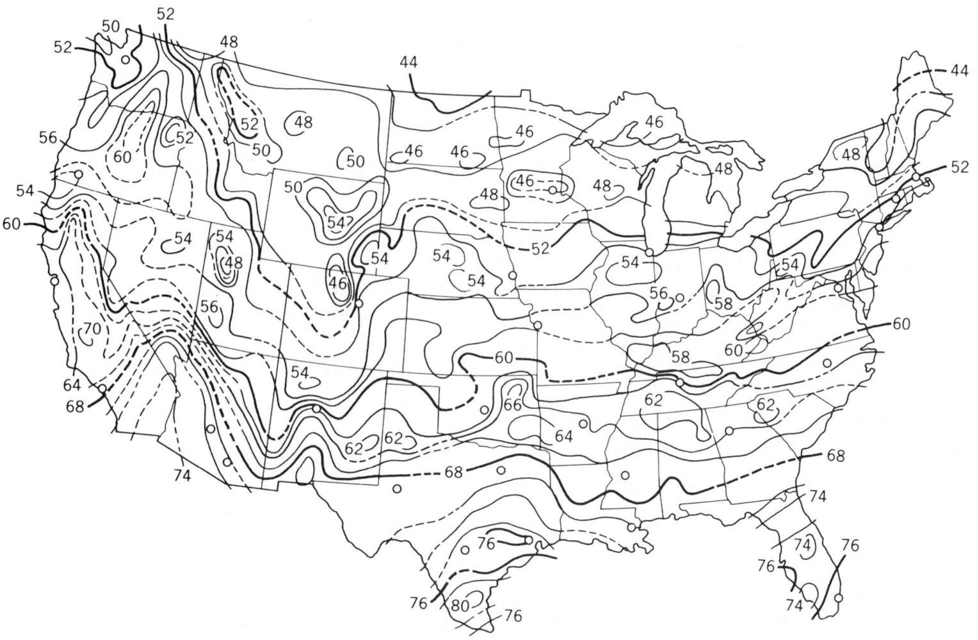

Fig. 5.25 Distribution of well water temperature in the United States, adapted from the National Water Well Association. These temperatures may be assumed to be the yearly average ground temperature. (Reprinted from Passive Cooling *by permission of the publisher, the American Solar Energy Society, Inc.)*

mains essentially the same. Considering this variation, estimate the average ground temperature outside the cooling wall, at its average depth below the surface.

4. Determine the temperature difference that makes cooling possible:

   ΔT = inside temperature − average depth summer ground temperature

5. Determine the long-term cooling rate:

   Btu/h ft^2 wall surface $= (\Delta T)(U \text{ wall})(C)$

   where C is a factor that accounts for the long-term effects of heat flow from the building to the soil through an uninsulated wall:

   $C = 0.11$ for dry soil (soil conductivity $k = 0.25$ Btu/ft-h-°F)

   $C = 0.28$ for average soil ($k = 0.75$)

   $C = 0.44$ for wet soil ($k = 1.5$)

A more thorough discussion of this procedure can be found in Watson and Labs (1983).

Earth tubes are devices for cooling the incoming ventilating air through earth contact, before it enters the building. In general, a small amount of air (perhaps equal to the minimum fresh air requirements—see Table 4.25) is brought in through several tubes. These tubes are usually 8 to 20 in. in diameter, are buried at a depth of 5 to 10 ft, and are up to 200 ft long.

The lowest temperature to which air can be cooled in such tubes will *approach* earth temperature; the slower the air moves through the tube, the more time there will be for cooling. The air temperature might come within 4 F° of the earth temperature, under good conditions.

To approximate earth tube performance:

Air velocity within each tube will be 500 fpm.

Then, air volume, cfm = (500 fpm) × (tube cross section, ft^2).

Sensible cooling will be about 1.3 Btu/h-°F per foot of length (maximum length about 300 ft), times the ΔT between earth temperature and outdoor air temperature.

For more detailed information, see C. E. Frances, "Earth Cooling Tubes, Case Studies . . ." in Bowen et al. (1981); see also Abrams (1980).

EXAMPLE **5.12.** A one-floor underground office building is being considered for Boise, Idaho. To maximize the use of the wall as a heat sink, the wall : floor ratio is set at 1 : 2 (as would be obtained in a cross section 10 ft high × 40 ft wide, with both walls underground). Uninsulated concrete walls 8 in. thick will be set at an average depth of 8 ft below ground level (including 3 ft soil depth over the roof). The U for these walls is the reciprocal of the resistances:

8-in. concrete R
= 0.08 × 8 in. = 0.64 (Table 4.2)

Inside air film R = 0.68 (Table 4.3)
Then, 0.64 + 0.68 = 1.32

$$U = \frac{1}{1.32} = 0.76$$

The skylights to be used are south-facing, triple-glazed, vertical monitors providing an average 3.0 DF. From Table 5.4,

$$\mathrm{DF}_{av}\ 3.0 = 0.2\left(\frac{\text{skylight area}}{\text{floor area}}\right)$$

skylight area = 3 ÷ 0.2
= 15% of the floor area

The triple-glazed south-facing clerestory skylights use ordinary glass, have a ⅜-in. airspace, are type C (commercial), and are set in a vinyl frame. From Table 4.16, the $U = 0.38$.

The sod roof's U value will be matched to the ASHRAE/IES Standard 90.1-1989, *Energy Efficient Design of New Buildings Except Low-Rise Residential Buildings,* requirement for Boise. From Table K.3, Boise is listed in ACP Table K.26. There we find the maximum U_o roof to be 0.051; assume that beneath the 3 ft of sod, enough insulation will be used to achieve U = 0.05. At 20 cfm per person, ventilation will meet the requirements of Table 4.25. Assuming (also from Table 4.25) that each seven persons occupy 1000 ft² of floor area, ventilation becomes (20 × 7)/1000 = 0.14 cfm/ft² floor.

The approximate heat gain is calculated as follows:

Internal (people, equip., lights) (Table 5.8)	= 7.9 Btu/h-ft²
From shaded skylights: 15% × (interpolated) 18	= 2.7 Btu/h-ft²
Through roof (at most, since sodded): 100% × 0.05 × (interpolated) 39	= 2.0 Btu/h-ft²
(No wall gains: these are *losses* to heat sink of earth.)	
Ventilation: 0.14 cfh/ft² × (interpolated) 20	= 2.8 Btu/h-ft²
Total gain	= 15.4 Btu/h-ft²

To find the cooling rate, begin with the average ground temperature, which, for Boise, is 55 F (from Fig. 5.25). Assume that the soil is dry outside the walls. The amplitude at the surface for Boise is about 20 F° (from Fig. 4.4), so in summer the surface temperature is 55 + 20 = 75 F. In dry soil, the temperature diminishes to the Boise average of 55 F at a depth of 14 ft. Since the average depth of the wall is 8 ft, assume that this outside-the-wall earth temperature in summer is about half-way between 75 and 55 F, or 65 F.

Consider an inside air temperature of 80 F, which would be quite comfortable when accompanied by cool wall surfaces; a lower MRT will allow a higher air temperature. The long-term cooling available through these walls is

Btu/h-ft² wall = (80 − 65)(0.76)
(0.11 long-term for dry soil)
= 1.25 Btu/h-ft² wall

Or, translated to floor area terms,

$$1.25 \times \frac{1 \text{ wall area}}{2 \text{ floor area}} = 0.68 \text{ Btu/h-ft}^2 \text{ floor}$$

or about 4% of the cooling requirements. Note that the wall's inner surface temperature will be approximately

$$80\text{ F} - \frac{(0.68R \text{ of air film})(15°\text{F } \Delta T)}{1.32 \text{ total } R} = 80 - 7.7 = 72.3 \text{ F}$$

(procedure from Fig. 4.5) Because 72.3 F is well

above the wet-bulb 64 F design outdoor temperature, summer condensation therefore would not seem to be a problem.

In *winter*, however, the heat gains from the envelope disappear; the internal gains at 7.9 Btu/h-ft^2 remain during occupied hours. To calculate envelope losses, assume an internal temperature of 75 F (remember those cold walls). At worst, winter conditions will be based on 75 inside − 10 outside = 65 F ΔT.

Skylights:	15% × 0.38 × 65	= 3.7 Btu/h-ft^2
Roof:	100% × 0.05 × 65	= 3.3 Btu/h-ft^2
Ventilation:	0.14 × 65	= 9.1 Btu/h-ft^2
Total		16.1 Btu/h-ft^2 loss, at worst

The earth temperature outside the walls now swings to the lower side of the average earth temperature. With the midpoint between the 55 F average ground temperature and the 35 F winter surface temperature being 45 F, wall heat losses become

$$(75 - 45)(0.76)(0.11) = 2.5 \text{ Btu/h-ft}^2 \text{ wall}$$

or

$$2.5\left(\frac{1 \text{ wall area}}{2 \text{ floor area}}\right) = 1.3 \text{ Btu/h-ft}^2 \text{ floor, average winter.}$$

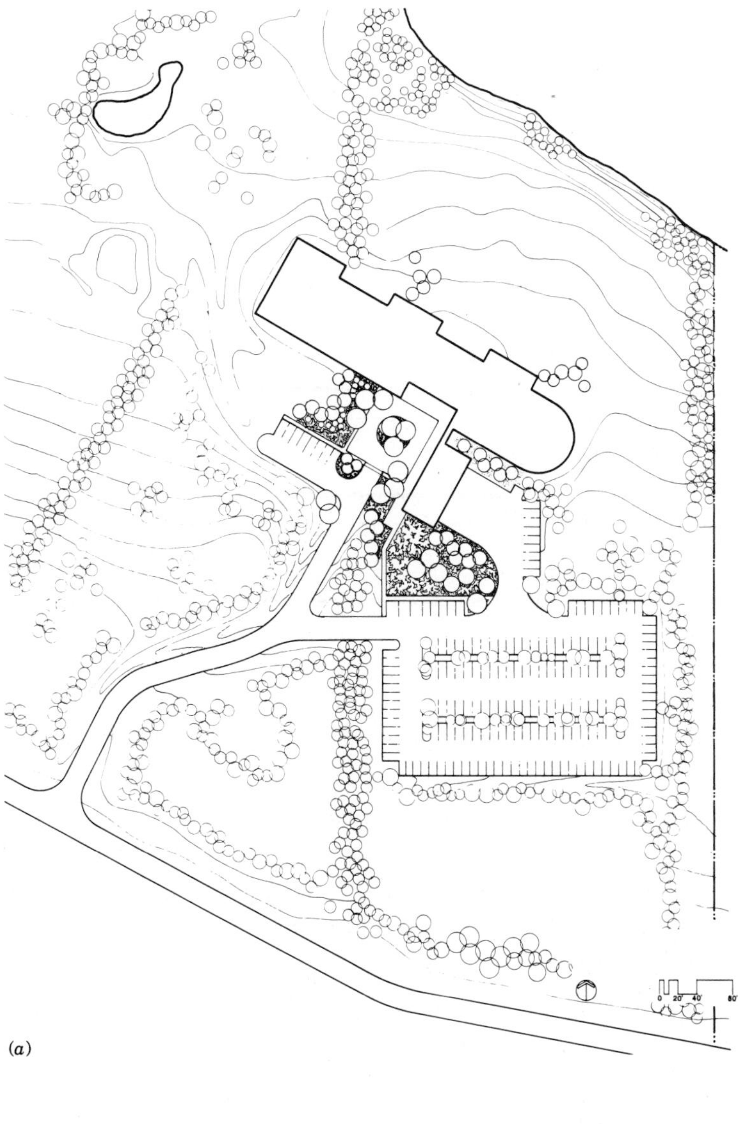

(*a*)

Fig. 5.26 *The suburban Buffalo, New York, office building for Ecology and Environment uses both stack effect and negative pressure at operable skylights to maintain indoor comfort for much of the summer.* (a) *The building is arranged perpendicular to the prevailing southwest winds.* (b) *Windows dominate the southwest (and northeast) facades; their top sections are operable.* (c) *Main floor plan; a central operable skylight and atrium provide daylight and allow heated air indoors to rise to the skylight.* (d) *Section shows relation of wing walls to skylight openings; wingwalls augment negative pressure at the skylight, helping the stack effect to exhaust air. [Courtesy of Cannon (Architecture Engineering Interiors Design), Grand Island, New York.]*

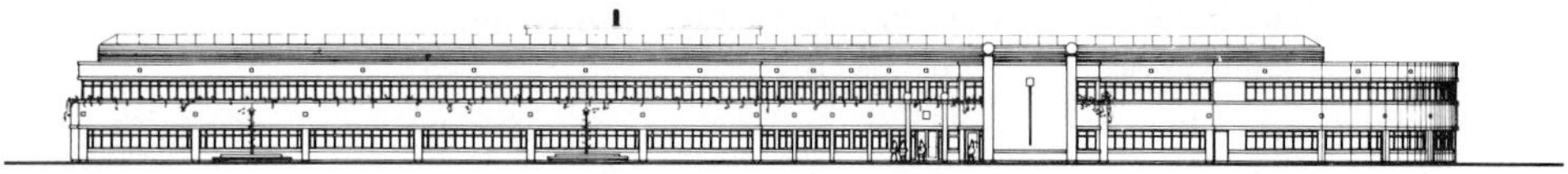

(b)

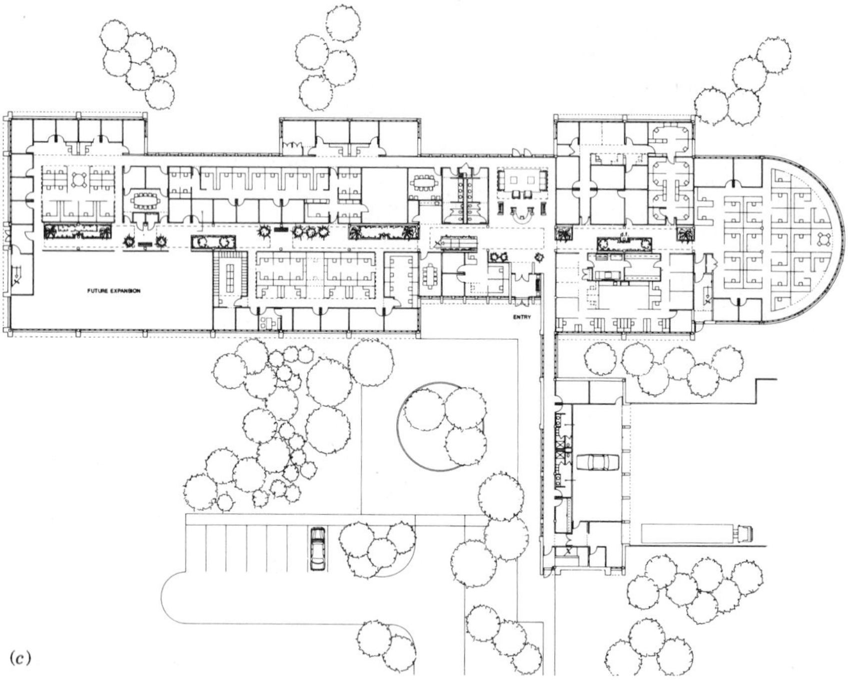

(c)

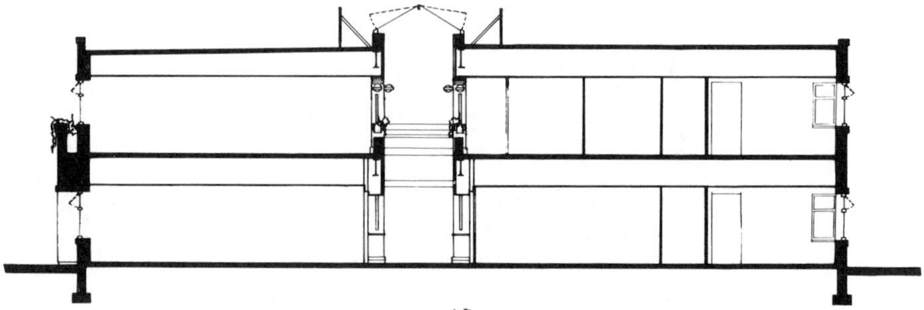

(d)

Fig. 5.26 *(continued)*

Total envelope losses therefore are 16.1 + 1.3 = 17.4 Btu/h-ft², under worst conditions. Comparison of the winter gain to the winter loss gives

9 h gain × 7.9 Btu/h-ft² = 71 Btu/day-ft²

24 h loss × 17.4 Btu/h-ft² = 418 Btu/day-ft²

347 Btu/day-ft² heat loss, at worst

Solar radiation through the south skylights will help:

Boise January insolation
= 927 Btu/ft²-day (Appendix B)

927 × 15% = 139 Btu/day-ft² floor, about 40% of the net daily heat loss under worst conditions

However, there may be condensation problems with the cold wall surfaces, as well as radiant heat loss discomfort.

Fig. 5.26 *(continued)*

(e) Natural Ventilation. The basic equations for calculating cross and stack ventilation were presented in Section 4.6*b* and *c*; rules of thumb were shown in Section 5.3*d* and *e*. To analyze natural ventilation in more detail, wind tunnel tests are more likely to be useful than are detailed calculations. A more detailed procedure for calculating cross ventilation in residences given in Chandra et al. (1986) accounts for such factors as neighboring buildings and terrain, height of opening above grade, insect screens, and framing members within openings. More details on some components associated with natural and stack ventilation, and examples of buildings using them, are given in Section 6.4*f*.

The office building for Ecology and Environment (a firm of international specialists in the environment) near Buffalo, New York, uses prevailing wind to augment the stack effect (Fig. 5.26). After wind tunnel tests of several roof vent configurations and building positions on the suburban site, the building was arranged with its long axis (and central skylight) perpendicular to the prevailing southwest winds. Wing walls were then placed along both sides of the skylight, about 4 ft away. Mechanically operable openings in the skylight thus face these wing walls. When the winds move over the roof, negative pressure areas are created at the skylight, thanks to the wing walls. The negative pressure helps draw out the hot air that has risen to the skylight via the stack effect. Fresh air enters the building readily through the windward (SW) side of the building. Fresh air will also enter the leeward side when the negative pressure at the skylight is strong enough. Wind tunnel testing and weather data indicate that this system will maintain indoor air temperatures about 4 F° above ambient. For about two-thirds of the occupied hours in summer, this natural ventilation will keep the building at or below 80 F.

REFERENCES

Abrams, D. W. (1986). *Low Energy Cooling,* Van Nostrand Reinhold, New York.

ASES (1982). *Progress in Passive Energy Systems,* Vol. 7, Proceedings of the 7th National Passive Solar Conference, American Solar Energy Society, Inc., Boulder, Colo.

Publications of the American Society of Heating, Refrigerating, and Air-Conditioning Engineers, Inc., Atlanta, Ga.: *Handbook of Fundamentals,* 1981, *Cooling and Heating Load Calculation Manual,* 1979.

Balcomb, J. D., Barley, D., McFarland, R., Perry, Jr., J., Wray, W., and Noll, S. (1980). *Passive Solar Design Handbook,* Vol. 2; *Passive Solar Design Analysis,* U.S. Department of Energy, Washington, D.C.

Balcomb, J. D., Jones, R. (ed.), Kosiewicz, C., Lazarus, G., McFarland, R., and Wray, W. (1983). *Passive Solar Design Handbook,* Vol. 3, American Solar Energy Society, Inc., Boulder, Colo.

Bowen, A., Clark, E., and Labs, K. (1981). *Passive Cooling,* American Solar Energy Society, Boulder, Colo.

Brown, G. Z., Reynolds, J. S., and Ubbelohde, S. (1982). *Inside/Out: Design Procedures for Passive Environmental Technologies,* Wiley, New York.

Buehrer, H. (1978). *Estimating Energy Usage,* developed for the AIA Energy Audit Seminars.

Chandra, S., Fairey, P. W., and Houston, M. (1986). *Cooling with Ventilation,* Solar Energy Research Institute, Golden, Colo.

THERMAL CONTROL

Egan, M. C. (1975). *Concepts in Thermal Comfort,* Prentice-Hall, Englewood Cliffs, N.J.

Johnson, T. (1981). *Solar Architecture, The Direct Gain Approach,* McGraw-Hill, New York.

Mazria, Edward (1979). *The Passive Solar Energy Book,* Rodale Press, Emmaus, Pa.

Miller, H., Editor (1980). *Proceedings of the Passive Cooling Workshop,* 5th National Passive Solar Conference, Amherst, published by American Solar Energy Society, Inc., Boulder, Colo.

Millet, M., and Bedrick, J. (1980). *Manual Graphic Daylighting Design Method,* Department of Architecture, University of Washington, Seattle, Wash.; sponsored by the U.S. Department of Energy.

Olgyay, V., and Olgyay, A. (1957). *Solar Control and Shading Devices,* Princeton University Press, Princeton, N.J.

Watson, D., and Labs, K. (1983). *Climatic Design: Energy Efficient Building Principles and Practices,* McGraw-Hill, New York.

6

SYSTEMS AND EQUIPMENT FOR HEATING AND COOLING

The preceding chapters have dealt with most of the early steps in the design process, such as the choice of thermal design strategy (Chapter 2), the choice of siting (Chapter 3), and the choice of the components and initial sizing of envelopes (Chapters 4 and 5). The next two chapters carry the design process into the specifics of mechanical systems and equipment.

There are various ways to organize this large collection of components and systems. The next two chapters are organized as follows:

Chapter 6:

1. A review of the relative thermal role of building envelopes versus their internal heating/cooling equipment.
2. A look at the heating/cooling *system design process* begun by the architect and completed by the engineer.
3. The basics of the process by which mechanical systems are integrated into building design.
4. Some details of *envelope components* typically used to reduce the need for heating and cooling.
5. Some details of *internal components*; the refrigeration cycles.
6. More details of small-building heating/cooling systems.

Chapter 7:

7. The four generic HVAC system types for large buildings.
8. Larger, centralized HVAC equipment.

9–12. Details of the four generic HVAC systems.

13, 14. Large-scale system opportunities.

Items 1, 2, 3, and 7 are a more theoretical introduction to systems and equipment; hence, rather than reading the chapter in order, you could read those items first, then go on to the component-and-equipment sections.

6.1 Review of the Need for Mechanical Equipment

One of the most basic functions of buildings is to provide shelter from weather. In a carefully designed building, the roofs, walls, windows, and interior surfaces alone can maintain comfortable interior temperatures for most of the year in most North American climates. With appropriate scheduling, the most uncomfortable hours within buildings can often be avoided; for example, the siesta avoids the hottest afternoon hours within stores and office buildings. Several aspects of comfort and climate, however, pose difficult challenges for ordinary building forms and materials.

A building surface primarily influences comfort through its *surface temperature*; secondarily, it slowly changes *air temperature* (as when cool air moves across a warm surface). Important as these two determinants of comfort are, they sometimes are not sufficient by themselves. Especially in cooling situations, *air motion* and *relative humidity* are significant comfort determinants; and for many indoor activities, *air quality* becomes an important issue in both heating and cooling.

Building form can work with climate to produce air motion for cooling, although the faster air speeds that extend the human comfort zone above 83 F (28°C) may be difficult to provide without mechanical assistance. Relative humidity is still controlled by mechanical (or chemical) means. Building form and materials may be able to keep spaces surprisingly cool, but without dehumidification, surfaces in many North American summer climates can become clammy and covered with mold.

It is difficult to filter air without a fan to force the air through the filtering media. Electrostatic filtering, of course, is done within mechanical equipment.

Thus, whereas the desired air and surface temperatures can often be achieved by passive means (a combination of building form, surface material, and occasional user response), the comfort determinants of air motion, relative humidity, and air quality often require mechanical devices. As the control of air properties—motion, moisture, particulate content—becomes more critical to comfort, the designer becomes more likely to respond with a sealed building, excluding outdoor air except through carefully controlled mechanical equipment intakes. In the recent past, this exclusion of outdoor air has often been accompanied by the exclusion of daylight, of view, of solar heat on cold days—in sum, by a general rejection of all aspects of the exterior environment. As designers come to terms with the role of mechanical equipment, we should also clarify the role of these devices in relation to the climate: Are they occasional modifiers, permanent interpreters, or permanent excluders of the outdoors?

6.2 Heating, Ventilating, and Air Conditioning (HVAC): Typical Design Processes

HVAC systems usually involve a minimum of three design stages. In the *preliminary design* phase, the most general combinations of comfort needs and climate characteristics are considered:

Activity comfort needs are listed.

Activity schedule is listed.

Site energy resources are analyzed.

Climate design strategies are listed.

Building form alternatives are considered.

Combinations of passive and active systems are considered.

One or several alternatives are sized by rule of thumb.

This level of analysis covers the process discussed in Chapters 1 to 4 and Sections 5.1 to 5.3. For smaller buildings, this analysis is often done by the architect alone. For innovative or unusual systems in smaller buildings, and especially for larger, multiple-zone buildings, consultants such as engineers and landscape architects often are included. The team approach is particularly valuable in assessing the strengths of various design alternatives. The architect and the consultants have very different perspectives, and when mutual goals can be clearly agreed upon early in the design process, these perspectives are not only mutually supporting but can produce striking innovations whose benefits extend far beyond services to the clients of a particular building. By setting an example, the team can make available better environments for less energy for hundreds of subsequent buildings. With inspired teamwork, the distribution of HVAC services can enhance building form, as many examples in this chapter show.

By the time the *design development* phase is reached, one of the design alternatives has probably been chosen as the most promising combination of esthetic, social, and technical solutions for the program. The consulting engineer (or architect, on a smaller job) is furnished with the latest set of drawings and the program. Typically, the architectural or mechanical engineer then:

1. Establishes design conditions.
   (a) By activity, lists the range of acceptable air and surface temperatures, air motions, relative humidities, lighting levels, and background noise levels.
   (b) Establishes the schedule of operations.
2. Determines the HVAC zones, considering:
   (a) Activities.
   (b) Schedule.
   (c) Orientation.
   (d) Internal heat gains.
3. Estimates the thermal loads on each zone.
   (a) For worst winter conditions.
   (b) For worst summer conditions.
   (c) For the average condition or conditions that represent the great ma-

jority of the building's operating hours.
   (d) Frequently, an estimate of annual energy consumption is made.
4. Selects the HVAC systems. Often, several systems will be used within one large building, since orientation, activity, or scheduling differences may dictate different mechanical solutions. Especially common is one system for the all-interior zones of large buildings and another system for the perimeter zones.
5. Identifies the HVAC components and their locations.
   (a) Mechanical rooms.
   (b) Distribution trees—vertical chases, horizontal runs.
   (c) Typical in-space components, such as underwindow fan-coil units, air grilles, and so on.
6. Sizes the components.
7. Lays out the system. At this stage, conflicts with other systems (structure, plumbing, fire safety, circulation, etc.) are most likely to become evident. Since insufficient vertical clearance is one of the most common building coordination problems with HVAC systems (especially all-air systems—see Sections 7.1 and 7.4), the layouts must include sections as well as plans. Opportunities for integration with other systems also become more apparent at this stage: air ducts can also help distribute daylighting, act as sunshading devices, or fulfill other functions.

After the architect and the other consultants hold conferences in which HVAC system layout drawings are compared to those for other primary systems (structure, plumbing, electrical, etc.), *design finalizing* occurs. At this final stage, the HVAC system designer verifies the match between the loads on each component and the component's capacity to meet the load. Final layout drawings then are completed.

6.3 HVAC and Building Organization

By this time, many decisions about a building design have been made: design strategies appropriate to the climate and the building's activities have been identified, and the basic siting and overall form of the building have been determined from daylighting and thermal considerations, among others. This section begins by considering the internal, yet broad issues of zoning and system choice and ends with a discussion of the more detailed consequences of system choice. A general guide to estimation of a building's thermal zone requirements was developed in Section 5.1, which discussed the importance of differences in function, schedule, and orientation.

(a) Zoning. The minimum number of thermal zones for a conventionally designed multipurpose building is shown in Fig. 6.1; more than these 16 zones could be produced by differences in scheduling within a zone, such as between "offices" and "stores." As is true of the other occupied floors, "apartments" have a minimum of five zones (based on orientation); however, the emphasis on individual controls—and the variation in usage patterns—often produces as many zones as there are apartments. When the details of zoning are added to the other preliminary design decisions, the details of HVAC systems can be considered.

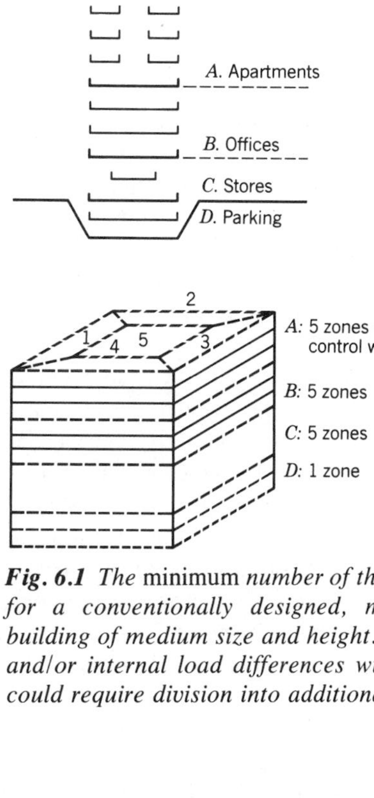

Fig. 6.1 *The* minimum *number of thermal zones for a conventionally designed, multipurpose building of medium size and height. Scheduling and/or internal load differences within a zone could require division into additional zones.*

TABLE 6.1 **Basic HVAC Systems: Tasks and Components**

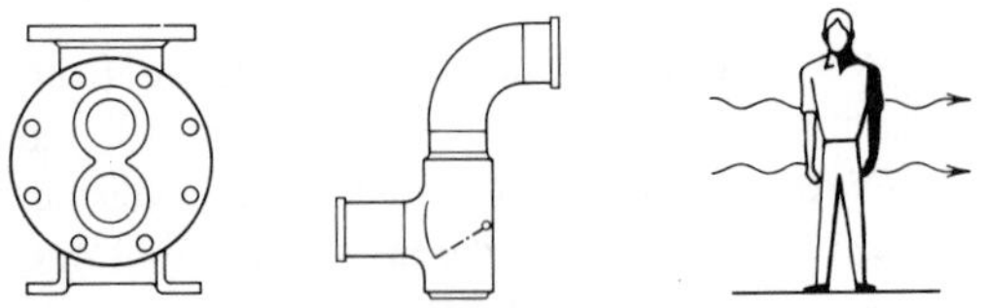

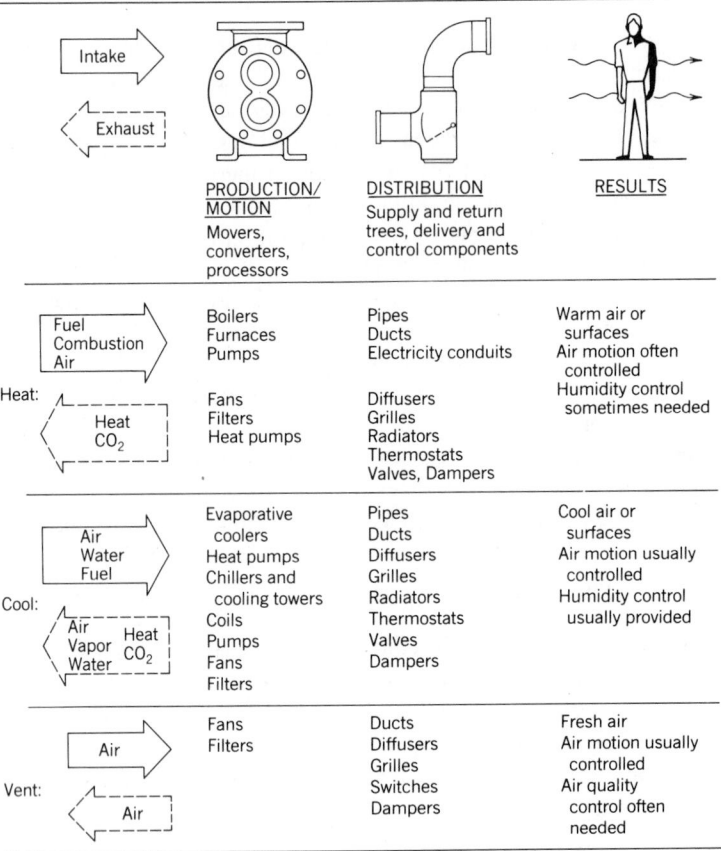

	Intake / Exhaust	PRODUCTION/ MOTION Movers, converters, processors	DISTRIBUTION Supply and return trees, delivery and control components	RESULTS
Heat:	Intake: Fuel, Combustion, Air Exhaust: Heat, CO_2	Boilers Furnaces Pumps Fans Filters Heat pumps	Pipes Ducts Electricity conduits Diffusers Grilles Radiators Thermostats Valves, Dampers	Warm air or surfaces Air motion often controlled Humidity control sometimes needed
Cool:	Intake: Air, Water, Fuel Exhaust: Air, Vapor, Water, Heat, CO_2	Evaporative coolers Heat pumps Chillers and cooling towers Coils Pumps Fans Filters	Pipes Ducts Diffusers Grilles Radiators Thermostats Valves Dampers	Cool air or surfaces Air motion usually controlled Humidity control usually provided
Vent:	Intake: Air Exhaust: Air	Fans Filters	Ducts Diffusers Grilles Switches Dampers	Fresh air Air motion usually controlled Air quality control often needed

Source: Class notes developed by G. Z. Brown, University of Oregon.

(b) System Anatomy. Table 6.1 describes the basic organization of any HVAC system. Three kinds of common tasks (heating, cooling, ventilating) are done by production components; usually, they require distribution and delivery components. Intake supplies and exhaust by-products accompany each task.

(c) Early Choices. Although the eventual choice of HVAC system should follow an analysis of the zone's needs, some early concepts underlie system choices. The first we will consider is the question of *central versus local* systems.

Central systems require one or several large mechanical spaces (often in basements and/or on roofs), sizable distribution trees, and complex control systems. The noise, heat, and other characteristics of such mechanical rooms can be controlled fairly easily, since the machinery is concentrated at a few locations. Similarly, maintenance is easy to perform, although breakdowns in central equipment can paralyze the entire building. Air quality can easily be controlled by location of the air intakes high above the pollution at street level and by regular maintenance of the centralized air filtering equipment. Longer equipment life can be expected, with regular maintenance. Energy conservation can be served by the recovery of one machine's heat by-product for a nearby machine's heat input. Although there are many ways to provide for the differing thermal needs of the many zones served by central systems, one important drawback of central systems is a difference in zone scheduling: when the entire system must be activated to serve one zone (such as computer operations in an office building on a weekend), energy is wasted.

Local systems therefore become increasingly attractive as scheduling differences multiply. Also, pronounced differences in other factors—function (with resulting comfort expectations) or placement within the building, for example—can lead to the choice of local systems. Large and centralized equipment spaces are not required with local systems; rather, production equipment is distributed throughout the building (or over the roof of low-rise structures). Dispersal of equipment minimizes the size of distribution trees and greatly simplifies control systems. Moreover, system breakdowns affect only small portions of the building. However, noise and other by-products of multiple machines pose numerous potential threats to occupied spaces, and maintenance is demanding, because access to so many locations is often not easy. Then, too, air quality depends on the regular cleaning of many filters scattered over the building, often within occupied spaces. The potential for energy conservation seems promising because heating or cooling is produced only as locally needed, but there is little chance to use one zone's waste heat as another's needed source.

Central heat/cool, local air distribution has become a popular way to take advantage of favorable characteristics of both central and local approaches. This was shown in Chapter 5 (Fig. 5.14*b*) with a central boiler/chiller space remotely located, and fan rooms on each floor. This minimizes the bulky distribution tree for air; although the distribution tree for heated and chilled water is extensive, it is also of much smaller diameter, therefore relatively easily accommodated. The central equipment room makes energy recovery systems more feasible and improves opportunities for pollution control.

Another basic question is that of *uniformity versus diversity* in the interior environments of buildings. This question encompasses not only thermal experiences, but visual and acoustical ones as well.

The advantages of uniformity are most evident in a rapidity of design and construction that, through mass production and speed, often brings lower first costs. As mentioned in Chapter 1, the flexibility in office arrangements that are accompanied by uniform ceiling heights, light placement, grille locations, and so on, can extend a building's usable life span. However, there are at least four types of offices, which may need to be interchangeable within such "flexible" space. The typical *enclosed office* has the privacy of four walls and a door. The *bullpen office* has repeated, identical workstations, with low dividers at about the height of the desk surface. The *uniform open plan office* resembles the bullpen, but with higher divider partitions for added privacy. The *free-form open plan office* has some individually designed workstations, with divider partitions of varying heights (sometimes reflecting varying status of workers). In the bullpen and uniform open plan office, the resulting uniformity is not always attractive to users, and diversity is often encouraged at a more personal level—with office furnishings, for example. A more thorough approach to diversity can provide stimulus to the user who spends many hours away from the variability of the exterior climate.

If offices must be uniform in ceiling lighting, air handling, and size, the corridors that connect them and the lounges, or other supporting service spaces, can deliberately be made different. Diversity requires a complete and detailed design of places; it gives the builder a more complex and interesting task; and it can provide orientation and interest to the users. The attractiveness of diversity is evident in most collections of retail shops, in which light and sound—and sometimes heat and aroma—are used to distinguish one shop from the next.

Diversity in the thermal conditions to be maintained, such as warmer offices and cooler circulation spaces in the winter, can be used to enhance the comfort of the office users. Designers have long recognized that a space can be made to seem brighter and higher if it is preceded by a dark, low transition space. Thermal comfort impressions can be manipulated similarly. Less-than-comfortable conditions in circulation spaces or other less critical zones not only make the critical spaces seem more comfortable by contrast, but also save significant amounts of energy over the life of a building. Furthermore, such conditions can make passive strategies more attractive.

A large-scale demonstration of diversity in thermal zones is shown in Fig. 6.2. Passive solar heating can make a significant contribution, even through a shallow-sloped, single-glazed cover in cloudy Glasgow, Scotland, largely because the mall area and leisure areas are allowed a much wider thermal range than would be permitted in stores and offices. The overcast skies are quite suitable for daylight, and the addition of summer sunshading makes natural ventilation (through stack effect, assisted by fans) possible during the cool summers. U.S. Pacific northwest climate conditions are similar.

(d) Comparing Systems and Zones. In the process of selecting systems from the wide variety available, it is helpful to consider the match between the zones' characteristics and those of various systems. Among the considerations are zone placement (near to or away from the building skin), the zone's thermal loads, the comfort determinants based on the zone's activ-

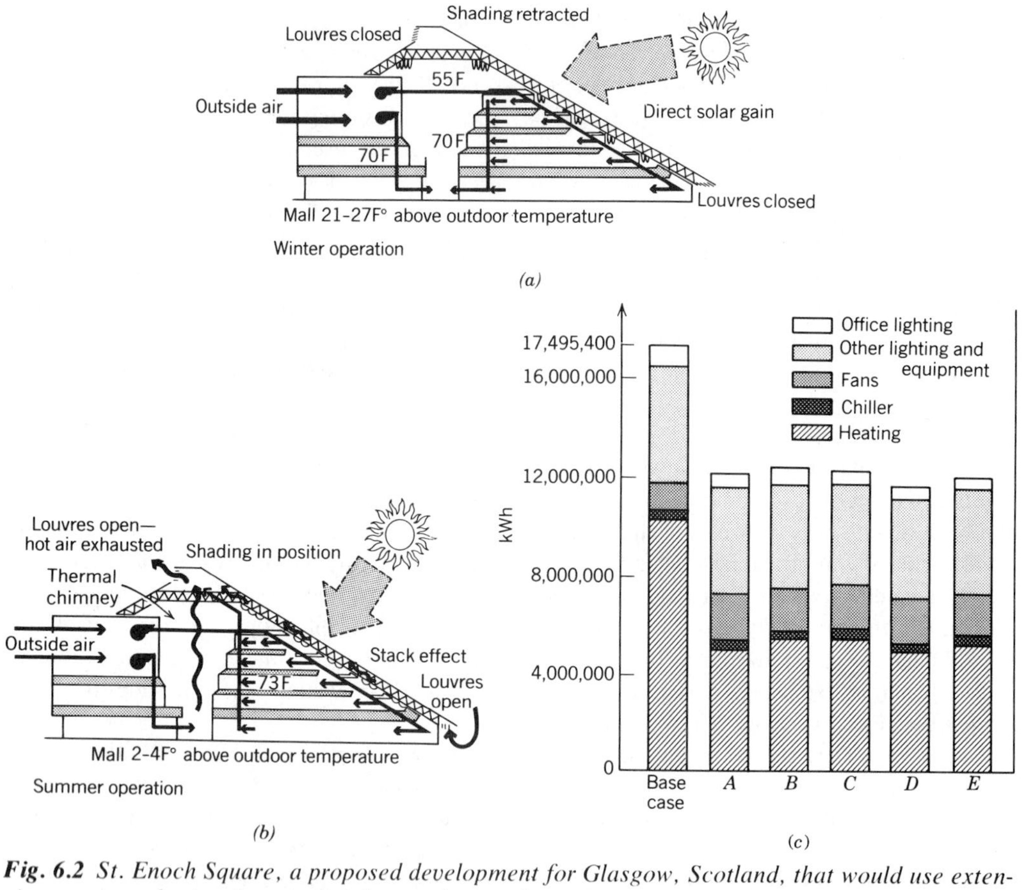

Fig. 6.2 *St. Enoch Square, a proposed development for Glasgow, Scotland, that would use extensive passive solar heating, daylighting, and natural ventilation. Reiach & Hall and GMW Partnership, architects (joint venture); Cosentini Associates, energy consultants; Princeton Energy Group, daylighting consultants.* (a) *Schematic section showing winter operation; the mall temperature varies around 63 F during operating hours, while offices are kept near 70 F.* (b) *Schematic section showing summer operation; the mall temperature swings from 68 to 74 F during operating hours.* (c) *Estimates of annual energy consumption, for a conventional-design "base case" and several alternative configurations of the proposed complex. Note the significantly lower heating energy requirements, resulting in part from the lower winter temperatures allowed for less-critical zones such as the mall and the leisure areas in configuration A to E.*

ities, the space available for system components within the zone, and the life-cycle costs of various system alternatives.

Zone placement will sometimes preclude local systems, which depend on easy access to outdoor air both for fresh air and for a heat source or sink. Local systems for interior (away-from-skin) zones are awkward. Relationships between zone placement and building forms are shown in Fig. 6.3.

The thermal loads on each zone will determine the extent to which heating or cooling is the dominant problem—which, in turn, can influence the choice of system. A zone with little

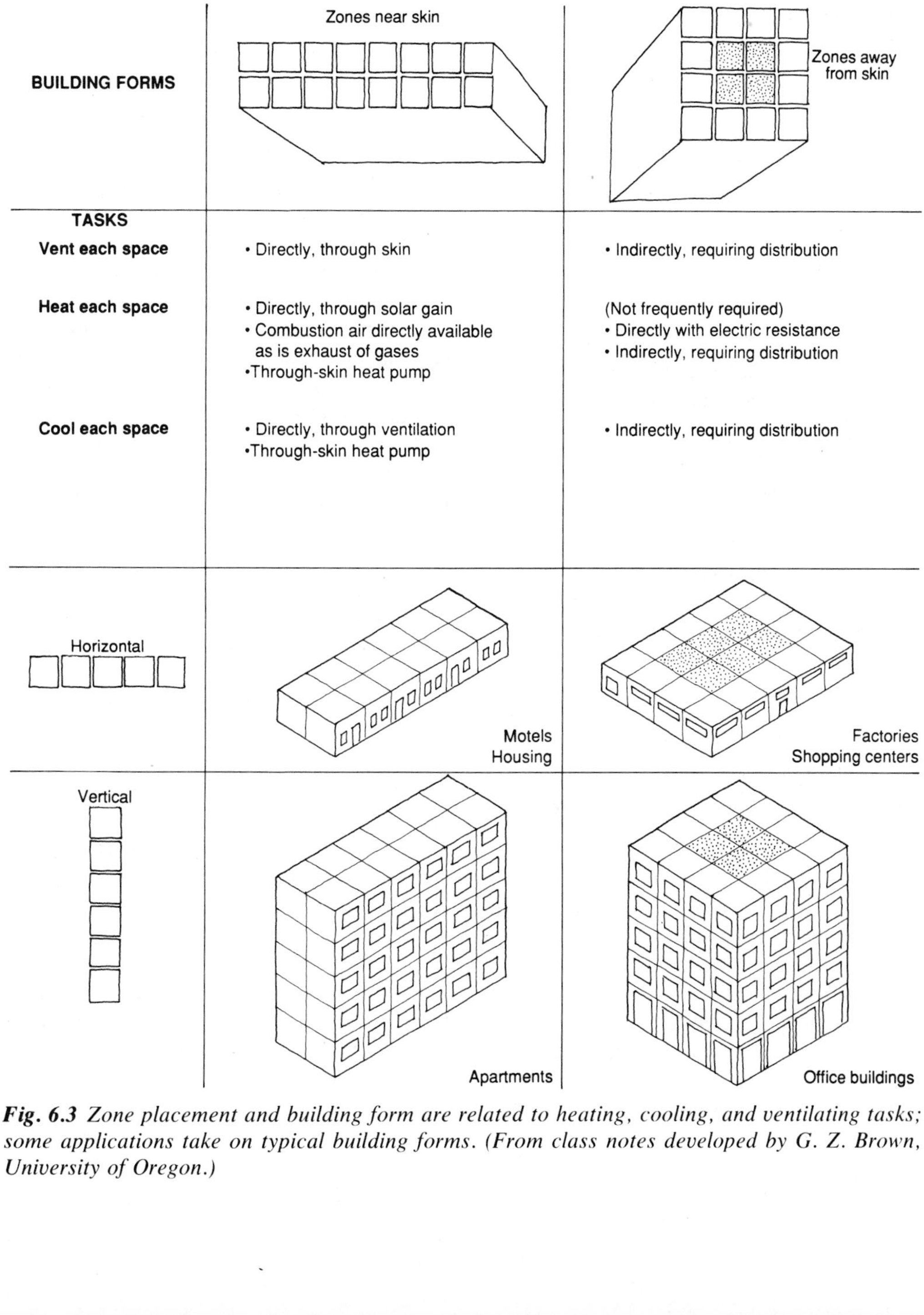

Fig. 6.3 *Zone placement and building form are related to heating, cooling, and ventilating tasks; some applications take on typical building forms. (From class notes developed by G. Z. Brown, University of Oregon.)*

cooling load and low moisture production may be well served by a simple system of fresh air plus heating, with no humidity control. Zones that require cooling will usually also require more complete control of air motion and relative humidity. Although it is risky to generalize about which comfort determinants are most important (given differences between activities and between individuals), it can generally be assumed that comfort and thermal tasks are related.

	For Heating of Spaces	*For Cooling of Spaces*
More important	Surface, air temperatures	Air motion
↓	Air motion	Relative humidity
	Relative humidity	Air, surface temperatures
Less important	Air quality	Air quality

Thus, the choice of systems can be based partly on whether the system provides good control of the more important comfort determinants.

Another important factor in system choice is the amount of space the system requires. In some cases, it is easy to provide small equipment rooms at regular intervals throughout a building, such that little or nothing in the way of a distribution tree will be required. In other cases, a network of distribution trees and central, large equipment spaces are easier to accommodate. These central systems typically fall into one of three classifications:

All-air (the largest distribution trees).

Air and water.

All-water (the smallest distribution trees, with local control of fresh air).

The details of these systems can be found in Sections 7.4 to 7.6 along with typical applications and space requirements. For now, a preliminary screening is necessary, to match zones and systems. A simplified procedure is shown in Table 6.2, in which preliminary system choices are made for a building such as the multipurpose structure shown in Fig. 6.1. In this process, the 16 zones first apparent are translated into three local systems and three central systems: one all-air, one air & water, and one all-water.

The Fox Plaza Building in San Francisco, which illustrate many of the matches between systems and zones, is shown in Fig. 6.4. This project includes four major building types in one structure:

1. Underground garage for the storage of cars.
2. Commercial center at ground level, including a bank, a women's specialty store, and other commercial establishments.
3. Ten floors of offices.
4. Sixteen floors of apartments.

The mechanical level is located between the office portion of the building and the apartments above. In this way, the main runs of all the equipment—heating, air conditioning, electrical, and so on—are directed both upward and downward, in the shortest possible distance. The spatial requirements of offices and the spatial requirements of apartments are quite different; thus, the total design layout of each of the two uses differs—floor-to-floor heights, window treatment, and the heating, cooling, electrical, elevators, and other services are all different. The placement of the mechanical level between the offices and the apartments not only facilitates an efficient operation but also provides for a definite visual separation between the two functions.

Quite unusual is the placement of the steam boilers on the thirteenth floor instead of in the conventional basement location. Adjunct to the general thirteenth-floor center, only a small amount of auxiliary equipment is located on the roof and in a small portion of the garage. Residential areas have hot-water heating, offices have dual-duct, high-velocity heating/cooling, and commercial tenants are supplied with hot and chilled water for individual climate control requirements.

(e) Central Equipment Location. The Fox Plaza Building exemplifies an intermediate location for the heating and cooling production equipment—one that separates floors of apartments from floors of offices. Other typical locations for central equipment are in the basement (where noise is most easily isolated, utilities are easily accessed, and machine weight is little

TABLE 6.2 **Procedure for Matching Zones and Systems**

Capsule Description

A multipurpose building (similar to that shown in Fig. 6.1) is situated in a cold winter–mild summer climate.

Apartments are on upper floors, surrounding an open-air central court; they have adequate daylight and cross ventilation. Floor heights are quite low.

Offices are rented to various tenants. Exterior offices have high ceilings, to facilitate daylighting, and therefore have low internal gains but restricted clearance for horizontal ducts. Interior offices have lower ceilings; vertical chase space is limited, since it reduces rentable area.

Shops are located around the perimeter on the high-ceilinged ground floor; some smaller shops are located on the mezzanine and ground floors in the interior zone. Space for vertical chases is severely limited on these highest rental floors.

The parking area is below grade, surrounded by air and light wells. Floor heights are very limited, to reduce ramp length.

Activities (Program)		*Offices*			*Shops*			
	Apartments	*Computer Center*					*Restaurant*	*Parking*
Schedule	24 h	24 h	9 hr		*9 h*		*12 h*	
Placement	Exterior	Exterior for access	Exterior	Interior	Interior	Exterior	Exterior for access	Entire floor
Internal gains	Vary	High	Low	Medium	High	Medium	High plus moisture	Low with exhaust gas
Dominant HVAC task(s)	Heat Ventilate	Cool	Heat/cool	Cool	Cool	Cool	Cool	Ventilate
HVAC space available								
Vert. (in plan)	Medium	Medium	Medium	Tight	Tight	Tight	Tight	Medium
Horiz. (in section)	Tight	Medium	Tight	Medium	Tight	Ample	Ample	Tight
System choices								
Local		*A*					*B*	*C*
Central								
All-air			*(D)*	*D*	*D*	*(D)*		
Air & water			*E*			*E*		
All-water	*F*							

TABLE 6.2 **Procedure for Matching Zones and Systems (*continued*)**

Summary

A. The computer center's unique schedule and rate of internal gain usually requires a separate system, equipped with humidity and air quality controls to protect the equipment. Some heat recovery for use in *E* and *F* seems advisable.

B. The restaurant's special problems of heat, moisture, and aroma, as well as its schedule, require a separate system.

C. The parking area needs only plenty of fresh air; it requires no tie with the other zones at all.

D. The always-cooling loads of interior zones are best served by all-air systems offering control of humidity and air quality. However, vertical chase size is tight, and high-velocity distribution may be required. (The exterior zones could also be served by all-air. But the need for heating, plus the likelihood of fresh air infiltration and the tight clearance for horizontal ductwork, all suggest that the system for exterior zones be separated from the system for interior zones.)

E. Quick changes from heating to cooling are best handled by water; some central air-quality control is offered by air & water systems.

F. A central all-water system offers energy conservation advantages, recovering waste heat from system *D* (and potentially from *A*). Fresh air is easily and cheaply handled on a local basis, which also provides cooling.

Mechanical Space

Probably best located on the top office floor, or on a floor of its own between offices and apartments. Distribution tree sizes will thereby be minimized on the high-rent ground floor.

problem) and on the roof, where access to air as a sink for reject heat is easiest of all and headroom is unlimited. Very tall buildings may require several intermediate mechanical floors. Examples of these approaches are found throughout the rest of Chapters 6 and 7.

These equipment spaces have special environmental needs: lots of fresh air, strong support, and high headroom. The equipment can generate considerable heat, moisture, air motion, noise, and vibration—to the potential annoyance of nearby floors (or neighboring buildings). As shown in the Fox Plaza example, the equipment can be strongly expressive of building services and can serve a useful demarcation role between vertical layers of high-rise buildings.

(f) Distribution Trees. Section 5.1 raised the preliminary questions about distribution trees: how many, what kind (air or water), and where to place them within buildings. Before carrying these questions further, we must consider two other basic concepts: the extent to which the distribution system should be architecturally expressed or concealed, and the extent to which the HVAC system might be integrated with structural elements.

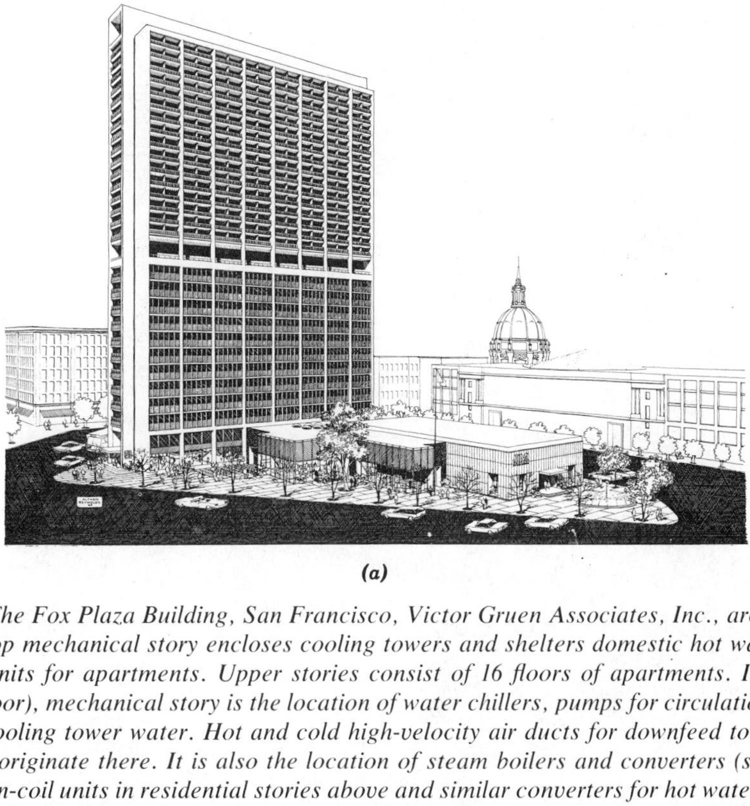

(a)

Fig. 6.4 (a) *The Fox Plaza Building, San Francisco, Victor Gruen Associates, Inc., architects and engineers. Top mechanical story encloses cooling towers and shelters domestic hot water generator-storage units for apartments. Upper stories consist of 16 floors of apartments. Intermediate (thirteenth-floor), mechanical story is the location of water chillers, pumps for circulation of chilled water, and cooling tower water. Hot and cold high-velocity air ducts for downfeed to lower story conditioning originate there. It is also the location of steam boilers and converters (steam to hot water), for fan-coil units in residential stories above and similar converters for hot water for coils in lower story office air units. Also on the thirteenth floor, adjacent is the domestic hot water generator-storage receiver for the offices below. Below the thirteenth floor, climate-making center are 10 stories of offices and two stories of commercial area. (Courtesy of* Progressive Architecture.*)* (b) *The Fox Plaza Building in construction. Clearly shown are the air-handling units on the thirteenth floor. Also visible are downfeed ducts that supply high-velocity hot and cold air from these units to the office stories below. Other ducts return the air at normal velocities to the air-handling units for reconditioning. Upper residential stories in this equable climate of San Francisco are supplied with heating but not cooling. Office cooling deals with people-concentration and lights. (Photo by Morley Baer.)*

Fig. 6.4 (continued)

Concealment and Exposure. The pipes, ducts, and conduits that take the necessary resources to and from the interior are often carried within a network of spaces unseen by anyone except builders and repair people. The advantages of concealment include less water and air noise, fewer surfaces requiring cleaning, less care necessary in construction (leaks, not looks, are important), and more control over the appearance of the interior ceiling and wall surfaces. Although maintenance access to such hidden supply lines is more difficult, a variety of readily removable covers is available, particularly in suspended ceilings.

On the other hand, the exposure of these supply networks provides an honest and direct source of visual (and occasionally acoustical) interest. Exposure in corridors and service areas and concealment in offices constitute an approach used in many office buildings. Flexibility is usually encouraged by exposure; changes can be easily made when there is no need for neatly cut holes in concealing surfaces. However, flexibility from full-height movable partitions requires constant ceiling heights—a feature of the suspended-ceiling approach.

One of the more spectacular examples of exposed mechanical (and structural) systems is shown in Fig. 6.5—the result of a design competition for a museum of modern art, reference library, center for industrial design, center for music and acoustic research, and supporting services, in downtown Paris.

When users are invited to play an active role in adjusting conditions inside, exposure of the switches they manipulate is helpful. Visible mechanisms not only remind users of their opportunities, but also encourage user interaction. In this way, adjustments are sometimes discovered that the designer had not anticipated.

Fig. 6.5 *Centre Georges Pompidou, Paris. A view of the mechanical support systems. The design competition for this complex was announced in 1971, but subsequent lawsuits and budget cuts delayed its opening until January 1977. Piano + Rogers architects. (Photo by John Tingley.)*

Mechanical-Structural Integration or Separation. The similarity of these two technical support systems—structures and environmental controls—has intrigued designers ever since mechanical systems began to require substantial volume for distribution, as in air-duct systems. As the complexity and size of the mechanical distribution systems was *increasing* with technological development (typically, more air is required to cool a space than simply to heat it), the increased strength of materials was *reducing* the size of the structural system. The "uncluttered" floor areas between the more widely spaced columns became desirable for flexibility in spatial layout. With the mechanical systems at or within these columns, floor areas remained clear, thus giving mechanical-structural integration further impetus. With the new expectations for cooling, the refrigeration cycle's cooling tower often moved to the roof, taking the air-handling machinery with it. This further encouraged the merging of systems, for one system was growing wider as the other diminished (see Fig. 6.6). Thus a fixed-column cross section, consisting mostly of the structural column at the base and the air duct at the top, became possible. (One of the responses to this opportunity, expressed as a dual-duct supply and return system, is shown in Fig. 7.53, page 445.)

Yet the functions of these systems differ widely: compared to the dynamic on-off air, water, and electrical distribution systems, the structural system is static—gravity never ceases. The moving parts in mechanical systems need maintenance far more frequently than the

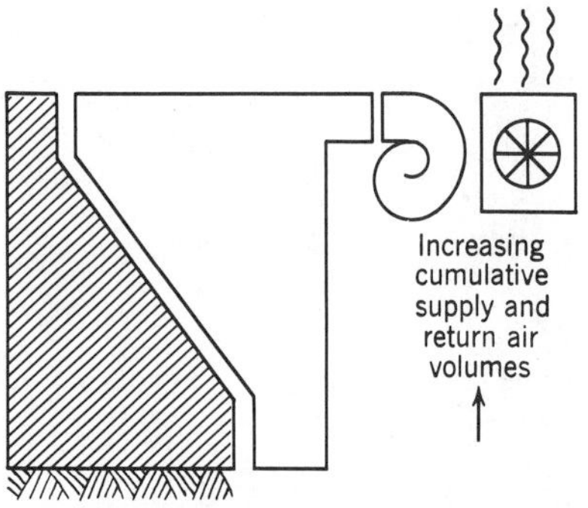

Fig. 6.6 Technical support system. With rooftop air handling, the total air duct size decreases toward the ground. Conversely, the total structural load increases toward the ground.

connections of structural components. Changes in occupancy can mean enormous changes in mechanical systems, requiring entirely different equipment; structural changes of such magnitude usually occur only at demolition. Mechanical systems can invite user adjustment; structural systems rarely do.

Thus, while it is possible to wrap the mechanical systems in a structural envelope, it is of questionable long-term value, given the differing life spans and characteristics of these systems. The probability of future change suggests that the *mechanical* system be the exposed one, despite the appeal to many designers of the structural system's cleaner lines.

Distribution Tree Placement Options. These options are summarized in Fig. 6.7. Vertical placement options are important because they affect floor space, influencing the flexibility of spatial layout and the availability of usable (or rentable) floor space. Horizontal placement options affect ceiling height—a particular issue in daylighting design, and sometimes a critical factor when overall height limits are imposed yet a maximum of usable floor space is desired. (In Washington, D.C., for instance, no building can rise higher than the Capitol.) Both vertical and horizontal distribution at the edges can have dramatic impact on building appearance.

The history of distribution trees and high-rise buildings is one of trends and countertrends. Initially, a multistory building relied on daylight and cross ventilation, so a thin, relatively high-ceiling plan with much perimeter was favored (refer to Fig. 3.29). The heat gain and loss was all at the perimeter, so perimeter distribution trees (carrying only steam or heated water, and of quite small diameter) were generally used. As electric lighting and thus the need for air conditioning increased, so did the thickness of plans; large central internal areas needed lots of forced, cooled air. Central boilers, chillers, and fan rooms were the norm. Thus, bulky air distribution trees appeared. At about the same time, the glass curtain wall and its slick, two-dimensional look of modernity became fashionable. The air distribution trees were so visually intrusive on the facades that they were pushed to the core, where cooling needs were relatively steady. However, the perimeter still experienced extreme needs for both heating and cooling; getting from vertical trees at the core to the perimeter required larger cavities above suspended ceilings. This pushed the ceiling in the offices down, to keep floor-to-floor distances economical. Vast office areas resulted that were visually dull, low ceilinged, and without daylight.

Now, countertrends include decentralized air handling, with small fan rooms on each floor. Vertical air distribution trees are shrinking; horizontal ones more common. At the same time, daylighting is pushing office ceilings higher; so is a preference for indirect lighting and its compatibility with computer screen visual comfort. Night cooling utilizing thermal mass is encouraging the exposure of concrete structure. A renewed interest in sun control is encouraging three-dimensional facades, replacing two-dimensional reflective glass facades (which merely redirect the sun toward someone else). (See later Fig. 6.12.) With increased three-dimensionality at the facade, perimeter distribution trees are once again conceivable.

It is logical to place at the perimeter the parts of the system that deal with the effects of sun, shade, and temperature change in the several perimeter zones, leaving at the core a separate network to handle the more stable interior areas. The disadvantages of perimeter distribution include (usually) higher construction costs and an

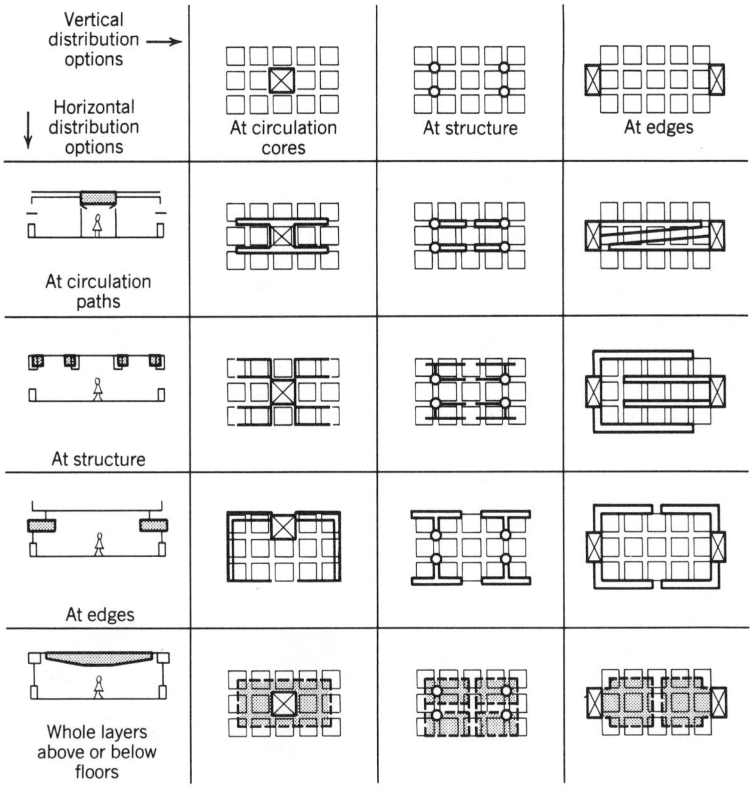

Fig. 6.7 *Distribution tree placement options, both vertical (impact on plan) and horizontal (impact on section). (Based on class notes developed by G. Z. Brown, University of Oregon.)*

environment that is more thermally hostile, due to the extremes of outdoor temperature.

Vertical distribution within *internal circulation cores* is very common, as it leaves a maximum of plan flexibility for the rest of each floor and does not disturb the prized floor areas nearest windows. However, one centralized vertical distribution trunk will require large horizontal branches near the core, so with this choice early thought must be given to the horizontal placement options.

An unusual example of vertical air distribution at the core is shown in Fig. 6.8. The Fox Plaza, Los Angeles, office building's unique features include both fan rooms on each floor *and* a large central vertical air shaft. This air shaft begins at bottom as a fresh air intake to each floor, and tapers to become, at top, an exhaust (heated) air outlet from each floor. Thus, the stack effect is utilized to supply fresh and exhaust stale air from a large building with very little need for fans.

Vertical distribution integrated with structure has some intriguing possibilities where the structure–HVAC integration concept is suitable. Because multiple HVAC trees are implied (since there are multiple columns with which they are integrated), the horizontal branches tend to be small. However, these branches often join the vertical trunk at the same place that critical column-to-girder structural connections need to be made; interference is common and can be costly to solve. Vertical distribution at the *edges* is potentially dramatic in form (see later Fig. 7.53) but costly to enclose (if outside) or wasteful of prime floor space (if inside).

Horizontal distribution above *corridors* is very common, since reduced headroom here is more acceptable than in the main activity areas. Furthermore, corridors tend to be away from windows, so their lower ceilings do not interfere with daylight penetration. Since corridors connect nearly all spaces, horizontal service distribution to such spaces is also provided. Furthermore, exposure of these services above corridors can heighten the contrast between such serving spaces and the uncluttered, higher ceiling offices that are served. Horizontal distribution at the *structure* is sometimes chosen, particularly where U-shaped beams or box beams

(a)

Fig. 6.8 The Fox Plaza, Los Angeles, office building is a 34-story, 800,000-ft² granite and glass tower (a) *with a very unusual vertical distribution tree.* (b) *Typical lower plan (floors 6 to 16) shows both a fan room and a large vertical air shaft. At this level most of the shaft area is supplying fresh air (from an intake in the bluff face below the building); the remainder is exhausting stale air toward the roof. Note the lack of columns between core and perimeter, contributing to office layout flexibility.* (c) *Typical upper plan (floors 31–33) shows fewer elevators; by this time most of the large vertical air shaft area is carrying exhaust air toward the roof; the remainder is supplying fresh air from the intake below.* (d) *Section shows the tapered central air shaft, which relies on the stack effect to help bring in cooler air below, exhaust-heated air above. (Courtesy of Johnson Fain Pereira Associates, Architects, Los Angeles; and Kim, Casey and Harase, Inc., Engineers, Los Angeles. Photo by Wolfgang Simon.)*

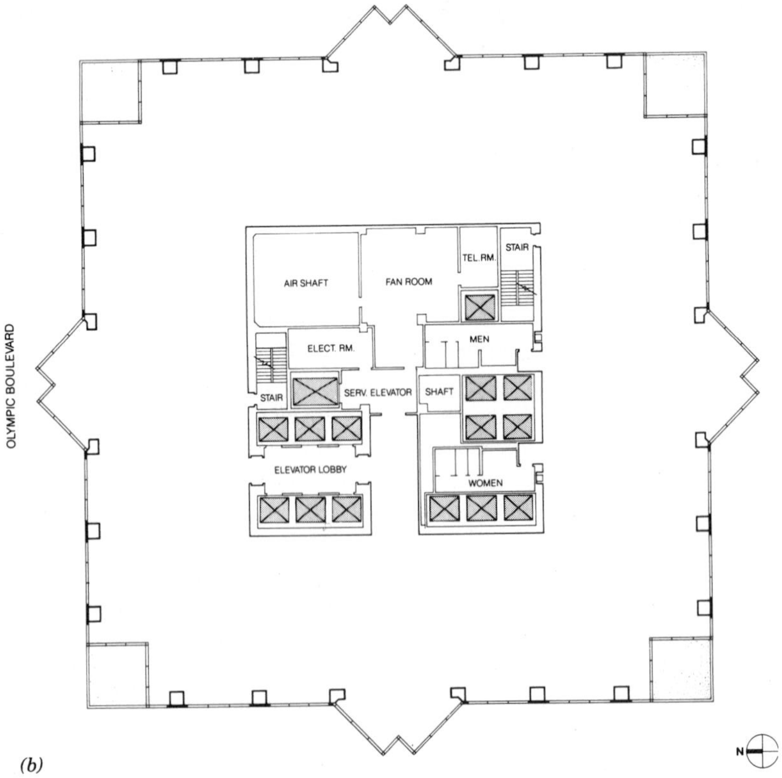

(b)

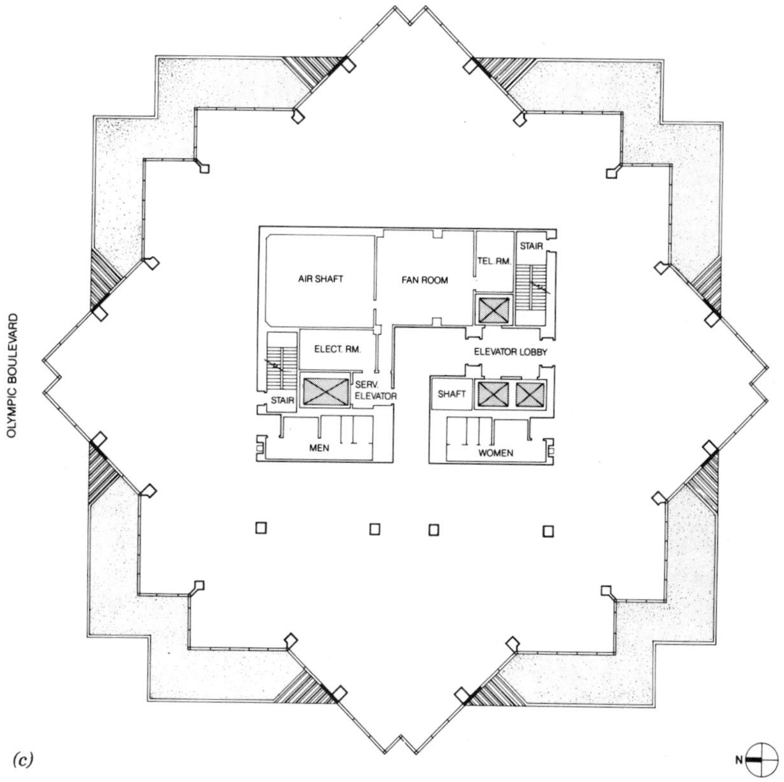
OLYMPIC BOULEVARD
AIR SHAFT
FAN ROOM
TEL. RM.
STAIR
ELECT. RM.
ELEVATOR LOBBY
SERV. ELEVATOR
STAIR
SHAFT
MEN
WOMEN
N

(c)

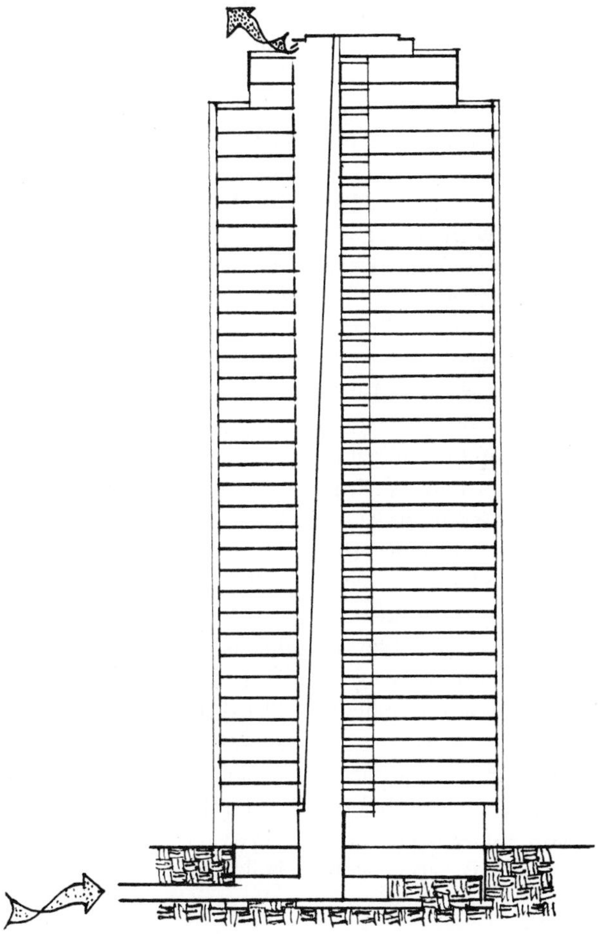

(d)

provide ready channels for HVAC distribution. However, the penetration of horizontal structure members by these continuous service runs must be coordinated. Horizontal distribution at the *edges* can be integrated usefully with sunshading devices and light shelves; it can also act as a spandrel element that contrasts with the window strips. Horizontal distribution within *whole layers* above or below floors is often utilized, particularly for return-air plenums (contrasted with supply air carried within ducts).

An example of supply at the edge for both vertical and horizontal distribution can be found in the International Building in San Francisco (Fig. 6.9). Here the vertical shafts are prominently exposed at the corners; these shafts carry supply and return ducts serving the four perimeter air conditioning zones. Air-handling equipment and a 750-ton refrigeration plant are located on the floors just below the terrace level (those least desirable for renting). Each corner duct branches to serve two zones, separately controlled. Pressure reduction and blending are done by equipment in the hung ceiling, and from these points air flows to strip-grille diffusers directly above the glass on the four sides of the building. Local controls assure comfort to personnel in each area.

Interior zones on each floor are supplied by a riser duct in the building core, which branches at each floor to a loop just outside the line of elevators. The loop serves ceiling diffusers.

Between the perimeter loop and the interior loop just described, a return loop collects air for return to the central station (second and third floors). These return loops on the eleventh to twenty-first floors are picked up by external return risers on alternate exterior corners. From the tenth floor on down, the loops are picked up (as shown in Fig. 6.9) by an interior return riser that extends down through the core in front of the blank faces of the high-rise elevators. To provide a clear space between the elevator banks on the main floor (fourth or terrace), the two core ducts risers are offset at the ceiling of that story.

In summary, perimeter air for all stories is supplied through corner ducts. Central air for all stories is supplied through a core duct. All return air above the tenth floor is carried down through the return ducts at the *other* two corners. Return air from the tenth floor and below is carried down by a return duct in the core.

Another example of vertical distribution at the edge, this time integrated with large hollow structural columns, is provided by the Chemistry Building at the University of California, Berkeley (Fig. 6.10). Because of the large number of distribution trees necessary for laboratory buildings, 16 hollow exterior reinforced-concrete structural columns were used to enclose pipes and ducts connected to short lateral branches, thereby minimizing crossovers and avoiding "spaghetti" patterns on the ceilings.

These are truly multipurpose columns. Together with the slimmer interior columns, they support the building. In the late afternoon, when the balconies (between adjacent columns) have ceased to exclude all direct sunlight, the deep exterior columns take over as totally effective sunshades. Finally, they enclose exhaust ducts from all exterior (and some interior) fume hoods (*FH* in Fig. 6.10). These ducts are exhausted by fans behind the masonry grill at the roof level. Four of the exterior columns are *pipe-station columns;* of the two shown in Fig. 6.10, one is detailed. The piping services include domestic hot and cold water, industrial hot and cold water, distilled water, demineralized water, natural gas, vacuum, compressed air, hydrogen sulfide, nitrogen, oxygen, steam, acid waste, vent pipes, and roof leader pipes. There is access to all exterior columns at all floors.

This is a once-through system; no air is recirculated. The pressure in the halls is greater than that in the rooms, preventing the flow of air or gases back out into the corridors.

Summary: All fresh air is brought in at ground level. It is partially conditioned and delivered to rooms by means of a central riser and ceiling ducts in corridors. After reheat and delivery to rooms, it is exhausted through fume hoods to ducts in hollow exterior columns; and in the case of interior rooms, to a central group of exhaust ducts. All exhaust fans are on the roof. Four exterior columns are also *pipe station* columns.

Further treatment of the relationships among HVAC systems, their distribution trees, and buildings is given in the examples that accom-

Fig. 6.9 The International Building, San Francisco. Anshen and Allen, architects; Eagelson, Engineers (Charles Krieger, E.E.), mechanical designers. (Courtesy of Progressive Architecture.*)* (a) *One of the four corner main ducts.* (b) *In this ingenious scheme, the major supply arteries for conditioned air are located in alternate corners. In diagonally opposite locations, supply risers are placed in large square enclosures. Although nonstructural, each enclosure is emphasized as a distinct vertical design element. Each encloses both hot- and cold-air ducts, which supply two separately controlled orientations on all 21 stories. Conditioned air originates at an intermediate floor, the third. In the opposite two corners, similar ducts return much of the air to the equipment story. The balance is returned through duct risers in the core.*

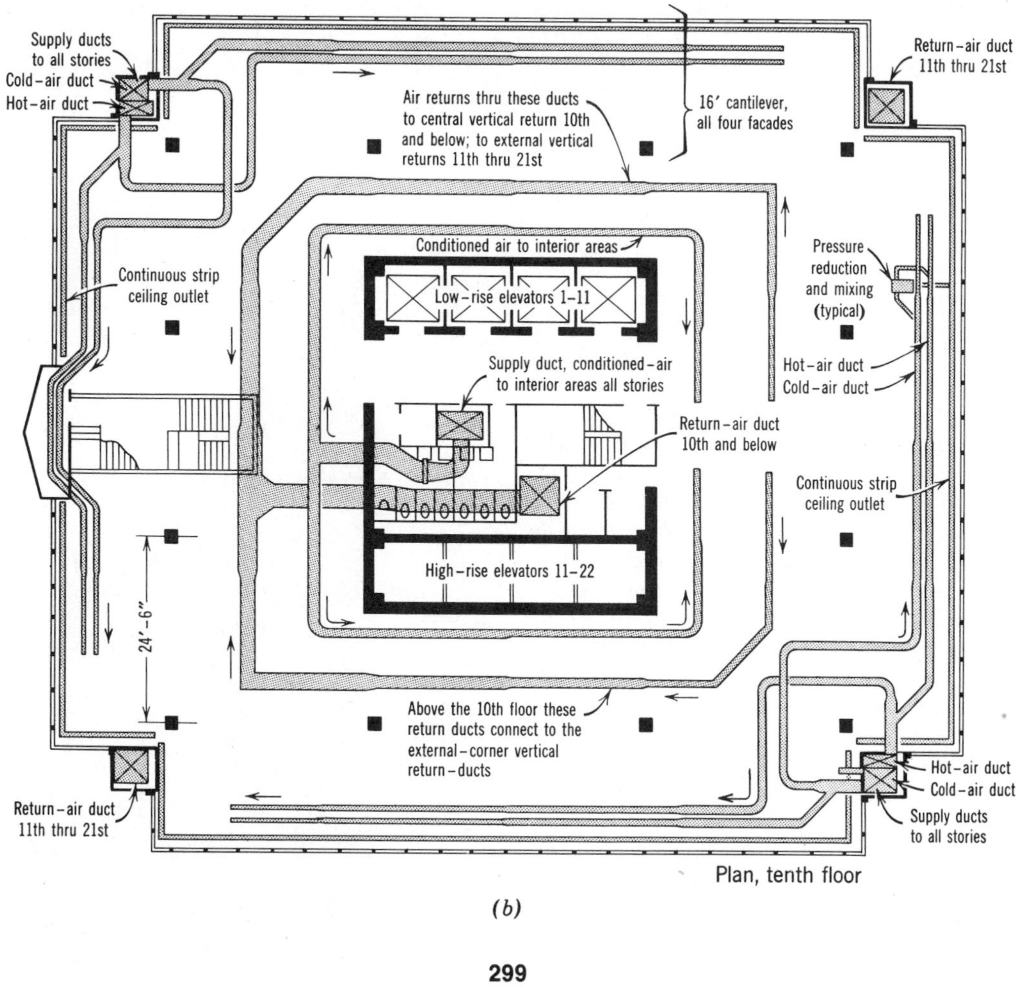

(b)

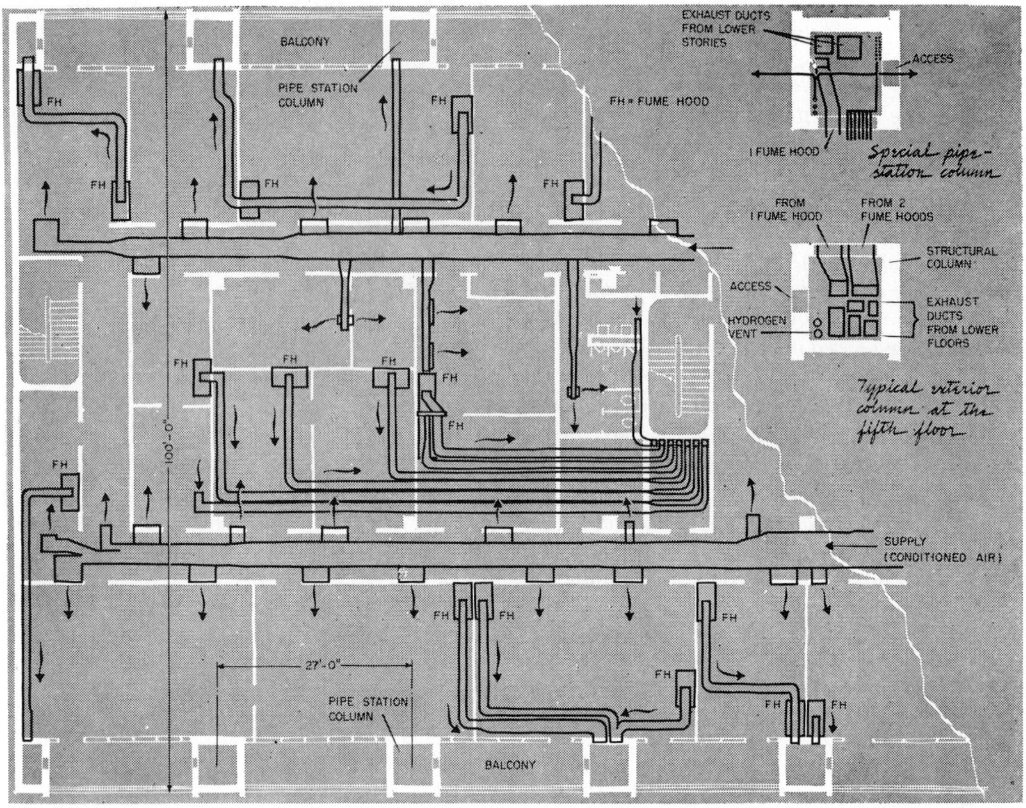

(a)

(b)

Fig. 6.10 *Chemistry Building, Unit 1, University of California at Berkeley. Anshen & Allen, architects. Mechanical design by Bayha, Weir & Finato.* (a) *Preliminary rendering and layout of supply and exhaust air shown in four of seven bays at typical floor, the fifth. The burden of exhaust from fume hoods is relieved by routing those located in outside rooms to exterior hollow structural columns. Hoods in* interior *rooms are ducted to exhaust stacks adjacent to the stairway. Hollow columns provide distribution space for piping. Lateral crossover of mechanical services at ceilings is minimized.* (b) *Behind the decorative masonry screen at the roof edge are exhaust fans that pick up the vertical ducts in the hollow columns to relieve the buildup of gases produced by a multitude of chemical research projects. At this hour the sun is intercepted by masonry walls that form an effective barrier against heat transmission to the indoors. This relieves a cooling system that must cope with the burden of people, lights, and heat-producing research.*

pany the more detailed descriptions of large-building HVAC systems in Sections 7.3 to 7.6.

6.4 Passive Heating and Cooling Components

This section precedes the discussion of conventional heating and cooling equipment and its related design process, because so many "passive" components *reduce the need* for conventional thermal devices.

At first glance, the office building in Fig. 6.11 appears to be a conventional glass—perhaps curtainwall—box. But passive approaches in this Niagara Falls, New York, example nearly eliminate infiltration, provide almost equal inside surface temperatures on all four exterior walls, and provide enough daylighting to eliminate daytime electric lights for almost half the office space. In this case, it all happens at the perimeter, where two vertical planes of glass are spaced 4 ft apart; in this void are movable horizontal sunshading louvres, which also help distribute daylight evenly and provide some night insulation. For cooling, natural ventilation of this glass cavity occurs through the stack effect.

At the end of this section is another example in which passive design principles more thoroughly organize the interior arrangement and overall building form. Before this example, an array of components is discussed, although we can only begin to cover the rapidly expanding range of products and processes available to the energy-conscious designer.

(a) Sunshading. Perhaps the single most important energy-related component for passively cooled buildings is the sunshade. (And since most solar-heated buildings will experience overheating in hot weather, sunshading is critical for passively heated buildings as well.) If correctly implemented, sunshading rejects most solar heat gains yet aids in distributing daylighting deep into buildings, where it replaces or reduces the internal gains due to electric lights.

If a building is arranged to intercept the intense rays of the sun before they pass through its glass walls instead of afterward, the air conditioning cooling load can often be cut in half. In approximate terms, external shading rejects about 80% of the fierce attack of solar energy, whereas internal shading accepts and reradiates 80% of it. Outside louvres have a chance to cool off in an occasional breeze, but inside drapes are part of a heat trap, and they constitute a system of hot-weather radiant heating that discomforts those who must work near perimeter surfaces.

Just how fierce is the sun? Compare it to the traditional heating effect of cast-iron steam radiation. Each square foot of such radiation produces 240 Btu/h. The amount of energy passing through 1 ft^2 of unshaded glass on an east wall on a summer morning is often evaluated at over 200 Btu/h. It can be frightening to think of an entire wall, blanketed *in summer* by a heat-producing source that is potentially almost as powerful as cast-iron radiation. The rejection of a maximum amount of this solar energy is obviously important for cooling performance. Exterior sunshades, then, represent the better initial design choice.

The Atlanta office building in Fig. 6.12 was designed with structural overhangs incorporating solar fins. The facades were chosen after computer-assisted evaluation of the shading effectiveness of several alternative designs. Sun penetration occurs after normal working hours; the wooded site helps reduce low-altitude solar gains, which can be especially troublesome. A clear-glazed window area equaling 50% of the wall area was utilized for perimeter daylighting and the unusually pleasant views of the site. The annual energy predictions show solar gains as but 21% of the cooling load; electric lights are the biggest item at 38%.

To properly reject direct sun, yet allow for view and daylight, many sunshades project out from the windows they protect. These exterior projections become highly visible elements of facades, and they tempt some designers into superimposing formal esthetic criteria that can be damaging to the solar control functions. A frequent example is the application of the same sunshade geometry to *all* elevations of a building. When the sunshades are fixed in position, this is usually counterproductive (see Fig. 3.16). Where they are movable, as in Fig. 6.11, this same sunshade approach is not so serious.

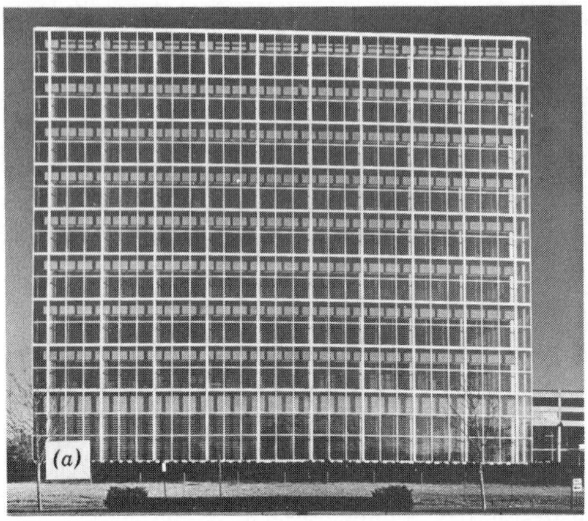

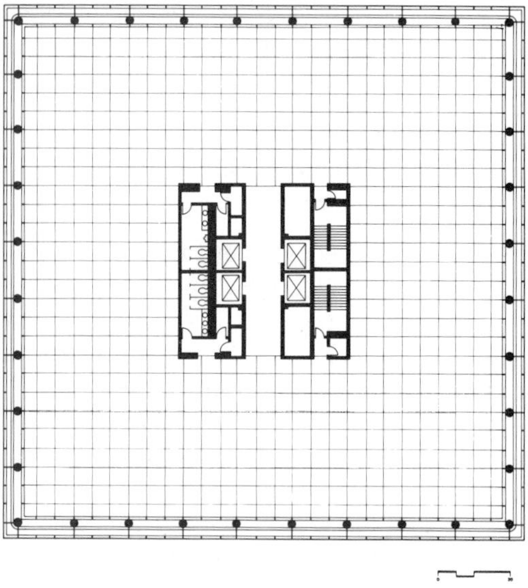

(b)

Fig. 6.11 The Occidental Chemical Corporate Office Building, Niagara Falls, New York. Architects: Cannon Design, Inc. (a) *Fully exposed to sun and wind despite its downtown site, the building appears as a conventional curtainwall box.* (b) *An ordinary-looking plan utilizes a central core, with suspended ceilings for horizontal service distribution.* (c) *A corner office displays the horizontal louvres, 15 ft long and 8 in. apart, painted white for best diffusion of daylight. They are motor-controlled by light sensors and can be closed on cold nights. Though all facades appear to be alike, the louvre positions will vary on sunny days.* (d) *The 4-ft-wide cavity allows for maintenance walkways and ample natural ventilation by stack effect. Internal and solar gains have made central heating unnecessary. Temperatures of the inside glass surface are nearly equal on all four sides; a perimeter heating/cooling system is therefore unnecessary, and the entire building is served by an all-air, variable-volume HVAC system.*

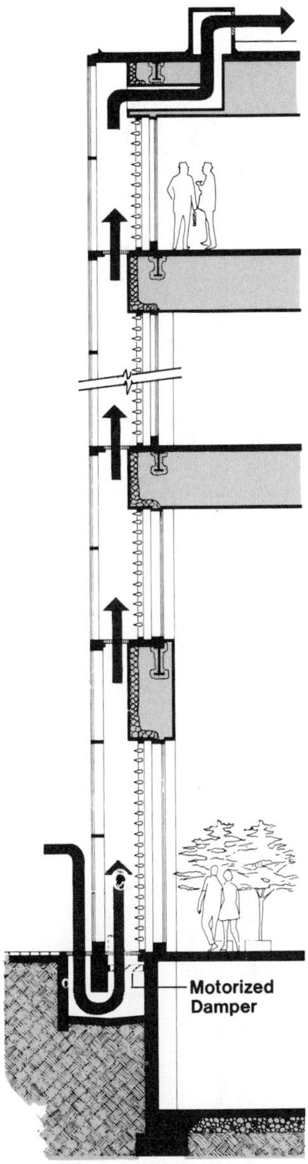

(d)

Fig. 6.11 (continued)

Fixed sunshading devices are very common, partly because they lack the moving parts and controls that can be expensive and troublesome. As explained in Section 3.4, they do not ordinarily serve very well, since in order to block the sun on any elevation in September, they must also block it in March. For many buildings in temperate climates, March is a heating-need month, whereas September is a cooling-need month. Because of this disadvantage, fixed sunscreens are not discussed further here; a procedure for evaluating their shading performance is presented in Appendix D.

Adjustable sunshading, once enormously common (it seemed as if every 1930s shop had an awning) but later considered "old-fashioned," has made a comeback with the advent of passive cooling. Some basic approaches to adjustable devices are shown in Table 6.3. Roll-down shades can be seen on Sacramento's Bateson Building (Fig. 5.20). Adjustable awnings, shown in Fig. 6.13, were also used on the buildings shown in Figs. 2.21 and 2.22 and in the Conservation Center for the Forest Society in Concord, New Hampshire (Fig. 6.34). Horizontal movable louvres enclose the Occidental Chemical offices (Fig. 6.11). Vertical movable louvres are particularly effective on east and west orientations, where they can block the sun from (for example) the southwest while maintaining a view out to the northwest.

How are movable shading devices adjusted to the desired position? Most manufacturers offer three types of controls: manual, motorized, and automatic. Manual systems are cheap and relatively trouble-free, but they require thoughtful, timely action on the part of the users of the building. So do motorized controls, but for large or heavy devices in remote places (clerestories, for example), the motorized control provides a steady hand for smooth operation. Automatic systems have the advantages of freeing the users from adjustment tasks and of taking into account the thermal needs of the building as a whole when setting the sunshade's position. With computerized controls becoming widespread for larger buildings (see later Fig. 7.40), the added costs of automatic sunshading control can be easily incorporated into the overall cost of controls.

(b) Glazing. Some of the most exciting recent technological developments in all of the architectural materials are those dealing with windows. Both energy conservation and postmodernism have contributed to these developments. The window is one of the most interesting components, combining unusually high degrees of view, light transmission, heat trans-

TABLE 6.3 **Adjustable Sunshading**

	Cooling and Daylighting Performance While Shading					
	Solar Heat Rejection	*Daylight Distribution*	*View*	*Ventilation*	*Typical Orientation*	*Heating: Night Insulation*
Roll-down shades						
Solid 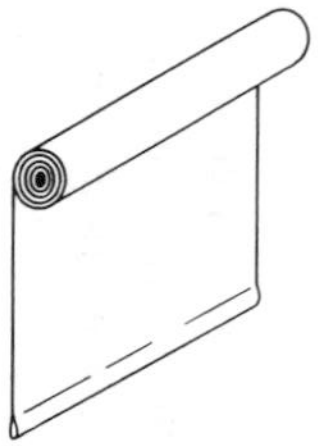	Blocks both direct and ground-reflected light. Heat builds up behind shade, unless vented.	Translucent shades diffuse sunlight evenly but may themselves become sources of glare, due to their brightness. If colored, they will color incoming diffuse light.	Blocks view.	Blocks ventilation.	East and west, where it is rolled down daily during hours of threat from direct sun.	Can be effective if infiltration above and around the shade is blocked.
Slatted 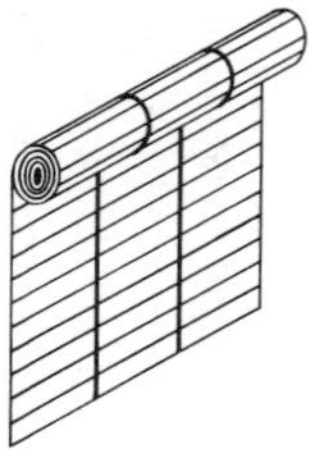	Blocks most direct sun and ground reflected light.	Thinly striped pattern of direct sun is admitted.	Heavily filters view.	Greatly reduces ventilation.		No value, unless slats can fit together to eliminate openings.

Fold-out Retractable Awning 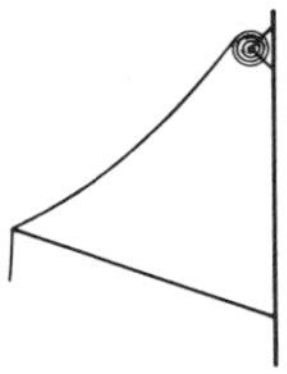	Blocks direct sun, admits ground reflected light. Heat on awning easily dissipated by breezes.	Translucent awnings diffuse sunlight evenly. (Colored translucent awnings color the diffused light.) Awnings admit ground-reflected light toward ceiling, where white surfaces diffuse it deeper into space.	Blocks sky view. Allows ground view.	Reduces ventilation somewhat, tends to direct wind upward into space.	Any orientation (on south, awnings can be deep and can be left in position for weeks at a time).	If awning folds back against window, it provides some insulation, but infiltration reduces its value.
Pivoting Louvres Horizontal 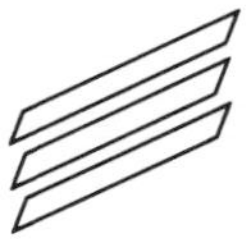	Blocks direct sun, admits ground-reflected light.	Sunlight and ground-reflected light directed toward ceiling, where white surfaces diffuse the light deeper into space.	Blocks sky view. Filters ground view.	Reduces ventilation somewhat; directs wind upward into space.	South; also used on other elevations.	Closed louvres provide some insulation, but infiltration reduces its value.
Vertical 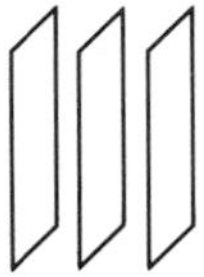	Blocks direct sun, admits some sky and ground-reflected light.	Diffuse and ground-reflected light directed sideways into space. Does not usually help with deeper daylight penetration.	Filters view.	Reduces ventilation somewhat; directs wind sideways into space.	East and west; sometimes south and north.	

(a)

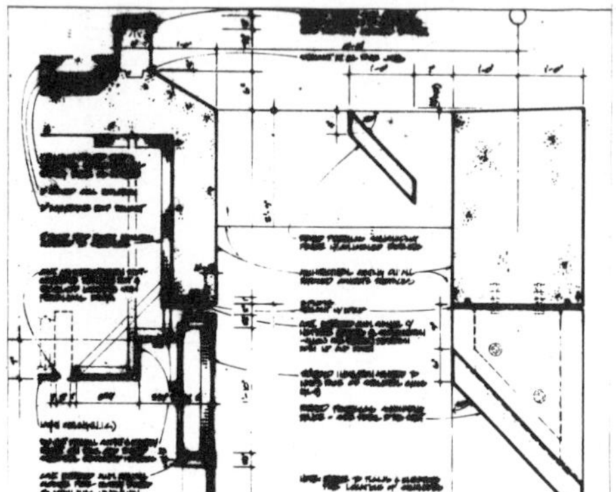

(b)

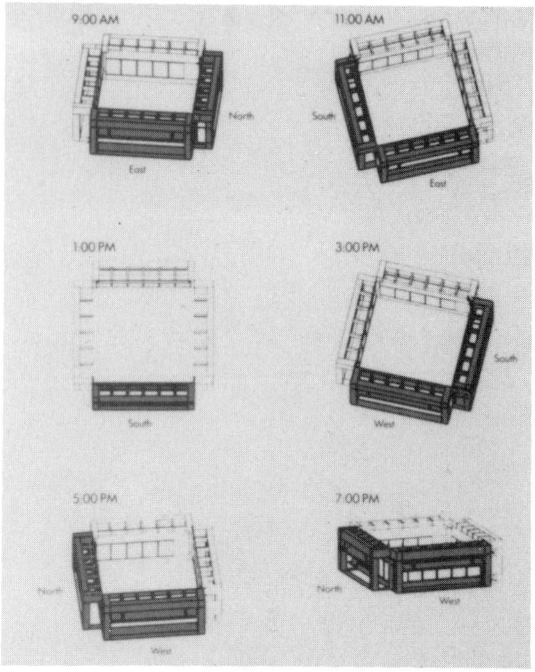

(c)

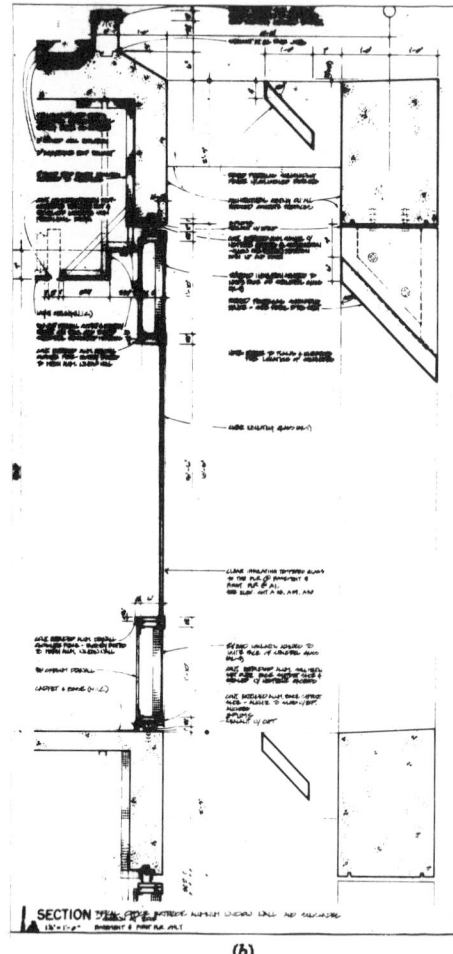

Fig. 6.12 *The 563,000-ft² Atlanta office building, North Terrace at Perimeter Center,* (a) *has facades incorporating a structural overhang with a solar fin* (b) *shown here in section.* (c) *Computer-generated summer views show shading effectiveness. (Courtesy of Skidmore, Owings and Merrill, Architects-Engineers, Chicago.)*

fer, ventilation, and sound transmission. It is also one of the elements that give a distinctive character to a facade. The search for thermally insulative windows that permit visual design freedom is leading in several directions.

A related component that can combine the functions of both glazing and sunshading is the *air-extract window* diagrammed in Fig. 6.14. Developed in Scandinavia in the 1950s and now utilized in the United States as well, this is a triple-glazed window that passes room air between a typical outer double-glazed window and an inner, single pane. The inner pane thus is kept at very nearly the same temperature as the room air, which greatly increases comfort near windows on very cold (or very hot) days. Venetian blinds are often inserted in this cavity, where they can intercept direct sun and redirect its light toward the ceiling (see also later Fig. 6.24). The solar heat intercepted by the blind is carried off by the room air to a plenum above the ceiling, where such air can be either exhausted or recirculated, and its heat content either reclaimed or rejected. The U value of these windows is dependent on the rate of airflow be-

Fig. 6.13 *Movable awnings can become an important design element, by helping to change a building's appearance with changing sun positions. (Courtesy of Levolor-Lorentzen, Inc.)*

tween the glazing (Fig. 6.14*c*); typical flow rates are 4 to 6 cfm per foot of window width. For an example of such windows in a large U.S. building, see Fig. 7.66, which shows the Comstock Center in Pittsburgh.

The thermal insulating properties of glass and plastics were detailed in Section 4.4 (see Table 4.16 in particular), and glazing's light transmission will be discussed in later chapters on lighting. A third important characteristic of glazing, for passive solar performance, is *solar transmission*. Several recently developed glazing products offer either increased solar transmittance or improved insulating performance. Either approach will aid in passive solar heating. Provided that careful attention is paid to glazing's role in daylighting year-round and that shading protection is provided in summer, good choices for all-season performance can be identified.

When cooling is the dominant concern, many designers use glazing that permanently rejects solar heat. The potential problems with reflective glazing were shown in Figs. 3.13, 3.14, and

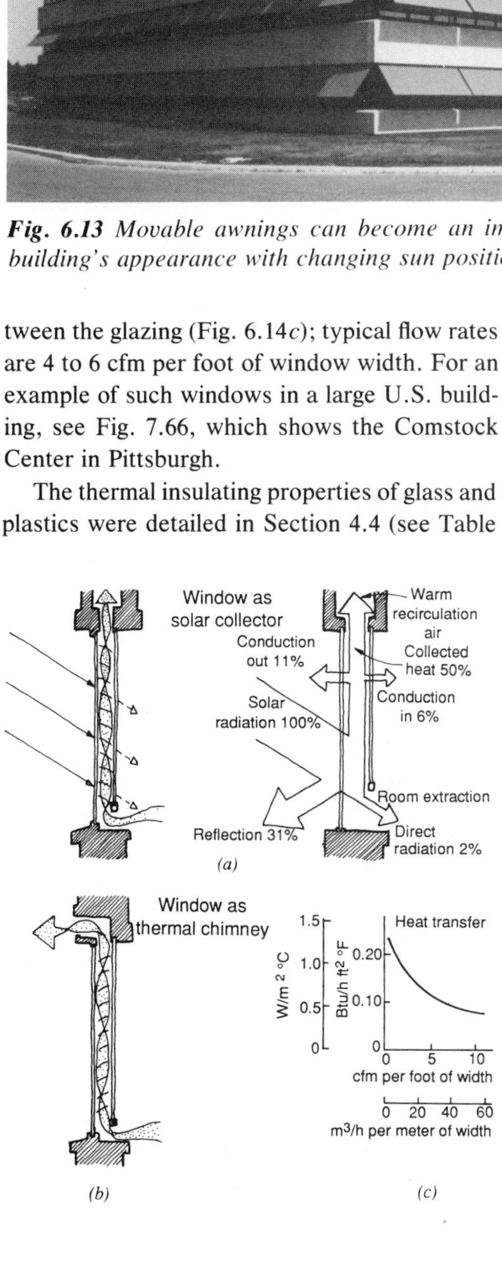

Fig. 6.14 *Air extract windows.* (a) *Cross section, showing the window as a solar collector.* (b) *Contrasts between the solar collector (winter) and the solar chimney (summer) modes.* (c) *The* U *value of these windows varies with the rate of airflow within the glazing; typical flow rate is 4 to 6 cfm per foot of window width. (From articles by D. Aitkin and O. Seppanen in* Proceedings of the Sixth National Passive Solar Conference. *© 1981 by the American Solar Energy Society, Inc.)*

TABLE 6.4 **Ratios of Average Solar Transmission to Heat Loss for Some Vertical Glazing Assemblies**[a]

1. Single-pane D.S. flat glass[b]	$\frac{84\%}{1.13 \text{ Btu/h F}} = 74.3$
2. Double glass (½″ air space)	$\frac{71\%}{0.58 \text{ Btu/h F}} = 121.6$
3. Triple glass (½″ air space)	$\frac{59\%}{0.36 \text{ Btu/h F}} = 164.6$
4. Quadruple glass (½″ air space)	$\frac{49\%}{0.28 \text{ Btu/h F}} = 175.0$
5. Selective (ITO) transmitter with D.S. glass spaced ½″ away[c]	$\frac{65\%}{0.29 \text{ Btu/h F}} = 224$
6. Double FRP (½″ air space)[d]	$\frac{66\%}{0.56 \text{ Btu/h F}} = 118$
7. Double 2-mil Teflon (½″ air space)[e]	$\frac{84\%}{0.61 \text{ Btu/h F}} = 138$
8. Quadruple Teflon (½″ air space)[e]	$\frac{71\%}{0.31 \text{ Btu/h F}} = 229$

Source: Johnson (1981).

[a]*U* values for glass systems are from the ASHRAE *1972 Handbook of Fundamentals.* Current window *U* values that account for edge and frame losses are considerably higher than these values; see Table 4.16.

[b]D.S. stands for double-strength glass.

[c]ITO stands for indium tin oxide coating.

[d]FRP stands for fiberglass-reinforced plastic.

[e]Teflon™ is a very-high-solar-transmittance plastic film.

3.15. However, heat-rejecting glass *is* appropriate when passive solar heating is *never* advantageous for a building.

Table 6.4 shows the ratios of solar transmission to heat loss for a variety of glazings used in passive solar heating applications. High ratios indicate materials that perform well. Glazings with lower ratios may have net daily heat losses; in colder climates, such glazings will require movable insulation at night, in order to admit more daytime solar heat than is lost through the glazing over 24 hours.

Recent improvements in the insulating performance of glazing commonly have followed two approaches: reduction of losses by convection and reduction of losses by radiation. *Convection losses* can be cut by the introduction of a screen of transparent cellular materials between the sheets of double glazing. This breaks up the convection currents that form in the glazing cavity, but it also tends to interfere with the view through the window.

Current research is focusing on transparent insulative materials that promise to give the window as high a thermal resistance as a well insulated wall. *R*-20 looks achievable with aerogel, a very porous glasslike substance that is both highly insulating (*R*-7 per inch) and highly transparent. The characteristics of this and several other new products are summarized in Table 6.5.

Alternatively, the glazing cavity can be filled with heavy gas (such as argon) and then sealed. The problem here is that the gas will leak out when there is any failure in the window seal; such leakage is usually undetectable.

TABLE 6.5 **Some Characteristics of Present and Future Glazing Products**

Product	*Approx. R Value*[a]	*Characteristics Likely to Be Advantageous*	*Characteristics Likely to Be Disadvantageous*
Low-emittance, or low-ε	*R*-3	Lower *U* value than ordinary double glazing; reduces fading; less condensation	Slightly higher cost than ordinary double glazing
Gas-entrained	*R*-3 to *R*-4	Lower *U* value than ordinary double glazing	Gas leaks likely over time, reducing *R* value
"Super windows"[b]	*R*-6 to *R*-10	May yield net solar gains even on north windows; will allow much more glazing area yet still meet energy codes	Needs high-quality sashes and frames; an inexpensive gas fill will help cost-competitiveness
Aerogels	*R*-7 to *R*-20	Per-inch resistance equal to best opaque insulations; sound transmission about one-tenth of ordinary window	Slight opacity or "frosted" (yellowish or bluish tint) appearance; degrades rapidly if exposed to moisture
Switchable glazings: passive[c]	*R* not affected	Low cost compared to active switchable glazing; responds to both temperature and intensity of sunlight; privacy; glare control; shading	Switches at preset combination of temperature and light intensity
Switchable glazings: active[d]	*R* not affected	Allows instantaneous control of natural light levels; privacy; glare control; shading	Very high cost; sometimes turns absorptive rather than reflective; currently degrades rapidly when exposed to ultraviolet radiation (therefore unstable in exterior applications)

Source: Adapted from *Inside Housing Newsletter,* Vol. 1, No. 1, December 1989; The Levy Partnership, Inc.

[a]*R* value at the center of the window glass, not averaged with edge and frame *R* values.

[b]Two low-ε coatings, krypton or argon gas filled, and a nonstructural center glazing layer.

[c]Such as Cloud Gel, manufactured by Suntek.

[d]Electrochromic devices allow much more control than do LCD (dispersed liquid crystal) devices, but require costly control systems.

Radiation losses can be greatly reduced by the introduction of *selective transmitter* (or "low-emittance") film somewhere within the glazing cavity. As shown in Fig. 6.15, these films admit most of the incoming solar radiation in both visible and near-infrared (short) wavelengths. As objects become warmed within the room, however, they begin to emit far-infrared (long wave) radiation that is reflected back into the room by the selective film. These selective

THERMAL CONTROL

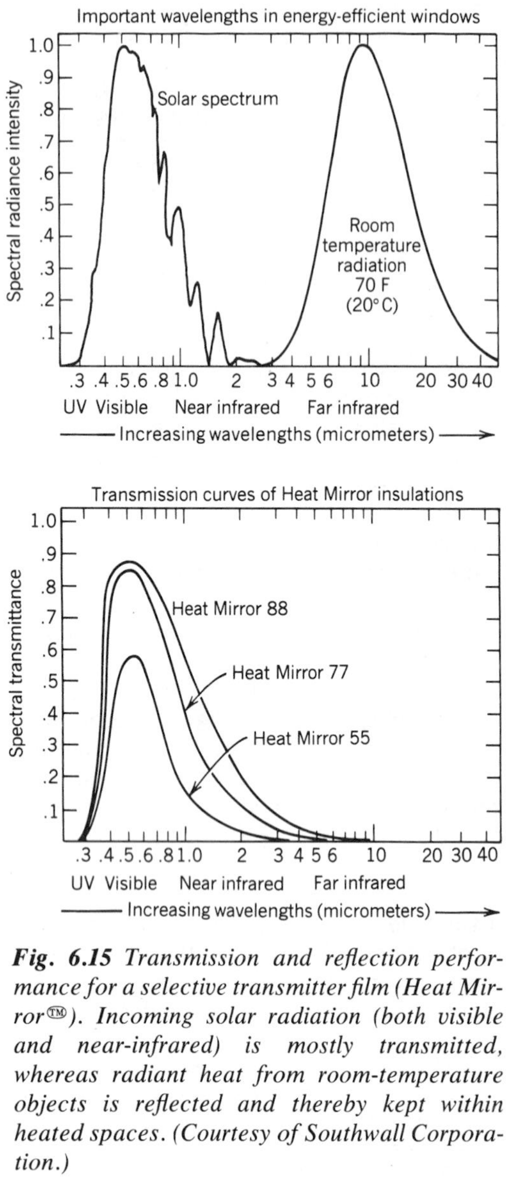

Fig. 6.15 Transmission and reflection performance for a selective transmitter film (Heat Mirror™). Incoming solar radiation (both visible and near-infrared) is mostly transmitted, whereas radiant heat from room-temperature objects is reflected and thereby kept within heated spaces. (Courtesy of Southwall Corporation.)

films typically are available either as separate sheets that can be inserted between sheets of glazing as a window is fabricated or as coatings that are applied to glass at the factory. As a separate sheet, a selective film could also be applied to existing windows—for instance, between storm windows and the ordinary windows they protect.

Two approaches to these low-ε films are *hard-coat* (durable, less expensive but less thermally effective) and soft-coat (better thermal performance but more expensive and subject to degradation by oxidation in the manufacturing stage). An important added benefit of these films is their reduction of ultraviolet transmission, thus reducing fading of objects and surface materials in rooms. Further, with their higher inner surface temperatures, condensation is much less likely on these windows.

Now that such products have gained acceptance, an increasing range of choices are offered. For example, a range from Heat Mirror 77 (low reflectance, recommended for vertical glass) to Heat Mirror 33 (low transmission, recommended for sloping glass) is available, and includes seven colors. (A clear Heat Mirror 88 has the lowest reflectance of all.) Some skylight and greenhouse manufacturers offer this low-emittance product as a standard option.

Selective films typically transmit slightly less solar energy than is transmitted by ordinary glass. Another approach to solar windows is to use a film that *increases* solar transmittance over ordinary glass (see Fig. 6.16). When such an assembly is quadruple glazed (item 8 in Table 6.4), it essentially matches the selective film's performance in triple glazing.

The high insulating value of these window assemblies is as attractive as the performance of ordinary double-glazed windows with movable night insulation, as indicated in Fig. 6.17. Given the dependence of movable insulation on either an automated control system or user response, the selective film assemblies may be the more attractive alternative in many cases.

"Super windows" is a term applied to products under development that combine several technical advances, in order to achieve thermal resistance values approaching that of well-insulated opaque walls. They combine two low-ε coatings, a gas fill (krypton or argon), and a nonstructural center glazing layer that can have various light transmittance characteristics (such as that of Fig. 6.16). All of these features are then compressed into a frame only 1 in. thick. With their combination of superior thermal resistance and solar gain potential, they can conceivably have a *better* heating season thermal performance than that of an insulated opaque wall. The potential design consequences are enormous.

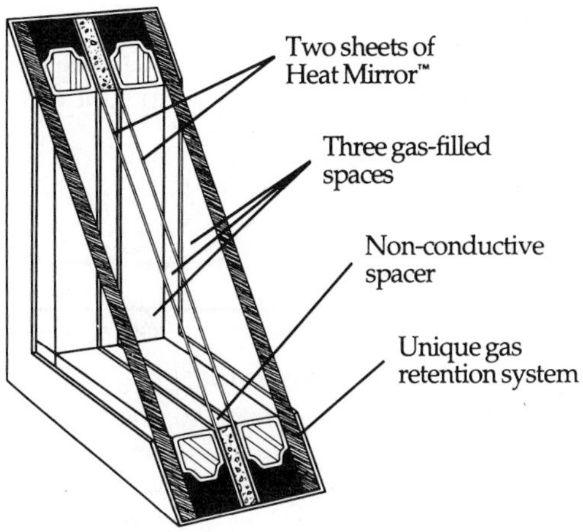

Fig. 6.16 *The* Superglass™ *window system utilizes two films of* Heat Mirror™ *between two outer panes of glass. Three air spaces are thus created, filled with a nonreactive gas mixture to retard convection. Non-metallic spacers reduce heat loss at edges. The resulting window has a center-of-glass insulating value of R-8.1. (ASHRAE overall window U-values are Type R, U = 0.22; Type C, U = 0.18.) Very low transmission of ultraviolet light is another characteristic. (Courtesy of Hurd Millwork Company, Medford, Wis.)*

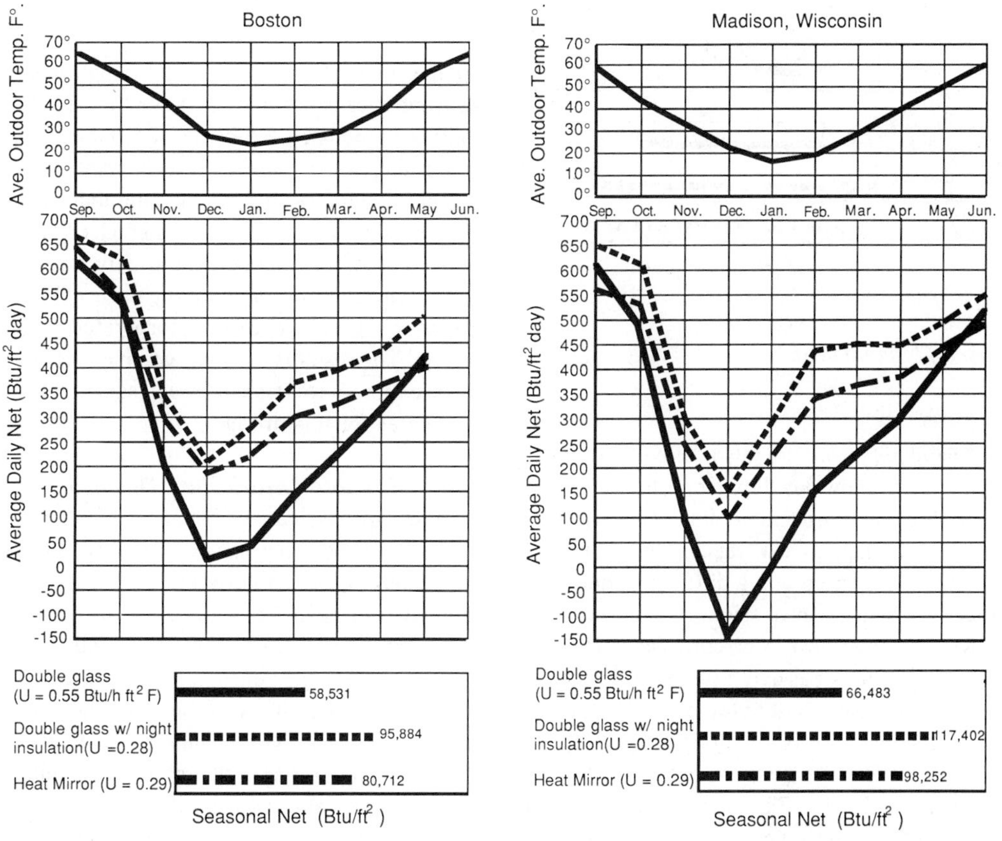

(a) *(b)*

Fig. 6.17 *Average daily net heating, per square foot of south window, for three glazing assemblies in three climates.* (a) *In Boston, ordinary double glazing is a slight energy plus even in December.* (b) *In Madison, ordinary double glazing is a net loser for almost two months.* (c) *In Seattle, ordinary double glazing is a net loser for more than two months. [From Johnson (1981).]*

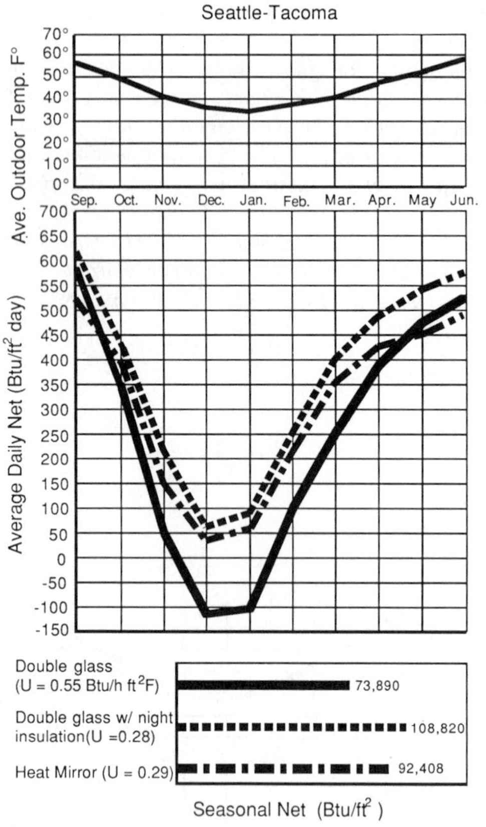

Fig. 6.17 (continued)

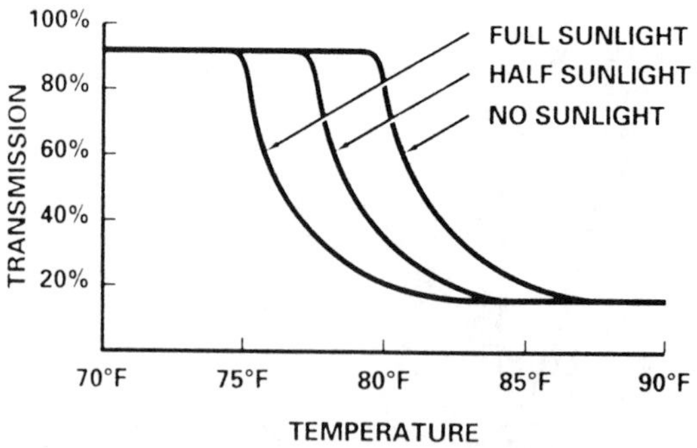

Fig. 6.18 *A passive switchable window film, Cloud Gel, changes from a transparent, highly transmissive film to a nearly opaque film under the combined influences of temperature and sunlight. (Courtesy of Suntek, Albuquerque, New Mexico.)*

Switchable windows (also called ''smart windows'') are a new development that has more to do with daylighting and cooling than with heating. These glazings can repeatedly change between a transparent and a diffuse/opaque state, affecting light transmission, solar gain, and view. A lower-cost, currently available film (Cloud Gel) acts passively (without electric input) in response to either temperature (for solar heating applications) or sunlight intensity (for daylighting applications), or both; see the performance curve in Fig. 6.18. Under development, and currently very expensive, are active switchable films. The more simple type uses *LCD (dispersed liquid crystal) devices* in a simple on–off, transparent-diffuse mode, and are most likely to be used in glare reduction, privacy, and skylight applications where high cost is not objectionable. *Electrochromic devices* are capable of electronic coloring through a large transmittance range and can have the added advantage of memory. Problems of turning absorptive rather than reflective, acceptable color, long-term stability of the switching unit, degradation under ultraviolet light, and high cost must be solved before they become widely available.

See-through photovoltaics have closely-spaced photovoltaic cells incorporated in the glazing. The result is a visual screen effect. Windows exposed to intense direct sun are likely candidates for this product.

(c) Movable Insulation. This component is unique to passive solar heating. Usually one of the costliest components, it is also the one most likely to be made by hand and on site. It demands the most from the users of the buildings. Living in passively heated and cooled buildings is often likened to ''thermal sailing'' (Figs. 2.19 and 2.20); movable insulation is the most visible sail. Advocates of conservation have pointed out that both architects and clients tend to undervalue conservation strategies because unlike passive or active solar heating, passive cooling, and mechanical equipment, the conservation features are almost always hidden away within the construction. Here is one that decidedly is not. Yet despite its promise of low-tech and individual design, movable insulation has not been used widely in the United States, and interest in more efficient glazings has nearly eclipsed movable insulation.

Many varieties of these thermal shades and shutters are available; they can be placed inside, within, or outside the glazing they protect. These options are summarized in Table 6.6. Typical U values for windows with movable insulation can be found in Table 4.17.

Along with movable insulation come some common problems. The first is control: Will the insulation be closed when it is cold and sunless and open when the sun should be admitted? As in the case of movable sunshading, the options here are manual, motorized, and automatic controls. For homes, the automatic option is certainly available but as yet is rarely installed. Another problem is the thickness needed to achieve good insulating performance. Bulky shades are awkward both in appearance and operation; thick shutters are less of a problem unless they stack when open (which rapidly increases the storage space they require). Unfortunately, some of the materials with the highest R value per inch emit toxic gases when burned, so their use within rooms requires costly fire-retardant covering materials. Another problem is infiltration into the airspace formed between the movable insulation and the glazing, which can reduce this assembly's thermal performance. Tight seals at the bottom and sides of thermal shades and shutters are particularly important for stopping the convection currents that would readily bring warmed room air into the top of the air space and send chilled air out of the bottom.

Available references that discuss the details of movable insulation options include Shurcliff (1980), Langdon (1980), and Niles and Haggard (1980); the latter is particularly rich in architectural detailing examples. Figure 6.19 shows one of the many approaches to movable insulation; they can also be found in the buildings shown in Figs. 1.10, 2.19, 5.15, and 6.11.

(d) Thermal Storage. The thermal storage properties of many common building materials—including density, conductivity, and specific heat—can be found in Table 4.2. The procedures for sizing of thermal storage for passive heating and cooling were outlined in Sections 5.3 (preliminary guidelines), 5.6 (passive solar heating), and 5.9 (passive cooling). The performance of varying degrees of thermal mass in passive solar applications can be found in Appendix C. Here, information about processes and products for thermal storage will be presented.

For *direct-gain* spaces, the thermal mass must be widely distributed around the room, so that direct sun can strike the mass surface and/or be reflected to the mass surface as soon as possible upon entering the window. Table 5.7 gave rules of thumb for mass; a common recommendation for direct-gain spaces above 50% SSF is for a thermal mass area five to seven times the area of glass (for no more than the optimum 4-in. masonry thickness). Appendix C shows how various ratios of mass to glass affect performance. There are many common ways to provide such mass surfaces, including brick veneer and clay tile over a bed of grout. These surfaces can be applied to frame construction.

For *thermal storage walls* (Trombé walls), previous design decisions have been made regarding size and thickness, to vent or not to vent, and movable insulation. Some detailing for vents and movable insulation is shown in Fig. 6.20. An unusually large Trombé wall application is shown in Fig. 6.21. This Vermont warehouse utilizes a vented Trombé wall and ceiling fans to prevent stratification of warm air at the ceiling. Daylighting is provided by roof monitors; the Trombé wall is vented to the outside in summer.

Water walls (Fig. 6.21, Table 6.7) also have been sized previously. They are commonly made of corrugated galvanized steel culverts, steel drums, or fiberglass-reinforced plastic tubes (for which manufacturers will provide suggested installation details). Within water wall containers, some air space should be provided, as water expands when it heats. A rust inhibitor should be added to the water within steel containers; algae growth in plastic tubes, encouraged by the daylight that passes through them, is usually controlled by the addition of an algicide to the water. Dyes may be added to change the color of daylight through the tubes. Water used as thermal mass in a direct-gain application is shown in Fig. 6.22.

TABLE 6.6 **Movable Window Insulation**

	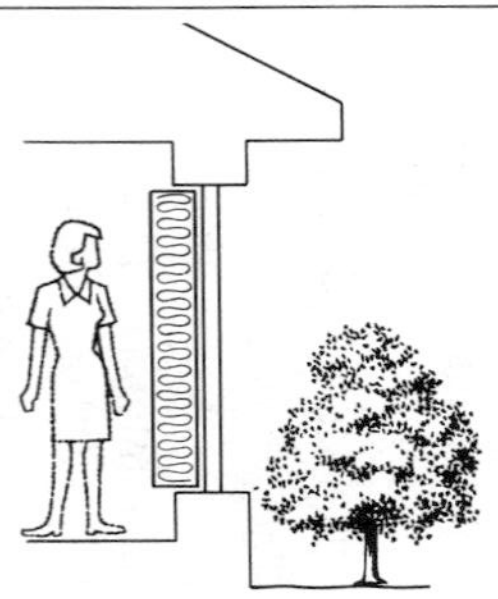 *Inside the Space*	 *Between Glazings (or with Trombe Wall)*	 *Outside*
General opportunities	Insulation unaffected by weather or exterior abuse.	Insulation unaffected by weather and exterior or interior abuse, and stays relatively dust-free.	Insulation protects window weathering or exterior abuse.
	Window surface changes texture when insulation covers it.		Most effective position to act as sunshade.
General problems	Least effective position to act as sunshade.	Between glazings, insufficient thickness inhibits developing adequate *R* value.	Wind and sleet can inhibit moving the insulation.
	Subject to interior abuse.	Difficult maintenance access.	Must be weatherproof construction and is subject to exterior abuse.
	Can affect furniture arrangement near window.	Storage is difficult.	Can affect landscaping materials near window.
	Some foams produce toxic gases when burned.		
How insulation moves			
Blow in/out 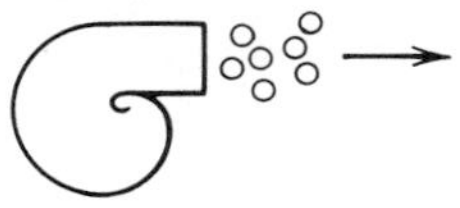	(Approach not developed.)	Bead-Wall™: Styrofoam beads fill glazing cavity. Vacuum-cleaner-type motor sucks beads out, blows beads in. Bead storage tanks required. Moderately poor visibility through window even without beads, although insulation disappears in open position.	(Approach not developed.)

Roll up/down 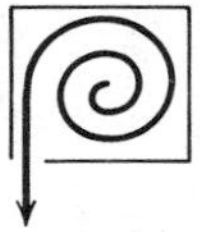 Insulation disappears in open position.	Many available products; often site-constructed. High R value usually means a bulky roller and awkward operation. Both automatic and manual operation are common. Little impact on furniture arrangement.	For Trombe walls, motorized and automated operation is common.	Widely used as sunshading and security, but seldom has high R value. Both automatic and manual operation are common.
Slide up/down/side 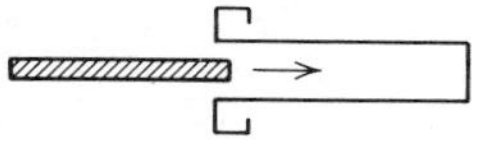 Insulation can disappear into pocket.	High R value can be obtained, if wide pocket is acceptable. Widely used, with or without pocket. Little impact on furniture arrangement. Usually manually operated.	Not frequently used.	High R value easily attained, since pocket usually omitted on exterior. Shutter therefore influences exterior appearance, either in open or closed position.
Fold up/down/side 	High R value easily obtained. Needs some clearance in front of window; can affect furniture. Usually manually operated.	Not frequently used due to insufficient clearance.	High R value easily obtained. Snow inhibits up/down operation. Some potential as reflector for added sun through glass. Needs some clear space outside window.
Swing up/down/side 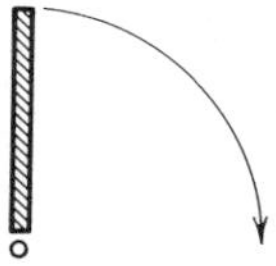	High R value easily obtained. Substantial visual impact, in any position. Needs considerable clearance in front of window, affecting furniture.	Rare: insufficient clearance.	High R value easily obtained. Snow inhibits up/down operation. Good potential as reflector for added sun through glass. Needs considerable clear space outside window.

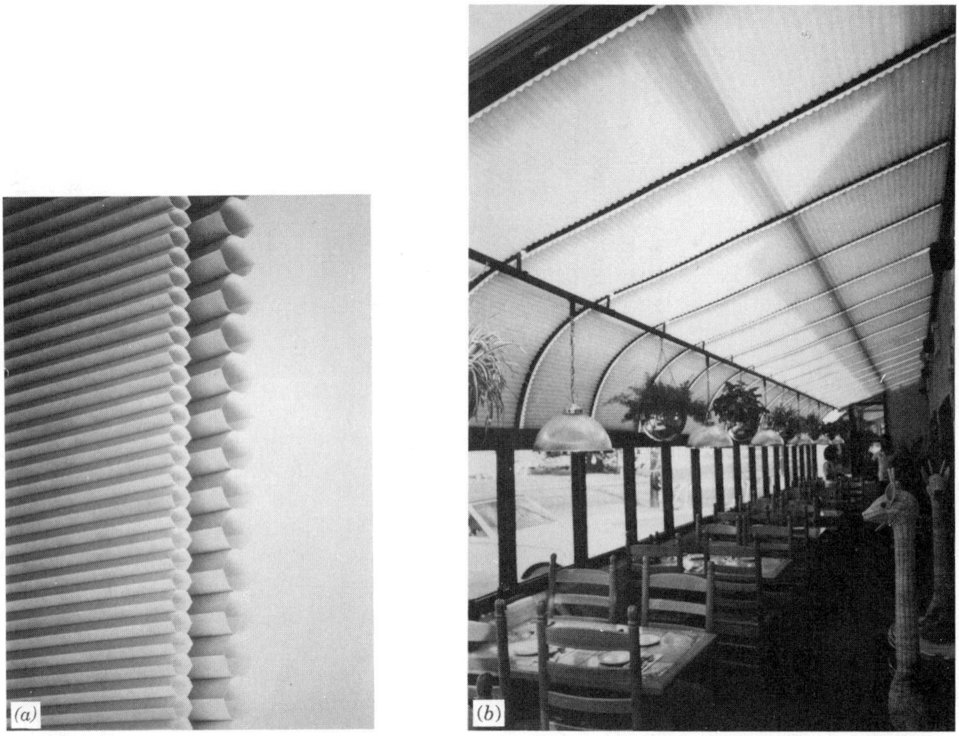

	Translucent		Opaque (With Film Interliner),
	3/8-in.[a]	3/4-in.	3/4-in.
Total *R* value, winter			
Over single glazing	2.22	2.38	3.1
Over double glazing	3.23	3.57	4.2
SC, summer			
With single clear glazing	0.49	0.45	(No daylight admitted)
With double clear glazing	0.45	0.44	

[a]A flame-resistant 3/8-in. version has slightly lower *R* values and SC.

Fig. 6.19 *One approach to movable window insulation is the Duette dual pleated shade.* (a). (b) *The shades can move either vertically or horizontally, cover both planar and curved openings, and be either motor or manually operated.* (c) *Available in either 3/8- or 3/4-in. thickness, the* R *value and light transmission vary with the materials of construction. (Courtesy of Hunter Douglas, Inc., Broomfield, Colorado.)*

Phase change materials, briefly described in Section 5.3, are available in configurations other than the flat bags of eutectic salts shown in Fig. 6.23. As flat tiles enclosing bags of salt, they can form the finished surface in any horizontal application—floors, ceilings, or counter and table tops, for example. Tubes and trays of salts can be arranged as desired (see Fig. 6.24). The function of these materials is to keep room temperatures steady; their performance in preventing overheating on a sunny winter day will also be appreciated on hot summer days, provided they are taken below the phase change or melting temperature at night. For U.S. locations with large daily temperature ranges in summer, thermal storage surfaces for passive solar heating in

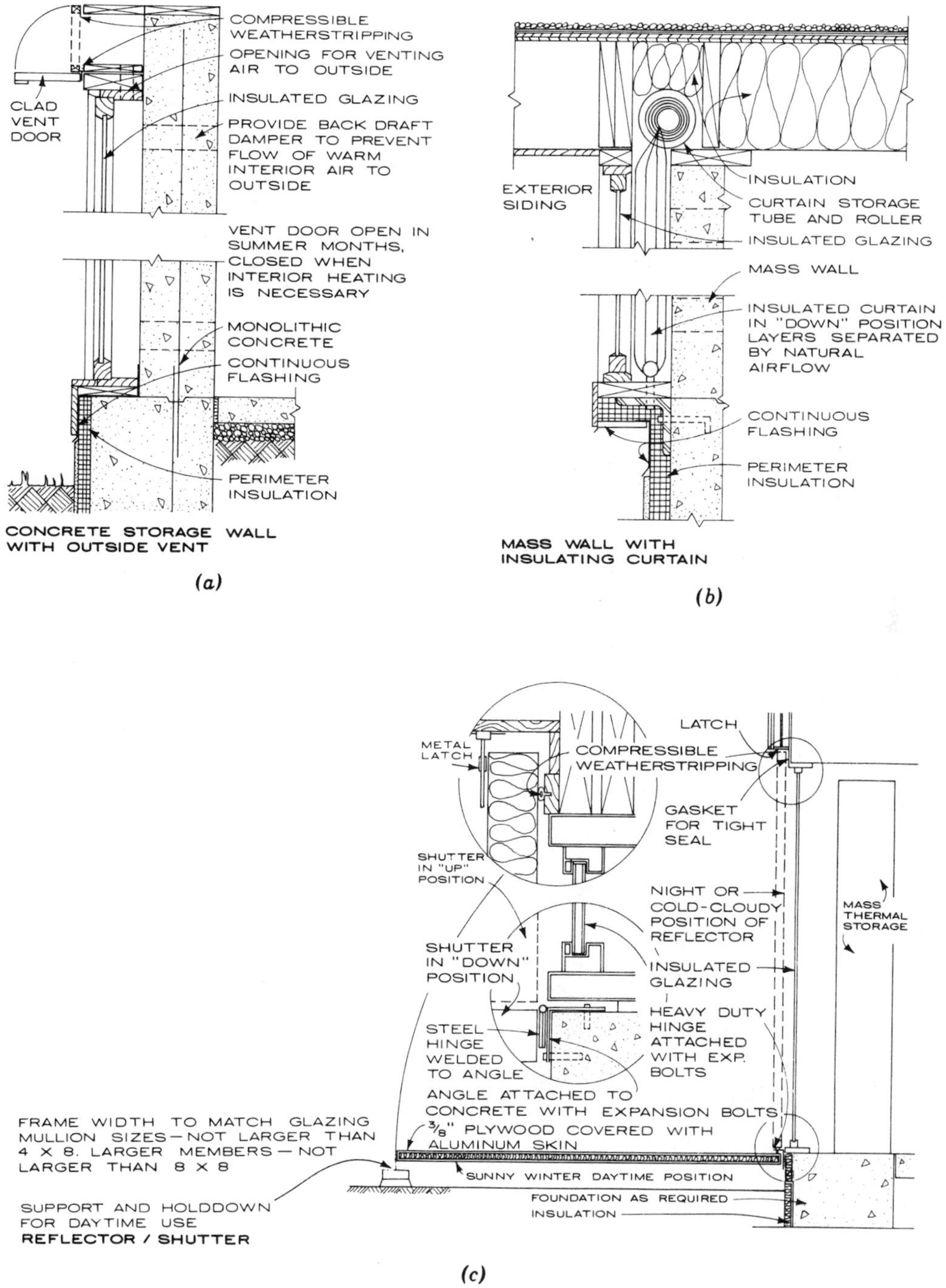

Fig. 6.20 *Detailing for thermal storage walls (Trombé walls).* (a) *If vented Trombé walls are used, a top vent to the exterior can provide low-velocity air exhaust from the space served by the Trombé wall.* (b) *Roll-down movable insulation such as this usually has motorized (frequently automated) controls. The cavity between wall and glass should be generous enough to avoid abrasion on the curtain as it moves.* (c) *Fold-down exterior insulating shutters are much easier to maintain than are curtains enclosed behind glass, but they are subject to severe weathering, and snow can inhibit operation. They have the significant added benefit of acting as reflectors to augment solar collection. (From AIA, Ramsey, and Sleeper,* Architectural Graphic Standards, *7th ed., © 1981 by John Wiley & Sons. Reprinted by permission.)*

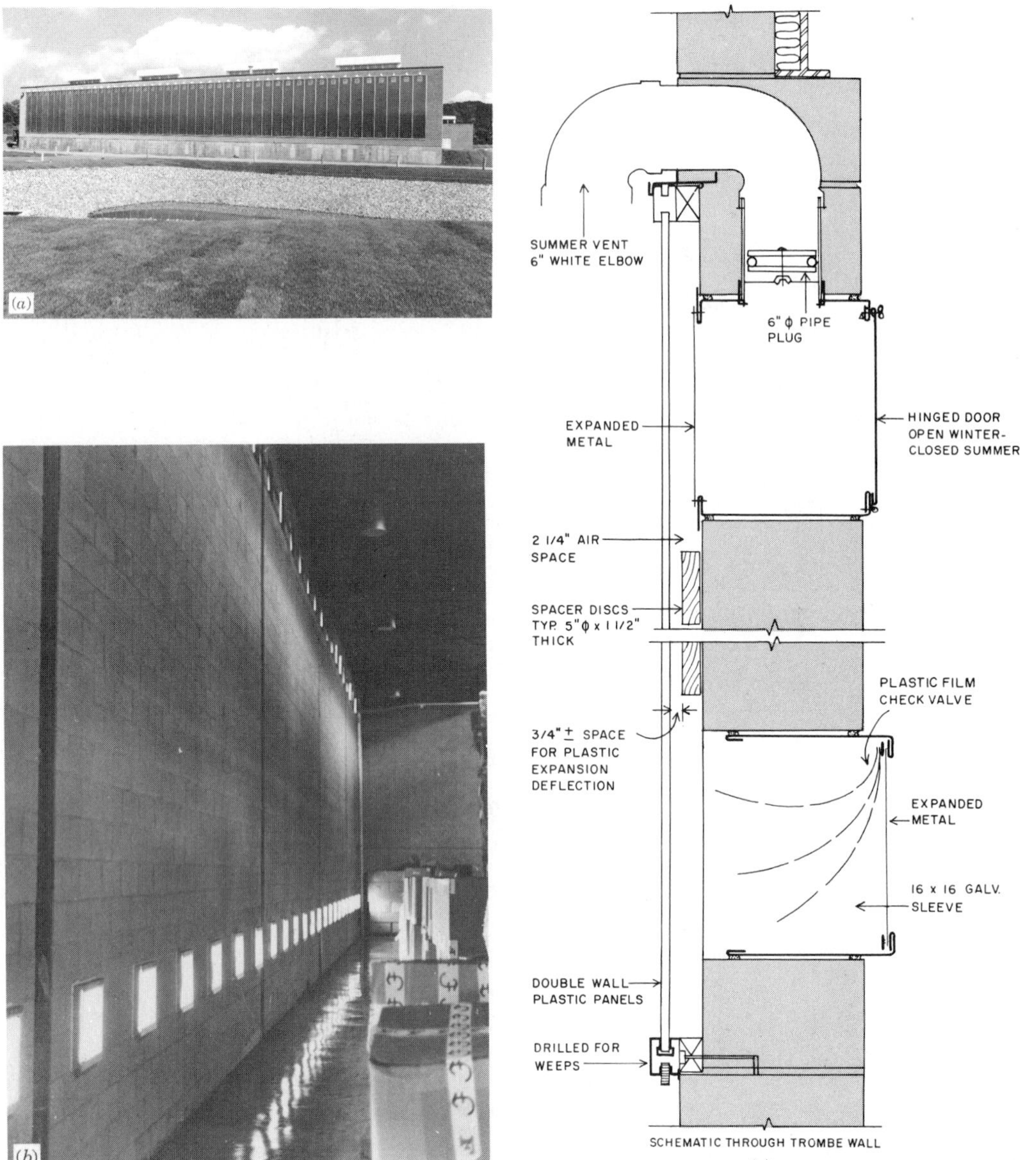

(c)

Fig. 6.21 *The warehouse for the Famolare shoe company in Brattleboro, VT uses a huge vented Trombé wall, 20 ft by 184 ft.* (a) *The Trombé wall faces 10° east of south. The pond in the foreground is an absorption basin for rainwater from roof and pavement. (Photo by Robert Perron.)* (b) *Interior shows the lower and upper vents.* (c) *Simple switches at bottom prevent reverse thermosiphoning on cold nights; at top, seasonal switches provide building ventilation in summer. Spacer disks prevent plastic double glazing from deflecting too far when its inner surface overheats. (Reprinted from* Architectural Record, *November 1979, © 1979 by McGraw-Hill, Inc. All rights reserved. Reproduced with the permission of the publisher. Courtesy of Banwell White Arnold Hemberger & Partners, Inc, Hanover, New Hampshire.)*

TABLE 6.7 **Water-Storage Tubes: Typical Data**

Diameter (in.)	*Height (ft)*	*Volume (gal)*	*Weight (lb) Full*	*Heat Capacity, 20F° Rise (Btu)*	*Floor Loading (lb)*	
					per lin. ft	*per ft²*
12	4	23.5	204	3,900	204	260
12	7	41	354	6,800	354	451
12	8	47	404	7,800	404	514
18	5	66	567	11,000	378	321
18	10	132	1122	22,000	748	635

winter are potentially useful for night ventilation cooling in summer.

The New Canaan Nature Center (Fig. 6.25) is an innovative New England example of many of the passive/low-energy components illustrated in this section. In addition to eutectic salts used in the south-facing railing on the upper level, the 4000-ft² building utilizes extensive clear, double-glazed south glass, movable insulating shades, operable vents at the skylight for stack effect ventilation, ceiling fans, a woodstove, solar collectors placed inside the skylight monitor (their water is used for warming planting beds), warm air heat recovery ducts, a well-insulated envelope, even a rainwater collection system. The manually operated switches for ventilation, insulation, and shading represent an unusual degree of user–building interaction. There is also unusual attention to microclimates and to transition between inside and outside in this award-winning building.

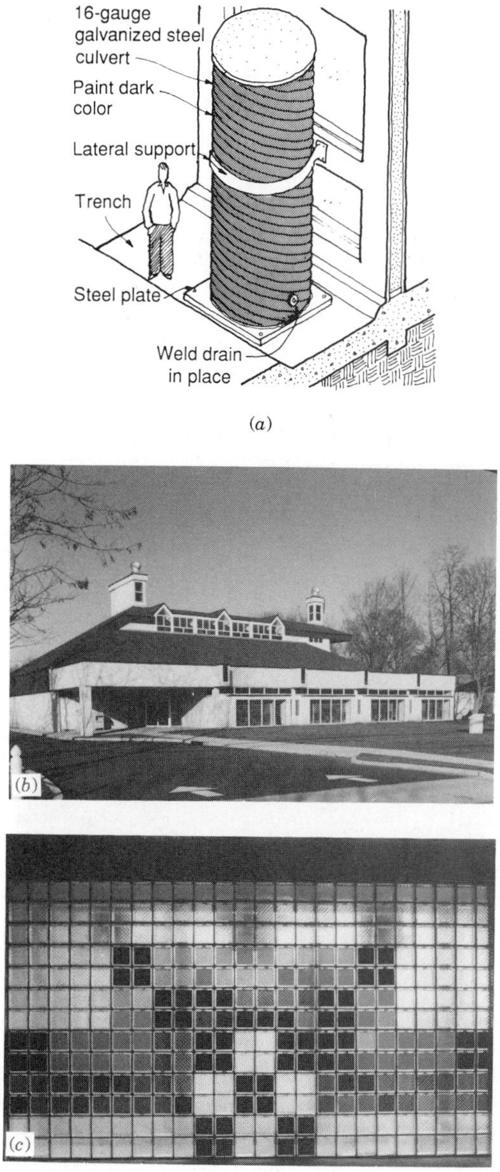

Fig. 6.22 *Water walls.* (a) *Detailing for a steel culvert water wall. The need for a cap (to prevent evaporation), lateral support at the top (to meet earthquake resistance requirements), and a trench for containment of accidental spillage is common to all water walls, regardless of container material.* (b) *The Center Moriches Free Public Library (New York) utilizes water-filled glass block* (c) *as a heat storage and daylight diffusing south wall. Dyes add color to selected blocks, forming the pattern shown. Lightshelves and stack ventilators add to its energy-conserving performance. (Courtesy of Banwell White Arnold Hemberger & Partners, Inc, Architects, Hanover, New Hampshire.)*

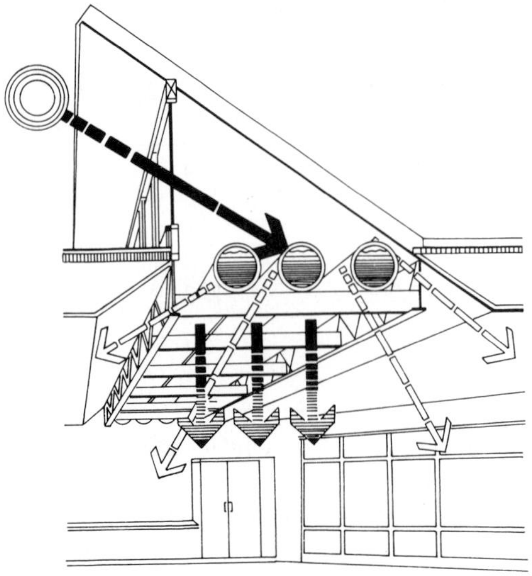

Fig. 6.23 The Abrams Elementary School in Bessemer, Alabama, uses plastic pipes filled with a total of 815 gallons of water beneath each double-glazed roof monitor. This system provides diffuse daylight as well as passive solar storage. Adams Peacher Keeton Cosby, Architects, Birmingham, Alabama. (Courtesy of the American Solar Energy Society, Inc., Boulder, Colorado.)

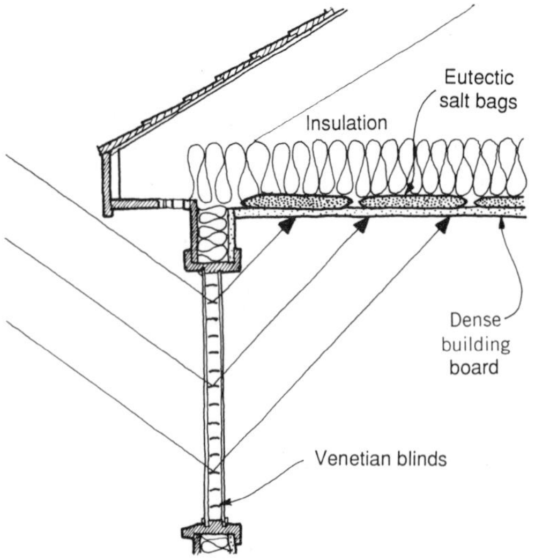

Fig. 6.24 Using bags of phase-change material (salt) for thermal storage above a flat ceiling. Mirrored-surface venetian blinds reflect direct sun to the ceiling, where it strikes a heavy (at least 90 lb/ft²) board product with high conductivity. Plaster is also suitable; ordinary gypsum board is not sufficiently conductive. Bags of salts weighing 5 lb/ft² are in contact with the ceiling board and must be installed in horizontal position. They are usually formulated to melt between 70 and 75 F. [Adapted by permission from Johnson (1981).]

Fig. 6.25 The New Canaan (Connecticut) Nature Center (a, b) *illustrates many passive and low-energy components.* (c) *Cross section illustrates the many design features that utilize renewable energy and enhance energy conservation. The "thermal storage elements" are filled with eutectic salts that absorb excess solar heat via phase change (from solid to liquid) by day, then again change phase (from liquid to solid) at night to release their stored heat to the space. (Buchanan/Watson, Architects. Photo by Robert Perron. Courtesy of Donald Watson FAIA, Troy, New York.)*

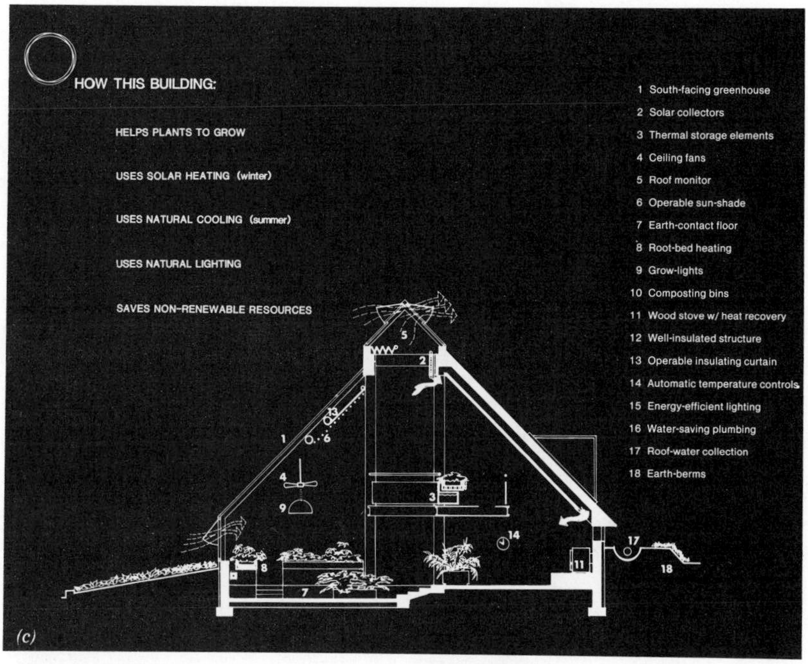

Fig. 6.25 *(continued)*

(e) Roof Ponds. Once the area and the depth of roof ponds have been determined (Sections 5.3 and 5.9), questions arise about the architectural integration of such large horizontal surfaces and their relative emphasis upon heating or cooling. Figure 6.26 shows variations in the treatment of containers and the insulating panels to match the most important thermal role of roof ponds; Fig. 6.27 shows three typical approaches to the placement of roof ponds on buildings. A valuable source of information about the container bags, sliding insulation, roof decks, and other components needed for roof ponds is the *California Passive Solar Handbook* by Phillip Niles and Ken Haggard, written for the California Energy Commission in 1980. (Publication No. P500-80-032, write to C.E.C. Publications Office, 1516 9th St., Sacramento CA 95814.) Commercially available products include the Skytherm® system, developed and pioneered by Harold R. Hay.

A variation of the roof pond (applying primarily to the cooling mode) is to move the water rather than move the insulation. Water-impervious insulation (such as Styrofoam) is secured in place above the metal ceiling, at a distance sufficient to be a reservoir for the cooling water. At night, this water is pumped onto the top of the insulation, where it cools by radiation and evaporation, then trickles slowly back into the reservoir. This variation is now being tested at the School of Engineering Technology, University of Nebraska at Omaha campus.

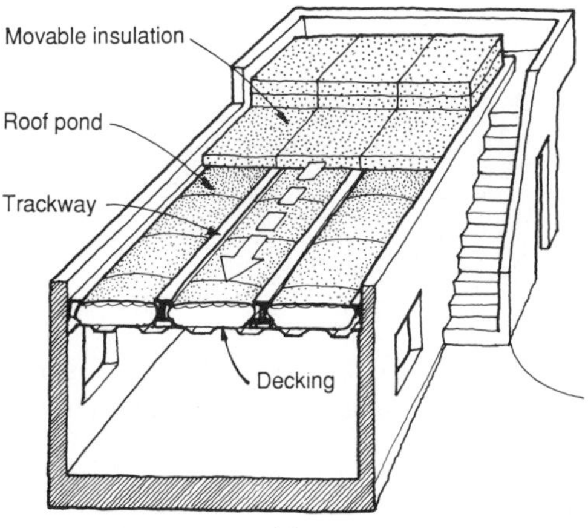

(*a*)

Fig. 6.26 (a) *Principal components of roof ponds.*

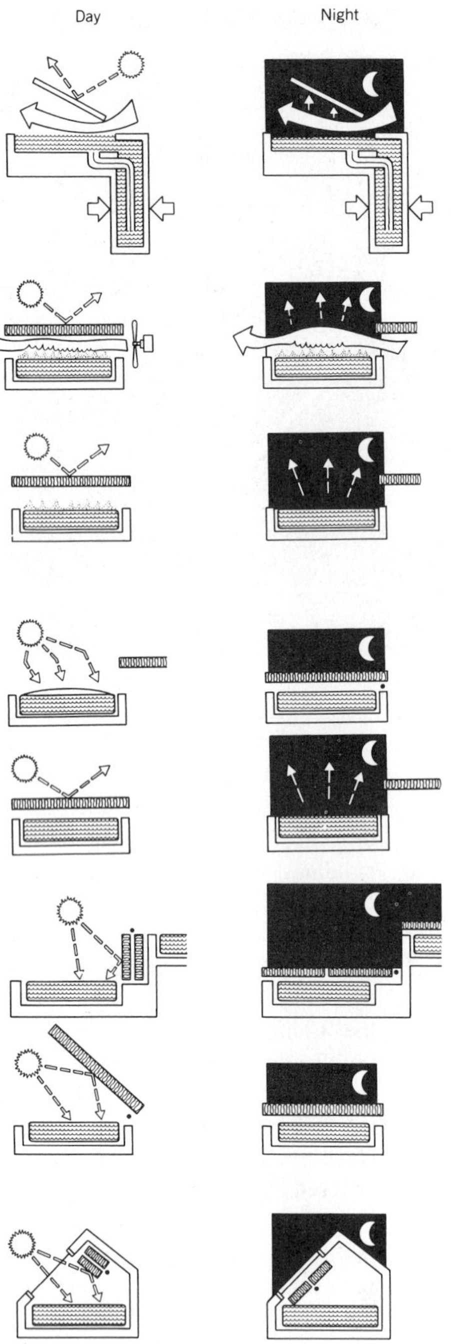

(b) Cooling Only

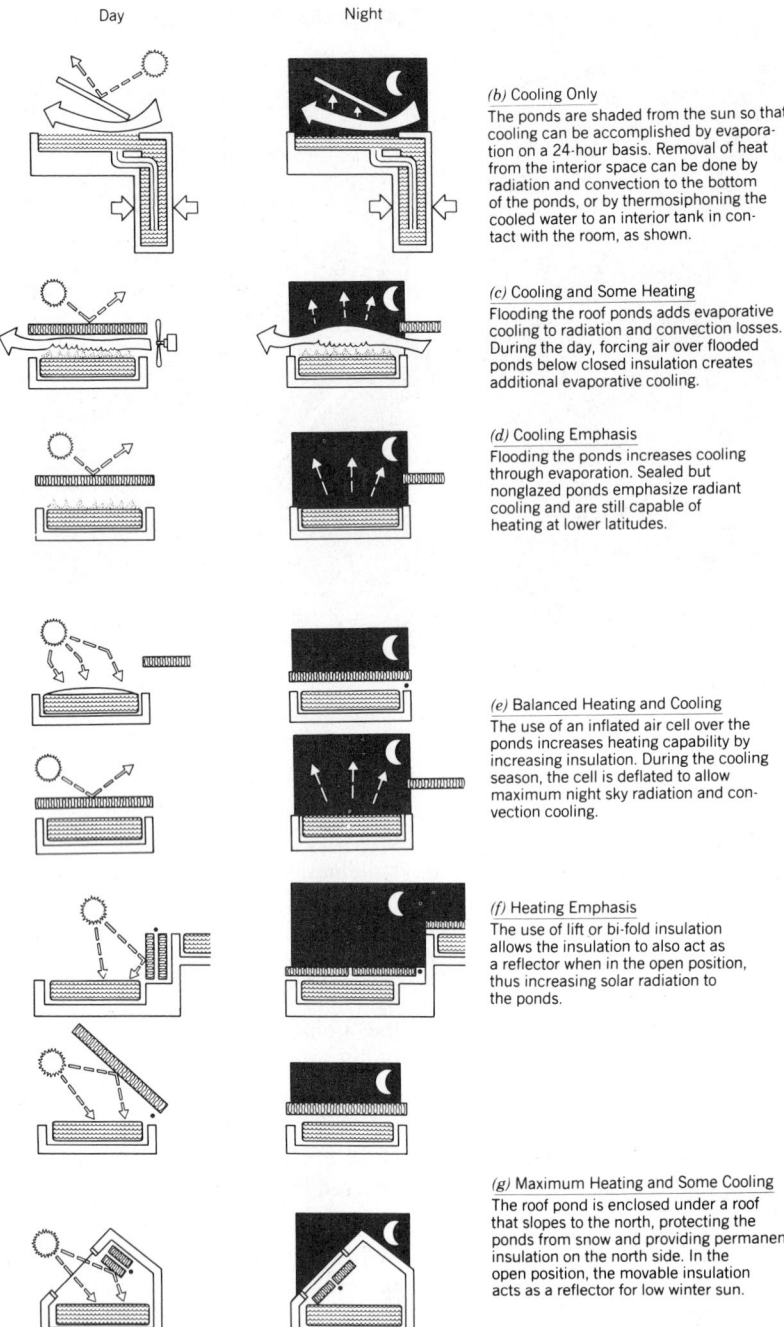

The ponds are shaded from the sun so that cooling can be accomplished by evaporation on a 24-hour basis. Removal of heat from the interior space can be done by radiation and convection to the bottom of the ponds, or by thermosiphoning the cooled water to an interior tank in contact with the room, as shown.

(c) Cooling and Some Heating

Flooding the roof ponds adds evaporative cooling to radiation and convection losses. During the day, forcing air over flooded ponds below closed insulation creates additional evaporative cooling.

(d) Cooling Emphasis

Flooding the ponds increases cooling through evaporation. Sealed but nonglazed ponds emphasize radiant cooling and are still capable of heating at lower latitudes.

(e) Balanced Heating and Cooling

The use of an inflated air cell over the ponds increases heating capability by increasing insulation. During the cooling season, the cell is deflated to allow maximum night sky radiation and convection cooling.

(f) Heating Emphasis

The use of lift or bi-fold insulation allows the insulation to also act as a reflector when in the open position, thus increasing solar radiation to the ponds.

(g) Maximum Heating and Some Cooling

The roof pond is enclosed under a roof that slopes to the north, protecting the ponds from snow and providing permanent insulation on the north side. In the open position, the movable insulation acts as a reflector for low winter sun.

Fig. 6.26 (b)–(g) *Variations on roof ponds for optimization of heating or cooling performance. [Reprinted from Niles and Haggard (1980) by permission of the California Energy Commission.)*

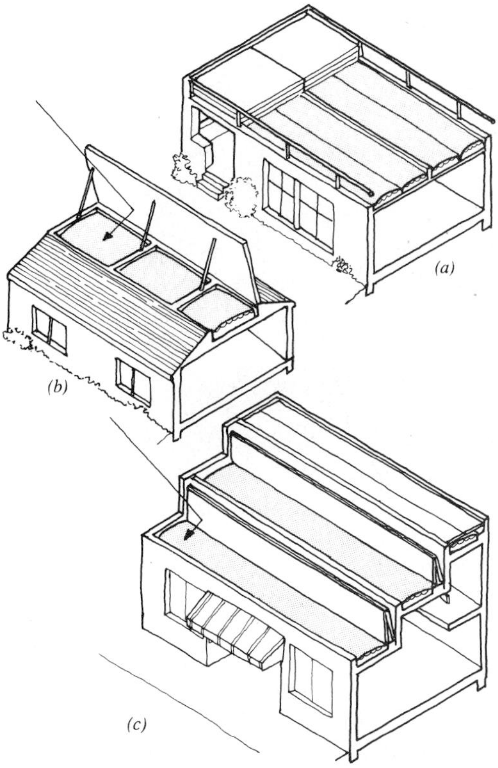

Fig. 6.27 Three typical roof pond integrations with building form. (a) *Insulating panels slide open to stack over exterior or untempered space. This is the most simple and common application.* (b) *Hinged insulating panels with reflective undersurfaces can act as reflectors to increase winter insulation. Snow and wind can inhibit operation.* (c) *Bifold insulating panels also act as reflectors, interfering less with diffuse radiation and allowing for exposure to night sky for summer cooling. [Adapted from Niles and Haggard (1980) by permission of the California Energy Commission.]*

Fig. 6.28 The addition to Barton Hall at the University of New Hampshire utilizes two towers with wind gravity ventilators (a) *to exhaust hot air at the south end of the auditorium; cool outdoor air is admitted through low windows and shutters at the north wall.* (b) *Below the sloping auditorium floor, sun warms a "greenhouse" space, then strikes manually operated louvres. These are adjusted by students as appropriate to provide diffused daylight to a multipurpose room beyond. (Courtesy of Banwell White Arnold Hemberger & Partners, Inc, Hanover, New Hampshire.)*

(f) Ventilation. Simple calculation procedures for cross ventilation and stack effect ventilation of buildings were detailed in Section 4.6. Several devices can be used to enhance the stack effect by creating suction when wind blows across the top of the stack. The most common such device (available in chain store catalogs) is shown in Figs. 6.28 and 6.29; its performance characteristics are listed in Table 6.8. This is probably not the most effective topping device, however. Figure 6.30 compares the volumetric airflow results for the turbine and sev-

Fig. 6.29 *Collection of passive heating and cooling components at the Cottage Restaurant, Cottage Grove, Oregon.* (a) *Northeast view, showing wind-gravity (turbine) ventilators (see Fig. 6.30) projecting above the roof; they provide night ventilation of thermal mass for summer cooling.*

eral other ventilators with those for a simple open stack. The tests were done in a wind tunnel at Virginia Polytechnic Institute.

Passive ventilation through windows and skylights is influenced by the position of the open window; if wind strikes the glass surface in its open position, it will be deflected. The direction of the wind approaching the window is generally unpredictable. Also, whereas for simple ventilation (without cooling) it is usually desirable to keep wind *away* from people, for cooling at temperatures above the standard comfort zone, wind *across* the body is helpful (Fig. 3.25). For these reasons, a window that can be opened in a variety of positions can be useful; some examples are shown in Fig. 6.31.

(g) Heat Exchangers. As attention to the design of energy-conserving envelopes increases, so does the proportion of winter heat loss (and summer heat gain) attributable to infiltration or ventilation. (In Example 5.8, 279 Btu/h-°F, or 33%, of the total UA_{ns} of 850 Btu/h-°F for an Omaha office building was attributable to infiltration.) As the tightness of construction increases, fewer air changes per hour (ACH) occur from natural infiltration; this makes forced ventilation attractive as a means of reducing indoor air pollution. If a heat exchanger is used, it is possible to maintain an adequate supply of fresh air without severe energy consumption consequences. Figure 6.32 illustrates the basic principle of the air-to-air heat exchanger that is becoming increasingly common for tightly built small buildings (large-building heat exchangers, often called heat recovery devices, are illustrated in Section 7.3).

Some commercially available heat exchangers are capable of extracting 70% or more of the heat from exhaust air. For the best diffusion of incoming fresh air through a building, the heat exchanger should be incorporated at the central forced-air fan (Fig. 6.32*b*). When a central forced-air system is not available, heat exchangers can be placed at various points in the building; typically, each heat exchanger is then equipped with its own fan.

Some cautions about air-to-air heat exchangers:

1. Avoid using them on exhaust airstreams that are contaminated with grease, lint, or excessive moisture (through cooking and clothes drying in particular) since clogging, frosting, and fire hazard problems can develop.
2. In colder winter conditions, a built-in defroster, which will require energy, will be needed.

Fig. 6.29 *(continued)* (b) *Southeast view, showing the solar-heating features: clerestory, protected by interior thermal shutters; and lower south windows, protected inside by automatic roll-down thermal shades and shaded outside in summer by deciduous vines.* (c) *Interior view, showing thermal mass in the concrete floor and in the many 30-gallon barrels of water placed throughout the interior. Barrels on suspended rack receive winter sun through clerestory and are cooled by summer night air drawn from below floor slab and out through turbine ventilators. (Courtesy of Equinox Design, Inc. Photos by G. Z. Brown.)*

3. Keep the outdoor fresh air intake as far as possible from the exhaust air outlet, to avoid drawing indoor-contaminated air back into the building.

Indoor pollution of air was discussed in Section 4.6, along with requirements for fresh air (Table 4.25). We usually refer to the quantity of fresh air assumed in heating or cooling calculations in terms of ACH. Although the guidelines given in Table 4.25 should provide for sufficient fresh air, the general question arises, What is the minimum ACH level compatible with adequate fresh air? A frequently cited (although often dis-

TABLE 6.8 **Turbine Ventilator Performance**

NOTE: The combination of wind suction and stack effect produces the following exhaust capacities (cfm) for various throat diameters and stack heights of turbine ventilators. Recommended minimum spacing between ventilators is 20 ft.

Outdoor Wind Velocity (mph):		2			4			5			6			8			10		
Temp. Diff. (F°):		*10*	*20*	*30*	*10*	*20*	*30*	*10*	*20*	*30*	*10*	*20*	*30*	*10*	*20*	*30*	*10*	*20*	*30*
Throat Diameter	*Ht. above Intake (ft.)*	*Exhaust Capacity (cfm)*																	
6″	10	114	125	130	210	221	226	266	277	282	314	325	330	426	437	442	534	545	550
	20	122	135	144	218	231	240	274	287	296	322	335	344	434	447	456	542	555	564
	30	129	144	156	225	240	252	281	296	308	329	344	356	441	456	468	549	564	576
	40	135	152	166	231	248	262	287	304	318	335	352	366	447	464	478	555	572	586
10″	10	209	222	274	370	383	435	463	476	528	545	558	610	728	741	793	915	928	980
	20	234	269	301	395	430	462	488	523	555	570	605	637	753	788	820	940	975	1007
	30	254	301	328	415	462	489	508	555	582	590	637	664	773	820	847	960	1007	1034
	40	269	318	355	430	479	516	523	572	609	605	654	691	788	837	874	975	1024	1061
14″	10	333	383	422	558	608	647	691	741	780	804	854	893	1062	1112	1151	1324	1374	1413
	20	376	444	496	601	669	721	734	802	854	847	915	967	1105	1173	1225	1367	1435	1487
	30	413	496	560	638	721	785	771	854	918	884	967	1031	1142	1225	1289	1404	1487	1551
	40	444	539	614	669	764	839	802	897	972	915	1010	1085	1173	1268	1343	1435	1530	1605
18″	10	476	564	623	755	843	902	924	1012	1071	1071	1159	1218	1399	1487	1546	1737	1825	1884
	20	549	662	747	828	941	1026	997	1110	1195	1144	1257	1342	1472	1585	1670	1810	1923	2008
	30	611	747	853	890	1026	1132	1059	1195	1301	1206	1342	1448	1534	1670	1776	1872	2008	2114
	40	662	819	941	941	1098	1220	1110	1267	1389	1257	1414	1536	1585	1742	1864	1923	2080	2202

24″	10	716	874	978	1101	1259	1363	1327	1485	1589	1522	1680	1784	1963	2121	2225	2412	2570	2674
	20	844	1046	1196	1229	1431	1581	1455	1657	1807	1650	1852	2002	2091	2293	2443	2540	2742	2892
	30	954	1196	1384	1339	1581	1769	1565	1807	1995	1760	2002	2190	2201	2443	2631	2650	2892	3080
	40	1046	1324	1542	1431	1709	1927	1657	1935	2153	1852	2130	2348	2293	2571	2789	2742	3020	3238
30″	10	1139	1385	1545	1719	1965	2125	2070	2316	2476	2379	2625	2785	3068	3314	3474	3769	4015	4175
	20	1342	1655	1890	1922	2235	2470	2273	2586	2821	2582	2895	3130	3271	3584	3819	3972	4285	4520
	30	1514	1890	2185	2094	2470	2765	2445	2821	3116	2754	3130	3425	3443	3819	4114	4144	4520	4815
	40	1655	2090	2430	2235	2670	3010	2586	3021	3361	2895	3330	3670	3584	4019	4359	4285	4720	5060
36″	10	1613	1967	2201	2475	2829	3063	2988	3342	3576	3418	3772	4006	4414	4768	5002	5428	5782	6016
	20	1901	2354	2692	2763	3216	3554	3276	3729	4067	3706	4159	4497	4702	5155	5493	5716	6169	6507
	30	2148	2692	3115	3010	3554	3977	3523	4067	4490	3953	4497	4920	4949	5493	5916	5963	6507	6930
	40	2354	2981	3470	3216	3843	4332	3729	4356	4845	4159	4786	5275	5155	5782	6271	6169	6796	7285
42″	10	2183	2663	2998	3350	3835	4170	4047	4527	4862	4645	5125	5460	6000	6480	6815	7365	7845	8180
	20	2588	3203	3668	3760	4375	4840	4452	5067	5532	5050	5665	6130	6405	7020	7485	7770	8385	8850
	30	2928	3668	4243	4100	4840	5415	4792	5532	6107	5390	6130	6705	6745	7485	8060	8110	8850	9425
	40	3203	4058	4723	4375	5230	5895	5067	5922	6587	5665	6520	7185	7020	7875	8540	8385	9240	9905
48″	10	2868	3500	3925	4412	5044	5469	5308	5940	6365	6078	6710	7135	7843	8475	8900	9638	10270	10695
	20	3378	4185	4785	4922	5729	6329	5818	6625	7225	6588	7395	7995	8353	9160	9760	10148	10955	11555
	30	3817	4785	5535	5361	6329	7079	6257	7225	7975	7027	7995	8745	8792	9760	10510	10587	11555	12305
	40	4185	5300	6175	5729	6844	7719	6625	7740	8615	7395	8510	9385	9160	10275	11150	10955	12070	12945

Source: Reprinted courtesy of Western Ventilating Equipment, Inc., Los Angeles, California.

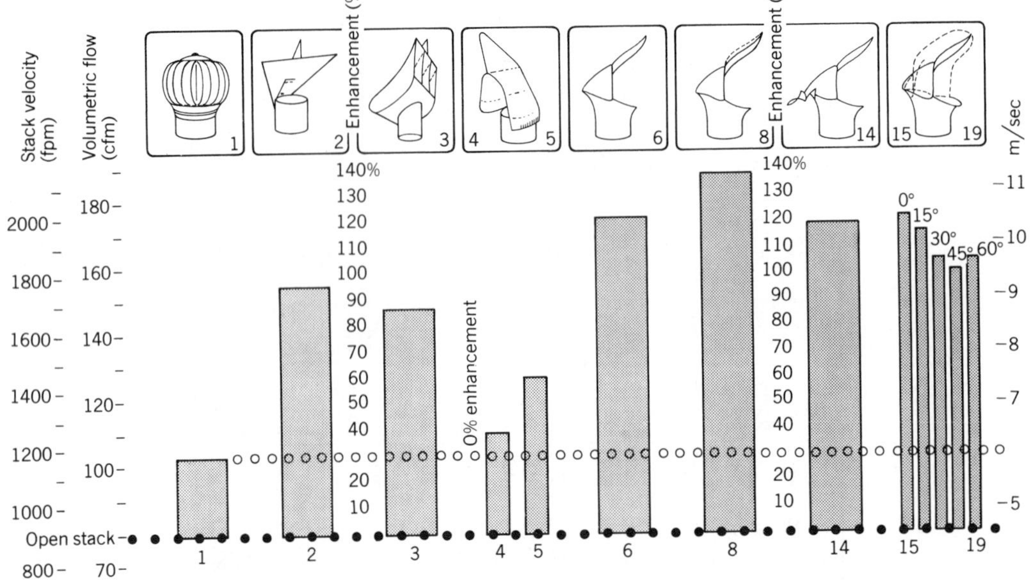

Fig. 6.30 Performance of some passive ventilators compared to a simple open stack. (From the work of R. P. Schubert and Philip Hahn; reprinted from Progress in Passive Solar Energy Systems, *© 1983, by permission of the publisher, the American Solar Energy Society, Inc.)*

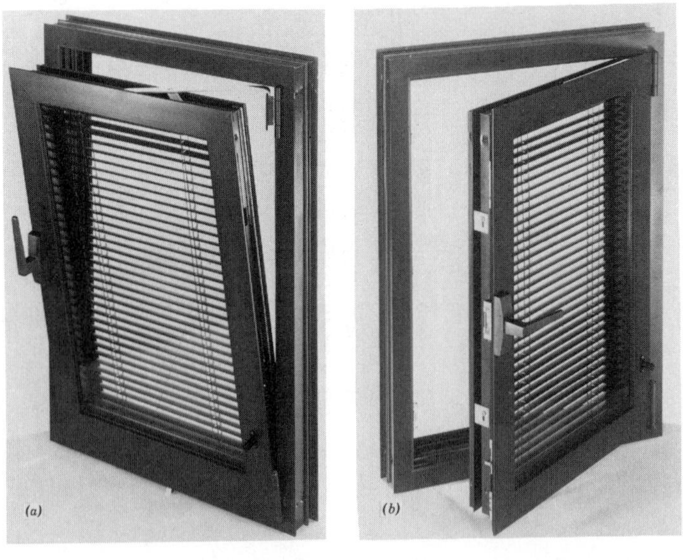

Fig. 6.31 *Window that can open in more than one position, providing flexibility for passive cooling performance.* (a) *The window tilts, directing the incoming air toward the ceiling for simple ventilation.* (b) *The window swings in, allowing incoming air to move across people and thus enhancing warm weather cooling. (Courtesy of Three Rivers Aluminum Company, Inc.)*

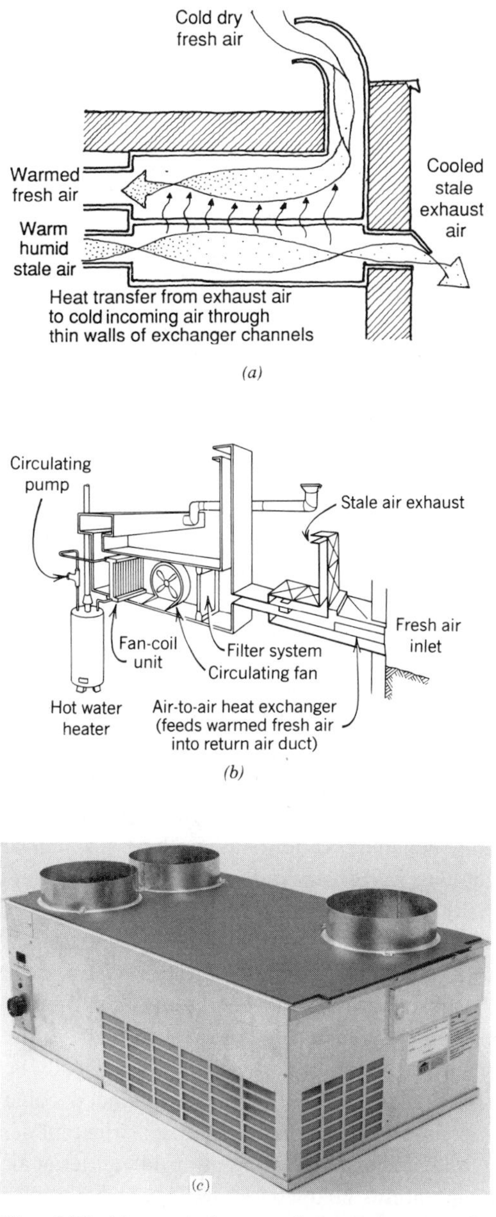

Fig. 6.32 *Air-to-air heat exchangers are particularly helpful for smaller buildings in cold climates.* (a) *The basic principle of operation.* (b) *Installation in a small hot water heating application. (Reprinted by permission of Alberta Agriculture, Home and Community Design Branch,* Low Energy Home Designs, *© 1983 by Brick House Publishing Co.)* (c) *Fan-powered heat exchanger, utilizing a heat/moisture exchange wheel (see also Fig. 7.29). This model is adjustable from 70 to 200 cfm, can be either ceiling, wall, or floor mount, and operates at 75 to 80% thermal efficiency. (Courtesy of Air-Xchange, Inc., Rockland, Massachusetts.)*

agreed with) minimum is 0.4 ACH. Table 6.9 shows that this 0.4 ACH rate of infiltration-plus-ventilation would produce *less* than 40% "new" air within a building after one hour; even after eight hours at this rate, not quite all the "old" air will be exhausted. There is, then, a difference between the fresh air input rate (ACH) and the *replacance*—the fraction of air molecules at one specified time that was *not* in the house at an earlier, reference time. This relationship, along with details of air pollutants and of heat exchanger design, is thoroughly discussed in Shurcliff (1981).

Water-to-water heat exchangers are also used to recover otherwise-wasted energy; since heat exchange is aided by steady flow, pumps are usually part of the water heat-exchange process. More information on these devices can be found in Section 9.4.

A new approach to both residential winter ventilation and heat exchange is the "breathable wall" combined with an exhaust air heat pump. This system depends on the house being under negative pressure, assured by installing a heat pump that takes heat from forced exhaust air and invests that heat either in space heating or domestic water heating. The fresh air to replace that being expelled is drawn in through the outside walls, in a unique combination of fiberglass lap siding board, fiberglass insulation batts, a breathable sheathing, and *no* vapor barrier. This allows a slow, steady stream of cold air to enter and be warmed by the insulation, then enter the house. More information is available from the Canadian National Research Council.

Another new development for winter preheating is the *unglazed transpired collector,* being studied at the National Renewable Energy Laboratory (formerly Solar Energy Research Institute). Installed on an unshaded exterior south wall, it is an enclosed airspace fronted by a perforated, heat absorbing panel. Air flows through these perforations, heated by the solar energy absorbed by the front panel. The air, pulled by a fan, then continues into the building.

(h) Dehumidification. Passive cooling techniques such as roof ponds and night ventilation of mass (discussed in Chapters 2 and 5) are *sensible* cooling processes: that is, they work by lowering surface temperatures, which lowers air

TABLE 6.9 **Air Replacance Compared to Input Air Changes per Hour (ACH)**

NOTE: Mixing is considered continuous and vigorous—as would be obtained in a forced-ventilation system.

Time from Start of Run (h)	*Rate of Air Input (ACH):*	*Replacance (%)*								
		0.06	*0.12*	*0.25*	*0.5*	*1*	*2*	*4*	*8*	*16*
1/16		0.4	0.8	1.6	3.1	6.1	11.7	22.1	39.3	63.2
1/8		0.8	1.6	3.1	6.1	11.7	22.1	39.3	63.2	86.5
1/4		1.6	3.1	6.1	11.7	22.1	39.3	63.2	86.5	98.2
1/2		3.1	6.1	11.7	22.1	39.3	63.2	86.5	98.2	99.9
1		6.1	11.7	22.1	39.3	63.2	86.5	98.2	99.9	100
2		11.7	22.1	39.3	63.2	86.5	98.2	99.9	100	100
4		22.1	39.3	63.2	86.5	90.2	99.9	100	100	100
8		39.3	63.2	86.5	98.2	99.9	100	100	100	100
16		63.2	86.5	98.2	99.9	100	100	100	100	100

Source: Shurcliff (1981). Reprinted by permission.

temperatures. In many hot and humid climates (and in tightly enclosed buildings that produce moisture, even in winter), *latent* cooling is also needed, to take excessive moisture from the indoor air. Either of two common processes is used for dehumidification: a chemical process of *absorption* (using desiccants) or a mechanical process of *refrigeration*. In absorption dehumidification, moisture is absorbed from the air by a chemical such as calcium chloride, silica gel, or lithium chloride. When these chemicals are saturated, they must be "regenerated" (relieved of excessive moisture) before they can again serve as dehumidifiers. (One absorption-regeneration process, applied to mechanical equipment, is illustrated in Fig. 6.36.) Passive techniques such as solar drying can be used for regeneration. The principal disadvantage of chemical dehumidification is that the process *raises the temperature* of the air while it lowers the relative humidity; it is the reverse of the evaporative cooling process, as was shown in Fig. 4.10. If desiccants are used, the passive cooling capacity must be great enough to further lower the indoor air temperature after it has been raised by dehumidification. Details of several chemical dehumidification-regeneration processes can be found in Bowen et al. (1981).

Refrigeration dehumidification is quite common; dehumidification is usually a part of the refrigeration cooling process, which is described in Section 6.5. For smaller buildings that need only dehumidification, rather than mechanical cooling, *refrigerant dehumidifiers* are commercially available (Fig. 6.33). Their advantage is that the air temperature remains essentially unchanged during dehumidification. However, these devices consume energy to run the refrigeration cycle, and this is added, as heat, to the space. Accumulated water must periodically be removed from the unit and if untended could become a source of disease. Most refrigerant dehumidifiers encounter operating difficulties at air temperatures below 65 F (18°C), at which point frost forms on their cooling coils. This could cause problems in a tightly enclosed residence in winter.

(i) Passive Components Summary. A look at a passively heated and cooled project will serve as a way of interrelating the array of components discussed above. The Conservation Center for the Society of the Protection of New Hampshire Forests, located near Concord, New Hampshire, is shown in Fig. 6.34. This 7000-ft^2 combination of office space, reception area, and lecture hall demonstrates the uses of New England forest products, as well as both direct-gain and sunspace heating strategies. Reflecting roofs and window louvres augment and direct sunlight, and water tubes, masonry walls, concrete floors, and phase-change materials (which fuse at 73 F) aid in thermal storage. Warm air is circulated behind and below spaces to distribute solar heat. Daylighting, summer shading, and clerestory ventilation are the passive cooling strategies. Although night-insulating thermal shutters were eliminated (due to budget considerations) the building apparently achieves over 60% solar heating; backup heat is provided by a wood-fired boiler whose hot water is distributed to radiant baseboard panels. The total south glass area is 2100 ft^2; the envelope heat loss rate is 6.9 Btu/DD-ft^2. The total auxiliary energy consumed is about 23,500 Btu/ft^2 per year (for lighting, heating, and office machines).

6.5 Refrigeration Cycles

Unlike the slow and diffuse heat transfer processes that characterize passive heating and cooling approaches, mechanical equipment can rapidly concentrate heating or cooling, on demand. The refrigeration cycle is a particularly useful mechanical process in heating as well as cooling applications. The two types of heat transfer process commonly used in mechanical equipment for buildings are the *compressive* and the *absorption* refrigeration cycles. (Equipment that utilizes these cycles is discussed in Section 6.7 and Chapter 7.)

(a) Compressive Refrigeration. As applied in Fig. 6.35, the compressive refrigeration cycle is a scheme for transferring heat from one circulated water system (chilled water) to another (condenser water). The means for doing this is the liquification and evaporation of a refrigerant, such as Freon, during which processes it gives off and takes on heat, respectively. The heat that it gives off must be disposed of (except in the heat pump), but the heat that it acquires is drawn out of the circulated water known as the chilled water, which is the medium for subsequent cooling processes.

Freon, a gas at normal temperatures and pressures, must be compressed and liquefied to be of service later as a heat absorber. To be liquefied (see Fig. 6.35), Freon must first be compressed to a high-pressure vapor; then, by means of cool water, latent heat is extracted from the Freon, which condenses it to a liquid. This product, high-pressure liquid Freon, is a potential heat absorber since, when it is released through an expansion valve, it springs back mechanically to gaseous form. In this change of state it must take on latent heat, by drawing heat

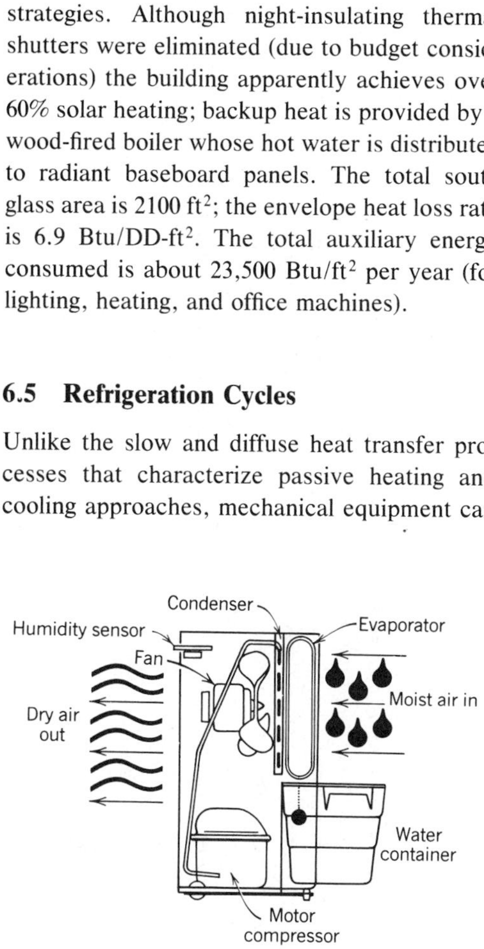

Fig. 6.33 Refrigerant dehumidifiers typically are installed as freestanding units in smaller buildings.

Typical Performance Data

Watts	290	450	465	630
Pints per day[a]	14	20	25	30
Unit dimensions (in.)				
Height	20½	20½	21½	21½
Width, 11¾				
Depth, 16¾				
"Average" room size served	20′ × 32′	35′ × 40′	40′ × 48′	40′ × 60′

[a]For air at 80 F, 60% rh.

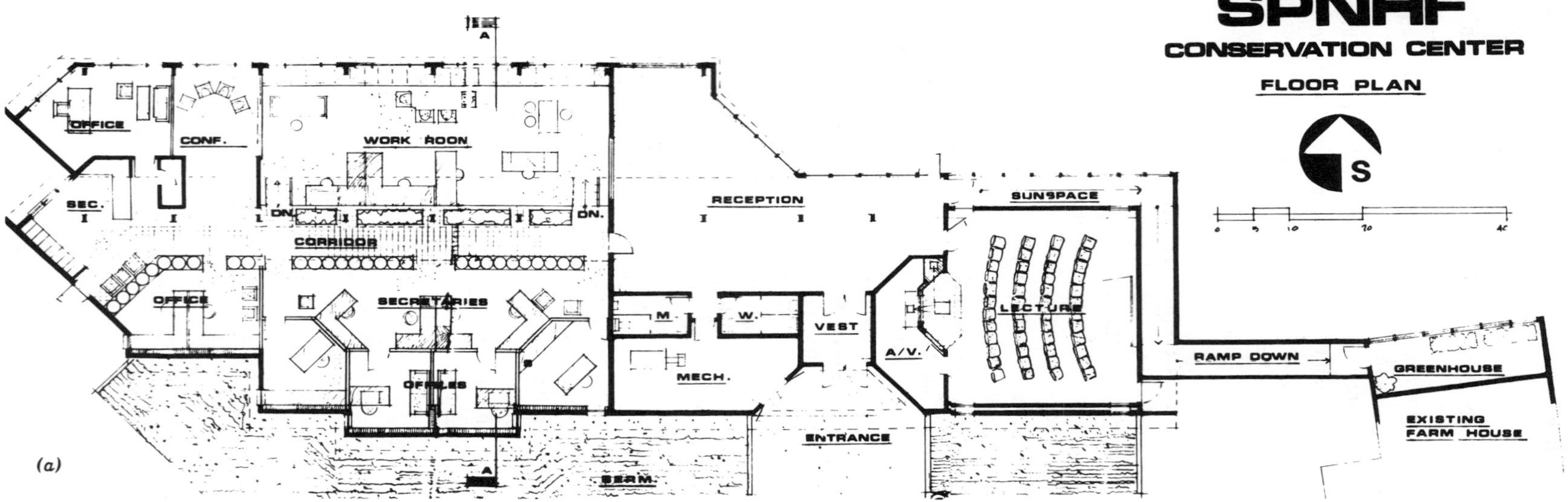

(a)

Fig. 6.34 The Conservation Center for the Society for the Protection of New Hampshire Forests, Concord, New Hampshire. Banwell White Arnold Hemberger & Partners, architects. (a) *The plan is elongated on the E–W axis, to maximize south solar exposure and facilitate daylighting. Direct gain serves the reception area, workroom, and offices, and a sunspace–double-envelope combination warms the lecture room. A wood-fired boiler provides backup heat.* (b) *This section relates south glazing to heat storage materials: translucent water tubes, masonry walls, and phase change materials both in ceiling and windowsill positions. The circulation of hot air that collects at the clerestory is also evident. In summer, the hot air is vented and an awning will shade the clerestory. Daylighting is diffused through the translucent tubes to the spaces beyond.* (c) *At the workroom, mirror-finish venetian blinds reflect solar energy to the phase-change materials in bags above the metal deck ceiling. (Photo by C. Stuart White, Jr.)*

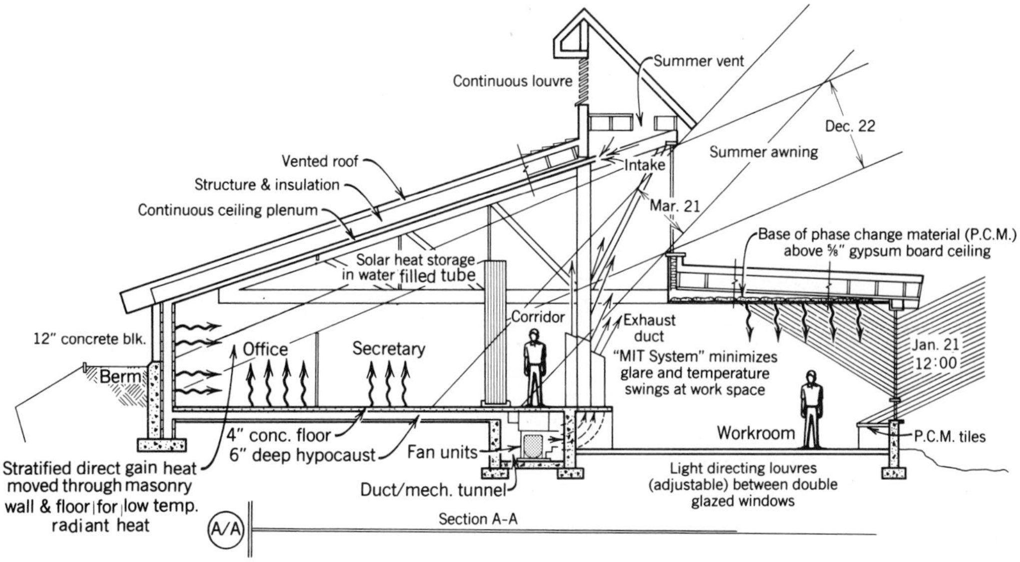

(*b*)

Fig. 6.34 *(continued)*

out of the circulated water of the chilled water system. It may be said that the refrigeration cycle pumps the heat out of the chilled water system into the condenser water system. Indeed, by special (reverse cycle) arrangements of the water systems, a *heat pump* is the result. (The compressive refrigeration cycle can be used to transfer heat between almost any media; whereas Fig. 6.35 illustrates a water–water cycle, Fig. 6.62 shows the heat pump in both air–air and water–air applications.)

Unfortunately, most common refrigerants are a threat to our atmosphere, because they can escape from the equipment as CFC gases. Considerable efforts are under way both to reduce the likelihood of escaping refrigerant and to develop non-CFC refrigerants that are as efficient as the CFC varieties now in use.

The piston-type compressor in Fig. 6.35 can instead be a *compliant scroll* compressor that promises higher efficiency, quieter operation, and longer life than today's piston-type compressors. As described by its developer, Copeland Corporation of Sidney, Ohio:

> Two spiral-shaped members fit together—forming crescent-shaped gas pockets. One member remains stationary, while the second member is allowed to orbit relative to the stationary member.
>
> This movement draws gas into the outer pocket created by the two members, sealing off the open passage. As the spiral motion

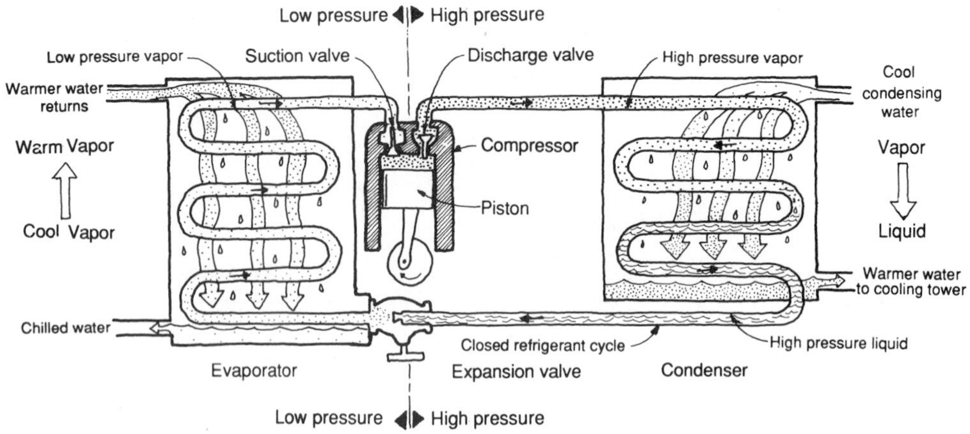

Fig. 6.35 Schematic arrangement of a compressive refrigeration cycle.

continues, the gas is forced toward the center of the scroll form as the pocket continuously becomes smaller in volume, creating increasingly higher gas pressures. At the center of the pocket, the high-pressure gas is discharged from the port of the fixed scroll member.

During the cycle, several pockets of gas are compressed simultaneously providing a smooth, nearly continuous compression cycle.

(b) Absorption Refrigeration Cycle. This process is illustrated in Fig. 6.36. In order to remove heat from chilled water, it uses still more heat in regenerating the salt solution and is typically somewhat less efficient than the simple compressive cycle. However, this heat for regeneration may be provided by solar energy, or by relatively high-temperature waste heat from another source. Because the high-grade energy (electricity) needed to run a compressor is replaced by the lower-grade heat needed to run the generator, the absorption cycle can enjoy an energy advantage over the compressive cycle.

6.6 Indoor Air Quality Control

Buildings can be designed for better indoor air quality (IAQ) and can then be equipped to maintain air quality. A discussion of design aspects is found in Sections 2.8, 4.6, and 5.2*g*.

Two new units have been proposed, to integrate the various indoor air pollutants in the same way as they are perceived by human beings. The *olf* is a unit of pollution (1 olf = the bioeffluents produced by the average person); the *decipol* is a unit of perceived air quality. These are related in this proposed comfort formula:

$$Q = 10 \frac{G}{C_i - C_o}$$

where

Q = ventilation rate, L/s
G = total pollution sources, olf
C_i = perceived indoor air quality, decipol
C_o = perceived outdoor air quality, decipol

At present, C_i is recommended to be set at 1.4 decipol, which represents an expectation of 80% of occupants satisfied with indoor air quality. C_o and G may be roughly estimated from Table 6.10. The greatly increased ventilation rates for existing buildings (compared to Table 4.25) suggest dealing with indoor pollution at its source rather than increasing outdoor air flushing—and related energy consumption—to such high levels. These proposals are discussed in more detail in Fanger (1989).

Indoor air pollution can be described both in terms of the types of contaminants (gaseous, organic, or particulate) and the types of effects (odors, irritants, toxic substances). Effects of air pollutants are also discussed in Section 3.5. For

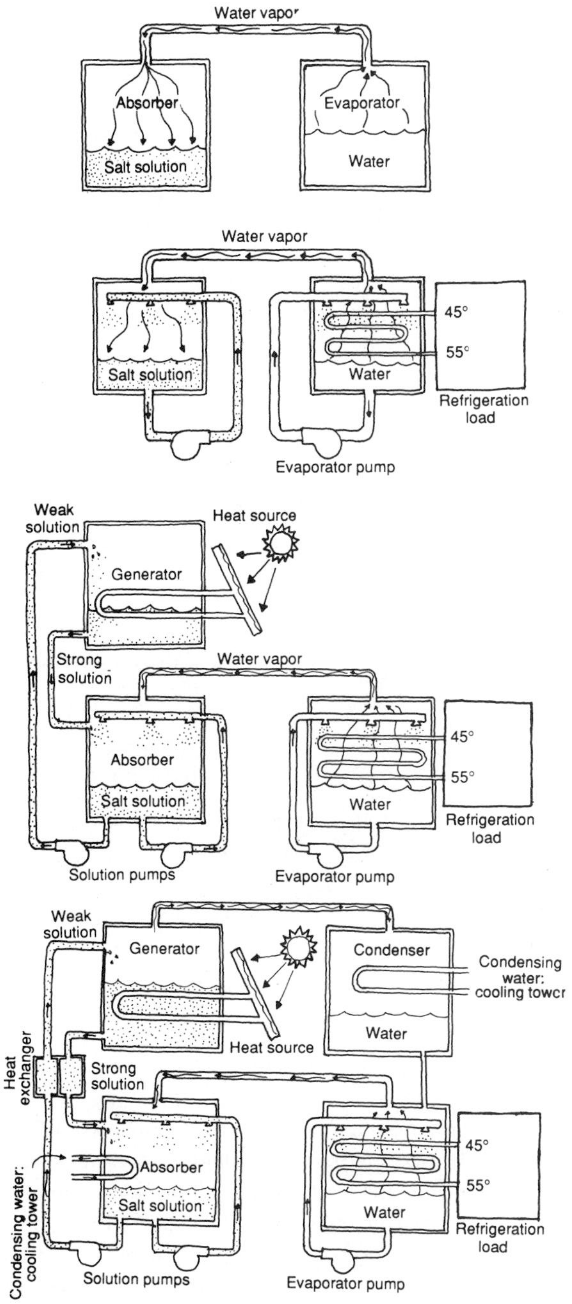

1 Evaporator and absorber

Consider two connected, closed tanks with a salt solution (lithium bromide) in one and water in the other. Just as common table salt absorbs water on a damp day, the salt solution in the absorber soaks up some of the water in the evaporator. The water remaining is thereby cooled by evaporation.

2 Evaporator coil and pump added

This refrigeration effect is utilized by putting a coil in the evaporator tank. Water from this tank is pumped to a spray header which wets the coil. The spray's evaporation chills water in the coil as it circulates to the refrigeration load. Solution pumped to spray in absorber raises efficiency.

3 Solution pumps and generator added

In an actual operating cycle, the salt solution is continuously absorbing water vapor. To keep the salt solution at proper concentration, part of it is pumped directly to a generator where excess water vapor is boiled off. The reconcentrated salt solution is returned to the absorber tank where it mixes with the solution sprayed to absorber in step 2.

4 Condenser and heat exchanger added

Water vapor boiled off from the weak solution is condensed and returned to the evaporator. A heat exchanger uses the hot, concentrated salt solution leaving the generator to preheat the cooler, weak solution coming from the absorber. Finally, condensing water circulating through the absorber and condenser coils removes the waste heat.

Fig. 6.36 *Absorption-refrigeration cycle. Connected systems are a heat source (such as steam, solar-heated water, or waste heat from an industrial process), condensing water, and chilled water (refrigeration load). (Adapted, courtesy of the Carrier Corporation.)*

some contaminants, the only method of avoidance is to design for their exclusion; equipment will not remove them, although increased ventilation can reduce their impacts. Examples are asbestos, radon, and pesticides. This section, based on SMACNA (1988), considers equipment to deal with odors, irritants, and some toxic particulate substances.

TABLE 6.10 **Estimating Indoor Air Quality**

Part A. ***Perceived Outdoor Air Quality, C_o***	
During smog episodes	>1 decipol
In cities with moderate air pollution	0.05–0.3 decipol
On mountains or at sea	0.01 decipol

Part B. ***Estimated Olf Loads in Offices per m² Floor Area***	
Pollution Source	*olf/m²*
Occupants (10 m² per person)	
Bioeffluents	0.1
Additional load from 20% smokers	0.1
40%	0.2
60%	0.3
Materials and ventilation system	
Average in existing buildings[a]	0.4
Low-olf buildings[b]	0.1
Total load in office buildings	
Average in existing buildings (40% smokers)	0.7
Low-olf buildings (non-smoking)	0.2

Part C. ***Ventilation Requirements in Office Buildings***[c]	
Based on 1.4 decipol indoor air quality	
Existing buildings[a] (0.7 olf/m²) smoking	5 L/s-m²
Low-olf buildings[b] (0.2 olf/m²)	1.4 L/s-m²
ASHRAE Standard 62-1989[d]	0.8 L/s-m²

Source: Fanger (1989). Reprinted by permission from the *ASHRAE Journal,* © 1989 by the American Society of Heating, Refrigerating, and Air-Conditioning Engineers, Inc., Atlanta, Ga.

[a]Based on field studies of 15 randomly selected buildings in Copenhagen.

[b]Designed to contain low-outgassing materials and a frequently maintained ventilating system.

[c]Assumes 10 m² per person and negligible outdoor air pollution.

[d]See Table 4.25.

(a) Odors. One of the most immediate indicators of IAQ problems is odor detection. People are sensitive to odors over an extraordinary range, while equipment to detect and classify odors is woefully lacking. Odors are perceived most strongly on initial encounter, then "fatigue" occurs and perception fades; thus visitors are more likely to detect odors than are the long-term inhabitants of a space. Odors may be simply unpleasant, or may be indicators of a more serious IAQ problem that has physiological consequences.

Sometimes odors are directly traceable to a source, but in office environments the odor is usually more complex. A typical blend for a office environment's odor may be tobacco smoke, body odor, grooming products (perfumes, colognes), copy machines, cleaning fluids, and outgassing from materials. This complexity produces an interesting reaction; people tend to become less sensitive to each of the component odors, with the result of overall masking. However, an architectural masking approach—the deliberate introduction of a "perfume" to cover offending odors—is rarely successful. Conversely, as the indoor environment is freed of multiple odors, people become more sensitive to the one or two odors that remain.

Sometimes a simple measurement of carbon dioxide concentration is used as a first indicator of potential body odor and/or inadequate ventilation problems, since the CO_2 concentration indoors is generally proportional to human concentration.

The most common remedy for odors is to increase the rate of outdoor air ventilation. Recommended rates of ventilation are found in Table 4.25. Human body odor control is usually achievable at a rate of from 6 to 9 cfm (3 to 4.5 L/s) of outdoor air per occupant. The less volume of space per occupant (i.e., lower ceilings), the higher the rate of ventilation required. An analogy: one fish will pollute a small pond more rapidly than it will a larger pond. Where tobacco smoking occurs, this recommended rate of outdoor air ventilation increases to 15 to 30 cfm (7.5 to 15 L/s), with unfortunate consequences for energy conservation in very hot or cold weather. In this instance, a fan-powered heat exchanger can be helpful, such as that shown in Fig. 6.32; see also Section 7.3.

Filtering out the odors from indoor air is usually achieved with electronic or with activated charcoal filters, described later in this section.

(b) Irritants. Unlike odors, which are immediately perceived and fade with prolonged exposure, irritants are often imperceptible at first and cause increasing distress over time. Symptoms of irritants include itching or burning eyes, sneezing, coughing, dry nose and throat, sore throat, and tightness of the chest. Most irritants are present in the form of particles and gas dispersoids, shown in Table 6.11.

Sources of irritants include the building itself and the equipment and occupants within. New and newly renovated buildings are particularly prone to problems from outgassing of paints, adhesives, sealants, office furniture, carpeting, and vinyl wall coverings. Formaldehyde and hydrocarbons such as styrene, benzene, and alcohols (to name only a few) are irritants found more frequently in these new spaces.

Long-term occupancy brings other irritants. Ozone, a friend in the upper atmosphere but a smog component below, is produced from copy machines, high-voltage electrical equipment, and—ironically—from electrostatic air cleaners. Mineral fibrous particles can result from the breakdown of duct insulation and fireproofing. Hydrocarbon compounds come from copy machines and their papers. Tobacco smoke is a mixture of gases and fine particles especially irritating to sensitive individuals. Low humidity can exacerbate problems with irritants, producing symptoms similar to those from chemicals. Carpet shampooing yields organic solvents and ammonia; nighttime cleaning coinciding with reduced or nonexistent ventilation is especially problematic. In contrast, night maintenance with *increased* ventilation rates—as with cooling by night ventilation of thermal mass—can reduce this threat.

As with odor control, the impacts of irritants can be reduced with increased outdoor air. However, care is needed when locating intakes, to avoid introducing irritants from motor vehicle exhausts. Filtering for the removal of irritants usually consists of particulate filters; less common are gaseous-removal filters, air washers, and electronic air cleaners. These are discussed later in this section.

(c) Toxic Particulate Substances. At the top of this list is asbestos, widely used in buildings until its toxicity was realized in the 1970s. We encounter asbestos in tightly bound form as in asbestos-cement and in vinyl-asbestos floor tiles, and in loosely bound form as in sprayed-on asbestos insulation. The latter form is particularly dangerous, readily releasing toxic asbestos fibers over the life of the material. With asbestos, neither increased ventilation nor filtering is acceptable; it must be removed under stringent controls to isolate the air within the affected space.

Some of the respirable particles (see Table 6.11) that result from incomplete combustion are toxic. Incomplete combustion can occur from smoking, in woodstoves, fireplaces, gas ranges, and unvented gas or kerosene space heaters. Lacking control of combustion at its source, the remedies are to isolate the source insofar as possible, exhaust air from the immediate vicinity, increase outdoor air to the area, and utilize particle filtering.

(d) Biological Contaminants. Because living things inhabit both buildings and outdoor air, there will be biological contaminants such as

TABLE 6.11 **Characteristics of Particles and Particle Dispersoids**

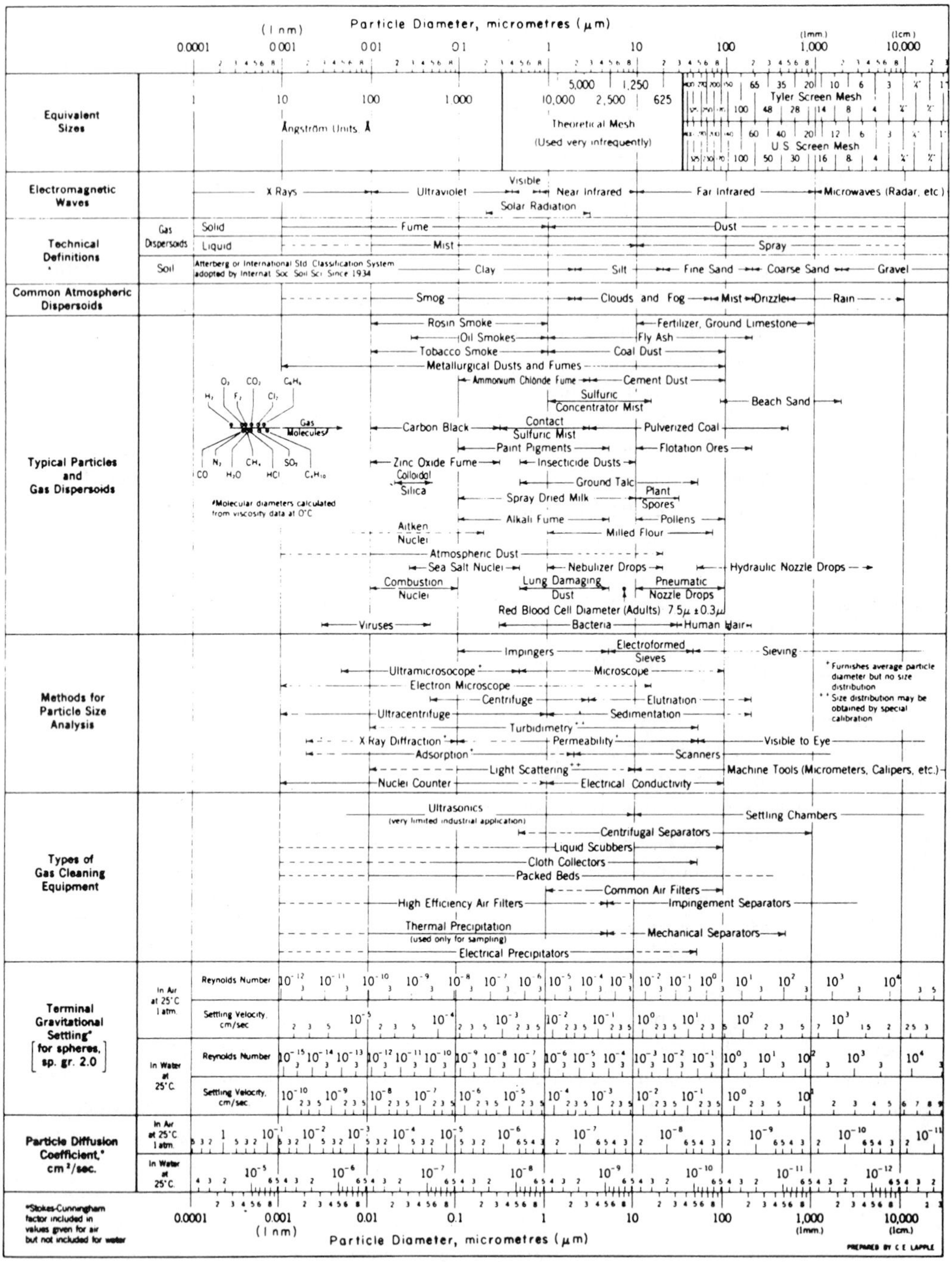

bacteria, fungi, viruses, algae, insect parts, and dust within buildings. Moisture encourages both the retention and growth of these contaminants; standing water (in HVAC system components) and moist interior surfaces are likely trouble sites. Allergic reactions and infectious and non-infectious diseases can result. Outbreaks of Legionnaire's disease occurred when improperly maintained HVAC systems within large buildings incubated, then distributed the disease-causing microorganisms. Now, residential humidifiers, dehumidifiers, and air conditioners are suspect.

Remedies for biological contaminants begin with good design and end with vigilant maintenance. Although exposure to ultraviolet light is sometimes employed, equipment such as filters are rarely an effective solution in this case.

(e) Equipment for Air Cleaning. Maintaining indoor air quality usually involves filters, of several types and in several places. Some typical filter characteristics are shown in Table 6.12 and Fig. 6.37.

Particulate filters are very common; *panel filters* are furnished with HVAC equipment and function mainly to protect the fans from large particles of lint or dust. Because they are relatively crude, they are not properly considered to be air cleaning equipment. *Media filters* are much finer, using highly efficient filter paper in pleats within a frame. They work both by straining and impaction. The larger particles are strained out by the closely spaced filter fibers, while some of the smaller particles that would otherwise pass through are pushed into the fibers due to the air turbulence. Particulate filters need regular maintenance, especially the media filters, which can become blocked and caused damage to HVAC equipment if not replaced frequently enough.

Media filters of high quality are expected to perform at an efficiency of 90% (minimum) and are typically at least 6 in. (150 mm) deep; this is for a 6-month minimum life cycle.

Adsorption filters are for gaseous removal and vary according to the pollutant in question. *Activated-charcoal filters* are the more common of these types, absorbing materials with high molecular weights but allowing those of lower weights to pass. Other adsorption filters use porous pellets impregnated with active chemicals such as potassium permanganate; the chemicals react with contaminants, reducing their harmful effects. Adsorption filters must regularly be regenerated or replaced.

High-quality adsorption filters contain gas adsorbers and/or oxidizers and are of sufficient capacity to remain active over a full service cycle of 6 months at 24 hours per day, 7 days per week. The air velocity should allow the air to remain in the filter about 0.06 second.

Air washers are sometimes used to control humidity and to control bacteria growth. The

TABLE 6.12 **Air Filter Characteristics**

Media and Type	Percent Efficiency Range: Atmospheric Dust	Percent Efficiency Range: Small Particles	Dust-Holding Capacity	Airflow Resistance (In. Water)[a]
Dry panel, throwaway	15–30	NA	Excellent	0.1–0.5
Viscous panel, throwaway	20–35	NA	Good	0.1–0.5
Dry panel, cleanable	15–20	NA	Superior	0.08–0.5
Viscous panel, cleanable	15–25	NA	Superior	0.08–0.5
Mat panel, renewable	10–90	0–60	Good to superior	0.15–1.0
Roll mat, renewable	10–90	0–55	Good to superior	0.15–0.65
Roll oil bath	15–25	NA	Superior	0.3–0.5
Close pleat mat panel	NA	85–95	Varies	0.4–1.0
High-efficiency particulate	NA	95–99.9	Varies	1.0–3.0
Membrane	NA	to 100	NA	NA
Electrostatic with mat	80–98	NA	Varies	0.15–1.25

Source: Reprinted by permission from AIA, Ramsey, and Sleeper, *Architectural Graphic Standards,* 7th ed., © 1981 by John Wiley & Sons.

[a]Higher airflow resistance values will require increased fan energy.

THERMAL CONTROL

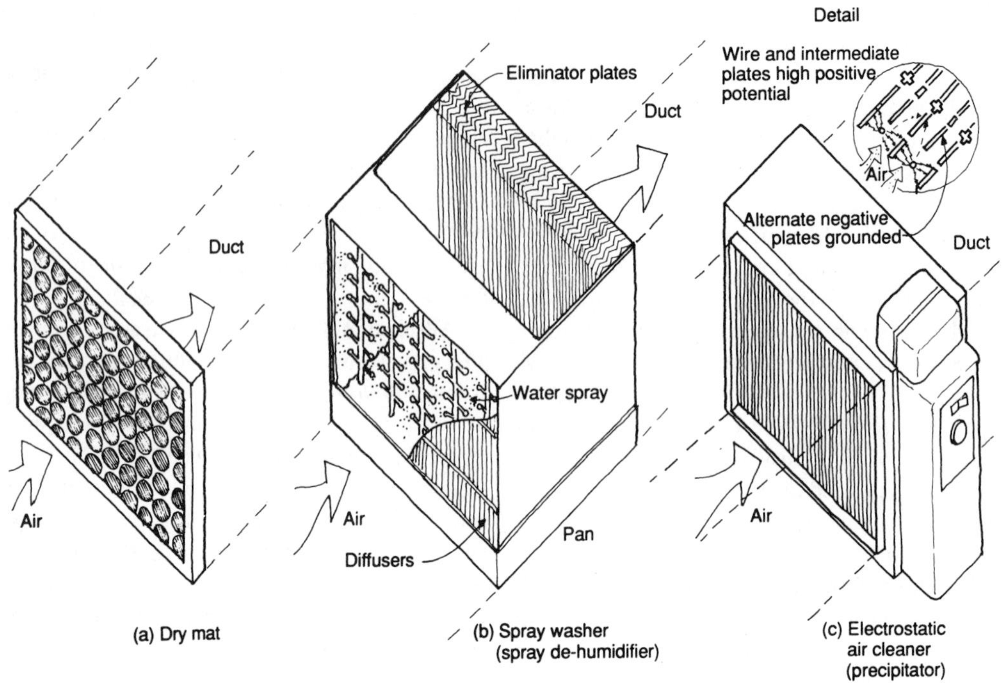

Fig. 6.37 *Air filter types. (Not shown are pleated and roll types.)*

moisture involved can pose a threat if these devices are not well maintained.

Electronic air cleaners can pose a different threat due to ozone production, but have the advantage of demanding less maintenance. Static electricity is produced in the *self-charging mechanical filter,* by air rushing through it; larger particles thus cling to the filter. The more humid and/or higher velocity the air, the less the filtering efficiency. In the *charged media filter,* an electrostatic field is created by applying a high dc voltage to the dielectric material of the filter. Many particles are not polarized, however, due to an insufficiently strong field. A *two-stage electronic air cleaner* first passes dirty air between ionizing wires of a high-voltage power supply. Electrons are stripped from the particulate contaminants, leaving them positively charged. Then, these ionized particles pass between collector plates that are closely spaced and oppositely charged. The particles are simultaneously repelled by the positive plates and attracted to the negative plates, where they are collected.

(f) Locating Air Cleaning Equipment. Before the advent of concern about IAQ, buildings were often designed with rather crude panel filters only at the HVAC equipment itself. In addition to these equipment-protecting devices, a building deserving a high IAQ will now have a combination of high-efficiency particle filters and adsorption filters.

The panel filters provided with the HVAC equipment are usually located *upstream* from the unit's fan. Subsequently, the high-efficiency particle and adsorption filtering system must be located *downstream* from the HVAC cooling coils and their drain pan, to ensure that any microbiological contaminants living on those wet surfaces are removed rather than being distributed throughout the building.

Energy conservation considerations have reduced the air circulation rate in today's central air handling systems. One result is a poor distribution efficiency, which means poorly mixed air within the occupied spaces. This discourages designers from integrating a high-efficiency filtering system with the central HVAC unit. In-

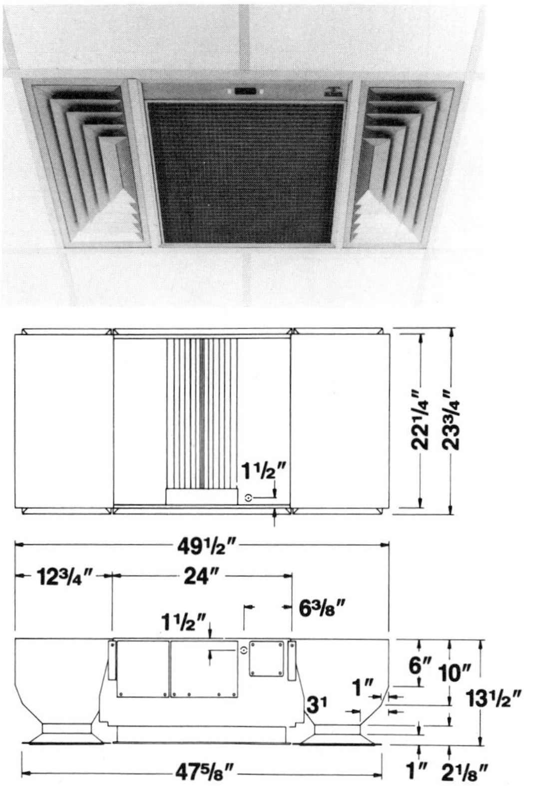

Fig. 6.38 A stand-alone electrostatic air filter operates independently of the central HVAC system; no ductwork is required. Airflow rates vary; the slower the flow, the more efficient the filtering. Flush mount (recessed in ceiling), ceiling surface mount, wall mount, and portable models are available. (Courtesy of Tectronic Products Co. Inc., East Syracuse, New York.)

stead, there is a trend toward *zoning* the air filtering equipment, to achieve both a high circulation rate and proper air mixing. Thus, each zone's filtering system has its own fan, which can operate either with or without the central HVAC fan. These zones should be as small as possible; the better the IAQ, the higher the cost of the many smaller filtering zones. An example of an independent electrostatic air filter is shown in Fig. 6.38.

Air circulation through these filters should occur at rates between six and 10 times per hour. The air is then ducted to diffusers, hence circulating in a sweeping pattern across a space to return air intakes on the *opposite side*. In buildings where the filtering system must be integrated with the central HVAC system (typical of small buildings), that system should be capable of continuously circulating at the rate of six to 10 times per hour, and of operating against the rather considerable static pressure that results from high-efficiency filters.

6.7 Heating and Cooling Systems for Small Buildings

Smaller buildings are typically skin load dominated: that is, for them the climate dictates whether heating or cooling is the major concern. In some climates, only heating systems are needed; the building can "keep itself cool" during hot weather without mechanical assistance. In other climates, only cooling is needed. In still others, both heating and cooling are required. In this section, heating-only mechanical systems will be treated first, then heating-and-cooling systems, then cooling-only systems.

(a) Central Heating or Cooling Equipment. In the early design stages, an approximate size for the largest equipment is sometimes useful. Once the heating or cooling capacities are known, manufacturers' catalogs can be consulted for the dimensions of the heating and cooling equipment.

The critical decision in sizing the heating equipment is the *design temperature:* What is the lowest reasonable outdoor temperature for which a heating device can be sized if the desired interior temperature is to be maintained? These winter design temperatures are listed in Appendix A. When this is known, the next step is

$$\text{design } \Delta T = \text{inside temperature} - \text{outside design temperature}$$

In Section 5.2, the criteria for Btu/DD-ft^2 were listed in Table 5.5. To convert to the required capacity of a building's heating equipment, calculate

$$\frac{\text{Btu/DD-ft}^2}{24\text{ h}} \times \Delta T \times \text{ft}^2\text{ floor area} = \text{Btu/h heating capacity}$$

Note: For passively solar heated buildings, the backup heating unit is sometimes sized for the heat loss of the *entire* envelope and sometimes for the envelope *minus the solar wall,* just as the criteria were defined in Table 5.5. Similarly, buildings with *reliable* internal gains (lights, equipment, people, etc.) are sometimes designed with smaller heating units, because internal gains supply a constant portion of the space heating needs. For a more complete discussion, see Section 5.4.

The sizing of cooling (mechanical refrigeration) units is not so straightforward, as were evident when detailed hourly heat gain procedures were presented (Sections 5.7 and 5.8). However, a *very approximate* early estimate can be obtained from the estimated hourly gains in Table 5.8. *Warning:* This estimate is likely to be *lower* than that obtained using the peak heat gain hour, for which cooling equipment is often sized.

$$\text{sensible Btu/h cooling capacity} = [\text{approx. heat gain (Btu/h ft}^2)][\text{floor area (ft}^2)]$$

Another common unit used for sizing mechanical refrigeration is "tons" of cooling capacity, one ton being equivalent to the useful cooling effect of a ton of ice, or 12,000 Btu/h (3516 W). Therefore, the required capacity in tons is

$$\frac{\text{heat gain (Btu/h)}}{12{,}000} = \text{tons of cooling}$$

(b) *Heating-Only Systems, For Rooms.* After the sun, the most ancient method of heating is the radiant effect of fire. With each step from campfire to fireplace to wood stove, more of the fuel's heat was captured for the room rather than wasted to the outdoors (Fig. 6.39). Although many people enjoy the sight, sound, and smell of the open fireplace, the tightly enclosed wood stove with catalytic combustor is a substantially more efficient and less polluting approach to heating.

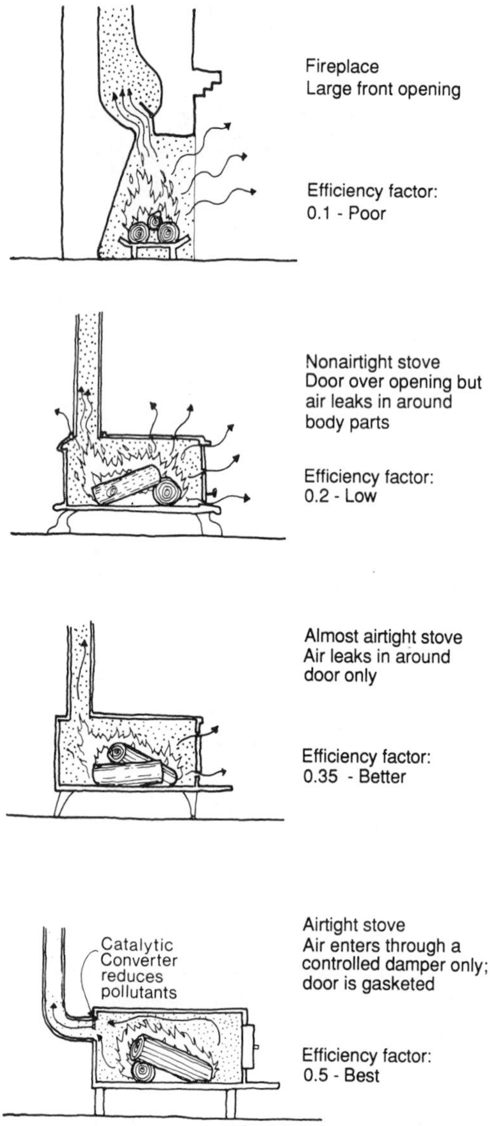

Fig. 6.39 *The efficiency of wood-burning devices has improved substantially. (Copyright © 1978 by* Alternative Sources of Energy. *Issue 35; 107 S. Central Ave., Milaca, MN 56353; adapted by permission.)*

Wood stoves are available in a wide variety of styles and are made of several materials. The sizes of such stoves are often difficult to determine. Manufacturers rarely specify the Btu/h output, which depends on the density, moisture content, and burn time of the wood fuel. Wood that has been split, loosely stacked, and covered from rain for at least 6 months should achieve a moisture content of about 20% by weight. The following sizing procedure assumes no more than this 20% moisture content. For more details, see *Alternative Sources of Energy Magazine* 35 (1978).

The formula for the hourly heat output to a room from a wood stove is

$$\text{Btu/h} = \frac{(V)(E)(D)(7000)}{T}$$

where

V = useful (loadable) volume of the stove (ft^3)
E = percent efficiency, expressed as a decimal (<1.0); see Fig. 6.39
D = density of the wood fuel (Table 6.13)
T = burn time (h) for a complete load of firewood; usually assumed at 8 h minimum
7000 = Btu/lb of firewood, 20% moisture content

("Bone-dry" wood can be assumed to have 8600 Btu/lb.) Note that a *drop* in burn time *increases* the Btu/h output; it is evident that when the air supply is increased to the stove, the fire burns hotter, consuming the wood more quickly. To meet the design heat loss (worst condition) for a room, burn times of 8 to 10 h should be assumed. Stoves rarely need relighting with a 10-h burn time.

Wood stoves are frequently used as the sole mechanical heat source for an entire building, such as a residence or a small commercial building that is passively solar heated. Since radiant heat is the dominant form of heat output, the areas that "see" the stove will get most of the benefit. However, "circulating" stoves convert a larger portion of their heat to convected heat, which produces a layer of hot air at ceiling level. By providing a path between rooms at the ceiling, this hot air will slowly spread throughout a building; it also easily finds its way upstairs, since warm air rises. The more thermally massive the ceiling construction, the longer it will store and reradiate the heat from this warm air mass.

The flue leading from a wood stove carries very hot gases, which are a potential source of heat (and pollution). The flue can be exposed to a space, making its radiant heat available, or simple heat exchangers can be constructed (such as the preheating of domestic hot water). More elaborate heat recovery devices—for boiler flue heat recovery—are discussed in Section 7.3.

Catalytic combustors, a recent development, reduce the air pollution from wood burning. These devices are honeycomb-shaped, chemically treated disks, as much as 6 in. in diameter and 3 in. thick. They are either inserted in the flue or built into the stove itself. When wood smoke passes through the combustor, it reacts with the chemical and ignites at a much lower temperature; this causes gases to burn that otherwise would have gone up the flue. The result is more heat produced, less creosote buildup in the flue, and fewer pollutants in the atmosphere. Like the catalytic converters in autos, these devices impose limits on the fuel: plastic, colored newsprint, metallic substances, and sulfur are ruinous to combustors, which means that the stove must be used as a wood burner, not a trash incinerator.

Wood stoves have a larger impact on building design than most other heating devices have. Either noncombustible materials must be placed below and around them, or minimum clearance to ordinary combustible building materials must be provided. Furniture arrangements and circulation paths must be designed with the very hot stove surfaces in mind. Hot spots occur near the stove; cold spots occur whenever visual access to the stove is blocked. Thermally massive ma-

TABLE 6.13 **Approximate Average Wood Density**

Type	*Density* (lb/ft^3)[a]
Shagbark hickory	40.5
White oak	37.4
Red oak	36.2
Beech	36.2
Sugar maple	34.9
Yellow birch	34.3
White ash	33.7
Black walnut	31.2
American elm	28.7
Spruce	25.6
Hemlock	23.7
Aspen	23.1
Red cedar	18.7
White pine	17.5

Source: American Sources of Energy Magazine, #35, © 1978. Reprinted by permission.

[a]These values are approximate; they vary a great deal.

terials near the stove are advantageous in leveling the large temperature swings that can accompany the on–off cycle of the stove; this affinity for thermal mass has made the wood stove a popular choice for backup heat in passively solar-heated buildings. Finally, the amount of space required for wood storage should not be overlooked; recall the impact of the wood storage space on the house shown in Fig. 1.9. A covered, well-ventilated, easily accessible, and quite large space is optimum.

Electric resistance heaters carry the disadvantage of using high-grade energy to do a low-grade task, as shown in Fig. 1.19. Their advantages, however, are impressive: low first cost and individual thermostatic control that can easily be used to make each room a separate heating zone. Thus, the energy wasted at the electricity-generating plant (usually, 60 to 70%) can be partially recovered at the building, where unused rooms can remain unheated. A few of the many types of electric resistance heaters are shown in Fig. 6.40. As in the case of wood stoves, surfaces can sometimes reach high temperatures, requiring care in the location of heaters relative to furniture placement and traf-

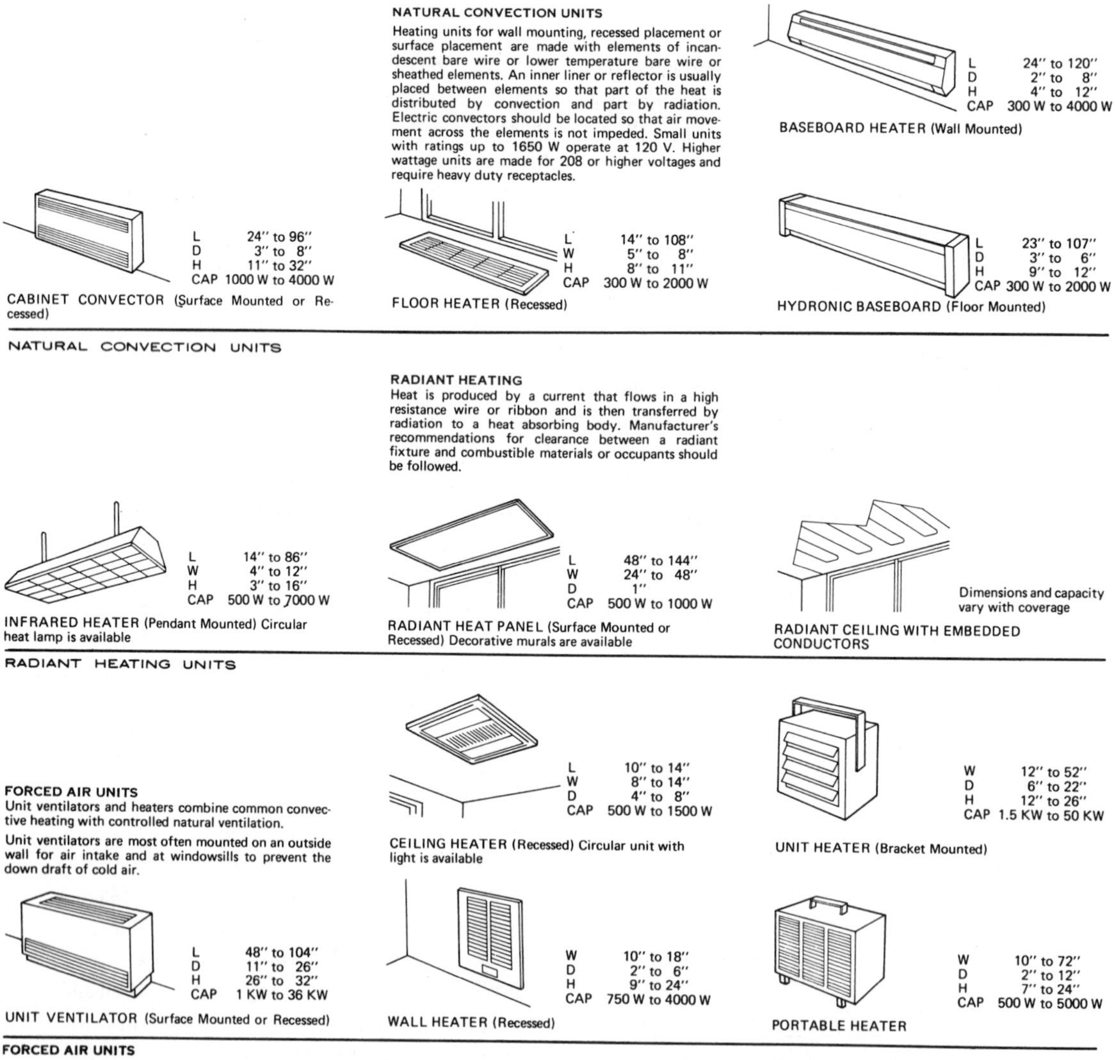

Fig. 6.40 *Varieties of electric resistance heating units. (From AIA, Ransey and Sleeper* Architectural Graphic Standards, *7th ed., © 1981 by John Wiley & Sons. Reprinted by permission.)*

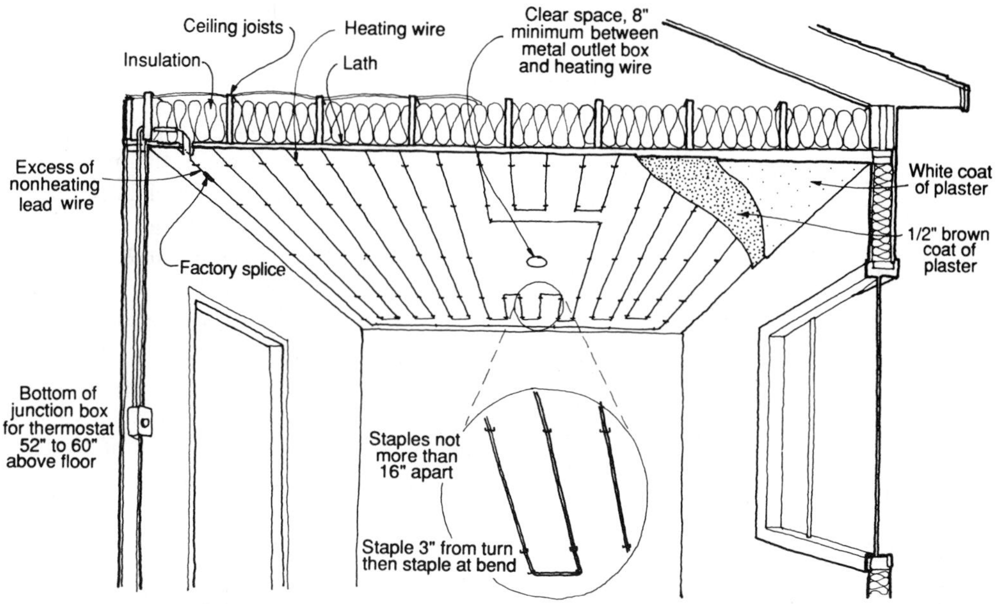

Fig. 6.41 *Typical installation of radiant-heating cable ready for completion of plaster ceiling.*

fic flow. Electric heaters are sized by their capacity, in kilowatts (1 kW = 3413 Btu/h).

Ceilings (or floors) can be constructed to include electric resistance in wiring (Fig. 6.41). The primary disadvantage of ceiling heat is that hot air stratifies just below the ceiling, so that air motion is discouraged.

Gas infrared heaters are often found in semi-outdoor locations such as loading docks and repair shops. When vented, they can be used in

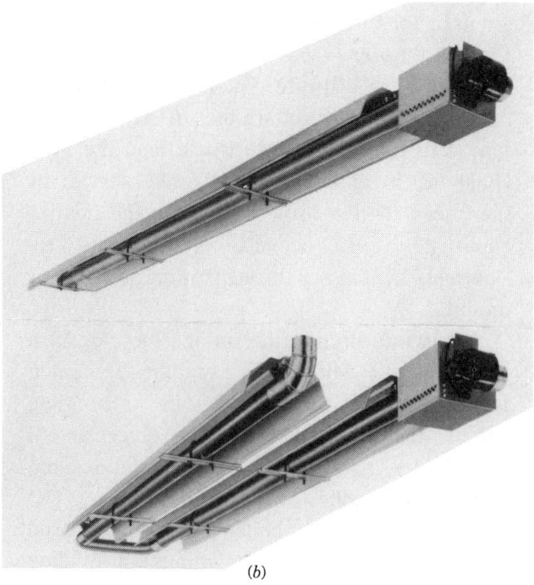

Fig. 6.42 (a) *A high-intensity infrared heater adds to the "historic" atmosphere of this retail store in an old warehouse. Exposed mechanical equipment includes the chains that operate the clerestory windows. The vented gas heater has an adjustable reflector that enables radiant heat to be directed. (Clark-Ditton, Architects, Eugene, Oregon.)* (b) *Vented gas high-intensity infrared heaters are available in both straight and U-shaped units. (Courtesy of Solaronics, Inc., Rochester, Michigan.)*

more traditional environments such as the retail store in a remodeled warehouse (Fig. 6.42). Their advantage is that they heat surfaces first rather than air, so that comfort is obtained without the need for such high air temperatures. When high rates of air exchange are expected, high-intensity radiant heaters are often used.

(c) Heating-Only Systems: Hot Water. The many remaining choices for whole-building heating systems will be discussed in the order (1) hot water and (2) forced air. Such systems will include a fuel, a heat source, a "mover" (such as pump or fan), a distribution system, a heat exchanger or terminal within the space, and a control system.

Hot water and steam boilers are rated according to heating capacity by several different categories. *Heating capacity, MBh,* is the rate of useful heat output with the boiler operating under steady-state conditions, expressed in 1000 Btu/h. This "useful heat" assumes that the boiler is within the heated envelope of the building; thus the heat that escapes from the boiler walls is available to help heat the building. *AFUE,* the annual fuel utilization efficiency, is defined as 100% minus the losses up the stack during both the on and off cycles, and the losses due to infiltration of outdoor air to replace the air used for combustion and for draft control. Finally, the *net I=B=R rating* is published by the Hydronics Institute. The I=B=R rating recommends both the square feet of stream radiation (1 ft^2 radiation = 240 Btu/h) and the MBh (1000 Btu/h) of heating load to be served by the boiler. This load is lower than the heating capacity rating, because it accounts for "normal" heat losses through piping and cycling on–off.

When interpreting the ratings, be careful to select a boiler whose rating matches the calculated critical heat loss of the house or building. When attempting to use AFUE to select an efficient boiler, take care to see that the assumptions about "inches of water draft" and % CO_2 are similar for the boilers being compared. (See later Table 6.17 for minimum efficiencies required for all heating and cooling equipment.)

Boilers and their accessories comprise a wide inventory. A few selected types are discussed here.

1. *Oil-fired steel boiler.* A refractory chamber receives the hot flame of the oil fire. Combustion continues within the chamber and the fire tubes. Smoke leaves through the breeching at the rear. Water, *outside* the chamber, receives the heat generated in the combustion chamber. If a domestic hot water coil is connected for use, a larger capacity boiler is selected. An aquastat (water thermostat) turns on the burner whenever the boiler water cools off, thereby maintaining a reservoir of hot water ready for heating the building.

2. *Gas-fired cast iron hot water boiler* (Fig. 6.43). Cast iron sections contain water that is heated by hot gases rising through these sections. Output is related to the *number* of sections. Additional heat is gained from a heat extractor in the flue. With induced draft combustion, condensing unit in the flue, and intermittent electronic ignition instead of a pilot light, up to 90% AFUE is attainable. The American Gas Association (AGA) sets standards for gas-fired equipment.

3. *Oil-fired, cast iron hot water boiler.* Primary and secondary air for combustion may be regulated at the burner unit. Flame enters the refractory chamber and continues around the outside of the water-filled cast iron sections.

Hot water heating circuits come in four principal arrangements. Figure 6.44*a* shows the series loop system, usually run at the building's perimeter. The water flows to and *through* each baseboard or fintube in turn. Obviously, the water at the end of the circuit is a little cooler, but since in all hot water systems the water temperature drop seldom exceeds 20 F° in residences, the *average* temperature can usually be used to select the baseboard or other elements. Valves at each heating element are not possible, since any valve would shut off the entire loop. Adjustment is by a damper at each baseboard, which reduces the natural convection of air over the fins. This is a "one-zone" system—all elements on, or off, together. There is no general rule

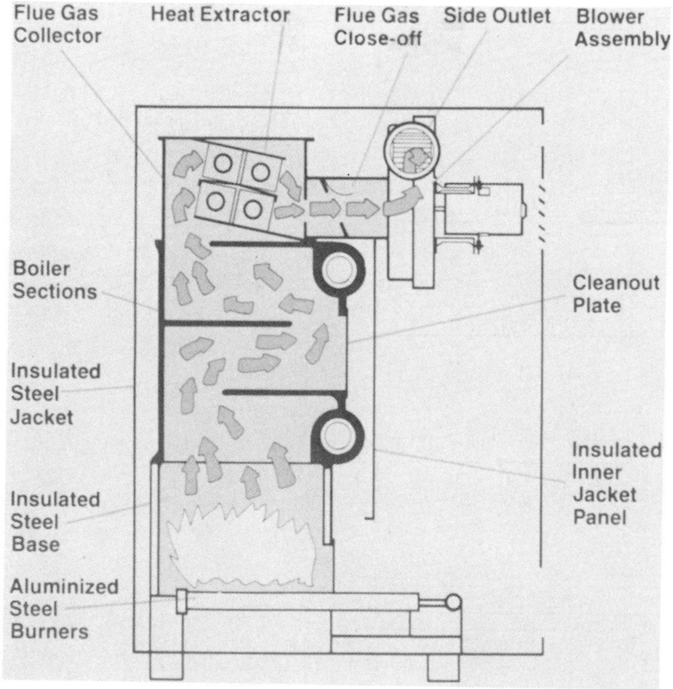

Fig. 6.43 *Gas-fired cast iron sectional boiler for hot water heating. Very high operating efficiency is possible; no chimney is required, as lower temperature exhaust gases can be vented through wall to exterior.*

about the maximum allowable length of water circuit, but for long runs, the pipe size can be increased or *several* loops used in parallel to create more than one thermal zone.

The one-pipe system shown in Figs. 6.44*b* and 6.45 is a very popular choice. Special fittings act to divert part of the flow into each baseboard. A valve may be used at each one to allow for reduced heat or for a complete shutoff, to conserve energy—an advantage that the loop system does not provide. The one-pipe system uses a little more piping and thus is not as economical to install as the loop system, in which piping is minimal.

The two-pipe reverse-return shown in Fig. 6.44*c* may be considered the classic piping arrangement. Water nearly at boiler temperature is supplied to each baseboard without being cooled by passing through a previous baseboard or accepting the cooler return water. Equal friction, resulting in equal flow, is achieved through all baseboards (nos. 1 to 5) by *reversing* the return instead of running it directly back to the boiler. This equality is effected by equal lengths of water flow through any baseboard together with its lengths of supply-and-return main. More pipe is required for this system than for the systems shown in Fig. 6.44*a* or *b*.

Figure 6.44*d* shows an arrangement that is not usually favored because the path of water through baseboard no. 1 is much shorter than that through the others, and especially no. 5. Baseboard no. 5 could easily be undesirably cool, since it is short-circuited by the others.

Pipe expansion requires expansion joints in long runs of pipe and clearance around all pipes passing through walls and floors. Each time a hydronic system changes from room temperature to a heated condition, the piping will undergo the following expansion (assuming 70 F initial temperature):

Water Temperature (F)	Pipe Expansion per 100 ft	
	Iron Pipe	Copper Tubing
160	0.7 inches	1.0 inches
180	0.9 inches	1.3 inches
200	1.0 inches	1.5 inches
220	1.2 inches	1.7 inches

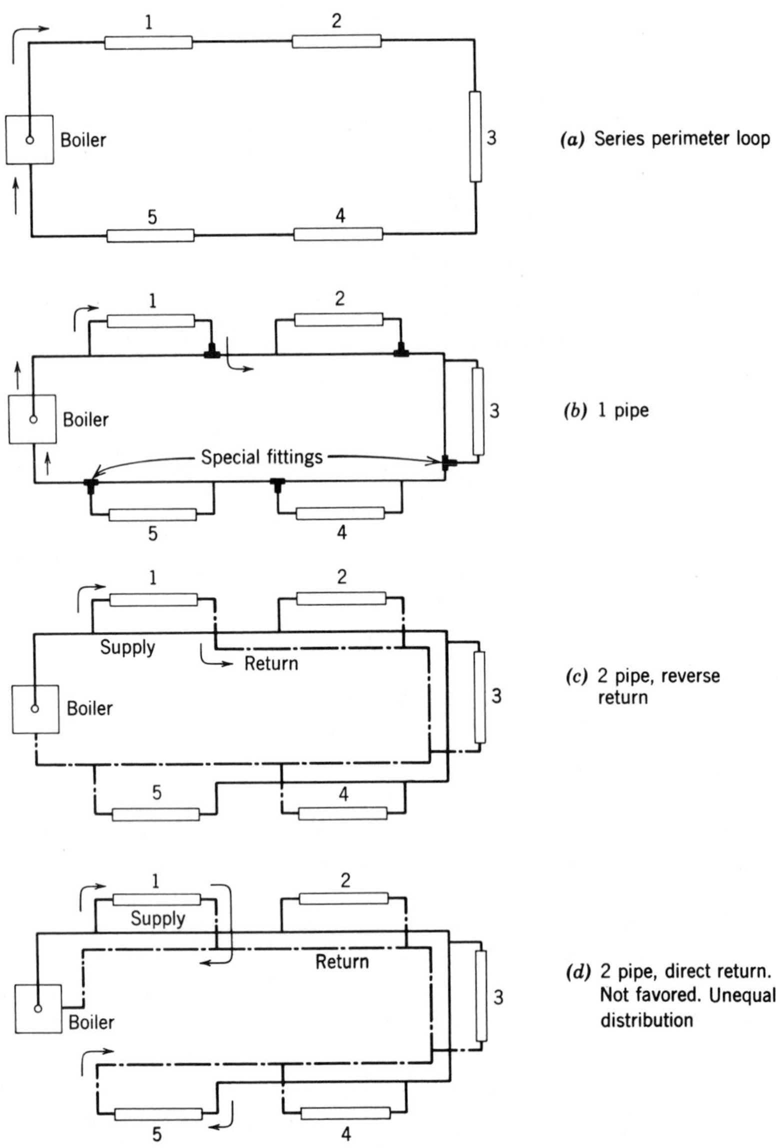

Fig. 6.44 *Plan views showing principles of piping for water distribution to heating elements (baseboard convectors shown here) in hot-water-heating systems. For simplicity, controls are not shown.*

An *air cushion tank,* compression tank, or expansion tank, is a closed tank containing air, usually located above the boiler. When the water in the system is heated, it expands, compressing the air trapped in the tank. This tank allows for the usual range of temperatures within the system, including temperatures above the usual boiling point of water, without the frequent opening of the pressure relief valve. One type of air cushion tank, called a *diaphragm tank,* separates the air and water with an inert flexible material; this prevents reabsorption of the air by the water.

For a conventional (unpressurized) air cushion tank, allow 1 gallon of tank capacity for every 5000 Btu/h of the total system heat loss. For a pressurized tank (at least 8 lb/in.^{2}) allow 1 gallon of tank capacity for each 7000 Btu/h (or, see manufacturer's recommendations).

Air vents and water drains are part of the distribution system. Except for the necessary air cushion in the upper part of the compression tank above the boiler, air must not be allowed to accumulate at high points in the piping or at the convector branches. Air vents relieve these possible air pockets, which would otherwise make the system air-bound and inoperative.

If a system is to be drained and left idle in a

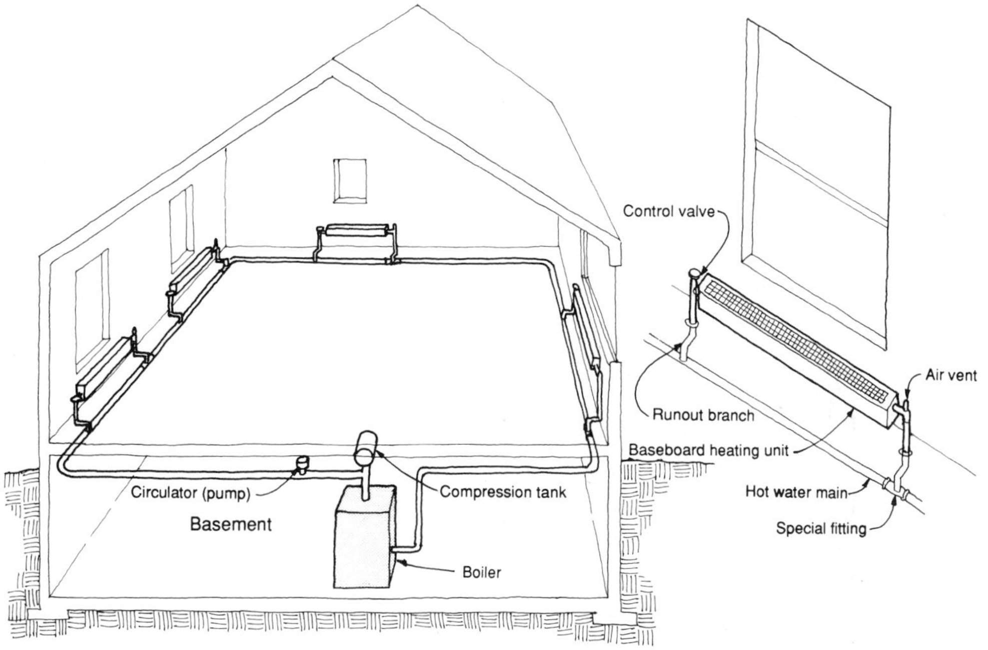

Fig. 6.45 *Hot water, one-pipe system—type* b *of Fig. 6.44.*

cold house, water trapped in low points could freeze and burst the tubing or fittings. Operable drain valves must be provided at such locations and, of course, at the bottom of the boiler, as shown in Fig. 6.46.

Hydronic and electrical controls allow automatic operation, described in Fig. 6.46. There are two options for system control:

1. As in Fig. 6.40, the thermostat controls the circulating pump and the boiler. In colder weather, the system operates almost continuously, and the average temperature in the system gradually rises.
2. The thermostat controls only the boiler, and the circulating pump operates continuously. This uses more energy for the pump but minimizes system temperature variation and thus the possibility of expansion noises.

Makeup water is added as required, the air level in the tank is regulated by the air control fittings, and the circulator and burner operate as controlled by the aquastat and thermostat. If air vents in the *piping* are not automatic, they will require periodic manual "bleeding" of unwanted air.

Circulating pumps are used to overcome the friction of flow in the piping and fittings and to deliver water at a rate sufficient to offset the hourly heat loss of the house or building.

(d) Hydronic Heating Design. The calculations for the sizing of a water distribution system are presented in Section 9.7; they are based upon the required flow and the friction in the piping. (For domestic water supply, another factor is the vertical distance through which the water must be raised. In these closed-loop heating systems, however, the weight of the cooler water falling back to the boiler essentially counterbalances the weight of the hot water being raised. Furthermore, gravity is helping the hot—lighter—water rise, and the cooler water fall.)

The key to pipe sizing is the overall required flow rate. Ordinarily, the temperature drop that occurs as the hot supply water gives up heat to the space (through the convector) is about 20 F° in residences; in commercial applications, 30, 40, and 50 F° are also common, as recommended

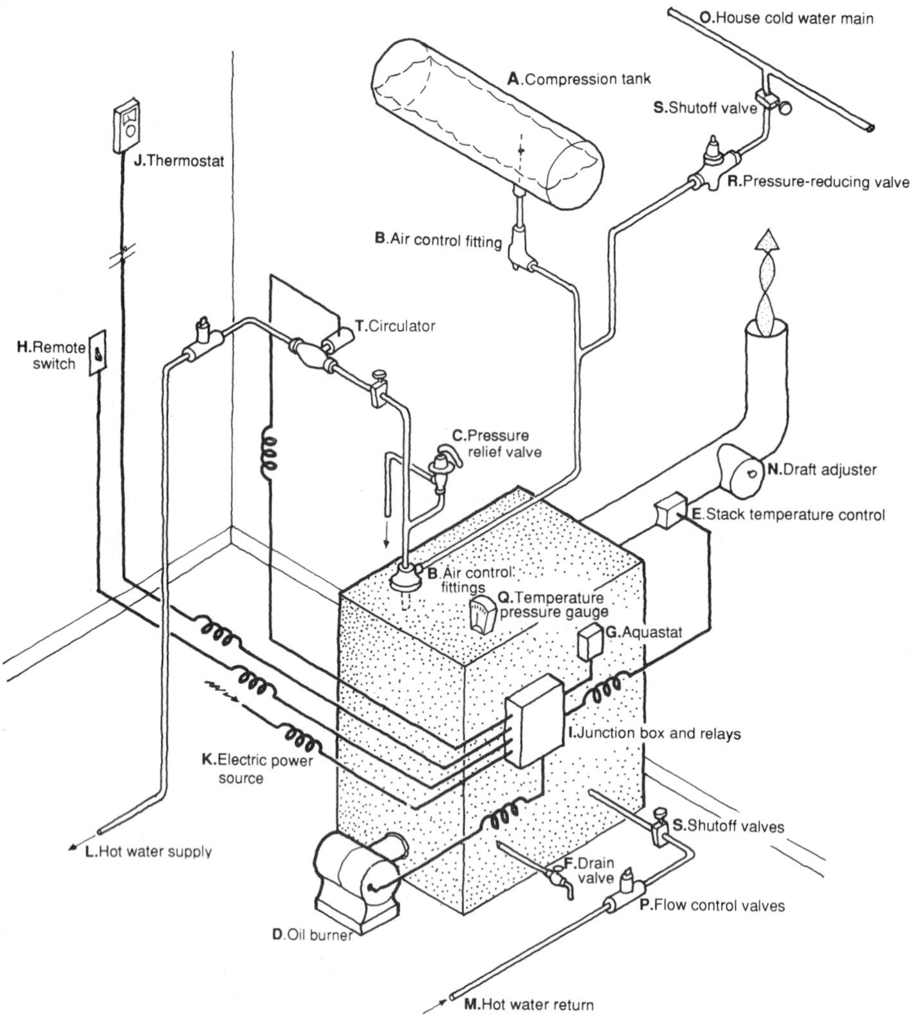

by the manufacturers of unit heaters and convectors. Because the entire building's design heat loss is to be overcome by this system,

overall flow rate, gpm

$$= \frac{\text{design heat loss, Btu/h}}{20\ \text{F}^\circ \times 60\ \text{min/h} \times 8\ \text{lb/gal}}$$

$$= \frac{\text{design heat loss}}{9600}$$

Then, by using Section 9.7, we can account for friction through piping and fittings and size the main supply and return pipes. The same procedure can be applied to branches, proportioned to the heat they must deliver.

With both the friction of the system (expressed in feet of water, or head) and the flow rate established, a pump can be selected. Figure 6.47 shows typical performance curves for four pumps. The designer enters the curve with desired flow rate, then selects the pump with a head capacity greater than or equal to the head required. Pump performance curves are provided by the manufacturer.

The critical choice, however, is not pipe size: it is relatively easy to distribute pipes within wall and floor/ceiling construction. Rather, the critical choice is *hot-water supply temperature:* the higher this temperature, the smaller the convector units that discharge the heat to each space. However, higher temperatures endanger occupants, who may suffer skin burns if they

Fig. 6.46 Hydronic and electrical controls; an oil-fired boiler for heating by hot water.

(A) *Compression tank.* Accommodates the expansion of the water in the system.

(B) *Air control fittings.* Vent out unwanted air in the boiler and maintain the level in the compression tank.

(C) *Pressure relief valve.* Usually set for 30 psi. Initial cold pressure about 12 psi. Relieves excessive system pressure.

(D) *Oil burner.* Responds to aquastat or thermostat.

(E) *Stack temperature control.* Senses stack temperature and stops oil injection if ignition has not occurred.

(F) *Drain valve.* At low point in the water system.

(G) *Aquastat.* Maintains temperature of boiler water by starting the oil burner when temperature of water drops below the aquastat's setting. Sometimes set at about 180 F.

(H) *Remote switch.* At a safe distance from the boiler so that the plant can be turned off in case of trouble during which the boiler cannot be approached.

(I) *Junction box and relays.* General control center.

(J) *Thermostat.* When the room temperature drops below its setting, it turns on both the oil fire and the circulating pump.

(K) *Electrical power source.* Operates from a separate individual circuit at the power panel.

(L) *Hot water supply.* Copper tubing to convectors or baseboards.

(M) *Hot water return.* Copper tubing from convectors or baseboards.

(N) *Draft adjuster.* Regulates the draft (combustion air) over the fire.

(O) *House cold water main.* From which water is fed automatically into boiler.

(P) *Flow control valves.* Prevent casual flow of water by gravity when the circulator is not running.

(Q) *Temperature pressure gauge.* Indicates water temperature and pressure. Sometimes supplemented by immersion thermometers in supply and return mains.

(R) *Pressure-reducing valve.* Admits water into the system when the pressure there drops below about 12 psi. Has a built-in check valve to prevent backflow of boiler water into the water main.

(S) *Shutoff valves.* Normally open. Can be closed to isolate the system and permit servicing of components.

(T) *Circulator.* Centrifugal circulating pump that moves the water through the tubing and heating elements.

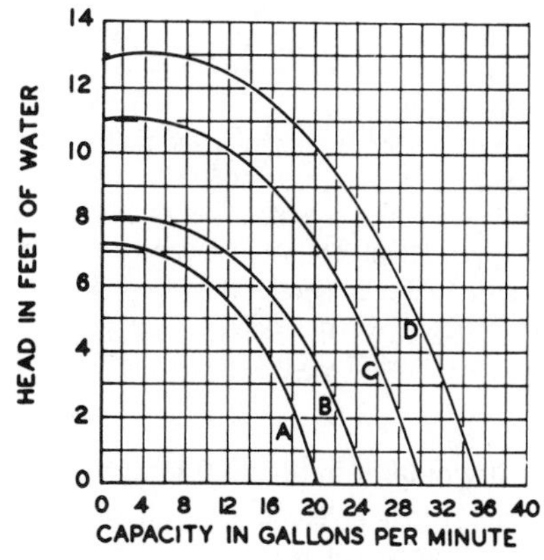

Fig. 6.47 Typical pump capacity (or performance) curves, for four pumps used in hydronic heating systems. (Courtesy of the Hydronics Institute, Inc., Berkeley Heights, New Jersey.)

touch exposed parts of the convectors or distribution tree. Higher temperatures also can lead to steam within the boiler/distribution tree. These systems are not designed to accommodate steam, and serious injury can sometimes result. A safer choice of average water distribution temperature is 170 to 180 F, even though these temperatures result in larger convectors.

Baseboard convector selection is then made from manufacturer's data tables, such as the one shown in Table 6.14. The two common baseboard types are *RC*, usually cast iron with a water-backed front surface and extended rear heating surface, and *finned tube*, a metal tube with extended surface in the form of fins, usually placed behind a metal enclosure.

EXAMPLE 6.1. Select a Heatrim R-750 baseboard convector (Table 6.14) for a 20-ft long living room wall. From heat loss calculation, the

TABLE 6.14 **Manufacturer's Data for Hot-Water Baseboard Convectors**[a]

Approved IBR water ratings[a]; model R-750 Heatrim—capacities Btu/h ft.

Number of Lineal Feet[b]	*Water Flow Rate 500 lb/h Average Water Temperature (F)*					
	170	*180*	*190*	*200*	*210*	*220*
1	500	560	630	690	760	820
2	1000	1120	1260	1380	1520	1640
3	1500	1680	1890	2070	2280	2460
4	2000	2240	2520	2760	3040	3280
5	2500	2800	3150	3450	3800	4100
6	3000	3360	3780	4140	4560	4920
7	3500	3920	4410	4830	5320	5740
8	4000	4480	5040	5520	6080	6560
9	4500	5040	5670	6210	6840	7380
10	5000	5600	6300	6900	7600	8200
11	5500	6160	6930	7590	8360	9020
12	6000	6720	7560	8280	9120	9840
13	6500	7280	8190	8970	9880	10660
14	7000	7840	8820	9660	10640	11480
15	7500	8400	9450	10350	11400	12300
16	8000	8960	10080	11040	12160	13120
17	8500	9520	10710	11730	12920	13940
18	9000	10080	11340	12420	13680	14760
19	9500	10640	11970	13110	14440	15580
20	10000	11200	12600	13800	15200	16400
21	10500	11760	13230	14490	15960	17220
22	11000	12320	13860	15180	16720	18040
23	11500	12880	14490	15870	17480	18860
24	12000	13440	15120	16560	18240	19680
25	12500	14000	15750	17250	19000	20500
26	13000	14560	16380	17940	19760	21320
27	13500	15120	17010	18630	20520	22140
28	14000	15680	17640	19320	21280	22960
29	14500	16240	18270	20010	22040	22780
30	15000	16800	18900	20700	22800	24600

Source: Courtesy of Burnham Corporation.

[a]Approved IBR water ratings shown above for American-Standard Heatrim Panels (with Model E-750 element) are based on a water flow of 500 lb/h with a pressure drop of 0.047 in. of water per lineal foot and a water flow of 2000 lb/h with a pressure drop of 0.530 in. of water per lineal foot. As allowed by the Institute of Boiler and Radiator Manufacturers (IBR) Testing and Rating Code for Baseboard Type of Radiation, 15% is added to water heat capacity.

The use of IBR ratings at water flow rates of 2000 lb/h is limited to installations

living room requires 9000 Btu/h. The average water temperature is 180 F, and water flow is 500 lb/h (equivalent to about 1 gpm).

SOLUTION. From the 180 F column of the left section of Table 6.14, choose a 16-ft-long baseboard convector for its capacity of 8960 Btu/h. It would fit in the 20-ft length of the wall, and 8960 Btu/h is close enough to the design figure (i.e., 9000 Btu/h).

Panel (radiant ceiling or floor) heating design usually depends on a water temperature of 120 F for heated floors, 140 F for heated ceilings. An

Water Flow Rate 2000 lb/h Average Water Temperature (F)					
170	180	190	200	210	220
530	590	670	730	800	870
1060	1180	1340	1460	1600	1740
1590	1770	2010	2190	2400	2610
2120	2360	2680	2920	3200	3480
2650	2950	3350	3650	4000	4350
3180	3540	4020	4380	4800	5220
3710	4130	4690	5110	5600	6090
4240	4720	5360	5840	6400	6960
4770	5310	6030	6570	7200	7830
5300	5900	6700	7300	8000	8700
5830	6490	7370	8030	8800	9570
6360	7080	8040	8760	9600	10440
6890	7670	8710	9490	10400	11310
7420	8260	9830	10220	11200	12180
7950	8850	10050	10950	12000	13050
8480	9440	10720	11680	12800	13920
9010	10030	11390	12410	13600	14790
9540	10620	12060	13140	14400	15660
10070	11210	12730	13870	15200	16530
10600	11800	13400	14600	16000	17400
11130	12390	14070	15330	16800	18270
11660	12980	14740	16060	17600	19140
12190	13570	15410	16790	18400	20010
12720	14160	16080	17520	19200	20880
13250	14750	16750	18250	20000	21750
13780	15340	17420	18980	20800	22620
14310	15930	18090	19710	21600	23490
14840	16520	18760	20440	22400	24360
15370	17110	19430	21170	23200	25230
15900	17700	20100	21900	24000	26100

where the water flow rate through the baseboard unit is equal to or greater than 2000 lb/h. Where the water flow rate through the baseboard is not known, the IBR rating at the standard water flow rate of 500 lb/h must be used.

[b]These ratings are based on active (finned) Heatrim lengths. Difference between active length and total length of the standard Heatrim heating elements is $2\frac{15}{32}$ in. Elements are unpainted.

Nonferrous fins on model E-750 elements measure $2\frac{1}{8} \times 2\frac{1}{8} \times 0.008$ in., spaced 52 fins per foot.

uncarpeted concrete floor slab using ¾-in.-diameter pipe or tube on 12-in. centers, with an average water temperature of 120 F, will deliver *50 Btu/h-ft² of floor panel.* A ceiling panel with nominal ⅜-in. tube on 6-in. centers, with an average water temperature of 140 F, will deliver *60 Btu/h-ft² of ceiling panel.* To determine the area of heated panel, divide the room's design heat loss by the rate of panel heat delivery. Although radiant ceilings deliver more heat per unit area, they also discourage air motion because the warmest air rises to lie against the warmest surface. In contrast, at a radiant floor the coolest air drops to contact the warmest surface, is then warmed, and rises to be cooled at the ceiling, drops to the floor, and repeats this cycle continually.

Each panel will contain one or more coils. In floor panels, each coil delivers 10,000 Btu/h and should be no longer than about 200 linear feet. In ceiling panels (with smaller-diameter tubes), each coil delivers 3000 Btu/h and should be no longer than about 100 linear feet.

Sizing of the main pipes and pump is based on the longest circuit, measured along the length of the supply pipe from the boiler to the coil, and back along the return line to the boiler. Two guidelines apply:

1. Do not include the length of the coil itself in this circuit length.
2. No section of a floor panel main should be less than ¾ in. in diameter.

EXAMPLE 6.2. Determine the panel area and coils required for a living room 15 ft × 25 ft (375 ft²) with a design heat loss of 16,000 Btu/h.

SOLUTION. Assume a radiant floor. The panel area required is

$$\frac{16{,}000 \text{ Btu/h}}{50 \text{ Btu/h-ft}^2} = 320 \text{ ft}^2 \text{ of floor panel}$$

The room has 375 ft² of floor available, so the heated panel is usually placed along the exterior walls where the room heat loss is greatest. The number of coils in the panel is

$$\frac{16{,}000 \text{ Btu/h}}{10{,}000 \text{ Btu/h}} = 1.6 \text{ coils}$$

Use two coils in this panel.

Zoning is relatively easy to accomplish with hydronic systems, as shown in Fig. 6.48. The installation shown in the figure is made up of three separately heated areas—the first, second, and third floors. Each can be heated to different temperatures as called for by thermostats in each separate apartment. For example, if the thermostat serving the second floor (zone *B*) calls for heat, it turns on pump *B*. Flow-control valves *B* open, admitting hot water from the boiler header to main *B*. Flow-control valves *A* and *C* remain closed, preventing flow in mains *A* and *C*. Any or all of the zones may operate at one time. The boiler keeps a supply of hot water continually ready to supply any zone on demand. This is achieved by an aquastat immersed in the boiler water. When the boiler water drops below the prescribed temperature, it turns on the firing device, such as an oil burner or a gas burner, which brings the water up to temperature. If an overhead main supplies downfeed, as in the first floor of this installation, special downfeed supply and return fittings are necessary. For the second- and third-floor zones, one special return tee is sufficient. If the designer also elects to use a special upfeed *supply* tee of the venturi type, higher outputs of the convectors will result.

Two of the more famous residences in the Midwest utilize hydronic systems. Frank Lloyd Wright's Robie House (Chicago) has wall radiators integrated below the north windows in the famous living room. Underfloor radiators with grilles in the floor were provided for below the full-height south windows but were apparently never installed. The boiler sits in a basement room. Mies van der Rohe's Farnsworth House (Fox River, Illinois) preserves its four walls of ceiling-to-floor glass by concealing radiant heating pipes in the floor slab. The boiler sits within the central utility "closet."

Today's radiators are designed to reflect the sleek and simple lines of contemporary architecture. See Section 7.6 for examples in larger buildings.

(e) Heating Equipment Efficiency, Combustion, and Fuel Storage. As fuels burn to produce heat, they require oxygen to support the combustion. Since oxygen constitutes only about one-fifth of the volume of air, reasonably large rates of airflow are required. The air should be drawn in from outdoors at a position close to the fuel burner or (preferably) led to this location by a duct. For residences and other small buildings a louver *about twice the cross-sectioned area of the flue* should prove satisfactory. It should be arranged to remain open at all times. This combustion air should *not* be drawn from the general building space—it is a waste of energy, and contemporary "tight" construction

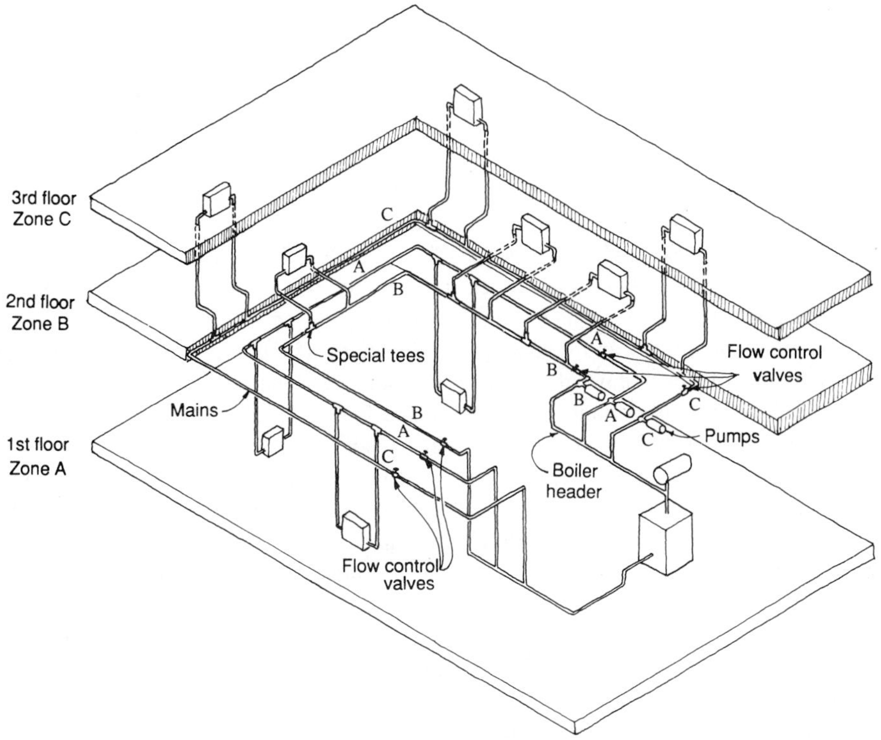

(a)

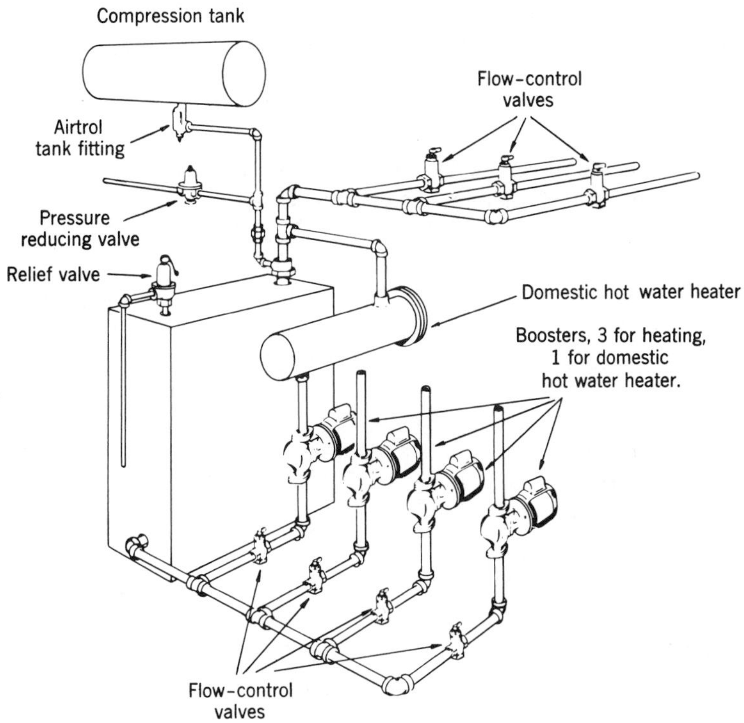

ITT Bell and Gossett

(b)

Fig. 6.48 *Three-zone, multicircuit, one-pipe system.* (a) *Each convector has connections to one pipe.* (b) *Boiler, piping, and water controls suitable for this three-zone, one-pipe system. Each one-pipe circuit should be provided with two flow-control valves and a circulator on the supply or return pipe. The terms* booster, pump, *and* circulator *are interchangeable.*

inhibits such airflow. A dangerous condition is created whenever stack flow is restricted.

The most important combustible element in the chemical makeup of fuels is carbon. It may be burned well or poorly. When burned poorly, it can cause great energy losses and sooty operation. For success, much depends on the proper selection of well-designed boilers, furnaces, and burners. Adjustments of primary and secondary air rates of flow and of draft (flow of air and gases through the boiler) are important responsibilities of the engineer and the heating contractor. Carbon may burn to carbon monoxide (CO) or (more completely) to carbon dioxide (CO_2) with greater heat production. Flue gases should be analyzed and the percentage of carbon dioxide measured. The best economy and the cleanest and most efficient combustion occur when the CO_2 content of the flue gases most nearly approaches the values given in Table 6.15 for maximum theoretical stoichiometric percent CO_2 (see also IBR standards, Table 6.16). The architect should require that these tests be made with the adjustments necessary to gain the most efficient performance. Flue gas temperatures should be taken. Temperatures lower than 600 F indicate that heat is properly retained in the boiler instead of escaping up the chimney.

It is important that chimneys, which carry high-temperature flue gases, be safely isolated from combustible construction, to prevent the possibility of fire. Conventional standards for houses call for a terra-cotta flue lining surrounded by 8 in. of brick, with an additional 2 in. of space between the brick and any wood. The space is usually filled with incombustible mineral wool. The size of flue will be dictated by the specification for the boiler or furnace selected for use. Flue height (see Fig. 6.49*a*) had traditionally been 35 to 40 ft. The function of providing a draft, for which chimney height was an important consideration, is no longer as necessary as it previously was, because fans are used now. For example, oil is injected under pressure, accompanied by air, and forced in by a fan. Often a draft adjuster in the breeching (smoke pipe) that carries the flue gases to the chimney is arranged to open slightly to *reduce* the normal stack draft. If increased draft should ever be required, an induced draft fan that puts a suction

TABLE 6.15 **Approximate Maximum Theoretical (Stoichiometric) CO_2 Values, and CO_2 Values for Various Fuels with Different Percentages of Excess Air**

Type of Fuel	*Theoretical or Maximum CO_2 (%)*	*Percent CO_2 at Given Excess Air Values*		
		20%	*40%*	*60%*
Gaseous fuels				
Natural gas	12.1	9.9	8.4	7.3
Propane gas (commercial)	13.9	11.4	9.6	8.4
Butane gas (commercial)	14.1	11.6	9.8	8.5
Mixed gas (natural and carbureted water gas)	11.2	12.5	10.5	9.1
Carbureted water gas	17.2	14.2	12.1	10.6
Coke oven gas	11.2	9.2	7.8	6.8
Liquid fuels				
No. 1 and 2 fuel oil	15.0	12.3	10.5	9.1
No. 6 fuel oil	16.5	13.6	11.6	10.1
Solid fuels				
Bituminous coal	18.2	15.1	12.9	11.3
Anthracite	20.2	16.8	14.4	12.6
Coke	21.0	17.5	15.0	13.0

Source: Copyright © by the American Society of Heating, Refrigeration and Air Conditioning Engineers, Inc., Atlanta, Ga. Reprinted by permission from *Handbook of Fundamentals,* 1989.

TABLE 6.16 **Conditions Set by the Hydronics Institute for I=B=R Ratings of Boilers**

Size	*Fuel*	*Minimum Efficiency*[a]	*Maximum Smoke*	*Maximum CO (%)*
Boilers under 300,000	Light oil	75% Steady state	No. 1	
Btu/h input	Natural gas	75% Steady state		0.04
Boilers 300,000 Btu/h	Light oil	75% Overall	No. 1	
input and over	Heavy oil	75% Overall	No. 4	
Power burner gas	Natural gas	75% Overall		0.04
Atmospheric gas (AGA)	Natural gas	75% Steady state		0.04

Source: Courtesy of the Hydronics Institute, Inc., Berkeley Heights, New Jersey.

[a]*Steady-state efficiency* is 100% minus the sensible and latent loses up the flue when operating at steady-state conditions. *Overall efficiency* is the measured heat output to the water or steam divided by the measured input to the boiler under steady-state conditions.

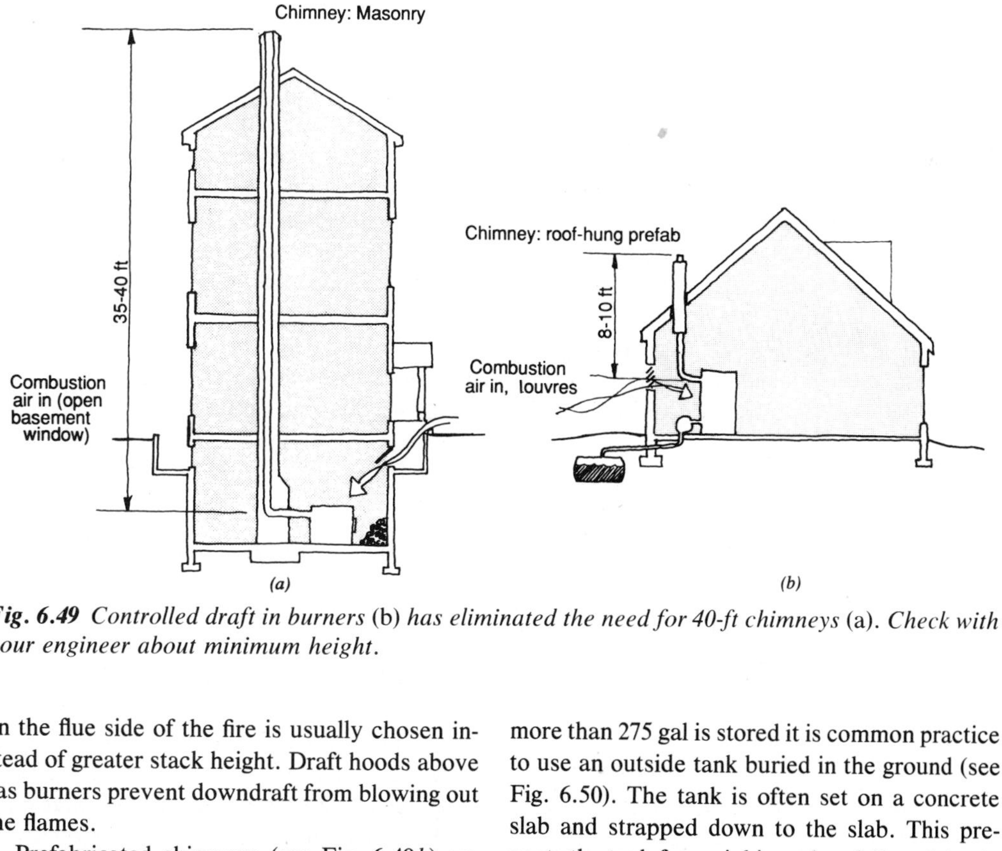

Fig. 6.49 *Controlled draft in burners* (b) *has eliminated the need for 40-ft chimneys* (a). *Check with your engineer about minimum height.*

on the flue side of the fire is usually chosen instead of greater stack height. Draft hoods above gas burners prevent downdraft from blowing out the flames.

Prefabricated chimneys (see Fig. 6.49*b*) are replacing with increasing frequency the bulkier and heavier field-built masonry. They offer a number of advantages and may be easily supported on a normal structure.

The storage space to be allowed for fuel oil depends on the proximity of the supplier and the space available at the building. For oil, when more than 275 gal is stored it is common practice to use an outside tank buried in the ground (see Fig. 6.50). The tank is often set on a concrete slab and strapped down to the slab. This prevents the tank from sinking when full or rising in flotation buoyancy that might be caused by adjacent groundwater when the tank is empty. The tank, usually made of steel, receives two coats of asphalt emulsion, to inhibit rust. Tubing for the gauge and for the supply and circulating lines are made of copper, and the fill and vent lines of wrought iron with swing joints to accommodate

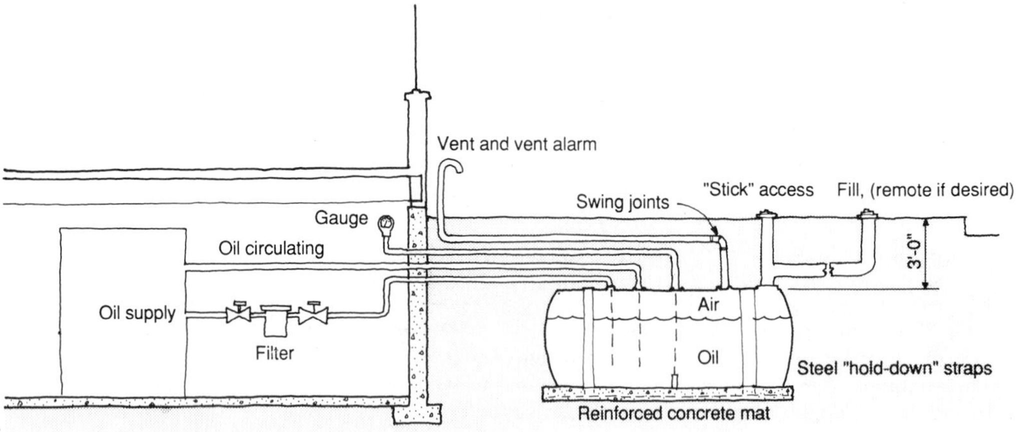

Fig. 6.50 Details of fuel oil storage tank.

possible slight settlement of the tank. Oil deliveries are often made on the basis of the degree days elapsed since the last fill-up of the tank. Thus, the customer is relieved of the chore of checking the gauge and ordering periodically.

The National Appliance Energy Conservation Act (1987) established minimum efficiency performance ratings for both small and large heating and cooling equipment. Table 7.3 (page 401) contains such standards for larger equipment; Table 6.17 contains the standards for smaller equipment.

AFUE, *annual fuel utilization efficiency,* is the ratio of annual fuel output energy to annual input energy, which includes any nonseason pilot input loss.

COP, *coefficient of performance,* is defined slightly differently depending on the task: for *cooling,* it is the ratio of the rate of heat removal to the rate of energy input in consistent units, for a complete cooling system (or factory-assembled equipment) as tested under a nationally recognized standard or designated operating conditions. For *heating* (heat pump) it is the ratio of the rate of heat delivered to the rate of energy input in consistent units, for a complete heat pump system as tested under designated operating conditions. Supplemental heat is not included in this definition.

EER, *energy efficiency ratio,* is the ratio of net equipment cooling capacity in Btu/h to the total rate of electric input in watts under designated operating conditions. (When consistent units are used, this ratio is the same as COP.)

IPLV, *integrated part load value,* is a single-number figure of merit based on part-load EER or COP expressing part-load efficiency for air-conditioning and heat pump equipment on the basis of weighted operation at various load capacities for the equipment.

SEER, *seasonal energy efficiency ratio,* is the total cooling output of an air conditioner during its normal annual usage period for cooling, in Btu/h, divided by the total electric energy input during the same period, in watt-hours.

(f) Warm Air Heating Systems. These systems began to supersede the open fireplace in about 1900. Originally, an iron furnace that stood in the middle of the basement was hand-fired by coal. Surrounding it was a sheet metal enclosure. An opening in its side near the bottom admitted cool air that gravitated to the basement. A short duct from the top of the enclosure delivered the warm air by gravity to a large grille in the middle of the floor of the parlor. Other rooms, including those in upper stories, shared a little of this warmth when doors were left open.

Very gradual changes had culminated by midcentury in systems essentially like the ones described in Fig. 6.51. The improvements included

Automatic firing of oil or gas.

TABLE 6.17 **Minimum Performance of Heating and Cooling Equipment of Smaller Size**

Part A. ***Boilers***

Category	*Rating Condition*	*Minimum Performance*[a]
Gas-fired <300,000 Btu/h	Seasonal rating	AFUE 80%[b]
Oil-fired <300,000 Btu/h	Seasonal rating	AFUE 80%

Part B. ***Warm-Air Furnaces and Combination Warm-Air Furnaces/Air-Conditioning Units***

Category	*Rating Condition*	*Minimum Performance*[a]
Gas-fired <225,000 Btu/h	Seasonal rating	AFUE 78%[c]
Oil-fired <225,000 Btu/h	Seasonal rating	AFUE 78%

Part C. ***Warm-Air Duct Furnaces and Unit Heaters***

Category	*Rating Conditions*[d]	*Minimum Performance*[e]
Duct furnaces, gas-fired	1. Max. rated capacity Steady-state	E_t[f] 78%
	2. Min. rated capacity Steady-state	E_t[f] 75%
Unit heaters, gas-fired	1. Max. rated capacity Steady-state	E_t[g] 78%
	2. Min. rated capacity Steady-state	E_t[g] 74%
Unit heaters, oil-fired	1. Max. rated capacity Steady-state	E_t[h] 81%
	2. Min. rated capacity Steady-state	E_t[h] 81%

Part D. ***Unitary Air Conditioners and Heat Pumps—Air Cooled, Electrically Operated <135,000 Btu/h Cooling Capacity—Except Packaged Terminal and Room Air Conditioners***

Category	*Subcategory and Rating Condition (Outdoor Temp., F)*		*Minimum Performance*[b]
<65,000 Btu/h Cooling capacity Cooling mode	1 ϕ	*Seasonal Rating* Split system Single package	 10.0 SEER 9.7 SEER[i]
<65,000 Btu/h Cooling capacity Cooling mode	3 ϕ	*Standard Rating (95 F DB)* Split system and single package *Integrated part-load value (80 F DB)* Split system and single package	 9.5 EER 8.5 IPLV
≥65,000 <135,000 Btu/h Cooling mode	All ϕ	*Standard Rating (95 F DB)* Integrated part-load value (80 F DB)	8.9 EER 8.3 IPLV
<65,000 Btu/h Cooling capacity Heating mode (heat pumps)	1 ϕ	*Seasonal Rating* Split-systems Single package	 6.8 HSPF 6.6 HSPF[i]

TABLE 6.17 **Minimum Performance of Heating and Cooling Equipment of Smaller Size *(continued)***

Part D. ***Unitary Air Conditioners and Heat Pumps—Air Cooled, Electrically Operated <135,000 Btu/h Cooling Capacity—Except Packaged Terminal and Room Air Conditioners (continued)***

Category	*Subcategory and Rating Condition (Outdoor Temp., F)*		*Minimum Performance[b]*
<65,000 Btu/h Cooling capacity Heating mode	3 φ	*Split-System and Single Package* High temp. rating (47 F DB/43 F WB) Low temp. rating (17 F DB/15 F WB)	 3.0 COP 2.0 COP
≥65,000 <135,000 Btu/h Cooling capacity Heating mode	All φ	*Split-System and Single Package* High-temp. rating (47 F DB/43 F WB) Low-temp. rating (17 F DB/15 F WB)	 3.0 COP 2.0 COP

Part E. ***Unitary Air Conditioners and Heat Pumps—Evaporatively Cooled, Electrically Operated —Cooling Mode <135,000 Btu/h Cooling Capacity—Except Packaged Terminal and Room Air Conditioners***

Category	*Rating Condition (F)*		*Minimum Performance[b]*
	Indoor Temp.	*Outdoor Temp.*	
<65,000 Btu/h Cooling capacity	*Standard Rating* 80 F DB/67 WB	 95 F DB/75 F WB	9.3 EER
<65,000 Btu/h	Integrated part-load value (80 F DB/67 F WB)		8.5 IPLV
≥65,000 < 135,000 Btu/h	*Standard Rating* 80 F DB/67 F WB	 95 F DB/75 F WB	10.5 EER
≤65,000 <135,000 Btu/h	Integrated part load value (80 F DB/67 F WB)		9.7 IPLV

Part E. ***Water-Cooled Air Conditioners and Heat Pumps—Cooling Mode <135,000 Btu/h Cooling Capacity—Electrically Operated***

Category	*Rating Condition (F)*		*Minimum Performance[b]*
	Indoor Air	*Entering Water*	
<65,000 Btu/h Cooling capacity	*Standard Rating* 80 F DB/67 F WB	 85	9.3 EER
	Low-Temperature Rating 80 F DB/67 F WB	 75	10.2 EER
<65,000 <135,000 Btu/h Cooling capacity	*Standard Rating* 80 F DB/67 F WB	 85	10.5 EER
<135,000 Btu/h	*Standard Rating*	70	11.0 EER
Cooling capacity	*Low-Temperature Rating*	50	11.5 EER
<65,000 Btu/h	*Standard Rating* 80 F DB/67 F WB	 85	9.3 EER
Cooling capacity	*Integrated Part Load Value*	75	8.3 IPLV
≥65,000 <135,000 Btu/h Cooling capacity	*Standard Rating* 80 F DB/67 F WB	 85	10.5 EER

TABLE 6.17 **Minimum Performance of Heating and Cooling Equipment of Smaller Size *(continued)***

Part F. ***Water-Source and Groundwater Source Heat Pumps—Electrically Operated <135,000 Btu/h Cooling Capacity***

Rating Condition F[j]	*Minimum Performance*[b]
Standard rating: 70 F entering water[k]	3.8 COP
High-temperature rating: 70 F entering water[k]	3.4 COP
Low-temperature rating: 50 F entering water[k]	3.0 COP

Part G. ***Packaged Terminal Air Conditioners and Heat Pumps—Air Cooled, Electrically Operated***[l]

Category	*Subcategory and Rating Condition (Outdoor Temp.)*	*Minimum Performance*[b]
PTAC's and PTAC H.P.'s[m]	Standard rating (95 F DB)	10.0 − 0.16 × Cap. (Btu/h/1000) EER
Cooling mode	Low-temp. rating (82 F DB)	12.2 − (0.20 × Cap. (Btu/h)/1000) EER
PTAC H.P.'s—Heating mode COP	Standard rating (47 F DB/43 F WB)	1.3 + 0.16 (EER above)

Part H. ***Room Air Conditioners and Room Air Conditioner Heat Pumps***

Category	*Minimum Performance*
Without reverse cycle and with louvered sides	
<6000 Btu/h	8.0 EER
≥6000, <8000 Btu/h	8.5 EER
≥8000, <14,000 Btu/h	9.0 EER
≥14,000, <20,000 Btu/h	8.8 EER
≥20,000 Btu/h	8.2 EER
Without reverse cycle and without louvered sides	
<6000 Btu/h	8.0 EER
≥6000, <20,000 Btu/h	8.5 EER
≥20,000 Btu/h	8.2 EER
With reverse cycle and with louvered sides	8.5 EER
With reverse cycle and without louvered sides	8.0 EER

Source: Adapted with permission from ASHRAE/IES Standard 90.1-1989, *Energy Efficient Design of New Buildings Except New Low-Rise Residential Buildings,* © 1989, American Society of Heating, Refrigerating, and Air-Conditioning Engineers, Inc., Atlanta, Ga. For larger equipment, see Table 7.4.

[a]As of Jan 1, 1992; National Appliance Energy Conservation Act of 1987.

[b]Except for gas-fired steam boilers for which minimum AFUE is 75%.

[c]For furnaces <45,000 Btu/h capacity, to be established by U.S. Department of Energy.

[d]Capacity provided and allowed by the controls.

[e]E_t, thermal efficiency; 100% flue losses.

[f]See ANSI Z83.9-86 for detailed definition.

[g]See ANSI Z83.8-85 for detailed definition.

[h]See U.L. 731-88 for detailed definition.

[i]Effective Jan. 1, 1993.

[j]Air entering indoor section at 70 F DB/60 F WB (maximum).

[k]Water flow rate per manufacturer's specifications.

[l]For multicapacity equipment the minimum performance shall apply to each capacity step provided and allowed by the controls.

[m]If the unit's capacity is less than 7000 Btu/h, use 7000 Btu/h in the calculation. If the unit's capacity is greater than 15,000 Btu/h, use 15,000 Btu/h in the calculation.

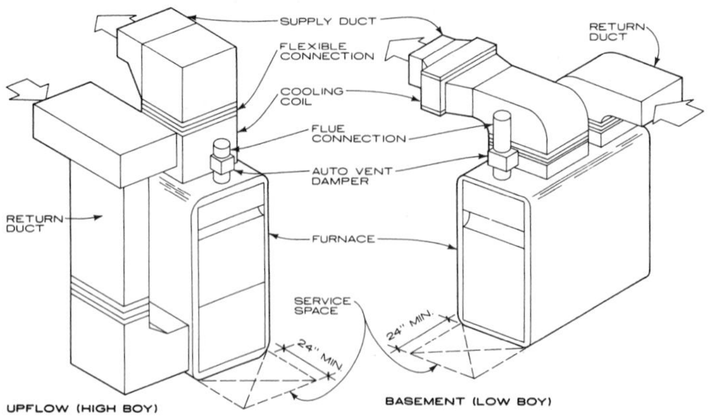

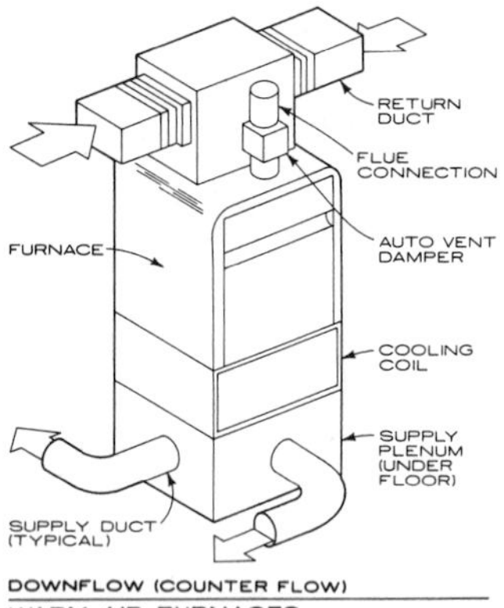

WARM AIR FURNACES

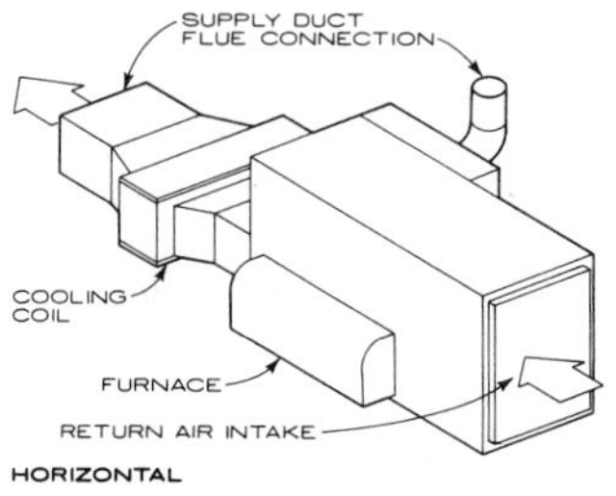

FLOOR AREA REQUIRED BY WARM AIR FURNACE

OUTPUT CAPACITY (BTU/HR)	FURNACE FLOOR AREA (SQ FT)*
Up to 52,000	2.4
52,000-84,000	4.2
84,000-120,000	6.6
120,000-200,000	13.1

*Based on net floor area occupied by the upflow or downflow furnace. Low boy unit requires 50% more floor area. Space for combustion air should be added as required by local codes. Adequate space should be provided for service.

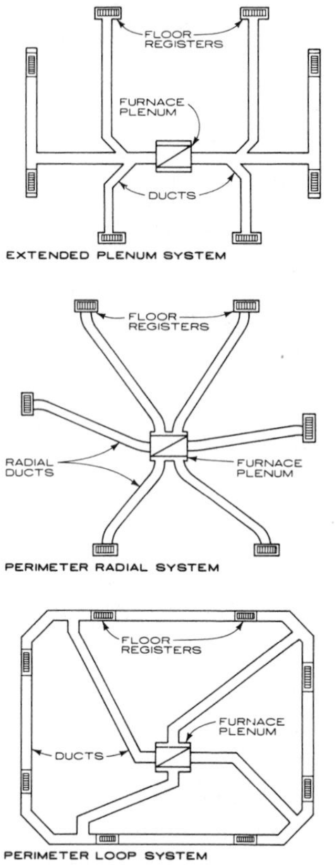

DUCT SYSTEMS

(a)

Fig. 6.51 (a) *Typical furnace types and duct distribution arrangements.* (b) *Solar collectors used as a source of warm air. (From AIA, Ramsey and Sleeper,* Architectural Graphic Standards, *7th ed., © 1981 by John Wiley & Sons. Reprinted by permission.)*

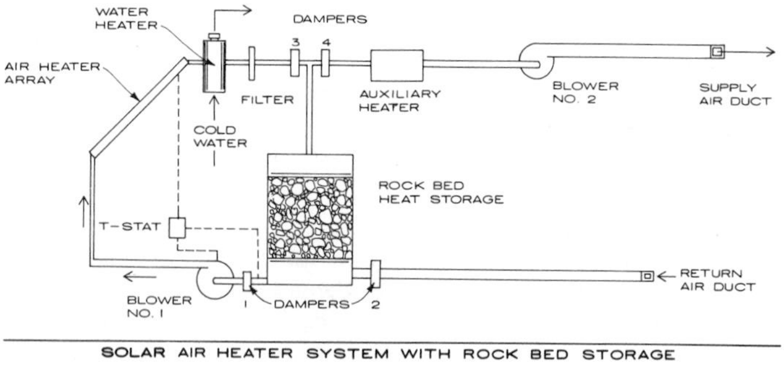

SOLAR AIR HEATER SYSTEM WITH ROCK BED STORAGE

(b)

Operational and safety controls.

Ducted air to and from each room.

Blowers to replace gravity.

Filters.

Adjustable registers.

By the 1960s it became apparent that the basement was beginning to disappear, as subslab perimeter systems became popular for basementless houses (see Fig. 6.52). In general, the features above were retained and air was delivered upward across glass, to be taken back at high-return grilles.

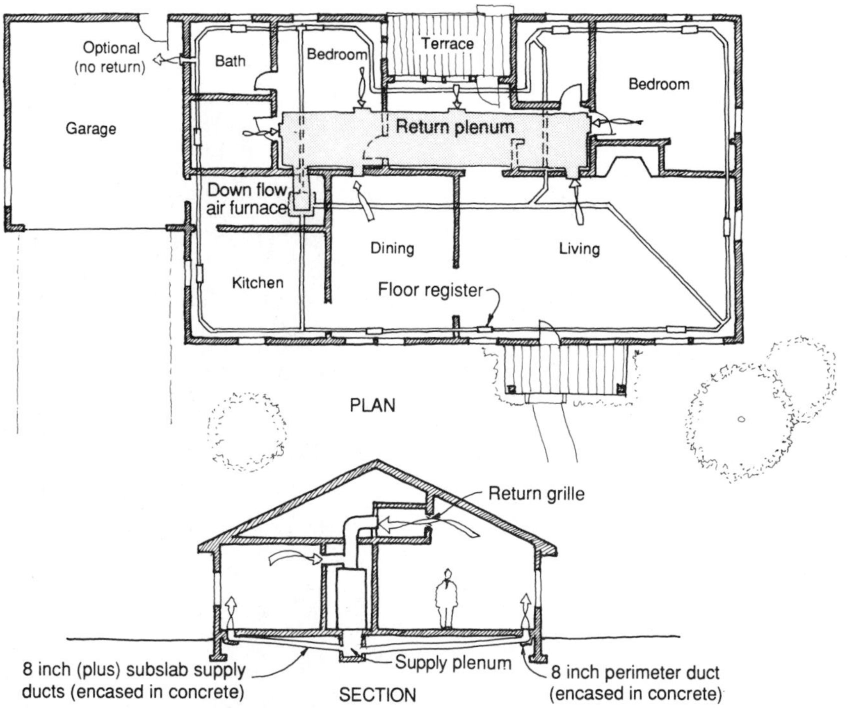

Fig. 6.52 *Forced-warm-air, perimeter loop system, adaptable for cooling. No returns from kitchen, baths, or garage.* (a) *Downflow air furnace.* (b) *Supply plenum.* (c) *Eight-inch (plus) subslab supply ducts (encased in concrete).* (d) *Eight-inch perimeter duct (encased in concrete).* (e) *Floor register, adjustable for direction and flow rate (Fig. 6.58).* (f) *Return grille.* (g) *Return plenum.*

As electricity began to replace oil and gas, former necessities such as combustion, chimneys, and fuel storage became nonessential. Horizontal electric furnaces began to appear in shallow attics or above furred ceilings. Air was delivered down from ceilings to warm exterior glass and taken back through door-grilles and open plenum space.

A new energy-saving trend appeared. The heat source (heat pump) was located centrally and fully *within* the insulated volume of the house. No stray heat escape from the unit was possible. Also, because of very small windows and double glazing, short ducts could deliver warm air from the inner side of each room, since warming of the double glass was less essential. Air returned to the unit through open grilles in doors and at the heat pump enclosure.

Today, solar energy can be used as a supplementary source of warm air (Fig. 6.51*b*).

Comfort is one of warm air heating's advantages. The motion of air in the space helps to assure uniform conditions and reasonably equal temperatures in all parts of a building. It is possible to clean both the recirculated air and the outdoor air by means of filters and other special air-cleaning equipment. Air may be circulated in nonheating seasons. Fresh air may be introduced to reduce odors and to make up the air exhausted by fans in kitchens, laundries, and bathrooms. Central cooling can be incorporated or introduced if ducts are designed originally to do so. Cooling sometimes calls for greater rates of air circulation. Humidification can be achieved by a humidifier in the air stream, and, if cooling is included in the design, dehumidification can be accomplished in summer. For both heating and cooling, a common arrangement is to place the supply registers in the floor, below areas of glass. This is important for winter operation. With adequate attention to supply register placement, return grilles can be located so as to minimize return air ductwork. High return grilles pick up the warmer air for reheating at the equipment. In many systems, air circulates at all times and is warmed or cooled as required.

Planning for warm air systems begins with the attempt to locate the furnace reasonably close to the center of the building. After the system is designed, a furnace must be selected. It should be capable of burning fuel at a rate suitable to make up the building's hourly heat loss. The rate of air delivery must be correct to transmit this heat at the air temperature *rise* that is planned. Finally, the motor and blower must be powerful enough to overcome the friction of air against metal in both the supply and return duct systems, as well as the friction of air flowing through the furnace, filters, registers, and grilles (see Fig. 6.57). Minor adjustments can be made at the furnace to adapt to the demands of the system and the building.

Some of the system components are discussed below.

Furnaces have become much more efficient in recent years, thanks to forced-draft chimneys and heat exchangers, as shown in Fig. 6.53. Seasonal efficiencies of up to 95% are possible, in contrast to about 62% for older furnaces.

The rating system for furnaces, like that for boilers, has been confused by a lack of standardized testing. As of 1992, AFUE ratings will be based on an "isolated combustion system" which requires that all combustion and dilution air be drawn from outside. (AFUE ratings were explained in Sections 6.7*b* and *e*.) Previously, by testing while using combustion air drawn from indoor spaces, manufacturers could show "higher" efficiencies, even though the furnace was clearly heating its own combustion air. The federal standard will require a minimum of 78% AFUE under this uniform testing procedure, as shown in Table 6.17.

Figure 6.54 shows the relationship between a furnace, the duct distribution tree, and some elements of the spaces they serve.

Ducts are constructed of sheet metal or glass fiber and are either round or rectangular. Ductwork will conduct noise unless these suggestions are followed:

Do not place the blower too close to a return grille.

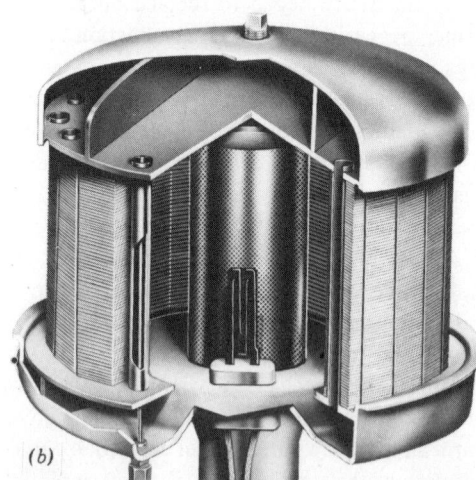

Fig. 6.53 *Furnaces with greatly increased operating efficiencies are now available.* (a) *An Amana gas furnace.* (b) *The small, high-efficiency heat exchanger utilized by the furnace, which recovers heat from exhaust gases. (Courtesy of Amana Refrigeration, Inc.)*

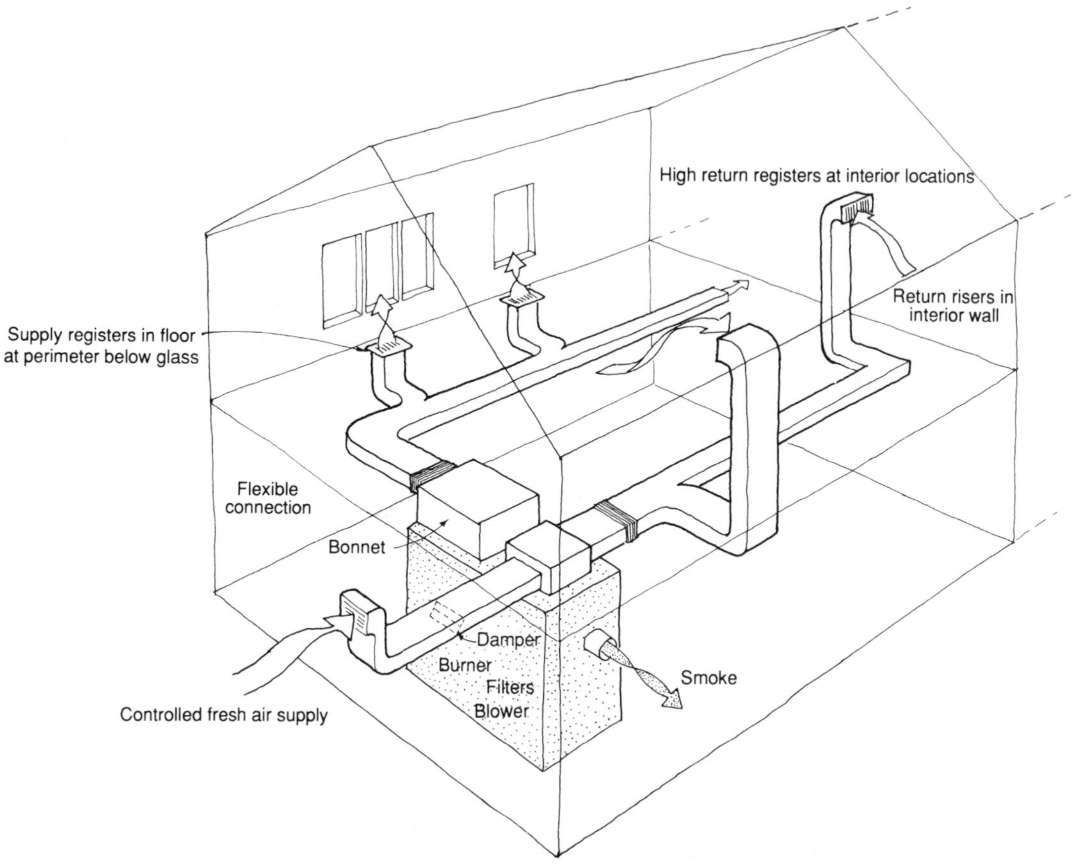

Fig. 6.54 *Conventional warm air furnace and ducts. This system, with supply registers in the floor under glass and interior, high-return registers, is suitable for heating or cooling. Furnace and ducts are in the basement of this basement-and-one-story house. In two-story houses, supply and return registers should be in the same relative positions in each story.*

Select quiet motors and cushioned mountings.

Do not permit connection or contact of conduits or water piping with the blower housing.

Use rubberized canvas flexible connection between bonnet and ductwork.

Ducts also can be lined with sound-absorbing material to further discourage noise transfer.

Duct sizes may be selected on the basis of permissible air velocity in the duct. See Section 7.4 for detailed duct sizing procedures.

EXAMPLE 6.3. The main duct in the low-velocity, warm air system of a residence delivers 1600 cfm. Select a size for this duct.

SOLUTION. Table 6.18 indicates that 800 fpm would be an acceptable velocity. The area of the duct in square inches would be

$$\frac{1600 \text{ cfm} \times 144 \text{ (in.}^2\text{/ft}^2\text{)}}{800 \text{ fpm}} = 288 \text{ in.}^2$$

A 20 × 14-in. (280-in.^{2}) duct is acceptable.

Figure 6.55 illustrates a device that simplifies duct sizing. At a glance, it will show many duct cross-sectional configurations that will satisfy the combined requirements of friction, airflow, and air velocity.

Dampers will be necessary to balance the system and adjust it to the desires of the occupants (see Fig. 6.56). Splitter dampers are used where branch ducts leave the larger trunk ducts. The flow of each riser can be controlled by an adjustable damper in the basement at the foot of

TABLE 6.18 **Recommended and Maximum Air Velocities for Ducts**

Designation	Residences	Schools, Theaters, Public Buildings	Industrial Buildings
	Recommended Velocities (fpm)		
Outdoor air intakes[a]	500	500	500
Filters[a]	250	300	350
Heating coils[a,b]	450	500	600
Cooling coils[a]	450	500	600
Air washers[a]	500	500	500
Fan outlets	1000–1600	1300–2000	1600–2400
Main ducts[b]	700–900	1000–1300	1200–1800
Branch ducts[b]	600	600–900	800–1000
Branch risers[b]	500	600–700	800
	Maximum Velocities (fpm)		
Outdoor air intakes[a]	800	900	1200
Filters[a]	300	350	350
Heating coils[a,b]	500	600	700
Cooling coils[a]	450	500	600
Air washers[a]	500	500	500
Fan outlets	1700	1500–2200	1700–2800
Main ducts[b]	800–1200	1100–1600	1300–2200
Branch ducts[b]	700–1000	800–1300	1000–1800
Branch risers[b]	650–800	800–1200	1000–1600

Source: Copyright © by the American Society of Heating, Refrigerating and Air-conditioning Engineers, Inc., Atlanta, GA. Reprinted by permission from *ASHRAE Handbook Systems and Equipment,* 1967.

[a]These velocities are for total face area, not the net free area; other velocities in table are for net free area.

[b]For low-velocity systems only.

the riser. Labels should indicate the rooms served. Some codes require dampers of fire-resistant material actuated by fusible links, to prevent the possible spread of fire through a duct system (see Chapter 13). Figure 6.56*d* shows how turning vanes can be used to assist airflow at sharp turns in ductwork. Such assistance reduces friction within the ductwork, thus reducing the total static head (Fig. 6.57) against which the supply fan must work.

Supply registers (Fig. 6.58) should be equipped with dampers, and their vanes should be arranged to disperse the air and to reduce its velocity as soon as possible after it enters the room. This is commonly done by providing vanes that divert the air, half to the right and half to the left. When a supply register is in the corner of a room, it is best if the vanes deflect all the air in one direction, away from the corner. Return grilles are of the slotted type in walls and of the grid type in floors. All registers and grills should be made tight at the duct connection. See Table 6.19 for selection of registers based on output and recommended face velocity.

Controls: The burner is started and stopped by a thermostat, which is placed in or near the living room at a thermally stable location that is protected from cold drafts, direct sunlight, and the warming effects of nearby warm air registers. A cut-in temperature of between 80 and 95 F is selected for the fan switch in the furnace bonnet. After the burner starts, the fan switch turns on the blower when the furnace air reaches

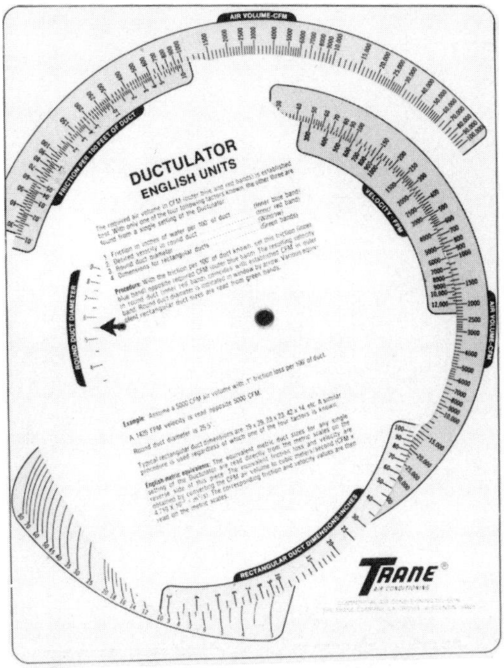

Fig. 6.55 The Ductulator is a duct-sizing device available from the Trane Company. The designer selects any two given factors (e.g., friction and air-flow volume) and the device yields all the other factors (e.g., air velocity, diameter of the round duct required, or combinations of rectangular-duct cross sections required).

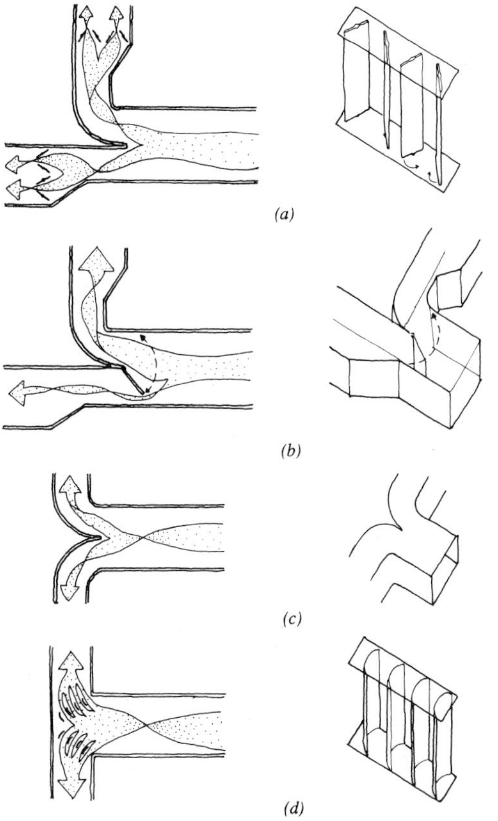

Fig. 6.56 Air controls in ducts. (a) *Air adjustment by opposed-blade dampers.* (b) *Air adjustment by splitter damper.* (c) *Conventional turns in ducts.* (d) *Right-angle turns with turning vanes— a more compact method.*

the selected cut-in temperature. Burner and blower then continue to run while heat is needed. When the burner turns off, the blower continues to run until the temperature in the furnace drops to a level a little below the cut-in temperature of the fan switch. If, during operation, the temperature unexpectedly exceeds 200 F, a high-limit switch turns off the burner in the interest of comfort and safety. As in all automatically fired heating units, a stack temperature control in the breeching cuts off the fuel if ignition fails.

(g) Heating and Cooling Systems. Mechanical cooling is a much more recent development than heating; in its early days, it was widely adopted as a retrofit to existing heating systems. The first two approaches to mechanical cooling reflect this early attitude.

Cooling coils added to warm air furnaces, which are commonplace, utilize a rather simple arrangement of the refrigeration cycle. Figure 6.35 illustrates the circuit of a refrigerant in compression, condensing, and evaporation, in which the condenser heat is carried away by water and the evaporation process draws heat out of water in another circuit to produce *chilled* water. Thus, the heat is *moved* to a heat rejection location outdoors. Figure 6.59 is a schematic diagram of an air-to-air (in contrast to a water-to-water) refrigeration device. Air instead of water can be used to cool the condenser, and indoor air can be cooled directly by being passed over the evaporator coil in which the refrigerant is expanding from a liquid to a gas. Thus, heat is moved from the indoor air to the outdoor air by the step-up action or heat-pumping nature of the

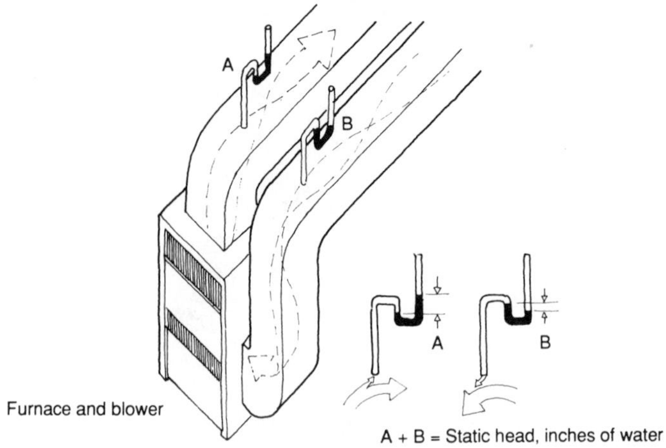

Fig. 6.57 The static head is the pressure, measured in inches of water, available to overcome friction in the entire system.

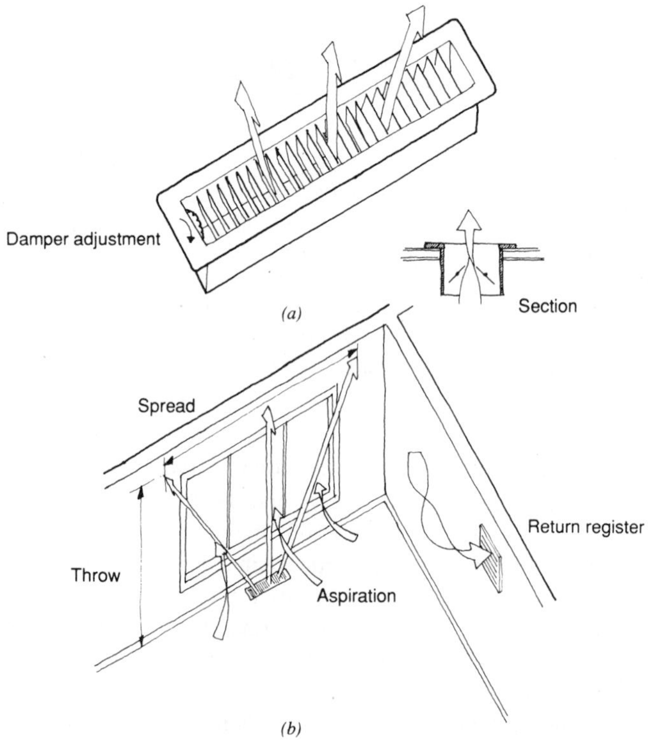

***Fig. 6.58** Floor registers and their action.* (a) *A 2¼-in. × 12-in. floor register (diffuser)—one of many sizes and shapes. It has diverting vanes for "spread" and an adjustable damper. See Table 6.19 for characteristics.* (b) *Concept of spread and throw. By aspiration (suction), cooler room air is induced to join the stream of warm air, resulting in a bland and pleasant airstream that crosses the room.*

refrigeration cycle. When indoor air is cooled directly in this manner, by the expanding refrigerant, the process is usually known as *direct expansion*. The cooling coils therefore are often referred to as "DX coils." Figure 6.60 shows another popular arrangement, in which the airflow through the furnace is horizontal.

Meanwhile, the compressor-condenser unit is placed outdoors, on a concrete slab or on the roof. The unit creates a noisy, hot microclimate in summer—an influence on both site and building planning.

Cooling by air/heating hydronically is the approach taken by, for example, Levitt & Sons in Levittown developments. This system combines a perimeter hot water heating pipe with an overhead air-handling system. A boiler with a tankless coil supplies domestic hot water. The heat

TABLE 6.19 **Forced-Air Registers**

*Part A. **Characteristics of 2¼ × 12–in. Floor Register***

Heating (Btu/h)	3045	4565	6090	7610	9515	11,415	13,320	15,220
Cooling (Btu/h)	855	1280	1710	2135	2670	3,200	3,735	4,270
Cfm	40	60	80	100	125	150	175	200
TP loss	0.009	0.015	0.027	0.037	0.050	0.080	0.105	0.134
Vertical throw (ft)	3	4	5	6	8	10	12	14
Vertical spread (ft)	6	8	10	11	14	17	22	25
Face velocity (fpm)	280	420	565	705	880	1,050	1,230	1,400

*Part B. **Recommended Delivery Face Velocities for Various Applications (Registers)***

Application	*Recommended Velocities (fpm)*
Broadcasting studios	500
Residences	500 to 750
Apartments	500 to 750
Churches	500 to 750
Hotel bedrooms	500 to 750
Legitimate theaters	500 to 1000
Private offices, acoustically treated	500 to 1000
Motion picture theaters	1000 to 1250
Private offices, not treated	1000 to 1250
General offices	1250 to 1500
Stores, upper floors	1500
Stores, main floors	1500
Industrial buildings	1500 to 2000

Source: The *Catalog* of the Lima Register Company.

NOTE: The sound caused by an air outlet in operation varies in direct proportion to the velocity of the air passing through it. The air velocity can be controlled by selecting outlets of proper sizes. The outlet velocities recommended above are within safe sound limits for most applications.

NOTE: For residences (500–750 fpm), any of four velocities of Part A qualify (280, 420, 565, 705 fpm). For stores (1500 fpm), a higher velocity (1400 fpm) would be permissible.

output supplies both the perimeter loop and a coil in the air-handling unit of the duct system. The total heating load is met by the combination of radiant heat generated by the perimeter loop and heated air from the overhead air-handling system.

The perimeter loop consists only of ½- or ¾-in. copper tubing embedded 4 in. below the top of the floor slab to kill the cold slab effect. It has the capacity to maintain a 35 F° differential between the inside and outside temperatures at the perimeter.

The air-handling unit and overhead duct system, incorporating supply outlets in each room and central return, is used throughout the year. Its cooling coil is connected to an adjacent outdoor condensing unit (see Fig. 6.61).

Standardization was one of the goals the Levitt organization aimed at with this approach to heating and cooling. Mechanical engineer John Liebl, the designer of the system, says that its major advantage is the fact that it provides year-round comfort that works with all types of slab, on any terrain, using any fuel. This standardization simplifies design and construction costs.

An important advantage of the system from a performance viewpoint is that problems of short cycling are minimized, as part of the heating load is carried by the loop and air discharge can

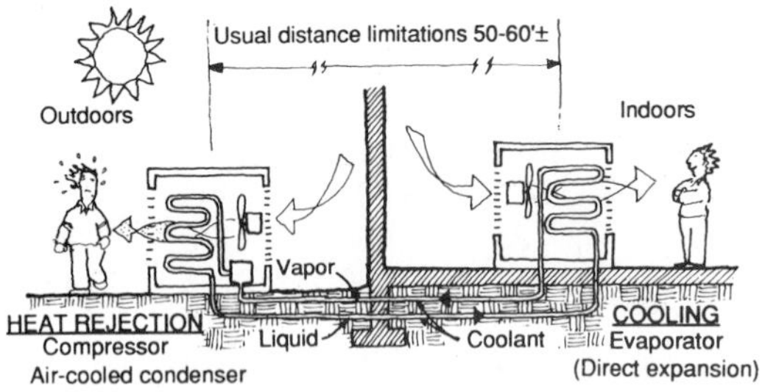

(*a*)

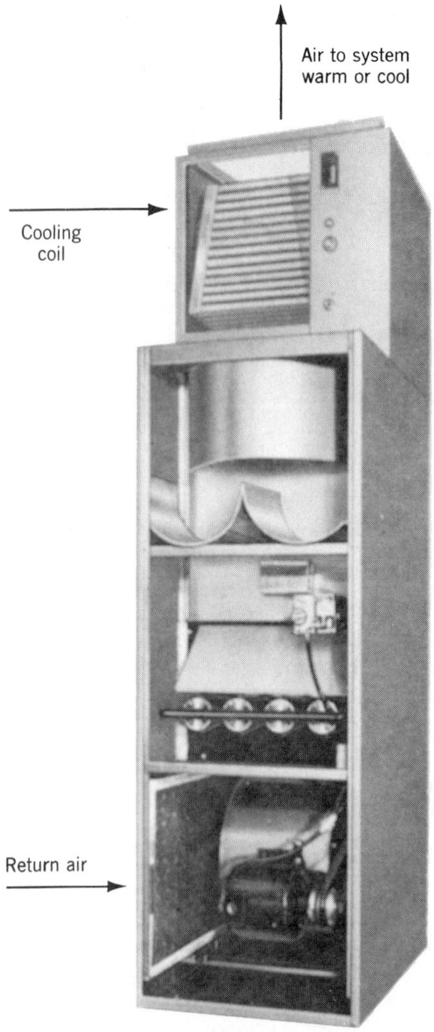

(*b*)

Fig. 6.59 (a) *Schematic diagram of an air-cooled conditioner combining remote outdoor heat rejection and indoor cooling coil suitable for placement in central airstream.* (b) *Cooling/heating air-handling unit. Removal of the front panel reveals upflow circuit. At lower left is air intake and filter; adjacent are fan and motor. Above these components are gas-burning elements. At the top is a direct expansion cooling coil. (Courtesy of American Furnace Division, Singer Company.)*

be kept at about 120 F, or 20 F° lower than with a conventionally ducted system.

With respect to comfort, it is claimed not only that the radiant effect eliminates any cold slab feeling, but also that the temperature variation from floor to ceiling in test houses has not exceeded 4 F°.

The *heat pump* is a single device that uses the refrigeration cycle to both heat *and* cool, thus eliminating the distinction between furnace (or boiler) and DX cooling coils. As shown in Fig. 6.62, heat is "pumped" from indoors to outdoors in summer (Fig. 6.62*a*) and from outdoors to indoors in winter (Fig. 6.62*b*). Heat pumps can transfer heat air–air, air–water, and water–water. The most common application for smaller buildings is the air–air heat pump, shown in Figs. 6.62 and 6.63. Although most of today's heat pumps utilize electricity to drive the compressive refrigeration cycle, there is a growing trend toward absorption cycle heat pumps that utilize natural gas or solar energy. Solar-driven refrigeration is a particularly elegant blend of energy source and task—the hotter the sun, the higher the cooling capacity.

One of the primary attractions of the heat pump is that in its heating mode it can give more energy than it receives (electrically). Although energy (usually electricity) is required to run the cycle, the pump draws "free" heat from a source such as outdoor air. The total heat delivered to the building is more than the heat (electricity) required to run the cycle. The measure of this heat advantage is called the *coefficient of performance* (COP), defined as

$$\text{COP} = \frac{\text{heat delivered to space}}{\text{necessary work input}}$$

(See related discussion in Section 6.7*d*.)

In typical space-heating applications, a seasonal COP of 2 or more is common in mild-winter areas. The energy advantage of heat pumps

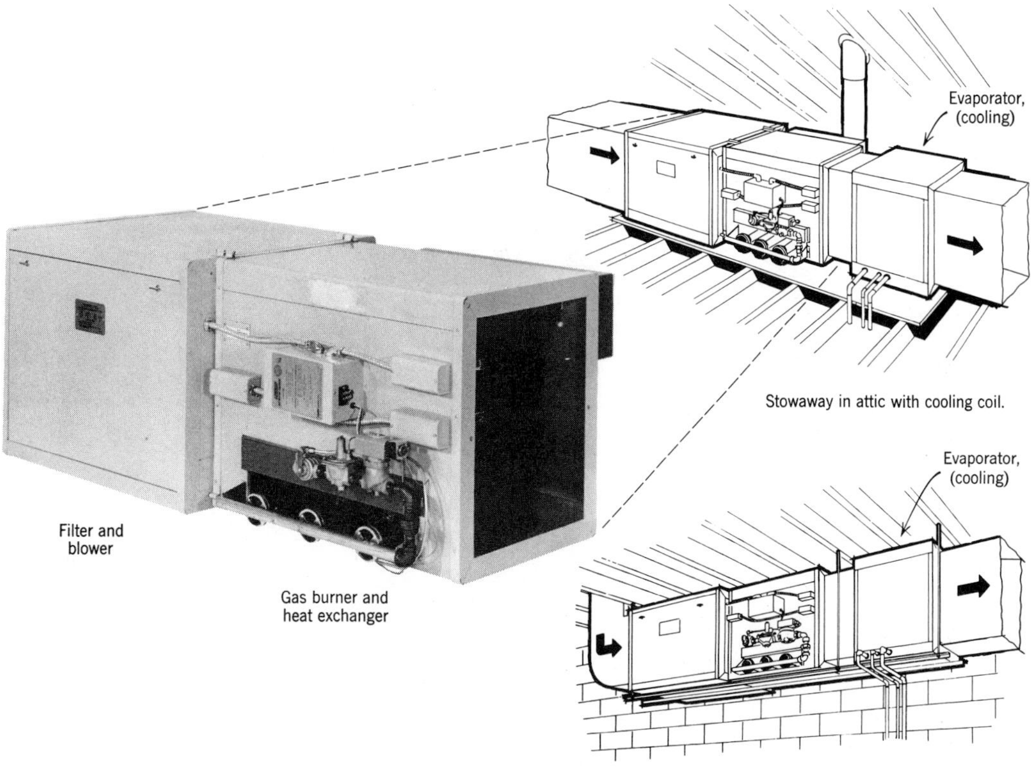

Fig. 6.60 *Compact, horizontal flow combinations for heating and cooling by air. The small pipe between the two* refrigerant *pipes of the cooling unit is a water drain that carries away the condensed moisture from the recirculated and outdoor air. Heating unit will require gas and flue connections. Refrigerant pipes connect to an outdoor compressor-condensor unit.*

that utilize electricity vis-à-vis simple electric-resistance heating coils, is shown in Fig. 6.64, in which a COP of 71/27 = 2.6 is demonstrated.

Because the COP changes with outdoor conditions and indoor load, a *seasonal energy efficiency ratio* (SEER) rating system has been established. SEER measures the number of Btu/h removed for each watt of energy input, averaged over the cooling season. The higher the SEER, the more efficient the heat pump's seasonal performance. SEER ranges are roughly as low as 5, as high as 15.

The heating cycle of the heat pump has a similar rating system, called the *heating seasonal performance factor* (HSPF). Federal minimum standards for SEER and HSPF take effect in 1992 and 1993, as shown in Table 6.17.

As might be expected from a device that draws heat from winter outdoor air. there are limitations to its heating performance. As outdoor temperatures approach 32 F, the COP drops and the outdoor coil tends to ice over. Built-in electric resistance coils must then be used; this, of course, ends the efficiency advantage that made the heat pump attractive. (See Fig. 6.65 for a demonstration of falling performance with falling temperatures.) Because of this characteristic, air–air heat pumps are less frequently used in cold-winter climates. They also generally make poor backup choices for passively solar-heated buildings, since backup sources are typically needed only in the coldest weather. Heat pumps that pump heat from *water* sources, such as wells or solar-heated storage tanks, or from the *ground,* are much more dependable cold weather performers.

Ground–air heat pumps, also called geothermal heat pumps, are being combined with "tri-

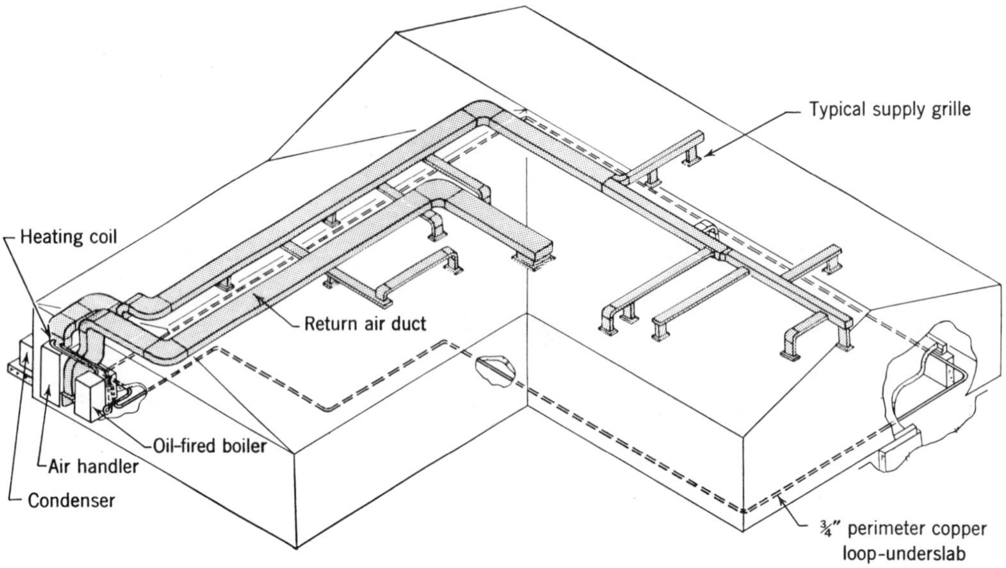

(a)

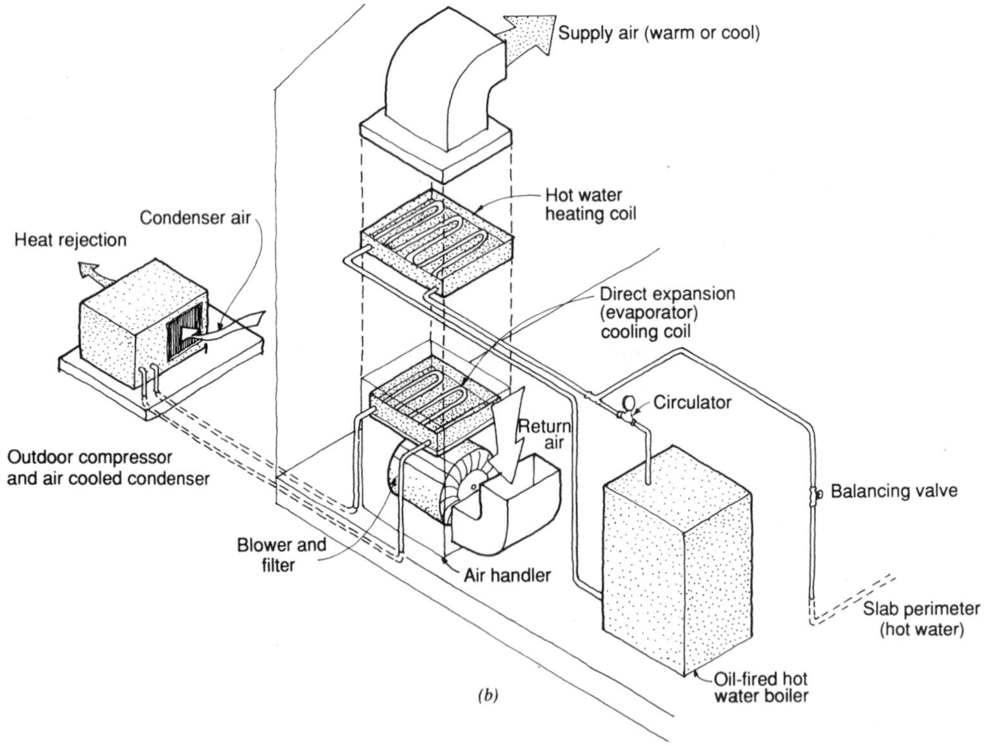

(b)

Fig. 6.61 (a) *Perimeter loop below slab, combined with overhead air system, provides heating and cooling. (Courtesy of Levitt & Sons. Design by mechanical engineer John Liebl. Reprinted from* Air Conditioning, Heating and Ventilating.*)* (b) *Expanded and partial view of the heating and cooling system, illustrating schematically the circulation of water, air, and refrigerant. Heating or cooling coils operate, as called for, to warm or cool the circulated air. This system combines air and water as thermal media with hydronic warming of the slab perimeter.*

ple function water furnaces'' to yield space heating, cooling, and domestic hot water. These units have four operating modes: standard space cooling, space heating, simultaneous space cooling and water heating, and water heating only. The water furnace is part of a closed-loop earth-coupled heat pump. The trenches in which the heat exchange pipes are laid are normally 4 to 6 ft deep; they are up to 400 ft long, depending on how many pipes are in a trench. A rule of thumb for earth heat exchanger size is 500 to 600 ft of pipe per ton (12,000 Btu/h) of heat pump capacity. The Mississippi Power Company reports a 3-ton triple function unit has SEER 13, in contrast to 3-ton air–air heat pumps with typical SEER ranging from 8 to 10.5. The HSPF for the triple function unit is 11.4, in contrast to 3-ton air–air heat pumps with typical HSPF ranging from 6.8 to 7.5.

Heat pumps have a high initial cost, and they have shown a relatively high frequency of compressor failure. Noise from the compressor may affect site planning, especially for residences. Section 7.6 provides numerous examples of multiple heat pumps for zoning within larger buildings, as well as systems that allow one pump's reject heat to become another's needed heat source.

Water–air heat pumps (shown in Fig. 6.62) and solar collectors (Fig. 6.66) make an effective team. As these rather high-COP heat pumps remove heat from the solar storage tank (to deliver it to the indoor air), the resulting lower temperature of the solar-heated water *increases* the solar collector's performance. Assume that on a cold, partly sunny day, the collector being fed water

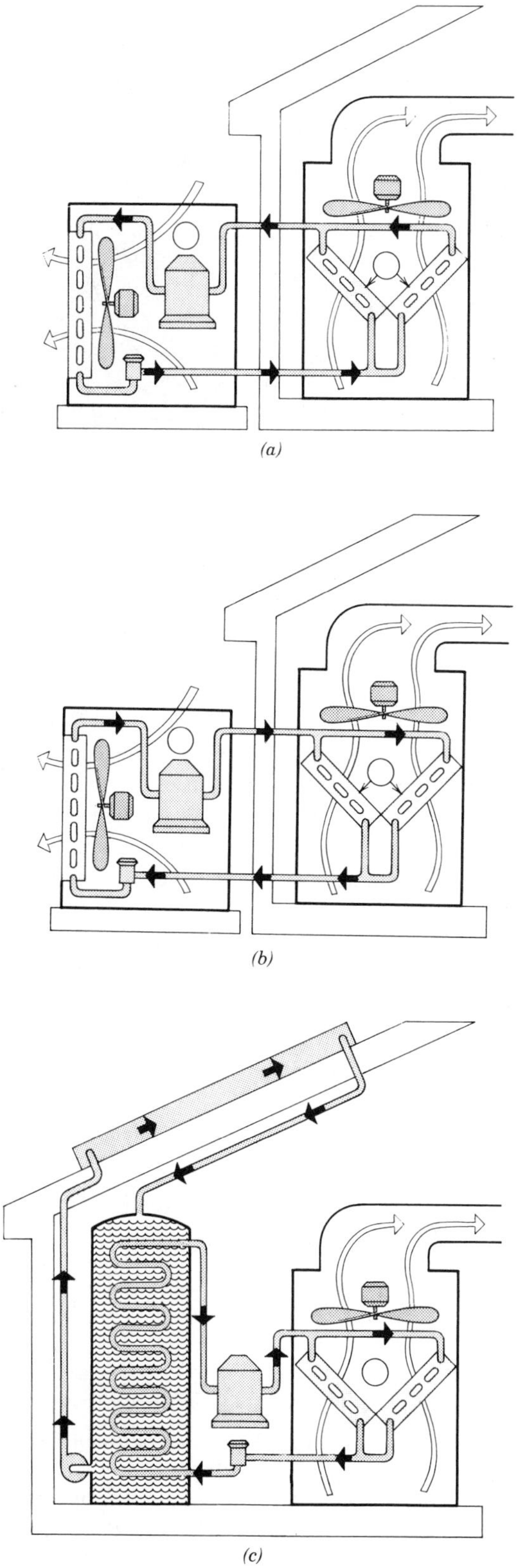

Fig. 6.62 *Applications of the compressive refrigeration cycle: as a simple heat pump, providing cooled air* (a) *or heated air* (b)*; and as* (c) *a device that increases solar collector efficiency by allowing the solar storage tank to operate at low temperatures (thus making it easily heated by the sun, even on cold days). To get usefully high temperatures from the low-temperature solar storage tank, the refrigeration cycle is utilized. (Reprinted with permission from* Popular Science *© 1978 by Times Mirror Magazines, Inc.)*

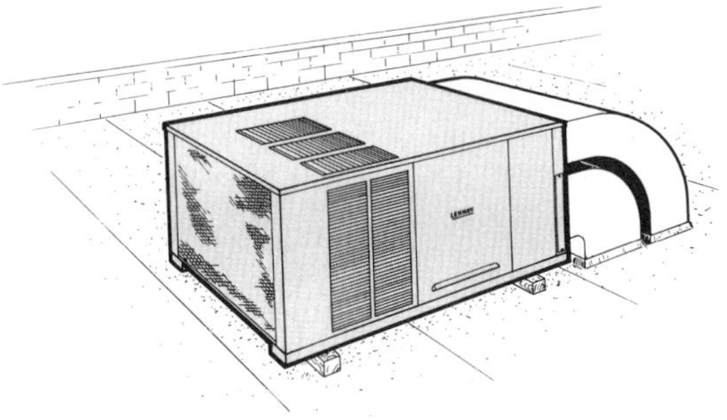

Rooftop installation

(a)

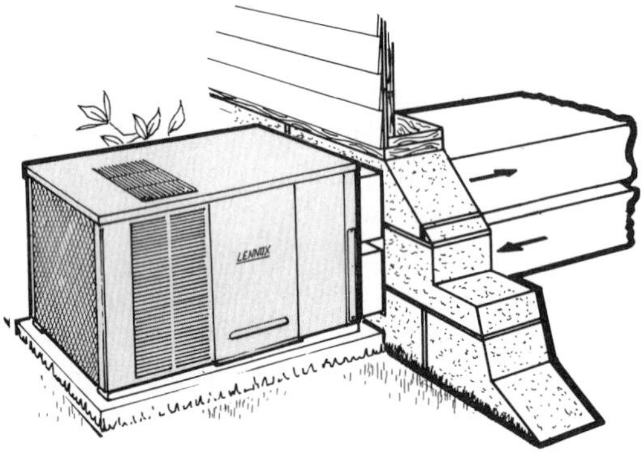

Unit on slab at grade level

(b)

(c)

Fig. 6.63 Heat pumps are available in either single-package or split systems. (a) *Single-package rooftop installation.* (b) *Single-package through-the-wall installation.* (c) *Cutaway view of the Lennox GCS16 5-ton gas heating/electric cooling single-package heat pump, featuring compact size and a "low-restriction" heat exchanger for higher efficiency.* (d) *Cutaway view of the Lennox HP22 outdoor unit of a split-system electric heat pump, with a scroll-compressor that contains far fewer moving parts than conventional compressors. (Courtesy of Lennox Industries Inc., Dallas.)*

(d)

from the tank at 90 F is able to raise its temperature to 94 F. This improvement is slight because of the rather high heat loss that a 94 F collector experiences when surrounded by cold air. If, however, the collector were to be fed 60 F water, its heat loss would be greatly reduced. The heat that the collector *does not* lose to the cold air can be invested in the 60 F water, which will leave at a considerably higher temperature than 64 F. Thus, more solar energy is collected, and

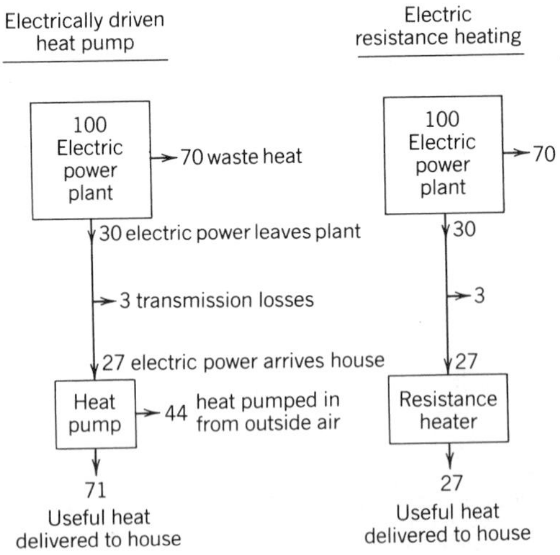

Fig. 6.64 Efficiency comparisons between simple electric resistance heating and the electrically driven heat pump under optimum operating conditions. See Fig. 1.19 for natural gas furnace comparison. [Reprinted by permission from Fisher (1974).]

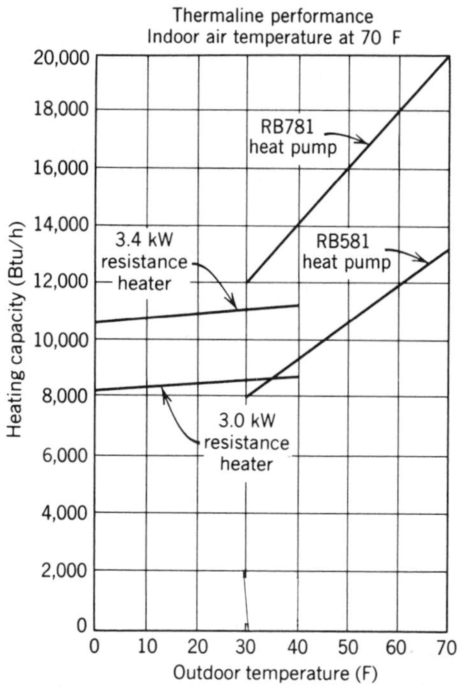

Fig. 6.65 *Typical air–air heat pump operating characteristics.*

more is available for transfer to the building via the heat pump.

(h) Cooling-Only Systems. Before the advent of mechanical air conditioning, cooling was commonly achieved with simple air motion provided by fans. The summer comfort chart shown in Fig. 2.3 encourages increased air motion as a way to extend comfort into air temperatures in the mid-80s F. As a general rule, people will perceive a 1 F° decrease in air temperature for every 15 fpm increase in that air's speed past the body. Ceiling fans are often installed and run at slow speed to destratify warm air at the ceiling in winter; they can be run at higher speed in summer to provide added comfort through increased air motion. The air motion produced by ceiling fans will vary with fan height above the floor, the number of fans in a space, and the fan's power, speed and blade size. Figure 6.67 shows expected air speeds with one 48-in. ceiling fan in a typical residential living room.

The device shown in Fig. 6.68 is perhaps the most commonly seen piece of mechanical equipment in the United States. Perched in windows in full view of passersby, these window-box air conditioners noisily remind us that most of our buildings still are *not* centrally mechanically

Fig. 6.66 (a) *Cutaway section of a typical solar collector.* (b) *Cross section of header.* (c) *Method of connecting manifolds of adjacent solar collectors.*

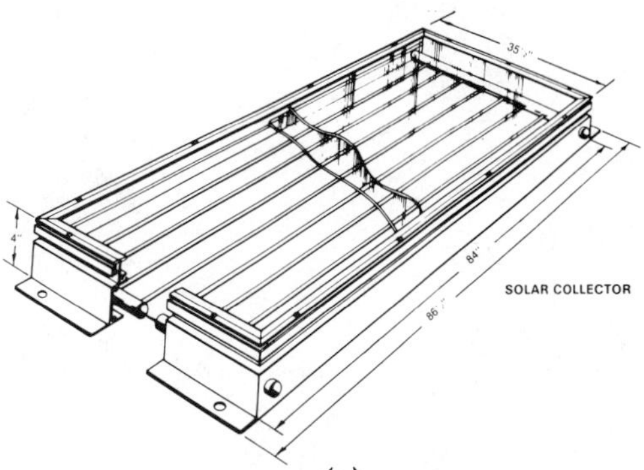

(a)

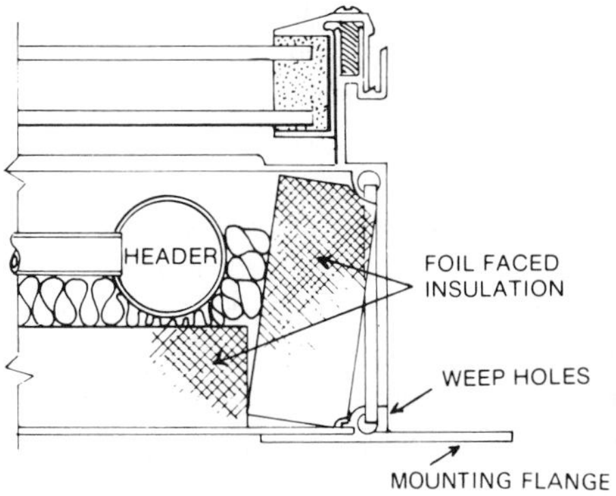

Cross section showing Mounting Flange on Top and Bottom Ends of Collectors.

(b)

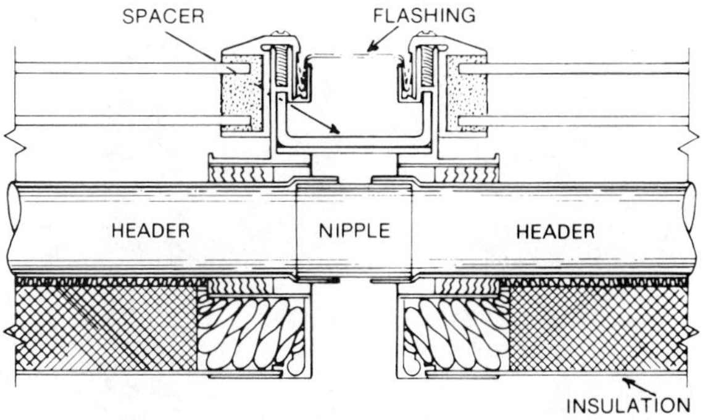

Cross Section Showing Connection of Internal Manifolds of Two Adjoining Collectors.

(c)

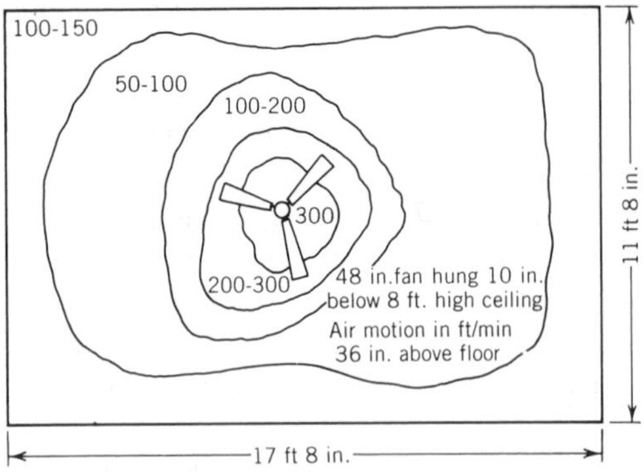

(a)

Fig. 6.67 Ceiling fans are useful at slow speeds in winter for destratifying warm air at the ceiling. In summer (a) *they extend the comfort zone by providing increased air motion at higher air temperatures. Room size is typical of residential living rooms. (Reprinted by permission from the 1989* Handbook of Fundamentals, *© 1989 by the American Society of Heating, Refrigerating, and Air-Conditioning Engineers, Inc., Atlanta, Georgia.)* (b) *and* (c) *A ceiling fan becomes a visual feature of the ceiling in this North Carolina residence. (Christopher C. Morgan, AIA, Architect, Charlotte, North Carolina.)*

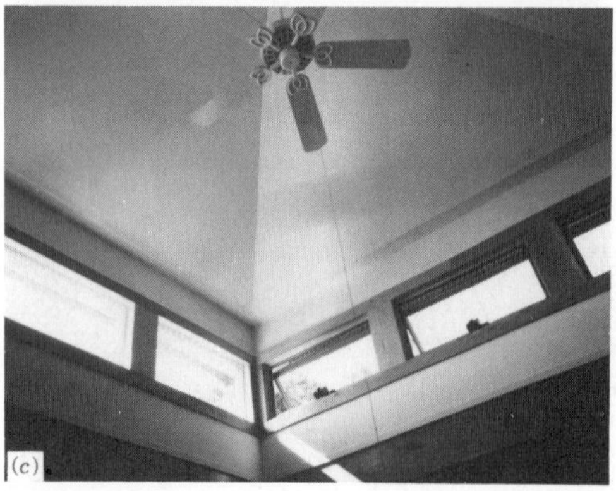

cooled. Mechanical cooling was considered a luxury until well after World War II. Built-in, through-wall air conditioners are still very popular; they offer a low-first-cost way to provide separate zones for individual apartments, motel rooms, and so on (see Section 7.2). In noisy cities, the drone of these units masks street noise for the interior, thus actually helping to promote relaxation. Unfortunately, such scattered units rarely afford the chance to conserve energy through the exchange of waste heat or the higher efficiencies that can accompany larger equipment. But if turned on only when cooling is needed (i.e., when people are present), they can provide substantial savings over the larger always-on systems.

Evaporative coolers (also affectionately termed ''swamp coolers'' and ''desert coolers'') are familiar devices in hot, arid climates (Fig. 6.69). (They are also used in other climates for special, high-heat applications such as restaurant kitchens.) They require a small amount of electricity to run a fan, and some water, to increase the relative humidity of the air they supply to the building. As explained in Section 4.7, the net effect of this device is *no total change* in heat content (enthalpy) of the indoor air; its DB temperature is *lowered,* but there is an *increase* in relative humidity. *People* feel cooler, although no change in total heat has occurred. However, as the air passes through the space, some ''actual'' cooling occurs, as explained in Section 5.9*b*.

The typical evaporative cooler shown in Fig. 6.69 needs full access to outdoor air and is thus

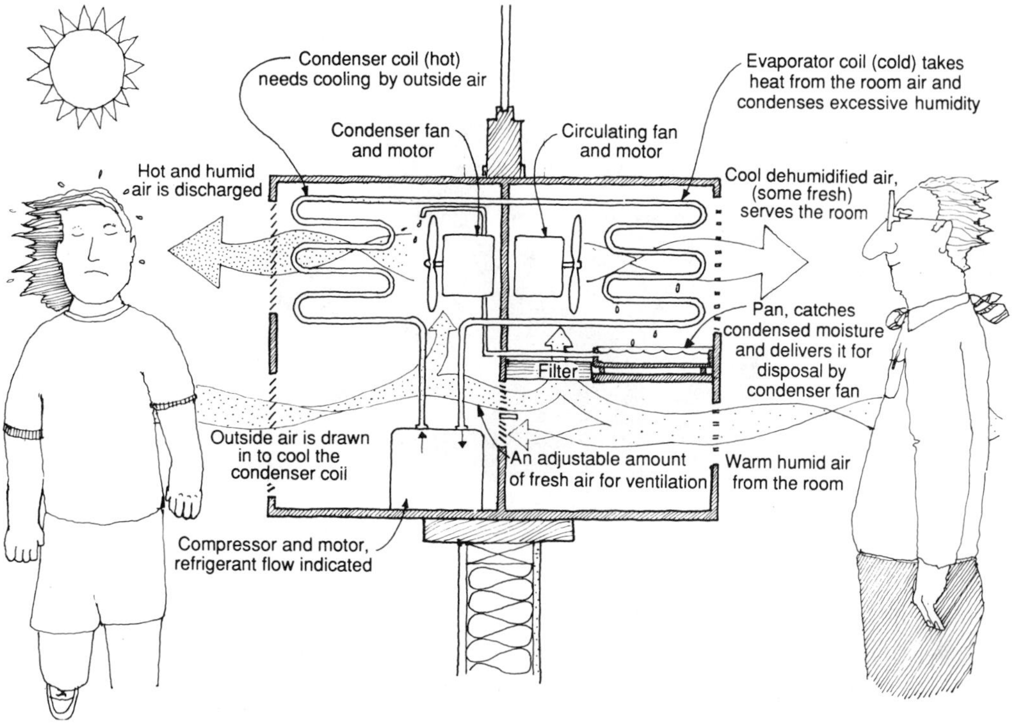

Fig. 6.68 *Schematic diagram of the operation of a through-the-wall, air-to-air conditioning (cooling) unit. Direct heat exchange occurs between air and the process of evaporation or condensation of the refrigerant. The unit is quite self-contained, requiring only access to outdoor air and an electrical connection that powers the motors of two fans and a compressor. The usual capacity is about 1 to 2 tons of refrigeration (12,000 to 24,000 Btu/h).*

often set on the roof; through-the-wall units are also available. Great quantities of dry, hot outdoor air are blown through pads kept moist by recirculated and makeup water. The "cooled" air is then delivered to the indoor space. The effect of the gently moving cool air is to cool the body and, additionally, to produce further cooling by evaporation of body moisture.

Air introduced into the indoor space must be exhausted for the system to operate properly. By selecting the room through which the air is exhausted, one can route cool air as desired in any chosen path from unit to relief opening.

The preceding discussion concerns the most simple application of evaporative cooling, known as the *direct* process (once-treated outdoor air directly introduced to the space). Unfortunately, some areas have such hot summer daytime conditions that such simple evaporative approaches cannot produce comfort indoors. Recently, *direct and indirect* processes have been combined to achieve more "real" cooling and better indoor comfort conditions. One of several such approaches is shown in Figs. 6.70 and 6.71. Warm, rather dry night outside air is evaporatively cooled and fed into a rock bed. The air's temperature is low enough to cool the rock bed, and its relative humidity is moderate. (At the same time, the house is directly evaporatively cooled by a second cooling unit.) Figure 6.71 traces the process by day. Extremely hot, dry outdoor air (*A*) is drawn into the rock bed, where it is cooled by contact (*D*). It can then be passed through an evaporative cooler, to achieve a better combination of rh and DB temperature (*E*). After picking up both sensible and latent heat, the air is exhausted (at approximately temperature *F*).

By comparison, simple direct evaporative cooling by day would have produced indoor sup-

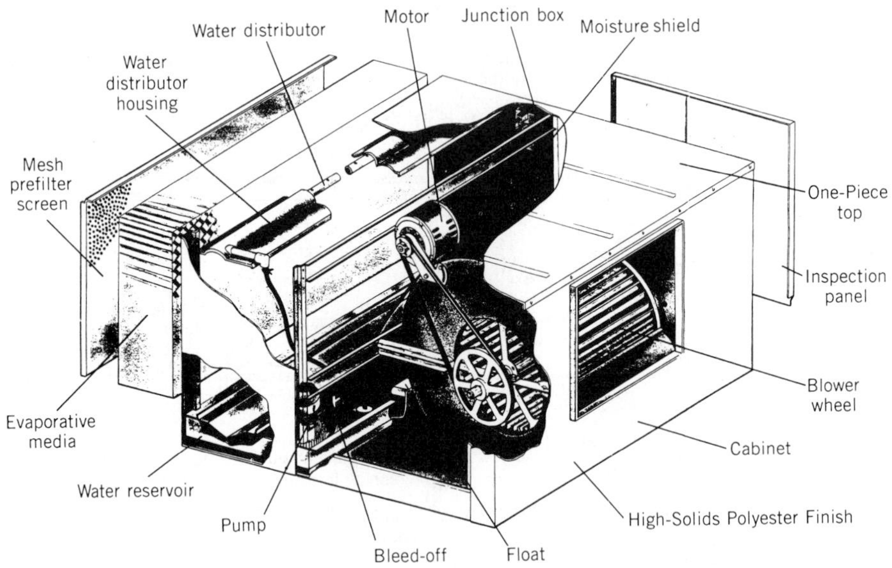

Model	Dimensions (in.) H	W	D	hp	Air Delivery cfm at 0.1 in.	cfm at 0.5 in.
AS50, AD50	28	42	45	1/3	2650	1600
				1/2	3110	2130
				3/4	3630	2806
AS70, AD70	35	42	48	1/2	3750	2610
				3/4	4350	3580
				1	4820	4100

Fig. 6.69 *The UltraCool evaporative cooler* (a) *has a lower profile than the traditional evaporative coolers; the inside components are plastic or stainless steel, to reduce corrosion.* (b) *Performance data; use 0.1 in. where ducts are short. (Courtesy of Champion Cooler Corporation, El Paso, Texas.)*

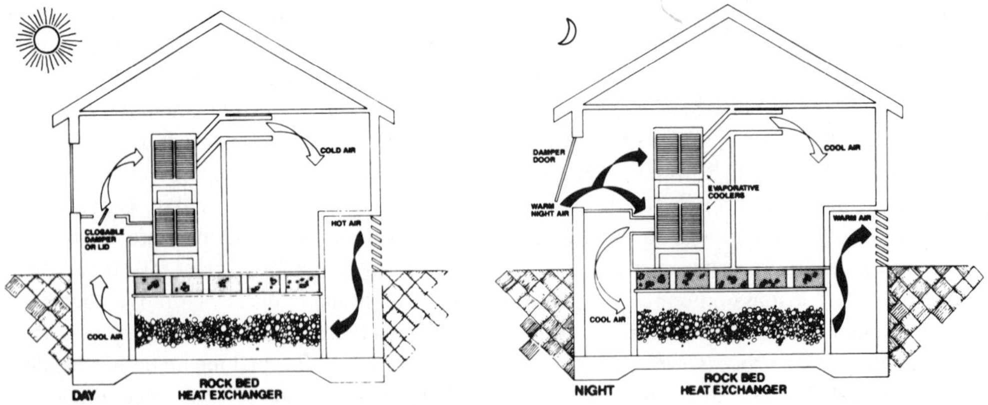

Fig. 6.70 *Direct–indirect evaporative cooling, utilizing a rock bed for storage and heat exchange. (Reprinted by permission of the Environmental Research Laboratory, University of Arizona.)*

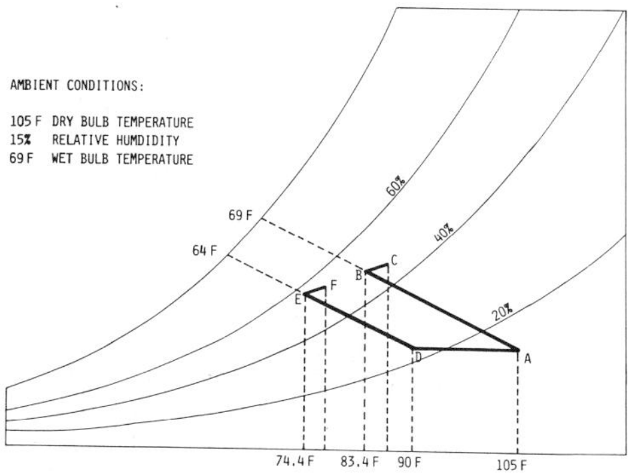

Fig. 6.71 Comparison of direct and indirect evaporative cooling on the psychrometric chart. Compare to the more simple direct process shown in Fig. 5.22. (Reprinted by permission of the Environmental Research Laboratory, University of Arizona.)

ply air too hot and humid for comfort (*B*). Further information on such processes is available from the University of Arizona Environmental Research Laboratory (Tucson International Airport, Tucson AZ 85706).

6.8 Psychrometrics and Refrigeration

The procedure for sizing cooling equipment is more complicated than for heating, because latent as well as sensible heat must be considered. The example following refers to the psychrometric chart, introduced in Fig. 4.8; a review of Figs. 4.8 to 4.16 may be helpful.

EXAMPLE 6.4. Find the total heat to be removed, and thus the refrigeration capacity required, for a dance hall. The design conditions are:

Room conditions (summer)	75 F DB, 50% RH
Number of occupants	80 people
Activity	Dancing
Ventilation provided (some smoking)	35 cfm per person (Table 4.25, Part A requires a minimum of 25 cfm per person)
Conditions, outdoor air	95 F DB, 75 F WB

Heat Gains in the Room	Sensible Heat, SH (Btu/h)	Latent Heat LH (Btu/h)
80 people dancing (see Table 5.22)		
80 × 405 Btu/h	32,400	
80 × 875 Btu/h		70,000
Total transmission and solar gain, and lights, equipment, etc.	67,600	(none)
	Room sensible heat (RSH) = 100,000	Room latent heat (RLH) = 70,000

Total heat gains in room: 170,000 Btu/h (RSH + RLH)

SOLUTION. First, determine the portion of the heat gain that is due to sensible heat gain, called the *sensible heat factor,* SHF:

$$\text{SHF} = \frac{\text{RSH}}{\text{RSH} + \text{RLH}} = \frac{100{,}000}{170{,}000} = 0.59$$

On the psychrometric chart (Fig. 6.72*a*), draw a line between the fixed "bull's-eye" (80 F DB, 50% rh) and the value of 0.59 on the SHF scale, at the upper right edge of the chart. This is called the "SHF line."

Point A is the condition of the "used" air within the dance hall, as it is returned for reprocessing: 75 F DB, 50% rh (62.5 F WB). Next, decide how much cooler the supply air should be than the return air. To avoid uncomfortable drafts, this supply temperature is usually 20 F° or less below the space's air temperature. In this case, choose 15 F°. Then the quantity of air required to cool the room will be:

$$\text{cfm} = \frac{\text{RSH}}{1.08\Delta t} = \frac{100{,}000 \text{ Btu/h}}{1.08(15 \text{ F}°)} = 6200 \text{ cfm}$$

(The factor 1.08 is the constant explained in the natural ventilation formulas in Section 4.6.)

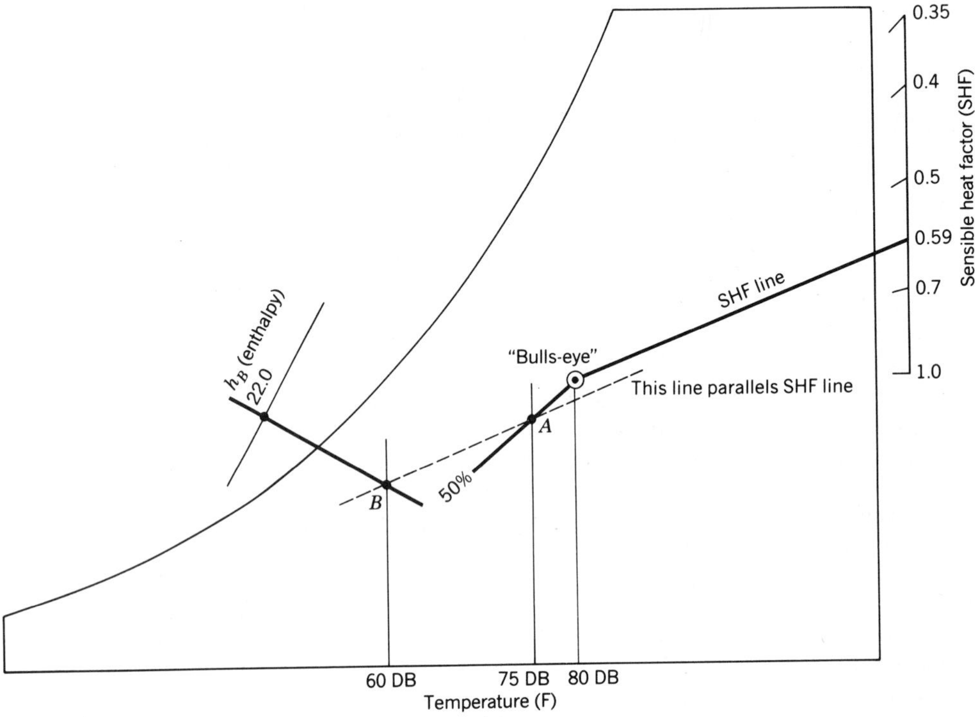

(a)

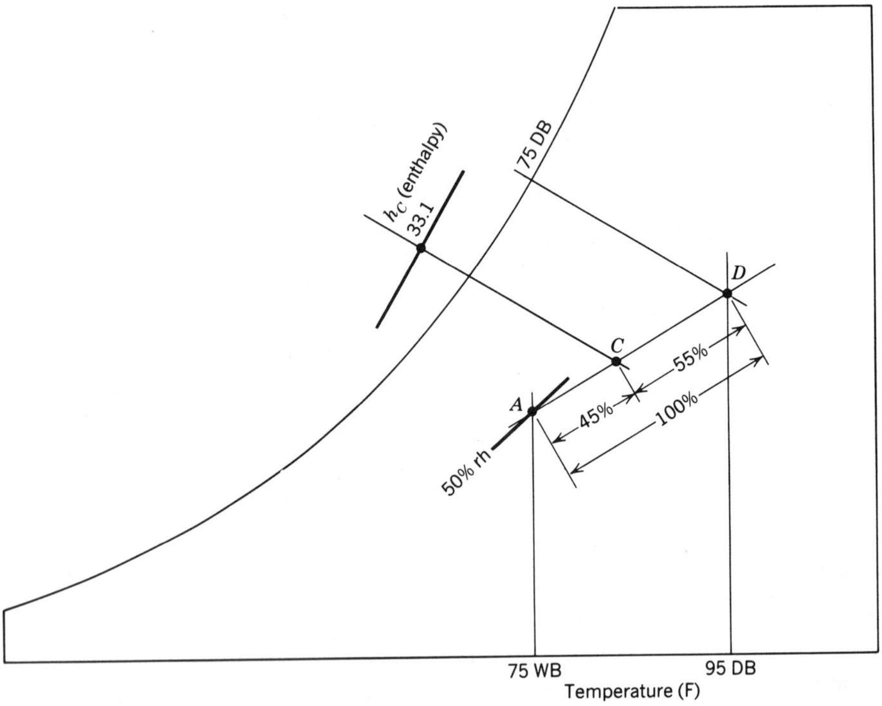

(b)

Fig. 6.72 *Sizing cooling equipment using the psychrometric chart.* (a) *Finding the conditions for the supply air.* (b) *Finding the conditions for the return air–outdoor air mixture.* (c) *Points* A, B, C, *and* D, *shown within the cooling equipment and the building.*

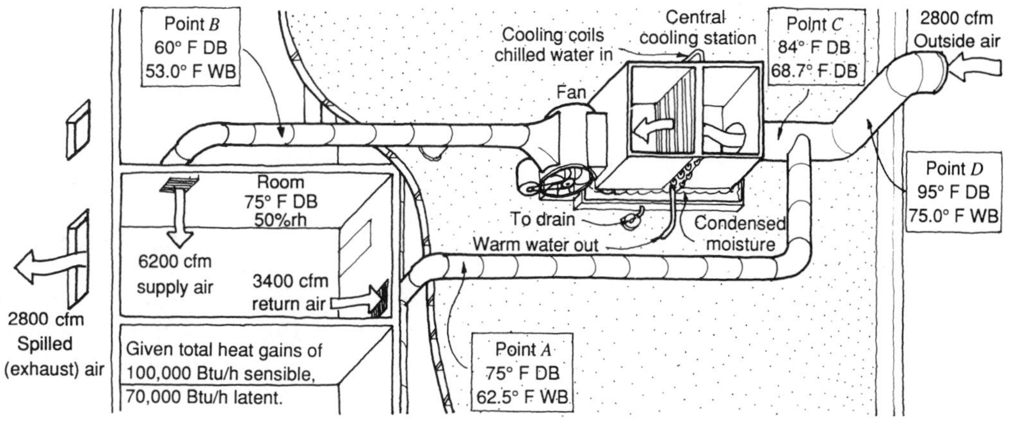

(*c*)

Fig. 6.72 *(continued)*

The portion of this supply air that is *outdoor* air is as follows:

$$80 \text{ people} \times 35 \text{ cfm/person} = 2800 \text{ cfm}$$

So the percentage of outdoor air is

$$\frac{2800}{6200} = 45\%$$

Now, several important points can be located on the chart. *Point B* is the condition of the air entering the rooms; it has been decided that it will be 15 F° cooler than 75 F, which places point *B* somewhere on the 60 F DB line. To determine exactly where, draw a dashed line through point *A*, parallel to the "SHF line," and extend it until it crosses the vertical 60 F DB line. This occurs at 60 F DB, 53 F WB; enthalpy (h_B) = 22.0 Btu/lb. *Point D* is the condition of the outdoor air, given at 95 F DB, 75 F WB.

Point C (Fig. 6.72*b*) represents the mixture of 45% outdoor air and 55% return air that is brought to the cooling equipment for treatment and distribution back to the dance hall. Connect points *A* (the return air) and *D* (outdoor air); then plot *C* at 45% of the distance from *A* to *D*. This occurs at 84 F DB, 68.7 F WB; enthalpy (h_C) = 33.1 Btu/lb.

The cooling equipment must remove the grand total heat (GTH) according to the formula

$$\text{GTH} = 4.5 \times \text{cfm} \times (h_C - h_B)$$

(where 4.5 is a constant = 60 min/h × 0.075 lb/ft^3 average air density). So, in this example:

$$\text{GTH} = 4.5 \times 6200 \text{ cfm} \times (33.1 - 22.0) = 309{,}690 \text{ Btu/h}$$

The size of the required refrigeration unit is specified in "tons," where 1 ton = 12,000 Btu/h.

refrigeration required

$$= \frac{309{,}690 \text{ Btu/h}}{12{,}000 \text{ Btu/h ton}} = 25.8 \text{ tons}$$

Note: If smoking were not permitted, the outdoor air requirements could drop to 25 cfm per person, the minimum required (Table 4.25).

$$80 \text{ people} \times 25 \text{ cfm/person} = 2000 \text{ cfm}$$

The percentage of outdoor air becomes 2000/6200 = 32%.

Point *C* then moves to about 81 F DB, 67 F WB, at which point h_C = about 31.5 Btu/lb.

The refrigeration required then becomes

$$4.5 \times 6200 \times (31.5 - 22.0)$$

$$= \frac{265{,}050}{12{,}000 \text{ Btu/h ton}} = 22 \text{ tons}$$

This represents a first-cost saving in equipment size, and of course energy savings over the life of the dance hall. (Dancers would smell less smoke, and more sweat.)

REFERENCES

American Institute of Architects, Ramsey, C. G., and Sleeper, H. R. (1981). *Architectural Graphic Standards,* 7th ed., Wiley, New York.

American Society of Heating, Refrigerating, and Air-Conditioning Engineers, Inc. (1981). *Handbook of Fundamentals,* and ASHRAE/IES 90.1-1989, *Energy Efficient Design of New Buildings Except New Low-Rise Residential Buildings,* ASHRAE, Atlanta, Ga.

Bowen, A., Clark, E., and Labs, K. (1981). *Passive Cooling,* American Solar Energy Society, Boulder, Colo.

Fanger, P. O. (1989). "The New Comfort Equation for Indoor Air Quality," *ASHRAE Journal,* October 1989, pp. 33–36.

Fisher, J. (1974). *Energy Crisis in Perspective,* New York.

Johnson, T. (1981). *Solar Architecture: The Direct Gain Approach,* McGraw-Hill, New York.

Langdon, W. K. (1980). *Movable Insulation,* Rodale Press, Emmaus, Pennsylvania.

Niles, P., and Haggard, K. (1980). *California Passive Solar Handbook,* California Energy Commission, Sacramento, Calif.

Shurcliff, W. A. (1980). *Thermal Shutters and Shades,* Brick House, Andover, Mass.

Shurcliff, W. A. (1981). *Air-to-Air Heat Exchanges for Houses,* Brick House, Andover, Mass.

SMACNA (1988). *Indoor Air Quality,* Sheet Metal and Air Conditioning Contractors National Association, Inc., Vienna, Va.

7

LARGE BUILDING HVAC SYSTEMS

This chapter continues the discussion that began in Chapter 6, with the focus now on more complex systems for larger buildings with many thermal zones.

Before selecting an HVAC system for a large building: For those using this book for reference more than for a text, you may have opened this book to this chapter. Your choice of an HVAC system should follow preliminary design decisions that are discussed in these preceeding sections:

6.1 Review of the Need for Mechanical Equipment reviews the relative thermal role of building envelopes compared to their internal heating, cooling, and ventilating systems.

6.2 Typical Design Process outlines the decisions made by the architect and the engineer.

6.3 HVAC and Building Organization is an especially important section that discusses zoning, basic components of systems, central versus local systems (also distinguishing between air and water distribution), thermal uniformity versus diversity, and presents a process for comparing HVAC systems and zones (Table 6.2, page 289). It then discusses the location of central HVAC equipment and distribution trees; Fig. 6.7 (page 295) may be especially useful. Several examples of high-rise buildings are presented.

6.4 Passive Heating and Cooling Components includes several approaches that are applicable to larger buildings, such as sunshading, glazing (including air extract windows and switchable windows), and thermal storage. Again several examples of larger buildings are presented.

6.5 Refrigeration Cycles reviews the compressive and the absorptive refrigeration processes.

6.6 Indoor Air Quality Control discusses the relationship between comfort, zoning, and equipment that helps maintain acceptable air quality in buildings.

7.1 Heating, Ventilating, Air-Conditioning (HVAC) System Types

Large buildings have so many thermal zones, and there are so many ways to move heat from one place to another, that hundreds of HVAC systems have been devised. A few of the most typical are introduced in this section; the following section will treat in detail the major components of HVAC production and delivery. Finally, some common variations on each of the four main system classifications will be presented.

One way to classify HVAC systems is by the media used to transfer heat. Although thousands of liquids and gases can be used as carriers of heat, the three most common in building applications are air, water, and refrigerant. Traditionally, there are four main system classifications:

Direct refrigerant systems.

All–air systems.

Air & water systems.

All-water systems.

In the last three cases, the heating/cooling production equipment typically is located centrally in a large building, often rather far from the thermal zones it serves. Distribution tree size and placement thus become important issues when those systems are selected. In direct refrigerant systems, the heating/cooling machine usually is located adjacent to the zone(s) it serves; thus, the machine's environment—the microclimate it creates, and its needs—relative to the zone's environment becomes an important consideration.

(a) Direct Refrigerant Systems. These systems nearly eliminate the distribution trees of air or water, relying instead on a heating/cooling device adjacent to or within the space to be served. The majority of such systems are air–air devices that can be located either on rooftops or on exterior walls—wherever a plentiful supply of outdoor air can be assured. In a *single-package* system (Fig. 7.1*a*), only one piece of equipment is involved. A single-package air–air heat pump, for example, moves heat between an outdoor airstream and an indoor airstream; although kept separate, both streams pass through a *single* outdoor unit. A system with both outside and inside components is called a *split system* (Fig. 7.1*b*). A split-system air–air heat pump moves heat between the outdoor unit (which also contains the compressor), through which outdoor air passes, and the indoor unit (which usually contains backup heating coils) for the treatment and circulation of indoor air. Further details and examples are presented in Section 7.2.

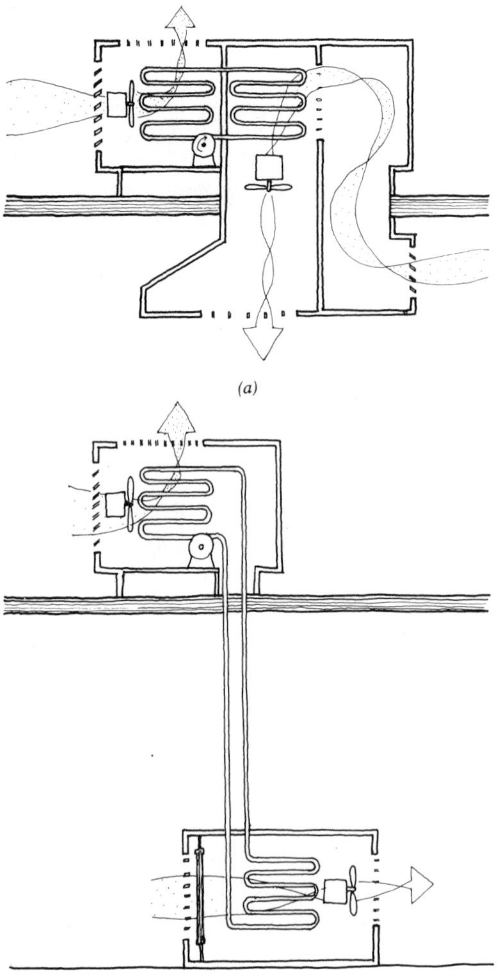

(a)

(b)

Fig. 7.1 *Typical arrangements for direct-refrigerant HVAC systems. In addition to the roof-mounted units shown, the equipment may also be through-the-wall.* (a) *Single package.* (b) *Split system.*

(b) All-Air Systems. The variations on all-air systems are shown in Fig. 7.2. Since air is the only heat transfer media used between the mechanical room (central station) and the zones it serves, and since air holds much less heat per unit volume than water, the distribution trees for this class are quite large. For comfort, however, these systems are, overall, the best. The quantities of air moved through the central station(s) are heated or cooled, humidity-controlled, filtered, and freshened with outdoor air—all under controlled conditions. Within the zones, supply registers and return grilles allow a well-planned stream of conditioned air to thoroughly permeate all work areas. More details on this HVAC class will be found in Section 7.4.

Single-Duct, Variable-Volume (VAV) Systems (Fig. 7.2a). Rapidly becoming the most popular of this class for large buildings, the single duct requires less building volume for distribution, and the variation of air *volume* flow rate (rather than of air temperature) saves energy relative to the single duct with reheat (Fig. 7.2*c*). Depending on outdoor conditions and prevailing indoor needs, the central station supplies at normal velocity either a heated or a cooled stream of air. Automatic volume controls (linked to each zone's thermostat) adjust the volume admitted to that zone, within an air terminal diffu-

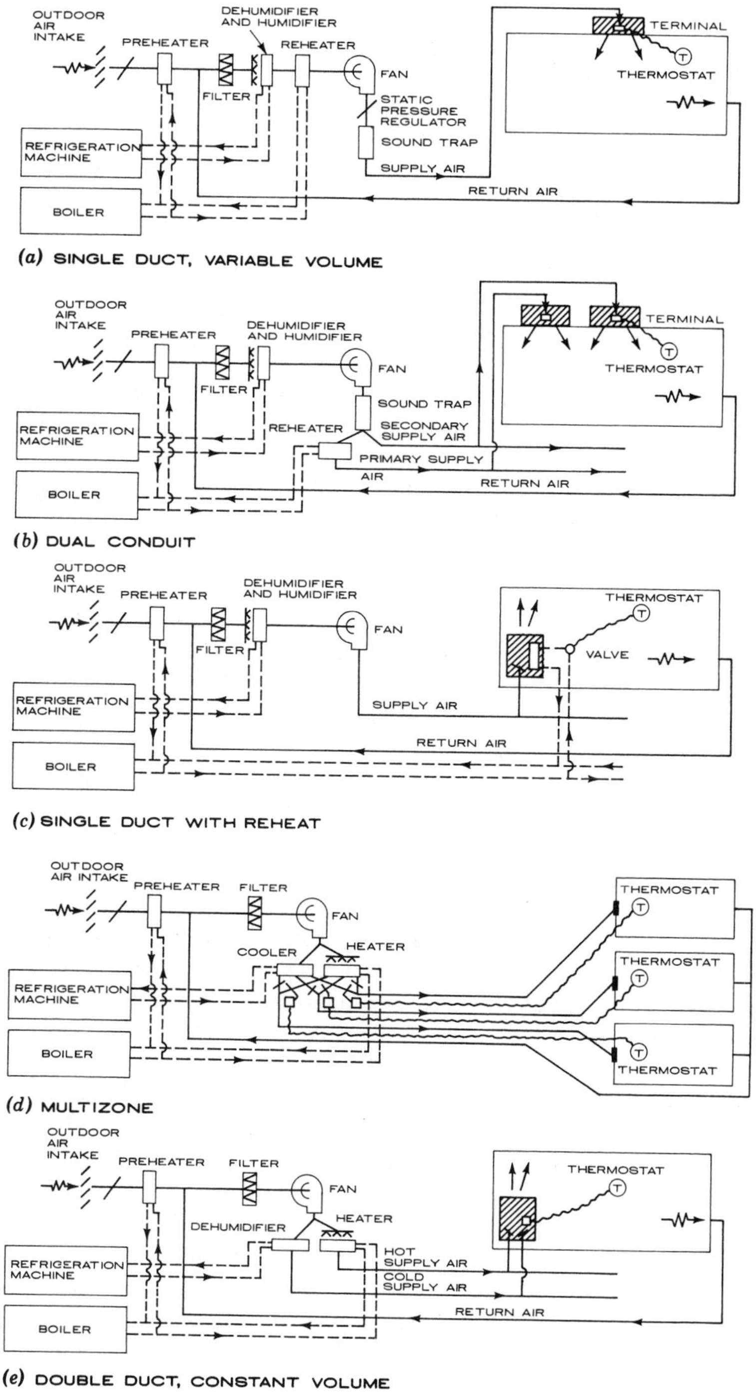

Fig. 7.2 *All-air HVAC system variations. Typical of these systems are large distribution trees, along with complete conditioning of the air. (Reprinted by permission from AIA, Ramsey, and Sleeper,* Architectural Graphic Standards, *7th ed., © 1981 by John Wiley & Sons.)*

ser (often located above a suspended ceiling). When the central station is supplying cold air, a zone that needs more cooling will get more air; an unoccupied room with no internal gains, or a space with heat loss through an exterior wall, will get less air. Clearly, such a system is well suited to serve the interior, always-hot zones of internal-load-dominated buildings. Less clear is its suitability for the perimeter zones of buildings in cold, cloudy conditions.

Dual-Conduit Systems (Fig. 7.2b). In these systems, the space-consuming disadvantages of two distribution trees are partially overcome by high velocity (reducing duct size, but increasing noise and friction). The *primary* air supply deals with heat gains or losses through the building's skin; accordingly, its temperature is variable but its volume flow rate is constant. The *secondary* air supply is as in the VAV constant temperature system—predetermined temperature, but variable volume. The two streams are mixed to order for each zone.

Single Duct with Reheat (Fig. 7.2c). This system (along with VAV) has the smallest distribution tree of this class, because at each zone the only object added to the duct is a small reheat coil (heat provided by steam, hot water, or electric resistance). (Technically, this could be also called an air & water system.) The central station provides a single stream of cold air that must be cold enough to meet the maximum cooling demand of any one zone. All other zones *reheat* this air as needed. In cold weather, outdoor air at temperatures as low as 38 F can be used; the colder this single central airstream, the less air need be circulated (and the smaller the ducts). For large buildings in most U.S. climates, however, the central airstream must be *cooled* most of the time; then, more energy must be spent to reheat the airstream at most zones. These systems thus are notorious for energy wastage, although careful engineering can make them attractive for some climates.

Multizone Systems (Fig. 7.2d). Because each zone has an individual, centrally conditioned airstream, the total distribution tree volume grows to astonishing size with only a few zones. The central station produces both warm and cool airstreams, which are mixed at the central location to suit each zone. These systems are more likely to be found on medium-sized buildings, or on larger buildings in which smaller central stations are located on each floor. The single-return airstream collects air from all zones (as is the case for the other systems in this class).

Double-Duct, Constant-Volume Systems (Fig. 7.2e). Two complete distribution trees are required: at the height of summer the cooling airstream does all the work, whereas in the coldest winter conditions the heating airstream carries the load. Most of the time, air from these two streams is mixed to order at each zone's air terminals. Because both temperature and volume can be controlled, this system offers better comfort under reduced load conditions than does the single-duct, variable-volume system. (Example: an only partially occupied room.) However, it is much more expensive to install, consumes much building volume for the two distribution trees, and is usually more energy consuming than the single-duct, variable-volume system that has largely replaced it.

(c) Air & Water Systems. Several variations on air & water systems are shown in Fig. 7.3. Most of the heating and cooling of each zone is accomplished via the water distribution tree, which is much smaller than that needed by air. For air quality—filtering, humidity, freshness—a small, centrally conditioned airstream, equal to the total fresh air required, is provided. Thus, several distribution trees are involved, yet the total space they require is almost always less than that required by all-air systems.

Exhaust air may be gathered in a return air duct system, making heat recovery possible. Or (a cheaper alternative) air can be exhausted locally, to avoid the construction of yet another distribution tree. If the water distribution provides either heating *or* cooling only, it is called a two-pipe system (shown throughout Fig. 7.3). If it provides simultaneous heating *and* cooling, it is either a three- or a four-pipe system, or yet another heating and cooling variation utilizing heat pumps (see Fig. 7.4). This class of system frequently serves the perimeter zones of large buildings, whereas all-air systems (commonly, single-duct, variable-volume) are used for the interior zones. More details on this HVAC class can be found in Section 7.5.

Induction Systems (Fig. 7.3a). This familiar system's air terminal may be found below windows throughout the United States. A high-velocity (and high-pressure), constant-volume fresh air supply is brought to each terminal,

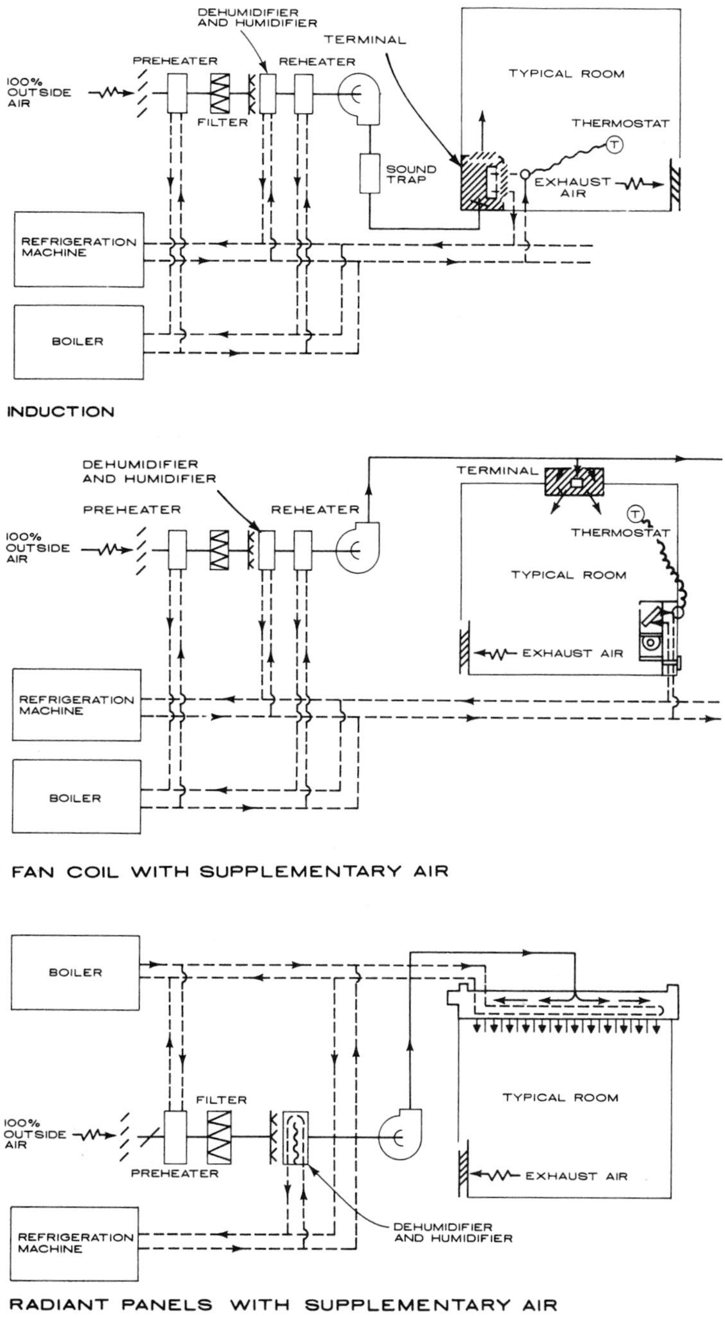

Fig. 7.3 *Air & water HVAC system variations. Although these systems are characterized by the use of several distribution trees, the volume they require is usually less than that required by all-air systems, and they condition the air nearly as thoroughly. (Reprinted by permission from AIA, Ramsey, and Sleeper,* Architectural Graphic Standards, *7th ed., © 1981 by John Wiley & Sons.)*

where it is forced through an opening in such a way that air already within the room (''bypass,'' or secondary, air) is *induced* to join the incoming jet of air. A fairly thorough circulation of room air is thus accomplished with only a little centrally treated air. Air then passes over finned tubes for heating or cooling. Thermostats control the unit's output by controlling either the flow of the water or the flow of secondary air.

Fan-Coil with Supplementary Air (Fig. 7.3b). Another familiar piece of below-window equipment is the fan coil, which moves the room air as it provides either heating or cooling. Cen-

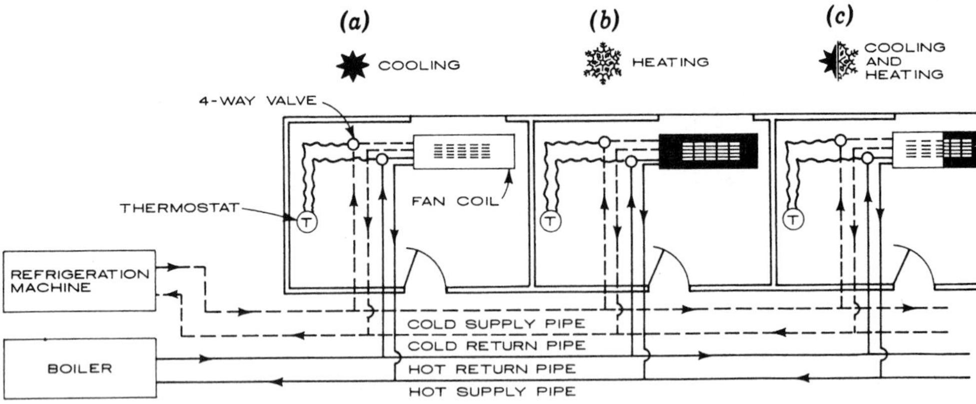

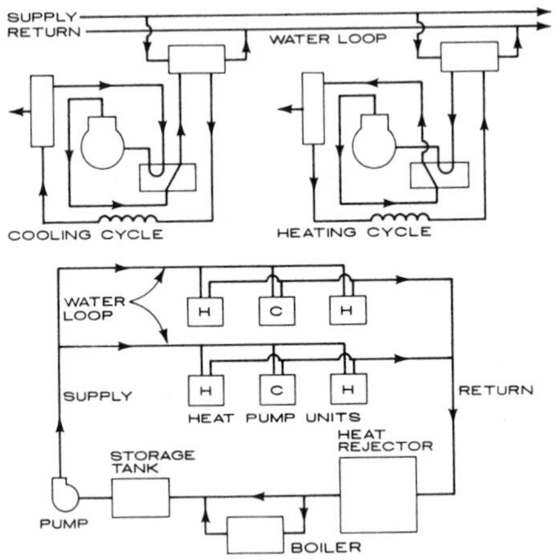

(d)

Fig. 7.4 All-water HVAC systems. (a, b, c) *Four-pipe distribution trees, which need smaller volumes than do those for air systems; however, less-thorough conditioning of the air is provided.* (d) *The water loop heat pump. This unit employs water–air units, allowing energy savings during periods of simultaneous heating and cooling. The heat pumps are usually acoustically isolated from the spaces they serve. (Reprinted by permission from AIA, Ramsey, and Sleeper,* Architectural Graphic Standards, *7th ed., © 1981 by John Wiley & Sons.)*

trally conditioned, tempered fresh air is brought to the space in a constant-volume stream; the fan moves both fresh and room air across a coil that either heats or cools the air, as required.

Radiant Panels with Supplementary Air (Fig. 7.3c). Either ceiling or wall panels contain the heated or cooled water, to provide a large surface for radiant heat exchange. Centrally conditioned, tempered fresh air is brought to the space in a constant-volume stream. The "piece of equipment" within the space is replaced by a large surface, which must be kept clear of obstructions to radiant heat exchange.

(d) All-Water Systems. The more simple-appearing all-water systems are shown in Fig. 7.4. These systems only heat and cool; air quality is dealt with elsewhere—either locally, by means of infiltration or windows; or by a separate fresh air supply system; or simply by fresh air from an adjacent system, such as a ventilated interior zone. A fan-coil terminal is again employed, so that air motion occurs along with heating or cooling. (Sometimes the fan-coil unit is located against the exterior wall, so that fresh air may be brought in and mixed with the room air through the fan.) Both baseboard and valence (above-window) units are also commonly available. More details on this HVAC class can be found in Section 7.6.

Two-pipe water distribution systems are shown throughout Fig. 7.3. They provide either heating or cooling. One pipe is for supply, the other for return.

Three-pipe systems are a first-cost-saving but energy-wasting alternative to four-pipe systems. They allow simultaneous heating and cooling (two supply pipes) but mix the very warm hot return and the mildly warm cold return in only one return pipe.

Four-pipe systems are shown in Fig. 7.4*a, b,* and *c*. They allow quick changeover heating and cooling, utilizing two supply and two return pipes. When simultaneous cooling and heating is required within the same fan-coil unit (Fig. 7.4*c*) a split or double coil is used.

A variation on the two-pipe distribution system is the *water loop heat pump,* shown in Fig. 7.4*d*. Heat pumps (water–air) either draw heat from the loop (heating mode) or discharge heat to it (cooling mode). The loop's temperature ranges between 60 and 90 F; in hot weather, a central cooling tower disposes of the loop's excess heat, while in cold weather a central boiler adds the loop's needed heat. Since individual heat pumps are used, this system is closely related to direct refrigerant systems.

(e) Typical Applications. Although each of these HVAC systems has been used in hundreds of different applications, the most common match between system and function is shown in Table 7.1. Note that several systems do not appear in this table: dual-duct, dual-conduit, four-pipe induction, and radiant panel with supplementary air. These systems are usually more expensive to install and/or operate, and thus less frequently used.

(f) Approximate Space Requirements. A general rule of thumb for the space taken by mechanical and electrical equipment was presented in Tables 5.9 and 5.10. A more detailed idea of space requirements can now be presented in Table 7.2. The percentage of the building's *gross floor area* is shown for equipment rooms, for vertical ductwork, and for the fan-coil units and other in-the-space equipment that typically occupy part of a building's floor area. For other parts of an HVAC system, the percentage of the building's *gross cubage* (volume) is indicated; horizontal ducts, air terminals, and outlets that typically are placed overhead and therefore do not require floor area allocations.

(g) Approximate Air Quantities. Especially for all-air HVAC systems, it is useful to get an early idea of the *total quantity* of air to be moved. Duct distribution trees often strongly influence floor-to-floor heights and the sizes of central core floor areas. Therefore, Table 7.5, in Section 7.4, presents some guidelines for the approximation of these air quantities and of the refrigeration equipment necessary to treat them. It is important to remember that these figures assume *all-air systems* (except as noted) and *normal percentage outdoor air* (ventilation rates).

7.2 Direct Refrigerant (Incremental) Systems

The most distinguishing feature of this approach to HVAC systems is decentralization. As shown in Fig. 7.1 and Section 7.1, these systems replace one or more central machines and associated distribution trees, with many dispersed machines requiring little, if anything, in the way of a distribution system. An analogy might be made between the first factories, which relied on a central power source (such as a water wheel) and distributed the motive power from that source to every machine within the factory via an elaborate system of belts and pulleys. Today, individually powered machines are far more common. And in buildings, dispersed machines are becoming more common as the savings (from elimination of distribution trees) and the convenience (of separate machines for separate zones) become increasingly attractive. Since these machines are easily added as zones proliferate, they are often called incremental units.

One of the simplest of these dispersed approaches is the *makeup air* approach, shown in Fig. 7.5. Especially common in factories or laboratory buildings with high exhaust air requirements, these simple devices either heat or cool the incoming fresh air that replaces air being exhausted. They often supplement the building's main heating/cooling system, which deals with heat gains/losses through the building's skin.

Incremental (or direct refrigerant) units are especially common in building types distinguished by all-perimeter spaces with varying orientations and numerous thermal zones. Motels are a prime example. In Fig. 7.6, separate air–air heat pumps serve each motel room; at best, their constant noise helps mask the intermittent sounds from the adjacent parking lot/circulation space. Opportunities for heat exchange between these heat pumps are scarce; if a central water

TABLE 7.1 **Typical HVAC System Applications**

NOTE: The more expensive systems within each classification—such as dual-duct and dual-conduit (for all-air) and four-pipe induction and radiant panel with supplementary air (for air & water)—may be substituted for those shown below.

	Individual Room or Zone Unit Systems				Central Station Apparatus Systems						
	Direct Refrigerant[a]		All-Water		All-Air				Air & Water		
			Room Fan-Coil		Single Air Stream				Primary Air Systems		
							Reheat				
Applications	Room 0.3–2 Tons	Area 2 Tons and Over	Recir. Air	With Outdoor Air	Variable Volume[b]	Bypass	At Terminal	Zone In Duct	Multizone Single Duct	Secondary Water H-V H-P Induction	Room Fan-Coil With O.A.
Single-purpose occupancies											
Residential:											
Medium	×										
Large		×	×						×		
Variety and specialty shops		×			×						
Restaurants											
Medium		×			×			×			
Large		×			×	×	×	×	×		
Bowling Alleys		×			×	×					
Radio and TV studios:											
Small		×			×	×		×	×		
Large		×			×	×		×	×		
Country clubs		×			×	×		×	×		
Funeral homes		×			×				×		

Beauty salons	×	×			×						
Barber shops	×	×			×						
Churches		×			×	×			×		
Theaters					×	×					
Auditoriums					×	×					
Dance and roller skating pavilions		×			×	×					
Factories (comfort)		×				×		×			
Multipurpose occupancies											
Office buildings					×				×	×	
Hotels, dormitories			×	×	×					×	×
Motels			×								
Apartment buildings				×						×	×
Hospitals				×	×			×		×	
Schools and colleges				×	×	×	×	×			
Museums					×			×	×		
Libraries:											
Standard		×			×	×		×	×		
Rare books		×			×			×			
Department stores					×	×					
Shopping centers		×			×	×			×		
Laboratories:											
Small		×			×			×	×		
Large					×		×			×	
Marine							×			×	

Source: Reprinted by permission from *The ABC's of Air Conditioning,* © 1975, Carrier Corporation.

[a]DX self-contained systems, either package or split-system.

[b]Conventional air outlets are not satisfactory when the variation in air quantity (volume) exceeds 20%, in which case self-controlled linear diffusers are used.

TABLE 7.2 **Typical HVAC System Space Requirements**

For equipment rooms, use the percentage of the entire building's gross area.
For distribution trees, use the percentage of the total served space's gross area (or cubage).

	Percent of Gross Area (or Cubage)					
	Equipment Room		*Distribution Tree*			
			Ductwork Distribution			
System	*Air Handling %*	*Refrig.*	*Riser*	*Horizontal*	*Piping*	*Outlets and Terminals*
Low-velocity conventional[a]	2.2–3.5	0.2–1.0	[b]	0.7–0.9	—	0.07–0.08[c]
High-velocity conventional	2.0–3.3	↓	↓	0.4–0.5	—	0.2–0.4[c]
Terminal reheat (hot water)	2.0–3.3	↓	↓	↓	0.03–0.04[c]	0.4–0.5[c]
Terminal reheat (elect.)	2.0–3.3	↓	↓	↓	—	0.4–0.5[c]
Variable volume		↓	↓	0.1–0.2	—	0.8–0.9[c]
Multizone unit		↓	↓	0.7–0.9	—	0.07–0.08[c]
Double-duct	2.2–3.5	↓	↓	0.6–0.8	—	0.4–0.5[c]
Dual-conduit	2.4–3.4	↓	↓	0.3–0.4	—	0.8–0.9[c]
All-air induction	2.0–3.3	0.2–1.0	↓	0.4–0.5	0.1–0.2	1.5–2.0
Air & water induction		↓				2.0–2.5
2-pipe	0.5–1.5	↓	0.25–0.35[d]		—	2.5–3.0
4-pipe	0.5–1.5	↓	0.3–0.4[d]		—	1.0–1.5
Fan-coil unit		↓		—	0.1–0.2	
2-pipe	—	↓	—	—	0.25–0.3	
4-pipe	—		—			

Source: Reprinted by permission from *The ABC's of Air Conditioning,* © 1975, Carrier Corporation.

[a] "Conventional" refers to simple single-zone systems.

[b] Included in the equipment room area.

[c] Percent of gross cubage (volume).

[d] Includes space for pipe risers.

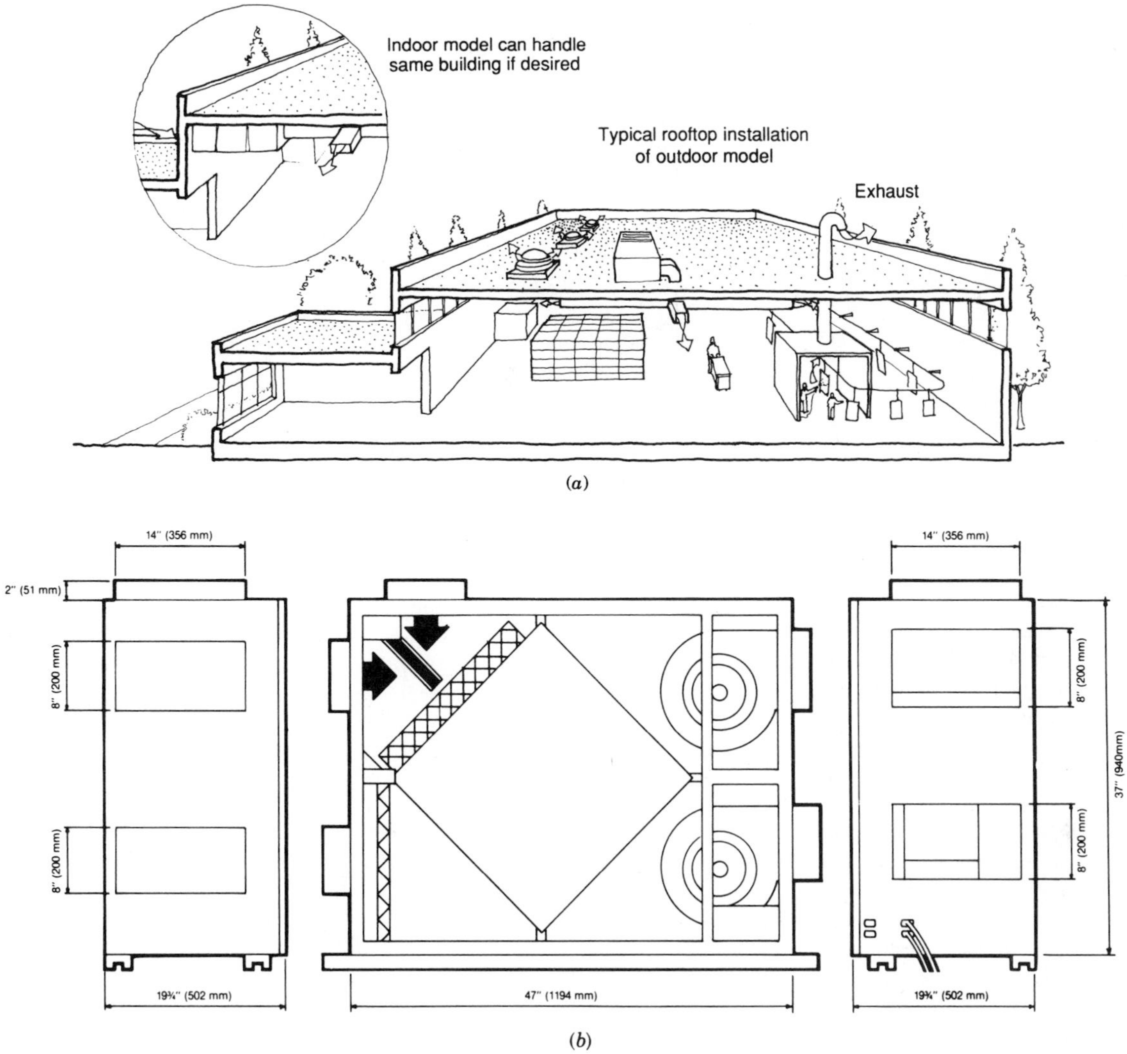

Fig. 7.5 *Makeup air unit heaters/coolers.* (a) *Distributed on a factory roof. These units prevent negative indoor air pressures by providing heated or cooled outdoor air to replace air being exhausted. Often, the building will have an additional (main) heating/cooling system, for overcoming heat losses/gains through the building's skin.* (b) *Heat recovery ventilator utilizing a cross-flow core, serves from 200 to 900 cfm, with supply air entering at 0.6 to 0.8 of exhaust temperature. (© Conservation Energy Systems, Inc., Minneapolis, Minnesota.)*

loop were substituted for outdoor air as the heat source/sink, energy costs would go down (although first cost would rise).

Figure 7.7 shows another incremental air conditioner. Note that the compressor is placed on the outdoor side of the baffle, which helps to muffle the indoor sound that it makes. These units have proven quite popular for apartment houses and motels and are being adapted for use in some office buildings. They are advantageous for several reasons. Control is in the hands of the occupant or tenant, relieving the management of complaints and making every room an individual zone. If a unit needs servicing, it is easy to remove the defective element and insert another. Cooling towers, central chillers, pumps, and piping for chilled and condenser water are all avoided, saving space and making un-

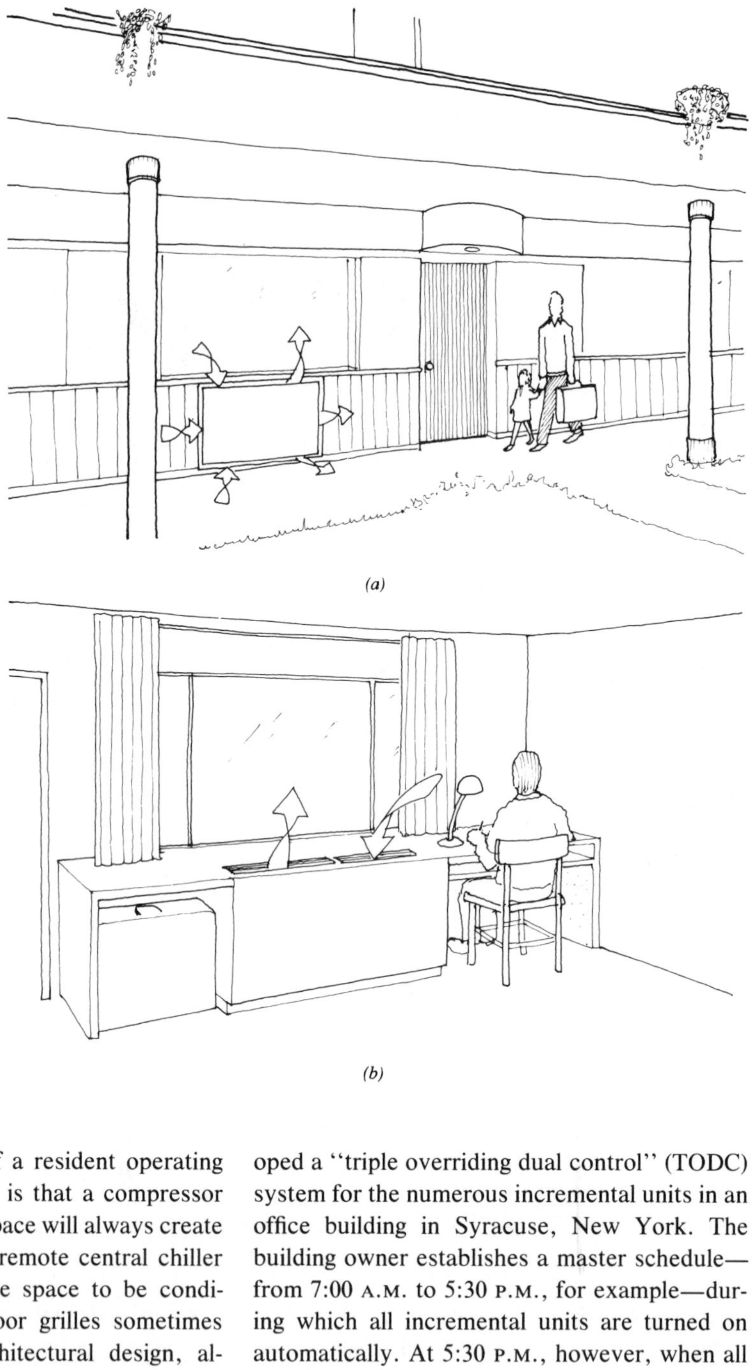

Fig. 7.6 (a) *In this application of the heat pump in a motel, a panel set forward of the exterior wall line allows the pump to inhale and exhale outdoor air around the panel edges. In summer it discharges warm air; in winter, cool air.* (b) *Interior cabinet detail and depth serve the heat pump. Room air is taken back as shown and discharged (after cooling or heating) upward across the glass surface.*

necessary the services of a resident operating engineer. A disadvantage is that a compressor so close to the occupied space will always create some sounds, whereas a remote central chiller would be inaudible in the space to be conditioned. Also, large outdoor grilles sometimes present problems in architectural design, although a repeating element such as three-dimensional grilles (and their shadow patterns cast by the louvres) can provide welcome relief in an otherwise flat facade.

Incremental units easily accommodate the varying schedules of different tenants. Some years ago, the Remington Corporation developed a "triple overriding dual control" (TODC) system for the numerous incremental units in an office building in Syracuse, New York. The building owner establishes a master schedule—from 7:00 A.M. to 5:30 P.M., for example—during which all incremental units are turned on automatically. At 5:30 P.M., however, when all units are turned off, they are all immediately reset by an electric impulse over the regular power wiring. No additional electric or pneumatic controls are required for this—an economy in installation cost. When units are thus reset through the special TODC panel within each unit, a single tenant may turn on his or her

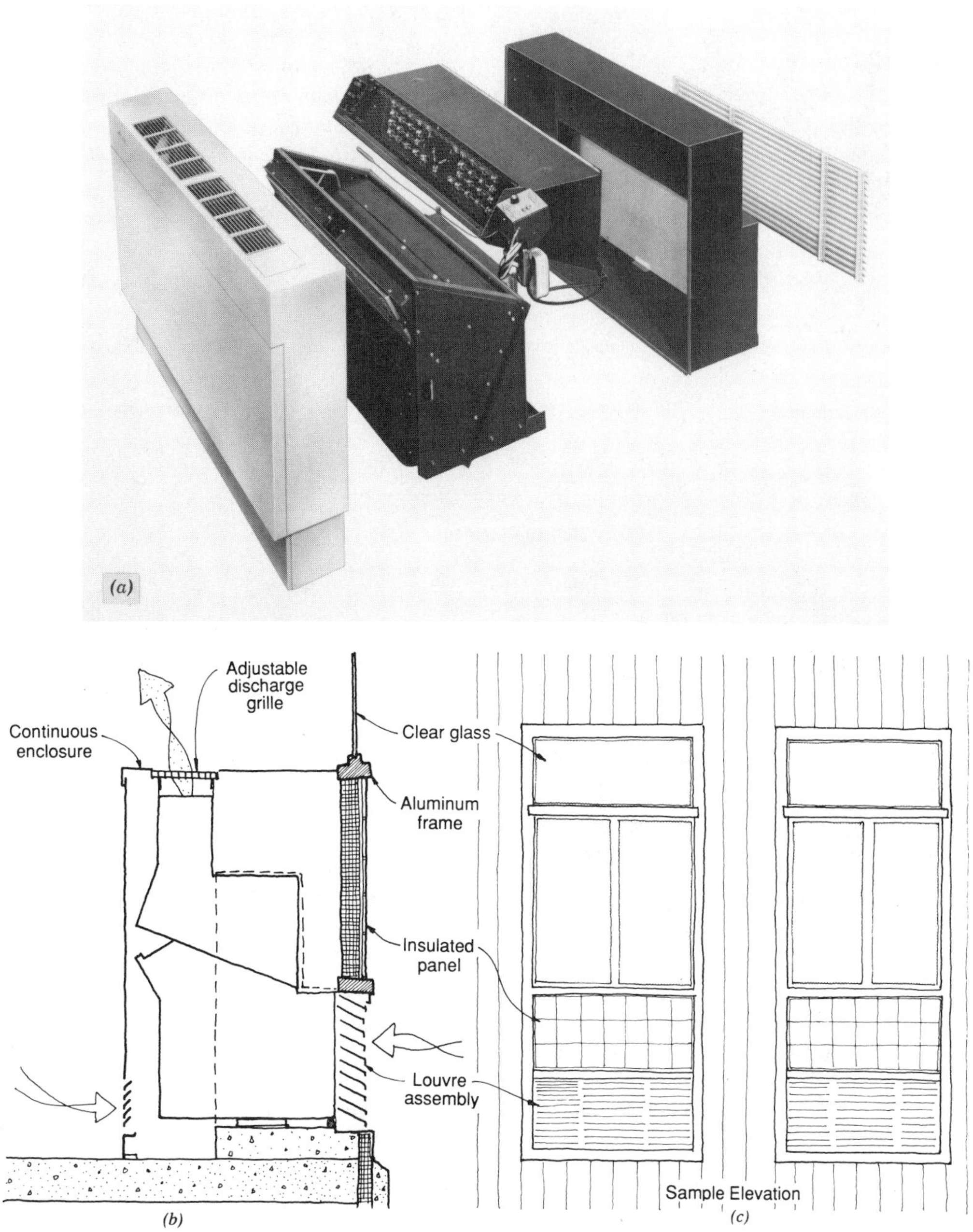

Fig. 7.7 *Incremental direct refrigerant unit.* (a) *Exploded view of the unit: left to right—room cabinet, electric-cooling chassis, room air circulation and electric-resistance heating element, wall box, and outside louver (Courtesy of Climate Control Division, the Singer Company.)* (b,c) *Facade treatment.*

conditioner for full operation by pressing a button. This will be the *only* conditioner operating in the building, unless others are similarly activated. At 9:00 P.M. and 12:00 midnight, impulses again turn off and reset all units then operating. The individual occupants may again press their buttons to continue the service, but if everyone has gone, the entire system shuts down. The

Fig. 7.8 The Vincent Smith School, Port Washington, New York. Budd Mogensen, AIA, architect. South elevation, construction phase. Passive solar heat in winter serves the four classrooms and (through clerestory glass) also warms the student commons area. Daylighting is also characteristic for these spaces, and for the two north classrooms.

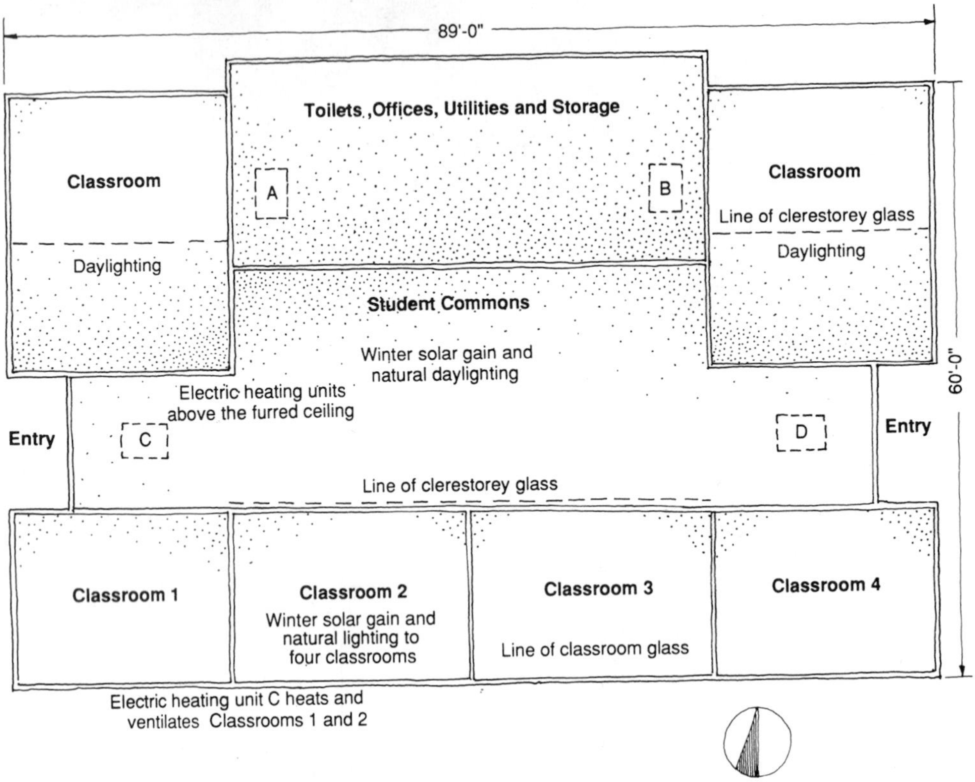

***Fig. 7.8** (continued)* (b) *Space allocations. The four south classrooms and the student commons receive solar warmth and daylighting. The two north classrooms are lighted from above by clerestory fenestration in addition to their first-story glass as shown in* (a). *Shown also are the four climate control units, each of which adds some fresh air to the circulated warm air.*

occupant is only slightly inconvenienced by the need to push a button at three-hour intervals during an evening or on a weekend or holiday.

Another common application for incremental units is in schools. In Fig. 7.8, a passively solar heated elementary school in New York combines several rooms with similar orientations and activities into a zone served by electric heating units above the ceilings. Fresh air is ducted to each unit; mechanical cooling is not provided.

Direct refrigerant (incremental) units are commonly located on roofs, where they have unlimited access to outdoor air and where their noise is less likely to annoy—provided they are sufficiently isolated from the building's structure. This approach is shown in the daylighted, passively solar heated Mount Airy, North Carolina, library (Fig. 7.9). This 14,000-ft^2 building

***Fig. 7.9** Mount Airy Library, Mount Airy, North Carolina. Mazria/Schiff & Associates, architects.* (a) *View from southwest. (Photo by Gordon H. Schenck, Jr.)* (b) *Thermal south–north sections showing natural flow of air and winter heating conditions. Incremental heat pumps are set on the flat-roof sections (left).* (c) *Plan.*

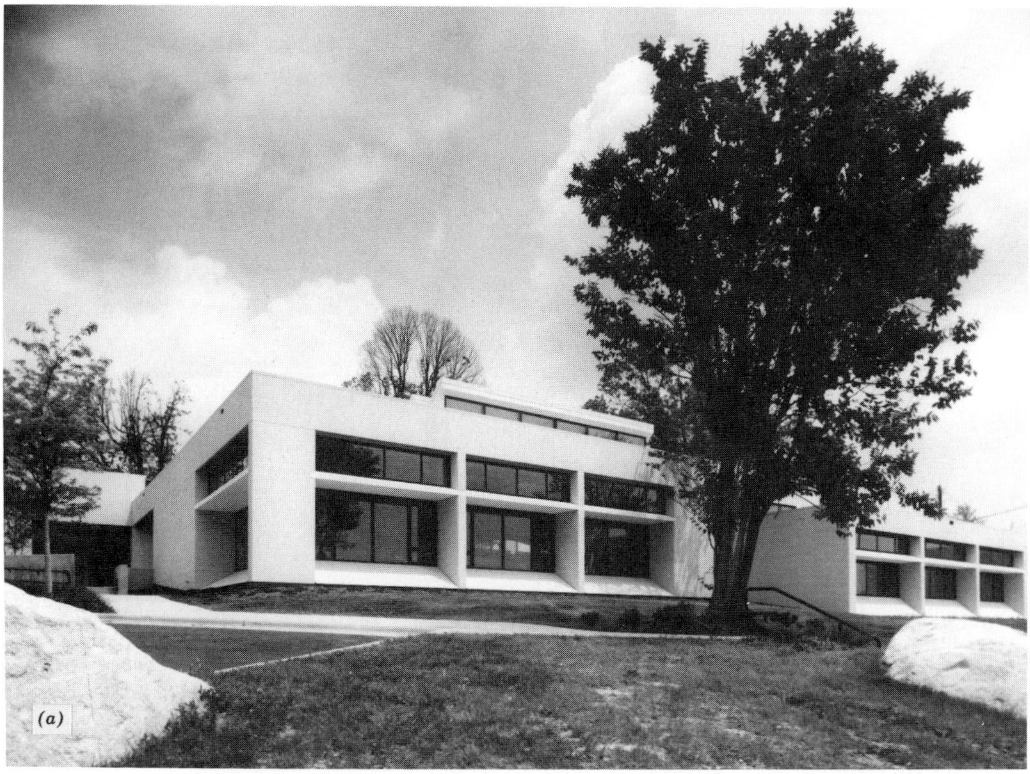
(a)

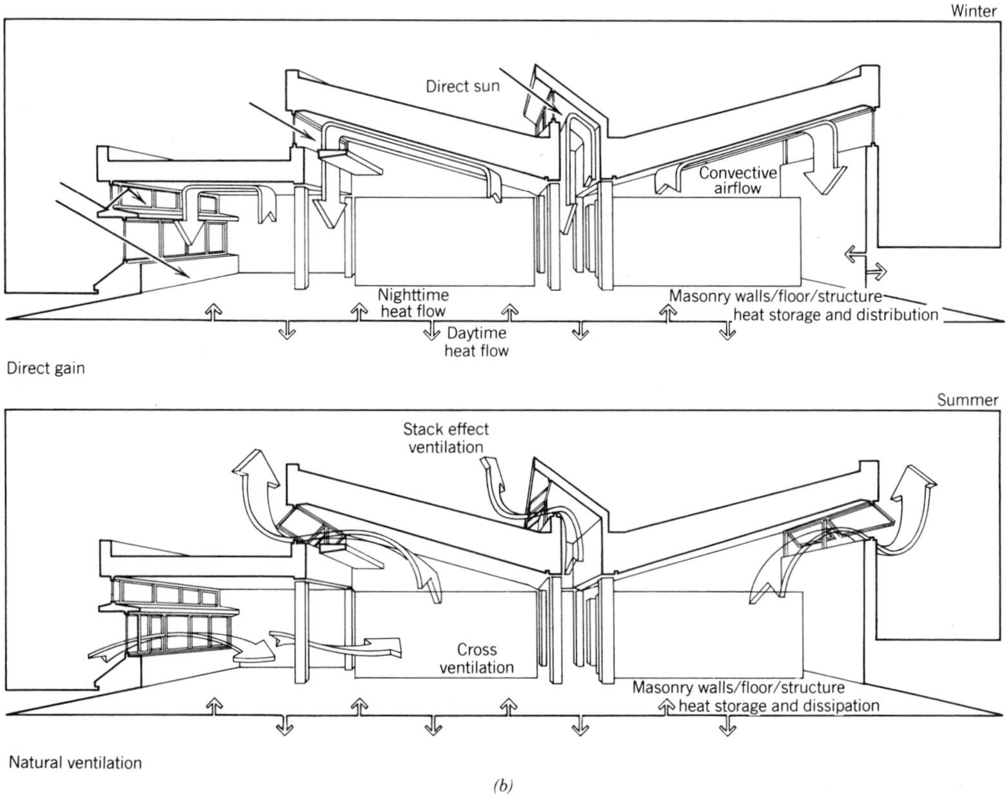
Winter
Direct sun
Convective airflow
Nighttime heat flow
Daytime heat flow
Masonry walls/floor/structure heat storage and distribution
Direct gain
Summer
Stack effect ventilation
Cross ventilation
Masonry walls/floor/structure heat storage and dissipation
Natural ventilation
(b)

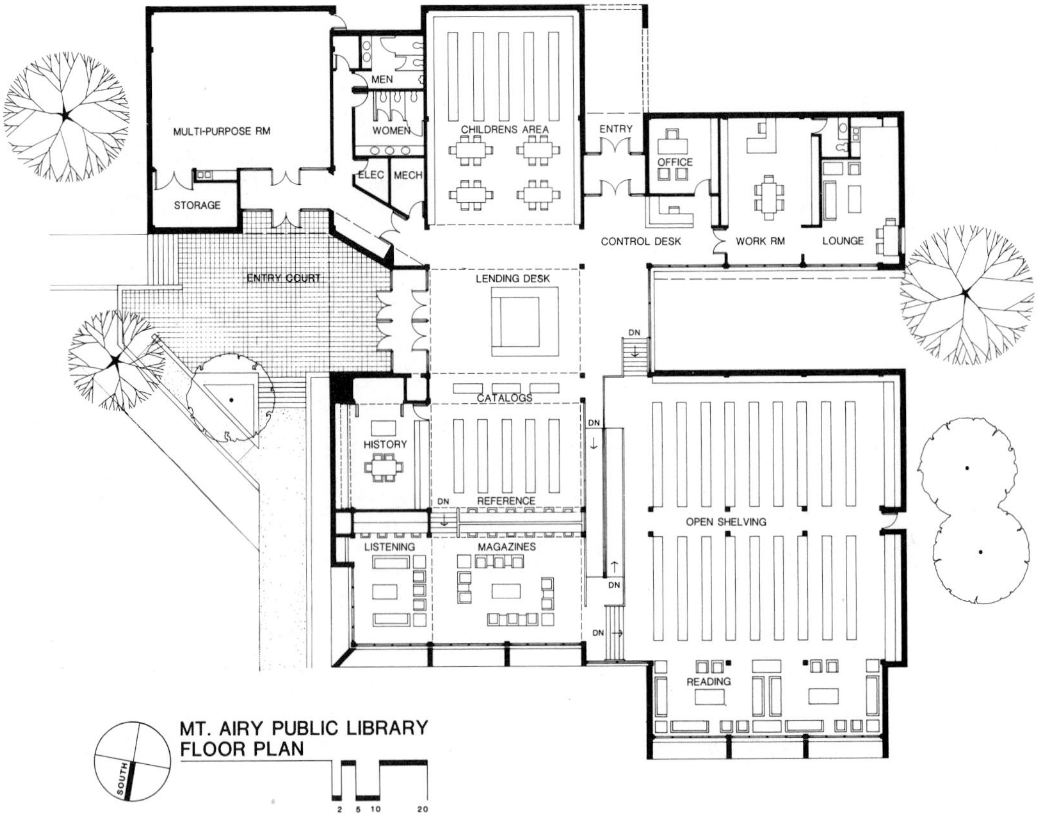

(c)

Fig. 7.9 (continued)

also has a solar preheating system for its hot water. The five heat pumps are air–air and utilize economizer cycles. The average annual building energy consumption has been monitored at about 17,000 Btu/ft^2—approximately one-third of nearby similar-function buildings.

The final example of a direct refrigerant system is provided by an innovative tent structure over a San Francisco department store (Fig. 7.10). Two layers of fiberglass, Teflon-coated fabric, separated by an average of 12 in., are supported by a network of cables hung from eight masts. The tent roof covers about 70,000 ft^2 of sales floor; its 7% translucency to sunlight provides 450 to 550 fc of daylight. This greatly reduces the need for electric lighting, although some, clipped onto exposed fire sprinkler pipes, is still used as accent lighting. About 3.5 W/ft^2 of solar gain, mostly in the form of this diffused daylight, penetrates the tent cover. When this solar gain is combined with heat gains from people and electric accent lights, a cooling load is always generated in San Francisco's mild climate, which rarely falls below 45 F (7°C). The roof's relatively high U value is thus advantageous in helping lose heat. Since San Francisco overheats even more rarely, such a low resistance to heat flow is not seriously disadvantageous in summer.

To remove this heat, four sets of direct refrigerant equipment are provided at the tent perimeter, on grade. These feed into a perimeter plenum, from which the entire store is supplied with cooled air. The exhaust air is collected at the center and returned to help with the task of cooling. This is possible because *indirect evaporative cooling* (also called sensible evaporative refrigeration) units, rather than conventional

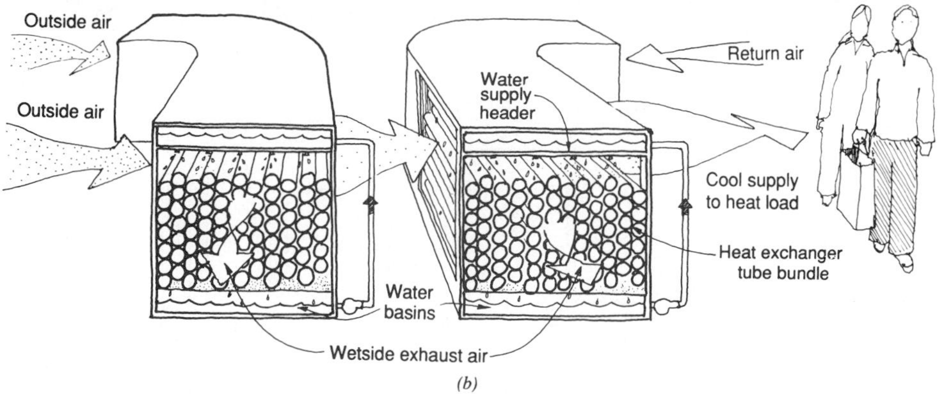

(b)

Fig. 7.10 *Bullock's Department Store at Fashion Island, San Mateo, California. Environmental Planning and Research, Inc., architects; Giampaolo and Associates, Inc., mechanical and electrical engineers.* (a) *The eight-masted white fabric roof, highly reflective to ward off solar heat, transmits some 7%, to provide ample diffuse daylight for sales areas. (Photo by Steve Proehl.)* (b) *Cooling is provided by a two-stage, indirect evaporative cooling system, which uses much less energy than does conventional compressive refrigeration.*

compressive refrigeration equipment, are used to cool air (see Fig. 7.10*b*). In this application, two units work in tandem. Air to be cooled (supply air) enters the heat exchanger of the first unit, which is cooled by evaporative cooling of outside air. As explained in Section 6.7*g*, this process depends on a WB temperature well under the DB temperature, a condition found throughout the West for most of the year. During peak temperature periods, the supply air is only somewhat cooled by this process; sufficiently low temperatures for use on the interior are obtained by passing it through the second unit's heat exchanger. This second unit is cooled by evaporative cooling of the *exhaust* air from the store, which is cooler than outside air under

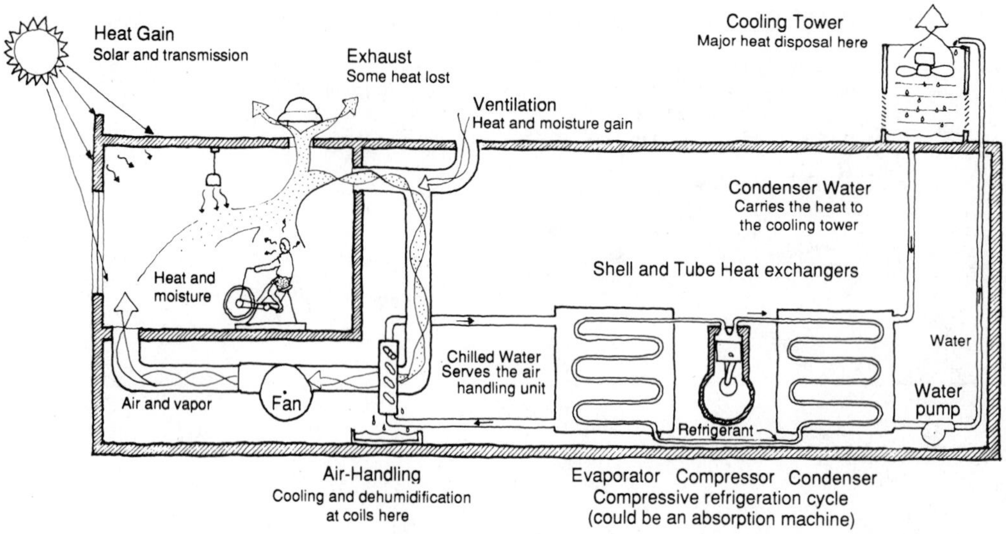

(a)

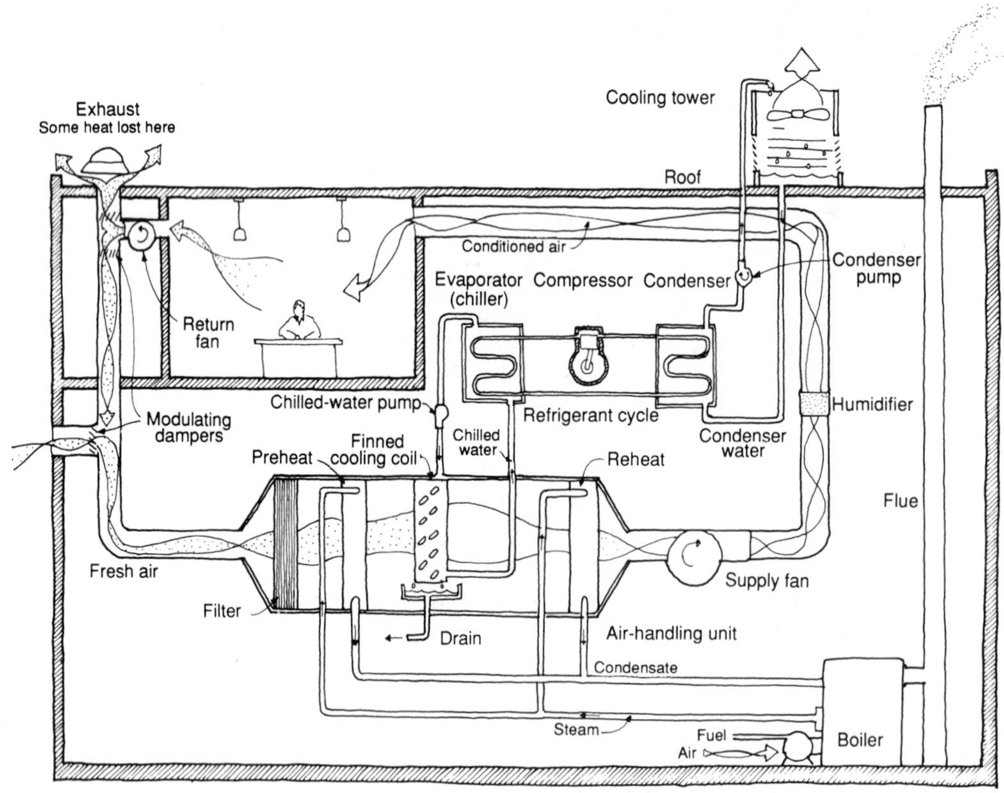

(b)

Fig. 7.11 *Some basic components of HVAC central equipment.* (a) *A simplified diagram of a cooling cycle in which chilled water is circulated to the air-handling coils and heat is disposed of through a cooling tower.* (b) *Schematic diagram of typical major components of central equipment for both heating and cooling.*

summer conditions. Thus the exhaust air does some work beyond the direct cooling of the tent's interior. This two-stage, indirect evaporative cooling process allows the supply air to be cooled without the relative humidity being raised, as would be the case if direct evaporative cooling were used. (Indirect evaporative cooling was also shown in Figs. 6.70 and 6.71.)

7.3 HVAC Central Equipment for Large Buildings

The many HVAC systems that are included in the categories of all-air, air & water, and all-water have in common a dependence on *central equipment* for the generation of heating and cooling, and/or air quality control. Figure 7.11 shows the basic relationships between some of the major pieces of central equipment and the spaces they serve. This section offers a general guide to some central equipment options and sizes.

The National Appliance Energy Conservation Act (1987) established minimum efficiency performance ratings for both small and large heating and cooling equipment. Table 6.17 (page 359) contains such standards for smaller equipment; see the related text for definitions of AFUE, COP, EER, IPLV, and SEER. Table 7.3 contains the standards for larger equipment.

TABLE 7.3 **Minimum Performance of Heating and Cooling Equipment of Larger Size**

Part A. ***Boilers: Gas- and Oil-Fired***

Category	*Rating Condition*[a]	*Minimum Performance*[b]
Gas-fired, ≥300,000 Btu/h	1. Max. rated capacity steady-state	E_c 80%[c]
	2. Min. rated capacity steady-state	E_c 80%[c]
Oil-fired, ≥300,000 Btu/h	1. Max. rated capacity steady-state	E_c 83%[d]
	2. Min. rated capacity steady-state	E_c 83%[d]
Oil-fired (residual),	1. Max. rated capacity steady-state	E_c 83%[d]
≥300,000 Btu/h	2. Min. rated capacity steady-state	E_c 83%[d]

Part B. ***Water Chilling Packages: Water- and Air-Cooled, Electrically Operated***

Category	*Efficiency Rating*	*Minimum Performance*[b]
Water-cooled		
≥300 tons	COP	5.2[e]
	IPLV	5.3[e]
≥150 Tons <300 Tons	COP	4.2
	IPLV	4.5
<150 tons	COP	3.8
	IPLV	3.9
Air cooled with condenser		
≥150 Tons	COP	2.5
	IPLV	2.5
<150 Tons	COP	2.7
	IPLV	2.8
Condenserless, air-cooled, all capacities	COP	3.1
	IPLV	3.2

TABLE 7.3 **Minimum Performance of Heating and Cooling Equipment of Larger Size (*continued*)**

Part C. ***Large Unitary Air Conditioners and Heat Pumps, Electrically Operated*** **>135,000 Btu/h Cooling Capacity**

Category	*Efficiency Rating*[f]	*Minimum Performance*[b,g]
Air conditioners		
Air-cooled	EER	≤760,000 Btu/h: 8.5; >760,000 Btu/h: 8.2
	IPLV	7.5
Water/evap.-cooled	EER	9.6
	IPLV	9.0
Heat pumps		
Air-cooled cooling	EER	<760,000 Btu/h: 8.5; ≥760,000 Btu/h: 8.7
	IPLV	7.5
Air-cooled heating	COP (47 F)	2.9
	COP (17 F)	2.0
Condensing units[h]		
Air-cooled	EER	9.9
	IPLV	11.0
Water/evap.-cooled	EER	12.9
	IPLV	12.9

Part D. ***Warm Air Furnaces and Combination Warm Air Furnaces/Air Conditioning Units***

Category	*Rating Condition*[a]	*Minimum Performance*[b]
Gas-fired ≥225,000 Btu/h	1. Max. rated capacity steady-state	E_t 80%[i]
	2. Min. rated capacity steady-state	E_t 78%[i]
Oil-fired ≥225,000 Btu/h	1. Max. rated capacity steady-state	E_t 81%[j]
	2. Min. rated capacity steady-state	E_t 81%[j]

Source: Adapted with permission from ASHRAE/IES Standard 90.1-1989, *Energy Efficient Design of New Buildings Except New Low-Rise Residential Buildings,* © 1989, American Society of Heating, Refrigerating, and Air-Conditioning Engineers, Inc., Atlanta, Ga. For smaller equipment, see Table 6.17.

[a]Provided and allowed by the controls.

[b]As of Jan 1, 1992; National Appliance Energy Conservation Act of 1987.

[c]E_c = combustion efficiency, 100%-flue losses; see ANSI Z21.13-87 for detailed definition.

[d]See U.L. 726-75, H.I. Heating Boiler Standard 86, and ASME PTC 4.1-64 for detailed definition.

[e]Where R-22, or CFC refrigerants with ozone depletion factors less than or equal to those for *R*-22 is used, these requirements are reduced to 4.7 COP and 4.8 IPLV.

[f]For units that have a heating section, deduct 0.2 from all required EERs and IPLVs.

[g]The levels are minimum performance levels. It is recognized that better energy efficiencies may be available and are encouraged.

[h]Condensing unit requirements are based on single-number ratings defined in paragraph 5.1.3.2 of ARI Standard 365.

[i]E_t, thermal efficiency, 100% flue losses. See ANSI 2.21.47-83 for detailed definition.

[j]See U.L. 727-86 for detailed definition.

(a) Boilers. These devices produce the heat for the recirculating hot water system used for building heating. The type of boiler selected depends on the size of the heating load, the heating fuels available, the desired efficiency of operation, and whether single or modular boilers are to be installed. Boiler sizes are commonly stated either in Btu/h of net output or in (gross) horsepower, where

$$\text{boiler horsepower} = \frac{\text{heating load (Btu/h)}}{\text{\% efficiency of boiler} \times 2247 \text{ Btu/h per horsepower}}$$

Efficiency depends partly on the number of passes that the hot gases make through the water—the more passes, the higher the efficiency. It also depends on burner efficiency and on regular maintenance. Finally, efficiency is best when the equipment is operating near its capacity. Table 5.12, page 230, can be used as a guide to the expected efficiency of boilers, according to fuel source. Figure 7.12 compares typical boiler types, including two- and three-pass boilers. Note that in "water tube" boilers (not shown), the water to be heated is taken through tubes that are surrounded by the boiler's fire. In "fire tube" boilers, the hot gases of the fire are taken in tubes that are surrounded by the water to be heated. In "wet-back" boilers, the boiler fire is always surrounded by water, beginning at the burner nozzle.

Fossil-fuel-burning boilers need flues for exhaust gases, fresh air for combustion, and required air pollution control equipment. The exhaust gas is usually first taken *horizontally* from the boiler; this horizontal enclosure, or flue, is called the breeching. The *vertical* flue section is called the stack. Guidelines for sizes and arrangements of breeching and stacks are shown in Fig. 7.13. Local codes determine the quantity of air required for combustion; local air pollution authorities set pollution control requirements. As a general rule, combustion air can be supplied in a duct to the boiler, at an average velocity of 1000 fpm. The duct should be large enough to carry at least 2 cfm per boiler horsepower. Furthermore, ventilation air to the boiler room should be provided; preferably, the inlet and outlet should be on opposite sides of the room. Minimum sizes: enough for 2 cfm per boiler horsepower, at a velocity of about 500 fpm.

Space requirements for boilers are summarized in Fig. 7.14, in which multiple boilers are shown. Note that clear space within the room must be provided, so that the tubes of the boiler can be pulled when they must be replaced.

Several types of single boilers are discussed below. The final boiler type discussed, the modular boiler, is preferred for energy conservation.

1. *High-output, package-type steel boiler.* For large buildings that use steam as a primary heating medium, one or several such boilers may be used. Direct use of steam can be seen in Fig. 7.11, supplying preheat and reheat coils and also a humidifying unit. The relative lightness of this boiler type, compared to the older styles with ponderous masonry bases (boiler settings), makes it suitable for use on upper floors of tall buildings. Figure 6.4 shows two such boilers on the thirteenth floor of the Fox Plaza Building.

2. *Converter, steam to hot water (Fig. 7.15).* When, in a building that uses primary steam boilers, secondary circuits that use hot water for heating are required, a converter is used. It is considered a heat exchanger. Figure 6.4 shows downfeed steam supply for the two boilers on the thirteenth floor to two such converters, one for hot water heating in the apartments and one below the garage ceiling for hot water heating in the commercial area. The secondary (hot water) circuits are not detailed in that illustration. A converter may also be used to transfer heat from steam to *domestic* (service) water. Converters are frequently used where central steam supply systems are available, as in large-city downtown areas. The easier, quieter distribution of heat by hot water has largely replaced steam heating distribution trees.

3. *Electric boilers (Fig. 7.16).* Where electric rates are competitive with those of fossil fuels, electric boilers are sometimes used. Both hot water and steam electric boilers are avail-

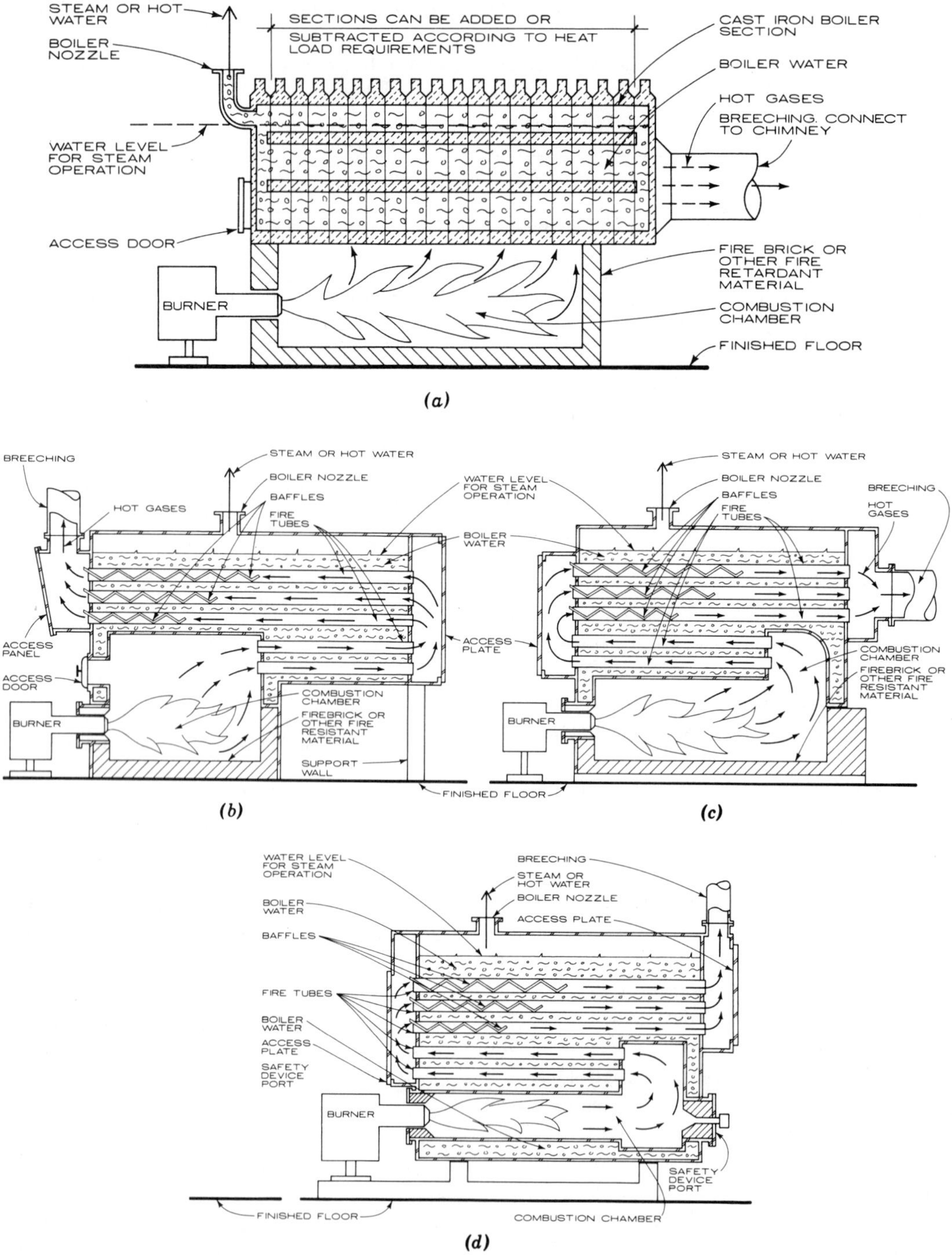

Fig. 7.12 *Comparisons of boiler types.* (a) *Cast iron sectional type.* (b) *Two-pass fire tube.* (c) *Three-pass fire tube.* (d) *Three-pass wet-back Scotch marine. (Reproduced by permission from AIA, Ramsey, and Sleeper,* Architectural Graphic Standards, *8th ed., © 1988 by John Wiley & Sons.)*

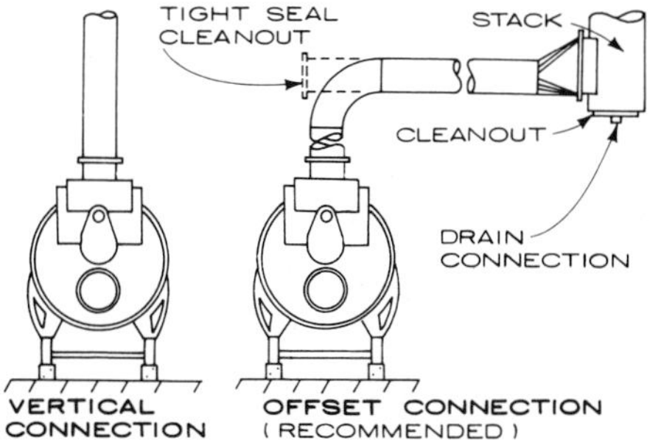

STACK DIAMETER—SINGLE BOILER VENT OR STACK

BOILER HORSE-POWER	STACK DIAMETER (IN.)	A (IN.)	B (IN.)	C (IN.)
15-20	6	15	15	12
25-40	8	20	20	16
50-60	10	25	25	20
70-100	12	30	30	24
125-200	16	40	40	32
250-350	20	50	50	40
400-800	24	60	60	48

STACK DIAMETER—MULTIPLE BOILERS: COMMON BREECHING AND STACK

BOILER HORSE-POWER	MINIMUM STACK DIAMETER (IN.) NUMBER OF BOILERS					
	2		3		4	
	100 FT	200 FT	100 FT	200 FT	100 FT	200 FT
25-40	11	12	13	14	14	16
50-60	13	14	15	16	17	18
70-100	16	17	19	20	21	23
125-200	21	22	24	26	28	30
250-350	26	28	32	34	34	40
400-600	32	34	38	40	42	46

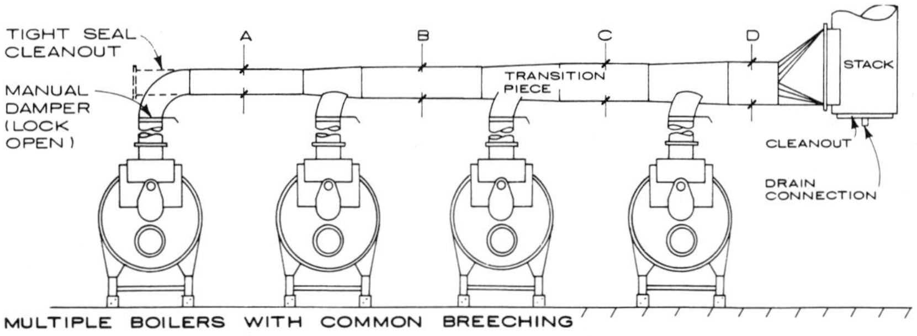

BREECHING DIAMETER—SINGLE AND MULTIPLE BOILERS

BOILER HORSE-POWER	MINIMUM BREECHING DIAMETER (IN. OD)			
	A (IN.) 1 BOIL-ER	B (IN.) 2 BOIL-ERS	C (IN.) 3 BOIL-ERS	D (IN.) 4 BOIL-ERS
15-20	6	8	9	9
25-40	8	10	11	12
50-60	10	12	14	15
70-100	12	15	17	18
125-200	16	20	22	24
250-350	20	25	28	30
400-600	24	30	33	36
700-800	24	34	38	42

Note: Stack diameter should be larger than breeching diameter.

***Fig. 7.13** Breeching and stack guidelines for fossil-fuel-fired boilers. (Reprinted by permission from AIA, Ramsey, and Sleeper,* Architectural Graphical Standards, *8th ed., © 1988 by John Wiley & Sons.)*

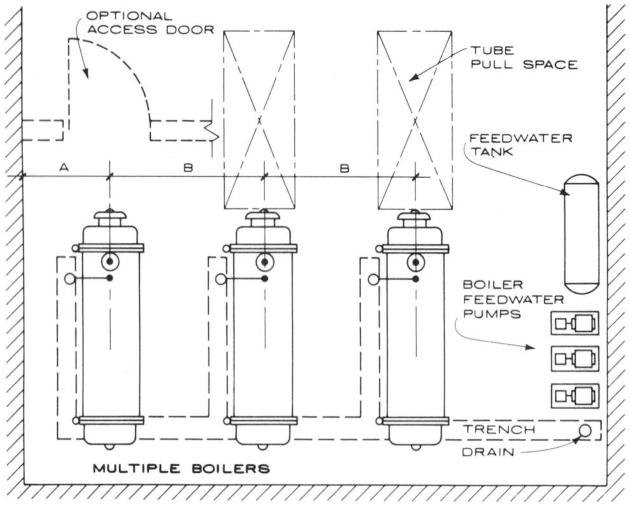

Fig. 7.14 Boiler room space requirements. Dimension A *includes an aisle of 3 ft 6 in. between the boiler and the wall. Dimension* B *between boilers, includes an aisle of at least 3 ft 6 in. (up to 5 ft for the largest boilers). (Reprinted by permission from AIA, Ramsey, and Sleeper,* Architectural Graphic Standards, *8th ed., © 1988 by John Wiley & Sons.)*

BOILER ROOM SPACE REQUIREMENTS

BOILER HP	15-40	50-100	125-200	250-350	400-800
Dimension A	5′-9″	6′-6″	6′-10″	7′-9″	8′-6″
Dimension B	7′-5″	8′-9″	9′-7″	11′-9″	14′-3″

able. The advantage of electric boilers is the elimination of combustion air, the flue, and air pollution at the building. The disadvantages are the use of a high-grade energy source for a relatively low-grade task and the pollution impact at the electric generating plant.

4. *Modular boilers (Fig. 7.17).* The primary advantage of modular boilers is efficiency. Boilers achieve maximum efficiency when they are operated continuously, at their full rated fuel input. The single boilers discussed above operate this way only under outside design conditions, which by definition occur, at most, during 5% of a normal winter. In a modular boiler design, each section is run independently. Therefore, only one section need be fired for the mildest heating needs; as the weather gets colder, more sections are gradually added. Because each section operates continuously at full rated fuel input, efficiency is greatly increased (see Fig. 7.18). Each module, being rather small, requires little time to reach useful temperature and (unlike the larger, single boilers) does not waste a lot of heat as it cools down. Thus, modular boilers usually produce a 15 to 20% fuel savings for the heating season, relative to single boilers. Their other advantages include ease of maintenance (one module can be cleaned while others carry the heating load) and small size (allowing easy installation and replacement in existing buildings).

Modular boilers also eliminate the cost of *oversizing* heating equipment. In cold climates, conventional boiler systems often use two or three large boilers to ensure that heat is available even if one large boiler fails. When two such boilers are used, it is common practice to size each boiler at two-thirds of the total heating load; an oversize of one-third results. When three such boilers are used, it is common practice to size each boiler at 40% of the total heating load; an oversize of 20% results. However, when a minimum of five modular boilers are used, oversizing can be eliminated since the failure of a single module will not have a significant impact on the overall heat output.

Gas-fired pulse boilers are an even smaller and more-energy-efficient choice for modular boilers. Pulse boilers utilize a series of 60 to 70 small explosions per second, making the hot flue gases pulse as they pass through the firetube. This makes for very efficient heat transfer. Pulse boilers are available up to about 300,000 Btu/h.

Pulse boilers operate with lower water temperatures, so water vapor in the flue gas can

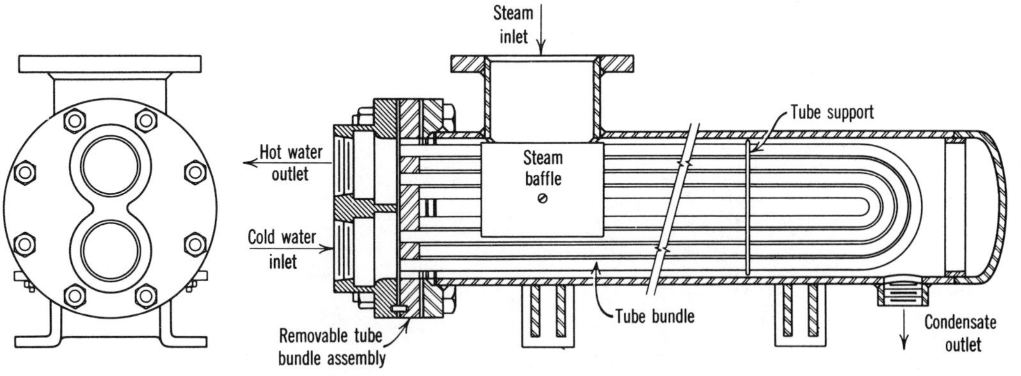

(a) Section illustrating the principle of heat transfer from steam to water.

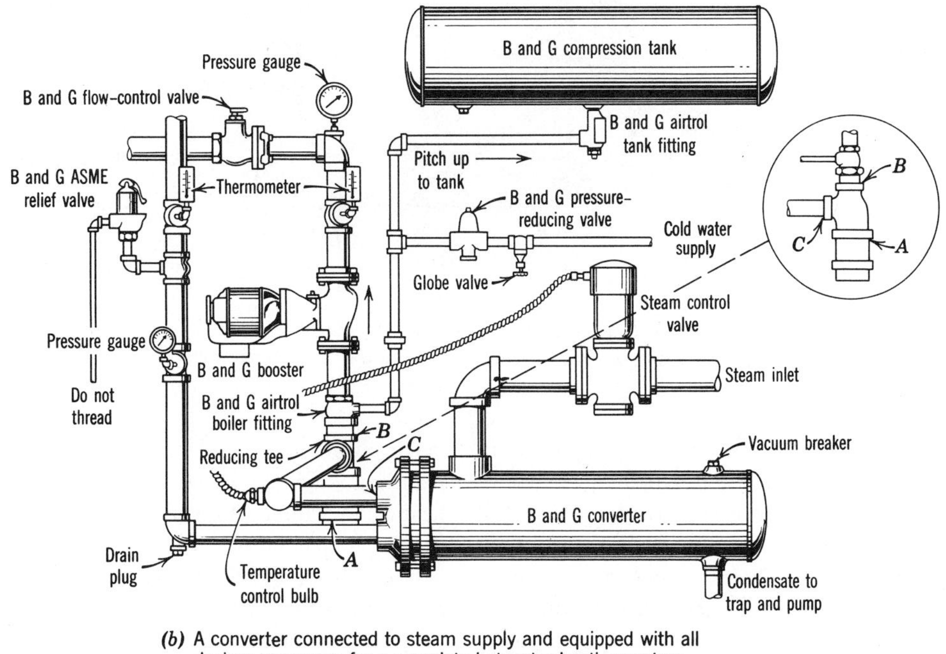

(b) A converter connected to steam supply and equipped with all devices necessary for a complete hot water heating system.

Fig. 7.15 *Conversion unit that transfers heat from steam to hot water. (Courtesy of ITT Bell and Gossett.)*

condense and drain. This condensation liberates additional heat via a change of state, allowing these pulse boilers to achieve efficiencies up to 90%. They exhaust moist air, not hot smoke, so flues can be small-diameter plastic pipe rather than large-diameter heat-resistant materials.

(b) Chillers. These devices remove the heat gathered by the recirculating chilled water system as it cools the building. See Table 7.4 for the minimum efficiency standards for chillers; Section 7.3*d* discusses dual-condenser chillers that conserve energy. The selection of chillers depends largely on the fuel source and the total cooling load. Chillers include absorption, or centrifugal, and reciprocating machines. Where central steam is available, or high-temperature water (from solar collectors, as waste heat from an industrial process, etc.), the *absorption chiller* (Fig. 7.19) is attractive. This device uses

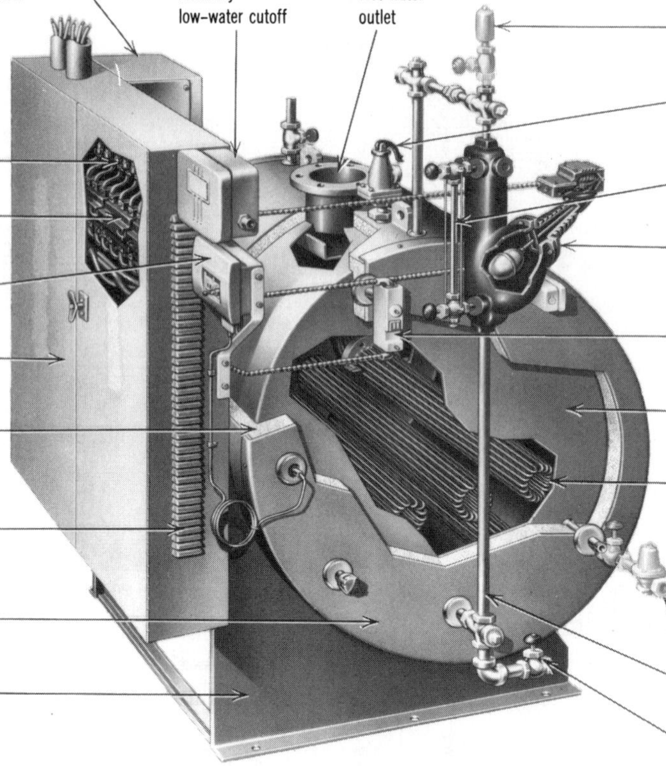

Fig. 7.16 *Electric hot water boiler, available in ratings from 180 to 1800 kW (614,160 to 6,141,600 Btu/h). Similar boilers with appropriate fittings are available for steam systems.*

the absorptive refrigeration cycle (explained in Fig. 6.36), requiring approximately 18 lb of steam at 14 lb of pressure to produce a ton (12,000 Btu/h) of cooling. In general, absorption equipment is less efficient than compressive refrigeration cycle equipment, although a "cheap" or even "free" heat source to power the cycle can rapidly overcome efficiency disadvantages. Absorption machines have fewer moving parts (and therefore require less maintenance) and are generally quieter than compressive cycle equipment.

The compressive refrigeration cycle (explained in Fig. 6.35) is used in the other two types of chillers. Larger units of this type are *centrifugal chillers* (Fig. 7.20), whose compressors either can be driven by an electric motor or can utilize a turbine driven by steam or gas. (When a steam-driven turbine is used, the exhaust steam is often used to run an auxiliary absorption cycle machine. These two devices make an efficient combination, and the steam plant that supplied them in summer can supply heating in winter.) Centrifugal chillers usually require about 1 hp/ton (0.75 kW, or 10 ft^3 gas, or about 15 lb of steam). These large chillers usually require a cooling tower.

Smaller compressive-cycle machines are called *reciprocating chillers* (Fig. 7.21). Usually electrically driven, they are often combined with an air-cooled heat rejection process, rather than a cooling tower. This makes them a closer relative of the smaller direct refrigerant machines discussed in Section 7.2.

Chilled water is usually supplied at between 40 and 55 F. When the chilled water is supplied cold and returns much warmer, the large rise in temperature reduces the initial size (cost) of equipment and increases the efficiency (thereby reducing operating cost as well). Water treatment may be needed for chilled water, to control corrosion or scaling.

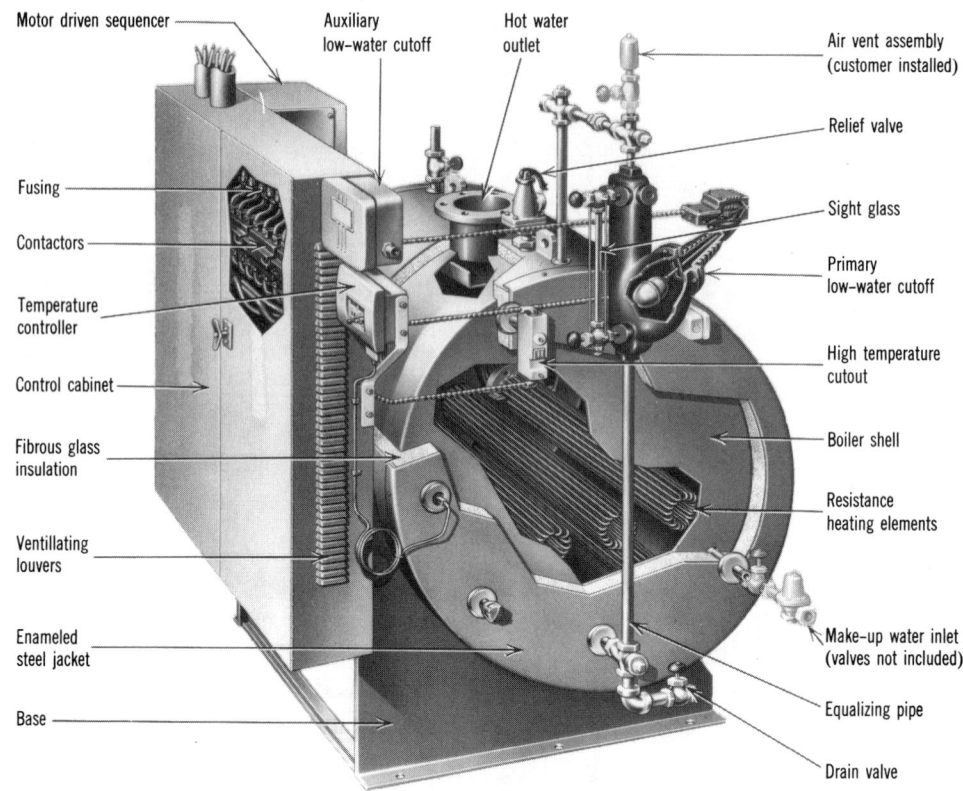

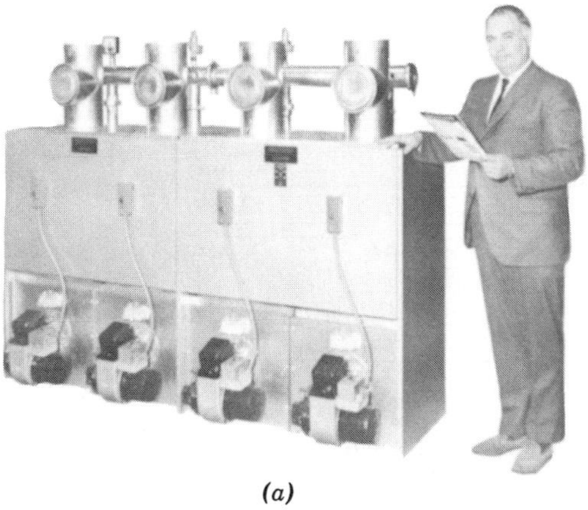

(a)

***Fig. 7.17** Modular boilers.* (a) *A bank of four modules—total input, 1.5 million Btu/h.* (b) *Details of one module.* (c) *Schematic of flow conditions in mild weather, with only one module in operation.*

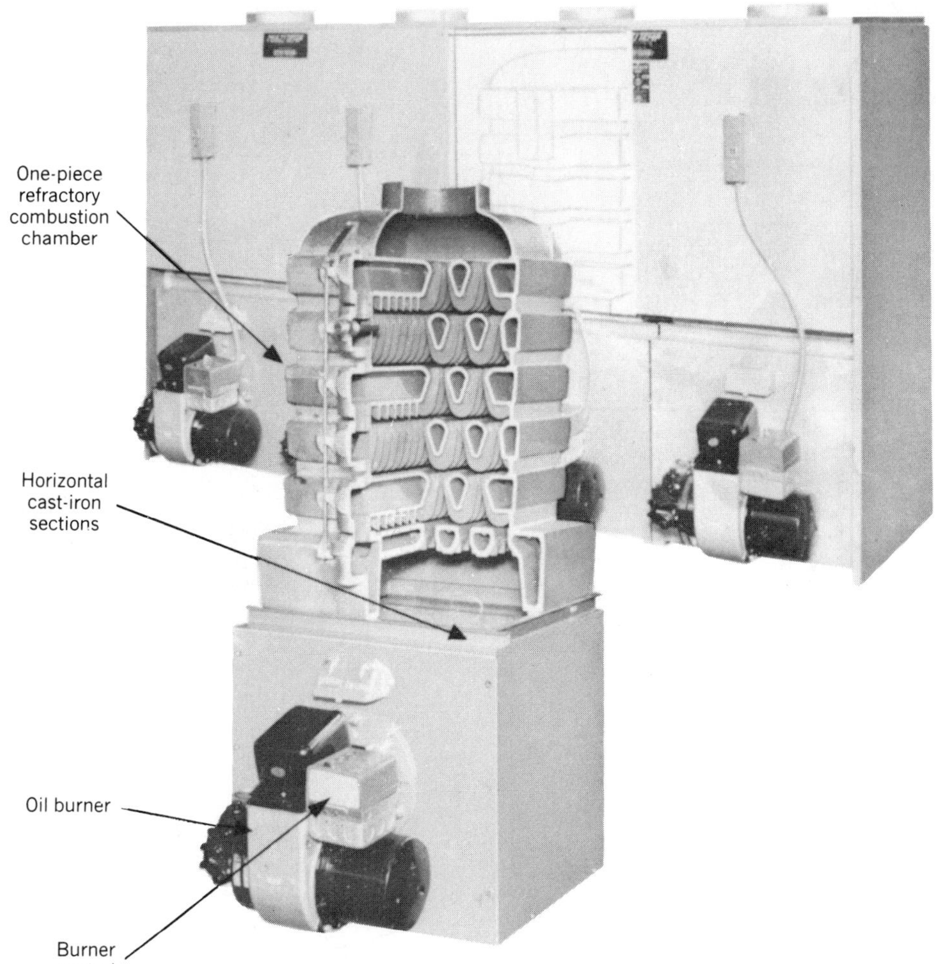

Cutaway of Heating Module
385,000 Btu/h input
Factory assembled with burners and all oil controls
20″ × 32″ × 48½″ high

Shipping weight:	644 lb	Floor loading:	179 lb/ft.^{2}
Water content:	7.25 gal	Pressure drop (ΔT = 20F°):	1.7′ WC
Fire side surface:	49 ft.^{2}	Water side surface:	42 ft.^{2}
Horsepower:	9.1 hp	Pressure rating:	100 psi ASME

(b)

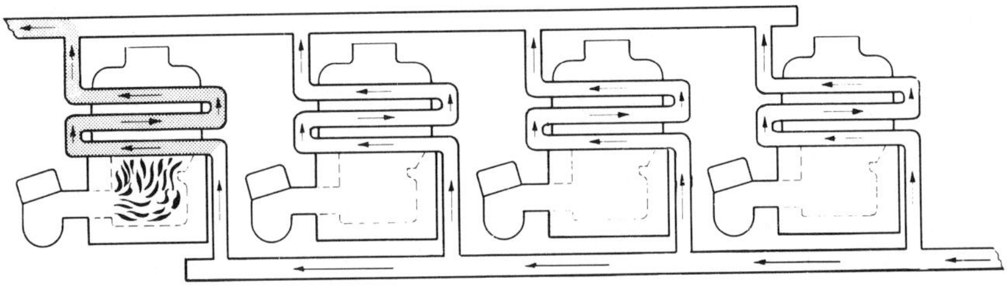

SCHEMATIC FLOW DIAGRAM

For 4-step 1,540,000 Btu/h Input Multi-Temp (MO-1540)

(c)

Fig. 7.17 (*continued*)

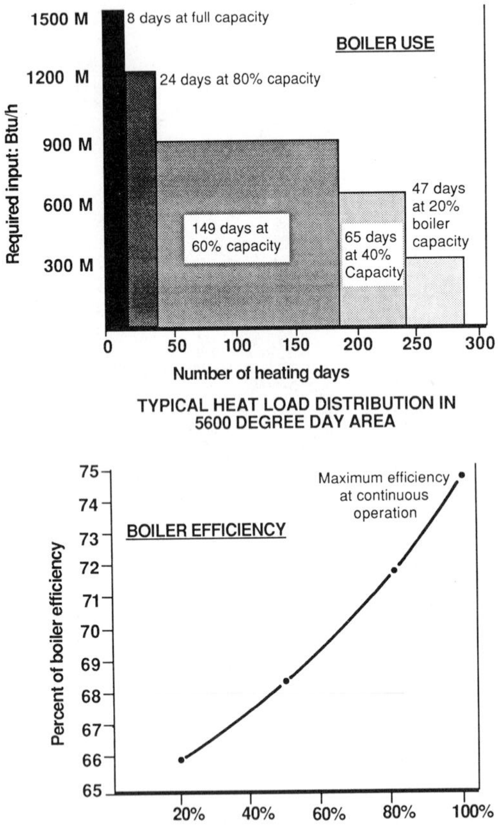

Fig. 7.18 *One large boiler versus many small ones. Whether in mild weather with few modules working, in cooler weather with several working, or in extreme winter weather with all working, the modules actually operating contribute with minimal short-cycling (lower diagram). Short-cycling of a single large boiler could drop the general efficiency to the lower levels of the 66 to 75% efficiency range.*

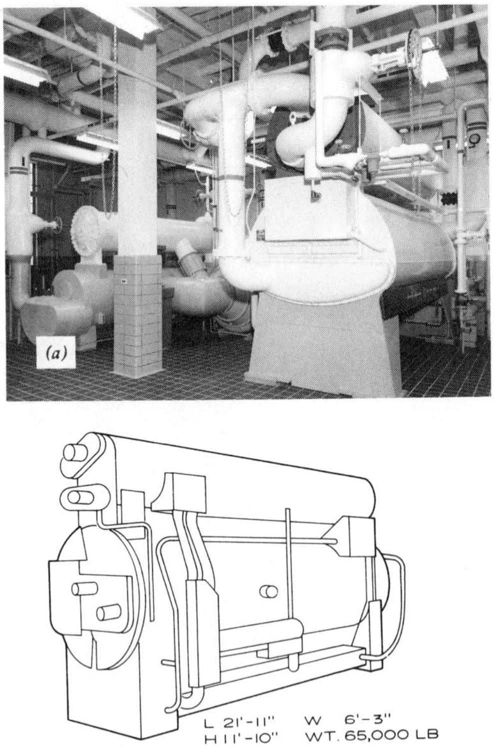

(b)

Fig. 7.19 (a) *An absorption chiller, used for producing chilled water. The Carrier Corporation. (Courtesy of Ingersoll-Rand.)* (b) *Two-stage absorption chiller utilizing steam, producing from 200 to 600 tons of cooling. (Reprinted by permission from AIA, Ramsey, and Sleeper,* Architectural Graphic Standards, *7th ed. © 1981 by John Wiley & Sons.)*

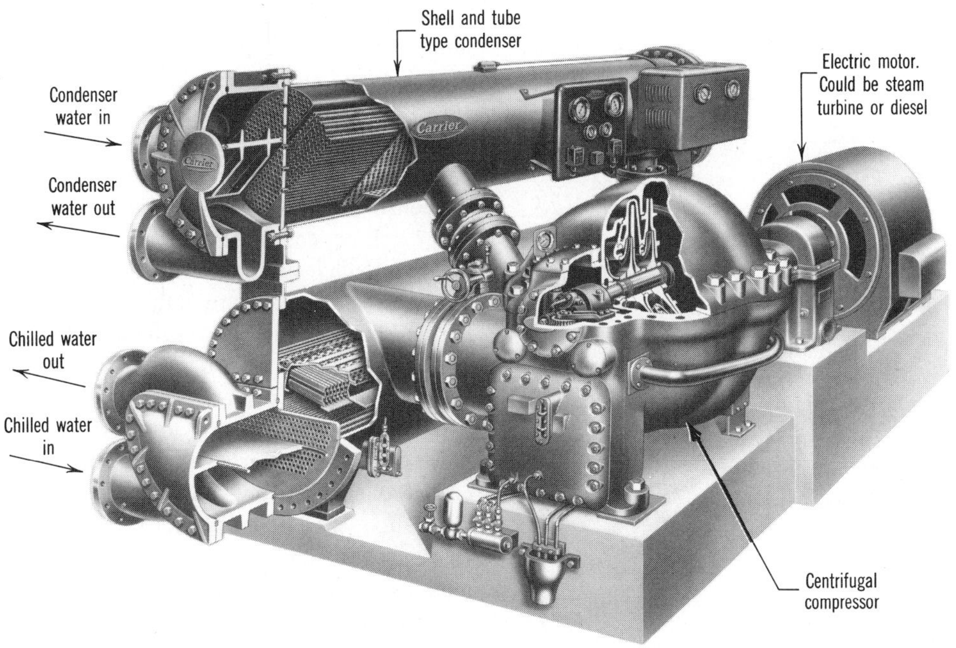

(*a*)

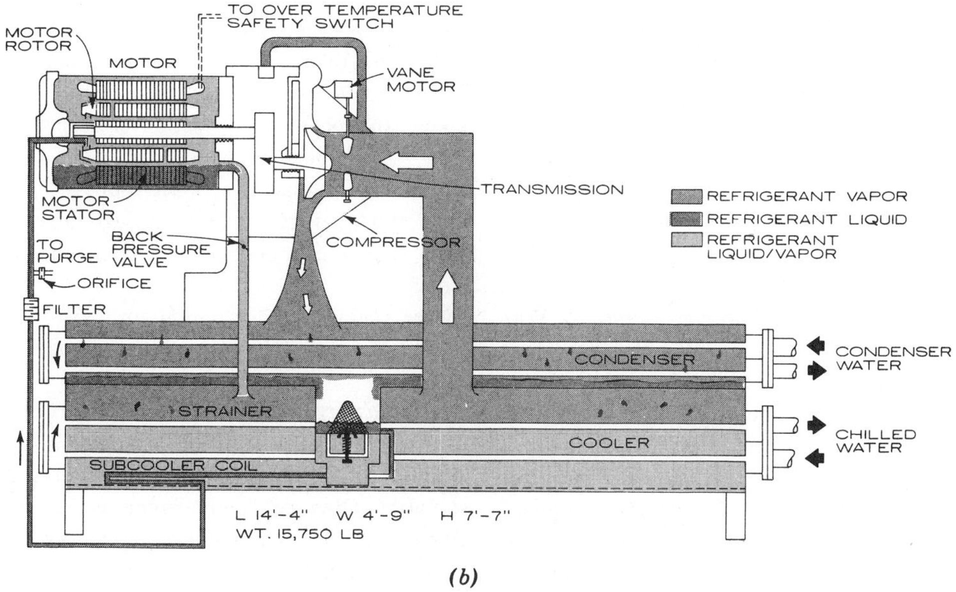

(b)

Fig. 7.20 (a) *A centrifugal chiller—a machine of large capacity that uses the compressive refrigeration cycle. (Courtesy of The Carrier Corporation.)* (b) *Centrifugal chiller with a flooded cooler and condenser within a single outer shell. This low-pressure unit typically produces 100 to 400 tons of cooling. (Reprinted by permission from AIA, Ramsey, and Sleeper,* Architectural Graphic Standards, *7th ed., © 1981 by John Wiley & Sons.)*

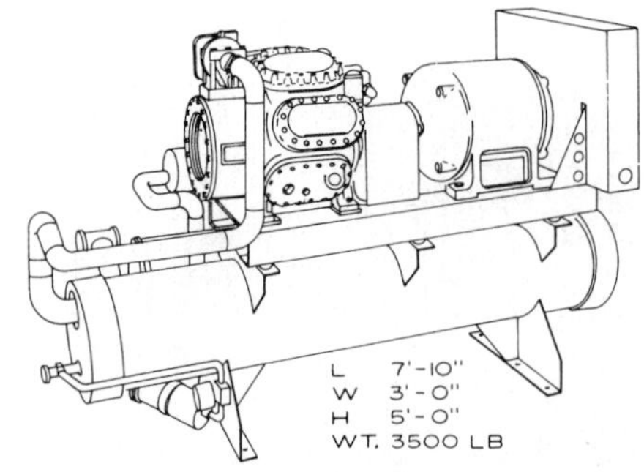

Fig. 7.21 *A reciprocating chiller—a small-capacity machine that uses the compressive refrigeration cycle. Typical at sizes of less than 200 tons of cooling. (Reprinted by permission from AIA, Ramsey, and Sleeper,* Architectural Graphic Standards, *7th ed., © 1981 by John Wiley & Sons.)*

Typical cooling capacities and space requirements of chillers are shown in Fig. 7.22. Each refrigeration machine in this illustration requires two pumps—one for the chilled water (to cool the building) and one for condenser water (to deal with reject heat). Typically, space is provided for future chiller additions, which may be required by building expansion and/or by higher internal gains from as-yet-uninstalled equipment, such as computer terminals within offices.

(c) Condensing Water Equipment. With chillers, there must be a way to reject the heat that is removed from the recirculating chilled water system. Reject heat is taken care of by the condensing water system, which serves the condensing process within refrigeration cycles. For larger buildings, the condensing water requirement is most likely to be met by a *cooling tower.*

The cooling tower's place within the overall equipment layout was shown in Fig. 7.11*b*; a more detailed guide to sizes and types is given in Fig. 7.23. The object is to maximize the surface area contact between outdoor air and the heat condensing water. In crossflow towers, fans move air horizontally through water droplets and wet layers of fill (or packing), while in counterflow towers, fans move the air up as the water moves down.

Cooling towers create a special—and usually unpleasant—microclimate. They demand huge quantities of outdoor air, which they make considerably more humid. In cold weather, they can produce fog. They are typically very noisy—a natural consequence of forced-air motion. The condensing water they evaporate (between 1.6 and 2 gph per ton of refrigeration) must be replaced, which is done automatically. The steady evaporation and exposure to the outdoors under hot and humid conditions spells trouble for the condensing water: controls for scaling, corrosion, and algae growth are especially important.

Although it is tempting to try to block the noise of cooling towers with solid barriers, it is critical that noise control not interfere with air circulation. The manufacturer's recommended clearances to solid objects near cooling towers must be consulted before a tower is enclosed in any way. The roof is a favorite location for cooling towers; at least nine such devices can be seen on the roof of the Chase Manhattan Building, shown in Fig. 9.35 (p. 569).

When fouling of the condensing water system cannot be tolerated, an alternative approach, called the *closed-circuit evarporative cooler,* is taken. Its schematic operation is described in Fig. 7.24. While the condenser water is protected within an always-closed loop, a separate body of water is recirculated through the cooler, with steady evaporation and attendant problems.

See Table 7.3 for minimum performance standards for condensing water equipment.

(d) Energy Conservation Equipment. One big advantage of central equipment rooms is the opportunity they present for energy conservation. Regular maintenance is simplified when all the equipment lives in a generous space kept at optimum conditions; with regular maintenance comes increased efficiency of opera-

Fig. 7.22 *Chiller room space requirements. Each refrigeration machine is served by two pumps (chilled water and condenser water). (Reprinted by permission from AIA, Ramsey, and Sleeper,* Architectural Graphic Standards, *8th ed., © 1988 by John Wiley & Sons.)*

Plan

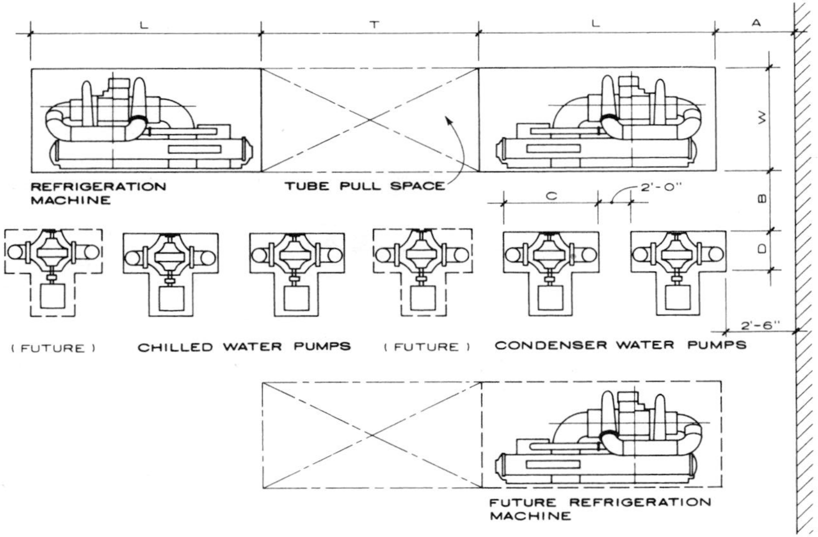

ROOM DIMENSIONS

Dimension A allows for a minimum 3 ft 6 in. aisle between the water column on the boiler and the wall. Dimension B between boilers allows for a clear aisle of:

3′6″— 15–200 hp
4′0″—250–350 hp
5′0″—400–800 hp

The shortest boiler room length is obtained by allowing for possible future tube replacement (from front or rear of boiler) through a window or doorway. Allowance is only made for minimum door swing at each end of the boiler.

AIR SUPPLY

Two permanent air supply openings on opposite walls of the boiler room are recommended. These openings should be located below a height of 7 ft with a total clear area of at least 1 sq ft. Air supply openings can be louvered for weather protection.

Size the openings by using the following formula:

$$\text{area (sq ft)} = \frac{\text{CFM}}{\text{FPM}}$$

Amount of air required (CFM):

Combustion air—max. boiler HP × 2 CMF/BHP
Ventilation air—max. boiler HP × 2 CFM/BHP

NOTE: a total of 10 CFM/BHP applies up to 1000 ft elevation. Add 3% more per 1000 ft of added elevation.

Air velocity required (FPM): Up to 7 ft height— 250 FPM
Above 7 ft height— 500 FPM
Supply air duct to boiler—1000 FPM

REFRIGERATION EQUIPMENT ROOM SPACE REQUIREMENTS

EQUIPMENT (TONS)	DIMENSIONS L	W	HEIGHT	T	A	B	C	D	MINIMUM ROOM HEIGHT
RECIPROCATING MACHINES									
Up to 50	10′-0″	3′-0″	6′-0″	8′-6″	3′-6″	3′-6″	4′-0″	3′-0″	11′-0″
50 to 100	12′-0″	3′-0″	6′-0″	9′-0″	3′-6″	3′-6″	4′-0″	3′-6″	11′-0″
CENTRIFUGAL MACHINES									
Up to 120	17′-0″	6′-0″	7′-0″	16′-6″	3′-6″	3′-6″	4′-6″	4′-0″	11′-6″
120 to 225	17′-0″	6′-0″	7′-0″	16′-6″	3′-6″	3′-6″	4′-6″	4′-0″	11′-6″
225 to 350	17′-0″	6′-6″	7′-6″	16′-6″	3′-6″	3′-6″	5′-0″	5′-0″	11′-6″
350 to 550	17′-0″	8′-0″	8′-0″	16′-6″	3′-6″	3′-6″	6′-0″	5′-6″	12′-0″
550 to 750	17′-6″	9′-0″	10′-6″	17′-0″	3′-6″	3′-6″	6′-0″	5′-6″	14′-0″
750 to 1500	21′-0″	15′-0″	11′-0″	20′-0″	3′-6″	3′-6″	7′-6″	6′-0″	15′-0″
ABSORPTION MACHINES									
Up to 200	18′-6″	9′-6″	12′-0″	18′-0″	3′-6″	3′-6″	4′-6″	4′-0″	15′-0″
200 to 450	21′-6″	9′-6″	12′-0″	21′-0″	3′-6″	3′-6″	5′-0″	5′-0″	15′-0″
450 to 550	23′-6″	9′-6″	12′-0″	23′-0″	3′-6″	3′-6″	6′-0″	5′-6″	15′-0″
550 to 750	26′-0″	10′-6″	13′-0″	25′-6″	3′-6″	3′-6″	6′-0″	5′-6″	16′-0″
750 to 1000	30′-0″	11′-0″	14′-0″	29′-6″	3′-6″	3′-6″	7′-0″	6′-0″	17′-6″

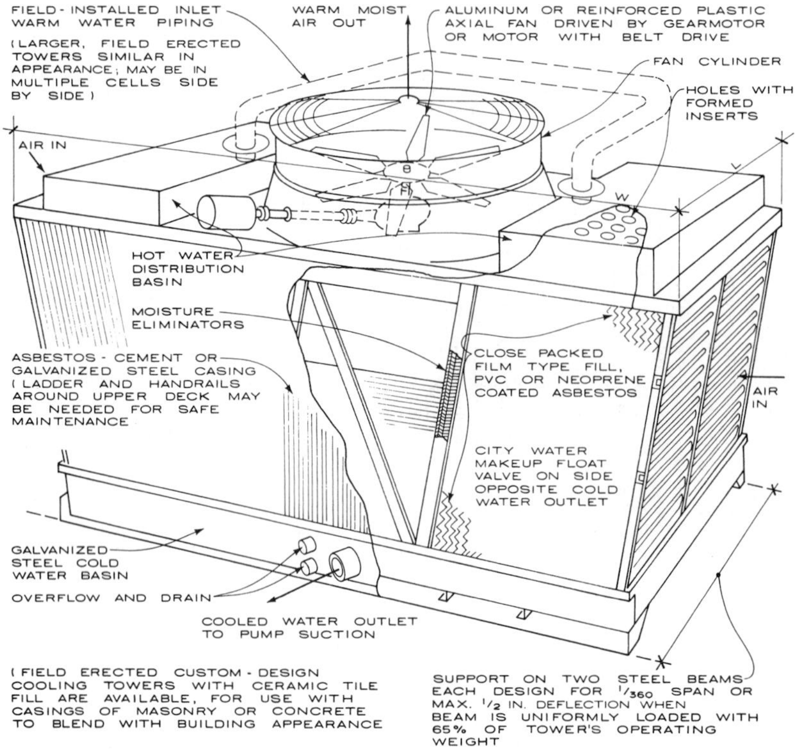

(a)

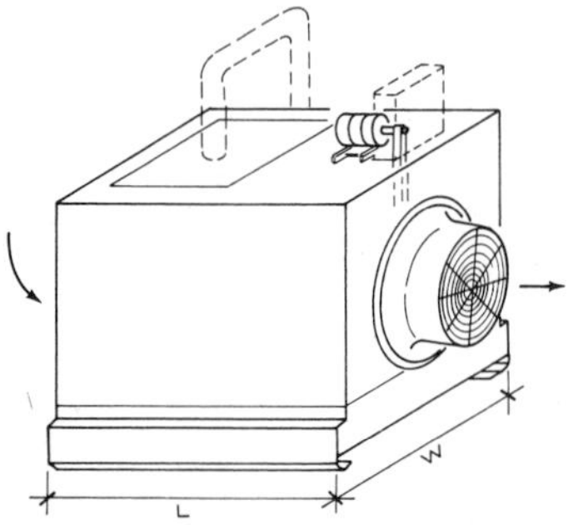

SMALL CROSSFLOW INDUCED DRAFT PACKAGE COOLING TOWER

TONS 3GPM/TON 95-85-78	OVERALL DIMENSIONS (IN.)			OPERATING WEIGHT (LB)	MOTOR (HP)
	L	W	HT.		
5	69	33	60	940	1/4
25	75	46	80	1600	1
50	84	64	92	2500	3
100	93	100	92	4200	5
150	100	144	112	8000	7 1/2

(b)

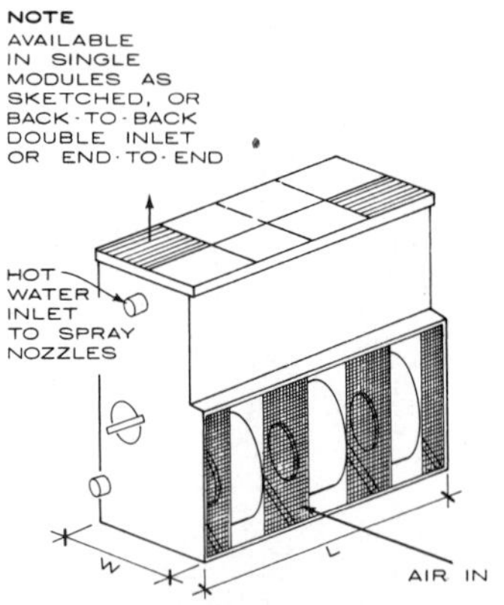

COUNTERFLOW FORCED DRAFT PACKAGE COOLING TOWER

TONS 3GPM/TON 95-85-78	OVERALL DIMENSIONS (IN.)			OPERATING WEIGHT (LB)	MOTOR (HP)
	L	W	HT.		
20	36	36	78	950	2
50	72	36	96	1700	7 1/2
150	144	56	122	4800	20
400	140	118	192	14,000	50

(c)

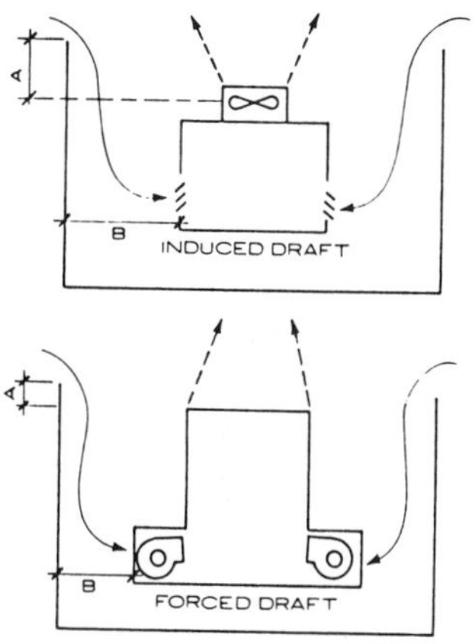

(d)

Fig. 7.23 Cooling towers that serve the condensing water system for large buildings. (a) *Cutaway view of a large-capacity (200 to 700 tons) crossflow induced-draft package cooling tower.* (b) *Size ranges for crossflow induced-draft package cooling towers.* (c) *Size ranges for counterflow forced-draft package cooling towers.* Note: *"3 gpm/ton 95-85-78" refers to the cooling capacity in tons, with condensing water flow at 3 gpm per ton of cooling, the condensing water entering the tower at 95 F and leaving at 85 F, and with outside air at no more than 78 F WB.* (d) *The more wall clearance, the better the operation. A = maximum height of enclosure above tower outlet; minimize this dimension. B = as large as possible, especially if walls have no air openings. (Consult manufacturer for minimum dimension.) (Reprinted by permission from AIA, Ramsey, and Sleeper,* Architectural Graphic Standards, *8th ed., © 1988 by John Wiley & Sons.)*

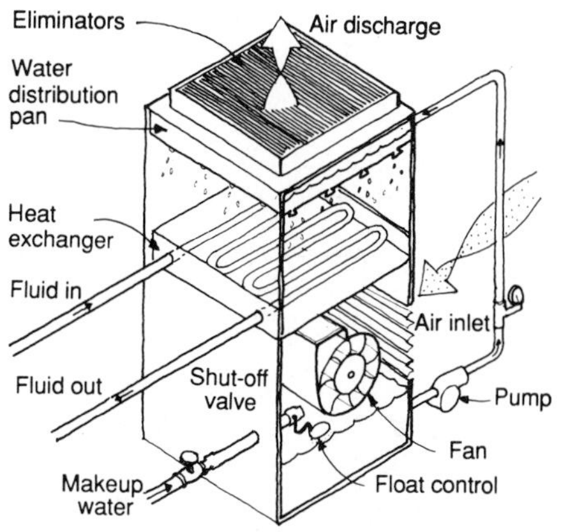

Fig. 7.24 Closed-circuit evaporative coolers, which cool the condensing water system while protecting it from contact with outside air. A self-contained water system is circulated through the evaporative cooler; steady evaporation is replaced by makeup water. (Based on AIA, Ramsey, and Sleeper, Architectural Graphic Standards, *8th ed., © 1988 by John Wiley & Sons.)*

tion. Another conservation opportunity is that of heat transfer between various machines, or between distribution trees, where one's waste meets another's need.

Within equipment, heat transfer can occur in a *boiler flue economizer,* through which the hot gases in a boiler's stack are passed for use in preheating of the incoming boiler water (Fig. 7.25). For cooling equipment, *dual-condenser chillers* (Fig. 7.26) can choose whether to reject their heat to a cooling tower (via the heat rejection condenser) or to building heating (via the heat recovery condenser).

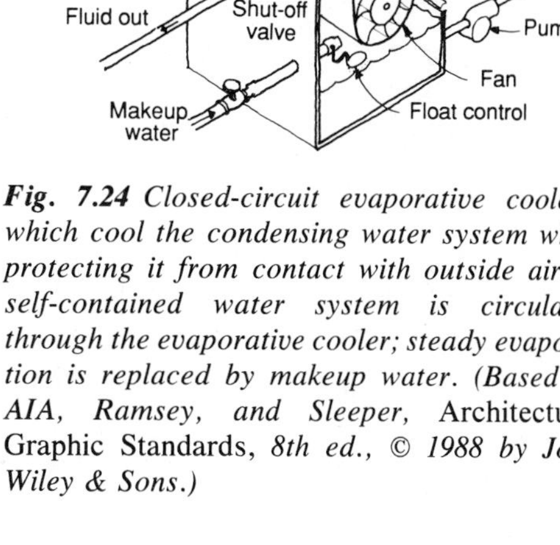

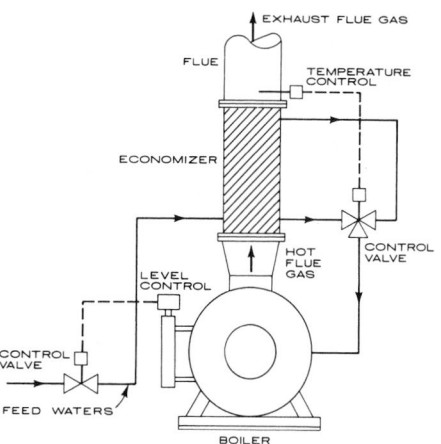

Fig. 7.25 Heat recovery for boilers. Flue gas entering at 500 F leaves the "economizer" at 325 F (a temperature still high enough to prevent condensation in the stack). The heat recovered here is added to incoming boiling water, raising its temperature from 200 F to 248 F. (Reprinted by permission from AIA, Ramsey, and Sleeper, Architectural Graphic Standards, *7th ed., © 1981 by John Wiley & Sons.)*

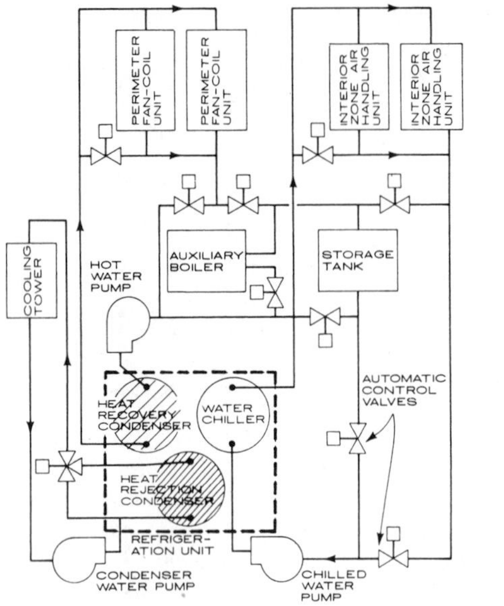

Fig. 7.26 *Dual-condenser chiller. Heat drawn from the chilled water system is either rejected to the cooling tower or recovered for use in building heating. (Reprinted by permission from AIA, Ramsey, and Sleeper,* Architectural Graphic Standards, *8th ed., © 1988 by John Wiley & Sons.)*

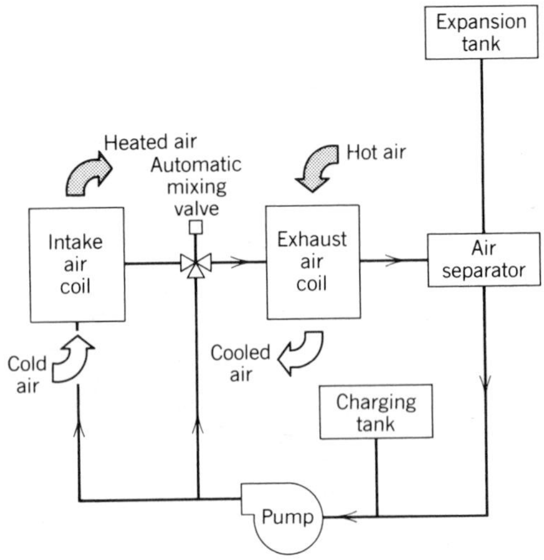

Fig. 7.27 *Runaround coils for heat transfer between fresh intake air and stale exhaust air, used where airstreams are in separate locations. (Reprinted by permission from AIA, Ramsey, and Sleeper,* Architectural Graphic Standards, *7th ed., © 1981 by John Wiley & Sons.)*

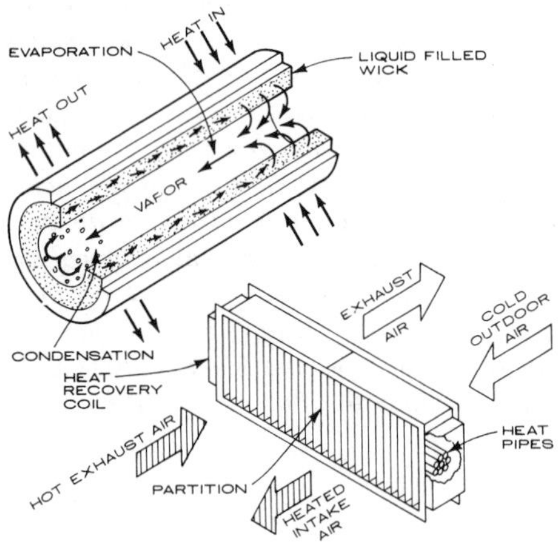

Fig. 7.28 *Heat pipe. These self-contained, no-moving-part devices have many applications. Shown here is sensible heat transfer between adjacent fresh-intake and stale exhaust airstreams. (Reprinted by permission from AIA, Ramsey, and Sleeper,* Architectural Graphic Standards, *8th ed., © 1988 by John Wiley & Sons.)*

There are numerous methods for heat transfer between distribution trees—especially between building exhaust air and fresh air. When these two airstreams are rather far apart, a set of *runaround coils* (Fig. 7.27) can be used. This circulating heat-transfer fluid usually contains antifreeze; it provides simple sensible heat transfer, with no restrictions on exhaust and intake location. No contamination of intake air by exhaust air is likely. The efficiency of such coils runs between 50 and 70%, and they are available in modular sizes up to 20,000 cfm.

The *heat pipe* (Fig. 7.28) also transfers sensible heat, but the airstreams must be adjacent. Within the heat pipe, a charge of refrigerant spends its life alternately evaporating, condensing, and migrating by capillary action through the porous wick. Since the only thing that moves is refrigerant, and it is self-contained, the ideal of no maintenance and long life is obtained. Efficiency is from 50 to 70%; modular sizes are available to 54 in. × 138 in. × 8 rows deep.

Energy transfer wheels (Fig. 7.29) go further than the two preceding devices, in that they transfer latent as well as sensible heat. In winter, they recover both sensible and latent heat

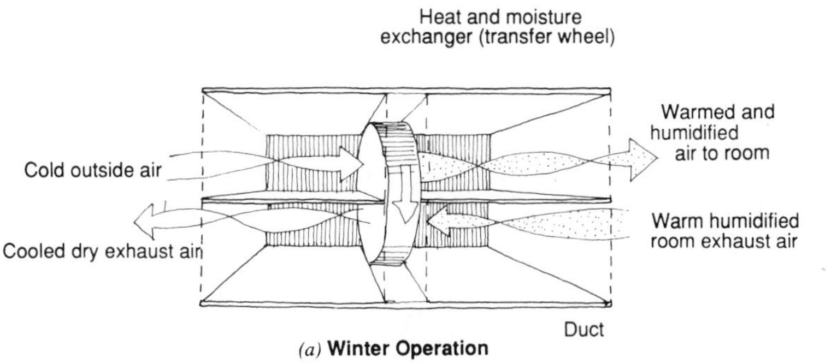

(*a*) **Winter Operation**

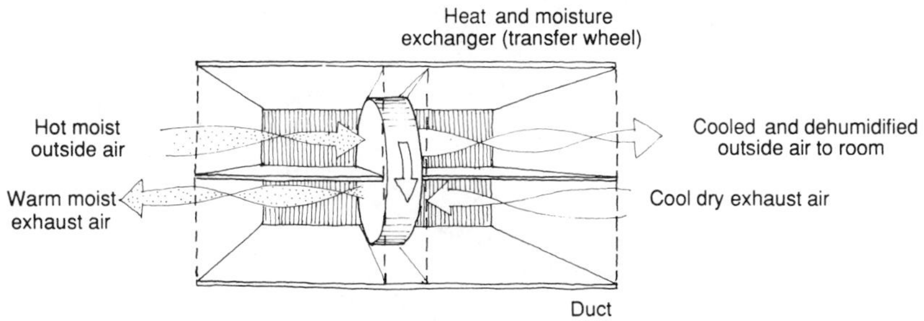

(*b*) **Summer Operation**

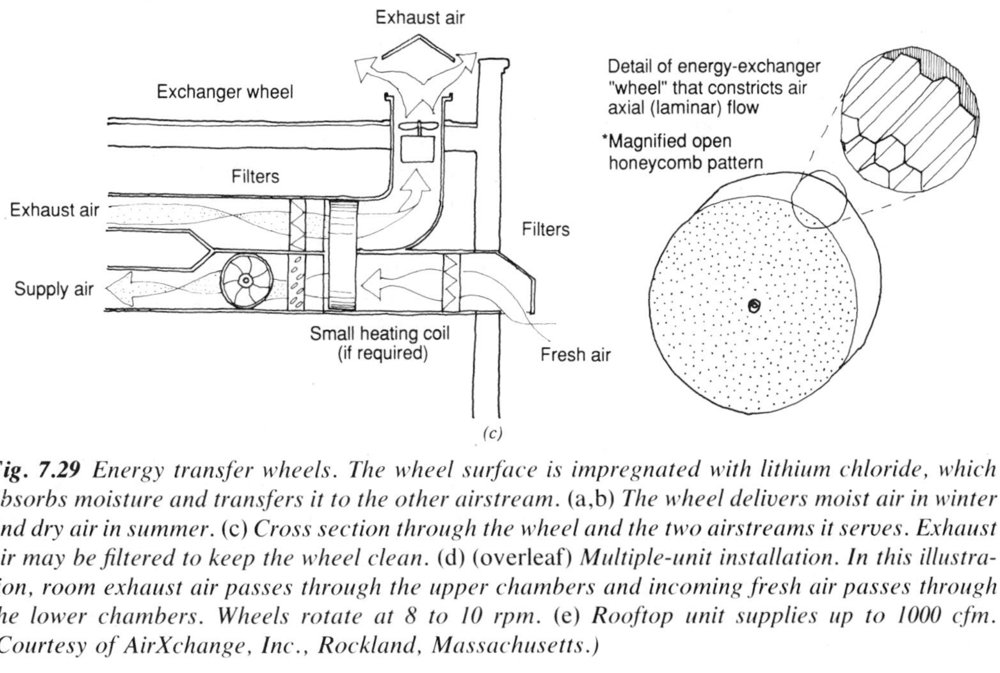

(*c*)

Fig. 7.29 *Energy transfer wheels. The wheel surface is impregnated with lithium chloride, which absorbs moisture and transfers it to the other airstream.* (a,b) *The wheel delivers moist air in winter and dry air in summer.* (c) *Cross section through the wheel and the two airstreams it serves. Exhaust air may be filtered to keep the wheel clean.* (d) (overleaf) *Multiple-unit installation. In this illustration, room exhaust air passes through the upper chambers and incoming fresh air passes through the lower chambers. Wheels rotate at 8 to 10 rpm.* (e) *Rooftop unit supplies up to 1000 cfm. (Courtesy of AirXchange, Inc., Rockland, Massachusetts.)*

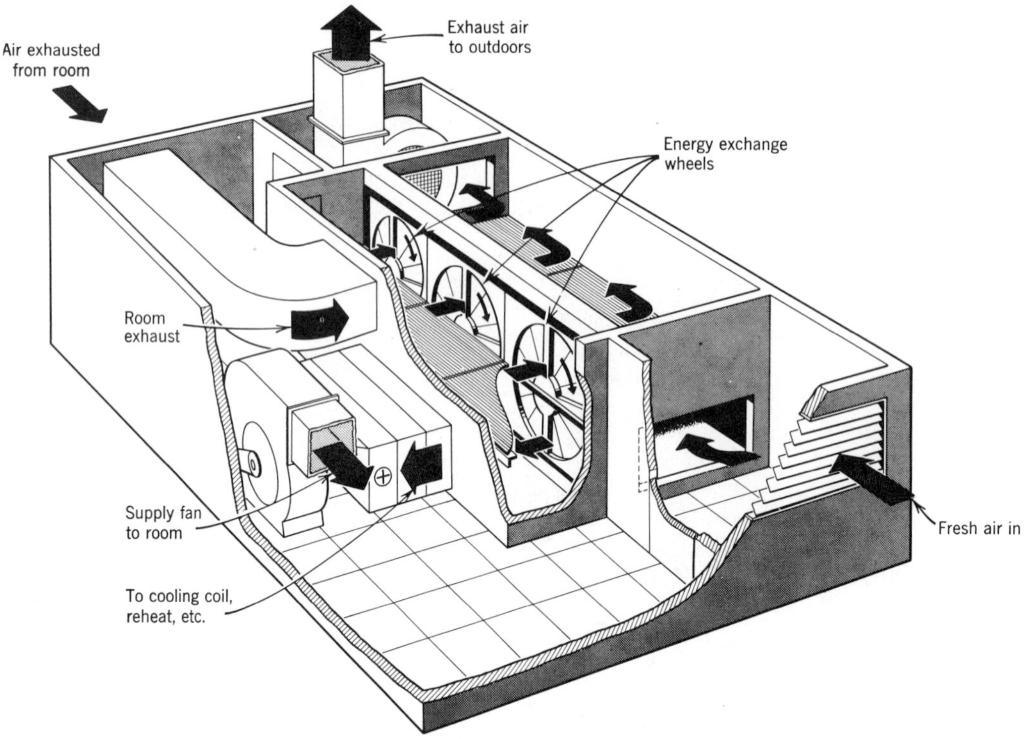

(*d*)

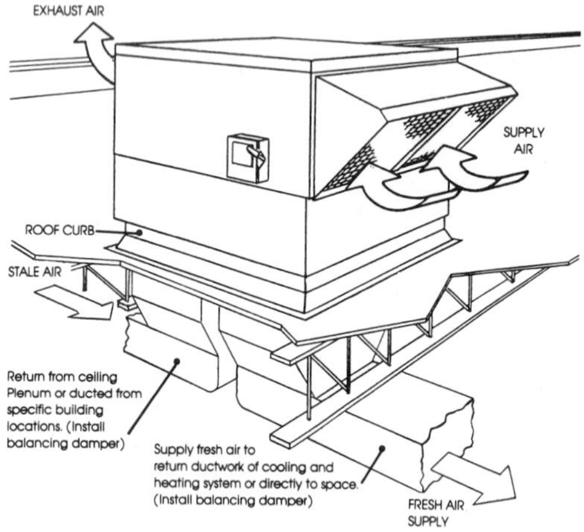

(*e*)

Fig. 7.29 *(continued)*

from exhaust air; in summer, they can both cool and dehumidify the incoming fresh air. Seals and laminar flow of air through the wheels prevent mixing of exhaust air and incoming air. A further precaution in the process purges each sector of the wheel briefly, using fresh air to blow away any unpleasant residual effects of the exhaust air on the wheel surfaces. Carryover of exhaust air qualities, except those of heat and moisture, is between 4 and 8% without purging, and less than 1% with purging. Efficiency is from 70 to 80%, and available sizes range up to 144 in. in diameter. An example of this device was shown in Fig. 6.32*c*.

A final conservation opportunity is offered by the *economizer cycle* (Fig. 7.30), which uses cool outdoor air, as available, to ease the burden on a refrigeration cycle as it cools the recirculated indoor air. The economizer cycle can thus be thought of as a central mechanical substitute for the open window; when it is cool enough (about 60 F or below), 100% outside air can be provided. Relative to open windows, this cycle has several advantages: energy-optimizing automatic thermal control, filtering of the fresh air, tempering of the cool outdoor air to avoid unpleasant drafts, and an orderly diffusion of fresh air throughout the building. Its disadvantages are the loss of personal control that windows offer and little awareness of exterior-interior in-

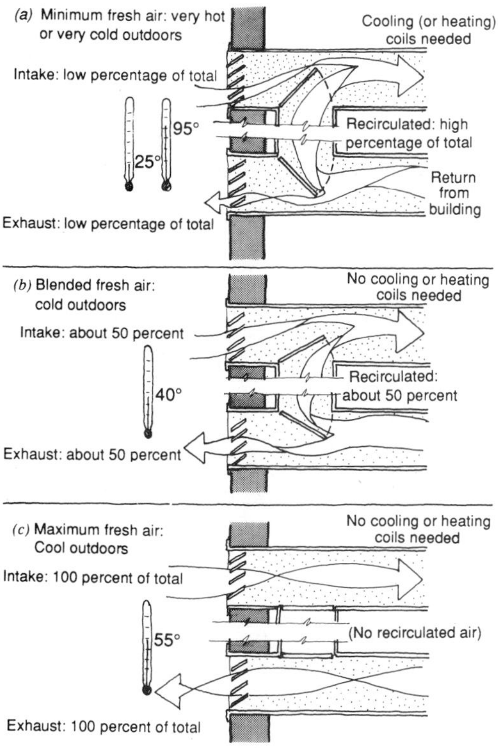

Fig. 7.30 *Economizer cycle controls the relationships between supply and return air: fresh, exhaust, and recirculated air.* (a) *When outside air is hot (or very cold), the economizer cycle is inactive.* (b) *As very cold outside air gets warmer, it can be blended with recirculated air, and neither heating nor cooling coils are needed.* (c) *When outside air is cool, it can completely replace recirculated air, making mechanical cooling inactive.*

teraction. Economizer cycles are available as options on most direct refrigerant machines (such as single-package rooftop units) and are typically installed for large-building central air supply systems. Buildings with high internal gains (internal load dominated) are particularly good targets for economizer cycles, since they need cooling even when the outside temperature is chilly. Economizer cycles lend themselves readily to a cooling strategy of night ventilation of thermally massive structures.

(e) Central Solar Energy Applications. In large buildings, mechanical equipment for solar energy is most commonly used for domestic (service) water heating and/or for powering the absorption refrigeration cycle. (Solar energy can also be passively collected in perimeter zones facing the sun and redistributed to the rest of the building by the central HVAC system.) Solar water heating is discussed in Chapter 9. Solar cooling, using moderately high-temperature (175 to 195 F) water provided by collectors, is shown in Fig. 7.31. Although the efficiency of such a

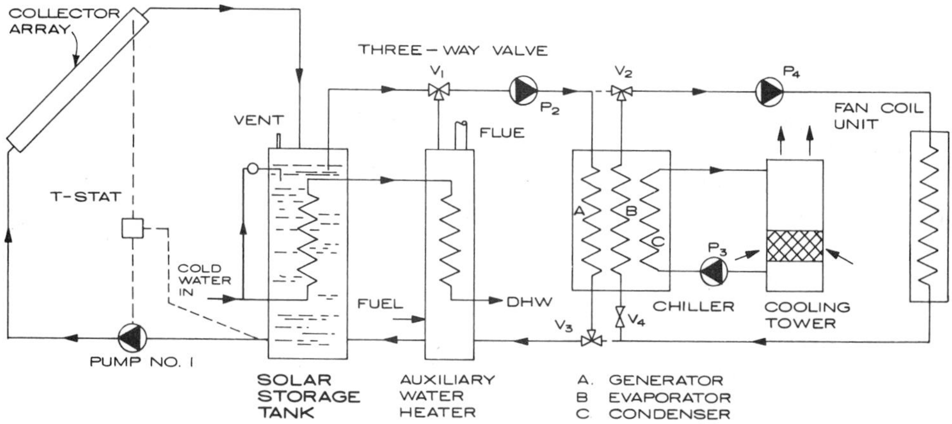

Fig. 7.31 *Solar-heated water for the absorption refrigeration cycle. Moderately hot water from the solar storage tank is supplied to the generator of the absorption cycle (explained in Fig. 6.36). When building heating is needed, solar-heated water (as available) can be supplied directly to the fan-coil unit, by reversing valves V_2 and V_3. (Reprinted by permission from AIA, Ramsey, and Sleeper,* Architectural Graphic Standards, *7th ed., © 1981 by John Wiley & Sons.)*

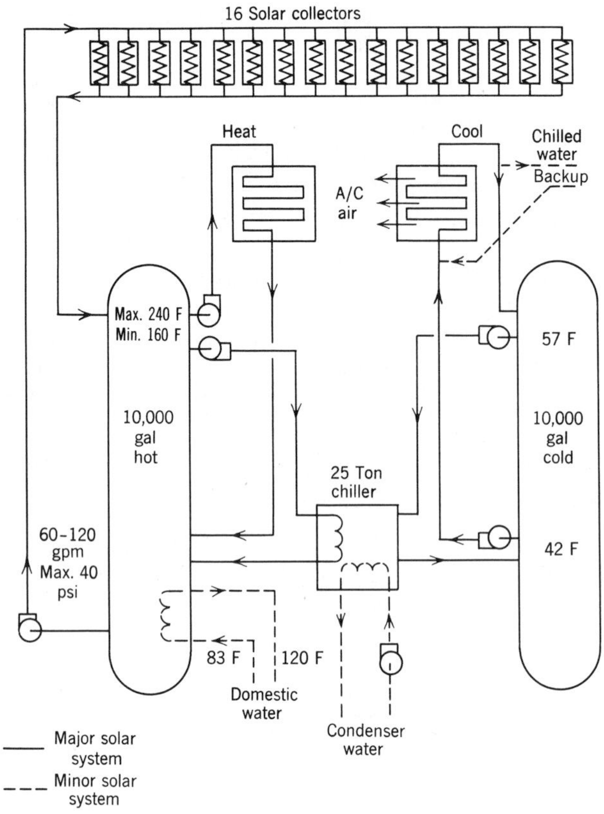

Application: Cooling, heating, hot water hydronic system.
Collector Type: Fixed horizontal parabolic trough mirrored roof with moving absorber.
Collector Manufacturer: AAI Corporation.

Collector Area: 3840 sq ft (356.75 m^2).
Storage Capacity: 10,000-gal, hot water steel tank (37,854 liters) plus 10,000-gal, chilled water steel tank (37,854 liters).
Total Btus: 400 x 10^6 Btu/yr.

(a)

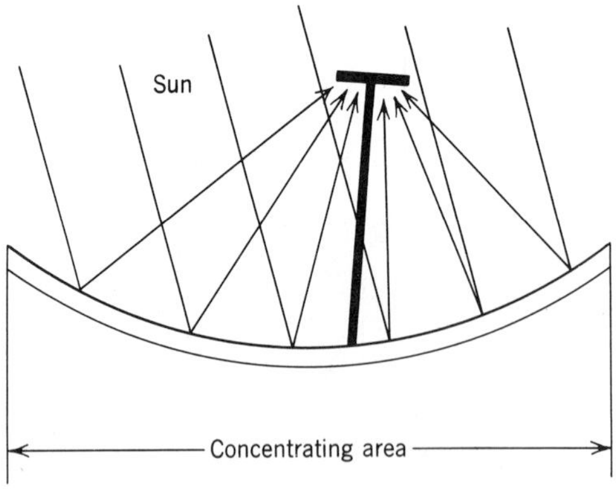

(b)

Fig. 7.32 *Solar-powered absorption cooling for an office building at Disney World, Orlando, Florida.* (a) *Schematic diagram of the system. For clarity, the piping at the 16 collector tubes is shown in simplified form. Actually the water passes through each collector tube twice before returning to the hot water storage tank.* (b) *The stationary mirrored panels concentrate the sun's heat on the absorber, which is an aluminum extrusion. Above this is swaged a U-shaped copper tube.*

moderate-temperature absorption unit is rather low (COP 0.5 to 0.7), the cost of the solar heat supplied may be less (particularly in summer!) than that of the alternative—using electricity to run higher efficiency compressive refrigeration cycles. Simple flat-plate solar collectors can easily deliver 160 F water under summer day conditions in most U.S. locations.

Parabolic concentrating collectors can deliver higher temperature water under clear sky conditions, producing somewhat higher efficiency in absorption chillers. A schematic of the parabolic collector-absorption chiller system for an office building at Disney World in Orlando, Florida, is shown in Fig. 7.32.

(f) Energy Storage. We commonly experience daily changes from warmer to colder conditions, even in summer. Central storage equipment for large buildings can take advantage of this cycle to increase operating efficiency and save energy.

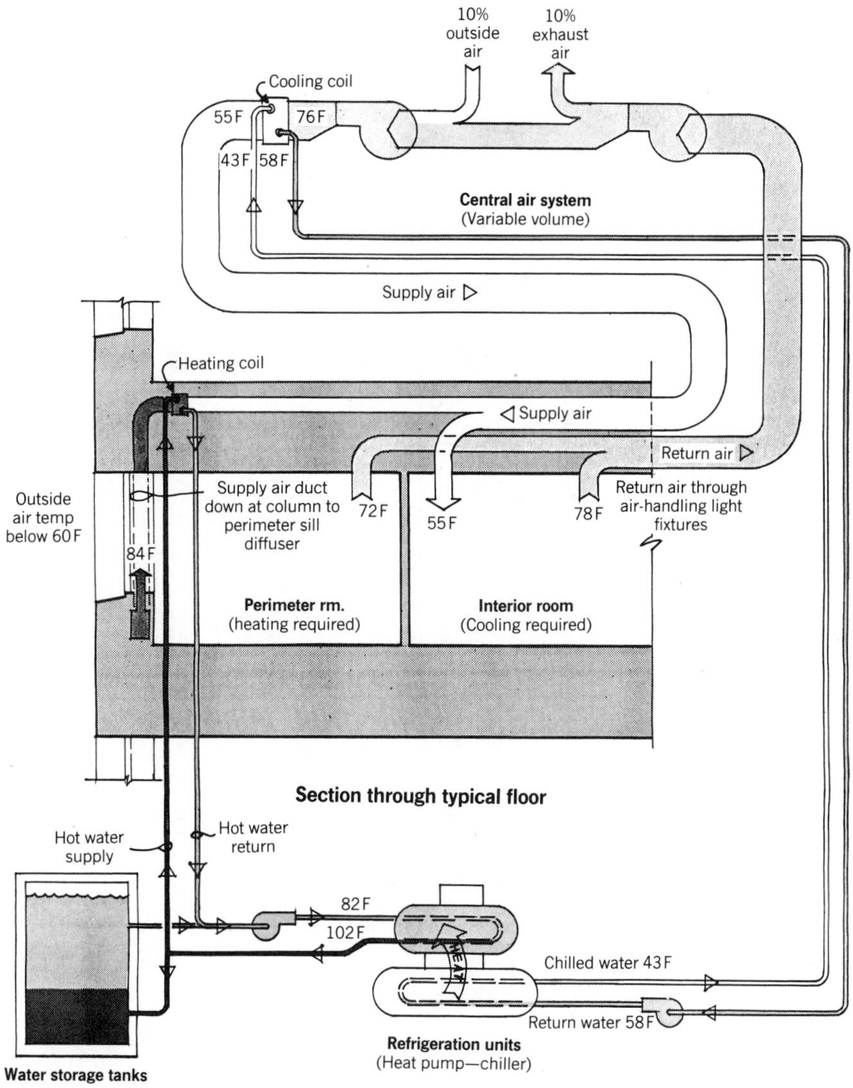

(a)

Fig. 7.33 *Water tank heat storage. The 870,000-ft² transportation (Park Plaza, Boston) office building, population 2000, uses a three-compartment insulated concrete tank storing 750,000 gal of water.* (a) *At outside air 40 to 50 F, the surplus internal heat is stored in the tanks, rather than rejected as exhaust air.* (b) (overleaf) *By the time outside air is about 50 F, the tanks are fully charged; up to outside 60 F, the economizer cycle (see Fig. 7.30) provides energy-conserving cooling.* (c) *At about 60 F outside, chillers must operate; but working all night when outside air is cooler, enables them to work less by day. Their night production is stored as cold water, available to help out at the following day's peak. Smaller machines and more efficient operation are the result. (Courtesy of Sooshanian Engineering Associates, Inc.)*

One common approach to storage is the use of *water storage tanks,* such as those shown in Fig. 7.33. On typical winter days, the total internal heat generated by a large building is somewhat greater than its total need for heating at the perimeter zones. Instead of being thrown away

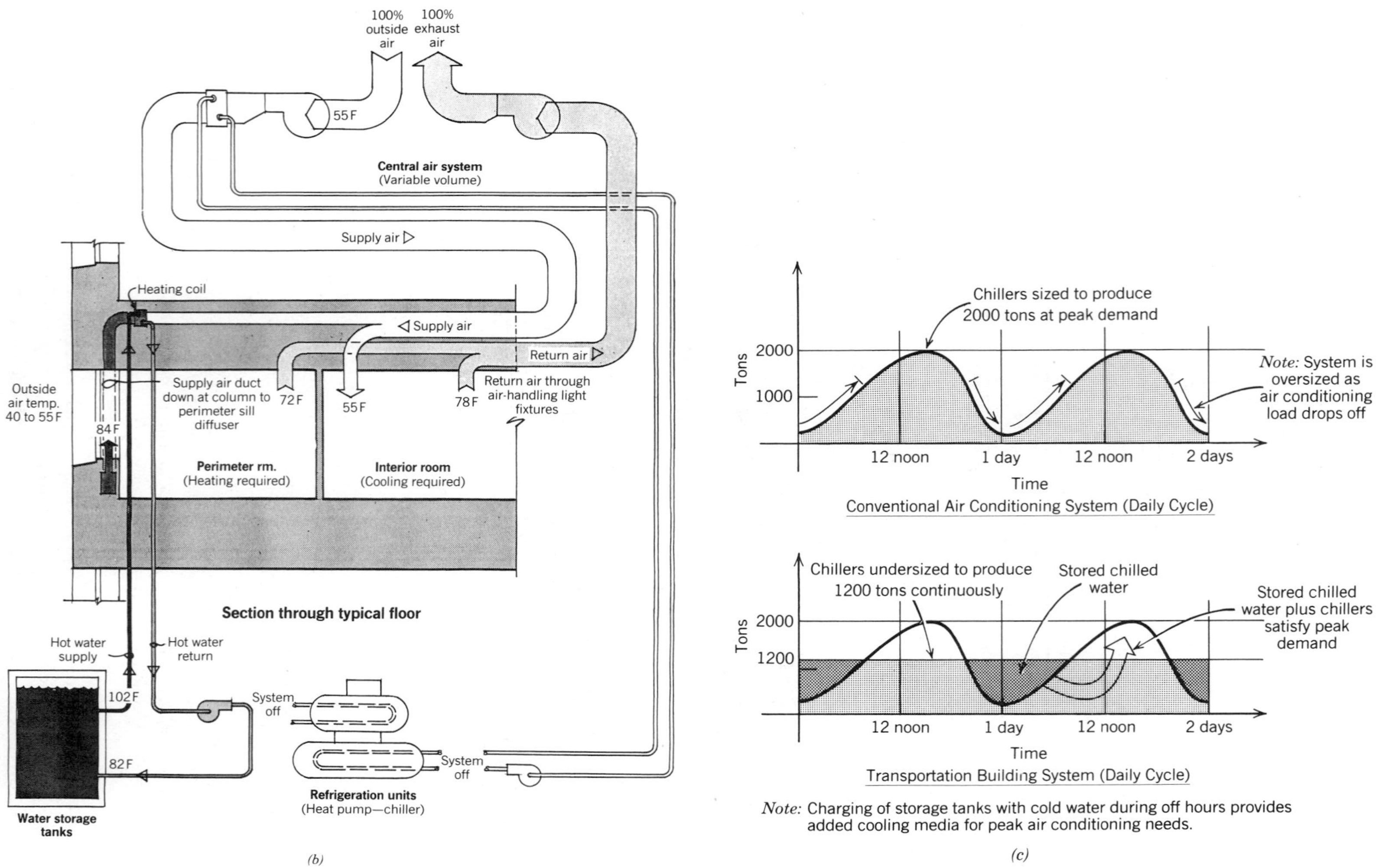
100% outside air
100% exhaust air
55F
Central air system
(Variable volume)
Supply air
Heating coil
Supply air
Return air
Outside air temp. 40 to 55F
84F
Supply air duct down at column to perimeter sill diffuser
72F
55F
78F
Return air through air-handling light fixtures
Perimeter rm.
(Heating required)
Interior room
(Cooling required)
Section through typical floor
Hot water supply
Hot water return
102F
82F
System off
System off
Water storage tanks
Refrigeration units
(Heat pump—chiller)
(b)
Tons
2000
1000
Chillers sized to produce 2000 tons at peak demand
Note: System is oversized as air conditioning load drops off
12 noon
1 day
12 noon
2 days
Time
Conventional Air Conditioning System (Daily Cycle)
Tons
2000
1200
Chillers undersized to produce 1200 tons continuously
Stored chilled water
Stored chilled water plus chillers satisfy peak demand
12 noon
1 day
12 noon
2 days
Time
Transportation Building System (Daily Cycle)
Note: Charging of storage tanks with cold water during off hours provides added cooling media for peak air conditioning needs.
(c)

Fig. 7.33 *(continued)*

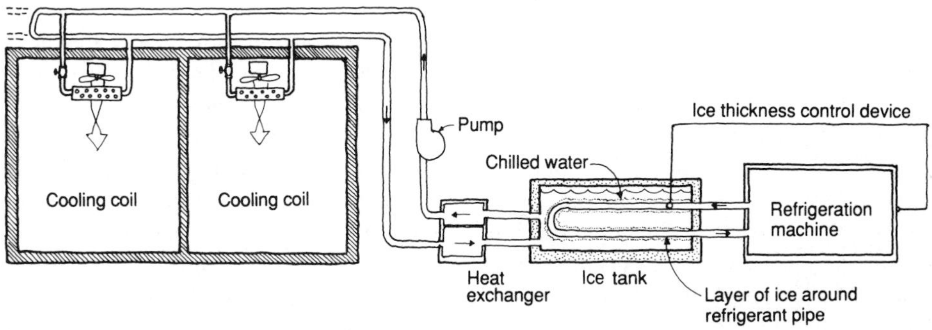

Fig. 7.34 *Ice tank heat storage. Chilled water in the ice tank forms a layer of ice around the refrigerant pipe (supplied by the refrigeration machine). As chilled water at about 35 F is removed from the tank, it enters a heat exchanger, so that its temperature is closer to 45 F for distribution to cooling coils throughout the building.*

as exhaust air, this surplus heat is captured and stored in large water tanks, from which it can be withdrawn and used on cold winter nights and weekends. In the summer, chillers can work at night, when efficiency is high because cool outdoor air helps the refrigeration cycle reject its heat. By storing the "coolth" produced, less work need be done by chillers during the next day's peak, when electric rates are highest and operating efficiency lowest. Daily cycles can also be used in rock bed storage, as we saw in Sacramento's Bateson Building (see Fig. 5.20).

Another storage approach is offered by *ice storage tanks,* as shown in Fig. 7.34. Because they take advantage of the latent heat of fusion (143.5 Btu/lb of water at 32 F), such units can store far more energy in a given-sized tank than can water (1 Btu/lb of water per °F of temperature change). A comparison of the relative sizes of storage tanks needed for an 18-story office building is shown in Table 7.4. Also, with ice storage there is less undesired mixing of hot and cold water within the tank.

Ice is usually made as a layer around a pipe that carries refrigerant in a closed circuit through the tank. Control of the thickness of this layer is important, because if too little ice is made at night, too little cooling will be available the following day. Considerable savings in electricity costs can be realized by this off-peak

TABLE 7.4 **Cooling Storage Comparison**

Eighteen-story office building: 375,000 gross ft^2, 800 tons peak load, 6250-ton h/day at design, 75 million Btu storage needed.

	Water Storage	*Ice Storage*
Btu/lb[a]	15	164
Pounds storage	5×10^6	0.457×10^6
Gallons	599,500	54,800
Storage efficiency	0.90	1.0
Percent ice		66%
Net gallons	666,000	83,030
Cubic feet	89,046	11,101
Size of tank approx. 8 ft deep	80- by 150-ft	30- by 50-ft
Storage ratio, water to ice	8 : 1	

Source: Reprinted by permission from *Specifying Engineer,* January 1983.

[a]Btu/lb based on ΔT of 15F° for water storage and on ΔT of 20.5F° in the melted ice water (added to 143.5 Btu/lb at fusion).

(nighttime) refrigeration operation. For a typical ice storage installation, see the Iowa Public Service Building (Fig. 7.59).

(g) Air-Handling Equipment. A given HVAC system may have many variations. In some variations, all the air is passed through one central equipment room. In others, air handling may be done in many separate, smaller rooms, whereas central heating and cooling will require only one equipment room. The guidelines for air-handling equipment for either case are shown in Fig. 7.35. Total air quantities may be estimated from Table 7.3 or obtained more precisely from cooling load calculations.

Air filtration and odor removal comprise an important part of the process; characteristics of various filters were discussed in Section 6.6. Filters are selected on the basis of the degree of air cleaning desired, first cost, and ease of maintenance.

(h) Controls. Almost all the above-discussed devices for heating, cooling, heat transfer, and so on depend on controls. Although the most obvious control function is to maintain desired comfort conditions, controls also increase fuel economy, by promoting optimum operation, and act as safety devices, limiting or overriding mechanical equipment. They also eliminate human error; controls fall asleep only during a power failure.

Although precise control of temperature and humidity everywhere in a building is a tempting thought, controls can usually maintain only a *range* of conditions, not a set point. This range (within which neither heating nor cooling is called for) is called the *deadband*. Temperature variations occur vertically within a space—the higher the space, the greater the variation. Variations between horizontal positions within a zone are highly likely, especially where one or more walls are exterior and where different rooms share a zone controlled by a single thermostat. Variations in time also occur: a building will always be warmer at the moment the furnace fan turns off than at the moment the fan turns on.

Individual controls can be classified as follows: *controllers,* which measure, analyze, and initiate action; *actuators,* which are the controller's slaves, and in turn become the masters of pieces of equipment; *limit and safety controls,* which may function only infrequently, preventing damage to equipment or buildings; and *accessories,* a miscellaneous collection.

Systems of controls can be classified in various ways. A common distinction is by power source: electric; pneumatic (in which compressed air is the motivating force); and self-contained, including "passive" controls such as those motivated by thermal expansion of liquids or metals. Another way to classify control systems is by the motion of the controller equipment: *two-position* systems are of the simple on–off type; *multiposition* systems have several varieties of ON position, commonly used for separate operation of more than one machine; *floating* controls can assume any position in the range between minimum and maximum; *central logic control* systems can be programmed to integrate the many aspects of building control into one decision-making unit.

Some control diagrams for common HVAC applications are shown in Fig. 7.36. Single-duct, variable-air-volume systems (Fig. 7.36*a*) are described in Section 7.4; with the constantly varying flow rate, the fan must be regulated so as to maintain the minimum pressure (and therefore, flow) needed at the most demanding outlet. This outlet may be either the one most remote from the fan or the one needing the greatest flow (because it has the highest gain) at the moment. Economizer cycles (Fig. 7.36*b*) compare outside to inside positions, and vary the proportion of fresh (outdoor) air to return air, to provide "free" cooling (see Fig. 7.30). Solar heating systems (Fig. 7.36*c* and *d*) compare solar collector temperatures to those of the storage tank, to regulate the pumped water to the collectors.

Today's buildings are increasingly regulated by central logic control systems, giving rise to the term *intelligent buildings*. In the broadest terms, an intelligent building provides a productive, cost-effective environment by optimizing the interrelationships between the building's structure, systems, services and management. In today's computer-controlled building, not only is the HVAC system centrally regulated, it is inter-tied with lighting, electric power (such as

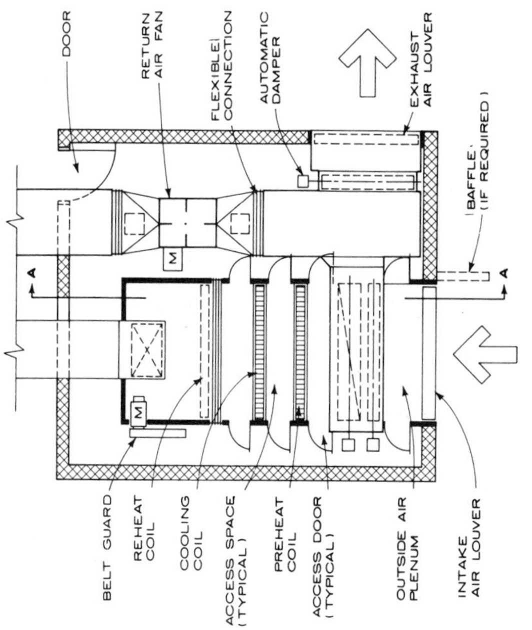

EQUIPMENT ROOM PLAN

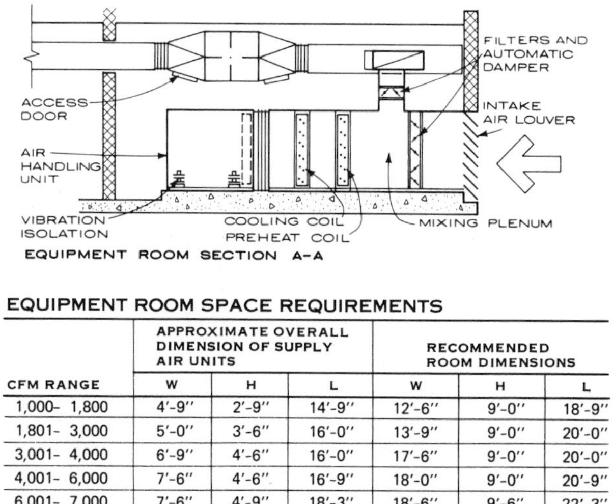

EQUIPMENT ROOM SECTION A–A

EQUIPMENT ROOM SPACE REQUIREMENTS

	APPROXIMATE OVERALL DIMENSION OF SUPPLY AIR UNITS			RECOMMENDED ROOM DIMENSIONS		
CFM RANGE	W	H	L	W	H	L
1,000– 1,800	4′-9″	2′-9″	14′-9″	12′-6″	9′-0″	18′-9″
1,801– 3,000	5′-0″	3′-6″	16′-0″	13′-9″	9′-0″	20′-0″
3,001– 4,000	6′-9″	4′-6″	16′-0″	17′-6″	9′-0″	20′-0″
4,001– 6,000	7′-6″	4′-6″	16′-9″	18′-0″	9′-0″	20′-9″
6,001– 7,000	7′-6″	4′-9″	18′-3″	18′-6″	9′-6″	22′-3″
7,001– 9,000	8′-0″	5′-0″	18′-9″	19′-0″	10′-0″	22′-9″
9,001–12,000	10′-0″	5′-6″	21′-0″	23′-0″	11′-0″	25′-0″
12,001–16,000	10′-3″	6′-0″	22′-0″	23′-6″	12′-6″	26′-0″
16,001–19,000	10′-6″	6′-6″	23′-9″	24′-0″	13′-0″	27′-9″
19,001–22,000	11′-9″	7′-3″	25′-0″	26′-9″	15′-0″	29′-0″
22,001–27,000	11′-9″	8′-6″	26′-0″	27′-0″	16′-0″	30′-0″
27,001–32,000	13′-0″	9′-9″	27′-9″	29′-0″	18′-0″	31′-9″

Fig. 7.35 *Air-handling equipment guidelines. Especially important are the separation of intake and exhaust locations, the isolation of noise and vibration from the fans, and access for maintenance of fans, filters, and coils. The return air fan and automatic damper assist in the operation of the economizer cycle (explained in Fig. 7.30).* Note: *Where higher spaces are available, air handlers may be vertical; this will shorten the length of the unit by between 2 ft and 3 ft 6 in. (Reprinted by permission from AIA, Ramsey, and Sleeper,* Architectural Graphic Standards, *7th ed., © 1981 by John Wiley & Sons.)*

load shedding), elevators, service hot water, access control and security, telecommunications, and information management. Such systems not only maintain comfort with energy conservation, they also sound the alarm about malfunctions, can learn from past practice, and keep records of performance. An example of a central processing system for building management is shown in Fig. 7.37.

This integration and automation of the many building control systems is made possible through *direct digital control* (DDC), which can

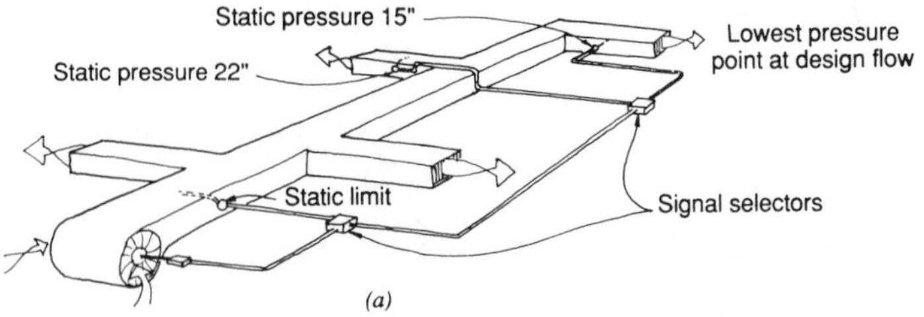

(a)

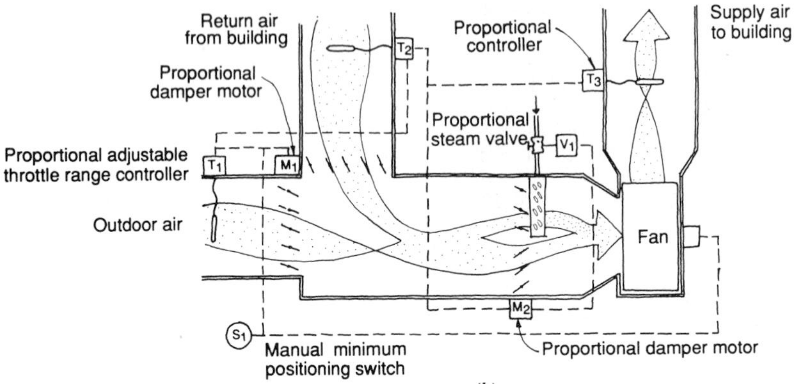

(b)

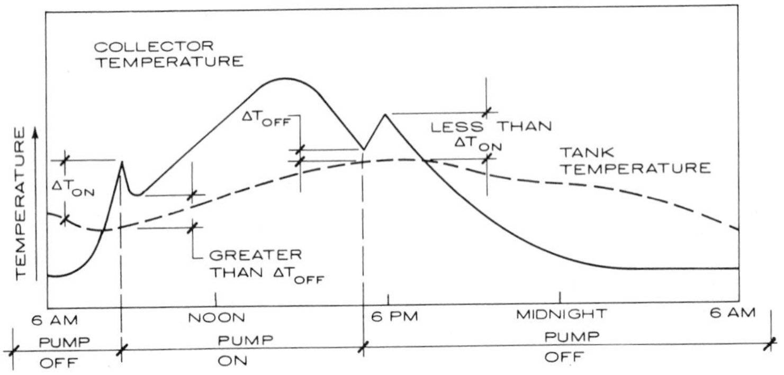

(c)

Fig. 7.36 *Control diagram for some common HVAC applications.* (a) *Single air duct with variable air volume (VAV).* (b) *Economizer cycle.* (c) *Solar water heating: the operating cycle.* (d) *Sensors within the solar water heating system. [*(a), (b), (d) *Adapted from, and* (c) *reprinted by permission from AIA, Ramsey, and Sleeper,* Architectural Graphic Standards, *7th ed.,* © *1981 by John Wiley & Sons.]*

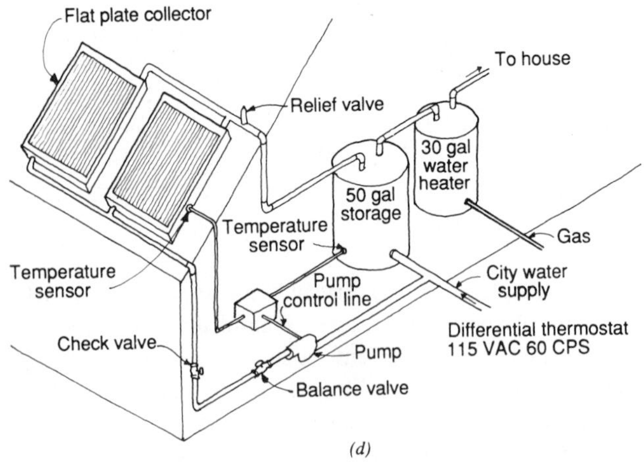

(d)

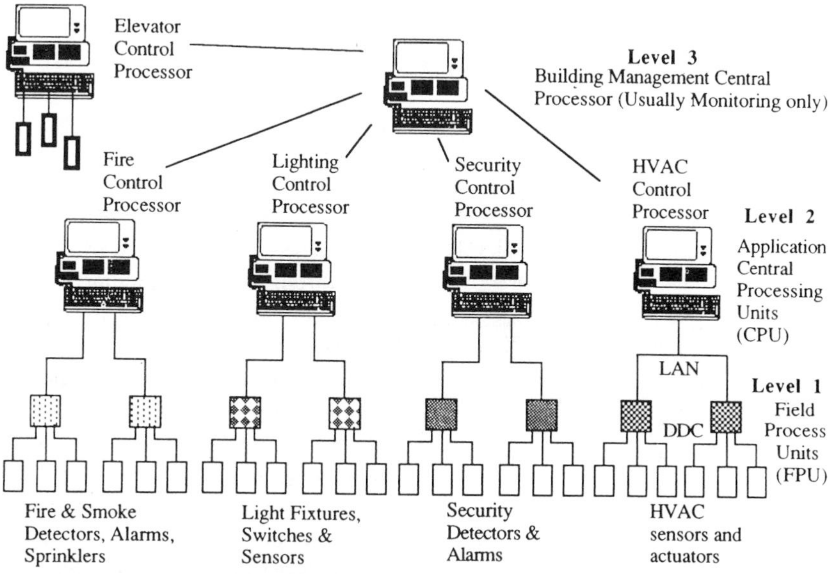

(a)

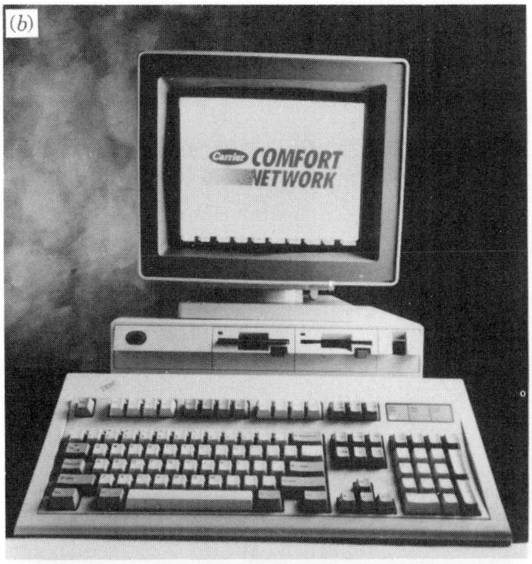

(b)

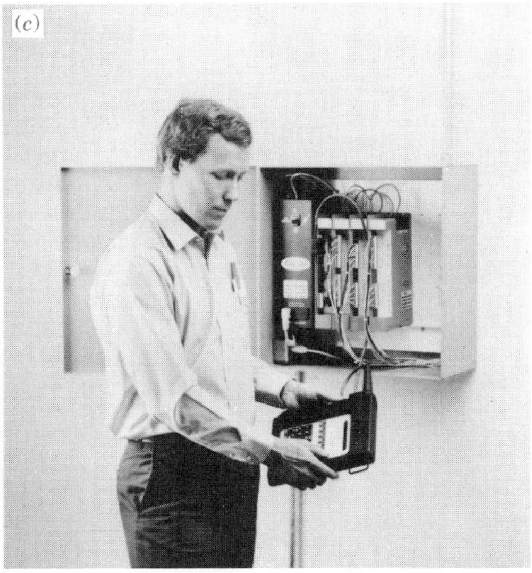

(c)

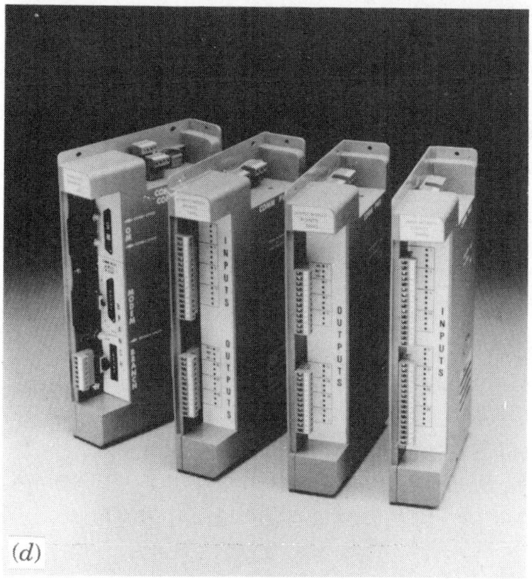

(d)

Fig. 7.37 (a) *Three-level building management system. (Reprinted by permission from ASHRAE Technical Data Bulletin* Intelligent Buildings, *© 1988, American Society of Heating, Refrigerating, and Air-Conditioning Engineers, Inc., Atlanta, Ga.)* (b) *The Carrier Comfort Network has a central control station, and a* (c) *portable owner's module that allows communication with the field-installed devices* (d) *which control individual pieces of equipment. (Courtesy of Carrier Corporation.)*

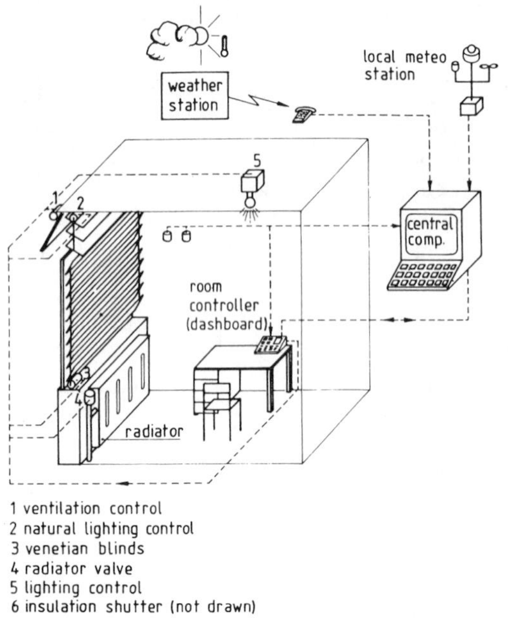

Fig. 7.38 *Control system for thermal and visual comfort in an office building. (Reprinted by permission from ASHRAE Technical Data Bulletin* Intelligent Buildings, *© 1988, American Society of Heating, Refrigerating, and Air-Conditioning Engineers, Inc., Atlanta, Ga.)*

be applied to a wide variety of elements. They provide "dynamic control" in the case of HVAC anticipating time-based changes in heat flow patterns or in occupancy schedules. These actions depend upon direct digital microcontrollers, located on each piece of regulated equipment throughout the building. Three building types show applications of this comprehensive automated control opportunity.

Offices might be provided with DDC not only for the variable-air-volume (VAV) supply units, but inter-tied DDCs for a ventilating window (preventing simultaneous open windows and treated forced-air delivery), daylight reflectors (mini-lightshelves), venetian blinds, radiant heater valve, electric light switch, and an insulating shade, as in Fig. 7.38. A control panel or "dashboard" gives the worker an opportunity to interact with the central control in operating these devices.

Laboratories have proven to be especially difficult HVAC control problems, due to their fume hoods. Conditioned air is provided from the central HVAC system, often by a VAV supply. Whenever a fume hood is exhausting air

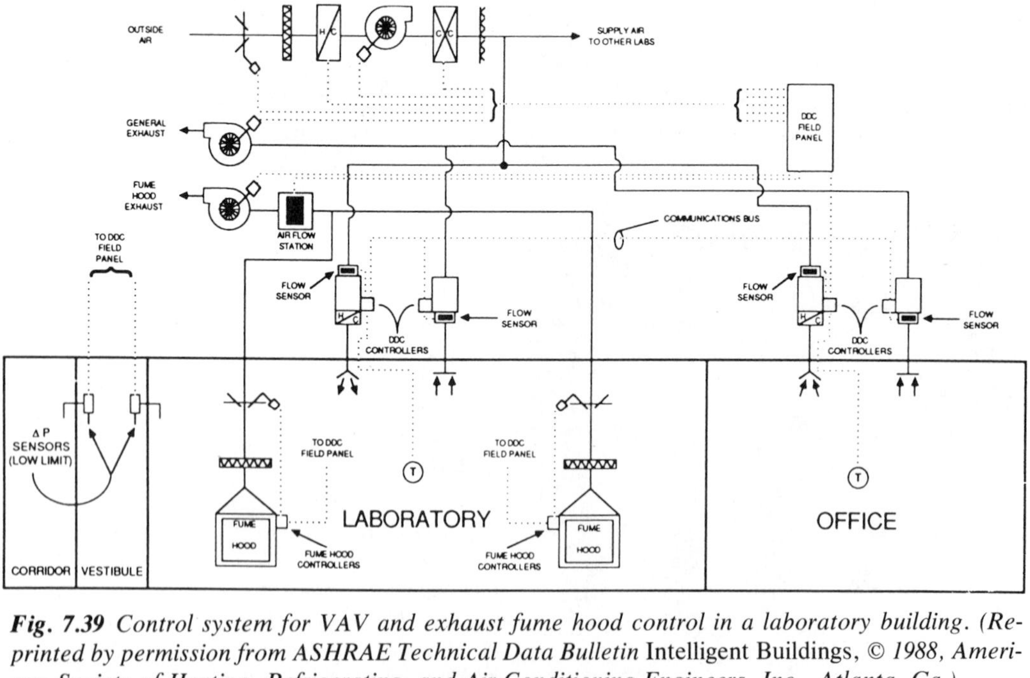

Fig. 7.39 *Control system for VAV and exhaust fume hood control in a laboratory building. (Reprinted by permission from ASHRAE Technical Data Bulletin* Intelligent Buildings, *© 1988, American Society of Heating, Refrigerating, and Air-Conditioning Engineers, Inc., Atlanta, Ga.)*

(frequently in huge quantities), the VAV supply and return systems are affected. Complicating this relationship is the nature of the laboratory work; where the experiment could be damaged by outside contaminants, the lab should be positively pressurized to minimize infiltration (and the VAV must therefore supply and return slightly *more* air than the fume hood exhausts). On the other hand, where the experiments involve diseased, toxic, or other hazardous substances, the lab must be negatively pressurized (and the VAV must supply and return slightly *less* air than the fume hood exhausts). DDCs tied to a central system can balance energy conservation, lab worker safety, and safety for the nonlab environment, as shown in Fig. 7.39.

Hotel rooms can present serious energy loss problems, from heating/cooling either an unoccupied room or a room with open windows. With DDCs inter-tied with the registration desk, an "unoccupied" mode of operating can be remotely controlled, with bare-minimum heating or cooling. When the room is occupied, the supply of either hot water heating or cooled air can be throttled back whenever the window is open. Also, a "purge" mode can enable a new arrival to select a greatly increased flow of outdoor air for a limited time period, to dilute cigarette smoke or other odors. Appendix J presents more detailed information on intelligent buildings.

The Albany, New York, County Airport uses automated controls to regulate solar gain through its large central skylight (Fig. 7.40). The computer monitors indoor and outdoor temperatures, keeps track of solar altitude and azimuth, then regulates the insulated shading louver position according to the building's need for heating or cooling. Solar gain is stored in the masonry wall that supports the skylight. A plenum behind the wall then heats air to be supplied to the vestibule areas of the airport, where winter heat losses are greatest. Photoelectric controls turn off electric lights when daylight is adequate. The skylight meets 40% of the lighting and 20% of the heating needs of this 57,000-ft^2 building.

Many examples of large buildings with central logic control have already appeared in this text—notably, the Bateson Building in Sacramento (Fig. 5.20), in which the decision as to whether to lower the east–west exterior sunshades depends partly on outdoor temperature; and the Occidental Chemical Building in Niagara Falls (Fig. 6.11), in which louvre positions on each of the four elevations are adjusted for a combination of sun position, sky condition (such as overcast), and outdoor temperature. Central logic control allows for a variety of switches that not only improve a building's performance, but also can enliven its appearance.

7.4 All-Air HVAC Systems

This large and complex family of systems was introduced in Section 7.1. All-air HVAC systems require large distribution trees but can promise comfortable results. With all-air systems, air quality can be manipulated through pressure control: *negative pressures* in odorous or excessively humid locations (kitchens, toilet rooms, pet shops in shopping centers, etc.), *positive pressures* in shopping malls, corridors of apartment houses, stair towers, and so on. The difference in pressure sets up an overall direction of air flow that helps prevent the spread of odorous or otherwise contaminated air and can even help manage smoke in a fire (see Chapter 13). Positive pressures in connecting spaces help each adjacent space keep its air to itself. All-air systems also offer an opportunity for increased electric lighting efficiency: because so much air is being moved, many supply air outlets and return air inlets are needed in each space. Return air can be channeled through lighting fixtures (luminaires), to serve two useful purposes: first, to lower the temperature of the air around the fluorescent tubes and thus (with most types of fluorescents) increase their light output (see Chapter 19); second, to whisk away much of the light's heat output *before* it can influence the temperature of the space. This heat is immediately available for use elsewhere, or for rejection to the outside, instead of increasing the need for cooling in the space. When the lighting heat is reused, the system is called a "heat-of-light" system.

(a) Component Sizing. Approximate sizes of mechanical equipment spaces were shown in

Fig. 7.40 The Albany, New York, County Airport features a central skylight (a) *that provides 40% of the light and 20% of the heat for the 57,000-ft² building.* (b) *The insulated louvers are controlled by computer to admit or block the sun, and to store winter night heat within the building. (Courtesy of Einhorn Yaffee Prescott, Architects, Albany, New York.)*

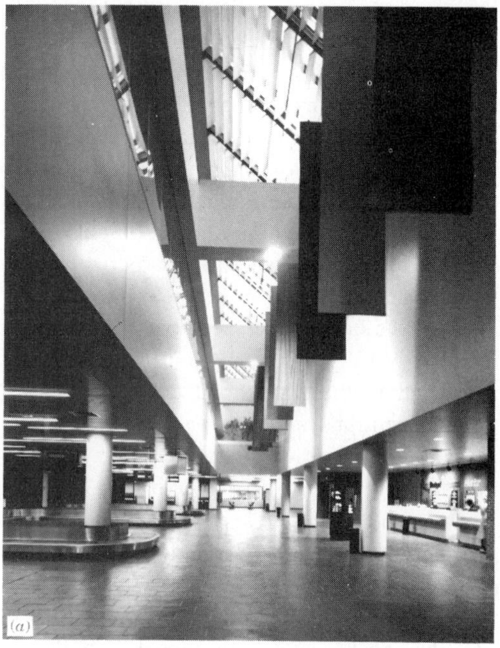

(*a*)

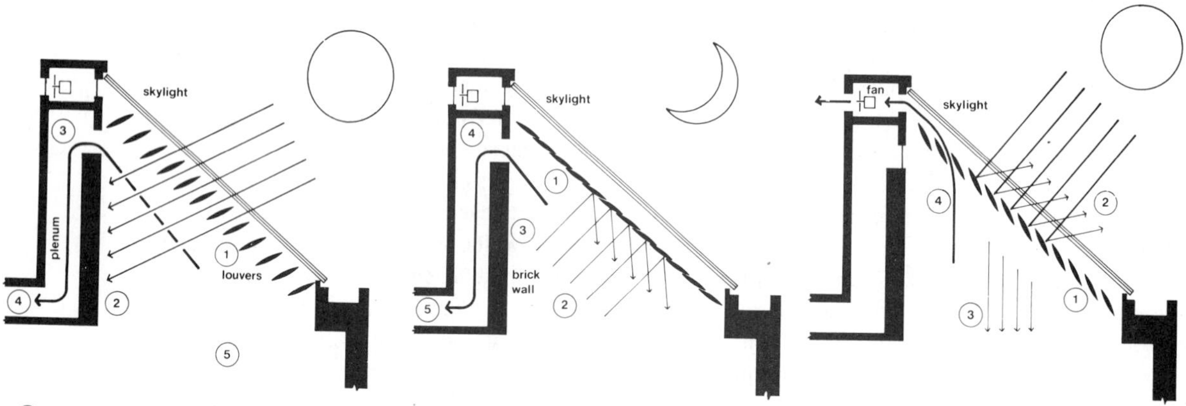

1. The operable louvers are open to allow sunlight into the building.
2. The sunlight warms a brick wall at the back of the skylight.
3. Air is drawn through a plenum behind the wall, and is heated during its passage over the warm bricks.
4. The heated air is drawn into the building heating system for distribution.
5. Natural light is provided to the space below.

Sunny Winter Day

The sun is used to provide heat and light.

1. The operable louvers are closed to provide an insulated "ceiling" below the glass and reduce heat loss through the skylight.
2. The louvers are insulated, blocking the flow of heat from the building.
3. The brick wall remains warm for some time after sunset due to its "thermal mass".
4. Air is drawn through a plenum behind the wall, and is heated during its passage over the warm bricks. When the heat stored in the brick wall is used up, air flow through the plenum is stopped.
5. The heated air is drawn into the building heating system for distribution.

Winter Night

Loss of heat through the skylight is reduced.

1. The operable louvers are partially open.
2. Direct sunlight is reflected by the louvers.
3. Diffuse northern light is provided to the spaces below.
4. The area below the skylight serves as a heat banking space. Warm air from the building below will collect beneath the skylight. Exhaust fans will then discharge the warm air from the building.

Sunny Summer Day

The sun is used to provide light, but is not allowed to penetrate into the building and generate excessive heat.

(*b*)

Table 5.9 and air-handling equipment in Table 5.10. At this stage, more detailed sizing guidelines may be useful. Table 7.5 shows approximate air quantities for various applications, depending upon internal gain and external orientation.

Air Ducts. Air duct sizes in cross section are frequently of interest early in the design process. Duct depths can help determine floor heights; duct cross sections influence the sizes and shapes of the vertical cores that serve multistory buildings. An approximation of duct size can be obtained as follows.

1. Determine the quantity of air to be distributed through the largest duct, using Table 7.5, or the air changes per hour, from a calculation of night ventilation of thermal mass (e.g., Example 5.1, Part M). This will usually be expressed in cubic feet per hour (cfh).
2. Convert cfh to cfm:

$$\frac{\text{cfh} \times 1\text{ h}}{60\text{ min}} = \text{cfm}$$

3. Find the recommended velocity of this air within the duct from Table 6.18 (page 366), expressed in feet per minute (fpm).
4. The approximate required cross-sectional area of the duct A is then

$$A_{\text{in.}^2} = \frac{\text{volume of air (cfm)}}{\text{velocity (fpm)}} \times 144\text{ in.}^2/\text{ft}^2 \times \text{friction allowance}$$

where the friction allowances are

round ducts = 1.0 (may be neglected)

nearly square ducts (ratio of width to depth, 1 : 1):

small (<1000 cfm) = 1.10

large (>1000 cfm) = 1.05

thin rectangular ducts (ratio of width to depth, 1 : 5) = 1.25

EXAMPLE 7.1. Approximate the largest duct sizes required to serve the Oregon office building shown in Fig. 5.4, page 191. On the building's upper floor are five bays, each 1440 ft^2, for a total of 7200 ft^2.

SOLUTION. To enter Table 7.5 for office buildings, this example's occupancy is 1440 ft^2/8 persons = 180 ft^2/person. Lights were shown (Example 5.1, Part C) to be 2 W/ft^2. Thus the "private offices," "low" range, applies. For zones of this office type, at any orientation, the "low" air quantity is 0.25 cfm/ft^2. The building's mechanical room is beyond the east end of this upper floor, so all the HVAC system supply air must enter through one end of the central duct. This duct must carry

$$0.25\text{ cfm/ft}^2 \times 7200\text{ ft}^2 = 1800\text{ cfm}$$

(Part of this 1800 cfm is the outdoor air provided, at 160 cfm/bay × 5 bays = 800 cfm.)

Within this duct, an air velocity must be established. From Table 6.18, the recommended velocity for main ducts in public buildings is 1000 to 1300 fpm; for branch ducts, 600 to 900 fpm. The lower the velocity, the larger, quieter, and less fan-energy intensive will be the duct; assume a lower velocity midway between "main" and "branch" in this example (800 fpm), because the main duct is so acoustically exposed to the space it serves. The friction allowance for the nearly square, large central duct can be assumed at 1.05. This supply duct will therefore be, at maximum:

$$A_{\text{in.}^2} = \frac{1800\text{ cfm} \times 144\text{ in.}^2/\text{ft}^2 \times 1.05}{800\text{ fpm}} = 340\text{ in.}^2$$

(The actual central supply duct size in this building's upper floor is 380 in.^{2} maximum. Compare the results above to the quicker sizing procedure of Example 5.1, Part I.)

(b) Accessories. Air diffusers and grilles (Fig. 7.41) are common to all these systems. The most popular locations for these accessories are ceilings, which are uncluttered by furniture and relatively unaffected by minor rearrangements in partitions. Also, the entire space above a suspended ceiling can be used as one huge return air duct; this arrangement is called plenum return. In a thermal zone where cooling is the prevalent condition, cool air introduced at the ceiling is ideal; being heavier, it sinks naturally upon entry. However, low wall or floor return air grilles will assure a better circulation of air. In a thermal zone where heating is the prevalent condition, the reverse locations are preferred—

TABLE 7.5 **All-Air HVAC Systems: Approximate Air Quantities and Refrigeration Sizes**

	Internal Gains						Approximate Requirements											
	Occupancy (ft^2/person)			Lights and Equipment[a] (W/ft^2)			Refrigeration[b] (ft^2/ton)			Air Quantities (cfm/ft^2)								
										East–South–West			North			Internal		
Applications	Low	Av.	High	Low	Av.	High	Low	Av.	High	Low	Av.	High	Low	Av.	High	Low	Av.	High
Apartment, high-rise	325	175	100	1.0	2.0	4.0	450	400	350	0.8	1.2	1.7	0.5	0.8	1.3	—	—	—
Auditoriums, churches, theaters	15	11	6	1.0	2.0	3.0	400	250	90	—	—	—	—	—	—	1.0	2.0	3.0
Educational facilities	30	25	20	2.0	4.0	6.0	240	185	150	1.0	1.6	2.2	0.9	1.3	2.0	0.8	1.2	1.9
Factories																		
Assembly areas	50	35	25	3.0[c]	4.5[c]	6.0[c]	240	150	90	—	—	—	—	—	—	2.0	3.6	5.5
Light manufacturing	200	150	100	9.0[c]	10.0[c]	12.0[c]	200	150	100	—	—	—	—	—	—	1.6	2.5	3.8
Heavy manufacturing[d]	300	250	200	15.0[c]	45.0[c]	60.0[c]	100	80	60	—	—	—	—	—	—	2.5	4.0	6.5
Hospitals																		
Patient rooms[e]	75	50	25	1.0	1.5	2.0	275	220	165	0.33	0.55	0.67	0.33	0.50	0.67	—	—	—
Public areas	100	80	50	1.0	1.5	2.0	175	140	110	1.0	1.25	1.45	1.0	1.1	1.2	0.95	1.0	1.1
Hotels, motels, dormitories	200	150	100	1.0	2.0	3.0	350	300	220	1.0	1.40	1.5	0.9	1.2	1.4	—	—	—
Libraries and museums	80	60	40	1.0	1.5	3.0	340	280	200	1.0	1.6	2.1	0.9	1.1	1.3	0.9	1.0	1.1
Office buildings[e]	130	110	80	4.0	6.0[c]	9.0[c]	360	280	190	0.25	0.5	0.9	0.25	0.5	0.8	0.8	1.1	1.8
Private offices[e]	150	125	100	2.0	5.8	8.0	—	—	—	0.25	0.5	0.9	0.25	0.5	0.8	—	—	—
Stenographic	100	85	70	5.0[c]	7.5[c]	10.0[c]	—	—	—	—	—	—	—	—	—	0.9	1.3	2.0
Residential																		
Large	600	400	200	1.0	2.0	4.0	600	500	380	0.8	1.2	1.6	0.5	0.8	1.3	—	—	—
Medium	600	360	200	0.7	1.5	3.0	700	550	400	0.7	1.1	1.4	0.5	0.7	1.2	—	—	—
Restaurants																		
Large	17	15	13	1.5	1.7	2.0	135	100	80	1.8	2.4	3.7	1.2	1.6	2.1	0.9	1.1	1.4
Medium							150	120	100	1.5	2.0	3.0	1.1	1.4	1.8	0.9	1.0	1.3

Shopping Centers																		
Beauty and barber shops	45	40	25	3.0[c]	5.0[c]	9.0[c]	240	160	105	1.5	2.6	4.2	1.1	1.7	2.6	0.9	1.3	2.0
Department stores																		
Basement	30	25	20	2.0	3.0	4.0	340	285	225	—	—	—	—	—	—	0.7	1.0	1.2
Main floor	45	25	16	3.5	6.0[c]	9.0[c]	350	245	150	—	—	—	—	—	—	0.9	1.4	2.0
Upper floors	75	55	40	2.0	2.5	3.5[c]	400	340	280	—	—	—	—	—	—	0.8	1.0	1.2
Dress shops	50	40	30	1.0	2.0	4.0	345	280	185	0.9	1.2	1.6	0.7	1.0	1.4	0.6	0.8	1.1
Drug stores	35	23	17	1.0	2.0	3.0	180	135	110	1.8	2.3	3.0	1.0	1.4	1.8	0.7	1.0	1.3
Variety stores	35	25	15	1.5	3.0	5.0	345	220	120	0.7	1.4	2.0	0.6	1.2	1.6	0.5	0.9	1.1
Hat shops	50	43	30	1.0	2.0	3.0	315	270	185	1.0	1.3	1.9	0.7	1.0	1.5	0.6	0.8	1.2
Shoe stores	50	30	20	1.0	2.0	3.0	300	220	150	1.2	1.6	2.1	1.0	1.4	1.8	0.8	1.0	1.2
Malls	100	75	50	1.0	1.5	2.0	365	230	160	—	—	—	—	—	—	1.1	1.8	2.5
Refrigeration for central heating and cooling plant																		
Urban districts							475	380	285									
College campuses							400	320	240									
Commercial centers							330	265	200									
Residential centers							625	500	375									

Source: Reprinted by permission from *The ABC's of Air Conditioning,* © 1975, Carrier Corporation.

[a]Codes limit installed electric lighting to lower levels in many cases. Equipment such as computer terminals may, in turn, increase these loads.

[b]Refrigeration loads are for entire application.

[c]Includes other loads that are ordinarily expressed in watts per square foot.

[d]Here, air quantities assume that supplementary means are used to remove excess heat.

[e]Induction (air & water) systems are assumed for hospital patient rooms and for the perimeter zones of office buildings.

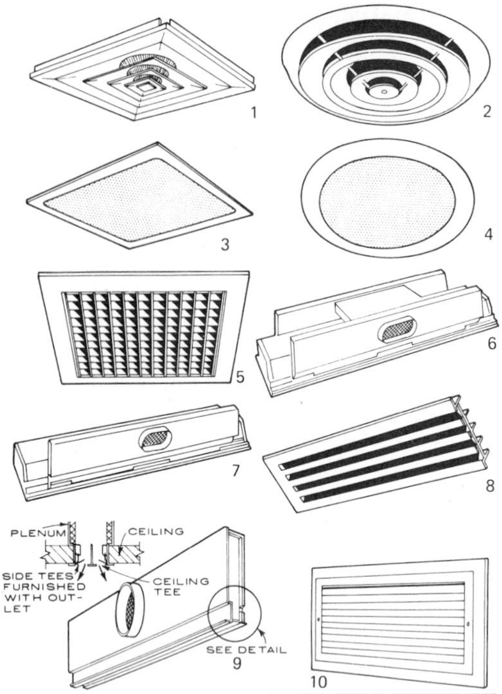

AIR DISTRIBUTION OUTLETS

KEY

1 RECTANGULAR LOUVERED FACE DIFFUSER: Available in 1, 2, 3, or 4-way pattern, steel or aluminum. Flanged overlap frame or inserted in 2 × 2 ft or 2 × 4 ft baked enamel steel panel to fit tile modules of lay-in ceilings. Supply or return.

2 ROUND LOUVERED FACE DIFFUSER: Normal 360° air pattern with blank-off plate for other air patterns. Surface mounting for all type ceilings. Normally of steel with baked enamel finish. Supply or return.

3 RECTANGULAR PERFORATED FACE DIFFUSER: Available in 1, 2, 3, or 4-way pattern, steel or aluminum. Flanged overlap frame or 2 × 2 ft and 2 × 4 ft for replacing tile of lay-in ceiling can be used for supply or return air.

4 ROUND PERFORATED FACE DIFFUSER: Normal 360° air pattern with blank-off plate for other air patterns. Steel or aluminum. Flanged overlap frame for all type ceilings. Can be used for supply or return air.

5 LATTICE TYPE RETURN: All aluminum square grid type return grille for ceiling installation with flanged overlap frame or of correct size to replace tile.

6 SADDLE TYPE LUMINAIRE AIR BOOT: Provides air supply from both sides of standard size luminaires. Maximum air delivery (total both sides) approximately 150 to 170 cfm for 4 ft long luminaire.

7 SINGLE SIDE TYPE LUMINAIRE AIR BOOT: Provides air supply from one side of standard size luminaires. Maximum air delivery approximately 75 cfm for 4 ft long luminaire.

8 LINEAR DIFFUSER: Extruded aluminum, anodized, duranodic, or special finishes, one way or opposite direction or vertical down air pattern. Any length with one to eight slots. Can be used for supply or return and for ceiling, sidewall, or cabinet top application.

9 INTEGRATED PLENUM TYPE OUTLET FOR "T" BAR CEILINGS: Slot type outlet, one way or two way opposite direction air pattern. Available in 24, 36, 48, and 60 in. lengths. Replaces or integrates with "T" bar. Approximately 150 to 175 cfm for 4 ft long, two-slot unit.

10 SIDEWALL OR DUCT MOUNTED REGISTER: Steel or aluminum for supply or return. Adjustable horizontal and vertical deflection. Plaster frame available. Suitable for long throw and high air volume.

(a)

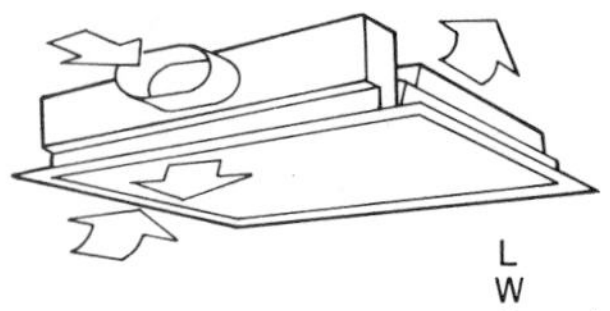

L 24" to 72"
W 24" to 72"
Capacity varies with air velocity

(b)

Fig. 7.41 *Common air distribution outlets.* (a) *Air diffusers of types 1 to 5 are for typical ceiling application; types 6 and 7 integrate with luminaires; types 8 and 10 may be used in other-than-ceiling applications; type 9 is a nearly "invisible" hung-ceiling adaptation.* (b) *Typical heat-of-light luminaires that integrate both supply and return air. (Reprinted by permission from AIA, Ramsey, and Sleeper,* Architectural Graphic Standards, *7th ed., © 1981 by John Wiley & Sons.)* (c) *Integral electric lighting, air distribution, and sound absorption, in an application at the* (d) *American Republic Insurance Building, Des Moines, Iowa; Skidmore Owings & Merrill, architects.*

supply low, return high. The final size of the supply registers depends heavily on the face velocity and the throw pattern of the air (see again Fig. 6.58 and Table 6.19) which are given in manufacturers' catalogs.

Noise within air ducts is commonly caused by excessive velocity and/or turbulence; this is not always avoidable. Techniques for noise suppression in ductwork are discussed in Chapter 27.

Where all-air systems are used throughout larger buildings, another accessory is the *draft barrier,* which can alleviate minor and temporary cold spots in an otherwise completely internal-load-dominated building. Large exterior glass areas cool the adjacent interior air in winter. The vertical layer of cool, denser air thus created then drops to the floor and blankets it like a chilly carpet. Unless one has witnessed tests using smoke and recording thermometers, the speed and resulting discomfort of this phenomenon are seldom fully comprehended.

Under the worst conditions, this cold layer must cross the space and reach a thermostat on an interior wall before a below-the-glass heating element takes over. When it does, the problem is a dual one—to reverse the downslip of air *and* to warm the space. The heating element is sel-

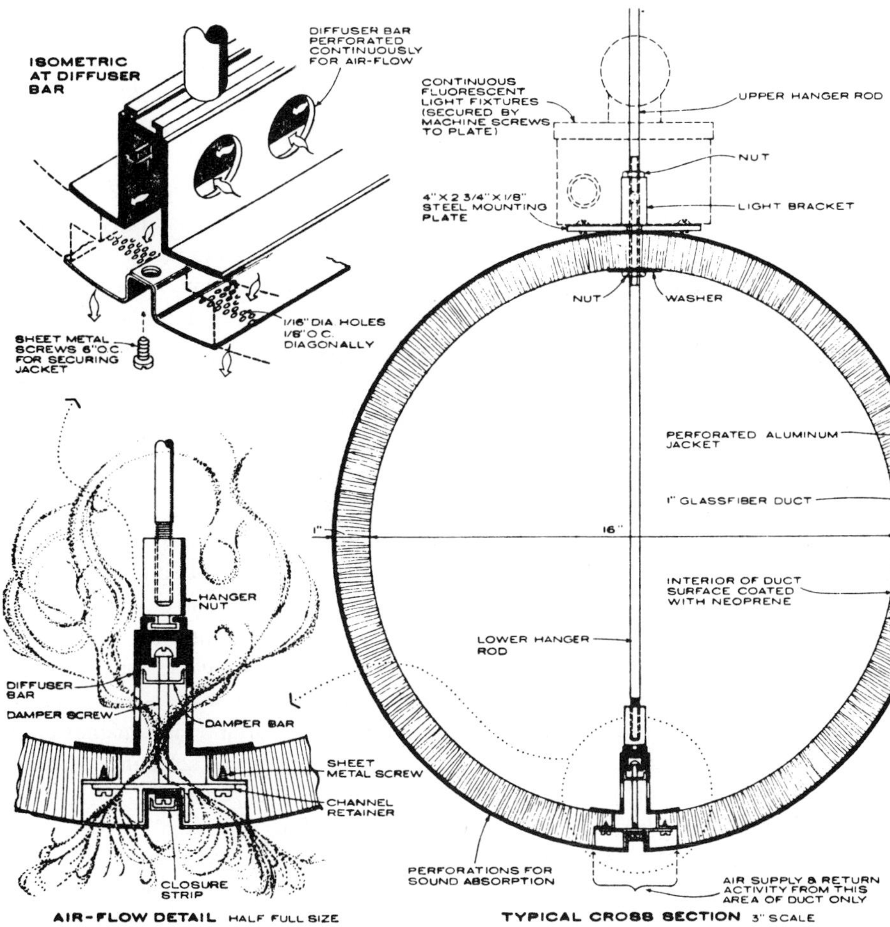
ISOMETRIC AT DIFFUSER BAR
DIFFUSER BAR PERFORATED CONTINUOUSLY FOR AIR-FLOW
SHEET METAL SCREWS 6" O.C. FOR SECURING JACKET
1/16" DIA HOLES 1/8" O.C. DIAGONALLY
CONTINUOUS FLUORESCENT LIGHT FIXTURES (SECURED BY MACHINE SCREWS TO PLATE)
UPPER HANGER ROD
NUT
4" X 2 3/4" X 1/8" STEEL MOUNTING PLATE
LIGHT BRACKET
NUT
WASHER
PERFORATED ALUMINUM JACKET
1" GLASSFIBER DUCT
1"
16"
INTERIOR OF DUCT SURFACE COATED WITH NEOPRENE
LOWER HANGER ROD
HANGER NUT
DIFFUSER BAR
DAMPER SCREW
DAMPER BAR
SHEET METAL SCREW
CHANNEL RETAINER
CLOSURE STRIP
PERFORATIONS FOR SOUND ABSORPTION
AIR SUPPLY & RETURN ACTIVITY FROM THIS AREA OF DUCT ONLY
AIR-FLOW DETAIL HALF FULL SIZE
TYPICAL CROSS SECTION 3" SCALE

(c)

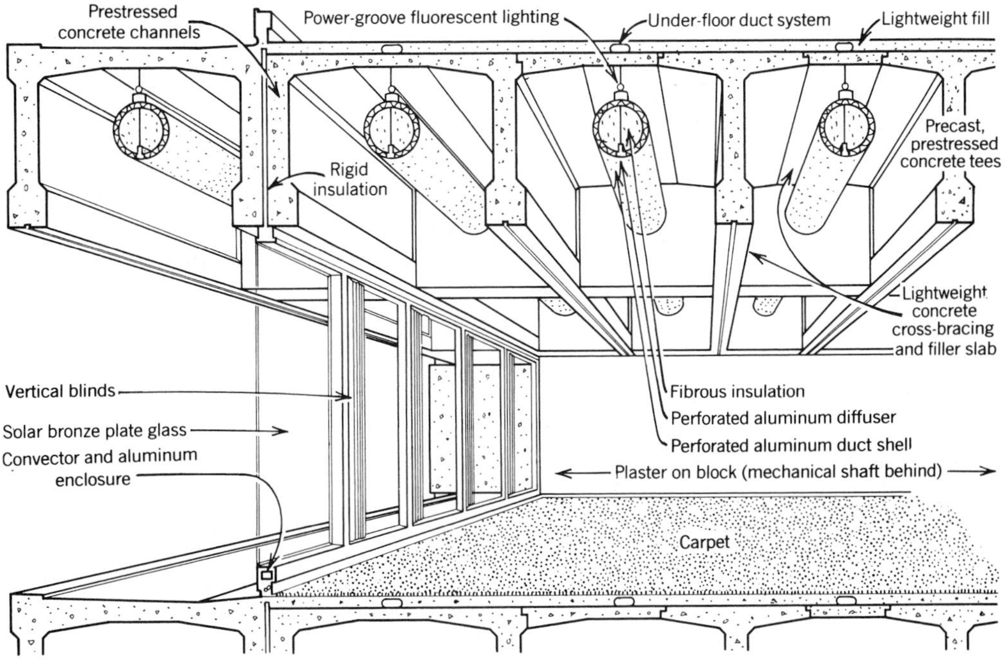
Prestressed concrete channels
Power-groove fluorescent lighting
Under-floor duct system
Lightweight fill
Precast, prestressed concrete tees
Rigid insulation
Lightweight concrete cross-bracing and filler slab
Vertical blinds
Solar bronze plate glass
Convector and aluminum enclosure
Fibrous insulation
Perforated aluminum diffuser
Perforated aluminum duct shell
Plaster on block (mechanical shaft behind)
Carpet

(d)

Fig. 7.41 *(continued)*

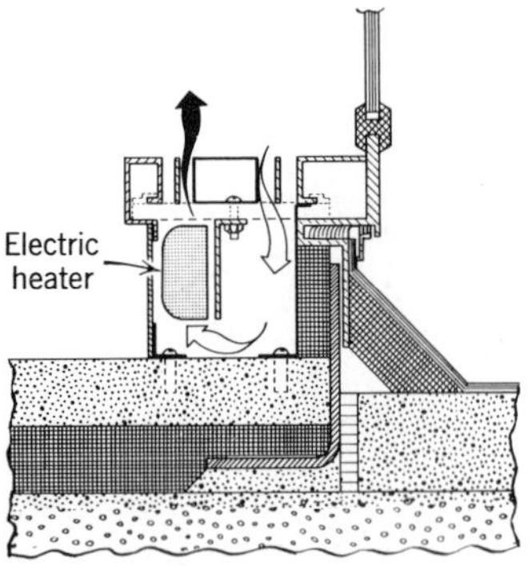

Fig. 7.42 Electric resistance draft barrier, for use below windows in buildings that generally require cooling.

dom properly adapted to meet both challenges correctly.

Admittedly, the use of a modulated heating medium—water or air—will provide a more continuous operation to maintain a warm upflow of air at glass, yet during even brief intervals between these heating periods, the cold air slides quickly past the nonoperative heaters. An electric resistance draft barrier for under-window application is shown in Fig. 7.42.

(c) Single-Zone Systems (Fig. 7.43). Smaller buildings are often served by this least-complicated all-air example (see examples in Section 6.7). Where an entire building is essentially one thermal zone, uncomplicated enough to be served by only one thermostat, choice of this system offers very low first cost. A closely related system is the multizone (see below), which is a combination of single zones.

(d) Single-Duct, Variable-Air Volume (VAV) Systems (Fig. 7.44). As described in Section 7.1, this system is an energy-saving option ideally suited to the internal zones of large buildings, where cooling is always needed. But it can also be adapted to serve the entire building, with important savings over the constant-volume (CV) systems described below [paragraphs (*e*) to (*g*)]. The air-handling units (fans, etc.) for each thermal zone are sized to meet the peak demand on that zone. In most CV systems, the fans always run at this peak condition speed, even though peaks are often of rather short duration. In VAV systems, fans only run at peak speed during peak hours. This obviously saves considerable energy needed to run fans. When the building uses only one or a few central fans, a VAV system will require smaller fans, because the fans need meet only one peak condition at a time and do not have to be sized for all zones' peak flows. The variation in demands on the fan can be met either by selecting variable-pitch blades or (less expensively) by varying the speed of the fan.

Where a VAV system is used, provision must be made for at least code-minimum fresh air levels, and noise may become a problem. Although VAV systems are typically less noisy than CV systems (because less air is moving), the air motion noise *varies* with the volume, and variable noise sources are inherently more noticeable than are steady ones.

Within (or near) the spaces it serves, the VAV system typically needs a mixing box or terminal; this is often placed above a suspended ceiling. Of the several terminal types available, the standard and most simple one is shown in

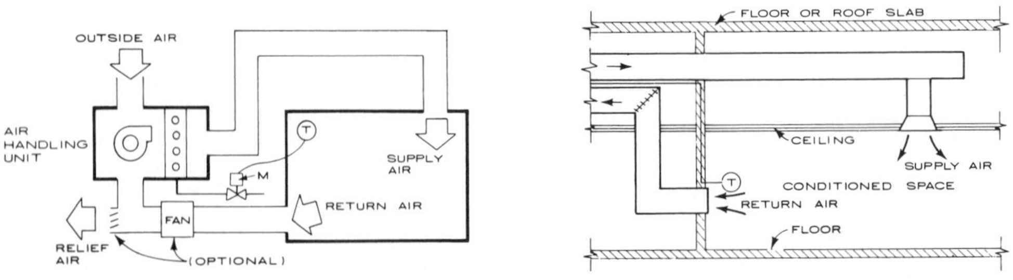

Fig. 7.43 Single-zone system. (Reprinted by permission from AIA, Ramsey, and Sleeper, Architectural Graphic Standards, *7th ed., © 1981 by John Wiley & Sons.)*

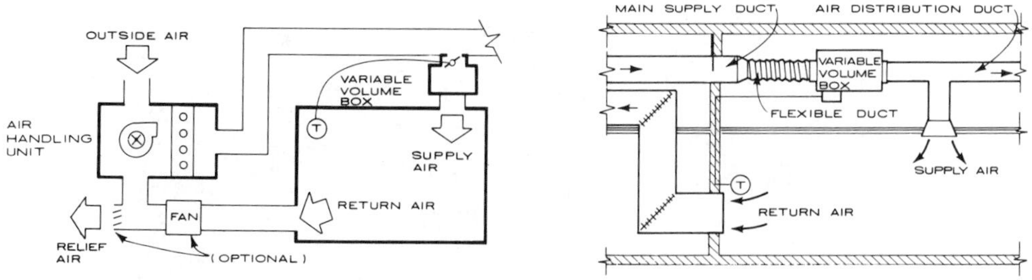

Fig. 7.44 Single-duct, variable-air-volume (VAV) system. (Reprinted by permission from AIA, Ramsey, and Sleeper, Architectural Graphic Standards, *7th ed., © 1981 by John Wiley & Sons.)*

Fig. 7.45. This terminal serves not only to vary the quantity of air, but to both attenuate the noise and reduce the velocity of air from the main trunk of the distribution tree. High velocity is commonly used in main ducts, because it reduces the size of these critical, large portions of the tree.

There are several variations on the basic VAV system, some of which respond to the problem of minimum airflows for rooms with little thermal load or to the desire to serve both interior and perimeter areas with the same HVAC system.

To maintain minimum fresh air, VAV terminals are often set so that they cannot be entirely closed off—a provision that ensures some outdoor air at all times. If this fresh air minimum, entering at low velocity, does not provide the desired air motion and mixing within the room, VAV terminals can be equipped with fans, which are activated as needed with decreasing incoming air volume. These self-contained fans mix room air with incoming air to provide an airstream of the right temperature and velocity to maintain comfort.

A variation of VAV shown in Fig. 7.46 allows much more individual control at each workstation. A mixture of outdoor and recirculated indoor air, called *primary air,* is brought from the main duct to each workstation's mixing box, in a duct carrying at maximum 150 cfm (typically, less). Each worker can adjust supply air temperature, the mixture of primary and locally recirculated air, air velocity (and therefore volume) and direction, radiant supplementary heat (below desk level), and even task lighting level and background (masking) sound level. At last, we have almost as much environmental control at our office workstation as we have in our automobile's front seat.

This approach to VAV is appropriate when simultaneous heating and cooling are needed. This is particularly true at perimeter zones, which can generate sizable heating needs while the rest of the building needs cooling. Another approach is to utilize an induction-type VAV terminal, with which air heated by electric lights

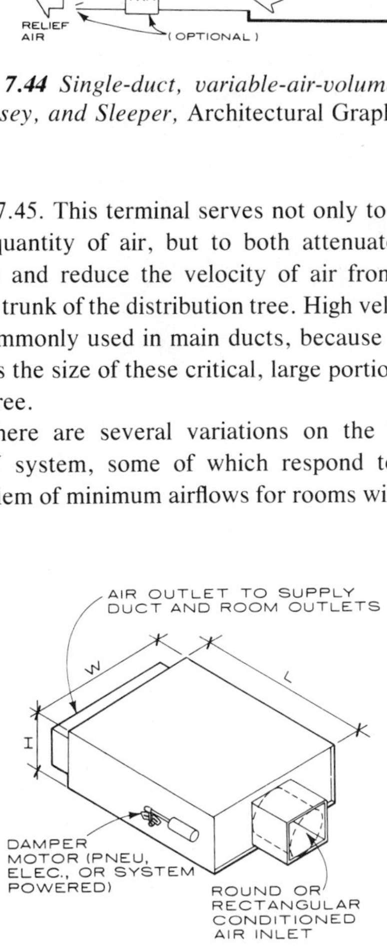

RANGE OF DIMENSIONS

CFM	HEIGHT	LENGTH	WIDTH
400	8″- 9″	24″-39″	14″-30″
800	10″-11″	24″-53″	18″-42″
1600	14″	30″-48″	22″-44″
2400	16″	42″-60″	26″-54″
3200	18″	42″-67″	33″-54″

Fig. 7.45 The most simple type of VAV terminal, which pinches back the volume of incoming air as thermal loads decrease. (Reprinted by permission from AIA, Ramsey, and Sleeper, Architectural Graphic Standards, *7th ed., © 1981 by John Wiley & Sons.)*

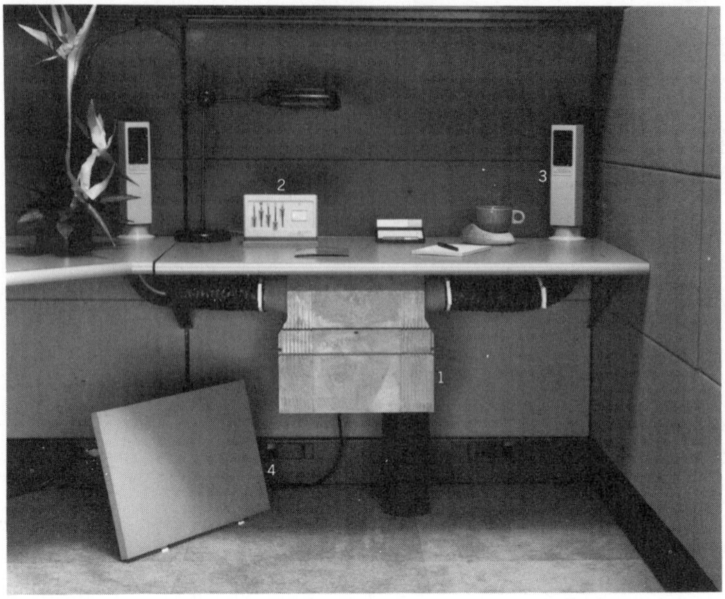

Fig. 7.46 *The Personal Environments workstation includes (1) fans, electrostatic air filters, air mixing box, and a background sound ("white noise") generator. (2) The control panel allows adjustment of task lighting, background sound, fan speed, primary/recirculated air mixture, and radiant heating. (3) Two diffusers (served by flexible ducts) distribute both air and background sound. Diffusers are variable about both horizontal and vertical axes. (4) The radiant heating panel warms the lower body. (Courtesy of Johnson Controls, Inc., Milwaukee, Wisconsin.)*

is induced to join the incoming cool airstream. Greater heating needs often require the use of reheat terminals supplied by a circulating hot water system or by electric resistance heating. In this reheat application (more energy efficient than the standard—CV—reheat system described in paragraph (*e*), because a much smaller volume of air is first cooled, then reheated), the water or electric coils can be incorporated either in the VAV terminal or in the ductwork between the terminal and the space it serves.

Another approach is to provide perimeter radiation via a water system, usually placed under the window while the VAV supply is in the ceiling above (see Fig. 7.47).

A final option for VAV at perimeters is the *dual-conduit* approach, diagrammed in Fig. 7.2. The VAV conduit works year-round, providing constant-temperature (cool) fresh air. The CV conduit carries hot air in very cold weather and cool air (supplementing the VAV conduit) in very hot weather or in the case of high solar gains.

An example of a VAV system used in a retrofit is shown in Fig. 7.48. The world headquarters of the Manufacturers Hanover Trust Company is a 52-story, 1.5-million-ft^2 building on New York's Park Avenue. Built over 20 years ago with a perimeter induction HVAC system and an interior double-duct system, it was recently refurbished at a cost of some \$60 million. Engineers Syska and Hennessy retained the perimeter induction units but replaced the double-duct interior system with VAV. With the aid of new air-handling equipment, rehabilitated chillers, and a more efficient lighting system (from 4 W/ft^2 to less than 2 W/ft^2), the annual energy budget is expected to drop from 123,000 Btu/ft^2 per year to between 80,000 and 90,000 Btu/ft^2 per year. Window glass constitutes almost 70% of the building's skin area, which helps explain such high energy consumption.

A final example is a floor-by-floor VAV system (Fig. 7.49) in a 1-million-ft^2 medium-rise (28-floor) Chicago office building designed (and occupied in part) by Skidmore Owings & Merrill.

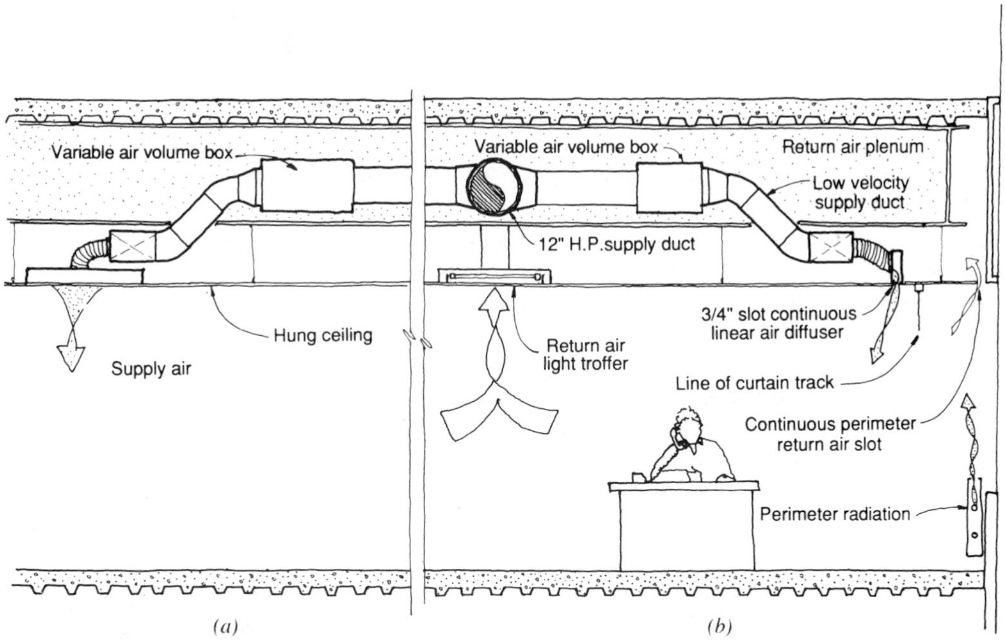

Fig. 7.47 *VAV in Denver's Anaconda Tower. Each VAV box (terminal) is controlled by a wall thermostat.* (a) *Interior zones use VAV boxes that induce low-pressure return air in the plenum to join the incoming high-pressure fresh air, for delivery to the space. Light troffers are used to admit return air to the plenum, as well as to supply air.* (b) *Perimeter zones receive only supply air, delivered through a continuous ¾-in. slot that skirts the building perimeter inside the curtain track. A hot water perimeter radiation system works on cold days. (Anaconda Tower: Skidmore Owings & Merrill, architects; Flack & Kurtz, engineers.)*

This lower, wider approach to office buildings utilizes three "stacked" atriums to relieve the monotony of the wide interior floors. Another result is lower structural and energy costs per square foot relative to conventional high-rise structures. The cubelike shape of the building exposes less skin area (38% of which is in insulating glass) to Chicago's cold winters; electric lighting at about 1.8 W/ft^2 holds down internal gains. To accommodate the differing schedules and comfort needs of a variety of tenants, each floor is provided with two VAV supply fans that can be operated nights and weekends, independently of the rest of the building. The mechanical core has one exterior wall (on an alley) to facilitate fresh air intake/stale air exhaust. The perimeter heating system is electric resistance fin radiation; an economizer cycle provides cooling with outdoor air below 55 F.

(e) Single Duct with Reheat (Fig. 7.50). Formerly a widespread system and usually supplied at constant volume (CV), single-duct reheat systems now are most often used when a constant air volume is important (as in hospitals and laboratories with large quantities of exhaust air) and the energy consumed in the reheating of cooled air is not a primary concern. For high-pressure and high-velocity main ducts, constant-volume reheat boxes (Fig. 7.50*b*) are used to control the noise, temperature, pressure, and velocity of the supply air. For more simple all-low-velocity systems, simple duct insert heaters (Fig. 7.50*c*) can be used. These are sized to fit the duct, which needs to be enlarged only slightly to accommodate them.

(f) Multizone Systems (Fig. 7.51). A multizone system is a collection of single-zone systems served by a single supply fan; such systems rarely exceed eight zones per air-handling unit. Simultaneous heating of some zones and cooling of others is possible, but leakage between zones at the decks of hot and cold coils is common. Return air from all zones is mixed within one return duct. One multizone system per floor of medium- to high-rise buildings is an increasingly common application.

(g) Double-Duct Systems (Fig. 7.52). Still considered the "Cadillac" of HVAC systems, because of its superior comfort control, the double-duct system is more rarely seen now, due to increased energy costs and the size of such systems. Building volumes needed for a double-duct system's three (two supply, one return) full-sized air distribution trees are harder to justify, given that VAV systems can provide acceptable comfort for most of the spaces.

A most expressive example of a double-duct system serving a building's perimeter zone is the Blue Cross–Blue Shield office building in Boston (Fig. 7.53). Here, each component of the all-air distribution tree can be seen on the facades. The closely spaced and strongly emphasized vertical elements are both structural and utilitarian. Two out of every three verticals are structural, and *all* the verticals enclose ducts. Each pair (one hot, one cold) of high-velocity, round air ducts at the corresponding structural columns constitutes a vertical air supply system. At each floor they feed an attenuation and mixing chamber. At these locations, temperature is

Fig. 7.48 *A retrofit application in which a new interior VAV system replaces a double-duct system.* (a) *Exterior of the world headquarters of the Manufacturers' Hanover Trust Company, New York City. Syska & Hennessy, engineers.* (b) *Detail of the retrofit energy-efficient electric lighting system. The new waffle grid luminous ceiling replaces the original flat solid vinyl panels, thus increasing lighting efficiency, diminishing acoustic problems, and conforming to then-current New York City code.*

Fig. 7.49 Floor-by-floor VAV for a Chicago office building. (a) *Exterior view of 33 West Monroe (photo by Merrick, Hedrick-Blessing.)* (b) *Section perspective showing "stacked atriums." Skidmore Owings & Merrill, architects and engineers.*

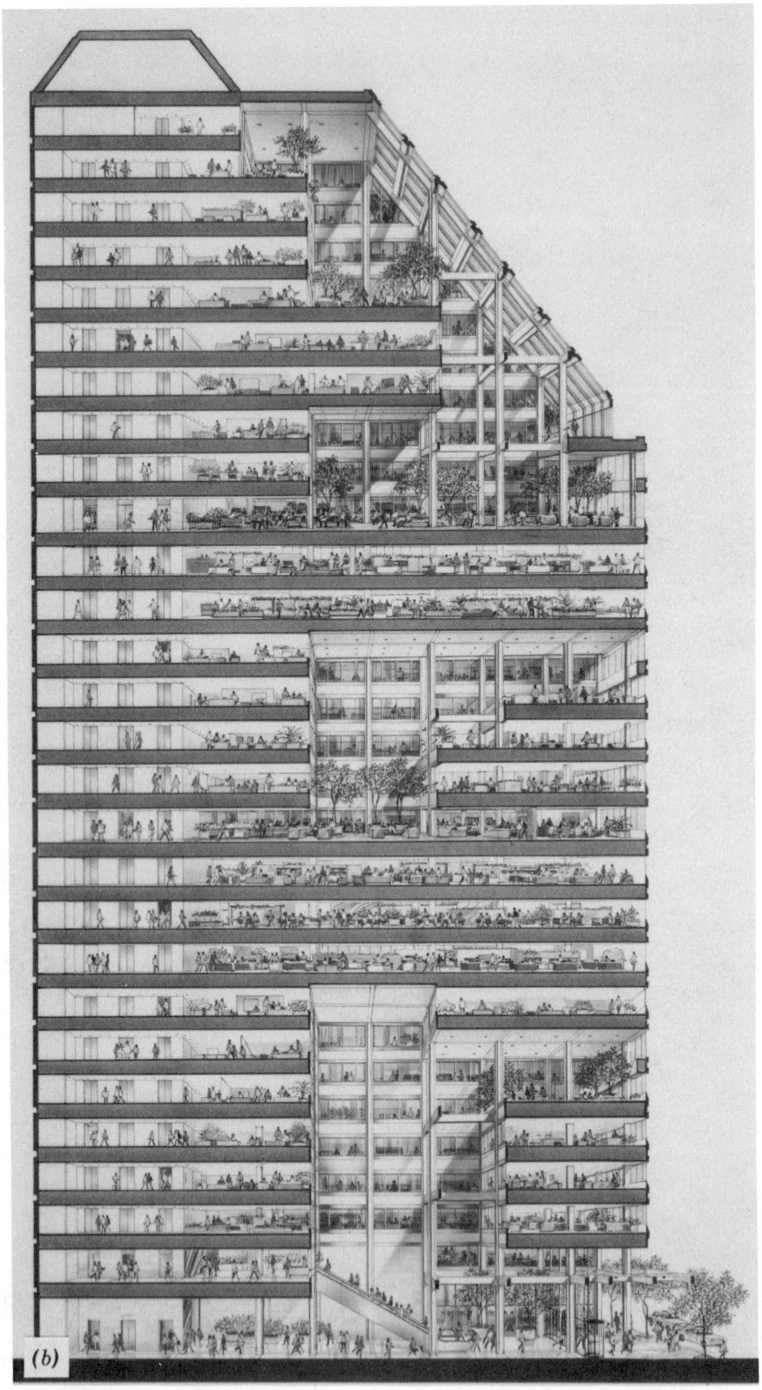

Fig. 7.49 (continued)

controlled as desired. The return ducts between pairs of hot-cold supplies complete the parallel, equally spaced pattern.

The mixing boxes (terminals) of double-duct systems are similar to those of other all-air systems (see Fig. 7.54). Although most double-duct

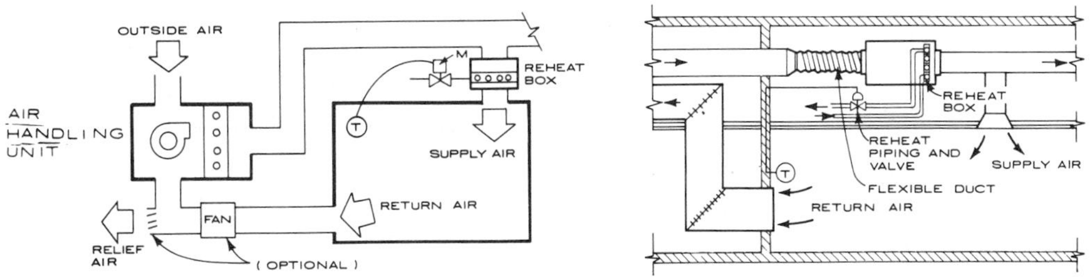

(a)

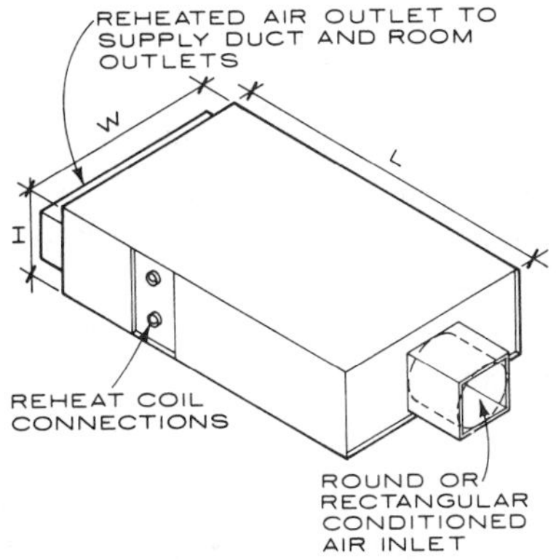

RANGE OF DIMENSIONS

CFM	HEIGHT	LENGTH	WIDTH
200	9″-11″	30″-50″	16″-22″
400	9″-11″	30″-51″	18″-30″
800	9″-11″	30″-51″	22″-42″
1600	14″-16″	48″-51″	40″-44″
2400	16″-18″	60″-55″	40″-54″
3200	16″-18″	60″-55″	16″-66″
5000	20″-18″	60″-55″	20″-80″

(b)

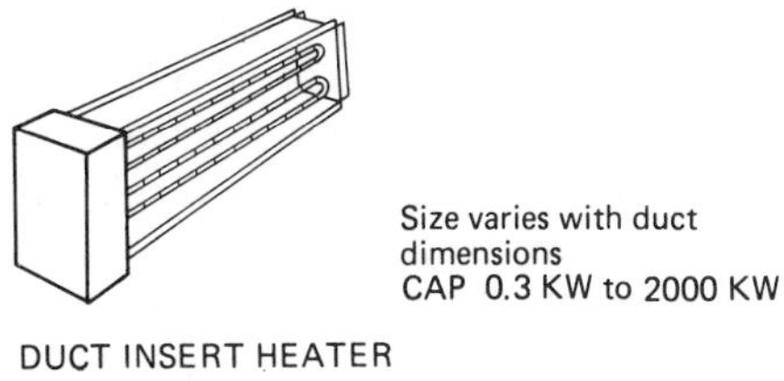

(c)

Fig. 7.50 *Single-duct with reheat system.* (a) *Diagrams of system and components.* (b) *Reheat box, or terminal, where velocity and pressure are reduced.* (c) *Simple duct heater. (Reprinted by permission from AIA, Ramsey, and Sleeper,* Architectural Graphic Standards, *7th ed., © 1981 by John Wiley & Sons.)*

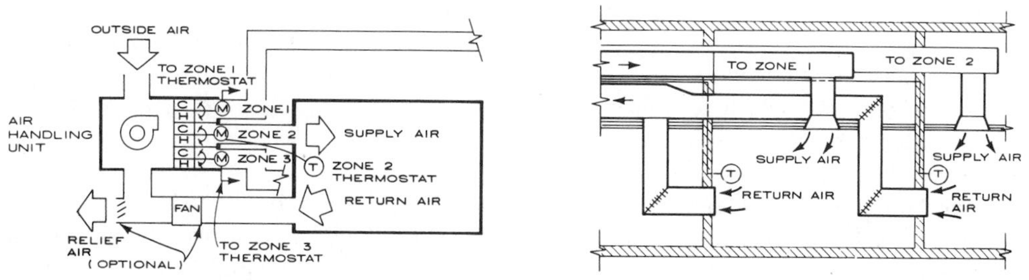

Fig. 7.51 *Multizone system. (Reprinted by permission from AIA, Ramsey, and Sleeper,* Architectural Graphic Standards, *7th ed., © 1981 by John Wiley & Sons.)*

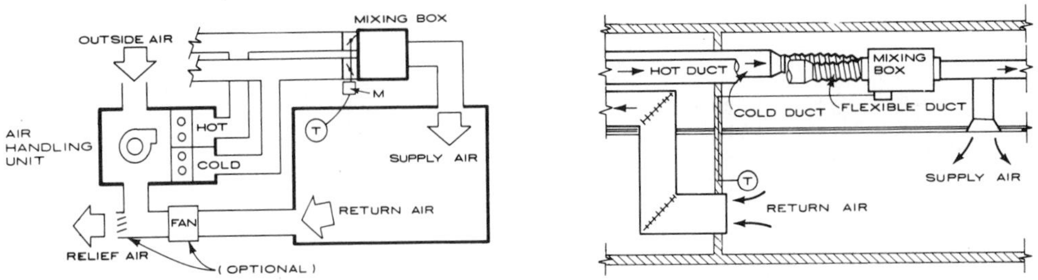

Fig. 7.52 Double-duct system. (Reprinted by permission from AIA, Ramsey, and Sleeper, Architectural Graphic Standards, *7th ed., © 1981 by John Wiley & Sons.)*

systems are CV, they can be VAV when the reduction in airflow is no more than 50% below the maximum.

7.5 Air & Water Systems

These systems, introduced in Fig. 7.3 and Section 7.1, have the design complexity—and first cost—of supply-and-return distribution trees for both water and air. This disadvantage is offset by the space-saving advantages of water trees and the superior comfort characteristics offered by air. Since only fresh air is centrally treated and distributed, only an equal quantity of exhaust air need be "returned." Because this exhaust air is not recirculated, these systems are attractive for hospitals and other buildings in which the mixing of air between zones is undesirable.

(a) Induction (Fig. 7.55). Centrally conditioned fresh air is supplied (at either high or medium pressures and velocities) to each induction terminal. Each terminal then mixes 20 to 40% incoming fresh air with 80 to 60% room air, passing it all over finned tubes for heating or cooling and circulating this mixture of air to the space.

In two-pipe systems, either hot or cold water—not both—is available to temper this air mixture ordered by the thermostat linked to each terminal. In four-pipe systems, the availability of both hot and chilled water makes it possible to switch instantly from heating to cooling, for excellent thermal control.

The induction terminals typically are located either above a space, as in Fig. 7.55, or below perimeter windows (Fig. 7.56). An unusually low-profile below-window induction unit can be seen in the Time–Life Chicago Subscription Office building (Fig. 7.57), where this perimeter system serves the outer 15-ft band of open-office work stations. The open-plan spaces are 30 ft wide, so the inner 15-ft band is served by a central system in the core, where all the return air is also gathered.

(b) Fan Coil with Supplementary Air (Fig. 7.58). This system is closely related to the preceding induction system; however, this system uses a fan at each unit, rather than high-velocity primary air, to move room air through the unit. Figure 7.59 shows a building that includes several variations on fan coils. The Iowa Public Service Building is an innovative 167,000-ft^2 utility office building that utilizes solar collectors and ice storage, supplemented by small backup boilers. Six ice-making machines serve a 75,000-gal ice storage pit with 90 million Btu capacity. The entire system, including building security and fire alarms, is controlled by computer. Heat exchange opportunities include reject heat from the ice-making machines, heat from the central toilet exhaust air, and heat from return air taken through luminaires. Solar collectors on the roof preheat the ventilation air, which is fed into the ceiling plenum at each floor. The fan-coil units then draw from this fresh air supply.

(c) Radiant Panels with Supplementary Air (Fig. 7.60). Large areas of radiant surface can be used to offset large losses of bodily radiant heat, as in the case of large areas of cold glass on a winter day, or when users are both

Fig. 7.53 *Blue Cross–Blue Shield Building, Boston. Anderson, Beckwith and Haible and Paul Rudolph, associated architects: Stressenger, Adams, Maguire, and Reidy, mechanical and electrical engineers. The two-story, Y-shaped forms are structural columns that divide at mezzanine level and continue to rise in pairs to form the exterior skeleton frame. Hollow channels on the exterior of each pair enclose, individually, a hot air supply duct and a cold air supply duct. These round, high-velocity ducts join for mixing and velocity reduction in attenuation boxes, located between columns at each floor. Conditioned air is discharged upward from a window sill grille above the box. Mullions between each pair of structural columns originate at the second-floor level and extend to the mechanical story at the roof. Each mullion encloses a return air duct, which draws air through grilles in the sills of the two adjacent windows on each story. Thus the air is delivered at the exterior, accomplishes its mission at that surface, and returns in the same vertical plane to the suction side of fans on the roof.*

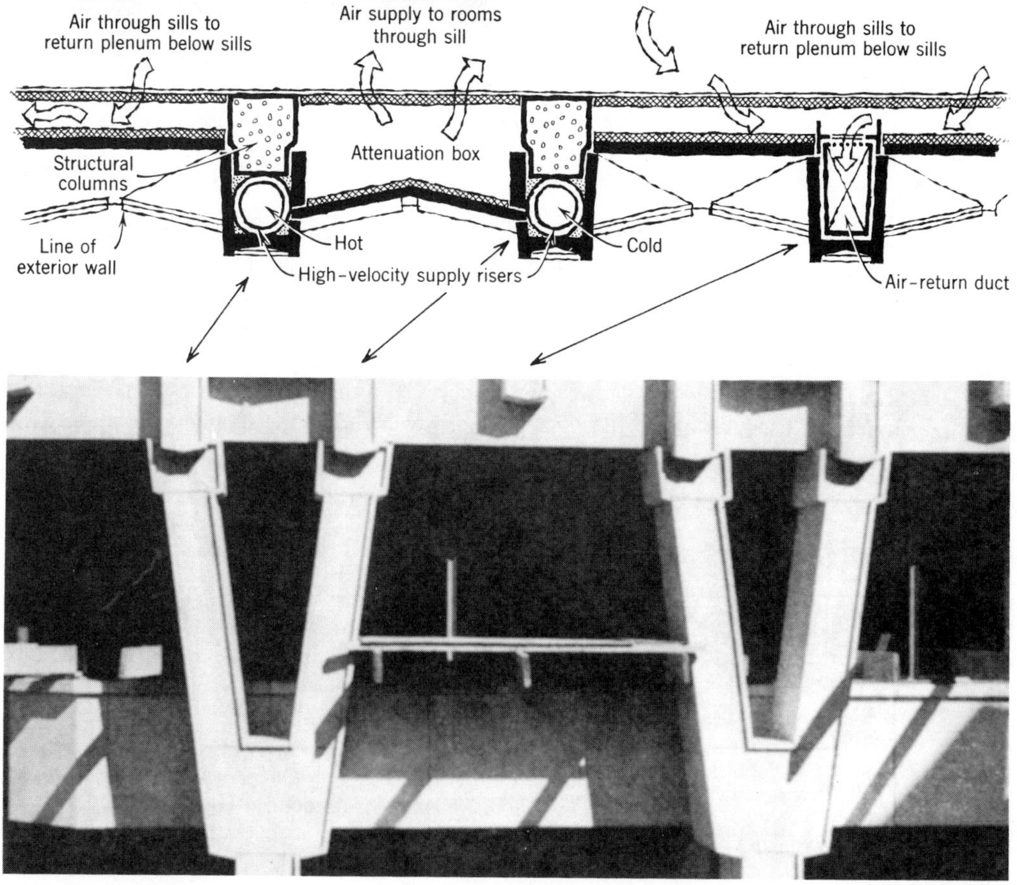

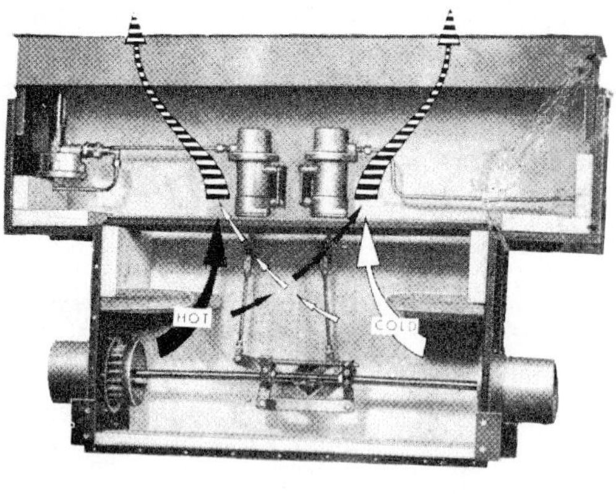

(a)

Fig. 7.54 *High-velocity double-duct unit, providing terminal mixing and attenuation (pressure and sound reduction).* (a) *Pneumatically controlled from a thermostat, the unit blends and delivers air at a selected temperature. These constant-volume units provide accurate, constant delivery at each outlet of the system, even though the pressure in the hot and cold ducts may vary widely (Courtesy of Anemostat).* (b) *Typical mixing box dimensions. (Reprinted by permission from AIA, Ramsey, and Sleeper,* Architectural Graphic Standards, *8th ed.,* © *1988 by John Wiley & Sons.)*

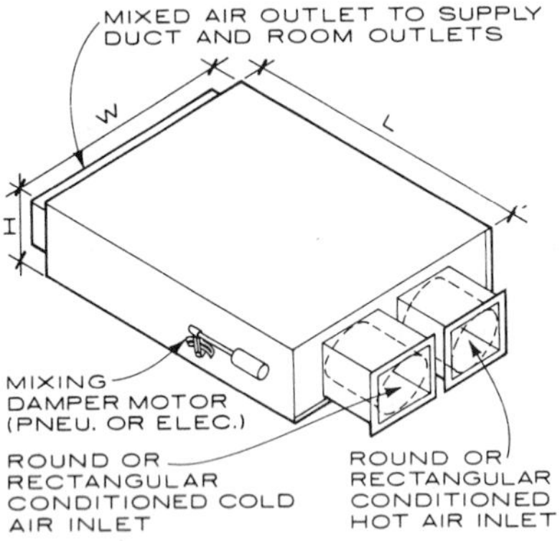

High, medium, or low velocity systems. Inlet pressure 1/4 to 1 1/2 in. W.C. Capacity range from 150 to 2000 cfm per box (low velocity) to 5000 cfm (high velocity). Box serves as converter from high to low velocity air system, noise attenuator, and control device by mixing hot and cold air streams.

RANGE OF DIMENSIONS

CFM	HEIGHT	LENGTH	WIDTH
400	6″-10″	40″-51″	30″-19″
800	8″-11″	50″-51″	42″-24″
1600	12″-14″	48″-51″	44″-40″
2400	14″-18″	60″-55″	54″-44″
3200	14″-18″	60″-55″	54″-44″
5000	16″-18″	60″-55″	54″-66″

(b)

scantily clothed and sedentary. In summer, such panels can help offset radiant gain from electric lights or large glass areas. The ceiling is often favored for the panel location, because it is uncluttered by the furniture, tackboards, and other items that cover floors and walls. However, floors are sometimes used, as in the Bleshman Regional Day School (Fig. 7.61). The designers of this school for the handicapped recognized that the majority of its users would spend much of their time quite close to the floor, and that the colder air near the floor could be uncomfortable, especially in the New Jersey winter. The entire floor is warmed by the supply (ventilation) air in

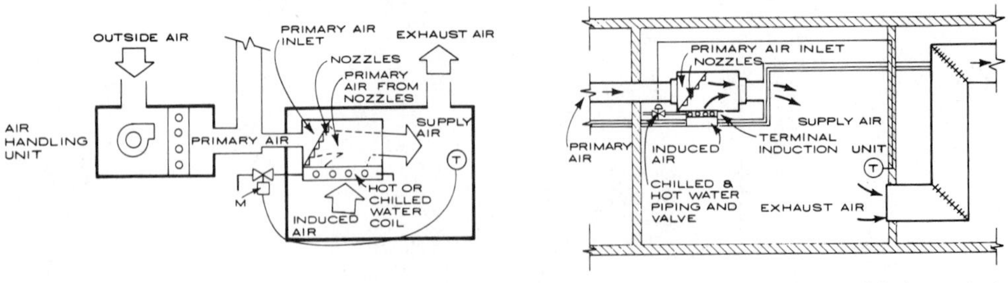

Fig. 7.55 *Induction system. Terminal induction units also are frequently located below windows. (Reprinted by permission from AIA, Ramsey, and Sleeper,* Architectural Graphic Standards, *7th ed.,* © *1981 by John Wiley & Sons.)*

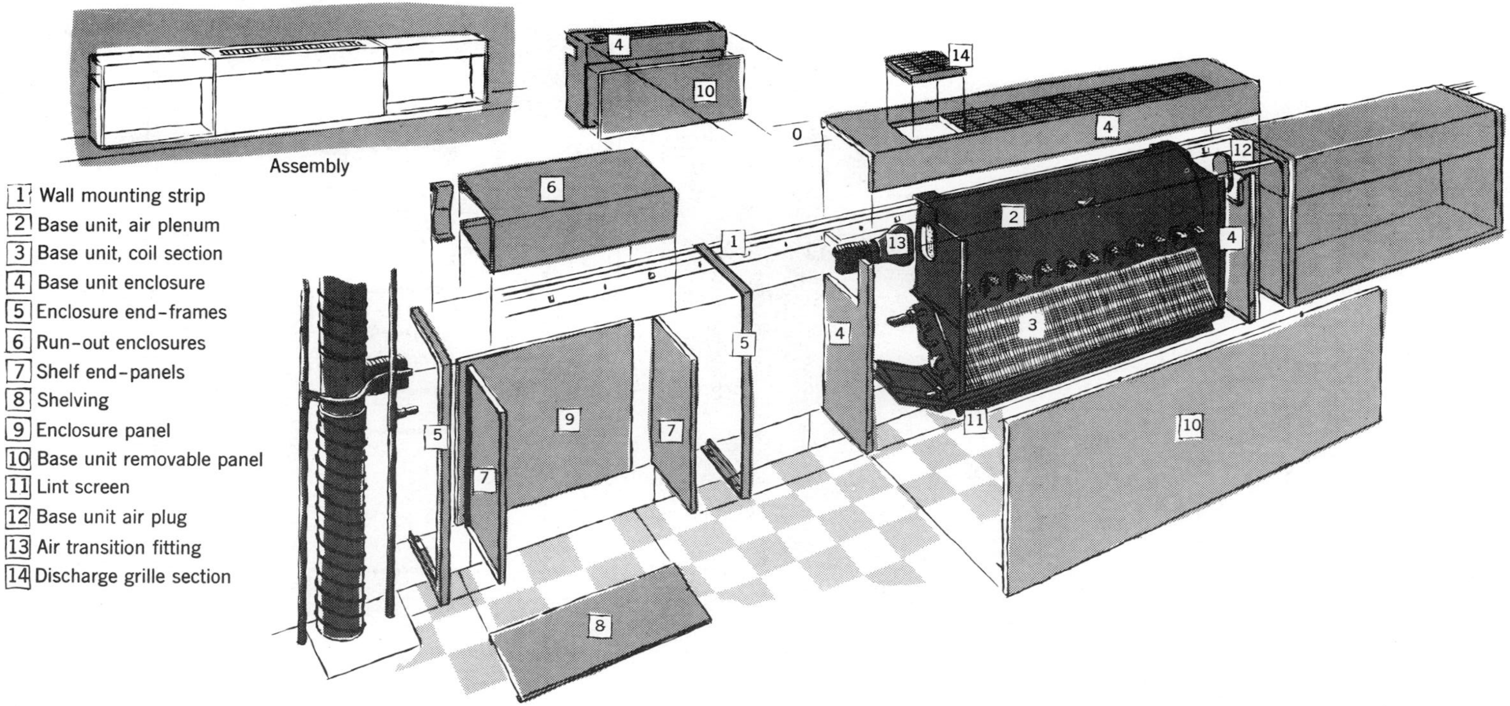

Fig. 7.56 *High-velocity induction system. Conditioned outdoor air for ventilation and to induce circulation of room air is brought in through a single high-velocity duct. It is attenuated and silenced in the chamber (2) and then, through jets in the front of this plenum, it induces flow of room air, which is heated or cooled at finned coil (3). The lint screen (11) requires periodic maintenance for proper airflow. (Courtesy of Carrier Corporation.)*

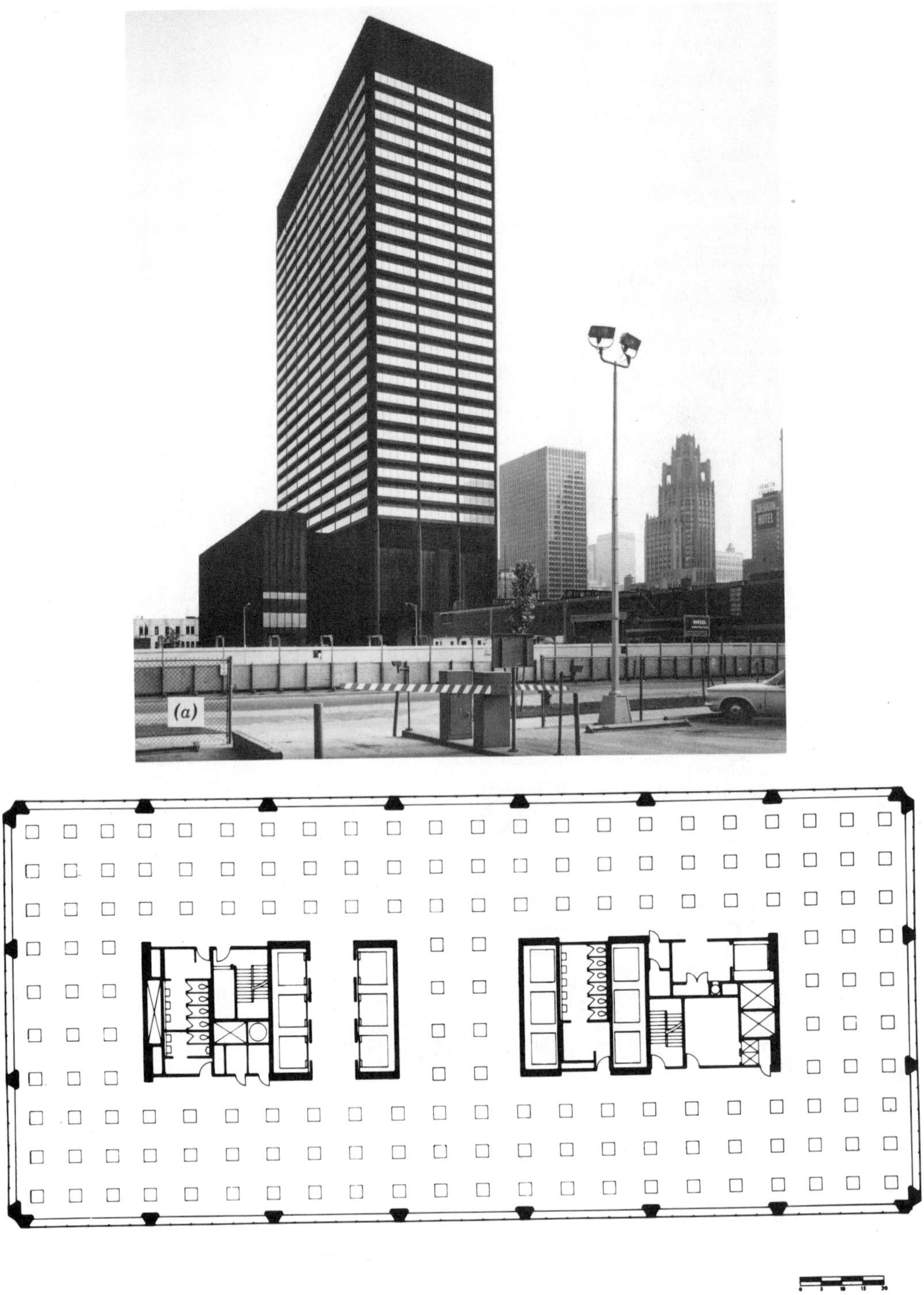

(b)

Fig. 7.57 *Induction units in the Time–Life Chicago Subscription Office building. Harry Weese and Associates, architects.* (a) *Exterior. (Photo, Daniel Bartush.)* (b) *Plan; all workstations are within 30 ft of a window.*

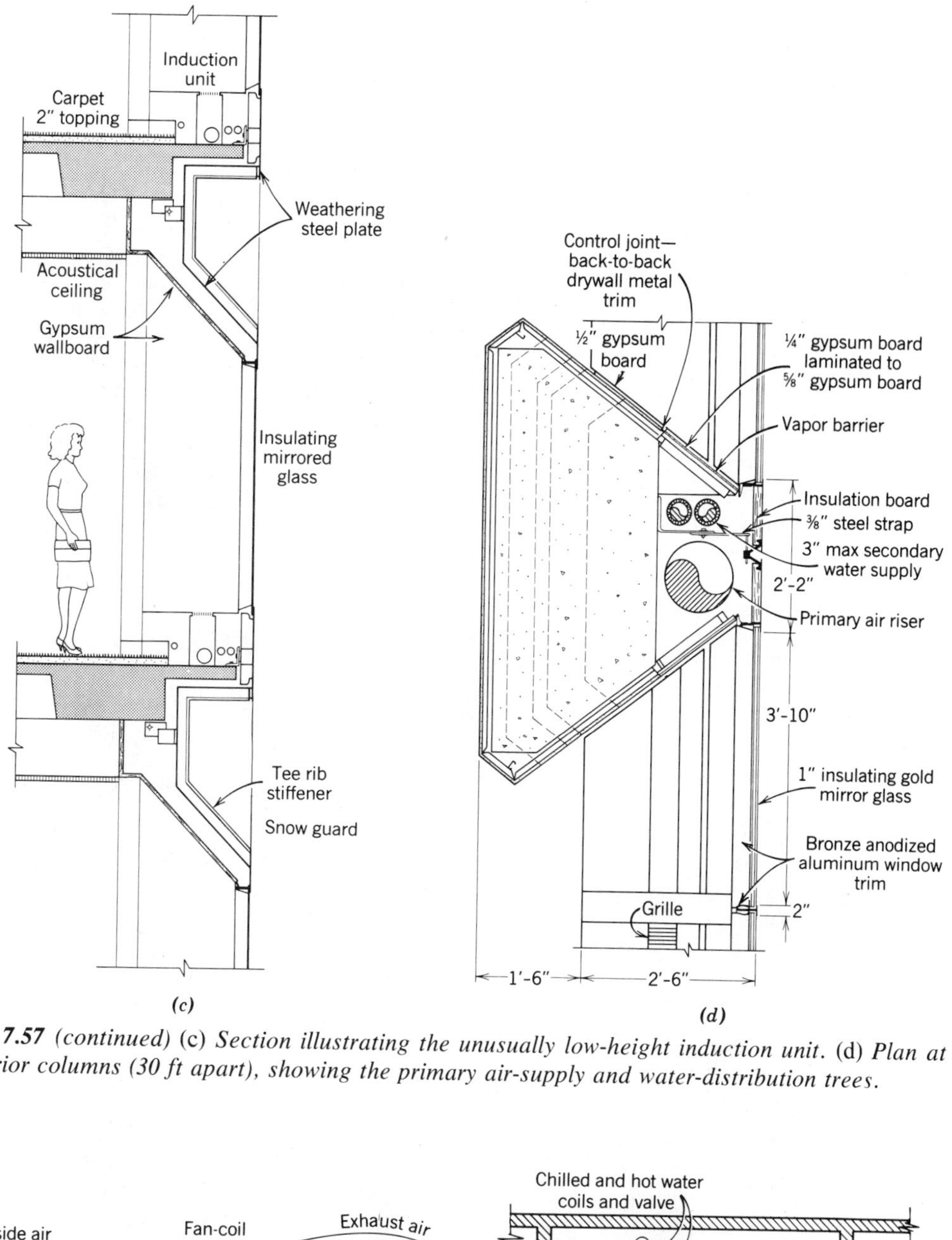

(c) (d)

Fig. 7.57 *(continued)* (c) *Section illustrating the unusually low-height induction unit.* (d) *Plan at exterior columns (30 ft apart), showing the primary air-supply and water-distribution trees.*

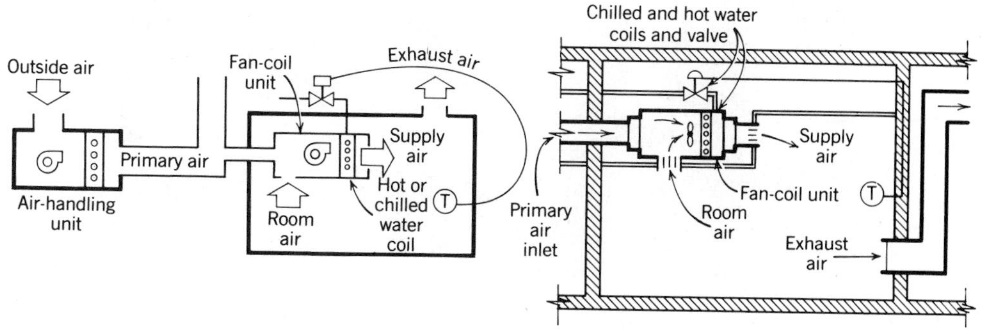

Fig. 7.58 *Fan-coil unit with supplementary air. These also are frequently located below perimeter windows.*

Fig. 7.59 Iowa Public Service Headquarters, Sioux City, Iowa. Joint-venture architects: Rossetti Associates and Foss, Engelstad, Heil Associates. (a) *Southeast elevation, with roof garden at the top floor.* (b) *Interior of the daylit atrium. (Photos by Balthazar Korab.)*

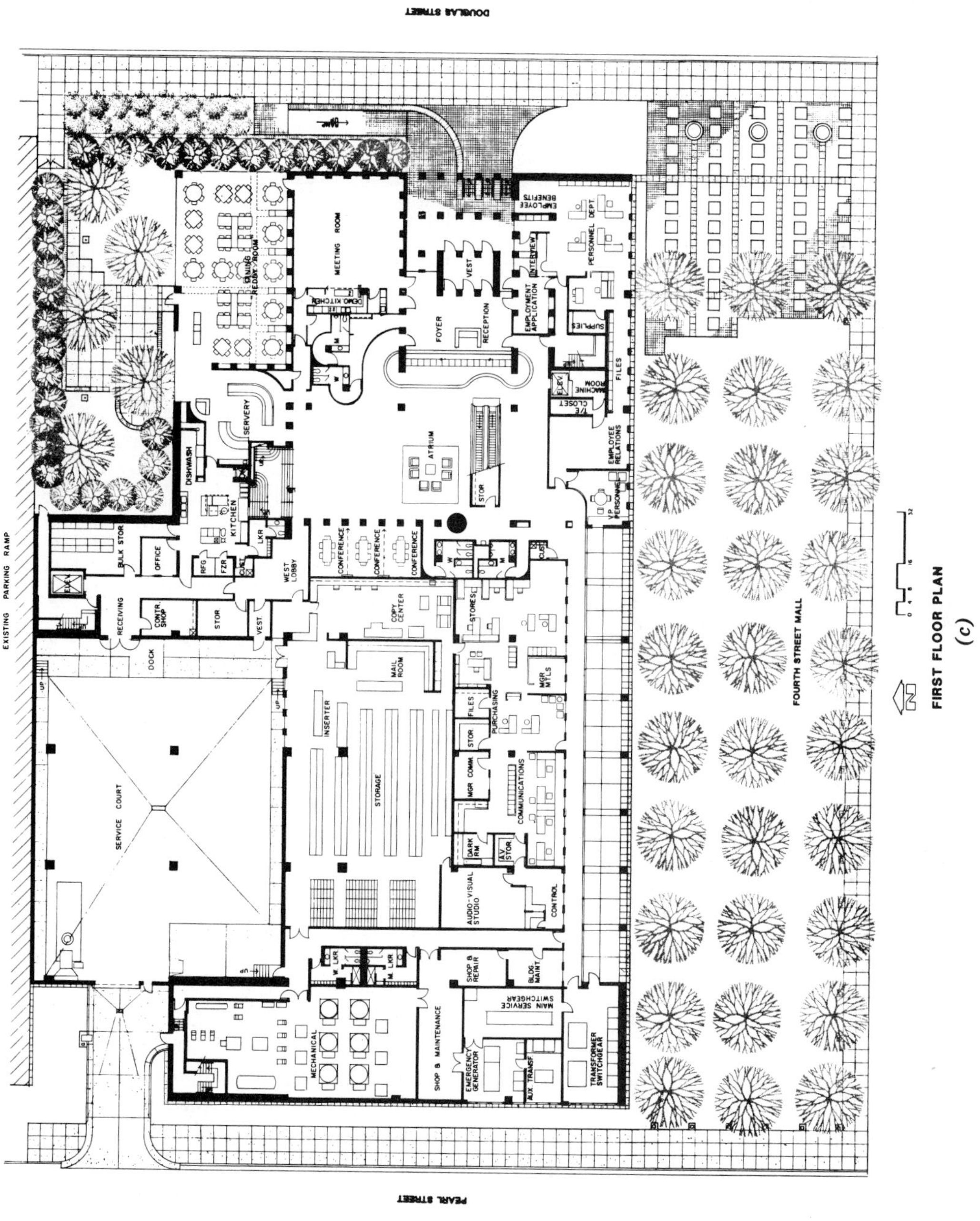

Fig. 7.59 (*continued*) (c) *Ground floor plan; six ice machines sit in mechanical room.*

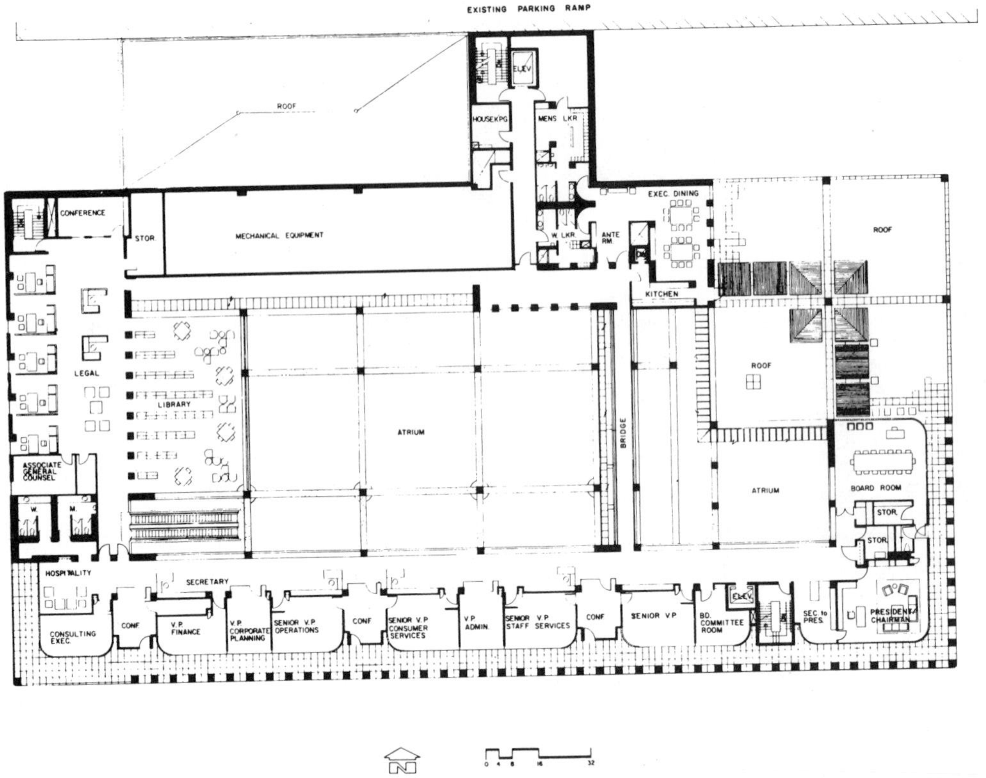

FIFTH FLOOR PLAN

(d)

Fig. 7.59 *(continued)* (d) *Top (fifth) floor plan; note the proportion given over to daylighting of lower floors.*

winter, which enters just below windows to counteract the downdraft off cold glass. In summer, the cool supply air first cools the floor, then cools the air in front of the warm glass. The concrete cellular "air floor" provides a thermal mass that helps maintain steady temperatures. Heated or cooled air is provided by rooftop air–air heat pumps, one of which is provided for each cluster of three to six classrooms. This HVAC example is therefore related to the direct refrigerant family, discussed in Section 7.2.

7.6 All-Water Systems

These systems (see Fig. 7.4 and Section 7.1) typically deal only with temperature control; air quality is left to separate systems. In its most simple application, the *two-pipe heating-only radiator* (discussed at length in Section 6.7) is getting new exposure with some colorful and pleasing products (Fig. 7.62). The Mayer Art Center at Phillips Exeter Academy in New Hampshire (Fig. 7.63), features new exposed radiators in some older buildings. These radiators are based on simple components (typically, 2¾ in. wide) that can be combined in many heights and widths, inviting the designer to feature them rather than to hide them in metal cabinets.

An especially familiar all-water application is the simple *fan-coil unit* (Fig. 7.64). These units, which may be found above ceilings, below windows, or in corners, simply control the temperature (and to an extent, the relative humidity) of the air already in the room, which is blown through the coils. Because water is often condensed from the room air when cooling is in progress, a drain line is required. Exterior air intake grilles can easily be added when fan-coil units

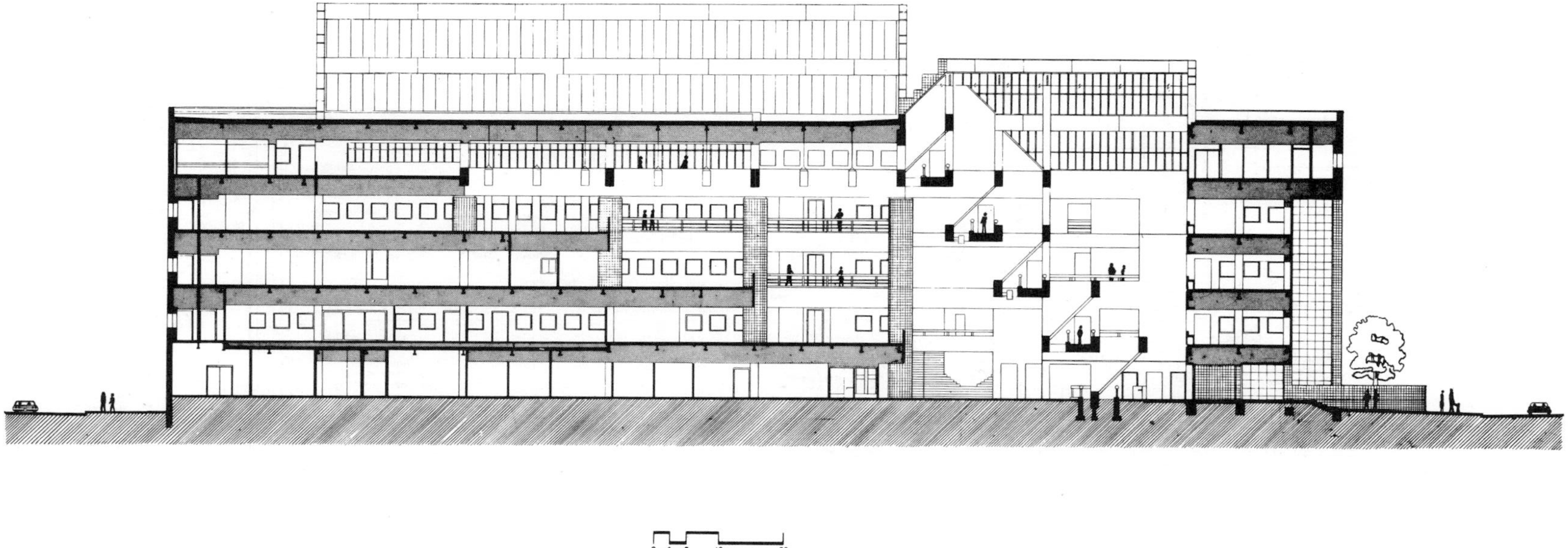

WEST - EAST SECTION LOOKING NORTH

(e)

Fig. 7.59 *(continued)* (e) *East–west section, looking north.*

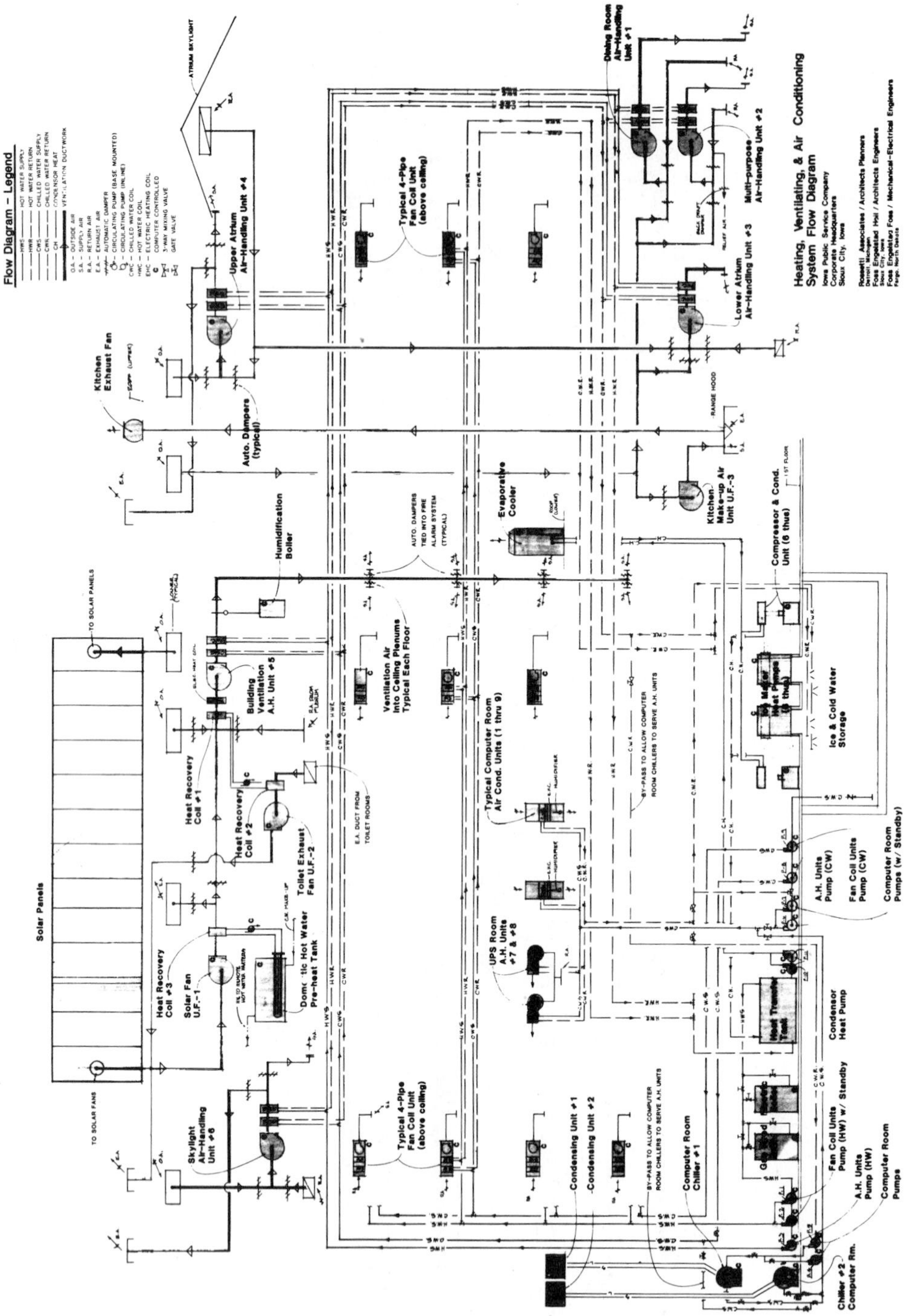

Fig. 7.59 (continued) (f) *System schematic, relating solar collectors, ice storage, and the four-pipe distribution system serving fan coil units (above ceilings) and air-handling units in many zones. Heat recovery units are also shown. (Courtesy of FEH Associates.)*

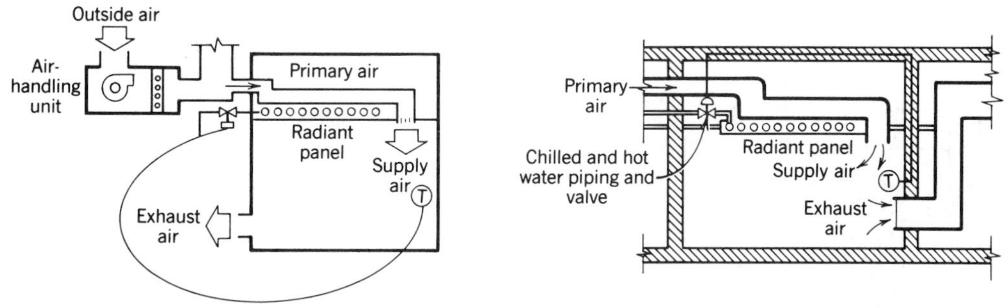

Fig. 7.60 *Radiant panel with supplementary air. Walls or floors are also locations for these large-area panels.*

are located below windows, to allow the local provision and tempering of ventilation air.

All-water perimeter systems with local fresh air can also take the simple form of *operable windows with hot water finned-tube radiation.* An expressive variation on this approach appears in the reading rooms and staff workrooms for the Seeley G. Mudd Library at Yale University (Fig. 7.65). Limestone spandrels are curved in to allow fresh air to enter these smaller perimeter rooms just below the windows, where hot water finned-tube radiators are available when

Fig. 7.61 *The Bleshman Regional Day School (for the multiply handicapped), Paramus, New Jersey. Rothe-Johnson Associates, architects.* (a) *Lobby. (Photo by Otto Baitz.)*

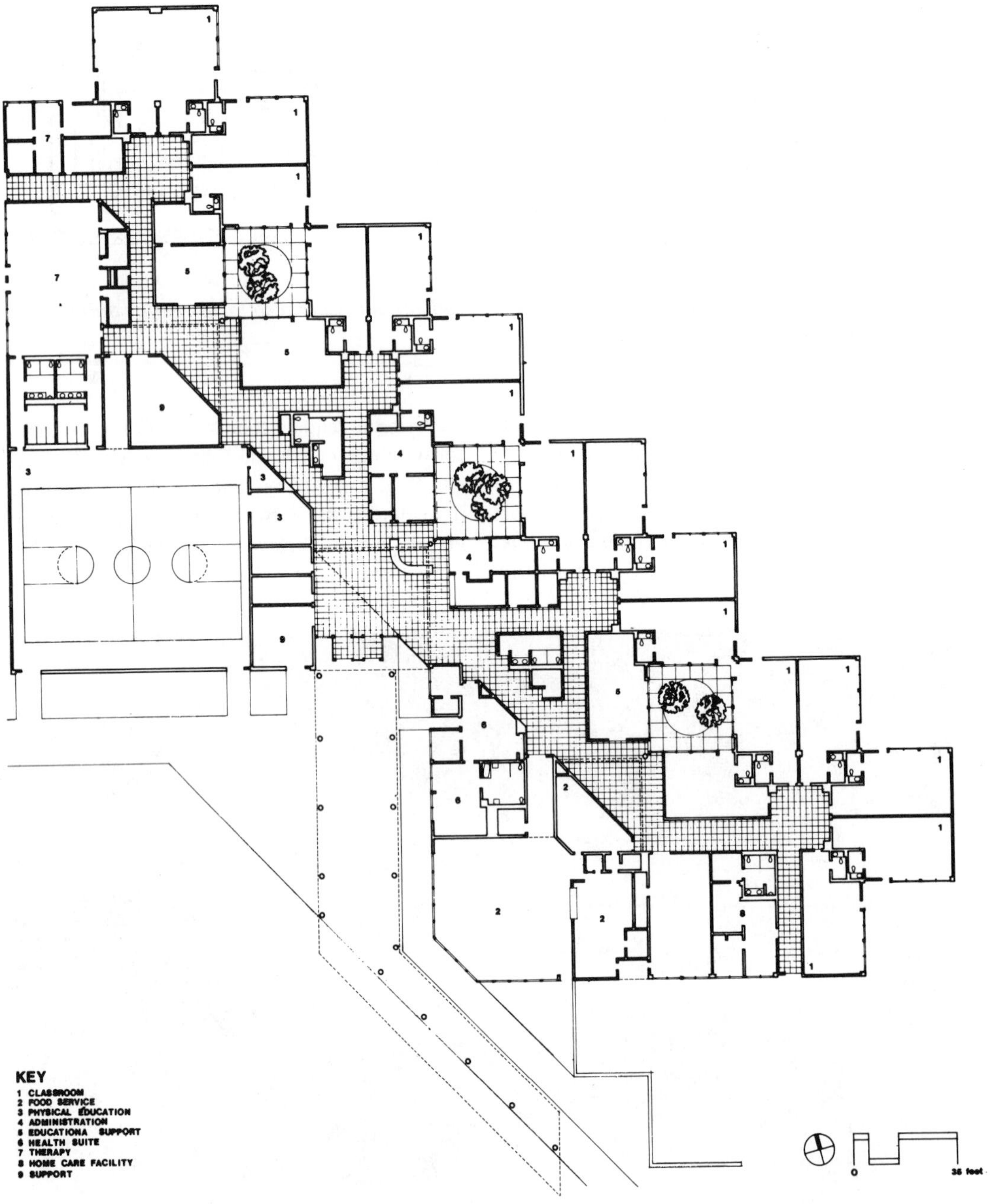

(b)

Fig. 7.61 *(continued)* (b) *Plan, showing courtyards that bring daylight between clusters of classrooms and into central circulation spine.*

needed. The incoming fresh air replaces exhaust air, which flows out the upper operable windows. The remainder of the building has conventional forced-air heating and cooling.

The last example of an all-water system involves the *water–air heat pump loop*. In the 175,000-ft^2 Comstock Center in Pittsburgh (Fig. 7.66), there are six to eight small heat pumps on

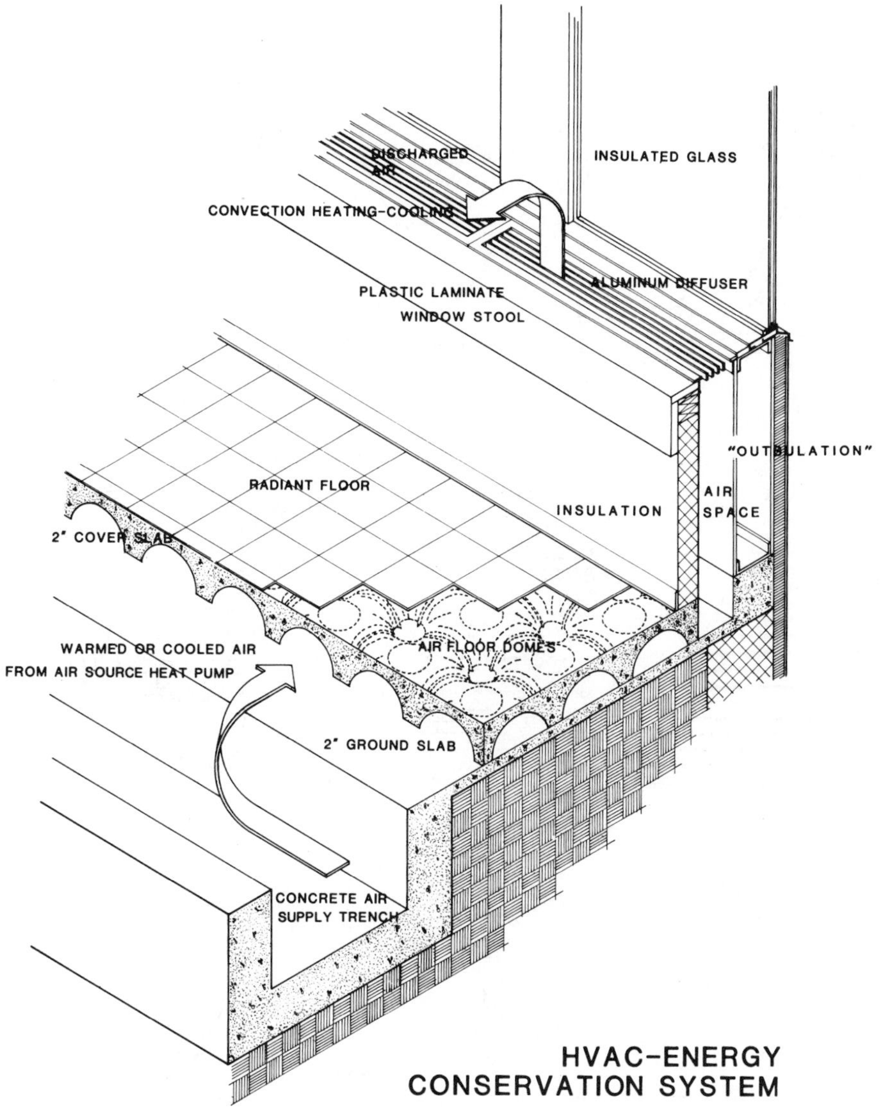

(c)

Fig. 7.61 *(continued)* (c) *Detail of the cellular concrete "air floor" and supply air diffuser below windows.*

each of 10 floors, located above the suspended ceiling. Their connecting loop doubles as the building's wet-pipe sprinkler system supply. (This is possible because the heat pumps keep the loop between 65 and 85 F.) Since neither hot nor chilled water-supply and water-return distribution trees are needed, there is a substantial first-cost saving, and relatively little building volume is consumed. A 23,000-gal water storage tank allows daytime heat, rejected to the loop by the heat pumps, to be stored and recalled for nighttime heating. When necessary, a small boiler will maintain the 65 F minimum temperature in the water loop.

Another feature of the Comstock Center is its use of *extract-air windows* to control infiltration and to moderate the perimeter zone's temperatures. In these triple-glazed windows, return air

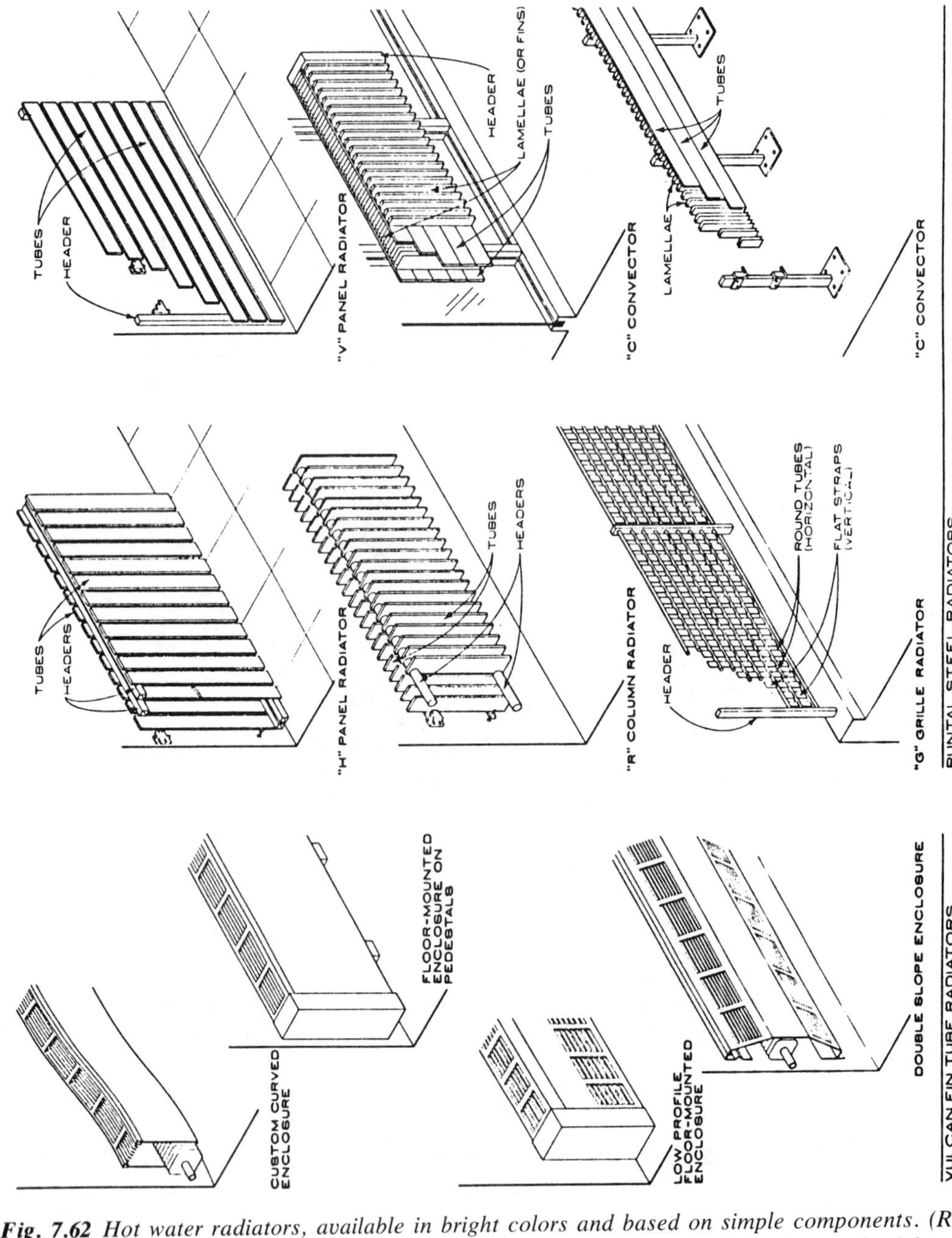

Fig. 7.62 *Hot water radiators, available in bright colors and based on simple components. (Reprinted by permission from AIA, Ramsey, and Sleeper,* Architectural Graphic Standards, *8th ed., © 1988 by John Wiley & Sons.)*

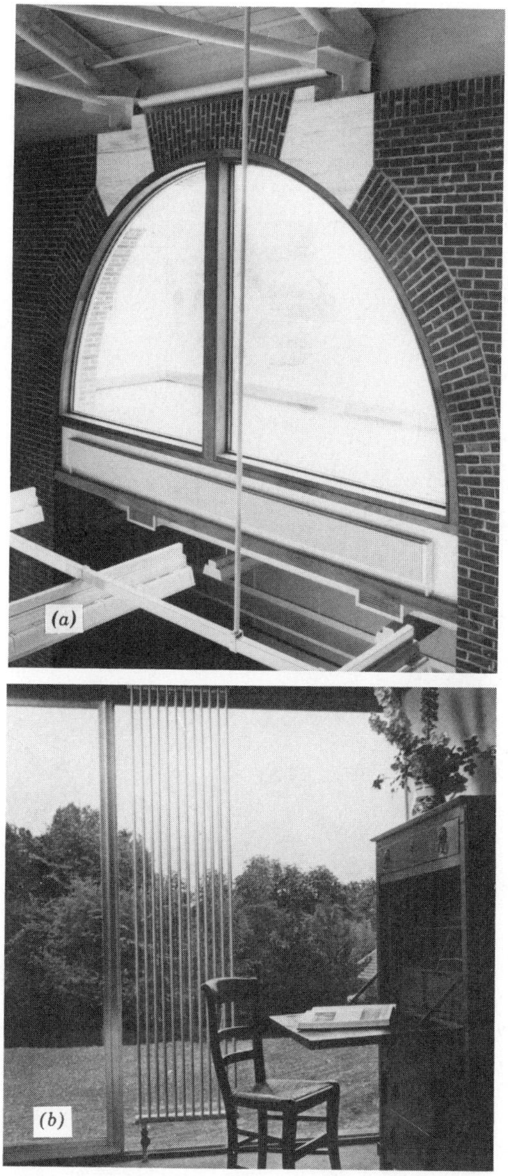

Fig. 7.63 *Exposed hot water radiators at the Mayer Art Center, Phillips Exeter Academy, Exeter, New Hampshire. Amsler Hagenah MacLean, architects.* (a) *Under-window panel radiator. (Photo by Alex Beatty.)* (b) *Radiator formed by repeating flat tubes. (Courtesy of Runtal/North American Energy Systems.)*

from the room is drawn up between the outer double-glazed window and an inner single sheet of glass. This air warms the inner pane in winter, substantially increasing radiant comfort and eliminating the need for a separate perimeter heating system. In summer, the same procedure cools the inner glass. After the return air arrives in the ceiling plenum, it is mixed with fresh air, then tempered and recirculated by the heat pumps.

The Comstock Center also utilizes a large daylighting atrium, whose temperature control is provided largely by exhaust air from the offices; the stack effect is utilized to provide natural ventilation on its west face in summer conditions.

The proliferation of individual air-handling units illustrated in this example, linked only by a common water loop, is a characteristic shared by the direct refrigerant systems, discussed in Section 7.2.

7.7 District Heating and Cooling

Often, large projects made up of many large buildings are well served by *one* central station heating/cooling plant. The familiar ''economies of scale'' apply here; very large, efficient, and well-maintained boilers and chillers encourage energy recovery through heat exchange, reduce air pollution, and remove the noise and other nuisances associated with heating and cooling from the other buildings. This approach is called *district heating and cooling.*

District heating for residences and small commercial buildings is common in northern Europe; district steam systems serve the central areas of many U.S. cities. Most often electrical generating plants are the heat (or steam) source; *cogeneration* (discussed in Section 7.8) allows the waste heat from the generating plant to be put to use as space heating (or steam absorption cooling). Ironically, trends toward better insulated buildings reduce the ''market'' for distributed heat, making the installation less cost-effective. This discourages the supplier both from investing in a district system, and from encouraging technical innovations in energy conservation on the part of their customers. Smaller, densely built-up central heating districts tend to be more successful than large, sprawling ones.

(a) High-Temperature Water and Chilled Water for Airports. Although long-distance

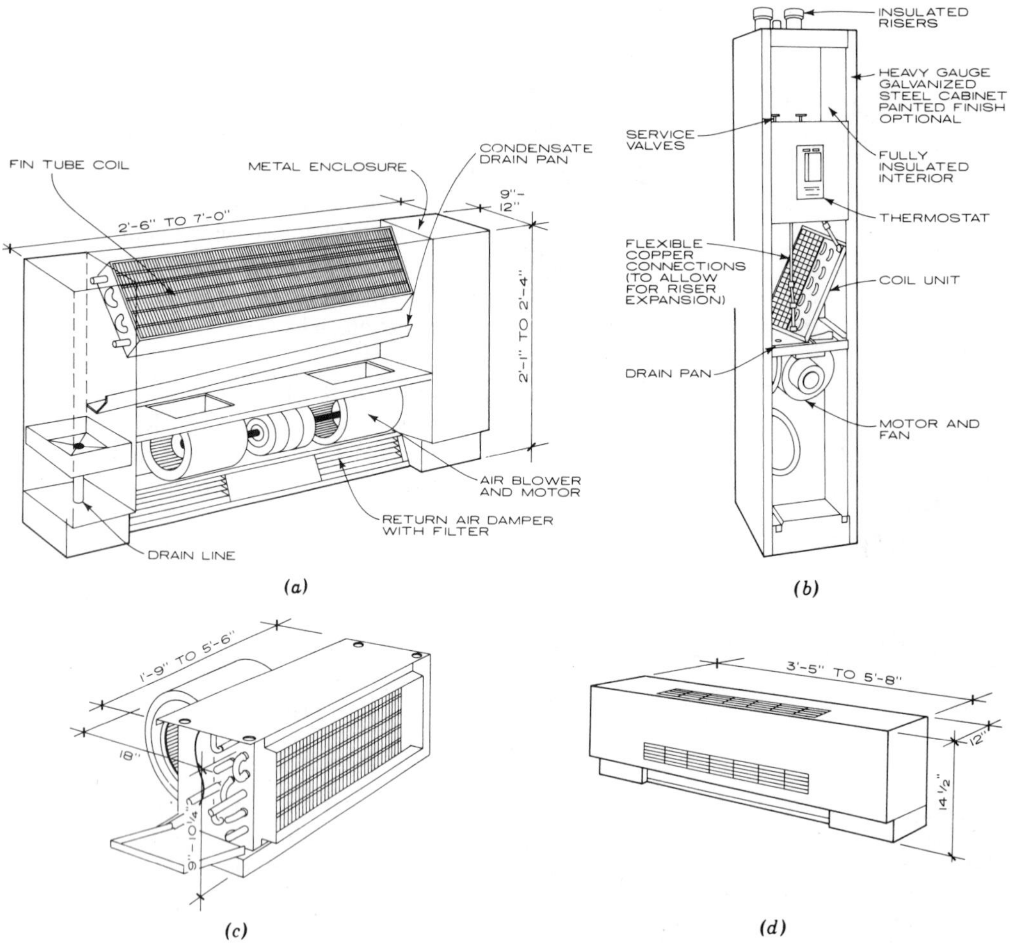

Fig. 7.64 *Simple fan-coil units, without fresh air.* (a) *Standard below-window unit.* (b) *''High-rise'' unit, for corners or cabinet locations.* (c) *Above-ceiling unit.* (d) *Low-profile unit (such as used in Time–Life Building, Fig. 7.57). (Reprinted by permission from AIA, Ramsey, and Sleeper,* Architectural Graphic Standards, *8th ed., © 1988 by John Wiley & Sons.)*

steam distribution has been used for more than a century, the development of the *high-temperature water* (HTW) principle is a product of recent decades. Offering many advantages (though steam is still frequently chosen for city distribution), circulated high-temperature, high-pressure hot water in closed systems has found great favor in Air Force bases and airports, and for groups of buildings such as hospital complexes and college campuses. Water will not flash into steam if kept at sufficiently high pressure. It may then be circulated by pumps through supply and return mains and through branches to heat exchangers, which operate conventional low-pressure hot water systems, generate steam, and perform numerous other thermal tasks. Pressures are on the order of 400 psig (pounds per square inch, gauge) and temperatures are about 300 F. During its circuit the water will sometimes lose about 150F° and 60 psig in pressure. A section shown in Fig. 7.67 illustrates a common arrangement.

High-temperature water has a number of advantages over steam for certain installations. It is a two-pipe system, and the temperature drop in the *supply* main is often as little as 10F°. With reasonably high water velocities, mains can be reduced to almost half the size of those required

Fig. 7.65 Local fresh air and finned-tube radiation at the perimeter. The Seeley G. Mudd Library at Yale University; Roth and Moore, architects. (a) *Exterior, showing curved limestone spandrels, along which flows incoming fresh air. (Photo © Steve Rosenthal.)* (b) *Section, showing the fresh air intake, finned-tube radiation, and upper operable sash for exhaust air. This system is used for the smaller reading and staff workrooms at the perimeter (at right end of plan.)* (c) (overleaf) *Plan.*

for steam distribution. Simplicity results from the lack of need for steam traps and pressure-reducing valves. The pipes need not pitch to low points, as in the case of steam, but may follow the contours of the ground. Although installation costs are greater, operational costs are less than for steam. Feed water treatment is negligible and corrosion is minimal. Underground problems of expansion and insulation are the same as in other subterranean systems. Large

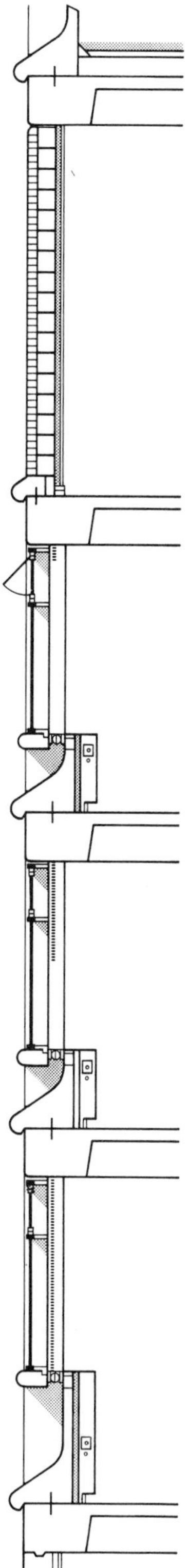

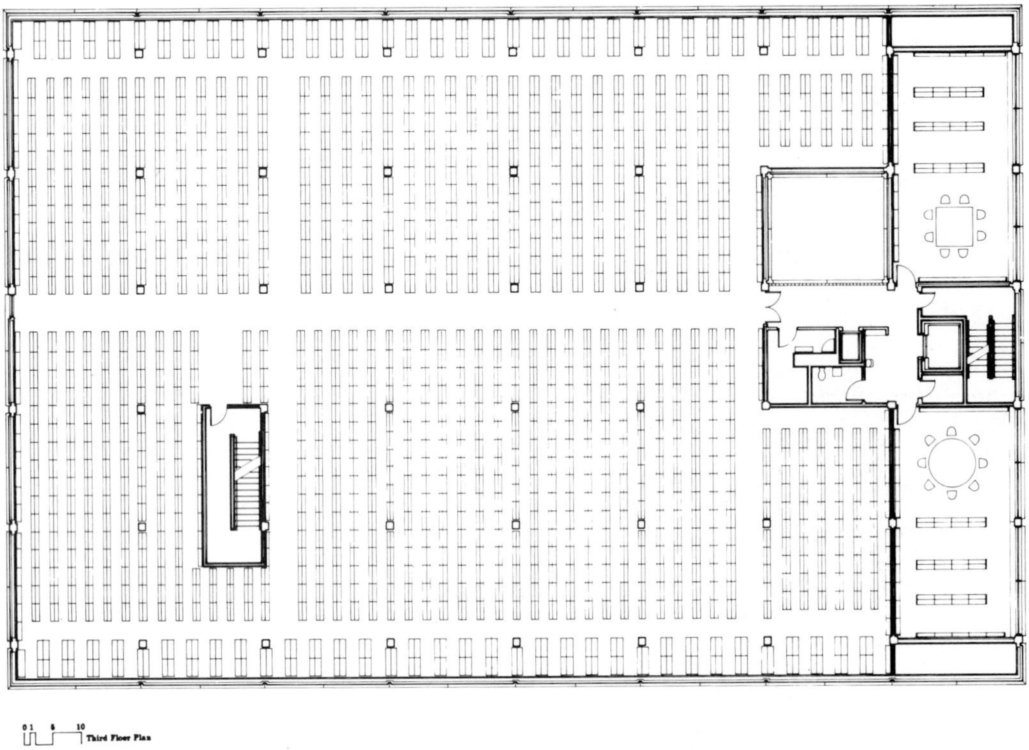

(c)

Fig. 7.65 (continued)

sweep-type loops accommodate expansion between fixed points, and underground piping is embedded in special, thermally efficient insulative fill.

(b) John F. Kennedy International Airport. During planning of the heating and cooling for complete air conditioning in the principal buildings of the Kennedy International (originally Idlewild) Airport, the initial concept was to have the facilities located in utility space in each of the buildings. However, the advantages of centralizing the basic equipment for both heating and cooling soon became apparent. A central heating/cooling plant was designed by the architects Skidmore Owings & Merrill and engineered by Seelye, Stevenson, Value, and Knecht. In the selected scheme (see Fig. 7.68), all facilities except air handling and ducts were assembled in the central plant. The elimination of a multitude of stacks, boilers, fuel storage, water chillers, and cooling towers at each building was considered to be a great step forward in the release of valuable space and for cleanliness, control, and architectural freedom. Moreover, it proved to be economically advantageous.

Four high-temperature water boilers supply the heating needs and also serve the absorption-refrigeration machines. These were specially adapted to use this high-temperature water instead of steam. Thus the boilers are active throughout the year, burning gas or oil. The latter fuel is stored in a 210,000-gal spheroid tank adjacent to the plant. Boilers, chillers, and cooling towers are all in close proximity.

Statistics for the initial stage of the plant are as follows.

1. *Hot water:* 160,000,000 Btu/h, 1,140,000 lb/h pump capacity, 160-ft head, 380 F supply, 240 F return.
2. *Chilled water:* 6210 tons, 16,800 gpm pump capacity, 150-ft head, 55 to 45 F cooling range.

GROUND FLOOR PLAN

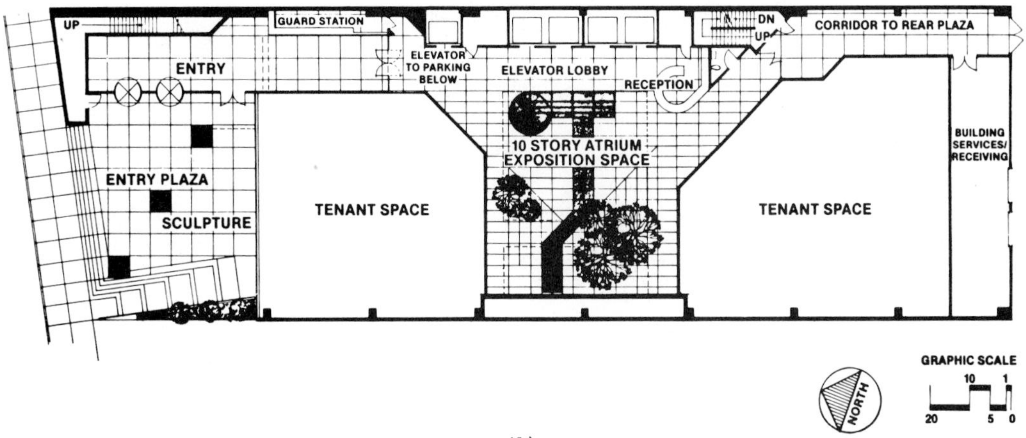

(*b*)

Fig. 7.66 *The Comstock Center, Pittsburgh. Burt Hill Kosar Rittelmann Associates, architects.* (a) *View of northwest corner.* (b) *Ground floor plan.*

3. *Condenser water:* 24,600 gpm pump capacity, cooling tower capacity for rated flow from 102.4 to 85 F with WB at 78 F.

At the various buildings of the airport group, chilled water and hot water are used in the coils of air-handling units from which conditioned air is circulated for complete climate control. Automatic electronic control and monitoring equipment was used. Checked periodically, this equipment is all but *completely* automatic in its regulation of the temperature and humidity throughout the building. (The initial stage as described here has been subject to expansion as air travel has increased.)

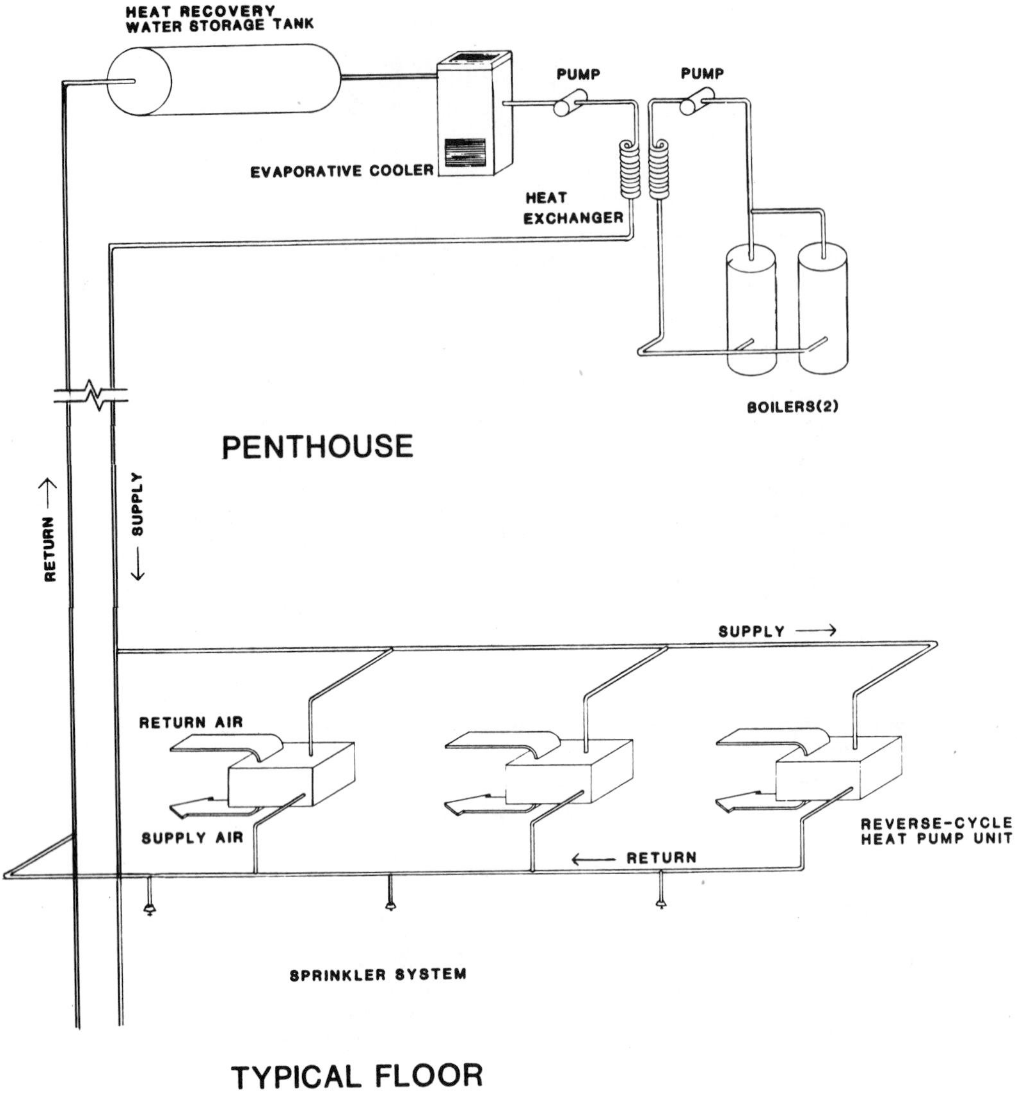

(c)

Fig. 7.66 *(continued)* (c) *Water loop heat pumps, combined with sprinkler system supply on each floor. Also called "tri-water" system."* (d) *Extract-air windows control infiltration and moderate perimeter zone temperatures. Fresh air is ducted to the plenum, where it mixes with return air before being treated and recirculated by the heat pumps. (See also Fig. 6.14.)* (e) *The central daylighting atrium is tempered by exhaust air from the offices; exhaust air then leaves via the stack effect.*

7.8 Cogeneration

In the course of these first seven chapters, electricity has repeatedly been called a "high-grade" source, in reference to the high temperatures needed to produce electricity by conventional (fossil fuel) means and to the large amount of waste heat (often twice the fraction of

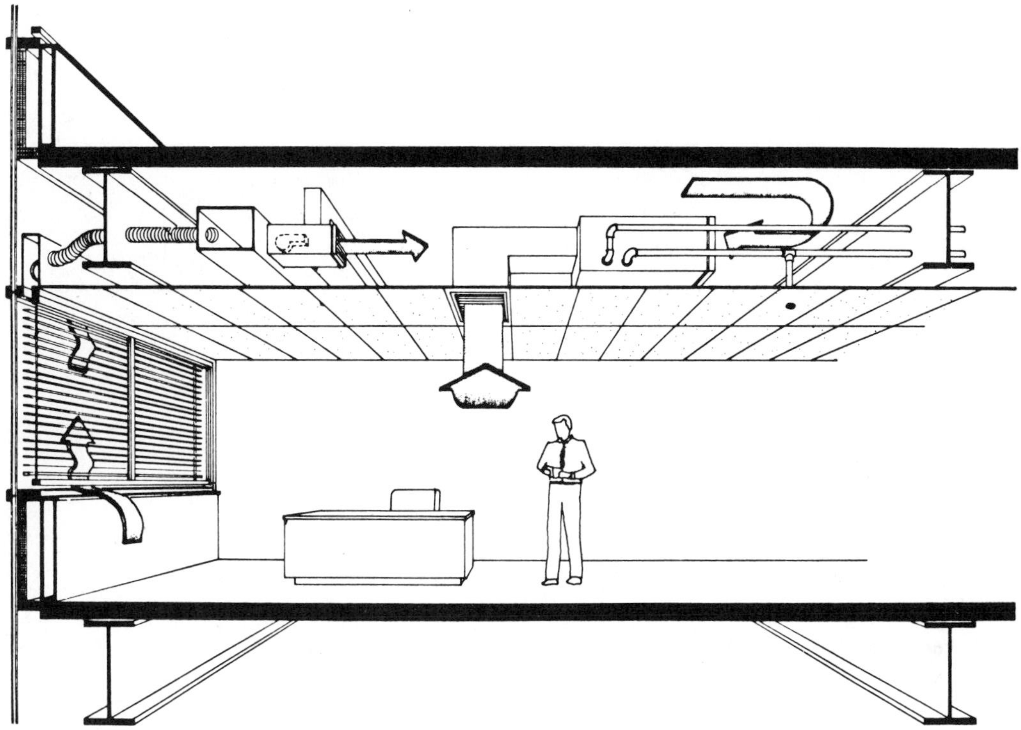

(d)

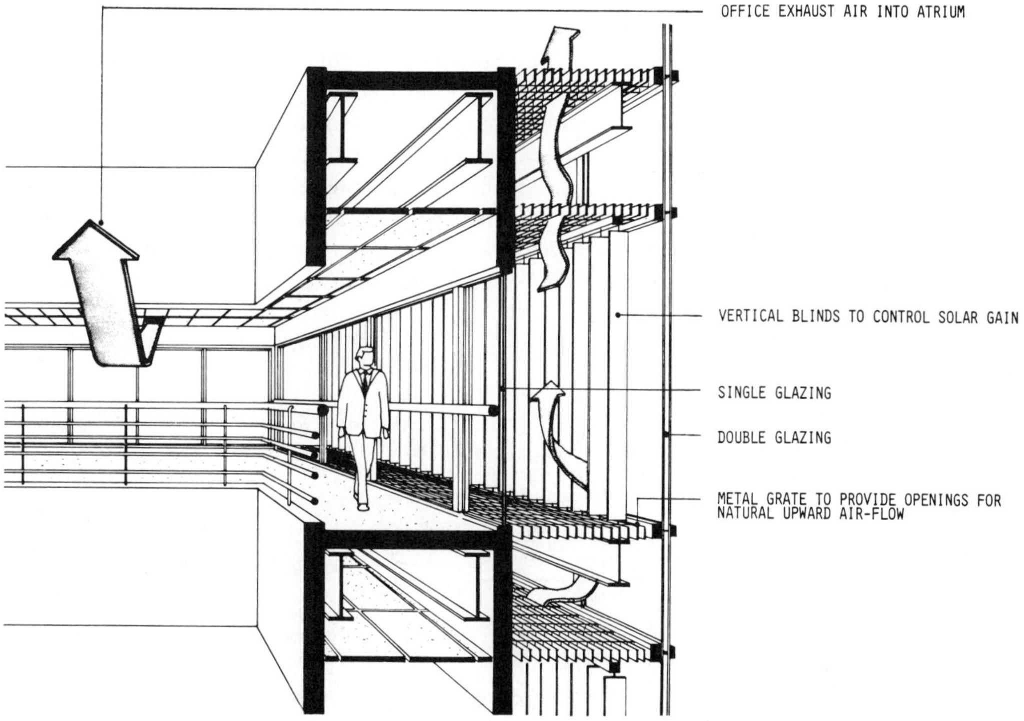

(e)

Fig. 7.66 *(continued)*

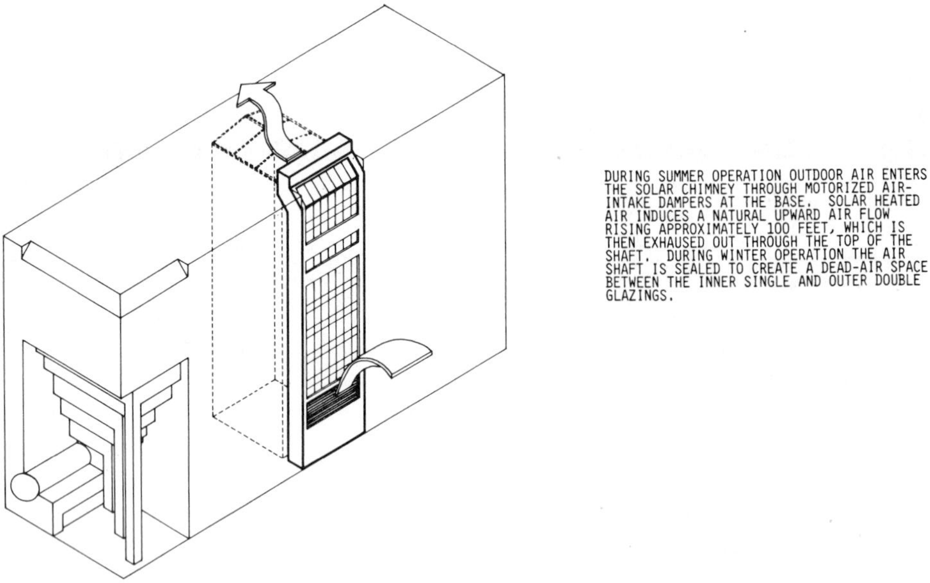
DURING SUMMER OPERATION OUTDOOR AIR ENTERS
THE SOLAR CHIMNEY THROUGH MOTORIZED AIR-
INTAKE DAMPERS AT THE BASE. SOLAR HEATED
AIR INDUCES A NATURAL UPWARD AIR FLOW
RISING APPROXIMATELY 100 FEET, WHICH IS
THEN EXHAUSED OUT THROUGH THE TOP OF THE
SHAFT. DURING WINTER OPERATION THE AIR
SHAFT IS SEALED TO CREATE A DEAD-AIR SPACE
BETWEEN THE INNER SINGLE AND OUTER DOUBLE
GLAZINGS.

(f)

Fig. 7.66 *(continued)* (f) *The stack effect also controls summer overheating.*

electricity) produced in the process. Cogeneration (also called "total energy") is an attempt to recover some of the otherwise-wasted, lower grade heat that accompanies the generation of electricity by steam turbines. Industrial co-generation facilities are especially cost-effective in the pulp and paper, petroleum and chemical industries. Building-scale co-generation is less obvious an energy bargain.

(a) Electrical Power Generation at the Site. It has been found that *where conditions are favorable,* electricity for power and light can be generated economically by a system that also supplies the building with heating in winter and cooling in summer. Such a system utilizes a fuel such as gas or oil and is often supplemental to the local electric utility company. Experience has indicated that although installation costs are greater than for the more conventional systems that use separate services of electricity and fuel, the savings in annual operating cost can sometimes pay for the excess installation cost in a reasonable time. Operational savings continue

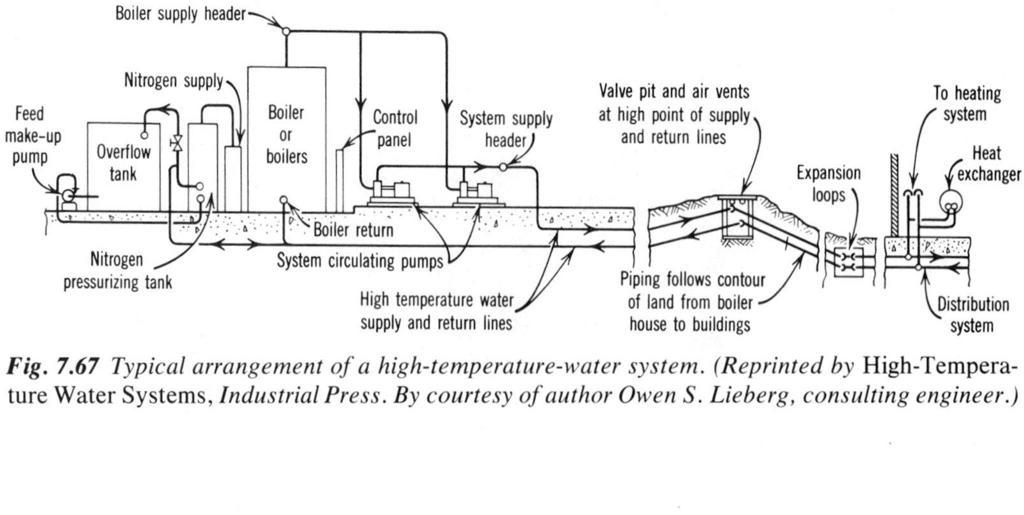

Fig. 7.67 *Typical arrangement of a high-temperature-water system. (Reprinted by* High-Temperature Water Systems, *Industrial Press. By courtesy of author Owen S. Lieberg, consulting engineer.)*

Fig. 7.68 (a) *John F. Kennedy International Airport. The central heating and cooling plant is to the left of the central pool. It is identified as an L-shaped building with a spherical oil-storage tank in its foreground. (Photos courtesy of Port Authority of New York and New Jersey.)*

thereafter. The success of this method, which developed largely in the 1960s, is evident by its use in hundreds of commercial and industrial buildings and in many schools.

Cogeneration is particularly attractive in district heating/cooling plants. This strongly suggests the need for expert engineering analysis of conditions before selection of the most appropriate method of providing the thermal and power services in a building. The conditions include

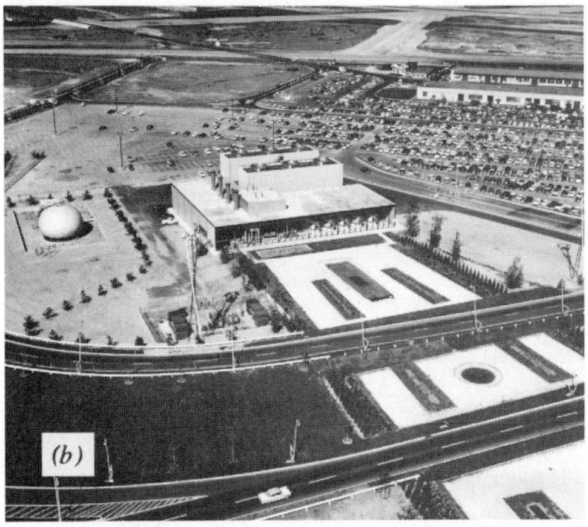

(b) *Heating and cooling plant, initial stage. At left, spherical oil storage tank. Four stacks serve the high-temperature water boilers. Five of the nine absorption refrigeration machines appear through glass at the right. Six cells identify the two cooling tower structures in the rear. Skidmore Owings & Merrill, architects. Seelye, Stevenson, Value, and Knecht, engineers.*

(c) *Night view of the refrigeration-absorption machines in the heating–cooling plant. Distinctive colors identify the piping connections to the machines. Chilled water, condenser water and, in this case, high-temperature water instead of steam, comprise the three circuits connected to each of the nine machines.*

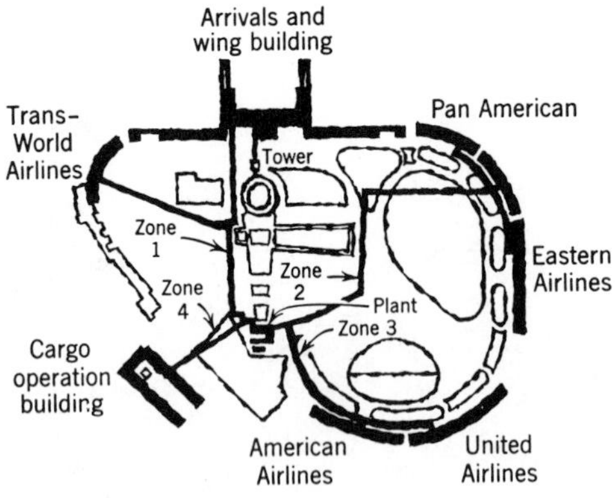

(d)

Fig. 7.68 *(continued)* (d) *Control terminal area, showing location of heating and cooling plant and the four zones of distribution for high-temperature water and chilled water. (Reprinted by permission from "Heating and Air Conditioning a Civilian Airport," by Charles Broder, M.E., in* High Temperature Water, *a Symposium bulletin published by the American Society of Heating, Refrigerating, and Air-Conditioning Engineers, Inc.)*

type and size of building, its geographic location, occupancy characteristics, relative costs of oil, gas, and electricity, and whether cooling as well as heating is essential.

(b) Early On-Site Power Generation. Before 1900 and in the early years of this century, nearly all large buildings and groups of buildings were supplied with direct current generated on or near the premises. The motive power was usually in the form of steam-driven reciprocating engines with belt connections to direct-current generators. Direct current cannot be transformed to different voltages and must be generated and distributed at the voltage used in the building. At these relatively low voltages, power loss in the distribution system is great, and distance adds greatly to the loss. With the development and use of alternating-current machines, utility companies were able to establish central power stations from which electricity could be transmitted great distances to the user

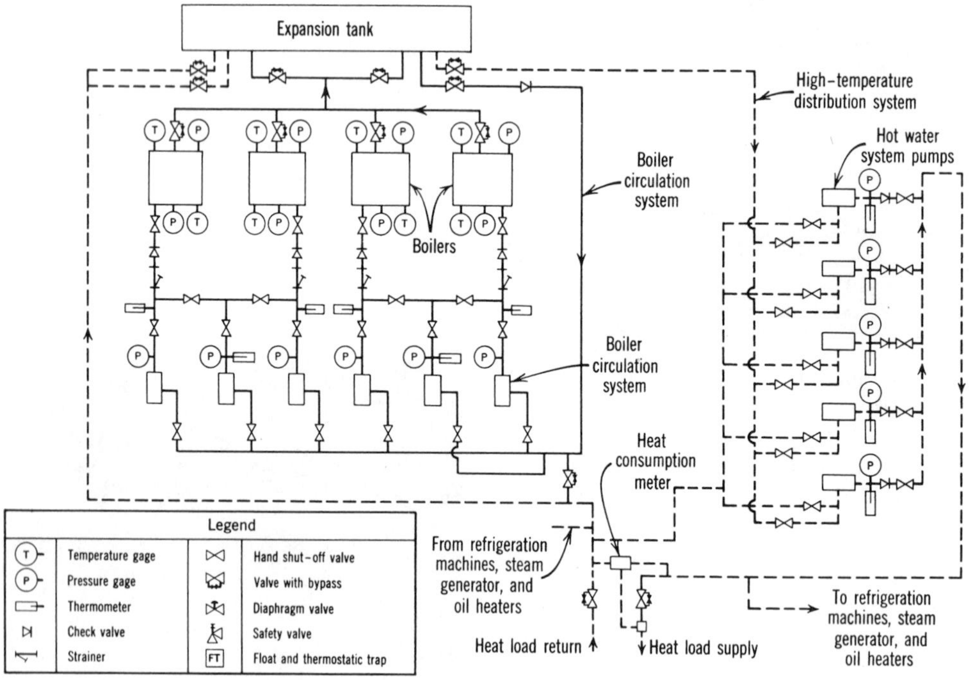

(e)

Fig. 7.68 *(continued)* (e) *Piping diagram for high-temperature water system, showing circulation through boilers and expansion tank and connection to heating load and refrigeration load. (From "Space Heating at Idlewild" by Charles Broder, M.E., in* Industry Power.*)*

at high voltage. There it was transformed down to domestic voltages for use. Since, at high voltage, power losses are very low, this system became universal. During the 1920s and 1930s, owners removed their private power generators and accepted utility service, with its savings in operating expense.

(c) How Cogeneration Developed. In earlier installations, in which steam, produced from coal-burning boilers, was used, there was little or no energy salvage from the steam, which was mostly wasted. The newer fuels as used directly in reciprocating engines or turbines have residual heat value that can be recovered for purposes of heating or cooling. The techniques for this recovery have been perfected by experience with nuclear power, rocketry, aviation, and other recent developments. For cogeneration to be successful, there has to be a reasonably steady demand in the building for the power generated and also for the heat recovered. It is easily understood that many buildings have this need. Lighting and the demand for power by computers, electrical business machines, electrical cooking, and the great multitude of electrically powered devices often make the call for power nearly constant throughout the year. Similarly, the exhaust heat recovery from the engines or turbines that power the generators is in demand for either heating or absorption cooling at any time of the year. Cooling, formerly something of a luxury, is now commonplace. It also fills the gap of energy demand that formerly existed in summer, when heating was not required.

(d) How It Works. Figure 7.69 shows the two principal systems for total energy, reprinted here through the courtesy of Educational Facilities Laboratories. In both systems—one using a turbine and the other a reciprocating engine to operate the generator that supplies electric power—heat is reclaimed to produce steam or hot water. The steam or hot water is then used for heating or, by use of an absorption chiller, for cooling. When a turbine is used, the fuels are natural gas or fuel oil. The heat is recovered by passing the hot turbine exhaust through a waste heat boiler to produce steam. Fuel for a reciprocating engine is natural gas or diesel fuel. Both the jacket cooling water and the hot engine exhaust are passed through heat exchangers that utilize the heat to produce steam or hot water for heating or cooling, as already described. In both systems, an auxiliary boiler, fired directly by gas or oil, stands ready to help maintain a balance in the system.

The unique feature of cogeneration is the use of the recovered heat. This is essentially free energy, since it would otherwise have to be purchased and paid for separately in the form of electricity or other fuels.

(e) Cogeneration for Schools. In the 1960s, changes in school buildings and in school programs prompted the Education Facilities Laboratories to support a study of the feasibility of cogeneration (total energy) for schools, methods of determining feasibility in specific instances, and guidelines for school administrators and engineers in developing solutions.

Schools in that era had large energy requirements; lighting was increased and cooling became as high a priority as heating. Electronic teaching aids added to these energy demands. Evening classes, multifunction buildings, and summer programs put many schools into a new category.

EFL's interest in sponsoring this work stemmed not only from the foregoing school demands but also from the rapidly increasing costs of mechanical and electrical systems in schools and, with regard to cogeneration, the less objective and inescapably conflicting commercial interests of electric utility companies and the manufacturers of electric generators. The report included a consideration of college buildings as well as of schools.

The foregoing conditions and criteria that evaluate schools and college buildings, as possibly suitable for the use of cogeneration, can also be applied to other building types. These can include commercial and industrial buildings as well as offices and motels.

(f) Cogeneration for Housing. The Harbortown apartment and townhouse complex in Detroit, MI utilizes a year-around natural-gas-fired primary energy and cogeneration/waste recovery system (Fig. 7.70). Each apartment has

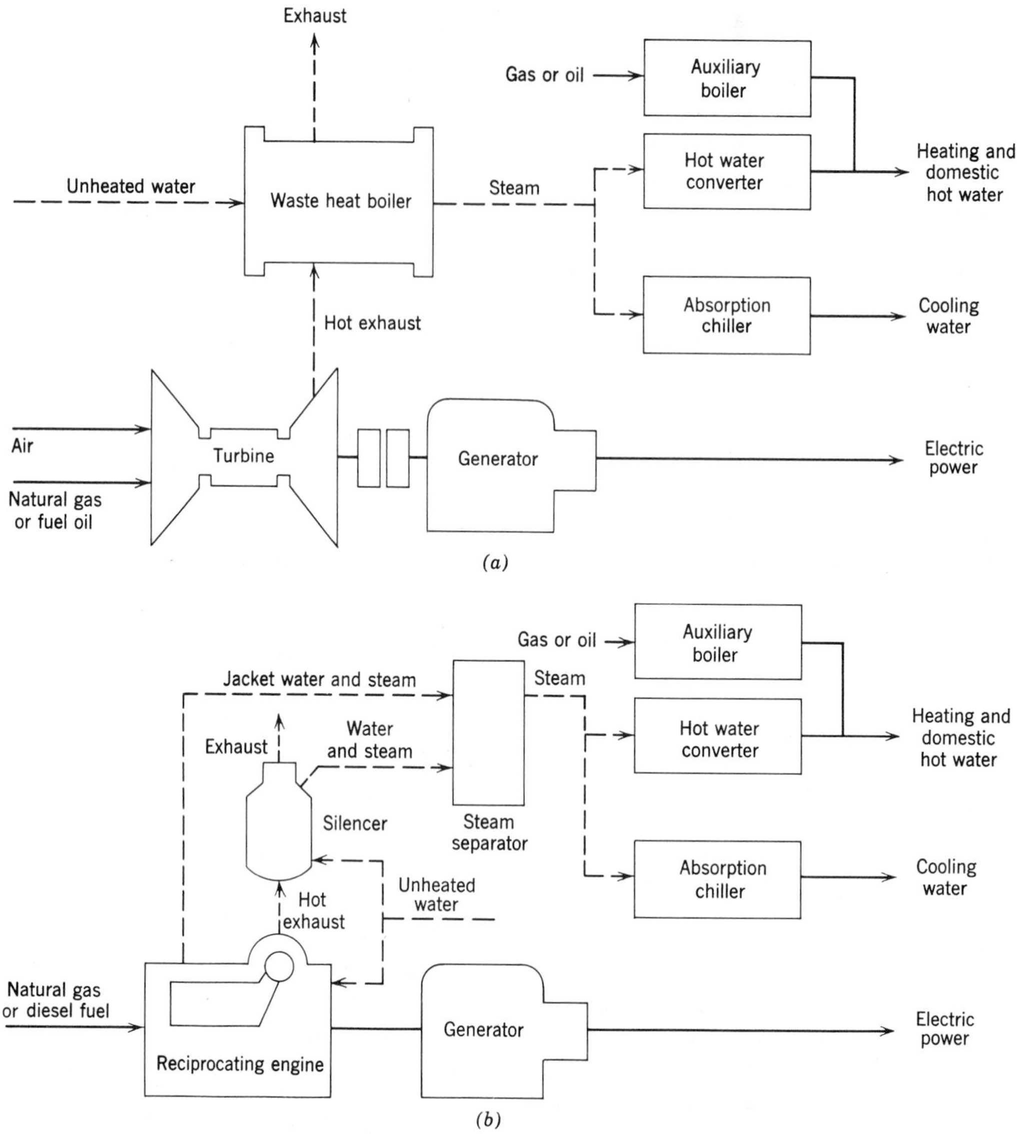

Fig. 7.69 *Cogeneration (total energy) systems.* (a) *Using turbine.* (b) *Using reciprocating engine. Reprinted by permission from* Total Energy, *Educational Facilities Laboratories.*

one or two stackable, upright fan coil units, using either chilled or heated water. This water is provided by two direct-fired chiller-heaters: each incorporates a natural gas boiler and a two-stage absorption chiller in the same unit, taking less space in the mechanical room. The natural gas-driven cogeneration electric plant serves rather constant loads such as corridor and outdoor lighting. The generator's waste heat is passed to a storage tank, where it preheats domestic hot water for the tenants. This meets about 87% of the hot water energy usage.

(g) System Components. The differences in the equipment needed for a cogeneration system and a conventional one are generally limited to the equipment room. No change is needed in the piping for chilled and hot water, and fan-coil units, the absorption machine, or the air-handling units. Figure 7.71 shows in schematic form

(a)

***Fig. 7.70** Harbortown is a 120,000-ft² apartment and townhouse complex* (a) *in Detroit that uses natural gas as a primary fuel year-round. Photo © 1989, William Kildow. Chiller–heaters (Hitachi) combine boiler and two-stage absorption chiller in one compact unit.* (b) *Comparison of headspace and* (c) *floor space required by chiller-heaters versus conventional units of similar capacities.* (d) *Cogeneration plant sends its waste heat to preheat domestic hot water.* (e) (overleaf) *Typical partial plan shows the space advantages of stackable two-pipe fan coil (FCR) units. (Courtesy of Skidmore, Owings and Merrill, Architects-Engineers, Chicago.)*

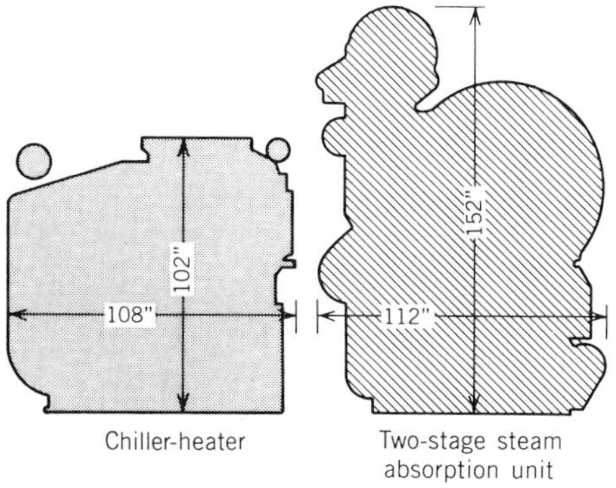

(b)

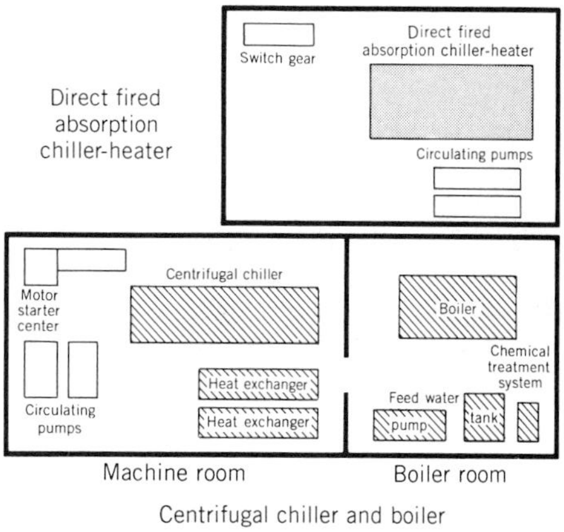

(c)

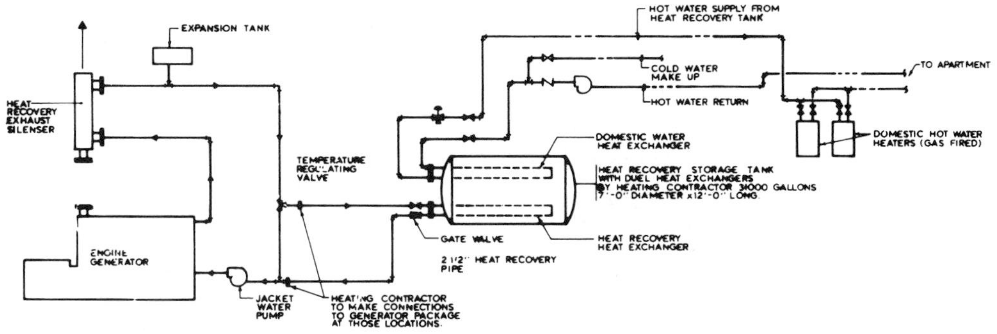

FLOW DIAGRAM OF HEAT RECOVERY SYSTEM

(d)

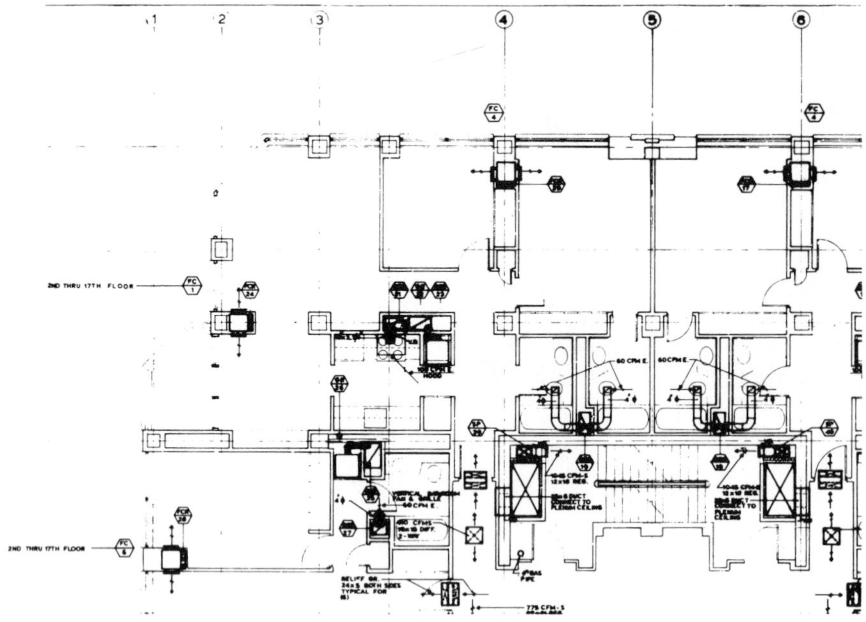

(*e*)

Fig. 7.70 *(continued)*

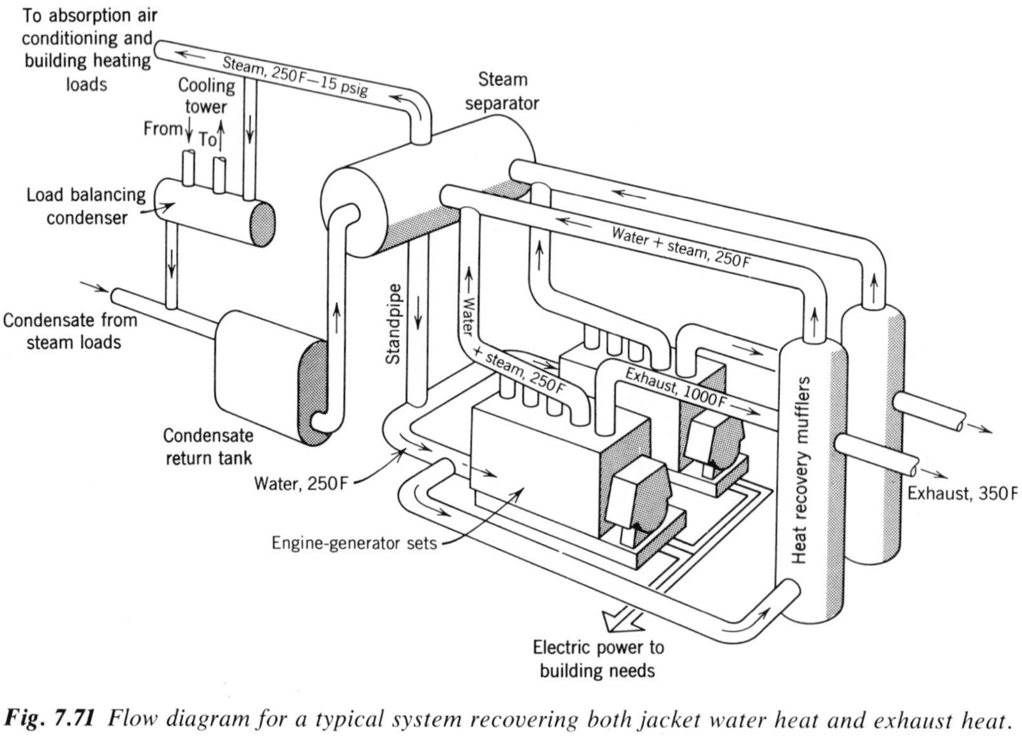

Fig. 7.71 *Flow diagram for a typical system recovering both jacket water heat and exhaust heat.*

the piping and a few additional devices commonly used. They comprise a steam separator, load-balancing condenser, and condensate return tank. After the steam has passed through the heat exchangers to produce hot water or through the absorption chiller for chilled water, the rest of the distribution system is as it would normally be in any design suitable for the specific building.

Although the terms just mentioned are typical, it must be understood that there are many reasonable or necessary variations within the design of an efficient cogeneration system.

REFERENCES

American Institute of Architects, Ramsey, C. G., and Sleeper, H. R. (1988). *Architectural Graphic Standards,* 8th ed., Wiley, New York.

American Society of Heating, Refrigerating, and Air-Conditioning Engineers, Inc. (1989). *Handbook of Fundamentals,* ASHRAE, Atlanta, Ga.

American Society of Heating, Refrigerating, and Air-Conditioning Engineers, Inc. (1989). ASHRAE Standard 90.1-1989: *Energy Efficient Design of New Buildings Except New Low-Rise Residential Buildings,* ASHRAE, Atlanta, Ga.

American Society of Heating, Refrigerating, and Air-Conditioning Engineers, Inc. (1988). *Technical Data Bulletin: Intelligent Buildings,* ASHRAE, Atlanta, Ga.

Fisher, J. (1974). *Energy Crisis in Perspective,* Wiley, New York.

PART III
WATER AND WASTE

Almost every building designed today in the United States is supplied with potable (drinkable) water. In the vast majority of buildings, most of this clean water is used to carry away organic waste. The result is a wide range of design impacts covering everything from the detailed arrangement of bathroom fixtures and interior surfaces to the overall plans of very large and complex water and sewage treatment facilities.

This section of the book presents the following topics: basic planning information and rainwater (Chapter 8); water supply (Chapter 9); organic waste disposal, both water and waterless (Chapter 10); the particular design issues of the bathroom (Chapter 11); and solid waste design issues (Chapter 12). Fire protection, which often requires the largest amount of water supply piping within a building, is covered in Chapter 13.

8

WATER AND BASIC DESIGN

8.1 Water in Architecture

Throughout history, in nearly all climates and cultures, the designer's major concern about water was how to keep it *out* of a building. Only within the last 100 years has water supply *within* a building become commonplace in industrialized countries. In the rest of the world today, running water is still not available within most buildings. Water's potential contributions to life-style and architecture are as numerous and varied as are appropriate design responses to the supply, usage, and return of such a versatile commodity.

(a) Nourishment. Much of the human body is water, the most abundant chemical in our bodies as well as in our diet. The amount of really "pure" (potable) water that we need for drinking and cooking is very small—only about 3 g/cd (gallons per capita per day) (11.4 L/cd) in the United States. The most common supply system throughout history has been the central municipal fountain or well (Fig. 8.1), whose technical importance to the community has often been emphasized by the esthetics of both the fountain's overall sculptural composition and the elegance of detail in water spouts, basins, and other elements. The social importance is evident from its location in the central plaza, from which site small amounts of water are carried daily by townspeople to homes or workplaces.

As daily potable water became available within buildings, water-related social opportunities diminished. However, employees still converse around the drinking fountain—for example, when it is located in a place that invites lingering or relaxing.

(b) Cleansing and Hygiene. Water is a nearly ideal medium for the dissolution and transport of organic waste, and its high heat-storage capacity makes the attainment of comfortable temperatures for bathing easy. Much larger quantities of water are used for cleaning than for nourishment: in the average U.S. home, about 14 g/cd (53 L/cd) is used for clothes washing and dishwashing and another 21 g/cd (79.5 L/cd) is used for bathing and personal hygiene.

In the past, water for cleaning was carried to the home less frequently; the Saturday night (only) bath was typical well into the 20th century in this country. Bathing vessels were usually portable and sometimes were combined with other pieces of furniture (a couch that sat over a tub, a metal tub that folded up inside a tall wooden cabinet, etc.) Thus, a bathplace rather than a bathroom was the common design response.

Although the physical constraints of water carrying were important design influences on bathing, cultural attitudes were at least as strong an influence throughout history. Perhaps the most startling contrast is between the medieval concept of bathing as an almost sinful indulgence and the earlier Roman attitude toward public baths as the social centers of cities.

Now, bathing facilities are commonly designed to be used on a very personal scale, in privacy. There also are welcome opportunities to design more-social bathing places, such as swimming pools, bath houses, and hot tubs. The characteristics of the water supply (spouts, jets, cascades) can be matched with those of the water body (mirror-smooth, gently flowing, rippled, rolling, foaming) to obtain the desired atmosphere.

(c) Ceremonial Uses. Largely through its associations with cleaning, water acquired a ceremonial significance that remains particularly evident in religious services. Examples of the ceremonial use of water include vessels containing holy water at entrances to Catholic

Fig. 8.1 The fountain in the central piazza at Tarquinia, Italy, still serves as a gathering place and as a symbol of the town's provision for a basic need.

churches, pools in the forecourts of mosques, and full-immersion baptismal fonts at the altars of some Protestant churches. The opportunities for esthetic expression are particularly rich in these ceremonial applications.

(d) Transportational Uses. In stark contrast to its uses in nourishing, cleansing, and celebrating, water is used in our buildings principally to transport organic waste. There is perhaps no more flagrant example of a mismatch in architecture than the high-grade resource of pure water being used for the low-grade task of carrying away a cigarette butt. The typical U.S. home uses 32 g/cd (121 L/cd) just to flush toilets.

In the past, table scraps were commonly fed to animals or composted, and human waste thrown out of windows (accompanied by warning cries) or deposited in holes below outhouses. Organic waste disposal was thus dependent on either portable vessels or special structures set apart from the typical building.

As water supplies were developed, water's advantages over the foul smell and inconvenience of these methods became irresistible. A typical sequence of events unfolded on Manhattan Island. In the 1700s, Manhattan was farm country that, like all other areas that later developed into large cities, had minimal water needs. Potable water was available in shallow wells and from some springs and streams. These sources were largely unaffected by the minor ground pollution from widely separated dry-pit privies (outhouses) that received human wastes. Paved city streets appeared in the 1800s, at which time the natural streams were enclosed in pipes, called *storm sewers;* these pipes led the rainfall to the many waterways surrounding the island. Then, in the later 1800s, flush toilets appeared. It seemed natural to connect the toilets to the already established "storm sewers" and to rename the pipeways "combined" sewers, which now carried both storm water and so-called "sanitary" drainage to the rivers (sanitary for the building, but not for the rivers). Fast-flowing rivers are natural sewage treatment plants, and, surprisingly, for many decades they did a fair job of keeping pollution reasonably in check. With the prospect of future sewage treatment plants, separate "sanitary" sewers were built. Also, there were some remaining (and some newly built) storm sewers that did not carry the wastes from toilets.

In cities where this confused pattern of sewer systems still exists (this includes most larger and older cities), it is now extremely difficult and expensive to sort out and reroute sewers so that *only* sanitary drainage goes to treatment plants and *all* storm drainage goes to waterways or into the ground. It seems particularly ironic that in most U.S. locations the rainwater that falls on a home's roof is adequate in both quality and quantity to supply a family's cleansing needs [21 + 14 = 35 g/cd (132 L/cd)]. In Chapters 8, 9, and 10, these possibilities for rainwater will be developed further.

As the human waste disposal place became a room within the building, the design issues grew more complex. Physically, there was the need for running water, and for large-diameter pipes that sloped downward continuously from the toilet to a sewer or septic tank. As sewer gas became a recognized problem, an elaborate system of traps and vents became necessary. Again, cultural attitudes were also influential: How private an activity was this elimination from the body? To what extent could one plumbing fixture accommodate both body cleansing and waste elimination? If males insisted on standing rather than sitting while urinating, how to devise a toilet that would also properly accommodate defecation, which requires a low seat? Some designer responses to these questions can be found in Chapter 11.

(e) Cooling. Water has a remarkable cooling potential: it stores heat readily, removes large quantities of heat when it evaporates, and vaporizes readily at temperatures commonly found at the human skin surface. In hot-dry climates, designers can place water surfaces (or sprays) upwind from the place to be cooled, or resort to the evaporative coolers discussed in Sections 2.4, 4.7, and 6.7. Cooling towers (Section 7.3) are familiar components of large-building cooling systems.

Because all of us have experienced the physical cooling of the skin by water, we all carry psychological associations between water and cooling that can enhance our comfort on hot days. The sight of sunlight reflected on a water surface, with its characteristic "dancing" quality, connotes coolness, as does the sound of running or splashing water. Thus, even when water does not physically cool people, it can make an important psychological contribution to human comfort (Fig. 8.2).

Fig. 8.2 *Water as an aesthetic feature of a courtyard in the hot-arid part of Mexico. The sound and sight of running water add to a psychological impression of coolness.*

(f) Ornamental Uses. In almost any landscaping application, indoors or out, water becomes a center of interest. Our association of water with nourishing, cleansing, and cooling make water a very powerful design element—a fact recognized by landscape designers throughout history. In arid regions, water is often used sparingly, in small, tightly controlled channels and at lower flow rates. The gardens of Islamic architecture are especially effective demonstrations of such design restraint. Where water is more plentiful, it has been used lavishly, as at the Villa d'Este in Tivoli, outside Rome.

Especially useful design characteristics of water include its *reflectivity,* which sets it apart from most plant and ground materials in a garden; its *liquidity,* which creates unique sounds wherever it is moved; and its *life-sustaining* potential, which allows the addition of both water plants and animals to a garden.

(g) Protective Uses. Every designer dreads water's ability to penetrate a roof and damage a building and its contents. But we all depend on water as the best fire protection available in most buildings. The vast quantities of water potentially required for firefighting must be delivered quickly; the result is pipes of enormous diameter regulated by very large valves. Because this system's distribution tree must be immediately obvious to firefighters, some degree of exposure is prudent. Despite its size and guarantee of at least partial exposure in public places, a fire protection water supply system is rarely treated as a visually integral design element. This mismatch of potential and actuality will be discussed further in Chapter 13.

8.2 The Hydrologic Cycle

There is a finite quantity of water in the earth and its atmosphere. The process whereby this water constantly circulates, powered by about one-fourth of earth's solar energy, is called the hydrologic cycle (Fig. 8.3). More than 99% of this water is "inaccessible"—either because it is salt water or because it is frozen in glaciers on polar ice caps. The most accessible sources of water are precipitation and runoff.

Precipitation has the advantage of relative purity, although acid rain is a growing threat in many parts of the world, including much of the United States and Canada. Like solar energy, precipitation is a very large but very thinly spread resource; its capture is therefore likely to take place on an individual basis. Until we experience a "water crisis" similar to the "energy crisis" that began in the 1970s, rainwater capture will remain a mostly untapped fresh water source in the United States.

Runoff enjoys the advantage of a concentrated flow of water, which permits easy capture of large quantities. Its most serious disadvantage is the possibility of pollution—organic, chemical, and radioactive—depending on what is upstream from the point of capture. In some regions of North America, river water is reused 50 times on its way to the ocean. Further discussion of water sources and treatments occurs in Chapters 8 and 9.

That part of daily precipitation which neither evaporates nor joins the runoff becomes part of *soil moisture*. Much soil moisture is used by growing plants and is soon transpired (evaporated) by the plant to the atmosphere. As water works downward below the root zone of plants, it eventually reaches a zone of saturation, where all voids in the earth's material are filled with water. This zone of saturation is called *groundwater;* the upper surface of groundwater is called the water table. Wells are commonly sunk to a point well below the water table, so that the latter's seasonal fluctuations will not interrupt the well's access to groundwater.

8.3 Basic Planning

After considering the relationship between a building and the roles that water plays within it, the designer must do basic sizing—of the quantities of water needed, of the areas in which water will be used, and of the areas and equipment associated with water's return to the hydrologic cycle.

But before discussing basic planning for the amounts of water used daily within buildings and for the treatment of that water, we should note that water is also an important component of building construction. The production of 1 ton (907 kg) of bricks requires 580 gal (2200 L) of water; 1 ton of steel requires 43,600 (165,000 L); 1 ton of plastic requires 348,750 gal (1.32 million L). Water is one of the main ingredients of concrete; a typical 94-lb (42.6 kg) bag of cement requires about 6 gal (23 L) of water.

Another way in which buildings contribute to water consumption is through electricity consumption. Most power plants require very large quantities of water, which they quickly return to the hydrologic cycle warmer in temperature (and perhaps as vapor). A large nuclear power plant that utilizes a cooling tower can evaporate daily the approximate equivalent transpiration from about 9 mi^2 (23 km^2) of forest.

(a) Water Supply. The task of rough sizing for water needs is complicated by the conflict between *current practice* and *conservation.* Current practice tends toward the use of large amounts of water for very low-grade tasks. Conservation reserves high-quality water for high-grade tasks and emphasizes recycling, as well as diminished overall usage, of water.

The supply of water often is first estimated in terms of gallons per capita per day (liters per capita per day). Table 8.1 shows some common terms used in water supply. Typical quantities were matched with nourishing, cleansing, and other usages in the United States in Section 8.1.

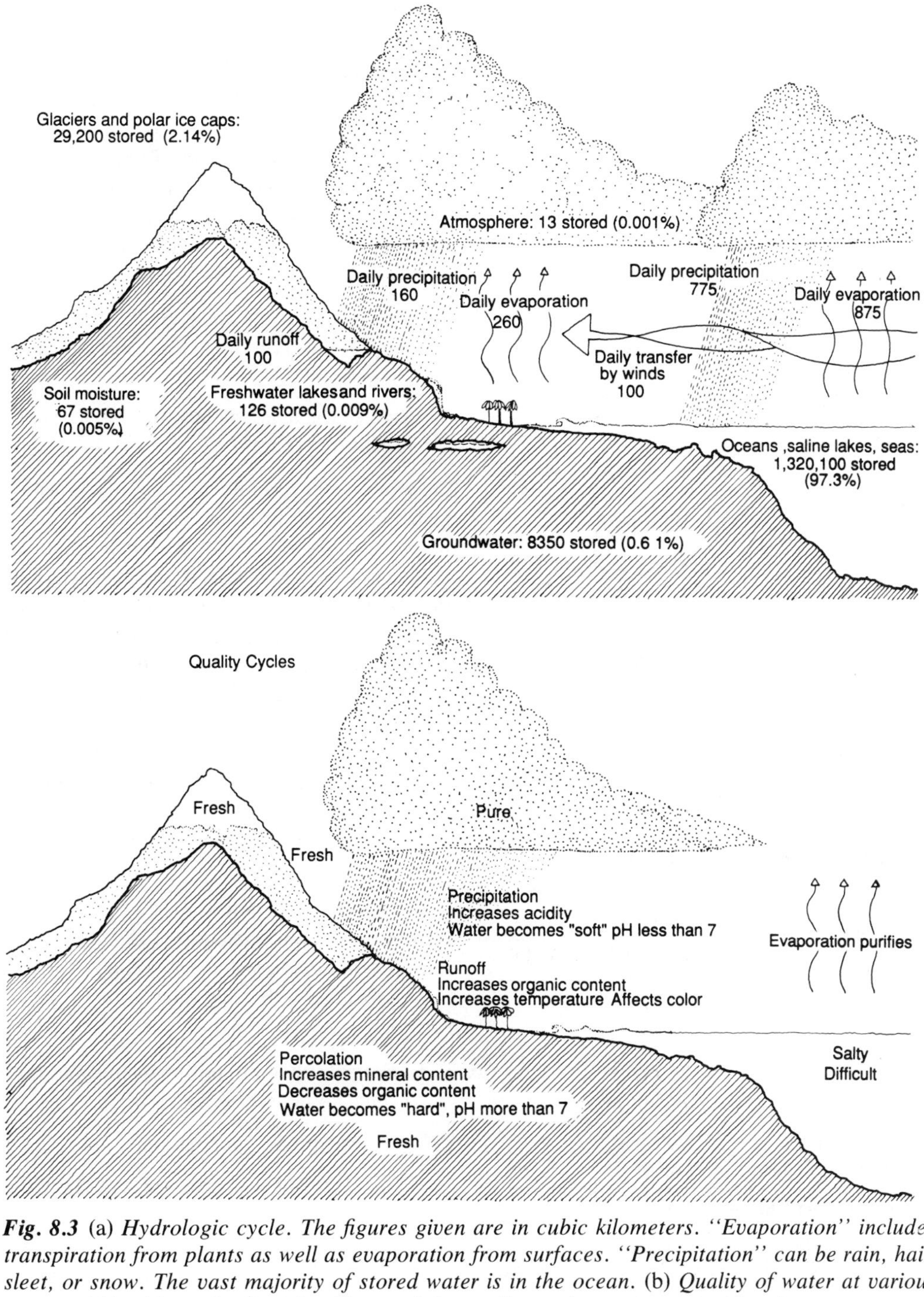

Fig. 8.3 (a) *Hydrologic cycle. The figures given are in cubic kilometers. "Evaporation" includes transpiration from plants as well as evaporation from surfaces. "Precipitation" can be rain, hail, sleet, or snow. The vast majority of stored water is in the ocean.* (b) *Quality of water at various stages within this cycle. See also Chapter 9 for quality and treatment considerations.*

TABLE 8.1 **Some Water Measurement Terms**

Quantity	
1 gal	= 8.33 lb
	= 231 in.^{3}
	= 0.134 ft^3
	= 3.785 L
1 liter	= 0.263 gal
1 cubic foot	= 7.48 gal
	= 6.24 lb
1 ton	= 240 gal
1 acre-foot (a-f)	= 325,851 gal
	= 43,560 ft^3
1 million gallons	= 3.07 a-f
Flow	
1 cfs	= 1.98 a-f/day
1 a-f/yr	= water supply for five for one year at 180 gal per person per day
1000 gpm	= 2.23 cfs
	= 4.42 a-f/day
1 million gpd	= 694.4 gpm
	= 1.55 cfs
	= 1,120 a-f/yr
Cost	
10¢ per 1000 gal	= 7.48¢ per 100 ft^3
	= $32.59 per a-f
10¢ per 100 ft^3	= $43.56 per a-f
	= $13.40 per 1000 gal
10¢ per ton	= $136 per a-f
Leaks	
Slow drip	= 170 gpd
	= 62,050 gal/yr
⅛″ (3.2-mm) diameter stream	= 3600 gpd
	= 4 a-f/yr

Source: Milne (1976).

To gain an appreciation of how the rate of urban water usage has changed over time, consider the following figures [from Milne (1976)]:

Imperial Rome	38 g/cd (144 L/cd)
London, 1912	40 g/cd (151 L/cd)
American cities just before World War II	115 g/cd (435 L/cd)
Los Angeles, mid-1970s	182 g/cd (689 L/cd)

Although the historical trend has clearly been toward higher per capita use of water, the recent emphasis on conservation has resulted in significant changes in this pattern. In Section 10.1, the influences of various conservation practices will be considered in more detail. Table 8.2 can be used to estimate the daily indoor usage of water in various facilities, if *current practices* are anticipated. For very rough approximations of *conservation* effects:

- Reduce Table 8.2 values by 25%, assuming simple conservation measures such as flow controls.
- Reduce Table 8.2 values by 50%, assuming partial recycling.

The more water used for flushing toilets within the building types listed in Table 8.2, the greater the potential for savings through conservation.

In urban areas, public water mains usually provide the necessary quantities of water at the pressures and rates of flow required to operate typical plumbing fixtures. For isolated systems, or those independent of public networks, water supply can come from individual sources: wells, springs, cisterns, lakes, and so forth. For the minimum pressures and flow rates necessary from these sources (and/or the storage vessels associated with them), see Table 9.16.

(b) Cisterns. Where rainwater is to be utilized, a rough approximation of catchment area and cistern storage volume is initially needed. This procedure is detailed later in this chapter. For now,

1. From Table 8.2, find the quantity of rainwater to be used daily:

   g/cd × population = gpd

   (L/cd × population = L/d)

2. Convert this quantity to the yearly need for water:

   gpd × 365 days = gal/yr

   (L/d × 365 days = L/yr)

3. Assume, conservatively, that a "dry" year will have two-thirds the precipitation of an average year; this measurement is the "design precipitation." (Average an-

TABLE 8.2 **Planning Guide for Water Supply**[a]

Building Usage	Per Capita (As Listed) Daily Usage Gallons	(Liters)
Airports (per passenger)	3–5	(11–19)
Apartments, multiple family (per resident)	60	(227)
Bath houses (per bather)	10	(38)
Camps		
Construction, semipermanent (per worker)	50	(189)
Day with no meals served (per camper)	15	(57)
Luxury (per camper)	100–150	(378–568)
Resorts, day and night, with limited plumbing (per camper)	50	(189)
Tourist, with central bath and toilet facilities (per person)	35	(132)
Cottages with seasonal occupancy (per resident)	50	(189)
Courts, tourist, with individual bath units (per person)	50	(189)
Clubs		
Country (per resident member)	100	(378)
Country (per nonresident member present)	25	(95)
Dwellings		
Boardinghouses (per boarder)	50	(189)
Additional kitchen requirements for nonresident boarders	10	(38)
Luxury (per person)	100–150	(378–568)
Multiple-family apartments (per resident)	40	(151)
Rooming houses (per resident)	60	(227)
Single family (per resident)	50–75	(189–284)
Estates (per resident)	100–150	(378–568)
Factories (per person per shift)	15–35	(57–132)
Highway rest area (per person)	5	(19)
Hotels with private baths (two persons per room)	60	(227)
Hotels without private baths (per person)	50	(189)
Institutions other than hospitals (per person)	75–125	(284–473)
Hospitals (per bed)	250–400	(946–1514)
Laundries, self-service (per washing)	50	(189)
Livestock (per animal)		
Cattle (drinking)	12	(45)
Dairy (drinking and servicing)	35	(132)
Goat (drinking)	2	(8)
Hog (drinking)	4	(15)
Horse (drinking)	12	(45)
Mule (drinking)	12	(45)
Sheep (drinking)	2	(8)
Steer (drinking)	12	(45)
Motels with bath, toilet, and kitchen facilities (per bed space)	50	(189)
With bed and toilet (per bed space)	40	(151)
Parks		
Overnight, with flush toilets (per camper)	25	(95)
Trailer, with individual bath units, no sewer connection (per trailer)	25	(95)
Trailer, with individual baths, connected to sewer (per person)	50	(189)
Picnic		
With bathhouses, showers, and flush toilets (per picnicker)	20	(76)
With toilet facilities only (per picnicker)	10	(38)
Poultry		
Chickens (per 100)	5–10	(19–38)
Turkeys (per 100)	10–18	(38–68)

WATER AND WASTE

TABLE 8.2 **Planning Guide for Water Supply** ***(continued)***

Building Usage	*Per Capita (As Listed) Daily Usage*	
	Gallons	*(Liters)*
Restaurants with toilet facilities (per patron)	7–10	(26–38)
Without toilet facilities (per patron)	2½–3	(9–11)
With bar/cocktail lounge (additional quantity per patron)	2	(8)
Schools		
Boarding (per pupil)	75–100	(284–378)
Day, with cafeteria, gymnasium, and showers (per pupil)	25	(95)
Day with cafeteria but no gymnasiums or showers (per pupil)	20	(76)
Day without cafeteria, gymnasiums, or showers (per pupil)	15	(57)
Service stations (per vehicle)	10	(38)
Stores (per toilet room)	400	(1514)
Swimming pools (per swimmer)	10	(38)
Theaters		
Drive-in (per car space)	5	(19)
Movie (per auditorium seat)	5	(19)
Workers		
Construction (per person per shift)	50	(189)
Day (school or office, per person per shift)	15	(57)

Source: Manual of Individual Water Supply Systems (1975).

[a] These values may be reduced as follows: with flow controls, up to 25% reduction
: with water recycling, up to 50% reduction.

nual precipitation is available from NOAA annual summaries.)

Average annual precipitation × ⅔ = design precipitation

4. From Fig. 8.4, determine the catchment area required.
5. Roughly size the cistern (storage) capacity by finding the longest dry period (in days of negligible rainfall, from NOAA local climatological data):

cistern capacity = gpd × days of dry period

6. Convert capacity to volume by the formula

$$1\ \text{ft}^3 \text{ stores } 7.48 \text{ gal of water}$$

$$(1\ \text{m}^3 \text{ stores } 1000 \text{ L water})$$

EXAMPLE 8.1, PART A. A 20,000-ft² one-story factory near Salem, Oregon, will use roof-collected rainwater to flush its toilets. Water-conserving toilets using 3 gal/flush and serving 20 workers are to be used. Table 8.2 shows that factory workers use between 15 and 35 g/cd. Since usage at this factory will be low—no showers, for example—assume 15 g/cd:

15 g/cd × 20 workers = 300 gpd, for *all* usages

Since low-flush toilets are to be used, reduce this figure by 25%:

$$0.75 \times 300 = 225 \text{ gpd}$$

Toilets will probably account for most of this 225 gpd: for example, at three flushes per day per worker,

3 flushes/day × 3 gal/flush × 20 workers = 180 gpd

Assume, then, that up to 200 gpd of rainwater will be utilized.

Catchment Area: Salem's average annual rainfall is 41 in. Design rainfall is ⅔ × 41 = 27.3 in., in a "dry" year; yearly need is 200 gpd × 365 days = 73,000 gal. The combination of 73,000 gal and 27.3 in. is off the chart in Fig. 8.4*a*, so divide 73,000 gal by 2, to obtain 36,500 gal, and

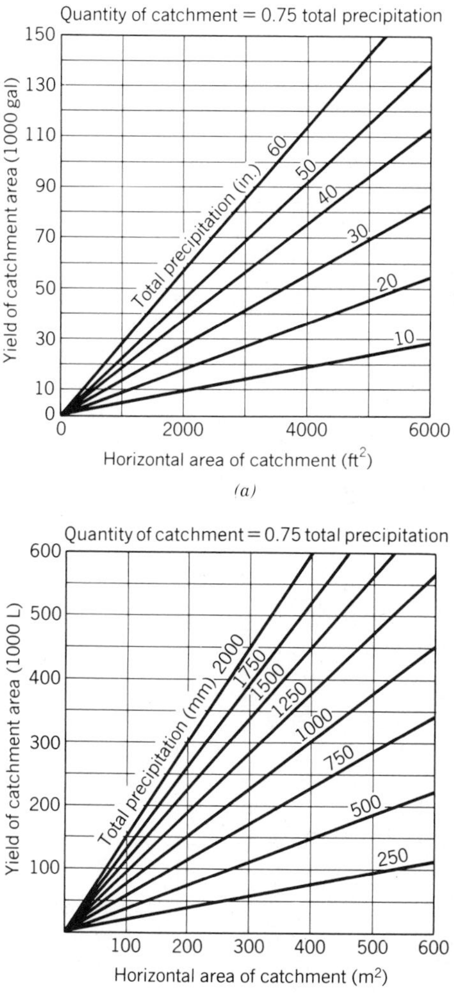

***Fig. 8.4** Yields of rainfall catchment areas (roofs), in terms of total precipitation. In these graphs, 75% of the total precipitation is assumed to be catchable; the remainder is lost to evaporation or spillage.* (a) *Conventional units, from the U.S. Environmental Protection Agency's* Manual of Individual Water Supply Systems, *1975.* (b) *SI units (1 m³ = 1000 L).*

then double the resulting catchment area. At 27.3 in. of precipitation, 36,500 gal will be caught by 2800 ft². The catchment area for 73,000 gal is, therefore, 2 × 2800 = 5600 ft². (So about 28% of the 20,000-ft² factory's roof area will suffice for a catchment area.)

Cistern Capacity: Salem normally has very dry summers; average monthly rainfall is as follows:

May	2.1 in.
June	1.4 in.
July	0.4 in.
August	0.6 in.
September	1.5 in.
October	4.0 in.

The dry period, then, runs from mid-June to mid-September—about 90 days. Thus

$$\text{capacity} = 200 \text{ gpd} \times 90 \text{ days} = 18{,}000 \text{ gal}$$

$$\text{volume} = \frac{18{,}000 \text{ gal}}{7.48 \text{ gal/ft}^3} = 2406 \text{ ft}^3$$

(For example, 5-ft deep × 22-ft square = 2420 ft³.) A more detailed sizing procedure for this building's cistern is presented later in this chapter.

(c) Required Facilities. Another important early design question is that of how many plumbing fixtures should be provided. Table 8.3 lists the minimum plumbing facilities required for various types and sizes of building occupancies. (Note that local requirements may differ somewhat from this particular guide, which is taken from the 1985 Uniform Plumbing Code. There are many plumbing codes in use in North America, and they sometimes disagree.) Since these are considered ''minimal'' requirements, more generous provisions are sometimes appropriate. Chapter 11 discusses the design of the spaces in which these facilities are used.

(d) Sewage. Where public sewers are available (as in most urban areas), the designer usually is not concerned with estimating the flow of sewage. However, where private or on-site sewage treatment is required, or where public sewage treatment facilities are overtaxed, total sewage flow is an early design concern. Table 8.4 lists these sewage flows, by type of occupancy, again in g/cd (L/cd). Note that sewage flow may differ from supply flow, especially where supply water is used for irrigation or in evaporative processes.

Once the daily flow of sewage is established, some early guidelines are needed for determining the suitability of and the area required by

TABLE 8.3 **Minimum Plumbing Facilities**[a]

Type of Building or Occupancy[b]	*Water Closets (Fixtures per Person)*	*Urinals*[c] *(Fixtures per Male)*	*Lavatories (Fixtures per Person)*	*Bathtubs or Showers (Fixtures per Person)*	*Drinking Fountains*[d,e] *(Fixtures per Person)*
Assembly places (theaters, auditoriums, convention halls, etc.)—for permanent employee use	*Male*: 1:1–15; 2:16–35; 3:36–55 *Female*: 1:1–15; 2:16–35; 3:36–55 Over 55, add 1 fixture for each additional 40 persons	1 per 50	*Male*: 1 per 40 *Female*: 1 per 40	—	—
Assembly places (theaters, auditoriums, convention halls, etc.)—for public use	*Male*: 1:1–100; 2:101–200; 3:201–400 *Female*: 3:1–100; 6:101–200; 8:201–400 Over 400, add 1 fixture for each additional 500 males and 2 for each 300 females	1:1–100 2:101–200 3:201–400 4:401–600 Over 600, add 1 fixture for each additional 300 males	*Male*: 1:1–200; 2:201–400; 3:401–750 *Female*: 1:1–200; 2:201–400; 3:401–750 Over 750, add 1 fixture for each additional 500 persons	—	1 per 75[f]
Dormitories[g]—school or labor	*Male*: 1 per 10 *Female*: 1 per 8 Add 1 fixture for each additional 25 males (over 10) and 1 for each additional 20 females (over 8)	1 per 25 Over 150, add 1 fixture for each additional 50 males	*Male*: 1 per 12 *Female*: 1 per 12 Over 12, add 1 fixture for each additional 20 males and 1 for each 15 females	1 per 8 For females, add 1 bathtub per 30; over 150, add 1 per 20	1 per 75[f]
Dormitories—for staff use	*Male*: 1:1–15; 2:16–35; 3:36–55 *Female*: 1:1–15; 2:16–35; 3:36–55 Over 55, add 1 fixture for each additional 40 persons	1 per 50	*Male*: 1 per 40 *Female*: 1 per 40	—	—

Dwellings[h] Single Multiple or apartment house	 1 per dwelling 1 per dwelling or apartment unit	 —	 1 per dwelling 1 per dwelling or apartment unit	 1 per dwelling 1 per dwelling or apartment unit	 —
Hospitals Waiting room For employee use	 1 per room *Male* / *Female* 1:1–15 / 1:1–15 2:16–35 / 2:16–35 3:36–55 / 3:36–55 Over 55, add 1 fixture for each additional 40 persons	 — 1 per 50	 1 per room *Male* / *Female* 1 per 40 / 1 per 40	 — —	 1 per 75[f]
Hospitals Individual room Ward room	 1 per room 1 per 8 patients	 — —	 1 per room 1 per 10 patients	 1 per room 1 per 20 patients	 — 1 per 75[f]
Industrial[i] warehouses, workshops, foundries and similar establishments (for employee use)	*Male* / *Female* 1:1–10 / 1:1–10 2:11–25 / 2:11–25 3:26–50 / 3:26–50 4:51–75 / 4:51–75 5:76–100 / 5:76–100 Over 100, add 1 fixture for each additional 30 persons		Up to 100, 1 per 10 persons Over 100, 1 per 15 persons[j,k]	1 shower for each 15 persons exposed to excessive heat or to skin contamination with poisonous, infectious, or irritating material.	1 per 75[f]
Institutional (other than hospitals or penal institutions), on each occupied floor	*Male* / *Female* 1 per 25 / 1 per 20	1 per 50	*Male* / *Female* 1 per 10 / 1 per 10	1 per 8	1 per 75[f]
Institutional (other than hospitals or penal institutions), on each occupied floor—for employee use	*Male* / *Female* 1:1–15 / 1:1–15 2:16–35 / 2:16–35 3:36–55 / 3:36–55 Over 55, add 1 fixture for each additional 40 persons	1 per 50	*Male* / *Female* 1 per 40 / 1 per 40	—	—

(continued)

WATER AND WASTE

TABLE 8.3 **Minimum Plumbing Facilities**[a] **(*continued*)**

Type of Building or Occupancy[b]	*Water Closets (Fixtures per Person)*		*Urinals*[c] *(Fixtures per Male)*	*Lavatories (Fixtures per Person)*		*Bathtubs or Showers (Fixtures per Person)*	*Drinking Fountains*[d,e] *(Fixtures per Person)*
	Male	*Female*		*Male*	*Female*		
Office or public building	1:1–15 2:16–35 3:36–55 4:56–80 5:81–110 6:111–150	1:1–15 2:16–35 3:36–55 4:56–80 5:81–110 6:111–150		1:1–15 2:16–35 3:36–60 4:61–90 5:91–125	1:1–15 2:16–35 3:36–60 4:61–90 5:91–125	—	1 per 75[f]
	Over 150, add 1 fixture for each additional 40 persons			Over 125, add 1 fixture for each additional 45 persons			
Office or public building—for employee use	1:1–15 2:16–35 3:36–55	1:1–15 2:16–35 3:36–55	1 per 50	1per 40	1 per 40	—	—
	Over 55, add 1 fixture for each additional 40 persons						
Penal institutions—for employee use	1:1–15 2:16–35 3:36–55	1:1–15 2:16–35 3:36–55	1 per 50	1 per 40	1 per 40	— —	1 per 75[f]
	Over 55, add 1 fixture for each additional 40 persons						
Penal institutions—for prisoner use							
Cell	1 per cell		—	1 per cell		—	1 per cell block floor
Exercise room	1 per exercise room		1 per exercise room	1 per exercise room		—	1 per exercise room

Restaurants, pubs and lounges[l]	*Male* *Female* 1:1–50 1:1–50 2:51–150 2:51–150 3:151–300 3:151–300 Over 300, add 1 fixture for each additional 200 persons	1:1–150 Over 150, add 1 fixture for each additional 150 males	*Male* *Female* 1:1–150 1:1–150 2:151–200 2:151–200 3:201–400 3:201–400 Over 400, add 1 fixture for each additional 400 persons		
Schools—for staff use	*Male* *Female* 1:1–15 1:1–15 2:16–35 2:16–35 3:36–55 3:36–55 Over 55, add 1 fixture for each additional 40 persons	1 per 50	*Male* *Female* 1 per 40 1 per 40	—	—
Schools[m]—for student use Nursery	*Male* *Female* 1:1–20 1:1–20 2:21–50 2:21–50 Over 50, add 1 fixture for each additional 50 persons	—	*Male* *Female* 1:1–25 1:1–25 2:26–50 2:26–50 Over 50, add 1 fixture for each additional 50 persons	—	1 per 75[f]
Elementary	*Male* *Female* 1 per 30 1 per 25	1 per 75	*Male* *Female* 1 per 35 1 per 35	—	1 per 75[f]
Secondary	*Male* *Female* 1 per 40 1 per 30	1 per 35	*Male* *Female* 1 per 40 1 per 40	—	1 per 75[f]
Others (colleges, universities, adult centers, etc.)	*Male* *Female* 1 per 40 1 per 30	1 per 35	*Male* *Female* 1 per 40 1 per 40	—	1 per 75[f]
Worship places—educational and activities unit	*Male* *Female* 1 per 250 1 per 125	1 per 250	1 per toilet room	—	1 per 75[f]
Worship places—principal assembly place	*Male* *Female* 1 per 300 1 per 150	1 per 300	1 per toilet room	—	1 per 75[f]

(continued)

TABLE 8.3 **Minimum Plumbing Facilities**[a] ***(continued)***

Whenever urinals are provided, one (1) water closet less than the number specified may be provided for each urinal installed, except that the number of water closets in such cases shall not be reduced to less than two-thirds (⅔) of the minimum specified.

Source: Uniform Plumbing Code, copyright © 1985. Printed by permission of the International Association of Plumbing and Mechanical Officials.

[a]The figures shown are based upon one (1) fixture being the minimum required for the number of persons indicated or any fraction thereof.

[b]Building categories not shown on this table shall be considered separately by the Administrative Authority.

[c]In applying this schedule of facilities, consideration must be given to the accessibility of the fixtures. Conformity purely on a numerical basis may not result in an installation suited to the needs of the individual establishment. For example, schools should be provided with toilet facilities on each floor having classrooms. Temporary workingmen facilities: one (1) water closet and one (1) urinal for each thirty (30) workmen.

a. Surrounding materials: wall and floor space to a point two (2) feet (0.6 m) in front of urinal lip and four (4) feet (1.2 m) above the floor, and at least two (2) feet (0.6 m) to each side of the urinal shall be lined with nonabsorbent material.

b. Trough urinals are prohibited.

[d]Drinking fountains shall not be installed in toilet rooms.

[e]There shall be a minimum of one (1) drinking fountain per occupied floor in schools, theaters, auditoriums, dormitories, offices or public buildings.

[f]Where food is consumed indoors, water stations may be substituted for drinking fountains. Theaters, auditoriums, dormitories, offices, or public buildings for use by more than six (6) persons shall have one (1) drinking fountain for the first seventy-five (75) persons and one (1) additional fountain for each one hundred and fifty (150) persons thereafter.

[g]Laundry trays: one (1) for each fifty (50) persons. Slop sinks: one (1) for each hundred (100) persons.

[h]Laundry trays: one (1) laundry tray or one (1) automatic washer standpipe for each dwelling unit or two (2) laundry trays or two (2) automatic washer standpipes, or combination thereof, for each ten (10) apartments. Kitchen sinks: one (1) for each dwelling or apartment unit.

[i]As required by ANSI Z4.1–1968, Sanitation in Places of Employment.

[j]Where there is exposure to skin contamination with poisonous, infectious, or irritating materials, provide one (1) lavatory for each five (5) persons.

[k]Twenty-four (24) lineal inches (609.6 mm) of wash sink or eighteen (18) inches (457.2 mm) of a circular basin, when provided with water outlets for such space, shall be considered equivalent to one (1) lavatory.

[l]A restaurant is defined as a business that sells food to be consumed on the premises.

a. The number of occupants for a drive-in restaurant shall be considered as equal to the number of parking stalls.

b. Employee toilet facilities are not to be included in the above restaurant requirements. Hand washing facilities must be available in the kitchen for employees.

[m]This schedule has been adopted by the National Council on Schoolhouse Construction.

TABLE 8.4 **Estimated Sewage Flow Rates**

Type of Occupancy	*Unit Gallons (Liters) per Day*
Airports	15 (56.8) per employee
	5 (18.9) per passenger
Auto washers	Check with equipment manufacturer
Bowling alleys (snack bar only)	75 (283.9) per lane
Camps, campgrounds with central comfort station	35 (132.5) per person
Campgrounds with flush toilets, no showers	25 (94.6) per person
Day camps (no meals served)	15 (56.8) per person
Summer and seasonal	50 (189.3) per person
Churches (sanctuary)	5 (18.9) per seat
With kitchen waste	7 (26.5) per seat
Dance halls	5 (18.9) per person
Factories, no showers	25 (94.6) per employee
With showers	35 (132.5) per employee
Cafeteria, add	5 (18.9) per employee
Hospitals	250 (946.3) per bed
Kitchen waste only	25 (94.6) per bed
Laundry waste only	40 (151.4) per bed
Hotels (no kitchen waste)	60 (227.1) per bed (2 person)
Institutions (resident)	75 (283.9) per person
Nursing home	125 (473.1) per person
Rest home	125 (473.1) per person
Laundries, self-service	50 (189.3) per wash cycle
Commercial	Per manufacturer's specifications
Motel	50 (189.3) per bed space
With kitchen	60 (227.1) per bed space
Offices	20 (75.7) per employee
Parks, mobile homes	250 (946.3) per space
Picnic parks (toilets only)	20 (75.7) per parking space
Recreational vehicles, without water hook-up	75 (283.9) per space
With water and sewer hook-up	100 (378.5) per space
Residences	75 (283.9) per person
Restaurants/cafeterias	20 (75.7) per employee
Toilet	7 (26.5) per customer
Kitchen waste	6 (22.7) per meal
Add for garbage disposal	1 (3.8) per meal
Add for cocktail lounge	2 (7.6) per customer
Kitchen waste—disposable service	2 (7.6) per meal
Schools—staff and office	20 (75.7) per person
Elementary students	15 (56.8) per person
Intermediate and high	20 (75.7) per student
With gym and showers, add	5 (75.7) per student
With cafeteria, add	3 (11.4) per student
Boarding, total waste	100 (378.5) per person
Service station, toilets	1000 (3785) for 1st bay
	500 (1892.5) for each additional bay
Stores	20 (75.7) per employee
Public restrooms, add (per unit of floor space)	1 (3.8) per 10 ft^2 (.9 m^2)
Swimming pools, public	10 (37.9) per person
Theaters, auditoriums	5 (18.9) per seat
Drive-in	10 (37.9) per space

Source: Uniform Plumbing Code, copyright © 1985. Printed by permission of the International Association of Plumbing and Mechanical Officials.

some treatment processes. One of the most common reasons for the rejection of a potential rural building site is a lack of suitability for sewage disposal. A geologic analysis of structural and sewage disposal potential is one of the first documents needed by the designer. Chapter 10 details the sizing of some common treatment methods. For this earlier stage, the following rules of thumb can be useful:

Septic Tank Drainfields. In conventional units (minimum 750-ft^2 area for any system):

- For shallow trenches in poorly draining soil: drainfield area = total sewage flow in gpd × 3.6 ft^2/gal.
- For deep trenches in well-draining soil: drainage area = total sewage flow in gpd × 0.4 ft^2/gal.

In S.I. units (minimum 70 m^2):

- For shallow trenches, poor soil: L/day × 0.087 m^2/L
- For deep trenches, good soil: L/day × 0.01 m^2/L

(These guidelines allow for an expansion area equal to the original size of the drainage field, in case of field failure.)

Mounds. These are built-up leaching fields on top of an existing grade (see Fig. 10.18). For a single-family dwelling, allow for a 4-ft-high mound whose bottom area is a square 44 ft (13.4 m) on each side and whose sides slope at a 1 : 3 vertical-to-horizontal ratio.

Package Sewage Plant Drainfields. In these, sewage is treated to a much greater extent than in septic tanks, and effluent is filtered:

- In poorly draining soil: total sewage flow in gpd × 0.49 ft^2/gal (L/day × 0.012 m^2/L).
- In well-draining soil: total sewage flow in gpd × 0.23 ft^2/gal (L/day × 0.006 m^2/L).

Sewage Lagoons. These consist of two open treatment ponds (primary and secondary) and are sized on the basis of pounds of biological oxygen demand (BOD) rather than gallons of sewage flow. A typical assumption for estimating BOD is

- 0.2 lb BOD/person for "ordinary" domestic sewage.
- 0.3 lb BOD/person where garbage grinders or other devices contribute added organic material to domestic sewage.

The *total* acreage (a) needed for the two ponds can be estimated as

- 20 lb BOD/a for the (colder) northern United States.
- 35 lb BOD/a for the (drier, warmer) southern United States.

(The primary pond is usually sized for 50 lb BOD/a.)

8.4 Rainwater

There is a striking similarity between rainwater and solar energy. Both are essential for agriculture, and both have been well understood and utilized by farmers since agrarian societies first emerged. Both can be very beneficial to architecture and were utilized as needed by anonymous builders for centuries. Both fell out of favor with designers as plentiful supplies of pure, centrally treated water and concentrated, centrally controlled fuels became commonplace. Both are thinly yet relatively evenly spread over the world's population, so they are at least seasonally available to help meet nearly every building's needs for water and heat. Yet both are difficult to utilize in industrialized societies, because they require *individual* expenditures.

Consider the typical public water and electric utility in a U.S. city. It can raise the funds to build large water treatment plants, electricity generating plants, and the network of pipes and wires that bring these commodities to every building. The utility's costs, including interest on its construction debts, will be passed on to its consumers, along with a margin of profit that is usually controlled by state governments. Thus, our society has a well-established method for encouraging central suppliers of water and heat.

Now consider the individual building owner.

To build a cistern and a solar-heated building, she must borrow money at an interest rate higher than that which the utility pays; and both options cost more initially than simple connections to the utility's pipes and lines. Even though she is willing to flush her toilets with rainwater rather than with chlorinated and filtered potable water from the utility, and even though she is willing to heat with lower-grade solar energy than with higher-grade electricity, she must pay a substantial first-cost penalty—with interest—to do so.

The overall public good could be well served by a mixture of public networks of pure water and electricity and individual cisterns and solar applications. The environmental benefits would be substantial—less water withdrawn from rivers, lakes, and underground aquifers; less stormwater discharged to pollute rivers; less fuel used to generate electricity; less environmental damage from power plants. Yet we continue to economically discourage the individual who uses the rain and the sun.

8.5 Collection and Storage

In terms of both quality and quantity, rainwater is an attractive alternative. Figure 8.3*b* showed that rainwater is near to the purest state in the hydrologic cycle. More recently, it is true, air pollution began to threaten the quality of rainwater in some areas, as "acid rain" has become widespread in the northeastern section of the North American continent and in Europe. In some particularly air-polluted locations, lead poses a threat to rainwater quality. Also, on any catchment surface, dust and bird droppings are common pollutants that must be considered. Other factors that bear on rainwater quality are roofing materials and the form of the roof. The appropriate health authority should be consulted for a list of roofing materials that will have no toxic effects on rainwater. Steeper roofs are scoured by winds and thus collect less dust and give cleaner runoffs. Devices to discourage roosting birds are strongly recommended, as are periodic checks of cistern water for bacteria. Fungicides (for moss control) should be scrupulously avoided, as should roofing paints containing lead. For these reasons, urban rainwater commonly is not used for drinking and cooking. (Bottled water or on-site water distilling can supply potable water.) For the typical residence, however, that still leaves about 95% of indoor water usage that could be provided by rainwater. And in those cities that have very "hard" public water supplies, rainwater's "soft" characteristics make it particularly attractive.

The quantity of rainwater available in most U.S. locations could meet a high percentage of typical home or business needs. Milne (1976) pointed out that the 42 in. of rain that falls annually on the streets and roofs of Manhattan Island could, if collected and stored, provide 148% of the residential needs of the 1.7 million inhabitants. For the typical U.S. suburban house (roof area of 1500 ft^2), the annual catchment can be estimated by combining the rainfall quantities (Fig. 8.5) with the resulting catchment yield (Fig. 8.4). Even at a "dry" rate of only 20 in. annually, a 1500-ft^2 roof would yield about 12,000 gal, or 33 gpd—nearly enough to meet the clothes cleaning and dishwashing needs of the family. In SI units, a 140-m^2 roof with annual 500 mm precipitation will yield 50,000 L, or 137 L/d.)

Unfiltered rainwater seems particularly well-suited to the irrigation of small lawns or gardens, both because it lacks additives unneeded by plants, such as chlorine or sodium fluoride, and because it can reduce the user's demand on the public water supply on the hottest summer days. Cisterns located above the irrigated area have the advantage of replacing a pump with simple gravity flow. The rainbarrel at the bottom of a downspout was an example of this approach—though one of limited capacity.

In many of the world's drier areas, smaller cisterns within the home are common. Such cisterns, which can be fed both by rainwater and by the public supply, are frequently used for all domestic purposes, including drinking. The presence of such a large water volume also can be advantageous in the event of fire. Sometimes these storage cisterns are required because the public supply is diminished or cut off at peak-usage hours, due to insufficient water main capacity. Figure 8.6 shows a cistern in the typical outdoor location, along with various options by

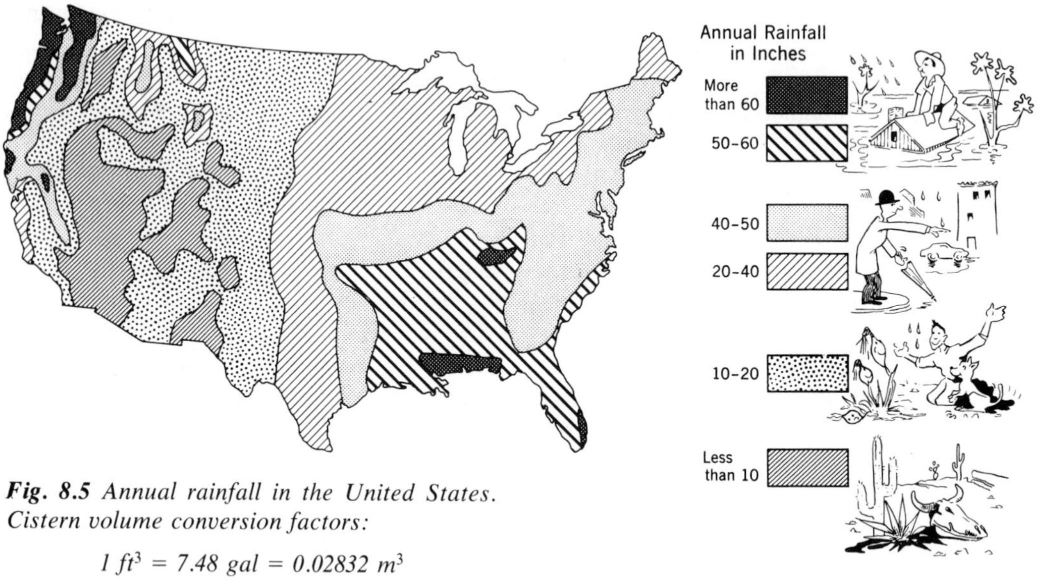

***Fig. 8.5** Annual rainfall in the United States. Cistern volume conversion factors:*

$$1\ ft^3 = 7.48\ gal = 0.02832\ m^3$$
$$1\ m^3 = 1000\ L = 264.2\ gal$$

which pollution from dirty roofs can be minimized.

When cisterns are taken seriously as water storage devices, their function and size can become strong design-form determinants. The country home of architect John Andrews in the dry ranchland of New South Wales, Australia, offers a particularly striking example of cisterns and form (Fig. 8.7). The corner rainwater collectors also help deflect cooling summer breezes into the house, along the diagonal walls at each corner. The central skylight then vents the breezes, along with the house's heat. From the corner cisterns, rainwater is pumped up to the central tower's storage tank. It then can be heated by solar collectors (to be installed on the sloping top of the tower), or used to supply the house's plumbing fixtures, or even used to sprinkle the metal roof surface, whose surface temperature could quickly be lowered by evaporation. Unevaporated roof water simply runs back into the cisterns. As a final design-with-water gesture, the shower (off bedroom 1) has been made into a true celebration place for cleansing and refreshing.

In this passively solar heated home, the living areas are placed on the elongated north side—the warmer side in the cold season of this Southern hemisphere house. Passive cooling is aided by pergolas on the west and south, which are covered with vines for hot season shading of windows.

Sizing. In Section 8.3, procedures were given for rough sizing of both the catchment area and the storage capacity for cisterns. When rainwater is to be a primary, as opposed to merely a supplementary source, a closer look must be taken at rainfall deposits and user withdrawals from a cistern.

***Fig. 8.6** Section through a typical outdoor cistern.* (a) *A "roof washer" gets the dirtiest, first runoff from the roof. It can later be emptied either by opening the faucet wide or by leaving the faucet slightly open so that it will slowly drain.* (b) *In place of the "roof washer," a sand filter may be used. The "flapper valve" is then used rarely, perhaps to divert the first rainfall after a prolonged dry spell. After this, the valve is left in a position to divert all rainwater to the sand filter. (Based on the U.S. Environmental Protection Agency's* Manual of Individual Water Supply Systems, *1975.)* (c) *Another roof-washing option is a "tipping valve," which dumps the first X gallons of each rainfall. After each rainfall, the valve must be manually (or spring) reset to "dry" position if it is to again intercept dirty water. In an extended wet period, it would probably be left in the "wet" position.*

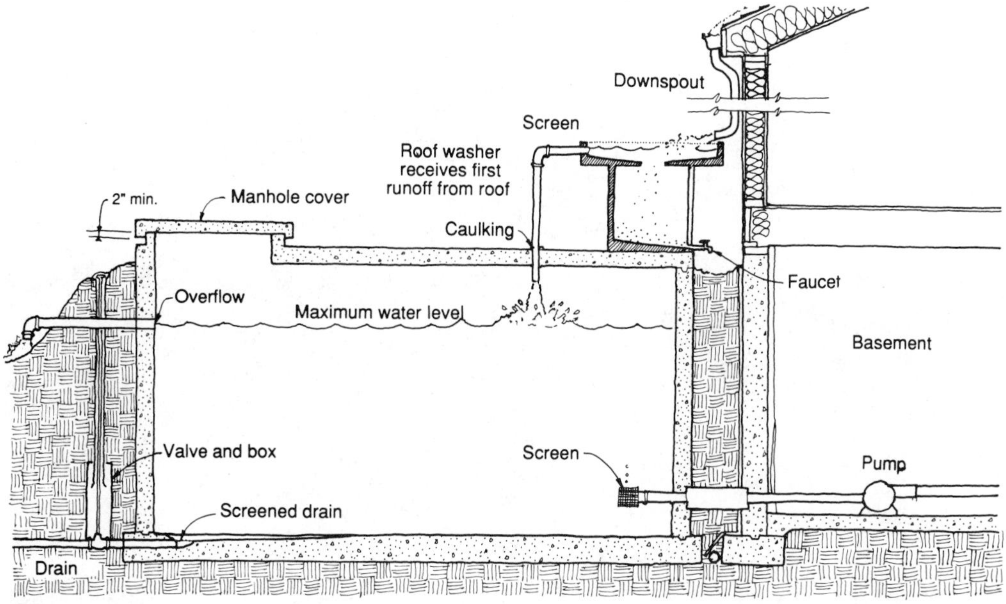
Downspout
Screen
Roof washer
receives first
runoff from roof
2" min.
Manhole cover
Caulking
Faucet
Overflow
Maximum water level
Basement
Valve and box
Screen
Pump
Screened drain
Drain

(a)

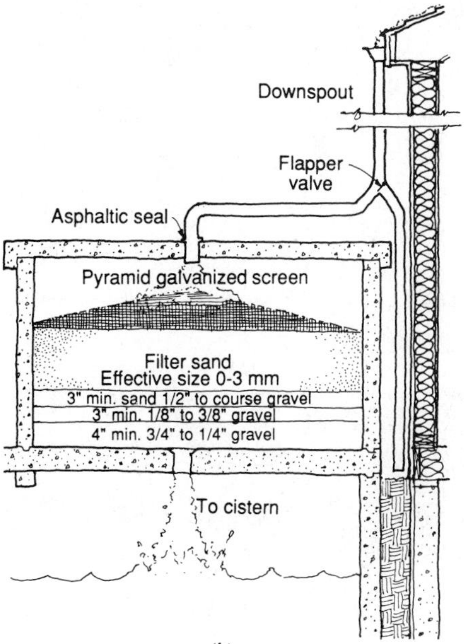
Downspout
Flapper
valve
Asphaltic seal
Pyramid galvanized screen
Filter sand
Effective size 0-3 mm
3" min. sand 1/2" to course gravel
3" min. 1/8" to 3/8" gravel
4" min. 3/4" to 1/4" gravel
To cistern

(b)

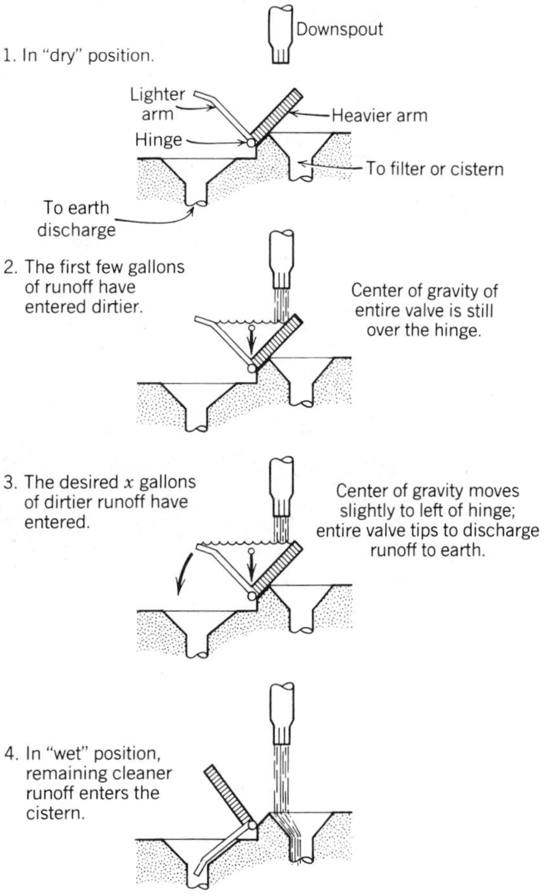
Downspout
1. In "dry" position.
Lighter
arm
Hinge
Heavier arm
To filter or cistern
To earth
discharge
2. The first few gallons
of runoff have
entered dirtier.
Center of gravity of
entire valve is still
over the hinge.
3. The desired x gallons
of dirtier runoff have
entered.
Center of gravity moves
slightly to left of hinge;
entire valve tips to discharge
runoff to earth.
4. In "wet" position,
remaining cleaner
runoff enters the
cistern.

(c)

Fig. 8.7 Country home of architect John Andrews, Eugowra, NSW, Australia. (a) *View from the southwest, showing corner cisterns, central daylight/ventilation barrel vault, and "energy tower" with water storage tank.* (b) *The view from the south, showing shading pergola to be covered by vines.* (c) *Floor plan, showing fireplace at centers and cisterns at all corners. Insect screens protect the openings to summer breezes.* (d) *Section. (Courtesy of John Andrews International Pty. Ltd.)*

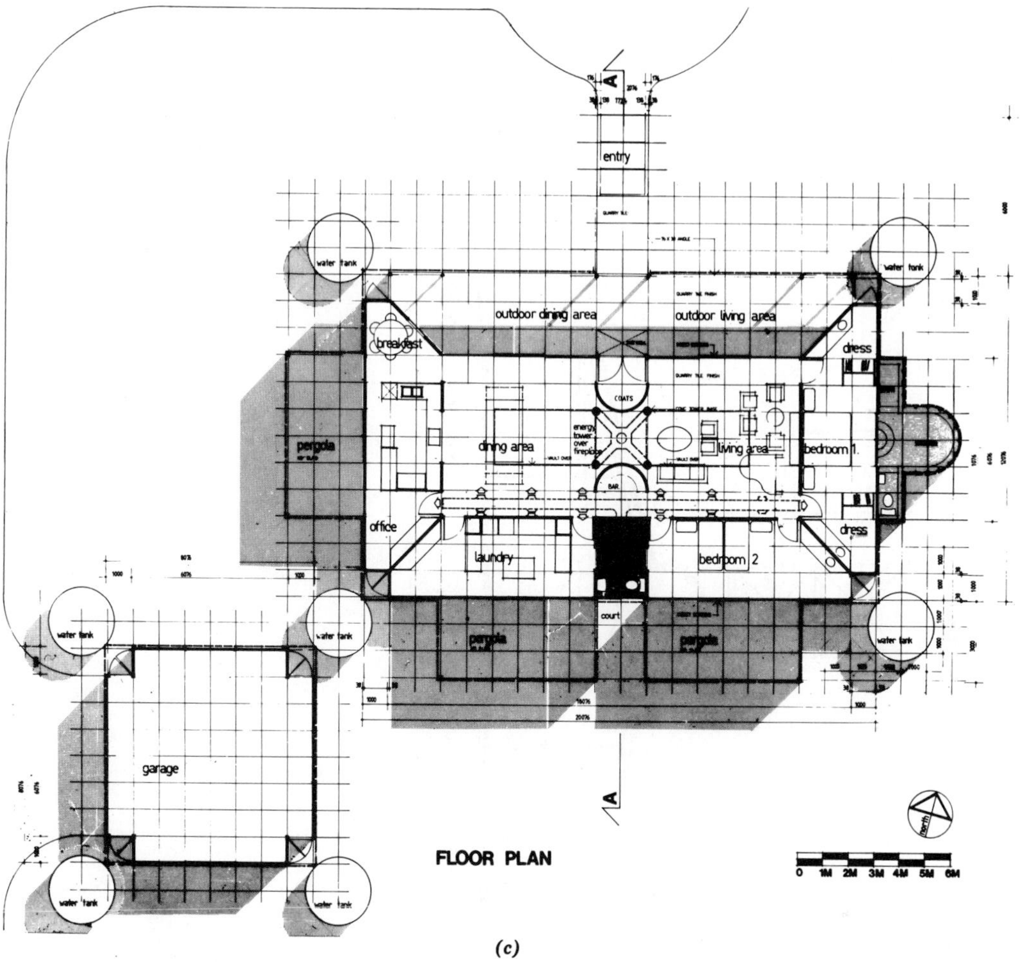

(c)

Fig. 8.7 *(continued)*

This procedure depends upon the monthly average rainfall (from NOAA Local Climatological Data), the monthly water usage, and the catchment area yield (from Fig. 8.4).

EXAMPLE 8.1, PART B. Take a closer look at the cistern that was approximately sized in Part A. For this cistern, to be used for toilet flushing in a factory near Salem, Oregon, daily usage was estimated at 200 gal and catchment area of 5600 ft^2. The cistern capacity was estimated at 18,000 gal. Begin the process in the midst of the wettest months.

From Table 8.5, the following conclusions can be drawn.

1. For an 18,000-gal cistern, when the end-of-December cumulative capacity is added to January's surplus, the cistern will be at capacity from November through April.
2. With no allowances for abnormally dry months, we could reduce the cistern size by about 3000 gal (the size of the smallest cumulative capacity, in September).
3. A larger cistern—one that could utilize everything from the catchment area—could be built. Its approximate size would be the year-end surplus of 38,510 gal plus the maximum spring monthly cumulative capacity of 25,410 gal, in April.
4. Or the surplus could be devoted to additional usage of rainwater, beyond mere toilet flushing, from November through April.

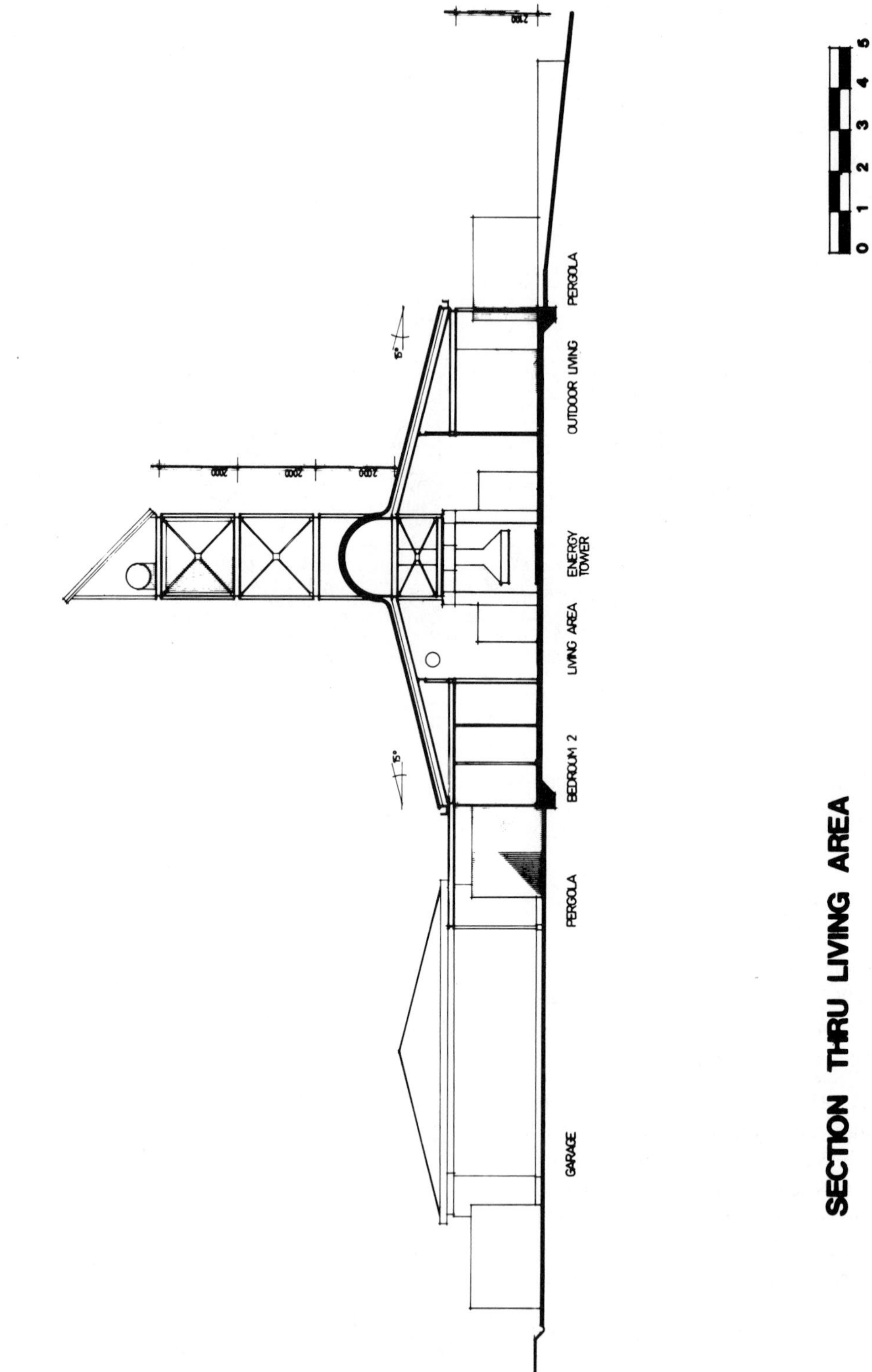

(d)

Fig. 8.7 (continued)

TABLE 8.5 **Rainfall Cistern Sizing Procedure**

I		II	III	IV	V	VI
Month and Rainfall[a] *(in.)*		*Catchment Yield*[b] *(gal)*	*Usage*[c] *(gal)*	*Net*[d] *(gal)*	*Cumulative Capacity*[e] *(gal)*	*Cumulative Capacity Adjusted for Actual Size*[f] *(gal)*
January	6.9	18,630	6,200	12,430	12,430	12,430
February	4.8	12,960	5,600	7,360	19,790[g]	18,000
March	4.3	11,610	6,200	5,410	25,200[g]	18,000
April	2.3	6,210	6,000	210	25,410[g]	18,000
May	2.1	5,670	6,200	−530	24,880[g]	17,470
June	1.4	3,780	6,000	−2,220	22,660[g]	15,250
July	0.4	1,080	6,200	−5,120	17,540	10,130
August	0.6	1,620	6,200	−4,580	12,960	5,550
September	1.5	4,050	6,000	−1,950	11,010	3,600
October	4.0	10,800	6,200	4,600	15,610	8,200
November	6.1	16,470	6,000	10,470	26,080[g]	18,000
December	6.9	18,630	6,200	12,430	38,510[g]	18,000

Source: Based upon procedure from Brown, Reynolds, Ubbelohde, *InsideOut: Design Procedures for Passive Environmental Technologies,* copyright © 1982 by John Wiley & Sons.

[a]NOAA data for Salem, Oregon.

[b]From Fig. 8.4, according to which 10 in. precipitation yields about 27,000 gallons for this 5600-ft^2 catchment area.

[c]The factory uses 200 gpd times days in each month.

[d]Col. II minus col. III.

[e]Values in col. IV, added month by month.

[f]Set at 18,000 gallons for the factory in Example 8.1.

[g]The cumulative capacity exceeds the actual storage capacity of 18,000 gallons.

8.6 Rainwater and Site Planning

Prior to the spread to rural areas of buildings, streets, roads, and paved parking lots, water from rainfall and melting snow found its own way to natural destinations. Surface flow to creeks, streams, and rivers accounted for part of this drainage. Underground flow aided the general runoff. Outcropping of flowing groundwater created springs and artesian wells. Low, dished areas formed lakes that in turn overflowed to outlet streams. Flat areas sometimes developed into swamps or marshes.

At a time when there was a choice of locations for towns and villages, sites next to rivers were usually chosen. The waterways provided transportation and water was supplied from the river or from adjacent wells. As streets and roads were built, slopes could be arranged whereby the rain falling on these areas and flowing onto them from roofs of buildings could run to the river. At interior parts of the country, high ground was favored for building sites and growing communities. Obviously, swampy or marshy ground would not be chosen, but it did provide terminal locations for the stormwater that ran off the high ground. In the course of this natural flow, much of the water was drawn by evaporation to the clouds. The rest, conforming to topographic river basins, continued along to seek its way to the sea.

As building increased, desirable locations grew scarce. The possibility of selecting high, dry ground diminished. Great areas, formerly low and marshy, were filled in and buildings constructed, often on piles. From such locations stormwater could not be disposed of by drainage to some adjacent lower area, or even recharged to the earth through dry wells. Moreover, extensive grids of paved streets and sidewalks in these level developments caught and held the water, resulting in considerable "ponding." Storm sewers had to be built and the water transported great distances, often having to be lifted at intermediate pumping stations before reaching its destination, which might be a remote river.

This emphasis on the removal of stormwater has led to an expensive and elaborate system based on quick "disposal" of rainfall. By decreasing the time between precipitation and runoff, quick disposal increases the peak flows within such systems, thereby increasing flooding of rivers during storms but reducing the rivers' flow between storms.

The overloading of storm sewers not only causes minor flooding, but also can influence building design. The designers of the New Orleans Convention and Exhibition Center, a building with 610,000 ft^2 of roof area, spared the city's storm sewers from the impact of a 14-acre runoff by carrying the rainwater over the roofs of adjoining wharfs, and thus discharging it directly into the Mississippi River.

As urban storm sewers reach capacity and suburban groundwater levels drop, designers have begun to emphasize "stormwater infiltration" (or recharge of groundwater), rather than "quick runoff." Three design strategies for encouraging such recharge have emerged: roofs that will retain water and slowly release it, porous pavement, and onsite infiltration of runoff.

(a) Roof Retention. If stormwater is to be sent to storm sewers or to soak into the ground, a *slow flow* from roofs will help by diminishing peak flows in sewers and giving soaked soil a longer time to absorb still more runoff. Nearly flat roofs with specially designed drains (see Fig. 8.18*c*) permit slower discharge yet eventually drain completely dry (to discourage mosquito breeding, etc.). (For cisterns, however, sloped roofs should be used, as they stay cleaner than flat roofs.) This temporary pond on top of a building will clearly add to structural requirements. Another problem could be posed by high winds blowing sheets of water onto people below. In summer, however, a flooded roof can provide the significant thermal advantage of greatly lowering daytime roof surface temperatures. Rainfall-retaining roofs are now required in some urban areas with overtaxed storm sewers.

(b) Porous Pavement. Where groundwater is in short supply, new building sites now are often required to retain rainfall on-site. To accomplish this, many builders use either porous concrete or "incremental" paving (alternation of paving materials with grass or ground-cover plants) for parking lots and roadways.

Porous concrete has been used for many years in building construction as a low-strength, high-porosity material that has some insulating properties (*R*-5 for 10-in. thickness). Patented porous concrete pavement now in use in Florida has a strength of over 3800 psi and a permeability of 2.3 gallons of water per minute per square foot. (A 2500-psi mix has a permeability of 18.5 gallons per minute per square foot.) In cold weather areas, the freeze–thaw cycle could be destructive to porous concrete.

Incremental paving, in which the many joints allow water to pass through, offers another possibility. So does the handsome, if more expensive, approach of alternating concrete paving with grass or ground cover, as shown in Fig. 8.8.

(c) Site Design for Recharging. This tactic is especially advisable for suburban-density developments in drier climates with ab-

Fig. 8.8 *An Oregon application of Grasscrete, a system in which concrete blocks are alternated with tufts of grass. Unthank Seder Poticha, Architects, Eugene Oregon. Photo by Jane Lidz.*

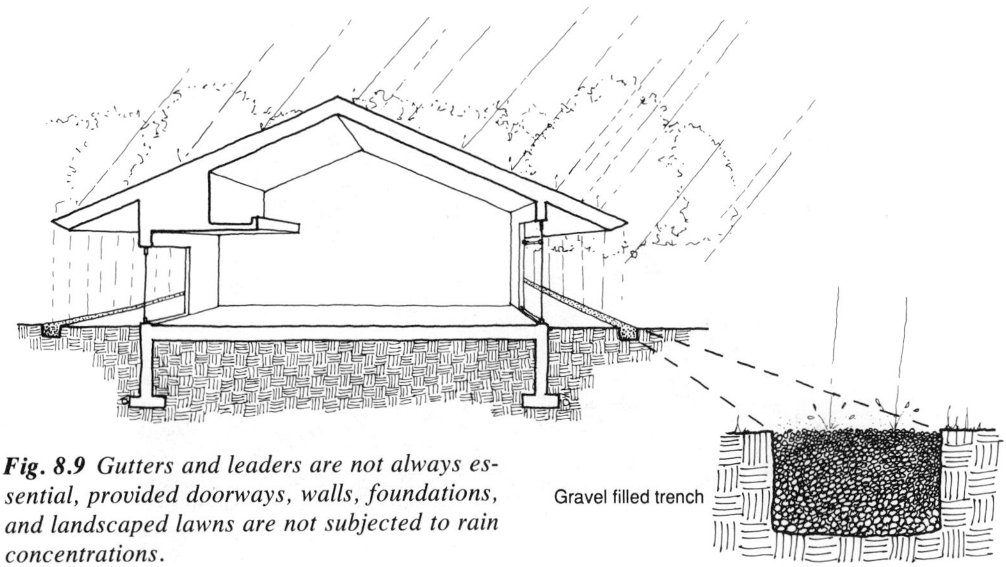

Fig. 8.9 *Gutters and leaders are not always essential, provided doorways, walls, foundations, and landscaped lawns are not subjected to rain concentrations.*

sorptive soil (sand, gravel, etc.). The first option is at each house; design of the entire subdivisions can also be influenced.

The simplest design approach is the gutterless sloped roof illustrated in Fig. 8.9, which is applicable to one-story, basementless homes with wide, overhanging roofs. A gravel-filled trench skirting the perimeter directly below the edge of the eaves catches the water flowing off the roof.

Some designers do not like the appearance of conventional gutters and leaders; other designers celebrate them (Fig. 8.10). There are many ingenious ways to avoid or modify their use and yet provide proper drainage. In many cases, however, gutters and leaders will be required, either to collect rainwater for cisterns or to control conditions around the perimeter of the house. Several options for stormwater recharge can be used with the gutter–leader combination.

A splash pan at the foot of each leader (Fig. 8.11*a*) offers the simplest method. It will lead the water a few feet from the house but will accommodate only a relatively low rate of flow. A gravel-filled pipe is somewhat more effective (Fig. 8.11*b*). When the soil is not very permeable (as, for instance, with clay), it is best to use a dry well with extended area and many perforations through which the water can be discharged to the ground.

Footing drains are often used to collect and lead away groundwater that accumulates around the foundations. This reduces the likelihood of basement wall leakage. These drains are most necessary when higher ground near the buildings increases the flow of groundwater against underground walls. Figure 8.12 illustrates this and also shows how stormwater from drains and roofs may be led to a surface absorption area of rock and gravel beyond a head wall where the general storm drain outcrops. This method can be chosen where there is sufficient property area and slope. It has the advantages of easy maintenance and service. Also, one can observe whether or not it is functioning correctly.

In new suburban developments, for which there are no storm sewers, recharge basins are sometimes required to deliver stormwater to the ground. Water from numerous roofs, paved areas, and curb catch basins is collected and piped to an open, unpaved pit, where it sinks into the earth. This method is not recommended in areas of dense, impervious clay soil (see Fig. 8.13). A particularly effective example of this approach is offered by the community of passively solar heated residences known as Village Homes, outside Davis, California. This area receives only about 20 in. (about 500 mm) of rain annually, so the recharge of groundwater seemed to this garden-oriented community to be a more attractive

Fig. 8.10 Falling rain is featured at this Oregon residence, where the gutter is held below the metal flashing at the roof's edge. Runoff is seen from indoors as it falls into the suspended gutter. Unthank Poticha Waterbury, Architects, Eugene, Oregon.

option than loss of the rainwater to the storm sewer. Stormwater flows from leaders to dry, rockbed channels, along which are gardens and bicycle paths (Figs. 8.14 and 8.15). Occasionally, small dams across these channels create temporary holding ponds, in case the runoff has not yet soaked through the channel bottom. In the event of extraordinarily heavy rainfall, an inlet to the public storm sewer is available beyond the final holding pond; this inlet is needed approximately once every five years.

The planning of the landscape around buildings may closely follow such considerations as irrigation. The ''hydrozone'' concept of landscape planning is shown in Fig. 8.16. To minimize water consumption, exotic plant species are kept to a minimum and located near the house, where storm runoff and irrigating water is readily available. Native and adapted plantings, which can survive on normal rainfall, are utilized elsewhere. (Systems for irrigation are discussed in Section 9.6.)

8.7 Components

The first stormwater system design decision to be made involves the establishment of ''watersheds'' on a building's roof. To what edges, or at what points, will runoff be directed? To what depths will it accumulate before it leaves the roof? Since the answers to these questions depend on the intensity of storms as well as on the roof's geometry, it is necessary to find the maximum hourly rainfall for each location. This figure is available from local building code officials, or from Fig. 8.17.

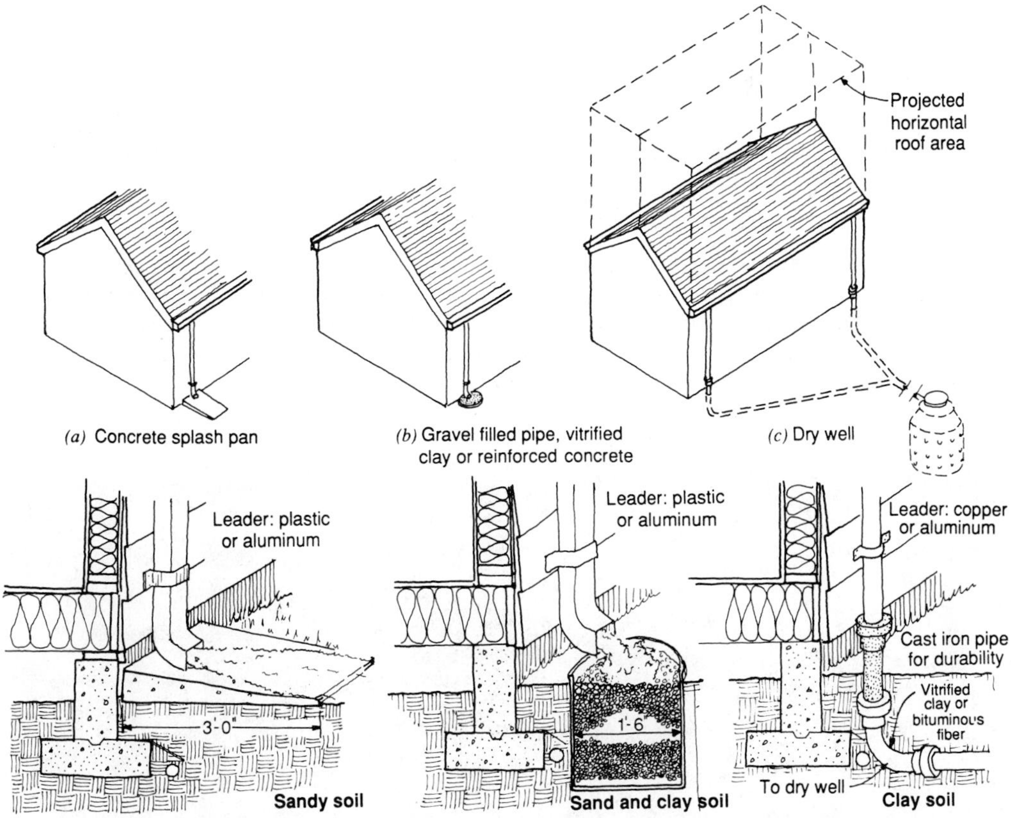

Fig. 8.11 *Roof drainage for houses. Method* (a) *is suitable for low rates of flow introduced into very pervious soil. When denser soil is encountered,* (b) *is used to get the water into the ground and thus avoid surface erosion. For heavy flow or to lead the water farther from the structure,* (c) *may be used with one or several dry wells.*

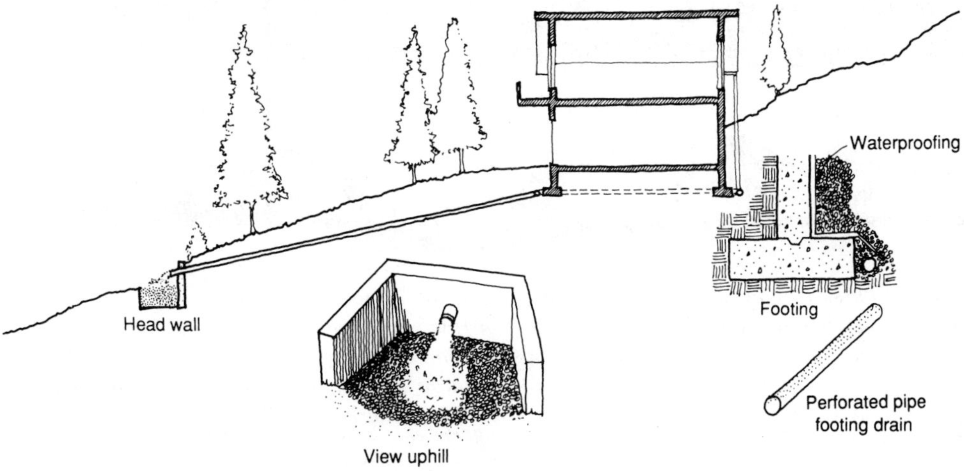

Fig. 8.12 *Disposal of stormwater on the site but remote from the house or building. When a wall is against a hill, it is usually subjected to pressure of groundwater during storms. Open-joint clay, plastic, or fiber tile accepts this water and carries it away. Footing drains are tight-joint clay tile or bituminous fiber pipe. Flow through stone and gravel returns the water to the earth. Head wall is appropriate in lieu of a dry well if the site permits.*

WATER AND WASTE

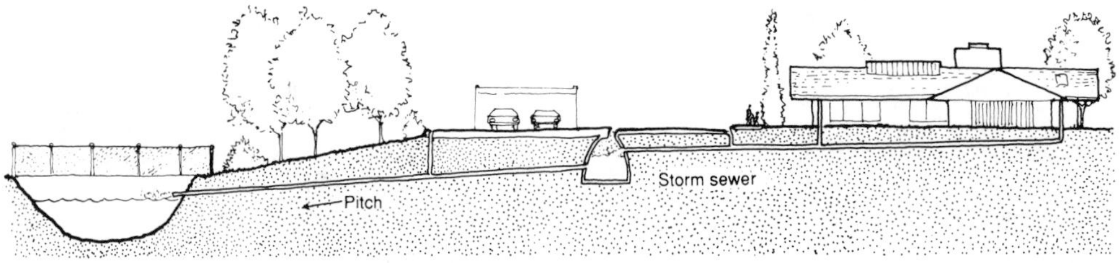

Fig. 8.13 *Recharge basins in suburban communities. When topography, groundwater level, and porosity of the soil permit, developers are sometimes required to install systems that collect stormwater and carry it from catch basins at street curbs, the roofs of all houses, and paved areas to a recharge basin that receives the water and returns it to the ground. For the safety of children, a fence is sometimes required to prevent unauthorized access to the basin.*

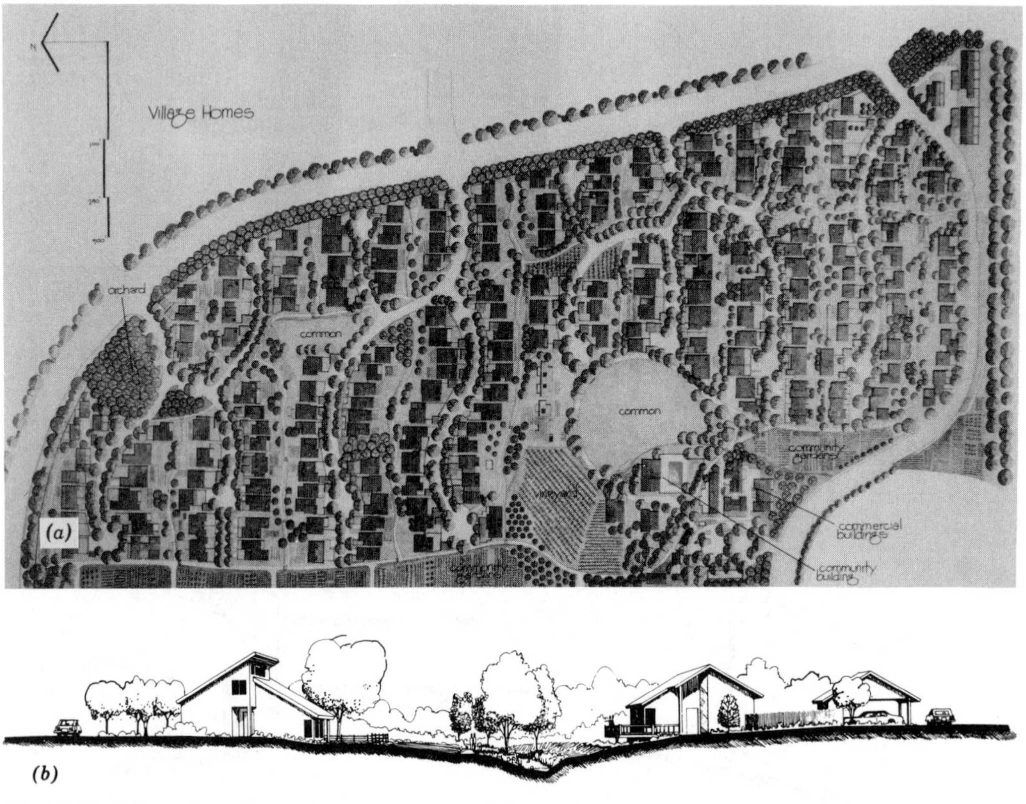

Fig. 8.14 *Village Homes, a California subdivision of solar homes and stormwater recharge areas.* (a) *Plan shows the emphasis on bicycle paths, narrow streets, and widespread community-maintained garden space through which the recharge streambeds are led.* (b) *Site section explains the gradual drainage from leaders at houses to recharge stream. (Reprinted by permission from M. Corbett,* A Better Place to Live, *copyright © 1983 by the Rodale Press.)*

Fig. 8.15 The open drainage ways and pedestrian paths of Village Homes, Photo by Alan Butler.

Gutters and *leaders* (downspouts) can be sized through the use of Tables 8.6 and 8.7. The sizing of both gutters and leaders depends both on the horizontal projected area of the roof, as shown in Fig. 8.11 and the maximum hourly rainfall.

EXAMPLE 8.2. Select a gutter and two leaders for the front half of a house, as shown in Fig. 8.11. Rainfall rate is 4 in./h. Projected roof area is 700 ft^2, and the slope of the gutter is 1/16 in. in 1 ft of length.

SOLUTION. From Table 8.6, choose a semicircular gutter with a 6-in. diameter. (Note that if a steeper gutter slope were designed, then at 1/2-in. slope per foot, only a 4-in.-diameter gutter would be required.)

Since two leaders will be used, each will drain 350 ft^2. Table 8.7 shows that a 2-in. leader can be used. For this gutter–leader combination, specify the detail of Fig. 8.18*a*.

Storm gutters and leaders can have an important impact on a building's appearance (Fig. 8.19). Alternatively, leaders can be set within buildings and gutters can be built into a roof's surface, to minimize the visual impact.

Where routing of stormwater inside a building is preferable, drains and leaders can be sized from Table 8.7, and *horizontal piping* can be sized from Table 8.8. Care should be taken to

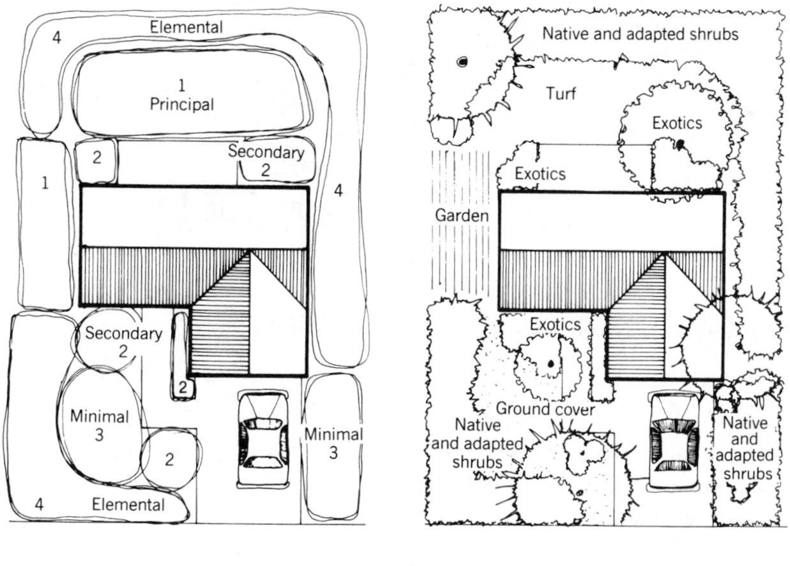

(a)

Fig. 8.16 The Hydrozone concept of landscape planning restricts exotic planting to special, easily watered areas. Native and adapted plantings are used elsewhere. (a) *Hydrozonics on a typical suburban lot.*

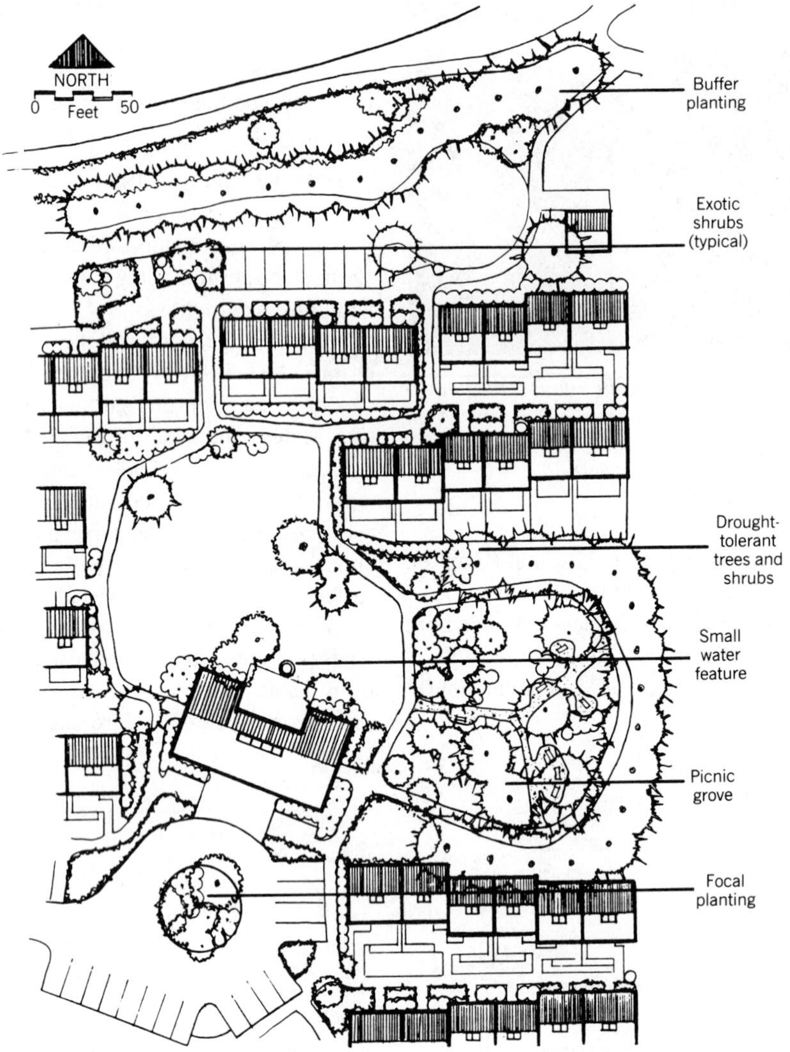

(b)

Fig. 8.16 *(continued)* (b) *Hydrozonics in a townhouse subdivision.*

insulate such lines; cold rainwater inside pipes can cause condensation to form on the outside, sometimes resulting in staining and other water damage.

EXAMPLE 8.3. Select sizes for vertical conductors and horizontal storm drains for the building shown in Fig. 8.20. Roof, balcony, and courtyard areas are as shown, rainfall rate is 4 in./h, and the pitch of horizontal drains is ¼-in. slope in 1 ft of run.

SOLUTION. Sizes selected and shown in Fig. 8.20 may be verified in Tables 8.7 and 8.8.

REFERENCES

Milne, M. (1976). *Residential Water Conservation,* U.S. Office of Water Research and Technology, Department of Commerce, NTIS.

U.S. Environmental Protection Agency (1975). *Manual of Individual Water Supply Systems,* U.S. EPA, Washington, D.C.

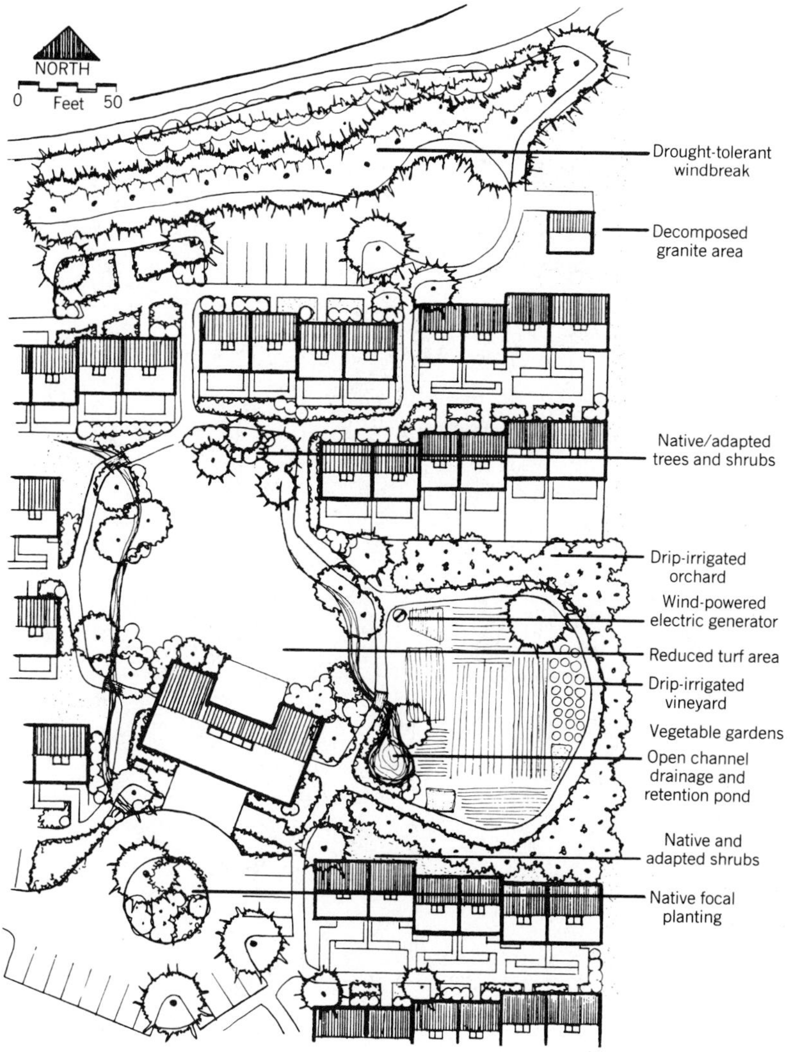

(c)

Fig. 8.16 *(continued)* (c) *Hydrozonics applied to more radical planning for water conservation. Reprinted by permission from* Energy Conserving Site Design, *E. G. McPherson, ed., copyright © 1984 by the American Society of Landscape Architects.*

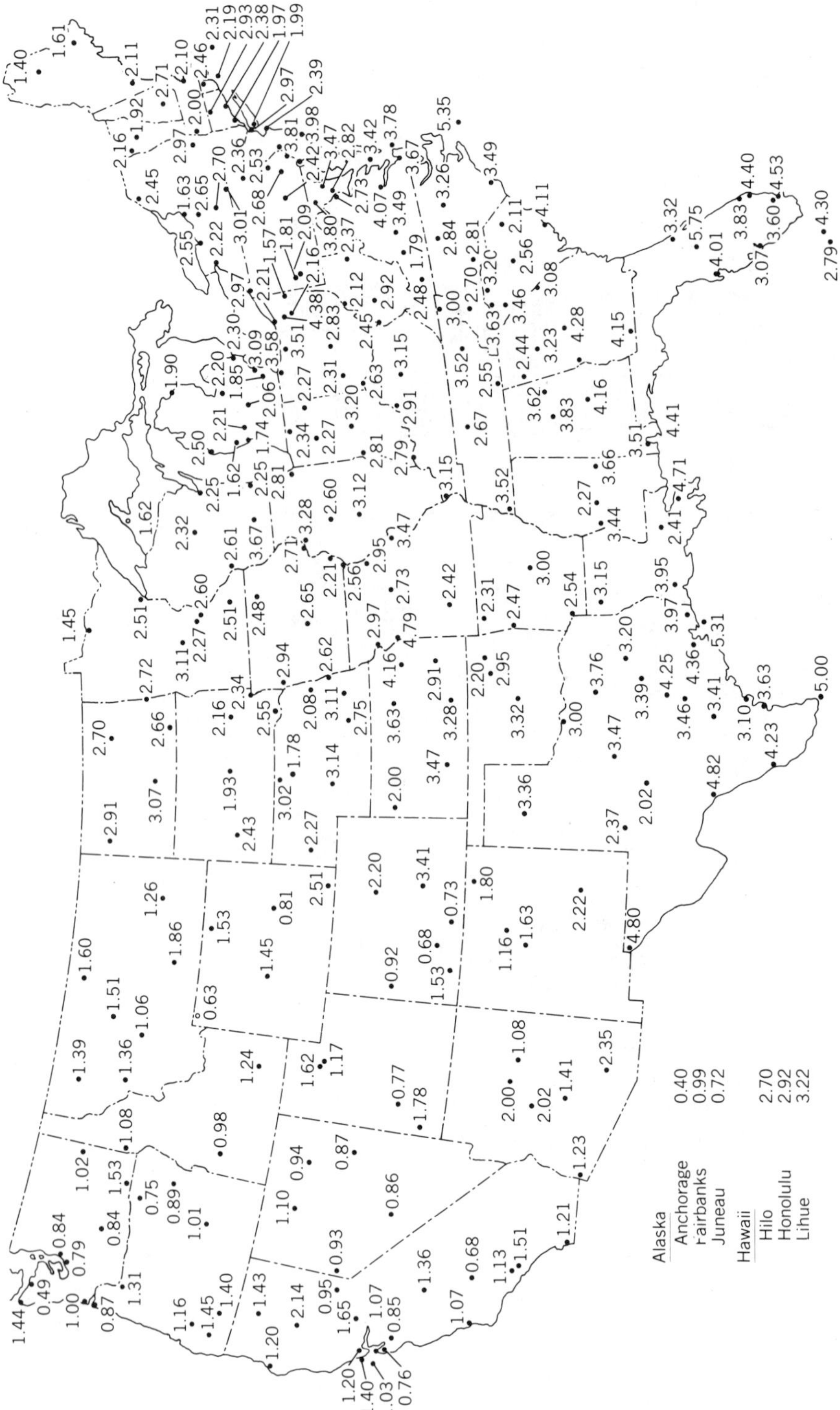

Fig. 8.17 *Maximum recorded hourly rainfall, in inches. (Latest data, 1961.) (From U.S. Weather Bureau Technical Paper 2, last issued in 1963.)*

TABLE 8.6 **Gutter Sizes**

Part A. ***Conventional Units: Maximum Horizontal Projected Roof Areas, Square Feet***

Diameter of Gutter 1/16″ Slope	Maximum Rainfall (in./hr)				
	2	3	4	5	6
3	340	226	170	136	113
4	720	480	360	288	240
5	1250	834	625	500	416
6	1920	—	960	768	640
7	2760	1840	1380	1100	918
8	3980	2655	1990	1590	1325
10	7200	4800	3600	2880	2400

Diameter of Gutter 1/8″ Slope	Maximum Rainfall (in./hr)				
	2	3	4	5	6
3	480	320	240	192	160
4	1020	681	510	408	340
5	1760	1172	880	704	587
6	2720	1815	1360	1085	905
7	3900	2600	1950	1560	1300
8	5600	3740	2800	2240	1870
10	10200	6800	5100	4080	3400

Diameter of Gutter 1/4″ Slope	Maximum Rainfall (in./hr)				
	2	3	4	5	6
3	680	454	340	272	226
4	1440	960	720	576	480
5	2500	1668	1250	1000	834
6	3840	2560	1920	1536	1280
7	5520	3680	2760	2205	1840
8	7960	5310	3980	3180	2655
10	14400	9600	7200	5750	4800

Diameter of Gutter 1/2″ Slope	Maximum Rainfall (in./hr)				
	2	3	4	5	6
3	960	640	480	384	320
4	2040	1360	1020	816	680
5	3540	2360	1770	1415	1180
6	5540	3695	2770	2220	1850
7	7800	5200	3900	3120	2600
8	11200	7460	5600	4480	3730
10	20000	13330	10000	8000	6660

(continued)

TABLE 8.6 **Gutter Sizes (*continued*)**

*Part B. **SI Units: Maximum Horizontal Projected Roof Areas, Square Meters***

Diameter of Gutter 5.2 mm/m Slope	*Maximum Rainfall (mm/hr)*				
	50.8	*76.2*	*101.6*	*127*	*152.4*
76.2	31.6	21	15.8	12.6	10.5
101.6	66.9	44.6	33.4	26.8	22.3
127	116.1	77.5	58.1	46.5	38.7
152.4	178.4	—	89.2	71.4	59.5
177.8	256.4	170.9	128.2	102.2	85.3
203.2	369.7	246.7	184.9	147.7	123.1
254	668.9	445.9	334.4	267.6	223

Diameter of Gutter 10.4 mm/m Slope	*Maximum Rainfall (mm/hr)*				
	50.8	*76.2*	*101.6*	*127*	*152.4*
76.2	44.6	29.7	22.3	17.8	14.9
101.6	94.8	63.3	47.4	37.9	31.6
127	163.5	108.9	81.8	65.4	54.5
152.4	252.7	168.6	126.3	100.8	84.1
177.8	362.3	241.5	181.2	144.9	120.8
203.2	520.2	347.5	260.1	208.1	173.7
254	947.6	631.7	473.8	379	315.9

Diameter of Gutter 20.9 mm/m Slope	*Maximum Rainfall (mm/hr)*				
	50.8	*76.2*	*101.6*	*127*	*152.4*
76.2	63.2	42.2	31.6	25.3	21
101.6	133.8	89.2	66.9	53.5	44.6
127	232.3	155	116.1	92.9	77.5
152.4	356.7	237.8	178.4	142.7	118.9
177.8	512.8	341.9	256.4	204.9	170.9
203.2	739.5	493.3	369.7	295.4	246.7
254	133.8	891.8	668.9	534.2	445.9

Diameter of Gutter 41.7 mm/m Slope	*Maximum Rainfall (mm/hr)*				
	50.8	*76.2*	*101.6*	*127*	*152.4*
76.2	89.2	59.5	44.6	35.7	29.7
101.6	189.5	126.3	94.8	75.8	63.2
127	328.9	219.2	164.4	131.5	109.6
152.4	514.7	343.3	257.3	206.2	171.9
177.8	724.6	483.1	362.3	289.9	241.4
203.2	1040.5	693	520.2	416.2	346.5
254	1858	1238.4	929	743.2	618.7

NOTE: Round, square, or rectangular gutters, leaders, or pipe may be used. They are considered equivalent when enclosing a scribed circle equivalent to the diameter listed above.

Source: Reprinted by permission from *Uniform Plumbing Code,* 17th ed., copyright © 1985 by the International Association of Plumbing and Mechanical Officials.

TABLE 8.7 **Roof Drain and Leader Sizes**

*Part A. **Conventional Units: Maximum Horizontal Projected Roof Area, (ft^2)***

Max. Hourly Rainfall (in.)	Size of Drain or Leader (in.)					
	2	3	4	5	6	8
1	2880	8800	18400	34600	54000	116000
2	1440	4400	9200	17300	27000	58000
3	960	2930	6130	11530	17995	38660
4	720	2200	4600	8650	13500	29000
5	575	1760	3680	6920	10800	23200
6	480	1470	3070	5765	9000	19315
7	410	1260	2630	4945	7715	16570
8	360	1100	2300	4325	6750	14500
9	320	980	2045	3845	6000	12890
10	290	880	1840	3460	5400	11600
11	260	800	1675	3145	4910	10545
12	240	730	1530	2880	4500	9660

*Part B. **SI Units: Maximum Horizontal Projected Roof Area, Square Meters***

Max. Hourly Rainfall (mm)	Size of Drain or Leader (mm)					
	50.8	76.2	101.6	127	152.4	203.2
25.4	267.6	817.5	1709.4	3214.3	5016.6	10776.4
50.8	133.8	408.8	854.7	1607.2	2508.3	5388.2
76.2	89.2	272.2	569.5	1071.1	1671.7	3591.5
101.6	66.9	204.4	427.3	803.6	1254.2	2694.1
127	53.4	163.5	341.8	642.9	1003.3	2155.3
152.4	44.6	136.6	285.2	535.6	836.1	1794.4
177.8	38.1	117.1	244.3	459.4	716.7	1539.4
203.2	33.4	102.2	213.7	401.8	627.1	1347.1
228.6	29.7	91	190	357.2	557.4	1197.5
254	26.9	81.8	170.9	321.4	501.7	1077.6
279.4	24.2	74.3	155.6	292.2	456.1	979.6
304.8	22.3	67.8	142.1	267.6	418.1	897.4

NOTE: Round, square, or rectangular gutters, leaders, or pipe may be used. They are considered equivalent when enclosing a scribed circle equivalent to the diamter listed above.

Source: Reprinted by permission from *Uniform Plumbing Code,* 17th ed., copyright © 1985 by the International Association of Plumbing and Mechanical Officials.

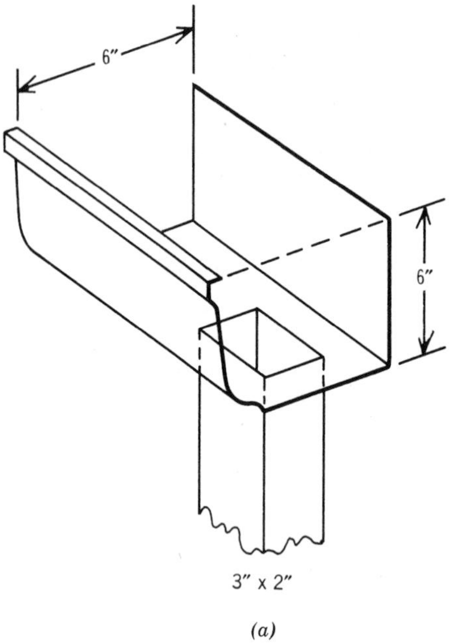

(a)

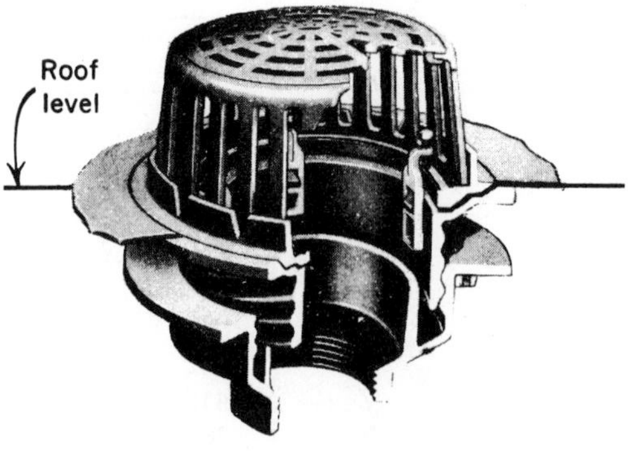

(b)

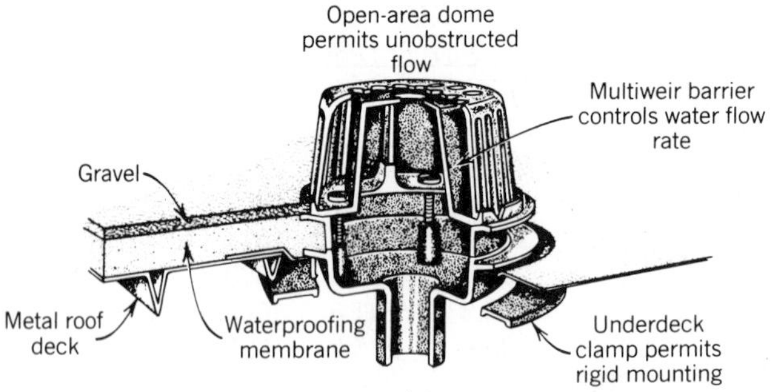

(c)

Fig. 8.18 *Storm drainage components.* (a) *Conventional gutter and leader for houses; sizes vary by manufacturer.* (b) *Ordinary roof drain (Josam Manufacturing Co.).* (c) *Roof drain for controlled flow. (From* Specifying Engineer, *November 1982.)*

Fig. 8.19 *Hayward Field West Grandstand at the Eugene Campus of the University of Oregon, built in 1976. The Amundson Associates, Architects, Springfield, Oregon. This is the track-and-field facility for the university.* (a) *In addition to the architectural and esthetic reasons for sloping the roof to the rear of this attractive and functional structure, the design allows for the collection of stormwater. The water collects at a horizontal gutter at the rear roof edge and drains through six leaders.* (b) *Portion of the roof gutter and four of the six leaders. Storm drainage from all downspouts is collected below ground and led to the street, where it empties into an 18-in. storm sewer.*

TABLE 8.8 **Horizontal Rainwater Piping Sizes**

Part A. Conventional Units: Maximum Horizontal Projected Roof Area, Square Feet

Size of Pipe (in.), 1/8" Slope	Maximum Rainfall (in./hr)				
	2	3	4	5	6
3	1644	1096	822	657	548
4	3760	2506	1880	1504	1253
5	6680	4453	3340	2672	2227
6	10700	7133	5350	4280	3566
8	23000	15330	11500	9200	7600
10	41400	27600	20700	16580	13800
12	66600	44400	33300	26650	22200
15	109000	72800	59500	47600	39650
Size of Pipe (in.), 1/4" Slope	**Maximum Rainfall (in./hr)**				
	2	3	4	5	6
3	2320	1546	1160	928	773
4	5300	3533	2650	2120	1766
5	9440	6293	4720	3776	3146
6	15100	10066	7550	6040	5033
8	32600	21733	16300	13040	10866
10	58400	38950	29200	23350	19450
12	94000	62600	47000	37600	31350
15	168000	112000	84000	67250	56000
Size of Pipe (in.), 1/2" Slope	**Maximum Rainfall (in./hr)**				
	2	3	4	5	6
3	3288	2295	1644	1310	1096
4	7520	5010	3760	3010	2500
5	13360	8900	6680	5320	4450
6	21400	13700	10700	8580	7140
8	46000	30650	23000	18400	15320
10	82800	55200	41400	33150	27600
12	133200	88800	66600	53200	44400
15	238000	158800	119000	95300	79250

TABLE 8.8 **Horizontal Rainwater Piping Sizes (*continued*)**

Part B. ***SI Units: Maximum Horizontal Projected Roof Area, Square Meters***

Size of Pipe (mm), 10.4 mm/m Slope	*Maximum Rainfall (mm/hr)*				
	50.8	*76.2*	*101.6*	*127*	*152.4*
76.2	152.7	101.8	76.4	61	50.9
101.6	349.3	232.8	174.7	139.7	116.4
127	620.6	413.7	310.3	248.2	206.9
152.4	994	662.7	497	397.6	331.3
203.2	2136.7	1424.2	1068.4	854.7	706
254	3846.1	2564	1923	1540.3	1282
279.4	6187.1	4124.8	3093.6	2475.8	2062.4
381	10126.1	6763.1	5527.6	4422	3683.5
Size of Pipe (mm), 20.9 mm/m Slope	*Maximum Rainfall (mm/hr)*				
	50.8	*76.2*	*101.6*	*127*	*152.4*
76.2	215.5	143.6	107.8	86.2	71.8
101.6	492.4	328.2	246.2	197	164.1
127	877	584.1	438.5	350.8	292.3
152.4	1402.8	935.1	701.4	561.1	467.6
203.2	3028.5	2019	1514.3	1211.4	1009.5
254	5425.4	3618.5	2712.7	2169.2	1806.9
304.8	8732.6	5815.5	4366.3	3493	2912.4
381	15607.2	10404.8	7803.6	6247.5	5202.4
Size of Pipe (mm), 41.7 mm/m Slope	*Maximum Rainfall (mm/hr)*				
	50.8	*76.2*	*101.6*	*127*	*152.4*
76.2	305.5	213.2	152.7	121.7	101.8
101.6	698.6	465.4	349.3	279.6	232.3
127	1241.1	826.8	620.6	494.2	413.4
152.4	1988.1	1272.3	994	797.1	663.3
203.2	4274.4	2847.4	2136.7	1709.4	1423.2
254	7692.1	5128.1	3846.1	3079.6	2564
304.8	12374.3	8249.5	6187.1	4942.3	4124.8
381	22110.2	14752.5	11055.1	8853.4	7362.3

Source: Reprinted by permission from *Uniform Plumbing Code,* 17th ed., copyright © 1985 by the International Association of Plumbing and Mechanical Officials.

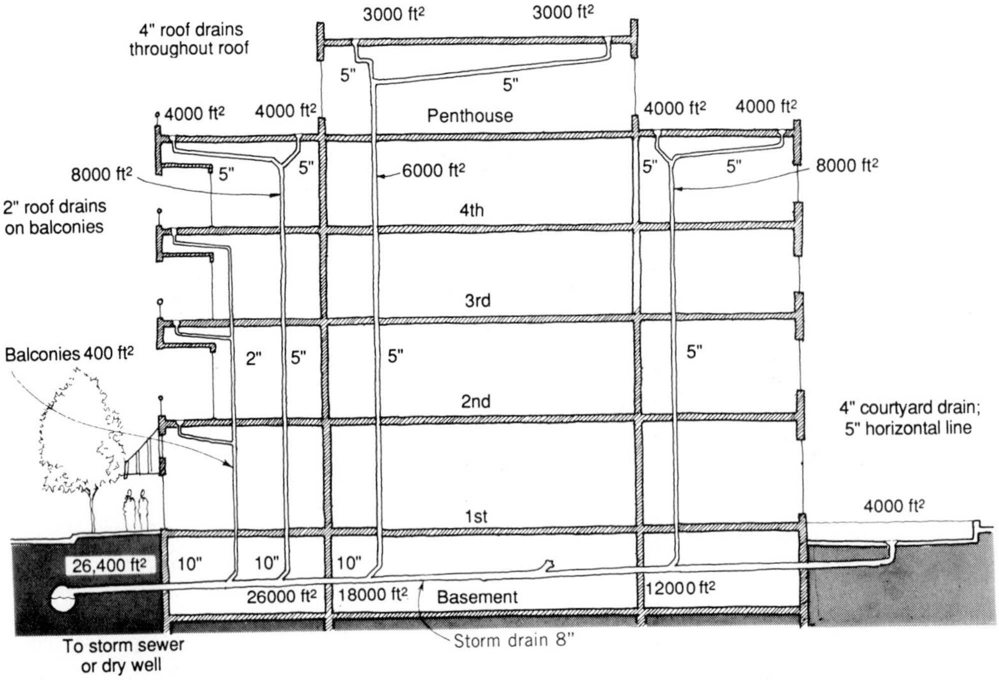

Fig. 8.20 *(Example 8.3.) Separate storm drainage. Areas drained and corresponding sizes of vertical leaders and horizontal drains are from Tables 8.7 and 8.8. Storm drain piping within a building needs insulative covering with a vapor barrier on the outside. This prevents condensation (sweating) on the pipes when in winter, warm, moisture-laden air in the building could otherwise reach the pipe surface (which would be cold from carrying icy water), condense there, and lead to wet, dripping conditions on the pipes. Each roof has two drains, in case one is temporarily blocked.*

9

WATER SUPPLY

With water, one of the designer's first concerns is to match the quality of the water to the task it performs. As this becomes a more serious design issue, designers will provide for the recycling of water within and around buildings, as well as to specify plumbing fixtures that use less water. Table 9.1 shows typical relationships between quality and usage. Water recycling and conservation are discussed in detail at the beginning of Chapter 10. In this chapter we deal primarily with *potable* (drinkable) water: first with issues of water quality, and then with the matter of assuring an adequate supply of water throughout the building.

9.1 Water Quality

A summary of water quality at the various stages of the hydrologic cycle was shown in Fig. 8.3. As precipitation, water contains few impurities: almost no bacterial content is present, and only small amounts of minerals and gases can be expected. To collect this nearly pure water, surfaces are needed—and at this point, foreign substances can readily contaminate the water. These pollutants can affect water's physical (mostly organic), chemical (mostly inorganic), biological, or radiological characteristics. Both surface water and groundwater are subject to pollution.

(a) Physical Characteristics. Some of the most noticeable alterations of water quality fall within this category. Water from surface sources (roof runoff, streams, rivers, lakes, ponds) is particularly subject to physical pollutants.

Turbidity is easy to see, and thus a likely source of dissatisfaction for the would-be consumer. It is caused by the presence of suspended material such as clay, silt, other inorganic material, plankton, or finely divided organic material. Even those materials that do not adversely affect health are usually esthetically objectionable.

TABLE 9.1 **Water Use and Quality in Buildings**

Use	*Desired Quality*
A. Consumed	
1. Drinking and cooking	Potable
2. Bathing	Potable
3. Laundering	Soft
4. Irrigation and watering of livestock	Unpolluted
5. Industrial processes	As required
6. Vapor to increase the relative humidity of air	
B. Circulated	
1. Hot water for heating	Note: Make-up water should be soft or neutral and, for swimming, potable
2. Chilled water for cooling	
3. Condenser cooling water	
4. Swimming pool water	
5. Steam for heating, later condensed	
C. Generally static	
1. Water stored for fire protection	No special requirement
2. Water in fire standpipes	
3. Water in sprinkler piping	
D. Controlled	
1. Vapor condensed to reduce relative humidity of air	

NOTE: For water uses in Section A, above, flow is often continuous. Section B comprises uses for which flow other than circulation is intermittent or at a relatively low rate, the water added to the systems being known as "make-up water." Items C2 and C3 call for piping to provide adequate though infrequent flow in emergencies. Obviously, item D1 relates only to moisture condensed out of the air and involves no design for supply.

Color, another visible alteration, is often caused by dissolved organic matter, as from decaying vegetation. Some inorganic materials also color water, as do growths of microorganisms. Like turbidity, such color changes usually do not threaten health but often are psychologically undesirable.

Taste and *odor* can be caused by organic compounds, inorganic salts, or dissolved gases. This condition can be treated only after a chemical analysis has identified which source is responsible.

Temperature is another characteristic of psychological importance—we expect drinking water to be cool. In general, water supplied between 50 and 60 F (10 to 16°C) is preferred.

Foamability is usually caused by concentrations of detergents. The foam itself does not pose a serious health threat, but it may indicate that other, more dangerous pollutants associated with domestic waste are also present. Because of increased foaming in water in the 1960s, today's detergents must use biodegradable LAS (linear alkylate sulfonate), which biodegrades rapidly—except in the absence of oxygen. Since this lack of oxygen is characteristic of some septic tank drainage fields, foam in drinking water should be promptly investigated.

TABLE 9.2 **Example of Chemical Analysis**

Quality		*Parts per Million (ppm)*[a]
Total hardness	as $CaCO_3$	30
Calcium hardness	as $CaCO_3$	20
Alkalinity (Methyl Orange)	as $CaCO_3$	27
Alkalinity (Phenolpntalein)	as $CaCO_3$	0
Free Carbon Dioxide	as CO_2	13.5
Chlorides	as Cl	6
Sulfates	as SO_4	4
Silica	as SiO_2	19
Phosphates—normal	as PO_4	0
Phosphates—total	as PO_4	0.5
Iron—total	as Fe	1.6
Total dissolved solids		66
Turbidity or sediment		present

Source: A report by Olin Water Service, for a private well in Virginia.

[a]Note that "ppm" and "mg/L" are essentially equivalent terms for concentration in water.

(b) Chemical Characteristics. Groundwater is particularly subject to chemical alteration, because as it moves downward from the surface it slowly dissolves some minerals contained in rocks and soils. A chemical analysis (such as that given in Table 9.2) is usually necessary for individual water supply sources. These analyses will indicate (1) the possible presence of harmful or objectionable substances, (2) the potential for corrosion within the water supply system, and (3) the tendency for the water to stain fixtures and clothing. Concentrations are expressed in mg/L (milligrams per liter), which is essentially equivalent to ppm (parts per million).

Some general terms commonly used to describe chemical characteristics of water are defined as follows.

Alkalinity is caused by bicarbonate, carbonate, or hydroxide components. Testing for these components of water's alkalinity is a key to determining which treatments to use.

Hardness is a relative term (see Fig. 9.1). "Hard" water inhibits the cleaning action of soaps and detergents, and it deposits scale on the inside of hot water pipes and cooking utensils, thus wasting heating fuel and making utensils unusable. Hardness, which is caused by calcium and magnesium salts, can be classified as temporary (carbonate) or permanent (noncarbonate). Temporary hardness is largely removed when the water is heated—it forms the scale just described. Permanent hardness cannot be removed by simple heating (see Section 9.2*g*).

pH is a measure of the water's hydrogen ion concentration, as well as its relative acidity or alkalinity (see Fig. 9.1). A pH of 7 is neutral. Measurements below 7 indicate increasing acidity (and corrosiveness); water in its natural state can have a pH as low as 5.5, with 0 the ultimate acidity. Measurements higher than 7 indicate increasing alkalinity; a pH as high as 9 can be found in water in its natural state, with 14 the ultimate alkalinity. The pH value is the starting point for determining treatments for corrosion control, chemical dosages, and disinfection.

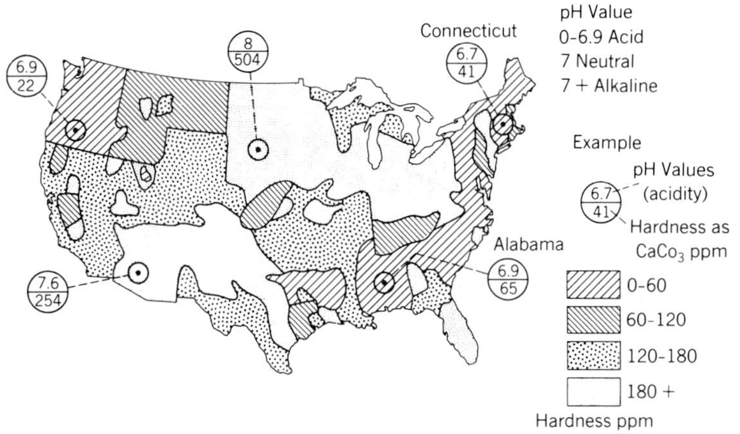

Fig. 9.1 Approximate groundwater chemical characteristics across the United States. Treatment may be needed when pH is less than 7.0 (acidity results in corrosion) or when hardness as $CaCo_3$ exceeds 65 ppm (essentially, ppm and mg/L are identical measures). (Courtesy of Progressive Architecture.)

Chemical additions to water supplies most commonly include the following elements.

Toxic substances are occasionally present in water supplies. Local health authorities can provide information about acceptable concentrations of such substances as arsenic (As), barium (Ba), cadmium (Cd), chromium (Cr^{6+}), cyanides (CN), fluoride (F), lead (Pb), selenium (Se), and silver (Ag). Although limited amounts of *fluoride* are frequently added to water supplies to help prevent tooth decay, fluorides in excess of such "optimum concentrations" can produce mottling of teeth. *Lead* poses a dangerous threat, even in relatively small amounts, because it is a cumulative poison. Lead in water usually comes from lead piping (in older buildings or cities) or corrosive water on lead-painted roofs. A maximum recommended concentration is 0.05 mg/L.

Unfortunately, the list of toxic chemicals is expanding, and as chemical waste dumps have been abandoned or mismanaged, groundwater has become contaminated. The U.S. Environmental Protection Agency has estimated that 75% of both active and abandoned chemical waste dumps—some 51,000 in all—are leaking. In addition to the following list of inorganic chemicals, we are becoming aware of many new organic chemicals as well, some of which are suspected of causing 5 to 20% of U.S. cancers. The threat to our groundwater supplies is illustrated in Fig. 9.2. Once polluted, aquifers are extremely difficult to clean.

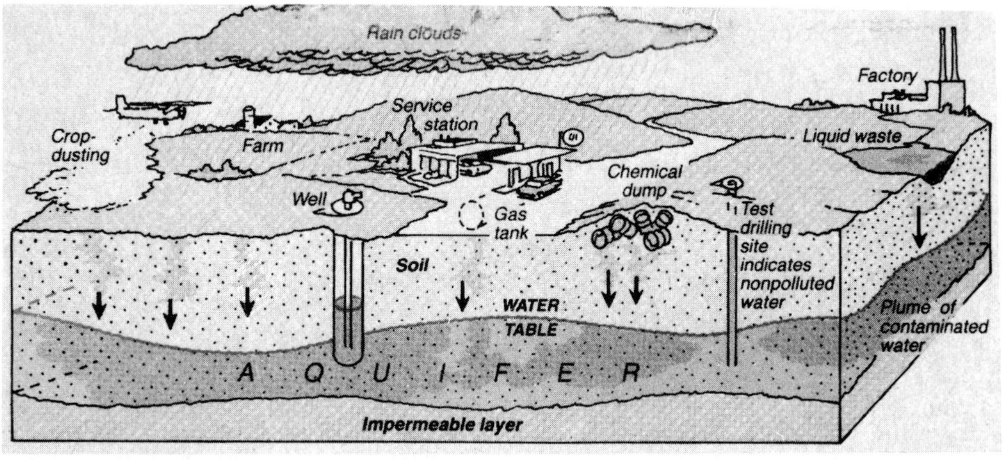

FPO

Fig. 9.2 How groundwater becomes contaminated. The "plume" formed by contaminants can often go undetected. (Copyright © 1982 by Newsweek, Inc., all rights reserved. Reprinted by permission.)

Chlorides can enter water as it passes through geological deposits formed by marine sediment, or because of pollution from sea water, brine, or industrial or domestic wastes. A noticeable taste will result from chloride in excess of 250 mg/L.

Copper can enter water from natural copper deposits or from copper piping that contains corrosive water. Concentrations of copper in excess of 1.0 mg/L can result in an undesirable taste.

Iron is frequently present in groundwater. Corrosive water in iron pipes will also add iron to water. At concentrations above 0.3 mg/L, iron can lend a brownish color in washed clothes and can affect the taste of the water.

Manganese can both pose a physiological threat (it is a natural laxative) and produce color and taste effects similar to those produced by iron. The recommended limit is 0.05 mg/L.

Nitrates in high concentrations pose a threat to infants, in whom it can cause "blue baby" disease. In shallow wells, nitrate concentrations can indicate seepage from deposits of livestock manure.

Pesticides, a growing threat to water supplies, are particularly common in wells near homes that have been treated for termite control. Avoid using pesticides near wells.

Sodium is primarily dangerous for people with heart, kidney, or circulatory ailments. For a low-sodium diet, the sodium in drinking water should not exceed 20 mg/L. Salts spread on roadways for ice control can leach into the soil and enter groundwater. Note that some "water softeners" (discussed below, in Section 9.2*g*) can raise sodium concentrations in water.

Sulfates, which have laxative effects, can enter groundwater from natural deposits of Epsom salts (magnesium sulfate) or Glauber's salt (sodium sulfate). Concentrations should not exceed 250 mg/L.

Zinc sometimes enters groundwater in areas where it is found in abundance. Although not a health threat, it can cause an undesirable taste at concentrations above 5 mg/L.

(c) Biological Characteristics. Potable water should be kept as free as possible from disease-producing organisms—bacteria, protozoa, and viruses. These organisms are not easily identified; a thorough biological water test is complex and time consuming. For this reason, the standard test is for *one* kind of bacteria—the coliform group (*E. coli*), which is always present in the fecal wastes of humans (as well as those of many animals and birds) and which outnumbers all other disease-producing organisms in water. The recommended maximum concentration for coliform bacteria is one organism per 100 mL (about ½ cup) water.

For biological activity to be kept to a minimum in drinking water, a water source should be chosen that does not normally support much plant or animal life; hence the popularity of groundwater, rather than surface water, as a source. In addition, the supply should be protected from subsequent biological contamination. Where cities depend on small lakes for water, humans are frequently excluded from the watersheds. Organic fertilizers and nutrient minerals should also be kept out of the water supply, to further discourage biological activity. For the same reason, stored water should be kept dark and at low temperatures. Finally, organisms (or their by-products) are commonly destroyed at treatment facilities.

(d) Radiological Characteristics. The mining of radioactive materials and the use of such materials in industry and power plants have produced radiological pollution in some water supplies. Since radiological effects are cumulative, concentrations of radioactive materials should be low indeed. "Safe" minimum concentrations have continually been revised downward for other radiation exposures; consult the local public health service for current recommendations.

9.2 Water Treatment

In the preceding section, water pollution was broken down into physical, chemical, biological, and radiological categories. The various forms of treatment for such pollutants do not necessarily fall into the same categories, as one treatment may be effective for several different polluted conditions. A general look at common domestic water quality problems is provided in Table 9.3. The treatment processes available range from the mild (sedimentation) to the extreme (distillation).

(a) Sedimentation. This process removes some suspended matter from water simply by allowing time and the inactivity of the water to do the work of settling out heavier suspended particles. Simple basins, ponds, or tanks constructed for this purpose are large enough to retain the water for at least 24 hours and are equipped with baffles to slow the water flow. To clean out the sediment, water usually is diverted to an identical second basin while the first is being cleaned.

(b) Coagulation (Flocculation). This process also removes suspended matter, along with some coloration. A chemical such as alum (hydrated aluminum sulfate) is added to turbulent water. The water is then held in a quiet condition, in which the suspended particles will combine with the alum to form floc. These heavy particles then settle out, in a process similar to sedimentation. Some adjustment of the pH may be necessary.

(c) Aeration (Oxidation). This simple process can improve the taste and color of water, help remove iron and manganese, and sometimes decrease corrosiveness. In aeration, as much of the water's surface as possible is exposed to air. The methods used are rich in esthetic possibilities—the spraying of water into air, the fall of a turbulent stream of water over a spillway, and *flowforms,* sculptural waterfalls designed to carry water in a rythmical, pulsating figure-eight pattern. To guard against contamination, these processes are often enclosed; if exposed, they must be kept clean. For aeration within tanks, water is passed through a series of perforated plates, in streams or droplets.

Ozonation is a less aesthetic but more certain oxidation process. It is most commonly used in cooling tower water treatment (see Fig. 9.3) and in addition ozonation has a very wide range of treatment applications.

Aeration improves the flat taste of distilled water and cistern water, by adding oxygen. It

TABLE 9.3 **Common Water Quality Problems and Treatment in Small Systems**

Item	*Cause*	*Bad Effect*	*Correction*
Hardness	Calcium and magnesium salts from underground flow	Clogging of pipes by scale, burning out of boilers, and impaired laundering and food preparation	Ion-exchanger (Zeolite process)
Corrosion	Acidity, entrained oxygen and carbon dioxide (low pH)	Closing of iron pipe by rust, leaking connections, destruction of brass pipe	Raising the alkaline content (Neutralizer)
Biological pollution	Contamination by organic matter or sewage	Disease	Chlorination by sodium hypochlorite or chlorine gas; or ozonation
Color	Iron and manganese	Discoloration of fixtures and laundry	Chlorination or ozonation and fine filtration
Taste and odor[a]	Organic matter	Unpleasantness	Filtration through activated carbon (Purifier); aeration
Turbidity[a]	Silt or suspended matter picked up in surface or near-surface flow	Unpleasantness	Filtration

[a]These problems are not common in private systems that use deep wells.

WATER AND WASTE

Fig. 9.3 *Recycled cooling tower water is treated by an ozonator, magnetic descaler, and a filtration system. This controls scale formation, algae and slime, corrosion, and sludge buildup. (Courtesy of Aqua-Flo, Inc., Baltimore.)*

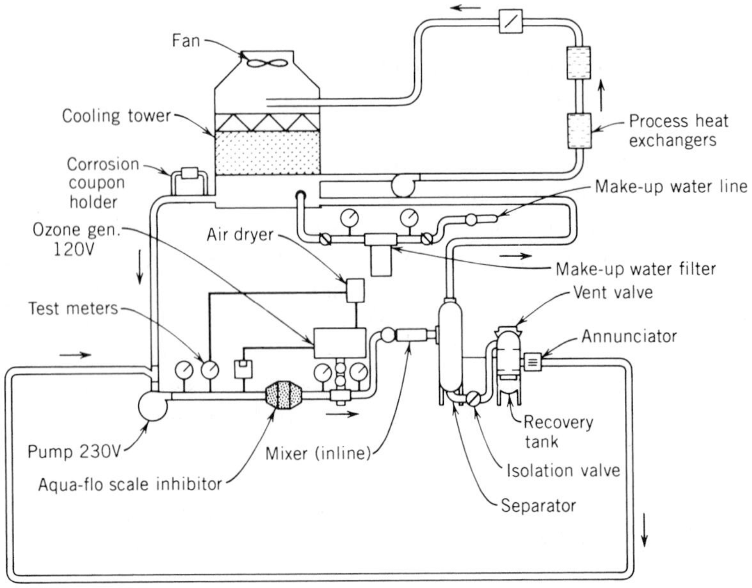

also oxidizes iron or manganese, which then can easily be removed by filtration, and removes odors caused by hydrogen sulfide and algae. Finally, it can reduce corrosiveness caused by carbon dioxide and other gases, although the increase in oxygen content will partially offset this process.

(d) Filtration. This very common treatment can remove suspended particles, some bacteria, and color or taste. The more common approaches are listed below, beginning with filtering to remove suspended particles then moving on to more specialized applications for the removal of iron and/or manganese, tastes, and odors.

Slow sand filters (see Fig. 8.6) for rainwater application) are low-maintenance, easily constructed devices that should be cleaned as often as the turbidity of the water demands—from once a day to perhaps once a month. They are cleaned by removal and replacement of the top 1 in. of sand, which is then either washed for reuse or discarded. The approximate rate of flow is between 60 and 180 gallons per day per square foot (2450 to 7350 liters per day per square meter) of filter bed surface. Overall thickness is usually 30 to 48 in. of sand over 6 to 8 in. of gravel (760 to 1220 mm of sand over 150 to 200 mm of gravel).

Pressure sand filters require control and the attention of an operator, and thus are rarely used for individual water systems.

Diatomaceous earth filters can be of either the vacuum or the pressure type. They require periodic attention to remain effective.

Porous stone, ceramic, or unglazed porcelain filters (also called *Pasteur filters*) are usually made in small sizes so that they can be attached to water faucets. They are used widely in some countries, such as Mexico, but poor maintenance or hairline cracks often lead to bacterial infiltration, complicating the filtration process. A more positive approach to the disinfection of drinking water thus is desirable.

Chlorination/fine filtration is a combined process that removes iron and/or manganese from water. The chlorine chemically oxidizes the iron or manganese, which forms a precipitate. Chlorine also kills iron bacteria (which can form a slimy mass when present) and disinfects. Then, the fine filter removes the precipitated iron or manganese. (*Ozonation/fine filtration* is an attractive alternative, whose process is similar.) Since water containing iron or manganese poses a threat to plumbing and other water treatment processes, this procedure should be the first one applied to water supplies that suffer from excess iron and/or manganese.

Alternative treatments for iron and/or manganese are aeration followed by filtration and treatment with potassium permanganate followed by filtration.

Activated carbon filters are particularly effective for removing tastes and odors. The water is passed through granular carbon, which attracts large quantities of dissolved gases, soluble organics, and fine solids.

Reverse osmosis is sometimes used to reduce the mineral content in water. An inert, semipermeable membrane has higher-pressure supply water on one side; as the pressure slowly forces water through this filtering membrane, most of the minerals (dissolved solids) are removed. Dissolved chemicals, however, remain. The principal disadvantage of this process is its enormous appetite for water; only a portion of the water supplied to a reverse-osmosis filtering unit is delivered to the user. The remainder must be used to flush the semipermeable membrane, so that mineral buildup is avoided.

Commercial reverse-osmosis units are available in sizes ranging from a 1 gpm (3.9 L/m) water delivery rate [using two membranes and a 3-hp motor, requiring 4.5 gpm (17 L/m) feedwater, for a 22% recovery rate] to a water delivery rate of 12.5 gpm (47.3 L/m) [using 12 membranes and a 15-hp motor, requiring 19.2 gpm (72.8 L/m), for a 65% recovery rate).

(e) Disinfection. This is the most important health-related water treatment. Although chlorination has become the standard approach to removing harmful organisms from water, there are alternatives: ultraviolet light (unsuitable for water with high turbidity, since the light cannot easily penetrate), bromine, iodine, ozone, and heat treatment, among others. An alternative is ozonation, which was first used in Nice, France, in 1908. It is used in municipal treatment plants for both bacterial disinfection and viral inactivation. Chlorine, however, continues to disinfect after the initial application. It is this continuing protection that has made it so universally relied on, despite dangers such as that posed by deadly chlorine gas. Although chlorine affects the taste and odor of water, it is also effective in removing less-desirable tastes and odors.

Factors that affect chlorine's ability to disinfect include

1. *Chlorine concentration*. The higher the concentration, the faster and more complete the rate of disinfection.
2. *Contact time*. The longer the chlorine contacts the organisms in water, the more complete the disinfection. At a minimum, 0.4 mg/L of chlorine should contact water 30 minutes before use.
3. *Water temperature*. The higher the temperature during contact, the more complete the disinfection.
4. *pH*. The lower the pH, the more effective the disinfection.

Chlorine can be added automatically to water supplies by *hypochlorinators*. These devices automatically pump (or inject) into water a chlorine solution formed from powder or tablets or from a prepared solution. They are usually no larger than the pumps used in small water systems (see Fig. 9.10). Some hypochlorinators are specially designed for low and fluctuating water pressures, or for use where electricity is not available.

(f) Corrosion Control. It is important to control corrosion in order both to keep water systems operating freely and to prevent corrosive water from increasing the concentration of hazardous materials (as from copper pipes). Corrosion is a slow degradation of a metal by a flow of electric current from the metal to its surroundings. The factors involved in corrosion control are:

1. *Acidity*. The more acid (low pH), the more corrosive the water.
2. *Conductivity*. As dissolved mineral salts increase the water's conductivity, they encourage the flow of the electrical current of corrosion.
3. *Oxygen content*. Dissolved oxygen destroys the thin protective hydrogen film on immersed metals, thus promoting corrosion.
4. *Carbon dioxide content*. Carbon dioxide forms carbonic acid, which attacks metal surfaces.
5. *Water temperature*. Increased temperature increases corrosion.

The products of corrosion often contribute to scale formation. Scale then lines surfaces, eventually clogging openings.

Acid neutralizers can be installed on water supplies with low pH; their function is often combined with those of hypochlorinators. Typically, neutralizing solutions are mixtures of soda ash and water.

Alternatives to corrosion control include feeding into the water small amounts of film-forming materials (polyphosphates or silicates), installing dielectric or insulating unions (to avoid complications from dissimilar pipe metals), and avoiding metal piping and fixtures altogether.

(g) Softening. Water "hardness" is caused primarily by calcium and magnesium deposits; when they are removed, water will be "soft." Where water hardness is mild enough to affect only laundering, cisterns may be used to collect "soft" rainwater to use for clothes washing. Where water hardness produces scale in pipes and within water heating appliances, and cisterns are not feasible, water softening equipment is used.

This equipment commonly uses zeolite, in an ion-exchange process. Water containing calcium and/or magnesium is passed through a bed of zeolite, which exchanges its sodium ions for those of calcium and/or magnesium. The water thus loses hardness, but it gains sodium content—an undesirable development for people on low-sodium diets. Periodically, as the zeolite loses its sodium ions, it must be regenerated. This is done automatically within the softening equipment, in which a brine solution is passed through the bed of zeolite. Alternatives to zeolite as the ion-exchange material include glauconite (or greensand) and a combination of resins [precipitated synthetic, organic (carbonaceous), and synthetic].

(h) Nuisance Control. Some organisms may not be injurious to health but can multiply so rapidly that piping or filters become clogged or the water's appearance, odor, and taste are affected. This process is most common in surface water sources, and it is within surface reservoirs that these treatments are most often applied. Algae growths, the most prevalent nuisance, can usually be controlled by applying copper sulfate (blue stone or blue vitriol) to the water body. Sudden and massive algae kills can have adverse impacts on other life forms within the water, because the decomposing algae rob the water of oxygen. As a further precaution, stored water should be shielded from sunlight whenever possible.

Cooling towers are an especially difficult water treatment problem. The murky water with high turbidity and bacterial count leads to clogged passages (thus inefficient performance), deteriorated surfaces, and to the growth of potentially lethal bacteria (*Legionella pneumophila*). As a result, enormous quantities of water are commonly used once-through the cooling tower, then dumped to storm sewers, rather than recycled in a closed system. (As late as 1989, San Francisco City Hall was reported to be using 96,000 gallons per summer day for once-through cooling—in a year of drought.) To treat cooling tower water successfully, a method is needed for microbial control, removing organics, and precipitating inorganics. Ozonation (Fig. 9.3) is one answer to this widespread treatment problem.

(i) Fluoridation. A heated controversy has developed over the addition of fluoride to drinking water. The advantage of fluoridation is that children who drink fluoridated water have lower rates of tooth decay. And since everyone drinks water, all children benefit, not just those who can afford fluoride pills. Its disadvantages are that only children need the fluoride, not adults, and that in amounts above those used in water treatment, fluoride is toxic and can cause mottled teeth. Opponents of fluoridation suggest that because sugar is a leading cause of tooth decay in children, sugar, rather than water, should be fluoridated.

Small water systems can be equipped with fluoridation units. However, fluoride levels in the water supply must be carefully monitored.

(j) Distillation. This is a simple, low-technology approach to purification that produces the equivalent of bottled water for drinking, cooking, and laboratory uses. In one process, it promises the removal of suspended solids, salts, bacteria, and (apparently) halogenated hydrocarbons. When water pollution is extreme, as in the case of sea (salt) water, distillation may be the best treatment. Water is heated, to encourage evaporation. As the water turns to vapor, virtually all pollutants are left behind. When this vapor encounters a cooler surface, it condenses, and pure water (although flat in taste) can be collected from this surface.

Any heat source can be used in the distilling of water; solar stills (Fig. 9.4) are gaining in popularity because the energy used is free. Solar distillation of cistern water is an autonomous approach. In semiarid, rather sunny climates, a solar still should produce about 1⁄12 gallon per square foot of collector surface area (4 liters per square meter per day). This rate of production suggests that only the water used directly for drinking or other specialized purposes is usually feasible for distillation. Another factor to consider is that the cleaning of the still is generally accomplished by flushing it with twice as much water as was delivered. If not excessively brackish, this flush water could be used for irrigation or other nonpotable-quality tasks. Other treatment processes also use water in treating water.

9.3 Water Sources

This section focuses on the equipment used to capture and store groundwater from wells. Other water sources are less often used for smaller systems, either because they require much more extensive treatment (surface water from lakes or rivers) or because they provide water intermittently (cisterns). The sequence of treatment for a city's water taken from a river is shown in Fig. 9.5; this multistage treatment is inappropriate for small water systems that receive only occasional maintenance. Cisterns were discussed in Chapter 8. Another increasingly important source of city water, the treated effluent of sewage treatment plants, is discussed in Chapter 10.

(a) Wells. Farms and remote housing developments usually have private water systems. In rural and suburban areas where the progress of building is faster than the development of municipal water supplies, private sources must be sought. The elaborate treatment necessary for the use of surface waters or those in dug wells of

Fig. 9.4 A solar still can be used to provide a small daily quantity of pure water; this installation serves a laboratory. (Courtesy of McCracken Solar Co., Alturas, California.)

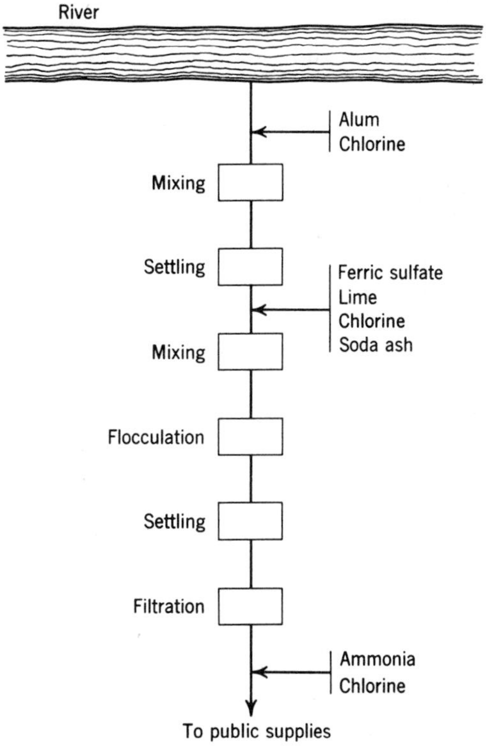

Fig. 9.5 When a city's water is taken from surface sources, such as rivers, multistage treatment is usually necessary. (From Water Quality and Treatment, *2nd ed., American Water Works Association, Inc.)*

the old type makes the use of driven or drilled wells preferable. Water from these sources usually has at least the advantages of purity, coolness, and freedom from turbidity, odor, and unpleasant taste—any of which *may* be encountered in addition to either acidity or hardness.

Bored wells, which are dug with earth augers, are usually less than 100 ft deep. They are used when the earth to be bored through is boulder-free and will not cave in. The diameter range is 2 to 30 in. The bored well is then cased with metal, vitrified tile, or concrete.

Driven wells are the simplest and, usually, the least expensive type. A steel drive-well point is fitted on the end of pipe sections and driven into the earth. The drive point is usually 1¼ to 2 in. (32 to 51 mm) in diameter. The materials and design of drive-well points vary according to the expected characteristics of the earth in which the well is driven. First, a pilot hole is bored (frequently, with a simple hand auger), and the drivewell point and pipe sections lowered into it. Then the well is driven to a point well below the water table.

The construction of *jetted wells* requires a source of water and a pressure pump. A washing well point is supplied with water under pressure; this loosens the earth and allows the point and pipe to penetrate.

Drilled wells require more elaborate equipment of several types, depending on the geology of the site. The percussion (or cable-tool) method involves the raising and dropping of a heavy drill bit and stem. Having thus been pulverized, the earth being drilled is mixed with water to form a slurry, which is periodically removed. As drilling proceeds, a casing is also lowered (except when drilling through rock).

Rotary drilling methods (either hydraulic or pneumatic) utilize a cutting bit at the lower end of the drill pipe; a drilling fluid (or pressurized air) is constantly pumped to the cutting bit to aid in the removal of particles of earth, which are then brought to the surface. After the drill pipe is withdrawn, a casing is lowered into position.

Another method is the down-the-hole pneumatic (air) hammer, which combines the percussion effect with the rotary drill bit.

Local well drillers, who usually use the method most suited to the prevailing geology of the region, can offer useful advice about well construction methods. When clients plan to build in a remote location, the architect and engineer will want to advise them about water problems. Quality-corrective measures can always be taken and pumping equipment purchased, but the amount of water that can be obtained from the ground and the depth and cost of wells are all-important considerations. There are some problem areas where wells several hundred feet in depth will yield as little as 5 gpm, or nothing. The cost of drilling a number of exploratory wells may sometimes be excessive. Unfortunately, when such difficulties occur, there is often no easy solution. Conferences with neighboring owners, state and federal geologists, and local well drillers are all helpful. Many regions, of course, yield plentiful water, but the cost of the probable depth of the well should be considered.

A low-yield well can be combined with storage tanks, so that the pump can run all night, slowly filling the tanks to meet the next day's demands.

(b) Pumps. Three common types of pumps used in well water supply are the positive displacement, the centrifugal, and the jet.

Positive Displacement Pumps. There are two principal types of positive displacement pump. In a *reciprocating pump,* a plunger moves back and forth within a cylinder equipped with check valves. The cylinder is best located near or below the groundwater level. Water enters the cylinder through an initial check valve (which allows flow in only one direction). As the plunger moves toward this check valve, the water is forced through the second check valve, located within the plunger itself. Then, as the piston returns to its original position, the water is forced upward toward the surface.

A *rotary pump* has a helical or spiral rotor—a turning vertical shaft within a rubber sleeve. As the rotor turns, it traps water between it and the sleeve, thus forcing the water to the upper end of the rotor.

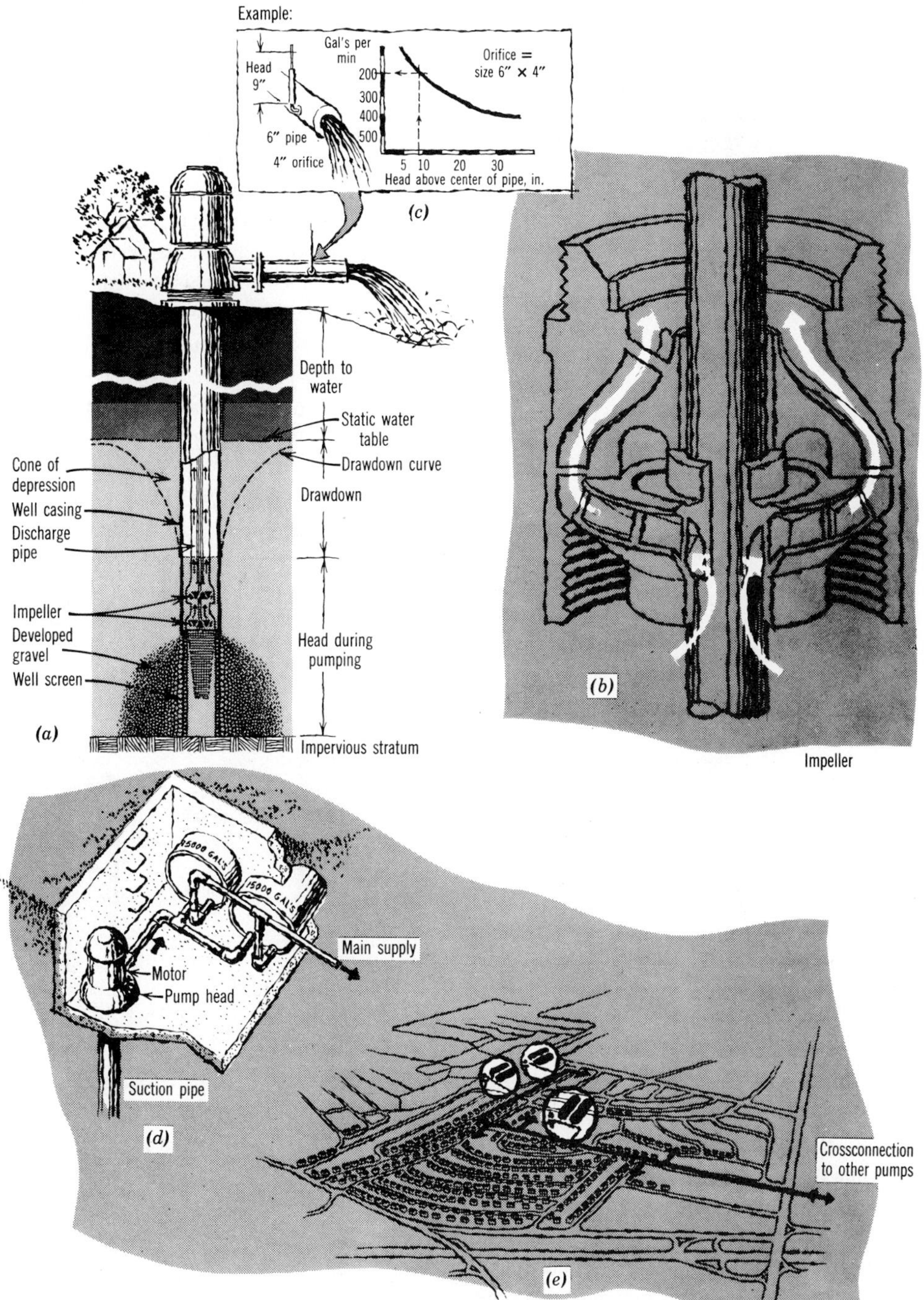

Fig. 9.6 (a) *A turbine well-pump.* (b) *Its operation.* (c) *Measurement of its capacity.* (d, e) *Its use in supplying a small community with groundwater.* (f) *A Jacuzzi multistage lineshaft turbine well-pump. (Jacuzzi Bros., Inc.) Capacities of turbine pumps range from 50 to 16,000 gpm. (By permission of* Progressive Architecture.*)*

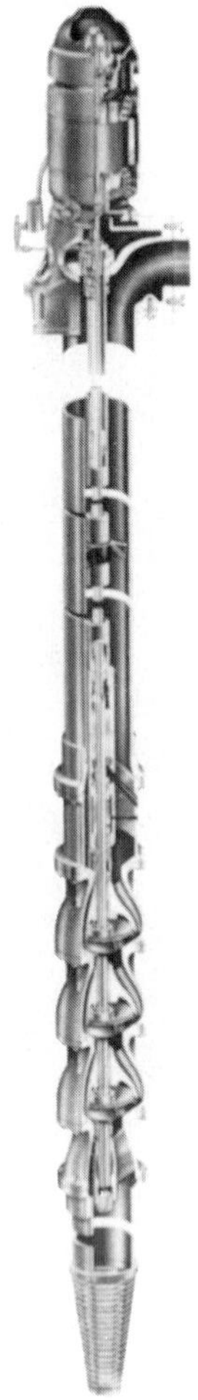

(f)

Fig. 9.6 *(continued)*

Centrifugal Pumps. This type of pump contains an impeller mounted on a rotating shaft. The rotating impeller increases the water's velocity while forcing the water into a casing, thus converting the water's velocity into higher pressure. Each impeller and casing is called a stage; many stages can be combined in a multistage pump. The number of stages depends upon the pressure needed to operate the water supply system, as well as the height to which the water must be raised. The most common centrifugal pumps are those used in deep wells.

The *turbine pump* has a vertical turbine located below groundwater levels and a driving motor located higher up, usually over the well casing at grade. A long shaft is thus required between the motor and the turbine. Substantial head clearances for this shaft's removal can be required. Figure 9.6 shows a turbine pump installation for a small community on Long Island, New York. The water is taken from the ground by multistage turbine pumps at depths of several hundred feet. The water is delivered to submerged hydropneumatic tanks at a pressure of about 80 psi. As water is demanded in the houses, the air under pressure in the upper part of the tanks forces it through the mains.

Submersible pumps are designed so that the motor can be submerged along with the turbine (Fig. 9.7). The lengthy pump shaft is thus eliminated.

Jet (or Ejector) Pumps. In a jet pump a venturi tube is added to the centrifugal pump. A portion of the water that is discharged from a centrifugal pump at the wellhead is forced down to a nozzle and venturi tube (Fig. 9.8). The

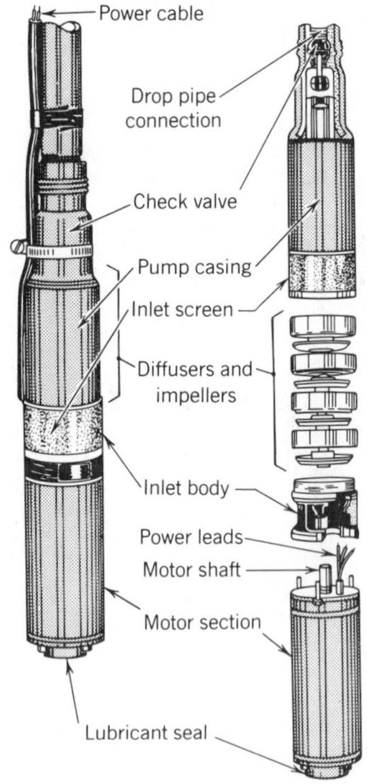

Fig. 9.7 *Submersible pump (centrifugal type), in exploded view. (From the U.S. Environmental Protection Agency's* Manual of Individual Water Supply Systems, *1975.)*

(a)

Fig. 9.8 *Details of the deep-well jet pump.* (a) *Photograph of a multistage jet pump housing and equipment. At the top is the on/off electrical switch activated by pressure settings. It controls the direct-connected electric motor at the left. Impellers are enclosed in the pump housing at the right. Circulating connections to and from the jet can be seen at the right, the pump discharge at the top. Note the pressure gauge readings up to 100 psi, within which range the pump can be set to operate.* (b) *Well casing and circulating lines. Jet element can be seen at the bottom of the left-hand (larger) pipe.* (c) *Cutaway section of the pump.* (d) *Pumping capacity gph under various conditions. Note the two pressure ranges, 20 to 50 psi and 30 to 60 psi. Jacuzzi Bros., Inc.* (e) *Jet-type (also known as venturi or ejector) deep-well pump and storage tank for a house or small building (for well lifts greater than 25 ft). Reduced pressure at* (f), *the jet nozzle, induces flow of groundwater into the circulated flow.*

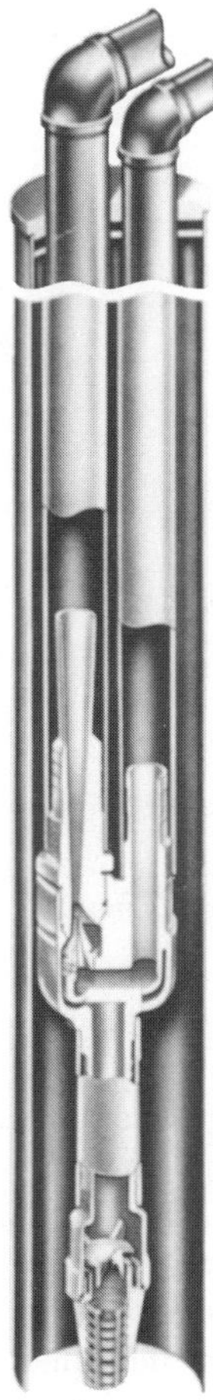

(b)

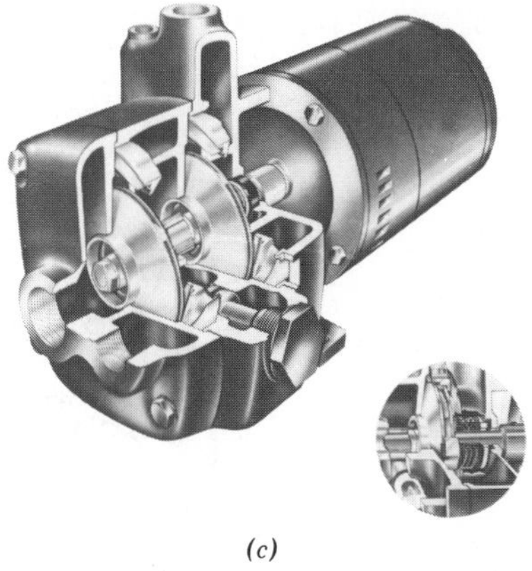

(c)

CHOOSE THE CORRECT JH FROM THESE CHARTS
DEEP WELL (Down to 120 feet)

The JH in the ½, ¾, 1 and 1½ horsepower rating, matched with the appropriate injector and pipe sizes, will pump this amount of water at the indicated lift or depth to the water in the well:

If suction lift or depth to water is	JH ½ HORSEPOWER	JH ¾ HORSEPOWER	JH 1 HORSEPOWER	JH 1½ HORSEPOWER
	Produces these gallons per hour between 20-50 lbs. discharge pressure		Produces these gallons per hour between 30-60 lbs. discharge pressure	
30 feet	795 G.P.H.	990 G.P.H.	1140 G.P.H.	1620 G.P.H.
40	680	875	1000	1470
50	575	735	875	1300
60	445	630	745	1200
70	360	495	620	920
80	310	385	530	820
90	255	315	435	700
100	220	275	340	590
110	195	240	295	540
120		205	250	480

(d)

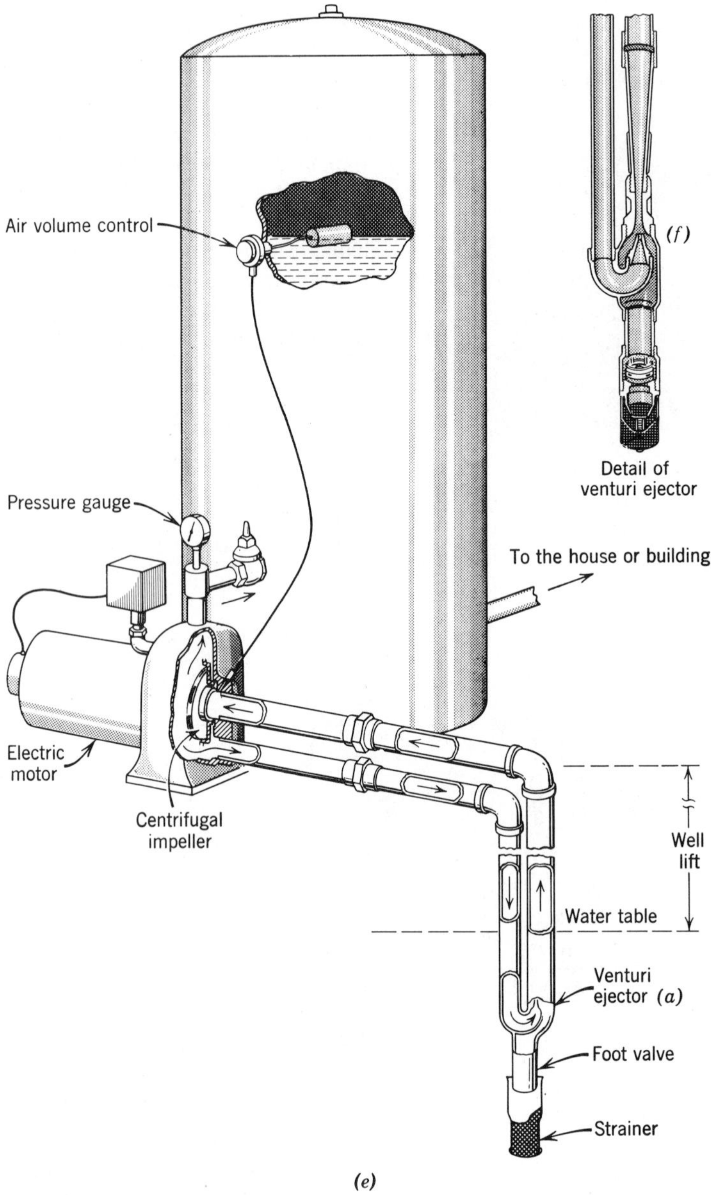

Fig. 9.8 (continued)

lower pressure within the venturi tube induces well water to flow in, and the velocity of the water from the nozzle pushes it upward toward the centrifugal pump, which can then more easily lift it by suction.

Pump Selection. Pump variations and characteristics are summarized in Table 9.4. The type of pump selected depends on many factors, including the rate of yield of the well, the daily flow (and maximum instantaneous flow rate) needed by the users, the size of storage or pressure tank used, and the total operating pressure against which the pump works (including the height to which water must be raised within the well). First cost, maintenance, and reliability are also factors, as is the energy used by the pump. In cold climates, the pump and the water supply system must be protected from freezing.

Of these factors, the two critical determinants are the flow rate (gallons per minute or per hour to be delivered) and the total pressure (or "head"). The flow rate depends on the number of fixtures to be served (Fig. 9.9), and the total pressure (Fig. 9.10) includes the suction lift, static head, and friction loss plus pressure head. This relationship will be explained in detail in Section 9.6.

Storage tanks are frequently used both to maintain a constant pressure on a pump-supplied water system and to allow for temporary peaks in water supply rates that exceed the capacity of the pump. Pressure tanks are most frequently used. Elevated tanks offer one alternative, cisterns another—although the latter usually are not located high enough to provide pressure to the supply system.

Pressure tanks are often housed in outbuildings, along with the pump and any water treatment equipment (Fig. 9.10). The temperature of the outbuilding must be kept above freezing, and its roof and walls should be removable, to allow

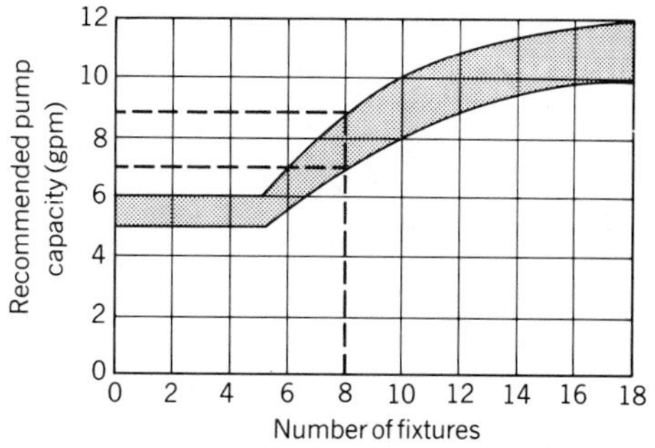

Fig. 9.9 *The relationship between the recommended flow-rate capacity of a pump and the number of domestic plumbing fixtures it supplies. See details in Section 9.6. (From the U.S. Environmental Protection Agency's* Manual of Individual Water Supply Systems, *1976.)*

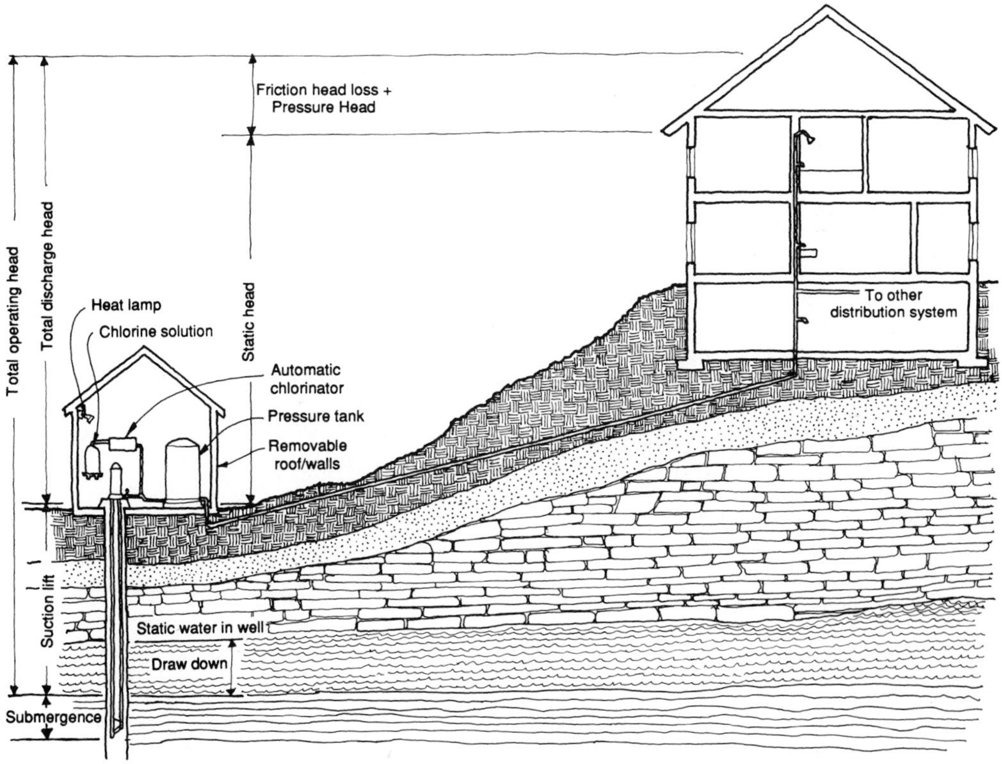

Fig. 9.10 *The components of the total operating pressure (or head), a critical determinant of pump size. (For details, see Section 9.6.) The pumphouse is usually a separate structure with water treatment and storage components. (From the U.S. Environmental Protection Agency's* Manual of Individual Water Supply Systems, *1976.)*

TABLE 9.4 **Pumps for Water Supply**

Type of Pump	*Practical Suction Lift*[a]	*Usual Well-Pumping Depth*	*Usual Pressure Heads*	*Advantages*	*Disadvantages*	*Remarks*
Positive Displacement						
1. Reciprocating						
(a) Shallow well	*22–25 ft*	*22–25 ft*	*100–200 ft*	1. Positive action. 2. Discharge against variable heads. 3. Pumps water containing sand and silt. 4. Especially adapted to low capacity and high lifts.	1. Pulsating discharge. 2. Subject to vibration and noise. 3. Maintenance cost may be high. 4. May cause destructive pressure if operated against closed valve.	1. Best suited for capacities of 5–25 gpm against moderate to high heads. 2. Adaptable to hand operation. 3. Can be installed in very small diameter wells (2″ casing). 4. Pump must be set directly over well (deep well only).
(b) Deep well	*22–25 ft*	*Up to 600 ft*	*Up to 600 ft above cylinder*			
2. Rotary						
(a) Shallow well (gear type)	*22 ft*	*22 ft*	*50–250 ft*	1. Positive action 2. Discharge constant under variable heads. 3. Efficient operation.	1. Subject to rapid wear if water contains sand or silt. 2. Wear of gears reduces efficiency.	
(b) Deep well (helical rotary type)	Usually submerged	50–500 ft	100–500 ft	1. Same as shallow well rotary. 2. Only one moving pump device in well.	1. Same as shallow well rotary except no gear wear.	1. A cutless rubber stator increases life of pump. Flexible drive coupling has been weak point in pump. Best adapted for low capacity and high heads.
Centrifugal						
1. Shallow well						
(a) Straight centrifugal (single stage)	*20 ft max.*	*10–20 ft*	*100–150 ft*	1. Smooth, even flow. 2. Pumps water containing sand and silt. 3. Pressure on system is even and free from shock. 4. Low-starting torque. 5. Usually reliable and good service life.	1. Loses prime easily. 2. Efficiency depends on operating under design heads and speed.	1. Very efficient pump for capacities above 60 gpm and heads up to about 150 ft.

(b) Regenerative vane turbine type (single impeller)	*28 ft max.*	*28 ft*	*100–200 ft*	1. Same as straight centrifugal except not suitable for pumping water containing sand or silt. 2. They are self-priming.	1. Same as straight centrifugal except maintains priming easily.	1. Reduction in pressure with increased capacity not as severe as straight centrifugal.
2. Deep well						
(a) Vertical line shaft turbine (multistage)	*Impellers submerged*	*50–300 ft*	*100–800 ft*	1. Same as shallow well turbine. 2. All electricai components are accessible, above ground.	1. Efficiency depends on operating under design head and speed. 2. Requires straight well large enough for turbine bowls and housing. 3. Lubrication and alignment of shaft critical. 4. Abrasion from sand.	
(b) Submersible turbine (multistage)	*Pump and motor submerged*	*50–400 ft*	*50–400 ft*	1. Same as shallow well turbine. 2. Easy to frost-proof installation. 3. Short pump shaft to motor. 4. Quiet operation. 5. Well straightness not critical.	1. Repair to motor or pump requires pulling from well. 2. Scaling of electrical equipment from water vapor critical. 3. Abrasion from sand.	1. 3500 RPM models, while popular because of smaller diameters or greater capacities, are more vulnerable to wear and failure from sand and other causes.
Jet						
1. Shallow well	*15–20 ft below ejector*	*Up to 15–20 ft below ejector*	*80–150 ft*	1. High capacity at low heads. 2. Simple in operation. 3. Does not have to be installed over the well. 4. No moving parts in the well.	1. Capacity reduces as lift increases. 2. Air in suction or return line will stop pumping.	
2. Deep well	*15–20 ft below ejector*	*25–120 ft 200 ft max.*	*80–150 ft*	1. Same as shallow well jet. 2. Well straightness not critical.	1. Same as shallow well jet. 2. Lower efficiency, especially at greater lifts.	1. The amount of water returned to ejector increases with increased lift—50% of total water pumped at 50-ft lift and 75% at 100-ft lift.

Source: U.S. Environmental Protection Agency, *Manual of Individual Water Supply Systems,* 1975.

[a]Practical suction lift at sea level. Reduce lift 1 ft for each 1000 ft above sea level.

for replacement of parts over time. One type of pressure-storage tank is shown in Fig. 9.11.

The capacity of pressure tanks usually is small in comparison to the daily *total* water consumption; they provide short-term responses to peak flow demands. As a general rule, the pressure tank should be sized to deliver about 10 times the pump's capacity in gpm (L/m). For a typical residence, allow 10–15 gal (38–57 L) tank capacity per person served.

For larger installations, the size of a pressure storage tank can be calculated by

$$Q = \frac{Qm}{1 - \dfrac{P_1}{P_2}}$$

where

Q = tank volume (gal)

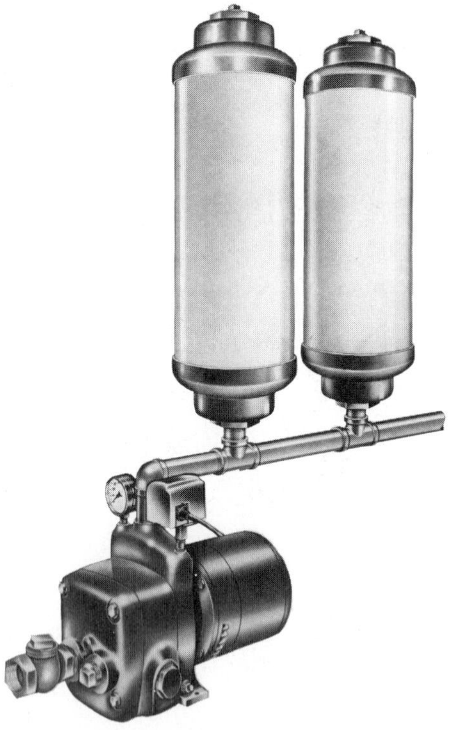

Fig. 9.11 Small pressure-storage tanks are installed primarily to keep a water supply system at constant pressure. These Hydrocel models are small (8½ in. in diameter, 27 in. long), so they can be installed almost anywhere along the supply system. (Courtesy of Jacuzzi Bros., Inc.)

Qm = 15 minutes of storage at peak usage rate (gal)

P_1 = minimum allowable operating pressure (psi) plus atmospheric pressure (14.7 psi)

P_2 = maximum allowable operating pressure (psi) plus atmospheric pressure (14.7 psi)

The ranges of allowable pressures are discussed in Section 9.6. (An alternative tank-sizing procedure is shown in Table 9.5.)

EXAMPLE 9.1. An office building in a remote location has a water supply system served by a pump and well. The peak demand is 50 gpm. The fixtures will operate at a minimum of 50 psi; a maximum of 70 psi should not be exceeded. Therefore,

$$Qm = 15 \text{ min} \times 50 \text{ gpm} = 750 \text{ gal}$$

and

$$Q = \frac{Qm}{1 - \dfrac{P_1}{P_2}} = \frac{750}{1 - \dfrac{50 + 14.7}{70 + 14.7}}$$

$$= \frac{750}{1 - 0.76} = 3125 \text{ gal}$$

The capacity of elevated tanks usually is equal to at least two days of average water usage. For firefighting or other special requirements, the capacity may have to be even greater.

A summary of the quality, treatment, and supply issues raised thus far is provided in the example of a rural estate shown in Fig. 9.12 and Table 9.5. On this large estate, water was required for an estimated demand of 30,000 gpd. This was for domestic use only, the irrigation needs being solved by a separate installation pumping from a lake. Wells were dug for the domestic supply. It was quickly evident that despite the great depths of the wells drilled, the available flow would be small. Four wells yielded a total rate of only 25 gpm.

Calculations for the amount of water to be stored to supplement this meager supply are shown in Table 9.5. The table shows that during the 14 daytime hours under conditions of peak demand, the well pumps would run continuously. Concurrently, an additional 9000 gal would be drawn from the tanks. At night, the

TABLE 9.5 **Calculations to Establish Capacity of Water Storage Tanks at Kinloch (Fig. 9.12)**

1. Potable domestic water demand per day	30,000 gpd
2. Assumed hours of use	7 A.M. to 9 P.M. 14 hours
3. Assumed hours of virtual nonuse	9 P.M. to 7 A.M. 10 hours
4. Well-yield rate per hour	25 gpm × 60 min = 1500 gph
5. Total well yield in 14 hours	15000 gph × 14 hr = 21,000 gal
6. Total water needed in 14 hours	30,000 gal
7. Well yield in 14 hours	21,000 gal (see 5 above)
8. Net-tank capacity required, minimum	30,000 − 21,000 = 9000 gal
9. Net-tank capacity as designed	
20,000 × 0.80 (80% full) = 16,000 (2 tanks at 10,000 gal each)	
2000 × 0.70 (70% full) = 1,400 (1 tank at 2,000 gal)	
17,400 gal, O.K., greater than 9000	
10. Well operation at night to restore 9000 gal to tanks	9000 ÷ 1500 = 6 hours of the 10 night hours available.

well pumps would run for 6 h to restore the 9000 gal drawn from the tanks during the day.

The operation of the system is illustrated in the diagram and notes of Fig. 9.12*a*. The pressure of 75 psi in the hydropneumatic tank is sufficient to raise the water to the greatest height in the distribution system, overcome friction in the piping, and leave a residual pressure available at each fixture of about 10 to 15 psi, as needed. Excessive pressure can always be moderated by a valve in the branch supply of the fixture.

Pressure in this tank is assured by the operation of the air compressor actuated by the pressure switch. The float switch starts the centrifugal pump, to deliver water from the storage tanks as needed. The storage tanks, piped together and acting as a single reservoir, are fed by the four wells. The wells operate in unison, singly, or in groups, depending on the level in the storage tanks. If the level drops rapidly, all well pumps can run. Although they are arranged to operate all day when needed, in periods of minimum demand only one or two may be called upon. Each has its own power supply, but controls emanate from panel *N*.

Swimming pools should be supplied with pure, potable water. Biological tests showed that this well water was safe. Thus, the fill-line to the swimming pool is connected from *this* control center rather than from the supply of lake water used for irrigation. Since swimming pool water is separately recirculated and provided with its own purification treatment at a location adjacent to the pool, this supply line for make-up water to the pool would have infrequent and off-hour use. Water for domestic use in the buildings is passed through one of three ion-exchangers to provide softening and to make the water more suitable for washing, bathing, and cooking. Periodically, the calcium precipitate can be flushed out and the tanks regenerated by sodium chloride.

9.4 Hot Water Systems and Equipment

There are many ways to provide the "domestic" or "service" hot water needs within a building—the hot water used *not* for space heating, but for bathing, clothes washing, dishwashing and other related functions. Whereas much of the world either heats such water on a cookstove or does without it, North Americans enjoy an array of choices for DHW (domestic hot water) supply systems. In this section, rough supply estimates are considered first, then basic design choices, conventional heater tanks, solar water heaters, and, finally, energy recovery heaters.

(a) Estimating the Demand. One of the more familiar applications for DHW is in the

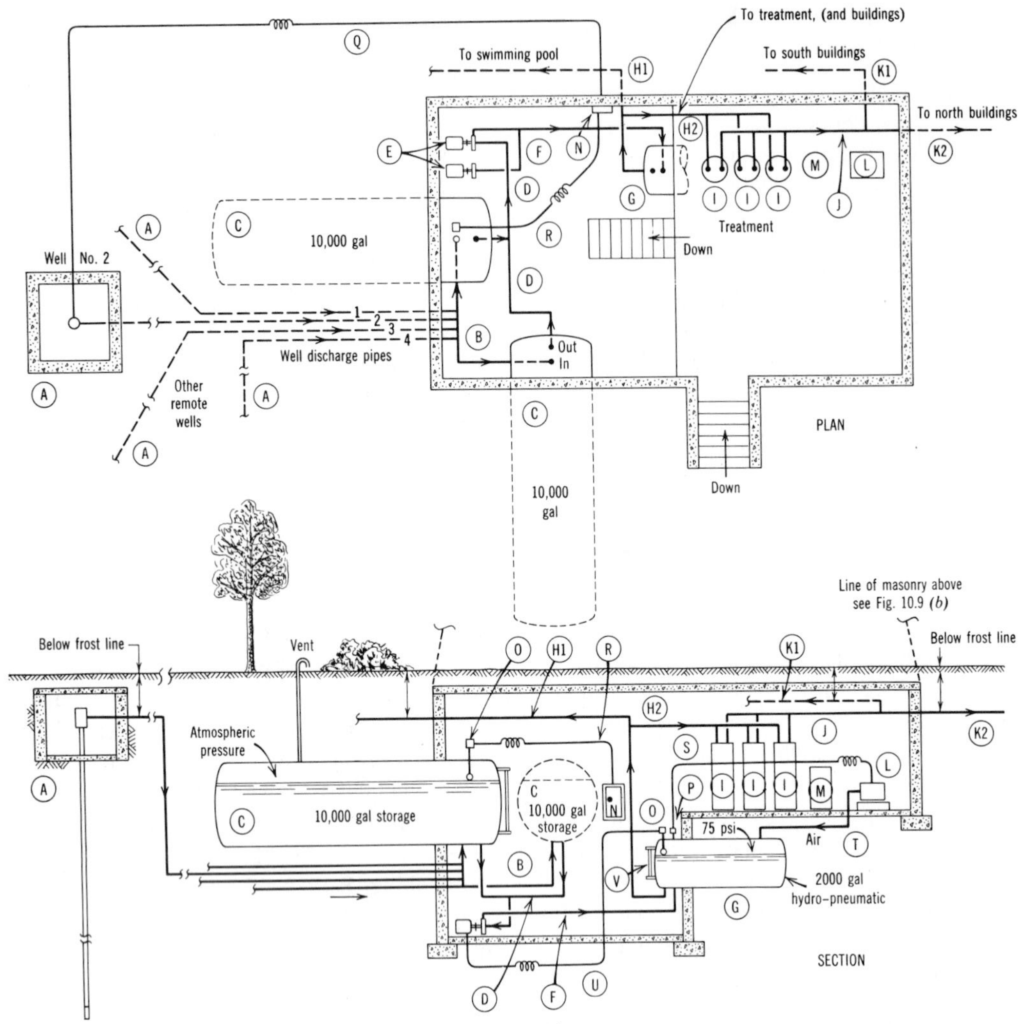

(a)

Legend

(A)	Wells
(B)	Well water header for 4 wells
(C)	Well water storage, vented tanks, (atmospheric pressure)
(D)	Pump supply pipe
(E)	Electric centrifugal pumps (one is a standby)
(F)	Pump discharge to hydro-pneumatic tank
(G)	Hydro-pneumatic tank, 75 lbs. per sq. in.
(H1) (H2)	Discharge pipes, pressurized well water
(I)	Water softeners, (ion exchangers)
(J)	Treated well water under pressure
(K1) (K2)	Branch mains to buildings (treated well water)
(L)	Air compressor maintains 75 psi in tank
(M)	Sodium chloride for regeneration
(N)	Electrical control panel
(O)	Float switches
(P)	Pressure switch (actuates air compressor)
(Q)	Control wiring to well pumps
(R)	Control wiring, float switch to panel
(S)	Control wiring, pressure switch to compressor
(T)	Compressed air piping
(U)	Control wiring, float switch to centrifugal pump
(V)	Gauge glass indicates water level

This schematic diagram shows system components and connections. For clarity, details (valves, drains, check-valves etc.) are not shown. All control wiring, here simply indicated, operates through control panel (N). See text for operation.

Fig. 9.12 *(continued)* (b) *One of two pavilions adjacent to the swimming pool that serves residents and guests at Kinloch. Below this pavilion is a separate story, with its own downhill access, that houses the control center. Exterior tubing to and from the center is all below grade.*

home. In the United States, it is often assumed that in a two-person family, each person will use 20 gal (76 L) of hot water per day, and that each additional family member will use another 15 gal (57 L) daily. Thus a family of four would require 70 gal (266 L) of hot water daily. To gain a better understanding of this quantity, see Table 9.6.

Estimation of the demand for commercial and institutional buildings is not so simple, as rules of thumb are less reliable. For larger buildings, there is a trade-off between quick recovery (high heating capacity) and storage size; big tanks have small heaters, and vice versa. The details of both residential and commercial water heater sizing will be explored in Section 9.4*c*.

An important variable is the temperature at which hot water is to be provided. Table 9.7 shows that although water becomes uncomfortably hot to the touch above 110 F (43.3°C), much higher temperatures are often used for some commercial processes. When determining the temperature at which DHW is to be supplied, consider the following factors.

Fig. 9.12 (a) *Water control and distribution center for the Kinloch Estate. (Bentel & Bentel, Architects, FAIA.) The plant is located in a hillside and forms a story below the bathing and dressing pavilion.*

High temperatures:

- Allow the installation of smaller storage tanks (less hot water is mixed with cold to achieve final usage temperature at the shower, sink, or lavatory)—but require larger heating units.
- Can be achieved at the point of use by heaters built into equipment such as dishwashers.
- Can cause scale to form on heating coils and within piping [above 140 F (60°C) in areas of ''hard'' water quality].
- May be required by code for some applications.
- Limit the potential for growth of *Legionella pneumophila* bacteria

Lower temperatures:

- Are less likely to cause burns, but may not achieve desired sanitation.
- Mean less energy consumed, because storage and pipe heat losses are lower.
- Allow the installation of smaller heating units—but require larger storage tanks.
- Make possible the use of lower-grade heat sources for DHW, such as solar energy or waste heat recovery.

Figure 1.12 showed that 4% of annual U.S. energy usage is for DHW, which requires the lowest-grade heat source of all the common usages.

(b) Basic System Choices. It is as difficult to categorize the array of available DHW systems as it was to categorize the many HVAC systems in Chapters 5 and 6. The most important variables are heat source, method of heat transfer to the water, central versus distributed heater/storage equipment, and related distribution trees.

Heat Sources. These include the familiar concentrated-energy, high-grade sources such as natural gas and electricity. Oil- and coal-fired boilers are frequently equipped with DHW coils as well. Buildings served by steam use

TABLE 9.6 **Domestic Hot Water Consumption—Residences**

	Gallons (Liters) per Use Hot Water Required	
	14-lb (6.4-kg) Machine	*18-lb (8.2-kg) Machine*
Clothes Washing Machine		
Hot wash/hot rinse	38 gal (144 L)	48 gal (182 L)
Hot wash/warm rinse	28 gal (106 L)	36 gal (136 L)
Hot wash/cold rinse	19 gal (72 L)	24 gal (91 L)
Warm wash/cold rinse	10 gal (38 L)	12 gal (45 L)
Dishwashing	*Small*	*Large*
Dishwashing machine	10 gal (38 L)	15 gal (57 L)
Sink washing	4–8 gal (15–30 L)	
Personal Hygiene		
Tub bathing	12–30 gal (45–134 L)	
Wet shaving/hair washing	2–4 gal (8–15 L)	
Showering	2–6 gpm (13–38 L/s)	

Source: Reprinted by permission from Russell Plante, *Solar Domestic Hot Water,* copyright © 1983 by John Wiley & Sons.

steam as a DHW heat source. Cogeneration provides heat for DHW, as in Fig. 7.70. Because of the relatively low-grade final temperatures needed, DHW can also be readily provided by wood-burning equipment, incinerators, solar energy equipment, heat pumps, and heat recovery devices (as in commercial ice-making machines whose discharged heat is contributed to DHW).

Heating Methods. There are two basic methods:

1. *Direct* heating brings water against directly heated surfaces: electric-resistance elements within tanks, or other electrically warmed surfaces, or surfaces directly exposed to fire or hot gases.

TABLE 9.7 **Temperatures at Point of Use, DHW**

Use	*Temp. F*	*(°C)*
Lavatory		
Hand washing	105	(40.6)
Shaving	115	(46.1)
Showers and tubs	110	(43.3)
Therapeutic baths	95	(35.0)
Commercial and institutional dishwashing		
Wash	140	(60.0)
Sanitizing rinse	180	(82.2)
Commercial and institutional laundry	180	(82.2)
Residential dishwashing and laundry	140	(60.0)
Surgical scrubbing	110	(43.3)

Source: Copyright © by the American Society of Heating, Refrigerating and Air Conditioning Engineers, Inc., Atlanta, GA. Reprinted by permission from the *ASHRAE Systems Handbook,* 1984.

2. *Indirect* heating can be accomplished in several ways. Coils containing steam or fluids can be submerged within water tanks or (Fig. 9.13) set within boilers, whose primary function usually is to provide space or industrial process heating. Alternatively, the coils containing DHW can be placed outside the boiler but within a casing containing steam, hot exhaust gases, or very hot water.

Both direct or indirect methods can be utilized in a variety of equipment:

Storage tank water heaters, the type most commonly used for residential and small commercial purposes (see Section 9.4*c*).

Circulating storage water heaters, in which the water is first heated by a coil, then circulated through the storage tank (as in some solar heaters—see Section 9.4*d*).

Tankless heaters, in which the water is very quickly raised to a desired temperature within a heating coil and immediately sent to the point of usage.

Tankless water heaters are available in larger sizes for central hot water systems (Fig. 9.13) or in smaller sizes for remote plumbing fixtures that occasionally need hot water or for isolated bathrooms, laundry rooms, and so forth. Decentralized tankless units can be as small as "instant hot water taps" for kitchen or bar sinks—electric-resistance heaters capable of generating perhaps ½ gph at up to 200 F (2 L/h at up to 93°C). Bathroom groups can require either electric resistance or gas-fired units (Fig. 9.14) producing up to 5 gpm (0.3 L/s) with a temperature rise of as much as 40F° (22C°). Because tankless heaters often consist of fairly long coils through which the water passes as it is heated, they may add considerable friction to the total for which the supply system is designed (see Section 9.7).

Central versus Distributed Equipment. This choice can be particularly complex. In the United States a central water-heater storage tank is standard equipment in homes and small stores. But the growing use of solar water heating, along with awareness of the heat losses from water-heater storage tanks, has created new interest in *combinations* of central and distributed DHW.

Consider a large residential DHW application such as that shown in Fig. 9.15. Here, two areas in which hot water is used are separated by some 50 ft. If the simple, centralized water-heater storage tank is used (Fig. 9.15*a*), it would probably be placed nearest the maximum-use fixtures—dishwashers and clothes washers. (Also, floor space for the water-heater storage tank is usually more plentiful there.) However, each time hot water is needed in the remote bathroom serving the bedrooms, the previously heated water in at least 50 ft of supply pipe—which, despite insulation, will almost certainly have cooled—must be drained off before hot water finally arrives. (A ¾-in. diameter pipe, for example, will hold 4.6 gal in 50 ft of pipe. If the water heater is set at an energy-conserving level of 120 F, and the incoming city water is at 50 F, the 4.6 gal of wasted water will also waste the 2682 Btu invested when it was heated.)

Another way to conserve water is with a recirculating hot water system (Fig. 9.15*b*). The primary disadvantages of this system are the increased heat loss through the hot water pipe (now kept at 120 F for 24 hours per day, rather than for just a few minutes) and the energy required to run the circulating pump (again for 24 hours daily).

A decentralizing option (Fig. 9.15*c*) is to install two water-heater storage tanks, one for each group. First cost probably will be greater, and more floor space required. The daily waste heat from *two* heaters is clearly a disadvantage that will at least partially offset the energy saved by eliminating the 50-ft run of pipe. (Further, in warm weather this waste heat will add to discomfort within the house.) A decentralized-centralized mix (Fig. 9.15*d*) will save energy, water, and floor space, but it will almost certainly be costlier to install. A central solar water heater—either passive (shown) or active—brings the water up to warm (winter) or very hot (summer). At each fixture group, a tankless heating coil instantaneously heats the water, as needed. Almost no water is wasted, and the energy lost in the 50 ft of pipe will be solar energy, not purchased energy from natural gas or electricity.

Another option would be to omit the solar water heater and simply install two tankless

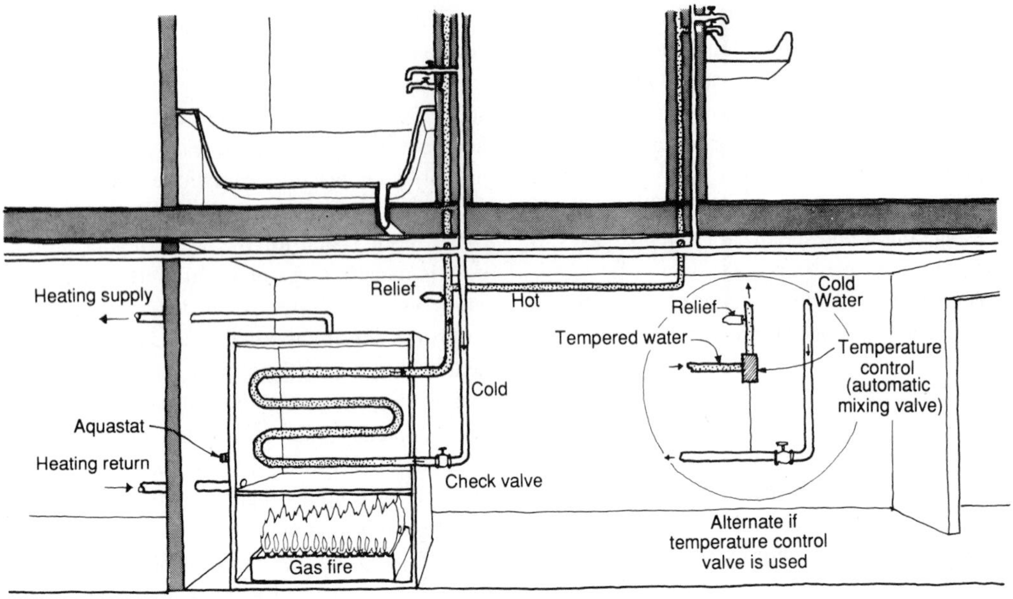

(a)

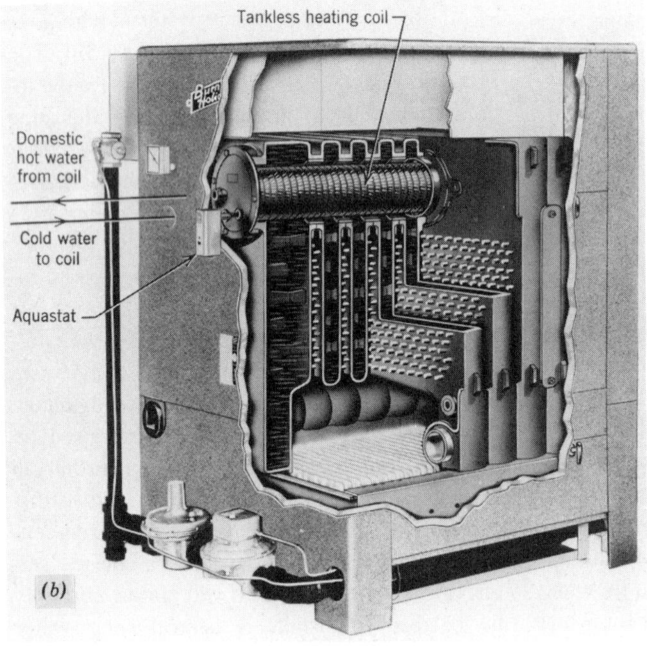

(b)

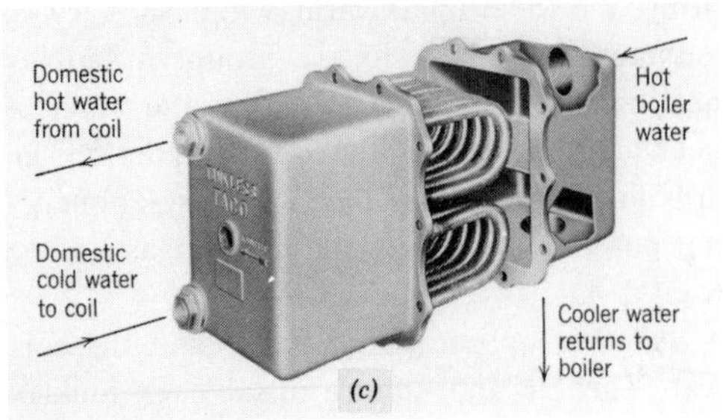

(c)

Fig. 9.13 (a) *Example of indirect, tankless heater for DHW, utilizing a boiler that provides hot water for space heating. Since no storage is used, the point of use for DHW should be very close to the boiler.* (b) *Internal tankless heating coil for domestic hot water immersed in the jacket water of a gas-fired hot water heating boiler. Approximate capacity range 3 to 15 gpm at 100F° rise.* (c) *External-type, tankless heater for domestic hot water. Boiler water is piped to the unit and circulates by gravity, transferring heat to the coil. Approximate capacity range 3 to 15 gpm at 100F° rise.* Note: *Because it can be highly inefficient to provide summertime DHW with a boiler designed for wintertime space-heating loads, codes may alter or prohibit this arrangement.*

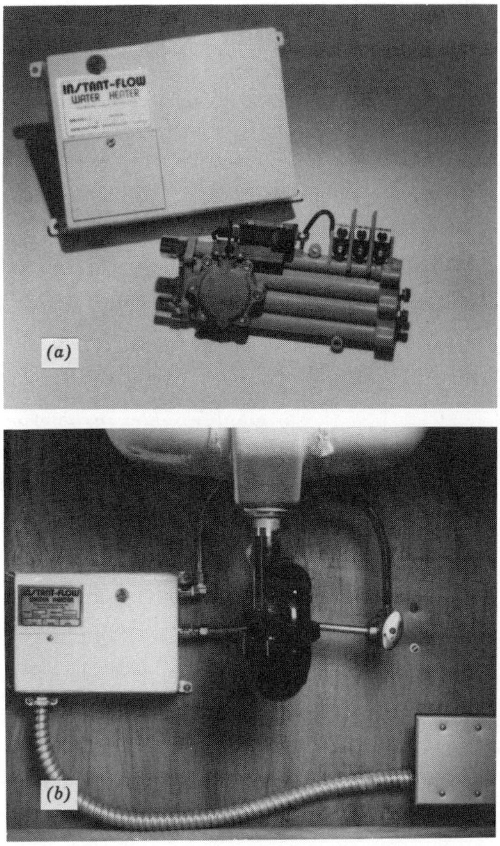

Fig. 9.14 *Instantaneous or tankless water heater.* (a) *In this "instant-flow" water heater, a series of coils heats water as it flows through.* (b) *Installation below a typical lavatory. A wide range of sizes is available; the 4.6 kW unit raises water by 31F° at 1 gpm; the 9 kW unit raises water by 61F° at 1 gpm. (Courtesy of Chronomite Laboratories, Inc.)*

heaters. The primary disadvantage of this setup would be the need for larger-capacity heaters, whose instantaneous demand for energy could lead to electric cost penalties in areas where peak-load rates are high.

Distribution Trees. The choice of a distribution tree associated with central water heaters must take into account the gradual cooling of water within the distribution system, once it has left the central heater. If the heat losses associated with constantly circulating hot water (looped trees) are preferable to the heat and water wastage associated with simple, single hot water distribution trees, such loop systems can be achieved in either of two ways.

Thermosiphon hot water circulation depends on the fact that water expands and becomes lighter when heated, as may be seen in Fig. 9.16*a*. If heat is applied to the lower loop of a glass tube, both ends of which have been inserted in an inverted bottle containing water, the water will move from *A* to *B* and will rise through the tube *BC* into the bottle. It here becomes cooled and drops through the tube *DA* to *A*, is again heated, and rises in the tube *BC*—thus completing the circulation. Since the movement depends on the difference in weight between the two columns of water, the velocity and consequent efficiency of the circulating system increase with both the temperature of the water and the height of the circuit. Hot water supply systems therefore usually consist of a heater with a storage tank, piping to carry the heated water to the farthest fixture, and a continuation of this piping to return the unused cooled water back to the heater. A constant circulation is thereby maintained, and hot water may be drawn at once from a fixture without first draining off through the faucet the cooled water that would be standing in the supply pipe if there were no recirculation. Because heat increases the corrosive action of water on metals, copper tubing is often chosen for use in hot water and hot water circulating systems. Thermosiphon circulating systems are particularly effective in multistory buildings, because of the beneficial effect of increased circuit height on circulation efficiency.

Forced circulation of hot water is often utilized where a height advantage is unavailable. Low, long, rambling buildings, such as some large one-story residences, schools, and factories, lack the height needed to set up good hot water circulation by gravity. Also, the flow is diminished by friction in long pipe runs. For such buildings the forced-circulation scheme shown in Fig. 9.17 offers one option. Three independent aquastats—devices that create an electrical signal impulse when water temperature drops—control this system. Aquastats *A, B,* and *C*, respectively, sense the temperatures of the water in the heater, the tank, and the end of the circulation-return main. As needed, they turn on

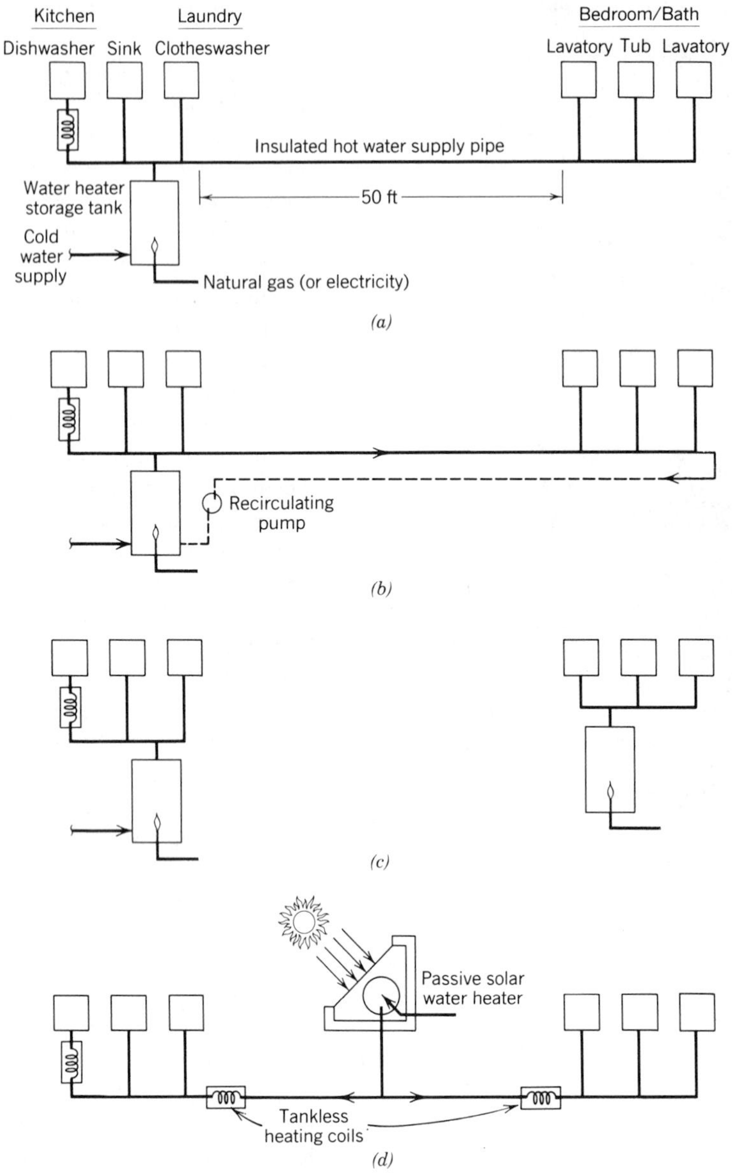

***Fig. 9.15** Options for DHW in a larger residence.* (a) *Typical centralized water-heater storage tank; water and energy is wasted in the 50-ft run.* (b) *Recirculating pump added; this saves water but requires continuous energy to operate pump and make up the pipe's heat losses.* (c) *Decentralized approach using two water-heater storage tanks; this roughly doubles the energy lost from storage but eliminates water and energy loss from the 50-ft run.* (d) *Central solar water heater/decentralized tankless heating coil combination; saves on everything except initial cost.*

the oil or gas burner, the tank-circulating pump, and the system-circulating pump. Fixtures remote from the tank are as close to hot water as the length of their hot water runout pipes. Water is usually available at full temperature in 5 to 10 seconds. Trial aquastat settings in °F could be *A* 180, *B* 160, *C* 120 (°C: *A* 82, *B* 71, *C* 49).

An energy-saving computer control is available for hotels, motels, apartments, and larger commercial buildings. This device (Fig. 9.18) *varies the supply temperatures* of hot water, so that the hottest water is supplied at the busiest hours. In hours of low usage, much lower supply temperatures mean that more "hot" water will be mixed with less "cold" water at showers, lavatories, and sinks. This increased "hot" water quantity poses no problem at off-peak hours, and the lower temperature means significant decreases in heat loss from the recirculating supply water system. A typical payback period is one year. The device stores a memory (adjusted weekly) of the typical daily patterns of usage, and varies the supply temperature accordingly.

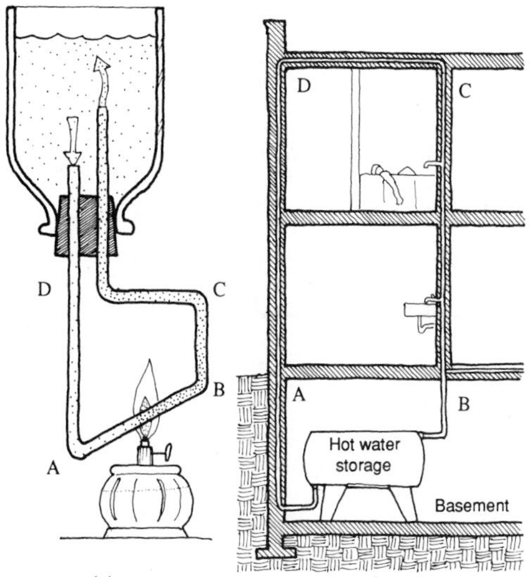

Fig. 9.16 (a) *Principle of hot water circulation by gravity (thermosiphon).* (b) *Its application to hot water service. During periods of no demand there is sufficient incidental cooling between* C *and* D *so that dense (less warm) water at* A *forces the lighter (hot) water at* B *to rise for speedy availability at each faucet.*

(c) Conventional Heat Sources. One of the most common appliances used in the United States is the water-heater storage tank, an energy-conserving model of which is shown in Fig. 9.19.

The National Appliance Energy Conservation Act (1987) specifies minimum performance standards for domestic (or service) water heating equipment; similar standards for smaller heating/cooling equipment were shown in Table 6.17; for larger heating/cooling equipment, in Table 7.3. The water heater equipment standards that take effect Jan. 1, 1992, are shown in Table 9.8.

For residences, the capacity of hot water heating/storage equipment can be taken from the values in Table 9.9. Note that for the familiar tank-type direct water heaters (gas electric, or oil), there is a stated relationship among storage size, rate of hourly heat input, rate of draw (demand) over one hour, and recovery rate. Example 9.2 uses Table 9.9 for the sizing of a residential water heater.

EXAMPLE 9.2. Select a natural gas water heater for a four-bedroom house with 2½ baths (two full baths and one powder room). The *minimum* requirements of HUD-FHA are shown in Table 9.10, column *A*. From Fig. 9.19, we select two trial units (40 gal and 50 gal) that *might* comply; see columns *B* and *C*.

SOLUTION. PCGS-50 is selected. (However, PGCS-40 is acceptable in all respects except that of Btu input.)

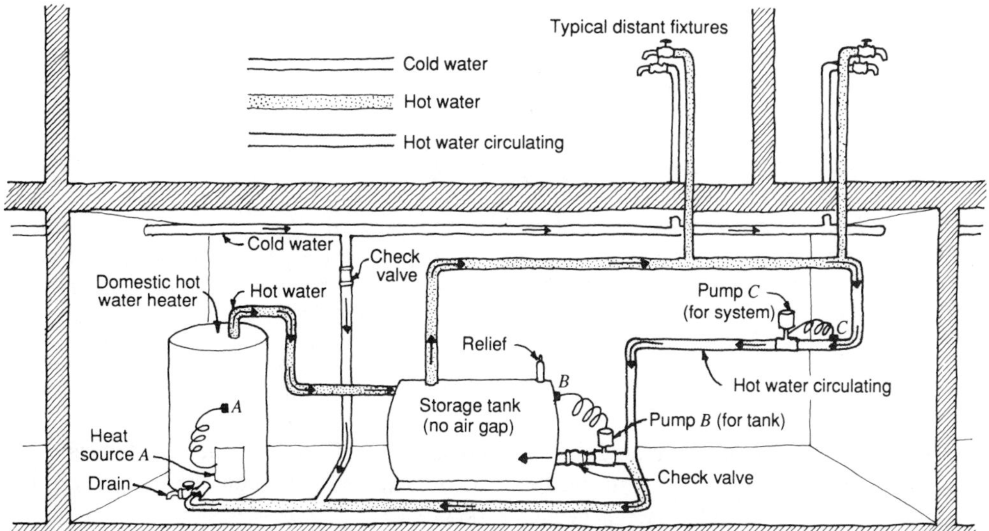

Fig. 9.17 Forced circulation of DHW, for a long, low building. (Note that the use of space-heating boilers for DHW may cause system inefficiencies in warm weather.)

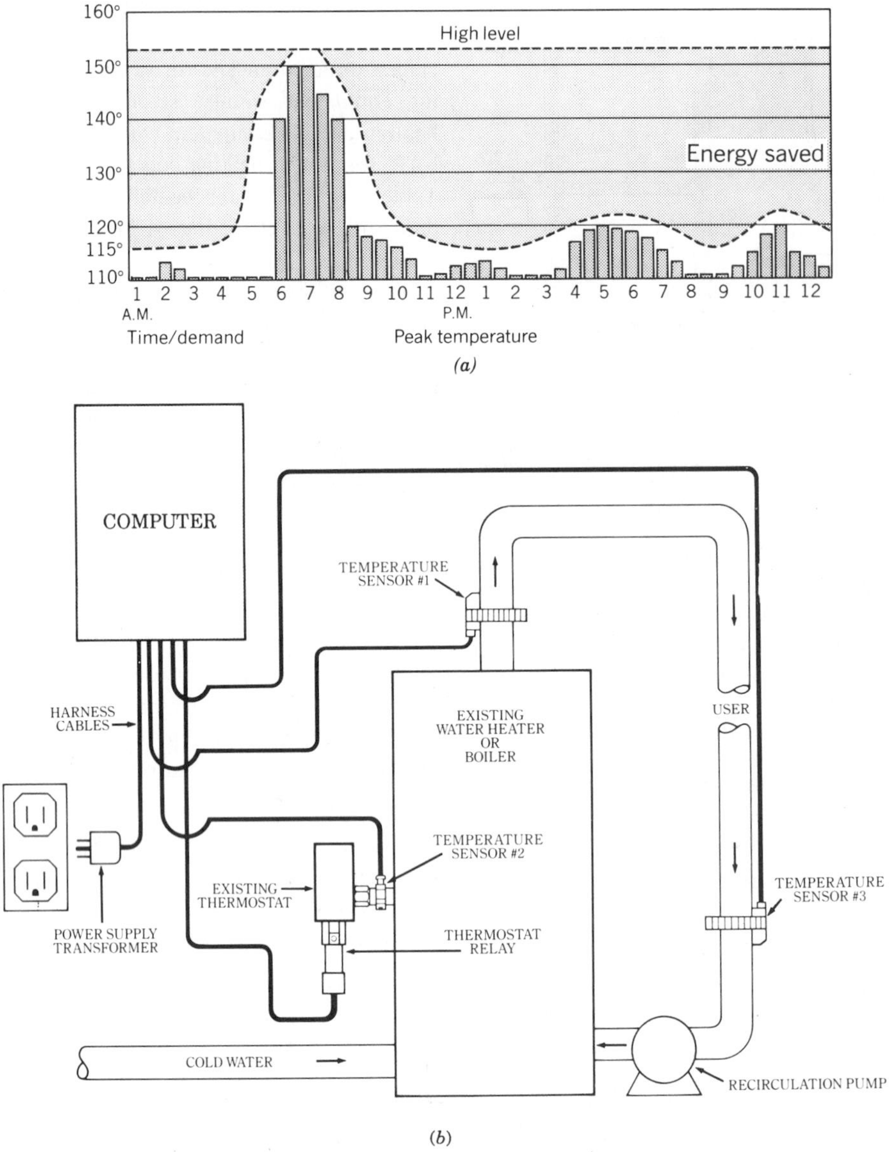

Fig. 9.18 *Variable temperature supply.* (a) *Energy savings possible when the supply temperature of hot water is varied. Heat losses from supply pipes are greatly reduced for most of the typical day. A computer controls the supply temperature.* (b) *Schematic of computer controlled variable temperature DHW system. (Courtesy of Fluidmaster, Inc., Anaheim, California.)*

Small shops and office buildings can be treated like residences for hot water sizing. For larger buildings, the trade-off between heaters and storage tanks should be explored. To begin this process, consult Table 9.11, which shows the maximum hourly and daily demands and contrasts them to the average day (which should be used to estimate monthly energy consumption).

The relationship between heater size and tank size is graphed in Fig. 9.20. The advantage of a larger heater is its smaller tank, which con-

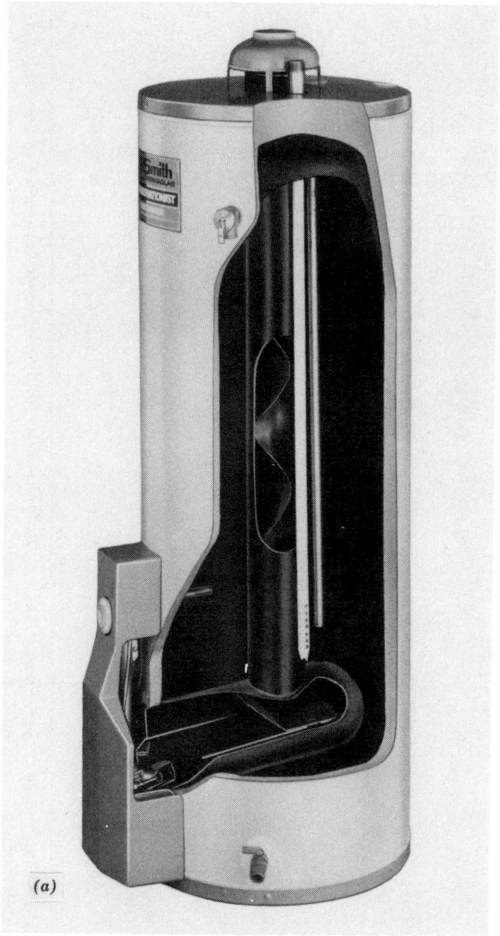

(a)

		All Dimensions in Inches				Natural Gas		
Model No.	Gal. Cap.	Vent Size	Height Plus Diverter	Height Less Diverter	Diameter	BTU Input Per Hr.	Gal./Hr.* Recovery 90° Rise	Approx. Ship Wt. (Lbs.)
PGCS-40	40	3	54	50 7/16	18 1/4	32,000	34.9	147
PGCS-50	50	3	57 1/4	53 11/16	20 3/16	40,000	43.1	156
PGCS-65	65	4	59	55 1/2	21 3/4	50,000	53.9	192

*Recovery capacities are based on DOE Method of Test (90° Rise).

Note: To compensate for the effects of high altitude areas above 2,000 feet, recovery capacity should be reduced approximately 4% for each 1,000 feet above sea level.

(b)

Fig. 9.19 (a) *Water-heater storage tanks, with such energy-conserving features as a gradually restricted flue outlet to maximize thermal efficiency and a combustion chamber that prevents heat from escaping directly up the flue (thus preventing continuing heat loss from the stack effect).* (b) *Numerical characteristics for such water heater storage tanks. (Courtesy of the A.O. Smith Corporation, Camden, South Carolina.)*

sumes less floor space and volume, weighs less, and probably has a lower first cost. The advantage of a large tank is that its smaller heater tends to work steadily rather than in spurts; this lower, steadier demand for energy will lead to lower utility rates in the case of electric heating, and thus to money savings over the life of the system. When the fuel supply is solar energy or waste heat, the larger tank again is usually better suited to the characteristics of the fuel supply.

TABLE 9.8 **Minimum Performance of Water Heating Equipment**[a]

Class	Fuel	Storage Capacity (gal)	Input Rating	Energy Factor,[b] EF	Minimum Performance: Efficiency	Minimum Performance: Heat Loss
Storage water heaters	Electric	≤120	≤12 kW	≥0.95 − 0.00132 V	—	—
		>120[c]	(or) >12 kW	—	—	SL <1.9 W/ft[2d]
	Gas	≤100	≤75,000 Btu/h	≥0.62 − 0.0019 V	—	—
		>100[c]	(or) >75,000 Btu/h	—	E_t 77%[e]	SL <1.3 + 38/V[f]
	Oil	≤50	≤75,000 Btu/h	EF ≥0.59 − 0.0019 V	—	—
		≤50	(or) ≤105,000 Btu/h	EF >0.59 − 0.0019V	—	—
		>50[c]	(or) >105,000 Btu/h	—	E_c 83%[g]	SL <1.3 + 38/V[f]
Unfired storage tanks	—	All Volumes	—	—	—	HL <6.5 Btu/h-ft[2h]
Nonstorage water heaters	Gas	—	All input ratings	—	E_t 80%[e]	—
	Distil oil	—	All input ratings	—	E_c 83%[g]	—
Pool heaters	Gas/oil	—	All input ratings	—	E_t 78%[e]	—

Source: Adapted with permission from ASHRAE/IES Standard 90.1-1989, *Energy Efficient Design of New Buildings Except Low-Rise Residential Buildings,* © 1989 by the American Society of Heating, Refrigerating, and Air-Conditioning Engineers, Inc., Atlanta, Ga.

[a] Minimum performance as of Jan. 1, 1992.

[b] EF, Energy factor, overall heater efficiency by DOE test procedure; as of 1990; V, storage volume in gallons.

[c] Storage water heaters and hot water storage tanks of more than 500 gallons storage capacity need not meet the SL or HL requirements above if the tank surface area is insulated to *R*-12.5 and if a standing pilot light is not used.

[d] SL, standby loss in W/ft² for electric water heaters, based on 80F° difference between water and air.

[e] E_t, thermal efficiency with 70F° difference between entering water and storage water.

[f] SL, standby loss in % per hour for electric fuel-fired heaters, based on 90F° difference between water and air; V, storage volume in gallons.

[g] E_c, combustion efficiency, 100% minus the flue loss when smoke = 0 (trace is permitted).

[h] HL, heat loss of tank surface area, Btu/h ft², based on 80F° difference between water and air.

[i] A nonstorage water heater has an input rating at least 4000 Btu/h per gallon of stored water, and a storage capacity of less than 10 gallons.

EXAMPLE 9.3. (Adapted from the *ASHRAE Systems Handbook,* 1984). A women's dormitory housing 300 students, with a cafeteria serving 300 meals in 1 h, is to be built. Find the required storage size for two conditions: (*a*) assuming a minimum recovery rate for both dorm and cafeteria; and (*b*) assuming a dorm recovery rate of 2.5 gph (2.6 mL/s), which is half the maximum hourly value given in Table 9.11, and a cafeteria recovery rate of 1.0 gph (1.1 mL/s), which is two-thirds the maximum hourly value given in Table 9.11.

SOLUTION. *Minimum Recovery:* From Fig. 9.20*a*, the minimum recovery rate for women's dormitories is 1.1 gph. For 300 students,

$$300 \times 1.1 = 330 \text{ gph recovery}$$

At this rate, again from Fig. 9.20*a*, the minimum usable storage capacity is 12 gal/student. Assume that 70% of the total size is this "usable" capacity. This means that after 70% of the stored hot water is withdrawn, the remaining water has been cooled (by incoming water) to an unusably low temperature. Storage size must be increased by

$$\frac{100\%}{70\%} = 1.43$$

Thus,

$$12 \times 300 \times 1.43 = 5150 \text{ gal storage}$$

From Fig. 9.20*e*, the minimum recovery rate for the cafeteria (serving full meals, type *A*) is 0.45 gph. For 300 meals,

$$300 \times 0.45 = 135 \text{ gph recovery}$$

At this rate, the minimum usable storage capacity is 7 gal/meal. Thus

$$300 \times 7 \times 1.43 = 3000 \text{ gal storage}$$

Combining these requirements for dorm and cafeteria,

$$\text{recovery} = 330 + 135 = 465 \text{ gph}$$

$$\text{storage} = 5150 + 3000 = 8150 \text{ gal}$$

Faster Recovery: At the specified dorm recovery rate of 2.5 gph,

$$300 \times 2.5 = 750 \text{ gph}$$

and the minimum usable storage required is 5 gal/student. Thus

$$300 \times 5 \times 1.43 = 2150 \text{ gal}$$

At the specified cafeteria recovery rate of 1.0 gph,

$$300 \times 1.0 = 300 \text{ gph}$$

and the minimum usable storage required is 2 gal/meal. Thus

$$300 \times 2.0 \times 1.43 = 860 \text{ gal}$$

Combining these requirements for dorm and cafeteria,

$$\text{recovery} = 750 + 300 = 1050 \text{ gph}$$

$$\text{storage} = 2150 + 860 = 3010 \text{ gal}$$

Note: This sizing procedure does not account for *system losses,* that is, heat lost from the storage tanks and from the hot water piping. Recovery capacities are usually increased because of these losses, which are simple to calculate when the U value of the tank and pipe insulation is known: $\text{Btu/h} = U \times A \times \Delta T$.

For this example, an increase in heater size of 225% for faster recovery allows the size of tank to be reduced to only 37% of original size. For these larger DHW applications, central steam is often used as a heat source. Figure 9.21 shows an indirect storage tank system. Such systems may cost more than tankless systems, but they are usually cheaper to operate, both because the peak prices for the instantaneous fuel demands of tankless heaters are avoided and because low-grade steam or waste high-temperature water sources can be utilized.

In any system that heats water under pressure, safety precautions are necessary. To minimize the dangers of excessive pressures (which can damage the system) and of superheated water (which can severely injure people), most codes require *pressure and temperature (P/T) relief valves* to be installed on top of all water heaters. Since these valves are designed to release hot water whenever dangerous pressure or temperature levels are reached, they should be attached to a length of pipe that will conduct the released water to a drain.

(d) Solar Water Heating. Solar energy is most attractive to a designer when it will do

TABLE 9.9 **Minimum Residential Water Heater Capacities**

Number of Baths	*1 to 1.5*			*2 to 2.5*				*3 to 3.5*			
Number of Bedrooms	*1*	*2*	*3*	*2*	*3*	*4*	*5*	*3*	*4*	*5*	*6*
Gas[a]											
Storage, gal (L)	20 (75.8)	30 (113.7)	30 (113.7)	30 (113.7)	40 (151.6)	40 (151.6)	50 (189.5)	40 (151.6)	50 (189.5)	50 (189.5)	50 (189.5)
1000 Btu/h (1 kW) input	27 (7.9)	36 (10.5)	36 (10.5)	36 (10.5)	36 (10.5)	38 (11.1)	47 (13.8)	38 (11.1)	38 (11.1)	47 (13.8)	50 (14.6)
1-h draw, gal (L)	43 (163.0)	60 (227.4)	60 (227.4)	60 (227.4)	70 (265.3)	72 (272.9)	90 (341.1)	72 (272.9)	82 (310.8)	90 (341.1)	92 (348.7)
Recovery, gph (mL/s)	23 (24.1)	30 (31.5)	30 (31.5)	30 (31.5)	30 (31.5)	32 (35.6)	40 (42.0)	32 (33.6)	32 (33.6)	40 (42.0)	42 (44.1)
Electric[a]											
Storage, gal (L)	20 (75.8)	30 (113.7)	40 (151.6)	40 (151.6)	50 (189.5)	50 (189.5)	66 (250.1)	50 (189.5)	66 (250.1)	66 (250.1)	80 (303.2)
kW input	2.5	3.5	4.5	4.5	5.5	5.5	5.5	5.5	5.5	5.5	5.5
1-h draw, gal (L)	30 (113.7)	44 (166.8)	58 (219.8)	58 (219.8)	72 (272.9)	72 (272.9)	88 (333.5)	72 (272.9)	88 (333.5)	88 (333.5)	102 (386.6)
Recovery, gph (mL/s)	10 (10.5)	14 (14.7)	18 (18.9)	18 (18.9)	22 (23.1)	22 (23.1)	22 (23.1)	22 (23.1)	22 (23.1)	22 (23.1)	22 (23.1)
Oil[a]											
Storage, gal (L)	30 (113.7)	30 (113.7)	30 (113.7)	30 (113.7)	30 (113.7)	30 (113.7)	30 (113.7)	30 (113.7)	30 (113.7)	30 (113.7)	40 (151.6)
1000 Btu/h (1 kW) input	70 (20.5)	70 (20.5)	70 (20.5)	70 (20.5)	70 (20.5)	70 (20.5)	70 (20.5)	70 (20.5)	70 (20.5)	70 (20.5)	70 (20.5)
1-h draw, gal (L)	89 (337.3)	89 (337.3)	89 (337.3)	89 (337.3)	89 (337.3)	89 (337.3)	89 (337.3)	89 (337.3)	89 (337.3)	89 (337.3)	89 (337.3)
Recovery, gph (mL/s)	59 (61.9)	59 (61.9)	59 (61.9)	59 (61.9)	59 (61.9)	59 (61.9)	59 (61.9)	59 (61.9)	59 (61.9)	59 (61.9)	59 (61.9)

Tank-type indirect[b,c]										
1-W-H rated draw, gal (L) in 3-h, 100 F° (55.6 C°) rise	40 (151.6)	40 (151.6)		66 (250.1)	66 (250.1)[d]	66 (250.1)	66 (250.1)	66 (250.1)	66 (250.1)	66 (250.1)
Manufacturer-rated draw, gal (L) in 3-h, 100 F° (55.6 C°) rise	49 (185.7)	49 (185.7)		75 (284.2)	75 (284.2)[d]	75 (284.2)	75 (284.2)	75 (284.2)	75 (284.2)	75 (284.2)
Tank capacity, gal (L)	66 (250.1)	66 (250.1)		66 (250.1)	66 (250.1)[d]	82 (310.8)	66 (250.1)	82 (310.8)	82 (310.8)	82 (310.8)
Tankless-type indirect[c,e]										
1-W-H-rated, gpm (L/s) 100 F° (55.6C°) rise	2.75 (0.17)	2.75 (0.17)		3.25 (0.205)	3.25 (0.205)[d]	3.75 (0.205)	3.25 (0.205)	3.75 (0.205)	3.75 (0.205)	3.75 (0.205)
Manufacturer-rated draw, gal (L) in 5 min, 100 F° (55.6C°) rise	15 (56.8)	15 (56.8)		25 (94.7)	25 (94.7)[d]	35 (132.6)	25 (94.7)	35 (132.6)	35 (132.6)	35 (132.6)

Source: U.S. HUD-FHA, for one- and two-family living units.

Copyright © by the American Society of Heating, Refrigerating and Air Conditioning Engineers, Inc., Atlanta, GA. Reprinted by permission from the *ASHRAE Systems Handbook*, 1984.

[a]Storage capacity, input, and recovery requirements indicated in the table are typical and may vary with each individual manufacturer. Any combination of these requirements to produce the stated 1-h draw will be satisfactory.

[b]Boiler-connected water heater capacities [180 F (82.2°C) boiler water, internal or external connection].

[c]Heater capacities and inputs are minimum allowable. Variations in tank size are permitted when recovery is based on 4 gph/kW (4.2 mL/s · kW) @ 100 F° (55.6C°) rise for electrical. A.G.A. recovery ratings for gas heaters, and IBR ratings for steam and hot water heaters.

[d]Also for 1 to 1.5 baths and 4 bedrooms for indirect water heaters.

[e]Boiler-connected heater capacities [200 F (93.3°C) boiler water, internal or external connection].

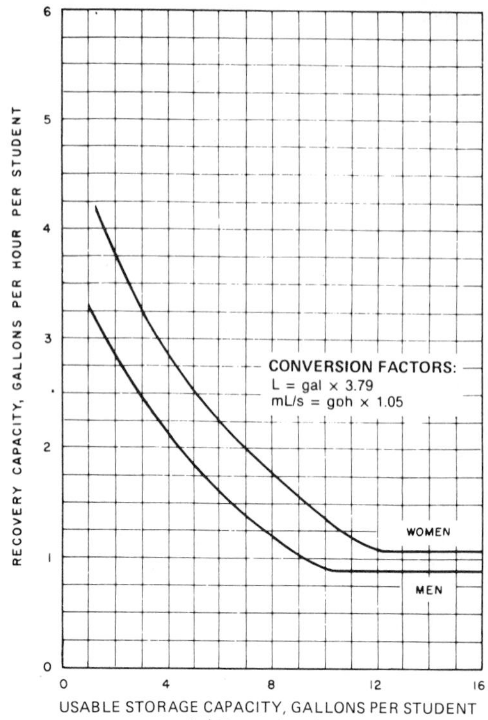

(*a*) Dormitories

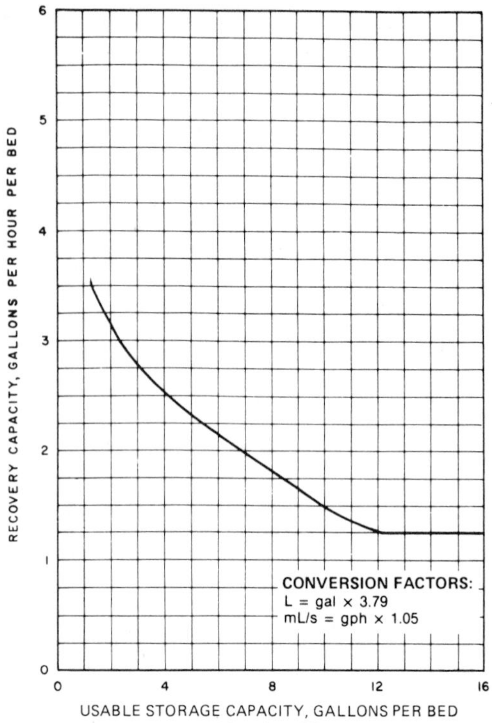

(*c*) Nursing Homes

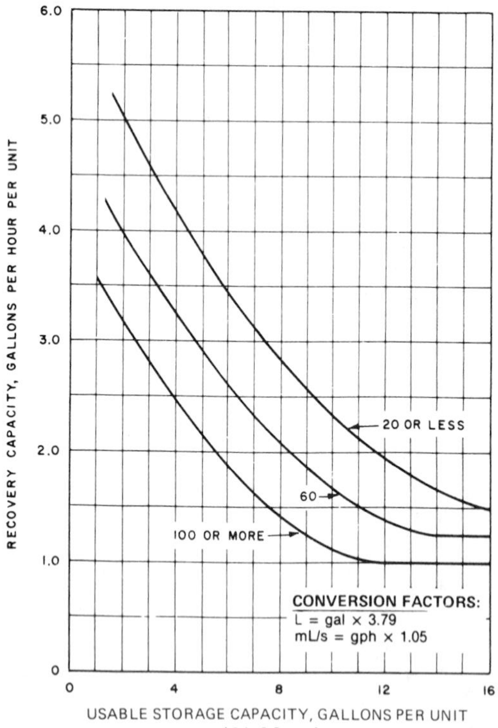

(*b*) Motels

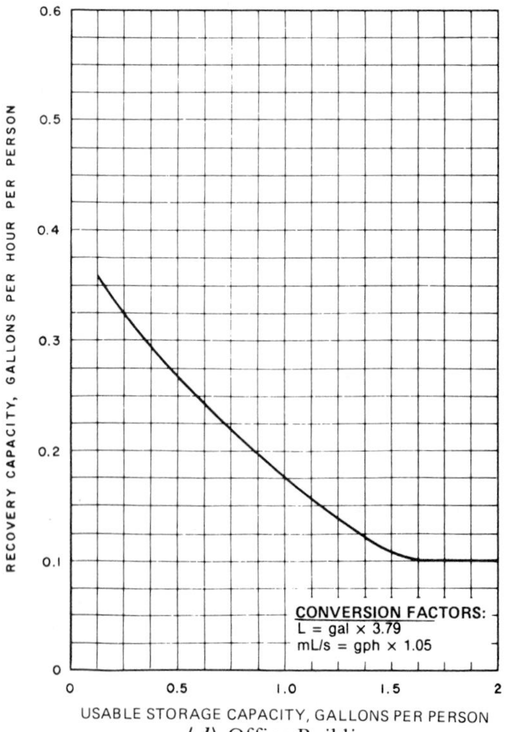

(*d*) Office Buildings

Fig. 9.20 *Domestic hot water sizing; the tradeoff between recovery (heater) capacity and storage capacity. "Usable" storage capacity is usually considered to be between 60 and 80% of the total tank capacity. (Copyright © by the American Society of Heating, Refrigerating, and Air-Conditioning Engineers, Inc., Atlanta, Ga. Reprinted by permission from the* ASHRAE Systems Handbook, *1984.)*

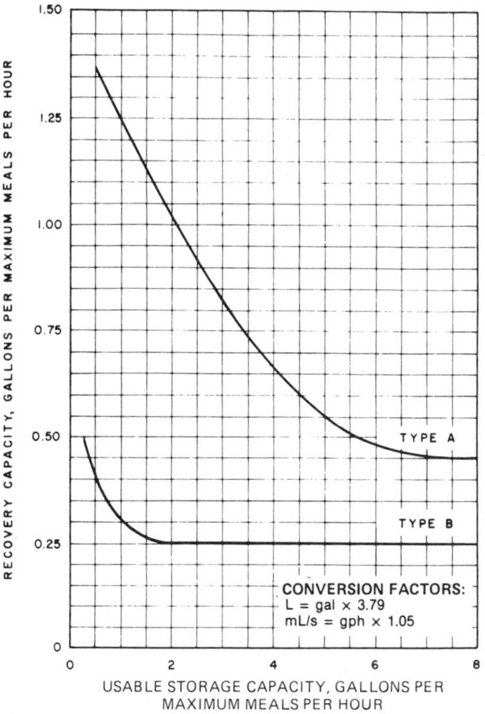

(*e*) Food Service

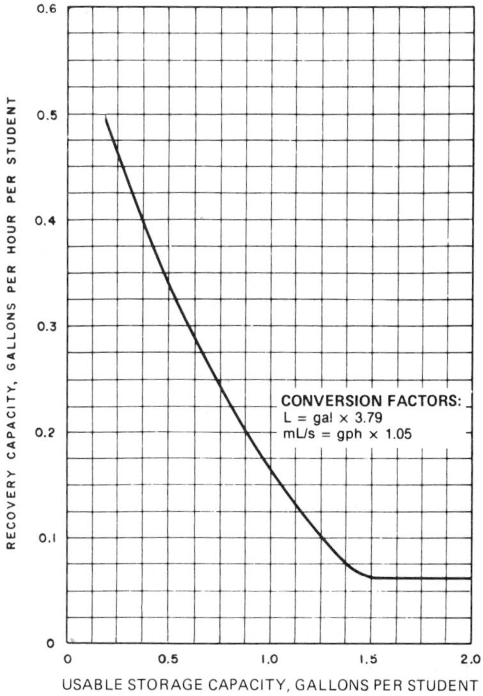

(*g*) Elementary Schools

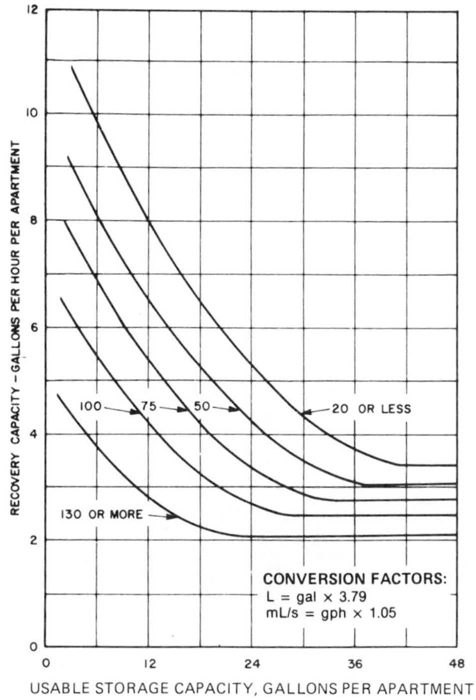

(*f*) Apartments

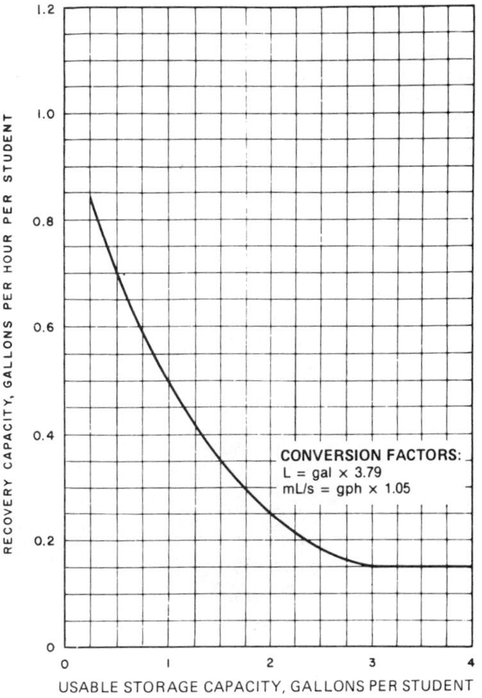

(*h*) High Schools

Fig. 9.20 *(continued)*

TABLE 9.10 **Procedure for Selecting a Residential Water Heater Storage Tank**

	Column A HUD-FHA Minimum (Table 9.9)	Column B Conservationist Water Heater, A.O. Smith (Figure 9.19) PGCS-40	Column C Conservationist Water Heater, A.O. Smith (Figure 9.19) PGCS-50
Storage (gal)	40	40	50
1000 Btu input	38	32.0	40
1-hr draw (gal)[a]	72	75.3	93.1
Recovery (gph)	32	35.3	43.1

[a]1-h draw = tank capacity + 1-h recovery.

most of its work during the time that it is most available. On a seasonal basis (Fig. 9.22), solar energy is especially attractive for outdoor swimming pool heating, as it will be used only in sunny, warm months. Throughout the United States, solar energy can easily meet most of the summertime demand for DHW as well. In the northern part of the country, a much smaller solar contribution can be expected in winter. An especially difficult problem is the winter space-heat mismatch between greatest need and least supply. For both solar DHW and space-heating systems, winter brings the added complication of the need to protect against freezing.

Again, there are many ways in which solar energy can be used to heat water. Solar water heating systems are usually classified by the means of fluid circulation ("passive" or "active"), the means by which heat from the collector piping is transferred to the DHW itself ("direct" or "indirect," or closed loop), and the means of protection against freezing.

Passive systems rely on gravity for circulation, as was explained in Section 9.4*b*. Hence, the storage tank usually must be placed above the collector and the number of bends in the collector's supply and return piping minimized, to minimize friction. Heavy storage tanks located high in a building can cause structural problems. The advantages of the passive ap-

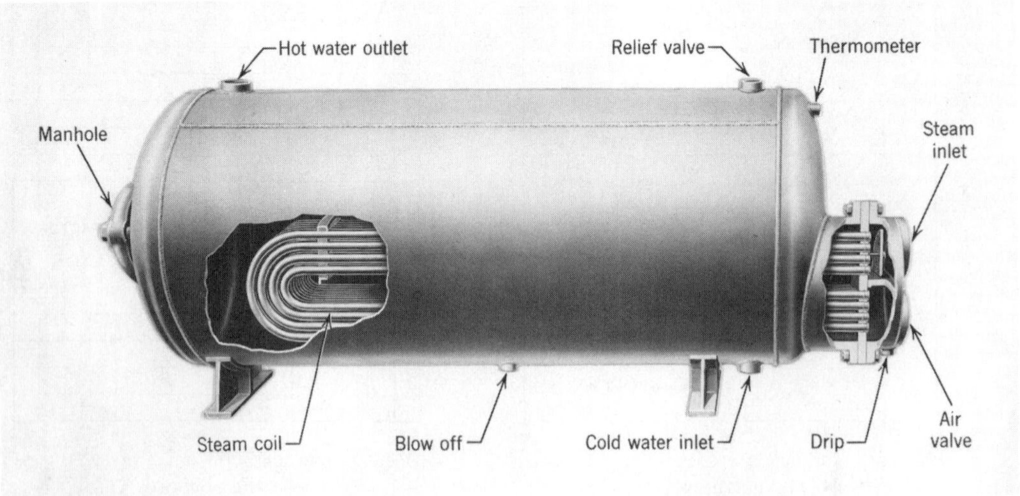

Fig. 9.21 *Storage tank heater and storage for domestic hot water for large-demand applications. Steam coil submerged in tank (see Fig. 9.36). Capacities 100 to 10,000 gph, varying by length of coil, for 140F° (40 to 180F) temperature rise.*

TABLE 9.11 **Domestic Hot Water, Commercial, Institutional**

Type of Building	Maximum Hour	Maximum Day	Average Day
Men's dormitories	3.8 gal (14.4 L)/student	22.0 gal (83.4 L)/student	13.1 gal (49.7 L)/student
Women's dormitories	5.0 gal (19 L)/student	26.5 gal (100.4 L)/student	12.3 gal (46.6 L)/student
Motels: no. of units[a]			
20 or less	6.0 gal (22.7 L)/unit	35.0 gal (132.6 L)/unit	20.0 gal (75.8 L)/unit
60	5.0 gal (19.7 L)/unit	25.0 gal (94.8 L)/unit	14.0 gal (53.1 L)/unit
100 or more	4.0 gal (15.2 L)/unit	15.0 gal (56.8 L)/unit	10.0 gal (37.9 L)/unit
Nursing homes	4.5 gal (17.1 L)/bed	30.0 (113.7 L)/bed	18.4 gal (69.7 L)/bed
Office buildings	0.4 gal (1.52 L)/person	2.0 gal (7.6 L)/person	1.0 gal (3.79 L)/person
Food service establishments:			
Type A—full meal restaurants and cafeterias	1.5 gal (5.7 L)/max meals/h	11.0 gal (41.7 L)/max meals/h	2.4 gal (9.1 L)/avg meals/day[b]
Type B—drive-ins, grilles, luncheonettes, sandwich and snack shops	0.7 gal (2.6 L)/max meals/h	6.0 gal (22.7 L)/max meals/h	0.7 gal (2.6 L)/avg meals/day[b]
Apartment houses: no. of apartments			
20 or less	12.0 gal (45.5 L)/apt.	80.0 gal (303.2 L)/apt.	42.0 gal (159.2 L)/apt.
50	10.0 gal (37.9 L)/apt.	73.0 gal (276.7 L)/apt.	40.0 gal (151.6 L)/apt.
75	8.5 gal (32.2 L)/apt.	66.0 gal (250 L)/apt.	38.0 gal (144 L)/apt.
100	7.0 gal (26.5 L)/apt.	60.0 gal (227.4 L)/apt.	37.0 gal (140.2 L)/apt.
200 or more	5.0 gal (19 L)	50.0 gal (195 L)/apt.	35.0 gal (132.7 L)/apt.
Elementary schools	0.6 gal (2.3 L)/student	1.5 gal (5.7 L)/student	0.6 gal (2.3 L)/student[b]
Junior and senior high schools	1.0 gal (3.8 L)/student	3.6 gal (13.6 L)/student	1.8 gal (6.8 L)/student[b]

[a] Interpolate for intermediate values.
[b] Per day of operation.

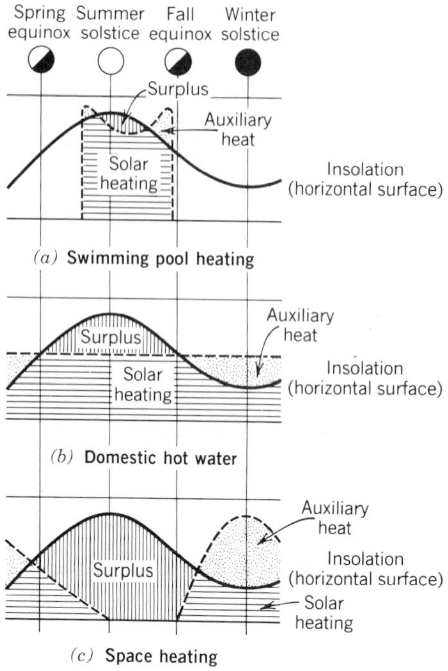

Fig. 9.22 *Comparison of the pattern of supply of solar energy (on a horizontal surface) to various patterns of heating needs.*

proach include lower cost of components, high mechanical reliability (no pump, etc.), and low operational costs.

Active systems use pumps to force the fluid to the collector. Although this setup allows the collector and storage to be located anywhere that is convenient for the designer, it introduces the complications of mechanical breakdown, increased maintenance, and the high cost of the energy needed to run the pump.

Direct systems utilize only one fluid: the water to be heated for use in the building is circulated through the solar collector. Such a system has the advantages of simplicity and efficiency, as it does not require a separate loop of fluid and the attendant piping complications and inefficiencies of heat exchange.

Indirect systems use a closed loop containing a fluid that circulates through the collector and storage tank. The fluid is not mixed with the DHW itself; rather, heat is passed from one fluid to the other through a heat exchanger. One advantage of this system is that it allows nonfreezing solutions to be used in the collector loop. It also allows collectors to be operated at low pressures, rather than at the high pressures typical of urban public water systems. The choice of fluid is determined by its freezing and boiling points, its specific heat, and its level of toxicity.

As Table 9.12 shows, these design elements (passive and active, direct and indirect) are combined in a number of typical solar DHW systems. Brief descriptions of some of the more common systems follow.

1. *Batch systems.* In these, the simplest of all such systems, a black-painted storage tank is exposed to the sun within a glazed collector box (Fig. 9.23). The price of simplicity is the inefficiency of exposing a tank of hot water to the cold nighttime conditions of the collector. Movable insulation can be used to reduce this problem. These heaters are also referred to as "breadbox" water heaters and "integral passive solar water heaters." The batch system was used at the Antelope Valley California Poppy Reserve (Fig. 5.15), as shown in Fig. 9.24. Because of their simplicity, they are favored by do-it-yourself builders, especially in mild-winter climates.

2. *Thermosiphon systems.* The sun acts as both the "pump" and the heat source for these systems (Fig. 9.25). With no moving parts, maintenance needs are low. The collector, however, must be lower than the tank, and piping must be kept as simple as possible. Since the coldest water remains in the lower collector at night, the hot water in the upper tank is not threatened with undue heat loss. However, freezing conditions pose a severe threat to the collector. Accordingly, indirect (closed-loop) systems containing a nonfreezing fluid are frequently used; *phase-change systems* are fast becoming the cold-winter option for this type of passive system.

3. *Passive geyser pumping systems.* The solar collector-tank system shown in Fig. 9.26 features no requirement for outside power (except supplementary water heating, as desired), no moving parts, and the freedom to locate the collector above the storage tank. It uses a methanol–water mixture rather than CFC refrigerant; it represents an environmentally attractive, au-

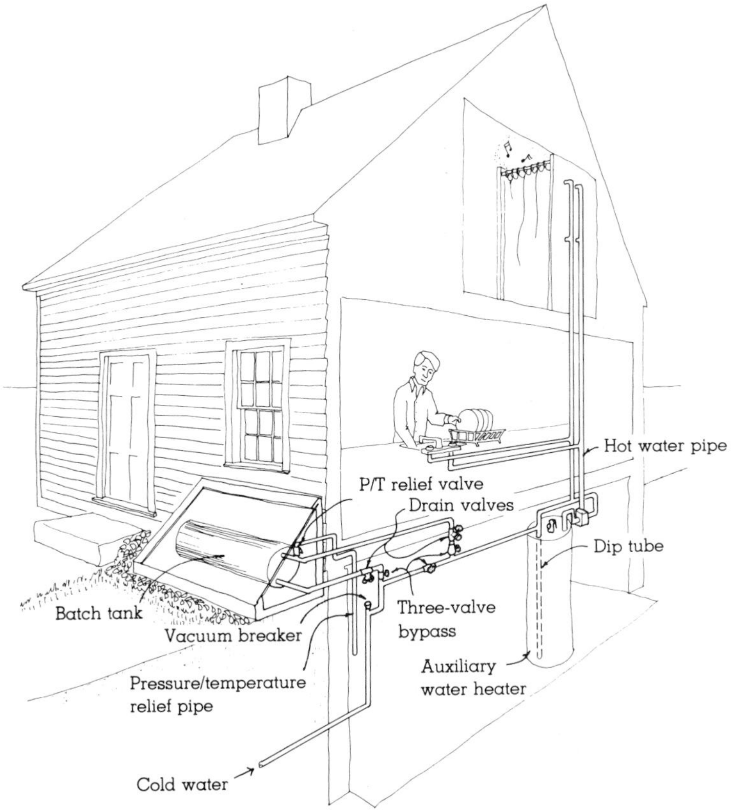

***Fig. 9.23** Batch solar water-heating system. (Reprinted by permission from Daniel K. Reif,* Passive Solar Water Heaters, *copyright © 1983 by Brick House Publishing Company, Inc.)*

tonomous water heating option. The methanol–water mixture boiling in the riser tubes of the solar collector drives the heated liquid up into the header, along with a burst of vapor. This action is called *geyser pumping*. Gravity then pulls the hot liquid down to the heat exchanger at the storage tank. This downward liquid flow pushes the cooled liquid out of the heat exchanger, back up to the vapor condenser. Here the vapor produced by the boiling meets the cooled liquid and changes state back to a liquid, which reenters the collector to repeat the cycle.

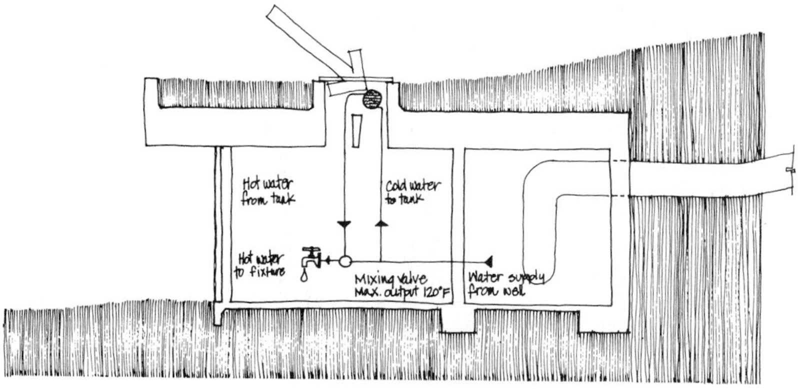

***Fig. 9.24** A batch solar water heater supplies the hot water for the Antelope Valley California Poppy Reserve (also described in Fig. 5.15). (Courtesy of the Colyer/Freeman Group, Architects.)*

TABLE 9.12 **Solar Domestic Hot Water (DHW) Systems**

Type	*Main Features*	*Advantages*	*Disadvantages*
Batch system	Batch tank inside collector box. Potable water within collector/tank.	No external power. Few components. Collector/tank at any location.	Seasonal; dependent on freezing locations. Heat loss at night from storage.
Thermosiphon system[a]	Flat plate liquid collectors. Normally open loop but no pump or external power (passive DHW system). Storage tank higher than collector.	No external power. Few components. High performance.	Seasonal; dependent on freezing locations (if water is collector fluid). Needs structural support for high storage tank.
Geyser pumping system	Flat-plate liquid collectors. Methanol-water solution passively pumped by solar heat to a heat exchanger below.	No freezing damage. No mechanical or electric parts. No liquid service.	Temporarily stops operation at subzero temperatures, as solution freezes to a slush.
Closed-loop freeze-resistant system[a]	Flat plate liquid collectors. Closed loop of piping from collectors to storage tank. Uses external energy (circulator and differential controller). Uses nonfreezing collector fluid. Pressurized stonelined storage tank.	Can be used in coldest climates. More and better-established competition. Circulator; small consumption of external energy. High performance.	Liquid; service/maintenance required. More components required.
Drain-back system[a]	Flat plate liquid collectors. Water is collector fluid (open loop). Potable water circulates through the heat exchanger in the storage tank (not through the collectors). Large heat exchanger. Pitched headers.	Can be used in the coldest climates. No antifreeze used. Most simple of active flat plate systems (no valves).	Larger pump; larger consumption of external energy. System must drain thoroughly. Use of corrosion inhibitor recommended.

Drain-down system[a]	Flat plate liquid collectors. Potable water circulated through collectors. Line pressure feeds collectors (open loop). Has automatic drainage valves. Pitched headers.	No heat exchanger or extra storage tank needed. High performance.	In some instances a larger pump; larger consumption of external energy. System must drain thoroughly. No corrosion inhibitor possible. Freeze danger with valve failure.
Air-to-liquid system[a]	Flat plate air collectors. Air-to-water heat exchanger. Ductwork and blower. Pipes and circulator. Larger collector area than liquid system.	Won't freeze (dependent on exchanger location). Air leaks won't cause damage. Integrates well with space heating.	Hard to detect leaks. More space required for ducts. Blower and circulator required. More carpentry involved. Less efficient than other systems.
Phase-change system[a]	Flat plate liquid collectors. Freon 114 or R12. Storage tank higher than collectors (passive type). Closed-loop refrigerant-grade piping from collectors to storage tank (passive type) or to condenser (subambient type).	No external power (passive type). Can be mounted at any location (subambient type).	Very hard to detect leaks. Special equipment to install. More components required (subambient type).

[a]Reprinted by permission from Russell H. Plante, *Solar Domestic Hot Water: A Practical Guide,* Copyright © 1983 by John Wiley & Sons.

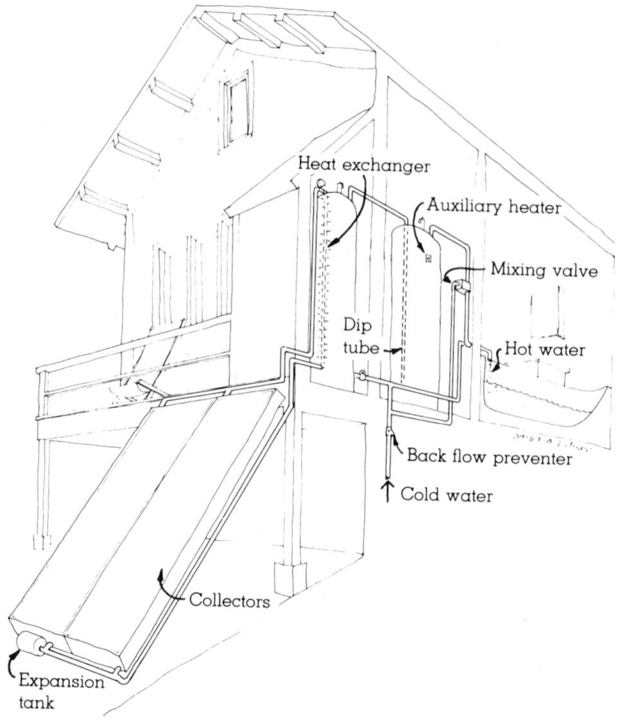

Fig. 9.25 Thermosiphon solar water-heating system. (Reprinted by permission from Daniel K. Reif, Passive Solar Water Heaters, *copyright © 1983 by Brick House Publishing Company, Inc.)*

4. *Closed-loop freeze-resistant systems.* In addition to being used in thermosiphoning systems, these are commonly used in active systems (Fig. 9.27). A small pump circulates nonfreezing fluid to the collector when there is sufficient sun. This process is governed by a "differential controller," a device that compares the temperature of the stored hot water to that of the collector. The price of this freeze protection is the inefficiency of the heat exchanger, used between the collectors' fluid and the DHW itself. Some codes require a *double* wall between any toxic nonfreezing fluid and the DHW, which further reduces efficiency.

5. *Drain-back systems* (Fig. 9.28*a*). Although these systems use water as the fluid pumped from tank to collector, the water is not

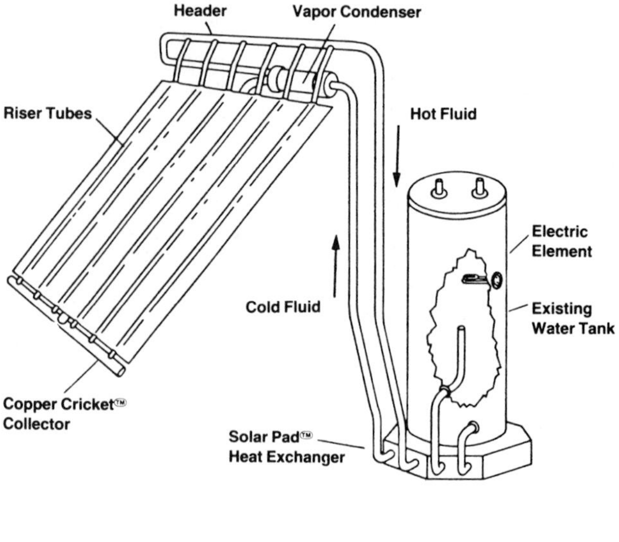

Fig. 9.26 The Copper Cricket™ is a passive pumping phase-change solar water heating system that allows the collector to be located above the storage tank, yet requires no outside power for water pumping to the collector. The methanol–water mixture passes through a heat exchanger at the storage tank. (Courtesy of Sage Advance Corp., Eugene, Oregon.)

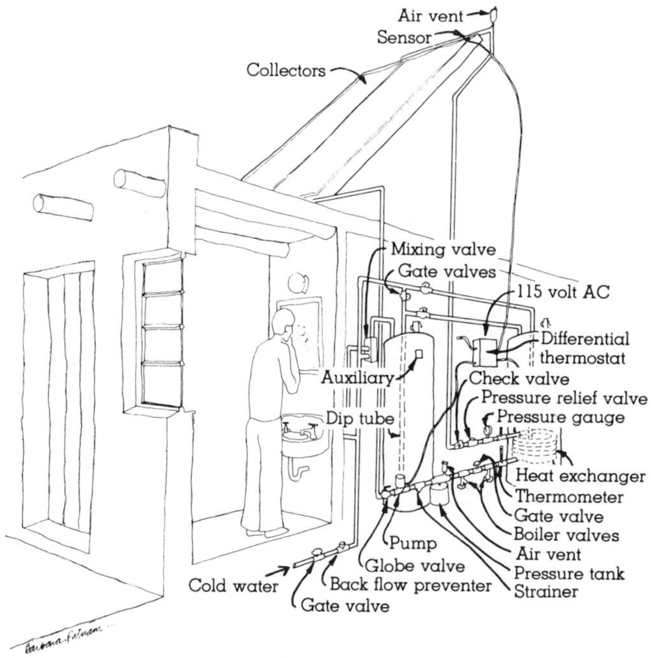

Fig. 9.27 *Closed-loop freeze-resistant solar water heating system. (Reprinted by permission from Daniel K. Reif, Passive Solar Water Heaters, copyright © 1983 by Brick House Publishing Company, Inc.)*

the DHW itself. Instead, the DHW passes through a heat exchanger in the solar storage tank. With the DHW kept out of the collector, the solar collector can operate at lower water pressure. This, however, requires a large heat exchanger, with attendant inefficiency. It also requires care in the design and installation of piping between collector and tank, so that the collector will drain thoroughly. When the controller senses that no solar energy can be gathered, it cuts off the pump, and water drains back into the tank. Therefore, the collector will be filled with air, not water, for all nighttime and cloudy-cold daytime hours. This suggests that corrosion inhibitors should be added to the collector/tank's water, since the piping is frequently exposed to air.

6. *Drain-down systems (*Fig. 9.28*b*). These are the only active systems that do not utilize heat exchangers; DHW is circulated directly through the collector. Both higher pressure and higher efficiency result. In this system, the collector is usually filled with water that moves only when the differential controller activates the pump. Whenever the outside temperature drops near freezing, the controller activates solenoid valves and the water in the collector is drained down (dumped). In cold-winter areas, this process could result in several gallons of water wasted per winter day. Although the lack of a heat exchanger is attractive from the standpoint of cost savings and thermal efficiency, the set of electrically controlled solenoid valves is not foolproof. When malfunctions occur in pressurized systems, the loss of water can be enormous, and there is great potential for water damage to the building.

7. *Air-to-liquid systems*. These systems use air collectors and rock storage beds; a heat exchanger transfers heat from the collector's air to the DHW. Much lower efficiencies result, and the ductwork is much more space consuming than are pipes to water collectors. Worse yet, leaks in air collectors or storage beds are very difficult to detect. However, air collectors are not damaged by freezing.

8. *Phase-change systems* (Fig. 9.29). In any of the above-discussed systems, in which DHW is kept out of the collector, the fluid within the collector not only can be freeze resisting, but also can take advantage of latent heat (the considerable amount of heat stored when this fluid

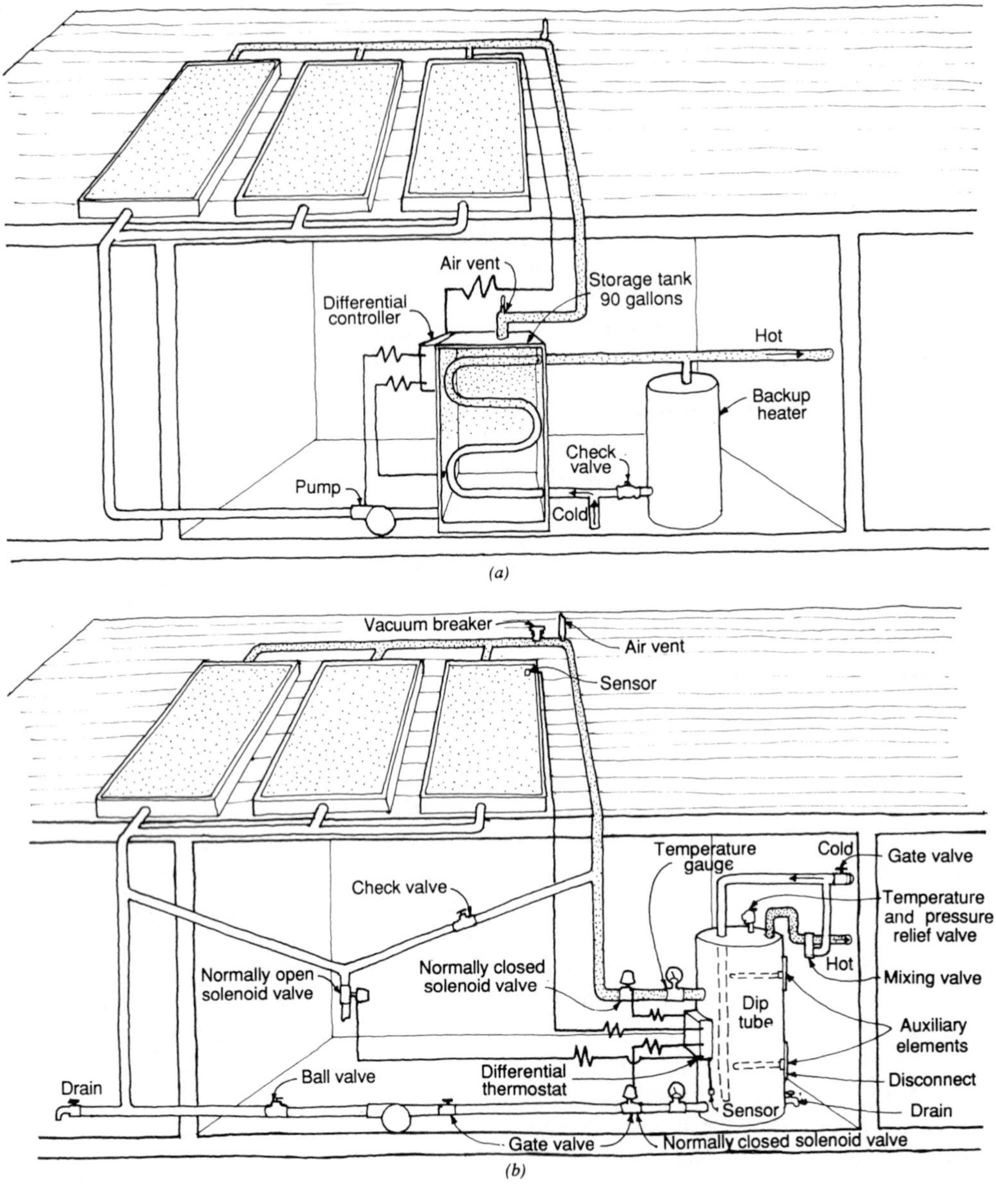

Fig. 9.28 *Comparison of* (a) *drain-back solar water-heating system with* (b) *drain-down system. The drain-back system's performance can be increased by the addition of the check valve that allows the colder water at the bottom of the storage tank to circulate into the heat exchanger. The drain-down system is best used in mild-winter climates with infrequent freezing temperatures, since the water in the collector is dumped with each freezing threat. (Adapted by permission from Russell Plante,* Solar Domestic Hot Water, *copyright © 1983 by John Wiley & Sons.)*

vaporizes and released when the fluid condenses). The primary disadvantages are the first cost, the difficulty in detecting leaks, and the resulting threat of some refrigerant fluids to the environment (such as the ozone layer).

Passive approaches to phase-change materials were outlined in the discussion of thermosiphoning systems. *Subambient* approaches utilize a heat pump and can glean heat even from cold-cloudy ("subambient") conditions. They

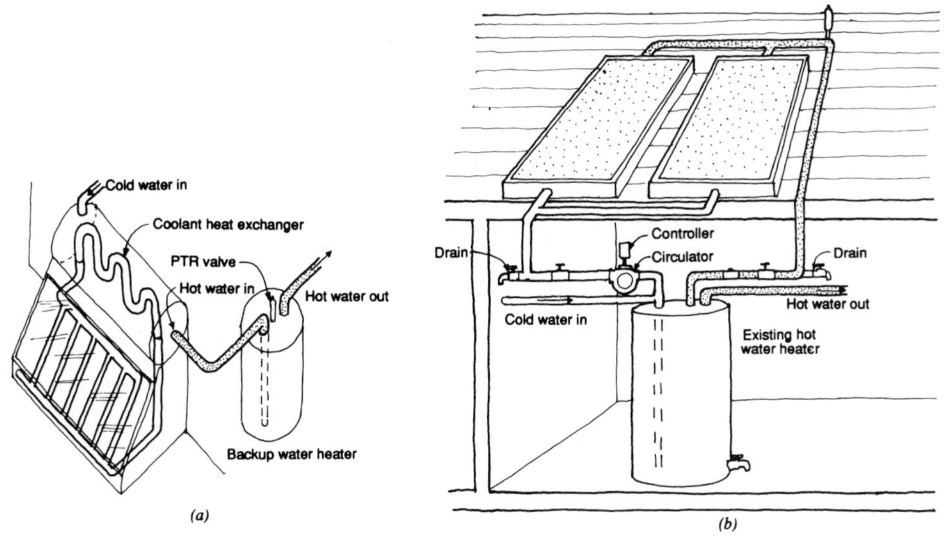

Fig. 9.29 Various approaches to phase-change solar water heating. (a) *A smaller solar-heated water tank can be placed at the top of the collector. The circulation of the phase change fluid within the collector is by thermosiphon. The exposed storage can cause problems in cold-winter areas.* (b) *Small heat exchangers can be built into the tops of thermosiphoning collectors; DHW is drained down when the threat of freezing arises. (Adapted by permission from* Solar Age, *November 1983.)*

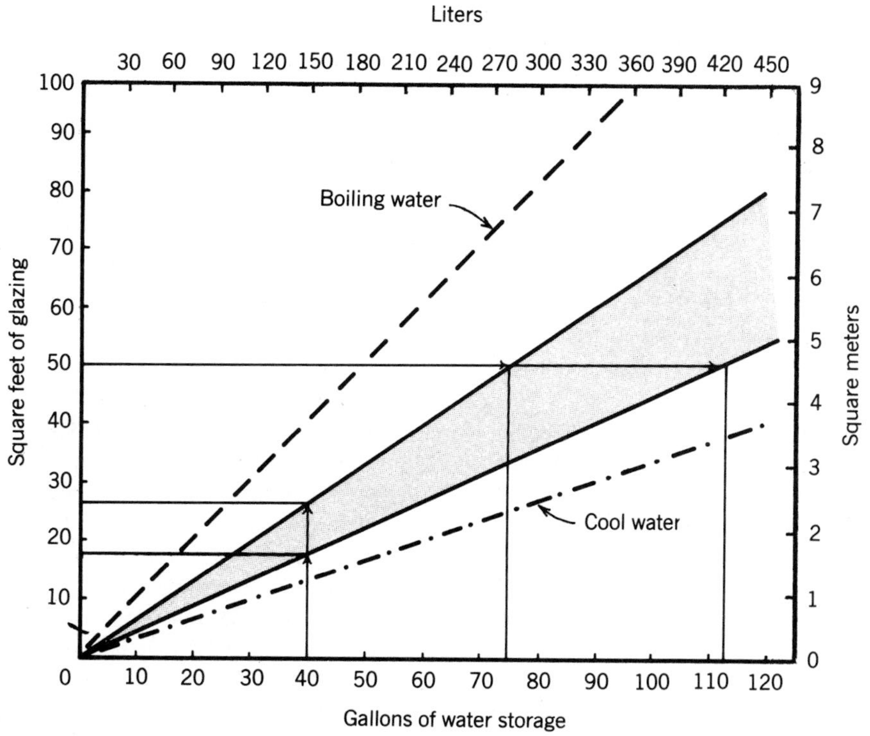

Fig. 9.30 Rule-of-thumb guide to the sizing of batch solar water heaters. In this example, a 40-gal storage tank should be contained within a collector that has between 18 and 27 ft^2 of south-facing glazing. Or a 50-ft^2 collector box should contain a tank of between 75 and 113 gal. (Adapted from Daniel K. Reif, Passive Solar Water Heaters, copyright *© 1983 by Brick House Publishing Co. Reprinted by permission.)*

thus represent a form of solar-assisted heat pump, closely related to those discussed in Section 9.4(*e*).

The *sizing* of solar water heaters usually begins with rules of thumb. For batch heaters, the sizing rule of thumb (Fig. 9.30) is

from 0.45 to 0.65 ft² glazing per gallon of water stored

For the flat-plate solar collectors used in all the other solar DHW systems listed above, the rules of thumb are:

12 to 25 ft² collector/person (residential)

optimum tilt (up from horizontal) equal to latitude (or less)

1 to 1.5 gal of storage per ft² collector area

This collector-sizing rule of thumb is for *residential* hot water usage, including cooking and bathing. For warmer climates with ample insolation, use the lower figure; this will supply about half of the hot water on an annual average basis. (For *nonresidential* DHW, adjust the rule of thumb by comparing gallons of hot water per person per day used in residences to those used in the nonresidential function for which you are designing—see Section 8.1.

A more detailed sizing procedure would consider the collector's expected heat contribution for some typical months. Such a procedure requires both weather data and assumptions about water temperatures and other values. As a result, detailed answers are best obtained from computer programs, such as those listed in Appendix H. For those interested in something more than rules of thumb but less than computer programs, however, the following approach to collector sizing (adapted from Brown, Reynolds, and Ubbelohde, *Inside/Out: Design Procedures for Passive Environmental Technologies,* copyright © 1982 by John Wiley & Sons) is provided.

STEP 1. *Select the collector tilt angle.* This can be done by consulting Tables B.11 to B.14, which present clear-day insolation values for various tilt angles. Appendix B, Table B.15, presents average insolation values on horizontal surfaces, which should also be checked. The tilt angle selected should be close to the optimum angle for the best month for total insolation.

STEP 2. *Check the collector efficiency.* This should be done at least twice: for the best insolation month and for the worst month. This step requires the following data:

Hourly insolation on tilted surface (Appendix B, Table B.11 to B.14).

Outdoor average temperature (Appendix B, Table B.15).

The hourly insolation values on tilted surfaces in Tables B.11 to B.14, are for *clear* days. For most locations; this should be adjusted for *average* conditions, which are shown for vertical (south) and horizontal surfaces in Appendix B, Table B.15. Since for most of North America the optimum yearly DHW tilt angle will be closer to horizontal than to vertical, a simple correction to insolation must be made:

$$\text{Hourly average day insolation on tilted surface} = \left[\begin{array}{l}\text{hourly clear-day}\\ \text{insolation on}\\ \text{tilted surface}\end{array}\right]\left[\frac{\text{average-day total horizontal insolation}}{\text{clear-day total horizontal insolation}}\right]$$

This step also requires assumptions about the input temperature of the water supplied to the collector, or T_i. In summer, this can be quite high—even higher than the thermostat set-point temperature of the hot water tank. In winter, it is likely to be lower, by perhaps 10 to 20F°, than the thermostat set-point temperature. Finally, this step requires a choice of collector type, so that efficiency can be determined. Figure 9.31 shows the efficiency curves for a variety of flat plate collectors. Note that the simplest of all, the unglazed flat black collectors, have the highest efficiency of all collectors at combinations of very low $T_i - T_a$ and very high I_o. These conditions are typical of summer days, on which the water is to be heated only a few degrees. The fact that this corresponds to swimming pool applications is one reason why such collectors are so popular for that purpose. Low cost is another reason. For the opposite conditions—those of

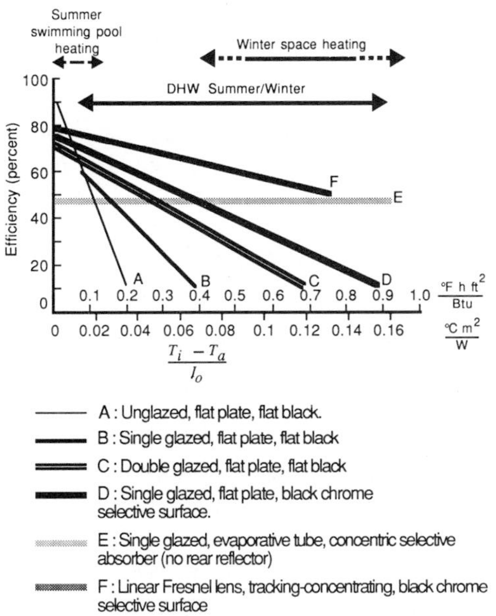

Fig. 9.31 *Collector efficiencies depend on insolation, water temperature, and ambient (air) temperature. (Based on the U.S. Dept. of Housing and Urban Development's* Intermediate Minimum Property Standards Supplement, *1977.)*

winter space heating—more-elaborate collectors are appropriate. Evacuated-tube collectors have the advantage of nearly eliminating collector heat loss by convection. The Fresnel lens and/or tracking-concentrating collectors increase the incoming solar energy per unit area, while heat loss increases only slightly (due to higher ΔT). Both these approaches are expensive, and tracking collectors require added maintenance for the additional motion. Selective surfaces represent a simple, relatively cheap improvement over flat black collectors; they absorb solar energy just as well but emit radiant energy at a vastly lower rate, thus cutting collector heat loss by radiation.

STEP 3. *Approximate system efficiency.* Once the hot water leaves the collector, it must be led back to storage. In indirect systems, it must go through a heat exchanger. The tank will lose some heat; heat losses also occur as the cooled water is led back to the collector. Another "loss" to be considered is noncollection time. Estimates of collector performance are based on total daily insolation, yet the first few and last few hours of daylight generally will *not* bring enough insolation to warm the collector. Therefore, some reduction should be made in daily insolation totals when performance is estimated. All these factors can be accounted for by assuming a lower system efficiency. A perhaps optimistic assumption is

$$\text{system efficiency} = 0.8\ (\text{collector efficiency})$$

STEP 4. *Determine the total heat needed for DHW.* Estimates of the quantity of hot water needed were made in Sections 9.4*a* and 9.4*c*. Once the total gallons (or liters) per day are known, and the desired storage temperature determined, the heat needed for the daily supply of DHW is easy to calculate. In conventional units,

$$Q = 8.33\ (\text{gpd})\ (T_s - T_g)$$

where

Q = daily heat need, Btu
8.33 = weight of water, pounds per gallon × Btu/lb-°F
T_s = storage temperature, °F (see Table 9.7)
T_g = groundwater temperature, °F (see Fig. 5.25, well-water temperatures)

In SI units,

$$Q = 1.16\ (\text{L/d})\ (T_s - T_g)$$

where

Q = daily heat need, W-h
1.16 = weight of water, × W-h/L-°C
T_s = storage temperature, °C
T_g = groundwater temperature, °C

STEP 5. *Determine the desired percentage of total heat to be solar, for DHW.* There are several approaches to this problem. One common strategy is to provide 100% of the hot water needs in the best insolation month. A closely related strategy is to choose the *yearly* percent solar desired, then provide that percentage in the "average" month (such as March). Whatever strategy is chosen, a month and its percent solar must be identified.

STEP 6. *Size the collector.* This is done by combining the preceding steps, as follows.

$$\text{collector area} = \frac{\text{daily heat need} \times \text{percent solar desired}}{\text{daily insolation} \times \text{system efficiency}}$$

This may be done for several months, as a check on optimum collector size.

EXAMPLE 9.4. Determine the approximate size of the solar collectors needed to serve the women's dormitory described in Example 9.3. Assume the location to be Springfield, Illinois.

STEP 1. *Tilt angle.* Refer to Appendix B (Table B.15). Springfield's latitude is 39.8°N. In Table B.15, average insolation is given for July. Use this month as the "best" month.

From Table B.12, the optimum tilt angle for July at 40°N latitude is horizontal. However, at 30° (latitude − 10°), much better performance will be obtained in "average" months, such as March and September. Choose a 30° tilt angle.

STEP 2. *Collector efficiency.* For the best month (July), the best hourly clear-day insolation is 307 Btu/h-ft^2, at 30° tilt, in the hour centered at noon. To adjust this for *average* hourly value, calculate

$$307 \times \frac{2058 \text{ Btu/day average horizontal (Table B.15)}}{2534 \text{ Btu/day clear horizontal (Table B12)}} = 249 \text{ Btu/h}$$

Outdoor average temperature is approximately obtainable from Table B.15; TA for July is 76 F. This is quite conservative, since it is the average daily temperature; the temperature at noon should be higher. From Appendix A, the mean daily range for Springfield is 21F°. Thus 76 + 21/2 = about 86 F. The thermostat setpoint temperature of the water storage tank is assumed at 115 F. This allows adequately hot water for most dormitory usages, although additional added heat will be required for the cafeteria's dishwashing.

Assume, since it's summer, that T_i is at the set-point of 115 F. (In winter, it would be lower). Therefore, the quantity $(T_i - T_a)/I_o$ can be calculated at

$$\frac{115 \text{ F} - 86 \text{ F}}{249 \text{ Btu/h}} = 0.116$$

Enter Fig. 9.31 with this number, and assume a single-glazed, flat plate selective surface collector (type D). The efficiency will be about 70% at these conditions.

STEP 3. *System efficiency.* Use the rule of thumb

$$0.8\ (70\% \text{ collector efficiency}) = 56\%$$

STEP 4. *Total heat needed.* The average usage in gpd can be estimated from Table 9.11. For women's dormitories, the rate is 12.3 gal/student × 300 students = 3690 gal. For the cafeteria, the average rate (type A) is 2.4 gal/meal = 2.4 × 300 = 720. Total gallons needed: 3690 + 720 = 4410.

Groundwater temperature for Springfield (from Fig. 5.25) is about 56 F. Assume that solar energy will be used to raise it to 115 F.

$$Q = 8.33\ (4410 \text{ gpd})\ (115 - 56) = 2{,}167{,}380 \text{ Btu}$$

STEP 5. *Desired percent solar.* For this best hour in July, assume 100% solar contribution.

STEP 6. *Collector size.* Average daily insolation can be obtained from Appendix B, as before:

clear day total, 30° tilt, July: 2409 (Table B.12)

clear day total, horizontal, July: 2534 (Table B.12)

average day total, horizontal, July: 2058 (Table B.15)

average July daily insolation, at 30° tilt

$$= \frac{2409 \times 2058}{2534} = 1956$$

collector area

$$= \frac{2{,}167{,}380 \text{ Btu} \times 100\% \text{ solar}}{1956 \text{ Btu/ft}^2 \times 56\% \text{ system efficiency}} = 1979 \text{ ft}^2$$

This size should be checked against average-month performance and adjusted as desired. Note that this is a ratio of about 1979/300 = 6.6

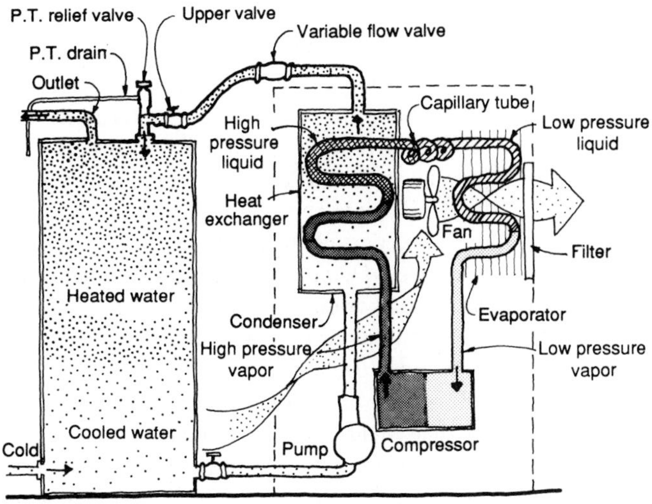

Fig. 9.32 *Heat pump (air–water) used for heating DHW. (Adapted by permission from* Specifying Engineer, *October 1983. Copyright © 1983 by Cahners Publishing Co.)*

WATER AND WASTE

ft^2 of collector per student—lower than the rule of thumb for typical residential DHW systems.

Note also that at the rule of thumb rate of 1 to 1.5 gal storage per ft^2 collector, a tank size of 2000 to 3000 gal is indicated. In Example 9.3, the minimum recovery rate needed a storage tank of 8150 gallons, and the fastest recovery rate needed 3010 gallons. Either seems sufficient for this size of collector.

Swimming pool heating is an especially attractive application for solar energy. A common rule of thumb for collector sizing is

collector area = 0.5 pool surface area

For summer ambient temperature operations, unglazed collectors are both the best performing and the least expensive, as explained for Fig. 9.31. Another important consideration is a pool cover, which will not only conserve the pool's heat, but also conserve water lost by evaporation. On very hot-dry days, water losses can reach 100 gpd from a 20 × 40-ft pool.

(e) Heat Pump Water Heaters. The heat pump's use of the compressive refrigeration cycle was explained in Section 6.7 (see Figs. 6.62 to 6.65). Air-to-water heat pumps are also used for DHW, as shown in Fig. 9.32. Since these devices remove heat from the air, they are usually installed either in normally overheated spaces (such as restaurant kitchens) or in unheated spaces such as garages or basements. (It can also remove heat from exhaust air, as explained in Section 6.4*g*.) The spaces that contain these units will be cooled and dehumidified by them. Heat pumps require only a little more space than that required by the simple hot-water storage tanks they serve. Some noise is created by the compressor and the fan that moves air across the evaporator.

Heat rejected from any refrigeration unit can be used for heating water, via the heat pump. Applications include ice making machines, refrigerated display units in grocery stores, walk-in freezers, and many others. Whenever constant refrigeration loads are present, there is the opportunity to utilize waste heat.

9.5 Water Distribution

This section looks at ways to supply water throughout buildings at pressures sufficient to operate plumbing fixtures. Smaller buildings may be served simply by the pressure available in water mains (or pressure tanks fed by pumped wells). This is called upfeed distribution, because the water rises directly from mains to the plumbing fixtures. For taller buildings, several other options are available: *pumped upfeed* (in which pumps supply the additional pressure needed); *hydro-pneumatic* (in which pumps force water into sealed tanks, compressing the air within; this maintains the needed water pressure); and *downfeed* (in which pumps raise the water to storage tanks at the top of the building, and water then drops down to the plumbing fixtures).

In cities with municipal water supply systems, water is distributed through street mains at pressures varying from about 50 psi (345 kPa) at the main to about 70 psi (483 kPa). For low-rise buildings of two or three stories, these pressures are adequate to act against the static pressure of water standing in the vertical piping, overcome the frictional resistance of water flow in the pipes, and still deliver water at the pressure required to operate plumbing fixtures. The flow-pressure required at the fixtures varies from 5 to 20 psi (35 to 138 kPa), depending on the type of fixture—for example, the basin faucet, showerhead and faucet, or water closet fixture. (Table 9.16 in Section 9.7 lists the minimum flow rates and pressures for typical plumbing fixtures).

(a) Static Pressure. The pressure exerted at the bottom of a stationary "head" of water is related directly to its height. One cubic foot of water weighs 62.4 lb. Consider a "cube" of water 1 ft square and 1 ft high. Its weight (62.4 lb) rests on a bottom area of 1 ft^2 (144 in.^{2}). The *static* pressure at the bottom is $62.4/144 = 0.433$ psi (3 kPa). Reciprocally, 1 psi of pressure will *sustain* a static (stationary) column of water $1/0.433 = 2.3$ ft in height. (In SI units, 1 meter of head = 10 kPa.) When fixture pressure and pressure lost in friction-of-flow in pipes are considered, the problem becomes more complex. Example 9.6 in Section 9.7 illustrates this problem in the calculation of pipe size. For upfeed and downfeed distribution, the relationship of heights and static pressure is one controlling factor.

(b) Upfeed Distribution. In small, low buildings of moderate water demand, it is seldom difficult to achieve the proper flow-pressure at fixtures by the use of an upfeed system. Pressure at the fixtures is usually more than required. When this causes an inconvenient splash, as at lavatory basins, a flow restrictor can be used in the faucet outlet.

Various parts of the typical upfeed system shown in Fig. 9.33 will now be considered, beginning at the point where supply water enters the building. In cold climates, water in the service entry pipe (or tube) must not freeze. The pipe must therefore be below the *frost line* of frozen ground. This could vary from 0 to 7 ft (0 to 2 m), depending on geographical location. The onset of winter in such climates requires the closing and draining of pipes supplying the hose bibbs (and other external piping) by means of a *stop-and-waste* valve. Houses left *unheated* in cold-winter weather must be entirely free of water that could freeze and burst the pipes. Note the drain valve at every low point in the system. House shutoff controls are usually located at the main, at the curb, and within the house.

Meters have recently taken on a new role: along with measuring the water quantity for which the occupant is to be charged, they now serve a restrictive function. During water shortages, they can be used to indicate water use in excess of established limits, beyond which limit fines are imposed and in some cases water supply reduced by valves controlled by the water company.

Treatment is most often performed to reduce hardness that could clog piping and equipment, or to neutralize acidity—a source of corrosion. During the short periods when the treatment tanks are valved off for backwashing or other servicing, the *bypass shutoff valve* is opened.

From this point on, the water continues under pressure to

1. Supply makeup water to the house-heating boiler, as required.
2. Supply water to and pressurize the cold water mains and branches, including the garden hose bibbs.
3. Supply water to and pressurize the domestic hot water system through the hot water heater, the hot water storage tank, and the mains, branches, and circulating lines.

The air-filled expansion chambers on cold water *runouts* absorb and reduce the shock of so-called "water hammer" when faucets are shut off. On hot water runouts they perform the same function, *plus* they allow for the expansion of the hot water as it increases in volume with in-

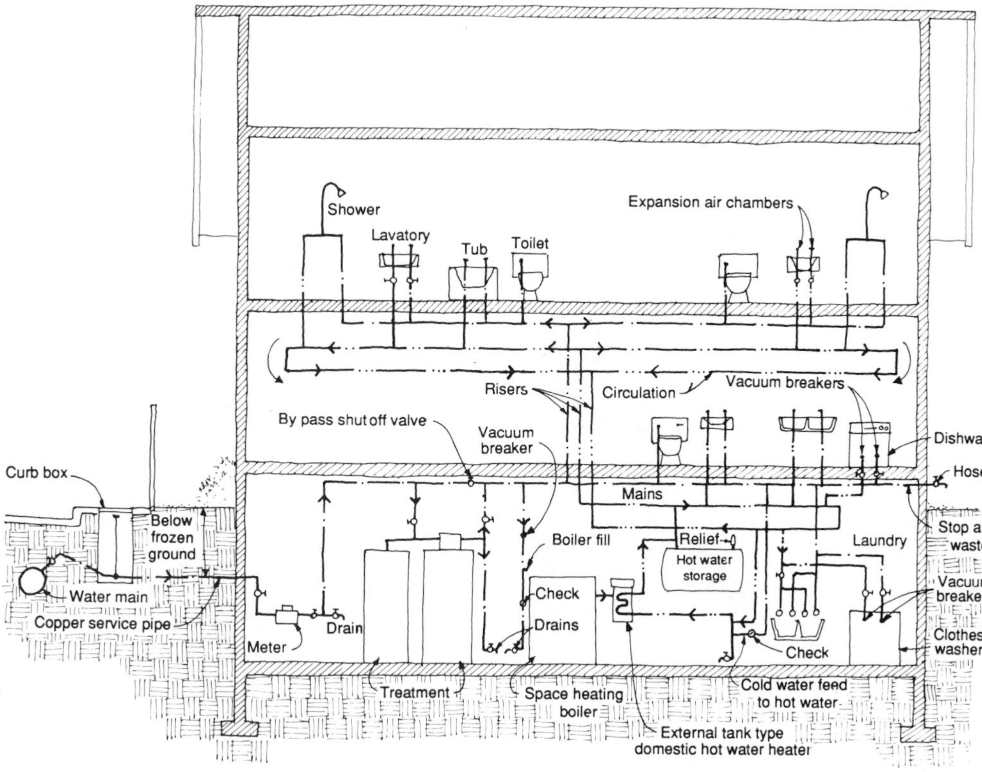

Fig. 9.33 Upfeed water distribution by pressure in street mains. A schematic section of the water services in a residence.

creasing temperature. Vacuum breakers prevent back-flow of polluted water into pipes carrying potable (hot *or* cold) water. Water from all fixtures and appliances, such as dishwashers, clothes washers, and boilers, is thus isolated.

In the legend for Fig. 9.33, the word *schematic* is significant. The entire diagram shows all parts lying in a two-dimensional *plane*. Obviously, in a real house a three-dimensional system would exist. For instance, economy of piping would suggest that, if possible, kitchen and lavatory, as well as the two upstairs bathrooms, be placed back-to-back.

(c) Principles of Downfeed Distribution. Water pumped directly from the street main (or from a basement "suction tank" filled by gravity from the main) is lifted to a roof-storage tank. In cold climates, the tank water in such exposed locations is kept at temperatures above freezing by heating coils in the tank. The fact that water pressure increases with distance below the tank water level is clearly shown by the construction of the tank (see Fig. 9.34). The iron rings, tensioned by adjustable threaded clamps, become more closely spaced toward the bottom of the tank, where the greater water pressure makes it increasingly difficult to restrain the vertical wooden staves of this cylindrical barrel.

Tanks like these add interest, though perhaps not beauty, to a building's silhouette. A "fence" enclosing this tank would have to be about 24 ft (7.3 m) high. Architecturally, this rectangular lump on the roof would not be much of an improvement. Yet in the century following the appearance of such structures as the Goodwill Building, our technology has become more complex. Presently, for most high-rise buildings, including many 60 stories and more in height, an entire rooftop crowded with equipment and technical facilities is needed to serve the stories

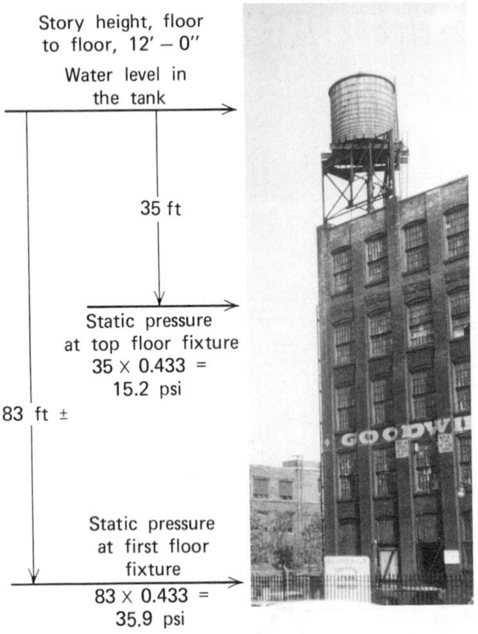

Fig. 9.34 *High-rise building of the 1870s, with downfeed distribution. Although upfeed distribution from a street main had been attempted in buildings higher than two or three stories, it was evident then (as it is now) that street main pressures that may drop below the usual 50 psi (345 kPa), together with heavy use, would result in very low pressure as fixtures in upper stories. Therefore, water is pumped from the main to elevated wooden roof tanks high enough to assure reasonable pressure at the top story and ample pressure at the bottom of the downfeed run.*

below, or the uppermost *zone*. The items could include

Water storage tanks.

Two-story penthouses over elevator banks.

Chimneys.

Numerous plumbing vents.

Exhaust blowers.

Air-conditioning cooling towers.

Cantilevered rolling rig to support scaffold for exterior window-washing.

Perimeter track for window washing rig.

Many of these units can be seen in the photographs of the roof of the Chase Manhattan Building shown in Fig. 9.35.

Thus, since the 1960s, a tall building commonly has a *band* or *screen* two stories (or more) high above the structural roof. It might be said that it all began with the need for elevated water-storage tanks.

(d) Tall Building Downfeed Distribution. Fig. 9.36*a* shows a medium-rise building, in which one elevated tank can serve all the lower floors. For taller buildings, it is advisable to separate groups of floors into *zones* with a maximum height (for plumbing pressure limits) of about 150 ft (about 45 m). This zone-height limitation is based on the height-to-static pressure relationship. At the top of the zone (about 35 ft (10 m) below the storage tank), the minimum desirable pressure is probably at least 15 psi (103 kPa). At the bottom of the zone, the maximum desirable pressure is perhaps 80 psi (552 kPa); above this pressure, damage to fixtures might occur.

$$80 \text{ psi} - 15 \text{ psi} = 65 \text{ psi difference}$$

$$65 \text{ psi} \times 2.3 \text{ ft/psi} = \text{about } 150 \text{ ft}$$

In SI units,

$$552 \text{ kPa} - 103 \text{ kPa} = 449 \text{ kPa difference}$$

$$449 \text{ kPa} \times 0.1 \text{ m/kPa} = \text{about } 45 \text{ m}$$

With pressure-reducing valves at lower floors, however, these zones can be much higher, as shown in Fig. 9.36*b*.

Consider the system described in Fig. 9.36*a*, beginning with the elevated tank. The lower part of the tank often serves as a reserve space to hold a supply of water for a fire hose system. In this case, only the water in the upper part is available for use as domestic (service) water. The amount stored must be enough to supplement what the pump will deliver during the several daily hours of high demand that occur in most buildings. The pump then continues, often for several hours, to replenish the house supply that had become partially depleted during the busy period. The suction tank is a buffer between the system and the street mains. It usually holds enough reserve to allow the pumps to make up the periodic depletion in the house tank. It refills automatically by gravity flow from the street main that, consequently, will not suf-

Fig. 9.35 (a) *High-rise building of the 1970s, with downfeed distribution. Uniquely, the elevated wooden outdoor water storage tank has not changed very much in 100 years (see Fig. 9.34). Its function is the same, and the tank's water level remains several stories above the top floor plumbing fixtures. As part of what might be called "Rooftop City," the tank has joined the contemporary equipment complex, surrounded by the visual barrier of a louvered "skirt."* (b) *The Chase Manhattan Building in New York City. Skidmore, Owings and Merrill, Architect. A helicopter view. Photo by Sky Service.*

fer as much of a drop in pressure as it would if it were connected directly to the suction side of the house pumps. Neighboring water users are protected from the adverse effects of sudden demands within adjacent large buildings. See Fig. 9.37.

House tanks and suction tanks are sometimes made of steel plate and divided vertically in half, each half having identical piping and controls. Hence, one half of the tank can be cleaned out at a time during hours of low demand without shutting down the entire system. One full-capacity pump is supplemented by an equivalent standby pump for alternate use. Since there is no suction lift below the pump or any fixture pressure at the top of the delivery leg (house tank supply), the head against which the pump works is the sum of the distance from the suction tank water level to the top of the house tank and the feet of head equivalent to the friction loss in the tank supply pipe. For this kind of service the vertical piping is on the order of 3 or 4 in. (76 or 102 mm) in diameter for large buildings. Sizes are established by a formal engineering design.

The house supply is fed by a short pipe from the house header to the cold water header that circles the top story and connects to many downfeed cold water risers. For simplicity, Fig. 9.36*a* shows only two risers and also omits many valves and controls.

Figure 9.36*b* is even more simplified. The hot water circulating originates as cold water at the house tank header and takes quite a long route. Descending to the bottom of the hot water

Fig. 9.36 (overleaf) (a) *Downfeed water distribution, a schematic section. Part of the water services for a 10-story building. Hot water circulation moves through the hot upfeed in two directions at the hot water heater and down to the tank through the two downfeed hot water risers. For details of one type of steel house tank, a typical centrifugal house pump, expansion joints, and expansion in hot water riser, see Fig. 9.37.* (b) *Downfeed water distribution, a schematic section. Part of the water services for a zoned building. Zone tanks include a fire reserve but standpipes are omitted from* this *drawing. For detail of steam-type domestic water heater, see Fig. 9.21.*

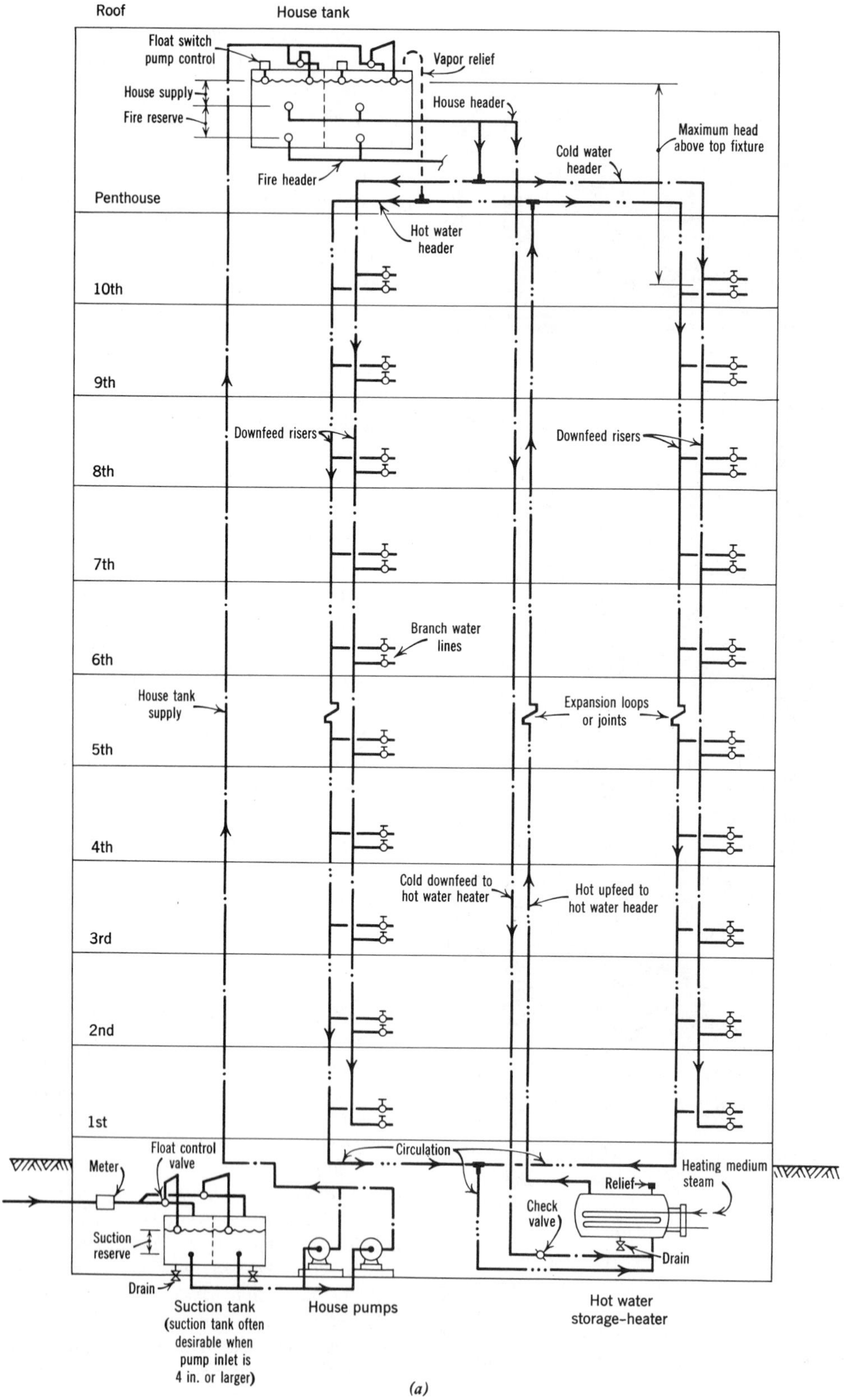
Roof
House tank
Float switch pump control
Vapor relief
House supply
Fire reserve
House header
Maximum head above top fixture
Cold water header
Fire header
Penthouse
Hot water header
10th
9th
Downfeed risers
Downfeed risers
8th
7th
Branch water lines
6th
House tank supply
Expansion loops or joints
5th
4th
Cold downfeed to hot water heater
Hot upfeed to hot water header
3rd
2nd
1st
Float control valve
Meter
Circulation
Heating medium steam
Relief
Check valve
Suction reserve
Drain
Drain
Suction tank (suction tank often desirable when pump inlet is 4 in. or larger)
House pumps
Hot water storage-heater

(a)

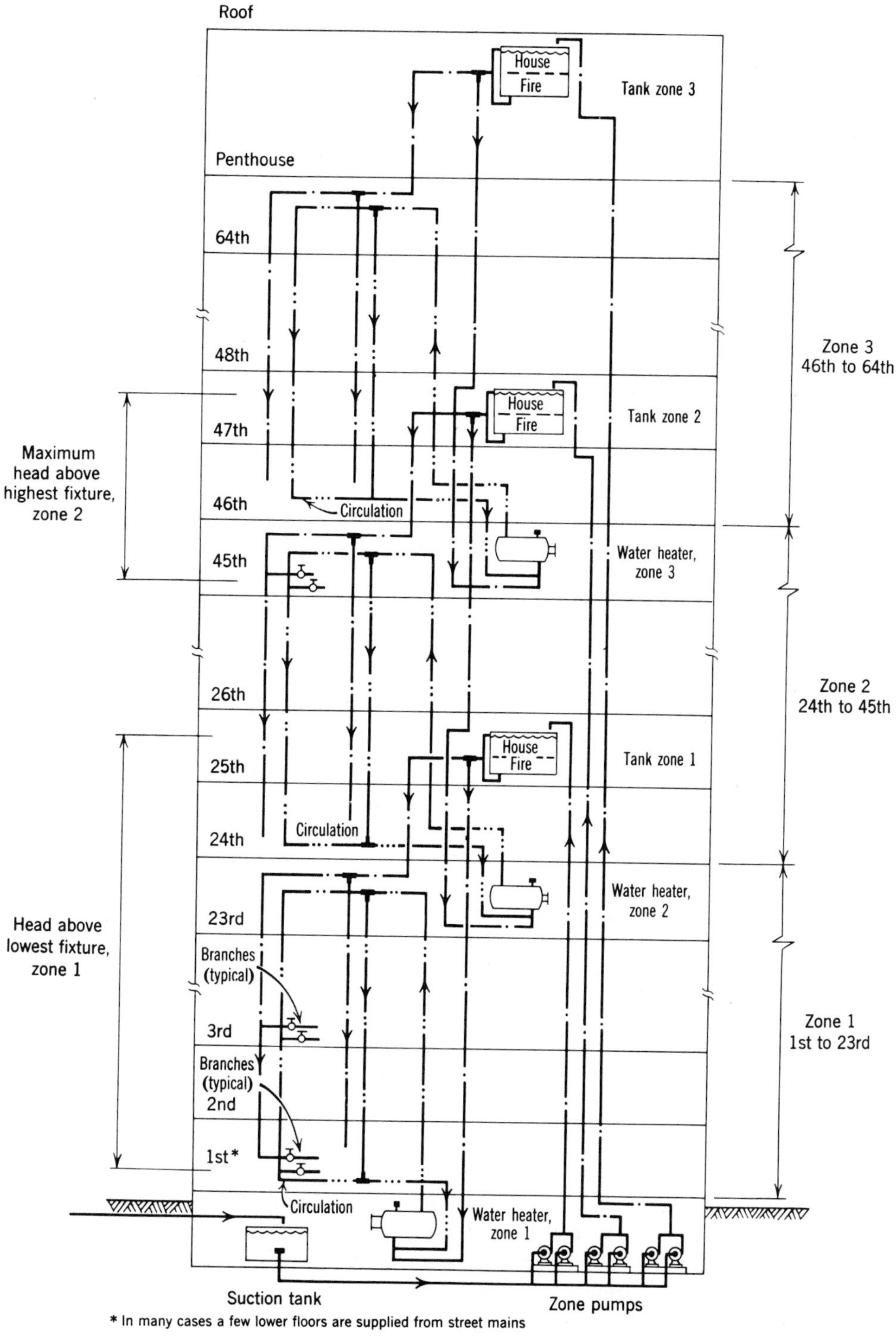

(b)

Fig. 9.36 (continued)

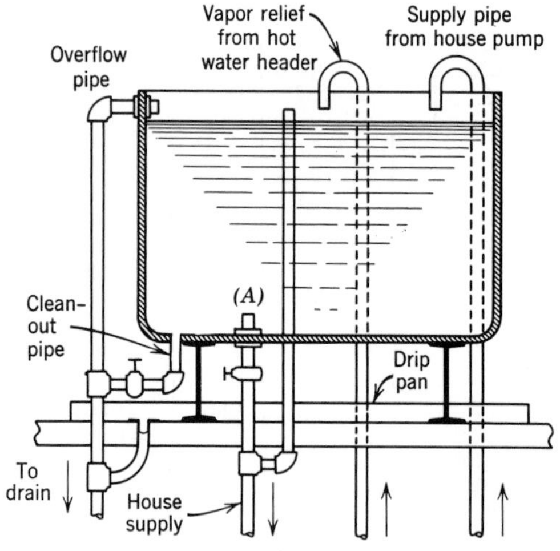

(a)

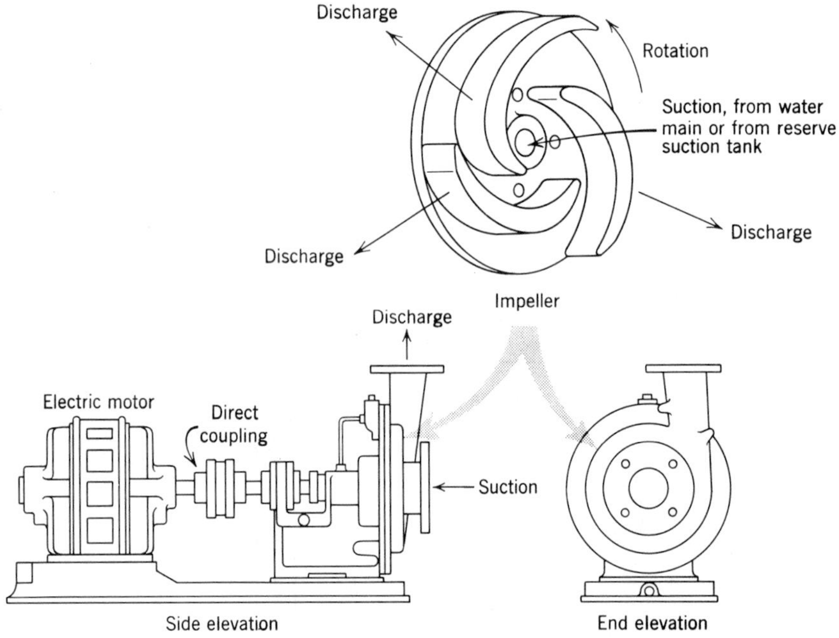

(b)

Fig. 9.37 (a) *House tank in elevated position for downfeed by gravity. Sediment in tank is drawn off through clean-out pipe and is prevented from entering house supply by pipe projection at* A. *Water for fire reserve could be provided by additional piping or by a separate tank.* (b) *Centrifugal house pump, commonly used to fill elevated house tank in top story or at intermediate mechanical floor. Electric motor responds to float switch pump control at the tank.*

heater, it rises to seek its own level at the hot water header, becoming available there for hot water downfeed on demand. All of this occurs as flow below the general pressurizing effect of the house tank. In effect, when there is a cold- and hot-water demand on a story near the top of the building, the cold water makes a short trip down to the faucet, while the hot water goes through three vertical pipes instead of one.

The general scheme just described, with tank above and heater below, is used in multiple forms for very tall buildings. The zones are quite independent; the only common service is that provided by the general suction tank. By this zoning method, problems of pipe expansion, excessive pipe sizes, and high pressures in lower stories are minimized. Commonly, 2½ stories, or about 35 ft (about 10 m), comprise the minimum pressure head above the top fixture served by any zone tank. The static pressure created at the fixture is thus 35 × 0.433 = 15 psi (10 × 10 = 100 kPa). If, during flow, not too much pressure is lost in friction, flushometers (flush valves) can be placed at this level, though flush tanks, because of their lower fixture-pressure demand, must often be accepted. Quite the opposite problem occurs at the bottom of the zones, where excessive pressures must be reduced at the fixtures. In zone I of Fig. 9.36*b*, first-floor fixtures are below a head of 24½ stories, or about 149 psi (1027 kPa) of static pressure. It is obvious that pressure-reducing valves must be used, and that fixture control valves must be throttled.

Pipe and Tube Expansion. The range of temperature from the normal indoor air temperature of about 70 F to that of service hot water (which often exceeds 160 F) is an index of the expansion of pipes and water as their temperatures rise from shut-down status (70 F) to operating status (160 F) (see Table 9.13). The longitudinal elongation of pipe, though negligibly small in houses, can be appreciable in a tall building. Two methods of allowing freedom for this longitudinal motion in long runs of expanding hot water piping are shown in Fig. 9.38*a*. The use of these devices precludes the buildup of excessive stresses in the metal of the pipes and the tendency of the pipes to buckle laterally.

TABLE 9.13 **Thermal Expansion of Pipe and Tubing**

Elongation in in./100 ft of pipe or tube for various increases in temperature. Multiply values below by 0.833 to obtain mm/m.

Increase in Temperature (F)	*(°C)*	*Steel Pipe*	*Copper Tubing*
20	11.1	0.149	0.222
40	22.2	0.299	0.444
60	33.3	0.449	0.668
80	44.4	0.601	0.893
100	55.5	0.755	1.119
120	66.7	0.909	1.346
140	77.8	1.066	1.575
160	88.9	1.224	1.805
180	100	1.384	2.035
200	111.1	1.545	2.268

NOTE: Special care must be taken to adapt for the expansion of *plastic* pipe, the elongation of which is about five times that of copper tubing under the same conditions.

EXAMPLE 9.5. A 20-story zone in a tall building has a height of 280 ft. What will be the increase in length of a copper tube carrying "service hot water" (domestic hot water), when its temperature increases from 70 to 160 F?

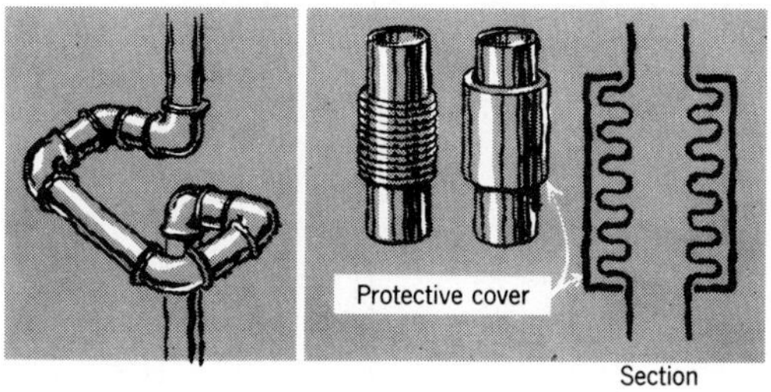

(a)

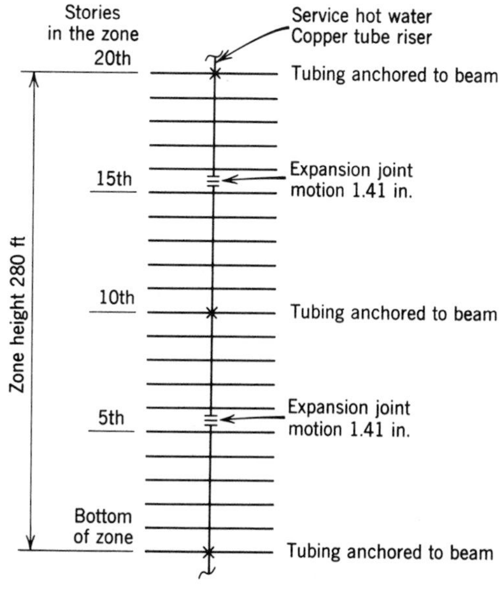

(b)

Fig. 9.38 (a) *Accommodation for the expansion of hot water piping or tubing. Left: Expansion joint of pipe and fittings. Right: A manufactured product.* (b) *(Example 9.5) Suggested scheme for location of the points of anchorage and expansion for service hot water tubing in a 20-story zone.*

SOLUTION. The difference in temperature is 90F°.

Elongation per 100 ft for 90F° increase is 1.01 in. (Refer to Table 9.13).

$$\text{expansion} = 2.8 \times 1.01 = 2.82 \text{ in.}$$

There are a number of ways of providing for this expansion. The one shown in Fig. 9.38*b* consists of accepting the motion at two locations, which would make the expansion in each case 1.41 in. Equidistant anchorage to fix the tubing is provided at the bottom, the tenth floor, and the twentieth floor. The support of the vertical riser at floors other than those at which it is anchored could consist of clamps of the type illustrated in Fig. 9.43, supported on springs.

(e) Pumped Upfeed Distribution. This distribution system (see Fig. 9.39) is for medium-size buildings—those too tall to rely on street main pressure but not so tall as to necessitate heavy storage tanks on the roof. This sophisticated equipment can deliver water at rates varying from what is needed for two or three faucets to full building demand, while maintaining at each outlet a pressure within 2 psi (14 kPa) of the design pressure for that outlet.

The installation shown in Fig. 9.39*a* and *b* is a triplex pump group. According to demand, one, two, or three pumps will operate. Since each pump is of the variable-speed type, virtually an infinite number of delivery rates can be achieved within the zero to maximum design rate. The pumps operate in sequence. When a very small rate of demand occurs, the smallest or "jockey" pump starts in response to a low voltage impressed on its motor. This and all other operations are triggered and adjusted by the pressure sensor at the base of the riser. The jockey pump continues to run until it has reached its maximum delivery rate, at which time the first of the larger pumps cuts in, joined by the other larger pump when required. Sequential operation of the three pumps, each increasing in delivery as called for by the sensor, meets the requirement for increasing supply at nearly constant pressure.

It would appear that the second of the two larger pumps would run less frequently and thus get the least wear. However, wear is equalized

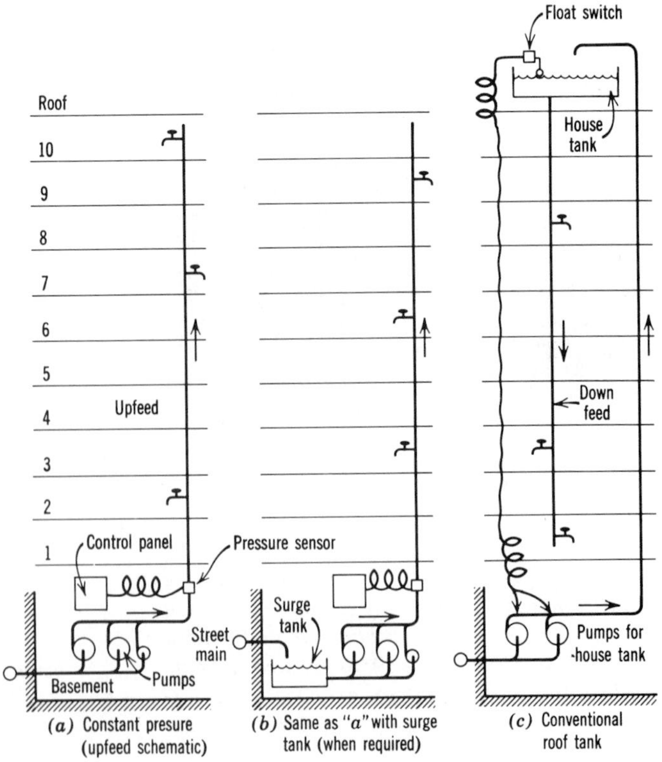

Fig. 9.39 Constant-pressure upfeed pumping (a) *and* (b), *compared to gravity downfeed from house tank on roof* (c). *(By permission of* Progressive Architecture.*)*

by the assigning of "lead" and "lag" positions. For a period of 24 hours, one of the large pumps holds the lead position and starts after the jockey pump, giving the other large pump a smaller burden. The next day the rested pump takes over the more active assignment. All of this occurs automatically.

At full operation this triplex unit can put a suction demand on the street main that could seriously reduce the available water pressure in the neighborhood (Fig. 9.39*a*). Therefore, city officials sometimes require that the system feed on a surge tank, filled by casual flow from the street main, independent of the building requirements (Fig. 9.39*b*). This requirement is often made when the design indicates a maximum building demand in excess of 400 gpm (25 L/s).

Obvious advantages of upfeed pumping are the elimination of the house tank and the heavy structure that transmits its weight down to the footings and, of course, the elimination of the necessary periodic cleaning of the tank. A shortcoming is the lack of reserve storage, which could cause a serious problem during an electrical power failure. However, minimum flow during this kind of emergency can be arranged if a diesel or other independent standby motor is available to drive one of the pumps.

9.6 Piping, Tubing, Fittings, and Controls

(a) Piping, Tubing, and Fittings. The conveyance of water through buildings to locations of use implies the design of a system of piping or tubing that efficiently fulfills its purpose, is easily maintained, and interferes as little as possible with the interior architectural form. It is common that except in basements, in utility rooms, and at points of access to controls, the system is concealed. Stud-and-joist construction provides space for concealment, but in concrete buildings, vertical and horizontal furred spaces must often be provided.

In general, public or private treatment should be provided to correct corrosive qualities. When this is done, it is theoretically suitable to use a cheaper piping material—steel. Yet prudence suggests that a better material be selected. In the nonferrous group, red brass and copper tubing are effective in corrosion resistance. Copper tubing is a very popular choice. It is less expensive than brass, assembles more easily, and is not subject to dezincification (attack by acids on the zinc in brass). For use in handling aggressive waters, plastic is often a good choice. Like copper, it is light in weight and assembles with great ease.

For ferrous pipes and "iron pipe size" brass, threaded connections are used. The external, tapered thread on the pipe is covered with pipe compound and screws tight against the internal tapered thread of the coupling or other fitting. The solder-joint connection in copper depends on capillary attraction that draws the solder into a cylinder of clearance between the mating surfaces of tube and fitting. This occurs after the surfaces are polished and fluxed and the parts placed in final position. They are then heated, and molten solder is applied to the circular opening where the fitting-edge surrounds the tube, with a small clearance. It is then drawn into the cylindrical connection. Solders are often tin-antimony alloys. This kind of joint permits the advantageous setting up of an entire tubing assembly without turning the parts (as in threaded installations), and before the soldering commences. For the same strength, copper tubing may have thinner walls because no threads need to be cut into it. Its smooth interior surface offers less friction to flowing water. While threaded- and solder-joint connections are the most common in small work, there are many other types. Ferrous pipes in the larger sizes are often welded or connected by bolted flanges (see Figs. 9.40 and 9.41 and Table 9.14).

(b) Plastic Pipe. Most of the plastic pipes and fittings now produced are synthetic resins derived from such materials as coal and petroleum.

Rapid increase in the development, acceptance, and use of plastics for water piping, fittings, and, indeed, drainage systems (see Chapter 10) suggests a separate discussion of this *family* of materials.

1. *Selection of material.* The chemistry of plastics is quite intricate. The material can ap-

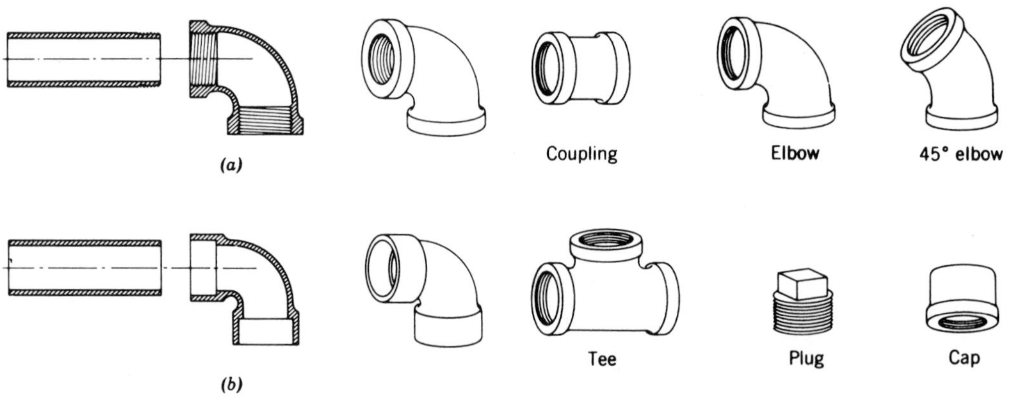

Fig. 9.40 Methods of connecting pipes and fittings, and tubes and fittings. (a) *Threaded: for ferrous pipe and fittings and for "iron pipe size" (IPS) brass.* (b) *Soldered: for copper tubing and fittings. A sliding fit similar to that of* (b) *is used for the solvent weld of plastic connections.*

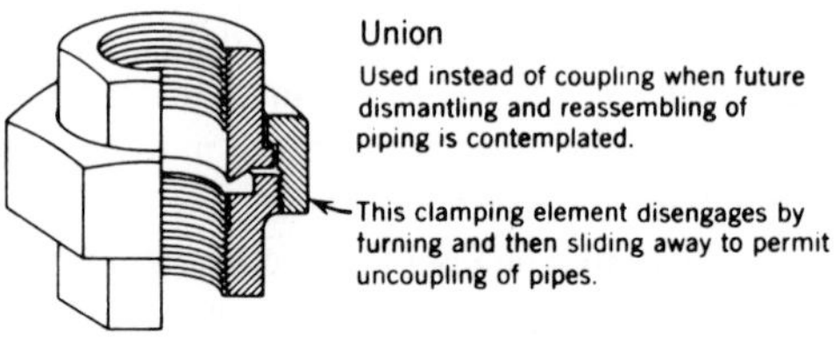

Union
Used instead of coupling when future dismantling and reassembling of piping is contemplated.
This clamping element disengages by turning and then sliding away to permit uncoupling of pipes.

Fig. 9.41 Examples of threaded *pipe fittings for ferrous or brass pipe. A few of the many fittings used in water piping. These and all common fittings are also available for solder-joint (copper) or solvent-weld (plastic) connections, and usually for transition from one system or material to another.*

pear in a great variety of forms, a few of which, especially suitable for water piping, are listed in Table 9.14.

2. *Sanitation.* Early concern about the possibility of adding odor, taste, or toxicity to potable water has long been dispelled. The National Sanitation Foundation tests and certifies plastic pipe. Their NSF seal must appear on pipes that are to carry potable water.

3. *History.* About 100 years ago the first plastic material, celluloid, appeared. Another product, Bakelite, was developed in 1905. Used for water piping, polyethylene has given successful service for more than 30 years. Intensive research and standardization beginning about 1945 has established many plastic materials as appropriate for plumbing.

4. *Code acceptance.* Codes usually govern which materials are acceptable to the local jurisdiction. Codes written by associations of plumbing officials or plumbing contractors are called *model codes.* The following model codes approve the use of plastic piping for all or a portion of the plumbing system.

- BOCA Basic Plumbing Code (Building Officials and Code Administrators International.)
- National Standard Plumbing Code (National Association of Plumbing, Heating, Cooling Contractors).
- Southern Standard Plumbing Code (Southern Building Code Congress).
- Uniform Plumbing Code (International Association of Plumbing and Mechanical Officials).

5. *Effects of temperature and vibration.* Most of the materials used for piping are thermoplastics and have the quality of repeated softening under the application of heat. PVDC material can carry water at 180 F (82°C), but plastic pipe should not be subjected to temperatures higher than this. Expansion (see note to Table 9.13) is great and affects the piping design. Quite shockproof, plastic is used in the plumbing systems of about 80% of mobile·homes.

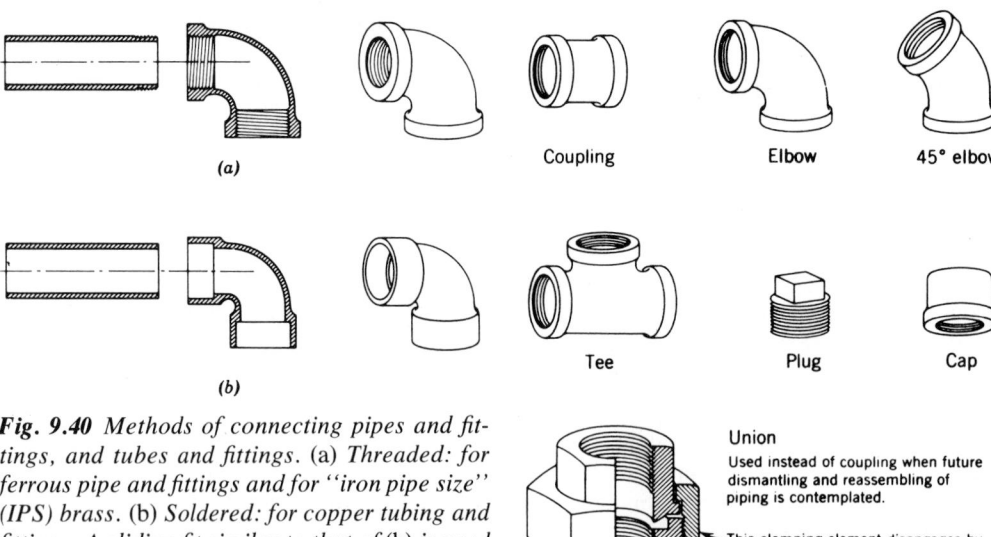

6. *Information source.* The Plastics Pipe Institute, a division of the Society of the Plastics

TABLE 9.14 **Pipe and Tubing for Water Services**

*Part A. **Characteristics of Materials***

Kind of Pipe	*Material or Manufacture*	*Connections*	*Qualities*	*Notes*
Steel	Butt welded to 2 in. (51) mm diameter, seamless, large sizes	Threaded	Basic	Should be used only when water is not corrosive
Brass, red	85% copper 15% zinc	Threaded, "IPS" (iron pipe size)	Corrosion-resistant	Bulky, because of the need for threading
Copper tube, type "K"	Seamless, hard or soft temper	Soldered fittings	Corrosion-resistant and easy to fabricate	Thinner-walled than brass; easy to put together and dismantle
Copper tube, type "L"	Seamless, thinner walls than type "K," hard or soft temper	Soldered fittings	Corrosion-resistant and easy to fabricate	Thinner-walled than brass; easy to put together and dismantle
Plastic[a]	See Part B	Solvent cement weld[b]	Very easy to fabricate	Not subject to electrolytic corrosion
Nickel, silver and chrome	Copper, nickel, and zinc, steel, and chromium	Threaded	Corrosion-resistant	Special applications
Galvanized steel	Zinc-coated steel	Threaded	Moderately corrosion-resistant	Suitable for mildly acid waters

*Part B. **Plastic Piping***

Symbol	*Material*	*Cold Water*	*Hot Water*
PE	Polyethylene	√	
ABS	Acrilylonitrile-Butadiene Styrene	√	
PVC	Polyvinyl Chloride	√	
PVDC	Polyvinyl Dichloride	√	√[c]

Source: Courtesy of the Plastics Pipe Institute.

[a]Upper limit of temperature, hot water, 180 F (82°C)

[b]For ABS and PVC (Part B).

[c]Developed for this special use. Most plastic materials not currently approved for hot water piping.

Industry, will furnish objective information about plastic pipe.

(c) Valves and Controls. A good system utilizes many valves. It is usually desirable to valve every riser, the branches that serve bathrooms or kitchens, and the runouts to individual fixtures. This setup facilitates repairs at any location with a minimum of shutdown within the system. Treatment equipment will have a valved bypass (see Fig. 9.33). Pumps and other devices that may need repair should be disconnectable by unions (Fig. 9.41) after valves are closed.

The gate valve (Fig. 9.42*a*), with its retractable leaf machined to seal tightly against two sloping metal surfaces when closed, offers the least resistance to water flow when open. It is usually chosen for locations where it is left completely open most of the time. The compression-type globe valve (Fig. 9.42*b*) is usually used for the closing or throttling of flow near a point of occasional use. Faucets are usually of the compression type, as are drain valves or hose connections. They are similar to the angle valve (Fig. 9.42*d*). When it is necessary to prevent flow in a direction opposite to that which is planned, a check valve (Fig. 9.42*c*) is introduced. The hinged leaf swings to permit flow in the direction of the arrow but closes against flow in the other direction.

(d) Pipe Support. If a piping system of conventional dimensions were to stand alone, without a building to rest on, it would quickly collapse. Quite heavy because of its water content, it needs closely spaced supports (Fig. 9.43). Vertical runs of 1-in. (25 mm) piping should be supported at every story, but larger sizes may extend for two stories. Horizontal pipes should be supported at intervals of 10 ft (3 m). Closer spacing (6 to 8 ft) (1.8 to 2.4 m) is preferred for sizes of ½ in. (12 mm) and smaller. Horizontal copper tubing should always be supported at closer spacing than steel. Adequate positioning of horizontal runs is important to assure correct pitch and drainage. Hangers are adjustable for this purpose.

(e) Shock and Hot Water Expansion. Water systems can be noisy. When faucets are shut off abruptly, the force exerted by the decelerated flowing water shakes and rattles the pipes. This phenomenon is called water hammer. Lengths of vertical pipe about 2 ft (0.6 m) long at the fixture branches (Fig. 9.44*a*) will usually solve this problem. They trap a certain amount of air, which absorbs the impact of the water with some resilience. A somewhat better and a more controllable device is a "rechargeable air chamber" (Fig. 9.44*b*). By closing the valve and draining the water through the hose bibb while the petcock above is open to admit air, the chamber may be refilled with air. Closing the petcock and hose bibb and opening the valve completes the service operation and reconnects the device with the water system. Rechargeable

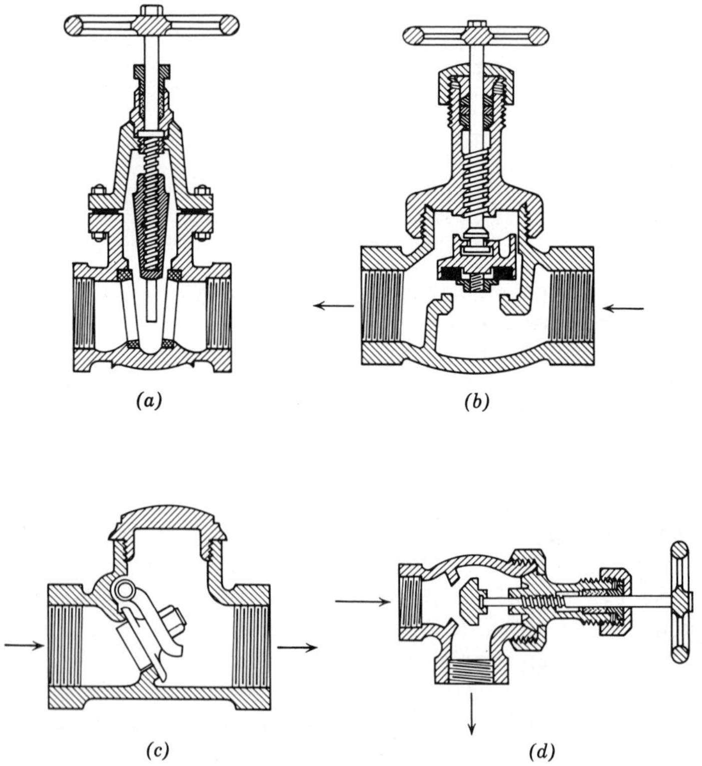

Fig. 9.42 *Typical valves for water systems.* (a) *Gate valve.* (b) *Globe valve.* (c) *Check valve.* (d) *Angle valve.*

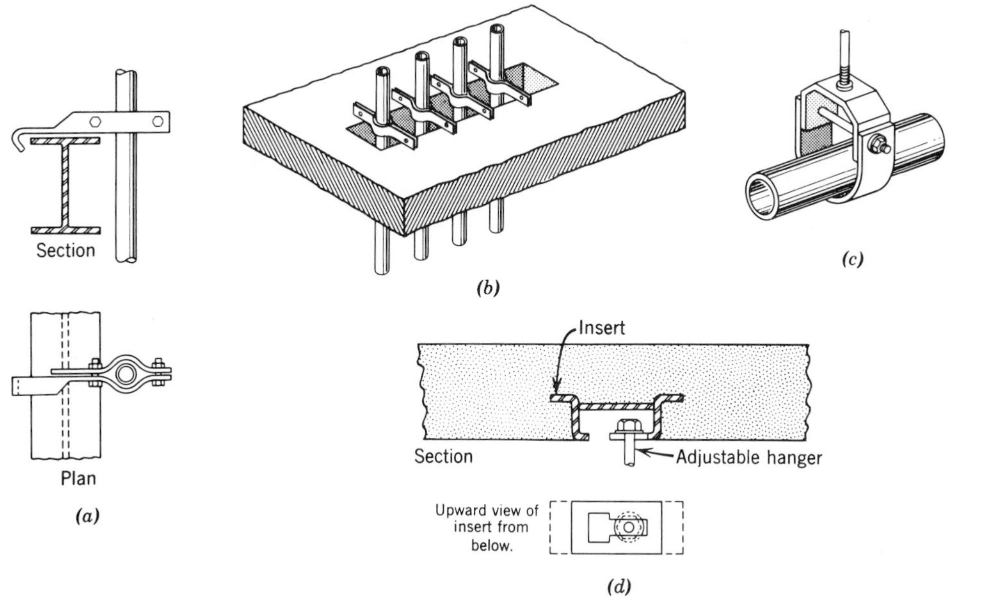

***Fig. 9.43** Pipe supports.* (a) *Vertical riser supported at steel beam.* (b) *Vertical riser group supported at slot in concrete slab.* (c) *Horizontal pipe hung from slab above by adjustable-length clevis hanger.* (d) *Typical metal insert in soffit of concrete to receive hanger rod.*

chambers are used on branch lines adjacent to groups of fixtures. Access for service must be provided. When this method is chosen, the smaller pipe extensions (Fig. 9.44*a*) are usually omitted. Perhaps the best method is the use of a special shock absorber (Fig. 9.44*c*).

Air cushions also protect the relief valve against frequent operation, with the resultant

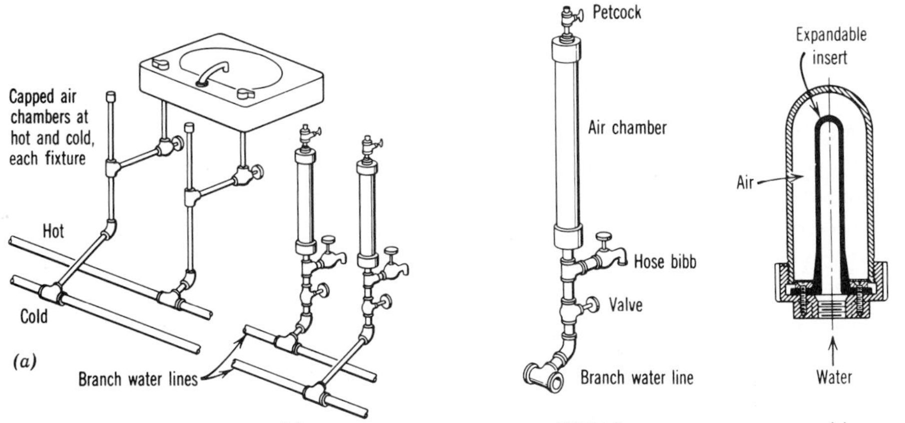

***Fig. 9.44** Shock relief and expansion chambers. Air chambers cushion the shock of the water hammer when the fixture faucets are shut off abruptly. They also permit hot water to expand, instead of periodically forcing open the hot-water-emergency pressure relief valve at the heater or tank.* (a) *Capped air chambers at each supply pipe of each fixture.* (b) *Rechargeable air chambers on hot and cold branch water lines. (Individual fixture chambers are omitted when these are used.)* (c) *Special shock absorber.*

leakage of hot water, as the hot water periodically expands and contracts in closed systems.

(f) Condensation, or "Sweating." The moisture that is always present in air often condenses on the exterior surface of cold pipes. Dropping off the pipes, it creates an unpleasantly wet condition and disfigures finished surfaces. Groundwater in some parts of the United States is 50 F (10°C) and colder. A pipe carrying such water might have a surface temperature of about 60 F (15°C). The psychrometric chart (page 130) indicates that at a summer air temperature of 85 F (29°C), condensation will occur on this pipe when the relative humidity exceeds 40%. All cold water piping and fittings should be covered. Glass fiber ½ or 1 in. (12 to 25 mm) thick is commonly chosen for this purpose. A tight vapor barrier on the exterior surface of the covering prevents the moisture-laden air from penetrating the insulation to reach the colder surface. The insulation provides another advantage of equal importance: it retards heat flow from the warmer air to the water, thus preventing the water from becoming disagreeably warm.

(g) Heat Conservation. Pipes carrying domestic hot water should be insulated to conserve the fuel used to heat the water and to assure a correct water temperature at the point of use. Table 9.15 lists minimum pipe insulation. Parallel hot and cold water piping, even though insulated, should be separated by 6 in. (150 mm) or more, to prevent heat interchange.

Storage tanks and water heaters are usually manufactured with integral insulation. Older devices, however, may have less insulation than today's energy supplies warrant. As a result, many of the older water heaters are retrofitted with added insulation, as shown in Fig. 9.45. Some electric utilities find the savings from conservation so attractive (relative to the cost of building new generating facilities) that they provide water-heater wrapping as a free service to customers.

(h) Irrigation. The highest priority for landscape watering is to design for optimum rainfall retention (see Figs. 8.8 to 8.16). In the past, provision for the watering of the landscape around buildings has frequently been limited to the installation of sill cocks (hose bibbs) at building exteriors. In many areas of North America, half the residential water usage is for outdoor purposes. With increasing demands on a finite water supply, water-conserving irrigation equipment has become available.

Fig. 9.45 *Retrofitting insulation on existing, minimally insulated water-heater storage tanks.*

Lawn sprinklers are relatively inefficient irrigating devices, as much of their water is lost to evaporation and to runoff. It is commonly estimated that lawn sprinkling will provide ½ in. of water per hour per square foot of lawn (or 0.3 gph per square foot) (0.005 L/s per m^2).

Landscape sprinkling is least efficient during the daytime, when hot sun and dry air combine to increase the rate of evaporation. However, nighttime sprinkling is rarely convenient for building custodians or homeowners. One solution to this problem is the use of *sprinkler timing devices* (Fig. 9.46*a*), which control electronic valves and a network of underground supply pipes and sprinkler heads that are permanently installed within landscaped areas. These timing devices typically include controls governing the length of the watering cycle, the time at which the cycle begins (usually before sunrise, when relative humidity is highest), and the number of days between cycles. A "rainy day switch" is often included, so that the irrigation can be discontinued during rainy weather. Rain sensors can be used to shut off irrigation automatically (Fig. 9.46*b*). Tighter control over the water and

TABLE 9.15 **Minimum Insulation for Pipes (Inches)**

Fluid Design Operating Temperature Range (°F)	*Insulation Conductivity*		*Nominal Pipe Diameter (in.)*					
	Conductivity Range [Btu-in./(h-ft³-°F)]	*Mean Rating Temperature (°F)*	*Runouts[a] up to 2*	*1 and less*	*1–1¼ to 2*	*2½ to 4*	*5 and 6*	*8 and up*
Heating Systems (Steam, Steam Condensate, and Hot Water)								
Above 350	0.32–0.34	250	1.5	2.5	2.5	3.0	3.5	3.5
251–350	0.29–0.31	200	1.5	2.0	2.5	2.5	3.5	3.5
201–250	0.27–0.30	150	1.0	1.5	1.5	2.0	2.0	3.5
141–200	0.25–0.29	125	0.5	1.5	1.5	1.5	1.5	1.5
105–140	0.24–0.28	100	0.5	1.0	1.0	1.0	1.5	1.5
Domestic and Service Hot Water Systems[b]								
105 and Greater	0.24–0.28	100	0.5	1.0	1.0	1.5	1.5	1.5
Cooling Systems (Chilled Water, Brine, and Refrigerant)[c]								
40–55	0.23–0.27	75	0.5	0.5	0.75	1.0	1.0	1.0
Below 40	0.23–0.27	75	1.0	1.0	1.5	1.5	1.5	1.5

Source: Reprinted by permission from ASHRAE/IES Standard 90.1-1989, *Energy Efficient Design of New Buildings Except Low-Rise Residential Buildings,* © 1989 by the American Society of Heating, Refrigerating, and Air-Conditioning Engineers, Inc., Atlanta, Ga.

[a]Runouts to individual terminal units not exceeding 12 ft in length.

[b]Applies to recirculating sections of service or domestic hot water systems and first 8 ft from storage tank for nonrecirculating systems.

[c]The required minimum thicknesses do not consider water vapor transmission and condensation. Additional insulation, vapor retarders, or both, may be required to limit water vapor transmission and condensation.

plant relationship can be obtained with *tensiometers* (Fig. 9.46*c*), which monitor the moisture content of the soil at the depth of the plants' root zone. These can be installed so as to override the automatic timing device, thus watering more—or less—frequently, depending on the plants' needs. Instead of sprinklers, *bubblers* (Fig. 9.46*d*) can be installed, with very low flow rates and less evaporative water losses.

Drip irrigation takes a very different approach to that of the flooding method typical of sprinklers. From a network of plastic tubes, either just underground or on the surface, *emitters* (Fig. 9.47) slowly and steadily drip water onto the ground surface at each needy plant. Most of this water soaks into the soil at a rate that is better for most plants than the sudden, short flooding of intermittent sprinkling. Two requirements are especially important for drip systems: the water must not contain materials that can clog the small holes of the emitters, and the pressure must be low. Pressure-reducing valves at the source of the drip system are advisable; if necessary, filters should also be installed there.

Drip irrigation is not a universal solution to landscape watering; it is best for individual plants such as shrubs and small trees, but is difficult to apply to large lawns. Where appropriate, it can achieve significant water conservation compared to sprinklers. A detailed discussion of soil texture, moisture, and plant growth is provided in Lowry (1988, Chap. 4).

9.7 Sizing of Water Pipes

There must be sufficient pressure at fixtures to assure the user of a prompt and generous flow of water. Municipal ordinances often state that the flow must be adequate to keep the fixtures clean and sanitary. The convenience of the user and the objectives of sanitation are consistent with each other and have resulted in prescribed pressures that must be maintained at the various fixtures to assure the proper flow rates. These pressures and flows are listed in Table 9.16.

Fixture pressures vary from 5 to 15 psi (34 to 103 kPa) for fixtures other than hose bibbs. Since the pressure in street mains is usually about 50 psi (345 kPa), it is possible to assure the correct fixture pressure, provided that the water does not have to be lifted to too great a height and that too much pressure is not lost by friction in distribution piping that is too long in *developed length* (actual distance of water flow), or that interposes too many fittings such as elbows and tees, or is too small in diameter.

The pressure components and their total in an upfeed system actuated by street main pressure are as follows.

Proper fixture flow pressure	A
Pressure lost because of height	B
Pressure lost by friction in piping	C
Pressure lost by flow through meter	D
Total street main pressure	E

In a design, items A, B, and E are known and are reasonably constant. The value of A can be found in Table 9.16. Street main pressure, E, is a characteristic of the local water supply. Item B, the pressure lost due to height, can be found by multiplying the height in feet by 0.433 (height in meters by 10) (see the discussion of static head, Section 9.5*a*). Item D, the pressure lost in flow through the water meter, depends on flow and pipe size (see Fig. 9.48), neither of which is yet known. Therefore, the value of item D is *estimated*. [For residences and small commercial buildings, meter size rarely exceeds 2 in. (51 mm)]. Later, it must be checked and a recalculation made if necessary.

The selection of a pipe size is facilitated by Fig. 9.49. Pipe diameter is determined by the point of intersection of a horizontal line representing flow and a vertical line expressing friction loss. To select a pipe size, one needs to know the probable flow and the *unit*-friction loss in the pipe and fittings. Also, the noise created by water flow must be considered. Above 10 fps (3 m/s) is usually too noisy; above 6 fps (1.8 m/s) may be too noisy in acoustical-critical locations.

Flow can be found by assigning the fixture units listed in Table 9.17, the sum of which is an index of the demand flow that can be found in Fig. 9.50. These curves, based on experience, indicate that flow does *not* increase in direct proportion to an increase in fixture units. In larger installations, there is less likelihood that many fixtures will be operating concurrently.

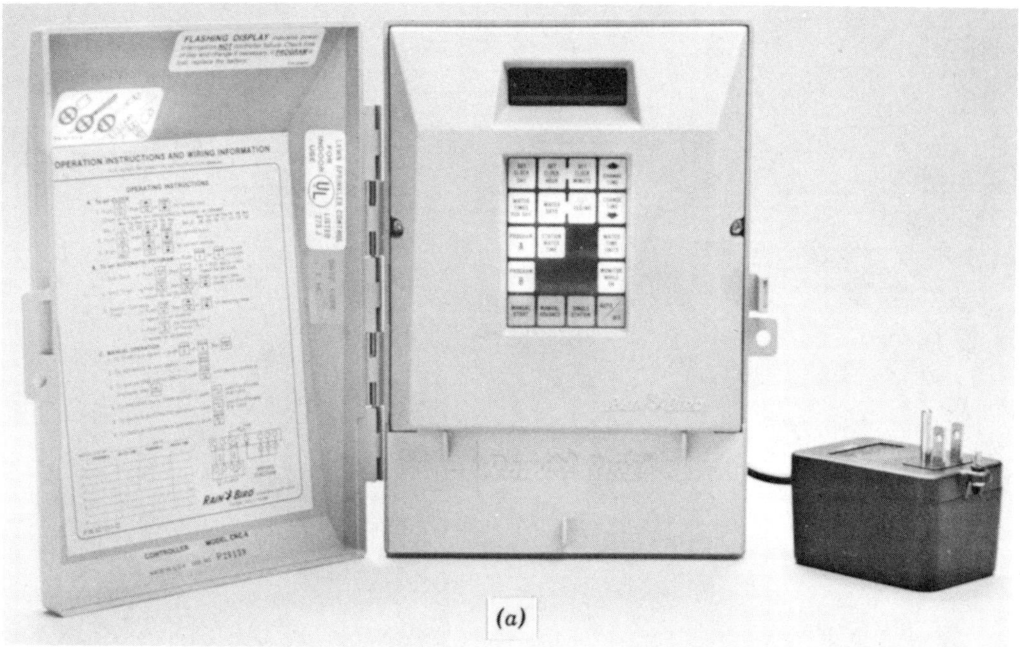

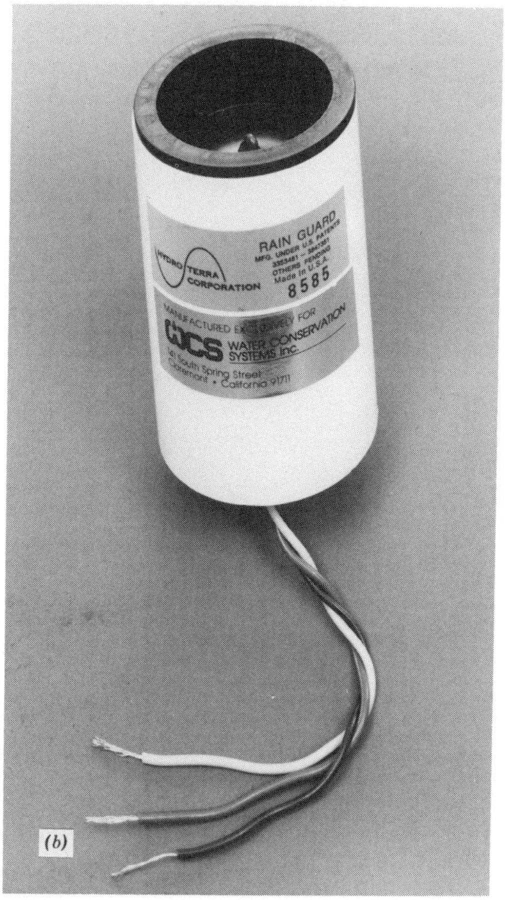

Fig. 9.46 *Components for automatic landscape sprinkling systems.* (a) *Programmable controllers feature multiple stations, up to 14-day sequences, and memory provisions in case of power failure. (Courtesy of Rain Bird Sales, Inc.)* (b) *A solid-state rain sensor automatically prevents needless watering during rainfall.*

To establish the desired friction loss, divide value C (pressure lost by friction in piping) by the *total equivalent length* (TEL) of the piping. This length is the sum of the *developed length* (DL) (total linear distance of water travel) and the length equivalent to the fittings. For instance, Table 9.18 shows that a 90° ell causes a friction loss equivalent to that of 3 ft of pipe in a 1-in.-diameter pipe run. Obviously the number and style of fittings must be estimated, and the *size* of fittings assumed. This is a puzzling but common engineering procedure that sometimes requires several recalculations.

EXAMPLE 9.6. Using the following data—some of which have been arrived at by the assumptions referred to above—find the proper size for a metered water supply main.

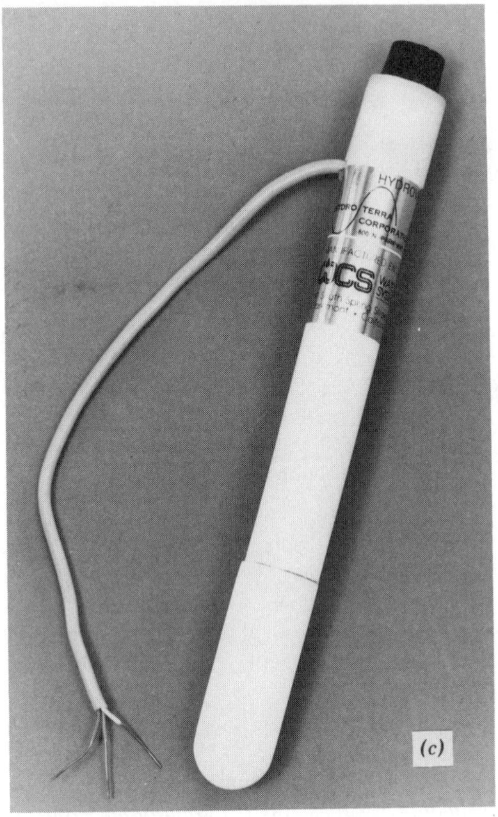

Fig. 9.46 *(continued)* (c) *A tensiometer controls irrigation by monitoring the moisture content of the soil. (Courtesy of Water Conservation Systems, Inc.)* (d) *Bubblers are low-flow substitutes for sprinklers—a step toward drip irrigation. These can be pressure-compensating, to permit constant flow. (Courtesy of Rain Bird Sales, Inc.)*

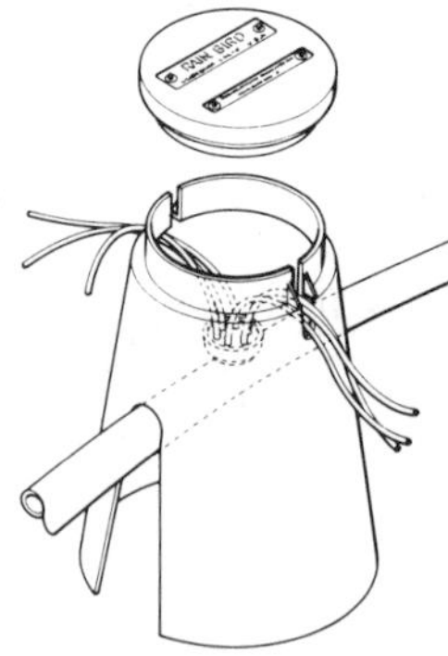

Fig. 9.47 *Equipment for drip irrigation.* (a) *On low-pressure lines, emitters are installed—one for each group of plants to be watered. The lines can be laid on the ground surface, or just under the surface.* (b) *Simple emitter boxes allow easy access. (Courtesy of Rain Bird Sales, Inc.)*

TABLE 9.16 **Minimum Flow and Pressure Required by Typical Plumbing Fixtures**

	Flow Pressure,[a]		*Flow Rate,*	
	(psi)	*(kPa)*	*(gpm)*	*(L/s)*
Ordinary basin faucet	8	55	2.0	0.13
Self-closing basin faucet	8	55	2.5	0.16
Sink faucet, 3/8 in. (9.5 mm)	8	55	4.5	0.28
Sink faucet, 1/2 in. (12.7 mm)	8	55	4.5	0.28
Bathtub faucet	8	55	6.0	0.38
Laundry tub faucet, 1/2 in. (12.7 mm)	8	55	5.0	0.32
Shower	8	55	5.0	0.32
Ball-cock for closet	8	55	3.0	0.19
Flush valve for closet	15	103	15–40[b]	0.95–2.52[b]
Flushometer valve for urinal	15	103	15.0	0.95
Garden hose (50 ft, 3/4-in. sill cock) (15 m, 19 mm)	30	207	5.0	0.32
Garden hose (50 ft, 5/8-in. outlet) (15 m, 16 mm)	15	103	3.33	0.21
Drinking fountains	15	103	0.75	0.05
Fire hose 1 1/2 in. (38 mm), 1/2-in. nozzle (12.7 mm)	30	207	40.0	2.52

Source: U.S. Environmental Protection Agency, *Manual of Individual Water Supply Systems,* 1975.

[a]Flow pressure is the pressure in the supply near the faucet or water outlet while faucet or water outlet is wide open and flowing.

[b]Wide range due to variation in design and type of closet flush valves.

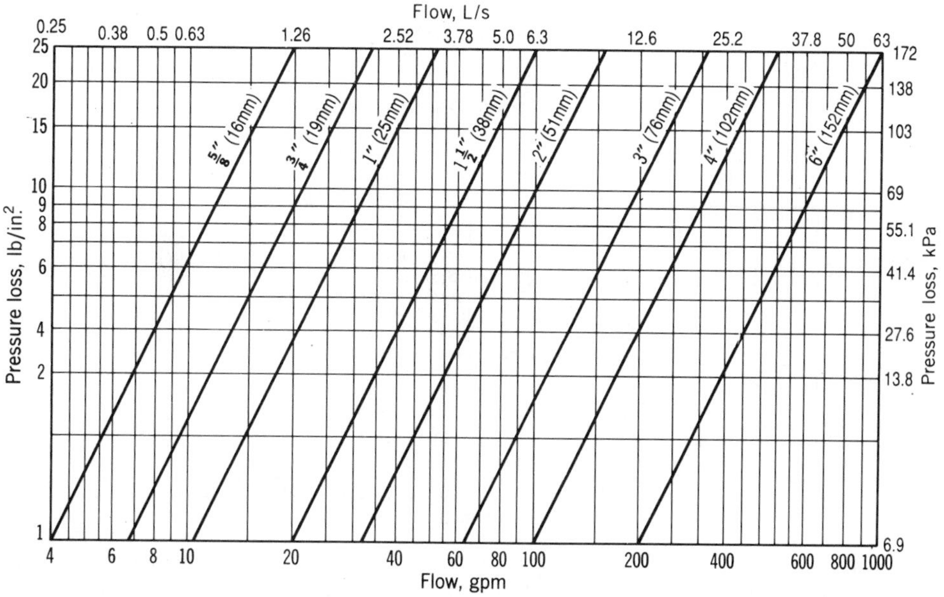

Fig. 9.48 *Pressure losses in water meters. (Adapted from the* ASHRAE Handbook of Fundamentals, *1989. Reprinted by permission.)*

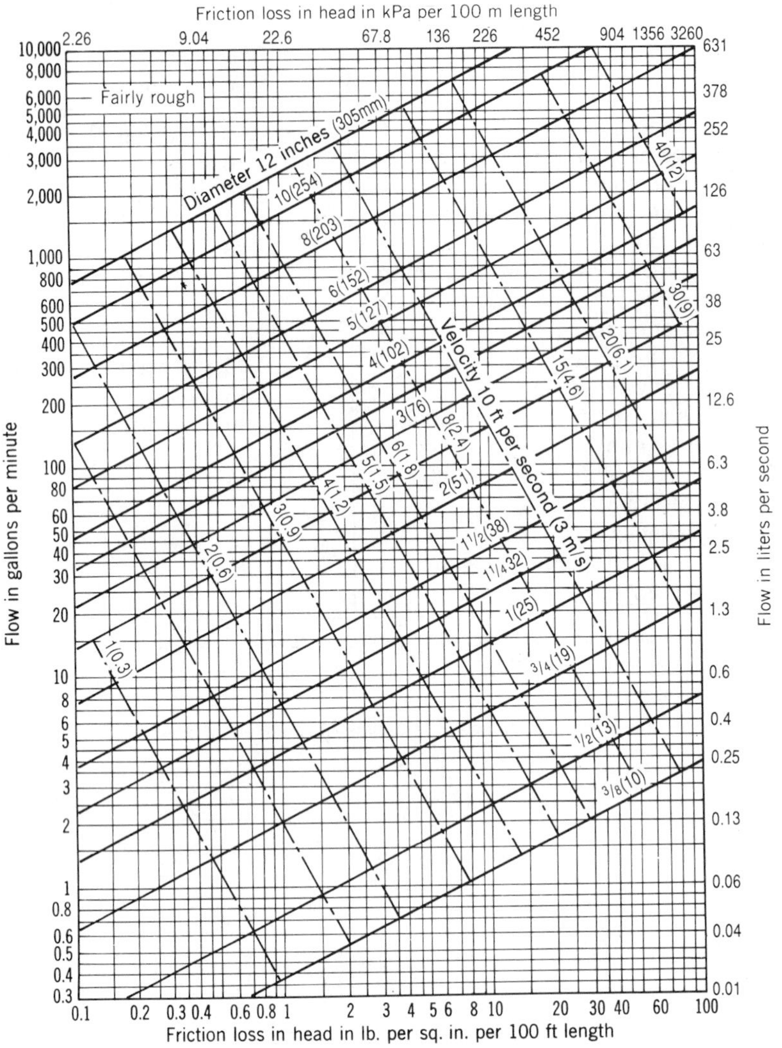

Fig. 9.49 *Flowchart for typical (fairly rough) pipe. Velocity is shown, as an aid in noise control: above 10 fps (3m/s), moving water can be heard within pipes. (Copyright © by the American Society of Heating, Refrigerating, and Air-Conditioning Engineers, Inc., Atlanta, Ga. Adapted by permission from* ASHRAE Handbook of Fundamentals, *1972.)*

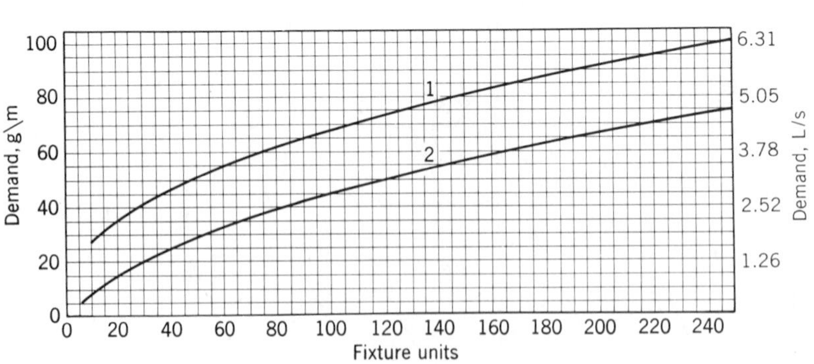

Fig. 9.50 *Estimate curves for demand load. Curve 1 is for a system of predominantly flush valves. Curve 2 is for a system of predominantly flush tanks. (Copyright © by the American Society of Heating, Refrigerating, and Air-Conditioning Engineers, Inc., Atlanta, Ga. Adapted by permission from* ASHRAE Handbook of Fundamentals, *1989.)*

TABLE 9.17 **Demand Weights of Fixtures in Supply Fixture Units**[a]

Fixture or Group[b]	*Occupancy*	*Type of Supply Control*	*Weight in Fixture Units*[c]
Water closet	Public	Flush valve	10
Water closet	Public	Flush tank	5
Pedestal urinal	Public	Flush valve	10
Stall or wall urinal	Public	Flush valve	5
Stall or wall urinal	Public	Flush tank	3
Lavatory	Public	Faucet	2
Bathtub	Public	Faucet	4
Shower head	Public	Mixing valve	4
Service sink	Office, etc.	Faucet	3
Kitchen sink	Hotel or restaurant	Faucet	4
Water closet	Private	Flush valve	6
Water closet	Private	Flush tank	3
Lavatory	Private	Faucet	1
Bathtub	Private	Faucet	2
Shower head	Private	Mixing valve	2
Bathroom group	Private	Flush valve for closet	8
Bathroom group	Private	Flush tank for closet	6
Separate shower	Private	Mixing valve	2
Kitchen sink	Private	Faucet	2
Laundry trays (1–3)	Private	Faucet	3
Combination fixture	Private	Faucet	3

Source: NBS Report BMS79, *Water-Distributing Systems for Buildings*. Copyright © by the American Society of Heating, Refrigerating and Air Conditioning Engineers, Inc., Atlanta, GA. Reprinted by permission from *ASHRAE Handbook of Fundamentals,* 1989.

[a]For supply outlets likely to impose continuous demands, estimate continuous supply separately, and add to total demand for fixtures.

[b]For fixtures not listed, weights may be assumed by comparing the fixture to a listed one using water in similar quantities and at similar rates. For example, a drinking fountain would be rated at 1 Fixture Unit, equal to that of a private faucet lavatory.

[c]The given weights are for total demand. For fixtures with both hot and cold water supplies, the weights for maximum separate demands may be taken as ¾ the listed demand for the supply.

Street main pressure (minimum)	50 psi
Height, topmost fixture above main	30 ft
Topmost fixture type	Water closet with flush valve
Fixture units in the system	85
Developed length (DL) of the piping to the highest and most remote fixture	100 ft
Pipe length equivalent to fittings (commonly estimated at 50% of the DL)	50 ft
System uses predominantly	Flush valves

TABLE 9.18 **Allowance in Equivalent Length of Pipe for Friction Loss in Valves and Threaded Flanges**[a]

	Equivalent Length of Pipe for Various Fittings						
Diameter of Fitting inches	*90-Deg Standard Ell, feet*	*45-Deg Standard Ell, feet*	*90-Deg Side Tee, feet*	*Coupling or Straight Run of Tee, feet*	*Gate Valve, feet*	*Globe Valve, feet*	*Angle Valve, feet*
$\frac{3}{8}$	1	0.6	1.5	0.3	0.2	8	4
$\frac{1}{2}$	2	1.2	3	0.6	0.4	15	8
$\frac{3}{4}$	2.5	1.5	4	0.8	0.5	20	12
1	3	1.8	5	0.9	0.6	25	15
$1\frac{1}{4}$	4	2.4	6	1.2	0.8	35	18
$1\frac{1}{2}$	5	3	7	1.5	1.0	45	22
2	7	4	10	2	1.3	55	28
$2\frac{1}{2}$	8	5	12	2.5	1.6	65	34
3	10	6	15	3	2	80	40
$3\frac{1}{2}$	12	7	18	3.6	2.4	100	50
4	14	8	21	4.0	2.7	125	55
5	17	10	25	5	3.3	140	70
6	20	12	30	6	4	165	80

[a]From NBS Report BMS66 *Plumbing Manual.*

SOLUTION. From the minimum street main pressure, subtract the sum of the fixture pressure, the static head, and the pressure lost in the meter. This sum is

	psi
A—fixture pressure (Table 9.16)	15
B—static head 30 × 0.433	13
D—pressure loss in meter (estimated, Fig. 9.48)	8
	36
E—pressure in street main	50.0
Subtract (A + B + D)	−36.0
	14.0

The pressure lost in 100 ft (DL) of piping plus the 50 ft of piping equivalent to the pressure lost by friction in the fittings therefore can total 14 psi. Total equivalent length (TEL) is 150 ft. This procedure assures 15 psi at the critical fixture. The unit-friction loss, psi/100 ft of pipe, will be

$$14 \text{ psi} \times 100/150 \text{ TEL} = 9.33 \text{ psi}/100 \text{ ft}$$

From Fig. 9.50, curve 1, a flush-valve system with 85 fixture units will have a probable flow (demand) of 64 gpm. Given this information, enter Fig. 9.49 horizontally at 64 gpm and vertically at 9.3 psi/100 ft. At the intersection of these lines, the pipe diameter and velocity are determined. Between 1½-in.- and 2-in.-diameter pipe.

velocity = 8 fps (less than 10, so OK)

Therefore, a 2-in.-diameter supply pipe will be chosen, with a 2-in. meter.

Now find the actual pressure loss in the 2-in. meter for a flow of 64 gpm. Figure 9.48 shows that this is 4 psi. Since this is *less* than the 8 psi estimated, the pressure at the *fixture* will be slightly higher than planned. When a final system layout is made, the fittings are tabulated and the length in piping equivalent to fittings is found. If this differs greatly from the 50 ft estimated in Example 9.6, a recalculation is made.

EXAMPLE 9.7. Solve the pipe sizing problem posed in the previous example now using SI units.

Street main pressure (minimum)	345 kPa
Height, topmost fixture above street	10 m
Topmost fixture type	WC with flush valve
Fixture units in the system	85
Developed length (DL) of the piping to the highest and most remote fixture	30 m
Pipe length equivalent to fittings (estimated at 50% of DL)	15 m
System uses predominantly	Flush valves

SOLUTION. From the minimum street main pressure, subtract the sum of the fixture pressure, the static head, and the pressure lost in the meter. This sum is

A—fixture pressure (Table 9.16)	103 kPa
B—static head 10 m × 10 kPa/m	100 kPa
D—pressure loss in meter (estimated, Fig. 9.48)	55 kPa
	258 kPa
E—pressure in street main	345 kPa
subtract (A + B + D)	−258
	87 kPa

The pressure lost in 30 m (DL) of piping plus the 15 m of piping equivalent to the pressure lost by friction in the fittings therefore can total 87 kPa. Total equivalent length (TEL) is 45 m. This procedure assures 103 kPa at the critical fixture. The unit-friction loss, kPa/100 m of pipe, will be

$$87 \text{ kPa} \times 100/45 \text{ (TEL)} = 193 \text{ kPa}/100 \text{ m}$$

From Fig. 9.50, curve 1, a flush-valve system with 85 fixture units will have a probable flow (demand) of about 4 L/s. Given this information, enter Fig. 9.49 horizontally at 4 L/s and vertically at 193 kPa/100 m. At the intersection of these lines, the pipe diameter and velocity are determined. Between 38-mm and 51-mm diameter pipe,

$$\text{velocity} = \text{about } 2.4 \text{ m/s (less than 3, so OK)}$$

Therefore, a 51-mm diameter supply pipe will be chosen, with a 51-mm meter.

As in the previous Example 9.6, the actual system pressures can be found, using the more exact pressure loss in the 51-mm meter.

REFERENCES

Lowry, William P. (1988). *Atmospheric Ecology for Designers and Planners,* Van Nostrand Reinhold, New York.

Milne, M. (1976). *Residential Water Conservation,* U.S. Dept. of Commerce, NTIS.

U.S. Environmental Protection Agency (1975). *Manual of Individual Water Supply Systems,* U.S. EPA, Washington, D.C.

10

WATER AND WASTE

The waste of resources inherent in the use of potable water to flush toilets was mentioned in Chapter 8. Here we will examine more closely this conventional approach to bodily waste removal, along with alternatives that use less water—or no water at all. For typical U.S. residences, the potential impact of such alternatives on water usage and treatment is shown in Fig. 10.1.

Almost every plumbing fixture within buildings is provided with both water supply and waste pipes. Since the toilet is usually the largest user of water (Fig. 10.1), as well as one of the worst polluters, this chapter's comparison of conventional, conserving, partial recycling, and full recycling/waterless alternatives is focused on toilets. Other water-conserving alternatives to conventional fixtures are included in this section.

10.1 Fixtures, Water, and Waste

A wide variety of water-saving fixtures is now available. For the toilet, or water closet (WC), there are four major categories of fixtures, depending on the amount of water used per flush. The *conventional* toilet uses 3.5 gal (13.2 L) or more per flush; the *watersaver* toilet uses from 1.7 to 3.5 gal (6.4 to 13.2 L) per flush; the *low-consumption* toilet (also called ultra-low-flush) uses 1.6 gal (6 L) or less per flush, and the *waterless* toilet uses no water at all.

(a) The Conventional Toilet. In the common toilet, a sudden deluge of water removes human waste, and simultaneously helps cleanse the toilet. Fast-moving water requires pressure and makes noise. The older flush-tank toilet (Fig. 10.2*a*) stores a smaller quantity of water (about 2.5 gal, or 9.5 L) at sufficient height above the toilet bowl to achieve fast flow. Although it uses minimal amounts of water, it is noisy and its elevated tank has a maximum visual impact. The more common (in North America) version is the two-piece toilet with the tank bolted to the bowl (Fig. 10.2*b*). With much less pressure available, this toilet requires 5 to 7 gal (19 to 26 L) per flush, but it is quieter. Newer "shallow trap" models reduce the quantity needed to 3.5 gal. (13.2 L). The maintenance problem posed by the seam between tank and toilet is eliminated in the one-piece toilet (Fig. 10.2*c*). The cost of its low profile is an even greater need for water: 6 to 8 gal (23 to 30 L) per flush.

The fourth common alternative is the flush-valve toilet, which depends on the building's water pressure rather than its own stored water. A noisy system requiring very high flow rates, it is seldom found in residences. However, it requires as little as 3 gal (11.4 L) per flush.

In addition to differing tank and bowl combinations, there are varieties of types of flushing actions. The four common flush toilets are compared in Fig. 10.3.

Washdown toilets are no longer made in the United States, although many are still in use. This is the noisiest toilet, and the most likely to become plugged, since it has the smallest-diameter trap. It was usually found in the two-piece flush tank toilet; with an elevated tank, the flush required was only about 2.5 gal (9.5 L).

Siphon jet toilets are in widespread use in North America, particularly in residences. A small priming jet hurries the bowl's contents along into the trap, and hastens the siphon action. With the elevated tank (two-piece) toilet, this process requires a flush of about 3.75 gal (14.2 L). With the more common close-coupled two-piece toilet, it requires a flush of from 5 to 7 gal (19 to 26.5 L). Siphon jet toilets are sometimes equipped with flush valves, which use less water (see the discussion of "Blowout," below).

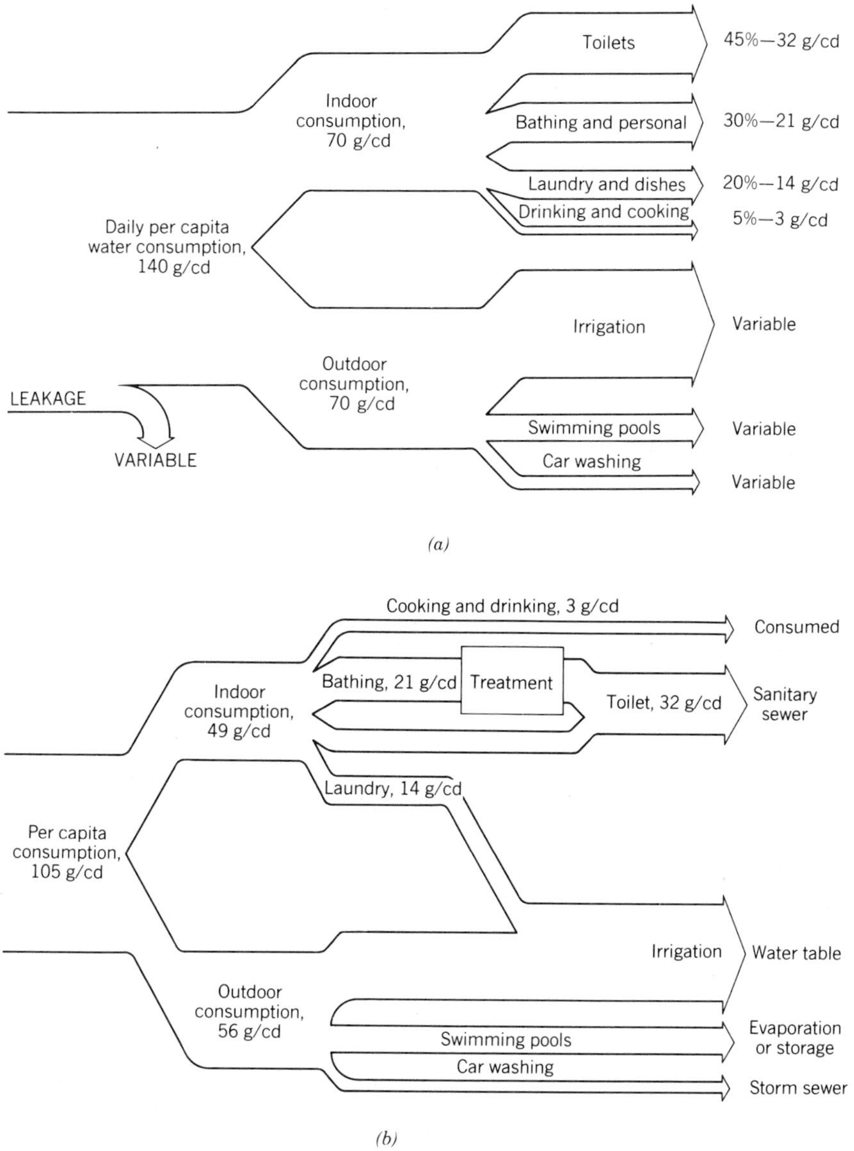

Fig. 10.1 *Opportunities for water conservation for the typical U.S. family.* (a) *At present the average U.S. residential usage is about 140 g/cd—all of it potable.* (b) *With attention to recycling and the matching of water quality to usage, potable water usage could be cut to 105 g/cd—a 25% reduction.* (c) (overleaf) *With further on-site treatment and recycling, potable water usage drops to 73 g/cd—a 40% reduction—and the need for public sewage treatment (for residences) disappears. [From Milne (1976).]*

Siphon vortex toilets are especially suitable for low-velocity water (often as a result of low pressure). They are therefore also the quietest, making them a favorite wherever bathrooms are adjacent to sleeping areas or other acoustically sensitive spaces. The water enters the bowl off-center in such a way as to form a vortex; this swirling action cleans the sides of the bowl and the trap, helping the siphon action in emptying both bowl and trap. The newer one-piece flush

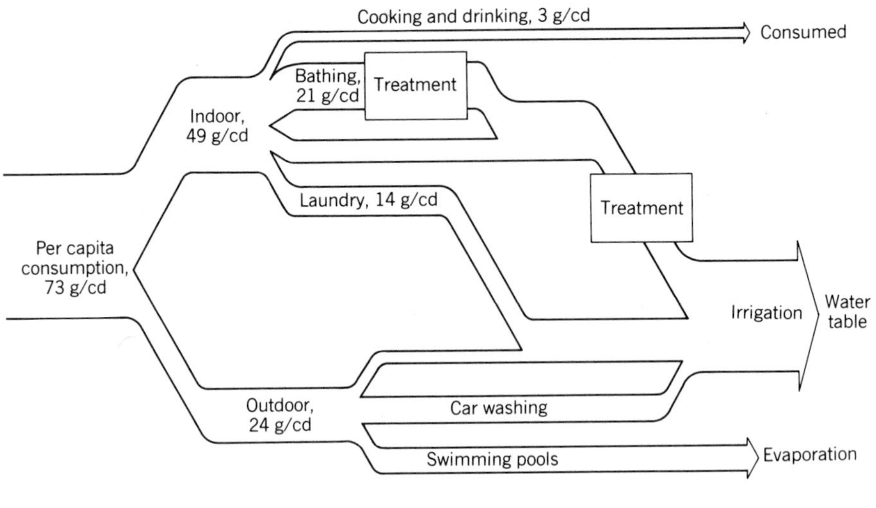

(c)

Fig. 10.1 (*continued*)

tank toilets usually have the siphon vortex flushing action, which typically requires 6 to 8 gal (22.7 to 30.3 L).

Blowout toilets combine very-high-velocity water and a simple trap to offer a noisy but very-low-maintenance toilet dependent on flush valves rather than tanks. They are very common in commercial and institutional toilet rooms, where large water supply lines and high pressures are available. The high velocity of the water lowers the quantity required from 3 to 4 gal (11.4 to 15.1 L) per flush.

Architects could easily specify toilets that require less water per flush and yet have very little impact on the user. With effort on the part of both users and architects, water can be recycled so as to supply conventional toilets with lower-quality water.

(b) Watersaver Toilets. Although these new toilets use less water than conventional ones, they may not meet the more strict water limits now established in water-conscious states and municipalities. Many of these toilets repre-

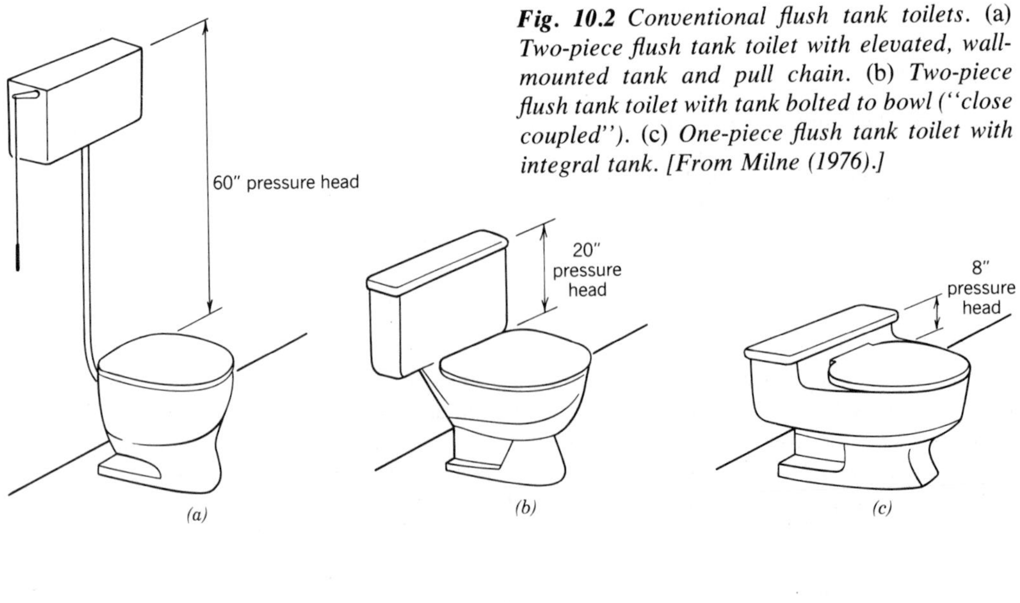

Fig. 10.2 *Conventional flush tank toilets.* (a) *Two-piece flush tank toilet with elevated, wall-mounted tank and pull chain.* (b) *Two-piece flush tank toilet with tank bolted to bowl ("close coupled").* (c) *One-piece flush tank toilet with integral tank. [From Milne (1976).]*

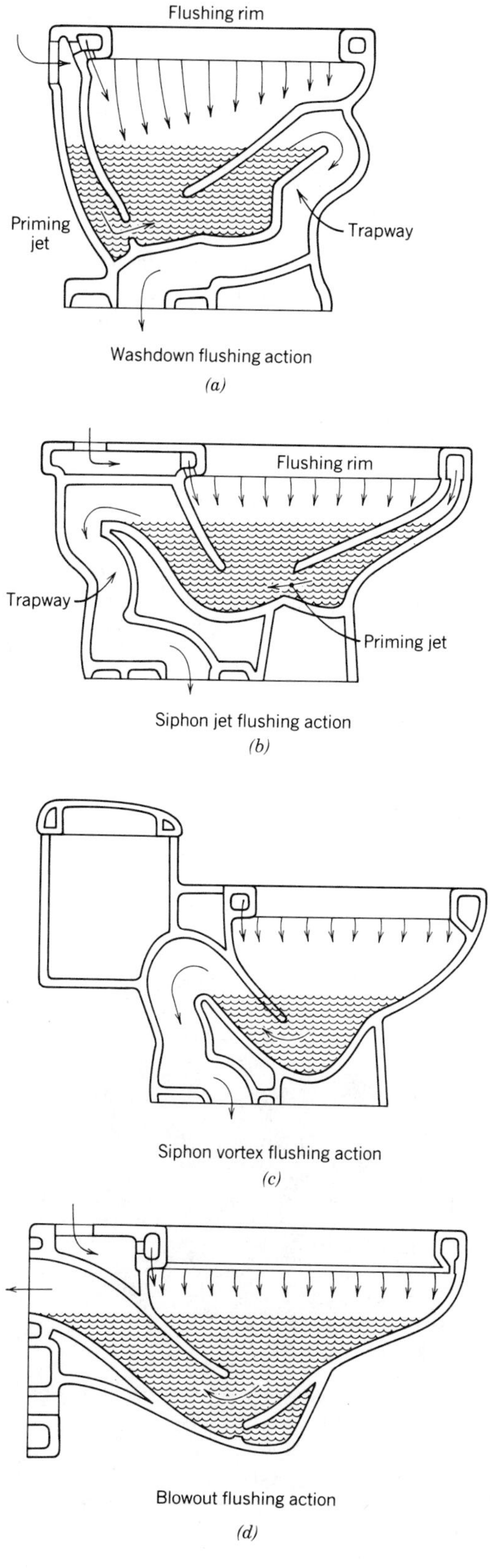

Fig. 10.3 *Four common flushing actions that are built into toilets. Newer flush toilets typically use either type* (b) *or type* (c)*; flush valve toilets use either type* (b) *or type* (d). *[From Milne (1976).]*

sent a compromise between the conventional "water waster" and the noisy operation and/or smaller water surface area that typifies so many low-consumption models. Watersaver toilets tend to use a conventional flushing action, hence need enough height between the tank and the bowl to provide sufficient water pressure within the bowl during the flush.

Toilets can be designed to use less water by varying several characteristics: the pressure behind the water entering the bowl (generated either by the height of the tank above the bowl, or by utilizing increased pressure from the water supply system or another source), the shape of the bowl, the way in which the bowl fills and empties, and the trap configuration. Watersaver toilets differ somewhat in most of these characteristics from the conventional toilet; more radical changes are needed to achieve the low-consumption toilet.

(c) Low-Consumption Toilets. Several of these toilets achieve water conservation by using a *flushometer tank* (Fig. 10.4). Satisfactory flushing can be achieved with much less water by flushing with a smaller amount of water entering at greater pressure. Instead of toilet tanks at ordinary air pressure, these tanks utilize water supply system pressure to compress the air trapped within the tank. Water enters the bowl with much greater force with this combination of water under system pressure and compressed air. The flush is thorough, quick, and noisy. If the water supply pressure is greater than about 65 psi (448 kPa), some problems with excessive tank pressures may be expected. A pressure-reducing valve would be helpful in such cases.

Another water-saving technique is to redesign the bowl to hold less water. The siphon jet, gravity-fed toilet whose tank is shown in Fig. 10.5 exposes a standing water surface of about 4½ by 6 in. (114 by 152 mm) (much smaller than that of Fig. 10.4). Along with a very low (1.4 gal)

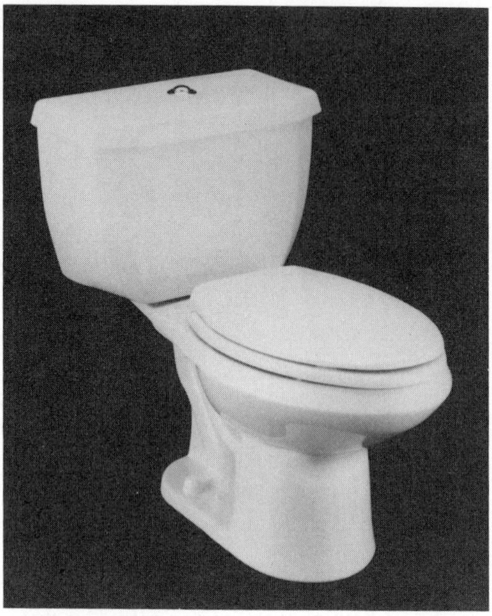

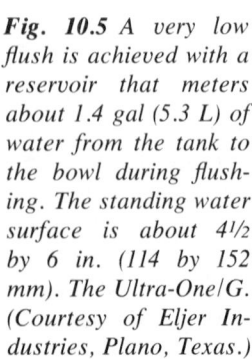

Fig. 10.4 *Example of a 1.5-gal-flush, pressure-assisted, direct-fed siphon jet flush action flushometer tank, yet with a larger 10 × 12 in. water surface area in the bowl: the Cadet Aquameter. (Courtesy of American Standard, Inc., Piscataway, New Jersey.)*

(5.3 L) flush, is trap is wide enough to pass a 2⅛-in. (54 mm) ball.

Even lower water consumption is achieved when a central compressed air system is combined with water supply system pressure. The Microphor flush toilet (Fig. 10.6) uses compressed air and 2 qt (1.9 L) of water per flush in a two-chambered toilet. When the flush lever is pressed, the water and waste in the bowl (upper chamber) is deposited into the secondary (lower) chamber, and 2 qt of water (direct from the supply pipe) washes down the sides of the bowl to await the next user. The now-closed secondary chamber then is pressurized with compressed air, and its contents deposited into the conventional sewer line.

Air compressors are needed, along with compressed air lines. A small compressor (¼ to ½ hp) (187 to 373 W) with accompanying air tank will operate up to three toilets. Although the compressor is noisy, the toilets themselves are no more noisy than conventional toilets; they use the same plumbing lines. With such low flows, the designer may choose to increase the slope of waste lines; however, the 2-qt quantity has proven sufficient to carry the waste in existing installations.

In contrast, the Envirovac flush toilet (Fig. 10.7) uses a vacuum and 1.5 qt (1.4 L) per flush. Because a central sewage tank (kept under vac-

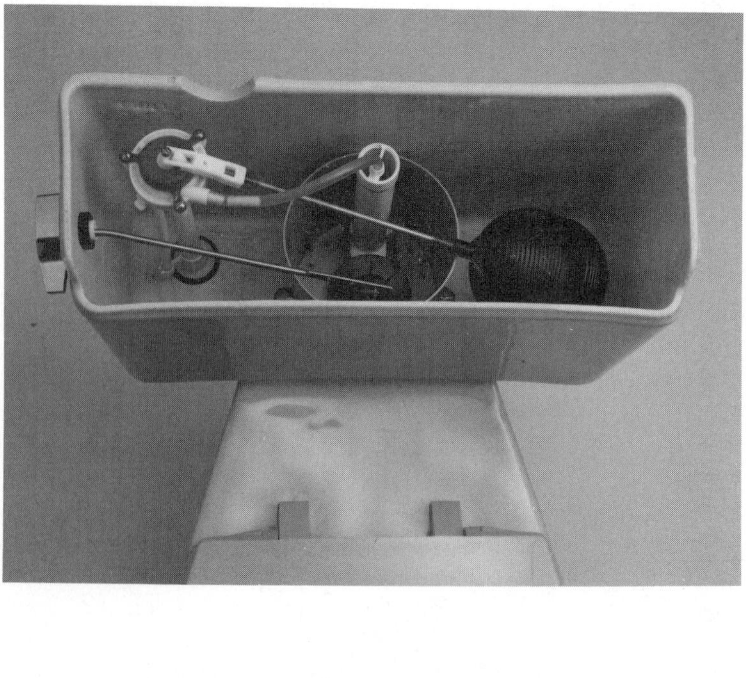

Fig. 10.5 *A very low flush is achieved with a reservoir that meters about 1.4 gal (5.3 L) of water from the tank to the bowl during flushing. The standing water surface is about 4½ by 6 in. (114 by 152 mm). The Ultra-One/G. (Courtesy of Eljer Industries, Plano, Texas.)*

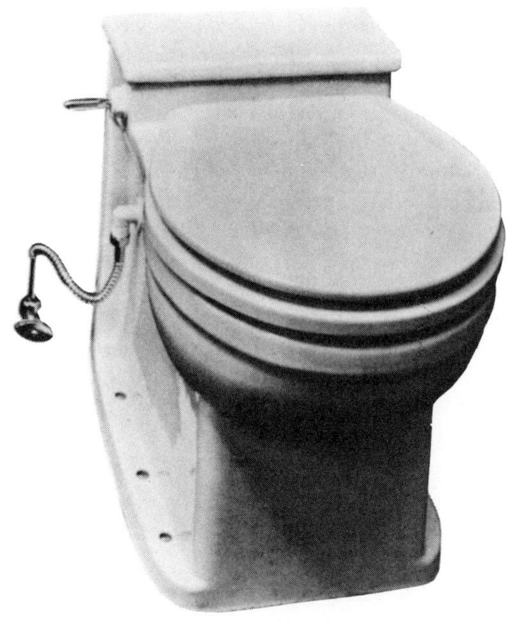

(a)

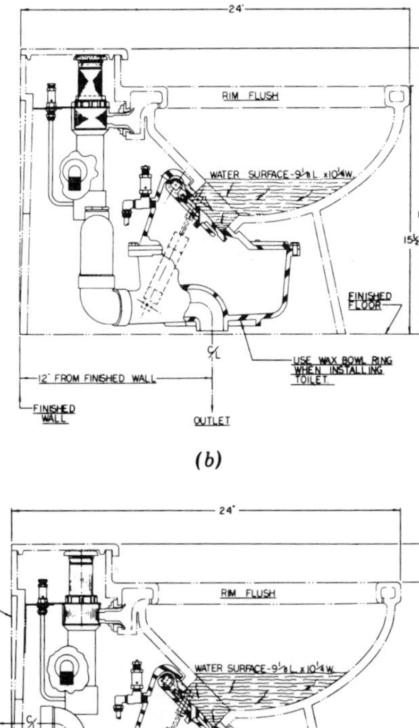

(b)

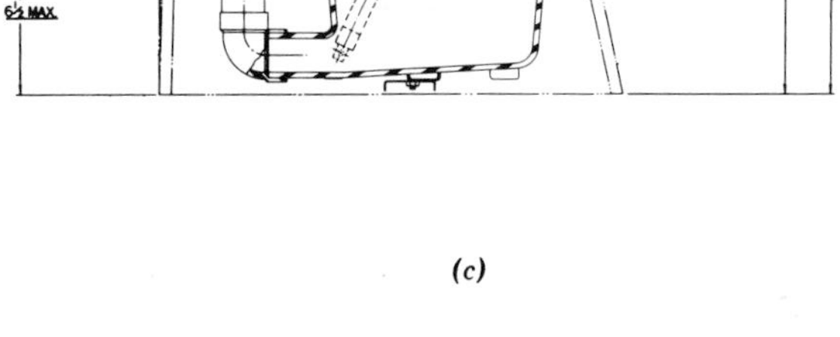

(c)

Fig. 10.6 (a) *Compressed air is combined with a 2-quart flush in the two-chambered Microphor toilet.* (b) *Section through the floor-discharge model.* (c) *Section through the wall-discharge model. (Courtesy of the Microphor Company, Willits, California.)*

uum by a pump) must be used, the sewer line from the toilet to this tank may run horizontally without a slope, or even run vertically upward from the toilet. This can have significant architectural advantages when vertical clearance is tight, toilets are being added within existing structures, or toilets must be located below the level of the public sewer, as in marinas or basements. In tall buildings, significant savings in power for water supply pumping are achieved. Easier approval for building permits, reduced water hookup fees, and lower monthly sewer charges can be substantial benefits. Further, the central sewage holding tank can be flushed into the sewer at off-peak times—a benefit to the treatment plant. These water-saving, architectural, and water treatment advantages must be weighed against the cost of the toilets and the tank/pump combinations, the space required for the tank/pump, the power used by the pump, and the possibility that a power failure will halt system operation.

Vacuum systems can be installed for groups of buildings, such as subdivisions. Small pipe sizes, freedom from the need for continuous-sloping sewer lines, and conservation advantages for both water supply and treatment are the benefits. Additionally, "graywater" (from kitchen, laundry, and bathing fixtures) can be kept separate from toilet water ("blackwater"), and thus made easier to recycle after moderate treatment.

(d) Flushing Controls. Water conservation is encouraged by the *dual cycle* toilet, whose flushing mechanism allows a choice of fewer gallons for liquid wastes, more gallons for solid wastes. More common outside the United States, this simple mechanism's handle is pushed up for liquid flushing and down for solids.

Another, newer development is the *auto-*

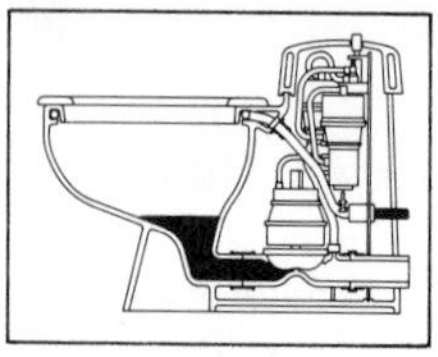

(a)

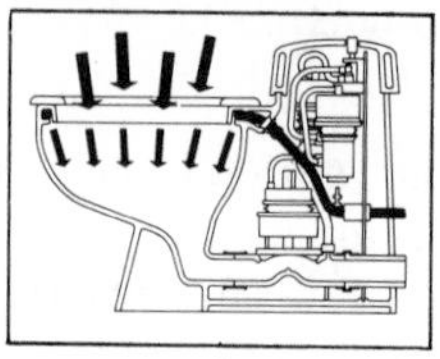

(b)

Fig. 10.7 (a) *The Envirovac system utilizes a vacuum and a 3-pt (1.4 L) flush. Less than 3 pints of water remain in the bowl until toilet is used.* (b) *When pushbutton flush is pressed, the discharge valve opens for about 3 seconds; a rush of air into the evacuated piping carries along sewage, odor, and airborne bacteria. Simultaneously, the washdown water flow begins. The flow continues for 4 seconds after the discharge valve closes, so that most of the 3-pt flush remains in the bowl.* (c) *Toilets are directly connected to a central holding tank, kept under a vacuum by a pump.* (d) *Central tank/pump combinations can serve one house or a large building.* (e) *The separation of graywater and blackwater is another water-conserving opportunity. (Courtesy of Envirovac, Inc., Rockford, Illinois.)*

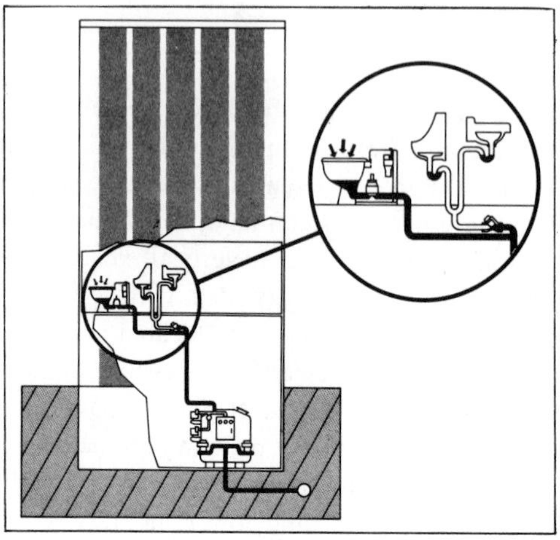

(c)

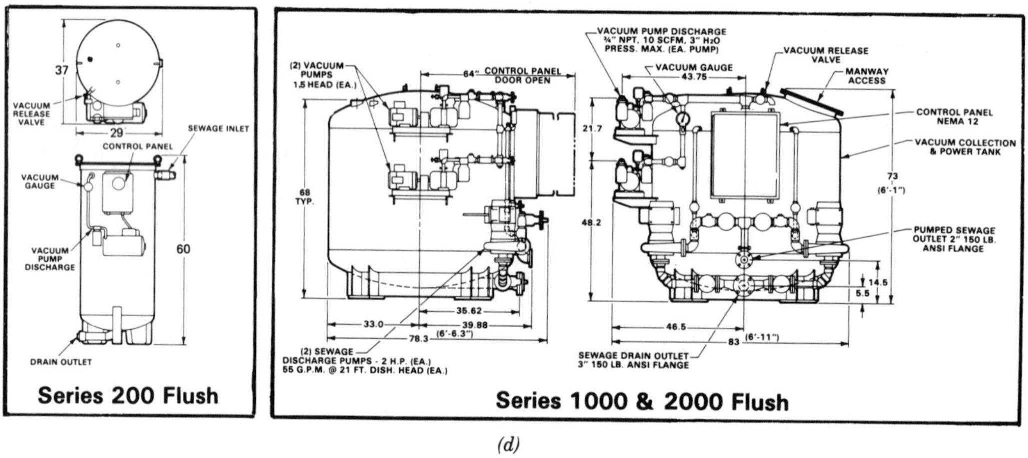

(d)

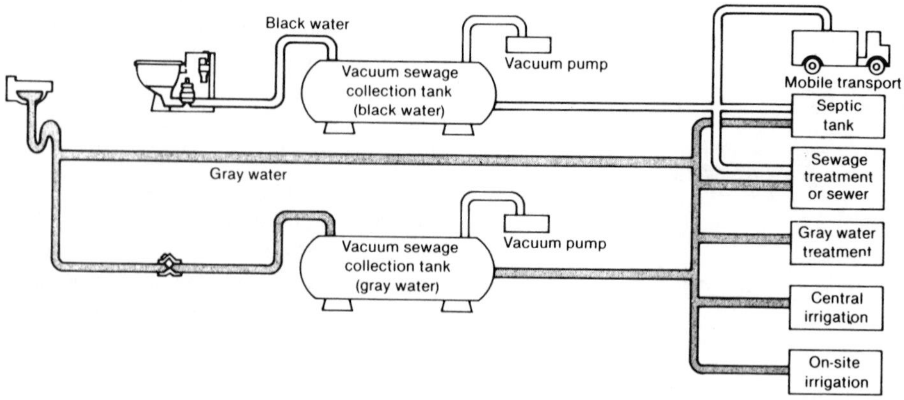

(e)

matic flush, triggered by radiant heat from the pressure of a body at the fixture or by light reflected off the user and back to the control. This "touchless" approach seems to promise more for hygiene than for water conservation, although it does prevent the flush valve from being held open too long by a careless user.

In general, plumbing fixtures will emit less water as the supply water pressure is reduced. For this reason, *pressure-reducing valves* are becoming popular as water conservation devices (when they would not otherwise be required to protect fixtures from overpressure). Installed on the supply line to a building, they can save water throughout the structure.

Several newer low consumption toilets use vertical flush valve sleeves, where the flush handle is in the center of the top of the toilet tank. The handle is lifted to initiate the flush. An advantage is that the less time the handle is raised, the less water enters the bowl. Again, less water may be admitted to flush away liquid waste, more water for solid waste.

(e) Waterless Toilets. The array of waterless alternatives include toilets in which chemicals or oil are substituted for water. These devices are commonly found in airplanes, vehicles, and boats, as well as in remote and environmentally sensitive areas. The chemicals must be frequently recharged, and the waste products removed. Other waterless toilets temporarily treat the waste awaiting discharge to a sewer by freezing it, burning it, or otherwise packaging it so as to remain unoffensive. Obviously, such devices can become energy-intensive solutions to waterless waste disposal. Also, they are still dependent upon public sewer systems.

Composting toilets offer a self-contained and much less energy-intensive strategy for waterless toilets. Such systems rely upon *aerobic* digestion of waste (i.e., that which occurs in the presence of oxygen). Aerobic systems usually are essentially odor-free, and the exhaust air is rich in CO_2 and water vapor. In contrast, *anaerobic* decomposition (that which occurs without oxygen) is malodorous and produces methane gas as an important by-product.

The *Clivus Multrum System* (Fig. 10.8) has a large decomposition chamber which must be below the toilet and the kitchen, from which it readily accepts organic waste. It must also be accessible to remove the humus—from 3 to 10 gal (11 to 38 L) of soil per person per year.

The relatively large chamber (about the size of a Volkswagen "bug") and its low position/access requirement can have a significant impact on design. Also of significance is the device's appetite for fresh air. The more air, the speedier the process of aerobic decomposition and, therefore, the less odor. In winter, this air could cause increased heat losses by infiltration in a building. The natural ventilation due to the stack effect (Section 4.6) is at least 15 cfm (7 L/s) in the Clivus Multrum. A ventilating fan in the stack will increase this rate significantly. On the other hand, if outdoor air is brought directly to the chamber, it could be too cold for proper decomposition. 98 F is optimum. Solar heating of such input air offers one alternative. However, adding heat to compost piles encourages oxidation rather than aerobic decomposition; inferior compost results.

(f) Shower Heads. These devices have been notorious for encouraging prodigal water usage; typical flow rates of 6 gpm (0.4 L/s) and maximum rates of 12 gpm (0.7 L/s) once were common. Even in a "short" (5-minute) shower, this rate of use could consume as much as 60 gal (227 L) of water, much of it heated. Many codes now require a limitation on shower-head flow; a fixture of 2.5 gpm (0.2 L/s) is common. These flows can be designed into the shower head, or they can be achieved by cheap, simple flow restrictors in retrofit applications. Some utilities distribute flow restrictors free of charge. Most bathers notice no difference, either in enjoyment or in cleansing, when using restricted-flow shower heads.

(g) Lavatory Faucets. At full flow, lavatory faucets typically deliver 4 to 5 gpm (0.25 to 0.3 L/s). Newer, low-flow faucets utilize a variety of devices to function as well (or better) with less water. Such devices include aerators (which add air bubbles to the stream, making it splash less and appear larger), flow restrictors, and mixing valves to control temperature. The lower

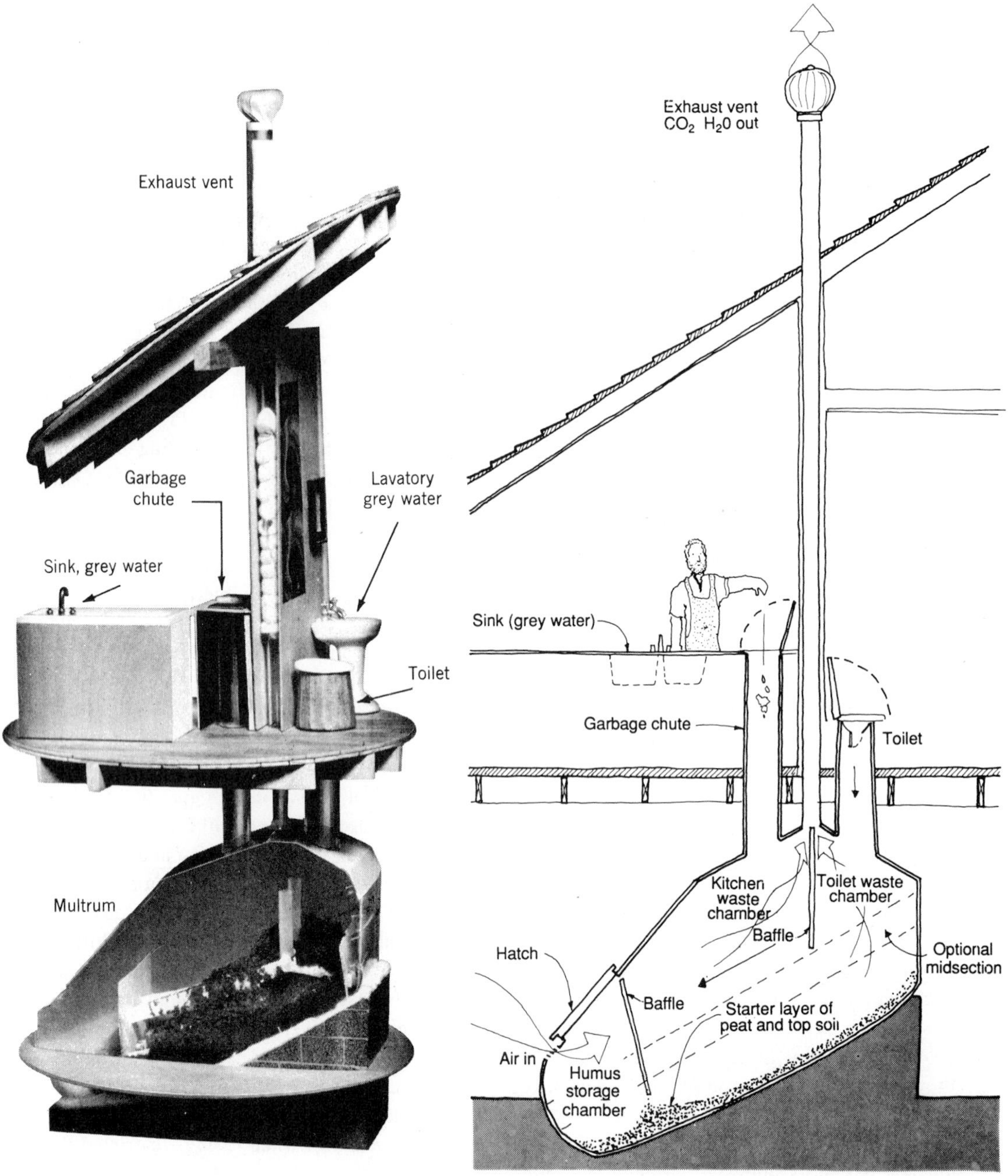

(a) (b)

Fig. 10.8 *Clivus Multrum System, with composting of both kitchen and toilet organic wastes.* (a) *The lower composting chamber requires 4 × 8 × 7 ft (1.2 × 2.4 × 2.1 m) of floor space, a generous supply of air (warm air speeds decomposition), and access for periodic removal of the garden-ready humus.* (b) *It must be arranged so that its upper end receives toilet wastes; as the bottom slopes downward, kitchen wastes are deposited on top of the decomposing toilet wastes. The vent stack assures continuous airflow, preventing odors from entering bathroom or kitchen.* (c) *The toilet seat and cover must be kept closed when not in use, so as to keep air flowing through the composting chamber and out the stack. The bowl (containing no water) swings open when the seat is occupied, exposing the composting chamber below. (Courtesy of Clivus Multrum USA, Inc.)*

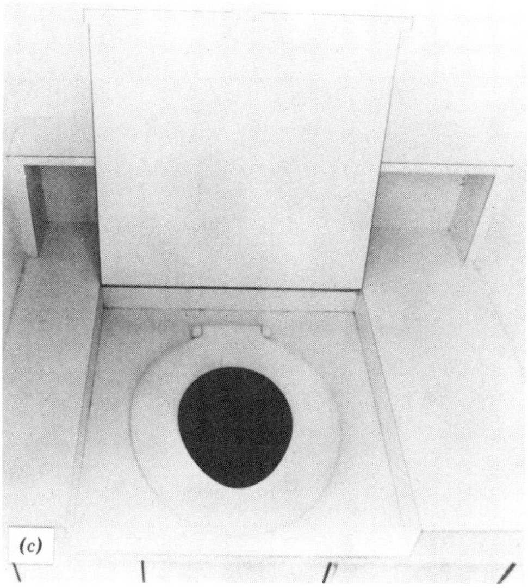

Fig. 10.8 (continued)

flows achieved range from ½ to 2½ gpm (0.03 to 0.16 L/s). Another promising development is the foot-operated faucet, which frees the hands from having to control water flow, thus saving a few seconds of flow during each lavatory usage. These devices could be particularly helpful at kitchen sinks, where extensive washing of objects takes place.

(h) Appliances. Dishwashers, washing machines, and other appliances are big users of water—and of the energy needed to heat it. Dishwashers use 12 to 18 gal (45 to 68 L) per cycle, heated well beyond the 120 F (48°C) typical of household hot water supply. Some models now allow shorter cycles, which can cut use to perhaps 7 gal (26.5 L).

Clothes washing machines use from 40 to 55 gal (151 to 208 L) for full-size loads. In the past, "suds saver" features allowed soapy, hot wash water to be reused. Many newer washers allow for a wider selection of water quantities and temperatures—a feature that can save considerable water and energy.

(i) Recycling of Water. These chapters on water and waste illustrate four common "grades" of water in buildings:

Potable water (usually treated, suitable for drinking).

Rainwater.

Graywater (wastewater not from toilets or urinals).

Blackwater (water containing toilet or urinal waste).

Many uses of water in and around buildings could be met satisfactorily with rainwater: bathing and laundry, irrigation, and toilet flushing, for example. Opportunities for the reuse of graywater and blackwater are listed in Table 10.1. In general, graywater must undergo treatment before reuse because it is likely to contain soap, hair, grease from the kitchen, and occasionally human waste (soiled clothes in the laundry). Considering these contaminants, "light graywater" might describe water from lavatories, tubs, and showers, and "dark graywater" that from washing machines, kitchen sinks, and dishwashers. More extensive treatment of blackwater is required, given the higher concentration of human waste in it.

The collection of these separate types of water poses a problem. In the past, rainwater, graywater, and blackwater were mixed together, which severely overloaded sewage treatment facilities in rainy weather. Today, we usually mix graywater and blackwater. There are precedents for keeping them separate, however (Fig. 10.9). The potential for graywater reuse with separate waste collection systems was shown in Fig. 10.1*b*: bathing's 21 g/cd (gallons per capita per day) can meet much of the conventional toilet's 32 g/cd, and laundry's 14 g/cd can help with irrigation. Another potential benefit of separate graywater collection is that the water's heat content can be partially reclaimed (Table 10.2).

Various stages of treatment are appropriate for different kinds of graywater (Fig. 10.10). Kitchen sink waste contains grease, which should be trapped (and periodically removed) before it can clog the filters and heat exchangers that serve graywater recycling systems. Similarly, lavatory showers and laundry waste contains lint and hair that must be intercepted quickly. Devices that do so, which are often called *interceptors*, are described in Fig. 10.39.

TABLE 10.1 **Potential for Domestic Water Recycling**

Original Use	Reuse: Toilet	Irrigation	Sprinkler	Kitchen Sink	Carwash	Laundry	Pool	Shower/Tub	Bathroom Sink	Dishwasher	Drinking	Cooking
1. Toilet[a]	2	–	–	–	–	–	–	–	–	–	–	–
2. Irrigation*[b]	1	1	1	–	1	–	–	–	–	–	–	–
3. Sprinkler*[c]	1	1	1	–	1	–	–	–	–	–	–	–
4. Kitchen sink with grinder	1	0	1	–	–	–	–	–	–	–	–	–
5. Carwash*	1	0°	1°	–	1	–	–	–	–	–	–	–
6. Laundry[d]	1	0°	1°	–	1	1	–	–	–	–	–	–
7. Pool (chlorinated)	1	–	–	–	1	1	2	–	–	–	–	–
8. Shower/tub	1	0°	1°	–	1	1	–	1	–	–	–	–
9. Bathroom sink[e]	1	0°	1°	–	–	–	–	–	–	–	–	–
10. Dishwasher	1	0°	1°	0	1	–	–	–	–	0	–	–
11. Drinking*	1	0	1	0	1	–	–	–	–	–	–	–
12. Cooking	1	0	1	0	1	–	–	–	–	0	0	0

Legend

0 Reusable directly, without treatment.
1 Reusable with settling and/or filtering (primary treatment).
2 Reusable with settling, filtering, and chemical treatment—usually chlorination (secondary treatment).
– Not reusable.

Source: Milne (1976).

*Very difficult to collect.

°Special soaps required.

[a]Small valves and underwater moving parts—clogging problem.

[b]Large orifice: unpressurized open hose or channel.

[c]Small orifice: pressurized.

[d]Assumes no diapers with fecal matter.

[e]Shaving and brushing teeth.

Various graywater-filtering devices are now available, including that shown in Fig. 10.11, whose effluent (outflowing water) could be used for either toilet flushing or irrigation.

In the blackwater-recycling toilet systems currently available, treatment is more extensive and first costs much higher. Although water is reused, the sewage sludge must be removed periodically from the treatment tank.

10.2 Waste Treatment

In addition to the self-contained composting toilets just described, there are numerous opportunities for the treatment of human and domestic wastes from buildings. In this section, individual building-scale systems will be considered first, then clusters of buildings, and finally municipal treatment plants.

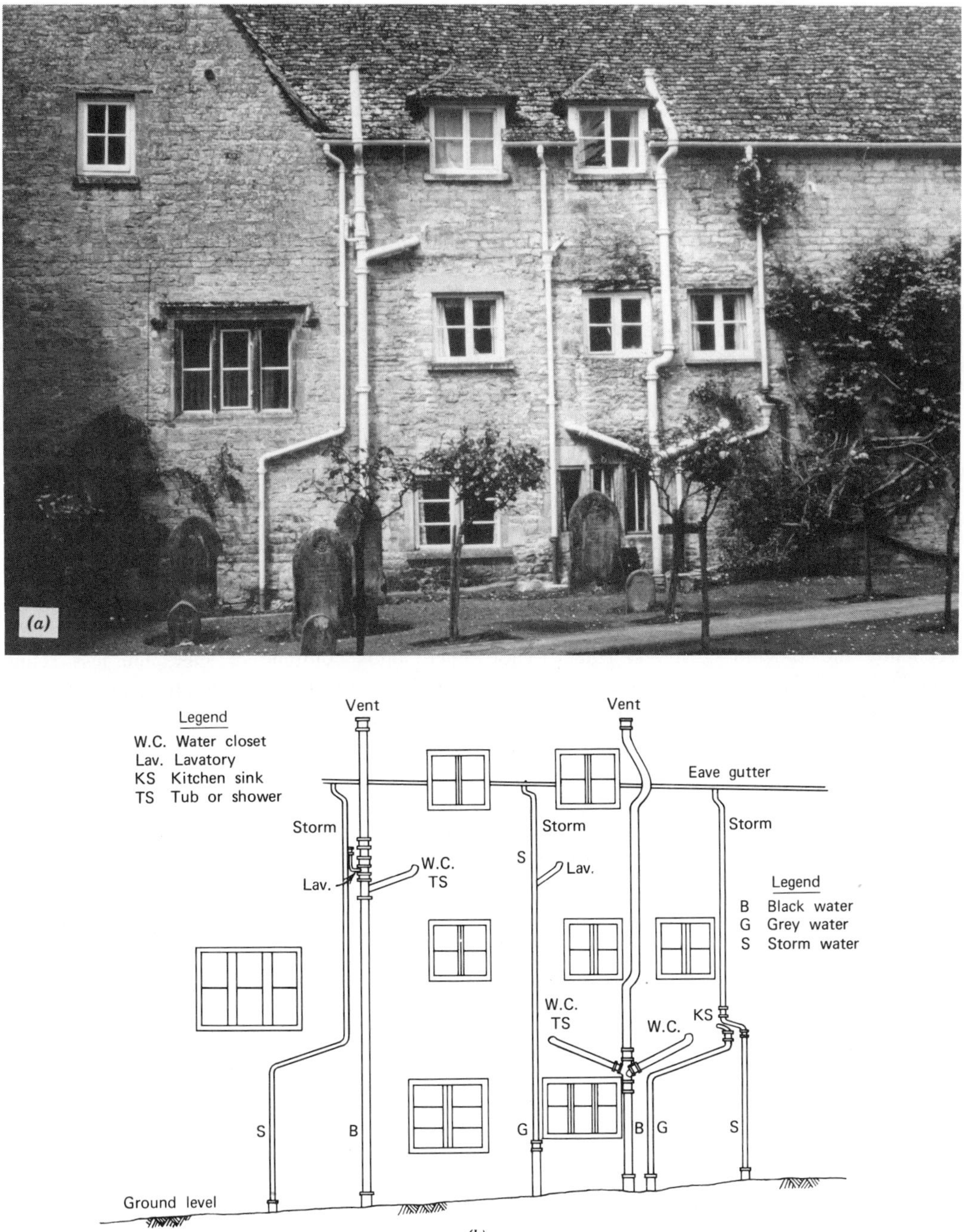

(b)

Fig. 10.9 *Residence of the 27th Vicar of the Parish Church at Bibury (established* A.D. *1086) in the English Cotswolds.* (a) *It is evident that the plumbing was added after the construction of the residence—perhaps several hundred years later. In England the drainage system of an older building often appears on the outside of the building. It seemed inappropriate to ask the Vicar's permission to inspect his indoor facilities, so we have taken the liberty of assigning probable uses and examining this conveniently visual "printed circuit" of drains. Note that (1) offsets of the vertical piping are made with "easy bends" and are pitched down in the direction of the flow; (2) vents are located at high outdoor points; and (3) stacks carrying so-called blackwater are of larger pipe size than those carrying graywater.* (b) *In this very sparsely settled region of the Cotswolds, the dispersal of stormwater to the ground and the private septic tank treatment of blackwater plus graywater can be satisfactory. In planning new systems the possible separate treatment of B, G, and S waters is becoming an important consideration.*

TABLE 10.2 **Heat Recovery from Graywater**

Source	*Volume/Use, Gal (L)*	*Flow Rate, Gal (L)*	*F*	*(°C)*	*Quality*	*Coincident*[a]
Kitchen sink	Up to 5 (1–20)	2.5 (10)/min	85	(30)	Poor	No
Lavatory	Up to 1.5 (1–5)	1.25 (5)/min	85	(30)	Fair	No
Bath	32 (120)	5 (20)/min	100	(37)	Good	No
Shower	15 (50)	2 (7)/min	100	(37)	Good	Yes
Washer	40 (150)	7 (25)/min	60	(15)	Moot	No

Source: Glenn Nelson, "Graywater Heat Recovery," © *Solar Age,* August 1981. Reprinted by permission.
[a]Coincident flow of warm waste water and input water to heater. If "no," a holding tank is needed for heat exchange. If "yes," a countercurrent heat exchanger can be built into the drain line. (See Fig. 10.10.)

(a) Individual Systems. The great majority of individual sewage treatment systems in North America use the *septic tank* (Fig. 10.12) as a *primary* treatment, where the settling of solids and anaerobic digestion take place. Subsequently, the effluent receives *secondary* treatment, which usually consists of a filtering process. Four common filtration systems are seepage pits, drainfields, mounds, and sand filters. Occasionally, a *tertiary* treatment (usually disinfection with chlorine) must be used.

Septic tanks (Fig. 10.13) are commonly constructed of precast concrete. The sewage enters the first chamber, where solids sink to the bottom as sludge, and scum forms on the surface. Anaerobic decomposition proceeds. The liquid moves through the submerged opening in the middle of the tank to the second chamber, where finer solids continue to sink, and less scum forms on the surface. Finally, the effluent, about 70% purified, leaves the septic tank for secondary treatment.

The longer that the sewage stays in the septic tank, the less polluted the effluent. This is why water conservation measures are so welcome with septic tank systems; the less the flow, the

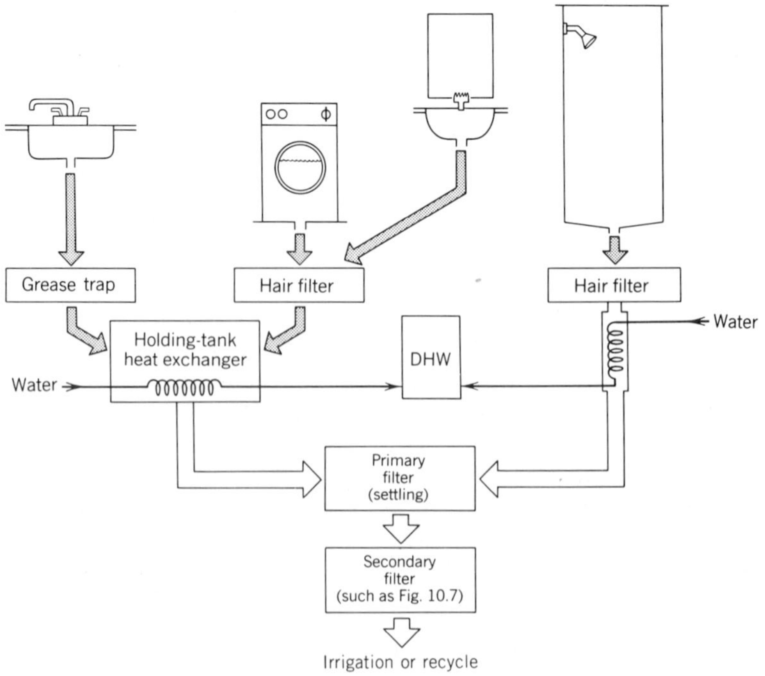

Fig. 10.10 *Sequence of water treatment and heat reclamation for domestic graywater.*

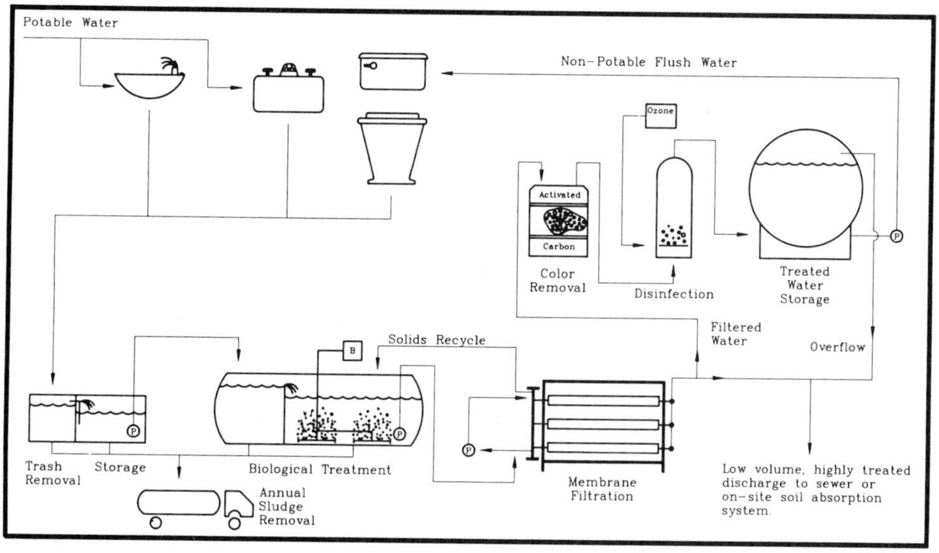

Fig. 10.11 *Cycle-Let on-site wastewater treatment and recycling system uses biological treatment, filtration, and final polishing/disinfection to provide an odor free supply of water, in turn reused for flushing toilets and urinals. Under some conditions, this water can also be used for irrigation. Residual stored solids are removed by hauling once per year. (Courtesy of Thetford Systems Inc., Ann Arbor, Michigan.)*

longer the water remains in the tank. The anaerobic decomposition process is so thorough that the sludge needs only occasional removal—once in several years is about average for residences.

Most systems with septic tanks will eventually experience "failure," usually due to a breakdown in the secondary treatment rather than in the septic tank. With periodic removal of sludge, the septic tank itself is a reliable and simple primary treatment device. However, some types of domestic waste can disrupt the anaerobic process within the tank: notably, towels, rags, insecticides, excessive paper, and hazardous materials.

Septic tank sizes are commonly based on code requirements that consider the number of bedrooms in residences (Table 10.3) or the number of waste fixture units served (see Table 10.9). (As shown in Table 10.4, maximum size may also be related to the soil condition.) Sewage flow rates are also considered. Oversized septic tanks are more expensive to install, but they release cleaner effluent and will prolong the life of the secondary treatment process.

The size of the secondary treatment system is usually based on the expected total flow over a 24-hour period. (This can be estimated from Table 8.4.)

Aerobic treatment units (Fig. 10.14) represent an increasingly popular alternative to septic tanks for primary sewage treatment. They depend upon air bubbled through the sewage to achieve aerobic digestion, which is faster than anaerobic digestion; hence, they can be smaller in size than septic tanks. They are energy-intensive and may require more maintenance than will the anaerobic tank. A secondary treatment process is required, as well. The effluent typically is less polluted than is that of septic tanks.

The sewage first enters an aeration chamber, where it is kept in turmoil so that air can continue to percolate through it. The air is forced by an air compressor; the sewage can be stirred by a variety of devices, depending upon the manufacturer. A second chamber is then used, to allow remaining solids to settle and to be filtered out before the effluent leaves the secondary treatment process.

Seepage pits (Figs. 10.13 and 10.15) are *secondary* treatment facilities that are only appropriate in very porous soil, where the water table

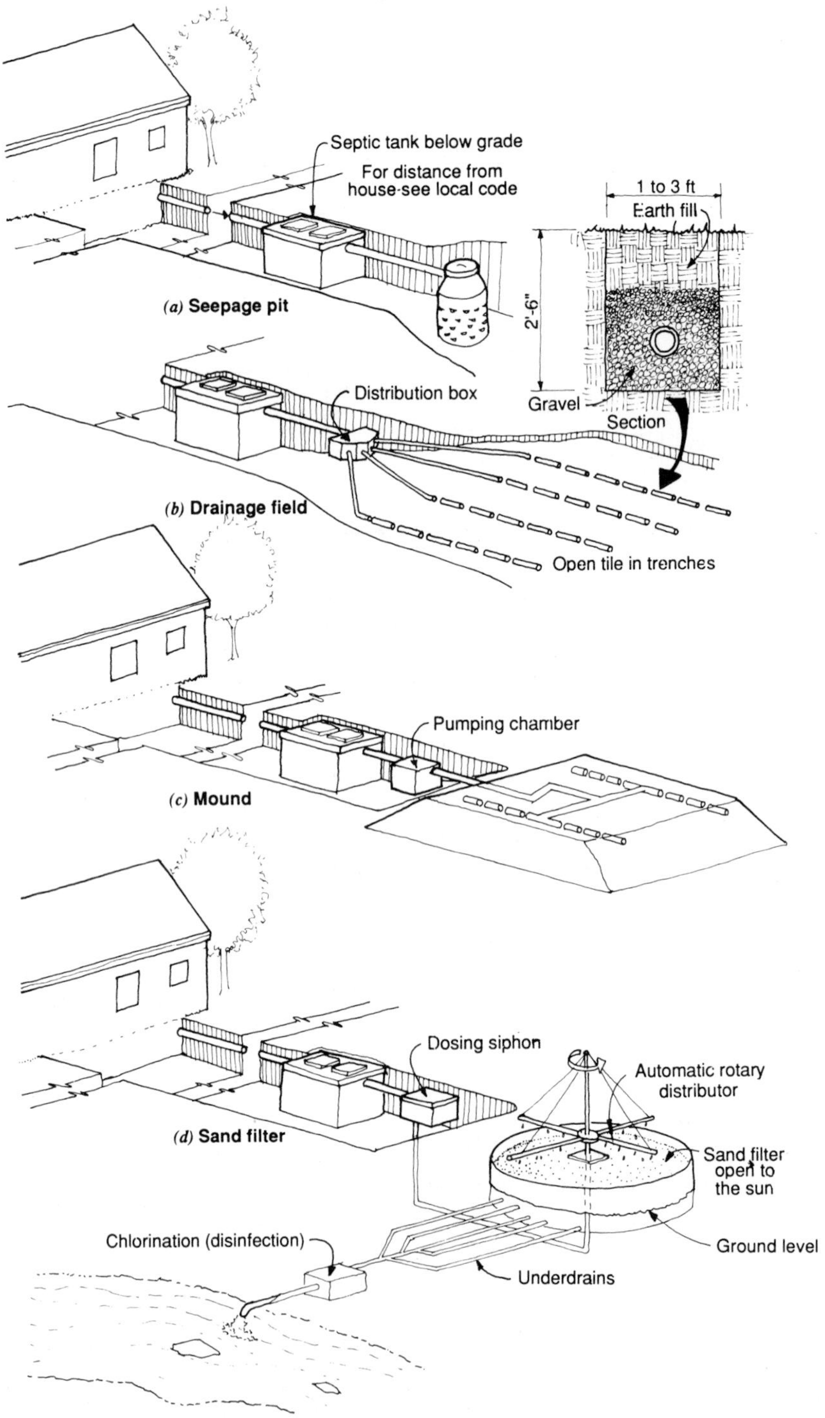

Fig. 10.12 In individual sewage treatment systems, septic tanks are commonly used for primary treatment. Four options for secondary treatment are shown here. Tertiary treatment usually is only required for effluent discharge into waterways. (a) *Seepage pits are not usually used.* (b) *Drainfields constitute the most commonly used options.* (c, d) *Mounds and sand filters are more expensive to construct and are used where high water tables preclude the use of option* (a) *or* (b).

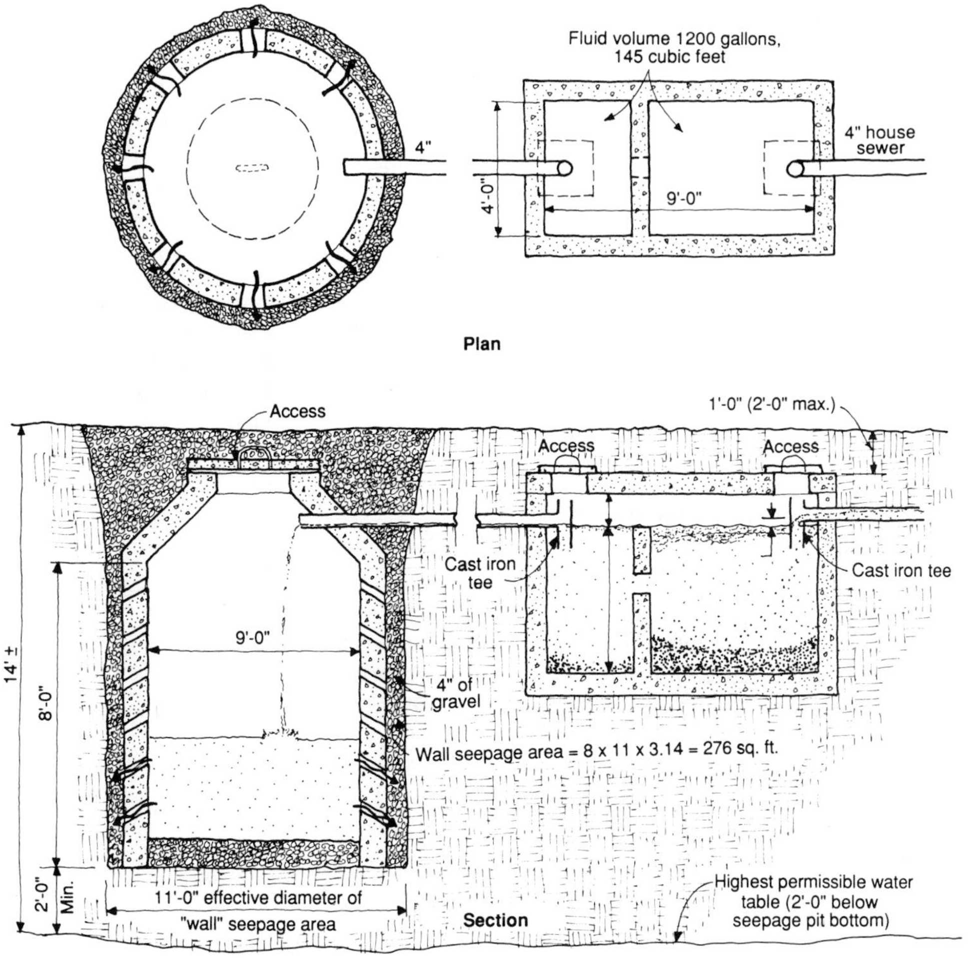

Fig. 10.13 *(Example 10.1) Plan and section of septic tank and seepage pit for four-bedroom house. Pit is suitable when the earth is absorbent and the water table low (below pit bottom). Drawing is not to scale.*

is at least 2 ft (0.6 m) below the bottom of the pit. (They should not be used as the only treatment process.) Since these pits are commonly 10 to 15 ft (3 to 4.6 m) below the earth surface, very low water table may be required. Another common usage of precast seepage pits is for "dry wells" that receive runoff from paved areas during rainstorms.

Seepage pits are sized by the square footage of wall area exposed to the earth—the "leaching area," as listed in Table 10.4. Placement of them relative to buildings, water sources, waterways, and property lines is strictly controlled—see Table 10.5 and Fig. 10.16.

EXAMPLE 10.1. (Fig. 10.13). Design a septic tank seepage pit system for a suburban residence under the following conditions.

Bedrooms	4
Occupants	8
Soil	Sandy loam
Depth to water table	14 ft

SOLUTION

Daily flow (Table 8.4)

$$= 75 \text{ gpd} \times 8 \text{ people} = 600 \text{ gpd}$$

Septic tank capacity (Table 10.3) = 1200 gal

TABLE 10.3 **Septic Tank Capacity**[a]

Single-Family Dwellings—Number of Bedrooms	*Multiple Dwelling Units or Apartments—One Bedroom Each*	*Other Uses: Maximum Fixture Units Served*[b,c]	*Minimum Septic Tank Capacity, Gal (L)*[c]
1 or 2		15	750 (2838)
3		20	1000 (3785)
4	2 units	25	1200 (4542)
5 or 6	3	33	1500 (5677.5)
	4	45	2000 (7570)
	5	55	2250 (8516.3)
	6	60	2500 (9462.5)
	7	70	2750 (10408.8)
	8	80	3000 (11355)
	9	90	3250 (12301.3)
	10	100	3500 (13247.5)

Extra bedroom: 150 gal (567.8 L) each.
Extra dwelling units over 10: 250 gal (946.3 L) each.
Extra fixture units over 100: 25 gal (94.6 L) per fixture unit.

Source: Reprinted by permission from the Uniform Plumbing Code, copyright © 1982 by the International Association of Plumbing and Mechanical Officials.

[a]Septic tank sizes in this table include sludge storage capacity and the connection disposal of domestic food waste units without further volume increase.

[b]See Table 10.9.

[c]For larger or nonresidential installations in which sewage flow rate is known, size the septic tank as follows.

1. Flow up to 1500 gpd (5677.5 L/d):

$$\text{flow} \times 1.5 = \text{septic tank capacity}$$

2. Flow over 1500 gpd (5677.5 L/d)

$$(\text{flow} \times 0.75) + 1125 = \text{septic tank capacity in gallons}$$

$$[(\text{flow} \times 0.75) + 4258 = \text{liters}]$$

TABLE 10.4 **Septic Tank and Leaching Area Design Criteria for Five Typical Soils**

Type of Soil	*Required ft²of Leaching Area/100 Gal (m²/L)*	*Maximum Absorption Capacity, Gal/ft² of Leaching Area for a 24-h Period (L/m²)*	*Maximum Septic Tank Size Allowable*	
			Gallons	*Liters*
1. Coarse sand or gravel	20 (0.005)	5 (203.7)	7500	(28387.5)
2. Fine sand	25 (0.006)	4 (162.9)	7500	(28387.5)
3. Sandy loam or sandy clay	40 (0.010)	2.5 (101.9)	5000	(18925)
4. Clay with considerable sand or gravel	90 (0.022)	1.10 (44.8)	3500	(13247.5)
5. Clay with small amount of sand or gravel	120 (0.029)	0.83 (33.8)	3000	(11355)

Source: Reprinted by permission from the *Uniform Plumbing Code,* copyright © 1982 by the International Association of Plumbing and Mechanical Officials.

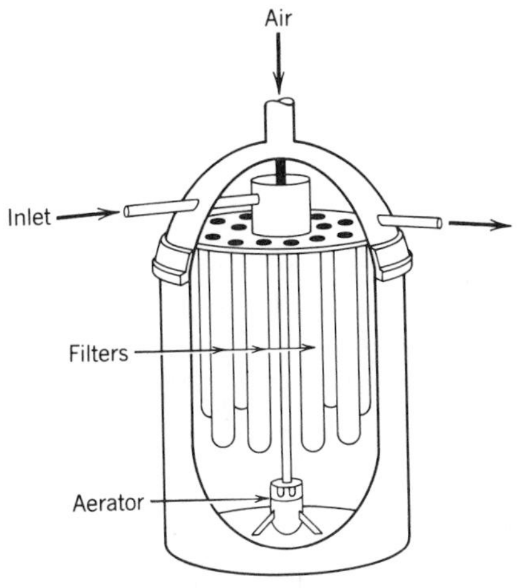

Fig. 10.14 Aerobic treatment unit, an alternative to the simple anaerobic septic tank. [From Milne (1976).]

Fig. 10.15 Precast elements used for recharge to the soil of the partially purified effluent of septic tanks or separately collected stormwater. In the former use, they are known as seepage pits; in the latter, dry wells. They are often made of prestressed concrete of high strength (4000 psi ultimate). Here, shown in the manufacturer's yard, are, right to left, two perforated rings with conical top, cylindrical extender and concrete cover for access, alternate perforated cone, and typical extenders for greater depth. Rings are usually available in 8- and 10-ft (2.4- and 3-m) diameters and in heights of 3, 4, and 5 ft (1, 1.2 and 1.5 m).

Fluid volume of tank

$$= 1200 \div 7.48 \text{ gal/ft}^3 = 160 \text{ ft}^3$$

Pit surface required (Table 10.4)

$$= 40 \times 600/100 = 240 \text{ ft}^2$$

Establish dimensions for the septic tank. Dimensions from the chosen tank (Fig. 10.13) are

$$4.5 \times 4.0 \times 9.0 = 162 \text{ ft}^3 \text{ volume}$$

OK, 162 > 160 (required)

Check the cylindrical effective seepage area of the selected pit (Fig. 10.13):

$$8 \times 11 \times 3.14 = 276 \text{ ft}^2$$

OK, 276 > 240 (required)

Tile drainage (disposal) fields (Fig. 10.17) are very commonly used as a secondary treatment method, since they are relatively inexpensive to build and do not require a water table so deep—or soil so permeable—as seepage pits require. For a tile field, drains of square-edge agricultural tile, 4 in. (102 mm) or more in diameter and placed in shallow trenches, are covered with gravel. The ends of the tiles are separated by ¼-in. (6-mm) openings. The effluent runs out of these spaces and stands in the interstices of the gravel until it seeps into the earth. In effect, the gravel provides spaces that act as a dry well to receive the fluids and accommodate them until they slowly sink into the ground.

Disposal fields are located according to regulations such as those given in Table 10.5 and sized in relation to total sewage flow and septic tank size (see Tables 10.4 and 10.6). Although Table 10.4 shows the maximum absorption capacities for soils based on gal/ft^2 over a 24-hour period, a common alternative is to use *percolation tests*. For such tests, a "test pit" is dug and water poured into it. The number of minutes it takes for the water level to drop 1 in. is then recorded. Some codes present design sizes based on percolation test data. In areas with poor soil or occasional flooding, codes often require that an area equal to the drainage field's required size be set aside for use in the event of failure in the original field.

The rules of thumb for sizing disposal fields were presented in Section 8.1. The trench width, depth, and spacing shown in Fig. 10.17 are typi-

TABLE 10.5 **Location of Sewage Disposal Systems**

Minimum Horizontal Distance Clear Required From:	*Building Sewer*		*Septic Tank*		*Disposal Field*		*Seepage Pit or Cesspool*	
Buildings or structures[a]	2 ft	(0.6 m)	5 ft	(1.5 m)	8 ft	(2.4 m)	8 ft	(2.4 m)
Property line adjoining private property	Clear		5 ft	(1.5 m)	5 ft	(1.5 m)	8 ft	(2.4 m)
Water supply wells	50 ft[b]	(15.2 m)	50 ft	(15.2 m)	100 ft	(30.5 m)	150 ft	(45.7 m)
Streams	50 ft	(15.2 m)	50 ft	(15.2 m)	50 ft	(15.2 m)	100 ft	(30.5 m)
Trees	—		10 ft	(3 m)	—		10 ft	(3 m)
Seepage pits or cesspools	—		5 ft	(1.5 m)	5 ft	(1.5 m)	12 ft	(3.7 m)
Disposal field	—		5 ft	(1.5 m)	4 ft[d]	(1.2 m)	5 ft	(1.5 m)
On-site domestic water service line	1 ft	(0.3 m)	5 ft	(1.5 m)	5 ft	(1.5 m)	5 ft	(1.5 m)
Distribution box	—		—		5 ft	(1.5 m)	5 ft	(1.5 m)
Pressure public water main	10 ft[c]	(3 m)	10 ft	(3 m)	10 ft	(3 m)	10 ft	(3 m)

NOTE: When disposal fields and/or seepage pits are installed in sloping ground, the minimum horizontal distance between any part of the leaching system and ground surface shall be fifteen (15) feet (4.6 m).

Source: Reprinted by permission from the *Uniform Plumbing Code,* copyright © 1982 by the International Association of Plumbing and Mechanical Officials.

[a]Including porches and steps, whether covered or uncovered, breezeways, roofed porte-cocheres, roofed patios, car ports, covered walks, covered driveways and similar structures or appurtenances.

[b]All drainage piping shall clear domestic water supply wells by at least fifty (50) feet (15.2 m). This distance may be reduced to not less than twenty-five (25) feet (7.6 m) when the drainage piping is constructed of materials approved for use within a building.

[c]For parallel construction—For crossings, approval by the Health Department shall be required.

[d]Plus two (2) feet (.6 m) for each additional foot (.3 m) of depth in excess of one (1) foot (.3 m) below the bottom of the drain line.

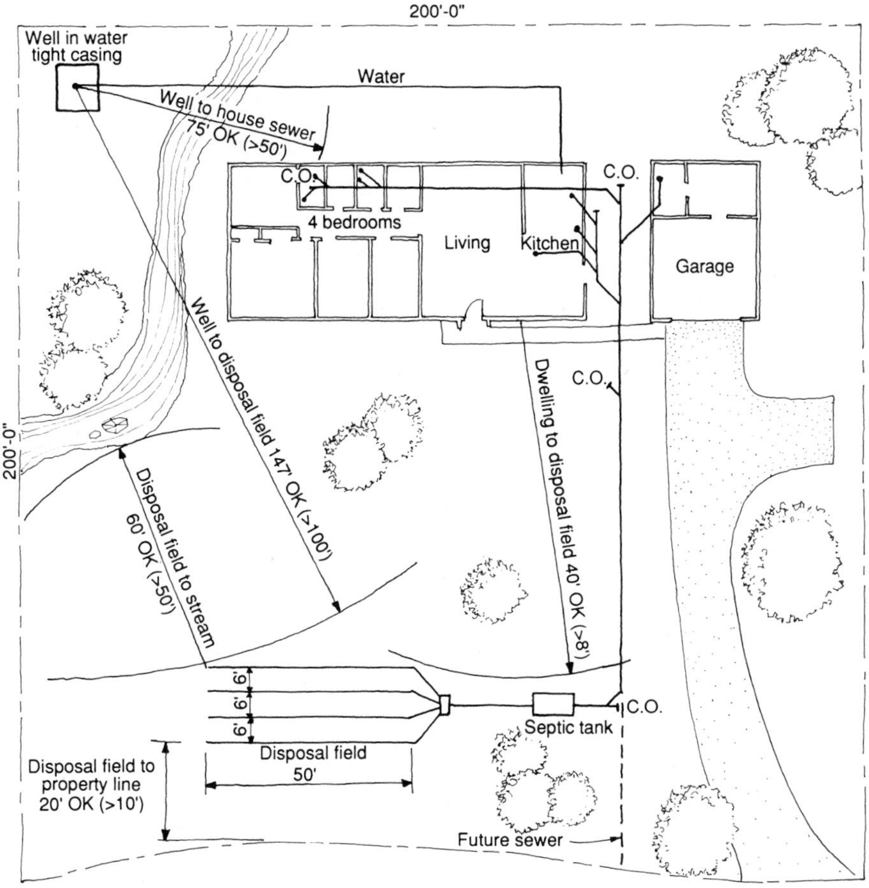

Fig. 10.16 Checking for required clearances on a suburban lot. Although the 40,000 ft² (3716 m²) lot is more than the minimum required, space should be left for a backup disposal field (of equal size) in case of failure of the original field. A seepage pit adjacent to this septic tank would barely meet the clearance criteria. With a stream nearby, the water table is probably too high for such a pit (see Table 10.5).

cal, but other combinations can be used, provided that they meet the requirements listed in Table 10.6.

EXAMPLE 10.2. (Fig. 10.17). Design a tile drainfield for the conditions given in Example 10.1.

SOLUTION. From the rule of thumb (Section 8.1), between 0.4 ft^2 and 3.6 ft^2 of drainfield area will be required per gallon of sewage; this will also allow for a second, backup drainfield.

$$600 \text{ gpd} \times 0.4 = 240 \text{ ft}^2$$

$$\times 3.6 = 2160 \text{ ft}^2$$

Since this is moderately well-draining soil, the total area will probably be about 1200 ft^2: 600 ft^2 of actual field and 600 ft^2 of backup area.

The effective absorption area of the typical trench depth and spacing shown in Fig. 10.17 is

$$\text{trench width } 2.0 \text{ ft}^2/\text{ft}$$

$$\text{trench sides } \underline{2.0} \text{ ft}^2/\text{ft } (12'' \text{ each side})$$

$$4.0 \text{ ft}^2/\text{ft}$$

From Table 10.4, the required square footage of leaching area for sandy loam is 40 ft/100 gal:

$$40 \text{ ft}^2/100 \text{ gal} \times 600 \text{ gpd} = 240 \text{ ft}^2$$

Since the trench has 4.0 ft^2/ft, the total trench length is

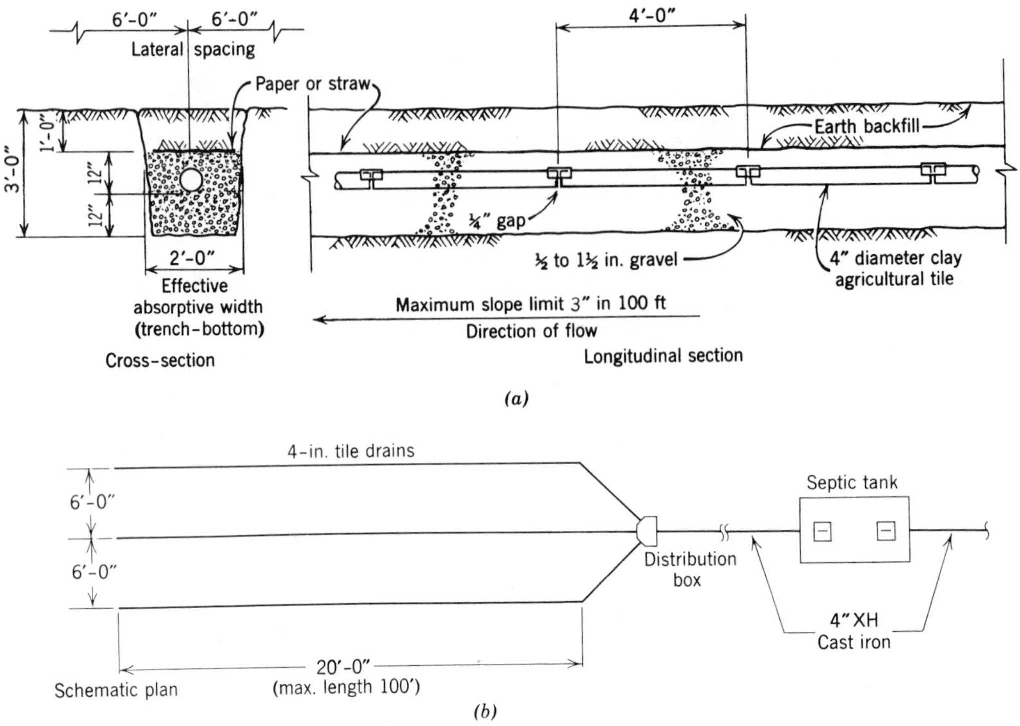

Fig. 10.17 *(Example 10.2) Tile drainfield for a four-bedroom, eight-person house. Although the drawings are not to scale, the dimensions would indicate a required area of about 20 × 70 ft (6 × 21 m) on the lot. When it is considered that it is best not to have the elements run below walks, drives, or other paved areas, sewage treatment on a small lot demands considerable space.* (a) *Transverse and longitudinal sections.* (b) *Schematic plan.*

$$240\ \text{ft}^2/4.0\ \text{ft}^2/\text{ft} = 60\ \text{ft}$$

A three-line disposal field is selected, so 60 ft/3 = 20 ft per line. Given the required clearances to the edges of the disposal field (Table 10.5) of 5 ft to property lines, the disposal field's area can be calculated as:

width: 5 ft + 12 ft (two spaces of 6 ft each) + 5 ft = 22 ft

length: 5 ft + 20-ft line + 5 ft = 30 ft

The area, then, is 22 × 30 = 660 ft^2 (about equal to the rule of thumb approximation).

Mounds with leaching beds (Fig. 10.18) represent a newer solution in the United States, and thus are likely to require special approval. The guidelines for leaching-bed sizing are essentially similar to those for drainage tile disposal fields. The absorption area for leaching beds, however, must be *50% greater* than that required for trenches. The bottom of the leaching bed generally must be at least 5 ft (1.5 m) above the water table, although in water-scarce areas, officials may reduce this requirement.

(b) Treatment Plants for Several Buildings. A variety of "package" treatment plants can be adapted for institutions or other groups of buildings such as subdivisions or shopping centers. One typical process is shown in simplified form in Fig. 10.19. Another example of the complexity of and the area required for this size of system is described here (Figs. 10.20 to 10.22).

A method of sewage treatment known as the Pasveer Oxidation Stream, which employs a principle similar to that followed by a natural stream, has been adopted at New York Institute of Technology. Serving the 450-acre campus at

TABLE 10.6 **Disposal Field Trenches**

*Part A. **Dimensions***

	Minimum	*Maximum*
Length of drain line(s)	—	100 ft (30.5 m)
Bottom width of trench	18 in. (457.2 mm)	36 in. (914.4 mm)
Spacing of lines, o.c.[a]	6 ft. (1.8 m)	—
Depth of earth cover over lines	12 in. (304.8 mm) [NOTE: 18 in. (457.2 mm) preferred]	—
Grade of lines	Level	3 in./100 ft (25 mm/m)
Filter material		
Over drain lines	2 in. (50.8 mm)	—
Under drain lines[a]	12 in. (304.8 mm)	—[d]

*Part B. **Leaching Areas***

Trench bottom[b]: minimum 150 ft² (14 m²) per system
Trench side wall: minimum[c] 2 ft²/ft
maximum[d] 6 ft²/ft

Source: Reprinted by permission from the *Uniform Plumbing Code,* copyright © by the International Association of Plumbing and Mechanical Officials.

[a]Minimum spacing of drain lines: 4 ft (1.2 m) plus 2 ft (0.6 m) for *each* additional foot (0.3 m) of depth *beyond* 1 ft (0.3 m) below the bottom of the drain line.

[b]Exclusive of rock, clay, or other impervious formations.

[c]Based on the minimum 12-in. trench depth below drain tile.

[d]Maximum of 36-in. trench depth below drain line can be counted when calculating required absorption area.

Old Westbury, Long Island, New York, this system provides an on-campus sewage treatment, which returns the purified effluent to the ground through 48 leaching wells located under the athletic field. The groundwater, thus restored, provides a contributing source of water for 400-ft-deep wells, distantly located, that furnish part of the water supply for the campus buildings (see Fig. 10.20).

During the development of the campus in the

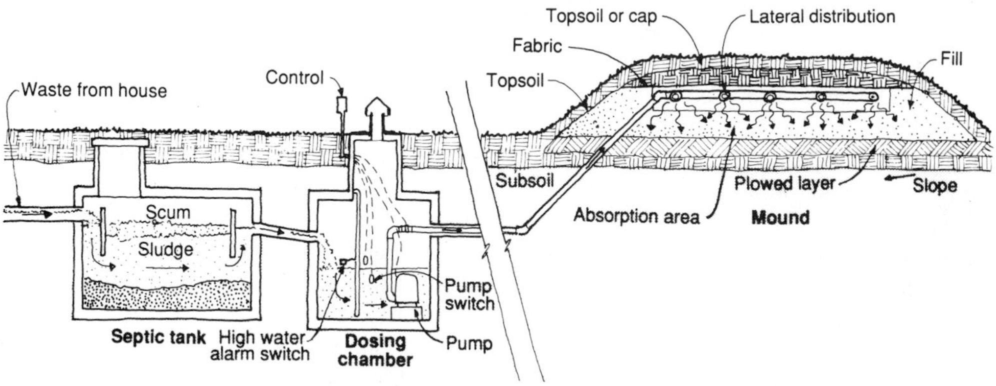

Fig. 10.18 *Mounds with leaching beds offer an option when the water table is high. The system serves a two- or three-bedroom home. [Adapted from Converse (1978).]*

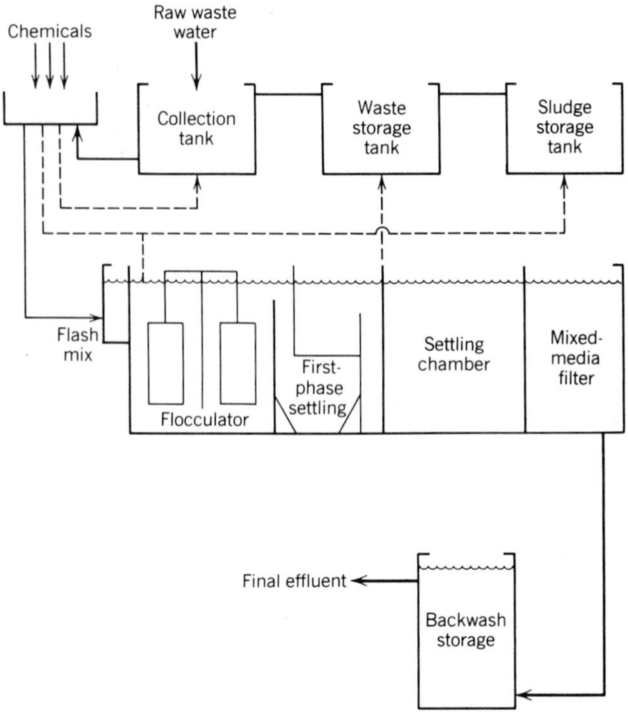

Fig. 10.19 Sewage treatment for groups of buildings. This is a combined physical–chemical treatment process: settling, anaerobic decomposition, filtering and chemical treatment from a primary–secondary–tertiary combination that results in a nearly pure effluent suitable for release to water bodies. [From Milne (1976).]

1960s and 1970s, studies were made of the sewage problem that would develop with the growth of the college. At the beginning, there were a few small buildings, later converted for administration offices and classrooms. These were, and still are, served by septic tanks and leaching fields. Presently, they constitute less than 20% of the sewage flow of the expanded building complex and may in future years be connected to the campus plant. There was no public sewer near the campus, and the health authorities ruled out the use of septic tanks for the numerous additional buildings that were contemplated.

The oxidation stream process applied here was developed by the Netherlands Research Institute for Public Health Engineering (T.N.) and is now in operation in many U.S. locations as well as in Europe. It is considered to be a modified form of the *activated sludge* process, with aerobic digestion and periodic sludge removal.

An important feature of the system is the mechanical aerator (Fig. 10.21), which keeps the stream of sewage moving and provides the oxidation necessary for aerobic digestion. In this design, sludge-drying beds are placed on an island, surrounded by the continuously moving oxidation stream. Another feature of the system is its low profile (Fig. 10.22), readily screened by trees. The aerobic digestion process is not malodorous. The sounds are produced by the waterwheel action of the mechanical aerators. No air compressor is required.

The plant has a full-time accredited operator; there are also two assistants and one relief operator. This design provides for a population of 4330, with a 340,000-gpd flow. It is presently used at about one-quarter of that capacity. The sludge has been removed from the drying beds twice in the first five years of operation.

(c) Large-Scale Sewage Treatment. Two examples of community-wide sewage treatment are presented here—one for a small, water-scarce community near San Diego, California, and one for the land-scarce conditions of New York City.

The *Padre Dam Municipal Water District* (Fig. 10.23) took brave steps toward water conservation and recycling during the 1960s and 1970s, with its Santee Water Reclamation Plant.

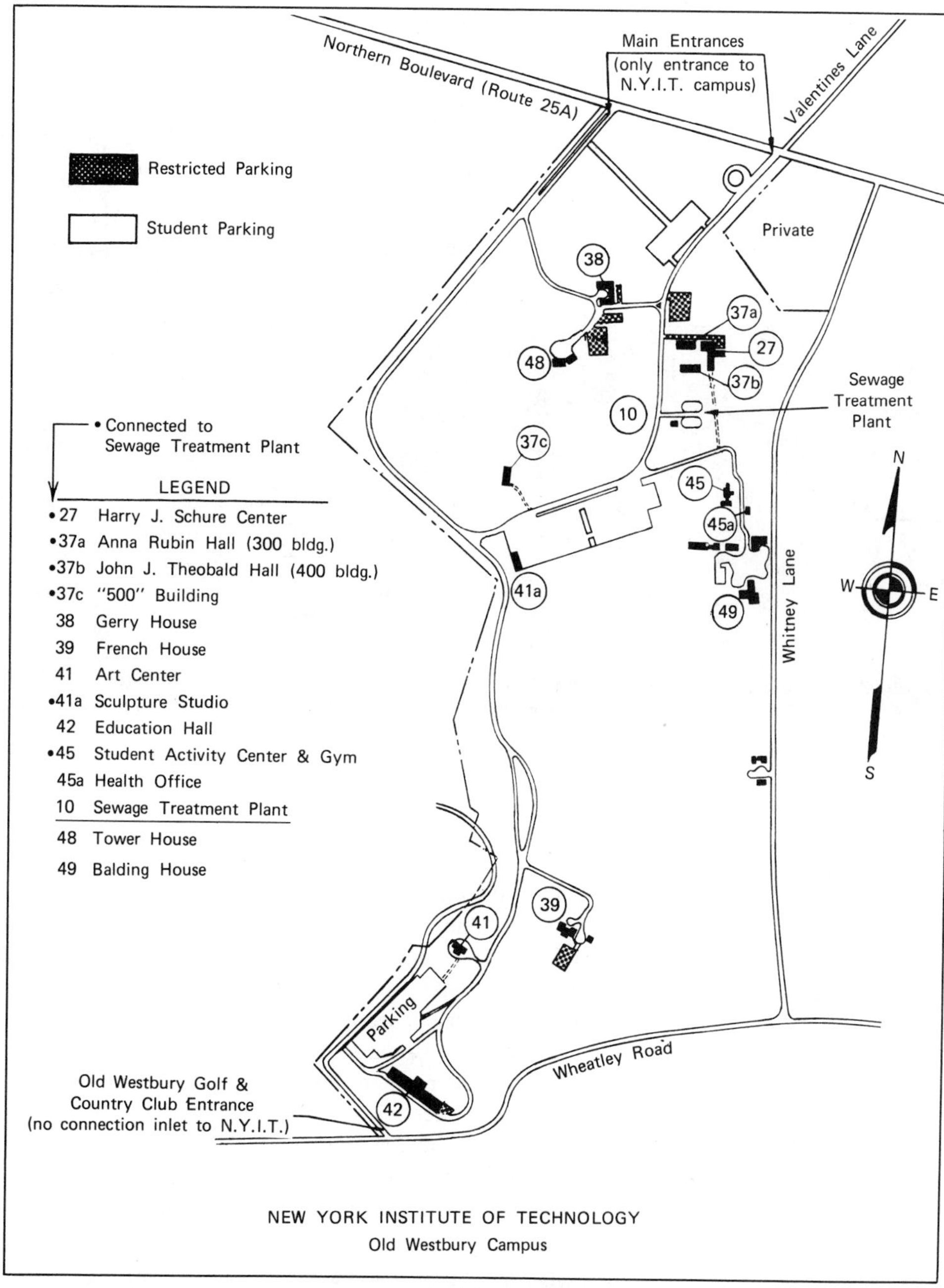

(*a*)

Fig. 10.20 *Sewage treatment plant, New York Institute of Technology (NYIT). The decision to locate the treatment plant at a position central to all the buildings it serves was a very appropriate choice for efficiency.* (a) *(opposite) Surrounded by heavily wooded areas (see Fig. 10.22), it is not a prominent landmark, though it offers a pleasant appearance to the occasional viewer. Although only half of the campus buildings are presently connected to the treatment plant, they are the largest ones and are responsible for 80% of the total campus sewage flow. A frequently used walkway (dotted lines on the map) connects the north academic center with the student activity and athletic center. This path passes just east of the woods that surround the plant. At certain points the plant can be seen through the trees, but no odor betrays its presence.* (b) *Layout of the treatment plant.* (c) *Plan of the two streams. Each of these units includes stream, rotor, clarifier with adjacent sludge pump pit, and three drying beds. Because of odorless operation, the proximity of the plant to the Student Activities Building poses no problem. (Courtesy of Bogen Jenal, Engineers, P.C.)*

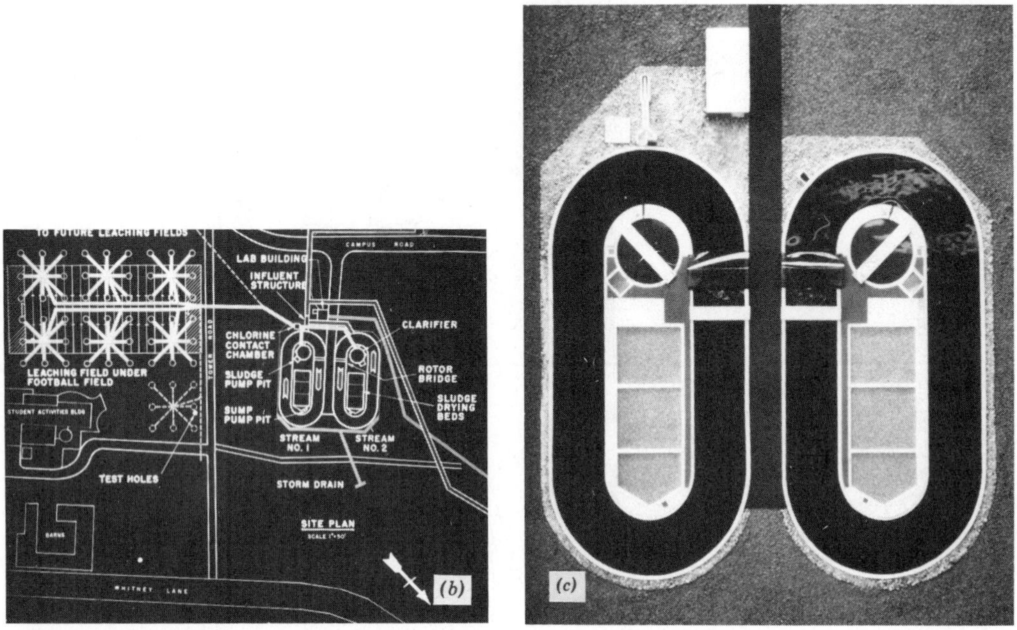

Fig. 10.20 *(continued)*

The Padre Dam Municipal Water District is a region where rainfall is less than 15 in. (380 mm) per year. It has no local water supplies available and water has to be imported 300 miles (480 km) from the Colorado River. At the same time it was contemplated that the wastewater resulting from this flow might have to be thrown away by discharging it into the Pacific Ocean. Adequate sewage treatment could process this fluid waste for secondary uses such as irrigation and recharging of groundwater. Considering further that many municipal supplies for potable water were taken from rivers often heavily polluted with sewage (see Fig. 9.5) and, after treatment, pumped into the domestic water mains, district officials decided to attempt to perfect a system that would make use of purified wastewater for many secondary uses.

Contrary to common opinion that such a concept might not be acceptable to the public, an effective public relations program concurrent with the technological advances has won over the citizens of the district.

The project involved building a sewage treat-

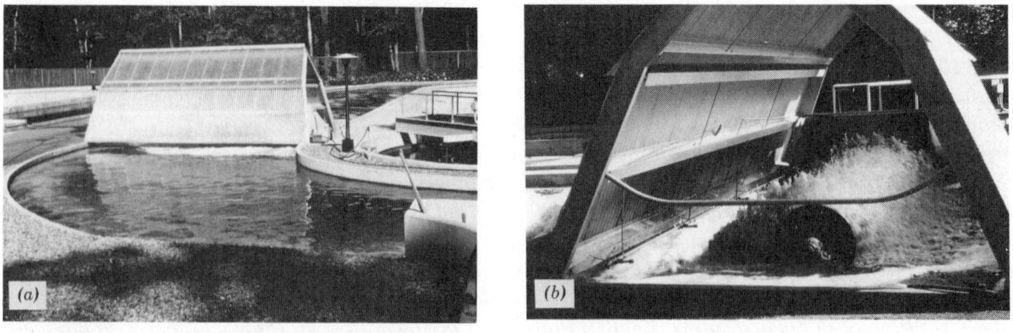

Fig. 10.21 (a) *Rotor with plastic "greenhouse" cover.* (b) *The rotor induces oxidation in this oval, circulating stream.*

Fig. 10.22 *NYIT sewage treatment plant at completion.* (a) *The only projections above ground level are the office/laboratory and the plastic covers over the rotors. Of the two lagoons, only the one in the foreground has been used, and only to a small percentage of its design capacity. The unused balance of its rating and the entire capacity of the second lagoon are available for future campus expansion.* (b) *The oxidation stream does not appear turbulent here, but it becomes so under the action of the mechanical aerator (rotor).*

ment plant and utilizing seven pits left over from a prior unrelated operation consisting of surface mining of sand and gravel. After partial purification of the sewage at the plant by the first two stages of the conventional activated sludge process, the effluent is discharged to form lakes adjacent to the plant. This provides tertiary (oxidation) treatment after which the purer effluent is pumped to a filter area at the north end of the complex where it is further purified by flow through sand and gravel. Chlorination is administered at one point. The water then flows through a series of seven lakes Nos. 7–6–5–4–3–2–1, in that order.

The appearance of new lakes in this semiarid region created an initial interest that was rapidly augmented by a very clever program. At first the lakes were fenced in. Then the fences were removed. The seven lakes were next made available for boating. They were stocked with fish and careful studies were made which indicated that the fish were healthy and flourishing. Fishing was permitted, but for a while all fish had to

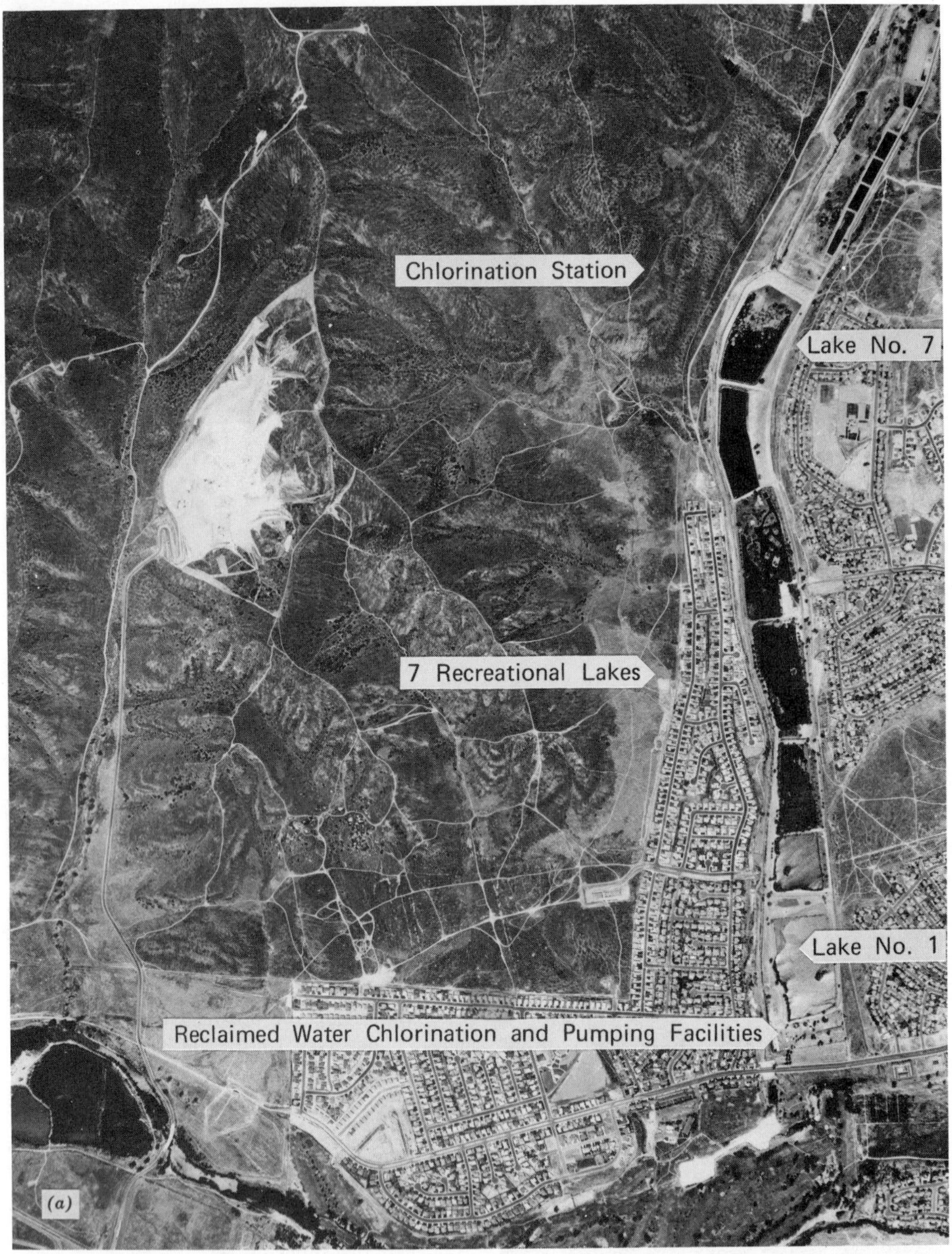

Fig. 10.23 *Santee Water Reclamation Plant and Santee Park and Recreational Facilities, Padre Dam Municipal Water District.* (a) *Aerial photograph of the Santee Plant, including the seven recreational lakes. Note the location of the San Diego River, at the bottom of the photograph. The water reclamation plant and the stabilization ponds do not appear in this aerial picture.* (b) *Raw sewage from the community of Santee enters the treatment plant, which is located at the top of this diagram. The process then proceeds southward to the point where reclaimed water is pumped to irrigation, or recharges ground water. Sludge does not enter the San Diego River, but is pumped to the San Diego Metro System.* (c) *This "laundry" process cleans up and makes reusable the valuable water that constitutes the major part of raw sewage.* (d) *Details of the reclamation plant. The item "Study Ponds," marked with an asterisk, indicates the continuing study and preparation for future distribution of drinking water that is reclaimed from sewage. The plant is not burdened by stormwater, which is recharged to the ground where it occurs locally in the community.*

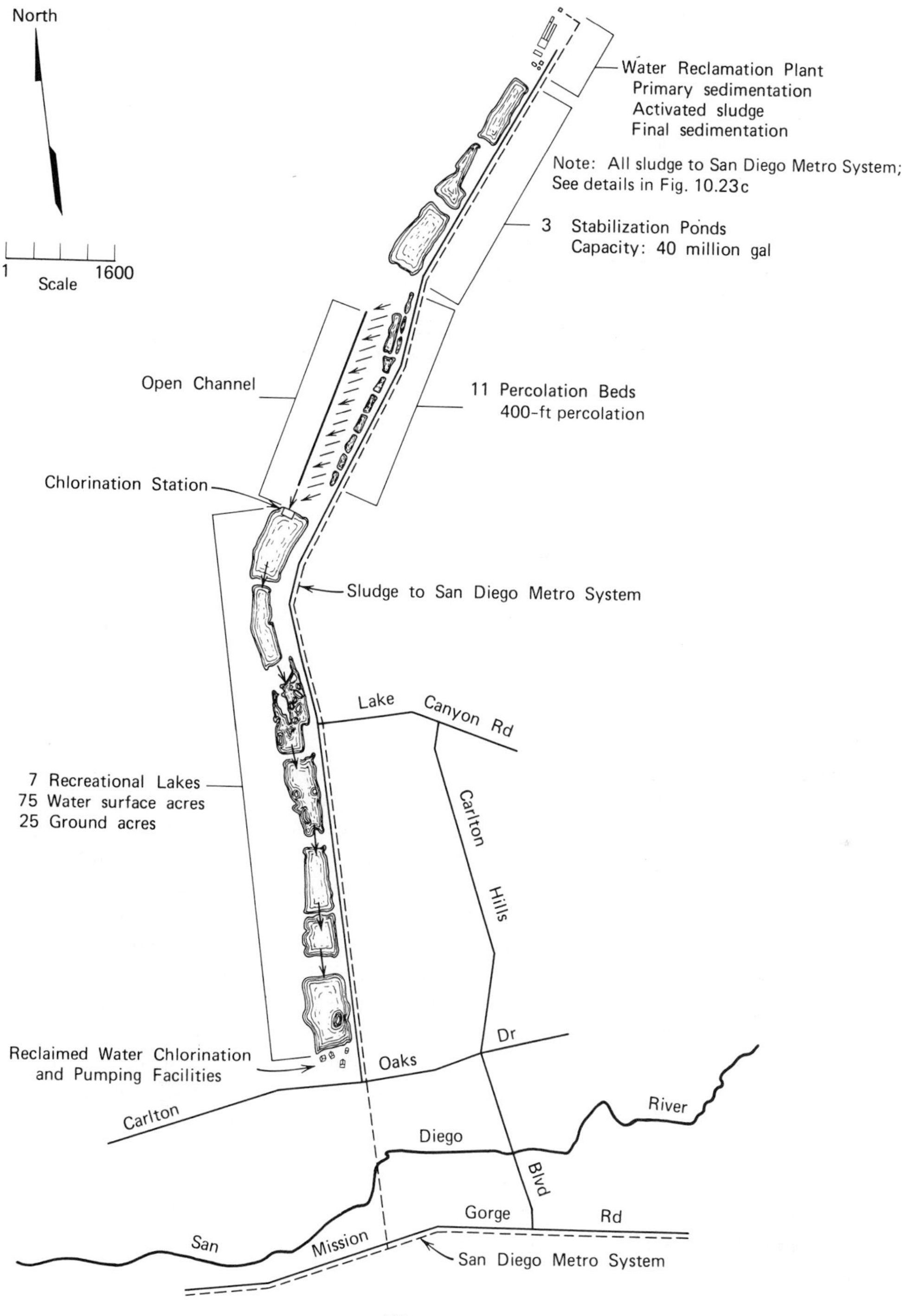
North
1
Scale
1600
Water Reclamation Plant
Primary sedimentation
Activated sludge
Final sedimentation
Note: All sludge to San Diego Metro System;
See details in Fig. 10.23c
3 Stabilization Ponds
Capacity: 40 million gal
Open Channel
11 Percolation Beds
400-ft percolation
Chlorination Station
Sludge to San Diego Metro System
Lake Canyon Rd
Carlton Hills
7 Recreational Lakes
75 Water surface acres
25 Ground acres
Dr
Reclaimed Water Chlorination
and Pumping Facilities
Oaks
Carlton
River
Diego
Blvd
Gorge
Rd
San
Mission
San Diego Metro System

(b)

Fig. 10.23 (continued)

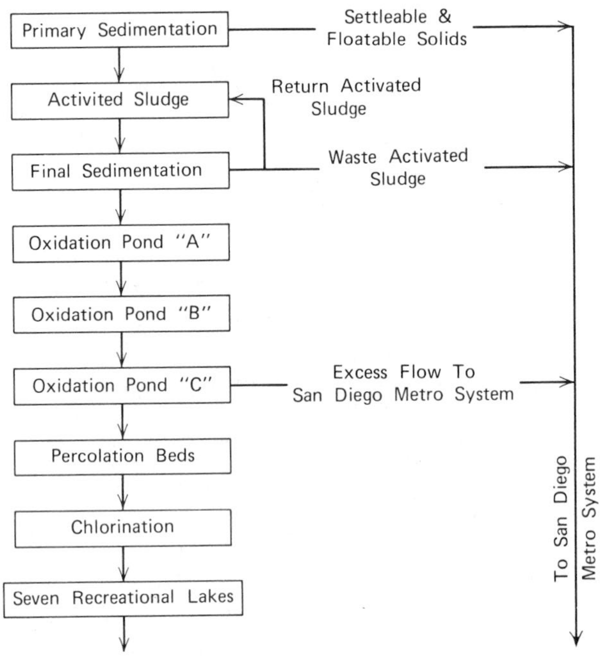

(*c*)

***Fig. 10.23** (continued)*

be returned to the water. Anglers were later permitted to keep and consume the fish. Finally, swimming was permitted, a use which connotes pure water that is in no way harmful to health. The overflow from the last lake, No. 1, discharges into Sycamore Canyon Creek and is used for irrigation of a golf course and the recharging of groundwater.

The *Ward's Island Sewage Treatment Works* (Fig. 10.24) in New York City provides a vastly larger example. Handling over 200 million gpd, it serves large parts of two city boroughs, Manhattan and the Bronx. Along the shorelines, "intercepting" sewers were built to head off the multitude of street sewers previously emptying into the rivers surrounding the city. They lead the sewage through deep under-river tunnels to the island plant.

Ward's Island uses the activated sludge process—one of the more efficient methods of treatment. Compressed air is bubbled through the sludge so that aerobic digestion can occur. On a much larger scale, it is the same process as described in Fig. 10.14. Grit settles out at chambers in Manhattan and the Bronx. Together with other intercepted solids, such as pieces of wood, it is trucked to outlying districts, where it is dumped as "garbage fill" along with garbage from other sources. In the preliminary settling tanks, other heavy solids drop during an hour's pause of the sewage flow. The remaining, highly polluted fluid flows through the aeration tanks for a 3-hour trip in the presence of a biologically active culture. As a result of bacterial action, accelerated by air pumped in along the way, digestion takes place and the fluid can be made 95% pure before it is discharged at two points to the river. The resulting floc, digested and much purified, is collected in the final settling tanks. The practice of dumping this floc at sea was stopped in 1981.

10.3 Principles of Drainage

Early in the history of indoor plumbing, waste drainage was a simple matter: a pipe containing the wastewater led to a sewer (Fig. 10.25*a*). Before long, the noxious gases that were created by the anaerobic conditions in the sewer became a threat to the health of those indoors. Thus, the *trap* was invented (Fig. 10.25*b*) to block the pipe so that gases could not pass. However, as moving water filled the pipe downstream from the traps, the mass of water acted as a plunger, creating higher pressures in front of it and negative pressures behind. The positive pressures might force sewer gas through the water in other traps; worse, the negative pressures could suck (or siphon) the water from the trap, leaving it open to gas passage. A way to deal with these pressures was through the installation of *vents,* so that the suction would draw air down through vents rather than water from the traps (Fig. 10.25*c*). A typical arrangement of fixtures, traps, and vents is shown in Fig. 10.26.

(a) Traps. The only separation between the unpleasant and dangerously unhealthy gases in a sanitary drainage system and the air breathed by room occupants is the water caught in the fixture trap after each discharge from a fixture. Sufficient water must flow, especially in water closets, to keep this residual water clean. Traps are made of steel, cast iron, copper, plastic, or brass—except those in water closets and urinals, which are often made of vitreous china cast integrally with the fixture. The deeper the seal, the more resistance to siphonage but the greater the fouling area; therefore, a minimum depth of 2 in. (51 mm) and a maximum depth of 4 in. (102 mm) are common standards. All traps should be self-cleaning: that is, capable of being completely flushed each time the trap operates, so that no sediment will remain inside to decompose.

There are a few exceptions to the rule that each fixture should have its own trap. Common exceptions include two laundry trays and a kitchen sink connected to a single trap, not more than three laundry trays using one trap, and three lavatories on a single trap. In the case of the laundry trays and sink, the sink is equipped with the trap and set nearest to the stack. (See stack *b* in Fig. 10.26).

Traps are usually placed within 2 ft (0.6 m) of the fixture and should be accessible for cleaning through a bottom opening that is otherwise closed by a plug. Overflow pipes from fixtures are connected into the inlet side of the trap. In long runs of horizontal pipe, so-called "running traps" are used only near the drains of floors, areas, or yards and should be provided with hand-hole cleanouts. "Island" sinks pose a special problem when the vent line cannot lead upward from such an exposed location. The sink's waste line can be taken to a distant sump, which is then itself trapped and vented. (See Fig. 10.28*c* for a similar process.)

When fixtures are used very infrequently, the water in traps can evaporate into the air, breaking the seal of the trap. In contemplating the possible frequency of use, this fact should be kept in mind by the designer. Unoccupied residences (such as weekend or vacation homes) are likely candidates for sewer gas penetration through traps emptied by evaporation. Otherwise, evaporation to a dangerous degree rarely occurs, except in the case of floor drains. Trapped drains of this type, employed to carry away the water used in washing floors or drained from heating equipment, may often lose the water seal between infrequent operations. Many authorities are reluctant to approve floor drains connected to the building's sewer, requiring instead that they be separately connected to a dry well. In either case the use of a special hose bibb, affording a source of water directly above the drain, is a wise precaution. It can easily be used to manually refill the trap of the drain, or, overflow lines from lavatories can lead to the floor drain trap.

(b) Vents. For the admission of air and the discharge of gases, soil, and waste, stacks are extended through roofs and a system of air vents, largely paralleling the drainage system, is provided. As in the case of drainage stacks, the ventilating stacks extend through the roof or vent through the drainage stack. The functions of venting are often misunderstood. It is true, of course, that one important purpose is to ventilate the system by allowing air from the fresh-air inlet (or from the sewer, if there is no house trap or fresh-air inlet) to rise through the system and carry away offensive gases. This provides some purification for the piping. However, several other purposes are served by the vent piping. The introduction of air near the fixture (and in

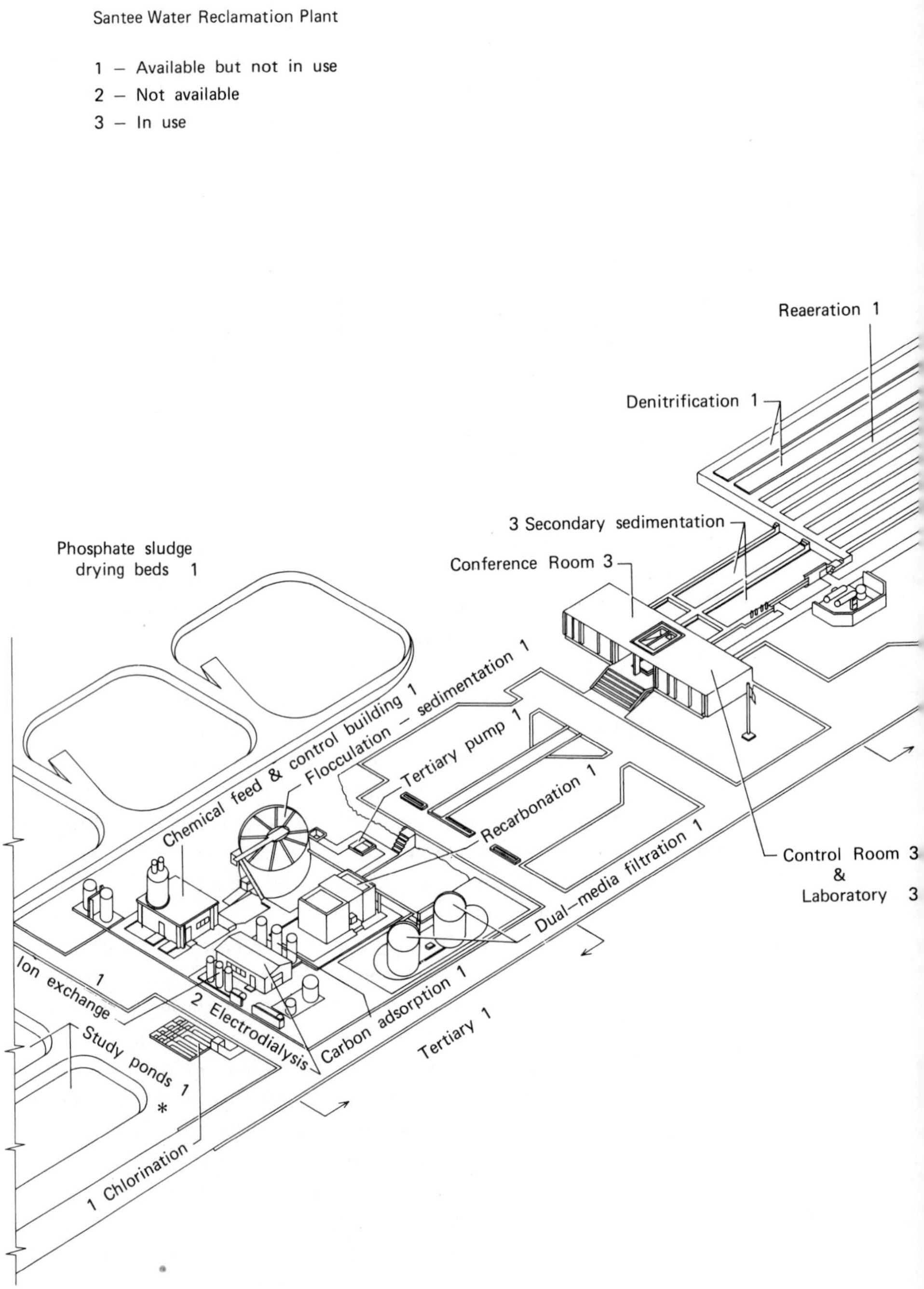

the case of circuit vents, at the branch soil line) breaks the possible siphonage of water out of the trap. Under other circumstances—namely, when drainage fluids descend to a fixture group through the soil stack—the foul gases would bubble through the trap-seals of that group. The vent system provides a local escape for these gases. Comprehensive experiments have shown

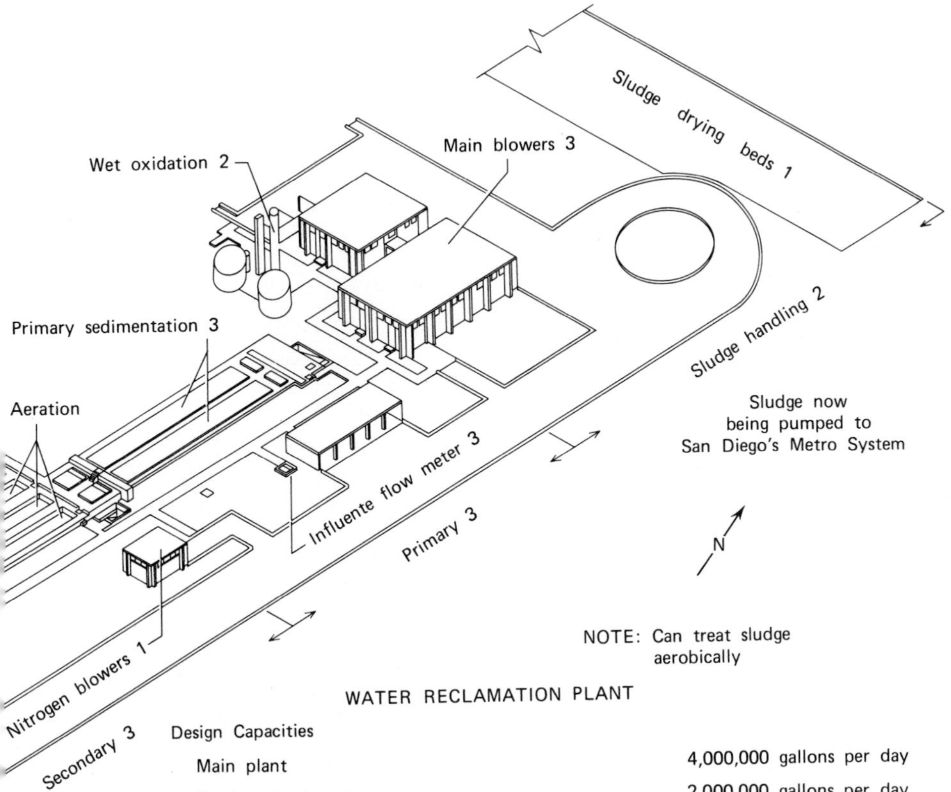

Design Capacities	
Main plant	4,000,000 gallons per day
Tertiary treatment	2,000,000 gallons per day
* Potable water study ponds	100,000 gallons per day
Total Construction Cost	$3,825,000
U.S. Government Construction Grants	$1,497,000
Joint Demonstration Research Project	$511,000
Santee County Water District	
Federal Water Quality Administration	

(d)

Fig. 10.23 *(continued)*

that circuit venting—which permits air and gases to pass in and out of the soil or waste branch instead of at each fixture, as in the case of continuous venting (individual fixture venting)—can fully prevent the siphonage of trapseals or their penetration by gases (see Fig. 10.27).

(c) Air Gaps and Vacuum Breakers. Every plumbing fixture is supplied with pure water at one point; most discharge contaminated fluids at another. The proximity of sewage to potable water at fixtures is inescapable. It is possible that sewage could accidentally be siphoned into a pipe carrying potable water. Consider an improperly placed faucet whose outlet is below the rim of a fixture. If the fixture overflow is plugged and the fixture bowl full, the faucet can easily project into the foul drainage water. If, in this circumstance, the water piping is drained while the faucet is open, contaminated water could be drawn by suction into the water piping.

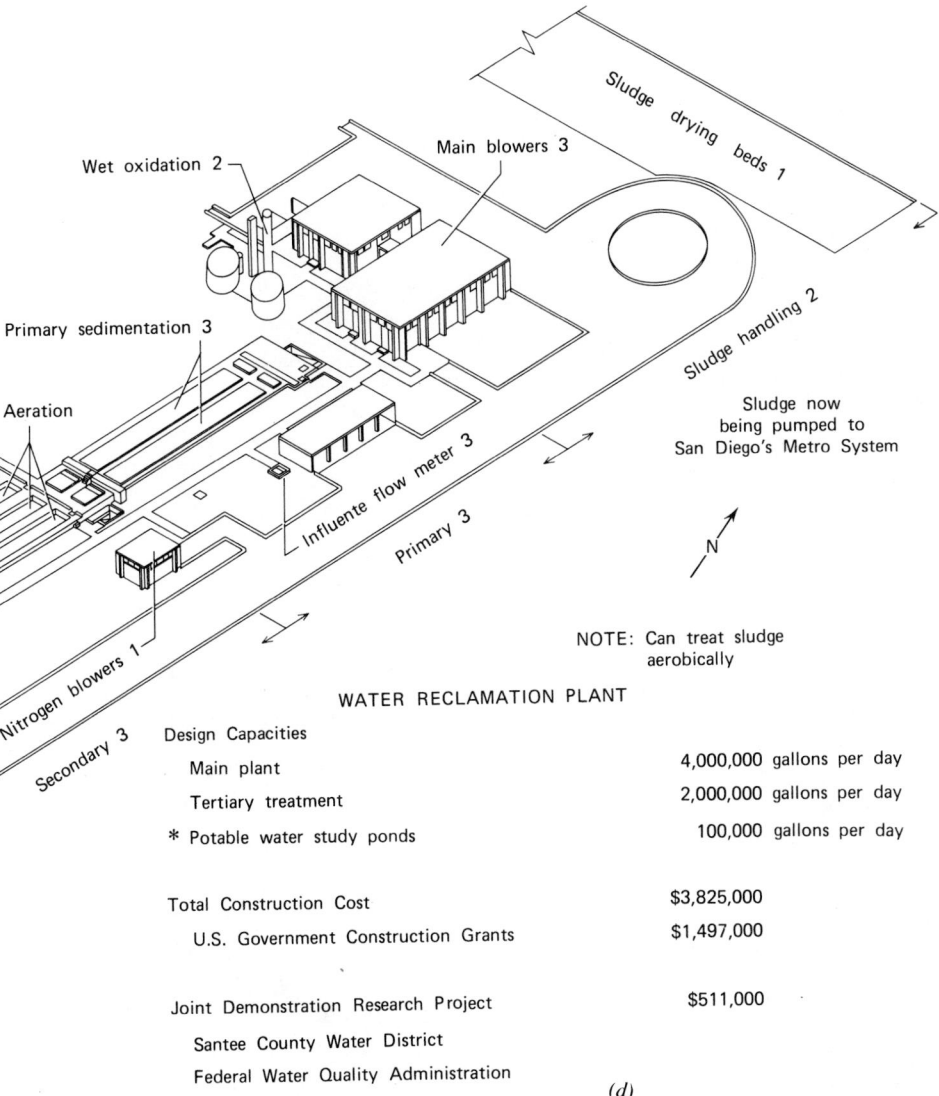

In water closets served by flushometers (flush valves), the water supply unavoidably enters the bowl below the rim. A vacuum breaker placed in the flushometer closes with water pres-

Fig. 10.24 *Ward's Island Sewage Treatment Works, New York City.* (a) *In the center of the aerial photograph is Manhattan Island, identified by Central Park. The island is flanked in the foreground by the Hudson River and, on its other side, by the East River, identified by Roosevelt Island.* (b) *Activated sludge process.* A, *Manhattan grit chamber;* B, *Manhattan sewage tunnel;* C, *Bronx sewage tunnel;* D, *laboratory and administration; E, power plant;* F, *pump and blower building;* G; *preliminary settling chambers;* H, *aeration chambers;* I, *final settling chambers;* J, *sludge storage building;* K *and* L, *95% pure water discharge;* M, *dock for sludge boats.*

sure but opens to admit air if there is suction in the water pipe. This prevents siphonage in much the same way that a vent prevents trap siphonage (see Fig. 10.28). The use of vacuum breakers at dishwashers and clothes washers is diagrammed in Figure 9.33. These are especially important locations, since in both these appliances pumps force the wastewater into the drainline.

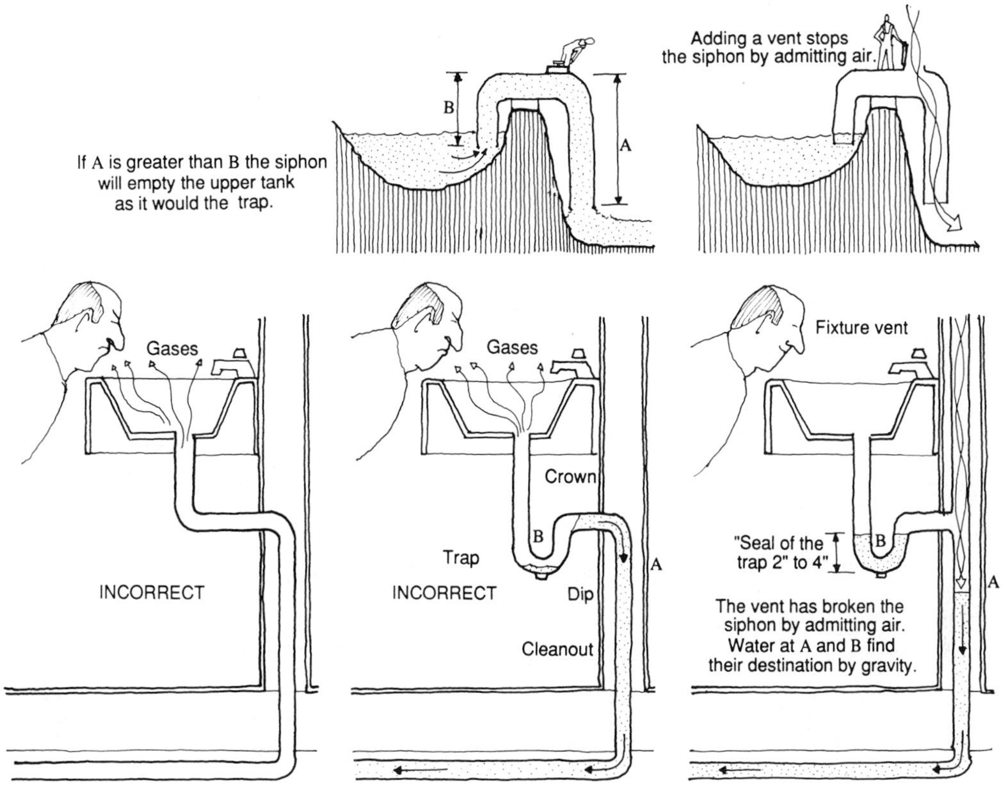

(*a*) Incorrect, no trap
Foul gases enter the room

(*b*) Incorrect, no vent
The water that should seal the trap is siphoned out

(*c*) Correct
Air entering from the vent breaks the siphonage

Fig. 10.25 *Function of a trap and one of the several functions of a vent (preventing siphonage).*

10.4 Piping, Fittings, and Accessories

(a) Piping and Fittings. The principal materials used for soil and waste piping and for venting are cast iron, copper, and plastic. Galvanized steel is sometimes chosen for vents and for tall stacks in high-rise structures (see Fig. 10.29). Sometimes, different materials are used in the same system. Where dissimilar metals are connected, dielectric unions are used to prevent corrosion due to electrolysis.

Cast Iron. Used first in Germany around 1562 and appearing in the United States about 1813, supplanting the tubing and culverts of early epochs that employed clay, lead, bronze, and wood, cast iron was the earliest of the modern materials used for piping. Its durability and resistance to corrosion makes it eminently suitable for the components of sanitary drainage systems. It is appropriate for a wide range of uses, from small residential work to the stacks and branches of tall buildings.

Typical fittings for sanitary drainage appear in Figs. 10.29 and 10.30. In sanitary flow systems composed of *any* material, changes in direction must be made with easy bends. To prevent clogging or fouling by the solid materials in the piping, right-angle connections are not used. Thus, the choices in Fig. 10.29 would be for ⅛ bend plus a 45° Y, or a ¼ bend long sweep. The top connection of the 90° T, in the position shown, would connect *only* to a vent.

The three cast-iron soil pipe joints shown in Fig. 10.31 are semirigid, watertight, and gastight connections of two or more pieces of pipe or fittings in a sanitary system. A special characteristic common to types *b* and *c* is that they provide a quieter plumbing system and slightly

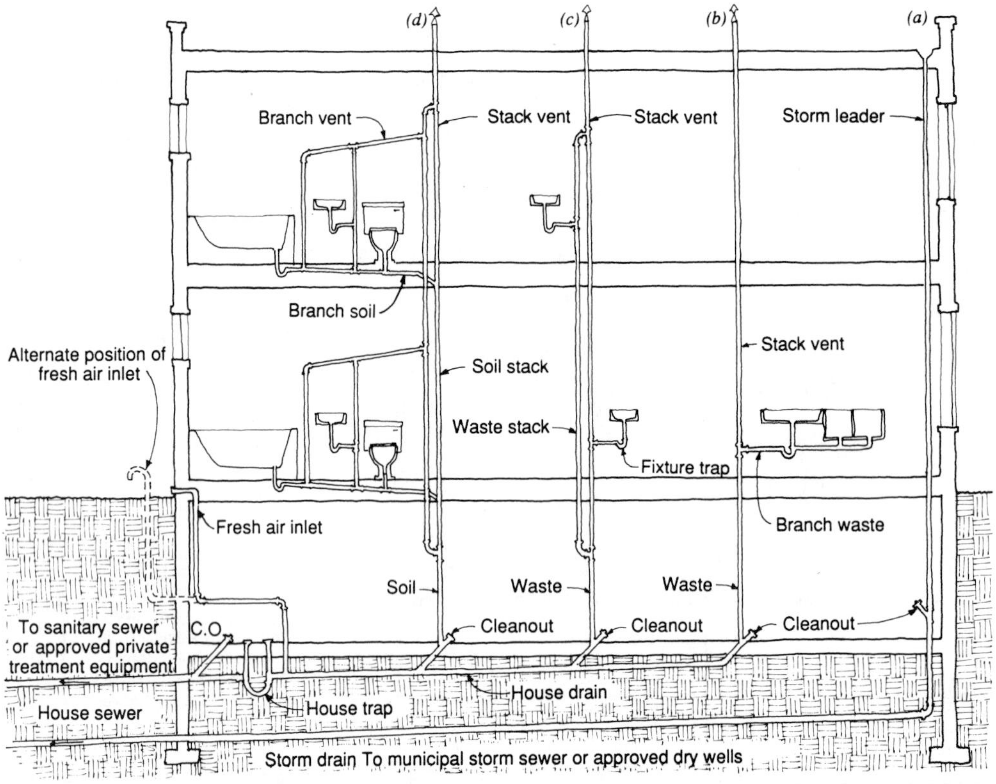

Fig. 10.26 *Typical sanitary drainage system, which separates stormwater from sewage. This installation combines blackwater and graywater, although their separation would encourage graywater recycling. The housetrap is optional under some plumbing codes, and illegal under others (since sanitary sewers could be vented at street level through the "fresh air inlet").*

more flexible joints. The use of cast-iron soil pipe and fittings in bathroom groups can be seen in Fig. 10.32.

Copper Tube and Fittings for DWV. There are several tube classifications for the copper products used in *plumbing* systems: K, L, and M are the choices for *water* systems, and DWV for *drainage, waste,* and *vent* installations (as the initials indicate). Connections between copper tubing and its couplings or fittings are made by a sliding fit (see Fig. 10.29*b*). Between the mating surfaces is a cylindrical capillary space filled with solder. The process of making the joint is a simple one, utilizing a "flux," heat, and solder. Properly made, the joint will be airtight and capable of withstanding high pressure. (High-temperature solder should be used for hot water lines, such as those to solar collectors.) To undo the joint for repair or renovation, it is simply reheated until the solder melts. Like cast iron, copper has a history of use in ancient installations. Updated and highly developed in recent decades, it is now in widespread use.

Plastic Materials for DWV. Along with copper and cast iron, plastics are very suitable for sanitary drainage systems. Table 10.7 lists the three kinds of plastics most suitable for drainage, waste, and vent. One of these materials, acrylonitrile–butadiene–styrene (ABS), is identified and further evaluated by the labeling shown in Fig. 10.33. One of several steps used in making a "solvent-weld" connection is seen in Fig. 10.33, as is a method of support in wood frame construction. Figure 10.34 shows assembled bathroom piping in place, and Fig. 10.35 illustrates the lightness of the material and its adaptability to prefabrication.

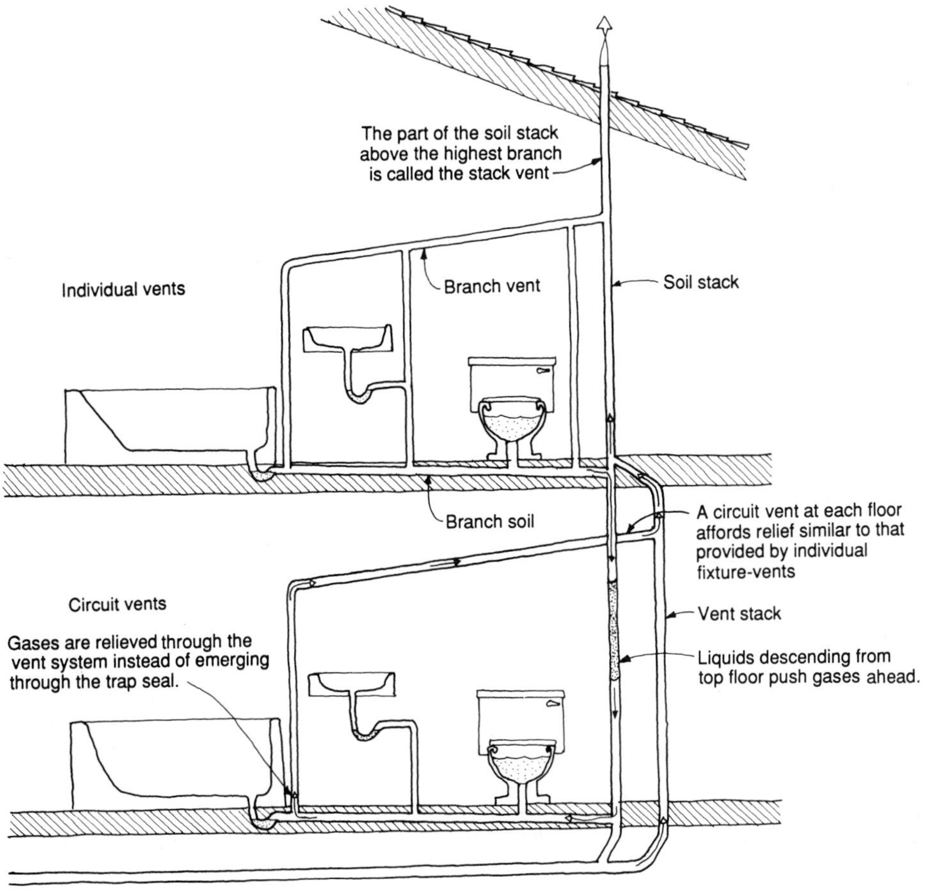
The part of the soil stack
above the highest branch
is called the stack vent
Individual vents
Branch vent
Soil stack
Branch soil
A circuit vent at each floor
affords relief similar to that
provided by individual
fixture-vents
Circuit vents
Vent stack
Gases are relieved through the
vent system instead of emerging
through the trap seal.
Liquids descending from
top floor push gases ahead.

Fig. 10.27 *Gas relief through vents. Gases pressurized by hydraulic action or by expansion due to putrefaction have an escape path through the vent system and will not enter the rooms.*

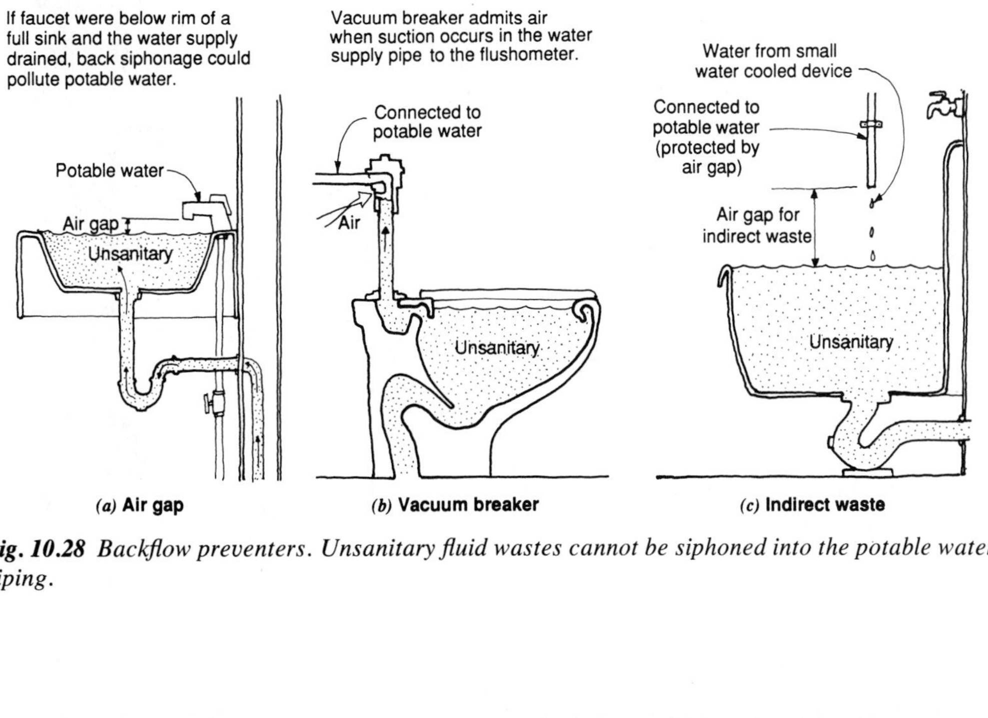
If faucet were below rim of a
full sink and the water supply
drained, back siphonage could
pollute potable water.
Vacuum breaker admits air
when suction occurs in the water
supply pipe to the flushometer.
Water from small
water cooled device
Connected to
potable water
Connected to
potable water
(protected by
air gap)
Potable water
Air gap
Unsanitary
Air
Air gap for
indirect waste
Unsanitary
Unsanitary
(a) Air gap
(b) Vacuum breaker
(c) Indirect waste

Fig. 10.28 *Backflow preventers. Unsanitary fluid wastes cannot be siphoned into the potable water piping.*

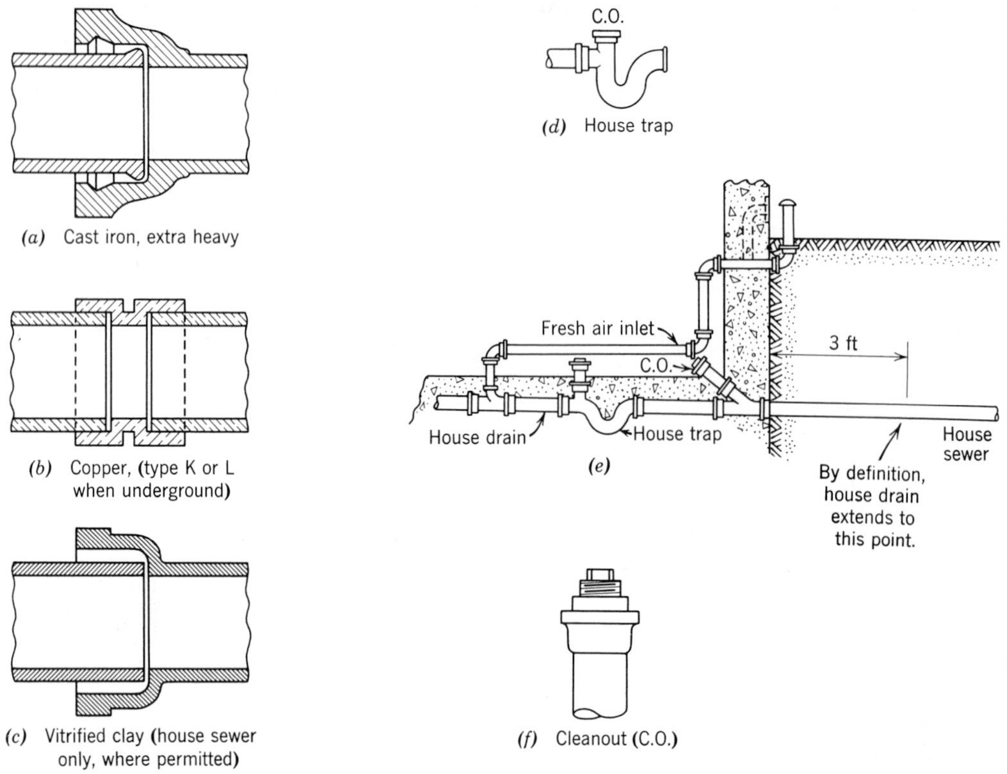

***Fig. 10.29** Piping and fittings.* (a) *Connection of cast iron piping.* (b) *Coupling to connect copper tubing.* (c) *Connection of vitrified clay piping.* (d) *Detail of house trap fitting.* (e) *House drain, house trap with cleanouts and vent (fresh air inlet), and house sewer.* (f) *Cleanout showing removable threaded plug. For large buildings the terms building drain, building sewer, and so on supplant "house." The inclusion or omission of the house trap depends on the local code.* Note: *Plastic connection is similar to* (b).

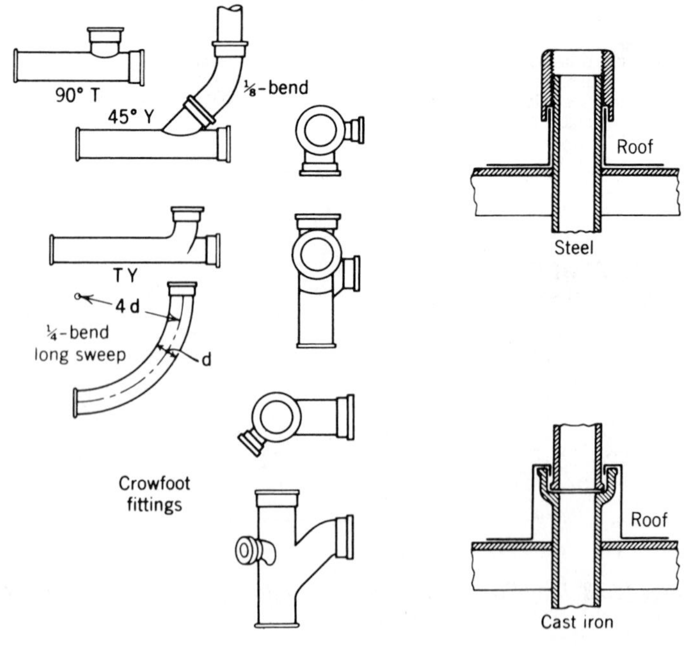

***Fig. 10.30** Cast-iron fittings, principal types, and method of flashing at roofs for steel and cast-iron.*

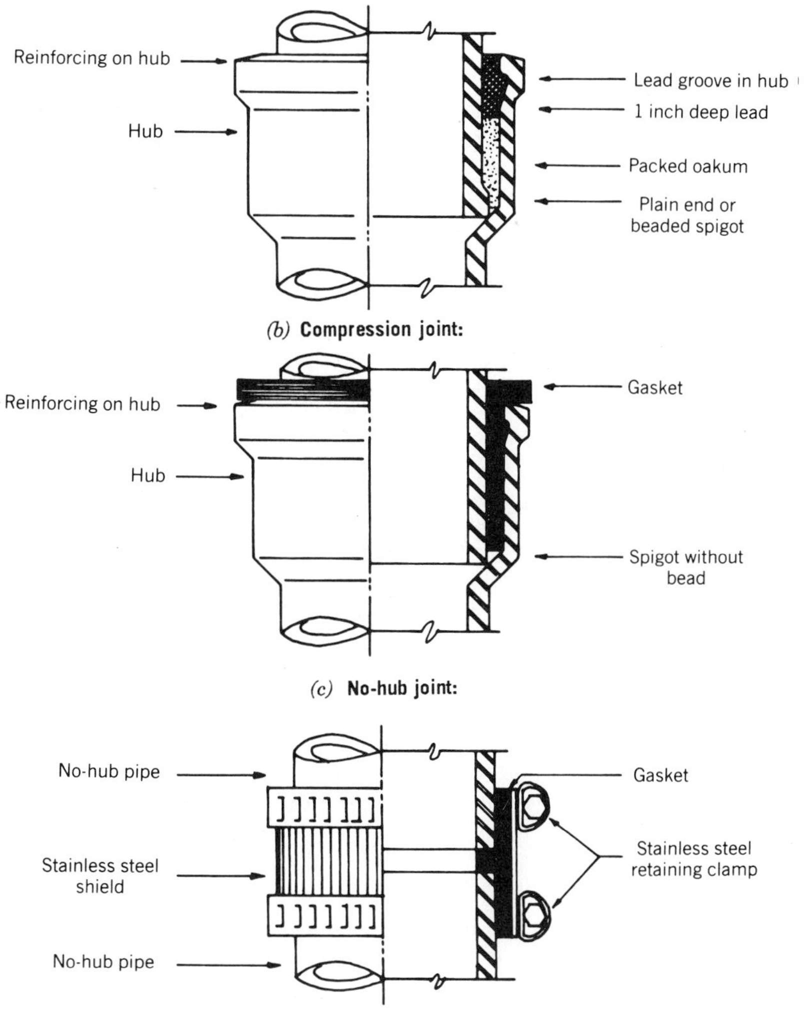

Fig. 10.31 Various joints used to connect cast-iron soil pipe and fittings. (Courtesy of Cast Iron Soil Pipe Institute.)

(b) Accessories. Among the many special devices that can form part of a plumbing system, a few are described here, including floor drains, backwater valves, ejectors, and interceptors.

Floor Drains. When floors in buildings must be washed down after such operations as food preparation and cooking, floor drains are usually necessary. Since they are sometimes connected to sanitary drainage systems and, in long periods of disuse, might lose their trap-seals by evaporation, special precautions are necessary to pre-

TABLE 10.7 **Suitable Choices of Material for Plastic Piping in DWV (Drainage, Waste, and Vent) and Sewer Systems**

Symbol	*Material*	*DWV*	*Sewer*
ABS	Acrilylonitrile-Butadiene Styrene	√	√
PVC	Polyvinyl Chloride	√	√
SRP	Styrene Rubber Plastic		√

Source: Reprinted by permission of *Progressive Architecture.*

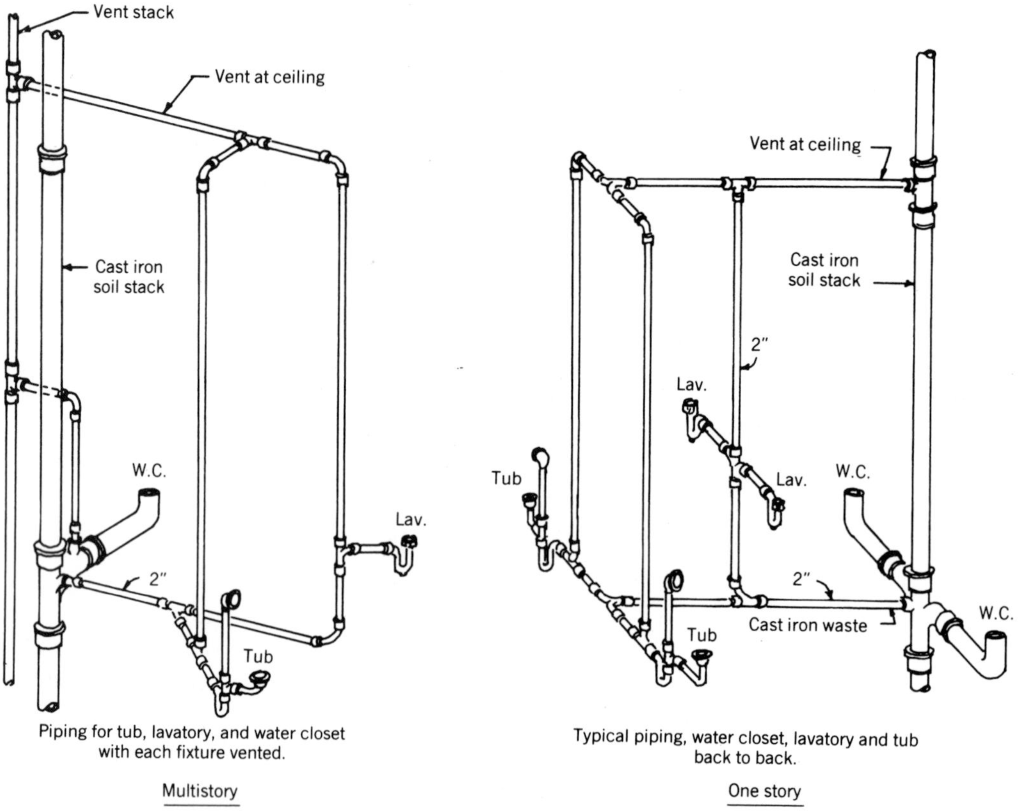

Fig. 10.32 *Two typical piping arrangements for water closet, lavatory, and tub. (Courtesy of Cast Iron Soil Pipe Institute.)*

serve the trap-seal and avoid odors and unsanitary conditions in the room (see Fig. 10.36).

Backwater Valves. These devices (Fig. 10.37) are sometimes used when plumbing fixtures are installed at low elevations, such as in basements, or in other locations that are near the level of the sewer. They cannot be used to protect the entire plumbing system, and should only be used when necessary. They must be accessible for maintenance. An alternative to the use of backwater valves is the sewage sump.

Sewage Sumps and Ejectors. Whenever subsoil drainage, fixtures, or other equipment are situated below the level of public sewers, a sump pit or receptacle must be installed. Into this pit the drainage from the low fixtures may flow by gravity, and from it the contents are then lifted up into the building sewer. Sewer ejectors may be motor-driven centrifugal pumps (see Fig. 10.38) or they may be operated by compressed air. The latter have no revolving parts within the receptacle. An air compressor is started when the float within the sump reaches a certain level, and air at a pressure greater than 0.433 psi for each foot (10 kPa for each m) of lift is delivered into the space above the liquid. The air pressure closes the inlet and opens the outlet check valves, expelling the contents of the sump and elevating it to the sewer.

Interceptors. Sanitary drainage installations ultimately discharge their waste matter into private or public sewage treatment plants that attempt to digest or cope with anything that may come through the pipes. Public plants are somewhat better equipped to handle this problem than private installations. From any plumbing fixture to the end of the disposal process, all

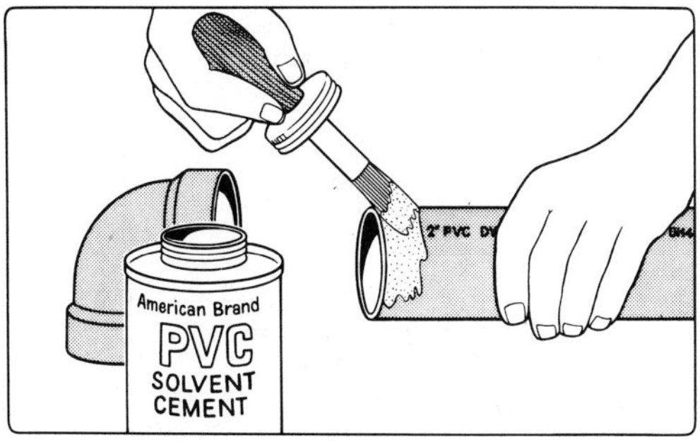

(a)

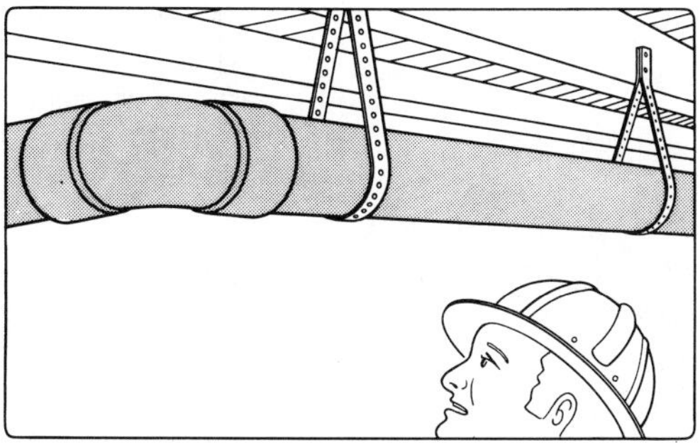

(b)

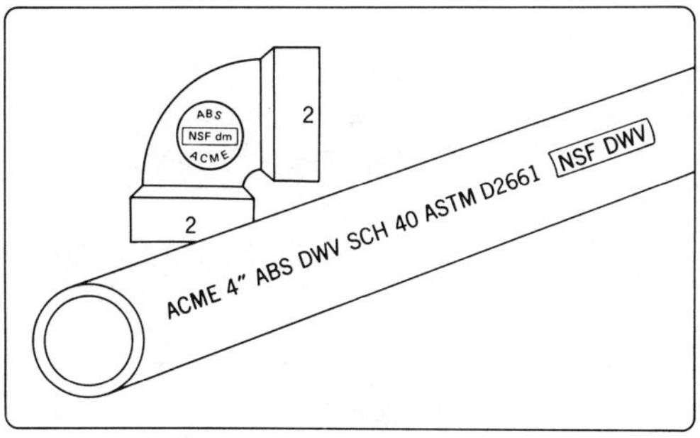

(c)

Fig. 10.33 Details in the use of plastic pipe. (a) *One of the steps in making a "solvent weld" of a plastic pipe to a plastic fitting.* (b) *In wood frame construction, plastic pipe assemblies can be supported by metal straps nailed to the wood joists. Flexibility of the plastic material suggests that the supports be more closely spaced than in the case of metal piping.* (c) *Typical identification symbols on plastic pipe. (Courtesy of Plastic Pipe Institute.)*

ACME	The name of the manufacturer.
4 in.	Diameter of the pipe.
ABS	Acrylonitrile-Butadiene-Styrene, the material.
DWV	Suitable for drainage waste and vent.
SCH 40	Schedule 40. This identifies the wall thickness of the pipe.
ASTM D2661	"Standards Number" assigned by the American Society for Testing and Materials.
NSF DWV	Tested by the National Sanitation Foundation Testing Laboratory. The pipe meets or exceeds the current standards for sanitary service.

parts of systems should be openable through cleanouts and other points of access, to relieve clogging that will often occur in the piping as well as in the septic tank or public disposal plant. Since it is impossible to control what will be discarded into the plumbing drains, trouble can usually be expected. The problem can be reduced somewhat by devices known as interceptors, which catch foreign matter before it travels too far into the system.

It is evident that interceptors require periodic servicing. Interceptors for as many as 25 different kinds of extraneous material are listed by some manufacturers. They include devices to catch hair, grease, plaster, lubricating oil, glass grindings, or troublesome unwanted material from many industrial processes. One of the few interceptors that is sometimes needed in homes, and more often in institutional kitchens, is the grease interceptor, or *grease trap*.

As the waste is passed from a kitchen sink through the circuitous path of the grease interceptor, the grease floats to the top, where it is trapped between baffles, while the more fluid wastes pass through at a lower level. There are special reasons for removing grease; for one thing, it congeals within piping and thus physically retards the sewage digestion process (see Fig. 10.39).

(c) Examples of Installations. In residential work, the piping assemblies may often be viewed as a "flag," as can be seen in Fig. 10.40. The mast is the soil stack, the horizontal top of the flag is the branch vent, the bottom is the soil or waste branch, and the outer edge is the vertical pipe of the last fixture. In frame construction, the flag usually fits into a 6-in. (152-mm)

WATER AND WASTE

Fig. 10.34 Plastic piping (solid lines) for water service, gas service, hot and cold water lines, and for drainage, waste, and vent. Gas service below grade can be PE, PB, or PVC. (Courtesy of Plastic Pipe Institute.)

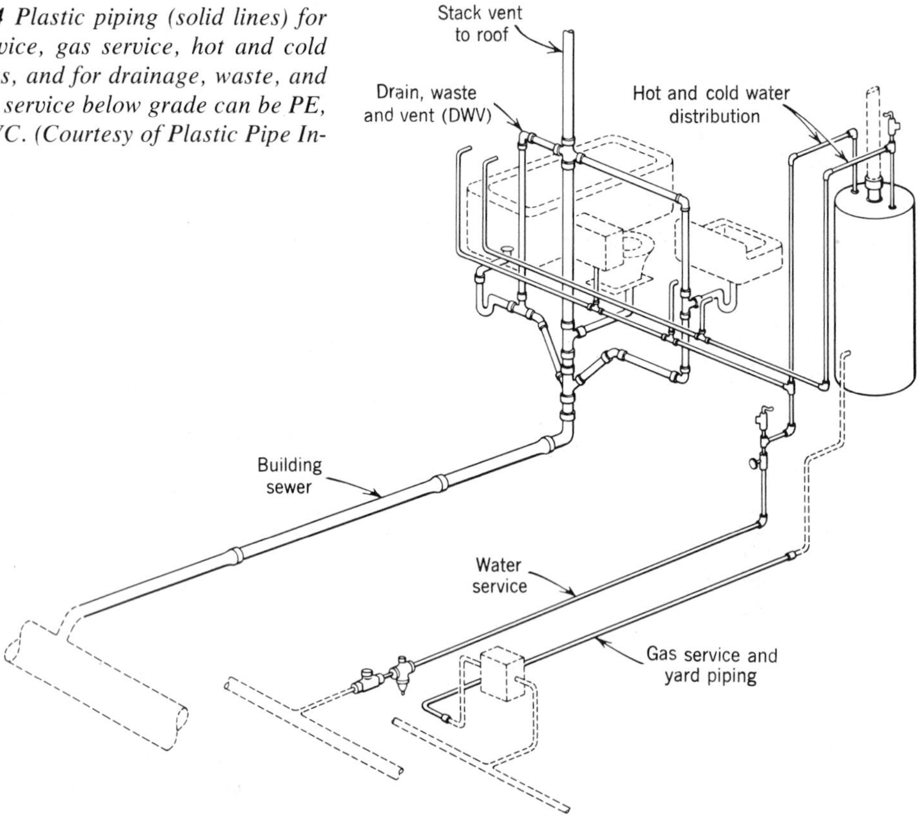

Fig. 10.35 Plastics lend themselves to preassembly of sections of DWV piping. Materials used for DWV are polyvinyl chloride (PVC), either tan or white, and acrylonitrile–butadiene–styrene (ABS), black. The architect or engineer should check structure that may be cut to accommodate piping. Notice in this picture that the notched joists are deeper and more closely spaced than other floor beams. (Courtesy of Plastic Pipe Institute. Photo by Richards Studio.)

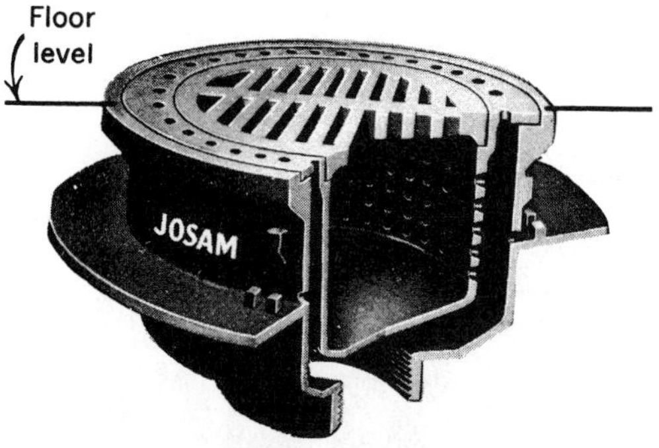

Fig. 10.36 *Floor drain. (Courtesy of Josam Manufacturing Company.)*

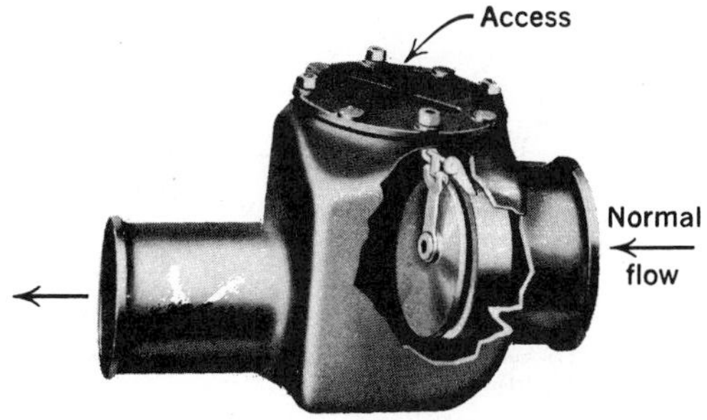

partition. Fixture branches project from the surface of the flag. There is considerable advantage in "back-to-back" planning of baths and kitchens; this allows the piping assembly to pick up the drainage of fixtures on both sides of it. When all the fixtures are on nearly the same level, as in Fig. 10.40, it is unnecessary to have a separate vent stack standing beside the soil stack, as is often the case in multistory construction. In this illustration, and generally in one-story construction, the upper part of the soil stack forms a vent called a *stack vent,* to which the branch vents connect. A separate major vertical vent would be called a *vent stack.*

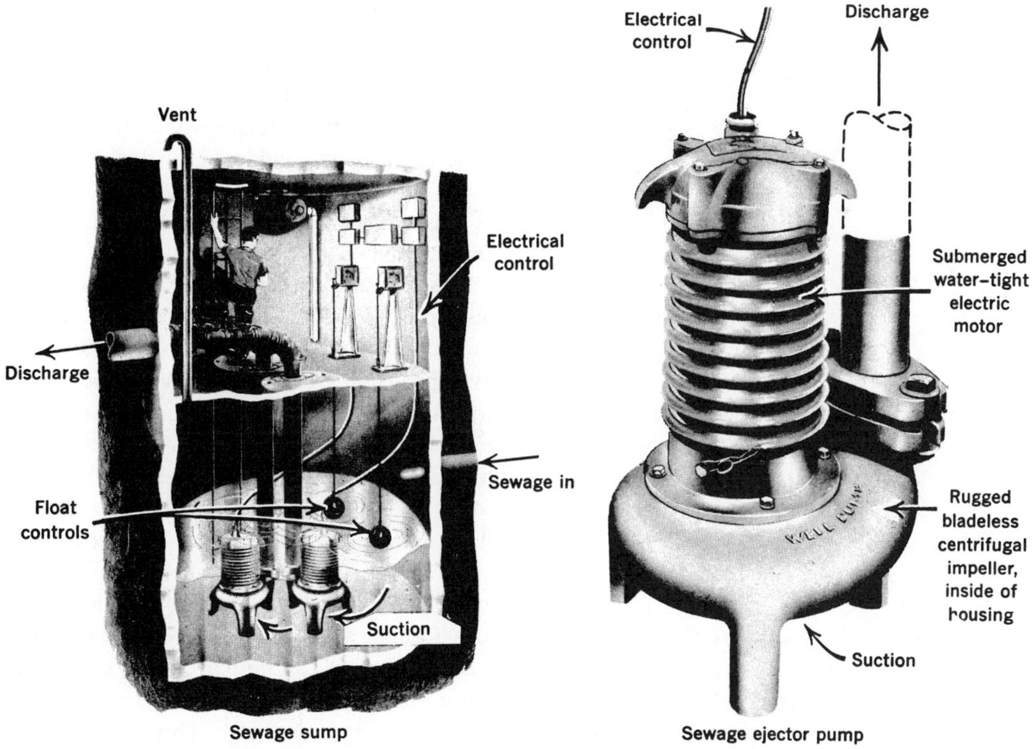

Fig. 10.38 *Sump and ejector: a submersible-type centrifugal pump for raising sewage to a higher level. This principle, shown here for an outdoor subgrade sewer installation, may be used in basement applications within buildings. Venting must be carried to roof. (Courtesy of Weil Pump Company.)*

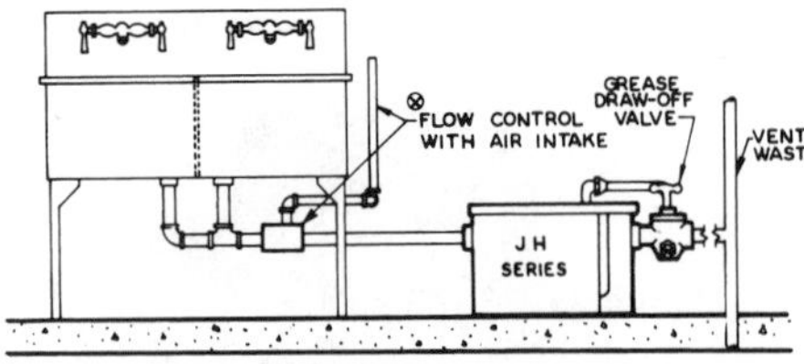

JH Series Semi-Automatic Interceptor installed on floor, servicing a double compartment sink.

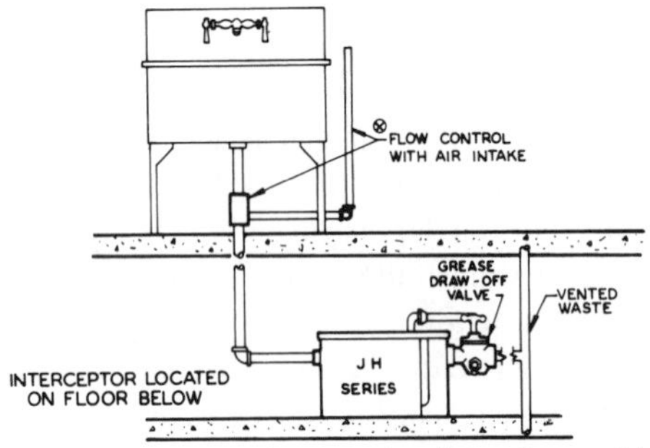

JH Series Semi-Automatic Interceptor installed on floor, servicing a sink on floor above.

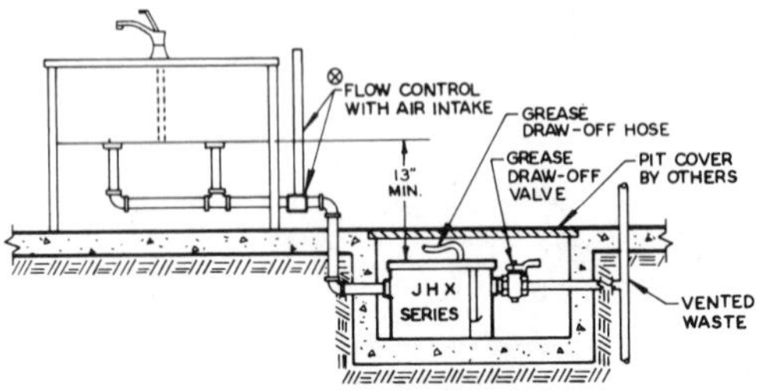

JHX Series Semi-Automatic Interceptor installed recessed below floor, in pit, servicing a double compartment sink.

(a)

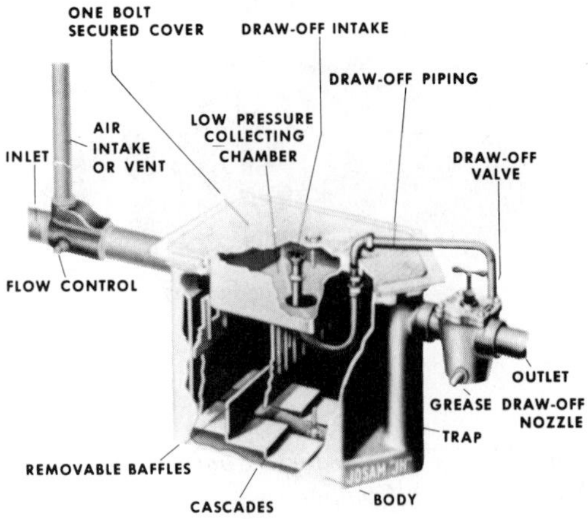

(b)

Fig. 10.39 (a) *One type of institutional interceptor, a grease trap. Choice of three locations—adjacent to sink, on floor below, or in a pit.* (b) *Cutaway view with identification of component parts. (Courtesy of Josam Manufacturing Company.)*

Multistory construction, especially in office buildings, is often designed to be flexible and free of random partitions that would interfere in the periodic replanning of interior spaces and the relocating of dividing partitions. The use of "cores" has offered a solution to this problem. Risers of the various systems are grouped in planes that coincide with permanent partitions of block masonry. This island of fixed construction is often placed at the central section of the building, freeing the surrounding areas for access to daylight. A hole in the floor for each pipe is often chosen in preference to a slot or shaft. This method interferes less with the floor construction (Fig. 10.41).

Offices often need a single lavatory or a complete toilet room for executives at locations away from the central core of the building. "Wet" columns with a full complement of plumbing pipes make this possible. If the pipes are to accompany a column in a steel building, structural coordination must be sought early in the planning if the pipes are to clear the structural framing of the floor (Fig. 10.42).

In some installations, the branch soil and waste piping perforates a floor and crosses below the slab to join the stack. Tubing has been developed, however, that sits above the structural slab, obviating the need for hung ceilings below (see Fig. 10.43). A lightweight concrete fill is cast to cover the tubing, raising the floor by 5 or 6 in. (127 or 152 mm). This can create a raised floor in the toilet room—with attendant access problems—so the higher floor level is usually carried throughout the floor of the entire story, forming a convenient space into which the electrical conduit can be placed at a time later than would have been required if it were to be placed in the structural slab. This affords some freedom in construction, because conduits in the structural slab conflict with reinforcement, and, furthermore, they must be planned and placed earlier.

Fig. 10.40 Drainage vent pipe in a frame residence. This installation in copper "drainage waste and vent tubing" (DWV) serves two bathrooms at the upper level behind the 6-in. stud partition and a kitchen sink and laundry tray at the lower level, which are on this side of the partition. In the bathrooms the roughing serves (from left to right) a lavatory, water closet, and bathtub and a lavatory, shower, and water closet. Bathtub and shower traps can usually be accommodated within the joist depth. The bend below the water closets, however, often leads to a horizontal branch exposed or furred-in below the joists. Some codes permit this branch from a water closet to be up to 6 to 10 ft long before it joins a vent. (Courtesy of Copper Development Association.)

10.5 Design of Residential Waste Piping

The sanitary drainage requirements for a residence (Fig. 10.44), provide the basis for an example.

EXAMPLE 10.3. Design, lay out, and size the piping for the sanitary drainage system for the house shown in Fig. 10.44.

SOLUTION. The first step is to identify the locations where hot and cold water is needed at fixtures and where soil or waste drains must be provided. Figure 10.45*a* illustrates how this is done. A plan layout for the drains in both levels follows (see Fig. 10.45*b*).

The task of fitting the two plumbing "distribution trees" (supply and waste) into available horizontal and vertical chases can be difficult. This is especially true of the waste system, which has larger pipes, because it is not under pressure, and must carry solids as well as liquid. To compound the problem, because of gravity flow requirements these larger pipes must slope continuously downward, from fixtures to the sewer.

Next comes the "plumbing section" (Fig. 10.46). The local administrative authority usu-

Fig. 10.41 Risers in a fireproof, multistory building. Pipes, tubes, conduits, and ducts virtually enclose toilet rooms and utility spaces. Ventilation ducts and a master 5-in. copper hot-water riser are just left of center. Soil and vent stacks with hot and cold water supplies, all of copper, are to the right of this group. At the left in lighter tone is the galvanized steel feeder conduit and distribution circuit conduits, of the same metal, for a local electrical control panel box. Note that some pipes and tubes are supported at this floor by bolted clamps. After testing and before pipes are enclosed, covering will be completed. (Courtesy of Copper Development Association.)

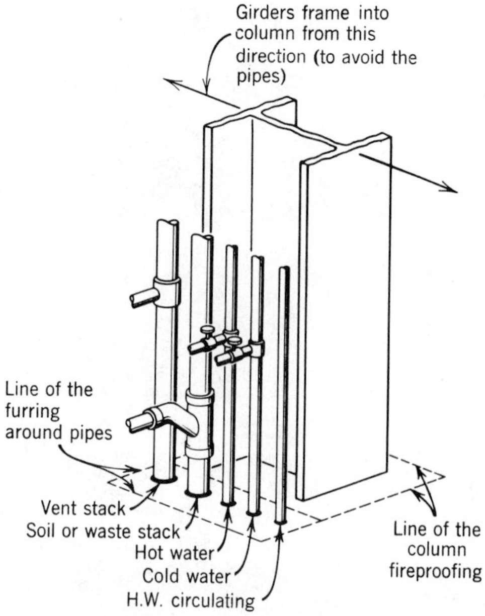

***Fig. 10.42** Piping at a "wet" column. In large office buildings, there are usually several of these, remote from the core and out in the general office area.*

ally requires this to be submitted for approval. Sizes of all piping are determined from Tables 10.8 through 10.12. Drainage fixture units (d.f.u.) for the system that we are designing are summarized in Table 10.13 from data given in Table 10.9. Runouts from *individual* fixtures should be the same size as the fixture trap. Branch runouts from *groups* of fixtures are sized by the drainage fixture units of the group. Although Table 10.11 permits a 3-in. branch for not more than two water closets, it is common practice to use a 4-in. runout for *every* water closet and not less than a 4-in. branch for every group of water closets. Since Table 10.11 allows 160 fixture units for a 4-in. branch, this size is acceptable for any branch to which a water closet is connected and, as can be seen, for any *stack*. The house drain cannot be less than 4 in., and because (at a ¼-in. fall per ft) a 4-in. house drain (Table 10.10) will carry 216 d.f.u., that size is chosen and is more than adequate for our 28 d.f.u. system (see Table 10.13). A water closet must have a 2-in. vent. Table 10.12 shows that a 2-in. vent, for vent-lengths not exceeding 150 ft, will serve 20 fixture units. This size is acceptable for branch vents and stack vents. For individual fixture vents, the vent size is usually the same as the size of the fixture runout. Under most codes, vertical vents that penetrate the roof must increase to a 4-in. size to prevent blocking by icing in freezing weather.

In residential applications and in other relatively small buildings certain fairly standard minimum sizes, such as a 4-in. soil stack and 2-in. vent, are usually adequate. Tables 10.8 and 10.12 are especially useful in the design of drainage systems for larger buildings.

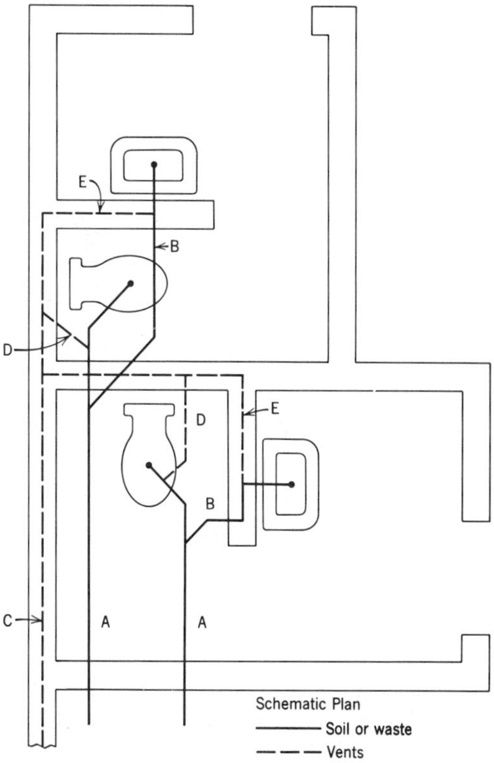

Fig. 10.43 *Horizontal waste piping* above *the structural slab, in an example of plumbing roughing for two lavatory rooms in a concrete office building. A lavatory and water closet in each room are served by soil and waste branches below and vent branches above. Hot and cold water tubing with air chambers can be seen. Although the extensions of the water tubing above the two flushometer connections appear to connect into the horizontal vent branches, they actually do not; they are capped and merely touch the bottoms of the vent branches. Note that soil branches are above the structural slab. A fill of 5 or 6 in. will be necessary to cover the tubing. All vertical tubing will be within the masonry block used to enclose the cubicles. (Courtesy of Copper Development Association.)*

Drainage	*Supply*
A Branch soil	F Hot
B Lavatory waste	G Cold
C Branch vent	H Flushometer supply (1¼-m.)
D WC vent	I Capped air chamber
E Lavatory vent	J Capped air chamber (flushometer)

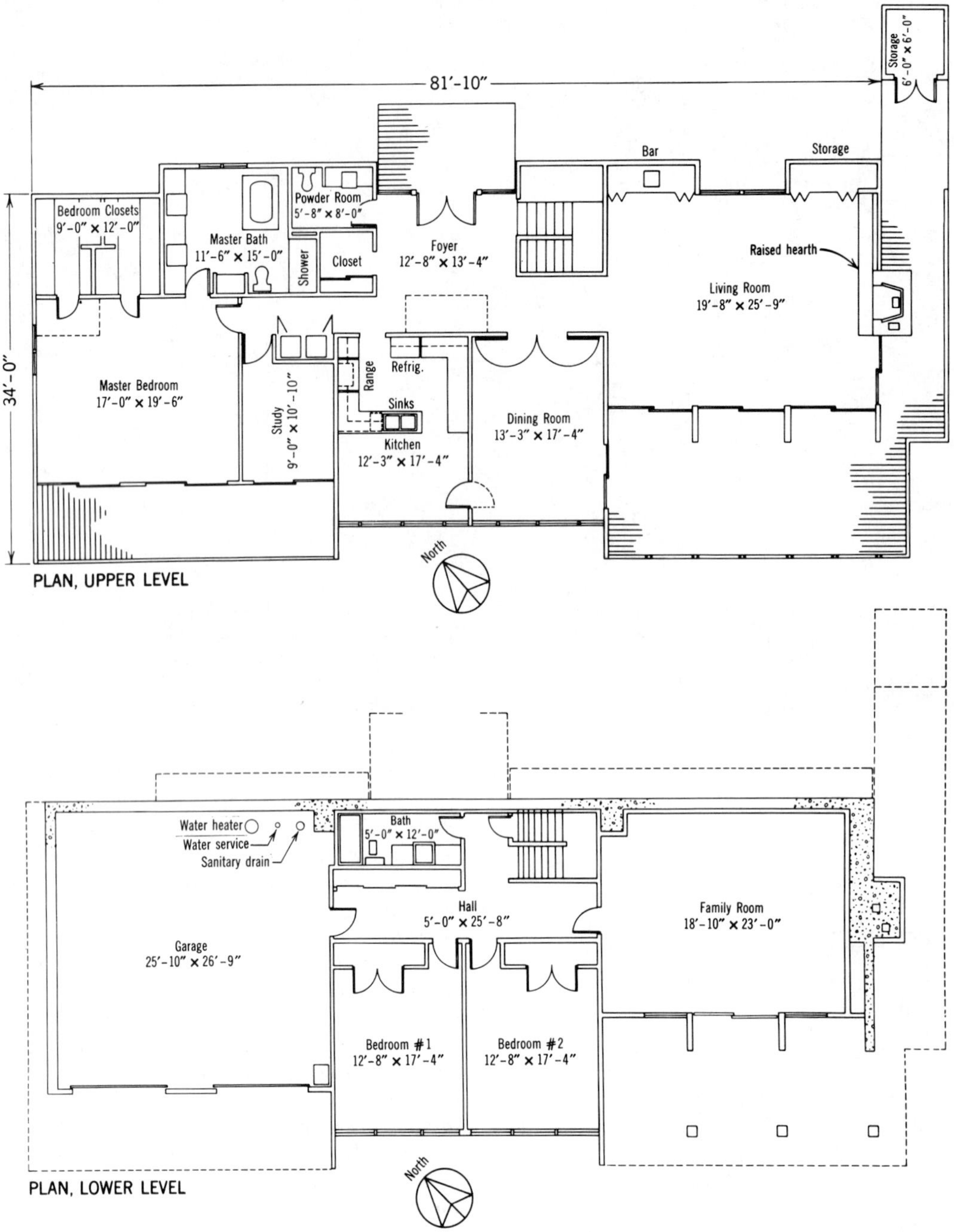

Fig. 10.44 *(Example 10.3) House on Long Island, Budd Mogensen, Architect and Planner. Floor plans to be used in the solution of the example.*

10.6 Design of Larger-Building Waste Piping

EXAMPLE 10.4. Select sizes for drainage and vent piping for the plumbing in an office building for which the fixtures are shown in Fig. 10.47.

SOLUTION. Individual fixture branches shall not be less than the size indicated in Table 10.8 for the minimum size of trap for each fixture. Selected fixture units from Table 10.9 are applied to each section of the piping and totaled for

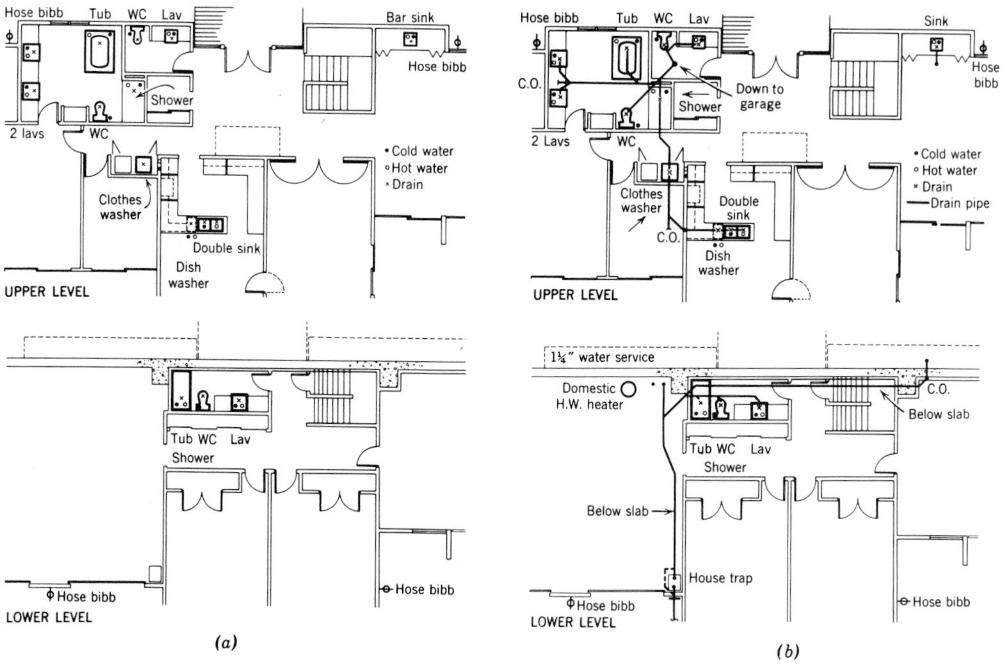

***Fig. 10.45** (Example 10.3)* (a) *Plumbing requirements, water supply, and partial sanitary drainage.* (b) *Sanitary drainage plan.*

each branch and stack, and for the building drain and the building sewer. An example of a fixture-unit summary and sample sizes of individual branches that connect into a typical branch of the men's toilet group on any floor are shown in the following table.

Features	Units per Fixture	Total Fixture Units	Diameter, Fixture Branch (in.)	(mm)
1 service sink	3	3	3	76
3 lavatories	1	3	1½	38
3 urinals, washout	4	12	2	51
3 water closets, valve operated	6	18	4	102
Total fixture units, men's toilet branch		36		

Table 10.11 indicates that a 3-in. horizontal fixture branch is inadequate for the above group because it will handle only 20 fixture units and not more than two water closets. Therefore, a 4-in. pipe is selected. Its capacity of 160 fixture units will be more than enough for the 36 needed here. The same table shows that the soil stack can be 4 in. in diameter (it is run thus for its entire height). Its capacity of 90 fixture units per story is sufficient for the 64 that connect in at each T-Y connection.

According to Table 10.10, the building drain and the building sewer at their pitch of ¼ in. in 1 ft should be 5 in. in diameter. Their capacity for 480 fixture units exceeds 350½ placed upon them. The vent stack at a 70 ft length and 338 fixture units would be 2½ in., but 3 in. is a better choice. This is increased to 4 in. as it passes through the roof.

Although opinion may vary about the relative merits of continuous or circuit venting, either system, properly designed, will effectively prevent the siphoning of traps or relieve air pressures that could cause foul gases to bubble through the traps into the occupied space. Another system, especially suitable for high-rise buildings, eliminates the vent stack completely with equal effectiveness.

(a) The Sovent System. This essentially ventless system changes the nature of the effluent (discharge of wastes and sewage from the

TABLE 10.8 **Size of Nonintegral Traps for Different-Type Plumbing Fixtures**

Plumbing Fixture	*Trap Size in Inches*	*Trap Size (mm)*
Bathtub (with or without overhead shower)	$1\frac{1}{2}$	38
Bidet	$1\frac{1}{4}$	32
Combination sink and wash (laundry) tray	$1\frac{1}{2}$	38
Combination sink and wash (laundry) tray with food waste grinder unit	$1\frac{1}{2}$[a]	38[a]
Combination kitchen sink, domestic, dishwasher, and food waste grinder	2	51
Dental unit or cuspidor	$1\frac{1}{4}$	32
Dental lavatory	$1\frac{1}{4}$	32
Drinking fountain	$1\frac{1}{4}$	32
Dishwasher, commercial	2	51
Dishwasher, domestic (nonintegral trap)	$1\frac{1}{2}$	38
Floor drain	2	51
Food waste grinder—commercial use	2	51
Food waste grinder—domestic use	$1\frac{1}{2}$	38
Kitchen sink, domestic, with food waste grinder unit	$1\frac{1}{2}$	38
Kitchen sink, domestic	$1\frac{1}{2}$	38
Kitchen sink, domestic, with dishwasher	$1\frac{1}{2}$	38
Lavatory, common	$1\frac{1}{4}$	32
Lavatory (barber shop, beauty parlor or surgeon's)	$1\frac{1}{2}$	38
Lavatory, multiple type (wash fountain or wash sink)	$1\frac{1}{2}$	38
Laundry tray (1 or 2 compartments)	$1\frac{1}{2}$	38
Shower stall or drain	2	51
Sink (surgeon's)	$1\frac{1}{2}$	38
Sink (flushing rim type, flush valve supplied)	3	76
Sink (service type with floor outlet trap standard)	3	76
Sink (service trap with P trap)	2	51
Sink, commercial (pot, scullery, or similar type)	2	51
Sink, commercial (with food grinder unit)	2	51

Source: National Standard Plumbing Code. (*Metric conversions by author.*)

[a]Separate traps required for wash tray for sink compartment with food waste grinder unit.

fixtures) instead of coping with the pressures and suctions that normal effluent would cause (see Figs. 10.48 and 10.49).

The "plunger" effect of a descending "slug" of water/waste within pipes was described in Section 10.3. If the effectiveness of the "plunger" can be reduced, the negative and positive pressures created by it will be also reduced. If their values can be brought down below the holding power of the several inches of water in the trap, no vents will be necessary. In the single-stack *Sovent* system illustrated in Fig. 10.49, this is done by dealing with the normal liquid effluent at each floor. Aeration there produces a *foam* that lacks the stack-filling tendency of the liquid effluent. Thus, through the creation of a *soft* plunger, pressure variations in the single stack are minimized.

Tests have shown that the positive and negative pressures produced by *normal* liquid effluent during its descent and relieved by the vent piping are often about 5 to 12 in. (127 to 305 mm) water gauge. Obviously, if the vents were not provided, the 2 to 4 in. (51 to 102 mm) of water seal in the traps would be vulnerable to penetration by gases from pipes under positive pressure or siphonage of water seals into pipes that may be under negative pressure.

TABLE 10.9 **Drainage Fixture Unit Values for Various Plumbing Fixtures**

Type of Fixture or Group of Fixtures	*Drainage Fixture Unit Value (d.f.u.)*
Automatic clothes washer (2-in. (51-mm) standpipe)	3
Bathroom group consisting of a water closet, lavatory, and bathtub or shower stall:	
Flushometer valve closet	8
Tank type closet	6
Bathtub[a] (with or without overhead shower)	2
Bidet	1
Clinic Sink	6
Combination sink-and-tray with food waste grinder	4
Combination sink-and-tray with one 1½-in. (38-mm) trap	2
Combination sink-and-tray with separate 1½-in. (38-mm) traps	3
Dental unit or cuspidor	1
Dental lavatory	1
Drinking fountain	½
Dishwasher, domestic	2
Floor drains with 2-in. (51-mm) waste	3
Kitchen sink, domestic, with one 1½-in. (38-mm) trap	2
Kitchen sink, domestic, with food waste grinder	2
Kitchen sink, domestic, with food waste grinder and dishwasher 2-in. (51-mm) trap	3
Kitchen sink, domestic, with dishwasher 1½-in. (38-mm) trap	3
Lavatory with 1¼-in. (32 mm) waste	1
Laundry tray (1 or 2 compartments)	2
Shower stall, domestic	2
Showers (group) per head	2
Sinks:	
Surgeon's	3
Flushing rim (with valve)	6
Service (trap standard)	3
Service (P trap)	2
Pot, scullery, etc.	4
Urinal, pedestal, syphon jet blowout	6
Urinal, wall lip	4
Urinal, stall, washout	4
Urinal trough (each 6-ft (1.8-m) section)	2
Wash sink (circular or multiple) each set of faucets	2
Water closet, tank-operated	4
Water closet, valve-operated	6
Fixtures not listed above:	
Trap size 1¼ in. (32 mm) or less	1
Trap size 1½ in. (38 mm)	2
Trap size 2 in. (51 mm)	3
Trap size 2½ in. (64 mm)	4
Trap size 3 in. (76 mm)	5
Trap size 4 in. (102 mm)	6

Source: National Standard Plumbing Code.

[a]A shower head over a bathtub does not increase the fixture unit value.

TABLE 10.10 **Building Drains and Sewers**[a]

Diameter of Pipe (in.)	(mm)	Maximum Number of Fixture Units That May Be Connected to Any Portion of the Building Drain or the Building Sewer Including Drain Branches — Fall per Foot (mm per m): 1/16 in. (5.2)	1/8 in. (10.4)	1/4 in. (20.9)	1/2 in. (41.7)
2	51			21	26
2½	64			24	31
3	76		36[b]	42[b]	50[b]
4	102		180	216	250
5	127		390	480	575
6	152		700	840	1,000
8	203	1,400	1,600	1,920	2,300
10	254	2,500	2,900	3,500	4,200
12	305	2,900	4,600	5,600	6,700
15	381	7,000	8,300	10,000	12,000

Source: National Standard Plumbing Code. (*Metric conversions by author.*)

[a]On-site sewers that serve more than one building may be sized according to the current standards and specifications of the Administrative Authority for public sewers.

[b]Not more than two water closets or two bathroom groups.

TABLE 10.11 **Horizontal Fixture Branches and Stacks**

Diameter of Pipe (in.)	(mm)	Maximum Number of Fixture Units that May Be Connected to: Any Horizontal Fixture Branch[a]	Stack Sizing for 3 Stories in Height or 3 Intervals	Stack Sizing for More than 3 Stories in Height: Total for Stack	Total at 1 Story or 1 Branch Interval
1½	38	3	4	8	2
2	51	6	10	24	6
2½	64	12	20	42	9
3	76	20[b]	48[b]	72[b]	20[b]
4	102	160	240	500	90
5	127	360	540	1,100	200
6	152	620	960	1,900	350
8	203	1,400	2,200	3,600	600
10	254	2,500	3,800	5,600	1,000
12	305	3,900	6,000	8,400	1,500
15	381	7,000			

NOTE: Stacks shall be sized according to the total accumulated connected load at each story or branch interval and may be reduced in size as this load decreases to a minimum diameter of ½ of the largest size required.

Source: National Standard Plumbing Code. (*Metric conversions by author.*)

[a]Does not include branches of the building drain.

[b]Not more than two water closets or bathroom groups within each branch interval or more than six water closets or bathroom groups on the stack.

TABLE 10.12 **Size and Length of Vents**
Part A, ***Conventional Units***

Size of Soil or Waste Stack	*Fixture Units Connected*	*Diameter of Vent Required (in.)* 1¼	1½	2	2½	3	4	5	6	8
		Maximum Length of Vent (ft)								
Inches										
1½	8	50	150							
1½	10	30	100							
2	12	30	75	200						
2	20	26	50	150						
2½	42		30	100	300					
3	10		30	100	200	600				
3	30			60	100	500				
3	60			50	80	400				
4	100			35	100	260	1000			
4	200			30	90	250	900			
4	500			20	70	180	700			
5	200				35	80	350	1000		
5	500				30	70	300	900		
5	1100				20	50	200	700		
6	350				25	50	200	400	1300	
6	620				15	30	125	300	1100	
6	960					24	100	250	1000	
6	1900					20	70	200	700	
8	600						50	150	500	1300
8	1400						40	100	400	1200
8	2200						30	80	350	1100
8	3600						25	60	250	800
10	1000							75	125	1000
10	2500							50	100	500
10	3800							30	80	350
10	5600							25	60	250

Source: National Standard Plumbing Code.

Figures 10.48, 10.49, and 10.50 illustrate the components and the action of the Sovent system. Effluent, already aerated and descending from upper stories, is diverted in the stack at each lower story. The aerator fitting there affords a passage for this diverted flow and also an air space into which the effluent from the local branch soil or waste can drop. Here it spatters, mixing with the air to form a rarified mixture of air and liquid. Tests show that this mixture does not produce pressures, positive or negative, of more than 1 in. (25 mm) water gauge. Thus a trap-seal of 2 in. (51 mm) or more is safe against siphonage or penetration.

At the foot of the single stack the aerated effluent is compacted—a process aided by a baffle in the path of the flow in the deaerator fitting (see Fig. 10.50). If not relieved, air piling up at this point could cause pressures in the stack at the first floor. An air-discharge pipe provides this relief of air from the deaerator fitting to the upper part of the building drain, above the liquid flow.

The Sovent system was invented by Fritz Sommer of Switzerland, who tested it in a 10-story drainage test tower. Since its introduction in 1962, it has been installed and used in hundreds of buildings in Europe and Africa. Canada used the Sovent method in the Habitat apartments at the 1967 Montreal expo. Sovent was

TABLE 10.12 ***Continued***
Part B. ***SI Units***

Size of Soil or Waste Stack (mm)	*Fixture Units Connected*	*Maximum Length of Vent (m) for Diameter of Vent Required (mm)*								
		32	*38*	*51*	*64*	*76*	*102*	*127*	*152*	*203*
38	8	15	46							
	10	9	31							
51	12	9	23	61						
	20	8	15	46						
64	42		9	31	92					
76	10		9	31	61	183				
	30			18	31	153				
	60			15	24	122				
102	100			11	31	79	305			
	200			9	27	76	275			
	500			6	21	55	214			
127	200				11	24	107	305		
	500				9	21	92	275		
	1100				6	15	61	214		
152	350				8	15	61	122	396	
	620				5	9	38	92	336	
	960					7	31	76	305	
	1900					6	21	61	214	
203	600						15	46	153	396
	1400						12	31	122	366
	2200						9	24	107	336
	3600						8	18	76	244
254	1000							23	38	305
	2500							15	31	153
	3800							9	24	107
	5600							8	18	76

Metric conversions by author, based on National Standard Plumbing Code.

TABLE 10.13 **Drainage Fixture Units (Example 10.3)**

	Units
Bar sink	2
Kitchen sink and dishwasher	3
Lavatory	1
Water closet	4
Clothes washer	3
Master bath, lavatory, water closet, tub	6
Extra lavatory	1
Shower	2
Lower bath, lavatory, water closet, tub and shower	6
	28

Values are from Table 10.9
Hose bibb drainage to ground
Roof drainage to dry wells

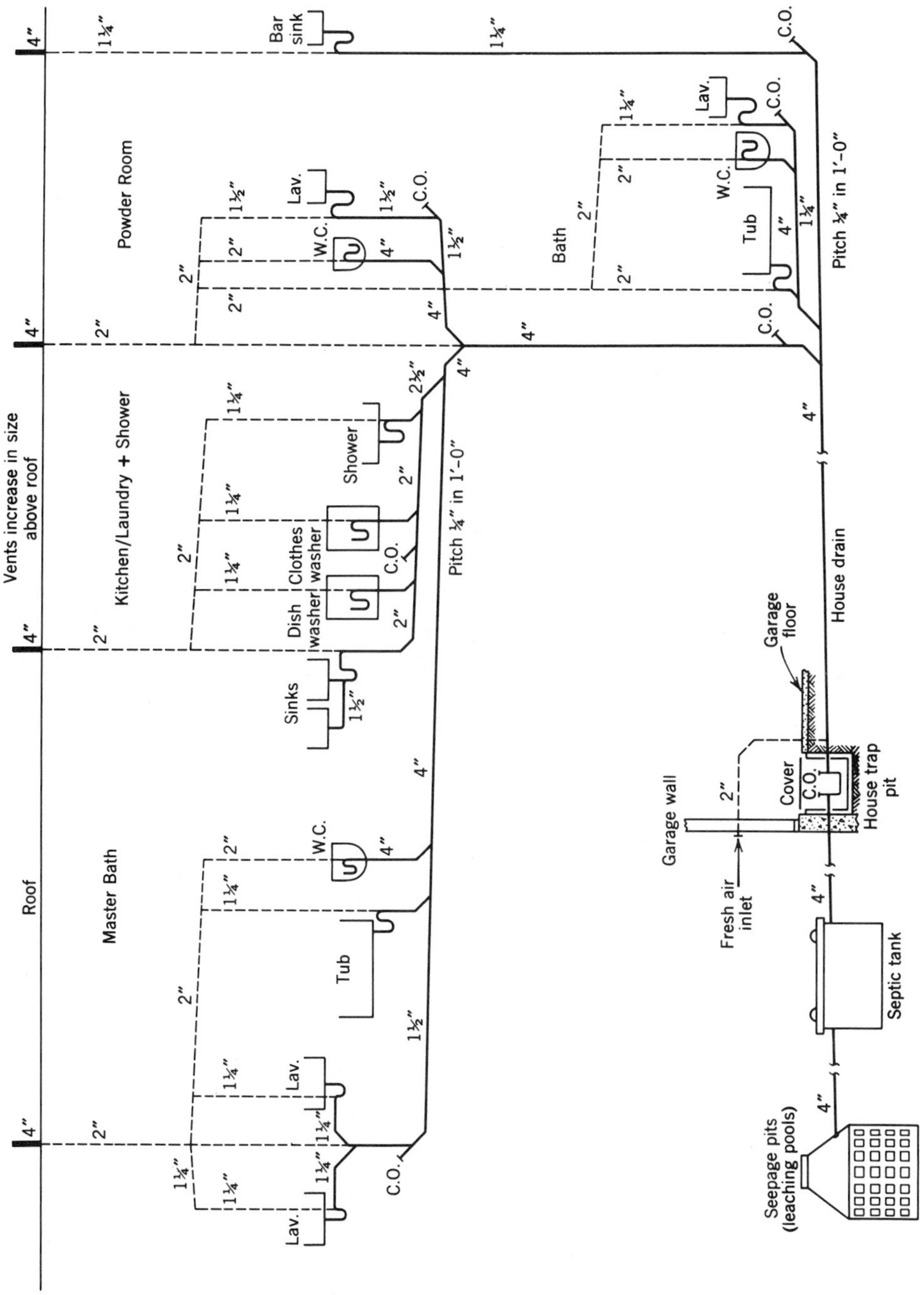

***Fig. 10.46** (Example 10.3) Plumbing section. When every fixture is vented individually, as in this example, the method is known as* continuous venting. *In larger systems, batteries of fixtures may be vented by a loop or circuit vent. This reduces the piping of the vent system (see Fig. 10.47).*

WATER AND WASTE

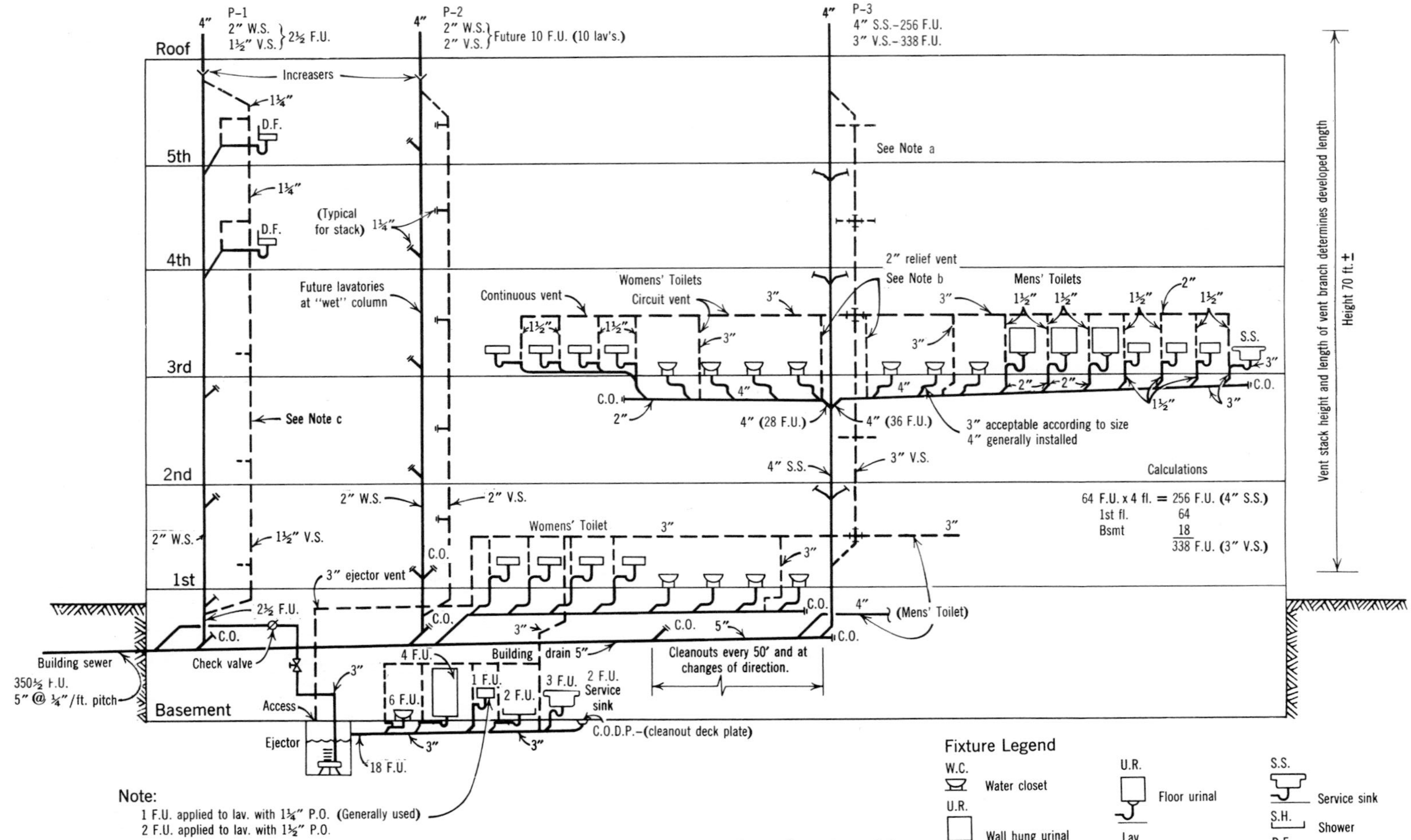

Fig. 10.47 *(Example 10.4) Plumbing section of an office building in general conformity with National Standard Plumbing Code. Circuit vents serve branch soil lines. The house trap and fresh-air inlet are omitted from building drain. Some codes require continuous venting (see Fig. 10.46). Note that* (a) *Relief vent not required on top floor.* (b) *Men's and women's toilets on 3rd floor are typical and would be repeated on 1st, 2nd, 4th, and 5th.* (c) *Drinking fountains on 5th and 4th floors would be repeated on 1st, 2nd, and 3rd.*

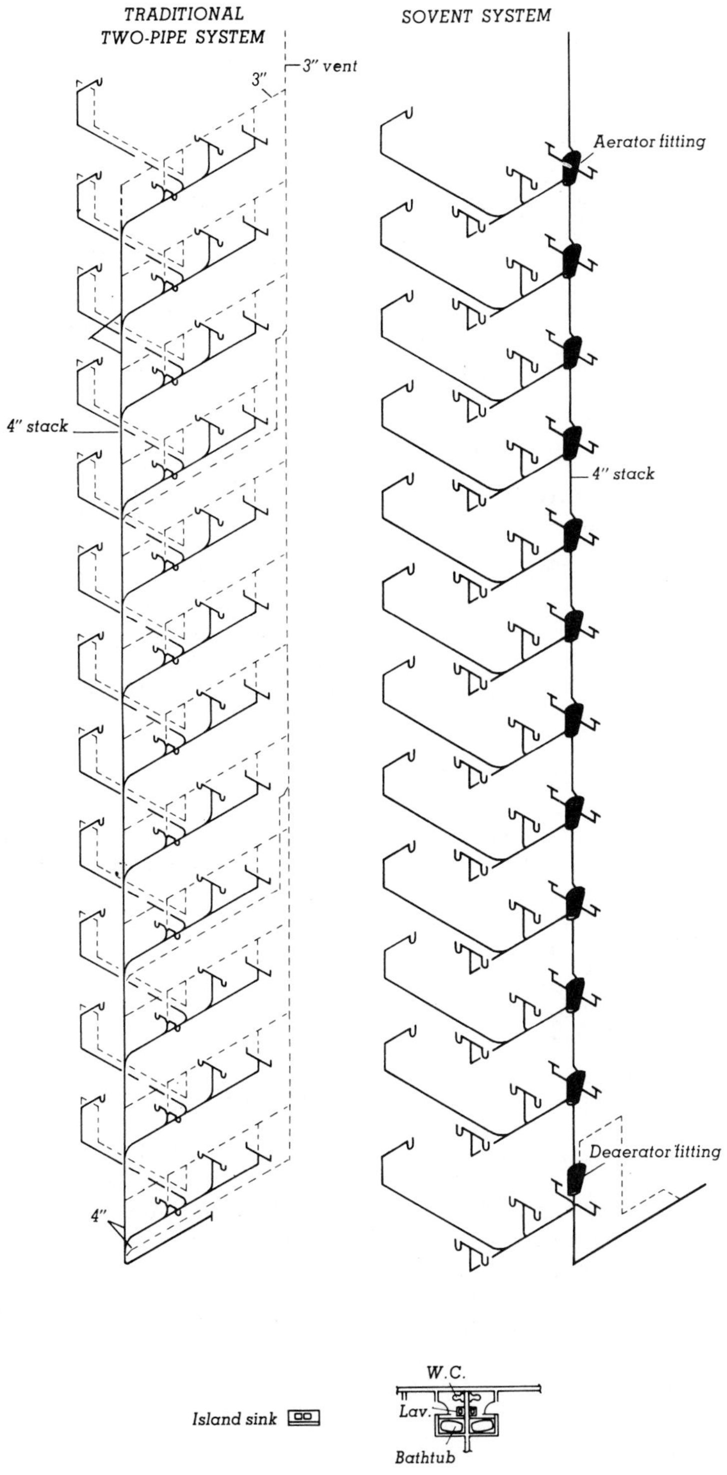

Fig. 10.48 *Cost-saving potential can be seen by the choice of the Sovent system over a two-pipe system for this 12-story stack serving an apartment grouping. (Courtesy of the Copper Development Association.)*

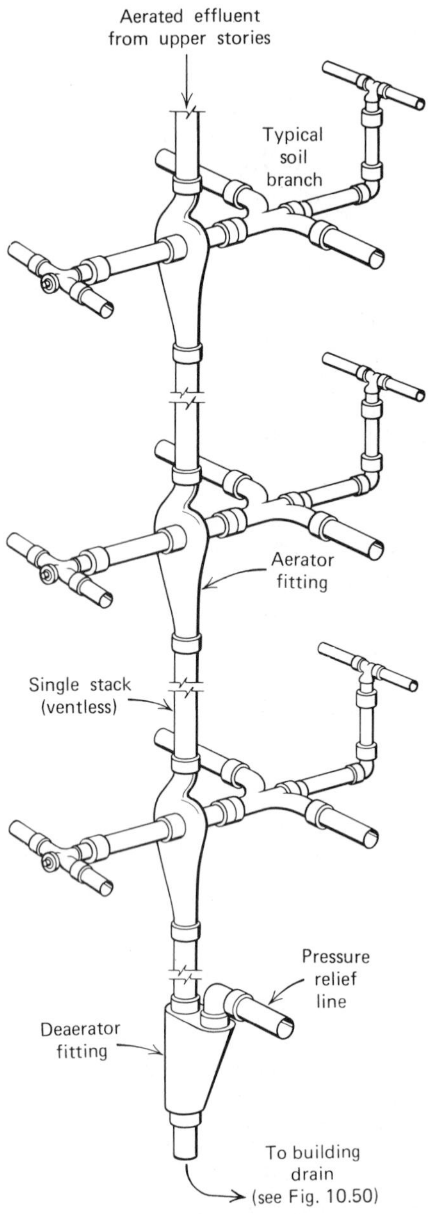

Fig. 10.49 The Sovent drainage stack consists of aerator fittings that join the horizontal branches to the stack at each floor level, and a deaerator fitting at the bottom of the stack. The stack is open to the atmosphere above the roof at the top. (Courtesy of the Copper Development Association.)

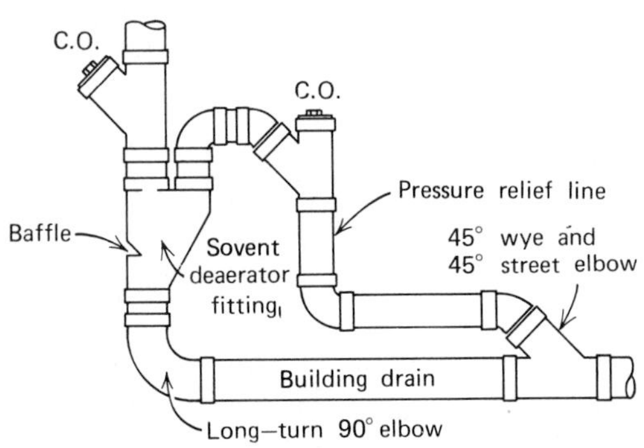

Fig. 10.50 The deaerator consists of an air separation chamber with an internal nosepiece, a stack inlet, a pressure relief outlet at the top, and a stack outlet at the bottom. The deaerator fitting at the bottom of the stack functions in combination with the aerator fittings above to make the single stack self-venting. The deaerator is designed to overcome the tendency for the falling waste to build up excessive back pressure at the bottom of the stack when the flow decelerates at the bend into the horizontal drain. (Courtesy of the Copper Development Association.)

first granted U.S. code acceptance in 1968 in Richmond, California. Following this success, its code acceptance grew rapidly during the early 1970s. Today, Sovent is accepted by some major model codes used nationwide. However, not all jurisdictions allow Sovent installations.

REFERENCES

Converse, J., et al. (1978). *Design and Construction Manual for Mounds . . .* , Small Scale Waste Management Project, University of Wisconsin Extension, Madison, Wis.

Milne, M. (1976). *Residential Water Conservation,* U.S. Office of Water Research and Technology, Department of Commerce, NTIS.

Stoner, C., ed. (1977). *Goodbye to the Flush Toilet,* Rodale Press, Emmaus, Pa.

11
BATHROOM DESIGN

More than any other familiar space, the bathroom reflects the interrelation of those factors—comfort, heating, cooling, ventilation, moisture control, water supply, and waste—that have provided the basis of the first 10 chapters. It also furnishes an excellent model through which to illustrate lighting considerations, sound isolation, and the special needs of the handicapped.

One of the more challenging aspects of bathroom design is the frequent conflict between physiological design criteria and such psychological influences as cultural attitudes about bathroom activities. The physiological criteria change as slowly as evolution. The cultural attitudes can change very rapidly, even within a generation. The bathroom supports two primary human activities: cleansing and elimination. Today's common attitudes about these closely linked activities are that one is "clean," the other "dirty." One might be discussed (though rarely) among friends; the other is simply not fit for conversation. This conflict in attitudes influences the design of our plumbing fixtures. The toilet is the fixture used for elimination; it is the logical place to provide for the cleansing of the perineal zone that should immediately follow elimination. (This is especially true of public toilets, where stalls close off toilets from lavatories.) Yet very few toilets presently incorporate the cleansing feature: the mixture of cleansing and elimination in the same fixture too often seems abhorrent.

Another reason to single out the bathroom for special attention is its role in the conservation of energy and water. Chapters 8 and 10 established the toilet as the key to significant water conservation in residences. Bathing—in particular, showering—represents another important target, one in which savings in water also mean savings in the energy used to heat water.

In discussing these topics, we will look first at the influences of physiology, psychology, and conservation on typical plumbing fixtures. Then we will consider how to put them together in various combinations, in spaces that are comfortable, pleasantly lighted, private, and fully accessible. Lastly, we will address some aspects of construction—the roughing in and maintenance of the fixtures.

11.1 Physiology, Psychology, and Fixtures

The brief summary presented in this section is based largely on a particularly revealing (and entertaining) study of the struggle between physiological and cultural criteria in fixture and bathroom design—Alexander Kira's *The Bathroom: Criteria for Design,* an expanded edition of which was published in 1976 by Viking Press.

(a) Cleansing. A key issue here is the contrast between running water and a standing water body. *Lavatories* are used primarily for the cleansing of hands, face, and teeth—activities done quickly with running water that is wasted directly, rather than being collected. Most lavatories are designed as collection bowls—perhaps a reflection of the days when washbasins of standing water, drawn and heated elsewhere, were brought to the bathing place. Most lavatories also have fittings that project out over the sink. Although these fittings have slowly evolved into very sleekly designed objects, most of them still dump running water directly into the drain, are hard to use as drinking fountains, and can be wounding to those who try to wash hair in the lavatory. In considering the prevalence of running as opposed to standing water, the role that running water could play in keeping lavatory surfaces clean, and the need for a drinking fountain when teeth are brushed, Kira proposed a quite different lavatory design (Fig.

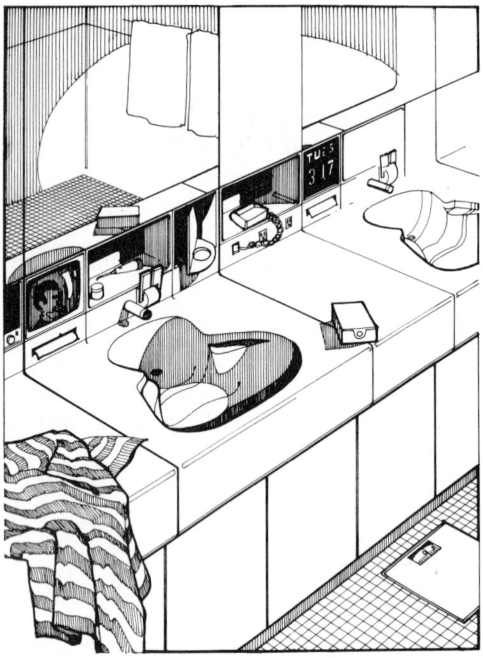

Fig. 11.1 Proposed lavatory that exploits running water. The water issues from a fountainlike stream, for ease in drinking and in hair and face washing. The stream strikes the lavatory bowl in such a way as to minimize splash outside the bowl, yet sets up a self-cleaning swirling action. A small repository at the back, over the drain, can serve for standing water when desired. (From The Bathroom: New and Expanded Edition, *copyright 1966, 1976 by Alexander Kira. By permission of Bantam Books, Inc. All rights reserved.)*

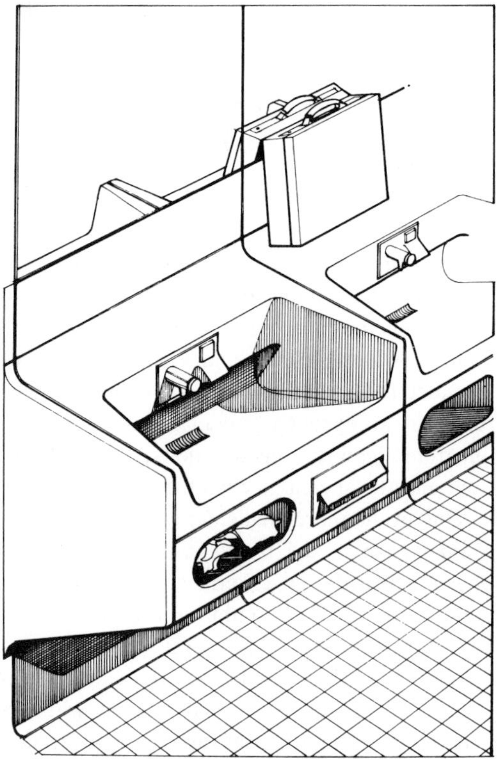

Fig. 11.2 Proposed public washroom lavatory in which hand washing is the main concern. A single source of tempered water is provided, again with an eye toward self-cleaning of the fixture. Facilities for soap and for hand drying are built directly into the lavatory. (From The Bathroom: New and Expanded Edition, *copyright © 1966, 1976 by Alexander Kira. By permission of Bantam Books, Inc. All rights reserved.)*

11.1). For public toilet rooms, used almost exclusively for hand washing, the lavatory can be kept very simple (Fig. 11.2).

Whole-body cleansing more clearly illustrates the running/standing water contrast. In cleansing, the ordinary sequence is *wet, soap, scrub, and rinse.* Standing water is good for wetting, and very good for soaping/scrubbing; running water is superior for rinsing. The psychological aspect enters here in common attitudes toward tubs and showers. Tubs are often seen as places to relax in, to spend more time in, perhaps to read in. Showers are viewed as quicker, "no-nonsense," stand-up places. Yet each could benefit from some features of the other.

Tubs should be designed so that the reclining body is supported at the back; this requires a contoured surface (Fig. 11.3) rather than the ordinary straight-line design. It also requires braces for one's feet, since the body will otherwise tend to float up and way from such a backrest. These tubs can be designed to accommodate persons of various leg lengths, as shown in Fig. 11.3*b*. There must also be a seat, to allow most of the body a chance to be out of the water and to facilitate safe entry/exit from the tub. Especially needed is a handheld shower for the final rinse; soapy standing water leaves a scummy film on both people and fixtures.

Showers may seem very "efficient," but

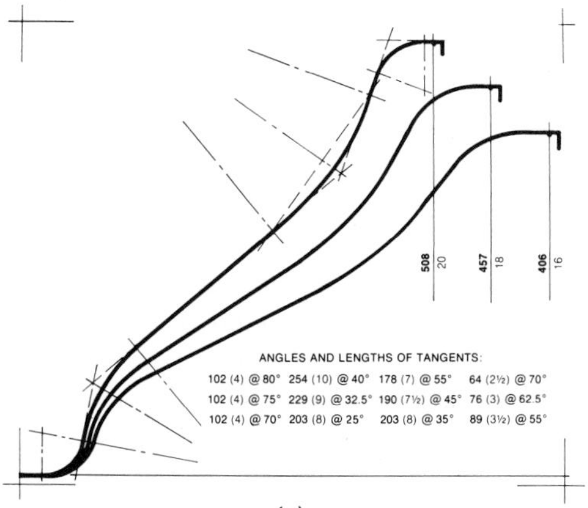

(a)

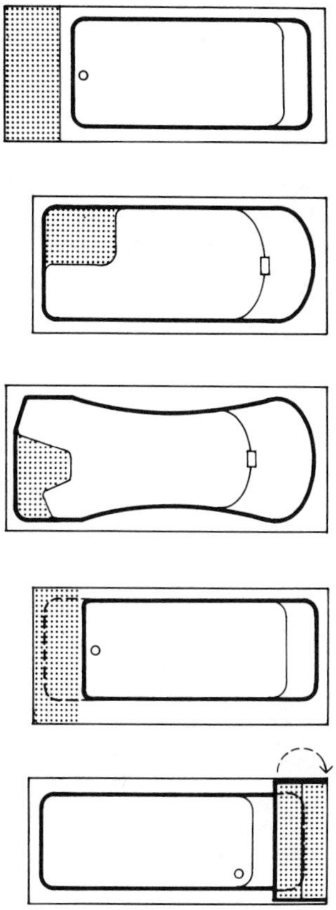

(b)

Fig. 11.3 *Some considerations for tub design, to facilitate relaxation during whole-body cleansing.* (a) *A contoured backrest allows comfortable reclining, with support to both the lower back and shoulder. Curve 2, with a median angle of 32.5°, promises comfort for most users.* (b) *Various approaches to tub plans, allowing long- and short-legged persons to brace their bodies against the contoured backrest. Raised seats are also provided, to facilitate entry/exit. (From* The Bathroom: New and Expanded Edition, *copyright © 1966, 1976 by Alexander Kira. By permission of Bantam Books, Inc. All rights reserved.)*

cleaning would be more thorough and safe if the bather could turn off the water and sit for at least part of the soap/scrub activity, especially for the lower legs and feet. Showers with integral seats (Fig. 11.4) are now common.

Another consideration is the location of water controls (fittings). The user should be able to reach them easily from outside the tub or shower without wetting his or her arm, but also able to manipulate them from within the tub/shower even if temporarily blinded by soap.

(b) Elimination. Two conflicts arise with toilets: the already-mentioned difficulty of combining cleansing with elimination, and the conflict over the height of the toilet. A *lower toilet* is

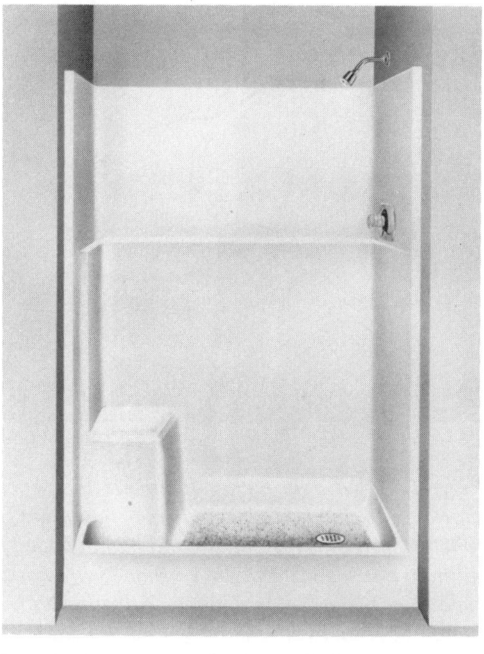

Fig. 11.4 *Many prefabricated showers now include seats, to encourage a safer posture while soaping and scrubbing the feet and legs.*

WATER AND WASTE

definitely of benefit to the average person, who will achieve far better bowel evacuation in a full squatting position. If combined with a toilet seat contoured to best support the body during defecation, the proposed toilet in Fig. 11.5*a* would be physiologically superior to most of today's fixtures. There are problems with low height. The typical male stands while urinating, and the lower toilet presents a more difficult target. This can have serious maintenance consequences unless males can be induced to sit or a separate urinal is provided—an unlikely option for residential bathrooms. (In our culture, urinals look "institutional"; they are also expensive.) Another problem with the lower toilet is that the elderly and some handicapped people will have difficulty getting on and off the seat (see also Fig. 11.17). Yet another problem arises from the fact that toilets are considered to be seats (for reading, toenail clipping, etc.). The low, squat-inducing toilet is decidedly not at a comfortable chair height.

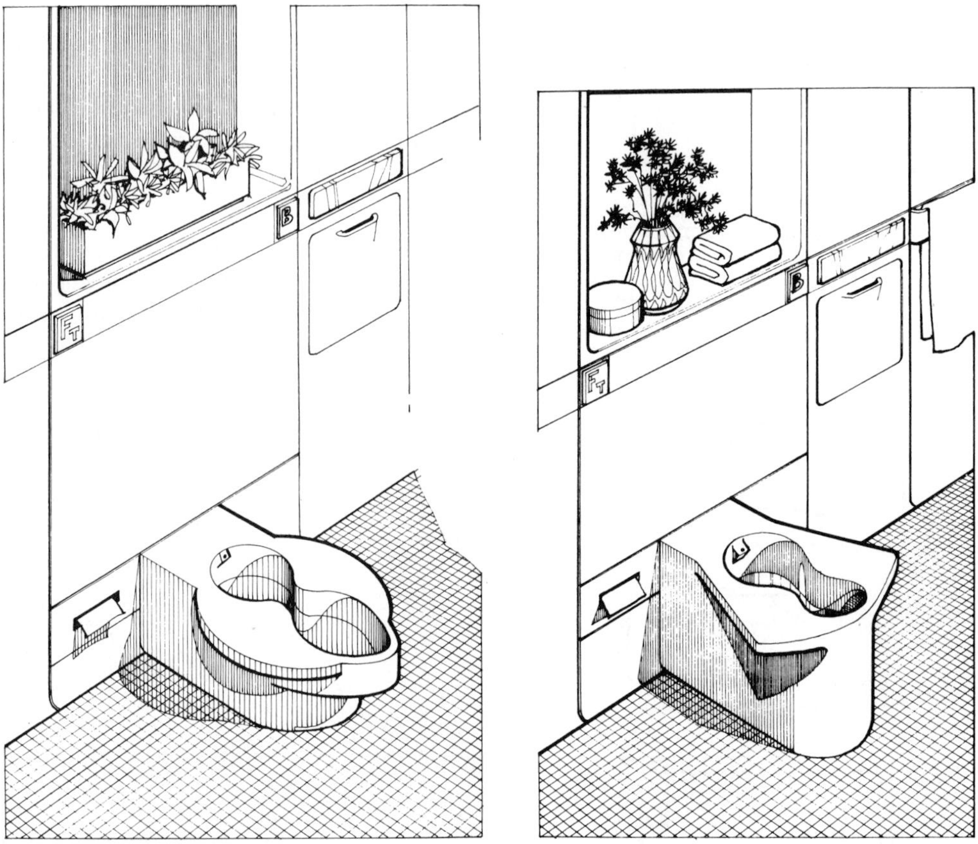

(a) *(b)*

Fig. 11.5 *Two approaches to toilets that encourage better posture during defecation and more thorough perineal cleansing.* (a) *A low toilet allows the best (full-squat) position.* (b) *The higher toilet is easier for the elderly and handicapped and better intercepts the urine from a standing male (although a separate foldout urinal is provided; see Fig. 11.8). Both toilets eliminate the separate toilet seat and its maintenance problems; both use small amounts of energy to warm the toilet at its point of contact with the body and to heat the water for perineal cleansing. Such cleansing is also the reason for larger front and back openings. The pushbuttons are labeled* F_T *for the flush, and B for the bidet (cleansing) function. (From* The Bathroom: New and Expanded Edition, *copyright © 1966, 1976 by Alexander Kira. By permission of Bantam Books, Inc. All rights reserved.)*

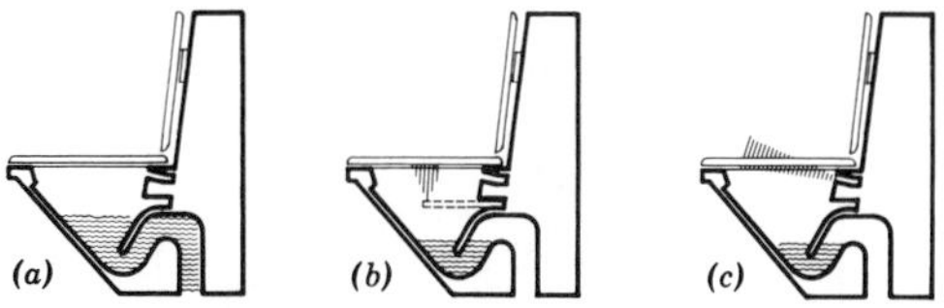

Fig. 11.7 *Provisions for perineal cleansing within toilets.* (a) *Section through toilet as it usually appears.* (b) *At the initiation of perineal cleansing, a pipe extends and emits a controlled spray of warmed water.* (c) *After pipe is retracted, a warm air jet issues from just below the seat. (From* The Bathroom: New and Expanded Edition, *copyright © 1966, 1976 by Alexander Kira. By permission of Bantam Books, Inc. All rights reserved.)*

As a result, a *higher toilet* with otherwise similar features is proposed (Fig. 11.5*b*). Note that both of these toilets feature openings whose shape differs from that of the conventional oval. This hourglass-shaped opening provides proper support for the body during defecation, and it allows a more generous opening, front and back, for proper perineal cleansing by hand. Another change is that there is no separate toilet seat; instead, these fixtures incorporate electric resistance heaters that warm the small portion of the toilet that contacts the body. Thus, the toilet becomes a consumer of energy—which is also used to heat the water that is provided for perineal cleansing. For those who insist on a separate seat, or for ordinary posture on conventional toilets, a physiologically sound contoured toilet seat is available (Fig. 11.6).

It is especially important that perineal cleansing be encouraged by the designers of bathrooms. The typical dry toilet tissue provided usually is not adequate for this task. In small, private bathrooms this problem is easily solved by placement of the toilet adjacent to the lavatory (or tub), where toilet tissue can be wetted with clean water. In public toilets separated by stalls, perineal cleansing requires either a clean water source built into the toilet, or a separate *bidet* within the stall—a highly unlikely provision, given the extra floor space and the cost of the bidet. Again, cultural issues arise: bidets are quite common in the private bathrooms of Europe, but a rarity in North America. Two common alternatives are (1) to build a cleansing source into the toilet itself (Fig. 11.7) or (2) to construct a toilet seat that provides for such cleansing. Several American manufacturers offer such seats, which can be easily adapted to existing toilets.

The *urinal* is an answer to the problem posed by lower toilets to males who stand while urinating. Although urinals are culturally acceptable in public toilet rooms, they have never achieved a place in private bathrooms. One solution proposed by Alexander Kira is to build the home urinal into the wall (Fig. 11.8*a*); it could be pulled out for use, at the optimum mounting height, and at an optimum receiving shape to eliminate backsplash onto the user. When pushed back into the wall, it would flush automatically. A less-satisfactory solution would be to modify the toilet by raising its back surface (Fig. 11.8*b*).

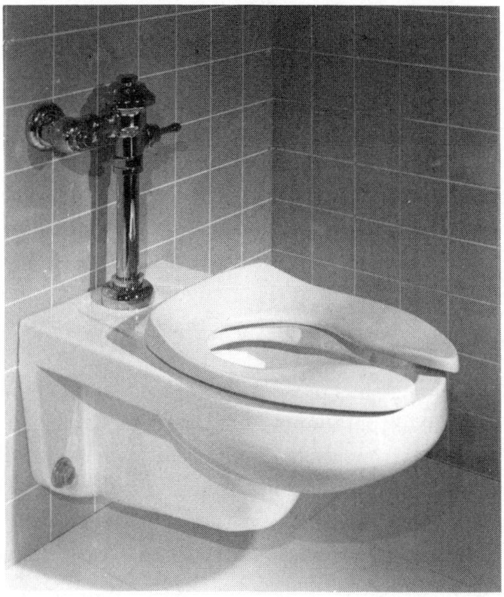

Fig. 11.6 *Toilet seat for the conventional WC, which encourages better posture for defecation and somewhat better access for perineal cleansing. (Courtesy of American Standard, Inc.)*

11.2 Design for Conservation

Several of the important options for conservation have included the substitution of rainwater

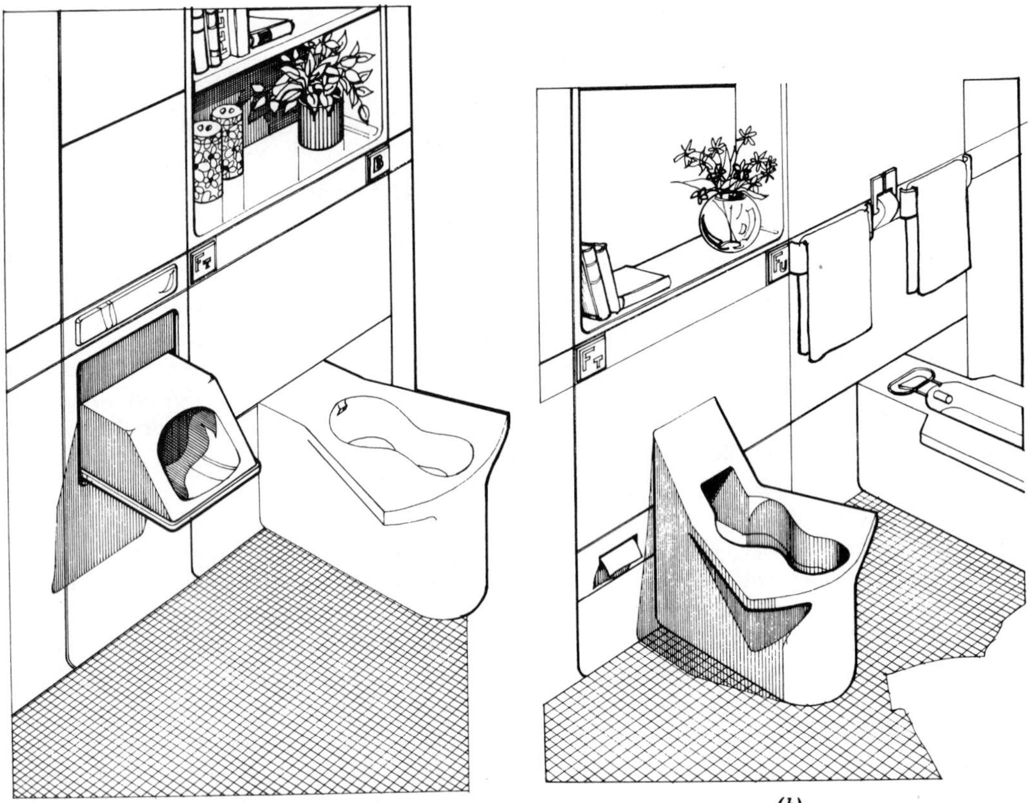

(a) (b)

Fig. 11.8 *Designs to accommodate male urination.* (a) *A built-in, tilt-out urinal for the home, visible only when in use.* (b) *A higher back for the toilet. A bidet for perineal cleansing is shown to the right of the toilet. (From* The Bathroom: New and Expanded Edition, *copyright © 1966, 1976 by Alexander Kira. By permission of Bantam Books, Inc. All rights reserved.)*

or graywater for potable water in fixtures such as toilets (Chapters 8 and 10), the use of less water in conventional fixtures, or the elimination of the use of water in toilets (Section 10.1). An obvious task for a designer is to specify water conserving fixtures, and water and energy conserving appliances. A less obvious and perhaps more interesting opportunity is the placement of the elements of the water and waste system in a relationship that encourages conservation.

Gravity flow is a key element when the use of rainwater and the recovery of graywater are desired. Vertical separation is desirable between points where rainwater is collected and stored, and the lavatories or toilets where it is used (Fig. 11.9). Vertical separation is also desirable between the fixtures that produce usable graywater (lavatories and showers, sometimes laundry rooms, perhaps dishwashers and kitchen sinks), the filtering system that renders the graywater useful, and the landscape that graywater irrigates.

Sloping sites present opportunities to exploit vertical separation, although the landscape downhill from a building has the irrigation advantage over landscape uphill. Access to graywater filtering systems placed at a lower elevation than bathrooms is much easier on sloping sites than on the flat land.

Even on a sloping site, the storage of rainwater above lavatories or toilets presents an unusual design challenge. Cisterns have traditionally been located underground and out of sight;

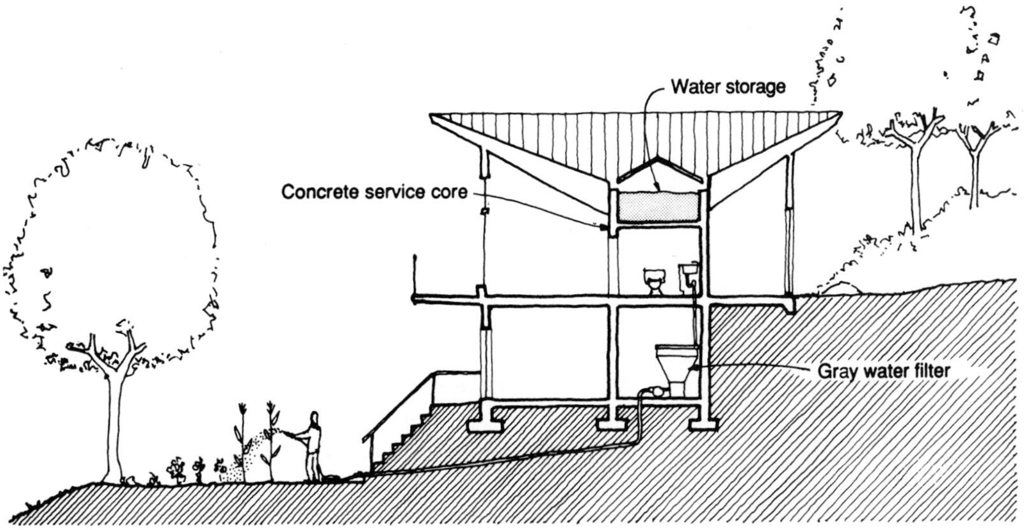

Fig. 11.9 Designing for water conservation and for the use of rainwater in gravity flow.

pumping energy is then necessary to use the water. In the Australian ranch house seen in Fig. 8.7, the above ground cistern is a highly visible feature. If rainwater is to be stored above points of use, water tanks become a prominent part of the form of roof eaves, affecting windows and upper walls particularly at bathrooms. This visual prominence of rainwater storage presents a second opportunity for the designer interested in promoting conservation.

Visible consumption is a strategy to encourage the users of fixtures to conserve water. We see, hear, feel and sometimes taste water as it issues from the fixture. If we could see *how much* water is involved in each use, we could more readily alter our pattern of use to conserve water where appropriate. Rainwater storage tanks outside bathroom windows might have a visible indicator of the water level; the lower this level, the more that conservation is an imperative. The water-level indicator might be as simple as visible condensation on the tank's outer surface. Small, transparent tanks of clean supply water could be placed above lavatories, filled only when empty by means of a simple float valve. The user might thus be encouraged to use only a portion of the water already in the tank. Sufficient pressure at the fixture is achieved by the tank's elevation above the fixture. The designer might also consider making the quantity of used (waste) water visible, but such water is often unsightly and the resulting quantity is "after the fact"—too late to conserve.

Audible consumption is another strategy, calling attention to the flow of supply (or waste) water. This is achieved most simply by slightly undersizing water supply pipes so that the water velocity becomes audible (about 10 fps, or 3 m/s). This strategy should be carefully considered; it is one thing to inform the water user, another to annoy. Audible flow is perhaps best justified on lines to exterior hose bibbs, where irrigation or car washing hoses may inadvertently be left on.

11.3 The Space Itself

After plumbing fixtures designed for the human physique and for water conservation have been chosen, they are placed in a space that has several unusual environmental needs. In this section, the first topic addressed is accessibility and privacy for the room and its fixtures; the emphasis here will be on the special problems of the handicapped. The next topic is the public bathroom, as typically designed into "cores" of multistory buildings. Finally, we will consider some special environmental control systems: thermal, luminous, and sonic.

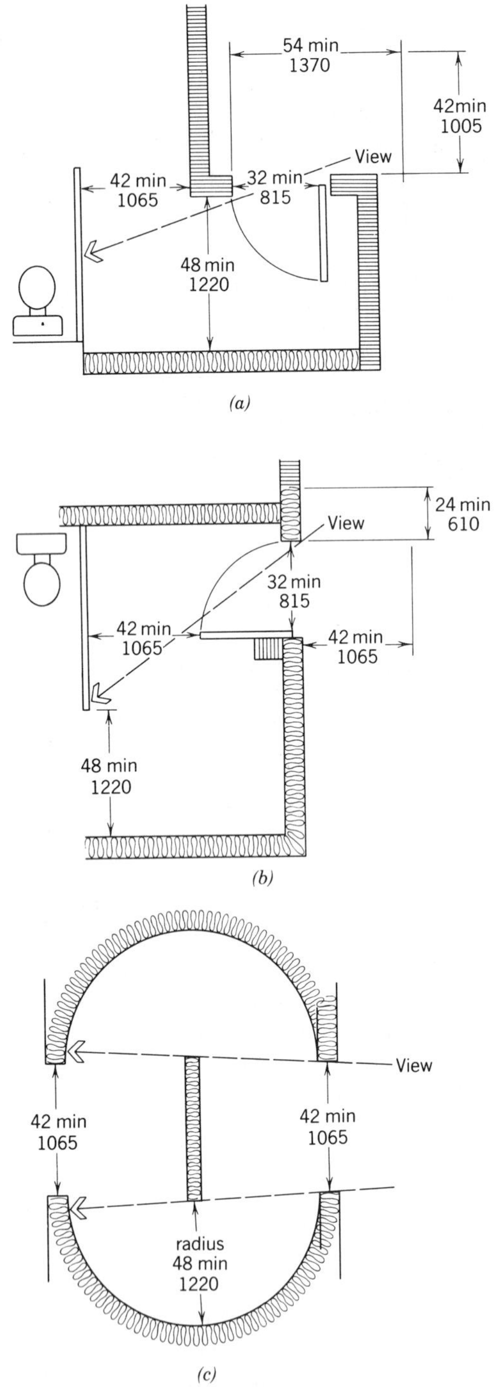

Fig. 11.10 *Entries to public toilet rooms should allow both visual privacy for those inside and easy access for the handicapped. All doorways must be at least 32 in (815 mm) in width and be clear of any obstruction, including protruding hardware. Designs* (a) *and* (b) *require the opening of doors, where design* (c) *allows rapid, easy passage of people and air—and sound.*

(a) Accessibility and Privacy. One of the most obvious requirements for bathroom design is that a person using any fixture within the bathroom should not be visible to someone outside the room. In private bathrooms, a simple closed door usually provides the solution. In public toilet rooms, however, entrances should be arranged so as to break the line of sight from hallways. The approaches shown in Fig. 11.10 show the amount of floor space required to balance privacy with adequate clearance for handicapped users in wheelchairs. (Much of the information in this section is reproduced with permission from *American National Standard for Buildings and Facilities A117.1-1986, Providing Accessibility and Usability for Physically Handicapped People,* copyright © 1986 by the American National Standards Institute (ANSI). Copies of this standard may be purchased from ANSI, 1430 Broadway, New York, NY 10018.)

Another accessibility issue is that of room for wheelchair maneuvering within such tight spaces as toilet rooms. The minimum clear floor area should be a circle 5 ft (1525 mm) in diameter. Minimum clearances for wheelchair maneuvering are shown in Fig. 11.11. Provisions for people in wheelchairs also influence the heights at which items such as light switches, electrical receptacles, paper towels, and water controls should be installed. The height limitations shown in Fig. 11.12 suggest that, in general, all objects to be reached by hand in toilet rooms be placed more than 15 in. (380 mm) and less than 48 in. (1220 mm) above the floor.

Floor space requirements and mounting-height limitations also apply to the various plumbing fixtures in bathrooms, at least one of which (in each case) should be accessible to people in wheelchairs in most buildings. These requirements are shown for drinking fountains (which often must be located outside toilet rooms) in Fig. 11.13, lavatories in Fig. 11.14, bathtubs in Fig. 11.15, and shower stalls in Fig. 11.16. The grab bars shown should have a diameter (or width) of 1¼ in. to 1½ in. (32 to 38 mm), with a clear space to the wall of 1½ in. (38 mm). They should be capable of resisting loads up to 250 lb-ft (1112 N), and not rotate within their fittings.

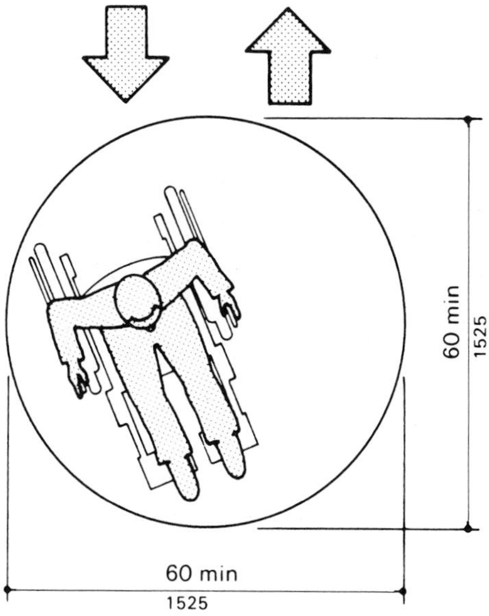

(a)

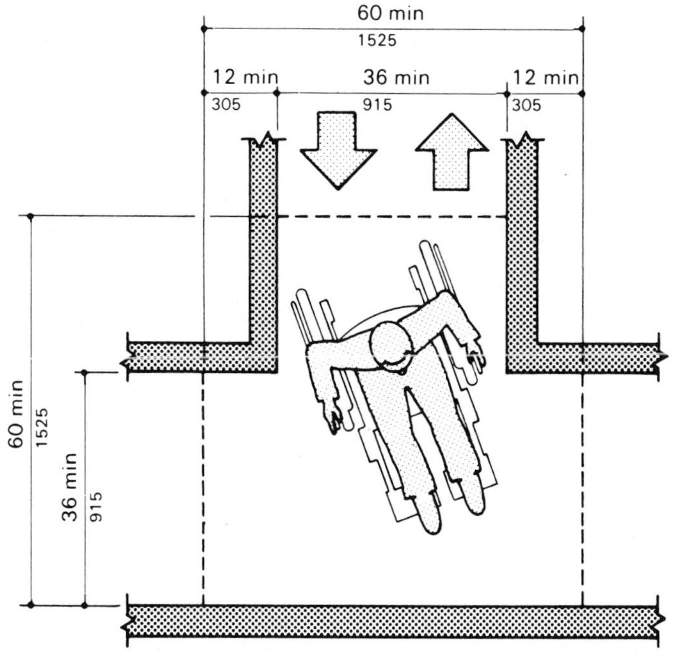

(b)

Fig. 11.11 *Wheelchair turning space, minimum requirements.* (a) *60-in. (1525-mm)-diameter space.* (b) *T-shaped for 180° turns. Dashed lines indicate minimum length of clear space required on each arm of the T-shaped space. (Reprinted by permission from American National Standard A117.1-1986, copyright © 1986 by the American National Standards Institute.)*

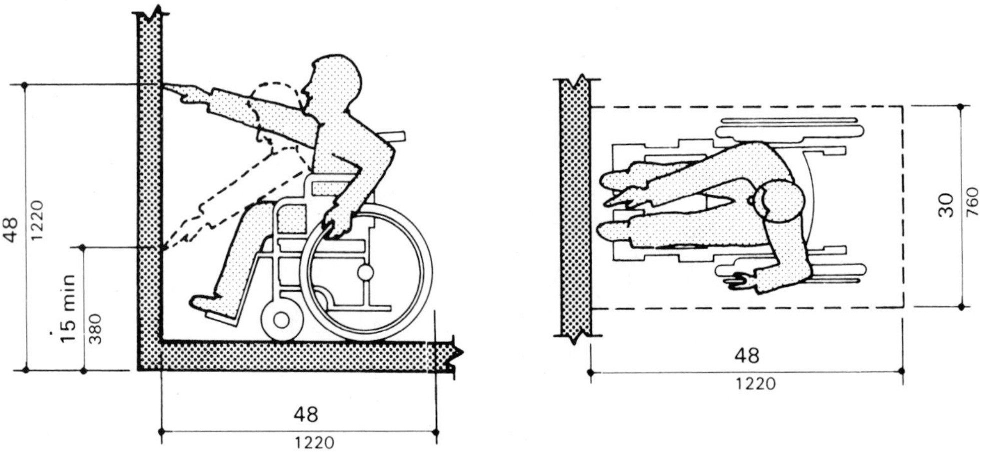

(*a*) *Forward Reach Limit*

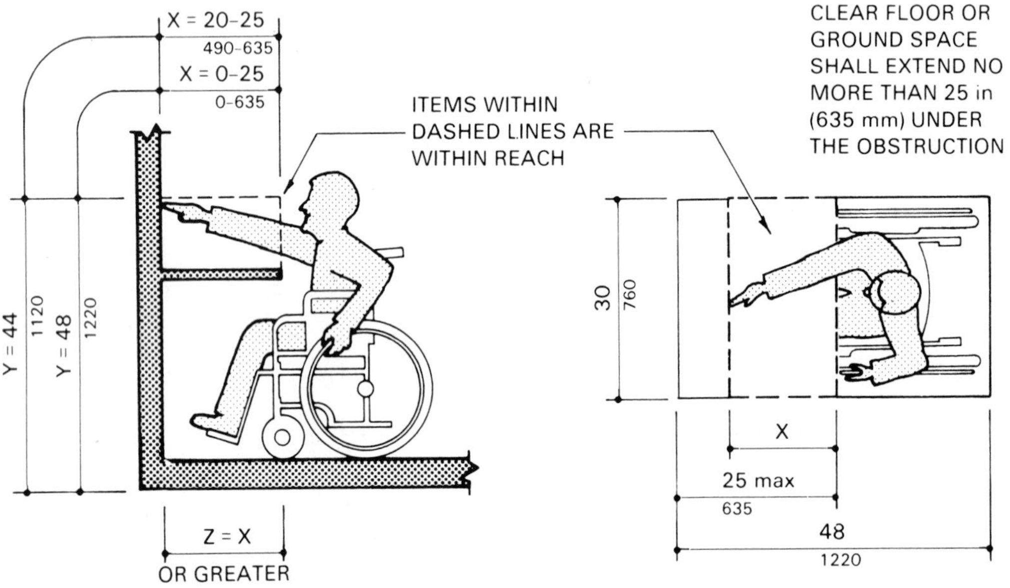

NOTE: x = Reach distance, y = Maximum height, z = Clear knee space. z is the clear space below the obstruction, which shall be at least as deep as the reach distance, x.

(*b*) *Maximum Forward Reach over an Obstruction*

Fig. 11.12 *Forward-reach and side-reach limits for people in wheelchairs. (Reprinted by permission from American National Standard A117.1-1986, copyright © 1986 by the American National Standards Institute.)*

Toilets (water closets) have particularly large floor space requirements, as shown in Fig. 11.17. There is a conflict between the lower mounting height needed for proper defecation, the conventional height of 15 in. (380 mm), and the recommended height for handicapped users of 17 to 19 in. (430 to 485 mm). The doors on stalls designated for wheelchair access should swing out rather than in. Provision for a wheelchair sitting beside the outswinging door must also be made (refer again to Fig. 11.10 for door clearances). No doors should swing into the clearance space required for any fixture.

Urinals for handicapped users should either

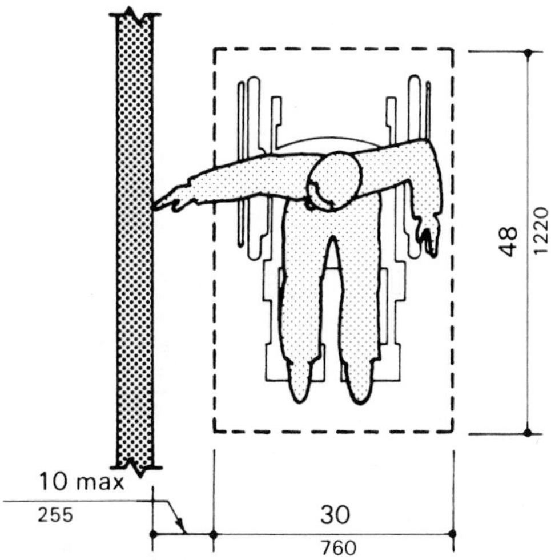

(c) *Clear Floor Space — Parallel Approach*

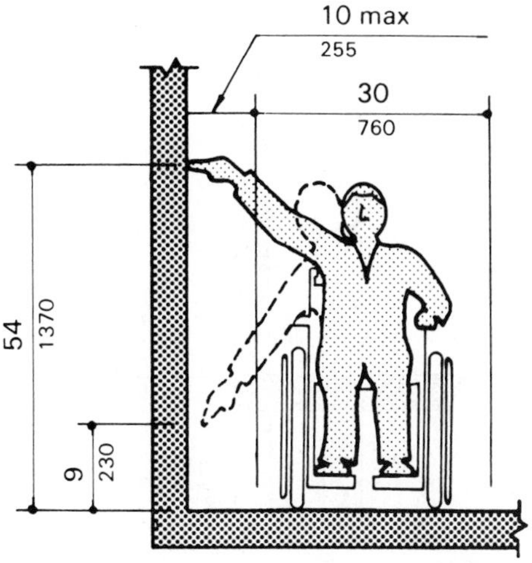

(d) *High and Low Side Reach Limits*

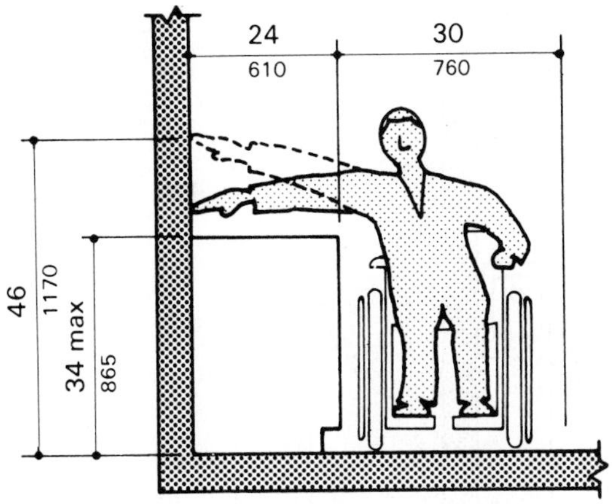

(e) *Maximum Side Reach over Obstruction*

Fig. 11.12 *(continued)*

be of the floor-mounted stall-type or, if wall-hung, have an elongated rim at a maximum of 17 in. (430 mm) above the floor. A clear floor space of 30 × 48 in. (760 × 1220 mm) is required in front of the urinal, and its flush controls should be no more than 44 in. (1120 mm) above the floor.

(b) Public Toilet Rooms. These spaces are especially challenging, because of several conflicting design criteria. They should be in easily accessible places, yet be as acoustically and visually private as possible. Unfortunately, they can easily become security problem areas in some buildings and cities. They generate no rental income (in contrast to office or commercial space around them), yet they cost significantly more per square foot to construct and require more-intensive maintenance. They need high rates of ventilation and could benefit from the psychologically purifying effects of plentiful sunlight—they even have high thermal mass for solar heat storage—yet they are usually placed in the innermost, lowest-rent zones of buildings (such as central service cores), far from wind or sun's benefits.

Further discussion of service cores and their related spaces can be found in Chapter 12. At

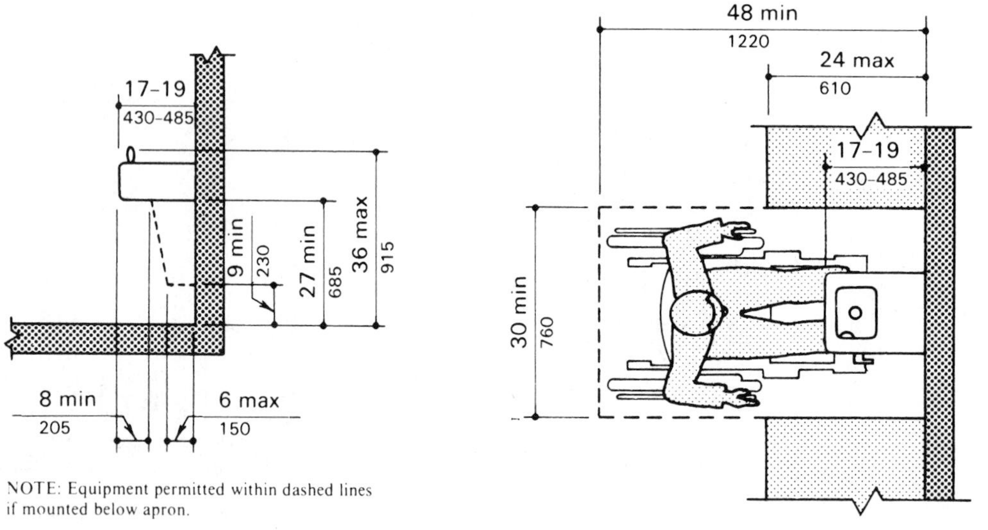

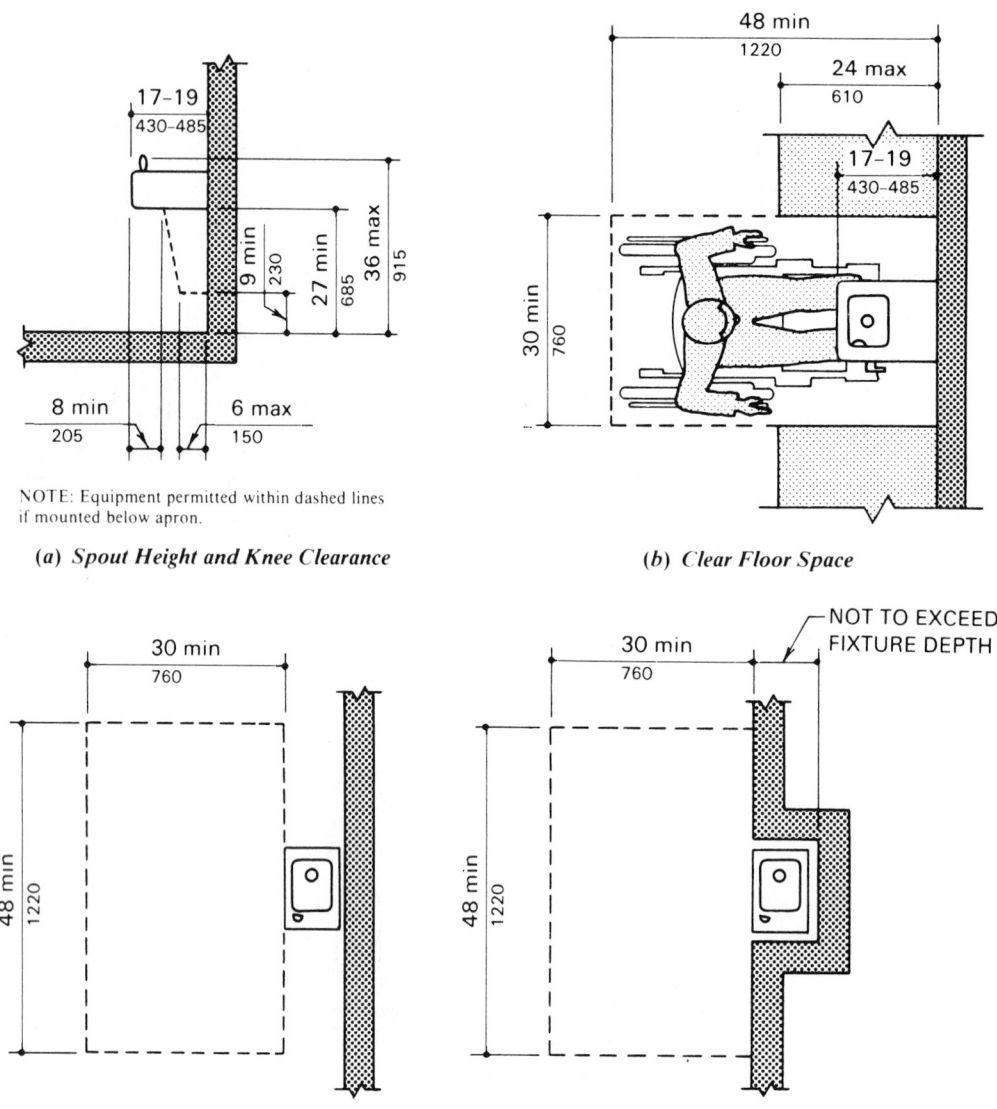

(*a*) *Spout Height and Knee Clearance* (*b*) *Clear Floor Space*

(*c*) *Free-Standing Fountain or Cooler* (*d*) *Built-In Fountain or Cooler*

Fig. 11.13 *Drinking fountain installation dimensions. (Reprinted by permission from American National Standard A117.1-1986, copyright © 1986 by the American National Standards Institute.)*

this point, some design considerations for public toilet rooms can be listed.

Location should usually be central without being "featured." They are necessary public services; people expect to find toilet rooms near reception areas, waiting rooms, eating places, elevators or stairs, coat or package-checking services, public telephones, drinking fountains, and major entrances.

Entries must balance accessibility and privacy (as in Fig. 11.10). Where possible, men's and women's toilet rooms should be located next to each other and both entries should be visible at a glance.

Fixtures are almost always grouped together for economy in both installation costs and floor area. A simple, clean counter of lavatories can be visually appealing; large mirrors add to a

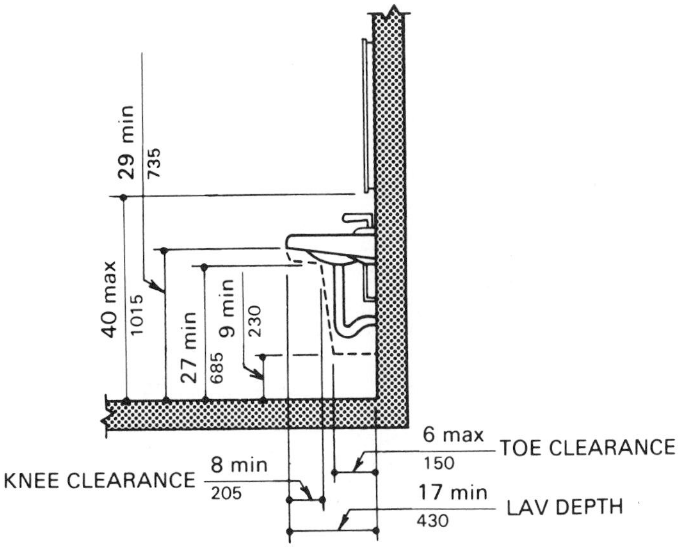

NOTE: Dashed line indicates dimensional clearance of optional underlavatory enclosure.

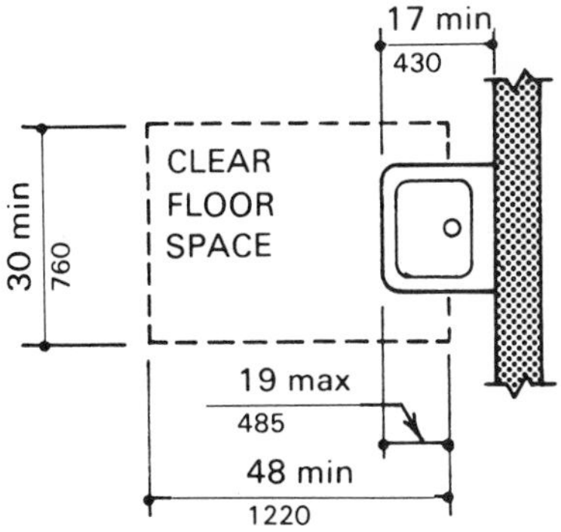

Fig. 11.14 *Clearances for lavatories. Hot water and drain pipes below lavatories must be insulated or otherwise covered. Mirrors should be mounted so that the lower edge is no more than 40 in. (1015 mm) above the floor. (Reprinted by permission from American National Standard A117.1-1986, copyright © 1986 by the American National Standards Institute.)* Note: *Standard height to lavatory counter is 31 in. (787 mm) (from* Architectural Graphic Standards, *7th ed.).*

sense of spaciousness in an otherwise crowded space. A row of toilet stalls, by contrast, rarely is visually inviting, and the separation of lavatories from toilets raises physiological problems, as discussed earlier. It is unfortunate that this stall, the most private place in public buildings, is usually the worst lighted, most crowded, worst smelling, and most poorly performing of all spaces. Some improvements for such toilet stalls are suggested in Fig. 11.18.

The required number of fixtures in such rooms is determined by codes, as was shown in Table 8.3. Although requirements differ by function, in general there will be somewhat fewer lavatories than urinals or water closets. Lavatories are usually closest to the toilet room entry, since some users need only cleansing facilities. Preferably, users of toilets/urinals will be brought next to lavatories before they leave.

Lighting can be of relatively low intensity compared to that required for work areas. Task lighting in toilet rooms should be concentrated on lavatories, urinals, and toilets. Other areas can be much less brightly lit. In keeping with the "restroom" concept of a refreshing space, designers might consider deliberately changing the lighting approach within toilet rooms from that of the surrounding work areas. If one is uniformly and diffused lighted, the other might consist of pools of bright light within a darker space. If one is primarily electrically lighted, the other might be dominated by daylight. These deliber-

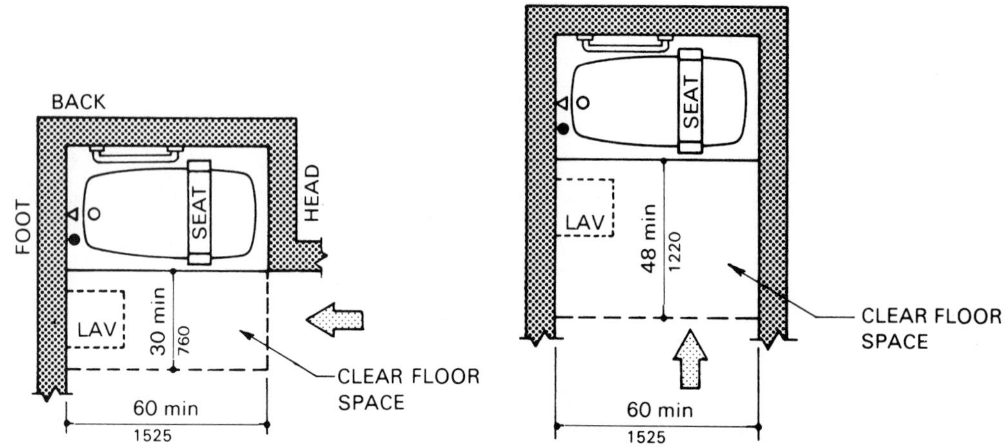

(*a*) *With Seat in Tub*

SYMBOL KEY:

● = Shower controls

◁ = Shower head

○ = Drain

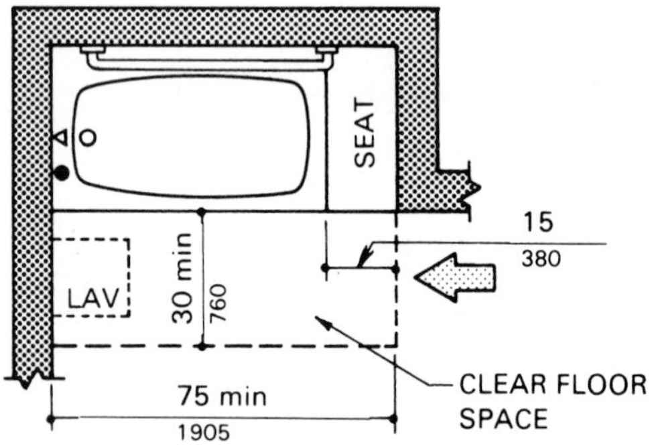

(*b*) *With Seat at Head of Tub*

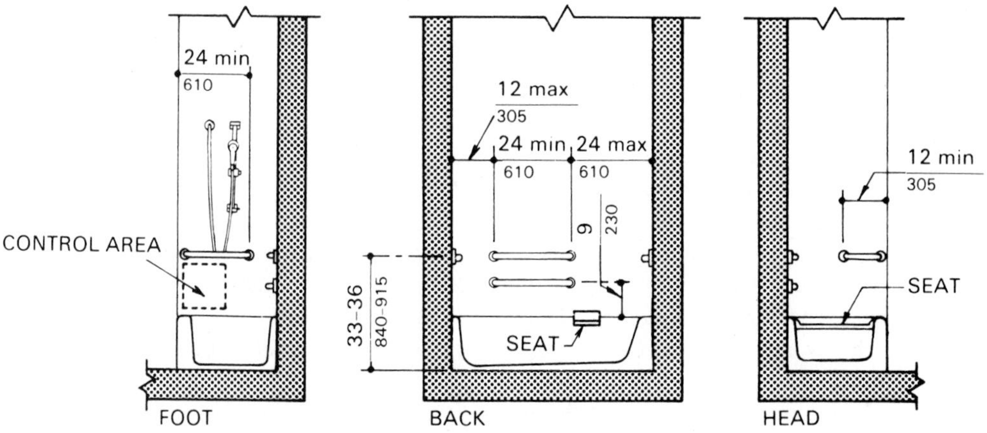

(c) *With Seat in Tub*

Fig. 11.15 *Clearances and grab bars for bathtubs;* (a) *and* (c) *are with seat in tub, whereas* (b) *and* (d) *are with seat at head of tub. (Reprinted by permission from American National Standard A117.1-1986, copyright © 1986 by the American National Standards Institute.)*

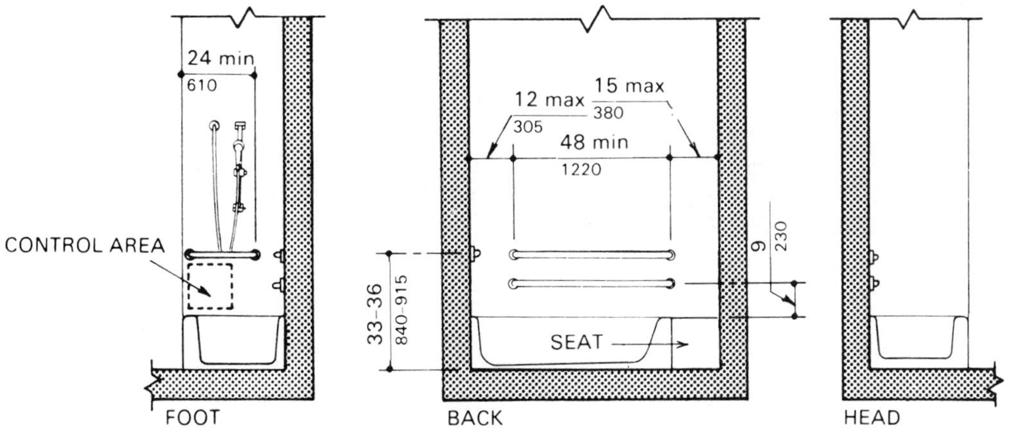

(*d*) *With Seat at Head of Tub*

Fig. 11.15 *(continued)*

ate contrasts can be extended to materials, ceiling heights, reverberance, temperature/air motion, and so on.

Ventilation is of particular importance. The primary things to remember are to keep the toilet downstream in the airflow and not to recycle air from toilet rooms. Accordingly, the air pressure within toilet rooms should be kept slightly below that of surrounding space, so that airflow is always *into* the toilet room, and exhaust vent intakes should be as close as possible to the toilet itself, yet kept above it to minimize maintenance problems and thermal discomfort due to drafts.

Although fresh air is sometimes brought directly to toilet rooms, a more common procedure is to supply slightly more air to surrounding office or commercial spaces than is returned; the resulting surplus is available to be drawn into toilet rooms, then exhausted.

Acoustic privacy is usually achieved by the placement of toilet rooms within service cores. However, ventilating openings and doorless-but-screened access ways present ready paths for sound. Some users of toilets respond to lack of acoustic privacy by repeatedly flushing the toilet to mask the sounds associated with elimination. This is a noisy and very wasteful practice, which could be minimized if the designer provides for acoustic privacy. Isolated toilet stalls usually are not practical—maintenance is complicated, and walls are far more expensive than simple visual-screen partitions. Instead, a steady masking sound is often provided. A higher-sound-level ventilating system is frequently chosen, since noisy fans are often cheaper than the quiet ones, and they reassure the user that polluted air is constantly being removed from the space.

Where office buildings have greater floor-to-floor heights (as for daylighting), it may be possible to locate the fan room that serves each floor over the toilet rooms. The rumbling of this air-handling equipment, normally an annoyance, is usually welcome in public toilet rooms.

(c) The Private Bathroom. Individual bathrooms within large buildings, such as those for hotel rooms, apartments, and hospital patients' rooms, commonly are subject to similar location constraints as the larger public toilet rooms: that is, a location away from the more desirable perimeter zone, with its daylight and fresh air. Accordingly, the ventilation strategy for such individual bathrooms is similar—an exhaust fan, drawing air from the spaces near the bathroom. Individual residences can often provide a happier environment for the bathroom, with daylight and fresh air available.

One of the first design considerations for private bathrooms is recognition that three distinct categories of activity are involved: partial

Fig. 11.16 Clearances and grab bars for shower stalls. (a) *and* (d) *show 36-in. by 36-in. (915-mm by 915-mm) stalls;* (b) *and* (e) *show 30-in. by 60-in. (760-mm by 1525-mm) stalls.* (c) *shows shower seat design. (Reprinted by permission from American National Standard A117.1-1986, copyright © 1986 by the American National Standards Institute.)*

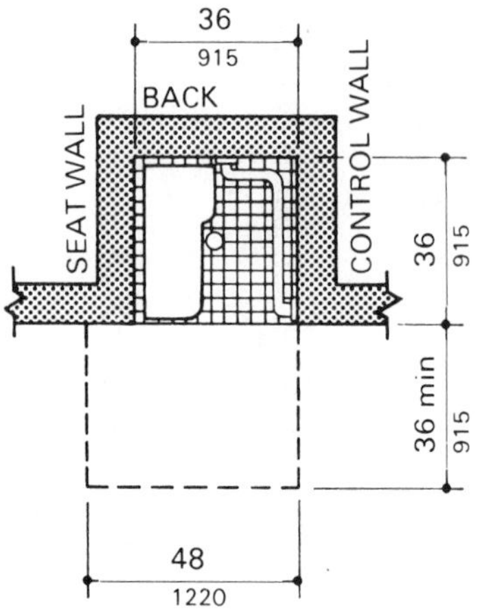

(a) 36-in by 36-in (915-mm by 915-mm) Stall

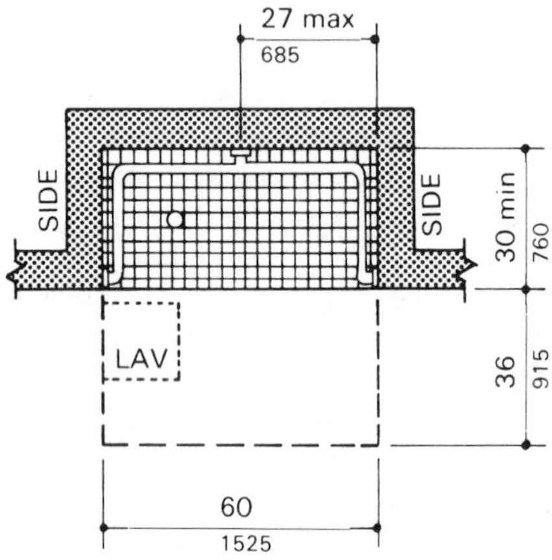

(b) 30-in by 60-in (760-mm by 1525-mm) Stall

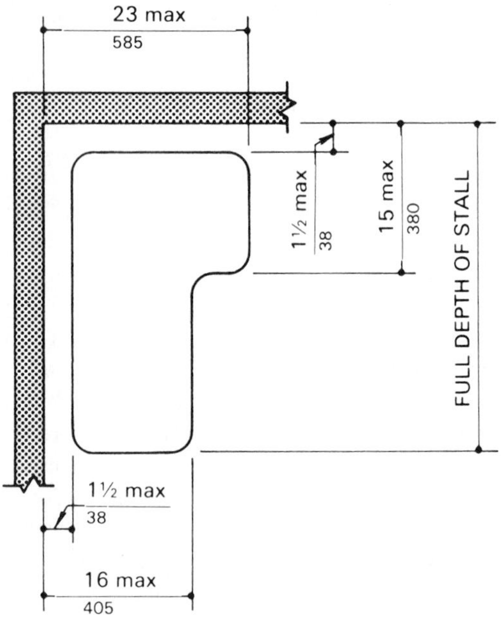

(c) Shower Seat Design

cleansing/grooming, full body cleansing, and elimination. A typical private bathroom is the *three-fixture type:* lavatory, toilet and tub/shower, arranged within one relatively small room [35 ft^2 (3.3 m^2) is minimal]. Advantages include a potentially efficient use of floor space and the convenience of lavatory/toilet adjacency for perineal cleansing. The great disadvantage is that the room is designed for one user at a time, so that others are deprived of access to fixtures for elimination or partial cleansing when one person is engaged in full-body cleansing.

As a result, there has been a trend toward *compartmented* bathrooms, in which the lavatory may be located in hallways, bedrooms, or small alcoves. This less-modest environment is usually suitable for the partial cleansing/grooming activities associated with lavatories. Also, lavatories are less expensive fixtures, require less floor area, and are somewhat less likely to do great damage if they malfunction. Most commonly, the toilet and tub/shower are then grouped in a more isolated, private space. Where many users of one such compartmented bathroom are expected, it is advisable also to separate the toilet (along with a smaller lavatory) from the tub/shower room.

Another typical approach to private bathrooms is the *guest bath,* in which a lavatory, a toilet, and a shower stall are included. The sub-

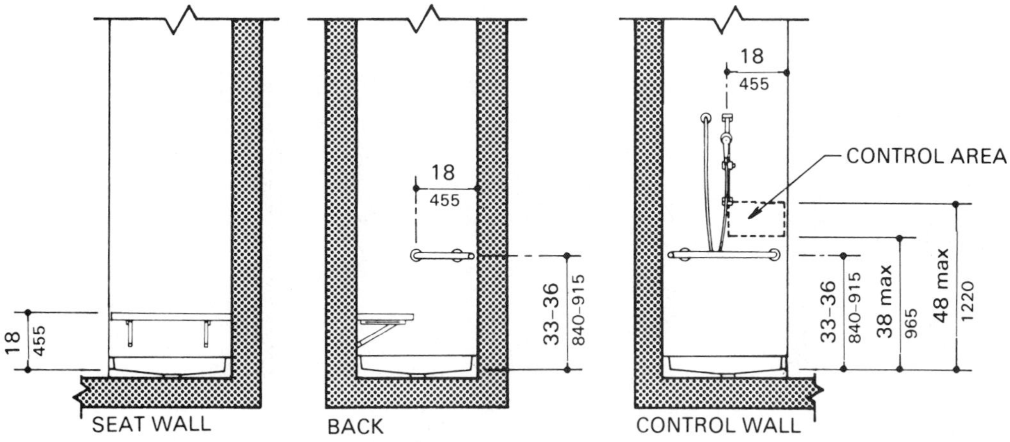

(*d*) ***36-in by 36-in (915-mm by 915-mm) Stall***

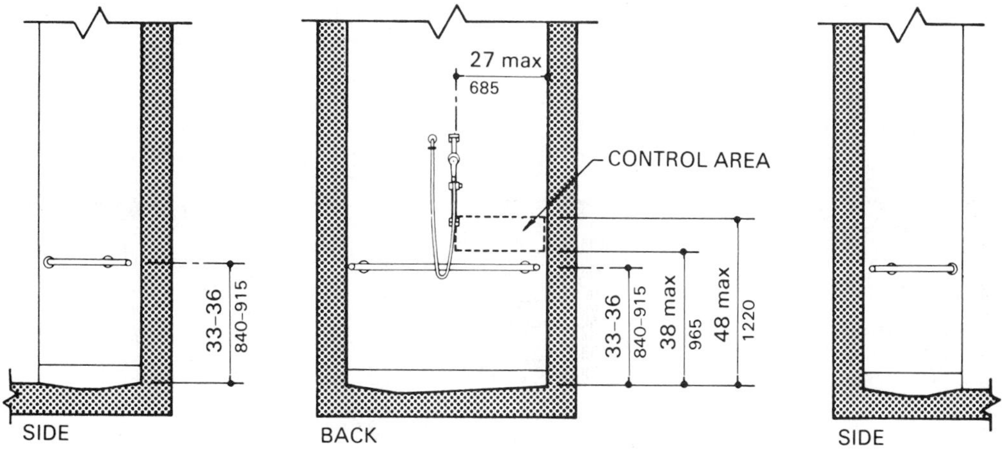

NOTE: Shower head and control area may be on back wall (as shown) or on either side wall.

(*e*) ***30-in by 60-in (760-mm by 1525-mm) Stall***

Fig. 11.16 *(continued)*

stitution of a shower stall for a tub cuts down the space needed to about 30 ft^2 (2.8 m^2). The guest bath is often the second bathroom in a residence, and sometimes is physically remote from the main one. A more limited and very common private bathroom is the *half-bath,* containing only a lavatory and a toilet. This room can be less than 25 ft^2 (2.2 m^2), and it is frequently found crouched below a stairway on the main floor of houses with upstairs bedrooms and main bath(s). The tight quarters and steeply sloped ceilings involved in such an arrangement offer a great opportunity for the use of mirrors as space expanders and light reflectors, as well as for aids to grooming.

Fixtures for private bathrooms commonly are more visually elegant, and somewhat more demanding of maintenance, than are their counterparts in public toilet rooms. The most dramatic difference is the presence of a tub, which can be large or small, raised or sunken, or built as a skylit greenhouse-like appendage to the bathroom (see Fig. 8.7, showing Andrews farmstead house). Special attention can easily be paid to accommodations such as pullout stepstools for small children at lavatories, and especially to storage—towels and soaps, at tubs/showers, grooming aids and medicines at the lavatory. It is common for as much design attention to be paid to these accessories in private bathrooms

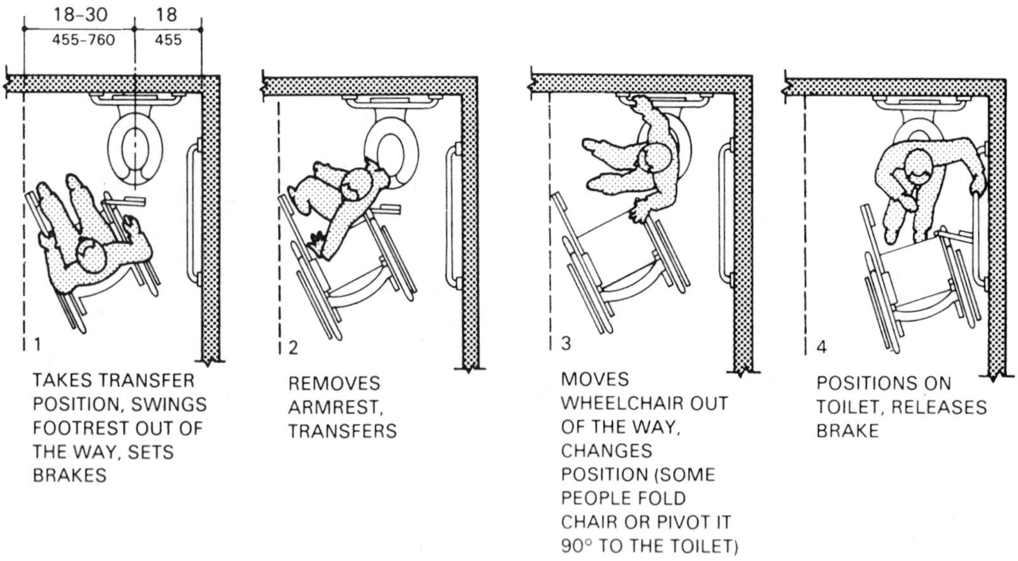

(*a*) *Diagonal Approach*

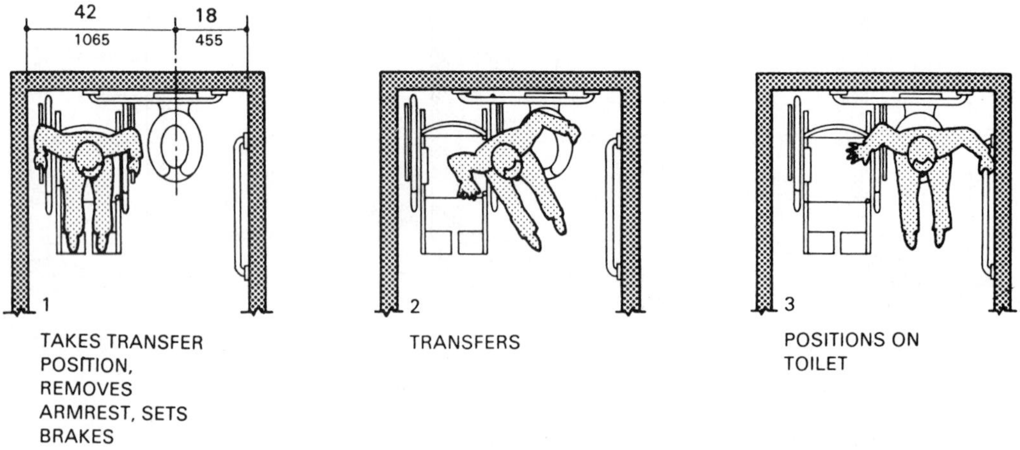

(*b*) *Side Approach*

Fig. 11.17 *Clearances and grab bars for toilets. Two typical approaches* (a, b) *necessitate the clearances of the stalls shown* (c, d, e, f), *as well as arrangements for toilets and urinals not within stalls* (g, h, i, j). *(Reprinted by permission from American National Standard A117.1-1986, copyright © 1986 by the American National Standards Institute.)*

as to the selection and placement of the fixtures themselves.

Lighting affords a special opportunity at the lavatory, because of the availability of natural light and the tradition of electric lights for grooming. There are several important points to remember here: that the objective is to light the human face, not the mirror surface; that excessively bright lights near the eyes can cause glare; and that the background seen behind the face in the mirror should be no lighter than the face itself. For these reasons, direct sun on the face at lavatories is usually avoided, as are windows directly opposite the mirror (unless they open into a low-daylight scene). Some possibilities for natural and electric lighting for lavatories are shown in Fig. 11.19.

Thermal considerations are likely to be domi-

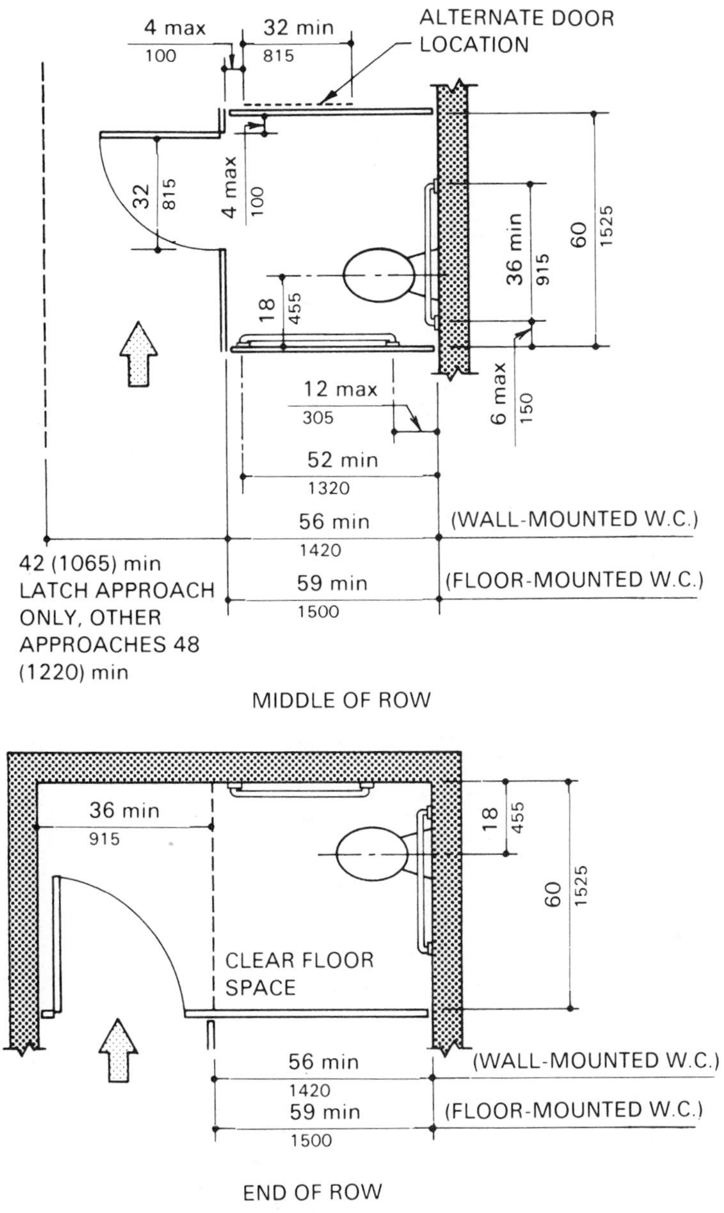

(*c*) *Standard Stalls*

Fig. 11.17 *(continued)*

nated by the ventilation needs, which are higher than those for any other residential space except for kitchens. If residences were to be designed as superinsulated buildings, a constant ventilating fan could be installed, pulling from the bathroom that amount of minimum fresh air required for the rest of the entire house. (However, if forced-air heat exchangers are provided for such residences, the bathroom fan should operate only as needed.) In hotels, this is a common way to assure fresh air to each room, and some multistory apartment houses also assure ventilation in this manner. Since most climates are not severe enough to warrant such tight regulation of

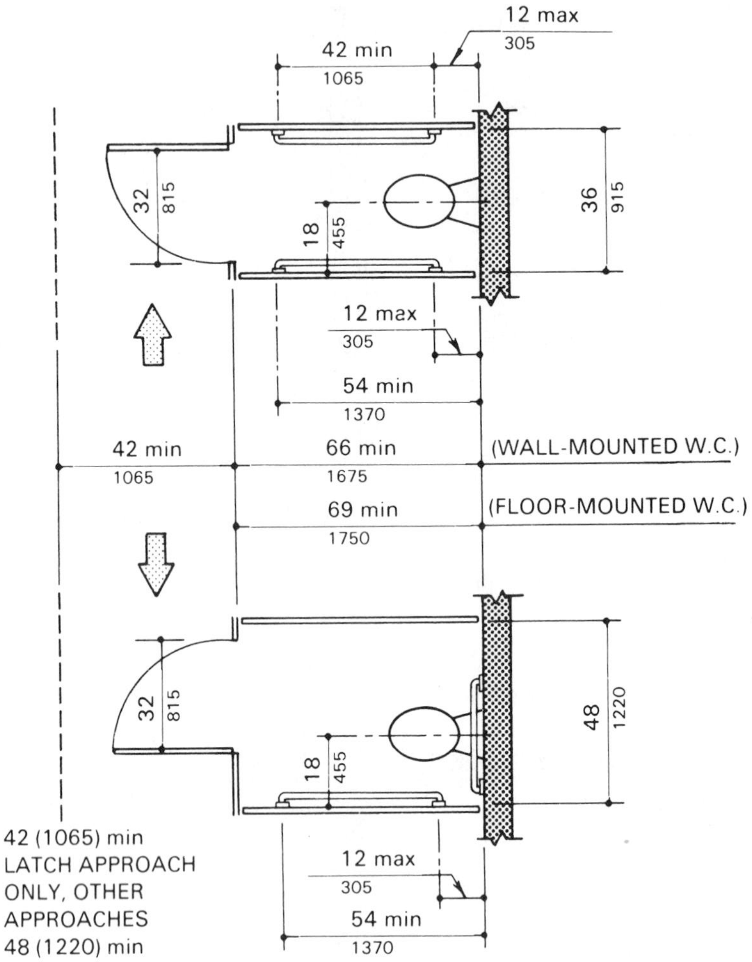

(*d*) *Alternate Stalls*

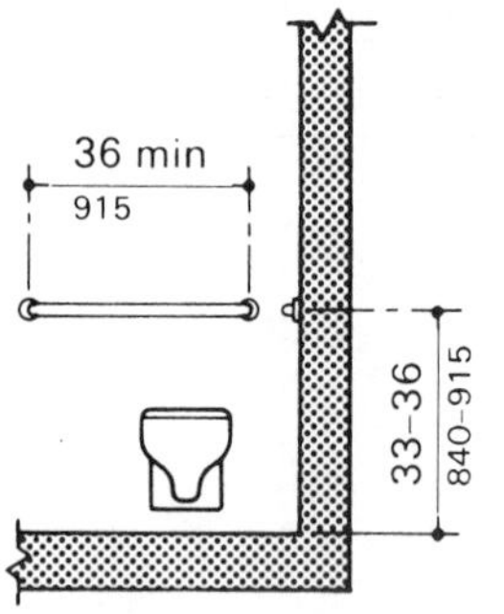

(*e*) ***Rear Wall of Standard Stalls***

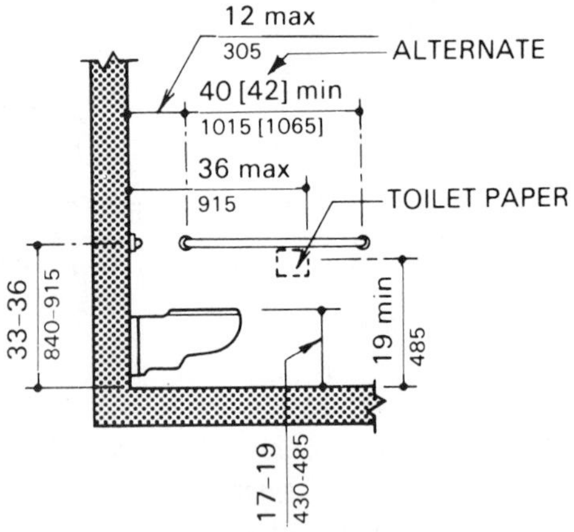

(*f*) ***Side Walls***

Fig. 11.17 *(continued)*

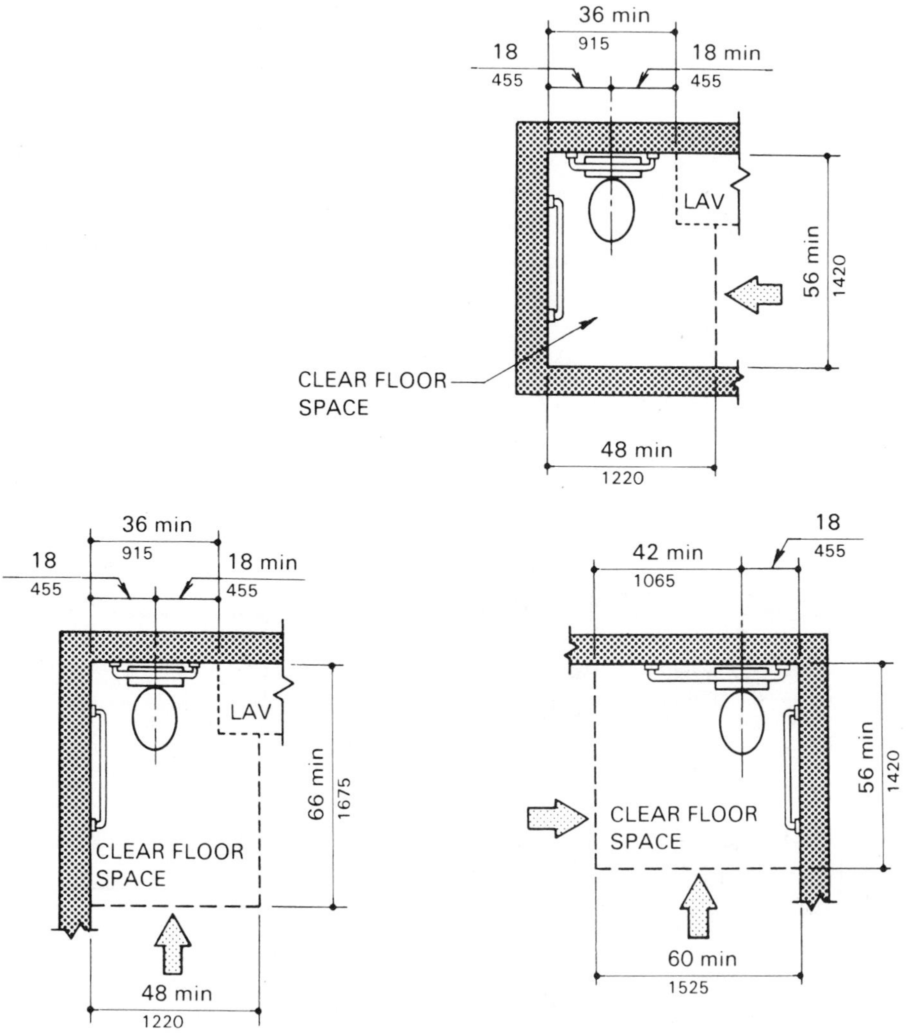

(*g*) Clear Floor Space at Water Closets

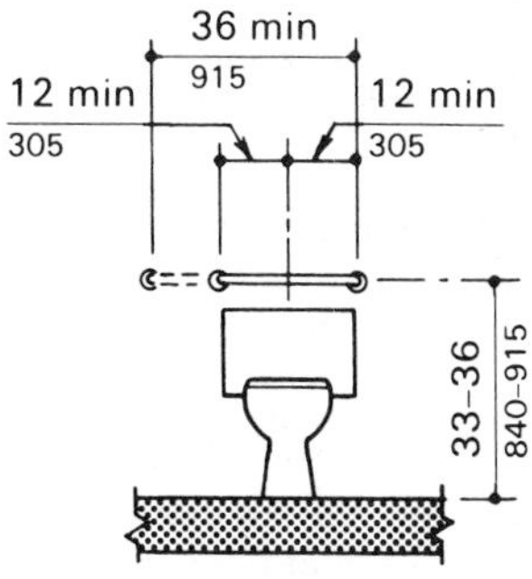

(*h*) ***Back Wall***

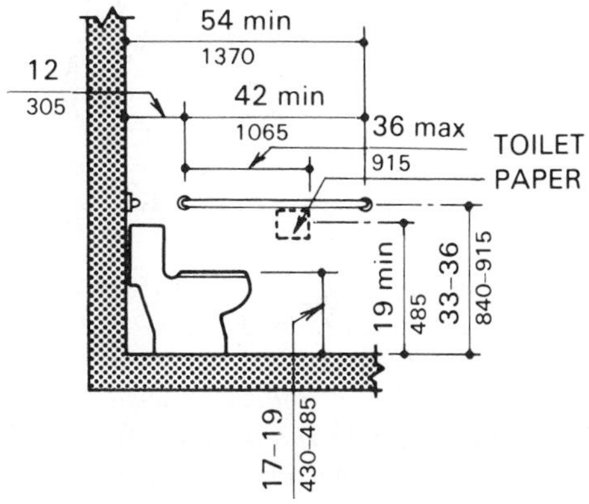

(*i*) ***Side Wall***

Fig. 11.17 *(continued)*

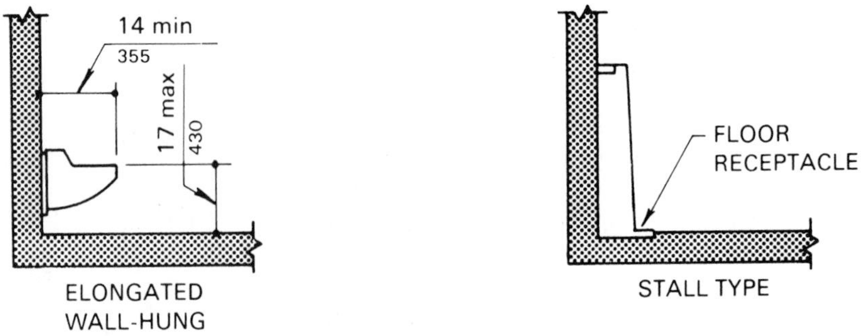

(*j*) ***Urinals***

Fig. 11.17 *(continued)*

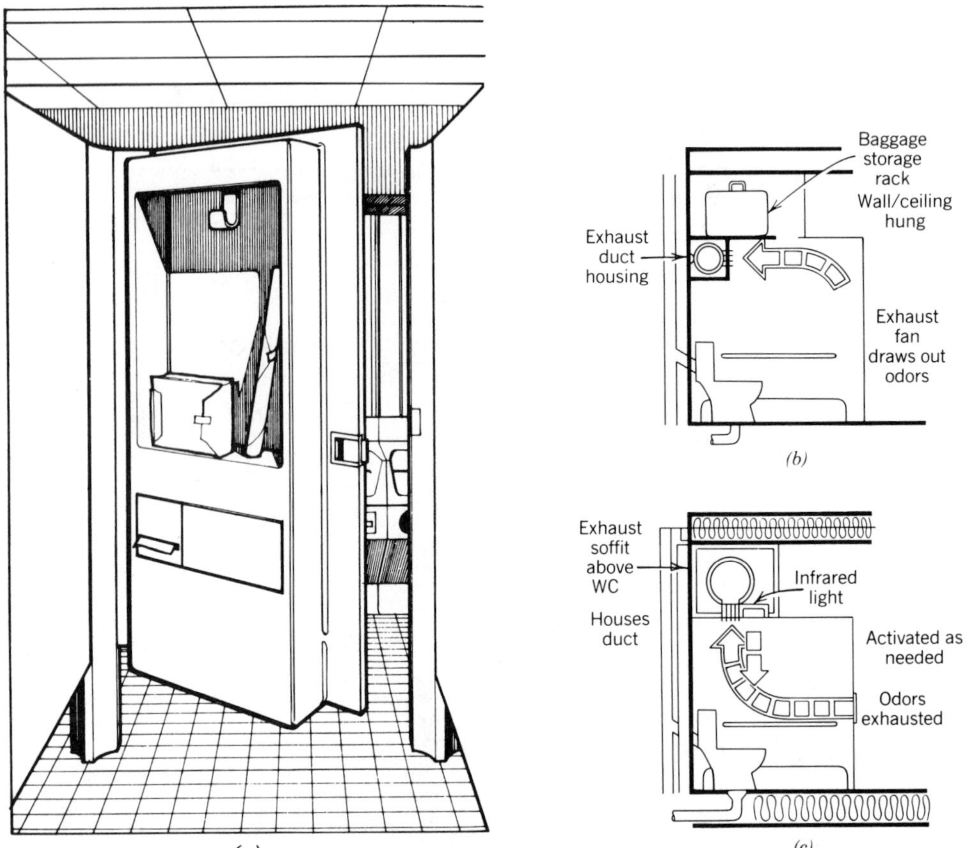

(*a*) (*b*) (*c*)

Fig. 11.18 *Improving the public toilet stall.* (a) *Stall doors could be designed to house small articles and outer clothing carried by the user. (From* The Bathroom: New and Expanded Edition, *copyright © 1966, 1976 by Alexander Kira. By permission of Bantam Books, Inc. All rights reserved.)* (b) *For transportation terminals, added provision for luggage, without an increase in floor area requirements, is desirable.* (c) *Exhaust ventilation directly above toilets is desirable. In cold weather, small radiant heat lamps with timer switches could provide task heating and lighting in an otherwise darker, colder environment. (From a study by Michael Bush, University of Oregon.)*

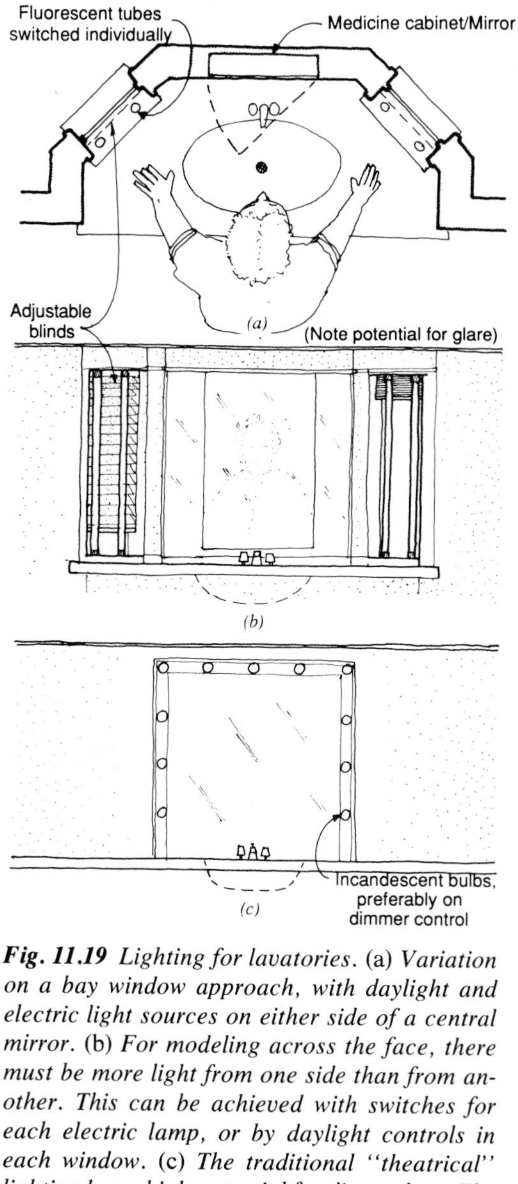

***Fig. 11.19** Lighting for lavatories.* (a) *Variation on a bay window approach, with daylight and electric light sources on either side of a central mirror.* (b) *For modeling across the face, there must be more light from one side than from another. This can be achieved with switches for each electric lamp, or by daylight controls in each window.* (c) *The traditional "theatrical" lighting has a high potential for direct glare. The larger the mirror, the farther the row of lamps from the face's image, and the less the glare problem. Less glare would be produced by a large, low-brightness overhead light source, with light-colored diffusing surfaces on walls and counters. (From a study by Michael Bush, University of Oregon.)*

outdoor air year-round, residential bathrooms are often ventilated, either naturally or with fans that are used only as needed.

Heating is the next most significant need, since a body that is naked and wet is one that is most susceptible to heat loss through both radiation and evaporation. Solar energy has much to offer the bathroom: strong radiant effects; the potential for some overheating, which may be appropriate in this more-needy space; and a psychological association between direct beams of sun and purification of the surfaces that it strikes. The bathroom contains fixtures and surfaces that are typically more thermally massive than those of most other spaces in houses. However, night insulation should cover windows and especially skylights, since radiant losses through unprotected glass are so great, and since the more humid air of the bathroom will readily condense on cold surfaces such as skylights exposed to the night sky.

Task heating is very popular and appropriate for the bathroom; radiant heat lamps provide the added heat for the body fresh from tub or shower, without overheating the entire bathroom. Timer switches are frequently used with these heat lamps. Radiant electric heaters are an opportunity to combine heating device and towel rack, shown in Fig. 11.20. If a forced-air heating system is used, an outlet near the lavatory is welcome for early morning grooming in winter. Air grilles should be installed on ceilings or on vertical surfaces in rooms that contain water, such as bathrooms and kitchens, since floors are frequently mopped and water spills are common. The toe space for cabinets beneath kitchen and bathroom sinks is a popular location for these supply air grilles.

When it is required, cooling is often linked with ventilation; the more cooling, the greater the ventilation rate. In very hot conditions, this is counterproductive, as it brings too much hot air into the house. A means of providing cool vertical surfaces (from night-ventilated thermal mass, for example), and a fan to stir the humid air, are a promising combination for hot-day comfort.

Some combinations of lighting and thermal considerations in bathrooms are shown in Fig. 11.21.

Acoustic considerations comprise another important aspect of bathroom design. Because of their nonabsorbent surfaces, bathrooms are frequently the most reverberant spaces in the

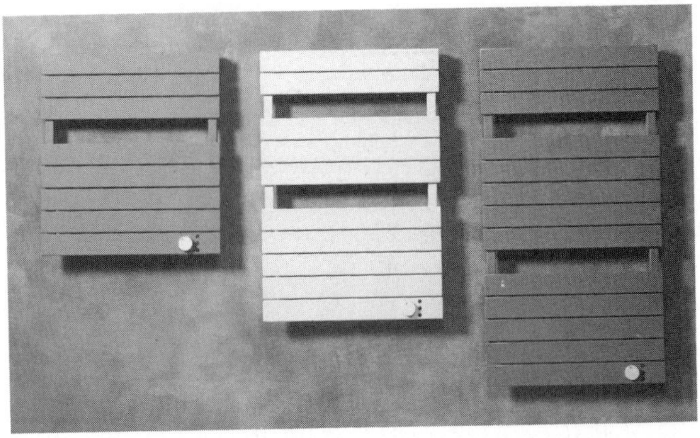

Fig. 11.20 *"Towel warmers" combine task electric heaters with a towel rack. They radiantly heat other bathroom surfaces as well as towels, and can provide local heat when a central heat system is not operating. (Courtesy of Runtal North America, Inc., Ward Hill, Massachusetts.)*

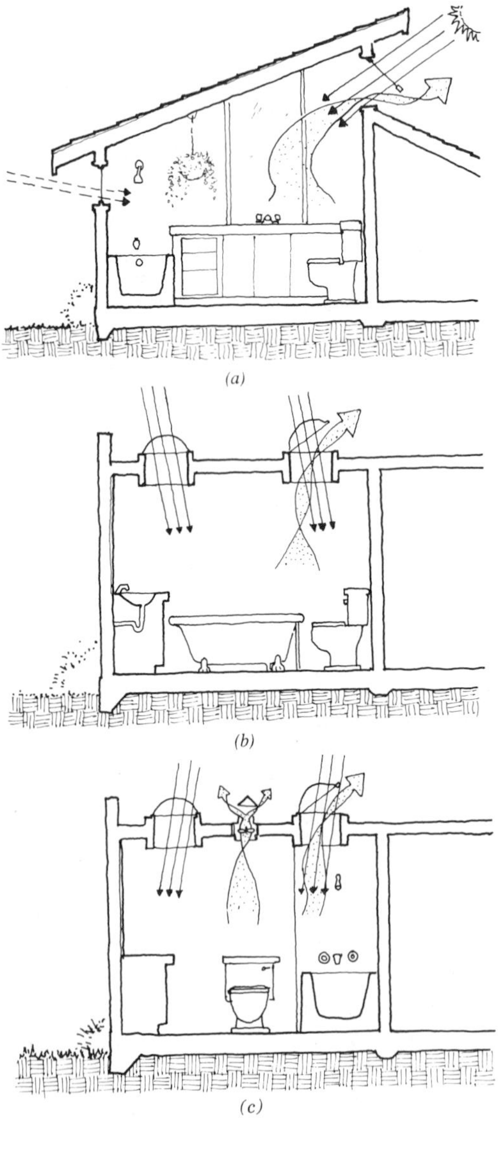

Fig. 11.21 *Light and air considerations in bathrooms.* (a) *An all-sidelight approach, in which south sun enters through clerestory and strikes the tub and floor. The lavatory is daylit from either side. Air past the toilet is drawn directly up and out of the clerestory, helped by the stack effect.* (b) *In this two-skylight approach, most light is given the lavatory and the toilet. An operable skylight over the toilet adds ventilation where it is needed. Without night insulation, these skylights could be the source of annoying droplets of condensation in winter. Horizontal skylights pose a problem for summer solar gains, however.* (c) *This two-skylight variation gives primary light and ventilation to the tub rather than to the toilet. Powered ventilating fans can assist when natural "stack" ventilation is insufficient. (From a study by Michael Bush, University of Oregon.)*

house. They encourage the singer or the whistler, and may even serve as practice rooms for the family musician. But they can also intimidate a person who prefers to be quiet, such as a guest using the toilet while the rest of the dinner party sits quietly in the adjacent room.

Two common design responses are isolation and masking sound. Bathrooms can be separated from acoustically sensitive spaces by closets or hallways. With careful attention to sufficiently massive construction and/or other construction details, the doors, walls, ceilings, and floors of bathrooms can be constructed so as to reduce the passage of sound to acceptably low levels. (These considerations are discussed in Chapters 26 and 27). Sound isolation also depends on attention to detail: no cracks around the bathroom door, no back-to-back electrical outlets between bathrooms and adjacent spaces, no air grilles into ducts that have other grilles nearby, and no other open window near the bathroom window. If such goals are unattainable, or if additional acoustic security is desired, masking sound can be added as needed within the bathroom. All too often, the masking noise is provided by repeated flushings of the toilet (itself an embarrassing sound for some) or the running of water in the lavatory. But a noisy ventilating fan would serve the purpose, as would a music source, such as a radio. An elegant, expensive touch would be a recirculating fountain gracing one corner; it could provide splashing sounds only when the light or fan is on.

The final consideration in bathroom acoustics is the sound of water moving within pipes. Where such pipes are exposed or barely concealed within, or firmly attached to, thin walls, perceptible sounds are highly likely. Although many of us may have experienced this sudden acoustic signal that a bathroom is in use, few have experienced it quite like the owner of a house on Long Island who was cited by Reyner Banham (1969). Designed at the height of the modernist expression of mechanical services, this house featured a dining table cantilevered from the exposed waste stack serving the bathroom above! Where water noise is undesirable, pipes should be wrapped, resiliently mounted, and/or located in a less acoustically critical wall.

11.4 Installation and Maintenance

The typical bathroom is a collection of individual fixtures, with fittings and accessories, combined with some custom cabinetry or shelving. The designer faces a wide range of styles offered by various manufacturers; the final array of components can be assembled from many different sources and chosen to create the most special effect or to meet the lowest budget. The preceding chapters have discussed fixture selection based on criteria such as physiology and water and energy consumption. To these must be added esthetics, durability and ease of cleaning, compatibility between fitting and fixture, and cost.

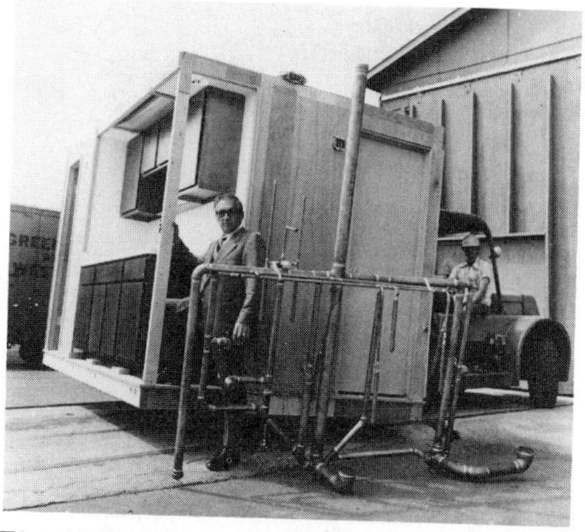

***Fig. 11.22** Preassembled plumbing "tree" for manufactured housing, along with the prefabricated kitchen/bath in which it is installed. (Courtesy of Wausau Homes, Inc., and the Copper Development Association.)*

There are prefabricated bathrooms, in which one manufacturer assembles the piping for preselected fixtures (Fig. 11.22). Going further, there are a few examples of entirely one-piece bathrooms, which incorporate the maintenance advantage of having no seams between fixtures, walls, and floors. This, of course, makes fixture replacement expensive and difficult.

For detail designers, however, the bathroom presents the demanding task of getting the right-sized pipes to appear in the right place. Figure 11.23 furnishes an example from a manufacturer's catalog, along with the information needed both for working drawings and for "roughing in"—the process of getting all pipes installed, capped, and pressure-tested before the actual fixtures are installed.

An example of roughing-in for an office building was shown in Fig. 10.43. The roughing-in of supply and waste piping for school lavatories is given in Fig. 11.24. Institutions such as schools have extensive requirements for durability and ease of maintenance. The fixtures are made of such resistant materials as stainless steel, chrome-plated cast brass, precast stone or terrazzo, or high-impact fiberglass. The fixture controls are designed to withstand heavy use—or misuse—and the fixtures are securely tied into the structure with concealed mounting

Fig. 11.23 Typical catalog description of a stainless steel sink. Included are fixture dimensions and roughing dimensions. (Courtesy of American Standard, Inc.)

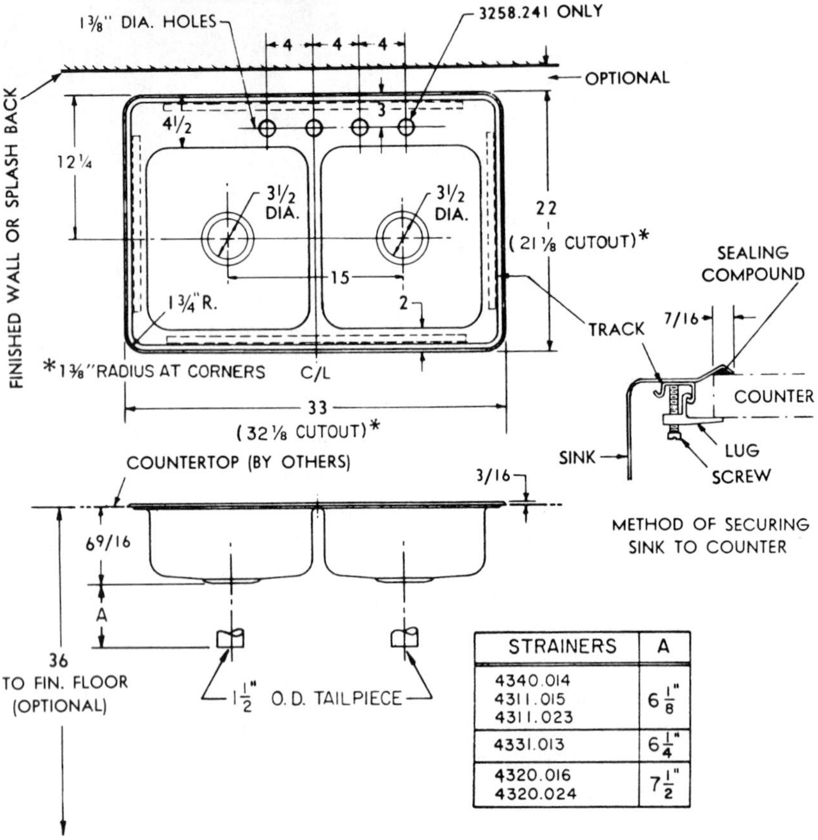

STRAINERS	A
4340.014 4311.015 4311.023	$6\frac{1}{8}$"
4331.013	$6\frac{1}{4}$"
4320.016 4320.024	$7\frac{1}{2}$"

hardware designed to resist extraordinary forces. (Some schools even move the lavatories into the hallway for better visual control of at least a part of the restroom facilities.)

In prisons, extreme measures are taken to prevent plumbing fixtures from becoming weapons. Heavy-gauge stainless steel fixtures with nonremovable fittings are provided, at very high cost for both the fixture and its tamper-proof installation.

For the more ordinary bathroom, the two maintenance questions of greatest concern are (1) how easy is cleaning around and within the fixture, and (2) how accessible are those parts of the fixture most likely to need repair or replacement? Ease of cleaning is often determined more by the space around the fixture than by the design of the fixture itself. This is particularly true of toilets, where a generous space of open floor to either side makes maintenance easy and therefore likely to be more frequent. Access to fixture parts may be more difficult. An access panel in the wall of the room behind the fixtures is often provided, encouraging speedy repair and replacement for fixtures at tubs, showers, and lavatories. Ideally, accessibility to all plumbing lines—whether via access panels in walls, trenches in concrete floors, exposed base-

Fig. 11.24 Roughing in place for four lavatories in the washroom of a school. At this stage, the waste branches have been capped and the system tested against possible leakage. The waste branches are made of cast iron, the vents of galvanized steel, and the water lines of copper with soldered fittings. Roughing dimensions have been followed. Vertical capped expansion and shock tubes serve as extensions of the entering water pipes that serve the hot and cold water branches.

ment ceilings, or adequately deep crawl spaces—should be provided. The bathroom is likely to undergo thorough "remodeling," including fixture replacement, as styles change and water/energy conservation becomes more important.

REFERENCES

ANSI (1986). *American National Standard for Buildings and Facilities A117.1-1986 Providing Accessibility and Usability for Physically Handicapped People,* American National Standards Institute, New York.

Banham, R. (1969). *The Architecture of the Well Tempered Environment,* Architectural Press, London.

Kira, A. (1966, 1976). *The Bathroom: New and Expanded Edition,* Bantam Books, New York.

Milne, M. (1976). *Residential Water Conservation,* U.S. Department of Commerce, NTIS.

12
SOLID WASTE

For some designers, the last building distribution system considered is the one involving the bulkiest items: the flow of supplies in and solid waste out. Since this system usually is not seen as consuming building energy or requiring specialized equipment, it ordinarily becomes a lower-priority system in the design process. Yet provisions for delivery of supplies, and especially for the collection and storage of solid wastes, can be more space-consuming than water/waste systems, can present a fire danger, and can create severe local environmental problems. The separation of solid waste for resource recovery involves significant energy and environmental consequences. Finally, mechanical equipment associated with solid waste is now more commonly installed.

12.1 Waste and Resources

In the past 40 years or so, there has been a marked increase in the amount of packaging material used for consumer products. Where shoppers once refilled reusable containers for bulk supplies at the market, for example, they now buy food in bags or cans that are discarded after use. This trend has increased spatial needs in the store, where shelves of cans are required instead of a bin of bulk products, and in the home, where such packaging soon turns into waste and must be stored until garbage collection day. Energy is required to make the boxes, bags, cans, and other containers; to transport them; and to collect them as trash. Devices such as trash compactors add both space and energy requirements to the process. Landfills for garbage disposal fill more rapidly as solid-waste flows increase.

There may not be much that building designers can do about these increased packaging trends. But the solid wastes from buildings do contain important resources, and a designer can help society to recover those resources, rather than to bury them in landfill or in the ocean. Further, the buildings themselves can be designed for materials recovery upon remodel or demolition, as was discussed in Section 1.7.

The resource within solid waste can be divided into the high-grade resources represented by recyclable materials and the low-grade resource of heat obtainable from the burning (incineration) of combustible solid wastes.

(a) High-Grade Resources. These include metals such as aluminum and steel, paper and paperboard, and some plastics. Glass is especially suitable for recycling when it can be reused after simple washing, as in the case of beer and soft drink bottles. "Returnable" bottles and cans have measurably decreased roadside litter in states that now require such containers. Newspaper is so readily stored and recycled that in many communities, charitable groups or service organizations recover newspapers for profit. Recycled paperboard can easily save 50% of the energy that would be required for pulp from virgin material.

The recovery of aluminum saves 96% of the energy necessary to produce it originally. Since the aluminum production process is dependent upon electricity, the recycling of aluminum is even more attractive. Energy conservation in aluminum production has been developed to the point that only 82% as much energy was needed to make a pound of aluminum in 1982 as in 1972. More than a third of that reduction was attributable to aluminum recycling programs; the remainder came from improvements in the production process. And as for steel, recycling can produce a 52% energy savings, compared to the use of virgin material.

Plastics are more difficult to recycle. Currently, USFDA regulations prevent the use of recycled plastics in food-related items. However, the plastics from items such as soft drink containers, margarine tubs and lids, and milk jugs can be reprocessed into plastic pellets, which are cheaper than virgin plastic pellets. These pellets are made into nonfood items—toys, building products, sports products, and other things.

(b) Low-Grade Resources. These resources include materials for which recycling is impractical, but which are combustible. Although to recover heat from such materials obviously seems better than to waste them altogether, the resulting additional problems of air pollution (and ash disposal) must be minimized. Low-grade resources include gaseous wastes, liquid and semiliquid wastes, and solid wastes. Industrial and commercial processes can generate wastes of all types, including some with very high heat content and some that are very toxic. In buildings, however, solid wastes are the ones most frequently subject to burning.

Some typical waste products are classified in Table 12.1; the higher the classification number, the more difficult the task of incineration. The

TABLE 12.1 **Types, Composition, Heating Values of Wastes**

Type	*Typical Mixture*	*Incombustible Solids (%)*	*Moisture (%)*	*Heating Value (Btu/lb)*
0	Trash, a mixture of highly combustible waste such as paper, cardboard, cartons, wood boxes, and combustible floor sweepings, from commercial and industrial activities. The mixtures contain up to 10% by weight of plastic bags, coated paper, laminated paper, treated corrugated cardboard, oily rags, and plastic or rubber scraps.	5	10	8500
1	Rubbish, a mixture of combustible waste such as paper, cardboard cartons, wood scrap, foliage, and combustible floor sweepings, from domestic, commercial, and industrial activities. The mixture contains up to 20% by weight of restaurant or cafeteria waste, but contains little or no treated papers, plastic, or rubber wastes.	10	25	6500
2	Refuse, consisting of an approximately even mixture of rubbish and garbage by weight. (Common to apartment and residential occupancy)	7	50	4300
3	Garbage, consisting of animal and vegetable wastes from restaurants, cafeterias, hotels, hospitals, markets, and like installations.	5	70	2500
4	Human and animal remains, consisting of carcasses, organs, and solid organic wastes from hospitals, laboratories, abattoirs, animal pounds, and similar sources.	5	85	1000
5	By-product waste—gaseous, liquid, or semi-liquid—such as tar, paints, solvents, sludge, fumes, etc., from industrial operations.	—	—	[a]
6	Solid by-product waste, such as rubber, plastics, wood waste, etc., from industrial operations.	—	—	[a]

Source: Incinerator Institute of America.

[a]Btu values must be determined by the individual materials incinerated.

heating values of these materials may be compared with those of the typical fuels listed in Table 5.12; this comparison shows that the most highly combustible grade of common solid waste has, per pound, better than one-half the heating value of anthracite coal. Air pollution regulations have severely restricted simple trash burning as a means of solid waste disposal; incinerators now must meet increasingly strict regulations.

Newer incinerators not only control the burning process to reduce pollutants, but also use the heat of combustion. The incinerator and waste heat recovery system shown in Fig. 12.1 has a boiler that can be fired either by a building's solid waste, or by a conventional fuel source when there is no waste to burn. The boiler gases can produce either hot water, or steam for an absorption chiller (see Fig. 7.19).

Where large quantities of mixed trash, garbage, and other refuse are collected, special resource-recovery plants can be built to recover materials, produce useful steam for electricity generation, and reduce the flow of waste to landfill (Fig. 12.2). In this energy-intensive process, the mixed garbage is shredded and blown through large "air classifiers" that separate the organic (burnable) wastes from metals and glass. The burnable wastes are then used, under controlled combustion, to generate electricity. The metals are further separated magnetically into ferrous and nonferrous classes; these and the glass are then recycled.

Ultimately, nonrecyclable materials are placed in a landfill. Over many years, anaerobic combustion in enclosed landfills generates methane gas, itself a usable resource. Figure 12.3 shows an architectural opportunity for such landfill sites: a greenhouse heated by landfill (methane) gas. Electricity generated by landfill gas supplies the lights for the greenhouse. As a rule of thumb, a site with 1 million tons of refuse in place should produce enough low-heat-value "landfill gas" to support a cogeneration system that includes a greenhouse.

12.2 Resource Recovery: Central or Local?

In general, the more thoroughly mixed are the different types of solid waste, the harder it is to recover their high- and low-grade resources. From an energy conservation viewpoint, solid wastes should be kept as separate as possible; glass bottles should be washed and reused rather than broken and recycled, and unrecyclable but burnable solid wastes should be kept clean and dry until they can be burned. The earlier that different types of metals are separated, for example, the less energy spent later on to separate aluminum from steel, and so forth. Organic food wastes could be composted for use on site, rather than ground up and added to the load on the sewage treatment system.

However, there are two disadvantages to "local" solid waste separation (at the point of their discard)—one cultural, one physical. Keeping solid wastes separate requires somewhat more effort and time on the part of the consumer. Rather than dumping everything in a garbage can, the resource-conscious consumer will wash metal cans, remove both ends and flatten them, then deposit them in a container reserved for ferrous metal. Newspaper and box cardboard are stacked separately; returnable bottles and cans are kept apart from glass; compostable kitchen wastes are kept separate from garbage. This disadvantage of slightly more work and time for waste separation is compounded by the physical disadvantage of the floor or cabinet space taken up by so many separate containers. In communities fortunate enough to have recycling-oriented garbage service, these containers can be carried out and lined up to await garbage collection. The garbage trucks used for this purpose are complex assemblages of various bins, rather than simple massive caverns. In many communities, however, each pile of recyclable materials at home must be taken to a different point—a disadvantage in terms of both the time and energy used in transportation.

The characteristics of local waste separation, then, are increased consumer time and building space requirements. Central waste separation is characterized by energy-intensive (and noisy) industrial processes. The next two sections offer a detailed look at the consequences of local waste separation on some common building types.

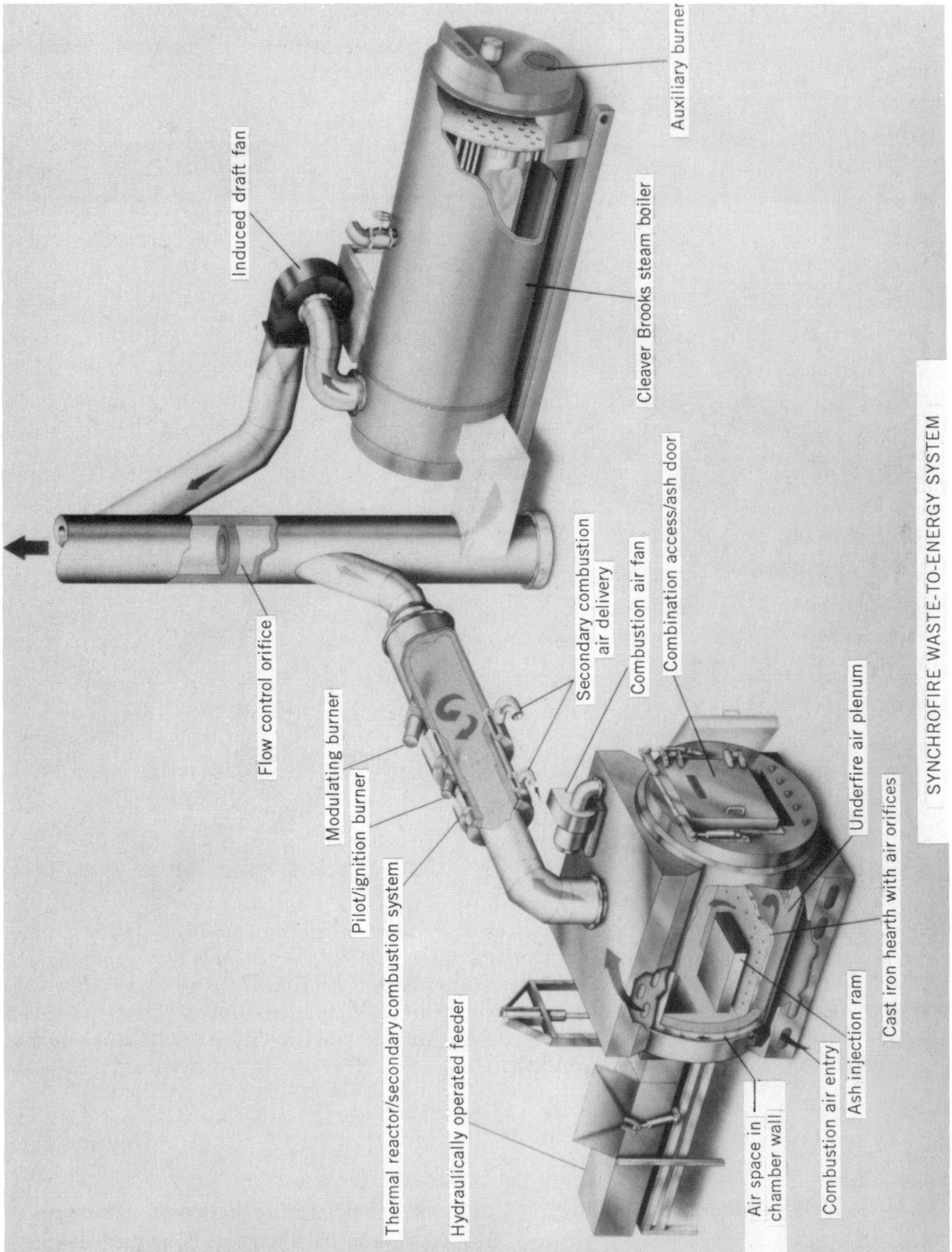

Fig. 12.1 *A pyrolytic incinerator controls the amount of combustion air to its firebed, producing temperatures of about 1200 F. At such temperatures in an oxygen-lean atmosphere, the waste thermally degrades, giving off combustion gases. These gases flow to a thermal reactor, where they are ignited at even higher (1800 to 2000 F) temperatures. The resulting flue gas is mostly carbon dioxide and water vapor, and thus meets most pollution standards. The incinerator gases are coupled to a heat recovery boiler which can be fired by conventional fuels when there are insufficient incinerator gases. (Courtesy of Cleaver-Brooks, Milwaukee, Wisconsin.)*

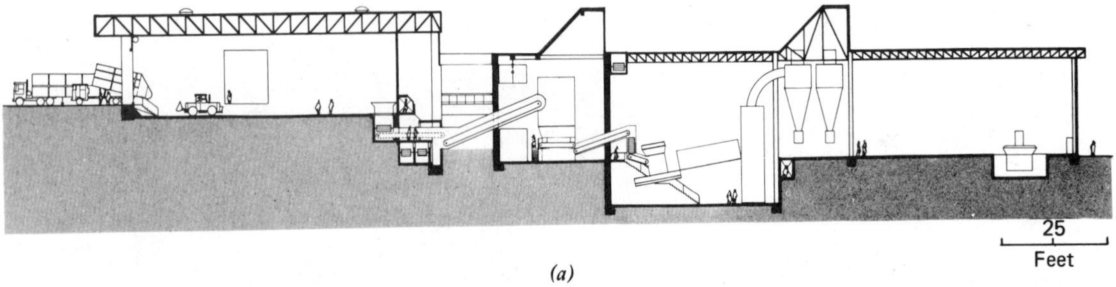

(a)

(b)

Fig. 12.2 Solid-waste resource recovery project. The flow of initially unseparated garbage through a proposed facility is shown in (a). *The rendering* (b) *shows the recovery plant in the foreground, the electric generating station beyond. Cone-shaped storage bins hold shredded refuse awaiting conversion to steam for the turbines. Clean Communities Corporation, Haverhill, Mass.; Camp Dresser & McKee, Inc., Engineers. (Reprinted from* Architectural Record, *June 1975. Copyright © 1975 by McGraw-Hill, Inc. All rights reserved.)*

12.3 Solid Waste in Small Buildings

The choice between separation of waste or the mixing of garbage in one can is most often enjoyed by the occupants of small buildings. Where food preparation is involved, as in residences or restaurants, the collection of special containers can reach surprising complexity (Fig. 12.4).

As the point of origin for so many types of solid waste, the kitchen is often the location of waste separation and storage. There is an inherent conflict, however, between the kitchen's frequently hot and humid environment and the need for a cool, dry, well-aired place for solid wastes. This suggests that a space—pantry, airlock entry, cabinet, closet—be provided that opens to the kitchen on one side and to the out-

Fig. 12.3 A 45,000-ft² greenhouse built over part of a 70-acre Michigan landfill utilizes "landfill gas" (including about a 65% methane component) both to power its space heaters and to generate electricity for lighting. Waste heat from the generator also contributes to space heating. (Photo by George P. Graff. Courtesy of Willow Run Farms, Inc., Ypsilanti, Michigan.)

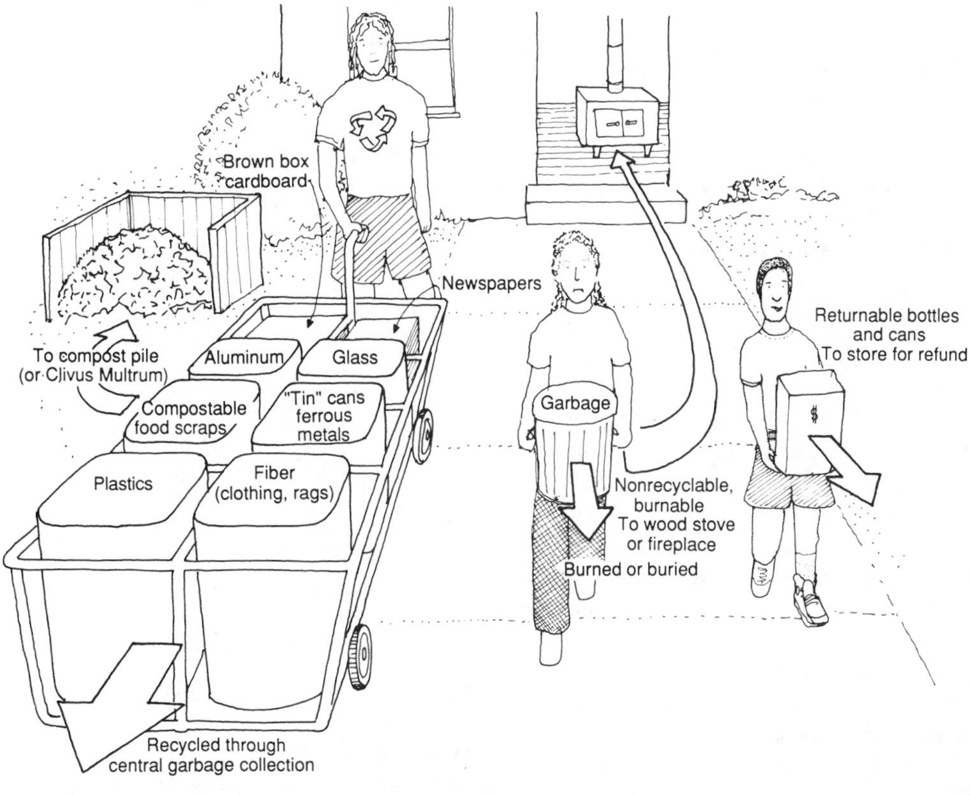

Fig. 12.4 Local separation of solid wastes can require many containers. In communities that offer recycling garbage collection service, the consumer can recycle without transporting individual containers to many different collection points. Some recycling services ask for glass to be separated into green, brown, and clear; some offer to recycle white paper.

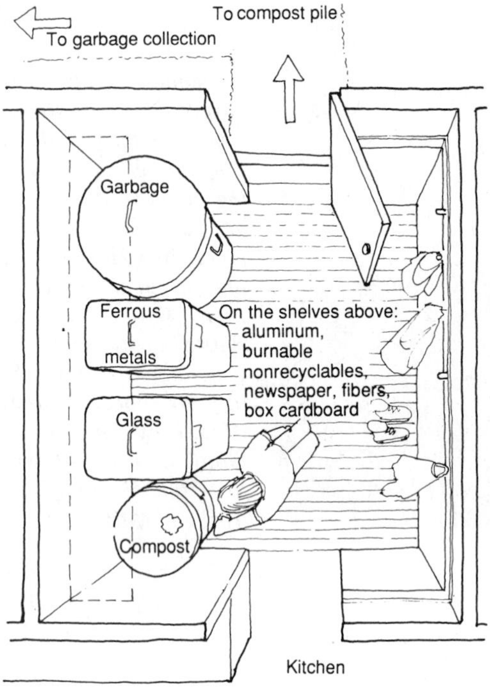

Fig. 12.5 *An entry vestibule (or airlock) to the kitchen can serve as a depository for solid wastes, as well as for coats, boots, and other items. Heavy, dirtier items belong on the floor, and many recyclables can be readily and conveniently stored on shelves. An outdoor hose bibb is useful for washing out soiled compost containers and garbage cans.*

side on the other (Fig. 12.5). Both the daily deposits to solid waste and the weekly waste removal are made easy, and the near-outdoor conditions are better for the waste in most U.S. climates. It is also important that cleaning of the waste storage area be easy.

Two mechanical devices increasingly found in homes and apartments are garbage compactors and garbage disposers, which usually are installed directly beneath the drain of the kitchen sink.

The *garbage compactor* allows a much less bulky storage arrangement. Used selectively, it can compact several of the stacks of items shown in Fig. 12.4, such as nonrecyclable burnable items, aluminum for recycling, ferrous metals, and box cardboard. Used indiscriminately, it can make the central process of garbage separation more difficult by crushing dissimilar items together. For a recycling-conscious household, it is questionable whether the compactor will save much more storage volume than it takes for itself.

The *garbage disposer* (Fig. 12.6) grinds up organic food scraps and sends them on through the sewer. This device is a boon to central garbage collection, since it lightens the weight of the garbage can and adds less moisture to the garbage (which thus can be burned more efficiently). Finely chopped organic matter has a better chance to biodegrade at the wastewater treatment plant than in a tightly packed landfill. However, garbage disposer units require both water and energy. Water must be kept running during the grinding process, to coagulate grease for chopping, to wash the blades and keep them free of debris, and to cool the grinder's motor. In all, from 2 to 4 gal (7.5 to 15 L) are required for 1 minute of operation. Because more water and solid waste is deposited in the sewer system, moreover, the central sewage treatment plant requires more energy to operate.

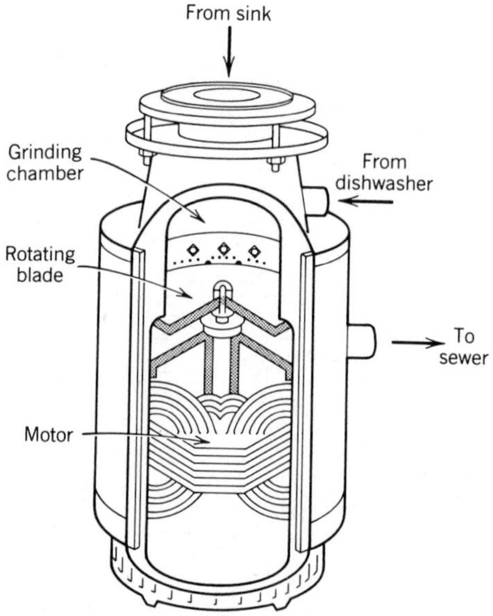

Fig. 12.6 *The garbage disposer (or grinder) diverts solid waste (food scraps) from landfills to sewage treatment plants. Occupying less than a cubic foot, it saves garbage storage space and some of the user's time, but it requires more energy and much more water than treatment of the food scraps as garbage. Composting is another alternative. [From Milne (1976).]*

The alternative is the *compost pile,* familiar to most home gardeners as a source of excellent soil conditioners. Urban opportunities for composting seem very limited; yet raised growing beds on a balcony or window flower boxes are logical recipients of a family's compost. The problem may not be so much what to do with the final product (humus), as where to locate the compost pile.

The outdoor compost pile has several characteristics that challenge the designer. At its best, it is a frequently turned, quite warm, damp, well-aired source of rich humus (and red worms) for gardens; odors are noticeable only while the pile is turned. At its worst, it is a source of unpleasant odors and a breeding place for vermin. Where odors are not objectionable, the heat generated in a frequently fed and tended compost pile might be welcome against the exterior walls of residences. Clearly, these walls must have nonorganic exterior materials!

As groups of residences are combined into large apartment complexes, the solid waste systems can make more significant demands on the designer. Where central storage compounds are provided, garbage cans should be fenced to ward off dogs and other marauders. Different bins for recyclable materials might be provided. Where space is limited, a single bin could accept specific materials on specified days—newspapers Mondays, metals Wednesdays—if collection schedules permit. A central compost pile could provide humus for the landscaping of the complex. An incinerator is frequently installed to recover heat from nonrecyclable burnable wastes. This combination of space and equipment has special environmental needs, including garbage truck access, noise control, and location of both incinerator stacks and compost pile with respect to prevailing winds.

12.4 Solid Waste in Large Buildings

As buildings become larger, it becomes more likely that solid wastes will be handled several times in the storage and collection process. This may inhibit the separated-waste approach, since not only the employees who generate it, but also the custodians who collect and store the waste, must understand what items go into which bins. However, larger buildings tend to generate large and concentrated kinds of wastes, which makes recycling more attractive.

Consider the multistory office building. The office floor operations are likely to waste large quantities of white paper and smaller quantities of newspaper, box cardboard, and unrecyclable burnable trash (including floor sweepings). Much smaller quantities of food scraps (coffee grounds), metals, and glass are also generated. Given the high cost of rental space, the pressures are very high for a simple mixed-garbage can (rather than multiple separate bins). However, the cost of hauling trash to increasingly scarce landfills is now so high that recycling programs pay for themselves more by avoided landfill fees than by the value of recycled materials. A building that utilizes an incinerator for heat recovery might provide for three waste categories: recyclables, burnable trash, and garbage (such as food scraps).

(a) The Collection Process. In larger buildings, the collection of solid waste typically involves a three-stage process (Fig. 12.7). The first stage is the generation of the waste itself, by employees who might be provided with one wastebasket apiece. If each employee is expected to separate wastes, this suggests a redesigned receptacle. At the typical desk in an office building, white paper, recyclables, burnable trash, and garbage would be deposited in separate compartments. Waste separation at the point of origin thus may not require much more floor area.

The second stage begins as custodians disconnect these individual baskets, dump them into separate bins on a collection cart, and reconnect the individual empty baskets for the next day's deposits. At various special-purpose stations, special wastebaskets can be supplied: white paper at the computer room and the copying machine, garbage at the employee lounge, burnable trash at the shipping/receiving station. Floor sweepings are added to burnable trash (or garbage, as is appropriate). When the cart is full, it is wheeled to the service closet, which is probably located within the core of the building. Here would be a container for each category of

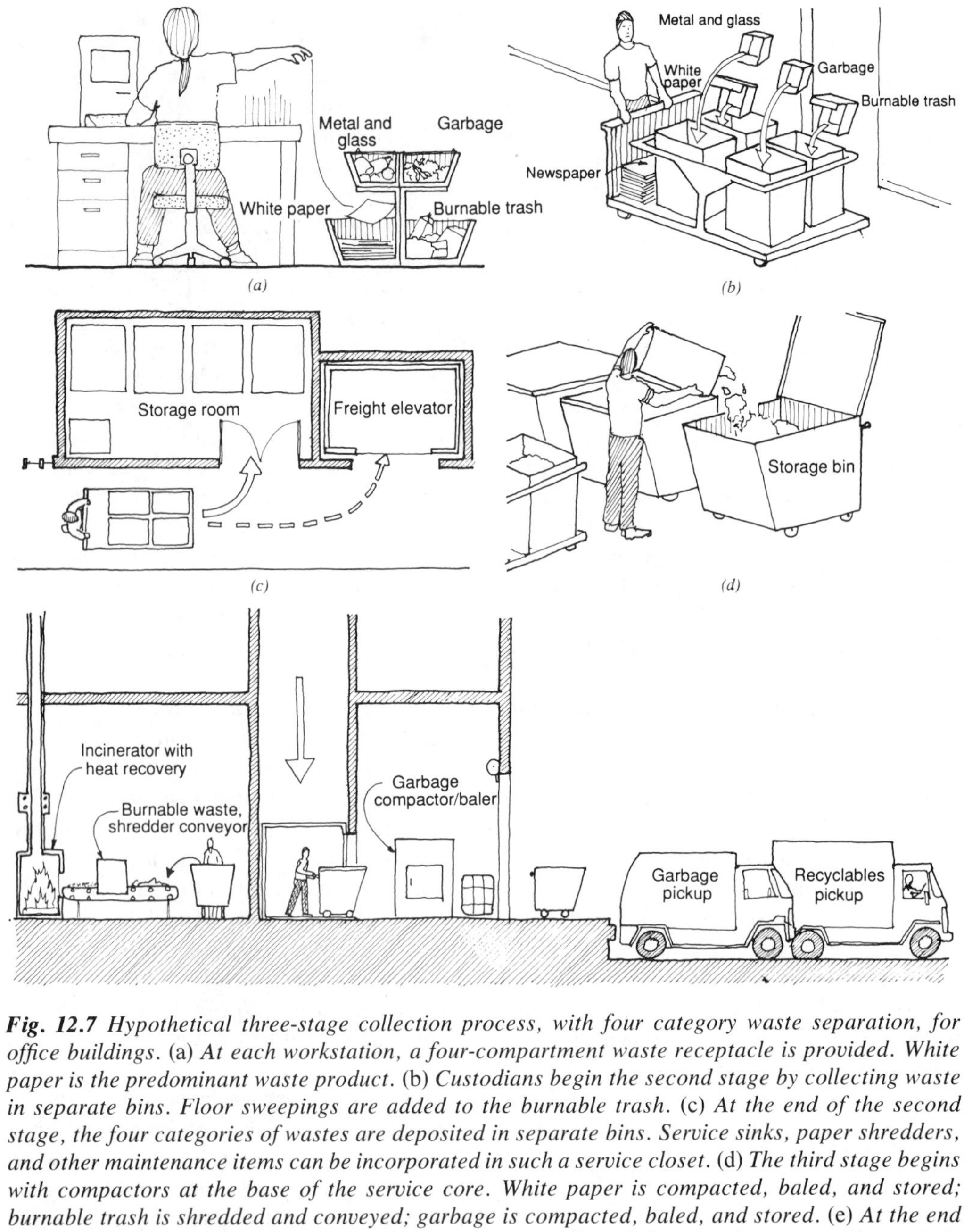

Fig. 12.7 *Hypothetical three-stage collection process, with four category waste separation, for office buildings.* (a) *At each workstation, a four-compartment waste receptacle is provided. White paper is the predominant waste product.* (b) *Custodians begin the second stage by collecting waste in separate bins. Floor sweepings are added to the burnable trash.* (c) *At the end of the second stage, the four categories of wastes are deposited in separate bins. Service sinks, paper shredders, and other maintenance items can be incorporated in such a service closet.* (d) *The third stage begins with compactors at the base of the service core. White paper is compacted, baled, and stored; burnable trash is shredded and conveyed; garbage is compacted, baled, and stored.* (e) *At the end of the third stage, white paper, metal, and glass is collected for recycling, burnable trash is incinerated and its heat recovered, and garbage is collected for separation at a central plant.*

waste, along with a service sink to wash out the garbage bin (and perhaps a paper shredder to be used selectively by employees).

The third stage begins at the ground floor of the service elevator where white paper and other recyclables are stored until collection, burnable trash shredded and conveyed toward the incinerator, and garbage compacted and bagged. In the storage space, cool, dry, and fresh air is desirable. A sprinkler fire protection system is advisable. The compactors and shredders are noisy, and must be vibration-isolated from the floor. At the end of the third stage, a truck or van from the recycling center collects recyclables, a garbage truck collects garbage bags, and the heat-recovery incinerator turns burnable trash into much heat and a little ash.

Where incinerators are impractical because of first-cost/space constraints or air quality considerations, the burnable trash can be combined with garbage throughout all three stages. In cities with resource recovery stations (see Fig. 12.2), the end result is similar, although more energy is consumed in the process.

The world headquarters of the Coca-Cola Company in Atlanta has an unusually thorough recycling program. Each office worker is given a legal-size plastic (from recycled plastics) file folder with printed instructions about what kinds of recycled papers this should contain. Periodically, as this fills, each employee then dumps high-grade office paper into containers, located at copy and computer centers on each floor; 40 to 60 employees use each such container. These containers are emptied nightly into wheeled 90-gallon (340-L) recycling carts, which are taken to a pickup area to await the waste paper buyer. About 120 tons (about 109,000 kg) of paper per year are thus recovered.

The company also has a "Recycling Works" recycling center, built of bright red recycled plastic "lumber," located in the employee garage. This roughly 500-ft^2 (46-m^2) area receives deposits brought from employees' homes; annually, about 14 tons of aluminum cans, 11 tons of glass, 6 tons of plastic bottles, and 26 tons of newspapers are recovered. The profits are donated to charities, and Coca-Cola saves about $33,000 annually on avoided costs of paper disposal.

(b) The Service Core. Medium- and high-rise buildings usually concentrate many building services within a "core," from which services can be distributed as needed throughout each floor. The core typically contains stairs; passenger and service elevators; toilet rooms; service closets; mechanical, plumbing, and electrical chases; electrical/telephone closets with local switching capabilities; fire protection equipment; and supply closets. The size of such cores varies widely; for example, the taller the building, the more elevators.

Often, these service cores are identical in plan from one floor to the next. Alternatively, the vertical services can be identical (stairs, elevators, chases) and the arrangement of other elements (toilet rooms, supply closets) allowed to vary. Whether or not minor floor-plan variations occur, the cores usually depart radically from the typical plan—and ceiling height—at both roof and ground floor (Fig. 12.8). (Middle interruptions also occur where intermediate mechanical floors are provided.) Depending upon the type of elevator machinery selected, the machine room ceiling height above the top floor served might be three times the ordinary floor height (see Chapter 23). On the delivery/loading floor, large trucks must be accommodated. For chillers, boilers, incinerators, and fans, higher ceilings may be required to accommodate the flues, ducts, and pipes to which they must be connected; see Section 7.3 for the dimensions required. Figure 12.9 offers an example of the way in which a building's service core can expand to nearly fill the ground floor.

A service core can be related to the remaining service floor area in any of several ways (Fig. 12.10). A single central core is a familiar arrangement (Occidental Chemical Offices, Fig.

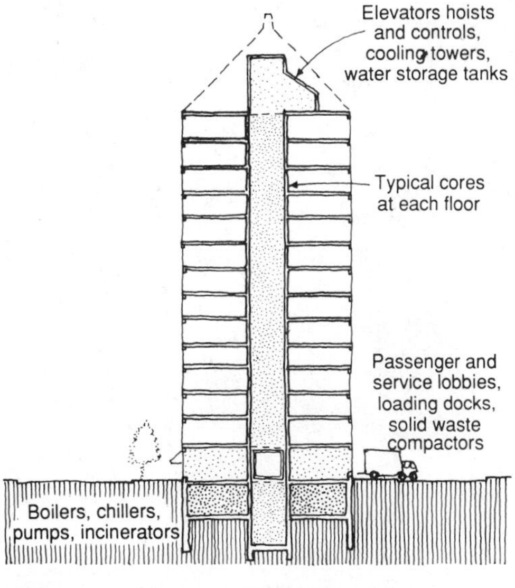

Fig. 12.8 Schematic section through a core of a multistory building. At the top and bottom, floor space and ceiling height requirements increase. The consequences on the ground (or delivery/collection) floor are particularly great.

WATER AND WASTE

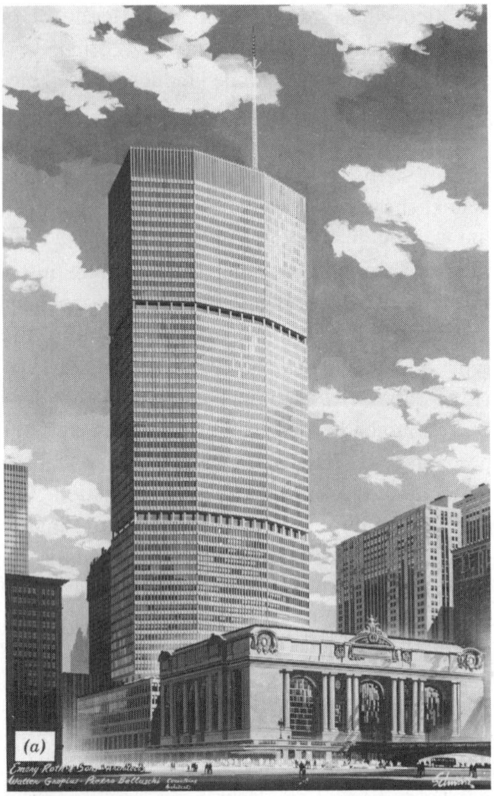

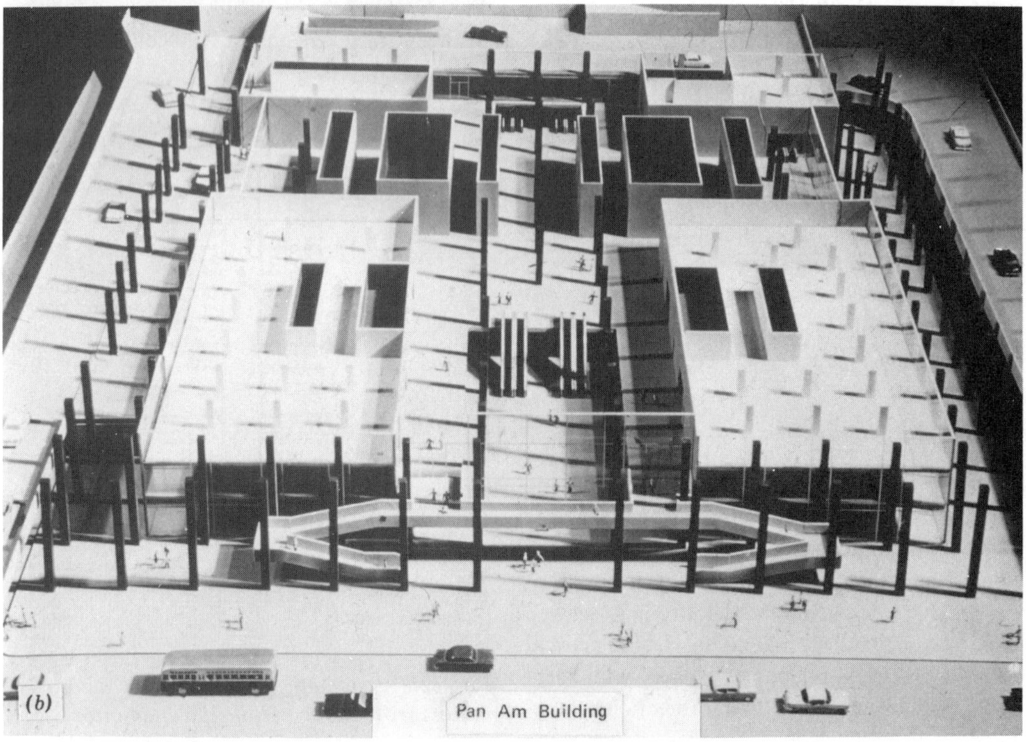

Fig. 12.9 *Pan Am Building, New York City. Emery Roth and Sons, architects; Walter Gropius, Pietro Belluschi, associated architects. The distribution principles of the building in Fig. 9.36 can result in the "mechanical stories" as used in the Pan Am Building. Stories of this type may also contain air-conditioning compressors, water chillers, and blowers.* (b) *Partial model, first floor. Typical of planning in many high-rise buildings, this is one of the early models used in developing space concepts for the Pan Am Building. The first floor and typical upper floor set the pattern. Space allocations must be made for sanitary and storm drainage, elevators, ducts, electrical and telephone risers, water risers, and fire stairs. Equipment planners must aid in establishing rooftop areas, mechanical stories, and the general building horizontal cross section. (Courtesy of Emery Roth & Sons, Architects.)*

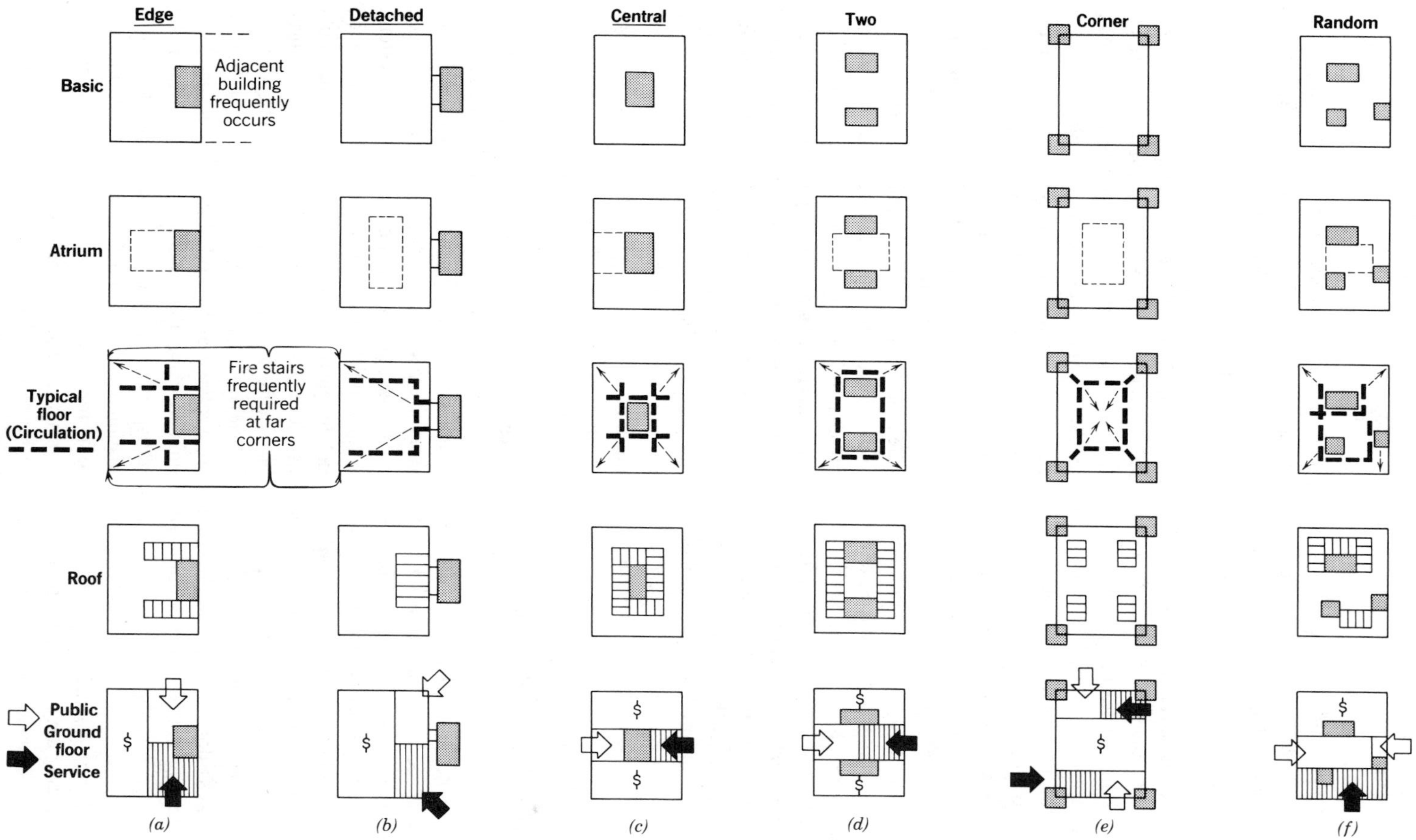

***Fig. 12.10** Comparison of core arrangements and building plans.* (a, b) *The edge and the detached cores give great flexibility of rental floor area, with light and view for core spaces.* (c) *The central core expands readily at the roof and the ground floor and has clear circulation and fairly flexible rental space.* (d) *Two-core is a popular, workable arrangement.* (e) *The corner cores give great flexibility to rental floors but are difficult at roof and ground floor.* (f) *Random cores generally occur in low-rise buildings, in which the benefits of repetitious plans are minimal. See also Table 12.2.*

6.11). Frequent variations include a core at the edge (Seeley Mudd Library, Fig. 7.65; Comstock Center, Fig. 7.66) or one that is detached, two cores symmetrically placed (Time-Life Building, Fig. 7.57), corner cores, or core services dispersed somewhat randomly (Bateson Building, Fig. 5.20; Iowa Public Service Building, Fig. 7.59).

Each core location has advantages and drawbacks. From the viewpoint of rentable space, important factors are the flexibility of the served (rental) floor area; the exposure of rental area to the perimeter, and thus to light, air, and view; and the extent to which ground floor high-rent space gets commercial exposure. For fire safety and occupant convenience, the distance of travel from the furthest rental floor area to the service core and the clarity of circulation within the rental area, are important. For the environment within the core—toilet rooms, stairs, elevator waiting areas—access to light, air, and view are desirable. For the mechanical services, ease of core area expansion on the roof, at the loading dock, and in the basement are important, as is the length of travel for ducts and other such elements within each floor. These factors are compared for various types of core designs in Table 12.2.

12.5 Equipment for the Handling of Solid Waste

Where large amounts of solid waste are generated, and little space is available in which to store it, various types of equipment can be used to change the volume or the composition of the waste, so as to facilitate its transportation and storage. Incinerators were discussed in Section 12.1; other approaches are considered here. The first step is to determine the probable daily flow of solid waste, as is done in Table 12.3. When the extent of the problem is known, equipment can be selected.

(a) Compactors. (Fig. 12.11). Of the wide variety of compacting devices, most are able to reduce the volume of solid waste to as little as 10% of the original volume. Among the many choices to be made are vertical versus horizontal compaction, automatic chute fed versus manual freestanding, whether wastes are to be bagged or baled, and the final size of each unit of compacted waste. Compactors can be noisy; they can also be prone to fires as heat is generated in the compaction process. Many compactors have built-in sprays for both fire control and disinfecting. Access to wash water and a floor drain are highly desirable.

(b) Pulping Systems. (Fig. 12.12). Rather than extend space-consuming chutes up through many floors, these systems grind waste into pulp in the presence of water, and thus make a readily transportable slurry. At the loading docks, this slurry enters a water press, where about 90% of the water is squeezed out, reducing the volume to about one-fifth that of the original wastes. This water is reused and replenished as required. Such pulping systems are used not only for general refuse, as illustrated, but also for the destruction of documents and for food service wastes. Pulping systems have limitations; they should not be used to handle metal or plastics, which are recoverable. Their use for paper wastes is questionable in light of paper recycling opportunities. Pulping systems are replacing incinerators in urban areas; their advantages in reduced air pollution must be weighed against the possible resource recovery, or energy recovery from incineration.

(c) Vacuum Systems. The primary advantage of the pulping system is that it reduces the building volume consumed by waste chutes or storage cans on each floor. A similar approach might be a grinder-plus-evacuated-tube system, similar to the vacuum sewage systems described in Section 10.1 (Fig. 10.6). Such a system would use less water, unless the recirculating slurry lines were omitted. Vacuum conveying systems are now used in several industrial applications, such as ash and wood-chip transporting. Vacuum systems that use air, not water, as the transport medium are frequently used for linens in hotels and for trash. Separate vacuum systems for trash and linens are desirable, both to lessen the soiling of the linens and to allow them to be delivered to a destination different from that for the trash.

The advantages of pressurized systems that use pulping or vacuum is that lines can be small and the contents can be moved horizontally or

TABLE 12.2 **Building Characteristics and Core Placement**

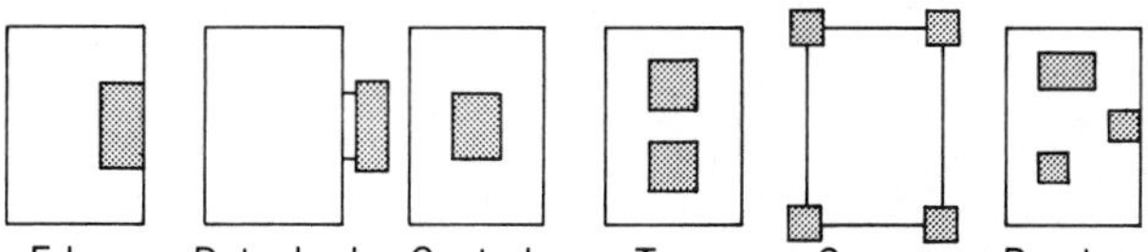

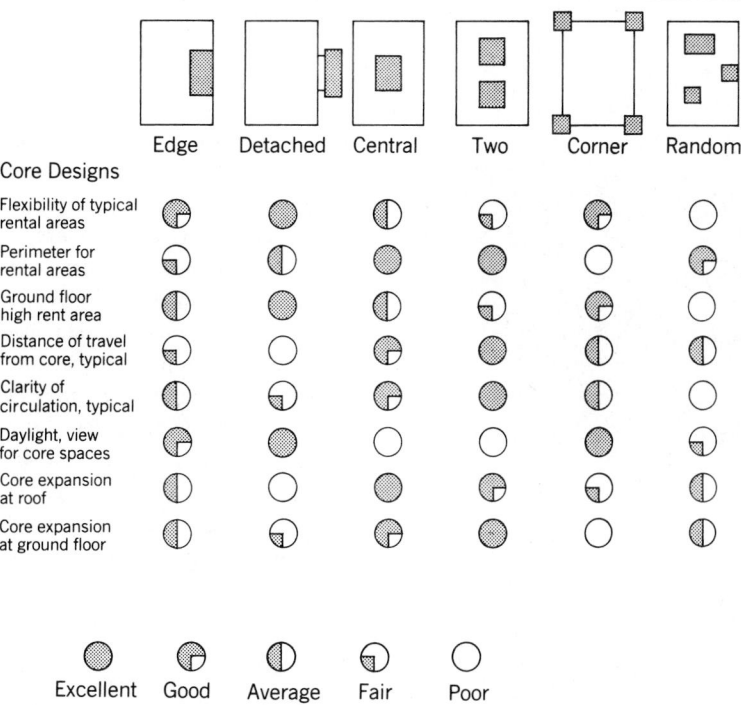

Core Designs	Edge	Detached	Central	Two	Corner	Random
Flexibility of typical rental areas	Good	Excellent	Average	Fair	Good	Poor
Perimeter for rental areas	Fair	Average	Excellent	Excellent	Poor	Good
Ground floor high rent area	Average	Excellent	Average	Fair	Good	Poor
Distance of travel from core, typical	Fair	Poor	Good	Excellent	Average	Average
Clarity of circulation, typical	Average	Fair	Good	Excellent	Average	Poor
Daylight, view for core spaces	Good	Excellent	Poor	Poor	Excellent	Fair
Core expansion at roof	Average	Poor	Excellent	Good	Fair	Average
Core expansion at ground floor	Average	Fair	Good	Excellent	Poor	Average

Excellent Good Average Fair Poor

TABLE 12.3 **Approximate Daily Flows of Solid Waste**

Classification	*Building Type*	*Amount of Waste*
Industrial buildings	Factories	Survey required
	Warehouses	1½ lb/100 ft^2-day (7.3 kg/m^2-day)
Commercial buildings	Office buildings	1 lb/100 ft^2-day (4.9 kg/m^2-day)
	Department stores	3 lb/100 ft^2-day (14.7 kg/m^2-day)
	Shopping centers	Survey required
	Supermarkets	7 lb/100 ft^2-day (34.2 kg/m^2-day)
	Restaurants	2 lb/meal (8.9 kg/meal)
	Drug stores	3 lb/100 ft^2-day (14.7 kg/m^2-day)
	Banks	Survey required
	Cafeterias	½–¾ lb/meal (0.2–0.3 kg/meal)
	Clubs	1½ lb/meal (0.7 kg/meal)
Residential buildings[a]	Private homes	3 lb (1.4 kg) basic + 1 lb/bedroom (0.5 kg/bedroom)
	Apartment buildings	3 lb/bedroom-day (1.4 kg/bedroom-day)
Schools	Grade, high, university	6 lb (2.7 kg)/room & ¼ lb/pupil-day (0.1 kg/pupil-day)
Institutions	Hospitals	20 lb (9.1 kg)/bed-day & 2 lb/meal (0.9 kg/meal)
	Nursing homes	4 lb/person-day (1.8 kg/person-day)
Hotels	First class	2 lb (0.9 kg)/room & 2 lb/meal-day (0.9 kg/meal-day)
	Medium class	1½ lb (0.7 kg)/room & 2 lb/meal-day (0.9 kg/meal-day)
	Motels	2 lb/room-day (0.9 kg/room-day)
	Trailer camps	6–10 lb/trailer-day (2.7–4.5 kg/trailer-day)

Source: Courtesy of International Dynetics Corporation.

[a]Residential wastes are typically 130 to 150 lb/yd^3, (77 to 89 kg/m^3) with a moisture content of 10 to 15%. On a per capita basis, it can average as high as 2.7 lb (1.2 kg) per person per day.

Fig. 12.11 *A refuse compactor suitable for apartment house or office building service. This will compact 54 yd³ (41.3 m³) of refuse per hour, packing a front- or rear-load commercial container. Overall size is 32 in. wide, 66 in. long (without container), and 67 in. high (813 by 1676 by 1702 mm); minimum room size is 6 ft wide by 13 ft long (1.8 by 4 m). (Courtesy of General Dynetics Corporation.)*

even upward. This allows far greater flexibility in design than is possible with gravity systems.

(d) Summary. There are many options to be considered within the present waste-handling process of grind, crush, burn, and bury. As earth's nonrenewable resources are stripped away, the recovery of materials from solid waste becomes increasingly attractive. The situation is similar to energy-and-design issues: after decades of energy-intensive building trends, designers are now giving space to daylighting atriums and allowing surfaces to act as thermal mass. A modest increase in floor space and equipment to encourage resource recovery from solid waste seems an increasingly worthwhile design investment.

REFERENCE

Milne, M. (1976). *Residential Water Consumption,* U.S. Department of Commerce, NTIS.

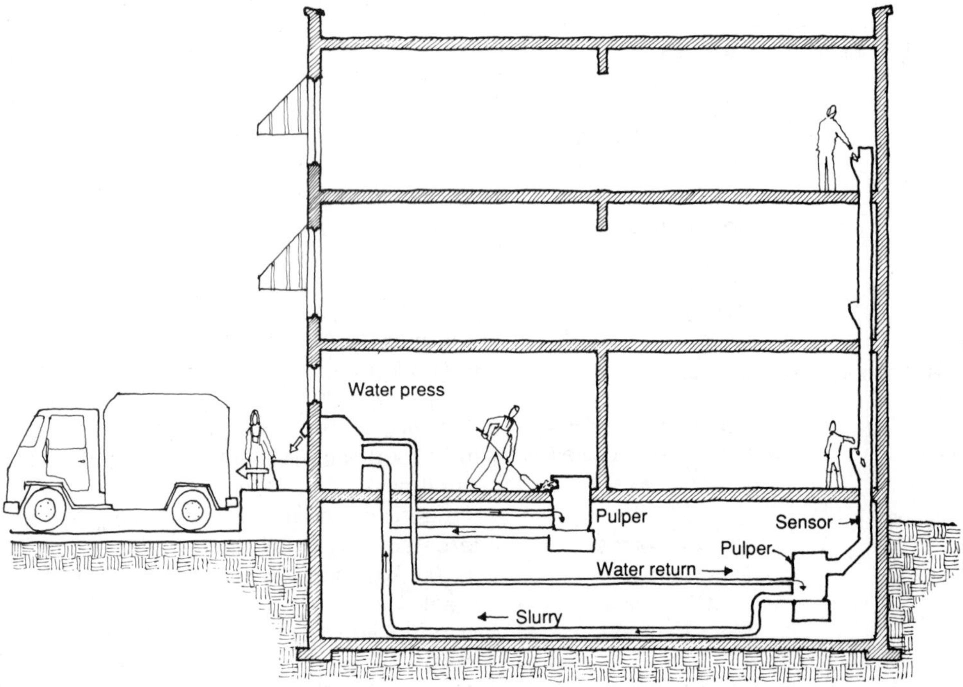

Fig. 12.12 *Pulping installation for apartment refuse. The pulper is located to deal directly with solid waste; the slurry then transports the waste to the press at the loading dock. The pulper can operate manually or automatically by means of a sensor as shown. One water press can serve two or more pulping units. (Courtesy of Wascon Systems, Inc.)*

PART IV
FIRE PROTECTION

One of the most challenging aspects of building design is that of resistance to fire. Designers find it easy to consider how their buildings can be made bearably cool on a stiflingly hot day or lighted and warm on a bitterly cold night. It is less pleasant to imagine a building burning, and to consider how first its occupants, then its contents—including the building itself—can be saved.

To complicate this process, there are inherent conflicts between some of the optimum features for fire resistance and some of those optimum features for passive strategies such as daylighting or passive cooling, or for forced-air systems. There are even potential design conflicts between systems for the safe evacuation of people and systems for the suppression of fire, since both people and fire thrive on oxygen.

This discussion of these topics in fire protection will begin with basic design considerations for fire resistance. Smoke management (for safe evacuation and for limited smoke damage) is next, followed by fire suppression systems such as sprinklers and non-water-based approaches. Lightning protection is then discussed, along with the many fire detection and alarm systems that are keyed to the four stages of the "typical" building fire.

Throughout Part IV, the influence of the National Fire Protection Association (NFPA) will be apparent. An especially useful source of information for diagnosis is the periodically updated *Fire Protection Handbook,* from which many illustrations have been reprinted in our text. Publications can be ordered from the NFPA, Batterymarch Park, Quincy, MA 02269.

13

FIRE PROTECTION

13.1 Design for Fire Resistance

Fire is a special kind of oxidation known as *combustion*. Oxidation, which has been discussed in the previous chapters in terms of rust within water supply equipment and aerobic digestion in waste disposal systems, is a process in which molecules of a fuel are combined with molecules of oxygen, producing a mixture of gases and energy. When this occurs rapidly, as in a fire, energy is released as heat and light, and some gases become visible as smoke.

Fire has a triangle of needs: fuel, high temperature, and oxygen. If deprived of any of these needs, building fires will be extinguished. In general, this triangle's influence on building design is as follows. The *fuel* is the building's structure and contents; the designer controls the choice of structural and finish materials, but rarely the final contents. The *temperatures* achieved in fires are well beyond the ability of building cooling systems to control, so special water systems (in the form of sprinklers) are often installed to deprive fire of the high temperatures it needs. *Oxygen* may be denied to a fire partly by limitations on ventilation, but these can have serious safety consequences. Another design response is to install fire suppression systems that either cover the fuel (foam, dry chemicals) or displace oxygen with another gas—for example, carbon dioxide or halogenated agents such as Halon 1301. The later agent inhibits the chemical action of the flame itself.

(a) Sources of Ignition. Buildings commonly contain three basic sources of ignition: chemical, electrical, and mechanical. Much more rarely, nuclear sources are present. In *chemical* combustion, most commonly known as "spontaneous combustion," some chemicals reach ignition at ordinary temperatures within buildings. Chemical combustion depends on the rate of heat generation (related to the degree of saturation of combustible products by the chemicals involved), the air supply (enough to supply oxygen but not enough to lower the temperature), and the insulation provided by the immediate surroundings (again—the more insulation, the easier the attainment of combustion temperatures).

Electrical heat energy is most commonly supplied by resistance heating, a familiar process in many appliances and in space-heating equipment. Less common are electric ignition by induction, dialectic process, arcing, and static electricity. An infrequent but enormously destructive electrical source is lightning.

Mechanical heat energy is produced by friction (including sparks), by overheating of machinery, and occasionally by the heat of compression.

(b) Products of Combustion. (Fig. 13.1). Everyone has experienced to some degree the dangers of the thermal products of combustion—flame and heat. These visible and tactile elements of fire can cause burns, dehydration, heat exhaustion, and fluid blockage of the respiratory tract, but they are responsible for only about one-quarter of the deaths resulting from building fires. The bulk of such deaths are caused by the nonthermal products—smoke (including gases). Smoke can usually be seen and smelled. Made up of droplets of flammable tars and small particles of carbon suspended in gases, it irritates the eyes and nasal passages, sometimes blinding and/or choking a person. Gases are especially dangerous because, without visible smoke, they are so often difficult to detect. Some gases are directly toxic, but all are dangerous because they displace oxygen. Many gases of combustion are discussed as air pollutants in Fig. 3.21. Common gases released in

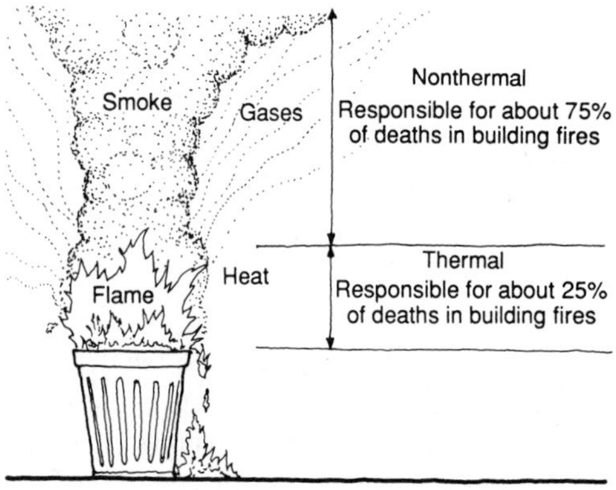

Fig. 13.1 Although fire's thermal products—flame and heat—make strong visible and tactile impressions, fire's nonthermal products—smoke and gases—pose the greater threat to life.

building fires include carbon monoxide and carbon dioxide. *Carbon monoxide* is not the most toxic product of combustion, but it is the most abundant. It is produced when insufficient oxygen is available to completely oxidize the burning material. Since *carbon dioxide* is produced in large quantities, it rapidly overstimulates breathing and thus causes the lungs to swell. Other dangerous and common building-fire gases include hydrogen sulfide, sulfur dioxide, ammonia, cyanide, and hydrogen chloride (from burning polyvinyl chloride).

In indoor fires oxygen commonly becomes insufficient, since the fire consumes it so rapidly. The normal concentration of oxygen in air is about 21%. At 15%, muscular skill is diminished; at 14% down to 10%, people remain conscious but judgment is faulty and fatigue is rapid; at 10% down to 6%, collapse occurs, but revival is possible when increased oxygen is supplied. The technique of starving the fire of oxygen can therefore pose a threat to humans, both by increasing the chances of carbon monoxide production and by depriving people of oxygen.

(c) Objectives in Fire Safety. Many older buildings were designed at a time when building fires usually resulted in the loss of several adjacent structures and, hence, the primary objective of fire fighting was to limit the conflagration to as few city blocks as possible. With the increased use of fire-resistant construction and code control of site and building planning, fires typically were confined to but one building at a time. As fire suppression systems came into common usage within buildings, it came to be expected that fires could be confined to one or two floors within a building. And now that automatic detection/suppression systems are technically advanced, fires can usually be confined to only one room, or to even smaller areas.

Three common objectives of building fire safety, in order of usual importance, are

- Protection of life.
- Protection of property.
- Continuity of operation.

Many of the elements of building fire safety are covered by building codes, but it is important to remember that codes specify the *minimum* acceptable protection. Designers can go much further than the codes require to enhance fire safety. It is also important to realize that codes *cannot* cover all aspects of fire safety, as there are too many variables to building design. Table 13.1 lists the stages of a fire and the factors that influence its growth. Many of these factors involve decisions made by building designers.

(d) Fire Safety and Other Environmental Control Systems. In some instances, the optimum design for fire safety will resemble the optimum design for lighting, thermal, acoustic, and water systems. One such matching characteristic is a high degree of thermal mass, which is useful for passive heating and cooling systems, for acoustic isolation of airborne sound, and for fire barriers (most thermally massive materials will not burn easily). Another such characteristic is the elevated water storage tank, which provides both adequate water pressure for plumbing fixtures and water for firefighting in the first few minutes of a fire, before firefighters arrive. Finally, solid (noncombustible) overhangs over windows not only provide sunshading, but also discourage the vertical spread of fire over the building face and can serve as emergency exterior places of refuge.

TABLE 13.1 **Major Factors Influencing Fire Growth**

Realm	*Approximate Ranges of Fire Sizes*	*Major Factors That Influence Growth*
1 (Preburning)	Overheat to ignition	Amount and duration of heat flux Surface area receiving heat from material ignitability
2 (Initial burning)	Ignition to radiation point [10 in. (254 mm) high flame]	Fuel continuity Material ignitability Thickness Surface roughness Thermal inertia of the fuel
3 (Vigorous burning)	Radiation point to enclosure point [10 in. to 5 ft high flame (254 mm to 1.5 m)]	Interior finish Fuel continuity Feedback Material ignitability Thermal inertia of the fuel Proximity of flames to walls
4 (Interactive burning)	Enclosure point to ceiling point [5 ft (1.5 m) high flame to flame touching ceiling]	Interior finish Fuel arrangement Feedback Tallness of fuels Proximity of flames to walls Ceiling height Room insulation Size and location of openings HVAC operation
5 (Remote burning)	Ceiling point to full room involvement	Fuel arrangement Ceiling height Length/width ratio Room insulation Size and location of openings HVAC operations

Source: Reprinted with permission from the *Fire Protection Handbook,* 16th ed., copyright © 1986, National Fire Protection Association, Quincy, Massachusetts. This reprinted material is not the complete and official position of the National Fire Protection Association on the referenced subject, which is represented only by the standard in its entirety.

In many other instances, there are potential design conflicts between building performance under ordinary conditions and building performance in the extraordinary event of a fire. Escalators invite shoppers to explore the upper levels of a department store; they are also efficient apertures for the vertical spread of fire and smoke. Special fire protection strategies are found in Figs. 25.8 to 25.10. Daylighting and natural ventilation are best served by high ceilings and low partitions, which encourage light and air from the perimeter to pervade building interiors. Fire and smoke can easily spread through such "open plans." (On the other hand, fire can build up much more rapidly in small, enclosed rooms, which retain heat.) Forced-air systems that can heat, ventilate, and cool are also potential pathways for smoke and fire; this hazard is especially serious when the systems penetrate floors, since vertically spreading fires are harder to fight than are horizontally spreading ones. (On the other hand, carefully designed forced-air systems can aid in smoke management.) "Windowless" buildings that rely on electric light in place of daylight are especially dangerous in fires, because firefighters cannot easily evacuate occupants or gain access to the building. Sunscreens that completely cover windows are disadvantageous for the same reason; nonoperable windows, while considered advantageous for tightly controlled air conditioning, must be broken for fire evacuation/access. Many excellent insulating materials will burn readily, and some will give off toxic gases. Some of the loveliest interior finishes are both flammable and deadly sources of gases in a fire. Table 13.2 summarizes the requirements for interior finishes.

This conflict, between interior comfort throughout a building's life and safety at the moment of its impending death by fire, is the principal reason why codes cannot alone assure fire safety. The dilemma of ordinary versus extraordinary performance is one that must be faced by the designer, owner, and occupants of buildings.

(e) Protection of Life. For most buildings, a reasonable goal is the evacuation of all occupants in the time interval between the detection of a fire and the arrival of the firefighters. Designers can provide clearly defined pathways to exits ("exit access") that can be kept relatively clear of smoke (Fig. 13.2). The absolute minimum width of an exit access is 28 in. (356 mm). "Exits" can take a variety of forms. Vertical exits (Figs. 13.3 and 13.4) include smokeproof towers, exterior and interior stairs and ramps, and escalators that meet specific requirements. *Vertical exits do not include elevators;* they are too easily stalled or, worse, opened at the floor of a fire by malfunctioning signal equipment. Exits in the horizontal plane include doors leading directly to the outside, two-hour fire-enclosed hallways, and moving walls. Special "horizontal exits" are provided by internal firewalls penetrated by two fire doors—one swinging open in each direction. "Exit discharge" is the actual point at which exit ways open to the outside.

TABLE 13.2 **Requirements for Interior Finish: Summary of Life Safety Code**

	Class of Interior Finish[a]		
Occupancy	*Exits*	*Access to Exits*	*Other Spaces*
Places of assembly[b]			
Class A	A	A	A or B
Class B	A	A	A or B
Class C	A	A	A, B, or C
Educational	A	A	A, B, or C
Educational—unsprinklered open plan buildings[c]	A	A	A or B; C or low-height partitions
Flexible plan buildings[d]	A	A	A or B; C or low-height partitions
Institutional, existing—hospitals, nursing homes, residential-custodial care	A or B	A or B	A or B
Completely sprinklered	A or B	A or B	A, B, or C
Institutional, new—hospitals, nursing homes, residential-custodial care	A	A	A; B in individual room with capacity not more than 4 persons
Residential, new—apartment houses	A or B	A or B	A, B, or C
Residential, existing—apartment houses	A or B	A, B, or C	A, B, or C
Residential—dormitories	A or B	A or B	A, B, or C
Residential, new—1- and 2-family, lodging or rooming houses		A, B, or C	

TABLE 13.2 **Requirements for Interior Finish: Summary of Life Safety Code (*continued*)**

Occupancy	*Class of Interior Finish*[a]		
	Exits	*Access to Exits*	*Other Spaces*
Residential, existing—1- and 2-family, lodging or rooming houses		A, B, or C	
Residential, new—hotels	A or B	A or B	A, B, or C
Residential, existing—hotels	A or B	A or B	A, B, or C
Mercantile[e]			
Class A	A or B	A or B	Ceilings—A or B Existing walls—A, B, or C
Class B	A or B	A or B	Ceilings—A or B Existing walls—A, B, or C
Class A or B, sprinklered	A, B, or C	A, B, or C	A, B, or C
Class C	A, B, or C	A, B, or C	A, B, or C
Business	A or B	A or B	A, B, or C
Business-sprinklered	A, B, or C	A, B, or C	A, B, or C
Industrial	A or B	A, B, or C	A, B, or C
Storage	A, B, or C	A, B, or C	A, B, or C
Unusual structures	A, B, or C	A, B, or C	A, B, or C

Source: Reprinted with permission from the *Fire Protection Handbook,* 16th ed., copyright © 1986, National Fire Protection Association, Quincy, Massachusetts. This reprinted material is not the complete and official position of the National Fire Protection Association on the referenced subject, which is represented only by the standard in its entirety.

[a]There are three classes of interior finish in the *Life Safety Code:* class A, flame spread 0–25, smoke developed 0–450; class B, flame spread 26–75, smoke developed 0–450; class C, flame spread 76–200, smoke developed 0–450. When a standard system of automatic sprinklers is installed, an interior finish with a flame spread rating not over class C may be used in any location where class B is normally specified, and with a rating of class B in any location where class A is normally specified, unless specifically prohibited elsewhere in the *Life Safety Code.*

[b]Class A, 1000 persons or more; class B, 300 to 1000 persons; class C, 100 to 300 persons.

[c]Open plan buildings—includes all buildings where no permanent partitions are provided between rooms or between rooms and corridors.

[d]Flexible plan buildings have movable corridor walls and movable partitions of full height construction with doors leading from rooms to corridors.

[e]Class A stores have aggregate gross area of 30,000 ft^2 (2800 m^2) or more, or utilizing more than 3 floor levels for sales purposes; class B, stores of less than 30,000 ft^2 (2800 m^2) aggregate gross area, but over 3000 ft^2 (280 m^2) or utilizing any floors above or below street floor level for sales purposes, except that if more than 3 floors are utilized, store shall be Class A; class C, stores of 3000 ft^2 (280 m^2) or less gross area, used for sales purposes on street level only. (A single balcony or mezzanine floor with less than half the area of the street level floor and which is used for sales purposes is not counted as another floor.)

Distances to and the sizes of exits are controlled by code minimums (Tables 13.3 and 13.4). When automatic fire suppression systems such as sprinklers are used, the allowable distances to exits are increased. Designers should remember, however, that at least 30% of building fire deaths result from fire cutting off the paths to exits. Another point to consider is that the building population as estimated for fire safety is usually much greater than the population for which HVAC, water, or elevator service is designed. The stairs must be designed so as to allow those already within the stairwell to continue down without interference from access doors on any floor. The minimum dimensions of stairs are specified by codes (Table 13.5). Stairs with direct access to outdoor air at each floor—so-called smokeproof towers—are the safest kind.

High-rise buildings present much more diffi-

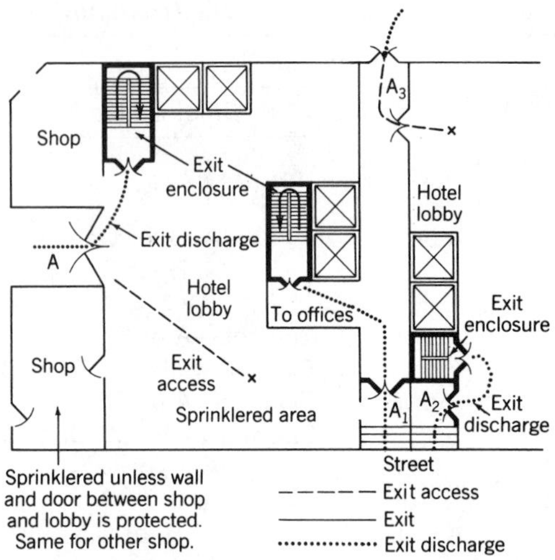

Fig. 13.2 Exit access, exit, exit enclosure, and exit discharge on the first floor of a multistory building. (Reprinted with permission from the Fire Protection Handbook, *16th ed., copyright © 1986, National Fire Protection Association, Quincy, Massachusetts. This reprinted material is not the complete and official position of the National Fire Protection Association on the referenced subject, which is represented only by the standard in its entirety.)*

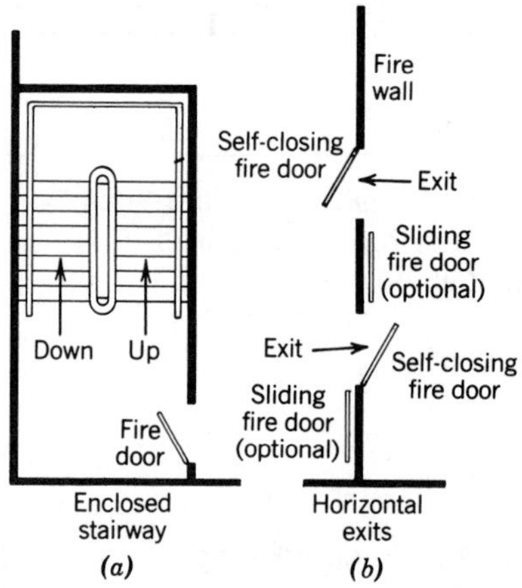

Fig. 13.3 Plan views of exits. (a) *Enclosed stairway allows occupants on any floor above a fire to escape.* (b) *Horizontal exit through an interior firewall provides a quick refuge and lessens the need of a hasty flight down the stairs. Two wall openings are needed to facilitate exit in either direction. The swinging doors are self-closing so as to discourage spreading smoke. Horizontal sliding fire doors, provided for safeguarding property, are normally open but close automatically in case of fire. (Reprinted with permission from the* Fire Protection Handbook, *16th ed., copyright © 1986, National Fire Protection Association, Quincy, Massachusetts. This reprinted material is not the complete and official position of the National Fire Protection Association on the referenced subject, which is represented only by the standard in its entirety.)*

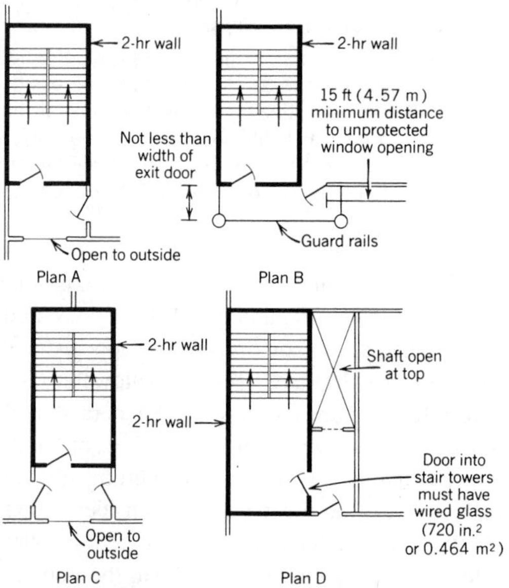

Fig. 13.4 Four variations of smokeproof towers. Plan A has a vestibule opening from a corridor. Plan B shows an entrance by way of an outside balcony. Plan C could provide a stair tower entrance common to two buildings. In plan D, smoke and gases entering the vestibule would be exhausted by natural or induced draft in the open air shaft. In each case a double entrance to the stair tower with at least one side open or vented is characteristic of this type of construction. Pressurization of the stair tower in the event of fire provides an attractive alternative for tall buildings and is a means of eliminating the entrance vestibule. (Reprinted with permission from the Fire Protection Handbook, *16th ed., copyright © 1986, National Fire Protection Association, Quincy, Massachusetts. This reprinted material is not the complete and official position of the National Fire Protection Association on the referenced subject, which is represented only by the standard in its entirety.)*

TABLE 13.3 **Summary of Life Safety Code Provisions for Travel Distances to Exits**

Occupancy	Dead-End[a] Limit, Ft (m)	Travel Limit to an Exit, ft (m) Unsprinklered	Sprinklered
Places of Assembly	20[b] (6.1)	150 (45.7)	200 (70)
Educational	20 (6.1)	150 (45.7)	200 (70)
Open plan	N.R.[c]	150 (45.7)	200 (70)
Flexible plan	N.R.	150 (45.7)	200 (70)
Health Care			
New	30 (9.1)	100 (30.5)	150 (45.7)
Existing	N.R.	100 (30.5)	150 (45.7)
Residential			
Hotels	35 (10.7)	100 (30.5)	150 (45.7)
Apartments	35 (10.7)	100 (30.5)	150 (45.7)
Dormitories	0	100 (30.5)	150 (45.7)
Lodging or rooming houses, 1- and 2-family dwellings	N.R.	N.R.	N.R.
Mercantile			
Class A, B, and C	50 (15.2)	100 (30.5)	150 (45.7)
Open Air	0	N.R.	N.R.
Covered Mall	50 (15.2)	200 (70)	300 (91.4)
Business	50 (15.2)	200 (70)	300 (91.4)
Industrial			
General, and special purpose	50 (15.2)	100 (30.5)	150[d] (45.7)
High hazard	0	75 (22.9)	75 (22.9)
Open structures	N.R.	N.R.	N.R.
Storage			
Low	N.R.	N.R.	N.R.
Ordinary hazard	N.R.	200 (70)	400 (121.9)
High hazard	0	75 (22.9)	100 (30.5)
Open parking garages	50 (15.2)	200 (70)	300 (91.4)
Enclosed parking garages	50 (15.2)	150 (45.7)	200 (70)
Aircraft hangars, ground floor	20 (6.1)	Varies[e]	Varies[e]
Aircraft hangars, mezzanine floor	N.R.	75 (22.9)	75 (22.9)
Grain elevators	N.R.	N.R.	N.R.
Miscellaneous occupancies	N.R.	100 (30.5)	150 (45.7)

Source: Reprinted with permission from the *Fire Protection Handbook,* 16th ed., copyright © 1986, National Fire Protection Association, Quincy, Massachusetts. This reprinted material is not the complete and official position of the National Fire Protection Association on the referenced subject, which is represented only by the standard in its entirety.

[a]Dead-end is an extension of a corridor or aisle beyond an exit (or exit access) that forms a pocket in which occupants may be trapped.

[b]In aisles.

[c]No requirement or not applicable.

[d]A special exception is made for one-story, sprinklered, industrial occupancies.

[e]See Paragraph 15–4.2 of Life Safety Code for special requirements.

cult problems, since firefighting equipment can ordinarily reach no higher than seven floors (about 90 ft) (27 m) and since, typically, only two exit stairways are provided. A fire stair must allow firefighters to move up while occupants are moving down. With a typical exit stair, a 15-story building housing 60 persons per floor per stair can be evacuated in about nine min-

TABLE 13.4 **Summary of Life Safety Code Provisions for Occupant Load and Capacity of Exits**

	Occupant[a] Load	*Capacity of Exits Number of Persons per Units of Exit Width[b]*					
Occupancy	*Ft² (m²) per Person*	*Doors[c] Outside*	*Horizontal Exit*	*Ramp Class A*	*Ramp Class B*	*Escalator*	*Stairs*
Places of Assembly	15 Net (1.4)	100	100	100	75	75	75
Areas of concentrated use without fixed seating	7 Net (0.7)						
Standing space	3 Net (0.3)						
Educational		100	100	100	60		60
Classroom area	20 Net (1.9)						
Shops and vocational	50 Net (4.6)						
Day Nurseries with sleeping facilities	35 Net (3.3)						
Health Care		30	30	30	30		22
Sleeping departments	120 Gross (11.1)						
Inpatient departments	240 Gross (22.3)						
Residential	200 Gross (18.6)	100	100	100	75	75	75
Mercantile		100	100			60	60
Street floor and sales basement	30 Gross (2.8)						
Other floors	60 Gross (5.6)						
Storage-shipping	300 Gross (27.9)						
Office areas	100 Gross (9.3)						
Business	100 Gross (9.3)	100	100	100	60	60	60
Industrial	100 Gross (9.3)	100	100	100	60	60	60
Detention and Correctional occupancies	120 Gross (11.1)	100	100	100	100[d] 60[e]		

Source: Reprinted with permission from the *Fire Protection Handbook, 16th ed.*, copyright © 1986, National Fire Protection Association, Quincy, Massachusetts. This reprinted material is not the complete and official position of the National Fire Protection Association on the referenced subject, which is represented only by the standard in its entirety.

[a]When actual maximum population count can be determined from plans, it can be subsututed for these figures.

[b]An exit width is defined as 22 in. (559 mm).

[c]Not more than three risers or 21 in. (533 mm) above or below grade.

[d]100 (down).

[e]60 (up).

TABLE 13.5 **Requirements for Exit Stairs**

		Existing Stairs[a]	
	New Stairs	*Class A*	*Class B*
Minimum width clear of all obstructions except projections not exceeding 3½ in. (89 mm) at and below handrail height on each side	44 in. (1.12 m) 36 in. (0.91 m) where total occupant load of all floors served by stairways is less than 50	44 in. (1.12 m) 36 in. (0.91 m) where total occupant load of all floors served by stairways is less than 50	44 in. (1.12 m)
Maximum height of risers	7 in. (178 mm)	10 in. (244 mm)	9 in. (229 mm)
Minimum height of risers	4 in. (102 mm)	7½ in. (191 mm)	8 in. (203 mm)
Minimum tread depth	11 in. (279 mm)		
Winders	See 5-2.2.2.5.	See 5-2.2.2.5.	See 5-2.2.2.5.
Minimum headroom	6 ft 8 in. (2.03 m)	6 ft 8 in. (2.03 m)	6 ft 8 in. (2.03 m)
Maximum height between landings	12 ft (3.7 m)	12 ft (3.7 m)	12 ft (3.7 m)
Minimum dimension of landings in direction of travel	Stairways and intermediate landings shall continue with no decrease in width along the direction of exit travel. In new buildings every landing shall have a dimension, measured in direction of travel, equal to the width of the stair. Such dimension need not exceed 4 ft (1.22 m) when the stair has a straight run.		
Doors opening immediately on stairs, without landing at least width of door	No	No	No

Source: Reprinted with permission from the *Fire Protection Handbook* 16 ed., copyright © 1986, National Fire Protection Association, Quincy, Massachusetts. This reprinted material is not the complete and official position of the National Fire Protection Association on the referenced subject, which is represented only by the standard in its entirety.

[a]Classes A and B are defined as stairs having these dimensional characteristics in *existing* buildings.

utes. However, with the same stair in a 50-story building housing 240 persons per floor per stair, evacuation will take at least 2 hours, 11 minutes! When doors are held open at each floor to admit occupants fleeing the fire, smoke can readily enter the stairwell. Because of the impracticality of rapid evacuation, larger buildings are required to provide "refuge areas" where smoke penetration is less likely. These details on smoke management are presented in the next section.

(f) Property Protection. One of the earliest design concerns in this category is that the site should permit access for firefighting equipment. Ideally, fire trucks should be able to pull alongside each exterior wall. If accessibility is limited by adjacent buildings, and roadways alongside buildings and other measures are impractical, more reliance must be placed on internal fire suppression systems. Another factor is the amount of time it will ordinarily take for firefighters to reach a site. In congested urban areas, or remote rural ones, time that elapses between the alarm and the arrival of firefighters can permit a fire to grow to unstoppable proportions. In these cases, emphasis again shifts to internal fire suppression systems.

Another design concern is for adequate water to fight the fire. Reliance on city water mains is not always a good solution. Elevated tanks on buildings can help in the early moments of firefighting, but their capacity is soon exhausted. Some buildings therefore rely on lakes or enclosed reservoirs for firefighting supply (see the discussion in Section 13.3).

Exposure protection is becoming common in areas where highly flammable surroundings pose a serious threat of fires originating *outside* a building. Candidates for exposure protection include buildings surrounded by older wooden buildings, bordered by lumberyards or other commercial activities that utilize highly flammable materials, or even bordered by open fields of dry grass or brush. Exposure protection guards against heat transfer by radiation and convective currents and against direct fire transfer via flying embers. Exposure protection begins with the use of nonflammable materials for the building's exterior. Erecting firewalls between the building and a fire-threatening neighbor is a more drastic, but sometimes necessary, step. Exposure protection sometimes includes external water sprinkler systems, in which a sprayhead is placed over (or under) each opening such as windows or doors. Sprinkler systems that soak the roof can play a cooling role on summer days, as well as an exposure protection role. Exterior doors can be chosen for their fire-delaying characteristics; fire-rated doors are shown in Table 13.6. Windows can also be protected by shutters with similar fire ratings that are designed to close automatically at high temperatures.

TABLE 13.6 **Fire-Rated Doors**[a]

Class	*Rating*	*Location*
A	3 h	Between buildings, or between separate fire areas in a single building
B	1 to 1½ h	Around vertical openings within buildings, such as stairs, elevator shafts, or mechanical chases
C	¾ h	In partitions between corridors and rooms
D	1½ h	Exterior walls with severe fire exposure from outside the building
E	¾ h	Exterior walls with light or moderate fire exposure from outside the building

Source: Excerpted from *Fire Protection Handbook,* NFPA (1976).

[a]Tight-fitting doors will help stop the spread of smoke.

Compartmentation has become increasingly important as buildings have become lightweight structures incorporating decreased fire resistance and open floor areas that encourage the spread of fires. Building codes establish the maximum floor areas permissible for various constructions and occupancies. If a building's floor area exceeds such limits, it must be subdivided by firewalls into areas that fall within the code limitations. Openings in the firewalls must be protected by fire doors (Table 13.6) and fire dampers in forced-air systems (Fig. 13.5). As important as compartmentation is in buildings with large floor areas, however, the vertical spread of fire poses a more serious problem. For this reason, compartmentation requirements around vertical openings are often especially strict.

Concealed spaces are found in many contemporary buildings, especially over suspended ceilings but also behind walls, within pipe chases, in attics, and in other places. All such spaces can offer paths for the spread of fire. The

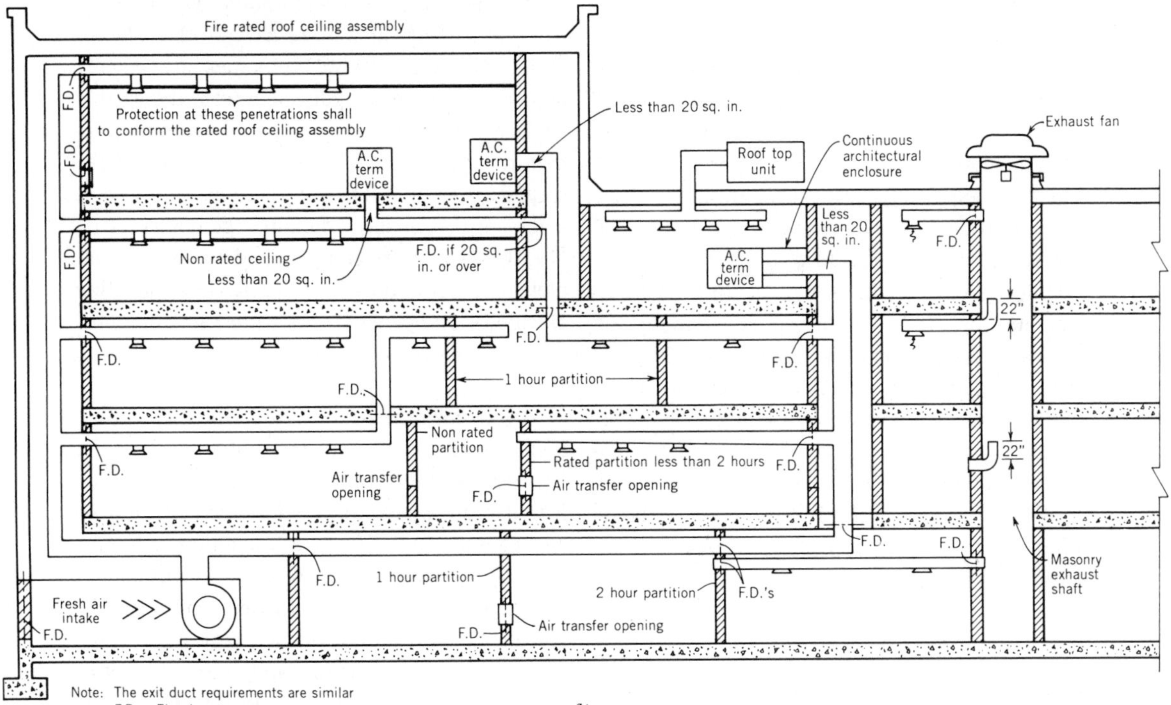

Fig. 13.5 *Openings in firewalls are protected by various devices.* (a) *Access doors to controls and other elements within firewalls can be of firerated construction.* (b) *Locations of fire dampers* (F.D.) *in a typical forced-air system—at firewalls, partitions, and shaft enclosures. [Part* (b) *reprinted with permission from the* Fire Protection Handbook, *16th ed., copyright © 1986, National Fire Protection Association, Quincy, Massachusetts. This reprinted material is not the complete and official position of the National Fire Protection Association on the referenced subject, which is represented only by the standard in its entirety.]*

designer can utilize noncombustible materials, insofar as possible, in such spaces. Another important step is to include automatic fire detection and suppression systems in these uninhabited spaces—and often, the use of oxygen-deprivation approaches to fire suppression. Another design response is compartmentation; using firestops (or firewalls) to break up otherwise continuous concealed spaces. Most codes, for example, require firestopping around each 1000 ft^2 (93 m^2) of suspended ceiling area, and every 2000 ft^2 (186 m^2) of attic floor area.

Structural protection, another important requirement, allows a building to continue to stand during a fire and enables the building to be salvaged rather than demolished after the fire. Codes require various protective layers for structural materials. In order of importance, it is usually most critical to protect columns, girders, beams, and, finally, the floor slab.

(g) Continuity of Operations. For most building functions, it is desirable to minimize the disruption of operation that fire will cause. Design strategies to encourage continuity of operations include special fire alarm/suppression systems for especially critical operations areas (control rooms, for example), the design of HVAC systems to allow for 100% outside air (to aid in purging a building of smoke after the fire is out), and provision for the speedy removal of the water dumped on a fire from a sprinkler system. The floors in sprinkler-served buildings should be waterproof; waterproofing should also be carried up walls, columns, pipes, and other elements to a height of 4 to 6 in. (102 to 152 mm) above the floor. These provisions will help to minimize the water damage associated with fires.

13.2 Smoke Management

As it has become increasingly evident that smoke kills more people in building fires than either heat or structural collapse, it has also become more common to design for smoke management as well as for resistance to fire and fire suppression. The objectives in smoke management are the same as those for fire resistance: to reduce deaths and property damage due to smoke, and to provide for continuity of operations with minimal smoke interference. Smoke management systems have been used for several generations; scientific laboratory buildings equipped with fume hoods (such as the Berkeley Chemistry Bldg., Fig. 6.10) offer a familiar example. Several options for smoke management are now available for ordinary building design.

(a) Factors in Smoke Management. The heat of a fire produces air pressure and buoyancy that aid the spread of smoke well beyond the scene of the fire itself. In low buildings, the smoke is spread primarily by heat-induced convective air motion and by the differential air pressures caused by the expansion of gases as the temperature increases. In tall buildings, the stack effect complicates this pattern of smoke spread, encouraging the rapid vertical rise of heated air within vertical shafts. Wind forces from outside are also more likely to be a factor in tall buildings. (Wind velocities are usually higher with increasing distance from the earth's surface). Forced-air systems, so common in high-rise buildings, can also contribute to the spread of smoke. The interactions among fire, stack effect, wind, building geometry, and HVAC systems are complex and not as yet completely understood. The ASHRAE publication *Design of Smoke Control Systems for Buildings* is a good source of more information.

(b) Confinement. The most passive design response to smoke is to try to confine it to the fire area itself. Another confinement technique is to exclude smoke from specially protected areas, called refuges. Compartmentation is important in these approaches: where firewalls cannot be used, special smoke barriers (often called curtain boards) are suspended from the ceiling (Fig. 13.6) in an effort to trap the initial layer of hot air and smoke, and therefore to set off the fire detection/suppression system more quickly. As useful as these partial barriers can be in the fire's early stages, they quickly lose effectiveness as the smoke layer thickens or as air pressures force the smoke below the barrier. Even in firewalls with fire door protection, small cracks around such openings provide smoke paths,

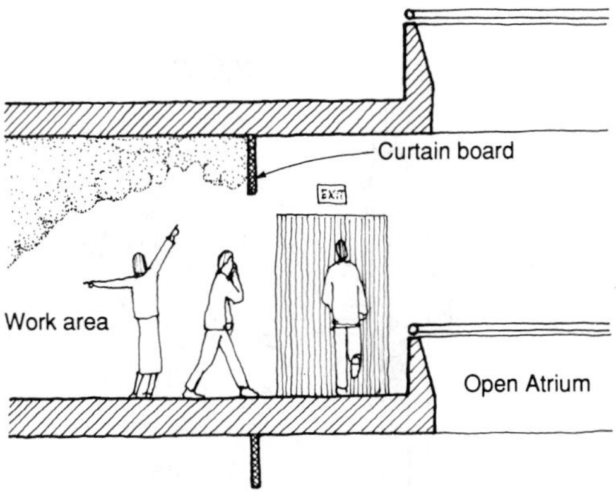

Fig. 13.6 Smoke barriers, often called curtain boards, are useful in a fire's early stages. By confining the initial layer of heated air and smoke produced by the fire, they help to slow the spread of smoke while making more likely the early detection and suppression of the fire.

aided by the air pressures and velocities that fires can quickly produce. Therefore, the typical "barrier" is not very resistant to spreading smoke, and smoke control based solely on confinement must be closely linked to an effective early detection/suppression system, such as a sprinkler system.

(c) Dilution. For a limited time in a fire's early stages, the dilution of smoke with 100% outdoor air (provided by the HVAC system) may make conditions bearable during occupant evacuation. However, such large quantities of fresh air are needed in so short a time that dilution alone is rarely sufficient for smoke control, particularly when the smoke contains toxic fumes. When the dilution system is combined with both confinement and early detection/suppression systems, it becomes more attractive. Dilution systems that dump large quantities of outdoor air into refuge areas can create such high pressures within the refuge that smoke cannot enter through cracks around doors. However, when such doors are opened (as when more people enter the refuge), dilution systems that rely on the conventional HVAC system can rarely provide sufficient pressure to exclude smoke through the open doorway.

(d) Exhaust. Special exhaust systems that function only in a fire are becoming more common. These systems employ both air velocity and air pressure to help control smoke. Although they involve a greater initial expense, because they require special fans and special smoke-exhaust shafts, exhaust systems have several advantages over simple confinement or dilution approaches.

1. They can remove toxic gases from refuge areas (particularly when there is a dilution or outdoor air supply to such areas).
2. They help firefighters by improving the air quality in the vicinity of the fire itself.
3. They can help control the direction that a fire takes, by creating air currents that a fire will follow.
4. They remove the unburned but combustible gases from a fire before the latter can cause a "backdraft" or "flashover" (smoke-explosion).
5. By creating higher air pressures in refuge areas and lower pressures in the fire zone, they keep smoke out of refuge areas, even when doors are temporarily open.
6. With them, the tall-building phenomenon of stack effect, complicated by buoyancy and wind, is less likely to overcome the smoke management system.

Two systems within the building must be closely coordinated with smoke exhaust systems: the HVAC system and the fire detection/suppression (usually sprinkler) system. As the fire detection system activates the smoke exhaust fans, it must also override the conventional HVAC system operation, usually by switching to 100% outside supply air and simultaneously blocking the return duct system. This process pressurizes each HVAC zone, to form a barrier against the smoke (Fig. 13.7), and also keeps smoke out of the return duct system. If the HVAC system is VAV, all supply control valves (dampers) must be moved to full open position. If additional protection is needed for refuge areas, additional air supply to such areas can be provided, perhaps by a separate (smoke dilution) system.

Sprinkler systems can hamper the functioning of smoke exhaust systems, both by creating

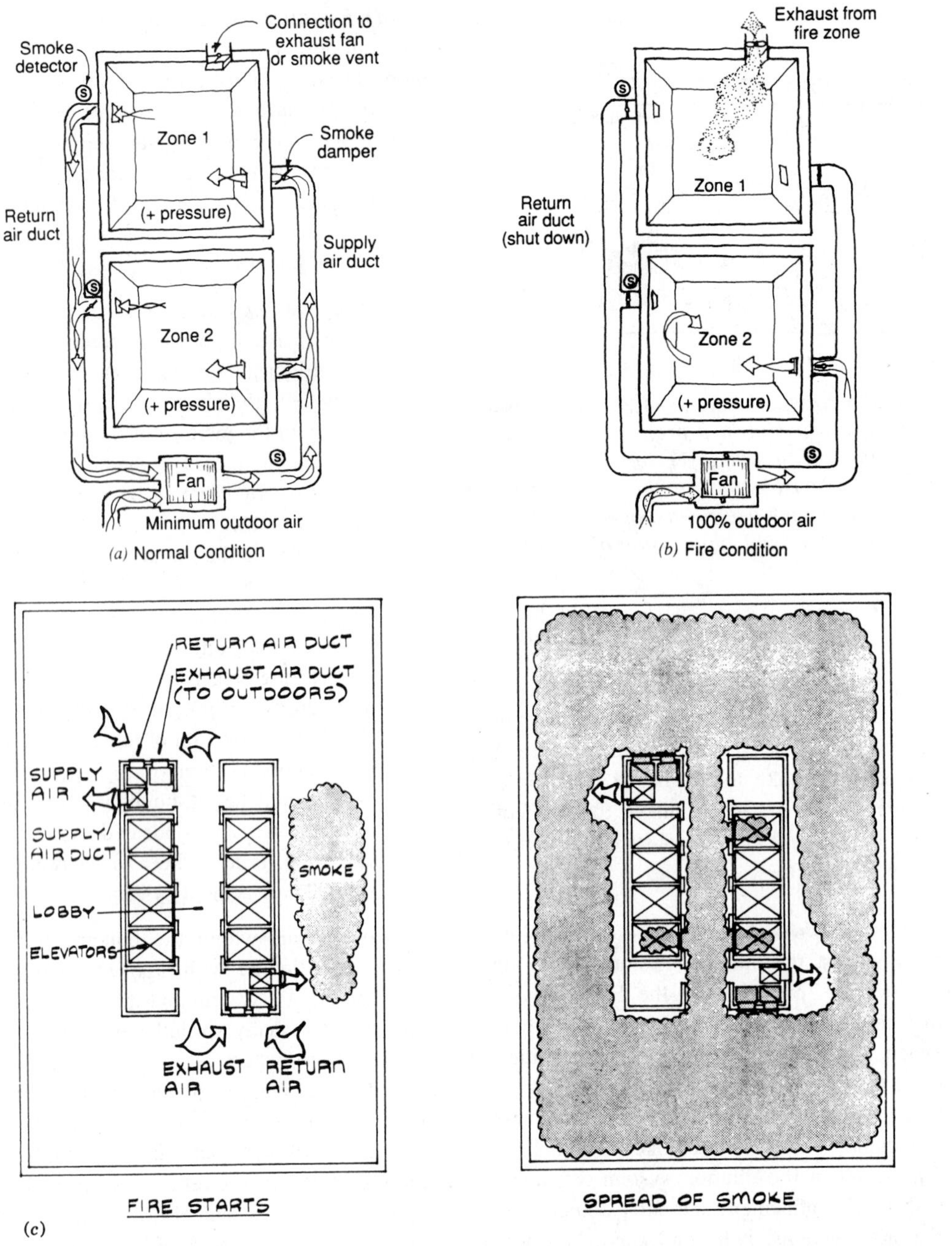

***Fig. 13.7** Smoke management by a smoke exhaust system.* (a) *The conventional HVAC system is shown operating at normal, minimum outdoor air supply conditions.* (b) *When smoke detectors trigger the smoke exhaust system, special fans begin to exhaust smoke, through a special duct, from the affected area only. Simultaneously, the conventional system undergoes two changes: All supply air becomes 100% outside air (for smoke dilution) and all return openings are closed off (so as to pressurize all nonfire areas, to keep smoke out of them and out of the return duct system). (Adapted from AIA; Ramsey and Sleeper,* Architectural Graphic Standards, *7th ed., copyright © 1981 by John Wiley & Sons. Reprinted by permission.)* (c) *Typical floor plans in tall buildings, showing fire and resulting spread of smoke in a conventionally zoned system.* (d) *Same plan, showing fire and smoke spread when smoke control zoning is designed. [*(c) *and* (d) *reprinted by permission from Egan,* Concepts in Building Firesafety, *copyright © 1978 by John Wiley & Sons.]*

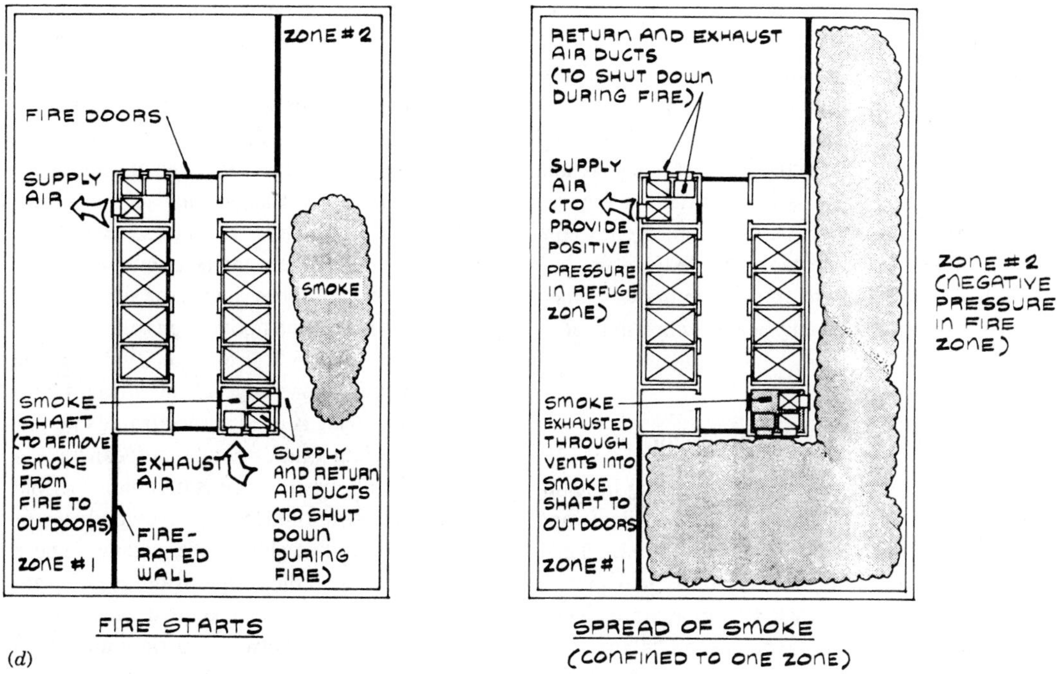

(*d*)

Fig. 13.7 *(continued)*

a curtain of water that inhibits the movement of smoke and by cooling the smoke and thus reducing its buoyancy. The buoyancy of smoke is the factor relied on by smoke exhaust shafts, whose intakes are located at ceiling level within each zone. Although these two systems thus appear to work at cross purposes, the suppression of the fire is obviously as important a goal as smoke management. Sprinkler systems suppress fires so quickly that the size of the accompanying smoke exhaust systems can generally be reduced. When the fire suppression system relies on oxygen displacement (such as carbon dioxide or halogenated agents), smoke exhaust systems clearly pose a threat; special attention is required if both systems are to be used.

Automatic heat-and-smoke ventilating hatches (without fans) are often installed (Fig. 13.8) in smaller buildings. These hatches open individually as heat or smoke from the fire trig-

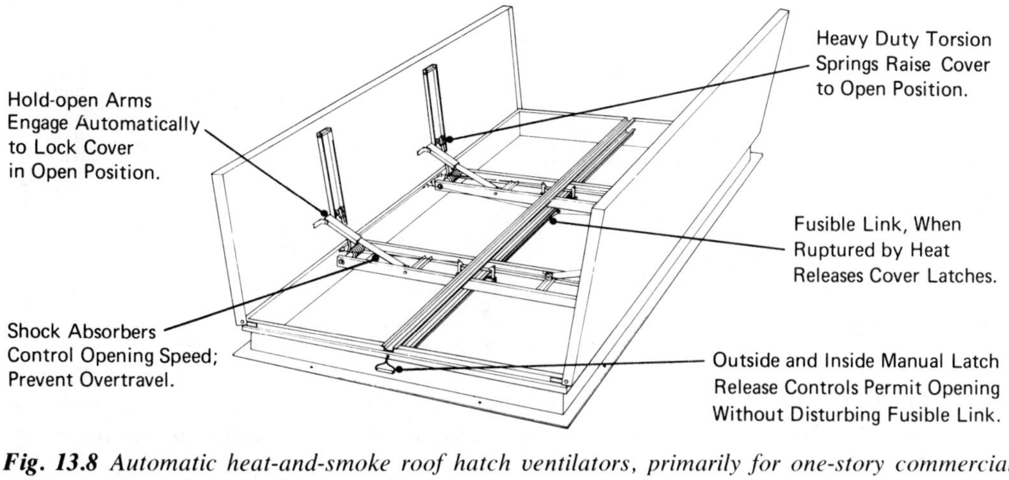

Fig. 13.8 *Automatic heat-and-smoke roof hatch ventilators, primarily for one-story commercial and industrial buildings. (Courtesy of Inryco, Inc.)*

FIRE PROTECTION

ger their control devices. As a result, indoor conditions near the fire are improved for firefighters, and firefighters on the roof are quickly alerted to the location of the fire indoors.

In summary, designers can provide several components that aid in the exhaust of smoke and the provision of access and ventilation for firefighters for almost any building: emergency controls on HVAC systems, operable windows and skylights (to facilitate ventilation and occupant escape), and smoke-and-heat hatches in roofs.

13.3 Water for Fire Suppression

The most popular medium for building fire suppression is water, which is readily available and relatively low in cost. Water cools, smothers, emulsifies, and dilutes. As it turns to vapor, it removes 970 Btu/lb (2256 kJ/kg) at atmospheric pressure and its volume increases 1700 times—a process that helps push away the oxygen needed by the fire. Water has several disadvantages, which sometimes preclude its use for fire suppression: it damages most contents of buildings, including interior surfaces; it conducts electricity readily as a stream (less readily as a spray); many flammable oils will float on the water's surface while continuing to burn; and as water vaporizes rapidly as steam, it can harm firefighters. Where these factors are major considerations, other suppression media can be considered (see Section 13.4).

Fire-suppressing systems are commonly combined with smoke management systems; Fig. 13.9 shows a typical combination of provisions in one-story buildings. Taller buildings, and especially those with multistory interior spaces (atriums), present a special challenge (Fig. 13.10). A typical approach is to construct an approximately 6-ft (1.8 m) deep curtain board (or fire spandrel) at the opening to the atrium on each floor. Smoke detectors, motorized dampers, and sprinklers can be combined in various ways. At the lobby level, the atrium floor level, and at any office floors open to the atrium, sprinklers at about 6-ft (1.8-m) o.c. will provide a water curtain at the edge of the atrium opening. Where glazing is used at the atrium's perimeter, the glazing frames sometimes can be designed to allow for considerable thermal expansion, and sprinklers can provide a water curtain on the office side of the glazing. Where balcony corridors adjoin the atrium, two sprinkler water curtains can be provided, one on each side of the partitions between corridor and office space. The smoke exhaust system often can provide six ACH (air changes per hour) for all spaces that open onto the atriums.

Sprinkler systems are widely relied on as proven, automatic fire-suppressors. However, designers must remember that the use of sprinklers does *not* give one license to ignore fire code limitations, even though many codes are more lenient for sprinklered buildings. In addition, provision must be made for adequate water supply, standby power for water pumping, and water drainage during and after the fire. Sprinklers require very large supply pipes and valves, and sometimes fire pumps—all of which are fre-

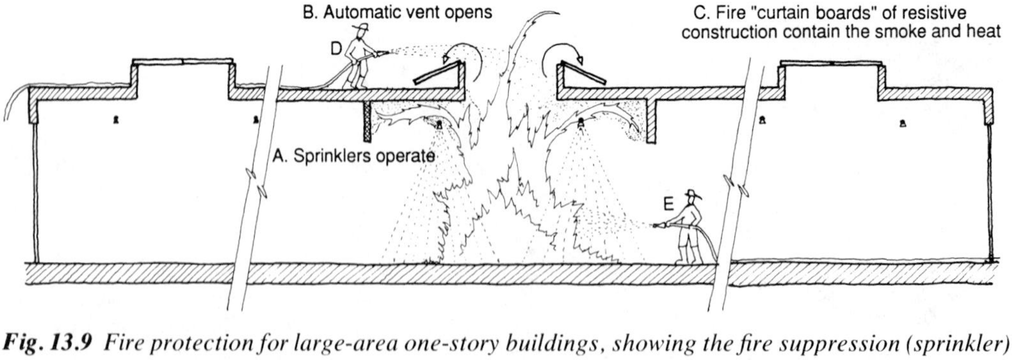

Fig. 13.9 Fire protection for large-area one-story buildings, showing the fire suppression (sprinkler) system, (a) *and the smoke management system,* (b *and* c). *Locations D and E are relatively safe positions for firefighters.*

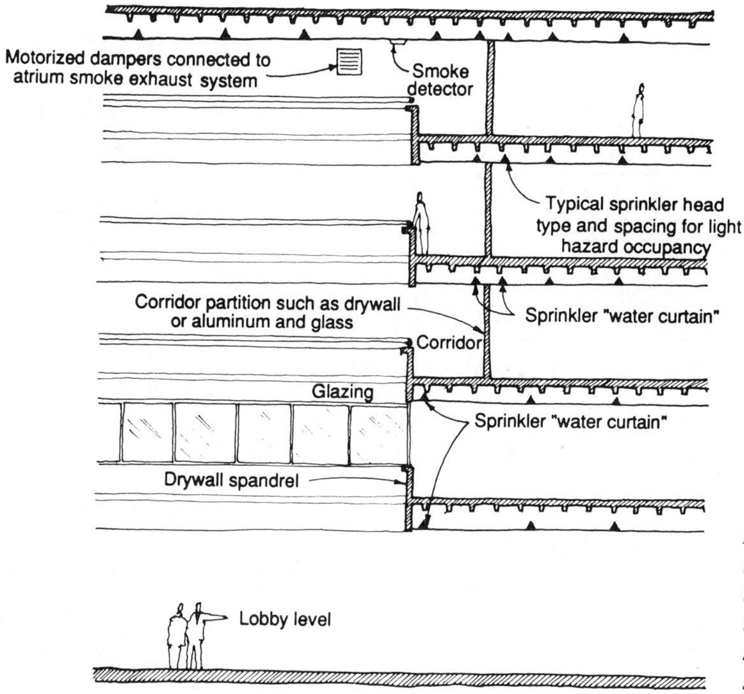

Fig. 13.10 *Fire protection in an atrium-type office building: section showing the detection/suppression system and provisions for smoke management.*

quently considered unsightly. On the other hand, sprinklers afford opportunities for integration with energy-conserving HVAC systems (see the Comstock Center, Fig. 7.66) and with display lighting (see Bullock's department store, Fig. 7.10). Also, the cost of sprinkler system installation can usually be recovered rather quickly through reductions in fire insurance premiums.

(a) Standpipes and Hoses. Standpipes and hoses with a separate water reserve or upfeed pumping are valuable in *any* building but become essential in tall buildings whose upper floors cannot be reached by firefighting equipment located at ground level. Figure 13.11 shows such a system, which is designed to be used by building personnel until the fire engines arrive, and thereafter by the firefighters. It is not practical to store enough water on the roof for a protracted firefighting period, and it is usually assumed that a half hour's supply will be more than enough to provide for the short period it takes the fire engines to arrive. When the system is used by the fire department, its pumps are attached to the street siamese (Fig. 13.12) to deliver water from street hydrants or the building's "secondary source." The check valve closest to the siamese in use opens and the check valves at the tank close, to prevent the water from rising uselessly in the tank. After the engines are disconnected from the siamese, the water between the siamese and the adjacent check valve drains out through the ball drip so that it does not freeze.

An overhead tank is considered a most dependable source, but the height required can be architecturally undesirable. In such a case, upfeed fire pumps operating automatically to deliver water to higher stories from lower suction reserve tanks may be used. Another option in this case is a pneumatic tank that delivers water by the power of the air that is compressed in the upper portion of the tank.

A *fire* standpipe zone will usually coincide with the service water zone (see Fig. 13.13, and also Fig. 9.36). Fire standpipes and their hoses (see Fig. 13.12) are usually located at or near fire stairs from which personnel or firefighters can approach a fire (see Fig. 13.17).

(b) Sprinklers. Automatic sprinkler systems usually consist of a horizontal pattern of pipes placed near the ceilings of industrial build-

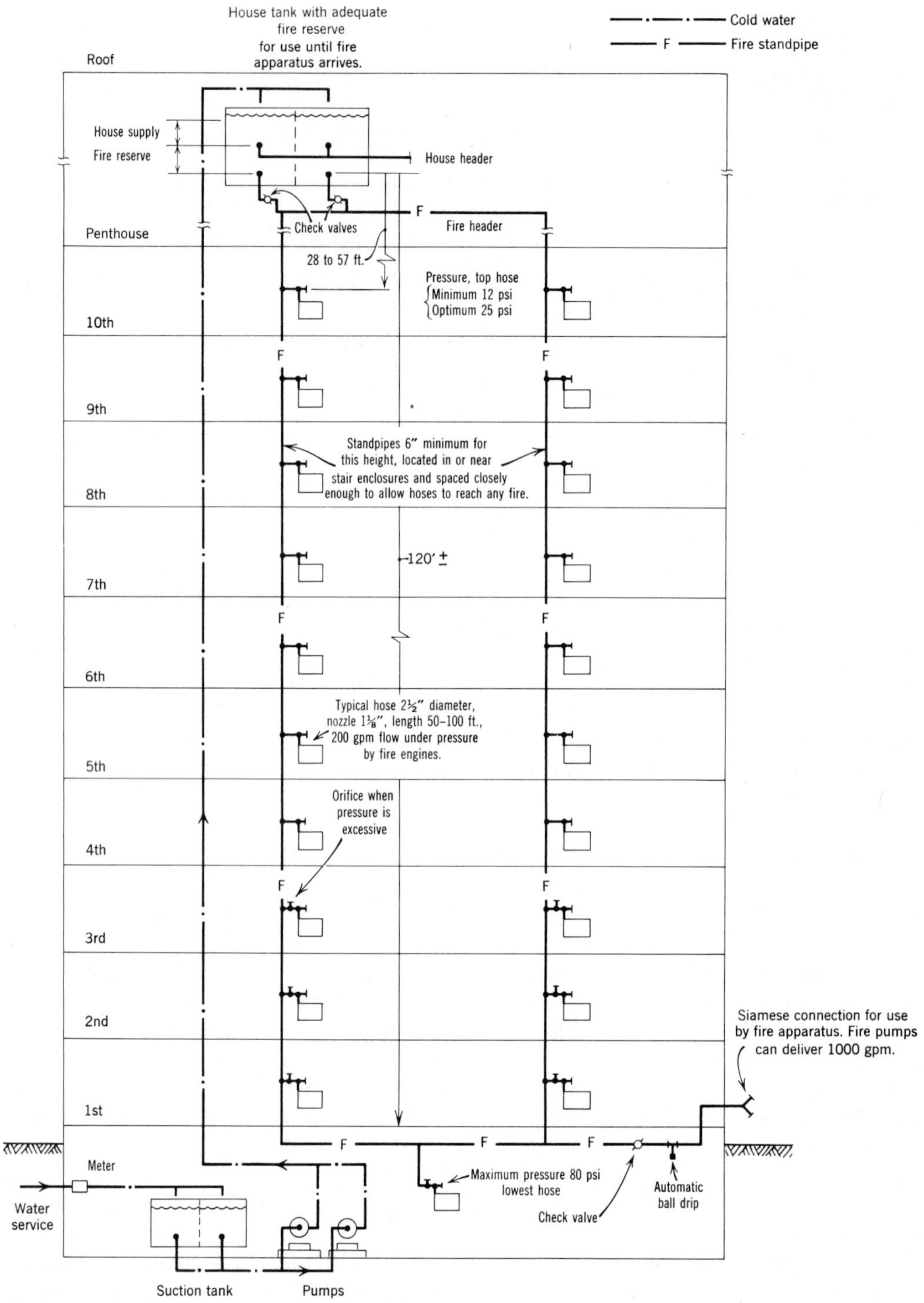

Fig. 13.11 *Schematic of a standpipe and hose system, which can sometimes feed sprinkler systems as well. Water is supplied initially by the roof tank's gravity downfeed, then by firefighters, through the siamese connection. Hoses at each floor are used first by building personnel, then by firefighters.*

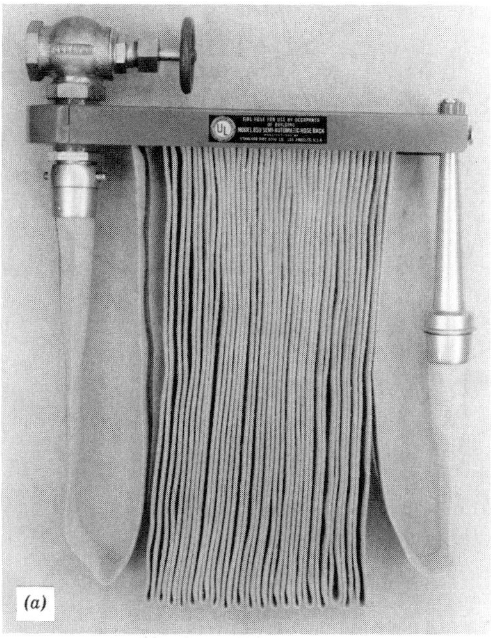

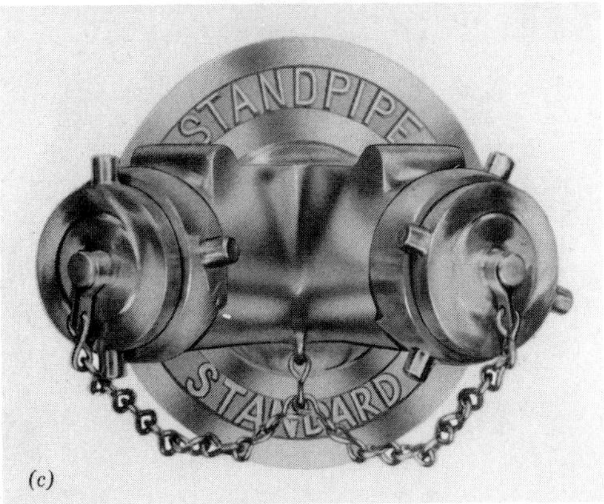

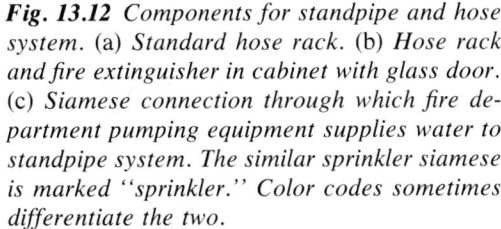

Fig. 13.12 *Components for standpipe and hose system.* (a) *Standard hose rack.* (b) *Hose rack and fire extinguisher in cabinet with glass door.* (c) *Siamese connection through which fire department pumping equipment supplies water to standpipe system. The similar sprinkler siamese is marked "sprinkler." Color codes sometimes differentiate the two.*

ings, warehouses, stores, theaters, offices, homes, and other structures in which a fire hazard requires their use. These pipes are provided with outlets and sprinkler heads constructed such that abnormally high temperatures will cause them to open automatically and emit a series of fine water sprays.

The temperature at which a sprinkler is triggered is usually specified; it should be at least 25F° (13.9C°) higher than the maximum ceiling temperature ordinarily expected. "Ordinary" sprinkler heads operate at between 135 and 170F. A sprinkler's heat-sensitive controls may be made of metal alloys that fuse, organic materials that soften, or organic liquids that expand and rupture their glass enclosure (Fig. 13.14). The sprinkler heads are usually upright, pendant, or sidewall types. Upright heads sit on top of the exposed supply piping. Pendant heads hang below piping, which thus can be concealed above suspended ceilings (Figs. 13.14*a* and 13.14*b*). The pendant heads themselves can be nearly concealed; low-profile, recessed, and concealed pendant heads are available (Fig. 13.4*c*). Sidewall sprinklers (Fig. 13.4*d*) are usually located adjacent to one wall of smaller rooms—hotels, apartments, etc.—and throw a spray of water entirely across such rooms. Therefore only one sidewall sprinkler head per small room often is used. A welcome development is the *flow control* sprinkler, which *closes* automatically once the ceiling temperatures are sufficiently reduced. Ordinary sprinklers, once activated, will continue to operate until a main valve is manually closed; this can waste large quantities of water and result in extensive water

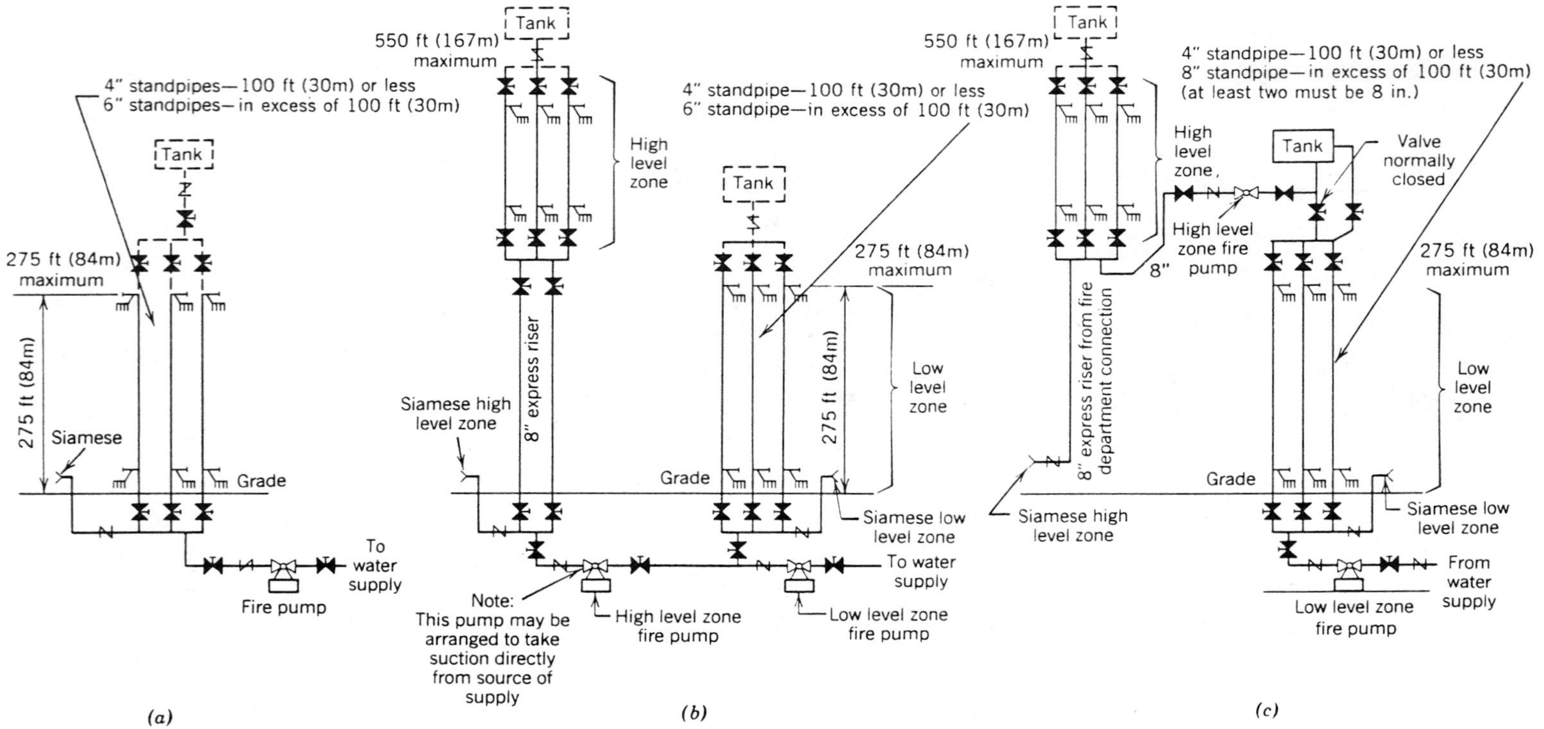

***Fig. 13.13** Zoning for standpipes in tall buildings.* (a) *A typical one-zone system will require a 1000-gpm (3785 L/min) fire pump.* (b) *An independent two-zone system for high-rise buildings, requiring two 1000-gpm (3785 L/min) fire pumps; the first pump provides high-pressure suction to the second, as well as supplying the lower zone.* (c) *An alternate two-zone high-rise system features an elevated tank, from which the upper zone's fire pump can draw; both pumps must be of 1000 gpm (3785 L/min) capacity. Reprinted with permission from* Fire Protection Handbook, *16th ed., copyright © 1986 by the National Fire Protection Association, Quincy, Massachusetts. This reprinted material is not the complete and official position of the National Fire Protection Association on the referenced subject, which is represented only by the standard in its entirety.*

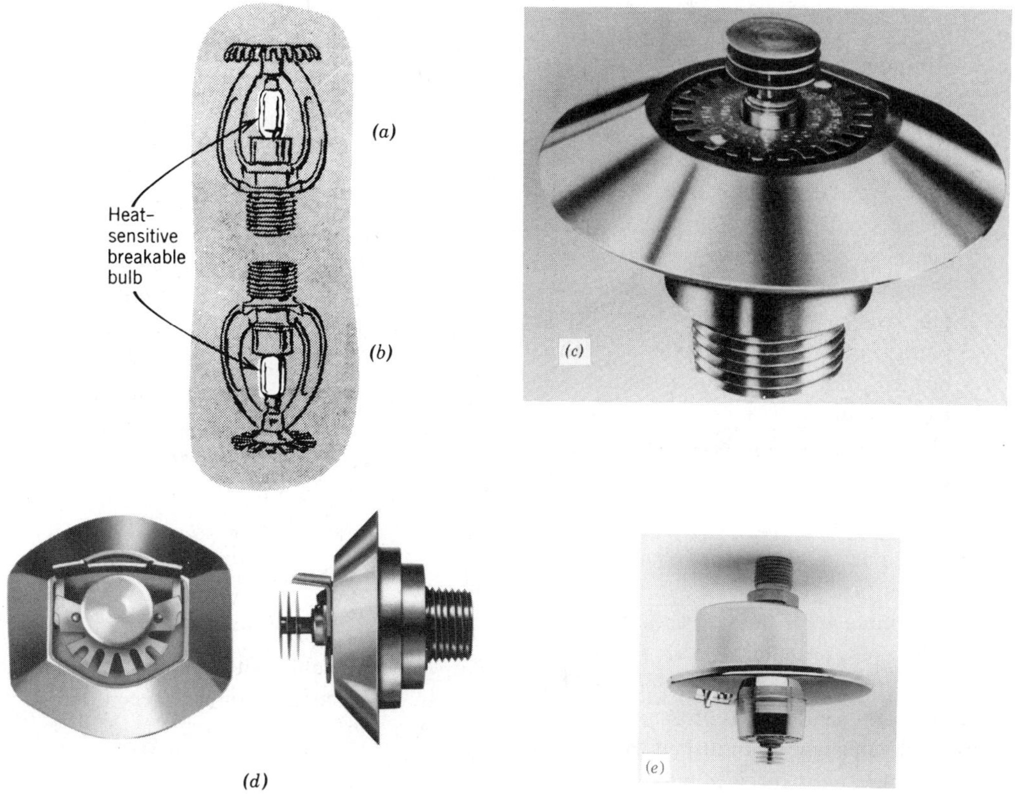

Fig. 13.14 *Sprinkler heads.* (a) *Upright; sits above exposed piping, just below the structural deck—where the hottest gases will first accumulate.* (b) *Pendant projecting through a suspended ceiling in most installations. Both* (a) *and* (b) *use the quartzoid bulb, a transparent bulb with a colored liquid that ruptures at a preset temperature to release the water stream; heads can be plated, polished, or colored—but never painted.* (c) *Pendant, styled for a more contemporary appearance.* (d) *A side-wall, front and side views.* (e) *Flow control pendant, which must be exposed within the space it serves. (Illustrations* (c), (d), *and* (e) *courtesy of Central Sprinkler Corporation.)*

damage. The flow control sprinkler shown in Fig. 13.14*e* is designed to operate individually, and in two stages. When the ambient temperature reaches 45F (60°C), a fusible alloy melts and the sprinkler is readied for action. Water discharge, however, does not occur until the second stage, when ambient temperatures reach 165F (74°C). Later, when the ambient temperature falls below 95F (35°C), the sprinkler returns to its "first stage" condition: readied for action, but no water discharge. Thus, if a fire reflash occurs, the sprinkler can again emit water, then again shut down as temperatures drop. Flow control sprinklers are not designed for use in dry-pipe systems, and require special detailing when used in preaction systems. They are not to be used in areas where nonfire ambient temperatures may exceed 100F (37°C). These sprinklers are more visible, since both the bottom and side elements must be exposed within a space.

Other special sprinkler models include *extra large orifice* (for delivering large water quantities where water pressures are relatively low) and *multilevel* sprinklers, for use where other sprinklers are at a higher plane within the same space (and these lower sprinklers may otherwise have their action retarded by water from the higher elevation heads).

All types of sprinkler heads must be replaced after use.

Sprinkler systems are of four common types.

1. *Wet-pipe* (most common), in which water is always standing under pressure in all pipes and mains.
2. *Dry-pipe* (used in unheated buildings), in which pipes are filled with compressed air (or nitrogen) until the opening of a sprinkler head permits water flow.
3. *Preaction,* which is very similar to dry pipe except that water is admitted to the pipes *before* any sprinkler head has opened, and which thus sounds a very early alarm.
4. *Deluge,* in which *all* sprinkler heads go off at once.

In the wet-pipe system, sprinklers in the area affected are opened by sensitive elements within the sprinkler heads themselves. In the dry-pipe systems, remote valves may be actuated by sensitive elements to admit water to sprinkler heads. Dry-pipe and preaction systems require a device to maintain design air pressure within the pipes; such air might be furnished from a compressor, from an air receiver tank, or from an existing pressurized air system (as in manufacturing operations). Dry-pipe systems also require a heated main control valve housing; and the pitching of all piping, to allow thorough drainage after usage. Preaction systems are especially popular where the building's contents are subject to water damage—computer rooms, retail stores, and so on—since the early alarm provided by water filling the piping often permits the fire to be found and extinguished manually, before any sprinklers open. Deluge systems are used where extremely rapid fire spread is expected—in aircraft hangars, for example, or other places where flammable liquid fires may break out.

The spacing of sprinkler heads is a complex matter, and sprinkler systems are usually designed by professionals working for sprinkler manufacturers. However, some guidelines for preliminary sprinkler location can be included here. The first consideration is the degree of hazard faced by the occupants, as listed in Table 13.7. Once this is known, sprinklers and pipes can be approximately located in plan with the aid of Table 13.8. Within each space, sprinklers should be located so as to detect a fire readily and to discharge water over the greatest area (considering obstacles such as beams, partial height partitions, etc.).

Piping for sprinkler supply can be hydraulically designed, as shown in Section 9.7. A complicating factor is the expectation that only a small percentage of the sprinklers will actually open; over 50% of the fires studied over a 49-year period were extinguished by two or fewer sprinklers. A detailed sizing procedure would consider both the available pressure at the highest sprinkler and the expected flow rate, which can vary from 500 to 5000 gpm (31.5 to 315 L/s). The sprinklers' actual performance is then determined by

$$Q = K\sqrt{p}$$

where

Q = flow rate, gpm
K = K factor, published for each sprinkler head by U.S. manufacturers (see Table 13.8, Part A)
P = pressure, psi

Sprinkler systems are usually designed for a maximum working pressure of 175 psi. Required flow rates are determined by the "density" of the fire hazard, which ranges from 0.1 gpm/ft^2 for light hazard to 0.5 gpm/ft^2 for extra hazard. In the preliminary design, however, pipe sizing may be approximated from the data given in Table 13.8. Reliable as automatic sprinklers are, no system is foolproof. Table 13.9 lists reasons for system failure: some due to occupant behavior, some to design deficiency.

Special installation requirements for sprinkler systems include: (1) at least one fire department connection on each frontage; (2) a master alarm valve control for all water supplies other than fire department connections; (3) special firewalls between protected areas and unprotected areas; and (4) sloping waterproof floors with drains or scuppers to carry away waste water.

When gravity tanks are used with sprinkler systems, they should provide an adequate reserve for this purpose and, in any case, enough to operate 25% of the sprinkler heads for 20 minutes. As in the case of standpipe and hose systems, this provision gives the fire company a chance to arrive and take over.

A typical automatic sprinkler system is shown in Fig. 13.15. The building, a printing and publishing plant, is in the category of "Ordinary Hazard, Group 2" (see Table 13.7). The sprinkler design results in a nozzle spacing such that one nozzle (sprinkler head) takes care of 125 ft^2 (11.6 m^2) of floor area. Standards of the National Fire Protection Association (NFPA) are constantly being reviewed and updated. NFPA 13-1989, *Installation of Sprinkler Systems,* lists the requirements (briefly referred to here), as well as numerous other regulations.

An alarm gong mounted on the outside of the building (see Fig. 13.15) warns of water flow through the alarm valve upon the actuating of a sprinkler head. This warning gives the building personnel an opportunity to make additional firefighting arrangements that can minimize loss and speed the termination of the fire; in this way, the sprinklers can be turned off, to prevent excess water damage to building contents after the fire is out. Sprinkler alarms commonly are also connected to private regional supervisory offices that communicate promptly with municipal fire departments upon the receipt of a signal. Siamese connections permit fire engines to pump into the sprinkler system in a manner similar to that used for standpipe systems.

All public buildings, and other buildings as required, should be provided with fire detection and alarm systems that indicate, in the custodian's office, the location of the fire.

In future sprinkler systems, different types of sprays may be ejected from a single sprinkler head: one spray of larger droplets to penetrate the fire plume and thereby cool the burning surfaces, as well as adjacent surfaces; and another, finer spray to cool the ceiling itself.

(c) Upfeed Pumping. In tall buildings, sprinklers can be supplied with water from elevated storage tanks. These tanks, rising above large buildings throughout the country, have become a traditional part of the American architectural scene. Now their use appears to be diminishing. Architects—such as Skidmore, Owings & Merrill, in their design for General Mills's central office building in suburban Minneapolis—are turning to automatically controlled pumping systems that obviate the need for elevated storage (Fig. 13.16).

The purpose of the familiar water tower has always been fourfold: to supply a constant pressure on the distribution lines, to store sufficient water to balance supply and demand, to prevent excessive starting and stopping of the pump, and to provide a dependable fire reserve. The last factor has been critical in the calculation of fire-insurance rates.

The principal objections to the use of tanks have been their unsightliness, the ever-increasing cost of steel and steel-construction work, the problem of freezing, and—in the case of large buildings—the tanks' tremendous weight. For the General Mills building, the fire underwriters possibly would have required at least 50,000 gal (189,250 L) of residual water in the tank for emergency purposes, which would mean about 100,000 gal (378,500 L) of elevated storage. The alternative chosen was a reinforced-concrete structure placed underground to one side of the building and covered with from 3 to 5 ft (0.9 to 1.5 m) of earth. Small vents rising from this reservoir blend in with the lawn and landscaped shrubbery above.

A comparison of costs would be difficult because of the number of factors involved. Generally, however, the saving in steel tended to make the overall cost comparable to that of an elevated tank of the same capacity. The savings were made possible chiefly by refined automatic-pump control. The underground reservoir eliminates the problem of appearance and weight, but the other usual advantages of an elevated tank—reliability in case of fire, a minimum starting and stopping of motors, and the maintenance of pressure while balancing supply and demand—must be equaled in the automatic circuitry. This is not so simple as it might at first seem. Factors such as fluctuations of demand, friction within the pipes, elevations, starting surges from the pumps, and pressure-flow characteristics of the pumps themselves must be met. Various combinations of these problems undoubtedly account to a great extent for the continued use of elevated tanks. The trend is toward the use of more-sophisticated pump control.

TABLE 13.7 **Relative Fire Hazard for Various Occupancies, as Related to Sprinkler Installations**

Classification of Occupancies

Occupancy classifications for this standard relate to sprinkler installations and their water supplies only. They are not intended to be a general classification of occupancy hazards. Examples listed below represent the norm for those occupancy types; unusual or abnormal fuel loadings or combustible characteristics may affect classification.

Light-Hazard Occupancies

Occupancies or portions of other occupancies where the quantity and/or combustibility of contents is low, and fires with relatively low rates of heat release are expected.

Light hazard occupancies include those having conditions similar to

Churches
Clubs
Eaves and overhangs (if combustible construction with no combustibles beneath)
Educational
Hospitals
Institutional
Libraries, except large stack rooms
Museums
Nursing or convalescent homes
Office, including data processing
Residential
Restaurant seating areas
Theaters and auditoriums excluding stages and prosceniums
Unused attics

Ordinary-Hazard Occupancies

Group 1—occupancies or portions of other occupancies where combustibility is low, quantity of combustibles is moderate, stock piles of combustibles do not exceed 8 ft (2.4 m), and fires with moderate rates of heat release are expected.

Group 1 ordinary-hazard occupancies include those having conditions similar to

Automobile parking garages
Bakeries
Beverage manufacturing
Canneries
Dairy products manufacturing and processing
Electronic plants
Glass and glass products manufacturing
Laundries
Restaurant service areas

Group 2—occupancies or portions of other occupancies where quantity and combustibility of contents is moderate, stock piles do not exceed 12 ft (3.7 m), and fires with moderate rate of heat release are expected.

Group 2 ordinary hazard occupancies include those having conditions similar to

Cereal mills
Chemical plants—ordinary
Cold storage warehouses
Confectionery products
Distilleries
Horse stables
Leather goods manufacturing
Libraries—large stack room areas
Machine shops
Metal working
Mercantiles
Printing and publishing
Textile manufacturing
Tobacco products manufacturing
Wood product assembly

TABLE 13.7 **Relative Fire Hazard for Various Occupancies, as Related to Sprinkler Installations (*continued*)**

Group 3—occupancies or portions of other occupancies where quantity and/or combustibility of contents is high and fires of high rate of heat release are expected.

Group 3 ordinary hazard occupancies include those having conditions similar to

- Feed mills
- Paper and pulp mills
- Paper process plants
- Piers and wharves
- Repair garages
- Tire manufacturing
- Warehouses (having moderate to higher combustibility of content, such as paper, household furniture, paint, general storage, whiskey, etc.)
- Wood machining

Extra-Hazard Occupancies

Occupancies or portions of other occupancies where quantity and combustibility of contents is very high and flammable liquids, dust, lint, or other materials are present, introducing the probability of rapidly developing fires with high rates of heat release.

Group 1—occupancies with little or no flammable or combustible liquids, such as

- Combustible hydraulic fluid use areas
- Die casting
- Metal extruding
- Plywood and particle board manufacturing
- Printing (using inks with below 100F [37.8°C] flash points)
- Rubber reclaiming, compounding, drying, milling, vulcanizing
- Saw mills
- Textile picking, opening, blending, garnetting, carding, combining of cotton, synthetics, wool shoddy or burlap
- Upholstering with plastic foams

Group 2—occupancies with moderate to substantial amounts of flammable or combustible liquids, or where shielding of combustibles is extensive, such as

- Asphalt saturating
- Flammable liquids spraying
- Flow coating
- Mobile home or modular building assemblies (where finished enclosure is present and has combustible interiors)
- Open oil quenching
- Solvent cleaning
- Varnish and paint dipping

Source: Reprinted with permission from NFPA 13-1989, *Installation of Sprinkler Systems,* copyright © 1989, National Fire Protection Association, Quincy, Massachusetts. This reprinted material is not the complete and official position of the National Fire Protection Association on the referenced subject, which is represented only by the standard in its entirety.

It can be seen from Fig. 13.16 that a continuous flow from the deep-well pumps, through both domestic and fire reservoirs, prevents the water from becoming stale and rancid. The fire reservoir is given the necessary priority over the domestic reservoir by means of a simple weir. Even if the domestic reservoir were completely empty, the fire reservoir would remain full.

TABLE 13.8 **Sprinkler Design Guidelines**

*Part A. **Sprinkler Heads***
Discharge Characteristics

Nominal Orifice Size (in.)	Orifice Type	K Factor[a]	Percent of Nominal 1/2-in. Discharge
1/4	Small	1.3–1.5	25
5/16	Small	1.8–2.0	33.3
3/8	Small	2.6–2.9	50
7/16	Small	4.0–4.4	75
1/2	Standard	5.3–5.8	100
17/32	Large	7.4–8.2	140
5/8	Extra large	11.0–11.5	200

Temperature Classifications

Max. Ceiling Temp.		Temperature Rating		Temperature Classification	Color Code	Glass Bulb Colors
F	°C	F	°C			
100	38	135 to 170	57 to 77	Ordinary	Uncolored or black	Orange or red
150	66	175 to 225	79 to 107	Intermediate	White	Yellow or green
225	107	250 to 300	121 to 149	High	Blue	Blue
300	149	325 to 375	163 to 191	Extra high	Red	Purple
375	191	400 to 475	204 to 246	Very extra high	Green	Black
475	246	500 to 575	260 to 302	Ultrahigh	Orange	Black
625	329	650	343	Ultrahigh	Orange	Black

Part B. ***Upright or Pendant Sprinkler Spacing***

Maximum Coverage per Sprinkler

Light hazard	• 200 ft^2 (18.6 m^2) under smooth ceiling, bar joist, and beam-and-girder construction [225 ft^2 (20.9 m^2) for hydraulically designed systems) • 130 ft^2 (12.1 m^2) under exposed construction such as wood joist, composite wood joist, and parallel chord wood truss of 16 in. (406 mm) nominal depth or less, and pitch chord wood truss of any depth[b] • 168 ft^2 (15.6 m^2) for other types of construction
Ordinary hazard	• 130 ft^2 (12.1 m^2) for all types of construction
High piled storage	• 100 ft^2 (9.3 m^2)[b]
Extra hazard	• 90 ft^2 (8.4 m^2) for all types of construction [100 ft^2 (9.3 m^2) for hydraulically designed systems)[b]

Maximum Spacing Between Sprinklers on Branch Lines, and Between Branch Lines

Light and ordinary hazard	15 ft (4.6 m)
High piled storage	12 ft (3.7 m)[b]
Extra hazard	12 ft (3.7 m)

Part C. ***Sidewall Sprinklers***

To be installed along walls, lintels, or soffits, where distance from ceiling to bottom of lintel or soffit is at least 2 in (51 mm) greater than the distances from ceiling to sidewall sprinkler deflector.

Maximum Coverage per Sprinkler, and Maximum Spacing Between Sprinklers

The dimensions shown below assume sprinklers along the length of one wall of a room or bay.[b]

	Light Hazard Occupancy			*Ordinary Hazard Occupancy*	
	Combustible Construction with Combustible Sheathing	*Combustible Construction Sheathed with Plaster-board, Metal, or Wood Lath and Plaster*	*Noncombustible Construction with Noncombustible Sheathing*	*Combustible Construction with Combustible Sheathing*	*Noncombustible Smooth Ceiling*
Maximum distance between sprinklers on branch line [ft (m)]	14 (4.3)	14 (4.3)	14 (4.3)	10 (3.0)	10 (3.0)
Maximum room width for single branch line along wall [ft (m)]	12 (3.7)	12 (3.7)	14 (4.3)	10 (3.0)	10 (3.0)
Maximum area coverage [(ft^2 (m^2)]	120 (11.1)	168 (15.6)	196 (18.2)	80 (7.4)	100 (9.3)

TABLE 13.8 **Sprinkler Design Guidelines (*continued*)**

Part D. ***Sprinkler Pipe Sizes***[c]

Light Hazard: Branch lines shall not exceed 8 sprinklers on either side of a cross main[d]

Number of Sprinklers at Ceiling				Number of Sprinklers Above and Below a Ceiling			
Steel		Copper		Steel		Copper	
1 in. pipe	2 sprinklers	1 in. tube	2 sprinklers	1 in.	2 sprinklers	1 in.	2 sprinklers
1¼ in. pipe	3 sprinklers	1¼ in. tube	3 sprinklers	1¼ in.	4 sprinklers	1¼ in.	4 sprinklers
1½ in. pipe	5 sprinklers	1½ in. tube	5 sprinklers	1½ in.	7 sprinklers	1½ in.	7 sprinklers
2 in. pipe	10 sprinklers	2 in. tube	12 sprinklers	2 in.	15 sprinklers	2 in.	18 sprinklers
2½ in. pipe	30 sprinklers	2½ in. tube	40 sprinklers	2½ in.	50 sprinklers	2½ in.	65 sprinklers
3 in. pipe	60 sprinklers	3 in. tube	65 sprinklers				
3½ in. pipe	100 sprinklers	3½ in. tube	115 sprinklers				
4 in. pipe	See Note *c*	4 in. tube	See Note *c*				

Ordinary Hazard: Branch lines shall not exceed 8 sprinklers on either side of a cross main.[b]

Number of Sprinklers at Ceiling				Number of Sprinklers Above and Below a Ceiling			
Steel		Copper		Steel		Copper	
1 in. pipe	2 sprinklers	1 in. tube	2 sprinklers	1 in.	2 sprinklers	1 in.	2 sprinklers
1¼ in. pipe	3 sprinklers	1¼ in. tube	3 sprinklers	1¼ in.	4 sprinklers	1¼ in.	4 sprinklers
1½ in. pipe	5 sprinklers	1½ in. tube	5 sprinklers	1½ in.	7 sprinklers	1½ in.	7 sprinklers
2 in. pipe	10 sprinklers	2 in. tube	12 sprinklers	2 in.	15 sprinklers	2 in.	18 sprinklers
2½ in. pipe	20 sprinklers	2½ in. tube	25 sprinklers	2½ in.	30 sprinklers	2½ in.	40 sprinklers
3 in. pipe	40 sprinklers	3 in. tube	45 sprinklers	3 in.	60 sprinklers	3 in.	65 sprinklers
3½ in. pipe	65 sprinklers	3½ in. tube	75 sprinklers				
4 in. pipe	100 sprinklers	4 in. tube	115 sprinklers				
5 in. pipe	160 sprinklers	5 in. tube	180 sprinklers				
6 in. pipe	275 sprinklers	6 in. tube	300 sprinklers				
8 in. pipe	See Note *c*	8 in. tube	See Note *c*				

Number of Sprinklers at Ceiling: Greater than 12 ft (3.7 m) Separations Between Sprinklers or Branch Lines

Steel		*Copper*	
2½ in. pipe	15 sprinklers	2½ in. tube	20 sprinklers
3 in. pipe	30 sprinklers	3 in. tube	35 sprinklers
3½ in. pipe	60 sprinklers	3½-in. tube	65 sprinklers

(for other pipe and tube sizes, see table immediately above for ordinary hazard)

Extra Hazard: Branch lines shall not exceed 6 sprinklers on either side of a cross main.

Number of Sprinklers at Ceiling			
Steel		*Copper*	
1 in. pipe	1 sprinkler	1 in. tube	1 sprinkler
1¼ in. pipe	2 sprinklers	1¼ in. tube	2 sprinklers
1½ in. pipe	5 sprinklers	1½ in. tube	5 sprinklers
2 in. pipe	8 sprinklers	2 in. tube	8 sprinklers
2½ in. pipe	15 sprinklers	2½ in. tube	20 sprinklers
3 in. pipe	27 sprinklers	3 in. tube	30 sprinklers
3½ in. pipe	40 sprinklers	3½ in. tube	45 sprinklers
4 in. pipe	55 sprinklers	4 in. tube	65 sprinklers
5 in. pipe	90 sprinklers	5 in. tube	100 sprinklers
6 in. pipe	150 sprinklers	6 in. tube	170 sprinklers
8 in. pipe	See Note *c*	8 in. tube	See Note *c*

Source: Reprinted with permission from NFPA 13-1989, *Installation of Sprinkler Systems*, copyright © 1989 National Fire Protection Association, Quincy, Massachusetts. This reprinted material is not the complete and official position of the NFPA on the referenced subject, which is represented only by the standard in its entirety.

[a]The *K* factor is explained on page 712.

[b]See NFPA 13-1989 for special requirements, or exceptions.

[c]These sizes do not apply to systems that are hydraulically designed.

[d]Supplied from a common set of branch lines, or separate branch lines from a common cross main.

TABLE 13.9 **Common Failure Modes for Automatic Sprinklers**

Failure Mode	*Potential Causes*
Water supply valves are closed when sprinkler fuses	Inadequate valve supervision Company attitude Maintenance policies
Water does not reach sprinkler	Dry pipe accelerator or exhauster malfunctions Pre-action system malfunctions Maintenance and inspection inadequate
Nozzle fails to open when expected	Fire rate of growth too fast Temperature of link inappropriate for the area protected Sprinkler link protected from heat Sprinkler link painted, taped, bagged or corroded Sprinkler skipping
Water cannot contact fuel (*Note:* The intent of this failure mode is to ensure that discharge is not interrupted in a manner that will prevent fire control by a sprinkler.)	Fuel is protected High piled storage is present New construction (walls, ductwork, ceilings) obstructs water spray
Water discharge density is not sufficient	Discharge needs are insufficient for the type of fire and the rate of heat release Change in combustible contents occurred Number of sprinklers open is too great for the water supply Water pressure too low Water droplet size is inappropriate for the fire size
Enough water does not continue to flow	Water supply is inadequate because of original deficiencies, changes in water supply, or changes in the combustible contents Pumps are inadequate or unreliable Power supply malfunctions System is disrupted

Source: Reprinted with permission from the *Fire Protection Handbook,* 16th ed., © 1986 by the National Fire Protection Association, Quincy, Massachusetts. This reprinted material is not the complete and official position of the National Fire Protection Association on the referenced subject, which is represented only by the standard in its entirety.

Pressure for the sprinkler is supplied by a small, 20-gpm (1.26 L/s) jockey pump. Signals from the sprinkler system bring in a 750-gpm (47.3 L/s) main pump. If this should fail, a diesel-engine-driven pump of equal capacity automatically takes over.

The circuitry of two 1000-gpm (63.1 L/s) deep-well pumps and two 200-gpm (12.6 L/s) domestic pumps was designed by engineers in the office of Skidmore, Owings & Merrill. Design of the controls was made by the Automatic Control Company of St. Paul. Three sensing units govern the operation of the pumps, bubble-control units in each of the two reservoirs, and a dual-control unit that regulates supply for the pressure tank. All three elements are connected to a large central cabinet located in an underground pump room next to the reservoir. Pumps are controlled through the cabinet.

The bubble control uses a small air compres-

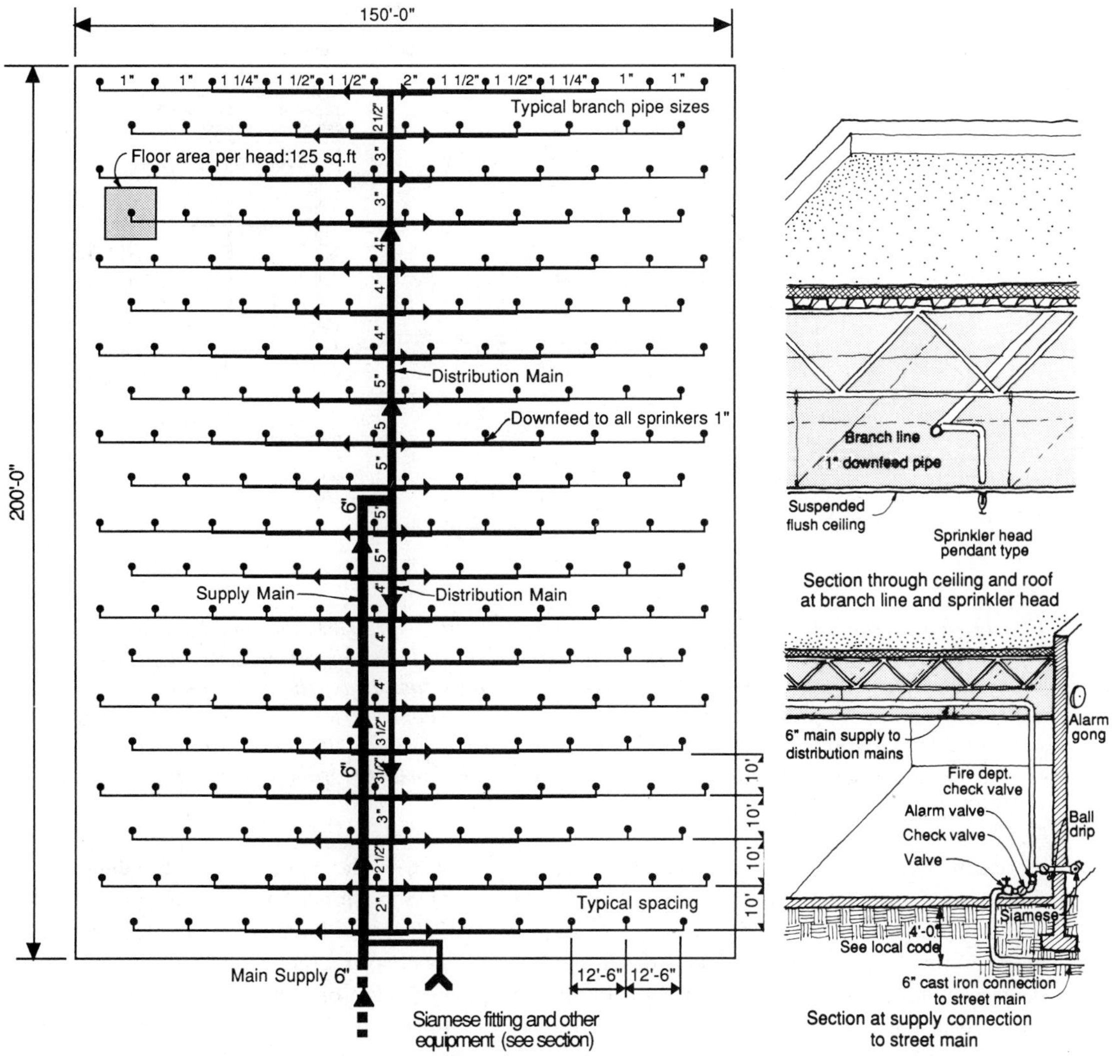

Fig. 13.15 *Plan of sprinklered industrial building, ordinary hazard occupancy, 125 ft² (11.6 m²) per sprinkler head. Sprinklers (and standpipes) may use water from street mains when pressure is adequate. Either system may use pneumatic or gravity tanks. When the latter is used to supply both systems, an independent sprinkler reserve occupies the bottom, and the fire standpipe supply occupies the top. Auxiliary fire-engine feed by siamese should be provided in all cases. Auxiliary sources and standby pressurization may be required if street main adequacy is questionable.*

sor within the central cabinet to send a flow of air through a ¼-in. (6.4-mm) tube to the reservoir. Back pressure on this flow, which varies with the level in the reservoir, operates pressure switches within the cabinet.

The hydropneumatic tank is used not for water storage, but to store air under pressure that will balance out surge from the two domestic pumps and reduce the number of times the pumps must be started and stopped. This is a hybrid of the closed system in which several pumps are sequenced automatically to supply even pressure. Its advantage is that only two pumps are used.

One of the disadvantages in the past has been the difficulty of maintaining the correct ratio of 60% air to 40% water. Tanks supplied with water from deep wells become air-bound as water

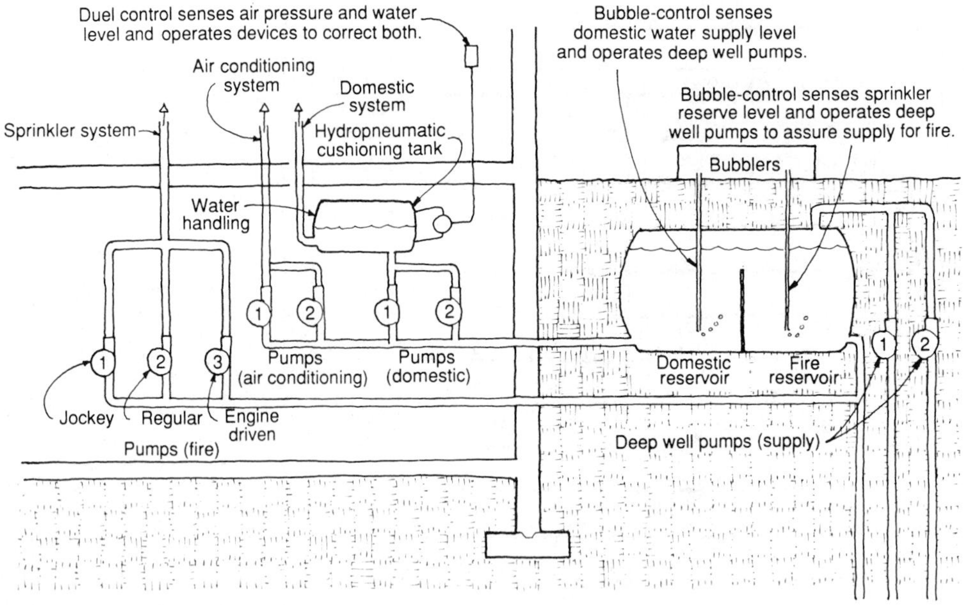

Fig. 13.16 Diagram of upfeed pumping system for supply of sprinklers and building demands for domestic water. In the General Mills Building, a large subsurface concrete water tank provides the secondary water source that fire insurance carriers look for. Thus, the wells and their pumps constitute a primary water source. If the well yield fails, the concrete tank supply can be tapped. An additional resource provides engine-driven pumping that operates in the event of a power outage at an electric utility company. Automatic controls and piping for the storage tank reserve system are not shown in this illustration.

stored in them gives up its absorbed air. The dual-control installation in the General Mills system eliminates the need for manual adjustment of this ratio by employing two sensing devices within a single control. A drop in air pressure in the tank sends signals to start the pump. A rise in water level sends signals to stop it.

The signals from the dual-control and the bubble-control units are processed in the central cabinet for correct time delay through motor-driven relays. The central control system also either alternates the pumps, to ensure even wear, or runs them together if the demand requires. In the event of low suction, it shuts the pumps off to prevent motor damage. At the same time, it sends an alarm to the office of the maintenance engineer, indicating the location of the trouble.

Normally, the system automatically satisfies the heavy demands of air conditioning, fire control, and domestic water supply in this modern, rurally isolated office building.

Upfeed pumping for an urban high-rise building is seen in the Transamerica Building (Fig. 13.17) in San Francisco. Two fire pumps can deliver 750 gpm at 275 psi (47.3 L/s at 1896 kPa) discharge pressure; they draw from city mains at 50 psi (345 kPa) or from a 5000-gal (18,925-L) closed tank in the basement. These pumps feed into two 6-in.-diameter (152-mm) "express" risers, one in each stair tower, that rise the full height of this 48-story office building. These risers serve both the sprinkler system and fire department hose lines. In each of three 16-story zones, "local" 6-in.-diameter risers branch off to feed a 2-in. (50.8-mm) looped main at each floor.

The sprinkler piping is carried above suspended ceilings; pendant sprinkler locations are coordinated with the modular grid that also locates partitions, air diffusers, and utility jacks. There are provisions for moving sprinkler heads as tenants change. At regular intervals, tees have been provided with one outlet stubbed and capped for future use. Typical office spaces are served by fully recessed, ½-in.-orifice, 165 F (12.7 mm, 74°C) pendant sprinklers. Exposed pendant sprinklers are used in toilet rooms and service areas.

(d) Provisions for Water Drainage. Sprinkler heads can release a great deal of water, most of which will remain unvaporized and quickly collect at floor level. In addition to waterproofing the floors and at lower walls, columns, and other elements, provision should be made, where possible, for gravity drainage of water. Scuppers in exterior walls are preferable to floor drains, which are more easily clogged by debris. Scuppers are provided with hoods that protect against control of infiltration, birds, or insects.

13.4 Other Fire Suppression Methods

When water poses almost as much of a threat as fire to a structure or its contents, a variety of other fire suppression methods are available. The most passive of such measures are intumescent materials, which expand rapidly as they are touched by fire. This process creates air pockets that insulate a surface from the fire or swell a material until it blocks openings through which fire (or smoke) could have passed. Intumescent paints, caulks, and putties are available. Some intumescent materials come in ½-in. (6.35-mm)-thick sheets, with various facing materials.

(a) Portable Fire Extinguishers. Most fires can be extinguished at an early stage with these common devices. Although some types of fire extinguishers do contain water or water mixtures, others contain dry chemicals and/or gases appropriate for various fire applications. A typical portable extinguisher was shown in Fig. 13.12*b*. These devices should be located in conspicuous places along ordinary paths of egress (see Fig. 13.18). Portable fire extinguishers are labeled as follows.

Fig. 13.17 *Transamerica Building, San Francisco, California; William Pereira & Associates, Los Angeles, California, architects.* (a) *The pyramid form provides rentable floor areas ranging from 22,000 to 3000 ft² (2044 to 279 m²).* (b) (overleaf) *Sixth-floor sprinkler plan, showing two risers, looped feed main, and branch lines. (Courtesy of Copper Development Association.)*

Class 1A to 40A: The numerals refer to the relative extinguishing potential (40A will extinguish 40 times as much as 1A). "A" refers to the contents: water or water-based agents, or multipurpose chemical agents. For use on ordinary combustibles such as wood, trash, paper, and textiles.

Class 5B to 40B: The numerals refer to the approximate square footage of deep-layer liquid fire that an inexperienced operator can extinguish. "B" refers to the contents: smothering or flame-interrupting chemicals such as CO_2, sodium, and potassium bicarbonate base dry chemicals, foam, or haloge-

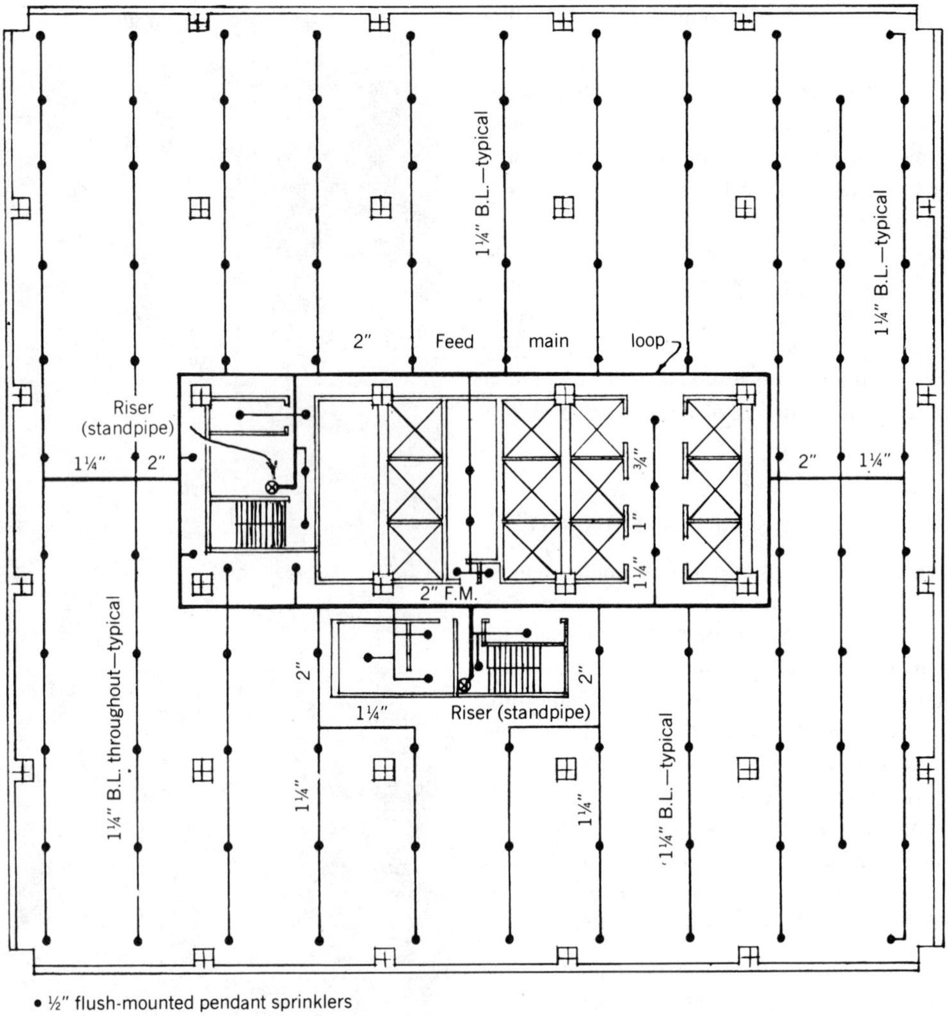

(b)

Fig. 13.17 (continued)

nated agents. For use on fires in liquid petroleum products or flammable liquids such as paint and solvents.

Class C: Contents are non-electrically-conducting, such as CO_2, sodium, and potassium bicarbonate base dry chemicals, or halogenated agents. For use on fires on or near electrical equipment.

Class A:B:C: "Multipurpose" dry chemical extinguishers filled primarily with ammonium phosphate. Although indicated for all three classes of use, ammonium phosphate is not ideal for electrical fires, because it leaves an especially hard residue.

Class D: Contents are dry powders, such as graphite or sodium chloride. For use on a variety of combustible metals; the specific combustible metal for which the extinguisher is designed is printed on the extinguisher's nameplate.

For the details of portable extinguisher types and location requirements, see NFPA 10, *Portable Fire Extinguishers,* available from the National Fire Protection Association.

(b) Carbon Dioxide (CO_2). This gas smothers a fire by displacing oxygen; therefore, CO_2 is usually used in tightly confined spaces that are free of people or animals—for example, display cases, mechanical or electrical chases, and unventilated areas above suspended ceilings. CO_2 is stored, under great pressure, as a liquid; when it is released as a gas, it absorbs about 120 Btu/lb, to provide cooling as well as smothering action. CO_2 is noncombustible and will not react with most substances. It spreads from discharge as a gas, but will stratify over time.

The main problem with CO_2 is that it must be used at concentrations of from 21% to 62% of that of the air, depending on the fire's fuel. However, at a CO_2 concentration of 9%, loss of consciousness will occur after a few minutes. This may allow enough time for an awake occupant to escape, but it doesn't help the firefighters' (or trapped persons') environment. Another potential problem is that after the initial smothering and subsequent dissipation of the CO_2, smoldering embers might reignite.

A typical CO_2 automatic detection/suppression system will have components similar to those of the Halon 1301 system shown later in Fig. 13.20.

(c) Foams. Because foams (masses of gas-filled bubbles) are lighter than water and flammable liquids, they float on the surfaces of burning liquids, smothering and cooling the fire. Foams can be designed to inundate a surface or to fill cavities; they can be thin and rapidly spread, or thick, tough, and heat-resistant. Both air-foaming and chemical-foaming methods can be employed. Because foam is so effective on flammable liquid fires, it is especially popular in airplane hangars (Fig. 13.19).

In Great Britain in the 1950s, a new high-expansion foam was introduced for firefighting in mines. Its introduction into the United States proved its effectiveness against fires of a similar nature in this country. Years of research and improvement improved the foaming agent and led to a wide variety of uses in many other fields of firefighting. The patents were acquired in 1963 by Walter Kidde & Company. Among the many unique features of this system is the expansion rate of the foaming process. One thousand gallons of bubbles are produced from 1 gal (3.785 L) of water. This 1000-to-1 rate is most dramatic when compared with the 10-to-1 rate of the fa-

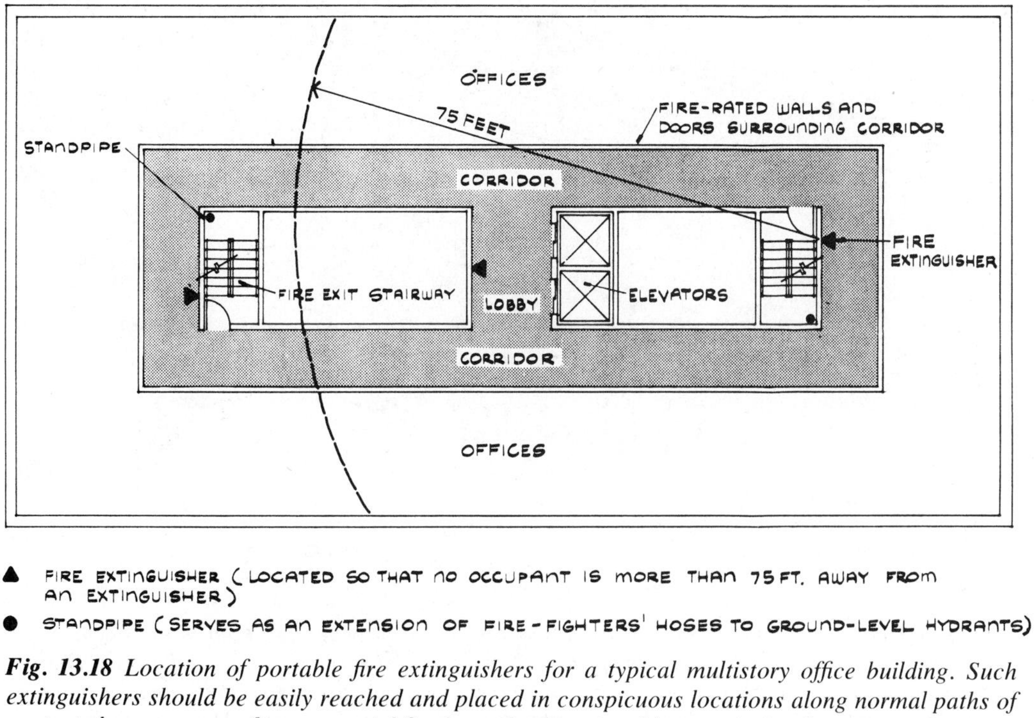

Fig. 13.18 *Location of portable fire extinguishers for a typical multistory office building. Such extinguishers should be easily reached and placed in conspicuous locations along normal paths of protected egress, away from potential fire hazards. (Reprinted by permission from Egan,* Concepts in Building Firesafety, *© 1978 by John Wiley & Sons.)*

miliar protein foam that firefighters have long used effectively, especially on oil fires.

Albert Kahn Associated Architects & Engineers, Inc., designed and installed a system of fire protection, using this method, in the North Central Airlines Hangar Building at the Metropolitan Airport in Detroit.

The foam-generating equipment in this installation can fill the 38,400-ft² (3567-m²) hangar with 1,400,000 ft³ (39,648 m³) of foam to a height of 36 ft (10.8 m) in less than 12 min. Automatic devices, upon sensing abnormal heat increase, operate the foam generators, open roof vents, start the smoke control exhaust fans in the vents and transmit a fire alarm signal to the Airport Fire Department. Foam discharge is delayed 30 seconds while evacuation sirens sound, to permit occupants to leave the fire area. Manual "override" controls allow personnel to start the system in the event of failure of the automatic controls, or to stop it if the fire is small and controllable by other methods.

The high-expansion foam is created by wetting a nylon net with a mixture of water and a special detergent soap concentrate. A large blower directs an air current through the net, producing an avalanche of foam.

Suds blanket the fire, attacking it in several ways. The water in the suds converts to steam, absorbing the heat of the fire. The expansion of the foam into steam reduces the oxygen content to about 7%, which is insufficient to support active combustion. A cooling effect is achieved by the wetting action of the breaking bubbles. The movement of air currents toward the fire to replace the rising hot gases draws the foam to the center of the fire. There it blocks the air flow and cuts off the supply of oxygen.

The fire, thus contained and diminished, can be approached by firefighters for further control.

Fig. 13.19 *Detergent foam discharged from four units, two of which are seen in the illustration, after 3 min of operation. This illustration at the North Central Airlines hanger building at Detroit Metropolitan Airport smothers fire and prevents its spread, and will not harm airplanes or machinery. Firefighters can breathe inside the foam. The system can handle combustible liquid fires. Note also the grid of upright sprinklers directly below the roof surface.*

Personnel advancing through the cooler sections of the foam are safe because the foam in these areas is 99% air and can support human life.

Aircraft are not harmed by the foam. Delicate machinery that might be injured by high velocity streams of water is undamaged and left quite clean when the foam is rinsed away. After all, it *is* a soap.

The structure—in this case, an open steel frame with a metal roof deck—is protected from excessive temperatures that might weaken it and cause it to collapse.

Perhaps the most serious potential disadvantage of foam is that the air turbulence created by a fire can, if violent enough, divert lightweight foam from its target.

(d) Halogenated Agents. Halogenated hydrocarbons, known commonly as "halons," are gases (stored as liquids) in which one or more hydrogen atoms have been replaced by halogen atoms. Although the original hydrocarbons are often highly flammable gases, the substitution of halogen not only makes the gas itself inflammable, but also gives it flame-extinguishing capabilities. The exact process in which halons extinguish fires is not known; apparently, unlike water or CO_2, they work primarily by inhibiting flames chemically rather than physically.

The most widely used of these agents is *Halon 1301;* the name signifies that its molecule contains one carbon atom, three fluorine atoms, no chlorine atoms, and one bromine atom. (When a fifth number is listed for halons, it refers to the number of iodine atoms.) Hydrogen atoms are not accounted for.

The advantages of Halon 1301 are that it can be released with relative safety in a flooding system in areas such as computer rooms, quickly extinguishing a fire with little harm to contents and—as yet—little demonstrated harm to people. It is also lightweight and space-saving relative to other fire suppressants. Halon 1301 is used on all commercial aircraft and in many special building applications, such as computer rooms, museums and libraries, telephone exchanges, and kitchens. Accordingly, it is effective against classes A, B, and C fires. It is commonly used in portable fire extinguishers.

Halon 1301 does have limitations: It will not extinguish burning metals or self-oxidizing materials such as gunpowder. For deep-seated class A fires, a higher concentration over a longer "soaking time" is required, necessitating much larger storage vessels and spaces that are relatively airtight. Halon is heavier than air, and it will slowly stratify in a steadily dropping layer as fresh air is introduced; also, it is more expensive than CO_2 or water. Finally, Halon 1301 at high concentrations will cause dizziness, impaired coordination, and reduced mental acuity. These are temporary effects, however; no residual toxic effects have yet been demonstrated. Ordinarily, Halon 1301 concentrations of 4% to 6% are sufficient to extinguish a fire. At concentrations below 7%, there is virtually no effect on humans. At concentrations of from 7% to 10%, the above-listed symptoms occur; at those above 10%, the toxic effects are potentially serious. When it is exposed to flame or very high surface temperatures, Halon 1301 decomposes to gases that are more dangerous; these have a sharp and acrid odor, which encourages occupants to leave but leaves a poor environment for firefighters. Halon 1301 is stored under high pressure in cylinders that should be kept from heat exposure. Halon is one of the CFC gases that threaten the ozone layer.

Halon 1301, unlike water, is really more a protector of building *contents* than of the building *structure*. It leaves no sticky residue, such as most dry chemicals produce, and is therefore a popular choice where a clean fire-suppressing agent is required, people are present, and there are objects or processes of high value. Occasionally, it is used where water availability is low, or where space cannot be found for systems that use other, bulkier fire suppressants. A simple diagram of a Halon 1301 installation is shown in Fig. 13.20.

Many other potential halogenated agents can by used in fire suppression. The prevalence of Halon 1301 seems attributable to its especially low corrosive effects on common building metals or plastics. For a more complete listing of halons, see the NFPA's *Fire Protection Handbook,* as well as specific NFPA publications on halon systems.

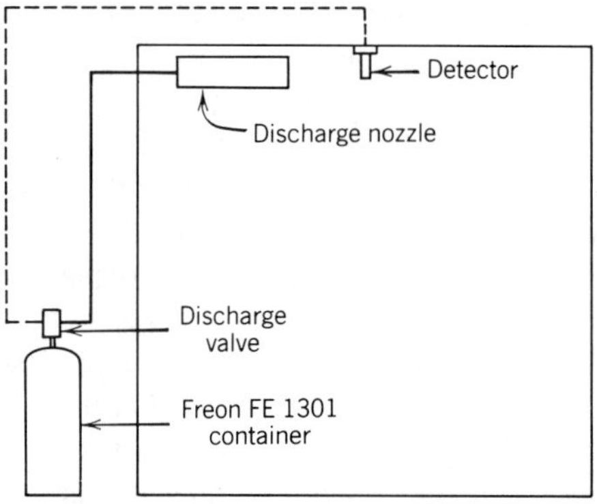

Fig. 13.20 Schematic of a Halon 1301 automatic fire detection/suppression system that provides total flooding of a space. Reprinted by permission from Jensen (ed.), Fire Protection for the Design Professional, *copyright © 1975 by Van Nostrand Reinhold Company, Inc.*

13.5 Lightning Protection

Lightning is nature's most destructive force: the average lightning discharge is estimated at 200 million V, 30,000 A, and courses through a grounded object in less than a thousandth of a second. "Cold" lightning bolts have ample current, voltage, and duration to shatter and kill, but not to ignite combustibles. "Hot" bolts will ignite combustibles as well.

The decision of whether or not to protect a structure depends on an evaluation of these factors:

1. Frequency and severity of thunder storms.
2. Value and nature of building and contents.
3. Hazard to building occupants.
4. Building exposure. Buildings in open and exposed areas are more susceptible to lightning than urban buildings.
5. Indirect effects. For example, loss of a water tower will seriously affect fire prevention and other services.

If a decision is reached to protect a building, it should be done completely and properly, with UL label equipment (Label A and B) and UL approved installation (Label C). A partially protected building is in reality an improperly protected building, which may well be worse than one with no protection at all.

The relevant standards are NFPA Bulletin No. 78, *Lightning Code* and Underwriters Laboratories Standard UL 96A. *Master Labeled Lightning Protection Systems.* The subject of lightning protection is a complex one, and the design of an adequate system is best left to specialists. Some of the design considerations and data on available materials are discussed briefly below.

The basic principle in lightning protection is to provide a continuous metallic path to solid (low resistance) ground for the lightning stroke, since *there is no known method of protection that will prevent the occurrence of a lightning stroke.* This will prevent the stroke from passing through the nonconducting portions of a building. Such passage is accompanied by great heat and mechanical forces caused by the high resistance of this path.

Any lightning protection should consist of at least two *air terminals;* at least two *ground terminals,* spaced as far apart as possible and terminating in always-moist soil; and the *down conductors* that connect air and ground terminals. The conductors may be exposed on the building exterior or concealed within a structure.

All metallic objects on the roof should be bonded to a looped conductor that joins the air terminals. These air terminals commonly are placed at both roof edges and ridges, at intervals of about 20 ft (6 m). The conductors are made usually of copper and are best housed in a plastic pipe conduit. Ground conductors, often called counterpoise conductors, are also installed to connect the earth terminals.

In reinforced-concrete buildings, column steel reinforcing can be used for lightning conduction *if* the column steel is welded, rather than merely tied. Where tied, and in buildings constructed with precast concrete panels, reinforcing steel (and other metal) within a few feet of lightning conductors should be bonded to them to avoid arcing. Such arcing, as noted above, is accompanied by great heat and mechanical forces which can produce major structural damage and fire. In steel buildings, the columns relatively easily become lightning conductors; care must be taken to adequately bond both tops and foundations of such columns to the air and ground terminals, respectively (see Fig. 13.21).

FIRE PROTECTION

A lightning arrester should be placed on all aerial service conductors, whether high or low voltage. Electrical and electronic equipment, which is sensitive to voltage surges such as those accompanying lightning strokes, should be individually protected by surge arresters (see Section 16.28).

Lightning protection systems are particularly important for very tall buildings. Figure 13.22 shows the schematic diagram for the John Hancock Center in Chicago, a structural steel building of unusual height. Intermediate loops at the mechanical equipment floors help to overcome the problems posed by the poorer conductivity of metal piping systems, which can become additional high-impedance paths for lightning, with resultant destructive effects. The intermediate loops are connected to each outside column, as well as to the main electrical, plumbing, air conditioning, and fire protection risers. These loops are also connected to both the roof conductor and ground counterpoise conductor.

FIRE ALARM SYSTEMS

13.6 General Considerations

A fire alarm system serves primarily to protect life and secondarily to prevent property loss. Since buildings vary in occupancy, flammability, type of construction, and value, the fire alarm system must be tailored to the needs of the specific facility. In schools for instance, where the paramount consideration is rapid, orderly evacuation, the type of system used will generally be the same although the means of automatic detection will vary with construction type, building height, specific area use, furnishings, and staffing. The fire alarm is part of the overall fire *protection* system design of the building. In particular, it overlaps with the design of safe egress and smoke/fire control in matters such as fan control and smoke venting, smoke door closers, rolling shutters, elevator capture, and the like. These automatic functions are initiated by operation of the alarm system but are designed in accordance with the overall fire protection plan.

As with other alarm systems, a fire alarm system has three basic parts: signal initiation, signal processing, and alarm indication. The signal initiation can be manual (pull stations or telephones) or automatic (fire and smoke detectors and/or waterflow switches). The alarm signal is processed by some sort of control equipment, which in turn activates audible and visible alarms and, in some cases, alerts a central fire station or municipal authorities.

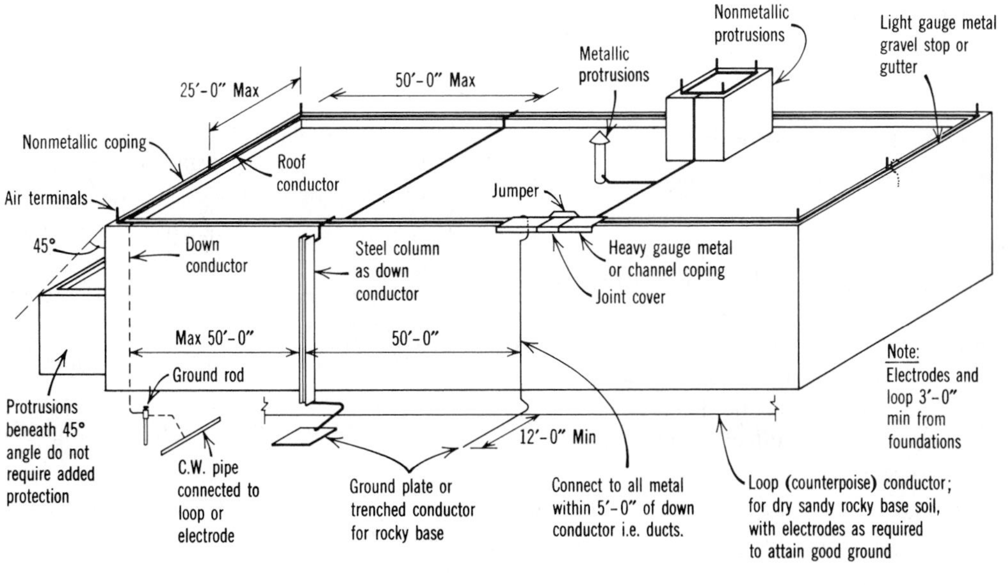

Fig. 13.21 *Diagram of a typical lightning-protection system.*

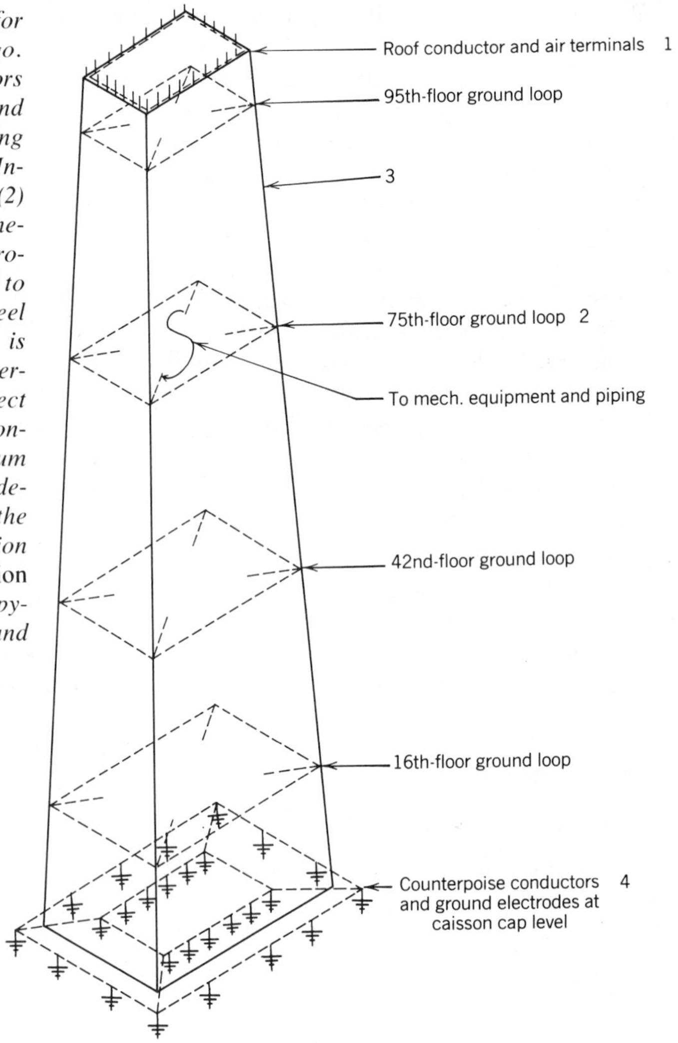

Fig. 13.22 *Lightning protection for the John Hancock Center, Chicago. Air terminals with roof conductors (1) are bonded to building steel and looped to assure that a single wiring break will not fault the system. Intermediate-floor ground loops (2) with connections to building mechanical equipment and piping protect against side-effect flashover to people or equipment. Building steel (3), properly bonded together, is used as a down conductor. Counterpoise conductors (4) cross-connect ground electrodes with ground conductors located below minimum groundwater level, to assure adequate and uniform dissipation of the charge. (Reprinted by permission from Jensen (ed.),* Fire Protection for the Design Professional, *copyright © 1975 by Van Nostrand Reinhold Company, Inc.)*

13.7 Fire Codes, Authorities, and Standards

There are probably more codes and standards in the area of fire protection than in any other area, with the possible exception of structural standards. Although these codes are devoted primarily to fire protection and life safety, they also govern the type of fire alarm system to be installed, its components and its installation. The following codes bear directly on fire alarm system arrangements.

NFPA (National Fire Protection Association) Life Safety Code 101—deals with municipal and auxiliary connections; auxiliary controls for elevators, door release, and so forth; type and location of alarm devices; circuit arrangement; alarm types and signals.

NFPA 70 National Electrical Code; particularly Art. 760—Fire Protective Signaling Systems.

NFPA 71 Signaling Systems for Central Station Service.

NFPA 72A Local Protective Signaling Systems.

NFPA 72B Auxiliary Protective Signaling Systems.

NFPA 72C Remote Station Protective Signaling Systems.

NFPA 72D Proprietary Protective Signaling Systems.

NFPA 72E Automatic Fire Detectors.

NFPA 72F Emergency Voice/Alarm Communication Systems.

NFPA 74 Household Fire Warning Equipment.

SBC (Standard Building Code), published by the Southern Building Code Congress—deals with auxiliary system controls such as door release, elevator capture, and so forth. In addition, SBC has a special section on high-rise buildings, covering fire communication, elevator control, and smoke alarms.

BOCA (Building Officials and Code Administrators International) Basic Fire Prevention Codes—cover smoke detection, alarm audibility, plus requirements for high-rise buildings.

HUD (U.S. Dept. of Housing and Urban Development)—covers requirements for residential buildings and care-type facilities. FHA minimum property standards are included here.

In addition, several insurance groups (e.g., Factory Mutual) issue standards that may apply to a specific facility, and there are local codes, standards, and fire marshal regulations. As with other aspects of construction—and perhaps more importantly than with most—the architect/designer must ascertain which regulations have jurisdiction before proceeding with the design. The specific recommendations given in Sections 13.16 through 13.20 are representative of present good practice. Actual design must be based on and meet the requirements of *current* NPFA standards plus other codes having jurisdiction in the particular locale of the project.

13.8 Fire Alarm Definitions and Terms

The following list can serve as a reference to terminology that the reader will encounter in fire alarm work.

Automatic System. A system in which an alarm-initiating device operates automatically to transmit or sound an alarm signal.

Auxiliary Fire Alarm System. See Section 13.13 and Fig. 13.34*b*.

Breakglass. Refers to a false-alarm deterrent available in fire alarm stations; a glass rod is placed across the pull-lever that breaks easily when the lever is pulled.

Coded Alarm Signal. An alarm signal that represents a 1-, 2-, 3-, or 4-digit number indicative of the location of the fire alarm station operated.

Coded System. One in which not less than three rounds of coded alarm signals are transmitted, after which the fire alarm system may be manually or automatically silenced.

Common Code. See *Dual-coded system, Master-coded system, Noncoded system, Selective-coded system, Zone-coded system.*

Continuous Ringing. Refers to a continuous alarm. In coded systems so arranged, it refers to the signal that sounds after the completion of the normal number of rounds (usually four) of identifying coded alarm signal.

Control Unit (Fire Alarm Panel). The controls, microprocessors, relays, switches, and associated circuits necessary to (1) furnish power to a fire alarm system, (2) receive and process signals from alarm-initiating devices and transmit them to indicating devices and accessory equipment, and (3) electrically supervise the system circuitry.

Dual-Coded System. See Section 13.15*d*.

Local Fire Alarm System. See Section 13.13 and Fig. 13.34*a*.

Local Noninterfering Coded Station. A fire alarm station that, once actuated, will transmit not less than four rounds of coded alarm signals and cannot be interfered with by any subsequent actuation of that station until it has transmitted its complete signal.

Manual System. One in which the alarm initiating device is operated manually to transmit or sound an alarm signal.

Master-Coded System. See Section 13.15*b*.

Noncoded System. See Section 13.15*a*.

Presignal System. See Section 13.15*f*.

Proprietary Fire Alarm System. See Section 13.13*d* and Fig. 13.34*d*.

Remote-Station Fire Alarm System. See Section 13.13*c* and Fig. 13.34*c*.

Selective-Coded System. See Section 13.15*e*.

Station, Fire Alarm. A manually operated alarm-initiating device; may be equipped to generate a continuous signal (noncoded station) or a series of coded pulses (coded station).

Supervised System. A system in which a break or ground in the wiring, that will prevent the transmission of an alarm signal will actuate a trouble signal.

Trouble Signal. A signal indicating trouble of any nature, such as a circuit break or ground, occurring in the device or wiring associated with a fire alarm system.

Zone-Coded System. See Section 13.15*c*.

13.9 Principles of Automatic Fire Detection

Fire authorities agree that *most* fires pass through four stages, the last of which is the visible, flaming fire. These stages are shown in Fig. 13.23 as a function of duration and degree of hazard. Also shown is the type of automatic detector recommended for detection at each stage.

(a) Incipient Stage. In this stage the combustion products comprise a significant quantity of microscopic particles (0.01 to 1.0 μm), which are best detected by *ionization-type* detectors (see Fig. 13.24). These detectors contain a small amount of radioactive material that serves to ionize the air between two charged surfaces, causing a current to flow. Combustion particles entering the detector chamber reduce air ion mobility, thus reducing current flow and increasing voltage. These changes are sensed and the alarm is set off. Response time of this type of detector depends on how rapidly the combustion particles can reach the detector—a factor that varies with room air currents and the type of material "burning." Once the particles reach the detector, response is instantaneous. Therefore, ionization detectors are best applied indoors, in spaces with stagnant air or low air velocity (below 50 fpm) and in which little visible smoke (large particles) is expected. (Some manufacturers make a special unit whose sensitivity *increases* with air velocity, for application in spaces with air velocity up to 500 fpm).

Ionization detectors should not be installed on warm or hot ceilings, or in any other location where hot air concentrates, since the hot air will prevent the combustion particles from reaching the detectors. As a corollary, ionization detector sensitivity is higher at low ambient temperature. Since particles tend to agglomerate after leaving the combustion area, because of mutual attraction, ionization detectors are most sensitive in low ceiling spaces.

Because ionization detectors react to minute particles of combustion, they should not be applied in, or close to, areas where normal activities produces such particles, since the obvious result will be nuisance alarms. Thus kitchens, bakeries, welding and brazing shops, workshops using open flames and burners, and areas with concentrated engine exhaust fumes are all spaces inappropriate for their application. Ionization detectors respond best to fast-burning flaming fires.

As can be seen from Fig. 13.23, the incipient stage of a fire also produces changes in the gas content of the environment. These combustion gases, which are not normally present in the air, are detectable by devices known as gas-sensing fire detectors. The principle of operation of these detectors depends on the type. The semiconductor type reacts to a change in conductivity due to the presence of combustion gases, while the catalytic element type detects a change in electrical resistance of the element caused by the presence of gases. Gas detectors are frequently used in conjunction with another type of detector.

Since dust settling in the ionization chambers reduces sensitivity, these units need periodic maintenance. Coverage of detectors varies from 150 to 750 ft^2 per unit, depending on the unit used, the type of combustible material in the space, and ambient conditions. Once the fire has passed the incipient stage and becomes smokey, or where even in the incipient stage the combustion products are the large particles typical of visible smoke, the ionization detector loses sen-

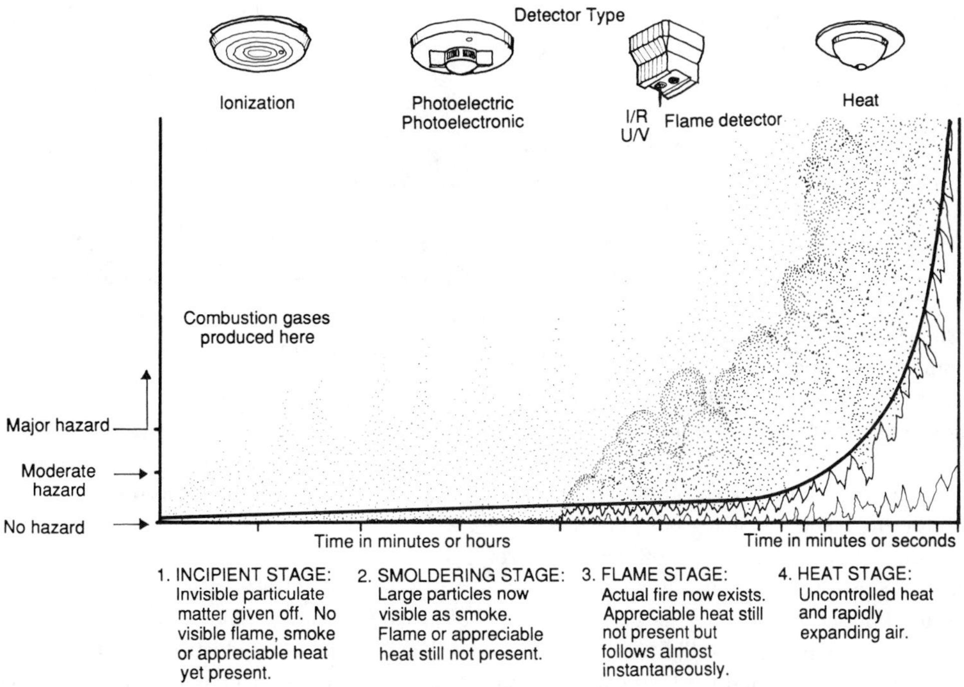

Fig. 13.23 *Four stages of a fire. Early detection of each stage requires a different type of detector that will respond to the fire's particular characteristics (fire signature) at that stage.*

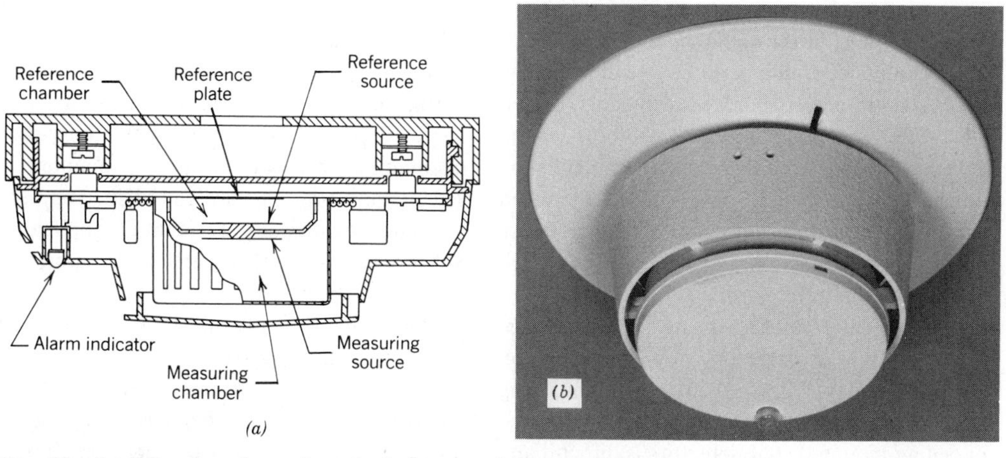

Fig. 13.24 (a) *Section through a dual-chamber ionization-type detector. The use of a reference chamber that is not exposed to combustion products reduces the influence of temperature and humidity, permitting sensitivity to be increased without triggering false alarms.* (b) *Photograph of a modern dual-chamber detector, with adjustable sensitivity and integral supervisory LED. Illustrated unit is 6 in. in diameter and 3 in. high. Ionization detectors are classified as early-warning units. (Courtesy of FENWAL, Division of Kidde.)*

sitivity, and photoelectric detectors are recommended. Ionization, gas, and photoelectric detectors are classified as early-warning types.

(b) Smoldering Stage. Refer again to Fig. 13.23. The smoldering stage of a fire is characterized by particles up to 10 μm in size. Such particles, although small, are visible to the naked eye as smoke and are best detected by photometric means. The simplest type of *photoelectric smoke detector* operates on the principle of beam obscuration, as shown schematically in Fig. 13.25. A beam of light is directed onto a photosensor and a steady-state, no-smoke, circuit condition is established. The presence of smoke in sufficient concentration will partially obscure the beam, changing current flow in the photocell circuit and setting off an alarm response. Sensitivity is typically set at 1 to 2% obscuration per foot.

In the older models of this type, the photosensor is a photocell and the light source a conventional lamp and a simple projection lens. Accumulation of dust and dirt and the presence of heavy fumes from industrial processes cause gradual obscuration of both the photo cell and the lamp. This, combined with lamp aging and consequent light reduction result in false alarms which are not only a nuisance (as will be discussed below) but also can constitute a hazard. To correct this in these older models, continuous maintenance and periodic recalibration is necessary. In newer short-throw models, pulsed LEDs are the light source, and photodiodes the detectors. Associated electronic circuitry is arranged to check that obscuration is indeed caused by smoke and not contaminants. These models require less maintenance and are much less likely to falsely alarm.

An application of long-throw beam-type photoelectric smoke detectors now approved by some fire codes is in areas such as high-ceiling atriums and similar spaces, where maintenance of ceiling-mounted spot detectors is difficult and expensive. Beam-type detectors with a throw of several hundred feet can be mounted on the sides of such spaces and readily maintained. They are normally quite effective since the line of sight between light source and photoreceptor in such areas is normally entirely unobstructed.

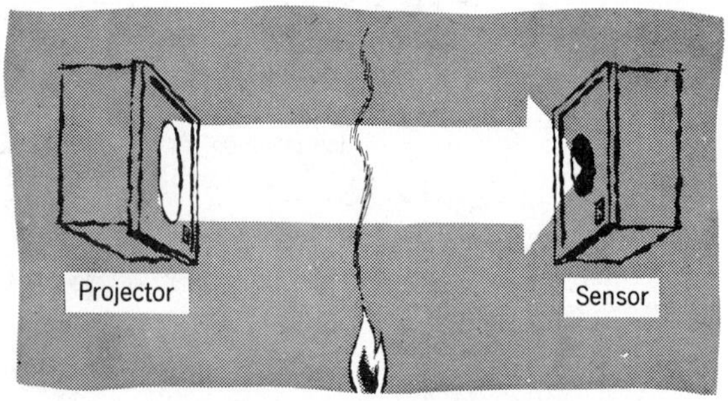

Fig. 13.25 Principle of smoke detection by beam obscuration. This design is used in two ways: as illustrated, with a projector in one location and a photocell at a remote location; or with the entire assembly in a single small housing, with light source and receiver on opposite sides of a 2- to 3-in.-wide smoke chamber.

A second design, usually referred to as a "photoelectronic," scattered-light, or "Tyndall-effect" detector, is illustrated in Fig. 13.26. In this design a beam of pulsed LED light is directed at a supervisory photocell, which serves to provide a baseline reference signal. The alarm photocell is shielded and normally receives no light. When smoke enters the unit, light is reflected from (scattered by) the smoke particles, and strikes the alarm cell. This changes the cell's resistance and the resultant signal is amplified electronically, causing an alarm. An alternative design depends on refraction of light to set off the alarm. These designs are not sensitive to normal dust and dirt accumulation or to light source depreciation, and high sensitivity can be maintained without continual maintenance. Thus, they normally are found in commercial use and high-quality residential applications. Sensitivity is usually set at 1% obscuration per foot.

Photoelectric smoke detectors of all designs will detect particles from about 0.2 μm (0.2×10^{-3} mm) to about 1000 μm (1.0 mm). Thus they are useful not only for smoldering fires but also for the smokey fires that characterize the burning of certain plastics and chemicals. Also, particle agglomeration, which increases with the distance the smoke travels, does not reduce their sensitivity, as it does with ionization types.

Maximum recommended spacing for photoelectric detectors, as for other types, is given by

FIRE PROTECTION

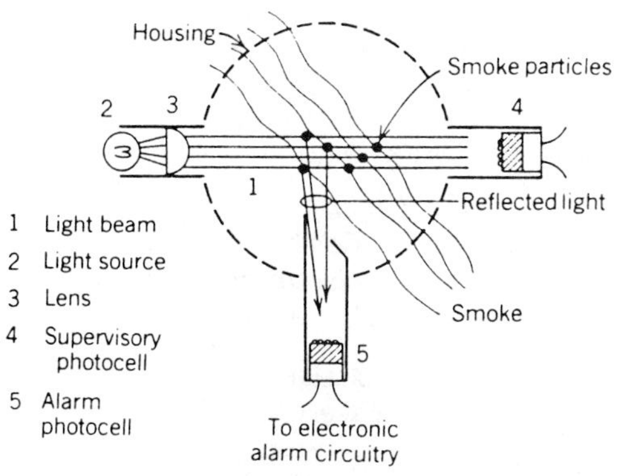

(a)

(b)

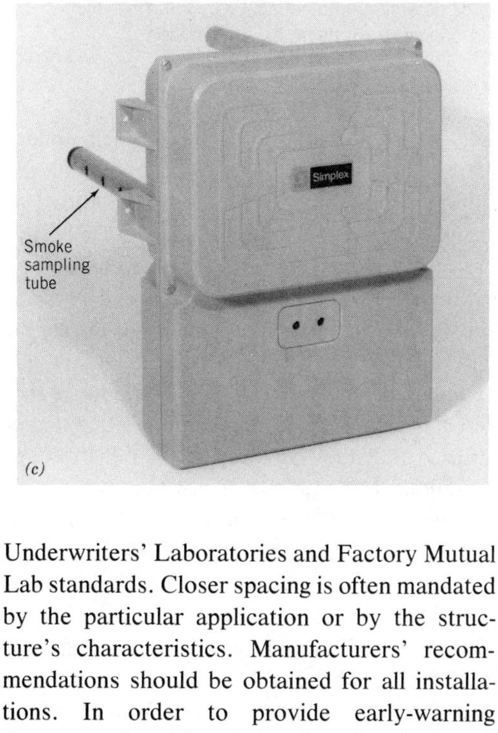

(c)

Fig. 13.26 (a) *Principle of operation of a scattered light photoelectronic spot smoke detector. A beam of light (1) from a source (2) is focused by a lens (3) on a supervisory cell (4). Smoke entering the unit reflects the light onto the alarm cell (5), changing its electrical characteristics and setting off an alarm. Photoelectric units with sensitivities up to 2% per foot are usually classified as early-warning units.* (b) *A typical "scattered light" spot smoke detector with integral 135°F (57°C) self-resetting temperature detector. The unit operates on 24 V dc and mounts on a standard 4-in. outlet box.* (c) *Smoke detector, intended for duct mounting, can utilize either a photoelectronic or ionization-type head. The unit, which is usually mounted in return-air ducts rather than supply ducts, to minimize smoke dilution, uses an air sampling tube that extends across the full width of the duct. Duct smoke detectors are intended primarily to prevent smoke dispersal throughout the building by causing shutdown of the air-handling system and by control of smoke damper operation. [*(b) *Courtesy of Protectowire;* (c) *courtesy of Simplex.]*

Underwriters' Laboratories and Factory Mutual Lab standards. Closer spacing is often mandated by the particular application or by the structure's characteristics. Manufacturers' recommendations should be obtained for all installations. In order to provide early-warning detection of a wider range of combustion products than is possible with either the photoelectronic- or ionization-type detectors individually, several manufacturers produce a unit that combines a multichamber ionization detector with a photoelectronic detector. Table 13.10 gives comparative detection characteristics of these three detector types.

Several other sophisticated detectors are available for early-warning detection in the incipient and smoldering fire stages. Probably the most sensitive unit is based on the operating principle of a *Wilson cloud chamber* (see Fig. 13.27). When microscopic particles, such as those produced by the early stages of fire, are introduced into a saturated atmosphere (cloud chamber), they act as nuclei around which water vapor condenses to form visible droplets. The detector operates by continuously sampling air from the protected space and setting off an alarm when the cloud chamber indicates the presence of particulate matter.

The continuous sampling procedure and the unit calibration make the system insensitive to dust and other noncombustion particulate matter and generally free of nuisance false alarms. The system's disadvantages are the need for piping and a high price in small installations. For large installations (30 or more detection points),

TABLE 13.10 **Detector Response Characteristics**

This chart lists the response characteristics of different types of smoke detectors to various smoldering and burning materials. This information should be used only as a general guideline. Actual detector performance of different manufacturers and in-field conditions may vary.

Material	*Ionization*	*Photoelectronic*	*Combination Ionization/ Photoelectronic*
Paper	Good	Fair	Excellent
Wood (Smoldering)	Fair	Good	Excellent
Wood (Flaming)	Very Good	Good	Excellent
Fabric	Fair	Good	Very Good
Upholstery	Fair	Good	Very Good
Polystyrene Foam	Very Good	Very Good	Very Good
Paint	Very Good	Good	Excellent
Oil	Good	Fair	Good
Polyethylene	Good	Good	Good
Isobutyl Rubber (Wire Insulation)	Good	Fair	Good
Phenol Formaldehyde (P.C. Boards)	*Fair	Good	Excellent
P.V.C.	Fair	Good	Good

Source: Courtesy of Cerberus Pyrotronics.

*Ionization detectors are not recommended for detecting smoldering P.C. Board type fires. An increased number (decreased spacing) of optical or combination detectors is recommended.

the price is competitive with those of comparable systems. This detector is particularly applicable to high-value installations, such as museums, data-processing spaces, and other high-value areas, because of its very high sensitivity.

Another "high tech" detector design reacts to the gases produced immediately after the incipient stage of a fire (see Fig. 13.23). A *laser-beam detector* measures the change in the index of refraction of air caused by combustion gases, heat, and smoke, and alarms when the change reaches a preset level. This unit, though sensitive, is expensive, not widely accepted, and heavily dependent on air currents, which limits its application.

(c) Flame Stage. As noted in Fig. 13.23, the appearance of flame is followed almost instantaneously by heat buildup and the rapid spread of flame, with a concomitant large increase in hazard. Detection of flame is no longer "early-warning," and the prime requirement for a detector at this stage is speed. It is self-evident that the actions taken as a result of a flame detection alarm, such as fire suppression and/or evacuation, must also be very rapid. This is not the case with early-warning equipment alarms, with which, several minutes are normally available to investigate the authenticity of the alarm, *if desired,* before taking appropriate action.

Flame detectors are of two types: those that detect ultraviolet radiation and those that detect infrared radiation. Both types of radiation are present at the beginning of the flame stage.

Ultraviolet (UV) radiation detectors operate by detecting the UV radiation produced by flames, which is typically in the 170- to 290-nm (nanometer) range. (The entire UV range is 40 to 390 nm.) Hydrocarbon (organic material) fires in particular produce strong radiation in this range. A quartz filter in front of the UV sensor tube limits the detector's reception range to between 200 and 250 nm, which is below the visual range. This desensitizes it to sunlight and to incandescent and fluorescent sources. The detector need

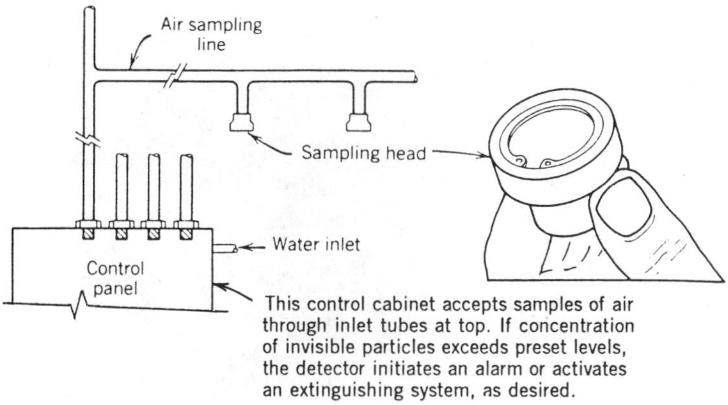
Air sampling line
Sampling head
Water inlet
Control panel
This control cabinet accepts samples of air through inlet tubes at top. If concentration of invisible particles exceeds preset levels, the detector initiates an alarm or activates an extinguishing system, as desired.

(*a*)

(*b*)

Tamper-proof sampling head

Fig. 13.27 Fire detection system based on the "cloud chamber" effect. (a) *Sampling heads are placed in all spaces to be monitored, including inside cabinets.* (b) *Tamperproof head housing, applicable to high-vandalism areas and correctional institutions.* (c) *Typical high-tech area (NASA Mission Control, Houston) installation. (Courtesy of Environment One and NASA.)*

Sampling heads
(*c*)

not "see" the flame directly; it can also detect UV radiation reflected from walls, ceilings, and the like. Obviously, the more direct the path, the stronger the radiation and the more rapidly the detector will respond. Since many devices—in particular, electric welders—produce UV radiation in sufficient quantity to activate the sensor, flame detectors are usually programmed to respond only to flickering sources at the usual flame flicker rate of 5 to 30 times per second.

In sum, UV detectors are very sensitive, react in seconds, and will respond to most types of fires. They are best applied in highly flammable or explosive storage and work areas, either open or closed. They are approved by major fire agencies.

Infrared radiation detectors are sensitive to radiation in the infrared region, between 650 and 6500 nm. Most commercial units are filtered to have maximum sensitivity at 4300 nm (4.3 μm), which corresponds to the radiation emitted by hot CO_2. They differ from ambient heat detectors in that they respond to radiant energy and are essentially optical detectors. Like the UV detectors, IR units are very sensitive, react in seconds, and must be programmed for flicker response to avoid false alarms, which can occur if they are exposed to other sources of IR, such as sunlight. Because of interference problems, IR units are normally applied to enclosed spaces such as sealed storage vaults and the like. A typical combination UV/IR unit is shown in Fig. 13.28.

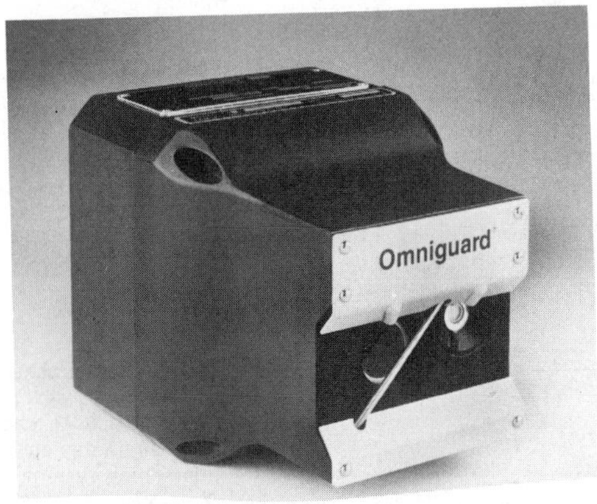

Fig. 13.28 *Spot fire detection unit containing both UV and IR detectors. This unit, which is explosion-proof, is about 5½ in. square and 7 in. deep without its swivel base. It is "tuned" to have peak response to hydrocarbon fires, such as petroleum and wood products. Built-in electronic circuitry plus combined peaked detectors minimize nuisance alarms. (Courtesy of Armtec/Meggit Industries.)*

(d) Heat Stage. Again referring to Fig. 13.23, the heat stage is the last and most hazardous stage, since by this time the fire is burning openly and producing great heat, incandescent air, and smoke. Spread of the fire depends on fuel, air currents, and the construction of the space in which the fire is burning. Detectors intended for use at this stage respond to heat and are referred to as heat-acuated, thermal, thermostatic, or simply temperature detectors. They act much like the fusible link in a sprinkler head. Effective application is restricted to locations where the subsequent alarm will permit adequate countermeasures to be taken in time to prevent injury and minimize loss. Keep in mind that the heat stage *follows* the smoke stage, and that smoke, not heat, is responsible for most casualties in fires.

Heat detectors are made in two designs: spot units, which are mounted in the center of the area to be protected; and linear units, which sense heat along their entire length. Both types respond to the high ambient temperatures caused by hot air convection from a fire. The linear type will also sense the overheating of an object or surface with which it is in contact, without the presence of fire.

Spot heat detectors are of two types: fixed-temperature units and rate-of-rise units. In the former, a set of contacts operates when a preset (nonadjustable) temperature is reached—usually 135°F or 185°F. The rate-of-rise type operates when the rate of ambient temperature change exceeds a predetermined amount, indicative of the heat stage of a fire. The rate-of-rise unit is normally combined with a fixed-temperature unit in a single housing. The fixed-temperature unit is available in either a one-time nonrenewable design that utilizes a low-melting-point alloy plug or an automatic resetting unit similar to a thermostat (hence the nomenclature). For most applications, the resettable unit is preferred. Two typical units are illustrated in Fig. 13.29. Spot units are best applied in areas subject to rapid-temperature-rise fires (e.g., basements; see Fig. 13.29*d*) that are separated from occupied areas, so that even at the heat stage, enough time remains for evacuation and fire suppression measures. Obviously, some property loss will occur, and this factor must be weighed against the higher cost of an early-warning system and the degree of hazard involved.

Linear (continuous-line) heat detectors are also of two major types. One type uses a pair of steel wires under tension, held apart by thermoplastic insulation (Fig. 13.30*a* and *b*). When exposed to its rated alarm temperature, the insulation melts and the wires come into contact, changing the current flow in the circuit to which they are connected and setting off an alarm. This type is available in a range of alarm temperatures (68°C/155°F, 88°C/190°F, 138°C/280°F) and can be used in open- or closed-circuit configurations (see Section 13.14) in lengths up to 2500 ft. Meters are available that will indicate the exact point of a fault, so that in long runs where overheating rather than fire is detected, countermeasures can be taken rapidly, without the delay caused by having to seek out the fault point. The major disadvantage to this type of detector is that it is nonrenewable; once having faulted, the melted section must be cut out and replaced.

The second type of linear detector (Fig. 13.30*c*) is basically a linear thermistor—that is, a device whose electrical resistance varies with temperature. The resistance is monitored at the control panel, and small changes noted. Depending on the thermistor material selected the detection range can be set anywhere from 70°F (21°C) to 1200°F (650°C). This type of linear detector is more sensitive, rapid, and expensive than the preceding design. When linear detectors are set to detect surface temperature rather than air temperature, they act more like early-warning devices. They are extensively used in this fashion to continuously monitor the temperature of such devices as transformers, cable trays, generators, and all sorts of hazardous equipment.

13.10 Detector Application Guidelines

Smoke detectors of all types (ionization, photo) are subject to false alarms activated by particulate matter in the air. Detectors are by nature threshold devices; raising sensitivity increases false alarming, decreasing sensitivity shortens the crucial early-warning period. False alarms are not merely nuisances; in facilities such as hospitals, and in places of public assembly such as theaters, public office buildings, and mass eating facilities, a false fire alarm can result in serious disruption, property loss, personal injury, and even death. As a result, conventional detectors in such buildings require continual maintenance and field sensitivity checks to minimize false alarms. In addition, most facilities of this type have fire/evacuation plans which call for some type of alarm verification before a general evacuation alarm is sounded. Since field experience has demonstrated that alarm verification can reduce false alarms by 80 to 90% without degrading the early warning performance of combustion product detectors, recent issues of some fire codes now permit, or even require, alarm verification. This verification can be accomplished in conventional (hard-wired, zoned, nonaddressable) systems by requiring activation of two cross-zoned detectors in a single area (since detectors report by zone). In addition, guidelines for the application of detectors have been established on the basis of field experience which will minimize the occurrence of false alarms. These guidelines, which are also applicable to the newer "smart" addressable detectors, which will be discussed below, include

1. Appropriate selection of smoke detectors. Choice of an ionization type where a photoelectric type is indicated, and vice versa, will almost guarantee false alarms.
2. Use of detectors with sensitivity-compensating circuitry to permit accommodating high temperature/humidity/ and air current ambient conditions, all of which increase false alarming.
3. Use, where necessary, of RFI (radio-frequency interference) filters on detector circuits since line radio noise (spikes, surges) can cause alarms. This factor

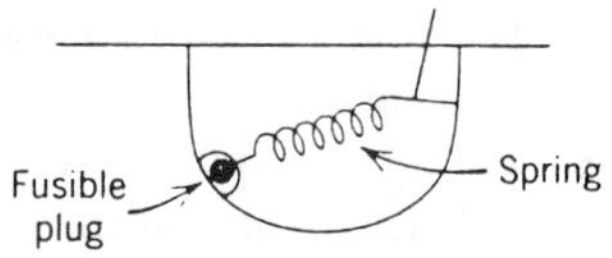

(a)

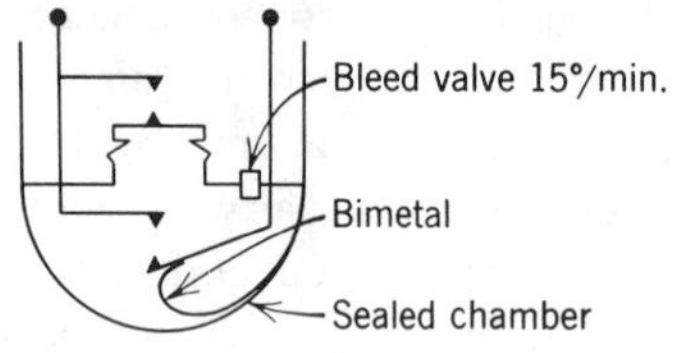

Rate of rise, fixed temp. (auto-reset)

(b-2)

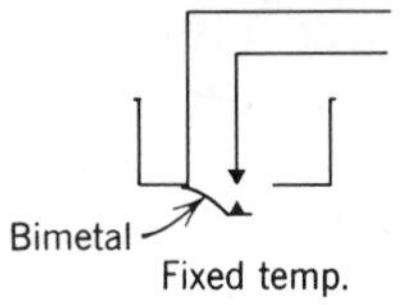

(c-2)

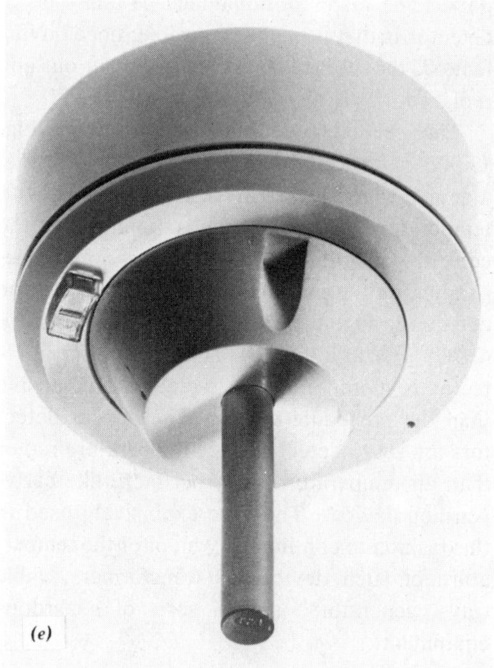

Fig. 13.29 Spot-type heat detectors. (a) *Fusible plug melts out at predetermined temperature, opening (or closing) the circuit and causing an alarm. Unit is indicating and nonrenewable.* (b) *Rate-of-rise unit consists of an air chamber with a restricted bleed valve. Rapid temperature rise causes the bellows to expand before air is lost by bleeding, thereby setting off the alarm. The unit illustrated is combined with a fixed-temperature unit, similar to* (c). (c) *Bimetallic unit action is similar to that of a thermostat, and is self-restoring.* (d) *Combination rate-of-rise and fixed-temperature unit installed in a wood-frame house basement adjacent to the furnace (See also Fig. 13.26*b*, which shows a photoelectronic detector with an integral, fixed-temperature detector. Such combined units cover a wider range of hazards than can either single unit.)* (e) *Commercial-grade surface-mount temperature detector of the self-resetting variety. This unit will alarm both at a preset fixed temperature of 135°F or 200°F and, regardless of temperature, when a rapid change in temperature occurs, as at the beginning of the heat stage of a fire. It operates on the principle of thermal lag: prevention of heat equalization on rapid ambient temperature changes sets off alarm. A separate contained element will alarm at a fixed temperature. The illustrated unit is 6 in. high overall, mounts on a standard 4-in. box, and is equipped with an indicating light. (Photo* (e) *courtesy of Cerberus Pyrotronics.)*

should be investigated in any installation with a heavy solid-state equipment load.

4. Use of detector covers in areas where any type of construction work is taking place. This will prevent fouling of detectors.
5. Establishment of a regular maintenance and testing program which includes field testing and recalibration of all detectors.
6. Avoidance of detector placement in areas where ambient conditions can cause problems. Where such placement is unavoidable, use compensating detectors, special maintenance, and appropriate verification procedures in the event of an alarm signal. These areas include:
   (a) Kitchens, laundries, boiler rooms, shower rooms, and other spaces with high humidity and steam.
   (b) Repair shops, laboratories, and other areas where open flames are used in normal work.
   (c) Garages and engine test facilities where exhaust gases are present.
   (d) Smoking rooms and areas near spaces designated as smoking areas.
   (e) Areas in which heavy accumulation of dust and dirt can be expected.
   (f) Areas of high air movement, such as near loading docks and exit doors and near the discharge from ducts or registers, since the air movement will act to dilute combustion products and thereby prevent false alarms.

Modern microelectronics has in large measure solved the problem of nuisance alarms by making possible the use of "smart" addressable detectors (i.e., detectors that report their condition *individually* to the control panel). These detectors are essentially identical to conventional detectors except that they are equipped with the appropriate electronic circuitry, usually mounted in a special base which makes each detector, in effect, a separate zone. Thus, in the event of an alarm, the control can require alarm confirmation from an adjacent detector or can require a repeat alarm from the same detector, after (remote) reset. These detectors can also be arranged to monitor their own actual field condition and report on degradation of sensitivity and maintenance requirements, thus further reducing conditions that cause false alarms.

Smoke detectors installed in air ducts are *not* substitutes for area coverage. Because of dilution, they will detect only a very heavy concentration of smoke in the occupied area; hence, they will not function as early-warning devices. Their purpose in ducts is normally to shut down the ventilation system in order to prevent smoke from circulating through a building, in the event that area detectors do not perform this function. Another reason for not using duct detectors for area protection is that ventilation systems are shut down for extended periods (nights, weekends)—which obviously disables the duct detectors during these periods. See NFPA Standard 90 for the application of smoke detectors to air conditioning and ventilating systems.

Heat detectors that react to ambient tempera-

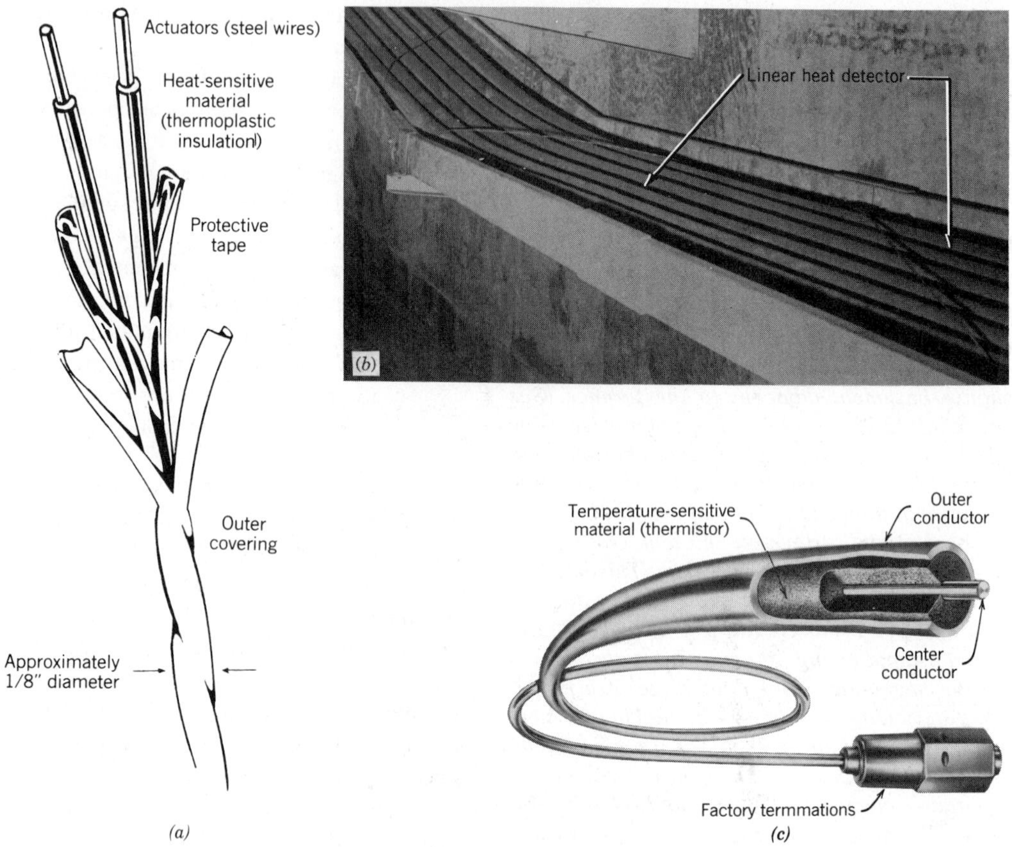

***Fig. 13.30** Linear heat detectors.* (a) *At a specified temperature the insulation melts, permitting the actuators (steel wires) to contact each other and cause an alarm.* (b) *Typical application of linear heat detector is to monitor cable temperatures in industrial cable tray.* (c) *The thermistor material is selected according to the alarm temperature required. Temperature change is detected as a change in the thermistor resistance (i.e., the resistance between the center and outer conductors). Assemblies are available in lengths of 1.5 to 15 ft and can be connected in series. [*(a), (b) *Courtesy of Protectowire;* (c) *courtesy of Fenwal, Division of Kidde.]*

ture are particularly sensitive to ceiling height and air movement. Rate-of-rise temperature detectors should not be employed where rapid temperature changes occur normally, as near an open furnace or in areas exposed to the sun. Where fires can occur within an enclosure such as switchgear, a linear detector in contact with the enclosure or a smoke detector *inside* the enclosure is indicated. Where fires of various types can occur, several types of detection can be applied simultaneously, on the same circuits, with indication or annunciation to indicate the alarming device.

When the manufacturer indicates a recommended area coverage for a detector, a square or a circumscribed circle is intended. Rectangular coverage will yield a smaller area, since the maximum safe distance between detector and fire is established by the diagonal of the square-area coverage, which is the diameter of the circumscribed circular coverage. For data on spacing, refer to NFPA Standard 72F and the UL and Factory Mutual standards cited earlier. The final decision on the type and placement of detectors is best made with the aid of an expert whose objectivity is unquestioned.

13.11 Manual Stations

In contrast to automatic detectors, the manual station is operated by hand (see Fig. 13.31).

Manual stations serve to spread the alarm that has already been detected by other means, either human or automatic. In conventional hard-wired systems, manual stations are either *coded* or *noncoded*. (See Section 13.15 for an explanation of coding.) If identification of the exact manual noncoded station operated is desirable, an annunciation panel can be added to the system; this is equivalent to using each station as a noncoded indicating zone. Because of wiring costs, annunciated systems become expensive, and beyond 10 stations, coding should be considered (see Fig. 13.32).

When the system design is such that immediate building-wide aural identification of the operated station is necessary, a coded station is used. The station code is received at the control panel, processed, and then transmitted audibly on the system gongs. Not less than three rounds of code, and normally four rounds, are transmitted.

The code usually comprises three or four digits, for example, 2-3-2 with a pause between the ringing groups and a longer pause between the rounds. The first number may identify the building floor, the second digit the wing, and the third digit the individual station. Establishment of codes is left to the user. Manual stations are placed in the normal path of egress from a building, so that an alarm may be turned in by a person as he or she exits. It is *imperative*, therefore, that stations be well marked and easily found. Architects who place fire alarm stations in nooks and corners and in camouflaged cabinets because they spoil the decor of the lobby are defeating the purpose of the system. Similarly, placement of bells *inside* hung ceilings because they are unattractive is not only foolish but dangerous, and should *never* be done, regardless of the circumstances. Loss of property and even of life may result from such ill-conceived aesthetic considerations.

Fig. 13.31 (a) *Manual fire alarm station with a break-glass rod. The unit is marked "local alarm," indicating that its operation will result in building-wide evacuation signal but not a city alarm (see Fig. 13.34*a). *Similar units are equipped with multiple sets of contacts for annunciation and other control functions, or with addressable system electronics. Other similar units use a lock-open design that requires a key to reset; this avoids the necessity of replacing the broken glass rod.* (b) *The manual station, which is normally surface mounted, can be enclosed in a well-marked recessed cabinet if desired for aesthetic reasons.*

13.12 Sprinkler Alarms

Water flow switches are placed in sprinkler pipelines and operate when a sprinkler head goes off (see Fig. 13.33). In electrical terms, a water flow switch is a set of contacts, similar to a temperature detector. It can be used to trip a coded transmitter, setting off a sprinkler code; to show up on a sprinkler annunciator board, called a sprinkler alarm panel; or to act as a zone in a noncoded system. Wiring of water flow switches or of switches activated by other extinguishing systems, is the same as that for a manual station.

13.13 Types of Fire Alarm Systems

Fire alarm systems are classified according to function and application and are covered in separate NPFA standards.

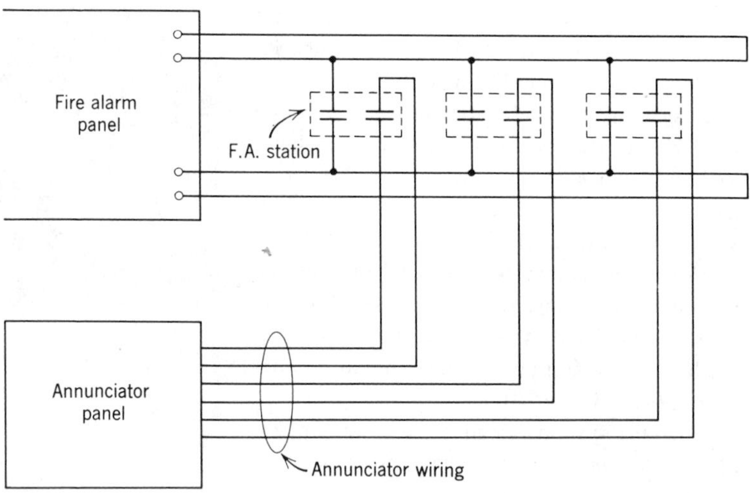

Fig. 13.32 *Wiring of noncoded fire alarm manual stations. An additional set of contacts in each station provides annunciation for that station. Note that the pair of annunciation wires required for each station increases the cost of such a system rapidly. Alternatively, electronic transmitters can be used on a single set of wires, thus reducing wiring costs but increasing equipment cost.*

(a) Local Protective Signaling System. See NFPA Standard 72A and Fig. 13.34*a*. This arrangement, as the name indicates, is intended to sound an alarm only in the protected premises. Any action must be taken locally, either manually or automatically. Thus, notification to the fire department must be manual, although fire suppression systems can be set into operation automatically. This arrangement is applicable to private residences, both single and multioccupancy, and to other private, relatively small facilities. When the building is unoccupied, notification to the fire department can come only incidentally, perhaps as a result of notification from a passerby.

(b) Auxiliary Protective Signaling System. See NFPA Standard 72B and Fig. 13.34*b*. This is simply a local system equipped with a direct connection to a municipal fire alarm box. The received alarm signal is identical to that resulting from a manual alarm at that city box. Since the fire department is aware of all direct connections, arriving firefighters would check the protected premises. This type of system is usually applied to public buildings such as schools, governmental offices, museums, and the like.

(c) Remote-Station Protective Signaling System. See NFPA Standard 72C and Fig. 13.34*c*. This system is similar to the auxiliary system, except that the alarm signal is transmitted via a leased telephone line to a remote location (a fire or police facility or a telephone answering service) that is manned 24 hours a day. The notice of alarm is then telephoned to the fire

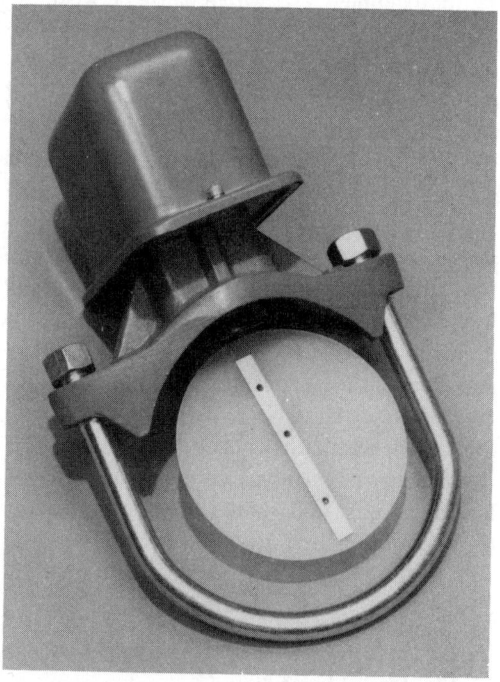

Fig. 13.33 *Typical water flow indicator. The unit bolts onto a sprinkler pipe with the paddle inside the pipe. Any water motion deflects the paddle, causing a signal to be transmitted from the microswitch mounted in the box on top of the pipe. (Courtesy of Notifier Company.)*

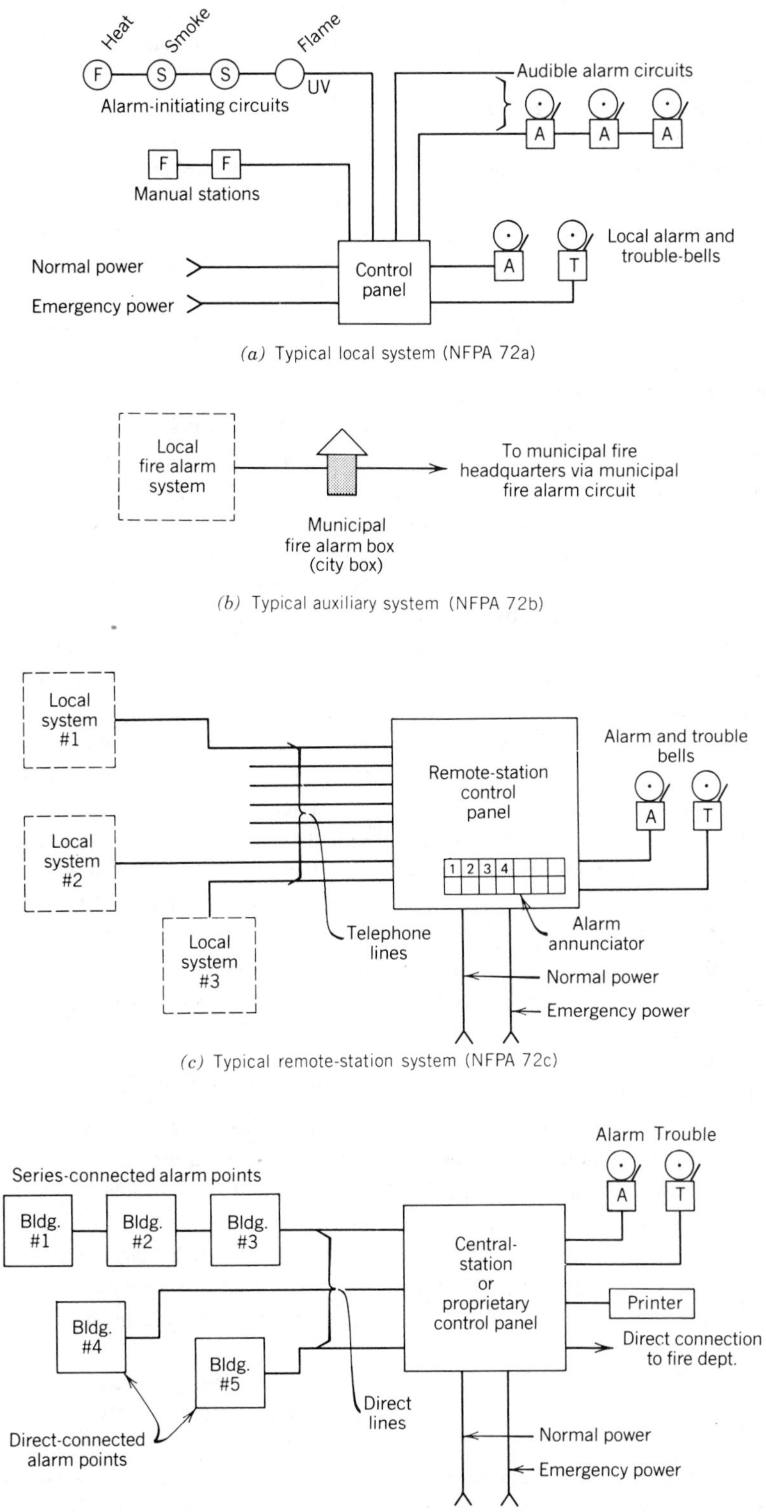

(a) Typical local system (NFPA 72a)

(b) Typical auxiliary system (NFPA 72b)

(c) Typical remote-station system (NFPA 72c)

(d) Typical proprietary or central-station system (NFPA 72d)

***Fig. 13.34** Fire alarm system arrangements.*

department. This arrangement is used in private buildings, such as stores and offices, that are unoccupied for considerable periods and for which reliance on passersby to turn in a fire signal is unacceptable. According to the NFPA definition, an audible alarm circuit extended from a local system to a nearby building—as, for instance, from a store to a nearby residence—does *not* constitute a remote-station system, unless all the requirements of Standard 72C are met.

(d) Proprietary Protective Signaling Systems. See NFPA 72D and Fig. 13.34*d*. This system, which is applicable to large, multibuilding facilities such as universities, manufacturing facilities, and the like, utilizes a dedicated central supervisory station to receive signals from all buildings. In a proprietary system, this station is on the site and is manned by persons associated with the facility. A common arrangement is for the station to be located in a guardhouse or similar supervisory location, from which point alarms are sent manually to fire stations. The central station system differs from the proprietary only in that the supervisory station is manned by persons unconnected with the facility, normally for a fee. The location may be on-site or off-site.

Information on the exact location or zone within each building at which the alarm occurred, is multiplexed or digitalized and transmitted over a single pair of wires to the central supervisory location. In older systems a separate pair of wires was required for each location. The central location has an audible alarm, some sort of visual display that indicates the alarm location, and a printer that makes a permanent record of each alarm. As has been mentioned, these central supervisory locations are frequently multipurpose, covering all aspects of facility security plus, frequently, control functions such as energy management.

13.14 Circuit Design

A system that is normally deenergized and carries no current except when functioning is called an open-circuit system (see Fig. 13.35*a*). Such a system is the simplest and most economical type but has the disadvantage of not indicating a broken wire or other malfunction that will render it inoperative.

Figure 13.35*b* shows a closed-circuit system. This arrangement will set off the alarm bells in the event of trouble in the equipment, but since this type of "false alarm" is to an extent undesirable, a further refinement can be added in the form of a trouble bell and/or light that will indicate an equipment failure to the occupants without ringing the fire alarm bells. This feature is known as *supervision,* and such a system is known as a supervised system. Supervised systems can utilize open- or closed-circuit devices, depending upon circuit arrangement. Furthermore, with special wiring and circuit design, the system can be arranged so that a single break or ground in the wiring to the devices will not prevent the operation of the system.

As a system becomes larger, a desirable feature to incorporate in the circuitry arrangement is the grouping of the devices into zones. Each zone is covered by a separate section or module in the control panel, or is annunciated, permitting extremely rapid identification of an alarm signal according to its zone. Of course, the individual alarming device remains unidentified; only the zone is known. Thus a balance must be struck during design between the economy of large zones with many devices, and the desirability of small zones from the operational and functional points of view.

13.15 System Coding

(a) Noncoded Systems. Noncoded systems are *continuous ringing* evacuation types using manual and automatic alarm initiation. If desired, the devices can be zoned and, if the system is sufficiently large, annunciation can be provided. Audible devices are continuous ringing vibrating bells and horns.

(b) Master-Coded Systems. This system, also called common-coded and fixed-coded, generates four rounds of code when any signal device operates. It utilizes a single code transmitter at the panel. Normally, the system stops after four rounds of code, although it can readily be arranged to sound continuously thereafter.

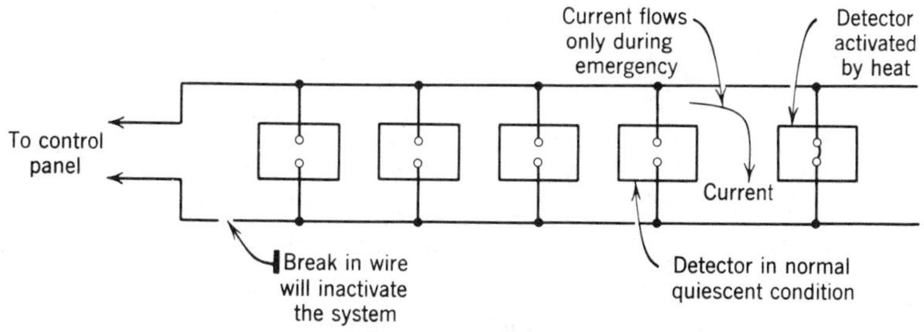

(a) Wiring of an open-circuit fire-alarm system

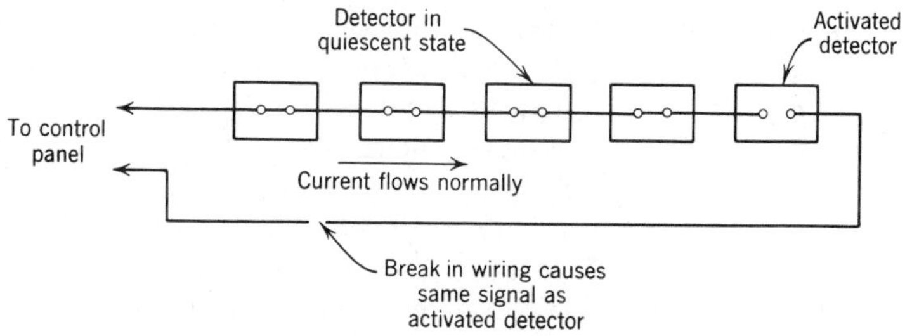

(b) Wiring of a closed-circuit fire-alarm system

Fig. 13.35 *Wiring of open- and closed-circuit unsupervised systems, showing the inherent advantage of closed-circuit wiring.*

When the code is set to ring the bells at an even 108 strokes per minute, it is known as "march time," because of the rhythmic cadence. This beat aids in the rapid panic-free evacuation of a building and therefore is frequently used in schools.

(c) Zone-Coded Systems. Identification of the alarmed zone in a system can be accomplished with zone lights, an annunciator, or by coding. In the first two cases, it is necessary to *go to the panel or annunciator* to determine the location of the operated device, which entails a possibly critical delay. All coded systems obviate this necessity by sounding the code on all the gongs in the building, thus immediately identifying the station and permitting the building staff to quickly investigate the cause of the alarm and take appropriate measures.

Therefore, if a *coded* system is desired, but by zone rather than by device (and this is less expensive by far), *noncoded* manual stations are used, along with automatic detectors, grouped by circuit into zones. These trip zone transmitters *in the panel,* which in turn ring the zone's code on the single stroke gongs or chimes. As with all coded systems, four rounds of coded signal are sounded, after which the system is silenced. In all coded systems it is advisable to include a device that records in plain English all alarms, including time of receipt and the code sounded.

(d) Dual-Coded Systems. This arrangement is a combination of noncoded and zone-coded systems. When an alarm device operates, it initiates two separate functions—an identifying coded alarm and a continuous ringing evacuation alarm. The alarms are sounded simultaneously; the coded alarm in the building's maintenance office and the evacuation alarm on separate audible devices throughout the building. A requisite to the application of this system is a continuously manned office in which the coded identifying signal can be received and acted upon.

(e) Selective-Coded Systems. These are fully coded systems in which all manual devices are coded and all automatic devices are arranged to trip code transmitters at the panel. Each manual station can be immediately identified by its distinctive code. Automatic devices may be grouped in any fashion desired, and annunciated if desired. The combinations and circuitry are entirely in the hands of the designer. In large systems, which fully selective-coded systems usually are, sprinkler transmitters and smoke detectors operate as integral subsystems of the main fire alarm panel.

(f) Presignaling. When it is desired to alert only key personnel, a system called presignaling is used. Small bells or chimes are activated only at their work locations. Since these systems are always selectively coded, the personnel alerted can immediately investigate and, if necessary, manually turn in a general alarm by key operation of a station. Because of the delay involved, this type of system is used only in buildings where evacuation is difficult and sufficient staff is available to immediately investigate the cause of an alarm.

13.16 Signal Processing

Once the hazard has been detected, a signal is transmitted to a control center and appropriate action taken. In conventional (at least for the present) systems, the alarm signal is transmitted over dedicated conductors ("hard wiring") to a control panel, consisting of either electromechanical relays or solid-state switching circuitry. These in turn actuate audible device circuits, illuminate annunciator panels, control fans and door releases, and so on, all via dedicated wiring. This arrangement has the advantages of reliability and simplicity. As the system grows, however, the wiring becomes heavy, complex, and expensive, panels become large and bulky, and changes become difficult. Also, troubleshooting of system faults becomes time consuming. Further, minimizing of false alarms is problematic because alarm verification in systems with large zones containing many alarm devices is very difficult within the extremely short time span permitted for this operation.

As a result of these problems, a panel design and wiring arrangement called multiplexing was introduced. This system uses a time-sharing electronic technique to transmit and receive multiple signals on a single 2-wire communications circuit. This has the great advantages of reducing the primary cost item in large hard-wired systems—individually wired circuits. The alarm and audible signal devices in standard multiplexed systems are the same as in conventional hard-wired systems. Only the system wiring and the panel architecture change. Zones, and even coding, are substantially unaltered, although the reduction in wiring costs permits the use of smaller zones. Control panel costs in such systems are considerably higher than in conventional systems, and the economic breakover point depends on the size of the system and the desired zoning. Some multiplexed systems are hybrid, using multiplexed detector circuits and hard-wired audible/visual alarm circuits. An additional advantage of multiplex panels is that they can readily be reprogrammed, which is emphatically not the case with hard wired relay panels.

The newest microprocessor-controlled addressable systems also use multiplexed transmission, but with the essential difference that much data processing is performed at each alarm-initiating device (i.e., each device is individually addressable and, to an extent controllable, from the central panel). Thus each device constitutes its own zone, as explained in Section 13.10. This arrangement, which also uses a 2-wire communications channel, has the additional advantage that it can:

1. Query and reset each device individually so that alarm verification is greatly simplified and false alarms in large measure eliminated.
2. Continuously scan and supervise all devices and circuits, thus anticipating faults by detecting changes and determine when maintenance is required.
3. Self-diagnose faults in all parts of the system.
4. Print out in plain language the results of all procedures and alarms.
5. In some cases, perform remote recalibration of devices.

6. Readily rezone, reprogram, and integrate with other building systems.

In new construction "intelligent" fire alarm systems are economically competitive only in large installations or in installations where the fire alarm system is coordinated with security and other control/signal systems and shares use of equipment. In retrofit work, however, in which hard-wiring costs can become prohibitively high, the increased cost of multiplex panels is more easily offset. Addressable alarm initiating devices are much more operationally desirable and their selection depends on the particular installation and the level of protection required. Typical system equipment is shown in Figs. 13.36 and 13.37. For major computer-controlled systems, of which the fire alarm is one aspect of an overall life safety and security system, see Appendix J.

The specific building system descriptions that follow are based on good present practice. Actual design must meet the requirements of the latest-edition NFPA standards, plus other codes having jurisdiction.

13.17 Residential Fire Alarms

Refer to NFPA 74, *Household Fire Warning Equipment,* and NFPA 101, Chapter 22, for detailed requirements. The system should provide sufficient time for the evacuation of the residents and for appropriate countermeasures to be initiated. The elements of the system are the various alarm-initiating devices, the wiring and control panel, and the audible alarm devices. The most basic system requires single-station (self-contained) *approved* smoke detectors with integral alarm in each sleeping area, at the top of stairs leading to occupied areas, and on each level of the structure. Some codes permit smoke detectors in existing construction to be battery powered provided the battery is monitored. In new construction, units must be powered by house current to avoid the common problem of dead batteries. The audible device is a bell in or near the sleeping rooms plus an exterior, weatherproof bell to alert neighbors.

An improved system would include one or more of the following:

1. Approved smoke detectors in each sleeping area and at the head of each stair, with at least one on every level. Combined smoke/heat detectors in boiler room, kitchen, garage. Heat detectors in the attic, kitchen and boiler room are frequently set at 185°F because of high ambient temperatures. Other units are set at 135°F.
2. Use of multiple-station smoke detectors—detectors capable of interconnection so that an alarm in any one unit will sound the alarm in all units (see Fig. 13.38).
3. Control unit (central panel) annunciated to show alarmed devices and arranged to shut off oil and gas lines and the attic fan (to prevent spread of smoke), and to turn on lights both inside and out. Also, an automatic dialer to ring a neighbor's phone and give a distinctive alarm sound when the phone is answered or some other device that will transmit an alarm outside the residence.
4. Backup power for the system—usually a large dry-cell, although a storage battery

Fig. 13.36 This control panel is a small microprocessor-based modular unit designed for use in a small (four zone maximum) building. It provides four device zones, two signal circuits, supervision of a-c line, ground connection, battery condition indication, annunciator connector, and system wiring and is field programmable for alarm verification, various coding arrangements, and alarm signal control. The cabinet is small and easily wall mounted. (Courtesy of Simplex.)

FIRE PROTECTION

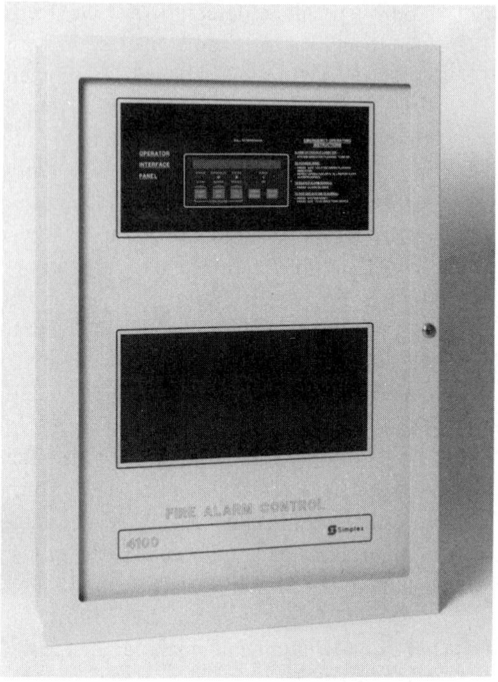

Fig. 13.37 Modular control panel that can be arranged to work with both hard-wired and addressable devices. The unit is field configurable, programmable, and editable. It can control up to 128 hard-wired or addressable devices, multiple signal circuits, and alarm relays in conventional or multiplex modes. The unit provides all necessary programming, supervision, and alarm functions, including variable coding, alarm verification, selective signaling, controllable alarm circuits, cross-zoning, and when so equipped, voice communication. The system is applicable to midsize buildings. The illustrated control cabinet measures 36 in. H × 28 in. W × 8 in. D. A similar-sized annunciator provides visual annunciation of the condition of 128 devices or points. (Courtesy of Simplex.)

with trickle charger is better. Power supply should be supervised.

5. Wiring of all devices connected to a system control unit to be on supervised circuits that will sound a trouble alarm in the event of a fault. Trouble alarm to be distinct from fire alarm.

Fig. 13.38 Typical residential smoke detector of the scattered-light photoelectronic type. The detector circuitry repeatedly checks the response of a photodiode to light from a pulsed infrared LED to minimize false alarms. This unit can also be equipped with a heat-sensing unit. These detectors are powered by line current and can be wired in tandem on a single circuit for housewide coverage. (Courtesy of Simplex.)

13.18 Multiple-Dwelling Fire Alarm Systems

See NFPA 101, Chapters 18 and 19. In addition to standard requirements, NFPA 101 lists many special requirements for tall multistory buildings, buildings intended for elderly residents, and structures with fire suppression systems.

In *all* multiple-dwelling structures, including apartment houses, dormitories, hotels, motels, and boarding houses, where occupants may be asleep, a general requirement exists that audible devices intended for general evacuation alarm be of such design that they be readily audible inside the individual rooms. The intent, of course, is that these bells or horns be loud enough to wake sleeping occupants. Most codes however, until very recently, did not specify required sound levels, relying instead on recommendations for minimum audibility. One recent code issue (BOCA, 1990) gives specific recommendations in dBA over ambient, plus minimum dBA levels for various spaces. It is therefore the responsibility of the fire alarm system designer to analyze the transmission loss (TL) of the construction in the various parts of the structure, calculate noise reduction (NR) between proposed audible device and sleeping room spaces, and only then both select, and finally locate, area alarm devices to satisfy code sound-pressure-level requirements. Data in Sections 27.12 to 27.21 and Appendix F will be useful in this regard.

(a) Apartment Houses. The basic protection scheme requires an approved single-station smoke detector in each apartment sleeping area, powered by house current. Battery units may not be used since, unlike homeowners, apartment dwellers rely on the building management for all building services, and periodic battery checks and replacement might not be carried out. Large buildings (see Code) require, in addition, a manual system with corridor and lobby manual stations, as well as loud bells in the corridors to signal general evacuation. The system is supervised and noncoded; the control panel is located in an attended location, such as the building superintendent's office. In tall multistory buildings, an annunciator is required, to indicate the floor on which the manual station was operated. A duplicate annunciator in the lobby will direct firefighters to the proper floor. Improvements in the basic protection scheme could include the following:

1. Annunciation of all the single-station apartment detectors at the system control panel with a duplicate annunciator in the lobby (see Fig. 13.39). A light over the apartment door, activated by the apartment alarm, to assist firefighters if full annunciation is not provided.
2. In large apartments, approved detectors in kitchens and in halls outside sleeping areas. All detectors in an apartment connected on a single alarm circuit.
3. Smoke detection in corridors, service and utility spaces, and storage areas, with annunciation.
4. Standby power to all fire-alarm system circuits. In all installations an auxiliary system is desirable, and in some local codes, one is required (see Section 13.13 and Fig. 13.34*b*).

(b) Dormitories. See NFPA 101 Chapter 16. Dormitories are classified as apartment houses when built in apartment style (i.e., a number of sleeping rooms around a living area). Otherwise, dormitories are classed as hotels. Basic protection requires:

1. A supervised, zoned, noncoded general evacuation alarm system (continuous ringing), initiated by manual stations.
2. Manual noncoded stations at each level, in the normal egress path and near each exit, plus a station at a continuously manned or monitored location such as a front desk or a security post.
3. Corridor smoke detection.
4. Sleeping room smoke detection; either single-station type or (preferably) connected as part of the central alarm system.
5. Appropriate detectors in storage rooms and mechanical equipment spaces.
6. Audible alarm devices throughout the building.
7. Alarm annunciator panel, located as directed by local fire authorities.
8. In high-rise buildings, an emergency voice/alarm communication system. (See Section 13.20, "Recommendations for Office Buildings," for a description of this system.)

An improved system might include:

9. Duplicate annunciator panel in the lobby if the principal annunciator is at the control panel.
10. Use of an approved method of alarm signal verification to avoid the major annoyance of evacuation in false alarm.
11. Means for operating the system in test mode, for evacuation drills.

(c) Hotels. Hotels are classified with dormitories and have the same minimum requirements as stated there.

A *hotel* of any magnitude would utilize a supervised dual-coded system with automatic stations in storerooms, kitchen, boiler room, and other unsupervised areas. A presignal system may be used where authorities permit it. Sprinkler transmitters and annunciation are common, as are smoke detectors in the ducts, activating an engineered smoke control system whose pri-

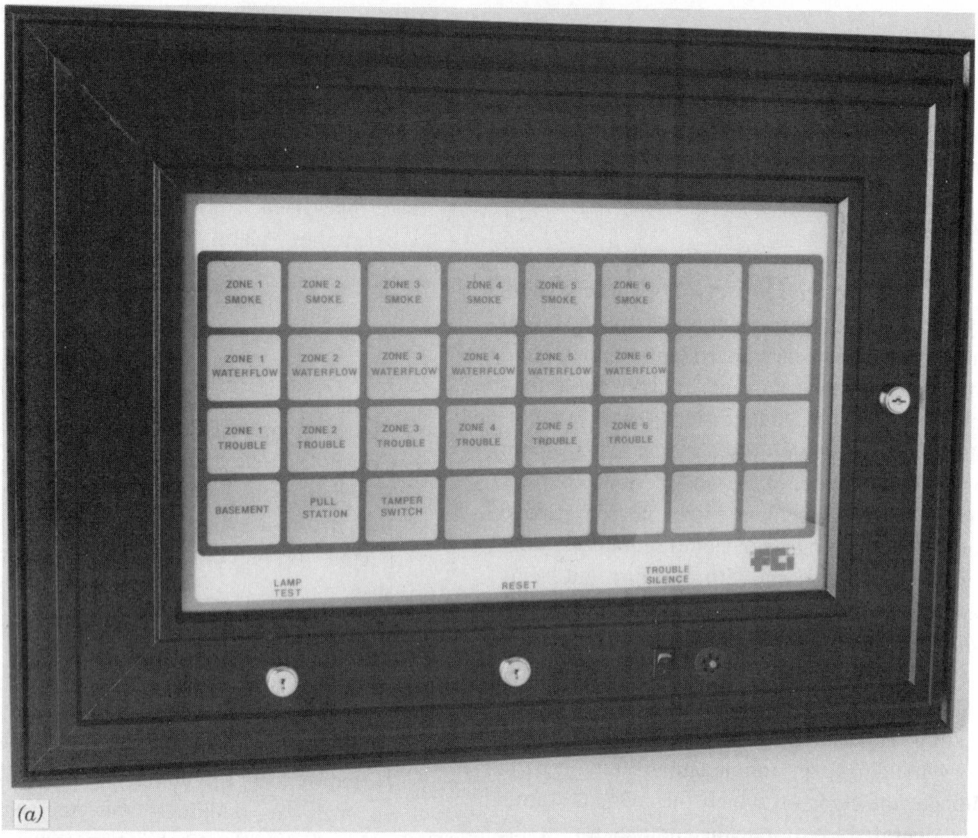

(a)

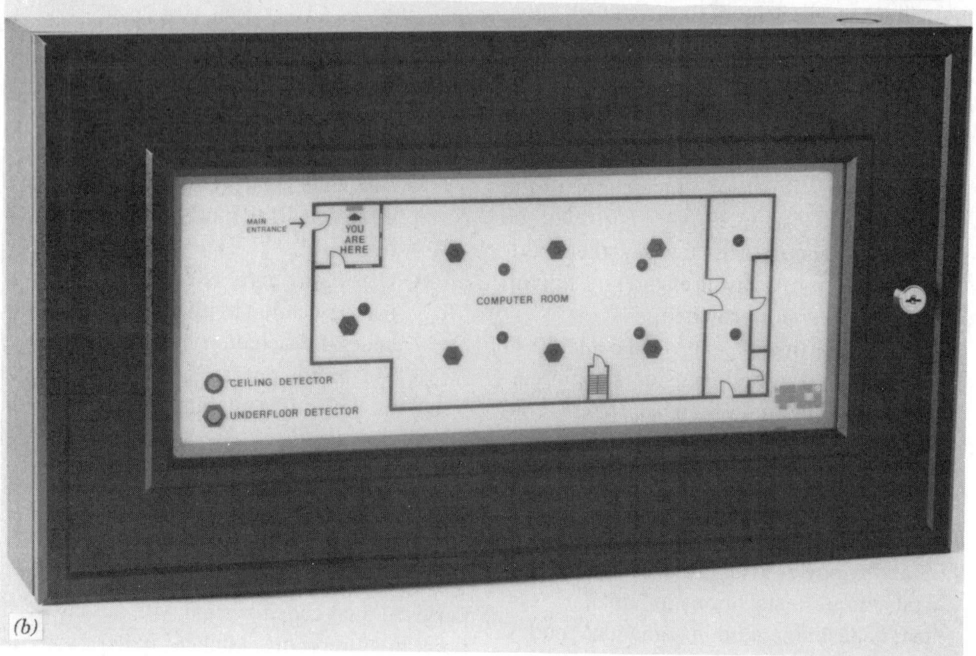

(b)

Fig. 13.39 *Typical annunciators, applicable to use with all types of fire alarm systems, can be either tabular type* (a) *or graphic* (b). *Both types can be furnished with lamp test devices and local/remote trouble indicators. The tabular type illustrated* (a) *has 32 zone sections and is available up to 128 zones (or devices). Graphic units* (b) *are custom designed for the specific installation and are available in large sizes to depict more extensive floor plans or riser diagrams. (Courtesy of Fire Control Instruments, Inc.)*

mary purpose is to prevent the spread of smoke through the building. Presignal or coded bells are installed in the office of the building engineer as well as in that of the hotel managers. (Where addressable detectors are employed, references to coding are inappropriate.) The fire alarm panel is most often placed in the mechanical equipment area. An auxiliary circuit, usually required by local codes, trips a city fire alarm box or, in large cities, connects to a fire supervision company, which in turn notifies the city fire department. All such connections are rigidly controlled by city ordinances. Figure 13.40 shows a typical hotel floor plan.

13.19 School Fire Alarm Systems

See NFPA 101 Chapter 10 for detailed requirements.

Although personal safety and prevention of property loss combine to form the purpose of all fire alarm systems, the former consideration far outweighs the latter in the instance of school buildings, particularly of the elementary grades. For this reason a general schoolwide alarm causing an immediate evacuation of the premises is the primary requirement of the system. Consideration also must be given to maintaining the uniqueness of the sound of the fire alarm gongs to allow no possibility of confusion with the program gongs, where the latter are used.

The system employed almost universally calls for a closed-circuit, supervised arrangement, noncoded or master-coded in the case of smaller schools and dual-coded in the instance of large or multibuilding institutions. For a multiple-building school, the circuitry is generally arranged to sound an evacuation alarm in the affected building only. The signal also is transmitted to administrative areas in other buildings, and almost always, via auxiliary circuit, to the municipal fire department.

Since regular fire alarm drills are mandatory in all schools, the system circuitry must be arranged to allow for this type of testing. As with other systems, manual stations are placed on each floor at exit points, such as stairways, with automatic stations in the boiler room, kitchen, some laboratories, shop classrooms, and selected storage areas (see Fig. 13.41). It is also advisable to connect any sprinkler flow switches to the alarm system to effect building evacuation while utilizing, in larger schools, a central sprinkler annunciator panel, which will indicate the particular water flow switch involved. This accomplishes the same purpose as coding of fire alarm stations.

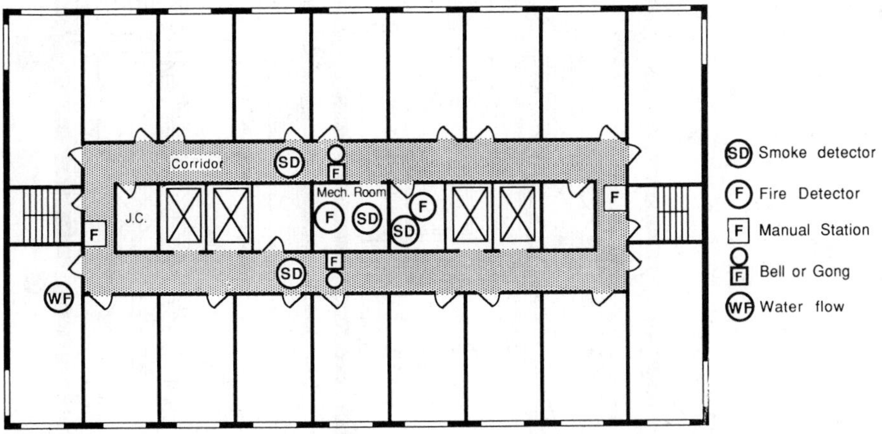

Fig. 13.40 Typical hotel floor plan. Bells are placed so that an alarm will be audible in all parts of the building. Manual stations are installed at all points of egress, such as stairwells and main floor exits. A typical location for sprinkler alarm transmitter would be in a janitor's closet (J.C.) or electric closet. Automatic smoke detectors are properly located in storage and mechanical spaces as well as in guest rooms.

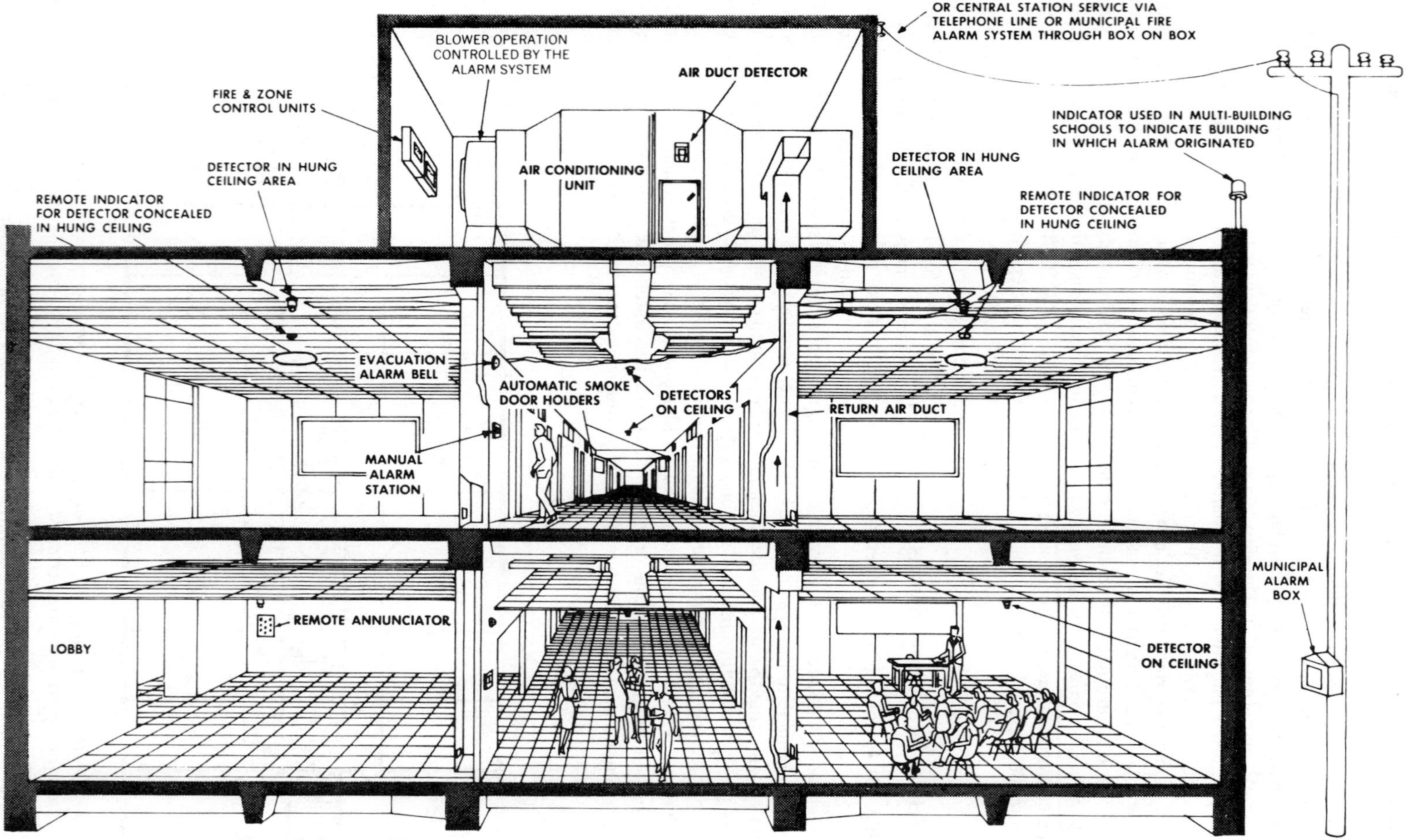

Fig. 13.41 *Section through school building showing typical fire system components. (Courtesy of Cerberus Pyrotronics.)*

Most states have adopted statutes requiring the application of automatic detection in addition to the previously universal requirement for manual systems. In an unoccupied building this will reduce property loss, and in an occupied building it may give the short alarm margin time required to avoid serious injury.

13.20 Office Building Fire Alarm Systems

See NFPA 101, Chapter 26.

(a) Small to Medium-Sized Buildings. The Life Safety Code (NFPA 101) requires that the fire alarm system sound a general audible alarm throughout the building plus a (separate) alarm in a continuously attended location in the building, for the purpose of initiating appropriate emergency action. This action almost undoubtedly includes a rapid verification of the validity of the alarm if this has not already been done automatically by the equipment, followed by an appropriate live voice announcement over a building-wide public address system, informing building occupants of the nature of the emergency (which has already been made known by the general audible alarm) and instructing them as to how to proceed, as necessary. This type of live voice announcement is imperative to avoid unnecessary building evacuation, and in the case of a real fire emergency to allay fear and to avoid confusion and even the risk of panic, since building tenants often, and transient occupants always, are unfamiliar with the building and its exit locations, regardless of the presence of signs and exit lights.

The fire alarm system itself, referring to conventional (not addressable) types, should be a supervised dual-coded type, with manual stations in the egress path and at exits, and automatic detectors in critical areas. Sprinklers are also used as a detection means.

Smoke detection in return-air ducts is often arranged to shut down all building supply and exhaust fans in order to avoid both feeding the fire and spreading smoke. A fan restart panel at an accessible location near the lobby should then be provided for selective restart of the *exhaust* fans only. Since this arrangement of fan shutdown and restart is not universal, it is well to consult local fire authorities and to follow their recommendations. Additional circuits can be arranged to release smoke control doors.

Water flow switches on the sprinkler system and smoke detectors in their system should run to individual annunciators, which in turn connect to the fire alarm control panel. This panel then may be auxiliarized for outside supervision if the building is unattended for periods of time. Where permitted by codes, presignal systems may be used in fully attended buildings. To avoid the possibility of the presignal going uninvestigated, a timing arrangement is possible that will turn in a general alarm if the alarm device causing the signal is not reset within a predetermined time.

(b) Large and High-Rise Office Buildings. Sad experience has demonstrated that high-rise buildings, once thought to be "fireproof," are emphatically not so. Indeed, due to their size, they have particularly severe fire protection problems, one of which is reliable communications during fire emergencies. As a result, fire codes now require that high-rise buildings be equipped with an emergency voice/alarm communications system that meets the requirements of NFPA 72F. The system provides full control of transmission and building-wide distribution of all tones, alarm signals, and voice announcements on a selective or all-call basis. Specifically, it makes possible:

1. Two-way active communications from the firefighters' control command post (usually in the lobby) to at least one fire station per floor, all mechanical equipment rooms, elevator machine rooms, and air-handling (fan) rooms.
2. Distribution to selected areas of alert tones, signals, and prerecorded messages on independent channels: one for paging and another for signal tones. (Obviously, a complete system of loudspeakers covering all areas of the building is an integral part of the system.)
3. Communication with fire department or central station.

In addition to communication functions, the (lobby) fire command post provides:

FIRE PROTECTION

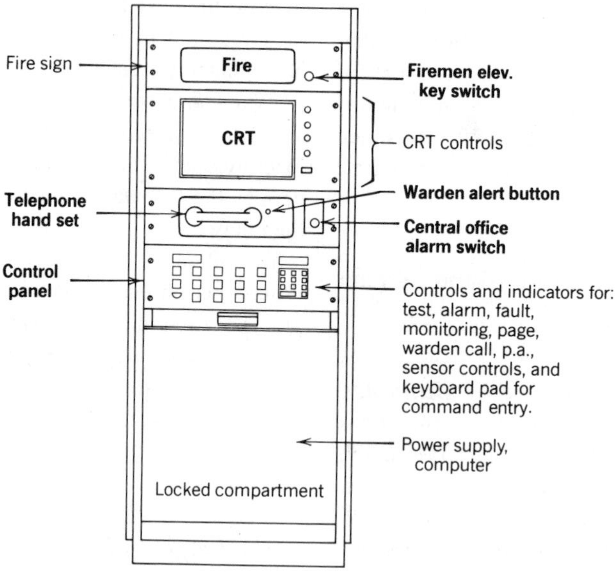

Fig. 13.42 *Front view of a computerized data acquisition and control center, which provides security system supervision and environmental monitoring, in addition to high-rise fire system control. The telephone handset provides communication with the fire stations on all floors. The condition of any device can be displayed on the CRT. All data are run via a single 11-conductor, multiplex cable. In addition to fire alarm communication, the unit provides elevator capture and control, control of all fans, vents, dampers, and smoke doors, and supervision of all fire devices and circuits. Other functions include control and monitoring of up to 2000 "addresses," which may include other building security systems. This unit is 2 ft square and 5 ft high. (Courtesy of Multiplex Electrical Services.)*

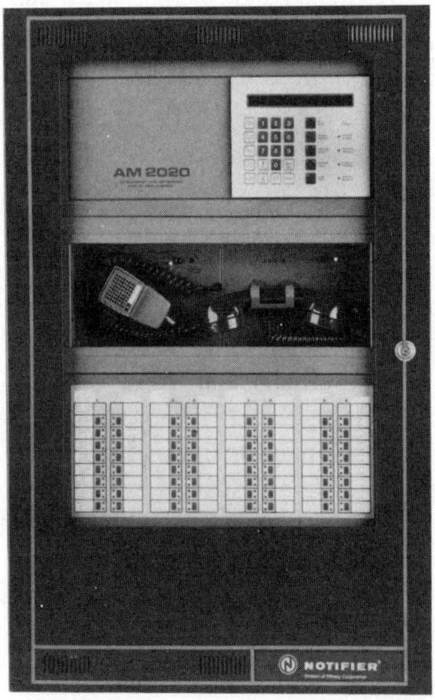

Fig. 13.43 *Control panel for a voice alarm multiplex fire alarm system. The basic unit is an intelligent addressable system with capacity for up to almost 2000 individually identified and controlled points (990 detectors and 990 monitor/control modules). The system is completely field programmable and uses multiple 2-wire communication loops and multiple microprocessors for fail-safe operation. The voice command section (center) is a multiplexed audio system which provides complete building intercom and emergency message dispersion. The bottom section of the panel contains zone annunciation and control modules, which provide alarm and trouble indication plus remote control of system relays, silence, reset, and other functions. (Courtesy of Notifier, a division of Pittway Corp.)*

4. Visual display (annunciation) of all fire alarm devices, including sprinkler valves and waterflow indicators.
5. Fire pump status indication.
6. Controls for any automatic stair door unlocking system (security access system).
7. Emergency generator status.
8. Elevator location indicators plus operation and capture controls.
9. Control of all air-handling units in the building.
10. Control of smoke control devices (doors, dampers, etc.).
11. Means for testing all circuits and devices.

The exact equipment supplied depends on the building and the local fire code. The outstanding characteristics of this system are the communications system, the visual display of alarm locations, and the remote control of air-handling equipment.

It should be readily apparent that in a large high-rise building, the requirements listed above are so complex and demanding that use of a conventional hard-wired system is not practicable. At the very least a multiplexed system is required, and in most modern buildings, particularly those being retrofitted with such a system, a computer-controlled, multiplexed, addressable-device system is used. The units shown in Figs. 13.42 and 13.43 provide a very sophisticated fire alarm and control system plus other control functions, as indicated.

13.21 Industrial Building Fire Alarm Systems

Industrial building systems are normally selective-coded and fully supervised. Presignaling is utilized in structures where for any reason an evacuation alarm is undesirable. In addition to manual stations at points of egress, the following devices may be used:

1. Temperature and smoke detectors in storage areas and laboratories.
2. Smoke and flame detectors in record rooms and continuous process laboratories.
3. Water flow switches on all sprinklers.

The annunciators, control panel, and alarm register are best placed in the guardroom. If none is available, an auxiliary or remote-station circuit should be added to allow remote monitoring.

Because of the high ambient noise level in many plants, horns are substituted in such areas for bells and gongs, which might be inaudible (see Figs. 22.15 and 22.16). If the building is sufficiently large, a computer-microprocessor-controlled system of the type shown in Fig. 13.42 can be used. As there, the unit can provide other control/monitoring functions as required.

In summary, the specific occupancy recommendations given above are representative of good practice at the present time. However, codes are constantly being changed in this particularly sensitive field, and they vary with locale. In actual design situations, all codes and regulations having jurisdiction must be complied with.

REFERENCES

Egan, M. David (1978). *Concepts in Building Firesafety,* Wiley, New York.

Jensen, Rolf, ed. (1975). *Fire Protection for the Design Professional,* Van Nostrand Reinhold, New York.

National Fire Protection Association, Batterymarch Park, Quincy, Mass. Latest issue of all relevant publications.

Fire Protection Handbook
National Fire Protection Association 13-1989 (or later update)
Installation of Sprinkler Systems

FIRE PROTECTION

PART V
ELECTRICITY

Electricity is the most prevalent form of energy in a modern building. It supplies not only electric outlets and electric lighting, but also the motive power for ventilation, heating, and cooling equipment; traction power for elevators and material transport; and power for all signal and communications equipment. An electric power failure can paralyze a facility. If properly designed, the facility will return to partial functioning by virtue of emergency equipment that will furnish some part of the facility's electricity needs for a limited time.

Given this complete dependence on electric power for normal functioning, it is apparent that designers must be familiar with normal electrical systems. Chapter 14 reviews basic relationships in electric circuits with emphasis on power, energy, energy costs, and methods of energy managements and electric load control. Chapter 15 introduces the reader to the concept and basis of electric equipment ratings and capacity, and continues with a description of modern wiring systems and their components. Chapter 16, essentially a continuation of the preceding chapter, describes service, utilization and emergency/standby power equipment. Also discussed are energy conservation considerations and economic factors. Chapter 17 draws on information given in the three preceding chapters to demonstrate straightforward design methods for building electrical systems.

14

PRINCIPLES OF ELECTRICITY

Historically, usable energy was most often produced by burning a fossil fuel such as coal or oil. The resultant heat energy was used directly as heat and light or converted by machines into motion. Only within the last century, however, has this heat in turn been used to create another form of usable energy—electricity. Even the recent partial substitution of nuclear for fossil fuels has affected only the heat production portion of this process. Beyond that point, the heat is utilized in the same manner to drive generators that produce electricity. It is well to remember, therefore, that in terms of natural resources electricity is an expensive form of energy, since the efficiency of the overall heat-to-electricity conversion, on a commercial scale, rarely exceeds 40%.

14.1 Electric Energy

Electricity constitutes a form of energy itself that occurs naturally only in unusable forms such as lightning and other static discharges or in natural galvanic cells, which cause corrosion. The primary problem in the utilization of electric energy is that, unlike fuels or even heat, it cannot be stored and therefore must be generated and utilized in the same instant. This requires an entirely different concept of utilization than, for example, a heating system with its burner, piping, and associated equipment.

The bulk of electric energy utilized today is in the form of alternating current (a-c), produced by a-c generators, commonly called alternators. Direct-current (d-c) generators are utilized for special applications requiring large quantities of d-c. In the building field such a requirement is found in elevator work. Smaller quantities of d-c, furnished either by batteries or by rectifiers, are utilized for telephone and signal equipment, controls, and other specialized uses.

14.2 Unit of Electric Current: The Ampere

Electricity flowing in a conductor is called current, or amperage, and is abbreviated *amp*, *amps*, or simply "A." When current is used in an equation, it is usually represented by the letter I or i.

It is convenient to establish an analogy between electric systems and mechanical systems as an aid to comprehension. Current is a measure of flow and, as such would correspond to water flow in a hydraulic system (see Fig. 14.1). The correspondence is not complete, however, since in the hydraulic system the velocity of water flow varies, whereas in the electric system the velocity of propagation is constant and may be considered instantaneous; hence the need to utilize the electric energy the instant it is produced.

14.3 Unit of Electric Potential: The Volt

The electron movement and its concomitant energy, which constitutes electricity, is caused by

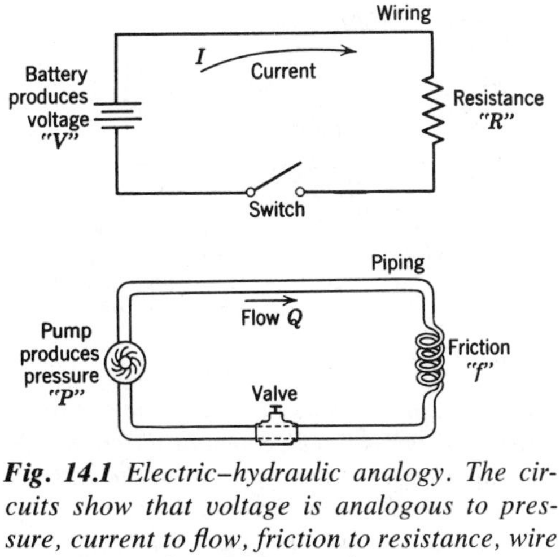

Fig. 14.1 *Electric–hydraulic analogy. The circuits show that voltage is analogous to pressure, current to flow, friction to resistance, wire to piping, and switches to valves.*

ELECTRICITY

creating a higher positive electric charge at one point on a conductor than exists at another point on that same conductor. This difference in charge can be created in a number of ways. The oldest and simplest method is by electrochemical action, as in the battery. In the ordinary dry cell, or in a storage battery, chemical action causes positive charges (+) to collect on the positive terminal and electrons or negative charges (−) to collect on the negative terminal. Assume for the moment that nothing is connected to the battery terminals. There is a tendency to flow between the electrified particles concentrated at the positive and negative terminals. *Potential difference* or *voltage* is the name given to this tendency, or force. It is analogous to pressure in a hydraulic or pneumatic system. Just as the pressure produced by a pump or blower causes water or air to flow in a connecting pipe, so too the potential (voltage) produced by a battery (or generator) causes current to flow in a conductor connecting the terminals between which a voltage exists. The higher the voltage (pressure), the higher the current (flow) for a given resistance (friction) (see Fig. 14.2). Other means of producing voltage, both direct (d-c) and alternating (a-c), are discussed in Section 14.8. The unit of voltage is the volt, abbreviated "V."

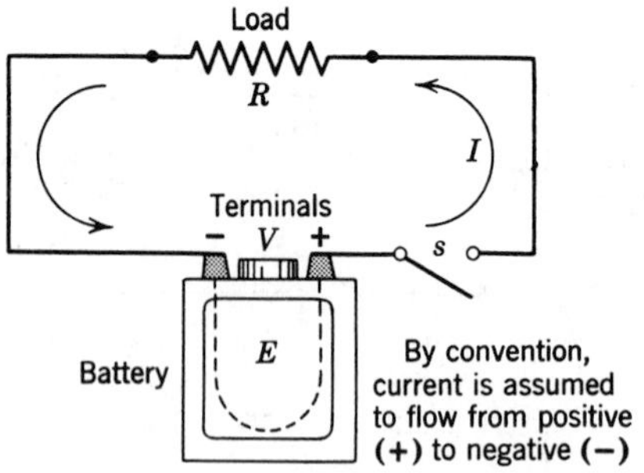

Fig. 14.2 *Current flows in the electric circuit as a result of the voltage (potential difference)* V *that exists between the terminals of the battery. By convention, current direction is from positive (+) to negative (−) in the circuit (and from − to + inside the battery).*

14.4 Unit of Electric Resistance: The Ohm

The flow of fluid in a hydraulic system is resisted by friction; the flow of current in an electric circuit is resisted by *resistance*, which is the electrical term for friction. In a d-c circuit this unit is called resistance and is abbreviated R; in an a-c circuit it is called impedance and is abbreviated Z. The unit of measurement is the *ohm*. (It is interesting to note that the scientific convention of naming units after persons whose work is closely related to the field is followed here also. Thus the units ampere, volt, and ohm are derived from André Ampere, Alessandro Volta, and Georg Ohm.)

Materials display different resistance to the flow of electric current. Metals generally have the least resistance and are therefore called conductors. The best conductors are the precious metals—silver, gold, and platinum—with copper and aluminum only slightly inferior. Conversely, materials that resist the flow of current are called insulators. Glass, mica, rubber, oil, distilled water, porcelain, and certain synthetics such as phenolic compounds exhibit this insulating property and are therefore used to insulate electric conductors. Common examples are rubber and plastic wire coverings, porcelain lamp sockets, and oil-immersed switches.

14.5 Ohm's Law

The current I that will flow in a d-c circuit is directly proportional to the voltage V and inversely proportional to the resistance R of the circuit. Expressed as an equation, we have the basic form of Ohm's law:

$$I = \frac{V}{R} \tag{14.1}$$

In a-c circuits, the same relation holds true except that instead of d-c resistance we use the a-c impedance. Ohm's law is frequently written in another form, that is,

$$V = IR \tag{14.2}$$

which states the mathematical relationship that volts = currents × resistance. This form has no logical basis; therefore, we recommend that the reader remember the form of equation 14.1, which clearly states the physical situation; that is, as a result of voltage V, a current I is produced that is proportional to the electric pressure V, and inversely proportional to the electric friction R.

An example will illustrate the application of Ohm's law. The example chosen is applicable to both a-c and d-c, since the device is purely resistive. This will be more fully explained in the subsequent discussion on alternating current.

EXAMPLE 14.1 An incandescent lamp having a hot resistance of 66 ohms is put into a socket that is connected to a 115-V supply. What current flows through the lamp (after it reaches operating temperature)?

SOLUTION

$$I = \frac{V}{R} = \frac{115}{66} = 1.74 \text{ A}$$

(These figures correspond to a normal 200-W general service incandescent lamp.)

We mentioned hot resistance in the above example since some materials exhibit higher resistance when hot than when cold. A typical example of this is the tungsten-filament lamp that when first turned on takes, for a fraction of a second, 10 to 15 times the steady-state hot filament current.

14.6 Circuit Arrangements

The two basic electric circuit arrangements are *series* and *parallel*. These concepts are the same for both d-c and a-c. As above, we will use purely resistive circuits so that circuit calculations will be applicable to both d-c and a-c. In other than purely resistive circuits, a-c circuit calculations are different, and much more complicated, than their d-c counterparts.

(a) Series Circuits. In a series arrangement the elements are connected one after the other—in series. Thus, resistances and voltages add. This is indicated graphically in Fig. 14.3*a*. An electric circuit may be defined as a complete conducting path that carries current from a source of electricity to and through some electrical device (or load) and back to the source. A current can never flow unless there is a complete (closed) circuit. It should be obvious that because of the arrangement of the components, *the current is the same in all parts of the circuit.* A somewhat more complicated circuit is shown in Fig. 14.3*b* and is analyzed in Example 14.2.

It is customary to refer to connection points on such wiring diagrams by letters as *a, b, c, d,* and so on. The battery voltage may then be called V_{ab} = 12 V; the voltage across the load resistance, V_{cd} = 11.5 V; the resistance of the two wires $r_{bc} + r_{da}$ = 0.04 ohm. The positive and negative terminals of the battery are shown.

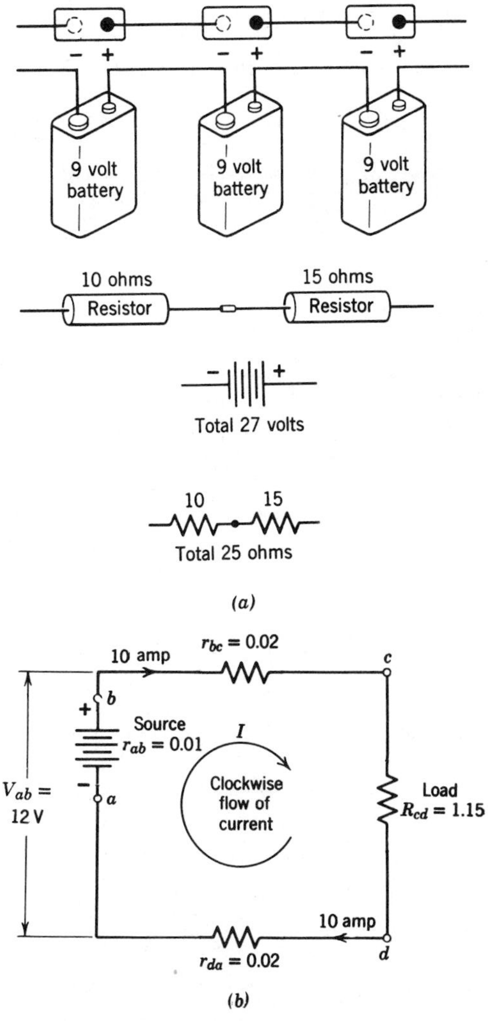

Fig. 14.3 (a) *Physical and graphic representation of series connection of batteries and resistors.* (b) *A simple series circuit.*

EXAMPLE 14.2. The battery in Fig. 14.3*b* is rated at 12 V, the line resistance (both wires) is 0.04 ohm, the battery internal resistance is 0.01 ohm, and the load resistance is 1.15 ohms. Determine (a) current flowing in the circuit, and (b) the voltage across the load (V_{cd}).

SOLUTION

(a) The current flowing is

$$I = \frac{V}{R} = \frac{V_{ab}}{r_{ab} + r_{bc} + r_{cd} + r_{da}}$$

$$= \frac{12}{0.01 + 0.02 + 1.15 + 0.02} = \frac{12}{1.2}$$

$$= 10 \text{ A}$$

(b) The voltage drop across the load is

$$V_{cd} = I \times R_{cd} = 10 \times 1.15 = 11.5 \text{ V}$$

Series circuits have very limited application in building wiring.

(b) Parallel Circuits. When two or more branches or loads in a circuit are connected between the same two points, they are said to be connected in *parallel* or *multiple*. Such an arrangement and its hydraulic equivalent are shown in Fig. 14.4*a*. From the circuit of Fig. 14.4*b* it should be apparent that multiple loads are across the same voltage and, in effect, constitute separate circuits. From this we conclude that in the multiple arrangement, the total current in the circuit is the sum of the individual currents flowing in the branches, that is,

$$I_T = I_1 + I_2 + I_3$$

Refer to Fig. 14.4*b*. Notice from the numbers shown there that the total current flowing in the

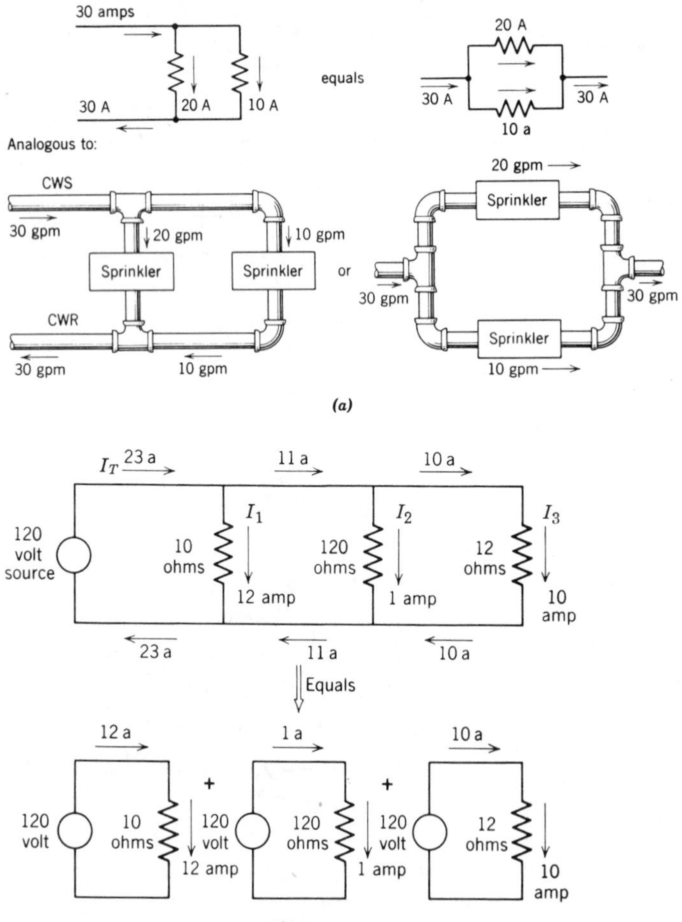

Fig. 14.4 (a) *In a parallel connection the flow divides between the branches, but the pressure is the same in each branch.* (b) *Note that loads connected in parallel are equivalent to separate circuits superimposed into a single connection. Each load acts as an independent circuit unrelated to, and unaffected by, the other circuits.*

circuit is the sum of all the branches, but that the current in each branch is the result of a separate Ohm's law calculation. Thus, in the 10-ohm load a 12-A current flows, and so forth.

The parallel connection is the standard arrangement in all building wiring. A typical lighting and receptacle arrangement for a large room is shown in Fig. 14.5. Here the lights constitute one parallel grouping, and the convenience wall outlets constitute a second parallel grouping. The fundamental principle to remember is that loads in parallel are additive for current, and that each has the same voltage imposed.

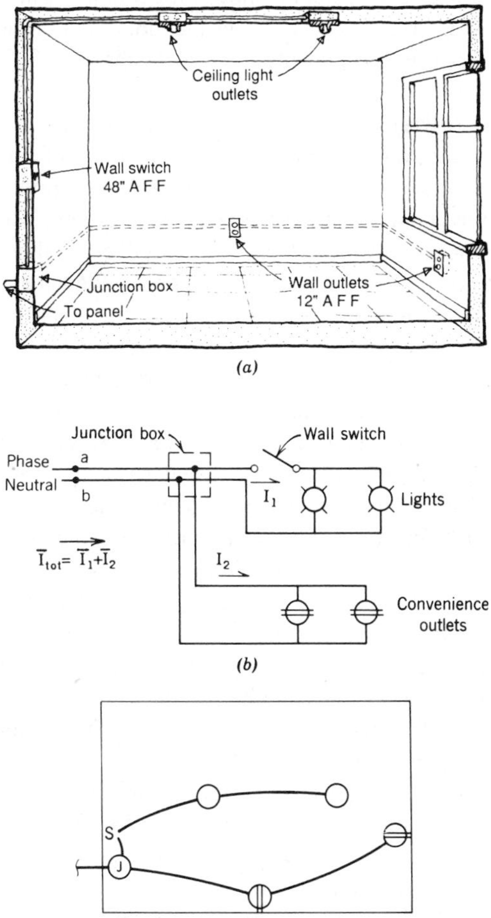

Fig. 14.5 Parallel groupings of ceiling light outlets and wall outlets are in turn connected in parallel to each other. A circuit is shown (a) *pictorially,* (b) *schematically, and* (c) *as on architectural plan. [The horizontal bar over the currents I_1 and I_2 in* (b) *signifies the use of vectorial addition. See Example 14.5]*

One additional point is important to appreciate. Remember, from Ohm's law, that current is inversely proportional to resistance. Thus as resistance drops, current rises. Now look at the circuit of Fig. 14.5. Normally that circuit will carry 10 amperes. But if by some mischance, such as deterioration of the wiring insulation, a connection appears between points *a* and *b*, the circuit is *shortened* so that there is no resistance in the circuit. The current rises instantly to a very high level, and we have a *short circuit.* If the circuit is properly protected, the fuse or circuit breaker will open, and the circuit will be deenergized. If not, excessive current will probably start a fire.

14.7 Direct Current and Alternating Current (d-c and a-c)

Whenever the flow of electric current takes place at a constant time rate, practically unvarying and in the same direction around the circuit, it is called d-c. The d-c voltages of 1.5 V positive polarity, and 1.0 V negative polarity are shown in Fig. 14.6*a*.

Whenever the flow of current is periodically varying in time and in direction, as indicated by the symmetrical positive and negative loops or sine waves in Fig. 14.6*c*, it is called an alternating current. The distance along the time axis spanned by a positive and a negative a-c loop is called one cycle. The number of such cycles occurring in one second is known as the *frequency* of the a-c current. Modern a-c systems in the United States operate at a frequency of 60 cycles per second, or 60 *hertz* (after H. Hertz). This means that current at 60 hertz (abbreviated Hz) is delivered to the consumer.

The a-c circuits differ from d-c circuits in a number of important respects and, since normal current supply is 60 Hz a-c, it is important to understand a-c terminology and usage. Instead of resistance the corresponding parameter in an a-c circuit is impedance, which is also measured in ohms. Depending on the device, it can be markedly different from the d-c resistance. For an a-c circuit, Ohm's law is

$$I = \frac{V}{Z} \tag{14.3}$$

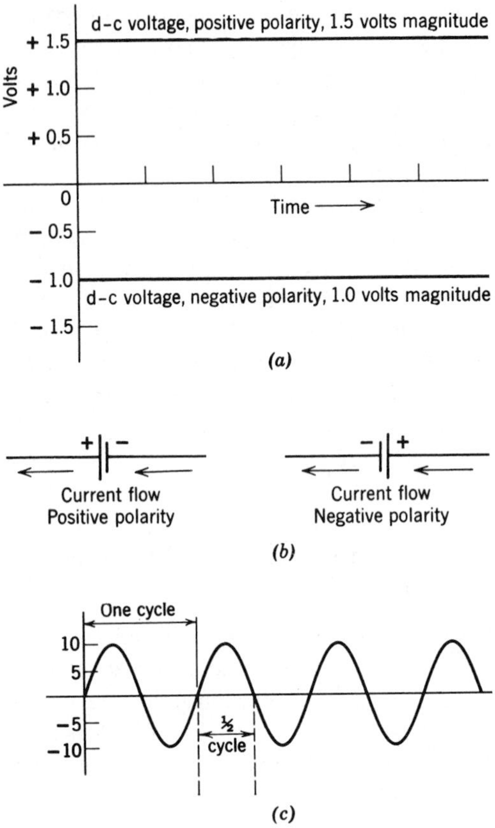

Fig. 14.6 (a) *Graphic representation of d-c voltages with positive and negative polarity.* (b) *Circuit symbol representation of a battery source. The longer bar is positive.* (c) *Alternating current (a-c).*

where Z is the symbol for impedance. We will not go into a-c circuit calculations primarily because such calculations are not especially useful to the reader. What *is* useful and important is an understanding of power and energy in both d-c and a-c circuits. This is discussed in the following sections.

14.8 Electric Power Generation

(a) Direct Current. With respect to generation of large amounts of power, photoelectric, piezoelectric, and thermoelectric effects including solar cells can be ignored, leaving the battery and the d-c generator as the sources of d-c current. Since the d-c generator is in reality an a-c generator with a device (commutator) attached that rectifies the a-c to d-c, the battery is still the only major direct source of direct current. (There are some special types of generators that produce direct current *directly*, but their use to date has been extremely limited.) A discussion of the application of batteries for emergency and standby power supply can be found in Section 16.31.

Another source of d-c power is rectification of a-c that can be accomplished on any desired scale to provide as much d-c power as there is available a-c power. The principal applications of d-c in buildings are for elevator power and to a lesser extent standby power for computers and other critical loads. Small amounts of d-c are also used for controls and telephones.

(b) Alternating Current. Alternating current is produced commercially by an a-c generator, generally called an alternator. The prime mover may be any type of engine or turbine. The process by which electricity is produced is illustrated in Fig. 14.7. It is based on the fundamental discovery in 1831 by Michael Faraday of the principle of electromagnetic induction. Put briefly, this principle states that when an electrical conductor is moved in a magnetic field, a voltage is induced in it (see Fig. 14.7*a*). The direction of movement determines the polarity of the induced voltage, as shown. If the conductor is formed into a coil and rotated in the magnetic field, a voltage of alternating polarity is produced, that is, alternating current. It does not matter whether the conductor or the magnet moves; the motion of the conductor and the field with respect to each other produces the voltage (see Fig. 14.7*b*). It is only one step (that of development) from this rudimentary a-c generator to the large, powerful alternator that produces a-c in a modern power plant. The frequency of the voltage generated is a function of the machine design (number of poles) and the speed at which it is driven. Normal generator frequency in the United States, as noted above, is 60 Hz, whereas in Europe and the Mideast it is 50 Hz.

14.9 Power and Energy

It is important, indeed imperative, that the distinction between power and energy be clearly understood, since all too frequently the terms

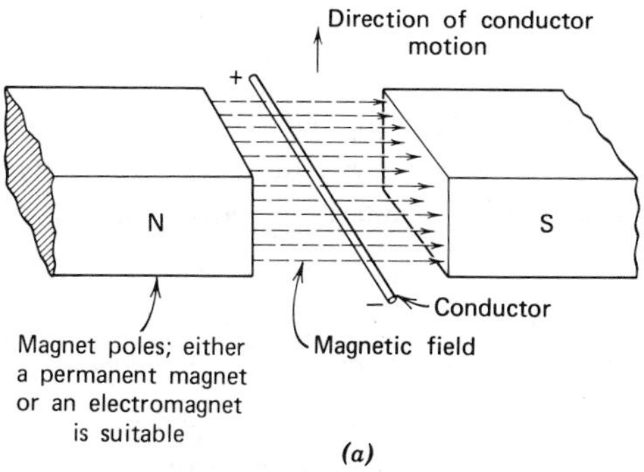

(a)

Fig. 14.7 (a) *The action fundamental to all generators is illustrated here. When a conductor of electricity moves through a magnetic field, a voltage is produced in the conductor, with polarity as shown.*

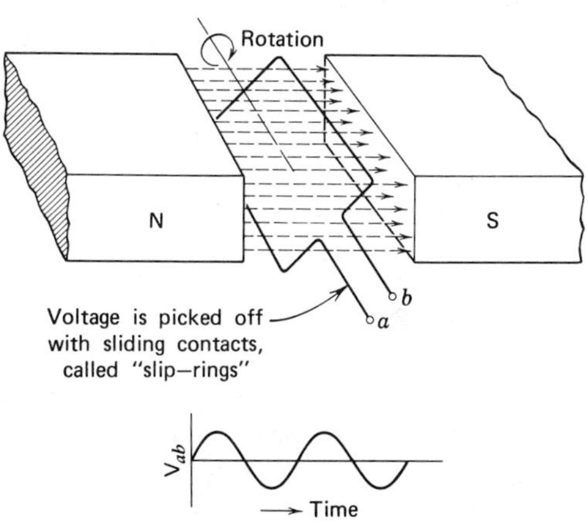

(b)

Fig. 14.7 (b) *Rotating a coil in a magnetic field produces an alternating sinusoidal voltage at terminals* a *and* b *because of the alternating polarity (see Fig. 14.7*a*).*

are incorrectly used interchangeably. *Energy* is the technical term for the more common expression—work. *Power* is the rate at which the energy is used. Stated otherwise and concretized, it takes a given amount of energy to lift a weight a given distance, but the faster it is lifted the more power is required. In terms of power, energy is the product of power and time, that is,

$$\text{energy or work} = \text{power} \times \text{time} \qquad (14.4)$$

In practical terms, energy is synonymous with fuel and therefore also cost. Thus energy can be expressed as barrels (tons) of oil, cubic feet (cubic meters) of gas, tons of coal, kilowatt hours of electricity usage, and dollars of operating cost. The concept of energy efficiency of structures can be stated in terms of annual usage of oil, gas, and electricity or alternatively in terms of dollars of total fuel cost. In technical terms, energy is expressed in units of Btu (calories), foot-pounds (joules), and kilowatt-hours. Conversion factors between these (and other) unit of energy are found in Appendix I. In terms of fuel, one kilowatt-hour (kWh) of energy is roughly equal to 0.5 lb of coal, 0.07 gallon of oil, 7.7 cubic feet of natural gas, and 8200 gallons of dam water.

Power is the rate at which energy is used or, alternatively, the rate at which work is done, since energy and work are synonymous. The term *power* implies continuity, that is, the use of energy at a particular rate, over a given, generally considerable, span of time. The concept of power necessarily involves the factor of time since it is, as stated above, the *rate* at which work is done. Thus multiplying power by time yields energy. Typical units of power in the English system are horse-power, Btu per hour, watt, and kilowatt. In the metric or SI system the corresponding units are joule per second, calorie per second, watt, and kilowatt. In physical terms, power is also the rate at which fuel (energy) is used. Thus power can also be expressed as gallons (liters) of oil per hour, cubic feet (cubic meters) of gas per minute, and tons of coal per day.

14.10 Power in Electric Circuits

The unit of electric power is the watt (W). A larger unit of 1000 watts is the kilowatt (kW). The power input in watts to any electrical device having a resistance R and in which the current is I is given by the equation

$$W = I^2R = I(IR) \qquad (14.5)$$

where W is wattage. This is true for both d-c and a-c circuits. However, since the resistance of an item is generally not known, but the circuit voltage and current *are* known, it would be preferable to be able to calculate power using these two quantities. This can be done, but is different for d-c and a-c.

In d-c circuits

by Ohm's law $V = IR$

and, since $W = I(IR)$,

$$W = VI \qquad (14.6)$$

where W is in watts, R in ohms, I in amperes, and V in volts.

In a-c circuits impedance is comprised of resistance and reactance (a-c resistance of inductance and capacitance) and causes a phase difference between voltage and current. This phase difference is represented by an angle, the cosine of which is called the power factor, abbreviated *pf*. This quantity is extremely important in that it enables us to calculate power in an a-c circuit. The a-c power equation is similar to that for d-c (see equation 14.6) with the addition of this special a-c term of power factor, that is

$$W = VI \times \text{pf} \qquad (14.7)$$

If power factor is not used in the equation, the product of voltage and current gives a quantity known as volt-amperes. In a purely resistive circuit, such as one with only electric heating elements, impedance equals resistance, power factor equals 1.0, and wattage equals volt-amperage. A few examples here should make applications of these equations clear.

EXAMPLE 14.3. Referring back to Example 14.1 in Section 14.5, calculate the power drawn using equations 14.5, 14.6, and 14.7.

SOLUTION. Since an incandescent lamp is purely resistive and therefore has unit (1.0) power factor, it does not matter where the circuit is a-c or d-c.

From Example 14.1, R = 66 ohms.

$$I = 1.74 \text{ A, and } V = 115 \text{ V}$$

1. In a d-c circuit, we would use equation 14.6:

$$W = VI = 115(1.74) = 200 \text{ W}$$

2. In an a-c circuit, we would use equation 14.7:

$$W = VI \times \text{pf} = 115 \times 1.74 \times 1.0 = 200 \text{ W}$$

3. In either a d-c or an a-c circuit, we can use equation 14.5:

$$W = I^2R = (1.74)^2 \times 66 = 200 \text{ W}$$

EXAMPLE 14.4. Using the data given in Example 14.2 and Fig. 14.3*b*, determine (a) the power lost in the wiring, and (b) the power input to the load.

SOLUTION

(a) The total line loss is

$$W = I^2R = I^2(r_{bc} + r_{da}) = (10)^2 \times 0.04 = 4 \text{ W}$$

(b) The power input to the load is

$$W = I^2R = I^2R_{cd} = (10)^2 \times 1.15 = 115 \text{ W or } 0.115 \text{ kW}$$

Alternatively, we can find this power by multiplying voltage and current. The voltage on the load is

$$IR = 10 \times 1.15 = 11.5 \text{ V}$$

and

$$W = VI = 11.5 \times 10 = 115 \text{ W} = 0.115 \text{ kW}$$

EXAMPLE 14.5. Refer to Fig. 14.5. Assume a 150-W incandescent lamp at each ceiling outlet. Also assume the load connected to one convenience outlet to be a 10-A hair dryer and blower, with a power factor of 0.80. Calculate the current and power in the two branches of the circuit, and the total circuit current, assuming a 120-V a-c source.

SOLUTION

(a) In the circuit branch feeding the lights we have (since the power factor is unity in resistive circuits)

$$\text{power} = VI$$

$$300 \text{ W} = 120 \text{ V} \times I$$

$$I = \frac{300}{120} = 2.5 \text{ A}$$

(b) In the second branch we have a 10-A, 0.8 pf load.

Power in watts = volts × amperes × power factor

$$W = 120 \times 10 \times 0.8 = 960 \text{ W}$$

But the circuit volt-amperes are

$$\text{V-A} = 120 \times 10 = 1200 \text{ V-A}$$

This latter figure is important for sizing equipment.

(c) To calculate the total current flowing from the panel to both branches of the circuit, we must combine a purely resistive current (lamp circuit) with a reactive one (dryer circuit). The exact value of current is the *vectorial* sum of the two branch currents, which calculates to be 12.1 amperes. This is close enough to the arithmetic sum of 12.5 amps to justify the use of the latter method except in cases where power factor is extremely low.

One further example at this point will demonstrate the importance of power factor in normal situations.

EXAMPLE 14.6. The nameplate of a (single-phase) motor shows the following data: 3 hp, 240 V, a-c, 17 A. Assume an efficiency of 90%. Calculate the motor (and, therefore, circuit) power factor.

SOLUTION

$$1 \text{ hp} = 746 \text{ W}$$

Therefore,

$$3 \text{ hp} = 3 \times 746 = 2238 \text{ W output}$$

$$\text{efficiency} = \frac{\text{output}}{\text{input}}$$

so

$$\text{power input} = \frac{2238 \text{ W}}{0.9} = 2487 \text{ W}$$

But for a-c,

$$\text{power} = \text{volts} \times \text{amperes} \times \text{power factor}$$

so

$$2487 = 240 \text{ V} \times 17 \text{ A} \times \text{power factor}$$

and

$$\text{power factor} = \frac{2487}{240 \times 17} = 0.61$$

Note the large difference between volt-amperes and watts.

$$VI = 240 \times 17 = 4080 \text{ VA}$$

$$P = \text{as above} = 2487 \text{ W}$$

where P designates power.

14.11 Energy in Electric Circuits

Since energy = power × time, the amount of energy used is directly proportional to the power of the system and to the length of time it is in operation. Since power is expressed in watts or kilowatts, and time in hours (seconds and minutes are too small for our use), we have for units of energy: watt-hours (Wh) or kilowatt-hours (kWh). Obviously, one watt-hour equals one watt of power in use for one hour, and one kilowatt-hour equals one kilowatt in use for one hour.

EXAMPLE 14.7

(a) Find the daily energy consumption of the appliances listed if they are used daily for the amount of time shown.

Toaster (1340 W)	15 min
Percolator (500 W)	2 h
Fryer (1560 W)	½ h
Iron (1400 W)	½ h

(b) If the average cost of energy is \$0.08 per kilowatt-hour, find the daily operating cost.

SOLUTION

(a) Toaster	1340 W = 1.34 × ¼ h	= 0.335 kWh
Percolator	500 W = 0.5 kW × 2 h	= 1.00 kWh
Fryer	1560 W = 1.56 kW × ½ h	= 0.78 kWh
Iron	1400 W = 1.4 kW × ½ h	= 0.70 kWh
	Total	2.815 kWh

(b) The cost is

2.815 kWh × \$0.08/kWh = \$0.225

or approximately 23 cents

Clearly the power being used at any specific time during the day by a residential household

ELECTRICITY

varies considerably. If we were to graph the power in use for a typical American household during a normal weekday, the plot might look something like that in Fig. 14.8. The average power demand of the household is obviously much lower than the maximum. The ratio between the two is called the overall *load factor* and runs between 20% and 30% for a typical household. The energy used by this household for the 24-h period shown is represented by the *area* under the curve of Fig. 14.8. This can be determined by integration only, since it varies continuously. That this is exactly what a kilowatt-hour meter does will be explained in Section 14.14, which deals with electrical measurements.

EXAMPLE 14.8. It has been estimated that the average power demand of an American household is 1.2 kW. Calculate the monthly electric bill of such a household, assuming a flat rate of \$0.08 per kilowatt-hour.

SOLUTION

Monthly energy consumption

$$= 1.2 \text{ kW} \times \frac{24 \text{ h}}{\text{day}} \times \frac{30 \text{ days}}{\text{month}} = 864 \text{ kWh}$$

Electric power bill

$$= 864 \text{ kWh} \times \$0.08/\text{kWh} = \$69.12$$

We stated in Example 14.8 above that the bill was based on a "flat" rate of 8 cents per kilowatt-hour. In actual fact the rate structure of most utilities is not so constructed. Generally, the tariff *decreases* with increasing use, thus encouraging larger use of electric power. This type of tariff was designed in the halcyon days of low-cost fuel, when the utility's costs were in large measure those of transmission, distribution, and administration. That this situation has changed

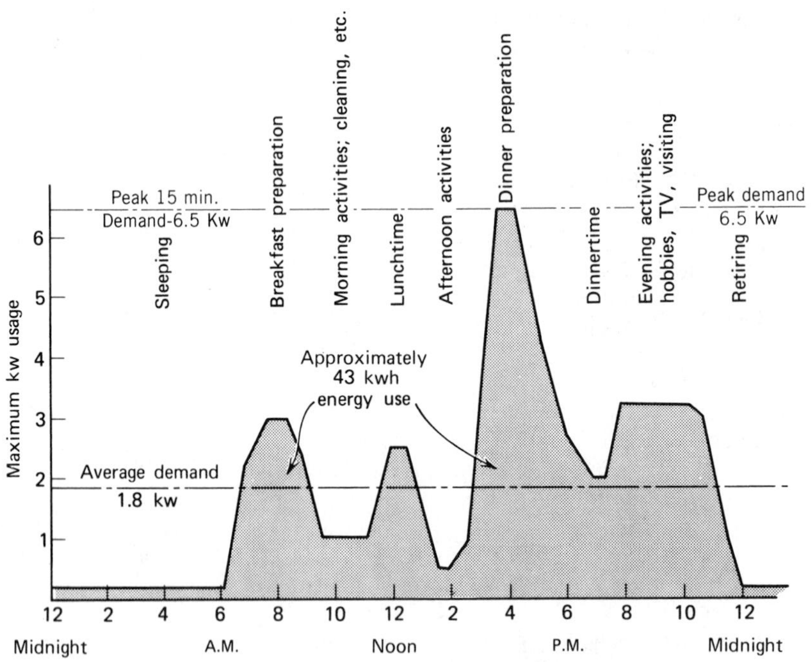

Fig. 14.8 *Hypothetical graph of power usage for a typical American household. Electric cooking is assumed. Area under the curve represents energy usage. Maximum kW demand (vertical axis) is based on a 15-min integrated demand, thus eliminating spikes in demand, such as those caused by starting a refrigeration (air-conditioning) compressor. This curve has a 24-h use of approximately 43 kWh, giving an average demand of 1.8 kW. The ratio between this average demand and the peak demand of 6.5 kW is called the load factor and is here 27.5%.*

is well known. Residential electric rates, with few exceptions, have not been restructured to encourage either efficient use of electric energy or reduction in energy usage. One technique, long standard in industrial and commercial user tariffs but almost never applied to residential users, is the levying of a charge for power (kW) in addition to the normal energy (kWh) charge. This *demand charge* is primarily useful in encouraging users to reduce their peak loads. In so doing, energy use is also reduced somewhat.

14.12 Electric Demand Charges

As stated, electric utility companies normally levy a kW demand charge on all but individual residential and a few special category customers. Varying with the individual utility company involved, this monthly charge runs between $2 and $15 per kilowatt of maximum average demand in any demand interval for that month. Demand intervals vary, being either 15 or 30 min (see Fig. 14.9).

Most utilities use a "sliding window" interval timing technique that starts a new interval every minute and updates the maximum interval demand accordingly. This enables them to find and bill for the maximum electric power demand in any 15- (or 30-) min period in the month. Some companies also include a "ratchet" clause that levies a demand charge for a number of months based on the maximum demand in any single month. This penalizes users with seasonal highs, that is, users with a low *yearly* load factor. The load factor is a measure of uniformity of power demand; a low load factor indicates short-time demand peaks for which the user is heavily charged. The reasoning behind the imposition of a demand charge and the significance of load factors can best be demonstrated by an example.

Assume that a pottery manufacturer, whose average 8-h daily load is 20 kW for lighting, pottery wheels, and the like, operates two 50-kW electric kilns twice a month for a 4-h period each

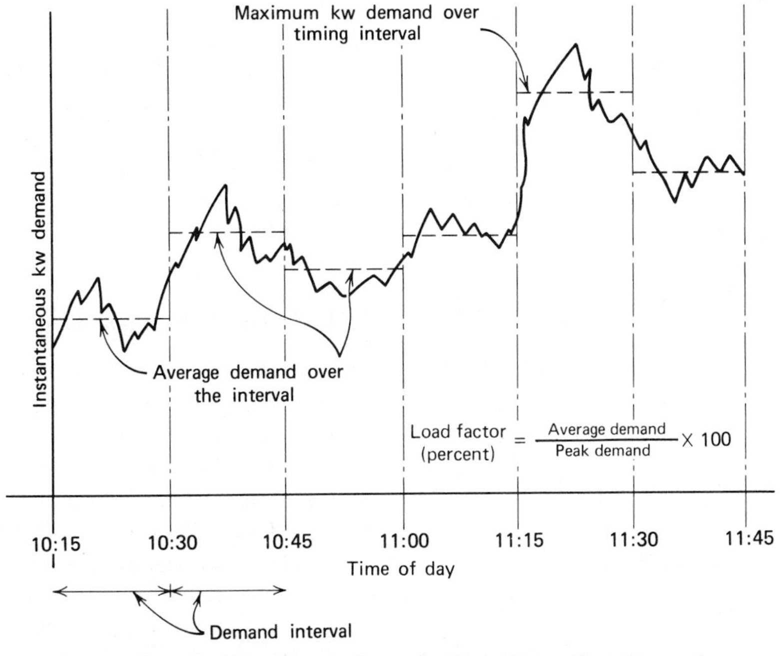

Fig. 14.9 *Typical instantaneous load curve for a facility. The utility demand meter records the average demand in each period (here 15 min). The maximum interval demand—in this case between 11:15 and 11:30—is used as the basis for monthly billing. A high load factor (utilization factor) indicate that little can be gained by demand control; a low load factor indicates the reverse.*

time. Further assume an energy charge of \$.05 per kilowatt-hour. The total monthly *energy* bill for the operation of the two kilns only would be

$$\text{cost} = 2 \times 50\text{ kW} \times 8\text{ h} \times \frac{\$0.05}{\text{kWh}} = \$40$$

Thus, were it not for the demand charge, the utility company (which is required by law to supply maximum customer demand) would have to provide and maintain 100 kW of generation, transmission, and distribution facilities in return for a payment that is the equivalent of 1.11 kW of average continuous load; that is,

$$\text{equivalent continuous load} = \frac{\$40}{720\text{ hr/month} \times 0.05} = 1.11\text{ kW}$$

The user's load factor can be calculated readily. By definition,

$$\text{load factor} = \frac{\text{average power demand}}{\text{maximum power demand}} \tag{14.8}$$

But, since for a given time interval the average power demand is equal to the energy used divided by the time involved, that is,

$$\text{average power demand} = \frac{\text{kWh energy use}}{\text{hours of use}} \tag{14.9}$$

we have the general expression

$$\text{load factor (LF)} = \frac{\text{kWh energy use}}{\text{maximum demand} \times \text{hours use}} \tag{14.10}$$

For the case under consideration, the *monthly* load factor is

$$\text{LF} = \frac{23\text{ days} \times 8\text{ h} \times 20\text{ kW} + 2\text{ days} \times 4\text{ h} \times 100\text{ kW}}{120\text{ kW} \times 720\text{ h/month}} = \frac{3680 + 800\text{ kWh/month}}{86{,}400\text{ kWh/month}} = 0.052 \text{ or } 5.2\%$$

This is obviously a very poor load factor, which results in a high demand charge. Assuming a \$5.00 per kW demand tariff, this pottery manufacturer would be billed, monthly, an additional:

$$\text{demand charge} = 120\text{ kW} \times \$5.00 = \$600$$

Since the energy bill is only

$$\text{energy cost} = 4480\text{ kWh} \times 0.05 = \$224.00$$

this manufacturer is paying heavily for highly peaked power use.

Although the illustration selected is somewhat extreme in its type of power use, it is not uncommon to find demand charges of the same order of magnitude as energy charges. Obviously, it is impossible to eliminate demand charges entirely, but it is certainly possible and frequently very simple to reduce them. Such a step is in the interest of the user, the utility, and the public at large: the user—for simple economic reasons; the utility—to permit more efficient use of its facilities; and the general public—by avoidance of unnecessary power plant construction and concomitant inefficient use of fuel, and overall reduction in fuel use. This last item is a secondary benefit of demand control.

In the next section we discuss electric demand control, whose primary function is to reduce electric *power* demand. This reduces demand charges and incidentally, secondarily, somewhat reduces energy consumption. Electric demand control is entirely different from *energy* management, which is primarily concerned with reduction of all types of energy use, including electric. Electric demand control is frequently included as one part of an overall energy management system.

14.13 Electric Demand Control

Electric demand control methods vary greatly in complexity and in degree of automation, but all basically perform the same task—efficient utilization of available energy to produce a high load factor, resulting in a lowering of demand charges. An ancillary, but important, benefit is the maximum utilization of building electrical power equipment, which normally runs underloaded. This results in smaller equipment, lower first cost, and less space utilization.

A further advantage to the user accrues from the fact that some utilities are offering their customers rebates that cover up to 40% of the cost of demand control and energy management equipment. This bonus serves to reduce by that amount the length of the payback period. (See Appendix E for a discussion of "payback period" calculation.) The reasoning of the utility is simple: it presently costs \$2000 ± per kilowatt of generating capacity for new power plant construction, whereas demand and energy management equipment costs only \$100 ± per kW demand reduction. Since a utility, by terms of its franchise, must supply all the power demanded, it is very much in the interest of any utility that is generating power near its equipment's maximum capacity, to reduce demand.

The oldest and simplest demand control scheme, still in use today, is the time-of-day dependent variable utility rate schedule. This scheme encourages off-peak use of electricity by offering a lower energy rate for off-peak hours use. Indeed, some utilities will install, at no charge to the user, a separate time-controlled circuit for use with time-deferrable loads such as water heaters, well pumps, battery chargers, and the like. The circuits are energized only during off-peak hours, which typically are mid-afternoon and after midnight. In the absence of a utility-supplied switch, the consumer can provide one of his own to accomplish his purpose (see Fig. 14.10).

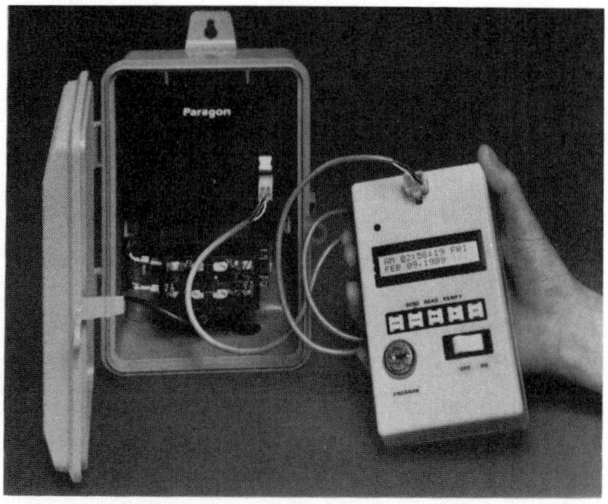

Fig. 14.10 *Programmable electronic time control switch designed for application with time-of-use (off-peak) utility price schedules. Programming device (shown hand-held) is separate from the switch and can be used by a utility to program many customer switches. The illustrated unit is arranged for 365-day scheduling, which permits full coordination with utility schedules that vary with the seasons of the year. Typical controlled loads are water heaters, thermal storage units, water and air accumulators, and any other load which either inherently or by design can be delayed for several hours. (Courtesy of Paragon Electric Co., Inc.)*

The basic technique of demand control is a simple one; electric loads are disconnected and reconnected in such a fashion that demand peaks are leveled off and load factor thereby improved. The extent to which a facility's electric loads can tolerate this type of switching is an indication of the potential effectiveness of a demand controller. An installation with a large *un*-interruptable load, such as computers or office lighting, will benefit minimally from demand control. Most industrial and commercial installations, however, contain a large percentage of interruptible loads (interruptions may be very short) and demand control systems frequently accomplish a 15 to 20% reduction in electric bills with a resultant short payback period on equipment investment.

The proliferation of demand control equipment has also produced a proliferation of nomenclature, including load shedding control, automated load control, peak demand control, and computerized load control. The last term refers to the type of equipment used to accomplish the intended control function. Descriptions that include the term *energy management* are inaccurate. Those devices are primarily concerned with control of *energy,* and are discussed in Section 21.2–21.4 for lighting and in Chap. 5 for HVAC. Demand control devices are intended to control *power*, which is *timed* energy use. Demand control produces energy savings as a secondary benefit. We will use the expression "electric demand control" for simplicity and avoid knowingly using a term that refers to a particular manufacturer's equipment.

We will discuss the various demand control schemes in some detail in order to enable the reader to differentiate between them by recognizing their unique characteristics. A manufacturer's descriptive literature usually comprises a lengthy list of its equipment's abilities, none of its shortcomings, and little of its specific or com-

ELECTRICITY

parative applicability. Furthermore, much of the equipment on the market today either overlaps a single control scheme or has additional abilities or functions not directly related to demand control. That being so, a good understanding of demand control schemes will greatly assist the prospective user.

(a) Level 1—Load Scheduling and Duty-Cycle Control. This level is the simplest and most obvious, and it is applicable to all types of facilities. The installation's electric loads are analyzed and then scheduled to restrict demand. Thus large loads can be shifted to off-peak hours and controlled to avoid coincident operation. The user can also take advantage of special night and weekend utility rates for loads that do not require immediate operation, such as battery charging and transfer pumping. Control can be entirely manual or automated by use of a duty-cycle controller. This device is essentially a programmable time switch (see Section 16.16) with switching for a number of circuits, or loads. Typical applications of this device are control of HVAC loads, lighting loads, and process loads in small commercial, institutional and industrial buildings.

The usefulness of these units lies in

1. Eliminating energy waste by shutting down units when not required.
2. Automatic control such as preheat and precool, which results in lower power and energy levels.

These devices are not, strictly speaking, demand controllers in that no cognizance is taken of the actual continuous electric loads. Instead, the devices operate on a preset duty cycle relying entirely on a prior analysis of the loads. Although such an analysis is a necessary first step in all levels of electric load control, its efficacy is limited since many of the loads are automatically controlled by pressure switches, thermostats, and the like. These loads cannot be scheduled with this type of controller and coincident operation cannot be prevented. Programmable temperature controllers (Fig. 9.18) and thermostats are primarily *energy* conservation devices, although appropriate programming is effective in reducing coincident demand loads and avoidance of full ON–OFF control.

(b) Level 2—Demand Metering Alarm. If in conjunction with a duty-cycle controller some type of continuous demand metering is installed that will go into alarm when a predetermined demand level is exceeded, a basic load control system will have been established. The load analysis discussed above would have to be extended to determine load priorities so that when the preset maximum demand load is exceeded and the alarm sounds, loads can be shed (disconnected) manually in a predetermined order of priority and, subsequently, reconnected also in order of priority. This type of control is practical only for a limited size installation inasmuch as most of the load switching activity is manual. A unit typical of this type is illustrated in Fig. 14.15*c*.

(c) Level 3—Automatic Instantaneous Demand Control. This type of control (also called *rate control*) is, in effect, an automated version of the level-2 system described above. The unit accepts instantaneous kW load information from the utility system, compares this information to the present kW limit (rate control), and acts automatically to disconnect and reconnect loads as required. These units *do not* recognize the utility's metering interval but act continuously on the basis of load comparison data. For this reason, these units are also referred to as load comparator controllers. Examination of Fig. 14.11 will make the unit's operation clear.

The first step in setting up this system is to separate the controllable (*sheddable*) loads from those that must remain uninterrupted. Depending on the type of facility, the two lists that follow are typical.

Sheddable Electric Loads

Nonessential lighting	Domestic hot water heating
Ventilation fans	Sewage ejectors with appropriate level controls
Space heating	Transfer pumps
Comfort cooling	Electric snow melting
Noncritical batch process equipment	Any device with flywheel effect
Electric boilers	

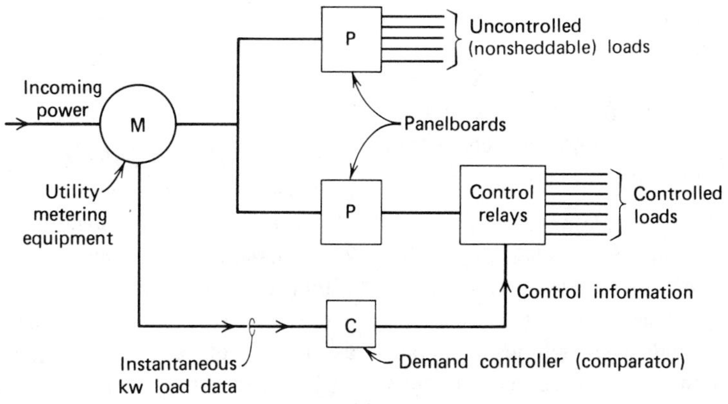

(a)

Fig. 14.11 (a) *Block diagram of a system of automatic electric power demand control. The demand controller receives instantaneous load data from the metering equipment, compares it to preset limits, and disconnects and reconnects controllable loads automatically to keep kW demand (load) within these limits.*

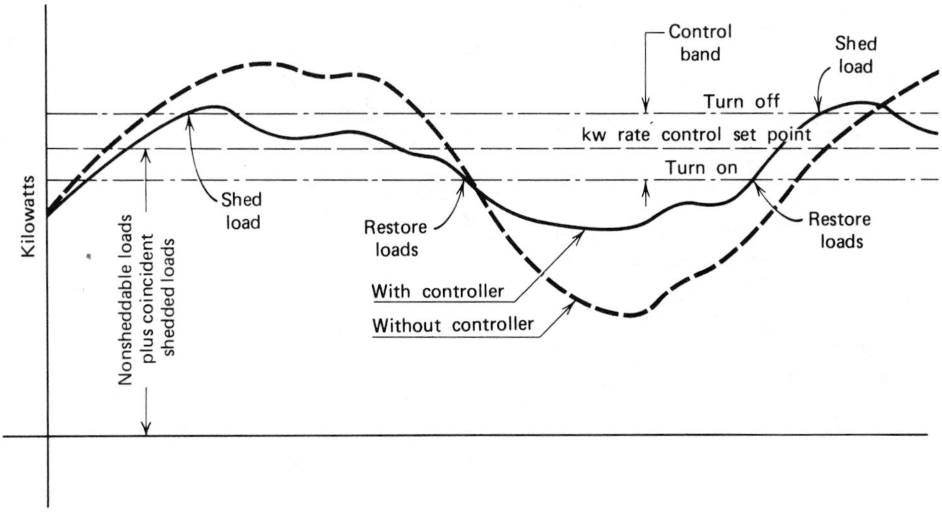

(b)

Fig. 14.11 (b) *Action of the demand controller is graphically illustrated. See the text, Section 14.13.*

Nonsheddable Electric Loads

Essential lighting	Process equipment
Elevators	Material handling equipment
Refrigeration compressors	Office machinery
Computers	
Emergency equipment	

The nonsheddable loads are fed directly from the power line. The sheddable loads are fed through a panel of control relays that respond to on/off instructions from the demand controller. Note that the controller acts to reduce maximum loads (peaks) and fill in low points (valleys). Although theoretically the energy use with or without the controller is the same, in actual practice energy savings of 15% and more are common. Some units have special provisions to overcome some of the limitations inherent in rate control systems. These are:

1. Excessive cycling.
2. Excessive off-time.
3. Inability to readily adapt to varying load patterns, resulting from variable production schedules, time schedules, changes in weather, and so on.

As a result of these limitations, this system is most useful in applications where operating modes do not change frequently and the facility is not very large. Thus stores, supermarkets, warehouses, small industrial facilities, and commercial installations are well served with this level system if they have at least 20% sheddable loads and their connected electric load is at least 150 kilovoltamperes (kVA).

(d) Level 4—Ideal Curve Control. This controller operates by comparing the actual rate of *energy* usage to the ideal rate, and controls kW demand by controlling the total *energy* used within a metering interval. A constant rate of energy use over a demand interval would show as the set of repeated straight lines in Fig. 14.12*a*. The utility company determines the demand over the demand interval by integrating

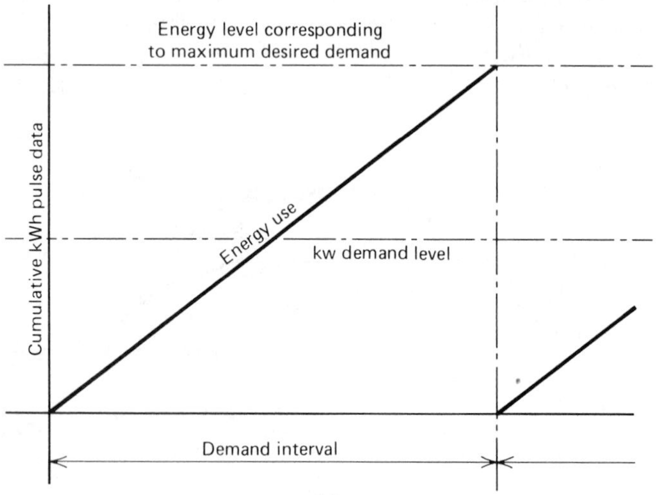

Fig. 14.12 (a) *Graph of cumulative energy use over a demand interval, corresponding to a constant kW demand. (Energy use is the time integral of power; i.e., kWh = kW × time.)*

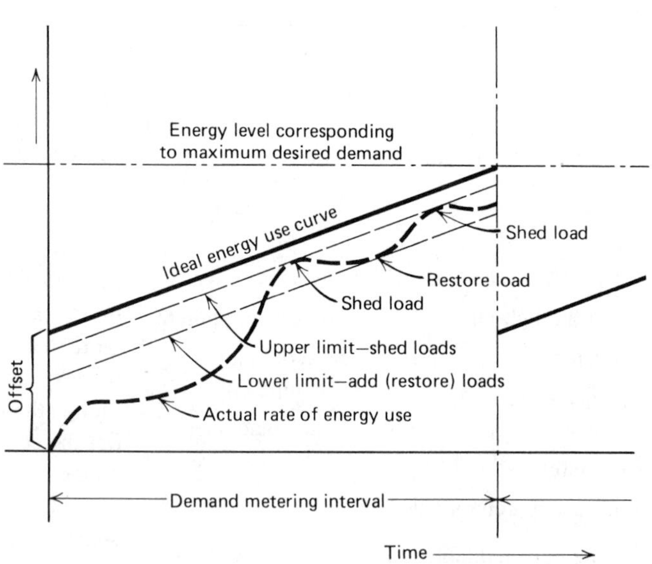

Fig. 14.12 (b) *Graph of action of an ideal rate controller. Note that toward the latter part of the period the pattern has been established, and the actual rate of energy use corresponds closely to the ideal.*

the kilowatt-hour energy over the interval and dividing by the interval time. Thus, the user is actually given a block of energy (kWh) that can be utilized at any desired rate, not necessarily at the constant rate of Fig. 14.12*a*. The desirable rate of energy use is shown as the "ideal curve" on the typical usage curve shown in Fig. 14.12*b*.

Shed points can be programmed for each load independently according to a predetermined priority, and priorities can be readily adjusted and rescheduled. Loads are normally shed only toward the end of the interval when the permissible energy total is approached and all loads are restored at the beginning of each interval. Thus during each interval loads are off for only a few minutes at most. Controllers operating on the "ideal curve" principle are considerably more flexible than the kilowatt rate controller described in the level-3 system above and are applicable to facilities of widely divergent load size, but with at least a 300-kW connected load. As with other controllers, the principal savings will be in demand charges, but almost always with considerable economies in energy billings.

An interesting application of these units is to utilize them to operate on-site electric generating equipment to supply peak power demands. This technique has come increasingly into use since recent rulings in many areas that not only permit on-site generation without utility approval but also require local electric utilities to purchase excess facility-generated power. These systems are generally referred to as *setpoint peak shaving* installations and are designed to operate in one of two modes: base-loaded peak shaving, in which the on-site generator produces a relatively constant output (Fig. 14.13*a*), and setpoint peak shaving, which operates the generator at an output level corresponding to specific power-level setpoints (Fig. 14.13*b*). These setpoints in turn are established at the beginning of each billing period and are changed automatically and continually to correspond to actual load conditions.

The advantage of the base-loaded arrangement is lower first cost for controls and efficient use of the equipment by operating at a relatively constant load. The disadvantage is the higher cost of generated electricity over purchased power. The ideal arrangement is to generate only, and exactly, the peak power required. This is what the setpoint arrangement attempts to do by utilizing microprocessor generator control to monitor power demand constantly and alter setpoints. Both of these arrangements are usable in facilities with minimal or zero sheddable loads since both always generate more power than is actually required. If some sheddable loads exist, a simpler (and cheaper) control system can be used which starts and stops the generator above a predetermined demand setpoint rather than utilizing a microprocessor to calculate the setpoint, relying on the demand controller to shed loads in order to minimize peak demand. Note, however, in Fig. 14.13*b* that the generating capacity utilized for load peaking is much smaller than the total load. Therefore, considerable idle standby power must be available to supply critical loads during power outages. This factor must be borne in mind when establishing system size since an auxiliary benefit of any peak shaving installation is the use of the generating equipment as part of the standby power plant.

The actual design of a peak shaving on-site generating system and the required load analyses are accomplished by the design and facility engineers and are considerably beyond our scope here. They are discussed briefly so that the architectural/engineering design group can be aware of the design alternatives and their physical and economic impact.

(e) Level 5—Forecasting Systems. These systems are by far the most sophisticated, the most expensive, and the most effective. They are best applied to large structures where the number of loads, load patterns, and complexity of operation preclude the effective use of the preceding systems. As a result of the large amount of load data required to be processed these systems frequently are installed as part of computerized central control facilities (see Appendix J). Details of operation are too complex for our needs, but the basic operation can be described. These units operate by continuously forecasting the amount of energy remaining in the demand interval, based on kilowatt-hour pulse data received. They then examine the status and priority of each of the connected loads

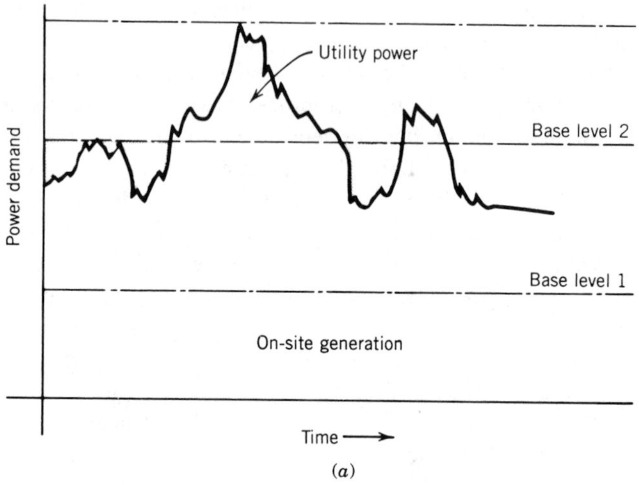

Fig. 14.13 (a) *With a base-loaded peak shaving system, selection of the base-load generation level determines all costs: initial, operating, and utility. Power below the base level is generated locally and above the base level by the utility. A low base level (1) yields low first cost and lowered peak demand but worse load factor. A high base level (2) means higher first cost, additional expensive power generated, including some sold at a low rate (shaded areas) but lowered demand charges.*

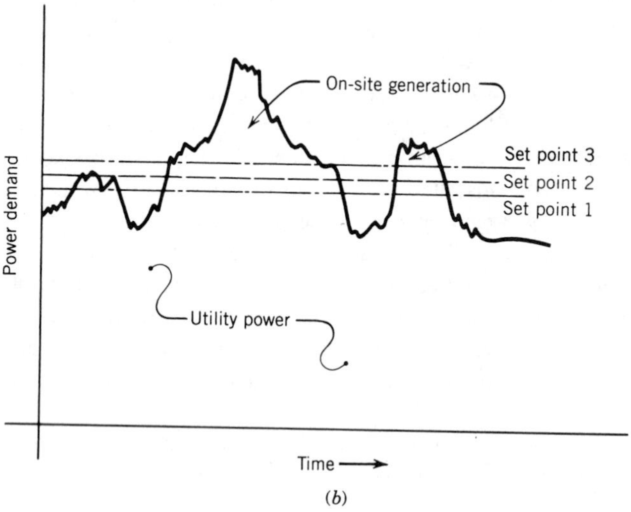

Fig. 14.13 (b) *A setpoint peak shaving generator supplies power* above *the setpoint (i.e., the power peaks). A simple modulating setpoint control would establish a setpoint (#1) and turn the local generator on and off to supply load peaks. These peaks may be reduced if loads are sheddable. With microprocessor control, the controller automatically alters the setpoint (#2,#3) to find the optimum combination of demand charges and generating cost for minimum total cost. Generator size is predetermined based on a detailed analysis of the load power profile.*

and decide on a course of action. Loads that in other systems are classified noncontrollable are in this system controlled, because of the accuracy and rapidity of the control function. A pneumatic compressor, for instance, that supplies process air might in other level control systems be classified as noncontrollable, despite the fact that it has long off periods. With computer control, such a load would be classified as "delayable" or "inhibited," since a 30-sec or 1-min delay in activation after the pressure switch closes its contact will normally be acceptable.

The advantage of these systems is that if programmed properly they can make small, accurate load changes throughout the interval, resulting in minimum load cycling and maximum efficiency.

14.14 Electrical Measurements

In the preceding sections we have explained the fundamental electric quantities of voltage and current and have defined the units involved as volts and amperes, respectively. As is true of all other physical quantities that are to be used in practical application, the need existed for a simple means of measuring these quantities. This need was met by the development of the meter movement illustrated in Fig. 14.14. Everyone at one time or another has felt the repulsion between like poles of two magnets held close together and, conversely, the attraction between opposite poles. This principle is used in the basic meter movement. It causes a deflection of the pointer as a result of the repulsion between the field of a permanent magnet and an electro-

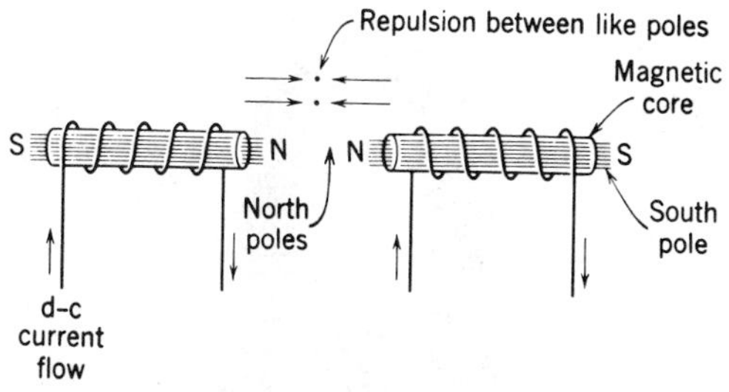

(a)

Fig. 14.14 (a) *Diagram showing basic electromagnetic principle and interaction between electromagnets. Any iron core becomes an electromagnet when current flows in a coil wound around it, as shown.*

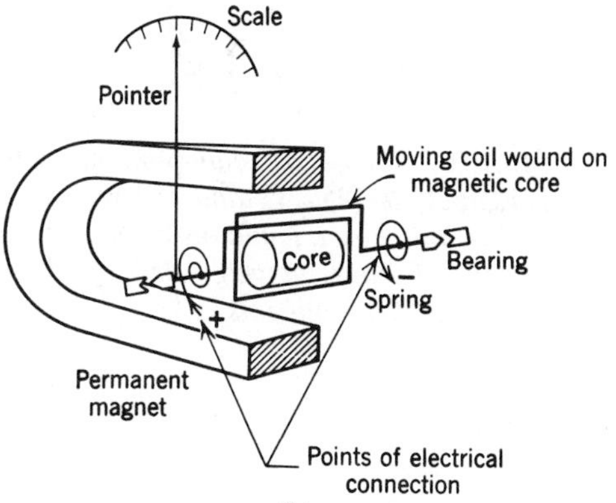

(b)

Fig. 14.14 (b) *The principle of the electromagnet is used in this basic meter movement. Current flowing in the movable coil forms an electromagnet whose field interacts (see part* a*) with the permanent magnet's field, causing a deflection proportional to the current flow. (Courtesy of Wechsler, Div. of Hughes Corp.)*

magnet. The electromagnet is formed when current flows in the coil, and its strength is proportional to the amount of current flowing. Thus, a strong current causes a larger deflection of the needle and, therefore, a higher reading on the dial. A spring (see Fig. 14.14*b*) provides restraining torque on the pointer. To make this very sensitive basic unit usable for large currents (it is intrinsically a microammeter, sensitive to millionths of an ampere), we simply divert, or shunt away, most of the current, allowing only a few microamperes to actually flow in the meter coil.

To use the same unit as a voltmeter, we put a large resistance (multiplier) in series with the meter, again limiting the current flowing to a few microamperes. The scale is then calibrated in volts. All d-c meters are made in this fashion. Most a-c meters operate on basically the same principle, except that instead of a permanent magnet an electromagnet is used. Thus when the polarity reverses, the deflecting force remains in the same direction. A d-c meter connected to an a-c circuit simply will not read, since inertia prevents the needle from bouncing up and down 60 times a second.

The meters described above are conventional analog units which read electrical units in proportion to mechanical forces exerted within the unit, that is, by analogy. Modern electronics has produced a complete line of solid-state electrical meters (see Fig. 14.15) which display the measured electrical values in either analog (dial) mode, digital mode, or both. They operate in a number of ways, all of which are different from that described above, and all of which are beyond the scope of this book. The fact that a meter reads digitally and does not rely on a reader to interpolate between lines or numbers on a printed scale, as is necessary with an analog meter, does not necessarily mean that it is highly accurate. Accuracy depends on the quality of the internal circuitry. Digital meters are, however, obviously easier to read since no visual interpolation is involved. This advantage, plus the constantly declining cost of sophisticated electronics which will increase the accuracy of low-cost meters will undoubtedly give solid-state digital meters a much larger market share in the immediate future.

The measurement of current and voltage in practical application is generally not as important as the measurement of power and energy, as the preceding section has made abundantly clear. To measure power, we utilize the fact learned earlier that power is proportional to the product of the voltage and current in the circuit. Although actual construction is complex, the theory of operationof a conventional coil-type wattmeter is simple. The meter has two coils; a current coil that is similar in connection to an ammeter, and a voltage coil that is similar in connection to a voltmeter. By means of the

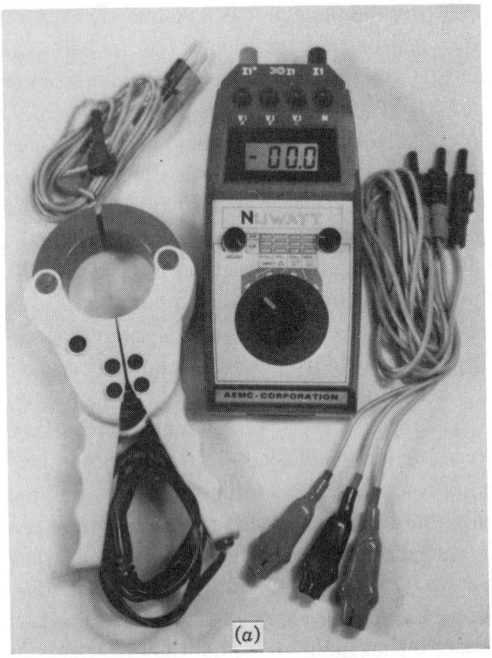

(a)

Fig. 14.15 *Several types of modern solid-state electrical meters.* (a) *Hand-held power meter capable of measuring 1/10 watt through 2000 kW of true rms (sinusoidal and nonsinusoidal) power. The importance of true rms measurement is discussed in Section 17.19. Connection to the electrical system is with clamp-on current probes, permitting measurement to be made without disturbing the system. Readout is digital. (Courtesy of AEMC Instruments.)*

(b)

Fig. 14.15 (b) *Solid-state clamp-on ammeter with 0–1250 A, 0–1250 V, and full resistance scales. The unit has an analog display, measures 8 × 3 × 1.5 in. and weighs less than 1 lb. (Courtesy of TIF Instruments.)*

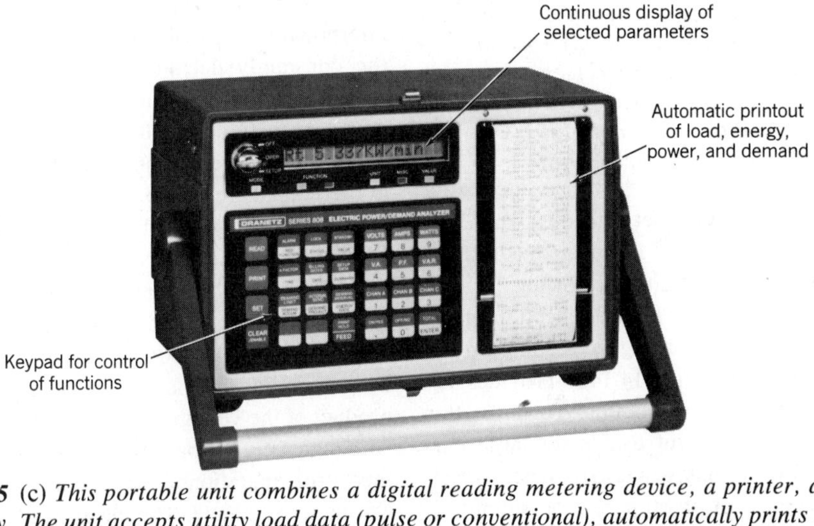

Fig. 14.15 (c) *This portable unit combines a digital reading metering device, a printer, and alarm capability. The unit accepts utility load data (pulse or conventional), automatically prints maximum demands per demand interval, day, billing period, or other desired period, and can readily be arranged to initiate an alarm signal when a preset power demand or rate of energy consumption is exceeded. The unit's probability permits analyzing separate parts of an electric power system. The unit is also arranged to display and print current, voltage, power factor, power, and energy use. (Courtesy of Dranetz Technologies.)*

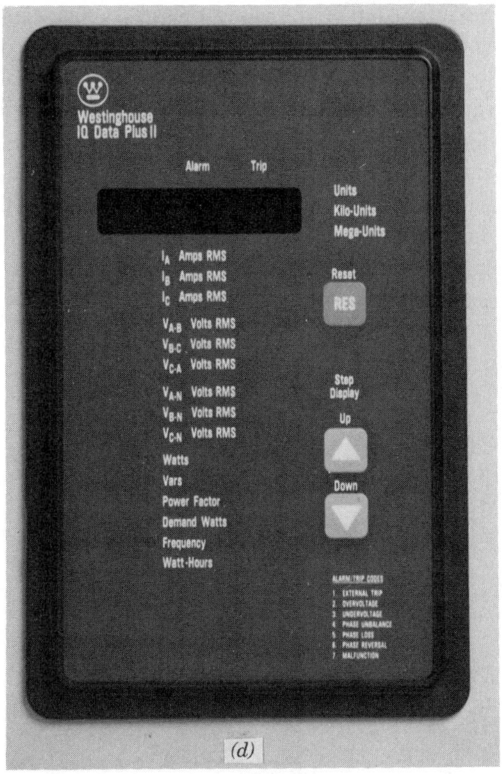

Fig. 14.15 (d) *This device is a microprocessor-based metering/protective device which measures current, voltage, frequency, power factor, power, energy, power demand, and reactive power. In addition, it can be programmed to alarm at a preset power usage rate, as part of a demand control scheme. It also provides monitoring of and protection from anomalies in the utility supply voltage. The unit is approximately 7 × 10 × 6 in. deep and weighs 6 lb. (Courtesy of Westinghouse Electric Corp.)*

physical coil arrangement, the meter deflection is proportional to the product of the two, and therefore to the circuit power. The meter can be calibrated as desired, depending on the size of the shunts and multipliers. To measure energy, the factor of time must be introduced, since as we know

$$\text{energy} = \text{power} \times \text{time}$$

D-c energy meters are available but are not of general interest because of the rarity of d-c power. A-c watt-hour meters are basically small motors, whose speed is proportional to the power being used. The number of rotations is counted on the dials, which are calibrated directly in kilowatt-hours. A diagram of the basic construction of an a-c kilowatt-hour meter is shown in Fig. 14.16. As can be seen, the kilowatt-hour energy consumption and the maximum interval kW demand can be read directly from the dials. (The illustrated unit is of the standard, not energy pulse, type.) If the numbers involved are too large or a meter is used with current transformers between it and the line, or for calibration reasons, a multiplying factor is required to arrive at the proper kilowatt-hour consumption. This number is written directly on the meter nameplate, and we multiply the meter reading by it to get the actual kilowatt-hours. In the absence of such a number, it can be assumed that the meter reads directly in kilowatt-hours.

Because manual reading of kWh meters of individual consumers is so labor intensive a task even without consideration of routine job difficulties such as inaccessible meters, dogs, inclement weather, and so on, utilities have long sought a more cost-effective means of determining consumer energy usage. To this end, with the benefit of modern microelectronics, a number of interesting, laborsaving kWh meters have been developed. One type is equipped with an electro-optical automatic meter reading system that is activated from a remote location. The meter data are transmitted electrically to a data-processing center, where they may be used by the utility to prepare subscribers' bills, to prepare customer load profiles, and to study, in combination with other such data, area load patterns, equipment loading, and so on. The customer can use instantaneous data to control loads, as explained at length in the preceding section.

Another meter type is equipped with a miniature radio transmitter which can be remotely activated to transmit the current kWh reading. The meter reader drives along the street and activates the meter from inside his vehicle by entering a customer code into a digital pad. A special receiver not only receives the transmitted kWh data but encodes and records the data automatically. These and other new kWh meters are obviously much more costly than the traditional units, but are finding increasing use, particularly in areas with high labor costs.

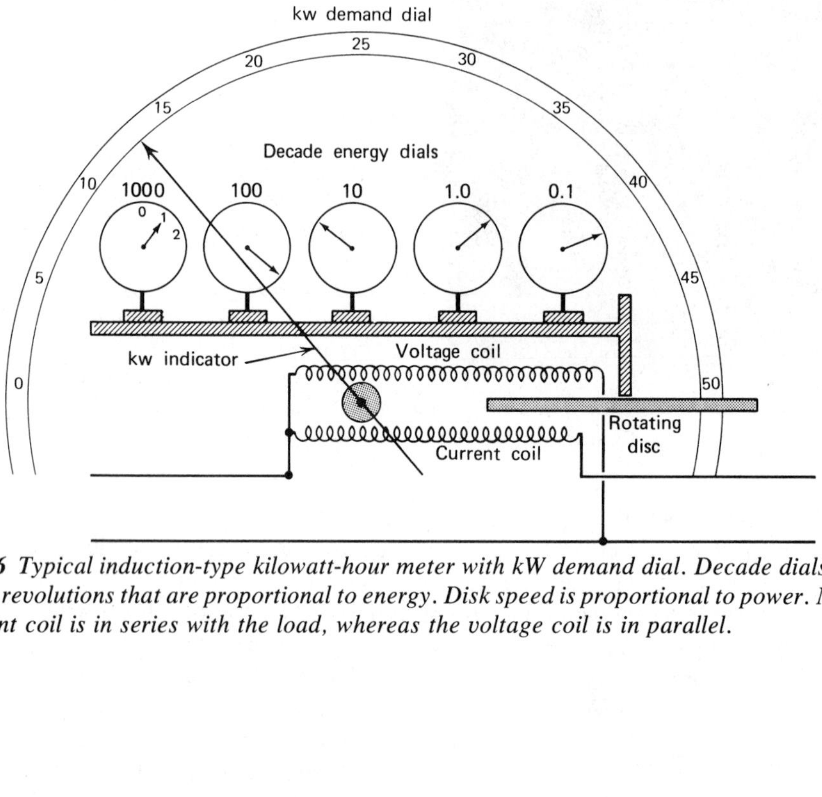

Fig. 14.16 Typical induction-type kilowatt-hour meter with kW demand dial. Decade dials register total disk revolutions that are proportional to energy. Disk speed is proportional to power. Note that the current coil is in series with the load, whereas the voltage coil is in parallel.

15

ELECTRICAL SYSTEMS AND MATERIALS: WIRING AND RACEWAYS

The major components of a building's electrical system can be arranged in three major categories; wiring and raceways, power-handling equipment, and control and utilization equipment. In the first category, we include conductors and raceways of all types; in the second, transformers, switchboards, panelboards, large switches, and circuit breakers; and in the last, actual utilization equipment such as lighting, motors, controls, and wiring devices. After some discussion that is applicable to all electrical materials, this chapter will discuss in detail the items in the first of these three categories, that is, the wiring and raceway system. Chapter 16 will cover most of the remaining items in the other two categories with the exception of lighting equipment, which will be discussed in the lighting section. Signal, control, and telecommunication equipment will be covered separately.

15.1 System Components

Referring to Fig. 15.1 note that the power system proceeds from the service point to the utilization point in a series of steps of decreasing circuit capacity. Although a normal single-line diagram does not differentiate by line weight between heavy and light (large and small) conductors (the heavier the conductor the greater the amount of power being carried), the single-line diagram of Fig. 15.1 does, in order to show relative power levels in the system. This ''size'' dif-

ELECTRICITY

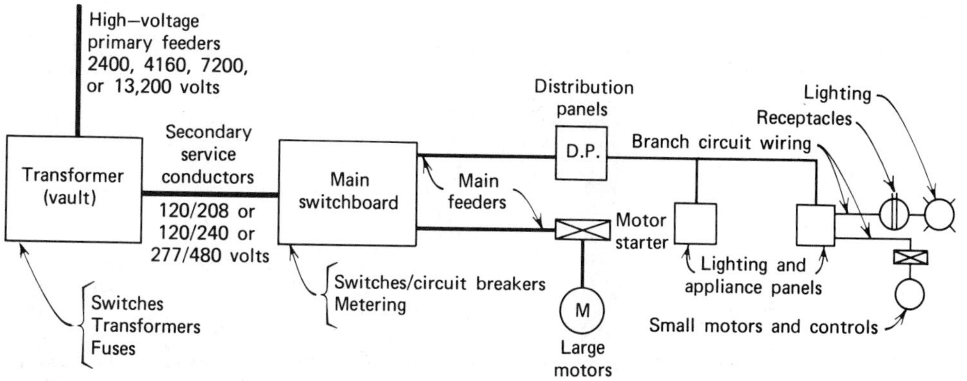

Fig. 15.1 Single-line diagram of a typical building electrical system, from the incoming service to the utilization items at the end of the system. This type of diagram is also referred to as a block diagram, *since the major components are shown as rectangles, or blocks. When this same type of information is presented showing the spatial relations between components, it is called a* riser diagram; *when electrical symbols are used in lieu of the blocks, it is called a* one-line *or* single-line diagram. *Were the connecting conductors between the major system components drawn to reflect size and thereby power-handling capacity, the system would appear as shown here.*

Typical Electrical Building Equipment

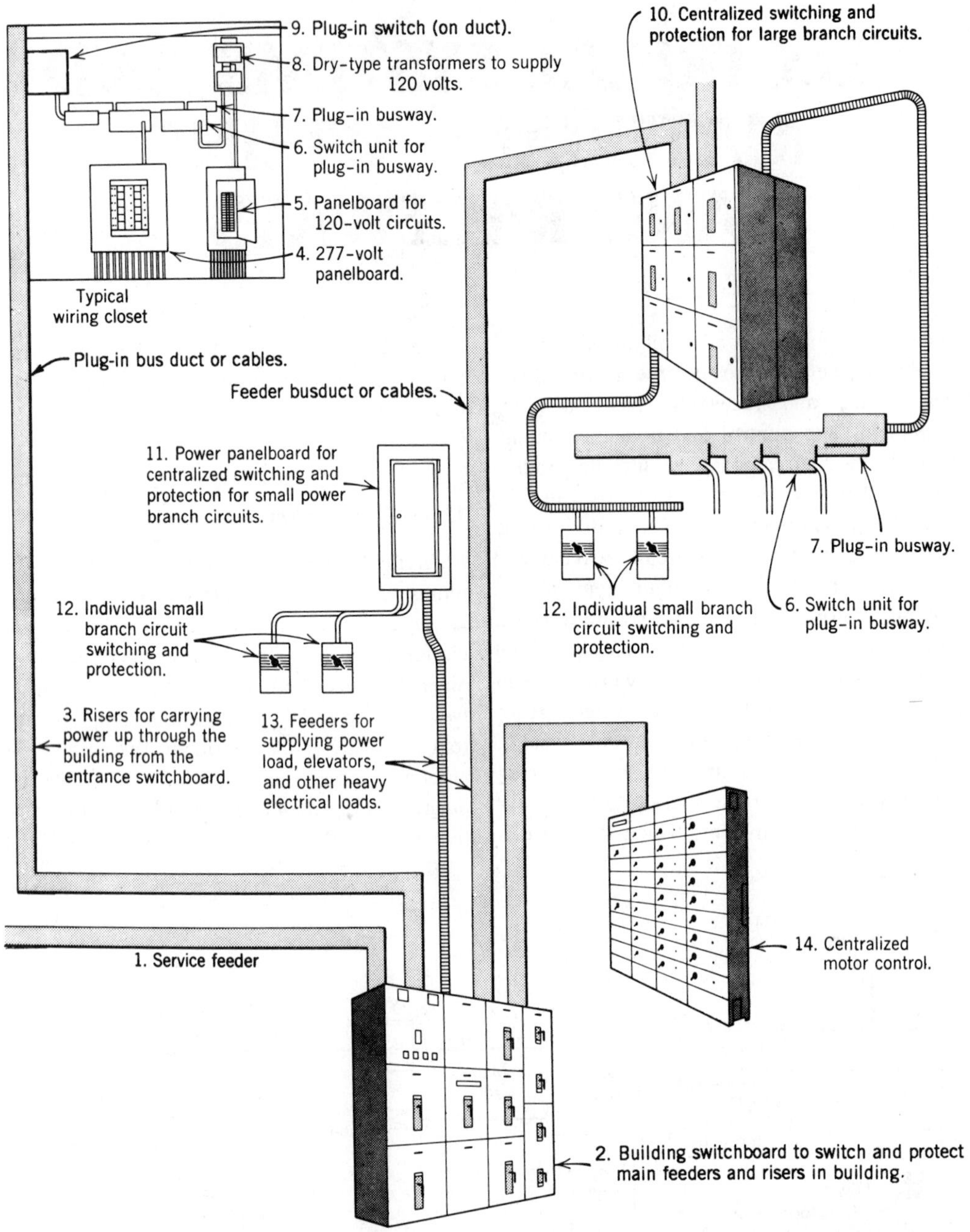

***Fig. 15.2** Building's basic electrical system shown pictorially, with the power capacity being indicated, approximately by the size of the conductors, Note that this diagram does not extend beyond the local panelboard, and includes only commonly used items. (Courtesy of General Electric.)*

ferentiation is more clearly shown in Fig. 15.2, which is a pictorial representation of a system similar to that of Fig. 15.1, but in somewhat greater detail and omitting items beyond the panel board. Note that only standard items are shown; special items, however common, such as emergency and uninterrupted power sources/supplies, energy controls, service transformers, and the like, are omitted for the sake of clarity.

15.2 National Electrical Code

The National Electrical Code (NEC) of the National Fire Protection Association (NFPA) defines the fundamental safety measures that must be followed in the selection, construction, and installation of all electrical equipment. This code is used by all inspectors, electrical designers, engineers, contractors, and the operating personnel charged with the responsibility for safe operation. Having been incorporated into OSHA (Occupational Safety and Health Act) it has, in effect, the force of law. The reader of this book should obtain for personal use the latest edition of the NEC from the NFPA at 80 Batterymarch Street, Boston, MA 02110. Frequent references will be made to this code.

In addition to the National Code, many large cities such as New York, Boston, and Washington, D.C., have their own electrical codes that, though similar to the NEC, contain numerous special requirements.

In order to assure a minimum standard of intrinsic electrical safety for electrical equipment, a single agency was needed to establish standards for, and to actually test and inspect, electrical equipment. Such an organization is Underwriters' Laboratories, Incorporated, which publishes lists of inspected and approved electrical equipment. These listings are universally accepted, and many local codes state that only electrical materials bearing the Underwriters' Laboratories (UL) label (of approval) will be acceptable.

15.3 Economics of Material Selection

The selection of electrical materials involves not only choosing a material or assembly that is functionally adequate and, where necessary, visually satisfactory, but also the consideration of economic factors. This is necessary since, in most instances, there is available a multiplicity of equipment that will fulfill the construction need. In such cases, economic factors often decide the issue. The decision is relatively simple when the various materials differ only slightly from each other and a straightforward first-cost comparison is all that is required. Often, however, the choice is not so simple, since the materials may vary considerably in characteristics other than functional suitability, and a more detailed cost study is required. As the reader is no doubt aware, such economic analyses are frequently of greater importance in comparisons between various HVAC systems. This is less so when dealing with electrical systems, since from an energy point of view electrical systems are more passive than HVAC systems, and it is energy costs that are often the decisive factor in economic analyses. We are speaking, of course, of life-cycle equipment costs (over the life of the *structure*) expressed in *present-value* dollars, or annual owning and operating costs, including equipment amortization costs. The type of analysis used depends on the situation. However, such comparisons are useful only when both the initial cost and the operating costs will be borne by the same individual, that is, an owner-operator. Obviously, in the case of a speculative building venture only first cost is considered.

It is often quite difficult to perform such cost analyses since accurate data on life and maintenance costs for electrical equipment may not be readily available. Still, even if a formal analysis cannot be done, the principle involved must be considered in the selection of all electrical material, bearing in mind its particular importance in the case of energy-utilizing equipment such as lighting and motors. Detailed economic discussions on these items will be found with the related technical discussion. See also Appendix E.

15.4 Energy Considerations

As mentioned above, energy costs are a major factor in economic analysis. However, energy considerations are at least as important, in and of themselves. This aspect will be examined in detail in the discussions both of building energy budgets and of their components, including lighting, elevators, and electric motors in all other systems.

15.5 Electrical Equipment Ratings

All electrical equipment is rated for the normal service it is intended to perform. These ratings may be in voltage, current, duty, horsepower, kW, kVA, temperature, enclosure, and so on. Ratings related to the specific equipment will be discussed in the sections below. The ratings that are specifically and characteristically electrical are those of voltage and current.

(a) Voltage. The voltage rating of an item of electrical equipment is the maximum voltage that can safely be applied to the unit continuously. It frequently, but not always, corresponds to the voltage applied in normal use. Thus, an ordinary wall electrical receptacle is rated at 250 V maximum, though in normal use only 120 V are applied to it. The rating is determined by the type and quantity of insulation used and the physical spacing between electrically energized parts.

(b) Current. The current rating of an item of electrical equipment is determined by the maximum operating temperature at which its components can operate properly continuously. That in turn depends on the type of insulation used. As a case in point, consider an electric motor. The current flowing in the motor windings causes a power loss (I^2R), which generates heat. If the windings are insulated with varnished cotton braid, with a maximum safe operating temperature of 65°C, the maximum permissible current (to which the horsepower rating of the motor is directly related) is that current which will produce this operating temperature. If these same windings are insulated with a silicone or glass compound with a maximum operating temperature of 150°C, obviously more current can be safely carried and the horsepower rating is consequently larger. Thus we see that, although a motor is rated in horsepower (or kW where SI units are used), a transformer is rated in kVA and a cable, as will be discussed below, is rated in amperes, the actual criterion on which all these ratings are based is maximum permissible operating temperature.

15.6 Interior Wiring Systems

At this point it will be helpful to survey the different types of interior wiring systems before commencing a discussion of components. When the primary purpose of the system is to distribute electrical energy, it is referred to as an electrical power system; when the purpose is to transmit information, it is referred to as an electrical signal or communication system. In this chapter we deal with electrical power systems except that the discussion on raceways will cover equipment used by communication systems as well.

Due to the nature of electricity, its distribution within a structure for power use poses basically a single problem: how to construct a distribution system that will *safely* provide the energy required at the location required. The safety consideration is all-important, since even the smallest interior system is connected to the utility's powerful network and the potential for damage, injury, and fire is always present. The solution to this problem is to isolate the electrical conductors from the structure except at those specific points, such as wall receptacles, where contact is desired. This isolation is generally accomplished by insulating the conductors and placing them in protective raceways. The principal types of interior wiring systems in use today are exposed insulated cables, insulated cables in open raceways, and insulated conductors in closed raceways.

(a) Exposed Insulated Cables. In this category would be included (using the NEC nomenclature) NM ("Romex") and AC ("BX") (see Sections 15.11 and 15.12).

Also included are other types where the cable construction itself provides the necessary electrical insulation and mechanical protection.

(b) Insulated Cables in Open Raceways (Trays). This system is specifically intended for industrial application, and it relies upon both the cable and the tray for safety.

(c) Insulated Conductors in Closed Raceways. This system is the most general type and is applicable to all types of facilities. In general, the raceway is installed first and the wiring pulled in or laid in later. The raceways themselves may be

1. Buried in the structure; for example, conduit in the floor slab or underfloor duct (see Sections 15.18, 15.19, 15.24, and 15.25).
2. Attached to the structure; for example, all types of surface raceways, including conduit and wireways suspended above hung ceilings (see Sections 15.23 and 15.30).
3. Part of the structure; for example, cellular concrete and cellular metal floors (see Sections 15.26 and 15.27).

(d) Combined Conductor and Enclosure. This category is intended to cover all types of factory-prepared and factory-constructed integral assemblies of conductor and enclosure. Included here are flat cable intended for under-carpet installation (Section 15.29), flat cable assemblies and lighting track (Section 15.16), manufactured wiring systems (Section 15.30), and all types of busway, busduct, and cablebus (Section 15.15).

15.7 Conductors

Electrical conductors (wiring) are the means by which the current is conducted through the electrical system, corresponding to the piping in the hydraulic analogy. By convention, a single insulated conductor No. 6 AWG (American Wire Gauge) or larger, or several conductors of any size assembled into a single unit, are referred to as cable. Single conductors No. 8 AWG and smaller are called wire.

The standard of the American wire and cable industry for round cross-section conductors is the American Wire Gauge. All wire sizes up to No. 0000 (also written No. 4/0) are expressed in AWG. The AWG numbers run in *reverse* order to the size of the wire, that is, the smaller the AWG number, the larger the size. Thus No. 10 is a heavier wire than No. 12 and lighter (thinner) than No. 8. The No. 4/0 size is the largest AWG designation, beyond which a different designation called MCM (thousand circular mil) is used. This unit is also referred to as kmil. In this designation, wire diameter *increases* with number; thus, 500 MCM (kmil) is a heavier wire (double the area) than 250 MCM (kmil).

Outside of the United States, where the metric system is in general use, conductor sizes are given simply as the diameter in millimeters (mm). Table 15.1 gives dimensional and stranding data for the commonest wire sizes and includes the millimeter equivalent of each size. This will prove useful in interfacing American gauges and metric sizes.

15.8 Conductor Ampacity

Conductor current-carrying capacity or *ampacity*, is determined as explained above, by the maximum operating insulation temperature. Heat generated as a result of the current flowing is dissipated into the environment. Thus, for a given environment (open-air, buried in earth, or enclosed), ampacity increases with increasing conductor size and maximum permissible insulation temperature. These facts are clearly shown in Table 15.2*a*. If more than three conductors are placed in a conduit, the increase in temperature requires that the conductors be derated by the amount shown in Table 15.2*b*.

Since heat dissipation from a conductor in free air is much greater than from the same conductor enclosed in conduit or direct buried, the corresponding allowable ampacity is also greater. Conversely, if the ambient temperature around the conductor is above 30°C, on which all the ampacity tables are based, the permissible ampacity must be reduced.

Ampacity tables for conductors in free air and cable types not shown in Table 15.2*a* and derating factors for high ambient temperature

TABLE 15.1 **Physical Properties of Bare Conductors**

Size AWG or MCM (kmil)	Area (Circular Mils)	Diameter (Inches)		Diameter (Millimeters)		d-c Resistance Ohms/1000 ft at 77° F, 25° C (Bare Copper)
		Solid	Stranded	Solid	Stranded	
16	2580	0.0508	—	1.29	—	4.10
14	4109	0.0641	—	1.63	—	2.57
12	6530	0.0808	—	2.05	—	1.62
10	10,380	0.1019	—	2.59	—	1.02
8	16,510	0.1285	—	3.26	—	0.64
6	26,240	0.162	0.184	4.11	4.67	0.41
4	41,740	0.204	0.232	5.18	5.89	0.26
2	66,360	0.258	0.292	6.55	7.42	0.16
1	83,690	0.289	0.332	7.34	8.43	0.13
0 (1/0)	105,600	0.325	0.373	8.26	9.47	0.10
00 (2/0)	133,100	0.365	0.418	9.27	10.62	0.081
000 (3/0)	167,800	0.410	0.470	10.41	11.94	0.064
0000 (4/0)	211,600	0.460	0.528	11.68	13.41	0.051
250 MCM (kmil)	250,000	0.500	0.575	12.70	14.61	0.043
300 MCM (kmil)	300,000	0.548	0.630	13.92	16.00	0.036
400 MCM (kmil)	400,000	0.632	0.728	16.05	18.49	0.027
500 MCM (kmil)	500,000	0.707	0.813	19.56	20.65	0.022

Source: Extracted from the National Electrical Code, except for millimeter dimensions.

are all found in the NEC. Typical ambient temperatures are given in Table 15.3.

15.9 Conductor Insulation and Jackets

Most conductors are covered with some type of insulation that provides electrical isolation and some physical protection. Additional physical shielding, where necessary, is provided by a jacket over the insulation. Insulation is rated by voltage. Ordinary building wiring is rated for 300 or 600 V. The common types of building wire insulation are listed in Table 15.4 with the associated trade names, code letters, maximum temperatures, and special provisions.

15.10 Copper and Aluminum Conductors

Aluminum has an inherent weight advantage over copper, with concomitant lower installation costs. Economy usually lies with copper in small- and medium-size cable, since weight is not a problem and the smaller conduit required for the smaller copper conductors generally make them cheaper. In the larger cable sizes, the aluminum weight advantage offsets the economy of smaller copper size and smaller conduit, and generally proves less expensive, particularly in areas of high labor cost such as urban areas.

Aluminum and copper both exhibit the low electrical resistivity necessary for a good electrical conductor. There are, however, difficulties inherent in splicing and terminating aluminum. These difficulties—which can be overcome with the use of proper equipment, techniques, and workmanship—stem from aluminum's cold-flow characteristic when under pressure (causing joints to loosen) and aluminum's oxide. This oxide, which forms within minutes on any exposed aluminum surface, is an adhesive, poorly conductive film that must be removed and prevented from reforming if a successful, long-life joint or termination is to be effected. If this is not done, the oxide causes a high-resistance joint with consequent excessive heat generation and possible incendiary effects. This oxide problem can in large measure be overcome by the

TABLE 15.2*a* **Allowable Ampacities of Enclosed Insulated Copper Conductors Rated 2000 V Maximum (not more than three conductors in raceway or directly burried in earth) Based on Ambient Temperature of 30°C (86°F)**

Size	Temperature Rating of Conductor[a]		
	60°C (140 F)	75°C (167 F)	90°C (194 F)
AWG MCM	Type UF	Types RHW, THW, THWN, XHHW[b]	Types THHN, XHHW[b]
14	20[c]	20[c]	25[c]
12	25[c]	25[c]	30[c]
10	30	35[c]	40[c]
8	40	50	55
6	55	65	75
4	70	85	95
3	80	100	110
2	95	115	130
1	110	130	150
0	125	150	170
00	145	175	195
000	165	200	225
0000	195	230	260
250	215	255	290
300	240	285	320
350	260	310	350
400	280	335	380
500	320	380	430

Source: Extracted from the National Electrical Code.

[a]These ampacities relate only to conductors described in Table 15.4.

[b]For dry and damp locations use 90°C rating; for wet locations use 75°C rating.

[c]Unless otherwise permitted by the NEC, the overcurrent protection for these conductors shall not exceed 15 A for 14 AWG, 20 A for 12 AWG, and 30 A for 10 AWG, after application of correction factors for ambient temperature and number of conductors in raceway.

use of copper-clad aluminum wire, but the cold-flow problem remains. Furthermore, when used in branch circuits, even if properly installed initially, aluminum can create problems when wir-

TABLE 15.2*b* **Current-Carrying Capacity Derating Factors**

Number of Conductors[a] in Raceway	Derating Factor
4 to 6	0.80
7 to 9	0.70
10 to 24[b]	0.70
25 to 42[b]	0.60
43 and above[b]	0.50

Source: Extracted from the National Electrical Code.

[a]Neutral conductors are not counted, but see Section 17.19.

[b]Includes a load diversity factor of 50%. For factors when diversity is not included, see NEC Table 310-16, note 8.

ing devices are replaced by unskilled homeowners.

As a result of a number of unfortunate incidents, some localities in the United States have banned the use of aluminum wire in branch circuitry. Heavy feeders are normally installed by experienced and skilled contractors, and the risk of a poor joint is minimized. We recommend that the use of aluminum conductors be restricted to sizes not smaller than No. 4 AWG, and that installation be permitted *only* by contractors who certify expertise in the specialized techniques involved. Also, local codes and the electrical inspectors should be consulted. All references in this text, including all tables and illustrations, are to copper conductors. The following sections are devoted to a brief description of the principal building wire types.

15.11 Flexible Armored Cable

Among the most common types of cable run without raceways is the NEC type AC armored cable, commonly known in the smaller sizes by the trade name "BX." It is an assembly of insulated wires, bound together and wrapped with a spiral-wound interlocking strip of steel tape (see Fig. 15.3). The cable is installed by simple U-clamps or staples holding it against beams, walls, and so on. This type of installation is frequently used in residences and in the rewiring of

TABLE 15.3 **Typical Ambient Temperatures**

Location	*Temperature*	*Minimum Rating of Required Conductor Insulation*
Well ventilated, normally heated buildings	30° C (86° F)	(See note below)
Buildings with such major heat sources as power stations or industrial processes	40° C (104° F)	75° C (167° F)
Poorly ventilated spaces such as attics	45° C (113° F)	
Furnaces and boiler rooms (min.)	40° C (104° F)	75° C (167° F)
(max.)	60° C (140° F)	90° C (194° F)
Outdoors in shade in air	40° C (104° F)	75° C (167° F)
In thermal insulation	45° C (113° F)	75° C (167° F)
Direct solar exposure	45° C (113° F)	75° C (167° F)
Places above 60° C (140° F)		110° C (230° F)

NOTE: 60°C for up to and including No. 8 AWG copper and 75°C for over No. 8 AWG copper.

TABLE 15.4 **Characteristics of Selected Insulated Conductors for General Wiring**

Trade Name	*Type Letter*	*Maximum Operating Temperature*	*Application Provisions*
Moisture- and heat-resistance rubber	RHW	75°C 167 F	Dry and wet locations
Heat-resistant thermoplastic	THHN	90°C 194 F	Dry and damp locations
Moisture- and heat-resistant thermoplastic	THW	75°C 167 F	Dry and wet locations
Moisture- and heat-resistant thermoplastic	THWN	75°C 167 F	Dry and wet locations
Moisture- and heat-resistant cross-linked thermosetting polyethylene	XHHW	90°C 194 F	Dry and damp locations
		75°C 167 F	Wet locations
Underground feeder; moisture resistant	UF	60°C 140 F	Underground, including direct burial

Source: Extracted from the NEC.

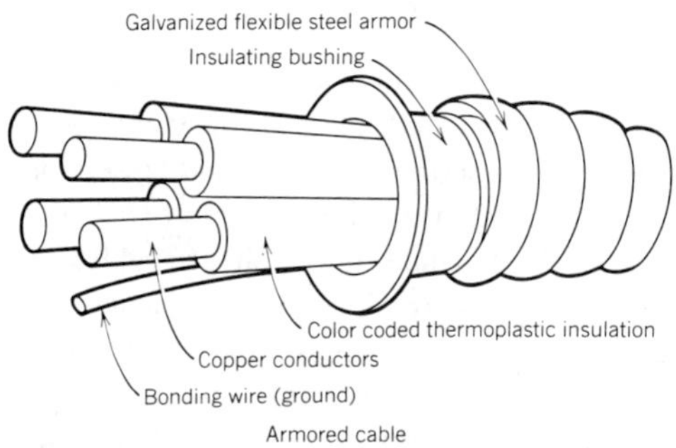

(a)

Fig. 15.3 (a) *Type AC flexible armored cable (BX). The bushing is installed to protect the wires from the sharp metal edges of the cut armor. (Courtesy of AFC/Nortek Co.)*

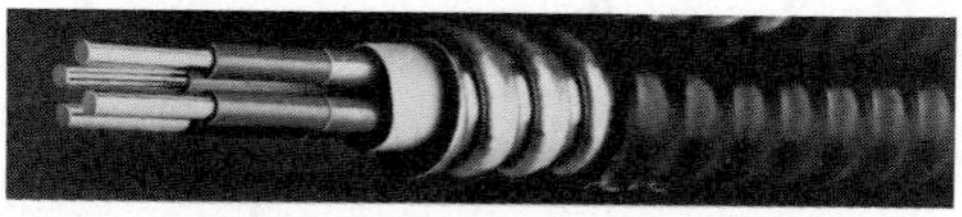

(b)

Fig. 15.3 (b) *Jacketed-type MC cable has a wide variety of applications, including wet locations, because of its water-impervious outer PVC jacket. Note nylon jacket over insulation and cable-tape binding conductors under the armor. (Courtesy of AFC/Nortek Co.)*

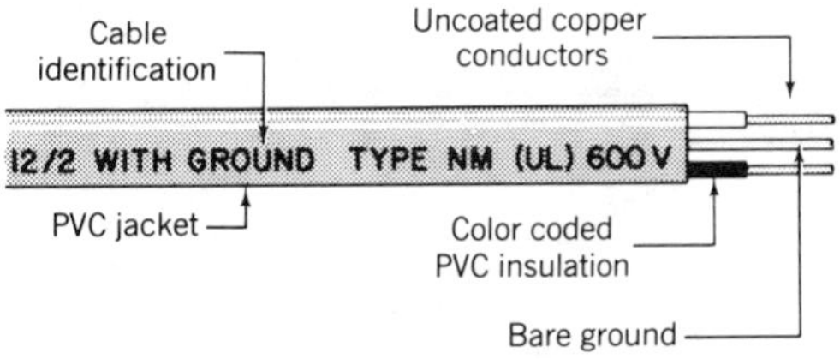

Fig. 15.4 *Construction of typical NEC type NM cable. The cable is a two-conductor, No. 12 AWG with ground, insulated for 600 V. Also normally shown are the manufacturer, cable trade name, and the letters (UL), which indicate listing of this product by Underwriters' Laboratories, Inc. The ground wire is bare or covered, and the entire cable may be obtained flat (illustrated), oval, or round.*

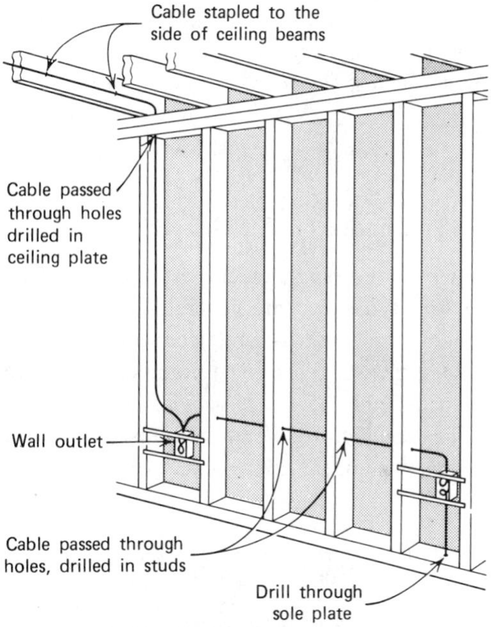

Fig. 15.5 *Typical wiring technique using types NM (Romex) or AC (BX), in wood stud construction. With metal stud construction BX cables are passed through precut openings in lieu of field-drilled holes.*

existing buildings. Use of type AC cable is generally restricted to dry locations. For specific application details see NEC Article 333, "Armored Cable."

A similar construction with much broader application, which is covered in NEC Article 334, is metal-clad (MC) cable. This cable may be used exposed or concealed and in cable trays, and when covered with a moisture-impervious jacket (see Fig. 15.3*b*) in wet and outdoor locations as well.

15.12 Nonmetallic Sheathed Cable (Romex)

In application, NEC types NM and NMC, also known by the trade name Romex, are restricted to small buildings, that is, residential and other structures not exceeding three floors above grade (see Fig. 15.4). The plastic outer jacket, unlike the armor on type AC, makes type NM easier to handle but more vulnerable to physical damage. For application details and restrictions see NEC Article 336, "Non-metallic Sheathed Cable." Typical installation technique is shown in Fig. 15.5.

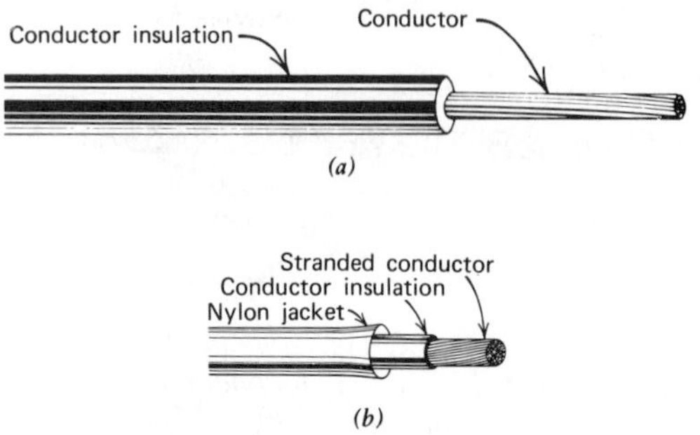

Fig. 15.6 (a) *Typical construction of unjacketed building wire such as type THW; see Table 15.4. Conductors normally are solid through No. 8 AWG, and stranded in sizes No. 6 AWG and larger; see Table 15.1.* (b) *The illustrated construction is typical for any nylon-jacketed cable such as THWN or THHN. (The first three letters indicate the type of insulation, and the final N indicates the nylon jacket.) [*(b) *Courtesy of ITT, Royal Electric Division.]*

15.13 Conductors for General Wiring

Under this heading (Article 310) the NEC lists the wire types that are generally used and installed in raceways and referred to by the term "building wire." The most common types are listed in Table 15.4. These wires consist of a copper conductor covered with insulation, and in some instances with a jacket (see Fig. 15.6).

15.14 Special Cable Types

Although most building wiring is accomplished with plastic-insulated 300- and 600-V conductors of the types described in the preceding sections, applications often require the use of special cables. These include high-voltage cables, armored cables, corrosion-resistant jacketed cables, underground cables, and so on. The reader is referred to manufacturers' catalogs and the NEC for construction and application details. Service entrance cables and installation are discussed in Sections 16.1 to 16.4, which cover electric service.

15.15 Busway/Busduct/Cablebus

A busway (busduct) is an assembly of copper or aluminum bars, in a rigid metallic housing. Its use is almost always preferable, from an economic viewpoint, in two instances: when it is necessary to carry large amounts of current (power), and when it is necessary to tap onto an electrical power conductor at frequent intervals along its length.

1. In the case of heavy current the alternative is to use paralleled sets of conductors or a single large conductor. This alternative becomes increasingly inefficient as cable size increases, because large round conductors require more cross section per ampere of current-carrying

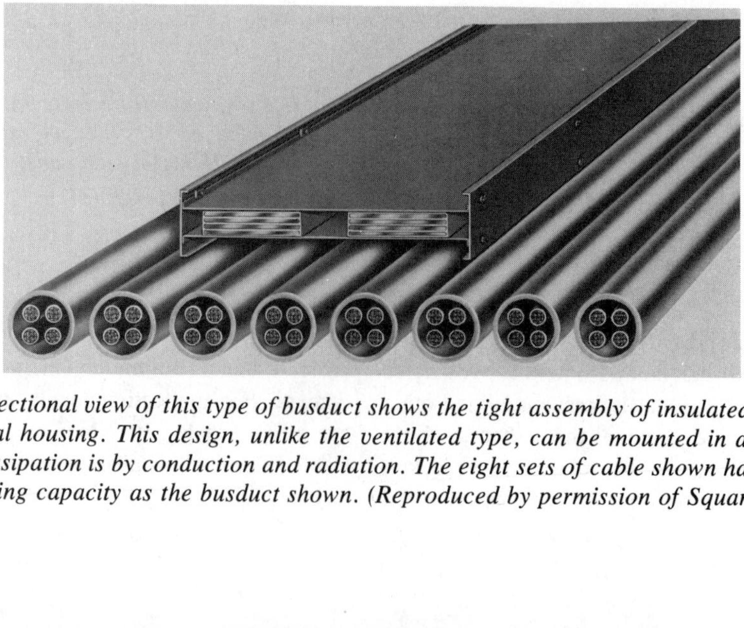

Fig. 15.7 A sectional view of this type of busduct shows the tight assembly of insulated conductors within a metal housing. This design, unlike the ventilated type, can be mounted in any position, since heat dissipation is by conduction and radiation. The eight sets of cable shown have the same current-carrying capacity as the busduct shown. (Reproduced by permission of Square D Co.)

capacity than small ones. This is not the case with the flat conductors (busbars) used in busduct (see Fig. 15.7).

2. Where many power tap-offs are required along an electrical feeder run, costs become very high because of the large amount of expensive hand labor involved, since a connection must be made to each conductor in the run. The preferable alternative is to use "plug-in" busway or busduct, to which connection can be made simply and rapidly with a plug-in device, in similar fashion to the insertion of a common plug into a wall receptacle. This has the additional advantage that connection and disconnection is simply a matter of inserting or withdrawing the plug-in device, whereas cable taps are permanent connections (see Fig. 15.8).

A typical application of heavy-duty busduct might be a vertical feeder in a high-rise building, from the basement switchboard to the penthouse machine room. The same structure could use heavy-duty plug-in busduct as vertical riser(s) with taps feeding individual floors (see Fig. 15.2). Typical applications for light-duty plug-in busduct (70 to 100 A) could be any machine shop or workshop. The electrical supply to individual machine tools is made very simply and flexibly with a tap-on device. See also the application in a typical wiring closet in Fig. 15.2.

Busduct is specified by type, material, number of buses, current capacity, and voltage: alu-

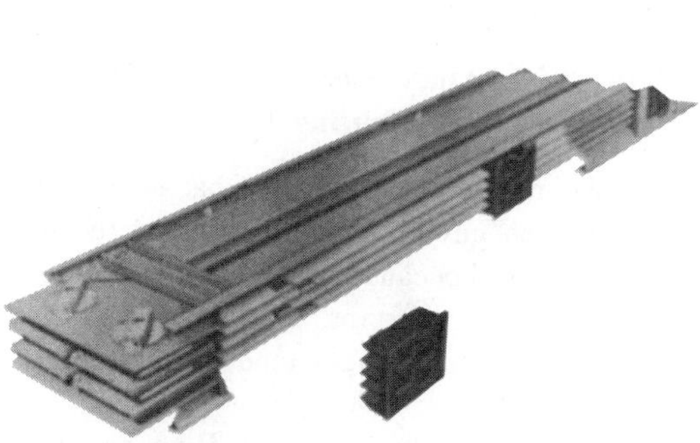

Fig. 15.8 *Construction of plug-in busduct. Plug-ins are spaced evenly on alternate sides to facilitate connection of plug-in breakers, switches, transformers, or cable taps. Housing is of sheet steel with openings for ventilation. Cover plate is not shown. (Reproduced by permission of Square D Co.)*

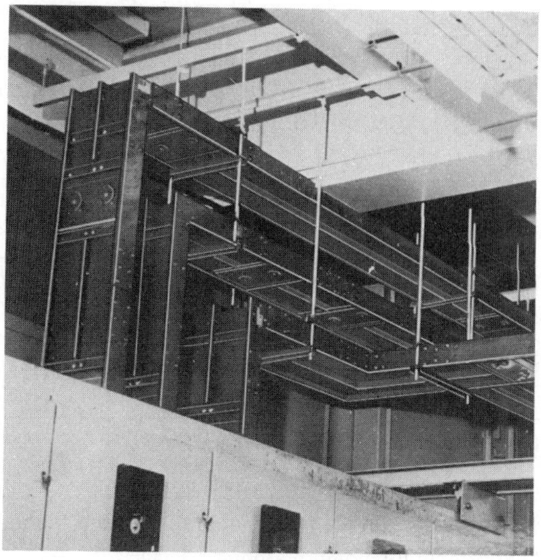

Fig. 15.9 *Typical installation of compact design busduct. Note that the individual busducts are supported by channels hung from the ceiling and that the same hangers support more than one level of bus. Turns are easily made in the same plane (horizontal angle) and in two planes (vertical to horizontal). (Reproduced by permission of Square D Co.)*

minum feeder busduct, 4-wire, 1000 A, 600 V; or copper plug-in busway, 100 A, 3-wire, 600 V. Feeder busduct (no plug-in capability) is available in ratings from 400 to 4000 A. Plug-in busway is available from 30 A for lighting or light machinery circuits (see the next section) to 3000 A. A wide variety of fittings and joints are available for all busways to permit easy installation (see Fig. 15.9). Designs are available for indoor and outdoor application.

Cablebus is similar to ventilated busduct, except that it uses insulated cables instead of busbars. These cables are rigidly mounted in an open space-frame. The advantage of this construction is that it carries the ampacity rating of its cables *in free air*, which is much higher than the conduit rating, thus giving a high amperes-per-dollar first-cost figure. Its principal disadvantage is bulkiness and the difficulty in making tap-offs.

As an example of the type of study that should be made when considering an item as relatively simple as an electrical feeder, refer to Table 15.5, which shows the results of such a

TABLE 15.5 **Life-Cycle Relative Cost Comparison of 2000-A, 208-V Feeder Installation**

Feeder System Description[a]	*Material Cost*	*Labor Cost*[b]	*Total First Cost*	*Power Loss per 100 ft*[c] *(kW)*	*Annual Energy Loss*[d] *(kWh)*	*Annual Energy Cost*[e]	*Life-Cycle Energy Cost*[f]	*Total Life-Cycle Cost*[g]
Cable tray	0.63	0.37	1.00	2.90	12,536	0.060	0.757	1.757 100%
Wire and conduit	0.68	0.61	1.29	2.90	12,536	0.060	0.757	2.047 117%
Busduct	1.14	0.24	1.38	5.60	24,207	0.116	1.461	2.851 162%
Cablebus	0.69	0.45	1.14	5.91	25,547	0.122	1.542	2.682 153%

[a]Equipment rating is 600 V, cable is copper, conduit is rigid steel, cable tray is aluminum.

[b]Labor cost includes overhead.

[c]Based on published resistivity data for cable and bus—assuming 80% demand (1600 A) and all conductors in the system equally loaded.

[d]Based on 80% demand, 12 hours per day, 360 days per year.

[e]Using $0.07 per kilowatt-hour as the combined net rate, including demand charges.

[f]Using 20-year life cycle, 8% fixed capital cost, and 3% annual escalation in energy cost.

[g]Sum of fourth and eighth columns.

study in relative cost terms. Note that when considering first cost alone, the advantage lies in cablebus; adding the energy-loss consideration shifts the advantage to cable tray and interlocked armor cables. No general conclusion should be drawn from Table 15.5 regarding costs. A change in feeder length, number of taps, hours of operation, energy rates, or any of the other factors can shift the advantage to a different system. The point of our study is to demonstrate clearly that life-cycle costs and first costs often yield entirely different results and, therefore, that this type of study is required before an intelligent decision can be made. (Life-cycle cost is taken to mean the present value of all costs over the installation's life cycle—in this case, 20 years for the system.)

Two additional items are worthy of note.

1. The very factors that yield lower first cost operate to yield higher operating cost. The smaller copper sizes in busduct and cablebus, permitted by high-temperature insulation and good ventilation, cause increased power loss because of their higher resistivity.
2. If the heat loss from the busduct or cablebus can be used to advantage, the related energy cost can be credited, instead of being considered a total loss, and life-cycle costs can be changed considerably. Conversely, it can also affect the building cooling load.

15.16 Light-Duty Busway, Flat Cable Assemblies, and Lighting Track

Special construction assemblies that act as light-duty (branch circuit) plug-in electrical feeders are widely used because of simplicity of installation and more important, the flexibility of use that derives from its plug-in mode of connection.

(a) Light-Duty Plug-In Busway. This construction, which may be used either for feeder or branch circuit application, is covered by the NEC general article on busways, with restrictions when applied as branch circuit wiring. Light-duty busways are rated from 20 to 60 A at 300 V, in 2- and 3-wire construction. A

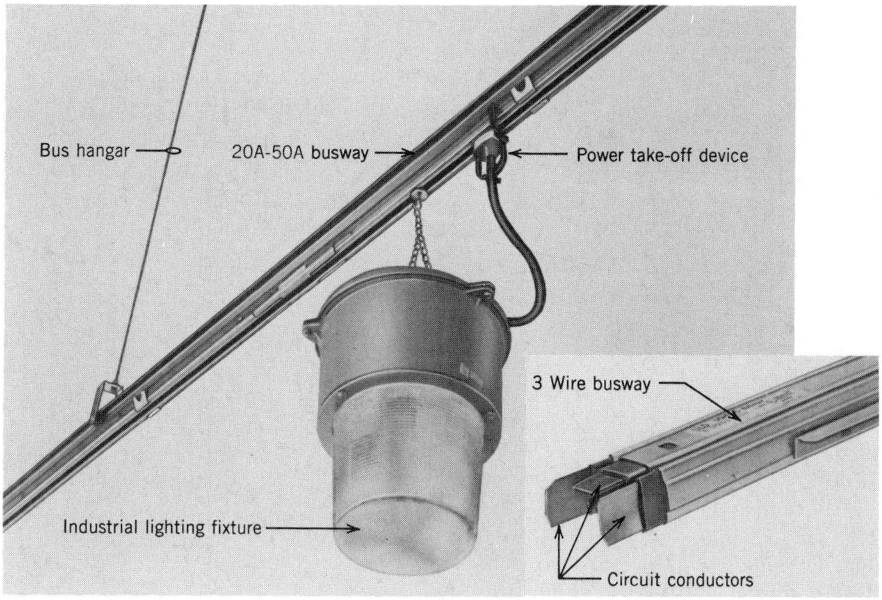

(a)

Fig. 15.10 (a) *Light-duty busway is rated 20 to 60 A, 300 V, and either 2- or 3-wire. Power-takeoff devices twist into the bus to make contact with the circuit conductors. Within NEC restrictions of overcurrent protection, this busway may be used for standard and heavy-duty lighting fixtures and for other electrical devices such as electrically powered tools. (Courtesy of Siemens Energy & Automation, Inc.)*

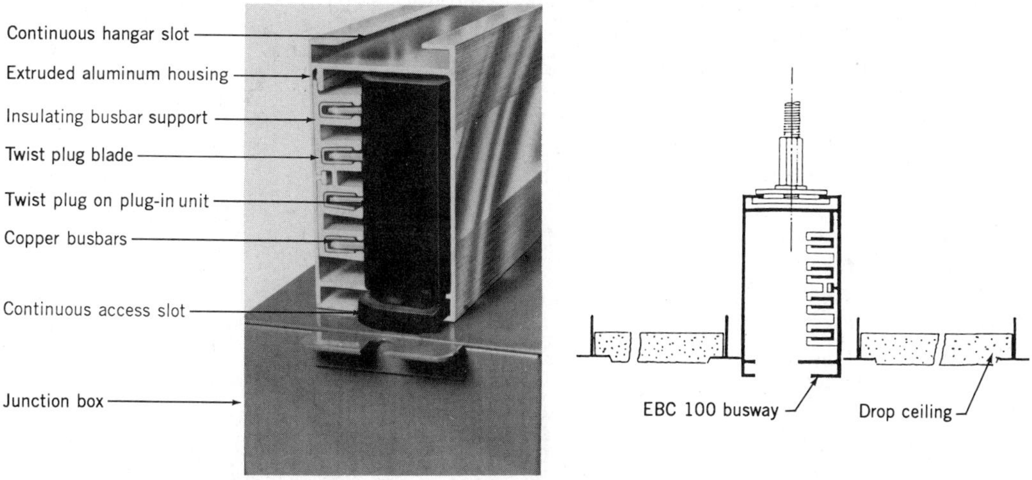

(b-1) (b-2)

Fig. 15.10 (b-1) *Plug-in busway rated 100 A, 3-phase, 4-wire, 600 V measures approximately 2½ in. W × 4½ in. H. The twist-in plug, which is integrally attached to a connection means (junction box in the illustration), is rated 30 to 100 A, single- or 3-phase, as required. The attached junction box (or receptacle, circuit breaker, or fuse box) then feeds the utilization device (e.g., heavy-duty lighting or machinery). This bus can also be installed in a hung ceiling* (b-2). *(Courtesy of Electric Busway Corp.)*

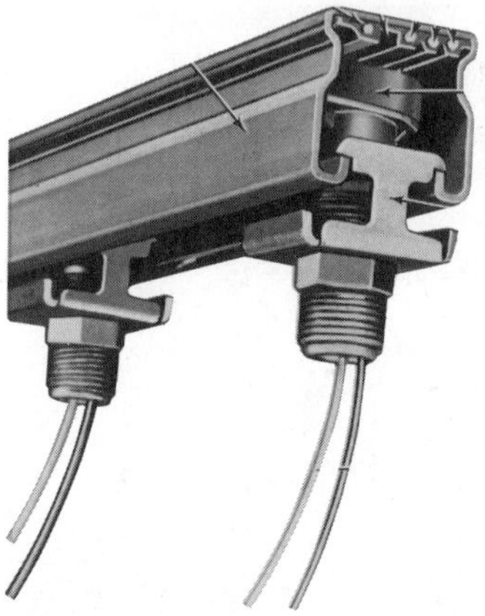

Fig. 15.11 Flat cable assembly installation. The unit illustrated is a four-conductor, 30-A, 600-V FC cable. Taps may be phase to neutral or phase to phase, that is, 120, 277, or 480 V. The tap can feed directly to a device or can energize a receptacle in an outlet box. A lighting fixture can be hung from the FC cable channel with a fixture hanger and hook. If the tap is removed, the pin holes in the PVC insulation "heal," maintaining the integrity of the insulation. (Courtesy of Chan-L-Wire®/Wiremold Co.)

somewhat heavier design rated 100 A at 600 V is available in 3- and 4-wire construction. Their application is principally for direct connection (with overcurrent protection as required by the NEC) of machine tools, light machinery, and industrial lighting (see Figs. 15.10 and 21.39).

(b) Flat Cable Assemblies (NEC Article 363; Type FC). A specially designed cable consisting of two, three, or four conductors, No. 10 AWG is field installed in a rigidly mounted standard 1⅝-in. square structural channel. Power tap devices, installed where required, puncture the insulation of one of the phase conductors and the neutral. Electrical connection is then made to the pigtail wires that extend from the tap devices. This connection can extend directly to the device or to an outlet box with a receptacle, which then acts as a disconnecting means for the electric device being served. In this fashion lights, small motors, unit heaters, and other single-phase, light-duty devices can be served without the necessity of "hard" (conduit and cable) wiring. Figure 15.11 illustrates this equipment.

(c) Lighting Track (NEC Article 410-S). This is a factory-assembled channel with conductors for one to four circuits *permanently* installed in the track. Power is taken from the track by special tap-off devices that contact the track's electrified conductors and carry the power to the attached lighting fixture, which can be positioned anywhere along the track. The tracks are generally rated at 20 A, and unlike FC cable assemblies, they are restricted to 120 V phase to neutral and may feed only lighting fixtures. Taps to feed convenience receptacles are not permitted. A typical design is shown in Fig.

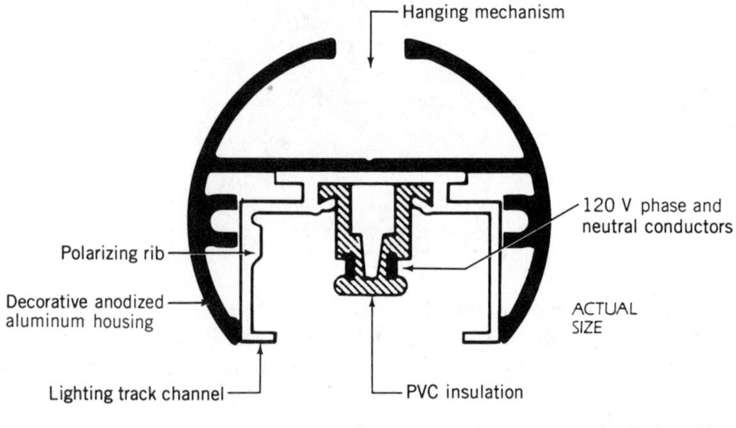

Fig. 15.12 Decorative lighting track. The circular housing can be obtained in a variety of finishes and hangar arrangements. The actual lighting track, shown full size, is readily available without the housing for direct surface mounting. Track is also available in a 4-conductor configuration (three 120-V conductors plus neutral). (Courtesy of Swivelier.)

15.12. An application of track lighting is shown in Fig. 21.18.

15.17 Cable Tray

This system, which is covered in NEC Article 318, is simply a continuous open support for approved cables. When used as a general wiring system, the cables must be self-protected. The advantages of this system are free-air-rated cables, easy installation and maintenance, and relatively low cost. The disadvantages are bulkiness and the required accessibility.

15.18 Design Considerations for Raceway Systems

The following sections deal with closed wiring raceways, which will complete our discussion of raceways. We will not go into the details of construction and application, because of space limitation, and because those data are readily available from manufacturers and the applicable NEC articles. We will, however, provide enough material for the reader to become familiar with the types of raceways, their common applications and limitations, and, where applicable, comparative characteristics.

Although our discussion in this chapter covers primarily equipment used in electrical power systems, empty raceways intended for signal and communications wiring (by others) are normally provided under the electrical section of the construction contract. The raceway's function for such systems is in large measure the same as it is for power wiring: protection and isolation of the wiring. Prior to the widespread use of computers in commercial establishments, the raceway space requirements for communication and signal wiring were easily established, since such wiring consisted of small telephone cables plus miscellaneous signal and alarm wires. These requirements were also easily satisfied with empty conduits and one or more cells in floor raceway systems. Today, when virtually every commercial establishment uses some type of data-processing equipment, and communication networking is becoming commonplace even in small facilities, raceways for communication cabling have become a major design item. They often far exceed in actual cross-sectional area the space required for power cabling and therefore must be considered early in the design process.

Sizing of raceways for power wiring is an exact process (see Section 17.17) based as it is on maximum permissible temperatures for specific materials in a given environment. Not so for communication cabling even when the system's present requirements are known, because of the extremely rapid growth of networking and of data interchange generally. The problem is all the more difficult when designing commercial space for rental to an unknown client. The advisable approach there is to provide a reasonable amount of floor-level raceway space for main cabling (Sections 15.24 and 15.25) and to rely on add-on systems such as under-carpet wiring (Section 15.29) and surface or ceiling raceways (Section 15.30) for additional raceway area, or alternatively, to use a structural system that provides virtually unlimited wiring space (Sections 15.26 to 15.28). Since the latter is a major structural/architectural decision, it must be taken at the preliminary stages of design. Design considerations for power system raceways will be discussed in the following sections and in Chapter 17. Design of raceways for data communication system wiring includes the following considerations:

1. Number type and location of data-processing terminals.
2. Networking requirements.
   a. Type of local area network (LAN): Token Ring (IBM), Ethernet (Xerox), ARCnet (Datapoint), or other. This will in large measure determine the communication media (i.e., coaxial cable, shielded and unshielded wire, and fiber optic cables), which affect the raceway space requirement. Cable type also determines the type of connectors needed (and their space requirements) and the type of floor outlets used for machine connection (see Fig. 15.27).
   b. Cable topology (i.e., interconnection

arrangements). This item is frequently not within the domain of the architectural designer, although the raceway space availability seriously affects cabling arrangement, and vice versa.

c. Requirement for interconnection of networks and connection to remote networks.

3. Number, location, and characteristics of major peripheral devices, such as mass storage, printing, and plotting.
4. Location and type of major subsystems, such as CAD/CAM design spaces.
5. Location of presentation spaces, which require interconnection to computer networks.

15.19 Steel Conduit

The purpose of conduit is to

1. Protect the enclosed wiring from mechanical injury and corrosion.
2. Provide a grounded metal enclosure for the wiring in order to avoid shock hazard.
3. Provide a system ground path.
4. Protect surroundings against fire hazard as a result of overheating or arcing of the enclosed conductors.
5. Support the conductors.

For these reasons the NEC generally requires that all wiring be enclosed in a rigid metallic conduit. Metal electrical raceways and associated fittings must be corrosion resistant. To this end, steel conduit is manufactured in several ways, among which are hot-dip galvanized, sherardized (coated with zinc dust), enameled, and plastic covered.

There are three types of steel conduit that differ basically only in wall thickness. They are, in order of decreasing weight:

1. Heavy-wall steel conduit, also referred to simply as "rigid steel conduit"; it is covered in NEC Article 346.
2. Intermediate metal conduit, usually referred to as IMC; this is covered in NEC Article 345.
3. Electric metallic tubing, normally known as EMT or thin-wall conduit; this is covered in NEC Article 348.

The differences are clearly shown in Table 15.6 and 15.7. Several types of heavy-wall conduit plus EMT are shown in Fig. 15.13. The equivalent millimeter (mm) sizes of conduits are given in Table 15.8.

Rigid conduit and IMC use the same fittings and are threaded alike. As a result of its thin wall, EMT does not lend itself to threading; instead, it uses set-screw and pressure fittings. The thinner walls of EMT and IMC yield a larger ID (inside diameter) and, therefore, easier wire pulling. This combination of lower weight and easier wire pulling gives EMT and IMC a distinct labor cost advantage over rigid conduit, which is enhanced further in jobs with a great deal of field bending and handling of conduit. Both, however, have application restrictions, which are detailed in the NEC.

Generally, no conduit smaller than ½ in., nominal trade diameter, is used. Ordinary steel pipe may not be used for electric purposes, and all electric steel conduit is distinctively marked as such.

When steel conduit is installed in direct contact with the earth, it is advisable to use hot-dip

TABLE 15.6 **Comparison of Steel Conduit Diameters**

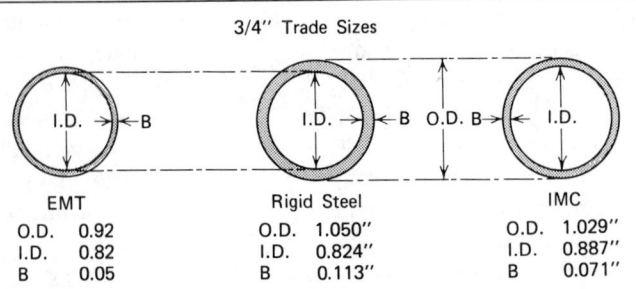

3/4" Trade Sizes

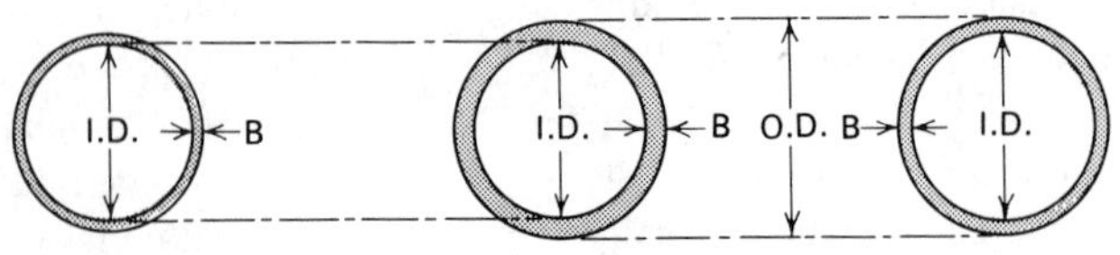

EMT		Rigid Steel		IMC	
O.D.	0.92	O.D.	1.050"	O.D.	1.029"
I.D.	0.82	I.D.	0.824"	I.D.	0.887"
B	0.05	B	0.113"	B	0.071"

TABLE 15.7 **Comparative Dimensions and Weights of Metallic Conduit**

Nominal or Trade Size	*Outside Diameter (Inches)*				*Weight per 10-ft Length (Pounds)*[a]			
	RS[b]	*IMC*[c]	*EMT*[d]	*AL*[e]	*RS*	*IMC*	*EMT*	*AL*
½	0.84	0.82	0.71	0.84	7.9	5.7	2.9	2.7
¾	1.05	1.03	0.92	1.05	10.5	7.8	4.4	3.6
1	1.32	1.29	1.16	1.32	15.3	11.2	6.4	5.3
1¼	1.66	1.64	1.51	1.66	20.1	14.4	9.5	7.0
1½	1.90	1.88	1.74	1.90	24.9	17.6	11.0	8.6
2	2.38	2.36	2.20	2.38	33.2	23.5	14.0	11.6
2½	2.88	2.86	2.88	2.88	52.7	39.3	20.5	18.3
3	3.50	3.48	3.50	3.50	68.3	48.3	25.0	23.9
3½	4.00	3.97	4.00	4.00	83.1	56.1	32.5	28.8
4	4.50	4.47	4.50	—	97.2	62.5	37.0	—

Source: All data courtesy of Allied Tube & Conduit Corp.
[a]Standard length including one coupling.
[b]Standard heavy-wall rigid steel conduit.
[c]Intermediate-weight steel conduit.
[d]Electric metallic tubing.
[e]Aluminum.

galvanized type and to coat the joints with asphaltum. If the earth is very wet, the complete conduit system should be coated with an asphalt compound. Conduit is fastened to the structure in much the same way as pipe; with pipe straps and clamps. Vertical load at floor opening is taken with special support clamps. Trapeze mounting is common for conduit banks hung from the ceiling, as in Fig. 15.14.

The selection of conduit size depends on the number and diameter of the wires that may be

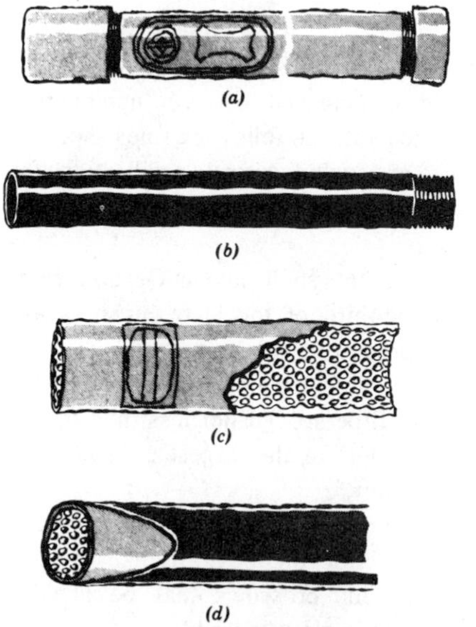

Fig. 15.13 *Steel conduits:* (a) *Galvanized, heavy wall, rigid;* (b) *black enameled;* (c) *EMT thin wall;* (d) *plastic-coated conduit for use in highly corrosive atmospheres.*

TABLE 15.8 **Metric Equivalents of American Conduit Sizes**

Conduit Size (Inches)	*Dimensions (Millimeters)*	
	OD	*ID*
$\frac{3}{8}$	16.5	9.5
$\frac{1}{2}$	20.5	12.7
$\frac{5}{8}$	23.0	15.9
$\frac{3}{4}$	26.5	19.0
1	33.0	25.4
$1\frac{1}{4}$	42.0	31.75
$1\frac{1}{2}$	48.0	38.1
$1\frac{3}{4}$	52.0	44.5
2	59.7	50.3
$2\frac{1}{2}$	76.0	63.5
3	89.0	76.2
$3\frac{1}{2}$	101.5	88.9
4	114.0	101.6

Fig. 15.14 *Typical overhead conduit bank installation. Note that due to field conditions the insert* (a) *for hanger rods was inadequate and an additional insert* (b) *was added. This conduit bank uses EMT, which has a pipe wall thickness approximately one-third that of heavy-wall rigid conduit. The resulting weight difference in a large bank such as this is very pronounced. EMT is suitable here since the overhead location protects the pipe from severe physical abuse. EMT joints are made with set-screw fittings* (c). *Note how individual conduits are fixed by clamps to the trapeze channel* (d). *(Courtesy of Republic Steel Corp.)*

drawn into the conduit without injuring the wire. The number and radius of bends in the conduit, as well as its total length, affect the degree of abrasion to the wiring insulation. No wires should be installed until the conduit system has been inspected and approved.

For structural reasons, conduits in concrete slabs are run close to the bottom surface (in the portion of the slab in tension) or near the central portion. If a great number of conduits must be embedded, it may be necessary to increase the slab thickness. In many instances, the structural slab is covered with a concrete topping, in which conduit may be installed without affecting slab integrity. In all cases, local building codes should be consulted for limitations on embedded conduits. In any event the top of any conduit shall be at least ¾ in. below the finished floor surface in order to prevent cracking. When heavy trucking is expected, this allowance should be increased to 1½ in. minimum.

In general, the following rules should be observed and included in all specifications for conduit work in concrete slabs:

1. Conduits shall have an OD no greater than one-third of the slab thickness as measured at its thinnest point.
2. Conduits running parallel to each other shall be spaced not less than three times the OD of the largest conduit center-to-center.
3. Conduits running parallel to beam axis shall not run above beams.
4. Conduit crossings shall be as near to a right angle as possible.
5. Minimum cover over conduits shall be ¾ in.

15.20 Aluminum Conduit

The use of aluminum conduit has increased greatly in recent years because of the weight advantage that aluminum has over steel, being even lighter than EMT. The savings in labor cost more than offsets the additional cost of the material itself. In addition, aluminum has better corrosion resistance in most atmospheres; it is nonmagnetic, giving lower voltage drop; it is nonsparking; and, generally, it does not require painting.

Its major drawback is its deleterious effect on many types of concrete, causing spalling and cracking when embedded. Although manufacturers can demonstrate many cases of embedding in concrete without harmful effect, it is a procedure that should be avoided unless the concrete additives are rigidly controlled. It is also inadvisable to bury aluminum in earth, with or without asphalt or other coating, because of the rapid corrosion often encountered. Other difficulties frequently encountered are freezing of threaded joints, because of thread deformation, and difficulty in obtaining electrical contact with grounding straps. With the exceptions noted above, aluminum conduit may be used in all locations where steel conduit is used.

Fig. 15.15 *This is a particularly good application of liquid-tight flexible conduit since it provides weatherproofing and acoustical isolation of the noise-producing transformer. (Courtesy of Electri-Flex Company.)*

15.21 Flexible Metal Conduit

This type of conduit construction—which consists of an empty spirally wound interlocked armor raceway—is known to the trade as *Greenfield* and is covered in NEC Article 350. It is used principally for motor connections or other locations where vibration is present, where movement is encountered, or where physical obstructions make its use necessary. The acoustic and vibration isolation provided by flexible conduit is one of its most important applications. It should always be used in connections to motors, transformers, ballasts, and the like. A typical applications is for wiring inside metal partitions. When covered with a liquid-tight plastic jacket, it is suitable for use in wet locations (see Fig. 15.15). In this construction it is most often known by the trade name Sealtite. Conduit material is either galvanized steel or aluminum.

15.22 Nonmetallic Conduit

A separate classification of rigid conduit (NEC Article 347) covers raceways that are formed from such materials as fiber, asbestos-cement, soapstone, rigid polyvinyl chloride (PVC), and high-density polyethylene.

For use above ground, this conduit must be flame retardant, tough, and resistant to heat distortion, sunlight, and low-temperature effects. For use underground the last two requirements are waived. Generally, nonmetallic conduit may be used without restriction in nonhazardous areas, within the physical limitations of the material involved. Thus plastic conduit has a temperature limitation, asbestos-cement has considerable physical strength limitations, and so on. As a result of these limitations, PVC conduit is the material of choice for indoor exposed use and asbestos-cement, fiber, and PVC plastic for outdoor and underground use. A separate ground wire *must* be provided, since the ground provided by a metallic conduit is absent.

15.23 Metal Surface Raceways

These raceways are covered in NEC Article 352. Surface metal raceways and multi-outlet assemblies may be utilized only in dry, nonhazardous, noncorrosive locations and may generally contain only wiring operating below 300 V. Such raceways are normally installed in exposed condition and in places not subject to physical injury.

The principal applications of surface metal raceways are:

1. Where the architecture does not permit recessing (see Fig. 15.16).
2. Where economy in construction weighs very heavily in favor of surface raceways and where expansion is anticipated (see Fig. 15.17).
3. Where outlets are required at frequent intervals, and where rewiring is required or anticipated (see Fig. 15.18).
4. Due to nature of wiring, and where access to equipment in the raceways is required (see Fig. 15.19).
5. Where rewiring existing installations, we wish to avoid the extensive and expensive cutting and patching required to "bury" a raceway (see Fig. 15.20).

15.24 Floor Raceways

The NEC recognizes three types of in-floor raceways:

Underfloor raceways—Article 354

Cellular metal floor raceways—Article 356

Cellular concrete floor raceways—Article 358

All three types are applicable to all types of structures and none may be used in corrosive or hazardous areas. The fundamental difference between them is that underfloor raceways are added on to the structure, whereas cellular floor raceways are part of the structure itself—and therefore have a pronounced effect on the building's architecture.

Fig. 15.16 The exposed wood members require the use of an unobtrusive surface raceway. A small flat raceway feeds receptacle outlets into which the elaborate hanging fixtures are plugged. (Courtesy of Wiremold Co.)

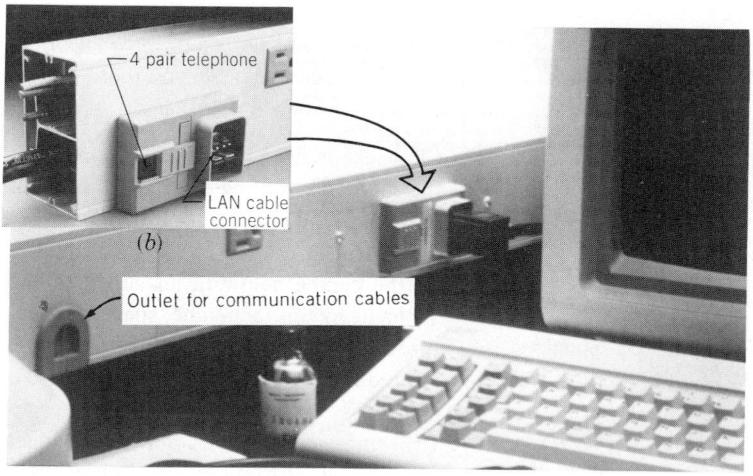

Fig. 15.17 *Large two-section surface raceway with a common cover handles wiring for power and communications.* (a) *The lower section has outlets for communication cables (dome-shaped outlet; see Fig. 15.19), plus a combination outlet for 4-pair telephone cables and for local area network (LAN) cable. This outlet is seen more clearly in the insert* (b). *The large dimensions of the raceway allows for anticipated future expansion. (Courtesy of Wiremold Co.)*

(*a*)

Fig. 15.18 *Multioutlet surface raceways find ready application in residences* (a) *and schools* (b). *Note the use of a surge and noise surpressor (foreground) to protect groups of computers from power line disturbances. See Section 16.29 for a discussion of power conditioning equipment.* (c) *Large-capacity surface metal raceways are particularly useful for wiring in full access floors (see Section 15.28) because of the heavy wiring (see wall box) and frequent rewiring. Note the perforated floor tile in the foreground, which is used to supply laminar airflow in this clean room at an integrated circuit manufacturer's laboratory facility. (Photos courtesy of Wiremold Co.)*

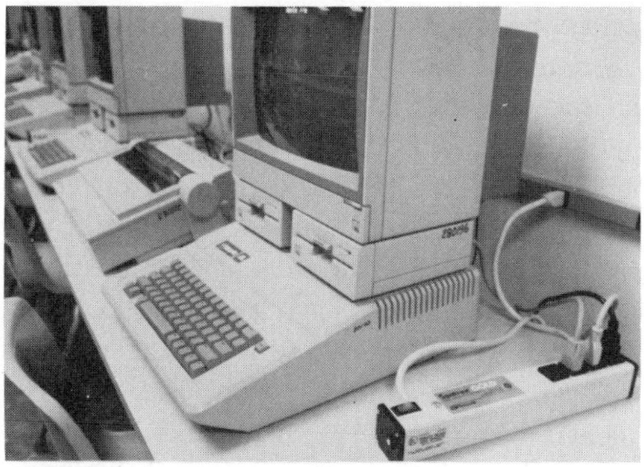

(*b*)

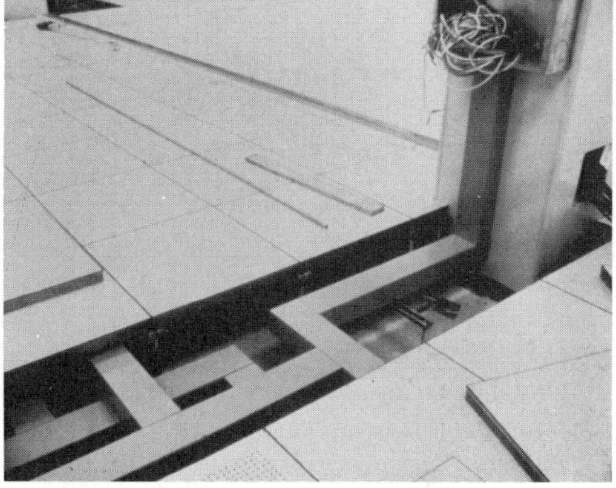

(*c*)

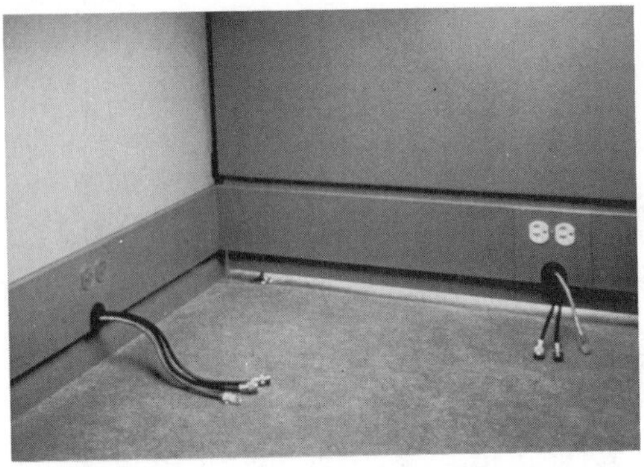

Fig. 15.19 *Since data cables are frequently preterminated, they must be laid into fully open raceways rather than pulled through. Further, low-tension compartments of surface raceways can contain terminal boards, strips, and the like, thus obviating the necessity for a separate cabinet. (Courtesy of Wiremold Co.)*

15.25 Underfloor Duct

These raceways may be installed beneath or flush with the floor. They find their widest application in office spaces, since their use permits placement of power, telecommunication, and signal outlets close to desks and other furniture, regardless of furniture layout. Where such underfloor raceways are not employed, and it is desired to place an outlet *on the floor,* one of the following methods is necessary:

1. Channel the floor and install a conduit in the chase, connecting it to the nearest wall outlet. Patch the chased portion.
2. Drill through the floor and run a conduit on the ceiling below to an outlet below. When using this technique, special "poke-through" fittings with adequate fire rating are required to restore the integrity of the slab.
3. Drill through the floor twice and connect the new outlet to an existing outlet via a conduit on the ceiling below. This, like method 2, is expensive and disturbs the occupant below.
4. Install a surface floor raceway.

Since all of these have obvious major disadvantages, underfloor (UF) duct systems were widely employed until the introduction of what may be called over-the-ceiling ducts (in apposition to under-the-floor ducts) and flat cable under-carpet wiring. These systems are discussed in Sections 15.29 and 15.30. The reason that alternate systems were developed is simply economic—underfloor duct systems are expensive, inflexible since they are literally cast in concrete, and frequently underutilized in one area and inadequate in another. However, before discussing the relative merits of systems, an understanding of what UF duct systems are, and how they are assembled and utilized, is in order.

An underfloor duct system is simply an arrangement of parallel rectangular metal raceways laid on the structural slab and covered with concrete fill. Access to the wiring in these *distribution ducts* is made via inserts that connect to openings in the ducts and terminate in floor fittings, for both power and signal/data wiring. Cable feeds to the distribution ducts are supplied by a second set of parallel rectangular metal raceways laid at right angles to the distribution ducts. In a *single-level* underfloor duct system the distribution and feeder ducts are on the same level, and the interwiring between them is accomplished in junction boxes. The advantage of a single-level system is shallow con-

Fig. 15.20 *The proliferation of signal, telephone, and data-processing equipment has necessitated extensive add-on wiring facilities. In many instances "burial" of raceways is either impractical, excessively time consuming, or prohibitively costly. Floor outlets, which are the most problematic when using surface raceways, may be utilized if properly positioned. (Courtesy of Wiremold Co.)*

ELECTRICITY

crete fill, normally 2½ to 3 in. The disadvantage (and bottleneck) of the single-level system is the junction box, which becomes more complex and multisectioned with increasing number of ducts and wires. To overcome the bottleneck and to decrease in some measure the high cost of underfloor duct systems generally, newer systems utilize a one-piece triple-cell duct for both distribution and feeder ducts, with factory set inserts every 24 in., which straddle all three cells at once (see Fig. 15.21). By placing distribution ducts on 5-ft centers, with adequate crosswise feeder ducts, and utilizing large flat junction boxes, a cost-effective installation can be accomplished which is adequate for all but the heaviest wiring demands. Distribution ducts can be equipped with special junction boxes which are designed to feed under-carpet cables that can take up additional load (see Sections 15.18 and 15.29).

However, because the initial cost of such a system is still high, an alternative arrangement utilizes *only* feeder ducts on approximately 25-ft centers, with flat (under-carpet) cable box connectors spaced approximately every 20 ft along the feeder ducts. The low-tension portion of this system relies completely on recently developed flat telephone and data transmission cables, including fiber optic cables. Since these cables are generally precut and factory terminated, the system requirements must be carefully analyzed (see Section 15.18) before committing to an complete under-carpet wiring system.

A two-level underfloor duct system is essentially the same as a single-level system with the exception that the distribution ducts and feeder ducts are on different levels (see Fig. 15.22*a*). This arrangement has the advantages of simplifying junction boxes and of giving the system unlimited feeder capacity but the distinct disadvantage of requiring a minimum of 3⅝ in. of concrete fill. This additional slab thickness can frequently be avoided by depressing part of the slab to accommodate feeder ducts run *under* the distribution ducts, as shown in Fig. 15.22*b*.

A modern two-level system is illustrated in Fig. 15.23. Here the feeder ducts run above the distribution ducts and intersect at an especially constructed junction fitting into which the distribution ducts partially recess, in order to reduce overall system height. The required concrete fill is either 3 or 4 in., depending on the depth of the distribution cells (2 or 3 in.).

In all underfloor duct systems, including one using only feeder ducts and under-carpet cable, the principal cable capacity bottleneck is usually the supply point to the *feeder* ducts. To overcome this problem, large floor areas can be subdivided and multiple feed points arranged in closets or at wall panels. In all such systems care must be taken to assure sufficient interconnection capacity between feed points since data networks are not only floor-wide but frequently building-wide.

Underfloor ducts may be cast into the structural slab in lieu of being in fill or topping, but the slab must be designed to accommodate them. The use of a fill or topping on the structural slab for underfloor duct has these advantages:

1. Ducts can be run in any direction, without conflict to structural elements.
2. Finishing is simplified.
3. Coordination is simplified.
4. Formwork and construction sequence are simplified.

The disadvantages are:

1. Additional concrete increases costs directly by increasing weight. This is particularly expensive in seismic designs.
2. Height of building may be increased.

In retrofit jobs where underfloor duct is decided upon rather than one of the other floor or ceiling raceway systems, the ducts will obviously be placed in a new floor fill.

In conclusion, some general comments on the application of underfloor duct systems are in order. Underfloor duct systems are *expensive*. They can add 50% to the building's electric system cost, without consideration of the construction costs involved. To justify their use, therefore, the building should meet these criteria:

1. Open floor areas, with a requirement for outlets at locations removed from walls and partitions.
2. Under-carpet wiring system inapplicable.

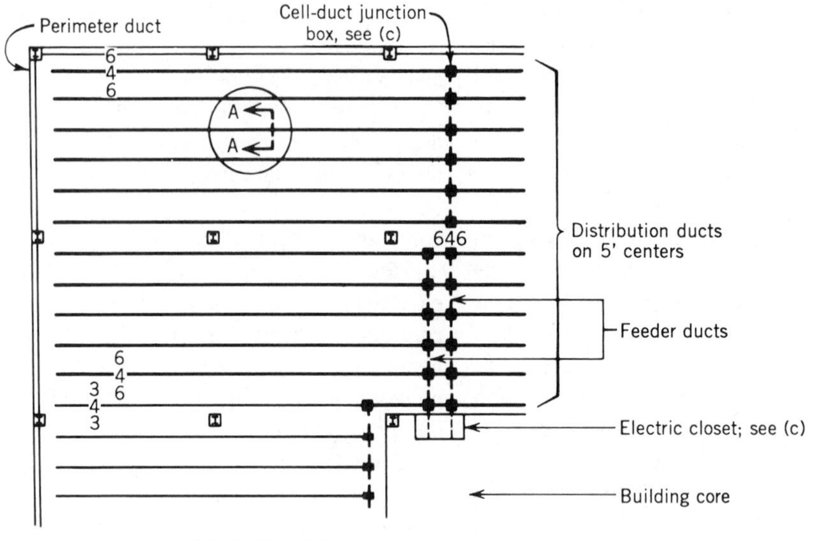

(*a*) Portion of floor plan

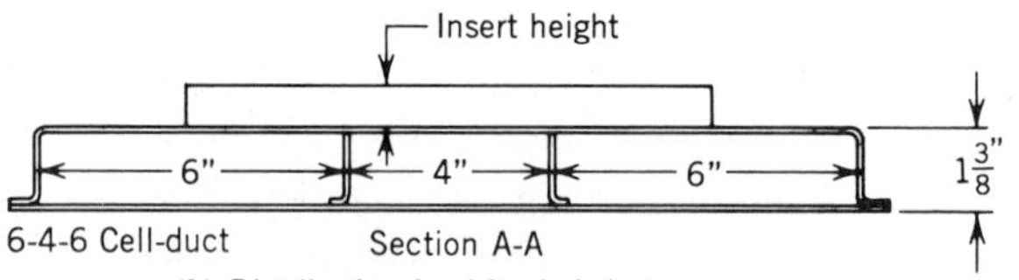

(*b*) Distribution (and feeder) duct

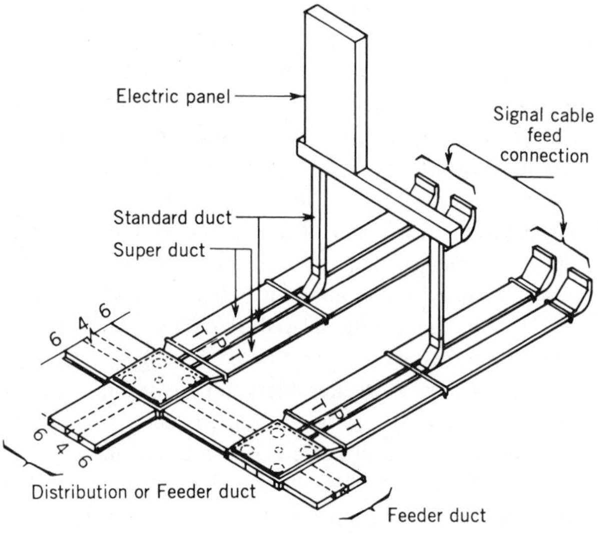

(*c*) Feeder details

Fig. 15.21 *Details of single-level underfloor duct system utilizing three-section cell duct for both distribution and feeder ducts.* (a) *Section of typical large open floor space in commercial facility. Distribution ducts may be placed as close as 5-ft centers to satisfy dense desk spacing.* (b) *Three-cell distribution duct utilizes a 4-in.-wide center cell (4.9 in.^{2}) for power and either 3-in.-wide (3.7 in.^{2}) or 6-in.-wide (7.4 in.^{2}) outer cells for signal and data cabling. Minimum concrete fill depth is 2½ in., resulting in a 1-in. cover over the distribution ducts. Service fittings are flush with the floor.* (c) *Due to the large capacity of both distribution ducts and feeder ducts, central cable feed points such as at electric closets can cause bottlenecks. Illustrated is one possible solution, consisting of a double-duct feed arrangement. Signal cable would feed in from cable boxes (not shown). (Diagrams courtesy of Square D Co.)*

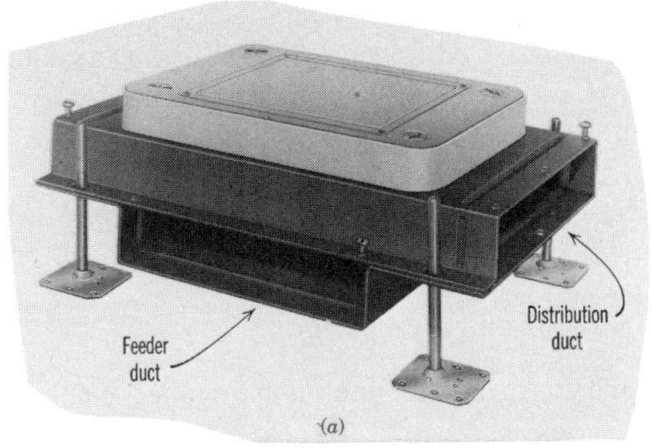

Fig. 15.22 (a) *Typical two-level junction box demonstrates the simplicity of the two-level system. (Courtesy of Square D Co.)*

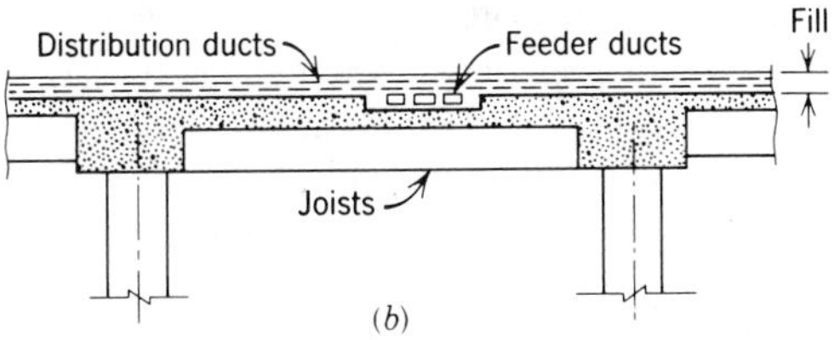

Fig. 15.22 (b) *Setting a two-level underfloor duct system. To avoid thickening fill, a depression in the slab can accept feeder ducts. Ducts would be run near the bay center to avoid the negative steel of joists, near columns.*

3. Outlets from ceiling systems unacceptable.
4. Frequent rearrangement of furniture and other items requiring electrical and signal service.

The facilities that may meet these criteria are prestige office buildings, museums, galleries and other display-case spaces, high-cost merchandising areas, and selected areas in industrial facilities. Bear in mind that even in high-cost office construction, underfloor duct systems are difficult to justify economically unless furniture layout will change. For fixed layouts, conduit-fed floor boxes are the economical choice. In doubtful cases, alternate arrangements can be planned and an intelligent choice made after costs and impact on the building structure are studied.

15.26 Cellular Metal Floor Raceway

The underfloor duct system described above is best applied to known furniture layouts and to

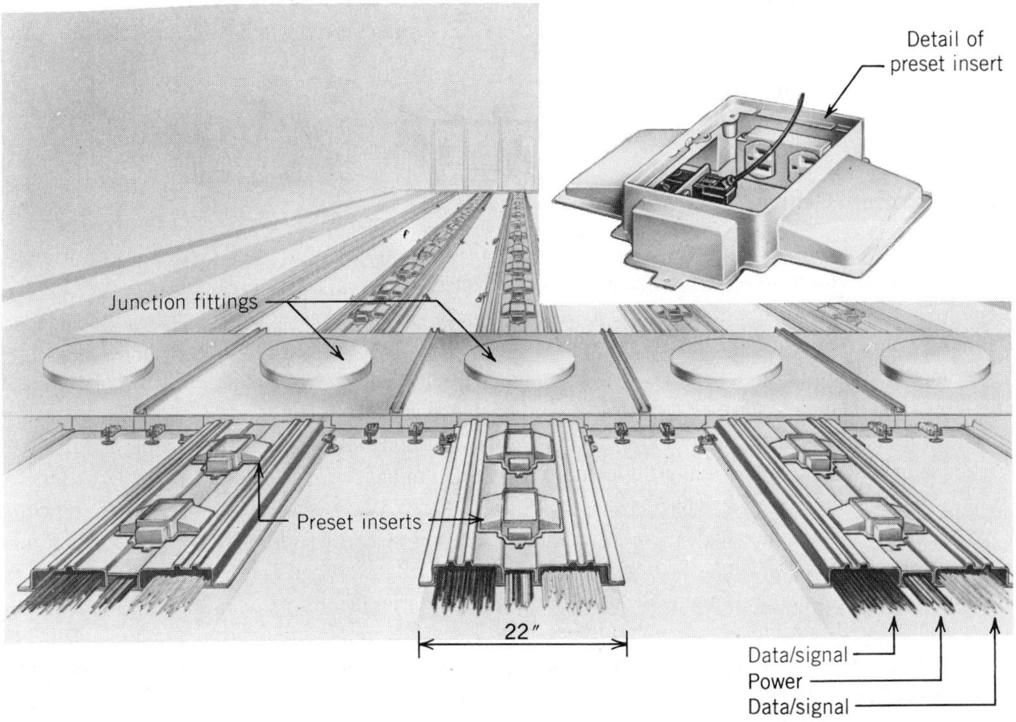

Fig. 15.23 Modified two-level underfloor duct system requires 3 in. of concrete fill for 2-in.-high ducts and 4 in. of fill for 3-in. ducts. Preset inserts straddle all three cells and provide two duplex power receptacles plus connection to both low-tension cells. Distribution ducts recess into feeder-header ducts at junction boxes to reduce overall height. See insert. (Courtesy of Walker, a division of Butler Manufacturing.)

rectilinear arrangements. Random arrangements, such as those found in office landscaping, require a fully accessible floor—if indeed the floor is to be used for electrification. This may be provided by a cellular (metal) floor that is an integrated structural/electrical system. The floor can be fully or partially electrified. A floor designed with two or three electrified cells adjacent to several cells of structural floor, as shown in Fig. 15.24, will give sufficient coverage for most purposes. One of the many structural element designs available is shown in that figure.

The cellular floor is part of the structural system and is designed accordingly. The electrical wiring is fed into the cells from header ducts and/or trenches that run perpendicular to the floor cells and constitute a system of underfloor ducts in themselves. The header ducts in turn are fed from lighting panels and signal data-transmission and telephone cabinets, in much the same manner as underfloor ducts.

Three types of wiring systems generally run in separate floor cells and header ducts: general lighting and appliances, data-transmission wiring, and telephone and signal systems. The latter two may be combined in a single cell only if the signal system voltage and power level are low. A complete range of outlets and fittings is available.

15.27 Precast Cellular Concrete Floor Raceways

This structural concrete system is similar to a cellular metal floor in application and has the same advantages: large capacity, versatility in that each cell is a potential raceway, and flexibility in outlet placement and movement. Here too, as with the metal cell constructions, first cost is higher than that of standard underfloor duct installation although life-cycle cost is frequently lower, depending on space use. A cell is defined in NEC Article 358 as a "single, enclosed, tubular space in a floor made of precast cellular concrete slabs, the direction of the cell being parallel to the direction of the floor member." Feed for these cells is provided, as with metal cellular floor construction, by header ducts. Although header ducts are normally installed in concrete fill above the hollow core structural slab, a header arrangement with feed from the ceiling below is also entirely practical. Like the metallic cellular floor, the cells can be used for air distribution and even for piping, although these items are generally installed in a hung ceiling.

15.28 Full Access Floor

This construction which is particularly applicable to spaces with very heavy cabling requirements, particularly if frequent recabling and reconnection is required, provides instant and complete access to an underfloor plenum. The system was originally developed for data processing areas that have a requirement for large, fully accessible cable spaces and for large quantities of conditioned air. The solution to both of these requirements is an infinite access floor consisting of lightweight die-cast aluminum panels supported on a network of adjustable steel or aluminum pedestals (see Figs. 15.25 and 15.18*c*). Panels are available from 18 × 18 in. to 36 × 36 in., and floor depth is normally 12 to 24 in., although taller pedestals are available. The construction is usually fireproof. Obviously, sufficient ceiling height is necessary to accommodate the raised floor.

15.29 Under-Carpet Wiring System

This system, which is covered in NEC Article 328, was originally developed as both an inexpensive alternative to an underfloor or cellular floor system and as a means for providing a flexible floor-level branch circuit wiring system. Essentially, the system consists of a factory assembled flat cable (NEC type FCC), approved for installation *only* under carpet squares, and the concomitant accessories necessary for connection to 120-V power outlets. The cable itself consists of three or more flat copper conductors, placed edge to edge and enclosed in an insulating material (see Fig. 15.26*a*). The entire assembly is covered with a grounded metal shield, which, like a metal conduit, provides both physical protection and a continuous electrical ground path. In addition, a bottom shield is required, which is usually heavy PVC or metal.

The cable, when properly installed on a hard flat surface, is approximately 0.03 in. high, and

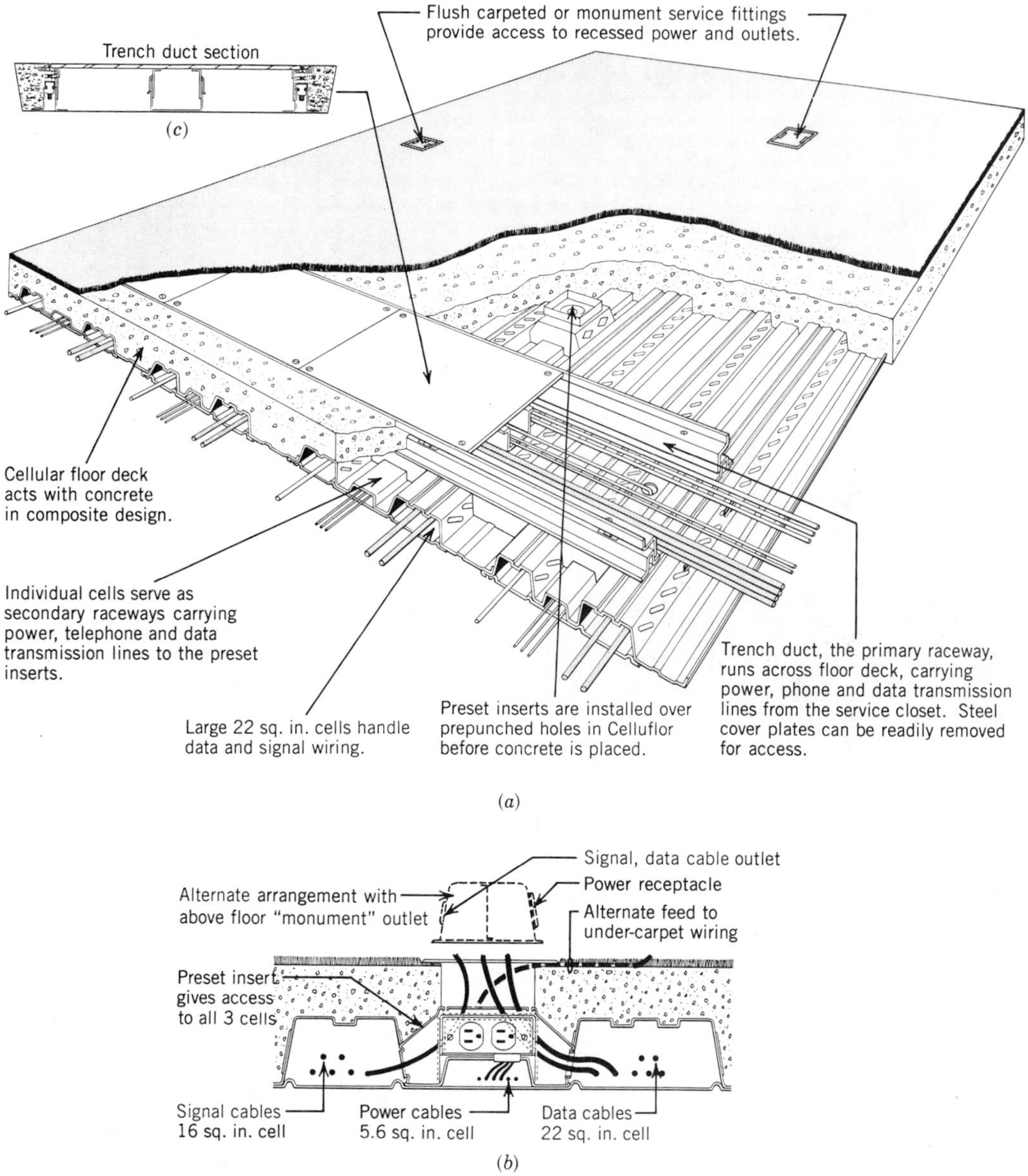
Trench duct section
(c)
Flush carpeted or monument service fittings provide access to recessed power and outlets.
Cellular floor deck acts with concrete in composite design.
Individual cells serve as secondary raceways carrying power, telephone and data transmission lines to the preset inserts.
Large 22 sq. in. cells handle data and signal wiring.
Preset inserts are installed over prepunched holes in Celluflor before concrete is placed.
Trench duct, the primary raceway, runs across floor deck, carrying power, phone and data transmission lines from the service closet. Steel cover plates can be readily removed for access.
(a)
Signal, data cable outlet
Power receptacle
Alternate arrangement with above floor "monument" outlet
Alternate feed to under-carpet wiring
Preset insert gives access to all 3 cells
Signal cables 16 sq. in. cell
Power cables 5.6 sq. in. cell
Data cables 22 sq. in. cell
(b)

Fig. 15.24 (a) *One of many designs for electrified cellular floor. The floor cells are available in many designs, depending primarily on the structural requirements. The trench* (c) *that straddles the cells provides the electrical feeds through precut holes in the cells. The trench itself is completely accessible from the top and, when opened, exposes all the wiring and the cells below.* (b) *Activated preset insert. Note that the insert straddles the center (power) cell and provides access to the two adjoining low-tension wiring cells. Power and signal wiring are completely separated at all times by metal barriers. If desired, a standard surface "monument" fitting can be mounted on the floor or connection can be made to undercarpet cables, in lieu of the flush plate shown. When an insert is to be deactivated, the flush cover plate is simply replaced with a blank plate.* (c) *Section through trench duct which acts as feeder for distribution ducts. Trench is available with or without bottom, in any required height, widths from 9 to 36 in., and one, two, or three compartments, depending on floor cell design and cabling requirements. (Courtesy of Walker, a division of Butler Manufacturing Corp.)*

(a)

(c)

(b)

Fig. 15.25 (a) *Typical access floor module. This unit is 24 in. square and of die-cast aluminum construction. Panels are available in a wide variety of floor finishes, including vinyl, laminates, conductive and antistatic compounds, and perforated for use in laminar airflow environments such as clean rooms.* (b) *Full access floor used for power, low-voltage, and data cabling and as a plenum for air distribution. Note flush-floor air diffuser. All cables in the plenum must be specifically rated for plenum installation.* (c) *Service box designed for installation in a full access floor. The box drops into a 6 × 9 in. hole in a floor panel and provides power, data, and communications connection from three separate internal compartments. [Photos courtesy of Tate Access Floors* (a), *USG Interiors* (b), *and Walker, a division of Butler Manufacturing* (c).*]*

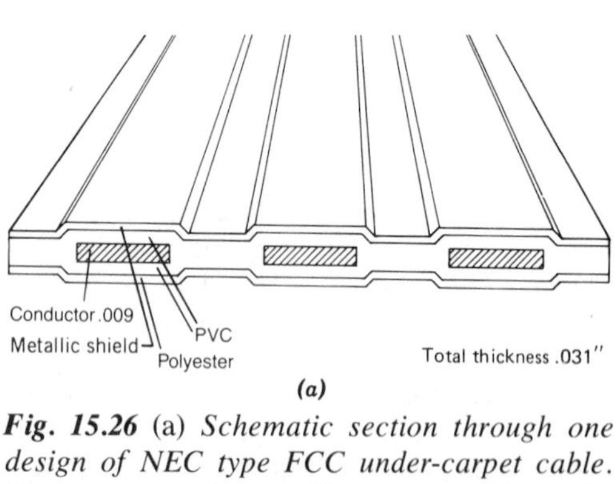

(a)

Fig. 15.26 (a) *Schematic section through one design of NEC type FCC under-carpet cable. The copper conductors illustrated are the equivalent of No. 12 AWG. The PVC acts as insulation, and the polyester, as both insulation and physical protection. All designs require a metallic top shield and a metallic or nonmetallic bottom shield for physical protection.*

(b)

Fig. 15.26 (b) *Preterminated 25-pair under-carpet telephone cable. These cables are commercially available in lengths from 5 to 50 ft in 5-ft increments. (Courtesy of AMP, Inc.)*

thus is essentially undetectable when covered with carpet. Since the carpet can readily be removed in sections, the entire system can be repositioned to meet changing furniture layout requirements, with a minimum of disruption and no structural work. The cable is designed to carry normal physical loads such as office traffic and furniture placement without affecting its electrical performance.

The attractiveness and simplicity of the system led to the development of similarly designed flat low-tension cables for signal and communication wiring, and more recently both electrical and fiber optic cables and accessories for data transmission (see Fig. 15.26*b*). (Nonpower cables are not limited to installation under carpet squares, although this ostensible advantage can detract from the flexibility of a combined power and signal cable installation.) Manufacturers also offer a complete line of junction fittings, connectors, adapters, and receptacles (see Fig. 15.27).

The problems inherent in this type of on-the-floor wiring system, such as cable crossings, splicing, interfacing with round cable systems, interconnections at floor boxes and fittings, feed connections from cabinets, underfloor ducts, floor cells, and through-the-floor fittings, have all been solved by a full line of factory-prepared devices designed for specific usages.

Since under-carpet wiring systems are separate and distinct from wire and conduit systems, they, like underfloor duct systems, are usually shown on a separate electrical plan. A small typical plan of this type is shown in Fig. 15.28*a*. Figures *b* and *c* are photographs of essential portions of such an installation. Note that a complex system such as that shown in Fig. 15.28*c* requires recessing a floor box into the slab, which, to an extent, defeats the essential simplicity and flexibility of an under-carpet system. Although these systems, at least in simplest form, are particularly applicable to retrofit work, their low cost, combined with the inherent advantages of a flexible *floor-level* wiring system, particularly in open office areas, has made them a widely used first choice in new construction as well.

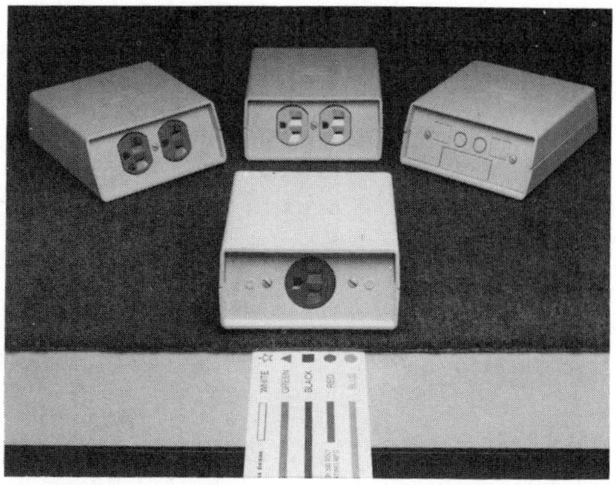

Fig. 15.27 *Typical components of an under-carpet wiring system. The under-carpet FCC power cable is shown without the metallic top shield required in actual installation. It is a color-coded 5-conductor cable (neutral, equipment ground, and three circuit conductors or two circuit conducts and an isolated ground conductor.) The floor outlets shown are: front, single power outlet; rear (left to right), duplex power outlet, one of which has isolated ground, standard duplex power outlet, and combination data cable, communications, and telephone outlet. (Courtesy of Hubbell, Inc.)*

15.30 Ceiling Raceways and Modular Wiring Systems

The need for electrical flexibility in facilities with limited budgets coupled with the high cost of underfloor electrical raceway systems encouraged the development of equivalent over-the-ceiling systems. These systems are actually more flexible than their underfloor counterparts, since they energize lighting as well as provide power and telephone facilities; furthermore, they permit very rapid changes in layouts at low cost. This last characteristic is particularly desirable in stores where frequent display changes necessitate corresponding electrical facility changes. In addition to the extreme flexibility of this electrical layout made possible by the ceiling raceway system, it has the additional advantage that it itself, not being cast in concrete like its underfloor counterpart, can be altered at will. Thus, not only layout changes (as mentioned above) but also changes in the utilization of existing spaces can readily be accommodated.

Fig. 15.28 (a) *Typical layout of undercarpet power and low-tension/data cabling for a small office. A power, phone, and data cable connection on the floor is provided under or immediately adjacent to each desk.* (b) *Power cables are run from a flush (recessed) wall box to a combination power/low-tension/data floor fitting. Data cables are connected to the combination floor fitting from their system boxes, either individually or via other floor fittings.* (c) *Four service floor junction boxes handle power, telephone, low-tension electric, and fiber optic cables. The floor box measures approximately 14 in. square and 2 in. deep. It can accommodate up to four 3-phase power circuits or eight 25-pair telephone/data cables or 24 coaxial cables or, as illustrated, a mixture of cable types. [Illustrations courtesy of AMP, Inc.* (a, c) *and Walker* (b).*]*

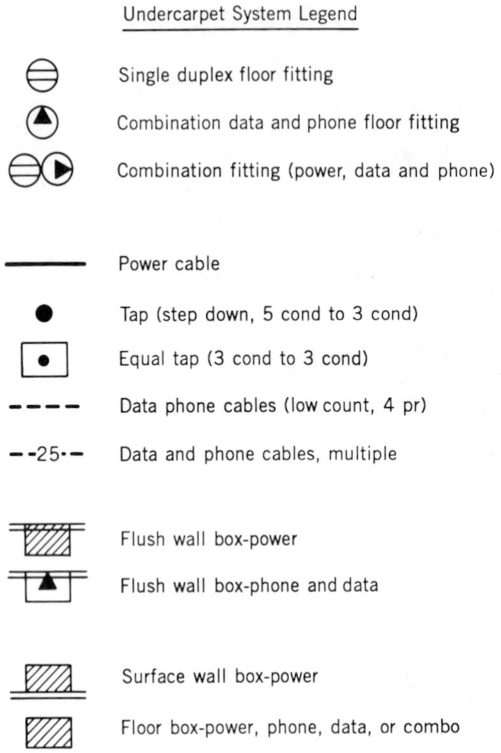

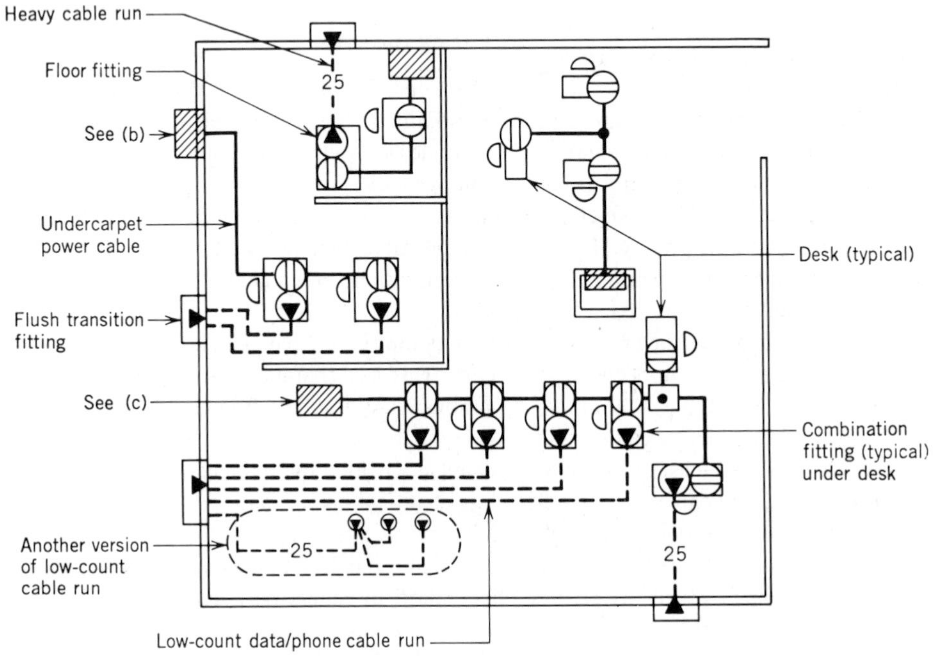

Typical Under-carpet Wiring Floor Plan Layout

(*a*)

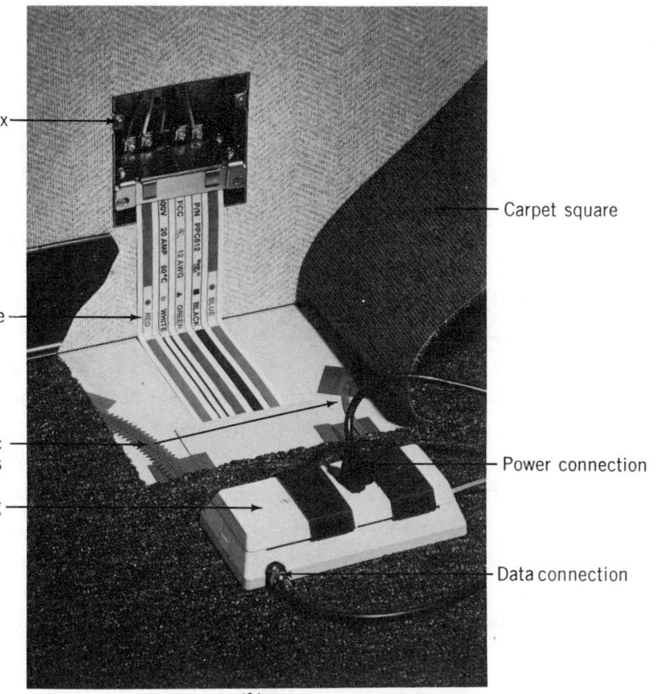
Flush wall box
Carpet square
Power cable
Fiber optic data cables
Power connection
Floor fitting
Data connection

(*b*)

Fig. 15.28 (b)

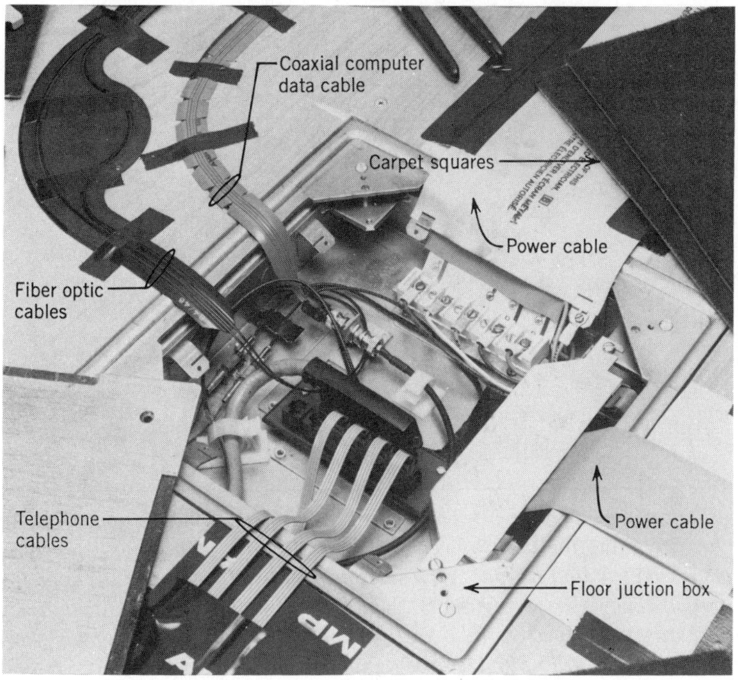
Coaxial computer data cable
Carpet squares
Power cable
Fiber optic cables
Telephone cables
Power cable
Floor juction box

(*c*)

Fig. 15.28 (c)

This is a particularly important characteristic in merchandising and educational facilities where spaces originally planned for one use have their application completely changed during the course of the building's life.

The systems vary among manufacturers but are essentially the same, and similar to underfloor systems. The entire system is shown clearly in Fig. 15.29. Header ducts (large area wireways) connect to electrical panels and signal, data, and telephone cabinets in the power and telephone closets, respectively. Data headers are normally of larger size than the power header and can carry other low-voltage, low-power signal equipment as well. Distribution ducts (laterals) tap onto the headers. These laterals may act as subdistribution raceways, feed lighting fixtures and data, signal, telephone, and power outlets directly, or do both.

The standard method for extending ceiling-level wiring to floor- or desk-level signal and power outlets is by means of a vertical multisection raceway fed from the top (see Fig. 15.30). These poles or posts are prewired with power wiring, contain several power outlets, a telephone connection, and in some models data cable outlets or connectors, and are simply and easily installed at any desired location. The end result in a hung ceiling office area or an exposed slab area (Fig. 15.31) is certainly less elegant than a floor-level wiring system, but for most users it is satisfactory, and its low cost as compared to any type of floor-duct system is a prime redeeming feature.

When power feeds to fixtures and receptacles are made with "hard" wiring, considerable field labor is involved with corresponding high cost. Furthermore, the permanent nature of such wiring lessens the inherent flexibility of the raceway system. To solve both of these problems a num-

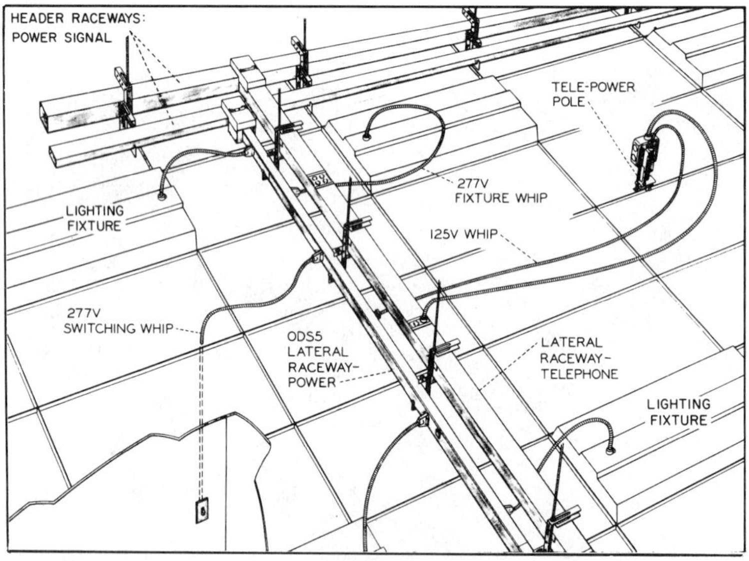

Fig. 15.29 *Overhead raceway system. The arrangement is a conventional "tree": large-cross-sectional-area header ducts for both power and signal/data/telephone are fed directly from the relevant cabinets in electric and telephone closets or in walls. These header ducts, in turn, subfeed to lateral distribution raceways (wiring ducts), which in turn either feed utilization devices directly, such as lighting fixtures, or feed vertical ceiling-to-floor raceways, known as power/telephone/communication poles (see Fig. 15.30). Depending on the size and layout of the system, laterals may serve as subdistribution only or as subdistribution plus feed-to-utilization devices. The "whips" shown are prewired cord sets that simply plug into the power raceway to feed lighting, switches, and receptacles in poles. Low-tension and communications raceways are provided with takeoff points for rapid connection. All numbers and nomenclature in this figure refer to Wiremold equipment. (Courtesy of Wiremold Co.)*

ELECTRICITY

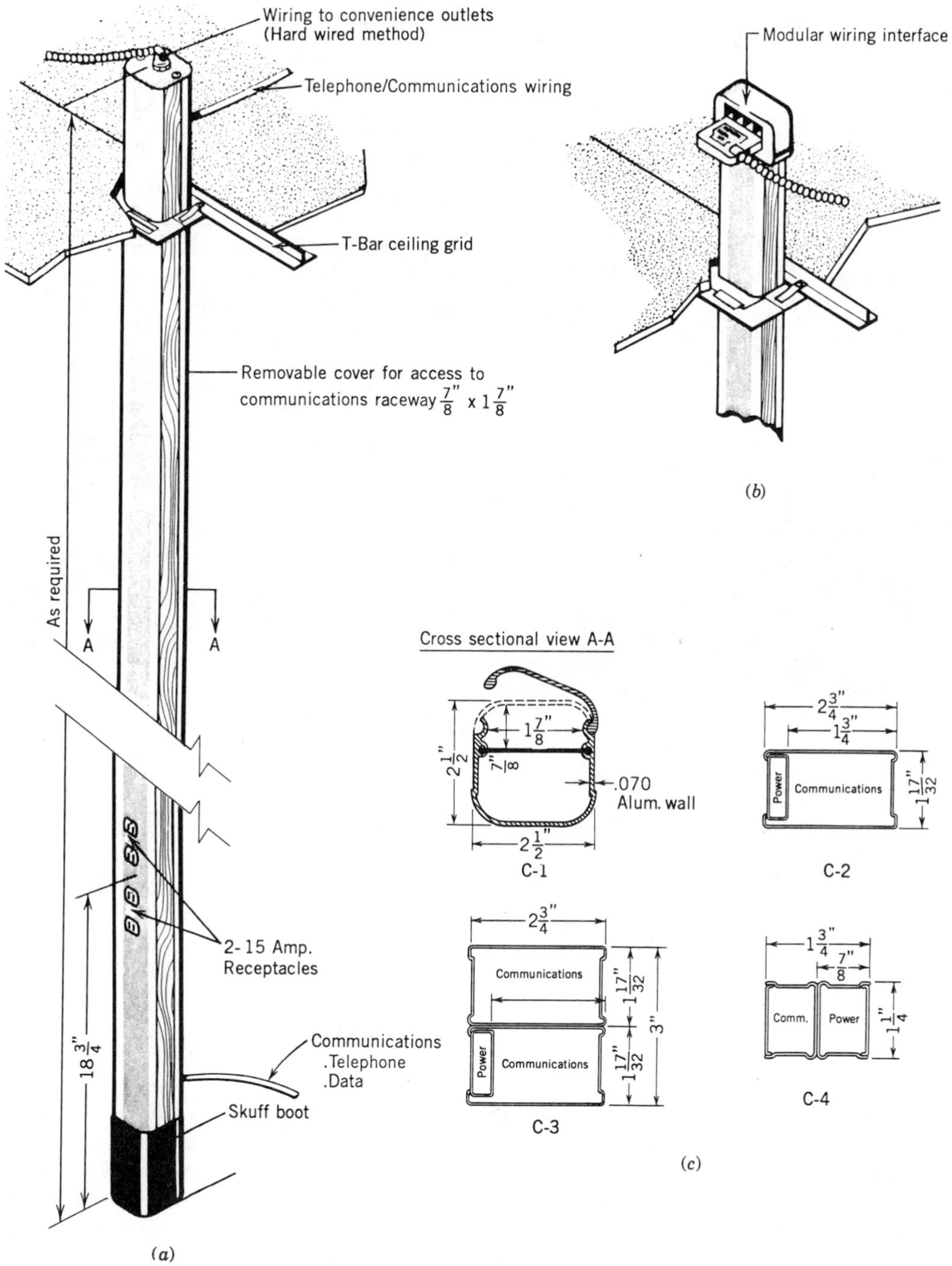

Fig. 15.30 (a) *A typical ceiling-to-floor "pole" raceway is fed at the top with both power and signal wiring from raceways in the hung ceiling (or suspended, exposed, below the ceiling slab). The feeds can either be conventional or by means of prefab modular wiring equipment as in* (b). *See also Fig. 15.32. The pole is available in various cross sections as shown in* (c-1) *through* (c-4). *[*(a, b, c-1) *Courtesy of Hubbell, Inc.;* (c-2, c-3, c-4 *courtesy of Wiremold Co.]*

(a)

Fig. 15.31 (a) *A "landscaped" open office plan utilizes power poles to bring power to the equipment level from an over-the-ceiling power distribution system. (Courtesy of Wiremold Co.)*

(b)

Fig. 15.31 (b) *In this economical installation the absence of a hung ceiling is obscured by painting the ceiling, fixture bodies, and raceways black. (Courtesy of Wiremold Co.)*

ber of manufacturers have developed a line of modular branch-circuit wiring elements. These, covered in NEC Article 604, consist of jacketed or armored cable sets terminating in polarized plugs. The polarization prevents accidental interconnection of low-voltage, 120- and 277-V systems. Ceiling raceways are equipped with matching receptacles, and connection to fixtures, poles, and other devices becomes a simple matter of insertion of plugs (see Fig. 15.32).

Disconnection is an equally simple action, resulting in a system of extreme flexibility. The additional cost of the manufactured wiring elements is frequently offset by labor savings even on initial installation and certainly after one or two field changes. Cable sets are available for power (120 V and 277 V), telephone, and all types of low-voltage signal equipment. The cables are approved for use in conditioned air plenums and hung ceilings.

To take full advantage of the potential labor-cost savings inherent in the system, field labor must be minimized. This is accomplished by factory equipping all utilization equipment, including lighting fixtures, with appropriate plug-in connectors.

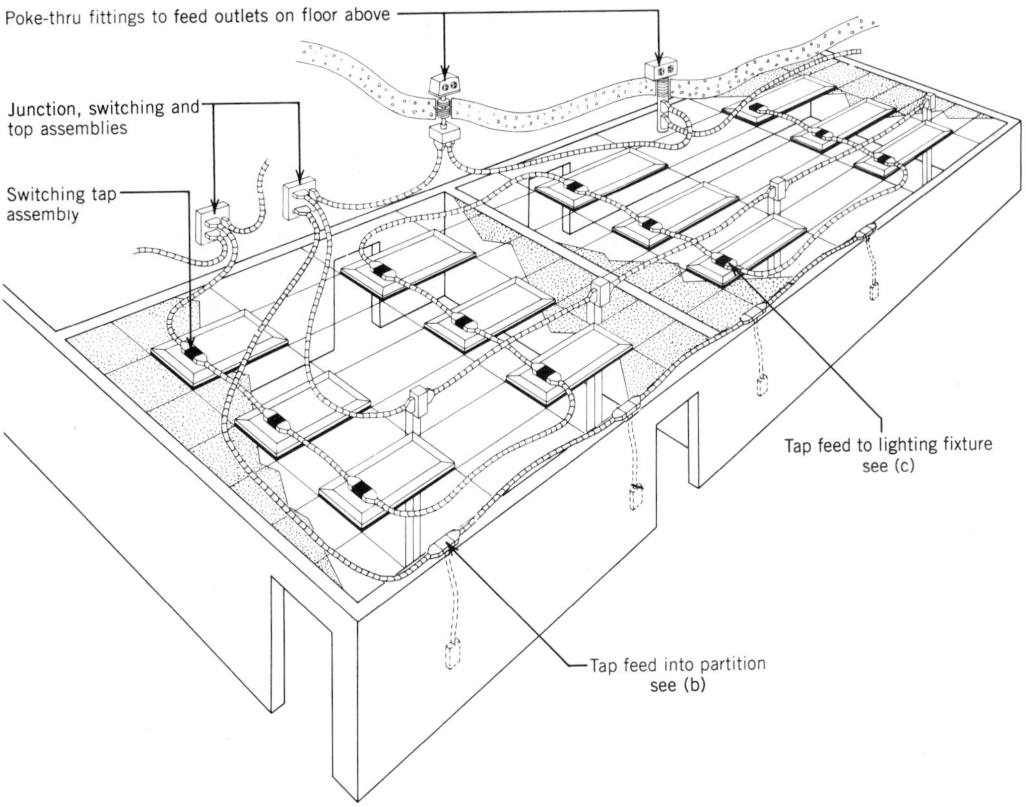

(*a*)

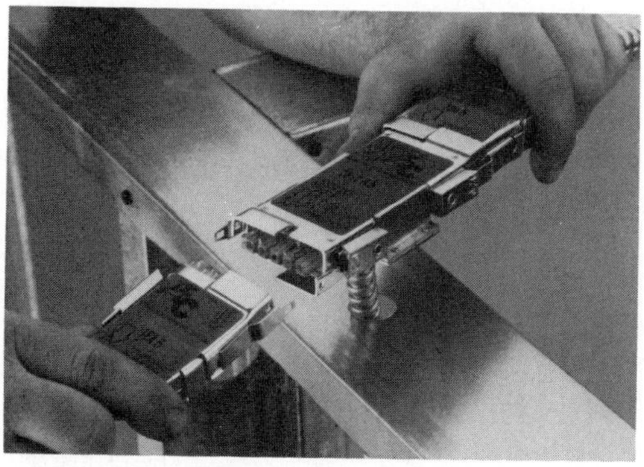

(*b*)

(c)

Fig. 15.32 (a) *Typical ceiling wired with modular elements. Manufactured modular wiring assemblies are used for tap connectoins* (b), *tap feed to fluorescent ceiling fixtures or any other ceiling lighting unit* (c), *and a complete range of junction, switching, tap and poke-through units. [*(a) *Courtesy of Walker, a division of Butler Manufacturing Co.;* (b, c) *courtesy of AFC/Nortek Co.]*

16

ELECTRICAL SYSTEMS AND MATERIALS: SERVICE AND UTILIZATION

ELECTRICITY

Having discussed in the preceding chapter conductors and raceways, the first of three major categories of electrical equipment, we turn now to the two remaining categories: power-handling and utilization equipment. The first step is to examine the means by which electric service is brought into a structure.

16.1 Electric Service

Public utility franchises require only that service be made available at the private property line. Thus service is normally tapped onto the utility lines at a mutually agreeable point at or beyond the property line. The service tap may be a connection on a pole with an *overhead service drop* or an *underground service lateral* to the building, or a connection to an underground utility line with a service lateral to the building. All electrical construction work on private property is normally at the owner's expense.

Under certain conditions the owner can influence the type of construction utilized by the electric utility company in conveying the electric service to the site. This is often the case in large tract developments and in places where owners are willing to share some of the cost of better grade construction. Also, in many areas, the utilities themselves have instituted "beautification" programs in an effort to decrease the objectionable appearance of much of their equipment. Service from the utility line to the building may be run overhead (OH) or underground (UG) depending on the following conditions:

1. Length of service run.
2. Type of terrain.
3. Budget limitations.
4. Utility company voltage.
5. Size and nature of electric load.
6. Importance of appearance.
7. Local practices and ordinances.
8. Maintenance and service continuity.
9. Weather conditions.

16.2 Overhead Service

The principal advantage of overhead electric lines is low cost. Depending on terrain and other factors, the cost of overhead as compared to underground installation has been in the range of 10 to 50% (the latter when compared to direct burial cable installation). This accounts for the majority of installations being overhead. In recent years special techniques in underground installation have lowered that cost, making it a feasible alternative. Where the service run is several hundred feet or more, voltages higher than utilization level may be involved. This weighs heavily in favor of overhead lines, particularly with voltages exceeding 5000 V. Similarly, when terrain is rocky and the electrical load is heavy, the cost of an underground installation rises sharply. Since overhead lines are easily maintained and repaired, and faults easily located, service continuity with overhead lines is usually quite good. In areas where there are extreme weather conditions, called heavy loading areas, where combinations of snow, wind, and ice increase the possibility of outages, un-

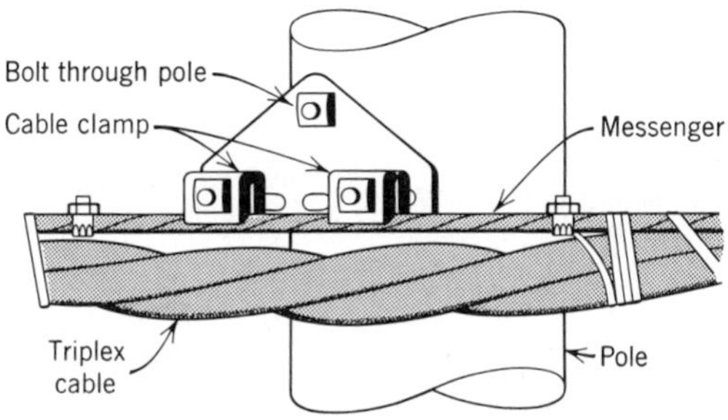

Fig. 16.1 *Preassembled aerial messenger cables are carried by steel messenger cables clamped to the poles.*

derground lines are preferred if service interruptions involve hardship or financial loss.

Overhead cables are of several types: bare, weatherproof, or preassembled aerial cable. Bare copper cable supported on porcelain or glass insulators on crossarms are normally used for high voltage (2.4 kV and higher) lines. Secondary circuits at 600 V and below are generally run on porcelain spool secondary racks with 1/c weatherproof cable as the conductor. Preassembled aerial cable consists of three or four insulated cables wrapped together with a metallic tape and suspended by hooks from the poles. This type of construction may be used up to 15 kV (see Fig. 16.1). It often proves to be more economical than crossarm or rack installation.

A typical detail of an overhead electric service entrance to a multiple residence is given in Fig. 16.2.

16.3 Underground Service

The advantages of underground electric service are attractiveness (lack of overhead visual clutter), service reliability, and long life. The principal disadvantage is high cost. To overcome this, utilities frequently use direct burial techniques which, by eliminating the raceway, reduce costs considerably. Since direct buried cable cannot be pulled out if it faults, as is the case with a raceway-installed cable, it is recommended that the decision of which technique is to be used be based on the consideration of these data:

1. Cost premium for underground raceway installation, including handholes if required (see Section 16.4).
2. Record of outages for direct burial installation.
3. Cost and availability of repair service (utilities frequently will repair customer-owned underground service laterals, *for a fee*).
4. Impact of electric service outage in terms of time delays, inconvenience, necessity to dig up lawns and paved areas, and cost impact in the case of a commercial facility.

16.4 Underground Wiring

Exterior installations are generally in connection with service—either directly from the utility or a subfeed between buildings. The methods available for underground wiring are:

1. Direct burial; see Fig. 16.3.
2. Installation in Type II, direct burial duct; see Fig. 16.4.
3. Installation in Type I, concrete encased duct, see Fig. 16.4.

The first alternative offers low cost and ease of installation, with the disadvantage stated above. The second offers median cost but little strength; therefore, only installations on undisturbed earth and/or under light paving are recommended. The last (3) offers high strength and permanence, but at the highest price of the three.

Nonmetallic duct (conduit) intended for underground electrical use is commercially available in two wall thicknesses. NEMA (National Electrical Manufacturers Association) type II with a heavy wall provides the physical protection required and is suitable for direct burial installation with no concrete encasement. Type I is manufactured with a thinner wall and is intended for encasement in a minimum of 2 in. of concrete. Common trade names for asbestos-cement and fiber ducts are Transite and Orangeburg. Plastic conduit is referred to as PVC, or simply as plastic. Non-metallic conduit is most frequently used without concrete encasement for low-voltage and signal wiring and with encasement for high-voltage wiring. It offers several advantages over steel conduit for under-

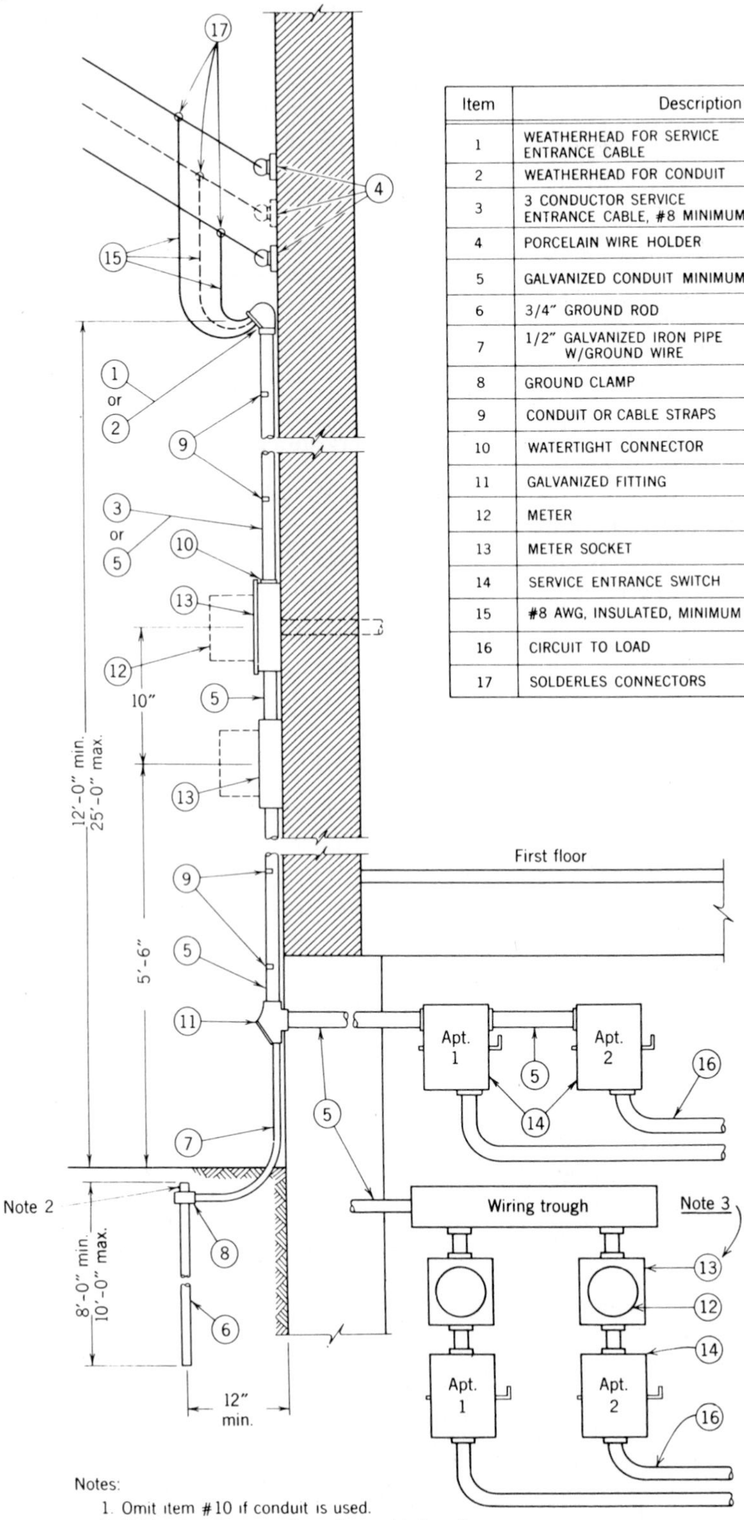

Item	Description
1	WEATHERHEAD FOR SERVICE ENTRANCE CABLE
2	WEATHERHEAD FOR CONDUIT
3	3 CONDUCTOR SERVICE ENTRANCE CABLE, #8 MINIMUM
4	PORCELAIN WIRE HOLDER
5	GALVANIZED CONDUIT MINIMUM 1″
6	3/4″ GROUND ROD
7	1/2″ GALVANIZED IRON PIPE W/GROUND WIRE
8	GROUND CLAMP
9	CONDUIT OR CABLE STRAPS
10	WATERTIGHT CONNECTOR
11	GALVANIZED FITTING
12	METER
13	METER SOCKET
14	SERVICE ENTRANCE SWITCH
15	#8 AWG, INSULATED, MINIMUM
16	CIRCUIT TO LOAD
17	SOLDERLES CONNECTORS

***Fig. 16.2** Detail of typical overhead electric service to a multiple residence. Note that meters are usually placed on the exterior of the buildings. If that is objectionable, they can alternatively be installed inside, provided that access is available to the utility's meter readers or that some type of remote reading system is installed. See Section 14.14.*

Notes:

1. Omit item #10 if conduit is used.
2. Cold water pipe ground may be used in lieu of ground rod.
3. Meters may alternatively be placed inside the building.

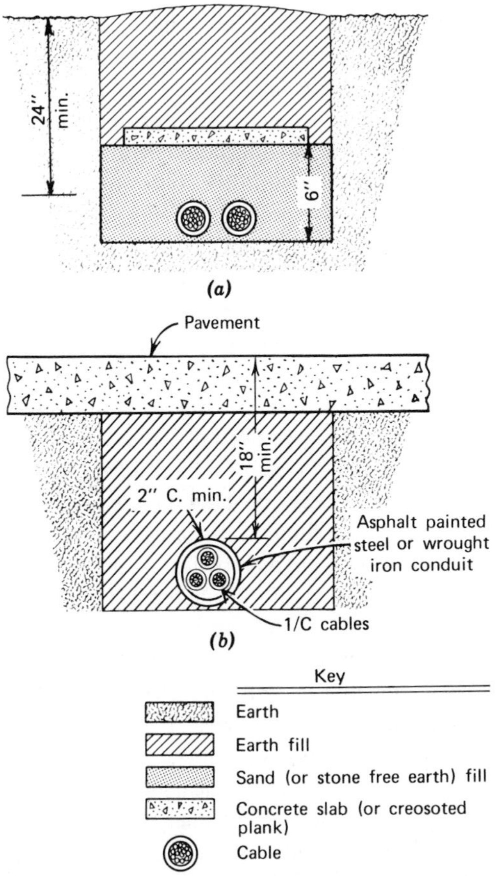

Fig. 16.3 (a) *Technique for installation of direct-burial cable.* (b) *Under highways, streets, and other high-load areas, cable should be installed in metal conduit.*

ground work, such as lower cost and freedom from corrosion.

When underground electric wiring is duct installed and the run extends over several hundred feet (the exact distance depending on pulling tension), a pulling handhole or manhole is necessary. Handholes are used for low-voltage power and signal cables, and runs with a small number of cables. Manholes are used for high-voltage cables and where large duct banks must be accommodated. Precast handholes and manholes are readily available in most standard sizes and are usually cheaper than field-formed and poured units.

Cable used in underground wiring must be specially manufactured and approved for that purpose. Type SE is the basic service entrance cable, constructed with moisture- and flame-

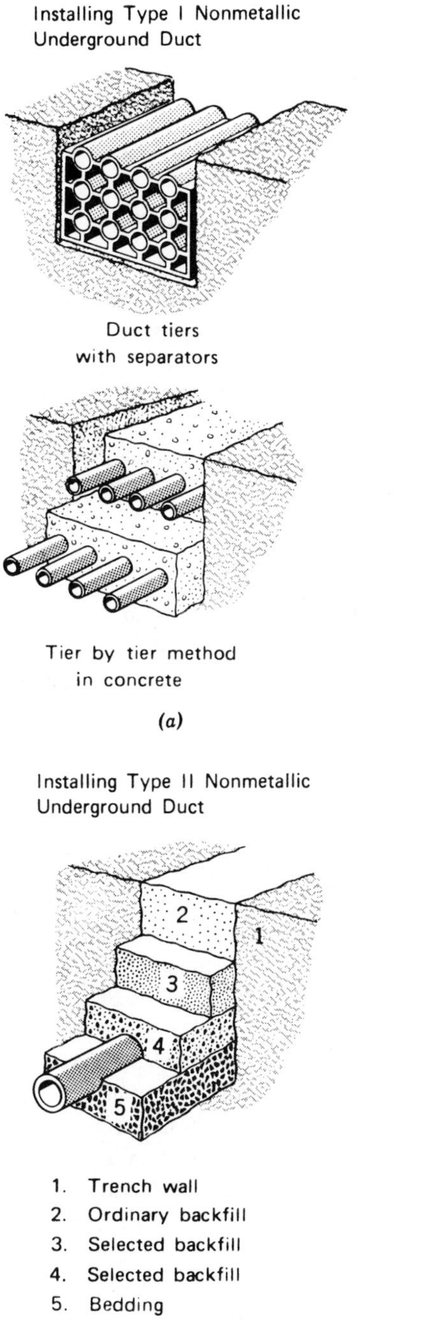

Fig. 16.4 *Underground duct installation.* (a) *Concrete duct bank should have minimum 6-in. earth cover.* (b) *Heavy-wall duct should be buried at least 24 in. in ordinary traffic ai eas and 36 in. where subject to heavy traffic. Each layer is about 8 in. thick.*

resistant covering. When provided with moisture proofing for underground use, the designation is SE type U, or simply USE. Underground cable for other than service runs is classified type UF (underground feeder).

16.5 Service Equipment

Referring back to Fig. 15.1, the reader will note that interposed between the high-voltage incoming utility lines and the secondary service conductors is a block labeled "transformer." This item is required whenever the building voltage is different from the utility voltage. It may be pole or pad mounted outside the building, or installed in a room or vault inside the building, as will be discussed below.

16.6 Transformers

A transformer is a device that changes or *transforms* alternating current of one voltage to alternating current of another voltage. Transformers cannot be used on d-c. A transformer would typically be used to step down an incoming 4160 V service to 480 V for distribution within a building; another transformer would be used in a local electric closet to step down the 480 V to 120 V for use on receptacle circuits. It is well to remember that ordinarily 120, 208, 240, 277, and 480 V are called secondary voltages, and 2400, 4160, 7200, 12,470, and 13,200 V are primary voltages.

Transformers are available in single-phase or three-phase construction. (See Section 17.3 for an explanation of these terms.) Transformer power capacity is rated in kVa (kilovolt-amperes). For a single-phase unit, this figure is the product of the full load current and the voltage. Since the voltages are different on primary and secondary, so are the currents—because the kVa remains constant. Thus a 100 kVa 2400/120 V transformer will carry at full load:

$$\text{primary current} = \frac{100{,}000\ \text{VA}}{2400\ \text{V}} = 41.6\ \text{A}$$

$$\text{secondary current} = \frac{100{,}000\ \text{VA}}{120\ \text{V}} = 832\ \text{A}$$

Cooling medium is a transformer characteristic of primary importance. Transformers are either dry (air cooled) or liquid filled. The choice depends upon required electrical characteristics, proposed physical location, and cost factors. Although detailed considerations are beyond the scope of this book, some selection criteria are presented in the following sections. In general, units rated above 5 kV are liquid filled and units in the 600-V class are dry. Units installed indoors, except in vaults, are normally dry type, and are intended for general purpose light and power circuits (see Table 16.1). Load center transformers are installed in unit substations, both indoor and outdoor. Distribution transformers are mounted on a pole or on a concrete pad outdoors. Substation transformers are large and are always concrete-pad mounted.

The cheapest liquid transformer coolant with good electrical characteristics is mineral oil. However, because of its flammability, its application is limited. Nonflammable liquid coolants, called *askarels*, generally containing polychlorinated biphenyl (PCB) were at one time very widely used. However, since 1979 PCB use in new equipment has been banned by federal regulation due to adverse ecological impact. Because of this ban, a number of other liquid coolants with low flammability and good electrical and physical characteristics have been developed. They are all more expensive than mineral oil (see Table 16.5).

The insulation class of a transformer affects its permissible temperature rise, operating temperature, physical size, electrical power losses, overload capacity, and life.

The physical size of a transformer of given kVA rating and voltage depends on the type of insulation used. In order of decreasing physical size and increasing operating temperature we have, for dry transformers, 105, 150, 185, and 220°C systems that represent organic, inorganic, and silicone insulating materials, respectively (see Table 16.2).

The dimensional data in Table 16.3 vary considerably from one manufacturer to another, and therefore the table is useful only for a concept of bulk volume and weight.

Although 220°C system insulation transformers can withstand 150°C rise, some users specify 220°C system insulation with 115°C rise or even 80°C rise; that is, a better grade of insulation is

TABLE 16.1 **Typical Transformer Data**

Transformer Type (Application)	*Maximum Capacity[a] (kVA)*	*Insulating (Cooling) Medium*	*Voltage Range*	
			Primary	*Secondary*
General purpose, dry type	1000	Air	120–600 V	120–600 V
Load center pad-mounted unit substation	3000	Air	2.5–15 kV	120–600 V
Load center pad-mounted unit substation	5000	Oil Silicone Fluid[b] Epoxy/resin[c]	2.5–34.5 kV	120–4800 V
Distribution system	10,000	Oil Silicone	2.5–67 kV	120–4800 V

[a]Three-phase bank.

[b]High-fire-point paraffinic hydrocarbon fluid, manufactured expressly as a transformer dielectric-coolant.

[c]Solid dielectrics used for special installation, such as high hazard areas.

used with an underrated transformer. The reason for this is fourfold:

1. Longer life.
2. Higher overload capacity.
3. Lower operating cost.
4. Capability to withstand heating effects of harmonics commonly found in commercial and industrial electrical systems (see Section 17.19).

A 220°C system transformer operated at full load *continuously* (an unusual situation) has a shortened life—estimated by some experts on the basis of accelerated aging tests to be between 3 and 10 years. The same class insulation transformer (220°C system) rated at 80°C has a life expectancy of over 100 years. With respect to operating cost, the same situations obtain here as with the high-temperature insulation discussed in Section 15.15 and studied in Table 15.5. The truism that one gets nothing for nothing can be expressed here as: In return for the smaller, lighter, cheaper, and hotter transformer (220°C system), we have higher losses and correspondingly higher operating cost, and shorter life.

ASHRAE/IES Standard 90.1-1989, Energy Efficient Design of New Buildings, which applies to most new construction, requires that for buildings with a total transformer capacity in ex-

TABLE 16.2 **Air-Cooled Transformer Electrical Insulation Temperature Ratings (Based on 40°C Ambient)**

Insulation Class (System)[a]	*Insulation Type*	*Average Conductor Temperature Rise*	*Ambient Temperature*	*Hot-Spot Temperature Gradient*	*Total Maximum Temperature*
105°C	Organic (A)	55°C	40°C	10°C	105°C
150°C	Mica, glass, resins (B)	80°C	40°C	30°C	150°C
185°C	Asbestos (F)	115°C	40°C	30°C	185°C
220°C	Silicones (H)	150°C	40°C	30°C	220°C

[a]The modern terminology for insulation class uses "system" in lieu of "class."

ELECTRICITY

TABLE 16.3 **Typical Dry-Type Transformer: Dimensions and Weights**[a]

kVa Output Continuous	Temp. Rise (°C)	Approximate Dimensions (Inches) H × W × D	Approx. Weight (Pounds)
Primary: 480 V Secondary 120/240 V			
Single-Phase 3	115	13 × 8 × 8	60
5	115	14 × 12 × 12	120
10	115	15 × 12 × 12	150
15	150	24 × 16 × 16	200
25	150	24 × 16 × 16	230
50	150	30 × 20 × 20	430
100	150	40 × 24 × 24	650
Primary: 480 V Secondary: 208Y/120 V			
Three-Phase 45	150	26 × 26 × 15	380
75	150	30 × 30 × 20	600
150	150	40 × 36 × 26	1100
225	150	44 × 36 × 26	1400

[a]Dimensions and weights vary among manufacturers. All figures increase with higher primary voltage and with lower temperature rise.

cess of 300 kVA, a calculation be made comparing the operating costs of various types of transformers and the results used to "balance energy costs with necessary operating flexibility, reliability and safety." The calculation should be based on projected transformer loading and loss data. Where this is not available, the standard provides a calculation form with average projected losses for an assumed loading. A typical calculation using the ASHRAE form is shown in Table 16.4, comparing two dry-type transformers, type H; one is designed for 80°C rise and the other for 150°C rise. The operating cycle is chosen to be representative of commercial use. With such a cycle, even the 150°C rise unit will probably last 30 years, the specified life cycle. The current different in first cost between the two units is less than $2000. Table 16.4 tells us that the lower energy waste of the 80°C unit will repay the first-cost differential in about 4 years. Obviously, a higher electric energy cost will reduce the payback time, and vice versa. Beyond the payback period the advantage is entirely on the side of the 80°C unit. Using a 30-year life, 8% fixed capital cost, and 3% annual cost escalation, the life-cycle cost of the two transformers is $47,688 and $57,504 respectively. Furthermore, with a total permissible hot-spot temperature of 220°C (428°F), the location of the 150°C rise unit must be very carefully chosen, since it can create a serious heat generation and radiation problem.

In summary, then, a transformer is specified by type, phase, voltages, kVA rating, sound level, and insulation class. Thus, 112.5 kVA, 3-phase, 480/120-208 V, air-cooled, indoor, dry-type transformer with 220°C insulation system and 115°C rise, 45 db (decibel) maximum sound level is an adequate transformer description. Sound ratings of transformers, as well as installation techniques and acoustical treatment, are discussed in Part IX, "Acoustics."

16.7 Transformers Outdoors

A service transformer bank is necessary, as explained above, when the facility utilization voltage is different from the utility voltage (see Fig. 16.5*a*). The designer occasionally opts for a step-up, step-down arrangement when the service run is so long that the conductor cost when run at a low voltage would be excessive because of large-sized cables. In such instances, the cost of the double transformer installation must be more than offset by the savings in feeder cost to be economically justifiable (see Fig. 16.5*b*).

The advantages of an outdoor transformer installation are:

1. No building space required.
2. Reduced noise problem within building.
3. Lower cost.
4. Ease of maintenance and replacement.
5. No interior heat problem.
6. Opportunity to use low-cost, long-life oil-filled units.

On the other hand, it is frequently easier to

TABLE 16.4 **Transformer Operating-Cost Comparison; 750 kVA, Dry Types**

ASHRAE 90.1—**Form 5-1 Transformer Loss Calculation Estimate**			
Transformer no.________		1	2
Rating________		750 kVA	750 kVA
Rated temperature rise____		80°	150°C
Cooling medium________		Air	Air
Rated no-load transformer Loss (%)________		0.293%—2.2 kW[a]	0.293%—2.2 kW
Rated full-load coil losses ________(kW)		9.0 kW[b]	12.0 kW
1. No-load transformer loss (%) manufacturers rating =________	%	0.293	0.293
2. No-load (%) loss × transformer full-load rating =________	kW	2.2	2.2
3. Annual no-load losses: 8760 × no-load kW (#2 above) =________	kWh	19,272	19,272
4. Annual hours of transformer operation from 10% to 50% load =________	hours	3650	3650
5. Use 10% of transformer full-load coil losses × #4 above =________	kWh	3285	4380
6. Annual hours of transformer operation from 50% to 80% load =________	hours	4380	4380
7. Use 40% of transformer full-load coil losses × #6 above = ________	kWh	15,768	21,024
8. Annual hours of transformer operation from 80% to 100% load =________	hours[a]	730	730
9. Use 80% of transformer full-load coil losses × #8 above =________	kWh	5256	7884
10. Total kWh (of #3, #5, #7, #9) =[c]________	kWh	43,581	52,560
11. Total kWh (#10 above) × average cost/[d] kWh for electricity =________	/Year	$3051	$3679

Source: ASHRAE 90.1 portion of this table reprinted with permission.

[a]Efficiency figures are averaged from manufacturer's published data.

[b]Loss at 0 load is core loss of 2200 W.

[c]Operating cycle, representative of a typical commercial building is full load—2 h, 75% load—8 h, 60% load—4 h, 40% load—2 h, 20% load—8 h.

[d]Based on electric energy cost of $0.07 per kWh, including demand.

find space indoors (preferably in a basement) than to find a suitable exterior spot; noise may be more disturbing from the available exterior spaces, such as courtyards, than from a basement. Costs can run high if long secondary runs are required; heat can often be handled by louvers or areaways adjoining a basement or even made use of profitably if the transformer load is fairly constant. Also, since exposure to direct sunlight decreases the unit rating by increasing its temperature, a shaded spot may be difficult to find. Furthermore, an exterior transformer, except where pole mounted, is a questionable choice in an area with a high incidence of vandalism, regardless of sturdiness of construction. Finally, appearance of an exterior unit may be objectionable. This latter point has received much attention from manufacturers, and numerous designs have been developed that minimize the appearance problem (see Fig. 16.6). The most popular type of exterior transformer installation is the pad mount. It has all of the advantages listed above in addition to extreme simplicity of installation—it is simply set

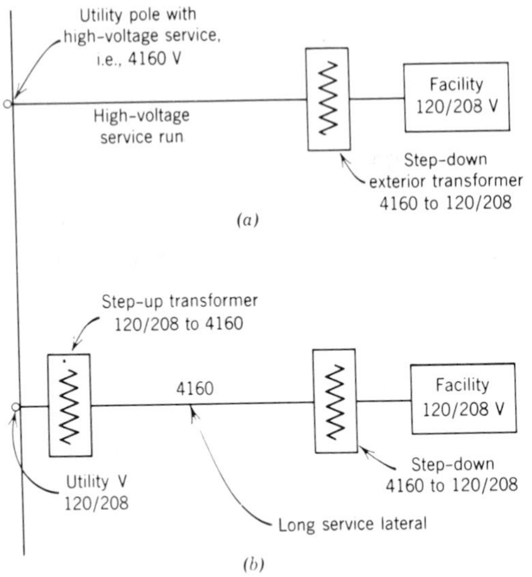

Fig. 16.5 Service transformer arrangements. (a) *High-voltage service with step-down service transformer at the facility, and* (b) *low-voltage service, with transformation at both ends of a long service run.*

on a concrete pad. Consult manufacturers' catalogs for dimensional data.

16.8 Transformers Indoors; Heat Loss

When an indoor transformer installation is indicated, special consideration must be given to the transformer's heat-generating properties. Between 1 and 1½% of the transformer's rating, depending on type, is converted to heat at full load. Table 16.4 shows that for a 750 kVA, dry-type, 150°C rise unit, 12 kW or 41,000 Btu/h of heat loss is generated at full load! Losses are lower for 80°C rise units, as indicated. Liquid-filled units have approximately the same losses as 80°C rise dry units. Unless the heat can be used, sufficient ventilation—either natural or forced—must be provided to keep the ambient temperature from exceeding 40°C.

Ventilation by natural convection is most desirable, necessitating the location of the transformer room on an exterior wall (with an areaway if the room is located below grade). The size of the *free* area required for a ventilation opening is 3 in.^{2}/kVA of capacity, plus an additional 1 in.^{2}/kVA for switchgear losses, if any. If

Fig. 16.6 (a) *Pad-mounted exterior transformers are neat, compact and if sited properly, unobtrusive. Large units can be partially screened by shrubbery.* (b) *When the size is such that visual screening becomes a problem, consideration should be given to a structural screen such as a decorative brick wall, which also provides some much needed sound screening. (Courtesy of General Electric.)*

a louver is used in the openings, the size of the opening must usually be doubled (total: 8 in.^{2}/kVA), since most louvers have a 50% free area. For good convection, it is desirable to divide the louvered areas in half, placing one-half near the ceiling, and the remaining half near the floor. To provide for equipment removal, if the areaway is large enough, it may be useful to add louver area between the upper and lower openings and make the full louver removable. A bird screen is also desirable.

Since outside air temperature will vary, it is advisable to use a temperature-controlled, adjustable louver. In severe cold climates, heat loss from the electrical equipment may not be sufficient in the winter to warm the room. In such instances, a unit heater should be installed in the room with a thermostat set at 55°F.

16.9 Transformers Indoors; Selection

When transformers are installed indoors, they are subject to stringent NEC regulations that are designed to make the installation intrinsically safe. These regulations are detailed in NEC Article 450 and no purpose is served in duplicating them here. Instead, we wish to present the reader with the essential considerations involved.

1. Oil-filled transformers present a fire hazard indoors, because flammable oil can spread from a tank leak or rupture. To prevent this, oil-filled transformers must be installed in a fire-resistant vault, the construction of which involves a heavy cost. Advantages offsetting this cost are the oil-filled transformer's small size, low weight, low first cost, low losses, long life, excellent electrical characteristics, low noise level, and high overload capacity. Despite these, the vault requirement has had the effect of restricting oil-filled units to industrial facilities and other structures where electrical considerations require its use.

2. Nonflammable liquid-filled units have most of the advantages of oil-filled units and do *not* require a vault unless voltage is very high. They do, however, require a sump or catch basin of sufficient capacity for all of the contained liquid. This and a relatively high first cost are the negative aspects.

3. Dry-type units are the transformers of choice in the majority of indoor installations, despite shorter life, higher losses, high noise level, greater weight, and larger size than the liquid-filled units. The principal advantage is ease of installation and almost unrestricted choice of location. As explained above, by using an underrated transformer (220°C system, 80°C rise), losses (heat) can be reduced and life extended. Also, for a price premium, noise level can be reduced. When a dry-type transformer is equipped and modified so that its characteristics are equal to that of liquid-filled units, its installed and owning costs are in the same range as the liquid units. Still, most users find the lack of restrictions on placement a sufficient reason for its choice. Table 16.5 gives a comparison of installed costs for these three classes of transformers.

16.10 Transformer Vaults

A transformer vault is basically a fire-rated enclosure, provided because of the possibility of transformer case rupture and an oil fire. However, this must not be construed as implying that transformers are hazardous or delicate devices prone to faults. On the contrary, transformers are extremely tough, sturdy, long-lived, capable of sustaining large and prolonged overloads, and indeed among the most reliable elements of an electrical system. However, faults do occur, and an oil-filled transformer is a potential fire hazard.

Transformer vaults should be located, if practicable, where they can be ventilated to the outside air without flues or ducts. The combined net area of all ventilating openings (gross area less screens, louvers, etc.) should be as explained in the preceding section, not less than 3 in.^{2}/kVA of transformer, but in no case less than 1 sq ft. Further ventilation recommendations plus details of enclosure construction materials, fire rating, door and sill details, and other relevant and important information are provided by NEC Article 450.

TABLE 16.5 **Relative Installation Costs for a 300 to 1000 kVA Transformer**

Transformer Type	*Temperature Rise*[a] *(°C)*	*Relative Transformer Cost*	*Construction Cost*	*Total Relative First Cost*
Oil filled	80	1.00	50–100[b]	1.50–2.00
Silicone, nonflammable fluids	80	1.40	20–40[c]	1.60–1.80
Dry, ventilated	80	1.65	—	1.65
	80	1.50	—	1.50
	150	1.35	—	1.35
	150	1.20	—	1.20

[a]Transformers of equal temperature rise have approximately equal life and equal losses.

[b]Cost of vault depends on local labor costs and size of transformer. The relative cost decreases with increasing transformer size.

[c]Cost of catch basin. As in the case of a vault, relative cost depends on labor rates and transformer size.

16.11 Service Equipment Arrangements and Metering

To summarize the preceding discussion and to proceed with our study of electric power-handling equipment, refer to Fig. 16.7. Electric service connection will be overhead or underground. Where building voltage is different from utility voltage, a transformer is required. This transformer may be pole or pad mounted outside the building, or installed in a room or vault inside or outside the building.

Metering must be provided at either the utility or the facility voltage, and at either the service point or inside the building. The choice is generally left to the owner with the understanding that inside meter equipment must be readily accessible to utility personnel. Although in increasingly large numbers of facilities, actual meter reading is accomplished remotely, the meter equipment must be available for inspection and service. If high-voltage service is purchased, then the transformers and all equipment beyond the service connection must be furnished by the owner. Conversely, if low voltage is purchased, all equipment necessary to provide low voltage is furnished by the utility. Obviously, the electric service rates for low-voltage service are *higher* than for high-voltage service in order to compensate the utility for the cost of providing and maintaining the step-down transformer and associated equipment. Therefore, it is often advisable for the owner of a facility (other than a small, residential facility) to investigate the economics of purchasing power at high voltage. Since many owners are not equipped to maintain high-voltage equipment, arrangements can sometimes be made to pay the utility to provide and maintain transformers, while taking the cost advantage of high-voltage service.

For a single-use building or a building where electric energy is included in the rental charge, only a single meter is necessary. Provision for such metering may be made in the main switchboard, or the meter may be independently mounted. In both cases, the meter is furnished and installed by the utility company. Where submetering is required, such as in apartment houses, banks of meter sockets are installed to accommodate the multiple meters. Federal regulations forbid master metering in new multiple-dwelling constructions, because it encourages energy waste. A low-voltage underground service detail as it would appear on a set of contract drawings, including relevant details, is given in Fig. 16.7. Note that here the service switch and meters are separately mounted. Meters are always installed *electrically* ahead of the service switch so that they cannot be disconnected.

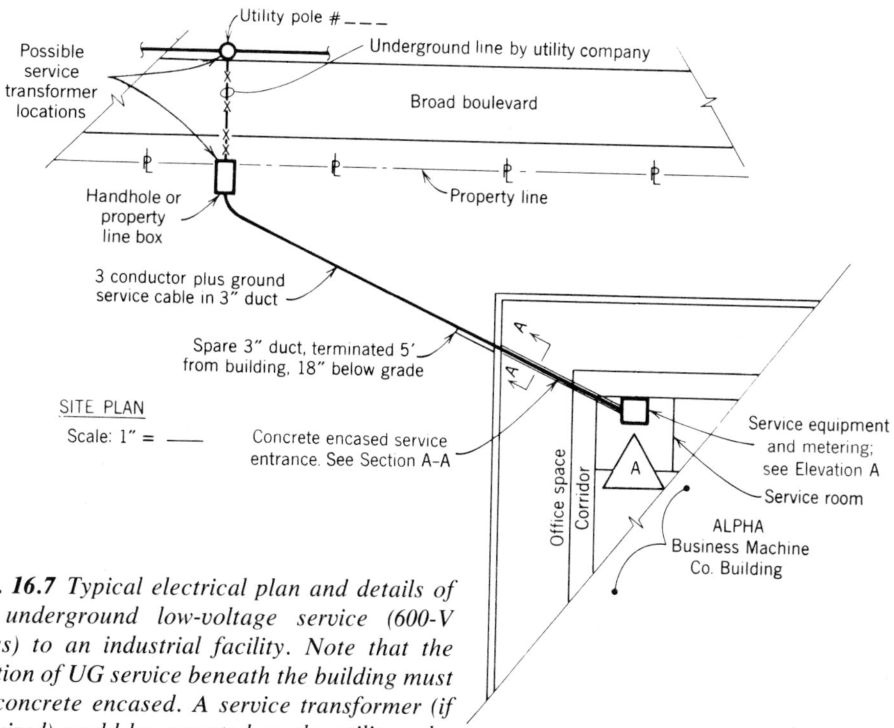

Fig. 16.7 Typical electrical plan and details of an underground low-voltage service (600-V class) to an industrial facility. Note that the portion of UG service beneath the building must be concrete encased. A service transformer (if required) could be mounted on the utility pole, at its base, or on a pad at the property line.

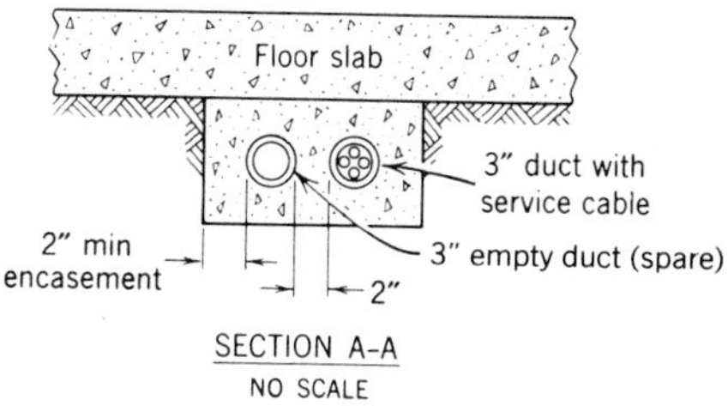

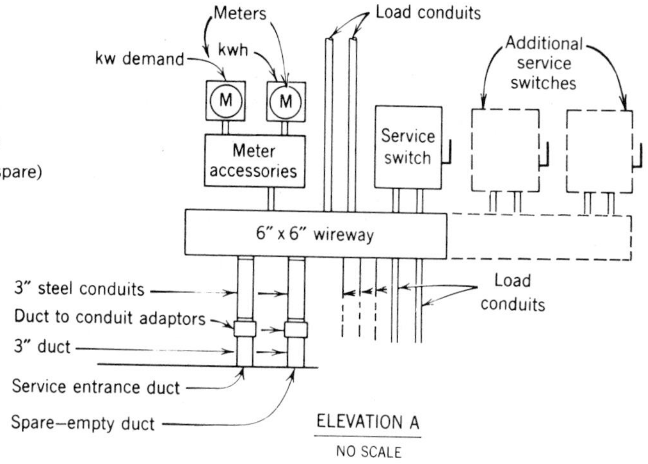

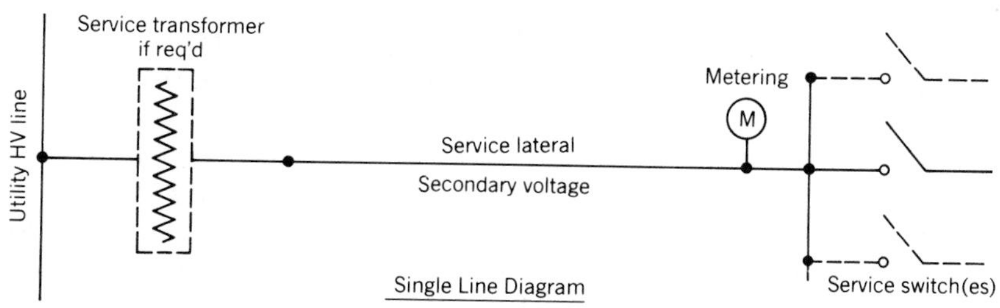

16.12 Service Switch

The purpose of the electric service switch is to disconnect all of the electric service in the building except emergency equipment. Thus, in the event of fire, no electrical hazard will face fire fighters. It is therefore obvious that this disconnecting apparatus must be located at a readily accessible spot near the point at which the service conductors enter the building. If such a location is not feasible, service conductors may be run in concrete encasement under the building and will be considered "outside the building" up to the point at which they emerge from the floor in the building (see Fig. 16.7). At that point, the service switch must be installed. The service switch or, more accurately, the "service disconnecting means," may comprise one to six properly rated switches. These are frequently assembled into a switchboard. Before discussing switchboards, a description of switches, circuit breakers, and fuses, the components out of which switchboards are constructed, is in order.

16.13 Switches

In this and the following two sections we discuss "traditional" electrical switching devices, which close and open an electric circuit by physically moving two electrical conductors into contact with each other to close the circuit, and physically separating them to open the circuit. The motive force can be supplied by hand, by an electrical coil, by a spring or by a motor, as discussed below, and the device name may change accordingly, but the action is the same. Solid-state switches perform the same electrical switching function but by a completely different process, without moving parts. They are discussed in Section 16.16.

An electrical switch is rated by current and voltage, duty, poles and throw, fusibility, and enclosure. The current rating is the amount of current that the switch can carry continuously and interrupt safely. Switches intended for motor control are also rated in horsepower. The voltage rating of a switch is, as for other electrical equipment, by voltage class. Thus a switch is rated 250 V, 600 V, or 5 kV as required. Switches intended for normal use in lighting and power circuits are called general duty safety switches. Switches intended for frequent interrupting, high fault currents, and ease of maintenance are rated HD for *heavy duty*.

An examination of Fig. 16.8 should clarify what is meant by the number of poles and throws of a switch. Unless otherwise noted, a switch is assumed to be single throw. Since the NEC states generally that the grounded neutral conductor of a circuit should not be broken, most switches carry the neutral through unbroken, by means of a solid link within the switch. This gives rise to the term *solid neutral* (SN) *switch*. Switches are available in 1-, 2-, 3-, 4-, and 5-pole construction. Poles are indicated by a "P"; thus 3-pole is written "3P," and so on.

A switch may be constructed with or without provision for fusing. If provided, the switch is fusible; if not, the switch is nonfusible. All separately enclosed switches must be in an appropriate cabinet. The National Electrical Manufacturers Association has standardized the nomenclature and application of enclosures for all electrical control equipment, of which switches are only one item. These are detailed in Section 16.15. Summarizing the above, then, we could adequately describe a switch thus: switch, HD, 3P & SN, 200A/150AF (fuse), 600 V, in NEMA 12 enclosure. Such a switch is illustrated in Fig. 16.9.

16.14 Contactors

A contactor is a switch. Instead of a handle-operated, movable blade, a contactor uses *contact* blocks of silver-coated copper, which are forced together to make (close) or are separated to break (open) the circuit. The common wall light switch is a small mechanically operated contactor. A relay is a small electrically operated contactor. Most contactors are operated by means of an electromagnet that causes the contacts to close. They open either by spring action or by gravity. Contactor terminology is somewhat different from that of a switch. Its condition when deenergized is its *normal* state. Thus, a contactor whose contacts are open when the coil is not energized is *normally open* (NO). One with normally *closed* contacts is NC. Units intended for motor control are called motor

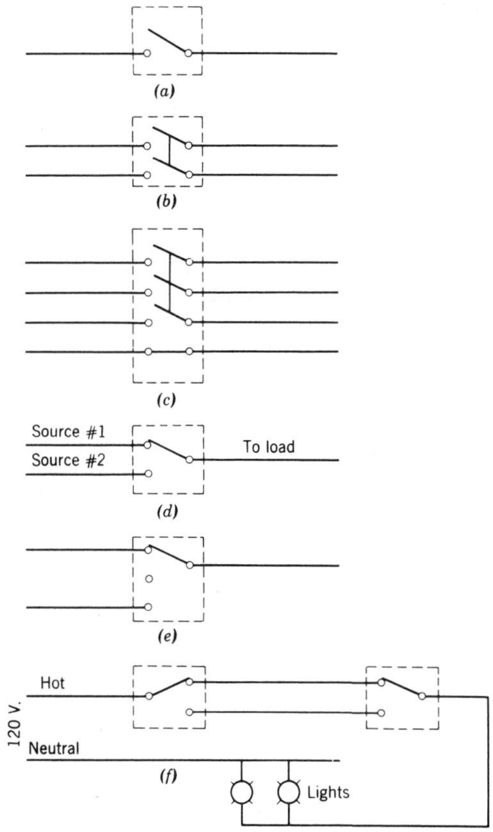

(a) Single-pole single-throw switch.

(b) Two-pole single-throw switch.

(c) Three-pole and solid-neutral (3P and SN) switch.

(d) Single-pole double-throw switch (also called, in small sizes, a 3-way switch).

(e) Single-pole double-throw switch with center "off" position (in control work called a hand-off-automatic switch).

(f) Use of two single-pole double-throw (3-way) switches for switching of a lighting circuit from two locations.

Fig. 16.8 *Typical switch configurations. Note that switches are always shown open.*

starters and are discussed in Section 16.23. Current, voltage, and number of poles have the same significance for contactors and relays as for switches.

The great advantage of contactors over switches is their facility for remote control. Switches must be manually thrown—or in very special cases with a motor. However, the magnetic contactor is inherently a remotely controlled device, making it ideal for a myriad of control functions. It can be controlled by pushbutton or automatic devices such as timers, float switches, thermostats, pressure switches, and so on. Since control can be both remote and automatic, the application of contactors is universal in control of lighting, heating, air conditioning, motors, and the like. Graphic representation of contactors are shown in Fig. 16.10.

16.15 Special Switches

Many special types of switches are available. Most of these types are beyond the scope of this book, except for the following, which are discussed briefly.

(a) Remote-Control (RC) Switches. A contactor that latches mechanically after being operated in known as a remote-control switch. It differs from a relay in that the latching operation performs the dual function of latching the contacts and disconnecting the electric circuit. To unlatch the contacts the electric control circuit must be reenergized. It is therefore also known as a mechanically held, electrically operated contactor. This characteristic is advantageous where the position of the contacts will be maintained for a long period (as in a lighting control

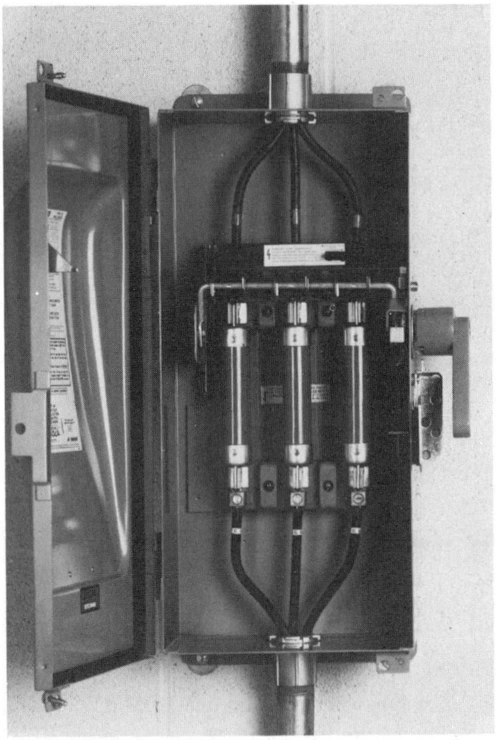

Fig. 16.9 The internal construction of a heavy-duty fused three-pole three-wire switch in an industrial NEMA 12 enclosure is clearly visible. Note the gasketing on the inside of the cover which seals the sheet steel box and makes it usable in atmospheres containing lint, dust, dirt, sawdust, and the like. The double lockable indicating type of external handle obviates the necessity to open the switch enclosure during normal use. (Courtesy of Siemens Energy & Automation, Inc.)

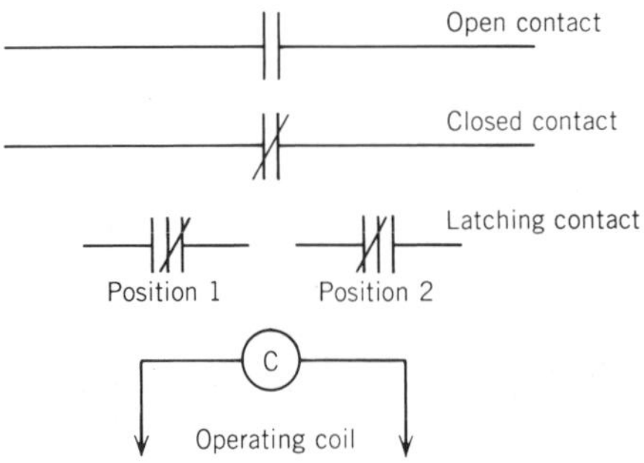

Fig. 16.10 *Graphic representation of switch (contact) position in a contactor or relay. Note that a latching contactor shows the switch in one of two positions, to indicate the latching nature of the contactor.*

circuit) and it is undesirable to hold the contacts by continuous energizing of a coil, as is required with an ordinary relay. A typical unit is shown in Fig. 16.11 and a typical application is shown in Fig. 17.12.

(b) Automatic Transfer Switch. This device, which is an essential part of all standby power arrangements, is basically a double throw switch—generally 3-pole—so arranged that on failure of normal service it *automatically transfers* to the emergency service. When normal service is restored, it automatically retransfers to it. The control devices are voltage sensors that sense the condition of the service and operate the switch accordingly. Auxiliary devices can be built on to the basic switch, the most common of which are emergency generator starting equipment. Figure 16.12 illustrates typical units. Transfer switches used in uninterrupted power supplies are solid-state since they involve no switching time lapse.

Fig. 16.11 *This remote control switch (mechanically held, electrically operated contactor) is a 12-pole, 20-A, 600-V unit measuring approximately 7 in. H × 6 in. W × 4 in. D, making it suitable for stud wall mounting. Compact solid-state control modules are available which mount directly onto the unit and convert it to a flush, wall-mounted multicircuit programmable controller. See Section 16.16. (Courtesy of Automatic Switch Co.)*

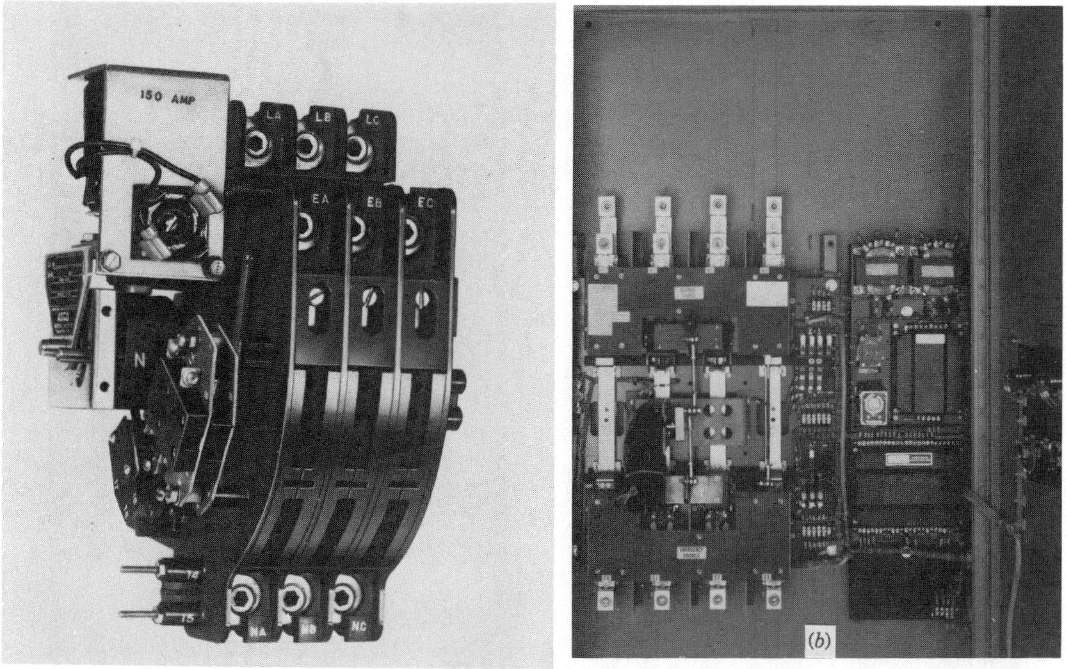

(a) (b)

Fig. 16.12 *Typical electromechanical automatic transfer switches. Switch* (a) *is a 150-A, 3-pole, 480-V unit arranged for connection of normal and emergency services at the top and load cable at the bottom. In a NEMA 1 general-purpose enclosure it measures approximately 17 in. W × 31 in. H × 12 in. D. Switch* (b) *is rated 800 A, 4-pole, 600 V a-c, which corresponds to approximately 300 kVA at 208 V, and 650 kVA at 480 V, 3-phase. The entire unit as shown in a NEMA 1 cabinet measures 37 in. W × 66 in. H × 14 in. D and weighs 400 lb. The subunit at the left contains both manual and motor operators and cable connections. The subunit at the right contains all sensing and control devices plus complete emergency generator starting and control equipment. [*(a) *Courtesy of Automatic Switch Co.,* (b) *courtesy of Russelectric, Inc.]*

(c) Time-Controlled Switches. In this category are included all switches whose operation is time based. The timer can be either the familiar electromechanical device consisting of a low-speed miniature drive motor to which some type of contact-making device is physically connected, or it can be a solid-state electronic timer, which in turn controls either a relay or a solid-state switch. The latter arrangement is the basis of all modern programmable time controls (see next section), which find wide application in lighting control (Section 21.3), energy management and automated building control (Sections 14.14 and 21.3), and institutional clock and program systems (Section 22.14).

Motor-operated timers depend for their accuracy on power line frequency, have moving parts that wear, must be reset after power outages (some units have spring-wound reserve power motors), and become cumbersome and expensive with increase in the number of controlled events. Electronic units are small, cool, quiet, independent of line frequency for accuracy (in many designs), carry through outages when standby-battery equipped, and have virtually unlimited event-control capacity. They will undoubtedly in large measure replace the older design except for the simplest applications. See Figs. 16.13 and 16.14 and Figs. 21.3 and 21.4.

16.16 Solid-State Switches, Programmable Switches, Microprocessors, and Programmable Controllers

A solid-state switch is an electronic device with a conducting state and a nonconducting state,

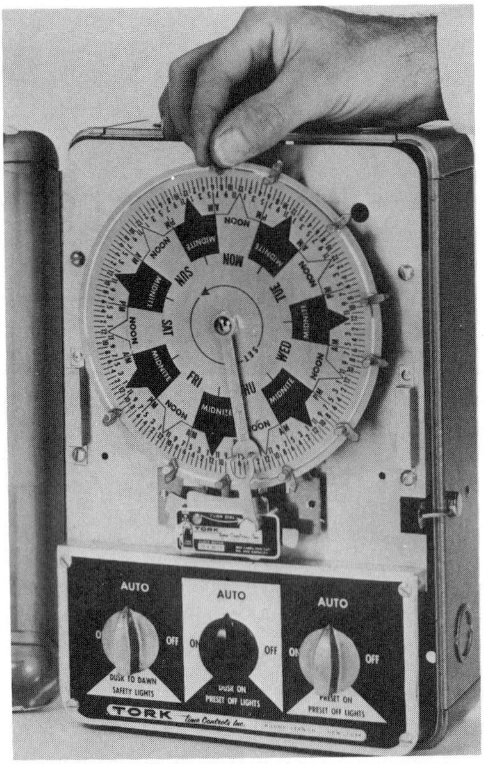

Fig. 16.13 *Time switch is arranged for three different types of control functions that are applicable to different types of load: (1) full photocell control—security lighting; (2) dusk ON, preset OFF—signs, parking lots; (3) preset ON, OFF—interiors. This unit's dial is a 7-day calendar type that permits setting a different schedule for every day of the week. The dial is not astronomical (solar), since all the turn-on functions are either photocell or time-of-the-day controlled. Solar dials are useful when photocell control is not used, since sunrise and sunset times vary with the season. (Courtesy of Tork, Inc.)*

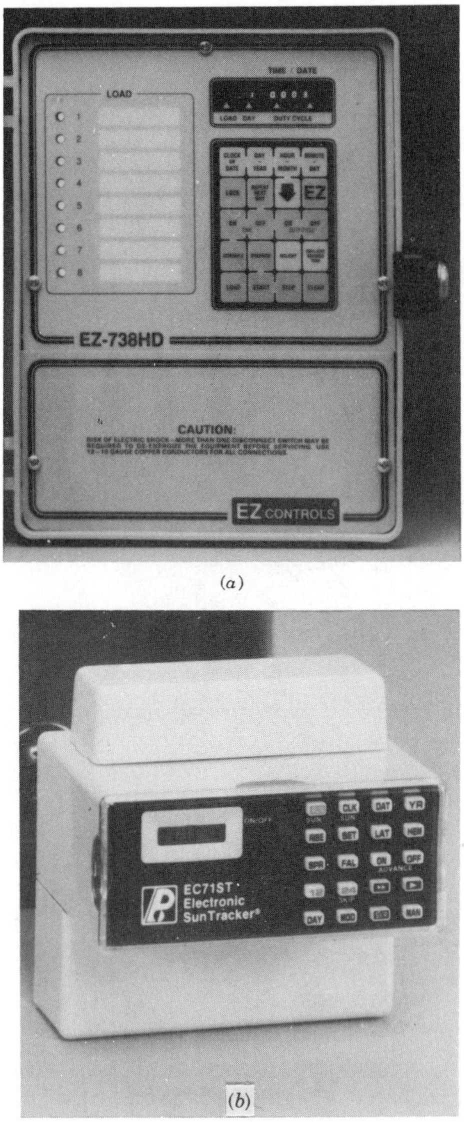

Fig. 16.14 *Programmable time switch.* (a) *This unit, which measures 9.5 in. H × 8 in. W × 4.5 in. D in its cabinet, permits independent programming of 8- to 20-A, 240-V circuits. Each circuit (channel) can be arranged for multiple ON–OFF daily operations, and can be programmed on a 7-day-week basis with special 365-day functions such as automatic daylight savings time adjustment.* (b) *This unit controls a single 15-A, 240-V circuit. The built-in astronomical feature automatically adjusts the ON–OFF setpoint according to sunset and sunrise throughout the year. [*(a) *Courtesy of E.Z. Controls;* (b) *courtesy of Paragon Electric Co.]*

corresponding to a conventional switch in its closed position and its open position. The change between the two states is accomplished by the application of a control signal (voltage). The change of state is instantaneous, noiseless, and arc-less. The current ratings of solid switches are approximately the same as those of their mechanical counterparts, ranging from fractions of an ampere to many hundreds of amperes. Voltage ratings are usually limited to secondary systems ratings (0 to 600 V), due to the nature of the semiconductors comprising the active elements of these switches, although higher-voltage equipment is available for special application.

If an electronic timing device is added to a solid-state switch, a time-base-controlled switch is created which will function exactly as the electromechanical time-based switches described in the preceding section. It is then only necessary to add a small programmable (electrically changeable) memory circuit in the form of an EPROM (*e*rasable *p*rogrammable *r*ead-*o*nly *m*emory) chip to create a programmable time switch. The simplest of these are single circuit 15-A devices intended to replace common wall switches (see Fig. 21.4). A more complex unit, corresponding roughly to the electromechanical unit shown in Fig. 16.13 but much more flexible, is shown in Fig. 16.14*a*. Addition of a ROM chip preprogrammed with astronomical information (sunrise–sunset) for the particular latitude involved converts the switch to a "sun-tracker." Such a switch is particularly applicable to control of circuits with year-round dusk-to-dawn control (see Fig. 16.14*b*).

If, in lieu of a programming device which until reprogrammed gives fixed invariable instructions to a (multichannel) switch, as is the case with a programmable time switch, we install a *microprocessor* [i.e., a logic/memory device programmed to *respond* according to a particular algorithm (control plan) to given input signals], we will have constructed a programmable controller (see Fig. 16.15). By definition, then, a programmable controller is an electronic device that uses a programmable memory for internal storage of instructions that implement specific functions, such as logic, timing, counting, and so on. The internal control/memory is frequently referred to as a microprocessor, whereas with respect to the overall programmable controller, it acts as a CPU (*c*entral *p*rocessing *u*nit). The programmable controller is therefore a special type of computer. It differs from a conventional computer in that its program is relatively short and exists in hardware (microprocessor), whereas a computer reads its program, which can be very lengthy, into its memory from software. Further, the programmable controller is designed for a specific type of function, whereas the conventional computer is a general-use calculating (computing) device that will process any type of software.

Programmable controllers have all but replaced the ubiquitous wired relay panel in applications such as industrial control, elevator control, process control, and the like, because of the ease with which such a controller can be reprogrammed (in lieu of rewiring a relay panel as was the case), the speed and accuracy of control and the vastly increased complexity possible. See Appendix J for definitions and for an explanation of application of these devices to building control.

16.17 Equipment Enclosures

Proper NEMA nomenclature, descriptions, and applications for the more common enclosures are found in Table 16.6. It is important to note that there is no enclosure described as WP or weatherproof. Equipment intended for outdoor use should be specified:

1. In a Type 3R enclosure to protect against rain (see Fig. 16.16).
2. In a Type 3S enclosure to protect against wind-driven rain and sleet.
3. In a type 4 enclosure to protect against the above plus splashing and condensation.

Note also that the type 12 industrial enclosure is similar to type 1, except that it is gasketed for dust and drip resistance and therefore is well applied in *all* "dirty" indoor environments, including commercial and institutional spaces (see Fig. 16.9).

16.18 Circuit-Protective Devices

To protect insulation, wiring, switches, and other apparatus from overload and short-circuit currents, it is necessary to provide automatic means for opening the circuit. The two most common devices employed to fulfill this function are the fuse and the circuit breaker (c/b).

(a) Fuses. The fuse is a simple device consisting of a *fusible* link or wire of low melting temperature that when enclosed in an insulating fiber tube is called a cartridge fuse, and when in a porcelain cup is known as a plug fuse. Figure 16.17 shows common types of fuses. Plug fuses such as those normally in residential use, are rated 5 to 30 A, 150 V max to ground. Cartridge

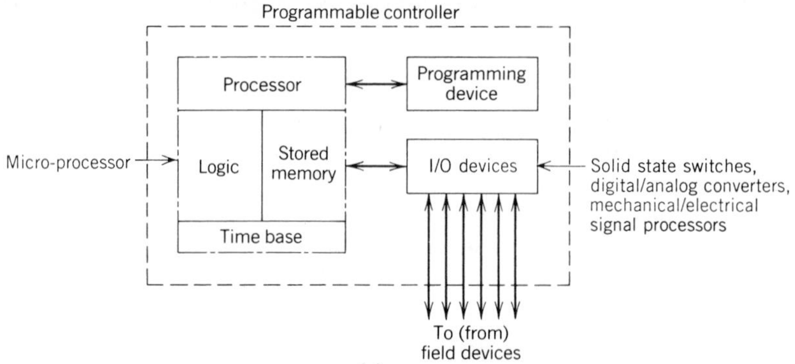

(*a*)

Fig. 16.15 (a) *Block diagram of a programmable controller. These devices are designed to supervise field conditions continuously and are therefore used in industrial applications as well as lighting and environmental system control.*

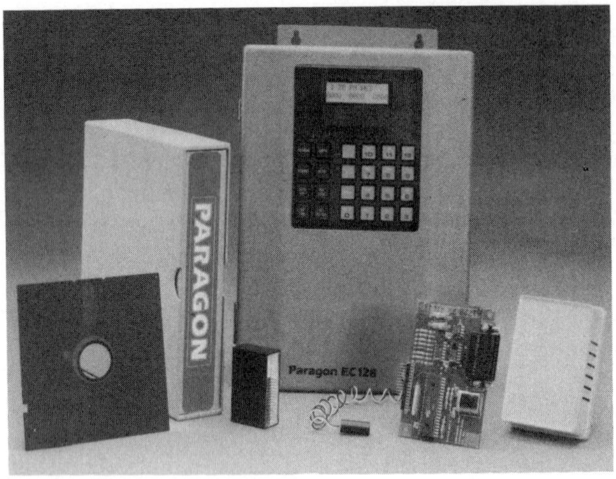

Fig. 16.15 (b) *Programmable controller designed for use as an energy monitoring and control device. The basic unit illustrated accepts up to eight analog or digital inputs, such as illuminance and temperature sensors, kW demand load, time signals, central control signals, and the like. It processes data according to its current program (note programming keypad) and can signal and control up to 12 outputs. When connected in multiplex with similar units the number of inputs and outputs increases accordingly. Accessories illustrated (from left to right) are a floppy disk and drive for program integration with a computer, a memory storage module, a communications module necessary for multiplexing in large systems, and a room-temperature sensor. (Courtesy of Paragon Electric Co.)*

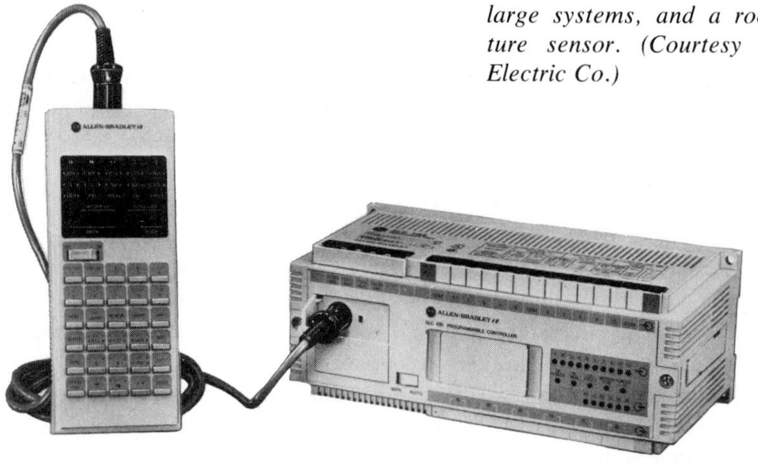

(*c*)

Fig. 16.15 (c) *General-purpose programmable controller. The processor unit contains the central processor (CPU), random access memory (RAM), and power supply. Peripherals include the illustrated hand-held programming unit and EEPROM (electrically erasable programmable read-only memory) module, and various timers, counters, and computer/software interfaces (not shown). The unit illustrated which measures approximately 10 × 9 × 3 in., has 16 I/O circuits and a 885-word memory. It is readily adaptable to a variety of automated control processes in manufacture, building control, and material handling. (Courtesy of Allen-Bradley Co.)*

TABLE 16.6 **Control Equipment Enclosures**

NEMA Designation: Type	Description	Application
1	General purpose	Dry, indoor use
2	Drip-proof	Indoor, subject to dripping
3	Dust-tight, rain-tight, and sleet-resistant	Indoor/outdoor, where subject to windblown dust and water
3R	Rain-proof and sleet-resistant	Outdoor, subject to falling rain, snow, and sleet
3S	Dust-tight, rain-tight, and sleet-proof	Outdoor, subject to windblown water, dust, and sleet; most severe exterior duty
4	Watertight and dust-tight	Indoor/outdoor, subject to water from all directions; not sleet-proof
7–9	Hazardous	Differing in application by class and group of hazardous use; see NEC
12	Industrial use, dust-tight and drip-tight	Indoor only, general use, industrial and other "dirty" environments

fuses of various designs are made up to 6000 A and 600 V.

(b) Circuit Breakers. A circuit breaker is an electromechanical device that performs the same protective function as a fuse and, in addition, acts as a switch. Thus it can be used in lieu of a switch-and-fuse combination to both protect and disconnect a circuit. Most circuit breakers are equipped with both thermal and magnetic trips. The thermal trip acts on overload, whereas the magnetic trip acts on short circuit. Both the thermal and the magnetic action have inverse time characteristics, that is, the heavier the overload, the faster the trip action. Modern circuit breakers in commercial and industrial applications are frequently equipped with electronic tripping control units, which provide fully adjustable overload, short-circuit, and ground-fault protection.

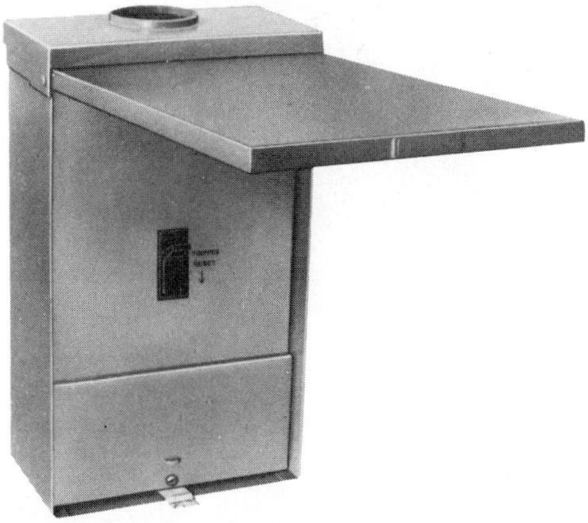

Raintight NEMA 3R circuit breaker enclosure.

Fig. 16.16 *Type 3R outdoor enclosure. This is the type usually intended when "weatherproof" is specified. (Courtesy of General Electric.)*

Air circuit breakers are available in two types: the molded-case breaker and the "large air breaker." Molded case breakers consist of a complete mechanism encased in a molded phenolic case. A light-duty molded case 50-A frame, plug-in circuit breaker, and a special type of circuit breaker designed to protect against ground faults in addition to overloads and short circuits are illustrated in Fig. 16.18. The large air circuit breaker is a more complicated and widely adjustable mechanism and can be used in applications that preclude use of molded case breakers. All breakers can be equipped with remote trip and auxiliary contacts, and all good breakers have trip-indicating handles and are *trip-free* (i.e., will trip out harmlessly if closed in on a short-circuited line). Low-voltage (600-V class)

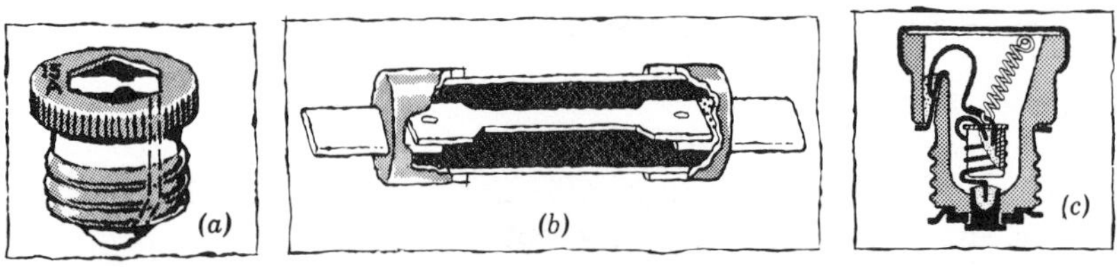

Fig. 16.17 Standard types of fuses are (a) *common household plug fuse;* (b) *single-element, knife-blade cartridge fuse;* (c) *dual-element, time-delay fuse with Edison base. Since fuses are inherently very fast-acting devices, time delay must be built into a fuse to prevent "blowing" on short-time overloads such as those caused by motor starting. A dual-element fuse such as that shown in* (c) *allows the heat generated by temporary overloads to be dissipated in the larger center metal element, preventing fuse blowing. If the overload reaches dangerous proportions, the metal will melt, releasing the spring and opening the circuit. The time to clear (blow) for fuses is inversely proportional to the amount of current. (Courtesy of Bussman Division, Cooper Industries.)*

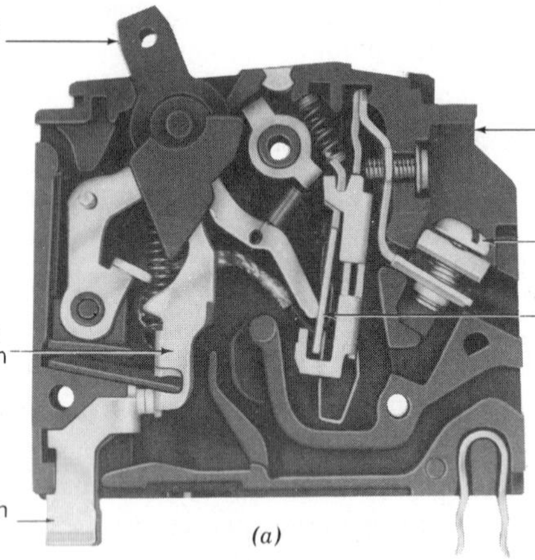

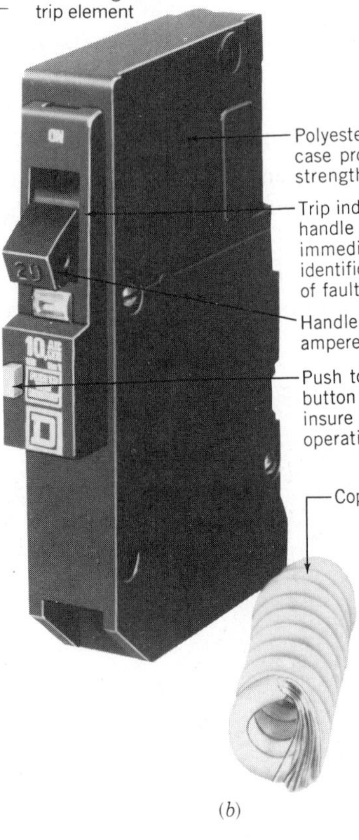

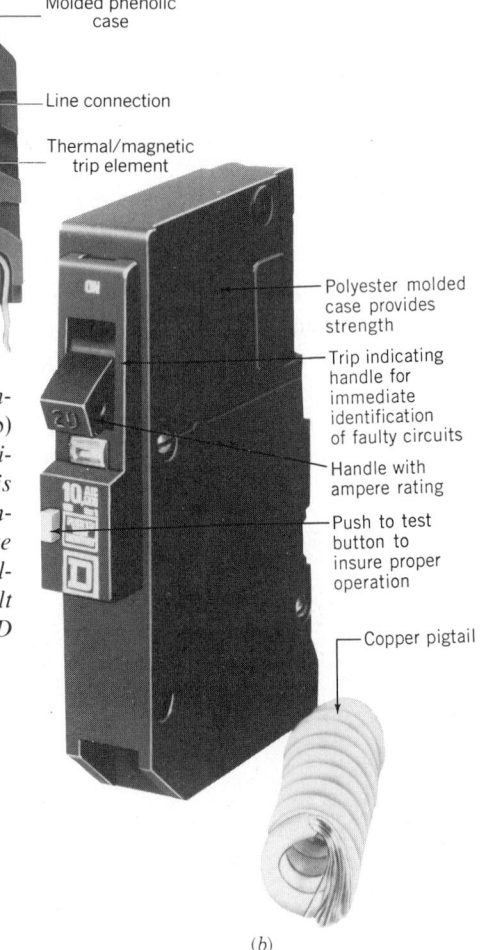

Fig. 16.18 Molded-case circuit breakers. (a) *Conventional single-pole, 50-A frame plug-in type c/b.* (b) *This unit will provide ground-fault protection in addition to functioning as an ordinary circuit breaker. It is designed to fit into the same physical space as a conventional molded-case circuit breaker and therefore can be used as a replacement in an existing panelboard. The units are referred to as GFCI (ground-fault circuit interrupters) or GFI. (Courtesy of Square D Co.)*

circuit breakers are available in frame sizes from 50 A to 2500 A and one to three poles. Characteristics vary widely and are beyond the scope of this work.

(c) Characteristics of Fuses and Circuit Breakers. Although both are circuit-protective devices, their characteristics differ markedly.

Fuses—Switch and Fuse Combination

Advantages	*Disadvantages*
Simple and foolproof	Single pole only
Constant characteristics	Requires switch
Initial economy	Necessity for storage of replacements
Very high IC (interrupting capacity)	Self-destructive (one-time operation)
No maintenance	Nonadjustable
Instantaneous, energy limiting	Nonindicating
	No electric or remote control
	Not trip free

Circuit Breakers

Advantages	*Disadvantages*
Usable as switch	Low to medium IC except for special units
Multipole	Periodic maintenance
No replacement storage	High initial cost
Resettable	Complex construction
Indicates trip	Aging
Trip free	
Remote control	
Adjustable	

The above lists demonstrate that there is no relevance to the oft-posed question "Which are preferable—fuses or breakers?" The answer depends on the specific application involved and is often based on highly technical factors, beyond this book's scope. Generally, breakers are used for all lighting and appliance panels.

16.19 Switchboards and Switchgear

Switchboards and switchgear are freestanding assemblies of switches, fuses, and/or circuit breakers, which normally provide switching and feeder protection to a number of circuits connected to a main source. A switchboard may be represented in a single-line diagram as in Fig. 16.19. It serves in an electrical system to distribute, with adequate protection, large bulk power into smaller "packages." Thus, by hydraulic analogy, the main buswork of the switchboard is equivalent to a main header, the switches to on/off valves, the fuses to flow-limiting devices, and the feeders to subheaders connected to the main header. Modern switchboards (Figs. 16.20 and 16.21) are all dead-front; that is, they have all circuit breakers, switches, fuses, and live parts completely enclosed in a metal structure. The operator controls all devices by means of push buttons and insulated handles in the front panel. When a switchboard has circuit breakers equipped with bayonet-type contacts and mounted in a movable drawer (like the drawers of a standard letter file), they are described as the *drawout* type. This drawout arrangement facilitates emergency replacements, inspection, and repairs.

There is no clear distinction made between the terms *switchboard* and *switchgear*, although often high-voltage equipment (above 600 V) is referred to as switchgear, as is equipment that

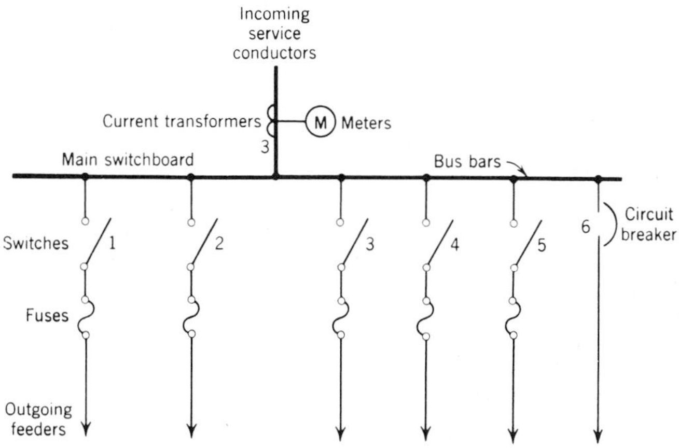

Fig. 16.19 *Typical switchboard. Switches are normally shown in the open position. Switches must be on the line (supply) side of fuses. Each line in a single-line diagram represents a 3-phase circuit. If circuit breakers were used, the entire board would be composed of units as illustrated in circuit 6.*

ELECTRICITY

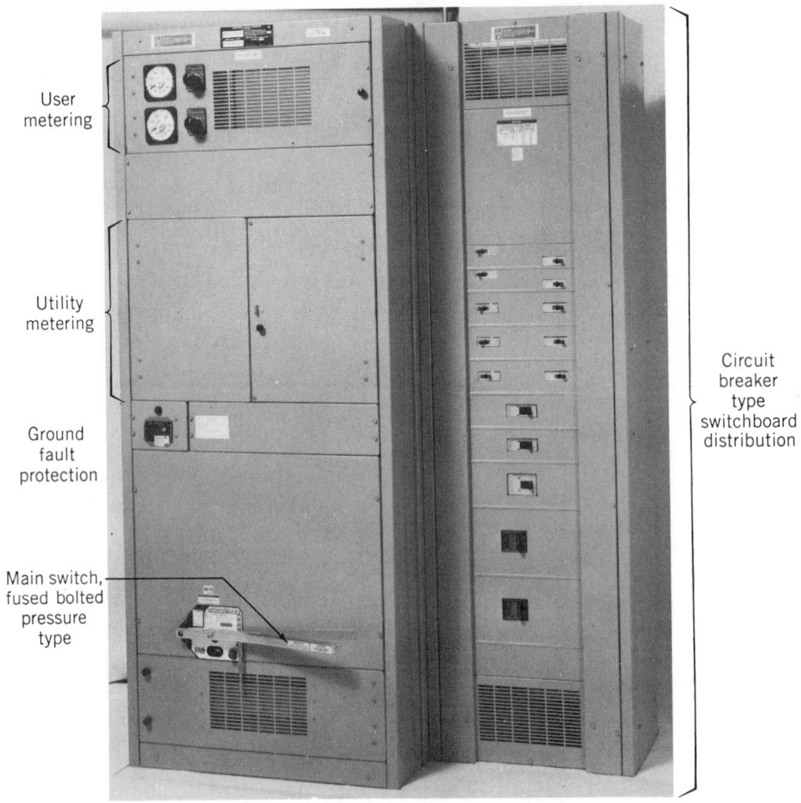

Fig. 16.20 Low-voltage (600-V class) switchboards are available in a number of design configurations. The illustrated unit is a main service switchboard for a small to medium-sized facility. The deeper section contains metering and service equipment rated up to 4000 A and varies in size from 26 to 45 in. wide and 18 to 66 in. deep, depending on equipment rating. The distribution section may contain 2- and 3-pole circuit breakers (or fused switches) rated 70 to 1200 A, and varies in size from 26 to 45 in. wide and 18 to 36 in. deep. Standard height is 90 in. Although the illustrated unit has conventional metering and circuit breakers, solid-state units are available in the same basic construction. (Courtesy of Westinghouse Electric Corp.)

comprises individual units rather than an assembly. When molded case circuit breakers are utilized in a switchboard it is often known as a building-type switchboard. Space requirements for various types of boards are shown in Fig. 16.22. These spaces, as well as all other electrical equipment spaces, must meet the requirements of the NEC. See NEC Article 110-16 for required clearances and access/entrance requirements. Main metal-clad switchgear for commercial, industrial, and public buildings is almost invariably located in the basement, and housed in separate well-ventilated electrical switchgear rooms. The designer must provide adequate lifting hooks, exits, hallways, and hatches for the entrance and exit of the equipment of largest dimensions and weight to be moved. Therefore, the specifications for switchgear should state the maximum number and overall maximum dimensions of sections to be bolted together as one portable section. Two, three, or four sections form the usual practical section. These sections may vary in length from about 8 to 12 ft. Smaller subdistribution switchboards require no special room enclosure, except to bar tampering or vandalism. In such instances a wire screen enclosure plus a large DANGER—HIGH VOLTAGE sign is usually adequate.

When switchgear is to be installed outdoors,

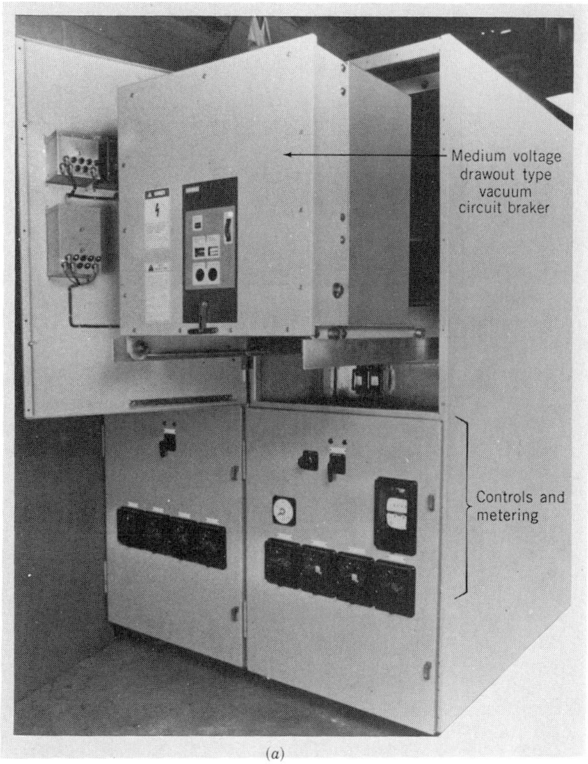

(a)

(b)

Fig. 16.21 (a, b) *Two sections of medium-voltage (15 kV class) metal-clad switchgear. Each section measures 36 in. W × 96 in. H × 94 in. D and can contain two drawout vacuum circuit breakers, rated up to 2000 A each. Due to the drawout characteristic of this gear, a 76-in. aisle in front of the gear is required.* (b) *Indoor switchgear if the type shown in* (a) *can be installed outdoors by using a weatherproof enclosure, as illustrated here. The enclosure is sized to provide sufficient front-aisle space to permit the drawout function of the gear. (Courtesy of Siemens.)*

Fig. 16.21 (c) *Four-bay (section) power-operated high-voltage service entrance switchgear rated 13.8 kV, consisting of (see single-line diagram inset, and reading left to right) a fused feeder section, two bays with parallel incoming lines and automatic transfer in the event of power failure, and a second switch and fuse feeder section. Fuses are electronic with circuitry designed to provide a trip signal when excessive current is sensed rather than simply relying on the thermal characteristic of a fusible element. This gear, which measures 93 in. H, 172 in. W, and 44 in. D and weighs 8000 lb, is suitable for interior or exterior installation. (Courtesy of S&C Electric Co.)*

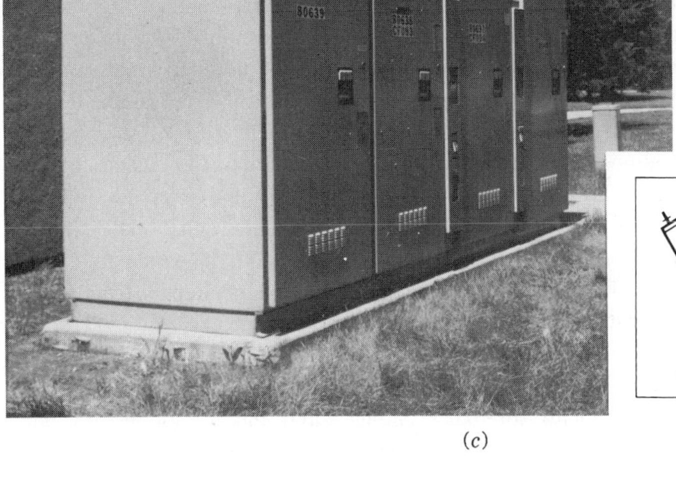

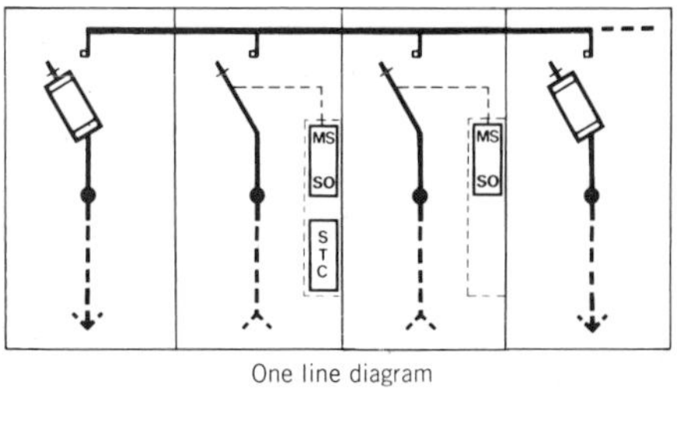

One line diagram

(c)

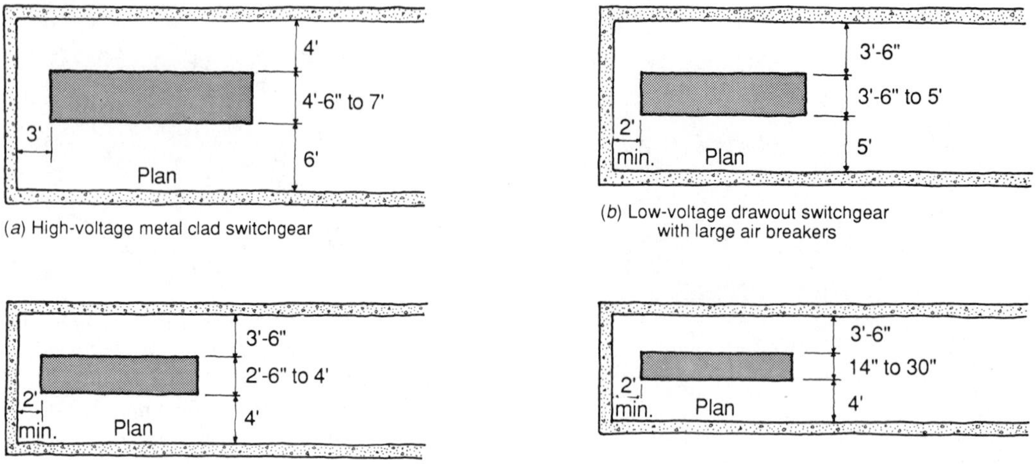

(*a*) High-voltage metal clad switchgear

(*b*) Low-voltage drawout switchgear with large air breakers

(*c*) Fixed low-voltage switchgear with large air breakers

(*d*) Building-type switchboard with molded case breakers

Fig. 16.22 *Minimum switchgear space requirements. Clearance to ceiling should not be less than 3 ft. Each room should have two doors when switchgear is connected to high-capacity systems. Switchgear is shown here in plan view.*

three methods may be employed: build a small house to enclose normal indoor gear, utilize weatherproof outdoor gear, or utilize switchgear which is built into its own metal house. Integral housings are equipped with heat and light and often prove the most economical choice.

16.20 Unit Substations (Transformer Load Centers)

An assembly of primary switch-and-fuse or breaker, step-down transformer, meters, controls, buswork, and secondary switchgear is called a unit substation, or a load-center substation. It is available for indoor or outdoor use, to supply power from a primary voltage line to any large facility. The location in the building of the unit substation is governed by the type of transformer utilized, as explained above in the discussion on indoor transformer installations. For this reason, almost all indoor unit substations utilize dry-type (air-filled) transformers. Unit substations are utilized to effect the economies inherent in prefabricated construction with coordinated components. A basement location is most often selected, with ventilation requirements as detailed above. Access should be restricted to authorized persons only. A typical unit is illustrated in Fig. 16.23.

16.21 Panelboards

A panel, or panelboard, serves basically the same function as a switchboard, except on a smaller scale; that is, it accepts a relatively large block of power and distributes it in smaller blocks. Like the switchboard, it comprises main buses to which are connected circuit-protective devices (breakers or fuses), which feed smaller circuits. The panelboard level of the system is usually the final distribution point, feeding out to the branch circuits that contain the electrical utilization apparatus and devices, such as lighting, motors, and so on.

The panel components—that is, the buses, breakers, and so on—are mounted on an insulating board that in turn is mounted inside an enclosing cabinet (see Fig. 16.24*a*). The line terminal of each circuit-protective device (breaker or fused switch) is connected to the busbars of the panelboard. The load terminal of the device then feeds the outgoing branch circuit. This is shown schematically in Fig. 16.24*b*. The busbars of the panelboard are energized by a feeder from some switchboard or load center.

Panelboards are described and specified by type, bus arrangement, branch breakers, main breaker, voltage, and mounting—though not necessarily in that order. A typical description

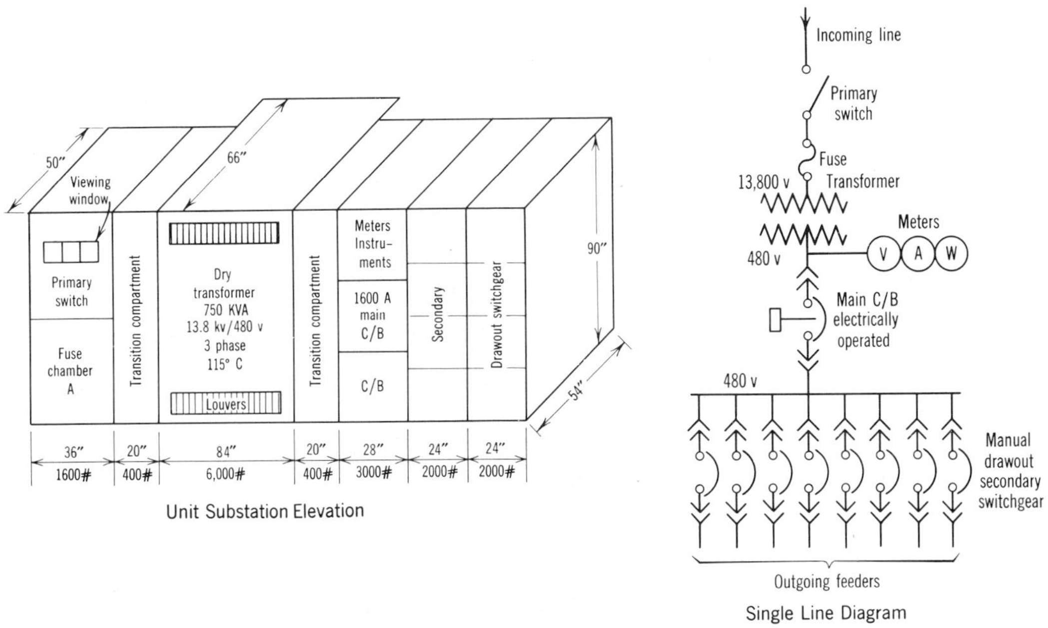

Fig. 16.23 Approximate sizes and weights of a typical large single-ended unit substation. Such a unit would supply a building with a maximum demand of 750 kVA. The incoming 13,800-V cables enter cubicle A and connect to the switch and fuses. The load side of the fuses connects to the transformer, which in turn connects to the secondary switchgear. This main secondary switchgear then feeds various switchboards and panelboards distributed through the building.

might be: Lighting and Appliance Panel, 3-phase, 4-wire; 200A mains; main c/b, 225A frame, 150A trip. Branch breakers—all 100A frame; 8 ea. SP-20A, 4 ea. 2P-20A, 4 ea. 3P-20A; flush with hinged locked door.

16.22 Electric Motors

Motors are very frequently supplied as adjuncts to specified driven equipment, such as fans, blowers, and so on, within the constraints of the voltage and enclosure. The actual choice of motor is left to the driven-equipment supplier, it being maintained that he is best qualified to select a motor that will optimally match the driven-equipment requirements and so supply a working whole for whose proper operation he is responsible. The supplier, however, is frequently guided primarily by the price motive. The specifier therefore should be sufficiently knowledgeable so that, within the stated constraints of applicability to the load, he can specify the particular motor desired. The following paragraphs are written with that purpose in mind and therefore concentrate on application data.

(a) Direct-Current Motors. As a result of the high cost and relative rarity of direct current, these motors are used only where continuous fine speed control is required, as in the case of elevator drives.

(b) Alternating-Current Motors. These motors fall into three general classifications: polyphase induction motors, polyphase synchronous motors, and single-phase motors. Within these categories there are further subdivisions. Of these many types, the vast majority are squirrel-cage induction machines; therefore, this type will be studied in some detail.

(c) Squirrel-Cage Induction Motors. This motor type owes its interesting name to an early design in which the rotor consisted of a group of bars welded together into a cylindrical cage-type shape. The design is basically unchanged today except for refinements. Squirrel-

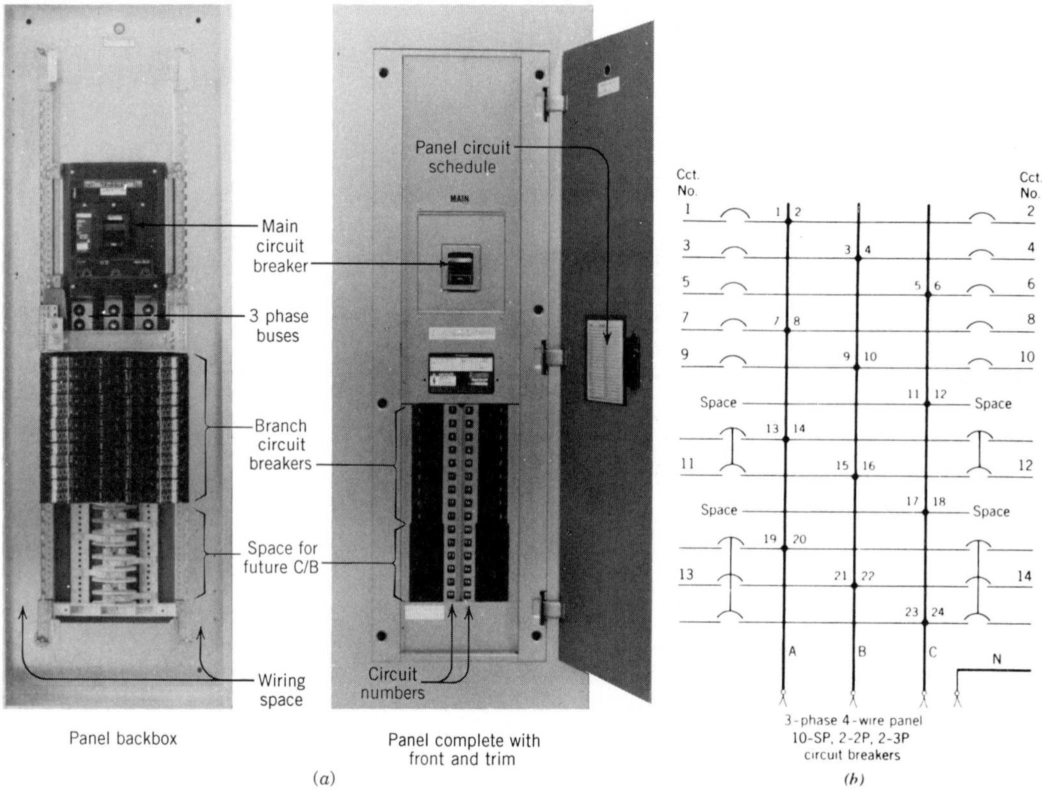

Fig. 16.24 (a) *Panelboards may be of the circuit breaker or switch-and-fuse type. The illustrated unit is a lighting and appliance panelboard by NEC definition (Article 384-14), and can contain 1-, 2-, and 3-pole circuit breakers. Such panels average approximately 6 in. D, 20 to 30 in. W, and a maximum of 62 in. H for a 42-pole panel. The 30-pole panel illustrated measures 20 in. W × 60 in. H × 5¾ in. D. (Courtesy of Siemens Energy & Automation, Inc.)*

Fig. 16.24 (b) *Typical schematic diagram for a panel. Note that single-pole, 2-pole, and 3-pole circuit breakers are connected to two and three buses respectively. With the neutral wire, they supply 120 V two-wire, 120/208 V 3-wire, and 120/208 V 3-phase 4-wire, respectively. See Figs. 17.1, 17.2, and 17.4.*

cage motors are manufactured in four different NEMA designs to meet different application requirements. Of these the most common are:

Type B—standard design, high efficiency and power factor, normal torque. Applicable to fans, blowers, and pumps

Type C—high starting torque, fair efficiency and power factor. Applicable to compressors, conveyors, and other devices that start under load.

A motor nameplate gives important data on the motor that are not self-evident:

1. *Type*—this is the manufacturer's designation and indicates primarily the enclosure. Common enclosures are open drip proof, totally enclosed, fan cooled, and weather-protected.
2. *Duty*—continuous or intermittent.
3. *Service factor*—permissible overload, generally 15%.
4. *kVA code*—indicates by a letter the maximum starting current per horsepower. This is useful in selecting motor protective devices.
5. *Frame*—a NEMA standard number that

indicates the motor's physical dimensions.

6. *Motor voltage*—the standard motor voltages are 208, 230/460, and 575 V. Induction motors will generally operate satisfactorily ±10% voltage. Only 208-V motors should be used on 208-V systems, since actual line voltage may be as low as 200 V. Using a 230-V motor will result in sharply reduced torque and increased temperature rise, and poor overload capacity.
7. Motor nominal full-load efficiency—a requirement of ASHRAE Standard 90.1-1989.

(d) Electric Motor Energy Considerations. Data compiled by the Federal Energy Administration indicate that more than one-half of all the electric energy generated in the United States is used to power electric motors. Thus an easily attainable overall decrease in motor circuit power of only 1% would result in an energy savings of 3+ billion kWh or 5+ million barrels of oil annually.

Energy losses associated with electric motors stem from two sources:

1. Losses in the motor itself, as indicated by the motor's efficiency.
2. Line losses caused by the motor's power factor.

Let us consider the latter item first.

The power factor (pf) of a device is an indication of the amount of reactive current that the device requires—the lower the power factor, the larger the reactive current. Reactive current does not draw power and does not therefore show up on the kilowatt-hour meter or the kilowatt-demand meter. It *does*, however, cause I^2R line (power) losses throughout the distribution system, and it does tie up generating capacity with no financial return to the utility. Recognizing this, utility tariffs almost invariably contain a low-power-factor penalty clause. This penalty revenue compensates the utility but does nothing to diminish the energy wasted in line losses. It is specifically this energy waste that a number of proposed and enacted, state and federal regulations aim to reduce. They generally state that utilization equipment rated greater than 1000 W and lighting equipment greater than 15 W, with an inductive reactance load component, shall have a power factor of not less than 85% under rated load conditions.

To gain an appreciation of the magnitude of losses involved, consider a small industrial facility with a continuous load of 100 kW at 60% power factor. This low figure is readily realized with low wattage HID lighting, and single-phase motors whose power factor is inherently low (of the order of 40 to 50%).

Consider now the effect of improving the power factor to 85%:

	60% pf	*85% pf*
Load	100 kW	100 kW
kVA (kW/pf)	166.67	117.65
Current (three phase, 208 V)	462 A	326 A

Losses are proportional to I^2. Assuming a 5% loss in the utility lines and another 5% loss in the building distribution system, the losses at the improved power factor are reduced to

$$\text{loss at 85\% pf} = 10\text{ kW}\left(\frac{326}{462}\right)^2 = 5\text{ kW}$$

That is, the loss is cut in half. This reduction is far in excess of the minimal 1% referred to above. Assuming an 8-h day, 5-day week, this amounts to 10,400 kWh/year, which is approximately twice an average residence's annual usage. Monetarily, the improvement is no gain to the utility, which loses its surcharge. For the industrial facility, the outlay needed for capacitors and other equipment (see below) to effect this improvement will generally pay for itself within a very few years, depending on local energy rates, electrical installation costs, and other factors. For the economy, however, the energy savings is pronounced.

Most motor manufacturers now produce a line of high power factor, high-efficiency motors. Using these in lieu of standard motors with power factor correction equipment will frequently reduce the pay-back period (refer to Fig. 16.25). Note the pronounced negative effect on motor efficiency and power factor that results from operation at partial load. To avoid this, the designer should match the motor as closely as

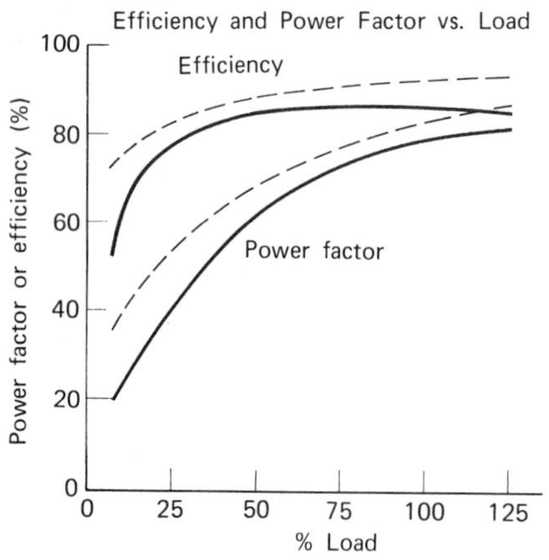

Fig. 16.25 Comparison of the efficiency and power factor characteristics of standard (solid line) and energy-efficient (dashed line) motors. Note that not only is full-load performance improved, but perhaps more important, the partial-load characteristic is much improved. (The data shown are based on industry averages and not any specific manufacturer.)

possible to the load requirement and avoid the oversizing so common in practice. This will do much to improve power factor and efficiency and concomitantly to conserve energy. Where this is impractical because of variable loads, the designer has two choices:

1. Use a standard motor, with an electronic control device that matches input to load. These devices, known as power factor controllers (*not* capacitors) act to reduce input when the motor is only partially loaded, thus reducing electrical and mechanical losses and increasing power factor.
2. Use an energy-efficient motor that has a better load versus efficiency and power factor curve than a standard machine.

ASHRAE Standard 90.1-1989 specifies minimum motor efficiency for permanently wired single-speed NEMA design B squirrel-cage induction motors, of common speeds and enclosure designs. (ASHRAE 90.1 Section 5.4.3.) The required efficiencies are listed in Table 16.7. A further requirement of the standard calls for the nominal full-load motor efficiency to appear on the motor nameplate.

With respect to motor efficiency itself, unlike power factor, any improvement is reflected in the facility's electric energy bill, in addition to effecting a savings in fuel. Unlike pf, efficiency cannot be improved by the addition of a device external to the motor, since it is inherent in the motor design.

Thus, for an *existing* facility a careful cost analysis would have to be made to determine whether replacing existing machines with

TABLE 16.7 **ASHRAE 90.1—(Table 5-1): Minimum Acceptable Nominal Full-Load Motor Efficiency for Single-Speed Polyphase Motors**[a]

hp	*Minimum Rated Efficiency, Percent*	*1992 Minimum Rated Efficiency, Percent*
1–4	77.0	78.5
5–9	82.5	84.0
10–19	84.0	85.5
20–49	87.5	88.5
50–99	89.5	90.2
100–124	91.0	91.7
125 or greater	91.7	92.4

Source: Reprinted with permission from ASHRAE/IES 90.1-1989; copyright © by American Society of Heating, Refrigeration, and Air-Conditioning Engineers.

[a]Motors operating more than 750 h per year are likely to be cost-effective with efficiencies greater than those listed. The more efficient motors are classified by most manufacturers as "high-efficiency" and are presently available for common applications with typical nominal efficiencies of: 5 hp, 89.5%; 10 hp, 91.0%, 50 hp, 94.1%; 100 hp, 95.1%; 200 hp, 96.2%. Guidance for evaluating the cost-effectiveness of high-efficiency motor applications is given in NEMA MG 10-1983.

TABLE 16.8 **Life-Cycle Cost Comparison for 5-hp Motor**

Motor Type	*Conventional Motor*	*High-Efficiency Motor*
Efficiency at full load	84%	89.5%
Power output	3730 W	3730 W
Power input	4.440 kW	4.168 kW
kwH per year (12 h/day, 5 day/wk, 52 wk/yr)	13,853	13,004
Annual kWh differential	849 kWh	
20-yr kW differential	16,980 kWh	
Annual energy cost at $.07 per kilowatt-hour	$970	$910
Initial motor cost	$310	$380
Motor life	10 yr	20 yr
Motor amortization cost	$22	$8
Total annual cost	$992	$918
Motor cost pay-back period	11 months	
20-yr life-cycle energy cost (8% capital cost, 3% annual price escalation)	$15,167	$14,229
Total 20-yr life-cycle cost	$15,510	$14,354

newer, more expensive, high-efficiency machines is justifiable. For a new facility, the payback time for the cost differential of the high-efficiency machines is short. An added advantage of these machines is that the lower losses result in lower temperature operation and therefore extended life. A life-cycle cost analysis of the type given in Tables 15.5 and 16.4 should be performed before purchase of motors. The results will probably indicate a distinct advantage to the high-efficiency designs. A typical analysis of a 5-hp machine is given in Table 16.8, based on the published manufacturer's data.

16.23 Motor Control

Since d-c controllers are highly specialized and infrequently encountered, we will confine our remarks to a-c motor control. A conventional a-c motor controller is basically a contactor (see Section 16.14) designed to handle the heavy in-rush currents encountered in an a-c motor starting. Its function is twofold: to start and stop the motor and to protect the machine from overload. These two separate and distinct functions are accomplished by combining a set of contacts for on/off control with a set of thermal overload elements for overload protection, in a single unit. When the contacts are operated by hand, the controller is called a manual starter; when the contacts are operated by a magnetic coil controlled by push buttons, thermostats, or other devices, the unit is known as a magnetic controller or simply and more commonly—a starter.

Recently electronic motor controllers have appeared on the market that not only provide the starting and protective functions of a conventional starter, but can also provide "soft" starts (i.e., starts that minimize the mechanical stresses caused by rapid application of accelerating torque). Additional advantages of these units are reduced size and weight, long life, more sophisticated motor protection, and additional operating functions such as jogging and reversal. Disadvantages are higher cost and possible radio frequency noise problems. Motors 1 hp or less are generally controlled by a manual switch that contains an overload protection device. It is advisable to utilize such a device for all fractional horsepower motors.

Starters are available in various sizes, voltages, and NEMA enclosures. Table 16.9 gives typical electrical and physical data for full-voltage conventional starters. Most starters are of the full-voltage across-the-line type; that is, the contacts place the motor directly onto the line and the motor starts up immediately. When such a procedure is undesirable because of voltage dip and flicker caused by the large inrush current or because of utility company limitations, a reduced voltage starter, sometimes called a compensator, is used. These units apply reduced voltage to the motor and thus reduce starting inrush current. Every motor controller is required by the NEC to have a disconnecting means within sight of the controller. Where convenient, this disconnect switch may be combined with the starter into a single unit, known as a combination starter. A circuit breaker or fused switch is often used in such an arrangement, which then constitutes the branch circuit protection and disconnecting means (see Fig. 16.26).

When starters are to be assembled for a group of motors, the motor starters, disconnect switches, motor controls, and indicating devices may be combined into a single large assembly for convenience and economy. Such an assembly is called a motor control center. Two typical motor control center construction types are shown in Fig. 16.27. A typical, brief description of a conventional motor controller would be similar to the following:

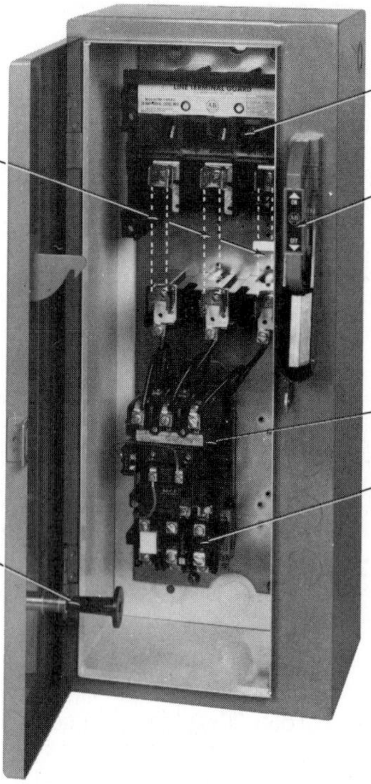

Fig. 16.26 *Interior of a combination fused-switch type, across-the-line motor controller. Note that the unit is essentially a switch and a starter wired together and installed in a single cabinet. (Courtesy of Allen-Bradley Co.)*

Combination circuit-breaker type, across-the-line motor controller, NEMA size 2, three O.L. elements, 208 V, in a NEMA 1 enclosure. Starter shall contain integral on/off push buttons.

TABLE 16.9 **Rating and Approximate Dimensions of a-c Full-Voltage Conventional Single-Speed Motor Controllers, Phase Combination Circuit Breaker Type**[a]

NEMA Size[b] *Designation*	*Maximum Horsepower*[b]	*Width*	*Height*	*Depth*
		Inches		
0	3	10	24	7
1	7½	10	24	7
2	15	10	24	7
3	30	20	24	9
4	50	20	48	9
5	100	20	56	11

[a]All starters are housed in a NEMA 1 indoor ventilated enclosure.

[b]Maximum hp that can be controlled at 208–230 V. Generally, when operating at 460 V, a starter one size smaller can be used.

Fig. 16.27 Typical low-voltage motor control centers. Back-to-back construction (b) *is space saving and adds only 5 in. to the basic 15 in. depth. Units are normally 90 in. high and 20 in. wide per section, although MCCs with especially large components are deeper and wide. [*(a) *Courtesy of Allen-Bradley Co.;* (b) *courtesy of Furnas.]*

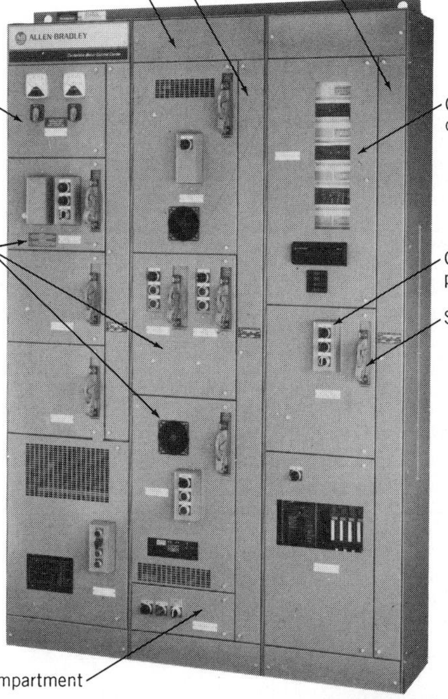

(*a*)

16.24 Wiring Devices

The general term *wiring devices* includes all devices that are normally installed in wall outlet boxes, including receptacles, switches, dimmers, and pilot lights. Attachment plugs, also called *caps*, and wall plates are also included in any discussion of wiring services. Devices are classified (by manufacturers) in descending order of quality, as premium or specification grade, intermediate grade, and standard or economy grade. These classifications are qualitative only, and one company's specification grade may only equal another's intermediate grade. To assure a certain quality level, NEMA and/or UL standards must be specified. In application, specification grade equipment is usually used in industrial and good commercial construction, intermediate grade in most educational and good residential buildings, and standard grade in low-cost construction of all types. The grade of wiring devices should, as with all electrical equipment, be consistent with the quality of construction in the entire facility. Wiring devices intended for hospital use are constructed to withstand heavy abuse. They are identified by a green dot on the device face, and must meet requirements of the UL standard for hospital-grade devices.

a. Control Unit
b. Vertical Wireway
d. Top Horizontal Wireway
e. Bottom Horizontal Wireway

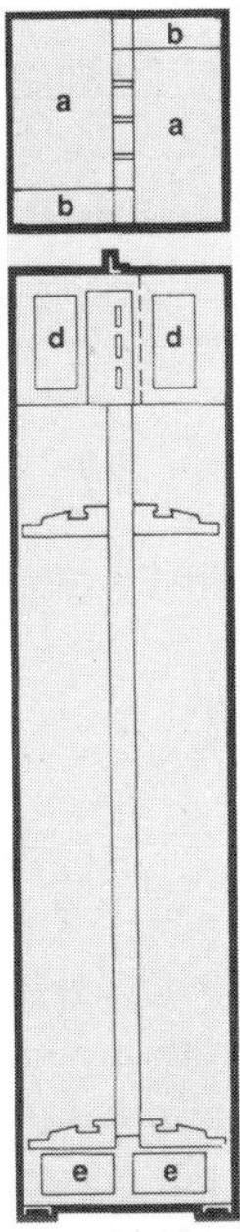

20" Back to back
Side view

(*b*-1)

(*b*-2)

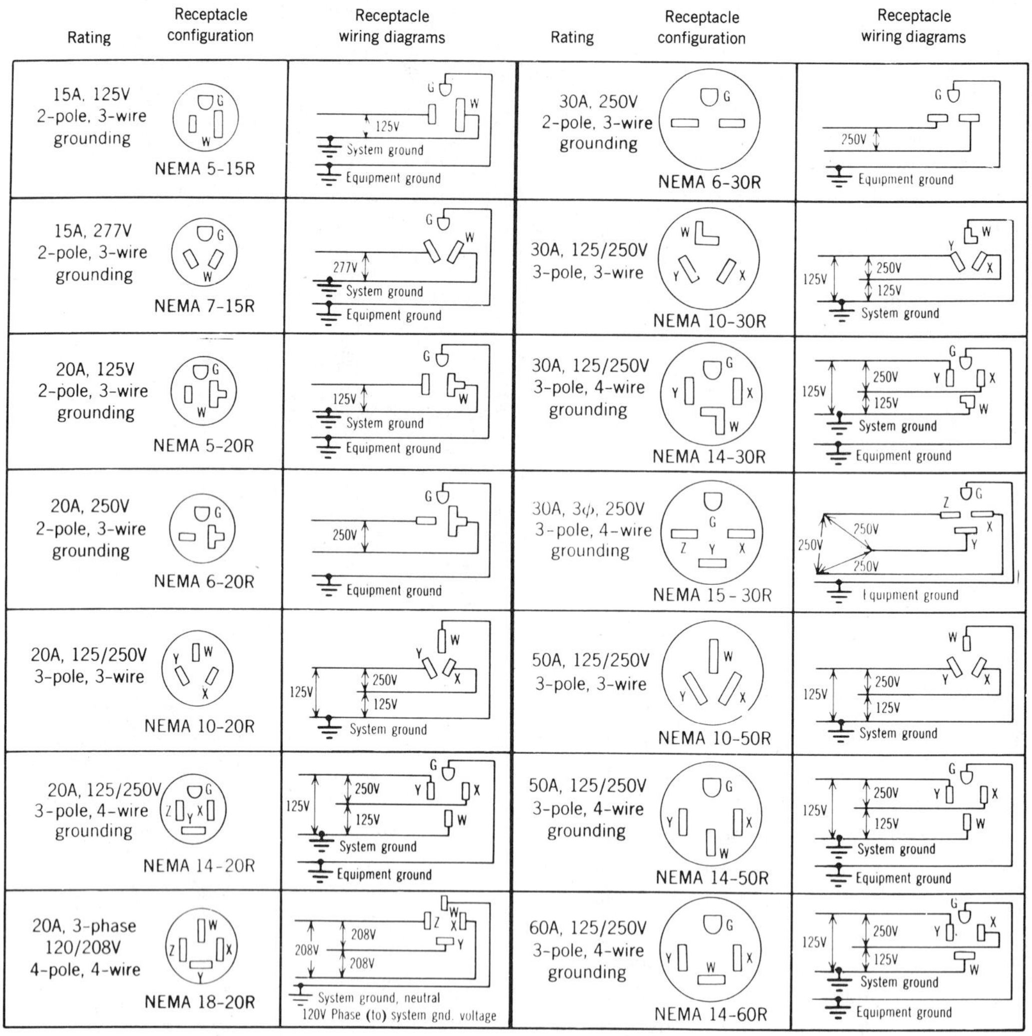

Note: "W" denotes white system ground or neutral
"G" denotes green equipment ground.

***Fig. 16.28** Receptacle configuration chart of selected common general-purpose, nonlocking devices with related NEMA designations.*

16.25 Receptacles

Receptacles are identified by number of poles and wires, and whether or not the device is designed for connection of a separate grounding wire. The grounding pole *is not counted* in the number of poles (see Fig. 16.28), but is counted in "wires." The grounding pole is connected to the green ground wire where this is run or to the conduit system where a ground wire is not run. The equipment grounding pole must not be confused with the system ground (neutral), nor must the wiring for the two be interchanged (see Section 17.3). In a typical application, a receptacle for an electric dryer with 4800 W, 208 V heating element, and ⅙ hp, 115 V motor would be NEMA 14-50R (see Fig. 16.28). The motor would connect across W and X, the heater across X and Y, and the appliance case to G.

Receptacles must be of the grounding type where installed on standard 15- or 20-A branch circuits. Receptacles connected to different voltages, frequencies, or current type (a-c or d-c) on the same premises must be polarized so that attachment plugs are not interchangeable. Figure 16.28 shows some of the standard receptacle configurations and their ratings.

Receptacles are regularly available from 10 to 400 A, 2 to 4 poles, and 125 to 600 V. In addition, special types are made such as locking and explosion proof, and units with decorative design. Also, specific usage units are available such as range receptacles. In addition to these types, several manufacturers produce a receptacle with built-in ground-fault protection, as illustrated in Fig. 16.29. For a discussion of these GFI (ground-fault interrupter) receptacles, see Section 17.4. All receptacles other than the normal 15/20 A, 3-wire, parallel-slot type should be specified to be furnished with the required number of matching caps (plugs).

Receptacles are normally mounted vertically between 12 and 18 in. from the floor, except that in shops, labs, and other areas, where tables are used against the walls, 42 in. is the usual mounting height. A typical receptacle specification would be: Receptacle, duplex, 2-pole, 3-wire, grounding type, 20 A, 250 V, specification grade, for indoor use.

Fig. 16.29 *Where it is desired to provide ground-fault protection at a single outlet only, and thereby localize power interruption (rather than using a GFCI circuit breaker on the entire circuit), a receptacle containing a ground-fault interrupter, as illustrated, may be used. This unit is also provided with an indicator light (lower right) to show whether the receptacle is energized. A test button is provided to permit periodic testing and a reset button reenergizes the receptacle after it has tripped on ground fault, and the fault has been cleared. (Courtesy of Leviton.)*

16.26 Switch Devices

Usually switches up to 30 A that can be outlet-box mounted fall into this category. Generally "a-c only" switches are preferable to the a-c/d-c type because of better construction. The usual a-c switch rating is 15, 20, or 30 A at 120 or 120/277 V. Normal constructions are single-pole, 2-pole, 3-way, 4-way, momentary-contact, 2-circuit, maintained-contact SPDT and DPDT. Operating handles are toggle-type, key, push, touch, rocker, rotary, and tap-plate types. The mercury and a-c quiet types are relatively noiseless; the toggle, tumbler, and a-c/d-c types are generally not. A typical switch specification would be: Switch, single-pole, a-c, quiet type, specification grade, 15 A, 125 V, with press handle lighted when OFF, suitable for back or side wiring.

A switch incorporating a solid-state rectifier is readily available, which will give high/off/low control for *incandescent lamps*, and costs very little more than an ordinary switch. Typical applications are areas where a lower illumination level is often acceptable and always desirable as an energy-conserving measure. In high-security areas, where the easily defeated normal key switch is inadequate, a tumbler lock controlled unit can be used. Loads that can be timed-out such as bathroom heaters and ventilating fans, can be controlled by a spring-wound timer as illustrated in Fig. 16.30. (A solid-state switch with preset time delay is shown in Fig. 21.3.) Other common switches are also illustrated in Fig. 16.30.

Thanks to miniature electronics, a programmable switch is available that fits into a wall outlet box in lieu of an ordinary switch. The unit acts as a solid state 15-A switch, and can be readily programmed to switch the controlled circuit or device at preset times (see Fig. 21.4).

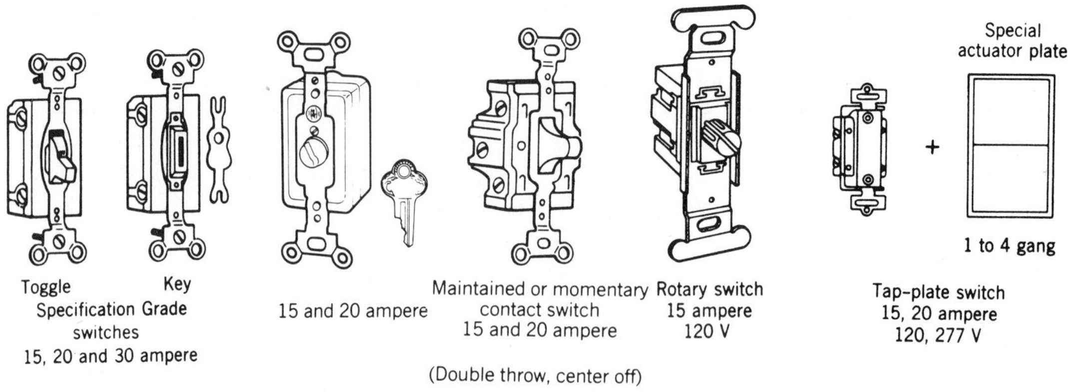

Lighting control, including dimming, is covered in Sections 21.2 and 21.3.

16.27 Low-Voltage Switching

The switches illustrated and discussed above are all full-voltage types; that is, they are wired directly into the load circuit and operate at line voltage and full current. A system of control that uses light-duty, low-voltage (24 V) switches to control line voltage relays, which in turn do the actual circuit switching, is illustrated in the wiring diagrams in Figs. 16.31 to 16.33. The system is variously referred to as low-voltage switching, remote control switching, and low-voltage control.

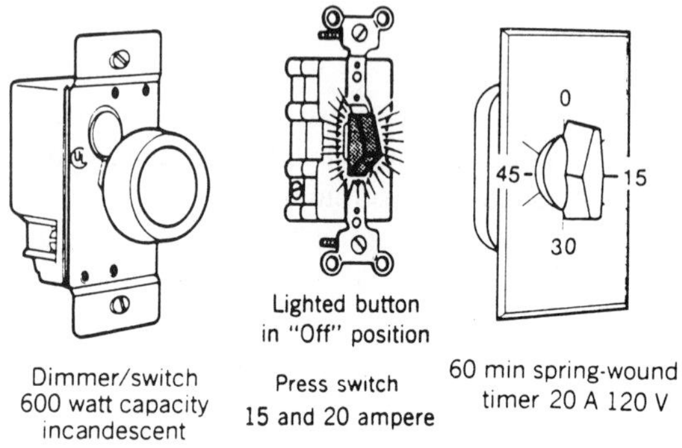

Fig. 16.30 Typical branch circuit switch types.

The basic control scheme is illustrated in Figs. 16.31 and 16.32. Since loads are relay controlled, any number of switches and control devices can be wired in parallel to operate the re-

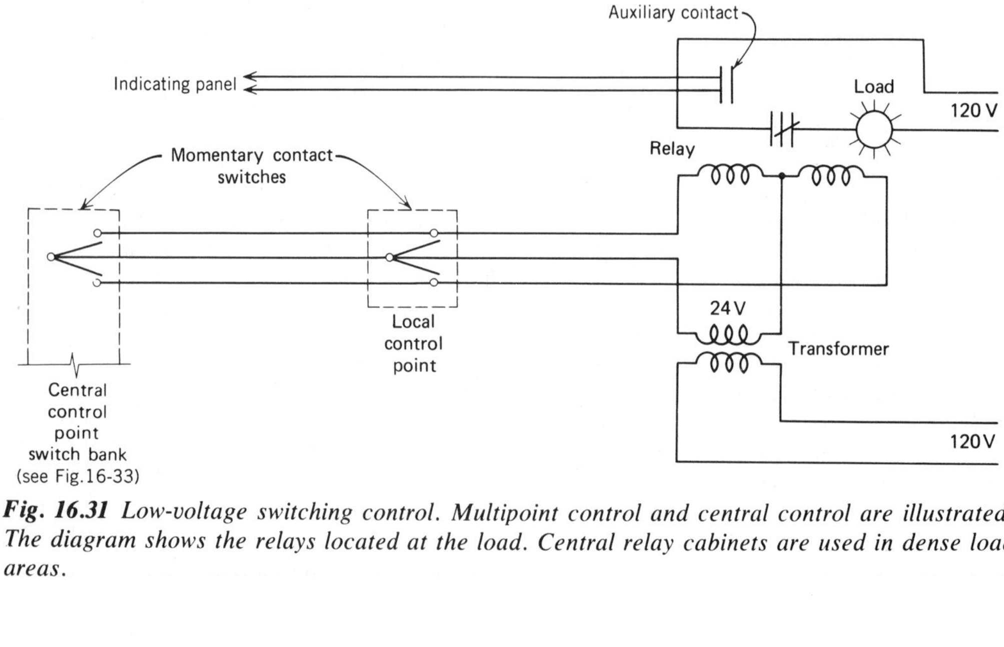

Fig. 16.31 Low-voltage switching control. Multipoint control and central control are illustrated. The diagram shows the relays located at the load. Central relay cabinets are used in dense load areas.

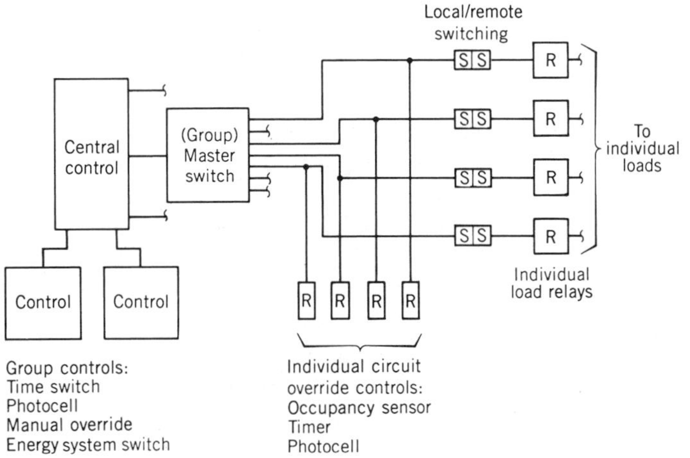

Fig. 16.32 *Schematic arrangement of a large low-voltage switching system. Individual load can be switch controlled from multiple locations, including a group (master) control point. In addition, local control devices such as photocells and occupancy sensors can override the local switches, to maximize energy conservation. Central (group) control devices perform the same override function at the group level.*

lay and thereby effect the switching. Thus the control can be local, remote, and master, with individual override by local control devices such as occupancy sensors and central control or override at the central controller by timers, daylight controllers, and so on. The advantages of this system of control over full-voltage switching are therefore.

1. Flexibility of control location.
2. Individual load override by local control devices (occupancy sensors, photocells; see Section 21.4).
3. Group load override by central control devices such as timers, daylight controllers, and energy management systems.
4. Low cost of low-voltage, low-current wiring.
5. Inherent system flexibility and simplification of alterations.
6. Monitoring of the status of individual loads at a centralized control panel via use of loads relay with auxiliary contacts (see Fig. 16.31).

(a)

(b)

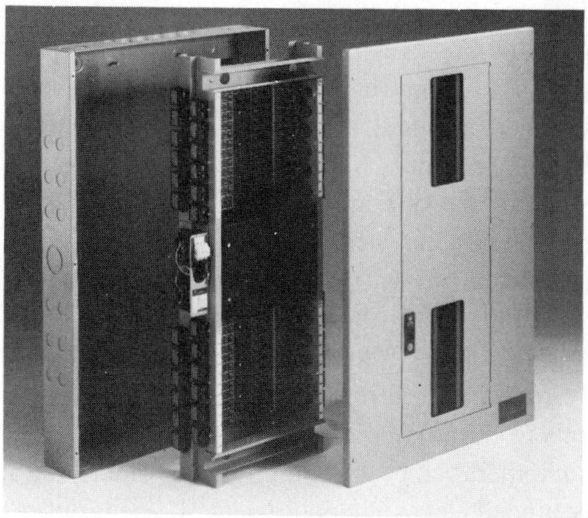

(c)

Fig. 16.33 *Components of a low-voltage switching and control system for lighting and other electric loads.* (a) *The basic switching device is a single-pole double-throw latching relay with a 24-V coil and 20-A 120- to 277-V contacts.* (b) *Low-voltage (24-V) wall switches can be mounted singly or ganged in groups to two to eight switches. Switches are available in single-button, two-button, rocker, key-operated, and lighted designs. Up to 48 relays can be centrally mounted in a single panel* (c) *with power supply and LED status lights. Control wiring from the panel extends to all of the control devices (switches, timing devices, automatic controls), and to the controlled loads (lighting, small motors, receptacles). (Courtesy of General Electric Wiring Devices.)*

ELECTRICITY

Points 2 and 3 are particularly important from the energy conservation point of view, as it has been amply demonstrated that reliance on manual control for energy usage control yields disappointing results. Where loads are concentrated, such as in commercial installations, relays are usually mounted in relay panels, along with the 24-V power supply. In residential application where loads are spread out, relays can be mounted adjacent to the controlled load.

Although the system described above is completely electromechanical, its flexibility can be enhanced by the addition of solid-state programmable switches and automation panels. Such a system has the additional advantage that any switch can be arranged to control any group of relays and the local *and* remote programming of all loads is possible. Fully programmable lighting (and other load) control systems are described in Section 21.3.

16.28 Power Line Carrier Systems

Although the advantages inherent in flexible, programmable load switching systems such as those described above are undeniable, particularly in connection with energy management systems, their high first cost frequently makes the entire project uneconomical. This is particularly true in retrofitting large complex facilities with energy management controls, where the cost of hard-wiring, even when using low-voltage control wires, can amount to more than half the total cost of the project. As a result of the need to minimize wiring costs, a system of control signal transmission that uses the electric power wiring (either new or existing) as its conductors was developed, thus obviating the need, and expense, of dedicated control wiring. The system is called a *p*ower *l*ine *c*arrier (PLC) system since the power line conducts the carrier control signal. An ancillary benefit of such an arrangement is that not only can it be added to an existing system, but it can also be removed, thus making possible and economical the leasing of energy management control systems.

The system operates by injecting into the power wiring a series of low-voltage high-frequency binary-coded control signals which then disperse over the entire power network. Only a receiver that is "tuned" to receive a particular code will react, generally by operating a local relay, which in turn connects or disconnects a particular load. In practice, the control signal generator can be a small manually programmed controller (Fig. 16.34) as in a residence, or a computer-operated energy management or lighting controller in a commercial facility. The receivers are normally of three types, most of which are designed to fit into an ordinary wiring device box (see Section 16.29):

1. A wall switch module, which acts as a normal manual wall light switch but also contains a coded receiver and fully rated (20 A, 125 V, or 277 V) relay.
2. A wall receptacle module, which contains a 20-A receptacle along with a receiver and relay.
3. A switching module, which can be connected to activate contactors, small motors, and the like.

Larger, multiple-circuit switching modules are also available for more extensive systems. Each miniature receiver can be manually tuned in the field [i.e., its code can be set (and reset) readily], giving the entire system extreme flexibility.

The system operates in the same manner as described above for a low-voltage switching system or for an energy management or lighting control system, with the essential difference that no dedicated wiring is required. In retrofit jobs, existing wall switches and receptacle outlets are replaced with receiver-controlled units; in new installations these devices are installed initially. Due to the fact that the control signals travel over the power wiring and are therefore attenuated by high-resistance connections, poor grounds, and faulty wire insulation, PLC installations require a high-quality power wiring installation to operate properly. In retrofit work preliminary tests occasionally indicate the impracticability of a PLC installation because of the condition of the power wiring system. Another problem frequently encountered in PLC systems is interference with the PLC control signals from radio noise generated by faulty power equipment and by improperly shielded or grounded electronic equipment. These problems can frequently be overcome by careful testing,

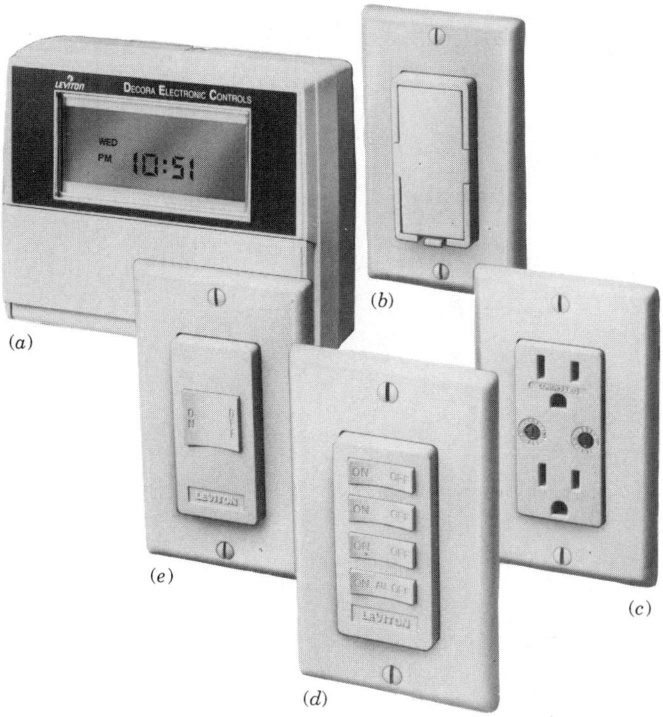

Fig. 16.34 *Components of a small residential-type PLC system.* (a) *Control unit, at which both manual and automatic (programmed) control of a maximum of 256 address codes is possible. The programming function is performed with control switches concealed under the cover at the base of the unit.* (b) *Wall switch module has local dimming and switching functions in addition to being remotely controlled from control center* (a). (c) *Duplex wall receptacle either split wired with local programming of one-half of the outlet or complete remote control.* (d) *Outlet box wall-mounted control switch that provides local group control (switching and/or dimming) for one to four devices (address codes). (e) Local switching module with remote control function for a single address. (Photo courtesy of Leviton.)*

installation, and repair of improperly operating equipment. In addition to energy measurement, PLC control can be applied to group switching of other loads, such as office machines, in addition to simplifying control of light and power for special events such as holiday festoon lighting.

16.29 Outlet and Device Boxes

These boxes are generally of galvanized stamped sheet metal. The most common sizes are the 4-in. square and 4-in. octagonal boxes used for fixtures, junctions, and devices, and the 4 × 2⅛ in. box used for single devices where no splicing is required. Box depths vary from 1½ to 3 in. Nonmetallic boxes may be used with NM and NMC cable and with nonmetallic conduit installations. In wet locations and for outdoor work, cast-iron or cast-aluminum boxes are recommended.

Floor boxes for floor outlets are usually of cast-metal construction and are installed directly in the floor slab.

As a result of a 1984 NEC (Article 300-21) requirement that electrical penetrations in fire-rated floors be designed to maintain the fire rating, electrical manufacturers have produced a line of poke-through fittings to meet this need. One such design is shown in Fig. 16.35. These electrical penetrations have become increasing prevalent in existing commercial spaces where the expanded need for desktop power and data wiring can be met most economically and rapidly by through-the-floor feeds from expanded wiring in the hung ceiling below. In addition, these fittings facilitate the electrical wiring relocations so common in rental office occupancies.

16.30 Power Conditioning

Power conditioning is a relatively new electrical term which describes the process of converting utility-supplied electrical power, which is frequently characterized by transients, spikes, radio-frequency noise, and voltage fluctuation, to a pure, accurately voltage-regulated sinusoidal waveform, normally referred to as *computer-grade power*. Some form of power conditioning is required by all data-processing and telecom-

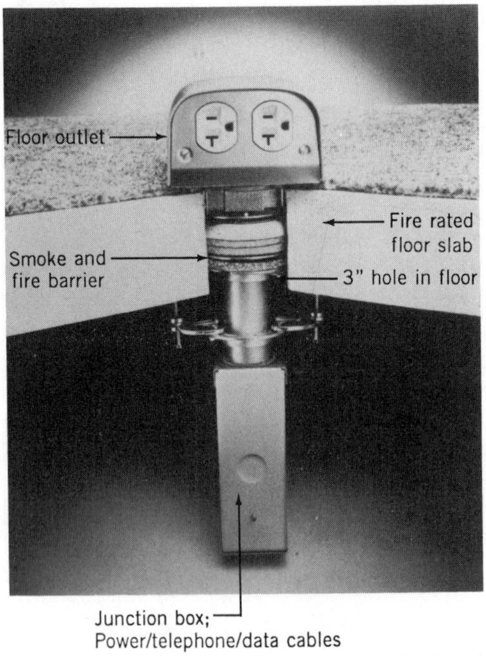

Fig. 16.35 Representative poke-through electrical floor fitting mounts in a 3-in. hole, is wired from underneath with power, telephone, and data cables and terminates in an appropriate floor fitting. The unit is adjustable to accommodate varying floor thicknesses. When abandoned, the floor fitting is replaced with a flat plate. (Courtesy of Hubbell, Inc.)

munication equipment, and by many other types of electronic equipment because of their extreme sensitivity to voltage fluctuations. In the case of data-processing equipment (computers and peripherals), radio noise can cause data errors, slow voltage fluctuations can cause overheating, data loss, and premature equipment failures, and large rapid voltage fluctuations (spikes, transients) can cause equipment burnout and system collapse. The economic ramifications of such effects on large commercial data-handling systems are so severe that power conditioning has become almost universal. [Power conditioners must not be confused with *u*ninterrupted *p*ower *s*upplies (UPS), which function to maintain power during utility *failure*, although in many commercial installations both functions are served by a single equipment package.]

(a) Source of Disturbances. Utility power systems are designed to maintain voltage and frequency within certain limits on an average, long-term basis. By its nature utility equipment cannot react to the short-time disturbances so harmful to electronic equipment. Furthermore, much of the noise in utility lines is introduced by users and is therefore beyond the utility's point of control. The impurities on the electrical power line can roughly be classified into three types: voltage variations, electrical noise, and transients.

1. *Voltage variations*. These can be caused by short-time high current drain, such as during motor starts, deliberate voltage reduction (brownouts), load switching of EMS systems, excessive line voltage drop due to heavy loading or current leakage, and similar problems.

2. *Electrical noise*. This is simply low-amplitude higher-frequency voltage which when superimposed on the power line results in a distorted waveform. Sources of such extraneous voltages are electronic equipment power supplies, lighting dimmers, solid-state motor controls, PLC systems, and small switching transients caused by branch circuit switches and relays.

3. *Transients*. These are large short-duration voltage variations, also known as surges or spikes. They are the principal culprits in major equipment failures. They are caused by lightning strikes to overhead lines (even many miles from the affected installation), utility switching of high-voltage lines, transformers, capacitor banks and generators, and electrical system faults.

(b) Treatment of Power Line Disturbances. Each of the three types of problems described above requires it own type of correction. Voltage variations are corrected with voltage regulators; noise problems by electrical isolation, filtering, and noise suppression; and transients are treated with surge suppressors. Each corrective device is a separate entity that can be applied individually to suit the particular problem encountered or in combination with another conditioning device. The choice in specific installations is a highly technical one and should be made by the facility's electrical engineer after careful analysis of the quality of the utility service. A few general recommendations can be made here.

1. All computer installations, including the smallest, should be protected from line transients with an appropriate surge suppressor. Multitap plug-in strips with integral surge suppressors are generally unsatisfactory unless they have surge current, clamping voltage, and surge-energy specifications, and these meet the installation requirements.

2. Major data-processing installations normally require all three types of line treatment. The most economical way to provide them is a single integrated power conditioning unit.

3. Considerable improvement in the quality of electrical power can frequently be achieved by the simple expedient of running separate feeders for sensitive loads. This too can be determined by the facility's electrical engineer.

Table 16.10 and Fig. 16.36 will give the reader an idea of the physical characteristics of power conditioning equipment. Dimensions are typical and are indicative only of equipment bulk. Designs vary widely among manufacturers as will therefore individual dimensions. Weights are not given, for the same reasons. Voltage-regulating equipment containing regulating transformers is very heavy. See also Fig. 15.18*b* for a typical application of a local power line surge suppressor.

(c) Uninterrupted Power Supply. As explained above, power conditioning equipment can supply clean utility power; it cannot supply any power during a utility outage. That eventuality is covered by an alternate supply of power. In a computer installation that alternate supply must be available instantaneously, hence the

TABLE 16.10 **Power Conditioning Equipment**

Equipment Functions				Dimensions (in.)		
Voltage Regulation	*Noise Suppression*	*Surge Suppression*	*kVA*[a]	H	W	D
×			5	15	20	20
			10	35	40	20
			20	40	40	30
×	×		0.25	6	6	10
			0.5	7	7	12
			1.0	8	10	15
			3.0	10	12	20
×	×		5	10	10	15
			10	20	30	15
			20	30	30	20
×	×	×	5	40	30	30
			10	50	35	30
			20	70	40	35
		×	NA[a]			
			15 A	3	3	16
			100 A	10	12	12
			500 A	15	15	12

[a]Surge surpressors are rated by transient energy-handling capability and therefore a kVA rating is not applicable. However, since transient energy increases with kVA size of protected equipment, a service ampere rating is given.

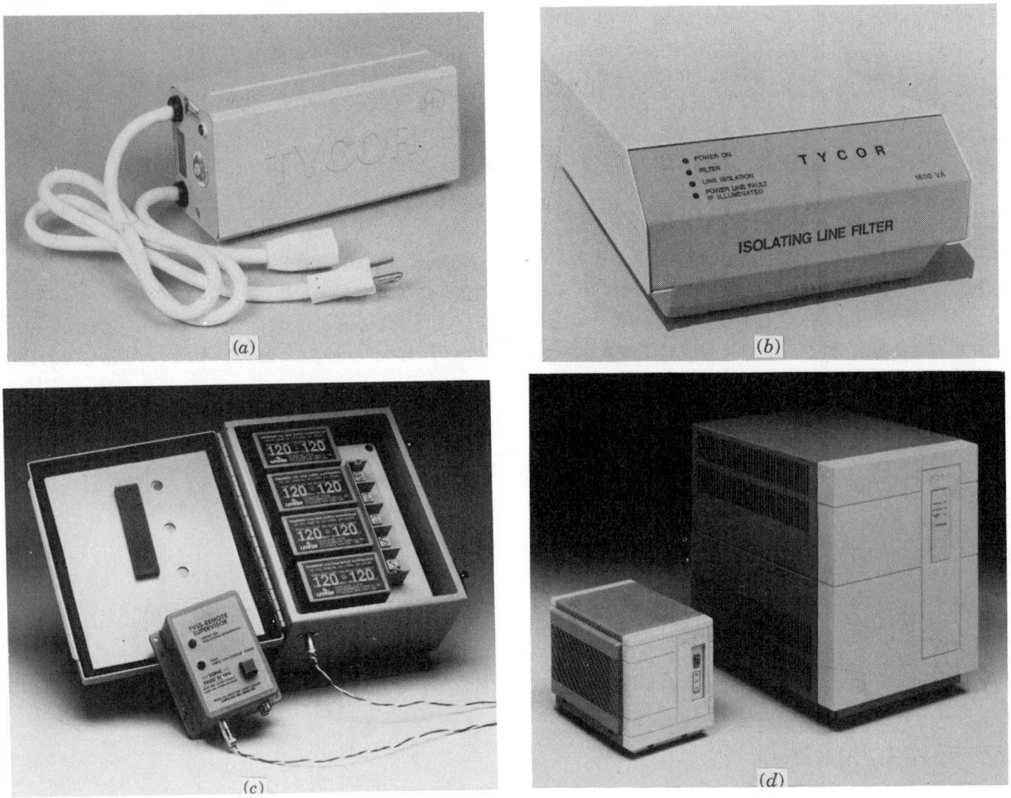

Fig. 16.36 (a, b, c) *Typical items of power conditioning equipment.* (a) *In-line (series) electrical noise filter and surge suppressor, intended for use with a single item of equipment. Rated 10 A, 120 V, the unit measures 8 × 3 × 5 in.* (b) *Isolating line filter contains an isolation transformer in addition to a line filter, providing isolation from spurious power line voltages, and noise filtering. Unit is rated 15 A, 120 V, and measures 17 × 10 × 5 in.* (c) *Transient voltage surge surpressor intended for connection at a panelboard and designed to provide both surge suppression and EMI/RFI noise attenuation to all circuits fed from the panel. The unit contains four single-phase modular-type surpressors (one for each phase line plus a spare) plus a diagnostic module (in front of open panel) which indicates the unit's condition. The cabinet measures approximately 12 × 8 × 4 in. and is applicable at most branch circuit panelboards.* (d) *Two commercial-grade power conditioning units which provide high-grade noise and surge suppression plus voltage regulation. The smaller unit measures approximately 12 × 9 × 9 in., weighs 60 lb, and is rated 2 kVA/240 V. The larger unit measure 20 × 15 × 20 in., weighs approximately 250 lb, and is rated 5 kVA/240 V. [*(a, b) *Courtesy of Tycor International;* (c) *courtesy of Leviton;* (d) *courtesy of SOLA.]*

name *uninterrupted power supply* (UPS). The selection of a UPS system for a computer installation is a complex process involving a choice between static and rotating systems and between off-line (switched in on utility failure) and on-line systems (connected and "floating" constantly). The selection procedure depends on the type of equipment in the computer/data-processing system and as such is well beyond the scope of this book. Suffice it to say that the choice must be made carefully, as UPS systems are large and very expensive. A rough idea of bulk for *static* UPS equipment including battery can be had from the following table.

Rating	*Type*	*Full-Load Capacity (min)*	*Dimensions (in.)*	*Weight (lbs)*
750 VA	On-line	10	8 × 12 × 20	70
1500 VA	On-line	10	9 × 12 × 24	100
3000 VA	On-line	10	12 × 20 × 24	250
10 kVA	On-line		40 × 25 × 45	700
10 kVA	Battery	10	40 × 10 × 45	600
		45	30 × 24 × 45	1000
		90	30 × 48 × 45	2000
20 kVA	Off-line	30	90 × 12 × 32	1400

16.31 Emergency/Standby Power Equipment

The NEC makes a clear distinction between emergency systems and standby systems, covering the former in Article 700 and the latter in Articles 701 and 702. The equipment, circuitry, and arrangement of both is similar; the purpose is somewhat different. The reader is also referred to NFPA Standard 110, Emergency and Standby Power Systems, which covers equipment requirements for these systems.

Emergency systems are intended to supply electric power to equipment essential for *safety* to human life, upon interruption of normal supply. Included in this classification would be illumination in areas of assembly, to permit safe exiting and panic prevention, and such other vital functions as fire detection and alarm systems, elevators, fire pumps, public address and communication, and orderly shutdown or maintenance of hazardous processes.

Standby systems are divided into two categories: those legally required (Article 701) and optional systems (Article 702). The former are intended to power processes and systems whose stoppage might create hazards or hamper fire-fighting operations. This classification is broader than the emergency system one and could include HVAC systems, water supply equipment, and industrial processes whose interruption could cause a safety or health hazard. It is intended primarily as a safety measure. *Optional* standby systems can cover any or all loads in a facility at the discretion of the owner, and are normally intended to protect property and prevent financial loss, in the event of a normal service interruption. A typical application would be a critical industrial process or an ongoing research project.

Health care facilities are covered by a separate set of regulations: NEC Article 517 and NFPA Standard 99, both of which are referenced as legally binding in the vast majority of jurisdictional codes. It is important to note that for both emergency and legally required standby systems, the *provision* of the system must be mandated by the authority having jurisdiction over construction, whether it be a local, state, or federal agency, or a combination of these. The NEC dictates how the system is to be designed and constructed; its existence depends on another authority. A case in point is the fundamental item of exit lights. These are mandated by the NFPA Life Safety Code 101, and Subpart E of OSHA regulations. They are not required by the NEC and reference to NEC will not assure their provision. The Code(s) having authority and jurisdiction must specifically require an emergency and/or standby system by that nomenclature, in order for NEC provisions to apply. The exact items of equipment to be powered are selected by the designer, keeping in mind the specific and general requirements of the codes having jurisdiction. The majority of codes make the provisions of the NEC and the relevant NFPA standards legally binding. Most codes require emergency systems; far fewer require standby systems, and then only for essential water and water-treatment systems and a few other essential uses. The designer must thoroughly investigate the matter of jurisdictional codes before proceeding with consideration of emergency electric power systems and equipment.

System design arrangements are discussed in Section 17.22. Emergency lighting equipment and system design is covered in Section 21.30. System equipment, which falls into two principal categories, that is, generator and battery installations, is discussed below. Optional standby systems normally use a fueled prime mover and not batteries.

(a) Engine-Generator Sets. An engine-generator set installation comprises basically three components: the fuel system including storage, if necessary; the set itself plus exhaust facilities; and the space housing the unit (see Fig. 16.37). The principal advantages of the engine-generator set are unlimited kVA capacity, duration of power limited only by size of fuel tank, use for peak-load shaving (see Section 14.13*d*), and, if properly maintained, indefinite life. Disadvantages are noise, vibration, nuisance of exhaust piping, need for constant maintenance and regular testing, and difficulties with fuel storage. Gasoline can be stored for only a year at most, and subsequent disposal is difficult. Diesel fuel keeps somewhat longer, but disposal is also difficult. Use of gas for a diesel or gas engine obviates the fuel storage problem but poses the alternate problem of availability of gas service during emergencies.

(b) Battery Equipment. Storage batteries are often used to supply limited amounts of emergency power, primarily for lighting (see Section 21.30). Such units are mounted in individual cabinets or in racks for larger installations and are always provided with automatic charging equipment. Installations of batteries and static inverters called UPS (uninterrupted power supply) systems are commonly found in computer installations where an outage, however slight, is unacceptable. Such installations are

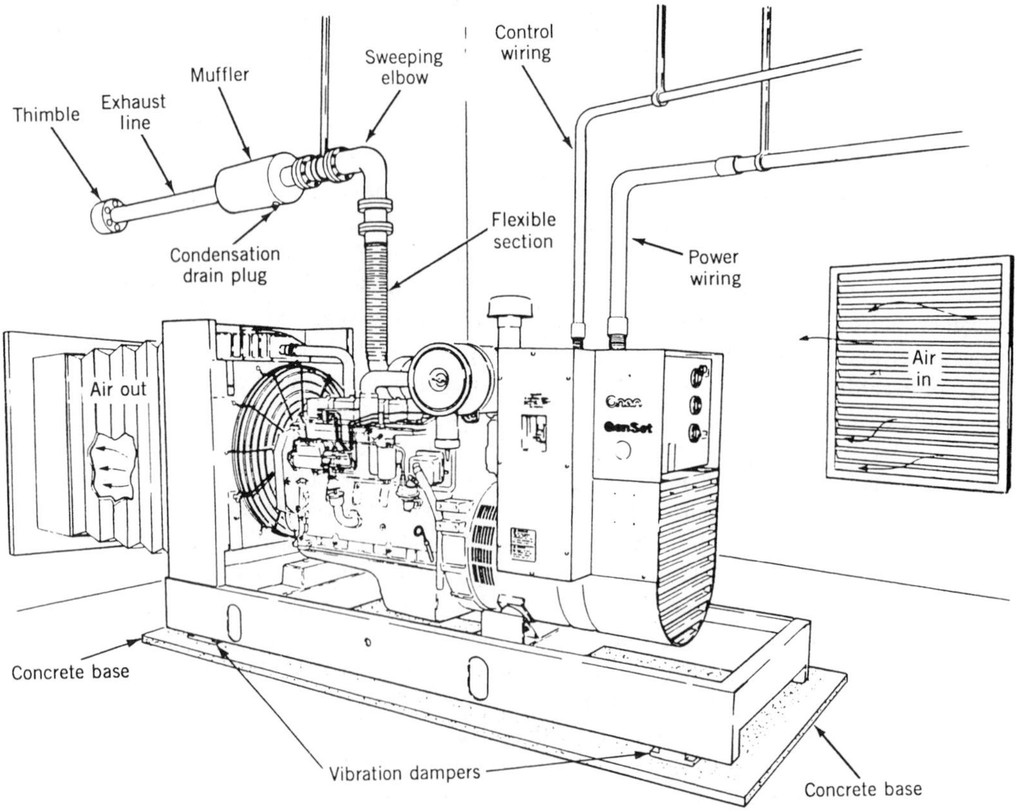

Fig. 16.37 Typical engine-generator unit and installation technique. Illustrated is a radiator-cooled diesel installation. (Courtesy of ONAN Corp.)

TABLE 16.11 **Comparative Characteristics of Engine-Generator Sets and Storage Batteries**

Characteristics	*Generator Set*	*Battery*
Capacity	Unlimited	Limited
Type of power	a-c or d-c	d-c[a]
Life	Unlimited	Limited
Relative size	Small	Large
Relative initial cost:		
(1) Below 50 kva	High	Medium
(2) Above 50 kva	Medium	High
Space Required	Medium	Large
Ventilation	Large	Small
Maintenance Cost	Medium	Low
Noise	Great	None
Vibration	Some	None
Fuel Storage	Yes[b]	None

[a]Alternating current can be provided, at high cost, by a static inverter.

[b]With natural gas engines, fuel storage is eliminated.

complex, expensive, and designed to meet the specific requirements of the installation.

Batteries need not be installed in separate battery rooms provided that live parts are guarded and sufficient ventilation is provided to prevent gas accumulation. Battery types are undergoing constant development. At this writing the types principally in use are lead acid, nickel cadmium, lead-antimony and lead-calcium cells. The choice depends on the application and is best left to a specialist. Batteries have the distinct advantage that they can be installed either in a central system with distribution by feeder of the battery power throughout the facility, or they can be installed in small package units around the building. Central systems are 24 to 125 V, d-c and normally feed nonfluorescent emergency lighting only. Individual packs are used often to supply a-c power via built-in inverters. The great disadvantage of battery systems is limited duration of power. The NEC requires that batteries maintain loads for a 1½ h *minimum*, but larger capacity is normally installed. Table 16.11 gives a brief comparison of engine-generator sets and batteries.

16.32 System Inspection

Each electric wiring system is inspected at least twice by the local inspection authorities; once after raceways (roughing) have been installed and before the wiring and closing in of walls, and once after the entire job is complete. The purpose of these inspections is to determine whether design, material, and installation techniques are meeting the national and local code requirements. Excellence of installation is the responsibility of the contractor. The designer, however, must be completely familiar with installation work and the equipment's physical characteristics in order to properly design an electrical system. Such a design will not present the contractor with unwarranted difficulties. The designer must also be very wary of equipment substitutions by the contractor who, having submitted a bid on the basis of plans and specifications, should be required to supply the specified equipment.

ELECTRICITY

17

ELECTRIC WIRING DESIGN

In wiring design, as in other design, there are numerous possible solutions to each problem—some good, some fair, and some poor. Experience guides the designer to a solution that best suits the job, since it is his or her responsibility to establish the most economical design within the framework of the design criteria. This chapter opens with a discussion of these criteria and continues on to actual wiring design. Actual electrical design intended for construction must meet all the requirements of local and national building codes. To our knowledge all building codes include NFPA 70, the National Electrical Code (NEC), as the minimum standard for all electrical design, and its requirements are legally binding. The electrical design data and procedures adduced in this chapter and in other sections of this book are based on NEC requirements and are to the best of our knowledge current. However, they are intended only as didactic materials. Code requirements change, as the NEC is revised every 3 years. The designer is therefore advised to obtain and study the latest edition of all codes having jurisdiction, including the NEC, and to assure that the electrical design is in accordance with the requirements of these codes.

17.1 General Considerations

(a) Flexibility. Every wiring system should incorporate sufficient flexibility in branch circuitry, feeders, and panels to accommodate all probable patterns, arrangements, and locations of electric loads. The degree of flexibility to be incorporated depends in large measure on the type of facility. Thus laboratories, research facilities, and small educational building require a great deal more flexibility than residential, office, and fixed-purpose industrial installations. As part of the design for flexibility, provision for expansion must be provided as experience has demonstrated that most facilities will grow, both physically and in electrical demand. Overdesign, however, is as bad as underdesign, being wasteful of money and resources both initially and in operation.

(b) Reliability. The reliability of the electrical power within a facility is determined by two factors: the utility's service and the building's electrical system. The service record of the utility should be studied along with the economic impact of a power outage to determine whether, and to what extent, standby power equipment is justified (see Sections 16.31 and 17.22). Emergency equipment required for the safety of a building's occupants is determined by local, state, and national building codes. Beyond the service point, the reliability of power is entirely dependent on the wiring system. Here, too, economic studies must be made to determine the quality of equipment and the amount of redundancy (duplicate equipment) to be installed. The subject of reliability is a complex one, and we can state only a few general principles here.

1. The reliability of an electric system is only as good as that of its weakest element. Therefore, it may be necessary to provide redundancy at anticipated weak points in the system.
2. The electrical service and the building's distribution system act together. An extremely reliable (and expensive) service is of little use if the power cannot reach the desired points.
3. Critical loads within the facility should be pinpointed to determine how best to reliably serve them, that is, by establishing reliable power paths to them or by furnishing individual standby power packages for them. The latter course is often chosen for healthcare and other critical loads.

(c) Safety. Although rigid adherence to the requirement of the NEC and other applicable NFPA codes will assure an initially safe electrical installation, the designer must be constantly alert to such factors as electric hazards caused by misuse or abuse of equipment or by equipment failure. Also, attention to the size of equipment used will eliminate the oft-encountered physical hazard caused by obstruction of access spaces, passages, closets, and walls with electric equipment. Finally, lightning protection can be subsumed under the heading of safety; this topic is discussed in Chapter 13.

(d) Economic Factors. This item can readily be divided into two frequently interrelated items: first cost and operating cost. All other factors being equal, the first cost depends in large measure on whether the constructor is interested in minimum first cost or minimum owning cost. We have demonstrated that these two costs frequently stand in inverse relationship to one another (exceptions are mentioned below). Low first-cost equipment generally results in higher energy cost, higher maintenance cost, and shorter life. The decision, however, is not purely an economic one, inasmuch as the electrical energy cost factor in the operating-cost equation is directly related to energy consumption, with one exception. That exception is the utility's demand charge, which is discussed in Section 14.12. Means for minimizing this cost are covered in detail in Section 14.13. Here there occurs a coincidence of reduction to both first cost and operating costs. Load-leveling equipment permits the electrical distribution system to be sized without consideration of coincident load peaks, thus resulting in smaller equipment, operating more efficiently—near its full-load capacity. All other reductions in electric energy cost flow directly from the corresponding reduction in energy consumption.

(e) Energy Considerations. This factor is a complex one involving considerations of energy codes and budgets, energy conservation techniques (see Section 17.5), and energy control.

Buildings constructed with governmental participation may be subject to energy budget limitations expressed in Btu/square foot/year. Although the lion's share of this budget is taken by heating/cooling and lighting systems, the electrical distribution system will also be subject to conformity to stated codes. Important among these is ASHRAE Standard 90. Detailed energy conservation considerations are discussed in detail in Section 17.5. A discussion of the IES Recommended Procedure for Lighting Power Limit Determination is found in Section 20.5.

(f) Space Allocations. The general impression that electrical equipment is small and easily concealed is accurate only for wire and conduit. Panels, motor control centers, busduct, distribution centers, switchboards, transformers, and so on can be large, bulky noisy, and highly sensitive to tampering and vandalism. Thus, space allocations must be concerned with maintenance ease, ventilation, expandability, centrality (to limit length of runs), limitation of access, and noise, in addition to the basic item of space adequacy.

(g) Special Considerations. These depend on the specialized nature of certain facilities and may include items such as security, central and/or remote controls, interconnection with other facilities, and the like.

17.2 Load Estimating

When initiating the wiring design of a building, it is important to be able to estimate the total building load in order to plan such spaces as transformer rooms, chases, and closets. This information is also required by the local power company well in advance of the start of construction. An exact load total can be made only after completing the design but, since this is often several months later, a good preliminary estimate is required. Such an estimate can be made from the figures given in Table 17.1. These figures are average. When it appears that the building will have heavier or lighter loads because of lighting levels or other factors, the figures should be modified accordingly. The figures having been established in the individual categories, they should be added together without ap-

TABLE 17.1 **Electric Load Estimating**[a]

I	II	III	IV	V	VI
	Volt-Amperes per Square Foot[b]				*Ten Year Percent Load Growth*
		Misc.	*Air Conditioning*[g]		
Type of Occupancy	*Lighting*[c,d]	*Power*[e]	*Electric*	*Nonelectric*	
Auditorium					
General	1.0–2.0	0	12–20	5–8	20–40
Stage	20–40	0.5			
Art gallery	2.0–4.0	0.5	5–7	2.0–3.2	20–40
Bank	1.5–2.5		5–7	2.0–3.2	30–50
Cafeteria	1.0–1.6	0.5	6–10	2.5–4.5	20–40
Church & synagogue	1.0–3.0	0.5	5–7	2.0–3.2	10–30
Computer area	1.2–2.1	2.5	12–20	5–8	50–200
Department store					
Basement	3–5	1.5			
Main floor	2.0–3.5	1.5	5–7	2.0–3.2	50–100
Upper floor	2.0–3.5	1.0			
Dwelling 0–3000 ft²	3.0	0.5	—	—	50–100
(not hotel) 3000–120,000	0.4	0.15	—	—	
above 120,000	1.5–2.5	2.0	—	—	
Garage (commercial)			—	—	10–30
Hospital	1.0–3.5	1.5	5–7	2.0–3.2	40–80
Hotel	1.0–2.0	0.5			
Lobby	1.0–1.5	1.0	5–8	2.0–3.5	30–60
Rooms (no cooking)	2.0–3.0	5–20	3–5	1.5–2.5	
Industrial loft building	1.2–2.2	0.5	—	—	50–100
Laboratories	1.5–3.0	1.5	6–10	2.5–4.5	100–300
Library	1.0–2.0	0.5	5–7	2.2–3.2	30–40
Medical center	1.5–3.5	2.5	4–7	1.5–3.2	50–80
Motel	1.2–2.5	0.5	—	—	30–60
Office building	1.5–2.8	2.5	4–7	1.5–3.2	40–80
Restaurant			6–10	2.5–4.5	20–40
School	2.0–2.5	2.0	3.5–5.0	1.5–2.2	50–80
Shops	2.0–3.5	0.5			
Barber & beauty	2.0–3.0	0.5	5–9	2–4	
Dress	2.0–3.0	0.5			
Drug	2.0–3.5	0.5	4–7	1.5–3.2	40–80
Five & ten	0.25–1.0	0.25			
Hat, shoe, specialty	0.3	—			
Warehouse (storage)	0.25	—	—	—	10–30
In the above except single dwellings:					
Halls, closets, corridors,	0.5	—	—	—	
storage spaces	0.25	—	—	—	

[a]Figures assume energy-conservation techniques applied.

[b]These figures do not include allowance for future loads.

[c]The figures given in Article 220 of the NEC are minimum figures for calculation of electric feeder sizes.

[d]See Section 20.5 for a detailed analysis of the IES lighting power limit determination.

[e]These figures are based on experience and must be applied judiciously.

[f]This figure does not include the power used by the computer.

[g]Includes the loads of air-handling equipment and pumps.

plication of demand or diversity factors in order to obtain the maximum demand load for which the building service equipment must be sized, in the absence of electric load-control (load-leveling) equipment. At this point an analysis must be made to determine the feasibility of incorporating such equipment into the facility (see Section 14.13). Input to this study includes the utility's complete rate schedule, including all penalty clauses, a detailed analysis of the building's equipment load patterns, and any external constraints such as maximum loads imposed by power and energy budgets.

Equipment load patterns must be carefully analyzed because they determine a load's "sheddability." Thus, for kitchen equipment, load interruption may be undesirable, but shifting of cooking time by a half hour is entirely feasible. On the other hand, for HVAC equipment, building thermal inertia and "stretching" maximum and minimum temperatures and humidities permit considerable load control without adverse effects. Also, as explained in Section 14.13, the degree and duration of load shedding is a function of the type of control equipment utilized. It is well to repeat here what is stated there—load control affects maximum demand, with only minor effect on total energy consumption.

The external constraints referred to are the energy budgets recommended or required by codes, legislation, and funding bodies. After the above load control analysis is completed, or simultaneously with it, a building energy consumption analysis must be performed. This may be done manually, although the numerous computer programs available are considerably more accurate. The results of this analysis will indicate whether the target electrical energy budget is being exceeded. If so, loads will have to be modified by reconsideration of projected systems and system criteria, by incorporating energy-conservation devices and techniques into the electrical system and by drawing up energy-use guidelines that will be applied when the building is occupied. Since this last item depends for its success on the day-to-day voluntary actions of the building's occupants, it should not be considered a major conservation source. Conservation measures are covered in Section 17.5.

The electrical loads in any facility can be categorized as follows:

1. Lighting.
2. Miscellaneous power, which includes data-processing equipment, convenience outlets, and small motors.
3. Heating, ventilating, and air conditioning (HVAC).
4. Plumbing or sanitary equipment.
5. Vertical transportation equipment.
6. Kitchen equipment.
7. Special equipment.

Category 1 is self-explanatory and is covered by column II of Table 17.1. Note carefully here however that lighting loads are carefully prescribed in ASHRAE/IES 90.1-1989 and that these ASHRAE/IES prescribed loads should be used where the ASHRAE standard has jurisdiction. Contradictions between recommendations in standards having jurisdiction must be adjudicated by local authorities before design is established. See Chapter 5, and Sections 20.5 and 21.2 for details of ASHRAE/IES prescribed loads.

Category 2, column III, includes, in addition to receptacles and small motors, such items as desktop computers and data-processing terminals, plug-in heaters, water fountains, and so on.

Category 3, column IV, includes all loads imposed by the HVAC equipment. Included therefore are fuel pumps, boiler motors, condensate pumps, all heaters, blower motors, exhaust fans, and so forth. Also included in column IV, for air conditioning loads, is refrigeration compressors. This item is omitted in column V since the air conditioning utilizes absorption machines that do not use electricity for primary power. When air conditioning is not anticipated, the HVAC load is still appreciable because of heating and ventilating (H and V) requirements. A rough estimate for this load would be two-thirds or 65% of the loads in column V.

Category 4 includes all loads associated with the water and sanitary system, including house water pumps, air compressors and vacuum pumps, sump pumps and ejectors, well pumps and fire pumps, water heaters and pneumatic tubes, plus such special items as display fountain pumps. Since these loads vary widely with local conditions and with facility design as much as type of facility, an estimate cannot be made on a volt-amperes per square foot basis, by type of building. For this reason, no figure is included in Table 17.1. If actual data cannot be used, it is helpful to remember that plumbing loads are relatively small, rarely exceeding 20% of the HVAC load, though, for the most part, they are unrelated to it.

Category 5, vertical transportation, is also obviously unrelated to square footage and therefore cannot be tabulated in Table 17.1. These loads, including elevators, moving stairs, and dumbwaiters, can be estimated from the data given in the vertical transportation chapters of this book.

Category 6, kitchen equipment, is also not included in Table 17.1, though obviously present in all restaurants, in most hospitals, and in some office and religious-use buildings. The reason for this omission is that the primary power for the major load, the cooking equipment, may be either gas or electric. Other large energy-use equipment such as dishwashers can be electric, gas, or steam fed. Furthermore, no correlation can be made between the facility type, area, and load, even if electrically powered, since population and schedule are also major factors. When kitchens are planned, a kitchen consultant or other experienced planner can usually supply an estimate of the electric power requirement, which can then be added to the other figures.

Category 7, special equipment, is so variegated that no figures can be listed. Under this title is subsumed such items as laboratory equipment, shop loads, display area loads, floodlighting, canopy heaters, display window lighting, and so on. This load data must be gathered for individual items of equipment and added to the foregoing totals.

Table 17.2 gives a tabulation of service entrance size in amperes, based on single- and three-phase service for typical occupancies. These figures are for quick estimate purposes and should be adjusted after the design is completed.

17.3 System Voltage

There are several systems of voltage commonly available in the United States and Canada. (See NEC Art. 230-SERVICES, for Code requirements.) The simplest of these is:

(a) 120 V, single-phase, 2-wire (see Fig. 17.1). This is used for the smallest of facilities such as out-buildings, and isolated small loads up to 6 kVA. Load is calculated by multiplying current and voltage. For 60-A service, which this type of service is normally limited to, no more than 50 A are usually drawn. Thus

$$VI = 120 \times 50 = 6000 \text{ V-A} = 6 \text{ kVA}$$

The nominal system voltage is 120 V, although it is also referred to as 110 V and 115 V (see Fig. 17.1).

TABLE 17.2 **Nominal Service Size in Amperes**

Nominal service sizes are 100 A, 150 A, 200 A, 400 A, 600 A.

	Area in Square Feet				
Facility	*1000*	*2000*	*5000*	*10000*	*Remarks*
Single-phase, 120/240 v, 3-wire					
Residence	100A	100A	150A	—	Minimum 100A
Store[a]	100A	150A	—	—	
School	100A	100A	150A	—	
Church[a]	100A	150A	—	—	
3-phase, 120/208 v, 4-wire					
Apartment House	—	—	100A	150A	
Hospital[a]	—	—	200A	400A	
Office[a]	—	100A	400A	600A	
Store[a]	—	100A	400A	600A	
School	—	100A	100A	200A	

[a]Fully air conditioned using electric-driven compressors. Based on figures in Table 17.1.

ELECTRICITY

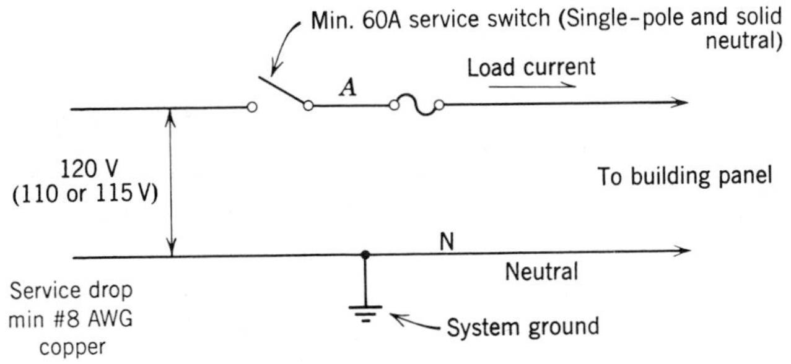

Fig. 17.1 *120-V, single-phase, 2-wire service. This is also the arrangement of the usual branch circuit.*

For somewhat heavier loads, the system normally used is:

(b) 120/240 V, single-phase, three-wire (see Fig. 17.2). The Code requires that all one-family residences with six or more 2-wire circuits have a minimum of 100 A, 3-wire service. Service disconnect for 100-A service would be a 100 A, 2-pole, and solid neutral switch, fused at no more than 80% of rating, or 80 A. This is usually written 100A, 2P & SN, 80AF. This service is used principally for residences, small stores, and other occupancies where the load does not exceed 80 A or 19.2 kVA. Load is calculated thus:

$$\text{kVA} = \frac{V \times 1}{1000} = \frac{240 \times 80}{1000} = 19.2$$

See Fig. 17.2. Although it may appear otherwise, the neutral carries no more than full-load current. Note that each "hot leg" of the 3-wire system carries line current. Thus, total load can also be calculated:

$$\text{load kVA} = \text{twice load on each line}$$

Assuming a balanced 80-A load:

$$\begin{aligned}\text{total kVA} &= 2 \times 80 \text{ A} \times 120 \text{ V}\\ &= 2 \times 9600 \text{ VA} = 19{,}200 \text{ VA}\\ &= 19.2 \text{ kVA}\end{aligned}$$

If load is unbalanced, with say 30 A in one line and 50 A in the second, total load is

$$\begin{aligned}120 \times 30 + 120 \times 50 &= 3600 + 6000\\ &= 9600 \text{ VA} = 9.6 \text{ kVA}\end{aligned}$$

Actual system voltages can be 120/240, 115/230, or 110/220, although 120/240 is the accepted industry standard. 120 V loads are connected between line and neutral and cause a current in only one line. 240 V loads such as a clothes dryer are connected between phase lines and cause current in both lines. For example, to find the line currents caused by the three loads shown on Fig. 17.3.

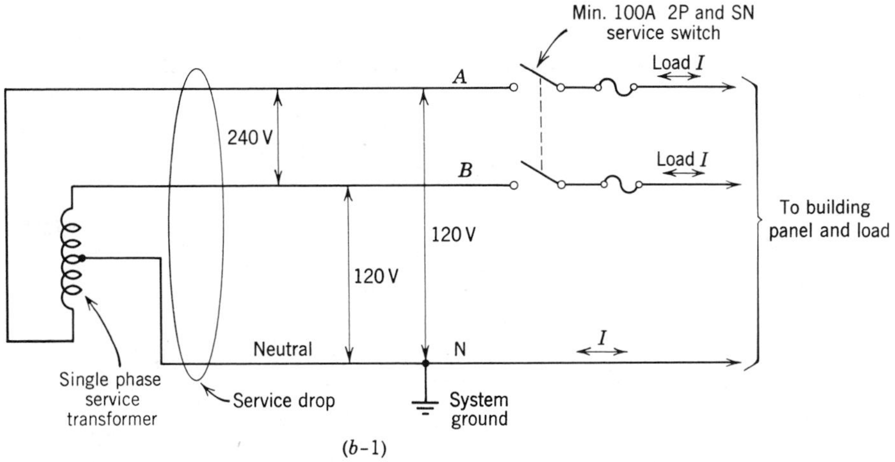

Fig. 17.2 *120/240-V, single-phase, 3-wire service. The single-phase transformer is center-tapped to establish a neutral. The neutral connection is always grounded.*

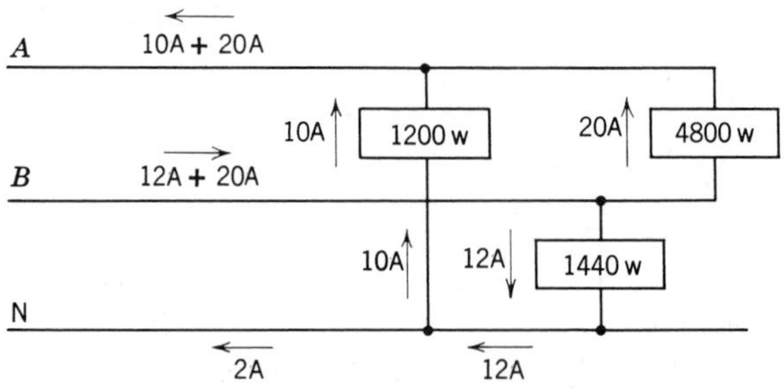

Fig. 17.3 *Note that the neutral carries the difference in current between the A and B legs and therefore a maximum that is equal to the current in one of the legs (when the other is zero).*

120-V, 1200-W iron on line *A*

120-V, 1440-W hair dryer on line *B*

240-V, 4800-W dryer on lines *A* and *B*

we calculate

$$I_A = \frac{1200}{120} = 10 \text{ A}$$

$$I_B = \frac{1440}{120} = 12 \text{ A}$$

$$I_{AB} = \frac{4800}{240} = 20 \text{ A}$$

Note that the neutral only carries the unbalance of 2 A.

Total current in $A = 20 + 10 = 30$ A
Total current in $B = 20 + 12 = 32$ A
Total load $= 120(30) + 120(32)$
$= 3600 + 3480 = 7440$ W
Loads are $1200 + 1440 + 4800 = 7440$ W

The 120/240-V, single-phase system comes from a center-tapped single transformer. When two hot legs and a neutral of a 3-phase system are taken, a similar system is obtained with a rating of:

(c) 120/208 V, single-phase, 3-wire (see Fig. 17.4). This system is really part of a 3-phase system and is most often found *within* a building with 3-phase distribution, and is used to serve a load that does not require 3-phase, 4-wire. Calculation of loads and line currents is considerably more complex than in (b) above because of the 120° phase displacement between phases *A* and *B*. Here, as before, the neutral will carry no more than line current whether the system is balanced or not.

When feeding large facilities, a 3-phase, 4-wire system is normally employed. This is usually:

(d) 120/208 V 3-phase, 4-wire (see Fig. 17.5). This system is the most widely used 3-phase ar-

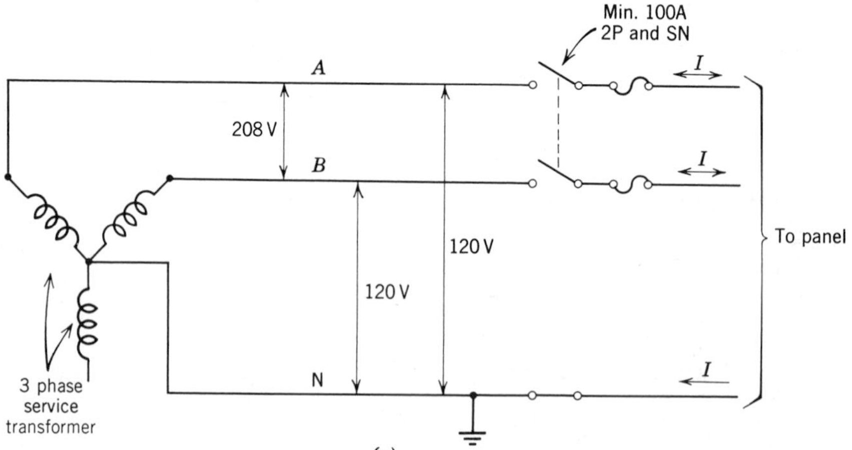

(c)

Fig. 17.4 *120/208-V, single-phase, 3-wire service. This arrangement comprises two-thirds of the full 120/208-V, 3-phase, 4-wire connection shown in Fig. 17.5.*

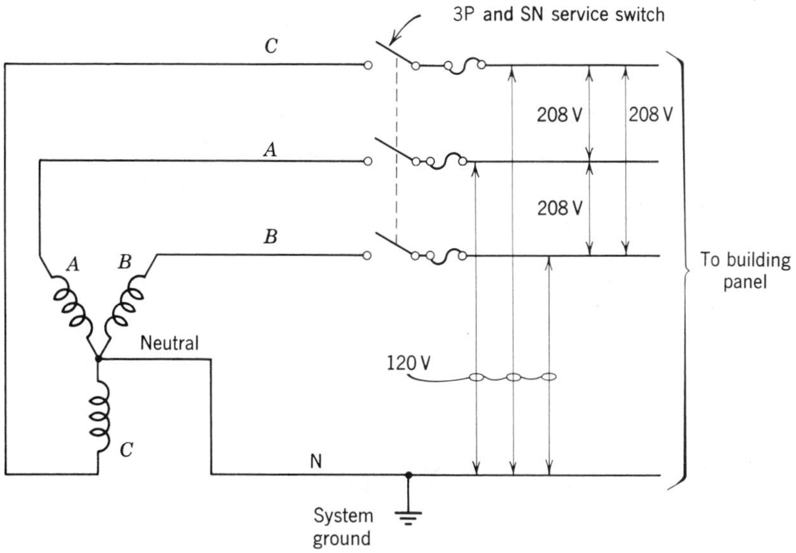

Fig. 17.5 *120/208-V, 3-phase, 4-wire system. The neutral connection is connected to the system ground and is not broken by the service switch.*

rangement and is applicable to all facilities except the very largest ones. In those, lengths of feeders and sizes of loads become so great that a higher voltage is required. In this system, 120-V loads such as lighting, computers and accessories, receptacles, and so on, are fed at 120 V by connection between each phase leg (see Fig. 17.6) and neutral. Motors larger than ½ hp and all 3-phase loads are fed at 208 V by connection between the 3-phase legs. Single-phase, 208 V loads such as heaters are accommodated as in (c) above, by connection between two phase legs. Such loads are often referred to as "2-pole" loads, alluding to the 2-pole current breakers used to feed them. Although this system is nominally 120/208 V, it is also used as 115/200. However, this is inadvisable because of nominal motor voltage, which will be discussed below.

Where buildings are larger, either horizontally or vertically, lighting is principally fluorescent, and the 120-V load does not exceed one-third of the total load, a more economical system of voltages is available. This system is:

(e) 277/480 V, 3-phase, 4-wire (see Fig.

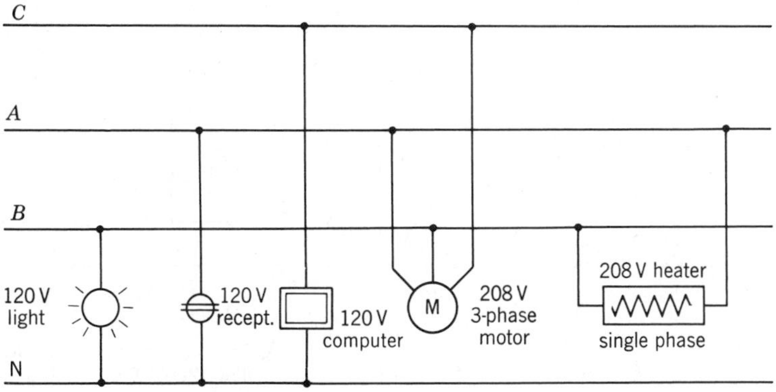

Fig. 17.6 *The flexibility of the 120/208-V system is illustrated and accounts for its wide application. Loads are shown schematically. In practice, the loads are fed via protective devices in the panel. These are omitted here for clarity. A, B, C, and N represent the panel buses.*

ELECTRICITY

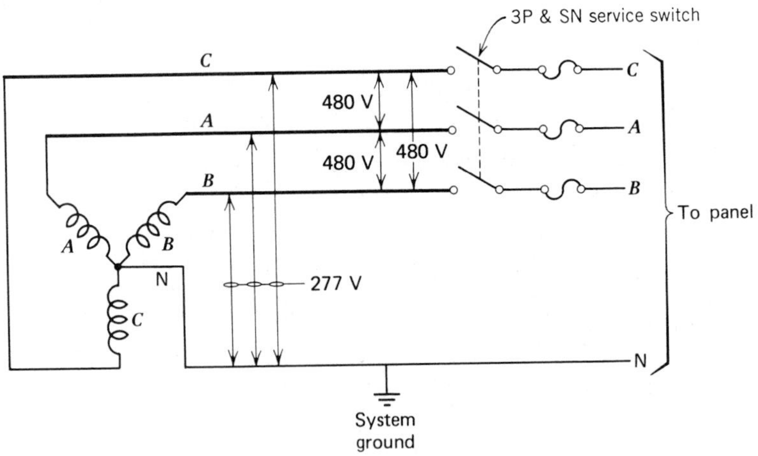

Fig. 17.7 (a) *277/480-V, 3-phase, 4-wire service system. The system is identical to the 120/280-V system shown in Fig. 17.5 except for voltages.*

17.7). It utilizes 277-V fluorescent lighting, 480-V machinery, and small (3 to 25 kVA) dry-type transformers to step down from 480 to 120 V for supplying receptacles and other 120-V loads. This system is ideally suited to multistory office buildings and large single-level or multilevel industrial buildings. Savings are generated by the smaller feeder and conduit sizes and smaller switchgear. which more than offset the additional cost of step-down transformers for the 120-V load.

As an example of the savings possible with this system, let us consider the wiring required for a 15 kW heater.

At 3-phase, 208 V:

$$I = \frac{15{,}000 \text{ W}}{\sqrt{3} \times 208 \text{ V}} = 42 \text{ A}$$

requiring No. 8 RHW wire (45-A capacity).

At 480 V:

$$I = \frac{15{,}000 \text{ W}}{\sqrt{3} \times 480 \text{ V}} = 18 \text{ A}$$

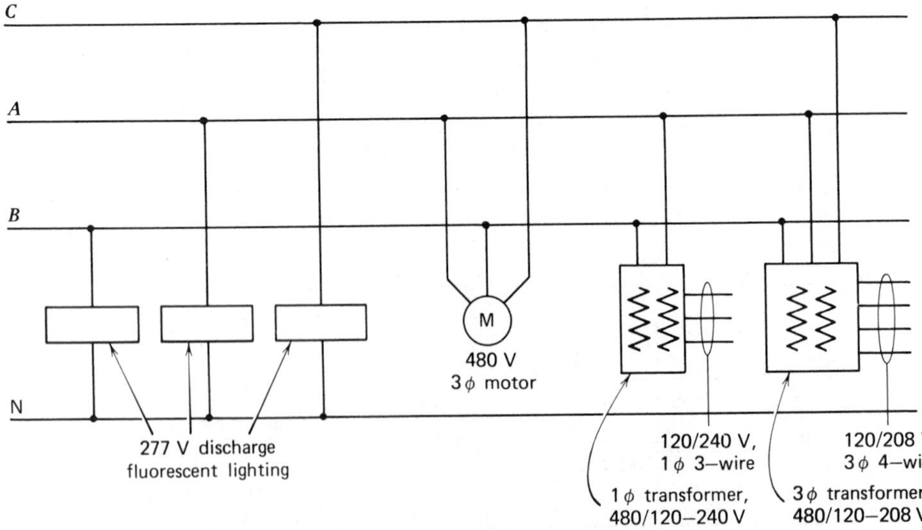

Fig. 17.7 (b) *Normal load arrangement for this system is illustrated. The lighting can be fluorescent or HID (mercury, metal-halide, sodium). Transformers, either single-phase or 3-phase, supply 120 V for receptacles and 208 V for loads requiring that voltage.*

requiring only No. 12 RHW wire (20-A capacity). This system is also frequently referred to as 265/460 V and 255/440 V.

Voltages above 150 V to ground are generally avoided in residential branch circuits, but may be used in commercial and industrial facilities, within the guidelines established by the NEC.

(f) 2400/4160 V, 3-phase, 4-wire systems are only used in very large commercial buildings or in industrial buildings with machinery requiring these voltages. The cost of running this voltage feeder within a building is high because of NEC requirements and the inherent higher cost of 5-kV equipment. A detailed cost and engineering analysis by a competent engineer is required for each case. Voltages above this level are widely used in large industrial plants and are beyond consideration in this book.

Reference was made above to the varied voltages assigned to the same voltage system. Thus, at the lowest level, we have 110, 115, and 120 V, at the next level 200, 208, 220, 230, and 240 V, at the next 255, 265, and 277 V, and finally 440, 465, and 480 V. These voltage differences arise because of the historical difference between transformer voltage standards, which establish the *system* voltage, and motor voltage standards, which govern *utilization* voltage level (see Table 17.3). Current motor voltage standards are established at a level that is consonant with the system voltage. Note the close correspondence between motor standard voltage and system voltage with a normal 4% feeder voltage drop. Thus we see that on 240- and 480-V systems, 230- and 460-V motors are completely suitable.

The difficulty that normally arises is in application of 230- and 240-V motors to the 208-V systems. Despite the fact that motors will operate at plus or minus 10% voltage, 230- and 240-V motors *should not* be used on 208-V systems. Instead, motors specifically wound for 200 V should be specified. A brief summary of the effects of undervoltage is given in Table 17.4.

TABLE 17.3 **System and Utilization Voltages**[a]

Standard Voltages[b]			
System Voltage (Transformers)		*Utilization Voltage*[c] *(Motors)*	
Nominal	*With 4% Drop*[c]	*Current Standard*[d]	*Obsolete Standard*
120	115.2	115	110
208[d]	199.7	200	208
240[d]	230.4	230	220
480	460.8	460	440
600	576.0	575	550

[a]To eliminate any confusion between system and utilization voltages, the current NEMA standards are tabulated above.

[b]When specifying transformers, use system voltages; for motors, use utilization voltage.

[c]Note that utilization voltage corresponds to a 4% drop from system voltage, well within the normal motor tolerance.

[d]Motors for 208-V systems are rated 200 V. Motors for 240-V systems are rated 230 V. They cannot be used interchangeably without seriously affecting motor performance.

17.4 Grounding and Ground-Fault Protection

The vast majority of secondary wiring systems are grounded. The reasons for this arrangement are several and varied. Among them are

TABLE 17.4 **Effects of Undervoltage on Utilization Equipment**[a]

Load	*10% Undervoltage*
Lighting	
Incandescent	Output reduced 30%
Fluorescent	Output reduced, poor start
Mercury	Low output, poor start
Motors	20% lower torque, hotter operation, reduced life, overloading
Heaters	20% reduction in output
Small tools	Stalling, low power

[a]Computers and peripherals are generally supplied with voltage-regulating power supplies which make them tolerant of ±10% variation in steady-state supply voltage. Conversely, they are highly intolerant of rapid supply voltage fluctuation.

1. To prevent sustained contact between the low-voltage secondary system and the high-voltage primary system in the event of an insulation failure. Such contact could cause a breakdown of the secondary system insulation and severely endanger the system's users.
2. To prevent single grounds from going unnoticed until a second ground occurs, which would extensively disable the secondary system.
3. To permit locating ground faults with ease.
4. To protect against voltage surges.
5. To establish a neutral at zero potential for safety and for reference.

Points (2) and (4) are quite technical and a full explanation is beyond the purpose of this book. Point (5) requires that the neutral in a single secondary system is

1. Never interrupted by switches or other devices.
2. Connected to ground only at one point—the service entrance.
3. Color coded white or natural gray for easy recognition.

A typical service-grounding diagram is given in Fig. 17.8. Universal acceptance and use of grounded secondary 120-V systems *introduces* another shock hazard while eliminating the dangers described above. This is shown in Fig. 17.9*a*. An accidental fault within an appliance can connect the metal case of the appliance to the line. This may readily occur with such common devices as an electric saw, clothes washer, dryer, or food mixer. A person contacting the appliance housing and simultaneously a ground, such as a water pipe, would receive a nasty 120-V shock. If the hands were wet, the shock could be fatal. Until such an incident occurred, however, the internal fault would remain an unnoticed but constant source of danger.

To eliminate this hazard, appliance manufacturers have always recommended that appliance housings be grounded to a cold water pipe and supply their appliances with 3-wire plugs: two wires connected to the appliance and the third wire to the housing. To accommodate such plugs and to provide a ground path, the National Electric Code requires all receptacles to be of the grounding type and all wiring systems to provide a ground path, separate and distinct from the neutral conductor (see Fig. 16.28). The result of such wiring is shown in Fig. 17.19*b*, where the ground current passes harmlessly through the internal fault, along the ground-wire path and back to the panel. A person contacting the appliance housing establishes a parallel ground path. However, since this path is usually of much higher resistance than the ground-conductor path, only a very small current will flow. Wet hands materially reduce contact resistance, and shock current can increase to a dangerous level. If the ground current is high, the branch circuit breaker or fuse will open, disconnecting the circuit.

When wiring systems are installed in metallic conduit, the conduit itself or the conduit plus a separate conductor within the conduit may be used as the grounding path. This latter method with the additional green ground wire is very much preferable, as explained in Fig. 17.9. When nonmetallic or flexible metallic wiring is used ("Romex" or BX), a separate grounding conductor run with the regular circuit conductors *must* be used. All insulated ground conductors must have their covering colored green for identification as a grounding conductor. Many industrial installations install complete "green-ground" systems in an attempt to eliminate shock hazard and to reduce insulation failures. This has not been entirely successful for the reason alluded to above; that is, in order to clear the ground fault, current must be high enough to trip the branch circuit protective device. Otherwise, the ground fault continues to "leak," unnoticed by the system's protective devices. Unfortunately, ground faults are by nature low-current, leak-type faults, because they result from weak spots in insulation, dirt accumulation, and so on. Therefore, although the shock hazard is greatly reduced by the green-ground path, the fault continues to leak and arc until it becomes large enough to cause a major breakdown, frequently accompanied by fire.

To eliminate this dangerous situation, which occurs anytime there is a leak of current to ground in an electric circuit, the ground-fault

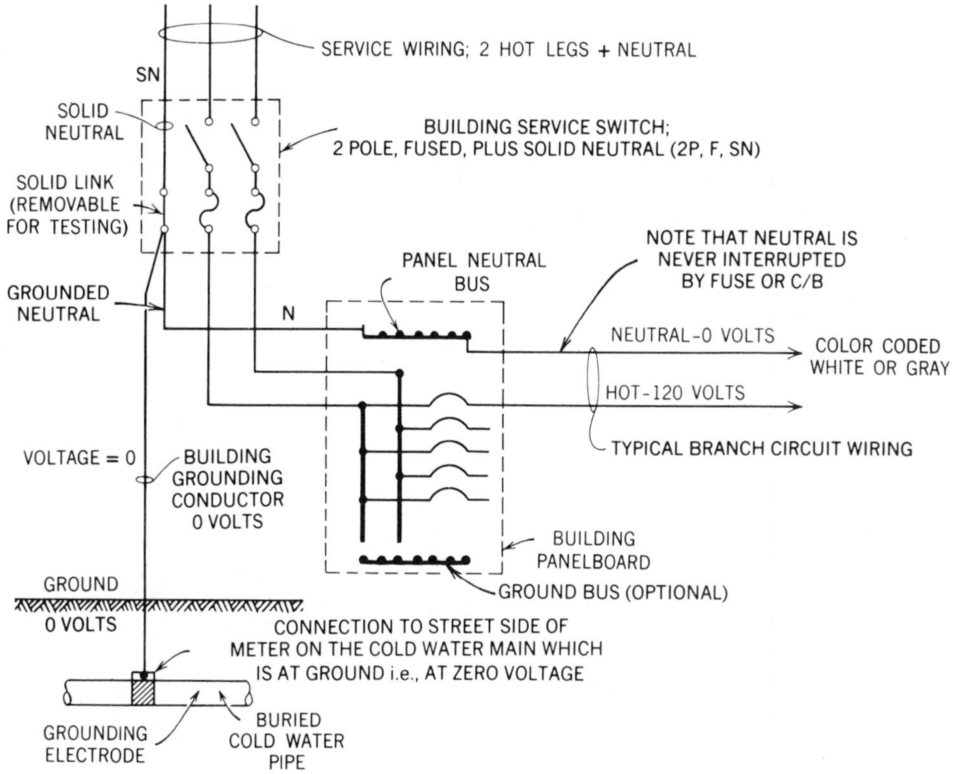

Fig. 17.8 *Typical service-grounding arrangement. Note that the grounded neutral is unbroken throughout. The ground bus, if present, is separate and distinct from the neutral bus, and both are grounded at the service entrance point. For the sake of clarity bonding connections, which are required to ground all non-current-carrying metal parts of the electrical system (boxes, enclosures, conduits, etc.) are not shown.*

circuit interrupter (GFCI or GFI) was developed (see Fig. 17.9*c*). This device compares, with extreme precision, the current flowing in the hot and neutral legs of a circuit; if there is a difference, it indicates a ground fault and the device trips out. The rapidity of this action—approximately half a second—eliminates the possibility of dangerous shock hazard, which exists even in a properly grounded circuit, as in Fig. 17.9*b*, and, all the more so, for the circuit arrangement of Fig. 17.9*a*. The separate ground path required by the NEC (bonded metallic conduit system, green ground or both) serves to minimize the shock current taken by a person before the GFCI operates. The device can be applied at the panel to replace a normal circuit breaker (see Fig. 16.7*b*) or at an individual outlet to replace a normal receptacle device (see Fig. 16.29). It is advisable to use GFI devices on all appliance circuits (see the NEC for locations where GFI use is mandatory, such as outdoors and in kitchens and bathrooms). Application to lighting circuits is generally not required, since fixtures are usually out of reach and are switch controlled. In mixed lighting and receptacle circuits, the GFCI is best applied at the outlet that is to be protected.

17.5 Energy Conservation Considerations

Before proceeding with a detailed description of design procedure, we will survey in this section many of the energy conservation ideas and techniques applicable to electrical distribution systems. This is done for ease of reference and crossreference, since the individual items appear throughout the lengthy design procedure. ASHRAE standard 90, latest issue, which is

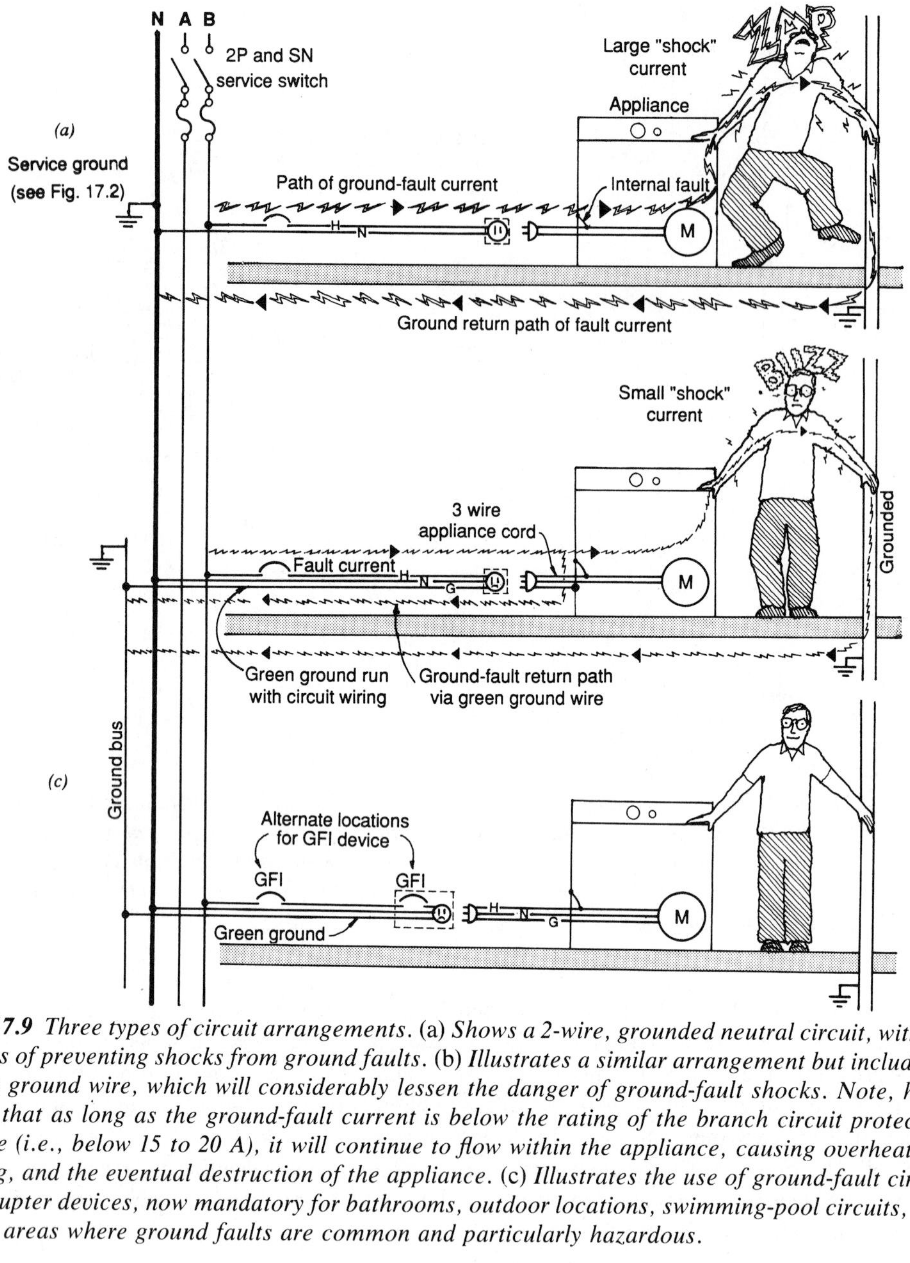

Fig. 17.9 *Three types of circuit arrangements.* (a) *Shows a 2-wire, grounded neutral circuit, with no means of preventing shocks from ground faults.* (b) *Illustrates a similar arrangement but includes a green ground wire, which will considerably lessen the danger of ground-fault shocks. Note, however, that as long as the ground-fault current is below the rating of the branch circuit protective device (i.e., below 15 to 20 A), it will continue to flow within the appliance, causing overheating, arcing, and the eventual destruction of the appliance.* (c) *Illustrates the use of ground-fault circuit interrupter devices, now mandatory for bathrooms, outdoor locations, swimming-pool circuits, and other areas where ground faults are common and particularly hazardous.*

mandated legally in most states, calls for a design of energy distribution systems "to conserve energy" (Section 8.0, Energy Distribution Systems). The following paragraphs are intended to fulfill that broad directive.

1. After establishing an energy budget based on projected loads and normal operation, set an energy reduction figure of 10 to 20% and accomplish it with the techniques given below. Annual energy consumption estimates are best made with the aid of one of the many computer programs available for that purpose.

2. Recognize the energy-use characteristics of all materials and systems specified (see Tables 15.5 and 16.4). In general, select high-efficiency equipment (motors, transformers, etc.).

If not available as such, use materials and equipment with the lowest temperature rise, since these have lowest losses. Economic justification can be established with life-cycle cost analysis. To avoid making a detailed cost analysis on every item, utilize one of the many available shortcut calculations for pay-back time (see Appendix E). This will generally indicate the material of choice. When comparisons are close, a detailed analysis may be necessary.

3. Provide electric load control equipment (demand control), either as part of an overall building control system or separately (see Sections 14.13 and 21.2).

4. In any multitenant residential building, provisions should be made to separately determine the energy consumed by each tenant. Where local codes and regulatory agencies permit, each tenant should be made financially responsible for the energy he uses. Exceptions to this rule would be made in the case of hotels, dormitories, and other transient facilities.

5. Where a choice of service voltages is available, select the highest available. Similarly, utilize the highest voltage in each class for interior distribution systems. This means 480 V in the 600-V class, 4 kV in the 5-kV class, and 13 kV in the 15-kV class. The result will be low line losses, small panelboards at the branch circuit level, and generally lower electrical contract cost (see Section 17.3).

6. In any building, the maximum total voltage drop *shall not* exceed 3% in branch circuits or feeders, for a total of 5% to the farthest outlet based on steady-state design load conditions.

7. Avoid the use of electric heating elements if alternatives are available. Electric heat is an inefficient use of national resources because of the low efficiency of fuel-to-electricity conversion.

8. Provide metering points (for fixed or plug-in meters) throughout the system to permit accurate analysis of power and energy use. Meters, both instantaneous reading and recording types, can provide data on equipment loading, load patterns, load coincidence, power factor, load voltage, power demand, and energy consumption. Analysis of these data will indicate how to program and shift loads for maximum operational efficiency. A flexible design that permits this load shifting is assumed (see item 10 below).

9. Include provisions for power factor correction in the system both at the device and at the feeder level. Then, if metering (see item 8 above) indicates the necessity, add capacitors as required. High power factor reduces line losses, permits maximum utilization of equipment capacity, and avoids utility penalty charges. Utilization equipment rated greater than 1000 W and lighting equipment greater than 15 W, with an inductive reactance load component, should have a power factor of not less than 85% under rated load conditions. Utilization equipment with a power factor of less than 85% should be corrected to at least 90% under rated load conditions.

10. Size equipment as close as possible to the load. This normally results in maximum efficiency and high power factor. Where load varies considerably, for example, between day and night or weekend, consider splitting the loads so that part of the equipment can be switched off when load is low, and the remaining load can be fed from a "night and weekend" feeder. Such a design permits one to shut down whole systems rather than operate at very low load with concomitant high losses and low power factor. The design must be sufficiently flexible to permit shifting of loads between feeders if measurements on the operational facility indicate that this is desirable (see item 8 above). The purpose would be to operate equipment near rated load and deenergize lightly loaded sections.

11. Use the most efficient type of control. This means solid-state control for motors, remote switch control for blocks of lighting, electronic control systems for elevators, and so on.

12. Arrange automatic time controls for 24-h loads such as vent fans, water coolers, vending machines, and calculators.

13. Seal all electric riser shafts to avoid heat loss by stack action.

14. Generally select the coolest possible lo-

cations for electric equipment. Low ambient temperature (below 40°C) permits use of smaller equipment for the same load with concomitant lower cost and losses. Thus, if below-grade space is available, it is well suited for this purpose.

15. Provision for future expansion should be made by means of a design that will accommodate additional equipment, in lieu of oversizing equipment initially. Here again the higher cost of two pieces of equipment can normally be justified by a detailed owning-and-operating cost analysis using realistic cost escalation and capital cost figures.

16. Energy-conservation techniques in lighting and lighting control are found in Sections 18.23, 18.24, 20.5, 20.7, 21.2, and 21.30.

17. Energy conservation techniques in vertical transportation are found in Sections 23.17, 24.1, 24.2, 24.3, and 25.12.

17.6 Terminology and Definitions

The succeeding discussions will utilize terms common in the electrical field. These terms are used in consonance with definitions given in Article 100 of the National Electrical Code. The reader is referred to that source for any necessary clarification.

17.7 Design Procedure

The steps involved in the electrical wiring design of any facility are outlined below. These may in some instances be performed in different order, or two or more steps may be combined, but the procedure normally used is that listed below.

1. Make an electrical load estimate based on areas involved, building data, and any other pertinent data (see Section 17.2).

2. In cooperation with the local electric utility, decide upon the point of service entrance, type of service run, service voltage, metering location, and building utilization voltage (see Sections 16.1 to 16.11 and Section 17.5, item 5).

3. Determine with the client the usage of all areas, and type and rating of all client-furnished equipment including their specific electric ratings and service connection requirements.

4. Determine from other consultants such as HVAC, plumbing, elevators, kitchen, and the like, the exact electrical rating of all the equipment in their designs. This determination will often be made after conferences during which the electrical consultant makes valuable recommendations to these other specialists about the comparative characteristics and costs of equipment (see Section 17.5, items 2, 7, 9, and 10).

5. Determine the location and estimate the size of all required electric equipment spaces including switchboard rooms, emergency equipment spaces, electric closets, and so forth. Panelboards are normally located in closets but may be located in corridor walls or elsewhere. This work is necessary at this point to enable the architect to reserve these spaces for electrical equipment. Once the design is accomplished in detail, the estimated space requirements can be checked and necessary adjustments made.

6. Design the lighting for the facility. This step, as will be discussed in detail in Chapters 18 to 21, is complex and involves a continued interaction between the architect and the lighting designer.

7. Depending on the type of facility, it may be necessary to separate the lighting layout from the receptacle and signal device layout for the sake of clarity of the plans. Once the decision has been made as to how this is to handled, the lighting fixture layout can be made.

8. On the same plan, or on a separate plan, as decided, locate all electrical apparatus including receptacles, switches, motors, and other power-consuming apparatus. Underfloor, under-carpet, and "over-ceiling" wiring and raceway systems would be shown at this stage. If extensive, a separate plan is made.

9. On the plans, locate data-processing and signal apparatus such as telecommunication outlets, network connections, phone outlets, speakers, microphones, TV outlets, fire and smoke detectors, and so on. At this stage provision is also made for load control wiring, building auto-

matic control wiring, programmable control and computer wiring, and the like. Since many of these systems are covered by special drawing or by specification alone, this step may be limited to showing outlets only. The material that follows deals with wiring design only; all further discussion of signal equipment will be reserved for Chapter 22.

10. Circuit all lighting, devices, and power equipment to the appropriate panels, and prepare the panel schedules. Include in this step the circuitry for emergency equipment.

11. Compute panel loads.

12. Prepare the riser diagram. This includes design of distribution panels, switchboards, and service equipment.

13. Compute feeder sizes and all protective equipment ratings.

14. Check the preceding work.

15. Coordinate the electrical work with the other trades and with the architectural plans. This is not really a separate step, but a continuing process starting at (9) above and covering all subsequent stages of the work.

The material for items 1, 2, 3, 4, 6, 7, and 9 is covered elsewhere in this book. The remaining steps, that is, 5, 8, 10, 11, 12, 13, and 15, will be discussed in order below.

17.8 Electric Spaces

The spaces required for electric equipment in a facility vary radically, depending on the design and nature of the building. The spaces required around major pieces of electrical switchgear and transformers are discussed above (see Fig. 16.22). The NEC in Article 110 further specifies the minimum working spaces required in front of electrical equipment. In general, a minimum of 42 in. of clear space should be maintained in front of panels, switches, and other electrical apparatus.

(a) Residences. In private residences, the service equipment and the building panelboard are generally incorporated into a single unit. The main disconnect(s) is usually installed as the main switch/breaker of the panel. A number of typical residential service-panel arrangements are shown in Fig. 17.10. The panel is normally placed in the garage, utility room, or basement. To minimize voltage drop, the panel should be placed as close to the major electrical loads as practicable, without sacrificing valuable space or making the panel inaccessible. Frequently a smaller panel can be subfed from the main panel to feed the kitchen and laundry loads. In apartments, panels are normally placed in the kitchen or the corridor immediately adjoining the kitchen. This location is chosen so that the panel circuit breaker may act as the required disconnecting means for fixed appliances rated at not more than 300 VA or ⅛ hp, as permitted by NEC.

(b) Commercial Spaces. The location of the required panelboards depends on their type and number, and on availability of space. In the research building of which Fig. 17.11 is a part plan, lighting panels are recessed into the corridor wall, since the building is only two stories high and the panels can be vertically stacked and fed by a single conduit. If this building were six or more stories high, an electric closet (see Section 17.9) would be advisable to accommodate the panel and riser conduits. Of course, when panels are installed in finished areas such as corridors, flush mounting is required.

To limit the voltage drop on a branch circuit in accordance with the Code requirements, panelboards should be located so that no circuit exceeds 100 ft in length. If circuits longer than this are unavoidable, No. 10 AWG wire should be used for runs of 100 to 150 ft and No. 8 AWG for longer circuits. These wire sizes apply to 15- or 20-A branch circuits, which are normally wired with No. 12 AWG wire.

The laboratory between the two offices of Fig. 17.11 is intended to function as a self-contained unit and is therefore equipped with its own panel. Multi-outlet assemblies, all wiring within the room, and the panel itself are surface-mounted to allow ready access to all components for the frequent rewiring encountered in laboratories. A main circuit breaker should be provided in such a panel to act as a main disconnect, whether required by Code or not. Where

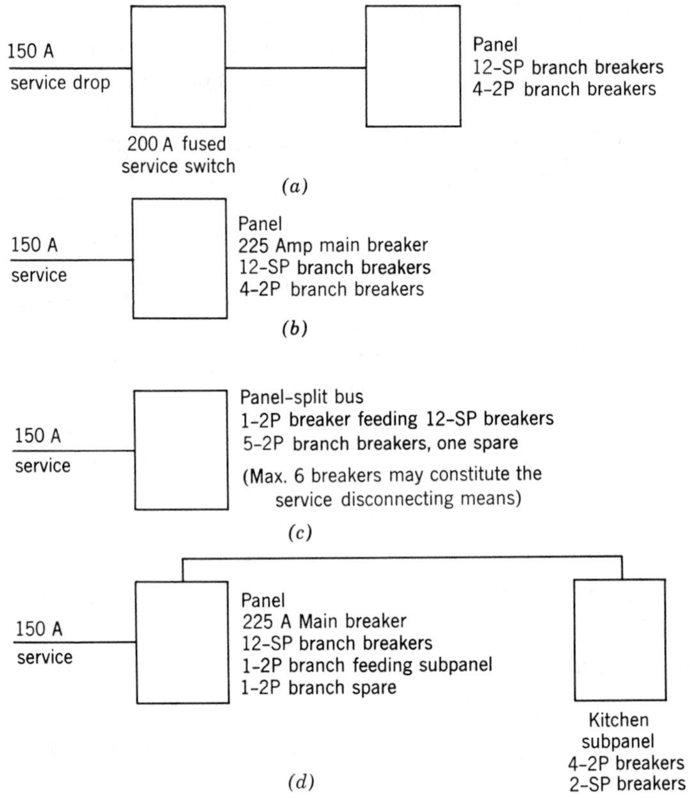

Fig. 17.10 *Typical 150-A service arrangements, applicable to residences and small commercial buildings. The service switch can be a separate unit* (a) *or combined with the branch circuit panelboard* (b–d). *The panel may be a single unit* (a, b), *two units in a single enclosure, fed by separate service switches* (c), *or a central panel and a subfed single use panel* (d), *in this case a kitchen subpanel.*

panels are convenient to the load controlled, the panel circuit breakers may be used for switching.

Panels supplying large blocks of load simultaneously switched, such as auditorium house lights, lobby lights, large single-use office areas, store lighting, and the like, can be constructed with built-in electrically operated, mechanically held contactors to switch the entire panel, with control at any desired remote location. These remote-control (RC) switches are discussed in Section 16.15. If only part of the panel's circuits is so arranged i.e., for remotely controlled block switching, a split bus panel can be used, with the RC switch controlling only that part of the panel's load (see Fig. 17.12).

Small offices, stores, and other small buildings frequently have lighting panels mounted in a convenient finished area and utilize the panels' circuit breakers for load switching. In large buildings, strategically located electric closets are provided to house all electrical supply equipment. Power panels and distribution panels are located as required by the loads fed through them.

In general, branch circuit panels, distribution panels, and switchboards are best located near the electrical load center. This minimizes feeder length and reduces voltage drop, making it the most economical arrangement.

Every completely enclosed switchgear room, emergency generator room, or transformer vault should be equipped with an emergency light source (see Section 21.30). In generator rooms these should be battery operated to give illumination for generator repairs in the event of generator failure during a power outage.

17.9 Electric Closets

In the design of a building electric system, particularly in multistory construction, it is often advantageous and convenient to group the electrical equipment in a small room called an electric closet (see Fig. 17.13). The shape of this space can be varied to fit the architectural and

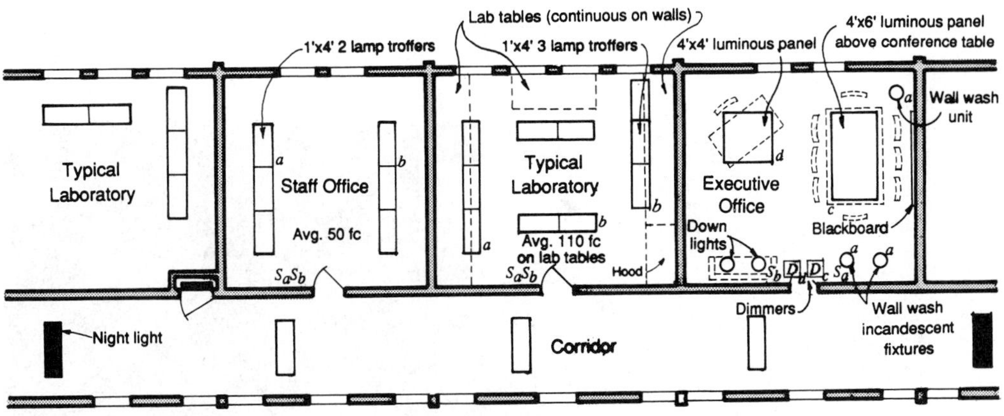

(a) LIGHTING PLAN

NOTES:

1. All receptacles in staff offices at 42" a f f to center line
2. All receptacles in executive offices at 12" a f f to c l
3. Surface raceway in labs mounted at 48" a f f to c l
4. Special outlets:

A 20 amp. 250 v 3 wire 2 pole grounded outlet (for 208 v 1/3 hp centrifuge)

B 30 amp. 250 v 3 wire 2 pole grounded outlet (for 208 v 6 kw kiln)

C 50 amp. 250 v 4 wire 3 pole grounded outlet (for 10 hp portable M-G set)

D 20 amp. 120 v/ 20 amp. 250 v 3 wire 2 pole grounded outlet (for cleaning and special use)

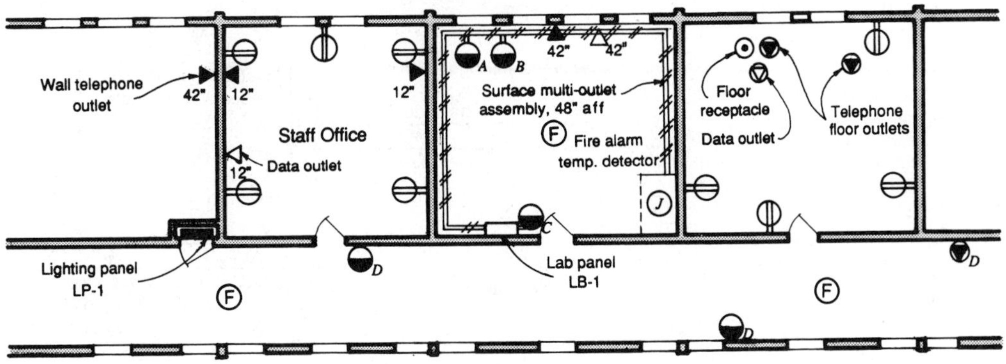

(b) POWER PLAN

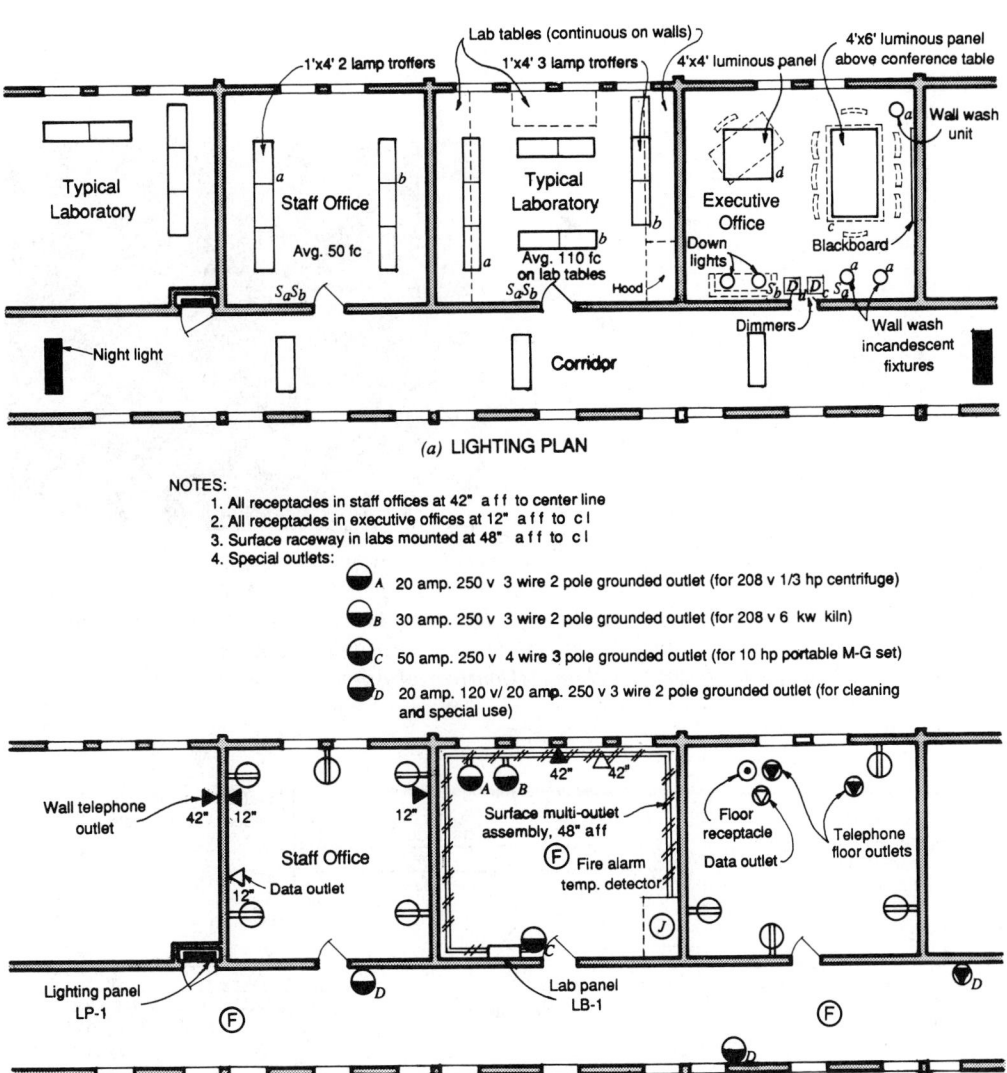

Fig. 17.11 Typical floor plans for lighting and power for a section of an office–laboratory building. Separate lighting and power plans are drawn for the sake of clarity. For circuited plan, see Fig. 17.21. Data outlets handle data processing and telecommunications cabling.

electrical demands, but it should provide the following:

1. One or more locking doors.
2. Vertical stacking, above and below other electric closets and located so as not to block conduits entering or leaving horizontally. Thus locations on outside walls and adjoining shafts, columns, and stairs are poor.
3. Space free of other utilities such as piping or duct passing through the closet, either horizontally or vertically.
4. Sufficient wall space to mount all requisite and future panels, switches, transformers, telephone cabinets, and communication equipment. Wall cabinet space must be coordinated with raceway connections to underfloor ducts and over-the-ceiling raceways systems (see Sections 15.25 to 15.30).

ELECTRICITY

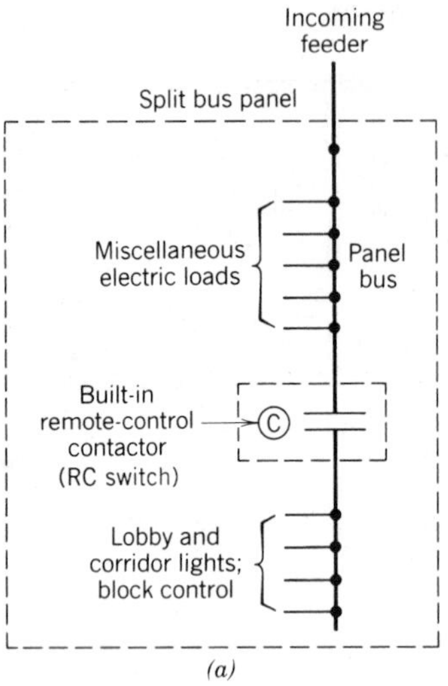

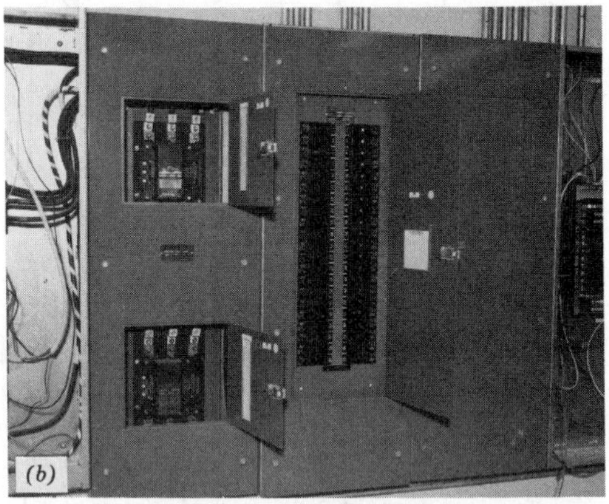

Fig. 17.12 (a) *Single-line diagram of a split bus panel. This design is used when a block of load is to be controlled as a unit. Control point is either remote, at the panel, or both. Contactor (RC switch) is electrically operated, mechanically latched.* (b) *RC switches (mechanically held contactors) are shown to the left of a bank of panels. The contactors were installed to accomplish photocell and timer control of a department store's lighting. (Courtesy of Automatic Switch Co.)*

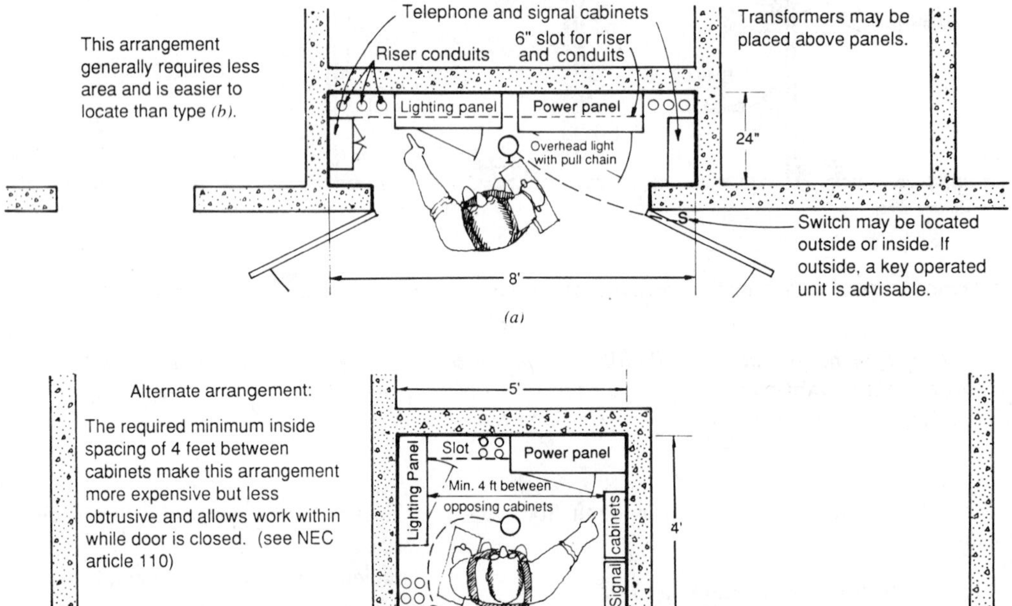

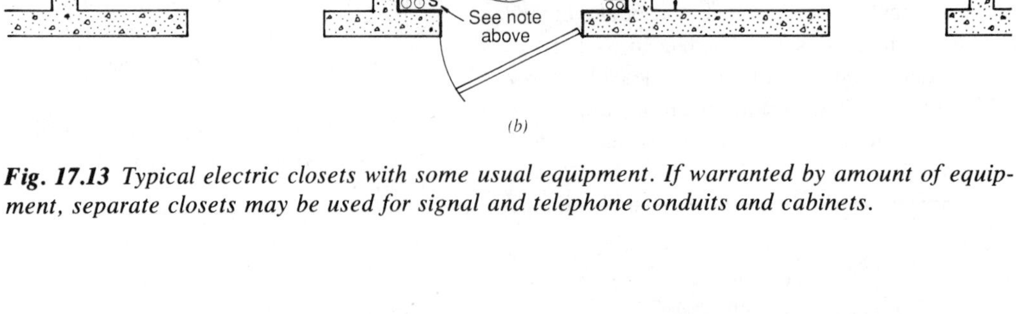

Fig. 17.13 Typical electric closets with some usual equipment. If warranted by amount of equipment, separate closets may be used for signal and telephone conduits and cabinets.

5. Floor slots or sleeves of sufficient size for all present and future conduit or bus risers.
6. Sufficient floor space so that an electrician can work comfortably and safely on initial installation and repair.
7. Adequate illumination and ventilation.

17.10 Equipment Layout

Wiring devices, principally comprising receptacles and switches, are located as required by the equipment to be served and by the anticipated area use.

Switches for control of lighting or receptacles are normally placed on the strike side of the door. Other devices such as plug-in strips on walls and special-purpose receptacles are shown and identified. Signal outlet locations are often noted but generally remain uncircuited on floor plans, a riser or floor raceway plan being utilized to show interconnections. These include fire-alarm equipment, telephone and intercom equipment, data and communications, radio and TV outlets, thermostats and so on. These devices may be identified by a special symbol or note where a standard symbol is not available.

As mentioned above, lighting fixture outlets are normally placed on the same drawing as wiring devices unless the large number of the latter precludes showing the lighting, without undue cluttering of the drawings. In such an event, the lighting is shown on one drawing and receptacles on another, with signals shown on the one least occupied. A ceiling or underfloor wiring system would probably necessitate such separation [see Fig. 15.28]. Motors, heaters, and other fixed and permanently wired equipment are shown and identified on the receptacle drawings (also called power drawings, in contradistinction to lighting drawings). Equipment furnished with a cord and plug is not normally shown. However, the receptacle intended for supplying such a device is shown and identified. A typical device layout is shown in Fig. 17.11. An abbreviated symbol list is given in Fig. 17.14. For a complete electrical symbol list, see McGuiness et al. (1980, pp. 665–672).

17.11 Application of Overcurrent Equipment

Before beginning an explanation of circuiting, it is necessary to explain the principles underlying overcurrent protection. As outlined in Chapter 16, the function of an overcurrent device is to open (interrupt) a circuit when the current rating of the equipment being protected is exceeded. These overcurrent devices are placed in circuits to protect wiring, transformers, lights, and all other equipment that can be damaged by excessive current. The following general rules govern the application of overcurrent protection:

1. Overcurrent devices must be placed on the line or supply side of the equipment being protected (see Fig. 17.15).
2. Overcurrent devices must be placed in all *ungrounded* conductors of the protected circuit (see Fig. 16.24*b*).
3. All equipment should be protected in accordance with its current-carrying capacity (see Fig. 17.16).
4. Conductor sizes shall not be reduced in a circuit or tap, unless the smallest size wire is protected by the circuit overcurrent devices (see Fig. 17.17).
5. Overcurrent devices shall be located so as to be readily accessible.

17.12 Branch Circuit Design

A branch circuit by NEC definition refers only to the circuit conductors, although for our purposes and in trade parlance it includes the protective device and the outlets served. Such circuits may be multi-outlet general-purpose type (Fig. 17.18*a*), multi-outlet appliance type (Fig. 17.18*b*), or single-outlet type intended for a specific piece of equipment (Fig. 17.18*c*). The multi-outlet types are limited to 50 A in capacity, while the single-outlet type is governed in size only by the requirements of the item being served, and may be 200 or 300 A in size.

In its simplest form, a branch circuit comprises only two wires. However, multiwire branch circuits carrying 2- or 3-phase wires plus a neutral are also widely used. Generally, each branch circuit should be sized for the load connected to it, plus the load expansion that is ex-

ELECTRICAL SYMBOL LIST

RACEWAYS

	CONDUIT AND WIRING CONCEALED IN CEILING OR WALLS TICS INDICATE NO. OF CONDUCTORS EXCLUDING GROUNDS; 2 #12, ¾" CONDUIT UON.
	CONDUIT AND WIRING CONCEALED IN OR UNDER FLOOR
	CONDUIT AND WIRING EXPOSED
F-6	FEEDER F-6, SEE SCHEDULE, DWG NO.
— BX —	BX WIRING
— NM —	NONMETALLIC CABLE (ROMEX) WIRING
2,4 2PLA 3# 12, ¾" C	HOME RUN TO PANEL 2PLA—NUMERALS INDICATE CIRCUITS, 3 #12 AWG, ¾" RIGID STEEL CONDUIT
TC2A	HOME RUN TO TELEPHONE CABINET TC2A
— EC_T —	EMPTY CONDUIT, SUBSCRIPT INDICATES INTENDED USE T—TELEPHONE, IC—INTERCOM, FA—FIRE ALARM, ETC.
4" × 4" A	SURFACE METAL RACEWAY, SEE NOTE ______, DWG ______ SIZE AND RECEPTACLES AS SHOWN

OUTLETS

CEILING	WALL	
(A) a	(A) 3	OUTLET AND LIGHTING FIXTURE: LETTER IN CIRCLE INDICATES TYPE—SEE SCHEDULE. SUPERSCRIPT NO. INDICATES CIRCUIT. SUBSCRIPT LETTER INDICATES SWITCH CONTROL. RECTANGLE INDICATES FLUORESCENT
A	B	OUTLET AND EXIT SIGN FIXTURE, UPPER CASE LETTER INDICATES TYPE, ARROWS INDICATE REQUIRED SIGN ARROWS
(E)	(E)	OUTLET BOX, BLANK COVER—NOTE 2
(J)	(J)	JUNCTION BOX, BLANK COVER

WIRING DEVICES

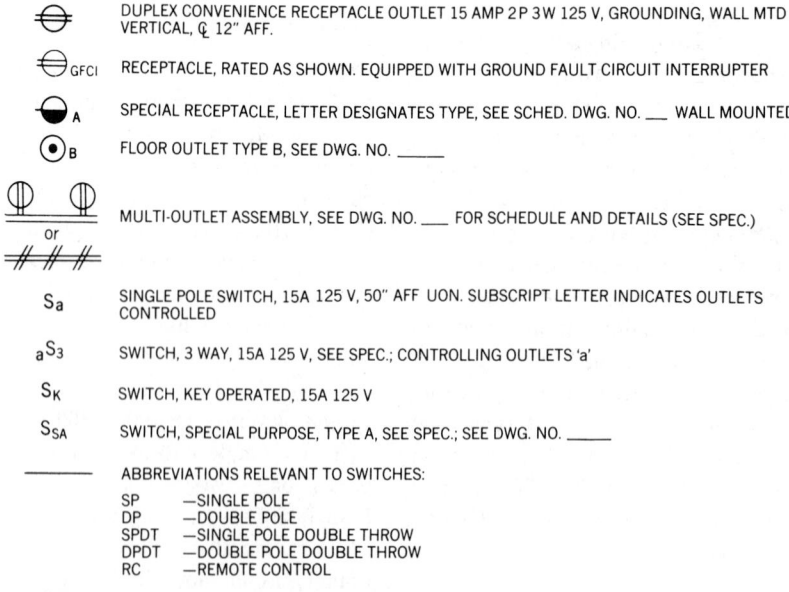

	DUPLEX CONVENIENCE RECEPTACLE OUTLET 15 AMP 2 P 3 W 125 V, GROUNDING, WALL MTD VERTICAL, ℄ 12" AFF.
GFCI	RECEPTACLE, RATED AS SHOWN. EQUIPPED WITH GROUND FAULT CIRCUIT INTERRUPTER
A	SPECIAL RECEPTACLE, LETTER DESIGNATES TYPE, SEE SCHED. DWG. NO. ___ WALL MOUNTED
B	FLOOR OUTLET TYPE B, SEE DWG. NO. _____
or	MULTI-OUTLET ASSEMBLY, SEE DWG. NO. ___ FOR SCHEDULE AND DETAILS (SEE SPEC.)
S_a	SINGLE POLE SWITCH, 15A 125 V, 50" AFF UON. SUBSCRIPT LETTER INDICATES OUTLETS CONTROLLED
$_aS_3$	SWITCH, 3 WAY, 15A 125 V, SEE SPEC.; CONTROLLING OUTLETS 'a'
S_K	SWITCH, KEY OPERATED, 15A 125 V
S_{SA}	SWITCH, SPECIAL PURPOSE, TYPE A, SEE SPEC.; SEE DWG. NO. _____
	ABBREVIATIONS RELEVANT TO SWITCHES: SP —SINGLE POLE; DP —DOUBLE POLE; SPDT —SINGLE POLE DOUBLE THROW; DPDT —DOUBLE POLE DOUBLE THROW; RC —REMOTE CONTROL
D	OUTLET-BOX-MOUNTED DIMMER

***Fig. 17.14** Abbreviated electrical symbol list. For complete list, see McGuinness (1980, pp. 665–672).*

LIST OF ABBREVIATIONS

A,a	AMPERES
AFF	ABOVE FINISHED FLOOR
C/B	CIRCUIT BREAKER
CCT	CIRCUIT
EM	EMERGENCY
EC	EMPTY CONDUIT
F	FUSE
FA	FIRE ALARM
GND	GROUND
GFCI	GROUND FAULT CCT-INTERRUPTER
HP	HORSEPOWER
LTG	LIGHTING
MH	MOUNTING HEIGHT, MANHOLE
N	NEUTRAL
NC	NORMALLY CLOSED
NO	NORMALLY OPEN
NIC	NOT IN CONTRACT
OH	OVERHEAD
OC	ON CENTER
SW	SWITCH
T	THERMOSTAT
TEL	TELEPHONE
TV	TELEVISION
UON	UNLESS OTHERWISE NOTED
UG	UNDERGROUND
WP	WEATHERPROOF

EQUIPMENT

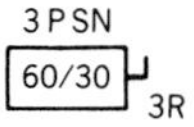

SW., 3 POLE AND SN., 60A/30 AF. NEMA 3R ENCL., ND (NORMAL DUTY) 250 V

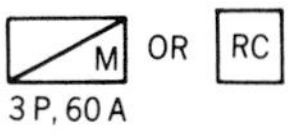

CONTACTOR, ENCLOSED, ELECTRICALLY OPERATED, MECHANICALLY HELD, 3 POLE 60 A

ITEM OF ELECTRICAL EQUIPMENT, AS INDICATED

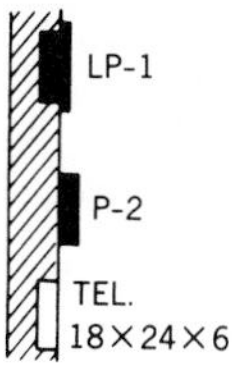

ELECTRIC PANELBOARD LP-1, RECESSED, SEE SCHEDULE ON DWG. ______

ELECTRIC PANEL (BOARD) P-2, SURFACE MTD

CABINET, MOUNTING, SIZE, AND PURPOSE AS SHOWN

SIGNALING DEVICES

BELL OR GONG, INSCRIBED LETTER INDICATES SYSTEM (SEE BELOW) AND SUBSCRIPT LETTER OR NUMBER INDICATES TYPE e.g. A—8" VIBRATING BELL, 12 V. DC; B—12" WEATHERPROOF SINGLE STROKE GONG, 120 V. DC, etc.

FIRE DETECTOR, TYPE 1
INTRUSION DETECTOR, TYPE 1 } ETC.
SMOKE DETECTOR, TYPE 2

MANUAL STATION—WATCHMEN TOUR, FIRE ALARM, ETC. LETTER INDICATES SYSTEM (SEE BELOW)

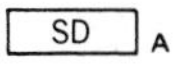

COMMUNICATIONS OUTLET, LETTER INDICATES TYPE
CABINET OR CONTROL PANEL, SMOKE DETECTION, USE IDENTIFYING TYPE LETTER IF MORE THAN ONE TYPE IS USED ON THE PROJECT

AUXILIARY DEVICE, AS INDICATED

HORN OR LOUDSPEAKER, TYPE A, TYPE 2

TELEPHONE OUTLET, TYPE B

COMMUNICATIONS OUTLET, LETTER INDICATES TYPE

CLOCK SYSTEM OUTLET, TYPE A

TV ANTENNA OUTLET

OUTLET, LETTER INDICATES TYPE

Fig. 17.14 (continued)

ELECTRICITY

SYSTEM TYPES

F, FA FIRE ALARM
S, SD SMOKE DETECTION
I, IA INTRUSION ALARM
T, TEL TELEPHONE
TV TELEVISION
IC INTERCOM
D DATA, TELECOMMUNICATION

AUXILIARY DEVICES

BATT BATTERY
S, SP SPEAKER, LOUDSPEAKER
TC TELEPHONE CABINET
WF WATERFLOW

MOTORS AND MOTOR CONTROL

COMBINATION TYPE MOTOR CONTROLLER; ATL STARTER PLUS FUSED DISCONNECT SWITCH, NEMA SIZE II, SEE SCHEDULE DWG.___ ___ ___

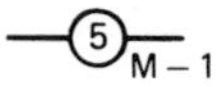

MOTOR #1, 5 HP, 3ϕ SQUIRREL CAGE UON

DEVICE 'T' (SEE LIST OF ABBREVIATIONS)

S_T MANUAL MOTOR CONTROLLER WITH THERMAL ELEMENT.

ATL ACROSS THE LINE STARTER—MAGNETIC

FS FUSED SWITCH

CB CIRCUIT BREAKER

FV FULL VOLTAGE

MCC MOTOR CONTROL CENTER

S START BUTTON—MOMENTARY CONTACT

ST STOP BUTTON—MOMENTARY CONTACT

CONTROL DIAGRAMS AND WIRING DIAGRAMS

MOMENTARY CONTACT PUSH BUTTON—NO—(START)

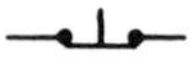
MOMENTARY CONTACT PUSH BUTTON—NC—(STOP)

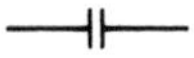
NORMALLY OPEN CONTACT—NO

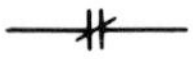
NORMALLY CLOSED CONTACT—NC

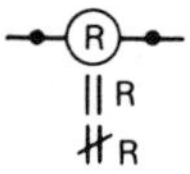

OPERATING COIL FOR RELAY OR OTHER MAGNETIC CONTROL DEVICE. WITH ONE NO AND ONE NC CONTACT. LETTERS NORMALLY USED ARE C, R FOR CONTROL COIL AND RELAY

POWER WIRING

CONTROL WIRING

ONE LINE DIAGRAMS

CONDUCTOR, SIZE, AND TYPE INDICATED

CONDUCTORS CROSSING, CONNECTED

GROUND CONNECTION

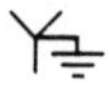
GROUNDED WYE

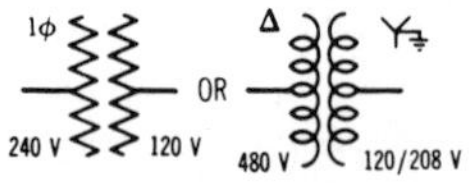

TRANSFORMER, 2 WINDING, 1ϕ INDICATES SINGLE PHASE; Δ-Y INDICATES 3 PHASE AND TYPE OF CONNECTION; NUMBERS INDICATE VOLTAGES.

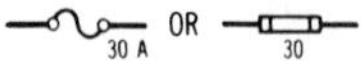

FUSE, 30 A TYPE - - -

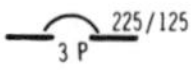

CIRCUIT BREAKER, MOLDED CASE, 3 POLE, 225 A FRAME/125 A TRIP

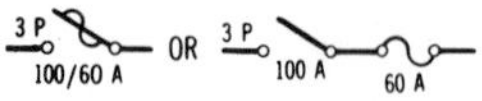

FUSED DISCONNECT SWITCH, 3 POLE, 100 A, 60 A FUSE

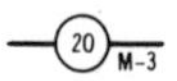

MOTOR, 3 PHASE SQUIRREL CAGE UON, 20 HP, MOTOR #3

Fig. 17.14 *(continued)*

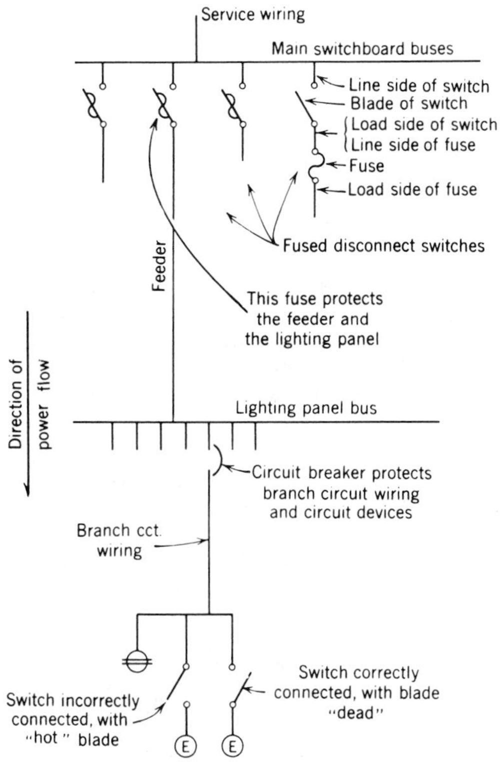

***Fig. 17.15** Location of overcurrent protective equipment. Protective equipment should always be located at the point where the conductor receives its source of supply so that when it operates the current supply is cut off.*

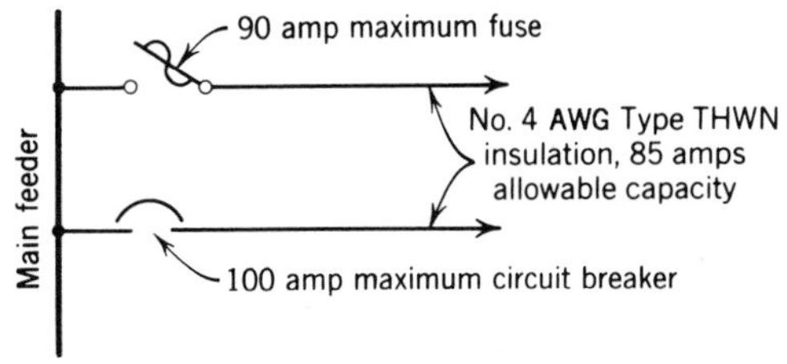

***Fig. 17.16** The overcurrent protection must correspond to the rating of the protected equipment. Where ratings do not correspond exactly, the next larger standard size for protective devices up to 800 A may be used. Above this, use the next* lower *rating. See NEC Article 240-3 for further restrictions.*

pected. These general rules of good practice should be followed:

1. In all but the smallest installations, connect lighting, convenience receptacles, and appliances on separate (groups of) circuits, although this is not an NEC requirement.

2. General-purpose branch circuits should be 20 A and wired with No. 12 AWG wire. Switch legs may be No. 14 AWG if the lighting load permits and the wiring meets the tap requirements of NEC Article 240-21.

3. Limit the circuit load on 15-A and 20-A circuits to the values shown in Table 17.5. This will provide the required building load expansion capability in the branch circuitry, that is, by bringing the loads on the branch circuits up to maximum, the additional building loads can be absorbed. However, since it is not always economical or feasible to expand existing circuits, panels are always equipped with spare circuit breakers and spaces for future circuit breakers. These then can alternatively be utilized to pick up building load expansion, or the user may use a combination of the two techniques, expanding some existing circuits and adding some new ones. (See Section 17.16 for a discussion of this subject.)

Since lighting and specific devices are circuited according to their nameplate rating, the only circuitry item left to the judgement of the designer is the number of convenience receptacles per circuit (cct). The NEC specifies that plug outlets (convenience receptacles) be counted, in totaling loads, at 1.5 A each unless included in the load for general lighting. This point requires some clarification. The receptacle load for general illumination in dwellings is included in the 3 /ft^2 specified by NEC Table 220-3(b) and found above in Table 17.1. This is because receptacle outlets on general illumination circuits are assumed to be for illumination (i.e., lamps), and therefore are necessarily included in the lighting load. Application of the 1.5 A (180 W) per outlet criterion to such outlets would, in the opinion of the NEC, unnecessarily limit the number of outlets permitted on such circuits. The same is true for appliance circuits. On both these types, according to the NEC, the number of receptacle outlets is not limited. On 15- and 20-A receptacle circuits which are neither general illumination nor small appliance branch cir-

ELECTRICITY

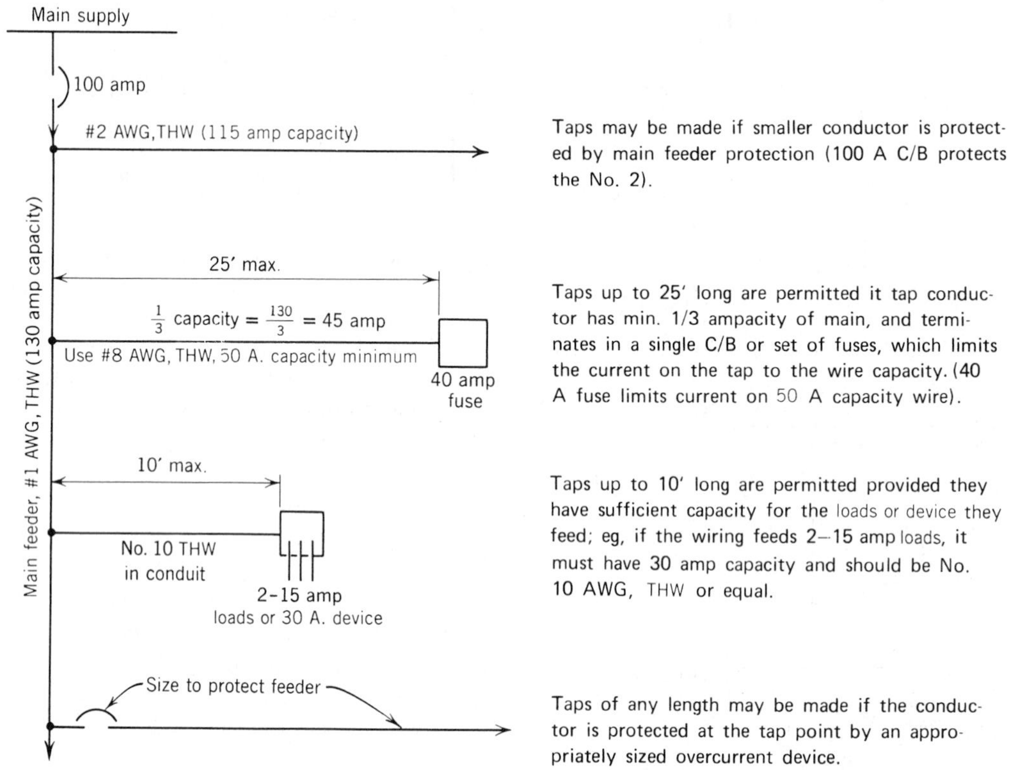

Fig. 17.17 *Permissible tap arrangements. See NEC Article 240-21 for additional restrictions on taps to those given in the figure.*

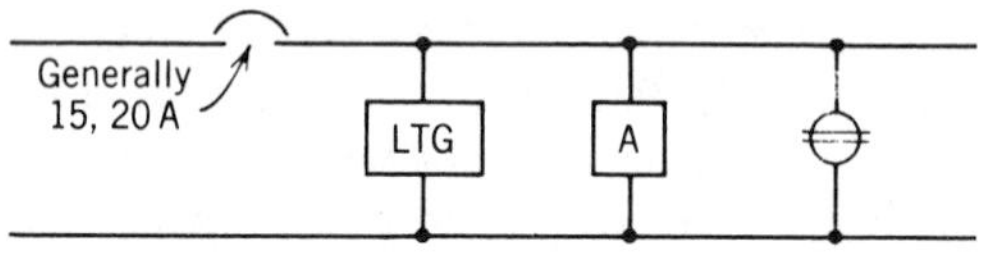

(a) General purpose branch circuit. Supplies outlets for lighting and appliances, including convenience receptacles.

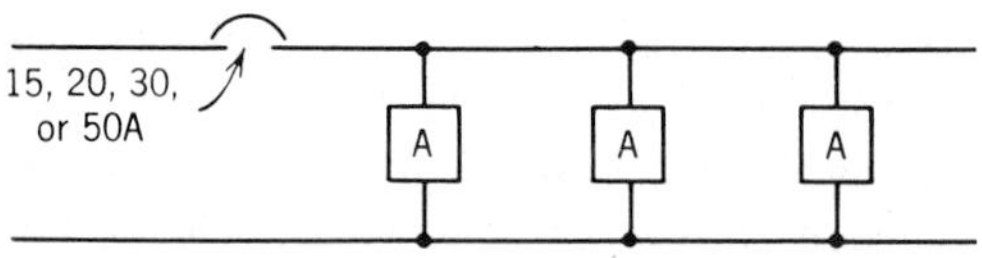

(b) Appliance branch circuit. Supplies outlets intended for feeding appliances. Fixed lighting not supplied.

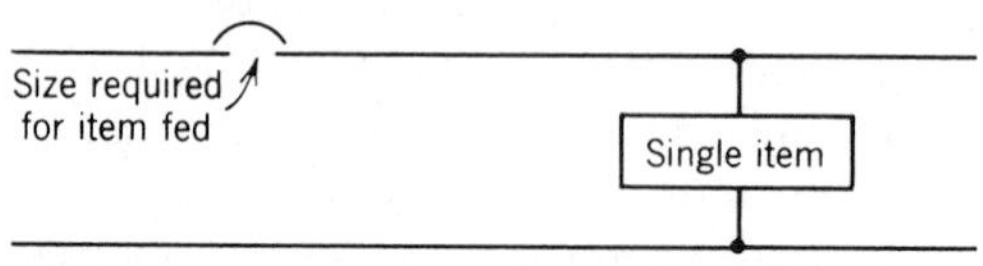

(c) Individual branch circuit, designed to supply a single, specific item.

Fig. 17.18 *Branch circuit types. See NEC Article 210.*

cuits, the number of outlets is limited to 10 (1800 W) on a 15-A circuit and 13 (2340 W) on a 20-A circuit.

It is our opinion that the distinction between circuit types is not completely clear and that good practice dictates the limitation of convenience receptacles per circuit to a considerably smaller number. Thus, following the guidelines stated above of 9- and 12-A loading on 15- and 20-A circuits, respectively, we would have by this method:

$$\text{15-A circuit } \frac{9}{1.5} = 6 \text{ outlets per cct}$$

$$\text{20-A circuit } \frac{12}{1.5} = 8 \text{ outlets per cct}$$

These figures must be used judiciously. If the devices to be energized are small, these quantities may be used. Such would be the case in a clerical-type business office where only an electric pencil sharpener or a desk lamp (100 VA) would be plugged in. However, for laboratories, modern computer equipped offices, service and repair facilities, and the like, no more than two

TABLE 17.5 **Recommended Branch Circuit Loads**[a]

Size Circuit	Circuit[d] Amperes	Volt-Amperes[d] at 120 V	Volt-Amperes at 277 V
		0% Expansion[b,c]	
15 A	12	1440	—
20 A	16	1920	4440
		25%[e] *Expansion*	
15 A	9.6	1150	—
20 A	12.8	1520	3600
		50%[f] *Expansion*	
15 A	8	960	—
20 A	11	1300	3000

[a]The loading shown will provide the needed expansion in the branch circuits. Where branch circuit expansion is not practical, the required expansion can be obtained from panel spares.

[b]See Table 17.1 for anticipated load growth.

[c]For 0% expansion, initial load = 80% of circuit rating, which is the maximum permissible for continuous loads. See Table 17.6, note 6.

[d]For branch circuits feeding utilization equipment such as air-conditioners in addition to convenience outlets, see NEC Article 210-23.

[e]To accomplish 25% expansion, the circuit is derated to 80%, i.e., 80% of 12 A = 9.6 A, etc.

[f]To accomplish 50% expansion, the circuit is derated by ⅓, i.e., ⅔ of 12 A = 8 A, etc. (Then 50% expansion on 8 A yields 12 A.)

ELECTRICITY

or three receptacles should be used or a 20-A circuit. Of course, diversity of use is an all-important factor and the more accurately it can be estimated, the better the design result.

A further note of caution is in order. Receptacles should be arranged, if at all possible, so that the loss of a single circuit does not deprive an entire area of power. That is, for the sake of reliability, circuitry should be alternated to give each space parts of different circuits.

NEC Table 210-24 specifies certain requirements for conductors, devices, and loads permissible on general-purpose branch circuits. These are summarized in Table 17.6.

17.13 Branch Circuit Design Guidelines; Residential

1. The NEC (1990 edition) requires for residences sufficient circuitry to supply a load of 3 W/ft^2 in the building, excluding unfinished spaces such as porches, garages, and basements. Using Table 17.5, this works out to 480 ft^2 on a 15-A circuit (1440 VA) and 640 ft^2 on a 20-A cct (1920 VA). Allowing for some expansion (better practice) reduces these figures to 400 ft^2 on a 15-A cct and 500 on a 20-A cct.

2. The NEC (Article 220-4) requires a minimum of two 20-A appliance branch circuits (see Fig. 17-18*b*) to feed all the receptacle outlets in the kitchen, pantry, breakfast, and/or dining room and similar areas, and *only* these outlets. This is because any receptacle in these areas is a potential appliance outlet and must be fed and circuited as such. An exception is made in that clock outlets and outdoor receptacles may also be fed from these circuits. Furthermore, the NEC requires that all kitchen outlets intended to serve countertop areas must be fed from at least two of these appliance circuits (which circuits may also feed appliance outlets in the other spaces specified above). Thus all countertop work space will not be deenergized by the failure of a single circuit. Although the Code does not limit the number of appliance outlets to be wired into each circuit, good practice dictates

TABLE 17.6 **Branch Circuit Requirements**

	Branch Circuit Size				
	15 A	20 A	30 A	40 A	50 A
Minimum size copper conductors	No. 14	12	10	8	6
Minimum size taps	No. 14	14	14	12	12
Overcurrent device rating	15 A	20	30	40	50
Lampholders permitted	Any type	Any type	Heavy duty	Heavy duty	Heavy duty
Receptacle rating permitted (see note 7)	15 A	15 or 20	30	40 or 50	50 A
Maximum load (see note 6)	15	20	30	40	50

NOTES:

1. Wiring shall be types RHW, RHH, T, THW, TW, THWN, THHN, XHHW in raceway or cable.
2. On 15-A circuit, maximum single appliance shall draw 12 A. On 20-A circuit, maximum single appliance shall draw 16 A. If combined with lighting or portable appliances, any fixed appliance shall not draw more than 7.5 A on a 15-A circuit, and 10 A on a 20-A circuit.
3. On a 30-A circuit, maximum single appliance draw shall be 24 A.
4. Heavy-duty lampholders are units rated not less than 750 W.
5. 30, 40, and 50-A circuits shall not be used for fixed lighting in residences.
6. When loads are connected for long periods (3 hours or more), actual load shall not exceed 80% of the branch circuit rating. Conversely, continuous type loads shall be figured at 125% of actual load in all load calculations.
7. A single receptacle on an individual branch circuit shall have a rating not less than the circuit, for example, 15 A on a 15-A circuit, etc. 15-A receptacles on 20-A circuit shall not supply a load greater than 12-A for appliances. 20-A receptacles on 20-A circuit shall be limited to a 16-A load.

that these receptacles be circuited with no more than four such outlets per 20-A circuit. This in turn will usually require more than the Code minimum of two appliance circuits. Additionally, the NEC requires that for kitchen and dining areas:

(a) No point on a wall behind a countertop be more than 24 in. from an outlet.

(b) All countertop convenience receptacles within 6 ft of a sink shall be of the ground-fault circuit interrupter (GFCI) type.

3. Locations utilized primarily or frequently for workshop-type activities, such as garages, utility rooms, and basements, should be provided with receptacles wired in appliance-type circuits (i.e., 20-A receptacles on 20-A circuits), with no more than four such receptacles to a circuit. Receptacles in garages, crawl space, below-grade finished or unfinished basements, and outdoors must be of the GFCI type. (For exceptions to this GFCI rule, see NEC Article 210-8.)

4. Additional circuits similar to appliance circuits (no fixed lighting outlets) should be furnished to supply one outlet in each bedroom of a house that is not centrally air conditioned. Such outlets are intended for window air conditioners. (Good architectural and HVAC design will provide window arrangement, attic ventilation, insulation, sun-screening, and the like to obviate the necessity for these noisy energy users.)

5. The NEC requires that at least one 20-A appliance circuit supply the laundry outlets only. This requirement satisfies good practice. If an electric clothes dryer is anticipated (and it should be unless it is definitely known that a gas dryer will be used), an individual branch circuit (distinct from the laundry outlets circuit and rated for the load) should be supplied to serve this load, via a heavy-duty receptacle. (Obviously, facilities for hanging clothes must be provided for those who prefer not to waste energy.)

6. Lay out convenience receptacles so that no point on a wall is more than 6 ft from an outlet. Use 20-A, grounding-type receptacles

only. Do not combine receptacles and switches into a single outlet except where convenience of use dictates high mounting of receptacles.

7. Circuit the lighting and receptacles so that each room has parts of at least two circuits. This includes basements and garages.

8. Avoid placing all the lighting in a building on a single circuit.

9. Supply at least one receptacle in the bathroom and one outside the house. Both must be GFCI types. This is an NEC requirement. An additional convenience is switch control of the outside receptacle from *inside* the house.

10. In rooms without overhead lights, provide switch control for one-half of a strategically located receptacle that is intended to supply a lamp. See Fig. 17.19 for the wiring arrangement in such a case.

11. Provide switch control for closet lights. Pull chains are a nuisance (but are considerably cheaper).

12. In bedrooms supply two duplex outlets at each side of the bed location to accommodate electric blanket, clocks, radios, lamps, and other such appliances.

13. A disconnecting means, readily accessible and within sight of the controlled item must be provided for electric ranges, cook tops, and ovens. It is good practice to utilize a small kitchen panel recessed into a corner wall to control the large kitchen appliances and to provide completely safe, accessible disconnecting means. Such an arrangement can also be cheaper if the length of run between the main panel and the kitchen is appreciable.

14. Perimeter lighting, inside controlled, can do much to lessen vandalism and discourage prowlers.

15. A tabulation of residential electrical equipment, including recommended circuits and receptacles, is shown in Table 17.7. A complete residential wiring plan for a small house is shown in Figure 17.20. Although residential plans are normally left uncircuited, a completely circuited design is shown here for didactic purposes.

17.14 Branch Circuit Design Guidelines; Nonresidential

(a) Schools. Since schools comprise an assembly of varied use spaces, including instruction, lab, shop, assembly, office, gymnasium, plus special areas such as swimming pools, photographic labs, and so on, it is not possible to generalize on branch circuit design considerations except for the following:

1. To accommodate the opaque, slide and video projectors frequently used in the classroom, 20-A outlets wired two to a circuit are placed at the front and back of each such room. A similar receptacle, wired six or eight to a circuit is placed on each remaining wall.

2. Light switching should provide:

(a) High–low levels for energy conservation and to permit low-level lighting for screen viewing. With fluorescent lighting this can be accomplished by alternate ballast wiring and switching, thus avoiding the high cost of dimming equipment.

(b) Separate switching of the lights on the window side of the room, which is often

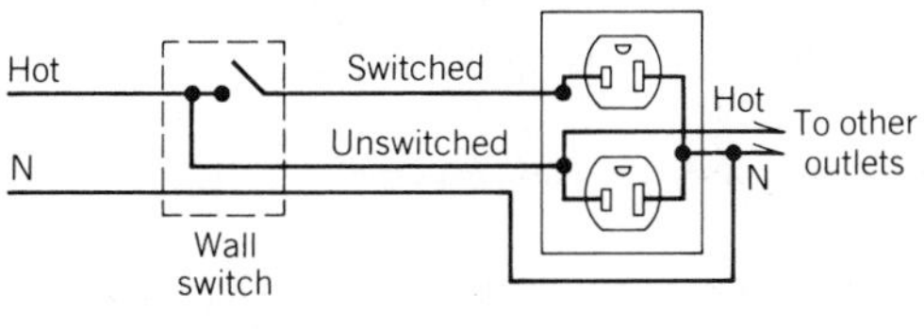

Fig. 17.19 *Split wiring of a duplex receptacle. Upper half is switch controlled; lower half is "hot" all the time. This allows wall switch control of a lamp or other device while maintaining part of receptacle live, for independent use. Notice that the receptacle is mounted with the grounding pole at the top. This is the safest way to install receptacles, since a metallic item such as a paper clip falling on to an inserted cap will contact the ground pole only.*

ELECTRICITY

TABLE 17.7 **Load, Circuit, and Receptacle Chart for Residential Electrical Equipment**

Appliance	*Typical Connected Volt-Amperes*[a]	*Volts*	*Wires*[b]	*Circuit Breaker or Fuse*[c]	*Outlets on Circuit*	*NEMA Device*[d] *and Configuration (See Fig. 16.28)*
				Kitchen		
Range[e]	12,000	115/230	3 #6	60 A	1	14–60R
Oven (built-in)[c]	4,500	115/230	3 #10	30 A	1	14–30R
Range top[c]	6,000	115/230	3 #10	30 A	1	14–30R
Dishwasher[c]	1,200	115	2 #12	20 A	1	5–20R
Waste disposer[c]	300	115	2 #12	20 A	1	5–20R
Broiler[e]	1,500	115	2 #12	20 A	1 or more	5–20R
Refrigerator[f]	300	115	2 #12	20 A	1 or more	5–20R
Freezer[f]	350	115	2 #12	20 A	1 or more	5–20R
				Laundry		
Washing machine	1,200	115	2 #12	20 A	1 or more	5–20R
Dryer[c]	5,000	115/230	3 #10	30 A	1	14–30R
Hand iron; Ironer	1,650	115	2 #12	20 A	1 or more	5–20R
				Living Areas		
Workshop	1,500	115	2 #12	20 A	1 or more	5–20R
Portable heater[g]	1,300	115	2 #12	20 A	1	5–20R
Television[g]	300	115	2 #12	20 A	1 or more	5–20R
				Fixed Utilities		
Fixed lighting	1,200	115	2 #12	20 A	1 or more	—
Air conditioner ¾ hp[h]	1,200	115	2 #12	20 A or 30 A	1	5–20R
Central air conditioner[i]	5,000	115/230	3 #10	40 A	1	—
Sump pump[i]	300	115	2 #12	20 A	1 or more	5–20R
Heating plant, i.e., forced-air furnace[h,j]	600	115	2 #12	20 A	1	—
Attic fan[i]	300	115	2 #12	20 A	1 or more	5–20R

[a]Wherever possible use actual equipment rating.

[b]Number of wires does not include equipment grounding wires. Ground wire is No. 12 AWG for 20-A circuit and No. 10 AWG for 30-A and 50-A circuits.

[c]May be direct connected. For a discussion of disconnect requirements, see NEC Article 422.

[d]Equipment ground is provided in each receptacle.

[e]Heavy-duty appliances regularly used at one location should have a separate circuit. Only one such unit should be attached to a single circuit at the same time.

[f]Separate circuit serving only one other outlet is recommended.

[g]Should not be connected to circuit with appliances or other heavy loads.

[h]Separate circuit recommended.

[i]Recommended that all motor-driven devices be protected by a local motor-protection element unless motor protection is built into the device.

[j]Connect through disconnect switch equipped with motor-protection element.

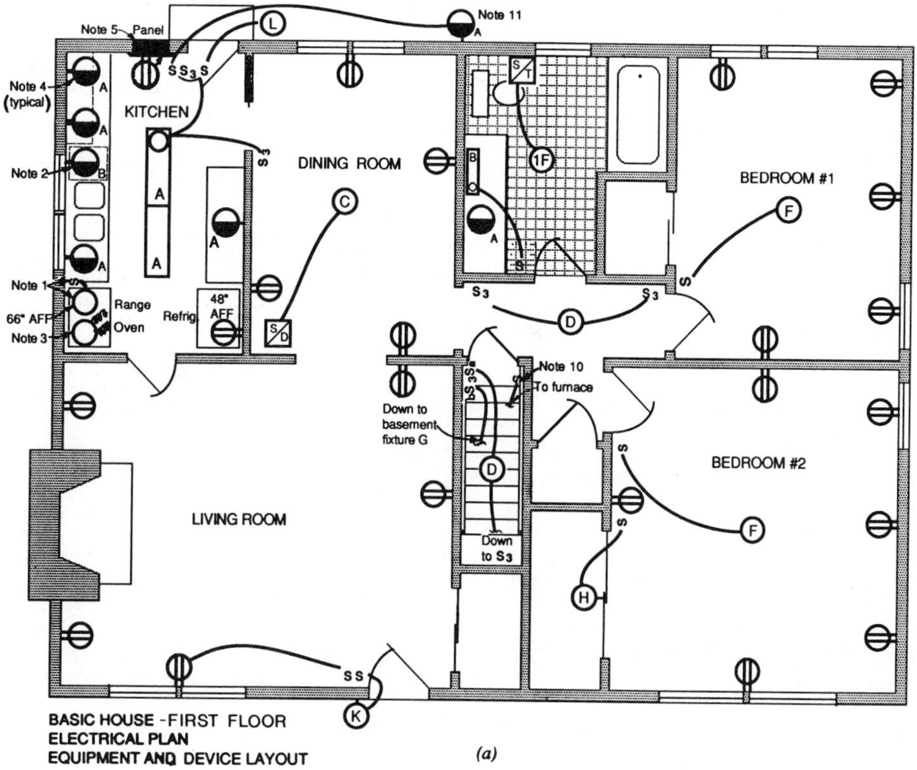

(a)

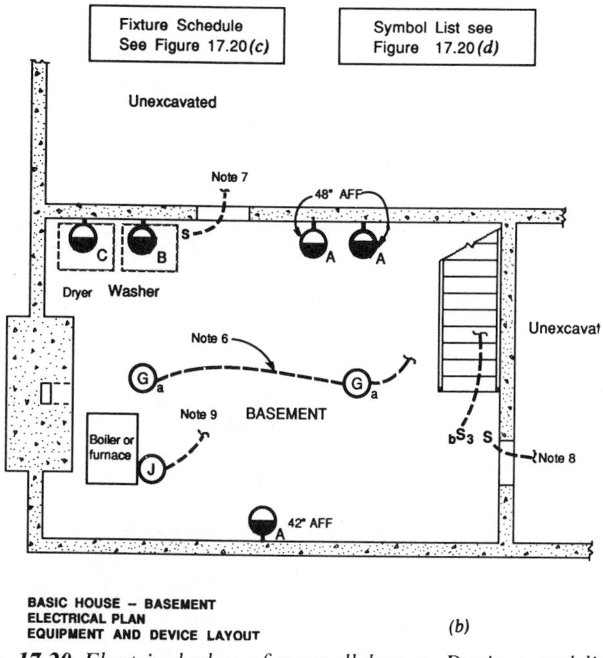

(b)

Notes:

1. Switch and outlet for exhaust fan. Switch wall mtd. above sink backsplash. Outlet with blank cover mounted adjacent to fan wall opening. Separate switch may be omitted if fan is supplied with integral switch.
2. Dishwasher receptacle wall mounted behind unit, 6" AFF.
3. Range and oven outlet boxes wall mtd., 36" AFF. Flexible connections to units.
4. Receptacles at countertop locations to be wall mounted 2" above backsplash.
5. Max. ht. of top c/b to be 78" AFF.
6. Wiring shown as run exposed indicates absence of finished ceiling in basement level. All BX to run through framing members. Attachment below ceiling joists not permitted.
7. Connect to 2-Type G fixtures ceiling mounted at 1/3 points.
8. Connect to 1-Type G fixture at center.
9. Connect to shutdown switch at top of stairs. Boiler control wiring by others see Note 10.
10. Boiler wiring safety disconnect. Provide RED wall plate, clearly marked "BOILER ON-OFF".
11. Equipped with self-closing gasketed WP cover.

Fig. 17.20 *Electrical plan of a small house. Devices and lighting are shown on main floor and basement plans* (a) *and* (b). *A detailed fixture schedule* (c) *and symbol list* (d) *is also prepared. If circuitry is added, the completed plans for both levels will appear as in* (e) *and* (f). *A detailed panel schedule is given in* (g).

LIGHTING FIXTURE SCHEDULE			
TYPE	DESCRIPTION	MANUFACTURER	REMARKS
A	48" L X 12" W X 4" DEEP NOMINAL, 2 LAMP/FLUORES-CENT, WRAP-AROUND ACRYLIC LENS, F 40 WW/LAMPS. SURFACE MTD.	BRITE-LITE CO. CAT. #2740/KFF OR EQUAL	4" DEPTH MAXIMUM
B	24" L, 1 LAMP 20W FLUOR. FIXTURE, WRAP-AROUND WHITE DIFFUSER, WITH SINGLE-SWITCHED RECEPTACLE. MOUNT ABOVE MEDICINE CABINET.	BRITE-LITE CO. CAT. #1/20/BFF OR EQUAL	MAX. MTG. HT. 78" TO ℄.
C	ADJUSTABLE HEIGHT PENDANT INCANDESCENT, 3-75W MAX., BUILT-IN 3-POSITION SWITCH.	HOMELAMP CO. CAT. #3/75/DRP OR EQUAL	———
D	10" D. DRUM-TYPE FIXTURE, WHITE GLASS DIFFUSER, CENTER LOCK-UP, 2-60W INCAND. MAX., SURF. MTD.	BRITELITE CO. CAT. #2/60/HF OR EQUAL	6" MAX. DEPTH.
F	12" D. DRUM FIXTURE, CONCEALED HINGE ON OPAL GLASS DIFFUSER FOR RELAMPING WITHOUT GLASS REMOVAL, 2-75W INCAND. MAX. SURFACE MTD.	DENMARK LIGHTING SPECIAL UNIT #374821	NO SUBSTITUTION WILL BE ACCEPTED.
G	PORCELAIN LAMPHOLDER, PULL CHAIN WITH WIRE GUARD, 100W. INCAND. SURF. MTD.	———	———
H	SAME AS TYPE G, EXCEPT W/O GUARD.	———	———
K	DECORATIVE OUTDOOR LANTERN, MAX. 150W INCAND., WALL MTD. 84" AFF TO ℄.	TO BE CHOSEN BY OWNER	———
L	UTILITY OUTDOOR LIGHT, ANODIZED ALUMINUM BODY AND CYLINDRICAL OPAL GLASS DIFFUSER. 1-100W INCAND. MAX. 84" AFF TO ℄.	UTIL-LITE CO. CAT. #1/100/BP OR EQUAL	IF VANDALISM IS OF CONCERN, SUBST. PLASTIC DIFFUSER.
IF	RECESSED FIXTURE WITH 150 W INFRARED HEAT LAMP	HEAT-LIGHT CO.	———

(c)

SYMBOLS AND ABBREVIATIONS

- BX CABLES RUN CONCEALED; TICS INDICATE NUMBER OF CONDUCTORS EXCLUDING GROUND WIRES. 2 #12 + BARE GROUND, UON
- SAME AS ABOVE EXCEPT RUN EXPOSED.
- WIRING RUN TURNING DOWN; WIRING TURNING UP
- 1 3 HOME RUN TO PANEL; ARROWS AND NUMERALS IDENTIFY CIRCUITS; TICS INDICATE WIRING – AS NOTED ABOVE.
- OUTLET BOX AND FINAL CONNECTION TO EQUIPMENT WITH FLEXIBLE CONDUIT (OR BX).
- OUTLET WITH SECTION OF SURFACE RACEWAY, 2 WIRE, SINGLE CIRCUIT, AND SEPARATE GREEN GND. 15A, 2P, 3 WIRE, RECEPTACLES ON 12" CENTERS.
- D a CLG. OUTLET WITH INCANDESCENT LTG. FIXTURE D, SWITCH CONTROL – 'a'.
- H WALL OUTLET W/INCAND. FIXT. 'H', M.HT. SHOWN.
- A CLG. OUTLET W/FLUOR. FIXT. 'A'.
- B WALL OUTLET W/FLUOR. FIXT. 'B', M.HT. SHOWN.
- J JUNCTION BOX
- DUPLEX CONVENIENCE RECEPTACLE, 20 A, 2 P, 3 W, 125 V. GROUNDING, WALL MTD., VERTICAL, ℄ 12" AFF NEMA 5-20 R.
- A DUPLEX CONVENIENCE RECEPTACLES, 15A, 2P, 3W, 125V, GROUNDING. W/INTEGRAL GFCI
- B SINGLE RECEPTACLE, 20 A, 2 P, 3 W, GND'G., NEMA 5-20 R.
- C SINGLE RECEPTACLE, 30 A. 125/250 V. 3 POLE-4 WIRE GROUNDING NEMA 14-30 R; SUPPLIED WITH MATCHING CAP
- S_a SINGLE POLE SWITCH, 15 A, 125 V, ℄ 50" AFF, UON, CONTROLLING OUTLET(S) 'a'.
- $_aS_3$ SWITCH, 3 WAY, 15 A, 125 V, ℄ 50" AFF, UON, CONTROLLING OUTLETS 'a'.
- S/T MANUAL TIMER SWITCH, 1 SET 15 AMP N.O. CONTACTS.
- S/D OUTLET BOX MTD. SWITCH AND DIMMER, INCAND. LOAD ONLY, 600 WATTS MAXIMUM. ℄ 50" AFF.
- FLUSH MTD. PANELBOARD;

AFF	ABOVE FINISHED FLOOR
MH	MOUNTING HEIGHT
T	THERMOSTAT
UON	UNLESS OTHERWISE NOTED
GFCI	GROUND FAULT CIRCUIT INTERRUPTER
WP	WEATHER-PROOF
NO	NORMALLY OPEN

(d)

Fig. 17.20 *(continued)*

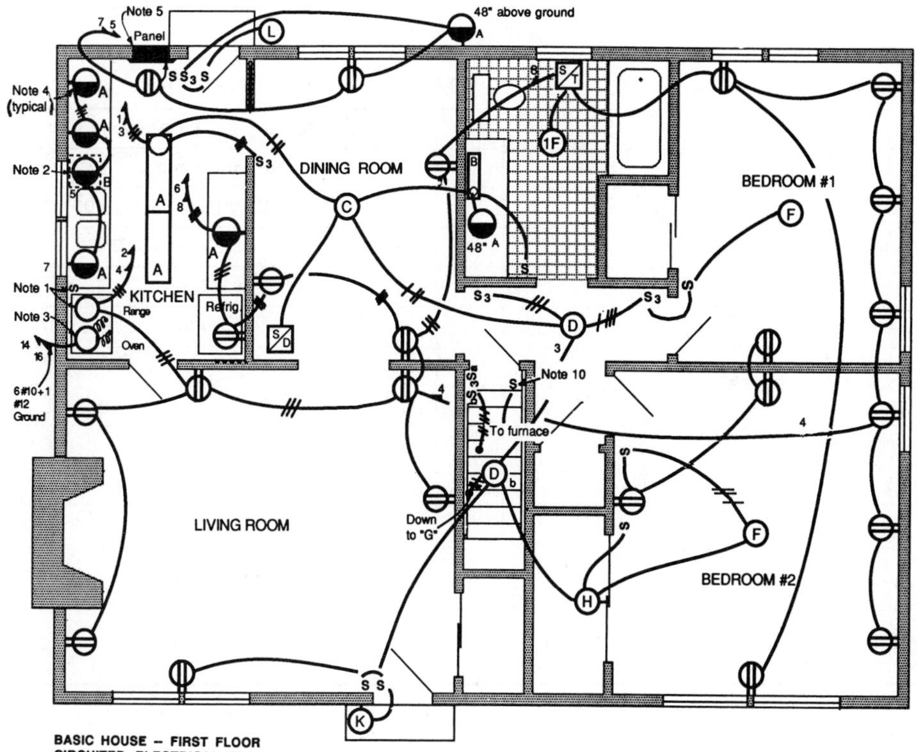

BASIC HOUSE – FIRST FLOOR
CIRCUITED ELECTRICAL PLAN

(e)

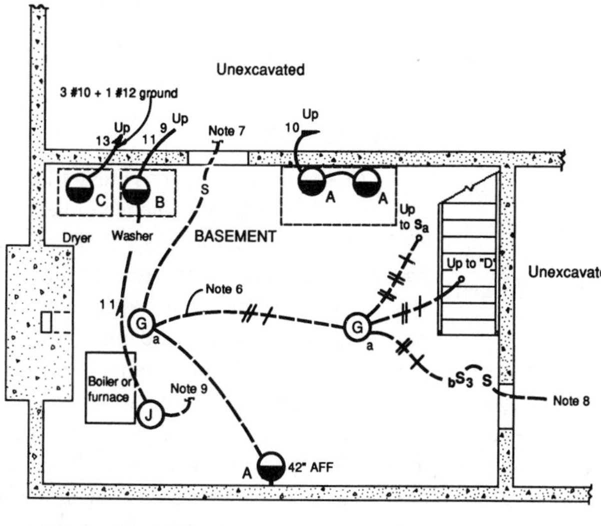

BASIC HOUSE – BASEMENT
CIRCUITED ELECTRICAL PLAN

(f)

Fig. 17.20 (continued)

PANEL SCHEDULE FOR BASIC HOUSE						
CIRC. NO.	DESCRIPTION	LOAD VA.	CIRCUIT BREAKERS	DESCRIPTION	LOAD VA.	CIRC. NO.
1	LTG – {KIT. DR., BR. #1, CORRIDOR / OUTSIDE, BATH + Φ}	720 1R	20 1 | 2 20	OUTLETS – LR. & DR. + EXH. FAN	30 5R	2
3	LTG – {OUTSIDE / LR., BR. #2, BSMT. + Φ}	720 5R	3 | 4	OUTLETS – BR. 1 & 2	6R	4
5	DISHWASHER	1500	5 | 6	APPLIANCE OUTLETS – KIT., DR.	—	6
7	APPLIANCE OUTLETS – KIT., DR., OUTSIDE	—	7 | 8	OUTLETS – BATH, BR's., CLG, HTR	150 2R	8
9	LAUNDRY OUTLET – BSMT.	—	9 | 10	OUTLETS – BSMT.	—	10
11	HOT WATER BOILER	1300	11 | 12	SPARE		12
13	ELECTRIC CLOTHES DRYER	5000	30 A 13 | 14 2P 15 | 16 30 A 2P	RANGE	6000	14
15	SPACE FOR 2 – 1P	—	17 | 18 30 A	OVEN	4800	16
	OR 1 – 2P	—	19 | 20 2P			

PANEL DATA
MAINS, GND. BUS: 150 A MNS., 60 A GND. BUS
MAIN C/B ~~OR SW/F~~ 150/100
BRANCH C/B INT. CAP. 5000 AMP.
MOUNTING – ~~SURF~~/RECESS
REMARKS: FRONT SUITABLE FOR PAINTING

VOLTAGE 120/240 1 PH.

(g)

lighted sufficiently by daylight. Control should be automatic; photocell initiated (see Section 21.3).

3. Provide appropriate outlets for all special equipment in labs, shops, cooking rooms, and the like.

4. Use heavy-duty devices and key-operated switches for public area lighting (corridors, etc.), plastic instead of glass in fixtures, and vandal-proof equipment wherever possible. All panels *must* be locked and should be in locked closets.

5. The NEC requires sufficient branch circuitry to provide a minimum of 3 W/ft^2 for general lighting in schools. Refer to the NEC Article 220. Unlike for residential occupancy, this figure does *not* include receptacles. Receptacles are calculated separately at 180 W each for ordinary convenience outlets.

6. Keep lighting and receptacles completely separate when circuiting.

7. Since most schools, including many at the elementary level, now use computer and video terminals for various learning facilities, their electrical and interconnection requirements must be determined and provision made accordingly.

(b) Office Space

1. In small office spaces (less than 400 ft^2) provide either one convenience outlet for every 40 ft^2, or one outlet for every 10 linear ft of wall space, whichever is greater. In larger office spaces, provide one outlet every 100 to 125 sq ft beyond the initial 400 ft^2 (10 outlets). These outlets are intended for miscellaneous electrical devices, in addition to anticipating the constantly increasing number of data system accessories. In addition, provide at *every* desk (on an adjacent wall, in a "power pole," or in the floor) a duplex receptacle intended for a computer terminal. These receptacles should be circuited at no more than six to a 20-A branch circuit, and less if the equipment to be fed so dictates. Figure 17.21 shows one possible circuiting arrangement for the room layouts shown in Fig. 17.11. Although other arrangements are possible, the

net result is the same. See Fig. 15.28 for a typical under-carpet wiring layout plan.

2. Corridors should have a 20-A, 120-V outlet every 50 ft, to supply cleaning and waxing machines.

3. As with all nonresidential buildings, convenience receptacles are figured at 180 W each.

4. Only specification grade equipment should be used.

(c) Industrial Spaces. These areas are so specialized that no meaningful guidelines can be given.

(d) Stores. In stores, good practice requires at least one convenience outlet receptacle for every 300 ft^2 in addition to outlets required for loads such as lamps, show windows, and demonstration appliances.

17.15 Load Tabulation

While circuiting the loads, a panel schedule is drawn up that lists the circuit numbers, load description, and wattage (actually volt-amperes), and the current rating and number of poles of the circuit-protective device feeding each circuit. Spare circuits are included to the extent that the

Fig. 17.21 (a) *Alternative methods of circuiting are shown. Room 205 shows the actual junction box location, with flexible connections to the box at each fixture. Room 207 shows circuit numbers and switch designations only; the placement of junction boxes is understood, and conduit runs are omitted for clarity and because they most often are not representative of actual installation. Room 209 shows an outlet box at each fixture, with schematic conduit connections. All of these systems are in common use.*

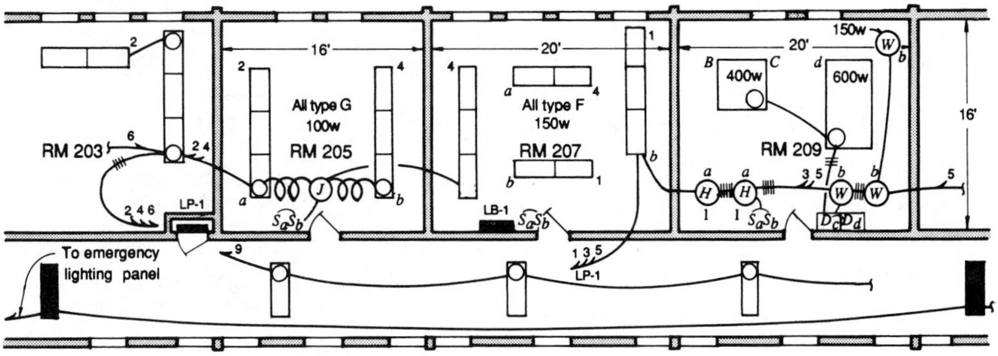

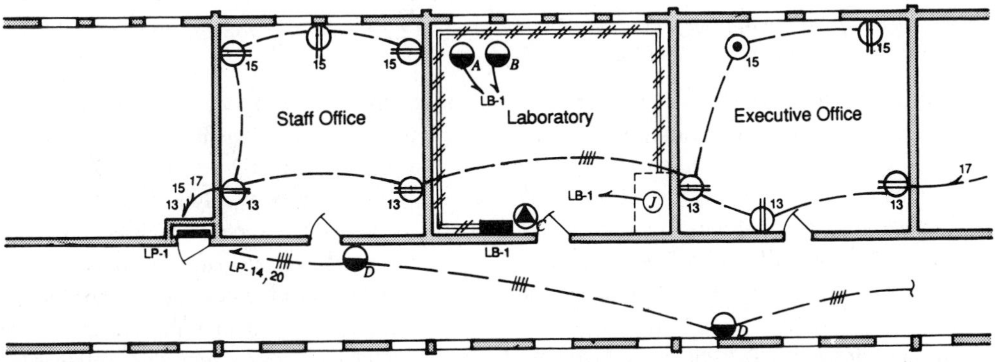

Fig. 17.21 (b) *Typical circuiting of several rooms in an office–lab building. Lighting and power (receptacles) are shown on separate plans* (a) *and* (b) *to avoid crowding (see Fig. 17.14 for symbols and Fig. 17.11 for notes). Lighting in offices is recessed; lighting in labs is surface mounted for flexibility. Note the double circuiting of the type D receptacles.*

designer considers them necessary and consonant with economy, but normally no less than 20% of the number of active circuits. Finally, spaces are left for future circuit breakers, in approximately the same quantity as the number of spare circuits, but always to round off the total number of circuits.

Panels are normally manufactured with an even number of poles. Thus if a panel had, with spares, 21 poles, the designer would probably require three spaces, to give a 24-circuit box. A typical panel schedule is shown in Fig. 17.22, which serves the laboratory of Fig. 17.21.

In calculating panel loads, the following rules apply:

1. Each specific appliance, device, lighting fixture, or other load is taken at its nameplate rating, except certain kitchen and laundry appliances for which the NEC allows a demand factor (see NEC Article 220).

2. Each convenience outlet, in other than residential spaces, is counted as 1.5 A (180 W).

3. Loads for special areas and devices such as show window lighting, heavy-duty lampholders, and multi-outlet assemblies are taken at the figures given in NEC Article 220.

4. Spare circuits are figured at approximately the same load as the average active circuits (1200 to 1500 W).

5. Spaces are not added into the load.

6. Continuous loads such as lighting are calculated at 125% of their actual value. See NEC Article 220-10(b), which applies to feeder calculations but applies equally to panel loading.

In calculating total panel loads, as shown in Fig. 17.22, no demand factor may be applied except as specifically stated in the NEC (see NEC Articles 220-11 and 220-13). This is true despite the knowledge that most often the usage will be such that average load will be lower than the maximum demand. (See Section 17.19 for multipanel feeders.) If it is known that certain loads will not or cannot be used simultaneously, the load total should reflect only the larger of the two. Thus, heating and cooling loads are *generally* not concurrent. Nor is building night floodlighting concurrent with the business office load, but it may be with general interior lighting. Note in Fig. 17.22 that 2-pole loads (208 single-phase) appear in two columns. Similarly, 3-phase loads would appear in three columns. Also note that the phase loads are *not equal*. It is the responsibility of the designer (or contractor) to circuit the loads so that the phases are as closely balanced in load as possible. If this is not done, one phase will carry considerably more current than the others. Since the panel feeder must be sized for the maximum phase current,

	ELEC. PANEL-LP-1[1]			
	120/208 V 3φ 4W	LOAD IN WATTS		
No.	LOAD	φA	φB	φC
1	Lighting	1050		
2	Lighting	1050		
3	Lighting		1450	
4	Lighting		1050	
5	Lighting			1100
6	Lighting			1200
7	Lighting	800		
8	Lighting	1100		
9	Lighting		700	
10	Lighting		1050	
11	Lighting			1000
12	Lighting			1200
13	Receptacle[2]	900		
14	Receptacle[3]	900		
15	Receptacle		900	
16	Receptacle		900	
17	Spare			1200
18	Spare			1200
19	Spare	1200		
20	Receptacle[4]		1000	
				1000
21	Spare	1200		
			1200	
22-26	Spaces only			
	25% addition for			
	continuos loads[5]	1000	1063	1125
	Phase totals	9300	9313	9025
	Panel total		27638	V-A
	Max. φ current		78 amp.	
	25% spare capacity	~	20 amp.	(Future)
	Max. Phase I		98 amps	

Main breaker 225A 3pole (see text example A.1, section 17.19)
Trip 100 A
Feeder size 4 #2 THW in 2"C.

NOTES:
1. All C/B 1P; 50 A frame, 20 A trip except ccts 20 and 21, which are 2P, 50 AF/20 AT.
2. 5 receptacles @ 1.5 A each.
3. Corridor receptacle; 120 V section.
4. Corridor receptacle; 208 V section.
5. See Section 17.15

***Fig. 17.22** Schedule for lighting panel LP-1.*

this may lead to an oversized feeder and therefore a waste of money.

Having tabulated and balanced the loads and totaled them by phase, the maximum current is calculated. A portion of the space capacity available in the branch circuits is added to the above total, as the basis for the calculation of the feeder load. This spare capacity, shown in Table 17.5, is something between 25 and 50%. The exact amount to be added initially in feeder sizing is developed below.

17.16 Spare Capacity

Load calculations for dwelling occupancies and other spaces are detailed in NEC Article 220 and examples given in Chapter 9 of the NEC. Since these calculations are specialized but routine, and are covered there in detail, they will not be repeated herein. Again it is emphasized that possession of a current copy of the National Electrical Code (NFPA Bulletin 70) is a sine qua non of proper electrical design.

Having arrived at the panel load totals as detailed above, the next step is to size the conductors feeding the panel. To do this, an examination of the spare capacity of the panel and of the feeders is necessary, in order that the system design be consistent, giving equal capacity for future growth in all its components. Considering the panel circuitry first, let us examine the effect of load expansion, including spares and spaces (see Table 17.8).

As noted in Section 17.12, spare capacity is build into the branch circuitry *and* into the panels. Most often expansion is accomplished by additional loading on some circuits and by adding new circuits via spare circuit breakers in the panel. Table 17.8 gives the *ultimate* capacity of the panel, that is, fully loaded circuits and fully utilized spares and spaces. Since this ultimate capacity is rarely achieved, panel feeders need only be sized for initial loads as detailed in Section 17.15, and provision made for rewiring to meet anticipated expansion, by one of the techniques listed in Section 17.17.

These results can be summarized as follows: For panels in buildings expecting limited expansion, for which branch circuits are loaded to 80% of capacity (25% *branch circuit* expansion, see Table 17.5, note e), the ultimate panel load without new conduit work, i.e., merely by filling out circuits, is 1.5 L, or 50% beyond the initial load. By adding breakers in the spaces, this load can be expanded to 75% beyond the initial load. The corresponding figures for panels that are lightly loaded (66% capacity, i.e., 50% branch circuit expansion as in Table 17.5 note f), in anticipation of considerable load growth, are 80 to 110%, respectively. These results are summarized in Table 17.8.

17.17 Feeder Capacity

To achieve economy, the panel feeder must accommodate the initial load plus some portion of

TABLE 17.8 **Panel Initial and Expanded Loads**

	Panels in Facilities Expecting Limited Expansion; Circuits Initially Loaded to Give 25% Expandability (See Table 17.5)	*Panels in Facilities Expecting Extensive Expansion; Circuits Initially Loaded to Give 50% Expandability (See Table 17.5)*
Initial load	100%	100%
Initial plus spares	120%	120%
Load after all circuits including spares are loaded to maximum allowable	150%	180%
Load after utilizing 20% spaces also	175%	210%

NOTE: For development of these figures, see McGuinness et al. (1980), pp. 690–691.

ELECTRICITY

the future load. Spare capacity in feeders (to accommodate a considerable portion of the panel spare capacity as shown in Table 17.8), is provided by one or more of the following procedures:

1. Provide feeder (and conduit) capacity initially, to handle the entire eventual load. This method is most expensive—requiring initial outlay for no return—and is rarely used.

2. Provide feeder for initial plus spare with properly sized conduit. Conduit is sized for type THW or RHW without covering. This method, as we shall see, yields very limited spare capacity.

3. Provide feeder for initial load plus spare, with conduit oversized by one size. Size conduit for type THW wire, which is very widely used because of attractive price and excellent electrical properties. Some additional costs are entailed here.

If the initial wiring is done with type TW wire, the effect is approximately the same as having oversized the conduit by one size and wiring initially with THW wire. This is caused by the lower ampacity rating of TW as compared to THW, resulting in a larger conduit size for the same initial ampacity. The reader can work out exact figures in each with the help of Tables 15.2*a* and 15.2*b*, and conduit capacity Table 17.9, or by using the NEC tables in Article 310 and Chapter 9.

4. Provide feeder for initial load plus spare and oversize conduit by *two* sizes. This will yield most of the capacity necessary in facilities anticipating large expansion.

5. Provide for initial load plus spare, with an empty conduit for future. This method is expensive because of high conduit cost, and it is infrequently advisable.

In procedures 2, 3, and 4 the future capacity beyond that initially supplied is handled by the use of larger gauge wire in the existing conduit. To examine exactly what these alternatives provide in spare capacity, we have tabulated in Table 17.10 the maximum ampacity of various size conduits and the future capacity obtainable. Table 17.9 is taken directly from the NEC.

Note from Table 17.10 that simply by rewiring we can obtain up to 37% additional capacity, whereas if the conduit had been oversized the additional capacity would be from 43 to 130%.

Returning now to the question of how large to make the feeder for a given panel load, we must balance future panel load, initial cost of feeder, and future capacity of existing conduit. It is best to avoid the installation of empty conduits since this is expensive. Rewiring, however, is relatively inexpensive and oversizing conduit is the method of choice if rapid expansion is anticipated.

Referring to Table 17.10, note that normal design uses THW cable. Design with T or TW is, in effect, a first step in oversizing conduit and is generally not economical. The second step is a deliberate oversizing of conduit that results in much increased conduit ampacity, as reflected by the figures. Using these figures in actual practice requires that the designer juggle cable and conduit cost against anticipated load growth to arrive at the most economical long-term solution. Applying these numbers to concepts previously developed we have:

1. Buildings designed for 25% branch circuit expansion (see Table 17.5) have a panel capacity of 1.75 times the load (Table 17.8). If it is desired to design the panel feeder to carry the full expansion possible, then: Calculate the feeder on the basis of panel load plus 20% spares, and oversize the feeder by 20%. This gives a feeder capacity of

$$1.20 \times 1.2 = 1.44 \times \text{initial load}$$

Table 17.10 indicates that rewiring will add another 15% on the average. This gives

$$1.15 \times 1.44 = 1.66 \times \text{initial load}$$

which corresponds sufficiently closely to the 1.75 desired.

2. For a building with 50% branch circuit expansion, utilizing full panel space capacity gives us an ultimate panel capacity of 2.1 times the initial load (see Table 17.8). This is accomplished in the following way: Oversize the feeder by 15% and oversize the corduit by one size. The latter step gives an expansion of approximately 50%. Therefore,

TABLE 17.9 **Maximum Number of Conductors in Trade Sizes of Conduit or Tubing**

Type Letters	Conductor Size AWG, MCM	Conduit Trade Size (Inches)									
		½	¾	1	1¼	1½	2	2½	3	3½	4
TW, T, THW, RHW, and RHH (without outer covering)	6	1	2	4	7	10	16				
	4	1	1	3	5	7	12	17			
	2	1	1	2	4	5	9	13			
	1		1	1	3	4	6	9			
	0		1	1	2	3	5	8	12		
	00		1	1	1	3	5	7	10		
	000		1	1	1	2	4	6	9	12	
	0000			1	1	1	3	5	7	10	
	250			1	1	1	2	4	6	8	10
	300			1	1	1	2	3	5	7	9
	350				1	1	1	3	4	6	8
	400				1	1	1	2	4	5	7
THHN	6	1	4	6	11	15	26				
	4	1	2	4	7	9	16				
	2	1	1	3	5	7	11	16	25		
	1		1	1	3	5	8	12	18		
XHHW (4 to 500 MCM)	0		1	1	3	4	7	10	15		
	00		1	1	2	3	6	8	13		
	000		1	1	1	3	5	7	11	14	
	0000		1	1	1	2	4	6	9	12	
	250			1	1	1	3	4	7	10	12
	300			1	1	1	3	4	6	8	11
	350			1	1	1	2	3	5	7	9
	400				1	1	1	3	5	6	8

Source: National Electrical Code.

TABLE 17.10 **Maximum Wire and Ampacity of a Conduit, and Ampacity Gain on Rewiring**

Initial Conduit Size (Inches)	Initial Installation THW Cable		Rewiring with THHN or THWN				
			Using Original Conduit			Capacity Increase Having Oversized Conduit	
	Max.[a] Wire	Max. Amperes	Max. Wire	Max. Amperes	Capacity Increase	One Size[b]	Two Sizes[b]
1½	1	130	1/0	150	15%	77%	119%
2	3/0	200	4/0	230	15%	43%	84%
2½	250	255	300	285	12%	44%	79%
3	400	335	500 2–4/0	380 368	13%	36%	60%
3½	600	420	700	460	10%	28%	—

[a]Assuming four single conductors in conduit; 3-phase conductors plus neutral.

[b]Using paralleled sets of feeders for conduits 3 in. and larger.

$$\text{feeder capacity} = 1.15 \times 1.2 \times 1.5 = 2.07$$

which is approximately the desired figure.

If, as in the case of laboratories, more than 100% expansion is anticipated (see Table 17.1), conduit should be oversized by two sizes and initial wiring oversized by approximately 25%. Feeders thus arranged will handle the new panels required to meet the anticipated expansion.

Two factors should be carefully noted here. First, note that the smaller conduits offer the largest expandability although, in dollars per amperes, they are more expensive. Second, in order to take advantage of spaces in a panel, conduit stubs should be taken from the panel and extended into hung ceilings or another procedure used to make the panel circuitry easily accessible in the future.

17.18 Panel Feeder Load Calculation

EXAMPLE 17.1. Refer to Fig. 17.22. The panel is for a laboratory/office area. Since large expansion is anticipated, circuitry follows the bottom section of Table 17.5. Ultimate panel load would be 26 cct at 1900 W = 50 kVA = 138 A. Thus the initial feeder is sized for 115 A (4 #2 THW, 1¼ in. C), but rewiring will allow as much as 230 A in a 2-in. conduit (oversized by two sizes; see Table 17.10). A 225-A frame c/b is chosen initially, since eventually the trip will be raised to 150 A.

The NEC in Article 220 specifies minimum watts-per-square foot (W/ft^2) figures for various occupancies, for lighting, and for miscellaneous power loads. Proper design procedure therefore requires that after detailed design of an area, the actual loading be compared to these minima, and the larger of the figures, as regards number of circuits and feeder load, be used. An example will help to make this clear.

EXAMPLE 17.2. Assume a single floor of an office building 100 × 200 ft. Assume also that 15% of the area is corridor and storage, equally divided between the two. Calculate the load and feeder size. Assume a good grade speculative construction venture.

SOLUTION

$$\text{Office space} = 85\% \text{ of } 20{,}000 \text{ ft}^2 = 17{,}000 \text{ ft}^2$$
$$\text{Corridor and storage} = 15\% \text{ of } 20{,}000 \text{ ft}^2 = 3000 \text{ ft}^2$$

The NEC specifies a minimum of 3½ W/ft^2 for lighting and 1 W/ft^2 for miscellaneous receptacles in office space. It further specifies ½ W/ft^2 minimum for corridors and ¼ W/ft^2 for storage. Therefore:

Lighting

Office load: $17{,}000 \times 3\tfrac{1}{2} = 59.5$ kW

Corridor: $1500 \times \tfrac{1}{2} = 0.75$ kW

Storage: $1500 \times \tfrac{1}{4} = 0.38$ kW

Minimum lighting load = 60.6 kW

Receptacles

$17{,}000 \times 1 \text{ W/ft}^2 = 17$ kW

These figures would then be compared to the actual design loads. Receptacles are counted at 180 W each, as noted in Section 17.15, item 2. If the Code minima exceed the design load (as it well may if lighting is properly designed), panels must be equipped with sufficient additional circuits to make up the difference. The number of such circuits is up to the designer, since circuit loading is not specified.

For instance, suppose that the lighting design was accomplished at 2 W/ft^2. Panel circuits would have to be provided for the additional 1½ W/ft^2, as follows:

$$17{,}000 \text{ ft}^2 \times 1.5 \text{ W/ft}^2 = 25.5 \text{ kW}$$

Assuming a 30 to 50% future load expansion, we would circuit the lighting loads at approximately 1300 W per circuit (see Table 17.5). Therefore, we would provide

$$\frac{25{,}500 \text{ W}}{1300 \text{ W}} = 20 \text{ additional circuits}$$

With respect to minimum feeder load, the NEC specifies that it be increased by 25% if loads are continuous (three or more hours). This requirement allows for breakers to heat up in panels while carrying continuous load, and is waived for circuit breakers that are ambient compensated, that is, are rated to carry 100% load. Since we have established 80% of the breaker rating as maximum load (see Table 17.5), *we have already accounted for this factor in circuitry but must utilize it in feeder calculation*. Assuming that the Code minima are the design loads, then, for feeder calculation:

Lighting load = 60.6 kW × 125%	= 75.75 kW
Receptacle load as above	= 17.0 kW
Minimum feeder load	= 92.75 kW
25% future load	= 23.2 kW
Design feeder load	= 116 kW

Since this load would be divided between several panels, the building electrical design might be such that the panels were not all fed by one feeder (see Fig. 17.23). However, assuming they *were*, the feeder would be calculated in terms of 3-phase current thus:

$$I = \frac{\text{kW}}{0.360} = \frac{116}{0.360} = 322 \text{ A}$$

Using THW cable, a minimum of 400 MCM would be required. Conduit would be a mini-

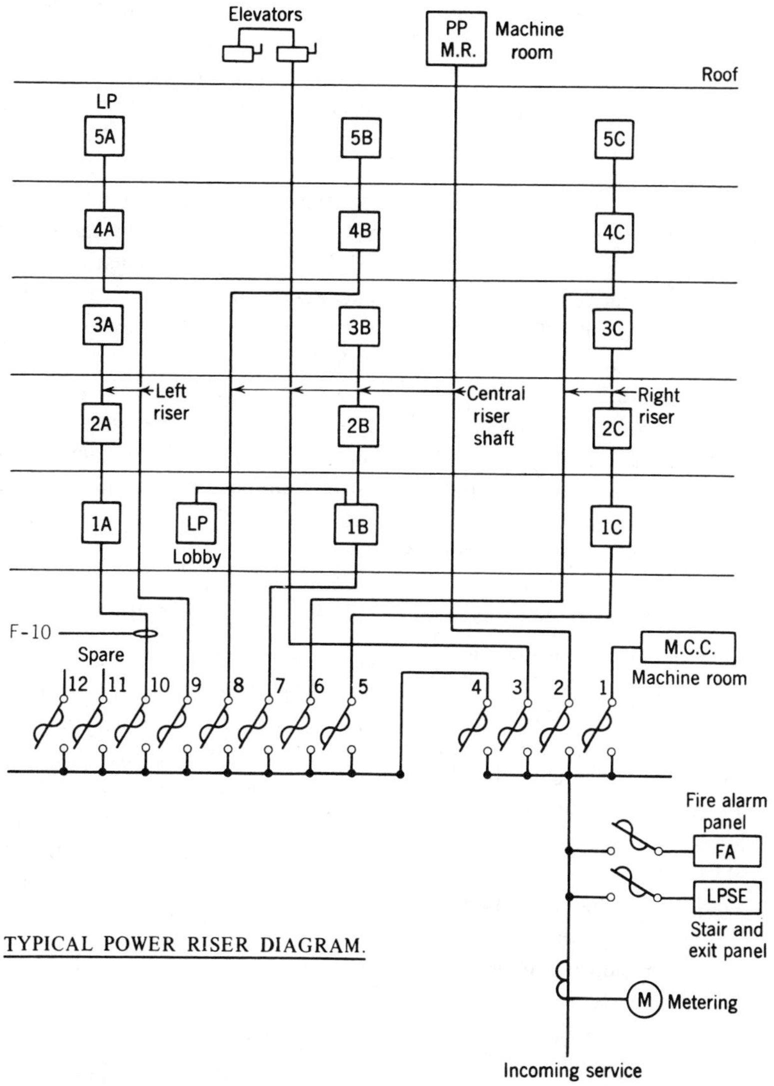

Fig. 17.23 *Typical power riser diagram. Ordinarily, the main switchboard would be shown as a large rectangle with the feeder emanating from it, and a switchboard schedule would detail the contents. Here, because of the unusual bus arrangement, the main switchboard appears as it would on a single-line diagram.*

TABLE 17.11 **Current and Wattage Relationships**[a]

120-V Single-Phase	*120/240-V 3-Wire*	*120/208-V Single-Phase, 3-Wire*	*120/208-V 3-Phase*	*277/480-V 3-Phase*	*277-V Single-Phase*
$I = \frac{W}{120}$	$I = \frac{W}{240}$	$I = \frac{W}{208}$	$I = \frac{W}{360}$	$I = \frac{W}{830}$	$I = \frac{W}{277}$

[a]Assuming 100% power factor.

mum of 3 in., but might be increased to 3½ or 4 in., according to the considerations of spare capacity discussed in the previous section.

As an aid in computing currents, Table 17.11 lists the relevant relations.

17.19 Harmonic Currents

A phenomenon of very recent vintage, that of large harmonic currents, has been the cause of considerable difficulty in modern electrical installations. Without going into detail in this highly technical subject, which is beyond the scope of our study here, a brief description of the problem and its causes can be given.

Conventional electrical loads such as lighting, resistive devices (heaters), motors, and the like are linear (i.e., the load impedance remains essentially constant regardless of instantaneous voltage). This is not the case with most electronic equipment. Computers, modems, printers, electronic lighting ballasts, variable-speed motor drives, and solid-state equipment of all types are essentially nonlinear loads. As such they produce harmonic currents, of which the odd-order ones are additive in the power system neutral conductor. The most troublesome of these are the third harmonic and its odd multiples (9th, 15th, 21st, . . .). These currents can become so large in a modern computerized office installation with electronic lighting ballasts that instead of the neutral conductors carrying the unbalanced current in a 3-phase system (zero in a balanced system), it actually carries *more* current than the phase wires. Another result of these currents can be severe overheating and even burnout of motors, generators, and transformers and malfunctioning of control and protective equipment.

There are at least partial solutions to this problem, and competent electrical engineers are aware of them. It is important that the architect, contractor, and building manager be at least apprised of the existence and potential seriousness of the problem, so that appropriate steps can be taken. This is particularly true for existing facilities which are adding computer networks, UPS systems, solid-state drives for pumps, fan, elevator motors, electronic lighting ballasts, and other solid-state equipment. In these instances changes to the existing electrical distribution system will probably have to be made to minimize the problem. Furthermore, manufacturers of solid-state equipment recognize the difficulties which their equipment can cause and some address the problem adequately. It is wise to write equipment acquisition specifications with this in mind.

17.20 Riser Diagrams

When all devices are circuited and panels are located and scheduled, we are ready to prepare a riser diagram. A typical diagram, shown in Fig. 17.23, represents a block version of a single-line diagram except that, as the name implies, vertical relationships are shown. All panels, feeders, switches, switchboards, and major components are shown up to, but not including, branch circuiting. This diagram is an electrical version of a vertical section taken through the building.

EXAMPLE 17.3. Feeder F10 of Fig. 17.23 serves lighting panels 1A, 2A, and 3A. Calculate the required feeder size, considering loads, future expansion, and voltage drop.

SOLUTION. The loads on these panels have

been computed in accordance with the above considerations and are:

Connected load:	LP-1A—100 A
	LP-2A—125 A
	LP-3A—110 A
	335 A

These figures include connected load, spares, and a 25% future factor. *The NEC requires that demand be taken per panel at 100% only for load calculated by square feet, which is lower than actual circuited loads.* If actual panel circuit loads are larger than the NEC minima, then they are utilized and any reasonable demand may be used, provided that at no time do the loads drop below the minima specified in the NEC. Therefore, we may apply diversity factors between panel loads in a judicious manner.

In office building work, typical diversity factors are

Lighting Panels Fed from a Single Feeder	*Diversity Factors*
1, 2	1.00
3, 4	1.09
5, 6, or 7	1.18
8, 9, or 10	1.33

Thus the load on feeder F-10, using 100% demand per panel and 1.09 diversity between panels, would be 335 × 1.0/1.09 = 307 A.

Methods for handling future expansion were discussed above. In this case the feeder, before voltage drop considerations, would be (from Tables 15.2 and 17.9) 4-350 MCM THW in 3 in. C. Note from Table 17.10, however, that 3 in. C gives a low percentage increase in rewiring. In this case, the wise choice would be to increase the feeder conduit to 3½ in., giving an eventual rewired capacity of 460 A; a 50% increase.

The final consideration in sizing a feeder is voltage drop. The recommended voltage drop from source to final panel should be no more than 1% for lighting and combination lighting and power feeders and 3% for power and heating load feeders. These figures should be adhered to as closely as possible. In instances of long runs where these restrictions will cause excessive cost, lighting feeders up to the branch panel may be run with 2% drop and power feeders up to 4% voltage drop. Total voltage drop to the last outlet in the circuit, including both feeder and branch circuit conductors should not exceed 5%. Many tables and curves are published by manufacturers from which voltage drop can be obtained. Such a set of curves is shown in Fig. 17.24, which shows maximum length of run for a 1% drop. Applying these curves to our last example:

Allowable voltage drop

= 1% of 208 V = 2.08 V

Distance—assume 80-ft run

From the curves, 307 A on 350 MCM cable will give a 1% drop in 50 ft. Therefore, the drop in 80 ft will be 1.6%, which is tolerable.

In summary, then, feeders are sized in accordance with load (actual or square feet, whichever is larger) and voltage drop. Conduit may be oversized for large future load expansion.

17.21 Service Equipment and Switchboard Design

The main switchboard shown in Fig. 17.23 constitutes a combination of service equipment and feeder switchboard. The service equipment portion of the board comprises the metering and the four main switches feeding risers, motor control center (MCC), roof machine room, and elevators. The feeder board comprises switches 5 through 12. Such an arrangement is permissible inasmuch as the NEC allows up to six fused switches or circuit breakers to serve as the service disconnect means. This arrangement was chosen in order to separate to the largest extent possible the motor loads (elevators, air-conditioning equipment, basement power, etc.) from the lighting. Such a procedure minimizes lighting fluctuations resulting from motor starting and yields simpler maintenance. Also, the size of the main switch is reduced. This switchboard would be of the metal-clad dead-front type with switches or circuit breakers, as desired.

Other considerations and general rules affecting service equipment are listed below.

1. A building may be supplied at one point by either a single set or parallel sets of service conductors.

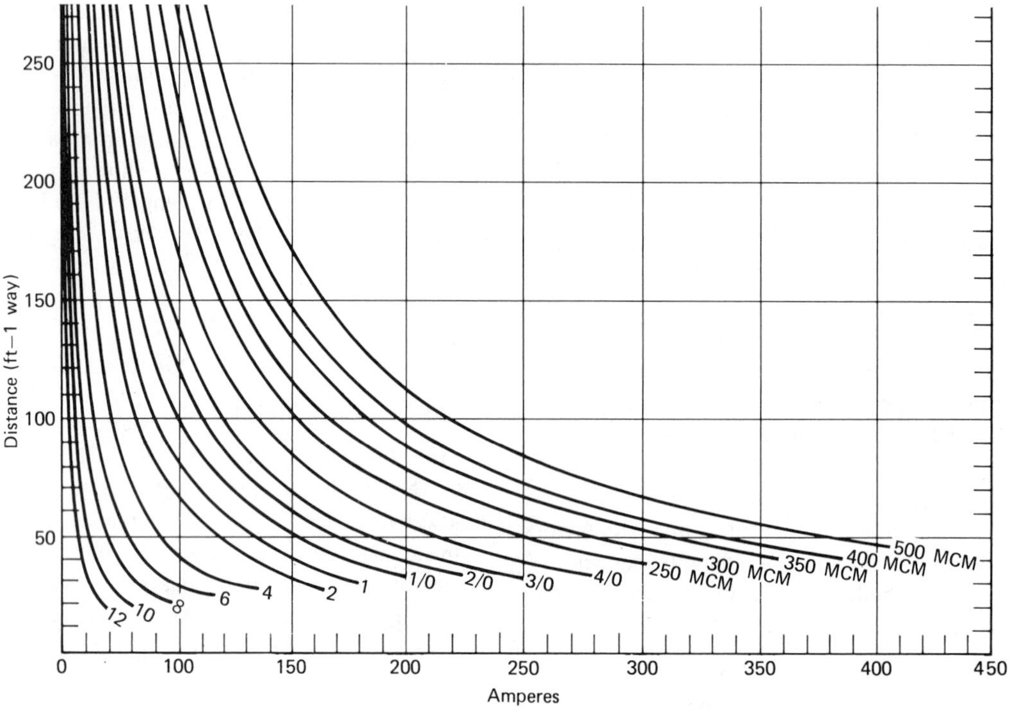

Fig. 17.24 *Curves for determining voltage drop in copper cables. Curves show maximum one-way circuit length for 1% voltage drop.*

2. Service drops may generally be not less than No. 8 AWG and service entrance conductors or underground service conductors not less than No. 6 AWG.

3. All equipment used for service including cable, switches, meters, and so on, shall be approved for that purpose.

4. It is recommended that a minimum of 100-A, 3-wire, 120/240-V service be provided for all individual residences.

5. No service switch smaller than 60 A or circuit breaker frame smaller than 50 A shall be used.

6. In multiple-occupancy buildings tenants must have access to their own disconnect means.

7. All building equipment shall be connected on the load side of the service equipment except that service fuses, metering, fire alarm and signal equipment, and equipment serving emergency systems may be connected ahead of the main disconnect (see Fig. 17.23).

8. For additional information on electric service, see Sections 16.1 to 16.12, Table 17.2, and NEC Article 230.

17.22 Emergency Systems

Some of the considerations relevant to power reliability were discussed in Section 17.1(*b*), and a brief review of the equipment available to supply emergency power was presented in Section 16.31. Emergency lighting equipment is covered in Section 21.30. In this section, we will discuss possible arrangements of emergency power supply. The choice of arrangement and the size and type of equipment depends in large measure on the requirements of local codes, which determine the loads to be fed from the emergency system. The reader should note that although we are using the term *emergency*, the concepts involved are equally applicable to standby systems.

In general, when emergency power is discussed, it is assumed to be replacing ''normal'' power; that is, the assumption underlying most codes and ordinances is that power must be supplied to selected loads within the building because of a utility power outage. Cognizance must also be taken of situations in which normal power has not failed and the outage is localized because of an equipment failure. That aspect of design—reliability—is left to the designer. Some of the arrangements that will be discussed below differentiate between the nature of outages, that is, a utility or general outage versus an equipment or local outage. An exception to the above generalization occurs with health-care facilities, where the NEC in Article 517 specifies an internal electrical design that will in large measure cover both types of outages, down to the distribution level. The interested reader should refer to the referenced NEC article for further reading.

The emergency system includes all devices, wiring, raceways, and other electrical equipment, including the emergency source that is intended to supply electric power to the selected loads. These loads normally include egress lighting (stair, corridor, and exit and lobby lights), signal equipment such as public address and fire alarm that must remain functional during an emergency, and one or more elevators as required by Code. The recognized arrangements are discussed below.

An important aspect of Code requirements permits the use of a single power source for (1) emergency, (2) legally required standby, and (3) optional standby systems provided it is equipped with automatic selective load pickup and load-shedding equipment that will assure adequate power to the three types of systems in the order of priority stated (1, 2, 3).

1. Where emergency loads are light, a storage battery arranged to be connected automatically on power outage may be used. Where all emergency loads can be operated on direct current, the arrangement of Fig. 17.25*a* is used. (Note that a-c lighting can accept d-c emergency power if equipped with a local inverter, as discussed in Section 21.30.) If a-c is required, the arrangement of Fig. 17.25*b* is utilized. When the emergency equipment is entirely separate from the normal equipment and is *normally de-energized,* the system of Fig. 17.25*c* is used. This arrangement is used in small facilities requiring egress lighting only, where it is found that supplying a completely separate emergency system is the preferred economic or engineering choice. Large battery installations are used where uninterrupted power is required, as is generally the case in computer installations where no power interruption, however short, can be tolerated. These systems are highly technical and beyond our scope here.

2. Where emergency loads are larger than can readily be practically supplied by a battery installation, and where start-up and power transfer time of up to 10 seconds is tolerable, a generator set is employed. Sufficient on-site fuel storage (gasolene, diesel fuel, bottled gas) must be provided for the prime mover to operate the generator at full emergency load for a minimum of 2 hours. In areas where the utility records indicate that simultaneous failure of electric service and utility-supplied petroleum-based fuels are rare, and where acceptable to the local jurisdictional authority, off-site fuel (utility gas and liquid petroleum products) may be used to fuel the prime mover. (It should be pointed out that a combination of sources can be used in a single building. For instance, a generator can supply bulk power loads and a battery installation selected lighting loads.)

The system can be arranged with a single transfer switch that senses normal power loss, as in Fig. 17.26*a*, or it can use multiple switches, each one of which will sense power loss *at its down-stream location*, as in Fig. 17.26*b*. The latter system provides greater power reliability, provided that the design is such that the emergency power uses an independent power path to the transfer switches. Otherwise, a faulted piece of equipment that will interrupt normal power downstream will also prevent emergency power from reaching that point.

3. Many codes permit the use of two separate electric services in lieu of a normal service plus an emergency source, provided that the two sources are independent, that is, come from different utility transformers or feeders, enter the building at different points and preferably from different directions, and use separate service drops or laterals. The point is, of course, that the type of reliability desired can be obtained only by minimizing the possibility of a single event interrupting both services. The usual ar-

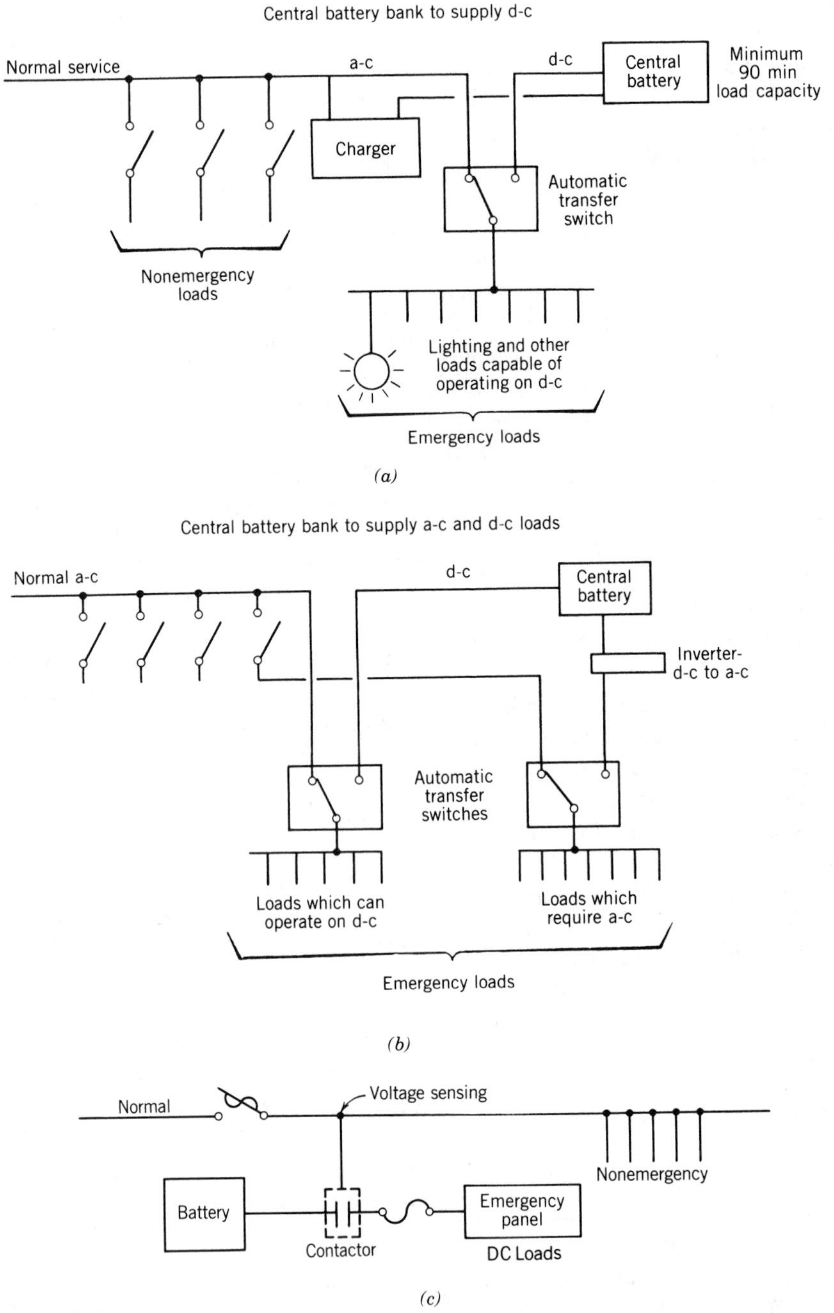

Fig. 17.25 *Use of a central battery for emergency power supply. Where all loads can be energized with d-c the arrangement of* (a) *is satisfactory. When a-c as well as d-c must be distributed, a central inverter is added as in* (b), *and the a-c and d-c emergency loads fed separately. In* (c) *the emergency loads are normally deenergized and activated through the contactor when it senses power loss.*

rangement is for one service to be "normal" and the other "standby."

4. The least reliable arrangement is one in which the emergency loads are connected ahead of the main disconnects and are so arranged that a downstream fault within the building will not affect these items. This situation is illustrated in the riser diagram of Fig. 17.18 where the stair

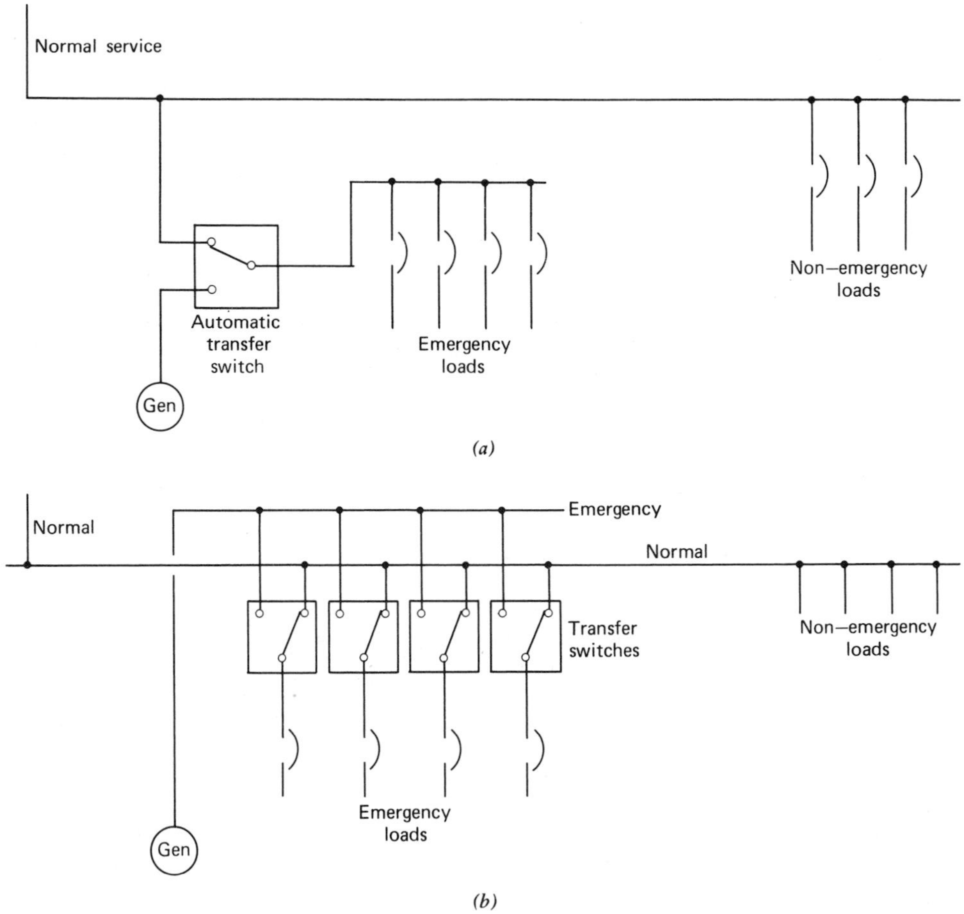

Fig. 17.26 Alternate arrangements of emergency/normal power feed. In (a) *a single transfer switch serves the normal power and transfers to the generator on power failure. In* (b) *the transfer switches are smaller, thus reducing the chance of a single equipment failure faulting out the entire emergency power system.*

and exit panel, which supplies egress lighting, and the fire alarm panel are connected ahead of the building main disconnect and protected with their own fuses. Such an arrangement obviously can do nothing in the event of a power outage and, although once very common, it is now falling into disuse as a result of more stringent codes.

5. The NEC recognizes a category of equipment for emergency illumination called "unit equipment." These devices, discussed and illustrated in Section 21.30, consist of individual self-contained packages with battery, charger, and light source *permanently* mounted and wired at required locations. The panel device feeding these units should be capable of being locked or so arranged as to be accessible to authorized personnel only.

6. Emergency system wiring (as the term is defined by Code) must be kept entirely independent of all other wiring and equipment and should not occupy the same enclosure or conduit as normal system wiring, except in dual-fed units such as transfer switches. Standby system wiring is not subject to this limitation, and may share raceways and other enclosures with general wiring.

REFERENCES

McGuinness, W. J., Stein, G., and Reynolds, J. S. (1980). *Mechanical Equipment for Buildings,* Wiley, New York.

PART VI
ILLUMINATION

It has been estimated that 90% of the information we obtain from all sources is received via our sense of sight. Vision, in turn, is made possible by the presence of light, the proper provision of which is discussed in this part of the book. Since architecture is a uniquely visually oriented profession, its practitioners must thoroughly understand the art and science of illumination in order to be able to integrate structure and light into a working whole that will appear, and function, according to the design intent.

Chapter 18, Lighting Fundamentals, introduces the subject with terminology, definitions, and basic characteristics and measures. A clear distinction is made between the study of light as a physical phenomenon and its study for visual-photometric effect. The former considers light as a form of energy whereas the latter views it as a visual phenomenon (i.e., the physiological response to a form of energy). Pursuing the subject of human response to light, the chapter continues with a discussion of factors in visual acuity. This subject is concluded with an explanation and tabulation of American and European lighting level recommendations.

The discussion continues with consideration of factors in lighting quality (as opposed to quantity), with particular emphasis on analysis and control of reflected glare. The chapter concludes with material on color vision, color in light, and color temperatures of light sources.

Chapter 19, Light Sources: Their Characteristics and Application, covers two basic subjects: daylighting and electric light sources. The day lighting section begins with a discussion of the characteristics of natural light and then proceeds to explain and demonstrate, with illustrative examples, a number of daylighting design methods. These are both graphic and analytic, American and international. The second subject, electric light sources, covers in detail incandescent, fluorescent, mercury, metal-halide, and sodium lamps. The discussions include a detailed description of construction and operating characteristics, including lamp accessories such as ballasts and lamp-holders. Emphasis is placed on luminous efficiency and overall life-cycle costs. The chapter concludes with a discussion of chromaticity of light sources, their spectral distribution, and color rendering index.

Chapter 20, Lighting Design, leads the reader step by step through the design procedure methodology. The first section, on preliminary design, covers cost factors, power budgets, energy considerations, and illumination methods. This is followed by a section on detailed design, which discusses lighting fixtures characteristics and illumination calculation techniques. The chapter concludes with a discussion of lighting design evaluation. Illustrative examples are used to demonstrate and compare calculation techniques.

Chapter 21, Lighting Application, applies the knowledge and techniques explicated in the three preceding chapters to specific occupancies, each of which has its special needs, namely: residential, educational, commercial, and industrial facilities. In addition, important special topics are discussed in detail, including emergency lighting, exterior lighting, and lighting for computer terminals. Finally, lighting control strategies and techniques are covered in detail from the points of view of energy conservation, economics, and automatic controls.

18

LIGHTING FUNDAMENTALS

Architecture is the masterly, correct and magnificent play of masses brought together in light. Our eyes are made to see forms in light; light and shade reveal these forms; cubes, cones, spheres, cylinders, or pyramids are the great primary forms which light reveals to advantage. *Le Corbusier*

18.1 Introductory Remarks

For many years an artificial division existed in the field of lighting design, dividing it into two disciplines: architectural lighting and utilitarian design. The former trend found expression in design that took little cognizance of visual task needs and displayed an inordinate penchant for incandescent wall washers and architectural lighting elements. The latter trend saw all spaces in terms of cavity ratios and performed its design function with foot-candles and dollars as the ruling considerations. That both these trends have in large measure been eliminated is due largely to the efforts of thoughtful architects, engineers, and lighting designers, assisted in part by the energy consciousness that followed the 1973 Arab oil embargo. The latter spurred research into satisfying real vision needs within a framework of minimal energy use.

Another positive factor in the rationalization of lighting design has been the work of the Illuminating Engineering Society of North America (hereafter referred to as IESNA). Its activities in research, standardization, and publication have done much to place lighting design on a stable scientific basis while taking full cognizance of its essential artistic aspects. It is precisely this combination of science and art that makes interior lighting design an architectural-type discipline. The responsible lighting designer must consider quantitatively:

1. Daylight—its introduction and integration with electric light.
2. The interrelationship between the energy aspects of electric and natural lighting, heating, and cooling.
3. The effect of lighting on interior space arrangement and vice versa.
4. The characteristics, means of generation, and utilization techniques of electric lighting.
5. Visual needs of specific tasks.
6. The effects of brightness patterns on visual acuity.

And qualitatively, the designer must consider:

7. The location, interrelationship, and psychological effects of light and shadow, that is, brightness patterns.
8. The use of color, both of light and of surfaces, and the effect of illuminant source on object color.
9. The artistic effects possible with color and patterns of light and shadow including the changes inherent in daylighting, and so on.

The list is almost endless because so much of the information we receive from our senses comes via our eyes, and what we see is a direct consequence of scene lighting.

As a result of the need to consider these and other interrelated factors, many of which are mutually incompatible, the lighting designer is faced with many difficult decisions. The purpose of the lighting chapters in this book is then twofold: to provide the background that will help the lighting designer make these decisions correctly, and to make him or her proficient in the use of lighting as a design material.

PHYSICS OF LIGHT

18.2 Light as Radiant Energy

The IESNA defines light as visually evaluated radiant energy or, more simply, a form of energy that permits us to see. If light is considered as a wave, similar to a radio wave or an alternating current wave, it has a frequency and a wave length. Figure 18.1 shows the position of light in the wave spectrum with relation to other wave phenomena of various frequencies.

From the chart we see that even the longest wavelength light (red) is a much higher frequency than radio and radar, and that visible light constitutes only a very small part of the wave energy spectrum.

Color is determined by wavelength. Starting the longest wavelengths with red, we proceed through the spectrum of orange, yellow, green, blue, indigo, and violet to arrive at the shortest visible wavelengths (highest frequency).

When a light source produces energy over the entire visible spectrum in *approximately* equal quantities, the combination appears white (as is the case with daylight), whereas a source producing energy over only a small section of the spectrum produces its characteristic colored light. Examples are the blue-green clear mercury lamp and the yellow sodium lamp. Chromaticity of light sources is discussed in Chapter 19.

18.3 Transmittance and Reflectance

Lighting design is possible because light is predictable, that is, it obeys certain laws and exhibits certain fixed characteristics. Although some of these are so well known as to appear self-evident, a review is in order.

The *luminous transmittance* of a material such as a fixture lens or diffuser is a measure of its capability to transmit incident light. By definition, this quantity, known variously as *transmittance*, *transmission factor*, or *coefficient of transmission*, is the ratio of the total transmitted light to the total incident light. In the case of incident light containing several spectral components passing through a material that displays selective absorption, this factor becomes an average of the individual transmittances for the various components and must be used cautiously. A piece of frosted glass and a piece of red glass may both have a 70% transmission factor but obviously affect the incident light differently. In general then, transmission factors should be used only when referring to materials

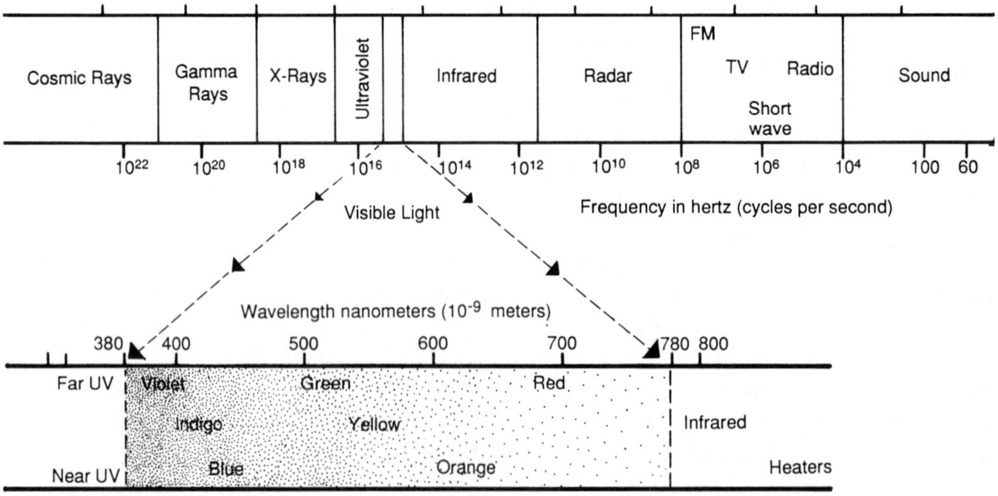

Fig. 18.1 Electromagnetic spectrum. See Chapter 19 for chromaticity diagrams of various light sources.

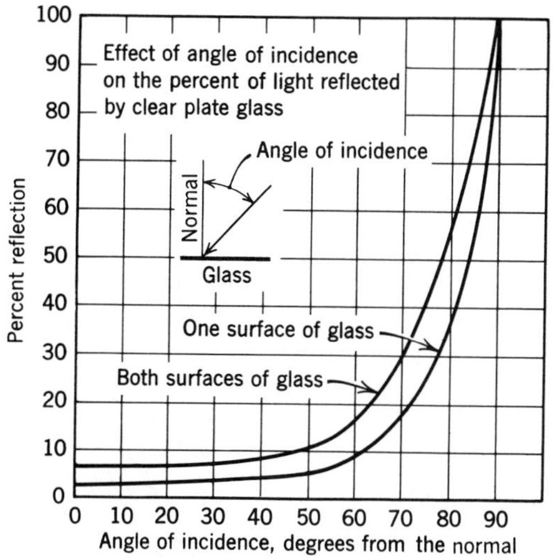

Fig. 18.2 Relation between angle of incidence and percent reflectance. This effect is important when considering the penetration of sunlight into interior spaces and, conversely, the exterior glare produced by reflection of the sun from building windows (see Fig. 3.11).

displaying nonselective absorption, that is, those that transmit the various component colors equally. Clear glass, for instance, displays a transmittance between 80 and 90%, frosted glass between 70 and 85%, and solid opal glass between 15 and 40%. The remainder is absorbed and reflected. See Table 19.6 for typical transmission factors.

Similarly, the ratio of reflected to incident light is variously called *reflectance, reflectance factor,* and *reflectance coefficient*. Thus if half the amount of light incident on a surface is bounced back, the reflectance is 50%, or 0.50. The remainder is absorbed, transmitted, or both. The amount of absorption and reflection depends on the type of material and the angle of light incidence (since light impinging upon a surface at grazing angles tends to be reflected rather than absorbed or transmitted; see Fig. 18.2). An example of almost perfect reflection from an opaque surface would be that from a well-silvered mirror, while almost complete absorption takes place on an object covered with lamp black or matte finish black paint. The effect of the materials surface finish on reflection is shown in Fig. 18.3. See Tables 19.14 and 20.3 for typical reflectance values. Reflectance measurement is discussed in Section 18.11.

If the reflection takes place on a smooth surface such as polished glass or stone, it is called specular reflection, as in Fig. 18.3*a*. If the surface is rough, multiple reflections take place on the many small projections on the surface, and the light is diffused as in Fig. 18.3*b*. Since the reflectance is a measure of total light reflected, it does not depend on whether the reflection is specular or diffuse, or a combination of both, as shown in Fig. 18.3*c*. Diffuse transmission takes place through any translucent material such as frosted glass, white glass, milky Plexiglas, tissue paper, and so on. This diffusing principle is widely employed in lighting fixtures to spread the light generated by the bulb or tube within the fixture. Diffuse and nondiffuse transmission are illustrated in Fig. 18.4*a* and *b*.

18.4 Terminology and Definitions

Before beginning any discussion of lighting studies, techniques, and effects, it is important to have a basic understanding of the physical concepts and terminology involved and their interrelations. We will use and explain both the conventional American Standard (AS) and the SI systems of units. The latter also called the metric system, is used as the basic system by the IESNA and in this book, whereas the lighting industry largely uses the AS system as its norm. To familiarize the reader with the SI system, we will frequently use duplicate units, with the second system's unit enclosed in square brackets [ ].

18.5 Luminous Intensity

The AS unit of *luminous intensity* is the candlepower [SI-candela], abbreviated cp [cd], and normally represented by the letter *I*. It is analogous to pressure in a hydraulic system and voltage in an electric system and represents the force that generates the light that we see. An ordinary wax candle has a luminous intensity horizontally of approximately one candlepower

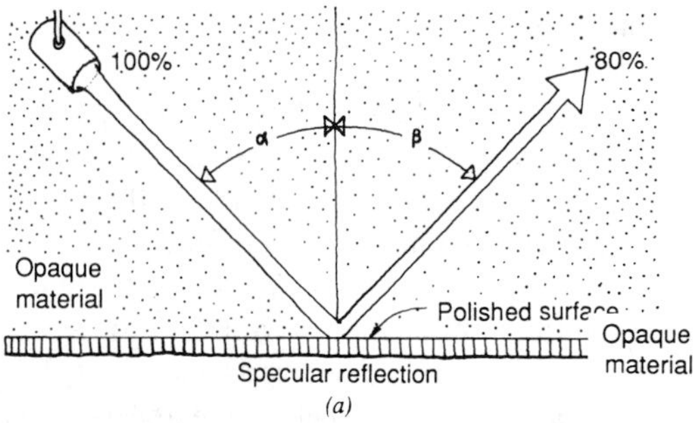

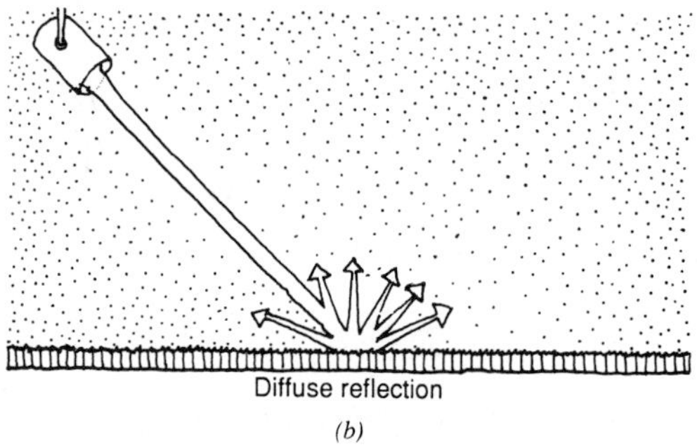

Fig. 18.3 Reflection characteristics. (a) *In specular reflection angle of incidence equals angle of reflection* ($\alpha = \beta$). *Since 80% of light is reflected, reflectance is 80%; 20% of light is absorbed.* (b) *In diffuse reflection, incident light is spread in all directions by multiple reflections on the unpolished surface. Such surfaces appear equally bright from all viewing angles.* (c) *Most materials exhibit a combination of specular and diffuse reflection. Such a surface will mirror the source while producing a bright background.*

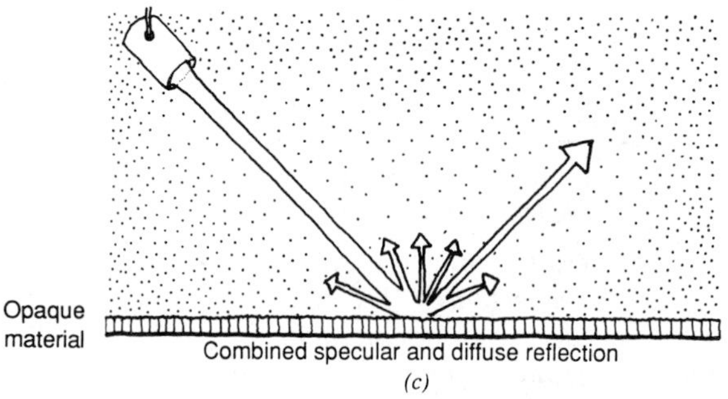

(candela), hence the name. The candela and candlepower have the same magnitude. Luminous intensity is a characteristic of the source only; it is independent of the visual sense.

18.6 Luminous Flux

The unit of luminous flux, in both SI and conventional units, is the lumen, abbreviated lm. If we take a one candlepower [candela] source that radiates light equally in all directions and surround it with a transparent sphere of one foot [meter] radius (see Fig. 18.5), then *by definition* the amount of luminous energy (flux) emanating from one square foot [meter] of surface on the sphere is one *lumen* [lumen]. Since there are 4π ft^2 [m^2] surface area in such a sphere, it follows that a source of one candlepower [candela] intensity produces 4π or 12.57 lm. The lumen, as luminous flux, or quantity of light, is analogous to flow in hydraulic systems and current in electric systems and is normally represented by the letter Φ.

In physical terms, the lumen is a unit of power, like the watt. However, unlike the watt, which is a radiometric unit directly convertible to other power units such as Btu/h, the lumen is a measure of *photometric* power. This means light power as perceived by the human eye and therefore a function of human physiology. Put another way, lumens or luminous flux is the time rate of flow of *perceived* luminous energy. Since the visual response of the eye is frequency dependent, the apprehended light power is also frequency dependent and therefore varies with the spectral content of the impinging light and the spectral sensitivity of the eye. Figure 18.6 should clarify this concept. Figure 18.6*a* shows the spectral content of the visible energy produced by a 500-W incandescent lamp. Measured radiometrically, it amounts to 45 W. However, when passed through a selective filter (Fig. 18.6*b*), which is effectively what happens when the light enters the eye, the resultant "understood" light power appears as in Fig. 18.6*c*, and therefore can no longer be measured in watts. Instead, we use a unit of eye-perceived, or photometric, power called the lumen. It should be obvious that if the spectral content curve in Fig. 18.6*a* were differently shaped, *even if the total radiometrically measured power were the same*, the resultant perceived power in Fig. 18.6*c* would be different.

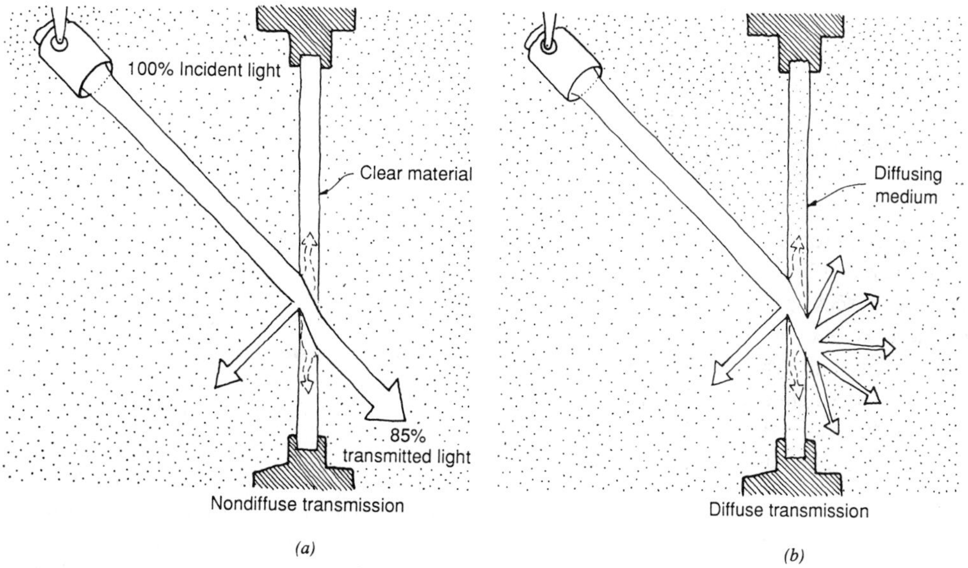

Fig. 18.4 *Transmission characteristics.* (a) *In nondiffuse transmission, the light is refracted (bent) but emerges in the same beam as it enters. Clear materials such as glass, water, and certain plastics exhibit this type of transmission. In the instance illustrated the transmittance is 85% (the remaining 15% is reflected and absorbed). The source of light is clearly visible through the transmitting medium.* (b) *With diffuse transmission, the source of light is not visible and, in the case of multiple sources, the diffusing surface will exhibit generally uniform brightness if the spacing between the light sources does not exceed 1½ times their distance from the material.*

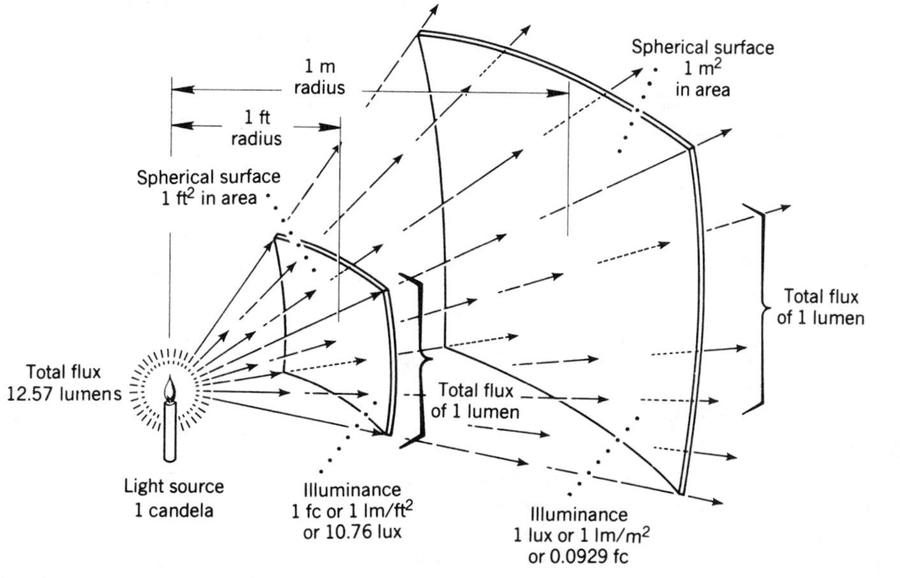

Fig. 18.5 *A source of one candela intensity will produce 4 π (12.57) lumens of light flux. Thus each square foot (square meter) of (spherical) surface surrounding such a source will receive one lumen of light flux. This quantity of light will produce an illuminance of one footcandle (lux) on the spherical surface.*

Refer to Fig. 18.6*b*. A correlation can be made between photometric and radiometric power at the point of maximum response of the eye, which occurs at 555 nanometers wavelength. [A nanometer is 10^{-9} m. The older unit, now obsolete, is the angstrom (Å), which is 10^{-8} cm, or 10^{-10} m (i.e., 1⁄10 of a nanometer).] One watt of monochromatic light at that wavelength will produce 683 lumens. However, since common light sources such as incandescent, fluorescent, mercury, etc., are not monochromatic, but produce light in many parts of the spectrum (see Fig. 19.60), no single conversion factor between watts and lumens exists. Each source has its own luminous efficiency (lumens/watt), determined by its spectrum. For the 500-W lamp used as an illustration in Fig. 18.6, its luminous efficiency (efficacy) is 10,000 lm/500 watts or 20 lumens per watt (lm/w or lpw).

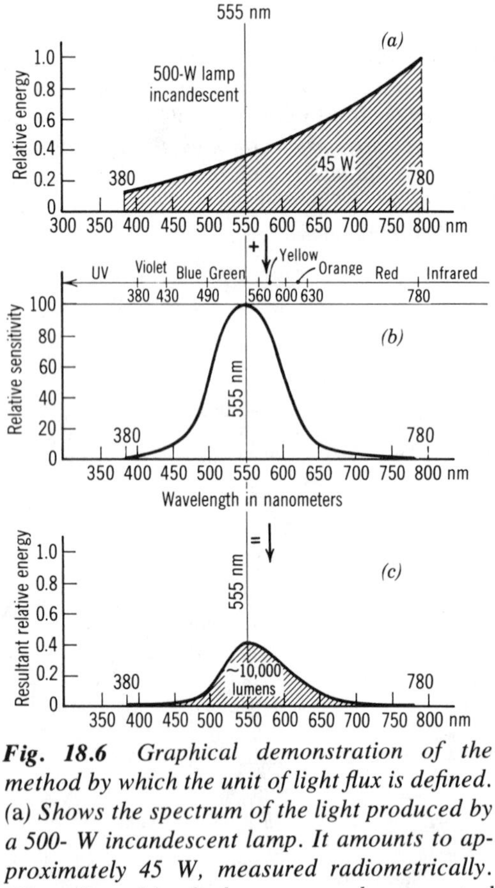

Fig. 18.6 *Graphical demonstration of the method by which the unit of light flux is defined.* (a) *Shows the spectrum of the light produced by a 500- W incandescent lamp. It amounts to approximately 45 W, measured radiometrically. When filtered by the human eye whose spectral sensitivity curve is given in* (b), *this light power is perceived as shown in* (c). *This new light power curve is expressed in lumens and indicates the quantity of light, as perceived by the eye.*

18.7 Illuminance

One lumen of luminous flux, uniformly incident on one square foot of area, produces an *illuminance* of one *footcandle* (fc). Illuminance is normally represented by the letter *E*. Restated, illuminance is the density of luminous power, expressed in terms of lumens per unit area. If we were to consider a light bulb to be analogous to a sprinkler head, then the rate of water flow would be the lumens and the amount of water per unit time per square foot of floor area would be the footcandles. The acoustic analog is intensity in watts per area. In SI units, the area is expressed in square meters and the illuminance (illumination) in lux (lx). Thus the SI unit, lux, is smaller than the corresponding unit, footcandles, by the ratio of square feet to square meters. That is,

$$10.764 \text{ lux} = \text{one footcandle}$$

or multiply footcandle by 10.764 to obtain lux. These relations are shown in Fig. 18.5. Restating the above mathematically yields

$$\text{footcandles} = \frac{\text{lumens}}{\text{square foot of area}}$$

$$\text{fc} = \frac{\text{lum}}{\text{ft}^2} \tag{18.1}$$

and

$$\text{lux} = \frac{\text{lumens}}{\text{square meters of area}}$$

$$\text{lx} = \frac{\text{lm}}{m^2} \tag{18.2}$$

As an *approximation* (8% error)

$$10 \text{ lx} \simeq 1 \text{ fc} \tag{18.3}$$

EXAMPLE 18.1. A 40-W, 430-mA (milliampere), 48-in. [122 cm] fluorescent tube produces 3200 lm. What is the illuminance on the floor of a 10-by-10 ft. room assuming 40% overall efficiency and uniform illumination?

SOLUTION

$$\text{Useful lumens} = 0.4 \times 3200 = 1280$$

$$\text{fc} = \frac{1280}{10 \times 10} = 12.8$$

$$\text{lx} = 12.8 \times 10.76 = 137.7$$

Calculating lux directly, we obtain

$$\text{lx} = \frac{1280}{10 \times 10} \times (3.28 \text{ ft/m})^2 = 137.7$$

By approximation:

$$\text{lx} \simeq 10 \cdot \text{fc} \simeq 128$$

Illuminance at a *point* can be computed from intensity, as shown in Section 18.12 below.

18.8 Luminance, Exitance, and Brightness

An object is perceived because light coming from it enters the eye. The impression received is one of object *brightness*. This brightness sensation, however, is subjective, and depends not only on the object *luminance* (L) but also on the state of adaptation of the eye (see Sections 18.18 and 18.19). For this reason the physiological sensation is generally referred to in the literature as subjective or apparent brightness or simply *brightness*, whereas the measurable, reproducible state of object luminosity is its *luminance* (formerly "photometric brightness"). Luminance is normally defined in terms of intensity; it is the luminous intensity per unit *apparent* (projected) area of a primary (emitting) or secondary (reflecting) light source. Thus its units are candela per area. Specifically, the SI unit of luminance is candela per square meter (cd/m^2), sometimes referred to as the nit. Another unit formerly in common use in the conventional (AS) system is the footlambert. Conversion factors for S1 and AS units (plus other, obsolete units, for the convenience of readers using older sources) are given in Table 18.1. Other luminance terms such as stilb, apostilb, blondel, millilambert, and candela per square in. are best avoided. In this book, the term *luminance* will be used except where it is specifically intended to refer to the physiological sensation involved, in which case the terms *brightness, subjective brightness,* or *apparent brightness* will be used. Luminance has no readily conceivable mechanical or electrical analogy.

Another concept that the lighting designer will encounter is known as *luminous exitance* or simply as *exitance* and as the name implies, describes the total luminous flux density leaving (exiting) a surface, irrespective of directivity or viewer position. For instance, if a surface one meter square emits one lumen, its luminous exitance is one lumen per square meter (1 lm/m^2) or 0.093 lm/ft^2. A surface that is a perfect diffuser, whether by emitting light diffusely or reflecting light diffusely, is known as a *Lambertian surface*. It is fairly simple to demonstrate mathematically that the luminance of such a surface equals $1/\pi$ times its exitance. The importance of this relationship is its usefulness as an approximation. Although very few surfaces are truly Lambertian, many are approximately so, and this relationship can be used as an engineering accuracy approximation in many such cases.

ILLUMINATION

TABLE 18.1 **Lighting Units—Conversion Factors**

Unit	*Multiply*	*By*	*To Obtain*
Illuminance (E)	Lux	0.0929	Footcandle
	Footcandle	10.764	Lux
Luminance (L)	cd/m^2	0.2919	Footlambert
	cd/cm^2	10000	cd/m^2
	$cd/in.^2$	1550	cd/m^2
	cd/ft^2	10.76	cd/m^2
	millilambert	3.183	cd/m^2
	Footlambert	3.4263	cd/m^2
Intensity (I)	Candela	1.0	Candlepower

The concept of exitance is important in detailed photometric calculations such as those involved in determining coefficients of utilization (Section 20.28), surface luminance coefficients (Section 20.43) and in detailed point illuminance calculations. All of these are beyond the scope of this book since they are not usually performed by the lighting designer. Use of the derived coefficients is demonstrated in the referenced sections.

Detailed point calculations are today almost universally performed by computer, and the necessary mathematics is built into the computer program. Readers interested in further background on luminous exitance are referred to reference Murdoch (1985) at the end of Chapter 21.

Since object luminance is that which is visually perceived and is a prime factor in visibility (and glare), it is important that the reader be able to perform basic luminance calculations. Although the eye does not differentiate between primary sources that generate and emit light, and secondary sources that derive their luminance from reflection or transmission, the differentiation is important in calculation procedures. See Fig. 18.7 for a graphic representation of the basic relationships.

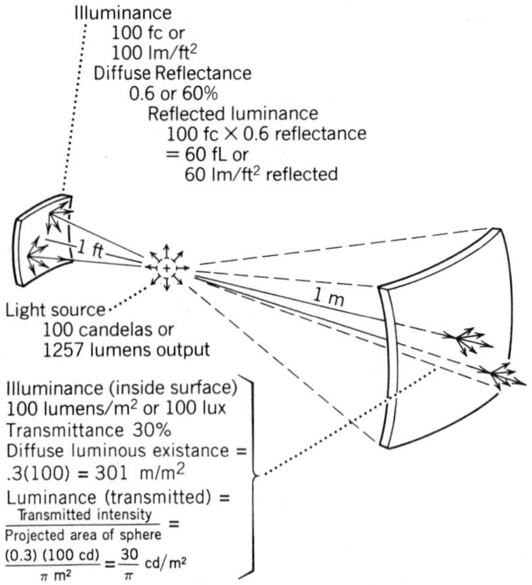

Fig. 18.7 Luminance may be either reflected or transmitted. In the former case it is calculated as the product of the incident lumens and the reflectance; in the latter, as the transmitted intensity divided by projected area.

Luminance of a light-emitting surface:

EXAMPLE 18.2

(a) Calculate the luminance of a standard inside-frosted, 100-W incandescent light bulb (an A-19 bulb—see Fig. 19.34 for characteristic) with a maintained output of 1700 lm. Assume (for simplicity sake) that the bulb is spherical.

(b) Assume that an opal glass globe of 8 in. diameter (20 cm) and a transmittance of 35% surrounds the above bulb. Calculate the luminance of the globe. Use SI units throughout.

SOLUTION

(a) Assume that the filament is a point source (an essentially valid assumption) and that the inside frosting of the glass does not reduce the output (it does by about 1%). The inside frosting serves to convert the point source filament to a uniformly emitting globe. The definition of a point source tells us that one candela produces 4π lumens, distributed spherically. Therefore,

$$\frac{\text{one candela}}{4\pi \text{ lm}} = \frac{I \text{ candela}}{1700 \text{ lm}}$$

and the intensity, I, of the filament, and therefore also of the bulb, since the frosting is assumed not to reduce output, is

$$I = 1700 \text{ lumens} \times \frac{\text{one candela}}{4\pi \text{ lm}} = 135 \text{ candela}$$

From Fig. 19.34, we find that the A-19 bulb has a diameter of 2⅜″ [6 cm]. The definition of luminance is

$$L = \frac{I}{A} \frac{\text{(intensity)}}{\text{(projected area)}}$$

The projected area becomes approximately a circle with a diameter of 6 cm (assuming a spherical bulb), whose area equals πr^2 or π (.03 m)^2. So

$$L = \frac{135 \text{ candela}}{\pi (0.03 \text{ m})^2} = 47{,}750 \frac{\text{candela}}{\text{m}^3}$$

This is a potential source of severe direct glare (see Section 18.25).

(b) We have already calculated the intensity of the source, but it is reduced to 35% by the globe. The projected area of the globe is larger than that of the bulb, that is,

$$A \text{ proj.} = \pi\left(\frac{0.20}{2}\text{ m}\right)^2$$

So the expression for luminance becomes

$$L = \frac{135 \text{ candela } (0.35)}{\pi(0.10 \text{ m})^2} = 1500 \frac{\text{candela}}{\text{m}^2}$$

This is no longer a potential source of direct glare.

Note two important principles demonstrated by this example:

1. Intensity is not area or distance dependent; it varies only with transmission factors.
2. Luminance varies with both intensity and area.

For example, if a white lamp whose inside coating reduces lamp output by about 10% had been used in lieu of an inside frost, both intensity and luminance would have been reduced by 10%. However, if a clear glass lamp had been used, intensity would remain the same but luminance would increase inversely with the ratio of filament area to bulb area. Therefore, a clear 100-W lamp, whose filament has an area of perhaps of 0.3 cm^2, would have a luminance of

$$\frac{135 \text{ candela}}{0.3E\text{-}4 \text{ m}^2} = 4{,}500{,}000 \text{ cd/m}^2$$

which is so severe a glare source as to be disabling when in the near field of vision.

EXAMPLE 18.3. Calculate the luminance of a standard 4-ft cool white fluorescent lamp, F40T12CW. Assume a viewing angle normal to the long axis of the lamp and that the lamp is a diffuse (Lambertian) emitter. Use SI units.

SOLUTION. This problem can be solved in two ways; using the relationship between exitance and luminance or by calculating intensity.

By Exitance/luminance: Use the 2770 lm at 40% life as an average condition of lamp output. The luminous length is 4 ft (120 cm) and the diameter is 12/8 in. (3.8 cm). The luminous surface area of the tube is then

$$\begin{aligned} A = L \times d \times \pi &= (120 \text{ cm}) \times (3.8 \text{ cm}) \times \pi \\ &= (1.2 \text{ m})(.038 \text{ m}) \\ &= 0.0456\pi \text{ m}^2 \end{aligned}$$

$$\text{exitance} = \frac{\text{luminous flux}}{\text{area}} = \frac{2770 \text{ lm}}{0.0456\pi \text{ m}^2}$$

$$\begin{aligned} \text{luminance} &= \frac{1}{\pi}\text{ (exitance)} \\ &= \frac{2770}{0.0456\pi^2} \text{ cd/m}^2 = 6155 \text{ cd/m}^2 \end{aligned}$$

By Intensity: The equivalent spherical intensity of the lamp can be calculated by using the relationship that one candela produces 4 π lumens. Therefore,

$$\text{equivalent intensity } I = \frac{2770/\text{m}}{4\pi} \text{ candela}$$

The radius of a sphere of equivalent surface area to the fluorescent tube would be

$$4\pi r^2 = 0.0456\ \pi \text{ m}^2$$

$$r^2 = \frac{0.0456}{4} \text{ m}^2 = 0.0114 \text{ m}$$

$$r = 0.1067 \text{ m}$$

The equivalent projected area of such a sphere is πr^2, or

$$A = 0.0114\pi$$

and its luminance is

$$\begin{aligned} L &= \frac{\text{intensity}}{\text{area}}\frac{\text{cd}}{\text{m}^2} = \frac{2770/4\pi}{0.0114\pi}\frac{\text{cd}}{\text{m}^2} \\ &= \frac{2770}{0.0456\pi^2}\frac{\text{cd}}{\text{m}^2} \\ &= 6155 \text{ cd/m}^2 \end{aligned}$$

EXAMPLE 18.4. To demonstrate the usefulness of the luminance/exitance approximation, calculate the luminance, in SI units, of the page that you are now reading. Assume a uniform illuminance of 500 lux, a diffuse reflectance of 0.77, and a viewing angle normal to the page.

SOLUTION. From the definition of illuminance, one lux is produced by one lumen falling on one square meter. Therefore, the exitance, or density of reflected lumens from the page is

$$\text{exitance} = 500 \frac{\text{lm}}{\text{m}^2} \times 0.77 = 385 \frac{\text{lm}}{\text{m}^2}$$

and

$$L = \frac{1}{\pi} \times 385 \frac{\text{lm}}{\text{m}^2} = 122.5 \text{ cd/m}^2$$

(Typical luminances are given in Table 18.2.)

18.9 Illuminance Measurement

Field measurements of illuminance levels are most commonly made with a portable illuminance meter, four of which are illustrated in Fig. 18.8. These devices comprise a photoelectric material connected to a microammeter via electronic control circuitry and are calibrated in footcandles or lux.

As explain in Section 18.6, and as shown in Fig. 18.6, the human eye is not equally sensitive to the various wavelengths (colors). Maximum sensitivity at high illuminance levels is in the yellow-green area (wavelength of 555 nm) while sensitivity at the red and blue ends of the spectrum is quite low. This effect is so pronounced that 10 units of blue energy are required to produce the same visual effect as 1 unit of yellow-green. Therefore, if a meter is to be useful, its inherent response, which is quite different from that of the human eye, must be corrected to correspond to the eye. For this reason meters are "color corrected."

The cells (meters) must also be corrected for light incident at oblique angles that does not reach the cell due to reflection from the surface glass and shielding of the light-sensitive cell by the meter housing. This correction is known as cosine correction. A good meter must therefore be (and it will plainly so indicate) color and cosine corrected.

Modern photometers may have considerable electronic circuitry, which provides such functions as automatic ranging, integration for flickering or time-varying sources, and connection facilities for data storage and transmission. For determining average room illuminance when using a conventional nonintegrating meter a number of readings should be taken and an average computed. Where no definite height is specified, readings are taken at 30 in. (75 cm) above the floor. The meter must always be held with the cell parallel to the plane of the test. Thus to measure wall illuminance, the meter must be held with the cell parallel to the wall. If electric lighting readings are desired and the test is being conducted during day-light hours, readings should be taken with and without the artificial illumination and the results subtracted. Detailed instructions for conducting field surveys are contained in the IES publication "How to Make a Lighting Survey." Briefly, a survey of an existing indoor lighting installation should establish:

1. Type, rating, and age of sources.
2. Type, design, and model of luminaires.
3. Maintenance schedule.

It should also measure:

1. Mounting height.
2. Spacing and pattern of luminaires.
3. Reflectances of walls, floor, ceiling, and major items of furniture and equipment.
4. Illuminance levels throughout the area plus levels at walls and columns, all at the working plane elevation. Additionally, vertical plane illuminance at walls and other major vertical planes should be measured. The significance of vertical surface luminance is discussed in detail in subsequent chapters.

18.10 Luminance Measurement

From the aspect of appreciation of the visual scene, including particularly considerations of glare, the measurement of luminance is more important and meaningful than that of illuminance (footcandles, lux). This is so since it is luminance, or more accurately subjective brightness and brightness contrasts caused by photometric luminance, that we see, not illuminance. That lux measurements are still more widely taken than luminance measurements and uti-

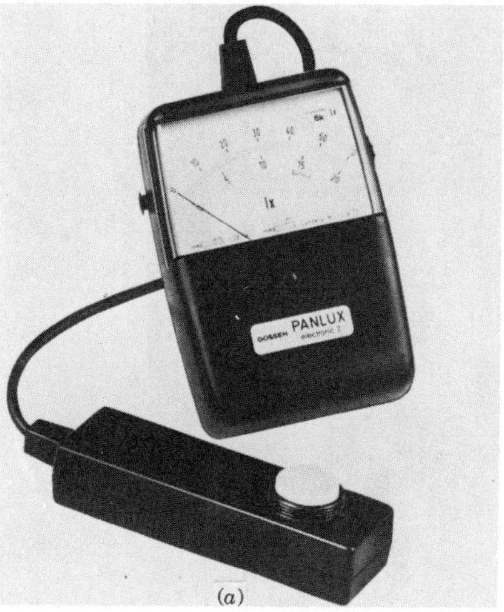

(a)

(c)

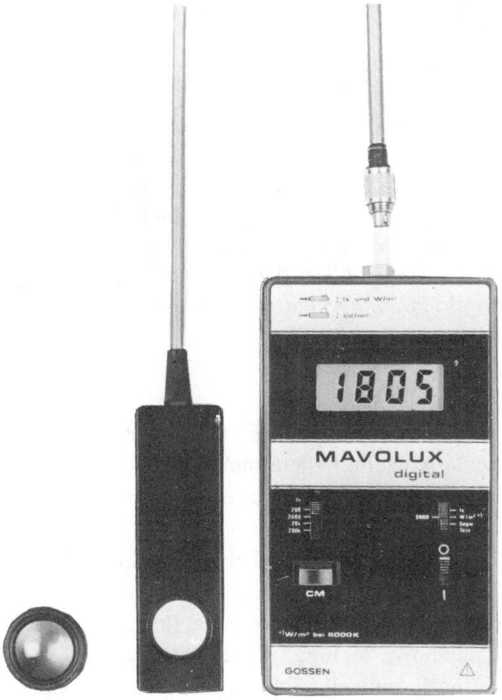

(b)

Fig. 18.8 (a) *Color and cosine corrected analog illuminance meter calibrated in lux, and equipped with a remote cable connected photocell. The unit has nine ranges covering 0 to 200,000 lux with a maximum ± 3.5% error. When equipped with the adapter shown in the lower left of* (b) *the unit becomes a wide angle (20°) aperture luminance meter with a range of 0 to 2,000,000 cd/m² in nine steps. The unit is battery powered, measures approximately 3 × 4 × 1½ in. and weighs 12 oz.* (b) *Electronic, digital, color- and cosine-corrected light meter which measures illuminance (0 to 200,000 lux) and luminance (0 to 2,000,000 cd/m²) in four ranges with a maximum 2½% error. The unit also measures solar radiation in W/m². It is equipped with a recorder output for use in extended-time monitoring applications. The meter measures 3½ × 6 × 1 in. thick and weighs under 1 lb.* (c) *Portable, highly accurate, autoranging digital illuminance meter has a range of 0.01 to 200,000 lux (20,000 fc) in five steps with a ±2% accuracy. The unit, which measures approximately 7¼ × 3 × 1 in. and weighs 10 oz, can measure flickering light sources, luminous intensity, comparative illuminances, and other convenient photometric measurements by means of a built-in microcomputer and available accessories. [(a,b) Courtesy of Gossen GmBH; (c) courtesy of TOPCON Instrument Corp. of America.]*

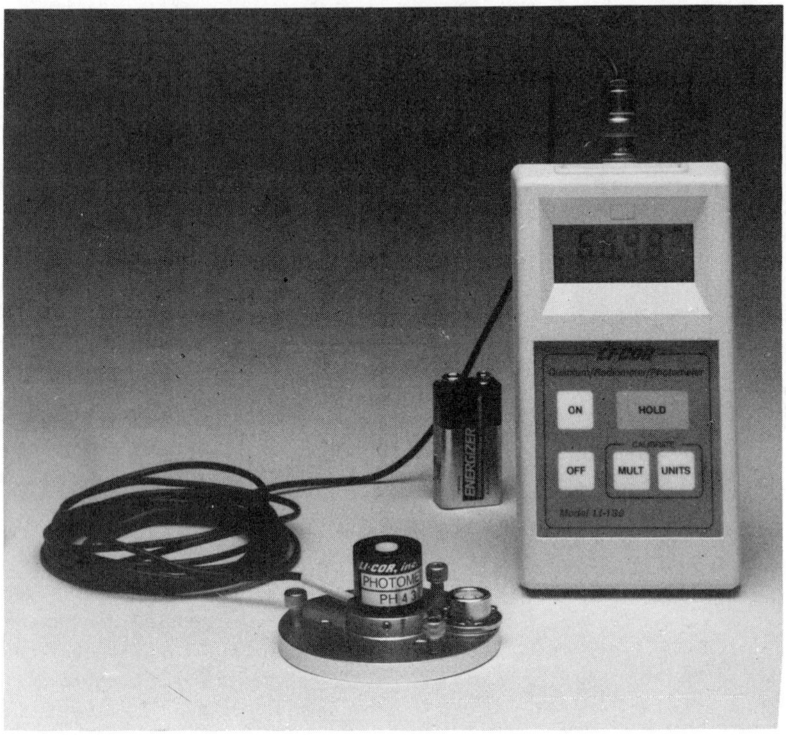

(*d*)

Fig. 18.8 (d) *Autoranging digital photometer with remote photometric sensor is ideal for measurements of light levels in architectural models as, for instance, in Fig. 19.24. The photometric sensor, which measures only 1¼ in. in diameter and 1¾ in. in height has minimum effect on light distributions in any but the smallest scale models. Illustrated is a mounting device which permits the sensor to be placed on any surface of a model. The meter has a range of 0 to 199,900 lux with better than 1% accuracy, is battery powered, measures 5½ × 3 × 1½ in., and weighs 9 oz. (Courtesy of Li-Cor, Inc.)*

lized as a gauge of the adequacy of a lighting installation is due to two factors:

1. Lux meters are cheaper and simpler to use than luminance meters.
2. Design recommendations for lighting levels are given in terms of illuminance (Section 18.24). Lux measurements can therefore be used as a rapid, simple method of determining whether a particular lighting installation meets these design requirements.

Luminance meters are available in a number of configurations, including a direct-reading narrow-angle (spot) type (Fig. 18.9*a*), an accessory attachment to an illuminance meter (Fig. 18.9*b*), and a laboratory-grade unit that reads both luminance *and* contrast (Fig. 18.9*c*). The last is particularly useful in determining the effect of veiling reflection or contrast reduction, discussed in Section 18.27.

An approximation of the luminance of a reflecting or luminous source can be obtained using an illuminance (footcandle, lux) meter of the type shown in Fig. 18.8. For diffuse *reflecting* surfaces, the cell of the meter is placed against the surface and then slowly retracted 2 to 4 in. (5–10 cm) until a constant reading is obtained. The luminance, in footlamberts, is then 1.25 times the reading in footcandles, the 1.25 factor compensating for wide-angle losses.

For a diffuse luminous source, the cell of an illuminance meter is placed directly against the surface (see Fig. 18.10); the source luminance in

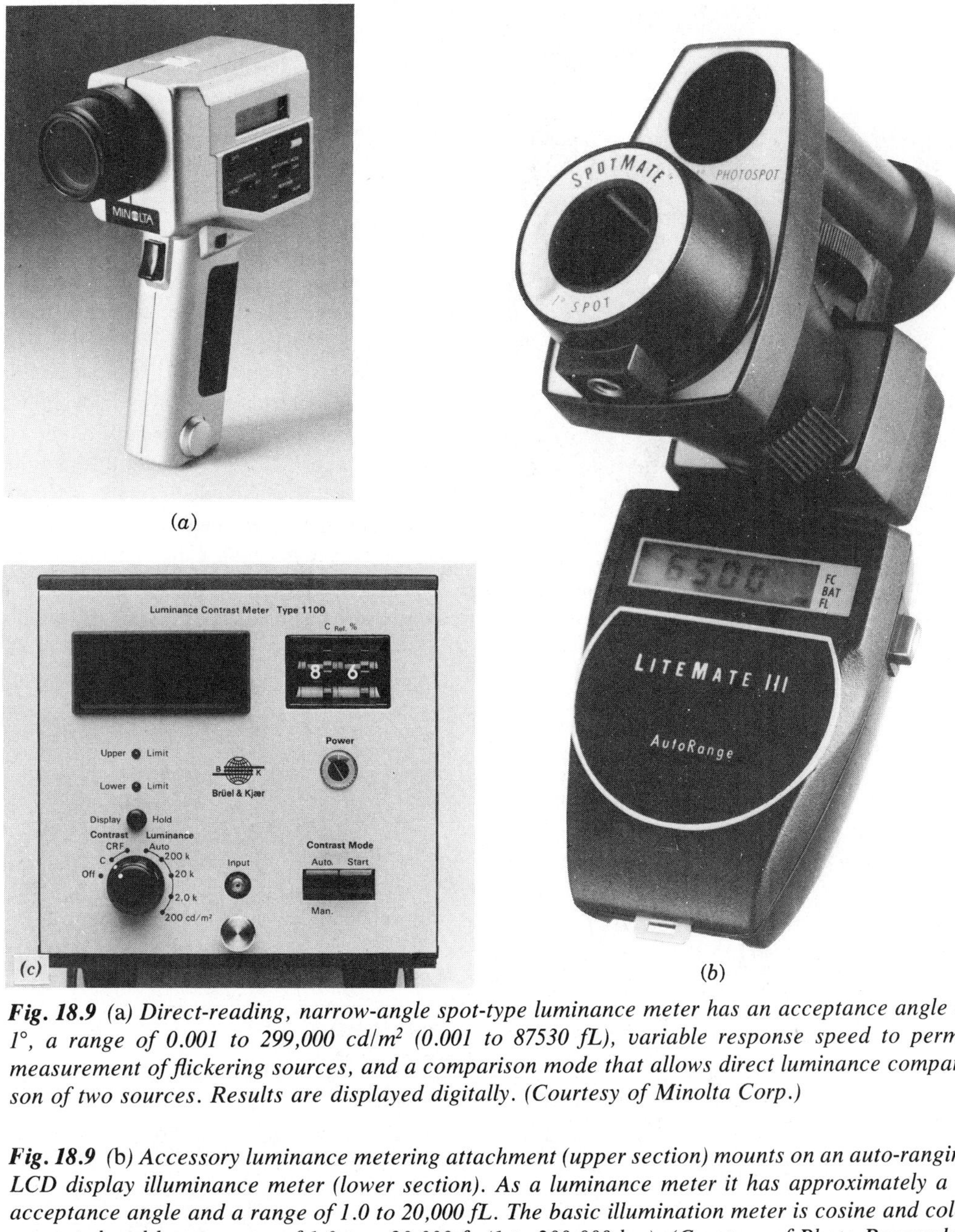

Fig. 18.9 (a) *Direct-reading, narrow-angle spot-type luminance meter has an acceptance angle of 1°, a range of 0.001 to 299,000 cd/m² (0.001 to 87530 fL), variable response speed to permit measurement of flickering sources, and a comparison mode that allows direct luminance comparison of two sources. Results are displayed digitally. (Courtesy of Minolta Corp.)*

Fig. 18.9 (b) *Accessory luminance metering attachment (upper section) mounts on an auto-ranging LCD display illuminance meter (lower section). As a luminance meter it has approximately a 1° acceptance angle and a range of 1.0 to 20,000 fL. The basic illumination meter is cosine and color corrected and has a range of 1.0 to ≈20,000 fc (1 to 200,000 lux). (Courtesy of Photo Research.)*

Fig. 18.9 (c) *This unit is a high-precision electronic digital instrument that measures luminance, luminance ratio, contrast, and contrast reduction. Use of the unit is explained in Fig. 18.30. Luminance measurement range is 0 to 200,000 cd/m² (56,400 fL) at high accuracy. (Courtesy of Brüel & Kjaer.)*

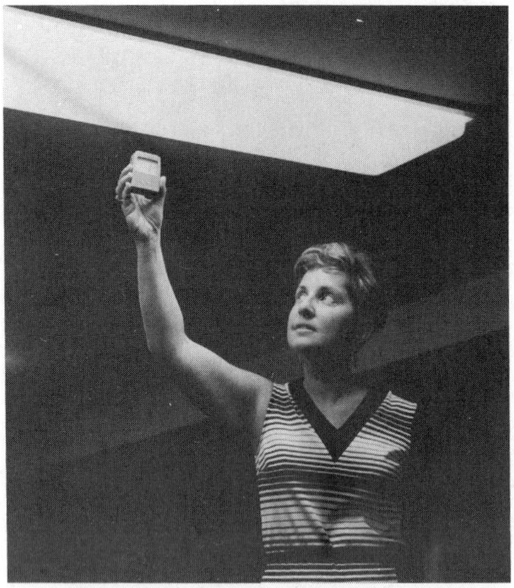

Fig. 18.10 When the cell of a direct-reading illuminance meter is held in contact with a luminous source, the surface luminance can be read directly or simply calculated. See the text.

footlamberts is equal to the reading on the meter in footcandles because footlamberts = lumens per area = footcandles. When using a meter calibrated in lux rather than footcandles, the readings must be divided by π to obtain the equivalent diffuse source luminance in cd/m^2.

18.11 Reflectance Measurements

It is often desirable to know the reflectance of a given surface since luminance can then be readily computed (see Fig. 18.7). Two methods of measuring diffuse (nonspecular) reflectance are shown in Fig. 18.11: the known-sample method and the light-ratio method. If a sample of known reflectance factor (RF) is available, this method should be used since it yields more accurate results than the ratio method. The sample should be no smaller than 8 in. by 8 in.

It is a good idea for an inexperienced lighting designer to determine the reflectances, illumi-

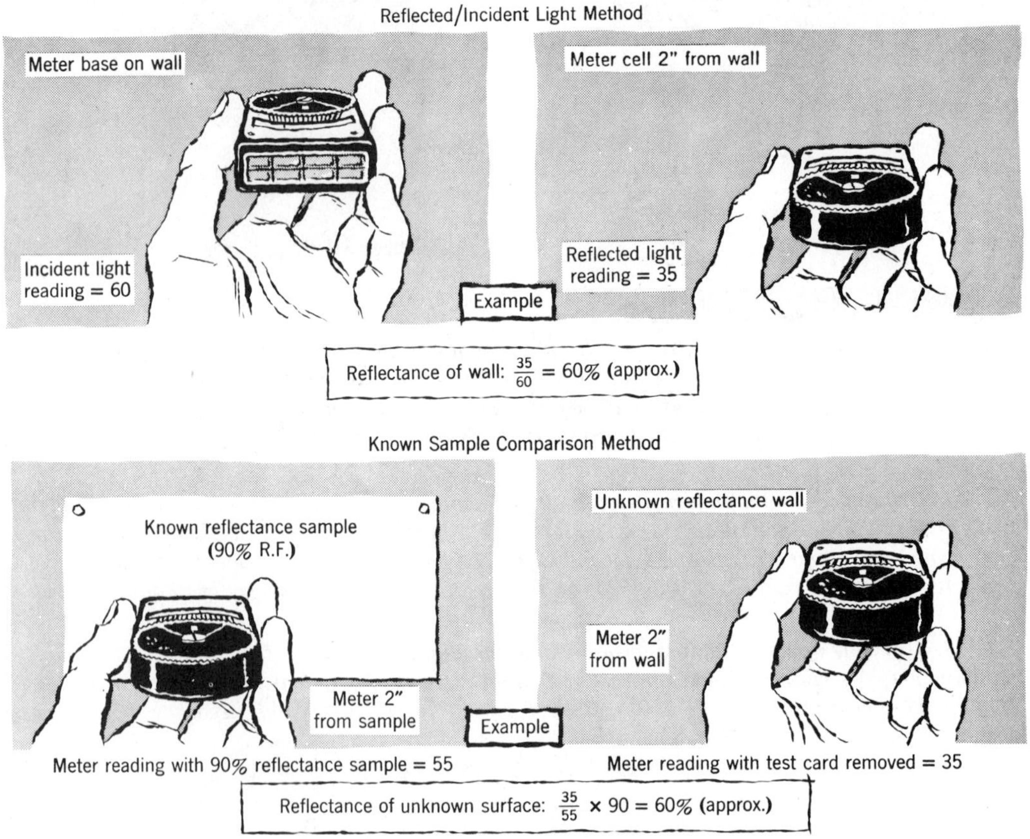

Fig. 18.11 Two simple methods of measuring diffuse reflectance of a surface.

nance, and luminance levels of spaces and surfaces familiar to him or her, such as his office desk, adjoining wall, and the like—even to the extent of marking these figures on the respective surfaces in order to develop an appreciation of and a memory for these parameters. This will enable the designer to visualize the result of a lighting design and should be of considerable assistance. See Table 19.14 for typical reflectance values.

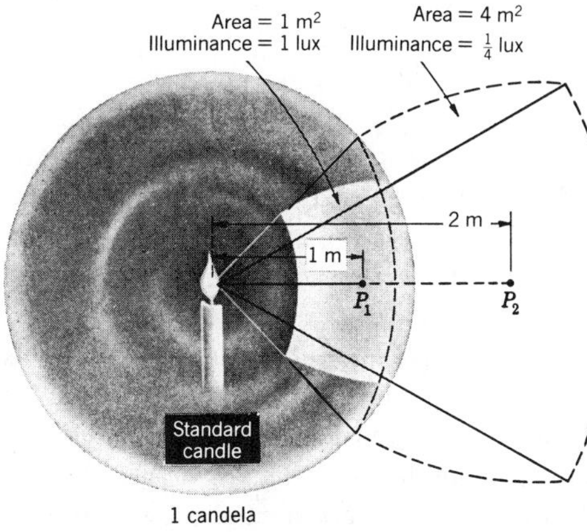

(a)

Fig. 18.12 (a) *Relations between candela, lumens, and lux defined with reference to a standard light source of one mean spherical candlepower (one candela) located at the center of a sphere of one meter radius.*

18.12 Inverse Square Law

We have already seen that, by definition, a point source of 1 cd intensity produces an illumination of 1 lux on the inside surface of a surrounding sphere of 1 m radius (r). Since the surface area of this sphere is 4π m^2, a 1-cd source produces 4π lm of luminous flux. Now assume a sphere of 2-m radius surrounding this same source (see Fig. 18.12*a*). Since the same amount of flux is spread over a larger area, the illumination on the larger sphere is inversely proportional to the ratio of the two-sphere areas, that is,

$$\text{lux}_2 = \text{lux}_1 \times \frac{\text{area}_1}{\text{area}_2} \qquad (18.4)$$

or

$$\text{lux}_2 = \text{lux}_1 \times \frac{4\pi r_1^2}{4\pi r_2^2}$$

$$= \text{lux}_1 \times \frac{r_1^2}{r_2^2} \qquad (18.5)$$

In other words, the illumination is inversely proportional to the square of the distance from the source. In general terms

$$\text{lux(fc)} = \frac{\text{cp intensity}}{\text{distance}^2} \qquad (18.6)$$

where distance is expressed in meters (feet). (This holds true for surfaces normal to a source. For other situations, see Section 20.38.)

This relationship can also readily be derived by using any solid angle and the area it intercepts, as in Fig. 18.12*b*. A glance at this figure shows clearly that the area intercepted is proportional to the square of the distance from the source and therefore the illumination is inversely proportional, as stated above.

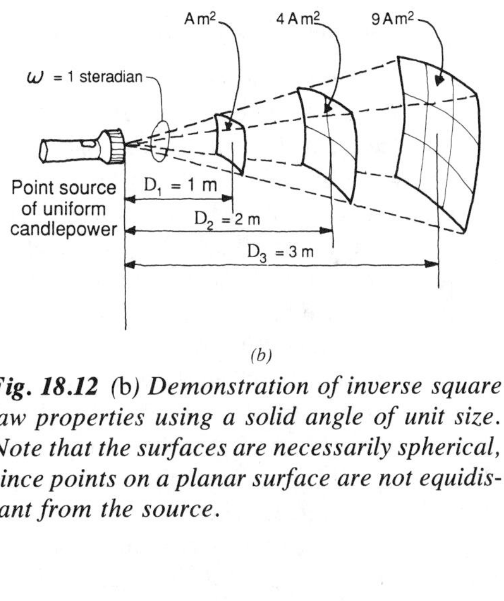

(b)

Fig. 18.12 (b) *Demonstration of inverse square law properties using a solid angle of unit size. Note that the surfaces are necessarily spherical, since points on a planar surface are not equidistant from the source.*

18.13 Luminous Intensity (Candlepower) Measurements

Luminous intensity (candela, candlepower) cannot be measured directly but must be computed from its illumination effects. The simplest way of doing this is to use the inverse square relationship developed in the preceding section. Measure the illumination produced on a plane at right angles to the source, at a known distance and apply equation 18.6. For accurate measurement, the distance should be at least five and preferably ten times the maximum dimension of the source, since for anything other than a point source the equation is an approximation. The candela (candlepower) thus calculated is the luminous intensity in the direction being viewed. Since candlepower is not uniform in all directions for anything except an ideal point source, and since a single intensity figure for a source is desirable for calculation purposes, the average of a number of intensity figures taken from several directions is used. This average figure is called the mean spherical candlepower (mscp), and represents an equivalent point source that will produce 4π lm for every candela. Thus a 10-cd lamp will exhibit an average intensity of $\pm$ 10 cd in all directions and will produce 40π lumens.

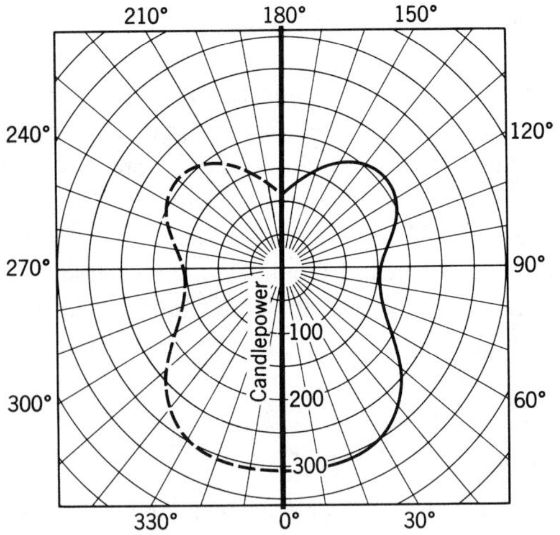

Typical candlepower distribution curve for a general diffuse type of luminaire

Fig. 18.13 *Typical candlepower distribution curve for a general diffuse-type luminaire. Since the unit is symmetrical about its vertical axis, only one curve need be shown. Normally, only the right side of even this single curve is shown, due to symmetry, as in Fig. 18.14.*

18.14 Candlepower Distribution Curves

If the candlepower figures calculated in the preceding section are plotted on polar coordinate axes, the resultant figure is called, logically, a candlepower distribution curve (CDC) for the particular source involved. The procedure for making this curve is straightforward. A photo cell is rotated around the source in a single plane, illuminance measured, and intensity (cd) calculated. Alternatively, the photo cell can be fixed and the source rotated. If the source's distribution is symmetrical, as shown in Fig. 18.13, then only a single set of values is required, and the resultant plot is valid in all vertical planes through the source. Thus for incandescent lamps, downlights, open circular reflectors, and the like, only a single CDC is required. For a nonsymmetric source such as a fluorescent luminaire, CDC curves in several planes are required to define the fixture's distribution characteristic. Normally, manufacturers will provide longitudinal and crosswise curves, plus a diagonal (45°) plane curve on request. This is illustrated in Fig. 18.14, where the three planes and typical resultant curves are shown. Most CDC plots are made on polar coordinates because such a plot clearly shows directions and magnitudes. Nevertheless, polar plots tend to crowd near the nadir and accurate magnitude readings at the cutoff angle are difficult to make. For this reason it is occasionally desirable to obtain a plot on rectangular coordinates. One such plot is shown in Fig. 18.15. The usefulness of candlepower distribution curves will become a clear in our subsequent discussions on fixture diffusers (Section 20.19), point-by-point calculations (Section 20.39), and direct and reflected glare (Sections 18.25 to 18.27). It should be noted that the area of the CDC curve is not a measure of the lumen output.

LIGHT AND SIGHT

18.15 The Eye

Since all discussion of light and lighting techniques is irrelevant to our purposes unless ultimately related to vision, we turn to a cursory examination of the human eye before proceeding further with discussions of lighting.

Light impinging upon the eye enters through the pupil, the size of which is controlled by the iris, thereby controlling the amount of light entering the eye. The lens focuses the image on the

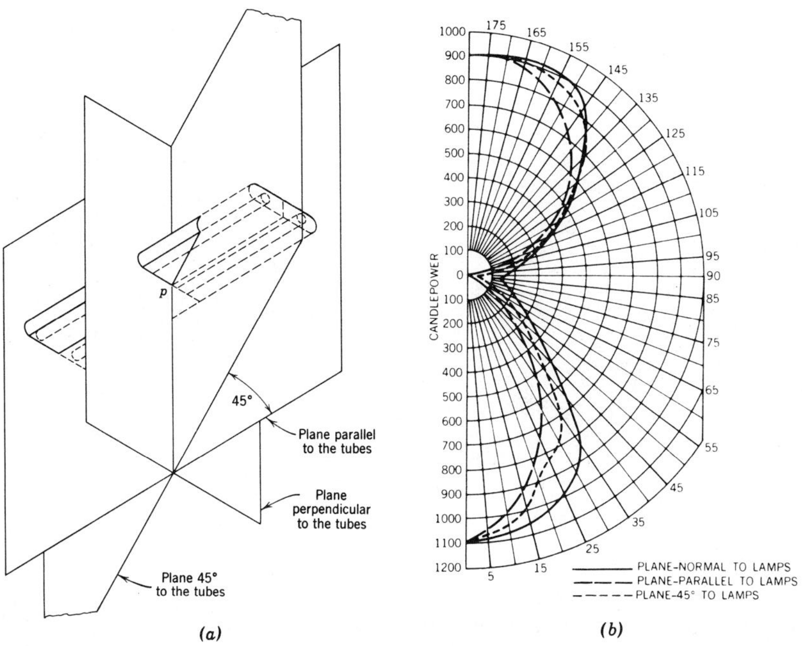

Fig. 18.14 Because of asymmetry of fluorescent fixtures, candlepower distribution curves in three planes are plotted, as shown.

retina, from which the optic nerve conveys the visual message by electric impulse to the brain. Figure 18.16 shows the structure of the eye and the parallel structure of a camera.

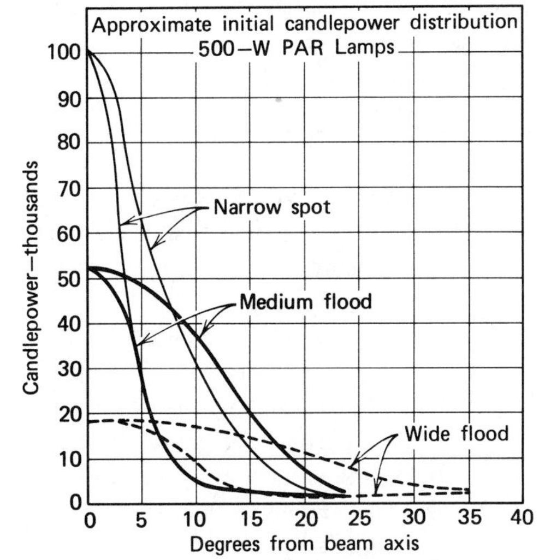

Fig. 18.15 Candlepower distribution curves plotted in rectangular coordinates. Note that candlepower values near the cutoff angles are easily read, which is not the case in polar plots.

The central portion of the eye, near the fovea, contains light-sensitive cells called "cones" because of their shape. The cones are responsible for the ability to discriminate detail and also give us our sensation of color. Proceeding outward from the fovea, a second type of cell is encountered called a "rod" cell, also named after its shape. These cells are extremely light sensitive, giving response to light 1/10,000 as bright as that required by cone cells. However, rod cells lack color sensitivity, thus accounting for the fact that in dim light (rod vision), we have no color perception and all colors appear as varying shades of gray. Rod cells also lack detail discrimination, making "night vision" quite coarse. Finally, rod cells are slower acting than cone cells and therefore have a low degree of flicker fusion, or, stated conversely, they are highly motion sensitive. Since these cells occur at the outer portions of the retina, their motion sensitivity results in our being best able to detect movement when looking out of the "corner of the eye." Looking at a fluorescent tube directly and then obliquely will demonstrate this effect.

Figure 18.17 is a sketch illustrating the angles involved in the field of vision. Of particular in-

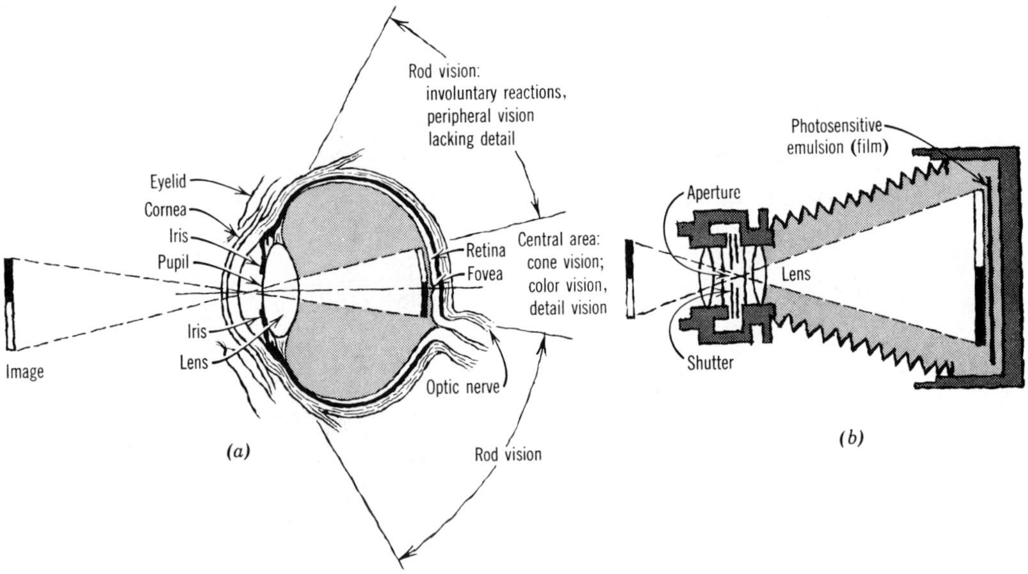

Fig. 18.16 The human eye and the camera operate on similar optic principles. The cornea acts as an outer refracting lens which introduces light into the iris. The iris and pupil control the f-stop, or opening of the eye, and correspond roughly to a range of f2.1 to f11. The lens, which acts as a perfectly smooth automatic zoom lens, can focus from about 2 in. to infinity. Light is focused on the retina, which contains in all some 150 million light-sensitive cells of two types: rod cells, which can detect luminances from 1/1000 cd/m² to approximately 120 cd/m² in black-and-white mode, and cone cells, which see luminances in the range 3 to 1,000,000 cd/m² in full-color mode. The fovea, which is the center of focus, is an area of pinhead size containing about 100,000 cone cells, which account for the extreme precision of foveal (center-focus) vision.

terest is the extreme narrowness of the cone of central (foveal) vision, in which acute perception of detail takes place. This area is so small that the eye must refocus on each dot in a colon (:) if you wish to examine each individually. Surrounding this central area is a cone of binocular vision of 30° half-angle, called the near field or surround, in which area most of coarser sight information is gathered. Beyond this cone we have far field and peripheral, primarily horizontal, monocular vision. It is the far field and peripheral areas that largely gives us our subjective, ambience-type reactions, which will be discussed below.

18.16 Factors in Visual Acuity

The three components of any seeing task are obviously the object or task, the lighting conditions, and the observer. Listed below are the variables affecting each of these three components. Based on the results of many investigations they can be categorized as of primary or secondary importance.

I. The Task
   Primary factors
   (*a*) Size
   (*b*) Luminance (brightness)
   (*c*) Contrast, including color contrast
   (*d*) Exposure time—needed or given
   Secondary Factors
   (*e*) Type of object—required mental activity; familiarity with the object (in reading, familiarity is so important as to become the primary factor)
   (*f*) Degree of accuracy required
   (*g*) Task; moving or stationary
   (*h*) Peripheral patterns
II. The Lighting Condition
   Primary Factors
   (*a*) Illumination level
   (*b*) Disability glare

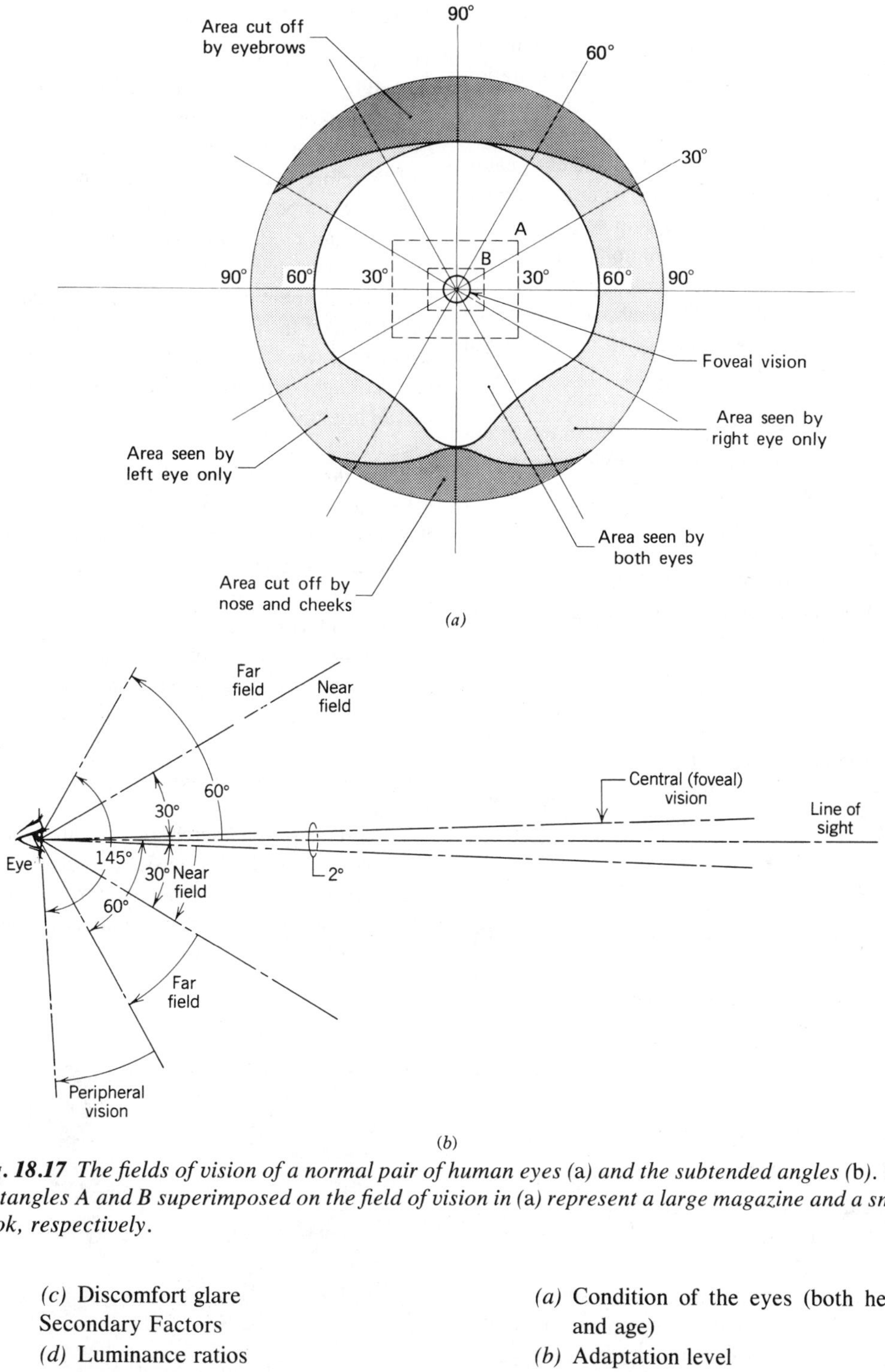

Fig. 18.17 *The fields of vision of a normal pair of human eyes* (a) *and the subtended angles* (b). *The rectangles A and B superimposed on the field of vision in* (a) *represent a large magazine and a small book, respectively.*

(*c*) Discomfort glare
Secondary Factors
(*d*) Luminance ratios
(*e*) Brightness patterns
(*f*) Chromaticity

III. The Observer
Primary Factors
(*a*) Condition of the eyes (both health and age)
(*b*) Adaptation level
(*c*) Fatigue level
Secondary Factors
(*d*) Subjective impressions; psychological reactions

Although in the following discussions these factors will be considered individually, many are interrelated. Thus luminance I(*b*) and adaptation III(*b*) result from the presence of illumination II(*a*); subjective impressions III(*d*) are dependent on brightness patterns II(*e*) and chromaticity II(*f*); fatigue III(*c*) results from a combination of many of the factors, and so on.

In the literature it is common to find reference to the quantity and quality of the lighting environment. In terms of the above factors, the quantity of light has reference to item II(*a*) and the quality to items II*(b)* through II(*f*).

The basic visual tasks are the perception of low contrast, fine detail, and brightness gradient. Assuming a good lighting environment, that is, low glare, acceptable luminance ratios, and white light plus a normal pair of unfatigued eyes, visual acuity is primarily dependent on items I(*a*) to I(*d*), the interrelated effects of which have been determined by a large number of field tests. Remember that the seeing task under discussion involves foveal vision, i.e. focusing and concentrating on small-area detail. This is a vastly different task than normal reading, where the eye rapidly scans familiar images without focusing on details and the brain immediately understands even when much of the information is missing, as in poor reading copy. The task discussed below is quite different and could be compared to studying mathematical equations or reading an unfamiliar language or even spelling proofreading (see Smith, 1978). All of these tasks necessitate detailed examination of each symbol individually.

18.17 Size of Visual Object

Visual acuity is generally proportional to the physical size of the object being viewed given fixed brightness, contrast, and exposure time. Since the actual parameter is not physical size but subtended visual angle, visual ability can be increased by bringing the object nearer the eye (see Fig. 18.18). It is assumed that we are dealing with a pair of *young* eyes, since at ages above 40 the accommodation ability of the eye becomes limited and bringing the object closer blurs the focus.

18.18 Subjective Brightness

The sensation of vision, as explained above, is caused by light entering the eye. This light may be thought of as a group of convergent rays, each ray coming from a different point in space and therefore carrying different visual information. The composite of these rays comprises the entire visual picture that the eye sees and the brain comprehends. The individual rays differ from each other in intensity and chromaticity depending on the part of the viewed object from which they were reflected. The intensity of these cones of light determines and describes the perceived brightness of the object being viewed (see Fig. 18.19).

The human eye detects luminance over an astonishing range of more than 100 million to 1, the lower levels being accomplished after an adjustment period, called adaptation time. This period varies from 2 min for cone vision to up to 40 min for rod vision, for dark adaptation, but is much faster for both types for light adaptation (going from dark to light). The effects of adaptation on apparent (photometric) brightness are discussed in the following section. Table 18.2 lists some measured luminances of everyday visual tasks.

An interesting characteristic of light-level adaptation is a shift in the sensitivity curve of the eye (see Fig. 18.6*b*). Whereas for the light-adapted eye (photopic vision), maximum sensitivity is at 555 nm in the yellow-green region, the

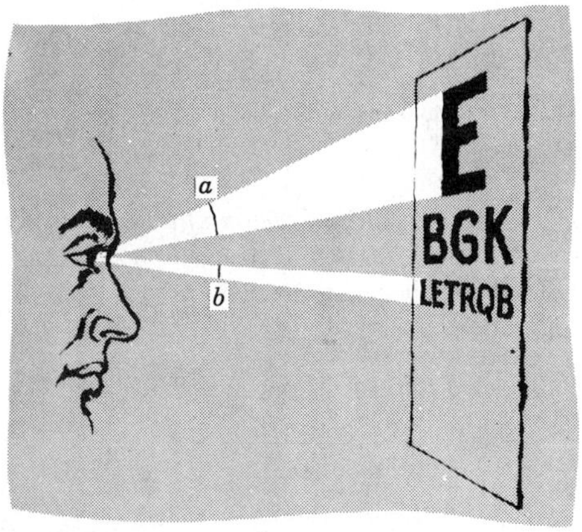

Fig. 18.18 *Relationship between object size and visibility is demonstrated by comparison of subtended angles* (a) *and* (b).

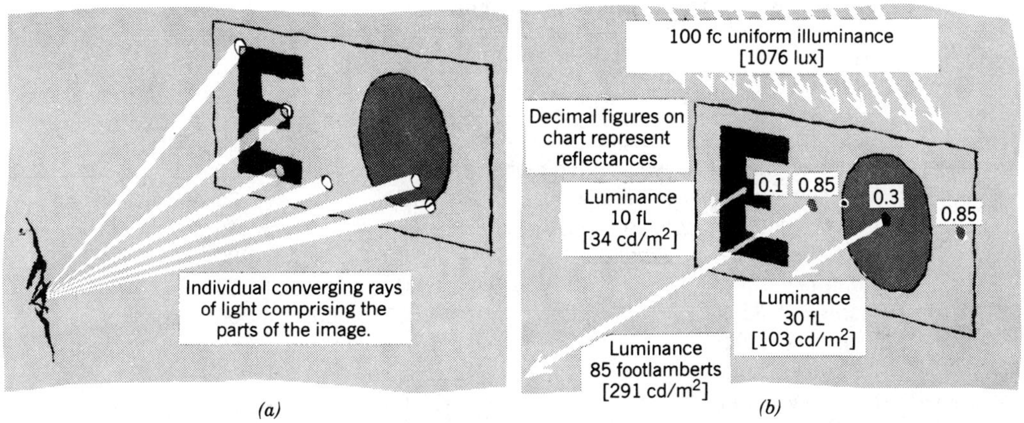

Fig. 18.19 (a) *Composition of the visual image.* (b) *Luminance of a nonuniform surface.*

dark-adapted eye (scotopic vision) peaks at 520 nm in the blue-green region. This means that as the light dims, the warm colors—yellow, orange, red—become grayed, and the blues and violets stand out. This phenomenon can be important in lighting design of restaurants, where light levels generally vary inversely with restaurant quality. Very few foods are blue or violet.

Returning then to the primary consideration of visual acuity as affected by luminance, we can state that, in general, visual performance increases with object luminance. However, a great deal depends on the background against which an object is viewed and the consequent contrast in brightness between the object being viewed and its surroundings.

18.19 Contrast

Many researchers in the area of visibility [see, e.g., Rea (1981)] have concluded that contrast is the single most important factor in visual acuity. This is actually self-evident when we realize that in point of fact the eye sees only contrast (and chromaticity). This can easily be demonstrated by viewing a large evenly lighted, monochromatic, diffuse finish surface (preferably white) which encompasses the entire visual field. The eye will be unable to focus on such a surface since it will see no contrast but only a single luminance. After a short while the sensation of vision will diminish and visual discomfort will increase. Thereafter the eye itself will attempt to provide the missing contrast by providing chromatic spots in the field of vision.

To properly evaluate the effect of contrast (luminance ratio) on visibility, we must first determine the nature of the visual task or, more simply, exactly what it is that we are trying to see. As stated before, the basic visual tasks are detail discrimination and detection of low contrast. Examples of the former are drafting, industrial product inspection, or something as simple as discriminating between the numbers 3, 6, and 8, which are similarly shaped. Detection of low contrast include reading faint copy, sewing a black fabric with black thread, and the like.

Contrast is a dimensionless ratio, defined as

$$C = \frac{L_T - L_B}{L_B} \text{ or } \frac{L_B - L_T}{L_B} \text{ or } \left|\frac{L_B - L_T}{L_B}\right| \tag{18.7}$$

Where L_T and L_B are the luminance of the task and background respectively, in any units. Thus, C varies from 0 for no contrast to 1.0 for maximum contrast. In most situations, the illumination on the task and background is the same. Therefore, since luminance is the product of illuminance (lux) and reflectance, contrast can also be expressed as

$$C = \left|\frac{R_B - R_T}{R_B}\right| \tag{18.8}$$

that is, *contrast is generally independent of illumination* (ignoring specularity). Thus, for the black on white lettering that you are now reading,

ILLUMINATION

TABLE 18.2*a* **Typical Luminance Values**[a]

Object	Luminance	
	cd/m²[b]	Footlamberts
Black glove on cloudy night	0.0003	0.0001
Wall brightness in a well-lighted office	100	30
This sheet of paper in an office	120	35
Green electroluminescent lamp	150	45
Asphalt paving—overcast day	1,300	380
North sky	3,500	1,000
Moon, candle flame	4,000–5,000	1,300
Fluorescent tube	6,000–8,000	2,200
Kerosene flame	8,500	2,500
Hazy sky or fog	15,000	4,400
Snow in sunlight	25,000	7,300
100-W inside frost incandescent lamp	50,000	14.600
Sun	2.3 E9	0.67 E9

[a]Values are rounded off.
[b]To obtain footlamberts; divide by 3.42.

TABLE 18.2*b* **Preferred and Permissible Luminances**

Item	Luminance in cd/m²
Recommended road luminance	1–2
Minimum discernible, chromatic	2–3
Clearly discernible human features	15–20
Preferred wall luminance	25–150
Preferred ceiling luminance	50–250
Preferred task luminance	100–500
Permissible luminaire luminance (depending on position in field of vision)	1000–7000

$$C = \frac{0.77 - 0.045}{0.77} = 0.94$$

which accounts for the excellent legibility. (We are assuming no specularity.) See Table 18.8 for reflectance figures.

It is obvious that high contrast is the critical factor in visual appreciation of outline, silhouette, and size, which is involved in the task of reading. Thus black-on-white print can be read with ease even in moonlight, which is at best 0.1 lux illuminance. This is because the contrast is so high (94%, as above). An important conclusion can then be drawn about a *reading* task: with high contrast (clear, legible print), visibility is essentially *independent* of illumination above a certain minimum. Indeed, high illuminance values can be detrimental, since they generally go hand in hand with high luminance sources, and these in turn can cause veiling reflections (see Section 18.27). This indeed is the conclusion of some recent studies [see Smith (1982)].

Now refer to Fig. 18.20. Note that as the contrast between the letters in the word "performance" and the background diminishes, the individual letters become harder to read. The end letters of the word require an illumination of up to 1000 lux, and that suffices only because we expect the letter "*e*" at the end. Were it an unknown sign, illumination of the magnitude of 10,000 lux + would be needed. These latter letters are an example of the second type of visual task mentioned above—low-contrast poor copy, or surface detail study.

The "*e*" at the end of "performance" in Fig. 18.20 is printed with the same density as the "*c*" in "contrast." It exists, and with enough lighting, the negative effect of lack of contrast can be overcome. This is not so with copy from a used-up printer ribbon or a washed-out photocopy. There, the data simply do not exist, and increased lighting will simply make this fact plainer.

High background luminance makes an object

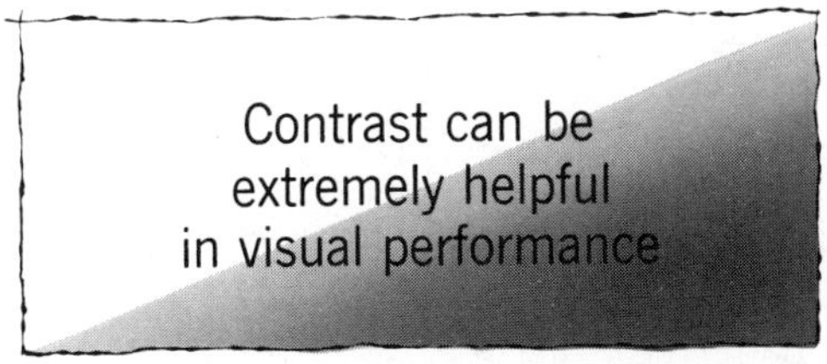

Fig. 18.20 *High contrast is helpful when the seeing task involves detection of silhouette detail.*

look darker and therefore assists in outline detail discrimination, which is precisely the visual task involved in reading. For this reason black on white is desirable for reading. (This is a special case of lateral adaptation which is discussed in Section 18.32). Conversely, high background luminance makes surface examination more difficult. A simple experiment will demonstrate this effect. Stand near a window and hold your hand in front of you with the floor as background. The skin surface detail will be perfectly clear—in rough proportion to its luminance. Now hold your hand up against the window with the daytime sky as background. The hand outline will be clear, but the skin surface will appear dark—the brighter the sky, the darker the skin surface. The reason for this is that the eye automatically adapts to the average brightness of the entire scene.

It is well known that when using an automatic exposure control camera to photograph a dark object on a light background such as a person in snow scene, it is necessary to manually increase the camera aperture in order to obtain additional light to photograph the detail of the darker object. (In doing this we overexpose the rest of the scene.) Since we cannot easily control the aperture of our eyes we must compensate for the detrimental effect of high background luminance in another way; for example, by increasing the surface luminance of the visual task. Indeed, this method is frequently employed [see Section 18.28(*b*)]. Limited visual compensation can be made by squinting; this reduces the field of vision and the overall scene brightness. For maximum visual acuity, *the luminance of a surface type task should be the same as, or slightly higher than, that of the background,* but ratios of 3 : 1 are acceptable in most circumstances.

Another way of understanding the above is to consider the adaptation characteristic of the human eye (refer to Fig. 18.21). As stated, the eye adapts to the brightness level of the *overall* scene and sees each object in the scene in the framework of that adaptation level. Thus at an adaptation level of 1 fL, a measured luminance ratio of 1 : 10 (horizontal scale on Fig. 18.21) appears to be only approximately 1 : 4 (vertical scale), that is, the apparent ratio is *smaller* than the actual. Put another way, the low level of eye adaptation causes the eye to diminish the difference between high brightnesses. This effect becomes smaller as the adaptation level rises, until at an adaptation level of 1000 fL (daylight conditions), the apparent and actual *ratios* correspond, that is, smaller ratios are recognizable. Since visual acuity is, by definition, the ability to distinguish between different levels of illuminance, we have in effect demonstrated that *visual acuity increases with increased adaptation level.* The second important conclusion that can be drawn is that at high adaptation levels apparent brightness is lower than actual and vice versa. Thus a shadowed object near a window will look *darker* than it actually is; contrary to first expectation, it must be better lighted than a similar object further inside the room for equal visibility.

That this effect (high level adaptation) is primarily important in daylight situations is also apparent from the curves, since at a 100-fL adaptation level, which is approximately the indoor condition, apparent and actual luminance levels coincide (within visual ability to recognize differences). However, the reverse effect, resulting from low adaptation levels, can be very important in design situations where low lighting levels are found, such as theaters, lecture halls, restaurants, and storage spaces. Sources of light that would be entirely acceptable at a higher adaptation level can easily become an annoying glare at a low level. Good examples are a theater usher's flashlight or the blinding glare of an oncoming car's headlights.

Reduction of contrast due to veiling reflections and the measurement of contrast and contrast reduction are discussed in Section 18.27.

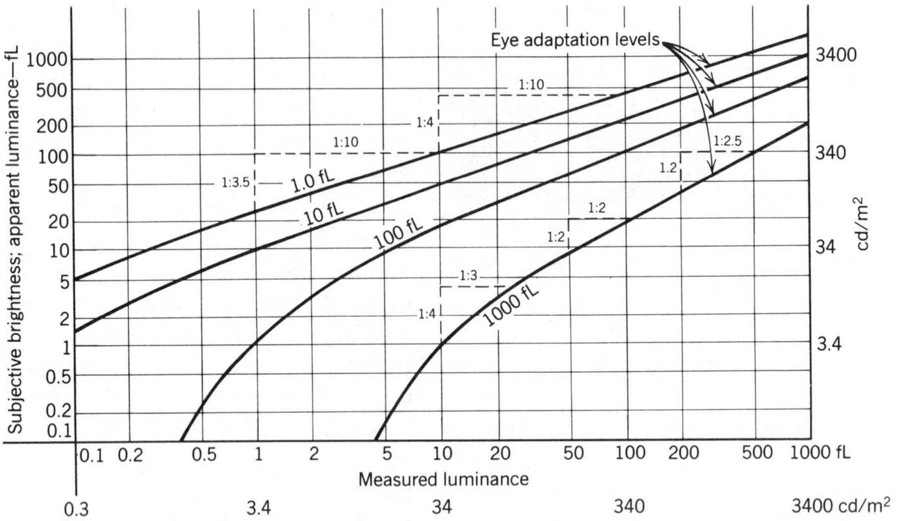

Fig. 18.21 Relation between subjective brightness and measured luminance levels. See the text. (Adapted from A. Cotton, Principles of Illumination, *Wiley, New York, 1960.)*

The foregoing discussion of contrast deliberately avoided any discussion of object colors and the effect of color contrast on visual acuity for several reasons:

1. Office-type tasks (paperwork) are most often black-on-white tasks.
2. Most work tasks involving colored objects deal with unsaturated colors, where the pronounced effects of color contrast are minimal.
3. The effect of object color on visual acuity is very complex since it involves the color characteristic not only of the object, but also of the background and the surround plus the chromaticity of the illuminant.

That being said, we must, however, mention a number of important color phenomena that bear on visual acuity.

1. The subjective brightness of a colored (heterochromatic) object is greater than that of an achromatic object for the same photometric luminance. This effect varies with hue and saturation. It is more pronounced with saturated colors than with those of low chroma and more in the blue-purple-red area than yellow-green (Alman, Breton, and Barbour, 1983).

2. Colored objects on a dark-to-black background appear light and desaturated. Conversely, colors on a light-to-white background appear darker and more saturated.

3. Adjacent complementary colors produce a pale-to-white border between them. The effect also appears when the task and background colors are complementary, and is most pronounced with saturated (high chroma) colors.

18.20 Exposure Time

Registering a meaningful visual image is not an instantaneous process, but one that requires finite amounts of time. Just as a photograph can be taken in dim light by using a longer exposure, so can the human eye better distinguish and discriminate fine detail in poor light given time (and neglecting eyestrain). Of course, the time needed depends on the type of task, but the principle of shorter time at higher illumination, within limits, remains the same. This is particularly true when the object being viewed is not static but in motion.

The phenomenon, however, is not linear. For one specific task tested, increasing the luminance by a factor of 6 halved the seeing time, whereas a further increase of sixfold in luminance reduced the time only approximately 20%. Thus, as in the case of improved contrast with increasing background brightness, we have a case of diminishing returns.

With the parameter of time, as with other parameters of visual acuity, the same qualification applies: When dealing with material that does not require detail discrimination, improved performance does not necessarily result from improved illumination. It has been amply demonstrated (*IES Lighting Handbook,* 1981) that speed of reading and comprehension are substantially independent of illumination levels above a minimum, but are very much dependent on the contrast quality of the material.

18.21 Secondary Task-Related Factors

Refer again to the list of factors in visual acuity in Section 18.16. We have discussed at some length the primary task-oriented factors I(*a*)–(*d*) and at this point wish to consider secondary factors (*e*)–(*h*) of that list. These items refer essentially to the level of concentration required. Thus spray painting a large metal object or packing fruit are very different from inspecting the painted object for defects or the fruit for bruises. The former tasks are largely mechanical and repetitious, whereas the latter require continuous judgmental decisions based on visual information. Since both the latter tasks are frequently moving (assembly-line type of work) and both involve penalties for inaccuracy (rejection at a later inspection state or negative feedback from the purchaser), the lighting required for these tasks is several orders of magnitude better than that for related but largely mechanical tasks. Indeed, establishing lighting recommendations for difficult visual tasks based on data extrapolated from laboratory condition tests may not be entirely accurate, and considerable field testing adjustment is often required. An observer performing a lab test has a different level of concentration and performance than that of a person at an 8-hour-a-day task. The latter will compensate for unsatisfactory seeing condition by:

1. Moving the work to a better viewing angle.
2. Moving head and eyes to a more comfortable position.
3. Reducing the distance between eyes and task, to the extent that the eyes accommodate.
4. Complaining about a poor contrast task so that something is done about it (such as fixing the photocopying machine).
5. Taking more time to perform the seeing task involved. (This item, if it affects production, will frequently spur management to make appropriate alterations in the work environment).

The last item in the task list—peripheral patterns—deals with the visual surround rather than the immediate area of the work. Other than glare sources, which will be discussed separately below, there are many items which, although outside the central field of vision, can disturb the viewer's concentration and therefore the task performance of a worker. These include movement (vehicles, machines, persons), to which peripheral vision is particularly sensitive, large variations in the brightness pattern of the background caused by such activities as periodic opening of an outside door or welding, and even nonvarying patterns, which are disturbing because of their very nature, such as checkerboard light–dark patterns, or devices upon which it is difficult to focus the eyes, such as crossed patterns of wires and bars. None of these items is strictly lighting oriented but they are noted here to demonstrate that an adequate lighting design necessarily includes adjustments for particular field conditions.

18.22 Observer-Related Visibility Factors

Refer to Section 18.16, category III. It is a well-documented fact that the visual performance of *healthy* eyes decreases with age. This reduction evinces itself principally in two areas: an increase in minimum focusing distance caused by increasing lens rigidity and a decrease in sensitivity caused by clouding of the cornea, lens, and vitreous humor. Both these effects can be compensated, the former with external lensing (eyeglasses) and the latter by increased task size, luminance, contrast, and exposure time, as explained above. The point, of course, is that the age of the worker is a primary parameter in the vision equation.

Furthermore, maximum performance may not be synonymous with maximum comfort or minimum fatigue. Indeed, the reverse may be true. This field, which is under active research at this writing, has tentatively established that what is normally referred to as "eyestrain" is a condition of the eye muscles resulting from extensive and intensive eye use (see Section 18.16, items I[*e, f*]). Thus, excellent performance under excellent lighting conditions can still produce fatigue because of the demanding nature of the task. Conditions will generally not produce fatigue if the observer is not "straining" to perform properly. In addition, as we discuss below, discomfort glare or even excessive lighting can cause fatigue without affecting performance. Thus the lighting designer who previously was concerned only with providing adequate uniform lighting levels should properly be concerned with all of the factors involved. Many are indeed beyond his or her control. Some, such as task contrast, previously thought to be outside the lighting designer's province should be examined by the designer in the framework of an overall lighting plus task-plus-observer problem, and recommendations made. Acceptance of these recommendations is a management decision.

Recent work in field testing of visibility situations in real work conditions (Rea, Pasini, and Jutras, 1990) considers all of these parameters in arriving at a visibility judgment, and in addition, goes far toward identifying any difficulty as lying with the task, the lighting condition, or the observer. This work is an extension of earlier research (Kambich and McClosky, 1984) which developed a new performance-related parameter for a specific visibility situation, called *relative visual performance* (RVP), which rates a vision situation in terms of speed and accuracy of performance. This parameter and the older contrast-related visibility criteria (ESI) are discussed below in the section on reflected glare. The entire field of visibility and visual performance is undergoing very active continuing research, and definitive answers to the elusive question of how to design for optimal viewing have not yet been found, nor indeed is there any assurance that they will be. What has emerged from the research to date is a confirmation of the complexity of human work-oriented visibility, plus our inability, to date, to adequately quantify the interrelationships between the factors involved—quantification being the basis of reliable, duplicatable design.

QUANTITY OF LIGHT

18.23 Illuminance Levels

Returning to the list of factors in Section 18.16, and having discussed the task-oriented and observed-oriented items [except for item III(*d*), psychological reactions, which is covered in Section 18.33], we turn now to item II, *the lighting condition*. This is frequently, if somewhat inaccurately, divided into two groups—quantity and quality of lighting—with item II(*a*) representing quantity and items II(*b*) to II(*f*) quality. That such a division is not accurate will become clear in our discussion of glare in Sections 18.25 to 18.28.

An understanding of the factors involved in visual acuity as discussed above, does not answer the most basic lighting design question, which is "How much light must I provide for the specific visual task at hand?" To provide an answer to that seemingly simple question, two approaches are possible, one pragmatic, one analytic. The former, which was the basic British approach, was to study specific tasks in actual and simulated field conditions. To these results, modifying factors of contrast, accuracy requirement, speed, and daylight provision were added. To a large extent this is still the basic British IES recommendation, as shown in Table 18.3. The disadvantage of this approach was the necessity to study a large number of visual tasks individually.

The American approach to the problem of determining required lighting levels was analytic and based principally on the work of Blackwell at Ohio University. The idea was to determine the conditions under which small differences in contrast could be detected for specific degrees of accuracy, with variable parameters of task size, luminance, and exposure time. These results were then related to field tasks by a process of extrapolation, using a device known as a *field task simulator*. The results obtained were in large measure adopted by the American IES and became the basis of most American lighting design. The levels were higher than their British counterparts for a number of reasons, among which were that the American recommendations were based on maximum performance whereas the British IES uses a graded system, starting with a median or average requirement and increasing it only for tasks with stringent requirements or poor visual conditions.

After a period of excesses spurred by cheap energy, a more-is-better philosophy, and interests of the not entirely objective lighting and electric power industries, the illuminance-level recommendation, which had reached 1000 fc (11,000 lux) for commercial offices, came back to earth. For more than a decade, illuminance recommendations have been based on a graded system somewhat similar to the British (explained in detail below), which takes into account not only physical task characteristics as previously, but also the worker's age and speed/accuracy parameters.

In addition, because of the moral and legal pressure (ASHRAE/IES Standard 90.1 is mandated by many codes) for energy conservation, the IESNA has taken additional salutary steps toward rationalization of its very influential illumination standards. They include recognition of fatigue, and task familiarity (e.g., reading) as factors in determining illuminance levels, establishment of lighting power budgets, and energy standards (see Section 20.5), which encourage use of daylight as a normal component of a space's illumination and use of task/ambient lighting design as the preferred technique where high levels of task lighting are required.

18.24 Illuminance Recommendations

Before discussing illuminance (lux) recommendations it is necessary to understand the basis of their derivation, applicability, and shortcomings. As noted above, most of the task illuminance recommendations are derived by extrapolation from threshold contrast visibility tests which yield a required task luminance. Assuming uniform, diffuse task reflectance and uniform illuminance, it is then a simple step to calculate required illuminance since luminance is simply the product of illuminance and reflectance. In AS units,

$$\text{fL} = \text{fc} \times \text{RF}$$

and in SI units

$$L(\text{cd/m}^2) = \frac{E(\text{lux}) \times \text{RF}}{\pi}$$

(see Example 18.4 and Fig. 18.7). This simple relationship is used to derive much of the current IESNA tables, with a modifying factor for varying background reflectance which affects contrast and therefore visibility, as explained above (see Table 18.6).

A number of reservations of this method have been voiced by respected authorities. One (Ngai 1981) is based on research which indicates that suprathreshold visibility requirements are more readily related to eye brightness adaptation levels than to threshold contrast luminance levels. Another (Rea 1982) indicates that deriving suprathreshold luminances from threshold values depends on applied criteria and can therefore vary considerably. Still another objection is based on the readily demonstrable fact that the sensation of vision is not mathematically related to photometric luminance. Thus a black surface of 10% reflectance illuminated with 900 lux and a white surface of 90% reflectance illuminated with 100 lux will both have exactly the same luminance, yet the eye will always see the white surface as lighter than the black by a large margin. All of these reservations add up to a caveat to the lighting designer to use the IESNA illuminance recommendations as one aspect of the overall design, not dogmatically.

Returning to recommendations, visual task studies indicate that assuming good contrast, the required luminances, categorized by type of task are as follows:

Category of Visual Task	*Required Luminance (fL)*
Casual	3–6
Ordinary	6–30
Moderate	30–60
Difficult	60–120
Severe	Above 120

The dependence of required illumination on task reflectance (RF) can be seen by a glance at the tabulation below, which shows quantitatively the illumination requirements in the above categories for tasks of radically different reflectance.

	*Required fc [lux]**	
	Reflectance	
Category of Visual Task	*50%*	*10%*
Casual	10 [100]	50 [550]
Ordinary	40 [400]	200 [2200]
Moderate	90 [100]	450 [4800]
Difficult	160 [1700]	900 [10000]
Severe	240 [2600]	1200+ [13000]

*Lux figures rounded.

This illustrates that a single illumination scheme is often inadequate for an area containing widely differing visual tasks. Note that a 10% RF makes all tasks difficult and that casual seeing would be only outline recognition.

(a) British IES. As stated above, the British IES recommendations consist of an extensive list of recommended "service illuminances" for specific tasks, modified by considerations of contrast, daylight availability, accuracy, and speed requirements. The method of use is to consult the detailed table of service illuminances (not reproduced here; see *British IES Code,* 1977) if the *specific* task is defined, or the flow chart (see Table 18.3) if only the type of task is known, in order to arrive at a recommended service illuminance. This is then amended by the self-explanatory technique shown on the flowchart, to arrive at a final illuminance recommendation. For example, if an illuminance for planing and benchwork in a woodworking shop is desired, the designer would consult the detailed list (*British IES Code,* 1977) and would find a recommendation of 500 lux. He then enters the flowchart (Table 18.3) at 500 lux (routine work) and, based on a knowledge of the task and work conditions, would either remain at 500 lux, decrease to 300 lux, or increase to 750 lux. These illuminance levels (300, 500, and 750 lux) are chosen specifically to correspond to minimum significant level change (see Table 18.4). If, alternatively, the designer knows only that the space under consideration is to be used as a woodworking shop, he could select the *standard service illuminance* directly from the general categories in the flowchart (Table 18.3).

The author has found this system of illuminance selection to be rapid, efficient, and reasonable and has used it effectively in many design projects.

(b) American IES. This (1981) system is patterned after the British system. In lieu of the single illuminance recommendation that characterized the early IES tables, this system provides a luminance range determined by task difficulty (contrast, size), within which range a specific illuminance is selected based on three weighting factors—age of the observer, the importance of speed and/or accuracy, and the reflectance of the background on which the task is seen. The procedure operates as follows:

1. An illuminance category (A through I) is initially selected, based on either a general description of the activity involved (see Table 18.5), or, if known, on a specific activity in a specific setting. (The extensive tables for this latter selection are found in the 1981 *IES Lighting Handbook* and subsequent editions and are not reproduced here.) The category selection gives a three-number range of illuminances that correspond to the minimum significant changes shown in Table 18.4. For instance, Category B is 50–75–100 lux and F is 1000–1500–2000 lux.

2. In the next step the three weighing factors are introduced. Age under 40 reduces the light requirement; unusual demand for speed or accuracy increases it, and particularly low or high background reflectance (below 30% or above 70%) increases or reduces it, respectively. This step is not a mechanical one in that at the higher luminances (categories D through I), the designer is expected to become thoroughly familiar with the importance, duration, and visual difficulty of the specific task involved, since the design aims at *task* lighting and not simply general room illumination (see Table 18.5). Having decided upon the weighting factor, the design can

TABLE 18.3 **Flowchart**

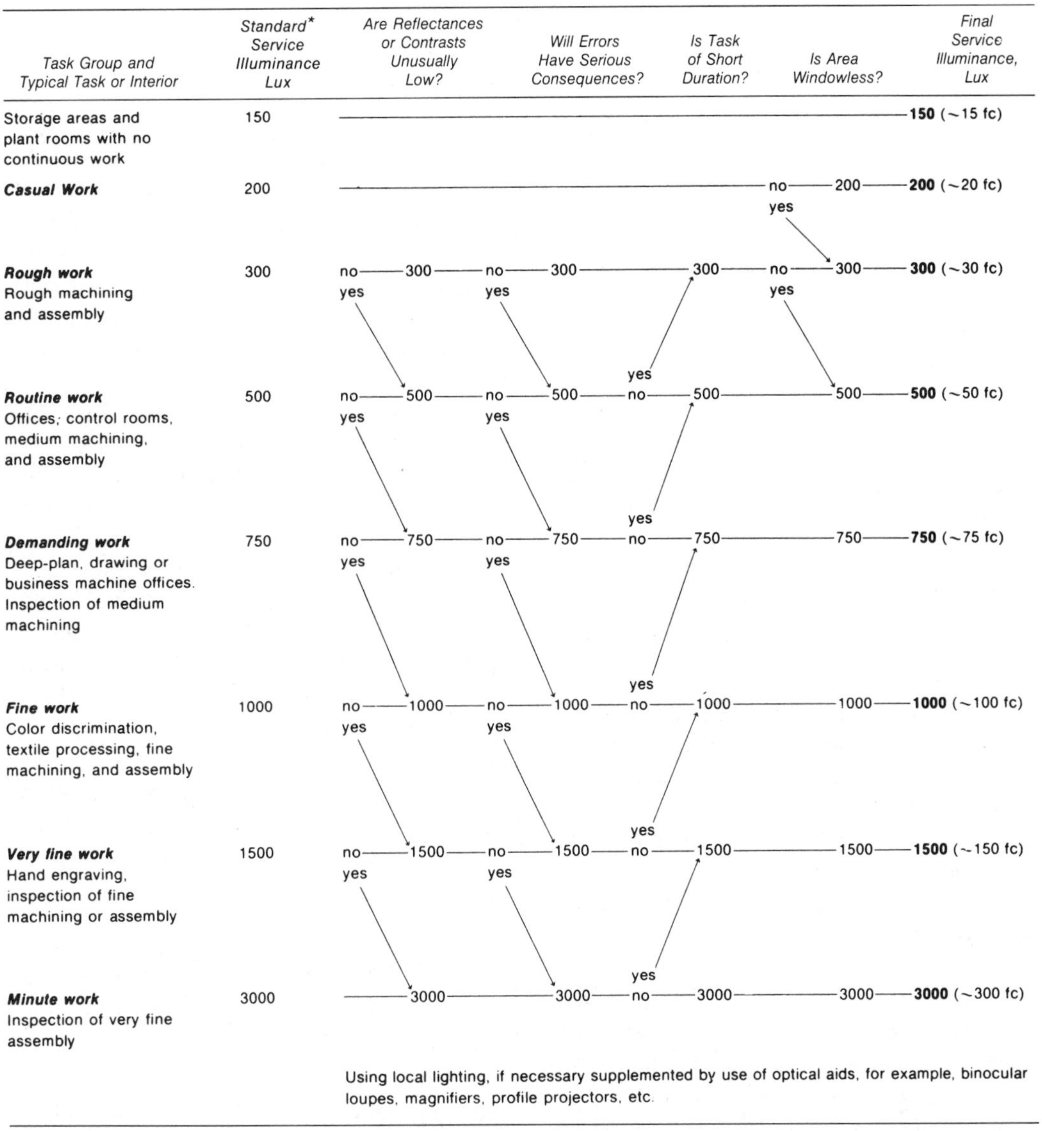
Task Group and Typical Task or Interior
Standard* Service Illuminance Lux
Are Reflectances or Contrasts Unusually Low?
Will Errors Have Serious Consequences?
Is Task of Short Duration?
Is Area Windowless?
Final Service Illuminance, Lux
Storage areas and plant rooms with no continuous work
150
150 (~15 fc)
Casual Work
200
no 200 200 (~20 fc)
yes
Rough work
Rough machining and assembly
300
no 300 no 300 300 no 300 300 (~30 fc)
yes yes yes
Routine work
Offices, control rooms, medium machining, and assembly
500
no 500 no 500 yes no 500 500 500 (~50 fc)
yes yes
Demanding work
Deep-plan, drawing or business machine offices. Inspection of medium machining
750
no 750 no 750 yes no 750 750 750 (~75 fc)
yes yes
Fine work
Color discrimination, textile processing, fine machining, and assembly
1000
no 1000 no 1000 yes no 1000 1000 1000 (~100 fc)
yes yes
Very fine work
Hand engraving, inspection of fine machining or assembly
1500
no 1500 no 1500 yes no 1500 1500 1500 (~150 fc)
yes yes
Minute work
Inspection of very fine assembly
3000
3000 3000 yes no 3000 3000 3000 (~300 fc)
Using local lighting, if necessary supplemented by use of optical aids, for example, binocular loupes, magnifiers, profile projectors, etc.

Source: British IES Code for Interior Lighting (1977).

NOTES

The standard* service illuminance recommended for an application should be increased if:

1. unusually serious consequences, in terms of cost or danger, could result from mistakes in perception, or
2. unusually low reflectances or contrasts are present in the particular task, or
3. tasks for which the recommended standard service illuminance is less than 500 lx are carried out in windowless interiors.

The recommended standard service illuminance may be decreased if, in the judgment of the designer, the duration of the task is unusually short. This flowchart gives the steps by which the recommended standard service illuminance should be modified when one or more of these conditions apply. The resulting final service illuminance derived from the flowchart should then be used as the design value.

*Author's note: standard service illuminance means the illuminance value recommended for use in actual design.

ILLUMINATION

TABLE 18.4 **Illumination Levels**

Each level represents a significant subjective change.

Footcandles	*Lux*
2.0	20
3.0	30
5.0	50
7.5	75
10.0	100
15.0	150
20.0	200
30.0	300
50.0	500
75.0	750
100.0	1000
150.0	1500
200.0	2000
300.0	3000

Source: Adapted from Note B of the "General Schedule" of the *British IES Code for Interior Lighting* (1977).

now select the recommended (target) illuminance from Table 18.6.

The resultant recommended illuminance is "raw" or conventional illumination, that is, average maintained lux (fc), *on the task* for categories D–I, and *in the room* for categories A–C. Therefore, the calculation procedure for the target illuminance is a lumen method for categories A–C and a point-by-point method for the *task* illumination of categories D–I.

For a space with several tasks of varying visual difficulty, the designer is expected to so design the lighting and controls so that task requirements are met without overlighting. A uniform layout keyed to the most severe task is tedious and energy wasteful and is to be discouraged.

The author selected three typical work environments and compared the illuminance recommendations of three authorities: British IES, IESNA, and a U.S. government agency (FEA). Results are tabulated in Table 18.7.

The IES recommended illuminance values are *not* applicable to installations where a visual task is not the deciding factor. Such installations include merchandising spaces, displays of all sorts, theatrical and artistic lighting, lighting for mood, lighting for safety, light used as part of an industrial process, and so on. Recommended illuminances for selected exterior spaces are given in Table 21.5.

QUALITY OF LIGHTING

18.25 Considerations of Lighting Quality

Quality of lighting is a term used to describe all of the factors in a lighting installation not directly connected with quantity of illumination. Certainly it is obvious that if two identical rooms are lighted to the same *average* illuminance, one with a single bare bulb and the second with a luminous ceiling, there is a vast difference in the two lighting systems. This difference is in the *quality* of the lighting, a term that describes the overall scene—that is, the luminances, diffusion, uniformity, and chromaticity of the lighting.

Excessive luminances and/or excessive luminance ratios in the field of vision are commonly referred to as glare. The quality of the lighting system must also include the visual comfort of the system, that is, the absence of glare. When the discomfort glare is caused by light sources in the field of vision, it is known as *direct* or *discomfort glare*. When the glare is caused by reflection of a light source in a viewed surface, it is known as *reflected glare* or *veiling reflection* (see Fig. 18.22).

18.26 Direct (Discomfort) Glare

The factors involved in producing discomfort glare are the luminance, size, and position of each light source in the vision field, plus the adaptation level of the eye. The discomfort of direct glare stems from two facts: first, that the eye adapts (rapidly) to the average brightness of the overall visual scene, and second, that the eye is attracted to the highest luminance in that scene. (The latter fact is used effectively in merchandising displays.) Thus if an area of high brightness such as a window or a lighting fixture exists in the visual scene and we are looking at an area of lower brightness such as a work task, three visually disturbing things occur:

1. The eye adapts to a higher luminance level, thus effectively reducing the subjective brightness of the task, or put more

TABLE 18.5 **Illuminance Categories and Illuminance Values for Generic Types of Activities in Interiors**

Type of Activity	*Illuminance Category*	*Ranges of Illuminances*	
		Lux	*Footcandles*
General lighting throughout spaces			
Public spaces with dark surroundings	A	20–30–50	2–3–5
Simple orientation for short temporary visits	B	50–75–100	5–7.5–10
Working spaces where visual tasks are only occasionally performed	C	100–150–200	10–15–20
Illuminance on task			
Performance of visual tasks of high contrast or large size	D	200–300–500	20–30–50
Performance of visual tasks of medium contrast or small size	E	500–750–1000	50–75–100
Performance of visual tasks of low contrast or very small size	F	1000–1500–2000	100–150–200
Illuminance on task, obtained by a combination of general and local (supplementary) lighting			
Performance of visual tasks of low contrast and very small size over a prolonged period	G	2000–3000–5000	200–300–500
Performance of very prolonged and exacting visual tasks	H	5000–7500–10000	500–750–1000
Performance of very special visual tasks of extremely low contrast and small size	I	10000–15000–20000	1000–1500–2000

Courtesy of Illuminating Engineering Society of North America.

ILLUMINATION

TABLE 18.6 **Illuminance Values, Maintained, in Lux, for a Combination of Illuminance Categories and User, Room, and Task Characteristics (for Illuminance in Footcandles, Divide by 10)**

Part A ***General Lighting Throughout Room***

Weighting Factors		*Illuminance Categories*		
Average of Occupants Ages	*Average Room Surface[1] Reflectance (%)*	*A*	*B*	*C*
Under 40	Over 70	20	50	100
	30–70	20	50	100
	Under 30	20	50	100
40–55	Over 70	20	50	100
	30–70	30	75	150
	Under 30	50	100	200
Over 55	Over 70	30	75	150
	30–70	50	100	200
	Under 30	50	100	200

TABLE 18.6 **(Continued)**

Part B ***Illuminance on Task***

Weighting Factors			*Illuminance Categories*					
Average of Workers' Ages	*Demand for Speed and/or Accuracy*[2]	*Task Background*[3] *Reflectance (%)*	*D*	*E*	*F*	*G*[b]	*H*[b]	*I*[b]
Under 40	*NI*[a]	Over 70	200	500	1000	2000	5000	10000
		30–70	200	500	1000	2000	5000	10000
		Under 30	300	750	1500	3000	7500	15000
	I[a]	Over 70	200	500	1000	2000	5000	10000
		30–70	300	750	1500	3000	7500	15000
		Under 30	300	750	1500	3000	7500	15000
	C[a]	Over 70	300	750	1500	3000	7500	15000
		30–70	300	750	1500	3000	7500	15000
		Under 30	300	750	1500	3000	7500	15000
40–55	NI	Over 70	200	500	1000	2000	5000	10000
		30–70	300	750	1500	3000	7500	15000
		Under 30	300	750	1500	3000	7500	15000
	I	Over 70	300	750	1500	3000	7500	15000
		30–70	300	750	1500	3000	7500	15000
		Under 30	300	750	1500	3000	7500	15000
	C	Over 70	300	750	1500	3000	7500	15000
		30–70	300	750	1500	3000	7500	15000
		Under 30	500	1000	2000	5000	10000	20000
Over 55	NI	Over 70	300	750	1500	3000	7500	15000
		30–70	300	750	1500	3000	7500	15000
		Under 30	300	750	1500	3000	7500	15000
	I	Over 70	300	750	1500	3000	7500	15000
		30–70	300	750	1500	3000	7500	15000
		Under 30	500	1000	2000	5000	10000	20000
	C	Over 70	300	750	1500	3000	7500	15000
		30–70	500	1000	2000	5000	10000	20000
		Under 30	500	1000	2000	5000	10000	20000

NOTES:

1. Average weighted surface reflectances, including wall, floor, and ceiling reflectances, if they encompass a large portion of the task area or visual surround. For instance, in an elevator lobby, where the ceiling height is 25 ft, neither the task nor the visual surround encompasses the ceiling, so only the floor and wall reflectances would be considered.
2. In determining whether speed and/or accuracy is not important, important, or critical, the following questions need to be answered: What are the time limitations? How important is it to perform the task rapidly? Will errors produce an unsafe condition or product? Will errors reduce productivity and be costly? For example, in reading for leisure there are no time limitations and it is not important to read rapidly. Errors will not be costly and will not be related to safety; thus, speed and/or accuracy is not important. If, however, prescription notes are to be read by a pharmacist, accuracy is critical because errors could produce an unsafe condition and time is important for customer relations.
3. The task background is that portion of the task upon which the meaningful visual display is exhibited. For example, on this page the meaningful visual display includes each letter, which combines with other letters to form words and phrases. The display medium, or task background, is the paper, which has a reflectance of approximately 85%.

[a]NI = not important, I = important, and C = critical.

[b]Obtained by a combination of general and supplementary lighting.

Source: Courtesy of Illuminating Engineering Society of North America.

Table 18.7 **Comparison of Lighting Level Recommendations (Lux)**

	Source of Recommendation		
Activity	*British IES*	*IESNA*	*U.S. Gov't Agency*
Classroom	500	200–500	400–600
Business office	500	300–750	500–750
Drafting room	750–1000	750–1500	800–1200

NOTES:
1. All levels are in lux. Divide by 10 for fc.
2. All levels refer to lighting on the task.

simply, making it harder to see what we are looking at (see Fig. 18.21). This is readily demonstrated by alternately blocking and unblocking a direct glare source with one's hand while trying to perform a moderately difficult visual task, and noting the immediate improvement in visibility when the glare source is obscured.

2. The eye is drawn simultaneously in two directions: automatically to the source of high luminance, and volitionally to the object we are looking at. The resultant tension causes considerable visual discomfort.
3. The adaptation level is continuously varying as the eye is drawn to the glare source and away again.

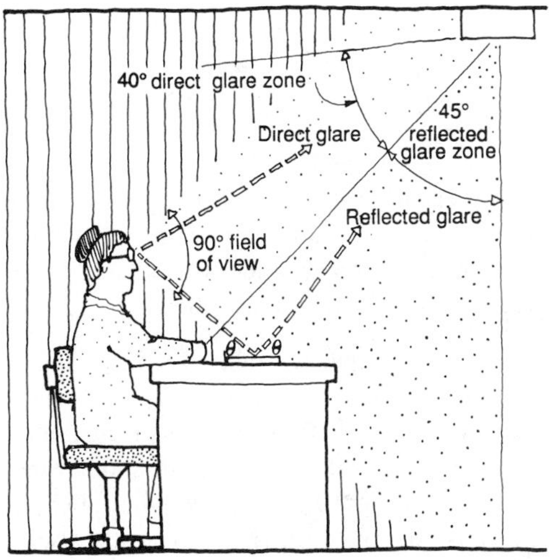

Fig. 18.22 *Glare zones. The direct and reflected glare light paths are delineated on the diagram. Direct glare presupposes a head-up position, whereas reflected glare assumes eyes down, at a reading angle. Other factors that affect the adaptation level of the eyes, the apprehended size of the glare source, and luminance ratios are room size and surface finishes, and size and position of lighting fixtures and windows. Light sources in the far-field and peripheral-vision areas beyond the central 90° cone are less troublesome as glare sources. See Fig. 18.17.*

Glare is proportional to a source's luminance and its apprehended *solid* angle. Therefore, a small bright source is usually not a problem, whereas a large, low brightness source (such as a luminous ceiling) may be. Indeed, a small bright source adds sparkle to the field of vision, and many observers find it a pleasant addition in a monotonous lighting environment. Although discomfort glare from a scene is cumulative, source luminance is more important than the number of sources. For instance, if the luminance of a number of sources is halved, the reduction in glare is greater than is achieved by reducing the number of such sources by half. Indeed the latter procedure will have little effect on discomfort glare.

The remaining two factors are less self-evident. Glare decreases rapidly as the brightness source is moved away from the direct line of vision, and thus the glare produced depends on its position in the field of view. The amount of discomfort glare produced by a source is inversely proportional to the background brightness (eye adaptation level). Thus a ceiling fixture with a luminance of 4000 cd/m^2 (1167/fL) at 65° might easily constitute a source of discomfort glare in a space with an eye adaptation level of 150 cd/m^2 (44/fL). The same fixture would not be objectionable in a daylight condition, where the eye adaptation level might be 1500 cd/m^2. A more striking example is that of an automobile's

ILLUMINATION

headlights, which at night are so severe a source of glare as to constitute "disabling glare," whereas in daylight, with its concomitant high eye adaptation level, these lights are noticeable, but not usually disturbing.

Keeping in mind the dependence of direct glare on eye adaptation level, a useful rule of thumb is that luminances of large sources should not exceed 2500 cd/m^2 (~700 fL) and small sources, 7500 cd/m^2 (~2200 fL). The former is about the luminance of blue sky, the latter of a fluorescent lamp. The terms large and small depend not only on actual physical dimensions of the source but also on the distance from the observer. That is, the actual criterion is apprehended size, or subtended visual angle, as shown in Fig. 18.18.

The glare effect of a number of individual glare source contributions in an interior space can be quantified by a criterion called "visual comfort probability," or VCP, which is defined as the percentage of normal-vision observers who will be *comfortable* in that specific visual environment. The IESNA has established a set of standard conditions for which VCP of sources can be calculated. These include a 1000-lux illuminance, representative room dimensions, fixture height and observer position, and a head-up field of view limited to 53° above and directly forward from the observer (see Fig. 18.23).

With these conditions, direct glare will not be a problem if all three of the following conditions are satisfied:

1. The VCP is 70 or more.
2. The ratio of maximum-to-average luminaire luminance does not exceed 5 : 1 (preferably 3 : 1) at 45, 55, 65, 75, and 85° from the nadir, crosswise and lengthwise.
3. Maximum luminaire luminances crosswise and lengthwise do not exceed:

Angle Above Nadir, Degrees	*Maximum Luminance*	
	cd/m^2	*fL*
45	7700	2250
55	5500	1605
65	3850	1125
75	2500	750
85	1700	495

A typical set of manufacturer's luminance and VCP data is shown in Fig. 18.24 for an actual ceiling-mounted fluorescent fixture with 4-40WT12 lamps. Note that all VCP values are considerably above the 70 minimum criteria. If full VCP data of this type are not available, an earlier criterion commonly called the "scissors curve criterion" can be used. This criterion, which is intended for application to a *single fixture*, by extrapolation can be applied to an entire space; that is, a room with fixtures that meet the scissors curve criterion will generally not fall below a VCP of 70. A luminance data check sheet using scissors curve criterion and typical fixture data is shown in Fig. 18.25. If we replace the scissors shown in the figure with the eye, this criterion tells us the limiting comfort brightnesses of the eye. That is:

Angle Above Nadir, Degrees	*Well-Tolerated Average Brightness, Footlamberts*
45	750
55	400
65	375
75	250
85	165

Despite the accuracy and excellence of the VCP criterion, it is inherently limited by its own standard conditions, which are not easily applied to other situations, as follows:

1. In small spaces, VCP has little significance.
2. Tabulated VCP figures are given for the worst viewing position in the room. Since VCP varies dramatically with observer position, the VCP values given are always lower than the space's average VCP.

In view of the above (and other) reservations, most of which tend to make the actual direct glare situation better than the VCP calculation would indicate, it is recommended that layouts giving a VCP somewhat below 70 not be discarded out of hand. Instead, they should be carefully examined and, if possible, several observer positions calculated using one of the many readily available PC-based computer pro-

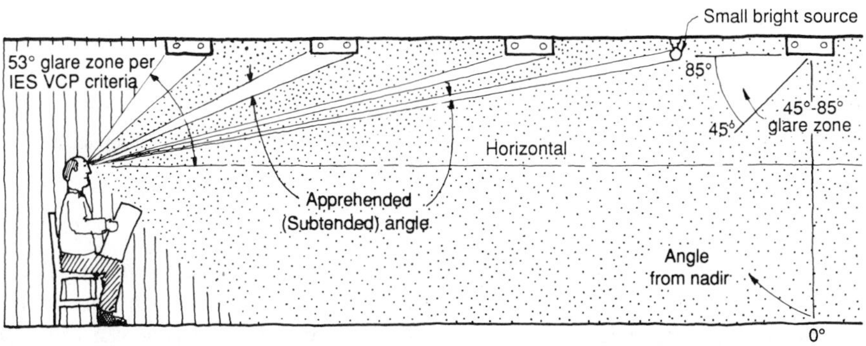

Fig. 18.23 Determination of direct glare. The glare contribution of each source depends on its size (subtended or apprehended solid *angle), luminance, and location in the field of view. Note that the apprehended* solid *angle of a small source is such that even with high brightness, it is not objectionable. Such sources are normally called* sparkle. *Glare will be much more objectionable with a dark background than with a light one; therefore, light-colored paints on ceilings and upper walls are recommended.*

4–40-W Lamps
Prismatic Lens Diffuser

Average Luminance Data (fL)

Vertical Angles	*Across Axes*	*Along Axes*
60°	587	512
65°	310	317
70°	149	165
75°	121	112
80°	140	100
85°	162	122

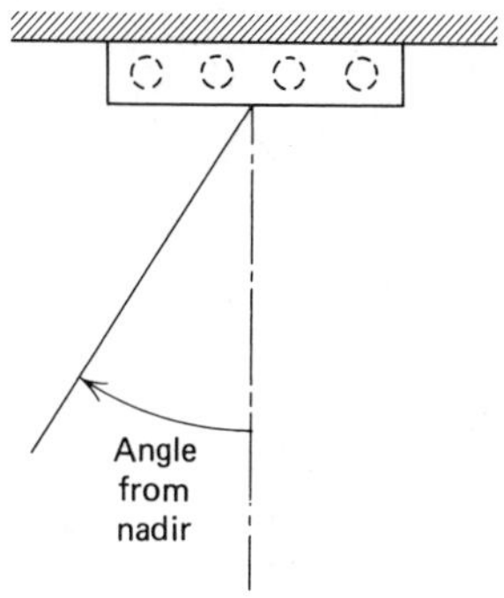

Reflectances:
- Wall 50%
- Ceiling cavity 80%
- Floor cavity 20%

Work plane illumination: 100 fc

IES Visual Comfort Probability Data

Room Size		*Luminaires Lengthwise*			*Luminaires Crosswise*		
W (ft)	*L (ft)*	*Ceiling Height (in ft)*					
		8.5	*10.0*	*13.0*	*8.5*	*10.0*	*13.0*
20 ×	20	81	77	76	79	75	72
	30	81	78	76	79	76	72
	40	82	79	77	79	77	74
	60	81	80	78	79	77	75
30 ×	20	84	80	77	82	78	74
	30	83	80	77	82	79	74
	40	83	81	78	81	79	75
	60	82	80	78	80	78	76
	80	82	80	79	80	78	76

Fig. 18.24 *A typical set of manufacturers' published VCP and luminance data. (Courtesy of Holophane.)*

Crosswise and Lengthwise

Angle	Limit Value[1]	Maximum	Average	Ratio[2] Max/Avg
45°	2250			
55°	1605			
65°	1125			
75°	750			
85°	495			

[1]Three times the value of the sloped limiting line, in footlamberts
[2]Should not exceed 3.

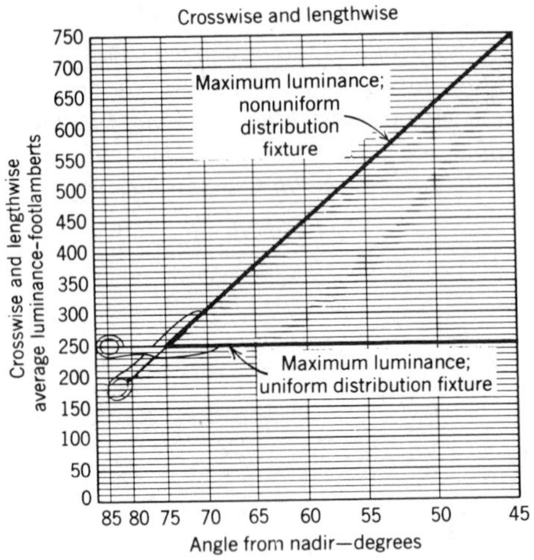

Fig. 18.25 Fixture luminance (brightness) check sheet. Values of average *luminance, in footlamberts, in both lengthwise and crosswise directions, are plotted on graphs of the type shown. All points must fall below the upper limit line for fixtures of nonuniform luminance and below the lower limit line (250 fL) for fixtures of uniform luminance (e.g., globes). Further, maximum fixture luminance should not exceed three times average at any angle, nor may the maximum exceed the limit values listed, which are three times the values of the sloping line.*

grams. With these as a guide, the designer can rearrange and substitute equipment to obtain the desired condition.

See Section 20.21 for a comparison of the direct glare characteristics of lighting fixture diffusers.

18.27 Veiling Reflections and Reflected Glare

Although there is no generally accepted convention with respect to nomenclature, many people refer to *reflected glare* when dealing with specular (polished or mirror) surfaces and to *veiling reflections* when considering source reflections in dull or semimatte finish surfaces, which always exhibit some degree of specularity. We use the terms interchangeably.

(a) Nature of the Problem. The problem of veiling reflections is much more complex than that of direct glare because it involves both the source *and* the task and is inherent in the act of seeing (refer to Fig. 18.26). Vision is produced by light being reflected from the object seen. The object viewed mirrors the source(s) of the light into the eye of the observer. Thus if the object being viewed were replaced by a mirror, we would see the source(s) of light clearly (see Fig. 18.26*a*). In a commercial space there are usually one or more lighting fixtures near the observer which furnish most of the light by which he sees. These *principal* sources are the main contributors to reflected glare. Other, more remote fixtures in the room are lesser sources of veiling reflections (see Fig. 18.26*b*). To the extent that the sources can be mirrored by the vision task, glare exists. It is imperative to an understanding of this problem to appreciate the importance of the reflection characteristic of the object being viewed. If the object were perfectly absorbent, that is, if it had a reflection coefficient of 0%, it would appear completely black as no light would be reflected into the eye (see Fig. 18.26*c*). Conversely, if the object were perfectly specular, like a clean mirror, and no light source were within the geometry of reflection, it too would appear black (see Fig. 18.26*d*). Thus if we took a mirror out on a cloudy night and shined a light on it from over our shoulder, it would be practically invisible since no light would be reflected back into the eye.

The reader might try this experiment: in an inside space with a single overhead luminaire, try to examine the surface of a very clean, dust-free mirror. You will find that the best angle to hold it is *almost* at the angle at which the light source is seen. This is because the mirror is *almost* completely specular, and it is the slight diffuse reflection near the viewing angle that permits us to see the surface. Thus we understand that reflected glare is due to task surface specularity, whereas object definition, that is, the ability to see the task itself, is due to task surface diffuseness. A corollary of this conclusion is that veiling reflections, which are caused by mirroring of a source in the task, are proportional to source luminance and substantially independent of illumination level. The brighter the source, the more troublesome its reflection. Glare sources within the geometry of reflected vision are shown in Fig. 18.27 and the effects are shown in Fig. 18.28. Figure 18.27 clearly shows that while large sources are difficult to avoid, small sources can usually be easily avoided by a small change in the source–task–eye geometry, as by moving the head or tilting the task. Table 18.8 lists a few sample reflectance figures to demonstrate that most materials exhibit both a specular and a diffuse reflectance. In studying Fig. 18.27 it is important to note that a majority of visual work is done in the zone of 20 to 40° from the vertical, below the eye, with a maximum at the 25° reading angle shown in Fig. 18.29.

(b) Contrast Reduction. The principal effect of the reflection of a light source in a visual object is to reduce contrast between the object and its background, thus reducing visibility. It is as if a bright veil were spread over the object being viewed, which accounts for the nomenclature "veiling reflection." As the angle of the incident light approaches the viewing angle, the specularly reflected component of this light becomes more and more pronounced, and task contrast drops. This is clearly visible in Figs. 18.28 and 18.29. The worst situation occurs when the incident angle equals the viewing angle. When the specular reflectance of the task and background is high, as with the glass screen of a video display terminal, for instance, an image of the source is superimposed on the object, making viewing impossible (see Fig. 18.29). However, even with the highly specular finish of "slick" magazine paper, vision is still possible because of the very high contrast between black ink and white paper, although with much reduced efficiency and considerable annoyance.

When considering specular *and* diffuse reflectance, the equation for contrast given in Section 18.19 must be rewritten as

$$C = \frac{(L_{BD} + L_{BS}) - (L_{TD} + L_{TS})}{L_{BD} + L_{BS}} \qquad (18.9)$$

where L_B and L_T are background and task luminances caused by diffuse (D) and specular (S) reflectance; that is, L_{BS} is the background luminance due to its specular reflectance, and so forth. If we were to rework the calculation of contrast of Section 18.19, including specularity, the result would be much different.

EXAMPLE 18.5. Assume an interior space lighted to an average illuminance of 75 fc (750 lux), using bare bulb fluorescent fixtures (luminance ≃ 7000 cd/m^2 or 2000 fL). The task is drafting with India ink on vellum. Reflectances are:

	Specular	*Diffuse*
Ink	0.021	0.038
Paper	0.018	0.71

Calculate the task contrast, without and with reflection of the 2000 fL source on the work.

SOLUTION

(a) Without specularity, using equation 18.6 for diffuse reflection only:

$$C = \frac{R_B - R_T}{R_B} = \frac{0.71 - 0.038}{0.71} = 0.946$$

(b) With specularity, using equation 18.9,

$$C = \frac{(75 \text{ fc} \times 0.71 + 2000 \text{ fL} \times 0.018)}{L_{BD} + L_{BS}}$$

$$- \frac{(75 + 0.038) + (2000 \times 0.021)}{L_{BD} + L_{BS}}$$

$$= \frac{(53.25 + 36) - (2.85 + 42)}{53.25 + 36} = 0.497$$

which is just over half of the previous contrast! If the contrast is normalized to the maximum contrast (as is usually done), the contrast reduction R can be expressed as

$$R = 1 - \frac{C}{C_{max}} \qquad (18.10)$$

Thus, in this case, contrast reduction would be

$$R = 1 - \frac{0.497}{0.946} = 0.47$$

That is, contrast reduction would be 47%.

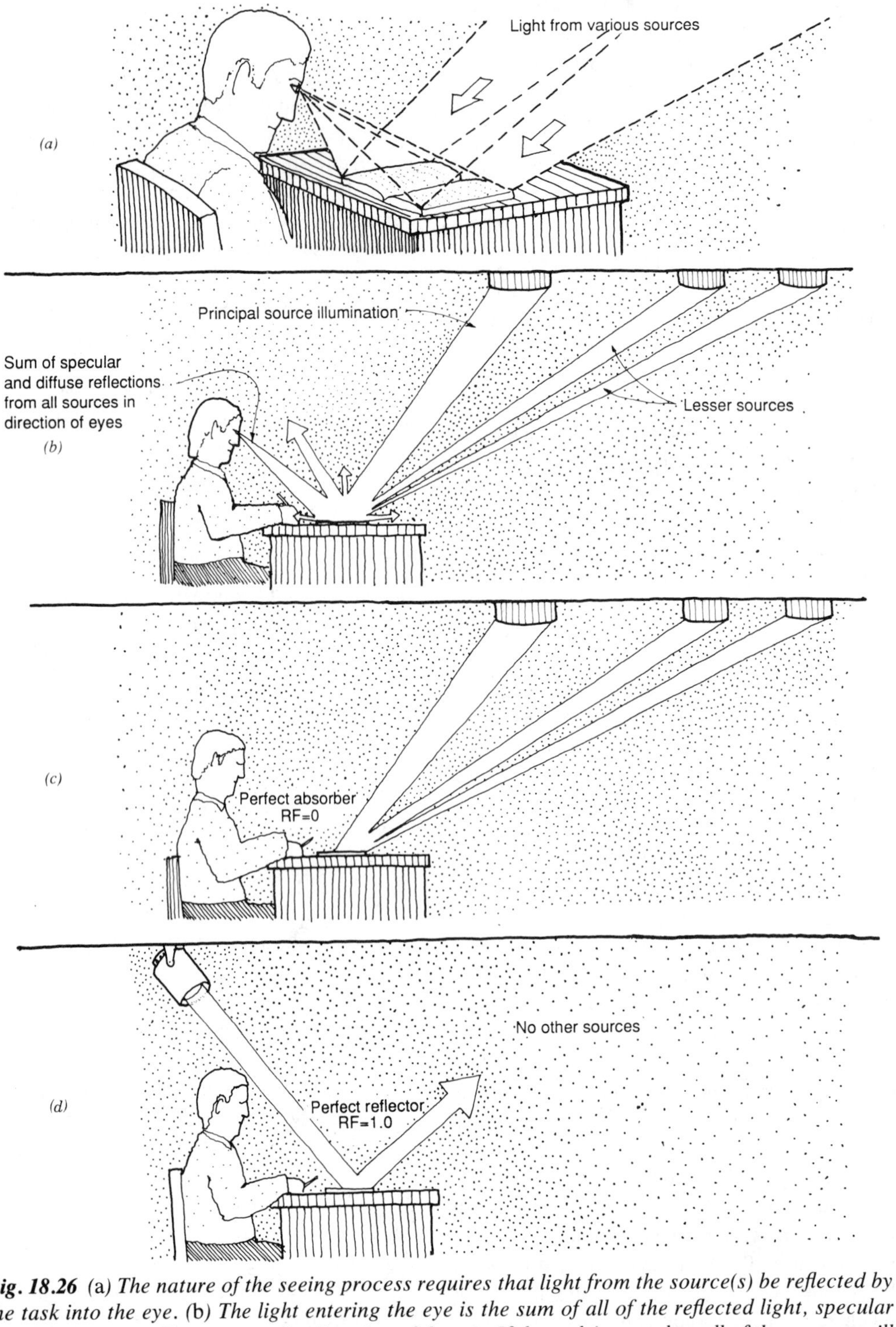

Fig. 18.26 (a) *The nature of the seeing process requires that light from the source(s) be reflected by the task into the eye.* (b) *The light entering the eye is the sum of all of the reflected light, specular and diffuse, from all sources, in the direction of the eye. If the task is specular, all of the sources will be seen reflected in the task.* (c) *A perfectly absorptive object is jet black since it reflects nothing.* (d) *A perfectly reflective object positioned as shown is also black since geometrically it cannot reflect light into the eyes.*

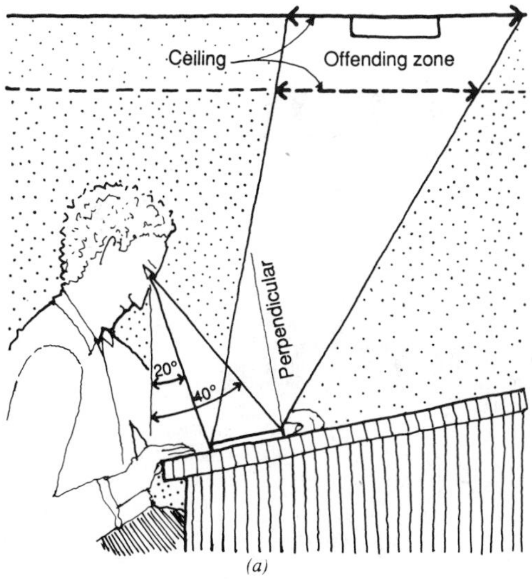

(a)

Fig. 18.27 Geometry of reflected glare. (a) *Since normal desktop, head-down viewing angles vary from 20 to 40° from the vertical, the offending zone is the area on the ceiling corresponding to specular reflection between these two angles. (In specular reflection angle of incidence = angle of reflection). Note that the higher the ceiling, the larger this area becomes. In an office situation the draftsman in* (b) *would see ceiling fixtures in the offending zone reflected in his instruments, parallel straightedge, and work. Note the important fact that the offending zone moves back and becomes smaller as the table tilts up (see Fig. 18.33).*

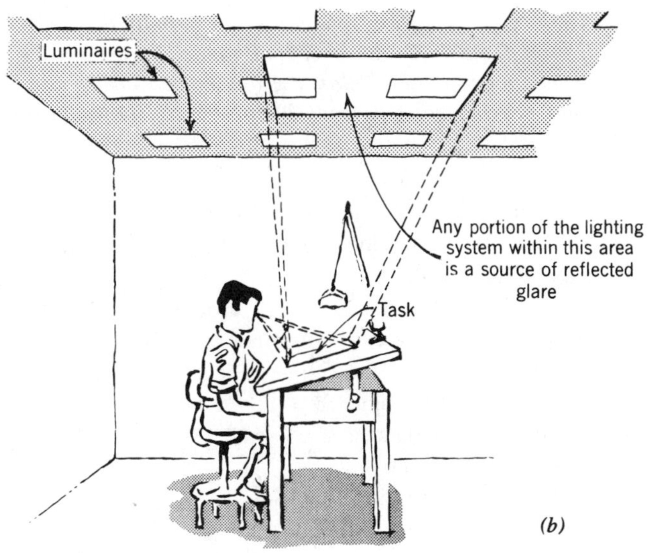

(b)

ILLUMINATION

A similar calculation for clear, black typewritten material on good white bond paper yields a contrast reduction from 94% to 77% or R = 18%. This 77% figure simply serves to emphasize the fact that with high task to background contrast, effective seeing is possible almost regardless of the lighting condition. However, this most emphatically does not relieve the lighting designer of the responsibility to provide a comfortable, efficient, lighting environment which is not the situation where pronounced veiling reflections exist. In general, any contrast reduction above 15% is undesirable.

Since both specular and diffuse reflectances frequently vary with the angle of view, and exact figures are rarely available, accurate calculation is difficult. If a lighting system exists, or a mock-up can be made, measurements of contrast reduction can be accurately made with a contrast/luminance meter of the type shown in Fig. 18.19*c*. The method of use is shown in Fig. 18.30. A standard contrast device that is designed to correspond to a normal office task (black typeface and white paper background) is positioned on the work surface and exposed to the ambient illumination. The task contrast is

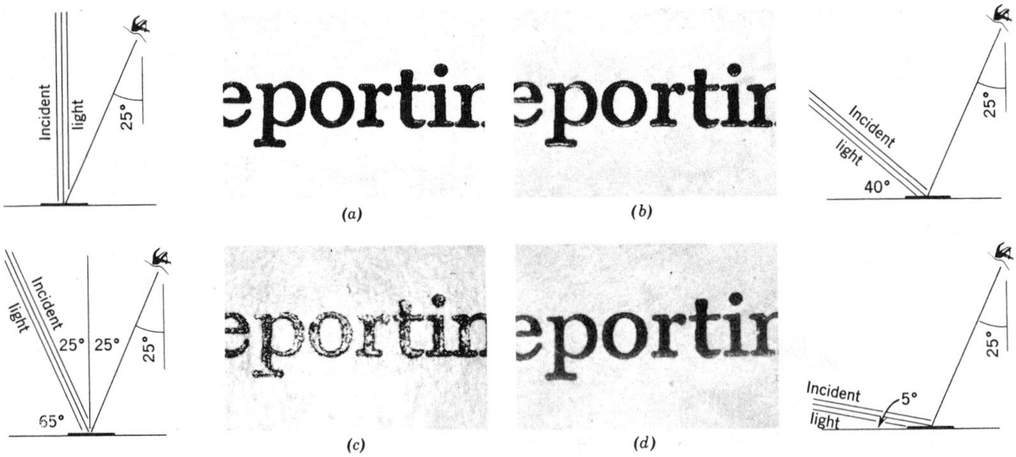

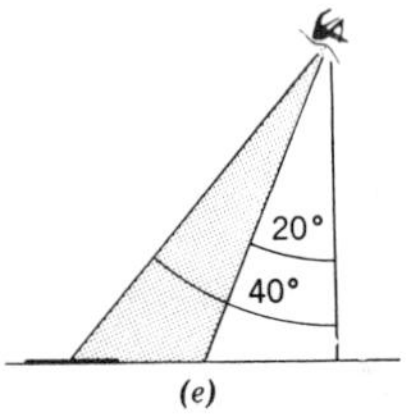

Fig. 18.28 Veiling reflection increases the difficulty of seeing the task. In (a), (b), and (d), it is diffuse reflection that permits us to see, since the specular reflection angle does not correspond to the viewing angle except in (c). Veiling is pronounced at the 5° glancing angle even on a matte-finish surface. The normal viewing zone is from 20 to 40°, as shown in (e). (Courtesy of the IESNA.)

then measured at the same angle at which it would normally be viewed. Contrast reduction is automatically calculated and displayed.

A contrast reduction map of the work surface can thus easily be made (Fig. 18.30*b*). Thereafter, changes can be made in position of the work, viewer, or illumination sources to minimize contrast reduction. Note the pronounced effect of simply shifting the source out of the offending zone (see Fig. 18.27). Contrast reduction in the primary work area should not exceed 15% for good task visibility with specular work items such as glossy papers.

Table 18.8 **Typical Reflectances**

Material	Reflectance: Specular	Reflectance: Diffuse
Matte black paper	0.0005	0.04
Matte white paper	0.0030	0.77
Newspaper	0.0065	0.68
Very glossy white photo paper	0.048	0.83
Metallic paper—copper	0.11	0.28
Dull black ink	0.006	0.045
Super gloss black ink	0.039	0.016

Source: Courtesy of the IESNA.

(c) Equivalent Spherical Illumination. Another way of approaching the problem of contrast reduction is to define a reference lighting

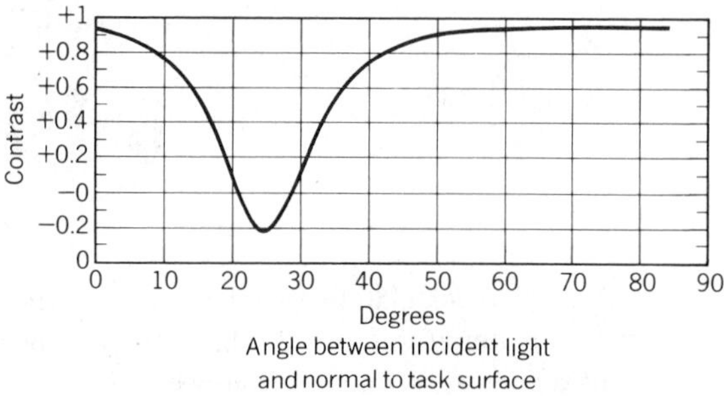

Fig. 18.29 Graph showing contrast reduction of a task with a specular background (such as a sheet of glossy paper) as a function of the angle between incident light and the normal to the task surface. Viewing angle is assumed to be 25° from normal (see Fig. 18.28). Note that between 22 and 27° the contrast is negative. This indicates that background luminance exceeds that of the task, making the task essentially invisible. What is visible, clearly, is a reflected (mirrored) image of the source.

Fig. 18.30 The luminance contrast meter measures contrast reduction at the viewing angle. (a) *Since the sample task can be positioned at any point on the table, a chart of contrast reduction can be made for any viewing position and any task location.* (b) *Note that with a lighting fixture directly in the "offending zone," that is, above and in front of the viewer, a severe loss of contrast occurs over much of the work surface (see Fig. 18.27).* (c) *By shifting the relative position of the viewer and the source so that no source exists in the offending zone, contrast reduction is held to 3 to 4% over most of the work surface. Contrast reduction in the normal work area should not exceed 15%. The contrast meter is shown in Fig. 18.9*(c). *(Courtesy of Brüel & Kjaer.)*

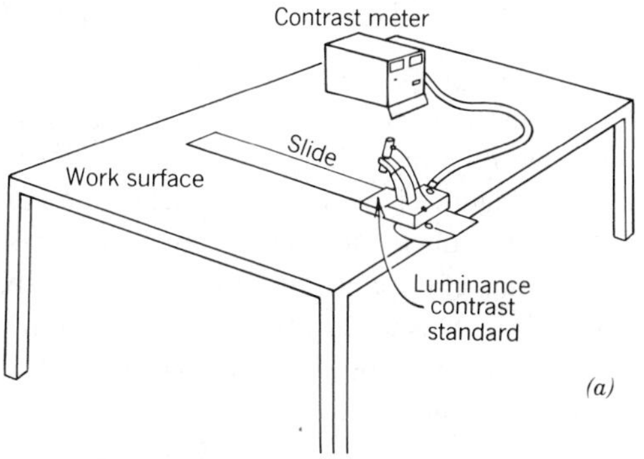

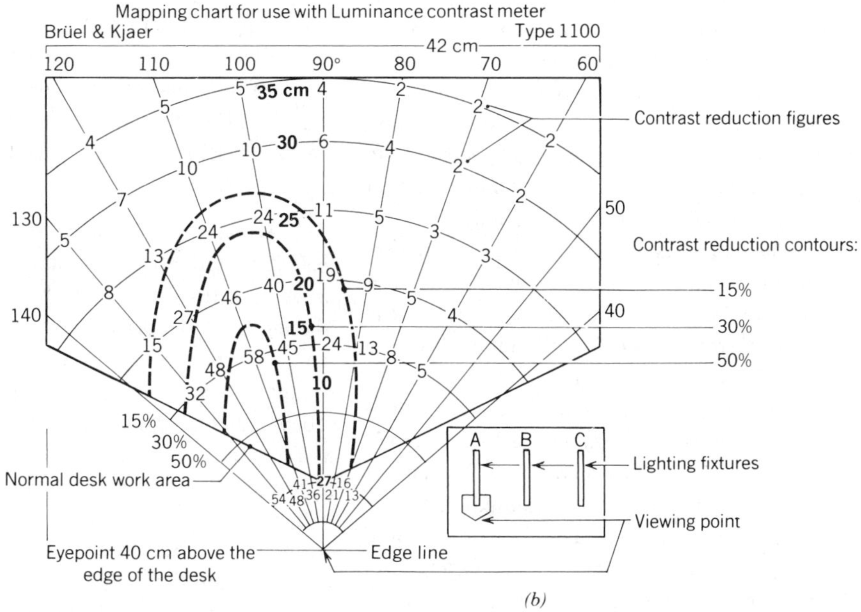

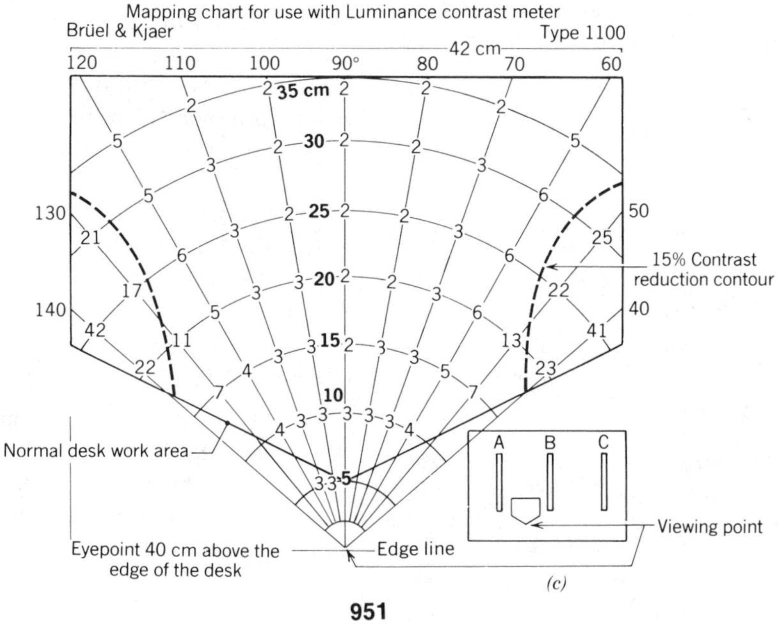

ILLUMINATION

system that is effectively free of veiling reflections, and then relate actual lighting systems to it with a figure of merit. Conversely, one can measure the effectiveness of a given lighting system in terms of the equivalent glare-free system. Both of these ideas, which are essentially the same, are the basis of the concept of equivalent spherical illumination, normally known simply as ESI.

In order to achieve a lighting system almost free of reflected glare, it is necessary to construct an enclosed volume whose surfaces are uniformly diffusely reflective and whose primary source is obscured to the maximum extent possible. As illustrated (see Fig. 18.31), the integrating sphere is such a device. Light is introduced from the outside, split by a deflector, and evenly distributed throughout the sphere by the multiple reflections from the white painted walls. The result is an evenly illuminated volume. When a task is introduced, the illumination falling on it is *entirely* uniform, that is, there are no high luminance sources reflected in it. It is therefore termed *spherically* illuminated. (Note the parallel to sky illumination.) The extent to which any other illumination system can duplicate this glare-free environment is that system's *equivalent* spherical illumination (ESI), and is simply the portion of its total illumination that is spherical, that is, diffuse, glare free. Determination of ESI is accomplished by comparing contrast rendition in the spherical and test systems.

A study of school lighting (Fig. 18.32) gave the illustrated results for four viewing positions

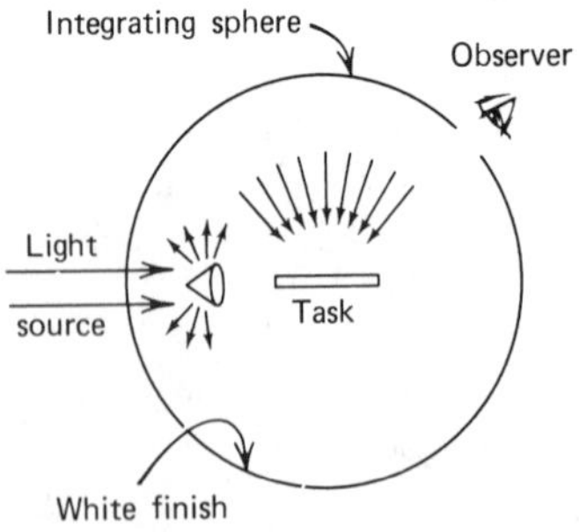

Fig. 18.31 *Sphere illumination is produced by illuminating an object by diffuse reflection from the inside walls of an integrating sphere. The light source and observer are normally external.*

in a classroom lighted with ceiling-mounted continuous rows of 2 by 4 ft, four-lamp, 40-W fluorescent fixtures with lens-type wraparound diffusers, on 10-ft centers. Carefully note that:

1. The ESI, depends entirely on viewing position and viewing angle, other factors in the space being equal.

2. In an ostensibly very well lighted (215 fc) position (M1), the glare-free illumination is only 28 fc! This does not mean that visual work in this position is impossible. It does mean, emphatically, that in position M1 a pronounced veiling reflection will exist on all specular objects. (Because of the size, orientation, and location of the glare source, this reflected glare will be difficult to avoid.) It further means that a large amount of energy is being utilized (effectively wasted) to produce essentially negative results.

The results could have been anticipated, at least qualitatively by examination of the observer positions via-à-vis the layout. Positions M1 and M3 have bright sources in the offending zone, M1 more so than M3, as is borne out by the results. M2 is an excellent position in that it receives light contributions from the two sides; its footcandle value being lower than the others is due to wide row spacing. M4 is ideally placed with no glare sources in the offending zone but with a row of fixtures positioned behind it, so that it is geometrically impossible to act as a glare source. The ESI analysis gives quantitative expression to our qualitative prejudgment and as such is a valuable design tool. Note particularly that the ESI results shown in Fig. 18.32 clearly correspond to the results of a similar test made with the contrast meter, as shown in the charts of Fig. 18.30.

As with the VCP criteria for direct glare, so with ESI; there are ameliorating factors that generally make a given lighting system better than these criteria figures would indicate. Some of these factors have already been mentioned but bear repetition.

1. ESI is critically dependent on observer position and viewing angle. Although position is generally fixed by chair location, observers can and do change viewing angle and head aspect to correct for glare situations.

2. The nature of the task (i.e., its specularity) is assumed to be fixed and unique. In some situations the task nature varies and thus also the contrast. When tasks are constant, severe veiling reflections will frequently lead to measures being taken that will improve the task, the lighting, or both.

3. The lighting distribution characteristics of the fixture involved is a critical factor in glare production. The characteristic of a wrap-around lens diffuser is such [see Section 18.28*c*(4), Fig. 20.24, and Section 20.19] that considerable light falls in the glare zone. Other diffuser characteristics will yield different results.

Calculation of ESI, like the contour plotting of Fig. 18.30, requires use of a computer. As with VCP, most of the widely used and readily available lighting calculation programs contain routines for ESI calculation and some, for contour plots. For designers who do not have access to such programs, most manufacturers will provide the service for their products.

(d) The Future of ESI. In 1970 the IES of North America introduced the concept of ESI to the lighting profession. The 5th edition of the IES handbook (1972) included in its extensive list of illuminance requirements for different tasks, a statement that for certain specified difficult specular tasks, the illuminance recommendation was specifically ESI and not, as for the other tasks not so specified "raw" footcandles. Beginning with the 6th edition of the IES handbook (1981) and in all subsequent editions, there appears the following note adjacent to certain highly specular tasks: "Task subject to veiling reflections. Illuminance listed is not an ESI value. Currently insufficient experience in the use of ESI target values precludes the direct use of ESI in the present consensus approach to recommended illuminance values. ESI may be used as a tool in determining the effectiveness of controlling veiling reflections and as a part of the evaluation of lighting systems."

Over the last decade ESI has come under considerable criticism in the professional lighting literature because it addresses contrast, which is not identical with visibility, because it requires recalculation for even the small geometry changes which can change contrast dramatically, because it tracks raw footcandles, and because it is based on extrapolation from threshold conditions (which procedure has been called into some doubt). These published reservations and criticisms of ESI have apparently sufficiently disturbed the lighting profession's consensus as to warrant the handbook note quoted above.

A textbook is not a proper forum for polemics. The author simply wishes to state that he has used the ESI concept and calculations many

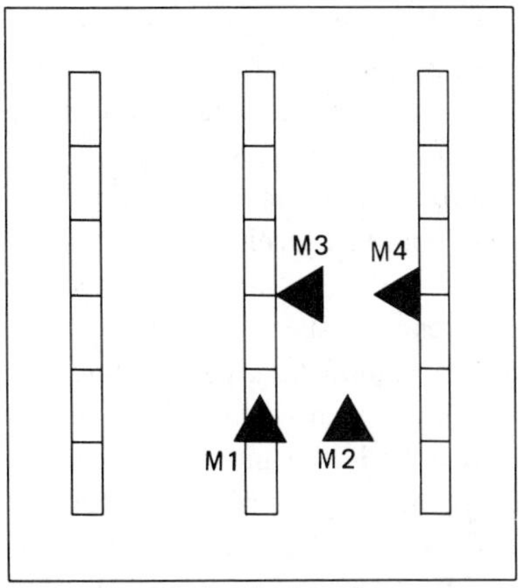

		Observer Position			
		M1	M2	M3	M4
TI	2L	108	92	125	118
	4L	215	185	250	235
CRF	2L	.75	1.00	.82	1.01
	4L	.76	1.00	.83	1.03
ESI	2L	17.8	91.9	31.5	127.8
	4L	28.4	185.3	58.1	308.3

TI–Task illumination
2L–2 lamps (inside pair)
4L–4 lamps
CRF–Contrast Rendition Factors

Fig. 18.32 *A test classroom illuminated by three widely spaced rows of four-lamp fixtures with lens-type wraparound diffusers. Observer positions are shown by arrows. The row of fixtures in front of position M4 is too far forward to be in the offending zone. [Reproduced from Sampson (1970).]*

ILLUMINATION

times since 1970 with excellent results, and will continue to do so. The procedure does exactly what it is designed to do: to point out locations of poor lighting geometry immediately and quantitatively, and to flag luminares with unsuitable distribution characteristics for the proposed use. That is, as stated in the IES handbook note, it is "used as a tool in determining the effectiveness of controlling veiling reflections and as a part of the evaluation of lighting systems."

(e) Relative Visual Performance. In recent years a metric called relative visual performance (RVP) has appeared in the literature and has merited considerable acceptance. It tests (also via computer calculation) the effectiveness i.e. the relative (to perfection) visual performance of a given visual environment as regards speed and accuracy of task accomplishment. Like ESI, it is based on luminance and contrast, but unlike ESI, it judges relative performance of a task rather than simply contrast reduction. It seems to ignore the discomfort of veiling reflections if the task can be performed efficiently. The author has computed RVP for a number of common lighting layouts with various types of lighting fixtures. The results give uniformly high values for RVP, varying between 0.95 and 0.99, which to our mind makes effective judgment of glare situations difficult. The reader is encouraged to use the available lighting programs to calculate ESI and RVP for proposed lighting layouts and to use wisely the results of both as tools in the design.

18.28 Control of Reflected Glare

Since the causes of veiling reflections are well understood, it would seem that a solution to the problem should by now have been adduced. Unfortunately, this is not the case. Although there is no known lighting method or material that will completely eliminate veiling reflections, there are a number of techniques that will minimize contrast loss due to veiling reflections while maintaining adequate illumination. These are:

Physical arrangement of sources, task, and observer so that reflected glare is minimal. See Section (*a*) below.

Adjusting brightness (eye adaptation level) so that objectionable brightness is minimized. See Section (*b*) below.

Design of the light source so that it causes minimal reflected glare. See Section (*c*) below.

Changing the task quality. See Section (*d*) below.

(a) Physical Arrangement of System Elements. The most effective technique is to arrange the lighting geometry so as to avoid sources of high luminance at reflection angles when dealing with specular tasks. This problem is compounded in modern offices, which frequently have both horizontal and vertical work surfaces, the latter being the specular screen surface of a VDT (video display terminal). The latter problem is so widespread that today, it effectively governs the design of office lighting. It is discussed in detail in Section 21.20. As should be clear from Figs. 18.30 and 18.32 and the related discussion, in a space using multiple sources, particularly in continuous rows, placing the work between rows with the line of sight parallel to the long axis of the units is a very effective technique (see Fig. 18.32, position M2). Position M4 is "dangerous" in that the center row can be source of reflections. In this case it is not, due to the 10-ft row spacing. Note that the offending zone for horizontal tasks is dependent on the tilt of the desk. Thus for a horizontal desk, the offending zone is forward of the desk, as in Fig. 18.33*a*; with an elevated table, the ceiling glare source zone may well be behind the source, as in Fig. 18.33*b*.

All of the above geometric solutions presuppose a known detail fixed-furniture layout, a situation that obtains in many but certainly not all cases. In the absence of such data, two alternatives are possible: a uniform layout and adjusting furniture to it, or vice versa. In practice, a combination of both is the most practical approach. Since low watts per square foot budgets have made ducted lighting fixture heat removal systems (air troffers) much less necessary, fixtures are easily shifted. This mobility is further enhanced by the extreme flexibility of lighting fixtures fed from ceiling plug-in raceways as dis-

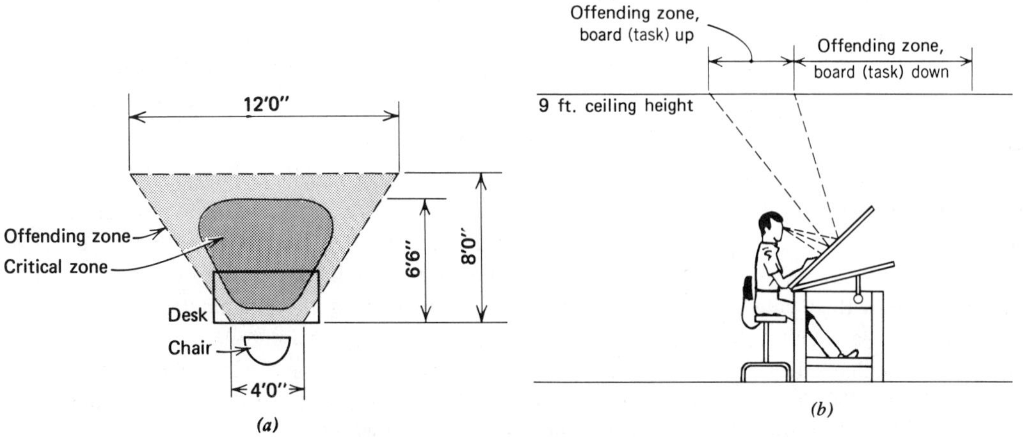

Fig. 18.33 (a) *If luminaires are kept out of the trapezoidal offending zone, contrast will be excellent. If the bulk of one or more luminaires projects into this zone and, in particular into the critical zone, contrast will drop sharply. The dimensions shown are for a flat desk 3 ft × 5 ft and a 9-ft ceiling height.* (b) *The dependence of glare zone on table tilt is illustrated. The offending zone becomes smaller as the table is raised, so that with a table near vertical position, glare is all but eliminated. See also Fig. 18.27. [*(a) *From Ross and Barruzini (1974).]*

cussed in Section 15.30. Figure 18.34 shows such a rearrangement, which results in saving five fixtures, a load reduction of 1 kW, and an *improvement* in visibility.

(b) Control of Area Brightness and Eye Adaptation Level. As discussed in Section 18.19, loss of contrast can be compensated for (and glare reduced) by increased overall non-glare illumination. We are simply making the task brighter to override the detrimental veiling reflection. The problem with this technique, however, is that a large increase in illuminance is required to overcome the glare. This required increase can, in many instances, be most practically accomplished not by increasing overall room illumination with the associated extremely high energy consumption but by adding a supplementary task lighting source so arranged as to be free of reflected glare. By making this supplementary source's position adjustable (as in Figure 18.27*b*), we accomplish three things.

1. Veiling reflection is overcome.
2. The high level of illumination needed for exacting tasks is provided, with minimum energy expenditure.
3. The observer is granted complete control

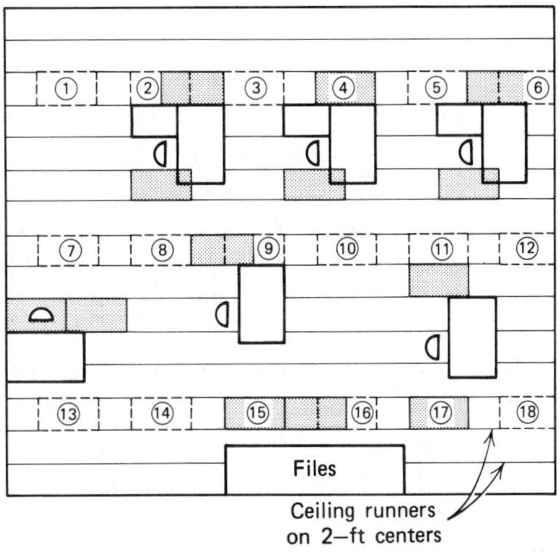

Fig. 18.34 The original uniform fixture layout utilized three rows of six 2 × 4, four-lamp fixtures, giving a total load of 3600 W, a load density of 3.3 W/ft², and a uniform illumination level of approximately 90 (raw) fc (900 lux). The original layout is shown dotted and numbered. The rearranged layout uses 13 fixtures (shown shaded) for a total 2600 W, a load density of 2.4 W/ft², and better than 100 ESI fc (1000 lux) on each work surface. In addition five fixtures are saved.

with resultant optimum lamp placement plus psychological satisfaction that will generally prevent worker complaints.

(Optimum position is generally to the left and slightly forward of the task).

We can demonstrate the effectiveness of a supplemental desk lamp by returning to Example 18.5.

EXAMPLE 18.6. Recalculate the contrast reduction of the ink on vellum visual task of Example 18.5, assuming that a desk lamp raises the illuminance to 200 fc (2000 lux) and is positioned so as to be glare free. (*Note:* An adjustable lamp with 2–15-W fluorescent tubes will produce about that illuminance on the task.)

SOLUTION. Contrast from equation 18.9:

$$C = \frac{(L_{BD} + L_{BS}) - (L_{TD} + L_{TS})}{L_{BD} + L_{BS}}$$

$$= 1 - \frac{200 \times 0.038 + 2000 \times 0.021}{200 \times 0.71 + 2000 \times 0.018}$$

$$= 0.72$$

With this contrast (0.72), the contrast reduction from the original no-glare situation has been improved from the original 47% reduction (0.946 to 0.497) to 24% reduction (0.946 to 0.72). Since even a contrast reduction of 24% is undesirable, a change in task-source geometry or a change in source luminance would be required, assuming that the task itself must remain unchanged.

(c) Control of Source Characteristics. The reflected luminance that causes loss of contrast is proportional to the luminaire luminance. It is apparent then that glare may be reduced by reducing luminaire luminance at the reflection angle. This can be accomplished in four ways:

Dimming or switching lamps. See number 1 below.

Using luminaires with lower overall luminance. See number 2 below.

Using the luminaire as a primary source to illuminate a large, low-brightness secondary source. See number 3 below.

Reduce the luminaire luminance *only at the offending angles*. See number 4 below.

1. Reducing the total output of a fixture will also reduce its output in the critical portion of the ceiling glare zone and can actually *increase* the ESI footcandles (i.e., improve task contrast).

2. In lieu of using a few small high-output sources, utilize larger-area, low-output sources (see Fig. 18.35). This has the effect of reducing the source luminance in the ceiling glare zone while increasing the illumination contribution from outside the glare zone, resulting in better contrast for the same or lower illumination level (lux). The disadvantage of this technique is increased fixture cost.

3. To overcome the economic disadvantage of multiple low-output, low-luminance sources, the ceiling can be used as a secondary source illuminated from high-output indirect or semi-indirect fixtures (see Section 20.11). These sources, which can be fluorescent or HID (metal-halide), have the advantage of high efficiency. The space's ceiling height must be sufficient to permit suspending the unit while avoiding "hot spots" on the ceiling. The minimum suspension length depends on the luminaire characteristic and is normally provided by the manufacturer. To assure high efficiency the ceiling should be painted with a high-reflectivity matte white paint and kept clean. The results obtained from a semi-indirect installation using 1500-mA, very high-output lamps are shown in Fig. 18.36.

4. Since most horizontal task vision takes place between 20 and 40° from the vertical (see Figs. 18.27 and 18.28), any fixture that emits little or no light below 40° from the horizontal *cannot* produce veiling reflection regardless of its position in the field of view (see Fig. 18.37). As a result, diffuser manufacturers designed and produce a prismatic diffuser whose output below 30° and above 60° are diminished, in order to minimize both reflected and direct glare. Due to the characteristic shape of the distribution curve, they are known industrywide as "batwing" diffusers. For observers positioned so that their sight lines are parallel to the longitudinal axis of the ceiling fixtures, lenses with linear (side-to-side) batwing characteristics will per-

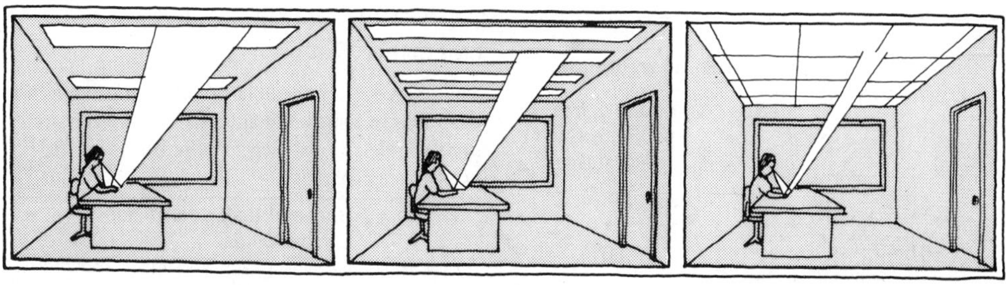

Fig. 18.35 A concentration of light in the glare zone (a) *produces the largest amount of reflected glare. As the number of light sources is increased* (b) *in the glare zone and luminance is decreased, reflected glare is decreased. Least glare is from an all-luminous ceiling, which also has the lowest luminance* (c)*. A similar result to* (c) *can be obtained with indirect lighting as in Fig. 18.36.*

form well. If the observing position varies in aspect with respect to the fixture, a radial batwing curve (in all directions) is required. Note carefully that these diffusers have only limited usefullness in reducing reflected glare in specular vertical surfaces (VDT screens). All types of diffusers and their characteristics are discussed in Section 20.21.

(d) Changing the Task Quality. At this point it should be abundantly clear that reducing the task specularity is at least as effective a means of reducing veiling reflections as changing the lighting system characteristics, if not more so. It is therefore recommended that task contrast and specularity be actively considered and recommendations made in a framework of energy effectiveness and cost effectiveness. Thus, to produce adequate visibility it will often be cheaper and always more energy-economical to upgrade the task (in the visibility sense) than to change the lighting system.

18.29 Luminance Ratios

As explained above, visual performance increases with contrast—that is, with difference in luminance between the object being viewed and

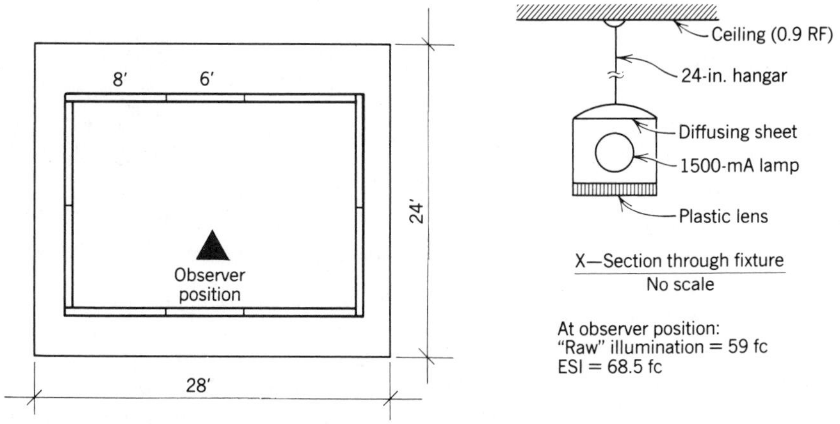

Fig. 18.36 *With a high-reflectance diffuse finish ceiling, this semidirect lighting installation yields a higher ESI illuminance than raw illuminance at the viewing location illustrated, indicating excellent contrast rendering. The plastic lens at the bottom of the fixture serves to provide perceived light source brightness and to avoid an impression of gloominess, despite the satisfactory overall luminance level. [Reproduced from Sampson (1970).]*

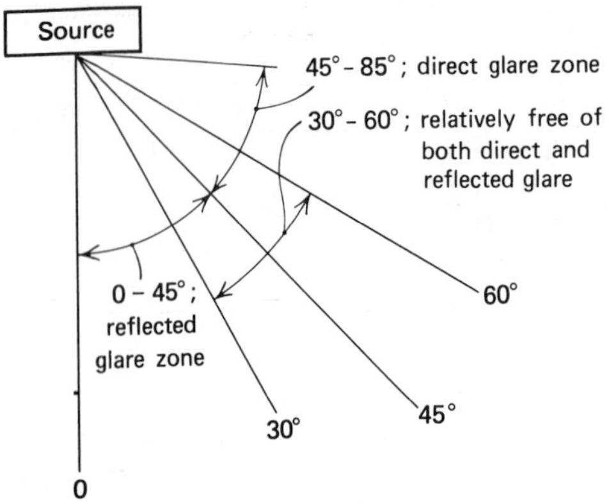

Fig. 18.37 Glare zones are 0 to 45° and 45 to 85° for reflected and direct glare, respectively. Therefore, a diffuser that emphasizes the 30 to 60° zone will be least objectionable on both counts.

TABLE 18.9 **Recommended Maximum Luminance Ratios**

To achieve a comfortable brightness balance, it is desirable to limit luminance ratios between areas *of appreciable size* as seen from normal viewing positions as follows:

1 to $\frac{1}{3}$	Between task and adjacent surroundings
1 to $\frac{1}{10}$	Between task and more remote darker surfaces
1 to 10	Between task and more remote lighter surfaces
20 to 1	Between luminaires (or fenestration) and surfaces adjacent to them
40 to 1	Anywhere within the normal field of view

These ratios are recommended as maximums; reductions are generally beneficial.

its immediate surroundings. Conversely, however, the difference between the average luminance of the visual field (task) and the remainder of the field of vision should be low to avoid the discomfort of large rapid changes in eye adaptation level. Restated, *contrast is desirable in the object of view but undesirable in the surrounding field of view*. Reflectances of 50, 30, and 80% for walls, floor, and ceiling, respectively, and 35% for furniture, will establish a fairly high eye adaptation level so that direct glare (that results from excessive luminances in the field of view) is minimized. Recommendations for maximum luminance ratios to achieve a *comfortable* environment are presented in Table 18.9. (Effective visual performance is entirely possible in environments with much higher ratios; It is simply not as visually comfortable and may be fatiguing.) To achieve these luminance ratios, it is obviously necessary to control carefully the reflectances of the major surfaces in a room. The reflectance figures given above are recommended averages for work-type commercial and educational spaces. The marked difference between a background with proper reflectance and one with excessive brightness ratios caused by the low surrounding reflectances is shown in Fig. 18.38.

18.30 Patterns of Luminance; Subjective Reactions To Lighting

In the list of characteristics in Section 18.16 we included among the secondary factors in illumination *patterns of luminance*, that is, the pat-

Fig. 18.38 The reflected glare from luminaires disappears when a piece of light, diffuse linoleum is placed over the dark, polished desktop. Light-colored desktops with a 35 to 50% reflectance result in task-to-background ratios within the 3 : 1 recommended range. Before the linoleum was placed, a reflection similar to the one seen above also existed on the left of the desk due to another luminaire. (Courtesy of IESNA.)

terns of light and shadow in a space resulting from the method of illumination. Thus a single source produces sharp shadows while a luminous ceiling or a completely indirect illumination system produces almost completely diffuse light (see Fig. 18.39). Diffusion is the degree to which light is shadowless and is therefore a function of the number of directions from which light impinges on a particular point and the relative intensities.

Perfect diffusion, rarely obtainable (or desirable), would have equal intensities of light impinging from all directions, therefore yielding no shadows. The only naturally occurring example of perfectly diffuse lighting is a daytime fog, which we know to be extremely disturbing to the eye, demonstrating that some directivity is desirable. Diffusion can be judged by the depth and sharpness of shadows. A room with well-diffused illumination resulting from multiple sources and high room surface reflectances yields soft multiple shadows that do not obscure the visual task. Since purely diffuse lighting in monotonous and therefore not conducive to extended periods of work effectiveness, some directional lighting is often introduced as an adjunct to diffuse general lighting to lend interest by producing shadows and high brightness variations. Where texture must be examined or surface imperfections detected by grazing angle reflections, highly directional lighting is required. Indeed, as seen in Figure 18.40, directional light is what creates shape and is precisely the characteristic best used to influence architectural space and form.

Sections 20.10 through 20.15, which deal with systems of lighting, illustrate a few of the light/dark patterns produced by different lighting arrangements. The combinations of uplighting and downlighting, perimeter and ceiling lighting, are legion; each produces its own shadows and modeling, and each has a quality of its own. It is very much in the interest of the lighting designer to be familiar with these effects so that he or she can mentally visualize them as the design progresses. Indeed it would be well for a designer to prepare a reference sketchbook of such shadow diagrams. It is these patterns of light and darkness that give the ambience and the subjective reactions of sociability/isolation, clarity/fuzziness, spaciousness/crampedness, simplicity/clutter, formality/informality, boredom/excitement, definition/shapelessness, and so on. (Color has a great deal to do with subjective reactions and is discussed separately below.) The subject of psychological reactions to lighting environment is extensive and complex and can be only touched upon here to the extent of mentioning a few of the salient lighting techniques and their usual subjective responses.

In addition to modeling and texture accent, small high brightness sources, usually called *sparkle* create points of interest and visual excitement. Lighting installations generally yield a sense of vividness or activity proportional to the level of illumination. This is not the case with very diffuse lighted areas, which even at high footcandle levels are tedious. This is particularly noticeable in large, luminous ceiling installations which are particularly oppressive when the ceil-

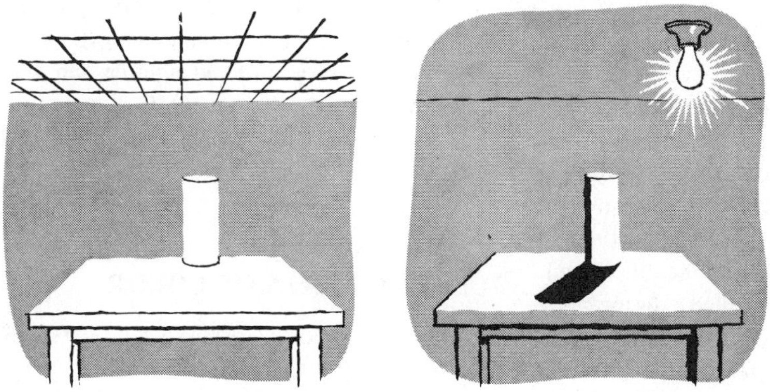

Fig. 18.39 *Diffuse illumination. The luminous ceiling installation provides shadowless, almost perfectly diffuse lighting. By comparison, the single ceiling bulb produces sharp shadows and very little light diffusion.*

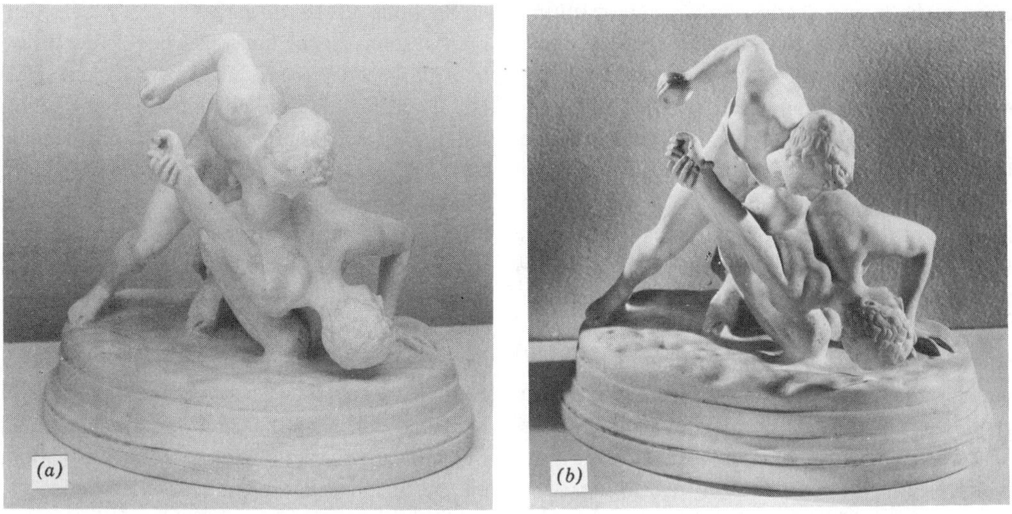

Fig. 18.40 Totally diffuse lighting (a) *destroys texture, whereas a combination of diffuse and directional lighting* (b) *produces the required modeling shadows. (Courtesy of Holophane.)*

ing is low. Small exposed incandescent lamps, a brightly lighted, rough-textured wall, and pendant fixtures with pierced reflectors are some of the techniques used to create this visual interest.

Visual attention can be drawn by high brightness. This well-known fact is used constantly in displaying merchandise. Note the following usual reactions:

A 3 : 1 luminance ratio between object and surround will be noticed but will usually not affect behavior or draw attention.

A 10 : 1 luminance ratio will attract attention and, if interesting, will hold it.

A 50 : 1 luminance ratio or larger will highlight the object thus illuminated, practically to the exclusion of all else in the field of view.

Since brightnesses draw the eye's attention, all of the individual brightness sources in the field of view produce an overall impression. If there is some form or order or pattern to them (as a pattern of lighting fixtures) then the overall impression is not disturbing—it can be thought of as visually harmonious. If, on the other hand, they are in disarray, they produce a discordancy precisely as sound produces discordancy in the ear. This visual "noise" is frequently referred to as visual clutter and can be very disturbing. The designer is well advised to keep this important fact in mind when arranging light sources that are the primary sources of luminance in an enclosed space. Aspects of fixture patterning are shown in Section 20.16.

Other subjective reactions to lighting, on which there is wide consensus are:

1. Bright walls (about 25% of horizontal light level) increase spaciousness. Conversely, dark walls diminish a space. As a corollary: high fixture luminance attracts the eye away from the walls and diminishes spaciousness.
2. Worker-adjustable task lights increase the feeling of control and therefore comfort.
3. Downlights (and color highlights) increase feelings of relaxation and comfort.
4. Hidden source indirect lighting and very low brightness lighting fixtures cause discomfort because of the inability to locate the source of light.

FUNDAMENTALS OF COLOR

The subject of color is vast. We can relate here only to a few aspects of color which it is imperative that the lighting designer understand. Further, since it is difficult to discuss color without actually using it, the coverage will be brief.

18.31 Color Temperature

A light source is often designated with a *color temperature,* such as 3400 K for quartz iodine lamps, 4200 K for cool white fluorescent tubes, and so on. This nomenclature derives from the fact that when a light-absorbing body (called a black body) is heated, it will first glow deep red, then cherry red, then orange until it finally becomes blue-white hot. The color of the light radiated is thus related to its temperature. Therefore, by developing a black-body color temperature scale, we can compare the color of a light source to this scale and assign to it an approximate "color temperature," that is, the temperature to which a black body must be heated to radiate a light approximating the color of the source in question. Temperature is measured in kelvin, which is a scale that has its zero point at minus 460°F. Figure 18.41 shows the assigned color temperature of some common light sources.

Strictly speaking, a color temperature can be assigned only to a light source that produces light by heating, such as the incandescent lamp. Other sources, such as fluorescent lamps, produce light by processes that will be detailed in the next chapter. Such sources are assigned a *correlated color temperature* (CCT), which is the temperature of a black body whose chromaticity most nearly matches that of the light source. For such sources there is *no relation whatever* between their operating temperature and the color of the light produced.

As is well known, any nonspectral color illuminant is composed of two or more component color illuminants. When such a composite light, for example, white, falls on a surface other than black or white, selective absorption occurs. The component colors are absorbed in different proportions so that the light reflected or transmitted is composed of a new combination of the same colors as had impinged on the surface. Thus a white light reflected from a red wall acquires a red tint since the component colors of the white light other than red were absorbed in greater proportion than the red. When reflected, the red light took prominence thus giving the reflected light a red tint. This is illustrated in Fig. 18.42*a*. Similarly, a white light when passed through a piece of red glass emerges as a reddish light since the other components were absorbed in much greater proportions than the red. This well-known phenomenon is illustrated in Fig. 18.42*b*.

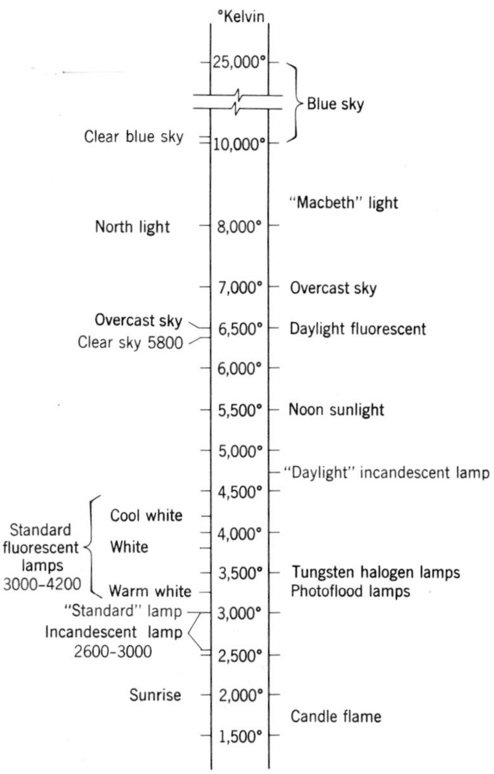

***Fig. 18.41** Approximate color temperatures of common illuminants.*

It is this phenomenon that allows us to see color at all; the individual object pigmentation absorbs all other colors of light and reflects or transmits to the eye only its own hue.

18.32 Object Color

The color of the illuminant (light) and correspondingly the coloration of the objects within a space constitute an important facet of the lighting quality. The two factors, however, must not be considered separately since by definition the color of an object is its ability to modify the color of light incident upon it by selective absorption. The color reflected or transmitted is apprehended by the eye as the color of the object. An object is technically said to be colorless (not transparent) when it does not exhibit selective absorption, reflecting and absorbing the various components of the incidenct light nonselectively. Thus, white, black, and all shades of gray are colorless, neutral, achromatic, or, more precisely, lack hue.

Hue is defined as that attribute by which we recognize and therefore describe colors as red, yellow, green, blue, and so on. Just as it is possible to form a series from white to black with the intermediate grays, so it is possible to do the same with a hue.

The difference between the resultant colors of the same hue so arranged is called *brilliance* or *value*. White is the most brilliant of the neutral colors and black the least, pink is a more brilliant red hue than ruby, and golden yellow a more brilliant (lighter) yellow hue than raw umber.

Colors of the same hue and brilliance may still differ from each other in saturation, which is an indication of the vividness of hue or the difference of the color from gray. Thus pure gray

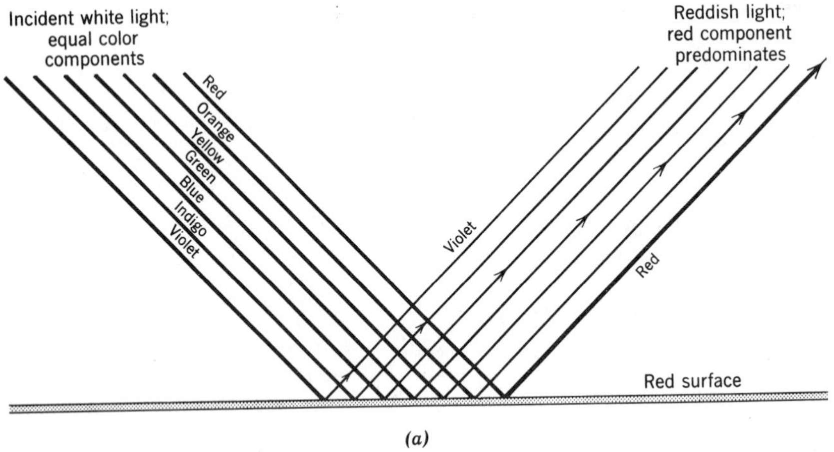

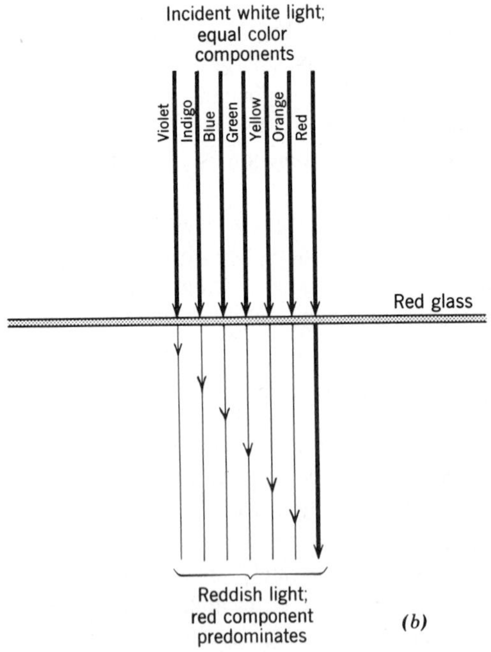

Fig. 18.42 Selective absorption of (a) *reflected light and* (b) *transmitted light.*

has no hue; as we add color we change the saturation without changing the brilliance. The three characteristics then that define a particular coloration are *hue, brilliance,* and *saturation.* Using these terms we may define "bay" as a color red-yellow in hue of low brilliance and low saturation; while carmine is a color red in hue, of low brilliance and very high saturation.

Various systems of color classification have been devised, including the ISCC- NBC color system, the Munsell Color System, the Ostwald Color System, C.I.E., and the Chromaticity Diagram. In the Munsell color system (see Fig. 18.43) brilliance is referred to as *value* and saturation as *chroma;* thus a color is defined by hue, value, and chroma. The brilliance (value) of a pigment or coloration is related to its reflectance to white light. The higher the brilliance or value, the higher the reflectance factor, as might be predicted when one considers that white and black are the poles of brilliance. Chroma or saturation may be thought of as either the difference from gray or the purity of the color. Spectral colors have 100% purity and therefore maximum chroma.

When white is added to a pigment, it produces a *tint;* adding black produces a *shade.* When pigments are mixed to produce a particular color, we create this color by a subtractive process. That is, each pigment absorbs certain proportions of white light; when mixed, the absorptions combine to subtract (absorb) various colors of the white spectra, and leave only those colors that finally constitute the hue, value, and chroma of the pigment. This subtractive effect is also utilized when producing colors by filtering white light. The filter selectively absorbs component colors, transmitting only the component desired. Thus a blue filter transmits only blue and so on (see Fig. 18.42 *b*).

Conversely, when light of the three primary colors of red, green, and blue are combined, they form white by an additive process (see Fig. 18.44).

The additive and subtractive primary colors are complementary; they combine to give a white or neutral gray, respectively. Thus, red and blue-green, blue and yellow, and green and magenta are complementary.

Therefore, if a red object is illuminated with blue-green light, the object color appears gray, since the red pigment absorbs the blue-green and reflects nothing; hence the gray. This accounts for the common "lost red car" in parking lots

ILLUMINATION

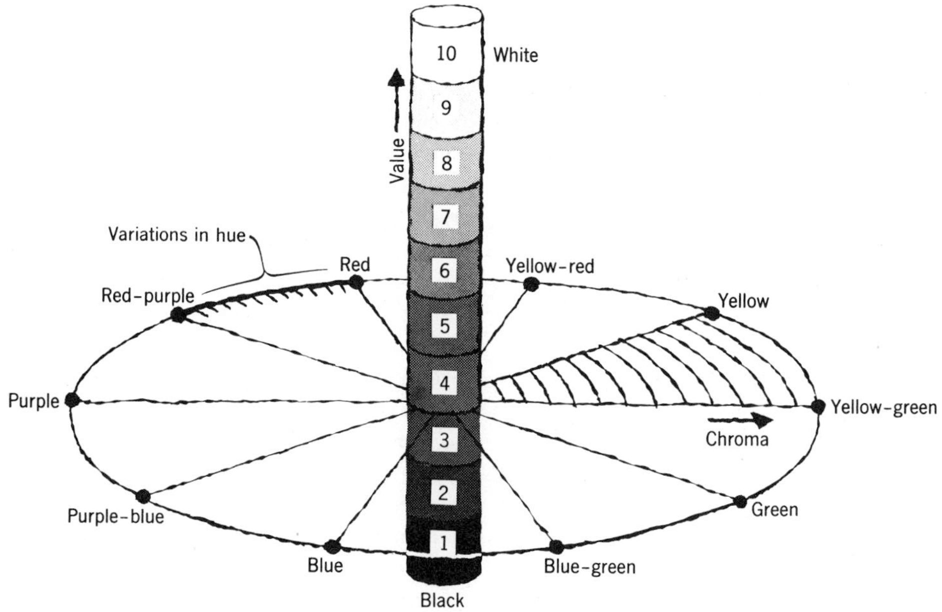

Fig. 18.43 *The Munsell color system defines a color by three characteristics: hue (color), chroma (saturation), and value (grayness).*

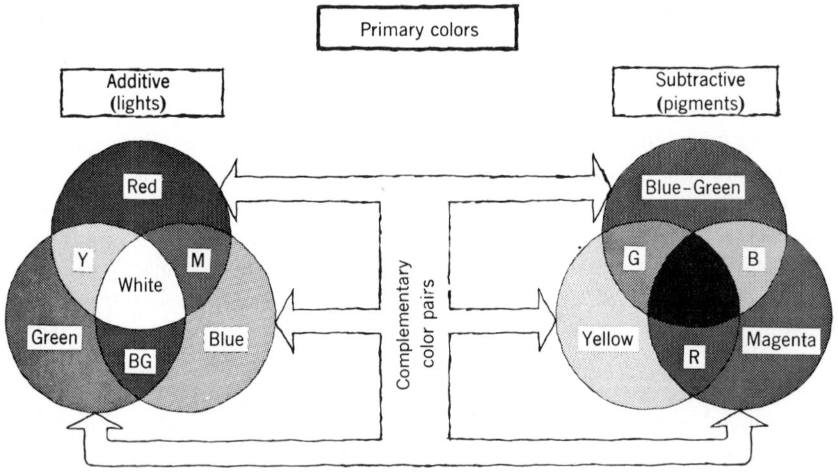

Fig. 18.44 Primary and complementary colors. Complementary color pairs are shown by arrows. Pigments form color by an absorptive (subtractive) process; colored lights form colors by an additive process.

illuminated with clear mercury lamps with their characteristic blue-green color. Similarly, a blue filter on a yellow light would transmit nothing.

18.33 Reactions to Color

Light of a particular hue (other than white) is rarely used for general illumination except to create a special atmosphere. When a space is lighted with colored light, the eye adapts by a phenomenon known as *color constancy* so that it can to a considerable degree, depending on the chromaticity of the light, recognize colors of objects despite the spectral quality of the illuminant. However, the eyes become more sensitive to the missing colors that would make up white light. This phenomenon could be used to make meat look redder on a butcher counter by using blue-rich, red-poor, cool white lighting in the remainder of the store.

A similar phenomenon occurs when the eye is exposed to a monochromatic scene where the chromaticity is due to coloration of the objects, rather than the illumination. The eye in such a situation becomes sensitized to the complementary color; thus if after looking at a green surface one shifts the gaze to a white surface, one sees the complementary red color. Returning to our meat market, the use of green paint on the walls also enhances the redness of the meat. This effect in reverse also partly accounts for the extensive use of green for paints, linens, and gowns, and so on, in operating rooms. The eyes of the surgeons and nurses when diverted from the redness of the surgical area will be more comfortable seeing green on a green background than on a white one.

By a process known as *lateral adaptation*, the apparent color of an object changes when background color is changed. Thus a green object looks somewhat blue-green on a yellow background because the eye is supplying the complementary color to yellow—that is, blue. Similarly, the same green object looks slightly yellow-green when on a blue background, the eye supplying the yellow.

Apparent brightness of a color is a function of its hue, in that light colors appear lighter than dark colors even when measured brightness is the same. Thus spaces may be defined by color within an area of equal illumination. Also, all colors tend to appear less saturated, that is, they appear "washed out" when illumination is high. Thus pigments of high saturation (chroma) must be used in well-lit spaces if they are to be effective, although extensive use of saturated colors is generally best avoided.

Other well-known psychological effects of colors are the coolness of blues and greens and the warmth of reds and yellows. Similarly, red and yellow are "advancing" colors because objects lit with them tend to "advance" toward the observer, giving the appearance of becoming larger. The opposite effect is noted with blue and green, accounting for their being known as receding colors. Thus cool colors might well be used in a fur salon and warm colors in a display of summer wear. A practical, energy-saving application of these color phenomena would be to use warm colors to compensate somewhat for lowered thermostats in the winter, and cool colors for the opposite effect in summer. How to accomplish this without the expense of repainting twice a year is left to the ingenuity of the architect and interior designer. In an atmosphere designed to be calm and restful, greens should generally predominate either in illuminant color, object color, or both, except in eating areas, which should be lighted with reds and yellows since cool colors are generally unappetizing. Yellows and browns emphasize motion sickness, whereas blues and greens tend to the reverse. Warm and saturated colors produce activity; conversely cool, unsaturated colors are conducive to meditation. Cool colors also seem to shorten time passage and are well applied in areas of dull repetitive work.

A further discussion of color control, illuminant colors, color measurement, and color matching will be found in Section 19.25 and 19.26, dealing with spectral energy distribution of sources and color rendering index.

REFERENCES

See listing at the end of Chapter 21.

19

LIGHT SOURCES: THEIR CHARACTERISTICS AND APPLICATION

19.1 General Remarks

Long before the dawn of recorded history, the double blessing of fire, heat and light, was discovered. Even today in our sophisticated space age, fire is still used almost universally as the source of heat and, in a large proportion of the world's dwellings, as the source of light. Electrical lighting had its real beginning in about 1870 with the development of commercially usable arc lamps and was given greater impetus nine years later by Edison's first practical incandescent lamp. Today's practical electric light sources fall into two generic classifications: the incandescent lamp, including tungsten-halogen types and the gaseous discharge lamp, which includes fluorescent, mercury, metal-halide, and sodium lamps.

The efficiency of a light source is termed its *efficacy* and is measured in lumens per watt (lm/W). Table 19.1 lists efficacies of modern light sources, including ballast losses where applicable. It is misleading to use the efficacy of the light sources alone, as is often done in the literature, because the ballasts are inseparable from the lamp. Note that Table 19.1 gives a range of efficacies for each lamp type. In general, efficacy increases with wattage, as is clear from Fig. 19.1. It is therefore energy-economical to use a small number of higher wattage lamps than the reverse. (It is also usually more economical with respect to fixtures.)

Since electric lighting in American nonresidential buildings consumes 25 to 60% of the electric energy utilized, any attempt to reduce this must necessarily include integration of the cheapest (insofar as energy is concerned), most abundant, and, in many ways, most desirable form of lighting available—daylight.

DAYLIGHTING

19.2 Daylighting as a Lighting Design Factor

The provision of daylight in structures in the United States had before the 1973 energy crisis largely been considered an amenity rather than a necessity. As such its provision had been the province of architecture rather than lighting design. The reasons for this are clear. Daylight is indeed an amenity. Windows provide visual contact with the outside and the resultant daylight provides a bright, pleasant, airy ambience. When daylight enters through windows (side lighting, as opposed to toplighting), its horizontal directivity provides good modeling shadows, minimal veiling reflections, and excellent vertical surface illumination. Furthermore, the continual variation of daylight, which is one of its prominent characteristics, provides a constantly changing pattern of space illumination—one that is unattainable with artificial light. Since these changes are gradual, the eyes adapt easily and the effect is one of visual interest. Undoubtedly, as a result of these effects, numerous studies have conclusively demonstrated a marked preference for daylight over any other form of lighting.

TABLE 19.1 **Efficacy of Various Light Sources**[a]

Source	*Efficacy (lm/W)*
Candle	0.1
Oil lamp	0.3
Original Edison lamp	1.4
1910 Edison lamp	4.5
Modern incandescent lamp	8–20
Tungsten halogen lamp	16–20
Fluorescent lamp[b]	32–102
Mercury lamp[b]	30–60
Metal-halide lamp[b]	70–100
High-pressure sodium[b]	50–130
Low-pressure sodium	120–140

[a]See Fig. 19.1.

[b]Including ballast losses.

On the other hand, although concern has been expressed about possible physical ill effects, none have *conclusively* been demonstrated to have been caused by lack of daylight, that is, by working in an electrically lighted space. Since an artificial lighting system had to be installed in any event, to furnish interior illumination during periods when daylight is insufficient, the practice arose in the United States of ignoring daylight and even of shutting it out deliberately! Careful design of an electric-lighting system can provide a satisfactory visual atmosphere, as these chapters explain. Furthermore, unlike daylight, control of such systems is relatively simple. Perhaps most important, an interior electric-lighting system has minimal impact on the building architecture, least of all on the all-important building facade. Finally, the energy to power electric-lighting systems *was* cheap.

The option of ignoring daylight in our high-energy-cost and energy-resource-poor society is no longer available. That being the case, the American designer must learn to cope with the special problems that daylight use presents in order to reap its benefits. Since daylight is variable, it creates special problems of glare control, direct sunlight control, and heat-gain limitation. In large measure the science

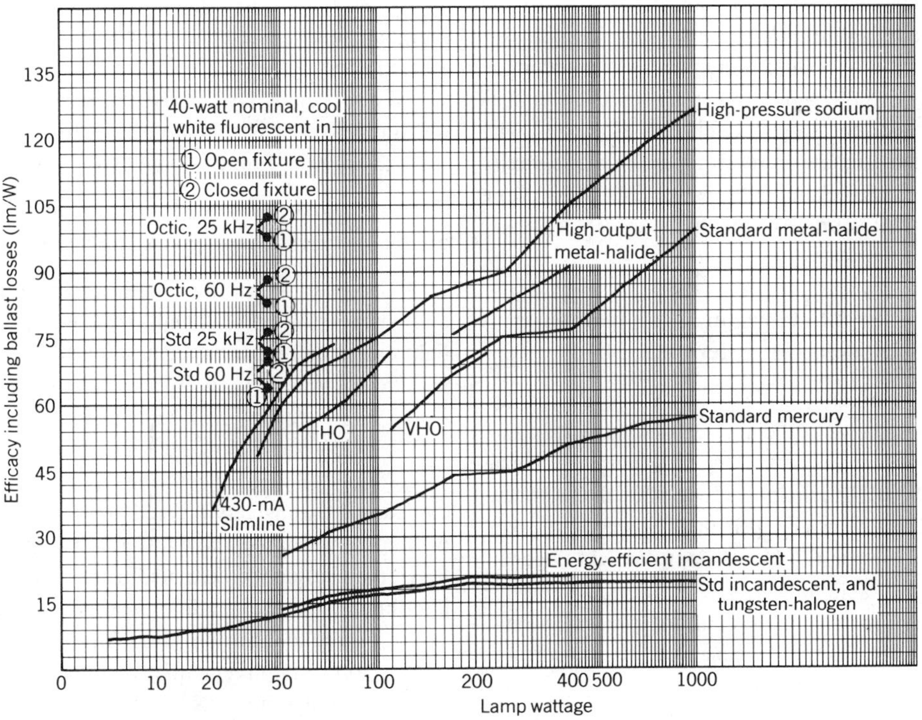

Fig. 19.1 *Average light source efficacies as a function of lamp size. Ballast losses are included. (Data from current published manufacturers' data.)*

(and art) of daylighting is not so much how to provide daylight as how to do so without the attendant undesirable effects. Put otherwise, the American designer must adapt and adopt the British technique of PSALI (Permanent Supplementary Artificial Lighting in Interiors), which is almost universally applied in Europe.

One caveat must be introduced immediately. Windows are a two-way portal. HVAC designers have long been aware of the major influence of windows on the thermal balance of a building and of the need for window controls. The same awareness is necessary for daylight design. Windows can be a daytime glare source and a nighttime light loss source unless appropriate controls are used. Furthermore, windows, particularly skylights, are a source of nighttime light pollution which affects not only astronomers but also ordinary citizens. This subject is discussed briefly in Section 21.33

The material below, which discusses PSALI, sky luminance, outdoor illumination, and calculation of interior daylight levels, is necessarily brief, because of space considerations. Probably no subject bearing on architectural engineering has received more attention in the past decade than daylight, its source, and its ramifications. A bibliography of books and papers bearing on the subject would alone cover more space than is allotted to the subject itself in these pages. The interested reader is referred to the bibliography at the end of Chapter 21, which lists a few of the important sources in this field.

19.3 PSALI (Permanent Supplementary Artificial Lighting, Interiors)

This technique, which is really a design approach, views artificial (electric) lighting as supplementary to daylighting and not vice versa. The PSALI technique recognizes that nonresidential buildings are principally used during daylight hours and that sufficient daylight is generally available during these hours to provide much of the structure's lighting needs. The success of the PSALI design approach is predicated on three assumptions:

1. The large variation of interior daylight illuminance levels during the course of the day will not adversely affect visual performance, even when the actual levels fall below accepted recommendations. Stated otherwise, the same visual performance can be achieved with less daylight than artificial light, when compared on a footcandle (lux) basis. The rationale for this conclusion is based on two well-known phenomena of human vision: lightness constancy and Weber's law of visual contrast. Lightness constancy is the ability of the eye to distinguish between light and dark objects over an extremely wide range of illumination. Thus, even though a lump of coal in sunlight is brighter than this piece of paper in moonlight, we will correctly identify the one as black and the other as white. This is because the eye appreciates lightness or brightness not only by actual luminance of the object being viewed but also by the surrounding area. Thus, in a space with *slowly* changing illumination levels, as is the case with daylighted spaces, the eye will adapt to the altered level and see basically the same scene, in terms of lightness (and darkness) of objects. That is, the eye's appreciation of the pattern of brightnesses in the room will remain constant.

   Weber's law states that within the very wide illuminance range of lightness constancy (above 10 lux), contrast ratios remain constant. Since visual performance is closely related to contrast discrimination, it is also largely unaffected over the wide absolute range of daylight illuminance.

2. Illuminance recommendations to be used are the current ones as given in Section 18.24.

3. Daylight and artificial light can be readily and successfully combined, that is, artificial light can supplement daylight when the latter is insufficient (see Fig. 19.2).

We made particular reference above to nonresidential buildings because residences are occupied and extensively utilized at night when no daylight is available and PSALI is inapplicable. Obviously residences are also used during daylight hours, but the conditions of fixed work location and long-duration visual tasks generally do not apply. Nonresidential buildings that are regularly used at night as well as during daylight hours should be designed for both conditions, with appropriate controls.

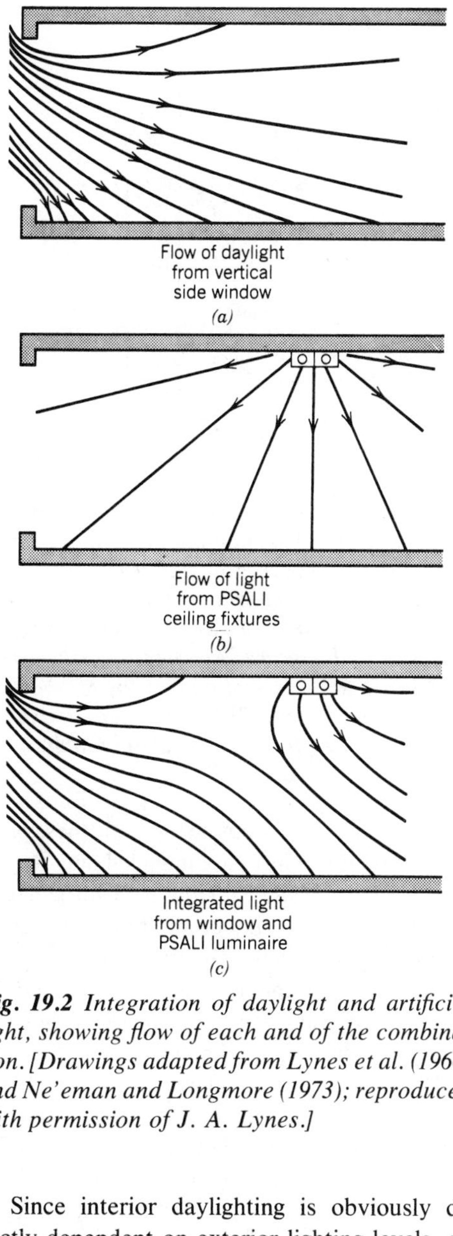

***Fig. 19.2** Integration of daylight and artificial light, showing flow of each and of the combination. [Drawings adapted from Lynes et al. (1966) and Ne'eman and Longmore (1973); reproduced with permission of J. A. Lynes.]*

Since interior daylighting is obviously directly dependent on exterior lighting levels, an understanding of the latter is a clear prerequisite to designing for the former. However, before beginning a discussion of exterior illuminance and the various methods by which the resultant interior daylight illuminance can be calculated, it is appropriate to ask several basic questions regarding daylight calculations in general:

1. Inasmuch as the majority of buildings must be designed for nighttime use as well, that is, with a complete system of electric lighting, what purpose does a knowledge of daylight distribution in such buildings serve?
2. Given the great variability of daylight with time of day, season, and weather, how can any meaningful calculations be made?
3. Furthermore, if daylight distribution is calculated for a variety of conditions, what use can be made of the results? What are the results we want, the definition of which will establish the exterior conditions that we wish to study?
4. What degree of accuracy in calculation is desirable? Obtainable?

The answers to these questions are fairly clear and will define the direction that our study of daylight will take.

The purpose of daylight analysis is manifold.

1. Granted the variability of daylight, it is possible to calculate maximum, minimum, and, if enough data are available, statistical averages. Furthermore, there are extensive areas of the United States where for long periods of time the weather is constant—clear blue skies in the Southwest, solid overcast in the Northwest—and for those areas fairly accurate calculations of interior daylight (±25%) can be made.

2. Maximum daylight data, including direct sunlight, are needed for design of glare controls, shading devices, light shelves, and other types of reflectors, and for zoning of automatic daylight compensation controls (see Section 21.4). These data are also important in establishing a maximum cooling load.

3. Minimum daylight data are necessary to establish maximum PSALI (see Fig. 19.3) and to design the controls that will differentiate PSALI from nighttime lighting. This also defines maximum electric lighting loads, which are needed for heating and cooling calculations.

4. Average daylight data are necessary to establish percentiles on the basis of which annual energy budgets can be determined. Furthermore, average data are important in designing fenestration, since use of minimums will probably result in excessive glass area with the concomitant thermal problems discussed in Chapter

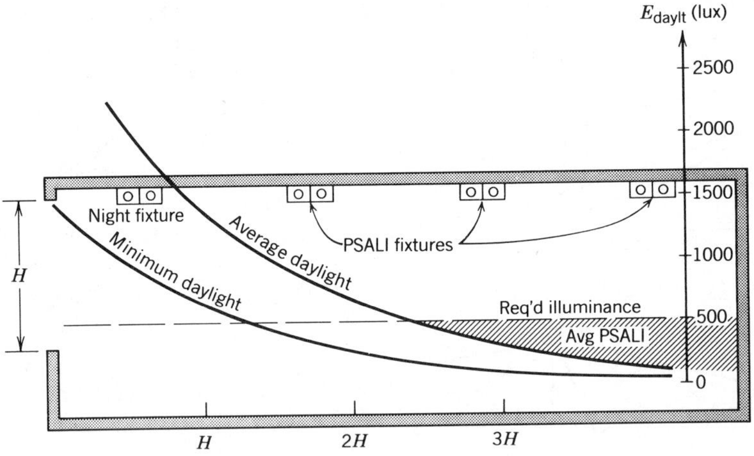

Fig. 19.3 *Section through a room with daylighting from one side window. The typical average daylight curve shows that for this building, to a depth of about 2½ times window height* H, *daylight provides the necessary illuminance. Beyond that, PSALI is required to maintain the required level. Even with minimum daylight conditions a depth of* H *into the room is daylighted, and PSALI lighting is required only beyond that depth. Thus the first (nearest the window) lighting fixture is used only in nighttime lighting, the second and third would be dimmed or multilevel switched, and the fourth (back wall) would be lighted during all hours of building use. Average daylight would represent the 85th percentile, that is, the level that (statistically) is maintained for 85% of the daylight hours. Setting the minimum is flexible and depends on the type of control system. It usually represents the 95% percentile, that is, the level that is maintained for 95% of daylight hours.*

5. Thus it is reasonable to design fenestration on the basis of a 50 to 60% figure, which states that a given daylight level will be maintained for 50 to 60% of the daylight hours. In the analyses that follow, it is assumed that fenestration has been established. Obviously, the procedures that will be presented can also be used to establish fenestration, by starting with rules of thumb and utilizing a modified trial-and-error procedure to improve the initial design.

5. The problem of (in)accuracy in interior lighting calculations can be divided into two: that caused by inaccurate input data, and that which is inherent in all calculations based on statistical averages. Referring to the first, we will see below that hand-type, pencil-and-paper methods, and use of sky luminance data based on season, latitude, and solar altitude may introduce an error of 25% or more, depending on the type of sky. Furthermore, the calculation procedure itself involves approximations of considerable magnitude, so that results may be 30 to 50% from actual measured levels. (That being so, the pointlessness of working to three decimal places should be obvious.) However, these *hand* calculation procedures are intended only to give the designer a "feel" for the daylight levels in the space and, more important, for the gradation of contours of daylight. Greater accuracy is possible only with computer-aided design procedures, which input more specific sky luminance data and reduce approximations.

For annual energy budget calculations, hour-by-hour daylight calculation is required, and, of course, this is done only by computer.

The second "inaccuracy" is a statistical deviation. Because of the infinite variability of weather in *temperate* climates, it is not possible to calculate the daylight condition for a specific time on a specific day, *nor is it necessary*. For this reason, the procedures shown below use single figures for whole seasons and rely on proper design of controls of both daylight and electric light, with PSALI, to furnish a well-lighted, low-energy-use interior.

We suggest that at this point the reader turn ahead to Section 21.4*e* and read through the section on daylight compensation, to understand how PSALI with proper controls will result in low energy use while maintaining the necessary lighting levels.

19.4 Characteristics of Outdoor Illumination

(a) Factors. The most prominent characteristic of daylight is its variability. Obviously the source of all daylight is the sun. The level of exterior illumination, at a particular place and time, depends on

1. Solar altitude which can be determined if latitude, date, and time of day are given. See Table 19.2.
2. Weather conditions (cloud cover, smog).
3. Effects of local terrain (natural and man made obstructions and reflections; see Fig 3.7).

The position of the sun in the sky is expressed in terms of its altitude above the horizon and its azimuth angle. The latter is defined as its horizontal position angle, measured from the *south*. Both are normally expressed in degrees (see Fig. 19.4*a*). Since the sun is due south at noon, azimuth angles are east of south before noon and west of south after noon. It is assumed that the reader is familiar with the basic astronomical phenomena governing the motion of the earth, which produce seasonal and latitudinal variations in the position of the sun. For our purpose, we will simply state the important facts, which are:

1. For all latitudes, the sun's altitude is highest in summer, lowest in winter, and in between in spring and fall (see Fig. 19.4*b* and Table 19.2). This fact, coupled with the shorter winter day, leads to the apparent rapid motion of the sun across the horizon in the winter and the apparent slow motion in the summer.

2. As a location approaches the equator (low latitude, either north or south), the sun's daily maximum altitude increases. The seasonal altitude *variation,* however, is the same for all latitudes (except at those extreme north and south latitudes where the sun is above or below the horizon for extended time periods; see Fig. 19.4*c*). This factor not only affects exterior illumination levels but also has a pronounced effect on the design and efficacy of sun-shading devices.

3. The sun's azimuth angle is entirely dictated by the time of day, since the sun by definition traverses the sky between sunrise and sunset. The principal significance of the azimuth angle is encountered when discussing building orientation, exposures, and shading angles (see Table 19.2).

The factor of cloud cover, unlike that of solar position, is predictable only statistically on the basis of extensive U.S. Weather Service observations at numerous weather stations throughout the United States. At locations other than those for which these data are available, an educated guess is necessary. Outside the United States, the designer must rely on locally available data, which are frequently meager or entirely unavailable. The third factor—that of local terrain and construction conditions that either reduce illumination by shadowing or increase it by reflection—is so particular and individual that it can be considered only on a case-by-case basis (see Fig. 3.7). Most of the calculation methods consider the effects of shadowing and obstruction and, to a lesser extent, the effects of reflections (see Sections 19.5 to 19.7).

(b) Sky Conditions. For manual calculation procedures, it is sufficient to establish four basic sky conditions. These are:

1. Completely overcast sky.
2. Clear sky, without sun (in the shade).
3. Clear sky, with sun.
4. Partly cloudy sky.

1. *Completely overcast sky.* This condition, which obtains for much of the year in northerly climates such as England, Scandinavia, and the Pacific Northwest, is also called the CIE sky, since it was adopted by the Commission Internationale de l'Eclairage (CIE) as the standard design sky for daylighting calculations (''Daylight,'' 1970). This sky as defined by the CIE has

TABLE 19.2 **Typical Solar Altitude and Azimuth Data as a Function of Date and Time of Day**

Latitude (°N)		Date	A.M.: 6 / P.M.: 6	7 / 5	8 / 4	9 / 3	10 / 2	11 / 1	Noon
			Solar Time[1]						
30	Altitude	June 21	12	24	37	50	63	75	83
		Mar.–Sept. 21	—	13	26	38	49	57	60
		Dec. 21	—	—	12	21	29	35	37
	Azimuth	June 21	111	104	99	92	84	67	0
		Mar.–Sept. 21	90	83	74	64	49	28	0
		Dec. 21	—	60	54	44	32	17	0
34	Altitude	June 21	13	25	37	50	62	74	79
		Mar.–Sept. 21	—	12	25	36	46	53	56
		Dec. 21	—	—	9	18	26	31	33
	Azimuth	June 21	110	103	95	90	78	58	0
		Mar.–Sept. 21	90	82	72	61	46	26	0
		Dec. 21	—	—	54	43	30	16	0
38	Altitude	June 21	14	26	37	49	61	71	75
		Mar.–Sept. 21	—	12	23	34	43	50	52
		Dec. 21	—	—	7	16	23	27	28
	Azimuth	June 21	109	101	90	83	70	46	0
		Mar.–Sept. 21	90	81	71	58	43	24	0
		Dec. 21	—	—	54	43	30	16	0
42	Altitude	June 21	16	26	38	49	60	68	71
		Mar.–Sept. 21	—	11	22	32	40	46	48
		Dec. 21	—	—	4	13	19	23	25
	Azimuth	June 21	108	99	89	78	63	39	0
		Mar.–Sept. 21	90	80	69	56	41	22	0
		Dec. 21	—	—	53	42	29	15	0
46	Altitude	June 21	17	27	37	48	57	65	67
		Mar.–Sept. 21	—	10	20	30	37	42	44
		Dec. 21	—	—	2	10	15	20	21
	Azimuth	June 21	107	97	88	74	58	34	0
		Mar.–Sept. 21	90	79	67	54	39	21	0
		Dec. 21	—	—	52	41	28	14	0
48	Altitude	June 21	17	27	37	47	56	63	65
		Mar.–Sept. 21	—	10	20	29	36	40	42
		Dec. 21	—	—	1	8	14	17	19
	Azimuth	June 21	106	95	85	72	55	31	0
		Mar.–Sept. 21	90	79	67	53	38	20	0
		Dec. 21	—	—	52	41	28	14	0

[1]NOTE: Solar time and clock time do not usually coincide. They are related by the equation of time. See, for instance, IESNA Recommended Practice RP-21, 1983.

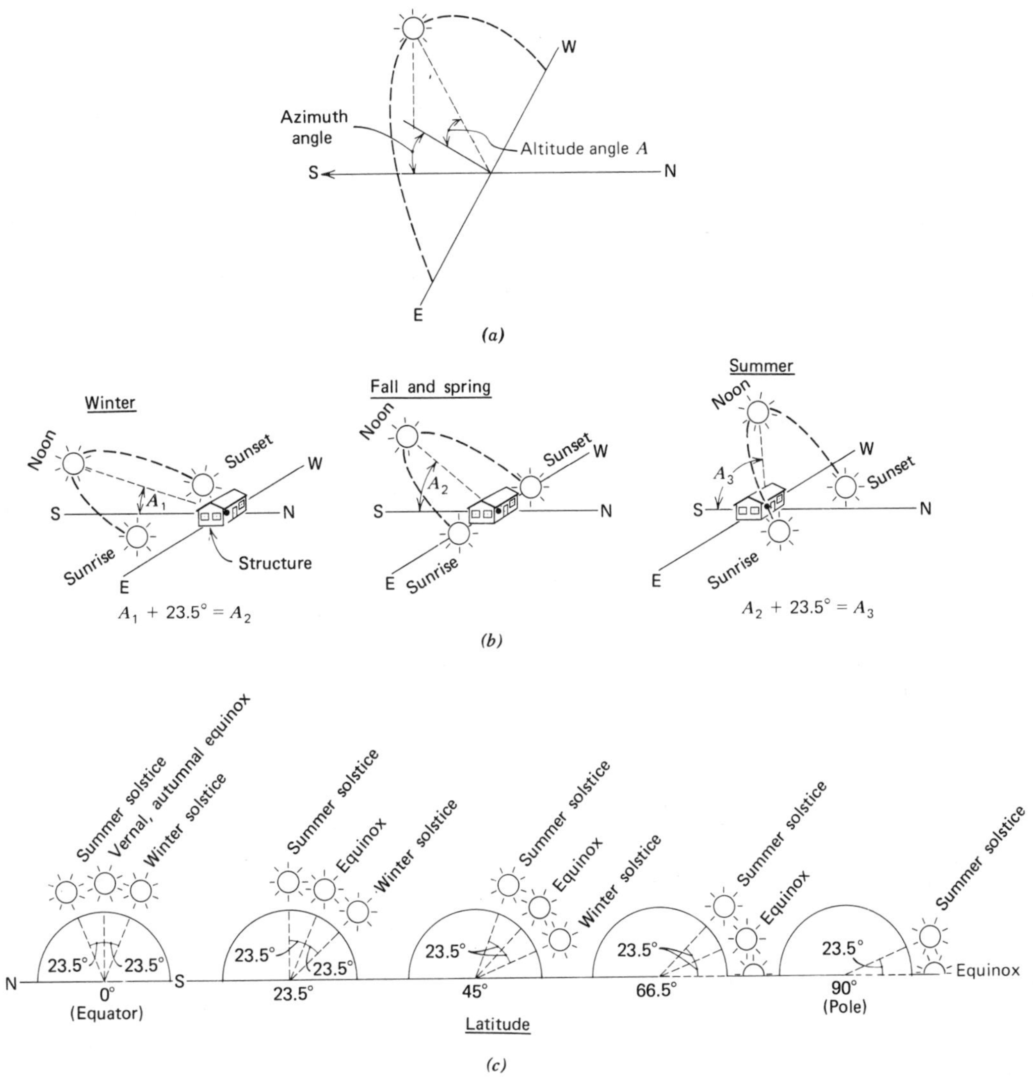

Fig. 19.4 (a) *The position of the sun is expressed in terms of vertical angle above the horizon (altitude) and horizontal angle, measured from the south (azimuth).* (b) *Approximate position of the sun in each of the seasons, at a midnorthern latitude (approximately 45°). Note that altitude angle is maximum in summer, minimum in winter, and in between in spring and fall. Note, too, the length of daylight hours; maximum in summer, minimum in winter, and in between in spring and fall.* (c) *Maximum sun altitude at various latitudes, for both solstices and equinoxes. Maximum summer sun altitude is 90° minus latitude plus 23.5°. Minimum winter sun altitude is 90° minus latitude minus 23.5°. Thus for all latitudes the yearly difference between maximum and minimum altitudes is twice 23.5°, or 47°, as shown.*

a nonuniform brightness distribution, increasing from horizon to zenith in approximately the ratio of 1 : 3. Sky luminance at any altitude angle above the horizon is defined as

$$L_A = L_z \frac{1 + 2 \sin A}{3} \qquad (19.1)$$

where L_A = luminance at $A°$ above horizon (in any direction)

L_z = luminance at zenith

Thus at the horizon, where $A = 0°$,

$$L_A = \frac{L_z}{3}$$

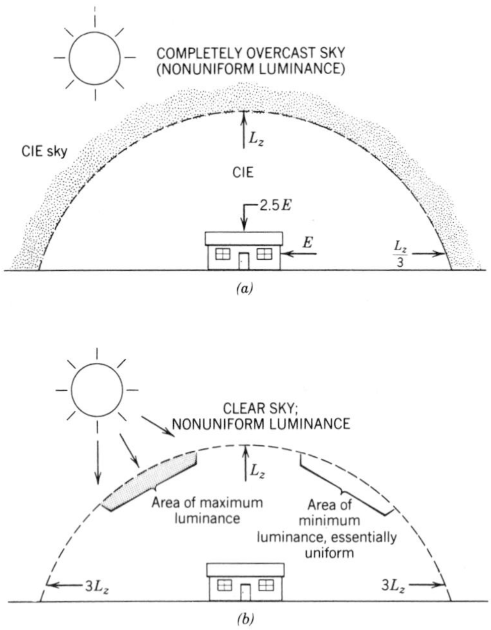

Fig. 19.5 (a) *The completely overcast sky has a zenith luminance* L_z*, which is three times the horizon luminance, according to the CIE formulation. With such a sky, illumination on unobstructed exterior horizontal surfaces* (E_H) *is about 2½ times that on a similar vertical surface* (E_V)*. This relation is also shown in Fig. 19.6.* (b) *One widely accepted formulation for clear sky luminance is simply an inversion of the overcast distribution* (a)*; horizon luminance is three times that of the zenith. Obviously, the area around the sun is brightest. The area opposite the sun is darkest and can be considered as essentially uniform at approximately 3500 cd/m² (1000 fL). Actual luminance varies from about 300 fL at the center to 2000 fL at the sides, averaging 1000 fL over the whole area.*

as stated above. The illuminance (lux, fc) produced by this distribution on unobstructed exterior horizontal and vertical surfaces stands in the approximate ratio of 2.5 : 1 (see Fig. 19.5*a*). This results from an integration of equation 19.1 over the whole sky.

There is, however, agreement among all sources that with an overcast sky, exterior horizontal illuminance varies directly with the sun's altitude, *irrespective of azimuth,* despite the overcast. Various formulations for the relation have been put forward. One formulation that gives good agreement with observations is given by Krochman (1963):

$$E_H = 300 + 21{,}000 \sin A \text{ (lux)} \qquad (19.2)$$

where E_H is exterior horizontal illuminance and A is the solar altitude, in degrees. Solar altitude (and azimuth) for various times of day can be obtained from Table 19.2. Figure 19.6 is a plot of year-round averages for both vertical and horizontal illuminance from an overcast sky as a function of solar altitude, based on U.S. Weather Service observations.

To manually perform accurate calculations of indoor daylight levels using a nonuniform sky luminance distribution, one requires some sort of sky protractor with which to measure sky angles from the interior observation point. With these angles, the sky luminance of each small patch of sky can be determined, and in turn the resultant interior daylight. Protractors are available from the Building Research Station, London, which simplify the work considerably, but it is still a time-consuming, laborious procedure. Today, the ready availability of computer programs that calculate interior daylight for all types of sky conditions makes such procedures substantially obsolete.

It is interesting to compare exterior horizontal illumination obtained from the two sources given: Krochman's formula (equation 19.2), and the observation based data of Fig. 19.6, for a few typical conditions. Solar altitude is obtained from Table 19.2.

Latitude: 38°N

Solar Time: 10 A.M.

Dates: Dec 21, March/Sept 21, June 21

	Eq. 19.2	Fig. 19.6
Dec. 21	790 fc	800 fc
Mar/Sept 21	1359 fc	1480 fc
June 21	1735 fc	2150 fc

The degree of agreement is generally satisfactory and either source will yield suitable results.

2. *Clear sky, horizontal illumination.* Exterior horizontal illumination on a cloudless day

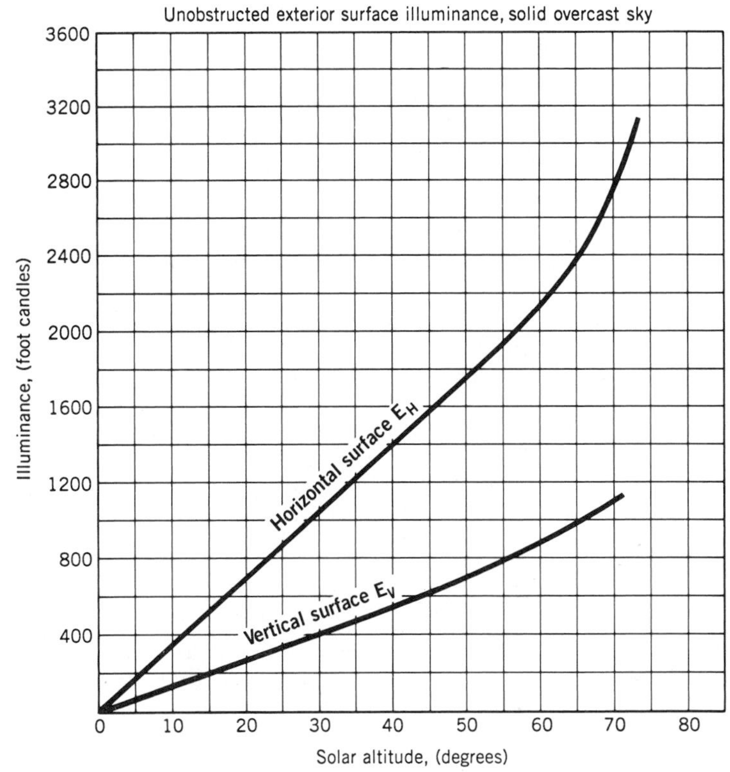

Fig. 19.6 *Curves giving unobstructed exterior surface illumination from an overcast sky, directly. To obtain illuminance in lux, multiply footcandles by 10.76 (or 10 as an approximation). The ratio between horizontal and vertical surface illumination is 2.5 : 1, as shown in Fig. 19.5*a. *(Data based on U.S. Weather Service observations; courtesy of Libbey-Owens-Ford.)*

consists of two components: diffuse illumination from the entire sky plus the much larger component of direct sunlight. As with overcast sky, various empirical formulas for both components have been proposed, and here too, all sources agree that the total illumination, diffuse plus direct, varies directly with solar altitude. [The most widely accepted empirical formulation for clear sky luminance distribution is that of Kittler (see *IES Handbook*, 1981, Ref. Volume, pp. 9–84)]. Figure 19.7 gives values for both components of exterior horizontal illumination, based on observations. The *sky only* values are used to determine shaded skylight illumination, or daylong ground illumination outside a shaded window, that is, a north-facing window, or an east/west window when the sun is on the opposite side of the building. In determining ground illumination, the values given in Fig. 19.7 must be reduced somewhat, as they represent unobstructed horizontal illumination, whereas the area outside a building window is partially obstructed from sky light by the building. If the building is so large that the ground outside the shaded window effectively receives diffuse radiation only from the half of the sky away from the sun, an average figure for E_H of 1000 fc can be used. This is because the luminance of the half of sky away from the sun varies from a minimum of approximately 300 fL for the deep blue patch directly opposite the sun to about 2000 fL at the sides, giving an average half-sky luminance of about 1000 fL. This, in turn, will give a horizontal illuminance E_H, diffuse, of about 1000 fc (see Fig. 19.5*b*).

Figure 19.7 also gives horizontal illuminance from the sun only, as a function of solar altitude. This value, when combined with the proper portion of diffuse illumination as discussed above, is useful in determining ground illumination outside a sunny building exposure, or illumination on an unshaded skylight. The light incident on an external reflector or light shelf at a window can also be determined from these figures.

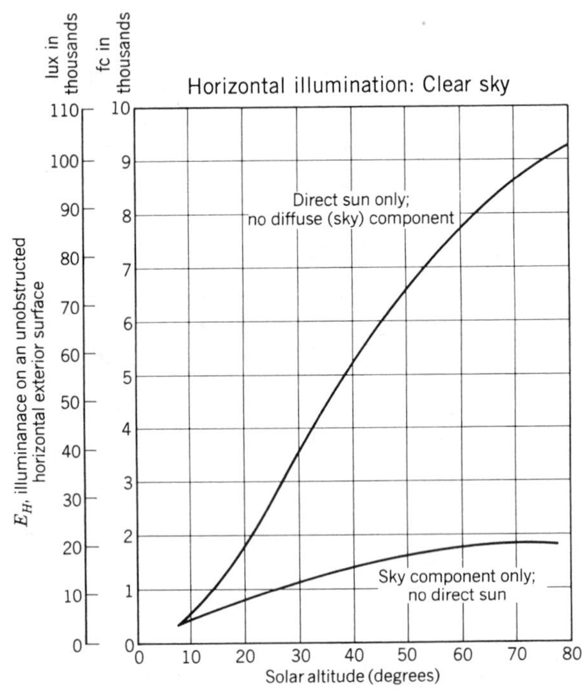

Fig. 19.7 *Components of the exterior horizontal illumination on an unobstructed surface, from a clear sky, as a function of solar altitude. Total illumination* E_H *is the sum of the two components. [From data in Rennhackkamp (1967).]*

3. *Clear, sky, vertical surface illumination.* Since most daylighting is accomplished via vertical fenestration, vertical surface illumination is the major component of interior daylight. It is also important for determining the daylight contribution of vertical elements in skylights. There is no simple relationship between horizontal and vertical illumination from a clear sky as there was for an overcast sky because the illumination on a vertical surface depends on solar azimuth as well as altitude. More specifically, it depends on the *bearing angle,* which is defined as the horizontal angle between vertical planes containing the sun and a perpendicular to the vertical surface in question. A bearing angle of 0° indicates that the sun plane is perpendicular to the vertical surface (see Fig. 19.8). As with E_H, E_V (vertical illuminance) is divided into two components: sky only, and direct sun only, which are plotted in Fig. 19.9*b–d* as a function of solar altitude and bearing angle. The *sky only* component is effectively for the half-sky, since a vertical surface can be exposed to a maximum of only half of the full sky.

The data are used as follows: Surfaces exposed to the sun are illuminated by the sun component at the relevant bearing angle (Fig. 19.9*a*) plus the sky component at the bearing angle (Fig. 19.9*b*–19.9*c*). Bearing angle for the sky component is important because of the nonuniform luminance of the clear sky—brightest near the sun and dullest opposite it. A vertical surface shaded from the sun (bearing angle >90°) is illuminated by the sky component, at the bearing angle, plus any reflected light from objects facing the window, at its bearing angle. For instance, a north-facing window at noon solar time has a bearing angle of 180°. It is illuminated by the sky component at 180° plus reflected light from south facing buildings *at 0° bearing angle.* These buildings (at 0°) are illuminated by both sun and sky (see Fig. 19.22).

The data presented in the curves of Fig. 19.9 are based on year-long averages of observations taken by the U.S. Weather Service, modified by later observations of Rennhackkamp (1967). When more accurate observation data are available, they obviously should be used in preference. In many areas of the world, data on solar radiation are available, but illumination data are not (see Treado and Gillette, 1983). To translate these radiation data to illumination, an "effi-

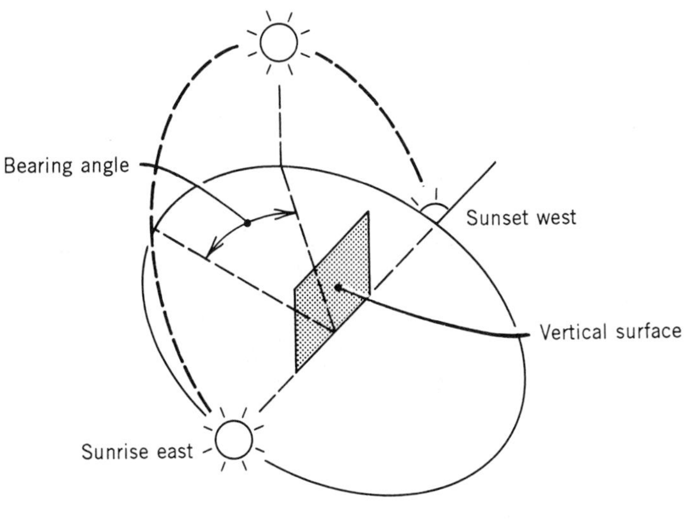

Fig. 19.8 Diagram showing the geometry of a vertical surface in relation to the sun position. The angle between a vertical plane perpendicular to the vertical surface (window) and a vertical plane containing the sun is defined as the bearing angle. (Other sources refer to this angle as the window-to-sun azimuth angle *or simply* azimuth angle.*)*

ciency'' figure is required for solar energy, in units of lumens per watt of received radiation. Reliable average figures are:

Total radiation, sun and sky = 119 lm/W

Diffuse (sky) radiation only) $\simeq$ 14 lm/W

Direct solar radiation only $\simeq$ 105 lm/W

Therefore, a surface receiving a total radiation of 750 W/m^2 is illuminated to

$$E = 750 \text{ W } (119 \text{ lm/W}) = 89{,}250 \text{ lux or } 8295 \text{ fc}$$

Obviously, the angle between the sun and the surface affects the amount of received radiation.

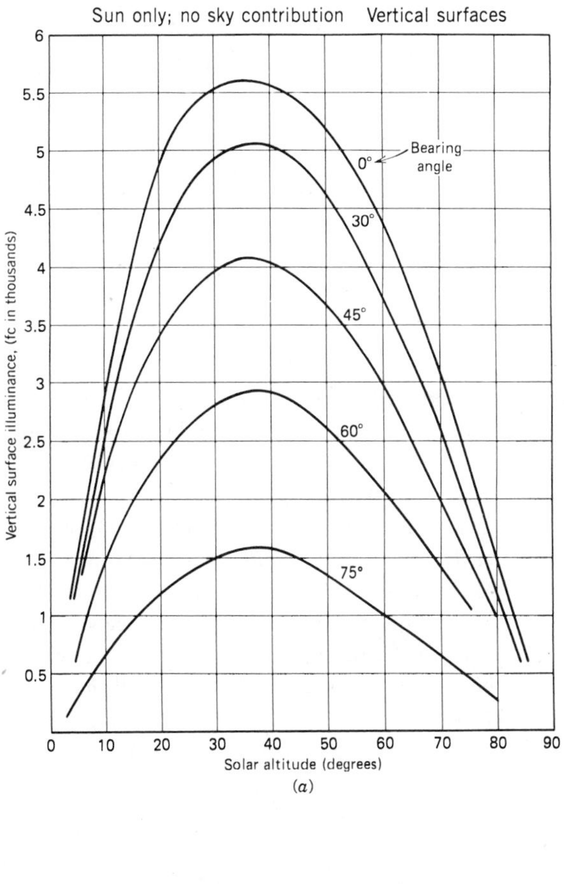

Fig. 19.9 (a) *Vertical surface illuminance, year-long average, sun only, no sky contribution.* (b, c) *Vertical surface illumination from a clear sky with no sun contribution for various seasons of the year. (Courtesy of Libbey, Owens, Ford.)*

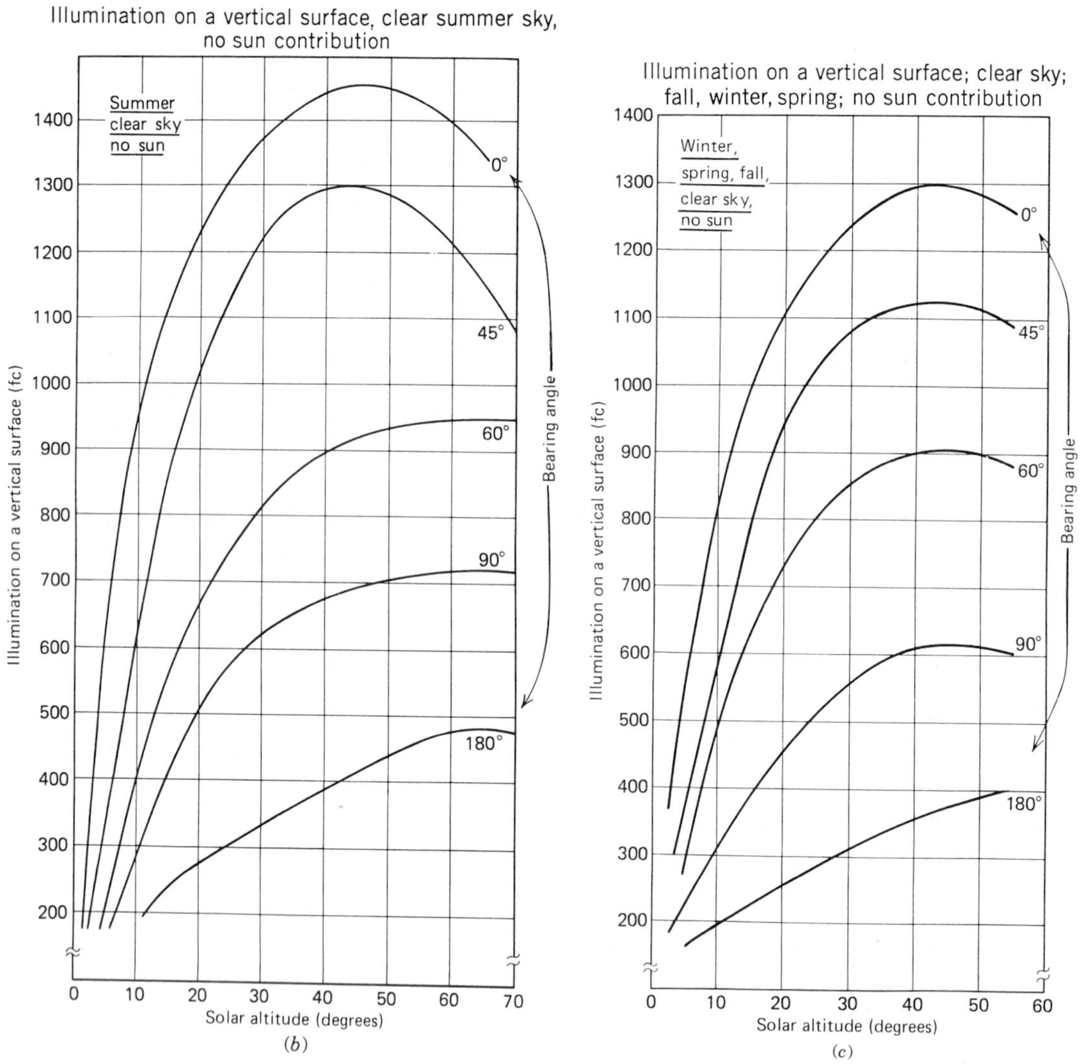

Fig. 19.9 *(continued)*

Therefore, if only horizontal or perpendicular-to-the-sun radiation data are available, somewhat complex trigonometric calculations will give the radiation on a surface at any bearing angle. The relevant equations are found in many sources; see, for instance, Wieder (1982, pp. 35–37).

4. *Partly cloudy sky*. There is no way of expressing mathematically the sky luminance of a partly cloudy sky because of its infinite variability. However, statistical data on cloud cover are available from observations at many weather stations, and these data should be used in computer-calculated, hour-by-hour energy programs. For the purpose of illumination, it is important to note that the illumination from a partly cloudy sky is *higher* than that from a clear sky by 10 to 15%, because of the additional reflected sunlight from cloud edges. Several attempts have been made recently to account for this type of sky—including a formulation (Nakamura and Oki, 1983) for an average intermediate sky between clear and solid overcast—

in terms of the effect on daylight factor within the room. As we will see in the next section, the concept of daylight factor was originally intended for overcast sky only, but, because of its great usefulness, it has recently been extended to clear sky (Bryan, 1980) and partly cloudy sky (Nakamura and Oki, 1983).

In the following sections several interior daylight analysis procedures will be presented. We specify "analysis" rather than design, because unlike the standard design procedure, we start with a daylight inlet (window, skylight, clerestory) and determine the daylight produced.

As discussed in detail in Section 19.3, hand-calculation methods (including small programmable calculators) address only maximum, minimum, and (possibly) average conditions. Microcomputer-aided daylight analyses can produce much more detailed and accurate results in both numerical and graphical form (two- and three-dimensional contours). These analysis techniques will also be demonstrated. It is our feeling that interior daylight contours (isolux lines) are frequently more meaningful than isolated individual absolute numbers, and for this reason we will also review "hand" graphic methods that produce such contours. Finally, because lighting design has as its final result a visual image, we will briefly discuss the use of physical models, which are particularly appropriate to the study of daylight.

19.5 Concept and Characteristics of Daylight Factor

Daylight factor (DF) is defined as

$$\mathrm{DF} = \frac{\text{indoor illuminance at a given point}}{\text{unobstructed exterior horizontal illuminance } (E_H)} \qquad (19.3)$$

expressed as a percentage. The concept is obviously applicable only where E_H can be calculated or measured, that is, where the sky luminance distribution is known or assumed. Thus it can be used with the CIE-defined overcast sky (equation 19.1) and clear sky whose distribution is fixed for the purpose of calculation. *Direct sun is excluded.* DF cannot readily be used for skies with constantly changing luminance (partly cloudy), since under such conditions the DF at a point also varies continuously, making the concept useless as a calculation tool for absolute daylight values. However, even with such a sky, plots of DF contours will show the *relative* daylight levels in an interior space. This is of great value to the lighting design in appreciating the patterns of brightness (luminance) within a room. As will become clear from an explanation of its components, a daylight factor associated with one type of sky luminance distribution cannot be used directly with another sky, without introducing considerable inaccuracy.

The reason for the primary reliance on overcast sky conditions is historical. The method was developed in England (see Hopkinson et al., 1966), where the solid overcast sky condition is predominant for much of the year. Considerable additional pioneer work on daylighting was done in Scandinavia, where the same sky condition prevails. As a result, the overcast sky became the basis for DF recommendations and remains so today. The validity of the DF concept as a method for expressing interior daylight conditions, both absolute and relative, should be obvious. With a given sky luminance distribution (which varies with solar altitude), variation in daylight *inside* will correspond exactly to variations *outside* for a given sky condition i.e., the ratio (DF) remains the same. This assumes minimal effect from obstructions and ground reflections, as is indeed the case with a dull overcast sky. Then, as explained in Section 19.3, the phenomenon of lightness constancy and Weber's law of contrast constancy will smooth out the variations, making the *apparent* brightness due to daylight appear fairly uniform. Thus the daylight factor method permits study of interior daylight distribution for varying fenestration, architectural arrangement, and building orientation.

The overcast sky in the DF method, as stated, represents minimum exterior illumination since the sky is frequently brighter. Hence actual interior daylight levels will frequently exceed the design minima. Since the overcast sky is *not* a glare source, the large window areas utilized in overcast sky design locales are not a glare source even when exterior levels are considerably above the design minima. However, such windows *are* a source of severe glare in clear sky conditions.

It is apparent that the DF varies within an interior space (see Figs. 19.12 and 19.13). This variation, however, is constant for a given architectural configuration. Therefore, knowing the DF variation for a given space and the exterior illumination as derived from sky luminance data, the actual interior daylight levels throughout the space can easily be calculated. Going one step further, minimum DF levels (as well as minimum artificial illumination levels) can be *specified* for different occupancies. These assume a minimum average daylight figure in a given geographical location, and the architecture is then designed to meet these requirements. Thus a design based on a combination of minimum exterior brightness and minimum DF requirements will result in sufficient daylight for a large part of the working day under almost all exterior conditions.

The daylight at any point within an enclosed space is comprised of three components (see Fig. 19.10):

1. Sky component (SC).
2. Externally reflected component (ERC).
3. Internally reflected components (IRC_1 + IRC_2).

The *sky component* (SC) is that portion of the total daylight at a point, which is received directly from the area of the sky visible *through* the window. Since the sky component is *received* light, it takes into account light reduction due to window obstructions (mullions, etc.) and losses in transmission; that is,

SC = incident skylight − window losses

The externally reflected component (ERC) is light reflected from exterior obstructions onto the point under consideration. This *does not include ground-reflected light*. ERC is of significance only in built-up areas (where there are structures opposite the window) and can be estimated as the portion of the sky component for that area of obstructed sky, reduced by the reflectance factor (RF) of the obstruction: that is,

ERC = sky component × RF (of obstruction)

Thus, if 25% of the sky is obstructed by a building with a 20% RF, we have

$$\text{ERC} = \text{SC} \times 0.25 \times 0.20$$

or

ERC = 5% of SC, to be added to the remaining 75% SC

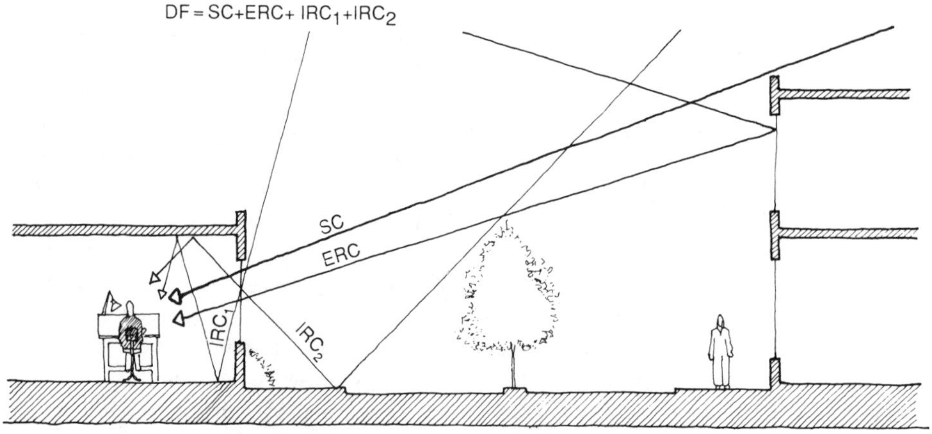

Fig. 19.10 *Total daylight factor (DF) is composed of SC (sky component), ERC (externally reflected component), and IRC (internally reflected component). The latter, in turn, is subdivided into reflected skylight and reflected groundlight. Note that surfaces deep in the room are lighted in large measure with rereflected light.*

TABLE 19.3 **Effect of Wall Reflectance Factor on Proportion of Internally Reflected Component (IRC) in Daylight Factor**

Distance from Window, Feet	*30% Wall Reflectance*		*60% Wall Reflectance*	
	Total DF	$\frac{IRC}{DF}$ %	*Total DF*	$\frac{IRC}{DF}$ %
0	30	1	31	3.5
5	16	1.9	17	6.5
10	5.5	5.5	6.3	16.9
15	2.1	14.3	2.9	37.9
20	1.3	23	2.1	52.4

ROOM DATA:
Room 24 × 28 ft. 70% ceiling RF
Window on 28-ft wall—one side only 20% floor RF
Window area = 20% of floor area

The *internally reflected component* (IRC) is the light received at the point under consideration that has been reflected from interior and exterior surfaces. Nevertheless, since IRC_2 is generally small, IRC ≅ IRC_1 (see Fig. 19.10). IRC is therefore entirely dependent on surface reflectances and on the amount of window glazing and becomes a large portion of DF deep within the interior space (see Table 19.3 and Fig. 19.11). IRC is normally calculated using publishes inter-

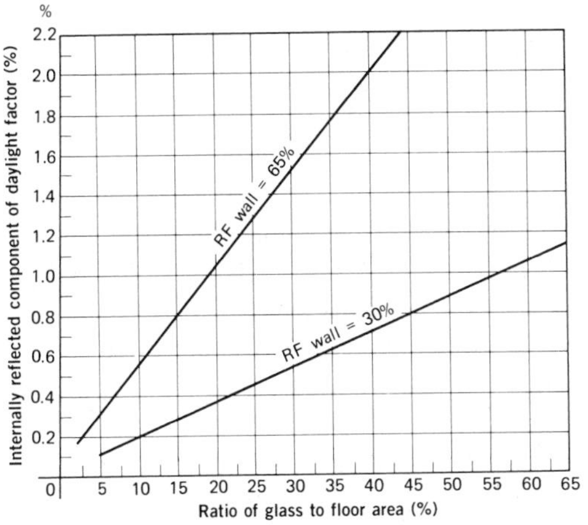

Fig. 19.11 Plot of internally reflected component (IRC) of daylight factor (DF) as a function of amount of glazing, expressed as DF, i.e., as a percentage of exterior illuminance. As expected, the effect of lighter wall finish becomes more pronounced as the fenestration area increases.

reflectance tables, as direct calculation is extremely complex. In summary, the DF then is the sum of the three components:

$$DF = SC + ERC + IRC$$

all calculated individually for each location being considered.

Typical curves for both horizontal and vertical daylight factor for a room with single sidelighting (windows on one side) are shown in Fig. 19.12. Note that DF_V is larger than DF_H, reversing the condition outside (see Fig. 19.5*a*). This is obviously due to the light coming from only the side. The reverse would be true for a toplighted space.

These curves are produced by a longhand daylight-protractor-aided technique (Building Research Station, London). The three components—SC, IRC, and ERC—are calculated separately and added to produce the total daylight factor. Any change in parameters, such as window dimensions or height above the working plane, ceiling height, surface reflectance, ground reflection, and obstructions, alters these curves and requires recalculation and replotting. Obviously, exact calculation of even a few variants in a space is a tedious, time-consuming procedure when done point by point, longhand. For this reason several alternative approaches are available:

1. Use of a computer program (see Section 19.9).

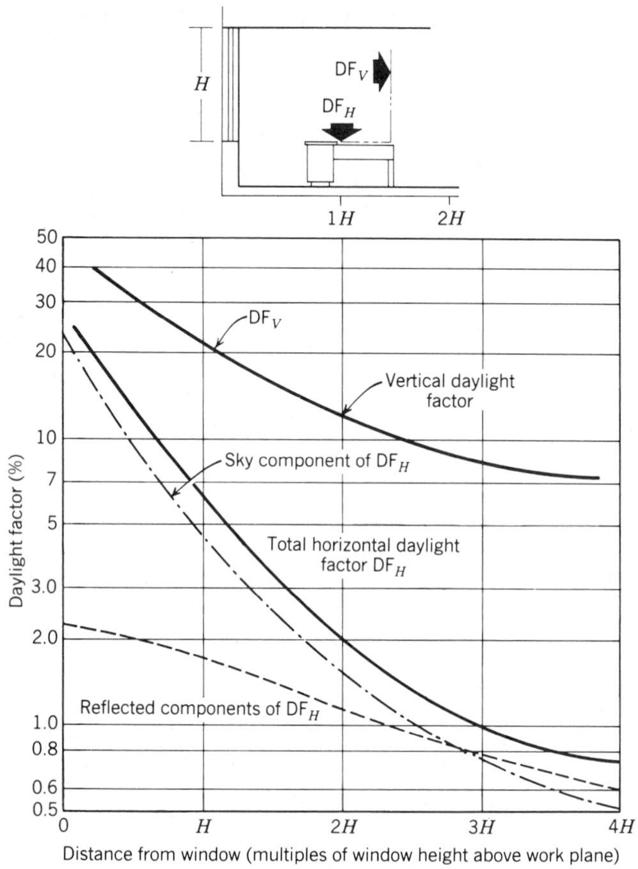

Fig. 19.12 Typical daylight factors curves for horizontal (DF_H) and vertical (DF_V) illumination for a room with large windows on one side only. Note that the sky component represents almost the entire daylight factor near the window, but reduces its proportion at greater depths. There, interreflected light constitutes 50% of the available daylight. See also Fig. 19.25.

2. Use of simplifications, such as standard curves, tabular data, or the CIE method described in Section 19.6.
3. Use of a library of graphic light distribution plots with varying parameters (see Section 19.7).
4. Use of a less laborious manual calculation procedure. One such technique, known commonly as the lumen method, or the IES method, is demonstrated in Section 19.8.

Space limitation prevents us from presenting complete data on each method. This is especially true with tabular data (item 2) and the library of graphic plots. We will therefore only survey those, while demonstrating in some detail the CIE and IES sidelighting methods and computer procedures. All will use the same example, for comparative purposes.

All the methods are primarily applicable to analysis rather than design, that is, calculating daylight, given the openings, not vice versa. The architectural designer, however, must establish an initial design, and this is done either intuitively or, lacking experience, by following some rules of thumb. A few such rules are given in Table 5.3 for overcast sky. Despite the fact the DF is a ratio and not an absolute value such as lux, it is applied as an absolute. That is, recommendations are made of minimum daylight factor for specific seeing tasks. Recommended minimum levels of daylight factor vary, but those listed in Table 5.2 are average.

To demonstrate the value of starting-point DF design figures such as those given in Table 5.2, we will translate them into actual illuminances in lux and compare the results to the recommendations in Tables 18.3 to 18.6. We will select two cities in the continental United States

TABLE 19.4 **Horizontal Illuminances (E_H) from Overcast Sky, at Selected Times, in Columbus, Ohio, and Seattle, Washington, from Overcast Sky, Corresponding to Recommended Daylight Factor**

Location	*10 A.M. Solar Altitude*[a]	*Horiz. Illum. E_H*[b] *(fc/lux)*	*DF Recomm.*[c]	*DF Recom. (lux)*	*Illuminance Recommendation*[d] *(lux)*
Columbus, Ohio	21 June 60°	2100/22,500	1.5–2.5%	338–563	
			2.5–4%	563–900	300–500
			4–8%	900–1800	
40° N latitude	21 Mar./Sept. 41%	1400/15,5000	1.5–2.5%	225–375	
			2.5–4%	375–600	500–750
			4–8%	600–1200	
	21 Dec. 21°	700/7500	1.5–2.5%	113–188	
			2.5–4%	188–300	750–1000
			4–8%	300–600	
Seattle, Washington	21 June 56°	1950/21,000	1.5–2.5%	315–525	
			2.5–4%	525–840	300–500
			4–8%	840–1680	
48° N latitude	21 Mar./Sept. 36°	1220/13,000	1.5–2.5%	195–325	
			2.5–4%	325–520	500–750
			4–8%	520–1040	
	21 Dec. 14°	500/5400	1.5–2.5%	81–135	
			2.5–4%	135–216	750–1000
			4–8%	216–532	

[a]From Table 19.2.
[b]From Fig. 19.6.
[c]From Table 5.2.
[d]From Tables 18.3 to 18.6.

that have overcast skies for appreciable portions of the year—Columbus and Seattle. Their approximate latitudes are 40°N and 48°N, respectively. Using Table 19.2 for solar altitude, Fig. 19.6 for overcast sky horizontal illuminance, and taking 10 A.M./3 P.M. for the average daily illuminance figures, we have tabulated the results in Table 19.4.

Note that only for midwinter (i.e., 21 December) are the actual values consistently below the recommendations of Tables 18.3 to 18.6. For most of the year daylight provides all the light necessary for the tasks at hand if the recommendations of Table 5.2 are followed. Obviously, then, where exterior illuminance is indeed as low as 5000 to 7000 lux, PSALI would be required for all areas beyond H feet, that is, one window height, from the window (see Fig. 19.12). In addition to the recommendations of Table 5.2, the designer should also see to it that the ratio between minimum and average DF, which relates to the contrast ratios in the space, should not be less than 30%; that is,

$$\frac{DF_{min}}{DF_{avg}} \geqslant 0.3$$

Finally, minimum DF in any portion of a space should not drop below 0.5%, which is sufficient for circulation.

19.6 Daylight Analysis: CIE Method

(a) General. This method was the result of a search for a simple, rapid, straightforward, and fairly accurate daylight calculation method that would yield reliable results without the time-consuming constructions and calculations

ILLUMINATION

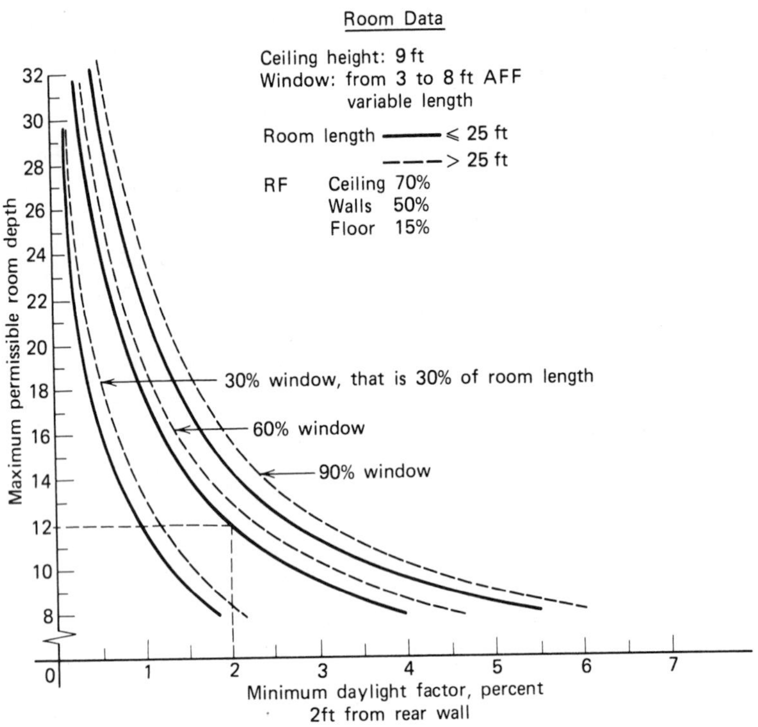

Fig. 19.13 Maximum room depth required to maintain minimum DF is proportional to window size. Thus for a room less than 25 ft long with a 5-ft-high window for 60% of the room's length, depth cannot exceed 12 ft if 2% DF is to be maintained at a point 2 ft from the rear wall. [From Daylight Design Diagrams *(1963).]*

necessitated by other manual methods. After a study of considerable length and intensity, the CIE adapted and adopted a system developed in Australia by Dr. A. Dresler (*Daylight Design Diagrams,* 1963). The current CIE recommendations were published in 1970 (*Daylight,* 1970).

This system is based on the daylight factor described above as applied to the *standard overcast CIE sky*. Dr. Dresler developed a set of more than 100 curves covering rooms of varying proportions and fenestration. One such typical curve is shown in Fig. 19.13. The curves relate minimum DF (at a point 2 ft from the wall opposite the window) to the maximum permissible room depth, for given reflectances and a standard window design, thus establishing the room's properties. Depth, or width, is the dimension at right angles to the window wall.

The curves were calculated using the techniques developed in England, as described in the preceding section. The published CIE recommendations make a clear, strong case for the minimalist approach in selecting sky illumination and in design. They state simply that the number of design variables is so large and daylight itself so variable, that a simple routine method can be based on only minimal conditions for a given (selected) percentile. Therefore, the diagrams will give the *lowest* level of daylight that can reliably be expected for a given percentage (percentile) of normal working hours in side-lighted rooms, and the *average* level in top-lighted spaces. Thus the designer can select the 95th percentile for minimum conditions, the 80th to 85th percentile for average conditions and the 60th percentile for maximum conditions. On the basis of these he can design maximum PSALI, estimate energy consumption, and establish the zones of lighting control.

(b) Characteristics of the Method. Advantages of the system are:

1. Consideration of obstructions, reflections, and interior reflections.
2. Use of overcast sky luminance that varies with latitude.
3. Applicability to a very wide range of side and top fenestration designs.
4. Establishment of required room proportions is architecturally more useful than solving for specific dimensions.

Limitations of the system are:

1. Inapplicable to clear-sky and sun conditions.
2. Inapplicable to other than rectangular-shaped rooms.
3. Unusable with sun-shading or high-reflectance ground.
4. Results give points of minimum, twice minimum, and four times minimum daylight only. Other points must be interpolated or extrapolated.
5. Window proportions and position in a wall are fixed.

Overall, the system accomplishes what it intended. The limitations listed are inherent in any rapid, simplified daylight calculation technique. They are listed here not as criticism, but to advise the reader of these limitations. The standard itself, which should be in the library of every daylight designer, also clearly states these limiting characteristics.

(c) Calculation Procedure. The CIE system is usable in two modes:

1. Given complete architectural data, find resultant daylight.
2. Given incomplete architectural data and required daylighting, find maximum room depth and/or other room proportions that will satisfy the daylighting requirement.

Obviously the former mode is simpler, since it leads directly to an answer. It will be demonstrated in an illustrative example below. The latter mode, since it has many answers, is more complex. That is, various combinations of fenestration and room dimensions will yield adequate daylight. For this reason, the designer should set room length (front to back) and percent fenestration of the window wall, leaving depth (width, side to side) as a variable, or alternatively, set room length and depth, with percentage of fenestration as the variable. Ceiling height is usually fixed by other considerations. See Fig. 19.14 for sketches showing room parameters. The procedure in mode 1 is as follows (the reader should work through the mode 2 procedure and apply it to a design project):

1. Express room depth in terms of window height. From Fig. 19.15 determine *design* daylight factor, that is, the daylight factor 2 ft from the wall opposite the window, on the window centerline.
2. The design daylight factor is larger than the actual service daylight factor because it is based on clear, clean glass and unobstructed horizon. Since this is rarely the case, the actual (service) daylight factor must be calculated using the correction factors of Fig. 19.16 and Table 19.5; that is,

$$DF_{service} = DF_{design} \times \text{correction factors}$$

3. Using the service daylight factor in equation 19.3

$$\text{required exterior illumination} = \frac{\text{required interior illumination}}{\text{service DF}}$$

obtain required exterior illumination.

4. From the curves in Fig. 19.17, obtain the percentage of hours between 0900 and 1700 that the required illumination is maintained. For time periods other than this, see Table 19.5*c*.
5. From Fig. 19.18 determine the room positions at which illumination is twice or four times minimum.

The curves given here for unilateral sidelighting are typical of those found in the standard (*Daylight*, 1970), which covers, in addition, bilateral sidelighting and toplighting of the three principal varieties, that is, skylights, sawtooth roofs, and monitor roof.

An example of this simple, rapid technique should make its use clear. We have selected a geographic location appropriate to this method.

ILLUMINATION

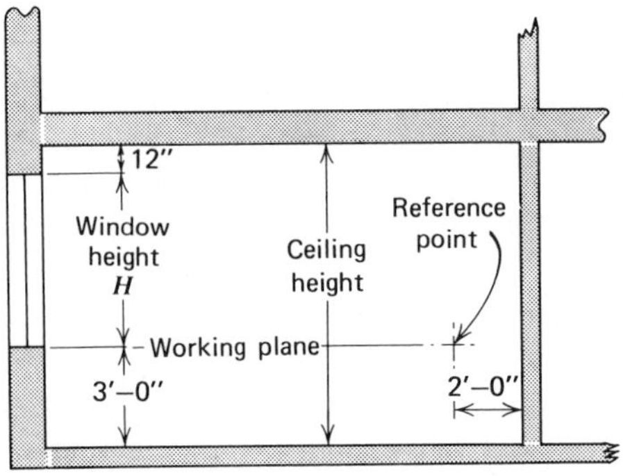

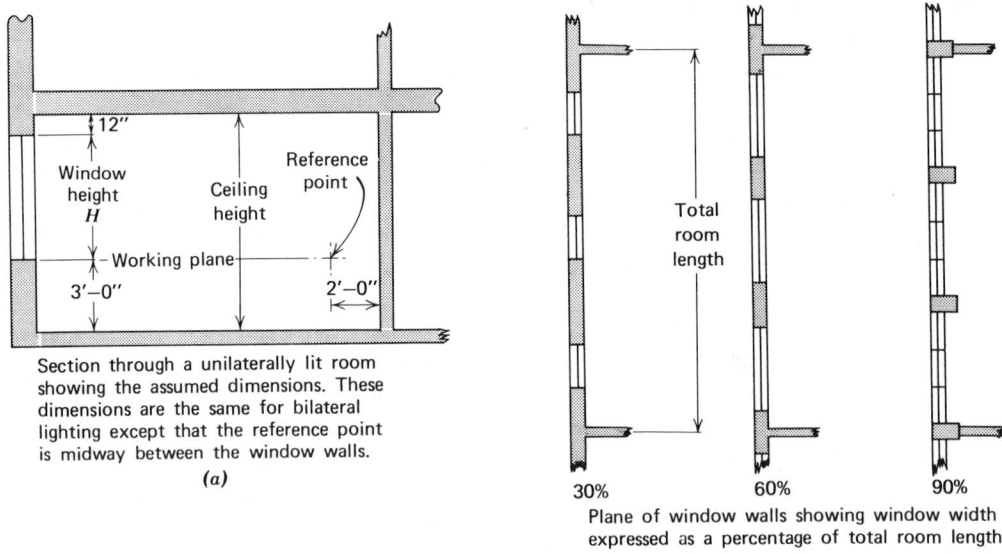

Section through a unilaterally lit room showing the assumed dimensions. These dimensions are the same for bilateral lighting except that the reference point is midway between the window walls.

(a)

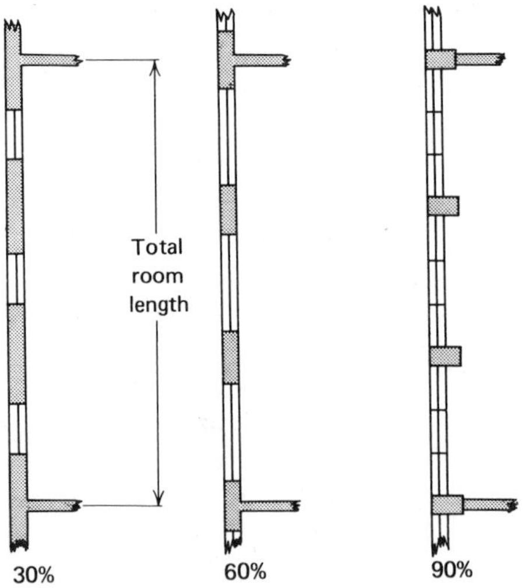

Plane of window walls showing window width expressed as a percentage of total room length.

(b)

Fig. 19.14 *Sketches indicating the parameters of the CIE calculation systems.* (a) *A vertical section through a room with dimensional data relevant to this system. Note that the sill height has been selected to coincide with a working plane at 90 cm (3 ft). Height of working plane varies between 30 and 36 in., the former being more common in American use, the latter in European use. A lower sill contributes only ground-reflected light at the working plane. Where window sill is much above the working plane (i.e., short windows, high on the wall), this design system is inapplicable.* (b) *Calculation of size (length) of windows with respect to overall room length. [From* Daylight, International Recommendations for the Calculation of Natural Daylight *(1970), Commission Internationale de l'Éclairage; reproduced with permission.]*

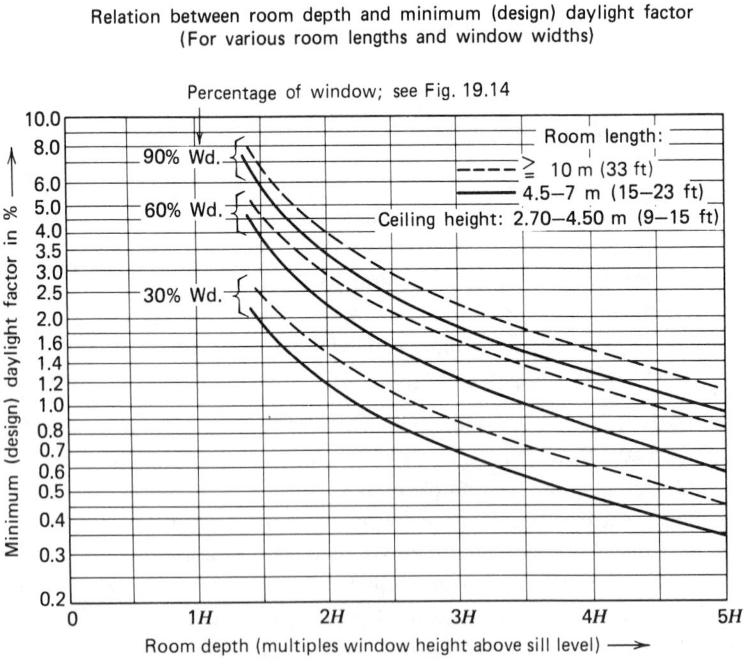

Fig. 19.15 *Basic design diagram that relates minimum daylight factor to room depth. Inasmuch as room depth is expressed in terms of window height, the curves effectively relate minimum daylight factor (2 ft from back wall) to room proportion. [From* Daylight, International Recommendations for the Calculation of Natural Daylight *(1970), Commission Internationale de l'Éclairage; reproduced with permission.]*

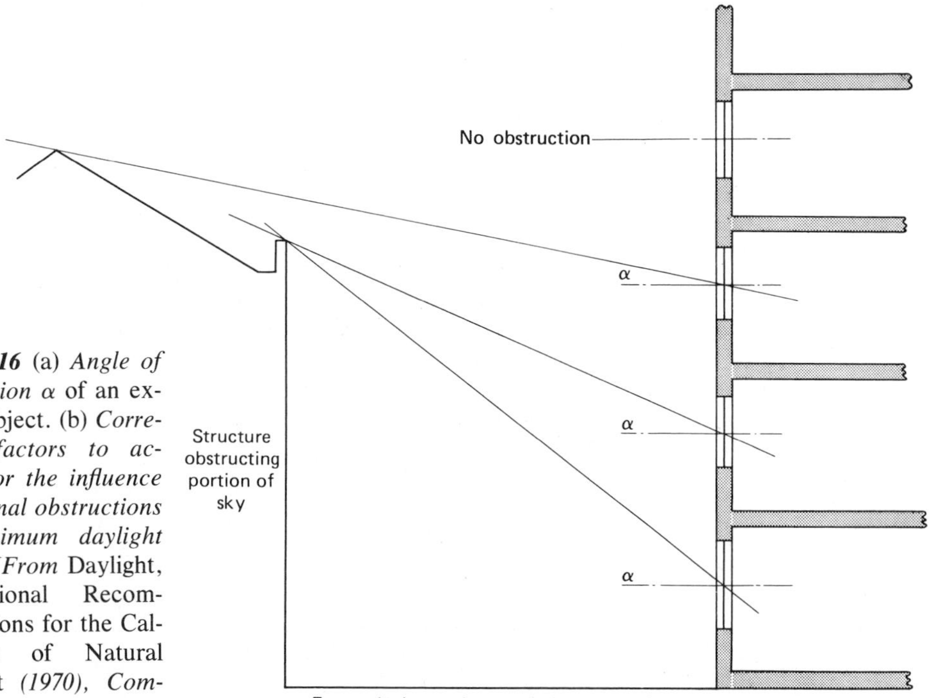

External obstructions and the angle of obstruction. The angle of obstruction α is measured from the center of the window to the outline of the opposite building.

(a)

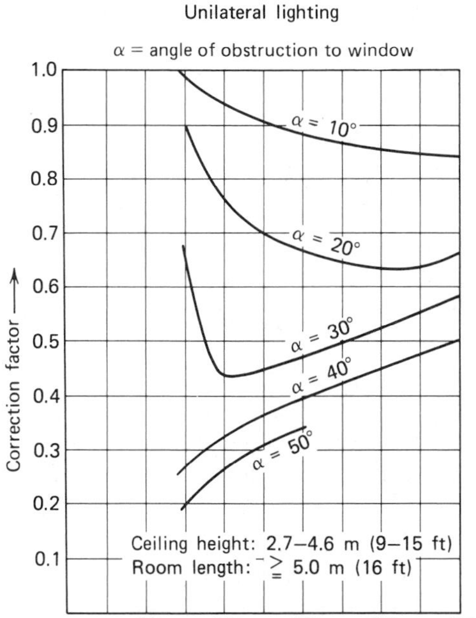

(b)

Fig. 19.16 (a) *Angle of obstruction* α of an external object. (b) *Correlation factors to account for the influence of external obstructions on minimum daylight factor. [From* Daylight, International Recommendations for the Calculation of Natural Daylight *(1970), Commission Internationale de l'Éclairage; reproduced with permission.]*

EXAMPLE 19.1. A classroom in a single-story Seattle elementary school is 25 ft long and 18 ft deep and has a 9 ft 6 in. ceiling. It receives daylight unilaterally from windows totally 18 ft in length. See room sketch, Fig. 19.19. Window glazing is wired glass having a transmittance of 80%. The school is situated in a built-up residential area with a satisfactory cleaning schedule. Determine the portion of the year during which tasks requiring a minimum illumination of 150 lux can be carried out by daylight throughout the room. Also, determine what levels can be maintained for 85% of daylight hours, and at what distances from the window. (Note that 150 lux corresponds to a DF of 2% applied to an E_H of 7500 lux.)

SOLUTION

General. The latitude of Seattle is 47.6° N. The design condition for Seattle is solid overcast sky for 85% of the hours between 0900 and 1700, producing an unobstructed exterior horizontal illuminance E_H of 7000 lux. (No azimuth for the window is specified because exterior horizontal

TABLE 19.5 **Correction Factors to Be Used in CIE Daylight Calculations**

Diffuse Transmittance of Glass	*Correction Factor*
80%	0.95
70%	0.8
60%	0.7
50%	0.6
40%	0.45
30%	0.35

(*b*) Correction Factors to Allow for Dirt Accumulation on Glass

Locality	*Class of Industry*	*Angle of Slope (Measured to the Horizontal)* 90–75°	60–45°	30–0°
Country or outer-suburban area	Clean	0.9	0.85	0.8
	Dirty	0.7	0.6	0.55
Built-up residential area	Clean	0.8	0.75	0.7
	Dirty	0.6	0.5	0.4
Built-up industrial area	Clean	0.7	0.6	0.55
	Dirty	0.5	0.35	0.25

(*c*) Percentages to Use When Figure 19.17 Curves Are Applied to Periods Other Than 09.00–1700

Curve in Figure 19.17	*95%*	*90%*	*85%*	*80%*	*70%*	*60%*
Alternative period	Percentage of alternative period					
07.00–15.00	95	90	85	80	70	60
08.00–16.00	100	100	95	85	70	60
07.00–17.00	95	85	75	65	55	45
06.00–18.00	75	70	65	60	50	40

Source: From *Daylight, International Recommendations for the Calculation of Natural Daylight,* Publication CIE No. 16 (E-3.2), 1970; Commission Internationale de l'Éclairage.

ILLUMINATION

illuminance is independent of azimuth according to the CIE formulation for overcast sky.)

1. Window height H is 5 ft (see Fig. 19.19). Room depth in terms of window height:

$$\frac{\text{18-ft depth}}{\text{5-ft window height}} = 3.6\ H$$

Window coverage:

$$\frac{3 \times 6 \text{ ft}}{25 \text{ ft}} = 72\%$$

From Fig. 19.15, for room length of 25 ft, ceiling height 9 ft 6 in., window coverage 72%:

$$\text{design DF at } 3.6\ H = 1.3$$

2. From Table 19.5 and Fig. 19.16*b*, we obtain correction factors:

Glass transmission	0.95
Glass cleanliness	0.8

Therefore,

$$\text{service DF} = 1.3 \times 0.95 \times 0.8 \cong 1.00$$

3. Required exterior illuminaton:

$$E_H = \frac{150 \text{ lux min.}}{1.00\ \text{DF}_{\text{min}}} \times 100 = 15{,}000 \text{ lux}$$

4. From Fig. 19.17 we see that with the given conditions, an illuminance of 150 lux will be maintained for less than 60% of the hours between 0900 and 1700. It was already apparent at step 2 that the desired 85th percentile of time would not be achieved, since the curves of Fig. 19.17 are based on a minimum DF of 2%, whereas our service DF is only 1.0, or half the required 2%. The level that *is* maintained for 85% of the hours is

$$E_{\text{min}} = 7000 \text{ lux} \times 1.0 \text{ DF} \simeq 70 \text{ lux}$$

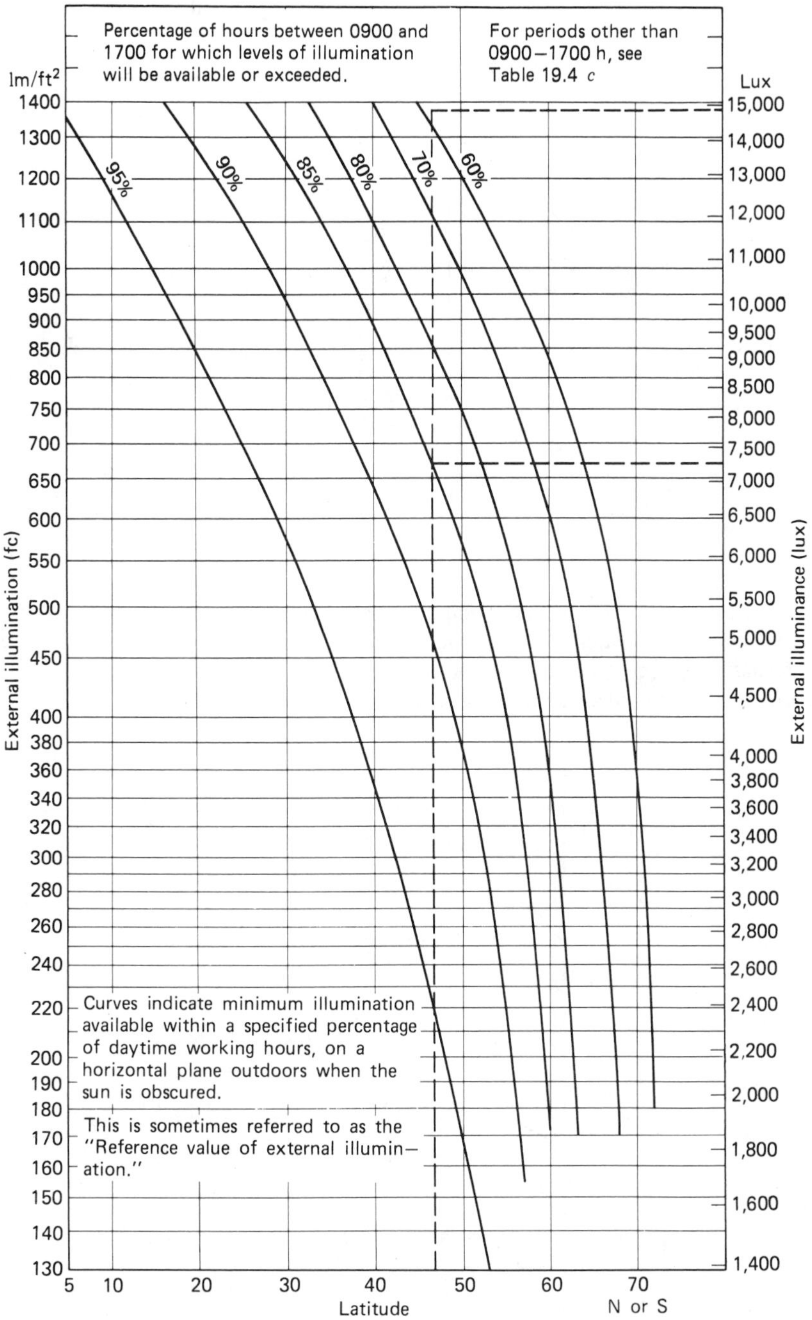

Fig. 19.17 *Minimum maintained external illumination as a function of latitude, for a given percentage of the normal working day. (From* Daylight International Recommendations for the Calculation of Natural Daylight *(1970), Commission Internationale de l'Éclairage; reproduced with permission.]*

5. To complete the picture, we have from Fig. 19.18
   (a) DF doubles at $1.8H$ corresponding to 140 lux at a depth of 9 ft (1.8×5 ft).
   (b) DF quadruples at $1.2H$, corresponding to 280 lux at a depth of 6 ft (1.2×5 ft). These values (a,b) are maintained for 85% of daylight hours.

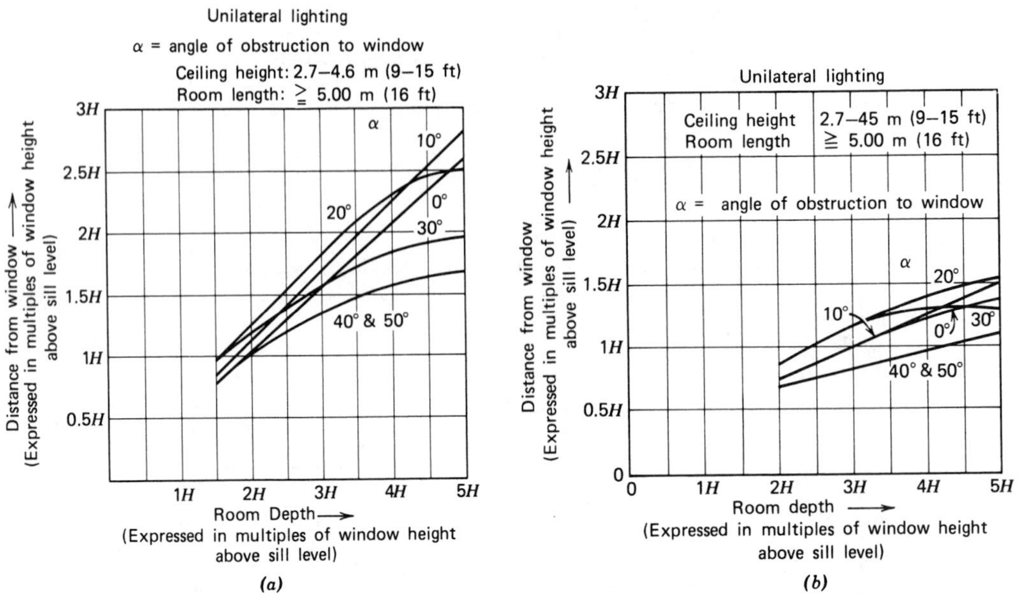

Fig. 19.18 (a) *Distance from window at which daylight factor is twice minimum daylight factor.* (b) *Distance from window at which daylight factor is four times minimum daylight factor. [From* Daylight, International Recommendations for the Calculation of Natural Daylight, *(1970), Commission Internationale de l'Éclairage; reproduced with permission.]*

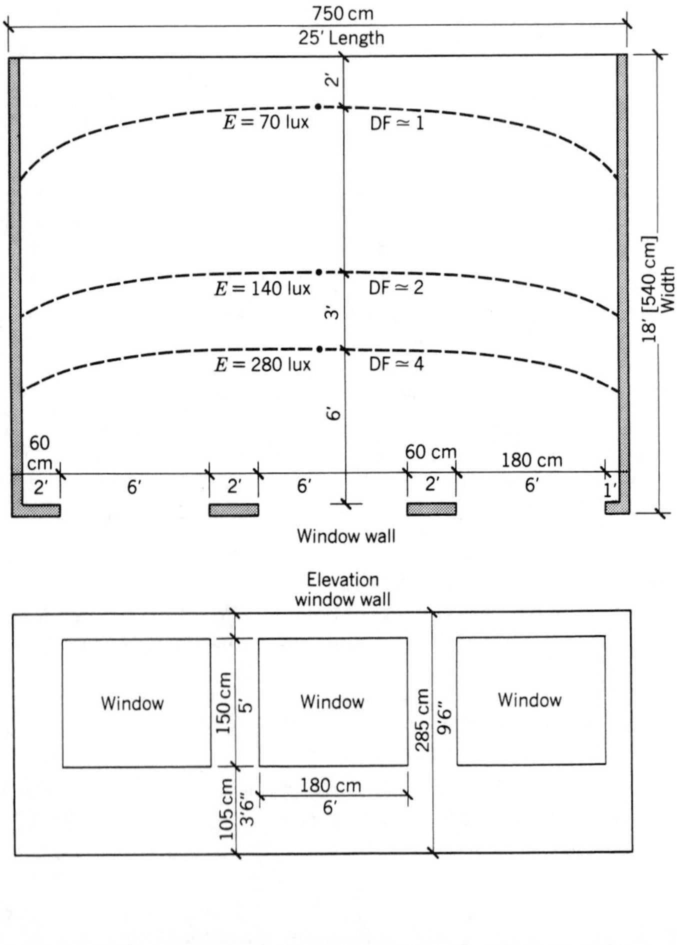

Fig. 19.19 *Plan and window wall elevation of room of Example 19.1. The three daylight contours are estimated, based on the calculated centerpoint. They represent the levels maintained for 85% of daylight hours. Levels twice as high are maintained for 60% of the daylight hours.*

6. The calculated illuminance levels are accurate only at the centerline of the window wall. By visual estimate and extrapolation, rough contours can be drawn. They are shown on Fig. 19.19, and also in the next section on Fig. 19.21*c* for the purpose of comparing results with the graphic method discussed there.

In summary then, the CIE method is simple, rapid, and flexible, but provides only limited data. It is usable in both the design and analysis mode. Its exterior illuminance data (Fig. 19.17) seem to agree well with actual measurement (7250 lux on the figure as compared to 7000 lux measured average for Seattle). If contours are roughed in as shown on Fig. 19.19, guidelines for PSALI design and lighting control strategy can be drawn.

19.7 Graphic Daylighting Design Method (GDDM)

A method that applies the principle of daylight factor to overcast sky and shows results as DF (isolux) contours within a room (rather than individual DF factors at specific points) was developed by M. S. Millet and J. R. Bedrick (1980). Its primary advantage over the CIE method is that its results are a family of DF contours that are more meaningful to a lighting designer than numerical output. Its shortcomings are that it, too, is not readily applied to clear-sky conditions and that it requires the designer to acquire a "library" of 200 or so patterns that cover most design situations. For this latter reason, only the outline of the method is presented here. We suggest that the interested reader acquire access to the catalog of patterns, so that he can apply this useful and relatively accurate method. (Application to clear sky is possible but not practical because of the size of the required pattern library.)

The authors of the method utilized a computer simulation program (UWLIGHT) to develop daylight distribution patterns in a room, resulting from either sidelights or skylights. To generalize the system, windows are identified by height to width proportion (H/W), and the position of the isolux contours on the plan is determined by the ratio of the elevation of the sill over the work plane to the window height. The latter factor is a considerable advantage over the CIE system which is restricted to a sill height at the work plane. Thus the GDDM system can account for high windows, clerestories, and other designs intended to introduce daylight deep into the space—something that the CIE method cannot do.

Figure 19.20 shows a typical isolux pattern for a window whose H/W ratio is 0.8 and whose sill is at the work plane. This particular window pattern was selected because it corresponds to the window of Example 19.1, Fig. 19.19, and will enable us to run through the same problem and compare results.

EXAMPLE 19.2. Repeat the design problem of Example 19.1 using the GDDM method.

STEP 1. Select the appropriate window pattern. Referring to Fig. 19.19, window proportion is

$$\frac{H}{W} = \frac{5}{6} = 0.83$$

We therefore select the pattern shown in Fig. 19.20 as the closest to our requirements.

STEP 2. On a plan of the room, trace the isolux pattern for each window. This is shown on Fig. 19.21*a*. $S/H = 0$ indicates that the pattern begins at the window wall, as shown. The patterns overlap, since the windows are close together. Where contours meet, the DF of the contours are added, producing points for new combined contours. The combined contours and their values, in DF, are shown in Fig. 19.21*b*.

STEP 3. In the final step the value of the combined contours is corrected to account for internally reflected components of daylight, plus light reduction due to glazing. The final contours are shown on Fig. 19.21*c*. Note that this diagram gives the designer a much more complete picture of the daylight contours than the results of the CIE method. For the purpose of comparison the three calculated DF factors of the CIE method are shown on Fig. 19.21*c*.

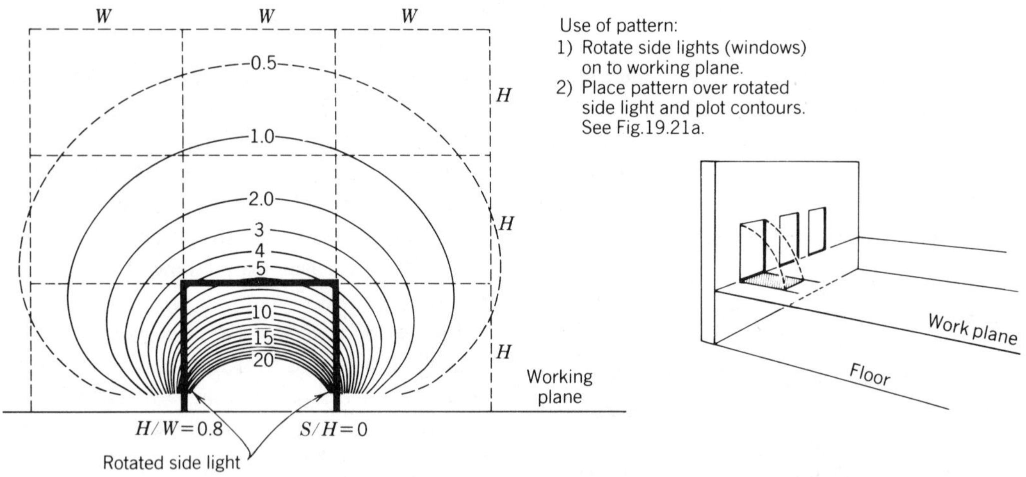

Fig. 19.20 Typical isolux contour map for a window with height-to-width ratio of 0.8, and sill at the working plane elevation. Numbers represent sky component of daylight factors, for overcast sky condition. The rectangles are the window outlines "rotated" (i.e., projected) onto the work plane. See insert. [From Millet and Bedrick (1980, p. 141).]

19.8 Daylight Design: IES/Lumen Method

A hand calculation procedure that addresses all types of sky conditions was developed in the early 1950s at Southern Methodist University (Dallas) by J. W. Griffith and Associates (Griffith, Arner, and Wenzler, 1957). The work was sponsored and published by the Libby-Owens-Ford (LOF) Glass Co. ("How to Predict. . . ," 1976). The procedure was then adopted by the IES of North America for its Recommended Practice of Daylighting in 1962. The current IES recommended practice (RP-23-1989), also frequently called the *lumen method,* is a modification of that method.

(a) General Remarks. This method is in large measure parallel to the IES method for calculating artificial lighting, in that it treats the window or toplight as a large area lighting source; essentially a transilluminated lighting fixture. It then applies coefficients to the light output of this "daylight fixture," similar in usage to the familiar coefficient of utilization and light loss factor of a lighting fixture, and arrives at an interior daylight level. For sidelighting, the lighting levels are calculated at five predetermined points in the room. For toplighting it is an average overall level, assuming proper spacing of toplights. The method is described completely in the relevant current IES publication (IES-RP23-1989). The reader is urged to obtain a copy since only selected portions will be reproduced in the discussion below, as applicable to the illustrative examples presented.

(b) Characteristics of the Method. Before beginning a description of the technique and demonstrating it with illustrative examples, a few general remarks are in order. The IES method is probably the most flexible manual technique available. It has three major characteristics for sidelighting:

1. It takes into account reflected light from the ground and adjacent structures as well as the reduction in sky light due to such structures.

2. It cannot accommodate direct sunlight, but conversely, it will readily accommodate the shading devices normally used to block direct insolation.

3. Provision is made for various types of glazing as well as common window controls such as horizontal and vertical blinds.

4. The principles of zonal cavity calculation of interior lighting (see Section 20.32) are ap-

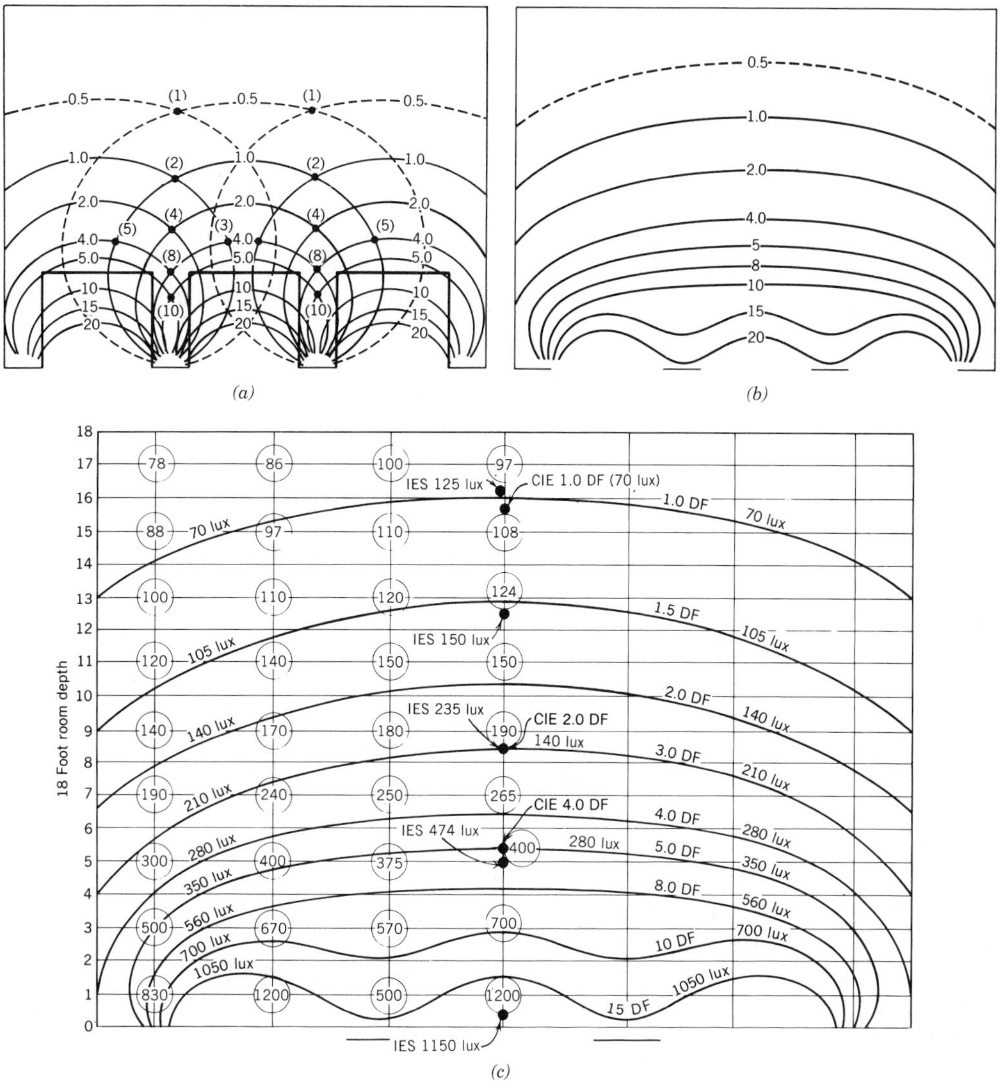

Fig. 19.21 (a) *Daylight contours for each window of Fig. 19.19 are plotted on the floor plan of the room being studied. Numbers in parentheses are combined sky components.* (b) *The isolux contours of* (a) *are combined to form new isolux contours, which represent the total sky component of daylight within the room.* (c) *The final isolux contours are calculated, including correction factors. The numbers represent daylight factors. Note the variance between these contours and the points calculated by the CIE method. The five design points calculated by the IES method, discussed in the next section, are also shown. Finally, the results of a computer calculation using the SUPERLITE program are shown in bubbles. This calculation procedure is discussed in Section 19.9. A comparison of the results of the four methods described indicates that on the room centerline for which location comparison of all methods is possible, agreement is within engineering accuracy. See the text discussion.*

plied. The window height determines the cavities (i.e., the floor cavity extends to the window sill height and the ceiling cavity from the top of the window to the ceiling). The room cavity is therefore the window height.

5. The working plane is always at the sill height of the window. Where this is decidedly not the case (a difference of up to a foot is usually negligible), such as when using a clerestory or a floor-to-ceiling window, working plane illuminance can be calculated by superposition. For instance, with a clerestory, we would subtract a working plane-to-celerestory sill height window from a working plane-to-top of clerestory window to obtain the desired result. Note, however, that a degree of inaccuracy is unavoidable when the working plane is *above* the sill.

6. Cavity reflectances are fixed (see Fig. 19.22) at 70–50–30 for ceiling, room, and floor cavities, respectively.

7. The system calculates only five points in a room, on the window centerline. As stated in reference to the three points calculated by the CIE method, this is not normally sufficient to give a picture of the interior daylight distribution.

8. The method is usable in only one mode; that is, given location and full architectural data, daylighting can be calculated. It cannot readily be used to determine room proportions, given the other data, as can the CIE method. Thus it is an *analysis* tool, not a *design* tool, except by trial and error.

(c) Calculation Procedure. The calculation procedure for sidelighting will be discussed, since it is more frequently encountered than toplighting. The reader is referred to the current *IES Recommended Practice* (RP-23-1989) for a discussion and presentation of the latter. It should present no difficulties since it is similar to, and simpler than, the sidelighting calculation procedure. In the discussion that follows we will use the same notation and terms as found in RP-23-89 *except* for bearing angle (see Fig. 19.8 and Section 19.2*b*(4)), which is there referred to as solar-window azimuth.

The method, as described in the referenced IES Recommended Practice, consists of four steps:

1. Determine the vertical illuminance on the *exterior* of the window.
2. Reducing this illuminance by the necessary factors (glazing, window controls, dirt) to obtain the transmitted illuminance.
3. Select the coefficients of utilization for the five calculation points on the basis of room dimensions and portion of sky seen by the window. The latter is determined by the ratio of vertical to horizontal illuminance at the window.
4. Calculate the illuminances at the five interior points by simple multiplication.

The terminology of the IES Recommended Practice is:

E_{xvk} = exterior vertical illuminance at window from half sky (since an unobstructed vertical window sees only half the sky)

E_{xHk} = exterior horizontal illuminance at window from full sky

E_{xhk} = exterior horizontal illuminance at window from half sky

E_{xvg} = exterior vertical illuminance at window from ground

τ = net transmittance of the window

CU_k = coefficient of utilization for sky light

CU_g = coefficient of utilization for ground light

E_i = interior illuminance at a specific reference point

In detail, then, the four-step procedure described above is as follows:

STEP 1. Determine vertical and horizontal surface illuminances from Figs. 19.6, 19.7, and 19.9. Calculate half-sky illuminance.

STEP 2. Determine net transmittance of window. This factor is the product of the glazing factor (Table 19.6), factors representing loss from glass, plus any complete uniform diffuser (*not* blinds, which are accounted for in the coefficient of utilization tables) and a light loss factor, LLF (Table 19.7), which represents the cleanliness status of the windows. This step also accounts for net glazing (i.e., gross window area less mullions, glazing bars, and so on).

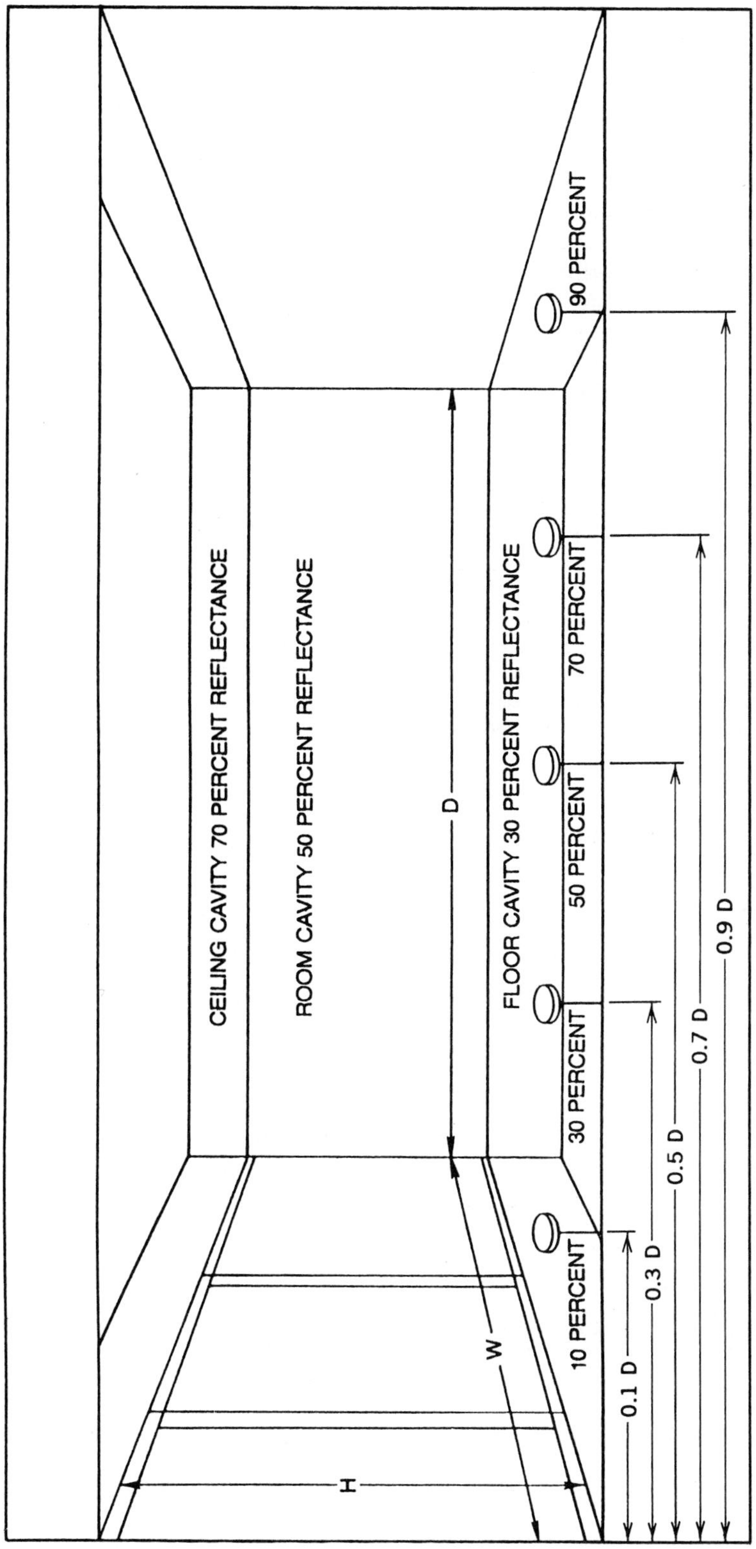

Fig. 19.22 *Standard conditions in a room for daylighting calculation: sidelighting. (From IES RP-23-1989; reprinted with permission.)*

TABLE 19.6 **Transmittance Data of Glass and Plastic Materials**

Material	*Approximate Transmittance (%)*
Polished plate/float glass	80–90
Sheet glass	85–91
Heat-absorbing plate glass	70–80
Heat-absorbing sheet glass	70–85
Tinted polished plate	40–50
Figure glass	70–90
Corrugated glass	80–85
Glass block	60–80
Clear plastic sheet	80–92
Tinted plastic sheet	9–42
Colorless patterned plastic	80–90
White translucent plastic	10–80
Glass-fiber-reinforced plastic	5–80
Double glazed—two lights clear glass	77
Tinted plus clear	37–45
Reflective glass[a]	5–60

Source: IES RP-23-1989; reprinted with permission.

[a]Includes single glass, double-glazed units, and laminated assemblies. Consult manufacturer's material for specific values.

STEP 3. Select coefficients of utilizations from Tables 19.8 to 19.13. These tables represent window without blinds, for clear and overcast sky conditions.

STEP 4. Calculate interior illuminance for each of the five reference points. The basic equation for each point is

$$E_i = \tau(E_{\text{xvk}} \times \text{CU}_k + E_{\text{xvg}} \times \text{CU}_g) \quad (19.4)$$

The ground component vertical illuminance E_{xvg} is the product of the ground reflectance (Table 19.14) and half the horizontal illuminance:

$$E_{\text{xvg}} = \text{RF}_g \times \frac{E_{\text{xHk}}}{2} \quad (19.5)$$

(For the derivation of this expression, see RP-23, page 8.) Due to space constraints, the coefficient tables for windows with venetian blinds are not presented here. For these, as well as all toplighting data, refer to IES *Recommended Practices,* RP-23-1989. To demonstrate the use of the IES method, we will apply it to the same example used in the CIE and GDDM methods, and compare results.

TABLE 19.7 **Typical Light Loss Factors for Daylighting Design**

	Light Loss Factor Glazing Position		
Location	*Vertical*	*Sloped*	*Horizontal*
Clean areas	0.9	0.8	0.7
Industrial areas	0.8	0.7	0.6
Very dirty areas	0.7	0.6	0.5

Source: IES RP-23-1989; reprinted with permission.

TABLE 19.8 **Coefficient of Utilization from Window Without Blinds; Sky Component $E_{xvk}/E_{xhk} = 0.75$**

Room Depth/ Window Height	Depth	Window Width/Window Height							
		.5	1	2	3	4	6	8	Infinite
1	10	.824	.864	.870	.873	.875	.879	.880	.883
	30	.547	.711	.777	.789	.793	.798	.799	.801
	50	.355	.526	.635	.659	.666	.669	.670	.672
	70	.243	.386	.505	.538	.548	.544	.545	.547
	90	.185	.304	.418	.451	.464	.444	.446	.447
2	10	.667	.781	.809	.812	.813	.815	.816	.824
	30	.269	.416	.519	.544	.551	.556	.557	.563
	50	.122	.204	.287	.319	.331	.339	.341	.345
	70	.068	.116	.173	.201	.214	.223	.226	.229
	90	.050	.084	.127	.151	.164	.167	.171	.172
3	10	.522	.681	.739	.746	.747	.749	.747	.766
	30	.139	.232	.320	.350	.360	.366	.364	.373
	50	.053	.092	.139	.163	.174	.183	.182	.187
	70	.031	.053	.081	.097	.106	.116	.116	.119
	90	.025	.041	.061	.074	.082	.089	.090	.092
4	10	.405	.576	.658	.670	.673	.675	.674	.707
	30	.075	.134	.197	.224	.235	.243	.243	.255
	50	.028	.050	.078	.094	.104	.112	.114	.119
	70	.018	.031	.048	.059	.065	.073	.074	.078
	90	.016	.026	.040	.048	.053	.059	.061	.064
6	10	.242	.392	.494	.516	.521	.524	.523	.588
	30	.027	.054	.086	.102	.111	.119	.120	.135
	50	.011	.023	.036	.044	.049	.055	.056	.063
	70	.009	.018	.027	.032	.035	.040	.041	.046
	90	.008	.016	.023	.028	.031	.034	.035	.040
8	10	.147	.257	.352	.380	.387	.391	.392	.482
	30	.012	.026	.043	.054	.060	.067	.070	.086
	50	.006	.013	.021	.026	.029	.033	.035	.043
	70	.005	.011	.017	.021	.023	.026	.027	.034
	90	.004	.010	.015	.019	.021	.023	.025	.030
10	10	.092	.168	.248	.275	.284	.290	.291	.395
	30	.006	.014	.026	.032	.036	.041	.044	.059
	50	.003	.008	.014	.017	.019	.022	.024	.032
	70	.003	.007	.012	.014	.016	.018	.019	.026
	90	.003	.006	.011	.013	.015	.016	.017	.024

Source: IES RP-23-1989; reprinted with permission.

TABLE 19.9 **Coefficient of Utilization from Window Without Blinds; Sky Component E_{xvk}/E_{xhk} = 1.00**

Room Depth/Window Height	Depth	Window Width/Window Height .5	1	2	3	4	6	8	Infinite
1	10	.671	.704	.711	.715	.717	.726	.726	.728
	30	.458	.595	.654	.668	.672	.682	.683	.685
	50	.313	.462	.563	.589	.598	.607	.608	.610
	70	.227	.362	.478	.515	.527	.530	.532	.534
	90	.186	.306	.424	.465	.481	.468	.471	.472
2	10	.545	.636	.658	.660	.661	.665	.666	.672
	30	.239	.367	.459	.484	.491	.499	.501	.506
	50	.121	.203	.286	.320	.335	.348	.351	.355
	70	.074	.128	.192	.226	.243	.259	.264	.267
	90	.058	.101	.156	.188	.207	.215	.221	.223
3	10	.431	.561	.607	.613	.614	.616	.615	.631
	30	.133	.223	.306	.337	.348	.357	.357	.366
	50	.058	.103	.155	.183	.197	.211	.213	.218
	70	.037	.064	.098	.119	.132	.147	.150	.154
	90	.030	.051	.079	.098	.110	.122	.126	.129
4	10	.339	.482	.549	.560	.563	.566	.565	.593
	30	.078	.139	.204	.234	.247	.258	.260	.272
	50	.033	.060	.094	.114	.126	.139	.143	.150
	70	.022	.039	.061	.074	.083	.095	.099	.104
	90	.019	.032	.050	.061	.070	.080	.084	.089
6	10	.211	.343	.433	.453	.458	.461	.461	.518
	30	.033	.065	.103	.123	.135	.145	.148	.167
	50	.015	.029	.047	.057	.064	.073	.077	.086
	70	.011	.021	.033	.040	.045	.051	.054	.060
	90	.010	.019	.028	.034	.038	.044	.046	.052
8	10	.135	.238	.326	.353	.362	.366	.367	.452
	30	.016	.034	.058	.072	.080	.090	.094	.116
	50	.008	.017	.027	.034	.039	.045	.048	.059
	70	.006	.013	.021	.026	.028	.032	.035	.043
	90	.005	.012	.019	.023	.025	.029	.031	.038
10	10	.090	.165	.244	.272	.283	.290	.291	.395
	30	.009	.020	.036	.045	.052	.060	.064	.087
	50	.005	.010	.019	.023	.026	.030	.033	.044
	70	.004	.009	.015	.018	.020	.023	.025	.033
	90	.003	.008	.014	.016	.018	.020	.022	.030

Source: IES RP-23-1989; reprinted with permission.

TABLE 19.10 **Coefficient of Utilization from Window Without Blinds; Sky Component $E_{xvk}/E_{xhk} = 1.25$**

Room Depth/ Window Height	*Depth*	*Window Width/Window Height* .5	1	2	3	4	6	8	Infinite
1	10	.578	.607	.614	.619	.621	.633	.634	.635
	30	.405	.525	.580	.594	.599	.612	.614	.615
	50	.287	.423	.519	.547	.556	.569	.571	.573
	70	.218	.347	.461	.501	.515	.522	.525	.526
	90	.186	.307	.428	.473	.491	.483	.486	.487
2	10	.472	.549	.566	.569	.570	.574	.575	.581
	30	.221	.337	.422	.447	.456	.465	.467	.472
	50	.120	.202	.285	.321	.337	.353	.357	.361
	70	.078	.136	.204	.242	.261	.281	.287	.290
	90	.064	.112	.174	.211	.233	.244	.251	.253
3	10	.377	.488	.527	.533	.534	.536	.536	.549
	30	.130	.217	.298	.329	.341	.352	.353	.362
	50	.062	.110	.165	.195	.211	.228	.231	.237
	70	.040	.070	.109	.132	.147	.166	.171	.175
	90	.033	.057	.090	.112	.127	.142	.148	.152
4	10	.300	.424	.484	.494	.497	.499	.499	.524
	30	.080	.143	.209	.240	.255	.267	.269	.283
	50	.036	.066	.104	.126	.140	.156	.160	.168
	70	.024	.043	.068	.083	.094	.109	.115	.120
	90	.021	.036	.056	.070	.080	.092	.099	.103
6	10	.193	.314	.395	.415	.420	.423	.423	.476
	30	.036	.071	.113	.136	.149	.161	.165	.186
	50	.017	.033	.053	.065	.074	.084	.089	.100
	70	.012	.024	.037	.045	.050	.058	.061	.069
	90	.011	.021	.031	.038	.043	.049	.053	.060
8	10	.128	.226	.310	.337	.346	.351	.352	.433
	30	.019	.039	.066	.082	.092	.104	.109	.134
	50	.009	.019	.031	.040	.045	.052	.056	.069
	70	.007	.015	.023	.029	.032	.037	.040	.049
	90	.006	.013	.021	.025	.028	.032	.035	.043
10	10	.088	.164	.241	.270	.282	.290	.291	.396
	30	.011	.024	.043	.054	.062	.071	.076	.103
	50	.005	.012	.022	.026	.030	.035	.038	.052
	70	.004	.010	.017	.020	.023	.026	.028	.038
	90	.004	.009	.016	.018	.020	.023	.025	.034

Source: IES RP-23-1989; reprinted with permission.

TABLE 19.11 **Coefficient of Utilization from Window Without Blinds; Sky Component $E_{xvk}/E_{xhk} = 1.50$**

Room Depth/ Window Height	*Depth*	*Window Width/Window Height*							
		.5	*1*	*2*	*3*	*4*	*6*	*8*	*Infinite*
1	10	.503	.528	.536	.541	.544	.557	.558	.559
	30	.359	.464	.514	.528	.534	.549	.550	.552
	50	.261	.384	.471	.499	.508	.524	.526	.527
	70	.204	.325	.432	.470	.485	.497	.499	.500
	90	.179	.295	.412	.456	.475	.474	.477	.478
2	10	.412	.477	.490	.492	.493	.498	.499	.505
	30	.201	.304	.379	.402	.410	.422	.424	.429
	50	.115	.192	.269	.304	.320	.339	.343	.347
	70	.078	.136	.204	.241	.261	.286	.292	.295
	90	.066	.117	.183	.221	.246	.262	.271	.273
3	10	.331	.426	.458	.461	.462	.465	.465	.477
	30	.121	.202	.275	.304	.316	.327	.329	.337
	50	.062	.109	.164	.193	.209	.228	.232	.238
	70	.041	.073	.114	.138	.154	.176	.183	.188
	90	.035	.062	.099	.123	.141	.159	.169	.173
4	10	.265	.372	.422	.430	.433	.435	.435	.456
	30	.077	.137	.199	.229	.243	.256	.259	.272
	50	.037	.069	.107	.130	.144	.161	.167	.175
	70	.026	.046	.073	.089	.101	.119	.126	.132
	90	.022	.039	.063	.078	.090	.106	.114	.120
6	10	.173	.281	.351	.368	.373	.375	.375	.422
	30	.037	.073	.115	.137	.151	.164	.168	.189
	50	.018	.036	.058	.071	.080	.092	.098	.110
	70	.013	.026	.040	.049	.056	.064	.069	.078
	90	.012	.023	.035	.043	.048	.057	.062	.070
8	10	.117	.207	.282	.305	.314	.319	.320	.393
	30	.020	.042	.071	.087	.098	.111	.116	.143
	50	.010	.021	.035	.044	.050	.058	.063	.078
	70	.007	.016	.026	.032	.036	.041	.045	.055
	90	.076	.014	.023	.028	.031	.036	.040	.049
10	10	.082	.153	.224	.250	.262	.269	.271	.368
	30	.012	.026	.047	.059	.068	.078	.084	.114
	50	.006	.014	.024	.030	.034	.040	.044	.060
	70	.005	.011	.019	.022	.025	.029	.032	.043
	90	.004	.010	.017	.020	.023	.026	.028	.038

Source: IES RP-23-1989; reprinted with permission.

TABLE 19.12 **Coefficient of Utilization from Window Without Blinds; Sky Component E_{xvk}/E_{xhk} = 1.75**

Room Depth/ Window Height	*Depth*	*Window Width/Window Height* .5	1	2	3	4	6	8	Infinite
1	10	.435	.457	.465	.471	.474	.486	.488	.489
	30	.317	.407	.452	.466	.471	.486	.488	.489
	50	.234	.343	.422	.447	.456	.472	.475	.476
	70	.187	.297	.395	.430	.445	.458	.461	.462
	90	.168	.276	.384	.426	.444	.447	.450	.451
2	10	.357	.412	.422	.424	.424	.430	.431	.436
	30	.180	.271	.335	.356	.363	.375	.378	.381
	50	.106	.177	.246	.278	.293	.313	.318	.321
	70	.074	.130	.194	.229	.249	.274	.282	.284
	90	.065	.116	.181	.219	.244	.264	.273	.276
3	10	.288	.369	.394	.397	.397	.400	.401	.411
	30	.110	.183	.247	.272	.282	.294	.296	.304
	50	.058	.104	.154	.181	.196	.215	.221	.226
	70	.040	.072	.112	.136	.152	.176	.184	.188
	90	.035	.063	.101	.126	.144	.166	.177	.182
4	10	.232	.324	.365	.371	.373	.375	.375	.394
	30	.071	.127	.183	.209	.222	.235	.238	.250
	50	.036	.067	.104	.125	.139	.157	.163	.171
	70	.025	.046	.072	.089	.101	.119	.127	.134
	90	.022	.041	.065	.082	.095	.114	.124	.130
6	10	.153	.247	.307	.320	.324	.326	.327	.367
	30	.035	.070	.109	.130	.143	.155	.160	.180
	50	.018	.036	.058	.071	.080	.091	.098	.110
	70	.013	.026	.041	.051	.058	.067	.073	.082
	90	.012	.023	.037	.046	.052	.062	.069	.078
8	10	.104	.184	.249	.269	.276	.281	.282	.346
	30	.020	.042	.070	.086	.096	.109	.115	.141
	50	.010	.022	.036	.046	.052	.060	.066	.081
	70	.008	.017	.027	.033	.038	.044	.048	.059
	90	.007	.015	.024	.030	.034	.040	.044	.054
10	10	.074	.138	.201	.223	.233	.240	.242	.328
	30	.012	.027	.048	.059	.067	.078	.084	.114
	50	.006	.014	.026	.032	.036	.043	.047	.064
	70	.005	.011	.020	.024	.027	.031	.034	.046
	90	.004	.010	.018	.022	.024	.028	.031	.042

Source: IES RP-23-1989; reprinted with permission.

TABLE 19.13 **Coefficient of Utilization from Window Without Blinds; Ground Component**

Room Depth/ Window Height	*Depth*	*Window Width/Window Height*							
		.5	*1*	*2*	*3*	*4*	*6*	*8*	*Infinite*
1	10	.105	.137	.177	.197	.207	.208	.210	.211
	30	.116	.157	.203	.225	.235	.241	.243	.244
	50	.110	.165	.217	.241	.252	.267	.269	.270
	70	.101	.162	.217	.243	.253	.283	.285	.286
	90	.091	.146	.199	.230	.239	.290	.292	.293
2	10	.095	.124	.160	.178	.186	.186	.189	.191
	30	.082	.132	.179	.201	.212	.219	.222	.225
	50	.062	.113	.165	.189	.202	.214	.218	.220
	70	.051	.093	.141	.165	.179	.194	.198	.200
	90	.045	.079	.118	.140	.153	.179	.183	.185
3	10	.088	.120	.157	.175	.183	.185	.163	.167
	30	.059	.107	.154	.176	.187	.198	.193	.198
	50	.039	.074	.114	.134	.146	.157	.163	.170
	70	.031	.055	.085	.101	.111	.122	.127	.130
	90	.028	.047	.070	.083	.092	.107	.113	.115
4	10	.073	.113	.154	.174	.183	.187	.176	.184
	30	.040	.082	.127	.148	.159	.170	.177	.185
	50	.025	.049	.078	.094	.103	.113	.117	.123
	70	.020	.036	.054	.065	.071	.079	.083	.087
	90	.019	.032	.046	.054	.060	.069	.073	.076
6	10	.056	.106	.143	.164	.175	.184	.173	.194
	30	.021	.050	.081	.098	.107	.117	.123	.138
	50	.013	.027	.041	.049	.054	.060	.064	.072
	70	.011	.021	.029	.033	.035	.039	.041	.046
	90	.011	.020	.026	.030	.032	.035	.037	.042
8	10	.036	.082	.122	.143	.156	.166	.170	.208
	30	.011	.029	.050	.062	.070	.078	.082	.101
	50	.007	.016	.024	.028	.031	.035	.038	.046
	70	.006	.013	.018	.020	.021	.023	.025	.030
	90	.006	.013	.017	.019	.020	.022	.023	.028
10	10	.024	.061	.109	.120	.131	.144	.147	.200
	30	.006	.017	.034	.040	.046	.053	.056	.076
	50	.004	.010	.016	.018	.020	.023	.024	.033
	70	.004	.009	.013	.014	.015	.016	.016	.022
	90	.004	.009	.013	.013	.014	.015	.016	.021

Source: IES RP-23-1989; reprinted with permission.

TABLE 19.14 **Reflectances of Building Materials and Outside Surfaces**

Material	*Reflectance (In percent)*	*Material*	*Reflectance (In percent)*
Bluestone, sandstone	18	Asphalt (free from dirt)	7
Brick		Earth (moist cultivated)	7
light buff	48		
dark buff	40		
dark red glazed	30	Granolite pavement	17
Cement	27	Grass (dark green)	6
Concrete	55	Gravel	13
Granite	40	Macadam	18
Marble (white)	45	Slate (dark clay)	8
Paint (white)		Snow	
new	75	new	74
old	55	old	64
		Vegetation (mean)	25

Source: IES RP-23-1989; reprinted with permission.

EXAMPLE 19.3. Using the building of Example 19.1, find E_i (1–5) for a spring day and a winter day, at 10 A.M. (2 P.M.). Assume the sky to be overcast so that a direct comparison with Examples 19.1 and 19.2 can be made. Reflectances are assumed to be 70% ceiling, 50% wall, 30% floor, to correspond to the IES method standard conditions.

SOLUTION. For completeness sake we will repeat conditions of the problem:

Location: Seattle, Washington

Latitude: 47.6° N

Room: 25 ft deep × 18 ft wide × 9 ft 6 in. high

Window: height 5 ft above 30 in. working plane, total length 18 ft

Transmittance: 85%; net glass area 92%

Ground Reflectance: not previously specified

Assume an extensive area of mixed grass and concrete walkways, with an average overall reflectance of 20%. See Table 19.14. (For more accurate calculation, see the method described in the IES Recommended Practice.)

STEP 1. Illuminances.

Solar altitude (Table 19.2).

Dec. 21: −14°

Mar. 21: −36°

Vertical window illuminance from Fig. 19.6.

$$\text{Dec. 21: } E_{xvk} = 200 \text{ fc} = 2150 \text{ lux}$$

$$\text{Mar. 21: } = 500 \text{ fc} = 5380 \text{ lux}$$

Horizontal illuminance from full sky from Fig. 19.6.

$$\text{Dec. 21: } E_{xHk} = 500 \text{ fc} = 5380 \text{ lux}$$

$$\text{Mar. 21: } = 1240 \text{ fc} = 13{,}340 \text{ lux}$$

Horizontal illuminance from half sky.

$$\text{Dec. 21: } E_{xhk} = 2690 \text{ lux}$$

$$\text{Mar. 21: } = 6670 \text{ lux}$$

From equation 19.5, the vertical illuminance from the ground is

$$E_{xvg} = \text{RF}_g \times \frac{E_{xHk}}{2} = 0.2\,(5380)$$

$$= 1076 \text{ lux on Dec. 21}$$

$$= 0.2(13{,}340)$$

$$= 2670 \text{ lux on Mar. 21}$$

STEP 2. Window transmittance (Tables 19.6 and 19.7). Glass transmittance 85%, light loss factor 0.9, net glass area 92%.

$$\tau = 0.85(0.9)(0.92) = 0.70$$

STEP 3. Selection of coefficients of utilizations (Tables 19.8 to 19.13).

Sky component E_{xvk}/E_{xhk}:

Dec. 21: 2150/2690 = 0.8 use Table 19.8

Mar. 21: 5380/6670 = 0.8 use Table 19.8

$$\frac{\text{Room depth}}{\text{window height}} = 18.5 = 3/6 \qquad \text{use } 4.0$$

$$\frac{\text{Window width}}{\text{window height}} = 18/5 = 3.6 \qquad \text{use } 4.0$$

Ground component: use Table 19.13

Point (%)	*CU_k*	*CU_g*
10	0.673	0.183
30	0.235	0.159
50	0.104	0.103
70	0.065	0.071
90	0.053	0.060

STEP 4. Calculation of illuminance (equation 19.4).

$$E_i = \tau(E_{xvk} \times CU_k + E_{xvg} \times CU_g)$$

E_i (%)	*Dec. 21*	*Mar. 21*
10	1151	2877
30	474	1182
50	234	585
70	151	378
90	125	312

To compare the results with those of the CIE and GDDM methods, the values calculated for December 21 are plotted on the room plan of Fig. 19.21*c*. The December values are selected since they are minimum and would correspond most closely to the 85th percentile figures of both previous methods. The agreement is excellent for the half of the room depth nearest the window. For the deeper half, where ground contribution is significant, the IES method yields higher figures because it considers the rereflected ground light contribution. Overall the agreement among the three methods is excellent. The figures for March 21 would correspond most closely to the 50th to 60th percentile of Fig. 19.17, that is, somewhat more than double the minimum figures—and here again the agreement is good.

To demonstrate the use of the IES method for clear sky conditions, we will work through a short example.

EXAMPLE 19.4. Find the five IES point illuminances for the same room as Example 19.3, on 21 June at 10 A.M. for clear sky conditions. Assume that the window faces southwest [azimuth angle (−)45°].

SOLUTION. Solar azimuth at 10 A.M., 21 June, 48° N latitude is 56° (Table 19.2). Bearing angle (Fig. 19.8) is therefore 45° + 56° or 101° [i.e., no direct sunlight enters the window (a necessary condition of the IES clear sky method)]. Solar altitude is 55°.

STEP 1. Horizontal illuminances (Fig. 19.7)

Sun only: 80,000 lux

Sky only: 18,000 lux, full sky

9,000 lux, half sky (E_{xhk})

Vertical illuminance (Fig. 19.9*b*): Bearing angle 101°; 700 fc = 7500 lux (E_{xvk})

From equation 19.5, vertical illuminance on the exterior of the window, resulting from ground light is

$$E_{xvg} = 0.2 \times \frac{80{,}000 + 9000}{2} = 8900 \text{ lux}$$

STEP 2. $\tau = 0.70$, as previously.

STEP 3. Selection of CU from Tables 19.8 to 19.13:

$$\frac{E_{xvk}}{E_{xhk}} = \frac{7500}{9000} = 0.83; \text{ Table 19.8}$$

Coefficients are the same as in Example 19.3.

STEP 4. Calculation of illuminance.

```
********************************************
* DAYLIGHT ILLUMINATION IN A RECTANGULAR ROOM *
********************************************

ROOM HAS A WIDTH (WINDOW WALL) OF 25.0 UNITS
A DEPTH OF 18.0 UNITS AND A HEIGHT OF   9.5 UNITS
ROOM ORIENTATION (FROM OUTWARD WINDOW NORMAL) = DUE SOUTH

ROOM HAS 3 WINDOW(S) IN THE FRONT WALL
WINDOWS ARE  6.00 UNITS WIDE BY   5.00 UNITS HIGH
    WINDOW TYPE IS "1" (CLEAR WINDOW)
    MAINTENANCE FACTOR      = .90, WINDOW TRANSMISSIVITY = .85
    WINDOW REFLECTANCE      =   .09

SUBJECT BUILDING..
      BUILDING WIDTH (FRONT   WALL)     =  25.0 UNITS
      BUILDING HEIGHT                   =   9.5 UNITS
      REFLECTANCE      =    .50

SKY LUMINANCE DETERMINED BY GEOGRAPHICAL DATA
      LATITUDE                        47.6 DEG.
      LONGITUDE                      122.0 DEG.
      ALTITUDE                          .0 METERS
      SKY MODEL                        1 (SOLID OVERCAST)
      GROUND REFLECTANCE               .20
      MONTH                           12(DEC)
      DAY OF MONTH                    21
      TIME OF DAY                     10.0 HRS.
      ====================================

ILLUMINANCE ON A HORIZONTAL SURFACE
      SOLAR COMPONENT=          0. FOOT-CANDLE
      SKY   COMPONENT=        525. FOOT-CANDLE
ZENITH LUMINANCE=      670. (FT.-LAMBERTS)

**************************************************************
* DATA FOR WORKING SURFACE NODES;    *         X,Y,Z=COORDINATES,
                                *
* I=TOTAL ILLUMINANCE (FOOT-CANDLE)  D=DAYLIGHT FACTOR (PERCENT) *
**************************************************************

  X = DISTANCE FROM FRONT WALL (PARALLEL TO FRONT WALL) FT
  Y = DISTANCE FROM ROOM CENTER LINE (PERPENDICULAR TO FRONT WALL) FT
      + VALUES (LEFT OF CENTER LINE), - VALUES (RIGHT OF CENTER LINE)
           HEIGHT OF WORK PLANE = 3.5 FT
```

X	-1.00	-1.00	-1.00	-1.00	-1.00	-1.00	-1.00
Y	10.71	7.14	3.57	.00	-3.57	-7.14	-10.71
I	77.53	110.80	44.92	111.75	44.92	110.80	77.53
D	14.78	21.12	8.56	21.30	8.56	21.12	14.78
X	-3.00	-3.00	-3.00	-3.00	-3.00	-3.00	-3.00
Y	10.71	7.14	3.57	.00	-3.57	-7.14	-10.71
I	46.14	62.29	53.08	65.21	53.09	62.30	46.14
D	8.79	11.87	10.12	12.43	10.12	11.87	8.80
X	-5.00	-5.00	-5.00	-5.00	-5.00	-5.00	-5.00
Y	10.71	7.14	3.57	.00	-3.57	-7.14	-10.71
I	27.71	35.52	35.39	36.55	35.39	35.53	27.72
D	5.28	6.77	6.75	6.97	6.75	6.77	5.28
X	-7.00	-7.00	-7.00	-7.00	-7.00	-7.00	-7.00
Y	10.71	7.14	3.57	.00	-3.57	-7.14	-10.71
I	18.27	22.24	23.53	24.64	23.53	22.25	18.28
D	3.48	4.24	4.48	4.70	4.49	4.24	3.48
X	-9.00	-9.00	-9.00	-9.00	-9.00	-9.00	-9.00
Y	10.71	7.14	3.57	.00	-3.57	-7.14	-10.71
I	13.30	15.87	16.80	17.45	16.80	15.87	13.31
D	2.54	3.02	3.20	3.33	3.20	3.03	2.54
X	-11.00	-11.00	-11.00	-11.00	-11.00	-11.00	-11.00
Y	10.71	7.14	3.57	.00	-3.57	-7.14	-10.71
I	11.16	12.88	13.74	14.24	13.74	12.89	11.17
D	2.13	2.45	2.62	2.71	2.62	2.46	2.13
X	-13.00	-13.00	-13.00	-13.00	-13.00	-13.00	-13.00
Y	10.71	7.14	3.57	.00	-3.57	-7.14	-10.71
I	9.17	10.43	11.19	11.48	11.19	10.44	9.19
D	1.75	1.99	2.13	2.19	2.13	1.99	1.75
X	-15.00	-15.00	-15.00	-15.00	-15.00	-15.00	-15.00
Y	10.71	7.14	3.57	.00	-3.57	-7.14	-10.71
I	8.13	9.20	9.88	10.06	9.88	9.20	8.14
D	1.55	1.75	1.88	1.92	1.88	1.75	1.55
X	-17.00	-17.00	-17.00	-17.00	-17.00	-17.00	-17.00
Y	10.71	7.14	3.57	.00	-3.57	-7.14	-10.71
I	7.36	8.33	9.02	9.14	9.02	8.33	7.36
D	1.40	1.59	1.72	1.74	1.72	1.59	1.40

(*a*)

Fig. 19.23 (a) *Output of SUPERLITE PC 1.01 program for the room of Examples 19.1 to 19.3 using the overcast sky condition prevalent in Seattle, Washington. The program output is in English units of footcandles on a room grid measured in feet. Daylight factor* D *is calculated along with the corresponding illuminance. The program measures distances from the window wall outward; hence the* X *values (parallel) to the window wall are negative. The* Y *values are measured on both sides of the room centerline. See* (b). *The results are shown in bubbles on Fig. 19.21c for comparison with results of the other calculation systems discussed.*

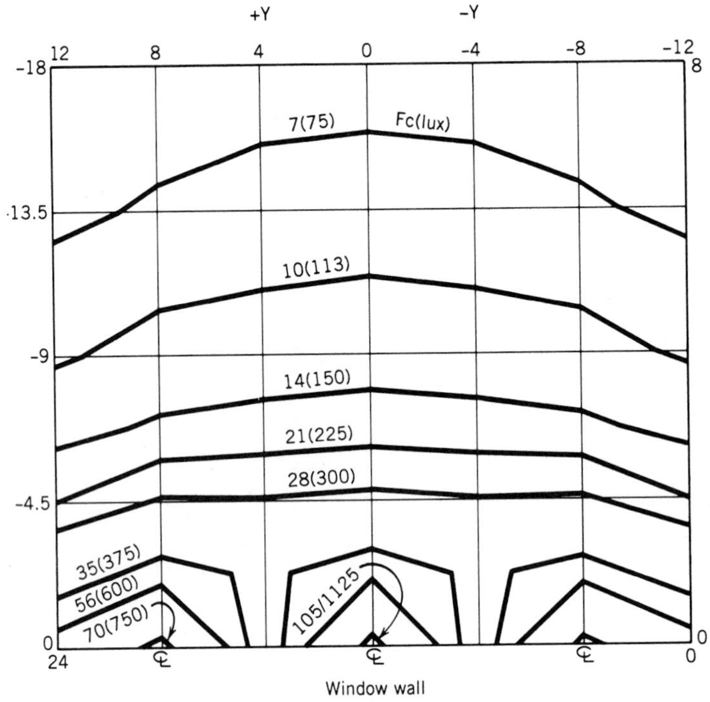

(*b*)

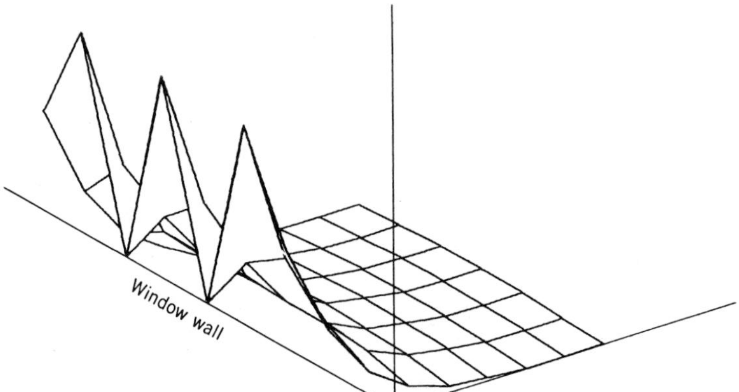

(*c*)

Fig. 19.23 *(continued)* (b) *Contour plot of illuminance values shown in* (a). *Compare to contours of Fig. 19.21*c. *Note that the principal difference occurs close to the window wall where the computer calculation shows a very sharp drop at the 2-ft wall space between windows. Contour plots were not produced by SUPERLITE.* (c) *A three-dimensional contour plot of illuminances in the room, as calculated and shown in* (a). *The sharp drop between windows is shown clearly as is the rapid falloff of daylight with distance from window wall (e.g., at the room centerline illuminance drops 75% within the first 25% of room depth).*

$$E_i = \tau(E_{xvk} \times CU_k + E_{xvg} \times CU_g)$$
$$= 0.7\ (7500 \times CU_k + 8900 \times CU_g)$$

Note that these results indicate that on a clear summer day, with no direct sunlight on the window, (bearing angle >90°) there is enough daylight throughout the space for routine work of moderate difficulty (500 lux requirement), since the calculation yields a minimum illuminance at 90% room depth of 652 lux.

19.9 Daylight Analysis: Computer Use

The most effective (and painless) method of performing the calculations necessary to produce meaningful numerical data is by use of one of the computer programs available. The advantage of such use is that data can be produced in any detail desired, for the entire room, with interior and exterior obstructions, for any configuration of windows and with varying room finishes. We used the SUPERLITE program version PC1.01 (see Appendix H) to produce the results shown in Fig. 19.23. A plotting routine is, at this writing, not available with the PC version of SUPERLITE. The plot in Fig. 19.23 was accomplished with one of the many extant plotting utilities. We ran the program with the same input data as in Examples 19.1 to 19.3 and showed the results on Fig 19.21*c* for comparison with the result of the other three methods. Agreement is well within engineering accuracy. This is to be expected, since all of the methods utilize the CIE overcast sky luminance distribution formulation. In summary, then, it is our recommendation that each daylighting problem be worked through assuming three different sky conditions:

1. *Completely overcast sky*. Use sky luminances that are maintained for 85% of the time, that is, minimum luminance. Use a computer program or the GDDM method if facilities are available. If not, use the CIE or IES methods and sketch in contours intuitively. As stated above, these results will establish maximum PSALI needed and extent/type of lighting control strategy.

2. *Average sky,* that is, overcast sky on March 21 or clear sky with no direct insolation either due to geometry or fixed shading, whichever is more applicable. This condition, in the design mode, will assist in determining glass area and room proportions. In the analysis mode the results are usable for HVAC design and energy calculations. Hand calculation is possible; computer assistance is highly desirable. For hour-by-hour energy calculations, computer calculation is required.

3. *Clear sky with sun*. These results will indicate need for shading and sun control devices and will give maximum HVAC loads.

In all studies, use the most accurate sky luminance data available; that is, if local data based on field observations are available, they are preferable to the generalized data presented here. Spaces lighted with multiple sources such as bilateral sidelighting, windows, and toplighting, or daylight plus artificial light are calculated by superposition (calculating the effect of each source separately and adding the results).

19.10 Physical Modeling in Daylight Design

As every architectural and engineering students is aware, physical models are a useful and occasionally indispensable tool for the investigation of complex phenomena. The variables in daylight design are so numerous and extensive that even simple physical models can give the lighting designer a visual understanding of a daylighted space that only the most experienced designer can mentally visualize on the basis of numerical results, and that, imperfectly. Physical models are an adjunct to calculation procedures and not a substitute (or vice versa) since each serves a different purpose. Modeling will give the designer a three-dimensional visual depiction of the space, with the lights and shadows produced by daylight. These can then be altered by changing window design and orientation, and addition of light shelves, reflectors, and window controls, to produce the visual image that corresponds to the intended design.

The principal problem in the use of models is how to expose them to the desired sky conditions. Thus if it is desired to view a particular space under both overcast and clear sky conditions, and the latter for different seasons of the

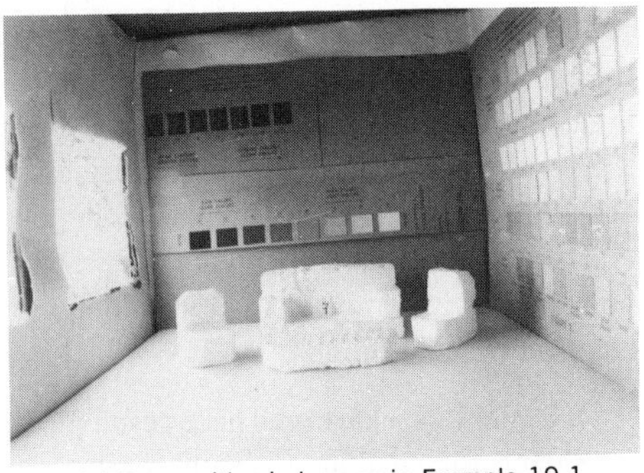

(*a*) Room with windows as in Example 19.1

Fig. 19.24 *These photographs illustrate the effectiveness of even a crude model, in daylight studies. The model was constructed out of a cardboard box, at a scale of 1 : 20, to represent a room of the same horizontal dimensions as the classroom of Example 19.1 but with a higher ceiling. A Munsell gray scale was mounted along the back wall on gray paper corresponding to the center chips of the gray scale. The upper left back wall has chips of varying chroma and the right wall has chips representing typical interior finishes. The bottom row shows accent colors.*

(*b*) Large high window, center of wall

(*c*) Two small high horizontal windows

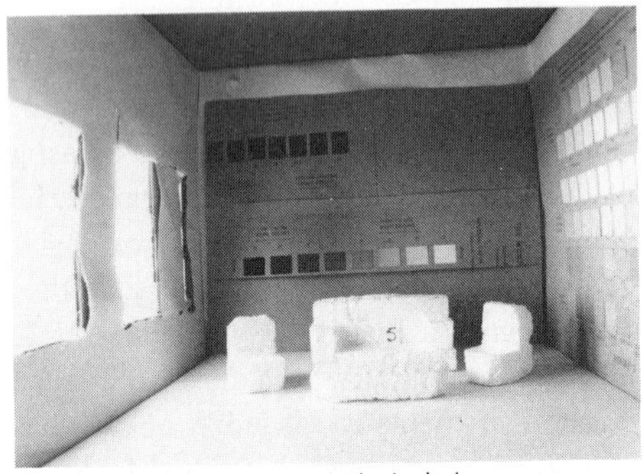

(*d*) Two small low vertical windows .....

(*e*) ...... with small skylight

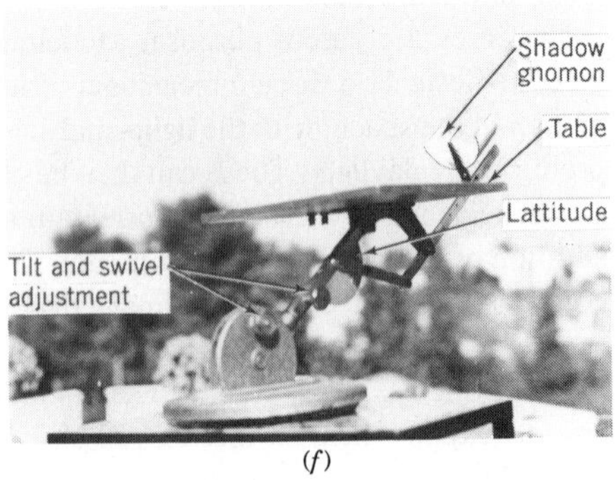

(*f*)

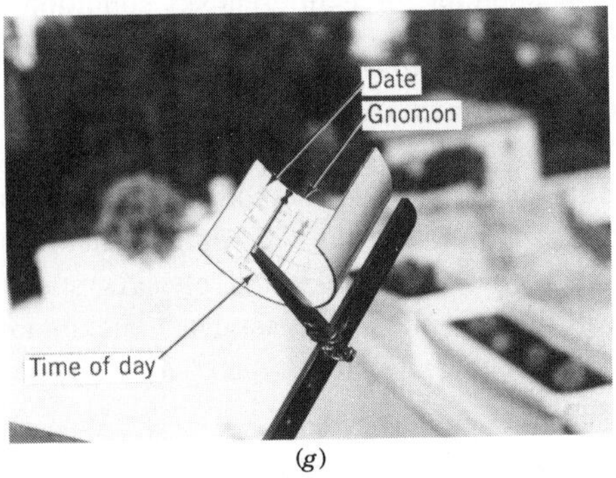

(*g*)

21 Jun

21 Mar/Sept

(*Fig. 19.24 continued*)

A reflex camera with an f2, 28-mm-wide angle was used for all photos. (a) *Overcast sky, 21 June, 6 PM. This photo represents the room of Example 19.1, with windows along one entire wall.* (b–e) *Overcast sky, 21 June 6 PM. These photos show the effect of window placement. The square-foot area of fenestration is equal in all the photos.* (f) *Table-type heliodon with adjustments for latitude, elevation, and tilt permit exposing the model to desired clear sky conditions.* (g) *Date and time of day are set by adjusting the position of the gnomons' shadow.* (h–j) *Clear sky, SW exposure, bright sun, 3 PM. The heliodon permits viewing afternoon sun penetration for the four annual seasons. Note the relative darkening of room surfaces because of eye adaptation to the extremely high luminance of the sun on the floor.* (k–m) *Clear sky, bright sun, 3 PM on 21 June. At the same time that window* (h) *is receiving direct sunlight, other exposures receive sky light only* (k, l), *and one exposure* (m) *receives reflected sunlight from an opposing structure.*

21 Dec

South

Southeast

Northeast

year (different times of day are obviously not a major problem), waiting for a real sky condition that corresponds to the one desired is obviously not a practical alternative. (Use of models in partly-cloudy-sky conditions is not useful because of the infinite variability of this condition: computer simulation is more practical.) Overcast skies are common in temperate climates, and therefore a model can be tested in real sky conditions without undue delay. In arid areas, where overcast skies are rare, the overcast sky condition would not constitute a design condition for local construction. The reasonable alternatives to real sky exposure are, in descending order of desirability:

Artificial sky

Heliodon (tilt-table)

Narrow-beam electric lights and diffuse light boxes

1. *Artificial sky*. Carefully designed and controlled artificial sky domes can duplicate overcast and clear sky conditions with a high degree of accuracy and thus are ideal for testing physical models. Unfortunately, to achieve an acceptable degree of accuracy, these domes are large, complex, and very expensive to construct. A number of units exist in major universities and test facilities around the country, and designers with access to them would do well to utilize them. Small domes and mirror boxes have much more limited usefulness and are applicable primarily to overcast sky conditions. For construction details, see Moore (1985).

2. *Heliodon*. The heliodon, or tilt-table (see Fig. 19.24) is a simple but very effective device that operates by rotating and tilting the building model with respect to the real sky until the desired solar altitude and azimuth is reached. Obviously, then, this technique is most effective with clear sky conditions. The objection often raised to this procedure is that in tilting the model a different portion of the sky and ground is "seen" through the window than would be the case in real sky conditions. Although this is obviously true, tests have shown the error to be minimal because:

(a) The solar component is much larger than the clear sky component when it is diffused by a window diffuser or even when reflected from facing structures and/or high reflectance ground (see Fig. 19.9).
(b) The clear sky component is roughly equal to the reflected component of low reflectance ground, so that a change in the portion of each "viewed" is not significant.

3. Use of a *narrow-beam electric light* source to simulate the sun and back-lighted translucent panels to simulate an overcast sky are crude and highly inaccurate methods which at best can be used to obtain some idea of interior light and shadow patterns.

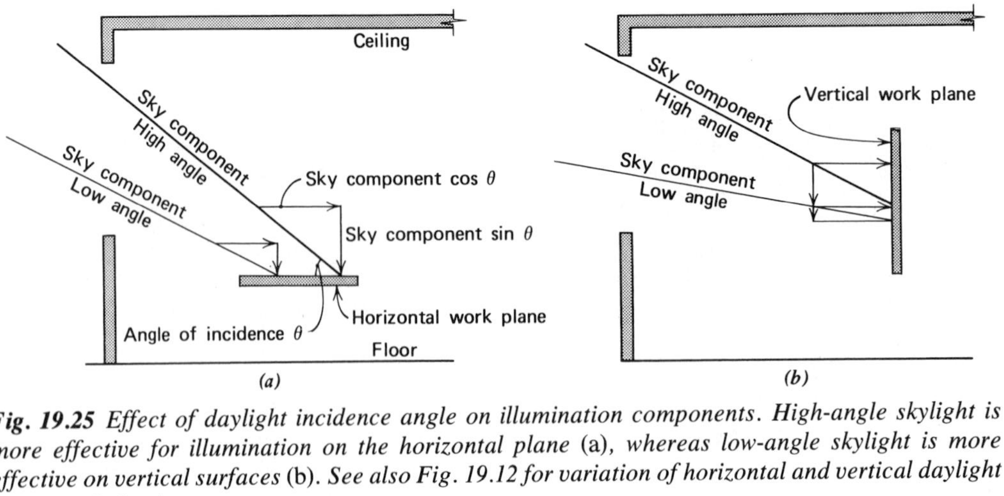

***Fig. 19.25** Effect of daylight incidence angle on illumination components. High-angle skylight is more effective for illumination on the horizontal plane* (a), *whereas low-angle skylight is more effective on vertical surfaces* (b). *See also Fig. 19.12 for variation of horizontal and vertical daylight factor with depth.*

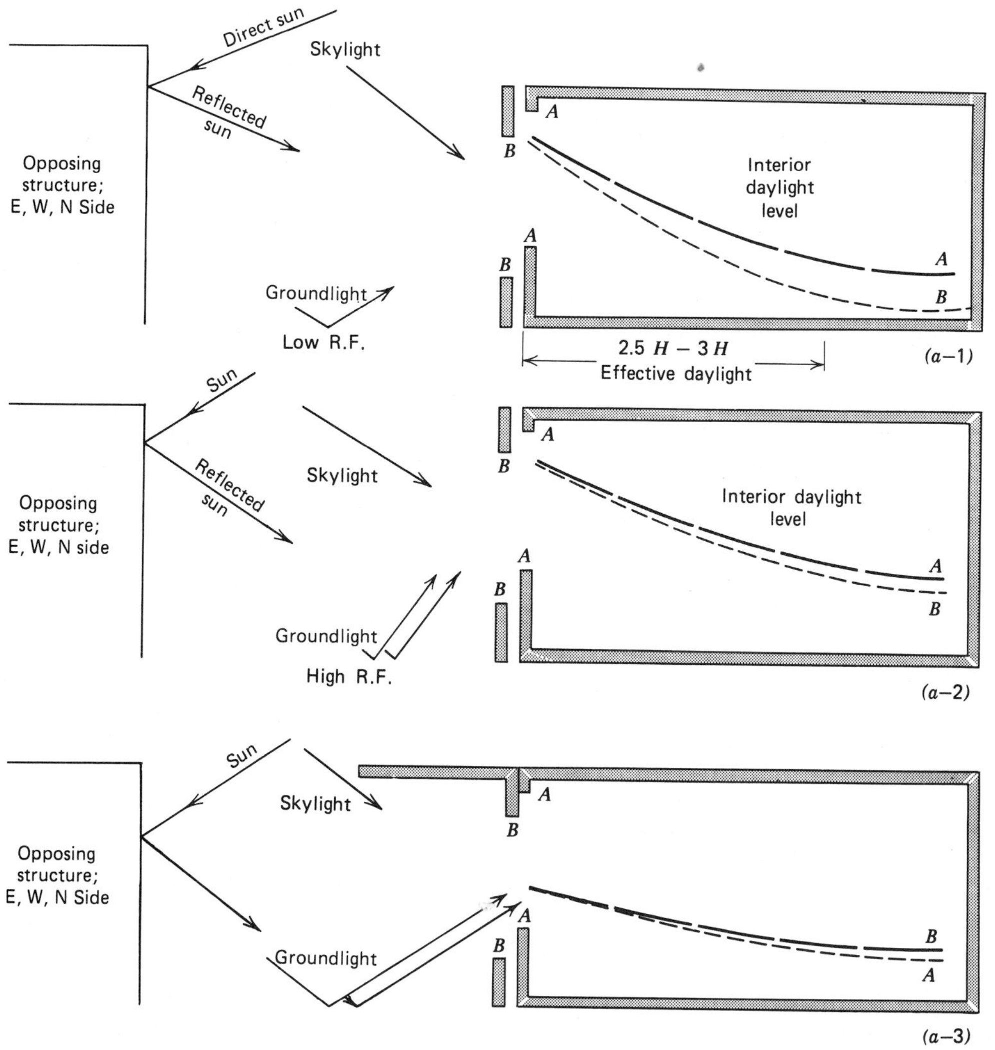

Fig. 19.26 *The effect of window placement on interior daylight. Where skylight predominates* (a-1) *because of high window/high sill/low ground reflectance/strong reflected component from opposite structure, illuminance drops off rapidly as depth increases. When reflected ground light is a major component, due to high ground reflectance* (a-2) *and/or strong rereflected sun from opposite structure, the window position on wall has minimum effect. Very low still height will bring curve B above curve A, as in* (a-3). *Note that where sky light is blocked* (a-3), *lower sill height obviously results in better penetration. Rereflected sun can be added to both skylight and groundlight, as shown.*

Measurements of light levels in scale models can be accomplished quite readily by using small photocells intended specifically for the purpose. The scale of the model must be large enough that the photocell(s) do not have a significant influence on the light distribution with the space. For further information on physical modeling, the reader is referred to Robbins (1986) and Moore (1985).

19.11 Additional Factors in Interior Daylighting

(a) Horizontal and Vertical Surfaces. Since the sky component of daylight enters side fenestration at an angle, it can be resolved into horizontal and vertical components, as shown in Fig. 19.25. The vertical component that illuminates horizontal surfaces is proportional to the sine of the angle of incidence, and the horizontal component that illuminates vertical surface is proportional to the cosine of this angle. Therefore, for horizontal tasks, windows should be *as high as possible* and, for vertical tasks, as low as possible. Since most tasks are horizontal, high windows will give better, deeper penetration than low windows of the same area.

(b) Window Details. The effect of window construction on total fenestration area reduction is often neglected. Even windows with narrow mullions and light metal frames have 8 to 10% obscuration; heavy window supports and small glass lights can result in 12 to 15% obscuration with proportional daylight reduction. Further obscuration readily results from dirt accumulation, wired glass, and mechanical system items such as pipes and ducts inside the room, adjacent to windows. The effect of window placement on interior light distribution is shown in Fig. 19.26.

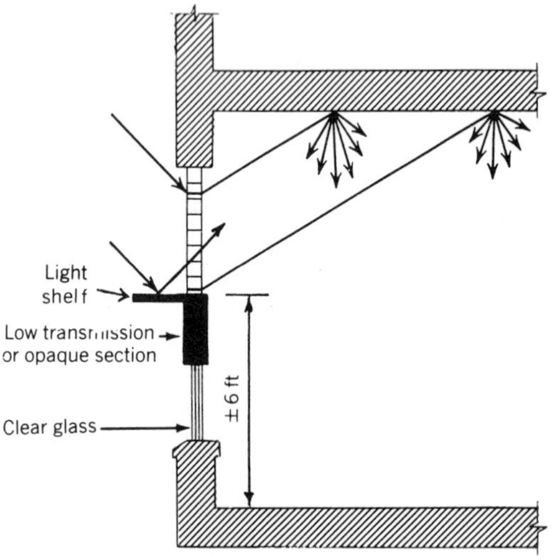

Fig. 19.27 *Arrows illustrate principal light paths into room with light-directing glass blocks above and clear glass below. Clear section permits unobstructed view outdoors. These blocks are particularly effective in low-ceilinged rooms, since the lower band of blocks near the 6-ft sill delivers most of the light to the deeper portions of the room. The use of a lightshelf below the blocks acts both as a reflector to increase the light transmission above and as a fixed horizontal sunshade for the window below.*

(c) Surface Reflectances. Interior reflectances are very important in daylight design. In addition to determining the magnitude of the internally reflected light component (IRC) within the room, they determine in large measure the eye adaptation level. A high adaptation level is desirable to avoid a sensation of glare when the window and its immediate surround are in the field of vision. Furthermore, the internally reflected light component contributes largely to the diffuseness of the room light. With low IRC, the sky component of daylight is the essential illuminant, and diffuseness and room penetration are reduced. Floors receive the sky component directly and should have at least 20% reflectance. Ceilings receive the ground component and should have at least 70% reflectance. Walls receive rereflected light. Since they are the surfaces seen at normal vision angles, they are responsible for eye adaptation levels and should have at least 50% reflectance.

Exterior surface reflection can provide the deep daylight penetration that is required for effective daylighting. Thus a concrete or light-painted ground surface (RF of 50 to 70%) will furnish one-quarter to two-thirds of the light incident on a window, depending on shading and orientation. When combined with a high-reflectance ceiling, good interior distribution can be achieved.

(d) Glare and Heat Control (Sunshading). These are among the most difficult problems to overcome in daylighting. All exposures except north are exposed to direct sun during clear-sky weather, with the attendant glare and heat problem. Although a northern exposure can receive direct sun during the summer, the early (and late) hour, oblique angle, and low sun altitude combine to essentially eliminate all problems except possibly glare. This subject was discussed at length in Chapters 5 and 6, primarily from the viewpoint of comfort. The reader is referred to Section 6.3*a* and to Table 6.6 for a description of adjustable sunshading devices. In addition to those devices, an effective technique that not only reduces glare, but also increases daylight penetration is the use of light-directing glass blocks (see Fig. 19.27).

We assume that it is understood that furni-

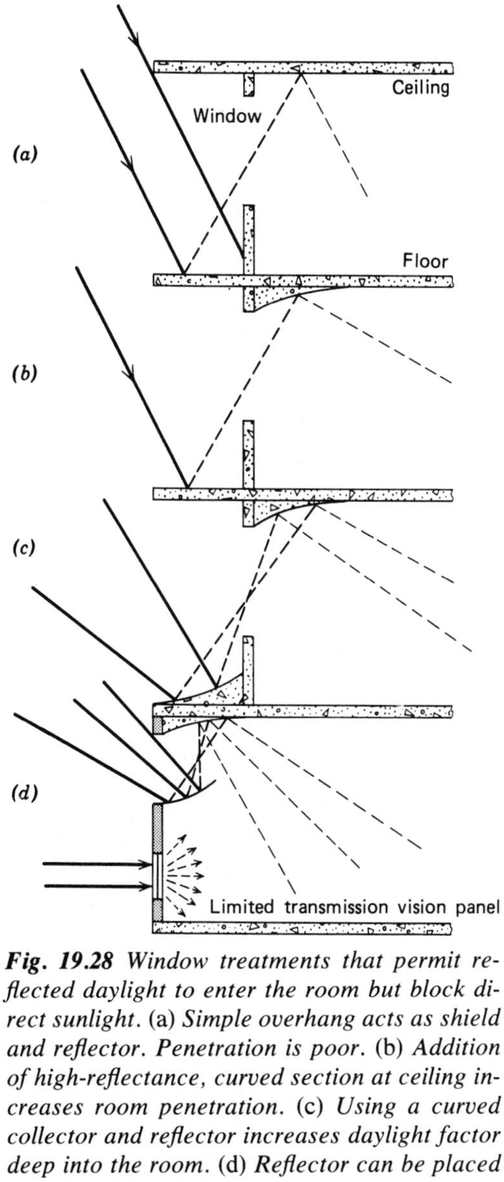

Fig. 19.28 Window treatments that permit reflected daylight to enter the room but block direct sunlight. (a) *Simple overhang acts as shield and reflector. Penetration is poor.* (b) *Addition of high-reflectance, curved section at ceiling increases room penetration.* (c) *Using a curved collector and reflector increases daylight factor deep into the room.* (d) *Reflector can be placed inside the building wall, using building wall for sun shield. Daylight penetration can be very deep, since sky factor angle is large.*

ture be oriented so that windows are to the side or, if not possible, to the rear. Windows in the line of sight must be sufficiently distant as not to be a glare source and must be provided with some type of glare control, such as blinds, curtains, shades, or the like.

(e) Fixed Sunshades and Lightshelves (See Appendix D). Fixed exterior sunshades,

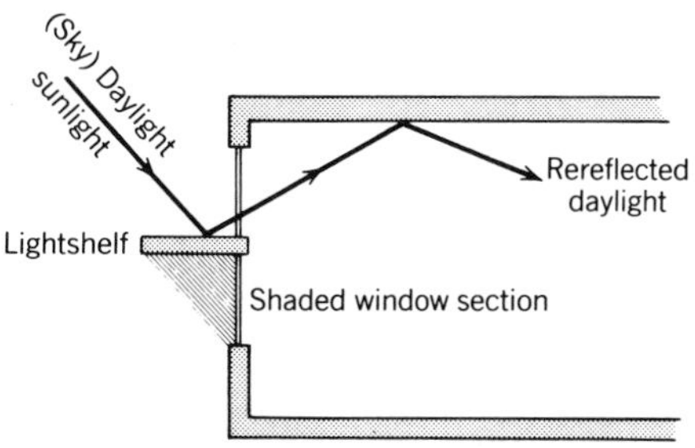

Fig. 19.29 Lightshelf acts to reduce daylight near the window and increases it at greater depth. Shelf material (opaque, translucent) and angle of installation (horizontal, sloped up) markedly affect performance.

particularly horizontal overhangs, can be arranged to function simultaneously as a light shelf, that is, as a reflector, to increase the flow of light to the *floor above* (see Fig. 19.28). If the overhang is moved down, as in Fig. 19.29, so that it shades the lower portion of the window, while reflecting light into the upper part of the same floor, it is usually referred to as a light shelf. Sunshades and light shelves have become common in recent architectural design because of the desire to increase the amount and penetration of daylight, while limiting the undesirable effects of direct insolation (see Fig. 19.26*a*-3). That being so, it is necessary to understand the effect of fixed sunshades and lightshelves on the amount of daylight introduced, and its distribution for all types of sky conditions including overcast. Beam sunlighting is not included in this section.

In one interesting study, Millet, Lakin, and Moore (1981) compared various fixed shading devices that have the same shading mask (see Fig. 19.30) as to their effect on interior daylight distribution, for both direct sun and overcast sky conditions. (Readers unfamiliar with shading masks are referred to Olgyay, and Olgyay, 1957; Ramsey and Sleeper, 1981; Lim et al., 1979; and Harkness and Mehta, 1978). The Millet et al. study conclusively indicates that devices with identical shading masks can yield vastly different interior daylight results. Therefore the lighting designer/architect would be well advised to test a model (either physically or mathematically) of any intended shading/reflecting device

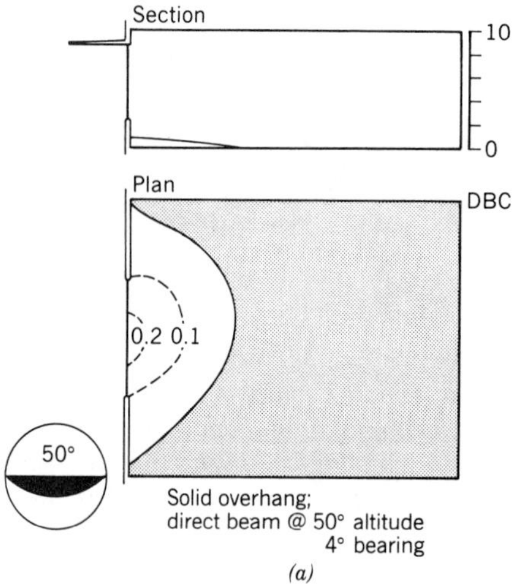

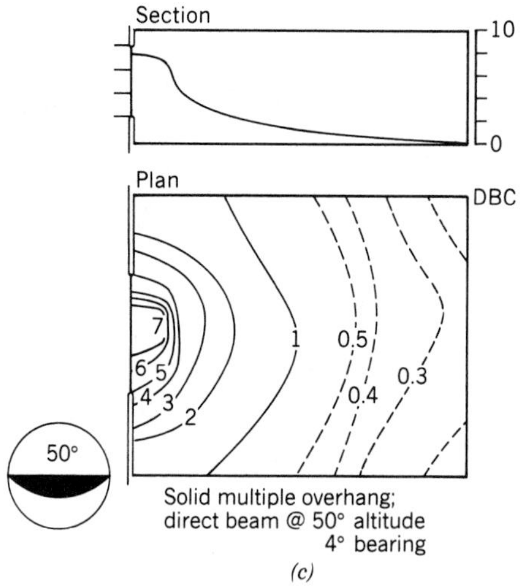

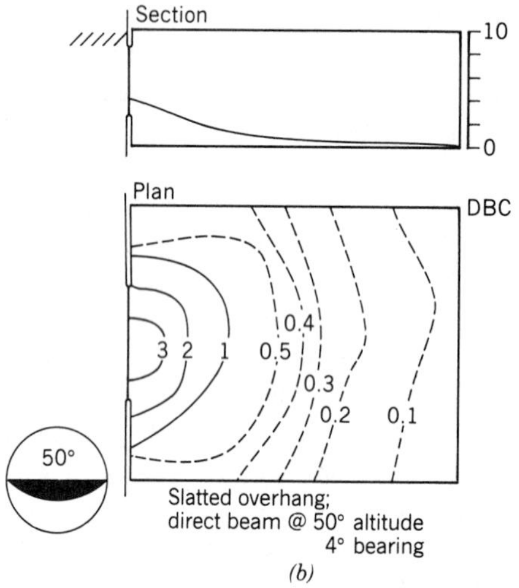

Fig. 19.30 *Three horizontal shading device designs with identical shading masks but resulting in entirely different interior daylight distributions under (simulated) direct sunlight conditions (labeled DBC). The solid overhang* (a) *provides complete (50°) window shading and results in a rapid drop-off of light in depth. The slanted slatted shade* (b) *acts both as shade and reflector, while the small stacked overhangs act as lightshelves, shading below and reflecting above. [Reproduced with permission from Millet, Lakin, and Moore (1981).]*

Fig. 19.31 (a) *When oriented east or west, expanses of glass must be provided with sun control devices. This unshielded, west-facing, 15-ft, glass-door facade creates intolerable disabling glare, the more so for the dimly lit interior corridor and associated low eye-adaptation level. (Photo by Stein.)* (b) *It is desirable from an architectural viewpoint to have the ceiling skylight area also incorporate artificial light sources, to be used when daylighting is insufficient. These alternative sources can be modulated automatically by the available daylight. This shopping mall uses suspended, acrylic-enclosed mercury luminaries that blend with the large skylight. They illuminate the skylight well with light of similar color to daylight, giving the ceiling area a daylight appearance at all times. Note, too, the stepped ceiling construction, which serves to reduce brightness gradually. (Photo courtesy of* LD&A Magazine, *from article by J. Wilson, May 1974.)* (c) *Daylight factor here has a large ground component due to cutoff of the sky component by deep overhang, which also acts as a reflector for the ground light. Low-reflectance surfaces are suitable for a corridor that has no demanding visual task. (Photo courtesy of Libbey-Owens-Ford.)* (d) *Full-height clear glazing with negligible direct sun shading is applicable only on north elevations or climates with continual overcast sky, as in this Pacific Northwest installation. Note that work positions are not*

Fig. 19.31 *(continued)*

placed facing the glass, to avoid excessive brightness ratios in the field of view. Exterior planting reduces any incipient glare. (Photo courtesy of Libbey-Owens-Ford.) (e) *Unusual application of skylighting provides daylighting to an inner court of the International Monetary Fund Headquarters, Washington, D.C. Translucent acrylic panels give sun control while limiting heating and glare. (Courtesy of Roper IBG.)*

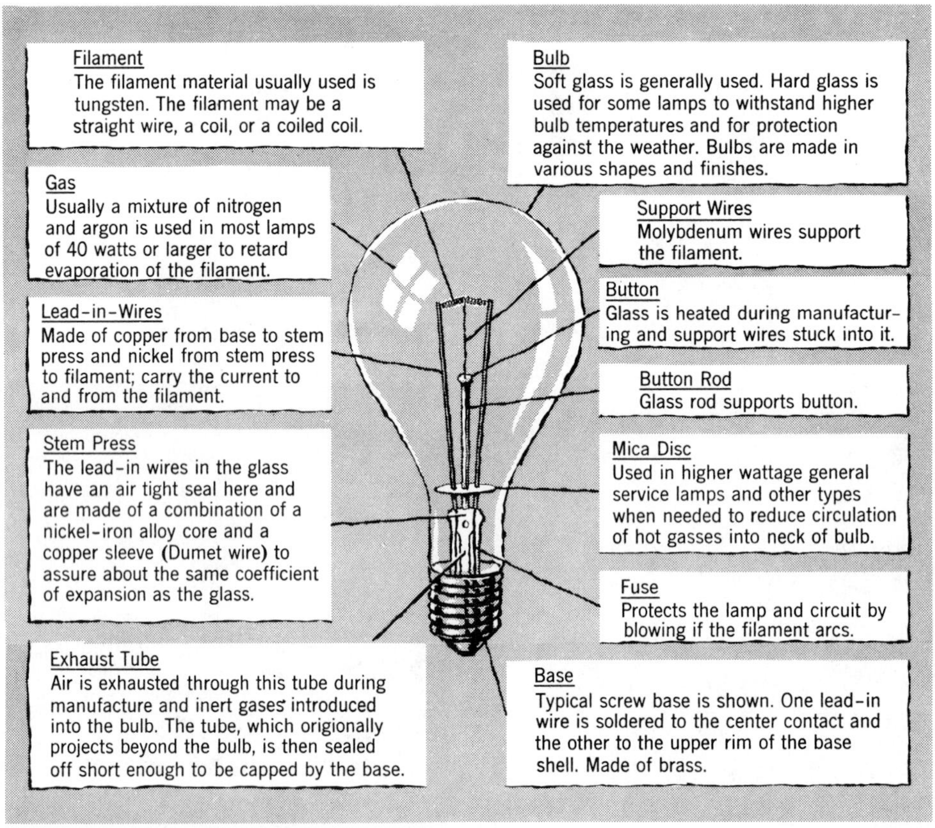
Filament
The filament material usually used is tungsten. The filament may be a straight wire, a coil, or a coiled coil.
Gas
Usually a mixture of nitrogen and argon is used in most lamps of 40 watts or larger to retard evaporation of the filament.
Lead-in-Wires
Made of copper from base to stem press and nickel from stem press to filament; carry the current to and from the filament.
Stem Press
The lead-in wires in the glass have an air tight seal here and are made of a combination of a nickel-iron alloy core and a copper sleeve (Dumet wire) to assure about the same coefficient of expansion as the glass.
Exhaust Tube
Air is exhausted through this tube during manufacture and inert gases introduced into the bulb. The tube, which origionally projects beyond the bulb, is then sealed off short enough to be capped by the base.
Bulb
Soft glass is generally used. Hard glass is used for some lamps to withstand higher bulb temperatures and for protection against the weather. Bulbs are made in various shapes and finishes.
Support Wires
Molybdenum wires support the filament.
Button
Glass is heated during manufacturing and support wires stuck into it.
Button Rod
Glass rod supports button.
Mica Disc
Used in higher wattage general service lamps and other types when needed to reduce circulation of hot gasses into neck of bulb.
Fuse
Protects the lamp and circuit by blowing if the filament arcs.
Base
Typical screw base is shown. One lead-in wire is soldered to the center contact and the other to the upper rim of the base shell. Made of brass.

Fig. 19.32 *Construction of a typical general-service incandescent lamp.*

under the anticipated sky conditions, before fixing his design in concrete.

Another study, Ander and Navvab (1983), used physical models to check the results of a computer program which calculates the daylighting impact of fenestration controls. Angled fins, lightshelves, and an overhang were tested under simulated clear and overcast sky conditions. The study developed a "fenestration coefficient" that indicates the effect of the device in question on the daylight factor of specific interior areas. Other studies by Modest (1983) and Selkowitz, Navvab, and Mathews (1983) confirmed (among other conclusions) the intuitive notion that a lightshelf's principal action is to project daylight deep into a space, and so even out the daylight level in the room. We trust that further studies will provide the designer with a useful "catalog" of exterior and interior shading devices and shelves, including reliable data on their daylighting (and thermal) impact.

(f) Suggested Daylighting Techniques. Some of the principles involved in the use of daylight in construction are illustrated in Fig. 19.31. The reader is referred to the references at the end of Chapter 21 for full coverage of the many aspects of daylighting that space limitations prevented our covering here. Among these are toplighting, treatment of lightshelves and overhangs, shading devices, daylight color, and similar topics.

INCANDESCENT LAMPS

19.12 The Incandescent Filament Lamp

(a) Construction. The standard lamp consists simply of a tungsten filament inside a gas-filled, sealed glass envelope (see Fig. 19.32). Current passing through the high-resistance filament heats it to incandescence, producing light. Gradual evaporation of the fila-

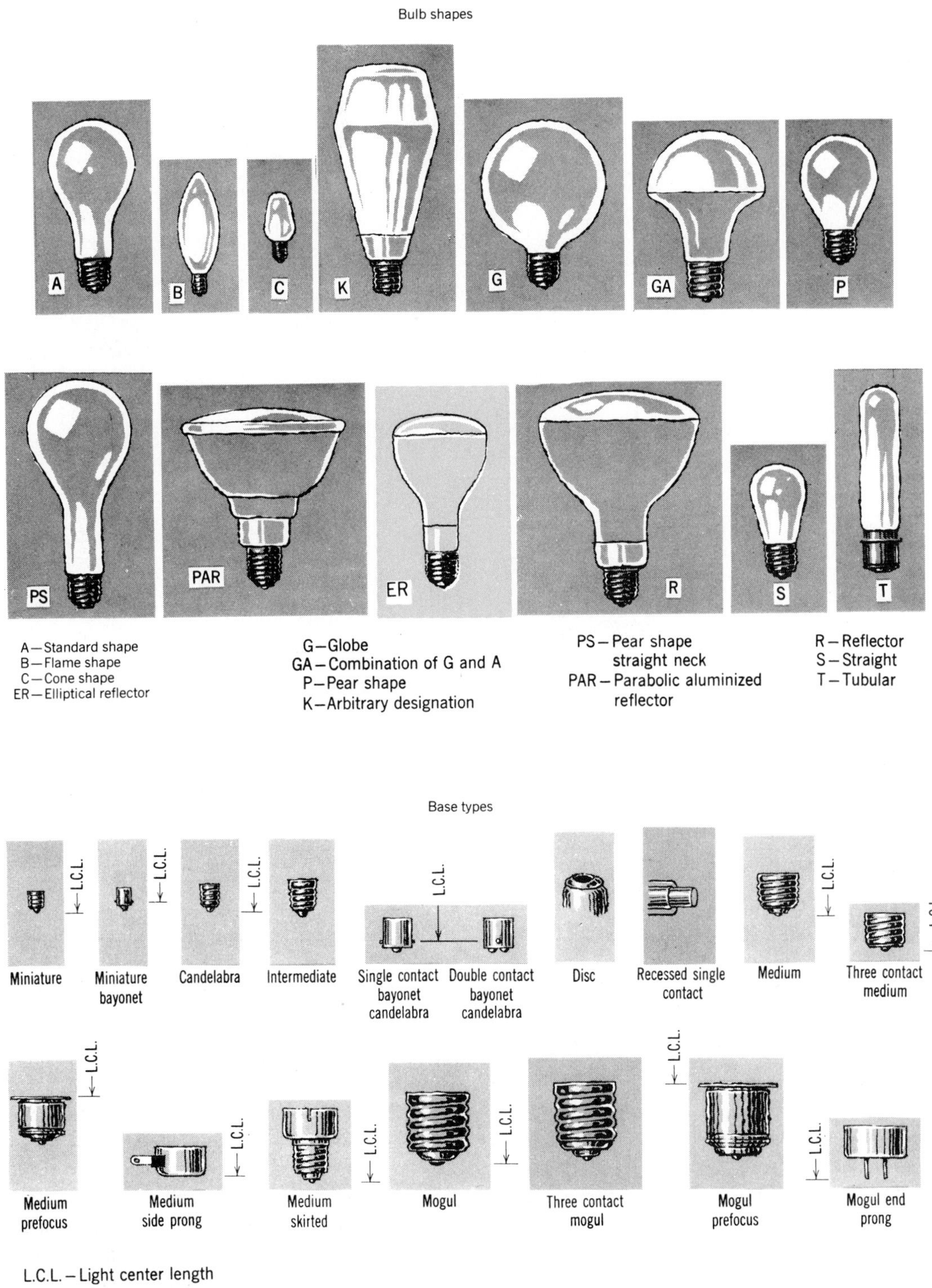

Fig. 19.33 *Common incandescent lamp bulb and base types with nomenclature. The bulb nomenclature indicates type and size; the letter being an abbreviation of the shape and the number equal to the maximum diameter in eighths of an inch. Thus a PS-52 is a pear-shaped bulb, 52/8 (6½) in. in diameter and an R-40 is a reflector lamp 40/8 (5) in. in diameter.*

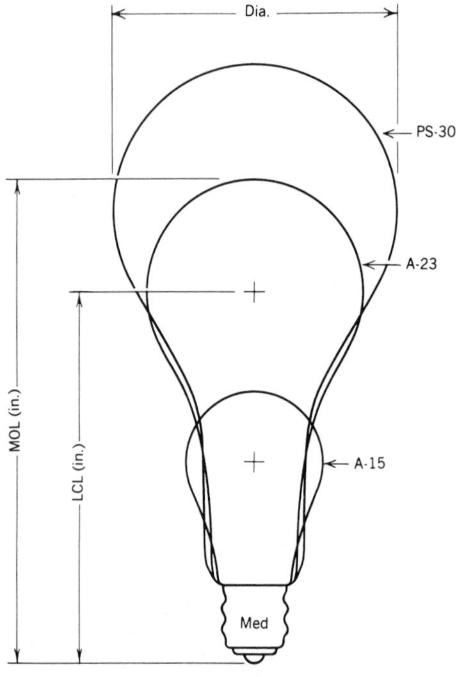
Dia.
PS-30
A-23
A-15
MOL (in.)
LCL (in.)
Med
Bulb diameter is given in 1/8 in. Example: An A-19 bulb has a diameter of 19/8 in. or 2 3/8 in.
MOL: Maximum overall length: This figure refers to the maximum length of the bulb.
LCL: Light center length: This dimension, important when designing reflectors, is measured from the filament to a point that varies with base type. See Fig. 19.31.

Lamps shown at slightly less than 1/2 actual size

	A – STANDARD SHAPE									PS – PEAR SHAPE							
WATTS	15	25	40	60	75	100	100	150	150	150	200	300	300	500	750	1000	1500
BULB	A–15	$A–19_1$	$A–19_2$	$A–19_3$	$A–19_3$	$A–19_3$	$A–21_1$	$A–21_2$	A–23	PS–25	PS–30	PS–30	PS–35	PS–40	PS–52	PS–52	PS–52
DIAMETER"	$1\frac{7}{8}$	$2\frac{3}{8}$	$2\frac{3}{8}$	$2\frac{3}{8}$	$2\frac{3}{8}$	$2\frac{3}{8}$	$2\frac{5}{8}$	$2\frac{5}{8}$	$2\frac{7}{8}$	$3\frac{1}{8}$	$3\frac{3}{4}$	$3\frac{3}{4}$	$4\frac{3}{8}$	5	$6\frac{1}{2}$	$6\frac{1}{2}$	$6\frac{1}{2}$
M.O.L."	$3\frac{1}{2}$	$3\frac{7}{8}$	$4\frac{1}{4}$	$4\frac{7}{16}$	$4\frac{7}{16}$	$4\frac{7}{16}$	$5\frac{1}{4}$	$5\frac{1}{2}$	$6\frac{3}{16}$	$6\frac{15}{16}$	$8\frac{1}{16}$	$8\frac{1}{16}$	$9\frac{3}{8}$	$9\frac{3}{4}$	13	13	13
L.C.L."	$2\frac{3}{8}$	$2\frac{1}{2}$	$2\frac{15}{16}$	$3\frac{1}{8}$	$3\frac{1}{8}$	$3\frac{1}{8}$	$3\frac{7}{8}$	4	$4\frac{5}{8}$	$5\frac{3}{16}$	6	6	7	7	$9\frac{1}{2}$	$9\frac{1}{2}$	$9\frac{1}{2}$
BASE	MED	MED	MED	MED	MED	MED	MED	MED	MED	MED	MED	MED	MOG	MOG	MOG	MOG	MOG
STANDARD FINISH	IF	IF	IF W	IF	IF	IF	IF	IF	CL IF	CL IF	IF	IF	IF	IF	CL IF	CL IF	CL IF

CL – CLEAR IF – INSIDE FROSTED

Fig. 19.34 *Typical dimensional data for common general-service incandescent lamps.*

ment causes the familiar blackening of the bulbs and eventual filament rupture and lamp failure. Incandescent lamps are available in many bulb and base types, with special designs for particular application (see Figs. 19.33 and 19.34). To diffuse the light, most bulbs are either etched on the inside (inside-frosted) or coated inside with white silica. The silica coating provides almost complete light diffusion at a cost of approximately 2 to 3% of the light output, whereas inside-frosted bulbs provide only partial diffusion but do not reduce light output. Inside-frosted bulbs are gradually being replaced by white lamps because of the ecological problems involved with the acid used in the etching process. Colored light is also readily available from either coated bulbs or bulbs of colored glass.

The lamp base is the means by which connection is made to the socket and thereby to the source of electric current. Most lamps are made with screw bases of various sizes, the most common being the medium screw base. General service lamps, of 300 W and larger, use the mogul screw base. Where exact positioning of the filament is important, as it is when lamps are placed in precise reflectors or in lens systems, a screw

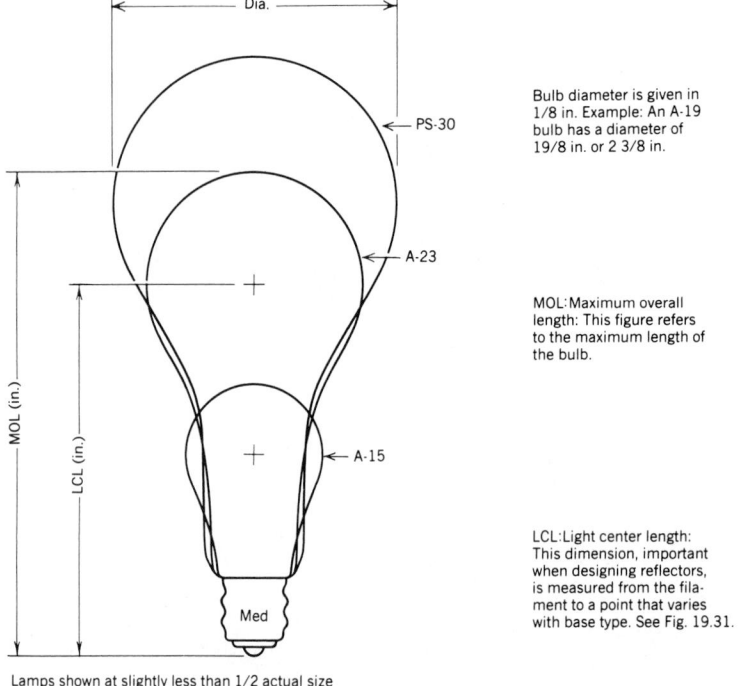

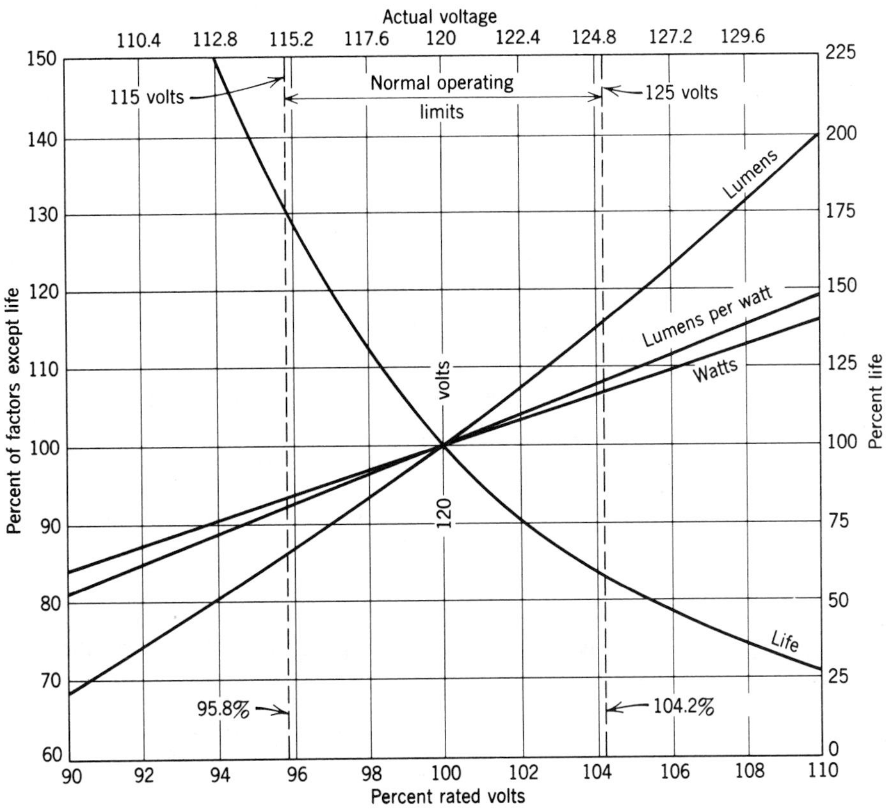

Fig. 19.35 *Characteristics of a standard 120-V general-service incandescent lamp as a function of voltage.*

base cannot be used. Lamps designed for such use are furnished with one of the special bases illustrated in Fig. 19.33.

(b) Operating Characteristics. These are critically dependent on the voltage at the lamps; therefore the life, output, and efficiency of a lamp can be markedly altered by even a small change in operating voltage, as illustrated by Fig. 19.35.

For example, burning a 120-V lamp at 125 V (104.2%) means approximately:

16% more light (lumens)
7% more power consumption (watts)
8% higher efficacy (lumens per watt)
42% less life (hours)

Burning a 120-V lamp at 115 V (95.8%) means approximately:

15% less light (lumens)
7% less power consumption (watts)
8% lower efficacy (lumens per watt)
72% more life (hours)

Particular note should be taken of the effect of voltage on lamp life. In installations where lamp replacement is difficult and/or expensive, lamps may be burned slightly under voltage and life prolonged, thereby decreasing the frequency of replacement. However, since efficiency is decreased by this procedure and since energy cost is normally a major cost in any lighting installation over the life of the installation, a detailed cost analysis should be made by the consulting engineer involved. Conversely, where lamps are replaced before burnout on a group replacement system and initial installation cost per footcandle and/or energy costs are high, lamps may be burned overvoltage, thereby increasing output and efficiency but shortening life. This procedure is normal in sports-lighting installations because of the high cost of tower-mounted floodlights, making it mandatory to extract the maximum light from each unit. In stadium installations that have yearly burning schedules averaging less than 200 hours, 10% overvoltage operation doubles the light output but still allows a once-a-year, off-season relamping and is therefore a highly economical procedure.

In general, however, it is advisable to operate incandescent lamps at rated voltage, accepting balanced efficiency, output, and life.

(c) Other Characteristics

1. *Lumen maintenance.* Light output decreases slowly with lamp life as the bulb blackens. Position during burning and bulb temperature affect this characteristic.

2. *Color.* White with a large yellow-red component and therefore highly flattering to the skin. As is explained in detail in Section 19.25, the spectral content of the light produced by a heated (incandescent) source depends on its temperature; high-wattage lamps are bluer, low-wattage lamps are yellower. Dimmed lamps give yellow-red light.

3. *Surroundings.* Generally, incandescent lamps are impervious to external heat, cold, or humidity. Starting of these lamps is completely unaffected by ambient conditions.

4. *Lamp efficiency.* Since incandescent lamps produce light as a by-product of heat, they are inherently inefficient. Luminous effi-

TABLE 19.15 **Typical Incandescent Lamp Data (Listing a Few of Many Sizes and Types of 115-, 120-, and 125-V Lamps)**

Watts and Life			Lumens		Physical Data		
Lamp Watts[a]	Average Rated Life (h)	Approx. Color Temp.[b] (K)	Initial Lumens	Lumens per Watt[c]	Shape of Bulb[d]	Base	Description
15	2500	—	126	8.4	A-15	Med	Long Life
25	1000	—	340	10.1	A-19	Med	Rough service
25	2500	2550	232	9.3	A-19	Med	—
40	1500	—	480	12.0	A-19	Med	—
50	2000	—	525	10.5	ER30	Med	See Fig. 20.23
60	1000	2800	890	14.8	A-19	Med	—
60	2500	—	750	12.5	A-19	Med	Long life
75	750	—	1220	16.3	A-19	Med	—
100	750	2870	1750	17.5	A-19	Med	—
100	750	—	1690	16.9	A-21	Med	—
100	2500	—	1460	14.6	A-19	Med	Long life
100	1000	—	1220	12.2	A-21	Med	Rough surface
150	750	2900	2810	18.7	A-21	Med	—
150	750	—	2680	17.9	PS-25	Med	—
200	750	2930	4000	20.0	A-23	Med	—
300	750	2940	6300	21.0	PS-25	Med	—
500	1000	3000	10850	21.7	PS-35	Mogul	—

[a]Figures in this column designate the input watts thus: 60 means 60 W. The letters identify the treatment of the glass bulb. All inside-frosted unless otherwise noted. Other letters have these meanings: W, white; SBIF, silver bowl, inside-frosted; Cl, clear; SW, soft white; Cer, ceramic.

[b]See Section 19.25.

[c]Efficacy (luminous efficiency), in lumens per watt, increases with filament temperature; therefore, with wattage.

[d]Bulb designations consist of a letter to indicate its shape and a figure to indicate the approximate maximum diameter in eighths of an inch (see Fig. 19.33).

ciency (efficacy) increases with wattage. Thus a 60-W general-service lamp produces 870 initial lumens or 14.5 lm/W, whereas a 100-W lamp produces slightly more output than two 60-W lamps, or put otherwise, it results in a 20% energy savings. Refer back to Section 18.6 for a discussion of photometric efficiency, or efficacy.

(d) Summary. The principal advantages of incandescent lamps are low cost, instant start and restart, simple inexpensive dimming, simple compact installation requiring no accessories, cheap fixtures, focusability as a point source, high power factor, life independent of number of starts, and good color.

The principal disadvantages are low efficacy (see Fig. 19.1), short lamp life, and critical voltage sensitivity. Low efficacy results in a large number of fixtures, high maintenance costs, and large heat gain. Short lamp life results in high replacement labor cost. Voltage sensitivity requires careful and and expensive circuit design. Also, light concentration of the filament (point source) requires careful fixture design in order to avoid glare and, if undesirable, sharp shadows. Because of the poor energy characteristic, incandescent lamp use should be limited to the following applications.

1. Infrequent or short-duration use.
2. Where low-cost dimming is required.
3. Where the point source characteristic of the lamp is important, as in focusing fixtures.
4. Where minimum initial cost is essential.
5. Where its characteristic color rendering is desired.

A brief list of conventional incandescent lamps and their physical and operating characteristics is given in Table 19.15. Lamp data for use in design should be taken from current manufacturers' literature. Data presented here are typical.

19.13 Special Incandescent Lamps, Including Energy-Saving Lamps

In the field of incandescent lamps other than the tungsten-halogen lamp, which is discussed sepa-

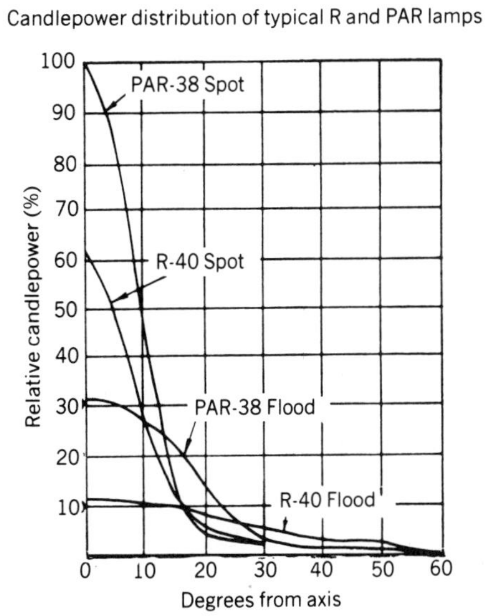

Fig. 19.36 *Typical photometric data for reflector spot and flood lamps. For dimensional data, see manufacturers' catalogs.*

rately, numerous special types are available. Some of the more important types are covered briefly in the following pages.

Rough service and *vibration* lamps are built to withstand rough handling and continuous vibration, respectively, both of which conditions are extremely hard on general-service lamp filaments. Neither type is intended for general use, and both types have lower luminous efficacy than a general-service lamp (see Table 19.15).

Extended-service lamps are designed for 2500-h life and are useful, as mentioned, in locations where maintenance is irregular and/or relamping is difficult. The lamp is really designed for slightly higher voltage than that at which it is applied, and therefore efficacy is reduced (see Table 19.15 and Fig. 19.35).

So-called long-life lamps, which are guaranteed to burn for 2, 3, or 5 years, are lamps designed for much higher voltages than that at which they operate. Since they normally sell at a high cost and are very inefficient, their use is seldom advisable. Before using such lamps, a cost comparison, including cost of lamps, energy, and relamping, should be made (see Appendix E).

(a) Reflector Lamps. These are made in "R" and "PAR" shapes (see Fig. 19.33) and

The beam of the PAR lamp is conical in shape. Each type of PAR lamp has a distinct illumination pattern which varies in size and light intensity — depending on the angle at which the lamp is aimed and on its distance from the area illuminated.

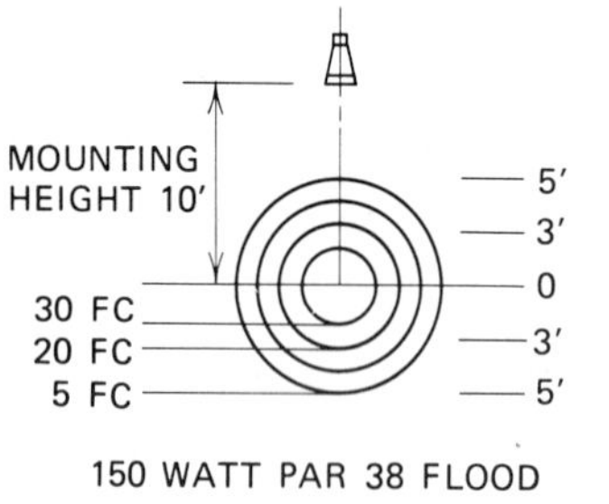

When centered directly on the surface to be lighted (at right angles or zero degrees) the small PAR 38 sizes give a round lighting pattern. The concentric rings show the amount of light measured in footcandles at various distances from the beam center. The round lighting pattern changes to oval or elliptical when the lamp is aimed at an angle.

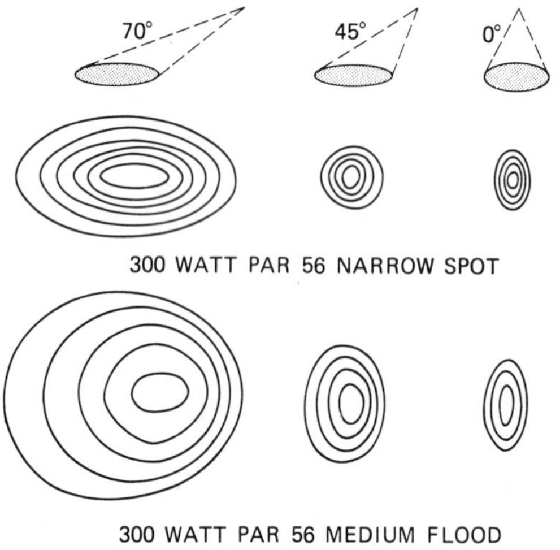

The lighting pattern of the larger PAR 46, 56, and 64 lamps is oval or elliptical, whether centered directly on the surface or aimed at an angle. As shown in the two diagrams, aiming PAR lamps at progressively greater angles — proportionately increases both the length and width of the area illuminated. In general, for spotlights the length of the lighting pattern becomes proportionately greater — and in the case of floodlights, the width.

Fig. 19.37 *Typical illumination patterns for PAR spot and flood lamps. (Courtesy of GTE/Sylvania, Inc.)*

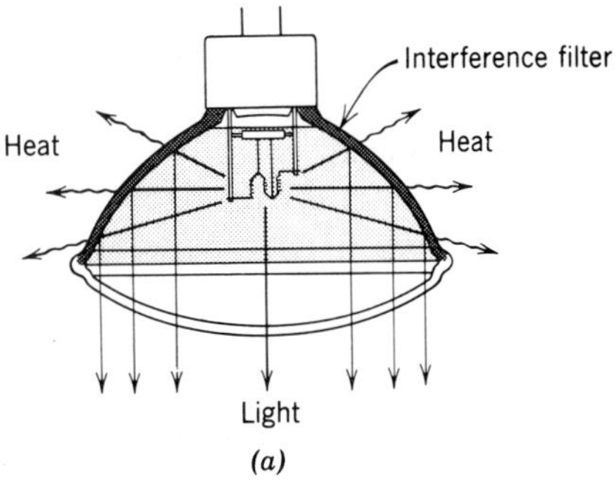

(a)

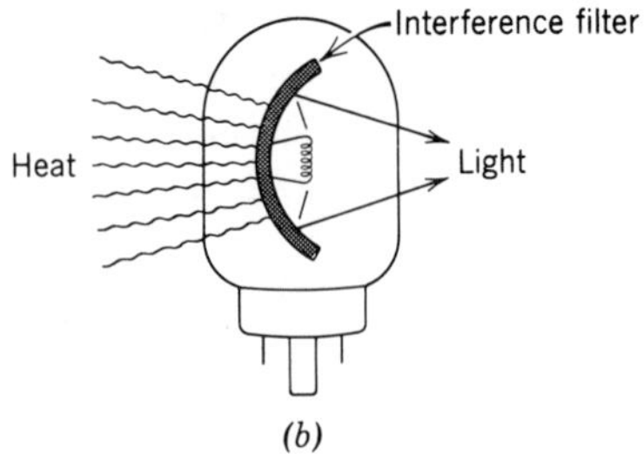

(b)

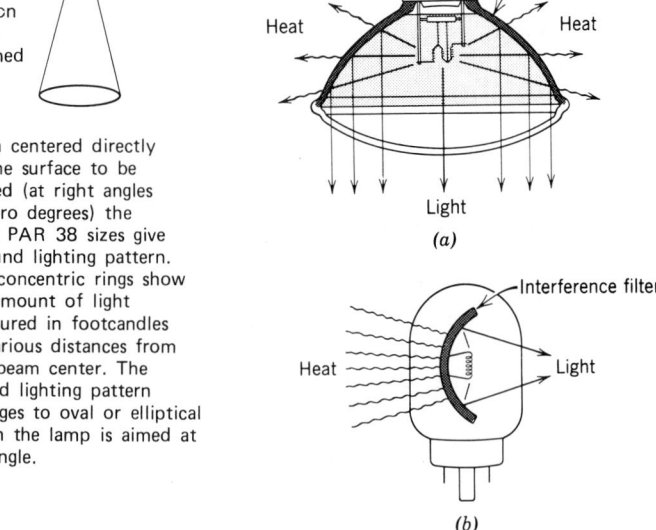

Fig. 19.38 *The reflectors in the cool-beam PAR lamp* (a) *and the projection lamp* (b) *reflect most of the visible light while transmitting the infrared energy, which constitutes two-thirds of the total heat energy. In an incandescent lamp, about 90% of the input energy is converted to heat. Provision for heat removal must be made in the fixture.*

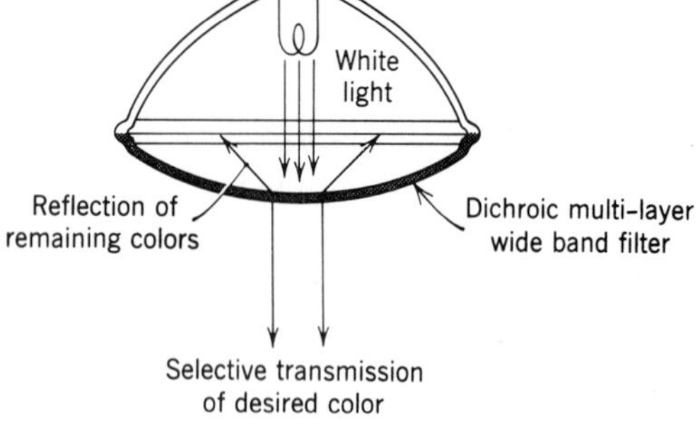

Fig. 19.39 *Action of dichroic filter is one of selective interference rather than absorption. Each film layer transmits one color while reflecting its complement. Desired color is obtainable by action of several films.*

contain a reflective coating on the inside of the glass envelope; this gives the entire lamp accurate light beam control. Both types are available in narrow or wide beam design, commonly called *spot* and *flood*, respectively. R lamps are generally made in soft glass envelopes for indoor use, whereas PAR lamps are hard glass, suitable for exterior application.

Typical reflector lamp photometric data are given in Fig. 19.36. Illumination patterns resulting from typical PAR spot and flood lamps are shown in Fig. 19.37. When using R and PAR lamps, the fixture acts principally as a lampholder, since beam control is built into the lamp.

(b) Interference (Dichroic) Filters. The basic filter is a thin film that operates on the interference principle rather than absorption. Thus the lamp surface remains relatively cool. In one design that is utilized to limit the heat in the light beam, the film is applied to the inside back of the lamp. It acts by transmitting infrared heat out the lamp back while reflecting light out the lamp front (see Fig. 19.38). Typical applications are in window displays, over food counters, and in any location where a "cool beam" is desirable. Obviously, provision must be made for removal of the heat from the fixture if the lamp is housed.

In a second design, multiple-layer filters are applied to the front of the lamp. Each film acts to transmit one color and reflect its complement (two-color, hence dichroic). These dichroic filter lamps produce a purer, more saturated color at high efficacy than is possible with selective absorption filters (see Fig. 19.39).

(c) Elliptic Reflector Lamps. These lamps, which are usually classed as energy-saving lamps, are simply an improved reflector design that increases the efficiency of the lamp–fixture combination over a similar installation using a conventional R reflector. The action of the reflector is shown in Figs. 20.18 and 20.23. This shape causes the beam to focus a few inches in front of the lamp, permitting high-efficiency application in pinhole downlights or deep baffle units, where use of ordinary R lamps causes trapping and loss of most of the lamps' output. Thus, although a 75-W ER (elliptical reflector) lamp will yield as much output from a pinhole downlight as a 150-W R lamp, it itself has no greater efficacy than the R lamp (see Table 19.15). In an open-type downlight or one designed for R-type reflectors, use of the (more expensive) ER lamp may actually *reduce* output.

(d) Energy-Saving Lamps. The requirement to conserve lighting energy has resulted in a line of energy-saving lamps being produced by every major manufacturer. These lamps, of all types, are known by trademarked names. Among them are Philips Econ-o-watt™ and GTE/Sylvania's Super-Saver™. These lamps are rated at a wattage lower than the standard lamps they are intended to replace, and their catalog listing indicates this. Thus a 60A-52A/EW—is a 52-W A-shape bulb Econ-o-watt™ lamp, intended to replace a 60-W lamp. Its 800-lm output gives it an efficacy of 15.38 *l*umens *p*er *w*att as compared to the standard lamp's 890 lm and 14.8/pw. Thus it has 10% lower output (800/890) but 13.3% lower energy consumption (52/60). Hence its energy-saving characteristic. However, not all reduced-wattage lamps have higher efficacy than the corresponding standard lamps. Some simply reduce light output proportionately to the wattage. In such instances, if light levels must be maintained, a *loss* is incurred, since more fixture bodies are required to hold the lower output lamps, for the same total output. Finally, because lamp life for energy-saving lamps is often different from that of a standard lamp, and first cost is usually higher, a simple comparison is not possible.

As has been repeatedly stated, any additional first cost should be analyzed by a life-cycle cost analysis and the payback period calculated. The designer will find that a proper control system is often more economically attractive than low-wattage lamps, and that energy-saving lamps are primarily useful in retrofit work.

19.14 Tungsten-Halogen (Quartz-Iodine) Lamps

This lamp type, illustrated in Fig. 19.40 is similar to the standard incandescent lamp in that it produces light by heating a filament. It differs in that a small amount of halogen gas (iodine or bromine) is added to the inert gas mixture that

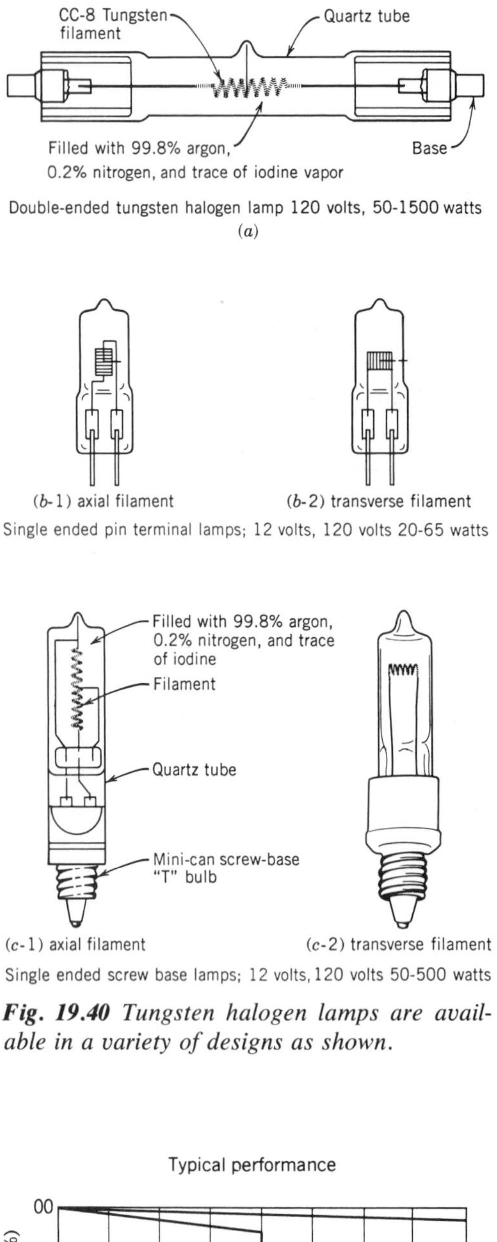

Fig. 19.40 *Tungsten halogen lamps are available in a variety of designs as shown.*

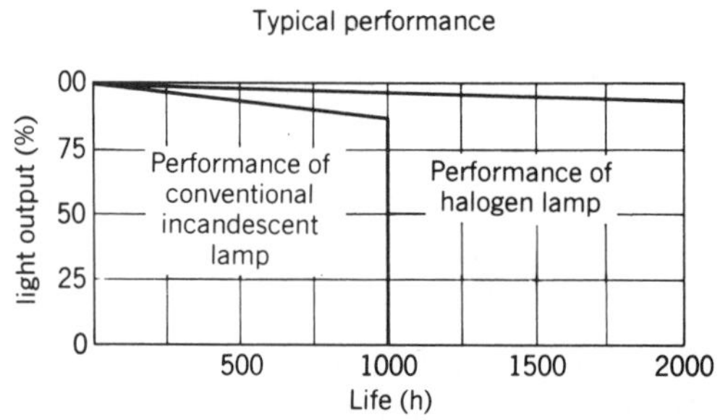

Fig. 19.41 *The self-regenerative halogen cycle results in slow evaporation of the tungsten filament with consequent lower light depreciation and longer life than those of a standard incandescent lamp.*

fills the bulb. This addition results in retardation of filament evaporation, which is the usual cause of lamp failure (see Fig. 19.41).

Although the lamp has only slightly higher efficacy than an equivalent common incandescent, it has the advantages of longer life, lower lumen depreciation (98% output at 90% life), and a smaller envelope for a given wattage (see Fig. 19.41). The latter characteristic is due to the high temperature required by the halogen cycle, which in turn requires a compact, high-temperature filament. As a result, the lamp is effectively a point source, making it ideal for use in precision reflectors. Indeed, it is in this area of compact precision reflectors, as discussed below, that most of the developments in tungsten-halogen lamp technology have occurred in the last four years. Due to the lamp's high filament temperature, the bulb envelope is made of quartz, which can withstand the temperature better than glass; this in turn gave rise to the name—quartz iodine—which is sometimes applied to this lamp type. Another result of the high filament temperature is that the gas pressure inside the quartz envelope is elevated and the lamps have been known to rupture violently, spraying hot quartz fragments over a wide area. As a consequence, all manufacturers now provide a cautionary notice with their lamps. The wording varies but essentially states that due to the possibility of rupture, lamps should be handled carefully, guarded against abrasion and overvoltage operation, and most important, adequately shielded or screened. The shield can be a reflector cover, a fixture lens, a screen, or other device that will contain hot flying fragments in the event the lamp shatters.

A number of standards organizations and professional societies have adopted and published cautionary notices, including the *IESNA Lighting Handbook,* ANSI Standard C78.1451.1988, Underwriters' Laboratories UL1591 (1990), Canadian Standards Association, and the International Electrotechnical Commission (IEC). Exceptions to these precautions obtain when a lamp is inherently protected by encapsulation inside a sealed envelope. This encapsulated construction is now manufactured in R, PAR, MR (multimirror reflector), and modified A-lamp shape. These encapsulated lamps are intended for direct replacement of standard incandescent lamps of the same type.

Other halogen lamp characteristics are similar in all respects to the incandescent lamp. Color temperature ranges between 2000 and 3400 K; spectral energy distribution is typical of a blackbody radiation (see Section 19.25), and dimming characteristics are similar as well.

(a) Lamp Types. The basic lamp is a small gas-filled quartz tube, as shown in Fig. 19.40. Because the lamp must be used with some sort of reflector, it is manufactured with different terminations to suit the fixture reflector or secondary lamp envelope into which it is placed. All the lamp types shown in Fig. 19.40 can be used where lamp-only replacement is intended, such as in floodlights or reflector fixtures by using appropriate bases—slide contacts for double-ended lamps, screw bases for screw-base lamps, and special ceramic pinhole bases for pin-type lamps.

(b) Encapsulated Lamps. See Fig. 19.42. These lamps are sealed units intended for direct replacement of either the corresponding distribution-type incandescent lamp in the case of reflector units, or general service incandescent lamps for reflector-less units. Being sealed units, the halogen lamp is not replaceable, and the entire unit is discarded on burnout. Reflector units are available in a wide variety of beam patterns which designers will find detailed in manufacturers' catalogs.

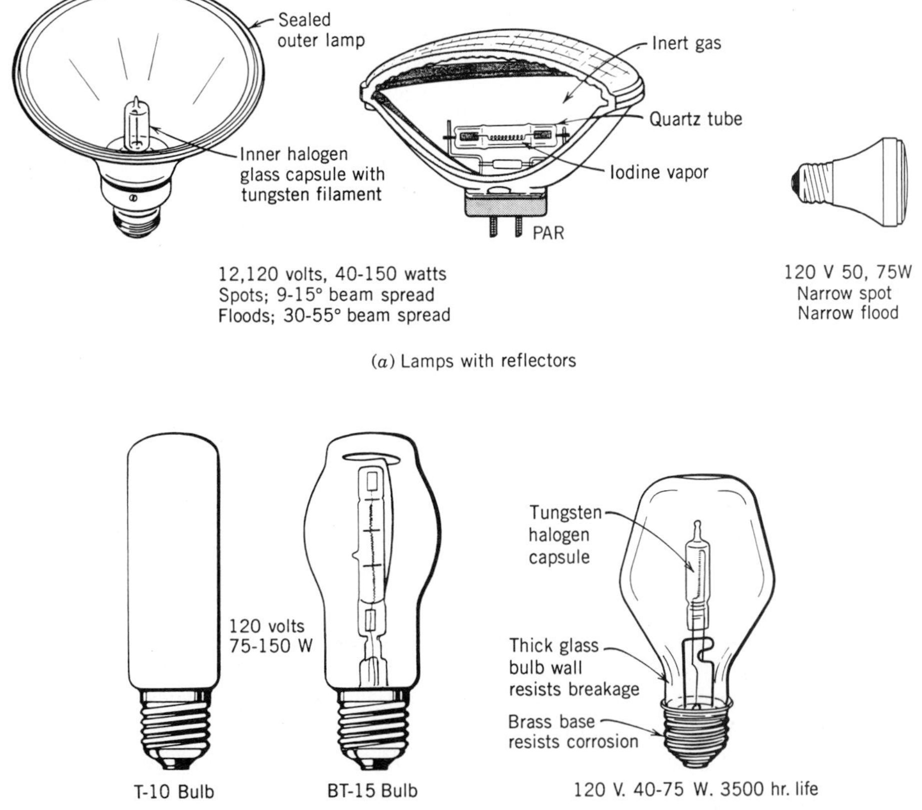

(b) General service encapsulated lamps

Fig. 19.42 *Tungsten halogen lamps can be mounted in a variety of ways inside an enclosing glass envelope.* (a) *Lamp is used either horizontally or vertically to suite reflector design or* (b) *used without a reflector in a protective glass envelope with Edison medium screw base.*

(c) Precision Reflector Units. Miniature single-ended 12-V lamps in 2-in.-diameter (and smaller) multifaceted dichroic (heat ejecting) reflectors with a bipin base have found very wide acceptance in all types of display and accent lighting applications. These units, illustrated in Fig. 19.43, are essentially an entire lighting fixture, like R and PAR lamps, requiring only a base for electrification. The lamps are known by the generic name of MR-16, after an early 2-in. diameter model, and although each major manufacturer utilizes its own trade name, equivalence can be established by referring to the ANSI Code. The multi-mirror-faceted dichroic reflectors produces a "cool" precision light beam by ejecting approximately two-thirds of the lamp heat through the back of the reflector. Means for this heat dissipation must be adequate, as in all installations, but particularly with tungsten-halogen lamps, whose life is sensitive to operating temperature. Beam characteristics of a few MR-16 lamps are given in Fig. 19.43.

FLUORESCENT LAMPS

19.15 The Fluorescent Lamp: Construction

The second major category of light sources is that of electric discharge lamps, of which the fluorescent lamp is the best known and most widely used type. It has become so popular since its major introduction in 1937 that it has almost completely supplanted the incandescent lamp in all fields except specialty lighting and residential use. The typical fluorescent lamp comprises a cylindrical glass tube sealed at both ends and containing a mixture of an inert gas, generally argon, and *low-pressure* mercury vapor. Built into each end is a cathode that supplies the electrons to start and maintain the mercury arc, or gaseous discharge. The short-wave ultraviolet light, which is produced by the mercury arc, is absorbed by the phosphors with which the inside of the tube is coated and is reradiated in the visible light range. The fluorescent lamp is so called because its phosphors fluoresce, or radiate light, when exposed to ultraviolet light. The particular mixture of phosphors used governs the quantity and spectral quality of the light output.

The descriptions that follow cover *standard* lamps and circuits, that is, lamps operating at power line frequency (60 Hz), with standard loading (200, 430, 800, 1500 mA), using core-and-coil magnetic ballasts, in common starting and operating arrangements. Special lamps, accessories, and circuits including low-wattage lamps and ballasts, high-frequency operation with electronic ballasts, 26-mm (T-8) tubes operating at 265 mA, modified waveform ballasts, triphospher lamps, and special-shape lamps are all discussed separately.

(a) Preheat Lamps. The original fluorescent lamp was a preheat design. Construction of a typical hot cathode lamp (preheat and rapid start) is shown in Fig. 19.44*a*; the basic preheat circuit is shown in Fig. 19.45*a*. The circuit utilizes a separate starter, which is a small cylindrical device that plugs into a preheat fixture. When the lamp circuit is closed, the starter energizes the cathodes; after a 2- to 5-s delay, it initiates a high-voltage arc across the lamp, causing it to start. Most starters are automatic, although in desk lamps the preheating is accomplished by depressing the start button for a few seconds and then releasing it. This closes the circuit and allows the heating current to flow; releasing the button causes the arc to strike.

Standard preheat lamps operate at 430 mA. All preheat lamps have bipin bases (see Fig. 19.44*b*). They range in wattage from 4 to 90 W and in length from 6 to 90 in. A typical ordering abbreviation for a preheat lamp would be F15T12WW. This translates: fluorescent lamp, 15 W, tubular-shaped bulb, 12/8-in. diameter (number represents diameter in eighths of an inch), warm white color (see Table 19.16). Preheat lamps are common in sizes up to 30 W; in larger sizes they have been supplanted by rapid-start and instant-start types.

(b) Rapid-Start Lamps. These are similar in construction to the preheat lamps; the basic difference is in the circuitry (see Fig. 19.45*b*). This circuit eliminates the delay inherent in preheat circuits by keeping the lamp cathodes constantly energized (preheated). When the lamp

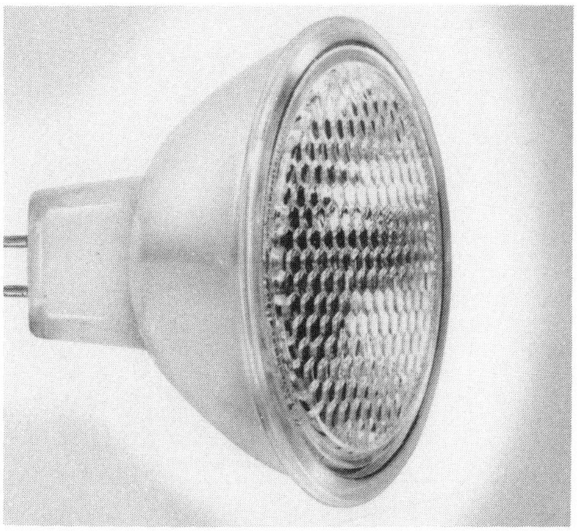

To find the MR 16 lamp appropriate for straight-on application (light source aimed perpendicular to object being illuminated), measure the distance between the fixture and object. Select the distance that is nearest this measurement from the chart below, i.e.from 2 to 12 feet, and find the lamp type that offers desired footcandle illumination level and beam size desired.

Beam pattern for projection perpendicular to surface is approximately circular.

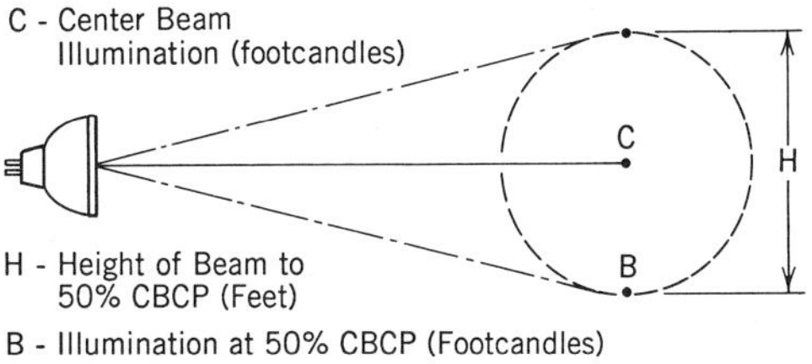

Watts	ANSI Code	Rated Life (Hrs.)	Color Temp. (°K)	CBCP (Candela)	Beam Spread to 50% CBCP	Distribution						
						0	2	4	6	8	10	12
Very Narrow and Narrow Spot												
20	EZX	2000	2925	10780	5.5°	C	2695	674	299	168	108	75
						H	.17	.35	.52	.70	.87	1.0
						B	1344	336	149	84	54	37
42	EZY	2500	3025	13500	7°	C	3375	844	375	211	135	94
						H	.24	.49	.73	.98	1.2	1.5
						B	1678	420	186	105	67	47
20	ESX	3000	2925	3300	12°	C	825	206	92	52	33	23
						H	0.42	0.84	1.3	1.7	2.1	2.5
						B	403	101	45	25	16	11
50	EXT	3000	3025	9150	13°	C	—	572	254	143	92	64
						H	—	0.84	1.3	1.7	2.1	2.5
						B	—	281	125	70	45	31
Spot and Narrow Flood												
42	EYS	3000	3075	2800	20°	C	700	175	78	44	28	—
						H	0.72	1.4	2.1	2.9	3.6	—
						B	334	83	37	21	13	—
50	EXZ	3000	3075	3000	26°	C	750	188	83	47	30	—
						H	0.92	1.9	2.8	3.7	4.6	—
						B	347	84	39	22	14	—
Flood												
20	BAB	3000	2925	460	36°	C	115	29	13	7.1	4.6	—
						H	1.3	2.6	3.9	5.2	6.5	—
						B	49	12	5.5	3.1	2.0	—
42	EYP	3000	3050	991	36°	C	248	62	28	15	9.9	—
						H	1.3	2.6	3.9	5.2	6.5	—
						B	106	27	12	6.6	4.3	—
65	FPB	3500	3050	2000	38°	C	500	125	56	31	20	—
						H	1.4	2.8	4.1	5.5	6.9	—
						B	211	53	23	13	8.5	—

Fig. 19.43 *The MR-16 lamp fixture comprises a single-ended tungsten halogen lamp in a 2-in.-diameter multimirrored precision reflector designed to produce a precise, relatively sharp cutoff beam. The table shows principal lamp characteristics and illuminances on a perpendicular surface at specified distance. On angled surfaces the beam produced is approximately elliptical. (Data extracted from GTE/Sylvania published material of 1988 is for reference only, and is subject to change. Photo of Sylvania Tru-Aim Professional lamp, courtesy of GTE Lighting, Europe.)*

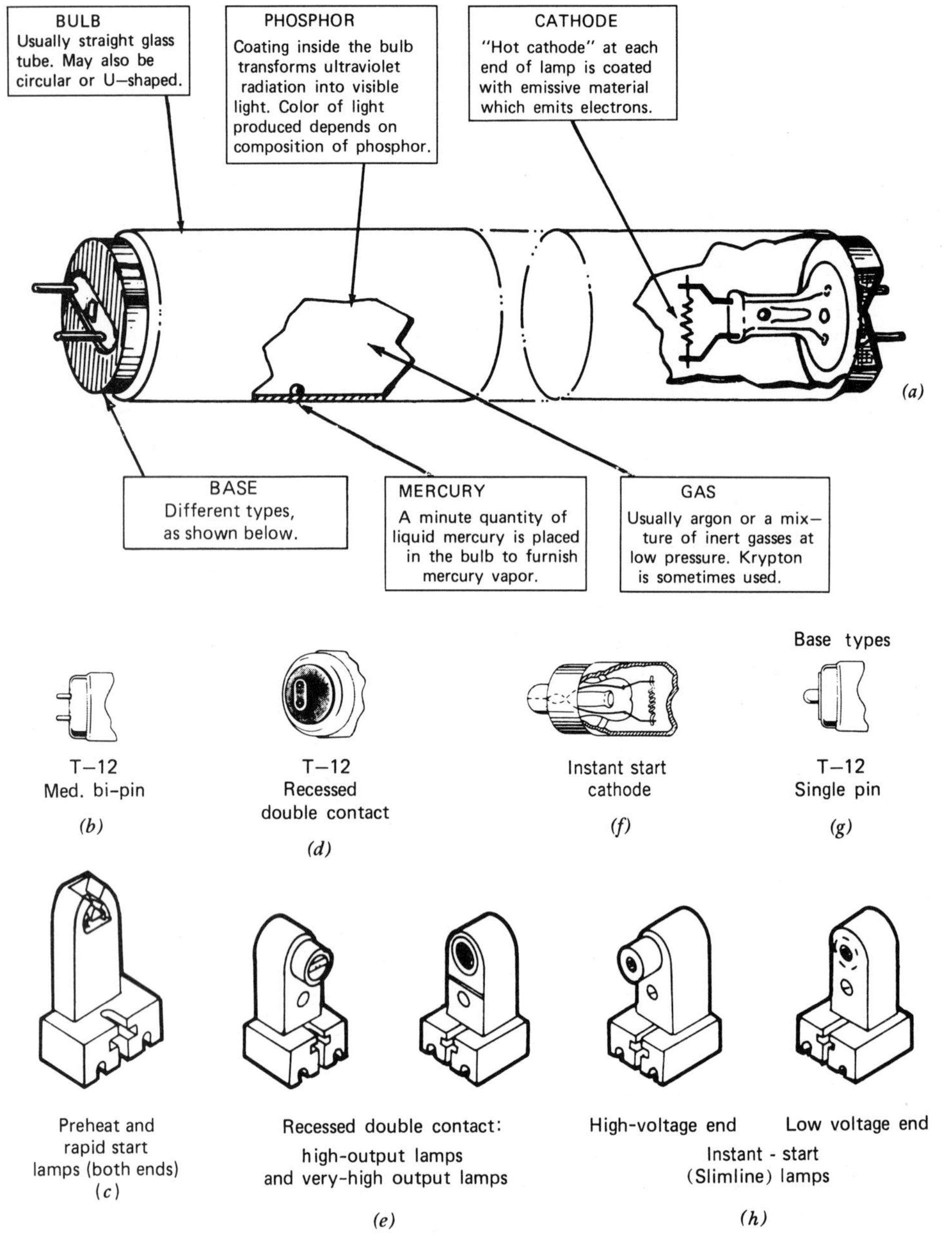

Fig. 19.44 Details of typical fluorescent lamps and associated lampholders. (a) *Construction of preheat/rapid-start bipin base lamp. (Courtesy of GTE/Sylvania, Inc.) This type of lamp has type* (b) *base and is held in type* (c) *lampholder. High-output HO and VHO rapid-start lamps use recessed d-c base* (d) *and lampholders* (e). *Instant-start lamps are similar in construction to* (a) *except with cathode construction* (f), *have a single pin base* (g), *and use single-pin lampholders* (h), *which are different for each end.*

circuit is energized, the arc is struck immediately. No external starter is required. Because of this similarity of operation, rapid-start lamps will operate satisfactorily in a preheat circuit. The reverse is not true, because the preheat requires more current to heat the cathode than the rapid-start ballast provides. (See Table 19.17 for interchangeability of lamps in the various circuits.) By far the most popular lamp is the 40-W T-12 lamp. A standard ordering abbreviation for

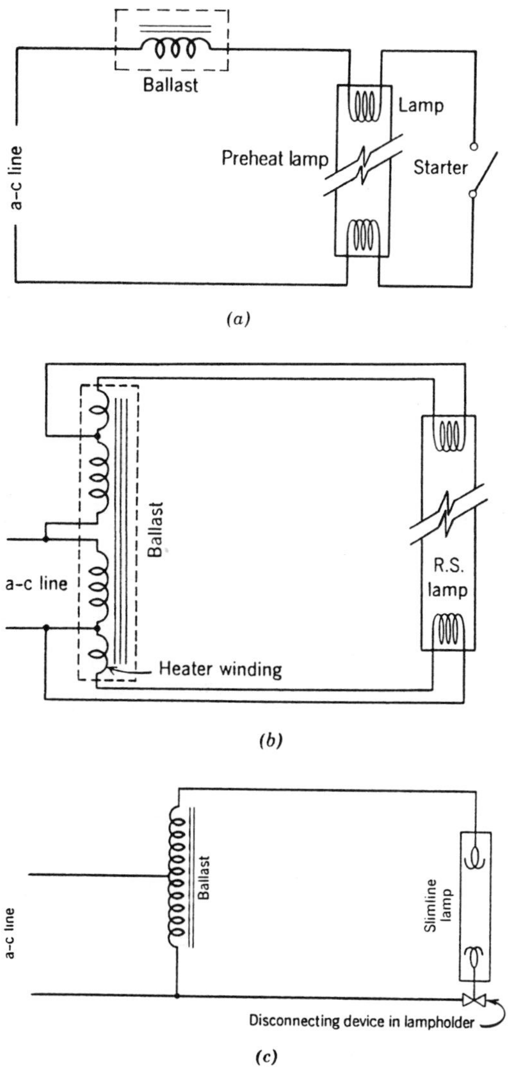

Fig. 19.45 *Basic circuits for preheat, rapid-start, and instant-start (Slimline). For the sake of clarity, only single-lamp circuits are shown, and power-factor correcting capacitors, autotransformers, and compensators are omitted.* (a) *Basic preheat circuit. Starter may be any of several types, manual or automatic. The circuit does not show compensators or other detailed elements, for the sake of clarity. Most preheat lamps are T-12 and operate at 430 mA.* (b) *Basic rapid-start circuit. Note the special end windings used to supply voltage to heat the cathode continuously. To assure proper starting, all standards RS lamps must be mounted within ½ in. of a grounded metal strip extending the full length of the lamp (1 in. for HO and VHO lamps). Normal output lamps operate at 430 mA; HO at 1500 mA.* (c) *Basic instant-start circuit. Voltage from ballast transformer is high enough to strike an arc directly. Note that unlike preheat and rapid-start lamps, these are single pin, since cathodes are not preheated. T-6 and T-8 lamps normally operate at 200 mA, T-12 lamps at 430 mA. Because of the high voltage involved, the lampholder at one end is a disconnecting device that opens the circuit when the lamp is removed.*

a lamp whould be F40T12WW/RS, which indicates fluorescent, 40 W, T-12 bulb, warm white color, rapid start. However, this size is so common that the tube size and rapid-start designations are omitted and the lamp is simply F40WW.

Standard rapid-start lamps operate at 430 mA. If this current is increased, the output of the lamp also increases. Two special types of higher output rapid-start lamps are available. One operates at 800 mA and is called simply high output (HO). The second, which operates at 1500 mA (1.5 A), is called by different manufacturers very high output (VHO), superhigh output, or simply 1500-mA, rapid-start lamp. There is also a 1500-mA special lamp that uses what looks like a dented or grooved glass tube. This lamp, called Power Groove by General Electric, has somewhat higher output than the standard VHO tube. All high-output lamps use double contact bases and special ballasts (see Fig. 19.44*d* and *e*). This lamp is used in applications where high output is required from a limited size source such as outdoor sign lighting, street lighting, and merchandise displays. Because of the serious heat problems involved, VHO lamps are frequently operated without enclosing fixtures. Conversely, HO and VHO lamps are frequently used in cold environments that would prevent proper operation of standard output 430-mA lamps.

Most HO and VHO lamps are slightly less

TABLE 19.16 **Typical Fluorescent Lamp Data: Standard Lamps 60 Hz Standard Ballasts**

Lamp Abbreviation	Lamp Data: Lamp (W)	Lamp Data: Diameter (in)	Lamp Data: Length (in)	Lamp Current (mA)	Ballast (W)[b,c]	Total Watts[c]	Lamp Life (h)[d]	Initial Output (lm)[e]	Lumens at 40% Life	Initial Actual Efficacy (lm/W)[f]	Remarks
Preheat lamps[a]											
F-15 T-8 CW	15	8/8	18	430	8	23	7,500	870	765	38	Cool white
F-20 T-12 WW	20	12/8	24	430	10	30	9,000	1,300	1,155	43	Warm white
Rapid-start—preheat lamps[g]											
F40 T-12 CW	40	12/8	48	430	7.5	46	20,000 +	3,150	2,770	68	
F40 T-12 WW	40	12/8	48	430	7.5	46	20,000 +	3,200	2,815	70	Warm white
F40 T-12 CWX	40	12/8	48	430	7.5	46	20,000 +	2,250	1,855	49	Cool white deluxe
F40 T-12 D	40	12/8	48	430	7.5	46	20,000 +	2,600	2,290	57	Daylight
F40 T-12/C50	40	12/8	48	430	7.5	46	20,000 +	2,200	1,890	48	5000 K color
F40 T-12/C75	40	12/8	48	430	7.5	46	20,000 +	2,000	1,720	44	7500 K color
F40 T-12/U	40	12/8	—	430	7.5	46	12,000	2,900	2,525	55	"U" shape[h]
Rapid start—high output											
F48 T-12 CW/HO	60	12/8	48	800	12.5	72.5	12,000	4,300	3,740	55	
F60 T-12 CW/HO	75	12/8	60	800	15	90	12,000	5,400	4,700	60	
F72 T-12 CW/HO	85	12/8	72	800	22.5	107.5	12,000	6,650	5,785	62	
F96 T-12 CW/HO	110	12/8	96	800	18.5	128.5	12,000	9,200	8,005	72	
Rapid-start—very high output											
F48T12 CW/VHO	110	12/8	48	1500	5	125	12,000	6,900	5,100	55	
F72T12 CW/VHO	165	12/8	72	1500	10	175	12,000	11,500	8,510	66	
F96T12 CW/VHO	215	12/8	96	1500	10	225	12,000	16,000	12,160	71	
Instant-start (Slimline) lamps											
F42 T-6 CW	25	6/8	42	200	10.5	35.5	7,500	1,750	1,490	49	
F64 T-6 CW	40	6/8	64	200	9	49	7,500	2,800	2,350	57	
F24 T-12 CW	20	12/8	24	430	14	34	7,500	1,150	1,035	34	
F36 T-12 CW	30	12/8	36	430	13	43	7,500	2,000	1,800	47	Warm white
F48 T-12 CW	40	12/8	48	430	12	52	9,000	3,000	2,760	58	
F72 T-12 CW	55	12/8	72	430	11	66	12,000	4,550	4,275	69	
F96 T-12 CW	75	12/8	96	430	13	85	12,000	6,300	5,800	74	

[a]Data given for a preheat circuit.

[b]Figures are for a two-lamp circuit.

[c]ANSI figures.

[d]Life figures are for 3-h burning per start.

[e]After 100-h burning.

[f]Includes ballast loss. (See Section 19.18 for explanation of Ballast Efficacy Factor.)

[g]Data given for lamps in a rapid-start circuit.

[h]U-shaped lamps available with 3 5/8- or 6-in. leg spacing; all other characteristics equal.

efficient than the standard 430-mA, rapid-start lamp and have considerably shorter life. Typical ordering abbreviations for high-output lamps are similar to the standard rapid-start lamps except that the number indicates length, not wattage. For instance, F72T12/CW/HO is fluorescent, 72 in. long, T-12 bulb, cool white, high output (800 mA). Similarly, F72T12/CW/VHO is fluorescent, 72 in. long, T-12 bulb, cool white, very high output (1500 mA). Typical characteristics for rapid-start lamps are given in Table 19.16.

A circuit known as trigger start uses a preheat lamp in a rapid-start type of circuit, but with higher strike voltage. It is used for lamps up to 20 W and does not use a starter. After the arc is struck, lamp voltage is reduced.

(c) Instant-Start Fluorescent Lamps. Slimline lamps are the best known variety of instant-start fluorescent lamps. They use a high-voltage transformer to strike the arc without any cathode preheating. These lamps have only a single pin at each end that also acts as a switch to break the ballast circuit when the lamp is removed, thus lessening the shock hazard (see Figs. 19.44*f*, *g*, and *h*, and 19.45*c*). The lamps are generally operated in two-lamp circuits at various currents; normal currents are 200 and 430 mA, and normal lengths are 24, 36, 42, 48, 60, 64, 72, 84, and 96 in. These lamps are actually hot cathode instant-start lamps, which differentiates from the high-voltage cold cathode type. The high-voltage starting characteristic of instant-start circuits lowers lamp life to about half that of the corresponding rapid-start lamp.

Slimline lamps and ballasts are more expensive than rapid-start and are somewhat less efficient. However, they are manufactured in certain sizes and currents not made in rapid-start (e.g., 96 in., 430 mA), and they have the additional advantage of being able to start in much lower ambient temperatures (below 50°F) than rapid-start circuits. This starting characteristic makes the instant-start circuit particularly applicable to outdoor use. A typical ordering description for such a lamp would be F42T6CW Slimline, which means fluorescent, 42-in. length, tubular, 6/8-in. diameter, cool white, instant start. Note also that in instant-start lamps the number following F indicates length, not wattage. This is true of all lamps that operate at other than 430 mA, which is the standard current. Typical characteristics appear in Table 19.16.

(d) Cold Cathode Tubes. The true cold cathode tube uses a large, thimble-shaped cathode and a high-voltage transformer that literally tears the electrons out of the large cathode to strike the arc. These lamps have a very long life that, in contradistinction to hot cathode lamps, is virtually unaffected by the number of starts. Cold cathode lamps have a lower overall efficiency than the hot cathode types and are nor-

TABLE 19.17 **Fluorescent Lamp Interchangeability: Standard Lamps and Ballasts Only**[a]

Lamp Type	*Ballast/Circuit Type*		
	Preheat	*Rapid-start*	*Instant-start*
Preheat	OK	Not good, poor starting	Not good, poor starting short life[b]
Instant-start (Slimline)	Won't start, not good[b]	Won't start, not good[b]	OK
Rapid-start	OK	OK	Not good, poor starting short life[b]
Preheat/rapid-start	OK	OK	Not good, poor starting short life[b]

[a]Special lamps such as low-wattage, high-wattage, 265-mA T-8 and so on must be used with matching ballast.
[b]Normally no possiblity of interchange. Instant-start lamp is single pin base; preheat/rapid-start lampholders are for bipin bases.

ILLUMINATION

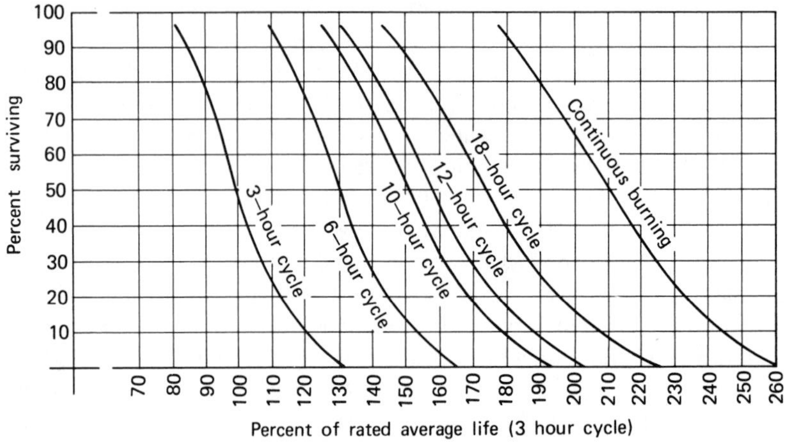

Fig. 19.46 Mortality curves of fluorescent lamps. (Data courtesy of GTE/Sylvania, Inc.; data subject to change.)

mally used where long continuous runs are required, as in architectural-type lighting rather than in lighting fixtures. Cold cathode lamps are readily dimmed and also operate well at varying ambient temperatures.

19.16 The Fluorescent Lamp: Characteristics of Operation

(a) Lamp Life. The lamp life of a standard fluorescent tube is greatly dependent on the burning hours per start. The figures listed in Table 19.16 and in the lamp catalogs for lamp life are based on a burning cycle of 3 h per start and represents the average life of a group of lamps; that is, half the lamps of any group will have burned out at this time. Typical lamp mortality curves are shown in Fig. 19.46, and the effect of burning hours per start is shown in Fig. 19.47.

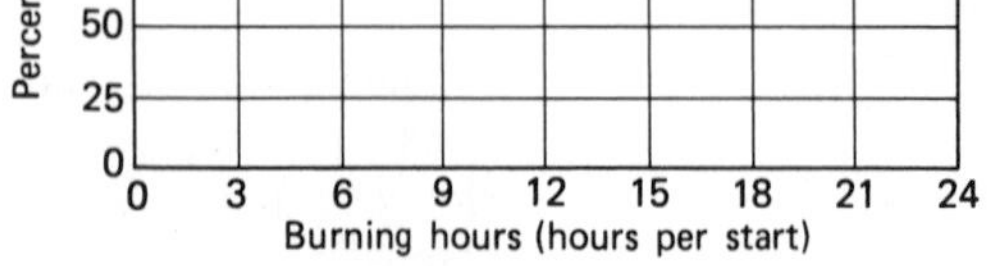

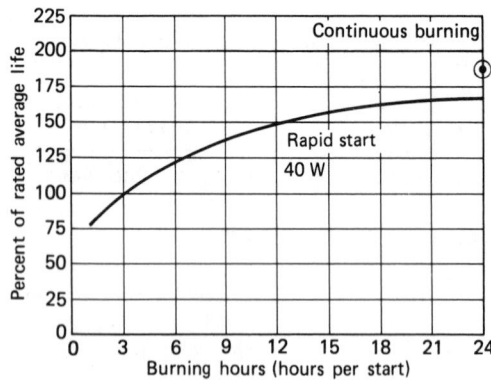

Fig. 19.47 Effect of burning hours on fluorescent lamp life. Note that at 3 burning hours per start, the average lamp life is 100% of the nominal catalog figure. (Data courtesy of GTE/Sylvania, Inc.; data subject to change.)

This item is significant in connection with energy costs and utilization. From an energy source viewpoint, if an area is not utilized for periods of 10 min or more, fluorescent lamps should be shut off. This takes into account the resource energy required to replace a tube as a result of shortening its life. From a cost viewpoint, the break-even point depends on these factors:

1. Lamp life reduction as a function of burning hours per cycle.
2. Cost of energy.
3. Cost of lamp and lamp replacement.
4. Amount of time lamp remains off when shut off.
5. Cost of switching equipment (if any).
6. Life of the building.

With this number of variables it is not possible to give general solutions, and an individual analysis is required. However, several analyses by the author have shown that assuming ordinary office conditions, using lamp life data as given in Fig. 19.47, 20-yr fixture life, \$0.085/

kWh energy cost escalating 3% annually, $1.25 lamp cost, 15-min relamping time, and $8 per lamp to provide the necessary switch (one switch per two 2-lamp fixtures), lamps should be switched off any time they are not in use for 5 to 8 min or more. (The spread is caused primarily by variation in local labor rates.) This figure should be used in setting timing for occupancy sensor switching (see Sections 21.2 to 21.4). It is thus clearly an economic fallacy to leave lamps burning to achieve longer lamp life.

(b) Effect of Temperature. The temperature of the coolest point on the lamp bulb wall determines the lamp's mercury vapor pressure, which in turn determines the lamp lumen output, wattage, and color. Maximum output occurs at a *bulb* temperature of 40°C (104°F). The bulb wall temperature itself is affected by room ambient temperature, airflow over the lamp (as in air-return fixture), and the temperature of adjacent surfaces such as the ballast enclosure. Thus it should be apparent that catalog data on lamp output and wattage, which are based on laboratory tests of bare bulbs at 25°C ambient temperature in still air (design conditions for maximum efficiency; see Fig. 19.48), may be very far from field operating conditions. For this reason, lamp (and ballast) manufacturers often list three values for lamp wattage in the catalog: laboratory conditions (ANSI), open luminaire (such as a parabolic reflector), and enclosed luminaire (such as a wraparound lens unit). Typical wattage/fixture values for two standard 40-W T-12 CW lamps with standard ballasts are 91, 88, and 78 W, respectively, that is, highest for laboratory conditions, with correspondingly high output, 3% less for an open lamp fixture, and 14% less for an enclosed fixture, because of its elevated lamp temperature. The well-documented effect of increased light output in air-return luminaires is due to the cooling action of the air on overheated fluorescent tubes. However, overcooling, which may occur with outdoor fixtures or open suspended fixtures in air-conditioned rooms, also reduces output, as can be seen from Fig. 19.48. Where knowledge of lamp wattage (and lumen output) is particularly important, as when selecting special ballasts for light reduction in retrofit work, field measurements on installed units must be made. For new construction, a full-size mockup with specific lamp and ballast combinations can be made and measurements of wattage and lumen output taken.

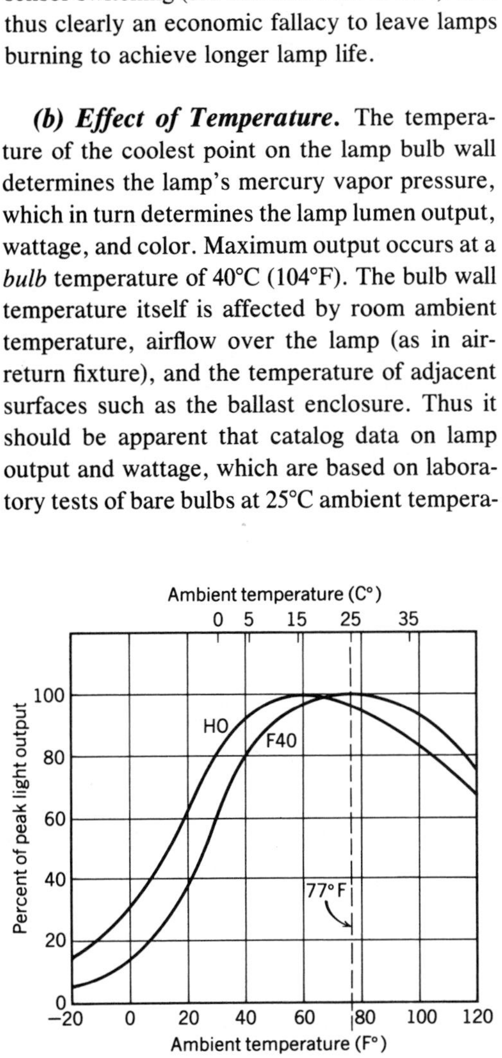

Fig. 19.48 *Relation of light output to ambient temperature immediately around a fluorescent tube. Light output drops 1.5% per degree C above and below 25°C ambient in still air. Since the HO lamps run hotter, they require a lower ambient to maintain proper bulb wall temperature. (Courtesy of GTE/Sylvania, Inc.; data subject to change.)*

Special all-weather and jacketed lamps are available that will maintain fairly constant lumen output over a wide ambient temperature range. For outdoor use where starting below 50°F is necessary, rapid-start lamps require special low-temperature ballasts. Slimline lamps with normal ballasts will start readily down to 20°F and, by using the next higher voltage ballast, starting down to −20°F is possible.

(c) Voltage Effects. Voltage either above or below rating adversely affects life, unlike the effect of low voltage on the incandescent lamp (see Fig. 19.49). Normal operating voltage range for ballasts is 110 to 125 V on 120-V circuits, 200 to 215 V on 208-V circuits, and 250 to 290 V on 277-V circuits.

(d) Lumen Maintenance. Lumen output of a fluorescent tube decreases rapidly during the first 100 h of burning and thereafter much

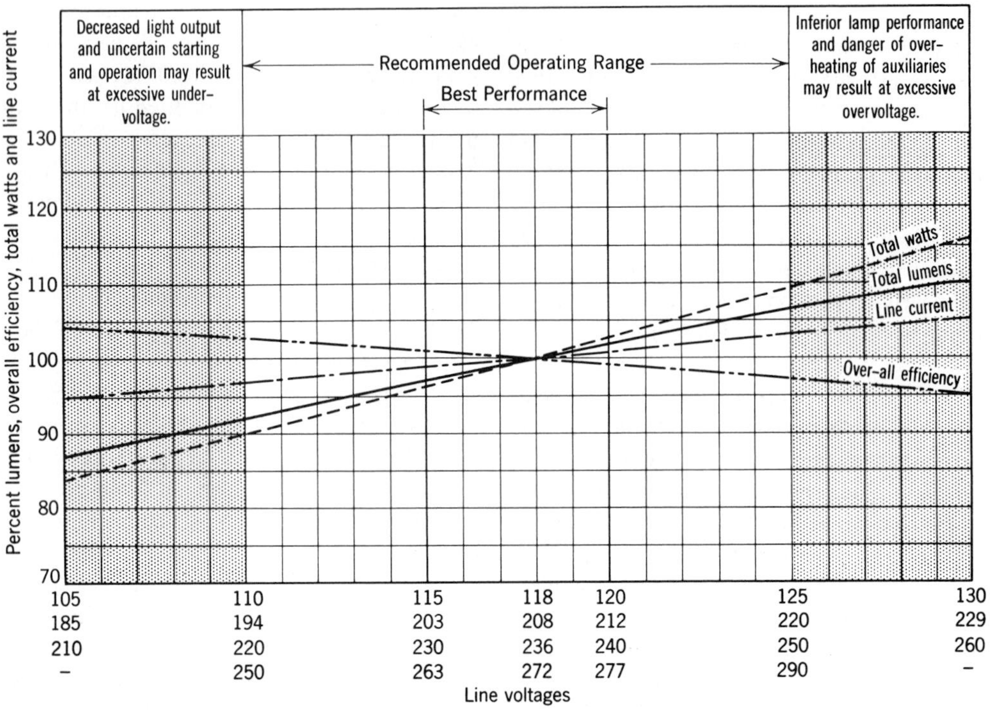

Fig. 19.49 Recommended operating ranges of circuit voltages for most satisfactory operation. The curves indicate the percentage changes in output lumens, efficiency, total watts, and current for line-voltage changes from the rated value. The nominal circuit voltages are 120, 208, and 277 V.

more slowly. For this reason the tabulated initial lumen figures represent output after 100 h of burning. Data are also generally published on the lumen output at 40% of average rated life. This figure is approximately 85 to 90% of the 100-h initial value (see Fig. 19.50). Lumen maintenance is *not* affected appreciably by the number of burning hours per starts. It is better for 200- and 430-mA lamps than for high-output (800- and 1500-mA) lamps.

(e) Efficacy. The design efficacy of a fluorescent lamp depends on operating current and the phosphors utilized. Figure 19.51 shows the energy distribution of a typical fluorescent lamp alone, not including ballast losses. All other conditions being equal, and assuming standard lamps and ballasts, warm white lamps are most efficient, followed closely by cool white, white, daylight, and colored lamps. Specialty colors or lamps designed to produce specific kelvin temperatures are low in output, with *lamp* efficacies

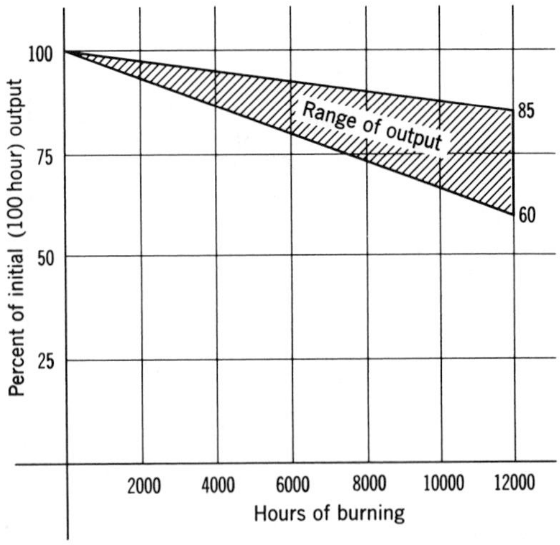

Fig. 19.50 Fluorescent lamp output depreciates with life. The majority of lamps fall in the upper part of the band. Lumen depreciation is unaffected by burning hours per start.

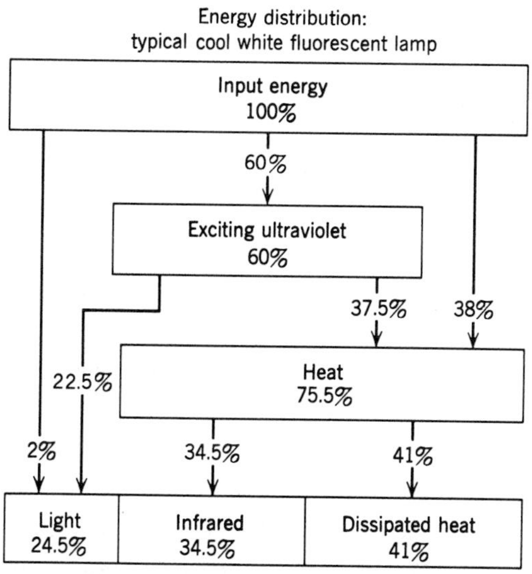

Fig. 19.51 Fluorescent lamps with efficacies of up to 85 lm/W are among the most efficient light sources available, yet still convert less than one-fourth of their energy to useful light. Ballast losses are not included in the percentages shown.

in the region 40 to 50 lm/W. Triphosphor lamps, discussed below, are an exception. The range of efficacy for standard lamps is 40 to 85 lm/W, *including* ballast losses in the wattage figure (see Table 19.16). This is important, since discharge lamps are inoperative without ballasts, and neglecting ballast losses yields an artificially high and therefore misleading efficacy. It is important that the designer appreciate the significance of efficacy as compared to lamp wattage. In a normal design situation, an interior space requires a calculated quantity of light, in lumens. The designer usually attempts to deliver that light at minimum overall cost, that is, a combination of first cost and operating cost. Minimum first cost usually means *maximum light output* per fixture, consistent with good lighting design. Minimum operating costs means maximum *lumens per watt*. These two demands are not normally consonant. Over the life of the building, operating cost dominates, as life-cycle cost analyses show, and the lamp–ballast combination of choice is normally the one with highest efficacy. *Wattage,* in and of itself, is a meaningless quantity unless associated with a lumen output figure. Thus low-wattage or high-wattage lamps with their special ballasts are rarely the indicated choice in new design, because their efficacy does not usually justify the cost premium. These special lamps are useful in retrofit work in which case, as noted above, field measurements of illuminance and lamp temperature are required before selecting the replacement lamp–ballast combination.

Generally, standard 430-mA lamps are most efficient, followed by HO 800-mA lamps, then VHO 1500-mA lamps. Specialty lamps such as reflector and low-wattage units are discussed in following paragraphs. Ballast losses, which constitute 5 to 16% of lamp wattage, depend on ballast type, circuit, manufacturer, type of fixture, ambient temperature, and number of lamps connected. Figures given in Table 19.16 are average. The amendment to Public Law 100-357, which mandates high-efficiency ballasts for certain lamp types is discussed in Section 19.18.

To make a proper comparison between lamp types, it is not sufficient to compare cost to produce a given quantity of lumens, since high-output lamp installations use a smaller number of fixtures. A meaningful cost comparison requires a full life cycle or annual owning cost analysis.

(f) Dimming and Reduced Output. It is often desirable to reduce the output of a fluorescent lamp in order to reduce energy consumption, correct overlighting, compensate for daylight, change an area's function, and so forth. If it is anticipated that full output will again be required, dimming is indicated. If the change is long-term or permanent, a stepped or fixed reduced-output arrangement is probably indicated. Output reduction is accomplished by reducing effective lamp current. This can be accomplished by lowering primary voltage, adding impedance to the circuit, or shortening the time of current flow in each half cycle. The last method describes the gating action of a modern solid-state thyristor dimmer (SCR, triac).

1. *Full-range dimming.* When dimming is required to levels as low as 5% of full output, each individual lamp must be provided with a special dimming ballast. (Electronic ballasts are discussed in Section 19.19.) This ballast uses "3-

wire'' control to provide high voltage for starting and restrike (even at low output levels) and continuous cathode heating. This latter is important to assure that even at low lamp arc current, the lamp does not flicker and that lamp life is unaffected, because only arc current is reduced while cathodes remain fully heated. Finally, the 3-wire control arrangement of individual dimming ballasts permits use of a central controller which makes presetting, zoning, and timing of local dimming functions possible.

The lamp of choice in dimming circuits is the standard 430-mA rapid start, although 800- and 1500-mA can also be dimmed. An important characteristic of this dimming system is the linear relation between light output and power input, such that lamp efficacy is maintained down to about a third of output, after which efficacy drops and color shifts toward the blue-green portion of the spectrum, because of bulb cooling.

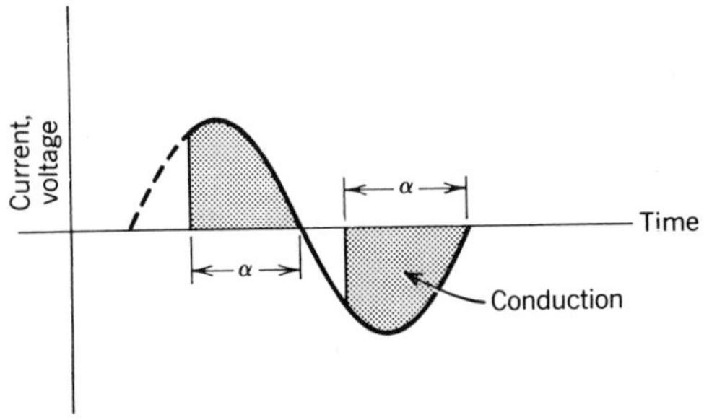

Fig. 19.52 Gating action of a solid-state dimmer. By controlling the gating, the conduction angle α can be adjusted, thus varying the effective voltage over the entire cycle.

2. *Partial dimming.* The thyristor action shown in Fig. 19.52 can be applied to control the effective input voltage to a fluorescent luminaire with *conventional* ballasts, and by so doing dim the lamp. This is commonly known as "2-wire" dimming, since the controlled voltage is applied to the two input wires of the luminaire. However, since the high strike voltage and cathode heating circuit necessary for full-range dimming is absent, this arrangement is only useful down to 40 to 50% of full output. Below this level, light output drops sharply, lamp color changes, starting becomes unstable, and lamp life is affected. However, within the 100 to 50% range, the energy-output curve is linear, that is, efficacy is maintained, making this arrangement very desirable as an economic lighting-energy management system. This is explained fully in Sections 21.2 to 21.4, which cover lighting control. A typical controller of this design for dimming an entire 20-A 277-V circuit is shown in Fig. 21.7.

Low-energy lamps are generally *not* suitable for dimming service. This is because they operate with special ballasts that reduce lamp current in various ways. Some ballast designs change current waveform, others disconnect the cathode circuit after the lamp starts (Sylvania's SuperSaver™ among others), and still others change the circuit impedance and current-voltage characteristics. Since most of these circuit alterations would interfere with normal dimmer functioning, their use is inadvisable without a prior detailed engineering analysis. Also, low-energy lamps themselves are not operable with dimming ballasts, since these ballasts are designed for the parameters of standard lamps only.

3. *Output reduction.* The cost of dimming equipment is usually justified only if smooth continuous-level changes are required (see Sections 21.3 and 21.4). Where discrete-level switching is acceptable, one solution is the use of multilevel ballasts. Such ballasts, once quite popular, are used less today because local codes generally prohibit use of wall switches or other simple switching devices to change levels, instead requiring that an electrician perform the function. Obviously, this is inconvenient, time-consuming, and expensive. The reason for the prohibition is that the level change involves switching of ballast capacitors, with the possibility of high voltages (about 300 V) and current surges (up to 50 A). However, circuiting can be arranged so that switching is performed when the unit is de-energized, and if that is acceptable to local code authorities, these ballasts can be very useful.

Where it is desired to reduce lamp output on a semipermanent basis while maintaining efficacy, impedance can be added to the circuit in various forms. When it is desired to restore full output, the device is simply removed. Alternatively, special lamps are available that contain integral impedances (Sylvania's Thrift-Mate™ is one) and that will operate at lowered output with standard ballasts. Both these solutions are best applied in enclosed fixtures, since the reduced lamp current may cause excessive lamp cooling in open fixtures with drop in efficacy and change in lamp color.

(g) Flashing. Fluorescent lamp flashing ballasts are similar to dimming ballasts in that the cathode circuit remains energized continuously and only the arc circuit is turned on and off. This permits flashing the lamps without affecting lamp life. Cold cathode, preheat, and rapid-start lamps are suitable for flashing service.

(h) High-Frequency Operation. It has long been known that operation of fluorescent lamps at frequencies above 60 Hz has many beneficial effects. Efficacy increases dramatically (see Fig. 19.53), the heavy and expensive magnetic core-and-coil ballast is replaced by small, light, quiet, and inexpensive solid-state components, heat loss is reduced by as much as 90%, maintenance is reduced, and dimming is greatly simplified. However, until recently, the drawback to all these desirable characteristics has been the problem of providing the requisite high-frequency power. Central static inverters are expensive, as is a high-frequency power distribution system. This situation has recently changed completely with the development of compact, *reliable*, and relatively inexpensive electronic ballasts. These devices, which are discussed further below in the ballast section, are actually small solid-state rectifier-inverter ballasts that change the incoming 60-Hz a-c to approximately 25 kHz. This local at-the-fixture conversion makes high-frequency lamp operation with all its advantages economically attractive; so much so that core-and-coil ballasts may soon be largely replaced for normal rapid-start installations.

(i) Other Characteristics. Fluorescent light color is discussed in Sections 19.26 and 19.27. Fluorescent lamps are large and therefore necessitate a relatively expensive fixture both to hold the lamps and to control the light. Since the

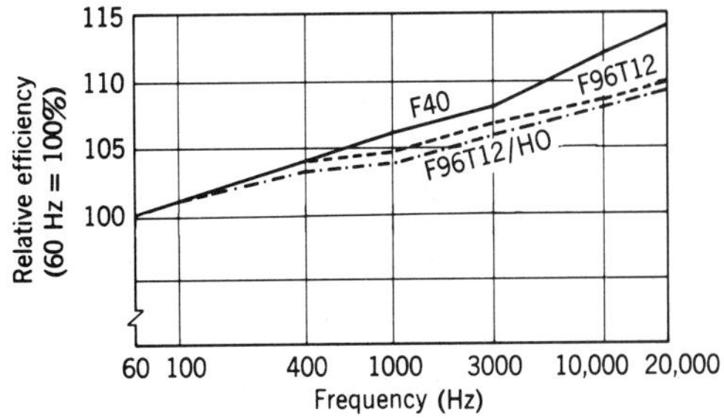

Fig. 19.53 *The efficiency of most fluorescent lamps increases with frequency increase. On this curve, the efficiency at 60 Hz is taken as the 100% value.*

tubes emit light throughout their considerable length, accurate beam control is difficult, making fluorescent units best applicable to area lighting.

In sum, then, the advantages of fluorescent lamps are long life, low cost, high output and efficacy, availability in an extremely wide range of sizes, colors, and brightnesses, and relative insensitivity to voltage fluctuation (important in brownout areas). Disadvantages are large size, which creates storage, handling, and relamping problems, and the fixture situation previously referred to.

19.17 Special Fluorescent Lamps

(a) Low-Energy Lamps. The need for energy conservation plus the desire to reduce lighting levels in existing overlighted spaced has resulted in a complete line of low-energy lamps. Their wattage ratings are lower than those of standard lamps because they are intended primarily as lower energy direct replacements for existing lamps. All such lamps are clearly marked by the manufacturer. They operate on lower current than standard lamps, achieved by one of the techniques discussed in section 19.16*f*(3), require special matching ballasts for maximum effectiveness, and have an efficacy equal to, or somewhat higher than, standard lamps and ballasts. They have the disadvantage of higher cost, need for special ballasts where maximum energy reduction is desired, generally

ILLUMINATION

shorter life, inapplicability to dimming circuits, and problems of inventory ad proper lamp replacement. To our mind, their use is indicated only where other light and wattage-reduction schemes such as circuit dimming, additional circuit impedance or reduced wattage ballasts are inapplicable. Here again, a proper economic analysis is indicated rather than relaying on oversimplified charts and graphs, which show savings as a function of energy cost for the life of the lamps or per year. Most of these charts ignore relamping costs and interest rates, which results in a grossly distorted cost picture. Cost analyses of new installations must also account for fixture cost, which is inapplicable in retrofit jobs.

(b) Reflector and Aperture Lamps. These lamps, which are available in 800- and 1500-mA sizes, contain an internal reflector that performs in the same fashion as the more common reflector in the incandescent R and PAR lamps. The reflector lamp is completely phosphor coated, while the aperture lamp has a 30° clear "window" resulting in very high luminance of this slot (see Fig. 19.54). Both types have lower efficacy than a normal tube and are generally applied where an enclosing fixture is uneconomical or impractical, as in handrails or for sign illumination.

(c) Triphosphor (Octic) Fluorescent Lamps. This is a high-efficiency, 1-in. (26-mm) diameter lamp whose high-output phosphors have the additional desirable characteristic of excellent color rendering index (CRI). (See Section 19.26 for an explanation of CRI.) The lamps are called triphosphor lamps because they use an additional inner coating of phosphors that produce light primarily at three points; about 450 nm (blue), 540 nm (green), and 610 nm (red). Among them are Sylvania's Octron™, and Philip's Octolume.™. They operate best with electronic ballasts at high frequency, although their performance with standard magnetic ballasts at 60 Hz is also an improvement on standard lamps. They are particularly temperature sensitive, making their use in dimming circuits inadvisable because of depreciation of output and color quality. Also, because of this characteristic, their performance in open-type luminaires such as parabolic reflectors should be carefully investigated before being specified. As with all special lamps they are restricted to use with special ballasts. However, unlike the reduced-energy lamp, their high efficacy, particular with electronic ballasts, makes their use in new construction attractive, if dimming is not intended.

The lamps are manufactured in straight lengths of 2, 3, 4, and 5 ft with corresponding nominal wattages (at 60 Hz) of 17, 25, 32, and 40 W, and in U-shaped lengths of 10, 16, and 22 in. with corresponding wattages of 16, 24, and 31 W at 60 Hz. All sizes are made in three color temperatures—3100, 3500, and 4100 K, each of which has a CRI of 75. Lumen output, color temperature, and color rendering index all remain unchanged when the lamps operate at high frequency on electronic ballasts, whereas lamp current and power drawn drop, accounting for

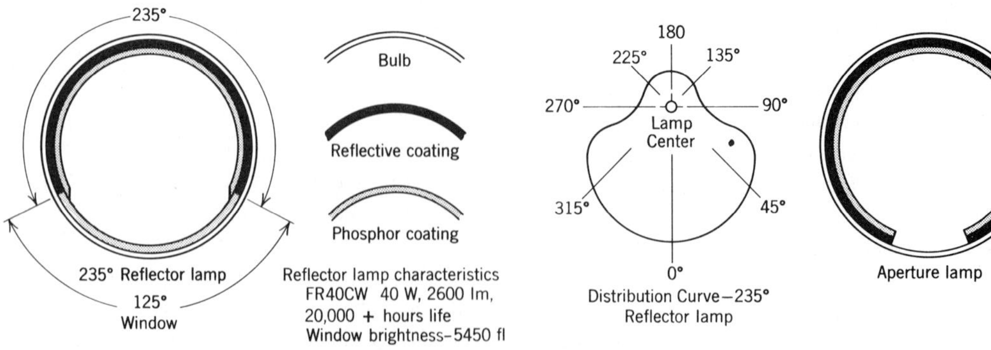

Fig. 19.54 Characteristics of reflector and aperture fluorescent lamps.

the increased efficacy shown in Table 19.18. Average field performance is shown in the table, as compared to a standard fluorescent tube, for operation both at 60 Hz and at 25 kHz.

(d) U-Shaped Lamps. U-shaped lamps were developed to answer the need for a high-efficacy fluorescent source that could be utilized in a square fixture, since the normal fluorescent lamp shape is frequently not architecturally suitable (see Fig. 19.55). The U lamp is basically a standard T12 fluorescent tube bent into a U shape and available with 3⅝- or 6-in. leg spacing; the former can be accommodated three to a 2-ft square fixture and the latter two to a 2-ft square fixture. (The narrower T-8 envelope of Octic lamps permits a tighter bend; those U lamps have 1⅝-in. leg-to-leg spacing.) U lamps operate on standard ballasts and have slightly lower output than the corresponding straight tube. In all other respects the U lamp has the same characteristics as the straight lamp of similar type. Because of the proliferation of lamp types and because, as pointed out previously, fluorescent lamps are designed to operate with a specific ballast, compatibility of lamps and ballast must be verified when relamping and retrofitting.

(e) Compact Fluorescent Lamps. The unwieldy linear shape of the standard fluorescent lamp, its double-ended connections, plus the need for a ballast have long restricted its use to specific commercial applications. Recently, however, new designs utilizing compact folded fluorescent tubes with integral ballast, terminating in a medium-screw Edison base, have appeared on the market as direct replacements for incandescent lamps and for fluorescents where the linear shape is objectionable. Although these lamps have lower efficacy than conventional fluorescents, it is still much higher than that of an incandescent lamp. This, coupled with long life (see Table 19.19) and excellent color rendering index, make these lamps attractive as direct replacements for low-efficiency incandescents. A large caveat is in order, however. These lamps

TABLE 19.18 **Comparative Average Characteristics: 4-ft Standard and Octic Lamps**

(a) 60-Hz Operation, Rapid-Start Ballast

Lamp and Diameter	W	Lm	I (mA)	CRI	Life (h)	Watts/Lamp[a]		Efficacy[a] (lm/W)	
						Open Luminaire	Closed Luminaire	Open Luminaire	Closed Luminaire
F-40 T-12 40 mm	40	3150	430	50–65	20k	47	43	64	70
Octic T-8 26 mm	32	2900	265	75	20k	35	33	83	88

(b) 25-kHz Operation, Instant-Start Ballast

Lamp and Diameter	W	Lm	I (mA)	CRI	Life (h)	Watts/Lamp[a]		Efficacy[a] (lm/W)	
						Open Luminaire	Closed Luminaire	Open Luminaire	Closed Luminaire
F-40 T-12 40 mm	40	3150	430	50–65	20k	39	37	72	76
Octic T-8 26 mm	32	2900	265	75	15k	29	28	98	102

[a]Including ballast loss.

ILLUMINATION

Fig. 19.55 Each 5-ft square module uses a 2-ft square fixture with two 40-W U-shaped fluorescent lamps, of the type shown in the insert. (Photo courtesy of GTE/Sylvania, Inc.)

are fluorescent, and like all fluorescents, life is proportional to burning hours per start, with catalog listed rated life based on 3 burning hours per start. They are therefore most definitely not applicable to such usages as closets, pantries, storage facilities, and the like where lamp on-time is short and on–off cycles are frequent. They are applicable to desk lamps, decorative fixtures, outdoor lighting in appropriate fixtures, and all applications where burning time per start is at least 3 h. Typical designs are shown in Fig. 19.56 and typical lamp data in Table 19.19.

19.18 Core-and-Coil Ballasts for Fluorescent Lamps

The conventional, magnetic core-and-coil fluorescent lamp ballast is an assembly of electrical components responsible for the proper starting and continued operation of the fluorescent lamps in a fixture. By nature of its construction, it also produces considerable heat and noise. Some of the important considerations in selection and application of ballasts are discussed below.

(a) Ballast Efficiency. Because of the considerable energy loss in inefficient ballasts, manufacturers and standards organizations have established two figures of merit by which ballast energy efficiency can be judged. *Ballast factor* (BF) is the ratio of light output of a lamp when operated on a test ballast to its output when operated on a standard reference ballast, using an ANSI test procedure. Minimum BF required for an F40 rapid-start lamp ballast to carry ANSI/CBM certification is 92.5%. (As pointed out above, only ballasts carrying CBM certification should be used where such certification is applicable.) The second figure of merit is the *ballast efficacy factor* (also called *ballast efficiency factor*). This rating is the ratio of ballast BF to its input watts. It is thus a measure of ballast efficiency. An amendment to the National Appliance Energy Conservation Law of 1988 mandates that after 1 January 1990 all ballasts for the three most common fluorescent lamp circuits—one- and two-lamp 40-W rapid start, two-lamp 96-in. instant start (Slimline), and two-lamp 96-in. high output (HO)—must have the minimum BEF specified in the law. Such ballasts are marked with a capital letter E in a circle. Exceptions to the law include special ballasts for these lamps (e.g., dimming), ballasts for other lamps, and ballasts used in residential buildings no higher than three stories above grade.

(b) Ballast Labels. There are several organizations involved with ballast standards and testing. They are:

CBM—Certified Ballast Manufacturers Association.

ANSI—American National Standards Institute. Originates standards on a national level.

ETL—Electrical Testing Laboratories, Inc. A private, independent organization and a recognized authority in measurements and testing of lamps and lighting equipment.

UL—Underwriters Laboratories, Inc. An independent, nonprofit organization testing for public safety.

CSA—Canadaian Standards Association.

Ballasts should be UL labeled and CBM/

TABLE 19.19 **Characteristics of Compact Fluorescent Lamps**[a]

Lamp Type	*Watts*	*Initial Lumens*	*Correlated Color Temperature, CCT (K)*	*MOL*[b] *(in)*
Folded[c]	5	250	2700	4.1–5.3
	7	400	2700	5.4–6.5
	9	600	2700/3500	6.6–7.5
	13	900	2700/3500	7.6–8.5
Double folded[c]	9	575	2700	4.4–5.5
	13	860	2700	4.9–6.0
Single PL[c]	5	250	2700	4.1
	7	400	2700/4100	5.2
	9	600	2700/4100	6.5
	13	900	2700/4100	7.4
	18	1250	2700/3500/4100	8.9
	24	1800	2700/3500/4100	12.6
	36	2900	2700/3500/4100	16.4
Double PL	10	600	2700	4.6
	13	900	2700	6.0
	18	1250	2700	6.8
	26	1800	2700	7.6

[a]Life of all lamps is 10,000 h at 3 burning hours per start. Color rendering index of all lamps is 81 to 82.

[b]Dimensions vary among manufacturers. A range is therefore given. Dimensions are for lamp alone, without adapters or base.

[c]Adapters with integral ballast and Edison screw base are available for all folded lamps and for single PL lamps in 5- to 13-W sizes.

ETL certified. The UL (Underwriters Laboratories) label assures intrinsic safety. CBM (Certified Ballast Manufacturers) establishes high-quality design criteria, and ETL (Electrical Testing Laboratories) tests the ballasts to determine that the design standards have been met.

(c) Ballast Protection. The National Electrical Code (NEC) requires that all ballasts for *indoor* fixtures be protected by an integral thermal-sensing device that will disconnect the ballast in the event of overheating. Overheating is caused by excessive voltage, excessive ambient temperature, or failure of a ballast component. These devices are either thermostatic (self-resetting) or fuse-type (self-destructive). Since two of the three conditions that cause overheating are usually correctable, our recommendation is to specify the self-resetting type of protector. Thermally protected ballasts are designated "type P" by the UL.

(d) Ballast Heat. Ballast heat is transferred to the fixture body by direct metal-to-metal contact (which must be unimpeded) and is then dissipated by radiation and convection from the fixture. Obviously, therefore, the location and method of fixture installation affect the heat transfer from the fixture and consequently the ballast temperature. Pendant fixtures (more than 6 in. below the ceiling) and fixtures recessed into ventilated suspended ceilings do not generally present a temperature problem. Fixtures mounted on insulating surfaces such as low-density acoustic tile, or into unventilated or heated ceiling spaces, or when boxed by a fire-rated enclosure and recessed into a fire-rated ceiling, *do* present serious heat-dissipation problems. When the material adjacent to the fixture is combustible, the installation may constitute a fire hazard. See National Electric Code Articles 410-73 and 410-76. Since each installation situation represents almost a unique case because of

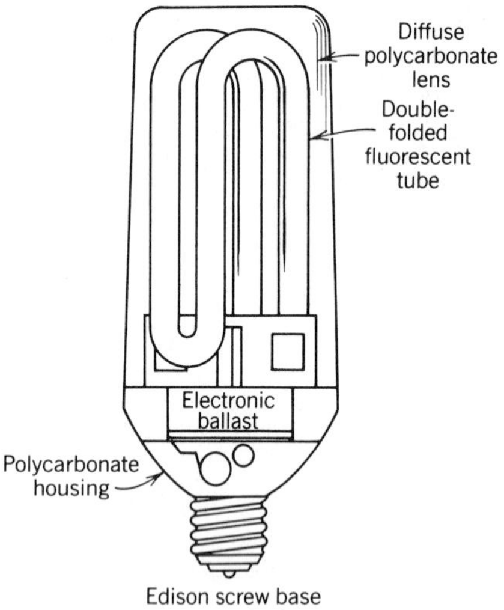

(*a*-1)

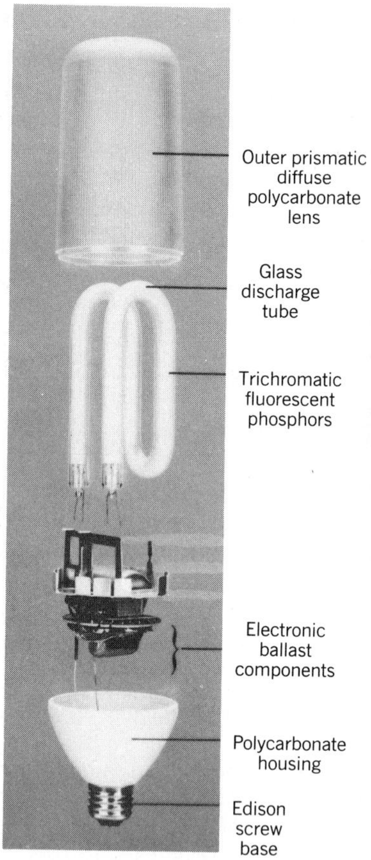

(*a*-2)

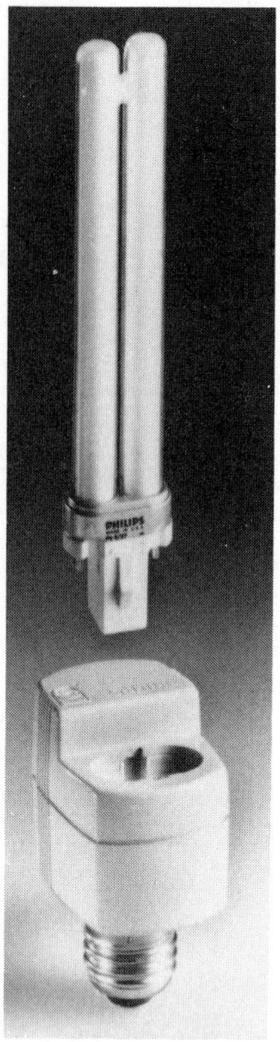

(*b*)

Fig. 19.56 (a) *The folded SL lamp is operated at high frequency to reduce heat buildup, improve output and efficacy, and produce light of acceptable color to serve as an incandescent replacement. Lamp characteristics: MOL, 7+ in.; diameter, 3 in.; wt, <10 oz; color temperature, 2700 K; CRI, 82; power input, 18 W; life, 10,000 h; efficacy, 61 lm/W; initial output, 1100 lm, which is the equivalent of a 75-W incandescent. The assembled lamp is shown in* (a-1) *and the components in* (a-2). *(Photos courtesy of North American Phillips Lighting Corp.)*

Fig. 19.56 (b) *Twin-tube bridged PL lamps are available in 5-, 7-, and 9-W sizes, with initial lumen output of 250, 400, and 600, respectively. All lamps utilize high-grade phosphors, giving a correlated color temperature of 2700 K (similar to incandescent) and a CRI of 82. All lamps have a 10,000-h life and when fitted with the illustrated ballast/base are usable in a standard Edison screw-base fixture. (Courtesy of Phillips Lighting.)*

the variables of fixture, ballast, ceiling material, and ambient temperature, the designer should require a temperature test of the specific fixture in the installation situation involved before acceptance of the unit. This can be specified along with shop drawings, photometric data, and a sample unit.

Conventional ballasts are designed to operate in the fixture at a maximum temperature of 90°C in a 25°C ambient. Ballast temperature is important, since ballast life is directly affected by it. At normal temperature a ballast life of 12 to 15 yr can be expected. A rule of thumb is that ballast life is halved for every 10°C above 90°C operation and conversely is doubled for every 10°C reduction in operating temperature. Ballast temperature rises 0.9°C per 1°C of ambient temperature above 25°C. Where ambient conditions above 25°C (77°F) are anticipated, special cool operation ballasts, should be specified, despite their higher cost.

Normal ballasts are designed to start fluorescent lamps in an ambient temperature range of 50 to 105 F. If conditions outside these limits are expected, special ballasts must be specified. For outdoor installations, ballasts suitable for starting at + 20, 0, and −20° F are available.

(e) Ballast Power Factor (pf). This factor is determined by unit design and is either high (above 0.9) or low (0.5 to 0.6). High-pf ballasts are more expensive than low-pf units, but the additional cost is readily repaid by lower line losses, smaller circuit conductors in long runs, and larger number of fixtures per circuit. Energy conservation and economic considerations dictate the use of power factor corrected (high power factor) ballasts.

(f) Radio Noise. Frequently referred to as radio frequency interference (RFI), this noise is *not* produced by the ballast, but by the arc discharge in the fluorescent tube. (Occasionally a defective ballast does cause RFI.) To minimize RFI, ballasts are available with integral RF noise suppressors. In extreme cases additional suppression can be obtained by installation of RF noise attenuators in the fixture.

(g) Stroboscopic Effect. This is no longer a problem with single-lamp ballasts due to the use of long-persistence phosphors in all lamps.

(h) Noise. See Section 27.31. Ballasts produce a hum or buzz when operating that is transmitted to the fixture because of the integral contact required for heat transfer. This contact generally amplifies the noise. Total noise depends on sound rating of the ballast, fixture design, and acoustical characteristics of the room.

Ballasts are sound rated by a letter that indicates, not actual sound developed, but performance in a space. The ballast selected should be suitable for the lowest sound level likely to be encountered in the subject space. Generally, A-rated ballasts are selected for offices and other normally quiet areas where ballast noise would be objectionable. Since no manufacturer's standardization of ratings exists, recommendations should be obtained from ballast and fixture manufacturers or from an acoustic consultant. Where ballast noise (or heat buildup) may present a problem, ballasts can be remotely mounted if provision is made for heat dissipation and noise control. See Table 27.11 for acoustic criteria for selection of ballasts.

(i) Special Ballasts. All nonstandard-current lamps, including low-energy and high-energy units, require matching ballasts to supply the required current, waveform, and circuitry. These ballasts, unlike standard-current units, are not interchangeable. Similarly, use of one maker's low-energy lamp with another's low-energy ballast is not suggested without prior testing or specific manufacturer's recommendation. There are four principal varieties of nonstandard ballast.

1. *Low-energy units* are intended to match specific low-energy lamps as explained above. Also, reduced current ballasts intended for use with *standard* lamps are readily available.
2. *High-energy units* are intended to be used with standard or high-output lamps. The purpose of this combination is either to increase output in an existing installation or to reduce the number of fixtures in a new installation.
3. *Multilevel ballasts* are useful when it is desired to change lighting level evenly. The usual unit is two-level, that is, full output and 50%, but three-level units are available for full, two-thirds, and one-third output. As explained above, switching may be a problem.
4. *Dimming ballasts* are made for standard lamps only and are normally applied one per lamp.

In addition to these types there are special units for low or high ambient temperature, weatherproof units, and low leakage-to-ground units for hospital applications.

(j) Lamp Removal. In general, two-lamp ballasts operate at their lowest temperature with two active lamps, other conditions being equal. Ballasts are affected by lamp failure or removal as follows:

1. In *rapid-start* circuits, operation of a two-lamp ballast with one lamp removed or burned out will not damage the ballast. Ballast power loss will remain approximately the same.
2. In two-lamp *preheat* circuits, one lamp operation will cause ballast overheating and shortened life.
3. In two-lamp *instant-start* circuits, deactivation of one lamp will normally deactivate the entire ballast because the lamps are usually connected in series.

19.19 Electronic Ballasts for Fluorescent Lamps

We have already noted in Section 19.16*b* the advantages to be gained in high-frequency operation of fluorescent lamps. The electronic ballast is primarily a generator of high-frequency a-c and secondarily a ballast, since at these frequencies (25–30 kHz) all that is required to "ballast" the circuit, once the arc is formed, is a capacitor. The following is a brief comparison of the characteristics of fluorescent lamp operation with an electronic ballast at 25 to 30 kHz versus operation with conventional core-and-coil magnetic ballast, at 60 Hz. For the purpose of comparison, and where applicable, two 40-W RS lamps are assumed *and light output equalized* for the two situations.

(a) Circuit Wattage, Losses, and Component Temperatures. All these factors are intimately interrelated. In a typical open two-lamp fixture, average power consumption with standard ballast is 88 W; with electronic ballast 70 W. This 18-W (20%) reduction is primarily due to reduction in ballast heating. As is characteristic of solid-state equipment, the electronic ballast is almost lossless, and as a result runs at 50 to 60°C, as compared to 85 to 90°C for the magnetic unit. This has several additional desirable effects; the bulb runs cooler because fixture ambient temperature is lowered, and the fixture mounting constraints that are directed at controlling fixture ambient temperature are almost completely eliminated.

(b) Flicker. Conventional ballast circuits have 29% flicker, which is visible in peripheral vision and can be annoying. Flicker is substantially reduced with all electronic ballasts (10% maximum) and with most is entirely absent. Users of video display units (computer terminals, monitors) report a sharp reduction in visual fatigue which some researchers attribute to elimination of fluorescent lamp flicker, or to elimination of the visual effect of lamp flicker *and* screen flicker.

(c) Dimming. This capability can be much more easily and cheaply incorporated into the electronic design without the appreciable increase in flicker which is characteristic of conventional circuits. Local control at the electronic ballast permits individual fixture level adjustment—a "tuning" aspect absent from conventional circuits. Also, advanced design electronic (EL) ballasts using integrated circuits can be remotely controlled in dimming, two-level switching modes, and on–off control modes.

(d) Noise. Solid-state ballasts are almost completely silent and readily meet A rating requirements.

(e) Other Characteristics. Both standard and electronic ballasts are available with power factor and ballast factor above 90%.

Standard ballasts will normally start a lamp at 50°F minimum. A special ballast is required for starting down to 0°F. Electronic ballasts normally start down to 0°F.

Standard circuits are voltage sensitive, as shown in Fig. 19.49. Electronic ballasts can easily be provided with a voltage-regulating circuit that will handle ±10% line voltage variation with no change in lamp performance.

Failure of one lamp in a two-lamp conventional circuit has the effects detailed in Section 19.18*j*. In electronic circuits, including Slimline, the lamps are operated in parallel. Failure of one lamp will cause the second either to be unaffected or to increase its output by 15 to 25%,depending on design.

Neither EMI (electromagnetic interference) nor RFI (radio-frequency interference) is a problem with properly designed, shielded, and grounded equipment.

Conventional circuits have 5 to 15% third-harmonic distortion, which appears as an undesirable current in the neutral of a 3-phase system. Electronic ballast circuits generally have higher third-harmonic currents, although stringent specifications can be met by proper design.

The capability to change lighting levels can readily be exploited by local control with a standard wall switch. Electromagnetic (EM) ballasts require special switching arrangements, as explained above.

EL ballasts readily meet ballast factor (BF) and ballast efficacy factor (BEF) standards because of their inherently low losses. However, care must be exercised in applying EL ballasts to low-energy lamps, particularly those with krypton gas filling. Integrated-circuit-controlled EL ballasts are available with integral wattage sensing, making them applicable to a range of different lamps. The normally higher price of EL ballasts can frequently be offset by the use of three- or four-lamp ballasts as opposed to one- and two-lamp EM ballasts. This advantage is particularly marked where remote modulating lighting control is utilized (see Sections 21.2 to 21.4).

Electronic ballasts are much lighter than their standard counterparts because of the absence of the heavy, magnetic, iron core. As a result, handling, maintenance, and storage are easier. For example, a standard EM two-lamp 40-W rapid-start ballast is approximately 10 × 1¾ × 2½ in. and weighs 3½ lb. The equivalent EL ballast for two *or three* lamps has the same physical dimensions (for complete replacement interchangeability) but weighs only 1.5 lb. It should be apparent from the above that the electronic ballast has so many advantages over the standard unit that its general acceptance is only a matter of time.

HIGH-INTENSITY DISCHARGE (HID) LAMPS

Subsumed under this heading are mercury, metal-halide, and sodium-vapor lamps. These lamps have inherently high efficacy and, with appropriate color correction, can be utilized in any application, indoor or outdoor, that does not have critical color criteria.

19.20 Mercury Lamps

These lamps operate by passing an arc through a *high-pressure* mercury vapor contained in an arc tube made of quartz (see Fig. 19.57). This action produces light in both the ultraviolet region (as in the low-pressure fluorescent lamp tube) and in the visible region, principally in the blue-green band. This color is characteristic of the clear mercury lamp. Details of spectral distribution are given in Section 19.26.

(a) Ultraviolet Radiation. A considerable portion of a mercury lamps' spectrum is in the ultraviolet range. This does not normally constitute a hazard to persons exposed to the light from even a clear lamp because the outer glass bulb absorbs most of the UV while transmitting the light in the visible range. However, if the outer bulb is broken, the quartz arc tube will continue to burn, and the UV emitted will constitute a safety hazard, particularly to the skin and eyes of persons exposed to the radiation. As a result of this situation, all manufacturers include a warning to the above effect with all mercury lamps sold. Users are also generally informed that a safety-type mercury lamp is readily available which will self-extinguish if the outer glass envelope is broken. Another alternative is, obviously, to use mercury lamps in an enclosing fixture that will both prevent lamp breakage from external sources such as vandalism and protect users of the light in the (unlikely) event of a spontaneous lamp fracture.

In connection with protection from the injurious effects of ultraviolet radiation, it is well to note two facts:

1. The shorter the UV wavelength, the more potentially irritating to skin and eyes. Ger-

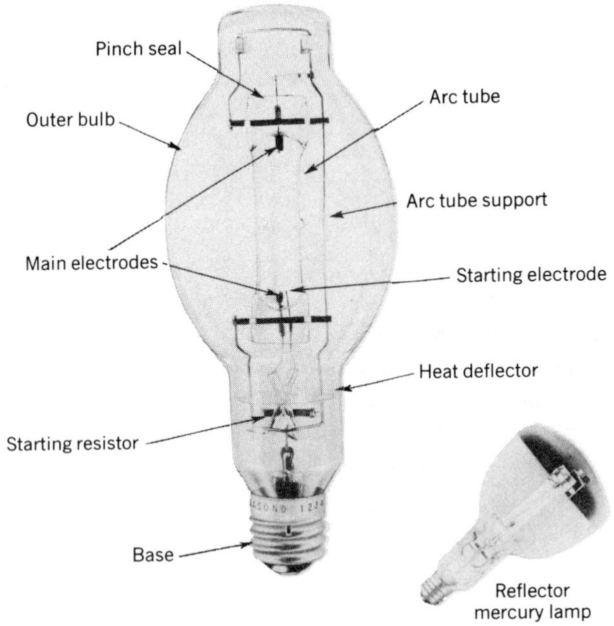

Fig. 19.57 Typical construction of a clear mercury lamp.

micidal UV is in the shortwave range (200 to 300 nm).

2. White plaster and polished metal are good reflectors of UV.

(b) Lamp Life. Lamp life is extremely long, averaging 24,000+ h based on 10 burning hours per start. Mercury lamps are not suitable for applications that are subject to constant switching; therefore, a relatively long period of burning per start was selected. Life is affected by ambient temperature, line voltage, and ballast design.

(c) Lumen Maintenance. This depends on the type of lamp and its burning position. Manufacturers publish data on *each* of their lamp types. In general, clear lamps have the best lumen maintenance, followed by color-improved and phosphor-coated units.

(d) Color Correction and Efficacy. Color correction is normally required because the blue-green light distorts almost all colors. The outer bulb is coated with phosphors that are excited by the UV light and reradiate generally in the red band, which is entirely absent in the basic lamp color. Depending on the arc tube design and the phosphors used, the color of the emitted light can be corrected to make it acceptable for general indoor use. Lamps are available in clear, white, color-corrected, and white-deluxe, in ascending order of color improvement. The deluxe lamp also uses a stain on the envelope to filter out some of the blue-green, which obviously reduces lamp output. Efficacy, including ballast loss, ranges from 25 lm/W for a 50-W lamp to a maximum of 55 lm/W for a 1000-W color-corrected lamp. Note that in general efficacy is lower than for fluorescent lamps.

(e) Ballasts and Lamp Starting. Ballasts are required, as with all arc discharge lamps, to start the lamp and thereafter to control the arc. The basic conventional ballast is simply a reactor that controls the arc after the discharge has been initiated. Three to six minutes is required for the lamp to reach full output, since heat must be generated by electron flow to vaporize the mercury in the arc tube before the arc will strike. Once extinguished, the lamp must cool and the pressure must be reduced before restrike is possible. This restart delay amounts to 3 to 8 min, depending on the ballast type, and is an important consideration in design, since a momentary outage will extinguish all lamps, leaving an interior area in the dark. Special fixtures now available utilize small quartz lamps to supply light during such outages. Alternatively, some incandescent lighting can be utilized that will maintain minimum illumination. Principal types of ballasts are reactor, regulating and electronic. Because lamp operating characteristics depend heavily on the type of ballast, the choice of which involves highly technical electrical considerations, we recommend that it be left to a electrical consultant. Mercury ballasts are normally quite noisy. Where this may be a problem, remote mounting should be considered.

(f) Dimming. Dimming of mercury lamps is possible and entirely practicable with the use of a dimming ballast and solid-state dimming control. These are available for 175-W through 1000-W units. Mercury lamps have so large an output that shutting off a unit creates an imbalance in the lighting coverage—a problem readily solved by dimming. Cost analyses indicate the economic feasibility of such control, including sensing equipment that will automatically maintain illumination levels at preset levels. The payback period depends on power rates and local labor rates. One analysis performed by the author for a large installation, with fifty 1000-W lamps, a $0.085/kWh energy rate, and 8% capital cost, indicated a 2½-yr payback period, assuming 4000-h annual operation. Where dimming is not required, two-level high–low operation is obtainable with a number of commercially available control systems.

(g) Application. Mercury-vapor lamps are applicable to indoor and outdoor use with proper attention to color and fixture brightness. The most common exterior application is for parking lots. Indoor application is generally limited to mounting 10-ft AFF or higher to avoid glare problems and to permit adequate area coverage. Use in industrial spaces and stores is common, as is discussed in detail in Chapter 21. However, because of relatively low efficacy, mercury installations are being replaced by the more efficient metal-halide and sodium lamps.

19.21 Metal-Halide Lamps

The metal-halide lamp is basically a mercury lamp that has been altered by the addition to the arc tube of halides of metals such as thallium, indium, or sodium. The addition of these salts causes light to be radiated at frequencies other than the basic mercury colors and increases efficacy, but reduces the life to 15,000 to 20,000 h and reduces lumen maintenance to about 60% at two-thirds life. Other characteristics, particularly as compared to mercury lamps, are discussed below.

(a) Safety. Being essentially a modified mercury lamp, metal-halide lamps carry the same safety warning as mercury lamps (Section 19.20*a*). An additional warning, however, refers to the fact that metal-halide lamp arc tubes have a tendency to rupture violently (explode) and therefore the lamp must be used in an approved enclosing fixture. Because this tendency to rupture is encouraged by continuous burning, users are instructed to turn off metal-halide lamps at least once daily for a period of time sufficient to allow the lamp to completely cool and stabilize. A line of tough plastic-coated lamps is available for use in open fixtures.

(b) Operating Characteristics. As can be seen from Table 19.20 efficacy of metal-halides in all wattages is considerably higher than that of mercury lamps, exceeding even that of standard fluorescents. Lamp color and color rendering are much better than those of mercury lamps because the spectrum is essentially continuous. They depend not only on the halide salt used in the arc tube but also on the type of phosphor used in coated lamps. Clear lamps have a correlated color temperature of 3500 to 4500 K and a color rendering index of about 65. Phosphor-coated lamps have a CCT of 3200 to 3800 K and a CRI of 70, making them suitable for almost all indoor applications. However, like mercury

ILLUMINATION

lamps, metal-halides are not instant starting, requiring approximately 2 to 3 min on initial startup and 8 to 10 min for restrike. As a result, when used for indoor installations a secondary instant-start source must be available. (Recently, a number of manufacturers have introduced special hot-restrike ballasts which provide immediate restrike of lamps on restoration of power after an outage. Lamp output on restrike is inversely proportional to the duration of the power outage.)

Alternatively, metal-halides can be successively mixed with fluorescent sources if the color characteristics of both types are carefully chosen. It is important to note here that the spectrum of light produced by metal-halides *changes* as the lamp ages. The change is gradual, definite, and usually unnoticed, and depends on the particular design of lamp. However, where color rendering is important, or where mixed with other light sources, the user must investigate the nature of these changes before final selection of a lamp.

One manufacturer of metal-halide lamps now produces lamps in a double-ended tubular design which looks very much like a tungsten-halogen lamp. These lamps are available in a range of ratings and color characteristics, including lamps with a CCT of 3000 K and a CRI of 80. Such lamps are applied where color rendering is important and where addition to, or replacement of, an incandescent lamp installation is contemplated. These lamps have lower efficacy and considerably shorter life than standard metal-halide designs.

One characteristic of metal-halides which is peculiar to them is their great sensitivity to burning position—so much so that lamps are designed for base-up, base-down, horizontal, and universal burning positions. If the burning position of a lamp is altered, its lumen output, lumen maintenance, and life are all reduced by factors up to 20% at best, and for some lamps much more severely. For this reason all metal-halide lamps are marked with the proper burning position.

Finally, dimming or reduced output operation of metal-halides is not recommended because of the drastic color shift that occurs when the lamp is dimmed. In effect, a dimmed halide lamp operates very much like a mercury lamp.

19.22 High-Pressure Sodium (HPS) Lamps

The highest efficacy general-purpose HID source is the high-pressure sodium lamp, marketed by its developer, General Electric, under the trade name *Lucalox®*, by Philips as Ceramalux™, and by Sylvania under the name *Lumalux®*.) Details of the lamp are shown in Fig. 19.58, and performance data are given in Table 19.21. Construction is quite different from that of mercury and metal-halide lamps and, although it operates as an arc discharge unit, its excellent characteristics stem from the spectral absorption phenomenon of the sodium contained under high pressure in the arc tube. The resultant light has a pronounced yellow tint. Typical characteristics are:

Lamp efficacy	80 to 140 lm/W
Efficacy, including ballast losses	57 to 128 lm/W
Life	24,000+ h
Lumen maintenance	80 to 90% at 50% life
Warm-up time	3 to 4 min
Restrike time	½ to 1½ min
Correlated color temp.	2100 K

Unlike the metal-halide lamp, the HPS unit is neither voltage nor burning-position sensitive and is color constant when operated undimmed. Its lumen maintenance is outstanding, as is its efficacy, both of which are the highest available for any general-use light source. Details of light color are discussed in Section 19.24, and comparative color characteristics with other HID sources and fluorescent lamps are given in Section 19.26. As with all discharge lamps, a ballast is required to supply the high voltage to strike the arc and to control the arc once struck. HPS ballasts are quite different from those for mercury or metal-halide lamps because of the high voltages necessary. In addition the ballast lamp power is a function of lamp voltage. As with other HID sources, electronic ballasts are now

TABLE 19.20 **Typical Metal-Halide Lamp Data**

Standard Design Lamps: rated life and mean lumens based on minimum of 10 h per start operation. Rated life and mean lumens are reduced for shorter burning cycles.

Watts	Bulb[a]	Base	Description	Average Rated Hours Life	Approx. Lumens Initial	Approx. Lumens Mean[b]	Efficacy[c] (lm/W)	CCT[d] (K)	CRI[e]
175[f]	BT-28	Mogul	Clear	7,500	14,500	10,800	71	4400	65
			Phosphor coated	7,500	14,500	10,200		3800	70
250[g]	BT-28	Mogul	Clear	10,000	20,000	17,000	73	4400	65
			Phosphor-coated Metalarc	10,000	20,000	16,000		3900	70
400[g]	BT-37	Mogul	Clear	15,000[h]	35,000	27,600	79	4100	65
			Phosphor coated	15,000[h]	35,000	28,600		3700	70
1000[g]	BT-56	Mogul	Clear	18,000	110,000	88,000	100	3800	67
			Phosphor coated	18,000	110,000	85,000		3400	72

High-Output Lamps: Lamps must be operated within ±15° of horizontal. Other lamps available for vertical operation.

Horizontal Operation (±15°)				Average Rated	Approx. Horiz. Lumens				
Watts	Bulb	Base[i]	Description	Hours Life	Initial	Mean[b]			
175	BT-28	Mogul	Clear	10,000	15,500	12,000	80	4600	65
			Phosphor coated		15,500	11,000		4100	70
175	BT-28	Mogul	Phosphor coated	10,000	14,000	10,000	69	3200	72
250	BT-28	Mogul	Phosphor coated	10,000	21,000	16,000	75	3200	72
400	BT-37	Mogul	Clear	20,000	39,000	31,000	89	3900	67
400	BT-37	Mogul	Phosphor coated	20,000	39,000	30,000	89	3500	72
400	BT-37	Mogul	Phosphor coated	15,000	37,000	28,000	84	3200	72

Source: Extracted from various manufacturers catalogs.

[a] For diagram of bulb shape, see drawings at right.

[b] Taken at 40% of rated life.

[c] Calculated using average ballast loss figures.

[d] Correlated color temperature.

[e] Color rendering index.

[f] Must be operated within ±15° of vertical.

[g] Any burning position.

[h] Life 20,000 h if operated within 15° of vertical.

[i] Position-oriented base.

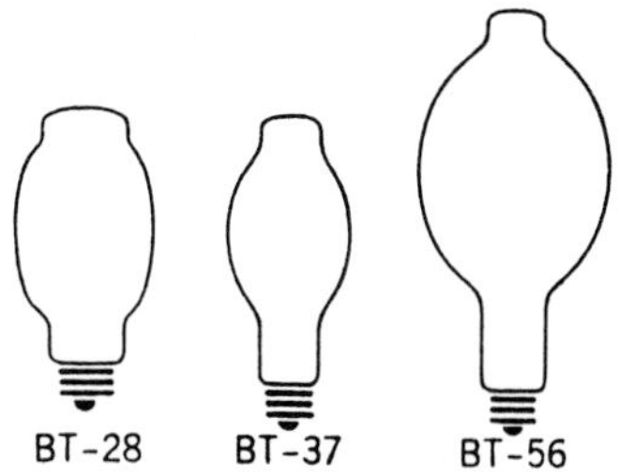

becoming available. As with halide lamps, ballasts are available for HPS lamps which provide instantaneous restrike after a power interruption. Light output on restrike is inversely proportional to the length of the outage; for example, after a 10-s outage, restrike light output level will be 85% of full capacity, whereas after a 2-min outage, lamp output will be only 10%. It will then take another minute (±) to regain full output.

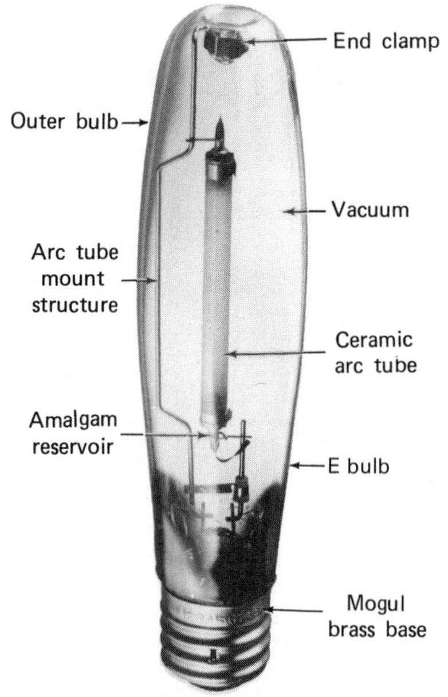

Fig. 19.58 *The main features of the HPS lamp are the alumina ceramic arc tube, amalgam reservoir, and rigid arc tube-mount structure. (Photo courtesy of General Electric Co., Lighting Business Group.)*

Although HPS lamps can be operated at reduced level or dimmed, this mode is not commonly used because the lamp color shifts to a deep yellow. In effect, the high-pressure sodium lamp becomes a low-pressure sodium lamp, whose characteristics are given in the following section. HPS lamps are available in clear and coated designs. The former is effectively a point source and, because of its extreme brightness, must be enclosed in a fixture. Table 19.21 gives characteristics of typical HPS lamps presently available. Because of high efficacy, their energy-saving implications are obvious. This lamp should be considered primarily, or as a substitute, for all existing or new HID applications. Its yellow color, which becomes less noticeable as the eye color adapts, can be made even more acceptable by using white sources in conjunction.

19.23 Low-Pressure Sodium (SOX) Lamps

This lamp, also referred to as SOX, produces light of sodium's characteristic monochromatic deep yellow color, making it inapplicable for general lighting. Because of its very high efficacy of over 150 lm/W *including* ballast loss, it can be applied wherever color is not an important criterion. Thus SOX is widely used for street, road, and area lighting, as well as for emergency or after-hours indoor lighting. Another desirable aspect of SOX lamps is their 100% lumen maintenance. This, coupled with the discharge lamp's typically long life (18,000+ h), makes SOX lamps the most economical source available today in terms of cost per million lumens produced.

19.24 Chromaticity

The CIE (Commission Internationale de l'Éclairage) color system is the internationally accepted standard for designating illuminant color. In this system the relative proportions of each of the three primary colors (red, green, blue) required to produce a given illuminant color are calculated, and the result is plotted on a standard chromaticity diagram (see Figure 19.60 in the next section). To calculate these proportions, measurements are made across the entire spectrum of the illuminant under test. These figures are then weighted in terms of the three primary colors. The resulting figures represent the proportions of red, green, and blue required to produce the spectrum color at that wavelength. These values are called the *tristimulus values* for that color and are designated by capital letters: X (red), Y (green), and Z (blue). The Y (green) value is also proportional to that color's luminosity.

19.25 Spectral Distribution of Light Sources

In addition to providing sufficient light of adequate quality, the lighting designer must also be concerned with the spectral content of the selected illuminant, since object color depends heavily on the illuminant. As shown earlier in Fig. 18.42*a*, perceived object color is the result of selective absorption and reflection of components of the illuminating light by the pigments of

TABLE 19.21 **Typical Data for High-Pressure Sodium Lamps**

Watts	Bulb	Base[a]	Average Rated Hours Life[b]	Approx. Lumens Initial	Approx. Lumens Mean	Description	Lamp Efficacy (lm/W)	Lamp and Ballast[c] Efficacy
35	B-17	Med	16,000	2,250	2,025	Clear	64	48
				2,150	1,935	Coated[d]	61	46
50	E-17	Med	24,000+	4,000	3,600	Clear	80	60
				3,800	3,420	Coated	76	57
70	E-17	Med	24,000+	5,800	5,200	Clear	83	63
				5,400	4,860	Coated	78	59
100	ED-17	Med	24,000+	9,500	8,850	Clear	95	74
				8,800	7,920	Coated	88	69
150	ED-17	Med	24,000+	16,000	14,400	Clear	106	85
				15,000	13,500	Coated	100	80
250	E18	Mogul	24,000+	27,500	24,750	Clear	110	90
	E-28			26,000	23,400	Coated	104	85
400	E-18	Mogul	24,000+	50,000	45,000	Clear	125	105
	E-37			47,500	42,750	Coated	119	100
1000	E-25	Mogul	24,000+	140,000	126,000	Clear	140	128

Source: Data extracted from various manufacturers catalogs.

[a]All bases are screw type. Lamps of 50-, 70-, 100-, and 150-W ratings are available in either medium or mogul screw base.

[b]Based on operation on proper auxiliary equipment for 10 h or more per start.

[c]Conventional core-and-coil ballast.

[d]Use in open-bottomed fixtures or where glare is a problem.

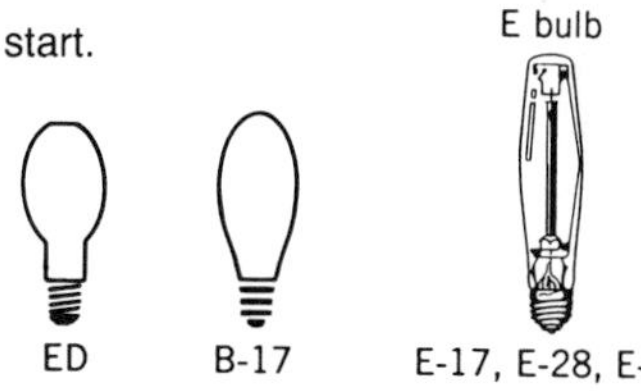

the object being viewed. It is therefore obvious that the *illuminant* must contain the color of the *object*, in order for us to see the object's color. It is not so obvious that the relative energy of an illuminant at a particular wavelength determines the saturation and brilliance with which we see a color. To understand this, refer to Fig. 19.59. In graphic form, the relative spectral energy of a few common light sources have been plotted, as a function of wavelength, that is, color. If we compare graphs 19.59*g* and 19.59*h*, which show the spectral content of two of the most common light sources, cool white and warm white fluorescent lamps, respectively, we note that the principal difference lies in the amount of blue and orange in their spectrum. Cool white has 1.3 blue units and 1.8 orange; warm white has 0.3 blue units and 2.7 orange. As a result, a blue object will be bright under cool white light and dull (grayed) under warm white, whereas the opposite will occur with an orange object. The situation is more pronounced with the clear mercury lamp (Fig. 19.59*k*) as compared to the clear metal-halide lamp (Fig. 19.59*m*). A red object will be gray under the mercury lamp and unrecognizable as red, whereas under the metalhalide its redness will show clearly, and so on.

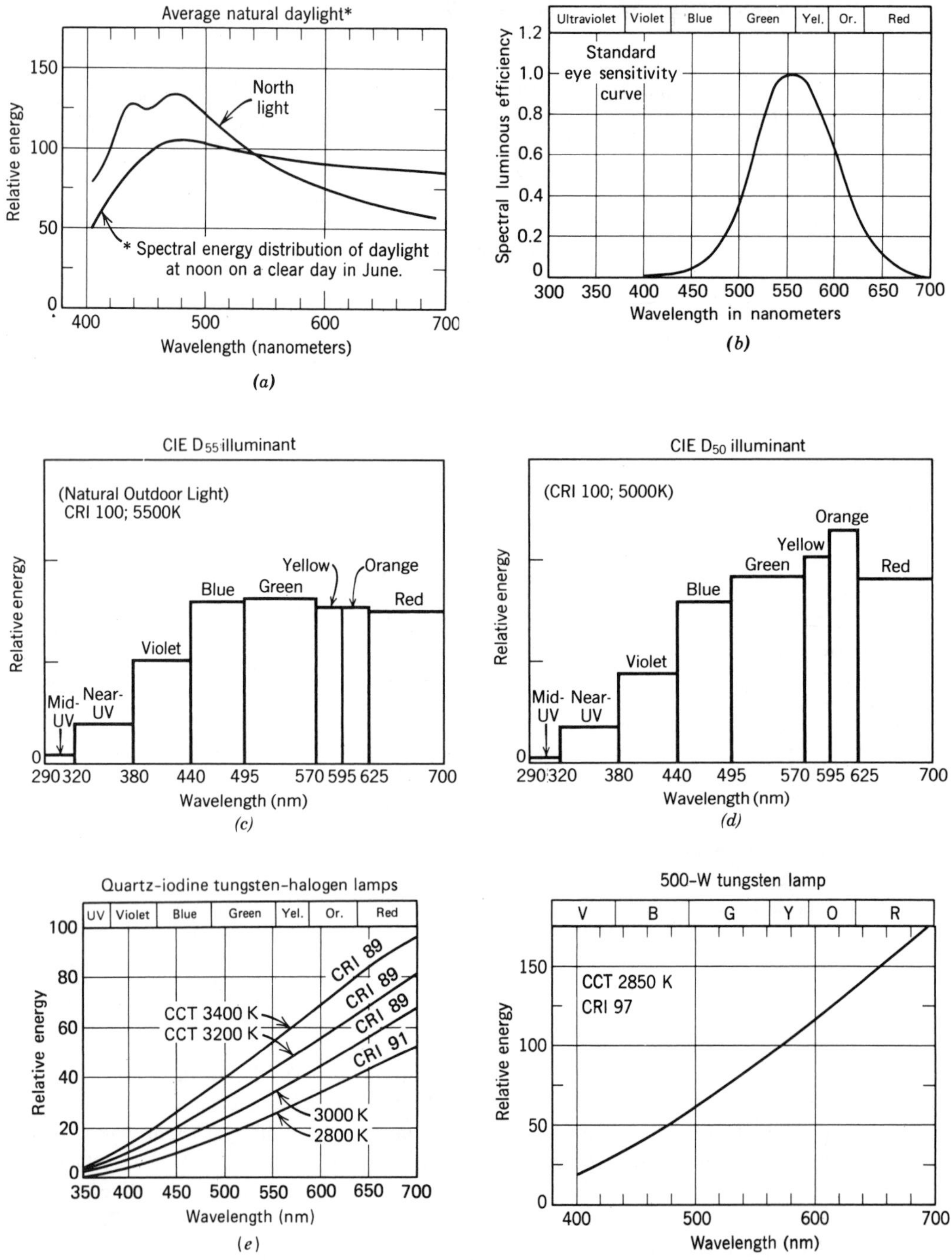

Fig. 19.59 *Spectral energy distribution curves for typical illuminants, plus their correlated color temperature (CCT) and color rendering indices (CRI). Since only a radiating blackbody has a true color temperature, a source with mixed color illuminants is assigned a correlated color temperature, which is the temperature of a blackbody radiator whose chromaticity most nearly matches that of the light source. See Section 18.31.*

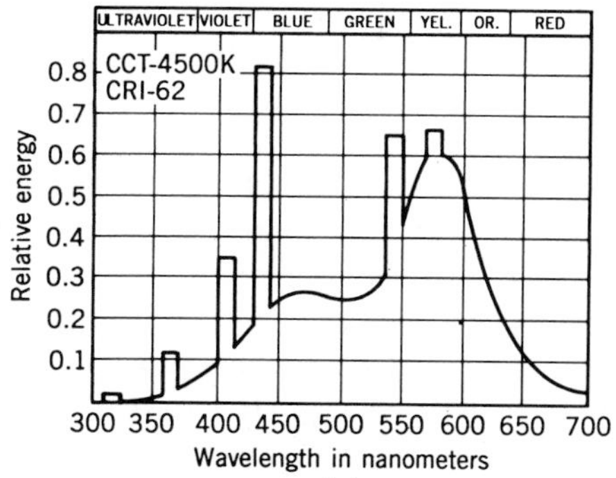

(g)

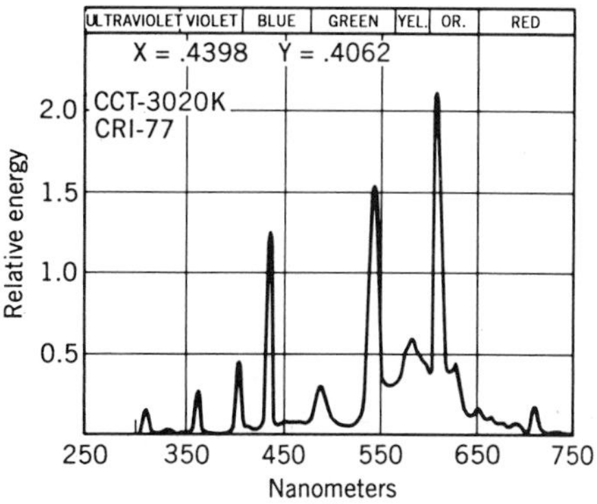

(h)

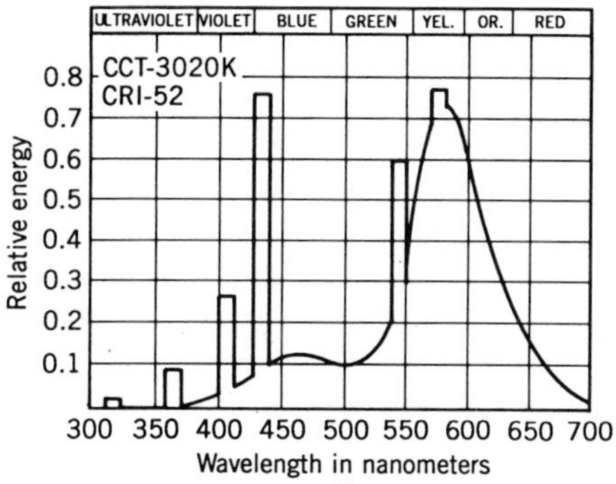

(i)

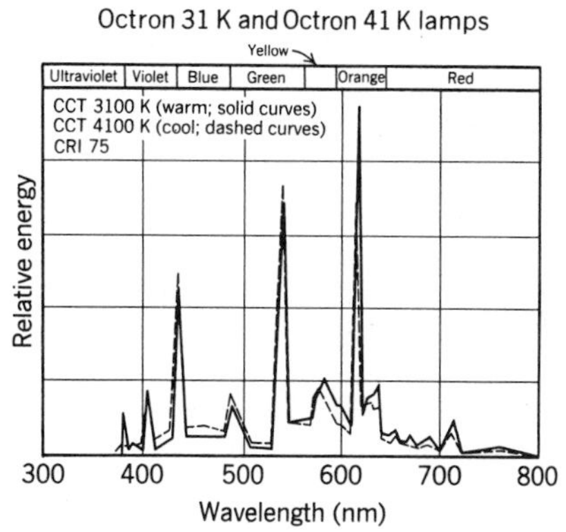

(j)

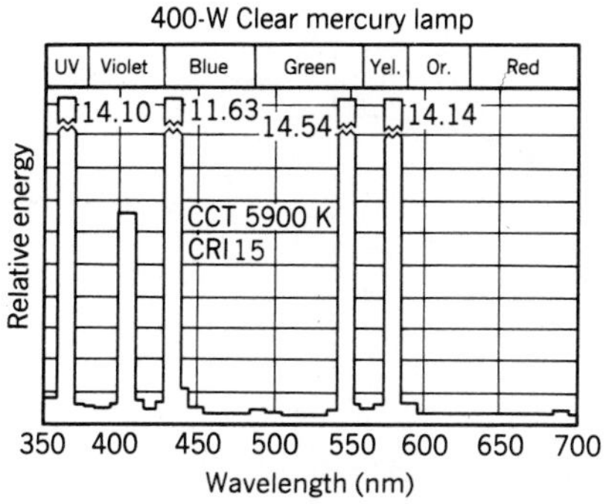

(k)

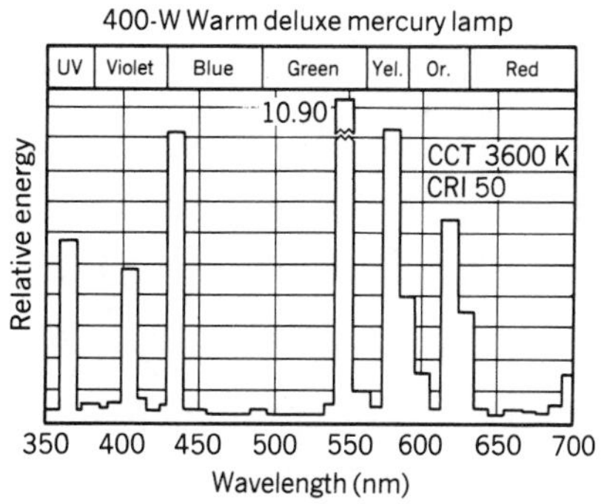

(l)

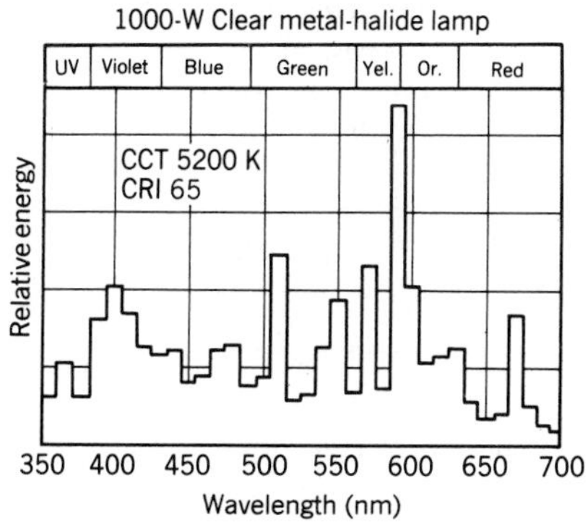

(m)

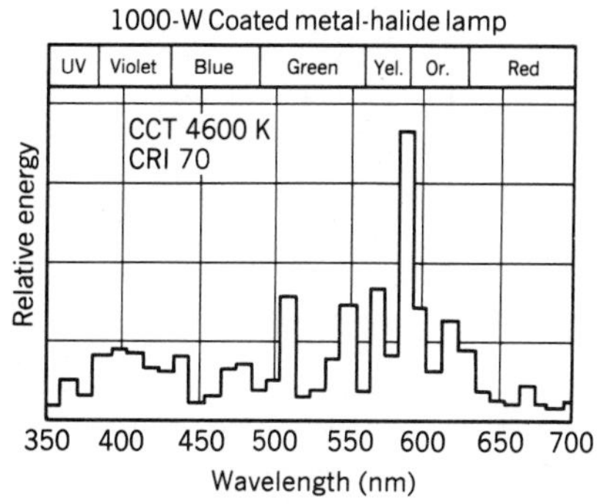

(n)

Fig. 19.59 *(continued)*

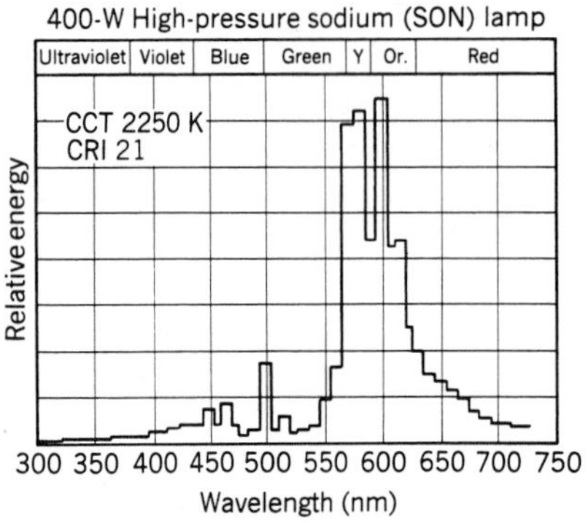

(o)

Fig. 19.59 *(continued)*

This concern for perceived object color, which relates not only to furnishings but also to paints and prefinished construction materials such as carpets or floor tile, is quite properly the province of the architect and lighting designer, who in turn must possess the necessary knowledge and information to make the appropriate choices of both illuminant *and* object color. Spectral composition graphs of the type shown in Fig. 19.59 are available from manufacturers for all light sources, and they should be examined when considering the characteristics of a particular light source.

One of the best ways to compare illuminants is first to expose a dull white surface to the illuminants, side by side but separated by an opaque divider, to get an impression of the illuminate color, and then to expose a series of colored chips, again side by side, to see which colors are brightened and which are grayed. The intensity of illumination also influences the appearance of colors, and it must be considered in choosing object colors. As intensity is increased, reflection increases, particularly with pale tints (high value) that contain much white

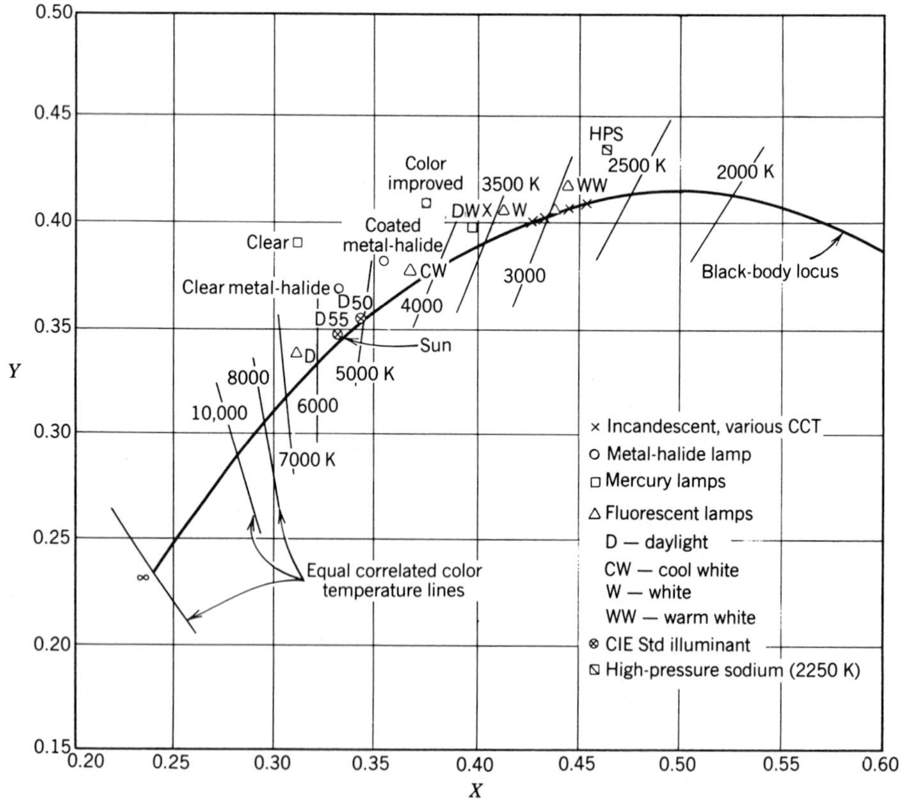

Fig. 19.60 *Basic chromaticity diagram showing relation of common illuminant chromaticities to that of the blackbody locus. Illuminants whose coordinates fall on the same line crossing the blackbody locus have the same correlated color temperature.*

TABLE 19.22 **Effect of Illuminant on Object Colors**

Lamp	*CRI (Approx.)*	*CCT (K)*	*Whiteness*	*Colors Enhanced*	*Colors Grayed*	*Notes*
Fluorescent Warm white	52	3000	Yellowish	Orange, yellow	Red, blue Green	
White	60	3450	Pale yellowish	Orange, yellow	Red, blue Green	
Cool white	62	4150	White White	Yellow Orange Blue	Red	
Warm white deluxe	77	3025	Yellowish	Red, green Orange Yellow	Blue	Simulates incandescent
Daylight	75	6250	Bluish	Green Blue	Red Orange	
Tri-phosphor	75	3000	Yellowish	Red Orange	Deep red Blue	T8 bulb
	75	3500	Pale yellowish	Red Orange Green	Deep red	T8 bulb
	75	4100	Pale greenish	Red, blue Orange Green	Deep red	T8 bulb
Mercury Clear	15	5710	Blue Green	Blue Green	Red Orange	Poor overall color rendering
Deluxe	50	3900	Pale purplish	Blue Red	Green	Shift over life to greenish
Warm deluxe	50	3300	Pinkish	Blue Red	Green	Shift over life to greenish
Metal halide clear	65	4100	White	Blue Green Yellow	Red	Shifts to pinkish over life
Phosphor coated	70	3900	White	Blue Green Yellow	Red	Shifts to pinkish over life
HPS standard	21	2100	Yellowish	Yellow	Red, blue	
HPS color corrected	65	2200	Yellowish white	Red, blue Green Yellow	Deep red Deep blue	CRI decreases slightly over life
Incandescent	99+	2900	Yellowish	Red, orange, yellow	Blue, green	

Source: Courtesy of General Electric Co., Lighting Business Group.

pigment and thus tend to wash out color. Therefore, with high-intensity lighting, saturation of colors should be high for true, brilliant color rendition.

Refer again to Fig. 19.59. Note from diagrams *a*, *e*, and *f* that the spectrum of a light source that produces light as a result of heating is continuous. Sunlight is equal in spectrum to a blackbody radiator at 5500 K; north light equal to one at about 8000 to 10,000 K; a 500-W incandescent lamp approximately equal to one at 2850 K; and so on. If the spectrum of a blackbody radiator is plotted on a chromaticity diagram, its locus is a continuous curved line as seen in Fig. 19.60. The chromaticity of all true blackbody radiators will fall exactly on this line, with the location depending on temperature. Daylight, for most purposes, falls on this locus, although because of selective atmospheric absorption and other phenomena, it is actually slightly off. Incandescent lamp chromaticity is very close to this locus because it is also a heat–light radiator.

A source that produces light by means of individual phosphors can also have chromaticity on this locus if the phosphors will, in effect, produce a continuous spectrum similar to that of a blackbody radiator. Thus we see that CIE standard illuminants (see Figs. 19.59*c* and *d*) have spectral components over the entire spectrum, yielding an exact equivalent of blackbody radiator at a particular color temperature. Therefore their color temperature and their correlated color temperature are the same. For other sources, their correlated color temperature is established by their chromaticity locus in relation to the diagonal lines crossing the blackbody locus, as in Fig. 19.60. Each of these lines is isothermal, that is, all chromaticities on it have the *same correlated color temperature*. Thus the reader can see readily that two sources with widely differing spectral content and therefore object color rendering can have the same correlated color temperature.

19.26 Color Rendering (Index), CRI

Color rendering is defined as the degree to which perceived colors of objects illuminated by a test source conform to the colors of the same objects as illuminated by a reference source. The *color rendering index* (CRI) of a source is a two-part concept, comprising a color temperature that establishes the reference standard and a number that indicates how closely the illuminant approaches the standard. *The standard is always daylight at that color temperature*. Therefore, the color rendering index of a light is really a measure of how closely it approximates daylight of the same color temperature. Two lights cannot be compared unless their color temperatures are equal or quite close. A color rendering index of 100 indicates an illuminant whose spectral content is equal to daylight of that temperature. Color rendering indices for typical common lamps are given in Fig. 19.59.

Table 19.22 lists the color characteristics of a few of the major sources. An illuminant's own color appearance on a neutral surface will depend on its own spectral content, but if the observer is placed in a space illuminated with this source, the eye after a short exposure time will become adapted to the source color and will detect only a degree of whiteness rather than an actual tint.

Where it is necessary to detect small color differences between two objects, a light poor in object color, or complementary to the object color, should be used, at a relatively high illumination level, followed by a light high in object color, at the same illumination level. If this is not possible, two widely different but broad-spectrum illuminants should be used, preferably at the same illumination level. Another technique is the use of a special, fixed color source. For a full discussion, see the *IES Handbook*. It should be remembered in all considerations of color comparison, matching, and rendering that object color depends on the spectral energy distribution of the light source (illuminant), and therefore any change in the spectral content will change the object appearance. Two sources of the same color temperature and therefore apparent whiteness can have quite different spectral content and will therefore render object colors differently. A case in point would be a 3000-K warm white fluorescent tube and an incandescent lamp (500-W photoflood) of approximately the same color temperature. Color temperature

Fig. 19.61 *A hand-held lightweight (≈ 8 oz) chroma meter for determining chromaticity and color temperature of a light source. The unit reads* X, Y *coordinates, color temperature in kelvin, and illuminance in lux. (Courtesy of Minolta Corp.)*

is an expression of dominant color, not spectral distribution.

A convenient hand-held chromaticity meter is illustrated in Fig. 19.61. This unit measures the X, Y coordinates of an illuminant, which can then be plotted on a standard CIE color diagram to determine absolute chromaticity. This is very useful in comparing illuminants to predict color response and to avoid color metamerisms. The meter also reads color temperature in kelvin and illuminance in lux.

REFERENCE

See the complete lighting bibliography at the end of Chapter 21.

20 LIGHTING DESIGN

20.1 General

Lighting design is a combination of applied art and applied science. There can be many solutions to the same lighting problem, all of which will satisfy the minimum requirements, yet some will be dull and pedestrian while others will display ingenuity and resourcefulness. The competent lighting designer approaches each problem afresh, bringing to it a knowledge of current technology and years of background and experience, yet rarely being satisfied with a carbon copy of a previous design. And it is these years of background with their successful and not-so-successful designs coupled with a constant striving for improvements that are the characteristics differentiating the competent lighting consultant, designer, or engineer from the person who attempts to force each new job into the unwilling mold of a previous design.

Because of the large number of interrelated factors in lighting, no single design is the correct one, and for this very reason it is not entirely desirable to solve a lighting problem with a step-by-step technique. However, since this technique is a good avenue of approach for the uninitiate who lacks the experience necessary to view an entire solution, we have adopted it.

20.2 Goals of a Lighting Design

Simply stated, the goal of lighting is to create an efficient and pleasing interior. These two requirements, that is, the utilitarian and esthetic, are not antithetical as is demonstrated by every good lighting design. Light can and should be used as a primary architectural material. We elaborate on these goals below.

1. Lighting levels should be adequate for efficient seeing of the particular task involved. Variations within acceptable luminance ratios in a given field of view are desirable to avoid monotony and to create perspective effects.

2. Lighting equipment should be unobtrusive, but not necessarily invisible. Fixtures can be chosen and arranged in various ways to complement the architecture or to create dominant or minor architectural features or patterns. Fixtures may also be decorative and thus enhance the interior design.

3. Lighting must have the proper quality as discussed previously. Accent lighting, directional lighting, and other highlighting techniques increase the utilitarian as well as architectural quality of a space.

4. The entire lighting design must be accomplished efficiently in terms of capital and energy resources, the former determined principally by life-cycle costs and the latter by operating energy costs and resource-energy usage. Both the capital and energy limitations are, to a large extent, outside the control of the designer, who works within constraints in these areas. Obviously, these constraints are maxima.

With these goals before us we can write a lighting design procedure, keeping in mind that the order of steps shown is not necessarily the same in each lighting problem and that, since all of the factors are closely interrelated, it is often necessary to apprehend several of the stages simultaneously before arriving at a decision.

It is appropriate to note at this point that the lighting design approach and procedure that we explicate in this chapter is primarily an analytic one; that is, the design procedure establishes requirements primarily in numerical form and then manipulates the variables of sources, fixtures, placement of units, and so on, to arrive at a design conclusion. There exists an alternative approach, frequently referred to as *brightness design*, in which the designer labels surfaces on a perspective plan of a space with desired luminances as established by a mental picture or by other means, and designs the lighting accordingly. This approach, which can be very effective, is highly intuitive (i.e., requires considerable prior experience on the part of the designer). For this reason, and because that approach requires a good deal of hands-on, field trial-and-error work, we feel that it is not appropriate to a textbook and is therefore not used here. The only exception to this statement occurs in our discussion of daylight models in Chapter 19. There, the entire purpose of the model is to give the designer a visual/mental picture of the space's brightness patterns, on the basis of which windows, window treatments, and possible exposures can be varied to achieve the desired light and shadow patterns (luminances and luminance ratios).

20.3 Lighting Design Procedure

For the benefit of the reader whose background and study framework is purely architectural, it may be helpful to draw parallels between the purely architectural approach to lighting design and the more structured approach described below. The terminology used in the text *Inside Out* (see Brown, Reynolds, and Ubbelohde, 1982) is typical of an architect's approach. There the authors speak of a four-stage plan: conceptualizing, scheming, developing, and finalizing. The *conceptualizing* stage is general and overall; it considers the occupant in the visual environment, and his visual relation to task, immediate environment, and surroundings. It corresponds, in the flowchart of Fig. 20.1, to the background thought and ideas for the two boxes labeled "task analysis" and "preliminary design." *Scheming* is the on-paper preliminary design stage; 1/16- or 1/32-in. scale, corresponding to parts of the "preliminary design" and "detailed design" stages. *Developing* is essentially the detailed design stage, with input from (the boxes on Fig. 20.1 labeled) costs, energy, codes, regulations, and other disciplines. *Finalizing* is comparing, checking, evaluating, and comparing design results to goals and requirements.

(a) Project Constraints. Refer to Fig. 20.1. This flowchart, which represents the design procedure and its interactions, should be referred to throughout the necessarily lengthy discussion that follows, in order to maintain perspective. It is important that the reader be aware of job constraints and of the interactions between the lighting designer and the remainder of the design group. We deliberately emphasize this to demonstrate the interdisciplinary nature of lighting design in general and its particular connection with HVAC and daylighting (fenestration). This approach, which is most often referred to as the systems design approach, will be followed throughout the discussion.

Item 4 of the list in Section 20.2 referred to constraints. These can alternatively be classified as the owner-architect-user team and the jurisdictional authorities. In some detail these are:

1. *Owner-designer-user group*. The owner establishes the cost framework, both initial and operating. A part of both of these may be a rent structure, which in turn determines and is determined by the space usage. If the owner is also the occupant, the cost factors change somewhat but remain in force. The architect determines the amount and quality of daylighting and the architectural nature of the space to be lighted. Many of these data are detailed in the building program. Obviously the architect and lighting designer (who may be one person) should interact in this aspect of building design.

2. The jurisdictional authorities *may* include:

FEA—Federal Energy Administration

GSA—General Services Administration

NCSBCS—National Conferences of States for Building Codes and Standards

ASHRAE—American Society of Heating, Refrigerating and Air Conditioning Engineers

IES—Illuminating Engineering Society

NBC—National Bureau of Standards

Most of these are jurisdictional by reference, that is, the actual authorities will specify that the lighting system meet the requirements of ASHRAE, IES, and so on. If public money is

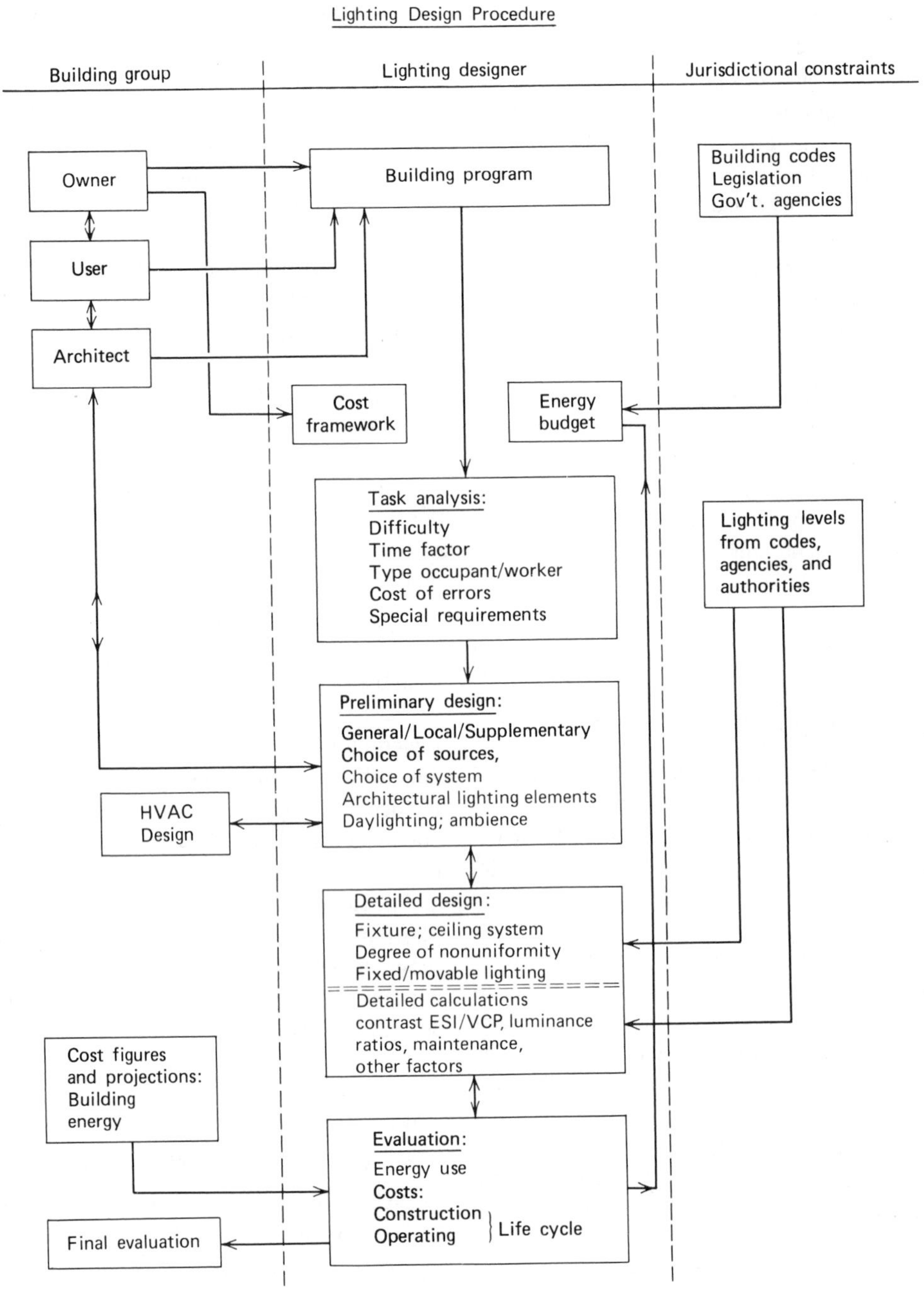

Fig. 20.1 *Lighting design procedure chart.*

involved, FEA/GSA standards will probably be involved. The principal area of involvement is that of energy budgets and lighting levels, both of which affect every aspect of lighting design including source type, fixture selection, lighting system, fixture placement, and even maintenance schedules. For this reason the first step in the lighting design procedure is to establish the *project lighting cost framework and the project energy budget.*

(b) Task Analysis. As shown in Fig. 20.1, this step essentially determines the needs of the task. Factors to be considered in addition to the nature of the task are its repetitiveness, variability, who is performing it (i.e., condition of the occupant's eyes), task duration, cost of errors, and special requirements. Several of these factors have been discussed in the preceding sections dealing with quality of light. The reader will be referred to the appropriate sections in the following analysis.

(c) Design Stage. This is the active consideration stage during which detailed suggestions will be raised, considered, modified, accepted, or rejected. This is also the most interactive stage as is clearly seen from Fig. 20.1. At its completion, a detailed, workable design is in hand. The critical interactions here are with the architect in daylighting and with the HVAC group in power loads. The former may result in relocating a space within the building, the latter in making a change in a lighting system or HVAC system. In brief, this stage consists of the following steps:

1. Select the lighting system. Select type of light source and distribution characteristic of fixture(s) or area source and consider effects of daylighting, economics, and electric loads.
2. Calculate the lighting requirements. Use the applicable calculation method and establish the fixture pattern, considering the architectural effects.
3. Design the supplemental decorative and architectural lighting.
4. Review the resultant design. Check the design for quality, quantity, esthetic effect, and originality.

(d) Evaluation Stage. With the design on paper, it can now be analyzed for conformance to the principal constraints of cost and energy. If the design stage has been carefully accomplished, with due attention to these factors, the result of the final evaluation should be gratifying. The results of this stage are fed to the architectural group for use in the final overall project evaluation. In the following sections we consider in detail each of the steps in the design procedure.

20.4 Cost Factors

This is a particularly difficult item for a novice lighting designer because it requires experience in the field and an acquaintance with commercially available equipment. Also, the inevitable tradeoffs between first cost and operating cost cannot intelligently be made unless the cost structure is clearly understood. The following guidelines should be of considerable assistance both in avoiding unpleasant surprises when a job is estimated and in preparing cost analyses:

1. Decide at the outset what cost criteria will be applied, that is, the relative importance of first cost, operating costs, annual owning costs, and life-cycle costs.

2. Trade-off decisions are required between first cost and operating costs. For example, incandescent lamps and fixtures are low in first cost and high in operating cost, and so on. Dimming and control equipment falls into this area of decisions.

3. Manufacturers' catalog items are *always* cheaper than specials, and can be priced more readily.

4. Compare annual owning costs of two systems or methods. Conversion of these data to life-cycle cost comparisons is straightforward. With the continued instability in energy costs, it is suggested that even in an *annual* owning cost comparison, two different energy costs be used and the effect of a sharp increase studied.

5. The impact of lighting energy on the operating cost of the entire building must be studied, and the apportionment of costs determined. The only practical means of accomplishing this is by computer program. Programs can be readily adjusted to reflect the effect of the lighting system on building costs, and in particular on HVAC first cost and operating costs. It is incorrect to artificially separate the lighting system from the HVAC system with which it intimately interacts.

The lower the lighting system's energy level, the lower the building's overall operating cost. The argument that heat from a lighting system is fully utilized to heat the building and is therefore not wasted is a specious one, which has been refuted on many counts:

HVAC system first cost is higher.

HVAC year-round cost is higher.

Lighting energy cost is higher.

Life-cycle costs are higher.

Energy resource use is higher.

This subject is discussed in detail in Section 21.23.

20.5 Power Budgets

The requirement to establish a project's lighting power budget in accordance with a specified procedure has now been incorporated into the building codes of many states.

The purpose of this budget determination procedure is not to dictate design procedure. Indeed all standards explicitly so state. Instead, the purpose is to develop an overall maximum power budget *within* which the designer is free to do as he or she wishes. Obviously, prodigality in one area is necessarily at the expense of another area, since maximum power is inflexible, and the entire budget is built on reasonable design techniques. Still, there is enough leeway in the budget and enough exceptions so that the designer is not overly restricted. It is strongly recommended that the reader obtain and scan through the ASHRAE/IES material at this point [see below], but reserve detailed study of the application guidelines, since they contain references to terminology and techniques with which he or she will not be familiar until completion of the chapter. However, this material is presented here to maintain the proper order of events in the design procedure.

The nationally accepted standard that defines the establishment of a lighting power budget is: ASHRAE/IES 90.1-1989, *Energy Efficient Design of New Buildings Except Low Rise Residential Buildings*; published by ASHRAE, 1791 Tullie Circle, N.E., Atlanta, GA 30329.

ASHRAE 90.1 sets forth design requirements for the efficient use of energy in new buildings intended for human occupancy, in the environmental system control sense. Specifically excepted are single- and multifamily residences of three or less stories above grade and buildings whose primary function is directed to a specific purpose, for which human occupancy may be required, but is secondary. These include industrial buildings, buildings with very low energy use, and very small buildings (<100 ft^2 area). The standard encompasses energy use requirements for all of a building's environmental systems (i.e., HVAC, lighting, electrical, power, water, envelope, and energy management). In addition, the standard lists suggestions for energy-efficient lighting design. These are discussed below in Section 20.7, along with a number of mandated lighting design items.

In contrast to its predecessor (ASHRAE 90A-1980), this standard recognizes the advances that have taken place in the field of lighting controls (see Sections 21.2 to 21.4) as a result of which the standard is able to focus on *energy use* limitation (which is its specific purpose) without necessarily limiting *power* use, as did the previous standard. It does this by a system of connected-lighting-power credits which permit increasing connected power load. These credits are given for automatic lighting energy conservation control techniques such as daylight compensation, occupancy sensing, and lumen maintenance techniques. An explanation of these control systems as well as the "credits" involved is given in the above-referenced section of Chapter 21. Additionally, the standard specifies the minimum number of control points required for lighting, in recognition of the fact that block control of area lighting leads to energy waste and that conversely localized lighting controls for small areas and limited circuit wattage makes energy use limitation at least possible. Since Standard 90.1-1989 involves all building systems, of which lighting is only one, and since space limitations prevent reproducing even the relevant parts of the standard here, we describe in the following sections how the standard is used in lighting design while referring the reader to various sources for further details.

Standard 90.1 is used to establish target light-

ILLUMINATION

ing power budgets for both interior lighting and exterior lighting, individually. These target budgets are the figures to which the final design's connected lighting load are compared to establish compliance. Trade-off between budgets is not permitted (i.e., the interior load must meet its budget and the exterior load its budget). Lack of compliance i.e. exceeding the target power budget would then necessitate a revision of the design to reduce the connected lighting load. Standard 90.1-1989 presents two methods (paths) for establishing the target lighting power budgets. [See flowchart, Appendix K, Fig. K-1 or Standard 90.1, page 9, for interior lighting power budget; exterior lighting budget is established in the same fashion.]

(a) Exterior Lighting Power Allowance (ELPA). This overall power budget figure is determined by multiplying the unit power allowance given in Table 21.9 (Standard 90.1, Table 6-1) by the appropriate corresponding linear distances or areas and summing the totals. As can be seen from Table 21.8, use of efficient sources such as high-pressure sodium and metal-halides is much more economical of power than tungsten-halogen or mercury lamps. Specifically excepted from the exterior lighting budget limitation is lighting for security, signs, store fronts, monuments, art displays, plant growth, research, outdoor athletics, and the visually impaired, normally OFF emergency lighting, and theatrical lighting.

(b) Prescriptive Criteria Path. This path, or method, for establishing an *interior* lighting power budget is applied when usage of interior spaces is not sufficiently defined to permit accurate application of unit power allowances. This situation exists:

1. At the preliminary design stage, before detailed space allocations have been determined, and/or
2. For speculative, core-and-shell type buildings, where the actual space usage is not determined until the space is rented

Accurate space allocation is defined in Standard 90.1 as allocation of at least 80% of the interior spaces by specific usage. Anything less than 80% prevents use of the more accurate system performance path described in the next section. The *prescriptive criteria path* classifies the entire building by usage type and gives unit (area) lighting power allowances accordingly (see Appendix K, Table K-2). This *unit lighting power allowance* (ULPA) is then multiplied by the gross lighted area of the building to arrive at the building's *interior lighting power allowance* (ILPA). Although theoretically the power credits for automatic controls described in the next section are applicable in this path, in actuality this is not the case since the connected lighting load to which these credits would apply is not known. This path is therefore simple to calculate, inflexible, and relatively inaccurate. The resultant lighting power budget figure is useful primarily in estimating lighting heat gain for building envelope energy calculations (Sections 5.7 and 5.8; Standard 90.1. Section 8), building energy calculations (Sections 5.4 and 5.5), and building energy cost calculation (Standard 90.1, Section 13). Compliance, as for exterior lighting, is determined by comparing actual design connected lighting load to the previously calculated budget (ILPA).

(c) System Performance Criteria Path. This method, as well as the *prescriptive path* described above, specifically excludes lighting for dwellings, industrial processes, theatrical, entertainment, and art display purposes, security, emergency (normally off), research and plant growth lighting, specialized medical lighting, and lighting for the visually impaired.

This path, as the name implies, is system performance oriented; that is, the building power budget (ILPA) is determined by a design approach to task-defined building spaces. The method requires that at least 80% of the building's areas be defined as to usage. Each such area is assigned a unit (area) lighting power density, which when multiplied by the space's physical area and an area factor AF which takes into account the space's physical proportions, including ceiling height, gives its power budget. The building's total power budget (ILPA) is simply the sum of these individual power budgets. Tables of unit power densities and charts of area factors are found in Appendix K, Table K-2, parts C through F, and in ASHRAIE 90.1 from which the Appendix K material is excerpted.

The system performance characteristic of this path is evinced primarily by the modification (reductions, credits) that can be made to the individual area budgets by the use of energy-saving controls, as noted above. Considering each space *individually*, the designer can reduce the space's effective (budget) connected load by 10 to 35% by use of automated lighting controls (see Table 21.1). The number of lighting control points for each space is mandated by Standard 90.1 on the basis of tasks, wattage, and physical area. As stated above, automated lighting energy controls not only reduce the required number of control points (a cost economy) but also the effective connected load. In this path, as in the prescriptive method, compliance is determined by comparing design connected lighting load to the building's total lighting power budget (ILPA) as adjusted for lighting energy control credits.

(d) Computer Assistance. As is undoubtedly obvious at this point, calculation of the interior lighting power budget for any but the smallest buildings by the system performance method is a time-consuming and painstaking job. To assist in this task, ASHRAE provides floppy diskettes with purchase of the standard which contain an interactive program that performs the tedious arithmetic calculation involved. (A second diskette performs the same function for the standard's Section 8, Envelope System Performance.) The program is sufficiently user-friendly as not to constitute a problem for even a computer novice, and we strongly suggest its use.

In addition to this ASHRAE/IES standard, there are a number of other IES publications which bear on the subject of lighting energy and which must be considered whether or not mandated by the jurisdictional codes. They are:

1. IES Standard LEM-3-1987, *IES Design Considerations for Effective Building Lighting Energy Use.*
2. IES LEM-1-1982, *IES Recommended Procedure for Lighting Power Limit Determination.*
3. IES LEM-2-1984, *IES Recommended Procedure for Lighting Power Limit Determination for Buildings.*
4. IES LEM-4-1984, *IES Recommended Procedure for Energy Analysis of Building Lighting Designs and Installations.*

The lighting design and power budget requirements of ASHRAE 90.1-1989 were published in the *Federal Register*, Vol. 54, No. 18, Monday, January 30 1989, Section 435.102 under the title "Principles of Effective Energy Building Design" but with one major and extremely important difference; that the unit power densities given in ASHRAE 90.1 would be cut sharply in 1993. Specifically, the prescriptive unit lighting power allowances (ULPA) of ASHRAE 90.1-1989, Table 6.5, and also found in Appendix K, Table K.2, would be reduced by the following multiplying factors:

Food service: fast-food	61%
Offices	74%
Retail	
general	82%
mall concourse	43%
service establishment	86%
Schools; all types	74%
Warehouse/storage	75%

Similarly, the unit power density (UPD) values used in the more detailed system performance criteria path calculation as listed in ASHRAE 90.1-1989, Table 6.6, and in Appendix K, Table K.2, parts C–F, are reduced by factors in the same range as those above but given in detail according to the task areas specified in the table.

Although no rationale is given in the *Federal Register* for these reduced allowances, which will apply to new construction where the jurisdictional authority so requires, it seems to us that the lighting design and manufacturing professions have been put on notice that there is considerable room for energy conservation in lighting, and that such (very desirable) reductions can be accomplished in a relatively short time (4 to 5 years) by improved equipment and design strategies.

20.6 Task Analysis

Refer to Fig. 20.1. This is the stage at which the quantity and quality of lighting required for the tasks are decided. The factors affecting this choice as shown in the Fig. 20.1 are difficulty, time factor, occupant, cost of errors, and special requirements.

(a) Difficulty. The components of visual difficulty were discussed at length in Sections 18.16 to 18.22, and the results in terms of lighting levels appear in Tables 18.3 to 18.7. Essentially, the designer will examine the type of task involved, and after determining the applicable authority he will select the required illuminance.

In the absence of specific instructions to the contrary, the designer will use IES recommendations. If there are several tasks to be performed at the same point, the most difficult, subject to time considerations, will be selected; that is, if the more difficult task occurs infrequently, it may be reasonable to provide supplementary portable lighting or even to suggest moving to another brighter location. If it is the major task, lighting should be based on it and provision made for intensity reduction for less demanding work.

Variation in task difficulty is particularly common in spaces in public buildings. Thus a school gym can be used for athletics, band concerts (despite the acoustics), and town meetings—three totally disparate lighting requirements. In these and similar instances it is common practice to treat the space as essentially three different spaces and to design lighting for each with a careful eye to maximum common equipment usage. Similar problems are encountered in basements, multipurpose rooms, and conference/meeting/lecture/exhibition rooms. Fortunately, most such spaces do not have severe seeing tasks.

The task variation referred to here is the variation that occurs in one very specific location and is not to be confused with task variation in an area, however restricted. Thus a small private office of, say, 8 × 8 ft has a desk, file cabinet, and circulation space, three tasks of differing but constant difficulty in one small space. The corresponding lighting for these is also fixed and varies with the task severity. The values listed in Table 18.3 to 18.7 represent the required illumination on the surface in question whether horizontal, vertical, or in between. Since the flux method of calculating illumination normally yields the 30-in., horizontal-plane illumination level, it is necessary to know the ratio of horizontal to vertical illumination for various lighting systems. This ratio is approximately:

narrow distribution	
direct and semi-direct	3 : 1
wide distribution	
direct and semi-direct	2.5 : 1
general diffuse	
indirect	1.5 : 1

Since the illumination values listed assume adherence to both recommended luminance ratios and reflectances (see Section 18.29), it is necessary to select, in conjunction with the interior designer, finishes and reflectances for surfaces within the area. If, for instance, in a private office a dark wall finish of 10% reflectance is chosen, it will be necessary for the lighting designer to compensate for this by additional wall lighting to maintain the recommended maximum 10 : 1 brightness ratio (see Table 18.9 and the discussion of point g in Section 20.7). The atmosphere created by vertical surface luminances will be discussed below. Table 20.1 lists reflectances of some common interior paint finishes.

(b) Time Factor. As discussed in Sections 18.20 and 18.23, the length of time in which the task must be accomplished is important in difficult work. Beginning with moderately difficult tasks, that is, luminance of about 50 fL (170 cd/m^2), prolonged intensive application, or rapidly changing tasks, would require illumination to be raised one (or more) levels. Alternatively, quality could be improved by increasing daylight or task contrast.

If, on the other hand, the worker readily learns the task and it becomes routine, illumination can be reduced one level because, effectively, the degree of difficulty has been reduced.

(c) Occupant. Since ordinarily the age and other specific characteristics of the worker are

TABLE 20.1 **Approximate Reflection Factors**

Medium Value Colors	
White	80–85
Light gray	45–70
Dark gray	20–25
Ivory white	70–80
Ivory	60–70
Pearl gray	70–75
Buff	40–70
Tan	30–50
Brown	20–40
Green	25–50
Olive	20–30
Azure blue	50–60
Sky blue	35–40
Pink	50–70
Cardinal red	20–25
Red	20–40

not known, a standard distribution is assumed and the recommendations as tabulated take account of this. On the other hand, if there is a high percentage of older workers, as is the case in certain industries, lighting should be raised one level. This compensates for inability of the eyes to accommodate and for the tendency to tire easily.

(d) Cost of Errors. This involves an economic tradeoff between cost of improving seeing accuracy as against the cost of the improved lighting. Performance can be brought close to perfection, but the cost of so doing increases much more rapidly than the proportional increase in performance. Standard lighting should provide 90%± work accuracy. Thus this step is basically an economic calculation, the criteria for which must come from the owner or user. Tasks in which this problem is encountered include all types of inspection, proofreading, textile matching, very fine machining, and jewelry manufacturing.

(e) Special Requirements. These include any nonstandard task lighting requirements. Some of these are specific illuminant color, directionality for shadowing, and reflections as required for inspection, polarization, and controlled variations, as required in a space with varied tasks or varying daylight factor. In addition to these the physical dimensions of the task often create special requirements of their own. We tend to assume a small object in the horizontal plane, since that is the normal office task. However, there are exceptions such as a drafting board, a large machine, an inspection bench, or a cutting table. Consequently, these special requirements arise:

1. *Large tasks.* In large tasks, the angle of seeing changes from 20 to 70° from the vertical, resulting in radically changed glare angles and reflection from the task.

2. *Three-dimensional tasks.* These tasks shadow themselves, particularly when containing undercuts and reveals. An architect's model shop presents such tasks. When it is necessary to see into an opening, an intense narrow beam is required.

3. *Tools.* Tools cast shadows below and in front when lighted from above and behind. A fabric cutter must see ahead of and below the cutting machine.

4. *Nonhorizontal tasks.* These must be calculated for the plane in which they stand. As noted in (*a*) above the ratio between horizontal and vertical illumination varies between 1.5 : 1 to 3 : 1, depending on the system. Task lighting requirements are stated in the plane of the task. This can have a pronounced effect on the lighting system selected and its arrangement.

5. *Task observed from various positions.* There are instances where a fixed task is observed from several angles, such as a drawing in a conference room or a wall display. Illumination must be adequate for all viewing angles.

PRELIMINARY DESIGN

20.7 Energy Considerations

Energy considerations must pervade every aspect of the design process. Some background material is in order here, to place the lighting energy subject in proper perspective. Best current estimates indicate that lighting consumes approximately 25% of the electric power generated in the United States. In terms of *resources* this amounts to approximately 4 million barrels (bbl) of oil per day. From these estimates, the usage by occupancy is approximately:

Residential	20%
Industrial	20%
Stores	20%
Schools and offices	15%
Outdoor and other	25%

In commercial buildings lighting consumes about 20 to 30% of the building's electric energy, more in residences and less in industrial facilities. By judicious design a reduction of 40 to 50% in lighting energy is attainable. Translated into resources, this reduction can readily amount to more than 1 million bbl of oil per day. Few will disagree that such a goal is well worth the effort. Every watt per square foot reduction in lighting energy results in at least 1.25 W/ft^2 (m^2) savings in air-conditioned buildings. It has been demonstrated by actual designs that offices and schools can be *well* lighted with less than 2.5 W/ft^2 (27 W/m^2). The question to be answered then is: What design guidelines can be followed to effect this energy-conscious design?

The two standards that provide the most authoritative and reliable guidelines are IES Standard LEM-3 (1987) and ASHRAE/IES Standard 90.1-1989. The latter contains two types of guidelines, mandatory and recommended. They will be so indicated where referenced.

1. IES Standard LEM-3 (1987), *Lighting Design Criteria for Effective Energy Utilization*, provides the reader with appropriate criteria for energy-conscious lighting design. (See also IES Recommended Procedures LEM-1, 2, 4.)
   (a) *Design lighting for expected activity.* This is the task lighting approach. It is wasteful of energy to light any surface to a higher level than it requires. Since many work spaces contain varied seeing tasks, nonuniform lighting is recommended where high levels are required. One way to accomplish this for areas where exact furniture layout is not available is to use readily movable fixtures. Providing overall high-level illumination with provision for switching to reduce levels is not advisable because of increased first cost and the psychological impetus to operate at maximum levels. Another solution is fixed fixtures for general low-level lighting and supplementary task lighting. Other factors and techniques to be borne in mind are:
   (1) Grouping of tasks with similar lighting requirements.
   (2) Place most severe seeing tasks at best daylight locations.
   (3) Improve the quality of difficult visual tasks. This is more energy-economical than providing additional light (see Section 18.21).
   (4) Heat-removal fixtures (air troffers) increase efficiency of the units 10 to 20% but make the fixtures immobile when ducted. Consider venting into ceiling plenum.
   (5) Advantages of nonuniform lighting increase as the space between work stations increases.
   (6) When using the task ambient design approach, keep in mind that a nonuniform ceiling layout gives a chaotic appearance to a space. Therefore, the preferred approach is uniform ambient lighting and local task lighting.
   (b) *Design with effective, high-quality, efficient, low-maintenance, thermally controlled luminaires. Effective* means providing useful light, with a high ESI component, and minimum direct glare. In cases where much of the viewer's time is spent in a head-up position, as in schools, or where the viewer can compensate for veiling reflections, the decision should lean toward high VCP (see Section 18.25). Where work and viewing position are fixed and most of the viewer's time is spent head down, the decision should lean toward low reflected glare (see Sections 18.27 and 18.28).
   (1) A *high-quality* luminaire is made with permanent finishes such as Alzac or multicoat baked enamel,

ILLUMINATION

so that its performance after 8 to 10 years of service will be comparable to the original.

(2) An *efficient* luminaire is one with a high CU (coefficient of utilization; see Section 20.27), that is, minimum light loss within the fixture.

(3) A *low-maintenance* luminaire remains clean for extended periods and is designed so that all reflecting surfaces can be easily and rapidly cleaned, without demounting. Enclosed fixtures should be gasketed. Nongasketed units collect and retain dust and cause rapid output depreciation. Relamping should be simple and rapid, to encourage group relamping programs that are energy efficient and cost effective. A 20% increase in maintained light is possible if lamps are replaced at the end of their *useful* life, that is, when output is down to 70% of initial maintained lumens, and fixtures are cleaned and maintained on a fixed schedule. No cost tradeoff is generally involved, since periodic maintenance and relamping is normally cheaper than one-at-a-time maintenance and burnout replacement, *and* yields 20% higher average lumens. A tradeoff is involved between the additional cost of more frequent maintenance and the energy savings produced. Fixtures in relatively inaccessible locations such as high ceilings must be designed for low maintenance, and maintenance should be on a fixed schedule. Tradeoff here is between higher cost of low-maintenance units, and high maintenance costs. Use of high-lumen-maintenance sources is assumed.

(4) A *thermally controlled luminaire* is one that controls the heat generated by the light source. This item depends to a large extent on the type of HVAC system, the lighting heat load, and the type of fixtures employed. Detailed analysis of this point involves HVAC considerations and the overall impact of lighting energy on the building. However, remember that the method of fixture installation controls, in large measure, the utilization of lighting heat (see Fig. 20.2). As noted above, return-air troffers raise light output by 10 to 15% (by lowering bulb temperature), in addition to reducing the a-c load.

(c) *Use efficient light sources and accessories*. This point is self-explanatory. Tradeoffs involved here are:

(1) Between first cost and life-cycle costs.

(2) Between desired illuminant color and efficiency. Some of the highest efficacy sources have less than ideal color but, since the eye adapts quickly, off-white color, as from HPS lamps, should be considered for indoor uses where color discrimination is not a factor in the work and will not create a negative response in transients.

(3) Between light control and efficiency. Fluorescent sources, which are highly efficient, do not lend themselves to good beam control and are principally useful for area coverage.

(4) Between architectural considerations and efficiency. Fluorescent sources are efficient and have good color. Linear lamps require a large fixture, which tends to dominate the space. Consider compact fluorescent lamps or U lamps in appropriate fixtures.

Figure 20.3 will assist the designer in determining roughly what the various sources represent in terms of watts per square floor (meter) load. See also Fig. 19.1 for a source efficacy comparison. Thus, with a target of 2 to 2.5 W/ft^2 for office-building lighting,

ILLUMINATION

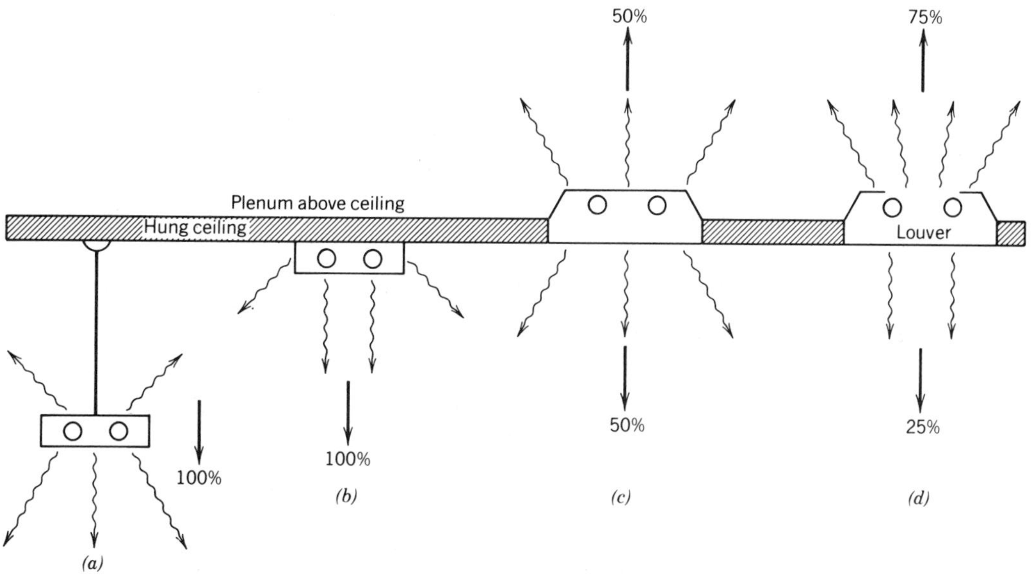

***Fig. 20.2** Method of fixture installation controls the transfer of lighting heat.* (a) *Suspended units contribute all their heat to the space, and themselves remain cool.* (b) *Surface-mounted fixtures also place all their heat in the space, but because of blocked transfer upward, run hot.* (c) *Static recessed units transfer about 50% of their heat to the plenum, whereas louvred units* (d) *transfer about 75%. When ducted, heat transfer up can be as high as 85%.*

the use of incandescent downlights is obviously severely restricted.

Spill light and borrowed light are often neglected sources. Glass in upper wall sections can provide sufficient corridor lighting from borrowed office lighting. Sources with high lumen *maintenance* such as tungsten-halogen and high-pressure sodium should be given preference.

In the category of accessories, ballasts are the largest energy users. For fluorescent lamps use electronic or low-loss (high BEF) conventional ballasts. Avoid single-lamp ballasts as they are less efficient than multilamp types. See Sections 19.18 and 19.19.

For HID sources, match ballast to the lamp, use solid-state types where available, and use the less efficient regulating ballast only where required.

For all sources, use high-power-factor ballasts only.

(d) *Select the appropriate lighting system.* As detailed in Sections 20.10 to 20.15, there are six types of lighting systems, each with its particular distribution characteristics and therefore applicability. In addition, all luminaires must be carefully located to provide uniformity of ambient lighting, and task lighting where applicable. When using indirect luminaires, correct spacing will avoid ceiling "hot spots," which can cause direct and reflected glare. Indirect lighting allows fully effective use of a space since users are not restricted to facing any particular direction to avoid direct or reflected glare from fixtures. Similarly, mounting height of suspended fixtures must be coordinated with cavity sizes and finishes to minimize light loss and maximize coefficient of utilization.

(e) *Use daylight, properly.* Daylight must

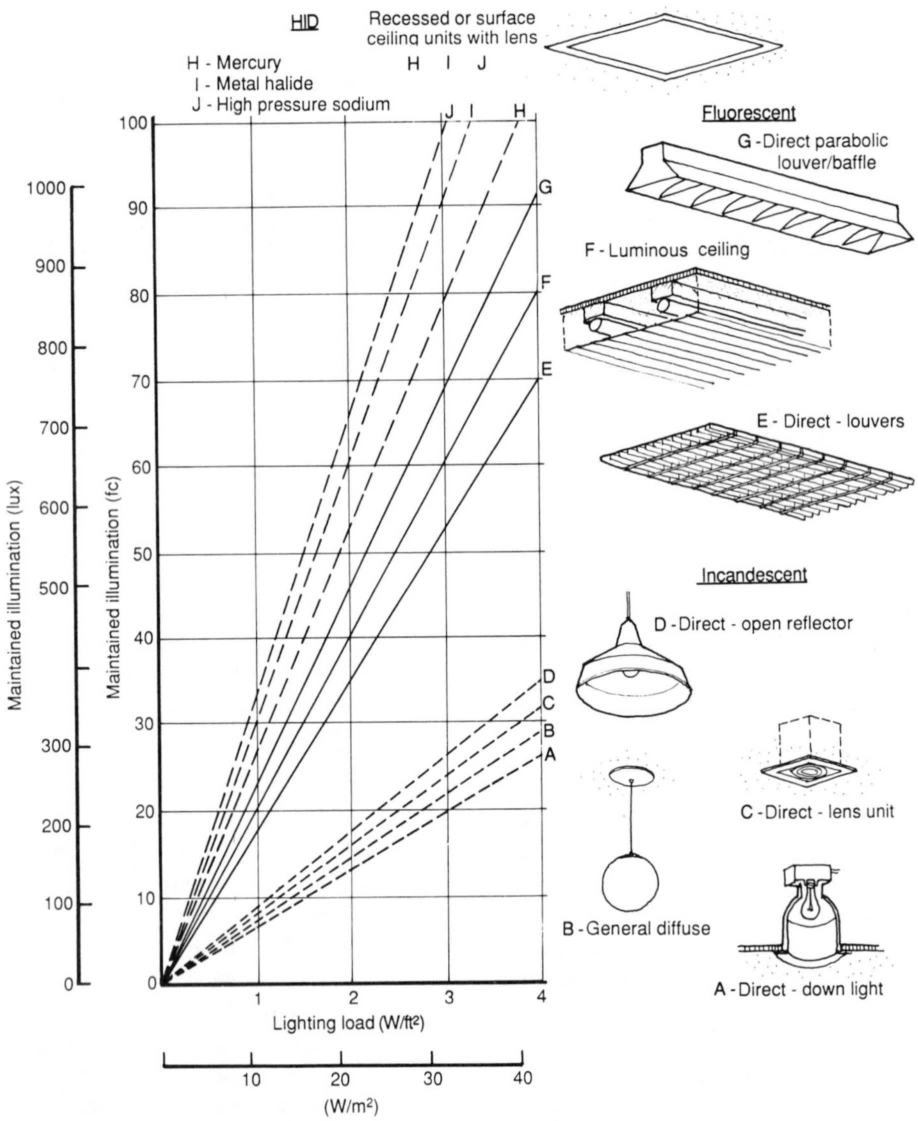

Fig. 20.3 Estimating chart for lighting load and related illumination levels for different light sources. The chart was calculated for a fairly large room, approximately classroom size. Although all the figures are necessarily approximate because of the variables involved, the chart gives figures close enough for a first approximation. Notice the increase in output as the sources change from incandescent to fluorescent to HID.

be considered as a normal light source subject to weather variations and time of building use. Obviously, a three-shift industrial plant cannot use daylight on all shifts, but it can for at least one shift, and design should reflect this fact. Part of proper daylight design is control of window brightness. Excessive brightness causes severe and even disabling glare. A corollary of excessive brightness is excessive heat gain. Both are manageable with common control devices, manual and automatic. Window control devices must also be designed to reflect light back into the space at night, to avoid light loss via windows.

(f) *Use energy-efficient lighting control strategies*. The subject of lighting control, including manual and automatic switching, dimming, sensing, and intensity control is covered in Sections 21.2 to 21.4, for both new installations and retrofit work. Proper design of controls can reduce energy consumption over a noncontrolled installation by as much as 60% without reducing lighting effectiveness. The economic trade-off in new construction is between the initial cost of control equipment on the one hand and the initial lighting equipment savings plus annual energy savings on the other hand. Proper decisions require life-cycle cost analysis. See Appendix E, Economic Analysis.

(g) *Use light finishes on ceilings, walls, floors, and furnishings*. This point is self-explanatory and is examined in a number of sections. A brief summary of recommendation ranges would be:

Ceilings—80 to 92%

Walls—40 to 60%

Furniture, office machines, and equipment—25 to 45%

Floors—20 to 40%

In addition to higher illumination levels in the room, high reflectances minimize uncomfortable luminance ratios, as between fixture and upper wall or task and background. With respect to these ratios (see Table 18.9):

(1) Between task and near surround—aim for 3 : 1, accept 5 : 1.
(2) Between task and immediate area—aim for 10 : 1, accept 20 : 1.
(3) Between luminaires and their background—aim for 20 : 1, accept 40 : 1.
(4) Anywhere in the normal field of view—aim for 40 : 1, accept 80 : 1.

Note that the targets themselves are maximums and values above maximum should be used only for very good reasons.

2. ASHRAE Standard 90.1 (1989)
   (a) *Mandatory requirements*
      (1) Design interior and exterior lighting to meet lighting power budgets and control point requirements as detailed in that standard.
      (2) Where applicable use only energy (E)-rated ballasts in fluorescent fixtures [see Section 19.8*a*]. For other installations use only high-power-factor, high-BEF ballasts.
   (b) *Recommendations*. Items not included in the LEM-3 listing are:
      (1) Where using task/ambient lighting, the ambient level should not be lower than a third of the task level, to avoid uncomfortable luminance ratios.
      (2) Accent lighting should not exceed five times the ambient level. Therefore, in merchandising areas where the contrast between ambient and task levels is critical, reduce ambient levels as much as practicable.

20.8 Preliminary Design

Again referring to Fig. 20.1, the preliminary design phase is the time during which ideas crystallize, but in terms of areas and patterns, as well as light and shadow; not yet in terms of hardware. At this stage the quality of the system is decided on, that is, the luminances, diffuseness, chromaticity, and proportion of vertical to horizontal lighting are determined. The latter establishes in large measure the room "mood" or lighting ambience. In preceding sections these items were discussed in some detail. In the sections that follow on lighting systems (direct, indirect, etc.), the quality of each will be considered and applications suggested. In the overall view, however, the ultimate quality of the lighting system, its visual pleasantness, centers of visual attention, high-lights and shadows, as well as texture and forms, will be a deft and perhaps artistic combination of the above considerations, and will establish, as the term implies, the quality of the lighting design. A few observations, not covered elsewhere, are mentioned below.

Planes other than the "working plane" must always be considered. The ratio of vertical to horizontal illumination of the chosen lighting system will determine wall luminance which greatly influences the overall impression of the space's brightness. (See Fig. 20.12.) The floor finish will have a pronounced effect on the ceiling illumination for direct lighting systems because in direct systems ceiling illumination derives only from room interreflectances, with floor reflectance being particularly important in large spaces. See Fig. 20.45.

The chromaticity of the room lighting depends primarily on the source but secondarily on the luminaire and surface finishes. A "white" source can be tinted slightly by the use of a colored reflector in the luminaire. Of course, the effect on luminaire output of such a change must be considered. In the case of semi-indirect and indirect lighting this same effect can be accomplished by the use of colored ceiling and upper wall surfaces, which serve as secondary reflectors and become the actual luminous source for the room. Recommendations in Table 19.22 cover choice of source.

20.9 Illumination Methods

There are three methods of illumination: general, local and supplementary, and combined general and local.

(a) General Lighting. This is a system designed to give uniform and generally, though not necessarily, diffuse lighting throughout the area under consideration. The method of accomplishing this result varies from the use of luminous ceiling to properly spaced and chosen downlights, but the resultant lighting on the *horizontal working plane* must be the same, that is, reasonably uniform. It may be, but is not necessarily, task lighting.

(b) Local and Supplementary Lighting. These are two terms that are used interchangeably but have slightly different meanings. By definition, *local lighting* provides a small, high-level area of lighting without contributing to the general lighting. *Supplementary lighting* also provides a restricted area of high intensity, but supplements the general lighting. In actual practice it is difficult to differentiate between the two. A desk lamp, a high-intensity downlight on a merchandising display, and a track light illuminating wall displays all seem to answer both definitions and in practice are referred to as local, supplementary, or local-supplementary lights. Typical of this genre are the units illustrated in Fig. 20.4.

(c) Combined General and Local Lighting. This illumination method is used in spaces where the general visual task is low, but supplementary lighting is required in a limited area for a particular task.

These three *methods* of illumination can be accomplished in many ways by the use of luminaires and luminous sources of different types, since the illumination method is a function of *both* luminaire arrangement and the luminaire's inherent lighting distribution. The term used to describe the effect of the combination of a particular fixture type applied in a particular way is the *lighting system*. Thus a reflector-type fixture when aimed down gives *direct* light. The same fixture beamed up at the ceiling gives *indirect* light. The following section describes the systems that constitute the vast majority of lighting installations.

20.10 Types of Lighting Systems

No one lighting system can be said to be the single choice in a given instance; on the contrary, the designer will normally have a choice of at least two systems that will, if utilized properly, yield illumination of adequate quantity and good quality. However, other factors, such as harmonization with the architecture and economics, will usually tip the balance in favor of one or the other. The five generic types of lighting systems are indirect, semi-indirect, diffuse or direct-indirect, semidirect, and direct.

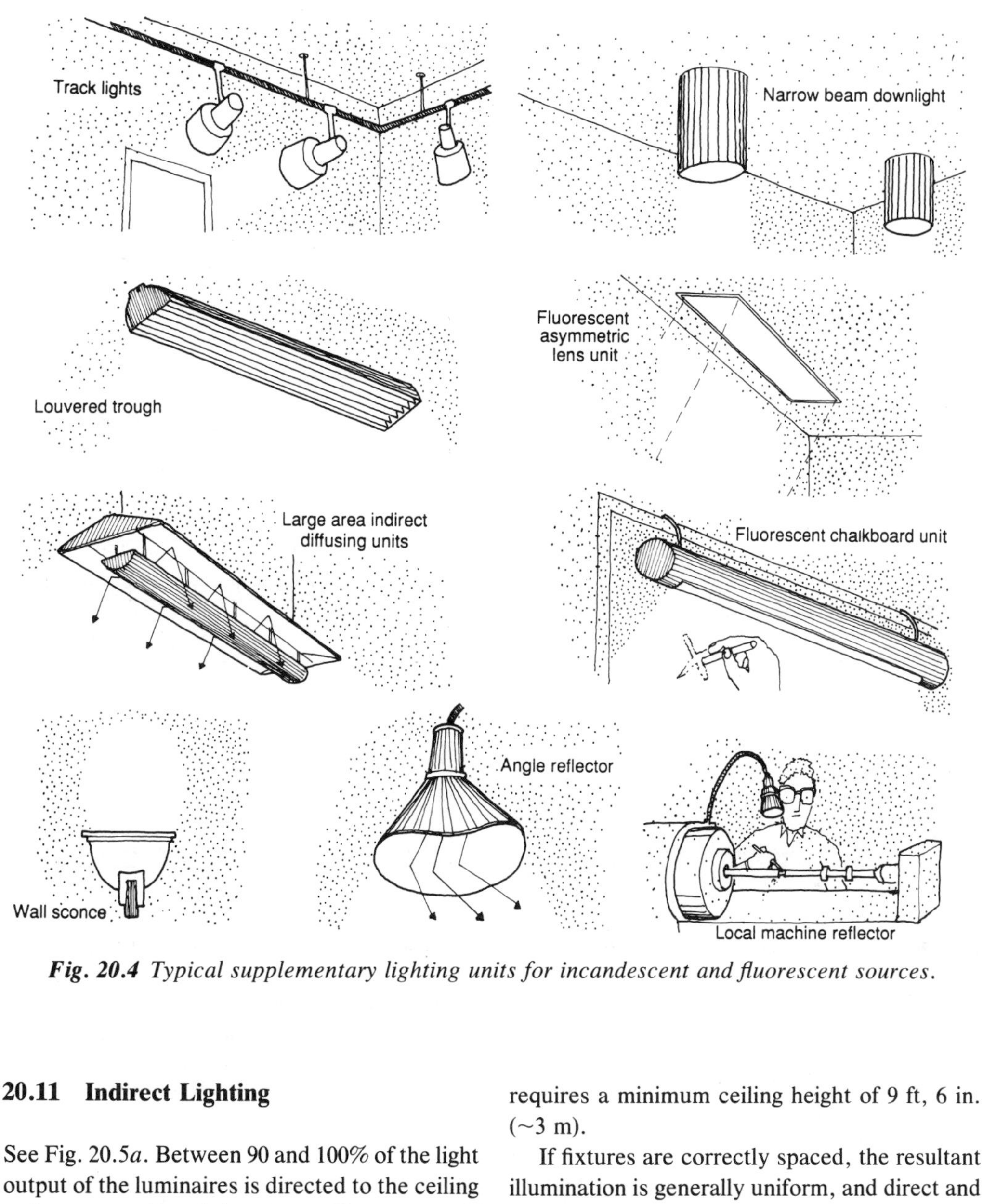

Fig. 20.4 *Typical supplementary lighting units for incandescent and fluorescent sources.*

ILLUMINATION

20.11 Indirect Lighting

See Fig. 20.5*a*. Between 90 and 100% of the light output of the luminaires is directed to the ceiling and upper walls of the room. The system is called indirect because practically all the light reaches the horizontal working plane indirectly, that is, via reflection from the ceiling and upper walls. *Therefore, the ceiling and upper walls in effect become the light source* and, if these surfaces have a high-reflectance finish, the room illumination is quite diffuse (shadowless). Since the source must be suspended at least 18 in. (45 cm) from the ceiling (depending on the unit's output) to avoid ceiling "hot spots," this system requires a minimum ceiling height of 9 ft, 6 in. (~3 m).

If fixtures are correctly spaced, the resultant illumination is generally uniform, and direct and reflected glare are low.

To avoid an excessive luminance ratio between the luminaire and its surrounding field, the luminaire can be made translucent, on the bottom, on the sides, or both. Approximately 750 lux (75 fc) is the maximum horizontal-plane illumination attainable without exceeding *overall* ceiling luminance of about 2500 cd/m^2 (730 fL) (see Section 18.26). With practically no veiling reflections, this is sufficient for all but the most difficult tasks. The lack of shadow, low

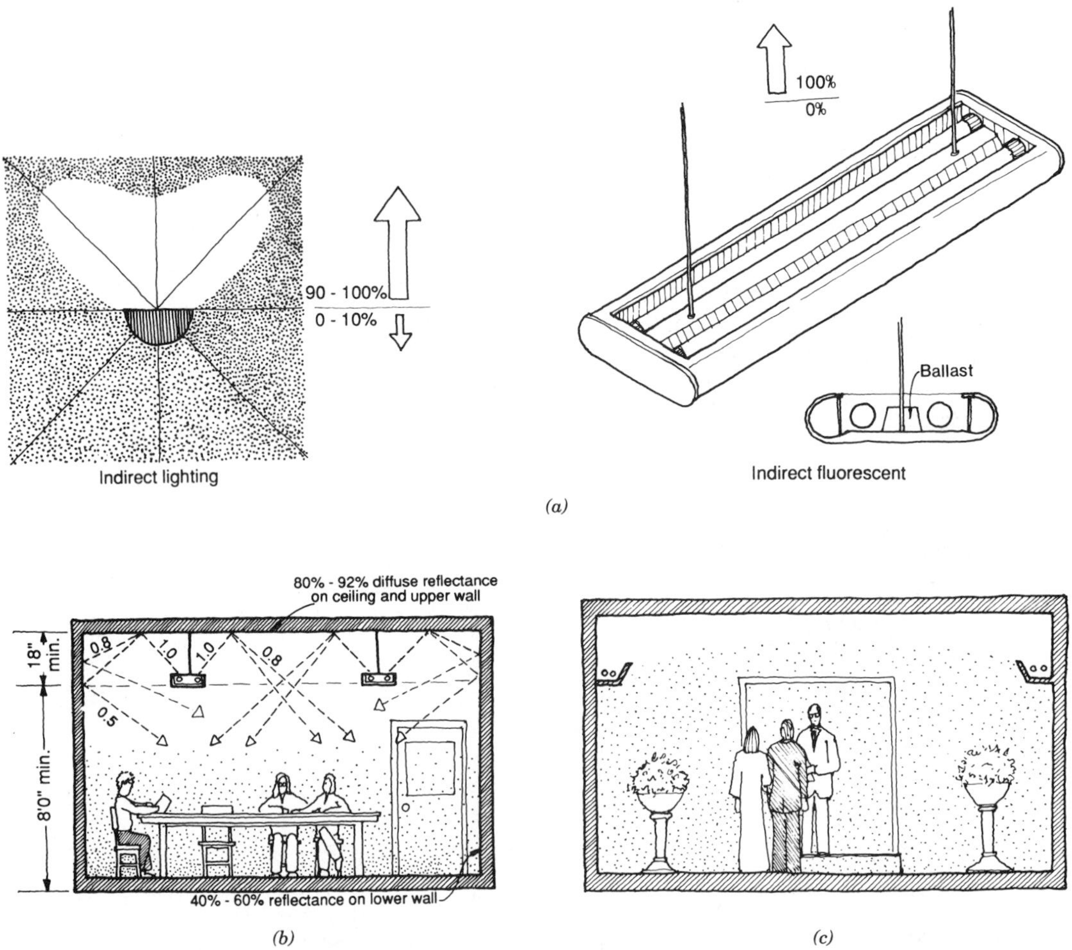

Fig. 20.5 Indirect Lighting (a) *The fixtures deliver 90–100% of their ouptut above the horizontal plane.* (b) *The ceiling and upper wall surfaces of the room are directly illuminated by the indirect fixtures; they then become secondary sources that in turn illuminate the room. When properly designed, this type of installation yields a substantially uniformly bright ceiling.* (c) *Use of architectural coves gives an acceptable luminance gradient on the ceiling and, if properly designed, nearly uniform, glareless illumination in the room. This system of illumination is particularly useful in VDT spaces.*

source brightness, and highly diffuse quality created by indirect lighting give a very quiet, cool ambience to this type of lighted space, suitable for private offices, lounges, and plush waiting areas. Areas having specular visual tasks such as office with VDT (video display terminals) use this system to great advantage. In such spaces, indirect fixtures *without* luminous bottoms or sides should be specified.

When properly designed, particularly when the source of light is architectural coves (see Fig. 20.5*c*), the ceiling has a floating, almost infinitely deep or skyline quality, which is pleasant and can be used to give an impression of height in a low ceilinged room. This system is not to be confused with a transilluminated ceiling, which is really a direct lighting system of entirely different quality and effect. A further characteristic of the indirect lighting system is loss of texture on vertical surfaces, as is common to all fully diffuse lighting.

Indirect lighting is inherently inefficient, since much of the useful light reaches the working plane only after double reflection—within the fixture and off the ceiling. Although to a considerable extent this is offset by the absence of veiling reflections, applications to difficult seeing tasks normally require an additional light source. This can be an additional array of ceiling mounted or suspended fixtures or, alternatively, a local light source. Thus an indirectly lighted drafting room having tables equipped with supplementary lamps would take advantage of both systems—the local high-intensity light at a maximum of 200 fc (~ 2000 lux) for the restricted area being worked on, and overall table lighting of 40 to 50 fc ESI. The latter would also solve any reflected glare problems arising from the many viewing angles required by large tasks such as drawings.

20.12 Semi-indirect Lighting

Between 60 and 90% of the light is directed upward to the ceiling and upper walls. This distribution is similar to that of indirect, except that it is somewhat more efficient and allows higher levels of illumination without undesirable brightness contrast between fixture and surroundings along with lower ceiling brightness. A typical fixture, illustrated in Fig. 20.6, employs a translucent diffusing element through which the downward component passes. The ceiling remains the principal radiating source and the diffuse character of room lighting remains. Direct and reflected glare are both very low, as with indirect lighting (see Fig. 18.36). In both indirect and semi-indirect systems, it is often desirable to add accent lighting or downlighting, to break the monotony inherent in these systems and to establish a visual point of interest, or to create required modeling shadows.

The quality of the semi-indirect system is somewhat different than indirect because attention is not drawn to the fixtures, since they exhibit less contrast with the background ceiling brightness.

In both indirect and semi-indirect lighting systems, the light undergoes a number of ceiling and wall reflections before reaching the horizontal working plane. The use of colored paints particularly on the ceiling, can serve to tint the room illumination slightly because of selective absorption.

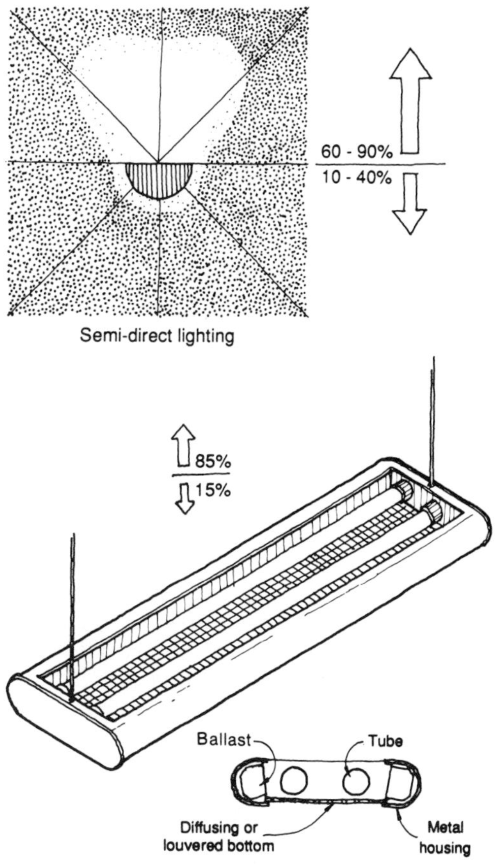

Fig. 20.6 Semi-indirect lighting.

20.13 Direct-Indirect and General Diffuse Lighting

Direct-indirect lighting provides approximately equal distribution of light upward and downward, resulting in a bright ceiling and upper wall (see Fig. 20.7). For this reason brightness ratios in the upper-vision zone are usually not a problem. Since the ceiling is a major though secondary source of room illumination, diffuseness will be good, with resultant satisfactory vertical-plane illumination.

Diffuse fixtures (See Fig. 20.8) gives light in all directions, whereas direct-indirect have little horizontal component. Stems for both types should be of sufficient length to avoid excessive ceiling brightness; generally not less than 12 in. [30 cm].

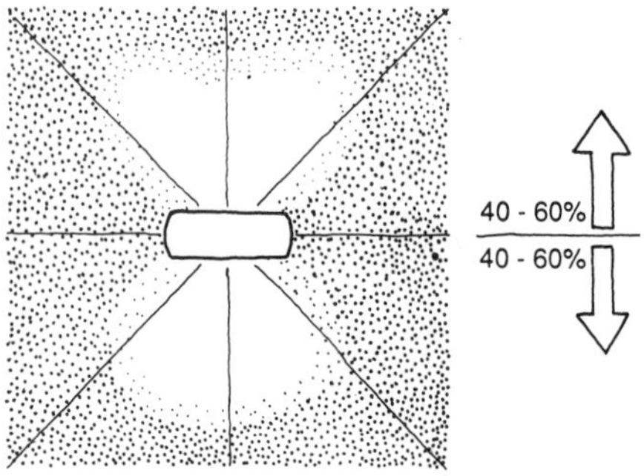

Direct-indirect

(a)

(b)

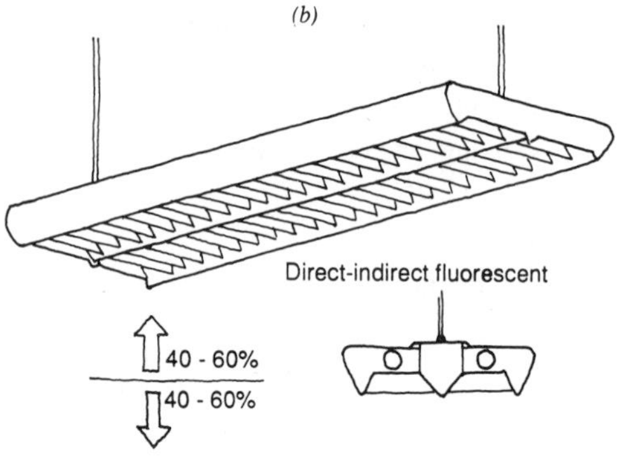

Direct-indirect fluorescent

(c)

Fig. 20.7 (a) *Direct-indirect lighting. Upper and lower room surfaces are luminous* (b), *but center of walls is not because of the lack of horizontal light from fixtures* (c). *Principal light on working plane comes directly from the luminaire.*

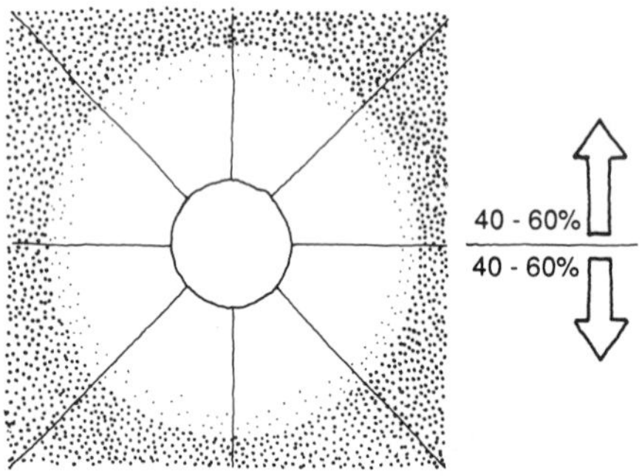

General diffuse

(a)

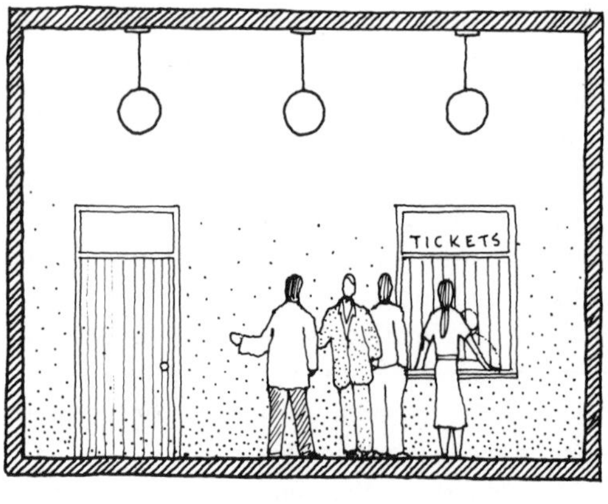

(b)

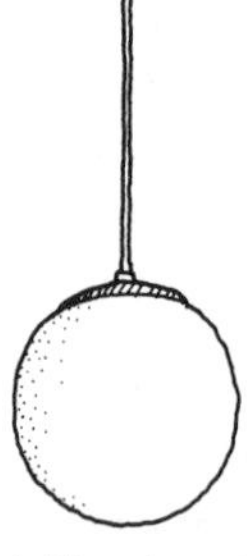

General diffuse incandescent

(c)

Fig. 20.8 (a) *General diffuse lighting. Note that all room surfaces are illuminated and become secondary sources, although those closest to the fixtures (ceiling, upper wall) are the brightest secondary sources* (b)*; the primary source of illumination is the direct radiation from the fixture* (c). *The floor contribution is low due to its normally low reflectance.*

Since the impression of illumination depends to a large extent on *wall brightness* because this is the surface we see most often, a space with general diffuse illumination will *appear* lighter than one with direct-indirect because of the darker walls in the latter (see Figs. 20.8*b* and 20.12).

By avoiding excessively bright luminaires and giving attention to positioning of sources and viewing angles, direct and reflected glare can both be kept low. Furthermore, since the luminaire (as any other source of brightness in the field of view which is much higher than the average scene) draws the eyes' attention, particular care must be taken to limit its brightness and to avoid disturbing fixture patterns (see Fig. 20.15).

Efficiency of these two systems is good. Both are well applied in spaces requiring overall uniform lighting at moderate levels such as classrooms, standard office work spaces, and merchandising areas.

20.14 Semidirect Lighting

With this type of lighting system 60 to 90% of the luminaire output is directed downward and the remaining upward component serves to illuminate the ceiling (see Fig. 20.9). If the ceiling has a high reflectance, this upward component will normally be sufficient to minimize direct glare from the luminaires, depending upon eye adaptation level. The degree of diffuseness will depend in large measure on the reflectances of room furnishings and of the floor. Shadowing should not be a problem when upward components are at least 25% and ceiling reflectance not less than 70%. With smaller upward components the system is essentially direct lighting (see Section 20.15). The system is inherently efficient. Reflected glare can be controlled by the methods discussed in Section 18.27. With adequate wall illumination, the quality of the lighting gives a pleasant working atmosphere. It is applicable to offices, classrooms, shops, and other working areas.

20.15 Direct Lighting

Since essentially all the light is directed downward, ceiling illumination is entirely due to light reflected from floor and room furnishings. This system then, more than any other, requires a light, high-reflectance, diffuse floor unless a dark ceiling is desired from an architectural or decorative viewpoint. Occasionally the ceilings are deliberately painted a dark color and pendant direct fixtures used to lower the apparent ceiling of a poorly proportioned room or to hide unsightly piping, ductwork, and so on.

The effect of direct lighting depends greatly on whether the luminaires are spread or concen-

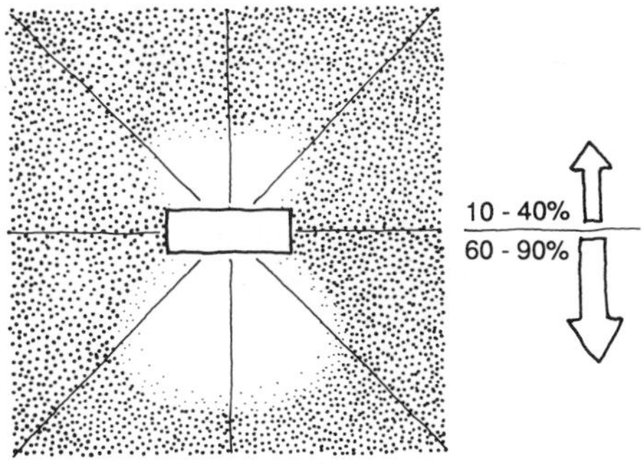

(a)

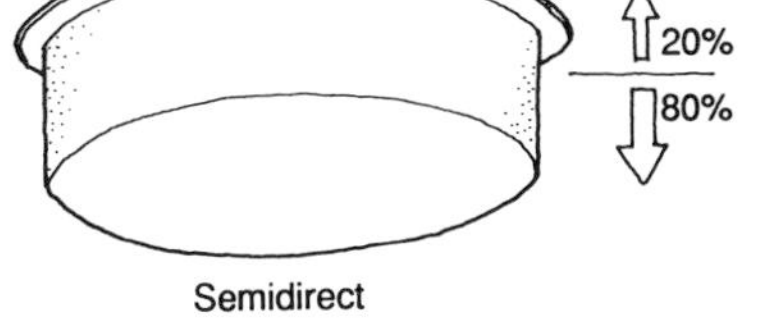

(b)

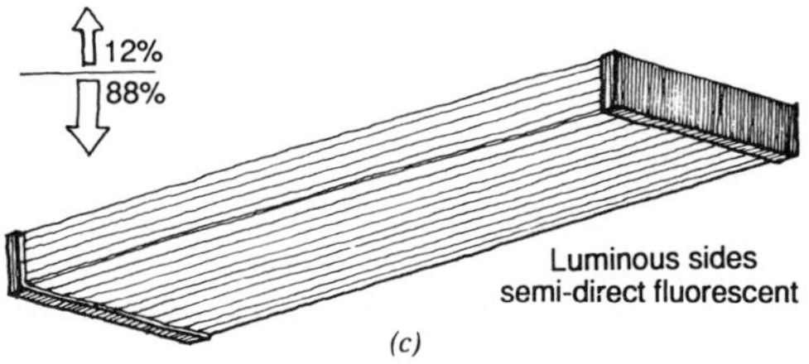

(c)

Fig. 20.9 *Semidirect lighting provides its own ceiling brightness* (a), *with surface-mounted fixtures* (b), *or pendant/surface units* (c). *Other characteristics are similar to direct lighting.*

trating (Figs. 20.10 and 20.11). In the former case considerable diffusion of light results from reflections on floor, furniture, and walls. The result is a working atmosphere with slightly darkened walls and ceiling. This type of lighting, which is most widely represented by the recessed fluorescent troffer in a suspended ceiling, is common for general office lighting. The fixtures themselves form a ceiling surface of light and dark areas, and the quality of the entire system is not unpleasant. Difficulties associated with direct glare and veiling reflections can be controlled by proper use of reflectances, use of low-brightness units, and judicious arrangement of viewing positions. When direct lighting units are used in a uniform pattern, this latter option disappears, and the need for particularly low-brightness units and high ceiling reflectivity, or specialty diffusers like the batwing design, is increased.

Direct lighting gives little vertical surface illumination, requiring the addition of perimeter lighting in business atmospheres (see Fig. 20.12).

Concentrating downlights create sharp shadows and a theatrical atmosphere, not appropriate to a working commercial space. When used alone, they can be appropriate in restaurants and other areas where the privacy type of atmosphere generated by limited-area horizontal illumination and minimal vertical-surface illumination is desired. When a fixture is designed with a black cone or baffle or other device that is nonreflecting at the viewing angle, the fixture appears dark. It is our opinion that installations providing high-horizontal surface illumination, yet with no apparent source of brightness, such as those using black-cone downlights, are disturbing to our normal bright-sun-and-sky orientation and should be used cautiously and only in limited areas. This same comment, but to a lesser extent, is applicable to very low-brightness diffusers such as the parabolic wedge type (see Fig. 21.33). (There, however, the unit has

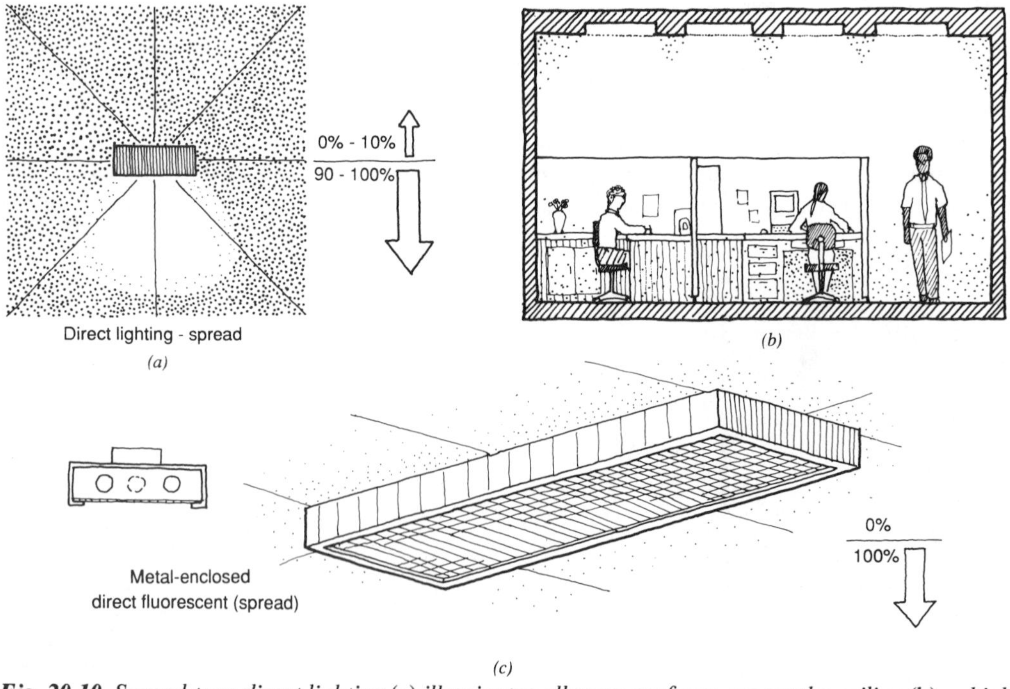

Fig. 20.10 *Spread-type direct lighting* (a) *illuminates all room surfaces except the ceiling* (b), *which is only illuminated by reflection from the floor. Some diffuseness is evident. The most common type of unit in this category is the direct fluorescent, either surface-mounted* (c) *or troffer type recessed in ceiling* (b).

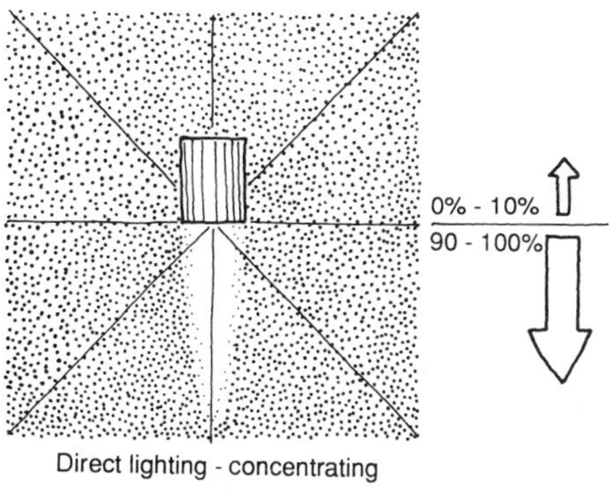

(a)

(b)

***Fig. 20.11** With concentrating direct distribution* (a), *the floor is the only luminous surface* (b) *other than the ceiling fixture. Diffuseness is absent. Walls are dark. Incandescent downlights* (c) *are of this type unless equipped with spread-type lenses.*

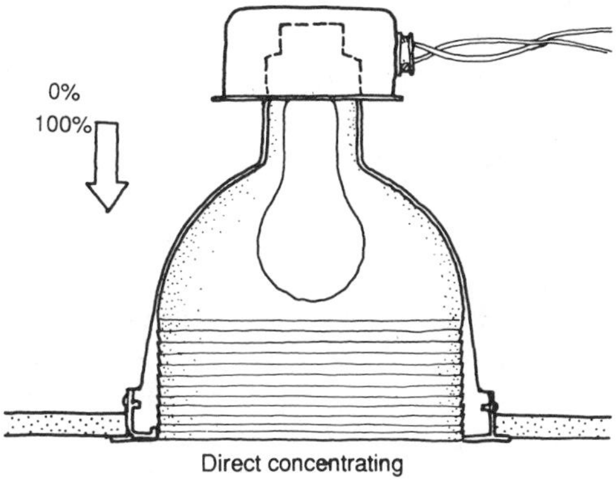

(c)

the redeeming characteristic of low reflected glare, which is not the cause with downlights.) Stated otherwise, the psychological impression of a space without a visible source of light is gloomy and cave-like. When using very low brightness sources, as in VDT areas, this negative impression can be alleviated by the addition of luminous surfaces in nonreflecting areas or points of light sparkle.

In summary then, spread direct lighting is suitable for general lighting while concentrated direct lighting, which reduces vertical illumination, is appropriate for highlights, local and supplementary lighting, and specialized viewing.

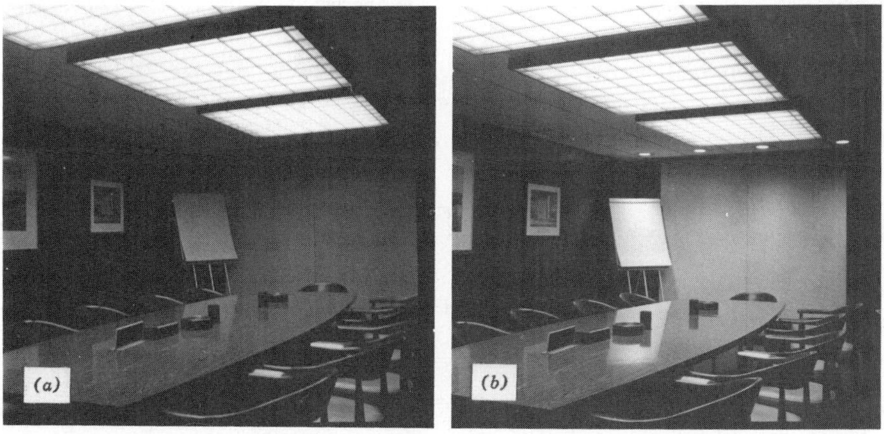

***Fig. 20.12** The use of large-dimension lighting fixtures in a low-ceiling room is possible if the apparent size of the unit is reduced. Here at a mounting height of 7 ft 6 in., 4 × 4 ft units are acceptable because the lattice on the face of each unit gives the impression of reduced fixture size. Note also that the apparent illumination in* (b) *is greater than in* (a) *[though both are exactly equal on the table surface] due to the wall wash in the background. The eye perceives vertical surface illumination more readily than horizontal and retains the impression for the entire space.*

20.16 Size and Pattern of Luminaires

Because of its luminance, each luminaire or other luminous source is a point of visual attention. To the extent that luminaires are numerous, large, very bright, or arranged in striking patterns, attention will be drawn to them and away from other surfaces. Furthermore, color elements or accent lighting can be added deliberately to draw attention. Rigid rules cannot be set down covering these criteria, but examples can demonstrate the principles.

Luminaire size should correlate with room size and ceiling height. Flourescent fixtures larger than 2 × 4 ft (60 × 120 cm) should not be used in ceilings below 10 ft unless their size is minimized by some sort of surface pattern (see Fig. 20.12). Transilluminated ceilings are *all* fixture and therefore require a minimum of 12 ft (3.66 m) mounting height. When installed below this level, particularly in large rooms, the effect is oppressive, as if the sky were lowered on us. To offset this effect, the use of colored, shaped, or dark panels is of some help. In place of a luminous ceiling (Fig. 20.13), a large-area, coffer-type fixture can be utilized, which gives the impression of great depth (see Fig. 20.14).

To achieve the uniformity of illumination necessary for general lighting, regular spacing is desirable. However, various effects may be obtained within the regularity, to accomplish an architectural purpose, as shown in Fig. 20.15. The pattern of lights must never be at cross-purposes with any dominant architectural pattern; rather it should either reinforce an architectural form or be neutral. If a strong architectural element is absent, a dominant lighting pattern may be desirable. Conversely, a strong architectural element can either be reinforced (Fig. 20.15*p*-1) or utilized to carry a neutral lighting pattern (Fig. 20.15*p*2, 3).

Continuous row installations eliminate the dominant checkerboard effect of closely spaced individual luminaires (and are cheaper). Coves and cornices give the ceiling a floating or light effect. Geometric patterns can be used to add interest or break monotony of large areas, such as department stores. Generally, incandescent downlights are not dominant, and regularity of placement is not essential. Nonuniform layouts with large sources create a distinct pattern problem inasmuch as they are too large to be neutral, and the nonuniformity can create visual confusion (see Fig. 20.16). The only cure for this problem is to minimize the source brightness by using low-brightness luminaires (see Fig. 21.29 and Section 20.28).

The incorporation of daylight into the luminous sources of a space should take cognizance of size and pattern as well. Large windows are not consonant with small ceiling sources, whereas skylights are readily integrated with other ceiling units. A frequently neglected consideration is the appearance of a source when de-energized. With proper daylight and energy-conserving design, many sources will be unlighted during the normal-use hours of the space. Obviously, low-brightness sources will

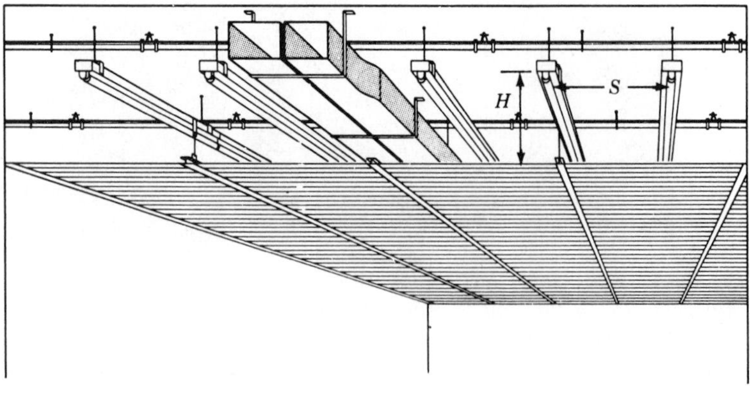

***Fig. 20.13** Luminous ceiling installation. Note that in a properly designed installation, piping and ductwork do not affect the light distribution. As a rule of thumb, fixture spacing* S *should not exceed 1.5 times fixture height* H *above the diffusing element, for uniform illumination. This type of installation is useful for specular tasks if supplemental lighting is impractical.*

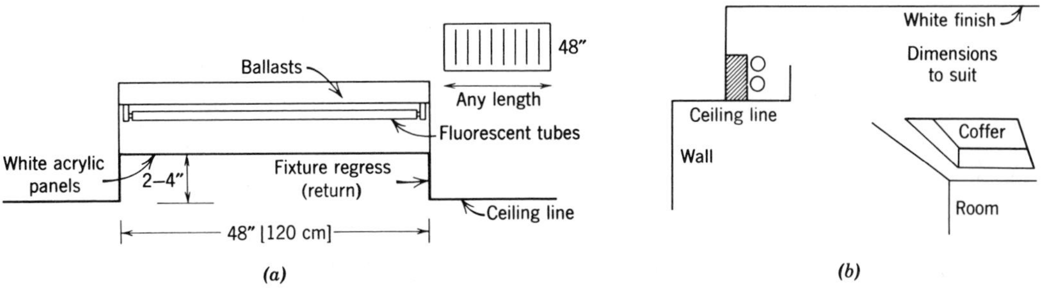

Fig. 20.14 *Types of coffer.* (a) *Coffer fixture, for direct diffuse lighting, generally available in 4 ft (120 cm) width and any length.* (b) *Architectural coffer, for indirect lighting, designed as required. (See Fig. 21.17 for dimensional data.) Both types give an illusion of great depth and soft, glare-free illumination. Coffers are also ideal daylight sources when designed in conjunction with skylights as in Fig. 19.31.*

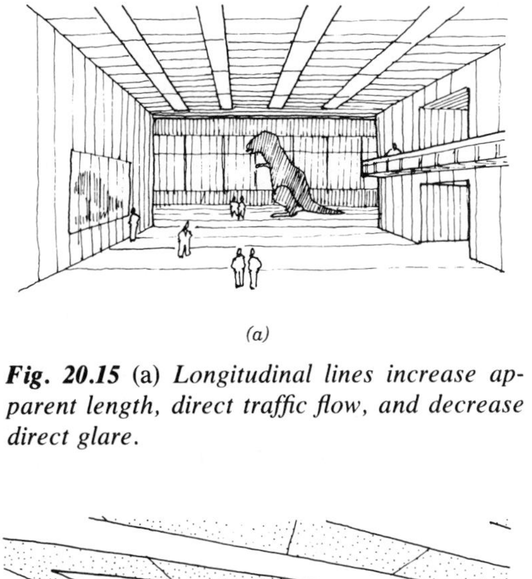

(a)

Fig. 20.15 (a) *Longitudinal lines increase apparent length, direct traffic flow, and decrease direct glare.*

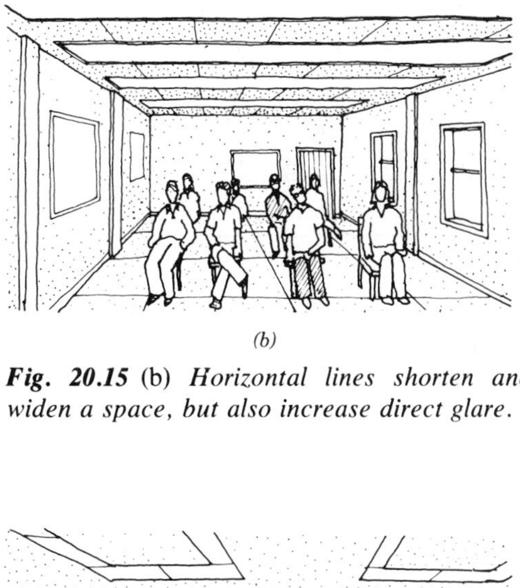

(b)

Fig. 20.15 (b) *Horizontal lines shorten and widen a space, but also increase direct glare.*

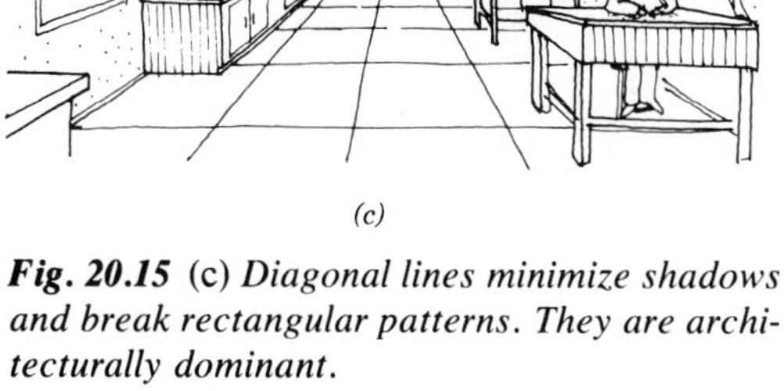

(c)

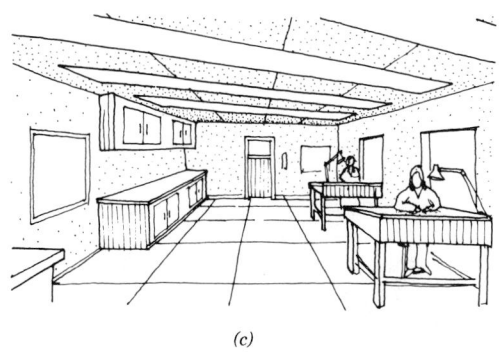

Fig. 20.15 (c) *Diagonal lines minimize shadows and break rectangular patterns. They are architecturally dominant.*

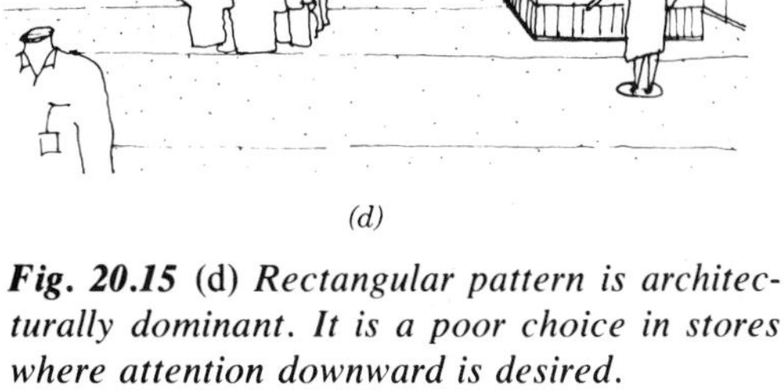

(d)

Fig. 20.15 (d) *Rectangular pattern is architecturally dominant. It is a poor choice in stores where attention downward is desired.*

ILLUMINATION

(e)

Fig. 20.15 (e) *Cornices, valances, and coves are luminous ceiling borders. In large rooms suspended coves achieve uniform ceiling brightness and when designed with a downward component or combined with local lighting, as illustrated, give a pleasant, intimate atmosphere.*

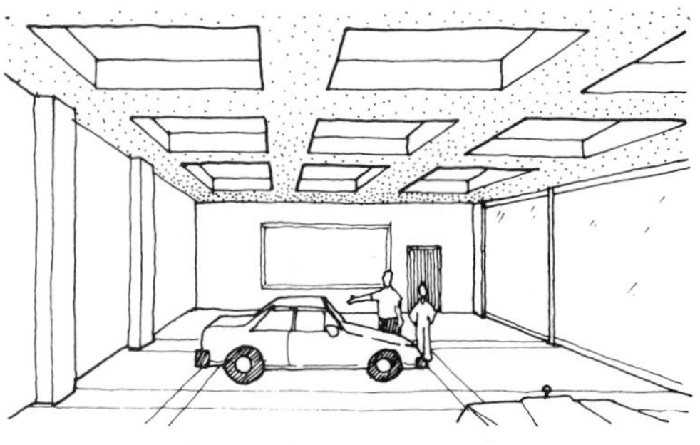

(f)

Fig. 20.15 (f) *Coffers create a decorative architectural effect and can be designed to resemble skylights or can be built into actual skylights.*

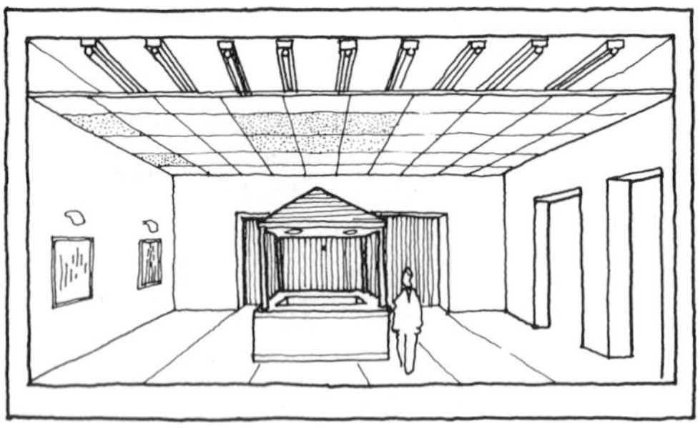

(g)

Fig. 20.15 (g) *Luminous ceiling system (see Fig. 20.13) provides high illumination, low brightness, and high diffusion. Some accent of either color or lighting is desirable to relieve monotony.*

change least in appearance, which is another factor in their favor.

20.17 Other Design Considerations

Since the phenomenon of vision as affected by lighting is the core of our discussion, it is appropriate to present here some miscellaneous, yet important observations which are not readily subsumed under any of the major headings in our discussion.

1. As shown in Fig. 20.15*a* and *b*, the impression of room length and width can be emphasized by the direction of lines of lighting. An even wall wash of light can also be used to shorten and widen a hallway or corridor.

2. As can be seen in Fig. 20.12, a light wall in the line of sight increases the room's size.

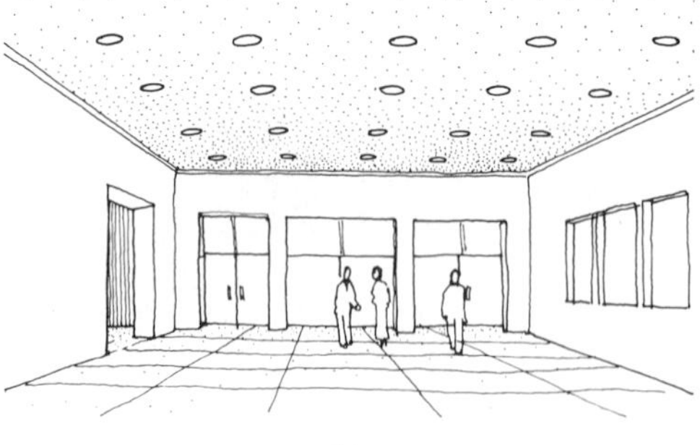

(h)

Fig. 20.15 (h) *Downlights are architecturally neutral and may therefore be spaced evenly* . . . .

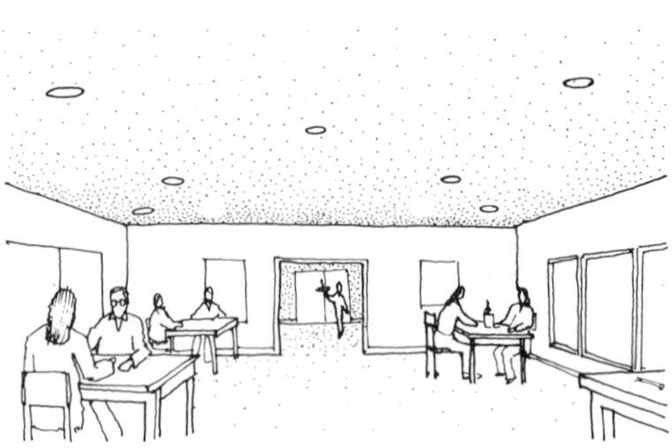

(j)

. . . *or unevenly.* ***Fig. 20.15*** (j)

Fig. 20.15 (k) *The large can-shaped surface-mounted downlights dominate the area's appearance despite the high ceiling. The pimpled look could be eliminated by recessing the units. Or, taking advantage of the prominence of these units, interest could be added by altering the regular pattern. [See* (l).*]*

Fig. 20.15 (l) *The lines of lights in the center converge optically to produce a directional flow of traffic toward the escalator. The remaining floor lighting is provided by visually pleasing circular patterns around columns (Penn Station, New York). (Photo by Stein.)*

Fig. 20.15 (m) *That patterns of lighting are plainly visible even during the daytime is apparent from this photo. The attractiveness of uniformity of fixture pattern can readily be seen. (Photo by Stein.)*

Fig. 20.15 (n) *Since fixtures are readily visible even when unlit during daylight hours, their outline can be accentuated and the resultant pattern utilized as an architectural motif. (Courtesy of Welton Becket & Assoc.)*

Fig. 20.15 (o) *Lighting can be utilized as a medium to connect the inside and outside of a building. The simple expedient of continuing the lighting pattern beyond the window or wall glass provides visibility from inside out as well as outside in. Care must be exercised to avoid fixture placement that will reflect in the glass.*

Fig. 20.15 (p-1) *Treatment of dominant architectural patterns. In each case the lighting designer was faced with essentially the same problem: a low-level seeing task in a large space with a dominant architectural ceiling. Different solutions were arrived at, each of which is consonant with the architectural ceiling. Designs* (p-1) *and* (p-2) *accomplish this by following the dominant line;* (p-3) *follows it with a neutral pattern. (Photo by M. B. Warren.)*

Fig. 20.15 (p-2) *Floor reflection and daylight provide the required ceiling and wall illumination. (Photo by L. Reens.)*

Fig. 20.15 (p-3) *The lighting in this extremely strong architectural pattern was harmonized deftly by recessing fixtures deeply into the lattice motif. Metal-halide HID units and tungsten-halogen units were used. (Courtesy of GTE/Sylvania, Inc.)*

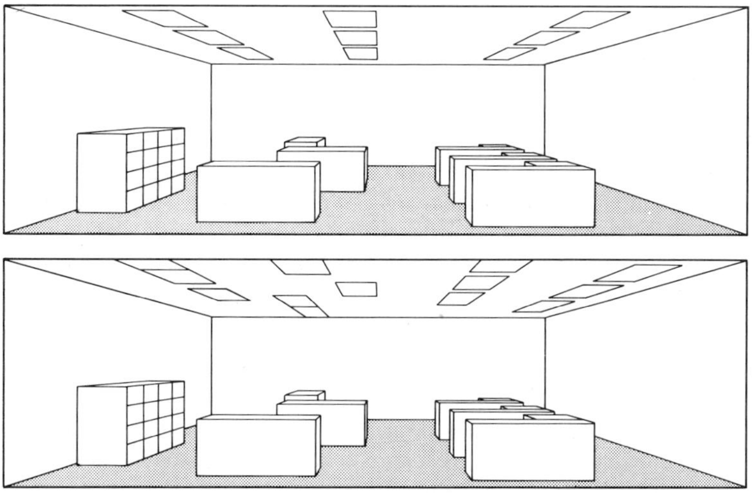

Fig. 20.16 (a) *The uniform layout of Fig. 18.34, shown here in perspective, is neutral in that it does not dominate the space or draw the eye.* (b) *The nonuniform layout can be dominant in the pejorative sense if the eye is drawn by the lack of pattern or symmetry.*

Therefore, the use of both light and low-chroma high-value paint hues can be used to expand a space (lighted walls, light colors) or contract it (unlightened walls, dark colors).

3. Vertical surface illuminance should be approximately 25 to 35% of horizontal for a space to appear dimensionally undistorted. Since high luminance attract the eye, fixtures with sparkle will draw the eye away from the walls and thereby shrink the space's horizontal dimensions.

4. As shown in Fig. 20.15*c, d, f, h,* and *j,* luminaire patterns can be dominant (i.e., can become a focus of attention) by virtue of their size or arrangement. The same is true of wall lighting patterns, and probably even to a greater extent, since walls are always in the direct line of sight. Therefore, wall lighting which creates meaningless scallops, spots, irregular gradients, and points of sparkle become a dominant visual element. In most cases this is neither intended nor desired. Unscalloped, even gradient wall illumination is readily accomplished with linear sources, with elliptical reflector lamps or fixtures, and with proper fixture spacing.

5. Concentrated pools of light in an overall low-ambient-light space effectively isolate the illuminated areas from each other (i.e., define a limited specific geographic space). This can be used to advantage in restaurants and in work or school areas where it is desired to define individual "territories" in a single large space.

6. For a discussion of behavioral and biological effects of lighting, see Corth (1986), Hughes and Neer (1981), Lewis (1987), Smith (1989), and Yorks and Ginthner (1987), among others.

LUMINAIRES

20.18 Design Considerations

(a) General. Refer again to Fig. 20.1. At this point in the design process the lighting hardware is chosen on the basis of the considerations adduced in the preliminary design stage and the appropriate calculations performed. Some spaces will require over-all, uniform illumination. These spaces are calculated by the lumen method, which yields average illumination. Other spaces will utilize local lighting alone, or local lighting in addition to general, requiring point-by-point illumination calculations or some other method for restricted-area calculation. Additional considerations at this design stage are the control strategy (see Sections 21.2 to 21.4), type of ceiling system, for example, modular, movable fixture, and integrated service, and ancillary considerations of ballast noise, fixture heat distribution, and maintenance. Also decided here is whether to utilize work-station-mounted or built-in lighting, both of which are principally applicable to open-plan spaces.

(b) Luminaire Characteristics. The purpose of a luminaire or lighting fixture (the terms are synonymous), are twofold: physically: to hold, protect, and electrify the light source, and photometrically: to control the lamp output (i.e., to redirect the light produced), since most common light sources emit light in substantially all directions. The means by which this beam shaping is accomplished are well known. The characteristics of reflectors, baffles, lenses, and louvres that perform these functions will be discussed in some detail in the following sections. However, the problem in luminaire selection is that the requirements are in many respects incompatible, and therefore a trade-off between various luminaire characteristics must be made. Thus, for instance, high efficiency usually means high fixture luminance with resultant glare. Desirable wall lighting means high-angle luminaire output with resultant direct glare (see Figs. 18.22 and 18.23). Low-angle light means minimum direct glare but possible veiling reflections. High shielding angle (>35°) means good visual comfort but reduced efficiency. An so on. It is therefore obvious that

1. No single luminaire design is ideal for even a majority of applications, and that
2. To make an intelligent selection among the literally hundreds of lighting fixtures commercially available, it is absolutely necessary that the specifier understand both the specific requirements of the application and the light control characteristics of the luminaire being considered.

Application requirements have been studied above; luminaire light control is discussed below. Readers interested in an in-depth study of light control design are referred to IESNA LM-17 (1958).

20.19 Lighting Fixture Distribution Characteristics

A review at this point of Section 18.14, by the reader, would be useful. The two distribution curves shown in Fig. 20.17 are actual test results of two 2-lamp, 1 ft wide by 4 ft long, semidirect fluorescent fixtures, with prismatic enclosure. The flat bottom of the curve in Fig. 20.17*a* indicates even illumination over a wide area, therefore permitting a high spacing-to-mounting height ratio (1.4) for uniform illumination. The rounded bottom of the curve in Fig. 20.17*b* indicates uneven illumination and closer required space for horizontal uniformity (spacing-to-mounting height ratio of 1.2). Uniformity is defined in Section 20.22. The straight sides of the curve in Fig. 20.17*a* show a fairly sharp cutoff, and the small amount of light above 45° means high efficiency, probably insufficient wall lighting, barely adequate diffuseness, and very little direct glare problem, but a distinct possibility of veiling reflections. Conversely, the curve in Fig. 20.17*b* shows a large amount of horizontal illumination (above 45°), with resultant direct glare, diffuseness, and relative inefficiency, since horizontal light is attenuated by multiple reflections before reaching the horizontal working plane. Here, however, low output below 45° minimizes reflected glare. The uplight component of fixture (*a*) is directed outward to cover the ceiling and will not cause hot spots; the corresponding light from fixture (*b*) is concentrated above the fixture and will give uneven illumination of the ceiling.

The conclusions reached above were based on the following observations:

1. *Uniformity of illumination* requires that the intensity at angles above the nadir (0° from vertical) be greater than those at 0° so that points distant from the fixture centerline obtain the same illumination as those below the fixture [since illuminance varies inversely with (square of) distance]. This is exactly the case with the flat bottom characteristics of Fig. 20.17*a*. Therefore, such fixtures can be spaced more widely than the units of Fig. 20.17*b*.

2. *High efficiency* is achieved by directing the luminaire output to the work plane (i.e.,

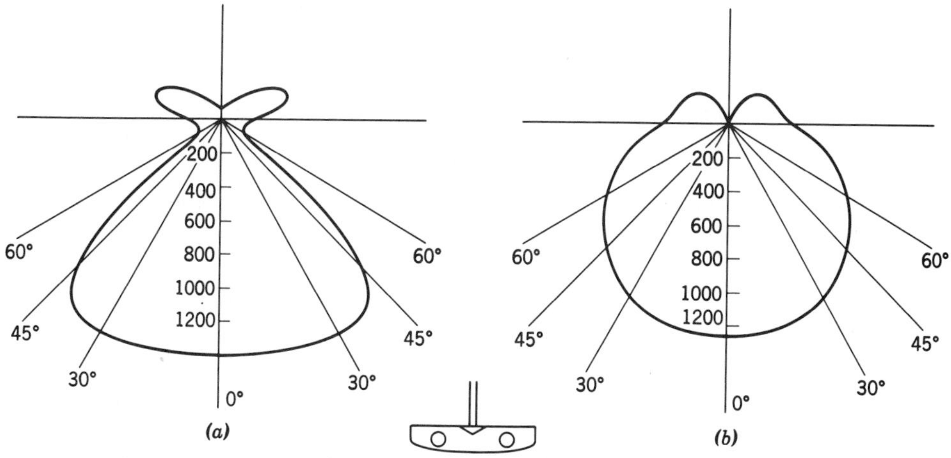

Fig. 20.17 Semidirect fluorescent fixture crosswise distribution (two lamps, 40 W, prismatic enclosure). Note the sharp cutoff and wide horizontally even distribution of (a) *in contrast to the diffuse, broad, and horizontally uneven distribution of* (b).

from 0 to 45° from the vertical). Light above 45° is directed to the walls and reaches the working plane only after multiple attenuating interreflections.

3. *Diffuseness* obtains when light reaches the work plane from multiple directions. This requires that light be reflected from walls and ceilings to the work plane, which in turn requires luminaire light output above 45° from the vertical.

4. *Direct glare* is caused by light output at high angles (i.e., above 45° from the vertical). Direct glare from linear fluorescent fixtures can be minimized by placing their long axis parallel to the line of sight, since such fixtures normally have low *endwise* high-angle output.

5. *Reflected glare* is caused by reflection of low-angle output from the task. Therefore, fixtures with control means that limit output between 0 and 45° (see Section 20.21*b*) will minimize the potential for veiling reflections.

6. *Shielding* is a function of the shape of the fixture housing plus any additional lamp concealment means, such as louvres or baffles. The shielding angle is defined as the angle between a horizontal plane through the louvres or baffles and the inclined plane at which the lamp first becomes visible (as one approaches the fixture; see Fig. 20.18). *Cutoff angle* is usually defined as being synonymous with shielding angle. However, because some sources define it as the complement of shielding angle, it is best to avoid the term and use only *shielding angle*.

7. *Ceiling illumination* is produced by light above the horizontal. As with light below the horizontal, a spread characteristic (Fig. 20.17*a* means good ceiling coverage, no hot spots, and ultimately good diffuseness. Concentrated uplight means a potential hot spot if the fixture suspension hanger is too short, and in any event uneven ceiling illumination.

Thus we see that a rapid inspection of a fixture curve performed by an informed person can yield a large amount of information on the fixture's performance. The reader is encouraged at this juncture to review our comments on the two distribution curves of Fig. 20.17 and then to analyze similarly other distribution curves in manufacturers' catalogs.

20.20 Luminaire Light Control

(a) Lamp Shielding. Except where it is desired to use a bare lamp as a source of sparkle, such as in chandeliers and other decorative fixtures, all lamps in interior fixtures should be shielded from normal sight lines (i.e., sight lines in a head-up, eyes-straight-ahead position (see Fig. 18.17)). The reason is obvious; bare lamps are so bright (see Table 18.2) that they usually constitute a source of direct glare or even disabling glare, depending on apprehended angle (closeness to the eye and size of lamp) and eye adaptation level. The range of permissible luminaire luminance listed in Table 18.2*b* of 1000 to 7000 cd/m^2 depends on these two variables (apprehended angle and adaptation level). Note that the upper limit corresponds to the luminance of a bare 40-W fluorescent tube, which accounts for the fact that bare-lamp fluorescent

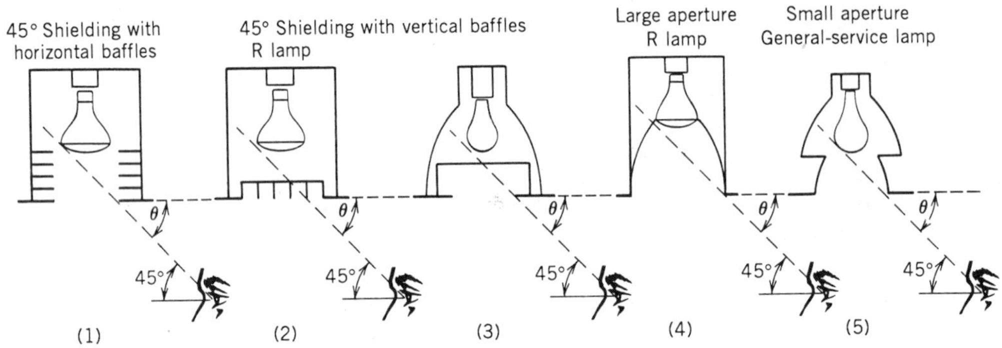

Baffled downlights (1), (2), (3) control unwanted high-angle light by cutoff as illustrated. Black baffles aid by absorbing and appearing dark. Other colors give a ring of light at the baffled edge.

Cones (4), (5) control brightness by cutoff and by redirection of light due to shape. They are either parabolic or elliptical. A light specular finish such as aluminum appears dull; a black specular finish appears unlighted.

(*a*)

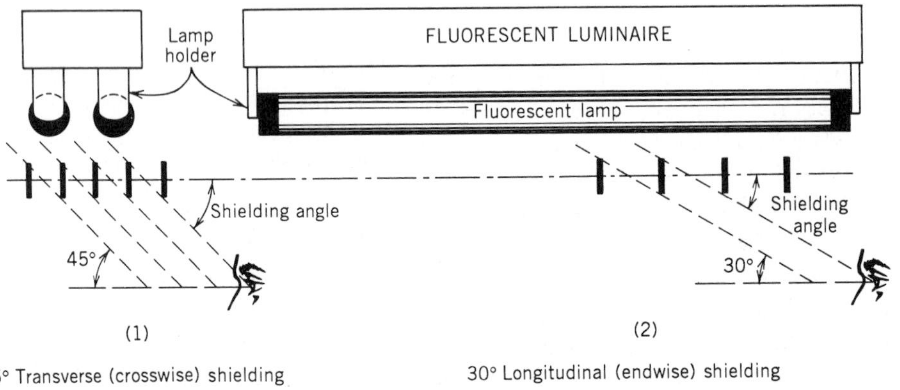

(*b*)

Fig. 20.18 *Methods for shielding light source.* (a) *Incandescent lamps use circular shields and have the same cutoff angle in all directions. A shielding angle of 45° minimum is recommended because of the extremely high luminance of incandescent lamps. Black finishes require high-quality maintenance since dust shows as a bright reflection.* (b) *Shielding of fluorescent lamp is less critical due to lower lamp luminance. 45° × 35° crosswise/lengthwise shielding as shown is excellent and 35° × 30° is satisfactory. The shielding elements may be louvres or baffles. Opaque shielding elements have a higher visual comfort rating than translucent plastic units. Since fixture luminance is higher in the transverse direction than lengthwise, a better cutoff angle is required.*

fixtures abound, with minimal user objection. This is definitely not the case with incandescent lamps.

Shielding of lamps is accomplished with the fixture housing/reflector, or with baffles and louvres, as mentioned above (see Fig. 20.18). Fluorescent fixtures require shielding most when placed crosswise to the line of sight, thus exposing the entire length of lamp to the field of view. Such fixtures require longitudinal baffles, deep housings, or louvres to provide the necessary shielding. Alternatively, when at all possible, place fixtures with their long axis in the direction of slight lines, as stated above.

(b) Reflectors. It is important to understand the action of luminaire reflectors. The basic shapes and beam patterns are illustrated in Figs. 20.19 to 20.22.

Note from Fig. 20.20*a* and *b* that the so-called pinhole downlight requires an elliptic reflector to focus the light through this hole at point "*x*" in order to maintain even minimal fixture efficiency. Elliptic reflectors are large and frequently space above the ceiling is too restricted for their use. A lamp with an integral elliptic reflector, which can therefore be utilized in a standard baffled reflector without severe losses, is illustrated in Fig. 20.21.

20.21 Luminaire Diffusers

Subsumed under this heading, in trade parlance, are all of the devices placed between the lamp(s) and the illuminated space, whose function it is to diffuse the light, control fixture brightness, redirect the light, obscure (hide), and shield the lamps. Since most of these devices perform mul-

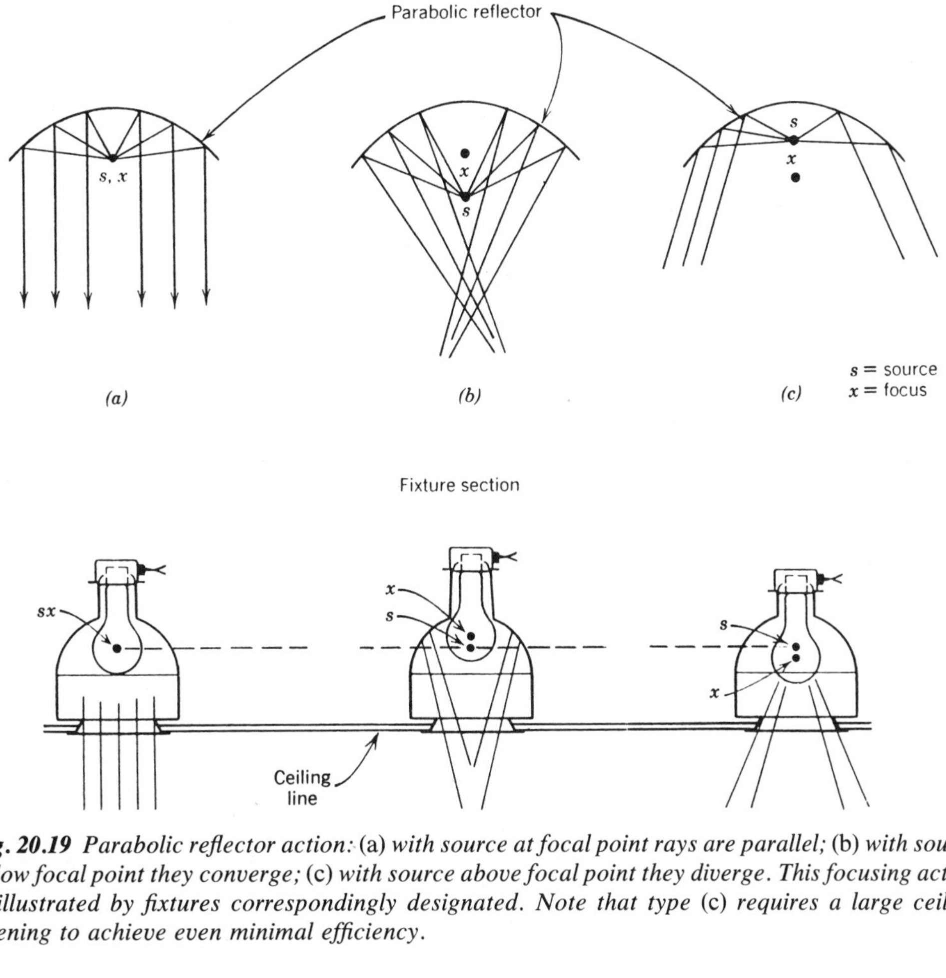

Fig. 20.19 *Parabolic reflector action:* (a) *with source at focal point rays are parallel;* (b) *with source below focal point they converge;* (c) *with source above focal point they diverge. This focusing action is illustrated by fixtures correspondingly designated. Note that type* (c) *requires a large ceiling opening to achieve even minimal efficiency.*

ILLUMINATION

tiple functions, they will be discussed individually.

(a) Translucent Diffusers. Because these do not redirect the light but merely diffuse it, the distribution characteristic is circular, as seen in Fig. 20.23*a*. Typical examples of this type are white opal glass, frosted glass, and white plastics such as Plexiglas, polystyrene, vinyl, and polycarbonates. The distribution is basically the same as it would be for bare lamps. Lamp hiding power is good. Depending on the material, direct glare can be a problem (VCP is poor). Veiling reflections are high, that is, ESI is low. Spacing-to-mounting-height ratio does not exceed 1.5. The fixture is generally inefficient. Wall illumination is good because of a large component of high-angle light (which reduces VCP). The net result of using this type of diffuser is to lower lamp luminance by distributing the lamp output over a larger diffusing area. Applications include corridors, stairwells, high ceiling spaces, and other areas without demanding visual tasks.

A special type of flat plastic panel that polarizes the transmitted light was introduced some years ago. These panels held great promise because they produce a marked decrease of veiling reflection at an angle of 60°, but much less at other angles. Since most viewing is in the range 20 to 40°, using these panels does not result in any appreciable reduction in reflected glare in normal office work situations. From experience in a drafting room equipped with luminaires utilizing high-efficiency multilayer polarizers, it can be stated that visual discomfort from reflected glare, as personally experienced and as reported by a large staff, is not noticeably reduced.

(b) Louvres and Baffles. These are generally rectangular section, metal or plastic, and serve primarily to shield the source (see Fig.

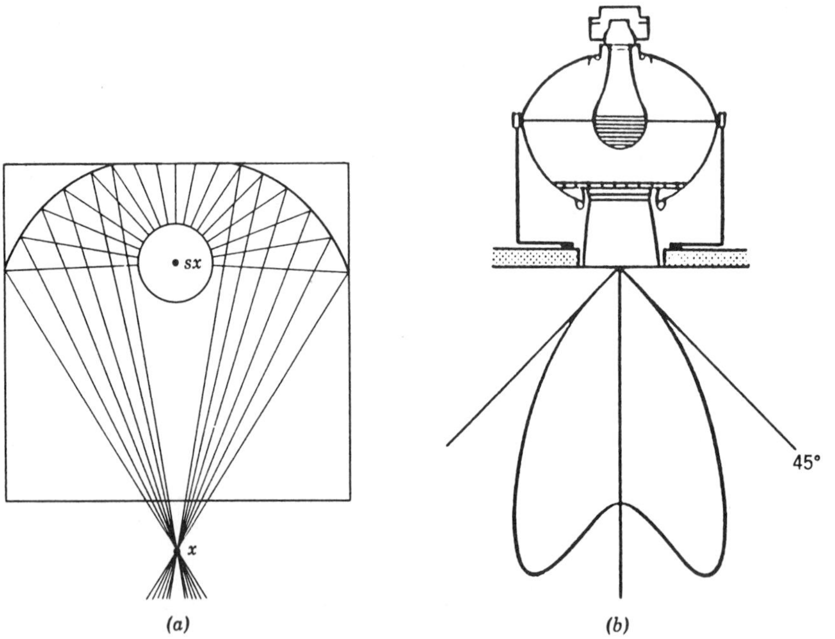

Fig. 20.20 (a) *Action of elliptical reflector section. With the light source at one focal point, the light converges at the other focal point. This effect is useful in fixture design as in* (b). *By projecting light up only, through use of silvered bowl lamp, the output light can be redirected through a constricted aperture at the other focal point, with little loss. Beyond the aperture, the light diverges. The cone in the bottom of the fixture provides cutoff at high angles of viewing (see Fig. 20.21).*

Fig. 20.21 Baffled downlights are low in brightness at normal viewing angles because the black baffles trap and absorb the lamp's output. The result is low efficiency, as in the fixture on the right. The lamp on the left contains an elliptical reflector that focuses the light below the fixture, resulting in even lower fixture brightness with high efficiency. This principle is useful in pinhole downlights, as illustrated in Fig. 20.20. (Photos courtesy of General Electric, Lighting Business Group.)

20.18) and also to diffuse the output, particularly when using plastic translucent louvres. Candlepower distribution curves are shown in Fig. 20.23*b*. The exact curve shape depends on the shielding angle, design of the louvre, and its finish. Louvres finished in specular Alzac or dark colors exhibit low direct glare. The large downward light component can cause serious veiling reflections. Overall fixture efficiency is average.

The spacing-to-mounting height ratio (S/MH), a luminaire metric which indicates the maximum spacing permissible for a given luminaire mounting height that will yield uniform illumination, and which is given as a dimensionless ratio, is fully explained in the next section. For this diffuser it does not exceed 1.5 and varies inversely with shielding angle. This is because the basic circular distribution is changed to egg shaped by cutoff and redirection, reducing the high-angle light. Thus a 45° shielding angle has lower direct glare but requires closer spacing.

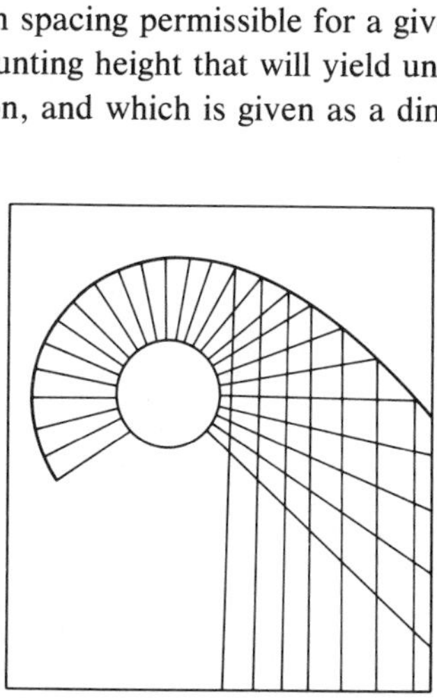

Fig. 20.22 The extended section reflector allows the source to be concealed (shielded) while projecting the light downward.

A special design of this category is the miniature egg-crate parabolic wedge louvre shown in Figs. 20.24 and 20.25. These units redirect a large portion of the light directly downward, and because of this redirection and their specular finish, they appear completely dark—darker, indeed, than the unlighted portion of the ceiling when viewed obliquely (see Figs. 20.25 and 21.33). Fixtures using these louvres have low efficiency due to trapped light, with a maximum coefficient of utilization of about 0.5 (see Section 20.32). VCP is very high but veiling reflections can be troublesome; S/MH is low, varying between 1.0 and 1.5. Shielding angle is usually 45°. The candlepower distribution curve is shown in Fig. 20.23*c*. When using these units, additional wall lighting is almost always required.

Because of the low efficiency of the parabolic louvre design shown in Fig. 20.24*a* caused by the wide light-trapping tops of the louvres, an improved version was developed with the tops

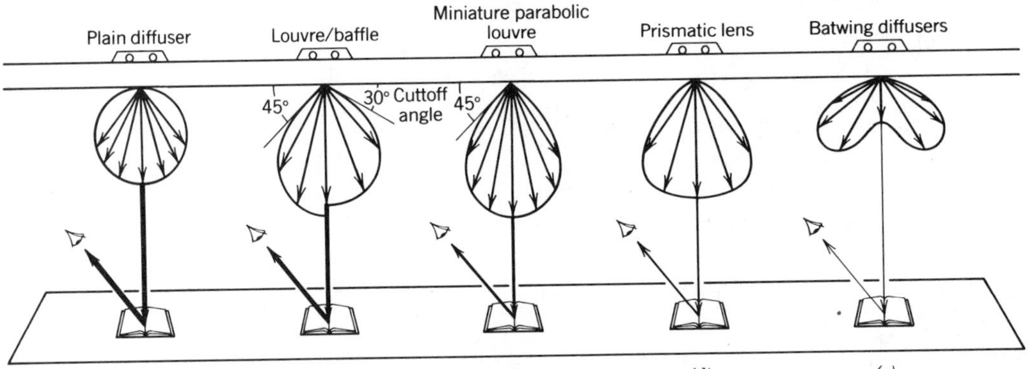

Fig. 20.23 *Comparison of typical candlepower distribution curves for common (fluorescent) luminaire diffuser elements. For a full description see the text. Note that for a given geometry of viewer and luminaire, the severity of veiling reflections depends entirely on the fixture's photometric characteristic. In the individual figures, the potential to produce reflected glare is indicated by weight of the line representing fixture output and reflectance from the work task. The batwing distribution* (e) *concentrates its output in the 30 to 60° range, which minimizes both direct and reflected glare.*

shaped to reflect incident light efficiently. This improved design, which is shown in Fig. 20.24*b*, increases luminaire efficiency by about 20%.

(c) Prismatic Lens. There are many designs available with varying distribution characteristics. Figures 20.23*d* and 20.17*a* can be taken as typical of this genre. They produce an efficient fixture (high coefficient of utilization), good diffusion, wide permissible spacing—S/MH as high as 2.0—and low direct glare (high VCP). Veiling reflections can be troublesome,

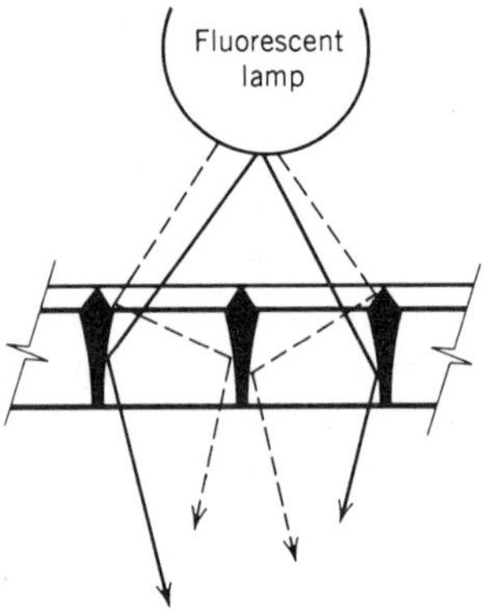

(b-1)

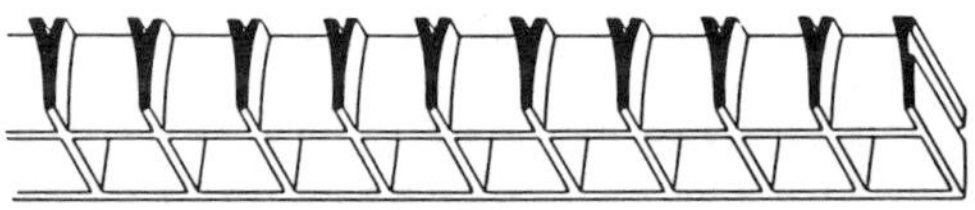

(a)

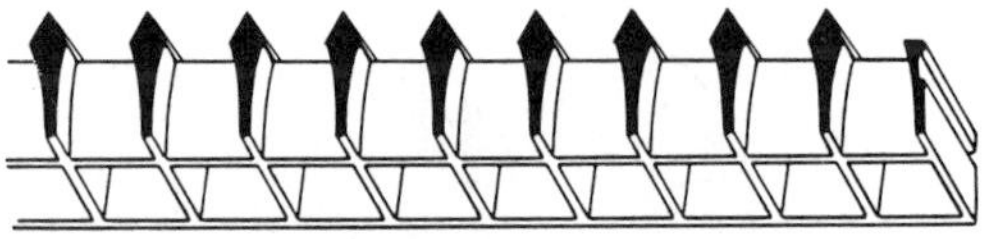

(b-2)

Fig. 20.24 (a) *Section through a conventional miniature parabolic wedge eggcrate type of louvre. These units give exceptionally low brightness when seen at normal viewing angle (see Fig. 20.25) and are therefore useful in VDT areas. Most such units are made of aluminized plastic. Fixtures equipped with these units exhibit low overall efficiency due to the large amount of light trapped by the broad top of each parabolic wedge.*

Fig. 20.24 (b) *A modified wedge design uses a curved top on each wedge to redirect and utilize light striking the top. Solid lines represent light rays redirected by the bottom curve, while dotted lines show light redirected by the top curve, which was lost in the design of* (a). *Typical louvre cell dimensions are ½-in. cube for design* (a) *with a consequent 45° shielding angle and ⅚ to ¾ in. square by ½ in. high for design* (b) *giving a 35 to 45° shielding angle.*

Fig. 20.25 In the illustration the author is checking the luminance of a fluorescent fixture equipped with a miniature parabolic wedge louvre, which has very low luminance above 45° (see Fig. 20.24). This characteristic is readily apparent in comparison to the adjacent fixture, which utilizes a prismatic lens diffuser.

depending on viewing angles and position (see Fig. 18.32).

(d) Fresnel Lens. The action of this lens is similar to that of a reflector. Lamp hiding power is poor but efficiency is high and visual comfort is good. Spacing-to-mounting height ratio (S/MH) is seldom above 1.5 (see Fig. 20.26).

(e) Batwing Diffusers. The theory of this type of diffuser is covered in paragraph 4 of Section 18.28*c*. A typical characteristic is shown in Fig. 20.23*e*. There are two fluorescent luminaire designs that produce the batwing shape distribution characteristic—a prismatic lens and parabolic reflectors and baffles.

1. *Prismatic batwing diffusers*. These are either linear or radial, that is, they produce the batwing distribution either in one direction or in all directions. Typical characteristics of both types are shown in Fig. 20.27. Note that the characteristic shape is more pronounced in the linear diffuser, which indicates better control of veiling reflections in that direction (normally crosswise). Fixtures equipped with these diffusers have good efficiency, low direct and reflected glare, and good diffusion. As with all enclosed ungasketed fixtures, the lens acts as a dust trap, necessitating frequent cleaning to maintain high output.

2. *Deep parabolic louvres and parabolic reflectors*. These luminaires, illustrated in Figs. 20.28 (deep parabolic louvre type) and 20.29 (parabolic reflector type), produce modified versions of the characteristic batwing distribution in the normal or crosswise direction. Distribution in the parallel or lengthwise direction is cir-

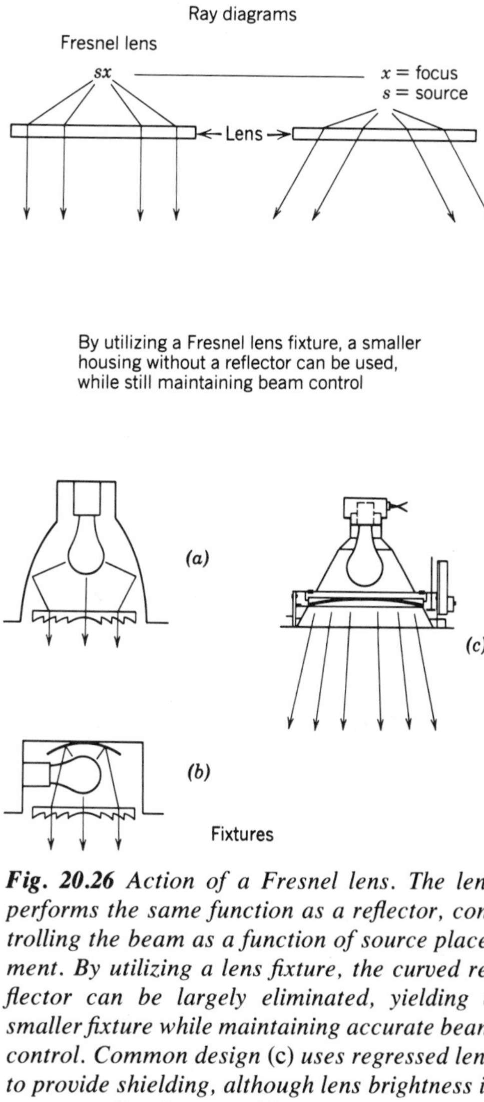

Fig. 20.26 Action of a Fresnel lens. The lens performs the same function as a reflector, controlling the beam as a function of source placement. By utilizing a lens fixture, the curved reflector can be largely eliminated, yielding a smaller fixture while maintaining accurate beam control. Common design (c) *uses regressed lens to provide shielding, although lens brightness is not normally objectionable.*

cular (diffuse) indicating minimum beam control in that direction. These fixtures, like the batwing lens-type diffuser units, have high efficiency, high S/MH, low reflected glare, and low to very low surface brightness, making them usable in VDT areas. They are normally applied with the long axis in the direction of sight lines.

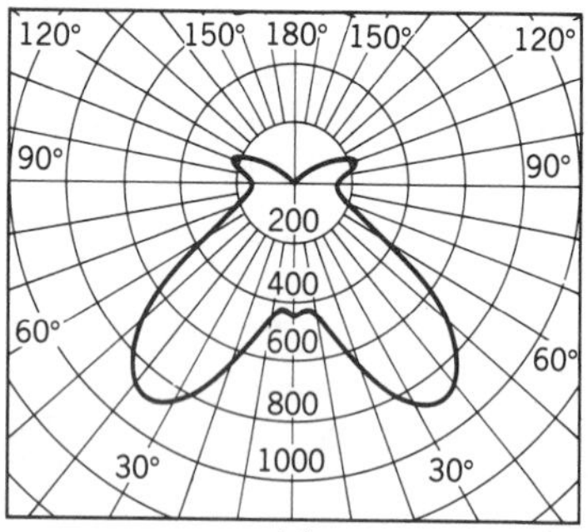

(a)

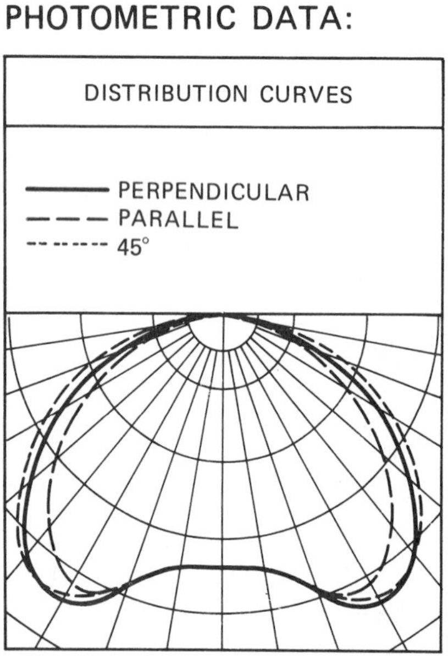

(b)

Fig. 20.27 (a) *Linear batwing distribution with extremely sharp cutoff in the upper and lower ranges. Curve taken across lamp axis for Holophane Percepta™ single-lamp unit.) (Courtesy of Holophane.* (b) *Distribution curves for a radial batwing distribution lens. Note that the perpendicular, parallel, and diagonal curves are almost identical. Zonal flux is maximum in the 30 to 60° range and drops off at both extremes, as desired.*

20.22 Uniformity of Illumination

In any space intended to be lighted uniformly with multiple, discrete, ceiling-mounted direct-lighting system light sources, it is necessary to establish a fixture spacing that will give acceptable uniformity of illumination. A ratio of maximum to minimum illumination on the working plane of 1.2 to 1.3 is readily acceptable since lesser ratios are not easily noticed (see Table 18.4). For general background or circulation lighting, up to 1.5 is acceptable. The recommended spacing-to-mounting-height ratio given by manufacturers (see figures immediately above the distribution curves for each fixture in Table 20.2) are generally based on a 1.2 figure (see Fig. 20.30).

When the luminaire's distribution characteristic is symmetrical in all directions, as is generally the case with point source lamps such as incandescent and HID, only a single S/MH figure is required. However, with the asymmetrical distribution of most fluorescent fixtures, a S/MH ratio is required both crosswise and lengthwise. Due to the characteristic of the lamp itself, the transverse (crosswise, perpendicular) ratio is almost always considerably higher than the longitudinal (parallel, endwise, lengthwise) ratio. See, for instance, fixtures 26, 28, and 42 of Table 20.2. The spacing-to-mounting-height ratio S/MH (also called *spacing criteria*, SC) for a specific luminaire is determined by measuring the distance between two test luminaires that yields the same illuminance midway between them as directly under each one. This ignores (deliberately) any contributions from other fixtures in a multifixture installation and from interreflections, accounting only for the direct component of illuminance from the two test fixtures. Therefore, in an actual installation with several rows of fixtures, the illuminance at point P_1 in Fig. 20.31 will be *higher* than the average by 20 to 30% because of the other fixtures and interreflections, and the illuminance at point P_2 will be approximately equal to the room average. As a result, the illuminance levels along the walls, assuming a distance between the last row and the wall equal to one-half of the side-to-side spacing, will range from 60% of average at point P_3 down to 50% at point P_4, assuming light-colored walls.

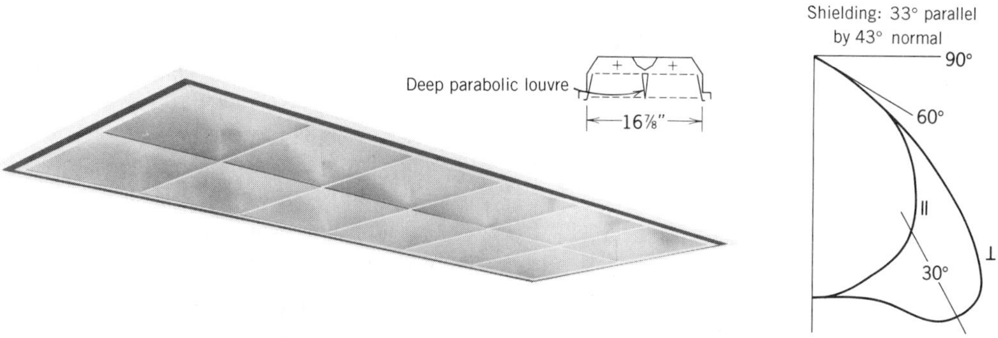

Fig. 20.28 Distribution characteristic of a typical two-lamp, parabolic louvre fixture. All surfaces are semispecular aluminum. The crosswise direction has a typical batwing characteristic, which gives good ESI and VCP. (Courtesy of Columbia Lighting.)

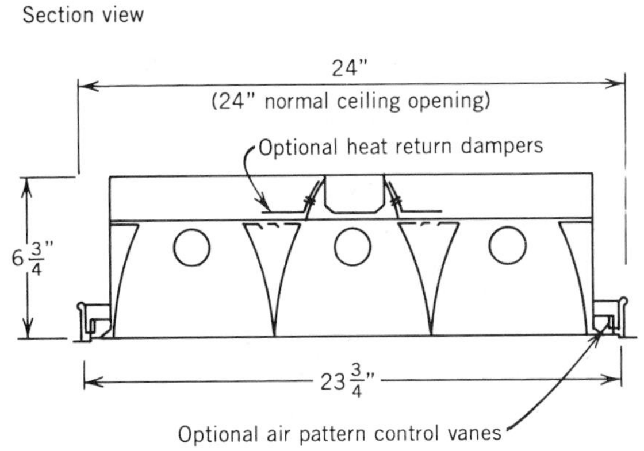

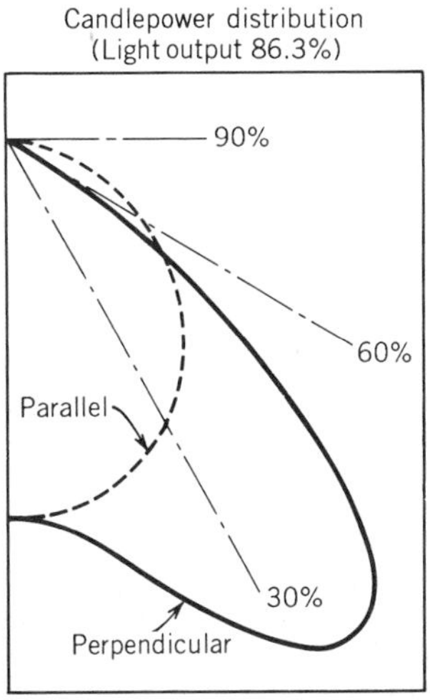

Coefficients of Utilization*
(Catalog No. XP 24G3R340)

Zonal Cavity Method						
RF	20					
RC	80			50		
RW	50	30	10	50	30	10
1	93	90	87	87	85	83
2	84	79	75	79	75	72
3	75	69	64	71	67	63
4	67	60	55	64	59	54
5	60	53	48	57	51	47
6	53	46	42	51	45	40
7	47	40	35	45	39	35
8	42	35	30	40	34	29
9	37	30	25	35	29	25
10	33	26	22	32	26	21

* Data is for 3 lamp unit utilizing lumen cool white lamps.

Luminaire Brightness Data

Brightness in Footlamberts		
	Average Brightness	
Angle	Crosswise	Lengthwise
45	2030	1000
55	610	960
65	70	850
75	0	630
85	0	110

S/MH = 1.7 perpendicular X 1.2 parallel

Fig. 20.29 Typical photometric data for a very efficient baffled, parabolic reflector linear fluorescent luminaire. Note the transverse (perpendicular) batwing distribution typical of this design.

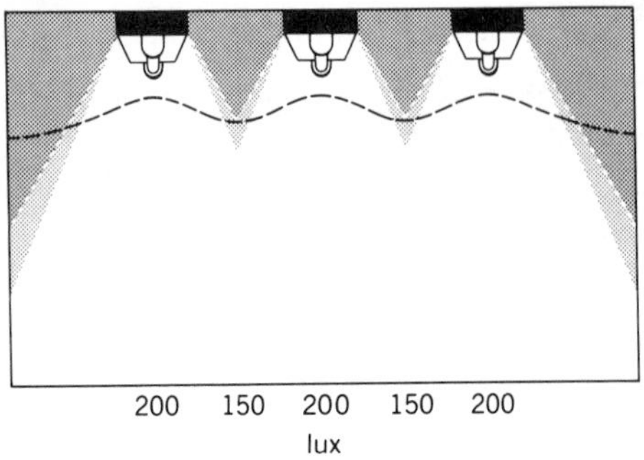

Fig. 20.30 The ratio of maximum to minimum illumination should not exceed 1.30 in areas requiring uniform illumination. See the text.

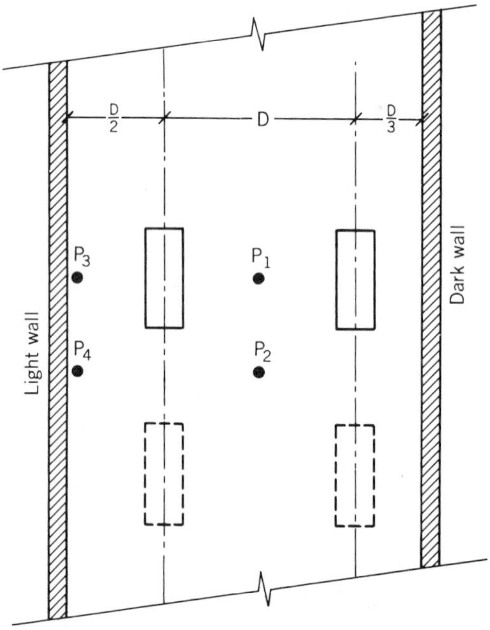

Fig. 20.31 Lighting fixtures installed according to the manufacturer's recommended spacing criteria (spacing to mounting ratio), with a row-to-row spacing D *and a row-to-wall spacing,* D/2, *as shown at the left wall. Assuming a high reflectance finish on the wall (light color), illuminances P_3 and P_4 are at least one half of the illuminance directly below a fixture. At a dark wall, as on the right, illuminance would fall below this value. As a consequence the designer would move the row of luminaires closer to the wall, as shown. For further explanation, see the text.*

When walls are dark due to paint aging, bookshelves, dark wood paneling, and the like, the illuminance levels drop to less than 50% of average, which is obviously insufficient as task lighting. To counteract this effect, particularly when placement of furniture is such that visual tasks *will* occur near walls, the designer has three choices:

1. Reduce the distance between the last row of lights and the wall to a third or less of the row-to-row spacing. This will also provide required wall lighting.
2. Provide some type of continuous perimeter lighting or wall wash units, both of which will increase illuminance levels at the walls.
3. A combination of the two methods above.

Because endwise illuminance from linear fluorescent fixtures is considerably lower than crosswise, end walls have lower illuminance then side walls and greater illuminance variation. It is therefore particularly important to provide some additional illumination as discussed above, and to terminate fixture rows no more than 1 ft from an end wall. This is all the more important where visual tasks without supplementary task lighting occur at these walls.

We mentioned above that the fixture of Fig. 20.17 had a high spacing-to-mounting-height ratio (S/MH) because of its flat-bottomed curved. This ratio, when not given by the manufacturer, may be approximated from Fig. 20.32. An accurate method of calculating maximum to minimum illumination ratios is available (see *IES Handbook,* 1981).

The foregoing discussion of illumination uniformity concerned itself with uniformity on a horizontal plane. Occasionally, it is necessary to know the degree of uniformity vertically, that is, on horizontal planes at different elevations *directly below the fixtures*. Four different lighting situations are normally encountered. They are point sources such as incandescent downlights, line sources such as continuous-row fluorescent fixtures, infinite sources such as luminous ceilings—whether transilluminated or indirect—and parabolic reflector beams such as from PAR lamps. The vertical uniformity of each type is shown graphically in Fig. 20.33.

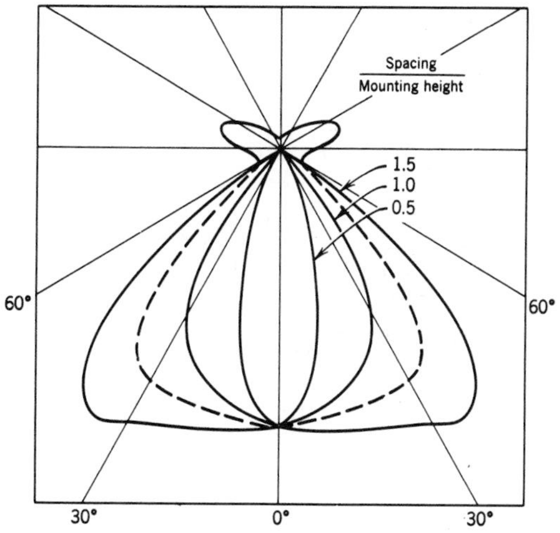

*Fig. 20.32 Typical distribution curves for approximating ratio of fixture spacing to the mounting height (S/MH) above the working plane, for direct lighting luminaires. To determine S/MH for a specific unit, compare its curve to those shown and interpolate. Although these curves were developed for direct lighting point sources such as incandescent (and HID), they can also be used with asymmetric distribution luminaires such as fluorescent. In such cases the curves will give the S/MH ratio in the plane whose characteristic is used, that is, lengthwise or crosswise. The dashed curve was transferred from Fig. 20.17*a. *Its permissible S/MH is somewhat higher than the curves indicate because it is a semidirect distribution, and its ceiling light component permits wider spacing between units. (After Odle and Smith, from* IESNA/Journal, *January 1963.)*

20.23 Luminaire Mounting Height

The mounting height of luminaires is normally established before their spacing. In arriving at a mounting height for fixtures with an upward component, a balance must be struck between low mounting, which will control ceiling brightness and give good utilization of light, and the reticence to dominate an area, particularly a

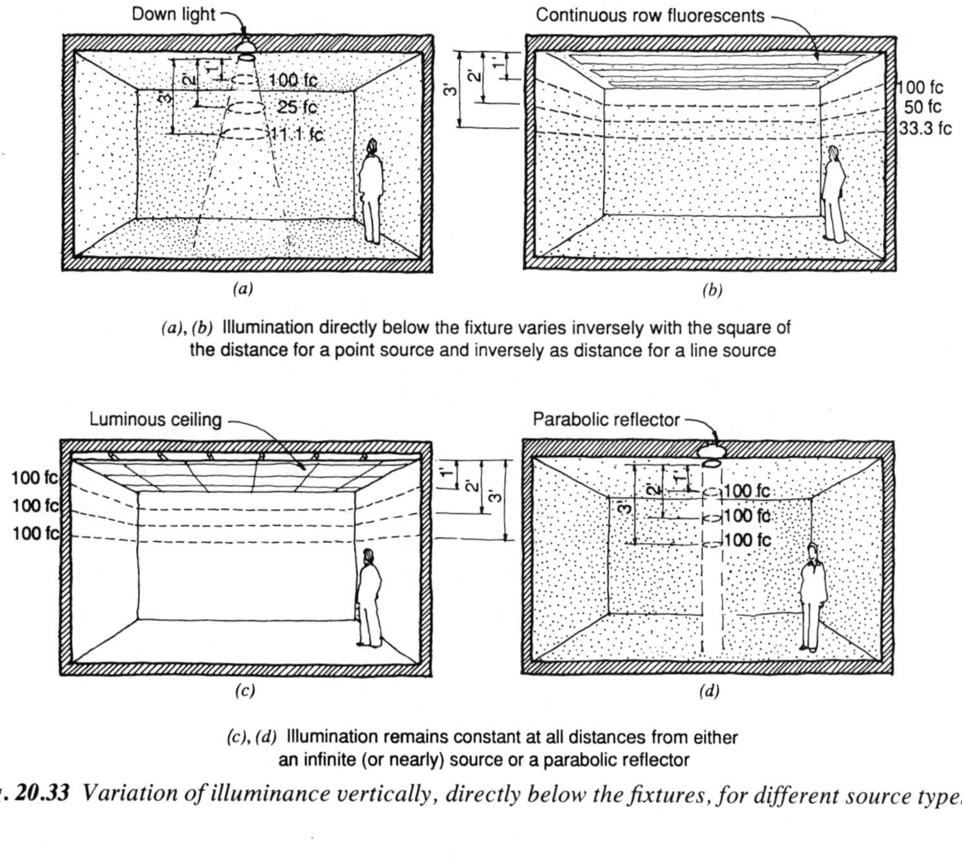

***Fig. 20.33** Variation of illuminance vertically, directly below the fixtures, for different source types.*

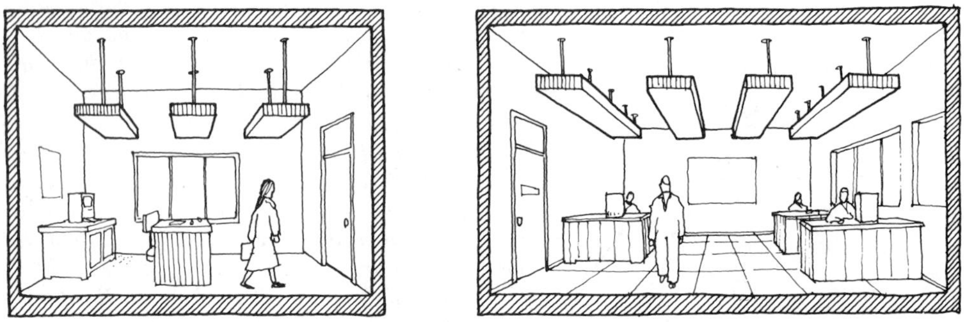

Fig. 20.34 Mounting height of fixtures may be lower in a small room than in a large room because of the illusion of lowness created in a large room.

large room, by using such a low mounting height that the apparent ceiling height is affected (see Fig. 20.34). Rules of thumb for mounting height are:

1. Indirect and semi-indirect luminaires should be suspended no less than 18 in. from ceiling, and preferably 24 to 36 in.
2. Direct-indirect and semidirect fluorescent fixtures should be suspended not less than 12 in. for two-lamp units and 18 in. for four-lamp units.

The effect of pendant length on coefficient of utilization (efficiency; see Section 20.28) is given in Fig. 20.35.

20.24 Lighting Fixtures

Before proceeding further with design, we wish to discuss the principal item of lighting hardware, the fixture itself. This and the following sections cover fixture construction, installation, and appraisal. The architect need simply stop to consider that lighting fixtures constitute 25 to 30% of the electrical budget or 4 to 5% of the overall building budget to appreciate their importance. Since the difference between a quality unit and an inferior one is often not readily visible to the casual observer, particular care must be taken in the specification of lighting fixtures and in examination of shop drawings *and* samples. All fixtures if applied properly will give a sufficient quantity of light, but only a good unit will combine quantity with good quality, ease of installation, facility of maintenance, and indefinite life. In addition, installation must be proper to ensure mechanical rigidity and safety, electrical safety, freedom from excessive temperatures, and requisite accessibility of component parts and of the fixture outlet box. The material below is a combination of NEC minimum requirements plus factors that the author has found important beyond these minima.

20.25 Lighting Fixture Construction

1. All fixtures should be wired and constructed to comply with local codes, NEC (Article 410) and Underwriters Laboratories Standard for Lighting Fixtures, and shall bear the Underwriters Label, where label service is available. Reflector Luminaire Manufacturers (RLM) standards should be adhered to for all porcelain enameled fixtures.

2. Fixtures should generally be constructed of 20 gauge (0.0359 in.) thick steel minimum. Cast portions of fixtures should be no less than 1/16 in. thick.

3. All metals should be coated. The final coat should be a baked-enamel white paint of minimum 85% reflectance, except for anodized or ''Alzac'' surfaces.

4. No point on the outside surface of any fixture should exceed 90°C after installation, and

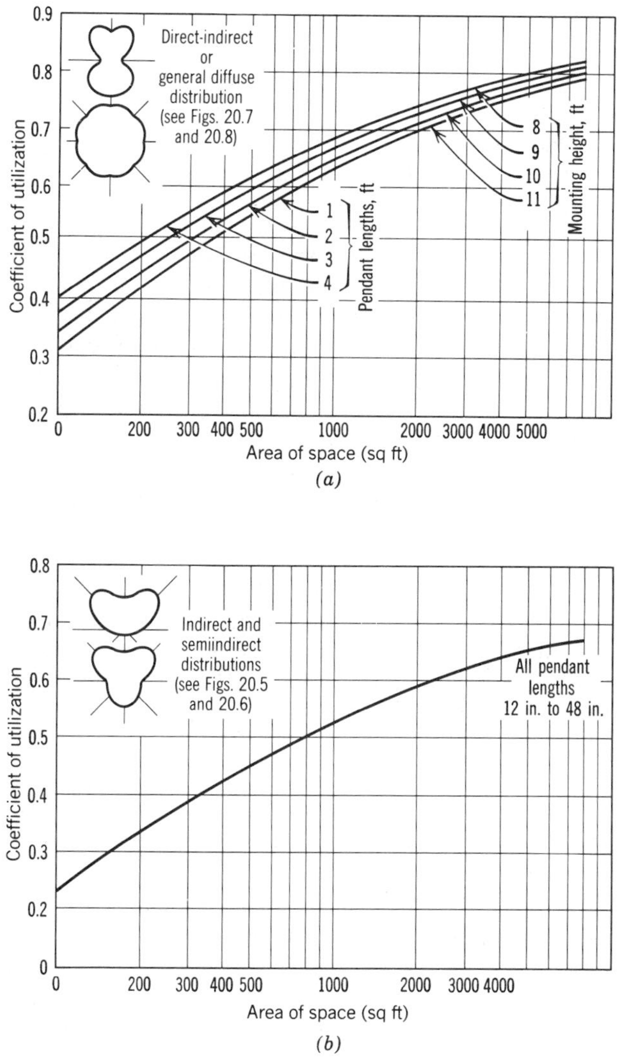

Fig. 20.35 *Coefficient of utilization (lighting system efficiency) as a function of pendant length for various distributions. With a substantial downward component as in direct-indirect or general diffuse* (a), *system efficiency rises slowly as the fixture descends (pendant length increases). Maximum differentials occur in small rooms and can reach 20%. Where the ceiling is the light source as in indirect and semi-indirect systems* (b), *the pendant length does not change the room illumination (see also Fig. 20.33*c*). This curve can be used to estimate CU for indirect and semi-indirect, where not found in Table 20.2 or manufacturers' catalogs.*

on continuous operation. For an exception, see NEC Article 410 M.

5. Each fixture should be identified by a label carrying the manufacturer's name and address and the fixture catalog number.

6. Glass diffuser panels in fluorescent fixtures should be mounted in a metal frame. Plastic diffusers should be suitably hinged. "Lay-in" plastic diffusers should not be used.

7. Plastic diffusers should be of the slow-burning or self-extinguishing type with low smoke-density rating and low heat-distortion temperatures. This latter should be low enough so that the plastic diffuser will distort sufficiently to drop out of the fixture before reaching ignition temperature.

8. It is *imperative* that plastics used in air-handling fixtures be of the noncombustible, low-smoke-density type. These requirements also apply to other nonmetallic components of such fixtures.

9. All plastic diffusers should be clearly marked with their composition material, trade name, and manufacturer's name and identifica-

tion number. Results of ASTM combustion tests should be submitted with fixture shop drawings. The characteristics of many plastic diffusers change radically with age and exposure to ultraviolet light. Glass and acrylic plastic are stable in color and strength. Other plastics may yellow and even turn brown, thus diminishing light transmission radically as well as changing the fixture appearance. Some plastics that are initially very tough and "vandalproof" embrittle with age and exposure to weather or the ultraviolet light of a mercury or fluorescent source. Thus the long-range as well as initial characteristics of all diffuser elements must be investigated before specification and approval.

10. Ballasts should be mounted in fixtures with captive screws on the fixture body, to allow ballast replacement without fixture removal.

11. All fixtures mounted outdoors, whether under canopies or directly exposed to the weather, should be constructed of appropriate weather-resistant materials and finishes, including gasketing to prevent entrance of water into wiring, and should be marked by the manufacturer, "Suitable for Outdoor Use."

20.26 Lighting Fixture Installation

1. Although some codes allow fluorescent fixtures weighing less than 40 lb to be mounted directly on the horizontal metal members of hung-ceiling systems, experience has shown that vibration, member deflection, routine maintenance operation on equipment in hung ceilings, and poor workmanship can cause such fixtures to fall, endangering life. It is therefore strongly recommended that all fixtures, surface, pendant, or recessed, whether mounted individually or in rows, be supported from the black channel iron supporting the ceiling system (purlins) or directly from the building structure, but in no case by the ceiling system itself. This is particularly important in the case of an exposed "Z" spline ceiling system.

2. Fixtures installed in bathrooms should *not* have an integral receptacle and when installed on walls, should have nonmetallic bodies. These are safety precautions.

20.27 Lighting Fixture Appraisal

As mentioned above, the intense competition in the lighting field necessitates close scrutiny of the characteristics of fixtures. To compare the relative merits of similar lighting fixtures as manufactured by different companies, complete test data plus a sample in a regular shipping carton from a normal manufacturing run are necessary.

The following list should be used as a basic guide, with additional items added according to job requirements:

1. *Photometric and design data.* Manufacturers should furnish complete test data, including candlepower distribution curve(s), coefficients of utilization (CU), wall and ceiling luminance coefficients, luminance data from 45 to 85°, a table of visual comfort probability (VCP), and recommended spacing-to-mounting-height ratio (*S*/*MH*). These data should come from a reliable, indepentent testing laboratory, and not from the manufacturer's test facilities. In addition, many manufacturers either regularly publish, or make available on request, various design aids such as isolux (isocandle) curves, ESI charts, and point-by-point computer printouts for different layouts. These very useful data and their application are covered in Section 20.38.

2. *Construction and installation.* The designer should check the sample for workmanship, rigidity, quality of materials and finish, and ease of installation, wiring, and leveling. Installation instruction sheets should be sufficiently detailed. Results of actual operating temperature tests in various installation modes should be included. Air-handling fixtures should be furnished with heat-removal data, pressure-drop curves, air-diffusion data, and noise-criteria (NC) data for different airflow rates.

3. *Maintenance.* Luminaires should be simply and quickly relampable, resistant to dirt col-

lection, and simple to clean. Replacement parts must be readily available.

20.28 Luminaire Efficiency: Coefficient of Utilization

Because of internal reflections inside a luminaire some of the generated lumen output of the lamp is lost. The ratio of output lumens to lamp (input) lumens, expressed as a percentage, represents the luminous efficiency of the fixture. This characteristic has little meaning by itself, however, since the actual overall efficiency of a luminaire depends on the space in which it is used.

To illustrate, let us consider the case of a large room with a high, dark ceiling. If we were to use a high-efficiency (say 80%) indirect lighting unit in such a room, most of the light directed upward would be lost (absorbed), and the actual lighting on the working plane would be very low. If, however, the same room were illuminated with 50% efficiency direct-lighting units utilizing the same wattage, the illumination on the working plane would be considerably higher than in the first case.

Similarly, if we consider a small room with dark walls and ceiling, lighted alternatively by diffuse lighting and by direct lighting units of the same wattage and unit efficiency, the horizontal-plane illumination will be higher in the case of the direct units because of the large loss of the horizontal and upward components of the diffuse lighting on the walls and ceiling. It should be obvious then that the fixture efficiency *alone* is not sufficient but *the overall luminous efficiency of a particular unit in a particular space* is the required figure of merit. This number, since it describes the utilization of the fixture output in a specific space, is known as the *coefficient of utilization* (CU). It is defined as the ratio between the lumens reaching the horizontal working plane and the generated lumens. Since each luminaire will have a different coefficient for every different space in which it is used, a system of standardization has been evolved utilizing room cavities (explained below) of certain proportions and various surface reflectances. The fixture coefficients are then computed and tabulated as shown in Table 20.2. It should be emphasized that the figures given in this table are for the generic fixture type only; in an actual job, actual fixture data as found in manufacturers' catalogs should be used. When fixture data are not available, as in preliminary estimates and calculations, use the coefficients listed in Section 20.5. To summarize, CU is a factor that combines fixture efficiency and distribution with room proportions, mounting height, and surface reflectances.

DETAILED DESIGN PROCEDURES

20.29 Calculation of Average Illuminance

Having selected a luminaire on the basis of all the foregoing criteria, it remains only to calculate the number of such units required in each space, for uniform *general* illumination, and to arrange them properly. Although a number of calculation methods are available, the lumen (flux) method is simplest and most applicable to our needs for area lighting calculations. Illuminance calculation from point, line, or area sources is covered in Sections 20.38 to 20.41. Luminance (photometric brightness) calculations are covered in Section 20.43.

Before beginning a detailed description of the zonal cavity calculation method, a general comment on precision and accuracy is in order. The degree of precision of any calculation need not exceed either the accuracy required or the precision of the data available. Thus there is no point in working to three decimal places if a ±10% rounding of the result is common and acceptable, or if the data available are only accurate to one decimal place. As the reader will see, the lumen (lighting flux) method of *average* illuminance calculation is replete with assumptions and estimates.

Among these are:

1. It is assumed that the space is empty. This is not normally the case.
2. It is assumed that all surfaces are perfect diffusers. This is not the case.
3. All surface reflectances are estimates, ±10%.

TABLE 20.2 **Coefficients of Utilization for Typical Luminaires with Suggested Maximum Spacing Ratios**

To obtain a coefficient of utilization:

1. Determine cavity ratios for the room, ceiling, and floor.
2. Determine the effective ceiling and floor cavity reflectances from Table 20.3. Use initial ceiling, floor, and wall reflectances.
3. Obtain coefficient of utilization (CU) for 20% effective floor cavity reflectance from appropriate table below for luminaire type to be used. Interpolate, when necessary, to obtain CU for exact room cavity ratio for nearest effective ceiling cavity reflectances above and below reflectance obtained in step 2; interpolate between these CUs to obtain CU for step 2 ceiling cavity reflectance.
4. If effective floor cavity reflectance differs significantly from 20%, obtain multiplier from Table 20.4 and apply this to the CU obtained in step 3.
5. To obtain CU for a ceiling cavity reflectance (ρ_{CC}) of 30 or 10%, multiply the figure for ρ_{CC} = 50% by 0.85 and 0.70, respectively. This is an approximation. For exact figures see *IES Handbook* (1987).
6. Use the figure in the last column (ρ_{CC} = 0; ρ_w = 0) for outdoor lighting, i.e., no walls or ceiling.
7. Legend:

ρ_{CC} = percent effective ceiling cavity reflectance

ρ_w = percent wall reflectance

RCR = room cavity ratio

Maximum *S/MH* guide = ratio of maximum luminaire spacing to mounting above work plane.

NOTE: In some cases, luminaire data in this table are based on an actual typical luminaire; in other cases, the data represent a composite of generic luminaire types. Therefore, whenever possible, specific luminaire data should be used in preference to this table of typical luminaires.

The polar intensity sketch (candlepower distribution curve) and the corresponding spacing-to-mounting height guide are representative of many luminaires of each type shown.

	Typical Distribution and Percent Lamp Lumens		ρ_{CC}: →	80			70			50			0
			ρ_w: →	50	30	10	50	30	10	50	30	10	0
Typical Luminaire	Maintenance Category	Maximum S/MH	RCR	Coefficients of Utilization for 20% Effective Floor Cavity Reflectance (ρ_{FC} = 20)									
1	V	1.5	0	.87	.87	.87	.81	.81	.81	.69	.69	.69	.44
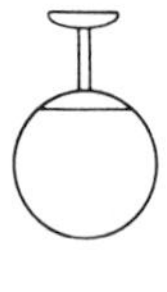	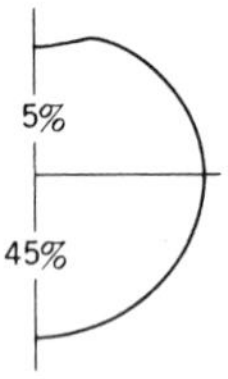 5% 45%		1	.71	.67	.63	.66	.62	.59	.56	.53	.50	.31
			2	.61	.54	.49	.56	.50	.46	.47	.43	.39	.23
			3	.52	.45	.39	.48	.42	.37	.41	.36	.31	.18
			4	.46	.38	.33	.42	.36	.30	.36	.30	.26	.15
			5	.40	.33	.27	.37	.30	.25	.32	.26	.22	.12
			6	.36	.28	.23	.33	.26	.21	.28	.23	.19	.10
			7	.32	.25	.20	.29	.23	.18	.25	.20	.16	.09
Pendant diffusing sphere			8	.29	.22	.17	.27	.20	.16	.23	.17	.14	.07
with incandescent lamp			9	.26	.19	.15	.24	.18	.14	.20	.15	.12	.06
3	IV	1.3	0	.99	.99	.99	.97	.97	.97	.92	.92	.92	.83
	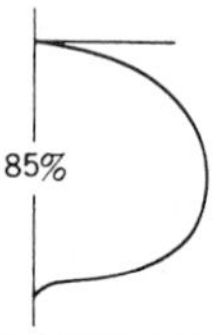 10% 85%		1	.88	.85	.82	.86	.83	.81	.83	.80	.78	.72
			2	.78	.73	.68	.76	.72	.67	.73	.69	.66	.61
			3	.69	.62	.57	.67	.61	.57	.65	.60	.56	.52
			4	.61	.54	.49	.60	.53	.48	.58	.52	.48	.45
			5	.54	.47	.41	.53	.46	.41	.51	.45	.41	.38
			6	.48	.41	.35	.47	.40	.35	.46	.39	.35	.32
Porcelain-enameled			7	.43	.35	.30	.42	.35	.30	.41	.34	.30	.28
ventilated standard dome			8	.38	.31	.26	.38	.31	.26	.37	.30	.26	.24
with incandescent lamp			9	.35	.28	.23	.34	.27	.23	.33	.27	.23	.21
7	IV	0.7	0	.52	.52	.52	.51	.51	.51	.48	.48	.48	.44
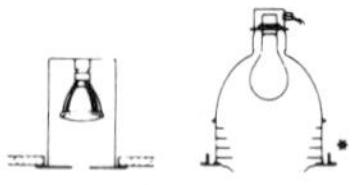	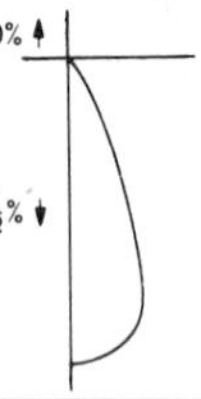 0% ↑ 43½% ↓		1	.49	.48	.48	.48	.48	.47	.47	.46	.46	.42
			2	.47	.46	.45	.46	.45	.44	.45	.44	.43	.41
			3	.45	.44	.43	.45	.43	.42	.44	.42	.42	.40
			4	.43	.42	.41	.43	.41	.40	.42	.41	.40	.38
EAR-38 lamp above 51-mm			5	.42	.40	.39	.41	.40	.38	.41	.39	.38	.37
(2 in.)-diameter aperture			6	.40	.39	.37	.40	.38	.37	.39	.38	.37	.36
(increase efficiency to			7	.39	.37	.36	.39	.37	.36	.38	.37	.35	.35
54½% for 76-mm (3 in.)-			8	.37	.36	.34	.37	.35	.34	.37	.35	.34	.33
diameter aperture)			9	.36	.34	.33	.36	.34	.33	.35	.34	.33	.32
18	III	1.5	0	.93	.93	.93	.91	.91	.91	.87	.87	.87	.78
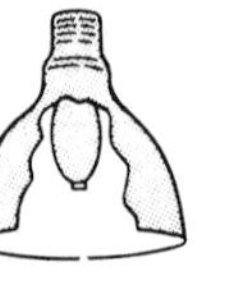	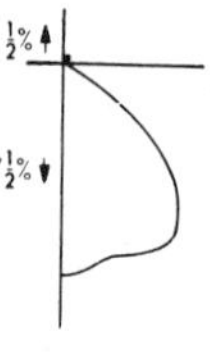 ½% ↑ 77½% ↓		1	.85	.82	.80	.83	.81	.79	.79	.78	.76	.70
			2	.77	.73	.70	.76	.72	.69	.73	.70	.67	.63
			3	.70	.65	.61	.68	.64	.60	.66	.62	.59	.56
			4	.63	.58	.53	.62	.57	.53	.60	.56	.52	.49
			5	.57	.51	.47	.56	.51	.47	.55	.50	.46	.44
			6	.51	.45	.41	.51	.45	.41	.49	.44	.40	.38
"High-bay" wide distribu-			7	.46	.40	.35	.45	.39	.35	.44	.39	.35	.33
tion ventilated reflector			8	.41	.35	.31	.41	.35	.31	.40	.34	.31	.29
with clear HID lamp			9	.37	.31	.27	.37	.31	.27	.36	.30	.27	.25

ILLUMINATION

TABLE 20.2 **Coefficients of Utilization for Typical Luminaires with Suggested Maximum Spacing Ratios (*continued*)**

	Typical Distribution and Percent Lamp Lumens		ρ_{CC}: →	80			70			50			0
			ρ_W: →	50	30	10	50	30	10	50	30	10	0
Typical Luminaire	Maintenance Category	Maximum S/MH	RCR	Coefficients of Utilization for 20% Effective Floor Cavity Reflectance (ρ_{FC} = 20)									
26	II	1.5/1.3	0	.95	.95	.95	.91	.91	.91	.83	.83	.83	.66
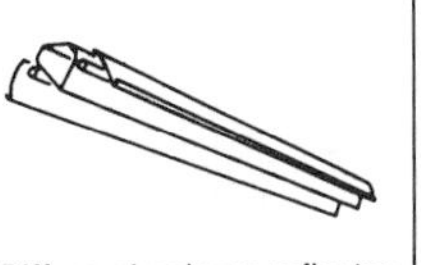	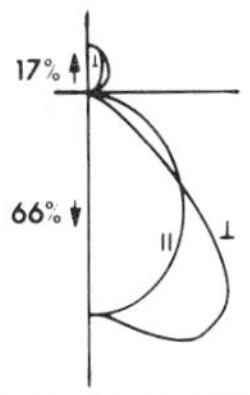		1	.85	.82	.80	.82	.79	.77	.75	.73	.72	.59
			2	.76	.72	.68	.74	.70	.66	.68	.65	.62	.52
			3	.69	.63	.59	.66	.61	.57	.62	.58	.54	.46
			4	.62	.56	.51	.60	.54	.50	.56	.51	.47	.41
			5	.55	.49	.44	.53	.48	.43	.50	.45	.41	.36
			6	.50	.43	.39	.48	.42	.38	.45	.40	.36	.31
Diffuse aluminum reflector with 35° crosswise shielding			7	.45	.38	.34	.43	.37	.33	.41	.36	.32	.27
			8	.40	.34	.29	.39	.33	.29	.37	.31	.28	.24
			9	.36	.30	.25	.35	.29	.25	.33	.28	.24	.20
28	II	1.5/1.1	0	.83	.83	.83	.79	.79	.79	.72	.72	.72	.56
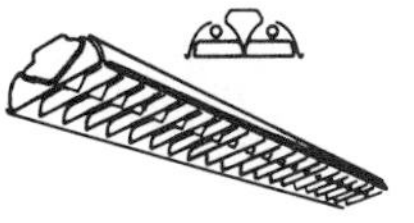	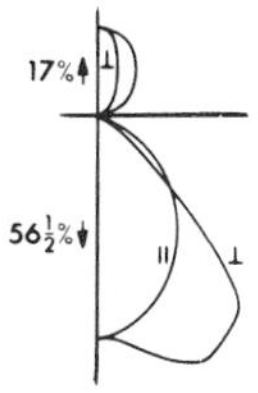		1	.75	.72	.70	.72	.69	.67	.65	.64	.62	.50
			2	.67	.63	.60	.65	.61	.58	.59	.57	.54	.45
			3	.61	.56	.52	.58	.54	.51	.54	.50	.48	.40
			4	.55	.49	.45	.53	.48	.44	.49	.45	.42	.36
			5	.49	.44	.40	.47	.42	.39	.44	.40	.37	.31
			6	.45	.39	.35	.43	.38	.34	.40	.36	.33	.28
Diffuse aluminum reflector with 35° crosswise × 35° lengthwise shielding			7	.40	.35	.31	.39	.34	.30	.36	.32	.29	.25
			8	.36	.31	.27	.35	.30	.26	.33	.28	.25	.21
			9	.33	.27	.23	.32	.26	.23	.29	.25	.22	.19
33	VI	N.A.	0	.77	.77	.77	.68	.68	.68	.50	.50	.50	.12
	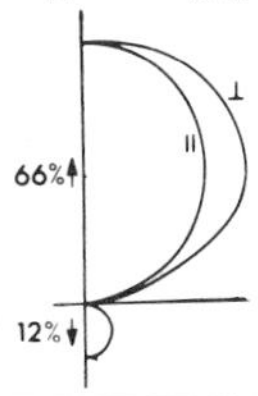		1	.67	.64	.62	.59	.57	.54	.44	.42	.41	.10
			2	.59	.54	.50	.52	.48	.45	.38	.36	.34	.09
			3	.51	.46	.42	.45	.41	.37	.34	.31	.28	.07
			4	.45	.40	.35	.40	.35	.31	.30	.27	.24	.06
			5	.40	.34	.30	.35	.30	.27	.26	.23	.20	.05
			6	.36	.30	.26	.32	.27	.23	.24	.20	.18	.05
Luminous bottom suspended unit with very-high-output lamp			7	.32	.26	.22	.28	.23	.20	.21	.18	.15	.04
			8	.29	.23	.19	.25	.21	.17	.19	.16	.13	.03
			9	.26	.20	.17	.23	.18	.15	.17	.14	.12	.03
35	V	1.5/1.2	0	.81	.81	.81	.78	.78	.78	.72	.72	.72	.59
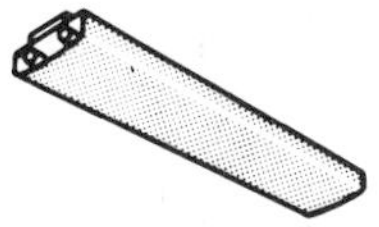	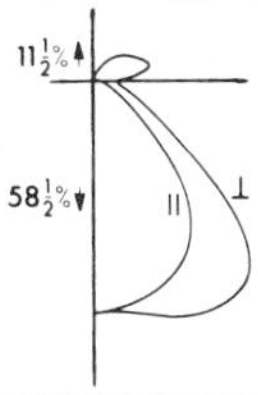		1	.71	.69	.66	.69	.66	.64	.64	.62	.60	.50
			2	.64	.59	.56	.61	.58	.54	.57	.54	.51	.44
			3	.57	.52	.48	.55	.50	.47	.51	.48	.45	.38
			4	.51	.46	.41	.49	.44	.41	.46	.42	.39	.34
			5	.46	.40	.36	.44	.39	.35	.41	.37	.34	.29
			6	.41	.35	.31	.40	.35	.31	.38	.33	.30	.26
Two-lamp prismatic wraparound; multiply by 0.95 for four lamps			7	.37	.31	.27	.36	.31	.27	.34	.29	.26	.23
			8	.33	.28	.24	.32	.27	.23	.30	.26	.22	.19
			9	.30	.24	.20	.29	.24	.20	.27	.23	.19	.17
38	IV	1.0	0	.60	.60	.60	.58	.58	.58	.56	.56	.56	.50
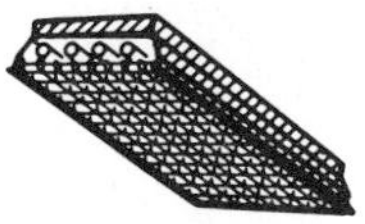	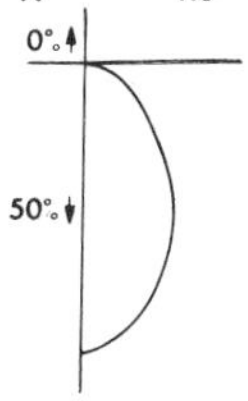		1	.54	.52	.50	.52	.51	.49	.50	.49	.48	.44
			2	.48	.45	.43	.47	.44	.42	.45	.43	.41	.39
			3	.43	.40	.37	.42	.39	.37	.41	.38	.36	.34
			4	.39	.35	.32	.38	.35	.32	.37	.34	.32	.30
			5	.35	.31	.28	.35	.31	.28	.34	.30	.28	.26
			6	.32	.28	.25	.32	.28	.25	.31	.27	.25	.23
Four-lamp, 610-mm (2 ft)-wide troffer with 45° plastic louvre			7	.29	.25	.22	.29	.25	.22	.28	.25	.22	.21
			8	.26	.22	.20	.26	.22	.20	.25	.22	.20	.18
			9	.24	.20	.17	.24	.20	.17	.23	.20	.17	.16
42	V	1.4/1.2	0	.75	.75	.75	.73	.73	.73	.70	.70	.70	.63
	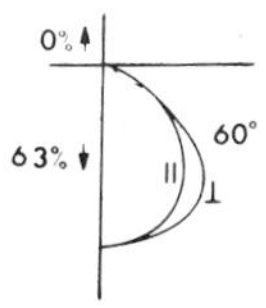		1	.67	.65	.63	.66	.64	.62	.63	.62	.60	.55
			2	.60	.57	.54	.59	.56	.53	.57	.54	.52	.49
			3	.54	.50	.47	.53	.49	.46	.52	.48	.45	.43
			4	.49	.44	.40	.48	.44	.40	.47	.43	.40	.37
			5	.44	.39	.35	.43	.38	.35	.42	.38	.34	.33
			6	.40	.34	.31	.39	.34	.31	.38	.34	.30	.29
Fluorescent unit with flat prismatic lens, four-lamp, 610 mm (2 ft) wide			7	.36	.30	.27	.35	.30	.27	.34	.30	.27	.25
			8	.32	.27	.23	.32	.27	.23	.31	.26	.23	.22
			9	.29	.24	.20	.28	.23	.20	.28	.23	.20	.19

TABLE 20.2 **Coefficients of Utilization for Typical Luminaires with Suggested Maximum Spacing Ratios (*continued*)**

Typical Luminaire	Typical Distribution and Percent Lamp Lumens: Maintenance Category	Maximum S/MH	RCR	ρ_{CC}: 80, ρ_W: 50	80, 30	80, 10	70, 50	70, 30	70, 10	50, 50	50, 30	50, 10	0, 0
				Coefficients of Utilization for 20% Effective Floor Cavity Reflectance (ρ_{FC} = 20)									
44	IV	N.A.	0	.71	.71	.71	.70	.70	.70	.66	.66	.66	.60
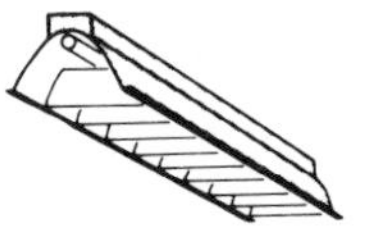	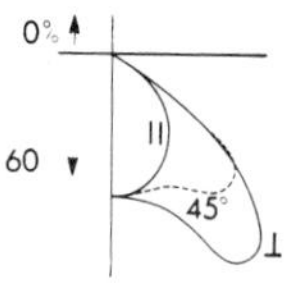		1	.65	.63	.61	.63	.62	.60	.61	.59	.58	.54
			2	.59	.55	.53	.58	.55	.52	.55	.53	.51	.48
			3	.53	.49	.46	.52	.48	.45	.50	.47	.45	.42
			4	.47	.43	.40	.47	.43	.40	.45	.42	.39	.37
			5	.42	.38	.34	.42	.37	.34	.41	.37	.34	.32
			6	.38	.33	.30	.38	.33	.30	.37	.33	.30	.28
Radial batwing distribution—louvred fluorescent unit			7	.34	.29	.26	.33	.29	.26	.33	.28	.25	.24
			8	.30	.25	.22	.30	.25	.22	.29	.25	.22	.20
			9	.27	.22	.18	.26	.22	.18	.26	.21	.18	.17
45	V	N.A.	0	.57	.57	.57	.56	.56	.56	.53	.53	.53	.48
	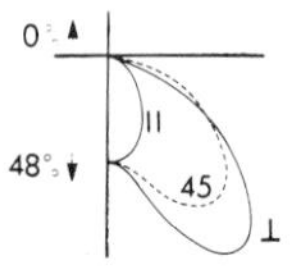		1	.50	.48	.47	.49	.47	.46	.47	.46	.44	.41
			2	.44	.41	.38	.43	.40	.38	.41	.39	.37	.34
			3	.39	.35	.32	.38	.34	.31	.37	.33	.31	.29
			4	.34	.30	.27	.33	.29	.26	.32	.29	.26	.24
Radial batwing distribution—four-lamp, 610-mm (2 ft)-wide fluorescent unit with flat prismatic lens and overlay			5	.30	.25	.22	.29	.25	.22	.28	.24	.22	.20
			6	.26	.22	.19	.26	.22	.18	.25	.21	.18	.17
			7	.23	.19	.16	.23	.19	.16	.22	.18	.16	.14
			8	.21	.16	.13	.20	.16	.13	.19	.16	.13	.12
			9	.18	.14	.11	.18	.14	.11	.17	.14	.11	.10
46	V	N.A.	0	.87	.87	.87	.84	.84	.84	.77	.77	.77	.64
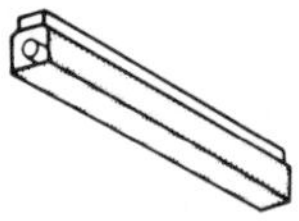	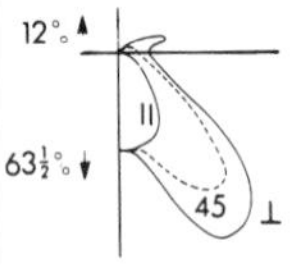		1	.76	.73	.70	.73	.70	.67	.67	.65	.63	.53
			2	.66	.61	.57	.64	.59	.56	.59	.56	.52	.44
			3	.59	.53	.48	.56	.51	.47	.53	.48	.44	.38
			4	.52	.45	.40	.50	.44	.40	.47	.42	.38	.32
Bilateral batwing distribution—one lamp, surface-mounted fluorescent with prismatic wraparound lens			5	.46	.39	.34	.44	.38	.33	.41	.36	.32	.27
			6	.41	.34	.29	.39	.33	.29	.37	.31	.27	.23
			7	.36	.30	.25	.35	.29	.24	.33	.27	.23	.20
			8	.32	.26	.21	.31	.25	.21	.29	.24	.20	.17
			9	.29	.22	.18	.28	.22	.18	.26	.21	.17	.14
47	V	1.7	0	.71	.71	.71	.69	.69	.69	.66	.66	.66	.60
	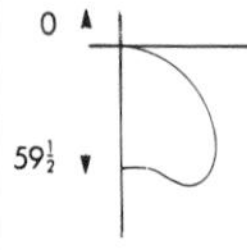		1	.62	.60	.58	.61	.59	.57	.59	.57	.55	.51
			2	.55	.51	.47	.53	.50	.47	.51	.48	.46	.42
			3	.48	.43	.39	.47	.43	.39	.45	.41	.38	.36
			4	.42	.37	.33	.41	.37	.33	.40	.36	.32	.30
			5	.37	.32	.27	.36	.31	.27	.35	.30	.27	.25
Radial batwing distribution—four-lamp, 610-mm (2ft)-wide fluorescent unit with flat prismatic lens			6	.33	.27	.23	.32	.27	.23	.31	.26	.23	.21
			7	.29	.24	.20	.29	.24	.20	.28	.23	.20	.18
			8	.26	.21	.17	.25	.20	.17	.25	.20	.17	.15
			9	.23	.18	.14	.23	.18	.14	.22	.17	.14	.13
48	I	1.6/1.2	0	1.01	1.01	1.01	.96	.96	.96	.87	.87	.87	.68
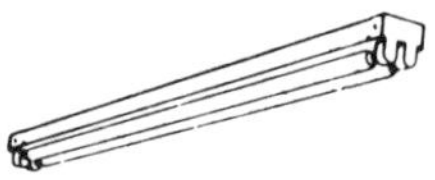	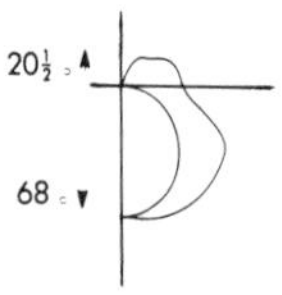		1	.85	.81	.77	.81	.77	.73	.73	.70	.67	.53
			2	.73	.66	.61	.69	.63	.58	.63	.58	.54	.42
			3	.63	.56	.50	.60	.53	.48	.55	.49	.44	.35
			4	.56	.47	.41	.53	.46	.40	.48	.42	.37	.29
			5	.49	.40	.34	.46	.39	.33	.42	.36	.31	.24
			6	.43	.35	.29	.41	.34	.28	.38	.31	.26	.20
			7	.39	.31	.25	.37	.29	.24	.34	.27	.23	.17
Two-lamp fluorescent strip unit			8	.34	.27	.21	.33	.26	.21	.30	.24	.19	.15
			9	.31	.23	.18	.30	.23	.18	.27	.21	.17	.12
49	I	1.4/1.2	0	1.13	1.13	1.13	1.09	1.09	1.09	1.01	1.01	1.01	.85
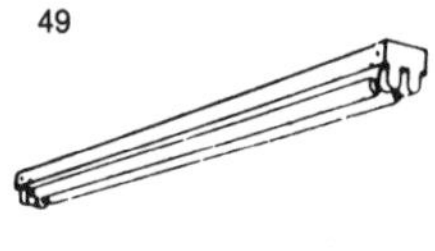	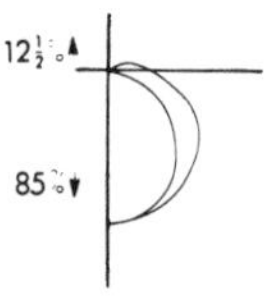		1	.96	.92	.88	.93	.89	.85	.87	.83	.80	.68
			2	.83	.76	.70	.80	.74	.68	.75	.69	.65	.55
			3	.73	.65	.58	.70	.63	.57	.66	.59	.54	.46
			4	.64	.55	.49	.62	.54	.48	.58	.51	.46	.39
			5	.56	.47	.41	.55	.46	.40	.51	.44	.38	.33
			6	.50	.41	.35	.49	.40	.34	.46	.38	.33	.28
Two-lamp fluorescent strip unit with 235° reflector fluorescent lamps			7	.45	.36	.30	.44	.35	.30	.41	.34	.28	.24
			8	.40	.32	.26	.39	.31	.25	.37	.30	.25	.21
			9	.36	.28	.22	.35	.27	.22	.33	.26	.21	.18

TABLE 20.2 **Coefficients of Utilization for Typical Luminaires with Suggested Maximum Spacing Ratios (*continued*)**

	Typical Distribution and Percent Lamp Lumens		ρ_{CC}: →	80			70			50			0
			ρ_W: →	50	30	10	50	30	10	50	30	10	0
Typical Luminaire	*Maintenance Category*	*Maximum S/MH*	*RCR*	*Coefficients of Utilization for 20% Effective Floor Cavity Reflectance (ρ_{FC} = 20)*									
50 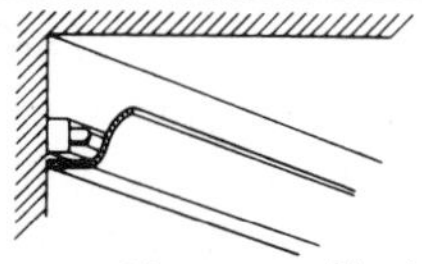			1	.42	.40	.39	.36	.35	.33	.25	.24	.23	Coves are
			2	.37	.34	.32	.32	.29	.27	.22	.20	.19	not rec-
			3	.32	.29	.26	.28	.25	.23	.19	.17	.16	om-
			4	.29	.25	.22	.25	.22	.19	.17	.15	.13	mended
			5	.25	.21	.18	.22	.19	.16	.15	.13	.11	for light-
			6	.23	.19	.16	.20	.16	.14	.14	.12	.10	ing areas
Single-row fluorescent lamp cove without reflector, multiplied by 0.93 for two rows and by 0.85 for three rows.			7	.20	.17	.14	.17	.14	.12	.12	.10	.09	having
			8	.18	.15	.12	.16	.13	.10	.11	.09	.08	low re-
			9	.17	.13	.10	.15	.11	.09	.10	.08	.07	flectances
													ρ_{CC} = 10%
													ρ_W = 50 30 10
53 ρ_{CC} from below ~45% 			1							.51	.49	.48	.47 .46 .45
			2							.46	.44	.42	.43 .42 .40
			3							.42	.39	.37	.39 .38 .36
			4							.38	.35	.33	.36 .34 .32
			5							.35	.32	.29	.33 .31 .29
			6							.32	.29	.26	.30 .28 .26
Louvred ceiling. Ceiling efficiency ~50%; 45° shielding opaque louvres of 80% reflectance. Cavity with minimum obstructions and painted with 80% reflectance paint—use ρ_{CC} = 50.			7							.29	.26	.23	.28 .25 .23
			8							.27	.23	.21	.26 .23 .21
			9							.24	.21	.19	.24 .21 .19
			10							.22	.19	.17	.22 .19 .17

NOTES:
1. Data extracted from *IES Handbook* (1981), Reference Volume, with permission.
2. Multiply by 1.05 for three lamps and 1.1 for two lamps.

4. Maintenance conditions are estimates, at best ±10%.
5. No allowance is made for deviation of the performance of an individual product from its specification.

Any attempt to account accurately for these approximations would enormously complicate the calculation and would serve no useful purpose for this type of *average* illuminance calculation. For this reason, the procedure presented below has introduced some approximations (which we feel are well justified), in the interest of simplicity. These approximations are noted wherever used.

20.30 Calculation of Horizontal Illuminance by the Lumen (Flux) Method

The lumen method of calculation is a procedure for determining the *average* maintained illuminance (footcandles, lux) on the working plane in a room. The method presupposes that luminaires will be spaced so that uniformity of illumination is provided in order that an *average* calculation have validity. The method is based on the definition of one foot-candle [lux] (of illuminance) as one lumen incident on one square foot [meter] of area, that is,

$$\text{footcandles (lux)} = \frac{\text{lumens}}{\text{area in ft}^2\ (\text{m}^2)}$$

As explained above, the ratio between the lumens reaching the working plane and the lumens generated is the coefficient of utilization, CU. Or

$$\text{lumens on the working plane} = \text{lamp lumens} \times \text{CU}$$

Therefore,

$$\text{illuminance } E = \frac{\text{lamp lumens} \times \text{CU}}{\text{area}}$$

ILLUMINATION

The coefficient CU is selected from Table 20.2 of generic fixture types or similar tables provided by the luminaire manufacturer, by a method known as the zonal cavity method. It is explained in Section 20.32.

The illuminance figure so calculated is *initial average* illumination. This initial level is reduced by the effect of temperature and voltage variations, dirt accumulation on luminaires and room surfaces, lamp output depreciation, and maintenance conditions. All of these effects are cumulatively referred to as the light loss factor (LLF):

$$\text{maintained } E = \text{initial } E \times \text{LLF}$$

(This factor was previously known as the maintenance factor, MF.) The procedure required to arrive at this factor is explained in the following section.

Our final expression for maintained illuminance E as calculated by the lumen method is, therefore,

$$E = \frac{\text{lamp lumens} \times \text{CU} \times \text{LLF}}{\text{area}} \tag{20.1}$$

where E is footcandles if area is expressed in square feet, or lux, where the area is in square meters. Lamp lumens is the total within the space, and equal to

no. of fixtures × lamps per fixture × lumens per lamp

Our formula then becomes

$$E = \frac{\text{luminaires} \times \text{lamps per luminaire} \times \text{lumens per lamp} \times \text{CU} \times \text{LLF}}{\text{area}} \tag{20.2}$$

or, conversely, solving for the number of luminaires required to achieve a target maintained luminance E,

$$\text{No. of luminaires} = \frac{E \times \text{area}}{\text{lamps per luminaire} \times \text{lumens per lamp} \times \text{CU} \times \text{LLF}} \tag{20.3}$$

If we wish to determine the area *per luminaire* that is required to give a specified illumination, we have

$$\text{area per luminaire} = \frac{\text{lamps per luminaire} \times \text{lumen per lamp} \times \text{CU} \times \text{LLF}}{E} \tag{20.4}$$

This is usually a much more useful procedure in large spaces than calculating for the entire room. For instance, it is much more convenient to know that to maintain, say, 60 fc with a given luminaire, 70 ft^2 per unit is required than it is to know that for an 18,000-ft^2 floor, 257 fixtures are necessary. The former figure allows us to establish a pattern, say 7 × 10; the latter figure is too large to be immediately useful. Therefore, for rooms requiring more than a small number of luminaires, the latter calculation should always be used.

20.31 Calculation of Light Loss Factor (LLF)

The light loss factor is composed of elements that can be categorized as recoverable and nonrecoverable. The former can be improved by maintenance; the latter cannot. The total LLF is the product of all the individual factors. The survey that follows includes approximations. For more precise data, see *IES Handbook* (1981).

Among the nonrecoverable loss factors are the following.

(a) Luminaire Ambient Temperature. Light output changes when a *fluorescent* fixture operates at other than its design temperature. With normal indoor installation use 1.0, that is, no depreciation. For air troffers *increase* output by 10%, that is, use a factor of 1.10. For other conditions, refer to technical data on the source involved.

(b) Voltage. When the lamp operates at rated voltage, use 1.0. Details of source sensitivity to voltage are given in Chapter 19.

(c) Luminaire Surface Depreciation. This factor is proportional to age and depends on type of surface involved. The designer must estimate this factor.

(d) Components. Losses can be due to use of nonstandard components such as ballasts, reflectors, louvres, and so on. When no exact data are available and no special conditions apply, use 0.9 for the product $a \times b \times c \times d$.

The factors below are *recoverable*, that is, can be improved to initial conditions by maintenance.

(e) Room Surface Dirt. This factor is self-explanatory. Obviously, lighting systems that depend heavily on surface reflections, such as indirect, are more seriously affected than systems that deliver most of their useful light directly. Assuming a 24-month cleaning cycle, and normal conditions of cleanliness, use the appropriate base factor below. Alter it for other conditions such as infrequent maintenance and unusual cleanliness or dirtiness.

Direct lighting: 0.92 ± 5%

Semidirect lighting: 0.87 ± 8%

Direct-indirect lighting: 0.82 ± 10%

Semi-indirect lighting: 0.77 ± 12%

Indirect lighting: 0.72 ± 17%

(f) Lamp Lumen Depreciation. This factor depends on the type of lamp and the replacement schedule. Use the following when exact data are unavailable:

	Group Replacement	*Replacement on Burnout*
Incandescent	0.94	0.88
Tungsten-halogen	0.98	0.94
Fluorescent	0.90	0.85
Mercury	0.82	0.74
Metal-halide	0.87	0.80
High-pressure sodium	0.94	0.88

(g) Burnouts. This self-explanatory factor depends on maintenance schedules and method of replacement. Use the following:

Group replacement procedures: 1.0

Individual replacement on burnout: 0.95

(h) Luminaire Dirt Depreciation (LDD). This factor depends on luminaire design, atmosphere conditions in the space, and maintenance schedule. The maintenance category is obtained from manufacturer's data or from Table 20.2. The type of atmosphere is determined by considering the space involved. Assuming a 12-month cleaning schedule and normal room cleanliness, use the base number in Fig. 20.36 and change it to match conditions of dirt and maintenance. The categories correspond to those of the IES.

Total LLF is the product of all the depreciation factors above, that is,

$$\text{LLF} = a \times b \times c \times d \times e \times f \times g \times h$$

For example, a fluorescent air troffer in a regularly maintained group-lamp-replacement, air-conditioned office might typically have an LLF of

$$\text{LLF} = 1.1 \times 1 \times 0.92 \times 1 \times 0.95 \times 0.9 \times 1.0 \times 0.93 = 0.80$$

The same fixture in the same office, but with walls and fixture cleaned only when replacing burned-out lamps would typically have an LLF of

$$\text{LLF} = 1.1 \times 1 \times 0.92 \times 1 \times 0.87 \times 0.85 \times 0.95 \times 0.78 = 0.55$$

Thus, if in the first case the maintained illumination is E fc, in the second case it is 0.55/0.80 or 0.69E fc, that is, a reduction of 31% as a result of poor maintenance. When a detailed determination of light loss factor is not possible, use the factors given in Section 20.33. They are somewhat more conservative than those given in Section 20.5*c*.

20.32 Determination of Coefficient of Utilization (CU) by the Zonal Cavity Method

The coefficient of utilization connects a particular fixture to a particular space, by relating the luminaire's light distribution characteristic to the room size and its surface reflectances. To account for the luminaire's mounting height and

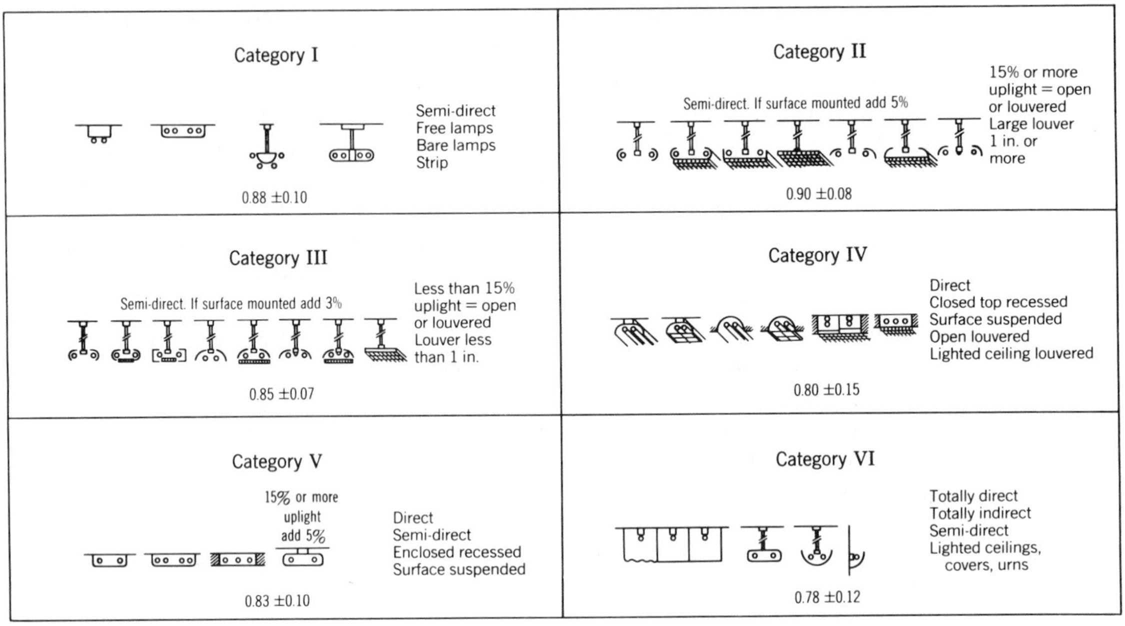

Fig. 20.36 *The LDD factor is determined from the category of luminaire, which is an indication of its proneness to dirt accumulation, plus a knowledge of room ambient conditions.*

its relationship to the working plane, the space is divided into three cavities: the ceiling cavity above the fixture, the floor cavity below the working plane, and the room cavity between the two (see Fig. 20.37). Given the surface reflectances, the effective reflectances of the floor and ceiling cavities can be obtained. With these, the CU can be selected from the tables (either Table 20.2 or manufacturer's data) and the lumen formula (equation 20.3 above) applied to arrive at average illuminance. A step-by-step explanation of the method plus illustrative examples will demonstrate the procedure. The reader should follow the steps with the flow chart in Fig. 20.38 and the calculation form in Fig. 20.39.

STEP 1. First, dimensional data are recorded. In offices, schools, and many other occupancies the work plane is 30 in. In drafting rooms it is 36 to 38 in., in shops 42 to 48 in., in carpet stores and sail-cutting rooms the work plane is at the floor level. The three "h" terms are the heights of the various cavities. In this step also, identify the initial reflectance of the room surfaces and fill in the sketch in Fig. 20.39. Similarly, establish wall and floor reflectances. Utilize the nearest reflectance to those given in Table 20.3. Interpolation between values is *not* required.

STEP 2. See Fig. 20.39. This step involves determining the cavity ratios of the room, by calculation. The basic expression for a cavity ratio (CR) is

$$\text{CR} = 2.5 \times \frac{\text{area of cavity wall}}{\text{area of work plane}} \qquad (20.5)$$

In a rectangular space the area of the cavity wall is $h \times (2l \times 2w)$ or $2h(l + w)$; therefore,

$$\text{CR} = \frac{2.5 \times 2h \times (l + w)}{\text{area of work plane}}$$

or

$$\text{CR} = 5h \times \frac{l + w}{l \times w} \qquad (20.6)$$

For other than rectangular rooms, the area can be calculated as required. For instance, in a circular room, the cavity wall area $= h \times 2\pi r$ and the work plane area is πr^2. Thus

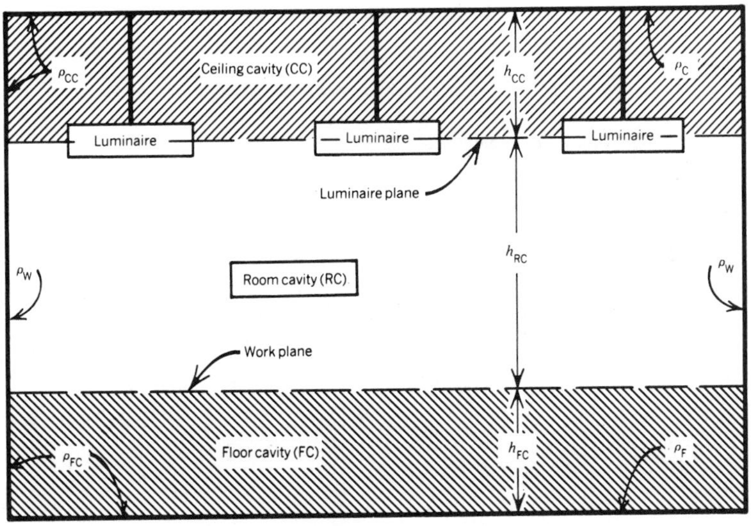

Notes:

1. Obtain effective ceiling cavity reflectance (ρ_{CC}) for combination of ceiling and wall reflectance from Table 20.3. Note that for surface-mounted or recessed luminaires, CCR = 0 and the ceiling reflectance may be used as the effective cavity reflectance. Since LLF accounts for room surface depreciation, use initial reflectances.
2. Obtain effective floor cavity reflectance (ρ_{FC}) from combination of floor and wall reflectances from Table 20.3.
3. Obtain coefficient of utilization for 20% effective floor cavity reflectance condition from table for luminaire, interpolating between tabulated values as required to match room size and ceiling and wall reflectance combinations. If effective floor cavity reflectance (ρ_{FC}) obtained in step 3 differes significantly from 20%, obtain the multiplier from Table 20.3. Multiply the coefficient of utilization by this multiplier.
4. If *initial illumination* is desired, CU selection should be obtained by above procedure and the light loss factor LLF omitted from the formula.
5. Legend:

ρ_C = ceiling reflectance
ρ_{CC} = ceiling cavity reflectance
ρ_W = wall reflectance
ρ_F = floor reflectance
ρ_{FC} = floor cavity reflectance
h = height in feet or meters
h_{RC} = height of room cavity

***Fig. 20.37** Room cavities as used in the zonal cavity method.*

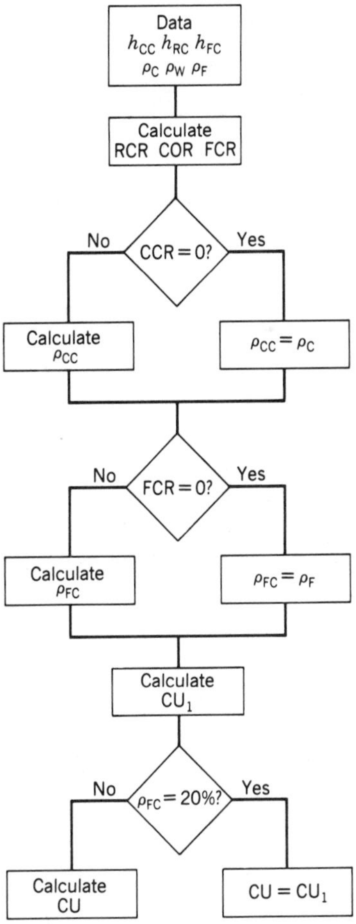

***Fig. 20.38** Zonal cavity method flowchart.*

GENERAL INFORMATION

1. Project identification: ______________________________________
   (Give name of area and/or building and room number)

2. Average maintained illumination for design: _____ lux [footcandles]

Luminaire data:
3. Manufacturer: ______________________
4. Catalog number: ______________________

Lamp data:
5. Type and color: ______________________
6. Number per luminaire: ______________________
7. Total lumens per luminaire: ______________________

SELECTION OF COEFFICIENT OF UTILIZATION

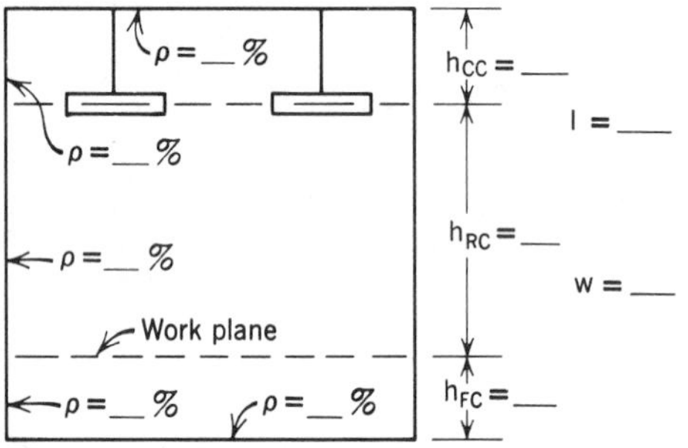

8. Step 1: Fill in sketch at right.

9. Step 2: Determine Cavity Ratios by formulas.

   9a. Room cavity ratio, RCR = ______________________
   9b. Ceiling cavity ratio, CCR = ______________________
   9c. Floor cavity ratio, FCR = ______________________

10. Step 3: Obtain effective ceiling cavity reflectance (ρ_{CC}) from Table 20.3. ρ_{CC} = ________

11. Step 4: Obtain effective floor cavity reflectance (ρ_{FC}) from Table 20.3. ρ_{FC} = ________

12. Step 5: Obtain coefficient of utilization (CU) from manufacturer's data. CU = ________

SELECTION OF LIGHT LOSS FACTORS

Unrecoverable
13. Luminaire ambient temperature ________
14. Voltage to luminaire ________
15. Luminaire surface depreciation ________
16. Other factors ________

Recoverable
17. Room surface dirt depreciation ________
18. Lamp lumen depreciation ________
19. Lamp burnouts factor ________
20. Luminaire dirt depreciation LDD ________

21. Total light loss factor, LLF (product of individual factors above): _____

CALCULATIONS

(Average maintained illumination level)

$$\text{Number of luminaires} = \frac{(\text{Illuminance}) \times (\text{Area})}{(\text{Lumens per luminaire}) \times (\text{CU}) \times (\text{LLF})}$$

22. = ______________________ =

$$\text{Lux [footcandles]} = \frac{(\text{number of luminaires}) \times (\text{lumens per luminaire}) \times (\text{CU}) \times (\text{LLF})}{(\text{area})}$$

23. = ______________________ =

24. Calculated by: ______________________ Date: __________

Fig. 20.39 *Zonal cavity method calculation form. (Courtesy of IESNA.)*

TABLE 20.3 **Percent Effective Ceiling or Floor Cavity Reflectance (ρ_{CC}, ρ_{FC}) for Various Reflectance Combinations**

Percent Ceiling ρ_C or Floor Reflectance ρ_F:	*90*				*80*				*70*			*50*			*30*				*10*		
Percent Wall Reflectance ρ_W:	*90*	*70*	*50*	*30*	*80*	*70*	*50*	*30*	*70*	*50*	*30*	*70*	*50*	*30*	*65*	*50*	*30*	*10*	*50*	*30*	*10*
Ceiling or Floor Cavity Rates CCR or FCR																					
0	90	90	90	90	80	80	80	80	70	70	70	50	50	50	30	30	30	30	10	10	10
0.2	89	88	86	85	79	78	77	76	68	67	66	49	48	47	30	29	29	28	10	10	9
0.4	88	86	83	81	78	76	74	72	67	65	63	48	46	45	30	29	27	26	11	10	9
0.6	88	84	80	76	77	75	71	68	65	62	59	47	45	43	29	28	26	25	11	10	9
0.8	87	82	77	73	75	73	69	65	64	60	56	47	43	41	29	27	25	23	11	10	8
1.0	86	80	74	69	74	71	66	61	63	58	53	46	42	39	29	27	24	22	11	9	8
1.2	86	78	72	65	73	70	64	58	61	56	50	45	41	37	29	26	23	20	12	9	7
1.4	85	77	69	62	72	68	62	55	60	54	48	45	40	35	28	26	22	19	12	9	7
1.6	85	75	66	59	71	67	60	53	59	52	45	44	39	33	28	25	21	18	12	9	7
1.8	84	73	64	56	70	65	58	50	57	50	43	43	37	32	28	25	21	17	12	9	6
2.0	83	72	62	53	69	64	56	48	56	48	41	43	37	30	28	24	20	16	12	9	6
2.2	83	70	60	51	68	63	54	45	55	46	39	42	36	29	28	24	19	15	13	9	6
2.4	82	68	58	48	67	61	52	43	54	45	37	42	35	27	28	24	19	14	13	9	6
2.6	82	67	56	46	66	60	50	41	53	43	35	41	34	26	27	23	18	13	13	9	5
2.8	81	66	54	44	66	59	48	39	52	42	33	41	33	25	27	23	18	13	13	9	5
3.0	81	64	52	42	65	58	47	38	51	40	32	40	32	24	27	22	17	12	13	8	5
3.5	79	61	48	37	63	55	43	33	48	38	29	39	30	22	26	22	16	11	13	8	5
4.0	78	58	44	33	61	52	40	30	46	35	26	38	29	20	26	21	15	9	13	8	4
4.5	77	55	41	30	59	50	37	27	45	33	24	37	27	19	25	20	14	8	14	8	4
5.0	76	53	38	27	57	48	35	25	43	32	22	36	26	17	25	19	13	7	14	8	4

Extracted from *IES Handbook;* reprinted with permission; for more complete data see *IES Handbook* (1981).

$$CR = \frac{2.5 \times h \times 2\pi r}{\pi r^2} = \frac{5h}{r} \quad (20.7)$$

For each of the cavities in a rectangular room we have:

Room Cavity Ratio

$$RCR = 5h_{RC} \frac{l + w}{l \times w} \quad (20.8)$$

Ceiling Cavity Ratio

$$CCR = 5h_{CC} \frac{l + w}{l \times w} \quad (20.9)$$

Floor Cavity Ratio

$$FCR = 5h_{FC} \frac{l + w}{l \times w} \quad (20.10)$$

For reference, since all cavity ratio figures are related, having determined one, the others are

$$CCR = RCR \frac{h_{CC}}{h_{RC}} \quad (20.11)$$

$$FCR = RCR \frac{h_{FC}}{h_{RC}} \quad (20.12)$$

and

$$CCR = FCR \frac{h_{CC}}{h_{FC}} \quad (20.13)$$

STEP 3. See Table 20.3 and Figs. 20.38 and 20.39. This step involves obtaining the effective ceiling reflectance (ρ_{CC}) from Table 20.3. Note that the wall reflectance remains as selected in step 1. If the fixtures are surface mounted or recessed, then CCR = 0 and ρ_{CC} = selected ceiling reflectance.

STEP 4. See Table 20.3 and Figs. 20.38 and 20.39. This step involves obtaining the effective floor reflectance ρ_{FC} as above in step 3, for ρ_{CC}. If the floor is the working plane, FCR = 0 and ρ_{FC} = selected floor reflectance.

STEP 5. Select CU from manufacturer's data. Note that interpolation may be necessary for ceiling cavity reflectance (ρ_{CC}), which is between the figures in the CU table. See Example 20.1 in the next section. Also, factors for CU correction for ρ_{FC} other than 20% (standard in CU tables) are given in Table 20.4.

TABLE 20.4[a] **Factors for Effective Floor Cavity Reflectances Other Than 20% (Any Wall Reflectance)[b]**

For 30% effective floor cavity reflectance, *multiply* by appropriate factor below.
For 10% effective floor cavity reflectance, *divide* by appropriate factor below.

Room cavity ratio	*Percent Effective Ceiling Cavity Reflectance, ρ_{CC}*			
	80	*70*	*50*	*10*
1	1.08	1.06	1.04	1.01
2	1.06	1.05	1.03	1.01
3	1.04	1.04	1.03	1.01
4	1.03	1.03	1.02	1.01
5	1.03	1.02	1.02	1.01
6	1.02	1.02	1.02	1.01
7	1.02	1.02	1.01	1.01
8	1.02	1.02	1.01	1.01
9	1.01	1.01	1.01	1.01
10	1.01	1.01	1.01	1.01

[a]Extracted from *IES Handbook* (1981), and reprinted with permission.
[b]For more precise data, for varying ρ_W, see *IES Handbook* (1981).

STEP 6. Calculate illuminance and number of fixtures or area per luminaire in the usual fashion.

Illustrative examples and shortcut methods are demonstrated below. CU coefficients are listed in Table 20.2.

20.33 Zonal Cavity Calculations: Illustrative Examples

EXAMPLE 20.1. It is suggested that the reader photocopy Fig. 20.39 and fill it in as the solution progresses.

Given. Classroom: 6 m × 8 m × 3.70 m height, elementary school. Initial reflectances: ceiling 80%, entire wall 50%, floor 20%. (Note that the sketch on Fig. 20.39 can accommodate different reflectances for the upper, center, and lower wall sections.) Provide adequate lighting using fluorescent lamp fixtures. Assume yearly maintenance, lamp replacement at burnout, proper voltage and ballasts, medium clean atmosphere.

SOLUTION

(a) Illuminance Level. Refer to Table 18.5. Depending on the class grade, the illuminance category could be either D or E. We will use E, which gives an illuminance range of 500–750–1000 lux. In Table 18.6 under category E we find: age—under 40, speed/accuracy are important, and background is over 70% (white paper). This yields a recommendation of 500 lux. Note that the British IES recommendation in Table 18.3 also gives 500 lux. The daylight contribution in much of the space is normally considerable in schools because of the hours of use. It will frequently exceed this 500-lux level, but to demonstrate the calculation procedure, we must assume an absence of daylight. Lines 1, 2, and 8 of Fig. 20.39 and the room sketch can be filled in.

(b) Luminaire Selection. The criteria we have previously developed for a classroom situation requires an installation that will yield

1. Low direct glare (high VCP), because students spend a large proportion of their time in a heads-up position.
2. Low veiling reflections because much of the seeing task involves high reflectance materials, occasionally specular.
3. High efficiency and low energy use to meet IES/ASHRAE and most governmental requirements.
4. Minimum required maintenance, in view of the poor cleaning and maintenance situation that obtains in many schools.

Although ceiling height is sufficient to permit use of indirect lighting (e.g., luminaire 33 in Table 20.2) with all its distinct advantages, it is not chosen, because

5. The luminaire maintenance category is bad, and given the type of maintenance expected, light reduction would be serious.
6. Indirect lighting depends on a highly reflective ceiling, requiring yearly cleaning and repainting at intervals not exceeding 5 years. This is not generally the case in public schools.

As a result we select a *two*-lamp version of luminaire no. 44 from Table 20.2. This unit is a parabolic aluminum reflector with louvres, and exhibits a 45° cutoff and crosswise batwing distribution. (Figures 20.28 and 20.29 are units of this design.) This unit meets requirements 1 through 4, above.

Although its distribution curve (see Table 20.2) shows no upward component, most commercial units of this basic design do have slots in the top of the reflector and show 5 to 10% uplight. We would select such a unit. This avoids an excessively dark ceiling and undesirable luminance ratios between fixture and background. Furthermore, fixtures with top slots stay cleaner because of the upward air movement through the fixture caused by the warm lamps and ballast. Mounting height should be about 270 cm [~9 ft] to permit easy maintenance, and good row spacing. For this luminaire the recommended maximum *S/MH* is between 1.5 and 2.0. Work plane height is 75 cm [30 in.].

(c) Calculations

STEP 1. The required data that should appear on the sketch in Fig. 20.39 are

$$h_{CC} = 1.0 \text{ m} \quad h_{RC} = 1.95 \text{ m}$$
$$h_{FC} = 0.75 \text{ m} \quad l = 8 \text{ m}$$

$$\rho_C = 80\% \quad \rho_w = 50\%$$
$$\rho_F = 20\% \quad w = 6 \text{ m}$$

STEP 2. From equations 20.8, 20.9, and 20.10

$$\text{RCR} = 5h_{RC}\frac{l + w}{l \times w} = 2.84$$

$$\text{CCR} = 5\,(1)(0.29) = 1.46$$

$$\text{FCR} = 5\,(0.75)(0.29) = 1.09$$

STEP 3. From Table 20.3 obtain effective reflectances:
For ρ_{CC} use $\rho_C = 0.8$, $\rho_w = 0.5$, and CCR = 1.46. Therefore,

$$\rho_{CC} = 0.61$$

STEP 4. For ρ_{FC} use $\rho_F = 0.2$, $\rho_w = 0.5$, and FCR = 1.09. Therefore,

ILLUMINATION

$\rho_{FC} = 0.18$ by interpolation

STEP 5. Interpolation between RCR of 2.8 and 3.0 is necessary. The coefficient of utilization for the selected luminaire—no. 44 of Table 20.2—can now be obtained by double interpolation. Required CU = 0.52. No correction from

$\rho_W = 0.50$		CU	
$\rho_{CC} \rightarrow$	0.70	0.61	0.50
RCR			
2.	0.58		0.55
2.84		?	
3.	0.52		0.50

Table 20.4 is required, as ρ_{FC} is close to 20%. At this stage lines 9 through 12 of Fig. 20.39 can be filled in.

STEP 6. The light loss factor (see Section 20.31 above) results from establishing items 13 to 20 of Fig. 20.39. These are:

Items 13–16	0.9
Item 17	0.95
Item 18	0.85
Item 19	0.95
Item 20	0.80

Item 21 : LLF =

$$(0.9)(0.95)(0.85)(0.95)(0.80) = 0.55$$

STEP 7. *Lumen calculation.* A typical classroom is normally divided into a large student seating area and a teacher's area. The illuminance requirement is approximately the same for both so that the room can be treated as a single unit for visual task purposes. A good lighting design will include chalkboard lighting, the spill light from which is usually sufficient to illuminate the very front of the room. Treating the entire space as one visual task area, and assuming two nominal 40-W cool white fluorescent lamps per fixture, the number N of luminaires required is, from equation 20.3,

$$N = \frac{(500 \text{ lux})(6 \times 8 \text{ m})}{2(3200 \text{ lm})(0.52)(0.55)} = 13.11$$

The remaining lines of Fig. 20.39 can now be completed.

Refer now to Fig. 20.40 for the layout. Two lengthwise rows of fixtures, spaced 300 cm apart, will give excellent coverage, as shown. The resultant spacing-to-mounting height ratio is only 1.1 (300 spacing/270 mounting height), well below the maximum of 1.5 to 2.0. The sixth fixture in the outside (window) row provides illumination for the teacher's desk. Two *single-lamp* batwing distribution or asymmetric distribution luminaires provide chalkboard and front-of-room lighting. The window-wall row is separately switched and is further from the wall than the inside row, due to the usual presence of

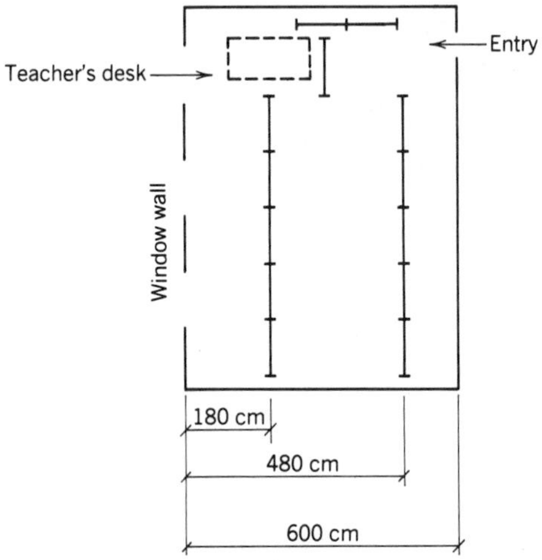

Fig. 20.40 *Layout of pendant parabolic aluminum reflector luminaires in a typical classroom. The units have a modified batwing distribution in the crosswise direction, which mandates their being hung with their long axis parallel to the line of sight, as shown. Further, the fixtures have lower brightness in that direction ie VCP for students in the head-up position is excellent. Note that the distance between the outside fixture row and the window wall is more than one-half of the side-to-side spacing because of daylight contribution during school hours, whereas the inside fixture row to inside wall spacing is less than one-half of the side-to-side spacing in order to maintain sufficient illuminance near that wall. The single fixture to the right of the teacher's desk (to the teacher's left) was so placed to avoid the veiling reflections that would result from a fixture directly above the desk, and also to take advantage of the fixture's transverse batwing distribution characteristic.*

daylight. Point-by-point illuminance including raw lux and ESI lux can be computed by one of the available computer programs. Daylight contribution can also be included. Actual *average* illuminance in the room would be simply the ratio of the number of fixtures used to the number required (calculated), times the target illuminance:

$$E = \frac{11 + \frac{1}{2} + \frac{1}{2}}{13.11}(500 \text{ lux}) \simeq 450 \text{ lux}$$

assuming similar characteristics for the one- and two-lamp fixtures.

EXAMPLE 20.2. Large modern business office, 60 × 100 × 8 ft hung ceiling height. Initial reflectances are 0.80, 0.50, 0.30. Provide general lighting using standard 4-ft fluorescent lamps with high-efficiency ballasts, in recessed troffer luminaires. The space is fully air conditioned. Lamps are replaced on a burnout basis and the fixture is then cleaned.

SOLUTION. From Tables 18.3 and 18.5 we obtain a recommended target illuminance for routine office tasks, of 500 lux. Since almost every desk in a modern business office is equipped with a visual display terminal (VDT), we must select a fixture that is designed to minimize the extremely disturbing reflection of luminaires in VDT screens. (See Section 21.20 for a full discussion of this ubiquitous problem.) Luminaires with parabolic diffusers, reflectors, and/or baffles, which exhibit very low high angle luminance, meet this requirement, as explained in Section 20.21. A second consideration is that in an office of this size, which may well be "landscaped" using half-height partitions, viewing directions will vary greatly, obviating any imperatives inherent in a luminaire's directionality. As a result we selected a specular parabolic baffled 2 ft square unit, which is directionally architecturally neutral and displays the required low-level high-angle luminance. The unit's characteristics are given in Fig. 20.41, when equipped with two 31-W, T8, U-shaped lamps (2800 lumens output; see Section 19.17*c*). Note that the luminaire exhibits a batwing intensity distribution in the transverse direction. This characteristic will be considered in deciding the fixture layout below.

Calculations. The reader should fill in a copy of Fig. 20.39 as we proceed. Working plane is taken as desktop height (i.e., 30 in.).

STEP 1. $h_{CC} = 0$; $h_{RC} = 5.5$; $h_{FC} = 2.5$

STEP 2. CCR = 0; RCR = 0.73; FCR = 0.33

STEP 3. $\rho_w = 50\%$; $\rho_{CC} = 80\%$ (for a recessed fixture we use the ceiling reflectance).

STEP 4. From Table 20.3, $\rho_{FC} = 29\%$.

STEP 5. By extrapolation, CU for $\rho_{FC} = 20\%$:

RCR	*CU*	
0.73	?	
1.0	0.57	CU = 0.59
2.0	0.50	

STEP 6. From Table 20.4, multiplier for $\rho_{FC} = 30\%$ (or 29%) is 1.085.

STEP 7. Final CU = 1.085(0.59) = 0.64.

STEP 8. Light loss factors, per Fig. 20.39:

1.1	0.92	
1.0	0.85	
0.9	0.95	LLF = 0.59
1.0	0.88	

STEP 9.

$$\text{Area/luminaire} = \frac{\text{lamp} \times \text{lumens/lamp} \times \text{CU} \times \text{LLF}}{\text{illuminance}}$$

$$= \frac{2(2800)(0.64)(0.59) \text{ lumens}}{500/10.76 \text{ fc}}$$

$$= 45.5 \text{ ft}^2$$

With a mounting height of 5.5 ft above the working plane, the spacing criteria of 1.8 transverse and 1.2 parallel yields maxima of 9.9 ft transverse and 6.6 ft parallel, on fixture centerlines. We would test grids of 9 × 5 ft and 8 × 6 ft on a computer to see which yielded better results for uniformity of illumination and lower reflected glare (i.e., higher ESI illuminance in *all* directions, since, as noted above, viewing directions

Photometrics

Luminaire type: RIK-D/REK-D 2/31-U

ZUMTOBEL

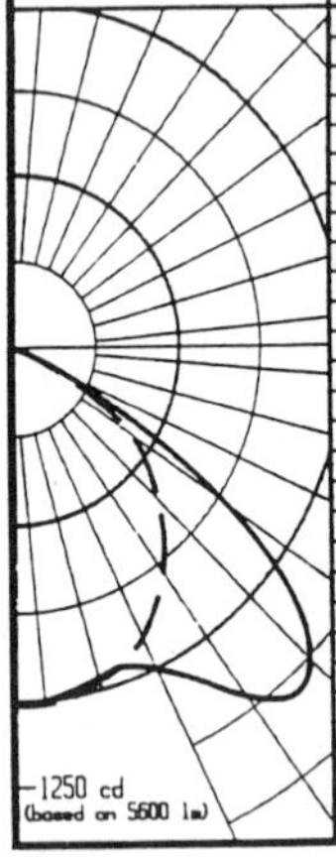

DEG.	Parl.	22.5	45	67.5	Norm.	Flux
0	1017	1017	1017	1017	1017	
5	1018	1017	1012	1011	1007	96
10	1008	1004	996	995	990	
15	982	978	966	975	969	275
20	952	945	938	964	960	
25	908	901	916	995	1003	438
30	847	852	923	1063	1085	
35	780	787	956	1166	1209	612
40	704	738	1002	1245	1300	
45	615	702	1024	1218	1261	733
50	531	672	933	1057	1096	
55	440	560	680	749	784	559
60	278	315	367	378	410	
65	94	92	72	102	145	136
70	10	15	8	19	16	
75	6	6	5	7	8	8
80	4	4	4	4	6	
85	3	3	2	3	4	3
90	2	2	2	2	3	
100	0	0	0	0	0	0
110	0	0	0	0	0	0
120	0	0	0	0	0	0
130	0	0	0	0	0	0
140	0	0	0	0	0	0
150	0	0	0	0	0	0
160	0	0	0	0	0	0
170	0	0	0	0	0	0
180	0	0	0	0	0	0

ZONAL SUMMARY				AVG [cd/m²]		
ZONE	LUMENS	LAMP	FIXT	DEG.	NORM.	PARL.
0 - 30	808	14.4	28.3	0	3022	3022
0 - 40	1420	25.4	49.7	45	3795	1852
0 - 50	2153	38.4	75.3	55	2375	1331
0 - 90	2859	51.0	100.0	65	443	288
90 -180	1	0.0	0.0	75	25	19
0 -180	2859	51.1	100.0	80	17	13

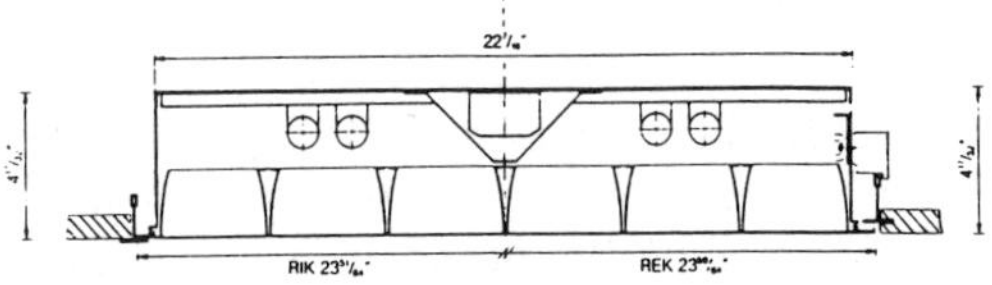

COEFFICIENTS OF UTILIZATION

LIGHT OUTPUT RATIO : 0.51

REFL. FACT. CEILING	80			70			50			30			10			0
WALLS	50	30	10	50	30	10	50	30	10	50	30	10	50	30	10	0
FLOOR	20	20	20	20	20	20	20	20	20	20	20	20	20	20	20	20
ROOM CAVITY RATIO - RCR 1	57	55	54	55	54	53	53	52	52	51	51	50	50	49	49	48
2	50	48	46	49	47	45	48	46	44	46	44	43	45	43	42	41
3	45	42	40	45	42	39	43	41	39	42	40	38	41	39	37	37
4	41	37	34	40	37	34	39	36	33	38	35	33	37	34	33	32
5	36	32	29	36	32	29	35	31	29	34	31	29	33	30	28	27
6	32	28	25	32	28	25	31	27	25	30	27	25	29	27	24	24
7	29	25	22	28	25	22	28	24	22	27	24	21	26	23	21	20
8	26	22	19	26	22	19	25	21	19	24	21	19	24	21	19	18
9	24	20	17	23	19	17	23	19	17	22	19	17	22	19	17	16
10	21	17	15	21	17	15	20	17	15	20	17	15	20	17	14	14

UNIFORMITY − $I_{min}:\bar{I}$ = 1:...

Transverse Spacing: MH×	1.1	1.2	1.3	1.4	1.5	1.6	1.8	2.0	2.2
Axial-Spacing MH × 0.7 :	1.3	1.4	1.4	1.4	1.4	1.3	1.3	1.2	1.3
MH × 1.0 :	1.3	1.3	1.3	1.3	1.3	1.3	1.3	1.2	1.2
MH × 1.5 :	1.5	1.5	1.5	1.5	1.5	1.5	1.4	1.3	1.3

S/MH NORM = 1.8 S/MH PARL = 1.2

Fig. 20.41 *Complete photometric data on a 2-ft² fixture with parabolic large square cell parabolic louvre. The unit uses 2-T12 U-shaped 31-W fluorescent lamps. Note that the transverse distribution characteristic has the batwing shape typical of this type of louvre. The louvre's specular finish yield a very low brightness at high angles, as can be seen from both the distribution curve and the luminance (cd/m²) data. Note also that the manufacturer supplies a table of uniformity versus spacing which assists the designer in deciding on a spacing grid in large installations. (Courtesy of Zumtobel.)*

will probably vary). In the absence of a computer program we would select the 9 × 6 grid to take advantage of the batwing characteristic. Note the important fact that a level of 500 lux (50 fc) of high-quality illumination for VDT areas is achieved with a power density of less than 1.6 W/ft² (17.54 W/m²).

20.34 Zonal Cavity Calculation by Approximation*

Although the foregoing zonal cavity calculations are straightforward and essentially simple, they can become tedious if more than a few areas are involved. Two alternatives exist: to utilize one of the many computer programs available or simply to do the calculations by approximations. Computer assistance is discussed and demonstrated in Section 20.37; an effective computational method using approximations is demonstrated in this section.

For a first approximation, make the following assumptions:

1. ρ_{CC} = 80 for a large room i.e. larger than 30 × 30 ft

   ρ_{CC} = 70 for a medium room i.e. between 12 × 12 ft and 30 by 30 ft

   ρ_{CC} = 60 for a small room less than 12 × 12 ft
2. ρ_{FC} = 20
3. Assume all rooms are square. To do this for a rectangle, take one-third the differ-

* Based on a method developed by B. F. Jones.

ence in dimensions and add to the smaller dimension to obtain equivalent width w. Then, for square rooms

$$\text{RCR} = \frac{10h_{\text{RC}}}{w}$$

4. Assume LLF = 0.65 for good conditions, 0.55 for average, and 0.45 for poor.

EXAMPLE 20.3 (by approximation). Classroom as in Example 20.1.

$\rho_{CC} = 70 \quad \rho_W = 50 \quad \rho_{FC} = 20$ (assumptions)

$$\text{RCR} = \frac{10 \times 1.95}{6 + 2/3} = \frac{19.5}{6.66} = 2.93$$

Obtain CU from Table 20.2, fixture no. 44.

$$\text{CU} = 0.52 \text{ by visual inspection}$$

$$= \frac{500(6 \times 8 \text{ m})}{2(3200)(0.52)(0.55)}$$

$$= 13.11 \text{ fixtures}$$

Thus, the result is the same as the accurate calculation. Let us check Example 20.2.

EXAMPLE 20.4 (by approximation)

$\rho_{CC} = 80 \quad \rho_W = 50 \quad \rho_{FC} = 20$

$$\text{RCR} = \frac{10 \times 5.5}{60 + 40/3} = \frac{55}{73} = 0.75$$

CU = 0.60 by inspection (mental interpolation)

LLF = 0.65

$$\text{area/luminaire} = \frac{2 \times 2800 \times 0.6 \times 0.65}{50}$$

$$= 43.7 \text{ sq ft}$$

This result is within 4% of the accurate calculation. We thus see that these simple approximations give answers sufficiently accurate for most uses and are therefore recommended.

In conclusion, then, with respect to zonal cavity calculations, we can make the following statements:

1. For everyday calculations of rectangular rooms, with assumed reflectances, use the assumptions listed above. A modified calculation form is provided in Fig. 20.42 to assist with this method.
2. For rooms where a high degree of accuracy is desired and actual reflectances are known, use the long method, with visual interpolation.
3. For rooms of unusual shape or rooms with special conditions such as coffered ceilings, mixed-material walls, and partial height partitions, use the long method, preferably with computer assistance. This should not be necessary for more than 5% of calculations.
4. For spaces in which a number of different solutions are to be tried, use a computer.

20.35 Effect of Cavity Reflectances on Illuminance

It should be obvious to every reader at this point in our study that the reflectances of the various room cavities have a marked effect on the CU, because of light reflections within the room. To demonstrate this graphically, we have plotted in Figs. 20.43, 20.44, and 20.45 the effect of varying cavity reflectances on the three principal types of fixture distribution: semi-indirect, direct-indirect, and direct-spread. Note that, as expected, the ceiling cavity reflectance has the most pronounced effect with indirect fixtures, and the floor reflectance with direct units. Since lighting costs amount to 3 to 5% of *total construction* cost for many types of buildings such as offices, a 20% differential in lighting units can amount to as much as 1% of the total cost of the facility. This amount would not only pay for the increased cost of higher reflectance finishes and materials but also would reduce both initial and operating costs. These data clearly indicate the necessity for the lighting designer to have considerable influence on the choice of room materials and finishes, a situation which unfortunately does not usually occur.

20.36 Modular Lighting Design

An increasingly large number of buildings are being designed on a modular system resulting in

Items 1–7, See Fig. 20.39.

8. Step 1: Fill in sketch at right.
9. Step 2: Determine room cavity.

9a: $w = W + \dfrac{L\text{-}W}{3}$

9b: $RCR = \dfrac{10\ hrc}{w}$

10. Step 3: Determine ceiling cavity ratio by equivalent square room size.

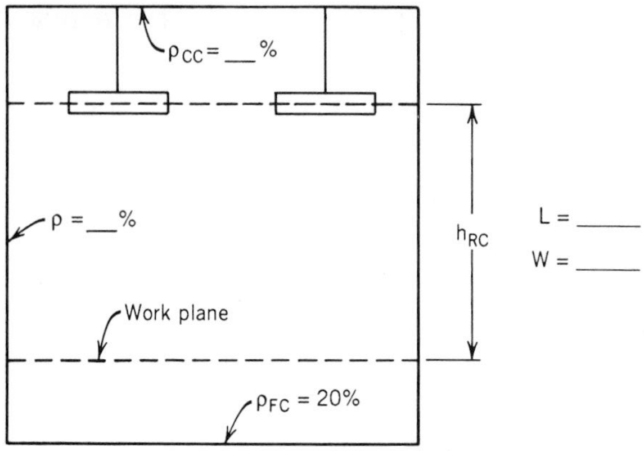

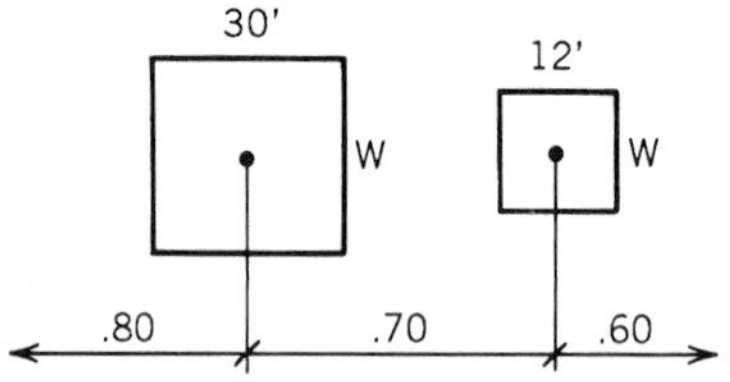

11. Step 4: Assume floor cavity reflectance ρ_{FC} = 20%.
12. Step 5: Obtain coefficient of utilization CU from Table 20.2 or from manufacturers' data: CU = ______
13. Step 6: Light loss factor LLF

Good conditions: 0.65
Average conditions: 0.55
Poor conditions: 0.45

14. Step 7

CALCULATIONS

$$\text{Number of luminaires} = \frac{\text{(Average maintained illumination level)}\ \text{(Illuminance} \times \text{(Area)}}{\text{(Lumens per luminaire)} \times \text{(CU)} \times \text{(LLF)}}$$

$$= \frac{\qquad\qquad}{\qquad\qquad} =$$

$$\text{Lux [footcandles]} = \frac{\text{(number of luminaires)} \times \text{(lumens per luminaire)} \times \text{(CU)} \times \text{(LLF)}}{\text{area m}^2\ (\text{ft}^2)}$$

Fig. 20.42 *Zonal cavity calculation by approximation.*

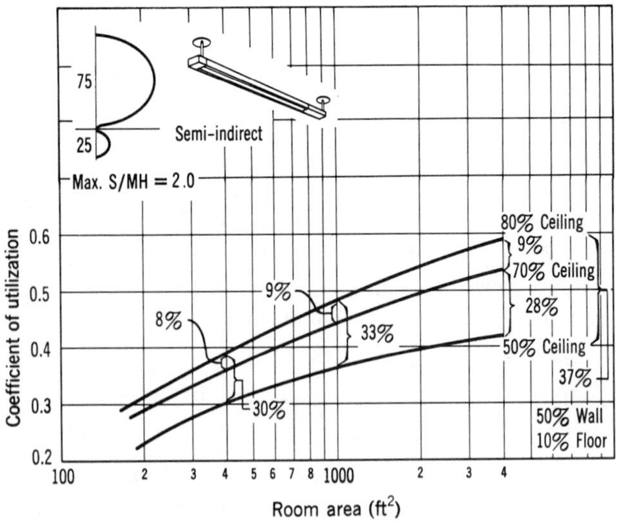

Fig. 20.43 *Effect of surface reflectances on the CU of a luminaire with semi-indirect distribution. As expected, since the ceiling becomes the light source, its reflectance has the most pronounced effect. With this particular unit having a 25% downward component, floor finish also has an appreciable effect, increasing CU by an average 10% for a 30% reflectance floor. The effect of wall reflectance naturally increases as rooms become smaller and the proportion of wall surface is larger. The change in CU between a 30% and a 50% reflectance wall varies from 15% for a 400-ft² room to 5% for a 4000-ft² room.*

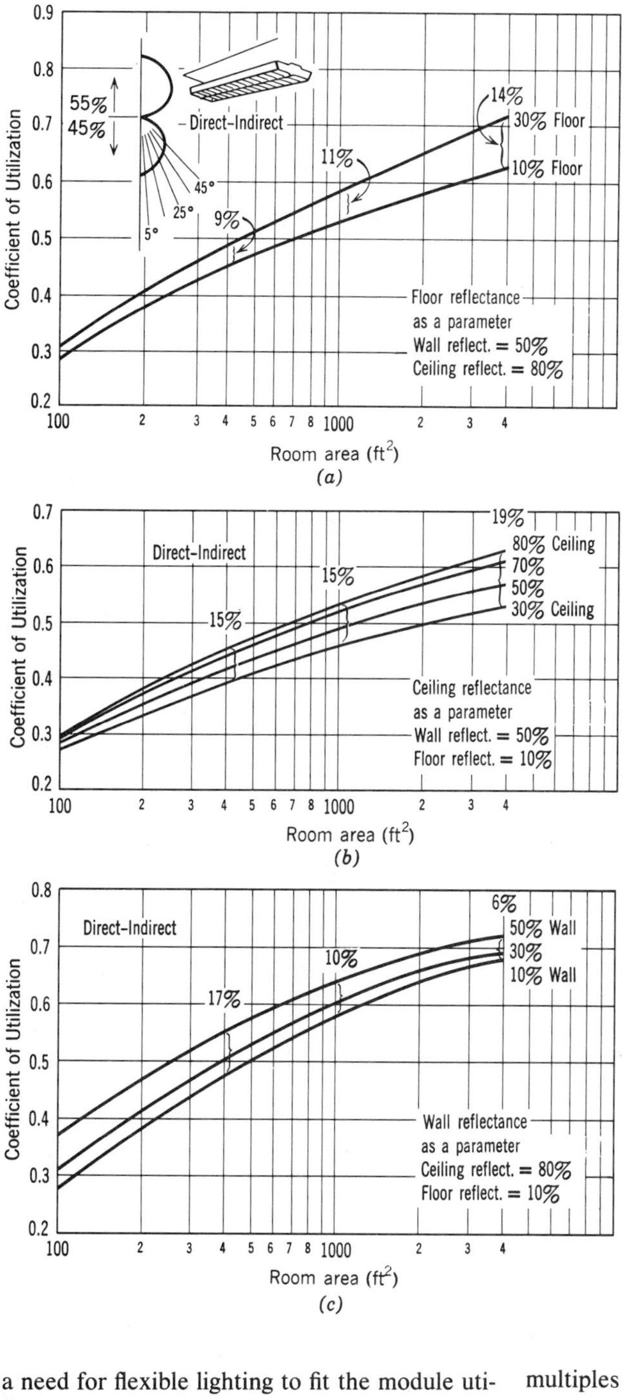

Fig. 20.44 *Effect of surface reflectances on the CU of a luminaire with direct-indirect distribution. With this distribution, the effects of ceiling and floor are most pronounced, with a wall effect only in small rooms.*

a need for flexible lighting to fit the module utilized. In such buildings, once the general lighting scheme and the fixture involved are established, it is convenient to draw a family of curves for the fixture chosen, thereby facilitating the utilization of the modular unit in various spaces. "Area" may readily be replaced with multiples of modular areas, as shown in Fig. 20.46.

20.37 Calculating Illuminance at a Point

The lumen (flux) method of horizontal illuminance calculation explained above is appropri-

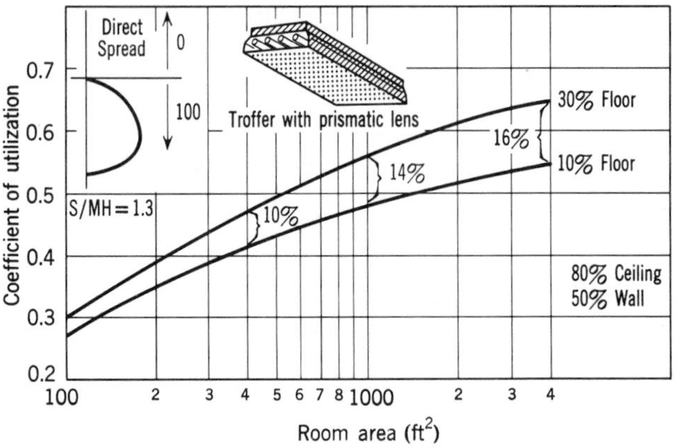

Fig. 20.45 Effect of surface reflectances on the CU of a luminaire with direct (spread) distribution. Floor finish is most important, with wall reflectance important only in small rooms. Since these fixtures have no upward component, all ceiling illumination is derived from reflection. Thus in a room with floor reflectance less than 20%, ceiling finish has no effect on room illumination.

ate for spaces in which illumination is essentially uniform throughout. However, even in such a space, illuminance varies at least ±10%, and near columns, walls, windows, bookcases, and the like, considerably more (see Fig. 20.53). Therefore, to answer the often asked question "How much light will I have on my desk?", the designer must turn to other methods. Three are available:

1. Calculation of illuminance at selected points by computer, as explained and demonstrated in Section 20.42.
2. Utilization of one of the design aids explained in the following section.
3. Longhand calculation by one of the methods presented in Sections 20.39 to 20.41.

These methods will also yield results where the lumen method is simply inapplicable. Among these applications are layouts that are intentionally nonuniform, calculation of illumination on planes other than horizontal (e.g., wallwashers), calculation of illuminance resulting from architectural lighting elements such as coves, valances, and the like, and illuminance calculations for nonstandard light sources for which coefficient data of the type given in Table 20.2 are not available.

20.38 Design Aids

By this term we mean any of the various curves, charts, plots, or tables either prepared by the designer or made available by luminaire manufacturers, the purpose of which is to simplify and speed lighting design when using that particular lighting fixture. The reliability of data so obtained depends entirely on that of the manufacturer involved, and their use should be governed accordingly. More recently, it has become customary for major manufacturers to provide

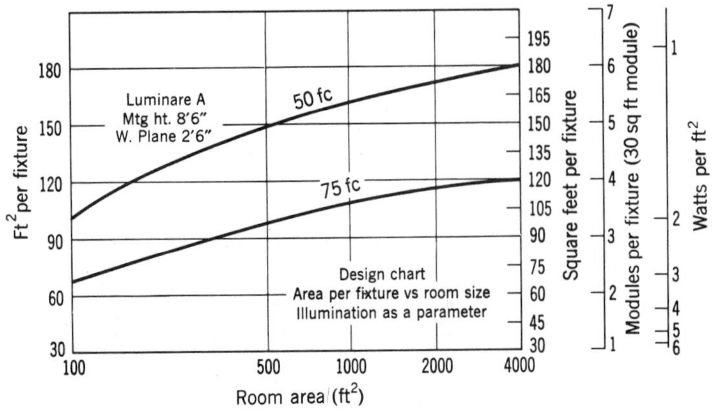

Fig. 20.46 Luminaire design chart. For frequently used fixtures, this type of chart gives a rapid design figure for various size rooms. As seen from the ordinates, the figures can be translated into number of modules and watts per square foot.

computer output charts and tables based on the designers' proposed layout(s). When using these, it behooves the designer to study the data carefully in the computer program being used, since the fixture supplier is certainly not a disinterested party, and the program may be "weighted" accordingly. A brief description of the common design aids follows.

(a) Isolux Charts. These charts, also called isofootcandle charts, are based on the type traditionally supplied by manufacturers of outdoor lighting equipment such as street lights and floodlights, but equally applicable to interior lighting. Their use is illustrated in Fig. 20.47. The basic tool is an isolux diagram for a single luminaire. This is either calculated (see Sections 20.39 to 20.42), measured from a full-scale mockup (the most accurate if not usually the most practical method), or obtained from the manufacturer. Since the relative positions of source and illuminated point are reversible, that is, if a source at *A* causes illuminance *E* at point *B*, then the source at *B* will cause the same illuminance *E* at point *A*, placing the center of the isolux chart at the point in question will permit direct reading of the illumination contribution of every other luminaire. It then remains simply to sum the individual contributions to obtain the (scalar) illumination at the desired point. An example is shown on Fig. 20.47.

(b) Illuminance "Cone" Charts. See Fig. 20.48. When the light distribution of a direct downlight is symmetrical, as is generally the case with recessed incandescent and HID downlights, a cone can be drawn showing illuminance directly under the fixture at various distances. The projected circles are defined by maximum illuminance at the center, and half of

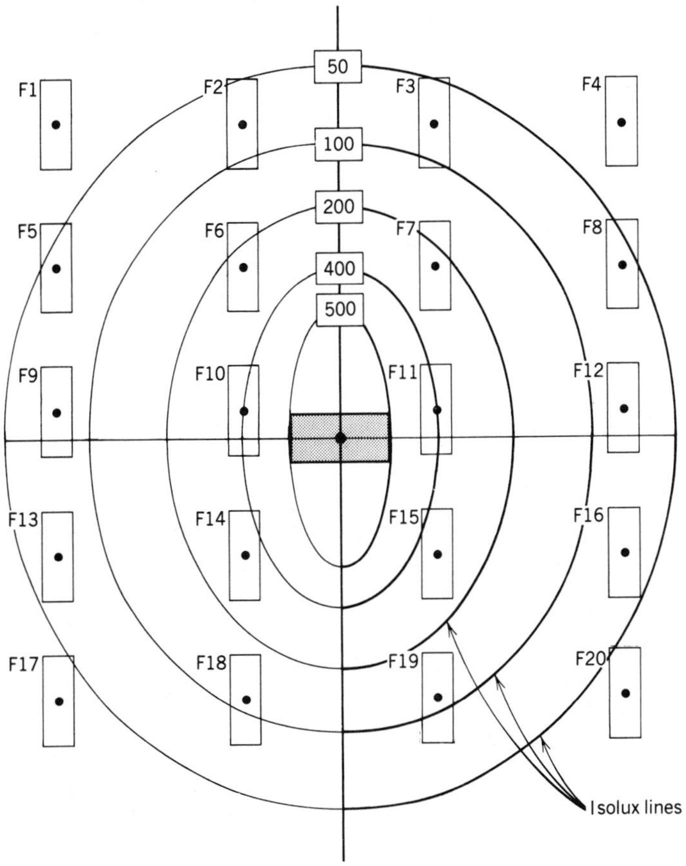

Luminaire	Contribution (Lux)
F2, F3	75 × 2
F5, F8	50 × 2
F6, F7	225 × 2
F9, F12	75 × 2
F10, F11	400 × 2
F13, F16	60 × 2
F14, F15	300 × 2
F18, F19	100 × 2
	2570 lux total

Fig. 20.47 *The ellipses represent isolux lines for a single luminaire at a given height above the work plane. They are centered on the point (the work area of a desk) for which it is desired to determine the illuminance. The total illuminance is the sum of the individual luminaire contributions. The center of the luminaire is the point of reference. Therefore, when two or more isolux lines pass through a fixture, its contribution is determined by the interpolated isolux line passing through its center. Note the symmetry around the vertical axis, necessitating a plot of only half the ellipses.*

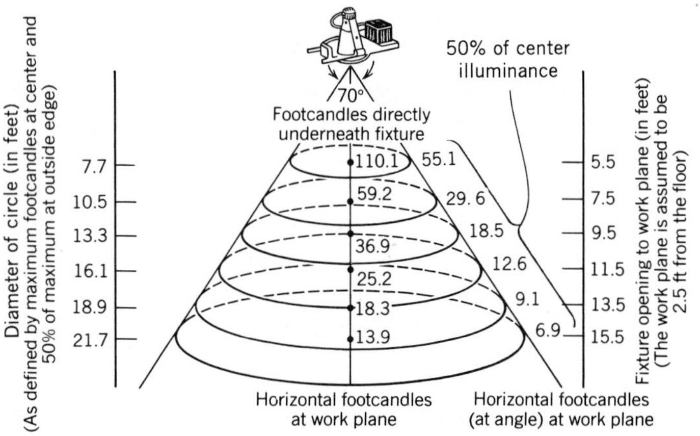

Fig. 20.48 *For downlights with symmetrical circular distribution, a "cone of light" as shown can be drawn, giving the illuminance directly below the luminaire, and a circle at the circumference of which illuminance is half this maximum. For more accurate work, additional horizontal circles can be added. (Courtesy of Halo Lighting.)*

this illuminance at the edge. This projected circle can be used in the same fashion as the isolux chart in the preceding paragraph, except that only two values are given—that at the center and that on the circumference.

(c) Illuminance Tables and Charts. These take various forms but all give specific illuminance data, in numerical form, for specific points. The values are usually obtained from computer printout and may be raw or ESI illuminance, or both.

20.39 Calculating Illuminance from a Point Source

It is well to note at the outset that all the methods discussed below calculate illumination at *a point*. The answer to the constant query, "How much light will I have on my desk from a luminaire at this location?" is arrived at by taking several points on the desk and calculating illumination at each one. Shortcuts can be made for symmetry and so on. The reader should also understand that most of these techniques are laborious and are generally performed by the consulting engineer or lighting specialist rather than the architect. They are presented here as source material for the technically oriented reader.

The basis of point source calculations is the inverse square law developed in Section 18.12:

$$fc = \frac{cp}{D^2} \qquad (18.6)$$

where *fc*, *cp*, and *d* are footcandle illumination, candlepower intensity, and distance, respectively. Refer to Fig. 20.49. The horizontal illumination at a point *P* as shown in Fig. 20.49*a* is

$$\text{horizontal } fc_P = \frac{cp}{D^2}\cos\theta \qquad (20.14)$$

and the vertical illumination at that same point is

$$\text{vertical } fc_P = \frac{cp}{D^2}\sin\theta \qquad (20.15)$$

However, since

$$\cos\theta = \frac{H}{D} \quad \text{and} \quad \sin\theta = \frac{R}{D}$$

we have then at point *P*

$$\text{horizontal illumination} = \frac{cp}{H^2}\cos^3\theta \qquad (20.16)$$

and

$$\text{vertical illumination} = \frac{cp}{R^2}\sin^3\theta \qquad (20.17)$$

Since the candlepower intensity in the direction of point *P* is normally taken from a candlepower distribution curve, θ is normally known, and these expressions are readily usable. Very few commercial sources are actually point sources. However, *when the maximum dimension of the source is less than five times the distance to point P*, the equations will give satisfactory results. Note that these equations can be used to calculate and plot isolux diagrams for point sources, of the type shown in Figs. 20.47 and 20.48.

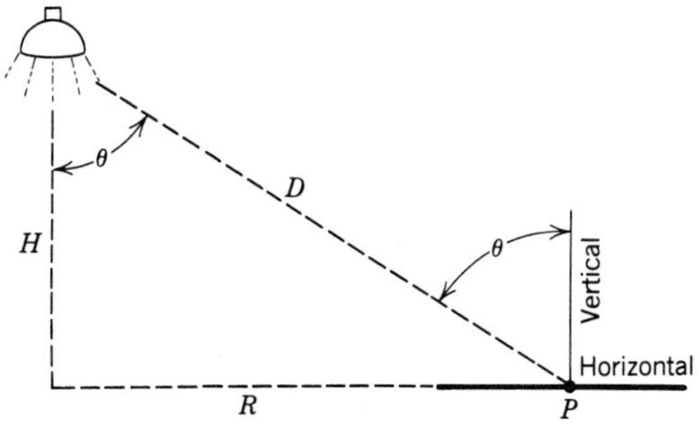

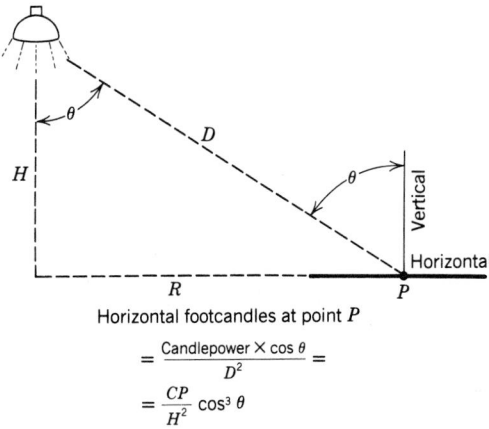

Horizontal footcandles at point P

$$= \frac{\text{Candlepower} \times \cos\theta}{D^2} = $$

$$= \frac{CP}{H^2}\cos^3\theta$$

Vertical footcandles at point P

$$= \frac{\text{Candlepower} \times \sin\theta}{D^2} = $$

$$= \frac{CP}{R^2}\sin^3\theta$$

Fig. 20.49 Relationship between intensity and illuminance when source can be considered a point source, that is, when the inverse square law applies. Source major dimension must not exceed 0.2D to be considered a point source.

EXAMPLE 20.5. Referring to Fig. 20.49 and the candlepower distribution curve of Fig. 20.50, find the horizontal and vertical illumination at point P, which is 10 ft below and 12 ft horizontally distant from the source.

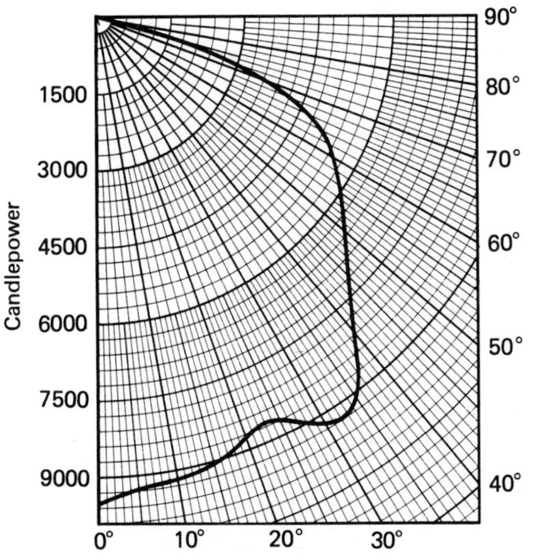

Fig. 20.50 Typical candlepower distribution plot for use in inverse square law calculation.

SOLUTION

$$H = 10 \text{ ft} \qquad R = 12 \text{ ft}$$

$$\theta = \tan^{-1}\frac{12}{10} = 50°$$

$$\sin\theta = 0.766 \qquad \cos\theta = 0.643$$

$$cp \text{ at } 50° = 6600$$

horizontal illumination =

$$\frac{6600}{10^2}(0.643)^3 = 17.5 \text{ fc}$$

$$\text{vertical illumination} = \frac{6600}{12^2}(0.766)^3 = 20.8 \text{ fc}$$

20.40 Calculating Illuminance from a Line Source

When the maximum source dimension exceeds 20% of the distance to the point illuminated, it can no longer be considered a point source. A source that is long and narrow such as a fluorescent tube can be considered as a line source. For a line source where length is much greater than width, the illumination varies *inversely with distance* as opposed to inversely with the square of the distance for a point source, that is,

$$fc = \text{constant} \times \frac{cp}{D} \qquad (20.18)$$

Thus if illumination at point P_1 is known, that at P_2 can be calculated as

$$fc_2 = fc_1 \times \frac{D_1}{D_2} \qquad (20.19)$$

When the point being considered is directly below the source and the source is very long, Equation 20.18 becomes

$$fc = \frac{fL \times W}{2D} \qquad (20.20)$$

where

fc = illumination at distance D in footcandles
fL = source brightness in footlamberts
D = distance in feet
W = width of source

When the point is not directly below the source, the equation becomes complex (see *IES Hand-*

TABLE 20.5 K_H and K_V Values

Part A. ***Horizontal*** $F_C = K_H \times$ ***Lamp Lumens per Foot***
Broad Distribution—White-Enameled Channel

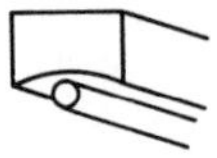

Vertical Distance from Centerline of Unit	K_H Values			
	Horizontal Distance from Centerline of Unit			
	0 ft	1 ft	2 ft	3 ft
1 ft	0.087	0.045	0.017	0.007
2 ft	0.040	0.033	0.020	0.011
3 ft	0.025	0.022	0.017	0.011
4 ft	0.018	0.015	0.014	0.009

Part B. ***Vertical*** $F_C = K_V \times$ ***Lamp Lumens per Foot***
Broad Distribution—White-Enameled Channel

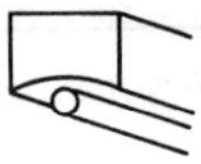

Vertical Distance from Centerline of Unit	K_V Values				
	Horizontal Distance from Centerline of Unit				
	3 in.	6 in.	9 in.	12 in.	18 in.
1 ft	0.014	0.030	0.037	0.043	0.041
2 ft	—	0.007	0.017	0.014	0.019
3 ft	—	—	—	0.005	0.008
4 ft	—	—	—	—	—

Part C. ***Horizontal*** $F_C = K_H \times$ ***Lamp Lumens per Foot***
Broad Distribution—White-Enameled Reflector

Vertical Distance from Centerline of Unit	K_H Values			
	Horizontal Distance from Centerline of Unit			
	0 ft	1 ft	2 ft	3 ft
1 ft	0.438	0.127	0.008	0.001
2 ft	0.233	0.150	0.061	0.017
3 ft	0.145	0.120	0.077	0.041
4 ft	0.106	0.095	0.072	0.048

Part D. ***Vertical*** $F_C = K_V \times$ ***Lamp Lumens per Foot***
Broad Distribution—White-Painted Cornice, No Reflector

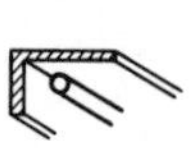

Vertical Distance from Centerline of Unit	K_V Values				
	Horizontal Distance from Centerline of Unit				
	3 in.	6 in.	9 in.	12 in.	18 in.
9 in.	0.185	0.159	0.175	0.165	0.129
1 ft, 9 in.	0.011	0.028	0.044	0.057	0.068
2 ft, 9 in.	0.004	0.010	0.017	0.023	0.032
3 ft, 9 in.	0.002	0.005	0.008	0.012	0.018

Source: Reproduced by permission of General Electric, Lighting Business Group.

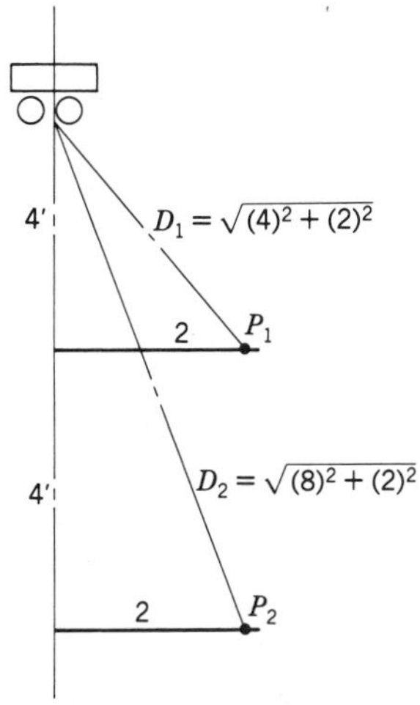

Fig. 20.51 For long linear sources, the ratio of illuminances is $E_{P_2}/E_{P_1} = D_1/D_2$.

book, 1987). When the source is very long with respect to width, the effect of longitudinal offset (in the same direction as the length of the source) diminishes and the only offset that need be considered is a lateral one—away from the axis of the source. Table 20.5 provides illumination values for this situation for several specific luminaire types. Obviously the longer the row, the more accurate the results. Lumens per foot data for fluorescent lamps are available from Table 19.16 by simply dividing lumen output by length.

EXAMPLE 20.6. Find the horizontal illumination on a line 8 ft below and 2 ft offset from a continuous row of two 800-mA HO lamps in a white-enameled reflector.

SOLUTION. Refer to Fig. 20.51 and Table 20.5, Part C. Note that the maximum vertical distance given in the table is 4 ft, whereas we require 8 ft. However, if we calculate the illuminance at 4 ft below and 2 ft offset (point P_1 in Fig. 20.51), then by equation 20.19 we can calculate the illuminance at point P_2. From Table 19.11, the output of an HO lamp is approximately 1100 lumens per foot.

Illuminance at point P_1 is

$$E_{P_1} = K \times \frac{\text{lumens}}{\text{foot}} =$$

$$0.072 \text{ (2 lamps) } (1100 \text{ lm/ft}) = 158.4 \text{ fc}$$

From Equation 20.19, the illuminance at point P_2 is related to that at P_1 inversely as their distance from the source, that is,

$$\frac{E_{P_2}}{E_{P_1}} = \frac{D_1}{D_2} = \frac{\sqrt{4_2 + 2^2}}{\sqrt{8^2 + 2^2}} = 0.542$$

and

$$E_{P_2} = 0.542\,(E_{P_1}) = 0.542(158.4 \text{ fc})$$
$$= 85.9 \text{ fc}$$

20.41 Calculating Illuminance from Linear and Area Sources

When the source is too wide to be considered a line source (the definition is relative and depends on distance to the illuminated surface), it is referred to as a linear source. The *direct component* of the illuminance at a point, resulting from such a source, or from a large area source, can be calculated by manual graphical or analytical means, both of which are based on an assumed distribution, generally lambertian (diffuse). Since most fixtures being applied today do *not* have lambertian characteristics (e.g., parabolic reflectors, prismatic diffusers), the results of such calculations are necessarily approximate. (Skylights and luminous ceilings *do* have a lambertian distribution, and there these calculation methods do give reliable results.) In addition to the direct component of illuminance, a *reflected component* must be added that depends on the point's location in the room and the room's characteristics. This calculation involves charts, diagrams, and tables. The interested reader will find a full description of these manual methods in Section 9, Reference Volume, 1987 *IES Lighting Handbook*.

However, because these manual methods are laborious, and frequently inapplicable, they are not presented here. The ready availability of desktop computers and computer programs that can handle detailed input for the specific light source involved, without broad approximations, has made these manual procedures obsolete. We recommend that when point-by-point illuminance calculation is desired, such a program be used. Alternatively, one of the design aids described above, based on a specific light source, can be used.

20.42 Computer-Aided Design

As pointed out repeatedly in previous sections, the use of a computer in lighting design is a practical nec̀essity if really useful results are to be obtained. Once the desired luminance patterns and illuminance levels have been established, and luminaires selected and located by using average illuminance (zonal cavity) calculations or by use of one of the design aids explained above and/or manufacturer's assistance, the responsible designer will confirm the preliminary design with accurate calculations. Furthermore, only a computer analysis will give useful point-by-point illuminance figures plus the necessary data on VCP and ESI for selected work locations and viewing directions. (A full-scale mockup with appropriate metering will also give these raw and ESI illuminance data. Such an alternative is obviously impractical except in major projects, and obviously much more expensive and time consuming).

In addition, the computer gives the designer a degree of flexibility not otherwise possible in that:

1. It performs the calculations accurately and rapidly.
2. It frees the designer for other, less routine work.
3. It gives the designer the ability to change parameters repeatedly, without making the amount of calculation excessive, as would be the case in hand calculation.

It is this last characteristic that gives the designer greatest flexibility. The ability to run a series of calculations for a pendant fixture with pendant lengths varying from 12 to 36 in. in 3-in. intervals, or to change paint colors and reflectance on various surfaces and to note the effect, or to test different layout patterns as mentioned above in Example 20.2, gives the designer a very powerful and urgently needed design tool. In addition, the computer can consider related items, such as first costs, energy use, operating costs, and impact on HVAC, items whose complexity because of interrelations put them well beyond the pencil-and-hand-calculator's ability. A few of the programs currently available are listed in Appendix H.

We have chosen to demonstrate the usefulness of computers in lighting design by analyzing Examples 20.1 and 20.2 with the help of a desktop microcomputer. The program used in Example 20.1 is LUMEN$ and that used in Example 20.2 is LUMEN-MICRO. Both are parts of the comprehensive LUMEN-MICRO series of sophisticated flexible lighting design programs produced by Lighting Technologies, 2540 Frontier, Suite 107, Boulder, CO 80301-2458. Both are used with permission. The computer we used is a 80386-based machine, although a standard 8086/8088 machine with mathematics co-processor will work as well, although more slowly.

(a) Computer Analysis of Example 20.1. See Fig. 20.52. Since the program was configured for English units (ft), we converted the metric measurements. The most time-consuming part of the analysis is entering the required data. The steps involved are:

1. Enter the room data on a input screen. This information includes room shape and dimensions, wall finishes, choice of luminaire, and its mounting arrangement. This will take no more than 3 to 5 min. If the program is to do a comparative analysis of lighting systems, with full economic analysis, data entry for up to four systems can take 15 to 20 min. The designer can manually select a CU from manufacturer's data after the program calculates all room ratios. If so, step 2 is unnecessary.

2. Enter fixture data into a database. This information, in the form of candela intensity photometric data as shown in Fig. 20.41, is sufficient for the Lumen$ program to calculate the luminaire's coefficient of utilization, given the room data of step 1. Once the data have been entered it will serve for all future use of this luminaire in both the Lumen$ program (zonal cavity calculation and economic analysis) and the Lumen-Micro Program, which calculates

```
                  LUMEN$ - ZONAL CAVITY ANALYSIS
                  ------------------------------
                        Meeb8 ex.20.1,run2


For:
 By:   Prof. B. Stein

         LUMEN$ © 1983 Lighting Technologies, Boulder, CO
------------------------------------------------------------------------
  Room Shape:    Rectangle                    Length:               26.30
  Reflectances:                               Width:                19.60
       Ceiling:              80.0 %           Height:               12.20
       Walls:                50.0 %
       Floor(actual):        20.0 %           Total Area:           515.5
       Floor(effective):     19.1 %           Work Plane Height:     2.50

  LUMINAIRE DESCRIPTION(S):
    #1:

                                              Parabolic
                                                Batwing
                                              ----------
                LUMINAIRE DATA:
                  Number of Lamps/Luminaire. .          2
                  Lumens/Lamp. . . . . . . . .       3200
                  Watts/Luminaire. . . . . . .         92
                  Suspension Length. . . . . .       3.30

                  Total Light Loss Factor(LLF)        .55

                COEFFICIENT OF UTILIZATION:
                  Room Cavity Ratio (RCR). . .       2.85
                  Effective Refl. of Ceiling .       60.7
                  Coefficient of Utilization .        .52

                SYSTEM PERFORMANCE:
                  Target Illuminance Level . .       46.5
                  Number of Luminaires Req'd .      13.10

                  Number of Luminaires Used. .         13
                  Average Illuminance (Maint.)       46.2

                  System Watts per Square Foot       2.32
```

Fig. 20.52 *Computer printout of zonal cavity analysis for Example 20.1. (Lumen$ © program courtesy of Lighting Technologies.)*

point illuminance and is demonstrated in (*b*) below. Entering a typical fixture's candela data will take less than 5 min.

3. Run the program. For zonal cavity calculations only, run time, including output printing, is 1 to 3 min. With a full comparative economic analysis of four different lighting systems, run time can be as much as 15 to 20 min. Hand calculation of the same type would take at least a day.

(b) Computer Analysis of Example 20.2. See Figs. 20.53 and 20.54. As explained above, we wish to know the point-by-point illuminance in the space, rather than simply the average illuminance as calculated by the zonal cavity method. This information will tell us whether and where we require supplementary lighting, whether the layout we have used yields the desired uniformity, and what the actual illuminance at walls (and partitions) will be. We are speaking here of a uniform layout. Obviously, if a nonuniform layout is intended, each fixture or group of fixtures must be considered individually and the resultant illuminance at a particular point calculated by superposition. The program

ILLUMINANCE

WORKING PLANE HEIGHT: 2.50

AVERAGE: 48.02 MINIMUM: 27.35 MAXIMUM: 55.56 MEAN DEVIATION: 4.73

ABS. Y COOR.	1.00	2.00	3.00	4.00	5.00	6.00	7.00	8.00	9.00	10.00	11.00	12.00	13.00	14.00	15.00	16.00	17.00	18.00	19.00	20.00
20.00	37.	40.	44.	47.	48.	49.	51.	53.	54.	53.	51.	50.	50.	50.	50.	51.	53.	54.	53.	51.
19.00	37.	41.	44.	47.	49.	50.	51.	53.	54.	54.	52.	51.	51.	51.	51.	52.	54.	55.	54.	52.
18.00	38.	41.	44.	47.	49.	50.	52.	54.	55.	54.	52.	51.	51.	51.	51.	52.	54.	56.	54.	52.
17.00	38.	41.	44.	47.	49.	50.	52.	54.	55.	54.	52.	51.	51.	51.	51.	52.	54.	56.	54.	52.
16.00	37.	41.	44.	47.	49.	50.	51.	53.	54.	54.	52.	51.	51.	51.	51.	52.	54.	55.	54.	52.
15.00	37.	40.	44.	47.	48.	49.	51.	53.	54.	53.	51.	50.	50.	50.	50.	51.	53.	54.	53.	51.
14.00	37.	41.	44.	47.	49.	50.	51.	53.	54.	54.	52.	51.	51.	51.	51.	52.	54.	55.	54.	52.
13.00	38.	41.	44.	47.	49.	50.	52.	54.	55.	54.	52.	51.	51.	51.	51.	52.	54.	56.	54.	52.
12.00	38.	41.	44.	47.	49.	50.	52.	54.	55.	54.	52.	51.	51.	51.	51.	52.	54.	56.	54.	52.
11.00	37.	41.	44.	47.	49.	50.	51.	53.	54.	53.	52.	51.	51.	51.	51.	52.	54.	55.	54.	52.
10.00	37.	40.	44.	46.	48.	49.	51.	53.	54.	53.	51.	50.	50.	50.	50.	51.	53.	54.	53.	51.
9.00	37.	40.	44.	47.	48.	49.	51.	53.	54.	53.	51.	51.	51.	51.	51.	51.	53.	54.	53.	51.
8.00	38.	41.	44.	47.	49.	50.	51.	54.	55.	54.	52.	51.	51.	51.	51.	52.	54.	55.	54.	52.
7.00	37.	40.	44.	47.	48.	49.	51.	53.	55.	53.	51.	50.	50.	50.	50.	51.	53.	55.	53.	51.
6.00	36.	39.	43.	45.	47.	48.	49.	52.	53.	52.	50.	49.	49.	49.	49.	50.	52.	53.	52.	50.
5.00	35.	38.	41.	44.	46.	47.	48.	50.	51.	50.	48.	48.	48.	48.	48.	48.	50.	51.	50.	48.
4.00	34.	37.	40.	43.	45.	45.	46.	48.	49.	48.	47.	46.	47.	47.	46.	47.	48.	49.	48.	47.
3.00	33.	36.	39.	41.	43.	43.	44.	46.	47.	46.	45.	44.	45.	45.	44.	45.	46.	47.	46.	45.
2.00	31.	33.	36.	39.	40.	40.	41.	42.	43.	43.	41.	41.	41.	41.	41.	41.	43.	43.	43.	41.
1.00	27.	29.	32.	34.	35.	35.	36.	37.	38.	37.	36.	36.	37.	37.	36.	36.	37.	38.	37.	36.

Desk — Desk — N — 5' — 5' — 2.5' — 4.5' — 9' — 9' — Wall — Wall — ℄ — ℄

```
***********
LUMEN-MICRO
***********
```

VERSION 4.1

meeb ex20.2 run5

DATE: 8/21/1990

(*a*)

Fig. 20.53 (a) *Computer printout of "raw" illuminance figure for a 20 × 20ft grid of 1-ft squares, for Example 20.2. (Lumen-Micro © program, courtesy of Lighting Technologies.)*

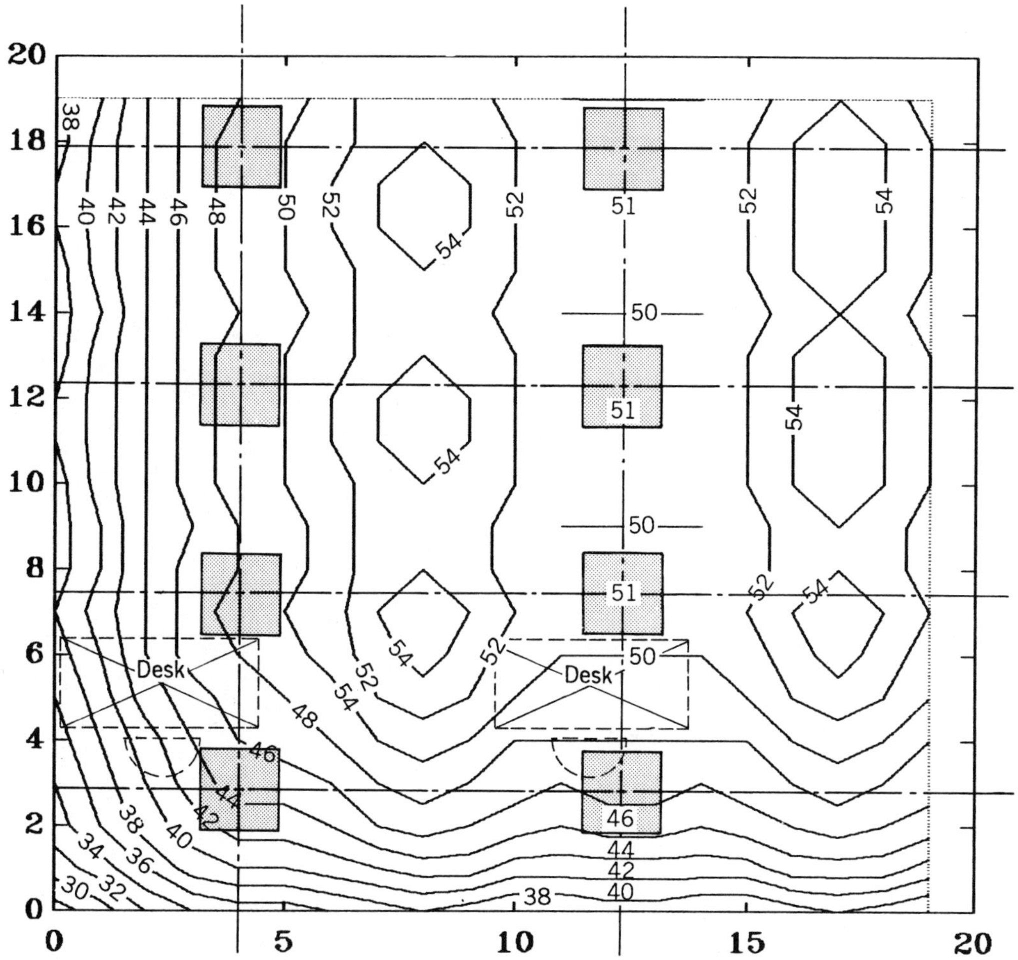

(b)

Fig. 20.53 (continued)

we used—Lumen Micro—has the capability of inputting multiple fixtures or groups of fixtures. The procedure for analyzing Example 20.2 would be:

1. Using a zonal cavity program, such as Lumen$, calculate the number of fixtures and the room floor area per luminaire.

2. Decide on a grid for the fixture layout. If more than one possibility exists, run the point-by-point calculation for both arrangements and select the more desirable.

3. Enter all of the relevant data in the input screens. These data include luminaire data, grid spacing, and information on the types of calculation desired. In our analysis using Lumen-Micro, we specified that the computer calculate:

(a) *"Raw" illuminance*. This is required, as noted above, to determine uniformity and compliance with the recommended illuminance for the visual task at hand. It is also required in order to know where, and in what quantity, to add supplemental lighting.
(b) *ESI illuminance*. This is useful in determining furniture arrangement (facing directions) which will result in minimum reflected glare.
(c) *Relative visual performance*. Since this

EQUIVALENT SPHERE ILLUMINATION
---------- ------ ------------

TARGET DESCRIPTION: PENCIL TARGET - CONCENTRIC RINGS @ 25 DEGREE VIEWING ANGLE
SPHERE CONTRAST: .1675

WORKING PLANE HEIGHT: 2.50

NORTH		EAST		SOUTH		WEST		TOTAL	
AVERAGE :	35.15	AVERAGE :	28.06	AVERAGE :	35.29	AVERAGE :	29.48	AVERAGE :	32.00
MINIMUM :	7.30	MINIMUM :	6.93	MINIMUM :	7.22	MINIMUM :	7.46	MINIMUM :	6.93
MAXIMUM :	65.84	MAXIMUM :	53.31	MAXIMUM :	65.72	MAXIMUM :	53.15	MAXIMUM :	65.84
MEAN DEV.:	14.90	MEAN DEV.:	10.25	MEAN DEV.:	14.38	MEAN DEV.:	9.29	MEAN DEV.:	12.41

ABS. Y- COOR.		ABSOLUTE X-COORDINATE(S) 1.00	2.00	3.00	4.00	5.00	6.00	7.00	8.00	9.00	10.00	11.00	12.00	13.00	14.00	15.00	16.00	17.00	18.00	19.00	20.00
20.00	N	34.	28.	16.	9.	10.	21.	41.	56.	65.	57.	42.	23.	12.	12.	23.	42.	57.	65.	57.	42.
	E	20.	21.	25.	32.	41.	49.	53.	50.	41.	32.	27.	28.	34.	41.	48.	53.	50.	41.	32.	27.
	S	35.	28.	16.	9.	10.	21.	41.	56.	66.	57.	42.	23.	12.	12.	23.	42.	57.	66.	57.	42.
	W	42.	45.	42.	36.	30.	26.	26.	31.	40.	50.	53.	48.	41.	34.	28.	27.	32.	41.	50.	53.
19.00	N	34.	28.	20.	13.	15.	26.	41.	55.	64.	56.	43.	28.	17.	17.	27.	42.	56.	64.	56.	42.
	E	16.	16.	19.	28.	39.	46.	47.	42.	33.	25.	20.	22.	29.	39.	46.	47.	42.	33.	25.	20.
	S	35.	28.	16.	9.	10.	21.	41.	56.	66.	57.	42.	22.	11.	11.	22.	42.	57.	66.	57.	42.
	W	36.	40.	39.	34.	26.	20.	19.	24.	32.	42.	47.	46.	39.	29.	22.	20.	25.	33.	42.	47.

PERCENT OF POINTS MINIMUM ESI

	N	E	S	W
95	11.2	11.7	10.6	12.1
90	13.0	12.5	13.3	13.9
85	15.8	14.2	15.9	16.1
80	17.4	16.0	19.0	19.1
75	20.4	17.7	21.5	21.3
70	22.1	19.7	23.0	22.9

(a)

***Fig. 20.54** Partial computer printout for lighting metrics as noted, for Example 20.2. (Lumen-Micro © program, courtesy of Lighting Technologies.)*

metric of lighting quality has been the subject of considerable recent interest (see Section 18.27*e*), we wish to see its applicability in this design.

(d) *Vertical illuminance*. This is helpful in determining the adequacy of lighting on vertical tasks, such as the copy-holders and other vertical desktop devices.

(e) *Visual comfort probability (VCP)*. Although we should be able to judge the VCP qualitatively by looking at the luminaire's distribution curves, an accurate analysis will pinpoint any problem areas in which direct glare might constitute a problem.

Because the room is so large (60 × 100 ft) an overall analysis is uselessly repetitive, wasteful of computer time, and insufficiently accurate if a large grid is used. We therefore chose to analyze a small 20 × 20 ft area containing eight luminaires. The 1 × 1 ft grid we used is small enough to show in detail the illuminance on desks and other areas.

4. This step involves entering luminaire candela distribution data into an appropriate database. If the luminaire already exists in a database, it is only necessary to make the data available to the program. To avoid the necessity of manually entering these data, some manufacturers are willing to furnish the designer with machine-readable intensity data in the standard IES format, which can then be read into the databases directly.

5. Run the program. The results, truncated in

RELATIVE VISUAL PERFORMANCE

-------- ------ -----------

TARGET DESCRIPTION: NUMBER 2 PENCIL ON HARD WHITE TABLET PAPER - FOR RVP CALCULATIONS
SPHERE CONTRAST: .6967

WORKING PLANE HEIGHT: 2.50

NORTH		EAST		SOUTH		WEST		TOTAL	
AVERAGE =	1.000	AVERAGE =	.999	AVERAGE =	.999	AVERAGE =	.999	AVERAGE =	.999
MINIMUM =	.995	MINIMUM =	.993	MINIMUM =	.994	MINIMUM =	.988	MINIMUM =	.988
MAXIMUM =	1.000	MAXIMUM =	1.000	MAXIMUM =	1.000	MAXIMUM =	1.000	MAXIMUM =	1.000
MEAN DEV.=	.001	MEAN DEV.=	.001	MEAN DEV.=	.001	MEAN DEV.=	.001	MEAN DEV.=	.001

ABS. Y COOR.		ABSOLUTE X-COORDINATE(S) 1.00	2.00	3.00	4.00	5.00	6.00	7.00	8.00	9.00	10.00	11.00	12.00	13.00	14.00	15.00	16.00	17.00	18.00	19.00	20.00
20.00	N	.999	1.000	.999	.997	.997	1.000	1.000	1.000	1.000	1.000	1.000	1.000	.998	.998	1.000	1.000	1.000	1.000	1.000	1.000
	E	.999	1.000	1.000	1.000	1.000	1.000	1.000	1.000	1.000	1.000	1.000	1.000	1.000	1.000	1.000	1.000	1.000	1.000	1.000	1.000
	S	.999	1.000	.999	.997	.997	1.000	1.000	1.000	1.000	1.000	1.000	1.000	.998	.998	1.000	1.000	1.000	1.000	1.000	1.000
	W	.999	1.000	1.000	1.000	1.000	1.000	1.000	1.000	1.000	1.000	1.000	1.000	1.000	1.000	1.000	1.000	1.000	1.000	1.000	1.000
19.00	N	.999	1.000	1.000	1.000	1.000	1.000	1.000	1.000	1.000	1.000	1.000	1.000	1.000	1.000	1.000	1.000	1.000	1.000	1.000	1.000
	E	.999	.999	1.000	1.000	1.000	1.000	1.000	1.000	1.000	1.000	1.000	1.000	1.000	1.000	1.000	1.000	1.000	1.000	1.000	1.000
	S	.999	1.000	.999	.996	.997	1.000	1.000	1.000	1.000	1.000	1.000	1.000	.998	.998	1.000	1.000	1.000	1.000	1.000	1.000
	W	.998	1.000	1.000	1.000	1.000	1.000	1.000	1.000	1.000	1.000	1.000	1.000	1.000	1.000	1.000	1.000	1.000	1.000	1.000	1.000

PERCENT OF POINTS MINIMUM RVP

	N	E	S	W
95	.997	.997	.997	.996
90	.998	.998	.998	.998
85	.999	.999	.999	.998
80	.999	.999	.999	.999
75	1.000	.999	.999	.999
70	1.000	1.000	1.000	1.000

(b)

Fig. 20.54 *(continued)*

the interest of space, are shown in Figs. 20.53 and 20.54. The results are analyzed below.

(c) Analysis of Computerized Solution of Example 20.2.

1. *Raw illuminance.* Refer to Fig. 20.53*a*. The computer output consists of a 20 × 20 ft grid showing the illuminance in footcandles in each 1-ft^2 space. On this grid we have superimposed the luminaires in a 5 × 9 ft grid, with half spacing to the walls, and have added two possible desk locations. Note the very high degree of uniformity and the expected drop-off at the walls, even with a 50% reflectance. Thus the desk adjacent to the wall might require a desk lamp, whereas other desks would not since the illumination over the entire desk surface is uniform and very close to target illuminance. In Fig. 20.53*b* we machine plotted an illuminance contour map with contour lines 2 fc apart. Isolux contour maps of this type are useful in obtaining an overall picture of the room illuminance and are frequently easier to understand than sheets of numbers as in Fig. 20.53*a*. In this case, however, illuminance is so uniform that such a map is not really necessary except to show how illuminance drops off at walls.

2. *Equivalent sphere illumination (ESI).* See Fig. 20.54*a*. We have reproduced only 2 ft from the top of the 20-ft grid, since each 1 × 1 square is analyzed for four viewing directions and the printout is therefore voluminous. Note the results. The highest values occur between fixtures (e.g., at 9 and 18 ft) in the north and south viewing directions (along the long axis of the lamps). This is as expected from the batwing

VERTICAL ILLUMINANCE

WORKING PLANE HEIGHT: 3.00

NORTH		EAST		SOUTH		WEST		TOTAL	
AVERAGE =	22.75	AVERAGE =	24.49	AVERAGE =	20.58	AVERAGE =	21.68	AVERAGE =	22.38
MINIMUM =	18.66	MINIMUM =	13.97	MINIMUM =	11.49	MINIMUM =	10.74	MINIMUM =	10.74
MAXIMUM =	24.85	MAXIMUM =	31.94	MAXIMUM =	24.80	MAXIMUM =	31.63	MAXIMUM =	31.94
MEAN DEV.=	1.11	MEAN DEV.=	4.56	MEAN DEV.=	2.89	MEAN DEV.=	5.94	MEAN DEV.=	3.68

ABS. Y COOR.		ABSOLUTE X-COORDINATE(S) 1.00	2.00	3.00	4.00	5.00	6.00	7.00	8.00	9.00	10.00	11.00	12.00	13.00	14.00	15.00	16.00	17.00	18.00	19.00	20.00
20.00	N	20.	21.	22.	23.	24.	24.	24.	24.	25.	25.	24.	24.	24.	24.	24.	24.	25.	25.	25.	24.
	E	31.	28.	24.	19.	17.	20.	24.	29.	31.	31.	28.	24.	19.	17.	20.	24.	29.	31.	31.	28.
	S	20.	21.	22.	23.	24.	24.	24.	25.	25.	25.	24.	24.	24.	24.	24.	24.	25.	25.	25.	24.
	W	12.	12.	12.	12.	16.	22.	27.	30.	31.	28.	24.	19.	17.	19.	24.	28.	31.	31.	28.	24.
19.00	N	20.	21.	22.	22.	22.	23.	23.	24.	24.	24.	24.	23.	23.	23.	23.	24.	24.	24.	24.	24.
	E	31.	29.	24.	19.	17.	20.	24.	29.	31.	31.	29.	24.	19.	17.	20.	24.	29.	32.	31.	29.
	S	20.	21.	22.	23.	23.	23.	24.	24.	24.	24.	24.	24.	24.	24.	24.	24.	24.	24.	24.	24.
	W	12.	12.	12.	12.	16.	22.	27.	31.	31.	28.	24.	19.	17.	19.	24.	28.	31.	31.	28.	24.

*

(c)

VISUAL COMFORT PROBABILITY

VISUAL PLANE HEIGHT: 4.00

NORTH		EAST		SOUTH		WEST		TOTAL	
AVERAGE =	95.40	AVERAGE =	92.81	AVERAGE =	95.39	AVERAGE =	90.33	AVERAGE =	93.64
MINIMUM =	90.82	MINIMUM =	76.42	MINIMUM =	88.49	MINIMUM =	74.22	MINIMUM =	74.22
MAXIMUM =	98.34	MAXIMUM =	99.99	MAXIMUM =	99.99	MAXIMUM =	99.99	MAXIMUM =	99.99
MEAN DEV.=	1.53	MEAN DEV.=	5.72	MEAN DEV.=	1.81	MEAN DEV.=	6.49	MEAN DEV.=	4.11

ABS. Y COOR.		ABSOLUTE X-COORDINATE(S) 1.00	2.00	3.00	4.00	5.00	6.00	7.00	8.00	9.00	10.00	11.00	12.00	13.00	14.00	15.00	16.00	17.00	18.00	19.00	20.00
20.00	N	98.	96.	97.	97.	97.	97.	96.	96.	97.	97.	97.	97.	98.	98.	97.	97.	97.	98.	97.	97.
	E	89.	97.	100.	100.	99.	95.	87.	80.	84.	89.	97.	100.	100.	99.	95.	87.	80.	84.	89.	97.
	S	98.	96.	97.	97.	97.	97.	96.	96.	97.	97.	97.	97.	98.	98.	97.	97.	97.	98.	97.	97.
	W	************************************						96.	86.	79.	80.	87.	95.	99.	100.	100.	97.	89.	84.	80.	87.
19.00	N	96.	94.	94.	95.	95.	94.	94.	93.	95.	95.	95.	95.	96.	96.	95.	95.	95.	95.	95.	95.
	E	90.	97.	100.	100.	99.	95.	87.	83.	83.	90.	97.	100.	100.	99.	95.	87.	83.	83.	90.	97.
	S	98.	98.	97.	97.	97.	96.	97.	98.	97.	98.	98.	97.	97.	97.	97.	98.	98.	97.	98.	98.
	W	************************************						97.	88.	78.	80.	87.	95.	99.	100.	100.	97.	90.	83.	83.	87.

(d)

Fig. 20.54 *(continued)*

distribution, which, as we have seen, must be used across the viewing direction. More important, however, is that for these locations ESI is *higher* than the raw footcandles by about 20%, indicating a viewing position and direction free of reflected glare. The situation is very different close to the fixture. Further, the E-W viewing direction which is transverse to the lamps has satisfactory ESI only when sufficiently distant from the luminaire so that the louvre's shielding angle prevents light from reaching that location at an angle whose reflective geometry would cause veiling reflections (see Section 18.27). Useful summaries are shown at the top of the sheet in Fig. 20.54*a*.

3. *Relative visual performance (RVP)*. See Fig. 20.54*b*. Comparison with the ESI chart in Fig. 20.54*a* shows that RVP is much less sensitive to situations where veiling reflections exist than ESI (see, for instance, 4-, 5-, 13-, and 14-ft north viewing) and as such in our opinion is a metric of only limited usefulness. See also the discussion in Section 18.27*e*.

4. *Vertical illuminance*. See Fig. 20.54*c*. Because the lighting system is direct but with a considerable element of diffusion from the reflective room surfaces, the vertical illuminance is uniform and relatively high, at about 50% of the horizontal illuminance. This should be enough for most routine vertical surface tasks, although reading small print or detailed text, particularly numbers, from vertical copy holders may require supplemental desk lamps. The electric distribution system (see Sections 15.18 to 15.30) should be so designed.

5. *Visual comfort probability (VCP)*. See Fig. 20.54*d*. As expected because of the specular, low-brightness parabolic louvres, VCP is very high. The areas on the truncated chart without numbers are locations from which no luminaire is seen.

It should be obvious from the foregoing discussion that the information made available by computer calculation is necessary to properly analyze a proposed lighting design and that simple average illuminance calculations are insufficient, particularly for the novice designer. Although veteran lighting designers can draw on their experience to predict fairly accurately the results that the computer presents numerically, we have found that these calculations are always useful and frequently point up lighting problems that might otherwise go unnoticed.

20.43 Average Luminance Calculations

The basic equations relating luminance to candela intensity and to illuminance are covered in Section 18.8. That section also deals with the calculation of source luminance, and with reflected luminance when the illuminance (lux) level is known. Thus, once horizontal illuminance has been calculated by any of the methods described above (Sections 20.29 through 20.42), and knowing the reflectance of an object, its horizontal luminance can readily be calculated (see Section 18.8). (Some of the computer lighting analysis programs, including Lighting Technologies' Lumen-Micro, which we used, will calculate surface luminances.) However, as explained in detail in Sections 18.29 and 18.30, which discuss lighting quality, and in Sections 20.10 through 20.16, which deal with lighting systems and patterns, the luminance impression of a visual environment is affected more by vertical surface luminance than by horizontal. (See also Table 18.2*b* and Fig. 20.12.) For this reason it is important to be able to calculate *average* vertical surface (wall) luminance in the same simple, straightforward fashion that *average* horizontal illuminance is calculated. In addition, it is useful to know the average luminance of the ceiling cavity in a space, in order to judge the contrast between all luminous objects, including luminaires, that have the ceiling cavity as background.

Straightforward calculation of both wall and ceiling cavity luminance (L_W and L_{CC}) is possible through the use of luminance coefficients that are similar in application and calculation to utilization coefficients. These coefficients are listed in Table 20.6 for some of the generic fixture types listed in Table 20.2. Others are listed in the *IES Lighting Handbook*. For actual installation calculations, it is preferable to obtain coefficients from luminaire manufacturers. The

TABLE 20.6 **Wall Luminance Coefficients and Ceiling Cavity Luminance Coefficients for Typical Luminaires**

To obtain a luminance coefficient, follow the procedure detailed at the head of Table 20.2 to find a coefficient of utilization. Data should be obtained from manufacturers of actual luminaires used.

	ρ_{cc}:	80			70			50			80		70		50	
	ρ_w:	50	30	10	50	30	10	50	30	10	50	30	50	30	50	30
Typical Luminaire	RCR	Wall Luminance Coefficients for $\rho_{FC} = 20$									Ceiling Cavity Luminance Coefficient $\rho_{FC} = 20$					
1 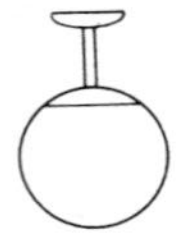											.42	.42	.36	.36	.25	.25
	1	.32	.18	.06	.30	.17	.05	.27	.15	.05	.42	.40	.36	.34	.25	.23
	2	.27	.15	.05	.25	.14	.04	.23	.13	.04	.42	.38	.36	.33	.24	.23
	3	.24	.13	.04	.22	.12	.04	.20	.11	.03	.41	.37	.35	.32	.24	.22
	4	.21	.11	.03	.20	.10	.03	.17	.09	.03	.41	.36	.35	.31	.24	.22
	5	.19	.10	.03	.18	.09	.03	.16	.08	.02	.40	.35	.34	.30	.24	.21
	6	.18	.09	.03	.16	.08	.02	.14	.07	.02	.39	.34	.34	.30	.23	.21
	7	.16	.08	.02	.15	.08	.02	.13	.07	.02	.39	.34	.33	.29	.23	.21
	8	.15	.07	.02	.14	.07	.02	.12	.06	.02	.38	.34	.33	.29	.23	.20
Pendant diffusing sphere	9	.14	.07	.02	.13	.06	.02	.12	.06	.02	.38	.33	.33	.29	.23	.20
with incandescent lamp	10	.13	.06	.02	.12	.06	.02	.11	.05	.01	.37	.33	.32	.29	.22	.20
3 											.15	.15	.13	.13	.09	.09
	1	.23	.13	.04	.23	.13	.04	.21	.12	.04	.15	.13	.13	.11	.09	.08
	2	.22	.12	.04	.22	.12	.04	.21	.11	.04	.14	.11	.12	.09	.08	.07
	3	.21	.11	.03	.21	.11	.03	.20	.11	.03	.13	.10	.12	.08	.08	.06
	4	.20	.10	.03	.19	.10	.03	.19	.10	.03	.13	.08	.11	.07	.08	.05
	5	.19	.09	.03	.18	.09	.03	.18	.09	.03	.12	.08	.11	.07	.07	.05
	6	.18	.09	.03	.17	.09	.03	.17	.09	.02	.12	.07	.10	.06	.07	.04
	7	.17	.08	.02	.16	.08	.02	.16	.08	.02	.11	.06	.10	.05	.07	.04
Porcelain-enameled	8	.16	.08	.02	.15	.08	.02	.15	.07	.02	.11	.06	.09	.05	.06	.04
ventilated standard dome	9	.15	.07	.02	.15	.07	.02	.14	.07	.02	.10	.05	.09	.05	.06	.03
with incandescent lamp	10	.14	.07	.02	.14	.07	.02	.13	.07	.02	.10	.05	.09	.04	.06	.03
7 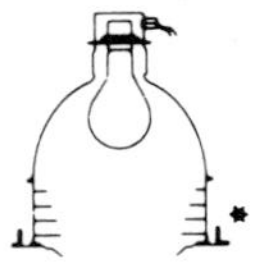											.08	.08	.07	.07	.04	.04
	1	.06	.03	.01	.06	.03	.01	.05	.03	.01	.08	.07	.06	.06	.04	.04
	2	.05	.03	.01	.05	.03	.01	.05	.03	.01	.07	.06	.06	.05	.04	.04
	3	.05	.03	.01	.05	.03	.01	.04	.02	.01	.06	.05	.05	.04	.04	.03
	4	.05	.02	.01	.05	.02	.01	.04	.02	.01	.05	.04	.05	.04	.03	.03
	5	.05	.02	.01	.04	.02	.01	.04	.02	.01	.05	.04	.04	.03	.03	.02
	6	.04	.02	.01	.04	.02	.01	.04	.02	.01	.05	.03	.04	.03	.03	.02
Reflector downlight with	7	.04	.02	.01	.04	.02	.01	.04	.02	.01	.04	.03	.04	.03	.03	.02
baffles and inside-frosted	8	.04	.02	.01	.04	.02	.01	.04	.02	.01	.04	.03	.03	.02	.02	.02
lamp (see note on this unit	9	.04	.02	.01	.04	.02	.01	.04	.02	.01	.04	.02	.03	.02	.02	.02
in Table 20.2)	10	.04	.02	.00	.04	.02	.01	.04	.02	.01	.04	.02	.03	.02	.02	.01

28											.27	.27	.23	.23	.16	.16
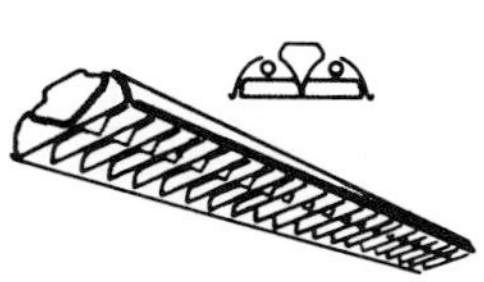	1	.17	.10	.03	.16	.09	.03	.14	.08	.03	.26	.24	.22	.21	.15	.14
	2	.16	.09	.03	.15	.09	.03	.14	.08	.02	.25	.23	.21	.20	.15	.14
	3	.15	.08	.02	.15	.08	.02	.13	.07	.02	.24	.21	.21	.19	.14	.13
	4	.15	.08	.02	.14	.07	.02	.13	.07	.02	.24	.21	.20	.18	.14	.12
	5	.14	.07	.02	.13	.07	.02	.12	.06	.02	.23	.20	.20	.17	.14	.12
	6	.13	.07	.02	.13	.06	.02	.11	.06	.02	.23	.19	.20	.17	.14	.12
	7	.13	.06	.02	.12	.06	.02	.11	.06	.02	.23	.19	.19	.16	.13	.11
Diffuse aluminum reflector with 35° crosswise and 35° lengthwise shielding	8	.12	.06	.02	.12	.06	.02	.11	.05	.02	.22	.18	.19	.16	.13	.11
	9	.12	.06	.02	.11	.05	.02	.10	.05	.01	.22	.18	.19	.16	.13	.11
	10	.11	.05	.01	.11	.05	.01	.10	.05	.01	.22	.18	.19	.15	.13	.11
33											.65	.65	.56	.56	.38	.38
	1	.20	.12	.04	.18	.10	.03	.13	.08	.02	.65	.63	.55	.54	.38	.37
	2	.19	.10	.03	.17	.09	.03	.12	.07	.02	.64	.61	.55	.53	.38	.37
	3	.17	.09	.03	.15	.08	.02	.11	.06	.02	.64	.60	.55	.52	.37	.36
	4	.16	.08	.02	.14	.07	.02	.11	.06	.02	.63	.59	.54	.51	.37	.36
	5	.15	.08	.02	.13	.07	.02	.10	.05	.02	.63	.59	.54	.51	.37	.36
	6	.14	.07	.02	.12	.06	.02	.09	.05	.01	.60	.58	.54	.50	.37	.35
	7	.13	.07	.02	.12	.06	.02	.09	.04	.01	.62	.58	.53	.50	.37	.35
Luminous bottom suspended unit with extra-high-output lamp	8	.12	.06	.02	.11	.05	.02	.08	.04	.01	.61	.57	.53	.50	.37	.35
	9	.12	.06	.02	.10	.05	.01	.08	.04	.01	.61	.57	.53	.50	.36	.35
	10	.11	.05	.01	.10	.05	.01	.07	.04	.01	.61	.57	.52	.49	.36	.35
35											.22	.22	.19	.19	.13	.13
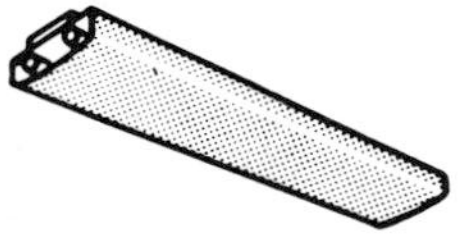	1	.19	.11	.03	.18	.10	.03	.17	.10	.03	.21	.20	.18	.17	.12	.12
	2	.18	.10	.03	.17	.09	.03	.15	.09	.03	.21	.18	.18	.16	.12	.11
	3	.16	.09	.03	.16	.08	.03	.14	.08	.02	.20	.17	.17	.15	.12	.10
	4	.15	.08	.02	.15	.08	.02	.14	.07	.02	.19	.16	.17	.14	.11	.10
	5	.14	.07	.02	.14	.07	.02	.13	.07	.02	.19	.15	.16	.13	.11	.09
	6	.14	.07	.02	.13	.07	.02	.12	.06	.02	.18	.15	.16	.13	.11	.09
	7	.13	.06	.02	.12	.06	.02	.12	.06	.02	.18	.14	.16	.12	.11	.09
Two-lamp prismatic wraparound—multiply by 0.95 for four lamps	8	.12	.06	.02	.12	.06	.02	.11	.05	.02	.18	.14	.15	.12	.11	.08
	9	.12	.06	.02	.11	.05	.02	.11	.05	.01	.17	.13	.15	.12	.10	.08
	10	.11	.05	.01	.11	.05	.01	.10	.05	.01	.17	.13	.15	.11	.10	.08
42											.12	.12	.10	.10	.07	.07
	1	.16	.09	.03	.16	.09	.03	.15	.09	.03	.11	.10	.09	.08	.06	.06
	2	.15	.08	.03	.15	.08	.03	.14	.08	.02	.10	.08	.09	.07	.06	.05
	3	.15	.08	.02	.14	.08	.02	.14	.07	.02	.10	.07	.08	.06	.06	.04
	4	.14	.07	.02	.13	.07	.02	.13	.07	.02	.09	.06	.08	.05	.05	.04
	5	.13	.07	.02	.13	.07	.02	.12	.06	.02	.09	.06	.08	.05	.05	.03
Fluorescent unit with flat prismatic lens; four-lamp, 2 ft wide—multiply by 1.05 for three lamps and 1.10 for two lamps	6	.12	.06	.02	.12	.06	.02	.12	.06	.02	.09	.05	.07	.04	.05	.03
	7	.12	.06	.02	.12	.06	.02	.11	.06	.02	.08	.05	.07	.04	.05	.03
	8	.11	.05	.02	.11	.05	.02	.11	.05	.02	.08	.04	.07	.04	.05	.03
	9	.11	.05	.01	.10	.05	.01	.10	.05	.01	.08	.04	.07	.03	.05	.02
	10	.10	.05	.01	.10	.05	.01	.10	.05	.01	.07	.04	.06	.03	.04	.02

TABLE 20.6 **Wall Luminance Coefficients and Ceiling Cavity Luminance Coefficients for Typical Luminaires (*continued*)**

Typical Luminaire	ρ_{cc}:	80			70			50			80		70		50	
	ρ_w:	50	30	10	50	30	10	50	30	10	50	30	50	30	50	30
	RCR	Wall Luminance Coefficients for $\rho_{FC} = 20$									Ceiling Cavity Luminance Coefficient $\rho_{FC} = 20$					
44	0										.114	.114	.097	.097	.066	.066
	1	.137	.078	.025	.133	.075	.024	.125	.072	.023	.105	.094	.089	.080	.061	.055
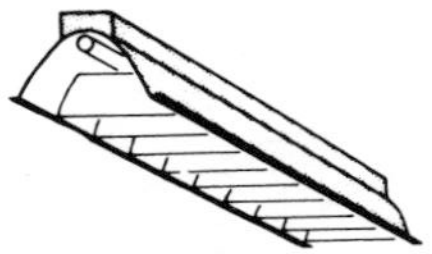	2	.131	.072	.022	.128	.070	.022	.121	.067	.021	.097	.079	.083	.068	.057	.047
	3	.127	.068	.020	.124	.066	.020	.118	.064	.019	.092	.068	.079	.059	.054	.041
	4	.123	.064	.019	.120	.063	.019	.115	.061	.018	.087	.060	.075	.052	.052	.036
	5	.119	.060	.018	.116	.060	.017	.112	.058	.017	.084	.053	.072	.046	.050	.032
	6	.114	.057	.016	.112	.056	.016	.108	.055	.016	.080	.048	.069	.042	.048	.029
	7	.110	.054	.015	.108	.054	.015	.104	.053	.015	.078	.044	.067	.038	.046	.027
Radial batwing	8	.106	.052	.015	.104	.051	.014	.101	.050	.014	.075	.041	.065	.036	.045	.025
distribution—louvred	9	.102	.049	.014	.100	.049	.014	.097	.048	.014	.073	.038	.063	.033	.043	.023
fluorescent, unit	10	.097	.047	.013	.096	.046	.013	.093	.046	.013	.070	.036	.060	.031	.042	.022
46	0										.236	.236	.201	.201	.138	.138
	1	.234	.133	.042	.225	.128	.041	.208	.119	.038	.229	.210	.196	.181	.134	.124
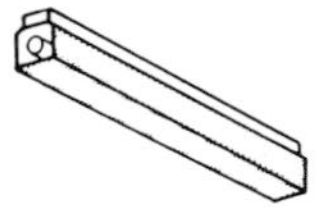	2	.213	.117	.036	.205	.113	.035	.190	.106	.033	.222	.193	.190	.166	.130	.115
	3	.195	.104	.031	.188	.101	.030	.175	.095	.029	.216	.180	.185	.155	.127	.108
	4	.181	.094	.028	.175	.091	.027	.162	.086	.026	.211	.170	.181	.147	.124	.102
	5	.170	.087	.025	.164	.084	.025	.153	.080	.023	.206	.163	.177	.141	.122	.098
Bilateral batwing	6	.159	.080	.023	.153	.078	.022	.143	.073	.021	.201	.157	.173	.136	.119	.095
distribution—one lamp,	7	.149	.074	.021	.144	.072	.020	.135	.068	.020	.197	.152	.169	.131	.117	.092
surface-mounted	8	.141	.069	.019	.137	.067	.019	.128	.064	.018	.193	.148	.166	.128	.115	.090
fluorescent, with prismatic	9	.134	.065	.018	.129	.063	.018	.121	.060	.017	.189	.144	.163	.125	.113	.088
wraparound lens	10	.126	.061	.017	.122	.059	.016	.115	.056	.016	.185	.141	.159	.122	.111	.086

47 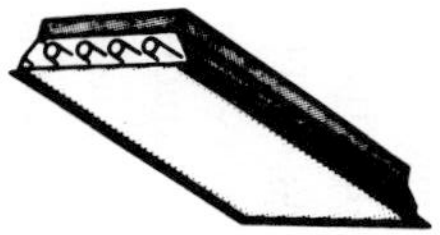	0										.114	.114	.097	.097	.066	.066
	1	.175	.100	.032	.171	.098	.031	.163	.094	.030	.107	.093	.091	.080	.063	.055
	2	.168	.092	.028	.164	.090	.028	.157	.087	.027	.102	.079	.087	.068	.060	.047
	3	.157	.083	.025	.153	.082	.025	.147	.080	.024	.097	.068	.083	.059	.057	.041
	4	.147	.076	.022	.144	.075	.022	.138	.073	.022	.093	.060	.080	.052	.055	.036
	5	.139	.071	.021	.136	.070	.020	.131	.068	.020	.090	.054	.077	.047	.053	.033
Radial batwing distribution—four-lamp, 610-mm (2 ft)-wide fluorescent unit with flat prismatic lens—see note 2	6	.130	.065	.019	.128	.065	.019	.123	.063	.018	.086	.049	.074	.043	.051	.030
	7	.122	.060	.017	.120	.060	.017	.116	.059	.017	.082	.045	.071	.039	.049	.027
	8	.115	.056	.016	.114	.056	.016	.110	.055	.016	.079	.042	.068	.036	.047	.025
	9	.109	.053	.015	.108	.052	.015	.104	.052	.015	.076	.039	.065	.034	.045	.024
	10	.103	.049	.014	.102	.049	.014	.099	.048	.014	.072	.037	.062	.032	.043	.022
48 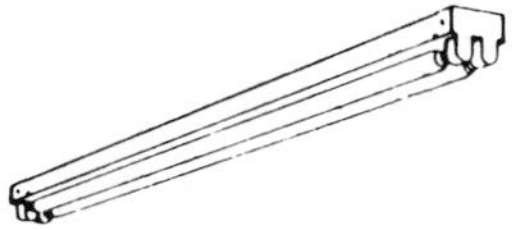	0										.325	.325	.278	.278	.189	.189
	1	.316	.180	.057	.304	.173	.055	.280	.161	.051	.320	.295	.274	.253	.187	.174
	2	.282	.155	.047	.271	.149	.046	.250	.139	.043	.314	.275	.269	.237	.184	.164
	3	.252	.134	.040	.242	.130	.039	.223	.121	.037	.307	.261	.264	.225	.181	.156
	4	.229	.119	.035	.220	.115	.034	.202	.107	.032	.301	.250	.258	.216	.178	.150
	5	.212	.108	.031	.203	.104	.030	.187	.098	.029	.295	.241	.253	.209	.175	.146
	6	.195	.098	.028	.188	.095	.027	.173	.089	.026	.289	.234	.248	.203	.171	.142
	7	.181	.090	.025	.174	.087	.025	.161	.081	.023	.283	.228	.244	.198	.168	.139
	8	.169	.083	.023	.163	.080	.023	.150	.075	.021	.278	.224	.239	.194	.165	.136
Two-lamp fluorescent strip unit	9	.159	.077	.021	.153	.074	.021	.141	.070	.020	.273	.220	.235	.191	.163	.134
	10	.149	.071	.020	.143	.069	.019	.132	.065	.018	.268	.216	.231	.188	.160	.132

NOTES:

1. Data extracted, with permission, from *IES Handbook* (1981), Reference Volume. Coefficients for fixtures 1, 3, 7, 28, 33, 35, and 42 have been rounded to two decimal places. For three decimal figures, see *IES Handbook* (1981).
2. Multiply coefficients by 1.05 for three lamps and 1.1 for two lamps.

average luminance calculations are parallel to those for illuminance.

Average initial wall luminance (cd/m²):

$$L_W = \frac{\text{lamp lumens} \times \text{wall luminance coefficient}}{\pi \times \text{floor area in m}^2} \tag{20.21}$$

and average initial ceiling cavity luminance in cd/m²:

$$L_{CC} = \frac{\text{lamp lumens} \times \text{ceiling cavity luminance coefficient}}{\pi \times \text{floor area in m}^2} \tag{20.22}$$

If area is expressed in square feet and π omitted from the equation, L is expressed in footlamberts.

To obtain *maintained* values, a light loss factor similar to that explained in Section 20.31 is introduced. It is calculated similarly except that item 17 (see Fig. 20.39), room surface dirt, is calculated using the following figures:

Lighting System	*Wall Luminance*	*Ceiling Luminance*
Direct	0.82 ± 10%	0.75 ± 10%
Semidirect	0.87 ± 7%	0.82 ± 10%
Direct-indirect	0.92 ± 5%	0.85 ± 8%
Semi-indirect	0.87 ± 7%	0.88 ± 7%
Indirect	0.82 ± 10%	0.90 ± 5%

For ceiling mounted or recessed luminaires, L_{CC} is the average luminance of the ceiling between luminaires. For pendant luminaires, the calculated L_{CC} is that of an imaginary plane at the height of the luminaires. It is useful in determining the brightness ratios when compared to luminaire luminance at the seeing angle involved. The ceiling cavity, as the wall, is assumed to have a lambertian characteristic, that is, perfect diffuseness, making luminance independent of viewing angle.

It would be instructive to calculate the wall luminance of the office of Example 20.2. The photometric data in Fig. 20.41 do not include the wall luminance coefficient, since luminaire coefficients are not normally published by luminaire manufacturers. However, based on other data available, a figure of 0.22 for wall luminance is a good estimate given RCR, ρ_{CC}, and ρ_W of 0.66, 80%, and 30%, respectively. Initial wall luminance is then

$$L_W = \frac{3200 \times 0.22}{32 \text{ ft}^2} = 22 \text{ fL} = 75.5 \text{ cd/m}^2$$

This is within the preferred range of 50 to 200 cd/m² (see Table 18.2*b*). In actuality the average wall luminance would probably be higher because of the practice of placing the last row of luminaires quite close to the wall.

EVALUATION STAGE

20.44 Design Evaluation

The design stage having been completed, the final step is evaluation of the design from the three aspects—lighting, costs, and energy. The lighting aspects include quantity, quality, ESI luminance ratios, mood, ambience, texture, color, variation, psychological impressions, orientations, and daylight use—in short, a review of all the lighting factors discussed in detail above. Since a good deal of experience is required to visualize the actual results from the drawings, the novice designer would do well to have the review done, at least in part, by one having such experience. The other two aspects of evaluation, cost and energy, can be performed readily with the aid of contractor's estimating figures for cost and a straightforward calculation for energy. The results are compared to the budget figures developed at the preliminary design stage. As we have repeatedly stressed, the useful cost figures are life cycle, annual operating and first cost, for economic comparisons, operating budgets, and construction budgets, respectively. In the following chapter we will present recommendations for specific occupancies, accompanied by actual cost studies and energy analyses. Detailed cost studies including the impact of lighting on air conditioning, the proportional cost of the wiring system, and the proper apportionment of costs involves the entire building and can be accurately performed only by computer. Studies of this type are generally made by consulting engineers rather than architects, and then only after initial, operating, and total costs have been set in proper perspective for the particular job, by the architect and client. This is necessary because often, as in the case of speculative construction, the client's overriding consideration is first cost, thereby rendering a complete cost analysis unnecessary. Any attempt to completely separate costs for lighting, HVAC, structure, and so on is arbitrary because of the intimate interaction between these factors. Lighting designers are well advised to keep themselves and the construction team aware of this, if they are to fully fulfill their responsibility.

REFERENCES

See complete listing for all lighting sections at the end of Part VI (page 1187).

21
LIGHTING APPLICATION

21.1 Introduction

In the preceding three chapters, we examined lighting fundamentals, sources, and design procedures. In this final lighting chapter we shall consider application of lighting principles to specific situations. The facilities covered in some detail include residential, educational, commercial, institutional, and industrial occupancies. Each is examined from the viewpoint of its special requirements and suggested approach. The latter includes lighting materials, sources, as well as comparative economics and energy considerations. The chapter concludes with consideration of special types of indoor lighting plus a short section on exterior lighting.

LIGHTING CONTROL

21.2 Requirement for Lighting Control

The term *lighting control* means all the techniques by which the lighting system will be operated and covers both manual and automatic controls. The control strategy must be decided on simultaneously with the lighting design because the control scheme must be appropriate to the light source. In turn, the system accessories and arrangement depend on the control scheme. For instance, if dimming is decided on, using a fluorescent light source, then the range of dimming determines the type of ballast, the ballast switching points, and the degree of dimming flexibility.

The primary purposes of lighting control are flexibility and economy: flexibility to provide the modifications of brightness and pattern desired by the designer, and economy of both energy resources and monetary resources. (See Appendix E for treatment of economic analyses.) A properly designed lighting control system will reduce energy usage over a noncontrolled installation by 10 to 60%. In addition, financial operating economies will result from:

1. Reduced energy use.
2. Reduced air conditioning costs as a result of lower lighting waste heat.
3. Longer lamp and ballast life due to lower operating temperatures and lower output.
4. Lower labor costs due to control automation.

Finally, as explained in detail in Section 20.5 (Power Budgets), lighting in most new construction will be designed with the energy constraints of ASHRAE/IES Standard 90.1-1989, *Energy Efficient Design of New Buildings*. This standard has a system of lighting power credits for lighting control systems designed with automatic energy-conserving controls. These credits, called *power adjustment factors*, permit the effective connected load to be reduced by factors as listed in Table 21.1. The reduction is in accordance with the following calculation:

$$\text{ALP} = \text{CLP} - \text{LPCC}$$

$$\text{LPCC} = \text{CLP} \times \text{PAD}$$

where

ALP = adjusted (connected) lighting power

CLP = connected lighting power for luminaires controlled by automatic control devices

PAF = power adjustment factor, as per Table 21.1

Thus, for example, circuits with a simple on–off mode initiated by a daylight sensor have a 10% power credit, whereas daylight sensing with continuous dimming has a credit of 30% because it is more energy economical (but much more expensive initially).

To avoid confusion, particularly in view of

TABLE 21.1 **ASHRAE 90.1 (Table 6.3): Power Adjustment Factor (PAF)**

Automatic Control Device(s)	*PAF*
(1) Daylight sensing controls (DS), continuous dimming	0.30
(2) DS, multiple-step dimming	0.20
(3) DS, On/Off	0.10
(4) DS continuous dimming and programmable timing	0.35
(5) DS multiple-step dimming and programmable timing	0.25
(6) DS on/off and programmable timing	0.15
(7) DS continuous dimming, programmable timing, and lumen maintenance	0.40
(8) DS multiple-step dimming, programmable timing, and lumen maintenance	0.30
(9) DS on/off, programmable timing, and lumen maintenance	0.20
(10) Lumen maintenance	0.10
(11) Lumen maintenance and programmable timing control	0.15
(12) Programmable timing control	0.15
(13) Occupancy sensor	0.30
(14) Occupancy sensor and DS continuous dimming	0.40
(15) Occupancy sensor and DS multiple-step dimming	0.35
(16) Occupancy sensor and DS On/Off	0.35
(17) Occupancy sensor, DS continuous dimming, and lumen maintenance	0.45
(18) Occupancy sensor, DS multiple-step dimming, and lumen maintenance	0.40
(19) Occupancy sensor, DS On/Off, and lumen maintenance	0.35
(20) Occupancy sensor and lumen maintenance	0.35
(21) Occupancy sensor and programmable timing control	0.35

Source: ASHRAE/IES 90.1-1989; reprinted with permission. Copyright © By American Society of Heating, Refrigeration, and Air-Conditioning Engineers, Atlanta, Ga.

overlapping and sometimes inaccurate terminology, it is necessary to differentiate between control functions, control devices, and control systems. For lighting, the only control *functions* are switching and dimming. The control *devices* are the means by which the switching and dimming functions are accomplished. They are numerous, ranging from a simple wall switch, through time switches and dimmers of all sorts. Generally also included in this category are control *initiation* devices, such as occupancy sensors and photocells. The control *system* is the entire assembly of control and signal initiating equipment together with their interconnections. Included here also are microprocessors and programmable controllers. The system can be a stand-alone arrangement, or alternatively, as is the case in large facilities, it may be part of an EMS (energy management system) or a BAS (building automation system) (Appendix J) or

ILLUMINATION

both. The difference in operation in these instances lies in the control algorithm, which is primarily energy oriented for an EMS and overall-building-function oriented for a BAS. In the discussion that follows the control criteria will be energy conservation, cost reduction, and operating flexibility.

21.3 Elements of Lighting Control

(a) Switching. There are two basic types of control functions—switching and dimming. Switching is an on–off function. By selecting the number of lighting elements to be switched in each switching action, the designer can establish the number of control levels. The more levels, the finer the control. Thus, in a space requiring several levels of uniform illumination for different functions, the designer has many control alternatives. He or she can switch entire fixtures, but this adversely affects uniformity. Taking three-lamp fluorescent fixtures as an example, the designer can obtain better uniformity and four levels of illumination by switching the ballasts (assuming one two-lamp and half of a two-lamp ballast per fixture):

All ballasts on	100% illumination
Two-lamp ballast on	66% illumination
½ of two-lamp ballast on	33% illumination
All ballasts off	0 illumination

This type of switching has the advantage of light reduction in relatively small steps, at low cost. A typical arrangement is shown in Fig. 21.1. Use of split-wired two-lamp or three-lamp ballasts rather than one-lamp units is advantageous from the cost and energy viewpoints. Even more uniform light reduction and finer control are possible with two-level ballasts (at increased cost). There, each lamp remains lighted but at either full or half output. Thus the designer could have 0, 50, and 100% levels, or by combining alternate ballast switching with two-level ballasts, could have 0, 17, 33, 50, 67, 83, and 100%. However, if that degree of control is desirable, dimming is probably preferable. The choice will depend on the type of space and on the situation economics, as discussed below. An alternative method of achieving lower lighting levels in discrete steps, by switching, is to introduce impedance into the lighting circuit. This acts to reduce circuit current and light output. Such devices are readily available for control of incandescent lamps (see Section 16.26) and fluorescent lamps (Fig. 21.6).

Recognition of the fact that an increased number of control (switch) points makes finer control and therefore energy conservation possible led to a new (and unusual) requirements in ASHRAE/IES 90.1 relating to the number of control points in a space and their types. As pointed out in Section 20.5 and above, manual

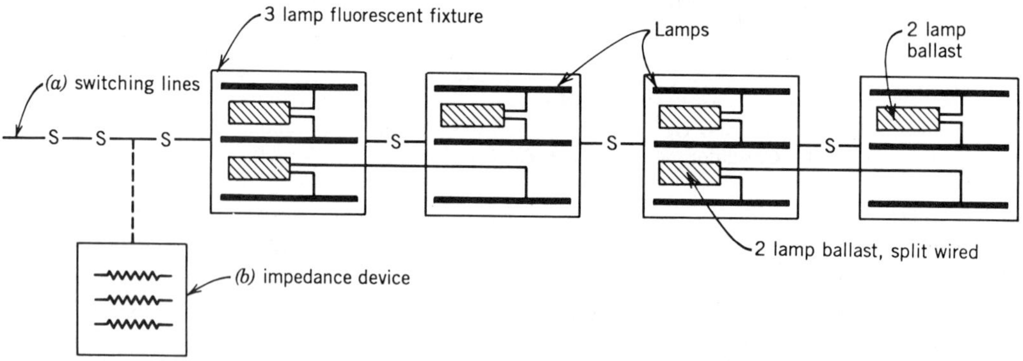

Fig. 21.1 *Schematic diagram of switching arrangements to achieve multiple discreet lighting levels with three-lamp fluorescent lighting fixtures. Two-lamp ballasts are used in the interest of energy conservation and financial economy. In scheme* (a) *ballasts are switched, thus removing either one or two lamps from service. Finer control is achieved by using two-level ballasts or by introducing impedance* (b) *into the circuit either in a block for an entire circuit or distributed in each fixture.*

TABLE 21.2 **ASHRAE 90.1 (Table 6.2): Control Types and Equivalent Control Points**

Type of Control	*Equivalent Number of Control Points*
Manually operated on–off switch	1
Occupancy sensor	2
Timer—programmable from the space being controlled	2
Three level, including off, step control or preset dimming	2
Four level, including off, step control, or preset dimming	3
Automatic or continuous dimming	3

Source: ASHRAE/IES 90.1-1989; reprinted with permission.

switching is ineffective as an energy-conserving means. Standard 90.1 therefore first specifies the minimum number of control points required in a space and then awards automatic switching or dimming with a higher number of equivalent control points (see Table 21.2). The basic requirement (see Standard 90.1, Section 6.4.2) is for one control point for every 450 ft^2 or fraction thereof of enclosed lighted space plus one control point for each task (or group of tasks) located in the space. Thus the classroom of Example 20.1 would require two control points for its 517 ft^2 of area plus one control point for all similar tasks grouped in the room, for a total of three control points. This requirement could be met with three wall switches or (see Table 21.2) one wall switch and an occupancy sensor, or an automatic dimming system probably initiated by daylight sensors. The point is, of course, to encourage use of automatic controls, which, as has been pointed out repeatedly, is the only proven method of attaining significant energy conservation.

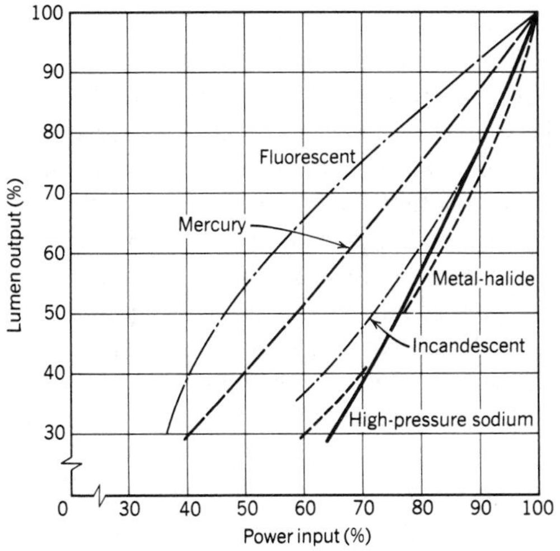

Fig. 21.2 *Typical dimming curves for generic light source types. Note that fluorescent is efficient and approximately linear, down to 40% output. All other sources have reduced efficacy when dimmed.*

(b) Dimming. The techniques and equipment required for dimming each of the different light sources, as well as the effect on the color of the light produced and on the lamp, are discussed in Chapter 19. Figure 21.2 shows typical lumen output versus power input curves for common light sources. Note that for fluorescent lamps, *even with conventional ballasts*, dimming down to approximately 40% of output is possible without reducing efficacy. This desirable characteristic can be exploited in control schemes where it is desired to gradually change light output without sacrificing efficiency. Since below 40% output efficacy drops off, an economical and efficient control scheme combines dimming and switching of multilamp fluorescent fixtures to yield an almost stepless output range of 13 to 100% output. Continuous dimming over a 10 to 100% range is practicable with special dimming ballasts or with electronic ballasts. As discussed in Chapter 19, electronic ballasts are much more energy efficient than conventional ones, and must be considered for all *new* installations, dimmed or not. For retrofit work, SCR (silicon-controlled rectifier) or triac dimmers will give excellent results with existing conventional core-and-coil ballasts, down to 40% output, as noted above.

(c) Control Initiation. The control operation is either manual or automatic. Manual operation is usually applicable only to a small number of simple functions such as on–off or level switching. Even then, however, the tendency is to leave lights on at the maximum level and not to shut them off when leaving a room. Numerous studies have demonstrated that no lasting appreciable energy economy is possible when the control initiation is entirely manual and relies on a facility's personnel, however well intentioned. Even a system of rewards for energy conservation rarely has long-term effects. A modicum of energy economy is possible when the turn-off function is automated by the use of "time-out" switches, which open after a preset interval (see Fig. 21.3). Significant long-term energy reduction is only achieved with automatic control initiation.

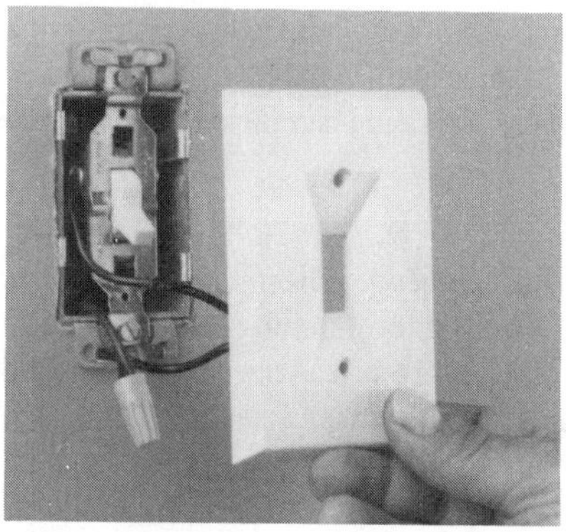

Fig. 21.3 Time-out switch replaces ordinary wall switch for control of incandescent and fluorescent lighting loads. Its solid-state timing mechanism, which is mounted on the back of the switch wall faceplate, is adjustable to turn off in 10 min to 12 h. Typical applications are closets, stock rooms, and other short-time occupancy spaces. (Courtesy of Paragon Electric Co., Inc.)

Automatic controls are of two types: an open-circuit type and a closed-loop feedback type, also known as static and dynamic control, respectively. The former initiates a control function that is independent of the actual lighting situation, whereas the latter reacts to the condition of the lighting situation it controls via a feedback loop.

(d) Static Control. The most common type of open-circuit lighting control is the programmable time controller. These vary from the small relatively simple unit of Fig. 21.4 to the more sophisticated units shown earlier in Fig. 16.13. These devices are available in a myriad of designs and capacities, but all perform the same basic function—(remote) control of loads and circuits on a preprogrammed time basis. The programming in turn is determined after analysis of operating schedules, task requirements, and field conditions. With "tight" programming, energy savings of up to 50% over an uncontrolled installation are possible.

Because these devices act only on a time base and are insensitive to actual field conditions, an override feature must be incorporated to permit accommodation of special conditions. Thus if the timer is arranged to shut off the row of lights adjacent to the windows between 10 A.M. and 3 P.M., local override must be provided to accommodate dark rainy days and the like. Similarly, if lights are shut off during nonworking hours, provision must be made for persons working overtime. The override arrangement can be entirely local, in which case it may lead to energy waste because it depends on local cancellation; it can be local with time-out, which can be a nuisance to a person working for an extended period; or it can incorporate an override feedback link to the controller, usually operated by telephone lines. In general, programmable time controls are best applied to facilities with regular, repetitive schedules and few exceptional situations.

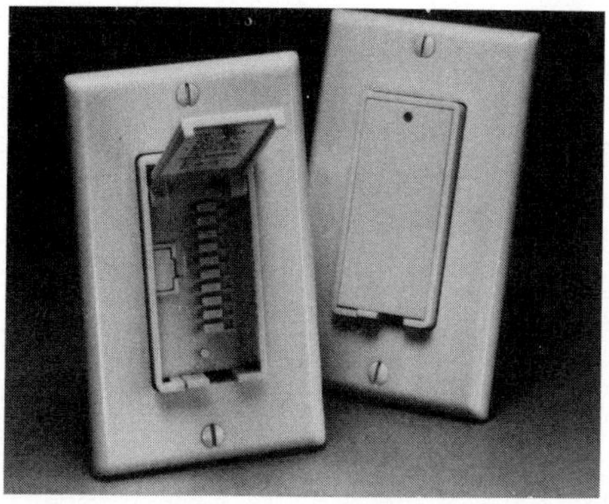

Fig. 21.4 Programmable lighting switch mounts in a standard wall device box. The unit can operate as a conventional on–off switch, and in automatic programmed mode with ten 1-h increments controlled by individual slide switches (see photo). (Courtesy of Leviton.)

(e) Dynamic Control. The second type of automatic control initiation responds to sensor-indicated field conditions via an information feedback loop. It is frequently referred to as dynamic control since the initiation of a control function depends not on a fixed programmed parameter such as time but on real-time field parameters (i.e., as measured at that instant). In the case of lighting control these parameters may be ambient illuminance, time, system kW demand, kWh usage in a time period, and space occupancy, singly or in a combination, depending on the programming algorithm (i.e., how the controller's microprocessor has been programmed). The control device in its entirety is called a *programmable controller* (see Section 16.16), which, in combination with the field sensors and the interconnecting wiring, constitute the control system. [Some systems are "wireless," using high-frequency signals impressed on the power wiring system to transmit control signals. This arrangement is known as a PLC (power line carrier) system and is discussed in Section 16.28.]

In addition to its CPU (microprocessor) the programmable controller contains input/output (I/O) interfaces, memory, and means for programming (and reprogramming). An operational block diagram is shown in Fig. 16.14*a*. The controller accepts not only the usual time-based signal function, but also information (feedback) from field devices via its I/O device. It then "processes" the signal in its CPU, which consists of logic and storage memory, and sends out a resultant processed control signal.

Large lighting control systems use a computer in lieu of a programmable controller and usually have additional control facilities, such as telephone interfaces and local control/relay/switching centers, which process local sensor input and control local lighting blocks (see Fig. 21.5). Identical systems are used for building automation (see Appendix J), HVAC control, energy management and the like, the difference being the type of sensing devices and the control algorithm.

21.4 Lighting Control Strategy

It is apparent that a good lighting control system will match the lighting supplied to the lighting required, *as the requirement varies*, and thus overlighting and underlighting are avoided. In addition, the control system must be capable of permitting initial adjustments and external, non-lighting-connected constraints such as commands from a peak-demand controller. The common lighting system situations addressed by the control system are listed below. Note the "value" of each strategy in terms of power credits, in Table 21.1.

(a) System "Tuning." In every lighting installation there is a difference between the design intent and the field result. This is due to assumptions and imprecision in calculation, differences between specified and installed equipment, equipment location changes, and so on. The responsible lighting designer will "tune" the lighting system in the field to attain the intended design. This usually means *reducing* levels in nontask areas, since spill light is frequently sufficient for circulation, rough material handling, and the like. This tuning can result in energy reduction of 20 to 30% depending on the control technique.

The smaller the group of light sources controlled, the more accurate the tuning and the larger the energy saving, proportionally. Lighting system retuning is also required when the function of an entire space is changed, or when a

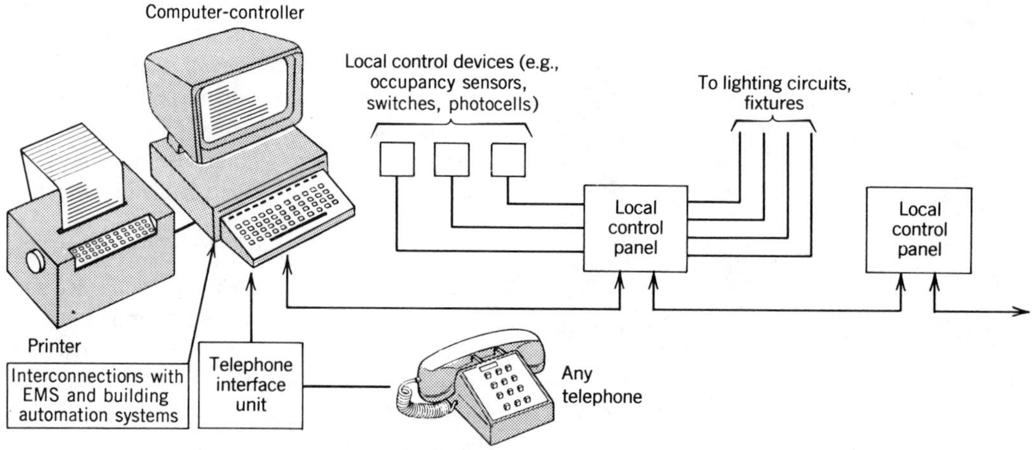

Fig. 21.5 *Schematic arrangement of a large lighting control system. The controller can schedule and supervise thousands of control points via the local control panels. The local panel accepts coded commands from the controller and operates individual devices and circuits via local panels. It also accepts local control signals from manual (switches) and automatic devices. Override is provided via telephone command to the controller, and all functions, including overrides, appear as "hard copy" on the printer.*

single area is altered by a furniture move or by a task change. An ancillary benefit of field tuning is glare reduction, which frequently improves task visibility dramatically. Tuning is a one-step function in the sense that once accomplished, it does not require change unless the space function changes. Therefore, it should also be reversible to accommodate such changes. Tuning should not be confused with *lumen maintenance*, described in (*d*) below, which is a control strategy designed to compensate for normal system output decline and its corollary—initial system overdesign.

As stated, the action most often required is a reduction in illuminance. This can be accomplished by:

1. Making appropriate field modifications to adjustable fixtures (aiming, lamp position, etc).
2. Replacing lamps with others of lower wattages, and replacing fluorescent tubes with low-wattage tubes or "phantom" tubes. These changes will reduce light output, reversibly.
3. Replacing fluorescent ballasts with low-current ballasts, thereby lowering output.
4. Adding current-limiting impedances to luminaires or lighting circuits as mentioned in Section 21.3*a* above (see Fig. 21.6).
5. Ballast switching or use of multilevel ballasts (see Section 21.3*a*.
6. Dimming by adjustment of potentiometer at individual fixtures so equipped.
7. Dimming of groups of fixtures (with standard ballasts) from centralized dimmers (see Fig. 21.7).
8. Replacing standard wall switches with time-out units (Fig. 21.3), programmable units (Fig. 21.4), or dimmer units (Fig. 16.30).

Items 5 and 7 can be accomplished locally or from a remote central controller.

(b) Variable Time Schedule. No normal task area has a constant 24-hour, 365-day, lighting requirement. In commercial and industrial spaces, work areas have regularly scheduled periods during which task lighting is not required. These include coffee and lunch breaks, cleaning periods, shift changes, and unoccupied periods. Programmed time controls can readily save 10 to 25% of the energy use as compared to relying on occupants to manually operate the controls. The action of such a controller in an actual installa-

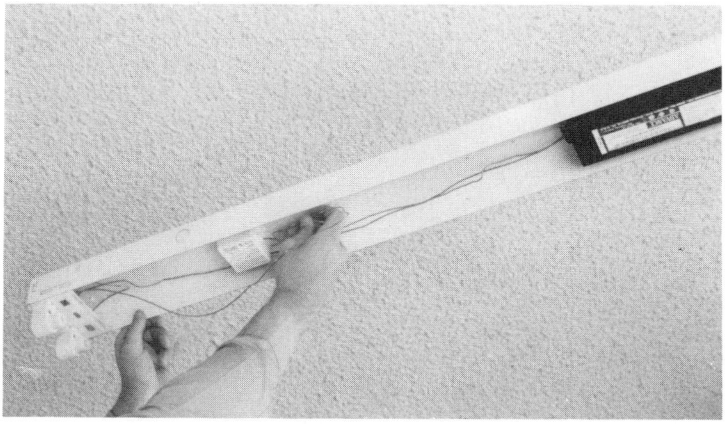

Fig. 21.6 This type of compact solid-state electronic circuit reduces ballast power input by approximately 30% and lamp output by about 28% (see the curve in Fig. 21.2), resulting in a net gain in efficacy. The power factor remains above 90%. Similar units are available for larger power decrease. They can be mounted on the lamp end or in the fixture channel, as shown. An ancillary benefit is cooler ballast operation, resulting in extended life. (Photo courtesy of Remtec Systems.)

tion is shown in Fig. 21.8. "Tight" scheduling took account of lunch hour and provided lighting only in restricted areas being cleaned rather than whole floors. Payback period for the investment in control equipment varies between 1½ and 5 years. Note that this type of static control (i.e., control that is insensitive to actual field conditions) gives only a 15% power credit (item 12, Table 21.1) because of its limited ability to conserve energy.

(c) Occupancy Sensing. Within a normal 9 A.M. to 5 P.M. working schedule, offices in commercial spaces are unoccupied for 30 to 60% of the time. The reasons are manifold—coffee break, conference, work assignment, illness, vacation, and reassignment to a different work location are a few. Occupancy sensors can operate relays to turn off lights after a preset minimum period of about 10 minutes or can dim the light level to a minimum in areas such as corridors, which always require some light. (They can also turn off other energy consumers

*Fig. 21.7 Typical single-circuit fluorescent dimmer. This unit is rated at 277 V and will adjust the energy input to all the fluorescent ballasts on a single 20-A circuit, from 50 to 100% of full capacity. (Dimming ballasts are not required; see Section 21.3*b*). Light output is reduced somewhat less than energy, improving efficacy. Lighting-level adjustment is either manual or automatic. Manual adjustment, at the dimmer or at a remote location, is smooth, uniform, and affects the entire circuit. Automatic dimming is initiated by the photosensor shown in the photograph, usually used as a daylight sensor to control dimming of perimeter lighting. Automatic dimming incorporates a 2-min fade time so that lighting-level changes will be smooth and will not disturb space occupants. Groups of such dimmers can be remotely adjusted from a central controller. The unit illustrated is 15 × 9⅝ × 8½ in. deep and weighs 13 lb. Similar units are available for other currents and voltages and for HID light sources. Their action is shown in Fig. 21.2. (Photo courtesy of Lutron.)*

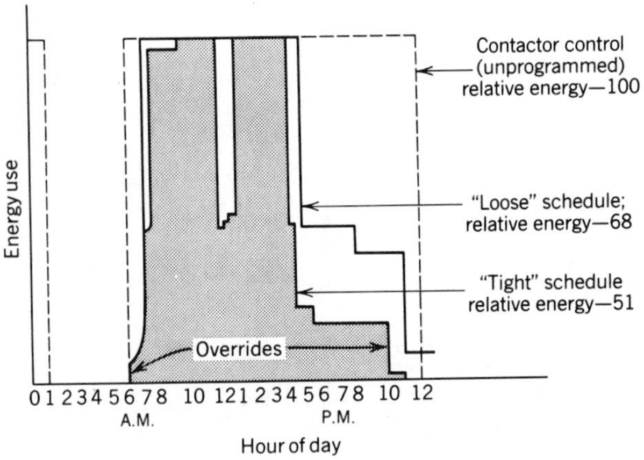

Fig. 21.8 *Actual plot of energy usage on the fifty-eighth floor of the World Trade Center in New York, with three degrees of lighting control. Note the importance of override by occupants when using "tight" scheduling. Override was provided on a 1000-ft² zone basis. [Data extracted from Peterson and Rubenstein (1982).]*

such as fan coil units, air conditioners, fans.) Reestablishment of the original lighting level can be instantaneous, delayed, or manual on the action of the occupant. Another useful function of an occupancy sensor is to provide automatic override in schedule systems, thus both relieving the occupant of the necessity of using a manual override and limiting the energy use to actual occupancy time. It can also light the occupant's way into a space, and shut off the system after he or she has left.

Occupancy sensors that react to presence are of two types—ultrasonic and passive infrared. Depending on type and mounting position, sensors cover a maximum area of 250 to 1000 ft^2 per unit. Payback for this equipment runs between 6 months and 3 years, depending on the type of space and the degree of control already existing. They are best applied in areas that are divided into individual rooms and work spaces. Sensors can be wall or ceiling mounted, or mounted on a wall outlet box in a combined sensor/wall switch configuration. A few designs are shown in Fig. 21.9. (See also Section 22.2, which discusses the security application of motion detectors and illustrates their operation.)

ASHRAE/IES 90.1 recognizes the great effectiveness of these devices in both a connected power adjustment credit of 30 to 40% (Table 21.1) and establishing a control point equivalency of 2 (Table 21.2).

(d) Lumen Maintenance. Referring to Section 20.31, we see that in order to maintain a minimum lighting level to the end of a maintenance period, we deliberately overdesign initially. The extent of the overdesign is the reciprocal of the light loss factor (LLF). With an average LLF of 0.60, this initial overdesign amounts to (1/0.60), or 66%. Assuming a linear light falloff over a 2-year maintenance period, this overdesign results in an average of 33% energy waste. (In the next 2-year maintenance cycle the savings will be slightly less due to a small amount of unrecoverable loss.) Because the light depreciation is a continuous and very gradual process over the maintenance period, the most appropriate control strategy is one that will reduce the initial overlighting by the required amount (as measured in the field) and gradually restore it, as the system ages. This control strategy, known as *lumen maintenance*, is best accomplished by a centrally controlled dimming system, operating in conjunction with local light sensors (photocells). The photocells measure ambient light, and in response to their signals the controller operates the dimming units to raise (or lower) the light output. Depending on the size of the installation, the dimmers (of the type shown in Fig. 21.7) can either be dispersed and controlled from a central point, or the entire system can be installed at one central location. The modulating action of such a system over the maintenance period is shown in Fig. 21.10. Note from Table 21.1 that this strategy in a purely manual mode (periodic maintenance, initial light reduction in accordance with the length of the maintenance period) gives only a 10% connected

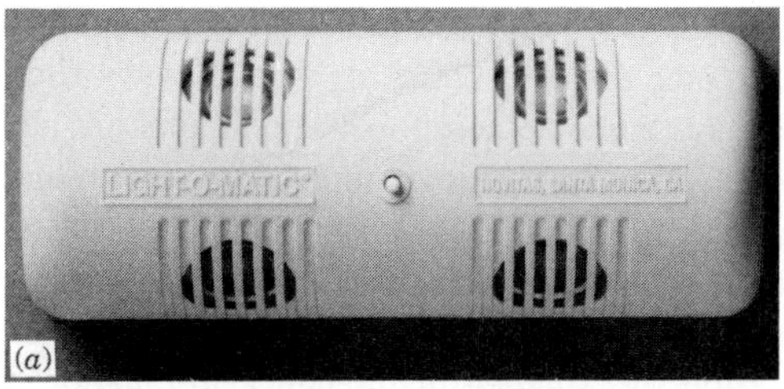

Fig. 21.9 (a) *Ultrasonic ceiling-mounted occupancy sensor. These units, which have field-adjustable sensitivity and time delay, are available with various coverage patterns to suit the installation. They initiate a control signal for relay control of lighting and, if desired, HVAC and energy control equipment as well. The unit illustrated measures approximately* $1\frac{1}{2} \times 2\frac{1}{4} \times 8$ *in. and weighs 5 oz. (Photo courtesy of Novitas.)*

Fig. 21.9 (b) *Combination ultrasonic occupancy sensor and manual lighting switch mounts on a standard wall junction box. The unit operates manual* ON, *automatic* OFF *with manual override. It is intended to control the lighting in a single room. The unit measures* $4\frac{1}{2}$ *in. square and weighs under 1 lb. (Photo courtesy of Novitas.)*

Fig. 21.9 (c) *Passive infrared motion sensor mounts on a standard wall outlet box and is designed, like* (b) *above, to control lighting in a single (large) enclosed space. It has a 170° horizontal coverage, manual* OFF *override, and can control fluorescent or incandescent lighting with its self-contained relay. The unit is approximately 3 in. W × 5 in. H and weighs 12 oz. (Photos courtesy of Hubbell, Wiring Device Div., Building Products Group.)*

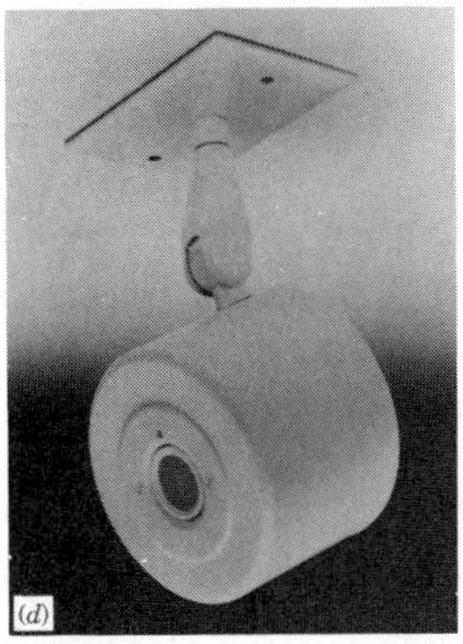

Fig. 21.9 (d) *Typical ceiling-mounted PIR occupancy sensor is available with narrow or wide coverage patterns to suit the application. This unit initiates control, which is actually performed by a remote relay at the luminaires, lighting panel, or in a relay panel. The PIR housing is 4 in. in diameter and 3 in. long. (Photo courtesy of Hubbell, Wiring Device Div., Building Products Group.)*

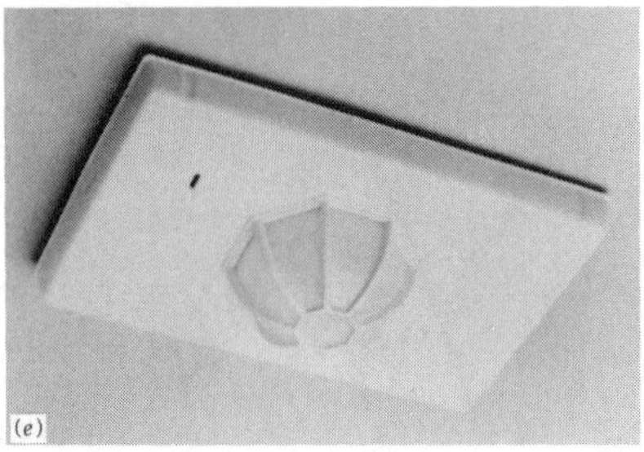

Fig. 21.9 (e) *PIR occupancy sensor with 360° coverage. Intended for ceiling mounting in large open areas. (Photo courtesy of Tork Time Controls.)*

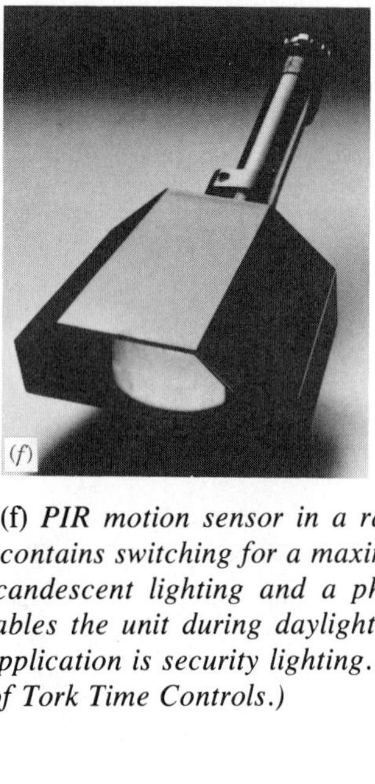

Fig. 21.9 (f) *PIR motion sensor in a raintight enclosure contains switching for a maximum of 500 W incandescent lighting and a photocell which disables the unit during daylight hours. Primary application is security lighting. (Photo courtesy of Tork Time Controls.)*

load reduction. When automated this "credit" can rise to 45%.

In a new installation, the choice of whether to use electronic ballasts and full-range dimming, conventional ballasts and partial dimming, or a system of multilevel switching is an economic one and depends on many factors, the most important of which is the cost of energy. Often a combined system is advisable. In interior zones initial lighting reduction does not exceed 40%, and full-range dimming is not required. In perimeter zones, daylight often provides all of the required light, and either full-range dimming, SCR dimming plus switching, or a multilevel switching system is required. (SCR dimmers are also referred to as lighting energy adjusters, reducers, or modulators in the manufacturers' literature.) See Section 21.3*a* above and the discussion of daylight compensation below. Payback period for a lumen maintenance installation of this type varies from 1 to 5 years. Shorter payback periods can be obtained by using multilevel switching rather than dimming, as discussed in Section 21.3*a*. Although dimming gives stepless control and therefore higher energy savings, the lower cost of switching reduces the payback period.

An additional favorable effect of initial light reduction is the lengthening of effective lamp life and a reduction in the rate of its lumen depreciation. When a lamp is operated at rated voltage, its lumen output drops during its life, according to the type of lamp. (See, for instance, Sections 19.12, 19.16, and 20.31). However, if lamps are operated at reduced output, as would be the case if lamps are dimmed to compensate for initial overlighting, the lamp life cycle is greatly extended, lumen depreciation is reduced, and lamp energy consumption is linearly reduced.

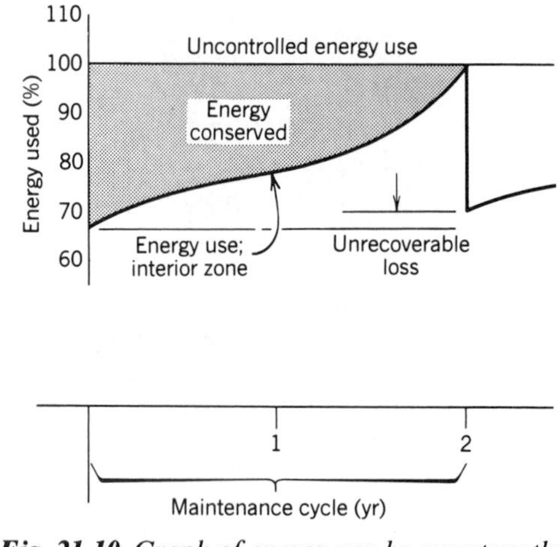

Fig. 21.10 Graph of energy use by a system that reduces initial lighting level to compensate for initial overdesign and gradually increases level as system output depreciates. In subsequent cycles the energy savings are reduced slightly because of unrecoverable output loss.

Typical life extension (to economic replacement) figures are:

Fluorescent	80%
Metal-halide	40%
High-pressure sodium	20%

These figures are for interior zones; for perimeter zones with ambient daylight compensation they are higher.

(e) Daylight Compensation. It should be apparent that a control system arranged for continuous ambient light compensation, as described in the preceding subsection, is automatically arranged to compensate for ambient daylight. The difference is that ambient compensation for lumen maintenance due to light loss factors is a very gradual process of *increasing* output, whereas daylight compensation can be a minute-by-minute variation, and generally in the direction of *decreased* artificial light. Because of these possible rapid variations, most switching systems are undesirable, as the constant on–off or level switching of lamps can be very annoying to occupants. A system that switches in small increments by use of multilevel ballasts in multiballast fixtures can be acceptable if combined with a fairly wide "dead band" between ON and OFF (as with a thermostat) to prevent "hunting" and excessive switching. (Indeed some variations in light levels as occur with daylight can be considered desirable. It is not necessary to compensate *exactly* for ambient daylight.)

Automatic dimming is the system of choice. Since in perimeter areas daylight often supplies all the required light, the system must either be full-range dimming with electronic ballasts (for fluorescent installations) or partial dimming with conventional core-and-coil ballast, plus switching. Here again the switching action can be disturbing and must be carefully designed. Figure 21.11 shows the action of both types of dimming system in a fluorescent installation with daylight compensation. The crucial design element in a daylight-compensating system is the establishment of zone areas. Depending on latitude and climate, the southern and possibly the east and west exposures can have an interior (second) perimeter zone that receives sufficient daylight for a large enough portion of the year to economically warrant dimming. The northern exposure has only a narrow perimeter zone (see Fig. 21.12). As a rule of thumb, the size of the zones is established by determining the maximum room depth that receives at least half its illuminance from daylight for several hours a day. A computerized economic study with perimeter zone depth as a variable is the proper method of determining zone size. See Matsuura (1979) and Rubenstein, Verderber, and Karayel (1985).

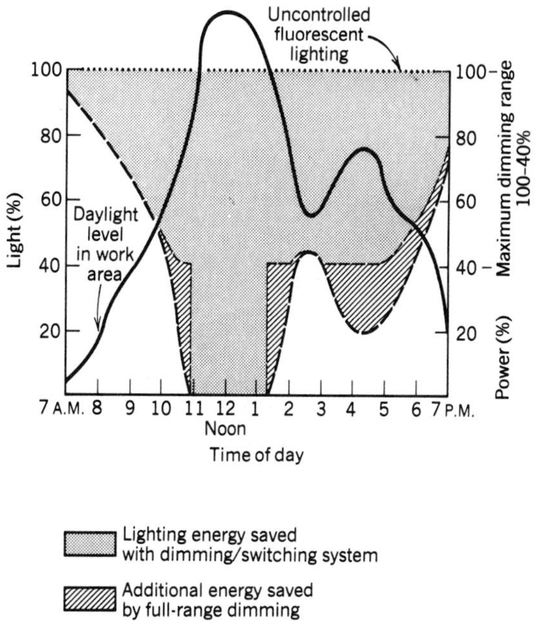

Fig. 21.11 *Typical graph of energy savings with daylight-compensating lighting control. A full-range dimming system is more effective than one that dims down only to 40%, since daylight often supplies most of the required lighting. An economic analysis is required to determine whether the additional cost of such a system is justified.*

Placement of the control photocells depends on the control system. Where daylight compensation is desired in conjunction with lumen maintenance (preceding subsection), area photocells are desirable, since they give a feedback control signal for the specific area involved. Alternatively, a daylight-factor (DF) map of a

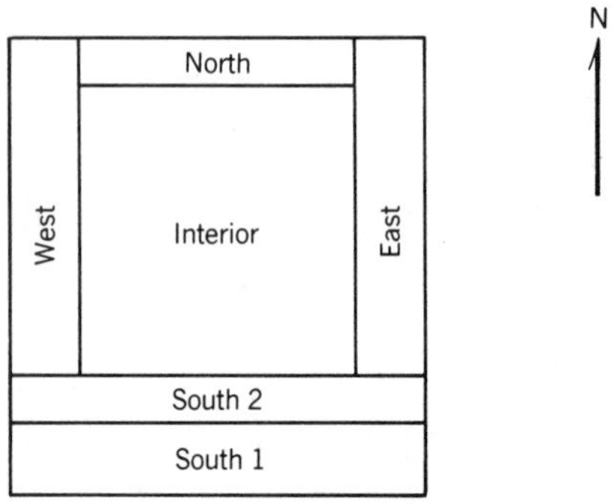

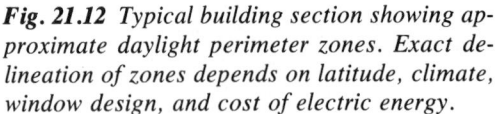

Fig. 21.12 *Typical building section showing approximate daylight perimeter zones. Exact delineation of zones depends on latitude, climate, window design, and cost of electric energy.*

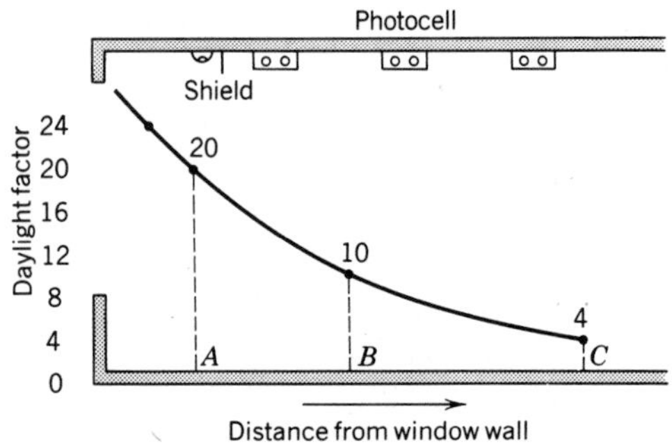

Fig. 21.13 *Typical daylight factor curve plotted on a room section with one side window. The photo cell control technique for daylight compensation described in the text assumes a constant DF distribution indoors, that is, fixed luminance sky and no direct sun.*

space can be made, preferably after the installation is complete and the furniture in place, and photocells located at one point, based on this map.

Refer to Fig. 21.13. We know from our study of daylight in Chapter 19 that the DF at any inside point is constant. Therefore, by measuring actual daylight level at a point either immediately inside or outside the window, shielded from direct sun and from inside artificial lighting, we can relate it to any area in the room by the ratio of daylight factors and establish a switching level at which the lighting for that inside area is switched, partially or fully. An example will make this clear. In Fig. 21.13, the photocells are mounted at point *A*. The ratio of daylight between points *A* and *B* is the ratio of their daylight factors, that is, 20/10 or 2.0. Therefore, if 500 lux of daylight is required at point *B* before switching lights in that area, 2.0 (500) or 1000 lux is required at point *A*. Similarly, a 500-lux requirement at point *C* corresponds to (20/4) (500) or 2500 lux at point *A*. Therefore, switching can be initiated by two single-level photocells or a dual-level unit at point *A*, with settings at 1000 and 2500 lux. Other switching arrangements can be made on the same principle. Dimming initiation can also be arranged in this fashion.

Daylight compensation can reduce energy use in perimeter areas by up to 60%, depending on latitude, climate, hours of building use, initial power density, and so on. The amount saved for this entire building obviously depends on the building configuration, that is, the ratio of perimeter to total area. Payback time is usually in the range of 3 months to 3 years. The power adjustment factor in Table 6.1 for this strategy varies from 10 to 30%, depending on the switching/dimming technique employed.

Summarizing, a well-designed lighting control system can provide energy conservation of up to 60%, extremely long lamp life, reduced cooling costs, extended ballast life, and reduced maintenance costs. Investment payback period (see Appendix E), when considering all these factors, is always short and therefore financially attractive. An aspect of a centralized lighting control system not discussed above, since it is not directly concerned with lighting, is its use in connection with peak demand reduction (see Section 14.13). When interconnected with a demand controller, this approach acts to reduce electric lighting load in accordance with a predetermined preferential-load schedule. Significant savings can thereby be achieved, as explained in Chapter 14. See also Rubenstein, Ward, and Verderber (1989) for notes on improving photocell-controlled systems.

RESIDENTIAL OCCUPANCIES

21.5 Residential Lighting

Residential lighting offers to the lighting designer great opportunity for originality and ingenuity, since a residence combines more functions and needs than any other building. Furthermore, it generally requires that all work be done at minimal cost, and that the end result please a range of tastes. The designer approaches the problem with a list of requirements, a perception of the space, and two basic tools: the lighting fixture and the architectural lighting element. The former was covered in Sections 20.18 through 20.28 and is augmented by the fixture data in Figs. 21.14 and 21.15. The latter is discussed below. The reader is referred to *Design Criteria* (1980) for additional details on residential lighting.

(a) Energy Considerations. In determining a power budget (see Sections 20.5 and 20.7), use the following unit power densities:

Bath	4.3 W/ft^2
Bedroom	1.4 W/ft^2
Finished spaces	2.2 W/ft^2
Kitchen	4.0 W/ft^2
Laundry	1.0 W/ft^2
Garage, unfinished spaces	0.5 W/ft^2

(Although residences are generally excluded from the requirements of ASHRAE/IES 90.1-1989, the reader is cautioned that there may be local jurisdictional standards whose requirements would alter the recommendations above.)

1. Provide means for reducing light levels in all areas. A kitchen during food preparation does not have the same lighting requirements as a kitchen being entered for a "refrigerator raid." Low-level lighting provision should be made in *all* rooms, including bathrooms. To accomplish this use high–low switches, simple dimmers, multilevel ballasts, and multilevel switching. An ancillary benefit is that ambience can be changed thereby in multiuse rooms such as dining rooms, family rooms, and finished basements.
2. Provide local task lighting for difficult tasks such as the location at which family accounts are handled.
3. Provide dimming and switching for accent lighting.
4. Use programmable timers for exterior lights.
5. In large residences consider programmable time switches and low-voltage control for ease of remote control and energy savings.
6. Use daylight in areas normally occupied during daylight hours such as kitchens and living rooms. Consider skylights with built-in artificial lighting for these areas.

(b) Sources. When using fluorescent, choose proper color for space (see Table 19.21). Despite their lower efficacy, use of 5000 K and daylight fluorescents as the artificial source in lighted skylights is very effective. Listed below are the appropriate sources to be used in different areas of a residence:

1. Work and utility areas, including kitchens, laundry, and workshop—fluorescent of appropriate color and shape.
2. Built-in architectural elements—fluorescent.
3. Bedrooms, portable lamps, accent lights—incandescent, tungsten-halogen, compact fluorescent, circular fluorescent.
4. Circulation areas, stairwells, closets—incandescent.
5. Exterior: for short periods—incandescents; for long periods—HID.
6. Bathrooms: general; incandescent or 3500 K fluorescent; mirror lighting—incandescent.
7. All rooms—daylight where possible.
8. All spaces—use incandescent when source is turned on and off frequently or lighted for short periods only.

(c) Recommendations. These design recommendations are applicable to residential occupancies of all types.

1. Use general/task-lighting concept with recommended levels as in Tables 21.3 and 21.4.

Wide Profile

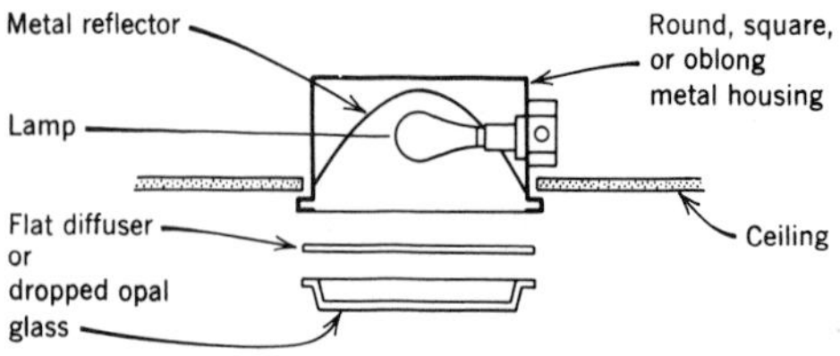

For general illumination (almost always used in multiple). Basement, reacreation rooms, kitchens, laundries, halls (service). Used singly or in small groups for small areas such as walk-in closets, garage, entry doors, overnangs in porches. Because of high luminance of diffuser, seldom used in living or social areas.

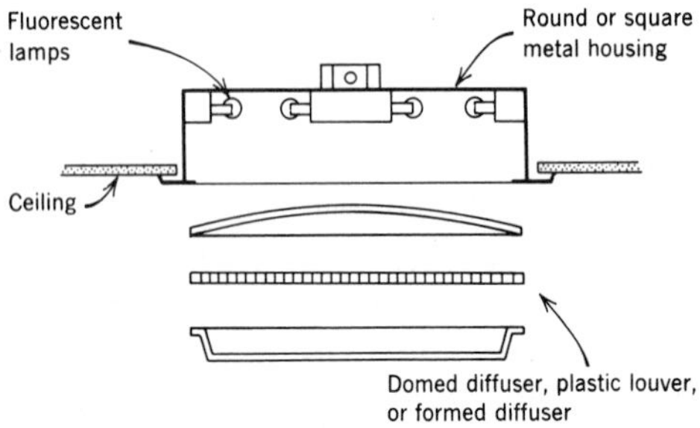

Very wide distribution, excellent as general lighting for kitchen, laundry, recreation room and bath. Because of large size and high lumen output, fewer units required. Often used signly or in pairs for entry halls and foyers and for skylight effect in interior halls and stairways.

Narrow Profile

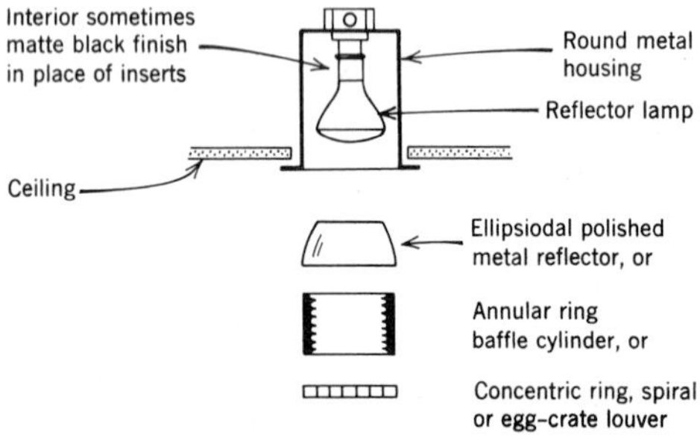

Accent lighting over plants, cocktail tables, etc. Wall lighting–mounted close to textured surfaces such as brick, stone, rough wood and fabrics. Task lighting–food preparation areas (may cause specular reflections), in multiple on quite close spacing for general lighting. Most effective when used near perimeter of room so that some light spills onto wall. Dramatic effects for family rooms, recreation rooms, and formal living areas. Supplementary stair lighting–shadow patterns define treads and risers. Dining tables–provide functional light on dining table to supplement decorative effect from hanging luminaire.

Medium Profile

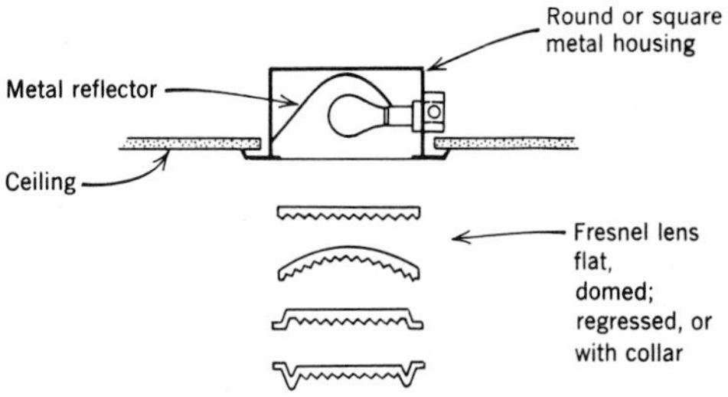

Used for specific task lighting where task area is large, such as kitchen sink, kitchen island counter or range, laundry tubs and ironing, game table, workbench, hobby area. Used for general lighting in restricted areas such as halls, entries and baths. Multiple groupings are satisfactory for the general lighting of kitchens and recreation rooms. If weatherproof, appropriate for outdoor uses, including overhangs, porches, entries.

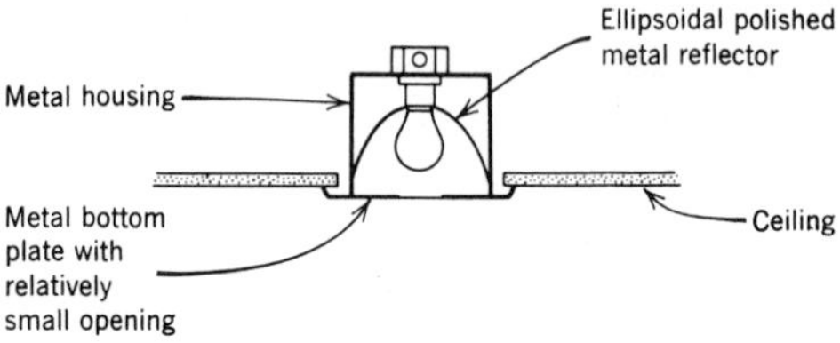

Uses basically the same as the Fresnel unit listed above, except that the lower luminance makes this type of equipment more usable in living and dining areas.

Special Asymmetric Profile

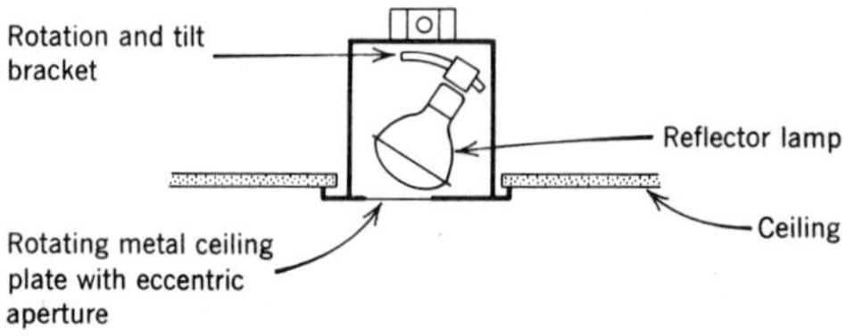

Useful for gallery or picture lighting and to light sculpture. If scalloping effect is acceptable can be used for wall lighting and to accent fireplace surfaces. Large size of bottom aperture sometimes make this unacceptable for highly styled interiors. May also be used for lighting piano music and sewing machines.

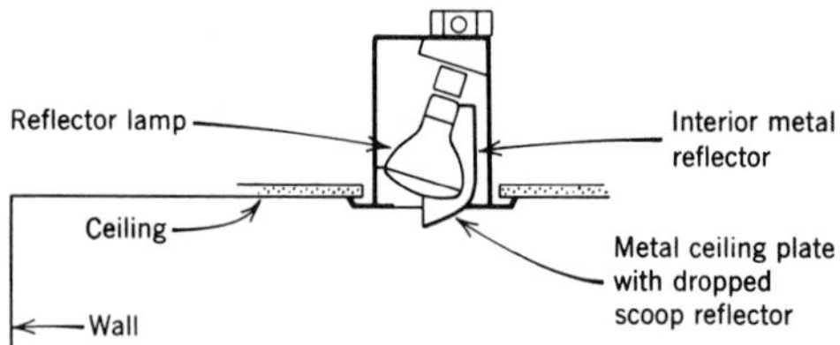

For uniform illumination of plane wall surfaces. Extremely effective for lighting murals and for minimizing wall imperfections. *Not* generally to be used for lighting textured wall surfaces because it directs no grazing light at wall. Spacing of these units is critical–follow manufacturer's recommendations closely.

Fig. 21.14 *Design features of recessed luminaires having wide, medium, narrow, and asymmetric profiles. See also Figs. 20.18 to 20.22 and 20.26. (Courtesy of IESNA.)*

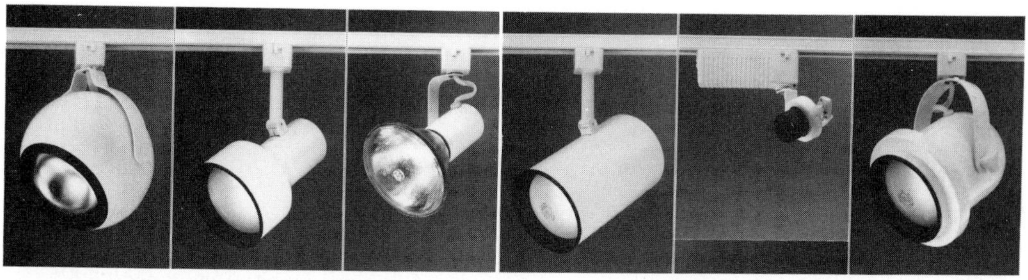

Fig. 21.15 Track lighting is available in a very wide variety of luminaire bodies. Those illustrated include (left to right) spheres, step cylinders, swivel reflector-lamp holders, flat-back cylinders, external transformer low-voltage spots, and gimbal-ring cylinders. Not illustrated are wall washers, adjustable and filtered spots, framing projectors, barndoor shutter units, and so on. In addition, a variety of single- and multicircuit track designs are made (see Fig. 15.12). (Photos courtesy of Rudd Lighting.)

2. Provide luminance ratios as in Fig. 21.16.
3. Provide general lighting in all spaces, sufficient for movement and casual seeing. Hallways require little lighting; stairs require more. Light stairs from directly above or ahead to create a shadow directly below the tread front. Lighting from the front eliminates shadows and can create a safety hazard.
4. Do *not* avoid ceiling lights as is so frequently done. Wide-profile ceiling fixtures provide general lighting; switch-controlled table lamps do not.

(d) Fixtures and Luminous Elements. See Figs. 21.14 and 21.15.

1. Utilize diffuse distribution for general lighting, narrow-distribution downlight for area and furniture accents, and narrow-distribution, ceiling-recessed incandescent wall-washers for accenting surface texture such as with brick walls.
2. Use built-in lighting to the extent possible, including architectural lighting elements (see next section). We believe that this demonstrates integrity of concept. For this reason, we recommend that the flexibility of track lighting be utilized for accent and task lighting but not for general lighting.
3. Private residences are the exception to the rule of selecting off-the-shelf items in preference to specials. The lighting should complement the architecture and furnishings, and frequently this can best be accomplished by original designs.

21.6 Architectural Lighting Elements

Architectural lighting elements are coves, cornices, valances, coffers, skylights, and other lu-

TABLE 21.3 **Current Illuminance Recommendations for General Lighting**

Activity or Area	Typical American Recommendation: Average Lux	Other Authorities: Average Lux
Conversation and relaxation	50–100[a]	50–100
Passage areas	50–100[a]	50–100
Areas other than kitchen	200–500	100–200
Kitchen	500–1000	300–500

[a]General lighting in these areas need not be uniform.

TABLE 21.4 **Current Illuminance Recommendations for Specific Visual Tasks**[a]

Seeing Task	*Typical American Recommendation: Average Lux*[b]	*Other Authorities: Average Lux*[b]
Dining	100–200	100–150
Grooming, makeup	200–500	500
Handcraft		
Ordinary seeing tasks	200–500	200–500
Difficult seeing tasks	500–1000	500–750
Critical seeing tasks	1000–2000	>1250
Kitchen duties		
Food preparation and cleaning involving difficult seeing tasks	500–1000	750–1000
Serving and other noncritical tasks	200–500	200–300
Laundry tasks	200–500	100–300
Reading and writing		
Handwriting, reproductions, and poor copies	500–1000	750
Books, magazines, and newspapers	200–500	300
Sewing, hand or machine		
Dark fabrics	1000–2000	>1250
Medium fabrics	500–1000	700–1000
Light fabrics	200–500	300–500
Table games	200–300	300

[a]Selection of illuminance within the given range is made on the basis of criteria given in Section 18.24.
[b]Divide by 10 to get footcandles. Due to the range of values, use of the exact 10.76 figure is unnecessary.

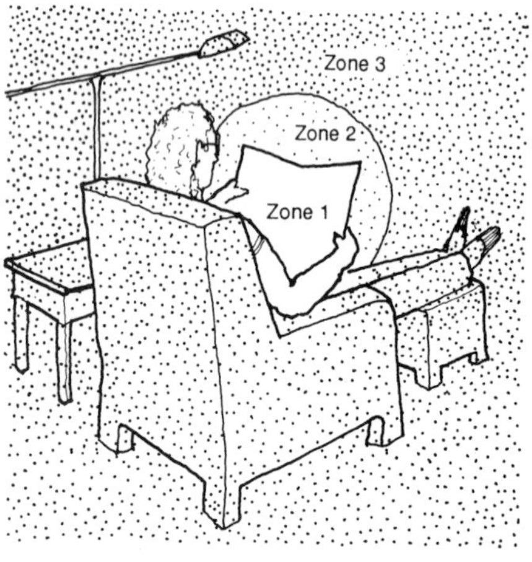

Zone 2 The immediate surroundings (area adjacent to the visual task)	
Desirable ratio	1/3 to equal to task*
Minimum acceptable ratio	1/5 to equal to task*
Zone 3 The general surroundings (not immediataly adjacent to task)	
Desirable ratio	1/5 to 5 times task*
Minimum acceptable ratio	1/10 to 10 times task*

*Typical task luminance range is 40 to 120 cd/m^2 [12–35 fL] and seldom exceeds 200 cd/m^2 [60 fL].

***Fig. 21.16** Seeing zones and recommended luminance ratios for residential visual tasks.*

minous constructions not normally comprising a lighting fixture. Although such units are normally less efficient than lighting fixtures, their use is often indicated by architectural considerations. Empirical design data are given in Fig. 21.17.

When using fluorescent tubes in these architectural lighting elements, it is possible to avoid dark spots between lamps by placing lamps at a slight angle rather than end-to-end, thus enabling ends to overlap. Reflectors increase installation efficiency. When used in coves, they should be aimed 15 to 25° above the horizontal and field-adjusted for best ceiling coverage. When using double strips, they should be stacked vertically as shown in Fig. 21.17*m*. Light output for double-lamp installation rarely exceeds 1.75 times the single-lamp output.

EDUCATIONAL FACILITIES

21.7 Institutional and Educational Buildings

The lighting requirements for some spaces in educational facilities coincide with requirements for commercial (office) and institutional buildings. To that extent the remarks herein are applicable there also. Generally, educational buildings (excluding private colleges and universities) are maintained from operating funds obtained from taxes, and the budget is *always* tight. Therefore, all equipment in these public buildings must be extremely hardy, vandal-proof, as maintenance-free as possible, and low in energy consumption. Maintenance in such buildings is generally poor and on a repair rather than preventive basis. With these overall criteria in mind, the following remarks apply to lighting equipment.

1. Use source with highest possible efficacy. Remember that daylight has the highest efficacy, followed by HPS, fluorescent, and metal-halide sources.

2. Where specific color lamps are called for, such as 3500 K, T8, the requirement should be permanently stenciled in large letters on the lighting fixture.

3. Long-life sources should always be given preference because of lower maintenance. Thus corridor and stair lighting should be fluorescent or HID. This is also important in locations where relamping is difficult, as in high-ceiling rooms such as gyms and assembly rooms. In such spaces, extended-life lamps are recommended, with preference to HID sources.

4. In calculating levels, low figures for LLF (light loss factors) should be used to allow for aging of paints and dirt accumulation. Cleaning of lighting fixtures in schools is virtually unknown. A figure of 0.5 to 0.6 is reasonable. This being so, provision should be made to reduce initial overlighting (see Section 21.4).

5. Most schools are *not* air conditioned. The masking air noise being absent, careful control must be exercised on noise and vibration from ballasts, diffusers, and so on. Ballast noise increases with current rating. Therefore, 800-mA high-output and 1500-mA very-high-output lamps must be used with caution, particularly in locations that amplify sounds, or where low NC (noise criteria; see Chapter 26) obtains.

6. Lighting equipment must be designed for an absolute minimum of maintenance. This means captive screws; rust-preventive plated parts; captive-hinged diffusers, whose cleaning requires only one person; ballast replacement without demounting fixtures (plug-in ballasts are available), nonyellowing plastics, and high-quality finish and assembly.

For recommended illuminances for the various visual tasks in an educational facility, see Section 18.24 and ANSI/IES RP-3 1988, *Guide for Educational Facilities Lighting*. In the following sections we discuss the lighting requirements of specific school building areas.

21.8 General Classrooms

The classroom is the basic space in a school. Unlike the classic schoolroom with fixed seats, a single viewing direction, and a fixed teacher location, many modern classrooms, at all grade levels, utilize multiple-student groups, teacher mobility, multiple tasks in the same overall space, and movable seating arrangements. Such spaces require:

1. Controls that permit subdivision of lighting and level control within the subdivision (see Sections 21.2 to 21.4).
2. A lighting system with a considerable indirect component and high diffusion to minimize the problem of veiling reflections due to viewing direction (see Sections 18.27 to 18.30).
3. Low-brightness luminaires with high VCP in all viewing directions, since a considerable portion of the students' time is spent in a head-up position (see Section 18.6).

In addition to these special recommendations, some lighting recommendations applicable to all classrooms are:

4. Classrooms using VDT units require special lighting (see Section 21.18).

ILLUMINATION

(a) Lighted Cornices

Cornices direct all their light downward to give dramatic interest to wall coverings, draperies, murals, etc. May also be used over windows where space above window does not permit valance lighting. Good for low-ceilinged rooms.

(b) Lighted Valances

Valances are always used at windows, usually with draperies They provide up-light which reflects off ceiling for general room lighting and down-light for drapery accent. When closer to ceiling than 10 inches use closed top to eliminate annoying ceiling brightness.

(c) Lighted Coves

Coves direct all light to the ceiling. Should be used only with white or near-white ceilings. Cove lighting is soft and uniform but lacks punch or emphasis. Best used to supplement other lighting. Suitable for high-ceilinged rooms and for places where ceiling heights abruptly change.

(d) Lighted High Wall Brackets

High wall brackets provide both up and down light for general room lighting. Used on interior walls to balance window valance both architecturally and in lighting distribution. Mounting height determined by window or door height.

(e) Lighted Low Wall Brackets

Low brackets are used for special wall emphasis or for lighting specific tasks such as sink, range, reading in bed, etc. Mounting height is determined by eye height of users, from both seated and standing positions. Length should relate to nearby furniture groupings and room scale.

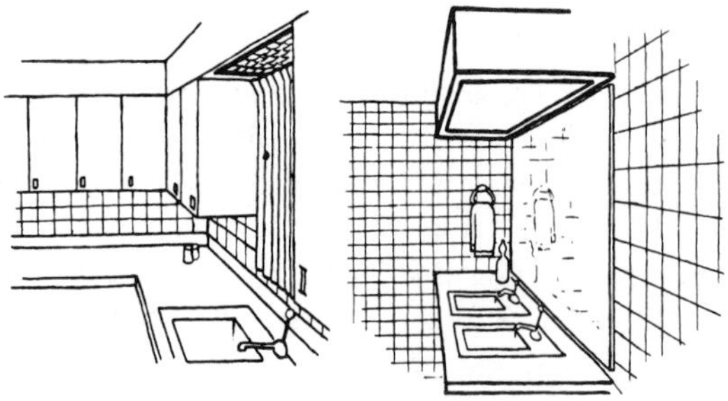

(f) Lighted Soffits

Soffits over work rease are designed to provide higher level of light directly below. Usually they are easily installed in furred-down area over sink in kitchen. Also are excellent for niches over sofas, pianos, built-in desks, etc.

Bath or dressing room soffits are designed to light user's face. They are almost always used with large mirrors and counter-top lavatories. Length usually tied to size of mirror. Add luxury touch with attractively decorated bottom diffuser.

Fig. 21.17 *Residential architectural lighting elements. (Courtesy of the IESNA except as otherwise noted.)*

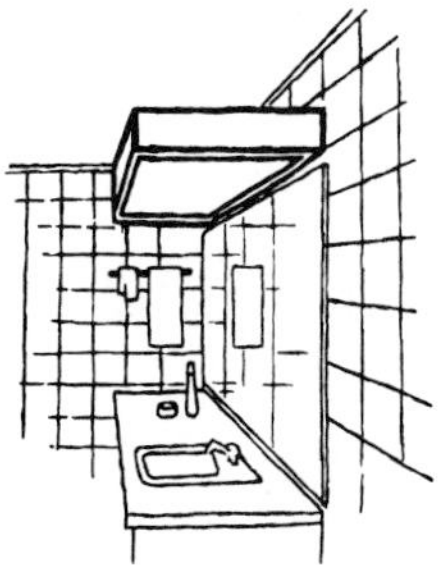

(g) Lighted Canopies

The canopy overhang is most applicable to bath or dressing room. It provides excellent general room illumination as well as light to the user's face.

(h) Luminous Ceilings

Totally luminous ceilings provide skylight effect very suitable for interior rooms or utility spaces, such as kitchens, baths, laundries. With attractive diffuser patterns, more decorative supports, and color accents they become acceptable for many other living spaces such as family rooms, dens, etc. Dimming controls desirable.

(j) Luminous Wall Panels

Luminous wall panels create pleasant vistas; are comfortable background for seeing tasks; add luxury touch in dining areas, family rooms and as room dividers. Wide variety of decorative materials available for diffusing covers.

(k) Typical Valance

This "typical" dimensional drawing applies only to commonly encountered window valance situations. Obviously, other window treatments could necessitate modifications in these critical dimensions; i.e., vertical blinds, double track situations, curved bay windows, etc.

The same "job-tailored" variations can occur in the design of any type of structural lighting device. Therefore no other dimensional drawings have been included here.

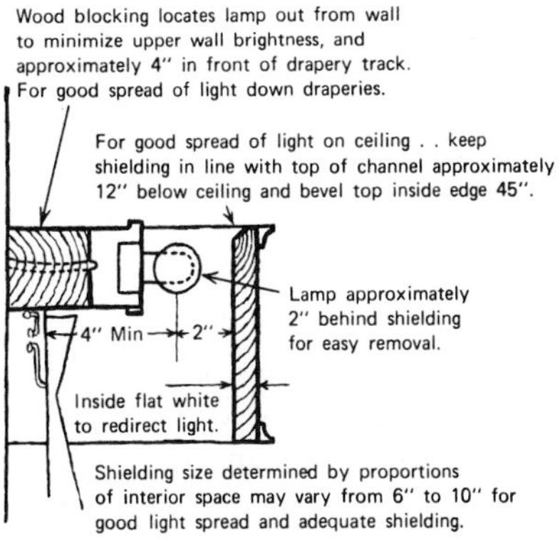
Wood blocking locates lamp out from wall to minimize upper wall brightness, and approximately 4" in front of drapery track. For good spread of light down draperies.
For good spread of light on ceiling . . keep shielding in line with top of channel approximately 12" below ceiling and bevel top inside edge 45".
Lamp approximately 2" behind shielding for easy removal.
4" Min
2"
Inside flat white to redirect light.
Shielding size determined by proportions of interior space may vary from 6" to 10" for good light spread and adequate shielding.

***Fig. 21.17** (continued)*

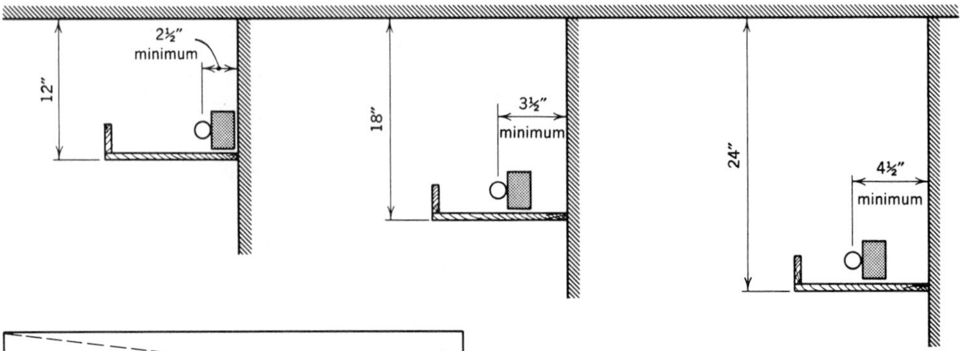

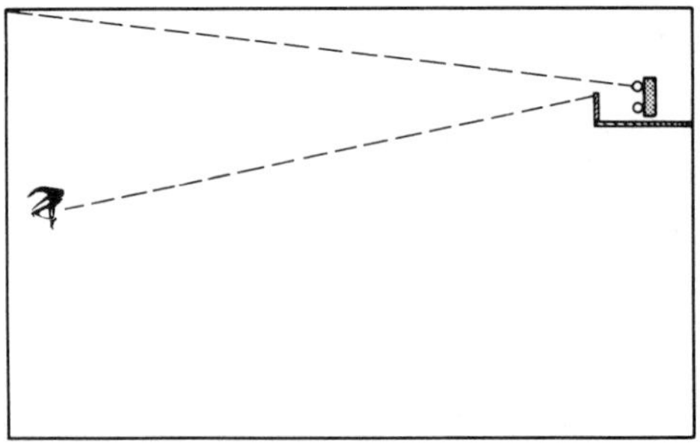

(l) and *(m)* Cove Installations

Proper cove proportions: Height of front lip of cove should shield cove from the eye yet expose entire ceiling to the lamp. Orientation of fluorescent strip as shown is preferable. Cove interiors should be painted with high reflectance matte-finish paint. *Westinghouse Lighting Handbook. (Out of print)*

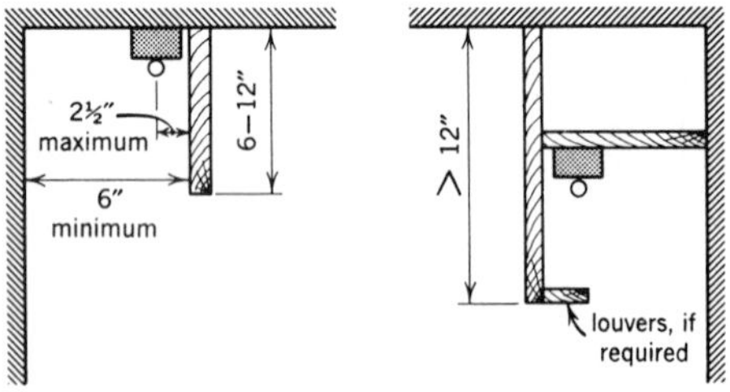

(n) Typical Cornices

Wall washing equipment mounted in valances and cornices provide improved brightness ratios and may be used for lighting desks against walls, or vertical illumination of walls and objects mounted thereon. *Westinghouse Lighting Handbook. (Out of print)*

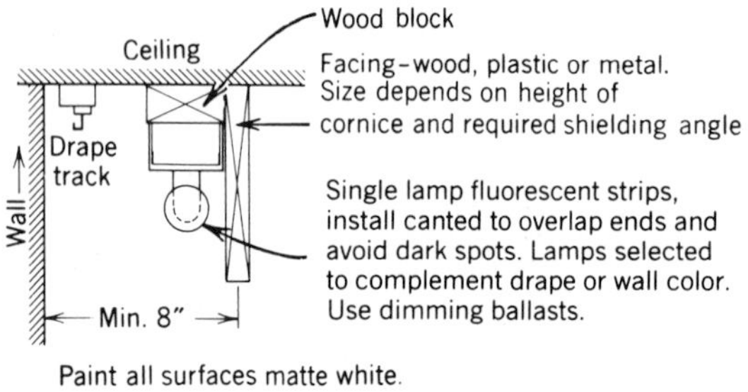

Fig. 21.17 *(continued)*

5. Use of standard or energy-efficient fluorescent lamps, cool white, or high-CRI white for direct and direct–indirect fixtures; 800- and 1500-mA lamps, same colors for semi-indirect and indirect lighting. Incandescent sources should not be used. Daylight, to the maximum extent, is desirable. Some form of daylight compensation is mandatory.
6. In comparing costs of alternative appropriate lighting systems, use life-cycle costing techniques only, since these facilities are nonprofit and long term. See Appendix E for an explanation of the principles of

economic analysis. Use a computer program such as LUMEN$ © to perform the economic analysis where possible (see Section 20.42). Do *not* compare costs on the basis of dollars per footcandle (i.e., by dividing maintained illumination by cost), since this leads to preference for higher illumination levels and consequently, higher wattage levels.

7. Illuminance category (see Section 18.24) and other lighting recommendations for the various types of classrooms and their activities are:
   (a) *Reading*. C, D, E, depending on the visibility characteristics of the reading material.
   (b) *Mechanical Drawing and Drafting*. E, F. Drafting tables should be equipped with an adjustable fluorescent lamp. Incandescents are not recommended because of heat, low efficacy, and both direct and reflected glare. Proper positioning of the drawing lamp will provide glareless illumination even when working with pencil on vellum.
   (c) *Typing*. Illuminance categories D, E. Where carbon paper is used, a desk lamp may be required. Since copyholders and paper in the machine are essentially vertical tasks, use a lighting system with a high vertical component, such as indirect and general diffuse. See also Section 21.28 on this point.
   (d) *Sight Saving*. These are general classrooms for visually impaired students. Illuminance category F is recommended, which indicates the need for supplemental lighting. These units should be individually adjustable by the user for both psychological and physiological reasons.

21.9 Special-Purpose Classrooms

(a) Shops. Illuminance categories for all types of bench and machine work would be

Rough	D
Medium	E
Fine	F, G
Extra fine	G, H

As for other tasks requiring levels above 500 lux, a task-ambient design involving an adjustable supplemental task light is recommended. As with drafting, adjustability of the task light is particularly important because of the frequently specular quality of the visual task, such as in a metal machining process. Another aspect of lighting design in shops that does not exist to the same extent in other school environments is that of safety lighting (see Table 21.5). Loss of illumination, which can be caused by a fault in the lighting circuitry only, will leave students in a darkened space with operating machinery. The recommendation is therefore to provide local sensing and an instantaneous uniform emergency lighting level of at least 10 to 20 lux without excess glare. Under normal shop lighting conditions the safety aspect of the lighting design should assure adequate lighting of all moving machine surfaces to which students may be exposed. This will probably entail field investigation and retrofit lighting work after machinery installation, possibly including portable lighting units.

(b) Music Rooms. The illuminance category here is D for clear printed music and E or F for older poor copy or handwritten scores. These requirements are for vertical surface illumination, as music scores are normally held in that position [see Section 21.8*g*(3) above]. For all spaces with vertical tasks a computer analysis of the design is recommended to determine compliance with recommendations (see also Section 21.28).

(c) Electronic Data Processing Rooms. Task light requirement is low. Tasks are generally reading vertical copy and working with a VDT. See Section 21.18 for a full discussion.

(d) Art Rooms. The primary requirement here is for constant color daylight. Thus north windows and skylights are highly desirable. For

TABLE 21.5 **Illuminance Levels for Safety**[a]

Hazard Requiring Visual Detection:	*Slight*	*Slight*	*High*	*High*
Normal Activity Level:	*Low*	*High*	*Low*	*High*
Areas	Normal classrooms Lounges Small offices Dorm rooms Washrooms	Shops Business classrooms Large offices Corridors Drafting rooms Lecture rooms Large classrooms Parking area Exterior walkways	Stairs Libraries Reading rooms Cafeteria Swimming pools Locker rooms Interior sports Bleachers	Boiler rooms Auditoriums Exits Exitways Laboratories Kitchens
Lux (Footcandles)	5.4 (0.5)	11 (1.0)	22 (2.0)	5.4 (0.5)

Source: ANSI/IES RP3-1988; reproduced with permission.

[a]Minimum illumination for safety of personnel, absolute minimum at any time and at any location where safety is related to seeing conditions.

artificial lighting, since color is so important, high-CRI fluorescent tubes are recommended. General illumination should be augmented by user-adjustable supplementary lighting. If use of models is anticipated, adjustable accent lights are advisable. For display of artwork, adjustable wall illumination is required. Ceiling track-mounted units are an excellent choice (see Figs. 21.15 and 21.18).

21.10 Assembly Rooms, Auditoriums, and Multipurpose Spaces

The varied activities in these rooms make flexible lighting imperative. For performances, low-level dimmed incandescent lighting is required. Here incandescent is the recommended source because of the lower cost of dimming and short burning periods. For assembly, this can be augmented by architectural elements along walls and drapes, and in the ceiling. For study, additional ceiling fluorescents or HID units can be switched on. The combinations are legion; the different usages are the critical consideration (see Figs. 21.19 and 21.20). Acoustic considerations are acute because of the low NC criteria. Thus the generally noisy ballasts of HID sources should be located with care. An additional consideration is step lighting. These units should be mounted to the side of, or in risers, to illuminate the tread, and particularly its leading edge. Stage lighting is too highly specialized to be discussed here. Obviously, though, some form of stage lighting is required and the building designer is well advised to consult an expert.

21.11 Gymnasium Lighting

Gyms present a situation similar to auditoriums in that they have widely varying usages. All fixtures should be sturdy and guarded. Phosphor-coated mercury, HPS, and high CRI metal-halide are excellent choices for color, life, control, and efficiency. Multiple levels should be available by switching or dimming. For dance and assembly use, other fixtures can be lamped with long-life incandescent or tungsten halogen, which provide good color for low-intensity lighting and also provide illumination during HID startup or restart after an outage. All fixtures should be designed for relamping from the floor. Locker rooms should use guarded-strip fluorescents.

21.12 Lecture Hall Lighting

Lecture hall lighting is similar, with respect to sources and other considerations, to illumina-

Fig. 21.18 Art exhibition room, illustrating good and bad lighting techniques. Upper wall fenestration is excellent for deep daylight penetration. Track lighting is ideal for display of art. The mixture of incandescent downlights for general lighting is excessive and an eyesore. Also, the positioning of the track lights can create both direct and reflected glare problems and annoying shadows unless the sources are selected properly and ceiling height is at least 10 ft.

Fig. 21.19 Schools frequently utilize spaces for multiple functions. This space, normally used as a dining area, doubles as an assembly room. The architecture did not lend itself to conventional fixtures. High-intensity, indirect tungsten-halogen units, in concrete beam junctures, provide sufficient light for both uses.

Fig. 21.20 Institutional cafeteria illuminated by cove lighting in deep pyramidal coffers. Lighting is even, glare free, soft in quality, and pleasant, yet of sufficient intensity to permit using the cafeteria as a working–meeting space. (Photo by Stein.)

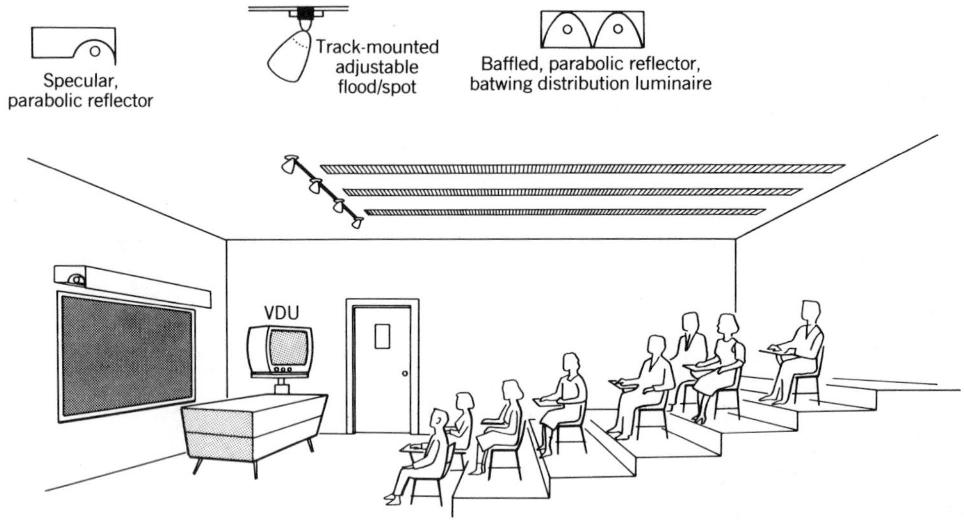

Fig. 21.21 *Typical lecture room lighting utilizes 45° cutoff baffled parabolic reflector troffers for minimum direct and reflected glare, adjustable track lights for demonstration table illumination, and an asymmetric reflector for chalkboard lighting. The large visual display unit in the room will not have veiling reflections with the illustrated lighting arrangement. See Section 21.18 for a discussion of lighting in areas using visual display devices.*

tion for classrooms. Adjustable-level fluorescent lighting is necessary for demonstrations, video, and the like. Auxiliary lighting for demonstration table and chalkboard completes the design. High-ceiling installations can utilize color-corrected mercury or metal-halide for general lighting. Controls for lighting should be at the demonstration table. See Fig. 21.21.

21.13 Laboratory Lighting

Laboratories differ from classrooms in that tables are fixed, bench surfaces are frequently very dark, many of the items used exhibit specular reflection, vertical surface illumination is important, and visual tasks are not normally prolonged or severe. With low ceilings, use direct fixtures with small uplight component, run crosswise to the tables. Luminaires with batwing distribution will minimize reflected glare from specular equipment. If ceiling height is sufficient, indirect lighting is highly desirable for the same reason. Indirect lighting will also provide a high degree of diffuseness necessary for vertical surface illumination. See Fig. 21.22 for suggested layouts.

21.14 Library Lighting

Libraries comprise several different seeing tasks, each of which requires its own lighting solution.

(a) General Reading Room. Here two solutions are possible and both are in common use. In the first, general lighting is supplied over the entire area, which is sufficient for reading tasks. For this purpose fluorescent or color-improved HID sources are normally applicable, the latter with ceiling heights of at least 10 ft. The long life and high efficacy of these sources are suited to the long burning hours found in libraries. The second, and more energy-efficient solution, involves low-level general lighting supplemented by local reading lighting on the tables or in carrels. This solution is consonant with task-lighting orientation and is to be preferred. Reading lights should be fluorescent, user adjustable if possible, and arranged to avoid veiling reflections when not user adjustable (see Section 21.21).

Wherever HID sources are used, an instant restart source must be available to supply mini-

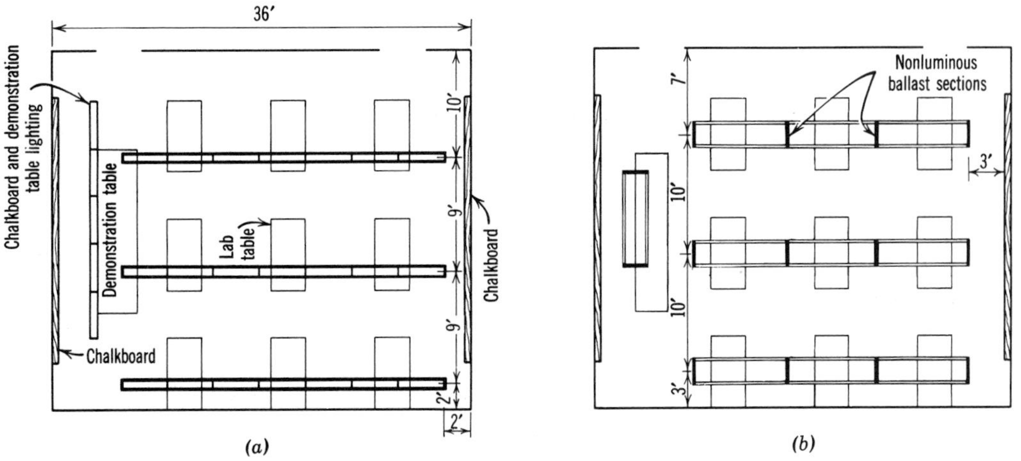

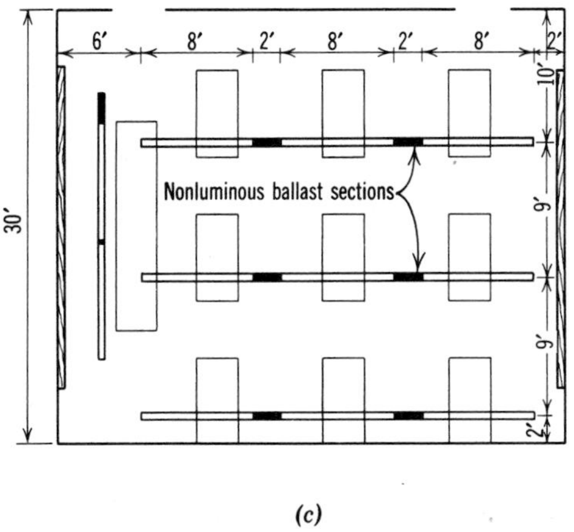

Fig. 21.22 Laboratory lighting schemes. Running lights across tables or in aisles is preferable from the aspect of reflected glare. (a) *Pendant direct–indirect units.* (b) *and* (c) *variations of the single, semi-direct HO and VHO unit design illustrated in Table 20.2, fixture 33. The relatively high noise level of ballasts is generally not objectionable in labs.*

mal lighting after an outage. Many commercial HID luminaires contain a small tungsten-halogen source for this purpose. Ballast noise can be a problem in low-NC-criteria spaces such as libraries. Special low-noise ballasts and enclosures are available and should be employed.

(b) Stack Areas. In stack areas the required vertical surface illumination is best supplied by one of the special fluorescent units designed for this purpose. These are mounted between stacks and no higher than 24 in. above them for best results (see Fig. 21.23).

21.15 Special Areas

Most schools contain areas devoted to functions not covered above. Some lighting recommendations for a few of these areas are:

1. *Spaces where color rendering and color matching are important*, such as sewing rooms and textile and art work spaces, must be lighted with particular attention to illuminant color. Consistent color rendering without disturbing metamerisms requires a continuous spectrum source such as incandescent (including tungsten halogen) and, of course, daylight. For lighting color matching, see Section 19.26.

2. *Food service areas* must be well lighted to emphasize cleanliness and food attractiveness. Color rendering of food is particularly important so as to enhance its appetizing appearance. Use incandescent in serving areas, 3100 K fluorescent or metal halide in preparation areas, and the latter two in eating areas, but at lower illuminance levels.

3. *Medical attention spaces* require high-

ILLUMINATION

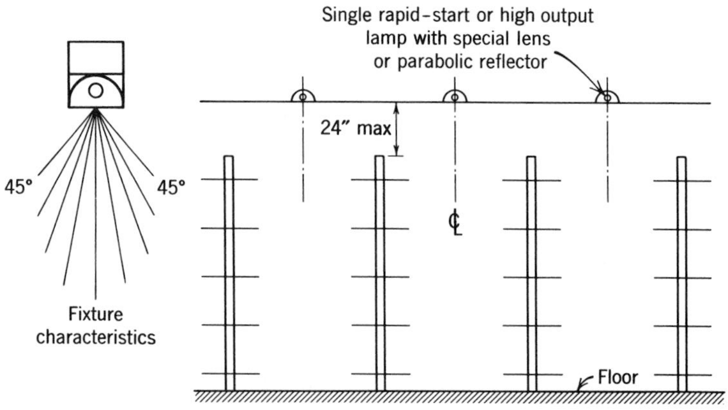

Fig. 21.23 Stack lighting is best accomplished by fixtures with lenses specifically designed for the purpose. Fixtures with baffles and plastic diffusers generally do not give adequate vertical surface illumination.

level, good color rendering light for diagnosis, and conventional office lighting for records and desk tasks.

4. *Offices, storage spaces, industrial spaces, and outdoor facilities* have the same requirements as similar spaces in other buildings, and are covered in the discussion below.

5. *Corridors and stairways in all types of buildings* require special lighting. Corridors intended only for circulation need only be lighted to ±10 fc unless a specific seeing task requires higher levels, for example, bulletin boards and lockers (see Fig. 21.24). Lighting can also be used to give direction by longitudinal arrangement. Wall-mounted or recessed wall lighting is particularly effective in corridors, giving walk illumination plus lighting for posters, bulletins, and so on. Fluorescent luminaires mounted across corridors, particularly when corridors are long, are effective in reducing the tunnel impression. Incandescent sources are not recommended, because of low efficacy, high maintenance, and frequency of relamping. Fluorescent and HID sources are also suggested for stairwells. Care must be exercised here, however, to avoid direct glare, which causes attention to shift from the stairs to the light and may thereby cause a hazard.

21.16 Other Considerations in School Lighting

(a) Controls. See Sections 21.2 to 21.4. Because schools operate for the most part on fixed time schedules and during daylight hours, lighting controls of the preprogrammed time-base type and daylight compensation controls are readily applicable. Other energy-conserving strategies, as discussed at length in the above-referenced sections, can be utilized, as applicable.

(b) Safety Lighting. See Table 21.5 for recommended illuminance levels. As pointed out, designers must be aware of requirements of all jurisdictional codes, including the NFPA 101: National Electrical Code for Safety to Life from Fire in Building and Structures, and all local municipal and state codes.

(c) Emergency Lighting. See Sections 21.29 and 16.31. Here too the local, state, and other NFPA codes establish minimum requirements. Exits must be clearly identified with lighted signs, and a lighted path to these exits provided. As with safety lighting, it is not sufficient that average illuminance meet requirements. Illuminance in any given area must be free of large level differences, which can cause disabling glare, particularly in view of the relatively high adaptation levels of occupants' eyes immediately before a lighting outage.

COMMERCIAL INTERIORS

21.17 Office Lighting—General

The following information applies primarily to offices in commercial buildings and secondarily

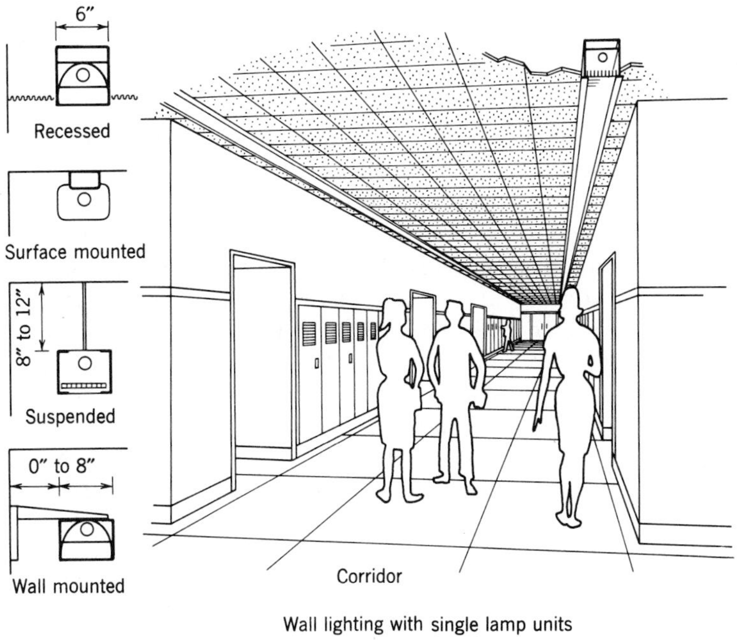

Fig. 21.24 *Lighting of school corridors. High-reflectance walls, floor, and ceiling improve utilization of light and increase the feeling of cheerfulness. The lighting technique illustrated is appropriate for school corridors. The rows of luminaires at each sidewall illuminate bulletin boards, special displays, and the faces and interiors of lockers more effectively than do units centered in the ceiling.*

to similar spaces in other occupancies, such as in educational and industrial buildings. In the latter cases, the general remarks applicable to facilities of those types take precedence. The special problems associated with VDTs-Visual Display Terminals (CRTs, computer screens, word processors) are discussed in Section 21.18. Task-ambient (nonuniform) lighting is covered in Sections 21.19 and 21.20. The reader is referred to *IES Standard Practice for Office Lighting* (1983) for a full discussion of office lighting.

(a) Light Sources. The most commonly used source is fluorescent. HID can be used to advantage in indirect lighting, in spaces with at least 10 ft (~3 m) ceiling height. When using HPS it is advisable to mix it with metal-halide to avoid possible objections to its yellow color. In most office applications, color is not critical but it is important. Use of T8 fluorescent lamps (at a small price premium) will assure high efficacy and good color since these lamps have high CRI.

Source color must be coordinated with the color scheme of room surfaces and furnishings. In areas with large daylight contribution, source correlated color temperature should be at least 4000 K. Incandescents may be used for storage areas, closets, and other short-burning-period uses. Incandescent and tungsten-halogen track lighting is used to advantage to illuminate displays of all sorts. Low-voltage units are useful in this application. See Section 19.14.

(b) Illumination Levels. These are to be found in Section 18.24 for the particular type of activity involved. See also *IES Standard Practice for Office Lighting* (1983). Recommended reflectances for room surfaces are:

Ceiling	80% minimum
Walls	50–70%
Partitions	50–70%
Floor	20–40%
Desktops, furniture	25–45%
Window blinds	40–60%

In landscaped offices, the half-height partitions block 30 to 80% of the light from ceiling lighting fixtures, depending on furniture arrangement. It is therefore all the more important that partition finish (including fabrics) be light colored. It is also desirable to have upper wall sections painted to match the ceiling, that is, with a lighter finish than the remainder of the wall. This serves the dual function of increasing ceiling cavity brightness, particularly with suspended fixtures, and increasing vertical illumination due to reflection from this surface.

(c) Vertical Surface Illumination. This is required for visual tasks in offices, such as files, desk drawers, card files, and copy stands. Large area luminaires and a high degree of diffuseness are desirable. This is especially true in large offices where wall reflections are absent. Light-finish furniture surfaces, fixtures that yield an illuminated ceiling, and high-reflectance floors will also assist in this.

21.18 Lighting for Areas with Visual Display Terminals (VDTs)

One result of the emergence of the computer era has been the explosive proliferation of visual display terminals in business offices. These devices, variously called VDTs, monitors or monitor screens, video displays, word processors, or simply computers, have become standard desktop items. Because of its specular surface, the VDT creates a special problem in office lighting that becomes the deciding consideration in selecting the lighting system. Simply stated, the primary problem is to avoid reflection in the screen of any luminous source in the area including luminaires, windows, illuminated walls, and even light-colored clothing. Any such reflection makes reading of the data on the screen always difficult, and sometimes impossible.

The problem of reflections in the VDT screen is not only one of work difficulty. With a *sharply defined* reflection in the screen, as is that of a lighting fixture, the problem becomes physiological. Refer to Fig. 21.25. As we already know, the eye is naturally and ineluctably drawn to any bright areas in the field of vision. On a VDT screen with a clear reflected luminaire image there are two bright items in the vision field: the luminaire reflection and the VDT text or graphics. These two images are separated in depth by the thickness of the glass on the face of the VDT. The eyes will be drawn to, and will attempt to focus on, both images but because of the difference in depth the eye focus will change continuously, causing eyestrain and severe fatigue. This effect is pronounced only with a clear reflected image; a fuzzy or blurred image, as in Fig. 21.26*a*, appears only as a bright veiling reflection. As a result of this and associated problems, a number of governmental agencies have already issued guidelines for employees using VDTs which limit work periods and monitor physical effects. The health department of the state of New Jersey, for instance, has issued guidelines for all state employees which call for 15-minute breaks every 2 hours and periodic eye examinations. In addition, these guidelines call for all VDTs to have tiltable swivel mounts, operator adjustable, and *separate* adjustable-angle keyboards.

Fixed-position keyboards such as those in Fig. 21.26*a* will not be permitted. The reasoning here is obviously that each operator's situation is unique because of differences in work position, posture, and the body's physical dimensions. Therefore, the VDT operator must be given the wherewithal to adjust and thereby maximize the visual (and physical) comfort at his or her working position.

A secondary problem is avoidance of reflections on the usually specular keyboard and other specular objects in the vicinity. Matte-finish keyboards are now available, as is semispecular

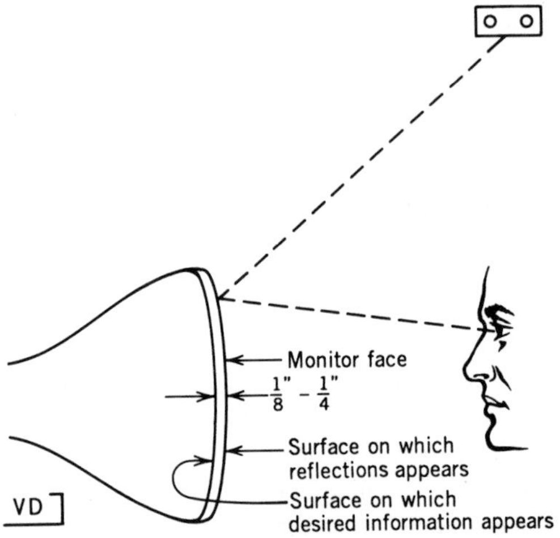

Fig. 21.25 *A clear reflection on the face of a VDT is physiologically disturbing because it causes the eye's focus to move continuously between the plane of the monitor image (inside) and the plane of the reflection (monitor face). This constant refocusing causes eye fatigue.*

glass. The problem with the latter is that it tends to dull and blur the monitor information, which, especially when working with complex graphics, is highly undesirable. This is even more true of retrofit monitor filters placed over the VDT screen.

It is obvious (see Fig. 21.26*b*) that the problem is one of geometry, complicated by the fact that the screen is often up to 20° from the vertical and the keyboard is up to 30° from the horizontal. A further difficulty lies in the contradictory nature of adjacent visual tasks: Viewing of the screen calls for a low (vertical) ambient light level (75 to 125 lux), whereas the reading and writing work done in conjunction with the screen data requires a much higher (horizontal) level (300 to 700 lux). A full solution to the problem therefore involves careful attention to the selection and location of the VDT equipment, and control of room surface reflectances, in addition to proper lighting design. The following recommendations refer to all these considerations, with the knowledge that some may not be in the purview of the architect/lighting designer.

(a) Equipment

1. The screen should be recessed as deeply as possible into the VDT. This will reduce the ambient light level at the screen and make reading easier. It will also reduce glare by restricting the room surface area that is reflected in the screen.
2. Convex screens reflect a larger ceiling area than flat screens and therefore should be avoided.
3. All parts of the VDT, including keys, should have a matte finish.

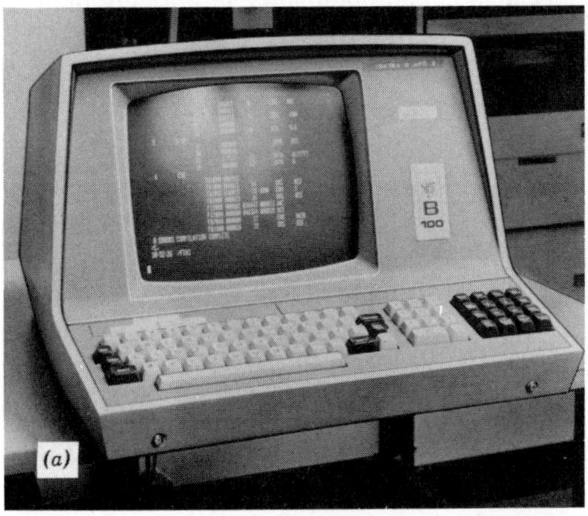

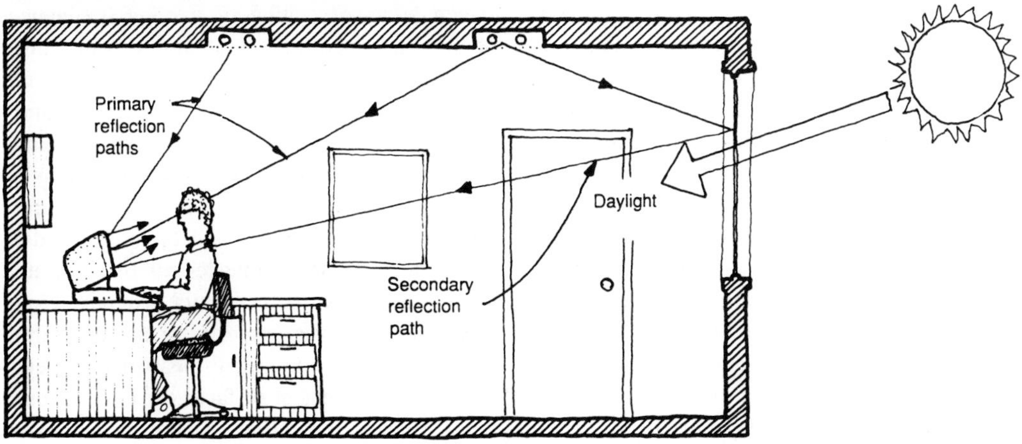

(b)

Fig. 21.26 (a) *Primary sources of reflection in the VDT screen and keyboard are luminous areas such lighting fixtures and windows. (Photo courtesy of Brüel and Kjaer.)* (b) *Specular surfaces such as window glass can become secondary reflection sources via rereflection of light from luminaires (or other light sources), as shown.*

4. The VDT should be mounted on a tilting swivel that will permit changing the geometry of reflection.
5. The keyboard should be separate from the monitor, with adjustable tilt angle.

(b) Location

1. Avoid locations where the VDT *viewer* faces a window or other large bright area, to avoid excessive luminance ratios.
2. Equip windows within the geometry of reflection with low-reflectance vertical blinds. Blackout curtains are not required. Remember that windows can become a secondary source of reflection, that is, luminaire to window to VDT screen. See Fig. 21.26*b*.
3. Test each VDT location by running a small mirror over the face of the screen to find glare sources, and correct each.

(c) Lighting. The lighting system needed is one that will prevent reflection of areas of high brightness on the face of the monitor, and provide sufficient task lighting for both horizontal and vertical tasks. There is no single solution to this difficult problem since the architectural aspects of VDT spaces vary widely. The designer must therefore adapt the solution to the constraints of the job A recently published IES standard, RP-24 (1990), should prove helpful.

1. *Indirect Lighting*. The luminance of an average monochrome monitor with text display is 15 to 20 fL (54 to 68 cd/m^2). [Graphics monitors, both monochromatic and color, may have higher luminance because the background is often bright and the contrast reversed (i.e., dark image on a light background). This situation is less troublesome than the reverse, standard, monochromatic text monitor, and therefore any lighting solution which satisfies the standard situation will also be acceptable for the graphics situation. Indeed, one approach to the entire problem is to use reverse contrast (i.e., dark text on a light background). However, since at this time the vast majority of VDTs are not thus arranged, we must address the more common light-on-dark situation.] Therefore, if we wish to permit a luminance ratio no greater than 10 : 1 to appear on the screen, maximum ceiling luminance must not exceed 150 to 200 fL (~500–700 cd/m^2). (IES standard RP-24 referenced above recommends a maximum of 850 cd/m^2 on any 60 cm square on the ceiling, measured at any angle). Higher ceilings can tolerate higher luminances. Luminance greater than this will cast a disturbing bright haze on the monitor screen (see Fig. 21.26*a*). With a ceiling height of 10 ft or more, an indirect lighting system with this luminance constraint can provide a uniform ambient illumination level of about 30 fc (300 lux) horizontal and approximately half that vertically. If these task levels are not sufficient, portable desktop task lights can be utilized. It is advisable to design for somewhat higher levels, and to incorporate switching or dimming controls for field "tuning" of the installation. A ceiling height of less than 10 ft would yield unacceptably high ceiling luminance in a totally indirect installation.

2. *Semi-indirect and Direct–indirect Lighting*. Spaces with a ceiling height of 9 to 11 ft can utilize these systems to advantage, using semi-indirect in the taller rooms and direct–indirect in lower ceiling areas. Uplight can vary between 40 and 80%, depending on the pendant stem length and the degree of spread of the upward component. Ceiling luminance should meet the criteria of item 1 above, and fixture luminance, the criteria that are adduced below for direct lighting systems. The ratio between fixture luminance and the background ceiling luminance, and between areas of ceiling of varying luminance, should not exceed 5 : 1 at any viewing angle within the reflection geometry of the screen. Since manufacturers provide only fixture luminance data, the only way to determine ceiling luminances with different pendant lengths is with a full-scale mock-up. A few manufacturers will provide *near-field* candela intensity data with which ceiling luminances can be calculated using a sophisticated computer program. However, since the vast majority of manufacturers use only far-field intensity measurement techniques, luminances of surfaces close to the luminaire, such as the ceiling, cannot be calculated

accurately, and only a mock-up will give reliable data. When designed properly, a direct–indirect system with approximately half the luminaire output in each direction can provide a very effective VDT environment. Figures 21.27 and 21.28 show a well-designed direct–indirect unit with its characteristics, and a typical installation.

3. *Direct Lighting.* In spaces with a ceiling height under 9 ft and where otherwise required, a low-brightness direct-lighting luminaire can be used. The CIE standard for direct luminaires in areas with heavy VDT use calls for a maximum luminance of 200 cd/m^2 (58 cd/m^2) at 50° from the vertical and above, and the same luminance at 60° and above in areas with moderate VDT use. Such a low luminance will assuredly prevent both annoying reflections on monitor screens and also direct glare, which can be problematic in VDT areas since the working posture is in a head-up rather than the traditional head-down position used for horizontal tasks. However, a glance at American manufacturers' catalogs will show that very few local manufacturers meet such stringent criteria. IESNA standard RP-24 recommends maxima of 850 cd/m^2 (~250 fL) at 55°, 350 cd/m^2 (~100 fL) at 65° and 175 cd/m^2 (~50 fL) at 75°, all measured from the vertical. Other American authorities suggest that a maximum fixture luminance of 850 cd/m^2 at 55° and above for dense VDT areas and the same luminance at 65° and above for light VDT use areas will suffice to limit both direct and reflected glare. Further, these luminances which can be met by many one- and two-lamp low-brightness fixtures will avoid the dingy, cavelike atmosphere created by very-low-brightness luminaires. The reader should note, however, that these luminances refer to the direction causing the reflection. Therefore, if viewing direction is not fixed, it is necessary to obtain and check parallel, normal, and 45° figures from the manufacturer.

To attain the required low luminance, fluorescent troffers can be equipped with miniature parabolic wedge louvres (Figs. 20.24*a* and 21.29) or the more efficient large-cell specular parabolic louvres (Fig. 21.30). One problem with the latter is that although a good shielding angle (minimum 35°, preferably 45°) will avoid lamp image reflection in the screen and will yield high VCP, (minimum of 80 recommended) the lamps will be reflected in the specular louvres, causing bright spots, which in turn can reflect in a monitor screen. To avoid this, some designers place a diffusing material (translucent overlay) above the louvres, which makes the louvre luminance uniform but reduces efficiency. Another alternative is to use semispecular louvres, which increase overall louvre luminance but without the highlights of specular louvres. A third alternative is to use specially made high-grade (and expensive) louvres whose curvature is carefully controlled to eliminate the rippled reflections seen in the photo of Fig. 21.30. Finally, specular louvres must be very well maintained since every speck of dust shows up as a bright spot on a dark background.

(d) Finishes. Walls, floors, and furniture should be finished in low-chroma, low-value colors, with a maximum reflectance of 50%. Very dark and very light desktops are to be avoided, because of excessive luminance ratios and the latter also because of reflections. Similarly, operators should be advised to avoid light-colored clothing and specular clothing accessories. In general, the standard recommended maximum luminance ratios of 1 : 3 in the near field and 1 : 10 in the far field should here too be the design goal.

21.19 Office Lighting Guidelines

(a) Private Offices. In these spaces, a task-ambient approach is generally indicated since there is usually only one primary visual task location, with the remainder of the space devoted to circulation and storage. Often, sufficient illumination for the latter is supplied by spill light from the task area, particularly when the task light is a pendant unit with uplight component, as seen in Fig. 21.31. In large rooms, provide downlighting in sitting areas and some type of wall illumination, using wallwashers, sconces, or a recessed perimeter unit, to brighten the of-

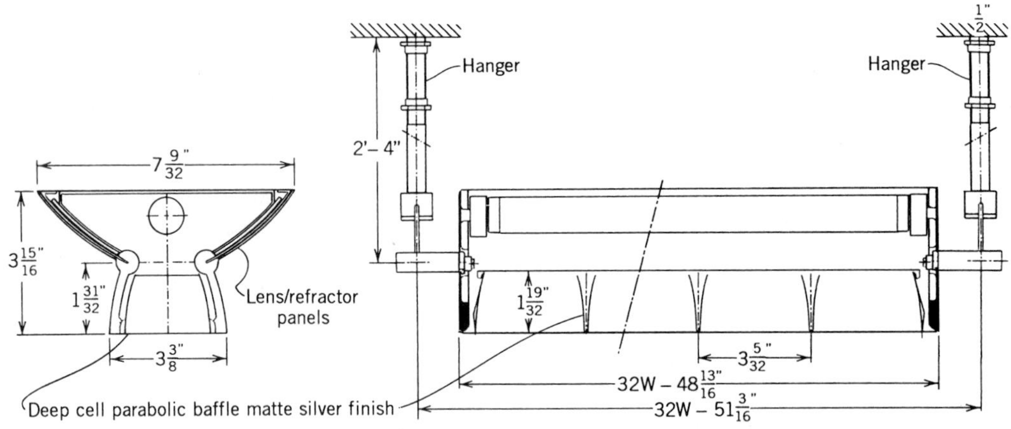

Photometrics

Luminaire type: ID-VMR 1/32

CANDLEPOWER

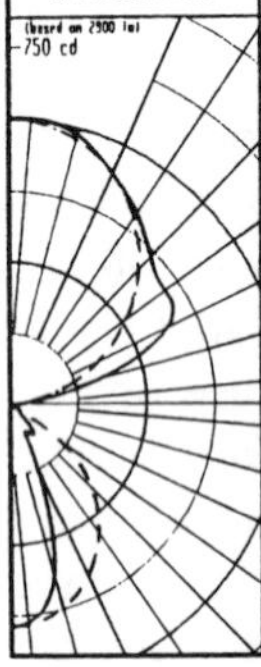

DEG.	Parl.	22.5	45	67.5	Norm.	Flux
0	473	473	473	473	473	
5	472	470	464	462	465	44
10	462	458	441	434	438	
15	447	437	407	374	375	111
20	428	409	336	263	254	
25	402	375	234	149	134	118
30	372	318	143	78	69	
35	337	251	81	69	75	97
40	297	181	59	77	85	
45	252	117	59	62	68	75
50	201	71	47	41	41	
55	135	38	29	29	30	40
60	65	19	18	24	26	
65	20	11	14	20	23	18
70	10	9	12	18	21	
75	7	8	10	16	18	12
80	6	6	8	13	16	
85	5	5	7	11	13	9
90	2	5	8	10	11	
100	54	101	140	113	93	51
110	164	184	285	322	305	195
120	265	268	318	386	413	292
130	360	363	383	405	419	318
140	440	448	451	460	462	324
150	507	519	518	512	512	303
160	557	566	575	567	561	250
170	588	593	598	599	600	165
180	602	602	602	602	602	57

LUMINANCE (fl)

DEG.	NORM.	PARL.
45	36	450
55	47	291
60	43	158
65	42	58
75	42	30
85	42	46

ZONAL SUMMARY

ZONE	LUMENS	LAMP	FIXT
0 - 30	273	9.4	11.0
0 - 40	371	12.8	15.0
0 - 60	485	16.7	19.6
0 - 90	524	18.1	21.1
90 - 180	1955	67.4	78.9
0 - 180	2479	85.5	100.0

COEFFICIENTS OF UTILIZATION

LUMINAIRE EFFICIENCY : 0.85

REFL. FACT. CEILING	80			70			50			30			10			0
WALLS	50	30	10	50	30	10	50	30	10	50	30	10	50	30	10	0
FLOOR	20	20	20	20	20	20	20	20	20	20	20	20	20	20	20	20
ROOM CAVITY RATIO - RCR 1	76	73	71	67	65	63	52	50	49	37	36	36	24	24	23	17
2	67	63	59	60	56	53	46	44	42	34	32	31	22	21	21	16
3	60	55	50	54	49	46	42	39	36	31	29	27	20	20	19	15
4	54	48	43	48	43	39	38	34	32	28	26	24	19	18	17	14
5	48	42	38	44	38	35	34	31	28	26	24	22	18	17	16	13
6	44	38	34	40	34	31	31	28	25	24	22	20	17	16	15	12
7	40	34	30	36	31	28	29	26	23	22	20	18	16	15	14	12
8	37	31	27	33	28	25	27	23	21	21	19	17	15	14	13	11
9	34	29	25	31	27	23	25	22	20	20	18	16	15	14	13	11
10	32	26	23	29	24	21	24	20	18	19	17	15	14	13	12	10

UNIFORMITY OF ILLUMINANCE - $I_{min}:I$ = 1:...

Transverse spacing: MH ×	1.1	1.2	1.3	1.4	1.5	1.6	1.8	2.0	2.2
Axial-Spacing Continuous:	1.3	1.3	1.4	1.3	1.4	1.3	1.3	1.4	1.4
MH × 1.0 :	1.3	1.3	1.3	1.3	1.3	1.3	1.3	1.4	1.4
MH × 1.5 :	1.3	1.3	1.3	1.3	1.3	1.3	1.3	1.3	1.4

S/MH = 1.9

Fig. 21.27 *Semi-indirect, low brightness, deep cell parabolic baffle fluorescent fixture. The parabolic baffle provides the required low luminance in the parallel direction (see the text), as does the fixture body in the normal plane, so that the fixtures should not constitute a glare problem. (See luminance figures in the photometric table.) Maximum ceiling luminance and ceiling luminance ratios will depend on hanger length, ceiling reflectance, and fixture spacing. Note also the very useful table of horizontal working plane illuminance uniformity as a function of both transverse spacing-to-mounting height ratio and axial spacing. (Courtesy of Zumtobel.)*

ten-dark wood paneled walls. Accent lights are required for pictures and other displays.

(b) General Offices. The two basic approaches to general office lighting are a uniform layout that provides task-level lighting in the entire area, and a task-ambient design. The former, discussed in this section, is most appropriate to speculative-type construction, where the furniture layout is unknown but a complete job is required. (Frequently, however, the construction contractor provides sufficient electric power only for lighting and miscellaneous power; the lighting of the space is designed *after* the space has been rented, that is, as "tenant" work. In such cases, a task-ambient design is possible since tenant work is tailored to the user's needs.)

As has been pointed out repeatedly, a task-level overall layout is wasteful of energy and will create problems with building energy budgets. Therefore, a uniform layout, when used, should

Fig. ***21.28*** *Semi-indirect lighting installation in a New York stock brokerage office, utilizing the fixture shown in Fig. 21.27. Every desk is equipped with a VDT. The translucent sides of the fixture are within the acceptable 5 : 1 luminance ratio to the ceiling recess luminance and therefore do not create a reflected glare problem. (Courtesy of Zumtobel.)*

Fig. ***21.29*** *The lighting system of choice uses 45° cutoff parabolic wedge louvres, shown here in a pyramidal 5-ft-square modular ceiling. The window would require treatment to reduce brightness except on a solidly overcast sky. Also, a darker desktop surface would be preferable. (Photo courtesy of Armstrong World Industries, which supplied the ceiling system.)*

Fig. 21.30 The illustrated installation shows a modern newspaper office with a VDT on every desk and a variety of viewing directions. The lighting fixtures utilized are deep cell specular parabolic louvre units, which typically have low surface brightness and crosswise batwing distribution. Note particularly that the VDT screens are deeply recessed into the monitor case to minimize glare and that each VDT is mounted with tilt and swivel capability. (Courtesy of Zumtobel.)

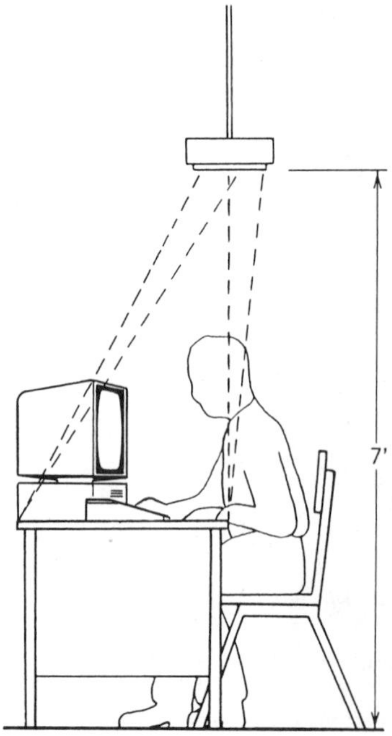

Fig. 21.31 Geometry of a pendant task-ambient fixture. This unit is intended not only to supply task illumination but also a measure of ambient lighting via its 20% uplight component. In a small room this component is sufficient for general circulation. When installed as shown, veiling reflections are minimized and glare on the VDT screen is essentially eliminated.

be designed so that levels can be lowered easily in areas not requiring task lighting (see Sections 21.2 to 21.4). Furthermore, since no control of the geometry of direct and reflected glare is possible in such layouts, the luminaire selected must give minimum average glare in *all* viewing directions. The system that will best meet these requirements is one that uses a large number of low-brightness luminaires in a dense layout. This is a very expensive solution, however, unless the ceiling is either modular or coffered.

Modular ceilings, with a single-lamp luminaire in each module, are a feasible, economical approach (see Fig. 15.31*a*). A structurally coffered ceiling with a luminaire in each coffer (see Fig. 21.32) is another feasible alternative, since the higher cost of luminaires is offset by the absence of a hung ceiling. For a flat ceiling, a low-brightness troffer would be selected. Batwing distribution is not indicated because of high direct glare (low VCP) and strong veiling reflections at work locations whose line of vision is crosswise to a lighting fixture. Where the viewing direction is established but exact furniture layout is unknown, a uniform layout with low-brightness parabolic reflectors with batwing characteristic can be used to advantage. Luminaire selection criteria for VDT areas were given in the preceding section. Where VDTs are not the overriding consideration, the following criteria will be helpful:

1. In terms of appearance, a 2 × 4 fixture is appropriate in a 2-ft^2 tile ceiling; a 1 × 4 fixture is most suitable in a 1 × 1 ceiling, and a square 2 × 2 fixture is suitable in all tile patterns.
2. Two-inch-deep parabolic louvres will permit a shallow fixture, but lamp shielding is poor. A 3-in. minimum louvre depth is suggested for all large-cell parabolic louvre luminaires.
3. Three-lamp parabolic louvre units, which are usually too bright for VDT areas, can be used to advantage. Two-lamp units will give better uniformity and VCP at a price premium.
4. Two foot square deep cell parabolics with standard U lamps, 31-W T8 U lamps, or biaxial compact lamps are an excellent choice from the aspects of VCP, uniformity of illumination, and appearance.
5. Premium-quality specular parabolic louvres are available that will eliminate hot spots (see Fig. 21.30).
6. In relatively small offices with 1 × 1 ceiling tiles, 1-ft^2 troffers with miniature parabolic wedge louvres and 16-W U lamps or compact fluorescent are an excellent all-purpose lighting solution.
7. Direct–indirect shallow luminaires with a wide, inverted batwing uplight distribution can be pendant mounted as low as 7 ft 6 in. AFF.

(c) Office Lighting Equipment. Office lighting equipment is generally not handled roughly. Fixtures may have touch latches, light hinges, and adjustable devices without fear of breakage or vandalism.

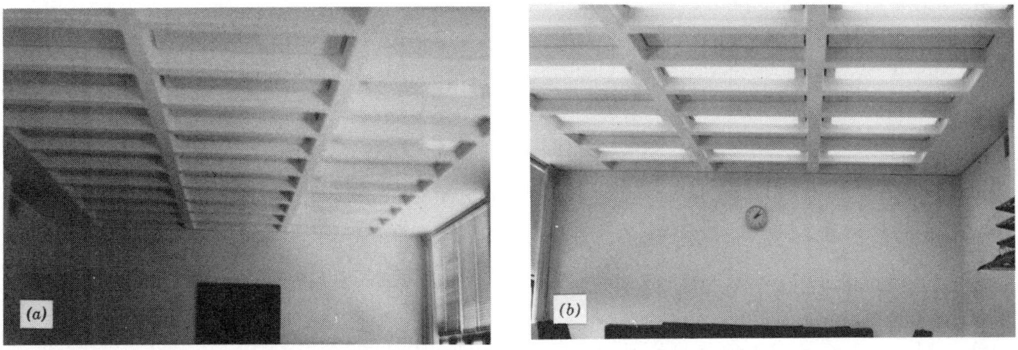

Fig. 21.32 *A coffer ceilinged room* (a) *is illuminated with 1 × 4 two-lamp units in alternate coffers. The appearance both unlighted* (a) *and lighted* (b) *is symmetrical and pleasing. A nonuniform layout would be less objectionable here than in a flat-hung ceiling. (Photos by Stein.)*

ILLUMINATION

(d) Maintenance. In most offices maintenance is provided on a trouble call basis. Lamps are replaced on burnout, and the fixture is then cleaned. Because of the long life of fluorescents and HID sources, this generally means a 3- to 5-year cleaning cycle. An LLF of 0.65 is reasonable in air-conditioned spaces, lower in open-window offices.

(e) Noise. Noise from air conditioning plus adjacent street, traffic, and process noise makes the use of higher noise level ballasts, as are found on HO and VHO lamps, frequently feasible. For private offices, A-rated ballasts are the best choice. (For more detail, see Chapter 27.)

(f) Fenestration. When fenestration is absent, a lighted valance is recommended around the room. This will remove the wall–ceiling line and will partially compensate for the lack of windows. It will also brighten the walls and increase illumination on desks placed adjacent to the walls.

(g) Supplementary Lighting. Supplementary lighting can be mounted on the ceiling or fed from poles of the type shown in Fig. 15.31.

(h) Control. Control strategy (see Sections 21.2 to 21.4) should *minimally* provide (after tuning):

1. Daylight compensation.
2. Small groups of lights to remain lighted while the remainder are off to permit off-hours work.
3. Path lighting through large spaces to permit traverse without turning on all lights.
4. Careful scheduling with supervised local override.

With these general guidelines in mind, the following sections discuss specific topics in office lighting.

21.20 Task-Ambient Office Lighting Design Using Ceiling-Mounted Units

Efficient modern office lighting, like other work area lighting, is often predicated on a task-ambient design. This approach has the advantages of minimum contrast reduction (high ESI) at the task, and minimum power use, the latter resulting in low operating cost. There are three applicable design methods:

1. Uniform *luminaire* layout with appropriate lamping and local controls that will result in the desired *task* lighting level. Since luminaire output will vary from one to another, only very low brightness units should be used. With such units, the differences in output will not be noticeable, and the desirable symmetry of ceiling appearances will be maintained. The disadvantage of this approach is the high first cost for luminaires and controls.

2. Uniform ceiling layout that provides low-level lighting for circulation and miscellaneous easy visual tasks, plus work-station-mounted task lighting. This latter can either be integral with the furniture (see the following section) or a separate, generally adjustable unit mounted on, or standing on, the desk. This latter is particularly useful for very severe seeing tasks when adjustability of angle and distance between light and work is vital. Integral furniture units are not adjustable, and are therefore limited to a maximum of about 750 lux on the task. Higher levels would generate excessive heat and noise and, almost certainly, glare.

3. Nonuniform (asymmetrical) luminaire layout, with luminaires located as required to suit the tasks. Obviously, a detailed furniture layout is required (see Fig. 18.34).

21.21 Task-Ambient Office Lighting Using Furniture-Integrated Luminaires

In lieu of ceiling-mounted fixtures, ambient and task lighting can be supplied by furniture-mounted units and freestanding, indirect HID units. Advantages include the following:

1. The problem of furniture layout and layout changes is eliminated.
2. Initial construction cost is reduced.
3. Energy requirements are lowered because of short distances between light source and task.

4. Each occupant has ON-OFF control of his or her task lighting including, in some designs, positioning control.
5. Maintenance is very much simplified, since fixtures are readily accessible from the floor.
6. Floor-to-floor height can frequently be reduced.
7. Tax advantages normally accrue due to higher depreciation rates on furniture than on the building.

Disadvantages include the following:

1. Difficulty in dissipating heat and minimizing ballast noise due to proximity of sources to user.
2. Veiling reflections are *always* present, and can be severe.
3. Luminance ratios in the near and far surround may exceed recommended levels.
4. Difficulty in lighting a free-standing open desk, since most of the fixture types are undercounter or sidewall mounted.
5. Difficulty in evenly lighting large table or L-shaped desk areas because of the concentrating nature of the lighting units.
6. Not readily applicable to automatic switching and dimming schemes.

Figure 21.33 graphically shows the problem of local desk lighting. The author's own experience has been that only systems that permit user positioning adjustment of the light source have a high degree of user acceptability.

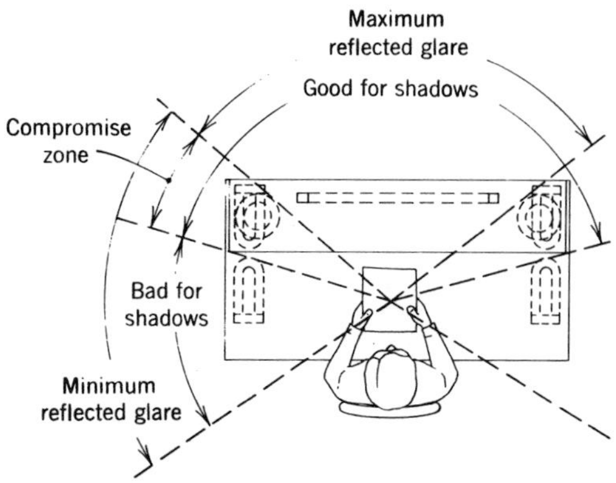

Fig. 21.33 *Furniture-mounted task lighting can create severe reflected glare or hand shadows if the luminaire location is not selected carefully. Further constraints are: location in elevation—for shielding (needs to be low) and for good distribution (needs to be high), appearance considerations, and compatibility with the worker's physical movements. Light source position adjustability removes most user objections to this type of task lighting.*

21.22 Integrated and Modular Ceilings

The cost, appearance, and design-flexibility advantages of an integrated ceiling design over field-assembled and coordinated systems have long been known. As a result, ceiling systems with integrated lighting, acoustic control, and air-handling capabilities are commercially available in modular sizes, among which are 60 in. square, 48 in. square, and 30 × 60 in. Modules are made in flat and pyramidal shapes, the latter having several distinct advantages over the flat:

1. More interesting and aesthetically pleasing.
2. More acoustic absorbency due to ceiling angles and large surface area.
3. Recessed center provides visual baffling, permitting use of higher brightness sources while maintaining high VCP.

A typically equipped pyramidal module is shown in Fig. 15.31. Possible luminaire arrangements for both flat and pyramidal shapes are given in Fig. 21.34, and examples of installation re shown in Figs. 21.29 and 21.35. In addition to the design flexibility available, electrified track can be integrated into the system runners, to supply both the lighting fixtures and power poles.

21.23 Lighting and Air Conditioning

The reduction of lighting power density levels to below 2 W/ft^2 in all but special areas has considerably reduced the impact of lighting-generated heat on a building's HVAC system. In non-air-conditioned buildings, the lighting heat contribution is partially applicable to building heating. Fixture efficiency is directly affected by its temperature. Fluorescent units operate at an opti-

ILLUMINATION

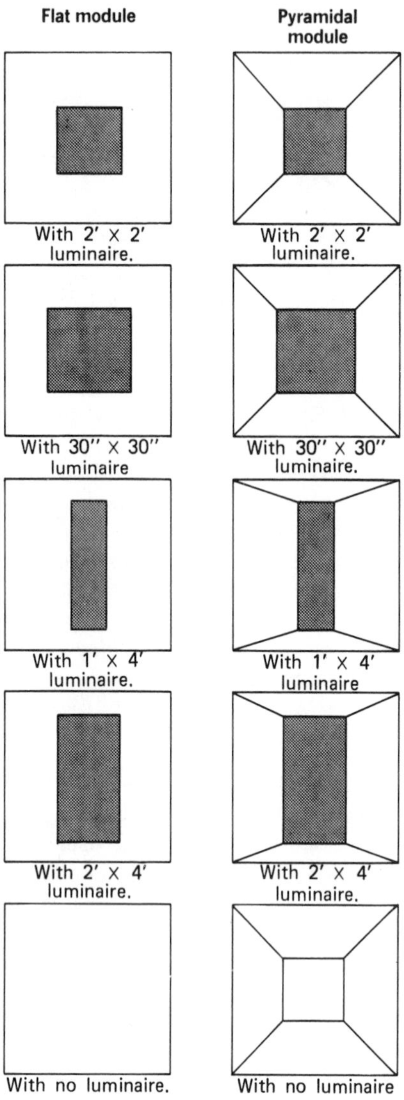

Fig. 21.34 *Various configurations of 5-ft² modules.*

mum temperature of 77°F. Temperatures above and below this decrease output and fixture efficiency. Thus heat removal from units is desirable even at low lighting-energy levels. The most effective method of fixture heat removal is by duct connection to the unit itself. This method, however, is relatively expensive and immobilizes the fixture. Alternatively the plenum can be exhausted with air passing through the fixtures, picking up excess heat. These details are essentially part of the HVAC design and are covered in Part II.

Fig. 21.35 *A square modular ceiling utilizing square lighting units demonstrates architectural symmetry, blending, and incidentally, that lighting is easily visible from the outside, even during the day. (Photos by Stein.)*

INDUSTRIAL LIGHTING

21.24 General

In industrial lighting the primary consideration is economic. Given acceptable standards of comfort and safety for the working staff, additional costs for lighting must be self-justifying economically. In one case a good lighting installation was improved at considerable cost. Production jumped 15%, of which 3% was sufficient to amortize the cost of the lighting alteration. In another case an outlay for new inspection lighting reduced product failures and proved economically sound. In a third, improved lighting reduced accidents, improved employee morale, and consequently improved production. The cases studied are far too numerous to mention; general principles will be adduced instead. For control strategies, see Sections 21.2 to 21.4.

21.25 Levels and Sources

Levels of illumination are detailed in the IES, *American National Standard Practice for Industrial Lighting* (RP-7, 1983). Other applicable recommendations are found in Section 18.24. Where levels above 50 to 75 fc are required; ambient illumination must be supplemented by local illumination. Industrial facilities lend themselves readily to daylighting, since many are one-story structures. Thus roof monitors, skylights, and clerestories are readily applicable and extremely desirable. However, since industrial facilities are frequently sited in industrial areas with attendant heavy atmospheric soot and dirt, a frequent cleaning and maintenance program is necessary if the LLF is to be kept at reasonable levels. This observation is obviously also applicable to indoor light facilities (to the extent that the indoor activity warrants).

Sources for industrial application should be high-efficacy, low-maintenance types, that is, HID and fluorescent. Where color is not critical, HPS is the recommended source. Adaptation to its warm yellow color is rapid and, as stated above, if it is mixed with metal-halides or mercury sources, no problem should be encountered. HID sources are easier to maintain, store, clean, and relamp than fluorescent and have equal or better efficacy, but have the disadvantage of delay and lowered output on restrike (see Section 19.22). Because of their concentrating nature, HID sources are more applicable to high-bay (>25 ft) and medium-bay (15 to 25 ft) installations, while fluorescent are suited for low-bay (<15 ft), although specially designed low-bay HID fixtures are available that minimize the inherent glare and distribution problems involved.

21.26 Industrial Luminance Ratios

For reasons explicated at length in preceding sections, luminance ratios in industrial situations must be controlled. Recommendations are given in Table 18.9. In many situations it is difficult to control the surrounding brightness. Ceilings, which so frequently are covered with piping, ducts, and other equipment, must be *light*. In other words, the above equipment must be painted with high-reflectance finishes, maintenance and cleaning must be good, and fixtures should have an upward component of light to avoid more than a 20 : 1 ratio of task to ceiling luminance.

Use of bright saturated colors for general surface painting should be avoided, however, since they draw attention and frequently have special significance. In addition to color-coded piping (banding is preferable), red frequently means fire equipment; green, first aid; orange, danger; and so on. White is also to be avoided, being excessively bright and susceptible to dirt. Light, unsaturated colors are preferable. Recommended minimum reflectances are:

Ceiling	75–85%
Walls	40–60%
Equipment	25–45%
Floors	20% minimum

21.27 Industrial Lighting Glare

The problem of direct glare can be acute in low-bay installations, and that of reflected glare in high-bay designs, when either uses a point source. One method of reducing direct glare is the use of low-brightness prismatic lens units that utilize a black reflector behind the prismatic lens. Methods of minimizing veiling reflection from all sources have been discussed previously.

21.28 Industrial Lighting Equipment

The cost of maintenance increases with labor rates. For this reason, high-quality lighting equipment will yield lowest owning and life-cycle costs. For instance, the cost of replacing a ballast for an HID lighting unit frequently *exceeds* the cost of the ballast. It is thus obvious that it is more economical to utilize long-life, hiqh-quality ballasts, particularly where luminaires are not readily accessible.

Other suggestions for lowering costs, both initial and operating, include using ventilated luminares that tend to be self-cleaning by convection (see Fig. 21.36) in addition to giving the needed upward light component, using bus-mounted fixtures for rapid installation and repair (see Fig. 15.10*a*), using lowering mechanisms on high-bay units to avoid catwalk or platform relamping with concomitant *extremely* high cost, using fixtures arranged for "stick" relamping from the floor in medium- and low-bay work, and generally incorporating the most modern equipment into the plant.

Proper maintenance is of paramount impor-

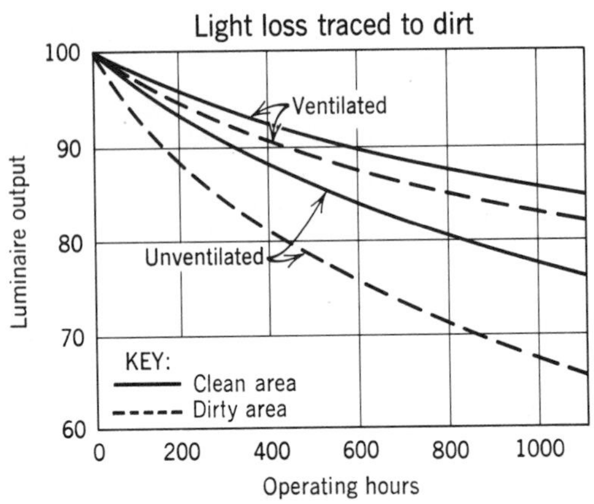

Fig. 21.36 Graph demonstrating the advantage of ventilated fixtures.

tance in industrial facilities because of the prevalence of dirt, vibration, and rough service. Under "maintenance" is subsumed cleaning, relamping, inspection, and preventive maintenance. Relamping on a burnout basis is extremely uneconomical because of disruption of production and lowered production due to lumen depreciation before burnout. Relamping should be done on a planned group basis. Similarly, if the specific facility has a high dirt accumulation rate, cleaning must also be done on a planned group basis, rather than only at relamping time.

Ballast noise, including the high levels of HID ballasts, is not usually a factor in industrial facilities because of high ambient noise. In relatively quiet installations and/or where fluorescent fixtures are mounted a short distance above the work bench as in inspection and fine assembly, this is not true and noise ratings and conditions must be examined carefully.

21.29 Vertical-Surface Illumination

In industrial facilities more than any other, the illumination of vertical surfaces is crucial. This is a result of the nature of work; machines, storage, gauges, and so on, all require high-level vertical-surface illumination. Refer to Fig. 21.37. The illumination on the vertical surface is the result of the horizontal component of the lighting. This is

$$\text{fc} = \frac{\text{cp}}{D^2}\sin\theta = \frac{\text{cp}\times R}{D^3} = \frac{\text{cp}\times\cos^2\theta\sin\theta}{H^2}$$

To maximize the horizontal component, we set to zero the derivative of fc with respect to θ. Thus

$$\frac{d\text{fc}}{d\theta} = \frac{\text{cp}}{H^2}(-2\cos\theta\sin^2\theta + \cos^3\theta) = \text{O}$$

or

$$2\sin^2\theta = \cos^2\theta \qquad \tan^2\theta = \tfrac{1}{2}$$

$$\tan\theta = 0.707 = \frac{R}{H} \qquad \theta \cong 35°$$

Therefore, maximum vertical illumination (illumination resulting from the horizontal-lighting component) is obtained when the angle between the fixture's vertical axis and the work is approximately 35°. Hence, we should select a fixture whose candlepower distribution curve demonstrates such a characteristic. Of course, the derivation is for a single location and fixture. For good vertical and angular illumination over a large area, arrange fixtures with considerable overlap.

SPECIAL LIGHTING APPLICATION TOPICS

21.30 Emergency Lighting

Emergency lighting is required when the normal lighting is extinguished for any of three reasons:

1. General power failure.
2. Failure of the building's electrical system.
3. Interruption of current flow to a lighting unit, even as a result of inadvertent or accidental operation of a switch or circuit disconnect.

As a result of the third requirement, sensors must be installed at the most localized level, i.e., at the lighting fixture (voltage sensor) or in the lighted space (photocell sensor).

(a) Codes and Standards. Since emergency lighting is a safety-related item, it is covered by various codes, all of which have jurisdiction. In addition, there are widely accepted technical society and industry standards whose

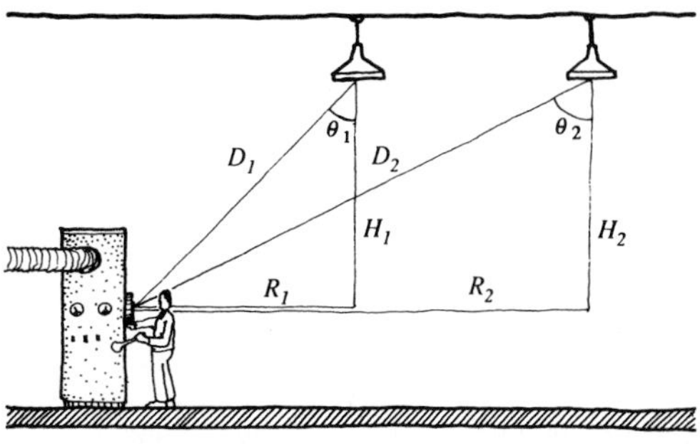

Fig. 21.37 Vertical surface illumination is maximum when θ is approximately 35° (see derivated in text).

recommendations normally exceed the minima legally required by the codes.

1. *Life Safety Code, ANSI/NFPA 101, current issue—1988.* This code defines the locations within specific types of structure requiring emergency lighting and specifies the level and duration of the lighting.

2. *National Electric Code, NFPA, 70, current issue—1984.* This code deals with system arrangements for emergency light (and power) circuits, including egress and exit lighting. It discusses power sources and system design.

3. *OSHA regulations.* These are primarily safety oriented, and in the area of emergency lighting discuss primarily exit and egress lighting requirements.

4. *Industry standards.* These include the publications of the IES and the IEEE (see Tables 21.6 ad 21.7). Because codes and standards are constantly being revised and updated, the designer in any actual construction project must determine which codes have jurisdiction, obtain current editions, and fulfill their requirements. The material presented below is in the nature of general information and good practice, but is not intended to take the place of construction and safety codes.

(b) Amount and Duration of Emergency Lighting. The codes, and many authorities, require a minimum of 1.0 fc (10.76 lux) at floor level throughout the means of egress, as sufficient to permit orderly egress. Other authorities recommend that escape routes be lighted to not less than 1% of their normal illuminance, but in no case less than 5 lux (0.5 fc). Where this latter recommendation falls below 1 fc (normal illumination less than 100 fc) we would abide by the 1 fc minimum required by the Life Safety Code.

Duration of the 1 fc level of emergency lighting is normally specified to be a minimum of 90 minutes (for egress) and 0.6 fc thereafter. Facilities that cannot readily be evacuated require higher levels for indefinite periods. Tables 21.6 and 21.7 give general criteria and typical recommendations for emergency lighting for egress as well as for other purposes.

(c) Central Battery Emergency Lighting Systems. Sections 16.30 and 17.22 cover emergency power systems, including generators and central battery systems. Generators are normally arranged to replace the utility during outages, using the building's normal electrical distribution system, with the difference that only essential loads remain connected. This obviously does not meet the requirement that emergency lighting be supplied on power failure, even if the fault is local (i.e., within the building).

Central battery systems are of three types:

1. Those supplying low-voltage dc (6 to 48 V) to dedicated emergency lighting units (see Fig. 21.38*a*). This central battery installation consists of a battery, charger, and switching equipment. The emergency lights are generally not part of the normal lighting system, consisting instead of dedicated incandescent fixtures. The system has the same limitation as that of a generator, and therefore is not applicable to most emergency lighting situations except when equipped with multiple downstream sensors. Since such sensing would be highly uneconomical, this system is limited to a single (large) space with common outage sensing. This arrangement has the further disadvantage that the emergency lighting units may obtrude on the building's decor and architecture since they are not part of the normal lighting system. Also, because the lighting units are isolated devices, it is difficult to obtain a satisfactory degree of emergency illumination uniformity [See (*e*) below].

ILLUMINATION

TABLE 21.6 **Condensed General Criteria for Preliminary Consideration for Emergency Lighting Applications**

Specific Need	*Maximum Tolerance Duration of Power Failure*	*Recommended Minimum Auxiliary Supply Time*	*Type of Auxiliary Power System*[a]	
			Emergency	*Standby*
Evacuation of personnel	Up to 10 s, preferably not more than 3 s	2 hours	×	
	System Justification: Prevention of panic, injury, loss of life. Compliance with building codes and local, state, and federal laws. Lower insurance rates. Prevention of property damage. Lessening of losses due to legal suits.			
Perimeter and security	10 seconds	10 to 12 hours during all dark hours	×	×
	System Justification: Lower losses from theft and property damage. Lower insurance rates. Prevention of injury.			
Warning	From 10 s up to 2 or 3 minutes	To return to prime power source	×	
	System Justification: Prevention or reduction of property loss. Compliance with building codes and local, state, and federal laws. Prevention of injury and loss of life.			
Restoration of normal power system	1 s to indefinite depending on available light	Until repairs completed and power restored	×	×
	System Justification: Risk of extended power and light outage due to a longer repair time.			
General lighting	Indefinite; depends on analysis and evaluation	Indefinite: depends on analysis and evaluation		×
	System Justification: Prevention of loss of sales. Reduction of production losses. Lower risk of theft. Lower insurance rates.			
Hospitals and medical areas	Uninterruptible to 10 s. ANSI/NFPA 99-1984 and 101-1988 allow 10 s for alternative power source to start and transfer	To return of prime mover	×	×
	System Justification: Facilitate continuous care to patients by surgeons, medical doctors, nurses, and aides. Compliance with all codes, standards, and laws. Prevention of injury or loss of life. Lessening of losses due to legal suits.			
Orderly shutdown time	0.1 s to 1 hour	10 minutes to several hours	×	
	System Justification: Prevention of injury or loss of life. Prevention of property loss by a more orderly and rapid shutdown of critical systems. Lower risk of theft. Lower insurance rates.			

Source: Extracted with permission from "Recommended Practice for Emergency and Standby Power Systems," ANSI/IEEE Standard 446-1987; © 1987 by the Institute of Electrical and Electronics Engineers, Inc.

[a] See Section 16.31 for definition of these terms.

TABLE 21.7 **Typical Emergency and Standby Lighting Recommendations**

Standby[a]	*Immediate, Short-Term*[b]	*Immediate, Long-Term*[c]
Security Lighting	Evacuation Lighting	Hazardous Areas
Outdoor perimeters	Exit signs	Laboratories
Closed circuit TV	Exit lights	Warning lights
Night lights	Stairwells	Storage areas
Guard stations	Open areas	Process areas
Entrance gates	Tunnels	
	Halls	Warning Lights
Production Lighting		Beacons
Machine areas	Miscellaneous	Hazardous areas
Raw materials storage	Standby generator areas	Traffic signals
Packaging	Hazardous machines	
Inspection		Health-Care Facilities
Warehousing		Operating rooms
Offices		Delivery rooms
		Intensive care areas
Commercial Lighting		Emergency treatment areas
Displays		
Product shelves		Miscellaneous
Sales counters		Switchgear rooms
Offices		Elevators
		Boiler rooms
Miscellaneous		
Switchgear rooms		
Landscape lighting		
Boiler rooms		
Computer rooms		

Source: Extracted with permission from "Recommended Practice for Emergency and Standby Power Systems," IEEE Standard 446-1987 © 1987, Institute of Electrical and Electronics Engineers, Inc.

[a]An example of a standby lighting system is an engine-driven generator.

[b]An example of an immediate short-term lighting system is the common unit battery equipment.

[c]An example of an immediate long-term lighting system is a central battery bank rated to handle the required lighting load only until a standby engine-driven generator is placed on line.

2. Those supplying 120-V d-c (see Fig. 21.38*b*). This arrangement supplies line voltage d-c to emergency lighting fixtures, which are also used as part of the normal system. HID and fluorescent luminaires are furnished with integral inverters to change the incoming d-c to a-c. This arrangement has the advantage of a central, well-maintained battery supply, but the disadvantage of requiring a full d-c distribution system.

3. Those supplying line voltage a-c (see Fig. 21.38*c*). Here the central unit contains a battery, charger, and central d-c/a-c inverter and switching equipment. Such a system can supply power to incandescent, fluorescent, and HID systems in spaces where a single voltage-loss sensing point is sufficient. The emergency lights are part of the normal lighting system. This arrangement is most frequently used where a distributed system with local batteries (discussed below) would not be readily applicable due to the size of the load as would be the case with high-wattage HID luminaires. A typical 10-kVA on-line UPS cabinet designed to supply emergency a-c power to HID lamps measures 42 in. W × 50 in. H × 19 in. D and weighs 1100 lb.

(d) Distributed (Local) Emergency Lighting Arrangements. As a result of the limitations of centralized systems, the most frequently used emergency lighting arrangement is one where the source of power as well as all the voltage sensing and switching equipment are installed at the emergency lighting fixture, which is usually part of the normal lighting system. The emergency pack consists of a rechargeable battery and charger, voltage sensing and switching equipment, and in the case of fluorescent lights, an electronic package (ballast) that will operate the lamp at high frequency and generally at reduced output (see Fig. 21.39). These packages, which are designed to supply the code required 90 min of emergency lighting, are maintenance free for periods of up to 5 years and use batteries that can withstand the high temperatures existing in fluorescent lamp ballast compartments. A second type of distributed emergency lighting unit is the familiar packaged unit with integral incandescent lights. Several designs are shown in Fig. 21.40.

(e) Emergency Lighting Design Considerations. The level of required emergency illumination in specific areas should be related to the area's level of normal illumination and the degree of hazard in the area. Therefore, we would provide:

Exit area	5 fc (50 lux)
Stairs	3.5–5 fc (35–50 lux)
Hazard areas, such as machinery room	2–5 fc (20–50 lux)
Other spaces	1.0 fc (10 lux)

These levels should be essentially uniform. When the illumination level in an interior space drops sharply from a level of 30–100 fc to 1–5 fc, the eyes require up to 5 min to fully accommodate. During this long period the space's occupants are essentially sightless—a condition that lends itself readily to panic. For this reason, bright, spotlight-type heads must be *very carefully arranged.* Otherwise they can create disabling glare and distorting shadows and impede eye accommodation.

Although it is customary to furnish the required emergency illumination from ceiling- or wall-mounted fixtures, consideration must be given to the code requirement that specified emergency lighting levels be maintained at floor level. Since heavy smoke can readily obscure light from overhead fixtures, it may be advisable

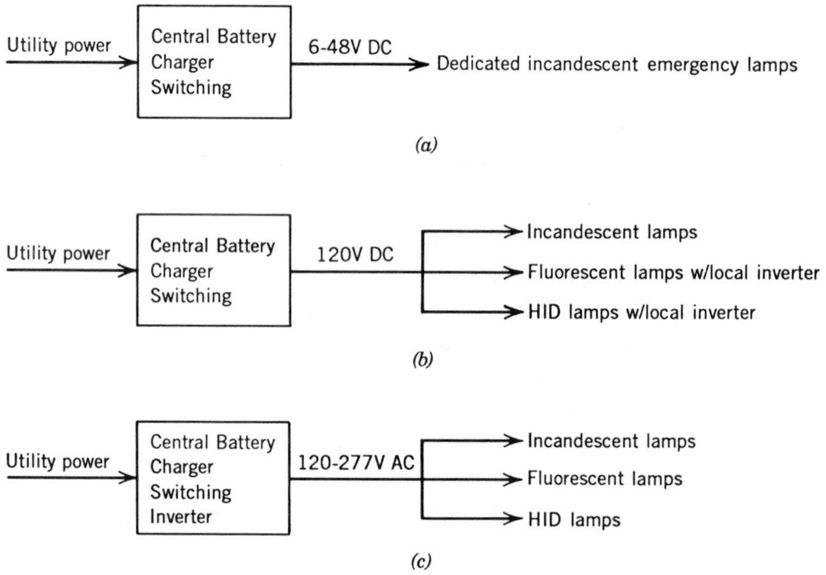

Fig. 21.38 *Schematic diagrams of emergency power system using central batteries. In* (a) *the emergency system is completely separate from the normal lighting system and utilizes dedicated lighting units.* (b) *This arrangement furnishes a d-c source of power to the luminaire, where it is inverted to a-c (except for incandescent lamps). The fixtures are utilized during both normal and emergency operation. Since switching is instantaneous, HID lamps are not extinguished.* (c) *Arrangement similar to* (b) *except that a central inverter supplies a-c continuously. Here too the lamps are used in both normal and emergency modes.*

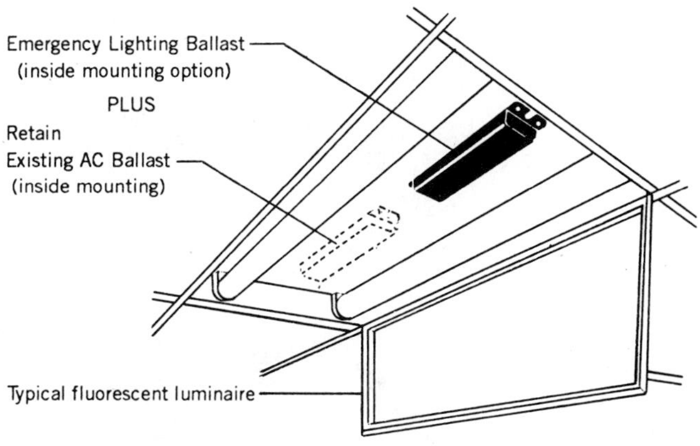

Fig. 21.39 *The emergency lighting ballast package contains a 90-min-capacity high-temperature battery, inverter, sensing and switching equipment, and electronic ballast circuitry to operate the lamp(s) at full or reduced output. The emergency ballast is energized on loss of normal power. The entire package measures 9½ × 2⅜ × 1½ in. (Courtesy of Teledyne Big Beam.)*

to install some emergency lighting fixtures near the floor level.

(f) Exit Lighting. Most codes require either self-illuminated signs or 5 fc (50 lux) on nonilluminated exit signs. Some exit signs are equipped with a battery and controls that will provide 1½ h of illumination on loss of utility power. Others are arranged to illuminate the exit area beneath the sign (see Fig. 21.41) and still others are equipped with a flasher and/or an audible beeper that will assist people to find the exit in a light-obscuring, smoke-filled room. Finally, some exit signs are nonelectrical self-illuminating, requiring no electrical wiring connection. All exit lights require continuous illumination and are part of the emergency lighting system.

21.31 Floodlighting

Floodlighting, both interior and exterior, is extensively used for such diverse locations as are listed in Table 21.8, in addition to the more common sports lighting, which is not listed. At the designer's disposal are a variety of sources with respect to output, color, life, efficiency, and wattage (see Chapter 19).

Although a detailed floodlighting design involves complex calculations beyond the scope of this work, it is often sufficient for the designer to utilize a watts per square foot table such as Table 21.8 to determine the approximate floodlighting requirements. Note that the unit power allowances for exterior lighting as specified in ASHRAE/IES 90.1-1989 and given in Table 21.9 weigh heavily in favor of HID sources, and that building facade lighting becomes a design challenge even when utilizing such sources.

Thus, if one is concerned with lighting a self-service parking lot at a neighborhood shopping center, and metal-halide is selected. Table 21.8 tells us that approximately 0.055 W/ft^2 will suffice. This is well within the UPA limits of Table 21.9. If the lot is 200 × 500 ft or 100,000 ft^2, then 0.055 × 100,000 = 5500 W is required.

Arrangement and choice of equipment remains then, before the problem can be considered solved. Considerable assistance on this score can be obtained from either the lighting engineer involved or from representatives of the equipment manufacturers.

Although most floodlight installations use a single type, the installation of Fig. 21.42*b* used a combination of metal-halide and mercury to obtain the desired effect.

21.32 Street Lighting

Although detailed street-lighting calculations and design considerations are beyond our scope (see appropriate IES standards), a few remarks are in order. New installations now use HID source almost exclusively. The low efficacy and short life of incandescent sources and the bulkiness of fluorescents make them obsolete. Furthermore, high street-lighting levels reduce vandalism and crime, improve night merchandising,

ILLUMINATION

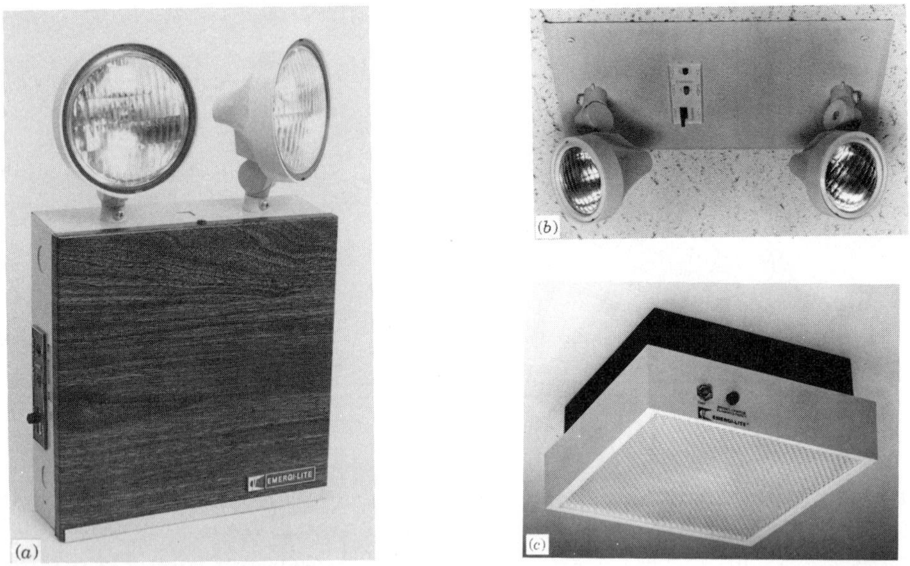

Fig. 21.40 *Various designs of self-contained, maintenance-free emergency lighting package units. Each package contains a 90-min-capacity battery, charger, inverter, and switching controls. Lamps are 6/12-Volt tungsten halogen.* (a) *Traditional type with sealed-beam-style heads.* (b) *Ceiling recessed unit.* (c) *Surface-mounted square-lensed unit.* (d) *Decoratively styled unit. (Courtesy of Emergi-Lite.)*

and add to an area's attractiveness. Some typical designs of street lighting and other outside luminaires are shown in Figs. 21.43 and 21.44.

21.33 Light Pollution

An oft-neglected corollary to the principle of placing light where it is required is not to place light where it is not required. *Light pollution* is frequently defined as unwanted light in public places, whereas *light trespass* is the intrusion of unwanted light onto private property. The latter would include light intrusion into windows and onto private property, whereas the former would cover excessive brightnesses everywhere, plus stray light that finds its way into the night sky to the distress not only of astronomers but also of anyone who simply wishes to enjoy the beauty of a star-filled sky. At this writing no standards exist which define the light levels that constitute an intrusion or a nuisance glare. However, a few simple guidelines will assist the designer in avoiding creation of a nuisance:

1. Light all exterior vertical surfaces from above, not below. This will reduce skylight pollution.
2. Use luminaires with sharp cutoff beyond the illuminated area. Shields can be added to standard luminaires to accomplish this.
3. After an installation is complete, inspect it at night to determine whether any nuisance has been created.

Further information on this subject can be found in IES publication CP-46-85, *Astronomical Light Pollution and Light Trespass*.

Fig. 21.41 Typical application of exit light with built-in battery, charger, and controls. Note that the bottom of the unit is designed to illuminate the area immediately in front of the exit. (Photo by Stein.)

REFERENCES AND BIBLIOGRAPHY

Alman, D. H., Breton, M. E., and Barbour, J. (1983). "New Results on the Brightness Matching of Heterochromatic Stimuli," *JIES*, July.

Ander, G. E., and Navvab, M. (1983). *Daylight Impacts of Fenestration Controls*, American Solar Energy Society, Washington, D.C.

ANSI/IEEE Standard 446, (1987). *Recommended Practice for Emergency and Standby Power*, IEEE/Wiley, New York.

Bernecker, C. A., and Mier, J. M. (1985). "The Effect of Source Luminance on the Perception of Environment Brightness," *JIES*, Fall.

Brown, G. Z., Reynolds, J. A., and Ubbelohde, M. S. (1982). *InsideOut*, Wiley, New York.

Bryan, H. J. (1980). "A Simplified Procedure for Calculating the Effects of Daylight from Clear Skies," *JIES*, April, pp. 142–151.

Corth, R. (1986). "Influence of Light Irradiance and Wavelength on the Primate Suprachiasmatic Nuclei," *LRI*, July.

Daylight Design Diagrams (1963). Service Division, Commonwealth Department of Labor and National Service, Melbourne, Australia.

Daylight, International Recommendations for the Calculation of Natural Daylight (1970). Publication CIE No. 16 (E-3.2), Commission Internationale de l'Eclairage, Paris.

Elmer, W. B. (1980). *The Optical Design of Reflectors*, Wiley, New York.

Griffith, J. W., Arner, W. J., and Wenzler, (1957). "Practical Daylighting Prediction," IES National Technical Conference, Atlanta, Ga.

Harkness, E., and Mehta, M. (1978). *Solar Radiation Control in Buildings*, Applied Science Publishers, London.

Hopkinson, R. G., Pethebridge, R., and Longmore, J. (1966). *Daylighting*, Heinemann, London.

"How to Predict Interior Daylight Illumination" (1976). Pamphlet, Libbey-Owens-Ford Co.

Hughes, P. C., and Neer, R. M. (1981). *Lighting for the Elderly: A Psychobiological Approach to Lighting*, The Human Factors Society, Inc.,

IES *Lighting Handbook* (1987). IESNA, New York.

IES RP1-1983. *American National Standard Practice for Office Lighting*, IESNA, New York.

IES RP2-1985. *Recommended Practice for Lighting Merchandising Areas*, IESNA, New York.

IES RP3-1988. *American National Standard Guide for Educational Facilities Lighting*, IESNA, New York.

IES RP4-1974. *Recommended Practice of Library Lighting*, IESNA, New York.

IES RP6-1988. *Recommended Practice for Sports and Recreational Area Lighting*, IESNA, New York.

IES RP7-1983. *American National Standard Practice for Industrial Lighting*, IESNA, New York.

IES RP11-1980. *Design Criteria for Lighting Interior Living Spaces*, IESNA, New York.

IES RP20-1984. *Lighting for Parking Facilities*, IESNA, New York.

IESNA RP-21, 1983. *Recommended Practice*

ILLUMINATION

TABLE 21.8 **Lighting Application Guide**

Application	Minimum Footcandles Maintained[a]	Watts per Square Foot Generally Required: Tungsten-Halogen	Mercury Units	Metal-Halide	High-Pressure Sodium
Automobile Parking					
Attendant parking	2	0.38	0.17	0.11	0.075
Industrial lots	1	0.13–0.15	0.06–0.07	0.037–0.044	0.026–0.03
Self-parking lots	1	0.13–0.15	0.06–0.07	0.037–0.044	0.026–0.03
Shopping Centers					
Neighborhood	1	0.13–0.19	0.06–0.09	0.037–0.055	0.026–0.038
Average commercial	2	0.26–0.3	0.12–0.135	0.075–0.087	0.052–0.06
Heavy traffic	5	0.65	0.29	0.19	0.13
Automobile Sales Lots					
Front row (Front 20 ft)	50	10.	4.5	2.9	2.0
Remainder	10	1.5–1.8	0.68–0.81	0.44–0.52	0.3–0.36
Building					
Construction	10	1.5–1.8	0.68–0.81	0.44–0.52	0.3–0.36
Excavation	2	0.26–0.3	0.12–0.14	0.075–0.09	0.052–0.06
Buildings up to 50 ft High	***Adj. Area*** ***Light Dark***				
Light surfaces	15 5	3.3 1.2	1.5 0.54	0.96 0.35	0.66 0.24
Medium light	20 10	4.3 2.2	1.94 1.0	1.25 0.64	0.86 0.44
Dark surfaces	50 20	10.0 4.3	4.5 1.94	2.9 1.25	2.0 0.86
Billboards and Signs	***Adj. Area*** ***Light Dark***				
Good contrast	50 20	10.0 4.3	4.5 1.94	2.9 1.25	2.0 0.86
Poor contrast	100 50	20.0 10.0	9.0 4.5	5.8 2.9	4.0 2.0
Protective Lighting					
Gates and vital area	5	1.2	0.54	0.35	0.24
Building surrounds	1	0.15–0.19	0.07–0.09	0.044–0.055	0.03–0.04
Roadways					
Along buildings	1	0.24	0.11	0.07	0.05
Open areas	0.5	0.08–0.1	0.036–0.045	0.023–0.029	0.02
Storage yards (active)	20	3.6–4.3	1.6–1.94	1.04–1.25	0.72–0.86
Storage yards (inactive)	1	0.15–0.19	0.07–0.09	0.044–0.055	0.03–0.04
Shopping Centers					
Parking areas (attraction)	5	0.65	0.29	0.19	0.13
Buildings (attraction)			(See Buildings)		
Used Car Lots			(See Automobile Parking)		

[a]All footcandle levels for ground area applications are *horizontal* values.

for the Calculation of Daylight Availability, IESNA, New York.

IES RP24-1990. Recommended Practice for Lighting Offices Containing Computer Visual Display Terminals, IESNA New York.

IESNA Recommended Practice for the Lumen Method of Daylight Calculations (1989). IESNA, New York.

IESNA CP29-1985. *Lighting for Health Care Facilities,* IESNA, New York.

IESNA LEM-1-1982. *IES Recommended Practice for Lighting Power Limit Determination,* IESNA, New York.

IESNA LEM-2-1984. *IES Recommended Procedure for Lighting Power Limit Determination for Buildings,* IESNA, New York.

IESNA LEM-3-1987. *IES Design Considerations for Effective Building Lighting Energy Use,* IESNA, New York.

IESNA LEM-4-1984. *IES Recommended Procedure for Energy Analysis of Building Lighting Designs and Installations,* IESNA, New York.

IESNA LM-17 (1958). *IES Guide to Design of Light Control,* IESNA, New York.

Jones, B. F. (1983). "Very Very Simple Hand

TABLE 21.9 **ASHRAE 90.1 (Table 6-1): Exterior Lighting Unit Power Allowances**

Area Description	*Allowances*
Exit (with or without canopy)	25 W/lin ft of door opening
Entrance (without canopy)	30 W/lin ft of door opening
Entrance (with canopy)	
High traffic (retail, hotel, airport, theater, etc.)	10 W/ft^2 of canopied area
Light traffic (hospital, office, school, etc.)	4 W/ft^2 of canopied area
Loading area	0.40 W/ft^2
Loading door	20 W/lin ft of door opening
Building exterior surfaces/facades	0.25 W/ft^2 of surface area to be illuminated
Storage and nonmanufacturing work areas	0.20 W/ft^2
Other activity areas for casual use such as picnic grounds, gardens, parks, and other landscaped areas	0.10 W/ft^2
Private driveways/walkways	0.10 w/ft^2
Public driveways/walkways	0.15 w/ft^2
Private parking lots	0.12 w/ft^2
Public parking lots	0.18 w/ft^2

Source: Copyright © by American Society of Heating, Refrigeration, and Air-Conditioning Engineers, Atlanta, Ga. ASHRAE/IES 90.1–1989; reprinted with permission.

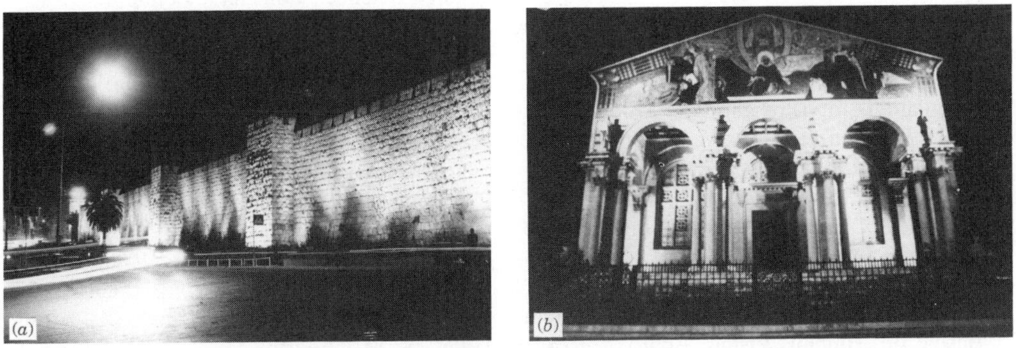

Fig. 21.42 (a) *Floodlighted section of wall surrounding the Old City of Jerusalem, Israel, adjacent to the Jaffa Gate. Light sources are 400-W HPS units, giving an average illumination level of 50 lux. Sodium source was chosen to enhance the yellow-red color of the stone.* (b) *Church of All Nations, Mount of Olives, Jerusalem, Israel. Floodlight sources are 250- and 400-W mercury and metal-halide units, giving an average illumination of 70 lux. Sources were selected to complement the colors in the mosaic at the top of the facade. (Photos courtesy of City of Jerusalem and J. Stroumsa, Chief Engineer.)*

Fig. 21.43 (a) *Concrete pole with aluminum arm and mercury luminaire.* (b) *Other concrete pole sections. (Photos courtesy of Union Metal Corp.)*

Calculations for Daylighting,'' *Proceedings, 1983 International Daylight Conference,* Phoenix, Ariz., pp. 87–92.

Kambich, D. G., and McCloskey, N. M. (1984). ''Practical Consequences of Using Relative Visual Performance to Analyze Lighting Systems,'' *JIES*, October.

Kaneko, T. (1982). ''An Alternative ESI System Based on Human Visual Acuity,'' *JIES,* April.

Kimball, H. H., and Associates (1923). ''Daylight Illumination on Horizontal, Vertical and Sloping Surfaces,'' *U.S. Weather Review,* Vol. 50, p. 615.

Krochman, J. (1963). ''Uber die horizontal Beleuchtungsstarke der Tagesbeleuxchtung,'' *Lichtechnik,* Vol. 15, No. 11.

Fig. 21.44 The ''lollypop'' fixture, even if aesthetically pleasing, gives poor illumination downward (note the large collar). The narrow pole is weakened by the large handhole. (Photo by Stein.)

Lewis, A. L. (1987). ''Light Affects Health: Light May Potentially Damage Health,'' *LD&A,* October.

Lim, Rao, et al. (1979). *Environmental Factors in the Design of Building Fenestration,* Applied Science Publishers, London.

Lynes, J. A., Burt, W., Jackson, G. D., and Cuttle, C. (1966). ''The Flow of Light into Buildings,'' *Trans. IES (London),* Vol. 31, No. 3, pp. 65–91.

Matsuura, K., 1979, ''Turning-Off Line in Perimeter Areas for Saving Lighting Energy in Side-Lit Offices,'' *Energy and Buildings,* Vol. 2.

Millet, M. S., and Bedrick, J. R. (1980). *Graphic Daylighting Design Method,* LBL/DOE, Washington, D.C.

Millet, S. M., Bedrick, J. R., and Adams, C. (1980). ''Graphic Daylight Design Method,'' paper presented at International Symposium on Daylighting, Berlin, July.

Millet, S. M., Lakin, J. E., and Moore, J. (1981). ''Rainy Day Shading; The Effect of Shading Devices on Daylighting and Thermal Performance,'' Department of Architecture, University of Washington, Seattle.

Modest, F. F. (1983). "Daylighting Calculations for Nonrectangular Interior Spaces with Shading Devices," *JIES,* July.

Moore, F. (1985). *Concepts and Practice of Architectural Daylighting,* Van Nostrand Reinhold, New York.

Murdoch, J. B. (1985). *Illumination Engineering,* Macmillan, New York.

Nagi, P. Y. (1981). "Some Issues on the Foundation of Visibility and Task Contrast," *JIES,* January.

Nakamura, H., and Oki, M. (1983). "Calculation of Daylight Factor Dominated by Intermediate Sky," *Proceedings, International Daylighting Conference,* Phoenix, Ariz., pp. 57–59.

Ne'eman, E., and Longmore, J. (1973). "The Integration of Daylight with Artificial Light," paper presented at CIE Symposium, Instanbul.

Olgyay, A., and Olgyay, V. (1957). *Solar Control and Shading Devices,* Princeton University Press, Princeton, N.J.

Peterson, D., and Rubenstein, F. (1982). *Energy Savings Through Effective Light Control,* Laurence Berkeley Laboratories, February.

Pierpoint, W. (1981). "Equi-visibility Lighting Control Principles," *JIES,* October.

Pierpoint, W. (1983). "A Simple Sky Model for Daylighting Calculations," *Proceedings, International Daylighting Conference,* Phoenix, Ariz., pp. 47–51.

Ramsey, C. G., Sleeper, H. R., and AIA (1981). *Architectural Graphic Standards,* 7th ed., Wiley, New York, pp. 76–86.

Rea, M. S. (1981). "Visual Performance with Realistic Methods of Changing Contrast," *JIES,* April.

Rea, M. S. (1982). "An Overview of Visual Performance," *LD&A,* November.

Rea, M. S., Pasini, I., and Jutras, L. (1990). "Lighting Performance Measured in a Commercial Building," *LD&A,* January.

Rennhackkamp, W. M. H. (1967). "Sky Luminance Distribution on Warm Arid Climates," *Proceedings, 16th International Conference on Illumination,* Washington, D.C.

Robbins, C. L. (1986). *Daylighting: Design and Analysis,* Van Nostrand Reinhold, New York.

Ross and Barruzini, (1975) Energy Conservation Applied to Office Lighting, Federal Energy Administration.

Rubenstein, F. (1981). *Energy Saving Benefits of Automatic Lighting Controls,* Laurence Berkeley Laboratories, September.

Rubenstein, F., Verderber, R., Karayel, M. (1985). "Determination of the Optimum Sector Size for Automatic Lighting Controls," *JIES,* Fall.

Rubenstein, F., Ward, G., and Verderber, R. (1989). "Improving the Performance of Photo-Electrically Controlled Lighting Systems," *JIES,* Winter.

Sampson, F. K. (1970). *Contrast Rendition in School Lighting,* Educational Facilities Laboratory, New York.

Selkowitz, S. (1982). *Daylight Design Overlays for Equidistant Sun-Path Projections,* LBL, Berkeley, CA.

Selkowitz, S., Navvab, M., and Mathews, S. (1983). "Design and Performance of Light Shelves," *Proceedings, International Daylighting Conference,* Phoenix, Ariz.

Smith, F. K. (1989). "Spaciousness, One of the Impressions Profoundly Affected by Light," *LD&A,* September.

Smith, S. W., and Rea, M. S. (1978). "Proofreading Under Different Levels of Illumination," *JIES,* October, p. 47.

Smith, S. W., and Rea, M. S. (1982). "Performance of a Reading Test Under Different Levels of Illumination," *JIES,* October, p. 29.

Szokolay, S. V. (1980). *Environmental Science Handbook,* Halsted/Wiley, New York, p. 100.

Thrun, E., and Jennings, R. (1981). "Levels of Performance Related Illumination Available from Daylight in Typical Office/School Interiors," *JIESNA,* October.

Treado, S., and Gillette, G. (1983). "Measurement of Sky Luminance, Sky Illuminance and Horizontal Solar Radiation," *JIES,* April, pp. 130–135.

Turner, W. C. (ed.) (1982). *Energy Management Handbook,* Wiley, New York.

Wieder, S. (1982). *An Introduction to Solar Energy for Scientists and Engineers,* Wiley, New York.

Worthey, J. A. (1989). "Effect of Veiling Reflections on Vision of Colored Objects," *JIES,* Summer.

Yorks, P., and Ginthner, D. (1987). "Wall Lighting Placement: Effect on Behavior in the Work Environment," *LD&A,* July.

PART VII
SIGNAL EQUIPMENT

Modern buildings, from the simplest residence to the most complex industrial facility, depend completely on electrical, signal, alarm, and communication systems for normal functioning. The systems discussed in this part of the book are security, music/sound, intercom, clock and program, and paging. Fire and smoke detection and alarms are covered completely in Part IV (Chapter 13), Fire Protection. Building automation is covered in Appendix J. This discussion is from the point of view of a user—that is, we emphasize system application rather than design. Operating principles are adduced, and application data are related to available equipment. Thus, although equipment is constantly being improved, operating and application principles remain substantially constant, thereby permitting the designer to readily adapt new hardware to existing system arrangements.

22
SIGNAL SYSTEMS

22.1 Introduction

No area of equipment design and application to buildings has seen such great and rapid changes as the field of signal equipment. Under this title is subsumed all signal, communication, and control equipment, the function of which is to assist in effecting proper building operation. Included are surveillance equipment such as fire and access control, audio and visual communication equipment such as telephone, intercom, and television, both public and closed circuit; and timing equipment such as clock and program. Specifically excepted is the entire area of data processing, which is not within the province of the building designer, except for equipment space allocation. These systems are no longer limited in application. Clock and program equipment, which once were the exclusive interest of schools and some industrial facilities, are now incorporated into building mechanical equipment control systems. Closed-circuit TV, which was once limited to classroom and college use, is commonplace in mercantile areas as part of surveillance systems. The hundreds of signals generated throughout a large facility are logged, channeled, and controlled by means of specially programmed computers and microprocessors. All the signal systems that once were separate and distinct are now frequently combined and serve multiple purposes.

Obviously a detailed study of the design and application of such equipment is beyond the scope of this book or, for that matter, of any single book. We shall attempt, however, to discuss the basic operation of the various systems, some of the equipment available, application to different types of facilities, and the impact of these systems on the spaces within a structure. The types of facility considered are single and multiple residences, schools, stores, office buildings, and industrial facilities. Hospitals and laboratories are combinations of the area types above plus facilities too highly specialized to be discussed herein.

The systems covered include surveillance, communication, and time-based signal arrangements. Antenna systems are discussed as a special case of a communication (reception) system.

22.2 Principles of Intrusion Detection

To understand the design of intrusion detection systems, it is necessary first to understand the characteristics of the commonly available intrusion detectors (sensors) on which these systems are based. Once the intrusion alarm has been given by a device (as with fire detection), the signal must be processed and appropriate measures taken. This may include sounding loud alarms, turning on lights, sending signals to private surveillance services or police, and so on.

(a) Normally Open (NO) Contact Devices. See Fig. 22.1*a*. Primary among these are switch mats. Contact is made as a result of pressure on the mat due to an intruder's weight. These devices, normally used in residences, are applied where only a single path of entry exists to the protected area, so that the mat cannot easily be circumvented, if it is not noticed. However, the device can be readily defeated by simply moving the mat or stepping over or around it. Other disadvantages are that the switch mat restricts free movement by authorized personnel (residents), and it can be set off by small animals. Finally, being an open-circuit device, its circuit is unsupervised.

(b) Normally Closed (NC) Contact Devices. See Fig. 22.1*b*. These come in a variety

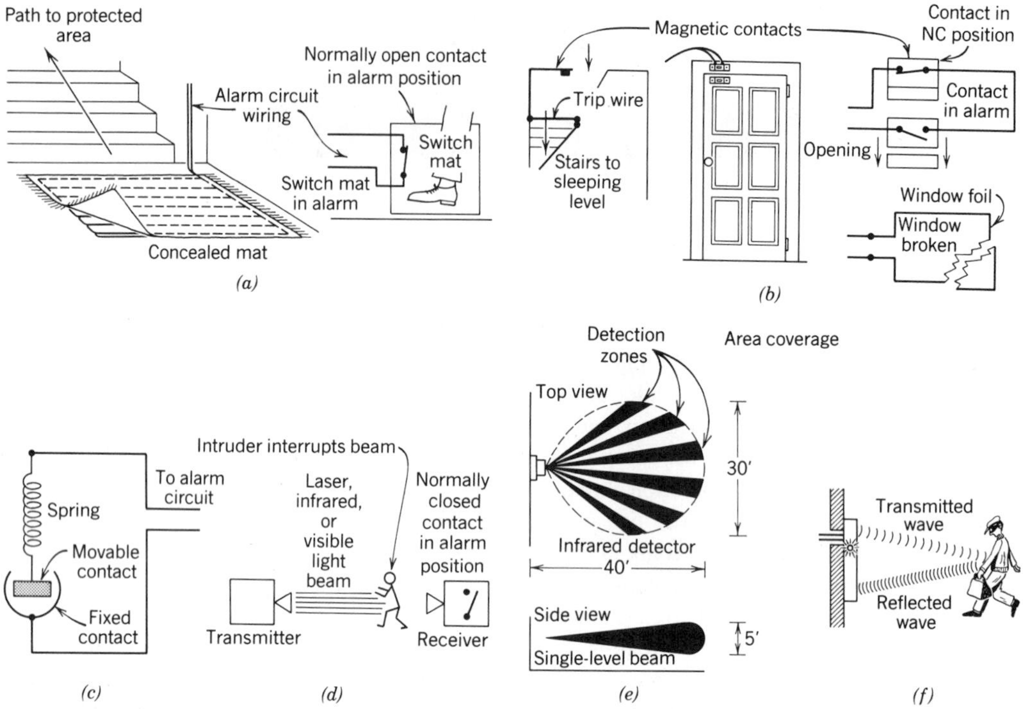

Fig. 22.1 *Typical intrusion detectors.* (a) *Normally open contact device such as a switch mat is operated by the weight of the intruder.* (b) *Magnetic door contacts are the first line of intrusion alarm at the house. The second may be a low-level trip wire at the base of the stairs.* (c) *Vibration detectors are very sensitive to motion and can be used on multipaned windows that cannot be conductive-strip foiled without spoiling their appearance.* (d) *Photoelectric beams form an effective intrusion barrier if placed so that their detection or avoidance is difficult.* (e) *Passive infrared detectors give basically a 30 × 40 ft oval protective zone, starting as a narrow beam and widening with distance. Focusability permits exact coverage of any area in a space. Units are also available with multilevel beams that give vertical as well as horizontal coverage.* (f) *Motion detectors depend on the Doppler effect. They detect changes in the frequency of a signal reflected from a moving object. Sensitivity is highest when relative motion is greatest, that is, when the intruder is moving directly toward (or away from) the detector.*

of designs, the most common of which are magnetic contacts for doors and windows, spring-loaded plunger contacts for doors and windows, window foil, and trip wires. These devices, though not easily defeated, also restrict use of the protected opening. This is not a problem with doors, but it can be inconvenient with windows in residences, where open windows are desired for ventilation. The problem with windows can be overcome by using multiple sets of contacts. Trip wires can be used only where traffic in the area is cut off during protected hours. (It is advisable to "exercise" magnetic contacts periodically to prevent their freezing in position after a long period of disuse.)

(c) Mechanical Motion Detectors. See Fig. 22.1*c*. Where window foil or fixed contacts are impractical, a mechanical motion detector can be used. This device is basically a spring-mounted contact suspended inside a second contact surface. Any appreciable motion of the surface on which the device is placed will cause the contacts to make *momentarily*, turning in an alarm. These devices are very sensitive and can be activated by sonic booms, wind, and even a heavy truck passing by. Because of this, most such units are provided with sensitivity adjustment.

(d) Photoelectric Devices. See Fig. 22.1*d*. These devices operate on the simple principle of beam interruption. When the beam is received, an alarm contact in the receiver is normally closed. Interruption of the beam causes the contact to open, setting off the alarm. Older devices of this design use a visible light beam, and rely, for concealment, on the fact that light is invisible except when reflected from an intervening object. This is quite effective indoors, but outside dust, insects, birds, and so on, will show the location of the beam, permitting it to be circumvented. Birds and small animals will set it off, too. Dispersion of visible light also limits the throw of the devices when used outside. Modern units use lasers or infrared beams, which are less easily detected and can be arranged to distinguish between intruder and other disturbance. These latter devices have a longer effective range for exterior use. When using a laser beam, the signal can be picked up, amplified, and retransmitted in a different direction, thus establishing a perimeter security "fence" from a single source.

(e) Passive Infrared (PIR) "Presence" Detector. See Fig. 22.1*e*. This device acts on the principle that all objects emit infrared radiation, or heat. The amount radiated depends primarily on the object's temperature and secondarily on its material, color, and texture. The IR passive sensor uses a lens or mirror that focuses on a small area and concentrates the IR radiation collected from that area into a sensor. Since IR radiation in an area that is undisturbed (not necessarily unoccupied) changes very slowly, because object temperature changes very slowly, any rapid change in the IR reading of that area indicates an object entering (or leaving) the space, causing an alarm. (We have already seen how these detectors can be used as occupancy sensors, to turn off lights when a space's occupant leaves; see Section 21.4.) The ability to focus on a particular area is utilized to cover areas both horizontally and vertically. Also, since the IR detector is not intrinsically sensitive to motion, but only to heat, it is usable where motion in the monitored area is unavoidable. The principal disadvantage to passive IR detectors is that rapid temperature changes caused by direct insolation, a cold breeze, a heater turning on, and the like can cause false alarms.

PIR detectors can be, and are, applied as sensitive motion detectors by using a multi-beam (zone) unit. Motion is detected as changes in the IR radiations of adjacent zones, or in the radiation of a zone (target) with respect to the background, both of which characterize motion.

(f) Motion Detectors. See Fig. 22.1*f*. These devices, which operate either at microwave frequencies or at ultrasonic frequencies, detect motion in the protected area by the Doppler effect. (This is the same effect that changes the perceived sound of a car horn or train whistle as the vehicle passes.) Any moving body will change the received frequency of the signal it reflects, and an alarm will sound. However, because the Doppler effect depends on relative motion between the source and the moving body, an intruder moving laterally may go undetected, if sensitivity has been reduced to avoid false alarms. Therefore, units should be located so that the path of an intruder is as nearly as possible directly toward or away from the detector. Ultrasonic units are cheaper than microwave but can be disturbed by strong air turbulence and very loud noises. Microwave units are undisturbed by air or noise, but because they penetrate solids (like ordinary TV signals), they can be affected by motion outside the protected area.

(g) Acoustic Detectors. These units alarm when the noise level exceeds a preset minimum. Alternatively, they can be arranged to respond to a particular range of frequencies, corresponding to the noise of breaking glass, forced entry, or whatever is desired. Although applied principally in security systems, they can also be used as occupancy sensors for switching of lighting. Such an application is described in Section 21.4*c* and illustrated in Fig. 21.11.

Fig. 22.2 *Typical intrusion detection equipment: (1) Passive infrared detector. Maximum sensitivity is 2°C differential between target and background, and target motion of 2½ in./s. Unit is approximately 6 in. high, 3½ in. wide, and 4 to 6 in. deep. They are available with different coverage patterns, with a maximum coverage of 2000 ft². (2) High-sensitivity multizone wide-angle passive IR detector gives coverage of 17 zones horizontally and vertically. (3) Ultrasonic detectors arranged to give broad coverage of approximately 200 × 50 ft. Each unit is 11 × 5½ × 3½ in. (4) High-sensitivity balanced signal type of ultrasonic detector differentiates between intrusion signal and random environmental signals, thus reducing false alarms. (5) Unobtrusive, button-type ultrasonic sensor intended for ceiling mounting. (6–8) System control panel, annunciator, and mechanical alarm interface unit. (Photos courtesy of Aritech Moose Security Products.)*

A variety of the above-described devices are illustrated in Fig. 22.2.

As with fire detectors, a balance must be struck between sensitivity of detectors and the nuisance of false alarms since unfortunately, increasing the former increases the latter as well. One very effective method of reducing nuisance alarms is to use multiple detectors of different technologies to, in effect, verify each other. Thus units are available that combine PIR and ultrasonic detectors in a single housing. Such a unit will not transmit an alarm until both detectors indicate intrusion. This dual technique is applicable to area and perimeter protection as well as portal surveillance.

The discussion above and that immediately below refer only to intrusion detection (i.e., indication of a human presence where it should not be). *Access* control, which is another area of security design entirely, deals with controlled entry to an area. It is generally more complex than intrusion control, and uses different technologies and a wholly different philosophy of design.

Elements of access control are discussed in Section 22.18. In both cases, the building designer is well advised to seek the advice of a security expert to assist first and foremost in developing an appropriate security philosophy for the facility, after which equipment selection and specification will follow.

PRIVATE RESIDENTIAL SYSTEMS

22.3 General

Modern private residences utilize a variety of signal apparatus that greatly enhance their functional value. Figure 22.3 shows a residence that has been provided with what would be considered adequate but by no means excessive sound and signal equipment for a house of its size. In general, all signal systems require a source of signal, equipment to process the signal, including transmitting it, and finally a means of indicating the signal, either audibly, visually, or on permanent record "hard copy." A complex system still falls into this threefold category except that the individual items of equipment and their functions become more sophisticated.

In connection with complexity of equipment, the reader is undoubtedly familiar with the sophisticated controls now on the residential market that will handle security, fire alarm, time functions, lighting, and so on. These are basically time-based programmable switches, of the type described in Section 16.16. They function on dedicated wiring or on power line carrier (PLC) signals, that is, high-frequency signals impressed on electric power wiring. Since it is not our purpose to discuss automation or central control, but rather the basic systems, the following descriptions consider the systems separately, despite the clear trend to consolidated control.

Residential fire alarm systems are covered in Section 13.17. The components and design of such a system, including detectors, circuit arrangement, and the like are also discussed in Chapter 13.

We have listed in Table 22.1 the systems and equipment found in the residence of Fig. 22.3, by the threefold classification. Note that the fire alarm, smoke detection, and intrusion alarm systems have been combined into a single system. This simplifies operation and avoids unnecessary equipment duplication. As we discuss the more complex systems, it will be seen that the basic functions remain unchanged.

As shown in Fig. 22.3, a single control panel can serve a multiplicity of residential systems. An annunciator, either integral with the panel or in a separate adjacent enclosure displays the nature and location of the alarm device that has "tripped." A riser diagram for this residence is shown in Fig. 22.3*b*. The alarm devices themselves are not shown, because they appear on the plans and duplication serves no useful purpose.

22.4 Residential Intrusion Alarm Systems

Although any or all of the devices described in Section 22.2 may be used, residences normally utilize door and window magnetic switches, and PIR and/or motion detectors, as shown in Fig. 22.3. A manual switch at the end of a long cord is also often provided so that the resident may at will set off the alarm if an intruder is heard. If the system employs the same audible signal devices as the fire system the sound should be distinctive so that the nature of the alarm can be discerned aurally. Intrusion alarm systems can be continuously supervised by connection with central stations of companies whose business is such supervision and who will either respond directly to an alarm call or notify local police authorities of any illegal entry.

22.5 Residential Television Antenna Systems

The large number of multiset American homes has made the central television antenna system a desirable feature of the modern residence. Systems with more than two outlets and not connected to commercial cable television services, generally require a booster amplifier (except in strong signal areas) and are known as amplified systems.

The function of the system is to supply a television signal at each wall outlet, so that a receiver may be operated at any location and so that two or more receivers may operate simultaneously.

The functioning of the system is simply to amplify the signal received by the antenna and by means of special cable to distribute these amplified signals in a concealed cable to the various wall outlets. The type and location of antenna, gain (amplification) of the amplifier, and type of cable are variables that, being dependent on the specific installation, are best left to a competent and reliable local television company or design engineer. If a dish-type or other antenna is considered, provision must be made for cable entry via wall sleeves, sealed inside and out, or empty conduit located as directed.

22.6 Residential Intercom Systems

The public demand for step-saving conveniences has resulted in the wide acceptance of the home "intercom." Although available with

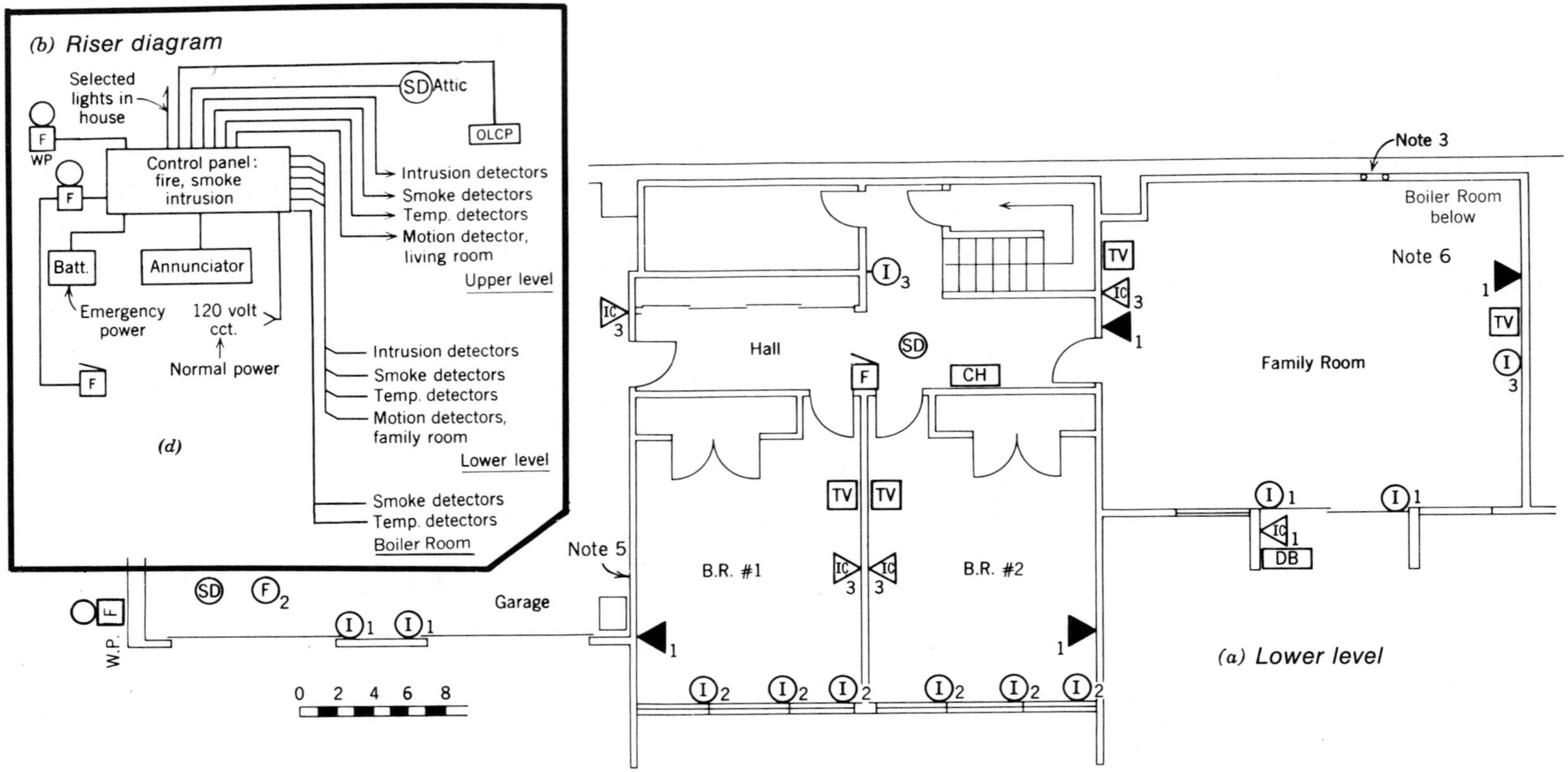

Symbols for signal equipment

○ F	6 in. a-c vibrating bell, concealed in recessed box, with grill cloth cover, 84 in. AFF.
WP ○ F	8 in. weatherproof bell or siren
F (buzzer)	Buzzer, a-c, similar installation to above
(F) 1	Temp. detector; rate-of-rise and fixed temp., resettable
2	Temp. detector; fixed temp., 190°C
(SD)	Smoke detector with resettable fixed temp. detector.
(I) 1	Intrusion detector, magnetic door switch.
2	Intrusion detector; magnetic window switch.
3	Intrusion detector; electronic, motion detector.
◇	Annunciator, custom design
CP	Central panel for fire alarm, smoke detector, and intrusion
DB	Door bell
CH	Chimes signal
◀ 1	Prewired phone outlet; jack 12 in. AFF.
2	Prewired phone outlet; fixed, 12 in. AFF.
3	Prewired phone outlet; fixed wall outlet 60 in. AFF.
IC 1	Intercom outlet, outdoor, W.P. 60 in. AFF.
2	Intercom outlet, master station 60 in. AFF.
3	Intercom outlet, remote station 60 in. AFF.
TV	Prewired TV antenna outlet, 12 in. AFF.

(c) Symbol list

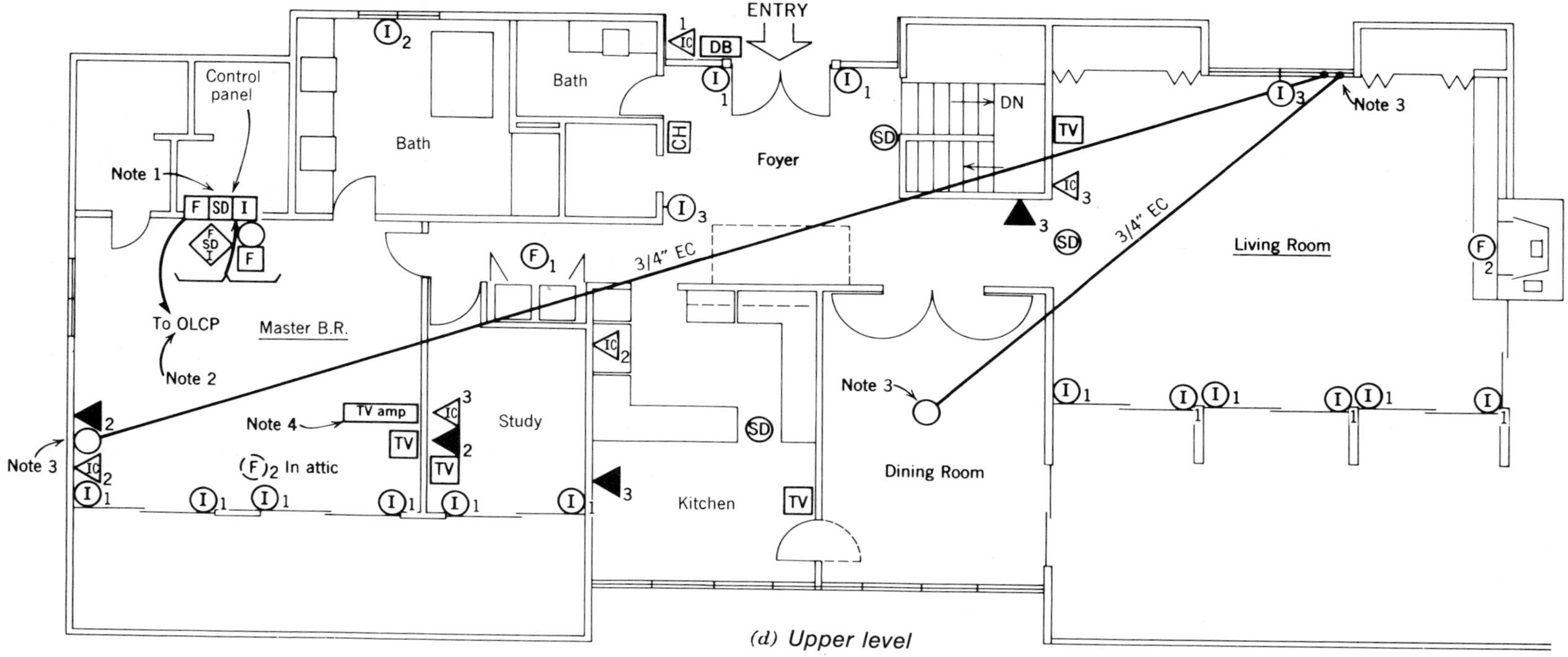

(d) Upper level

Fig. 22.3 *Signal plans for typical one-family residence.*

Notes:

1. *The fire detection, smoke detection, and intrusion alarm devices all operate from a single control panel, see* (d). *The alarm bell is common. The annunciator indicates the device operated and its location.*
2. *The connection between the signal control panel and OLCP (outside lighting control panel) activates all outside lights when a signal device trips. Selected lights inside the house can also be connected to go on. See riser diagram* (d).
3. *Two ¾-in. empty plastic conduits, extending from 4-in. boxes in living room wall down to family room and terminating in 4-in. flush boxes. Boxes to be 18 in. AFF and fitted with blank covers. Also, extend a ¾-in. plastic EC from one 4-in. box in living room to 12-in speaker backbox recessed in dining room ceiling. Locate in the field. From the second 4-in. box in living room extend a ¾-in. empty plastic conduit to an empty 4-in. box in the master bedroom, 18-in. AFF. Finish with blank cover.*
4. *Provide television antenna amplifier, recessed in wall box, with hinged ventilated cover, 18-in. AFF. Connections to antenna and to all television outlets by television antenna subcontractor. Provide 120-V outlet at the amplifier, with switch to disconnect.*
5. *Provide ¾-in. EC from TV amplifier to an exterior location, as directed, for future satellite dish cable.*
6. *Boiler room to contain smoke detector, fixed 190 F heat detector, and remote station intercom outlet.*

TABLE 22.1 **Elements of Residential Signal Systems**

System Type	*Signal Generator*	*Signal Processor*[a]	*Signal Transducer*
Fire alarm	Temperature and smoke detectors	Control cabinet(s)	Bells, annunciator, buzzer, siren
Intrusion alarm	Door and window switches, motion detectors	Control cabinet	Bells, buzzer, annunciator, siren
Emergency call system	Pull, push button	Control cabinet	Bells, annunciator, corridor lights
Door bell	Push button	Transformer	Buzzer, chime
TV antenna	TV station and house antenna	Amplifier	TV set
Intercom	Microphone, speaker—mike	Amplifier	Speakers in various stations

[a]The proper wiring and switching is included under this title in all cases.

various features, the basic system comprises one or more masters and several remote stations, one of which monitors the front door, allowing it to be answered from various points within the home. Where desired, a closed-circuit TV system can be added so that visual identification at entrances, in addition to voice communication, is effected. In general, master stations allow selective calling, whereas remote stations operating through the masters are nonselective. The systems are particularly useful when left in the open (monitor) position for remote "baby-sitting." The applicability of such systems to residences with outbuildings should be immediately apparent. Since wiring is low voltage and low power, multiconductor color-coded intercom cable is generally used, run concealed within walls, attics, and basements.

Systems are also available that impose the signals onto the house power wiring. This has the advantage of eliminating separate wiring and making remote stations portable—they are connected simply by plugging into a power outlet.

22.7 Residential Telephone Systems

Prior to court decisions permitting users to install their own telephone equipment, the actual wiring within a structure was done only by the utility, in the user's raceway system. For some years now, work beyond the service entrance may be done by the owner, in fashion similar to other signal work. In residential work the telephone company normally follows the route of the electric service, entering the building overhead or underground as desired. In both cases a separate service entrance means must be provided: if aerial, a sleeve through the wall; if underground, a separate entrance conduit. Unless a residence has many entering lines, no source of power is required for the telephone equipment.

Wiring of telephone instruments when installed *after* completion of the residence consists of a single surface-mounted cable that, even if skillfully installed, is unsightly at best and completely objectionable at worst. Prewiring consists of running the cables on the wall framing and into empty device boxes to which instruments are later connected.

MULTIPLE-DWELLING SYSTEMS

See Section 13.18 for a description of multiple-occupancy residential fire detection and alarm systems.

22.8 Multiple-Dwelling Lobby Intercom and Security Systems

Apartment houses, residences, and hotels combine the functions of the intrusion alarm and doorbell systems in the familiar lobby-to-apartment communication system. The most basic system is a series of pushbuttons in the lobby and an intercom speaker or telephone with which to communicate with residents. At the other end, the tenant has a speaker microphone plus a lobby door-opener button (see Fig. 22.4). This system can also be arranged to utilize the tenants' regular telephones. When the number of tenants is large, an alphabetical roster is added to the apartment-button panel to avoid the nuisance of scanning all the apartment names when the sought party's apartment number is not known. When the number is larger yet, a simple pushbutton per apartment arrangement becomes cumbersome, and is usually replaced by an alphabetical tenant register plus a dial or button phone. Closed-circuit television can be added to the lobby–tenant system, enabling the occupant not only to converse with, but actually to see, the caller. Such a system will increase (by 7 to 10%) the electrical contract cost for an average apartment house.

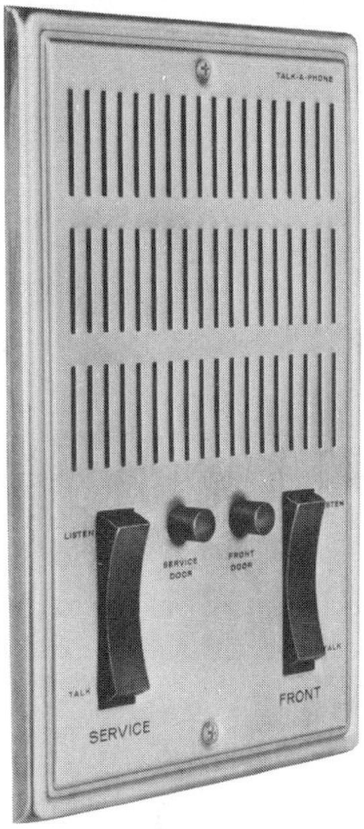

Fig. 22.4 Apartment house communication-security equipment. Typical apartment unit contains intercom controls and door-opening buttons for main and auxiliary entrances. (Photo courtesy of Talk-a-Phone Co.)

In addition to the security provisions provided by the apartment-to-lobby audio and video connections, additional security and alarm devices have been used such as emergency call buttons within the apartment at desired locations. These will light alarm lights, ring bells, and perform any other alarm functions required, to cover the situation of an intruder who manages to bypass the lobby security check. In geriatric designs, these buttons serve to *unlock* the apartment door to allow aid to enter, if summoned by lights and alarms. In luxury apartments, apartment doors can be monitored from a central security desk and any unscheduled door movement subjected to immediate investigation. These systems are custom designed to the needs and requirements of the owner.

A security problem applicable to all facilities, including residential ones, deals with the problem of limiting entry in unsupervised areas to authorized persons (i.e., unsupervised access control). The problem of keys and locks is well known, despite advances in that field. More sophisticated means include magnetic cards and electronic combination locks, which, because of ease of code change, are particularly useful in residential facilities that cater to transients.

Another aspect of security that is particularly appropriate to housing for the elderly and handicapped, although it is applicable to all housing installations, is the emergency call system. The purpose of this system is to alert *outsiders* to an emergency situation *inside* a closed apartment. This alarm system is essentially a way to call for help in time of illness or other distress. Many construction and housing codes include descriptions of the required equipment. Most often they prescribe a call initiation button in each bedroom *and* bathroom, which will register an audible (alarm) and visible (annunciated) signal at a location that is monitored, locally or remotely, 24 hours a day. Additional signals are required in the floor corridor and at the apartment, the purpose of which is to alert immediate neighbors to the distress call.

22.9 Multiple-Dwelling Television Antenna Systems

All modern multiple residences supply each room with one or more TV/FM jack outlets. The master antenna systems feeding these are similar to the residential type except for size and electronic design of the components. The antenna should be placed only after a survey has been made to determine signal strength patterns on the roof. Space below the roof and a single 15-A circuit should be allowed for the amplifier equipment. Coaxial cables are run through floor sleeves and tapped at various points to provide good signals in each apartment.

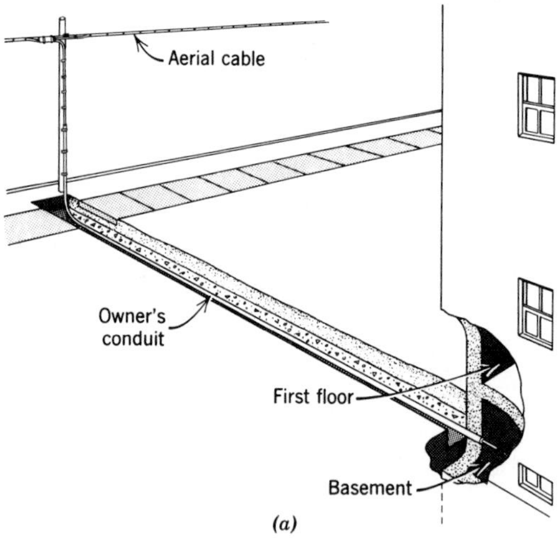

(a)

22.10 Multiple-Dwelling Telephone Systems

As in the small residence, the telephone service normally follows the same entrance path and method of entrance as the electric power service. For the sake of economy in underground construction, the two services often share the same trench, albeit in different raceways, and utilize twin manholes where such are required. Typical entrance arrangements for any large building, residential or other, are shown in Fig. 22.5.

The service entrance space requirements vary with the size of the building and telephone capacity. For a small apartment house of the garden or three-story type, a clear wall space of 2 to 4 ft is sufficient. A terminal room is required only in very large buildings.

Apartment buildings and dormitories differ from commercial structures in that the floor plans of all floors are similar, so that the arrangement of risers is relatively simple. It is common practice to utilize cable only, in risers that extend through vertically aligned closets in apartments. To accommodate these cables, a sleeve through the floor between closets is necessary. If a riser is located in a shaft other than a closet, conduit is normally utilized to allow for easy installation, protection, and repair. If the location is accessible, as in an alcove, only a sleeve is provided. When the riser is located outside the apartment, each dwelling unit is connected to the riser by a conduit with a junction box at either end. These conditions are illustrated in Fig. 22.6.

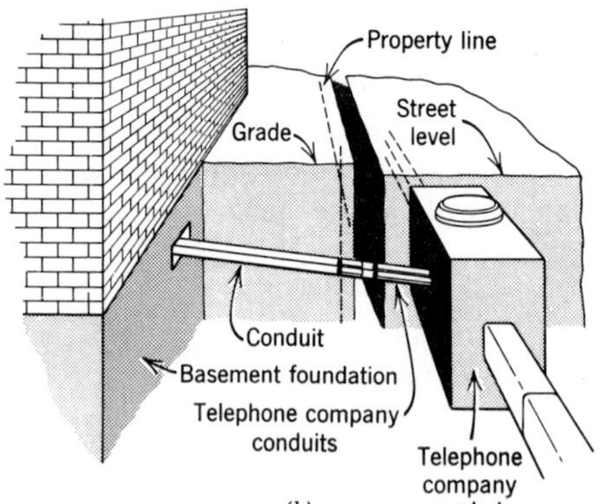

(b)

Fig. 22.5 *Telephone cable may enter a building underground, originating on overhead lines* (a) *or in manholes* (b).

Beyond the apartment service point, the individual rooms can normally be prewired entirely without conduit, or with only a few short sleeves.

22.11 Hotels and Motels

The distinction that once existed between hotel and motel generally resided in the vertical dimension; hotels were multistory and motels sin-

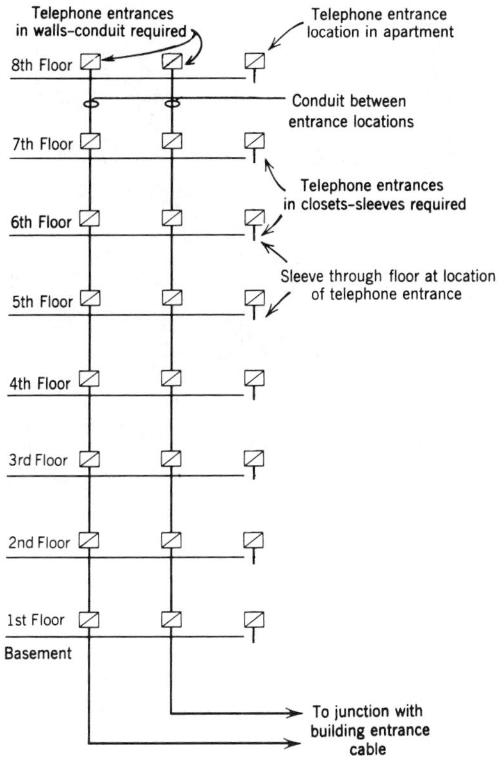

Fig. 22.6 Typical telephone riser diagram. Note the need for conduit between apartments when installation is made inaccessible, as in a wall.

gle story. That distinction has disappeared, so that motels are subsumed under the hotel category.

(a) Security. Because of the transient population of these facilities and the need to provide maximum service and security to the guests during their stay, the two principal problems that must be addressed in this area are room access security and hotel equipment security (i.e., prevention of what is euphemistically known as *shrinkage*).

1. *Room Access Security.* The ineffectiveness of key locks in preventing undesired entry even in private homes is well known; in hotels, where the key is passed from occupant to occupant, the lock is little more than a psychological barrier. As a result, most modern hotels have installed (electronic) room locks whose opening device code is changed with every new guest. These locks may be of the coded pushbutton type, whose coding is changeable only from a central lock-security console; a magnetically or punched-hole coded card type (Fig. 22.7*a*); or a programmable electronic lock and coded key type (Fig. 22.7*b*). All of these systems have relative advantages and disadvantages, depending on the number of rooms, whether the installation is new construction or retrofit, and average length of guest stay.

2. *Equipment Security.* The need to provide service requires that guest rooms be equipped with television and possibly a VCR, meeting rooms with TV, VCR, projectors, and computer terminals, and that these and other devices be made available for guest use on a rental basis. Theft control is a consultant specialty on its own and far beyond the scope of this book. However, we will mention one system that is equally applicable to equipment in hotel, schools, office buildings, and industrial facilities. It senses the disconnection of the equipment from its power connection (wall plug) and transmits an alarm over the power lines (see Section 16.28) to an annunciator at a selected control location (see Fig. 22.8). This arrangement has the advantage of alarming immediately on equipment removal, thus normally permitting appropriate retrieval action to be taken in time.

(b) Telephone, Data. The telephone and data communication system in large modern hotels is among the most important and complex. Hotels catering to business persons must make available computer terminals and modems that can be used at the very least in rooms dedicated to that purpose, and in high-quality hotels, in (selected) guest rooms as well. This requires that the building designer must provide adequate raceway (and cabling) facilities, as discussed in Section 15.18. The current practice of holding business meetings and technical conferences in hotels also requires that conference and meeting rooms be arranged for very heavy concentrations of electronic equipment that can be installed and rearranged in short order. This normally calls for some sort of access floor and modular cabling, as described in Sections 15.25 to 15.30.

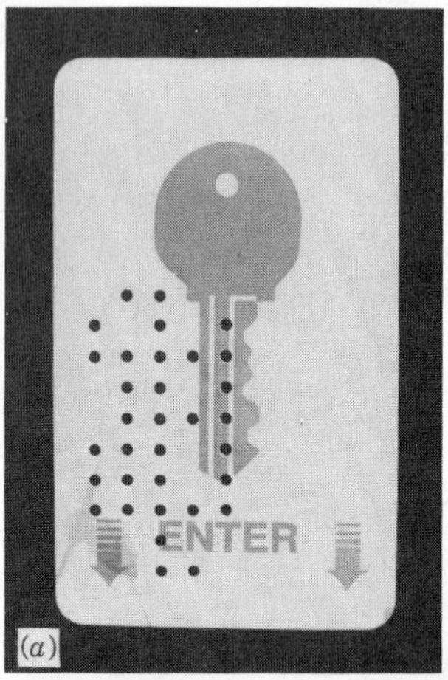

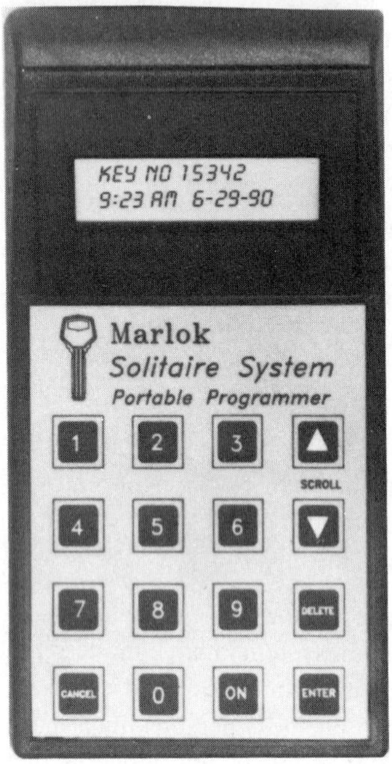

(b-2)

(b-1)

Fig. 22.7 (a) *Punched-card coded type of "key" in common use in hotels. The punched hole coding is changed after each room use. The drawing of a key on the card is the size of an ordinary door key. (Photo by Stein.)*

Fig. 22.7 (b) *This electronic door locking system uses a nonduplicatable fixed-code metal key to operate a programmable electronic lock* (b-1). *The lock, which can be reprogrammed by use of the portable programmer* (b-2) *with upward of 10 million different codes, reads the key code with an internal infrared reader. Various system configurations, including central room status reporting, are readily accomplished. (Courtesy of Marlok Co.)*

SCHOOL SYSTEMS

22.12 General

The proper operation of a modern school requires that flexible and efficient signal and communications equipment be available to the administrative and teaching staff. Such equipment, engineered to meet the needs of the individual institutions, will do much toward optimum utilization of staff and student time. School fire detection and alarm systems are discussed in Section 13.19.

22.13 School Security Systems

Although intrusion alarms and security systems were not historically normal school requirements, this situation has unfortunately changed. Sensing devices on doors and windows can be arranged both to trip local alarm devices and, via auxiliary circuits, to notify police headquarters. Often, vandals can be frightened off by having the alarm system actuate a protective lighting system that will illuminate the building exterior and any interior areas desired, such as record rooms. A perimeter alarm detection sys-

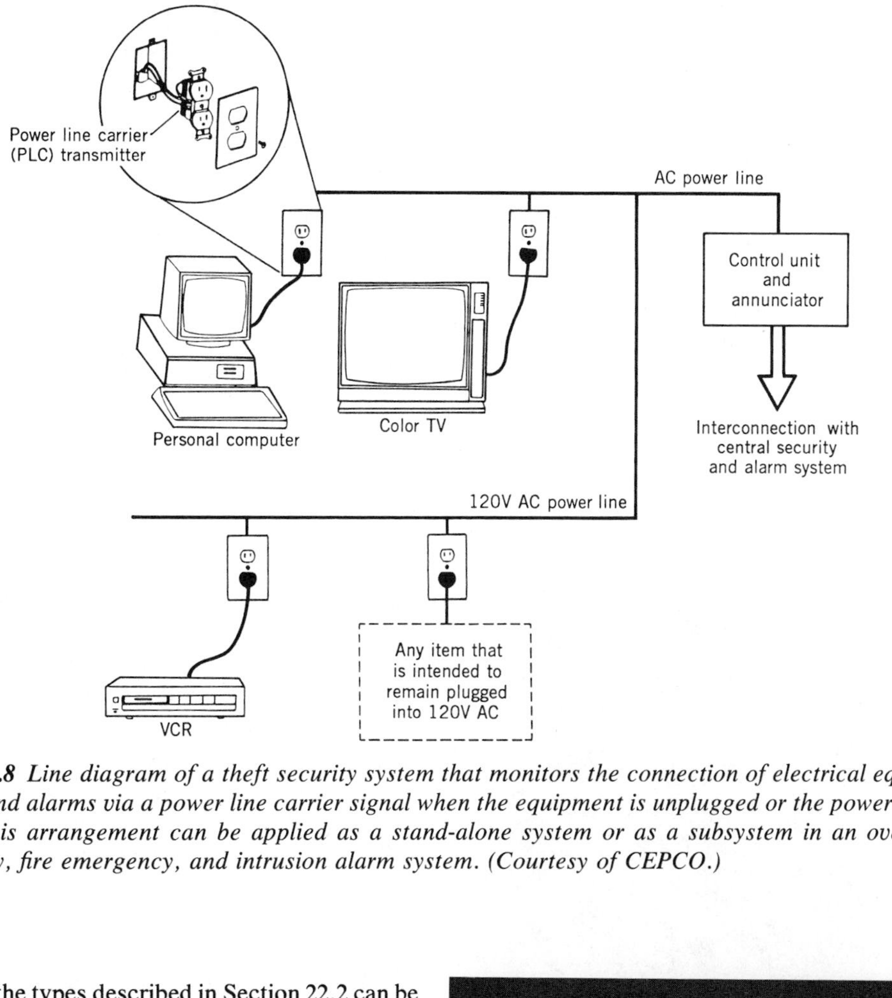

Fig. 22.8 Line diagram of a theft security system that monitors the connection of electrical equipment and alarms via a power line carrier signal when the equipment is unplugged or the power line cut. This arrangement can be applied as a stand-alone system or as a subsystem in an overall security, fire emergency, and intrusion alarm system. (Courtesy of CEPCO.)

tem of the types described in Section 22.2 can be installed in particularly vandalism-prone areas to assist in preventing entry to the school premises after hours. Though expensive, they are very frequently cost effective.

A type of exterior door security device applicable to schools and all other facilities with doors that are locked from the outside but which must be openable from the inside in the event of emergency is the exit control alarm. This device (see Fig. 22.9) alarms as desired (i.e., audibly or visually, locally or remotely) when the controlled door is opened. It is normally equipped with a timed bypass that will allow authorized personnel to key-operate the bypass but will prevent the door being held open without alarming. One model of door opener alarm requires a continuous pressure on the door opening bar of 10 to 20 s before the door will open, with the alarm

Fig. 22.9 Alarming-type door control device permits door to be opened from one side, with immediate alarm. The alarm of the illustrated device is locally audible but can also be arranged to be for remote and visible signals. The alarm can be bypassed and reset by key only. This bypass is timed to prevent the door from being propped open. (Courtesy of Detex Corp.)

SIGNAL EQUIPMENT

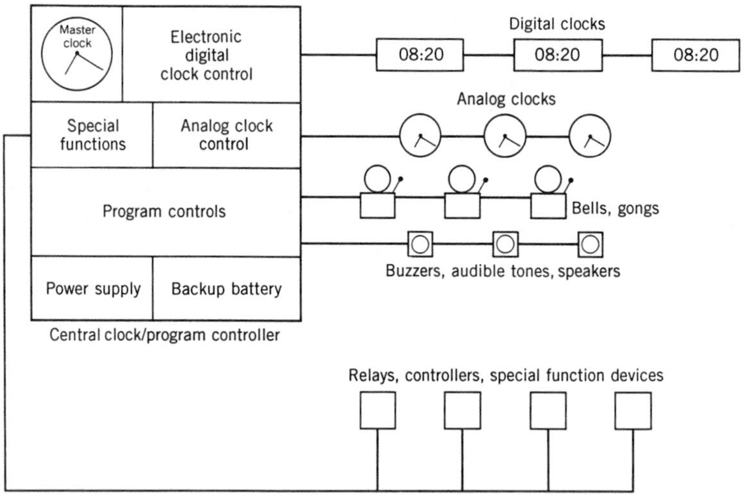

(*a*)

Fig. 22.10 (a) *The central clock and program device uses its electronic time base to control clocks of all types, time-based audible signals, and event controllers such as relays and switches.*

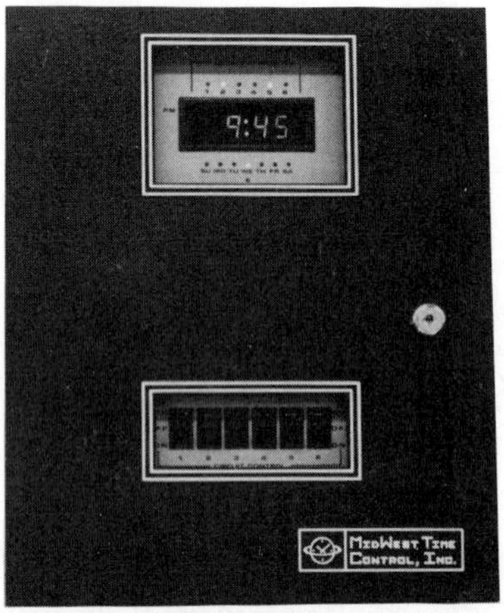

(*b*)

Fig. 22.10 (b) *Typical microprocessor-based master time clock and program device. This unit, which measures 14 × 12 × 5 in. deep, provides time correction signals to synchronous wired clocks of the digital or analog type and can be adapted to control minute impulse clocks. Programming of four output circuits can be daily or weekly, and can accommodate daylight savings, holidays, and leap-year update with a total program capacity in excess of 600 events. Programming is done locally or via a modem, from a remote computer. This type of device is applicable to any facility requiring a clock and/or program system in addition to its primary application in schools. (Courtesy of Midwest Time Control Inc.)*

being activated immediately on application of pressure. This arrangement (usable only where fire codes permit) allows sufficient time for the facility's staff to investigate the attempted exit before the door will open, and is applicable only where exit control is an overriding consideration. As mentioned in Section 13.20*b*, stair and exit door exit locks can be arranged to be centrally released from the fire control center.

22.14 School Clock and Program Systems

The clock system and the program system were at one time separate and distinct, sharing only the timekeeping facilities provided by the master clock. Now that electromechanical programming devices are effectively obsolete, the two systems are actually one, but the traditional two-system name remains. Referring to Fig. 22.10, we see that the heart of the system is the time base (electronic clock). This is precisely the same device that provides timing for all programmable switches and controllers. In a clock and program device it controls clock signals, audible devices, and if desired, other switching functions.

The clock system may use analog clocks (with hands) or digital units. The former are usually locally powered, and a correction signal is periodically transmitted via dedicated wiring from the master clock at the controller. This same controller continuously transmits a binary-coded signal to digital "slave" clocks, which can be either of the self-illuminated LED (light-emitted diode) type or of the LCD (liquid-crystal display) type. LCD units are easily visible only in high ambient illumination and when viewed directly (not at an acute angle). LED units are best applied in areas of low ambient illuminance. Conventional largeface clocks are easily visible in all ambient light situations.

The programming function of the controller serves to delineate audibly the various time periods into which the school day and week are divided. A single-circuit unit is utilized in the instance of an institution that operates entirely on one schedule, such as an elementary school on a morning period–lunch–afternoon period regimen. For a school employing different schedules for its various parts, the controller can provide multiple program schedules on different circuits, depending on its design. Controllers are user programmable and are normally provided with a local, master clock, crystal control to assure accurate timekeeping regardless of line frequency variation, backup power source to maintain user programming and master clock local display, and various conveniences such as daylight saving time correction, security arrangements, even timers, and so on. If desired, the clock and program controllers can also be used for mechanical system control, by the simple expedient of adding relays and switching devices, as shown in Fig. 22.10.

The audible devices in a program system may be bells, gongs, buzzers, horns, or a tone reproduced on a classroom loudspeaker. This last system has the following advantages:

1. Clear audibility in each classroom, with adjustable volume to accommodate both quiet and noisy areas.
2. No possibility of confusion between program tone and other signals such as fire alarm gongs.
3. Multiple use of the speaker unit, for classroom sound as well as program tone.
4. Complete flexibility of programming that is not possible with hall gongs. This is particularly desirable in schools with special programs for special groups of students.

Fig. 22.11 *Administrative telephone units of a modern school telephone/intercom system. The system, which is entirely solid state and requires no operator or switchboard, provides complete internal communications plus internal connection for selected administrative units, selective paging, and emergency call signals. Class-change signals on classroom intercom system speakers are also possible using time signals via interconnection with the school's master clock system. (Photo courtesy of Rauland-Borg.)*

22.15 School Intercom Systems

Various types of intercom systems are available depending on the needs of the school building involved. In small schools, a simple wired intercom system connecting the various offices is usually sufficient. This is supplemented by outside telephones in the administrative offices and a functional paging system that is normally part of the school's sound system arrangement. With larger buildings and correspondingly larger numbers of extensions and multiple-function demands, more sophisticated equipment is required. The unit illustrated in Fig. 22.11 is typical of modern school intercom equipment, and is actually a private telephone system of considerable flexibility. Such a system is generally interfaced with the school sound system, and provides these functions:

1. Intercom between staff members and offices.
2. Direct communication with classrooms, including selective and all-call capability.

3. Zone, group call, and conference call functions.
4. Interconnection with the outside phone system.
5. Class-change signals via interface with master clock system.
6. School intercomputer communication via modems.

These systems use direct pushbutton dialing and programming, thus eliminating the necessity for switchboards and operators. All stations are coded with three-digit codes, and all switching is solid state, minimizing maintenance problems. Such systems are adequate for all but the largest institutions.

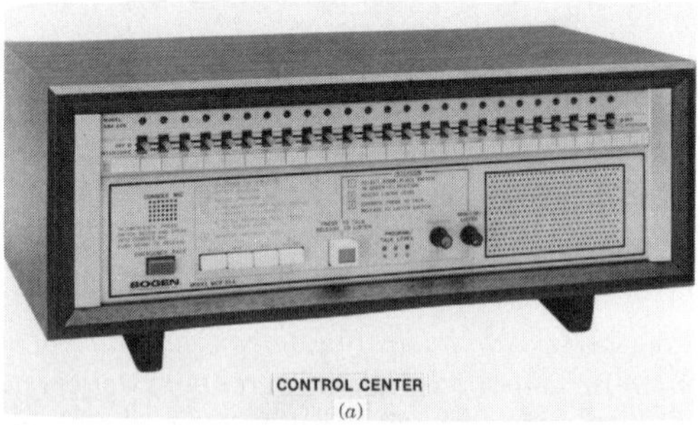

Fig. 22.12 (a) *Compact program/intercom control center. This unit, which measures 20.5 × 11 × 8.5 in. H has separate power amplifiers for intercom and program material. It distributes program material or provides intercom with 25 loudspeaker locations, and has additional capabilities, such as emergency/all-call paging, tone calling, time signal, night bell, and telephone paging. Program material derives from separate external sources. (Courtesy of Bogen Communications Inc.)*

22.16 School Sound Systems

An integrated sound-paging-radio system designed for school use offers several modes of operation and considerable flexibility. Its function is to provide a means for distributing recorded (records, tapes), broadcast (AM/FM), or live sound to preselected areas of the school. Thus, a simple system might provide a record player and single microphone input, and a single channel to all the speakers in the school, whereas a complex system can be arranged to operate with three simultaneous input signals distributed to six different areas of the school (see Fig. 22.12).

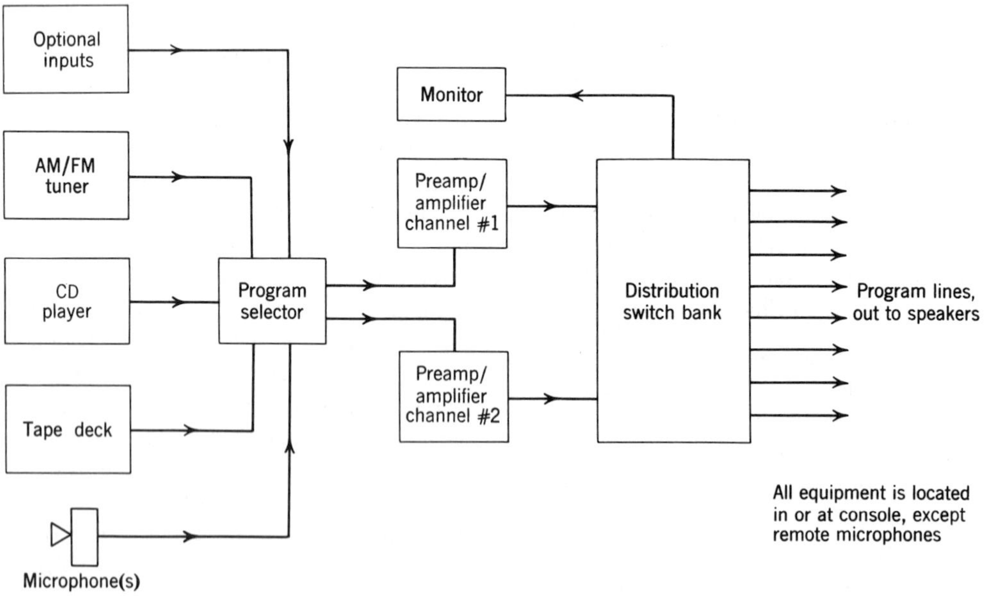

Fig. 22.12 (b) *Block diagram of a two-channel sound system. Optional facilities that can be added to the console are tape deck, private telephone communications, and equipment for rebroadcast of signals between areas. Intercom and master clock and program can also be incorporated.*

A conventional system consists of a desk-size control console containing most of the input units, amplifiers, and switching devices and connections to the remote loudspeakers. The input units may comprise one or more AM/FM tuners, turntables, CD player, tape deck, and microphones. One microphone is normally located at the console with others in the principal's office, auditorium, school office, or other selected location. If desired, mike outlets can be spotted around the school and a spare mike and stand supplied that can be plugged in at any of these points.

Loudspeakers, located in classrooms, gymnasium, auditorium, cafeteria, and outdoors receive the amplified signal through the switching mechanisms located in the console. It is the function of these switches to deliver the program material to the various loudspeaker circuits, which are also called program lines. Thus, using a system with multiple amplifiers, music can be piped to the cafeteria, an important radio program to senior classrooms and teachers' lounge, and instructions to an outdoor gym class or team practice. An all-call feature also allows announcements to reach all speakers in the system simultaneously. The intercom system discussed above can be interconnected with the sound system to allow conversation between classrooms and the console or other points. A small system can be contained in a compact desktop console (see Fig. 22.12*a*). A large system frequently requires a console. A console is usually built in a desk arrangement, and it is advisable to provide sufficient space for it and for the person who operates it. Often an alcove of 30 to 50 ft^2 is reserved for it and a library of recordings.

Loudspeakers may be in flush or surface baffles, at the discretion of the designer. Gymnasium, cafeteria, and auditorium units are normally flush-mounted in the ceiling. For large areas such as these it is well to provide a volume control, enclosed in a recessed wall box with a locking cover. A common variation of the above-described system uses separate subsystems for the cafeteria, auditorium, ball field, or other areas utilizing sound equipment frequently. These smaller systems have their own input, amplification, and control devices but utilize speakers in common with the central console. Normally, the console has an override feature that allows it to override local systems.

For a discussion of high-quality sound systems required for recital halls in music schools and the like, refer to Sections 26.27 to 26.29.

22.17 School Electronic Teaching Equipment

This area, as so many others that utilize computers, is growing and changing so rapidly that almost anything written on it is immediately out of date. We will therefore simply outline the present and projected near-future uses of electronic teaching equipment.

(a) Passive-Mode Usage. In this category we can include all recorded material, in whatever form, which is available to the student via some form of information retrieval technique. This would include printed, audio, and video material, in conventional and electronic library forms.

(b) Interactive Mode. Here students use a computer teaching terminal interactively to study at their own pace on a one-to-one basis, with the computer acting as tutor. Modern teaching programs sense the students' weak points (as do good teachers) and emphasize them in teaching. The building designer must be aware of the rapid developments in this field at all educational levels, including elementary school, in order to make adequate electric power, cable raceway, lighting, and HVAC provisions.

OFFICE BUILDING SYSTEMS

22.18 General

Under this category we include systems found in all office, professional, and sales-type buildings. Such buildings house tenants with varying schedules and requirements. This must be considered in the design of the signal systems for such buildings.

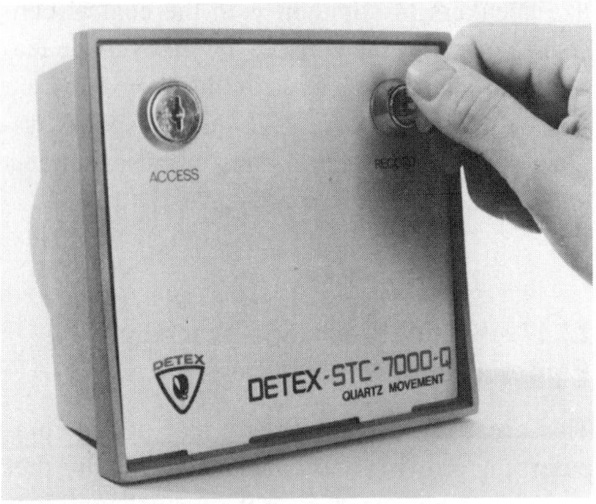

Fig. 22.13 Watchman's tour station. Here the clock and recording tape are inside. The station is operated by simply inserting a key as shown and giving it a quarter-turn. (Courtesy of Detex Corp.)

Although in many of the medium-sized and large buildings control, alarm, and security functions are combined in multiuse apparatus and consoles, which we will discuss, the basis systems are essentially separate despite shared-use equipment. We discuss them individually to demonstrate function and equipment, and combined to show economies and modern practice.

22.19 Office Building Security Systems

Although automatic surveillance systems *are* applicable to office and mercantile occupancies, they are more frequently found in industrial facilities and are discussed under that heading. Office buildings normally utilize some type of manual watchman's tour system so that surveillance of unoccupied areas is conducted on some regular basis.

The simplest type is nonelectric and comprises a number of small cabinets, each containing a key, placed at intervals around the interior and exterior of the building. The watchman uses these keys to operate a special clock that he carries about, thus recording the exact time at which he "clocked in" at any specific location. Alternatively, the clock is wall mounted and the guard carries only a key (see Fig. 22.13). A computerized version of this system is available which simplifies station check-in, automatically records guard visit data, and provides hard-copy printout.

Guard tour systems are also available that permit constant supervision and are particularly effective where more than one person is on duty. Such systems show on a panel the location and progress of the watchman, by means of lights that glow when the device at each location is operated. Since part of the effectiveness of these systems lies in the timing of the tour, a system can be arranged to sound an alarm if a particular station is not operated within a specific time period. Telephone jacks spaced at points along the guard's route allow him to communicate with the supervising office or other point, without interrupting the scheduled tour. For protection of areas housing extremely valuable items or documents, an intrusion alarm system may be employed.

22.20 Office Building Communications Systems

This planning item is composed of three parts, which are frequently melded into a single network:

1. Intraoffice voice communication, or intercom (see Fig. 22.14).
2. Interoffice and intraoffice data communication using telephone and communication cabling.
3. Outside-the-building communication via phone company lines.
4. Paging function. This item is discussed in Section 22.25 under industrial facilities.

Today, in most locations, the user has the choice of purchasing or leasing as much of the system as he desires. In other words, the communication system including cables, instruments, switching equipment, and so on can either be all privately owned, all supplied by the local telephone company, or almost any division desired. However, except for a small interoffice intercom, duplication is eliminated. That is, the same instruments and switching equipment are used for both intercom and outside connection.

Fig. 22.14 This desktop multifunction intercom unit is usable in either handset or hands-free mode. In addition to standard intercom functions, each handset, which is connected to its central exchange with a four-conductor cable, can be used for such additional facilities as automatic redial, call conferencing, searching, all-call, group call, all transfer, call registration (store incoming call numbers), privacy mode, special meaning tones, and networking possibility. In addition, the unit can be used for voice and "beeper" paging. Exchanges, which can handle up to 4000 stations, are microcomputer-controlled solid-state assemblies which vary in physical size according to the number of stations served. A typical large exchange might measure 60 in. W × 24 in. H × 20 in. D. (Courtesy of Dukane Corp., Communications Systems.)

What to buy, rent, or lease constitutes an economic decision, since the required functions are satisfied by either private or telephone company equipment.

The item of data handling is highly specialized and was discussed in Section 15.18 with respect to the responsibility of the building architect and designers.

22.21 Office Building Communications Planning

Planning for the telephone and other communications equipment in an office building is of prime importance because of the large amounts and critical locations of required space. Therefore it must be done simultaneously with other space planning. Exact requirements for office space are generally unknown at design time and, even if they were known, planning would have to account for changes in space usage as well as increased communications and data transmission services. For this reason, all planning is based on square foot areas. Planning is essentially for spaces only, from incoming service to final instrument, since cabling and equipment are furnished and installed either by a private telephone equipment supplier or by the telephone company.

The planning information that follows is applicable to office buildings with average telephone and data transmission loads. Buildings whose tenants include brokerage houses, insurance companies, headquarters of multibranch operations, and other heavy telephone and data transmission users may need more space. On the other hand, advances in technology, including the use of fiber optic cables, act to reduce equipment size and space requirements. As a result, the best course of action is to design for current requirements, with a reasonable estimate for the future based on advice from experts in the field.

Space is required for:

1. Service entrance including terminal space, cabinet, or room.
2. Riser spaces, shafts, conduits, and cabinets.
3. Apparatus closets for equipment.
4. Satellite locations for interconnections.
5. Equipment rooms for specific-use equipment.
6. Distribution system including conduits, boxes and underfloor duct, and raceway systems.

(a) Service Entrance. Adjacent to the service conduit a spare sleeve placed in the foundation wall and sealed will provide for future service expansion. Inside the building the telephone cable is terminated in a wall box, cabinet, or terminal room depending on the cable size. For buildings up to about 70,000 ft^2 of rentable area (depending on the type of anticipated tenant use), wall-mounted terminal cabinets are normally sufficient. In any case, the area should be dry, and it requires light, ventilation, one or more 120-V outlets on a separate circuit, and a good ground connection. In special cases more power is required.

Surface-type wall installations comprise simply a sheet of ¾-in. marine plywood on to which the phone company mounts its cabinet. Flush installations in finished areas require a cabinet with ¾-in. plywood back.

(b) Riser Shafts. These can accept the cables extending beyond the terminal room and carry them vertically through the building. Connection between terminal space and risers is preferably in conduit.

The riser shafts provide means for the cables to extend vertically and to terminate at each floor. Ideally, the risers comprise a series of vertically aligned closets connected by 3½- or 4-in. sleeves set in the floors and extending 1 in. above the floor. It is preferable to separate communications closets from electrical power closets. Where multiple risers are used, shafts should be interconnected by several 2-in. conduits to allow for interconnection of systems.

(c) Riser Closets. Here, cables from the riser system are interconnected to switching and power equipment, as well as to the cables that radiate from the closet to station locations throughout the floor. These closets may also be called *zone closets* or *apparatus closets*, particularly if they function with an underfloor raceway system. Walls of the closet should be lined with plywood at least ¾-in. thick to support the weight of switching and connection panels, power equipment, terminals, connecting blocks, and other hardware. Each riser (apparatus) closet must be provided with a switched ceiling light, and a separate 20-A, 120-V circuit with two duplex receptacles. A source of emergency power is desirable to avoid curtailment of telephone service during outages.

(d) Satellite Closets. Unlike riser (apparatus) closets, satellite closets do not contain switching and power equipment. Their primary use is to provide cable-connecting and -terminating facilities in large complex facilities, where riser closet space is insufficient.

(e) Equipment Rooms. Where extensive crossconnection is required or where tenants utilize PBX (private switchboard) equipment, the required equipment is placed in equipment rooms. These spaces, which are actually small closets or alcoves, should contain a 20- to 30-A, 2P, 120/208-V circuit, a 20-A, 120-V outlet, a grounding point, good lighting, and, obviously, sufficient equipment space. Space requirements vary with each installation and are obtainable from the equipment supplier. Ventilation is essential, as is absorptive acoustic material on the ceiling and at least one wall. Connections between these spaces and other communication equipment closets should be via floor or ceiling ducts or multiple 3½-in. conduits.

(f) Horizontal Distribution. Cabling from riser, satellite, apparatus, and equipment closets to individual outlets and instruments can be underfloor (Sections 15.24 to 15.28), in ceilings and plenum spaces (Section 15.30), under carpet (Section 15.29), or in surface raceways (Section 15.23). Because of the large raceway volumes required, conduit is infrequently used. When using underfloor raceways a header capacity of 2 $in.^2$ per workstation is reasonable, based on one multiline telephone and one video display unit per station.

(g) Fiber Optic Cables. In installations with very heavy data transmission loads, in those using video systems, and/or in applications for which the high-security, low-noise, and broad-bandwidth characteristics of optic fiber cables are desirable, they are used in lieu of copper cabling. Space requirements and connection accessories are obtainable from manufacturers. Since cables are often supplied with factory-applied terminations, raceways must be of the large capacity full access type.

22.22 Office Building Supervisory Control Center

As the modern office building's mechanical and electrical systems increased in complexity, the need arose for a central point of supervision, control, and data collection from which to survey and control an entire building's functioning. From such a point the water, air conditioning, heating, ventilating, electrical, and other systems could be controlled manually and automatically with much greater accuracy than if no such central control point were available. Data on temperatures, pressures, flow, current, voltage, and all of the many parameters of mechanical systems could be made instantly available, so that operational decisions could be made more accurately. Also, all systems could be monitored here, and all alarms instantly acted upon, automatically or manually.

Such control centers, generally called *supervisory control and data centers*, are now installed as a matter of course in office buildings. They are equipped with microprocessors or computers that process the huge amount of data received to arrive at operational decisions intended to optimize system performance. As such, they result in a considerable savings in operating costs, in addition to the work force savings generated by their remote monitoring and control functions. These units normally perform these functions:

1. Remote monitoring, recording, logging.
2. Remote start–stop.
3. Remote controls, resets, changes.
4. Alarm functions.

The system for which the supervisory control center is the control, calculation, data logging, and building systems connection point is called a *building automation system* (BAS). The format and functioning of a generalized version of a BAS is discussed in Appendix J, to which the reader is referred.

Architecturally, the BAS supervisory control center requires good lighting and ventilation, extensive raceway space, but little area. Because systems are tailored to the building, no guidelines can be stated for space requirements.

INDUSTRIAL BUILDING SYSTEMS

22.23 General

All industrial facilities ranging from the taxpayer loft, housing a small hand assembly plant, to the immense steel manufacturing plant, require a variety of signal and alarm equipment. Although, as in the case of commercial structures, a detailed analysis of the equipment is out of place here, it behooves the building designer to know generally the function, operation, and availability of this type of equipment.

Fire alarm systems for industrial buildings are discussed in Section 13.21. Audible alarms in industrial facilities for any of the building security systems must be selected with the usually high ambient noise level in mind. See Figs. 22.15 and 22.16 for recommendations.

22.24 Industrial Building Security Systems

Among the most important signal functions in this type of facility is protection. Although they may use a common control console, these systems are varied and perform separate functions.

(a) Door and Exit Controls. Outside doors and doors to restricted areas are supervised by electrified security hardware that triggers an alarm when a door is opened without authorization. The alarm mechanism may be concealed, or openly displayed, to act as a deterrent (see Fig. 22.9). Annunciation can be provided by a separate panel or tied into a central alarm panel.

(b) Personnel Entry Control. Several levels of security are available and can be applied in accordance with security needs. Beyond a simple lock, the first level at an unmanned portal is a card reader that grants entry to the card holder. The breaches at this level, however, are numerous. With the assistance of the card holder, more than one person can enter during a single portal opening (tailgating), or the card can be transferred to an outsider after the authorized person enters (passback). Finally, cards can be

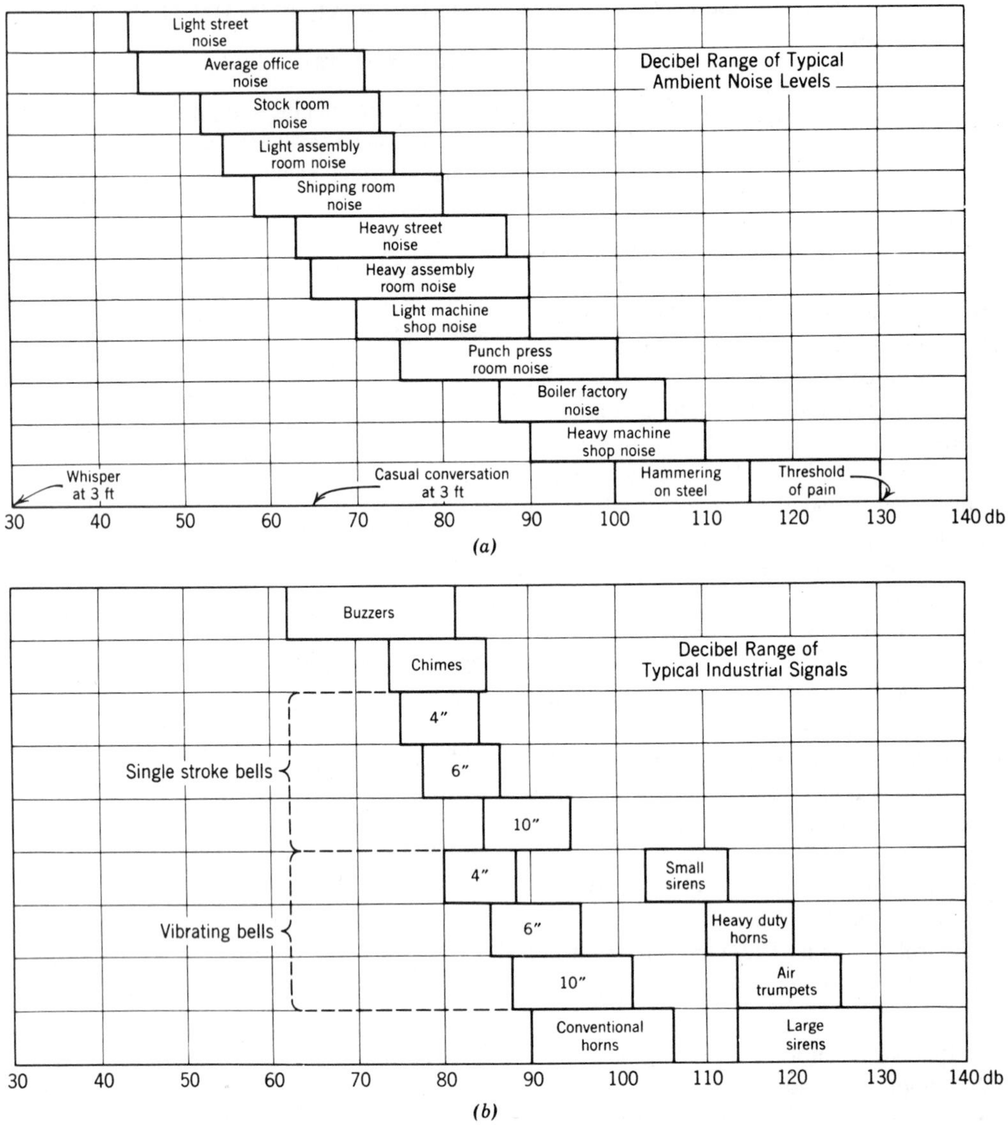

Fig. 22.15 (a) *Decibel range of typical industrial noise levels.* (b) *Decibel rating of typical industrial signals may be compared with ambient noise levels* (a) *to facilitate selection.*

found, forged, or stolen. The next level requires the encoding of a number simultaneously with presentation of a card, thus barring entry to holders of found or stolen cards. Beyond this level there are many techniques, including use of a computer to track card location so that a card already inside the building cannot be reused without exiting, thus eliminating passback. Also, portals can be arranged so that tailgating is extremely difficult.

When entry security levels are high, personal identification techniques that cannot be forged are usable. Among them are voice prints, hand geometry, and retinal scans (see Fig. 22.17). Finally, an access control procedure involving a control card, visual identification, and visual tracking of the entry procedure through a controlled, physically limiting access portal can be used where a high risk of unauthorized entry justifies it (see Fig. 22.18).

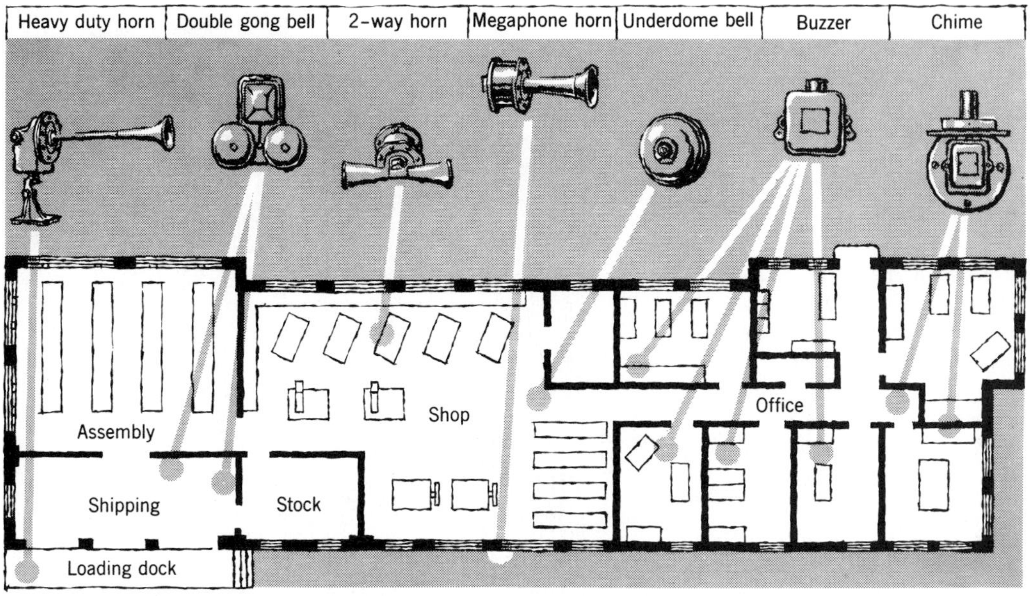

Fig. 22.16 Plan of a small industrial facility, showing suggested locations of typical signals.

Fig. 22.17 This foolproof identification procedure relies on the fact that no two persons have identical retinal eye patterns. The illustrated device, usually placed at an unmanned controlled access portal, scans the entrant's retina, compares it to the previously recorded pattern, and then either permits or denies entry. (Photo courtesy of Eyedentify Inc.)

(c) Watchman's Tour Equipment. When tour equipment is used, it is frequently of the combination-alarm type. This type of station allows guards to call in, alerts them to waiting calls, permits a general alarm to be turned in by key operation, and is usable as a manual fire alarm station. The station may be coded or noncoded as desired. The tour is normally timed so that any delay automatically sends an alarm and summons help.

(d) Intrusion Security Systems. A combination of the systems described in Section 22.2 is normally employed. Perimeter systems are normally arranged to sound an audible alarm and to energize all security lighting.

All the alarms above, plus others not discussed, can be connected to a central monitoring console. If the console is unattended, it is normally auxiliarized, so as to alert security personnel. Usually, however, the alarm signal is transmitted to a central security station—normally a constantly attended guard post. In all cases, a permanent printed record of every alarm should be made. See Appendix J for a diagram of centralized security control in a multibuilding facility.

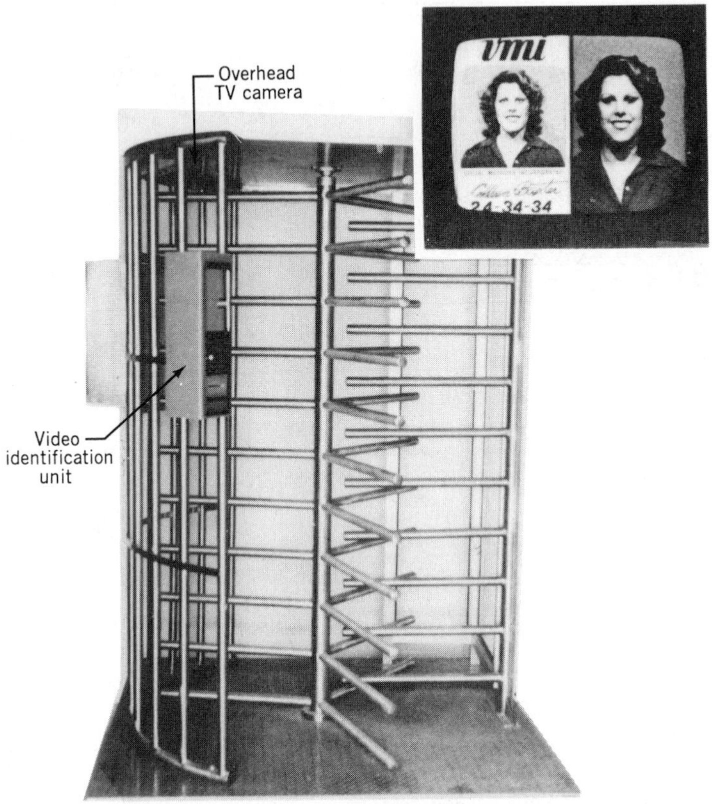

Fig. 22.18 *A person desiring entry at this high-security portal inserts her identification card and is viewed by a camera in the entrance device. At a remote location, images of the subject and her ID card are displayed on a screen (inset), while ID data from a memory bank are displayed simultaneously for comparison. Actual physical entry through the turnstile is controlled electrically and monitored by an overhead camera. (Photos courtesy of Visual Methods, Inc.)*

22.25 Industrial Building Paging Systems

All the time saving and efficiency potential that results from the proper use of the various signal and communication equipment available can be lost if there is no recipient for the information transmitted. Furthermore, sometimes a decision must be reached quickly to avoid costly delay, reruns, and so on. These factors combine to make rapid and accurate paging an extremely important function in a manufacturing operation, particularly of the larger variety.

Paging systems fall into two general categories and several subcategories: they are either visual, audible, or both, and are either common or selective. The simplest visual and audiovisual types comprise flashing lights, which may be combined with buzzers or bells, either or both of which are generally coded. Such systems are nonselective in that they impinge on the senses of all the building occupants—an obvious disadvantage. Furthermore, they must override ambient noise levels, which may be high (see Figs. 22.15 and 22.16).

More sophisticated systems utilize a small pocket device that is carried by each person likely to be paged—maintenance personnel, plant engineers, executives, and so on. By means of either direct radio transmission or of electric fields induced by induction loops installed throughout the building, an individual pocket device can be alerted by a beep. In some systems, the alerted person then listens to the message directly. On others, it is necessary for a person, once having been paged, to go to a phone and call in to a central paging desk to receive the message. Others utilize small hand-held, two-way radio transmitters with paging, to enable conversation between the page originator and the recipient.

In older paging systems it was usual to have a dedicated paging console and in some cases even a separate paging operator. Modern systems generally incorporate the paging function into one of the facility's systems: telephone, intercom (see Fig. 22.14), or sound and music. Paging calls can originate at the paging control

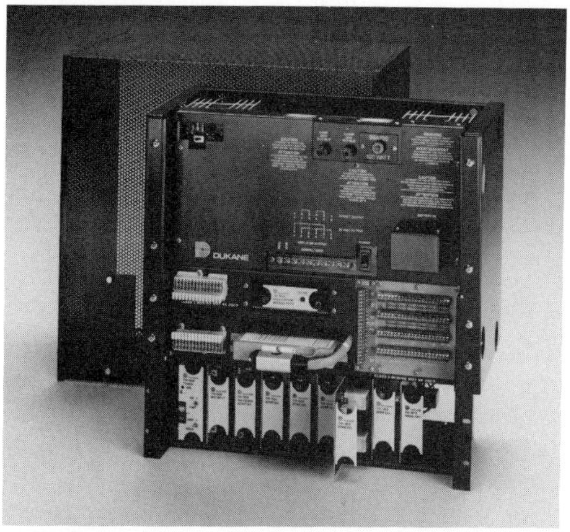

Fig. 22.19 Voice paging equipment, designed to interconnect with a telephone system so that paging calls can be placed at any phone and distributed over the facility's loudspeaker system. Modular electronic construction permits addition of special paging features such as zone paging, talk-back via speakers (two-way communication), priority access, all call, and others. The illustrated wall-mounting cabinet, which measures approximately 24 in. W × 24 in. H and weighs 66 lb, contains all of the electronic equipment required, including an audio power amplifier. (Courtesy of Dukane Corp., Communications Systems.)

console, at an intercom station, or at an internal or external phone and can be placed by anyone entering the appropriate access code (see Fig. 22.19). Voice paging via speakers requires audio equipment only; paging via pocket pagers or belt-beepers requires radio frequency transmitters and antenna systems as well.

REFERENCES

Bukowski, R., O'Laughlin, R., and Zimmerman, C. (1987). *Fire Alarm Signaling Systems Handbook*, NFPA & SFPE, Boston.

Fire Protection Handbook, 16th ed. (1986).

National Fire Protection Association Codes, 1990. NFPA, Boston.

Life Safety Code, NFPA 101, 1988

PART VIII
TRANSPORTATION

No facilities in a multistory structure are taken for granted more consistently than elevators and escalators, which are relied on to move people quickly and safely under all conditions, including emergencies. This movement should be quiet, trouble free, and economical. Furthermore, since vertical transportation accounts for 10 to 15% of a construction budget, somewhat less of the building area, and somewhat more of the operating cost, and is a determining factor in building shape, core layout, and lobby design, we see that elevator and escalator selection is a major task for the architectural designer.

Chapter 23, Passenger Elevators, introduces the subject with a description of the components of such facilities, including traction equipment, cars, safety devices, and systems of control and supervision. Special topics such as codes and standards, and requirements for the handicapped are also covered. The discussion then turns to actual selection of units. Criteria are established for elevator performance, including interval, waiting time, handling capacity, average trip time, and round-trip time. Extensive car performance data (curves) are presented, which can be used with the other criteria to select the size, number, and speed of elevators to fulfill the requirements. The discussion continues with principles of elevator zoning. The chapter concludes with an analysis of spatial requirements for elevators, including shafts, lobbies, and core arrangement. Throughout, application of principles and data is demonstrated through the use of examples.

Chapter 24, Special Topics, addresses important passenger elevator considerations other than elevator selection. These include power and energy requirements, with emphasis on energy conservation, fire-emergency operation, emergency power requirements, and passenger security. The next part of the chapter is devoted to special cases: freight elevators, hydraulic elevators, nonvertical designs, residential elevators, and chair lifts, with discussion of equipment, capacities, selection, and economy, and comparison (where applicable) to conventional design. The chapter concludes with a section on material-handling devices: dumbwaiters, horizontal and vertical conveyors, pneumatic systems, container delivery systems, and automated self-propelled vehicles. This section answers the need to plan for material handling in commercial and institutional buildings rather than treating the subject as an afterthought, with resultant inefficiency, overloading of freight facilities, and misappropriation of passenger elevators.

Chapter 25, Moving Stairways and Walks, treats these subjects with a discussion similar to the coverage of passenger elevators. Descriptions of equipment components, capacities, sizes, speeds, and methods of selection are followed by special topics including lighting, fire protection, power and energy requirements, and budget estimating. Information on inclined walks (ramps) is also presented, as is a section on shuttle-type people movers, which also affect building design.

23

VERTICAL TRANSPORTATION: PASSENGER ELEVATORS

GENERAL INFORMATION

23.1 Introduction

Among the many decisions that must be reached by the designer of a multistory building, probably none is more important than the selection of the vertical transportation equipment, that is, the passenger, service, and freight elevators and the escalators. Not only do these items represent a major building expense, being in the case of a 25-story office building as much as 10% of the construction cost, but also the quality of elevator service is an important factor in a tenant's choice of space in competing buildings.

Although the final decision as to the type of equipment rests with the architect, the factors affecting it are so numerous that it behooves the building designer to consult with an elevator expert. Such consultation service is readily available from consultants in the field and to an extent from the major elevator and escalator manufacturers. It is the function of this chapter to familiarize the architect and engineer with the nature and application of vertical transportation equipment, to enable them to make preliminary design decisions before turning to the consultants.

23.2 Passenger Elevators

Our discussion is principally concerned with the general-purpose traction elevator. Hydraulic elevators are covered in Section 24.10.

Ideal performance of an elevator installation provides minimum waiting time for a car at any floor level, comfortable acceleration, rapid transportation, smooth and rapid retardation, accurate automatic leveling at landings, and rapid loading and unloading at all stops. Furthermore, the system must provide quick and quiet power operation of doors, good visual floor and travel direction indication both in the cars and at landings, easily operated car and landing call buttons (or other devices), smooth, quiet, and safe operation of all mechanical equipment for all conditions of loading, comfortable lighting, reliable emergency and security equipment, and generally pleasant car atmosphere.

In addition to the passenger-oriented service considerations above, the elevators have architectural aspects as well. The cars and shaftway doors must be treated in a manner consonant with the architectural unity of the building. More important, though, the shaftways are major space elements whose integration into the building is a prime factor in composition.

23.3 Codes and Standards

Perhaps more than any other item of construction, elevators are governed by strict installation codes. The "bible" of the industry is the American National Standards Institute (ANSI/ASME) code A 17.1, *Safety Code for Elevators, Dumbwaiters, Escalators and Moving Walks,* an up-to-date copy of which should be an integral part of every architect's and engineer's working library. This code has legal force in most parts of the United States. In addition, Standards 17.3 and 17.4 cover existing elevators and escalators, and emergency evacuation of elevators, respectively. As with other items, some states and municipalities have their own elevator codes (Massachusetts, Wisconsin, Pennsylvania, New York City, Seattle, Boston, among others) that are generally based on, and more stringent than, the ANSI code.

In addition to the elevator code, other construction and installation codes have an influence on elevator work. Thus NFPA No. 101, *Life Safety Code*, states certain fire safety requirements, NFPA No. 70 (the National Electric Code) governs some of the electrical aspects of elevator construction, and state and local laws add a multitude of requirements and restrictions bearing on fire safety, emergency power, security regulations, and special accommodations for handicapped persons. Provisions for the handicapped are covered by ANSI A 117.1 (Barrier-free), a special industry code, and in most locations by local law. As with most large industries, the elevator industry is self-regulating and standardized. The National Elevator Industry, Inc. publishes standard elevator layouts for traction and hydraulic installations, as well as its own elevator standard for the handicapped, "Suggested Minimum Passenger Requirements for the Handicapped" (NEII, 185 Bridge Plaza North, Ft. Lee, NJ 07024). Elevator consultants and elevator companies are normally knowledgeable as to all the codes and standards in force, but this does not relieve the architect-engineer of legal responsibility for the installation. Therefore, we strongly recommend that in the preliminary planning stage all pertinent regulations concerning vertical transportation be acquired and studied.

ELEVATOR EQUIPMENT

23.4 Arrangement of Principal Parts

The car, cables, elevator machine, control equipment, counterweights, hoistway, rails, penthouse, and pit make up the principal parts in any traction elevator installation. An idea of the functioning and orientation of these units of equipment can be obtained from an inspection of Fig. 23.1. Specific installations will vary somewhat; in newer systems microprocessor logic modules are frequently used in lieu of the older electromechanical, relay-operated control panels, and solid-state equipment may be used in lieu of a motor-generator (m-g) set. On the whole, however, the basic functional components remain the same for all installations.

The cars, with their equipment for safety, convenience, comfort, and finish, are the only items with which the average passenger is familiar. Indeed, some of the building's prestige depends on proper design of the car. Essentially, the cab is a cage of some fire-resistant material supported on a structural frame, to the top member of which the lifting cables are fastened. By means of guide shoes on the side members, the car is guided in its vertical travel in the shaft. The car is provided with safety doors, operating-control equipment, floor-level indicators, illumination, emergency exits, and ventilation. It is designed for long life, quiet operation, and low maintenance.

The cables (ropes) that are connected to the cross-head (top beam of the elevator) and carry the weight of the car and its live load are made of groups of traction steel wires especially designed for this application. Four to eight cables, depending on car speed and capacity, are placed in parallel. Although multiple ropes are used primarily to increase traction area on the sheaves, they also increase the elevator safety factor, since each rope is normally capable of supporting the entire load. The minimum factor of safety varies from 7.6 to 12 for passenger elevators and 6.6 to 11 for freight elevators. The cables from the top of the car pass over a motor-driven cylindrical sheave at the traction machine (grooved for the cables) and then downward to the counterweight.

The counterweight is made up of cut steel plates stacked in a frame that is attached to the opposite ends of the cables to which the car is fastened. It is guided in its travel up and down the shaft by two guide rails typically installed on the back wall of the shaft. (Obviously the counterweight travels in the reverse direction to the car.) See Fig. 23.1.

The weight of the counterweight is equal to the weight of the empty car plus 40% of the live load. It has several purposes: to provide adequate traction at the sheave for car lifting, to reduce the size of the traction machine, and to reduce power demand and energy cost. Approximately 75% of the energy expended in lifting a load is returned to the system by regeneration when the load is lowered. (Regeneration is the process in which the traction motor becomes a

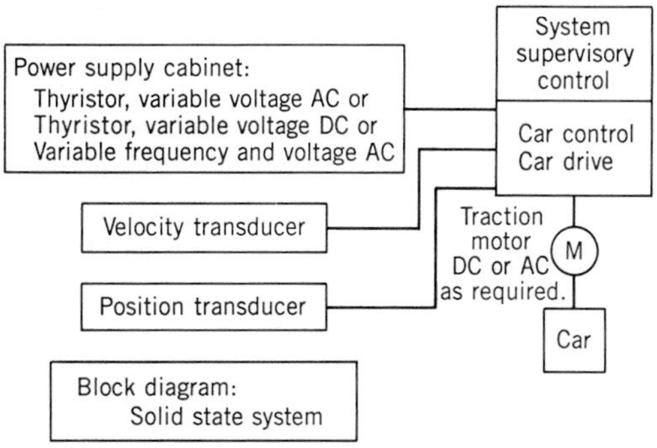

STEIN: Equipment 8E
Fig.23-1
LEFT SIDE

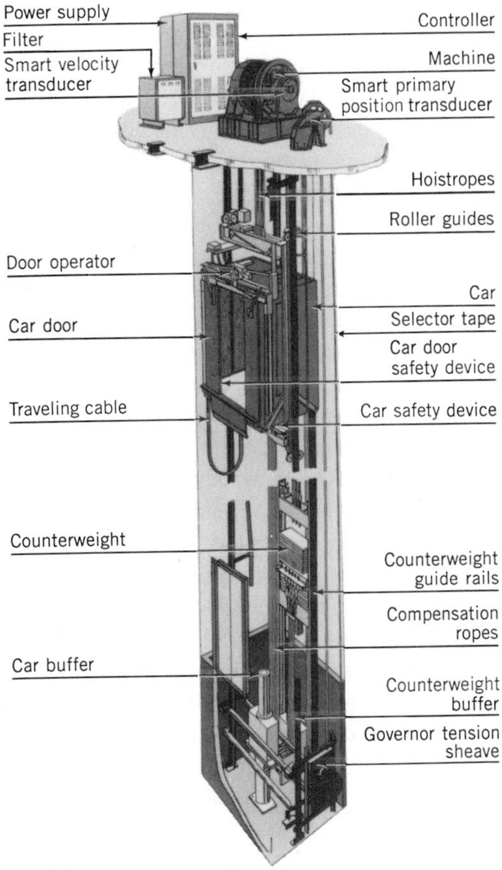

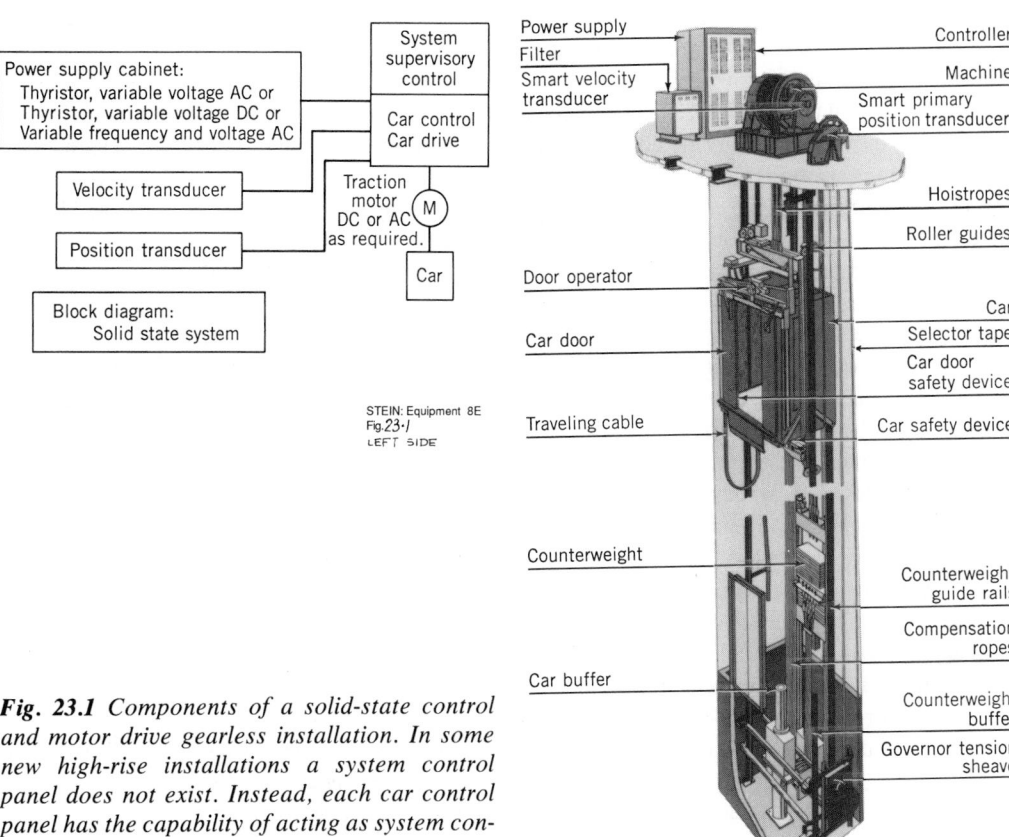

Fig. 23.1 *Components of a solid-state control and motor drive gearless installation. In some new high-rise installations a system control panel does not exist. Instead, each car control panel has the capability of acting as system controller (see Section 23.22). See block diagram insert. (Courtesy of Otis Elevator Co.)*

generator when the car is lowered, and pumps power back into the electrical system.) The "lost" energy appears as heat, primarily in the machine room. See Section 24.2 for a discussion of system energy requirements.

To compensate for hoist rope weight, which in high-rise elevators becomes an important factor, cables are attached to the bottom of the car and the counterweight, thus equalizing loads regardless of the cab position. These cables can be seen in Fig. 23.1.

The elevator machine turns the sheave and lifts or lowers the car. It consists of a heavy structural frame on which are mounted the sheave and driving motor, the gears (if any), the brakes, the magnetic safety brake, and certain other auxiliaries. In most existing installations the elevator-driving (traction) motor receives its energy from a separate m-g set, which is in operation during the period that the particular elevator is available for handling traffic. This m-g set is properly considered a part of the elevator machine, although it may be located some distance from it. In new installations, solid-state power and control equipment replaces the m-g set, as discussed in Section 23.17. The governor that limits the car to safe speeds is mounted on or near the elevator machine.

The control equipment is usually divided into three groups:

1. *Drive control* is concerned with the velocity, acceleration, position determination, and leveling of the car.
2. *Operating control* covers the area of car door operation and functioning of car signals, including floor call buttons and all indicating devices.
3. *Supervisory control* is concerned with group operation of multiple-car installations.

The actual physical devices in these control systems were electromechanical in the past but are principally solid state in new installations. The indicating and control devices that are seen and used by the elevator user, including car and hallway buttons, lanterns, and audible devices, are all coordinated into the overall operational control scheme, which produces rapid, safe, and comfortable vertical transportation.

The shaft, or hoistway, is the vertical passageway for the car and counterweights. On its sidewalls are the car guide rails and certain mechanical and electrical auxiliaries of the control apparatus. At the bottom of the shaft are the car and counterweight buffers. At the top is the structural platform on which the elevator machine rests. The elevator machine room (which may be on one or two levels) is usually directly above the hoistway. It contains the traction motor, the m-g set or solid-state control that supplies energy to the elevator machine, the control board, and other control equipment. All machinery and control equipment are designed for quiet, vibration-free operation.

23.5 Gearless Traction Machines

A gearless traction machine consists of a d-c or a-c motor, the shaft of which is directly connected to the brake wheel and driving sheave. The elevator hoist ropes are placed around this sheave. The absence of gears means that the motor must run at the same relatively low speed as the driving sheave. Since it is not economically practical to build motors for operation at very low speeds, this type of machine is utilized for medium- and high-speed elevators, that is, 500 fpm and above. The motors range from 20 to 375 hp. Gearless machines are generally utilized for passenger service, with usual car capacities of 2000 to 4000 lb, although specials of up to 10,000 lb, such as at the World Trade Center, have been built. Below 500 fpm, geared machines are used. In the range of 500 to 700 fpm, a 2 : 1 roping arrangement (see Section 23.7) is generally used. This reduces motor size and increases sheave speed, thus reducing the cost. Above 800 fpm, motor speed is high enough for 1 : 1 roping to be applied economically. At this writing maximum car speeds are 1600 fpm, although 2000-fpm systems and faster have already been developed and will undoubtedly be an important factor in the development of ever taller, practical, workable buildings.

The gearless traction machine is generally considered superior to the geared machine. It is more efficient, quieter in operation, requires less maintenance, and has longer life. The decision as to whether these advantages are worth the additional cost involved is made only after a careful analysis. Generally, a gearless machine is chosen where rise is more than 250 ft and smooth, high-speed operation is desired. In the intermediate range of rise and speeds, that is, 150 to 250 ft height and 400 to 500 fpm, excellent equipment, both geared and gearless, is available. The choice depends on performance characteristics and cost.

Because virtually every major elevator company operates internationally, and because outside the United States, elevator size and speed are specified in kilograms and meters per second, it is useful to have approximate conversion factors in mind. Since 1 kg is 2.2 lb and 1 m/s is 196.86 fpm, the factors of 2 and 200 will give approximate American equivalents. Thus:

Weight		*Speed*	
Standard Metric (kg)	*Approximate U.S. (lb)*	*Standard Metric (m/s)*	*Approximate U.S. (fpm)*
1000	200	1	200
1250	2500–3000	1.5	300
1600	3500	2	400
2000	4000	2.5	500
		4	800
		6	1200

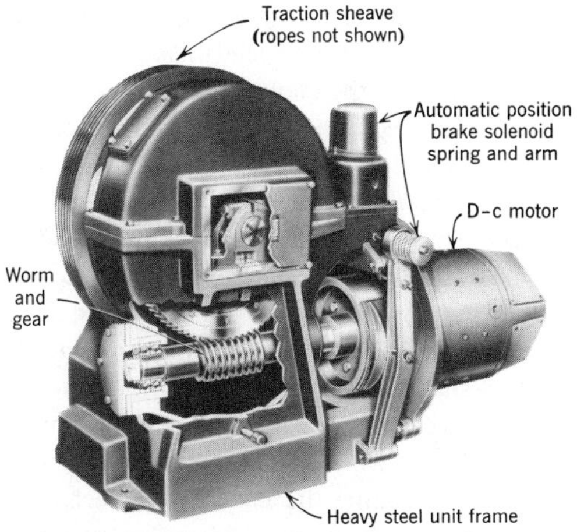

Fig. 23.2 *Cutaway of a typical d-c geared traction machine. Note the grooves for multiple ropes in the traction sheave. (Photo courtesy of Schindler Elevator Corp.)*

23.6 Geared Traction Machines

The geared traction machine (Fig. 23.2) employs a worm and gear interposed between the driving motor and the hoisting sheave. The driving motor can therefore be a smaller, cheaper high-speed unit rather than the large low-speed unit that a gearless installation would require.

The motor used in a geared installation, as in gearless, depends on the type of drive system, and may be either d-c or a-c. The characteristics of the various drive systems and their applicability is discussed in detail in Sections 23.12 to 23.17. Geared machines are used for car speeds up to 450 fpm and a maximum rise of about 300 ft. With an appropriate drive and control system a geared traction machine can give the same high-quality, accurate, smooth ride as is available from a gearless installation.

23.7 Arrangement of Elevator Machines, Sheaves, and Ropes

The simplest method of arranging vertical travel of a car is to pass a rope over a sheave and counterbalance the weight of the car by a counterweight. Then, by rotating the sheave, the car will move up or down and require very little energy to do it. This is essentially the scheme that is used on a majority of high-speed passenger elevators, as illustrated in Fig. 23.3*a*.

When the four or more supporting ropes merely pass over the sheave *T* and connect directly to the counterweights, the lifting power is exerted by the sheave through the traction of the ropes in the parallel grooves on the sheave. This system is referred to as the single-wrap traction elevator machine. The function of sheave *S* is merely that of a guide pulley; it is called the deflector sheave.

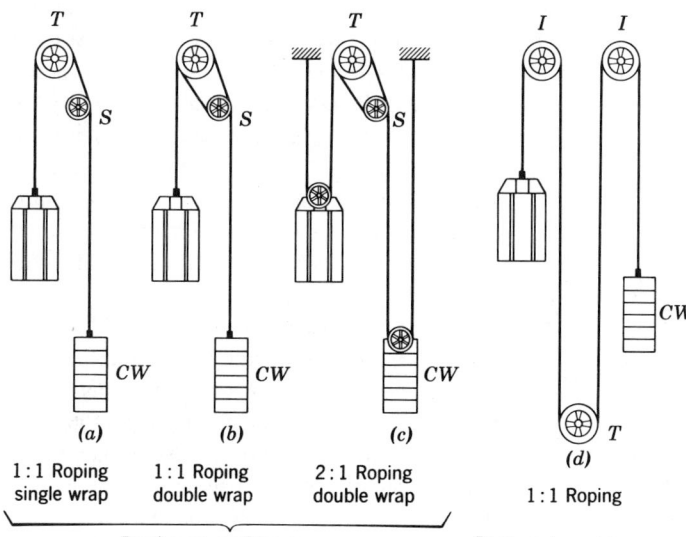

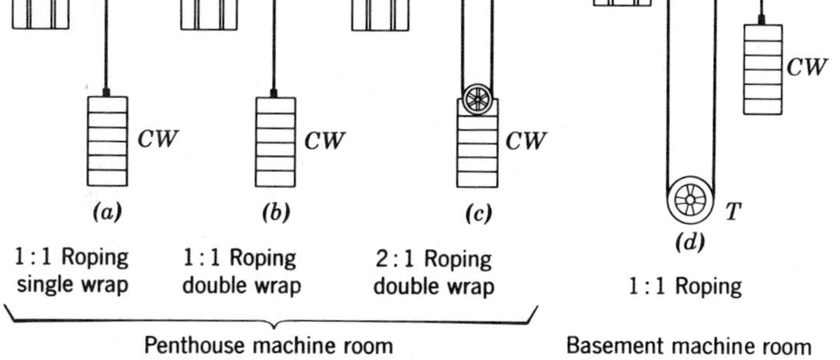

Fig. 23.3 *Elevator roping and sheave arrangement.* T, *traction sheave;* S, *secondary sheave;* I, *idler sheave;* CW, *counterweight.*

TRANSPORTATION

In Fig. 23.3*b* the ropes from the car are first wrapped over the traction sheave *T*, then around the secondary or idler sheave *S*, once more around sheave *T*, and back over *S* to the counterweights. This arrangement is characteristic of the one-to-one, double-wrap traction machine. It provides greater traction than the single-wrap machine and is used in many automatic high-speed installations.

A 1 : 1 roping arrangement (Fig. 23.3*a*, *b*, and *d*) gives no mechanical advantage; that is, the drive must supply sufficient power to move the unbalanced load. The 2 : 1 roping (Fig. 23.3*c*) has a mechanical advantage of 2, which results in a high-speed, low-power, and, therefore, low-cost traction machine. This arrangement is used for a wide variety of installations varying from medium-speed (500–700 fpm) gearless passenger elevators to low-speed, heavy-duty freight units.

In types *a*, *b*, and *c* in Fig. 23.3, the elevator machines are located at the top of the hoistway. When the elevator machines are placed in the basement, a very different arrangement of cables and sheaves must be utilized to secure the same results (see Fig. 23.3*d*). Much more rope is required in such an arrangement, and consequently the problems of rope maintenance are increased. These systems, however, obviate the necessity for a tall penthouse; and where this is desirable for architectural or other reasons, a basement machine is used. This arrangement uses geared traction equipment, with speeds up to 400 fpm. All the illustrated ropings are applicable to the full range of car capacities up to 4000 lb.

23.8 Safety Devices

The main brake of an elevator is mounted directly on the shaft of the elevator machine (see Fig. 23.2). The elevator is first slowed by dynamic braking of the motor, and the brake then operates to clamp the brake drum, thus holding the car still at the floor.

A dual safety system, designed to stop an elevator car automatically before the car's speed becomes excessive, is normally used. The device that acts first is a centrifugal governor, which is independent of the other elevator machinery, and at normal speeds has no effect on the operation of the elevator. In the event of a limited overspeed, the governor will cut off the power to the d-c motor and set the brake. This usually stops the car, but should the speed still increase, the governor actuates the two safety rail clamps, which are mounted at the bottom of the car, one on each side. These devices clamp the guide rails by wedging action, bringing the car to a smooth stop (see Fig. 23.4). Solid-state controllers can be arranged with sensing devices (transducers) which detect car overspeed, thereby eliminating the need for a mechanical governor or tachometer. The action on power control and brake activation is essentially the same.

Oil or spring buffers are usually placed in the elevator pit. Their purpose is not to stop a falling car but to bring it to a partially cushioned stop if it should overtravel the lower terminal. Electrical final-limit switches are located a few feet below and above the safe travel limits of the elevator car. If the car overtravels (down or up), these switches de-energize the traction motor and set the main brake. Safety arrangements under emergency condition of fire or power failure are discussed in Section 24.4.

23.9 Elevator Doors

The choice of car and hoistway door affects the speed and quality of elevator service considerably. Doors for passenger elevators are power operated and are synchronized with the leveling controls so that the doors are fully opened by the time a cab comes to a complete stop at the landing. The closing time, however, varies with the type of door and size of opening. For safety reasons, the kinetic energy of an automatic door is limited to 7 ft-lb, and its closing pressure to 30 lb. To provide fastest closing within this energy limitation, a center-opening door is used. Also, to reduce passenger transfer time and avoid discomfort, a clear opening of 3 ft, 6 in. (106.7 cm) is used in most commercial installations, which permits simultaneous loading and unloading without undue passenger contact. [Some consultants feel that simultaneous passenger trans-

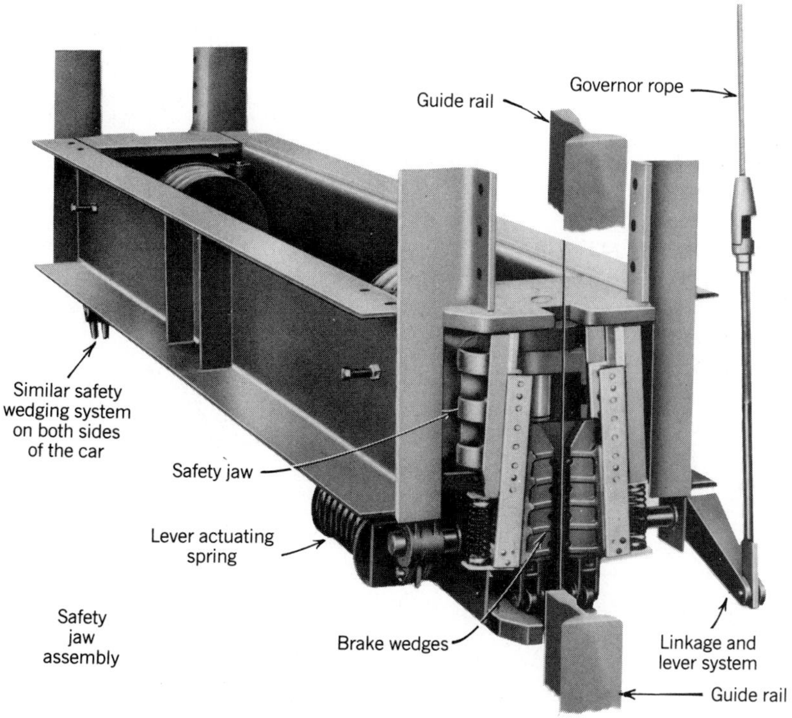

Fig. 23.4 *Elevator safety devices. The governor trips, clamping the governor rope and releasing the safety jaws, which exert a constant retarding force on the car rails, thus bringing the car to a gradual and safe stop. (Photo courtesy of Schindler Elevator Corp.)*

fer is practical only with a 48-in. clear opening; see Strakosch (1983).] See Fig. 23.5. When an opening narrower than 36 in. is used, loading will be delayed until unloading is complete; therefore, speed and quality of service will be markedly reduced. Such small doors are applied only in residential or small, light-traffic buildings. The available types are shown in Fig. 23.6.

A two-speed door design is used where space conditions dictate or where a wide opening is required. The two-speed nomenclature reflects the fact that the two halves of the door must travel at different speeds to complete their travel simultaneously (see Fig. 23.6*c*).

Installations can be equipped with an electronic sensing device that detects passengers in a wide area on the landing in front of the car door, rather than only directly in the door path. Such detection, often accompanied by an audible signal, causes the car door to remain open for a predetermined length of time, or a closing door to reverse. These devices are particularly useful in installations where passengers cannot approach the entrance or cannot enter the car quickly, for example, riders with baggage or holding children, people in wheelchairs, and employees moving bulky objects such as beds or carts in hospitals or wheeled objects in office and industrial facilities.

All automatic elevators, whether equipped with detection beams or not, are required by ANSI to have a safety edge device on the car doors that will cause the car and hoistway doors, which operate in synchronism, to reopen when the safety edge meets any obstruction.

23.10 Cabs and Signals

Possibly the only area in which the architect has a free hand in selection of equipment is in the decor of the cabs and the styling of hallway and cab signals. A normal elevator specification is a functional one that describes the intended oper-

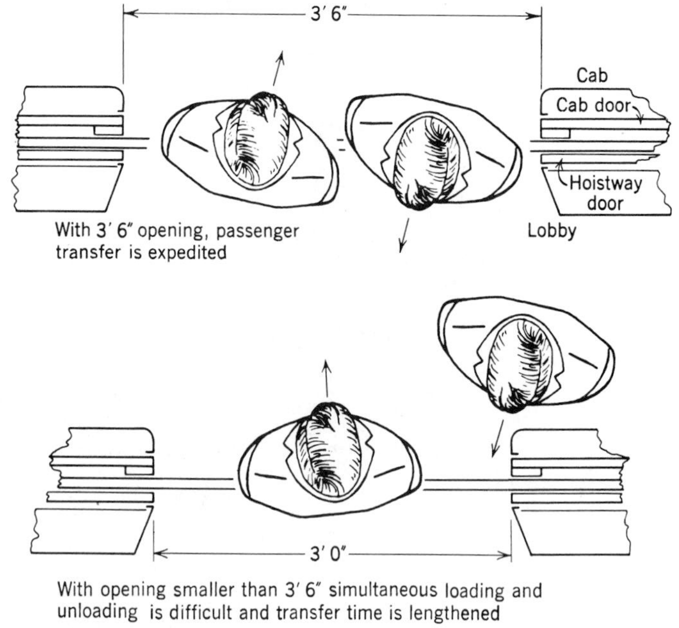
3′ 6″
Cab
Cab door
Hoistway door
Lobby
With 3′ 6″ opening, passenger transfer is expedited
3′ 0″
With opening smaller than 3′ 6″ simultaneous loading and unloading is difficult and transfer time is lengthened

Fig. 23.5 Transfer of passengers with door openings of different widths. With openings smaller than 3 ft 6 in. (42 in.), simultaneous loading and unloading is difficult and transfer time is lengthened. With a 42-in. opening, large people or people with bulky outerwear may brush against each other in passing. For complete isolation of passengers, a 48-in. opening may be necessary.

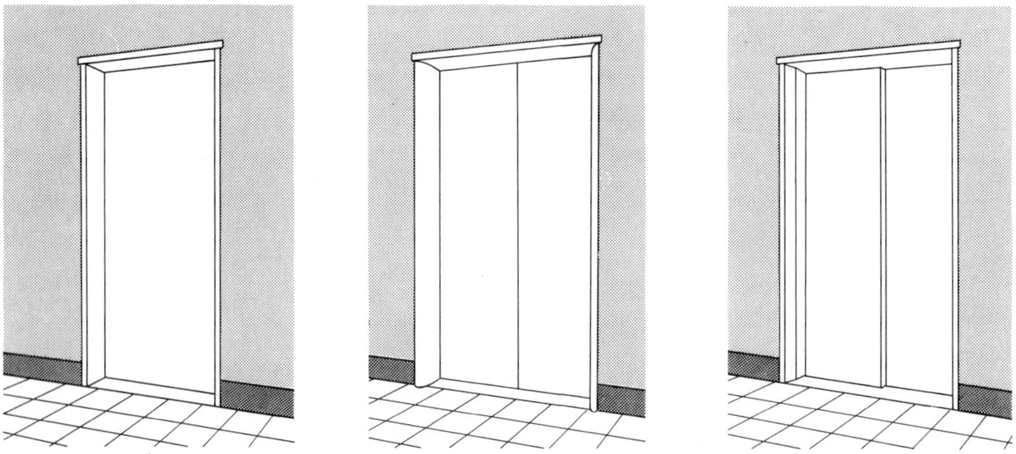

(a) *(b)* *(c)*

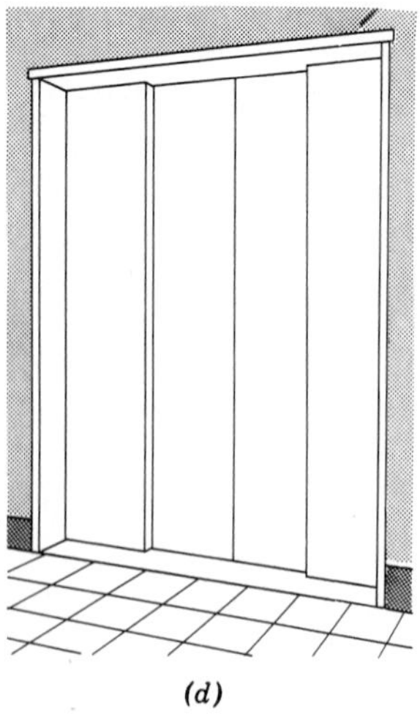

(d)

***Fig. 23.6** Typical hoistway doors and applications.* (a) *Single slide door, 24 to 36 in. wide, for small commercial building or residential use.* (b) *Standard commercial door, 42-in. center opening, for office buildings use, or 48-in. to 60-in. center opening, for hospital or service car.* (c) *Two-speed 42-in. general commercial use door.* (d) *Two-speed, center-opening, 60-in. department store door, freight and passenger, nonautomatic service.*

ation of the equipment and normally includes an amount to cover optional decor of the cabs. The type and functioning of signal equipment is also specified, but finish and styling are optional. Cab interiors may be finished in wood paneling, plastic (Micarta or Formica), stainless steel, or almost any material desired. Floors may be tile, wood, or carpeting as selected. Illumination may be from ceiling fixtures, coves, or completely illuminated luminous ceiling, of standard or special design. For each bank of elevators, it is wise to furnish at least one set of wall mats, to protect wall finishes when cars are being used to move tenant furniture. This is especially important where no separate service car has been provided.

Car and hallway signals and lanterns should be designed to fulfill their basic functions, take proper consideration of the needs of the handicapped, and be consonant with the decor of the cabs and the corridors. (For a discussion of cab control panel, see Section 23.26.) The hall buttons should indicate desired direction of travel and by visual means confirm that a call has been placed. The hall lantern located at each car entrance must visually indicate the direction of travel of an arriving elevator and preferably its present location. An audible signal should signal its imminent arrival. The latter feature allows waiting passengers to move to the next arriving car in a bank and thereby speeds service. Hall stations can be equipped with special switches for fire, priority, and limited access service, as required (see Fig. 23.7).

Within the cab, indication of travel direction and present location can be accomplished either with separate fixtures or by indicators built into the car panel. Several manufacturers now market voice synthesizers built into the car panel that announce the floor, direction of travel, and any other desired message such as safety or emergency messages.

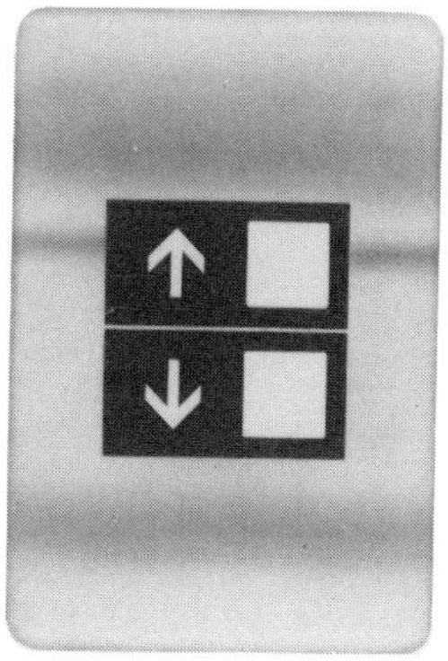

(a) Basic hall station

(b) Combination hall lantern and car position display

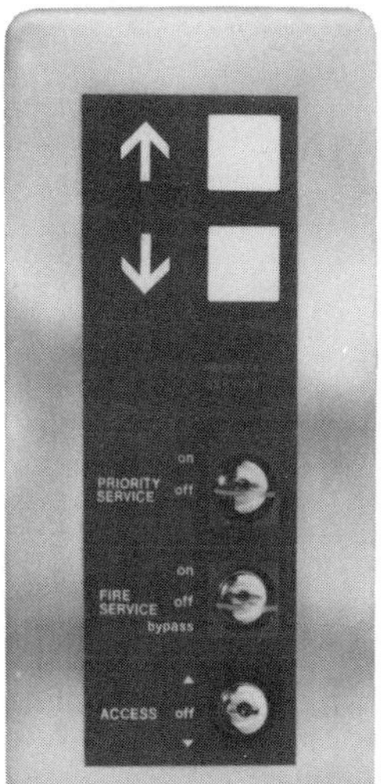

(c) Expanded hall station

Fig. 23.7 *Typical hall fixtures. Buttons* (a) *must show direction and have visual indication. Lantern* (b) *must show travel direction, audibly announce imminent car arrival, and preferably also show present car location. Special hall stations* (c) *may incorporate priority, emergency, and security features. (Photos courtesy of Otis Elevator Co.)*

23.11 Requirements for the Handicapped

Although the recommendations of the NEII special standard "Suggested Minimum Passenger Elevator Requirements for the Handicapped" are not mandatory except where local building

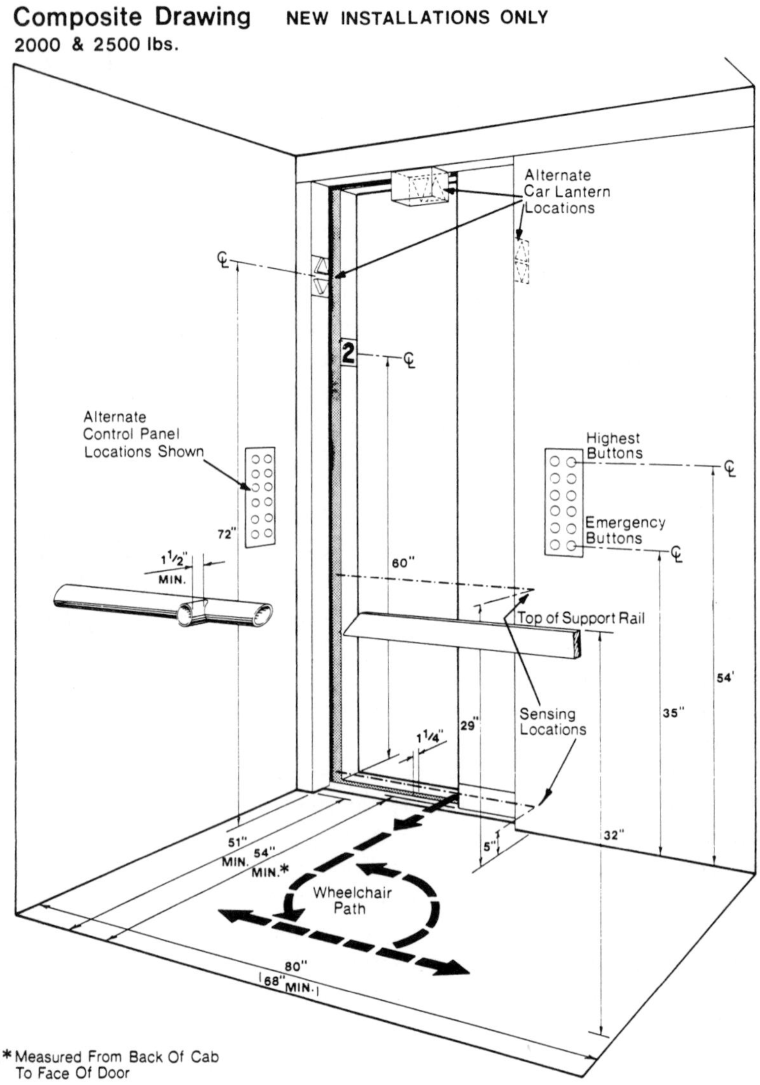

Fig. 23.8 Composite sketch embodying NEII recommendations for 2000- and 2500-lb elevator car sizes, arrangements, and appointments to accommodate the handicapped. (Reproduced with permission of National Elevator Industry, Inc., 185 Bridge Plaza North, Fort Lee, NJ 07024. Copyrighted July 1985, Elevator Requirements for the Handicapped, *4th ed.)*

codes make them so, it is certainly advisable to adopt them in consonance with the modern concept of barrier-free buildings.

The basic physical limitations addressed are those of ambulation and sight. Thus to ease access for passengers in wheelchairs (or with walking aids), the standard requires excellent car leveling, 42-in. minimum clear door opening, delayed door closing, detection beams that will reopen the door *without contact* on sensing a passenger, inside car dimensions that will permit turning a wheelchair, buttons and emergency controls within easy reach, and appropriate car furnishings. For those with sight impairment, the NEII standard calls for audible signals in addition to easily seen and recognized visual ones, both in the car and at landings, to indicate call registration, car approach, car landing, direction of travel, floor, car position, and so on. In this connection, a voice synthesizer is of invaluable assistance. Additionally, Braille plates adjacent to car floor buttons and large, easily recognizable symbols adjacent to passenger-controlled emergency controls are to be used.

Many of these ideas will also make life easier for passengers with full faculties. A note of caution, is in order, however. Delayed door closing may increase travel time appreciably. In buildings with traffic peaks, it may be necessary to designate one or more specific elevators for use by the handicapped during these periods. Figure 23.8 is a composite sketch showing the application of some of these recommendations.

In addition to the NEII standard's requirements, peripheral facilities should meet the requirements of ANSI standard A117.1 (latest issue) *Buildings and Facilities—Providing Accessibility and Usability for Physically Handicapped People.*

ELEVATOR CAR CONTROL

23.12 Drive Control

Elevator drive control, or motion control, comprises the arrangement of hardware and in most modern installations, software, which effects the controlled motion of a car. The car's acceleration, velocity, braking, leveling, and all aspects of power control, including regenerative braking, are subsumed under this heading. The other aspect of car control is usually known as operating control, and deals with the operation of the car doors and the integration of car buttons, lanterns, and other passenger-operated devices into the overall control and indicating system. Elevator *car* control is separate and distinct from the control system that governs the functioning of a group of cars acting as a unit building system. That control arrangement is generally designated as the supervisory system, and is discussed separately below. It retains its separate identity as a supervisory system, albeit a relatively simple one, even in the case of a single-car system.

Elevator car acceleration and deceleration are accomplished by controlling the speed of the motor that drives the elevator traction machine. This speed control can be accomplished in a number of ways, all of which are in use in elevator installations. They are:

1. Rheostatic control of a single-speed a-c traction motor.
2. Rheostatic control of a two-speed a-c traction motor.
3. Thyristor control of an asynchronous (squirrel cage) a-c traction motor.
4. Thyristor control of a d-c traction motor.
5. Motor-generator set control of a d-c traction motor (Ward–Leonard system).
6. Variable-voltage, variable-frequency control of an asynchronous a-c traction motor.

We will describe each system briefly and note its applicability, advantages, and disadvantages.

23.13 Rheostatic a-c Motor Control

See Fig. 23.9*a*. Speed control is accomplished by introducing resistance into the motor circuit. This results in a relatively rough ride with poor leveling. Rise is limited to 150 ft and speed to 200 fpm (1 m/s). Use of a two-speed motor improves the ride and leveling by using the motor's low-speed winding for landing the car, but the ride quality is still not smooth. Other disadvantages of this arrangement are high starting currents, which cause line disturbances and require heavy feeders, and poor operating efficiency. The system is used where economic considerations are paramount and where a hydraulic unit is impractical.

23.14 Thyristor Control, a-c and d-c

(a) A-c. See Fig. 23.9*b*. The development of high power transistors in the last decade made possible and practical accurate speed control of inexpensive a-c squirrel-cage motors by the simple expedient of supplying the motor carefully regulated variable voltage. The resultant speed control is smooth and stepless, making it applicable to passenger elevator installations. However, the high motor slip caused by the constant frequency of the variable-voltage a-c supply causes large thermal losses with resultant low operating efficiency. Too, the "chopper" which is used to accomplish the necessary voltage control introduces undesirable harmonics into the power system in considerable quantity. These adversely affect all of the electrical systems components and can cause equipment overheating and overload (see Section 17.19). Furthermore, these same harmonics appear as line

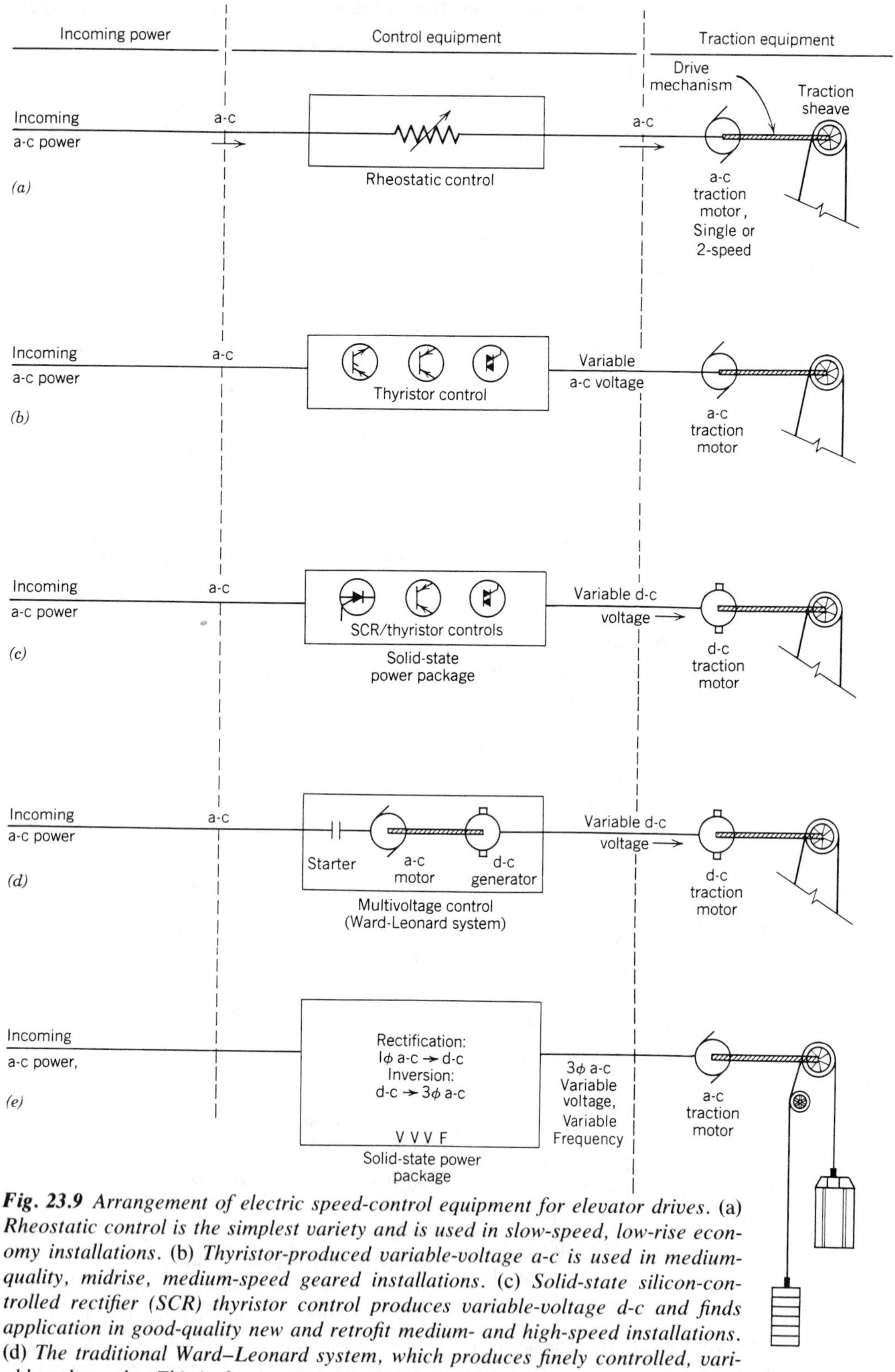

Fig. 23.9 *Arrangement of electric speed-control equipment for elevator drives.* (a) *Rheostatic control is the simplest variety and is used in slow-speed, low-rise economy installations.* (b) *Thyristor-produced variable-voltage a-c is used in medium-quality, midrise, medium-speed geared installations.* (c) *Solid-state silicon-controlled rectifier (SCR) thyristor control produces variable-voltage d-c and finds application in good-quality new and retrofit medium- and high-speed installations.* (d) *The traditional Ward–Leonard system, which produces finely controlled, variable-voltage d-c. This is the system in use today in the vast majority of existing high quality installations.* (e) *Variable-frequency a-c is energy efficient, highly accurate, and applicable to all types of installations.*

"noise," which can also adversely affect computer and data communications. Finally, the system's low power factor increases line losses and necessitates increased feeder sizes. Despite excellent control electronics, the ride and leveling are not of the highest quality. This option is applicable to low and midrise residential and commercial passenger service, using geared machines, where economic factors offset the disadvantages. Maximum rise is approximately 250 ft and maximum car speed is 350 fpm (1.7 m/s).

(b) D-c. See Fig. 23.9*c*. A large improvement in ride quality can be had by utilizing the d-c version of thyristor control to supply variable voltage to a d-c traction motor. This arrangement provides the high-quality ride and leveling characteristic of d-c drives and removes the rise and speed limitations of its a-c equivalent, but not the system's low power factor, heavy line harmonics, and high machine room thermal losses. Line harmonics can be sharply reduced by the use of power line filters, which are supplied as a matter of course in good-quality installations. This control system is widely used today, with both geared and gearless traction machines as applicable, for installations of all rises and car speeds. Its extremely smooth ride makes it one of the three systems of choice in high-quality, center-city prestige buildings. The other two systems used for such installations are a Ward–Leonard variable-voltage d-c system and the relatively new variable-voltage, variable-frequency a-c system.

23.15 Variable-Voltage d-c Motor Control

See Fig. 23.9*d*. Before the development of electronic motor control, the only practical method of obtaining the precise motor speed control necessary for smooth stepless elevator operation was to impress a variable d-c voltage on a d-c traction motor. The variable d-c voltage required to drive the d-c traction machine was traditionally obtained from an auxiliary m-g set comprising an a-c motor and a d-c generator. This arrangement is known as a *Ward–Leonard system* or as a *unit multivoltage* (UMV) drive. Although this arrangement has disadvantages, it is the classic high-quality elevator drive arrangement and is found in the vast majority of existing high-quality geared and gearless installations. The disadvantages of this system—low overall efficiency, expensive machines, high thermal losses in machine rooms, high maintenance costs for three rotating machines, and high noise levels—are all accepted as the price of a system that operates faultlessly, with supreme accuracy at all speeds, for long periods of time given proper maintenance. As noted above, it is today still one of the three principal drive control systems being used in new installations.

In modernization work, buildings with existing Ward–Leonard systems are frequently converted to solid-state d-c thyristor control. This change maintains the excellent ride characteristics of variable-voltage d-c traction machine installations while improving system performance by replacing the auxiliary m-g set with a low-loss, low-maintenance, quiet solid-state thyristor drive control.

23.16 Variable-Voltage, Variable-Frequency (VVVF) a-c Motor Control

See Fig. 23.9*e*. This system, which is entirely solid state like the thyristor controls described above, is in the opinion of many elevator experts the system of choice for all high-quality new installations of any speed or rise. The system consists of a rectifier, which changes the incoming 60 Hz a-c to d-c, and an inverter, which creates variable-voltage, infinitely variable frequency, three-phase a-c from the (rectified) d-c. This a-c is then applied to a standard squirrel-cage a-c motor, which operates essentially at the speed corresponding to the frequency of the input. By controlling both the voltage and frequency input to the traction motor, it is possible to provide continuously variable, highly accurate speed control at extremely high efficiencies throughout the full speed range of the motor. This arrangement is precisely what is required for high-quality geared and gearless installations, without the energy losses associated with the auxiliary m-g set of UMV control. As a result, it is coming into increasing use in new installations, particularly in areas where high electric utility rates provide a short payback period for the higher initial cost of the system. This arrangement eliminates most of the disadvantages of the solid-state thyristor control system and has these characteristics:

1. Overall system efficiency is high at all motor speeds.
2. Traction motors are economical single-speed squirrel-cage a-c.
3. Motors are 10 to 15% smaller than equivalent drive systems.
4. System power factor is close to unity.
5. Line harmonics are lower than with thyristor controls.
6. Speed control and leveling are very high quality; equal to d-c traction motor control.
7. Equipment is 98% solid state, thus requiring less maintenance.
8. Starting currents are low.
9. System is applicable to all rises and speeds.
10. Machine rooms are smaller, cooler, and quieter than those using Ward–Leonard UMV controls.
11. First cost is higher than the competing alternatives.

23.17 Comparative Drive System Characteristics

Solid-state equipment will in time undoubtedly replace the traditional m-g set and its associated controls, because of inherent advantages. The principal drawback at present is the higher initial cost of solid-state equipment. However, when considering life-cycle cost, the relative amounts may be reversed.

(a) Energy and Cost Considerations. One of the principal advantages of solid-state elevator control lies in the energy savings possible. If we were to tabulate energy use of the various systems of motion control discussed above, on a relative scale of 1 to 10, the comparison would appear as in Table 23.1.

Considering that in many parts of the United States energy costs are well above 10 cents/kWh, an economic analysis (see Section 24.2 and Appendix E) may well demonstrate that the higher first cost of the electronic motion control system is economically justifiable, particularly when combined with lower maintenance costs.

Contract maintenance costs in major U.S. cities for high-rise gearless elevators exceed $1000 per month per car. Even a 15% reduction in this cost over an estimated 25- to 30-year equipment life gives a present value (PV) for life-cycle maintenance cost reduction of $25,000 to $35,000 per car, depending on cost of capital.

An ancillary cost benefit accrues due to the sharp reduction in the size of the forced-ventilation or air-conditioning equipment in the machine room, because of the lower heat loss of solid-state motion control equipment as compared to UMV equipment. This economic benefit is appreciable since it affects both first cost and operating cost of the cooling equipment.

(b) Physical Considerations. Although elimination of the m-g set in the solid-state drive system does save space, the corresponding cabinets for car controller and motor drive are larger than their UMV counterparts. Thus the machine room for a solid-state design is only slightly smaller than in the UMV arrangement. A typical solid-state car control panel for a 3500-lb, 800-fpm gearless elevator measures 80 in. H × 48 in. W × 24 in. D and weighs 800 lb.

(c) Car Operation. Somewhat better passenger service is obtainable when using electronic equipment because of better car control. This in turn is due to faster response of components and finer control, particularly of jerk (rate of change of acceleration). Jerk is the principal factor in passenger discomfort.

(d) Other Factors. Elimination of the m-g set and use of electronic motion control have other desirable effects:

TABLE 23.1 **Comparative Energy Use of Various Motion Control Arrangements**

Energy Use	*Relative Amount*
Hydraulic elevators (discussed in Section 24.10)	10
Rheostatic control; geared a-c traction motors	7.5–8.5
Thyristor control; geared a-c traction motors	7–7.5
UMV motion control; geared d-c traction motor	7–7.5
UMV motion control; gearless d-c traction motor	6–6.5
Thyristor control; geared d-c traction motor	5–5.5
Thyristor control; gearless d-c traction motor	4–4.5
Variable-voltage variable-frequency control, a-c traction motor	2.5–3.0

1. Elimination of current and kilowatt demand peaks, which are characteristics of motor starting operation. This will result in a reduction in electric bills. The overall reduction, including energy and demand changes, can reach 35%.
2. Reduction in noise and vibration in the machine room, which frequently is transmitted to other parts of the building.
3. Reduction in maintenance costs, because the periodic maintenance required for continuously rotating and sliding equipment is eliminated. Solid-state equipment maintenance is minimal.
4. Possible reduction in the size of the emergency generator required to operate an elevator, with concomitant cost reduction. This reduction is possible since the very high current inrush taken on m-g starting is eliminated. The generator must, however, be large enough to handle the regenerative braking current produced by a full car traveling down.

At this point we can summarize the discussion above with a tabular comparison of the various drive system in use today (Table 23.2). The exact degree of improvement possible when changing to solid-state equipment in a retrofit job depends on the quality of both the old and the new systems. It is greatest when a microcomputer group control system replaces a relay-logic terminal dispatch system.

23.18 Elevator Operating Control

Assuming the system to be energized and at rest, registration of a call from a station in the

TABLE 23.2 **Comparative Characteristics of Elevator Drive Systems**

Type	*Rise*	*Speed*	*Control*	*Initial Cost*	*Operating Cost*	*Performance*
Geared a-c	150	50–200	Rheostat	Low	Medium	Poor
	250	150–350	Thyristor	Medium	Medium	Fair
	300	150–450	VVVF	High	Low	Excellent
Geared d-c	175	50–450	UMV	Medium	High	Excellent
	250	50–450	Thyristor	Medium	Low	Very good
Gearless d-c	Unlimited	400–1200	UMV	Medium	High	Excellent
			Thyristor	Medium	Low	Excellent
Gearless a-c	Unlimited	400–2000	VVVF	High	Low	Excellent

Note: Life of equipment is generally indefinite except for worm and gear of geared unit, which has a 30- to 40-year life.

TRANSPORTATION

lobby or an upper-floor corridor will activate the system. The particular elevator that will answer the call is selected by the supervisory system. In a UMV system, the m-g set related to that elevator car is started and comes up to speed slowly. In the case of solid-state motion control, power is available immediately, and the m-g set starting time is eliminated. This difference between m-g set and solid-state equipment disappears in the more usual case of an m-g set that is already running. With voltage available the elevator control is cut in and the car is put into operation. Switching devices in the car control panel release the brake, energize the elevator motor, and accelerate the car to its rated speed. Reverse operations are initiated when decelerating and stopping (landing) the car. When the car stops, the brake holds the sheave and elevator stationary.

The motion of a single car is determined by the action of three principal items of equipment: the car controller, the motor controls, and the system supervisory equipment. The action of this latter equipment is discussed in Sections 23.19 to 23.23. The function of the car controller panel is generally to duplicate the action of the elevators and to anticipate hall calls. This is accomplished in older controllers by the action of a sort of miniature elevator system that contains sets of contacts over which small sliding brushes move in synchronism with the corresponding car movements in the hoistway. Most recent systems utilize an electronic or optical position transducer, which is at least as accurate as its mechanical forerunner but does not require periodic alignment to maintain accuracy. This information, that is, the car position and both waiting hall calls and car calls, is fed into the group supervisory system. This system may reside in a separate panel or, as described below, in each car controller panel. It acts to initiate all the procedures necessary to answer all calls via the individual car controller panels. In almost all new systems the switching and control devices are solid state, although small low-rise, low-speed, single-car installations such as are found in small apartment houses and loft buildings may still use electromechanical switches (relays) and position devices to good effect.

SYSTEMS OF ELEVATOR OPERATION AND SUPERVISION

23.19 System Control Requirements

An operating system is that control, as stated above, that governs the automatic (operatorless) response of a single car or a group of cars, to calls for service. An effective system therefore must take cognizance of all hall calls and car calls, car travel directions, and car positions—in relation to each other and in relation to the call requirements, plus the relationships between up and down traffic and the trends of traffic. This last is required in order that the system anticipate demand rather than react to it, since only by anticipation can the system operate at maximum efficiency. The operating system considers all the above data and dispatches cars accordingly. Obviously, since traffic and calls are never static, the control that will satisfy all these demands in a large elevator system is an extremely sophisticated one. Without such control, however, large elevator systems will render poor service, to the dissatisfaction of owner and rider. On small systems the operating control is much simpler, as described below. Throughout the discussion that follows, reference to m-g sets applies, obviously, to UMV motion control. In solid-state systems, as explained above, drive voltage is available instantaneously.

23.20 Single Automatic Push-button Control

This system is the simplest of the passenger-operated automatic control schemes. It handles only one call at a time, providing an uninterrupted trip for each call. A single corridor button at each level can register a call only when the car is not in motion. To indicate availability of the car, an "in use" light is placed over the hall call button. Calls can be placed when the light is off. This control scheme is applicable only to a short-rise, inactive elevator, that is, one making five or fewer trips per hour. Such elevators are found in small apartment houses, residences, small professional buildings, and industrial buildings using freight elevators.

23.21 Collective Control

This system is no longer used in new installations in the United States, although it is common in other countries. With only a slight increase beyond very light traffic single automatic push-button control becomes unsatisfactory, because no call storage provision is made. Waiting periods can become extremely long, particularly with slow cars and moderately high rise. A control system therefore evolved that provides a single call button at each landing. The elevator stops at each floor that has registered a call to "collect" the waiting passenger. Hence the "collective" nomenclature. The car cannot differentiate between up and down calls since only a single button is provided. A car direction light is frequently placed adjacent to the call button. This will tell the prospective passengers whether the car that stops in response to their call is going in the right direction. As can readily be seen, this type of operating system is acceptable only for light residential or industrial service.

23.22 Selective Collective Operation

This type of collective operation is "selective" in that it is arranged to collect all waiting "up" calls on the trip up and all hall "down" calls on the trip down. The control system "stores" all calls until they are answered, and automatically reverses the direction of travel at the highest and lowest calls. When all calls have been cleared, the car will remain at the floor of its last stop awaiting the next call, and its m-g set will stop after several minutes. Pressure on any "up" or "down" button at any landing, or on any floor button in the car, will start the m-g set and set the car into operation.

Selective collective control is standard in locations where service requirements are moderate such as in apartment houses, small offices, and hospitals. Since these locations often require more than one car, a control scheme is available for groups of one to three cars. It automatically assigns each landing call to the car best situated to answer it, prevents more than one car from answering a call, allows one car to be detached for freight duty while others serve passengers, and automatically parks all but one car at the ground floor, that car acting as a free car until service calls require the use of the parked cars.

Although selective collective control is standard for most residential and other light-to-moderate service requirement buildings, its inherent characteristics frequently result in long waiting periods for elevators. These characteristics are:

1. Highest call reversal.
2. Shutdown of the m-g set (where used) after calls have been answered.
3. Strong tendency toward bunching of cars.

The last characteristic is particularly annoying in groups of three cars. Frequently a passenger will arrive at a landing to find that all three cars have just passed, going in the same direction. The result is that service is only slightly better than that which would be rendered by a single car, except that load (handling) capacity is greater. For this reason, operation of more than two cars with this system is not recommended and operation of more than three cars is not feasible.

Although collective control furnishes adequate elevator service where requirements are light to moderate, it is still basically a signal-controlled system that weighs all calls equally and takes no cognizance of traffic patterns.

To overcome this shortcoming, elevator engineers over the years 1950 to 1980 developed numerous supervisory control systems, among which were multiple zoning arrangements and programmed traffic patterns. The former divided the building into zones and dispatched elevators within these zones, thus reducing waiting time. The latter recognized that most buildings exhibit repetitive traffic patterns—such as morning up-peak, evening down-peak, lunch hour up and down peaks, and so on—and arranged car travel accordingly to give optimum service. Choice of the mode of operation was generally left to the lobby attendant, whose judgment was relied on to select the mode appropriate to the need. This was frequently less than satisfactory and was replaced by a system in which the operational mode was selected on a clock-time basis. The basic problem with all these older preprogrammed operating systems is that they can react in only a few modes because of physical limitations. The control system consists of relays, contacts, and hardwiring. As the system becomes more complex, the number of these devices increases enormously and places a practical limit on the sophistication of the system. This limitation disappeared with the advent of the computerized, microprocessor-controlled operating system.

23.23 Computerized System Control

The most advanced type of control system continuously monitors demand and controls each car's motion in response to demand only; that is, it analyzes all the possibilities and answers each call in optimum fashion. The term "optimum" depends, of course, on the system design strategy, and this varies between manufacturers. Such a system is possible only with the aid of a central computer combined with programmable microprocessor-controlled peripherals, since the amount of data that must be collected and processed is enormous.

All manufacturers attempt to optimize the parameters by which system quality is measured: specifically, to minimize interval, hall waiting time, and average trip time. How this is accomplished depends on the relative importance or weight assigned to each item of input data in an extremely complex computer algorithm. One manufacturer (Schindler's Miconic V) uses an algorithm that calculates a figure of merit for each car to answer each waiting hall call. The number arrived at represents the weighted sum of prospective passenger hall waiting time (interval and hall wait time) and traveling passenger delay (average trip time). The best figure of merit (minimum time) is selected by the group supervisory system and that car is assigned to the landing call. Another manufacturer (Armor/Kone, TM5 516) uses a somewhat simpler dynamic call allocation algorithm (see Fig. 23.10) which relies on the high-speed calculation capabilities of a central computer to make a last-moment decision as to which car answers which call. Another algorithm (Armor/Kone TMS 900), in addition to analyzing each car's capability to answer a call, calculates the effect of the decision on the overall elevator service quality for the building and uses this factor as well in the final decision.

Still another algorithm (Otis, Elevonic 411) uses a car dispatch program based on preprogrammed and learned traffic patterns modified by the history of the previous few minutes of operation. In the basic program, and within the constraints of a particular mode, each car computes its own response time to a waiting hall call, considering its own position, velocity, and car calls, and compares it to that of all other cars in the group. The car that will provide optimum hall call service time (waiting time plus trip time) answers the hall call. To facilitate this type of car "bidding" for calls, some systems are arranged so that each car controller is, in effect, a master group controller. At any one time, one car in the group acts as master, but if taken out of service, another car controller becomes the master.

In addition, this program (and that of several other manufacturers) uses an artificial intelligence module to learn traffic data and history and continuously change the car dispatching program. These changes consist in part of sectoring or zoning the building so that cars are dispatched and grouped to provide optimal service. This type of system is particularly useful in buildings with single- or multifloor use occupants, with repetitive traffic patterns. Another program option provides variable grouped or sectored floor assignments for cars leaving the lobby during periods of peak up traffic. The purpose of this function is to reduce the number of stops in a typical round trip, thereby reducing lobby waiting time.

Most algorithms permit overriding the present program mode in response to crowd sensors or to particularly heavy, concentrated, service demand such as might be the case if occupants of an entire floor left the building at an unusual hour. The algorithm possibilities are le-

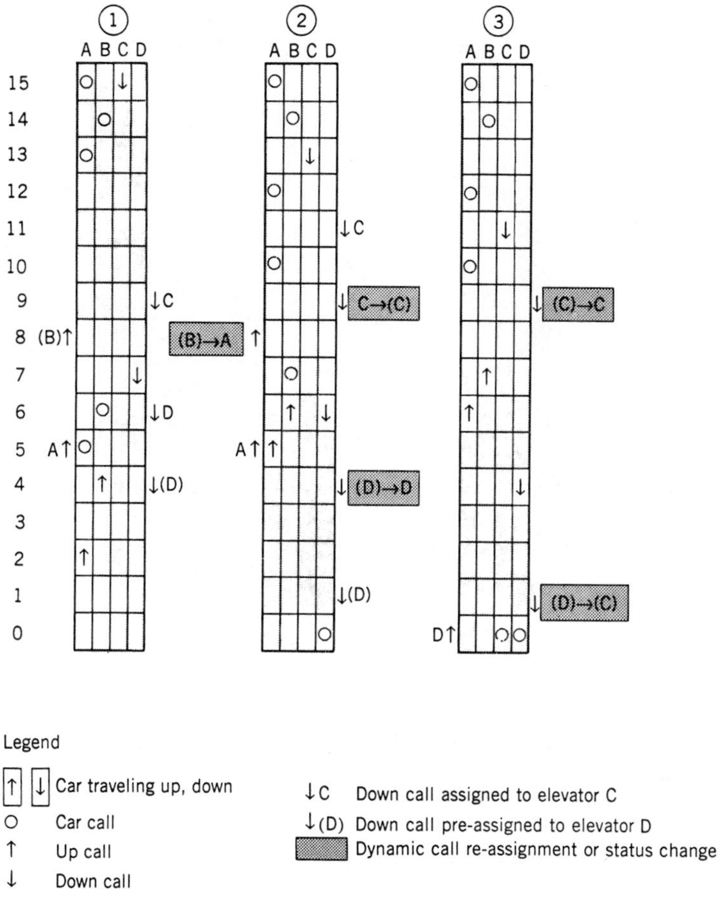

***Fig. 23.10** Simplified explanation of the algorithm logic for a group supervisory control system of medium complexity. Myriad timing calculations of acceleration, deceleration, and door operation go into each decision. These calculations are continuously repeated and updated, to provide a second-by-second picture of the entire system. (Courtesy of Kone Elevator Co.)*

TMS 516 Dynamic Call Allocation

The sequence on the left demonstrates the call assignment effect of the dynamic last moment decision principle of the TMS 516 in intensive day traffic with incoming, outgoing and interfloor traffice elements.

Stage 1: Due to coinciding car call to floor 5, A is assigned the up call on that floor, despite B being closer. Up call on floor 8 is pre-assigned to B. Vacant C is assigned the down call on floor 9. D is assigned floor 6 down call, and pre-assigned down call on floor 4.

Stage 2: A few seconds later, the up call on floor 8 is dynamically re-assigned to A as B has to stop for the intermediate car call to 7. C is assigned the new down call on floor 11, and the status of the down call on floor 9 is changed from assignment to pre-assignment. Opposite change occurs on floor 4. A new down call on floor 1 is pre-assigned to D.

Stage 3: Several seconds later, the pre-assignment of the down call on floor 1 is transferred to C, to enable fast service of the new main floor priority up call by D. Floor 9 down call pre-assignment is confirmed.

gion because of the large number of factors involved. These factors include:

1. Condition of each car, including its load; registered calls; priority status, if any; and status of its m-g set, if any.
2. Distance of each car from a registered hall call.
3. Car unloading calls ahead of and behind each hall call.
4. Special conditions, including delays, priority calls, and timing of registered lobby calls.

On the basis of all these data, the computer calculates not only which car is in the best position to answer a call but also travel time for the hall call *and* the calls already registered in that car. The computer also analyzes the traffic trends many times each second and reevaluates the importance of all factors in order to anticipate service requirements. It does this by recognizing traffic patterns as they begin to form and then comparing them to preprogrammed pattern responses for such traffic patterns as peak-up, peak-down, light, and balanced traffic. This recognition function permits anticipating the ser-

Fig. 23.11 *Building elevator traffic studies can be accomplished quickly and accurately using a traffic analyzer with multiple inputs to feed data via appropriate software into a conventional laptop computer* (a). *The unit illustrated has 64 inputs, all of which are scanned once per second, and is capable of unattended monitoring for a week. The accumulated data on call registration, response time, waiting time, maximum interval, cars in service, and so on, can be viewed visually on the computer's LCD screen or printed out as in* (b). *(Courtesy of Delta Elevator Equipment Corp.)*

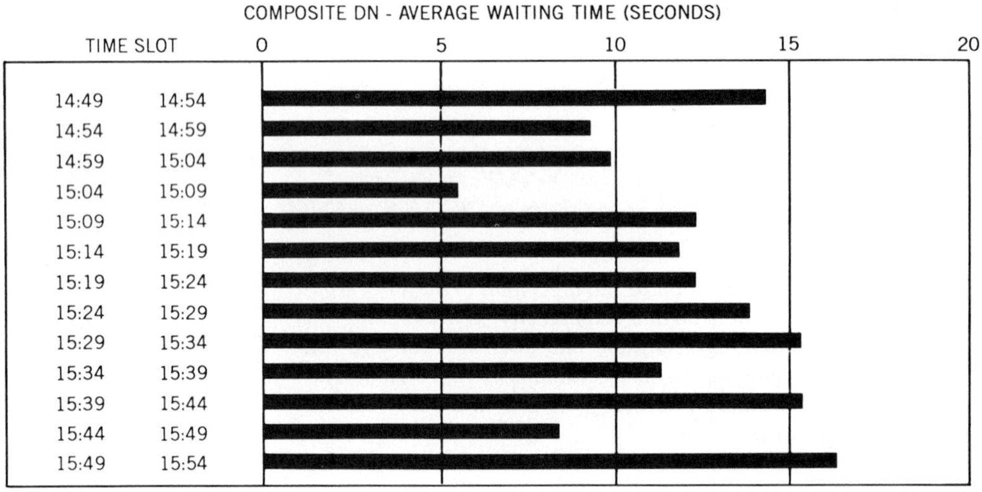

(b)

vice requirements, rather than simply reacting to them, thus improving actual service.

The actual program logic for even a small group supervisory traffic control system is beyond the scope of this book, but its guidelines are not. An adequate system should:

1. Be programmed initially for the anticipated service needs. These needs can be analyzed in existing buildings with computerized traffic analyzers (see Fig. 23.11) and in planned buildings by detailed occupancy analysis using computerized elevator system simulation (see Fig. 23.12).
2. Be reprogrammable to meet changes in building needs at nominal cost and with minimum shutdown time. It must be possible to detach cars from the system during testing, reprogramming, and routine maintenance to provide minimal service even during off-hours.
3. In case of retrofit work, be adaptable not only to existing traction equipment but also to anticipated modernization, again

***Fig. 23.12** Elevator system simulation. A computerized mathematic model of an automated elevator system permits the operator to enter building data and vary elevator control strategy. The results, displayed graphically, show the system response. (Courtesy of Otis Elevator Co.)*

without extensive outlay. When an existing group operating system is modernized by the installation of a microcomputer group controller, it is almost always also necessary to change the car controls to programmable solid-state equipment.

4. Provide for priority call calculation (based on landing waiting times), statistical analysis of traffic in order to anticipate patterns, adaptive (zoned) car parking to meet specific needs, adjustable door timing based on type of call (lobby, landing, car), backup dispatch means (in case of dispatch system failure), and automatic call cutout for constant (stuck) signals.
5. Provide means for lobby and management office viewing (monitor) and obtaining hard copy (printer output) of elevator traffic information in both real time and stored (see Fig. 23.13). This data-handling equipment must provide a fault mode that will store, display, and diagnose system operating faults. The owner/manager unit may have on-site control and reprogramming capabilities, if so required.
6. Provide additional functions such as emergency power elevator selection and control (see Section 24.4), priority service, selective hall/car call cutout, swing (separate) car operation, and hoistway access controls for maintenance.
7. Provide, where specifically required by the building management, a riot control feature (access limitation at entrance levels), crossover floor operation in a zoned system, convention service (intense short-time usage at selected levels), and controlled access at given floors and for specific occupants.
8. Be fully coordinated with the fire protection system in accordance with the local fire regulations (see Section 24.5).
9. Act in consonance with the elevator security equipment in accordance with predetermined security measures so that operation of security/alarm devices will initiate automatic elevator motion control procedures. This too must be coordinated with the local security authorities, and the automatic procedures must be subject to manual override (see Section 24.6).

Proper operation of the system should also result in:

1. All floors getting equal service, including the basement, if desired.
2. Proper handling of multifloor tenants in office buildings, to permit efficient interfloor traffic.
3. Appropriate action in emergencies, such as general or local circuit power failure or any type of abnormal car or signal operation.

23.24 Rehabilitation Work: Performance Prediction

The primary reason for rehabilitating an existing elevator system is to improve its operating performance. As mentioned above, the traction machinery has extremely long life, particularly in the gearless configuration. Thus the usual rehabilitation project consists of replacing the m-g set, car control panel, and group controller with solid-state, microprocessor-controlled programmable equipment, while retaining the original

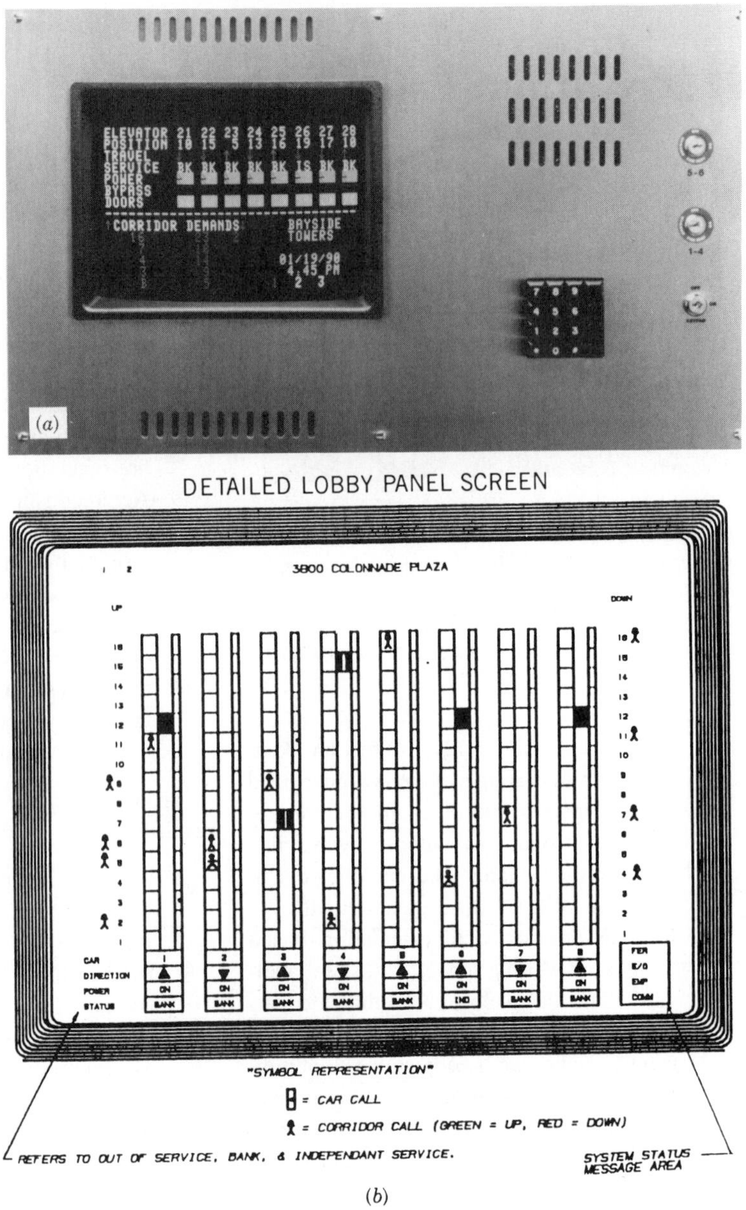

(b)

Fig. 23.13 *Elevator system lobby panel, also known as a traffic director's station, or as in illustration* (a)*; Schindler's Building Management Interface unit. The keypad at the lower right of the photo permits the operator to select the desired display. He can display elevator system status of selected banks in tabular form* (a) *or graphically* (b). *Displayed data included car status, all waiting calls, special car modes, and system modes and messages. From the unit illustrated he can enable and disable various system modes and identify any malfunctions. Other building management units (not illustrated) permit the operator to alter system operation by changing its programming. (Illustrations courtesy of Schindler Elevator Corp.)*

traction equipment. The car door operator, which controls door action, is usually also replaced, since door opening and closing characteristics are an important factor in overall trip time, and therefore system performance.

Using computers, major manufacturers have developed elevator system simulators. These devices enable architects and owners to input proposed (or existing) building data, and to receive visual and printed readout on the performance of proposed systems (see Fig. 23.12). Since such programs are interactive, the operator can change the input data and the characteristics of the proposed equipment until the desired performance level is reached. These devices are particularly useful in modernization work, since the owner will see in advance the operation of a proposed system *in his building* and can make the required decisions on this basis.

23.25 Lobby Control Panel

A lobby control panel (also called a traffic director's panel or an information control center) is furnished with each bank of elevators and is usually mounted in a readily accessible and easily visible position in the building lobby.

The contents of the panel are generally these:

1. An information module that gives information on corridor calls, special status of cars, car movement direction and location of cars. The information can be displayed in tabular form (Fig. 23.13*a*) or graphically (Fig. 23.13*b*).
2. A communications compartment that permits two-way communications with cars and other selected locations.
3. A switch section containing a "power-on" and "in-service" switch for each car in the bank, plus special switches.

The panel provides for manual intervention to permit special types of operation. Thus a car can be:

1. Arranged to travel without operating the usual audible and visual signals (*inconspicuous riser*).
2. Taken out of supervisory control and operated manually (*attendant or independent service*).
3. Selected for night or weekend service while the other cars are shut down.
4. Assigned to a particular floor on a fixed- or priority-basis call (*convention feature or priority*).

Other operational features that can be provided are concerned with emergency service, including the "fireman's return" feature required by ANSI and many local fire codes (Section 24.5) and the controls related to switching of power between cars in the event that operation of an emergency generator is necessary.

Fig. 23.14 *Car panel ergonometrically designed for convenient use by normal as well as handicapped passengers. Each floor button is lighted and adjoins a Braille plate (raised numbers). Four special buttons—door-open, door-close, alarm, and emergency stop—are placed at the bottom of the panel and marked with internationally recognized symbols. Firemen's controls and special access switches are at the top of the panel, along with speaker/microphone. This panel is also equipped with a voice synthesizer and an upper panel that can display any desired information, in addition to showing the car location. Note also the arrow on the door jamb, which clearly indicates the direction of car travel. (Photo courtesy of Otis Elevator Co.)*

23.26 Car Panel

A typical modern car panel is illustrated in Fig. 23.14. Every car panel (station) is equipped with full-access buttons for call registry, door-open and alarm, and switches for emergency stop and fireman control. Also always provided behind an unlocked door is a telephone or intercom device that permits communication with the lobby panel. A door-close button is sometimes provided if extensive hand operation of the car is anticipated. It is activated only when the car is under manual control. Controls that do not concern the normal passenger are grouped in a locked compartment in the car station. These include hand-operation switch, light, fan and power switches, and any special controls such as security and emergency devices. Finally, a compartment accessible only to technicians contains the devices controlling door motion, car signals, door and car position transducers, load-weighing control, door and platform detection beam equipment, speech synthesizer (if used), and visual displays.

ELEVATOR SELECTION

23.27 General Considerations

The selection of elevators for any but the simplest buildings requires the simultaneous consideration of several factors: adequate elevator service for the intended building usage, the economics of elevator selection, and the architectural integration of spaces assigned to elevators, including lobbies, shafts, and machine rooms. As must be obvious, these factors are interdependent; therefore, in large complex buildings many combinations are possible. The selection of a single optimum system for such cases is most practical with the aid of a computer, or simulator. For most buildings, however, certain guidelines can yield entirely satisfactory results with hand computation. These guidelines will be developed and explained below. The criteria of elevator service quality are:

1. Interval and average waiting time.
2. Handling capacity.
3. Travel time.

23.28 Definitions

A clear definition of terms, including variant usages, is imperative for the proper study of a subject. To that end, an abbreviated list of important definitions for our study follows.

Interval (I) or Lobby Dispatch Time. Average time between departure of cars from the lobby.

Average Lobby Time or Average Lobby Waiting Time. Average time spent by a passenger between arriving in the lobby and leaving the lobby in a car.

Registration Time. Waiting time at an upper floor after registering a call.

Round-Trip Time (RT). Average time required for a car to make a round trip, starting from the lower terminal and returning to it. The time includes a statistically determined number of upper-floor stops in one direction and, *when calculating elevator requirements based on up-peak traffic,* an express return trip.

Travel Time or Average Trip Time (AVTRP). Average time spent by passengers from the moment they arrive in the lobby to leaving the car at an upper floor.

Handling Capacity (HC). Figure of merit for an elevator system, indicating the maximum number of passengers that can be handled in a given period, usually 5 min; thus "5-min handling capacity." It is expressed as a percentage of the building population.

Zone. Group of floors in a building that is considered as a group, with respect to elevator service. It may consist of a physical entity—a group of upper floors above and below which are blind shafts—or it may be a product of the elevator group control system, changing with system needs.

23.29 Interval or Lobby Dispatch Time, and Lobby Waiting Time

In an ideal installation, at least from the riding public's point of view, a car would be waiting at the lower terminal on the rider's arrival or would be available after a short wait. Since cars leave the lobby separated in time by the *interval (I)* and passengers arrive at the lobby in random fashion, the average waiting time in the lobby should be half the interval. Field tests show, however, that it is actually higher than this, and the figure most used in the industry is 60%; that is,

$$\text{average lobby waiting time} = 0.6I$$

Excellent office building design provides an average lobby wait of 15 to 18 s during up peaks, with 22 s considered good and 26 borderline.

Since some control systems zone the building in such a way that some cars do not return to the lobby, the interval as a figure of merit can be somewhat misleading. As a basis of comparison, however, a 30-s interval is excellent, 35 s borderline, and 40 s just acceptable in other than center-city office spaces. Table 23.3 lists acceptable intervals for various building types.

With intervals in this range, riders will not be conscious of any irksome delay in elevator service. Consciousness of delay is considered a major drawback in rental desirability and should be sedulously avoided for all conditions of traffic except during morning and evening peaks, when a certain delay is expected and therefore tolerated, however grudgingly. Even in both these cases, a modern group supervisory system will recognize any *timed-out* call, that is, a call with registration time exceeding 50 s, as a priority call. Priority calls are answered by the first available car, usually within 15 s. If a considerable amount of interfloor traffic is expected during peak periods, as may be the case when a large company occupies several upper floors, elevator service must be increased by 20 to 40% over that calculated to maintain proper intervals.

23.30 Handling Capacity

The frequency, or interval, with which a car appears at the lobby is one of the two factors that determine the passenger capacity of an elevator system. The other is obviously the size or capacity of the elevator car. The systems's *handling capacity* is completely determined by these two factors—car size and interval, and is independent of the number of cars. This can be best understood by visualizing the system as a single set of doors that open periodically (interval) to remove a given number of passengers (car capacity) from the waiting group. Whether that set of doors represents a single car or many cars that take turns is immaterial. The only factors that fix handling capacity are passenger load (car capacity) and frequency of loading (interval) (see Table 23.4).

Note that cognizance is taken of the fact that during peak traffic periods cars are not loaded to maximum capacity, but only to about 80%—a figure determined by actual count in many existing installations.

As a convenient measure of capacity, the handling capacity of a system for 5 min is taken as a standard. This is because a 5-min rush period is used as a measure of a system's ability to handle traffic. This may be expressed thus:

TABLE 23.3 **Suggested Elevator Intervals**

Facility	Interval (s)
Office Buildings	
Center city	25–30
Investment	28–32
Suburban	32–35
Residential	
Prestige apartments	50–70
Middle-income apartments	60–80
Low-income apartments	80–120
Dormitories	60–80
Hotels—1st quality	30–50
Hotels—2nd quality	50–70

TABLE 23.4 **Car Passenger Capacity (p)**

Elevator Capacity (lb)	Maximum Passenger Capacity	Normal Passenger[a] Load per Trip
2000	12	10
2500	17	13
3000	20	16
3500	23	19
4000	28	22

[a]The number of passengers carried on a trip during peak conditions is approximately 80% of the car capacity.

TABLE 23.5 **Minimum Handling Capacities (HC)**

Facility	*Percent of Population to be Carried in 5 Min*
Office Buildings	
Center city	12–14
Investment	11.5–13
Single purpose	14–16
Residential	
Prestige	5–7
Other	6–8[a]
Dormitories	10–11
Hotels—1st quality	12–15
Hotels–2nd quality	10–12

[a]Due to more urgent traffic demands, particularly at the school and work exodus.

TABLE 23.6 **Population of Typical Buildings for Estimating Elevator and Escalator Requirements**

Building Type	*Net Area*
Office Buildings	*Square feet per person*
Diversified	
Large lower floors	140–160[a]
Upper floors	160–180
Average use	160
Single purpose	150
Hotels	*Persons per sleeping room*
Normal use	1.3
Conventions	1.9
Hospitals	*Visitors and staff per bed*[b]
General private	3
General public (large wards)	3–4
Apartment Houses	*Persons per bedroom*
High-rental housing	1.5
Moderate-rental housing	2.0
Low-cost housing	2.5–3.0

[a]Density may vary for different floors. Clerical and stenographic area may have a population density as high as 70 square feet per person.

[b]If visiting hours are restricted, visitor population will determine elevator requirements. If visiting is not restricted to a certain few hours, staff requirements may determine elevator design. Where traffic is heavy, a combination of passenger cars and larger "hospital" cars should be used to provide optimum service.

TABLE 23.7 **Office Building Efficiency**

Building Height	*Net Usable Area as Percentage of Gross Area*
0–10 floors	Approximately 80%
0–20 floors	Floors 1–10 approximately 75%
	11–20 approximately 80%
0–30 floors	Floors 1–10 approximately 70%
	11–20 approximately 75%
	21–30 approximately 80%
0–40 floors	Floors 1–10 approximately 70%
	11–20 approximately 75%
	21–30 approximately 80%
	31–40 approximately 85%

NOTE: Applicable to buildings with 15,000 to 20,000 gross square feet per floor.

Source: Reprinted from G. R. Strakosch, *Vertical Transportation, Elevators and Escalators,* 2nd ed. New York: Wiley, 1983.

handling capacity (HC)

$$= \frac{300 \text{ s} \times \text{passengers/car}}{\text{interval}}$$

or

$$HC = \frac{300p}{I} \tag{23.1}$$

where p is individual car loading. When the interval is 30 s, the system handling capacity is $10p$, a convenient figure to remember.

To establish a figure of merit for building service, system capacity HC must be related to building size. This is normally done by establishing a minimum percentage of building population that the system will handle in 5 min. A good system for a diversified office building will handle no less than 12% of the building population. Similar figures are shown in Table 23.5 for various types of facilities.

In planning a building the population must, of course, be estimated. This is particularly difficult in speculative-type, diversified-use buildings. However, based on rental cost, area, and building type, a fair estimate can be made. Population estimates for office buildings are based on net area, that is, actual available area for tenancy. Table 23.6 gives suggested density figures, while Table 23.7 gives average office building efficiency figures for use in calculating net area.

23.31 Travel Time or Average Trip Time

The average trip time or time to destination is the sum of the lobby waiting time plus travel time to the median floor stop. Car round-trip time is also used as a criterion but is not as relevant or meaningful as trip time. In a commercial atmosphere a trip of less than 1 min is highly desirable, a 75-s trip acceptable, a 90-s trip annoying, and a 120-s trip the limit of toleration. Obviously, in the more relaxed atmosphere of a residence, where interval alone can account for a minute or more of trip time, these maxima are revised upward. From Fig. 23.15 we see that the 2000-lb and 2500-lb cars used in residential buildings can have a 17-story rise, even with 60-s interval, without excessive trip time. On the other hand, the 3500-lb car, which is almost universal in office buildings (Fig. 23.16), is limited to a maximum of 16 floors local run before exceeding the 90-s limit and about 6 to 8 floors to stay within the 75-s criterion.

An important reservation on the foregoing statements must be noted. The curves of Figs. 23.15 and 23.16 are based on statistical calculations, empirical data, and field observations, as discussed in the next section. This being so, the average values that these figures give should be considered to be ±15%, and borderline cases can be shifted either way. Designs that show high travel time on paper frequently work out well in the field, because lobby loading is often less than 80%, upper floor stops are less than the statistical figure due to groups of people going to the same floor, and staggered working hours relieve traffic peaks. Also, a feature called high-call reversal takes account of the fact that cars do not travel to the top on each trip, but reverse at the top call. This can reduce average trip time by 5 to 10%. Finally, sophisticated solid-state traffic controls allow more rapid acceleration and deceleration without discomfort, variable door closing time, and more efficient selection of landing call answers, all of which can further reduce trip time another 5%.

23.32 Round-Trip Time

The figure for round-trip time during up-peak traffic conditions, which are those used for calculating elevator requirements, is composed of the sum of four factors. They are: time to accelerate and decelerate, time to open and close doors at all stops, time to load and unload, and running time (see Figs. 23.17 to 23.19). It is physically the time consumed by a car from door opening at the lower terminal to door opening at the same terminal at the end of a round trip. Since the actual number of stops made by a car is unknown, a statistical probability figure is used, based on the passenger capacity of the car and number of local floors above the lower terminal. In calculating this round-trip time (*RT*), it is assumed that a car will depart the lower terminal when loaded. No intentional delay is included at either lower or upper terminal. The *RT* thus calculated is a median figure, with any single actual round trip taking more or less time. In detail, round-trip time consists of the time expended in:

1. Loading at the lobby.
2. Door closing at the lobby.
3. Accelerating from the terminal and from each stop.
4. Decelerating at each stop.
5. Passenger transfer at each stop.
6. Door operations at each stop.
7. Running time at rated speed between stops.
8. Return express run from the last stop.

These figures are obtained as follows:

1. *Field observations:*—items 1 and 5 are based on a 3 ft, 6 in. door opening. A smaller door opening increases passenger transfer time.
2. *Calculations:*—items 2, 3, 4, 6, 7, and 8.

Door-closing time is based on a 3 ft, 6 in. center-opening door with adjustable speed.

Acceleration and deceleration times are calculated with a maximum of 4 ft/s/s because anything beyond that results in physical discomfort to the passengers.

Running time at rated speed takes place after the car has accelerated and before it begins to decelerate. If we consider that it takes between 20 and 30 ft to accelerate to 700 fpm, depending on rate of acceleration, we see that in local runs the car never gets to rated speed. It simply

TRANSPORTATION

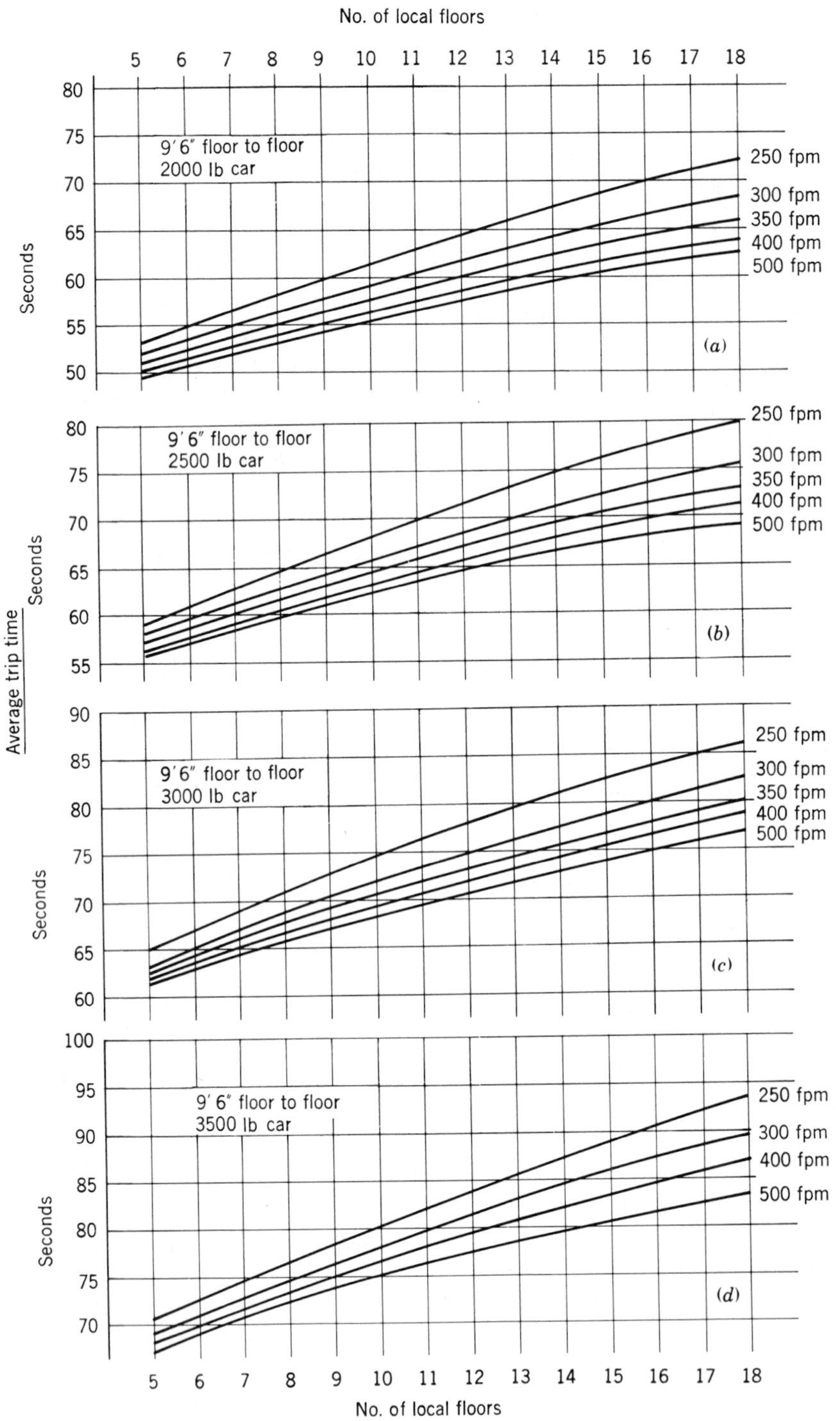

Fig. 23.15 Plots of average trip time for various car speeds and capacities with 9-ft 6-in. floor height and 30-s interval.

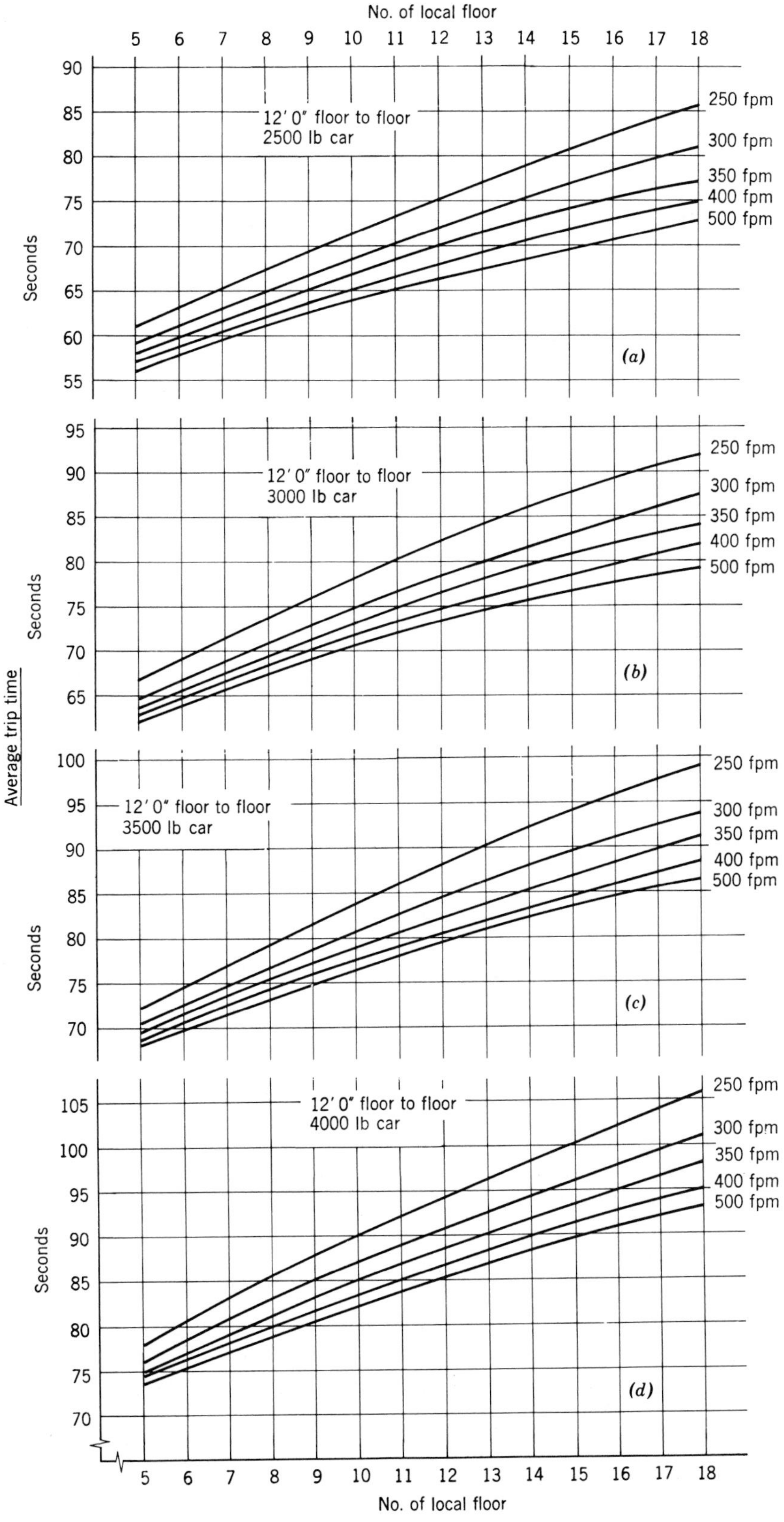

Fig. 23.16 Plots of average trip time for various car speeds and capacities for 12-ft floor height and 30-s interval.

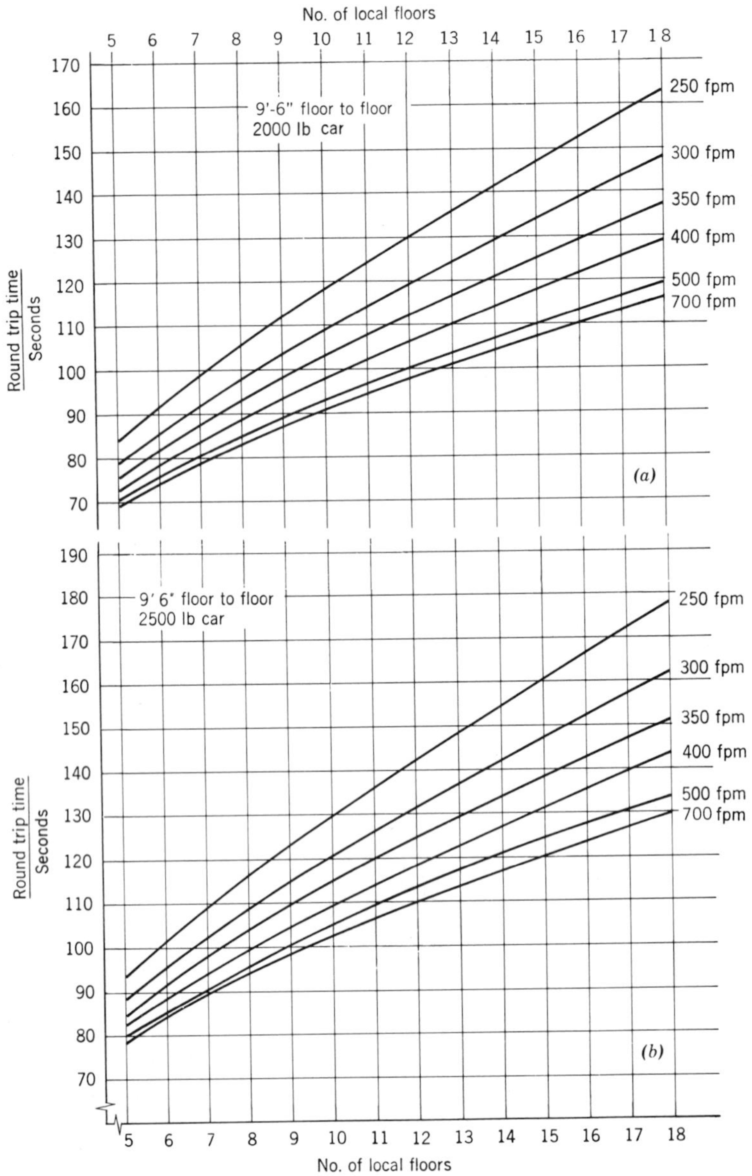

Fig. 23.17 *Plots of round-trip time for various car speeds and capacities with 9-ft 6-in. floor heights and 30-s interval.*

accelerates and decelerates. The higher speed equipment with its larger motor accelerates more quickly and gives some time advantage on the return express run, but has no great time advantage overall. This accounts for the small reduction above 500 fpm in Figs. 23.17 and 23.18.

In calculating round-trip time for cars in upper zones, it is necessary to know the time required to traverse the express floors. This may be obtained from Fig. 23.19. The times given therein are for *one-way* express runs. Thus to calculate *RT* for an upper zone car, take the *RT* corresponding to the upper local floors and add *twice* the figure obtained for express run time from Fig. 23.19.

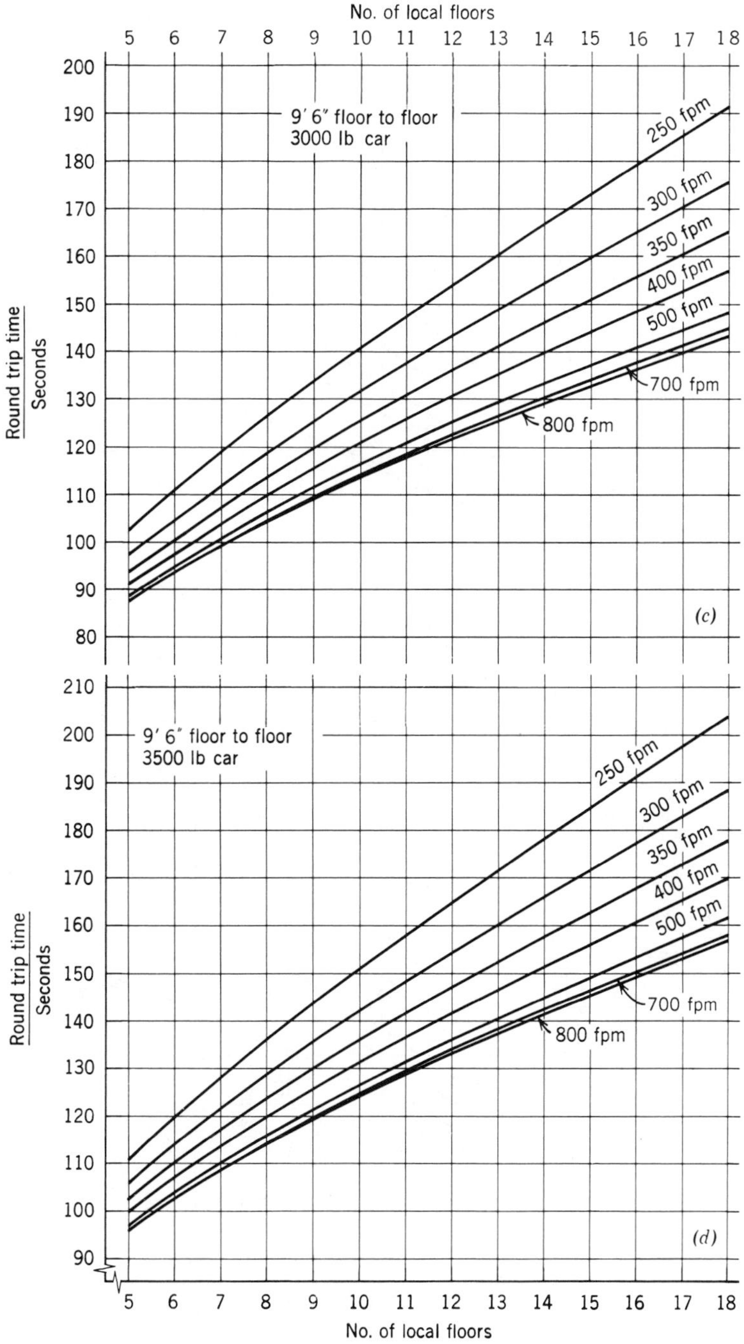

Fig. 23.17 *(continued)*

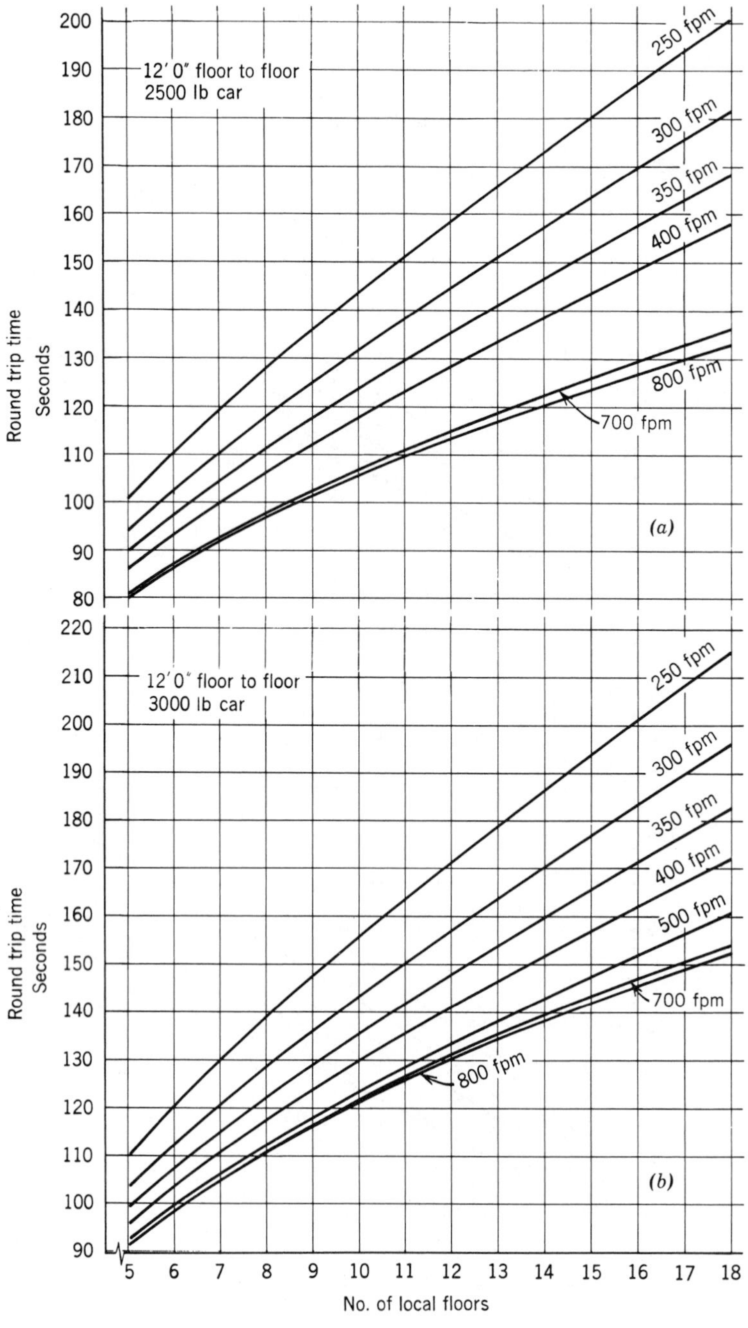

Fig. 23.18 Plots of round-trip time for various car speeds and capacities with 12-ft floor height and 30-s interval.

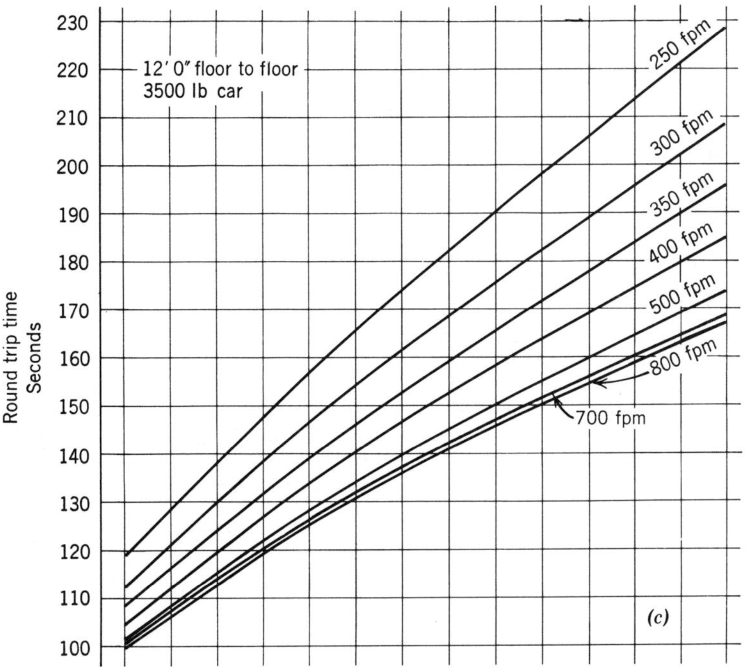

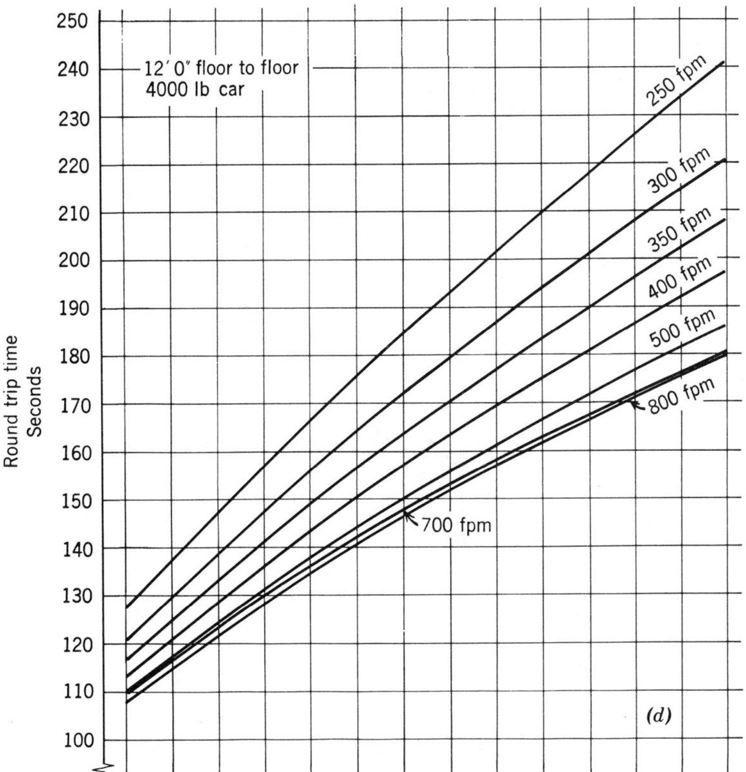

Fig. 23.18 *(continued)*

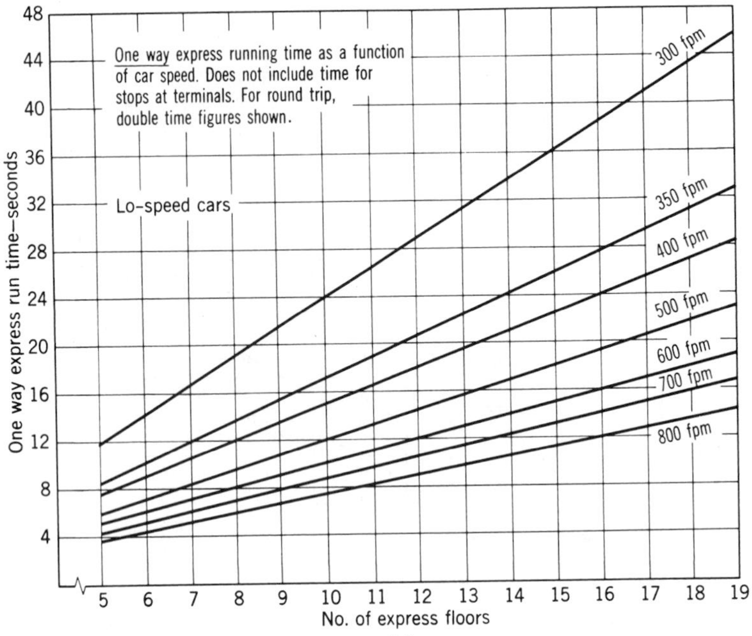

(a)

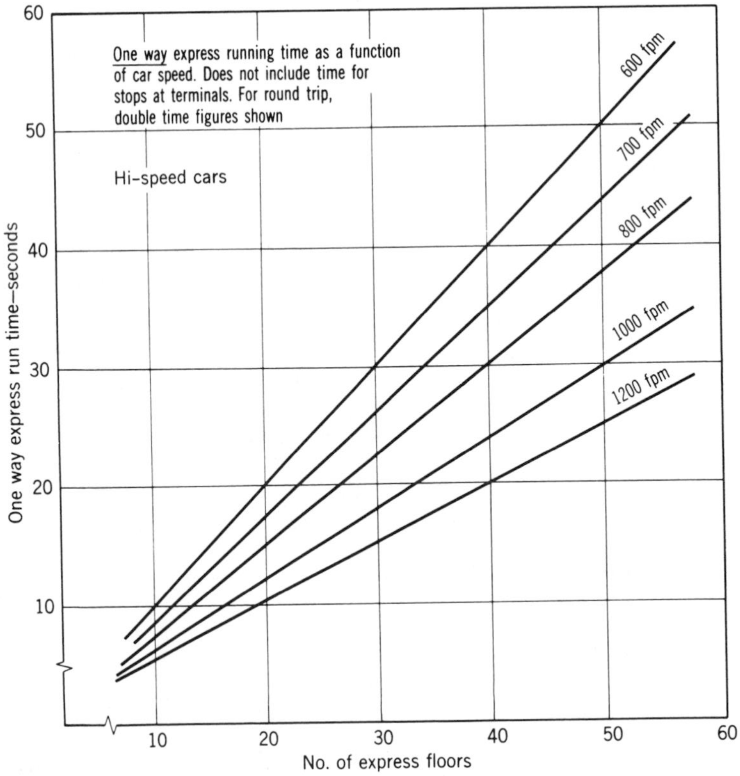

(b)

Fig. 23.19 *One-way express running time, not including terminal time.*

23.33 System Relationships

The symbols that will be used in all elevator discussions are:

p	Individual car capacity, equal to 80% of maximum during peak hours.
h	5-min capacity of a single car.
N	Number of cars in a system.
HC	System 5-min handling capacity, expressed in number of persons.
RT	Round-trip time, in seconds.
$AVTRP$	Average trip time, in seconds.
I	Interval, in seconds.
D	Population density, in square feet per person.
PHC	Percent of population to be moved in 5 min, expressed as a percentage figure.

Having considered the definition of interval, handling capacity, average and round trip time, it is well to demonstrate at this point the interrelationships between these quantities, and the other equations governing the remaining factors that define elevator systems. We have previously established that handling capacity is determined by car capacity p and interval I:

$$HC = \frac{300p}{I} \tag{23.1}$$

In a system comprising a single car the interval (I) is obviously equal to the round trip time (RT). In a system with more than one car, the interval will be reduced in proportion to the number of cars. Thus,

$$I = \frac{RT}{N} \tag{23.2}$$

The 5-min handling capacity (h) of a single car is then

$$h = \frac{300p}{RT} \tag{23.3}$$

if we remember that for a single car, its interval is its round trip time. It follows then that if the handling capacity of a single car is h, then the handling capacity of N cars is N times as much. Thus

$$HC = N \times h$$

or

$$N = \frac{HC}{h} \tag{23.4}$$

23.34 Car Speed

The selection of the car speed to be used is a matter of trial and error, the final selection being that required to give a round-trip time (RT) which will in turn give an acceptable interval. However, in order to establish a starting point, it has been found that a minimum car speed corresponding to a given building height, or in elevator parlance—rise, can be established. Similarly, although car size can be selected at any value, it has been shown that for certain facility types, specific-size cars are indicated. These recommendations are tabulated in Table 23.8.

It should also be borne in mind that elevator equipment falls into distinct speed categories. Thus most manufacturers will use geared equipment through 400 fpm and gearless thereafter. The next category is 500 fpm gearless, followed by 600 to 700 fpm, and so on. Therefore, it is wise to avoid moving into the next higher and more expensive equipment category, if possible. This may mean exceeding recommended interval or dropping slightly low in handling capacity. It will be found, however, that this can be done without injury to the elevator system, provided a good group supervisory control system is employed.

23.35 Single-Zone Systems

Having established the relationships that govern the design and performances of an elevator system, comprising a single zone, it would be helpful to follow through an illustrative example.

EXAMPLE 23.1. Office building, downtown, diversified use, 14 rentable floors above lobby, each 12,000 ft net. Floor-to-floor height—12 ft. Determine a workable elevator system.

SOLUTION. From Table 23.5, average HC is 13%. From Table 23.3, maximum recommended

TABLE 23.8 **Elevator Equipment Recommendations**

Usage		Car Capacity (lb)	Minimum Car Speed (fpm)	Car Travel (feet)
Office buildings			350–400	0–125
	Small building	2500	500–600	126–225
	Medium building	3000	700	226–275
	Large building	3500	800	276–375
			1000	Above 375
Hotels		2500		
		3000	As above	As above
Hospitals			150	0–60
			200–250	61–100
		3500	250–300	101–125
		4000	350–400	126–175
			500–600	176–250
			700	Above 250
Apartment houses[a]			100	0–75
		2000	200	76–125
		2500	250–300	126–200
			350–400	Above 200
Retail stores			200	0–100
		3500	250–300	101–150
		4000	350–400	151–200
		5000	500	Above 200

[a]FHA minimum requirements call for full-collective variable voltage control, minimum of two cars, and approximately 120 bedrooms per car, for all buildings exceeding seven stories in height.

interval is 30 s. From Table 23.6, average population density is 160 ft^2 per person.

Building population:

$$\frac{14 \text{ floors at } 12{,}000 \text{ ft}^2}{160 \text{ ft}^2 \text{ per person}} = 1050 \text{ persons}$$

Suggested minimum handling capacity:

$$PHC = 13\%$$
$$HC = 0.13 \times 1050 = 137 \text{ persons}$$
$$\text{rise} = 14 \text{ floors at } 12 \text{ ft} = 168 \text{ ft}$$

From Table 23.8 we select a recommended car size of 3000 lb at 500 fpm.

3000 lb
500 fpm

Then from Figs. 23.18*b* and 23.16*b*:

$$RT = 143 \text{ s} \qquad AVTRP = 76 \text{ s}$$

Single-car capacity $h = 300p/RT$ (see Table 23.3 for p):

$$h = \frac{300(16)}{143} = 33.5 \text{ persons}$$

$$N = \frac{HC}{h} = \frac{137}{33.5} = 4.1 \text{ cars, say } 5$$

(four cars would give an excessive interval)

$$I = \frac{RT}{N} = \frac{143}{5} = 28.7 \text{ s}$$

$$\text{actual } PHC = \frac{5(13\%)}{4.1} = 15.9\%$$

These figures are excellent.

We should try faster cars, to reduce the number of cars, or slower cars, to reduce car cost. We select 700 fpm and 400 fpm.

3000 lb
700 fpm

$$RT = 140 \qquad AVTRP = 73\text{ s}$$

$$h = \frac{300(16)}{140} = 34.3\text{ persons}$$

$$N = \frac{137}{34.3} = 4\text{ cars}$$

$$I = \frac{RT}{N} = \frac{140}{4} = 35$$

$$\text{actual } PHC = \frac{4(13\%)}{4} = 13\%$$

This solution, that is, 4 cars at 700 fpm, 3000 lb, is marginally acceptable. The interval is somewhat high, but the handling capacity is acceptable. Now, trying the slower car:

3000 lb
400 fpm

$$RT = 152, \qquad AVTRP = 77\text{ s}$$

$$h = \frac{300(16)}{152} = 31.6\text{ persons}$$

$$N = \frac{137}{31.6} = 4.34\text{ say 5 cars}$$

$$I = \frac{152}{5} = 30.4\text{ s}$$

$$\text{actual } PHC = \frac{5(13\%)}{4.34} = 15\%$$

This is also an acceptable arrangement. Thus, of the three arrangements, the two most likely choices are

4 cars, 3000 lb, 700 fpm

5 cars, 3000 lb, 400 fpm

To reduce the interval of the four-car solution, we can try a high-speed smaller car, since a larger car will *increase* the interval.

2500 lb
700 fpm

$$RT = 122\text{ s} \qquad AVTRP = 66\text{ s}$$

$$h = \frac{300(13)}{122} = 32\text{ persons}$$

$$N = \frac{137}{32} = 4.28\text{ cars}$$

Using four cars:

$$I = \frac{122}{4} = 30.5\text{ s}$$

$$PHC = \frac{4}{4.28}(13) = 12.1\%$$

Handling capacity is insufficient. Using five cars:

$$I = \frac{122}{5} = 24.4\text{ s}$$

$$PHC = \frac{5}{4.28}(13\%) = 15.19\%$$

Of these two solutions only the five-car arrangement is satisfactory, since four cars yield insufficient *PHC*. However, because interval is low and *PHC* high, we can drop the speed and still obtain a workable arrangement:

2500 lb
400 fpm

$$RT = 139\text{ s} \qquad AVTRP = 71\text{ s}$$

$$h = \frac{300(13)}{139} = 28.0\text{ persons}$$

$$N = \frac{137}{28.0} = 4.9,\text{ say 5 cars}$$

$$I = \frac{139}{5} = 27.8\text{ s}$$

$$PHC = \frac{5}{4.9}(13\%) = 13.3\%$$

which is acceptable, and obviously cheaper than five cars of 3000 lb capacity and 400 fpm speed.

We thus have two solutions to compare:

1. Four 3000-lb cars at 700 fpm, gearless, giving an interval of 35 s and a handling capacity of 13%.
2. Five 2500-lb cars at 400 fpm, geared, giving an interval of 27.8 s and a handling capacity of 13%.

The final selection should at this point be made on the basis of cost and core layout. When considering cost, note that first cost is the governing factor only in a speculative venture. With an owner-operator building, the cost comparison should be on a life-cycle basis. Cost figures must reflect the impact of elevator space requirements on net rentable area in the building.

TABLE 23.9 **Relative First Cost[a] Figures for Passenger Elevators at Various Speeds (fpm) (for a Given Drive System)**

Car Size	*Hydraulic*	*Geared Traction*			*Gearless Traction*			
(lb)	*100*	*200*	*350*	*500*	*500*	*700*	*1000*	*1200*
2000	40	80	100	130	165	170	220	235
2500	43	85	115	145	175	180	235	250
3000	50	90	120	150	180	185	250	265
3500	58	95	125	155	190	195	265	275
4000	60	100	135	165	200	205	280	300
4500[b]	70	120	150	185	225	230	300	325
5000[b]	75	130	160	200	240	250	330	350

[a]Costs are ±10%; based on standard fixtures, cabs, and entrances, and average rise for the speed indicated.

[b]Service elevator or hospital elevator.

NOTE: See Table 23.8 for speed/rise recommendation.

Comparative relative cost figures are given in Table 23.9. Utilizing these data, we have:

1. Four 3000-lb cars at 700 fpm = 4 × 185 = 740 cost units.
2. Five 2500-lb cars at 400 fpm = 5 × 125 = 625 cost units.

The choice, *on a first-cost basis*, is clearly in favor of the geared 400-fpm units. However, the costs are sufficiently close that in a life-cycle cost analysis, even without considering the comparative space cost, which favors the four-car bank, the lower maintenance and energy costs of the gearless machine might reverse the results. Furthermore, 2500 lb cars are generally considered too small for office building use except for small suburban buildings not exceeding 10 floors as seen in Table 23.8. The recommendations in Table 23.8 are based on extensive field surveys and experience.

As mentioned in Section 23.23 and shown in Fig. 23.12, planners of new buildings today generally take advantage of elevator selection software provided by consultants, manufacturers, or by in-house capabilities. The results of one such program as applied to the problem just solved manually are shown in Fig. 23.20. This particular program was prepared by Otis Elevator Co. Note that the results in Fig. 23.20*a* correspond exactly to the results obtained manually (compare with the first trial calculation in the solution above). The correspondence is not surprising since the round-trip curves in Fig. 23.18 are based on a 3.3-s door time and a 4.0-ft/s^2 car acceleration, as shown in the figure. Note that high call reversal occurs at the top floor [13.5 (i.e., 14)] and that the number of up stops is 10 (9.7 from statistical calculations). The up-peak calculation assumes no counterflow traffic (i.e., an express down run and no interfloor traffic), as shown. These can be added, however, yielding much different results, as shown in Fig. 23.20*b*. We see there that counterflow traffic (down stops) of only 3% and interfloor traffic of 1%, both of which are reasonable, change the interval and handling capacity to borderline acceptability.

23.36 Multizone Systems

In general, buildings with less than 15 stories are elevatored with a single zone (i.e., all cars serve all floors), and buildings with more than 20 stories are split into two or more zones. The area in between—16 to 19 stories—can go either way, depending on population density and required interval. Actually, a modern group supervisory system can automatically zone a building when traffic requires it. However, such an arrangement, although efficient, is expensive both as to

```
OtisPlan Single Group Performance

Single Deck Up Peak

name: Sample problem
floors above lobby: 14                 population per floor:    75people
lobby height:   12.0ft                 % counterflow: 0.0
average floor height: 12.0ft           % interfloor:   0.0
express zone height:     0.0ft         number of cars: 5
door time:   3.3sec                    car speed:     500.0fpm
12ft flight time:   4.3sec             car capacity: 3000lbs
car acceleration: 4.00ft/sec/sec       up car loading: 16people/car
high call reversal:   0.0
up probable stops:   0.0               added trip time:   0.0sec

(Use 0 to calculate high call
 reversal and probable stops)
                              *** RESULTS ***
round trip time:          144.0sec     high call reversal:      13.5
interval:                  28.8sec     up probable stops:        9.7
up handling capacity:     15.9%/5 min
```

(*a*)

```
OtisPlan Single Group Performance

Single Deck Up Peak

name: Sample problem
floors above lobby: 14                 population per floor:    75people
lobby height:   12.0ft                 % counterflow: 3.0
average floor height: 12.0ft           % interfloor:   1.0
express zone height:     0.0ft         number of cars: 5
door time:   3.3sec                    car speed:     500.0fpm
12ft flight time:   4.3sec             car capacity: 3000lbs
car acceleration: 4.00ft/sec/sec       up car loading: 16people/car
high call reversal:   0.0
up probable stops:   0.0               added trip time:   0.0sec

(Use 0 to calculate high call
 reversal and probable stops)
                              *** RESULTS ***
round trip time:          176.1sec     high call reversal:      13.5
interval:                  35.2sec     up probable stops:        9.7
up handling capacity:     13.0%/5 min

                     Alt-H - Help Information
```

(*b*)

Fig. 23.20 *Printout of the results of a computerized elevator selection program.* (a) *Up-peak conditions, assuming no down stops (0% counterflow) and no interfloor traffic. The results correspond very closely to those obtained manually in Example 23.1 (see text).* (b) *Addition of even light counterflow and interfloor traffic can seriously affect the system's performance, as can be seen from the round trip, interval, and handling capacity figures. (Courtesy of Otis Elevator Co.)*

equipment and from a construction viewpoint since it does not take advantage of the considerable savings engendered by blind lower shaftways for upper-zone elevators. Analysis of multizone systems is complex and today is rarely done by hand. The interested reader is referred to Stein et al. (1986) for a detailed explanation of the technique involved. Most designers and consultants use one of the many extant computerized simulation and selection programs. These have the advantage that in addition to using basic criteria and building parameters, they can also consider the effect of variations in traffic control. Furthermore, the best of these programs can evaluate the engineering and economic impact of such factors as varying rental rates for different floors, rental space, machine room and hoistway space, core layout, and the structural ramifications of the elevator system.

23.37 Other Elevator Selection Considerations

For commercial buildings rough costs can be estimated from Tables 23.9 and 23.10. Special considerations for various building types are given below.

(a) Office Buildings. All necessary design criteria can be selected from Tables 23.3 to 23.6. Supervisory group control is normally microprocessor based. Approximately one service car per 10 passenger cars should be provided, or alternatively one car for every 300,000 ft^2 of net area. Service cars should be 4000 lb or larger without dropped ceiling and, if also used for passenger service, equipped with wall pads. An oversized door (e.g., 4 ft, 0 in. or 4 ft, 6 in.) is particularly useful in handling furniture. Service elevators should have a shaftway door at every level plus easy access to the truck dock or other freight entrance as well as the lobby. These cars operate as service cars normally but can serve as passenger cars in peak periods to alleviate congestion and delay. This fact is particularly useful in marginal service designs.

TABLE 23.10 **Office Buildings: Cost of Elevator and Electric Work**

Item	*Number of Stories* 20	35	60
Elevator work	10.9%	11.9%	12.2%
Electric work	13.3%	12.6%	12.2%

(b) Apartment Houses. Studies indicate that apartment house traffic depends not only on population but also on location and type of tenant. Houses with many children experience a school-hour peak; houses in midtown with predominantly adult tenancy exhibit evening peaks due to the homecoming working group and outgoing amusement traffic. Normally, a single elevator will suffice, although a second car functioning as a service and/or passenger car is sometimes indicated, particularly in buildings taller than six stories. The cars may be banked or separated, as desired. If a single car is used, it should be service elevator size.

Self-service collective control is the general choice, with provision for attendant control in prestige buildings. With smaller cars and short rise, a swing-type manual corridor door is acceptable; in larger installations both the cab and corridor door should be the power-operated sliding type.

Service elevators must be large enough to handle bulky furniture and should therefore be at least 3000 lb and preferably 3500 lb, with a 48-in. door and high ceiling. Hoistways must be isolated from sleeping rooms by lobbies or other space. Similarly, machine rooms must be isolated, since the starting and stopping of motors and other machine room noises are an effective barrier to sound sleep. Security arrangements are discussed in Section 24.6.

(c) Hospitals. As mentioned in Table 23.6, the governing factor in the determination of elevator requirements may be either normal hospital traffic or visitor traffic, depending on the visiting-hour schedule. Due to the large amount of vehicular traffic such as stretcher carts, wheelchairs, beds, linen carts, and laundry trucks, the elevator cabs are much deeper than the normal passenger type. This type of car when used for passenger service holds more than 20 persons and therefore gives slow service. For this reason, it is occasionally advisable to utilize some normal passenger cars in addition to hospital-size cars, particularly in large hospitals.

The use of tray and bulk carts in food service imposes a considerable load on the elevators before, during, and after meals, and passenger service is seriously disrupted. To alleviate this congestion and delay, many architects and hospital administrators prefer the use of dumbwaiter cars or another of the many types of materials-handling system that will handle a 15½ × 20 in. food tray. These systems can also be used for transporting pharmaceuticals and other items and are discussed in Sections 24.19 and 24.20.

Elevators should be grouped centrally, although separated by type of use. Car control is normally self-service collective.

Population of the hospital may be estimated from Table 23.6. Experience has shown that a carrying capacity of 45 passengers in a 5-min period is adequate (estimating each vehicle as equivalent to 9 passengers).

Intervals should not exceed 1 min. All recommendations for service to the handicapped should be adopted (see Section 23.11).

(d) Retail Stores. Retail stores present a unique problem in vertical transportation inasmuch as the objective is partially to transport persons to specific levels and partially to expose the passengers to displayed merchandise. For this reason modern stores rely almost exclusively on escalators, with one or two small elevators intended for use by staff and handicapped persons. When for some reason it is desired to equip a store with elevators, use the recommendations shown in Table 23.8, calculated for a load of between 10 and 20% of store population. Control should be automatic, selective collective. Cars are arranged in a straight line to facilitate loading and waiting.

THE PHYSICAL PROPERTIES AND SPATIAL REQUIREMENTS OF ELEVATORS

23.38 Shafts and Lobbies

The elevator lobbies and shafts form one of the major space factors with which the architect is concerned. The elevator lobby on each floor is the focal point from which the corridors radiate for access to all rooms, stairways, service rooms, and so forth. Such lobbies obviously must be located one above the other. The ground-floor elevator lobby (also called the lower terminal) must be conveniently located with respect to main entrances. The equipment within or adjacent to this area should include public telephones, building directory, elevator indicators, and control panels.

All lobbies should be adequate in area for the peak-load gathering of passengers to ensure rapid and comfortable service to all. The number of people contributing to the period of peak load (15- to 20-min peak) determines the required lobby area on the floor.

Approximately 5 ft^2 of floor space per person should be provided at peak periods for waiting passengers at a given elevator or bank of elevators. The hallways leading to such lobbies should also provide about 5 ft^2 per person, approaching the lobby. Under self-adjusting relaxed conditions, density is about 7 ft^2 per person. However, in peak periods, crowding occurs, reducing this to 3 to 4 ft^2. An acceptable compromise is 5 ft^2 per person.

The main lower terminal of elevator banks is generally on the street-floor level, although it may be on the mezzanine level, when the elevations of the street entrances vary so that one side of the building is at mezzanine level, while another entrance is lower. Such a situation is ideal for the use of escalators, which will economically and rapidly carry large numbers of persons between levels, thus making practical and efficient a single main lower elevator terminal. The upper terminal is usually the top floor of the building. Typical dimensional data and lobby arrangements are shown in Figs. 23.21 to 23.23.

23.39 Dimensions and Weights

Manufacturers and elevator consultants will supply standard layouts for elevators including dimensions, weights, and structural loads. Furthermore, to assist in preliminary design, major manufacturers have agreed upon, and publish, a set of Standard Elevator Layouts via their trade organization, the National Elevator Industry, Inc. (See Section 23.11 for NEII recommendation for the handicapped.) One such standard,

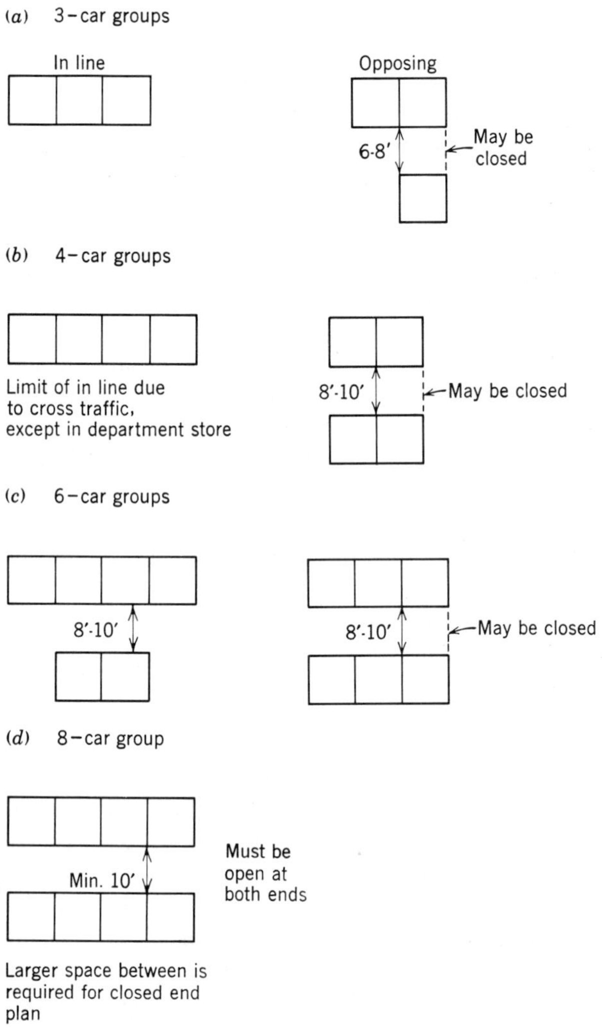

Fig. 23.22 Lobby groupings for single zones: three-, four-, six-, and eight-car groups.

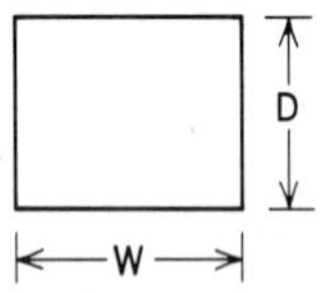

Car size	D	W
2500 lb	70″	8′ 6″
3000 lb	7′ 6″	8′ 6″
3500 lb	8′ 2″	8′ 6″
4000 lb	8′ 2″	9′ 6″

Fig. 23.21 Rough hoistway dimensional data to be used in architectural single-line planning stage.

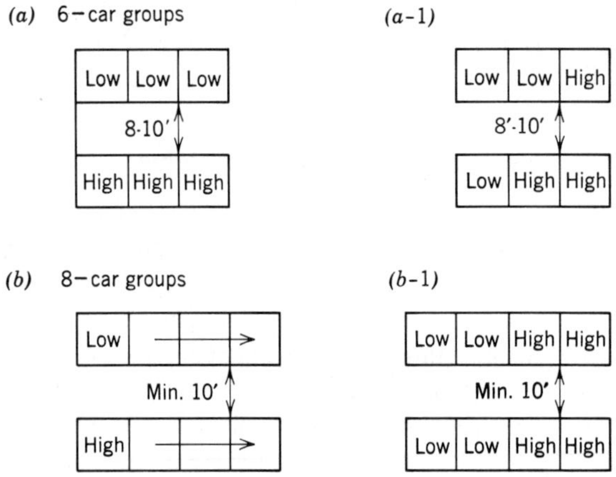

Fig. 23.23 Lobby groupings for multiple zones. Arrangement (a) is preferable to (a-1), and (b) to (b-1). Groups with more than four cars in a row are not used because end-to-end walking time would excessively lengthen landing stops and hence total travel time.

with applications, is reproduced in Fig. 23.24 for 500- to 700-fpm gearless units in the full range of car capacities. These standards are available from the NEII.

As may be seen from Fig. 23.24, it is necessary, in providing for an elevator installation, to consider such factors as the depth of the pit, the dimensions of the hoistway, the clearance from the top of the hoistway to the floor of the penthouse, the size of the penthouse, and the loads that must be carried by the supporting beams.

The penthouse floor and secondary level floor are located above the shaft of each elevator and need approximately 1½ stories of additional height above the top of the support beam of a given elevator when it is standing at its top-floor location. The actual floor area required by the elevator traction machine and its controls is roughly two times the area of the elevator shaft itself. The required area of the floor of the secondary level is no larger than the elevator shaft it serves. The machine room contains the bulk of the elevator machinery. Since some of this equipment will have to be moved for maintenance, it is advisable to furnish an overhead trolley beam that can be used during installation as well. The maximum beam load will be supplied by the elevator manufacturer.

Some typical machine room dimensional data follows, taken from actual installations. Because of the multiple drive options and the flexibility in equipment arrangement, no general conclusions can be drawn from these figures; they are listed simply to give a conceptual picture.

Bank Description	*Machine Room Dimensions (Rounded to Nearest Foot)*
2—2500 lb 125 fpm	16 × 18
3—2500 lb 350 fpm	19 × 26
3—2500 lb 700 fpm	21 × 26
6—3500 lb 500 fpm	26 × 29
6—3500 lb 700 fpm	27 × 27
6—3500 lb 1000 fpm	27 × 27
6—5000 lb 450 fpm	31 × 33

A manufacturer's layout giving dimensional data for hoistway and machine room is shown in Fig. 23.25.

When penthouse space is not available and a hydraulic unit is not desired, a basement traction unit, also referred to as an underslung arrangement, can be used. These units are low speed (100 to 350 fpm) and are therefore applicable where rise is limited and traffic light to medium. Figure 23.26 shows a shaft section with car and dimensional data.

23.40 Structural Stresses

For the purpose of structural design it is necessary to know the overhead load that must be supported by the foundations, by structural columns extending upward to the penthouse, and by the main beams that support the penthouse floor and subfloor. These loads (reactions) are supplied by manufacturers and usually include the actual dead weights of equipment when the elevator is not in motion, plus the added weight caused by the momentum of all moving parts and passengers when the elevator is at top speed and is suddenly caused to stop rapidly by the safety devices.

23.41 Linear Elevator Motor Drive

A revolutionary elevator dreive design developed by OTIS utilizes a linear induction motor build into the counterweight frame. Motive power is therefore supplied at the counterweight, thus entirely eliminating the overhead traction machine and machine room. Trial installation is in progress at this writing in several locations outside the USA. The inherent architectural, construction, energy usage, maintenance and cost advantages of this elegant design will make it an extremely attractive option for American building designers once it has achieved the required Code and jurisdictional approvals. However, because the Code is predicated on the use of a traction machine, which in this design is entirely absent, an essentially new section of Code will have to be prepared to

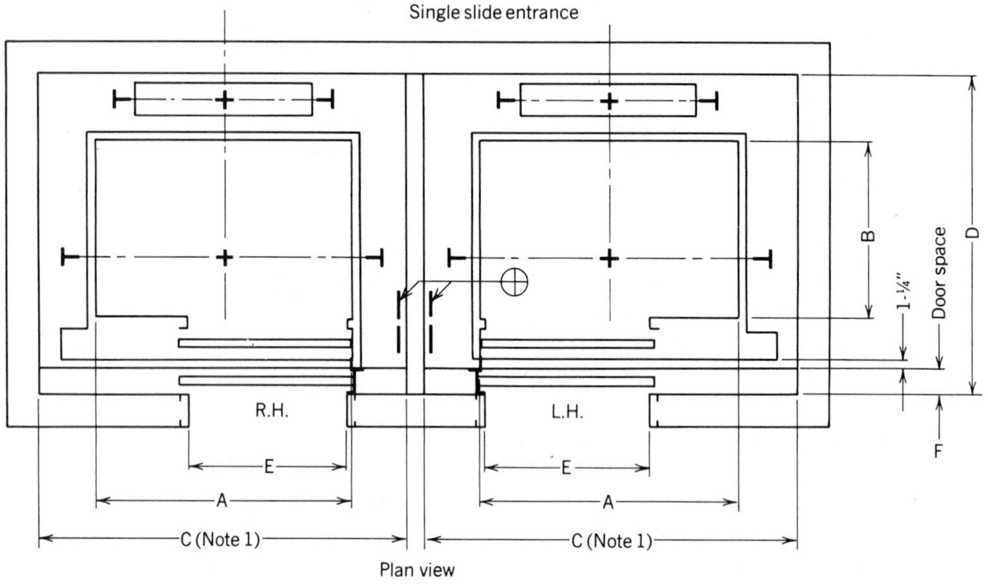

Note 1. "C" dimension does not include any allowance for rail backing.

Rated Speeds 500–700 fpm

Rated Load (lb)	Area[a]	+A	+B	C	D	E	Entrances Type		F min (in.)
Application: Office Bldg. Apartment Houses, Hotels, Banks, Etc.									
2000	24.2	5–8	4–3	7–5	6–11	3–0	Single slide[b]	Center opening	5
2500	29.1	6–8	4–3	8–6	6–11	3–6	Single slide[b]	Center opening	5
3000	33.7	6–8	4–7	8–6	7–4	3–6	Single slide[b]	Center opening	5
3500	38.0	6–8	5–3	8–6	8–0	3–6	Single slide[b]	Center opening	5
4000	42.2	7–8	5–3	9–6	8–0	4–0	Center opening[b]		5

Dimensions (ft–in.)

[a]Maximum allowable inside car area in square feet.

[b]These car sizes and entrance types comply with NEII suggested minimum passenger elevator requirements for the handicapped; 2000 lb rated load will not accommodate ambulance-type stretcher.

Fig. 23.24 *Typical elevator installation dimensional data. (Reproduced with permission of National Elevator Industry Inc., 185 Bridge Plaza North, Fort Lee, NJ 07024. Copyrighted 1983,* Vertical Transportation Standards.*)*

Vertical Transportation Standards
Electric Passenger Elevators

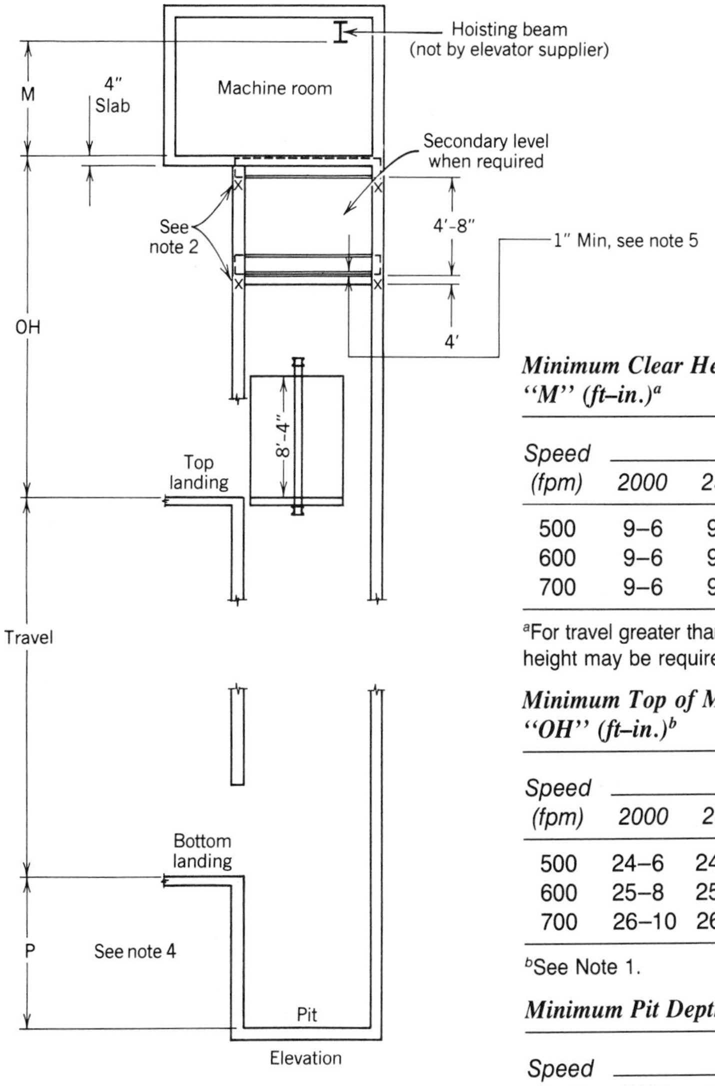

Minimum Clear Height of Machine Room "M" (ft–in.)[a]

Speed (fpm)	Rated Load (lb) 2000	2500	3000	3500	4000
500	9–6	9–6	9–6	9–6	9–6
600	9–6	9–6	10–0	10–6	10–6
700	9–6	9–6	10–6	10–6	10–6

[a]For travel greater than 500 ft, increased machine room height may be required.

Minimum Top of Machine Room Floor "OH" (ft–in.)[b]

Speed (fpm)	Rated Load (lb) 2000	2500	3000	3500	4000
500	24–6	24–6	24–6	24–6	24–6
600	25–8	25–8	25–8	25–8	25–8
700	26–10	26–10	26–10	26–10	26–10

[b]See Note 1.

Minimum Pit Depth "P" (ft–in.)[c]

Speed (fpm)	Rated Load (lb) 2000	2500	3000	3500	4000
500	11–0	11–0	11–0	11–0	11–0
600	12–9	12–9	12–9	12–9	12–9
700	12–9	12–9	12–9	12–9	12–9

[c]See Note 4.

***Fig. 23.24** (continued)*

Notes

1. "OH" dimensions are based on an 8 ft, 4 in. overall car height.
2. Supports for elevator machine beams at X-X in elevation, not by elevator supplier.
3. Machine room floor slab is flush with or above top of machine beams.
4. For travel greater than 400 ft, increased pit depth may be required. Consult elevator supplier.
5. Top of support to be 1 in. above top of slab.

TRANSPORTATION

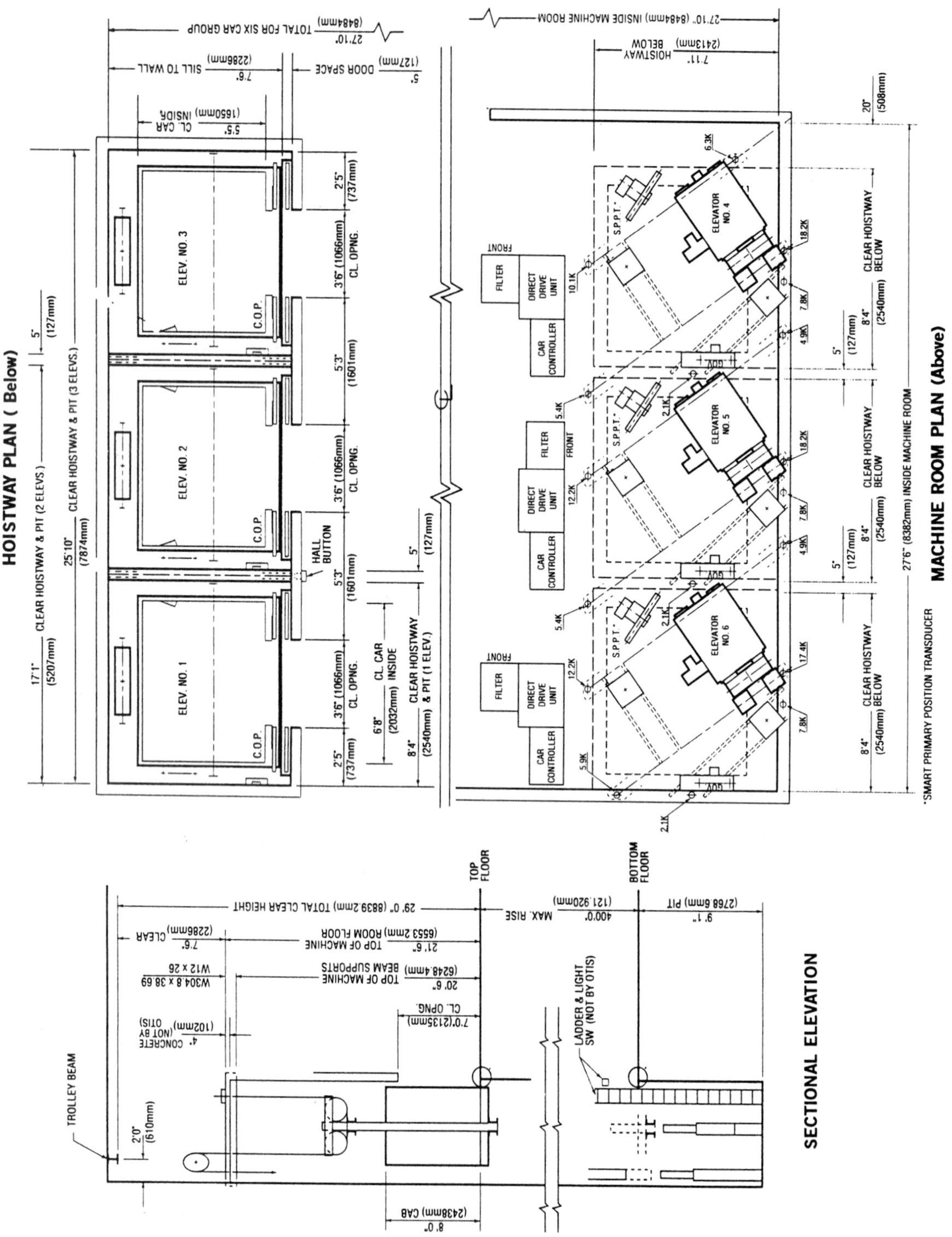
HOISTWAY PLAN (Below)
ELEV. NO. 1
ELEV. NO. 2
ELEV. NO. 3
C.O.P.
HALL BUTTON
CL. OPNG.
CL. CAR INSIDE
CLEAR HOISTWAY & PIT (1 ELEV.)
CLEAR HOISTWAY & PIT (2 ELEVS.)
CLEAR HOISTWAY & PIT (3 ELEVS.)
SILL TO WALL
DOOR SPACE
TOTAL FOR SIX CAR GROUP
MACHINE ROOM PLAN (Above)
ELEVATOR NO. 4
ELEVATOR NO. 5
ELEVATOR NO. 6
S.P.P.T.*
FILTER
DIRECT DRIVE UNIT
CAR CONTROLLER
FRONT
GOV.
CLEAR HOISTWAY BELOW
INSIDE MACHINE ROOM
*SMART PRIMARY POSITION TRANSDUCER
SECTIONAL ELEVATION
TROLLEY BEAM
TOP FLOOR
BOTTOM FLOOR
MAX. RISE
PIT
TOP OF MACHINE ROOM FLOOR
TOP OF MACHINE BEAM SUPPORTS
TOTAL CLEAR HEIGHT
CLEAR
4" CONCRETE (NOT BY OTIS)
LADDER & LIGHT SW (NOT BY OTIS)
CAB

Fig. 23.25

Basement Traction – medium and low speed

Basement traction elevators are utilized for limited overhead conditions in new and existing buildings. This type of elevator facilitates future floor expansion.

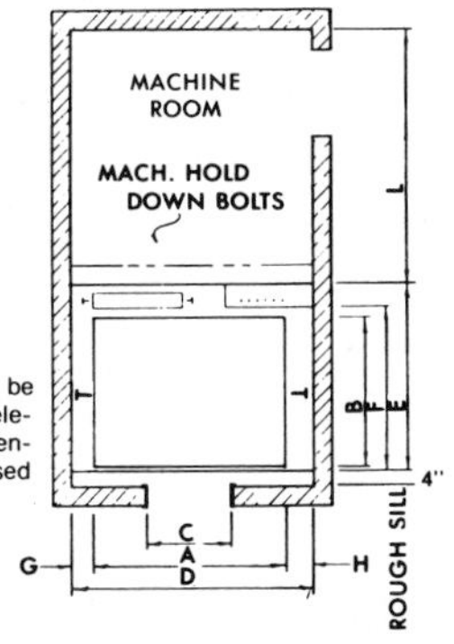

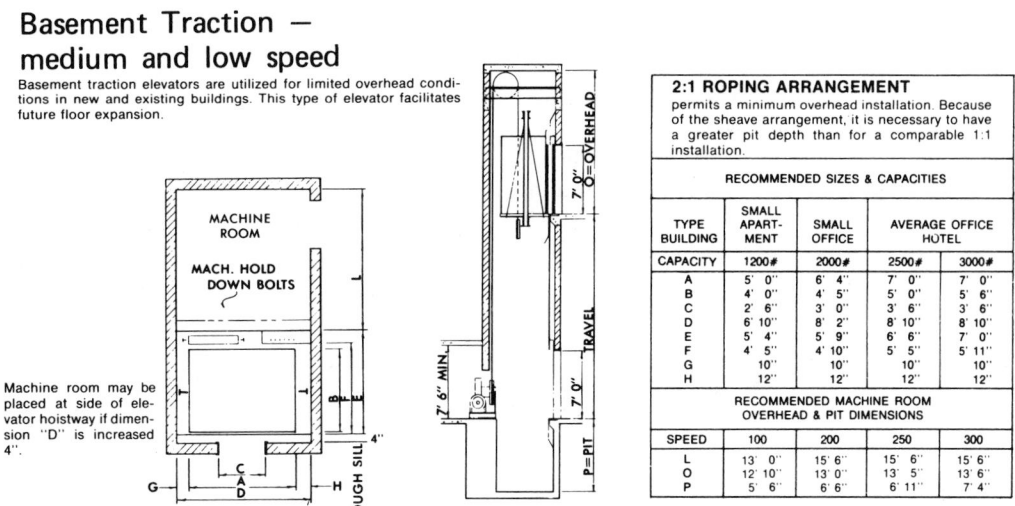

Machine room may be placed at side of elevator hoistway if dimension "D" is increased 4".

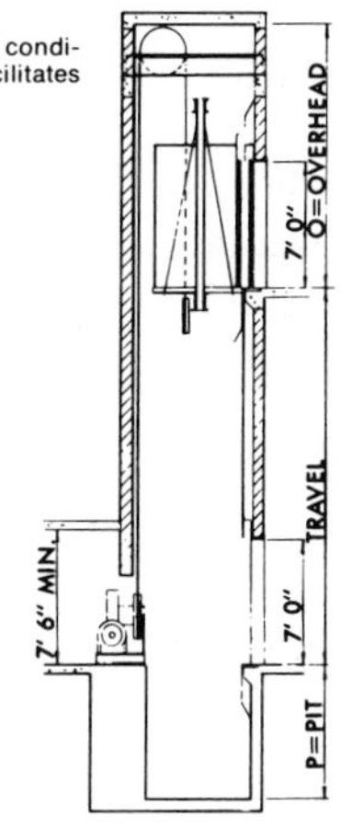

2:1 ROPING ARRANGEMENT
permits a minimum overhead installation. Because of the sheave arrangement, it is necessary to have a greater pit depth than for a comparable 1:1 installation.

RECOMMENDED SIZES & CAPACITIES				
TYPE BUILDING	SMALL APARTMENT	SMALL OFFICE	AVERAGE OFFICE HOTEL	
CAPACITY	1200#	2000#	2500#	3000#
A	5' 0"	6' 4"	7' 0"	7' 0"
B	4' 0"	4' 5"	5' 0"	5' 6"
C	2' 6"	3' 0"	3' 6"	3' 6"
D	6' 10"	8' 2"	8' 10"	8' 10"
E	5' 4"	5' 9"	6' 6"	7' 0"
F	4' 5"	4' 10"	5' 5"	5' 11"
G	10"	10"	10"	10"
H	12"	12"	12"	12"
RECOMMENDED MACHINE ROOM OVERHEAD & PIT DIMENSIONS				
SPEED	100	200	250	300
L	13' 0"	15' 6"	15' 6"	15' 6"
O	12' 10"	13' 0"	13' 5"	13' 6"
P	5' 6"	6' 6"	6' 11"	7' 4"

Fig. 23.26 *Typical data for basement traction machine (underslug) arrangement, used where penthouse is unavailable or undesirable. (Courtesy of Montgomery Elevator Co.)*

LINEAR INDUCTION MOTOR ELEVATOR SYSTEM

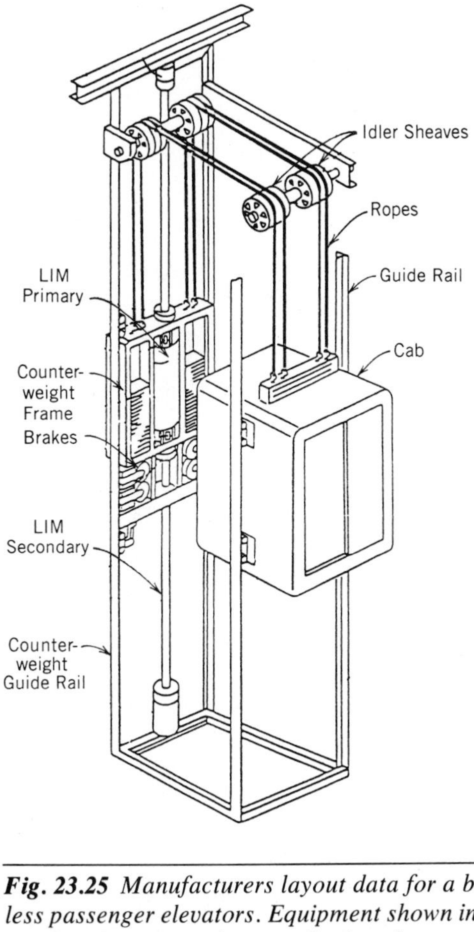

Fig. 23.27 *Schematic drawing of OTIS' new Linear Induction Motor (LIM) elevator design. Because the motor is built into the counterweight and supplies motive power linearly (vertically), the conventional overhead rotary traction machine is entirely eliminated, as is also therefore, the machine room. The system operates more efficiently, quietly and smoothly than the conventional design. (Courtesy of OTIS Elevator).*

Fig. 23.25 *Manufacturers layout data for a bank of six 3500-lb (1588-kg), 700-fpm (3.56-m/s) gearless passenger elevators. Equipment shown in the machine room is thyristor control for d-c traction machines. Since each controller has the capability of group supervisory control in this design (Otis Elevonic 411), no separate group supervisory equipment is shown. No additional space would be required if a UMV drive with m-g sets were selected rather than thyristor control. (Courtesy of Otis Elevator Co.)*

cover this design, and in particular its safety aspects. It should not be expected that so extensive a Code revision will be a short-term project. A schematic diagram of the design is shown in Fig. 23.37.

REFERENCES

Strakosch, G. R. (1983). *Vertical Transportation, Elevators and Escalators,* 2nd ed., Wiley, New York.

Stein, B., Reynolds, J., and McGuinness, W. (1986). *Mechanical and Electrical Equipment for Buildings,* 7th ed., Wiley, New York.

24

VERTICAL TRANSPORTATION: SPECIAL TOPICS

POWER AND ENERGY

24.1 Power Requirements

The power required by an elevator is that amount necessary to perform the necessary traction work and to overcome friction. Since power is equal to the *rate* at which work is done the elevator motor size is directly proportional to the speed of the system. In other words, it requires proportionately more power to lift a 3000-lb car at 700 fpm than at 200 fpm. This relationship is shown in Fig. 24.1, which shows the minimum size of a d-c elevator traction motor as a function of speed, for different capacity cars. This motor supplies the system friction in addition to the required traction power. (For power data on hydraulic elevators, see Section 24.10.) Since friction is higher in geared machines than in gearless units, the traction motor must be larger. The m-g set motor size (where used) is approximately 20% larger than the value shown, to compensate for the set losses.

When solid-state control is used (and the m-g set eliminated), the values shown in Fig. 24.1 are still valid inasmuch as the size of the traction motor is unaffected by the type of variable-voltage or -frequency supply.

An elevator moves only about 50% of the time, the remainder being spent standing at various landings. As the number of cars in a bank increases, the probability of *all* the cars being in operation simultaneously decreases, resulting in a system demand factor less than 1.0. The factor is shown directly in Fig. 24.1.

As an example of the use of the curves, consider a bank of five 3500-lb, 600-fpm units. From Fig. 24.1, each car requires 48 hp:

$$\text{group demand factor} = 0.67$$

$$\text{total instantaneous power required} = 5 \times 48 \times 0.67 = 160 \text{ hp}$$

Note that this is the *traction motor* power requirements. Assuming an overall efficiency (eff) of 80% for an m-g set, the elevator system power requirement is

$$\text{system power} = \frac{160 \text{ hp}}{80\% \text{ eff}} = 200 \text{ hp}$$

which would have to be provided by the building power system. When using solid-state control, with an efficiency in excess of 90%, system power requirements would be reduced proportionately.

24.2 Energy Requirements

The energy used by an elevator is essentially the system friction, including the heat generated by the brakes plus the electrical losses in the traction motor and power supply equipment—rotary or solid state. The energy expended in raising the car and its passengers is simply stored as potential energy. It is *returned to the power system* when the car and passengers descend, via the system of regenerative braking in use in almost all elevator systems. Refer to Fig. 24.2, which shows the approximate efficiencies of the components of a typical system. With these data, we are able to calculate a system's energy consumption.

EXAMPLE 24.1. Given a system of five 3500-lb, 600-fpm (gearless) cars calculate:

(a) Heat generated in the machine room during peak periods. Assume solid-state control.

TRANSPORTATION

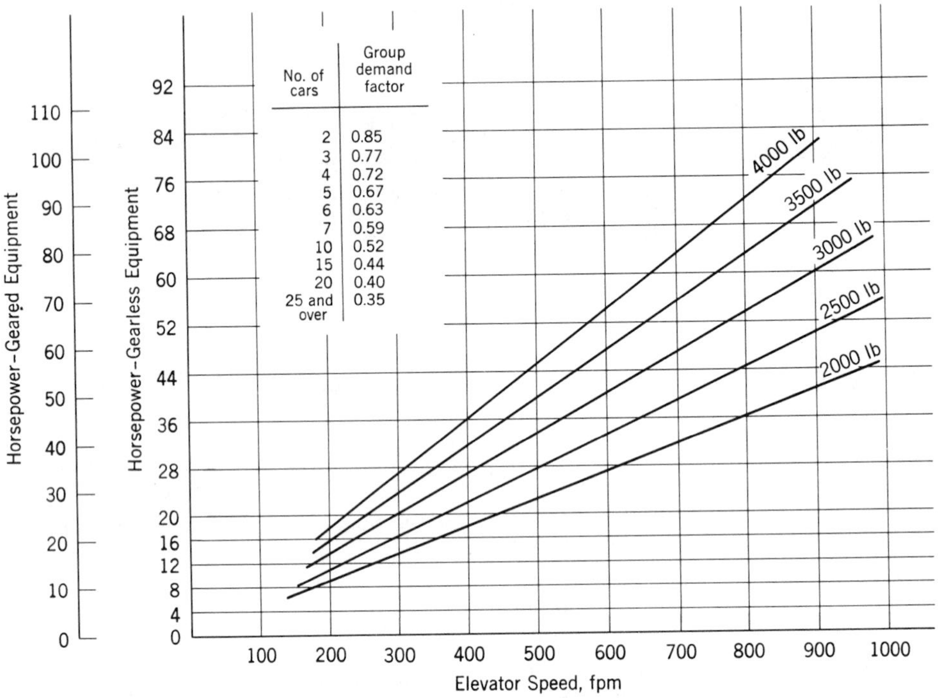

Fig. 24.1 Elevator traction motor power requirements per car. The m-g set motor (if used) is approximately 20% larger than the traction machine.

(b) Approximate monthly energy cost using a combined demand/energy rate of \$0.08/kWh.

SOLUTION. (a) During peak periods the traction motor operates approximately 50% of the time and is at standstill the other half. Assume that while operating it draws 90% of full load. Therefore, for one car, from Fig. 24.1,

traction motor = 48 hp

Total loss per machine:
In controls:

$$\frac{48 \text{ hp}}{0.9 \text{ eff}} \times 90\% \text{ load} \times 50\% \text{ operation} \times 10\% \text{ loss} = 2.4 \text{ hp}$$

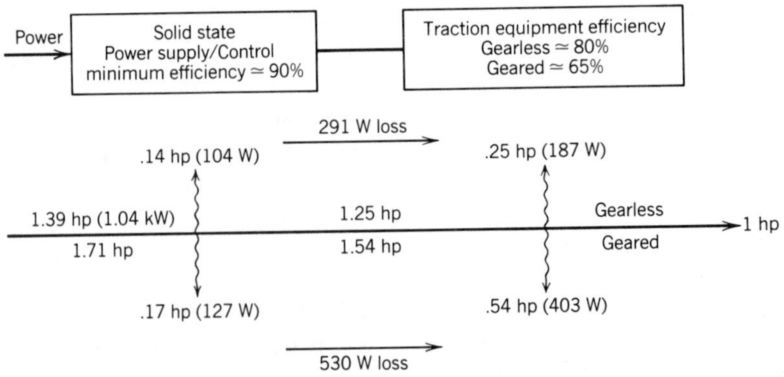

Fig. 24.2 Block diagram showing losses in the system per horsepower delivered to the elevator car, and the equivalent wattages. Note that the losses in a geared system are almost double those of a gearless one. Figures shown are for solid-state controls.

In traction motor:

$$48 \text{ hp} \times 90\% \text{ load} \times 50\% \text{ operation} \times 20\% \text{ loss} = 4.32 \text{ hp}$$

$$\text{Total} = 6.72 \text{ hp} = 17{,}100 \text{ Btu/h}$$

Since there are five elevators operating, the total heat generated is

$$5 \times 17{,}000 \text{ Btu/h} = 85{,}500 \text{ Btu/h}$$

This is the rating of a home furnace. As a result, machine room temperature in warm climates frequently reach 120°F! (No diversity is taken since all the machines are operating and the heating is additive; diversity is applicable only in calculating instantaneous load.)

Since the solid-state elevator equipment now used is much less tolerant of high ambient temperatures than were the old electromechanical switches and relays that preceded them, the machine room should be held to a maximum dry bulb temperature of 90°F. (Temperatures above 90°F can result in unreliable elevator system performance.) This can sometimes be accomplished by thermostatically controlled forced ventilation, particularly if spill air from an air-conditioning system is present. However, since machine rooms are frequently on the building roof and exposed on all surfaces to direct solar radiation, air conditioning may be necessary. It is also important to prevent machine room temperature from dropping below 55°F, for the same reason as mentioned above. This can usually be done with one or more unit heaters, which will normally operate only during the winter, and then only nights and weekends.

In actual design situations, accurate heat loss figures, which are available from manufacturers, should be used, along with accurate heat gain and heat loss calculations for the specific machine room being designed. Frequently, use of thermal insulation, thermostatic louvers, and sunshading can ease the thermal load and result in appreciable financial and energy savings. Because of the very high initial and operating cost of air conditioning, an effort is now being made to produce solid-state elevator controls using components that are specifically designed to operate at elevated temperature. As of this writing such equipment is not commercially available.

(b) To calculate the monthly energy cost, an estimate must be made of the total usage of the system. Assuming the system to be in an office building, a reasonable breakdown of operation during a 24-h day would be

2 h peak use
2 h 70% of peak
6 h 50% of peak
14 h 10% of peak

This gives an average of 30% of peak load for the bank. Therefore, per car

$$\begin{aligned} \text{energy} &= 30\% \times \text{total losses} \times 24 \text{ h} \\ &= 0.3 \times 6.72 \text{ hp} \times 24 \text{ h} \\ &= 48 \text{ hp-h} = 36 \text{ kWh/day/car} \end{aligned}$$

Monthly cost would be

$$\begin{aligned} 36 \frac{\text{kWh}}{\text{day}} \times 25 \text{ days} \times \$0.08 & \\ &\simeq \$72/\text{month/car} \\ &\simeq \$360/\text{month for the bank} \end{aligned}$$

In a design using m-g sets in lieu of the solid-state controls, the monthly cost would be \$630 for the bank.

24.3 Energy Conservation

A reduction in energy consumption can be accomplished by implementing the following recommendations:

For Existing Elevators

1. Increase interval during nonpeak hours.
2. Replace m-g sets with thyristor controls. This conserves energy not only due to higher efficiency of solid-state gear but also because energy consumption of idling machines is eliminated.
3. Recycle machine room waste heat.
4. Shut down some units completely during off hours.

For Building in the Planning Stage

1. Design for maximum trip time.
2. Use the lowest speeds possible, within a type—that is, geared and gearless.
3. Use gearless equipment whenever possible (see Fig. 24.2).
4. After construction, implement the recommendations for existing elevators above.

Since elevator shafts have a powerful stack effect, measures should be taken to counteract this during the heating season.

24.4 Emergency Power

Major power failures and local "brownouts" have demonstrated forcefully the need for a standby or emergency power source of adequate size to operate a building's elevators. Few experiences are so harrowing as being trapped in the crowded confines of a small box suspended in a long vertical shaft, with little or no light, and complete strangers for companions.

A common misconception relative to elevators is that on failure of power, the cars will automatically descend to the nearest landing where exit will be possible. In actuality, the brake is set immediately on power outage and the car remains stationary. Hydraulic cars can be lowered by operation of a manual valve; *small* traction cars can be cranked to a landing by hand; but large cars are fixed in position. This is particularly bad for cars in blind shafts, that is, express shafts with no shaftway doors. In such cases escape from the cars via hatchway is not practicable and, when emergency power is not available, breaking through the shaftway walls is the only recourse.

In addition to simple inconvenience, loss of elevator service in facilities such as hospitals, mental institutions, and jails constitutes a danger to life. For this reason most codes require that emergency power be available in public buildings to operate at least one elevator at a time, and for lighting and communications. Most installations separate the emergency power functions, providing a diesel generator for traction power and separate individual elevator battery packs for communications, lighting, and preferably, also the car fan. The latter two items can be furnished by the elevator manufacturers with the cars, as an option.

The generator is normally sized to supply one elevator motor at a time, with manual or automatic switching arranged between unit controllers. Thus each car in turn can be brought to a landing and thereafter a single car retained in service. Obviously, if it is desired to operate more than one car, a larger generator can be installed. This might well be the case in a multiwing building with critical service requirements, such as a hospital.

The amount of power required, the size of the emergency generator, and the equipment size necessary to absorb regenerative power are all data that can be furnished by the consulting engineer and the elevator manufacturer.

OTHER CONSIDERATIONS

24.5 Fire Safety

Most codes specify the procedures that the elevator control equipment must implement once a fire emergency has been initiated. Details vary somewhat, but in general the actions are these:

1. All cars will close their doors and return nonstop to the lobby or other designated floor, where they will park with doors open. Thereafter they will be operable in manual mode only, by use of the firemen's key in the car panel.
2. All car and hall calls are canceled and call registered lights and directional arrows extinguished.
3. The fire emergency light or message panel in each car is activated to inform passengers of the nature of the alert and that cars are returning to a designated terminal.
4. Door sensors and in-car emergency stop switches are deactivated.
5. Traveling cars will stop at the next landing without opening doors and then proceed to the designated terminal.

The cars can then be used by trained personnel to transport fire personnel and equipment, and for evacuation. In the event of a false alarm, the emergency procedure can be overidden at the lobby panel and the system returned to normal while the source of the alarm is located. (This is a particularly important feature in large buildings with automatic fire alarm systems containing hundreds of fire, smoke, and water-flow detectors. See Section 13.20.)

24.6 Elevator Security

Elevator security has two aspects: physical security of riders, and consideration of the elevator as a portal in a building-access security system.

(a) Rider Security. This problem is particularly difficult inasmuch as the traveling elevator is an enclosed space that can be rendered inaccessible simply by pressing the emergency stop button. Thereafter, an attacker can escape at the floor of his choice. To ameliorate this danger to an extent, elevators are equipped with alarm buttons, which alert residents and security personnel, if any. Every elevator, by code, must be equipped with communication equipment, but a hand-held phone is not practical in a security-emergency situation. More effective is an open two-way communication system with "no-hands" operation, in the car. When a closed-circuit TV monitor is added, utilizing a wide-angle camera in each car (see Fig. 24.3), the security problem will have been addressed to a considerable extent. Obviously, using a communication and TV system presupposes continuous manning of the building security desk.

At least one major manufacturer now markets a device that will alarm automatically on detecting sudden violent motions or a sharp pointed instrument. The matter of handling the alarm is problematic, since an automatically locked door can be forced open manually, and, furthermore, the advisability of locking a violent person *in* the elevator with potential victims is questionable.

Man Trying To Hide In Corner Under Camera

Fig. 24.3 *Wide-angle TV camera intended for elevator cab surveillance. A prominent, printed warning in the cab is an integral part of the system's effectiveness. (Photo courtesy of Visual Methods, Inc.)*

(b) Access Control. This is often a matter of restricting access to (and from) a floor or car. This can be accomplished by pushbutton combination locks or coded cards, the proper use of which will permit access (see Chapter 22). However, if a second person happens to accompany the authorized person, the effectiveness of this type of access barrier is seriously compromised. In sum, the most effective security system is a combination of automatic monitoring and access devices, coupled with continuous supervision by persons who know the appropriate action to take in an emergency.

24.7 Elevator Noise

As already noted, elevator operation, with its rotating, sliding, and vibrating masses, can be a cause of serious noise disturbance to quiet areas such as sleeping rooms, libraries, and certain types of office space. Noise can be reduced by the appropriate application of vibration isolators (e.g., between guide rails and the structure) and by proper control, but primarily by placing noise-sensitive areas away from shafts and machine rooms. Furthermore, the clatter and whirring sound of the older machine room, caused by relays, step switches, m-g sets, and sliding contacts, can be entirely eliminated by the use of solid-state equipment.

24.8 Elevator Specifications

Two basic types of specification for elevator equipment, as for other types of equipment, are utilized. The performance specifications describe job conditions and invite contractors to submit detailed proposals including full engineering. The burden of comparing proposals then falls on the owner, who—if competent to properly perform such an evaluation—would most probably do better to utilize the equipment-type specification.

In recent years the use of performance specifications has increased because of the advent of preengineered, premanufactured systems. These are supplied by the major manufacturers and have these advantages:

1. Approximately 10% lower cost than a custom-designed system.
2. A complete engineered and tested system whose performance and cost are known exactly.
3. Rapid delivery.
4. Minimum supervision required by the owner and architect.

If architects decide to use a custom-designed system, they must prepare detailed drawings and specifications. The specifications must include:

Elevator type, rated load and speed.

Maximum travel.

Number of landings and openings.

Type of control and supervisory system.

Details of car and shaft doors.

Signal equipment.

Characteristics of power supply.

Finishes.

This last item can be left as a dollar allowance for architectural treatment of the car interior. Since the selection of, and technical specifications for, elevators are specialized and complex, the services of an elevator consultant are usually required.

In addition to the technical portions of the specifications, it is imperative that the following items be covered in detail.

(a) Owner's Responsibility. The *construction* contractor normally provides:

1. The hoistway, including properly designed, lighted, drained, waterproofed, and ventilated machine room and pit.
2. Access doors, ladders, and required guards.
3. Guide rail bracket supports, and support for machine and sheave beams.
4. Electric feeder terminating in switch in machine room.
5. Hoistway outlets for light, power, and telephone.
6. Temporary light and power during construction.
7. Concrete machine foundations.
8. Vents, holes, and other work to satisfy fire codes.
9. All cutting, patching, and chasing of walls, beams, masonry, and so on.
10. Coordination of all work.
11. Any special work, as negotiated.

(b) Elevator Contractor's Responsibility. Provide complete, working, tested, and approved system in accordance with specification, plus any special work such as painting, special tests, scheduling of work, and temporary elevator service.

(c) Special Job Conditions. These include work restrictions, scheduling, penalties or bonuses, test reports, and the like.

In alteration and modernization work, the problems of coordination are more complex, and an elevator contractor experienced in this type of work must be selected. To this end, in all elevator contract work, bids should be solicited from parties named on qualified bidder lists. Part of an elevator contract comprises maintenance of the installation for a specific period after completion. Contractors with poor maintenance facilities should be avoided.

SPECIAL ELEVATORS

24.9 Unique Traction Designs

Elevator engineers have always provided interesting and original solutions to unusual problems. A detailed analysis of these solutions is beyond our scope here; however, a rapid review is of interest. For more details see Strakosch (1983).

(a) Sky-Plaza System. For skyscraper buildings such as the World Trade Center in New York and multiple-use buildings such as the John Hancock Center in Chicago—both of which are, in effect, stacked multiple buildings—the elevator solution involves transporting large groups of people from the street lobby to an upper lobby called a sky plaza. At this point the passengers transfer to another elevator to continue their upward journey. The advantages of this system are

1. Reduction in the space occupied by elevators, since the shafts do not extend the entire height of the buildings.
2. Interrupting the otherwise lengthy vertical trip by the horizontal break at the sky lobby.

(b) Double-Deck Elevators. This is an old technique, recently revised to answer the needs of tall buildings such as the Sears and Citicorp Towers. The principal purpose is to limit the otherwise prohibitively large amount of space occupied by elevator shafts (see Fig. 24.4). The double-deck car increases shaft capacity, decreases the number of local stops, and increases the rental area available. This technique can also

Fig. 24.4 (a) *The double-deck car serves to increase cab capacity and decrease shaft space. Coincidence of calls in the upper and lower cabs reduces the number of local stops made by the double car.* (b) *Graphical representation of the space saved in a 40-story building by the use of double-deck elevators. (Courtesy of Otis Elevator Co.)*

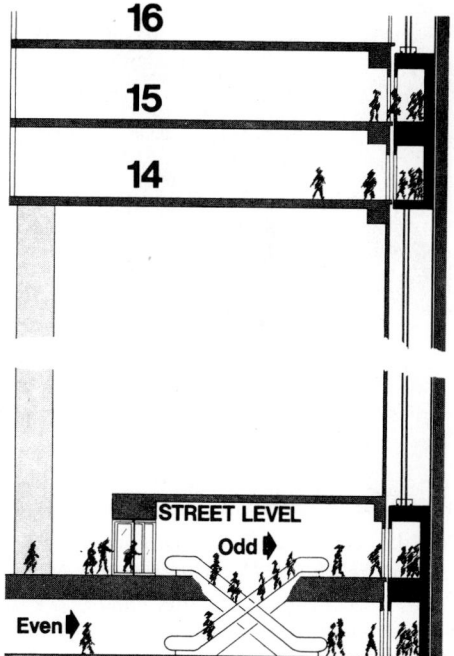

(a)

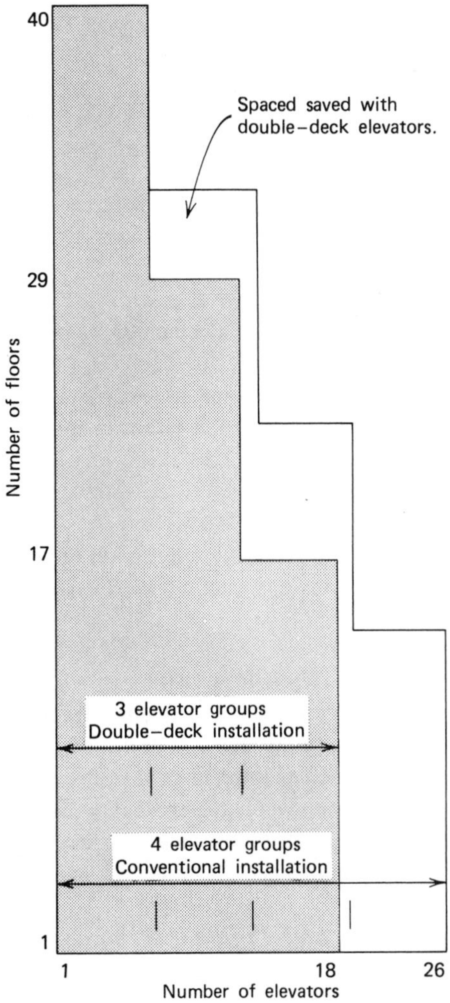

(b)

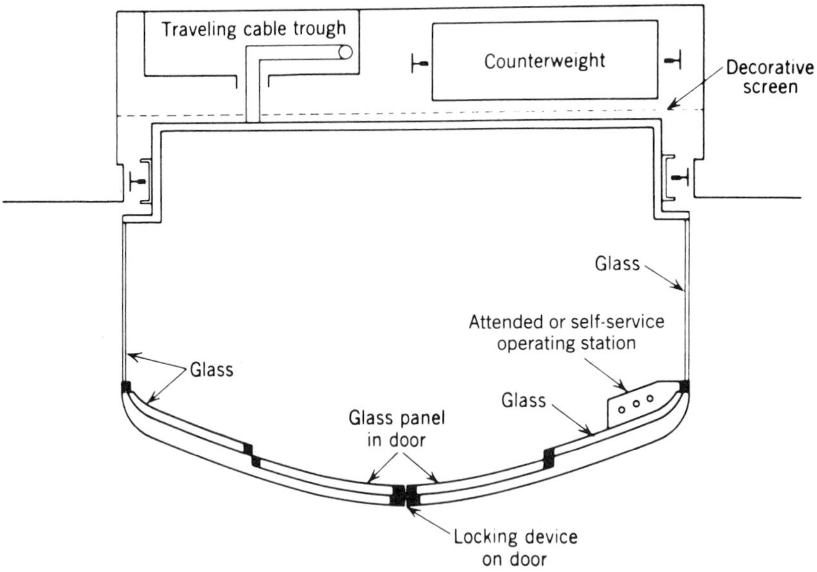

Fig. 24.5 Basic design of a traction-type observation elevator. Cars are small to permit all the occupants to enjoy the view through the cab's glass walls. [From Strakosch (1983); reprinted with permission of John Wiley & Sons.]

be combined with sky lobbies for further space economy, as was done in the Sears Tower.

(c) Observation Car Elevators. By placing the traction lifting mechanism *behind* the car, attaching the car at the back, and using a glass-enclosed, observation-style car, a spectacular unit can be constructed that becomes an attraction in itself. The basic construction is shown in Fig. 24.5, and a well-designed example is seen in Fig. 24.6. If the back screen is treated properly, the car give the impression of movement without any apparent motive force or machinery.

(d) Slant Elevators. Although elevators are normally conceived as traveling vertically, this is not necessarily so. Slant or inclined elevators have been constructed in numerous locations. The design varies depending on the angle of incline. Figure 24.7 shows a well-known application of an inclined elevator.

(e) Rack and Pinion Elevators. These operate on the straightforward principle of a rotating cogged wheel pinion attached to a vertical

Fig. 24.6 Twin installation with unusual car design, at the Hilton Hotel in Atlanta. Note the effectiveness of the screening, making the traction equipment barely visible. (Photo courtesy of Hauenstein and Burmeister, Inc., which supplied the cars for this installation.)

Fig. 24.7 The St. Louis Gateway Arch has a 10-passenger, inclined elevator in each leg. Placement of doors, arrangement of counterweight, and size of shaft all depend on the angle of incline. The car moves 82 ft horizontally during its 386 ft of total travel, at an incline of approximately 12°. (Photo courtesy of Bethlehem Steel Corporation, which supplied the elevator rope for this installation.)

rack. Rotation of the pinion forces the attached cab up and down along the rack (see Fig. 24.8). The advantages of this system are its inherent simplicity and safety, unlimited rise, low maintenance and operating costs, plus minimum space requirements. This last characteristic was primarily responsible for the selection of a rack and pinion design for a 210-ft-rise emergency elevator installed as part of the renovation and rehabilitation of the Statue of Liberty in 1986. The elevator is used primarily to evacuate victims of heart attacks, which average one a month. Carrying them down the 171-step spiral staircase which connects the main upper landing to the observation platform in the crown is obviously not practical. The tight headroom available—less than 9 ft—ruled out a traction unit, as did the extremely tight shaftway, which measures only 2 ft ½ in. × 4 ft 10 in. The car itself (see Fig. 24.9) measures only 1 ft 11 in. × 3 ft 10 in. outside dimensions. Space was so tight that the drive was divided between two motors, both of which are mounted under the passenger compartment. Rack and pinion lifts are used both indoors and outdoors primarily in industrial environments for both passengers and material vertical transport.

Fig. 24.8 The basis of the rack and pinion drive is simply a driven pinion on a stationery rack; rotation of the pinion forces the structure attached to the pinion to move along the rack. (Courtesy of Alimak Elevator Co.)

24.10 Hydraulic Elevators

Most of the foregoing elevators are traction types; that is, they are raised and lowered as a result of the tractive force of cables attached to, or passing under, the car. In contrast to these, the conventional hydraulic or plunger elevator is raised and lowered quite simply, by means of a movable rod (plunger) rigidly fixed to the bottom of the elevator car. The absence of cables, drums, m-g sets, elaborate controllers and safety devices, and penthouse equipment makes this system inherently inexpensive and often the

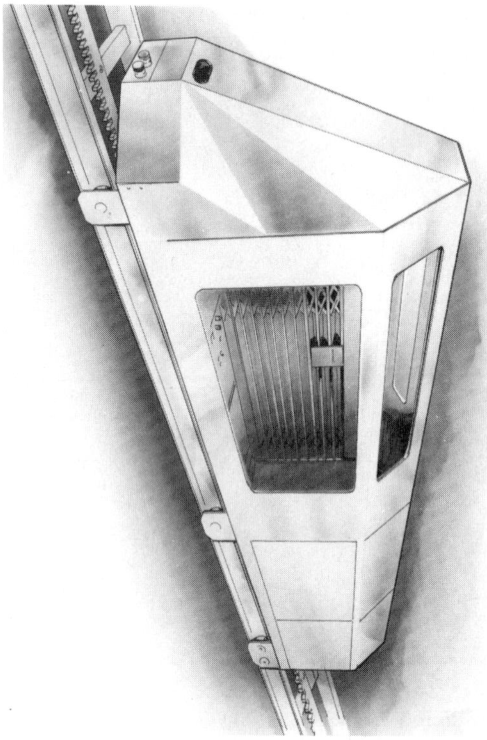

High-speed emergency elevator in Statue of Liberty.

Fig. 24.9 *The rack and pinion principle is used in the very small Statue of Liberty emergency elevator. The rack extends the entire length of the shaft and the car moves along the rack, driven by two motorized pinions below the passenger compartment. (Courtesy of Alimak Elevator Co.)*

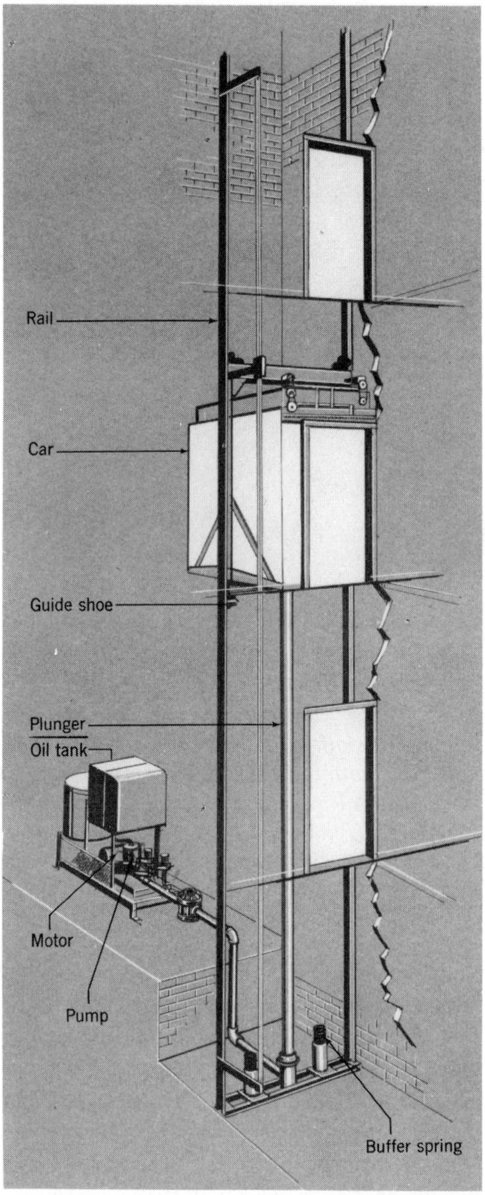

Fig. 24.10 *Phantom view of a typical plunger-type hydraulic elevator installation. (The oil pump is actually inside the tank, but it is shown outside for clarity.) (Courtesy of Schindler Elevator Corp.)*

indicated choice for low-speed (up to 200 fpm), low-rise (up to 65 ft) applications, where construction of the plunger pit does not present difficulties and absence of a penthouse is desirable.

The components of a typical hydraulic unit are shown in Fig. 24.10. This system operates very much the same way as a hydraulic automobile jack. Oil from a reservoir is pumped under the plunger thereby raising it and the car. The pump is stopped during downward motion, the car being lowered by gravity and controlled by the action of bypass valves, which also control the positioning of the car during upward motion. Control systems normally used are similar to those for traction types, for example, single pushbutton, collective, and selective collective. Similarly, door arrangements are the same as in traction types, that is, single slide, center opening, and two speed. Automatic leveling is readily available and is standard on all automatic units. Typical layout and dimensional data for standard plunger units are given in Fig. 24.11, along with capacities and application recommendations.

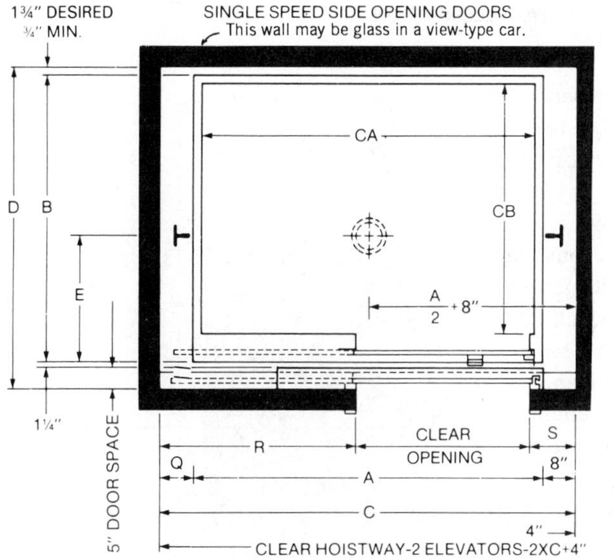

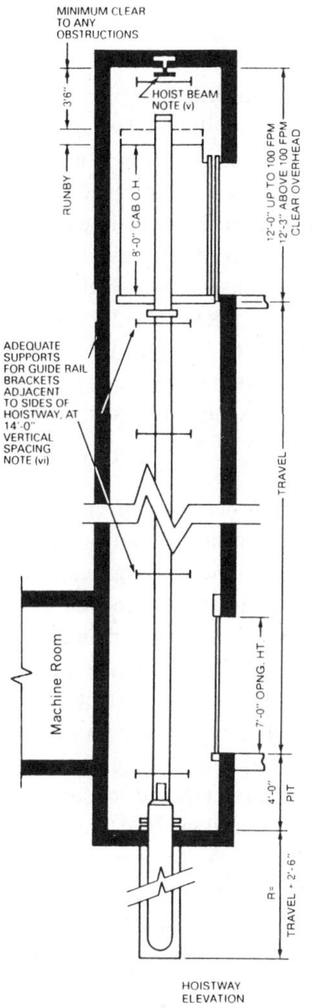

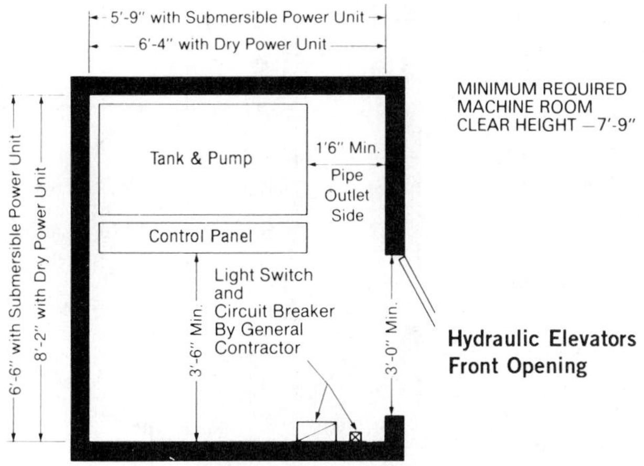

Hydraulic Elevators Front Opening

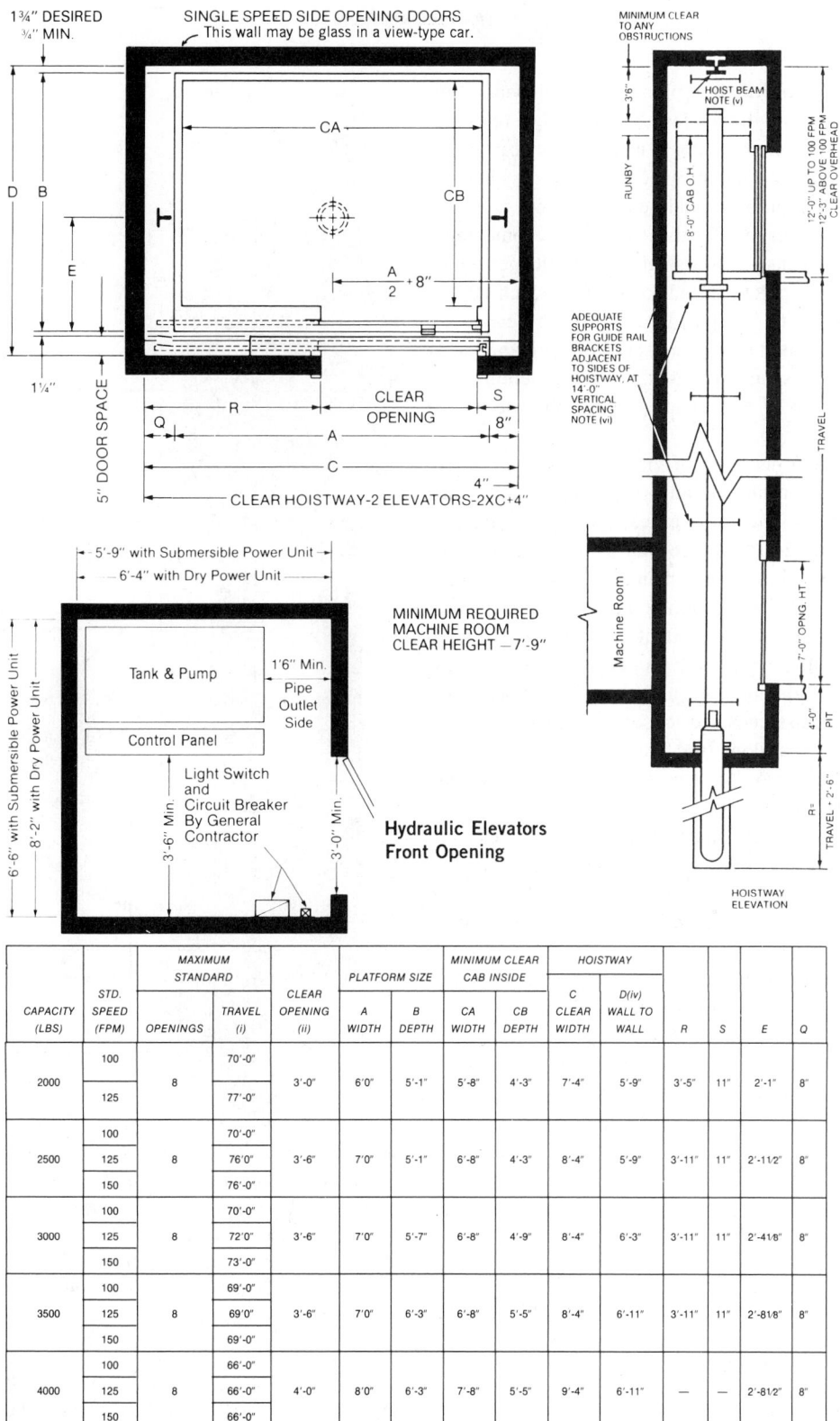

CAPACITY (LBS)	STD. SPEED (FPM)	MAXIMUM STANDARD OPENINGS	MAXIMUM STANDARD TRAVEL (i)	CLEAR OPENING (ii)	PLATFORM SIZE A WIDTH	PLATFORM SIZE B DEPTH	MINIMUM CLEAR CAB INSIDE CA WIDTH	MINIMUM CLEAR CAB INSIDE CB DEPTH	HOISTWAY C CLEAR WIDTH	HOISTWAY D(iv) WALL TO WALL	R	S	E	Q
2000	100	8	70'-0"	3'-0"	6'0"	5'-1"	5'-8"	4'-3"	7'-4"	5'-9"	3'-5"	11"	2'-1"	8"
	125		77'-0"											
2500	100		70'-0"											
	125	8	76'0"	3'-6"	7'0"	5'-1"	6'-8"	4'-3"	8'-4"	5'-9"	3'-11"	11"	2'-1 1/2"	8"
	150		76'-0"											
3000	100		70'-0"											
	125	8	72'0"	3'-6"	7'0"	5'-7"	6'-8"	4'-9"	8'-4"	6'-3"	3'-11"	11"	2'-4 1/8"	8"
	150		73'-0"											
3500	100		69'-0"											
	125	8	69'0"	3'-6"	7'0"	6'-3"	6'-8"	5'-5"	8'-4"	6'-11"	3'-11"	11"	2'-8 1/8"	8"
	150		69'-0"											
4000	100		66'-0"											
	125	8	66'-0"	4'-0"	8'0"	6'-3"	7'-8"	5'-5"	9'-4"	6'-11"	—	—	2'-8 1/2"	8"
	150		66'-0"											

***Fig. 24.11** Typical dimensional, capacity, and layout data for conventional plunger-type hydraulic elevator. Door systems are single slide (SS) or center opening (CO). The car can be used as a viewing type unit by utilizing a glass wall at the back of the car. (Extracted with permission from published data of Schindler Elevator Corp.)*

Where drilling a plunger hole presents difficulties, a hydraulic installation using a telescoping plunger or a roping arrangement can be installed. The latter, which is simpler, cheaper, and less prone to maintenance than a telescoping plunger, has for these reasons become popular. It consists of a single-action plunger which lifts a rope crosshead, or pulley, which is then roped to the car (see Fig. 24.15). These cars are very limited in rise and speed and are applicable only to small two- to three-story buildings. A cutaway of a twin telescoping plunger unit is given in Fig. 24.12.

From the point of view of the construction, the major inherent advantage of hydraulic units is the absence of overhead traction equipment. In Fig. 24.10 we see that only the guide rails project above the car and, if these are camouflaged, the impression of a freestanding elevator car is given. This effect can be used to good advantage *inside* large, open spaces such as exist in shopping malls and stores; when combined with glass-enclosed, observation-type cabs, the effect is striking (see Figs. 24.13 and 25.3).

The major inherent *disadvantage* of the standard hydraulic elevator is its operating expense. Since it is not counterweighted, it requires a relatively large motor to drive the oil pump, and *all* the energy is lost in heat (see Table 23.1). As an example of the operating cost, consider a 3500-lb, 125-fpm hydraulic unit in a department store. Such a unit requires a 40-hp motor. Assuming the unit to be in operation 10 h/day, 6 days a week, and assuming a normal 60% time-in-operation figure, we have (remembering that the motor operates only in the up direction)

$$\text{energy used/day} = \frac{40\text{ hp}}{0.82\text{ eff}} \times \frac{0.746\text{ kW}}{\text{hp}} \times 60\% \times 10\text{ h} \times \tfrac{1}{2} = 110\text{ kWh/day}$$

At \$0.08/kWh we have

$$\text{monthly energy cost} = 110 \times \frac{6\text{ days}}{\text{week}} \times \frac{4.33\text{ weeks}}{\text{month}} \times \$0.08 = \$229/\text{month}$$

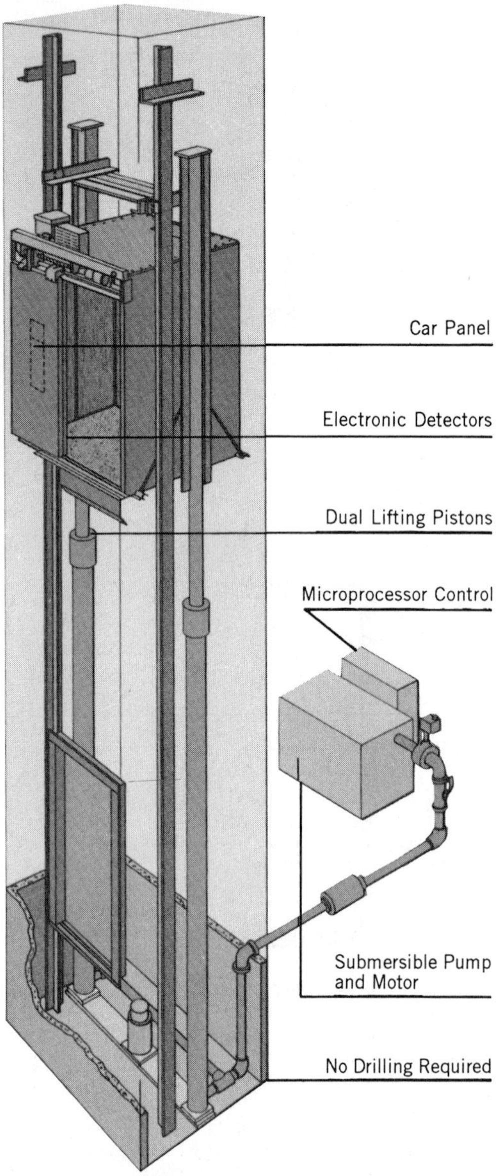

Fig. 24.12 *The illustrated holeless hydraulic elevator is similar in construction to standard plunger-type except that it uses two telescoping jacks to lift the car. Service for this unit is limited to two stops, a maximum rise of 15 ft, and a car rating of 2000 or 2500 lb. (Courtesy of Otis Elevator Co.)*

Compare this to the previously calculated (Section 24.2) monthly energy cost of \$72 when using solid-state equipment and \$126 with m-g set power supply, for a 3500-lb, 600-fpm traction car, for an appreciation of the value of a counterweight.

Fig. 24.13 This four-story-high, exterior-plunger elevator in the Los Angeles rapid transit system is practical and fulfills viewing function. (Photo courtesy of Montgomery Elevator Co.)

24.11 Residential Elevators and Chair Lifts

Although the special needs of the handicapped have only within the last decade been widely recognized officially in legislated requirements, the elevator industry has been providing for the handicapped for years, on a voluntary basis. Small private residence elevators can double as wheelchair lifts, and are available in a wide range of designs, including winding-drum units (Figs. 24.14 and 24.15), roped hydraulics (Fig. 24.16), and worm and screw units. Standard traction designs are used infrequently due to the overhead space requirement. Similarly, standard hydraulics requiring a plunger bore hole are rarely used. Chair lifts and wheelchair platform lifts also come in a variety of designs. The screw worm and cog design chair lift shown in Fig. 24.17 is also utilized as a wheelchair lift that can negotiate stair turns, by adding a second upper screw worm to support a chair platform. The hydraulic scissor-jack type of wheelchair platform lift is most frequently found in public and institutional buildings because it is particularly applicable to short-rise, heavy-load, frequent-duty applications, and requires minimal maintenance.

All of these units are covered by various sections of the elevator code and must be installed in accordance with code requirements, including safeties and controls. Residential units are normally arranged to operate on 120 V a-c, although heavy-duty units may require 240-V service.

FREIGHT ELEVATORS

24.12 General

The preceding material, which dealt with passenger traffic, had as its prime consideration the most effective solution to the problem of vertically transporting a given number of persons. The problem with respect to freight elevators is similar: to transport a given tonnage of freight efficiently, economically, and quickly. The service car in a facility can be considered to be a freight car but, if utilized for passenger duty at all, it must meet passenger service requirements. If passenger duty is not required, or if much freight is to be handled, a car designed specifically for freight is used.

Factors to be considered in freight elevator selection, in addition to tonnage movement per hour, are size of load, method of loading, travel, type of load, type of doors, and speed and capacity of cars. Since these factors are interrelated, the actual process of selection involves making assumptions on the basis of recommendation and then arriving at a solution, very much as was done for passenger elevators.

It is beyond the scope of this book to discuss in detail the selection of material-handling elevators because of the large number of considerations involved. Therefore we shall restrict ourselves in the following paragraphs to descriptive material and recommendations. Also, since freight elevators form such an important link in industrial processes, a careful and detailed material-flow study should be made before freight elevators are selected. Elevator manufacturer's representatives and materials-handling consultants can be very helpful in this regard.

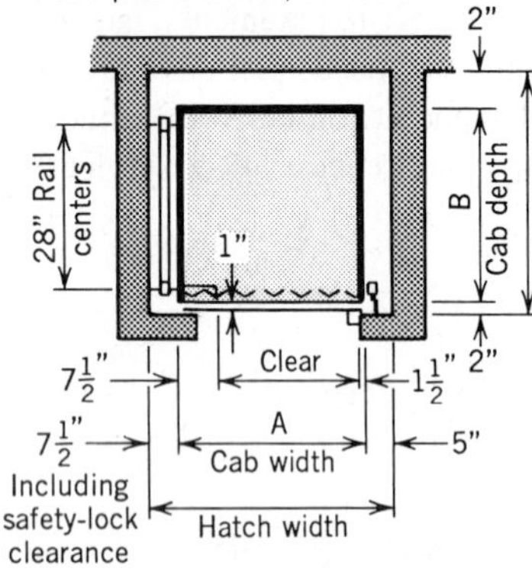

Cab sizes		
	A (width)	B (depth)
3 x 3	3 ft.	3 ft.
3 x 4	3 ft.	4 ft.
Special	To Suit (12 sq. ft. max. platform area)	

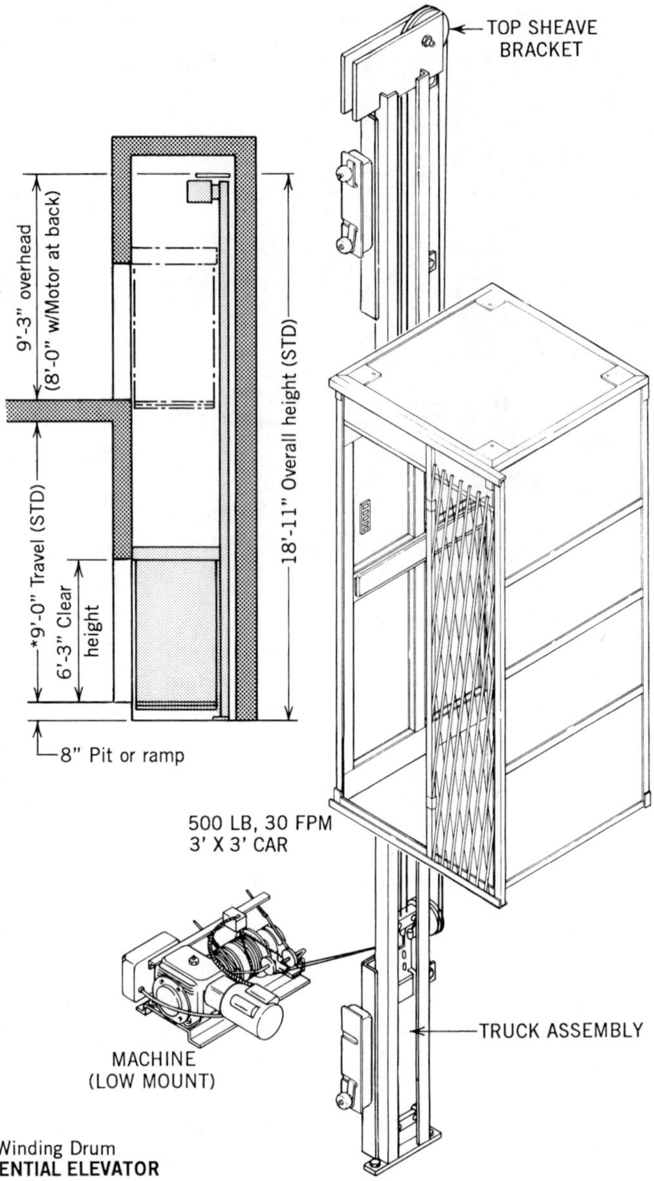

Fig. 24.14 *Isometric drawing of a typical residential elevator of the winding drum design. Units of this type are usually limited to 500-lb car capacity, 40-ft rise, and 30-fpm speed. The cab is rigidly attached in cantilever fashion to a rolling truck, which is raised and lowered by cables attached to the winding drum. The driving motor and winding drum shown here at the base of the assembly can be installed at any floor stop along the hoistway, or overhead. Control is normally pushbutton automatic with limit-switch leveling. A 6-in. pit is required at the bottom of the hoistway. Door interlocks, cable failure, and overrun safeties are standard items in such installations, which must meet elevator code requirements. (Courtesy of Waupaca Elevator Co., Inc.)*

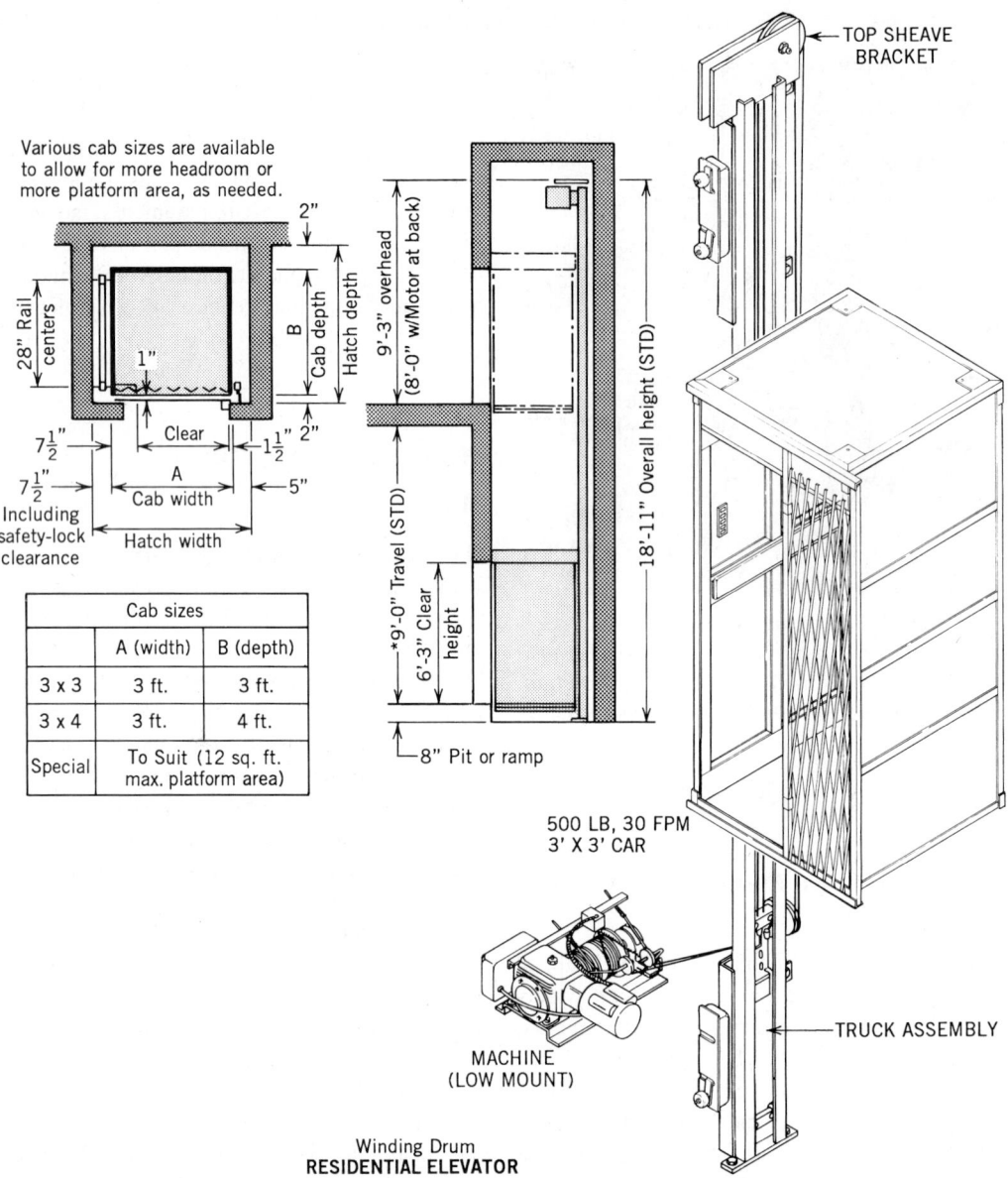

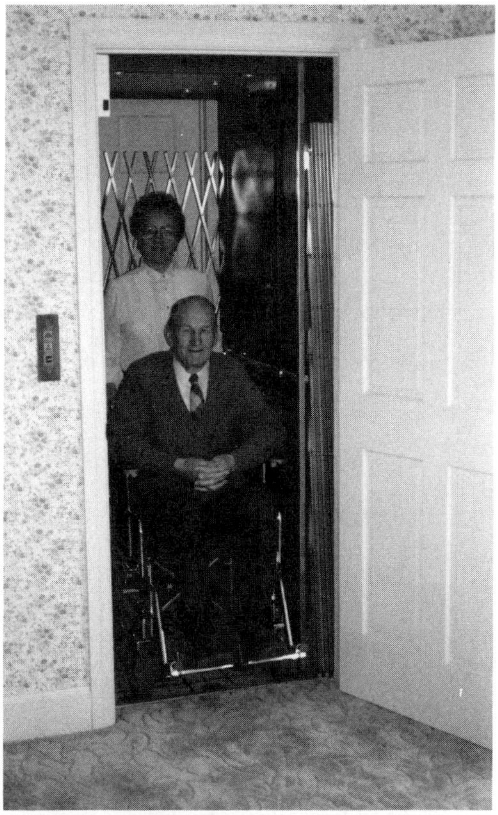

Fig. 24.15 Winding-drum residential elevator specially designed and adapted for use by handicapped. Car floor area is 12 ft² to accommodate a wheelchair and another passenger. Note also that the car is arranged with front and back access for maximum flexibility. (Courtesy of Waupaca Elevator Co., Inc.)

24.13 Freight Car Capacity

Figure 24.18 is a section through a typical traction type freight car shaft. Capacities corresponding to a specific platform size are due to the varying square-foot loads that are permissible. Cognizance of this is taken by the ANSI Code (A17.1) for Elevators, which has established five load classifications for freight elevators.

Class A. General Freight Loading, by hand truck. Single items may not exceed 25% of the car-rated load. Rated load is based on 50 pounds per square foot (psf) of net inside platform area.

Class B. Motor Vehicle Loading. Car will carry automobiles or automobile trucks. Rating is based on a load of 30 psf of net inside platform area.

Class C1. Industrial truck loading; truck carried.

Class C2. Industrial truck loading; truck not carried.

Class C3. Concentrated loading; no truck used; increments greater than 25% rated capacity.

For classes C1, C2, C3-rated load based on 50 psf; cars have automatic leveling.

24.14 Freight Elevator Description

Since speeds are generally between 50 and 200 fpm, a geared-type traction machine is used almost universally.

The preferred system of control is collective, with variable-voltage, d-c supply. However, if the car is used infrequently (fewer than five trips a day), economy is very important, accurate leveling is not essential, and a rougher ride is tolerable, then a-c rheostatic control may be used. For low-rise jobs, a hydraulic unit is most often employed. These, as with the variable-voltage d-c traction units, provide accurate control, smooth operation, and very accurate automatic leveling. Hydraulic units rarely exceed 60 ft in height, and operate at speeds up to 125 fpm. Accessories such as governors, safeties, and brakes are similar to those for passenger elevators previously described.

General-purpose freight elevators, whether traction (Fig. 24.19) or hydraulic (Fig. 24.20), in load ranges up to 20,000 lb, are standard design items, applicable to all types of commercial and industrial buildings. Heavier units are individually engineered. Units 20,000 lb and over require special safeties. As with passenger elevators, structural reactions are supplied by the manufacturer to the architect, who is responsible for adequate structional supports. This item is of great importance in larger car installations, since traction unit rails must be supported every few feet and additional steel provided to accomplish this.

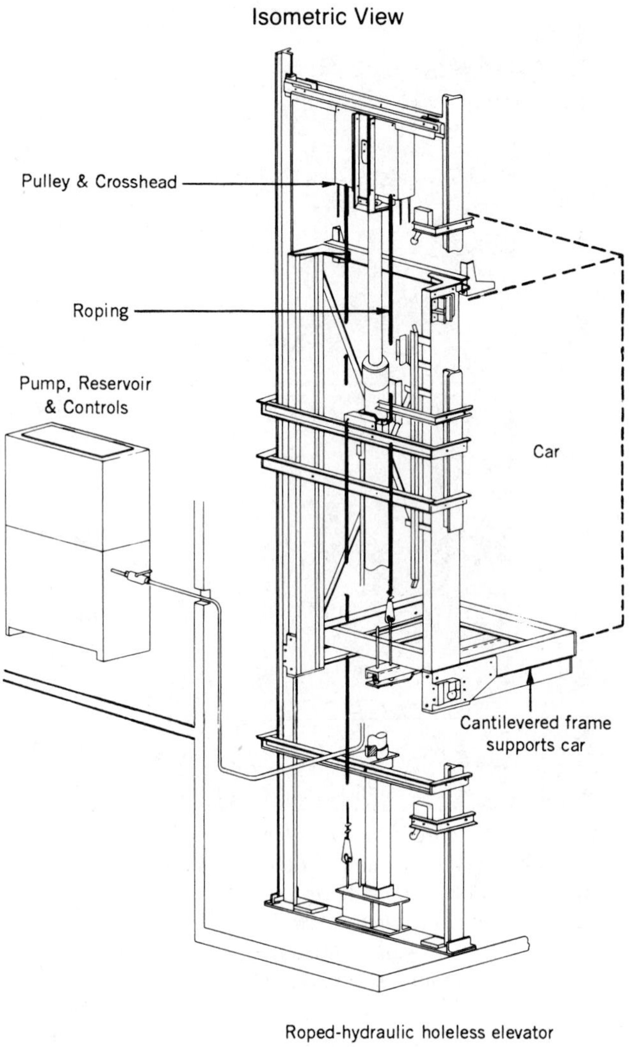

Fig. 24.16 Low-rise residential-type elevator of the roped hydraulic design. The cantilevered car is lifted by cables attached to the cable crosshead, which is in turn lifted (and lowered) by the single-section telescoping piston (jack). The power unit, which includes the oil tank, pumps, and control, is usually mounted at the lower level. Control is automatic, including automatic leveling. Depending on the specific design, the car is a 700- to 750-lb, 30- to 36-fpm unit, normally with a single-story rise. The hatchway is 16 to 26 ft² with the larger size used for a car intended to accommodate a wheelchair. (Courtesy of CEMCO, Corbett Elevator Co.)

24.15 Freight Elevator Cabs, Gates, and Doors

Cabs for freight service are normally built of heavy-gauge steel with a multilayer wooden floor, the entire unit being designed for hard service. Guarded ceiling light fixtures are required. Cab gates slide up vertically and are a minimum of 6 ft high. Hoistway doors are normally vertical lift, center-opening, manual or power operated. Both cab gate and hoistway doors are counterweighted and open fully to give complete floor and head clearance (see Fig. 24.21).

24.16 Freight Elevator Cost Data

The cost of a freight elevator installation, as with passenger elevator installation, is dependent on many factors, principally: capacity, type of control, use, and type of door operation.

Since exact pricing, like actual selection, is outside our scope, it is our recommendation that a reputable manufacturer or elevator consultant be consulted for such information. We can, however, make some general remarks on pricing, as follows:

1. Variable-voltage controlled equipment, depending on type, costs 20 to 50% more than rheostatically controlled equipment.
2. Above a basic two-floor rise, the cost increases linearly with rise.
3. Electric door operation can increase the cost of a car installation 10 to 25%.

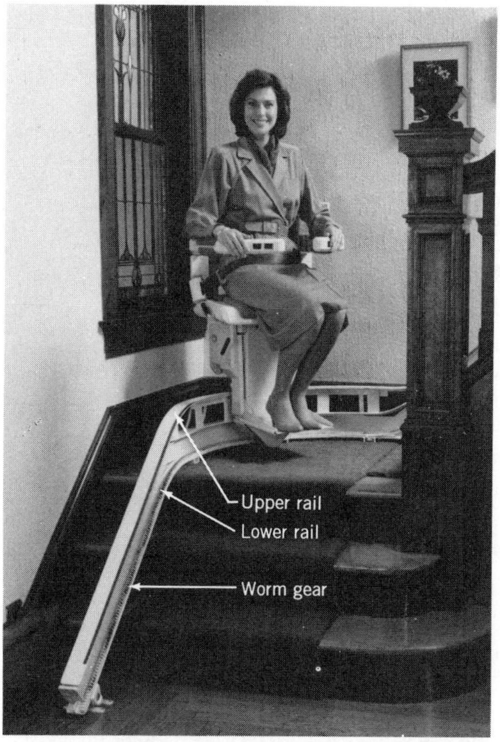

Fig. 24.17 This chair lift is rated 300 lb and travels at 25 fpm. The chair itself, which is of the folding seat and arm design for space economy when not in use, is rigidly attached to a roller carriage that rides on a double-rail section. Motive power is provided by a motor-driven rotating cog which engages a fixed worm in the lower rail section. The worm is clearly visible in the photo. This type of positive-engagement gear drive gives the lift the distinct advantage over traction cable designs of being able to negotiate curves, as illustrated. Safeties include obstruction sensors and an overspeed brake. (Courtesy of Cheney.)

As an example, of *comparative* pricing, using a nominal 100 for an 8000-lb, 75-fpm, four-floor, manual door car with a-c rheostatic control and automatic leveling, the same car with variable-voltage control, 150 fpm, and electrically operated doors would cost approximately 180.

24.17 Garage Car-Parking Facility

Multistory automobile parking facilities that increase parking space by replacing interfloor ramps with freight elevators have the characteristic of very slow service. Since this is obviously unacceptable in a downtown commercial application, an alternative arrangement using a materials-handling approach is available. One such system consists of a central structural tower that supports a continuous loop of car-carrying "baskets," that is, a sort of Ferris wheel. An automobile enters a cage at a terminal, which can be anywhere in the loop—top, bottom, center—and is "parked" by the simple expedient of operating the controls at the terminal. The loop rotates, the car remains in its cage, and an empty cage is presented at the terminal. A car is retrieved by calling back the appropriate cage. In principle, the system is similar to the vertical conveyor described in Section 24.22.

The advantages of this parking system are as follows.

1. Very high space utilization efficiency; approximately 40 m^3 (1400 ft^3) per midsize car.
2. Minimal horizontal space requirement.
3. Simplicity of operation with minimum labor.
4. High degree of security to parked vehicles and to drivers.
5. Rapid operation.

Where height of the structure is limited but horizontal site is not, a similar arrangement can be constructed horizontally. In effect the vertical structure described above is lying on its side and is two "stories" high, and as wide as necessary to accommodate the required number of vehicles.

This parking system is used principally overseas because of the prevalence of small cars, which readily fit into the carrying cage. With the increasing number of compact cars in the United States, it may find increasing application in congested areas of American cities, as well.

MATERIAL-HANDLING EQUIPMENT

24.18 General

The material-handling equipment discussed briefly here is that which finds application in commercial and institutional buildings. Industrial materials handling is an entirely separate

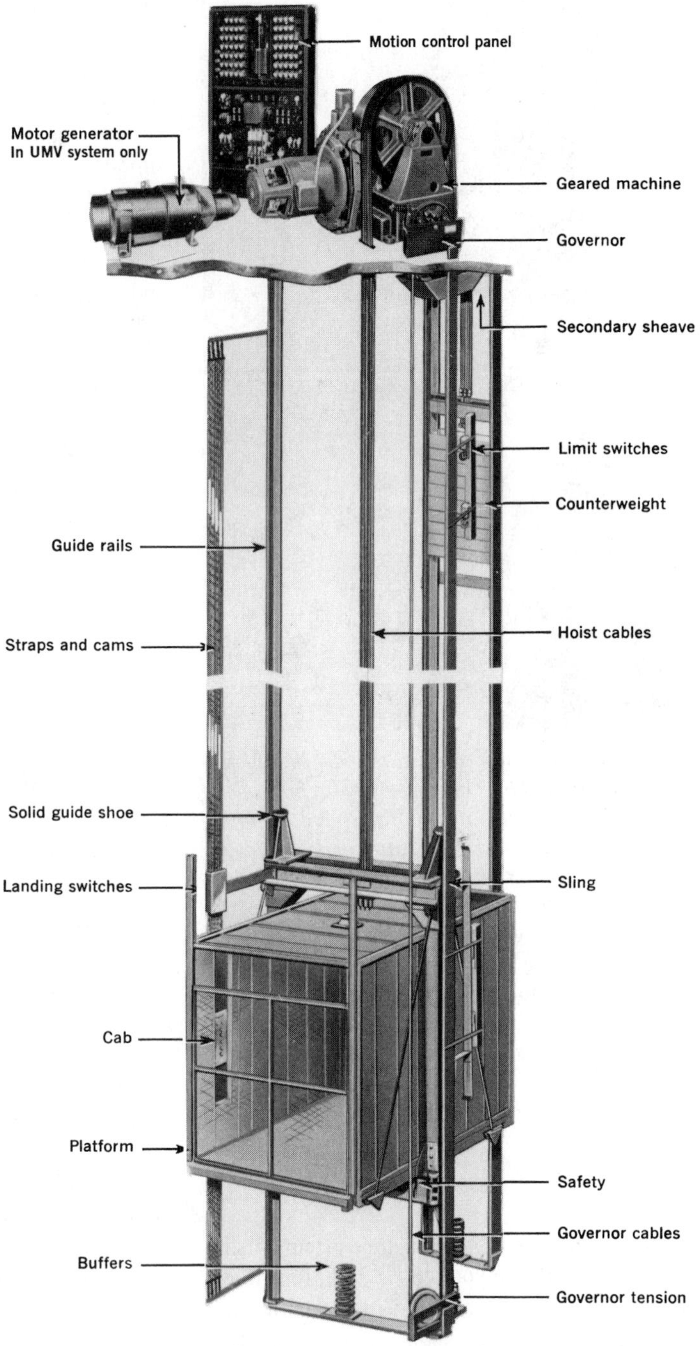

Fig. 24.18 Components of a typical freight elevator installation utilizing a variable-voltage-controlled geared traction machine. The sling that lifts the car is frequently arranged with double sheaves over which the hoisting ropes pass. This roping arrangement increases the mechanical advantage of the lifting ropes.

subject not germane to our purpose. The need to transport material within a building has always existed and until approximately a decade ago was done largely manually, with mechanical assistance. Thus offices used messengers; hospitals used dumbwaiters, service elevators, conveyors, and chutes. The single exception to this situation was (and still is) the extensive use of pneumatic tube systems in large stores, although in new installations their use is declining. Today's systems accomplish the same end—that is, the transfer of materials—but automatically

Freight Elevators — Traction

LIGHT AND MEDIUM DUTY FREIGHT ELEVATORS					
CAPACITY	2500#	3000#	4000#	6000#	8000#
A	5′-4″	6′-4″	6′-4″	8′-4″	8′-4″
B	7′-0″	8′-0″	8′-0″	10′-0″	10′-0″
C	5′-0″	6′-0″	6′-0″	8′-0″	8′-0″
D	7′-10″	8′-10″	8′-10″	10′-10″	10′-10″
L	13′-0″	14′-0″	14′-0″	14′-0″	14′-0″

HEAVY DUTY POWER TRUCK FREIGHT ELEVATORS					
CAPACITY	10,000#	12,000#	16,000#	18,000#	20,000#
A	8′-4″	10′-4″	10′-4″	10′-4″	12′-4″
B	12′-0″	14′-0″	14′-0″	16′-0″	20′-4″
C	8′-0″	10′-0″	10′-0″	10′-0″	12′-0″
D	11′-4″	13′-6″	14′-0″	14′-2″	16′-6″
L	14′-0″	15′-0″	15′-0″	17′-0″	21′-0″

MINIMUM PIT & OVERHEAD DIMENSIONS FOR LIGHT & MEDIUM DUTY FREIGHT ELEVATORS				
CAR SPEED	50	75	100	200
O	16′-0″	16′-0″	16′-0″	16′-0″
P	5′-6″	5′-6″	5′-6″	6′-0″

H = 7′8″ for light end medium duty doors
8′0″ for heavy duty doors

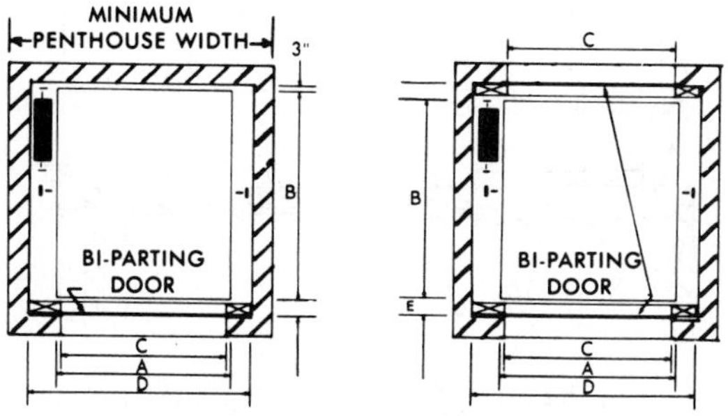

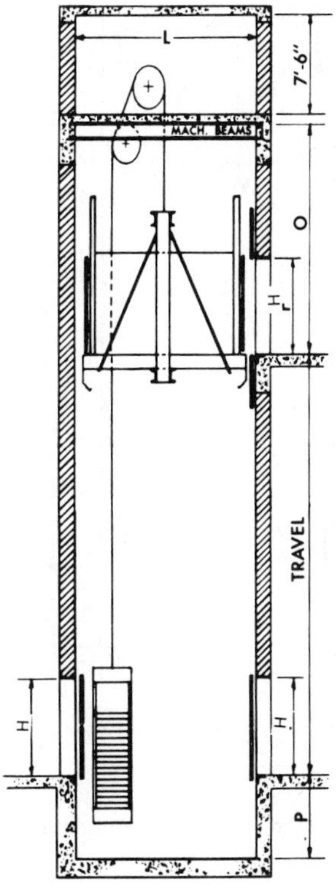

Fig. 24.19 Typical dimensional data for traction-type freight elevators. (Courtesy of Montgomery Elevator Co.)

and, in general, much more rapidly. First cost of these systems is frequently high, but the reduction in labor and increase in speed generally yield a short payback period combined with a marked rise in efficiency.

Modern, commercial material-handling systems can be grouped roughly into four categories:

1. *Elevator-type systems.* These are vertical lift, car-type systems including the common dumbwaiter and ejection lifts, which are basically automated dumbwaiters.
2. *Conveyor-type systems.* These include horizontal and vertical conveyors.
3. *Pneumatic systems.* These include sophisticated pneumatic tube systems and pneumatic trash and linen systems.
4. *Other systems.* Systems that do not fit easily into any of the above categories include automated messenger carts and automatic track-type container delivery systems.

24.19 Manual Load/Unload Dumbwaiters

Dumbwaiters often provide the most convenient and economical means of transporting relatively small articles between levels. In department stores, such units transport merchandise from stock areas to selling or pickup counters; in hospitals dumbwaiters often transport food, drugs, linens, and other necessary small items. In multilevel restaurants, office dining rooms, and the like, dumbwaiters are almost always used

Freight Elevators — Oil Hydraulic

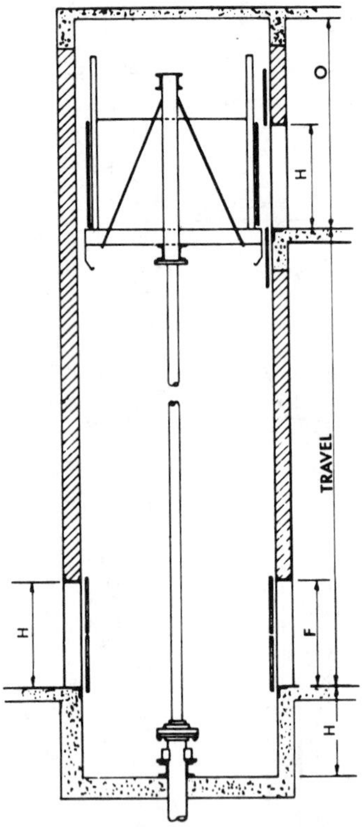

LIGHT AND MEDIUM DUTY HYDRAULIC FREIGHT ELEVATORS							
CAPACITY	2000#	3000#	4000#	5000#	6000#	8000#	
A	5'-0"	5'-6"	6'-6"	8'-6"	8'-6"	8'-6"	
B	6'-0"	7'-0"	8'-0"	10'-0"	12'-0"	12'-0"	
C	4'-8"	5'-2"	6'-2"	8'-2"	8'-2"	8'-2"	
D-manual doors	6'-4"	6'-10"	7'-10"	9'-10"	10'-6"	10'-6"	
D-power doors	6'-10"	7'-4"	8'-4"	10'-4"	10'-6"	10'-6"	
P	4'-6"	4'-6"	4'-6"	4'-6"	4'-6"	5'	

HEAVY DUTY POWER TRUCK HYDRAULIC FREIGHT ELEVATORS					
CAPACITY	10.000#	12.000#	16.000#	18 000#	20 000#
A	10'-6"	10'-6"	10'-6"	10'-6"	12'-6"
B	14'-0"	14'-0"	16'-0"	16'-0"	20'-0"
C	10'-2"	10'-2"	10'-2"	10'-2"	12'-2"
D-manual doors	12'-6"	12'-6"	12'-6"	12'-6"	14'-6"
D-power doors	12'-6"	12'-6"	12'-6"	12'-6"	14'-6"
P	6'	6'	6'	6'	6'

PIT AND OVERHEAD DIMENSIONS				
SPEED FPM	25	50	75	100
P	4'-6"	4'-6"	5'-0"	5'-0"
0 (7'-0" Door)	13'-2"	13'-2"	13'-2"	13'-2"
0 (8'-0" Door)	14'-2"	14'-2"	14'-2"	14'-2"

H = 7' for light duty; 8' for heavy duty

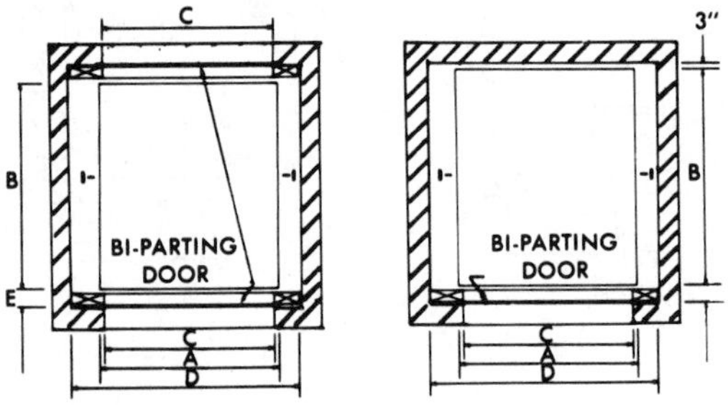

Fig. 24.20 *Typical dimensional data for hydraulic-type freight elevators. The machine room, preferably at the lowest landing level, should be arranged to maintain a temperature between 65 and 100°F for proper elevator operation. The dimensions shown are for biparting car doors. (Courtesy of Montgomery Elevator Co.)*

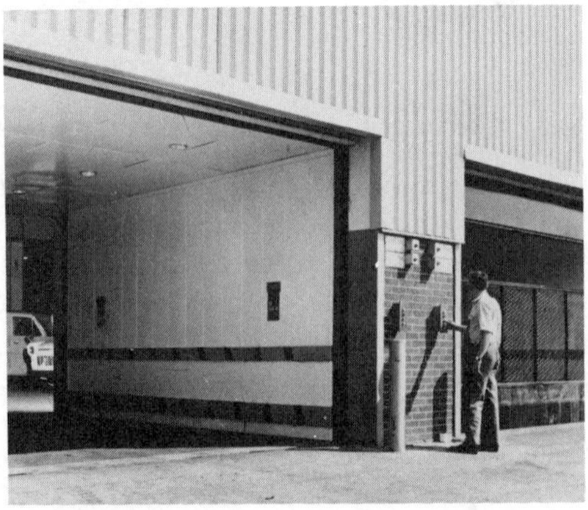

Fig. 24.21 *A pair of large hydraulic freight elevators for automotive use. These units are automatic, self-leveling, and equipped with power-operated biparting shaftway doors and cab gate. As with many freight installations, both ends of the cab are open. Note that two control stations are provided in the cab, one at each gate. (Photo courtesy of Harris-Preble Co., which supplied the doors in this installation. Hydraulic lifts are by Becker.)*

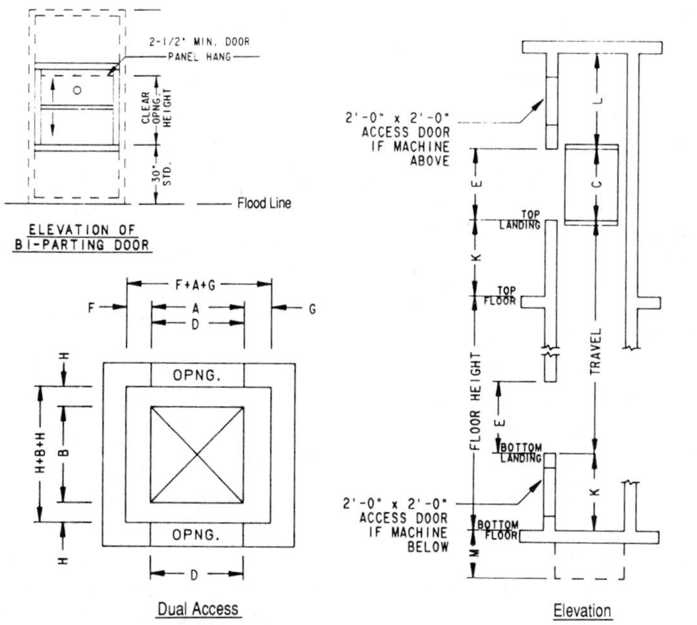

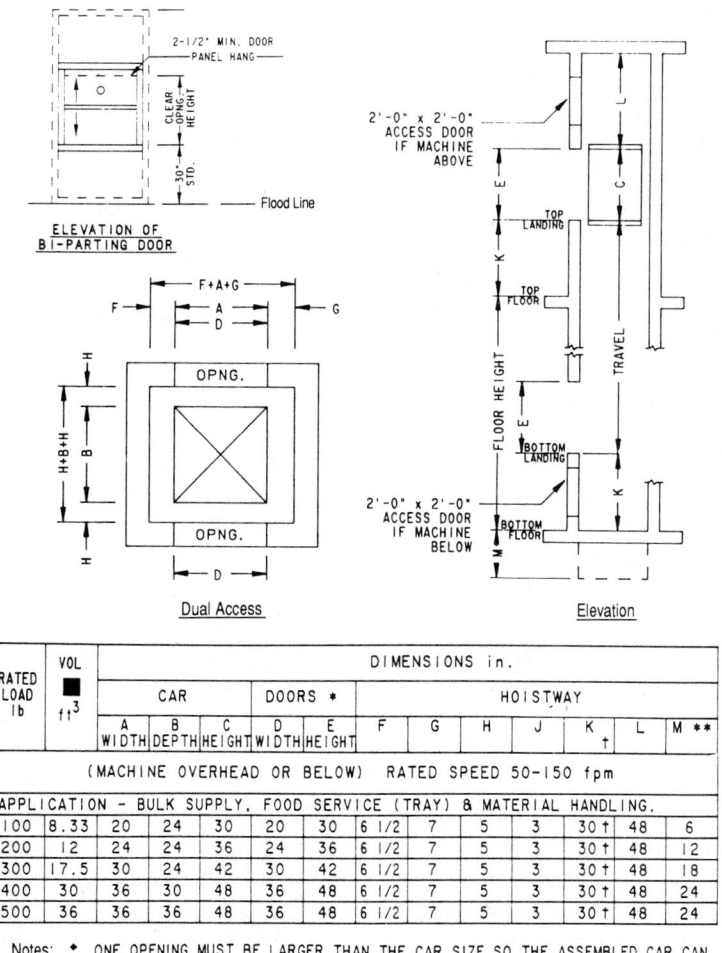

RATED LOAD lb	VOL ft³	DIMENSIONS in.											
		CAR			DOORS *		HOISTWAY						
		A WIDTH	B DEPTH	C HEIGHT	D WIDTH	E HEIGHT	F	G	H	J	K †	L	M **
(MACHINE OVERHEAD OR BELOW) RATED SPEED 50–150 fpm													
APPLICATION – BULK SUPPLY, FOOD SERVICE (TRAY) & MATERIAL HANDLING.													
100	8.33	20	24	30	20	30	6 1/2	7	5	3	30 †	48	6
200	12	24	24	36	24	36	6 1/2	7	5	3	30 †	48	12
300	17.5	30	24	42	30	42	6 1/2	7	5	3	30 †	48	18
400	30	36	30	48	36	48	6 1/2	7	5	3	30 †	48	24
500	36	36	36	48	36	48	6 1/2	7	5	3	30 †	48	24

Notes: * ONE OPENING MUST BE LARGER THAN THE CAR SIZE SO THE ASSEMBLED CAR CAN BE PLACED INTO, OR REMOVED FROM THE DUMBWAITER HOISTWAY.
** DEPTH OF PIT REQUIRED IF SLIDE DOWN TYPE DOORS ARE USED.
† IF BASEMENT MACHINE (DRUM OR TRACTION) 42 in.

Fig. 24.22 *Typical layout of counter-height dumbwaiter. The shaft openings may be single (one side) or dual access. Doors can be biparting, slide up or slide down. In addition to the sizes shown in the table, light-load cars rated 25 and 50 lb at 50 fpm are available in standard designs. (Reproduced with permission of National Elevator Industry, Inc., 185 Bridge Plaza N., Fort Lee, NJ 07024; copyrighted 1983, Vertical Transportation Standards.)*

for delivery of food from the kitchen and for return of soiled dishes.

Dumbwaiter cars are limited to a platform area of 9 sq ft and a maximum height of 4 ft. The car may be, and frequently is, compartmented by shelves. Normal speed ratings are 50 to 150 fpm, with a capacity of up to 500 lb. Cars may be of the traction (counterweighted) or drum (direct pickup) type. Control is normally "call and send" between two floors, although multibutton selector switch or central dispatching arrangements are available for applications with more than two floors. Loading may be floor, counter, or any other specified height. (See Fig. 24.22 for typical layouts.)

24.20 Automated Dumbwaiters

These units are also known as ejection lifts because of the method of delivery (see Fig. 24.23). They find their best application in institutions and other facilities that require rapid vertical movement of relatively large items. Thus, this device is ideally suited for delivery of food carts, linens, dishes, bulk liquids containers, and so on. The load can be a cart (Fig. 24.24) or a basket (Fig. 24.23) containing the items being transported. At the delivery terminal the item must be picked up and transferred horizontally to its final destination if remote from the delivery point. Loading can be manual or automatic. Sophisticated ejection lifts use programmable controllers for automated loading, dispatch, and ejection; electronic sensors to determine whether space is available for a load; and automated return of the unloaded cart.

Payload capacity for cart systems is available up to 1000 lb and car speeds up to 350 fpm. Maximum cart size is approximately 32 in. W × 68 in. L × 70 in. H. Round-trip time for a 200-fpm unit with 5 loading stations is approximately 2 min; with 10 stations about 2.5 min. Major considerations for these units are their relatively high cost and the large shaft area required.

24.21 Horizontal Conveyors

Although horizontal conveyors find their best application in industrial facilities, they are also usable in commercial buildings such as mail-order houses, which require a continuous flow of material. Restrictions in application stem from inflexible right-of-way requirements, noise generation, and a degree of danger if left unprotected or exposed to unauthorized persons. Cost is relatively low, and capacity is virtually unlimited.

24.22 Selective Vertical Conveyors

The action of this system is similar to the automated dumbwaiter in that the system transfers vertically and automatically loads and unloads, but the similarity ends there. Vertical conveyors are constructed with a moving continuous-loop chain to which are attached carriages that pick up and deliver tote boxes. At sending and receiving stations the operator places the items to be moved (up to 60 lb) in the tote box, "addresses" the box in one of several ways depending on the system, and places it at a pickup point (see Fig. 24.25). The first empty carriage on the chain will pick up the box and deliver it to its address. Functions in modern vertical conveyors are monitored by a microprocessor, which tracks system operation and furnishes maintenance data and operational diagnostics. Drawbacks of this system are the large shaft required, noise, and cumbersome arrangements when interfacing with horizontal conveyors. Cost is moderate.

24.23 Pneumatic Tubes

This well-tried system will undoubtedly continue in use where physical transfer of an item is required. Where information must be moved, electronic data reproduction will completely replace the transfer of pieces of paper between two points. Pneumatic tube systems are available with 2¼- to 6-in. ranges of tube diameters (special shapes are also used) and with single or multiple loops.

Whereas older systems were generally pressurized using a single large noisy compressor, newer systems are computer controlled, utilize a small blower in each of the zones, operate basically on vacuum but also on pressure, are relatively quiet, and are capable of being constructed in unlimited system length. Carriers travel at 25 fps. The computerized control center provides information and control of all system components, including status, traffic data, station assignments, scheduling, and the like. Overall, pneumatic tube systems perform their task reliably, rapidly, and efficiently at relatively low cost if installed during initial building construction. Typical system components are shown in Fig. 24.26.

24.24 Pneumatic Trash and Linen Systems

Rapid movement of bagged or packaged trash and linen from numerous outlying stations to a central collecting point is usually handled by this system. (Health codes require using separate tubes for trash and linen.) Linen systems are found generally in hospitals; trash systems in various facilities, frequently in conjunction with compactors. The system is basically a network of large pipes, negatively pressurized, with numerous loading stations throughout the building. Pipes are 16-, 18-, or 20-in. sizes, operating at

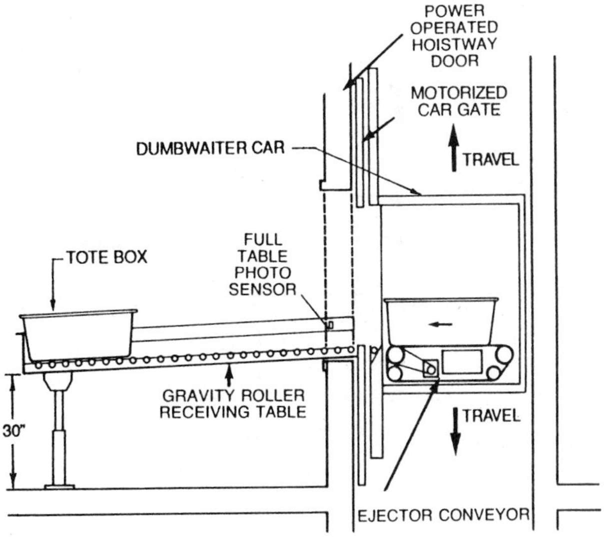

(a-1)

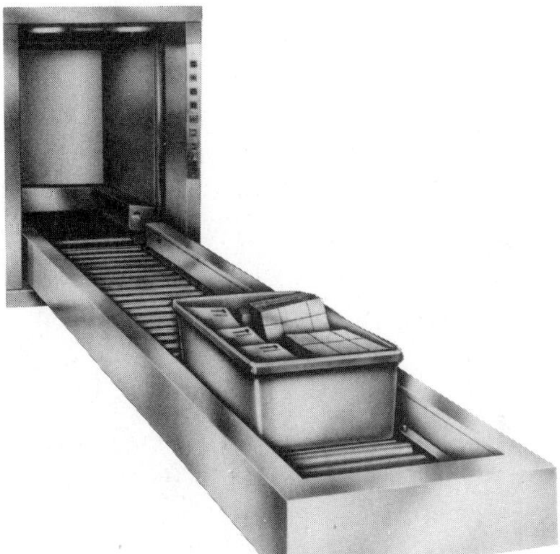

(a-2)

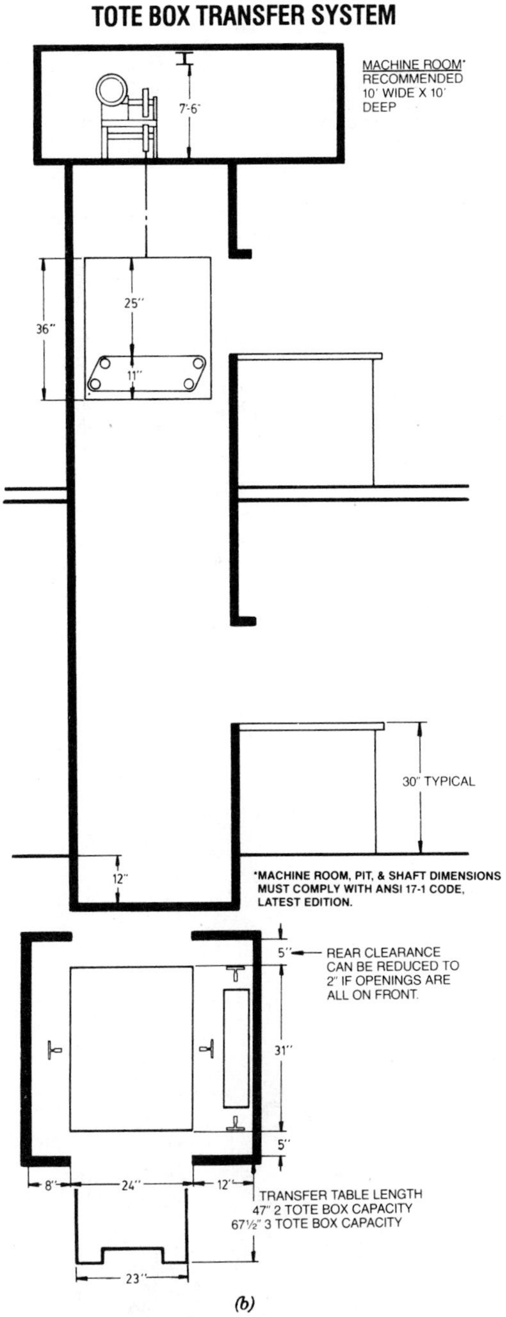

(b)

Fig. 24.23 This automated dumbwaiter or vertical ejection lift is designed to eject (unload) a container automatically at a preselected station, after manual loading and dispatching of the container (tote box) at another station. (Systems are available with automatic loading as well.) The tote box is carried on an ejection conveyor (a), *which electrically senses arrival at its destination and effects ejection. Station doors are automatically electrically operated. The dumbwaiter car drive mechanism is a counterweighted traction drive. Car capacity is 300 lb, although containers are normally limited to 50 lb for ease of handling. Car speeds are normally 50 to 150 fpm, with higher speeds available for large rise. A typical hoistway section is shown in* (b). *Vertical travel is not limited. (Courtesy of D. A. Matot.)*

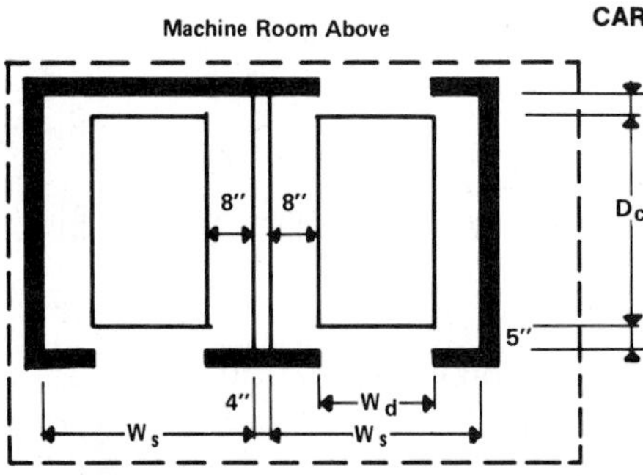

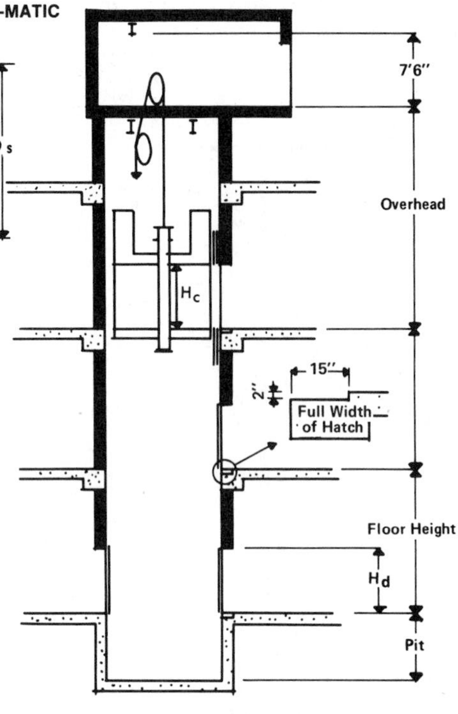

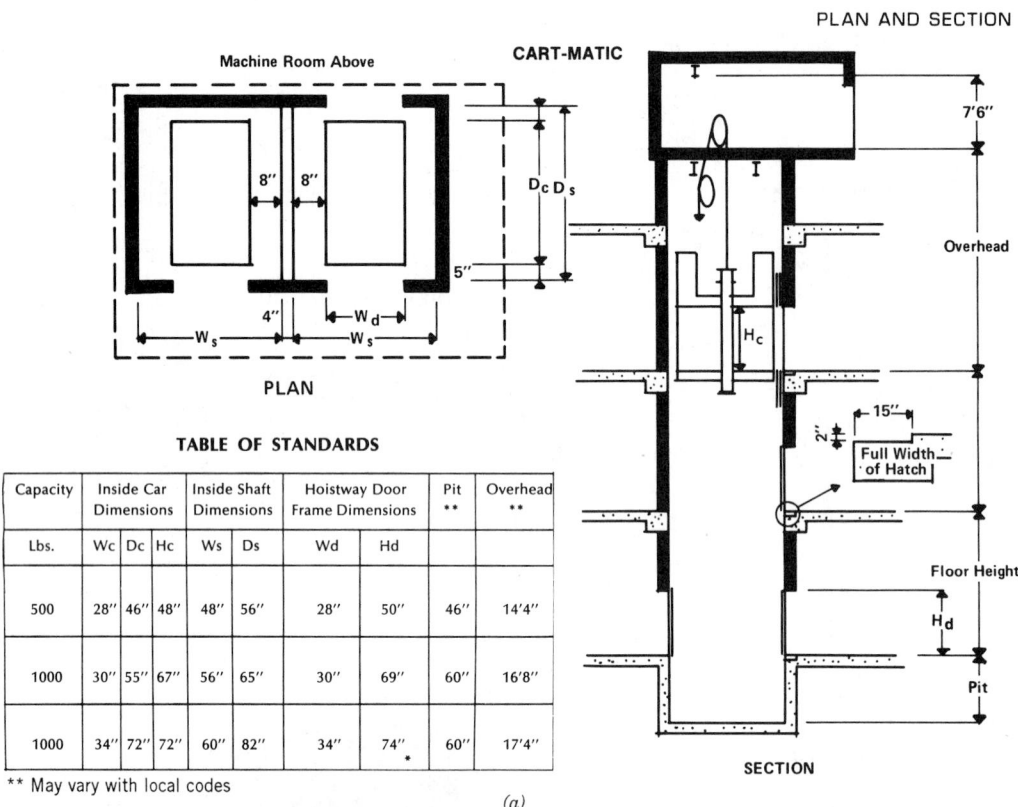

TABLE OF STANDARDS

Capacity	Inside Car Dimensions			Inside Shaft Dimensions		Hoistway Door Frame Dimensions		Pit **	Overhead **
Lbs.	Wc	Dc	Hc	Ws	Ds	Wd	Hd		
500	28"	46"	48"	48"	56"	28"	50"	46"	14'4"
1000	30"	55"	67"	56"	65"	30"	69"	60"	16'8"
1000	34"	72"	72"	60"	82"	34"	74"*	60"	17'4"

** May vary with local codes

(a)

***Fig. 24.24** (a) Plan, section, and dimensions for vertical ejection lift system (automated dumbwaiter). (b) Open ejection-lift unit showing cart ejection mechanism. Shaftway doors are vertical biparting. (c) The same unit being loaded with food carts and dispatched to the various floors of the hospital. At the upper floors the carts are rolled away by the attendants. Later the lifts are used to return soiled dishes and trays. (Courtesy of Courion Industries, Inc.)*

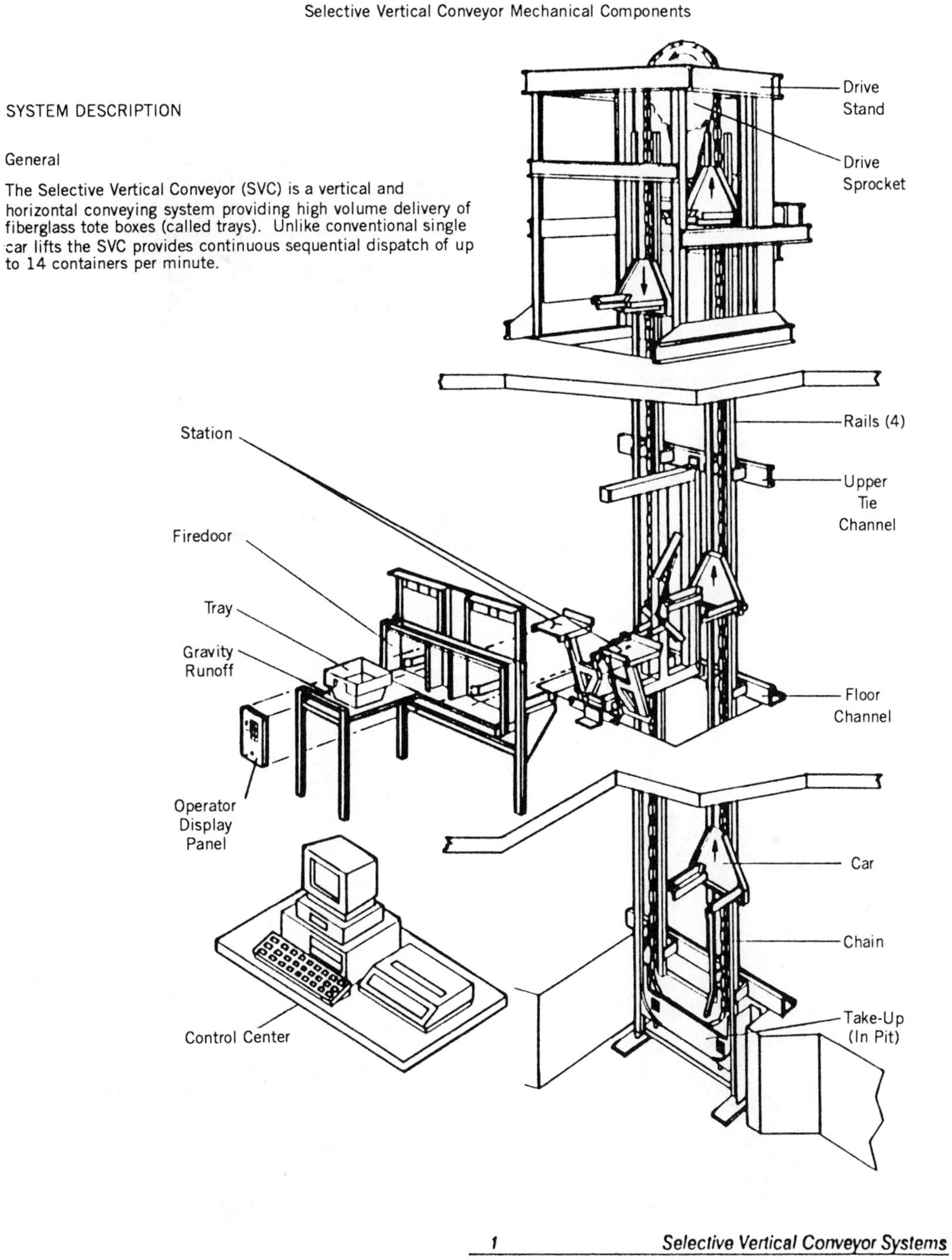

(*a*)

Fig. 24.25 (a) *System diagram showing operation of a selective vertical conveyor. Tote-box addressing and dispatching can be accomplished remotely from the control center.* (b) *Typical sending and receiving terminal. Stations are arranged vertically in a common shaftway. Each station contains a keypad for addressing and dispatching boxes. (Courtesy of TransLogic Corp.)*

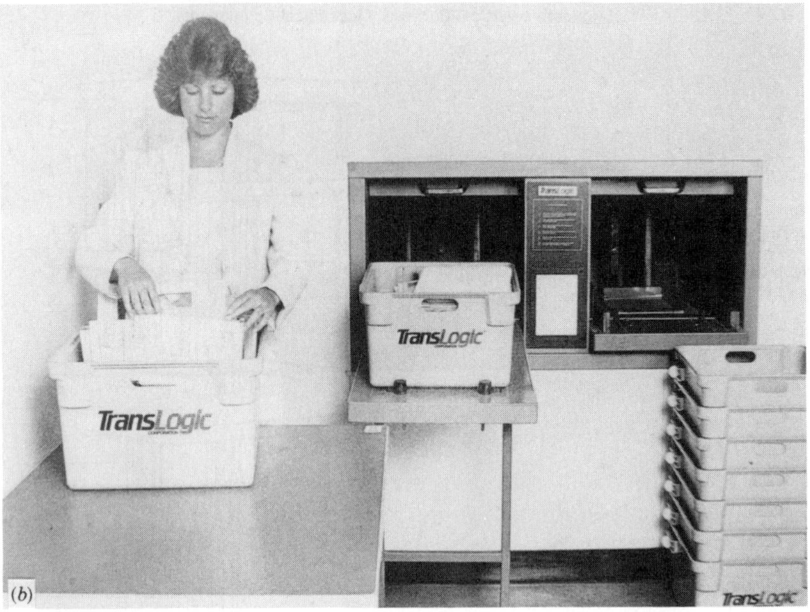

Fig. 24.25 (continued)

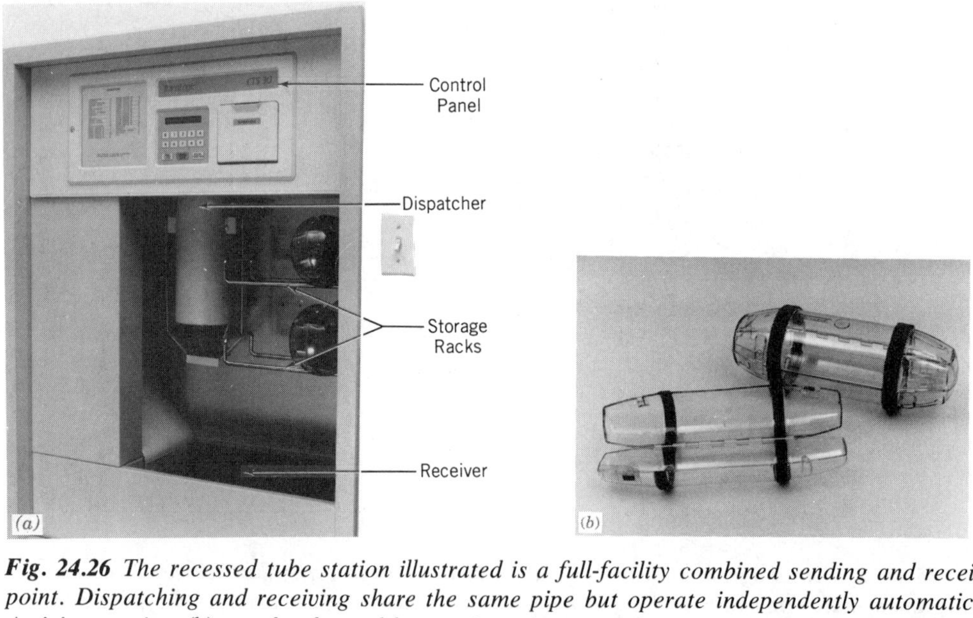

Fig. 24.26 *The recessed tube station illustrated is a full-facility combined sending and receiving point. Dispatching and receiving share the same pipe but operate independently automatically. Arriving carriers* (b) *are decelerated by an air cushion and drop into the front bin on arrival.* (b) *Carriers (3¾ in. D × 15½ in. L and 5¾ in. D × 16 in. L) are made of impact-resistant transparent polycarbonate plastic. Liners are used with fragile items. The bands around the carriers that maintain the air seal in the transport tubes are replaceable. (Courtesy of TransLogic Corp.)*

high static pressure. A system normally can handle only one unit load at a time, but moves it so quickly (20 to 30 fps) that system capacity is large and delays are not encountered. Material placed into a loading station is picked up as soon as the previous load clears. Compressors are large and very noisy, requiring considerable space allocation and acoustical isolation. In addition to the main vacuum system, a high-pressure air line is required to operate the doors, and sprinkler heads must be installed every few floors. Overall costs are low to moderate. For the specific task performed, a cheaper and more efficient transfer technique is difficult to find.

24.25 Automated Container Delivery Systems

This ingenious arrangement employs captive and secure containers locked onto a motorized carriage that, in turn, is locked onto the track system. Power for the motor in the carriage is picked off a third rail at 24 V d-c. The entire assembly moves horizontally or vertically with equal ease. Containers move at a constant 120 fpm horizontally but slower on rises, depending on container load and the steepness of the rise. Containers are available in a number of shapes and volumes to suit the particular installation. Two standard sizes are approximately 18 in. × 6 in. × 13 in. narrow profile and 18 in. × 12 in. × 8 in. low profile. Normal payload is 20 lbs, although for horizontal runs only considerably heavier loads can be carried.

In simple single- or dual-track point-to-point or loop systems, right-of-way conflicts cannot occur and routing is simply a matter of addressing the car. For complex systems involving loops and branches, routing is decided by a central computerized controller that finds the shortest route for each car (up to 250 cars per system), side-tracks, parks cars, and in effect, operates a miniature railroad type system. Addressing the cars is done at each station simply by dialing an address code and dispatching the car. Cars move quietly, unobtrusively, and rapidly throughout the system. The system is easily added onto a structure as a retrofit operation because of track flexibility and small size. Its major drawback is its high cost. Components are shown in Fig. 24.27.

24.26 Automated Self-Propelled Vehicles

These robot battery-powered vehicles follow a route determined by a passive guidance floor tape. They can be arranged to interface automatically with vertical transport means (elevators) so that the route can cover various levels in a facility. The floor tape, which can be installed below carpets, tile, and wood and directly on concrete, is entirely passive. It determines only the path that the vehicle will follow. All sensing, instruction, motive power, and vehicle control are located on the vehicle itself, which can carry approximately a 300-lb payload and can operate a full 8-h day without recharging its on-board batteries. Vehicle speeds are variable, ranging from 20 to 120 fpm. Programming of route, stops, vertical interfaces, changing of cycle, and other functions is accomplished with an external programming device and an on-board programmable controller. Applications for an automated vehicle of this type are limitless: parts delivery and pickup in industrial facilities, food and supplies distribution in hospitals, and mail and document pickup and delivery in offices are among its basic functions. Typical units are shown in Fig. 24.28.

24.27 Conclusion

The foregoing very brief summary simply describes the types of equipment available. For each facility being planned, the architect must study the material transfer problems, remembering that buildings not only handle and process but also generate material; an office building generates about a pound of waste per 100 ft^2 per day—a prodigious amount in today's large office structures. This type of dry waste can be compacted, bailed, and sold, unlike garbage and wet waste. In addition to considerations of the type of material being handled, there are factors of speed, scheduling, location of stations, labor and material costs, space requirements, noise generation, and energy requirements. To consider and evaluate all these factors in a large and complex facility is generally beyond the ability of the architect alone. Thus expert advice from consultants who specialize in materials handling and from manufacturers' representatives must be sought.

REFERENCE

Strakosch, G. R. (1983). *Vertical Transportation, Elevators and Escalators*, 2nd ed., Wiley, New York.

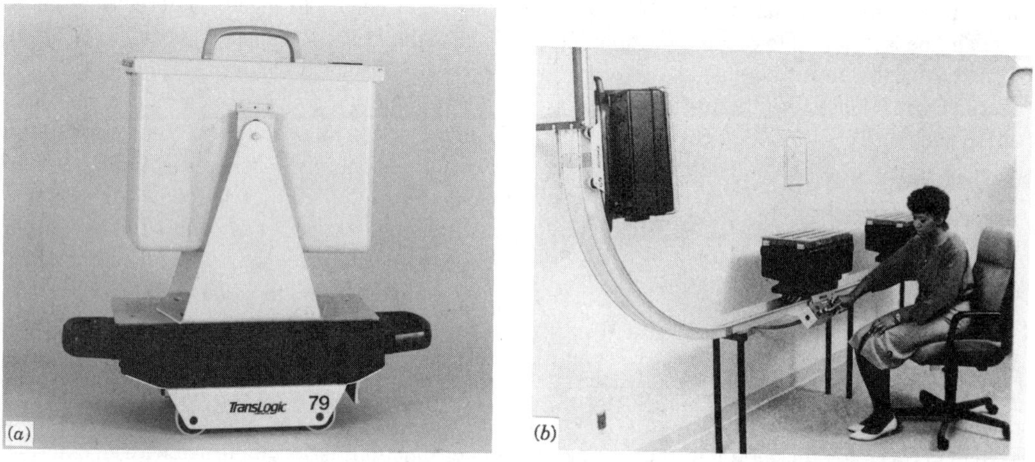

Fig. 24.27 (a) *One of many designs of carriage and container for an electric track vehicle system. This design uses a pivoted container that is maintained in a vertical position regardless of the carriage position, which may be horizontal, vertical, sloped, or even inverted.* (b) *Dispatching and receiving station for a track vehicle system. (Courtesy of TransLogic Corp.)*

1 Pneumatic Safety Bumper

2 Sensor Panel
Contains metal detection sensors which confirm presence of the guide tape to the vehicle guidance system.

3 Steering System
Tricycle wheel design.

4 Batteries

5 Coupling Unit
Transcar tows trolleys. This device is activated to couple Transcar with trolley for transport. When trolley is delivered to destination, coupling unit retracts and vehicle drives from underneath trolley.

6 Drive Motors
Each drive wheel is powered by a 24 VDC motor.

7 Control Unit
Includes all electronic equipment

8 Charging System

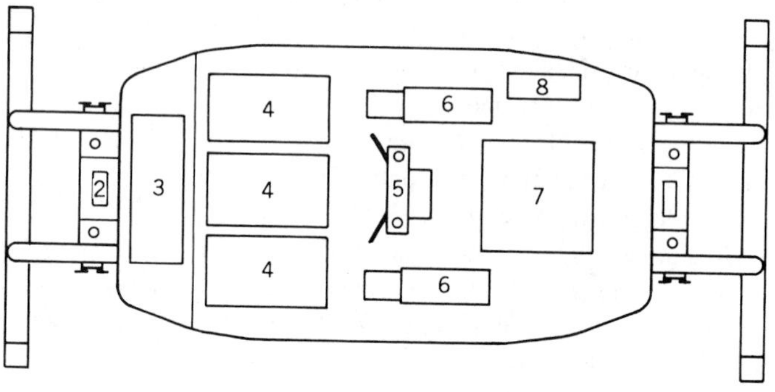

(a)

Fig. 24.28 *Components of the automated vehicle are shown in this block diagram* (a). *The vehicle fits underneath its load, as can be seen in* (b) *and* (d), *which show two typical applications; automated food cart delivery in a hospital and automated mail service in a large office. The external programming device is seen in* (c). *(Courtesy of TransLogic Corp.)*

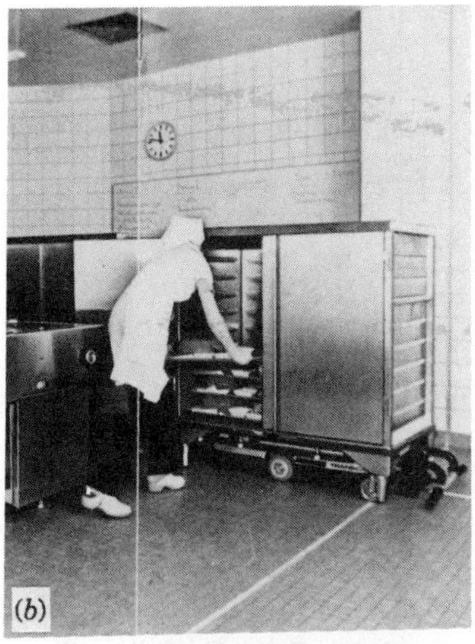

(b)

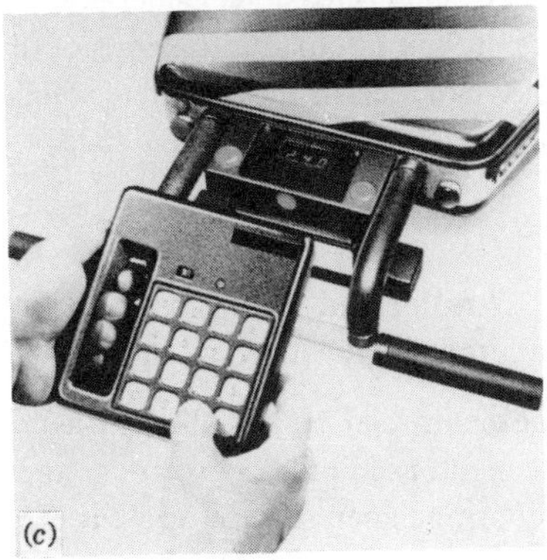

(c)

(d)

Fig. 24.28 (*continued*)

25
MOVING STAIRWAYS AND WALKS

MOVING ELECTRIC STAIRWAYS

25.1 General

The moving stairway is also referred to as an escalator or as an electric stairway. This section uses all three names. The escalator was first operated at the Paris Exposition in 1900. Its successors not only deliver passengers comfortably, rapidly, and safely, but also continuously receive and discharge their live loads at a constant speed with practically no delay at any landing. The annoyance of waiting for elevators is not present. Also, time is not lost by acceleration, retardation, leveling, or door operation nor by pressing hall buttons, by passenger interferences in getting in or out of the cars, and so on. One seldom sees a waiting passenger or congestion of passengers at the lighted combplate of an escalator.

Instead of formal lobbies and hallways leading to a bank of elevators on each floor, the electric stairway is always in motion, inviting passengers to ride. The corridors, aisles, and other passageways in existing buildings usually provide space for floor openings adequate for the installation of escalators. In contrast, it would in most cases be almost impossible to install an adequate bank of elevators in an existing building to meet the need for vertical transportation. Elevator hoistways must be vertical from bottom to top floors; an escalator installation can be "staggered" at various appropriate locations (see Fig. 25.1).

A standard stairway is assembled from three separate sections of structural truss—an upper section, a middle section, and a lower section similar to the upper one (see Fig. 25.7*a*). The middle straight section may be any desired length to provide rises for floor heights from 10 to 30 ft, for example. When the rise exceeds approximately 25 ft, depending on the unit's width and type, an intermediate support is located between the two end supports of the stairway where such is feasible. Generally, the upper corners of the bottom and top ends of the truss, after assembly, carry the complete weight of the stairway mechanism and its live load.

25.2 Parallel and Crisscross Arrangements

Moving stairs can be installed in a crisscross arrangement, as in Fig. 25.1*a*, or in either of two parallel arrangements, shown in Fig. 25.1*b* and *c*. The essential difference between the two plans is that in the crisscross arrangement the entrances (exit) to the up and down escalators are separated by the horizontal length of an escalator (see Fig. 25.2), whereas in either of the parallel arrangements the two escalators face in the same direction (see Fig. 25.12).

Thus in the crisscross plan little possibility of confusion exists since at each escalator location only one choice exists. This is considered by most designers to be a definite advantage, particularly in crowded buildings such as stores and where riders are generally unfamiliar with the layout. Furthermore, the crisscross arrangement is more economical of space than the standard parallel arrangement (Fig. 25.1*c*), as can be seen clearly from the plan. The stacked parallel arrangement (see Figs. 25.1*b* and 25.3) must be used with caution because of the inconvenience to the rider of an enforced long walk-around to continue the trip. This arrangement is found most often in mass-purchase-type facilities and in malls, since its raison d'être is to expose the

TRANSPORTATION

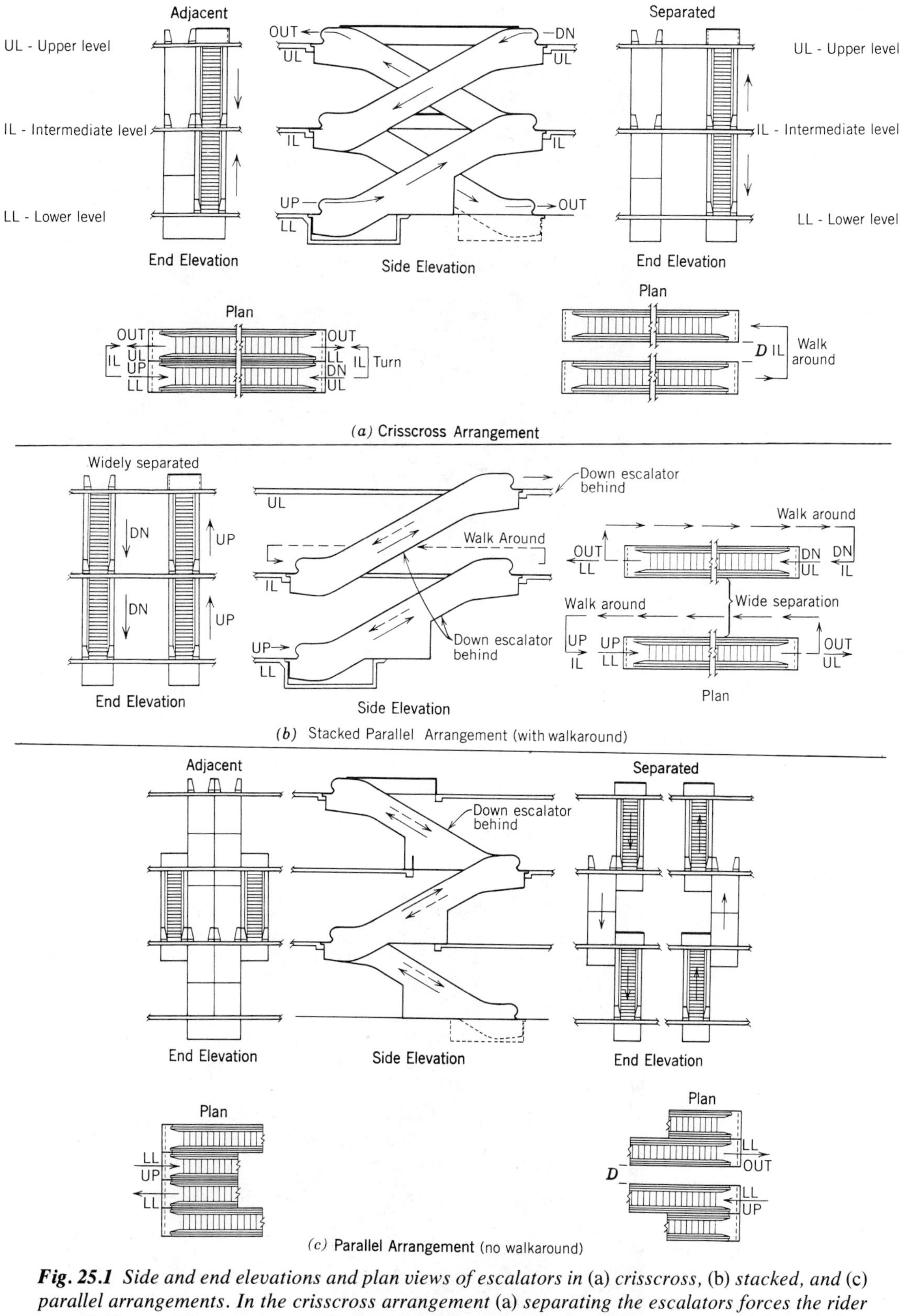

(a) Crisscross Arrangement

(b) Stacked Parallel Arrangement (with walkaround)

(c) Parallel Arrangement (no walkaround)

***Fig. 25.1** Side and end elevations and plan views of escalators in* (a) *crisscross,* (b) *stacked, and* (c) *parallel arrangements. In the crisscross arrangement* (a) *separating the escalators forces the rider to walk distance* D *to continue a trip. In the parallel arrangement* (c) *the separation* D *has no effect on the rider, who simply makes a turn to continue in the same up or down direction. In the stacked arrangement* (b) *the rider must traverse the entire length of the escalator at each level in order to continue since the escalators are stacked vertically on each floor.*

Fig. 25.2 *Upper terminal of two elegant glass balustrade escalators in crisscross arrangement. The large horizontal distance between the two is considered an advantage of the crisscross plan since it effectively eliminates confusion for riders. (Photo courtesy of O&K Rolltreppen, West Germany.)*

rider to attractive special sale and impulse purchase items which are deliberately placed along the length of the walk-around in a single store. In malls (see Fig. 25.3) this arrangement is much less objectionable since many of the people are there specifically to browse and window-shop, rather than to purchase and leave, as is the case in a single store. In a store, however, the inconvenience of this walk, which is frequently compounded by the crowds of people that normally congregate at special sale merchandise displays and counters, rapidly engenders annoyance. Thus the stacked arrangement is rarely used above two floors (one such walk-around), and

Fig. 25.3 *The stacked parallel arrangement of escalators at Stix, Baer & Fuller's Independence Mall store (Kansas City, Mo.), requires riders to walk around when traveling more than one floor. To avoid the annoyance of excessively long trips, riders are provided with the alternative of glass-cab hydraulic elevators. The glass escalator balustrades are architecturally consonant with the glass cab and the glass-protective barriers around the center well of the store. (Photo courtesy of Montgomery Elevator Co.)*

when it is, riders can be expected to gravitate to elevators. A similar, but generally much shorter and therefore less objectionable walk-around exists in the case of the crisscross plan, where the escalators are separated by distance D (see Fig. 25.1*a*). This distance is usually limited to about 10 ft because any greater distance places the trip-continuation escalator out of sight and causes not only annoyance at the enforced walk but also confusion and consequently resentment. Separation of the escalators is frequently an architectural consideration and does have the advantage of easier mixing of riders entering at the various levels with riders making a continuous trip. Designers must, however, be continuously aware of the possibility of a negative reaction to separation of escalators, which can be further reinforced when:

1. Insufficient floor space is provided for the transit between escalators, causing crowding, pushing, and delay.
2. Insufficient elevator service is provided for passengers wishing to travel at least three floors. This forces people to make a multistory escalator trip, which in itself can be wearying, particularly when carrying parcels. If such a trip is further lengthened by an enforced walk-around at each floor, it becomes a source of severe irritation, often sufficient to keep customers away from the store.

The consideration of division of rider traffic between elevators and escalators in a store is an important one. The general philosophy of store owners is to make escalators the primary means of vertical transportation, for the obvious reason of merchandise exposure. As a result, elevators are frequently placed at one or both ends of a store while the escalator banks are central. However, care must be taken to provide sufficient elevatoring in stores exceeding three floors, particularly in stores using the stacked or separated crisscross plans, because of the inconvenience of multifloor escalator trips.

25.3 Location

Because escalators are constantly moving and are generally part of a horizontal and vertical trip, they must be placed directly in the main line of circulation. This is in contrast to the elevator bank, which, being a vertical transportation unit, can be set off as an element on its own, for people to approach and utilize. Escalators must therefore be placed in the area served, and with a total and even dominating view of it. This allows potential riders to immediately:

1. Locate the escalators.
2. Recognize the individual escalator's destination.
3. Easily and comfortably move toward the escalator.

One of the most effective ways to disorient traffic movement is to inadequately or poorly mark escalator destinations. The resultant milling about, false starts, and general unhappiness can be observed in numerous otherwise well-designed buildings.

Sufficient lobby space must be provided at the base for queuing where anticipated, and most particularly at discharge points. A restricted or poorly marked area here will cause passenger hesitation and traffic backup. Since the escalator discharges continuously, backup of traffic is dangerous and therefore intolerable. This is particularly crucial in theaters and stadiums, where even momentary hesitation during peak traffic periods could be disastrous. To avoid this, four design steps, in descending order of importance, are taken:

1. Provide well-marked escalators with sufficient traffic-carrying capacity.
2. Provide collecting space at intermediate landings so that pressure can be relieved.
3. Provide a physical divider at intermediate landing turnaround points which will guide riders away from the discharge point and provide adequate space (and time) for riders either to leave at that level or to follow the guide around and continue the trip (see Fig. 25.4).
4. Provide a slight setback for the next escalator so that the necessary 180° turn can readily be negotiated (see Fig. 25.5).

At the landing, an escalator should discharge into an open area with no turns or choice of direction necessary. Where such is unavoidable,

Fig. 25.4 *Bank of glass balustrade escalators with mirrored sides, in a crisscross plan. Note the dividers at the turn-around at the intermediate level whose function is to guide traffic smoothly either away from the escalator or to the turn point. This effectively eliminates undesirable bunching and crowding. (Photo courtesy of O&K Rolltreppen, Germany.)*

large clear signs should make hesitation unnecessary. Landing space beyond the escalator newels should be 8 ft for a 32-in. unit and 10 ft for 40- and 48-in. units *minimum* for a standard 100-fpm speed. For escalators that will be reversed to accommodate change in traffic direction, this landing space must be provided at top and bottom.

The crisscross arrangement is generally favored because of lower cost, minimum floor space occupied, and lowest structural requirements. The parallel arrangement, being less efficient and more expensive, has as its only virtue a very impressive appearance that strongly draws people to it. For this reason it is frequently employed, particularly in multiple banks of three or four, in transportation terminals (see Fig. 25.6). In such installations, flexibility is maintained by operating all but one escalator in the direction of heaviest traffic. Reversibility of escalators provides this most desirable feature.

Fig. 25.5 *Single, 32-in. crisscross escalator in Lord & Taylor's Oakbrook Center store (Chicago). Note the setback of the descending escalator, which is helpful in making a smooth turn to the next escalator. By separating the two escalators horizontally, an enforced walk-through can be created. (Photo courtesy of Montgomery Elevator Co.)*

Fig. 25.6 The largest moving stairways in Western Europe are installed in the Stockholm subway system. This parallel bank of three escalators is 70 m (230 ft) long with a 33-m (108-ft) rise. Unlike American standards of 30° incline, 100-fpm speed, and 32- or 48-in. size, these units are at 27.3° incline, 45-m/min (147.6-fpm) speed, and 1000-m (40-in.) width. (Photo courtesy of O&K Rolltreppen, Germany.)

25.4 Size, Speed, Capacity, and Rise

Moving stairs are built according to manufacturers' and industry standards and are therefore available in standard designs. However, all major manufacturers will also produce special designs for particular applications. The data that follow refers to standard designs.

All escalators in the United States are installed at an angle of 30° from the horizontal, with a minimum vertical clearance of 7 ft for escalator passengers (see Fig. 25.17). The 30° inclination means that the rise is equal to 57% of the unit's projected floor area for its inclined portion. The length of the horizontal portions of the stairway depends on the specific design. Elongated newels (see Figs. 25.2 and 25.17) require at least two horizontal treads before the landing plate, whereas short newels (Fig. 25.5) use only a single horizontal tread at the landing. In older designs standard units were available in either 90- or 120-fpm speeds, with the latter requiring a longer horizontal entrance and exit to permit passengers to use the stair safely. Today, although the maximum linear speed permitted by the safety code (ANSI/ASME 17.1) is still 125 fpm, the industry has standardized on a single speed of 100 fpm. Sports facilities and transportation terminals which experience very heavy traffic for short periods are frequently equipped with two speeds (100 and 120 fpm), with the higher speed used only during heavy traffic since the 120-fpm speed presents some difficulty to the less agile or hesitant passenger.

A recent change in the code has redefined the size designation of escalators. The previous *width* designations of 32, 40, and 48 in. referred to the distance between balustrades at the hip level (i.e., between handrails). Since this was not a particularly useful figure, and one that varied slightly with unit design, the new *width* designation refers to the width of the stair tread, in inches. The old width dimension now is referred to as *size*. The standard sizes and widths are:

Escalator Size (in.)	*Tread Width (in.)*
32	24
40	32
48	40

Treads are nominally 16 in. deep with an 8-in. rise.

Although units are available in all three sizes

TABLE 25.1 **Escalator Passenger Capacity**

Size (in.)	Tread Width (in.)	Speed (fpm)	Passengers per Hour Maximum[a]	Design[b]	Observed[c]
32	24	100	5200	4000	2300
40	32	100	7300	5300	2900
48	40	100	9000	6750	4500

[a]Theoretical maximum (see text).
[b]Heavy loading (see text).
[c]Average long-period loading.

(32, 40, 48 in.), the bulk of manufacture is concentrated in the 32- and 48-in. sizes (24- and 40-in. tread width) because of carrying capacity. Although theoretically the intermediate-size 40-in. escalator (32-in. tread) can carry 40 to 50% more passengers than the 32-in. size (24-in. tread width), in actual practice this is not the case and little advantage is gained from the additional cost of the 40-in. size over the 32-in. size moving stair. As a result, the 40-in. size escalator is something of a novelty. Table 25.1 lists theoretical, nominal (design), and observed average escalator passenger capacities. Maximum loads assume approximately 1¼ persons per tread for a 32-in. unit, 1½ persons per tread for a 40-in. unit, and almost 2 persons per tread for a 48-in. unit. In actuality, maximum capacity is approached only during peak-load periods in transportation terminals and stadiums. At other times full (heavy) load is represented by the nominal capacity figure and average (long period) load by the observed capacity figure. Although a 40-in. tread can indeed carry 2 persons, psychological factors plus physical ones, such as bulky clothing, packages, purses, and briefcases, mitigate against such loading. As a result, on 40-in.-wide treads one person uses each step in a diagonal pattern and on 24-in. stairs one person occupies every other tread. The 32-in. tread has slightly larger capacity than the 24-in. unit.

Some major manufacturers make two standard models (not styles) of moving stairs: a standard-duty unit intended for general indoor use, which provides low to medium rise, and a heavy-duty, sturdier unit intended for all-weather heavy-traffic use such as at transport terminals. The latter, because of stiffer heavier construction, can provide a greater maximum rise than that of standard units. Maximum rise for off-the-shelf design units varies among manufacturers. Approximate maximum figures are given in Table 25.2.

Specially designed units are available with rises of up to about 60 ft. In escalator design all the motive power is delivered at one point; that is, the drive motor drives the main chain, which drives the top sprocket, which drives the step chain, which pulls up the steps, causing the entire assembly to move. This arrangement is suitable for moderate rises of up to approximately 25 ft. Beyond that, the design becomes increas-

TABLE 25.2 **Approximate Maximum Escalator Rise**

Unit Size (in.)	Type	Supports	Maximum Rise (ft)
32			24
40	Standard	Ends	20
48			16
32		Ends	30
40	Standard	and	24
48		center	20
32			24
40	Heavy	Ends	20
48			18
32		Ends	40
40	Heavy	and	30
48		center	20

ingly inefficient. As the rise increases, the loads on all the drive components, including chains and sprockets, increase sharply. Furthermore, to accommodate the heavier equipment necessitated, truss width increases, as does wellway size and balustrade decks. For rises above 25 to 35 ft (depending on unit width) the drive motor is too large to fit inside the truss and requires a separate machine room below the truss, with attendant cost. All these factors combine to limit escalators to a maximum rise of 60 ft (varies slightly between manufacturers).

25.5 Components

The major components of a standard escalator installation are shown in Fig. 25.7. Safety devices are discussed in Section 25.6.

The truss is a welded steel frame that supports the entire apparatus. The tracks are steel angles attached to the truss on which the step rollers are guided, thus controlling the motion of the steps. The sprocket assemblies, chains, and machine provide the motive power for the unit, somewhat similar to the chain drive of a bicycle. The control cabinet, which is normally located near the drive machine, contains malfunction indicators in addition to the drive controls. The cabinet may also contain a microprocessor malfunction analyzer and communication means for transmitting escalator operating conditions to a central control point. An emergency stop button wired to the controller and placed near or on the escalator housing at both ends will stop the drive machine and apply the brake (see lower left of escalator, Fig. 25.5).

Key-operated control switches at the top and bottom newels will start, stop, and reverse the stairway. The handrail is driven by sheaves powered from the top sprocket assembly. It is synchronized with the tread motion to provide stability to riding passengers and a support for entering and leaving passengers. Handrails disappear at inaccessible points at newels. The balustrade assembly is designed for maximum safety of persons stepping on or off the escalators.

A particularly attractive design utilizing a transparent balustrade made of tempered glass is illustrated in Fig. 25.2. They are frequently referred to as crystal balustrades. In these units, the handrail is pinch-driven within the truss. In addition to metal and glass as balustrade materials, fiberglass, wood, and various plastic materials are also used.

25.6 Safety Features

Protection to passengers during normal operation is assured by a number of safety features associated with moving stairways:

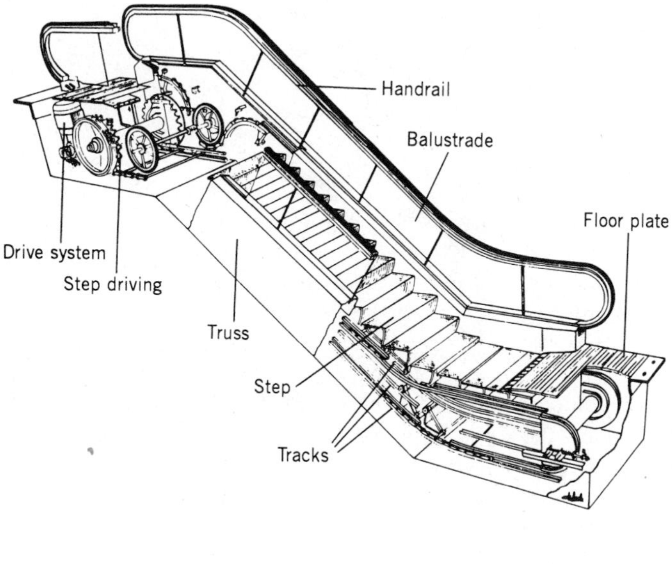

Fig. 25.7 *Cutaway view of a standard escalator showing principal parts. (Drawing courtesy of Kone.)*

TRANSPORTATION

1. Handrails and steps travel at exactly the same speed (100 fpm) to assure steadiness and balance and to aid stepping on or off the combplates.

2. The steps are large and steady, and are designed to prevent slipping.

3. Step design and step leveling with the combplates at each landing prevent tripping upon entering or leaving the escalator. This is accomplished with two or three (depending on manufacturer) horizontal steps at each end of the escalator.

4. The balustrade is designed to prevent the catching of passengers' clothing or packages. Close clearances provide safety near the combplates and step treads.

5. Adequate illumination is provided by the building at all landings, at the combplates, and completely down all stairways. Some escalator designs provide built-in lighting, as discussed in Section 25.8.

6. An automatic service brake will bring the stairway to a smooth stop if:

(a) The drive chain or the step chain is broken or abnormally stretched.
(b) A foreign object is jammed into the handrail inlet, between the skirt guard and step or between steps, causing them to separate.
(c) A power failure occurs.
(d) The emergency stop button is operated (one is located at each end of the escalator).
(e) Any of the fire safety system devices operates (see Section 25.7).
(f) A tread sags, rises, or breaks.
(g) A drive motor malfunction occurs.

7. In case of overspeed or underspeed, an automatic governor shuts down the escalator, prevents reversal of direction (up or down), and operates the service brake.

If the escalator is stopped by operation of a safety device, passengers can then walk the steps as they would any stationary stairway.

25.7 Fire Protection

Four methods of affording protection in case of fire near escalators are available: the rolling shutter, the smoke guard, the spray-nozzle curtain, and the sprinkler vent. One of these methods is required by code when more than two floors are pierced. Figure 25.8 illustrates clearly how the wellway at a given floor level may be entirely closed off by the fire shutter, thus preventing draft and the spread of fire upward through escalator wells. The movement is actuated by temperature and smoke relays that automatically actuate the motor-driven shutters. The shutter in Fig. 25.8 is shown at the third floor level, but other shutters may be installed at the tops of horizontal wellway openings at any floor.

Figure 25.9 illustrates the smoke-guard method of protection. It consists of fireproof baffles surrounding the wellway and extending downward about 20 in. below the ceiling level. Smoke and flames rising upward to the escalator floor opening meet a curtain of water automatically released from sprinkler heads of the usual type, shown at the ceiling level. The baffle is a

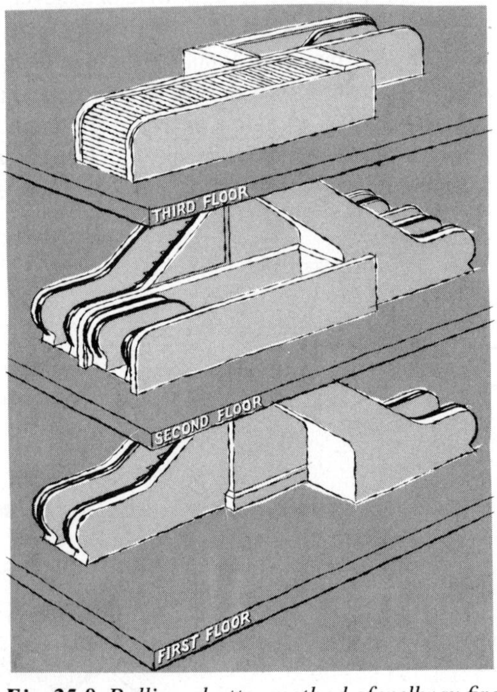

Fig. 25.8 Rolling-shutter method of wellway fire protection.

TRANSPORTATION

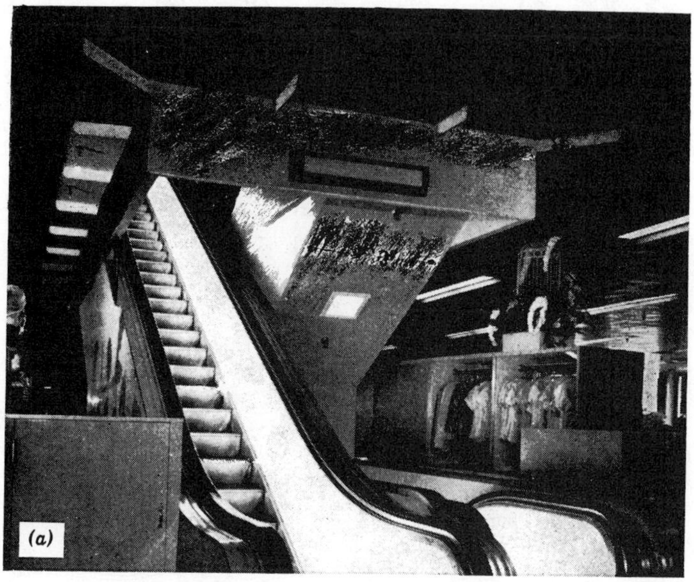

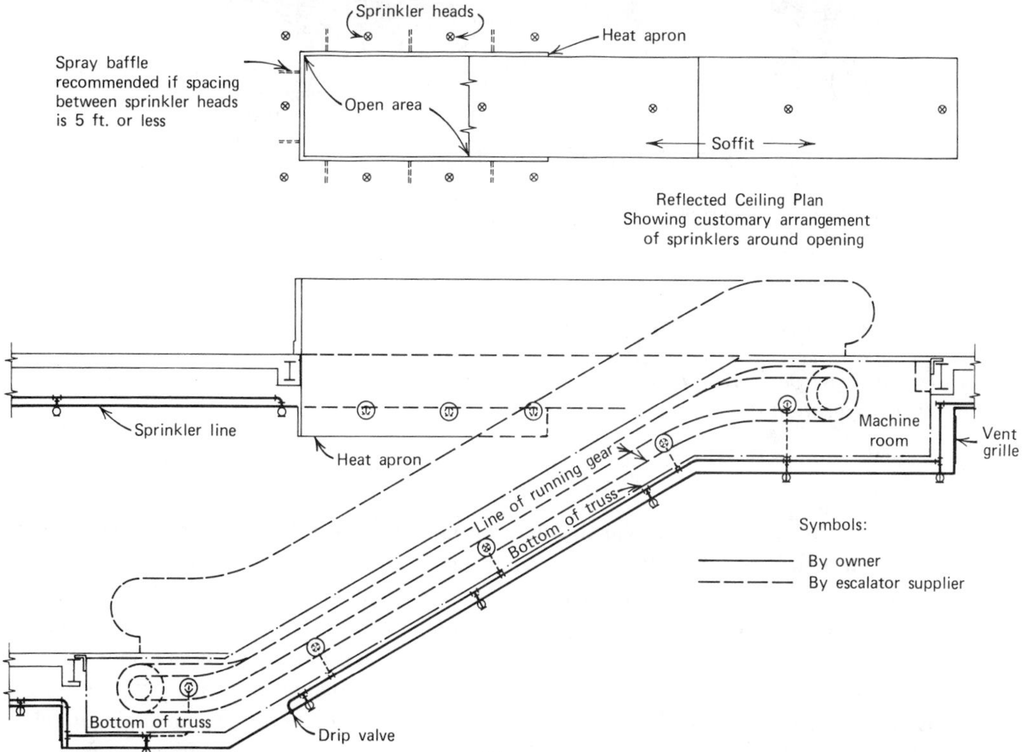

Fig. 25.9 (a) *Smoke-guard method of fire protection for a 32-in. moving stairway, crisscross type. The escalator floor opening (per floor) is approximately 4 ft 4 in. by 14 ft 8 in.* (b) *Reflected ceiling plan and section showing baffle and sprinkler layout. (Courtesy of Otis Elevator Co.)*

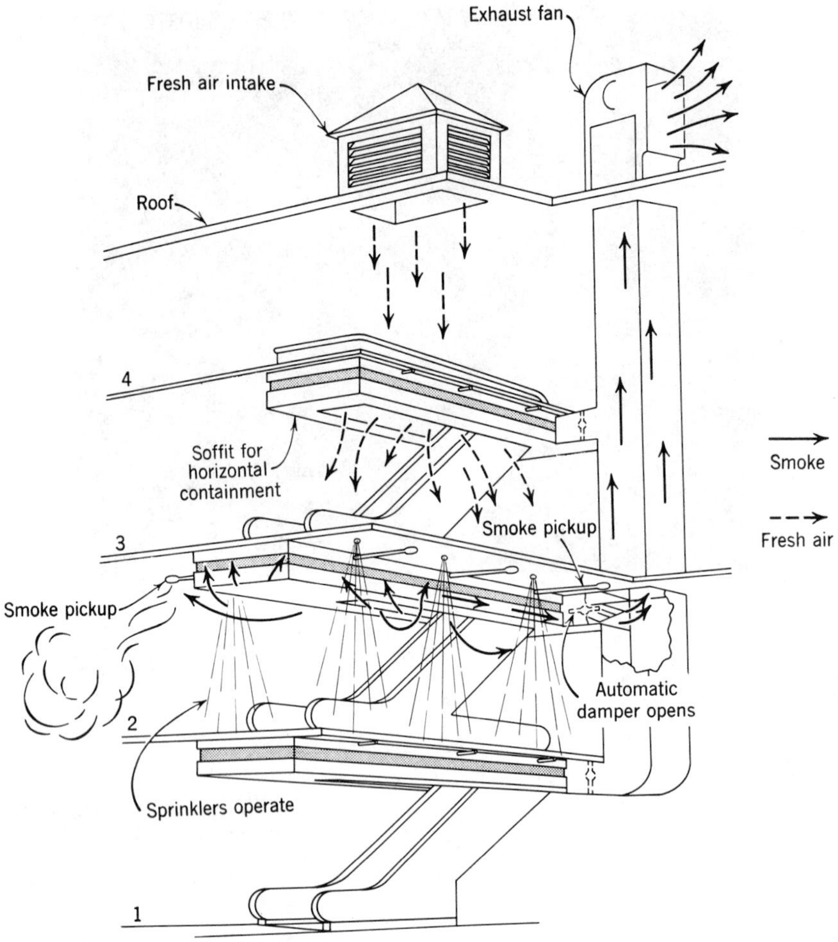

Fig. 25.10 Sprinkler vent fire protection for escalator openings. An exception (with control) to the rule against perforation in floors.

smoke and flame deflector. The vertical shields between adjacent sprinklers ensure that the spray from one will not cool the nearby thermal fuses, preventing the opening of adjacent sprinklers.

The spray-nozzle curtain of water (not shown) is quite similar to the smoke-guard protection. Here closely spaced, high-velocity water nozzles form a compact water curtain to prevent smoke and flames from rising through the wellways. Automatic thermal or smoke relays open all nozzles simultaneously.

The sprinkler-vent fire control is shown in Fig. 25.10. The fresh air intake housed on the roof contains a blower to drive air downward through escalator floor openings, while the exhaust fan on the roof creates a strong draft upward through an exhaust duct; this duct in turn draws air from the separate ducts just under the ceiling of each moving stairway floor opening. Three such separate wellway ducts are shown. Each duct has a number of smoke-pickup relays that automatically start the fresh air fans. The usual spray nozzles on the ceiling near the stairways aid in quenching the fire.

25.8 Lighting

Adequate illumination of a moving stairway, particularly at the landings, is important from

***Fig. 25.11** The balustrade section of an escalator can be used to supply the required task (stair) lighting.* (a) *A continuous flourescent strip is installed at the balustrade base at Fox Hills Mall (Culver City, Calif.). (Courtesy of Montgomery Elevator Co.)* (b) *A similar source is placed under the handrail of the crystal balustrade in this covered mall in West Germany. Note the additional light at the base for illuminating the combplate. (Courtesy of O&K Rolltreppen, Germany.)*

decorative as well as safety standpoints. In a stairwell-type installation, where general-area lighting does not provide sufficient illumination for the escalator, lighting consonant with the adjacent illumination is installed on the ceiling above the stairway, with special emphasis on lighting the combplate. Thus for instance in Fig. 25.5 the general illumination is supplemented by down-lights above the stairs. In Fig. 25.6 banks of fluorescent lights are placed across the escalator bank at frequent intervals along the rise. Note the additional concentrated light at the combplate. Two different lighting treatments of similar installations are shown in Fig. 25.11.

25.9 Application

1. Main floor locations should be chosen in the direct flow of traffic to assure maximum use.

2. Vertical arrangements should be made to accomplish specific purposes, such as exposure of merchandise, maximum passenger capacity, and maximum accessibility to various areas.

3. The aspect of reversibility of an electric stairway should be considered in applications where major traffic flow is unidirectional. Light traffic in the reverse direction can be handled by a normal fixed stair, adjacent to the escalator (see Fig. 25.12). Similarly, a bank of two escalators can operate either both up, both down, or one up and one down to handle variable traffic conditions in such areas as office buildings and transportation terminals.

4. Exterior escalators can provide an attractive, interesting, and economical solution to transporting people to selected entry points in a building without the necessity of extending the building to cover the entrance (see Fig. 25.13).

25.10 Elevators and Escalators

The use of elevators and escalators should be considered together, as a single problem in vertical transportation as applied to the particular facility being designed. In this connection, and particularly in case of modernization, Fig. 25.14 provides an interesting comparison.

In certain facilities there are times during which demand for vertical transportation is so large that elevators are not a feasible solution. A

Fig. 25.12 *A conventional arrangement provides a parallel bank of two electric stairways separated by a fixed stair in this Canadian application (Place des Jardins, Montreal). Reversibility of the escalators provides the desired flexibility, although normal operation is one up and one down. The fixed stair accommodates those who cannot or will not use a moving stair, as well as reverse traffic when both escalators are operating in the same direction. (Photo courtesy of Montgomery Elevator Co.)*

prime example is the school building. During class change, virtually the entire building population moves, with as much as 80% moving between floors. Since class change time is at most 10 min, the only reasonable solution is the combined use of fixed and moving stairs. In other buildings such as multifloor stores, the escalator provides for short trips of one or two floors, and the elevator generally transports passengers traveling three or more stories.

A comparison of travel time between escalator and elevator is of interest. Using a normal speed of 100 fpm, a 12-ft floor requires 14.4 s for travel plus about 5 to 6 s to turn, for a total of

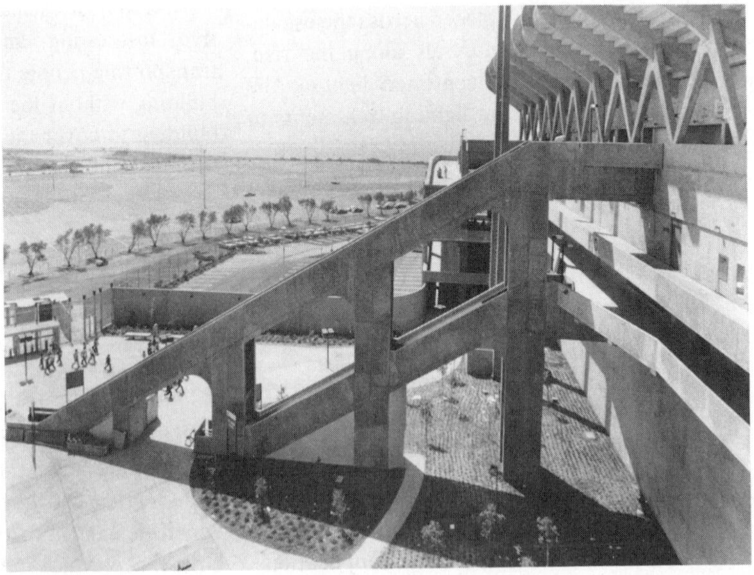

Fig. 25.13 *An attractive exterior escalator installation avoids the necessity of interior stairs and escalators, while providing an item of architectural interest at San Francisco's Candlestick Park. (Courtesy of Montgomery Elevator Co.)*

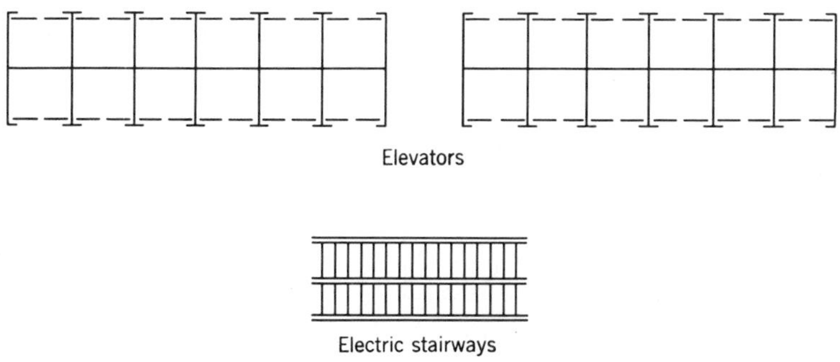

Fig. 25.14 Comparative space requirements for equivalent passenger-handling capacity. Note the marked space savings offered by escalators.

approximately 20 s. Thus a four-story trip would take approximately 80 s. A similar elevator trip would take at most 60 s. The additional escalator time is not noticed in the activity of boarding, turning, and riding. However, a trip of more than four stories becomes tiresome, and all the more so with the enforced ''walkaround'' at each floor in the separated arrangement (see Section 25.2).

25.11 Electric Power Requirements

Standard American electric stairways are driven by three-phase, 60-Hz, a-c induction motors at standard voltages (208, 230, and 460 V). Horsepowers of driving motors are shown in Table 25.3.

It is recommended that no more than four escalators be served by a single electric feeder, and further that not all the escalators of an installation, whatever the number, be served from the same feeder.

Since obviously one cannot be trapped on an escalator, emergency power is rarely required. Ventilation for the machinery should be supplied for approximately 40% of the power to be dissipated as heat. Thus a 10-hp motor would require the dissipation of 0.40 × 10 × 2500 Btu/hp or approximately 10,000 Btu/h.

TABLE 25.3 **Typical Standard-Duty Escalator Motor Sizes**

Escalator Size (in.)	*Maximum Rise (ft)*	*Size of Motor (hp)*
32	14	5
	22	7½
	30	10
48	10	5
	15	10
	20	15

25.12 Special-Design Escalators

As with elevators, escalators of nonstandard design are available on special order. These include units with slopes other than 30°, speeds other than 100 fpm, and rises beyond normal. The most unusual of these special designs, illustrated in Fig. 25.15, is the curved escalator, which made its debut overseas and is now available in the United States. This design, which solves the very common problem of directing

Fig. 25.15 Exposition model of a curved escalator which makes a 90° turn in it rise. These unconventional units offer an elegant (and expensive) solution to the problem of directing traffic to turn after arriving at the escalator's upper terminal. (Photo courtesy of Otis Elevator Co.)

Fig. 25.16 Escalator dimensional data in a range representing most of the major manufacturers. These data can be used for preliminary planning only.

		Range From	Range To
A		13'-2"	16'-10"
B		7'-5"	9'-4"
C		5'-9"	7'-6"
D		3'-3"	4'-2"
E		3'-3"	3'-9"
F		12'-5"	14'-10"
G		2'-11"	3'-1"
H		2'-7"	2'-9"
I	32"	3'-11"	4'-6"
I	48"	5'-3"	5'-8"
J		0	1"
K		2'-11"	3'-2"
L	32"	4'	4'-4"
L	48"	5'-3"	5'-10"
M	32"	2'	2'
M	48"	3'-4"	3'-4"
N		1'-9"	2'-2"
O		3'-6"	4'-0"

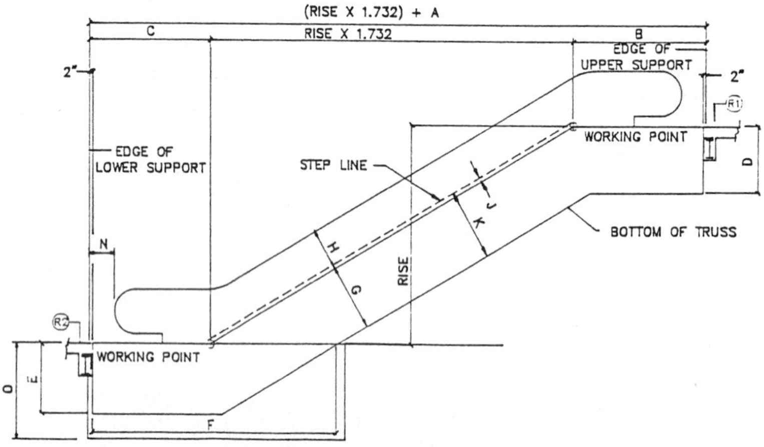

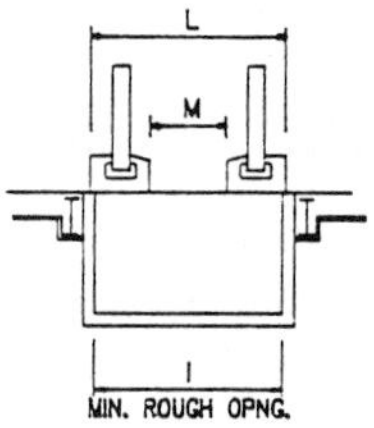

Glass Balustrade

PRELIMINARY ESCALATOR SPACE REQUIREMENTS

passengers to make a 90° turn as they leave the escalator, is an engineering tour-de-force. The entire drive mechanism had to be designed anew because tread speed along the inside curve must be lower than that on the outside curve. Handrail drive was less of a problem since the handrail is narrow and therefore speed variation within it is minimal. As of this writing, these units are coming on the market in the 90° turn design, but the drive problem having been solved, other angles will be available as well.

25.13 Preliminary Design Data and Installation Drawings

At the preliminary design stage, the architect requires rough dimensional and structural data for escalators. Since at this stage of design a specific manufacturer for the stairs has not yet been selected, and because these data vary from one supplier to another, the information in Fig. 25.16 is given as a range that covers most of the major manufacturers. Similar data are available from the NEII. Once construction contracts have been awarded and an escalator supplier selected, exact data on space requirements and structural reactions, similar to that shown in Fig. 25.17, are made available. It will be seen that two "working points" are identified, between which a very strong steel wire is tightly stretched. From these two points all other measurements are made, that is, locating the centerline of the truss sections, placing the lower and upper landing truss support beams, and so on.

25.14 Budget Estimating for Escalators

The cost of an escalator includes the cost of the associated mechanical and electrical equipment, plus the shipping installation charges. The manufacturer provides expert engineering and a field erector who supervises the installation.

A tabulation of *relative* escalator prices on a base of 100, is given in Table 25.4. Prices for units with rise above 35 ft rise very rapidly and depend on the type of unit used. The designer is referred to the suppliers for quotes on all units. To these figures must be added the cost of builder's work, wellway protection, lighting, outside balustrades, and plaster.

TABLE 25.4 **Relative[a] Escalator Prices**

Rise (ft)	*Width*	
	32 in.	*48 in.*
14	100[a,b]	118
16	103	121
18	105	124
20	108	127
22	111	130
24	113	133
26	115	135
28	118	138
30	120	141

[a]Base price figure is 100 for the shortest, narrowest, slowest unit (i.e., 14-ft rise, 32 in. wide). All units are 100 fpm. Add 5% for glass balustrade for any unit.

MOVING WALKS AND RAMPS

25.15 General

Moving walks and ramps are different from moving stairways in application, function, construction, and capacity. Escalators have as their primary function the movement of large numbers of people vertically, when such vertical distance does not exceed approximately five stories, as noted above. This specific transportation function the moving stair performs extremely well with minimum cost, space, and maintenance.

When vertical transportation of wheeled vehicles and large parcels is required, the use of an electric stairway is awkward, if not entirely impossible. For such functions and others discussed below, the moving ramp may best be utilized.

Unlike the elevator and escalator, the moving walk or ramp serves a dual function, that is, horizontal transportation only, or a combined function of horizontal and vertical transportation. For the purpose of our discussions, we will define a moving *walk* as one with an incline not exceeding 5° where the principal function is horizontal motion and inclined motion is incidental to the horizontal. A moving *ramp* is a device with an incline limited by code to 15°, where vertical motion is as important as or more important than the horizontal component. It should be understood that the walk and ramp are physically the same device, differently applied.

25.16 Application of Moving Walks

The principal uses of moving walks or moving sidewalks, as they are sometimes called, are to:

1. Eliminate and/or accelerate burdensome walking.
2. Eliminate congestion.
3. Force movement.
4. Easily transport large, bulky objects.

Anyone who has walked the seemingly endless distances in a major airport, carrying a heavy suitcase, can appreciate the near-absolute necessity for a moving walkway. It is for this reason that transportation terminals have become

ESCALATOR/ESCAL-AIRE
506 SERIES
LAYOUT DETAILS
ARR. TWO FLAT STEPS
ELONGATED NEWELS

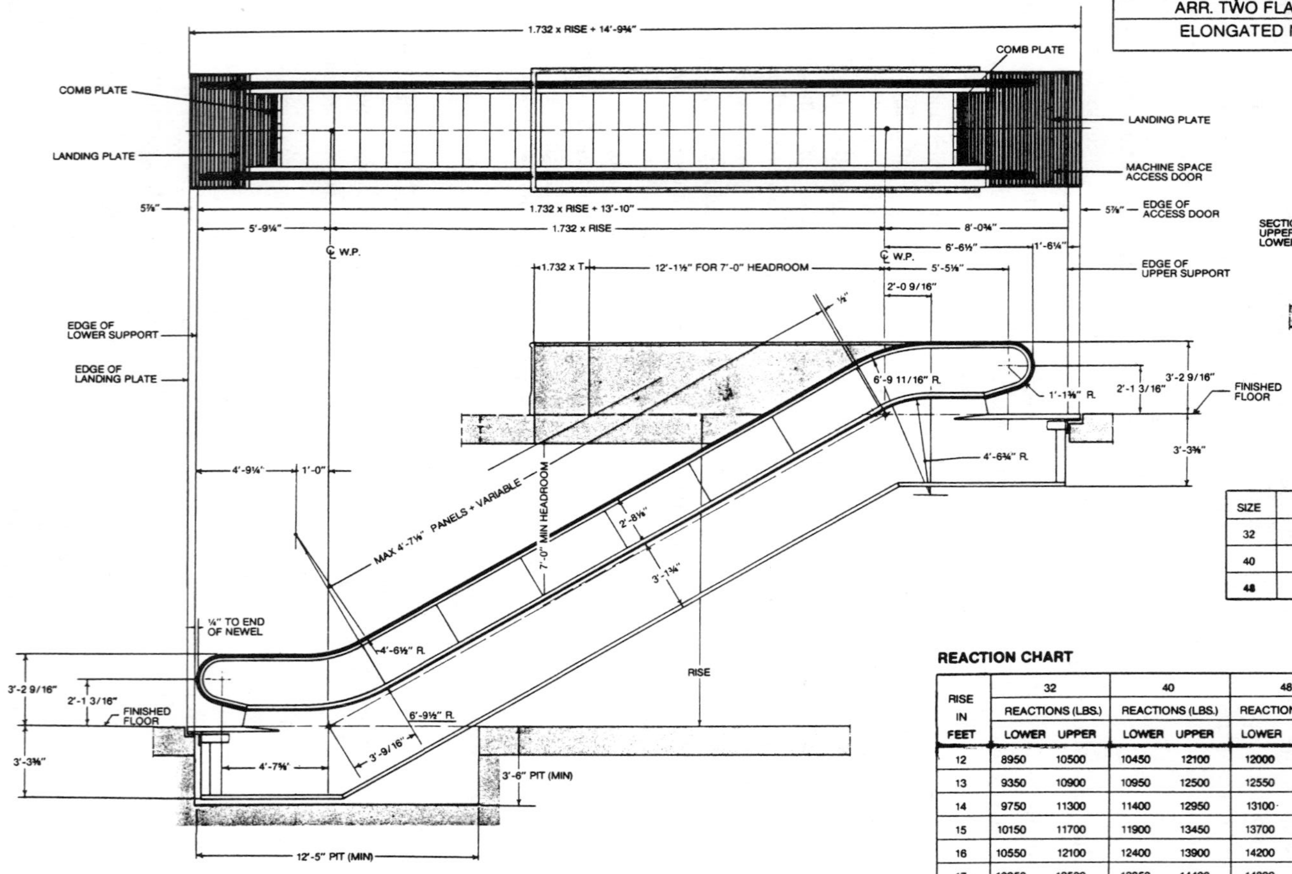

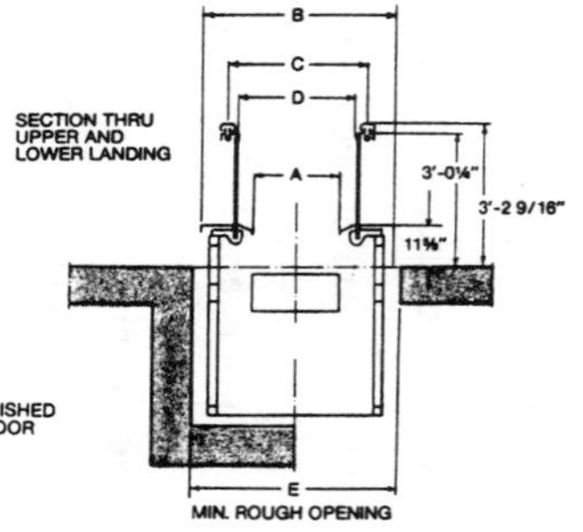

SIZE	A	B	C	D	E	G
32	2'-0"	3'-11½"	3'-0¼"	2'-8"	4'-1½"	1'-11⅛"
40	2'-8"	4'-7½"	3'-8¼"	3'-4"	4'-9½"	2'-7⅛"
48	3'-4"	5'-3½"	4'-4¼"	4'-0"	5'-5½"	3'-3⅛"

REACTION CHART

RISE IN FEET	32 REACTIONS (LBS.) LOWER	32 REACTIONS (LBS.) UPPER	40 REACTIONS (LBS.) LOWER	40 REACTIONS (LBS.) UPPER	48 REACTIONS (LBS.) LOWER	48 REACTIONS (LBS.) UPPER
12	8950	10500	10450	12100	12000	13550
13	9350	10900	10950	12500	12550	14100
14	9750	11300	11400	12950	13100	14700
15	10150	11700	11900	13450	13700	15200
16	10550	12100	12400	13900	14200	15800
17	10950	12500	12850	14400	14800	16300
18	11300	12900	13300	14900	15300	16900
19	11750	13300	13800	15350	15900	17400
20	12050	13600	14100	15700	16250	17800

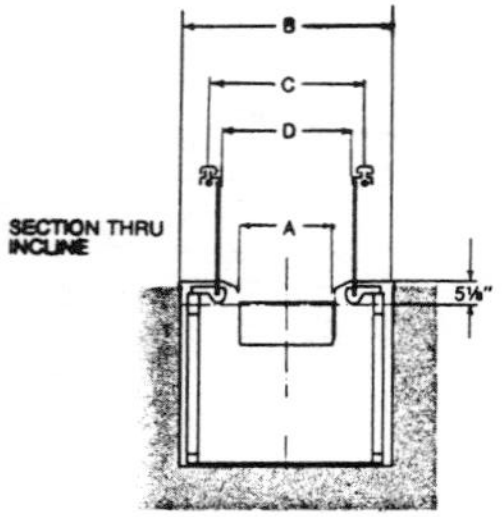

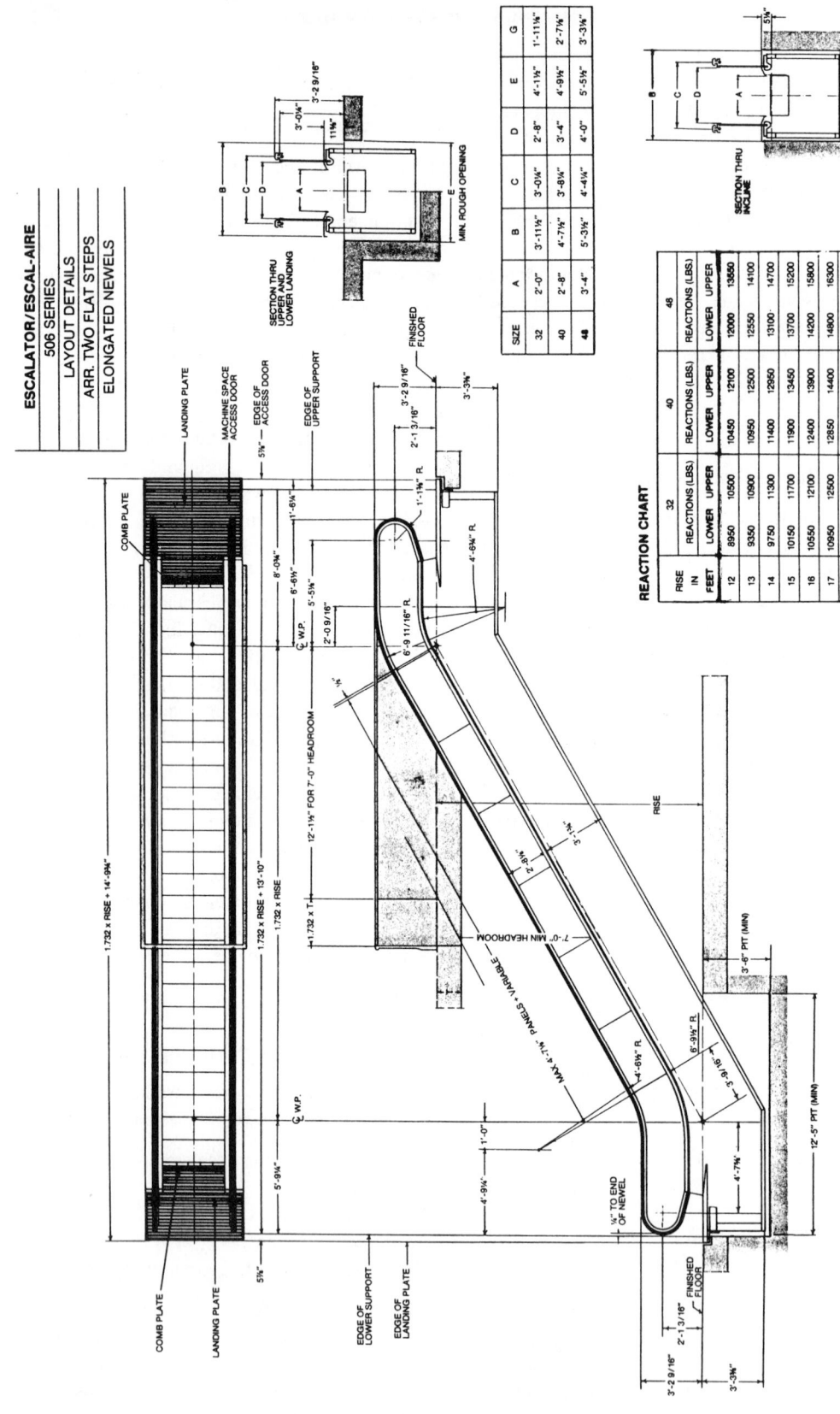

Fig. 25.17 Typical detailed space requirements and structural information as supplied by a manufacturer for a specific design of escalator. (Courtesy of Otis Elevator Co.)

Fig. 25.18 *Twin autowalks at Manchester Airport, England. These are the pallet type with glass balustrades and continuous built-in fluorescent lighting. These walks, which are each 1 m (40 in.) wide and 55 m (180 ft) long, travel in opposite directions. Because a single-loop drive runs both walks, they cannot both be operated in the same direction. (Photo courtesy of O&K Rolltreppen, Germany.)*

major users of this item (see Fig. 25.18). Other transportation terminals, such as rail and ship terminals, also can often find excellent applications for the moving walk, since much heavy and bulky luggage is moved in these areas.

Application of the apparent distance compression that moving walks provide permits placement of parking areas remote from the pedestrians' destination. Thus, a store can extend its parking area with no annoyance to patrons who must make the long trip to their cars with bulky packages or shopping carts. These advantages are all the more appreciated by persons with a walking impediment.

A second application of walks, as noted above, is the routing of traffic to avoid congestion, milling about, and lost time and motion. This is particularly applicable in transportation terminals, where persons are always traveling in opposite directions through the same and often restricted area, such as in "fingers" leading from airplanes to the main air terminal.

Moving walks are also useful to move people past a display window or some other point where congestion caused by stopping, is undesirable. This "movement of objects" application demonstrates clearly that the moving walk is in function simply a large conveyor belt, regardless of its construction.

25.17 Application of Moving Ramps

The moving ramp that combines horizontal and vertical movement is principally applicable as follows:

1. To move persons and wheeled vehicles vertically.
2. To move persons who lack the agility required to use an escalator.
3. To vertically move large, bulky objects.

Ramps have found a fertile field of application in multilevel stores where escalators are not feasible for shopping-cart users. Such stores may also utilize rooftop parking that is made accessible to cart users via a moving ramp. Since luggage carriers do not easily adapt to usage on escalators, transportation terminals, which are almost always multilevel, also find extensive application for the moving ramps (see Fig. 25.19).

25.18 Size, Capacity, and Speed

The speed, physical dimensions, and therefore passenger capacity of walks and ramps are not as extensively standardized as is the case with escalators. Manufacturers utilize several different tread widths, combined with various speeds and ramp angles of incline. The combinations are designed to suit the situation. Furthermore, since the maximum permissible walk ramp speed varies with angle of slope, and with design of entering point, passenger capacity ratings vary with each design. Higher speeds are allowed by code for level entrance than with sloping entrance, for the obvious reason that the level entrance is easier to board. Tables 25.5 and 25.6 and Fig. 25.20 give data for typical units.

Since capacity varies with width, speed, and type of entrance, exact capacity figures must be obtained for each specific design. Maximum

Fig. 25.19 (a) *Moving ramps in weatherproof design provide direct access to upper floors. This 12° ramp leads to the second floor of a store in Copenhagen. Note the glass balustrade and continuous flourescent lighting fixtures below the handrail. (Photo courtesy of O&K Rolltreppen, Germany.)*

TABLE 25.5 **Maximum Permissible Operating Speeds of Moving Ramps**

	Maximum Speed [fpm (m/s)]	
Angle of Incline	*Level Entrance*	*Sloping Entrance*
0–3°	180 (0.9)	180 (0.9)
3–5°	180 (0.9)	160 (0.8)
5–8°	180 (0.9)	140 (0.7)
8–12°	140 (0.7)	130 (0.65)
12–15°	140 (0.7)	125 (0.625)

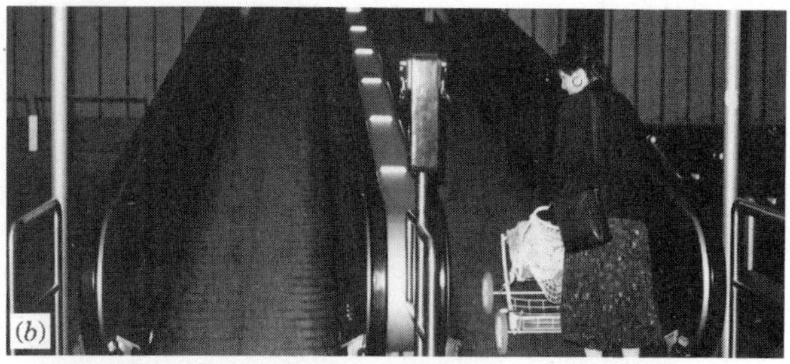

Fig. 25.19 (b) *Palletized inclined moving ramps are particularly useful in installations that do not lend themselves to the use of escalators because of the need to transport wheeled carts in addition to passenger traffic. (Photo courtesy of Schindler Wertheim.)*

practical lengths at present are approximately 1000 ft, with longer units in design.

25.19 Components

Moving walks are manufactured in only one design: a derivative of the escalator, which uses a flattened pallet in place of a step. In all other respects—the drive mechanism, safeties, brake, handrails, and so on—the unit is similar to an escalator. Preliminary layout data and detailed plans are, as with moving stairs, available from vendors, who generally are the same vendors as supply escalators.

25.20 Horizontal Shuttles

The last item in our discussion is not strictly a building construction item since the bulk of its

TABLE 25.6 **Moving Walks/Ramps: Commercial Sizes**

Angle (deg)	*Pallet Width [in. (mm)]*	*Ramp Speed [fpm (m/s)]*
0–3	32 (800)	100 (0.5), 130 (0.65), 150 (0.75)
Walk	40 (1000)	100 (0.5), 130 (0.65), 150 (0.75)
	62 (1400)	90 (0.45), 100 (0.5), 130 (0.65)
10, 11, 12	32 (800)	100 (0.5), 130 (0.65)
Ramp	40 (1000)	90 (0.45), 100 (0.5), 130 (0.65)

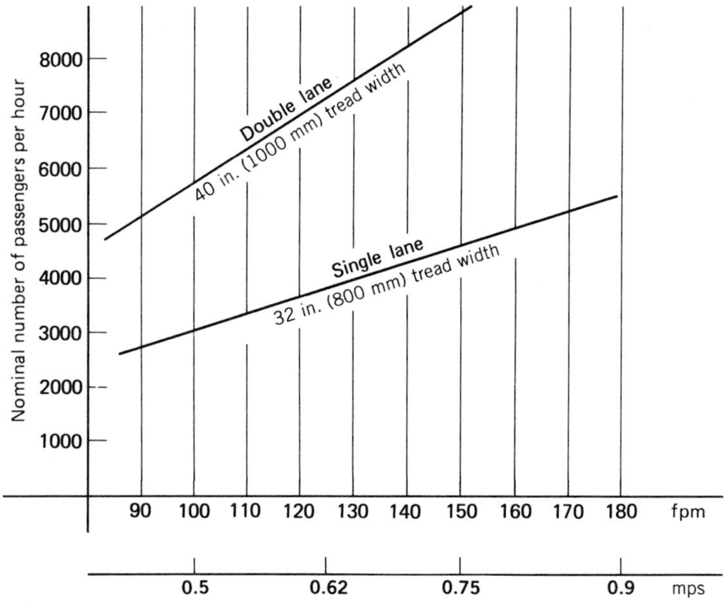

Fig. 25.20 *The capacity of moving walks varies with speed, angle of incline, and tread width. Capacity shown is for maximum incline permitted at that walk speed. Because of the requirement for handrail support, tread widths greater than double lane are not often utilized.*

construction generally exists outside the facility. Horizontal shuttles, also called horizontal elevators, people movers, transit shuttles, and the like are a relatively new item whose use and technology are burgeoning rapidly. The essential construction consists of a car or gondola, generally completely enclosed, which transports groups or batches of riders between two or more terminals. The greater similarity to an elevator than to a moving walk is obvious since passengers are grouped rather than transported continuously, and entry and exit are door and stop dependent rather than continuous. Technologies vary among manufacturers. One supplier uses a hovercraft design with the car supported on a blower-controlled air cushion. Another supplier uses the magnetic levitation principle combined with a linear drive motor, and so on.

The building designer must be concerned with terminal space requirements, effect of the vehicle entry on the HVAC system since the vehicle frequently enters from outdoors, and the movement of batch traffic from the shuttle terminal to other parts of the facility. Shuttles find application in airports, large recreational facilities such as holiday villages, and multibuilding industrial and educational campuses. Details of the vehicles and terminals are far too variegated to be detailed here, but it behooves the modern architect to be fully aware of even the more unusual people transport facilities and their impact on building design.

REFERENCE

Vertical Transportation Standards (1983). National Elevator Industry, Inc. (NEII).

PART IX
ACOUSTICS

The best (and simplest) distinction between the terms ''sound'' and ''noise'' is a subjective one—the former is desirable, the latter is not. This definition does not consider the content of the sound. For example, speech, which is a desirable sound in most instances, can become a ''noise'' when emanating from a neighbor's apartment or an adjoining office. In such a case an air conditioner, which is most often considered to be a ''noise,'' can provide a desired ''sound'' by masking the intruding speech. The function of architectural acoustics is simply to reinforce this definition; that is, to enhance desired sound and to maximally attenuate noise.

Chapter 26, Fundamentals of Architectural Acoustics, introduces the subject with a discussion of physical sound theory and physiological hearing phenomena. The latter include the negative effects of noise, which are primarily psychological (annoyance) at low levels but become physical and can result in hearing damage at high levels. Criteria of the U.S. Occupational Safety and Health Administration (OSHA) are given. The discussion moves to the core of the subject—room acoustics—with an explanation of absorption and reverberation, and develops acoustic design criteria for various indoor activities. The chapter concludes with a description of high-grade sound reinforcement systems.

Chapter 27 is devoted to noise control. Noise within a space is controlled by absorption, whereas control of noise transference between spaces is a matter of isolation. The chapter begins with a discussion of the application of absorptive materials for room noise reduction. It treats the problem of interspace noise conduction, dividing the problem into two parts (airborne noise and structure-borne noise), since the solutions are different. Relevant criteria, including sound transmission class (STC) and impact isolation class (IIC), are adduced and explained, and solutions to noise transference problems of different types are suggested. The chapter offers sections on mechanical system noise control and on acoustic recommendations and criteria, ending with a section on exterior acoustics.

26

FUNDAMENTALS OF ARCHITECTURAL ACOUSTICS

SOUND THEORY AND HEARING PHENOMENA

26.1 General

Architectural acoustics may be defined as the technology of designing spaces, structures, and mechanical systems to meet hearing needs. With proper design, "wanted" sounds can be heard properly and "unwanted" sounds, or "noise," can be attenuated or masked to the point where it does not cause annoyance. However, achieving good acoustics has become increasingly more difficult for a variety of reasons. To cut costs, the weight of construction materials used in many of today's buildings is reduced. Since light structures generally transmit sound more readily than heavy ones, this practice poses major acoustical problems. Forty percent or more of a building budget may be allocated for mechanical systems—most of which make noise. Outside noise sources such as cars, trucks, trains, and airplanes present problems in isolating interior spaces from exterior sound.

Building owners and tenants are aware that good acoustic environments in buildings are possible, and the architect is expected to provide them. A clear understanding of the principles explained in this and the following chapter will assist the architect in accomplishing straightforward designs alone and, in more complex instances, cooperating knowledgeably with the project's acoustic consultant. The importance of proper acoustic design is all the more critical, since after-the-fact acoustic "repair" is often difficult and frequently impossible without substantial structural alterations.

All acoustics situations have three common elements—source, transmission path, and receiver. The source can be made louder or quieter and the path can be made to transmit more or less sound. The listener's reception of sound also may be influenced. This chapter presents essential aspects of architectural acoustics to assist a designer in defining acoustic goals. Moreover, it describes the achievable goals as well as various methods for reaching them through design.

26.2 Sound: Definition and Generation

Sound can be defined in a number of different ways depending on the aspect we desire to study. Thus sound is a physical wave, or a mechanical vibration, or simply a series of pressure variations, in an elastic medium. For airborne sound, the medium is air. For structure-borne sound, the medium is concrete, steel, wood, glass, and combinations of all these materials. A much more limited definition, more appropriate to our study, is to define sound simply as an audible signal. This means that the science of architectural acoustics is concerned with the building occupant, and sounds that he or she cannot detect are generally not our concern. To further clear the air, it is always assumed that the hearer has a pair of healthy young ears with a detection range of 20 to 20,000 Hz. With these givens, it is probably best to view sound as a series of pressure variations. In air, these pressure variations take the form of periodic compressions and rarefactions. Refer to Fig. 26.1. The bell radiates a pure tone in all directions equally, that is, it creates a circular wave front.

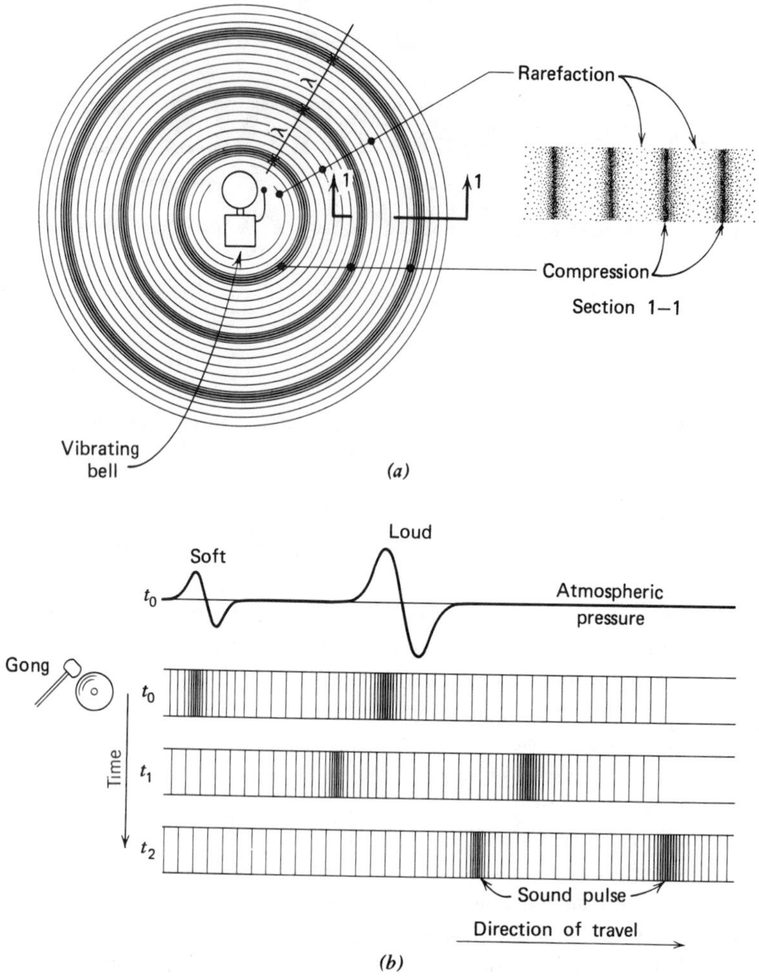

Fig. 26.1 *Sound pressure waves.* (a) *The continuous vibration from the bell causes a series of compressions and rarefractions of the air to travel outward longitudinally from the source. Amplitude information is carried by pressure; that is, greater amplitude means greater compression and greater rarefraction. (Compression and rarefraction are shown diagrammatically as line density, although they are actually molecular phenomena, as shown in the upper drawing).* (b) *Two single impulses of different magnitude (amplitude) traveling away from the source. Note how amplitude information is carried by difference in pressure.*

As the bell vibrates, it sets up vibrations in the air, of the same frequency, which can best be visualized in the form of a sectional view. Notice that the pressure changes containing the sound information travel in the same direction as the wavefront—longitudinally. This is unlike a radio signal, for instance, in which the wave travels longitudinally but the information—that is, the modulation, is expressed in wave height and shape—is transverse. Sound is therefore *longitudinal* mechanical wave motion.

26.3 Frequency

The number of times the cycle of compression and rarefaction of air occurs in a given unit of time is described as the frequency of a sound. For example, if there are 1000 such cycles in 1 second, the frequency of the sound is 1000 cps or 1000 hertz (Hz) in the standard nomenclature. Thus in Fig. 26.1*a*, higher frequencies would be shown by compressions and rarefactions that are closer together and lower frequencies by those that are further apart. In sound, the concept of frequency is often referred to by a term borrowed from music—*pitch*. The higher the frequency, the higher the pitch, and vice versa. As stated, the approximate frequency range of a healthy young person's hearing is 20 to 20,0000 Hz. The upper limit decreases with age as a result of a process called presbycusis. Recognition of this phenomenon can be of importance in schools, since very high-pitched sounds that are inaudible to most adults (the loss is more pronounced in men than women) can be a source of extreme annoyance to students. For example, dentists report that high-speed turbine drills and teeth-cleaning devices cause extreme auditory discomfort to many young patients. These devices produce sounds in the 15- to 20-kilohertz (kHz) Range (see Fig. 26.2).

The human speaking voice has a range of approximately 100 to 600 Hz in *fundamentals*, but harmonics (overtones) reach to approximately 7500 Hz. Most speech information is carried in the upper frequencies, while most *acoustic energy* exists in the lower frequencies. The critical frequency range for speech communication is 300 to 4000 Hz. Overtones outside these frequencies give the voice its characteristic sound and specific identity. Telephone and radio communication are accomplished in a considerably narrower frequency band by sacrificing some voice quality and intelligibility. (See Figs. 26.14 and 26.20.)

A sound composed of only one frequency is a pure tone. Except for the sound generated by a tuning fork, few sounds are truly pure. Musical sounds (tones) are composed of a fundamental frequency and integral multiples of the fundamental frequency (harmonics). Most common sounds are complex combinations of frequencies. Figure 26.3 shows examples of pure tones, musical notes, and common sounds; Fig. 26.4 shows the frequency ranges of some common devices and phenomena. The frequencies in the scale of Fig. 26.4 all stand in the ratio of 2 : 1 to each other, that is, 16 : 32 : 63 : 125 : 250, and so on. Borrowing again from musical terminology,

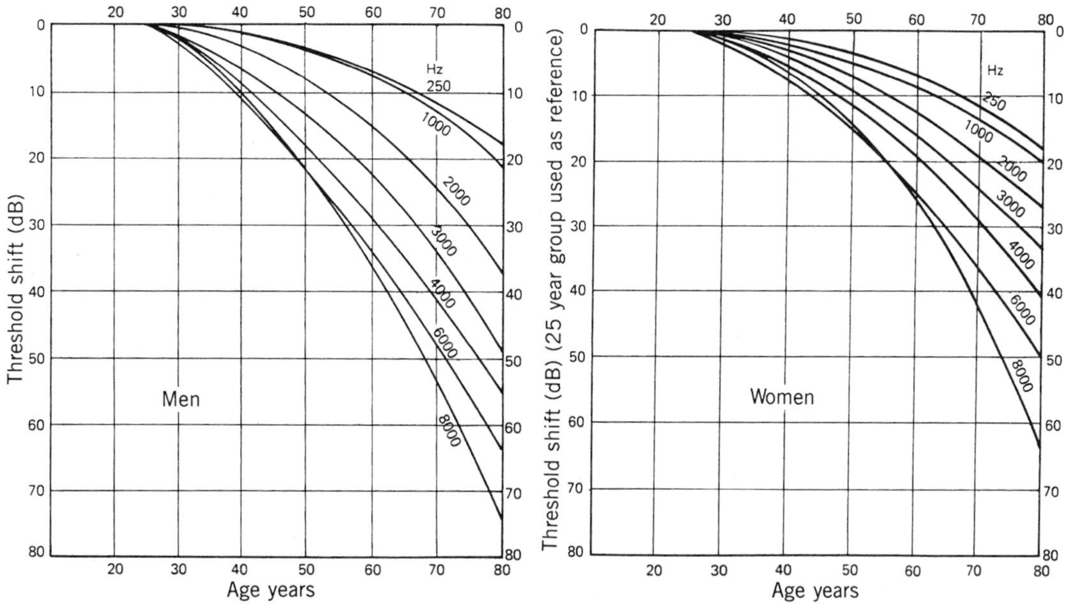

Fig. 26.2 *The curves illustrate the average hearing threshold shift for men and women, with increasing age relative to 25-year-old reference group. Note that at age 60 the male threshold shift at upper speech frequencies (4000 Hz) is 30 db as compared to 20 db for women, a difference of 10 db or one-half the subjective loudness (see Table 26.3). For men older than 50 and women older than 60, frequencies above 10 kHz are effectively inaudible, and even at normal speech frequencies of 1000 to 2000 Hz subjective loudness is reduced to one-half of that of a 25-year-old listener. (Reproduced with permission from F. A. White,* Our Acoustic Environment, *Wiley, New York, 1975.)*

ACOUSTICS

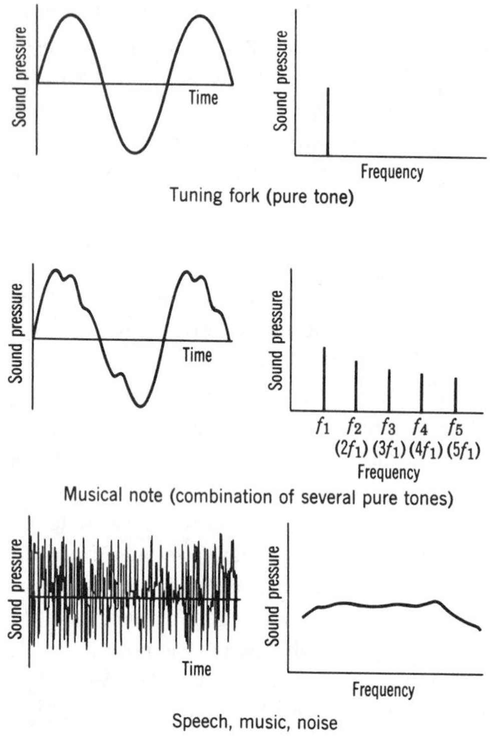

***Fig. 26.3** Schematic representations of a pure tone, a musical note, and more complex sounds (speech, music, and noise), showing the variation of sound pressure with time and frequency.*

they are one *octave* apart. These particular frequencies are also accepted internationally as the center frequencies of octave bands used for the purpose of sound specification. For technical reasons, a geometric mean is used. Thus 250 is the center frequency of an octave band ranging from $250/\sqrt{2}$ to $250\sqrt{2}$ and that particular octave is known as the 250-Hz octave. If finer division is required, ½-octave and ⅓-octave bands are used.

26.4 Velocity of Propagation

Sound travels at different velocities depending on the medium. In air, at sea level, sound velocity is 344 m/s or 1130 fps. This corresponds to 770 miles per hour or 1239 kilometers per hour—slow indeed when compared to light, which has a velocity of 186,000 miles per *second*. Since sound travels not only in air but also through parts of a structure, it is of interest to know the velocities in other media (see Table 26.1). For our purposes, velocity changes due to temperature and altitude (atmospheric pressure) may be ignored and, for rough calculation, 1100 fps and 350 m/s may be used as velocity in air, since both are within 3% error.

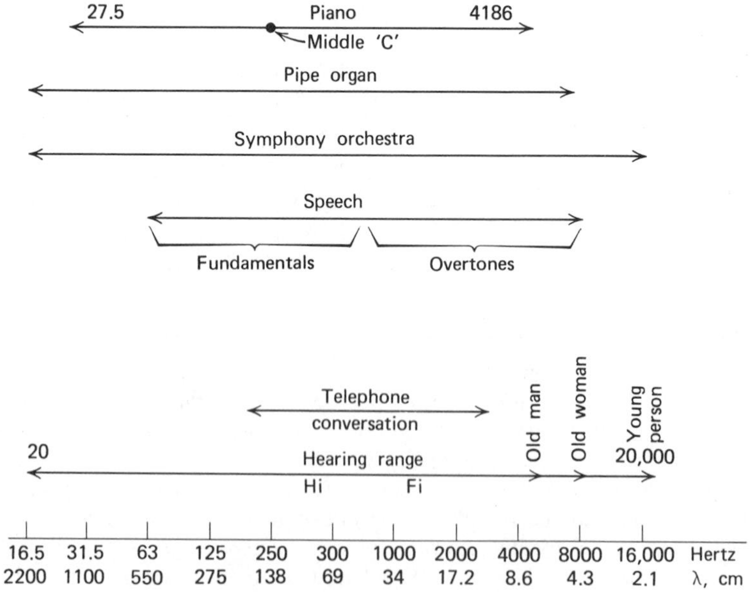

***Fig. 26.4** Frequency ranges of common instruments. Wavelength (λ) is calculated on an assumed propagation velocity of 344 m/s.*

TABLE 26.1 **Sound Propagation Velocity in Various Media**

Medium	Velocity: Meters per Second	Velocity: Feet per Second
Air	344	1130
Water	1410	4625
Wood	3300	10,825
Brick	3600	11,800
Concrete	3700	12,100
Steel	4900	16,000
Glass	5000	16,400
Aluminum	5800	19,000

NOTE: These figures are approximate, since the listed materials vary in density. Average frequency is used.

26.5 Wavelength and Types of Propagation

The wavelength of a sound may be defined as the distance between similar points on successive waves or the distance the sound travels in one cycle. The relationships between wavelength, frequency, and velocity of sound is expressed as

$$\lambda = \frac{c}{f} \tag{26.1}$$

where

λ = wavelength, ft (m)
c = velocity of sound, fps (m/s)
f = frequency of sound, Hz

Low-frequency sounds are characterized by long wavelengths and high-frequency sounds by short wavelengths. Sounds with wavelengths ranging from ½ in. to 50 ft can be heard by humans. A simple nomograph presented later in this chapter (see Fig. 26.29) permits rapid determination of wavelength, given the frequency, and vice versa.

26.6 Sound Magnitude

The magnitude of a sound signal is a much more complex subject because of the different terms in use and because of the great range of numbers involved. When we speak of sound magnitude, we think of loudness, which is a subjective, ear-oriented reaction not linearly related to the sound power (watts). The strength of sound is described variously as sound power, sound pressure, sound pressure level (SPL), sound intensity, and sound intensity level (IL), all of which differ from each other, and from subjective loudness. To clearly understand these concepts, and it is imperative that they be so understood, a comprehension of how we hear and how sound is propagated in free space is necessary.

26.7 Sound Intensity; Free Field Propagation

A point sound source of constant power P (watts) radiating in free space, that is, at a distance far from the effects of any reflecting surface, is represented in the drawing of Fig. 26.5. The *sound intensity* at any distance from the source is expressed as

$$I = \frac{P}{A} \tag{26.2}$$

where

I = sound (power) intensity, W/cm^2 (W/m^2)
P = acoustic power, Watts
A = area, cm^2 (m^2)

Since it is traditional in architectural acoustics to express area in square centimeters although the MKS (SI) system would require area to be stated in square meters we will use square centimeters. Conversion data are supplied in Table 27.15. See also Table 27.16 for a listing of sym-

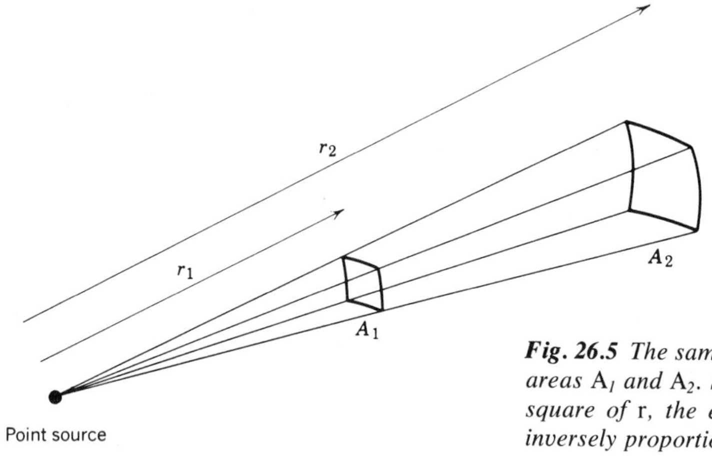

Fig. 26.5 The same total energy passes through areas A_1 *and* A_2*. Since A is proportional to the square of* r*, the energy density or intensity is inversely proportional to* r^2.

bols and abbreviations. Since the sound radiates freely in all directions,

$$I = \frac{P}{4\pi r^2} \quad \text{W/cm}^2 \tag{26.3}$$

where r is the radius of an imaginary enclosing sphere, which is a form of the classic *inverse square law*, stating that *intensity is inversely proportional to (distance)² from the source*. [In English units this is

$$I = \frac{P}{930 \times 4\pi r^2} \quad \text{W/ft}^2 \tag{26.4}$$

since 1 ft² = 930 cm².] Using equation 26.3 to determine the intensities I_1 and I_2 at distances r_1 and r_2 from point source P, and dividing the equations, we find that the intensities at distances r_1 and r_2 from the source stand in the ratio

$$\frac{I_1}{I_2} = \frac{r_2^2}{r_1^2} \tag{26.5}$$

Note the exact correspondence of these relations to the derivations for illumination from a point source, found in Section 18.12 and Fig. 18.12. Figure 26.6 shows graphically how a sound pulse is attenuated in *strength* (but not in waveform) as it travels outward from the source by action of distance.

The preceding derivation is based on a *point* source, that is, a source that is small relative to the wavelength produced. This type of source produces spherical waves. Line sources such as strings produce cylindrical waves. Large vibrating surfaces such as walls produce plane waves. The importance of these facts will become clear in the discussions on barriers and diffraction in Chapter 27.

The threshold of hearing, that is, the minimum sound power intensity (I) that a normal ear can detect, is 10^{-16} W/cm². (Actually, the ear responds to pressure, as explained below.) The maximum sound intensity that the ear can accept without damage is approximately 10^{-3} W/cm². This gives a range of 10^{13}, or 10 million million to 1 (10,000,000,000,000 : 1). Table 26.2 gives an idea of the physical significance of these numbers. Two problems arise immediately when dealing with quantities of this type; the numbers themselves are very small and the ratios are very large. Furthermore, the human ear responds logarithmically, not arithmetically to sound pressure (and intensity); that is, doubling

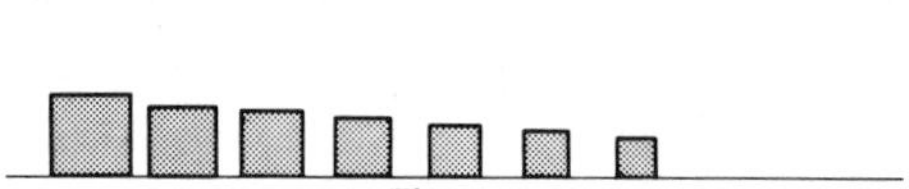

Fig. 26.6 Attenuation of a sound signal in air as it travels away from the source. The shape (information) remains constant when traveling in a nondispersive medium such as air. This is not the case with travel in solids, where different frequencies travel at different velocities, and therefore cause a wave-shape change with time and distance. Since velocity of propagation is constant, time and distance are linearly related and can therefore be plotted on the same axis.

TABLE 26.2 **Comparison of Decimal, Exponential, and Logarithmic Statements of Various Acoustic Intensities**

Intensity (W/cm²)		*Intensity Level—Logarithmic Notation*	*Examples*
Decimal Notation	*Exponential Notation*		
0.001	10^{-3}	130 db	Painful
0.0001	10^{-4}	120 db	
0.00001	10^{-5}	110 db	75-piece orchestra
0.000001	10^{-6}	100 db	
0.0000001	10^{-7}	90 db	Shouting at 5 ft
0.000000001	10^{-9}	70 db	Speech at 3 ft
0.00000000001	10^{-11}	50 db	Average office
0.0000000000001	10^{-13}	30 db	Quiet unoccupied office
0.00000000000001	10^{-14}	20 db	Rural ambient
0.000000000000001	10^{-15}	10 db	
0.0000000000000001	10^{-16}	0 db	Threshold of hearing

the intensity of a sound does not double its loudness—the change is barely perceptible. To solve these problems it would be much more convenient if we were to construct a scale that:

1. Started at zero for the minimum sound (intensity or pressure) we can hear.
2. Used whole numbers rather than negative powers of 10.
3. Had some fixed relationships between an arithmetic difference and a loudness change; say, 10 units equals a doubling (or halving) of loudness. Thus, on such a scale, the difference between 20 and 30, 60 and 70, would always be a doubling of loudness.

Such a scale exists. It is the decibel scale.

26.8 Intensity Level (IL) and the Decibel (db)

The word "level" indicates a quantity relative to a base quantity. Intensity *level* is the ratio between a given intensity and a base intensity. If we express intensity level as

$$IL = 10 \log \frac{I}{I_0} \qquad (26.6)$$

where

IL = intensity level, db
I = intensity, W/cm²
I_0 = base (i.e., 10^{-16} W/cm², the threshold of hearing)
log = logarithm to base 10

then we have established a scale that satisfies the three conditions set forth above. The quantity IL, intensity level, is dimensionless, since it indicates simply a ratio between two numbers. It is measured in decibels (db) for convenience in expressing the large numbers involved. Table 26.2 shows the great convenience of using the logarithmic decibel scale as compared to either decimal notation or exponential notation. Table 26.3 gives a short listing of subjective loudness changes expressed in decibels. Note that *10 db indicates a doubling of loudness,* and 20 db is loudness doubled twice, that is, four times as loud. The *difference* (Δ) between any two intensity levels

TABLE 26.3 **Subjective Loudness Changes and Corresponding Intensity Level Changes**

Change in Level (db)	*Subjective Change in Loudness*
3	Barely perceptible
6[a]	Perceptible
7	Clearly perceptible
10	Twice or half as loud
20	Four times or one-quarter as loud

[a]Six decibels corresponds to the change encountered when distance to the source in a free field is doubled (halved).

ACOUSTICS

$$\Delta IL = IL_2 - IL_1 = 10 \log \frac{I_2}{I_0} - 10 \log \frac{I_1}{I_0}$$

Therefore,

$$\Delta IL = 10 \log \frac{I_2}{I_1} \text{ db} \qquad (26.7)$$

A few examples using decibel notation and logarithmic calculations should help the reader become familiar with this useful system.

EXAMPLE 26.1. Two sound sources produce intensity levels of 50 and 60 db, respectively, at a point. When these are functioning simultaneously, what is the total sound intensity level? (We assume identical frequency content and random phase; that is, the phase relationship between the two sources changes in a random manner.)

SOLUTION. Note that we are dealing with intensity level, not intensity, since intensity itself has little significance for us. The technique involved in adding two sound intensity levels has three steps:

1. Convert both to actual intensity.

$$IL = 10 \log \frac{I}{I_0}$$

so

$$60 = 10 \log \frac{I_1}{10^{-16}}$$

or

$$6.0 = \log \frac{I_1}{10^{-16}}$$

Then, using the definition of a base 10 logarithm:

$$10^6 = \frac{I_1}{10^{-16}}$$

$$I_1 = (10^{-16})\, 10^6 = 10^{-10} \text{ W/cm}^2$$

By similar calculation,

$$I_2 = 10^{-11} \text{ W/cm}^2$$

2. Add arithmetically.

$$I_1 + I_2 = 10^{-10} + 10^{-11}$$

$$= (10 \times 10^{-11}) + 10^{-11}$$

$$I_{tot} = 11 \times 10^{-11} \text{ W/cm}^2$$

3. Reconvert to decibels. To find the intensity level IL corresponding to the new total intensity $I_1 + I_2$, we simply apply equation 26.6:

$$IL_{tot} = 10 \log \frac{I_{tot}}{I_0}$$

$$= 10 \log \frac{11 \times 10^{-11}}{10^{-16}}$$

$$= 10(\log 11 + \log 10^5)$$

$$= 10(1.04 + 5)$$

$$= 60.4 \text{ db}$$

which is only a fraction larger than the original 60 db of the stronger sound.

EXAMPLE 26.2. Assume two noise signals of 60 db each. What is the combined strength in decibels?

SOLUTION. One method would be to calculate levels as in Example 26.1. A shorter method is to find the difference between the sum and either signal and to add it to either original signal. Using equation 26.7,

$$\Delta IL = IL_{comb} - IL_1 = 10 \log \frac{I_{comb}}{I_1}$$

$$= 10 \log \frac{2I_1}{I_1}$$

$$= 10 \log 2$$

$$= 10(0.30)$$

$$= 3 \text{ db}$$

This answer (which is independent of any particular level) gives us the extremely important fact that *doubling a signal intensity raises the intensity level by 3 db*. (In our case, the combined intensity level would obviously be 60 db + 3 db or 63 db). Similarly, quadrupling a signal's intensity raises the received level by 6 db. This is because

$$\Delta IL = 10 \log \frac{4I}{I}$$

$$= 10 \log 4$$

$$= 10(0.60)$$

$$= 6 \text{ db}$$

Therefore, quadrupling 60 db gives 66 db, or 60

db + 60 db = 63 db and 63 db + 63 db = 66 db. This technique is very useful when combining a large number of identical levels, as in the following example.

EXAMPLE 26.3. A factory will contain 20 identical machines, each of which generates a noise level of 80 db. What will be the combined noise level? (Ignore problems of frequency content, phase, and fields.)

SOLUTION

$$\begin{aligned}\Delta IL &= IL_{tot} - IL_{single} \\ &= 10 \log \frac{I_{tot}}{I_{single}} \\ &= 10 \log \frac{20I_{single}}{I_{single}} \\ &= 10 \log 20 \\ &= 10(1.3) = 13 \text{ db}\end{aligned}$$

Therefore the total noise level will be

$$IL_{tot} = 80 \text{ db} + 13 \text{ db} = 93 \text{ db}$$

A chart for combining decibel levels of two sources is given below in Fig. 26.7 which eliminates the somewhat lengthy procedure detailed in Example 26.1. Referring to Table 26.3, we note that the human ear is not responsive to fractional db changes; indeed even a 3-db change is barely perceptible. This being so, we recommend that the detailed calculation and the chart be reserved for situations where a high degree of accuracy is required. For everyday calculations, the following rules of thumb may be used to combine db levels of two sources:

1. When the difference between two sources is 1 db or less, add 3 db to the higher level to obtain the total.
2. When the difference is 2 to 3 db, add 2 db.
3. When the difference is 4 to 8 db, add 1 db.
4. When the difference is 9 db or more, ignore the lower source.

A comparison of addition by these rules of thumb and by an accurate method shows that at usual levels, the error involved is always less than 1%.

db Level		*Sum*	
Lower	*Higher*	*Approximate*	*Accurate*
60	60	63	63
60	62	64	64.4
60	64	65	65.5
60	66	67	67
60	68	69	68.7
60	70	70	70.5

Returning to the inverse square law expressed in equation 26.5, we are now in a position to determine the effect on sound intensity level of moving away from the sound source.

EXAMPLE 26.4. Given a sound source that produces sound intensity *IL* at a distance d_l from the source (the reader can substitute any numbers desired, or follow the problem with symbols), what are the intensities at twice the distance? Four times?

SOLUTION. We know from equation 26.7 that

$$\Delta IL = \frac{I_2}{I_1}$$

and we also know from equation 26.5 that

$$\frac{I_1}{I_2} = \frac{d_2^2}{d_1^2}$$

therefore, since $d_2 = 2d_1$

$$\frac{I_2}{I_1} = \frac{(d_1)^2}{(2d_1)^2} = \frac{1}{4}$$

Substituting in equation 26.7, we have

$$\begin{aligned}\Delta IL &= 10 \log \frac{I_2}{I_1} \\ &= 10 \log 1/4 \\ &= 10(-0.6) \\ &= -6 \text{ db}\end{aligned}$$

which tells us that sound intensity level (not pressure) is reduced by 6 db. Similarly, when distance is quadrupled, it is reduced by 12 db.

To summarize, then, *intensity level changes 3 db with every doubling or halving of power and changes 6 db with every doubling or halving of*

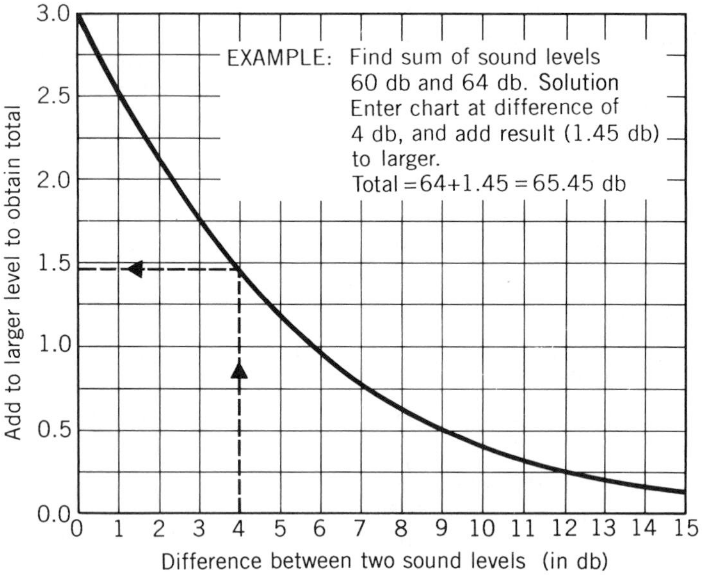

Fig. 26.7 *Chart for adding two uncorrelated sound pressure levels when both are expressed in db. (reprinted by permission, from E. B. Magrab,* Environmental Noise Control, *Wiley, New York 1975.)*

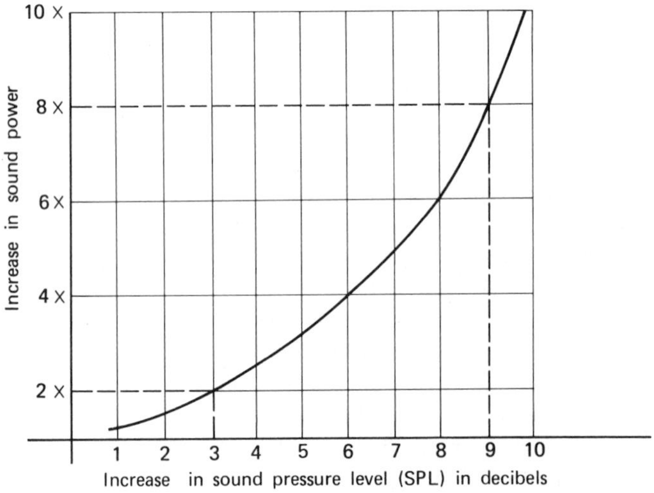

Fig. 26.8 *Decibel level increase as a function of power (intensity) increase.*

distance from a point source. Figures 26.7 through 26.9 illustrate these relationships.

The ear responds to sound pressure, not intensity. As explained below, sound *pressure level* (SPL) is equal numerically to *intensity level* (IL) (at least for normal temperature and pressure, i.e., for our use in architectural acoustics). Therefore, the foregoing examples and manipulations of intensity level are equally applicable to sound pressure level.

As explained in some detail in the following section, the combined effect of two sounds depends on their frequency content. In the examples above we assumed signals either of identical frequency and random phase or of very-wide-frequency spectrum—so wide that phase phenomena are not significant. In architectural acoustics work, such an assumption is generally valid.

26.9 Human Hearing

Refer to Fig. 26.10. The outer ear is funnel shaped and serves as a sound-gathering input terminal to the auditory system. Sound energy

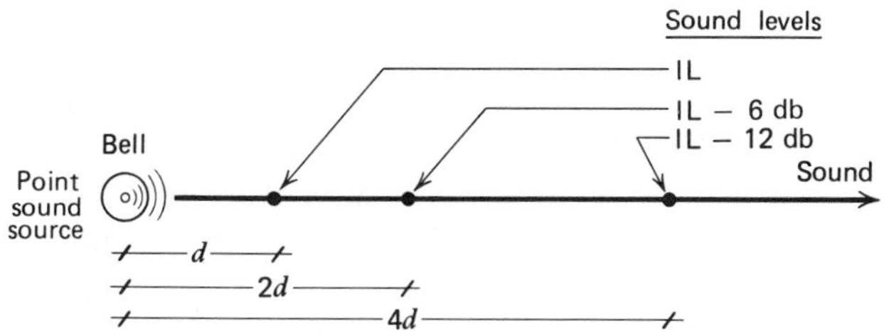
Sound levels
IL
IL − 6 db
IL − 12 db
Sound
Bell
Point sound source
d
2d
4d

Fig. 26.9 *Sound energy levels at varying distances from the source. Each doubling of distance reduces the level of 6 dB. These relationships hold true only in a free field.*

travels through the auditory canal (outer ear) and sets in motion the components of the middle ear, comprising the eardrum, hammer, anvil, and stirrup. The stirrup acts as a piston to transmit vibrations into the fluid of the inner ear. This fluid motion causes movement of hair cells in the cochlea, which in turn stimulates nerves at their bases, which in turn transmit electrical impulses along the eighth cranial nerve, to the brain. These impulses we understand as sound.

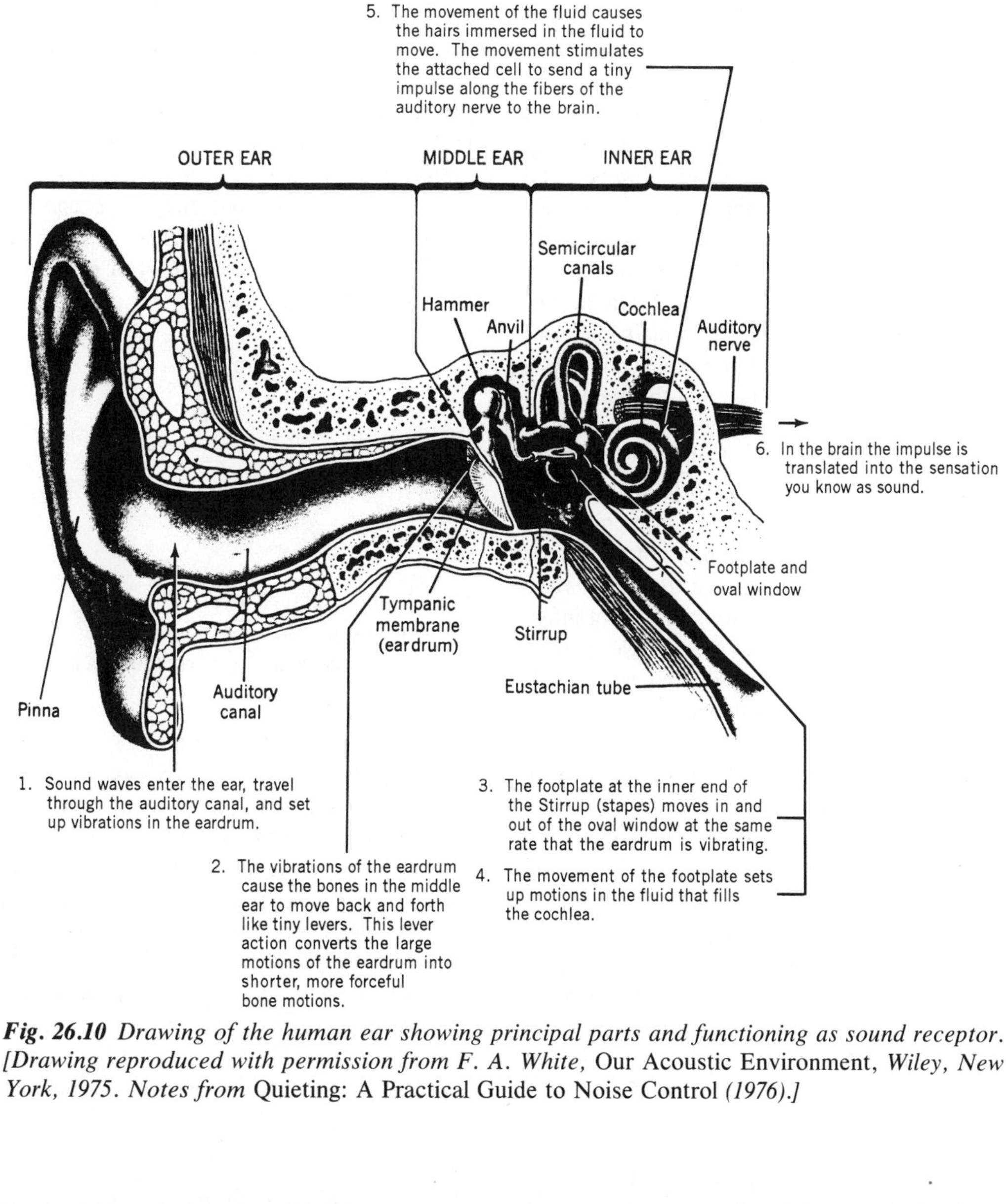
5. The movement of the fluid causes the hairs immersed in the fluid to move. The movement stimulates the attached cell to send a tiny impulse along the fibers of the auditory nerve to the brain.
OUTER EAR
MIDDLE EAR
INNER EAR
Semicircular canals
Hammer
Anvil
Cochlea
Auditory nerve
6. In the brain the impulse is translated into the sensation you know as sound.
Footplate and oval window
Tympanic membrane (eardrum)
Stirrup
Eustachian tube
Pinna
Auditory canal
1. Sound waves enter the ear, travel through the auditory canal, and set up vibrations in the eardrum.
2. The vibrations of the eardrum cause the bones in the middle ear to move back and forth like tiny levers. This lever action converts the large motions of the eardrum into shorter, more forceful bone motions.
3. The footplate at the inner end of the Stirrup (stapes) moves in and out of the oval window at the same rate that the eardrum is vibrating.
4. The movement of the footplate sets up motions in the fluid that fills the cochlea.

Fig. 26.10 *Drawing of the human ear showing principal parts and functioning as sound receptor. [Drawing reproduced with permission from F. A. White,* Our Acoustic Environment, *Wiley, New York, 1975. Notes from* Quieting: A Practical Guide to Noise Control *(1976).]*

Frequency recognition is accomplished in the cochlea, by the basilar membrane. This membrane resonates at one end at about 20 Hz and at the other at 20 kHz, giving the ear its frequency range. The ear hears and recognizes distinct frequencies, yet the hearing mechanism has the ability, apparently as directed by the brain, either to hear individual frequencies *or* to combine them into a single sound. Thus when we hear a string quartet we can, *at will,* hear either the entire quartet or each instrument individually. With concentration (vision helps in this), a "trained" ear can pick out a single instrument in an orchestra of 120 pieces, even if there is more than one such instrument in the group. Conductors do this regularly. Similarly, the ear can perform the selection known as the "cocktail party effect," that is, pick out one voice in background noise that may be 20 db louder than the wanted signal. In effect, the ear is attenuating the unwanted signals. Normally, however, the ear does precisely the opposite. It combines sounds that are clearly distinct from each other in frequency and phase. The three tones in a chord struck on a piano are different in frequency and, if played as a very rapid triplet, out of phase. Yet the ear combines them and hears a single sound, despite the fact that the maxima of the three tones do not occur simultaneously. In the preceding decibel calculations (Section 26.8), as well as those that will follow, it is assumed that the ear ignores phase differences and combines frequencies. This may not always be the case, particularly when the frequencies are very far apart. For this reason the type of information found in Table 26.4 (i.e., single-number representations of complex sounds) can be misleading and must be used with caution. This problem is discussed in the next section and in Section 26.12, which explains measurements and weighting networks.

At the threshold of hearing (approximately 0 db), the displacement of air molecules impinging on the eardrum, and the eardrum excursion, are approximately one angstrom unit (1 Å = 10^{-8} cm), which is approximately the diameter of an *atom*. Were the ear an order of magnitude more sensitive, it would hear thermal noise. The human ear is thus close to the practical limit of sensitivity. At the other end of the noise spectrum, the threshold of pain corresponds to a sound level of 130 db, and to an eardrum motion of approximately 0.25 mm. An astonishing range indeed!

As is obvious from the brief description of the hearing process above, movement of the eardrum is caused by air pressure. Therefore the quantities that interest us more than intensity (*power density*) are pressure, which is *force density,* and pressure level, which is the ratio of sound pressure to a base level, expressed in db.

26.10 Sound Pressure and Sound Pressure Level (SPL)

The usual base or reference pressure corresponds to the threshold of hearing and is taken to be 20 μPa or 2 × 10^{-4} microbar (μbar) (see Table 27.15). As with intensity, this pressure is established as 0 db for the purpose of sound pressure *level* calculation. Since the ear responds logarithmically to intensity and since pressure varies as the square root of intensity, we can write the expression

$$\text{SPL} = 10 \log \frac{p^2}{p_0^2}$$

or

$$\text{SPL} = 20 \log \frac{p}{p_0} \qquad (26.8)$$

where

SPL = sound pressure level, db
p = pressure, Pa or μbar
p_0 = reference base pressure, (20 μPa or $2E^{-4}$ μbar)

Since for both intensity level and sound pressure level, the 0-db base corresponds to the hearing threshold, we can have equalized the db scales for sound pressure level (SPL) and intensity level (IL) *and the db values of the two can be used interchangeably.* Obviously, though, the actual intensity and the actual pressure corresponding to a particular decibel level are different—completely different, in magnitude and units. For instance, 70 db equals 10^{-9} W/cm intensity, and 0.063 Pa pressure. However, the important fact is that 70 db corresponds approximately to a particular noise level. We say "approximately" because assigning a single-number decibel level to a sound presents two difficulties:

1. The sound pressure level varies with time, except for a pure steady tone.
2. The different components of most common sounds vary in pressure level.

To overcome this problem two techniques are used. If a sound has a dominant frequency, that frequency's level can be used (see Fig. 26.11). This would be the case for a relatively constant sound such as a motor, or a fan or blower. Other sounds that vary widely in level and frequency can be plotted on an octave-band chart using maximum level for minimum percent of time (see Fig. 26.12). Where the position of the listener is not specified in the table, it is assumed to be at normal *close* distances, that is, 10 to 20 ft from a train, 3 to 5 ft from a radio, and the like.

26.11 Loudness Level: The Phon Scale

The human ear is not uniformly sensitive over its entire frequency range of 20 Hz to 20 kHz. The 120- to 130-db upper limit pain threshold

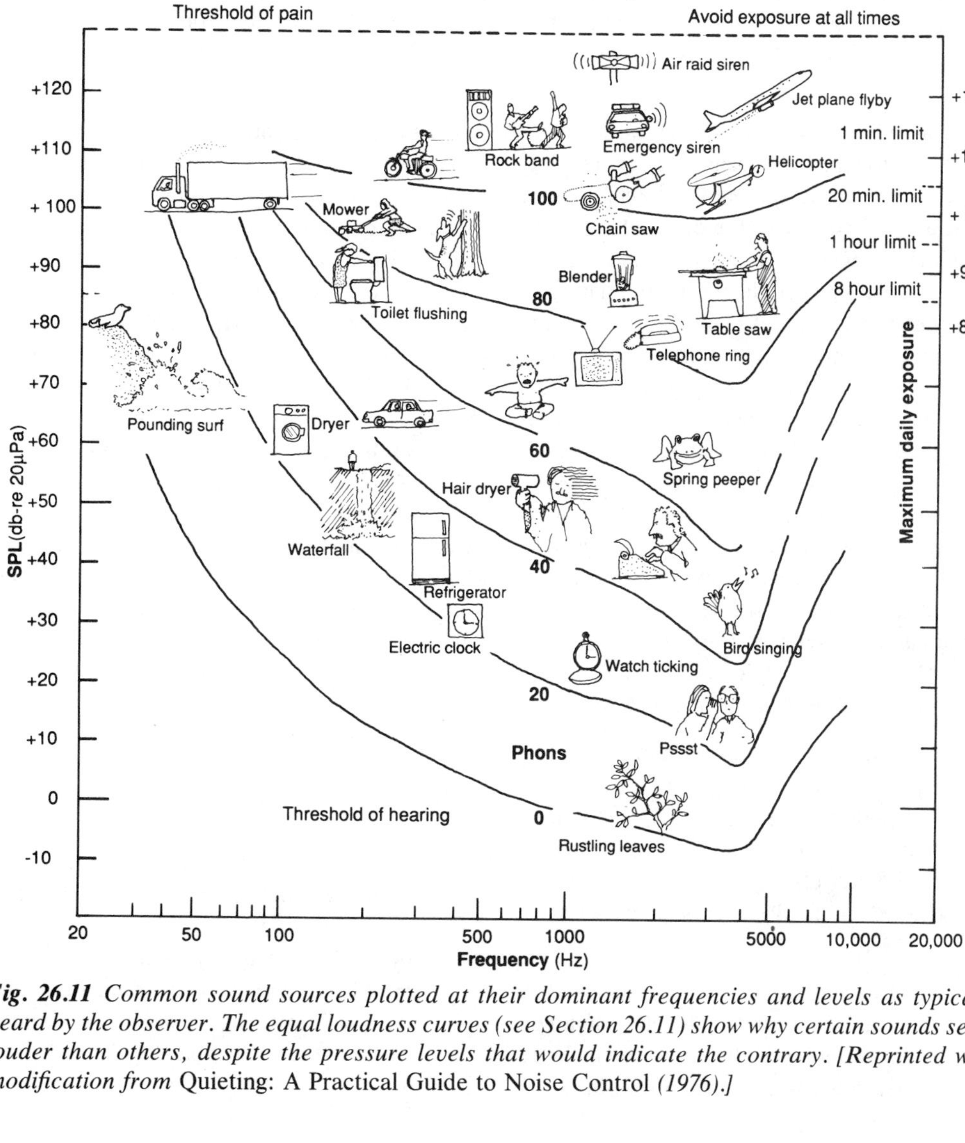

Fig. 26.11 *Common sound sources plotted at their dominant frequencies and levels as typically heard by the observer. The equal loudness curves (see Section 26.11) show why certain sounds seem louder than others, despite the pressure levels that would indicate the contrary. [Reprinted with modification from* Quieting: A Practical Guide to Noise Control *(1976).]*

ACOUSTICS

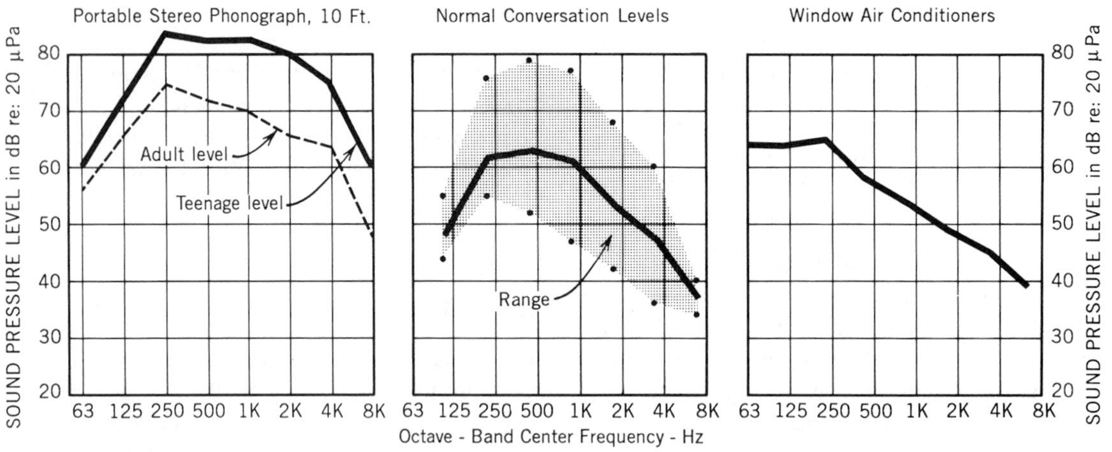

***Fig. 26.12** Curves of SPL of common noise sources plotted according to octave band frequency content. Values shown are averages of multiple measurements. [Reprinted from* A Guide to Airborne, Impact, and Structure-Borne Noise Control in Multifamily Dwellings *(1968).]*

occurs at all frequencies. However, at the lower limit, the 0-db threshold occurs only at 1000 Hz. The ear is in fact most sensitive at 3000 to 4000 Hz, at which frequencies the threshold is about −5 db. This nonlinear response exists throughout the ear's range. To determine the nature of this nonlinearity, a great number of tests were conducted with pure tones at different frequencies, in which listeners were asked to equate subjective loudness of signals. These tests resulted in a family of curves called *equal loudness level contours* or, alternatively, *Fletcher-Munson equal loudness contours* (after two of the principal researchers). These curves (see Fig. 26.13) are internationally recognized and standardized, and are used as the reference for normal hearing response. They are also used to "weight" measuring devices, as is explained in Section 26.12. Note that *by definition:*

1. All points on a single contour have the same *subjective* sensation of loudness.
2. The loudness level in *phons* (number shown in the center of each curve in Fig. 26.13) of the *entire* contour is defined by the db level of that contour at 1000 Hz.

The phon scale was constructed as a way of defining perceived sound (subjective loudness impression) in terms of the ear's nonlinear response. Assigning a single number—in a unit called the *phon*—to an equal loudness contour makes it possible to compare subjective loudness impressions of two sounds regardless of frequency or actual sound pressure level. Thus in Fig. 26.11, the subjective loudness of a whisper in the ear (20 phons) is the same as that of the distant pounding surf, despite the fact that the SPL of the former is 20 db and of the latter, 80 db. The 60-db differential is due to the ear's sharp drop in sensitivity at low frequencies. However, because the phon scale is nonlinear in terms of perceived loudness (a given phon differential does not correspond to the same perceived loudness change throughout the phon scale) and because phons cannot be combined arithmetically to give a resultant subjective loudness (60 phons plus 50 phons is not 110 phons), the scale has found only limited use.

The Fletcher–Munson contours demonstrate some interesting phenomena:

1. Sensitivity drops off sharply at low frequencies, particularly at low-db levels. (For this reason most high-fidelity amplifiers provide automatic bass boost at low-volume levels.)
2. Maximum sensitivity occurs between 3 and 4 kHz—precisely the frequencies that contain most information in human speech. (See Section 26.17 and Fig. 26.14.)

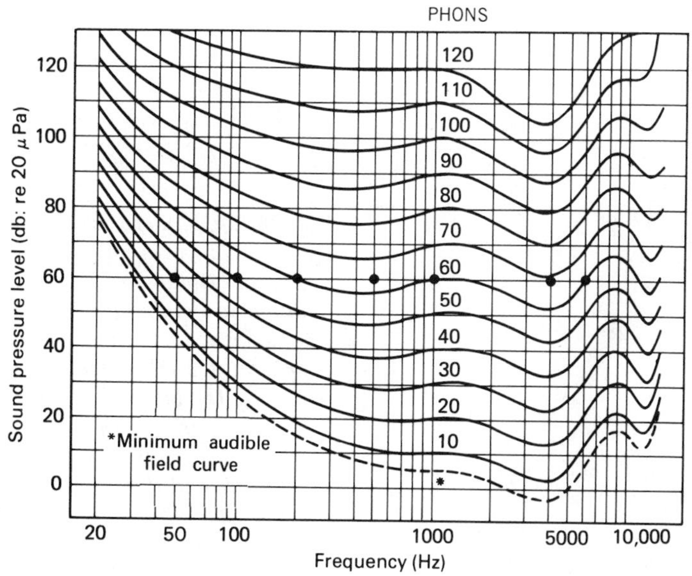

Fig. 26.13 Standard equal loudness contours. These curves are accurate for a listener with normal binaural hearing, situated in the near field of a source producing pure tones directly ahead of the listener. The subjective loudness of the perceived sound of each contour is quantified in a unit called the phon. *The phon value of each curve is shown in the center of the figure. Note that an SPL of 60 db corresponds to 30 phons at 50 Hz, 50 phons at 100 Hz, 63 phons at 500 Hz, 60 phons at 1 kHz, 68 phons at 4 kHz, and 60 phons at 6 kHz. This indicates the relative flatness of the ear's response in the central frequency-loudness range, and the sharp drop at low frequencies.*

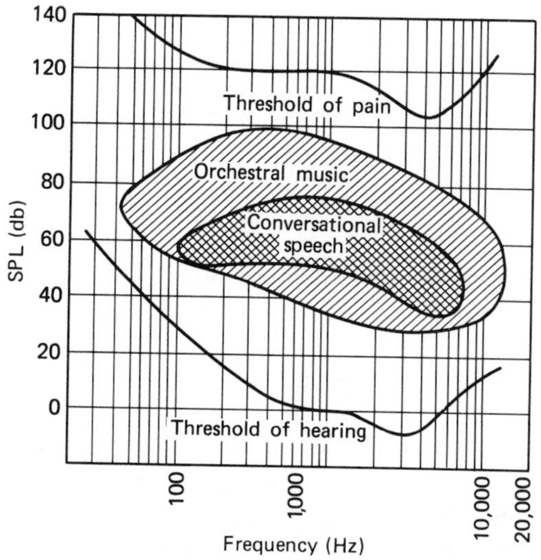

Fig. 26.14 The positions of speech and wide-range music in the human ear's aural field are illustrated. Speech is in the nominally linear response area of the ear as is most music (see Fig. 26.13). Beyond these frequencies, the ear's action is effectively to attenuate the signal.

3. In the normal listening range of 45 to 85 db, and the most used frequency range of 150 Hz to 6 kHz, the contour is substantially flat; that is, the ear's response is effectively linear. It is only at extremes of sound level and frequency that nonlinearity occurs.

26.12 Measurement of Sound Pressure Level

The need for a means of measuring sound levels is obvious. One such instrument is the integrating sound-level meter illustrated in Fig. 26.15*a*. To correlate meter readings with subjective loudness impressions, most such single-reading instruments are furnished with three weighing networks, the characteristics of which are given in Fig. 26.15*b*. The A network corresponds to the 40-phon contour and discriminates against low frequencies (see Fig. 26.13). The B network corresponds similarly to the 70-phon contour. The C network gives substantially flat uniform

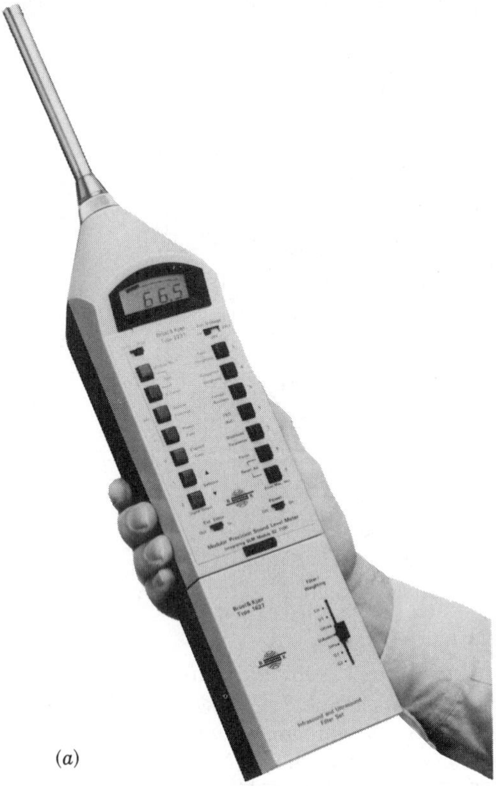

(a)

Fig. 26.15a Modern precision sound level meter. This unit with its accessories will measure and record isolated nonrepetitive noise events in addition to continuous, fluctuating and impulsive sounds, and will give both instantaneous sound level and equivalent continuous sound level L_{eq}*. The latter is important when sound at a location varies with time, and an equivalent level for design purposes is required. The unit takes level and duration into account and displays* L_{eq} *directly. It is arranged to measure dbA or linear (dbC), fast, or slow time constant. It adapts with filter networks to read octave or ⅓-octave bands, and reads out digitally. It also has peak-hold settings, which measure maximum sound during a time exposure. (Courtesy of Brüel and Kjaer.)*

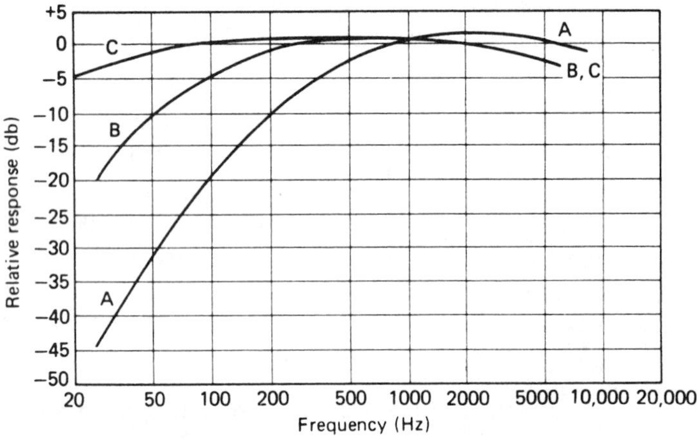

(b)

***Fig. 26.15b** Internationally standardized A, B and C weighting curves.*

response. In practice it was found that the B and C networks did not correspond well to subjective loudness reports. This was because the loudness contours on which they were based were taken for pure tones, whereas most field measurements are of very complex sounds. The A network, however, did correspond well, and has therefore become the standard single-number measuring scale for loudness of sounds at any frequency and intensity. Measurements taken in this way should be labeled "dbA" to identify the scale. The sound meter is equipped with two different damping characteristics—fast and slow. The former will follow reasonably rapid fluctuations. The latter is more heavily damped and will give an average reading of sounds with rapid changes of more than 4 db. Extremely rapid noise pulses are impulse noises, requiring special instruments.

It is known that the ear "averages" sounds over a minimum period, and that sound impulses shorter than this period sound quieter than they would sound as steady-state noise. This minimum sound length i.e. the time required for the sound to achieve full loudness is taken to be between 50 and 200 milliseconds (1/20–1/5 s) and depends to an extent on the sounds frequency content. A figure of 70 ms (1/14 s) is frequently found in the literature. This figure becomes important when considering the effect of echoes (see Sections 26.23 through 26.25).

More accurate measurements of complex sounds than are possible with a standard meter are accomplished with sophisticated instruments that measure intensity in octave bands and plot the results as in Fig. 26.12. Such measurements are necessary for accurate application of sound absorption and attenuation materials whose characteristics are also nonlinear over the frequency spectrum. Single-number dbA readings are known as *overall* levels and are useful as preliminary data and for broad–spectrum design.

Table 26.4 shows a list of common noise levels as measured by the dbA scale. Such single-value numbers are useful to establish a mental-aural comparison base, and to use in maximum exposure calculations, as discussed in Section 26.21.

26.13 Sound Absorption

When sound energy impinges on a material, part is reflected and the remainder is absorbed, in the sense that it is not reflected. (Some of the energy is transmitted, as discussed in the next chapter; see Fig. 27.2.) Most materials are neither perfect reflectors nor perfect absorbers. The coefficient of absorption (α) is defined as

$$\alpha = \frac{I_a}{I_i} \qquad (26.9)$$

where

I_a = sound power density (intensity) absorbed by the material, W/cm^2
I_i = intensity impinging on the material, W/cm^2
α = absorption coefficient (no units)

Thus α is a measure of absorption efficiency. When $\alpha = 1.0$ all the impinging energy is absorbed. Since open space has this characteristic, α has also been defined as the ratio between the absorption of a given material and that of an *open window* of the same area. Obviously, then, for an opening (open window, open door, etc.)

$$\alpha = 1.0$$

The total absorption (A) provided by a surface is proportional to its area and its absorption coefficient, that is,

$$A = S\alpha \qquad (26.10)$$

where

A = total absorption, *sabins*
S = surface area, square feet or meters
α = coefficient of absorption

Since α is a ratio and thus unitless, and S is a unit of area, $S\alpha$ should be area as well. Instead, sound absorption units are called sabins in honor of W. C. Sabine, a pioneer in architectural acoustics. One sabin (m^2) is the sound absorption equivalent to an open window one square meter in area. Similarly, one sabin (ft^2) is equivalent to 1 square foot of open window. Obviously, 1 sabin (m^2) equals 10.76 sabin (ft^2). When not specified, a sabin (ft^2) is intended.

Most rooms are constructed of several materials, each having a different absorption coefficient α. Thus to determine the total absorption of the room, it is necessary to sum the component absorptions, that is,

$$\Sigma S\alpha = S_1\alpha_1 + S_2\alpha_2 + \cdots + S_n\alpha_n$$

or

$$\Sigma A = A_1 + A_2 + \cdots + A_n \qquad (26.11)$$

where

$\Sigma S\alpha$ = total absorption in the room, sabins
S_1, S_2, etc. = area of each material
α_1, α_2, etc. = coefficient of each material
A_1, A_2, etc. = total absorption of each different material

If S is expressed in square feet, then A will be in sabins (ft^2); if S is in square meters then A is sabins (m^2).

TABLE 26.4 **Common Noise Levels**

Sound Pressure Level, SPL (dbA)	Typical Sound	Subjective Impression
150		(Short exposure can cause hearing loss)
140	Jet plane takeoff	
130	Artillery fire, riveting, machine gun	(Threshold of pain)
120	Siren at 100 ft, jet plane (passenger ramp), thunder, sonic boom	Deafening
110	Woodworking shop, hard-rock band, accelerating motorcycle	Sound can be felt (Threshold of discomfort)
100	Subway (steel wheels), loud street noise, power lawnmower, outboard motor	Very loud, conversation difficult; ear protection required for sustained occupancy
90	Noisy factory, truck unmuffled, train whistle, machine shop, kitchen blender, pneumatic jackhammer	
80	Printing press, subway (rubber wheels), noisy office, supermarket, average factory	(Intolerable for phone use)
70	Average street noise, quiet typewriter, freight train at 100 ft, average radio, department store	Loud, noisy; voice must be raised to be understood
60	Noisy home, hotel lobby, average office, restaurant, normal conversation	Usual background; normal conversation easily understood
50	General office, hospital, quiet radio, average home, bank, quiet street	
40	Private office, quiet home	Noticeably quiet
30	Quiet conversation, broadcast studio	
20	Empty auditorium, whisper	Very quiet
10	Rustling leaves, soundproof room, human breathing	
0 db		Intolerably quiet Threshold of audibility

The mechanics of absorption and the use of absorptive materials are discussed in Sections 27.3 and 27.4 in connection with noise reduction techniques. Absorption coefficients for common materials, acoustic materials, and auditorium furnishings are tabulated in Table 27.1. It is important to note that for most common materials, absorption (and therefore α) varies with frequency. Thus absorption calculations must be made individually for the frequencies being studied.

26.14 Reverberation

Reverberation is the persistence of sound after the source of sound has ceased a result of repeated reflections. Reverbation time (T_R) describes the period required for the sound level to decrease 60 db after the sound source has stopped producing sound. For rooms of usual size and shape, the reverberation time at a specific frequency may be found by the formula:

$$T_R = K \times \frac{V}{\Sigma A} \quad \text{seconds} \qquad (26.12)$$

where

K = a constant, equal to 0.05 when measurements are in feet and 0.16 when in meters
V = room volume, ft^3 (m^3)
ΣA = total absorption, sabins (ft^2 or m^2) at that frequency

(For spaces of unusual shapes, see Beranek, 1988; Embelton, 1971; Young, 1959.) Subjectively, reverberation is one of the most pronounced hearing reactions in an enclosed space. Simply, it is the ear's reaction to echoes, giving an impression of ''liveness'' or ''deadness of the space.''

''Liveness'' and ''deadness'' are terms that describe the subjective impression of sound reflection in a space. A space with highly reflective surfaces, and therefore low absorption ($\bar{\alpha} < 0.2$), sound live. Conversely, a highly absorptive nonreflective environment ($\bar{\alpha} > 0.4$) sounds dead. ($\bar{\alpha}$ is the average absorption of the entire space, as explained in detail in the next section.) Since room absorption is related to total surface area, which in turn is related by room proportions to room volume, it is possible to relate all three factors in a single diagram. Figure 26.16 is drawn using room proportions of 1.5 : 2 : 1 for L : W : H, based on Fig. 26.34 for preferred room proportions. These proportions (1.5 : 2 : 1) are the average of the extremes of Fig. 26.34 (i.e., 1.5 : 2.4 : 1, 1.8 : 2.2 : 1 and 1.3 : 1.6 : 1). The maximum differential introduced by using average proportions is 2½%—well within engineering accuracy.

The ultimate dead space is one where there is no reflection at all—as is the case outdoors. Indoors, we receive an auditory cue in the form of reflected sound (feedback) which helps us regulate our sound power output (voice level). Outdoors, the absence of this cue, to which we are so accustomed by our largely indoor existence, automatically causes us to raise our output. This is the reason then that most people tend to speak loudly and even to shout when outdoors. Conversely, and correctly, a highly reflective indoor condition gives a large feedback signal, which usually results in a lowered vocal output.

In most room acoustics studies, reverberation times are calculated at 125, 500, 1000, and 2000 Hz. The midfrequency (500 to 1000 Hz) range is generally the reference used in specifying the reverberation time of a room when studying the speech characteristic of the space.

Reverberation can be considered as a mixture of previous and more recent sounds. The converse of reverberation or reverberance is articulation. An articulate environment keeps each sound event separate rather than running them together. Spaces for speech activities should be less reverberant—more articulate—than those designed for performance of music. See Sections 26.23 and 26.24 for reverberation criteria for speech and music rooms, respectively.

26.15 Sound Fields in an Enclosed Space

The inverse square law described in Section 26.7 holds true for the acoustic *far field,* which is the sound field sufficiently far from the source that intensity is proportional to pressure squared. This type of acoustic field is developed in open, obstruction-free circumstances. Propagation in an enclosed space is quite different. There, when a sound reaches a wall or other large (with respect to wavelength) obstruction, part of the sound energy is reflected and part absorbed. Thus the sound at any point in the room is a combination of direct sound from the source plus reflected sound from walls and other obstructions. If the reflections are so large that the sound level becomes uniform throughout the room, the field within the room is termed a *diffuse* one (no shadows), and intensity measurements with respect to a specific source are meaningless. Of course, if it is our intention to measure sound pressure level at a specific point, such as a seat in an auditorium, the type of acoustic field in the room is irrelevant.

Most rooms do not have such a high level of reflection that a diffuse field is created. Instead, there is a *near field* near the source, a *free field* beyond the near field, and a *reverberant field* near the walls (see Fig. 26.17). They can be recognized as follows:

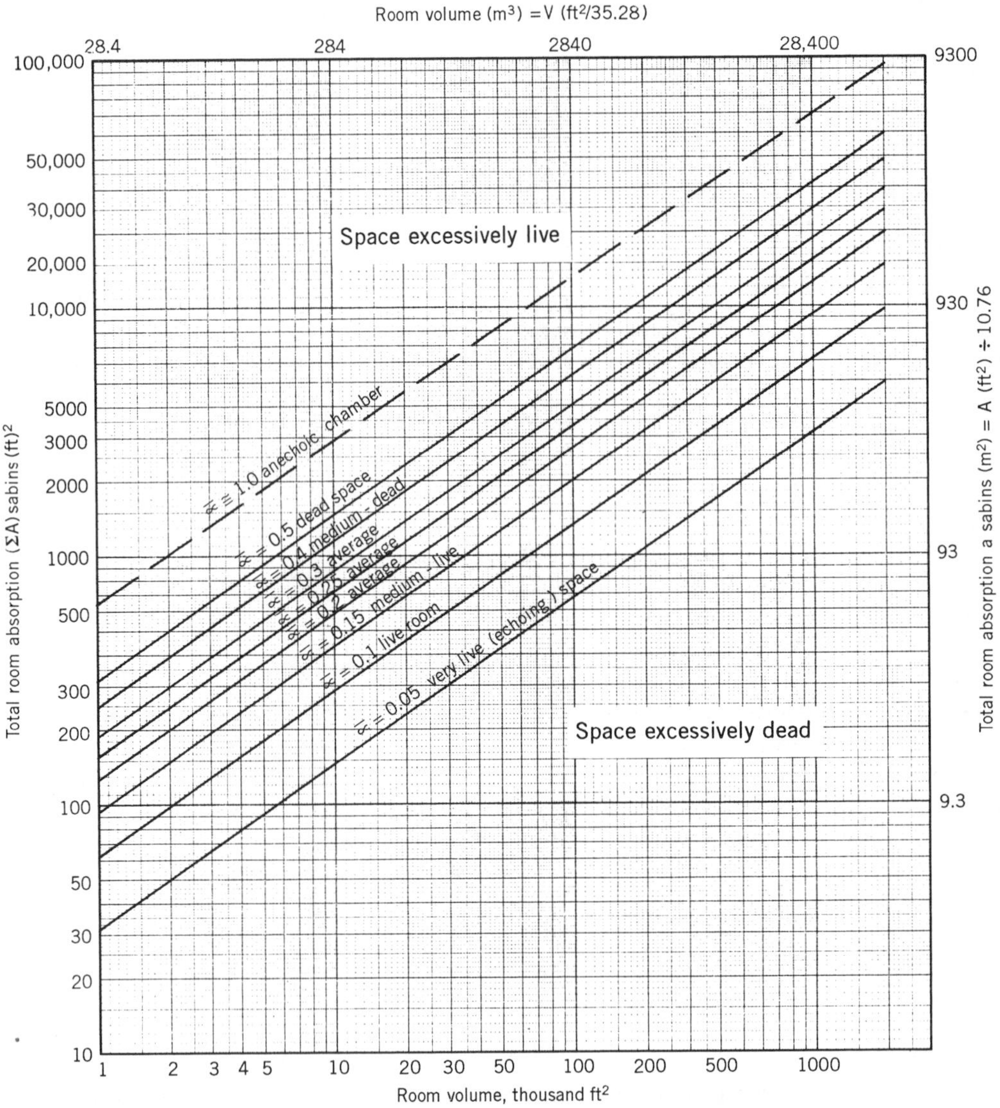

Fig. 26.16 *Chart indicating a room's "liveness" as a function of its volume and total Sabin absorption. The L-W-H room proportions chosen*—2H : 1.5H : H—*represent the average of the three extreme points of the recommended room proportion triangle in Fig. 26.34. For these proportions,* $S = 6.25V^{2/3}$ *and therefore* $A = S\bar{\alpha} = 6.25\bar{\alpha}V^{2/3}$.

1. The near field is generally within one wavelength of the lowest frequency of sound produced by the source. Within this distance SPL measurements vary widely and are not meaningful. (The maximum wavelength for the human male voice is about 11 ft.)

2. Near large obstructions such as walls, the *reverberant field* is dominant and approaches a diffuse condition. In well-designed music auditoriums the reverberant (diffuse) field predominates and sound pressure level remains relatively constant beyond the free field area.

3. The *free* (far) *field* exists between the near and reverberant fields, and there intensity varies as pressure and inversely with distance squared. In this field, sound pressure level drops 6 db

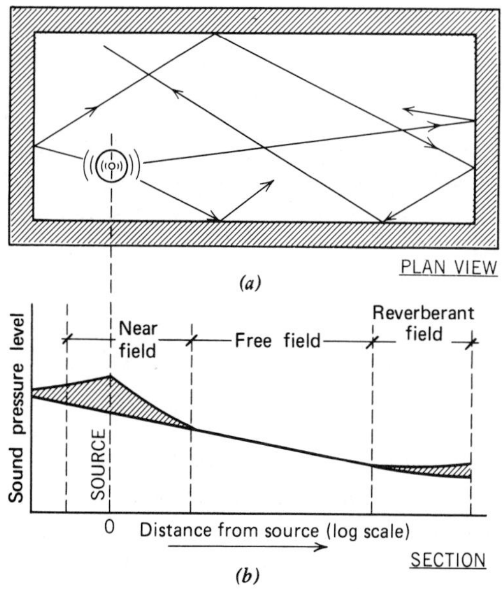

Fig. 26.17 *Type of sound field in an enclosed space depends in large measure on reflections (reverberation) and absorption. In a typical room there is a near field adjacent to the sound, a free field beyond that, and a reverberant field adjacent to the walls. In a large hall or auditorium the reverberant field dominates and level remains approximately constant.*

with each doubling of distance from the source, and it is in this field that meaningful SPL measurements can be made, with respect to a specific small source.

Since it is desirable to be able to predict the sound level (SPL) in a space during the design stage, and also because of the difficulty of SPL measurement in rooms with various types of sound field, as explained above, it is useful to have equations that relate sound *power* level (PWL) to sound *pressure* level. Sound power level is supplied by manufacturers with each item of equipment, at either octave or ⅓-octave points.

To avoid confusion, it will be helpful to the reader to think of sound pressure level (SPL) as the resultant "noise" or sound in an enclosed space, resulting from a source in that space, and affected by the characteristics of the space and the position of the listener. It is thus an end effect. The sound power level (PWL) is a measure of the amount of sound generated by a source, independent of its environment. In free space the two quantities are simply related by the inverse square law, whereas in enclosed spaces the room characteristics come into play. Roughly speaking, an analogy to lighting can be drawn: SPL corresponds to room illumination, that is, footcandles, and PWL corresponds to the lumen output of the source. The basic relationship between sound pressure level and sound power level for a single, small, non-directive source in a room large enough to have both free and reverberant fields is:

English units:

$$SPL = PWL + 10 \log \left(\frac{Q}{4\pi r^2} + \frac{4}{R} \right) + 10.5 \tag{26.13}$$

Metric units:

$$SPL = PWL + 10 \log \left(\frac{Q}{4\pi r^2} + \frac{4}{R} \right) + 0.2 \tag{26.14}$$

where Q is a directivity factor and

SPL = sound pressure level, db
PWL = sound power level, db
r = distance from source, ft (m)
R = room constant, ft^2 (m^2)

The factor R can be calculated from

$$R = \frac{S\bar{\alpha}}{1 - \bar{\alpha}}$$

where

S = total room surface area, m^2
$\bar{\alpha}$ = average absorption coefficient of all materials in the room

that is,

$$\bar{\alpha} = \frac{\Sigma A \text{ (total room absorption)}}{S \text{ (total room surface area)}} \tag{26.15}$$

or

$$\bar{\alpha} = \frac{S_1\alpha_1 + S_2\alpha_2 + \cdots + S_n\alpha_n}{S_1 + S_2 + \cdots + S_n} \tag{26.16}$$

The directional constant Q is either inherent in the sound source, and as such will be part of the given data, or can be obtained from Fig. 26.18

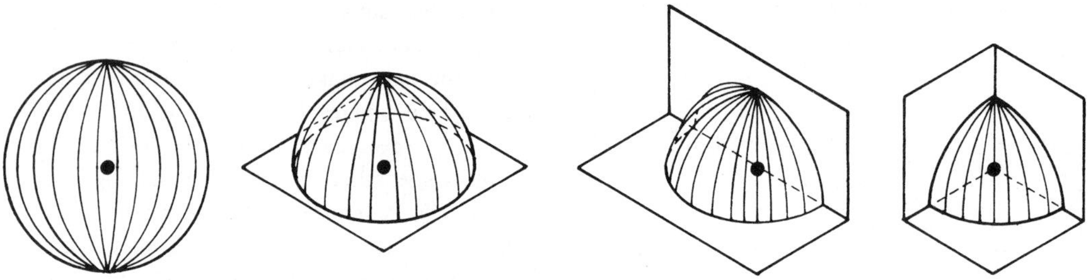

Fig. 26.18 Diagrams illustrating directivity factors for either inherently directive sources or nondirective sources placed adjacent to large reflecting surfaces. (Courtesy of Barry Blower Co.)

for a nondirectional source made directional by adjacent reflecting surfaces.

Thus for a source suspended or supported at least ½ wavelength (at its lowest frequency) from the ceiling or floor, $Q = 1$; a source near the floor and distant from the walls uses $Q = 2$; and so on. In a space containing more than one sound source, the sound pressure levels can be combined as explained in Section 26.8. In the two extreme cases [i.e., rooms with only far (direct) field and rooms with only reverberant (diffuse) fields, (corresponding to dead and live rooms, respectively)] for a nondirectional source located on the floor and away from walls ($Q = 2$), equations 26.13 and 26.14 reduce to:

Dead room, direct field:

English units:

$$\begin{aligned} \text{SPL} &= \text{PWL} - 20 \log r + Q - 0.6 \\ &= \text{PWL} - 20 \log r + 1.4 \end{aligned} \qquad (26.17)$$

Metric units:

$$\begin{aligned} \text{SPL} &= \text{PWL} - 20 \log r + Q - 10.9 \\ &= \text{PWL} - 20 \log r - 8.9 \end{aligned} \qquad (26.18)$$

Live room, diffuse-reverberent field:

English units:

$$\text{SPL} = \text{PWL} - 10 \log \Sigma A + 16.3 \qquad (26.19)$$

Metric units:

$$\text{SPL} = \text{PWL} - 10 \log \Sigma A + 6.0 \qquad (26.20)$$

To understand the application of these equations, and the effect of absorptive material in a space, see Example 27.1 in Section 27.7, Noise Reduction by Absorption.

To understand the effect of room absorption on the development of the reverberant field in a large space, refer to Fig. 26.19. The field is plotted from equation 26.13, for a room 80 ft × 120 ft × 30 ft high, using average absorptions $\bar{\alpha}$ ranging from 0.05 to 0.7, that is, from very live to very dead. Note particularly that the addition of sufficient absorbent material can entirely prevent the development of a reverberant (diffuse) field. This is particularly important in noisy industrial interiors where the building materials commonly employed are generally nonabsorbent, and the noise sources are both numerous and loud. For further discussion, see Sections 27.2 through 27.8.

26.16 Other Factors in Hearing

(a) Masking. When two separate sources of sound are perceived simultaneously, the perception of each is made more difficult by the presence of the other. This effect is known as masking, and is defined as the number of decibels by which the threshold of audibility of a sound is raised by the presence of another sound. Masking is greatest when two sounds are close in frequency or frequency content, since the ear has greater difficulty separating them. Also, a low frequency will more effectively mask a high frequency than the reverse, for the same decibel levels. With broad-frequency

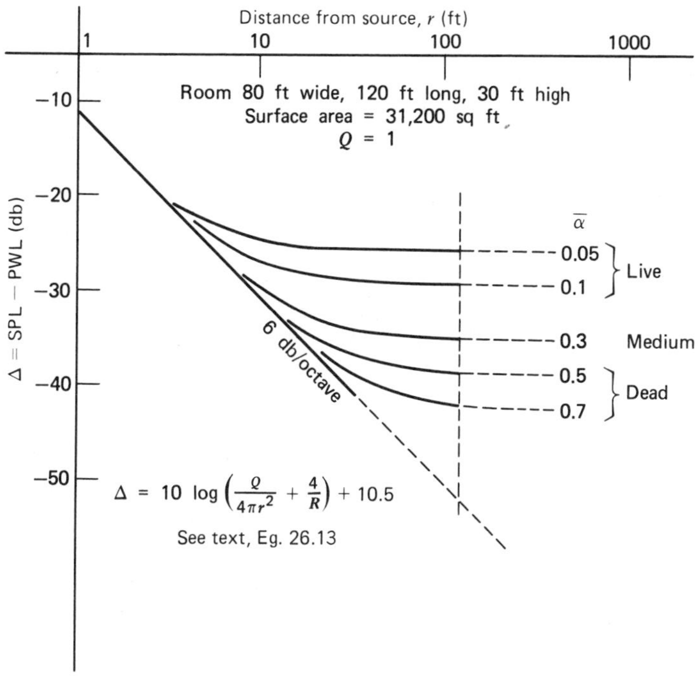

Fig. 26.19 *Curves show that development of reverberant field (constant SPL, or flat curve) is dependent on the room's absorption characteristics. In a live room ($\bar{\alpha} = 0.1$) the reverberant field begins 20 ft from the source. In a dead room ($\bar{\alpha} = 0.5$) it begins at 120 ft from the source, that is, at the back wall. Effectively, then, there is no reverberant field in such a room. Slope of asymptote is 6 dB per octave, that is, inverse square attenuation.*

sounds the masking effect is difficult to predict, since it depends in part on how "hard" the listener is listening. Masking is an extremely important technique in noise control where background noise levels are deliberately manipulated to mask unwanted sounds (see Section 27.22). The background noises used for this purpose are of the broadband continuous variety, such as "pink" noise, which are themselves noninformation bearing. They serve to obliterate lower level, information-bearing sounds that would cause annoyance and are therefore particularly useful in open (landscaped) office plans. Frequently, mechanical equipment noise such as from air diffusers and fans can be utilized to good advantage as masking noise.

(b) Directivity. The exact mechanism by which the binaural aspect of hearing detects direction is not entirely understood. It is clear, however, that in enclosed space, reverberation will blur most directivity and "stereo" information will be almost completely dependent on high frequencies in the near field.

SOUND SOURCES

26.17 Speech

As can be seen from Fig. 26.14, the ear's sensitivity is maximum in the speech frequency and normal energy range. Speech sounds vary in time length between 30 and 300 ms and the ear normally perceives them individually and clearly. Speech is comprised of *phonemes*, which are individual and distinctive sounds that to an extent vary from language to language. That is, certain ones exist in one language and not in another. Since certain phonemes carry more information than others, it is these that good architectural acoustics must be particularly careful to preserve intact, to preserve intelligibility. In English, consonants carry much more information than vowels, as can readily be demonstrated by writing a sentence first without consonants and then without vowels:

Most speech energy is concentrated in the 100- to 600-Hz range.

o ee ee i oeae i e 100–600 e ae

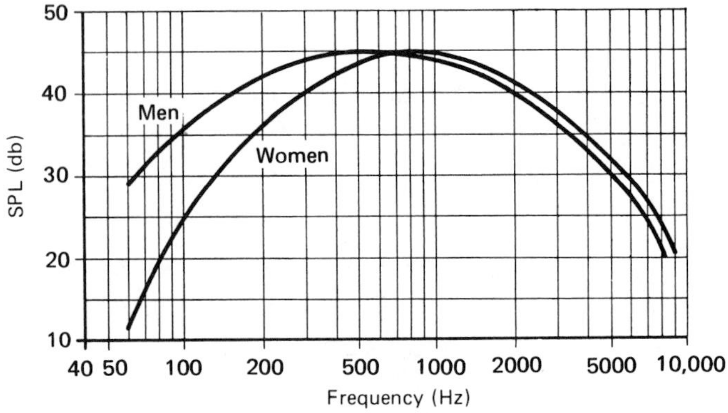

Fig. 26.20 Frequency distribution of typical male and female voices. [From Brüel and Kjaer (1963).]

Mst spch nrgy s cncntrtd n th 100–600 Hrtz rng.

The male voice centers its energy around 500 Hz; the female about 900 Hz (see Fig. 26.20). It is, however, in the high frequencies that consonants have most of their energy. Phonemes such as *s* and *sh* have most of their energy above 2 kHz and both are particularly important in conveying intelligence.

Normal speech averages between 40 and 50 dbA sound pressure level at 3 to 4 ft, with a dynamic range of from about 30 db for soft speech to about 65 db for loud speech at the same distance. Extremes of speech are 10 db for a soft whisper and 80 db for a shout, but in both these instances intelligibility is sharply reduced because of lack of consonant power. Indeed, in shouting, emphasis is necessarily on vowels, so that it is generally accepted that 70-db SPL is about the upper limit of fully intelligible human speech. (Singers who frequently exceed 90 db do so at great loss of intelligibility.) Another result of the high-frequency content of consonants, and therefore intelligibility, is its directiveness. The higher the frequency, the greater the sound's directivity and the less its diffraction (ability to be heard beyond a barrier). Therefore, intelligibility of speech is greatest directly in front of the speaker and least behind him. High-frequency tones are most easily absorbed and least diffracted.

26.18 Other Sounds

Music is much broader and complex than speech in frequency and dynamic range. It has no direct parallel to intelligibility. "Reception" of music is a combination of physiological and psychological phenomena. As such, it is beyond most of the purposes of this study but is briefly examined in our discussion of room acoustics, auditoriums, and halls. *Noise* is variously defined as unwanted sound, sound with no intelligence content, and broadband sound, depending on the listener and the situation. All three definitions are correct at various times and situations. For our purpose we must assume that any sound can be referred to as noise.

NOISE CRITERIA

26.19 Negative Effects of Noise

Although noise effects and their control are the subject of Chapter 27, noise criteria and their background are discussed here as part of our overall study of hearing and sound sources. There are two basic approaches to the negative effects of noise, a pyschological-practical one and a purely physiological one. The latter is concerned with the physical impact of noise on the body including hearing loss and other deleterious conditions (see Section 26.22). The former is concerned with noise levels that cause annoyance and disturbance to daily activities including work, relaxation, and rest, and is discussed in the following sections.

26.20 Noise and Annoyance

Research has developed accurate data on loudness. The concept of annoyance, however, being primarily subjective and psychological, is much more elusive. Tests have shown that *in general,* annoyance as a result of noise is:

1. Proportional to the loudness of the noise.
2. Greater for high-frequency than low-frequency noise.
3. Greater for intermittent than continuous noise.
4. Greater for pure-tone than for broad-band noise.
5. Greater for moving or unlocatable (reverberant) noise than for a fixed-location sound.
6. Much greater for an information bearing noise (neighbor's radio) than for a nonsense noise.

To establish criteria for *acceptable background noise* (i.e., noise whose annoyance level is acceptable), certain of these effects must be neglected for the sake of simplicity. [They can be and are considered in construction design and in establishing levels of masking noise (see Chapter 27)]. We thus ignore:

Factor 3, since design is based on continuous sounds.

Factor 4, since broadband noise is assumed.

Factor 5, since the noise source is assumed to be fixed in location.

Factor 6, since we can only consider noise level, not content.

Thus the particularly and special characteristics of noises such as a barking dog (3), a whistle (4), a single passing vehicle (5), and intelligible sounds (6) are not considered. In order to quantify the concept of acceptable background noise, it was necessary to remove it from the purely psychological arena and relate it to a physical phenomenon. This was done by studying the effect of the two remaining annoyance factors (1 and 2) on speech communication. This study resulted in two concepts: the Articulation Index (AI) and the Speech Interference Level (SIL). They are determined by reading a carefully selected set of phonetically balanced nonsense syllables to a test audience in the presence of different levels and compositions of background noise. The ratio of correct answers to total syllables is the Articulation Index. An Articulation Index (Beranek, 1949) in excess of 0.5 indicated a condition in which perfect intelligibility could be expected. A simplified version of the AI called the Speech Interference Level (SIL) was devised by Beranek. The SIL consists simply of the arithmetic average in decibels of the background-noise sound levels in three octave bands, 600 to 1200, 1200 to 2400, and 2400 to 4800 Hz, since it was found that correlation between intelligibility and the sound power in these three bands could be established.

Beranek developed the well-known and widely accepted noise criteria (NC) curves shown in Fig. 26.21. The NC curves take cognizance of the field-determined fact that most people prefer to speak at a level no greater than 22 phons above the background noise. By combining the SIL levels in decibels with this fact, the NC curves are derived, that is, they represent a loudness level 22 phons higher than the SIL in db. These contours represent then the maximum *continuous* background noise that will be considered acceptable in the environment specified and correspond fairly accurately to background noise level in commercial environments. A similar set of curves called noise rating (NR) curves, proposed by the International Standards Organization (ISO), finds considerable application outside the United States. They are less stringent than NC curves in the low frequencies and more stringent in the high frequencies.

In application, the octave-band spectrum of the noise being studied, over the range of 63 to 8000 Hz, is measured and plotted on an NC curve sheet. The lowest NC curve that is *not* exceeded by any portion of the plot is the NC rating of the noise. Thus specifying a maximum noise level of NC-30 for a space means that no portion of the sound pressure level curve of any continuous background noise in the area may cross the NC-30 contour. Conversely, a piece of equipment rated NC-35 has an octave-band spectrum completely below NC-35. A fan rating of NC-53 would indicate that at some point on its frequency-spectrum plot the fan exceeded NC-50 by 3 db (see Fig. 26.21).

The NC number of a noise varies between 5 and 10 db below the measured dbA. NR numbers correspond closely to NC numbers. The

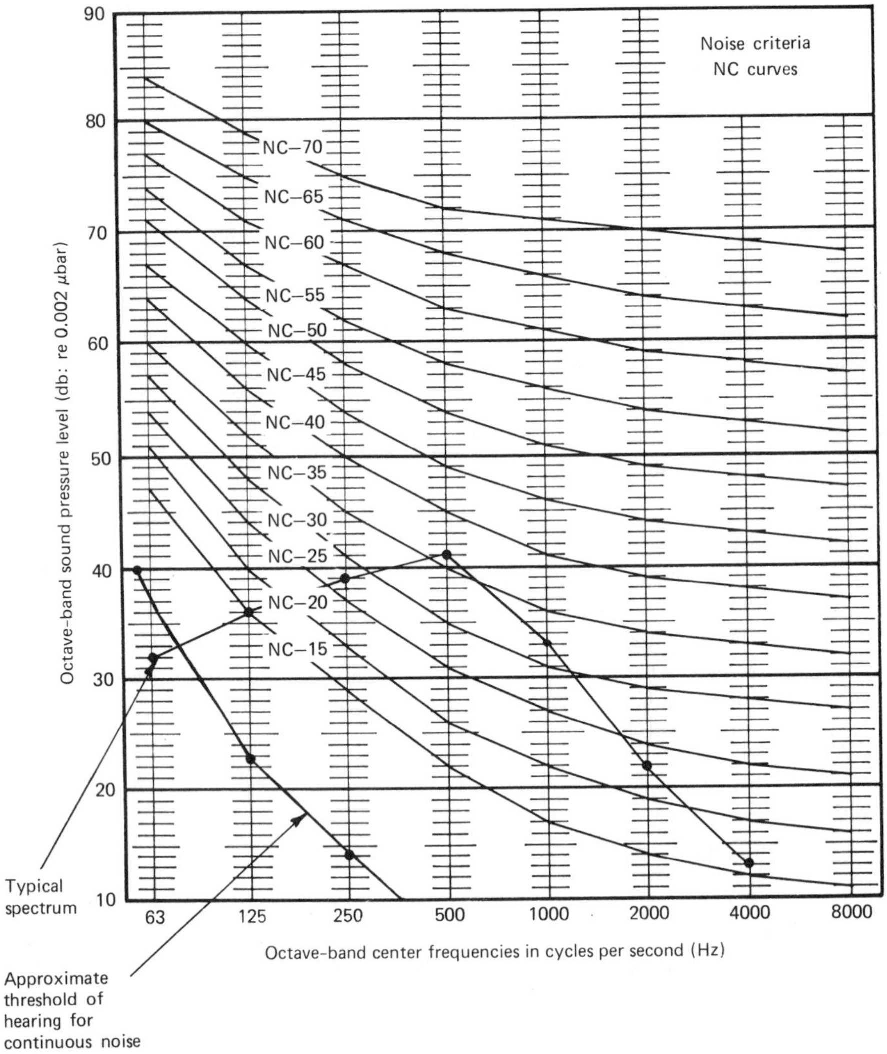

Fig. 26.21 *Noise criteria curves. Typical spectrum plotted would be rated NC-36, as it exceeds NC-35 at 500 Hz by 1 db. (See Table 27.8 for specific recommendations for interior spaces.)*

virtue of the NC curve system is that it is a single-number specification for an entire frequency spectrum. Its disadvantage is that it was derived for, and is primarily accurate for speech conversation against a continuous, non-intelligence-bearing noise. This is not the situation that is most troublesome in an office; indeed, continuous equipment noise may be *helpful* in masking unwanted speech interference, as mentioned above. See Sections 27.21 and 27.22, and Fig. 27.33. Nevertheless, NC curves remain the commonly used criteria for continuous *nonintentional* background noise acceptability.

26.21 High Noise Levels; Hearing Protection; OSHA

It has long been recognized that continuous exposure to high noise levels causes a degree of temporary deafness in a majority of people and that long periods of such exposure, even on an intermittent 8-h workday basis, appears to produce permanent hearing impairment. Many experts place the safe upper limit at 85 db. In addition, studies have indicated that continual exposure to noise levels as low as 75 to 85 dbA level can produce or contribute to numerous

§ 1910.95 Occupational noise exposure.

(a) Protection against the effects of noise exposure shall be provided when the sound levels exceed those shown in Table G-16 when measured on the A scale of a standard sound level meter at slow response. When noise levels are determined by octave band analysis, the equivalent A-weighted sound level may be determined as follows:

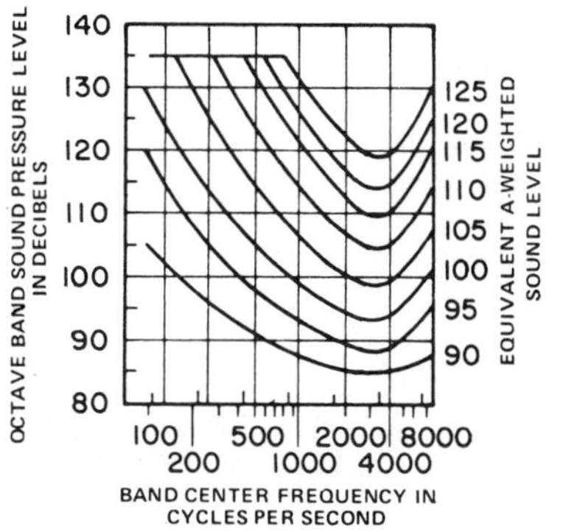

Equivalent sound level contours. Octave band sound pressure levels may be converted to the equivalent A-weighted sound level by plotting them on this graph and noting the A-weighted sound level corresponding to the point of highest penetration into the sound level contours. This equivalent A-weighted sound level, which may differ from the actual A-weighted sound level of the noise, is used to determine exposure limits from Table G-16.

[1910.95 amended at 39 FR 19468, June 3, 1974]

(b)(1) When employees are subjected to sound exceeding those listed in Table G-16, feasible administrative or engineering controls shall be utilized. If such controls fail to reduce sound levels within the levels of Table G-16, personal protective equipment shall be provided and used to reduce sound levels within the levels of the table.

(2) If the variations in noise level involve maxima at intervals of 1 second or less, it is to be considered continuous.

(3) In all cases where the sound levels exceed the values shown herein, a continuing, effective hearing conservation program shall be administered.

TABLE G-16—PERMISSIBLE NOISE EXPOSURES[1]

Duration per day, hours	Sound level dBA slow response
8	90
6	92
4	95
3	97
2	100
1½	102
1	105
½	110
¼ or less	115

[1] When the daily noise exposure is composed of two or more periods of noise exposure of different levels, their combined effect should be considered, rather than the individual effect of each. If the sum of the following fractions: $C_1/T_1 + C_2/T_2 + \cdots C_n/T_n$ exceeds unity, then, the mixed exposure should be considered to exceed the limit value. Cn indicates the total time of exposure at a specified noise level, and Tn indicates the total time of exposure permitted at that level.

[1910.95 Table G-16 amended at 39 FR 19468, June 3, 1974]

Exposure to impulsive or impact noise should not exceed 140 dB peak sound pressure level.

Fig. 26.22 *The standard for exposure to noise in the workplace. (From OSHA Rev., July 1988.)*

physical and psychological ailments including headache, digestive problems, tachycardia, high blood pressure, anxiety, and nervousness—an almost complete catalog of human illnesses. Since continuous noises are most severe in industry, regulatory legislation in the United States has been most stringent and effective in this area. In 1969 the Walsh-Healy Public Contracts Act was passed, and thereafter its provision for maximum permissible exposure to noise levels was incorporated into the Occupational Safety and Health Act. Both the act and the associated regulatory agents, the Occupational Safety and Health Administration, are known as OSHA. The relevant provisions of this act are reproduced in Fig. 26.22. To avoid complex regulations, limitations of exposure are given in terms of single-number dbA values. Since workers rarely remain in a single acoustic environment for 8 h, their total daily exposure, or *time-*

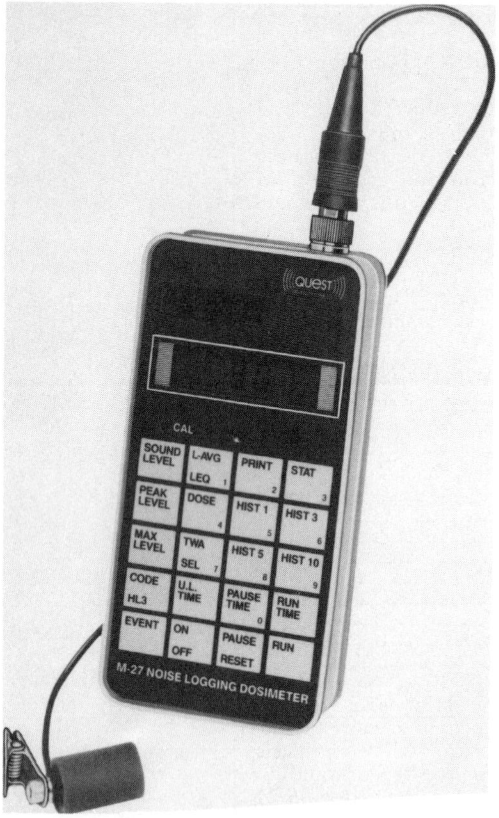

Fig. 26.23 This small ($5\frac{1}{2}$ in. × $2\frac{3}{4}$ in.) multifunction noise meter can be worn by the user throughout the workday. It measures noise dose, time-weighted average (TWA) noise level, and other noise levels, in addition to functioning as a normal sound-level meter. It can also be connected to a printer to provide timed histograms for statistical noise analyses. (Courtesy of Quest Electronics.)

weighted average (TWA) exposure can be calculated from timed dbA measurements using formulas and tables given in the OSHA code and then compared to permissible levels. Alternatively, one of the newer dose meters (see Fig. 26.23) can be used, which automatically integrates the noise to which a person is exposed over a given time period and reads out the TWA exposure directly.

Typical industrial noise levels are given in Table 26.5. When these levels are exceeded, management must take steps to reduce the exposure, either by reducing the noise or by providing hearing protectors. Typical characteristics of a few types are given in Fig. 26.24.

TABLE 26.5 **Typical Industrial Noise Levels**[a]

Equipment	*dbA*
Printing press plant (medium size automatic)	86
Heavy diesel-propelled vehicle (about 25 ft away)	92
Heavy-duty grinder	93
Air compressor	94
Plastic chipper	96
Cutoff saw	97
Multiple spot welder	98
Turbine condenser	98
15 cu ft air compressor	100
Drive gear	103
Banging of steel plate	104
Magnetic drill press	106
Air chisel	106
Positive displacement blower	107
Air hammer	107
Vacuum pump	108
Jolt squeeze hammer	122

[a]These are approximate values for typical generic equipment types and should not be used as design values.

OSHA does not deal extensively with impulse noises, except to state that they shall not exceed 140-dB peak sound pressure level. The area of impulse noise is quite different from continuous noise, since apprehended levels depend on duration, and specification is difficult. Much work has been done in this area by the military for obvious reasons. The interested reader is referred to the literature since this subject is substantially outside the area of architectural acoustics.

ROOM ACOUSTICS

26.22 Sound in Enclosures

When a continuous sound is generated in an enclosure, fields are set up as described in Section 26.15. When the sound is not a continuous tone or noise but a series of discrete sounds, following one upon the other and containing information as in speech or music, the room must be designed to maintain and enhance the information's intelligibility. That is what is meant by design of room acoustics.

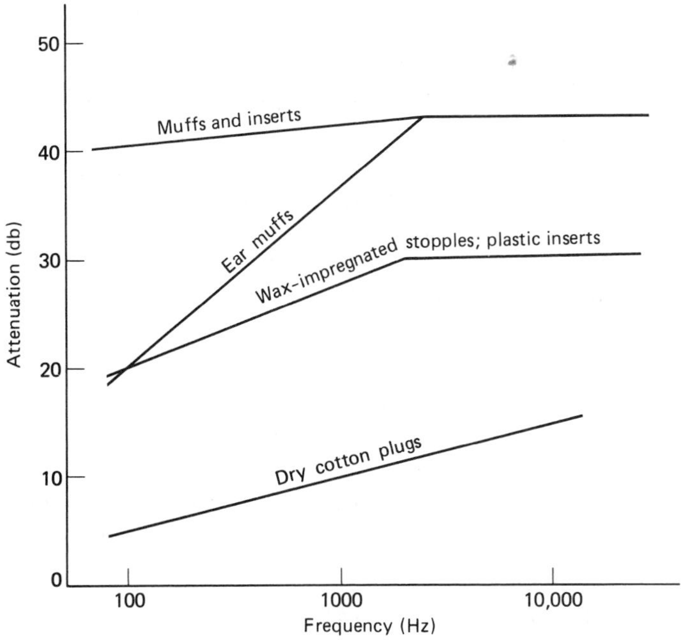

Fig. 26.24 *Sound attenuation characteristics of various types of ear protectors. [From* Quieting: A Practical Guide to Noise Control *(1976).]*

The generated sound radiates out from the source until it strikes a room boundary or other large surface. Before reaching this surface the sound intensity is attenuated by distance (inverse square law) and by absorption in the air. This latter is appreciable only in large rooms, and at frequencies above 2000 Hz. [For detailed data on air absorption, see Magrab (1975). In the absence of exact data, assume a figure of 0.40 sabin (m^2)/100 m^3 and see Table 27.1.) When the sound reaches the wall, it is partially reflected and partially absorbed. A small portion is transmitted into adjoining spaces. The energy transmitted is usually so small that it has little effect on the space within which the sound originates although, as discussed in Chapter 27, it may be very important in the surrounding spaces.

The ratio between the energy absorbed and the energy reflected will significantly affect what one hears within the space. Specifically, if little energy is absorbed and much is reflected, two effects will be noticeable. Intermittent sounds will be mixed together (which may make speech *less* intelligible or music *more* pleasant), and steady sounds will accumulate into a reverberant field, making the space "noisy." Conversely, if much energy is absorbed and little reflected, the room will sound quiet to speech and "dead" to music. At this point it may be helpful to the reader to review the technicalities of absorption and reverberation as discussed in Sections 26.13 through 26.15 above.

26.23 Criteria for Speech Rooms

The overriding criterion for speech is intelligibility. Since speech consists of short disconnected sounds 30 to 300 ms in length (see Section 26.17), among which are high-frequency, low-energy phonemes, the ideal room must assure the ear's undistorted reception of these phonemes. This requires keeping reverberation to a minimum. We can obtain a good approximation of the subjective feeling of liveness of a room, for purposes of *speech,* from the relation

$$T_R = 0.3 \log \frac{V}{10} \qquad (26.21)$$

where

T_R = optimum reverberation time in seconds, for speech

V = room volume, m^3

For instance, a typical classroom might have a volume of 150 m^3 (5300 ft^3). Optimum reverberation time is

$$T = 0.3 \log 15 = 0.35 \text{ s}$$

Reverberation times longer than this would sound live, shorter ones dead and flat. Indeed an increase of 20% in reverberation time would make the room excessively live and boomy and would negatively affect speech intelligibility. Figure 26.25 gives optimum mid-frequency reverberation times as a function of room size and use, since experience has shown these figures to be a good indication of the space's reverberation characteristic over the entire frequency band. For good speech intelligibility, reverberation time should remain flat down to 100 Hz.

Figure 26.26 gives *maximum* reverberation time figures for speech, in large rooms. The reflected sounds associated with reverberation have either a salutary or a deleterious effect. The ear cannot distinguish between sounds that arrive within a maximum of 50 ms (1/20 s) of each other. (Some authorities use 40 ms, i.e., 1/25 s.) Sounds arriving within this time *reinforce* the direct path signal and appear to come from the source. Sounds arriving after this time are apprehended as a fuzzy echo or elongation of the sound, reducing intelligibility and directiveness. Since the range of 40 to 50 ms corresponds at 344 m/s to 13.76 to 17.2 m, a speech room must be so arranged that the difference between the first reflection path and the direct path is no greater than 17 m and preferably 14 m or less (see Fig. 26.27). For more detail concerning this and related factors on reflected paths, refer to Section 26.25.

Too *low* a reverberation time (very high absorption, minimum reflection) is also undesirable because:

1. It limits the size of the room to that which can be covered by direct sound only.
2. It is disturbing to the speaker, since absence of reflection prevents him or her from gauging proper voice level and tends to cause excessive effort (shouting, as when outdoors).

Thus proper design of a room for speech is a compromise between the need for some reflection and the desire to minimize reflection to preserve intelligibility.

26.24 Criteria for Music Performance

Adequate design for a music space requires recognition of the following:

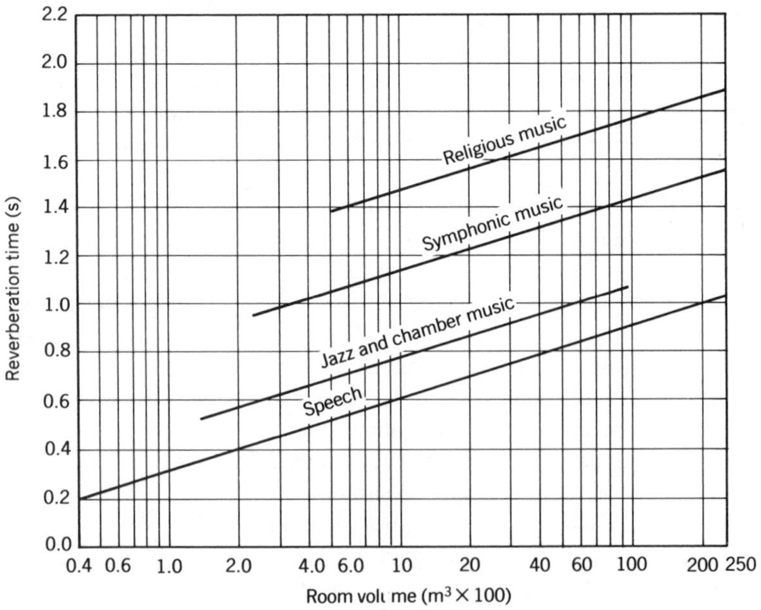

Fig. 26.25 Optimum reverberation times for the frequency range 500 to 1000 Hz. (Reprinted by permission from E. B. Magrab, Environmental Noise Control, *Wiley, New York, 1975, p. 206.)*

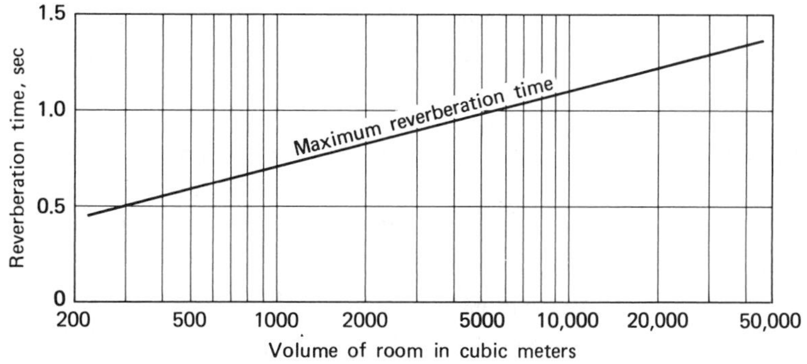

Fig. 26.26 Maximum recommended reverberation time for speech in auditoriums and lecture halls. [From P. V. Brüel, Sound Insulation and Room Acoustics, *Chapman & Hall, London, 1951, as reprinted in Brüel and Kjaer (1963).]*

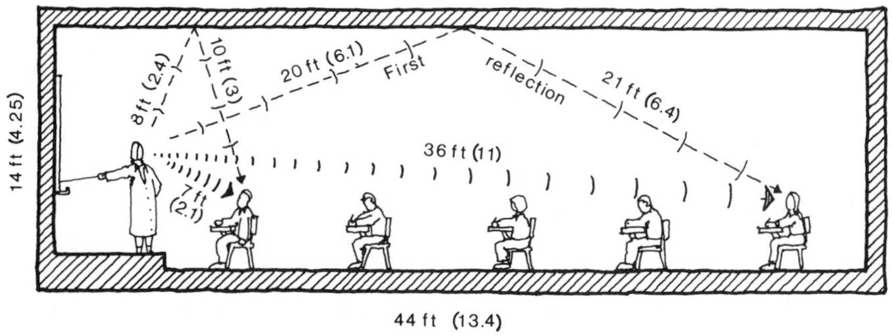

Fig. 26.27 Sound paths in a typical medium-sized lecture room. Note that for both extremes of listener position, the maximum path-length difference between direct and first reflection is 11 ft. Thus signal is reinforced, and intelligibility should be excellent if room absorption is provided to limit reverberation time to about ½ s maximum (see Fig. 26.26). Numbers in parentheses are dimensions in meters.

1. Large-volume spaces require direct-path sound reinforcement by reflection.

2. Relatively long reverberation time is needed to enhance the music—the exact amount depending on the type of music (see Fig. 26.25). Designers should keep in mind that recommendations vary as much as 100% between respected sources. In addition to an optimum reverberation time in the central 500-Hz to 1-kHz range, the reverberations at other frequencies should exhibit a slight drop in the higher frequencies to compensate for the sharp drop in ear sensitivity. Thus T_R at 100 Hz should be, according to most researchers, between 25 and 50% longer than T_R at the center frequencies. As a matter of interest, concert halls judged to be excellent by musical experts have a center frequency reverberation time between 1.6 and 1.8 s.

3. Directivity declines if the reinforcing signal is excessively delayed. With large ensembles, directivity gives the sense of depth and instrument location necessary for proper appreciation. This is often referred to as clarity or definition in music. With a solo instrument this problem is diminished.

4. Brilliance of tone is primarily a function of high-frequency content. Since these frequencies are most readily absorbed, a good direct path must exist between sound source and listener. Since our eyes and ears are close together, a

ACOUSTICS

good sound path exists when a good vision path exists. At the other end of the spectrum, lack of sufficient bass expresses itself as a loss of "fullness," which is often caused by resonant absorption (see Section 27.3).

The actual design of music performance space is a very complex procedure involving extensive calculations of absorption, reverberation time and ray diagramming, and juggling of materials, dimensions, and wall angles. Simulation techniques and acoustic models are also employed. Most modern design also uses movable reflector panels and other active variables. After construction is completed, extensive tests are conducted and field adjustments are made.

26.25 Ray Diagrams and Sound Paths

Ideally, every listener in a lecture hall, theater, or concert hall should hear the speaker or performer with the same degree of loudness and clarity. Since this is obviously impossible by direct-path sound, the essential design task is to plan methods for reinforcing desirable reflections and minimizing and controlling undesirable ones. Normally only the first reflection is considered in ray diagraming, since it is strongest. Second and subsequent reflections are usually attenuated to the point that they need not be considered except for the special situations of flutter, echoes, and standing waves discussed below.

(a) Specular Reflection. Specular reflection occurs when sound reflects off a hard polished surface. This characteristic can be used to good advantage to create an effective image source. In ancient Greek and Roman theaters, seats were arranged on a steep, conical surface around the performers. The virtue of the arrangement (see Fig. 26.28*a*) is that the sound power travels to each location with minimal attenuation. The same effect can be accomplished by placing the sound source above the seats. This is not practical physically, but it can be accomplished effectively by the use of a reflecting panel (see Fig. 26.28*b*). Panel dimension must be at least one wavelength at the lowest frequency under consideration. Figure 26.29 is a conversion chart from frequency to wavelength in feet and meters.

(b) Ray Diagrams. Ray diagramming is a design procedure for analyzing reflected sound distribution throughout a hall, using the first reflection only. Figure 26.30 shows a ray diagram. The rays are drawn normal (perpendicular) to the spherically propagating sound waves. Specular reflection is assumed. That is, the angles between the reflecting panel and the incident and reflected rays are always equal. Thus, in addition to the direct sound, each listener is receiving reflected sound energy. It is as though there were additional sound sources, the *real* one and numerous image sources. Figure 26.30 shows the application of a ray diagram to a lecture hall. In Fig. 26.30*a* the stage height and seating slope are arranged to provide good sight lines, and the ceiling height is established by reverberation requirements esthetics, cost, and so on. It can be seen that less than half the ceiling is providing useful reflection. By dividing the ceiling into two panels (Fig. 26.30*b*), people in the rear of the room perceive the direct source plus two image sources, and the useful reflecting area is increased by 50%. In Fig. 26.30*c*, the shape has been further refined to include a lighting slot and a loudspeaker grille.

Although they are a useful design tool, ray diagrams have certain restrictions. For example, the hall is three dimensional and the diagrams two dimensional, that is, sectional. To properly ray diagram, depth (width) must also be considered, and this unduly complicates the diagrams. Also, design must always be a compromise between ray diagrams for various "speaking positions" on the stage. Thus a paraboloid may be a perfect shape for one source position but will be very poor for other positions.

(c) Echoes. As explained in Section 26.23, a clear echo is caused when reflected sound *at sufficient intensity* reaches a listener more than 50 ms after he has heard the direct sound. Echoes, even if not distinctly discernible, are undesirable in rooms. They are annoying and make speech less intelligible. The relative annoyance is dependent on the time delay and loudness rel-

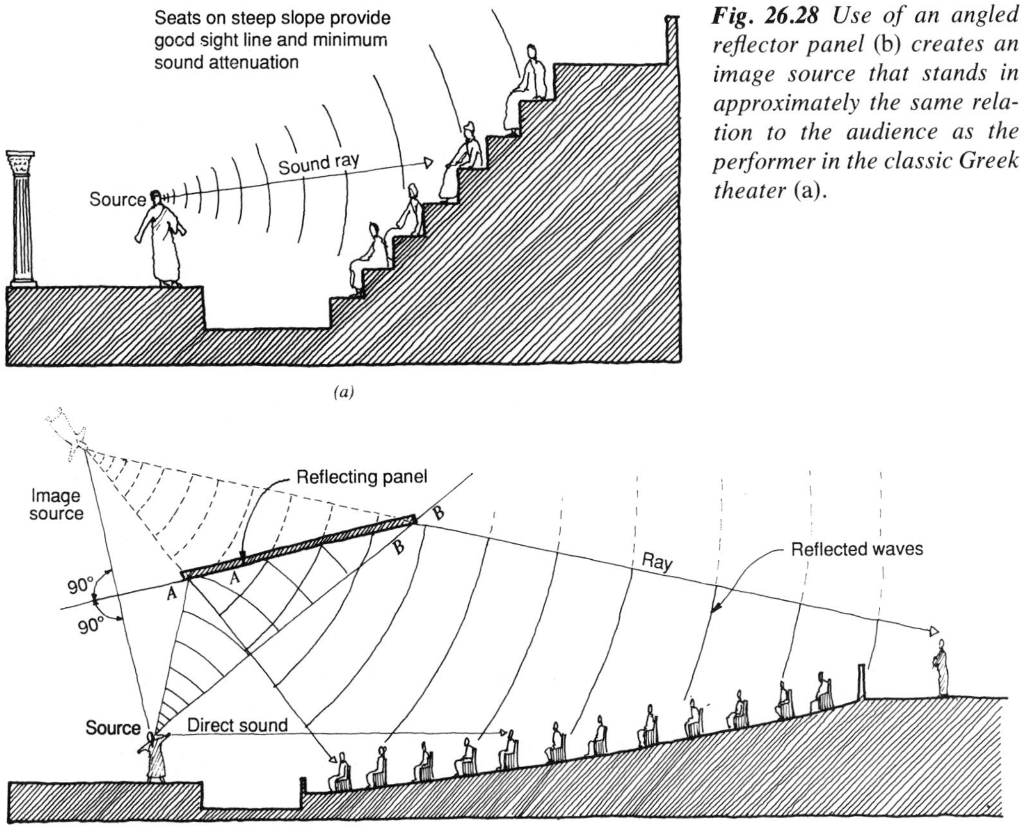

Fig. 26.28 Use of an angled reflector panel (b) *creates an image source that stands in approximately the same relation to the audience as the performer in the classic Greek theater* (a).

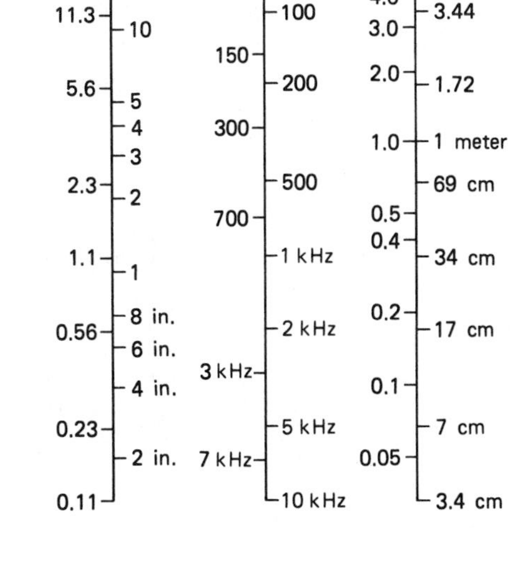

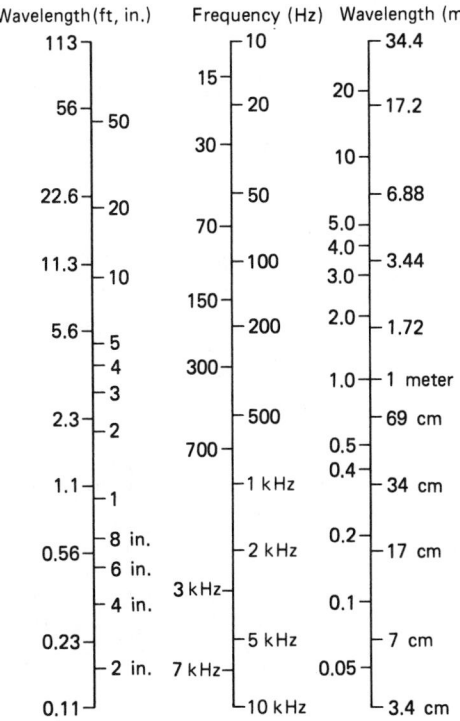

Fig. 26.29 Nomograph for determining wavelength in feet or meters, given frequency in hertz, or vice versa. Speed of sound is taken as 344 m/s and 1128 ft/s. To use, hold a straightedge horizontally across the nomograph, and read the figures directly.

ative to the direct sound, which, in turn, are dependent on the size, position, shape, and absorption of the reflecting surface.

Typical echo-producing surfaces in an auditorium are the back wall and the ceiling above the proscenium. Figure 26.31 shows these problems and suggests remedies. Note that the energy that produced the echoes can be redirected to places where it becomes useful reinforcement. If echo control by absorption alone were used on the ceiling and back wall, that energy would be wasted. The rear wall, since its area cannot be reduced too far, may have to be made more sound absorptive to reduce the loudness of the reflected sound.

Fig. 26.30 Section through a typical lecture room showing use of ray diagrams.

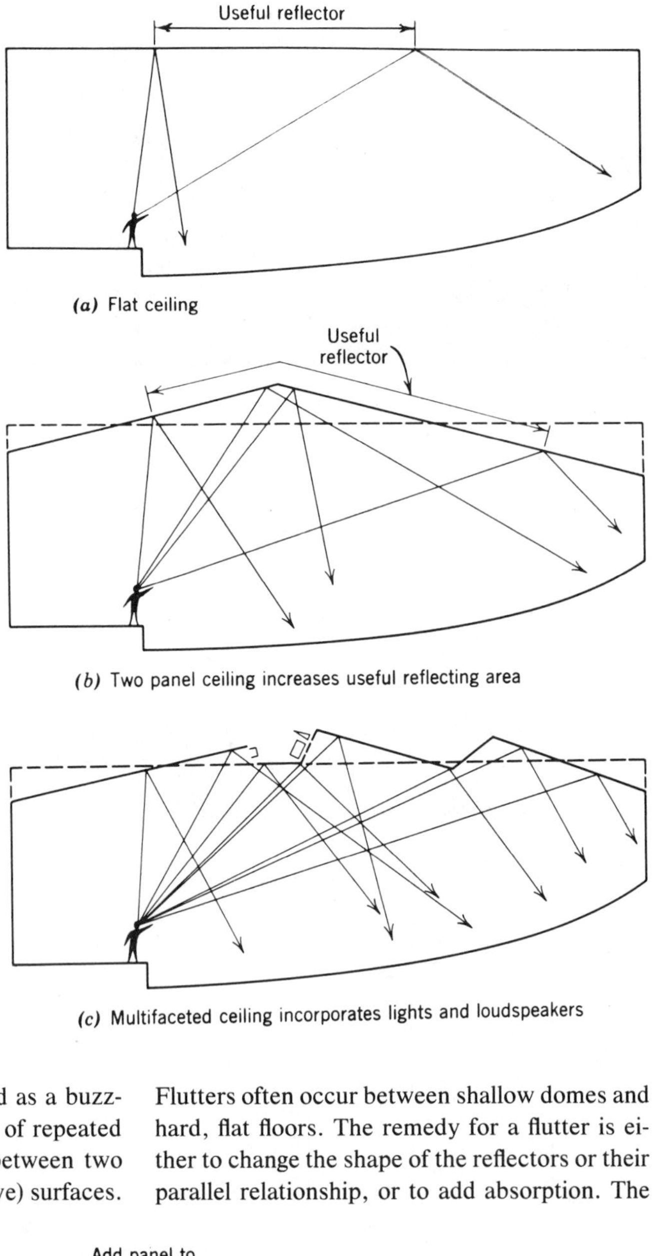

(d) Flutter. A flutter, perceived as a buzzing or clicking sound, is comprised of repeated echoes traversing back and forth between two nonabsorbing parallel (flat or concave) surfaces. Flutters often occur between shallow domes and hard, flat floors. The remedy for a flutter is either to change the shape of the reflectors or their parallel relationship, or to add absorption. The

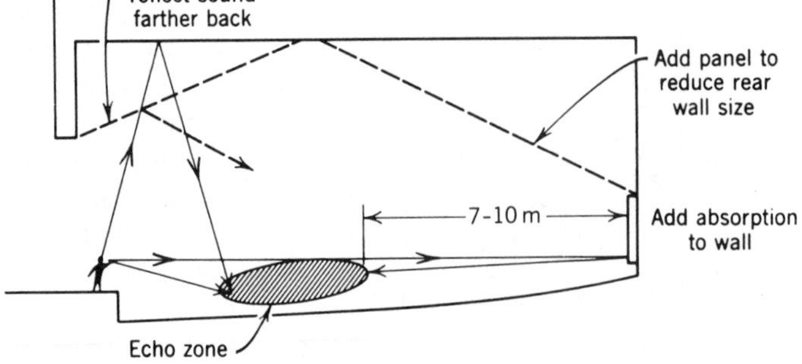

Fig. 26.31 Auditorium section showing the causes and remedies for two typical echoes.

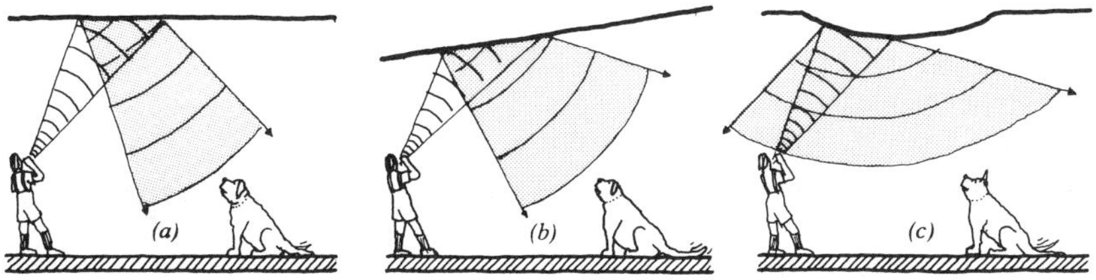

***Fig. 26.32** Sound diffusion can be created with reflectors of different shapes, ranging from the horizontal flat* (a), *inclined flat* (b), *or convex* (c). *Diffusion improves from* (a) *to* (c).

solution chosen will depend on reverberation requirements, cost, or esthetics.

(e) Diffusion. This is the converse of focusing and occurs primarily when sound is reflected from convex surfaces. A degree of diffusion is also provided by flat horizontal and inclined reflectors (see Fig. 26.32). In a diffuse sound field the sound level remains relatively constant throughout the space; an extremely desirable property for musical performances.

(f) Focusing. Concave domes, vaults, or walls will focus reflected sound into certain areas of rooms. This has several disadvantages. For example, it will deprive some listeners of useful sound reflections and cause hot spots at other audience positions (see Fig. 26.33*a*).

(g) Creep. This describes the reflection of sound along a curved surface from a source near the surface. Although the sound can be heart at points along the surface, it is unaudible away from the surface. Creep is illustrated in Fig. 26.33*b*.

(h) Standing Waves. Standing waves and flutters are very similar in principle and cause, but are heard quite differently. When an impulse (such as a hand clap) is the energy source, a flutter will occur between two highly reflective parallel walls. It is perceived as a slowly decaying buzz. When a steady, pure tone is the source, a standing wave will occur, but only when the parallel walls are spaced apart at some integral multiple of a half-wavelength.

When the parallel walls are exactly one-half wavelength apart, the tone will sound very loud near the walls and very quiet halfway between them. This is because at the center, the reflected waves traveling in one direction are exactly one-half wavelength away from those traveling in the other, and thus equal *but opposite in pressure,* which results in total cancellation. In other rooms standing waves are noted as points of quiet and maximum sound in the room. Standing waves are important only in rooms small with respect to the wavelengths generated (smallest room dimension <30 ft for music or <15 ft for speech).

Another effect of standing waves, *resonance,* is the accentuation of the particular frequency

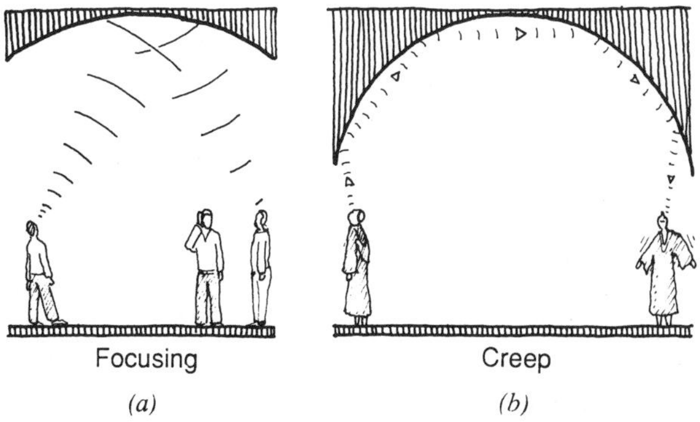

***Fig. 26.33** Two undesirable phenomena in room acoustics.*

ACOUSTICS

that will cause a standing wave in a room of that dimension. Thus, if one speaks (or plays a musical instrument) standing near a wall of a room, about 8 ft by 8 ft, one will notice an abnormal and sometimes unpleasant loudness in the sound at about 280 Hz.

Thus, when a musician plays a scale, one note may seem far louder than the adjacent ones, and listeners in one section of the room will hear a quality of sound different from that heard by those in other sections. This effect *must* be avoided for music performance but is merely an annoyance in rooms designed for speech use. This is one of the reasons that one finds music rehearsal rooms, broadcast studios, and so on with nonparallel walls and undulating ceilings. These irregularities prevent most of the undesirable effects described above from occurring. However, since rooms with nonparallel walls and undulating ceilings are for most applications unacceptable, and since standing waves and other resonant phenomena are related to room geometry, it is possible to calculate room proportions that will minimize these effects. Figure 26.34 shows such room proportions, which are applicable to bass frequencies (i.e., below 100 Hz). These are the problematic frequencies, since at higher frequencies (and in large rooms) many standing waves occur and the total effect is much less disturbing and frequently hardly noticeable.

26.26 Auditorium Design

This is a general term used to describe a space where people sit and listen to speech or music. Acoustical design of an auditorium includes room acoustics, noise control, and sound system design. Noise control is covered in Chapter 27. The first step in the acoustical and architectural design must be determining the program. If the program for the auditorium includes activities that need different acoustical environments, it must be decided early whether the acoustics will be a compromise between the program extremes or adjustable for various activities. Acoustical environments can be altered by changing volume, moving reflecting surfaces, and adding or subtracting sound-absorbing treatment. Figure 26.35 illustrates several examples of acoustical adjustability.

Factors that influence the acoustical design include audience size, range of performance activities, and sophistication of the potential audience. A small school auditorium and a professional theater will have widely divergent demands from both audience and performers.

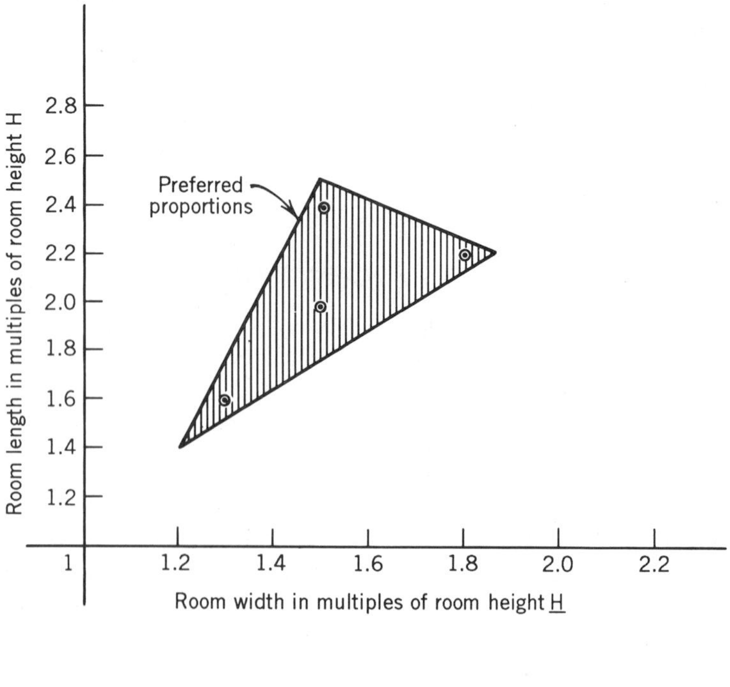

Fig. 26.34 Triangular figure contains the room proportions recommended, for avoidance of disturbing low-frequency acoustic phenomena such as flutter, standing waves, and resonances, particularly at frequencies below 100 Hz. The points shown at the three apexes were averaged to obtain the center, average proportion of 2H : 1.5H : H. *See also Fig. 26.16 for application of these proportions.*

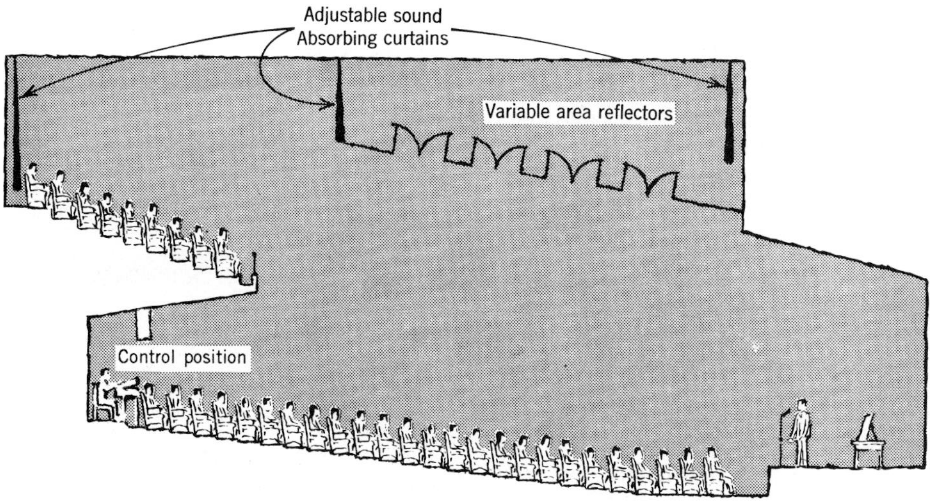

Fig. 26.35 Adjustable acoustic elements in an auditorium.

The audience size determines the basic floor area of an auditorium. Once this area has been fixed, the volume of the room is developed according to reverberation requirements of the space.

Figure 26.36 shows a typical auditorium in plan and section. The shape of wall and ceiling surfaces is developed to give proper distribution of sound and eliminate focusing or echoes. Essential characteristics of the design include:

1. Ceiling and side walls at the front of the auditorium distribute sound to the audience. These surfaces must be close enough to the performers to minimize time delays between direct sound and reflected sound.
2. Ceiling and side walls provide diffusion.

Acoustics must be considered in selection of materials used in an auditorium. All auditoriums use both sound-reflecting and sound-absorbing materials. Since the largest area of sound-absorbing material in any auditorium is the audience, the difference in acoustical characteristics that occur without an audience may be minimized by using fully upholstered seating.

Chairs with fully upholstered seats and backs, covered in an open-weave material, will have absorption characteristics closely approximating an audience. Using the auditorium in Figure 26.36 as an example, the reverberation characteristics of an auditorium with various materials may be examined. In the first example, the room use is assumed to be for music performance. The only sound absorption is that provided by the audience and seating. In the second set of calculations absorptive curtains were installed along the rear wall and a portion of the side wall. This configuration might be used for lectures in a room that is adjustable between speech and music configurations. A third configuration might use permanent sound-absorbing treatment installed on the ceiling and rear and side walls. Because of its low reverberation time, this configuration would be appropriate only for movies and lectures, not for music activities.

These simple examples indicate the effect of changes in the amount of absorption on the characteristics of a room. Adjustable treatments permit the characteristics of the room to be modified to any point between the extremes to meet the program acoustic requirements of a multipurpose hall.

Existing spaces may require remedial treatment to eliminate unwanted phenomena such as focusing and echoes, as shown in Fig. 26.37. In the first example, the surface of the dome was covered with sound-absorbing material to eliminate focusing; in the second, sound-absorbing treatment was applied to a curved rear wall to

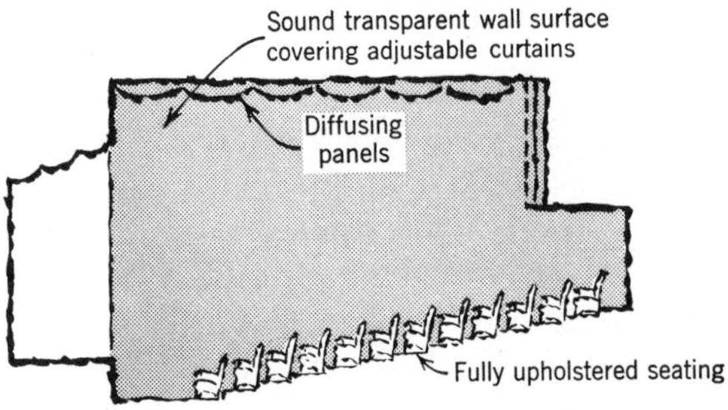

Section

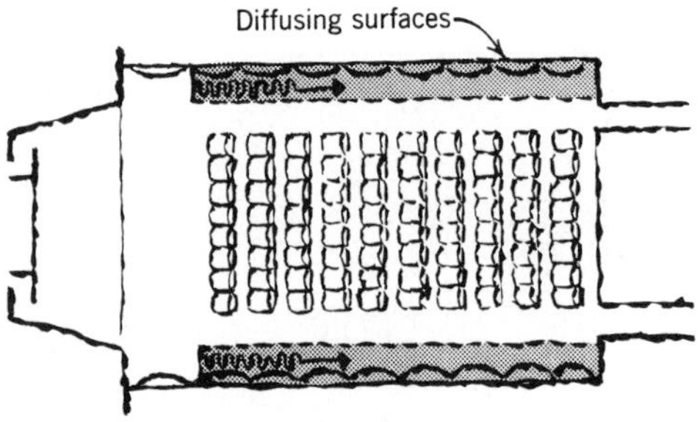

Plan

Simplified Calculations of Midfrequency (500 and 1000 Hz) Average Reverberation Times

$$\text{reverberation time } (RT) = \frac{0.05 \times \text{volume (cu ft)}}{\text{total absorption (sabins)}}$$

$$\text{volume} = 155{,}500 \text{ cu ft}$$

More Reverberant Condition
(curtains retracted)

	Area	*a*	*Absorption*
Seating and stage (with audience and performers)	3323	.92	3060
Wall area Concrete block	8000	.2	1600
Lower rear wall Permanent sound-absorbing treatment	450	.88	396
Total absorption More reverberant condition			5056

$$RT = \frac{0.05 \times 155{,}500}{5056} = 1.5 \text{ s}$$

Less Reverberant Condition
(curtains exposed)

	Area (sq ft =)	*a*	*Absorption*
Seating and stage (with audience and performers)	3323	.92	3060
Wall area Concrete block (balance covered by curtains)	3600	.2	720
Curtains	4400	.45	1970
Lower rear wall Permanent sound-absorbing treatment	450	.88	396
Total absorption Less reverberant condition			6146

$$RT = \frac{0.05 \times 155{,}500}{6146} = 1.2 \text{ s}$$

Fig. 26.36 *Auditorium with surface treatments for control of reflections and reverberation.*

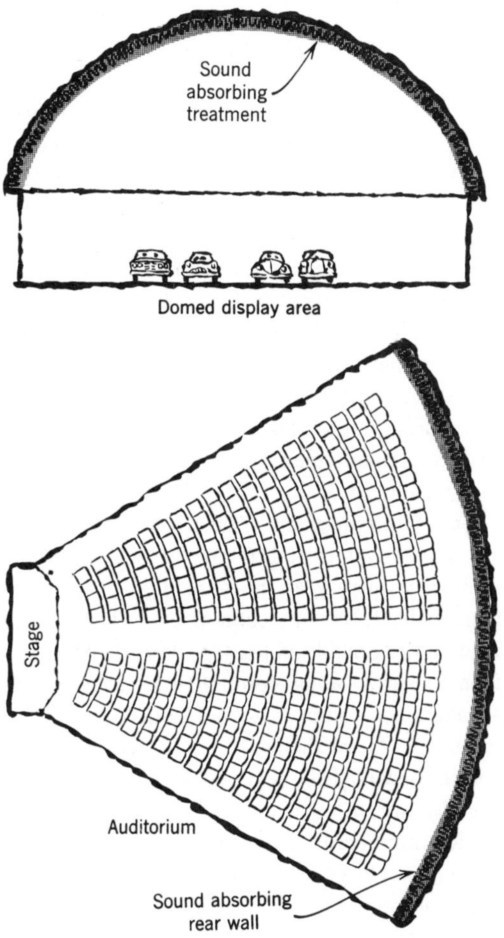

Fig. 26.37 Sound-absorbing treatment used to eliminate focusing from dome and curved auditorium wall.

eliminate an echo. Such treatment also will affect the reverberation characteristics.

SOUND REINFORCEMENT SYSTEMS

26.27 Objectives and Criteria

The purpose of a sound reinforcement system is just what the name indicates—to reinforce the sound, which would otherwise be inadequate. Thus an ideal sound system will give the listener the same loudness, quality, directivity, and intelligibility as if the source of sound were immediately adjacent to him—a distance of 2 to 3 ft for speech and farther for music, depending on type and number of instruments. This situation must obtain for every position in the space except that a variation of ±3 db for loudness is tolerable. The other factors should remain constant. Of these factors, loudness and intelligibility have been discussed at length and should be well understood. By quality we mean that frequency response should be linear so that reproduced sound bears the same relation between its frequency components as the original sound. (Quality is then field-adjusted by "voicing" or "equalization," as discussed below.)

Directivity is the characteristic whereby the sound appears to be coming from the originating source, that is, the loudspeakers should be directionally "invisible," and the listener must have the impression of actually hearing the source. It should be emphasized that sound systems cannot correct poor acoustic design completely, although they can improve a bad situation.

Generally, sound systems will be required in spaces larger than 50,000 ft^3 (~1400 m^3). In terms of population, this volume translates as 550 persons in lecture rooms (15 ft average ceiling height and 6 ft^2 per person) and 325 persons in theaters (20 ft average ceiling height and 7.5 ft^2 per person). In such a room (50,000 ft^3) a normal speaking voice can maintain a volume level of only 55 to 60 db, depending on room design and voice strength. With background noise at NC 30 (see Table 27.8), the speaker will be heard; at higher noise levels intelligibility will suffer.

26.28 Components and Specifications

All sound systems consist of three basic elements: input devices, amplifier(s), and loudspeaker systems.

(a) Input. Input is usually a microphone, a radio source, or some form of prerecorded material.

(b) Amplifier and Controls. Amplifiers must be rated to deliver sufficient power to produce intensity levels of 80 db for speech, 95 db for light music, and 105 db for symphonic music.

This assumes a *maximum* background noise level of 60 dbA. Thus 80-db speech intensity will be 20 db higher, or four times as loud as the noise level. If noise level is *known* to be below 60 db maximum, amplifier and loudspeaker power ratings can be reduced accordingly. The amplifier should carry technical specifications for signal-to-noise ratio, linearity, and distortion. Exact figures here depend on application and are left to the acoustics specialist or sound engineer to supply.

In addition to the usual volume, tone mixing, and input–output selector controls, the amplifier *must* contain special equalization controls for signal shaping. These are highly critical filter networks which by selective amplification and attenuation of portions of the overall audio frequency spectrum, *voice* or *equalize* a system after installation. Equalization is a sine qua non of a good sound system; without it the system will howl, sound rough, give insufficient and poorly distributed gain and sound level, and generally sound poor. Essentially, voicing tailors the system to the acoustic properties of the space. A system not equipped for equalization is not a professional system, and results will verify it. Furthermore, the specification must provide for the services of a competent sound engineer to perform the equalization after installation and construction is complete.

Another control frequently required in theater systems is a delay mechanism or circuit that can introduce a time delay into a signal being fed to a loudspeaker. Figure 26.38 shows a sound system that covers a majority of an auditorium from a central-loudspeaker cluster. The underbalcony seating areas are hidden from the central cluster and receive the reinforced sound from distributed loudspeakers in the underbalcony soffit. To provide directional realism, the signal to the underbalcony loudspeakers must be delayed to allow the weaker signal from the central speakers to arrive first. Delay is necessary because electrical signals travel at the speed of light, whereas sound is much slower (one-millionth the speed, approximately). With this arrangement, sound will seem to come from the source, and the directivity so necessary to realism is maintained.

(c) Loudspeakers. These are the heart of any sound system and obviously must be of the same high quality as the remainder of the system. Indeed, system economies will show up much more quickly in loudspeaker performance than in any other component. Selection of speakers is a complex, technical task not within our scope. Nevertheless, a few general remarks are in order. The best systems use central-speaker arrays consisting of high-quality, sectional (multicell), directional, high-frequency horns, and large-cone woofers. These assemblies are frequently very large, and the architect should be aware of the dimensions that must be accommodated. Smaller units with folded horns can be used, at a sacrifice in low-frequency response. If only speech is to be reproduced, these will perform adequately. Distributed systems

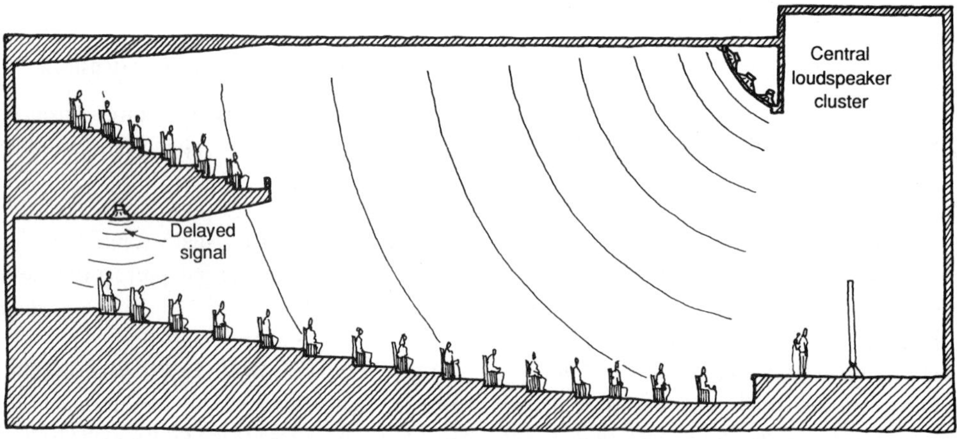

Fig. 26.38 *Loudspeaker system using delayed signal to under-balcony area.*

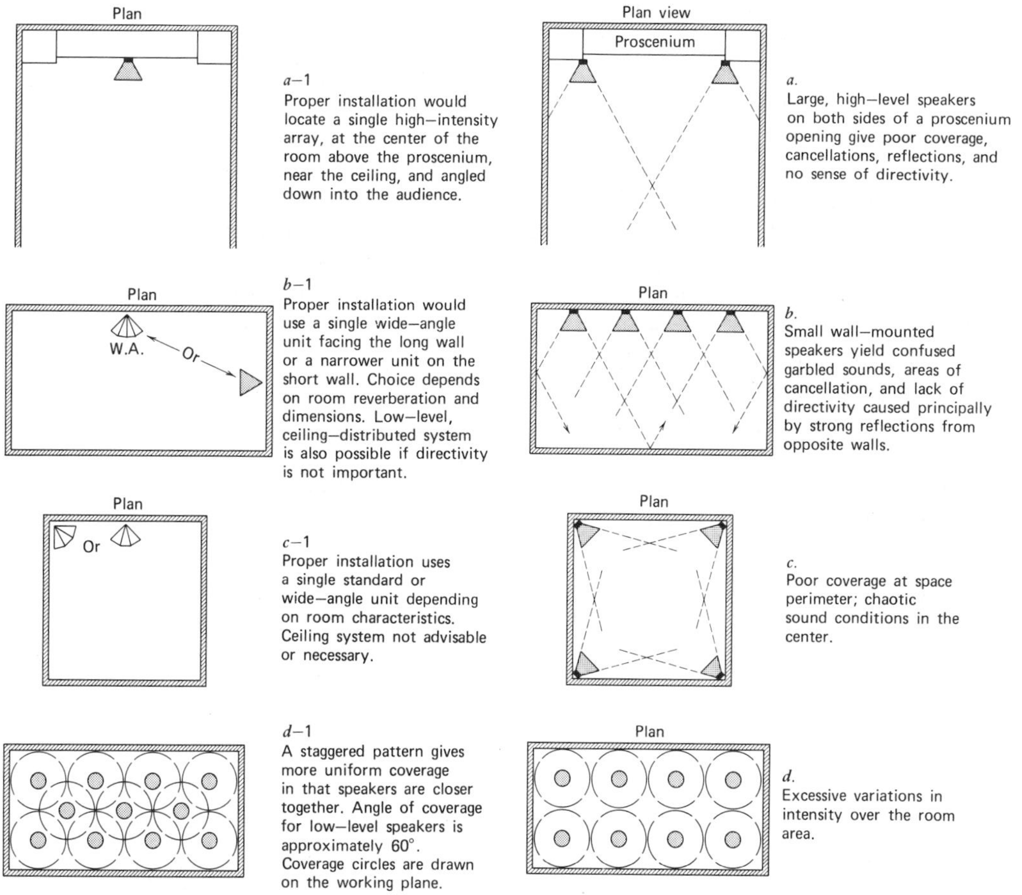

Fig. 26.39 *Poor layouts* (a) *to* (d) *and good layouts* (a-*1*) *to* (d-*1*). *Wide-angle speakers are available to fit most needs. Ceiling speakers give effective coverage on a 60° cone. Working plane is taken to be 4 ft AFF for seated audience and 6 ft AFF for standing listeners.*

use small (4 to 12 in. diameter) low-level speakers, ceiling mounted, and firing directly down.

26.29 Loudspeaker Considerations

Loudspeaker system design and placement must be coordinated with the architectural design. The two principal types of loudspeaker system are central and distributed. The loudspeakers in a central system are a carefully designed array of directional high-frequency units combined with less directional low-frequency units, placed above and slightly in front of the primary speaking position. In most theaters, this location is just above the proscenium on the centerline of the room. Located in this position, the system provides directional realism and is simple in its design.

The distributed loudspeaker system consists of a series of low-level loudspeakers located overhead throughout the space. Each loudspeaker covers a small area, in a manner similar to downlights. This type of system is used in low-ceilinged areas where a central-loudspeaker cluster could not provide proper coverage. It also can be used for public address functions if directional realism is not essential, in spaces such as exhibition areas, airline terminals, and offices. In addition, distributed loudspeaker systems provide flexibility for use in spaces where source and listener locations vary according to the use of the space, since loudspeakers can easily be switched to provide proper coverage. In general, a listening position should receive sound from only one loudspeaker. Systems that cover seating areas with signals from several scattered loudspeakers usually increase the loudness of the sound but tend to produce garbled speech. This rule is the principal reason that the arrangements shown in Fig. 26.39*a–d* will guarantee a bad job. The common practice of placing one loudspeaker on either side of a proscenium opening (Fig. 26.39*a*), or rows of speakers on one or both sides of a room (Fig. 26.39*b*), is particularly to be deplored.

Essential parameters for location and design of the control position create problems for the architect. The sound system operator must be within the coverage pattern of the loudspeakers. For proper operation, he or she should be able to hear the sound as it is heard by the audience. Some current auditorium designs locate sound system controls within the audience seating pattern (see Fig. 26.35). Others place a control room with a completely open wall or a large window at the rear of the auditorium. Monitor loudspeakers and earphones are inadequate substitutes for actual listening within the auditorium. In churches the simple control equipment can be located at the rear of the congregation area.

REFERENCES

See complete list of acoustics references at the end of Chapter 27.

27

BUILDING NOISE CONTROL

27.1 Introduction

Noise control in buildings is comprised of three components:

1. Reduction of noise generation at the source by proper selection and installation of equipment.
2. Reduction of noise transmission from point to point (along the transmission path) by proper selection of construction materials and appropriate construction techniques.
3. Reduction of noise at the receiver through acoustical treatment of the relevant spaces to meet NC criteria developed in Chapter 26.

Assurance of speech privacy is achieved by manipulation of all of the above plus the use of masking noise where necessary.

NOISE REDUCTION

27.2 Principles of Noise Reduction

Noise reduction is essentially the science of converting acoustical energy into another, less disturbing form of energy—heat. Since the amounts of energy involved are minute—130 db corresponds to 1/1000 of a watt or 0.003 Btu/h—the heat produced is completely negligible. This energy conversion is accomplished by absorption of the sound energy by the room contents and wall coverings, and by the structure itself. The former controls noise levels *within* a space and the latter noise transmission between spaces. The reasons for this will become clear as our discussion proceeds. Noise control treatment of a room will affect the reverberant noise level *within* that room but will have minimal effect on the noise level in adjoining spaces. Refer to Fig. 27.1 for a graphic presentation of this fundamental fact. The best that can be accomplished with acoustic room treatment is elimination of the reverberant field, that is, to make the intensity at the room boundaries what it would have been in free space, as in Fig. 27.1*d*. (Even this is extremely difficult; the actual field at the wall would be above 72 db, except in a completely anechoic chamber.) Adding further wall or other acoustic absorbent as in Fig. 27.1*e* does nothing in the room itself and has minimal effect on the overall transmission loss, since the transmission loss in the acoustic material itself is very low, as can be seen in Fig. 27.2*b*.

ABSORPTION

27.3 Mechanics of Absorption

We have already learned in Sections 26.13 and 26.14 the definition of sound absorption and its relevance to room reverberation characteristics. At this point it is appropriate to reexamine absorption as an acoustic phenomenon so that we may understand the application of absorptive material. Refer to Fig. 27.2*a*. In an untreated room of normal construction, when the sound waves strike the walls or ceiling, a small portion is transmitted, a small portion is absorbed, and most of the sound is reflected. The exact proportions obviously depend on the nature of construction. When acoustical treatment is applied to the room surfaces as in Fig. 27.2*c*, some of the energy in the sound waves is dissipated before the sound reaches the wall. The transmitted portion is slightly reduced, but the reflection is greatly reduced. The difference between the two situations is shown graphically in Fig. 27.3. Refer back to Fig. 26.19, where the result of adding absorptive material to a room is shown in greater detail. The difference between a room with average absorption of 0.1 and the same

(a) TV set in free space produces 75-db sound level, which drops 6 db for each doubling of distance. Attenuation by inverse square law (see Section 26.7).

(b) TV still produces 75 db. In the free field, sound drops to 72 db but builds up to 74 db at the wall due to reverberant field reinforcement (see Figs. 26.16 and 26.18). Wall attenuation is 30 db. Sound on other side of the wall is 74 − 30 = 44 db.

(c) Acoustic tile ceiling acts to reduce room reverberant field. Free field is extended. Level at wall is 73 db. Level in second space is 73 − 30 = 43 db.

(d) Entire room is acoustically treated, effectively eliminating reverberant field. Room is "dead." Level on second side of wall is 72 db less acoustic tile loss, less wall loss (that is, 72 − 2 − 30 = 40 db).

(e) Add another 2¼ in. of acoustic wall treatment. Room is "dead." Level at wall 72 db. Level in second space = 72 − 4 − 30 = 38 db.

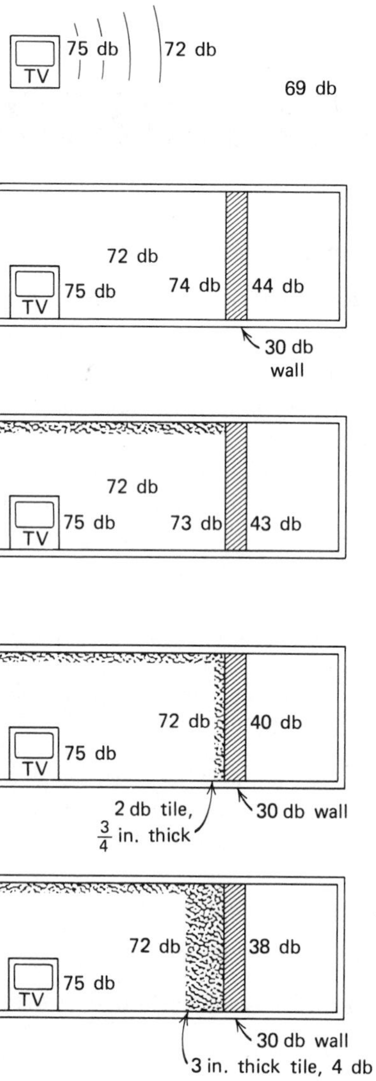

Fig. 27.1 Action of acoustic absorbent material.

room with $\bar{\alpha}$ of 0.7 is 15 db, which is a reduction in *loudness* of approximately 1.5 times (see Table 26.3). We will now examine the acoustic materials themselves and the effect of varying type, quantity, thickness, and installation methods.

27.4 Absorptive Materials

There are three families of devices for sound absorption—fibrous materials, panel resonators, and volume resonators. All types absorb sound by changing sound energy into heat energy. Only fibrous materials and panel resonators are used commonly in buildings. Volume resonators, also known as Helmholtz resonators, after their originator, are used principally as enclosures for absorbing a narrow band of frequencies. The following discussion refers to fibrous absorbers—the other two types are discussed separately in Section 27.8.

Fibrous or porous materials absorb acoustic energy by the frictional drag produced by moving the air in small spaces within the material. The absorption provided by a specific material depends on its thickness, density, porosity, and

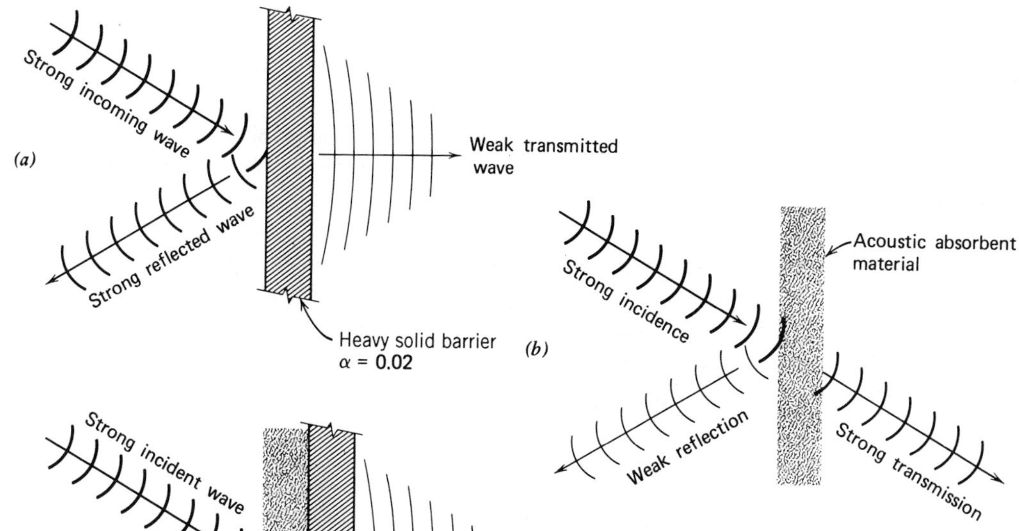

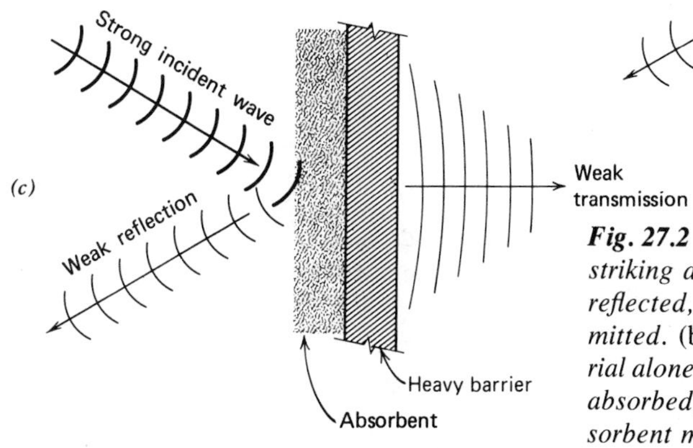

Fig. 27.2 (a) *Action of an incoming sound wave striking a heavy barrier. Much of the energy is reflected, some is absorbed, and a little is transmitted.* (b) *Action of acoustic absorbent material alone. Very little energy is reflected, some is absorbed, and most is transmitted.* (c) *When absorbent material is applied to the heavy wall, it "traps" sound, preventing reflection, while wall mass acts to reduce transmission.*

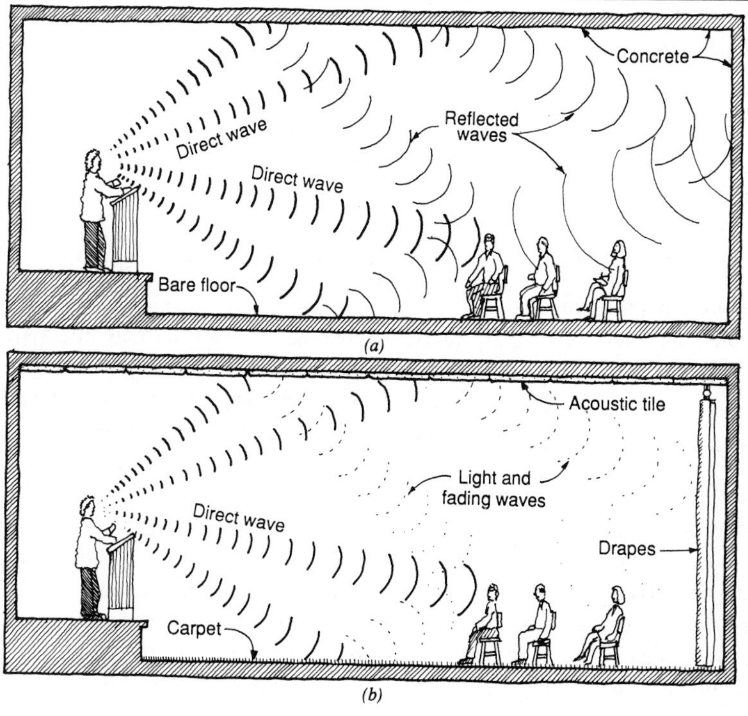

Fig. 27.3 *In the untreated space* (a) *reverberant (reflected) sound constitutes a large portion of received sound in much of the room. These reflections are largely eliminated in* (b) *by wall and ceiling absorption. Note that direct wave is completely unaffected.*

ACOUSTICS

resistance to airflow. For example, materials must be thick to absorb low-frequency sounds effectively. Since the action depends on absorbing energy by "pumping" air through the material, the air paths *must extend from one side to the other*. A fibrous material with sealed pores is almost useless as an acoustic absorbent. (Therefore, painting will generally ruin a porous absorber.) A simple test is to blow smoke through the material. If the smoke passes through freely and the material is porous, fibrous, and thick, it should be a good sound absorbent. Absorbency increases with increasing material porosity up to approximately 70% porosity; above that figure absorbency remains fairly constant.

Table 27.1 gives absorption coefficients for fibrous absorbent materials and for building materials and furnishings. Several important conclusions can be drawn from examination of this table.

1. For absorbent materials, absorption is normally higher at high frequencies than at low.

2. Absorption is not always proportional to thickness, but depends on the type material being used and the method of installation (see Fig. 27.4). It is clear from this figure that beyond a nominal thickness, little is to be gained by adding thickness except at very low frequencies, or when installed discontinuously, as in item 3, below.

3. It is possible to obtain an α greater than 1.0 by using very thick blocks. See "Sound blocks" in Table 27.1. These are installed at a distance from each other. Edge absorption is very large, particularly at high frequencies.

4. Installation methods have a pronounced effect as discussed in the following section.

27.5 Installation of Absorptive Materials

Coefficient ratings for absorptive materials are always given with mountings corresponding to ASTM requirements. The most common standardized mounting methods are given in Fig. 27.5. Installation of absorbent directly on a wall or ceiling is the least effective means, since exposure to sound energy is minimal. When an air gap is left between the porous layer and the rigid surface, the combination acts almost as well, in midfrequencies, as a layer of absorbent equivalent in thickness to the air plus porous material (see Fig. 27.6). One problem with this technique is that at the $\lambda/2$ node of a standing wave there is a severe drop in absorption, as can be seen from Fig. 27.6*c*. Thus, at 1000 Hz, one-half wavelength is approximately 7 in. At that distance, α drops severely, but is a maximum at $\lambda/4$ or 3½ in. For ceiling tile hung at 16 in. below the slab (Fig. 27.5, method E405), the drop in absorption occurs at

$$\lambda/2 = 16 \text{ in.}$$

$$\lambda = 32 \text{ in.} = 2.67 \text{ ft}$$

$$f = \frac{1128}{2.67} = 422 \text{ Hz}$$

which is midfrequency. This factor should be considered in applying absorptive material; that is, avoid a spacing corresponding to a drop in absorption at a sensitive frequency. To obtain good low-frequency absorption, it is essential that a deep air space be provided behind the porous absorbent material, and that walls be treated in addition to the ceiling.

In increasing order of effectiveness, absorbent material can be applied:

1. Directly to room surface.
2. Hung below ceiling and supported away from the walls.
3. Hung from the ceiling as louvres or baffles.
4. Made up into shapes such as cubes, or tetrahedrons, and suspended from the ceiling.

The last two techniques are extremely effective because they expose a very large surface of porous material—much larger than could be accomplished with wall or ceiling covering. Of course, these suspended objects become architectural elements and must be handled accordingly. In contrast, surface coverings are relatively neutral. In general, treatment should not be limited to one room surface such as the ceiling. All three principal surfaces in the direction of sound propagation, that is, ceiling, floor, and back wall, should be treated *approximately*

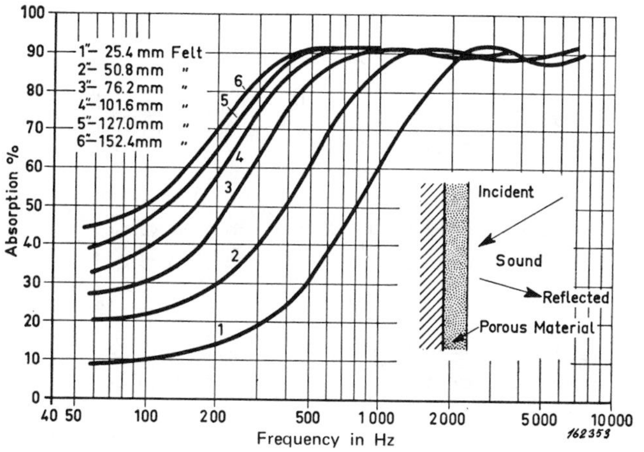

Fig. 27.4 Variation of absorption coefficient with thickness of felt absorbent. Note particularly that beyond 1 kHz, all thicknesses give the same α, whereas at low frequencies the absorption is proportional to thickness. Furthermore, it requires a very heavy layer to give appreciable absorption at low frequency. [Courtesy of Brüel and Kjaer (1951), as reprinted in Brüel and Kjaer (1963).]

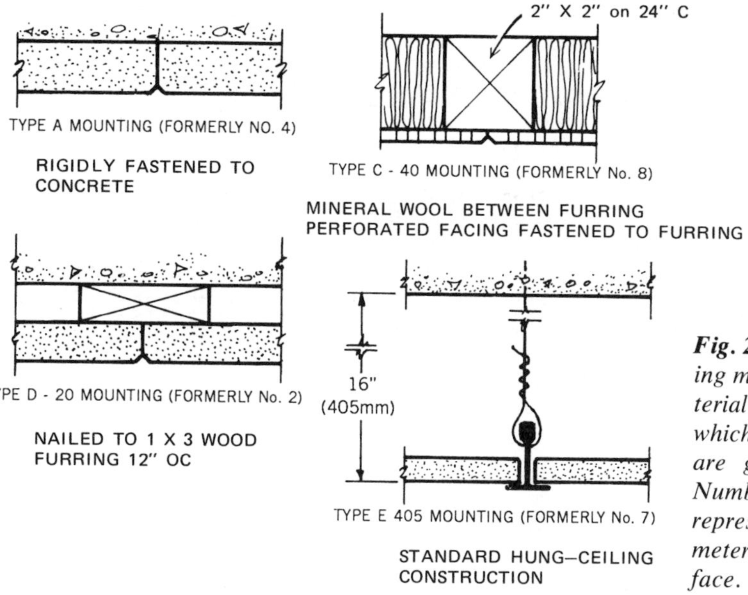

Fig. 27.5 Standard test-mounting methods for absorptive material, in accordance with which absorptive coefficients are given by manufacturers. Number following the letter E represents the distance in millimeters from the mounting surface.

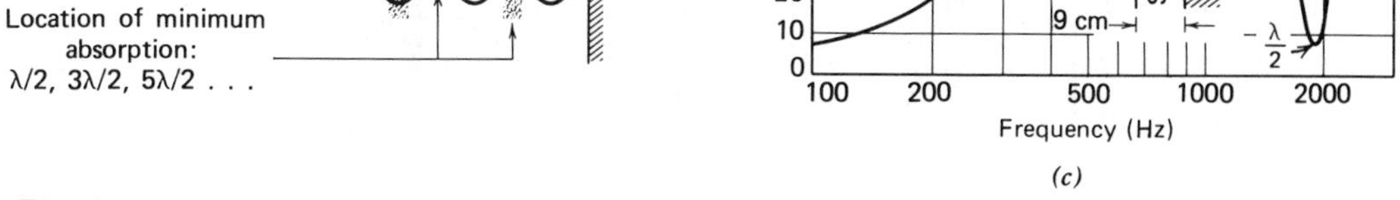

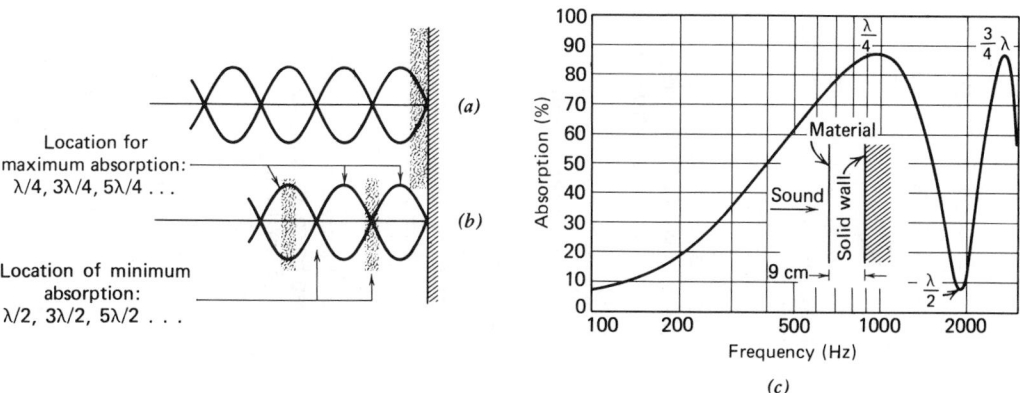

***Fig. 27.6** Sound waves striking a surface of large mass will create standing waves at certain frequencies, depending on room dimension* (a). *If insulation is placed at the nodes of these waves* (b), *attenuation is drastically reduced* (c). *[Illustration* (c) *from Brüel and Kjaer (1951), as reprinted in Brüel and Kjaer (1963).]*

TABLE 27.1 **Coefficients of Absorption[a]—α**

	Absorption Coefficients						
General Building Materials and Furnishings[b]	*125 Hz*	*250 Hz*	*500 Hz*	*1000 Hz*	*2000 Hz*	*4000 Hz*	*NRC[c]*
Air, sabins per 1000 ft³ at 50% rh				0.9	2.3	7.2	—
Brick, unglazed	0.03	0.03	0.03	0.04	0.05	0.07	0.005
Brick, unglazed, painted	0.01	0.01	0.02	0.02	0.02	0.03	0.00
Carpet, heavy, on concrete	0.02	0.06	0.14	0.37	0.60	0.65	0.29
Carpet, heavy, on 40-oz hairfelt or foam rubber	0.08	0.24	0.57	0.69	0.71	0.73	0.55
Concrete block, coarse	0.36	0.44	0.31	0.29	0.39	0.25	0.35
Concrete block, painted	0.10	0.05	0.06	0.07	0.09	0.08	0.05
Fabrics							
Light velour, 10 oz/yd², hung straight, in contact with wall	0.03	0.04	0.11	0.17	0.24	0.35	0.15
Medium velour, 14 oz/yd², draped to half area	0.07	0.31	0.49	0.75	0.70	0.60	0.55
Heavy velour, 18 oz/yd², draped to half area	0.14	0.35	0.55	0.72	0.70	0.65	0.60
Floors							
Concrete or terrazzo	0.01	0.01	0.015	0.02	0.02	0.02	0.00
Linoleum, asphalt, rubber, or cork tile on concrete	0.02	0.03	0.03	0.03	0.03	0.02	0.05
Wood	0.15	0.11	0.10	0.07	0.06	0.07	0.10
Glass							
Large panes of heavy plate glass	0.18	0.06	0.04	0.03	0.02	0.02	0.05
Ordinary window glass	0.35	0.25	0.18	0.12	0.07	0.04	0.15
Gypsum board, ½ in. nailed to 2 × 4's 16 in. o.c.	0.10	0.08	0.05	0.03	0.03	0.03	0.05
Marble or glazed tile	0.01	0.01	0.01	0.01	0.02	0.02	0.00
Openings							
Stage, depending on furnishings			0.25–0.75				
Deep balcony, upholstered seats			0.50–1.00				
Grilles, ventilating			0.15–0.50				
Plaster, gypsum or lime, smooth finish on tile or brick	0.013	0.015	0.02	0.03	0.04	0.05	0.05
Plaster, gypsum or lime, on lath	0.14	0.10	0.06	0.05	0.04	0.03	0.05
Plywood paneling, ⅜ in. thick	0.28	0.22	0.17	0.09	0.10	0.11	0.15
Rough wood as tongue and groove cedar	0.24	0.19	0.14	0.08	0.13	0.10	0.14
Water surface, as in a swimming pool	0.008	0.008	0.013	0.015	0.020	0.025	0.00
Absorption of Seats and Audience[d]	*125 Hz*	*250 Hz*	*500 Hz*	*1000 Hz*	*2000 Hz*	*4000 Hz*	*NRC[c]*
Audience, in upholstered seats, per ft² of floor area	0.60	0.74	0.88	0.96	0.93	0.85	—
Unoccupied cloth upholstered seats, per ft² of floor area	0.49	0.66	0.80	0.88	0.82	0.70	—
Wooden pews, occupied, per ft² of floor area	0.57	0.61	0.75	0.86	0.91	0.86	—
Students in tablet-arm chairs, per ft² of floor area	0.30	0.42	0.50	0.85	0.85	0.84	—

Acoustic Absorptive Materials	Mtg.[e]	Absorption Coefficients 125 Hz	250 Hz	500 Hz	1000 Hz	2000 Hz	4000 Hz	NRC[c]
Armstrong[f]								
High-performance vinyl-faced fiberglass ceiling panels:								
1 in. thick	E405	0.82	0.86	0.72	0.96	0.96	0.77	0.90
1.5 in. thick	E405	0.79	0.98	0.83	1.03	0.98	0.80	0.95
Painted nubby glass cloth panels								
3/4 in. thick	E405	0.81	0.94	0.65	0.87	1.00	0.96	0.85
1 in. thick	E405	0.78	0.92	0.79	1.00	1.03	1.10	0.95
Random fissured 5/8 in. thick panels								
Perforated	E405	0.66	0.76	0.68	0.80	0.89	0.80	0.75
Nonperforated	E405	0.44	0.29	0.28	0.77	0.83	0.76	0.55
Typical averages, mineral fiber tiles and panels								
3/4 in. fissured	E405	0.47	0.50	0.52	0.76	0.86	0.81	0.65
3/4 in. textured	E405	0.49	0.55	0.53	0.80	0.94	0.83	0.70
5/8 in. fissured	E405	0.28	0.33	0.66	0.73	0.74	0.75	0.60
5/8 in. textured	E405	0.29	0.35	0.66	0.63	0.44	0.34	0.50
5/8 in. perforated	E405	0.27	0.29	0.55	0.78	0.69	0.53	0.60
Gold Bond, National Gypsum[g] Quiet Touch 5/8-in. wall panel fabric-covered glass fiberboard on honeycomb core	A	0.00	0.07	0.37	0.78	0.91	0.94	0.55
	E60	0.29	0.65	0.91	0.89	0.87	0.92	0.85
	E405	0.61	0.70	0.65	0.81	0.96	0.96	0.80
Sound Touch 5/8-in. wall panel fabric-covered glass fiberboard on gypsumboard base	A	0.01	0.06	0.20	0.46	0.68	0.84	0.35
	E60	0.41	0.18	0.25	0.43	0.64	0.86	0.40
	E405	0.17	0.15	0.20	0.46	0.76	0.91	0.40
Tectum[h] Sound Blocks 3 in. thick × 15 in. square on 24-in. centers	A	0.36	0.53	1.81	2.65	2.75	2.73	

[a]Complete tables of coefficients of the various materials that normally constitute the interior finish of rooms may be found in books on architectural acoustics. This table will be useful in making simple calculations.

[b]Selected data courtesy of Owens-Corning Fiberglass.

[c]Noise reduction coefficient: the arithmetic average of the α values at 250, 500, 1000, and 2000 Hz.

[d]When the audience is randomly spaced, use an average of 5.0 sabins (ft^2) per person.

[e]See Fig. 27.5.

[f]Courtesy of Armstrong World Industries.

[g]Courtesy of Gold Bond Building Products, Division of National Gypsum Co.

[h]Courtesy of Tectum, Inc.

[i]Figures are sabins (ft^2) per unit.

equally for best results. The common practice of treating the ceiling only is generally inadvisable, since high frequencies are highly directive and may not reach the ceiling until the third reflection. See specific recommendations in Section 27.35.

27.6 Noise Reduction Coefficient

The last column in Table 27.1 is labeled NRC—noise reduction coefficient. This figure is the arithmetic average of the absorption coefficients at 250, 500, 1000, and 2000 Hz. The name is ill chosen inasmuch as it cannot be used as the words would seem to imply. It is useful as a single-number criterion for the *midband* effectiveness of a porous absorber. Obviously, if careful design is required or high- and low-end frequencies are of interest, NRC is nearly useless, and detailed calculations over the entire frequency range must be made. Similarly, two materials with the same NRC can perform quite differently, since the NRC is an average and few materials have a flat characteristic.

27.7 Noise Reduction by Absorption

The reader is referred back to Sections 26.13 through 26.15 and Fig. 26.19 for a review of the characteristics of absorption, reverberation, and fields in an enclosed space, as background material for a discussion of noise reduction by absorption. Equation 26.14 relates sound pressure level (SPL) and sound power level (PWL) as a function of distance from the source and room absorption. The following example shows the application of this equation and the result of adding absorption to the space.

EXAMPLE 27.1 An open blower is installed on the floor, away from the walls, in a large enclosed space that is 6 m long, 12 m wide, and 4 m high. The floor is concrete ($\alpha = 0.01$); walls and ceiling painted block ($\alpha = 0.07$). The sound power levels (PWL) supplied by the manufacturer are 90 db at 500 Hz and 87 db at 2000 Hz. Calculate the noise level SPL at distances of 5 m and 10 m from the blower outlet: (*a*) in the original room; (*b*) with double the absorption; (*c*) with quadruple the absorption.

SOLUTION

1. From Equation 6.14, we have the expression

$$SPL = PWL + 10 \log_{10}\left(\frac{Q}{4\pi r^2} + \frac{4}{R}\right)$$

In our example, $Q = 2$ (see Fig. 26.18).

2. The first step is to calculate room factor R for the three situations of absorpton.

$$R = \frac{S\bar{\alpha}}{1 - \bar{\alpha}}$$

$$S = 2(48) + 2(24) + 2(72) = 288 \text{ m}^2$$

from equation 26.16,

$\bar{\alpha}$ original

$$= \frac{3(48)(0.07) + 1(72)(0.07) + (72)(0.01)}{288}$$

$$= 0.055$$

$$\bar{\alpha}_a = 0.055 \qquad R_a = 16.76$$

$$\bar{\alpha}_b = 2(0.055) = 0.11 \qquad R_b = 35.6$$

$$\bar{\alpha}_c = 4(0.055) = 0.22 \qquad R_c = 81.23$$

3. Calculating SPL values and tabulating, we obtain

	500 Hz		*2000 Hz*	
SPL for:	*5 m*	*10 m*	*5 m*	*10 m*
Original room	83.9	83.8	80.9	80.8
Double $\bar{\alpha}$	80.8	80.6	77.8	77.6
Quadruple $\bar{\alpha}$	77.5	77.1	74.5	74.1

The results indicate the very important fact that *doubling the absorption decreases the noise level by only 3 db*. Therefore it requires a quadrupling of absorption to make the decrease noticeable (see Table 26.3). This is obviously an expensive procedure with diminishing returns.

If the entire space is considered to be a reverberant field, then from equation 26.19 the noise (sound) intensity level (IL) can be expressed as follows:

$$IL = PWL - 10 \log \Sigma A + 16.3 \text{ db} \qquad (26.19)$$

where

ΣA = total absorption in room, sabins (ft^2)

IL = intensity level, db

PWL = sound power level, db

Although increasing absorption decreases the noise level, the level never is reduced below the free field level for that distance from the source (see Fig. 27.1).

The amount of noise reduction provided by additional absorption may be determined by

$$\text{noise reduction} = IL_1 - IL_2$$
$$= 10 \log \Sigma A_2 - 10 \log \Sigma A_1$$

Therefore,

$$NR = 10 \log \frac{\Sigma A_2}{\Sigma A_1} \qquad (27.1)$$

where

ΣA_1 = total absorption, initial condition

ΣA_2 = total absorption, final condition

From equation 27.1 it is obvious that doubling the absorption results in a noise reduction of 3 db, since $10 \log_{10} 2 = 3$ (db). It is of interest to work out a practical example using equations 26.19 and 27.1 and comparing the results with the more precise relation given in Equation 26.13.

EXAMPLE 27.2. Referring to Fig. 27.7, calculate the original noise level and the subsequent noise reduction by three steps of sound absorption treatment, assuming a completely reverberant field in the space, that is, SPL independent of location.

SOLUTION. *(a) Original Condition.* Painted concrete block chamber, 10 × 10 × 10 ft.

Fan power level:

At 500 Hz = 88 db

At 2000 Hz = 78 db

Absorption	*Area*	α	*Total Absorption* ($\Sigma S\alpha$)
500 Hz	600 ft^2	0.06	36 sabins (ft^2)
2000 Hz	600 ft^2	0.09	54 sabins (ft^2)

Sound intensity level before treatment:

At 500 Hz:

$$IL = \text{sound power} - 10 \log \Sigma S\alpha + 16.3 \text{ db}$$
$$= 88 \text{ db} - 10 \log 36 + 16.3 \text{ db}$$
$$= 88 \text{ db} - 15.6 \text{ db} + 16.3 \text{ db}$$
$$= 88.7 \text{ db}$$

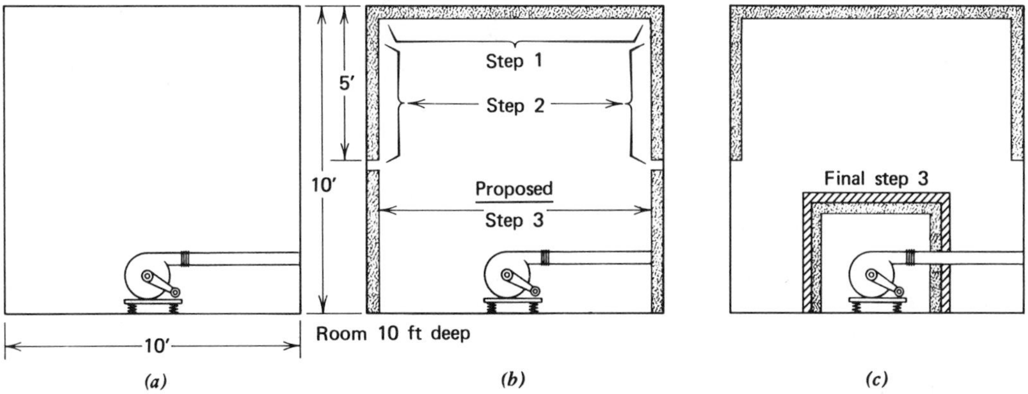

Fig. 27.7 *Quieting the room* (a) *by addition of absorptive material is cost effective only through step 2* (b). *Further quieting would be accomplished locally* (c), *which might obviate the necessity for wall treatment (step 2). Ceiling treatment should remain.*

At 2000 Hz:

$$IL = 78 \text{ db} - 10 \log 54 + 16.3 \text{ db}$$
$$= 78 \text{ db} - 17.3 \text{ db} + 16.3 \text{ db}$$
$$= 77 \text{ db}$$

(b) Ceiling Treatment Only

At 500 Hz:

$$\alpha = 0.82$$
$$\text{additional absorption} = 100(0.82 - 0.06)$$
$$= 76 \text{ sabins}$$
$$NR = 10 \log \frac{76 + 36}{36}$$
$$= 4.9 \text{ db}$$

At 2000 Hz:

$$\alpha = 0.94$$
$$\Delta A = 100(0.94 - 0.09) = 85$$
$$NR = 10 \log \frac{85 + 54}{54}$$
$$= 4.1 \text{ db}$$

(c) Ceiling and Half-Wall Treated

At 500 Hz:

$$\text{added absorption} = 300(0.82 - 0.06)$$
$$= 228 \text{ sabins}$$
$$NR = 10 \log \frac{228 + 36}{36}$$
$$= 8.7 \text{ db}$$

At 2000 Hz:

$$\text{added absorption} = 300(0.94 - 0.09)$$
$$= 255 \text{ sabins}$$
$$NR = 10 \log \frac{255 + 54}{54}$$
$$= 7.5 \text{ db}$$

(d) Ceiling and Full-Wall Treatment

At 500 Hz:

$$\Delta A = 500(0.82 - 0.06)$$
$$= 380 \text{ sabins}$$
$$NR = 10 \log \frac{380 + 36}{36}$$
$$= 10.6 \text{ db}$$

At 2000 Hz:

$$\Delta A = 500(0.94 - 0.09)$$
$$= 425 \text{ sabins}$$
$$NR = 10 \log \frac{425 + 54}{54}$$
$$= 9.5 \text{ db}$$

Summary

	IL (SPL)	
	500 Hz	*2000 Hz*
Bare room	88.7	77
Ceiling treated	−4.9 db (−4.9)	−4.1 db (−4.2)
Half-wall treatment	−8.7 db (−8.7)	−7.5 db (−7.7)
Full-wall treatment	−10.6 db (−10.7)	−9.5 db (−9.6)

The numbers in parentheses are the SPL differences as calculated from equation 26.13, using $Q = 2$ and $r = 5$ ft.

We would conclude here that the third step is really not worthwhile, since only a negligible additional decibel of quieting is accomplished. This example is intended to indicate the law of diminishing returns in quieting by absorption. Starting with a live room, the initial application is effective. Beyond that, additional quieting by absorption is not economical, and the same outlay would be better applied in quieting the machine itself, probably with a machine enclosure as indicated in Fig. 27.7*c*.

27.8 Panel and Cavity Resonators

Panel resonators are built with a membrane such as thin plywood or linoleum in front of a sealed air space generally containing absorbent material. The panel is set in motion by the alternating pressure of the impinging sound wave.

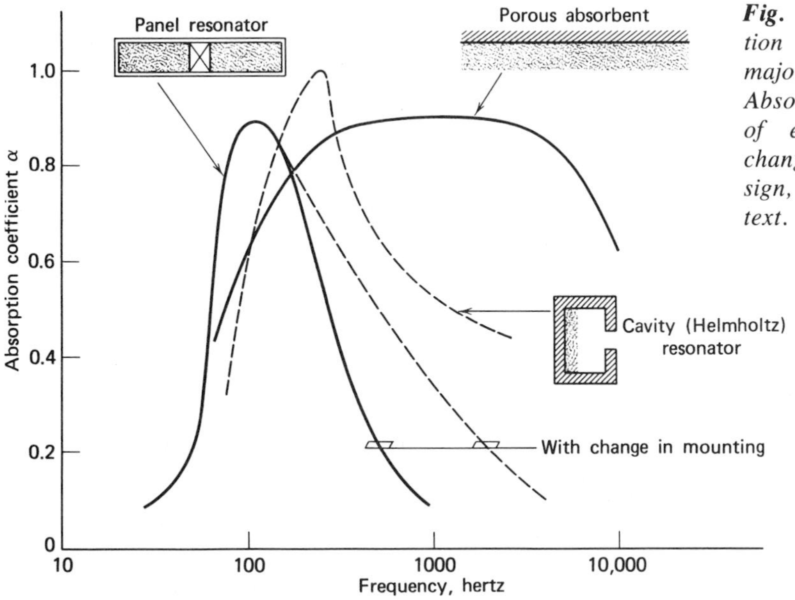

Fig. 27.8 *Typical absorption curves for the three major types of absorbers. Absorption characteristic of each type can be changed by varying design, as discussed in the text.*

The sound energy is converted into heat through internal viscous damping. Panel resonators are used where efficient low-frequency absorption is required and middle- and high-frequency absorption is unwanted or provided by another treatment (see Fig. 27.8). Panel resonators are often used in recording studios.

A *volume or cavity resonator* (Helmholtz resonator) is an air cavity within a massive enclosure, connected to the surroundings by a narrow neck opening. The impinging sound causes the air in the neck to vibrate, and the air mass behind causes the entire construction to resonate at a particular frequency. At that frequency absorption is very great—approaching unity, and dropping sharply above and below this frequency (see Fig. 27.8). By adjusting neck opening and cavity dimensions, the unit can be tuned to resonate at any desired frequency. This makes it extremely useful when a major single frequency is present, as with 120-Hz transformer hum. Such an installation is shown in Fig. 27.9 along with a typical concrete block resonator. Fibrous filler can be used in the block to increase high-frequency absorption.

Fig. 27.9 *Typical application of cavity resonators to absorb transformer hum in a large high-voltage substation. Wall is built of masonry units constructed to act as resonators (see insert). (This installation is made with SOUNDBLOX as manufactured by the Proudfoot Co., through whose courtesy photo is supplied.)*

27.9 Acoustically Transparent Surfaces

The soft porous material of which most absorbents are constructed is covered with perforated

ACOUSTICS

metal or wood panels, which act as physical protection and as stiffeners. These coverings are acoustically transparent except at higher frequencies. The frequency at which a noticeable reduction in absorption occurs can be estimated for circular holes as

$$f = \frac{40p}{d} \tag{27.2}$$

where

f = frequency, Hz

p = percentage of open area

d = diameter of holes, in.

Thus for ¼-in. holes and 60% open area, which is a typical commercial construction,

$$f = \frac{40(6)}{0.25} = 9600 \text{ Hz}$$

which is very high and generally not of major concern. It is always preferable, given a fixed percentage of open area, to use small holes, rather than large ones since, as is obvious from the formula, this raises the frequency at which absorption drops. It is also desirable to stagger the holes as this improves absorption. An open-weave fabric is almost completely transparent to sound and can be used as a decorative cover on absorbent wall coverings (see Fig. 27.10).

27.10 Absorption Recommendations

To summarize the discussion above, absorption techniques are useful and effective:

1. To change room reverberation characteristics.
2. In spaces with distributed noise sources such as offices, schools, restaurants, and machine shops.
3. In spaces with hard surfaces and little absorptive content.
4. Where listeners are in the reverberant field. (No amount of absorptive material can reduce intensity levels in the direct field.)

Concentrated noise sources are better handled by individual equipment enclosures than by room treatment, since enclosures reduce direct field noise, which, as stated above, room surface treatment cannot do. In addition to ceiling tiles and wall panels, acoustic absorbent material can be applied in the form of a sprayed-on finish (acoustic plaster), suspended unit absorbers, carpets, draperies, and other surface treatments. See Table 27.2 for characteristics of the most common of these elements.

SOUND ISOLATION

27.11 Airborne and Structure-Borne Sound

In contradistinction to the preceding sections, which were concerned with the phenomenon of in-room sound reduction by absorption, the following sections will discuss the characteristics of sound *transmission* between enclosed spaces. The distinction is often made between airborne and structure-borne sound, although in reality they differ only in sound origin. *Airborne sound* originates in a space with any sound-producing source and although it changes to *structure-borne* sound when the sound wave strikes the room boundaries it is still referred to as airborne because it originated in the air. *Structure-borne* sound is generally understood as *direct impact* caused by a vibrating or impacting source directly contacting the structure. Hence a child crying in the adjoining apartment is contributing airborne sound; the same child bouncing a ball on the floor is creating structure-borne sound, in this case by impact. The building pumps that were installed without proper damping mounts create structure-borne sound by vibration.

In reality, all sound transmission is both airborne and structure-borne since, once having entered the structure, the sound travels along the structure and causes the structure to vibrate, generating airborne sound. Figure 27.11 should assist in understanding this action. In Fig. 27.11*a* the sound is airborne, originating in the air at one side of the partition. The incidence of sound energy *causes the partition to vibrate*, generating sound on the other side. Sound does not "pass through"—it causes the structure to become a secondary source. The partition vi-

Fig. 27.10 *Acoustic ceiling tiles and wall panels are available in a wide variety of materials, designs, and finishes. Ceiling tiles:* (a) *Pressed organic fiber and inorganic binder, textured finish;* (b), (c) *mineral fiberboard, carved pattern and vinyl paint finish;* (d) *random fissured rectangular fiberglass;* (e) *painted mineral fiberboard specially treated for high-humidity areas;* (f) *mineral fiberboard covered with brushed finish aluminum strips: 2 ft × 4 ft panel with routings between strips painted black. Wall panels:* (g) *Cloth-covered panel of wood fibers and inorganic binder;* (h) *perforated mineral fiberboard with cloth covering;* (i) *fiberglass panel covered with a variety of surfacing materials and finished in fabric. [Photos courtesy of Tectum* (a), (g), *and Armstrong World Industries* (b–f,h,i).*]*

TABLE 27.2 **Typical Recommended Acoustical Treatment**

Material Characteristics	*Acoustical Tile*	*Metal-Faced Acoustical Units*	*Acoustical Plaster*	*Acoustical Form Board*
Size and form	Square tiles 12 in. × 12 in.; some up to 48 in. × 96 in. Roof deck 2 ft × 8 ft—most common thicknesses, 5/8 and 3/4 in.	12 in. × 24 in. up to 24 in. × 120 in. Thickness is controlled by acoustical pad backing. Either paper-wrapped mineral wool or cut or roll glass fiber. Units run 2 to 3 in. total thickness.	A plasterlike material of special fibrous or particulate aggregate, 1/2 to 1 1/2 in. thick.	Sizes and thickness vary to fit structural requirements (floor and roof slabs only) 1 to 2 1/2 in. for wall application.
Surface	Wide variety of perforated, textured, and sculptured surfaces with white, painted, vinyl, and glass cloth finishes.	Baked-enamel finish usually white, but available in color, perforated or slotted.	Fine-grain white texture, but may be spray painted.	"Shredded Wheat" or smooth pattern.
Method of installation	Adhesive, nailing, or stapling to wood furring; lay-in grids; concealed spline-grid.	Attached to metal supports nailed to wood furring or hung in a proprietary metal suspension system.	Trowel or spray.	According to structural design (floor and roof slabs only). Adhesive or mechanical application on walls.
Major area of application	All interior applications. Check specifications for application in high-humidity areas.	All interior applications. Check specifications for application in high-humidity areas.	Most interior applications. Especially useful for large curved surfaces. Not satisfactory in high humidity.	Floor and roof decks wherever applicable. Wall surfaces in most interior spaces.
Advantages	Provide widest range of finishes in high-absorption units. Mineral fiber tiles are rated incombustible. Lay-in units provide access to plenum above.	Incombustible. Easily maintained. Can be washed or painted; replacement units will match original job. Permits easy access to plenum.	Rapid, low-cost installation on large irregular or curved surface. Fine-grain texture. Incombustible—can be applied with (or as part of) fire protection.	Combines form board with thermal insulation and acoustical absorption.

Limitations	Allows transmission of sound over partition in hung ceiling unless selected for high-attenuation factor.	Allows transmission of sound over partition in hung ceiling unless backed by impervious layer.	Easily abused. Difficult to match finish in patching or repairing. Does not always perform as advertised. Performance limited to high-frequency absorption. Dusty in application.	Not advisable for high-humidity areas.

Material Characteristics	*Unit Sound Absorbers*	*Carpeting*	*Drapery*
Size and form	From 12 in. × 12 in. × 2 in. units to 24 in. × 48 in. plastic coated glass fiber baffles.	Sized or seamed to fit any floor or wall area.	Acoustically transparent fabrics ranging from opaque velours to transparent/translucent glass fibers.
Surface	12 in. × 12 in. units in smooth to rough surface. Baffles are vinyl wrapped.	Cut, looped, or combination to achieve any degree of "fuzzing."	Velveteen to bouclé (rough).
Method of installation	Units may be applied using special wall clips, pendents (ceiling only), or adhesive. Baffles are wire supported.	Tacking (floor only) or adhesive.	According to function and esthetics.
Major area of application	12 in. × 12 in. units are useful in all interior areas. Baffles in industrial areas.	Floors and walls.	Window or opaque wall areas and room dividers.
Advantages	Permit maximum flexibility. Hung units add sound absorption without requiring lowering lights or sprinkler heads.	Provides a relative degree of mid- and high-frequency absorption and luxurious appeal.	Provides a relative degree of overall sound absorption and luxurious appeal.
Limitations	Each application must be designed individually.	Acoustical absorption increases with pile height, pad, and fiber density.	Acoustical absorption increases with fabric density percent, fold when drawn, and air space behind.

Source: Reprinted with permission of the authors from Goodfriend and Sulewsky, "A Guide to Acoustic Materials," *Architectural and Engineering News*, February 1970.

Fig. 27.11 (a) *Airborne sound is so called because it originates in the air.* (b) *Structure-borne sound originates from mechanical contact between structure and vibrating or impacting sources. As such its energy level is usually much higher than airborne sound.* (c) *Some techniques for controlling airborne sound. [Data from* A Guide to Airborne, Impact, and Structure-Borne Noise Control in Multifamily Dwellings *(1968).]*

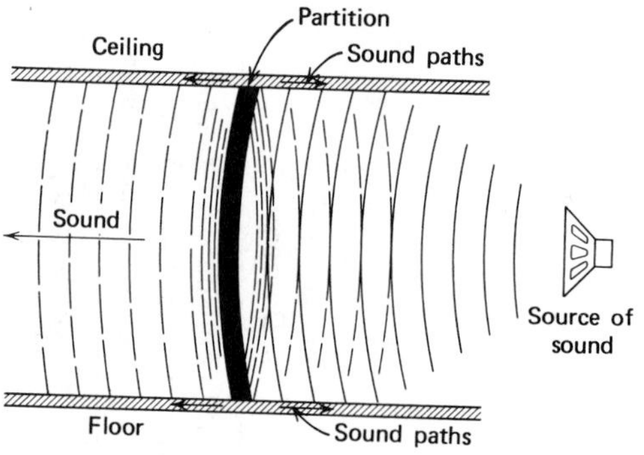

(a)

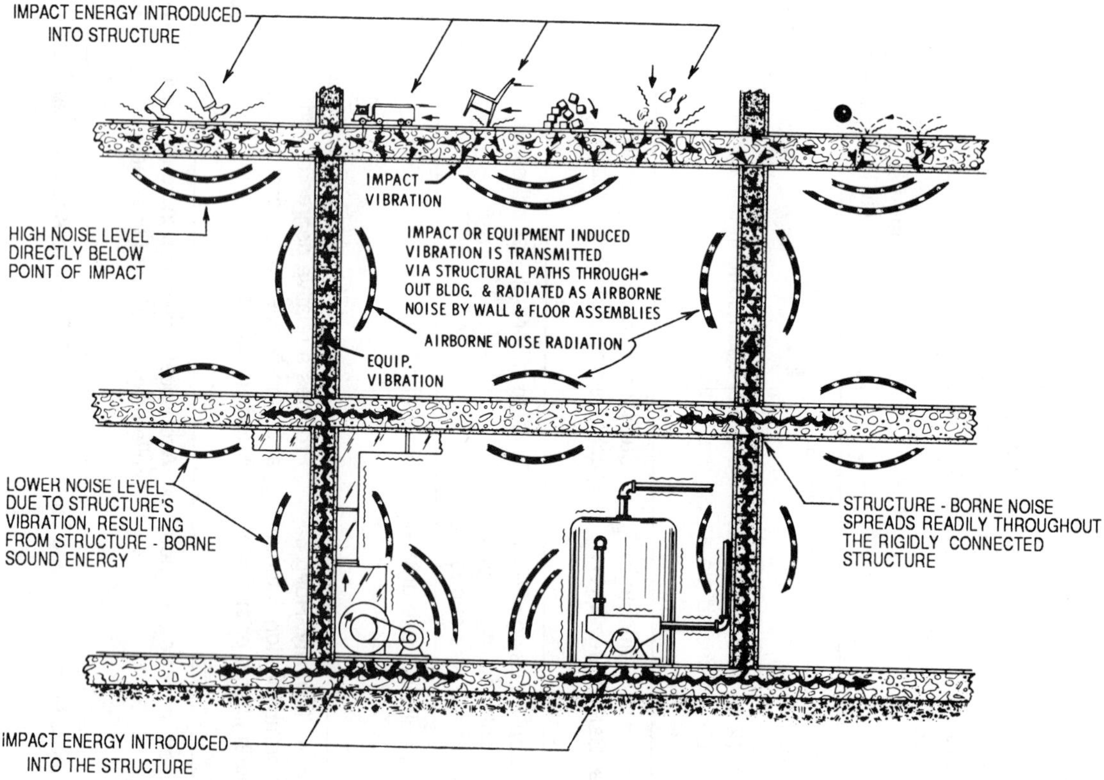

Transmission of Impact and Structure - Borne Noise.

(b)

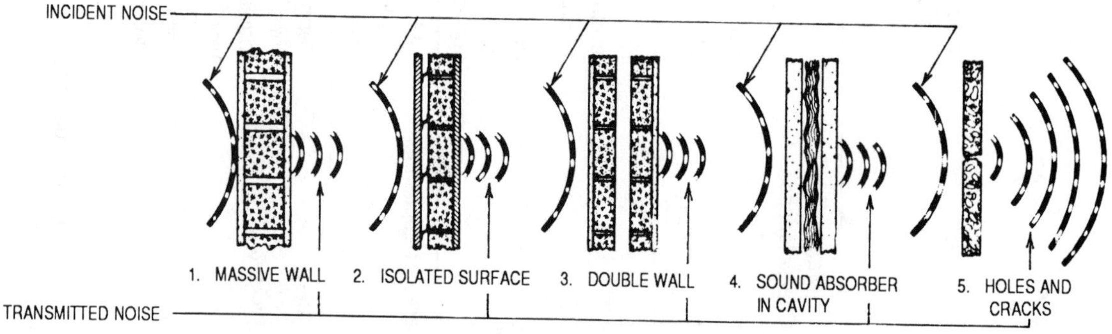

Constructions 1-4 reduce transmitted noise; holes and cracks (5) can drastically reduce a wall's transmission loss and thereby its effectiveness as a sound barrier

(c)

brates primarily in the direction of the sound, that is, in the vertical plane. It also vibrates in other modes causing some sound energy to pass into the floor and ceiling, depending on the mode of attachment. This energy becomes structure-borne sound.

In Fig. 27.11*b* the process is similar, but reversed. Energy is introduced into the structure directly and efficiently by mechanical contact, that is, vibration and impact. Sound travels along the structure, as shown and by causing the structure to vibrate, creates airborne sound. In a structure with rigid wall-to-floor connections, these sounds are clearly heard throughout the structure. Airborne sound (originating in the air) is generally much less disturbing than structure borne, since its initial power is very low and it attenuates rapidly at boundaries. Structure-borne sound is generally at a much higher energy level initially and attenuates slowly as it travels through the structure, thereby causing disturbance over large sections of a building. This disturbance is magnified by the "sounding board effect."

We are all familiar with the fact that a tuning fork must be held up to the ear to be heard directly, but if its handle is placed on a table the sound is amplified. This action is not really amplification but an increase in the efficiency of energy transfer. In general, the efficiency of a radiator is proportional to the ratio of its surface dimensions to the sound wavelength. A tuning fork vibrating at concert A (440 Hz) of wavelength 2½ *feet* cannot couple its energy into the air. It is simply too small. By placing the instrument on a table whose dimensions are approximately one wavelength, we permit it to transfer its energy efficiently; hence the "amplification." The same effect can be extremely troublesome in structure-borne sound. A vibrating pump itself makes little sound. However, it transfers a large amount of energy into the structure, which will appear as audible sound at each partition, floor, and wall that is rigidly coupled to the structure. Soft (damping) connections prevent energy transfer, thereby greatly attenuating the transmission of sound energy into connecting efficient radiating surfaces; hence the desirability of such flexible connections.

Airborne sound changes direction easily (diffracts), with low frequencies being most flexible in this regard. Structure-borne sound travels much more rapidly than airborne sound (see Table 26.1) and with attenuation as low as 1 db per kilometer. Mass of the structure is an effective attenuator only in the direction of radiation. That is, a sound traveling along a massive structure will radiate little outward from the structure (although enough to be very annoying) because the large mass minimizes vibration in that direction. Thus in Fig. 27.11*b*, noise from impact on the floor above will be louder (for equal impact energy) than noise from machines below because the former generates sound directly downward while the latter introduces energy into the entire network of parallel paths.

The sections immediately following will deal with airborne sound and the means for controlling it (see Fig. 27.11*c*). Impact noise (structure-borne sound) is covered in Sections 27.23 to 27.25.

AIRBORNE SOUND

27.12 Transmission Loss (TL) and Noise Reduction (NR)

The transmission loss (TL) of a barrier is the ratio, expressed in decibels, of the acoustic energy reradiated by the barrier to the acoustic energy incident on it. This number is a figure of merit for the sound-isolating quality of the wall itself and is obtained from controlled laboratory tests. (In Europe transmission loss is referred to as sound reduction index, *R*.) However, the number that is of greater importance to the building designer is the actual noise reduction between two spaces separated by a barrier, that is, the action of the barrier in situ. This noise reduction is defined as the difference between the intensity levels in the two rooms, that is,

$$NR = IL_{\text{room 1}} - IL_{\text{room 2}} \qquad (27.3)$$

and is related to the TL of the barrier by the expression

$$NR = TL - 10 \log \frac{S}{A_R} \qquad (27.4)$$

ACOUSTICS

where

NR = noise reduction, db

TL = barrier transmission loss, db

S = area of barrier wall, ft² (m²)

A_R = total absorption of *receiving* room, sabins, ft² (m²)

We see, therefore, that noise reduction and transmission loss are not equal but are related by the size of the dividing wall S and the absorption characteristic of the receiving room, A_R. A moment's thought will confirm the logic of this relation. When sound energy impinges on the wall, the wall in turn becomes the sound source, radiating into the receiving room. Therefore, the amount of sound energy transferred is proportional to the area S of the common wall between the two spaces.

The sound level in the receiving room is related to its own reverberance (absorption characteristic A) as we have seen repeatedly. (Refer back to equation 26.13, for example.) Thus if the receiving room is a reverberant, live space, A is low and NR is less than TL. Conversely, if the receiving room is dead, A is large and noise reduction can be greater than transmission loss, depending on the ratio of barrier wall size to room area (see Fig. 27.12). In lieu of calculating, the following rule of thumb can be used:

1. For a live receiving room,

$$NR = TL - 1 \text{ db}$$

2. For a medium receiving room,

$$NR = TL + 4 \text{ db}$$

3. For a dead receiving room,

$$NR = TL + 7 \text{ db}$$

The extreme case of "deadness" of the receiving room is one with no walls, that is, sound transmission from inside to the exterior. In such cases, NR exceeds TL by 10 to 15 db, depending on the size of the exterior opening and the point outside where IL is measured. To acquire facility with sound-isolation techniques, the reader must become familiar with the relationship of transmission loss to the barrier wall's physical characteristics, its mass, ridigity, material of construction, and method of construction and attachment. These considerations are the subject of the following sections.

Note: The reader is cautioned to be careful with the term *noise reduction* (NR), since a similar term—noise reduction coefficient (NRC)—also

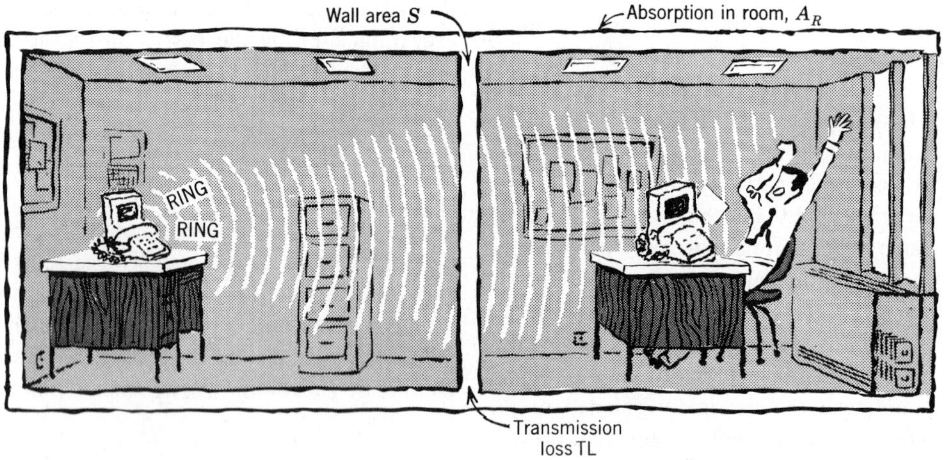

Fig. 27.12 *A simple case of airborne sound transmission between adjacent rooms through a common barrier. With a sound source in one room, the transmitted sound level is dependent not only on the transmission loss of the barrier but also on the area of the barrier and the receiving-room absorption. The background noise level in the receiving room determines whether the transmitted sound will be heard. See Section 27.21 and Figs. 27.32, 27.33, and 27.35.*

exists, which is completely unrelated. The latter, as pointed out in Section 27.6, is very poorly named.

27.13 Barrier Mass

Sound transmission requires that the barrier be set into vibration by the incident sound energy. Although this was stated above, we repeat it here to emphasize the fundamental importance of this simple statement. [We are of course referring to a barrier that is impervious to air (i.e., a solid barrier). Otherwise, the moving air molecules carrying the sound will simply pass through with minimal transmission loss.] The impinging energy acts as a force on the wall. Since $F = MA$, the larger the mass, the less it will vibrate. When other factors are taken into account, the resultant relationship is known in acoustics as the *mass law*. It states that for a nonporous, homogeneous structure of low stiffness, the sound transmission loss is proportional to the logarithm of surface mass (the weight of the wall per square foot of surface area) and to the frequency of vibration. Thus doubling the mass or frequency will, theoretically, cause an increase of 6 db in the transmission loss or, stated otherwise, the slope of *TL* versus the frequency times mass (*fM*) curve is 6 db (see Fig. 27.13). Table 27.3 lists the average transmission loss of some common barriers, confirming the overall operation of the mass law. Figure 27.13 is a graphic representation of mass law operation. With sound incident at 90°, maximum energy is imparted to the barrier and the entire mass resists, resulting in maximum transmission loss. In practice, however, sound is incident from 0 to 80° (called field or random incidence) reducing the mass effect, but keeping the slope at 6 db per octave. Due to nonhomogeneity, porousness, and stiffness, actual field results indi-

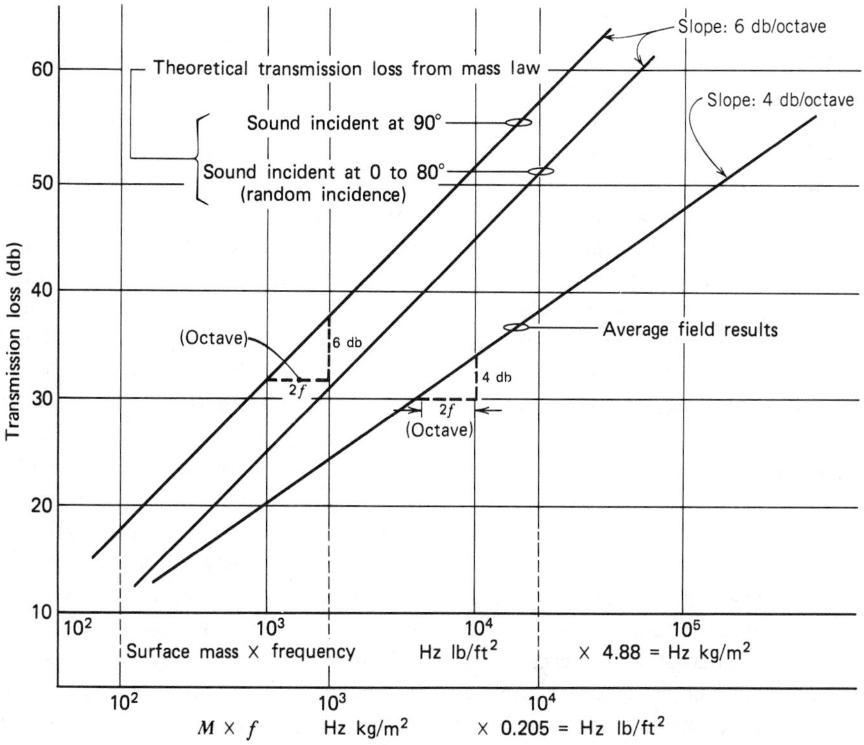

Fig. 27.13 *Graphic representation of mass law action in attenuation of transmitted sound. Perpendicular (90°) sound incidence results in maximum transmission loss. Field or random incidence (0 to 80°) is approximately 6 db lower, but maintains the 6 db per octave slope. Field results are much lower due to flanking and stiffness effects, averaging 4 db per octave (frequency doubling).*

ACOUSTICS

TABLE 27.3 **Typical Average[a] Transmission Loss**

Barrier Construction	Surface Weight (lb/ft^2)	Transmission Loss (db)
3⁄16-in. plate glass	2.5	20
1⁄4-in. asbestos—cement sheet	3.0	21
2-in. plaster on wire lath	15	34
3-in. cinder block	25	39
4-in. concrete	50	45
8-in. brick, plastered	80	50
12-in. brick, plastered	120	53

[a]Over the frequency range of 150 to 3000 Hz.

cate transmission losses nearer 4 db per octave, as shown by the lower curve in the figure.

27.14 Stiffness and Resonance

The stiffness of a barrier is a function of its material composition and the rigidity of its mounting. The former is dependent on its internal cohesiveness, that is, its modulus of elasticity, and the latter depends on its boundary restraints—whether the barrier is tightly or loosely held. A homogeneous material of high Young's modulus (such as steel) has great cohesiveness between its molecules. As soon as one molecule is set in motion by the incident sound, the motion is passed to the next, and so on, making the material an excellent sound conductor. Homogeneous materials with low modulus of elasticity have high internal damping (motion of molecules is not transmitted well), and they are good sound insulators. Composite materials such as concrete and organic materials such as wood do not conform to these general rules.

The rigidity of mounting can be likened to a drumhead—the tighter it is stretched, the better it resounds. Ridigity (stiffness) in a panel barrier resists damping and assists vibrations, making it a poor noise transmission insulator. To recapitulate, stiffness, both internal and external, is an undesirable characteristic in sound isolation. A material such as lead, which has high mass and low modulus of elasticity, and resists rigid mounting, is an excellent sound attenuator.

The effects of stiffness and mass both vary with frequency, unfortunately in opposite directions. Stiffness acts to reduce transmission loss as frequency increases while, as we have seen, mass acts to increase it; the combined effect is shown in Fig. 27.14. Therefore, stiffness is most effective at low frequencies and mass at high frequencies. At very low frequencies the mass and stiffness effects negate each other, giving the resonance dips shown. Beyond approximately 200 Hz most common wall construction enters the mass law range and continues with it until the critical frequency. Deviations from a smooth 4 to 6 db per octave slope are due to the nonhomogeneous nature of most wall construction. At the critical frequency the phase of incident sound waves corresponds or "coincides" with the phase of vibration (shear wave) of the barrier in such a way as to pass a large portion of the incident energy. See insert on Fig. 27.14. This shows in Fig. 27.14 as the "coincidence dip." This effect is most pronounced in thin homogeneous partitions and light stiff ones.

Critical frequency f_c, as a function of panel thickness for common materials, is plotted in Fig. 27.15. To avoid a coincidence dip in the audible range, partitions can be either very heavy and/or very stiff, which greatly decreases the critical frequency, or heavy and limp (resilient), which greatly increases the critical frequency. In practical terms, cost effectiveness is heavily in favor of the latter alternative. Thus, for instance, the TL of a wood partition can be improved by grooving it to increase flexibility thereby increasing the critical frequency. The dramatic improvement in TL, resonances, and coincidence dip achieved by use of resilient mounting of a simple masonry partition is shown

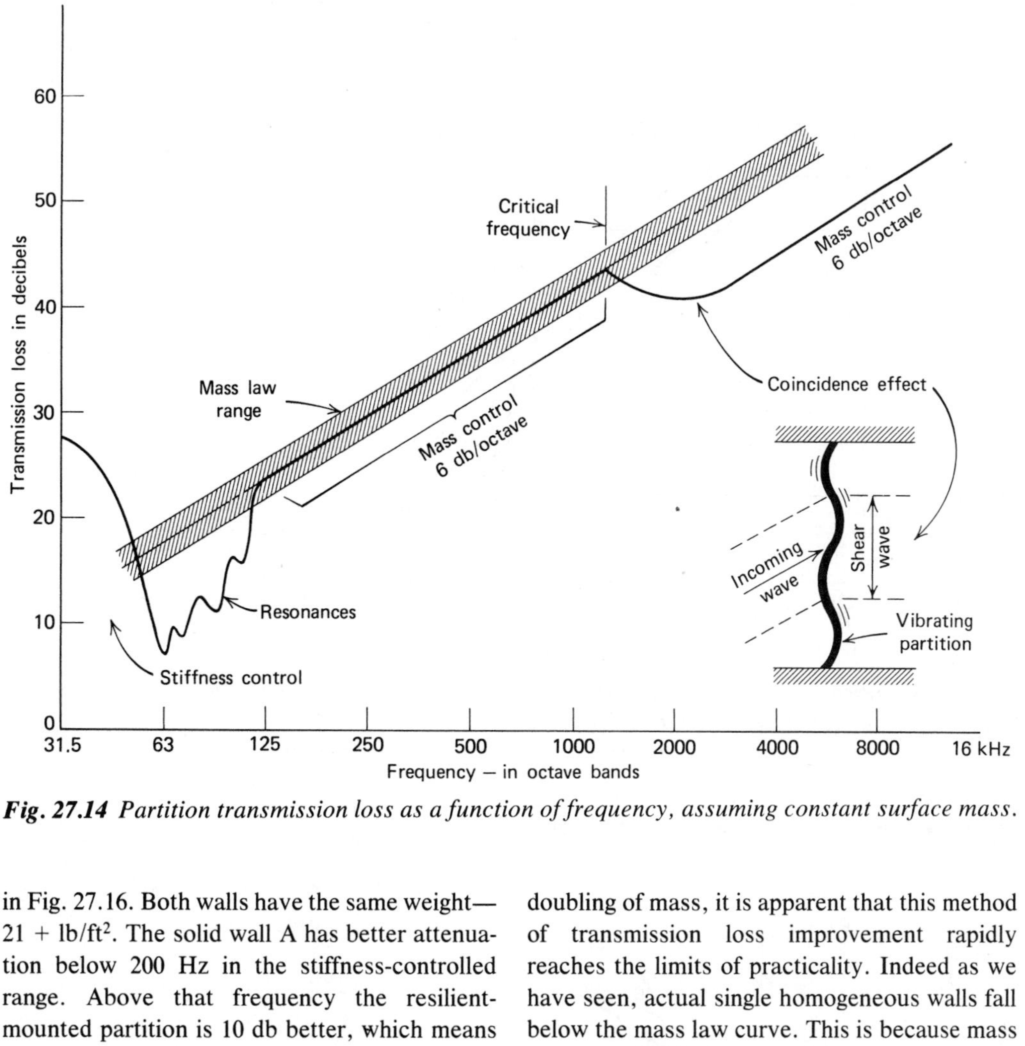

Fig. 27.14 *Partition transmission loss as a function of frequency, assuming constant surface mass.*

in Fig. 27.16. Both walls have the same weight—21 + lb/ft^2. The solid wall A has better attenuation below 200 Hz in the stiffness-controlled range. Above that frequency the resilient-mounted partition is 10 db better, which means that the transmitted noise is only one-half as loud. The sound transmission class (STC) (which is a figure of merit for partition sound transmission, as explained in Section 27.16) of wall A is 40; that of wall B is 51. Both show a coincidence dip at approximately 250 Hz (see Fig. 27.15) but that of wall B is shallow, that of A deep and wide. Furthermore, wall B is consistently better than mass law attenuation; wall A consistently worse (due to stiffness).

27.15 Compound Barriers (Cavity Walls)

Since the maximum theoretical increase in transmission loss with mass increase is 6 db per doubling of mass, it is apparent that this method of transmission loss improvement rapidly reaches the limits of practicality. Indeed as we have seen, actual single homogeneous walls fall below the mass law curve. This is because mass increase brings with it stiffness increase, which as we have seen acts to *reduce* transmission loss. If, however, a barrier is constructed of two separate layers without rigid interconnection, its performance is better than the calculated transmission loss based on mass alone. Note that even the nonrigid wire ties of wall B in Fig. 27.17 lower its STC by five points. At low frequencies, where stiffness controls the transmission loss (see Fig. 27.14), the cavity in *C* acts as a rigid connection between the layers, adding stiffness and increasing transmission loss. At higher frequencies, in the mass law range, the air in the cavity acts as a damping coupling to reduce stiffness. The net result is an improvement in performance throughout the frequency range.

Transmission loss for the entire cavity wall

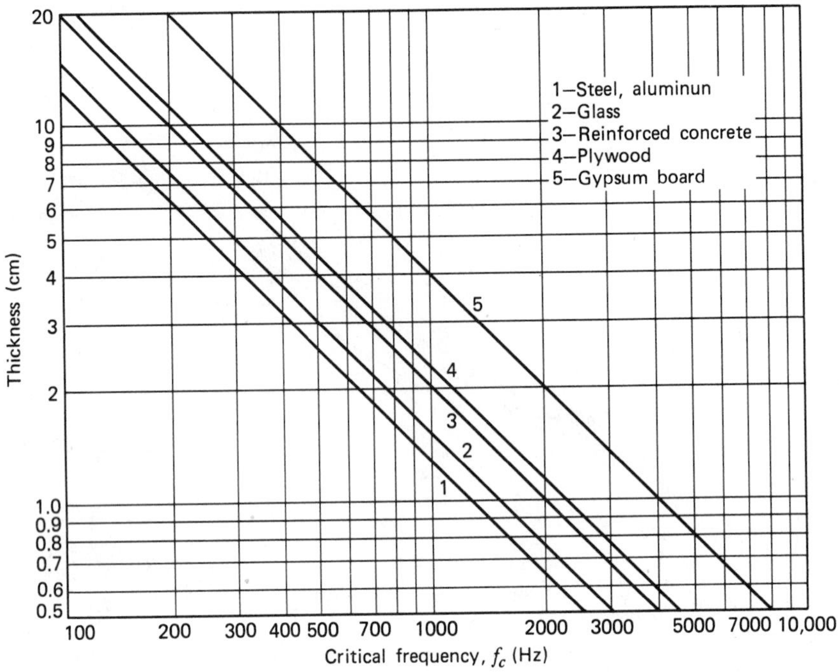

Fig. 27.15 *Critical frequency as a function of thickness for several common materials. (Reprinted with permission from E. B. Magrab,* Environmental Noise Control, *Wiley, New York, 1975, p. 262.)*

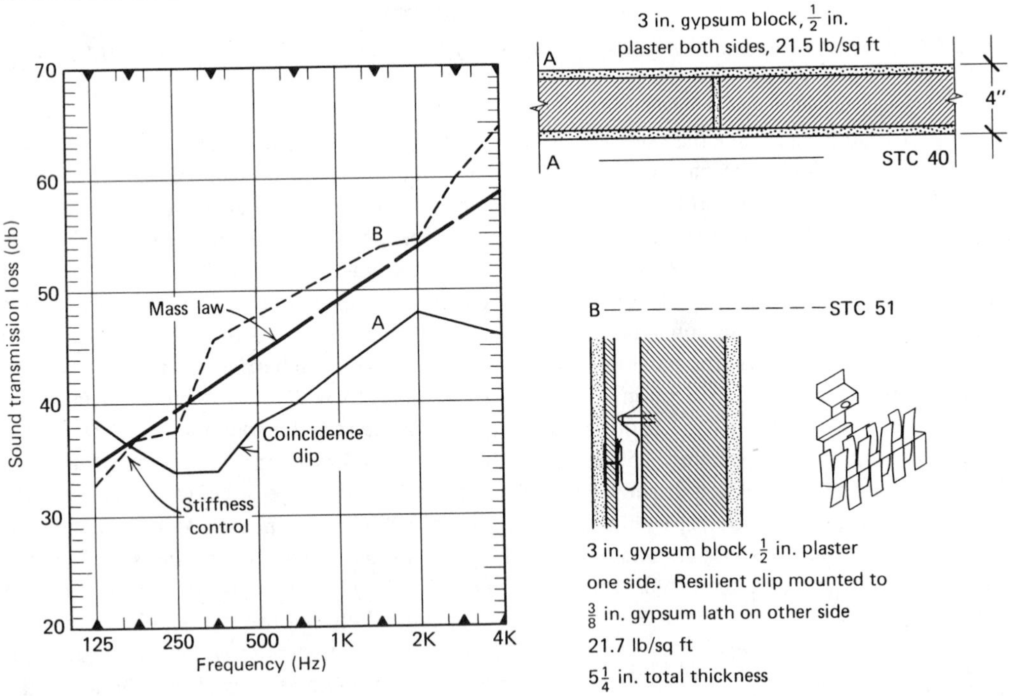

Fig. 27.16 *Transmission loss characteristics of two equal-weight partitions with similar boundary constraints. The solid partition A is worse than the mass law due to stiffness. The resilient-mounted wall B performs better than the mass law and much better than wall A, except at the lowest frequencies. [Data extracted from* A Guide to Airborne, Impact, and Structure-Borne Noise Control in Multifamily Dwellings *(1968).]*

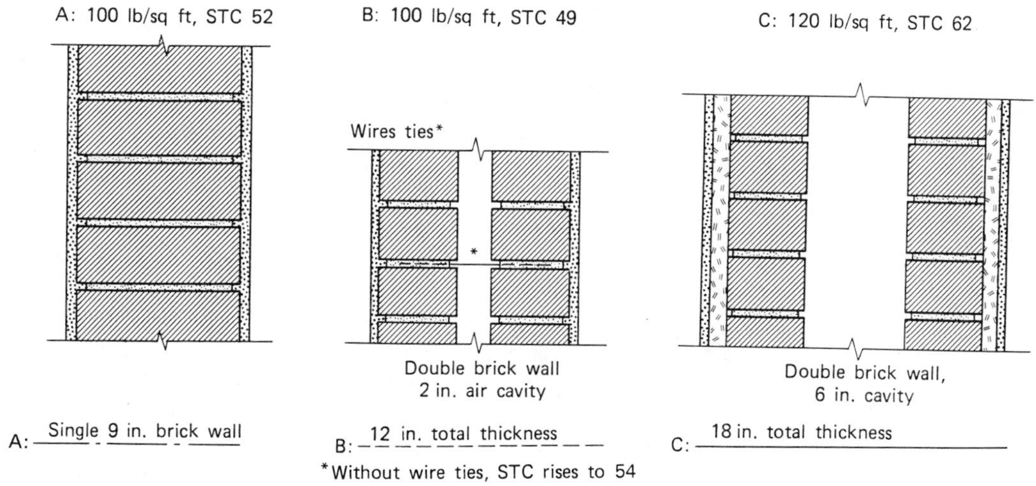

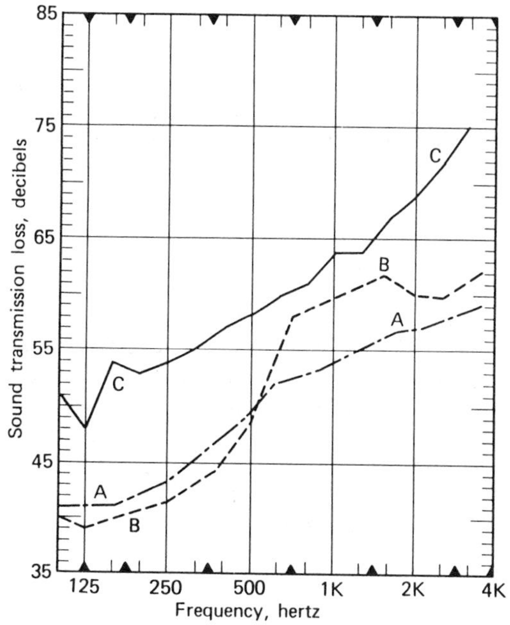

Fig. 27.17 Transmission loss curves showing the effect of air space on heavy wall construction. All three walls are approximately the same mass. The 2-in. airspace in wall B is not significant until the higher frequencies, whereas the large, 6-in. airspace of wall C is effective throughout the frequency spectrum. [Data from A Guide to Airborne, Impact, and Structure-Borne Noise Control in Multifamily Dwellings *(1968), pp. W6, 22, 23.]*

increases with the width of the air space at the rate of approximately 5 db per doubling. Performance can be improved still further by filling the void with porous, sound-absorbent material. This acts to further decrease the stiffness of the compound structure *and* to absorb sound energy that reflects back and forth between the two inside surfaces. The performance of cavity walls is reduced by any rigid interconnections between leaves. Thus a common stud wall with frequent rigid interconnections acts little better than a single homogeneous wall. However, a stud wall with staggered studs exhibits much improved performance over a single-material wall or a common stud wall. The above effects are illustrated in Figs. 27.17 and 27.18 (see also Appendix F).

27.16 Average Transmission Loss (TL); Sound Transmission Class (STC)

Various attempts at using a single-number average transmission loss to describe a barrier's characteristics have been tried but with little success. Indeed these averages can be misleading, since they ignore both deficiencies and proficiencies at particular frequencies. Their use, therefore, in all but rough work is to be discouraged.

To avoid the shortcomings of averages and yet to benefit from the indisputable convenience of single-number ratings, a system of standard contours was developed, called sound transmission class (STC) contours. Actual test results for a given construction, measured in a series of sixteen ⅓-octave bands, are compared to the

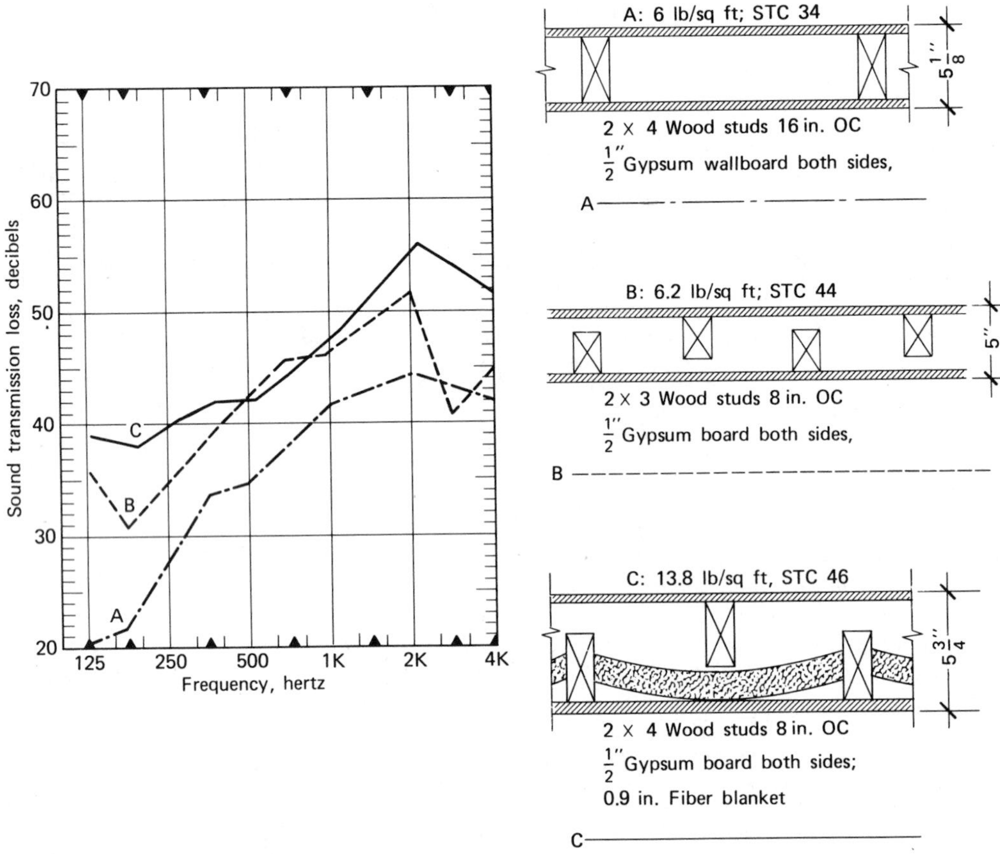

Fig. 27.18 *Transmission loss curves illustrating the effect on lightweight walls of stiffness reduction and addition of absorptive material in the cavity. Curve A is a standard stud wall as found in frame construction. Curve B shows the advantage of staggered studs over the entire frequency range. The dip at 3 kHz is a coincidence dip for a single leaf. Addition of absorption material C improves the attenuation characteristic at both ends and is particularly useful for its low end improvement. [Data from* A Guide to Airborne, Impact, and Structure-Borne Noise Control in Multifamily Dwellings *(1968), pp. W20, 34, 37.]*

standard STC contours according to a fixed procedure, and the STC number for that barrier is derived. The technique is illustrated in Fig. 27.19. Figure 27.20 shows two *TL* curves and the STC ratings of each. The nine-frequency average *TL* for each is approximately 40. However, there is a major dip in transmission loss of curve *B*. The STC of *A* is 42 and of *B*, 33. Thus, the STC approach considers a flaw in performance that is ignored by average *TL*. Nevertheless, since STC fails to give credit for performance *above* the established requirements, octave-band transmission-loss data, rather than STC ratings, should be used in all critical work such as music rooms or mechanical rooms where certain particular frequencies may be dominant.

Figure 27.21 gives three standard STC contours that are of interest because they are used by the FHA to specify grades of construction. The criteria for their application are found in Section 27.33. An appreciation of the degree of speech sound isolation provided by walls with different STC ratings is given in Section 27.21 (see Table 27.7). Since the subjective reaction on the quiet side depends on the background noise level there, the table gives this reaction for two NC curve levels. To assist the designer, extensive sound transmission testing has been performed on most types of standard wall and parti-

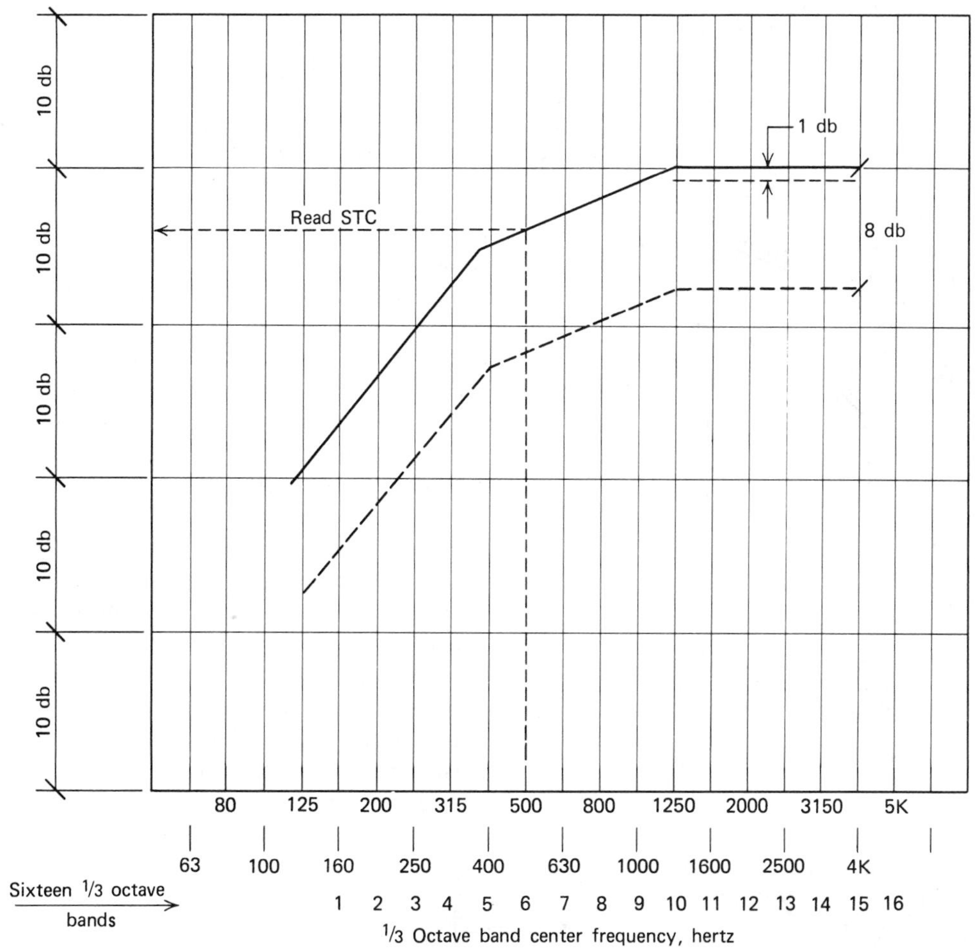

The STC is determined by comparison with a transparent overlay of Figure 27.19 on which the STC contour is drawn. The STC contour is shifted vertically, relative to the test curve, until some of the measured TL values for the test specimen fall below those of the STC contour (the solid line) and the following conditions are fulfilled:

1. The sum of the deficiencies (i.e., the deviations below the contour) shall not be greater than 32 db.
2. The maximum deficiency at a single test point shall not exceed 8 db [the broken (dashed) line beneath the STC contour].

When the contour is adjusted to the highest value (in integral db) that meets the above requirements, the sound transmission class for the specimen is the TL value corresponding to the intersection of the contour and the 500-Hz ordinate.

Fig. 27.19 *Overlay from which sound transmission class (STC) is determined graphically.*

tion construction and the results published. Table 27.4 and Appendix F give descriptions of construction with typical details, transmission loss data, STC ratings, and other pertinent data.

27.17 Composite Walls and Leaks

It is frequently necessary to determine the transmission loss of a composite wall, that is, a wall with a window, door, vent opening, and the like. It should be clearly appreciated that the two materials are "in parallel," to borrow an electrical concept, and the behavior is similar to that situation. That is, the overall performance will be strongly affected by the poorer of the two, with some tempering of the degradation when the poorer barrier is much smaller in area than the other material. Figure 27.22 enables us to calculate situations of this type.

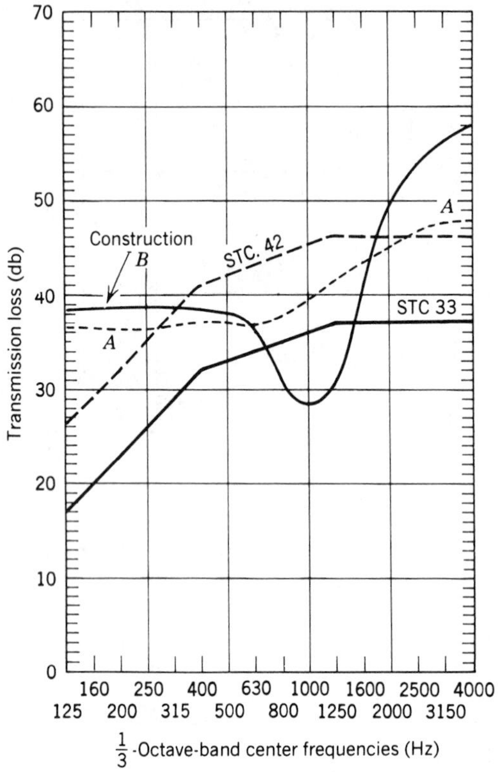

Fig. 27.20 *Curves A and B are two different construction types with the same average TL. However, application of the STC curve criteria given in Fig. 27.19 yields an STC of only 33 for construction B because of its deep center dip, as compared to STC of 42 for construction A.*

Since an opening in a wall is effectively a second material of *TL* = 0, the curves of Fig. 27.22 can be replotted as in Fig. 27.23. Note carefully that the curves very rapidly flatten out; thus any wall with 1% open area will have a *maximum* transmission loss of 20 db, which is all but useless as sound insulation. For this reason it is absolutely imperative that all openings be completely sealed, particularly those around doors and windows. A hairline crack degrades a wall 6 db; a keyhole will degrade a door 3 db, and so on. Special considerations for doors and windows are discussed below. Care must also be taken with such common acoustic leaks as back-to-back electric outlets, pipes passing through walls, and medicine cabinets—in fact any break in the integrity of a partition. All such openings must be caulked to make an airtight joint if any degree of sound isolation is to be maintained.

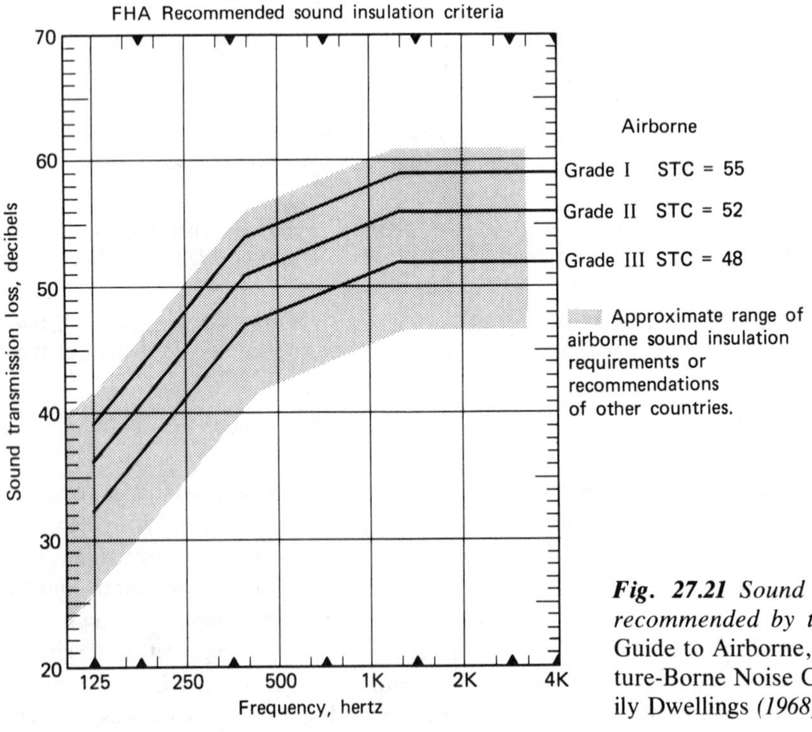

Fig. 27.21 *Sound insulation criteria recommended by the FHA. [From* A Guide to Airborne, Impact, and Structure-Borne Noise Control in Multifamily Dwellings *(1968), pp. 10–11.]*

ACOUSTICS

TABLE 27.4a **Improvements in STC Rating of Stud[a] Partitions[b]**

Description	*STC*[c]
Basic partition: single wood studs, 16 in. on centers, ½-in. gypsum board each side, air cavity	35
Add to basic partition	
Double gypsum board, one side	+2
Double gypsum board, both sides	+4
Single-thickness absorbent material in air cavity	+3
Double-thickness insulation	+6
Resilient channel supports for gypsum board	+5
Staggered studs	+9
Double studs	+13

[a]For application to metal stud partitions, use adders as in note *b*, but begin with STC = 40 for 3⅝-in. basic partition.

[b]When using two improvements add additional +2; for three improvements add +3.

EXAMPLE: Improvements to 35 STC basic partition:

Staggered wood studs	+9
Double gypsum board, one side	+2
Single insulation	+3
Adder	+3
Total	+17
Total STC	35 + 17 = 52

[c]The STC figures are conservative. Other sources list the same constructions with 1 to 5 points higher STC.

TABLE 27.4b **STC Ratings of Masonry Walls**

Description	*STC*[a]
4-in. lightweight[b] hollow block	36
4 in. dense hollow block	38
6-in. lightweight hollow block	41
6-in. dense hollow block	43
8-in. lightweight hollow block	46
8-in. dense hollow block	48
12-in. lightweight hollow block	51
12-in. dense hollow block	53
4-in. brick	41
6-in. brick	45
8-in. brick	49
12-in. brick	54
6-in. solid concrete	47
8-in. solid concrete	50
10-in. solid concrete	53
12-in. solid concrete	56

[a]See note *c*, Table 27.4*a*.

[b]All ratings of lightweight block assume sealing with paint. Note that this reduces absorption.

MODIFICATIONS

Add sand to cores of hollow blocks		+3
Add plaster to one side		+2
Add plaster to both sides		+4
Add furring strips, lath and plaster:	1 side	+6
	2 sides	+10
Add plaster via resilient mounting:	1 side	+10
	2 sides	+15

Examples 27.3 and 27.4 show how barriers of high *TL* are degraded by standard openings. To maintain the integrity of a barrier, special care must be taken in the design of windows and doors, as explained below.

EXAMPLE 27.3. Given a wall 9 ft × 18 ft with a transmission loss of 52 db at 1000 Hz, containing a 3 ft × 7 ft 6-in. hollow core door of 22-db transmission loss. Find the overall *TL* of the composite wall.

SOLUTION. Refer to Fig. 27.22.

$$TL_1 - TL_2 = 30 \text{ db}$$

$$\frac{S_2}{S_1} = \frac{3 \times 7\frac{1}{2}}{9 \times 18} \times 100 = 13.9\%$$

From curves:

$$TL_1 - TL_c = 21.5$$

$$TL_c = 52 - 21.5 = 30.5 \text{ db}$$

That is, a door with an area of only 14% of the entire wall reduces the *TL* of the structure from 52 to 30.5 db—that is, from excellent to *very poor*.

EXAMPLE 27.4. An exterior brick/frame wall having a *TL* of 54 at 1000 Hz, measuring 8 ft × 16 ft, is pierced by two wood frame windows, each of area 3 ft × 4 ft, with single ⅛-in. glass, and has a *TL* of 34 at 1000 Hz. Find the combined transmission loss.

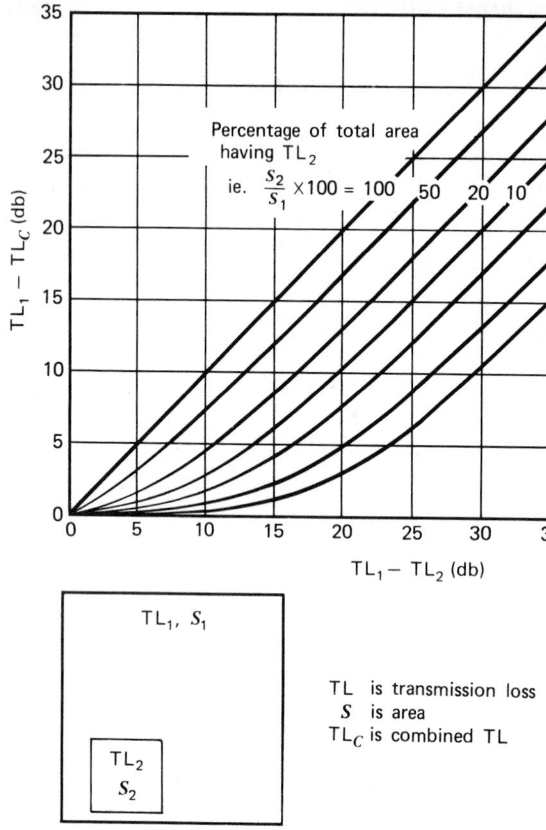

***Fig. 27.22** Transmission loss of a two-element composite barrier as a function of the relative transmission loss of the components. (From E. B. Magrab,* Environmental Noise Control, *Wiley, New York, 1975, p. 266, Fig. 7-45.)*

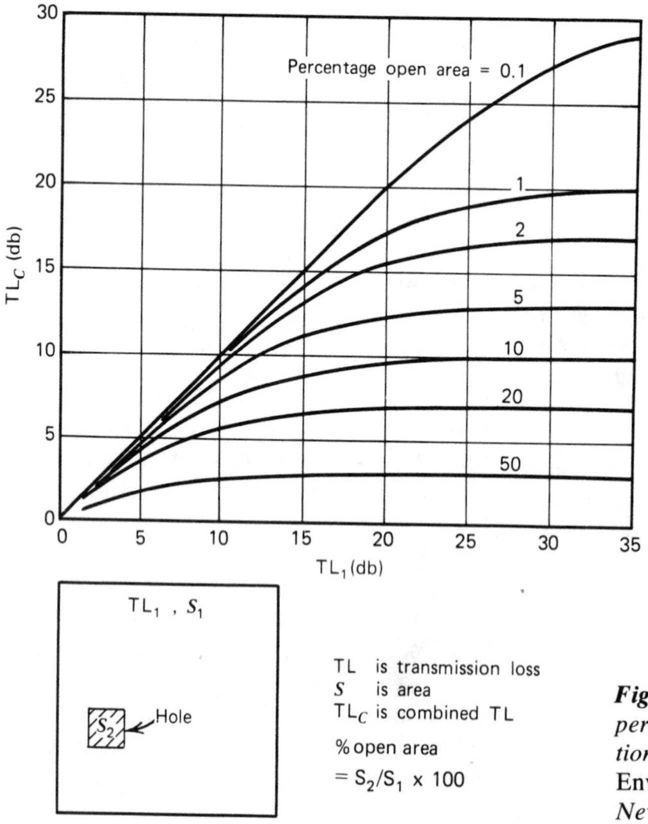

***Fig. 27.23** Effect of a hole of a given percentage of the total area on a partition of TL, db. (From E. B. Magrab,* Environmental Noise Control, *Wiley, New York, 1975, p. 268, Fig. 7-47.)*

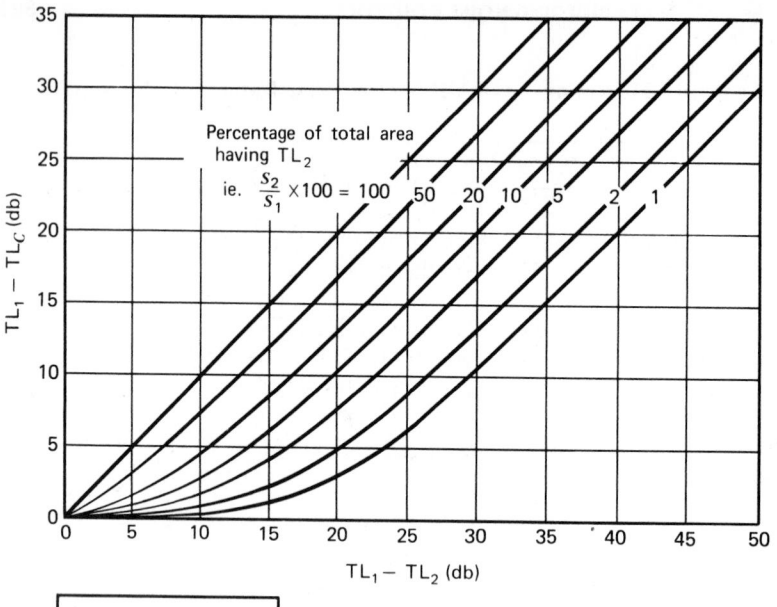

SOLUTION

$$TL_1 - TL_2 = 54 - 34 = 20 \text{ db}$$

$$\frac{S_2}{S_1} = \frac{2 \times 3 \times 4}{8 \times 16} \times 100 = 18.75\%$$

From Fig. 27.22,

$$TL_1 - TL_c = 12.5 \text{ db}$$

$$TL_c = 41.5$$

Again the result is a reduction from an excellent wall to a poor one.

27.18 Doors and Windows

As can be appreciated from the preceding section, doors and windows in large measure determine the overall transmission loss of a wall. Since in almost every instance doors and windows are of lower acoustic rating than that of the wall in which they are mounted, particular care must be taken not to further degrade performance with leaks.

(a) Doors. Table 27.5 gives average *TL* figures for common wooden doors and for doors arranged in a sound lock, that is, two doors with a sufficient space between to allow door swing. As with all transmission loss averages, the number is usable only as an approximate figure of merit. STC is a better criterion; a frequency analysis is best. The important conclusions to be drawn from this table are:

1. Louvered doors and doors undercut to permit air movement are useless as sound barriers.
2. The most important step in sound-proofing doors is complete sealing around the opening. The door in the closed position should exert pressure on these gaskets, making the joints airtight.
3. Doors constructed of two leaves separated by a sound-absorbing material (e.g., steel sheets with fiber batting inside) are more effective than a single solid material door of the same weight.

Commercial products are available with STC rating up to 65 for application in sensitive locations.

(b) Windows. Windows are critically important to block exterior noise, and all the more so, since exterior wall construction is generally of high STC, making the window the deciding factor in the composite transmission loss. Sound leaks through cracks in operable windows will normally establish the windows' rating regardless of type of glazing. Fortunately, the attention now given to the sealing of windows for thermal purposes has had a salutary effect on their acoustic properties. As with doors, the importance of proper gasketing and sealing cannot be overstressed. Double glazing is effective only when the two panes are separated by a wide air gap (see Table 27.6). A narrow air gap acts as a stiff spring between the panes. The resultant TL

TABLE 27.5 **Average Sound Transmission Loss of Doors**

Construction	*Average TL (db)*	*STC*
Louvered door	10	15
Any door, 2-in. undercut	12	17
$1\frac{1}{2}$ in. hollow core door, no gasketing	15	22
$1\frac{1}{2}$ in. hollow core door, gaskets and drop closure	20	25
$1\frac{3}{4}$ in. solid wood door, no gasketing	25	30
$1\frac{3}{4}$ in. solid wood door, gaskets and drop closure	30	35
2 hollow core doors, gasketed all around with sound lock	38	45
2 solid core doors, gasketed all around, with sound lock	48	55
Special commercial construction, with lead lining and full sealing		45–65

ACOUSTICS

TABLE 27.6 **Sound Transmission Class (STC) of Window Construction**

Window Construction	*STC*
Operable wood sash, ⅛-in. glass, unsealed	25
Operable wood sash, ¼-in. glass, unsealed	25
Operable wood sash, ¼-in. glass, gasketed	30
Operable wood sash, laminated glass, unsealed	28
Operable wood sash, double-glazed, ⅛-in. panes, ⅜-in. air space, gasketed	29
Fixed sash, double ⅛-in. panes, 3-in. air space, gasketed	44
Fixed sash, double ¼-in. panes, 4-in. air space, gasketed	48

is approximately that of a single pane of double weight.

In addition to a window's sound transmission characteristic when closed, it is important to consider the TL when open, because of ventilation and natural cooling requirements. The sound attenuation between the center of a room with a clear-through open window and a point some distance outside, is 5 to 15 db. This drops to ±3 db as the receiver–observer approaches the open element. By making the path from inside to outside indirect, the "open" window attenuation can be increased to as much as 25 db, but with considerable reduction of airflow, hence ventilation. Several possible arrangements with approximate TL figures are given in Fig. 27.24. This principle can be applied advantageously when exterior noise reduction is important, but sealed windows are undesirable. Window opening style and placement can also have an effect on the amount of exterior noise admitted, as shown in Figs. 27.25 and 27.26.

27.19 Diffraction; Barriers

The physical process by which sound passes around obstructions and through very small openings is called diffraction. Simply stated, diffraction is a process whereby any point on a wave front establishes a new wave front when passing an obstacle. Thus, although much of a sound wave is blocked by a small opening, the portion that does get through establishes a new wave front [see Fig. 27.11*c*(5)]. The *amplitude* of the diffracted wave is determined by the relationship between the size of the opening and the

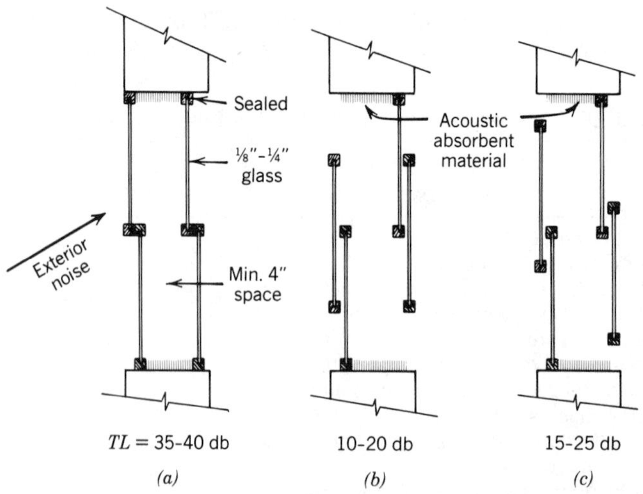

Fig. 27.24 *The degree of attenuation of external noise can be regulated when using pairs of double-hung (or horizontally sliding) windows with acoustic sealant and absorbent materials. Ventilation airflow varies inversely with transmission loss.*

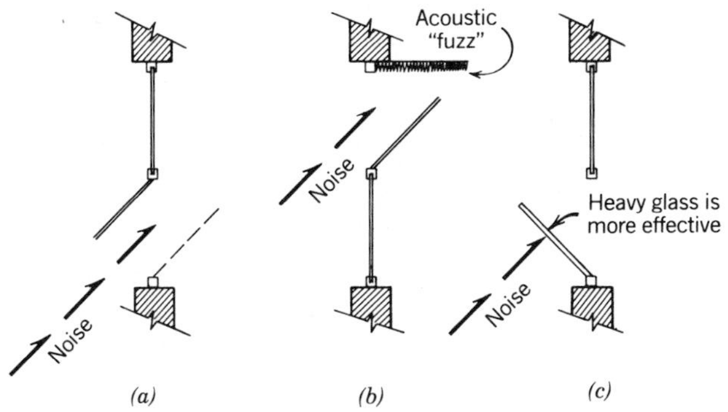

Fig. 27.25 Alternative arrangements of the same basic "hopper" window design can yield results differing by up to 10 db. Design (a) *is entirely open, and the noise path is unobstructed deep into the room. Design* (b) *is about 5 db better than* (a) *at frequencies above 1 kHz because of higher absorption and less diffraction. Lower frequencies diffract readily around the window leaf and are less affected by absorptive material. Design* (c) *can be 10 db better than* (a) *because it interposes a rigid barrier into the noise path, particularly at high frequencies. In this arrangement, the glass thickness is important.*

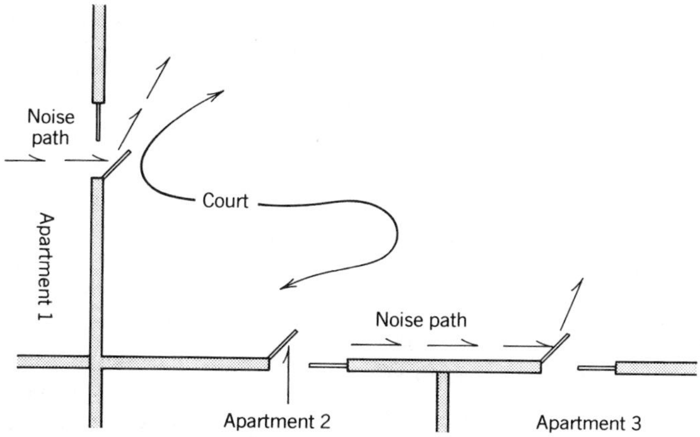

Fig. 27.26 Noise transfer between contiguous corner spaces, as in apartments 1 and 2, can be particularly severe if windows are improperly designed. Swinging windows, as shown, are preferable to double-hung or hopper, because they act to reflect sound away from the adjacent space. Similarly, adjacent spaces on the same wall, as for apartments 2 and 3, can benefit from this type of swinging arrangement, which is preferable to sliding or double-hung designs.

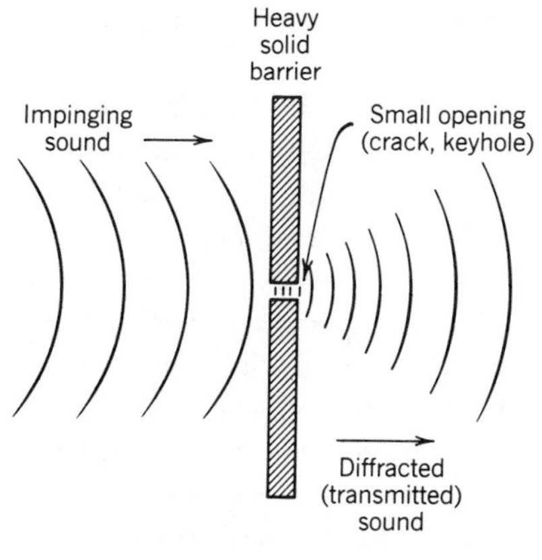

Fig. 27.27 Sound passes through small openings by diffraction. The intensity of the transmitted sound is proportional both to frequency and to the size of the opening. It is always less than the intensity of the impinging sound [see also Fig. 27.11c(5).]

ACOUSTICS

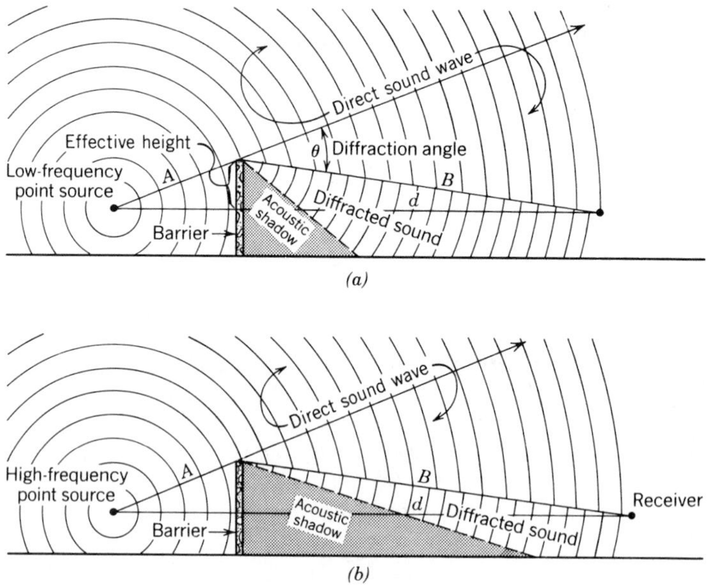

Fig. 27.28 *Comparison of the effect of a barrier on sources of different frequencies. The low-frequency sound* (a) *diffracts more readily over the barrier than the high-frequency sound* (b) *because of its longer wavelength. Thus the lower the frequency, the smaller the "acoustic shadow" and the lower the barrier attenuation. Of course, the "shadow" is not sharply defined; it represents increasing attenuation as the observer approaches the barrier. See Fig. 27.29, for evaluation of barrier attenuation.*

wavelength. For a small hole, short wavelengths (high frequencies) are attenuated less than long wavelengths (low frequencies). See Fig. 27.27.

When sound encounters a finite-length barrier, it diffracts around and over it, approximately as shown in Fig. 27.28. The attenuation of the diffracted sound depends on frequency, type of source, and the dimensions of the barrier. For a point source, with only a single *practical* path around the obstruction (barrier) the noise reduction to db can be calculated from Maekawa's equation:

$$NR = 20 \log \left[\frac{\sqrt{2\pi N}}{\tanh \sqrt{2\pi N}} \right] + 5 \text{ db} \quad (27.5)$$

where

$N = (F/565)(A + B - d)$

NR = noise reduction, db

F = frequency, Hz

$A + B$ = shortest path length around the barrier, ft (over or around)

d = straight-line distance, source-to-receiver, ft

Note that this equation:

1. Is applicable only to exterior barriers where sound passing over the barrier is partially diffracted and partially attenuated by distance. In an interior situation, sound passing over a partial height barrier (see Fig. 27.35) strikes the ceiling and is reflected down, increasing the received sound and effectively reducing barrier attenuation. Maximum exterior barrier attenuation is 24 db as compared to about 15 db for a partial height interior partition.
2. Assumes that the barrier is very long, (or very high) so that only one sound path exists. In practice, a barrier whose length (height) is at least four times the distance between the source and the wall is sufficient, if *the barrier is close to the source.* If the barrier is close to the receiver, it must be longer (higher) still.
3. Assumes a point source. Line sources (such as traffic) show 20 to 25% less attenuation for the same barrier.

The equation will, however, give reliable, usable results when the dimensions of the source are small with respect to the barrier, as is the case for speech; individual motors, fans, engines, and other mechanical devices; and individual motor vehicles. The chart in Fig. 3.19 relates barrier dimensions and position to traffic noise reduction. Note that there frequency is not a variable, since the chart has been plotted for an average attenuation at 220 Hz, which is traffic center frequency.

It should be apparent that the best location for a barrier is either very close to the source or very close to the receiver. The worst position for attenuation is halfway between. All effective barriers are assumed to be opaque and to have a minimum surface density of 5 lb/ft^2 ($\sim$20 kg/m^2). The inherent TL of the barrier need not be very high; a massively thick barrier has only marginally higher attenuation than one with the minimum surface weight specified above. Absorptive material placed on the source side of a barrier will reduce the noise reflected back toward the source but will not effectively increase the barrier's attenuation with respect to the receiver. Although maximum noise reduction of an *exterior* barrier is about 24 db, in practice it rarely exceeds 20 db. Figure 27.29 is a nomograph based on equation 27.5.

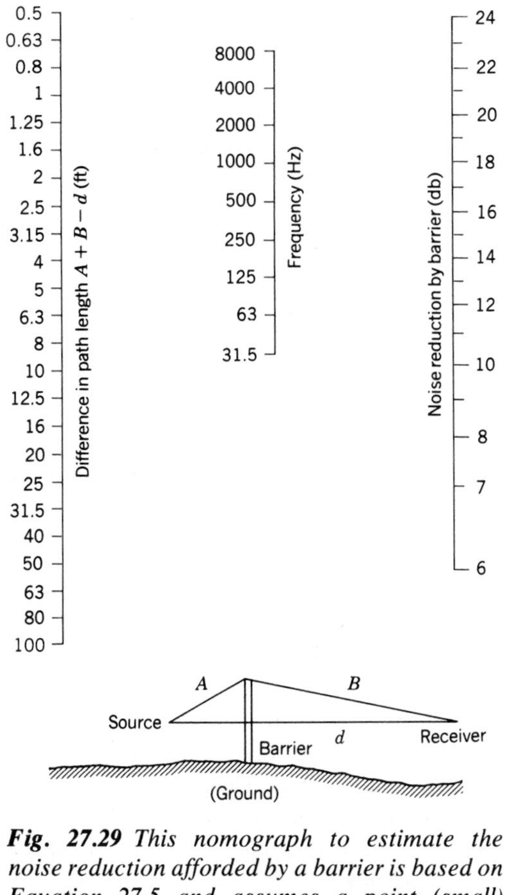

Fig. 27.29 *This nomograph to estimate the noise reduction afforded by a barrier is based on Equation 27.5 and assumes a point (small) source and only a single path around the barrier. The dimensions* A, B, *and* d *are taken from the insert sketch.* A *plus* B *represent the shortest path around the barrier—which may be over or around it. (Reprinted with permission from B. Fader,* Industrial Noise Control, *Wiley, New York, 1981, pp. 148–149.)*

27.20 Flanking

Just as sound will pass through the acoustically weakest part of a composite wall, so it will also find parallel or flanking paths, that is, an acoustic short-circuit. Proper design of window locations to avoid flanking paths has already been shown in Fig. 27.26. The same situation obtains with respect to doors and any other openings between spaces. Thus in Fig. 27.30 a high STC wall between the two spaces is in large measure defeated by flanking paths F5, F6, and F7. In other spaces the most common flanking path is via the plenum, as in Fig. 27.30, path F1, and Fig. 27.31*b* and *d*. Ductwork with registers or grilles in various rooms acts as an excellent intercom system unless completely lined with sound-absorptive material (see Section 27.29). Even then, low-frequency sound is only minimally attenuated, and special measures must be employed if good transmission loss is required. This subject is discussed further below, under, Mechanical System Noise Control.

27.21 Noise Reduction and Background Noise

Referring back to Section 27.12, we saw that

$$IL_2 = IL_1 - NR \qquad (27.3)$$

where NR is noise reduction and IL_2 and IL_1 are sound intensity levels in the receiving and source rooms, respectively. If the resulting IL_2 is below the background noise level, it will not be a source of annoyance; conversely, if it is

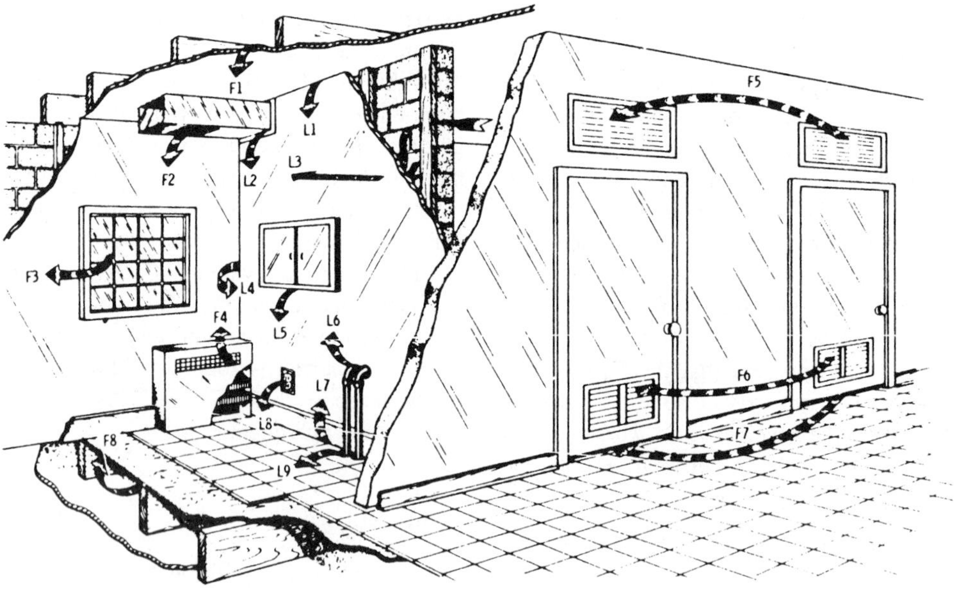

FLANKING NOISE PATHS

F1 OPEN PLENUMS OVER WALLS, FALSE CEILINGS
F2 UNBAFFLED DUCT RUNS
F3 OUTDOOR PATH, WINDOW TO WINDOW
F4 CONTINUOUS UNBAFFLED INDUCTOR UNITS
F5 HALL PATH, OPEN VENTS
F6 HALL PATH, LOUVERED DOORS
F7 HALL PATH, OPENINGS UNDER DOORS
F8 OPEN TROUGHS IN FLOOR-CEILING STRUCTURE

NOISE LEAKS

L1 POOR SEAL AT CEILING EDGES
L2 POOR SEAL AROUND DUCT PENETRATIONS
L3 POOR MORTAR JOINTS, POROUS MASONRY BLK
L4 POOR SEAL AT SIDEWALL, FILLER PANEL ETC.
L5 BACK TO BACK CABINETS, POOR WORKMANSHIP
L6 HOLES, GAPS AT WALL PENETRATIONS
L7 POOR SEAL AT FLOOR EDGES
L8 BACK TO BACK ELECTRICAL OUTLETS
L9 HOLES, GAPS AT FLOOR PENETRATIONS

OTHER POINTS TO CONSIDER, RE: LEAKS ARE (A) BATTEN STRIP A/O POST CONNECTIONS OF PREFABRICATED WALLS, (B) UNDER FLOOR PIPE OR SERVICE CHASES, (C) RECESSED, SPANNING LIGHT FIXTURES, (D) CEILING & FLOOR COVER PLATES OF MOVABLE WALLS, (E) UNSUPPORTED A/O UNBACKED WALL BOARD JOINTS (F) EDGES & BACKING OF BUILT-IN CABINETS & APPLIANCES, (G) PREFABRICATED, HOLLOW METAL, EXTERIOR CURTAIN WALLS.

Fig. 27.30 *Flanking transmission of airborne noise. [Reprinted from* A Guide to Airborne, Impact, and Structure-Borne Noise Control in Multifamily Dwellings *(1968).]*

greater than the background noise it will be heard and, depending onintensity and content, could be a source of disturbance. See Fig. 27.32 for a graphic representation of this idea.

With this in mind, we can also tabulate the performance of a sound barrier, in terms of receiving-room background levels (see Table 27.7). Stated otherwise, the apparent isolation provided by a barrier may be greater than the actual noise reduction. Figure 27.33 shows two conditions of adjacent spaces. Although the source room level is uniform and partitions on both sides of the source room are identical, the background noise in the two receiving rooms is different. In *A*, the background is NC-35; in *B*, it is NC-25. The occupant of room *A* hears nothing from the source room. The occupant of room *B* hears clearly. Occupant *A* probably will praise the partition while occupant *B* will complain. Although the levels of reradiated sound are identical in the two receiving rooms, the intruding signal is masked by the background noise in *A* but it is clearly audible in *B*. Thus the apparent noise reduction is substantially higher in *A* than in *B*. Section 27.22 investigates this phenomenon of speech privacy.

Sound isolation may be achieved very economically by careful planning. Storage and circulation can serve as buffers for noise-sensitive areas. Physical separation of noisy areas from quiet ones often eliminates the need for complicated and expensive compound barriers.

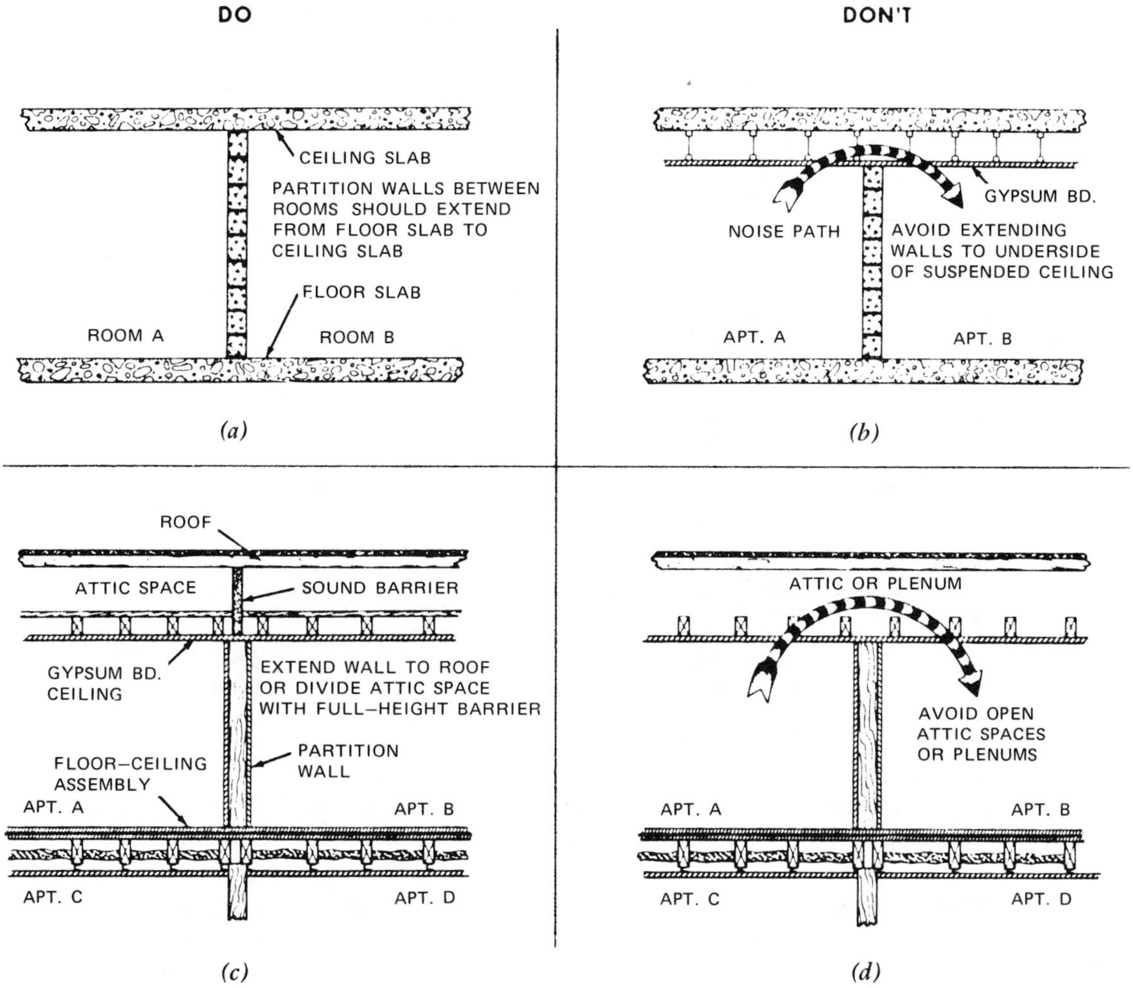

Fig. 27.31 *Construction techniques to avoid flanking paths. [Reprinted from* A Guide to Airborne, Impact, and Structure-Borne Noise Control in Multifamily Dwellings *(1968).]*

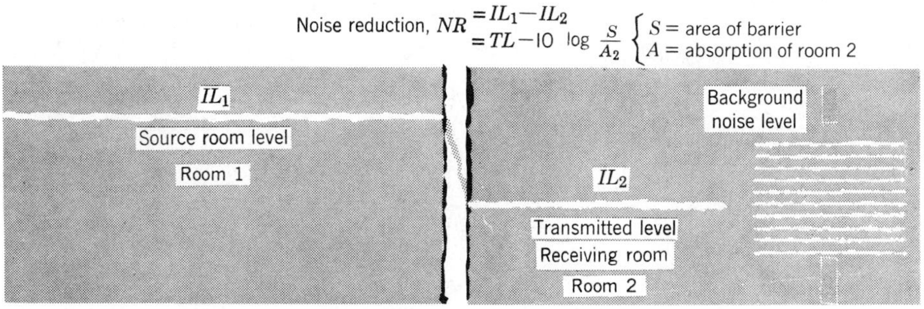

Fig. 27.32 *The background noise level determines whether the transmitted sound will actually be heard.*

TABLE 27.7 **Relations Between Barrier STC Rating and Hearing Condition on the Receiving Side**

Barrier STC[a]	Hearing Condition	Description	Application
25	Normal speech can be understood quite easily and distinctly through the wall.	Poor	Space divider
30	Loud speech can be understood fairly well. Normal speech can be heard but not easily understood.	Fair	Room divider where concentration not essential
35	Loud speech can be heard, but is not easily intelligible. Normal speech can be heard only faintly, if at all.	Good	Suitable for offices next to quiet spaces
42–45	Loud speech can be faintly heard but not understood. Normal speech is inaudible.	Very good	For dividing noisy and quiet areas; party wall between apartments
46–50	Very loud sounds, such as loud singing, brass musical instruments, or a radio at full volume can be heard only faintly or not at all.	Excellent	Music room, practice room, sound studio, bedrooms adjacent to noisy areas

[a]Assuming a background noise level of NC-25. With higher background, (e.g., NC-35), STC can be degraded one step.

27.22 Speech Privacy

The subject of speech privacy received considerable additional study and emphasis with the advent of open-plan offices (office landscaping), although the same problem prevails in all spaces. Essentially the purpose of the investigations was to determine the factors affecting speech privacy and to quantify them with a degree of accuracy sufficient for design purposes.

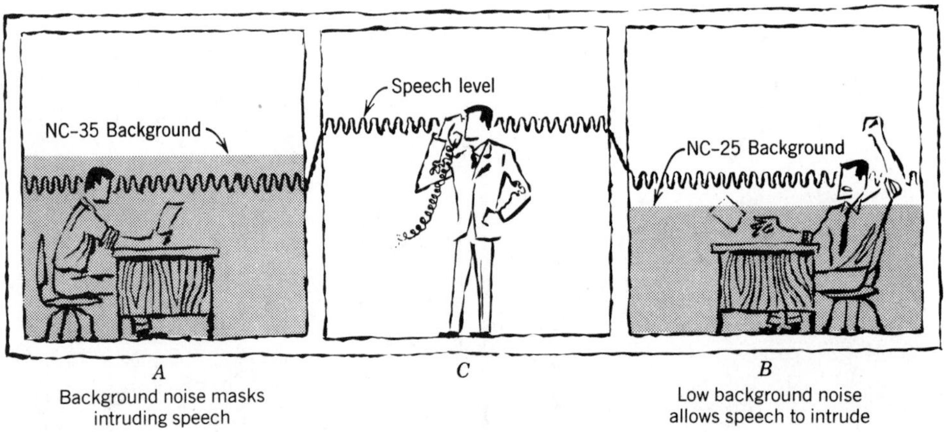

Fig. 27.33 The occupant in room A with background noise NC-35 (≈45 dbA) is unaware of the noise that is to disturbing to occupant B, whose NC-25 (≈36 dbA) is insufficient to mask C's loud speech.

These studies indicate that there are six factors involved, which can be subsumed under two headings:

1. Speech rating (source room, No. 1)
   a. Speech effort—a measure of the loudness of speech.
   b. Source room factor—gives the approximate effect of room absorption, on speech level in source room. Scale in Fig. 27.34 drawn for average normal absorption. For live rooms raise factor by two points; for dead rooms lower it by two points. Factors $a + b$ give approximate source room voice level.
   c. Privacy allowance—What is the measure of privacy required?
2. Isolation rating (receiving room, No. 2)
   d. STC—rating of the barrier.
   e. Noise reduction factor A_2/S is an indication of receiving-room absorption, that is, the difference between *NR* and *TL*; A_2 is area of receiving room; *S* the area of the barrier between rooms. Absorption is assumed to be average. For live rooms lower this factor two points; for a dead room raise two points.
   f. Recommended background noise level—in receiving room from Table 27.8.

An analysis sheet for enclosed spaces is provided in Fig. 27.34. See also Cavanaugh et al. (1962) and Young (1965). The two examples of this analysis that follow should clarify its use. The reader should follow the analysis with Fig. 27.34 in hand. The numbered steps in the examples correspond to the numbers in the figure.

EXAMPLE 27.5

(*a*) Source room:

General clerical office, 40 × 60 × 9 ft, average $\bar{\alpha}$
16-ft partition, STC 40

(*b*) Receiving room:

Conference rooms, 16 × 24 ft, medium-dead room
Background noise level, 40 dbA (NC-30), from Table 27.8

Privacy analysis (see Fig. 27.34)

(*a*)	1. Speech effort: raised	66
	2. Area $A_1 > 1000$ ft^2	0
	3. Privacy—normal	9
	Speech rating:	75
(*b*)	4. STC (given)	40
	5. A_2/S	3
	[(16 × 24)/(16 × 9) = 2.6, corresponding to 2.0; add 1.0 for higher than average $\bar{\alpha}$	
	6. Noise level	40
	Isolation rating:	83
	$a - b =$	−8

Therefore the STC rating of the partition can be reduced to 32 without affecting speech privacy.

EXAMPLE 27.6

(*a*) Source Room:

Drafting room 20 × 30 ft, medium-live
Common wall 12 × 8 ft high
STC: 26 (half glass, with door)

(*b*) Receiving room:

Supervisor's office, 12 ×14 ×8 ft, average absorption
Background noise level, 35 dbA

Privacy analysis:

(*a*)	Speech effort—conversational	60
	Source room factor, 2 + 1	3
	Privacy—confidential	13
	Speech rating:	76
(*b*)	STC	26
	A_2/S	2
	[(12 × 14)(12 × 8) = 1.8, corresponds to 1.6; use 2.0, since system uses whole numbers only.]	
	Noise level	35
	Isolation rating:	63
	$a - b =$	13

or strong dissatisfaction

The suggested corrections here are to increase STC to 36 by gasketing the door and increasing background noise level in the receiving room to

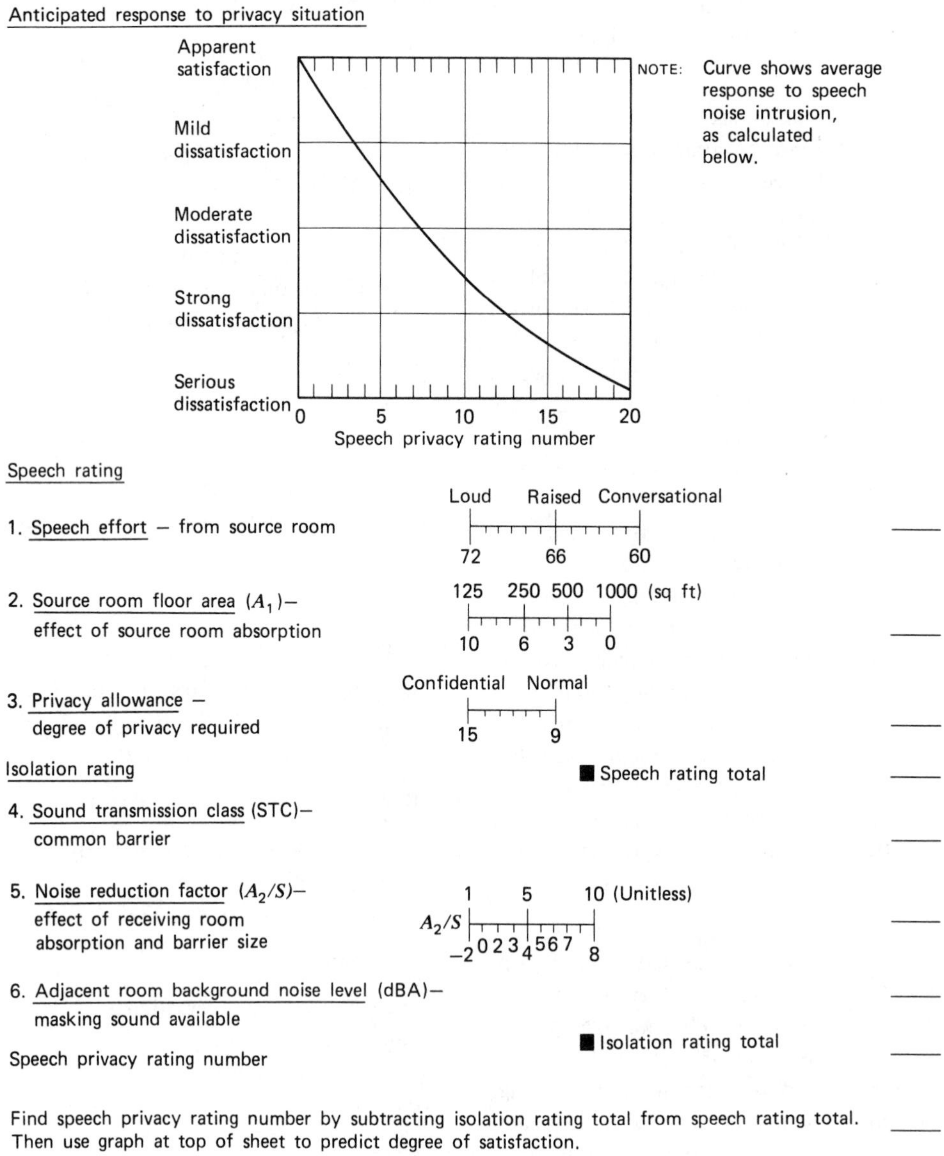

Fig. 27.34 *Speech privacy analysis sheet. [Based on Cavanaugh et al. (1962).]*

40 dbA (NC-30). This should result in a satisfactory condition. If still unsatisfactory, the glazing could be doubled.

A form similar to Fig. 27.34 is available for open plan offices. We have not included it because these spaces require critically accurate design best left to an acoustics expert. In these open spaces where partial height partitions can provide a *maximum* of 15 db isolation, masking noise of fairly high level must be introduced. Too high a level will result in the noise itself being disturbing; too low a level results in speech interference, extreme dissatisfaction, and impediments to work. Since in such areas, control of background noises is extremely difficult, success depends in large measure on proper physical placement of noise sources and grouping of common acoustic need areas. The indiscriminate reliance on noise generators is to be discouraged (see Fig. 27.35). Generally

TABLE 27.8 **Recommended Category Classification and Suggested Noise Criteria Range for Steady Background Noise**

Type of Space (and Acoustical Requirements)	*NC Curve*	*Equivalent[a] dBA*
Concert halls, opera houses, and recital halls (for listening to faint musical sounds).	10–20	20–30
Broadcast and recording studios (distant microphone pickup used).	15–20	25–30
Large auditoriums, large drama theatres and houses of worship (for excellent listening conditions).	20–25	30–35
Broadcast, television, and recording studios (close microphone pickup only).	20–25	30–35
Small auditoriums, small theatres, small churches, music rehearsal rooms, large meeting and conference rooms (for good listening), or executive offices and conference rooms for 50 people (no amplification).	25–30	35–40
Bedrooms, sleeping quarters, hospitals, residences, apartments, hotels, motels, and so forth (for sleeping, resting, relaxing).	25–35	35–45
Private or semiprivate offices, small conference rooms, classrooms, libraries, and so forth (for good listening conditions).	30–35	40–45
Living rooms and similar spaces in dwellings (for conversing or listening to radio and TV).	35–45	45–55
Large offices, reception areas, retail shops and stores, cafeterias, restaurants, and so forth (for moderately good listening conditions).	35–50	45–60
Lobbies, laboratory work spaces, drafting and engineering rooms, general secretarial areas (for fair listening conditions).	40–45	50–55
Light maintenance shops, office and computer equipment rooms, kitchens, and laundries (for moderately fair listening conditions).	45–60	55–70
Shops, garages, power-plant control rooms, and so forth (for just acceptable speech and telephone communication). Levels above PNC-60 are not recommended for any office or communication situation.	—	—
For work spaces where speech or telephone communication is not required, but where there must be no risk of hearing damage.	—	—

[a]For information only.

Source: Extracted with permission from E. B. Magrab, *Environmental Noise Control,* Wiley, New York, 1975.

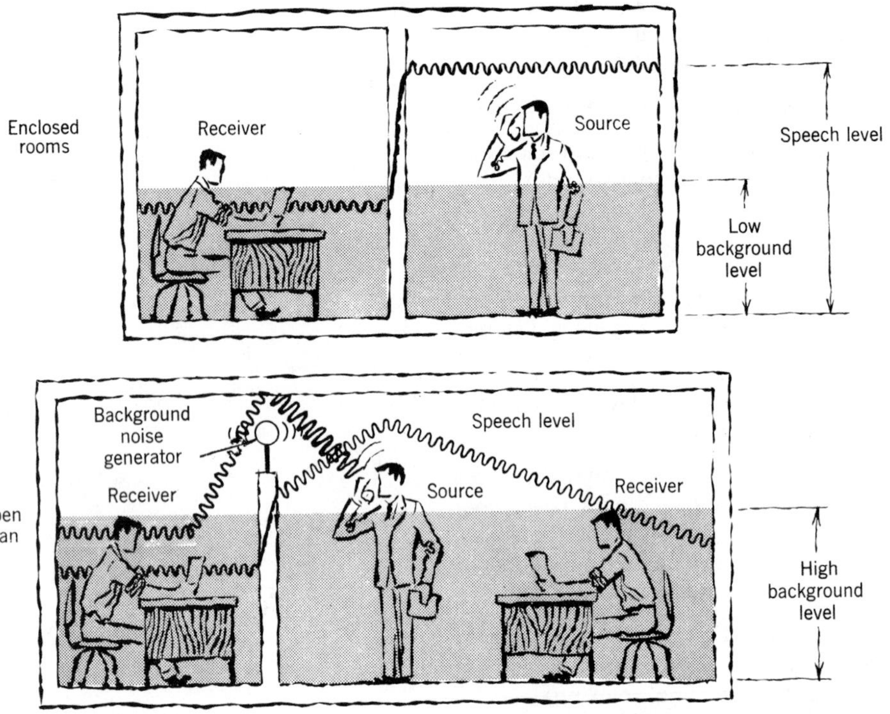

Fig. 27.35 *The open-plan background noise level must be high enough to mask adjacent speech but not so loud that it constitutes a disturbance itself. Received noise is a combination of partition-transmitted and ceiling reflected noise.*

speaking, open areas should use barriers at least 5 ft high × 8 ft wide of very high absorbency, and acoustic ceilings no higher than 9 ft with a minimum absorption coefficient of 0.7, backed by a plenum at least 30 in. deep. The partial height barriers must be set so that the listener is shielded from line-of-sight noise, either direct or on first reflection. Obviously the taller the barrier, the better.

It was found, when introducing background noise to provide masking, that a noise with a frequency spectrum conforming in shape to an NC curve was neither pleasant nor unnoticed (neutral), but instead sounded emphasized at both ends; that is, it rumbled and hissed. Similarly, it was found that neither white noise (a signal with equal energy per frequency band width) nor pink noise (a signal with equal energy per octave) were satisfactory. Instead, a signal with a drooping energy per octave characteristic was found to provide the greatest masking satisfaction in field tests, and is now generally used, although older installations still use pink noise.

STRUCTURE-BORNE SOUND

27.23 Structure-Borne Impact Noise

As stated in Section 27.11, structure-borne noise is at least as serious a problem as airborne noise because:

1. There is no air cushion between the source and the structure. Thus high-intensity energy is introduced into the structure, through which it travels with minimum attenuation and at great speed.

2. Sound, once introduced into the structure, is attenuated well only by discontinuities in the structure. Since the structure must have integrity to carry the loads, discontinuities of the type that will stop noise are complex and expensive.

3. The entire structure constitutes a network of parallel paths for sound. Therefore, partial solutions are useless, since sound will find flanking paths. The entire structure must be soundproofed to yield good results.

4. Unlike the case of airborne noise, additional mass does not usually alleviate floor-borne noise, particularly in long spans where the floor acts as a diaphragm.

5. The increasing use of exposed structural ceiling eliminates the attenuation that can be introduced into a plenum above a hung ceiling. This is particularly bad, since most structure-borne noise is carried by floor structures (rather than walls), which radiate sound up and down.

The discussion that follows will be limited to impact noise. Refer to Section 27.27 for a brief treatment of vibration that is felt rather than heard, and is in effect a very low-frequency noise. Many of the recommendations and techniques that will minimize impact noise will also reduce vibration.

27.24 Control of Impact Noise

Impact noise problems can be controlled in two ways—by preventing or minimizing the impact, and by attenuating it once it has occurred. Since we are concerned with structure, the latter problem is discussed first. The former problem is covered in Section 27.26. Impact on floors is more serious than wall impact because the latter will be partially attenuated at the wall/floor joint, whereas the former is introduced directly into the building framework.

The discussion below addresses each of the solutions shown in Fig. 27.36.

(a) Cushion the Impact. See Fig. 27.36*a*. This obvious solution will frequently eliminate all but severe problems. The resilient materials in common use are floor tiles of vinyl, rubber, and cork, or carpeting on pads, in ascending order of impact insulation. See Section 27.25 and Appendix G for quantitative data on impact isolation.

(b) Float the Floor. See Fig. 27.36*b*. Since the key to elimination of structure-borne sound is *isolation,* separating the impacted floor from the structure floor by a resilient element is extremely effective. This element can be rubber or mineral wool pads, or blankets, or special spring metal sleepers. The effectiveness depends on the mass of the floating floor, compliance of the resilient support, and degree of isolation of the floating floor. This last element is extremely important, since flanking paths via end contacts with walls can short-circuit the floating element's sound impedance and defeat the system. With floating floors it is important that:

1. Mass of the floating floor be large enough to properly spread the loads. Otherwise, the pad will compress and deform sufficiently to transmit the impact.
2. Total construction must be airtight. Airtight is soundtight.
3. Particular care be exercised where partitions rest on the floating floor (see Fig. 27.37).
4. Short-circuits at walls or by penetrations be avoided; see Fig. 27.11*b*. Details of proper construction techniques are given in *A Guide to Airborne, Impact and Structure-Borne Noise Control in Multi-family Dwellings* (see References.)

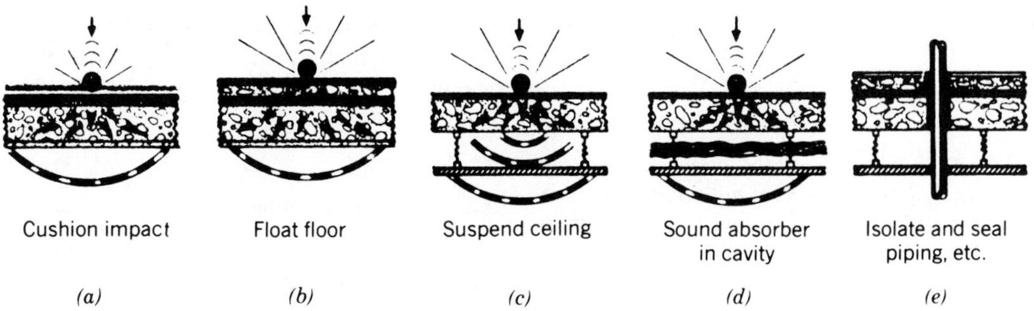

Fig. 27.36 *Methods of controlling impact sound transmission through floors. [Reprinted from* Quieting: A Practical Guide to Noise Control *(1976), p. 50.]*

ACOUSTICS

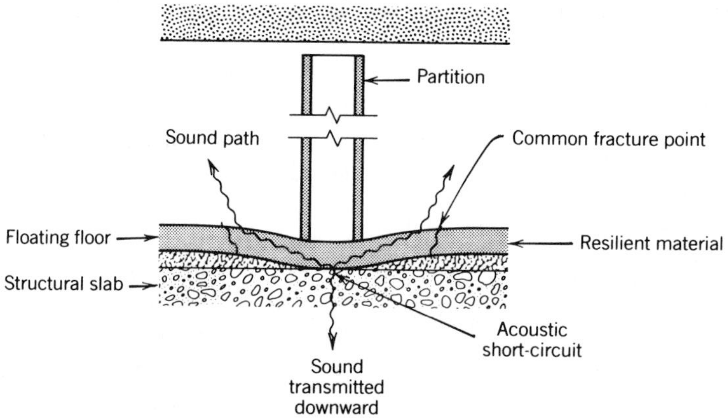

Fig. 27.37 Caution must be exercised when supporting partitions on floating floors to prevent structural failures or short-circuiting of the floating element, as illustrated. [Reprinted from A Guide to Airborne, Impact, and Structure-Borne Noise Control in Multifamily Dwellings *(1968).]*

5. Construction throughout be consistent. Mixed construction types invite flanking noise paths. (See Fig. 27.38).

(c) Suspend the Ceiling—and Use Absorber in Cavity. See Fig. 27.36(*c,d*). As stated, the most disturbing noise is radiated down from the ceiling. A flexibly suspended ceiling with an acoustic absorbent layer suspended in it can be very effective if not flanked by paths leading into the walls and from there reradiating into the space below. It is imperative that the entire floor slab above be decoupled from the walls below by resilient separators.

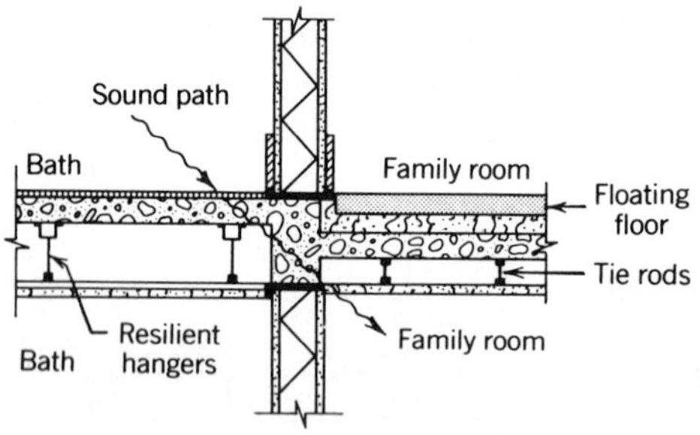

Fig. 27.38 Flanking paths in mixed-construction type floors. FHA does not recommend mixing construction types unless provisions (e.g., expansion joints or breaks in all structural paths between each space) have been made to prevent flanking. [From A Guide to Airborne, Impact, and Structure-Borne Noise Control in Multifamily Dwellings *(1968).]*

(d) Isolate All Piping. See Fig. 27.36*e*. All rigid structures such as piping must be isolated so as not to form a flanking path, and caulked with resilient sealing so as not to constitute an air-sound leak.

27.25 Impact Isolation Class (IIC)

This is a single-number, impact isolation rating for floor construction, similar in intent and derivation to STC wall ratings. Tests are made with a standard tapping machine and the noise levels measured in ⅓-octave bands. These are plotted and compared to a standard contour, exactly as with sound transmission class. Details of typical floor constructions along with IIC ratings are given in Appendix G. Resilient floor finishes on any of the floor constructions not provided with them will add to the IIC ratings approximately as follows:

1/16-in. vinyl tile	0
⅛-in. linoleum or rubber tile	4 ± 1
¼-in. cork tile	10 ± 2
Low-pile carpet on fiber pad	12 ± 2
Low-pile carpet on foam rubber pad	18 ± 3
High-pile carpet on foam rubber pad	24 ± 3

MECHANICAL SYSTEM NOISE CONTROL

27.26 Mechanical Noise Sources

Mechanical devices obviously make noise. And, generally, the more power they consume, the more noise they make. In many of today's buildings, 40% of the total cost is spent on mechanical systems. These systems are located throughout a building.

In most buildings, the primary sources of mechanical noise are the components of the air conditioning and air-handling systems such as fans, compressors, cooling towers, condensers, ductwork, dampers, mixing boxes, induction units, and diffusers. The curve of Fig. 27.39 depicts typical air-handling system noise and indicates the portions of the spectrum produced by each group of components.

Pumps are another source of mechanical noise. Pump noise is frequently transmitted along pipes to remote points.

Elevators, escalators, and freight elevators also introduce mechanical noise into buildings. Escalators and freight elevators pose few problems, since they are localized in a specific area and have low operation speeds. However, elevator car operation is rapid, and it affects large areas. In addition, the motors and controls are located on or above the prime upper floors of a building. Motor, shaftway, and other equipment noise must be properly controlled to prevent annoyance to building tenants located near the shaftways or mechanical penthouses. Vibration isolation of these major components is a specialized problem beyond the scope of this text.

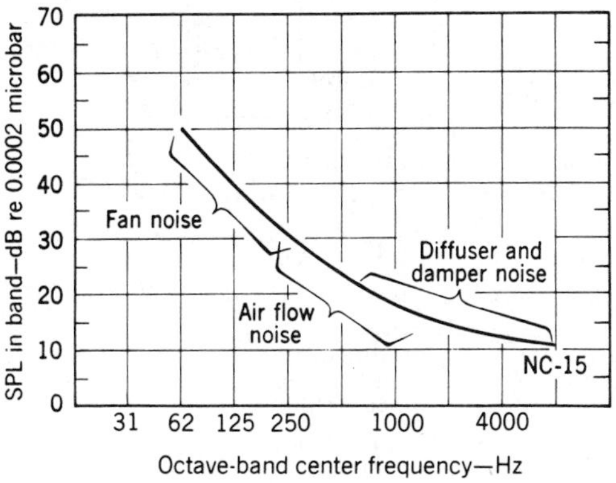

Fig. 27.39 *Frequency spectrum of noise of HVAC system components.*

27.27 Quieting of Machines

Machines cause noise by vibration. This noise is imparted directly to the surrounding air and by vibrational contact to the surrounding structure. Therefore, there are three ways to reduce this noise:

1. Reduce the vibration itself.
2. Reduce the airborne noise by decoupling the vibration from efficient radiating sources.
3. Decouple the vibrating source from the structure.

Refer to Fig. 27.40: Items 1, 3, and 4 reduce vibration; items 4, 5, 6, and 7 reduce and decouple the vibration from the radiating cabinet; and items 2 and 8 decouple the vibrating source from the structure. Once the noise becomes airborne or structure borne, the isolation techniques studied above are employed.

Vibration reduction takes two forms, damping and isolation. Damping is accomplished by rigidly coupling the vibrating source to a large mass, frequently called an inertia block. Much of the energy is absorbed and dissipated as friction; the remainder results in lower-amplitude vibration (see Fig. 27.41). *Isolation* is accomplished by supporting the vibrating mass on resilient supports. These take many forms and are used in tandem. Thus machines are supported on fibrous, rubber, or spring steel vibration isolators, and the entire mass can be supported on a floating floor, which in turn rests on resilient vibration isolators as in Fig. 27.41. Large machines are supported on special commercial "sandwiches" of asbestos, lead, cork, and other strong resilient materials. Piping is supported on cork pads and hung on resilient hangers.

The extensive use of emergency electric generators has caused a serious noise and vibration problem due to the large mass and extremely high noise levels. For such units, complete enclosures are frequently the best approach. Flexible joints in all pipes and ducts connected to vibrating machines are mandatory. This in-

Fig. 27.40 Techniques to reduce the generation of airborne and structure-borne noise in machines and appliances. (Reprinted from A Guide to Airborne, Impact, and Structure-Borne Noise Control in Multifamily Dwellings *(1968).]*

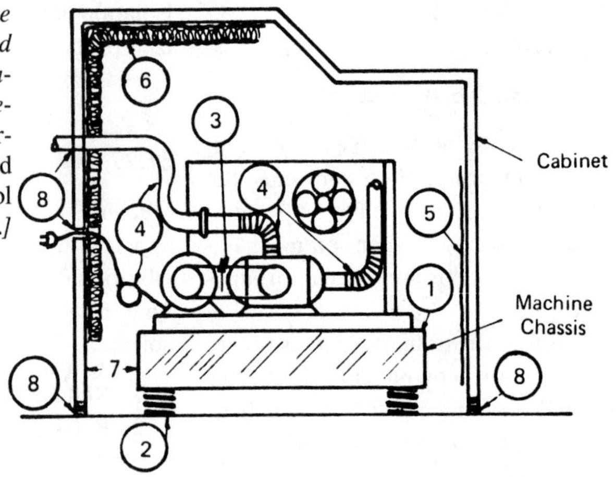

1. Install motors, pumps, fans, etc. on most massive part of the machine.
2. Install such components on resilient mounts or vibration isolators.
3. Use belt drive or roller drive systems in place of gear trains.
4. Use flexible hoses and wiring instead of rigid piping and stiff wiring.
5. Apply vibration damping materials to surfaces undergoing most vibration.
6. Install acoustical lining to reduce noise buildup inside machine.
7. Minimize mechanical contact between the cabinet and the machine chassis.
8. Seal openings at the base and other parts of the cabinet to prevent noise leakage.

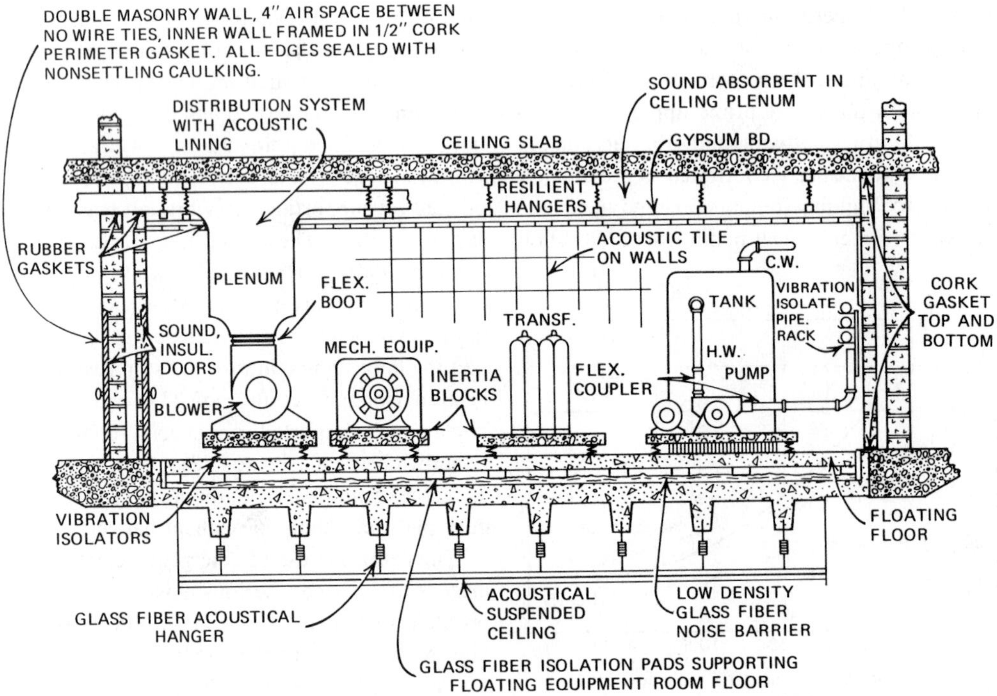

Fig. 27.41 Soundproofing a mechanical equipment room. (Reprinted from A Guide to Airborne, Impact, and Structure-Borne Noise Control in Multifamily Dwellings *(1968).]*

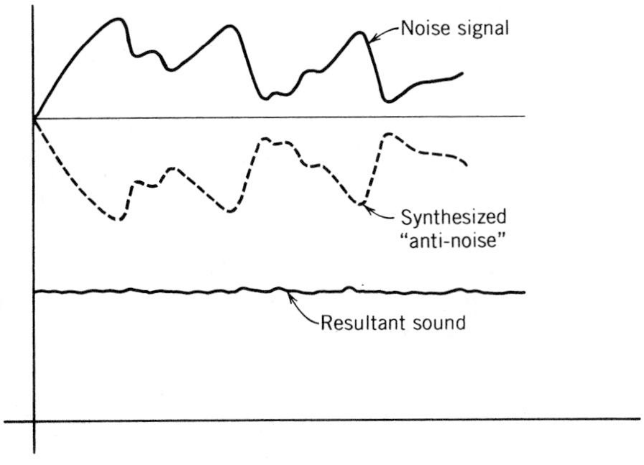

Fig. 27.42 Silencing of noise by introduction of a synthesized noise signal exactly out of phase with the original signal. The resultant sound is effectively zero: that is, silence.

cludes flexible conduit connectors to all motors, transformers, and lighting fixtures using ballasts.

27.28 Active Noise Cancellation

Since noise is a phenomenon consisting of acoustic wave energy being transmitted at certain frequencies, it is theoretically possible to *eliminate* (not mask) noise by simultaneous transmission of identical wave energy exactly out of phase with the noise. The sum of the two energy waves is zero—hence silence (see Fig. 27.42). The problem with this simple solution always has been (and still is) that even with the fastest computers, it still takes a finite amount of time to analyze the noise waveform, synthesize the antinoise, and inject it into the acoustic environment. Since the out-of-phase generated signal lags behind the noise, it cannot completely cancel it. Worse still, if the original noise is of random (i.e., nonpredictable) waveform, it will have changed in content during the synthesizing time lag so that the result of the combined waves is far from silent. Obviously, the faster the noise analysis and antinoise synthesis is accomplished, the more effective will be the partial silencing. Fortunately, however, most mechanical noise is repetitive or periodic because it is caused by repetitive actions—the pulsing of pistons in engines and compressors, the rotation of rotors and fan blădes, the noise caused by mechanically produced fluid movement, the hum of transformers, and so on. That being so, it is a fairly simple (though expensive) procedure to determine the periodicity of the noise, analyze it, synthesize the antiphase waveform, and inject the ''antinoise'' into the *next* cycle, thus eliminating the time lag and effectively canceling the noise and creating silence. A block diagram of such a system is shown in Fig. 27.43.

A particularly important application of active noise cancellation, and one already in use, is in the area of hearing conservation for people exposed to high noise levels at work. Here the out-of-phase noise is introduced into earphones in headsets. This allows the wearer to hear random noise such as speech clearly, while repetitive cyclic noise from engines and the like is can-

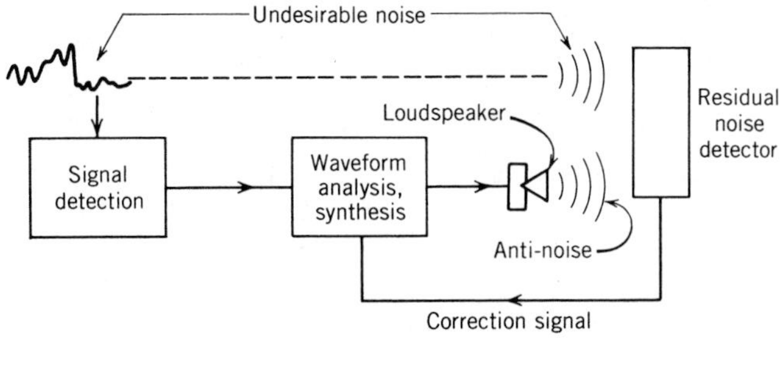

Fig. 27.43 Block diagram of active noise cancellation. The noise signal is detected, its periodicity determined and its waveform analyzed, and an out-of-phase noise is synthesized and injected into the acoustic environment. A residual noise detector provides a feedback signal, which acts to improve noise cancellation.

celed. Another important application of active noise cancellation is in the area of vibration control, since vibration is no more than very low frequency noise.

Active noise canceling systems are commercially available and are coming into increasing use at this writing. Their principal drawback presently is their high cost, but as with most high-tech equipment, costs will probably drop with increasing commercial acceptance.

27.29 Duct System Noise Reduction

Design of quiet duct system operation entails more than specification of duct lining. Air turbulence generates noise. Turbulence and noise levels increase as the velocity of airflow increases. Table 27.9 gives recommended design velocities for various distances from the terminal devices. Velocities should be as low as possible, since sound increases exponentially with velocity. Allowable velocities increase as the distance from the terminal device becomes greater. Return velocities may be slightly higher than supply.

Sound travels as easily against as with the airflow in ductwork. Therefore, both supply and return systems must be lined to control transmission of fan noise. Maximum noise reduction occurs at bends in the ducts. For maximum noise reduction in short runs, a pair of 90° bends often is designed into a system, since lining is not effective in runs shorter than 50 to 60 ft.

Other design approaches that are part of a quiet system include smooth transitions at changes of duct size and large radius bends with turning vanes. Attenuation drops rapidly with increasing size of duct, and therefore ducts should not be deliberately oversized. Crosstalk between rooms and between ducts can be minimized by using lined ducts, separating adjacent ducts as much as possible, and gluing damping material on the outside and lining on the inside. Damping material is particularly effective in preventing the thin metal walls of the ducts from resonating. Mufflers and silencers are effective in reducing fan noise. The pressure drop they introduce must be compensated for in the fan. The *ASHRAE Guide and Data Book*'s chapter on noise control in fans, plenums, housings, and ducts should be consulted for their recommendations. Figure 27.44 shows some of the ways by which cross-talk and flanking noises can be reduced. Figure 27.45 shows some of the techniques employed for quieting duct noise. Active noise cancellation is particularly useful in duct systems since it does not reduce air flow as do liners, baffles, and other mechanical silencing devices.

The increased use of variable air volume (VAV) systems has introduced some noise problems which should not be neglected. VAV system noise can be minimized by following a few basic design rules. Maintain minimum system static pressure since fan noise increases exponentially with static pressure. Select the air volume modulating device at the fan with care as it can be a noise source. Since outlet air volume control involves duct area restriction with attendant velocity increase and resultant noise, the design must include some sort of downstream silencing equipment. Ceiling diffuser acoustic

TABLE 27.9 **Air Speeds in Ducts to Yield NC-15 or NC-25 Background Levels**[a]

	Supply		*Return*	
Location	*NC-15*	*NC-25*	*NC-15*	*NC-25*
Slot speed at min. $\frac{1}{2}$ in. opening	250 fpm	350 fpm	300 fpm	420 fpm
10 ft of duct before opening	300	420	350	490
Next 20 ft	400	560	450	630
Next 20 ft	500	700	570	800
Next 20 ft	640	900	700	980
Next 20 ft	800	1120	900	1260
Next 20 ft	1000	1400	1100	1540
Next 20 ft	1300	1820	1450	2030
Next 20 ft	1600	2240	1800	2520

[a]Ducts with 1- to 2-in. inside duct lining, all duct sizes.

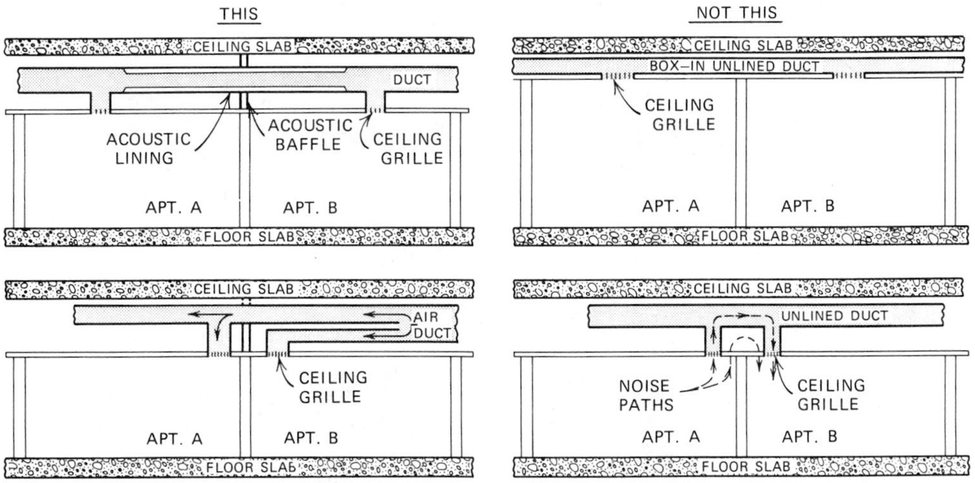

Fig. 27.44 Since ducts are extremely efficient sound transmission paths, considerable precaution must be taken to avoid crosstalk and ventilation, combustion, and equipment noise. Avoid running ducts as a common supply or return between rooms unless they are properly baffled and lined with sound-absorbing material. The common practice, in wood frame structures, of using troughs between joists as a common return duct between dwellings results in very serious noise transmission problems. Caulk or seal around ducts at all points of penetrations through partitions. Use double-wall ducts, acoustical lining, flexible boots, and resilient hangers where required. Dwelling units should be serviced by separate supply and return ducts, which branch off a main duct system. (Reprinted from A Guide to Airborne, Impact, and Structure-Borne Noise Control in Multifamily Dwellings *(1968).]*

characteristics must be coordinated with design air velocity and with any requirement for masking noise. Finally, avoid the use of throttling dampers on ceiling diffusers since a partially closed damper can generate very high noise levels.

27.30 Piping System Noise Reduction

As with airflow, noise increases exponentially with velocity. Piping is not a major noise source normally, since the radiating diameter is small, although flow velocities much in excess of 8 fps, where the pipe is in contact with the structure, can create noise problems, particularly when passing through NC-15 to NC-25 areas (see Table 27.8). Domestic water system mains should be limited to 50 psi in other than tall buildings and pressure in branches to 35 psi. In high-rise structures pressure-reducing valves will be required in high-pressure mains to meet these recommendations. Obviously piping must be designed to prevent water hammer, and noise sources must be located away from quiet areas.

Pumps, as with all rotating equipment, are sources of vibration and noise and should be treated as described in Section 27.27. Figure 27.46 shows a typical pump installation with appropriate noise reduction measures. For at least a distance of 100 pipe diameters beyond the pump, resilient pipe hangers should be used. With centrifugal pumps as with fan and blowers, machine sound concentrates in narrow bands and, if extremely disturbing, can be attenuated with resonant filters. Reciprocating pumps are more difficult to control as the pulsations are more vibration than noise. Flexible connections in the piping and U-joints in the piping will absorb much of this vibration.

27.31 Electrical Equipment Noise

Electrical equipment is generally overlooked as a noise source and this is unwise. Most electrical noise is 120-Hz hum. This can be very disturbing because it is so low a frequency and, as we have noted repeatedly, low-frequency noise is diffi-

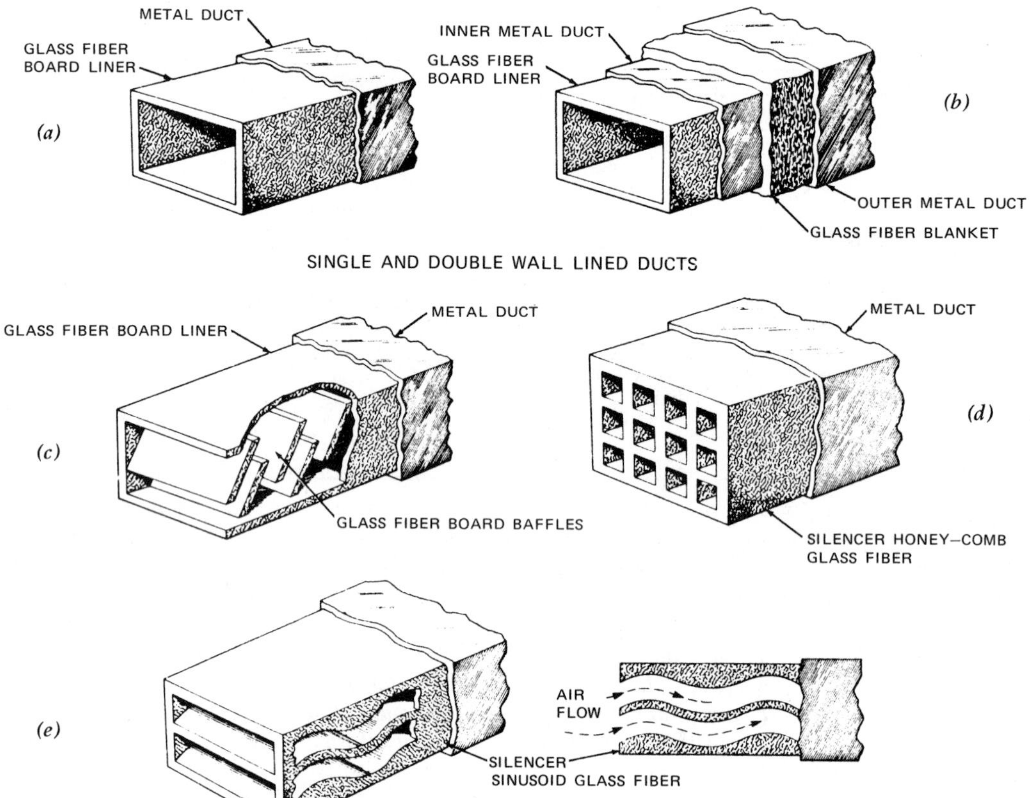

Fig. 27.45 *Unlined duct has negligible sound attenuation and acts as an excellent speaking tube. Inside lining* (a) *gives 2 to 3 db attenuation per foot in the range 1 to 2 kHz, dropping rapidly above and below those frequencies, and giving negligible low-frequency attenuation.* (b) *Double lining gives higher attenuation and reduces crosstalk between ducts. Duct silencers and baffles* (c–e) *give high broadband attenuation: a maximum of 10 to 12 db/ft in the range 1 to 2 kHz, and lower above and below. They are useful to reduce fan noise in short runs but cause considerable pressure drop. [Reprinted from* A Guide to Airborne, Impact, and Structure-Borne Noise Control in Multifamily Dwellings *(1968), Fig. 8-66.]*

cult to attenuate passively. Transformer noise levels are dictated by NEMA and ANSI standards. For a premium price lower noise units are obtainable. Table 27.10 lists maximum noise levels for dry-type units. Most manufacturers warranty noise levels below these. Oil- and silicone-filled units are normally quieter than dry-type transformers, as are units designed for lower temperature rise. Transformer noise can be minimized by these steps:

1. Mount unit on vibration isolators.
2. If transformer is wall hung, use resilient hanging. If it is floor mounted, place on as massive a slab as possible.
3. Locate the unit so that reflections do not amplify the sound. Sound-absorbent material on the walls behind the units is not useful at 120 Hz. Only cavity resonators will absorb appreciable amounts of sound at that frequency.
4. Use only flexible conduit connections.
5. Avoid locating transformers adjacent to, or immediately outside, quiet areas. A common error in this regard is placing a transformer pad immediately below the window of an NC 15–25 area.

The second major source of 120-Hz hum is conventional, core-and-coil discharge light fixture

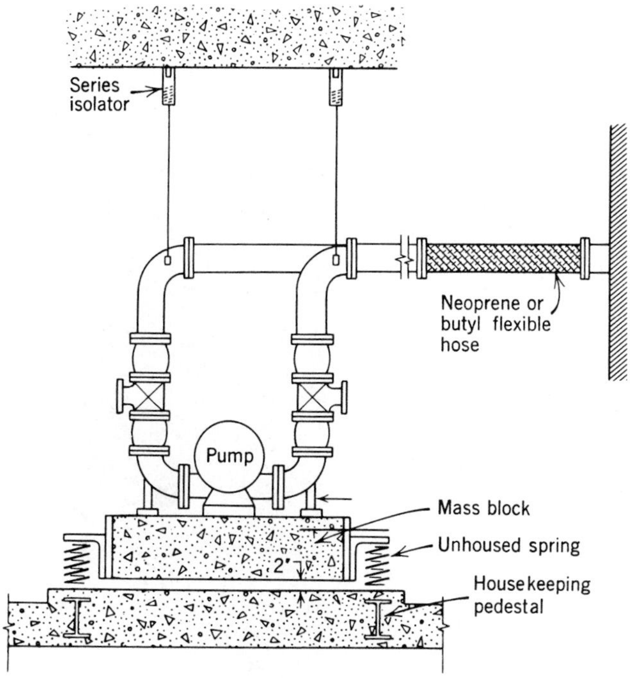

Fig. 27.46 Typical pump installation with appropriate isolation and damping measures.

ballasts. (Electronic ballasts are practically noiseless.) This includes fluorescent plus all the HID sources. Table 27.11 lists the recommended application of conventional fluorescent ballasts. HID ballasts were not a problem until recently when metal–halide, and sodium lamps moved from their former noisy industrial surroundings into quiet office spaces. These ballasts can be very noisy, and care must be exercised in their placement. With all ballasts, the method of mounting has a marked effect on the radiated noise. As pointed out earlier, when a small vibrating source is coupled rigidly to a larger body, noise is amplified because of increased source-to-air coupling. Since fluorescent ballasts for linear lamps are necessarily closely coupled to large metal fixtures for heat dissipation purposes, the sound radiation is much amplified. A large number of fluorescent fixtures mounted in a plenum can create a serious problem. Solution of the problem lies in use of absorptive material in plenums, flexible con-

TABLE 27.10 **Maximum Sound Levels: Dry-Type Transformers**

kVA	*Decibels (NEMA Standard)*
0–9	40
10–50	45
51–150	50
151–300	55
301–500	60

TABLE 27.11 **Acoustic Criteria for Selection of Conventional Core-and-Coil Fluorescent Lamp Ballasts**

For an Installation in	*Use of Ballasts with This Rating Will Usually Be Satisfactory*
TV or radio station, church, synagogue	A
Offices, residence, night school	B
Library, reception or reading rooms, school study hall	C
Noisy office, doctor's or dentist's office, classroom	D

ACOUSTICS

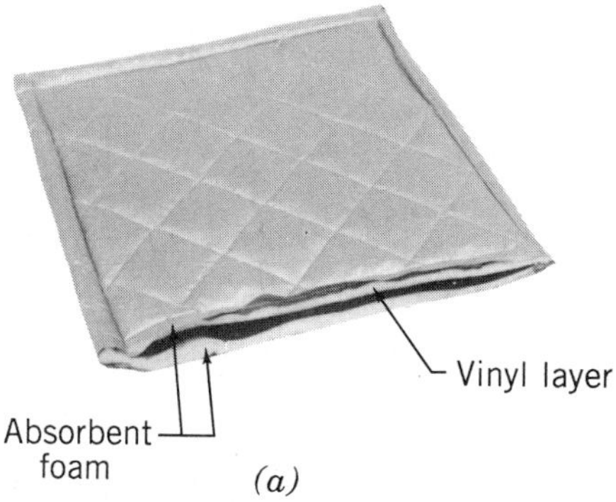

(a)

Fig. 27.47 (a) *Typical flexible nonmetallic composite acoustical barrier material which can be used as a free-hanging curtain or applied to a rigid surface. It consists of a layer of flexible vinyl sandwiched between two layers of fiberglass or other acoustic foam. It is most effective when applied with an airspace between it and any rigid support barrier. See Fig. 27.6 and Section 27.5. (Courtesy of EAR, division of Cabot Corp.)*

(b)

Fig. 27.47 (b) *Composite semirigid acoustical barrier material as used in enclosure panels or as an enclosure lining. The material comprises a thick (1- or 2-in.) foam layer installed facing the noise source, a 1-lb/ft² lead or steel layer (septum), and an outer ¼-in. foam layer, whose function is to isolate the inner foam and the metal from the supporting structure or wall when the material functions as an inner liner. See Fig. 27.47c.*

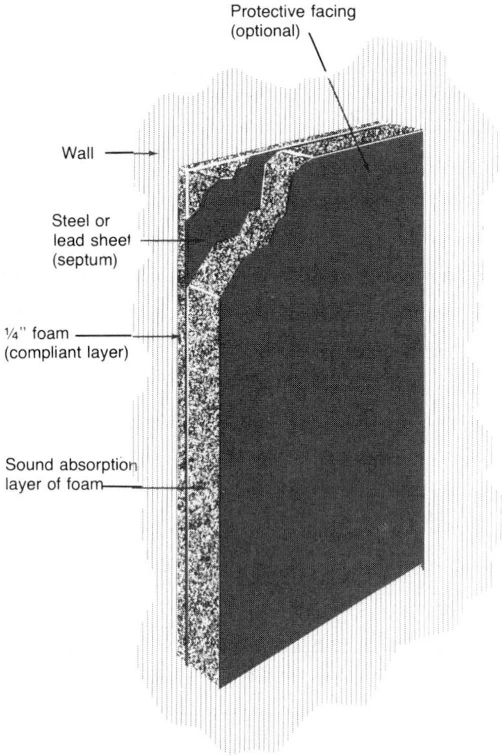

Foam/Metal/Foam

(c)

Fig. 27.47 (c) *Application of composite sound barrier material as an acoustic liner on a nonacoustic partition. [(b, c) Courtesy of United McGill Corporation.]*

duit connection to fixtures, and resilient fixture hanging. In severe cases ballasts can be remote-mounted. HID ballasts are inherently noisier than fluorescent but like ballasts for nonlinear fluorescent lamps are generally less troublesome, being coupled to small radiating bodies.

27.32 Noise Problems Due to Equipment Location

Roof-mounted HVAC units have proven themselves very economical and very noisy. The vibration, short duct runs, and sound reflections are serious problems that can be solved with vibration isolators, sound mufflers, and careful location of equipment. Roof-mounted cooling towers are a particular problem when located adjacent to a taller building. That problem has led to a spate of lawsuits and noise control legislation in many cities. For this reason particular attention should be paid in design to all exterior equipment.

In high-rise buildings, problems are caused by the conflicts between the stringent noise requirements of the prime upper floor space and the near presence of elevator machine rooms, mechanical equipment rooms, and cooling towers. These problems are almost impossible to solve after construction and require the services of an acoustics expert during design.

27.33 Sound Isolation Enclosures and Barriers

In buildings with concentrated high-level noise sources such as certain types of machinery, it is always more desirable to reduce the noise at its source rather than attempt to treat the larger enclosing space. This is most effectively accomplished by enclosing the noise source with materials that will furnish a combination of reverberant noise reduction by absorption as explained in Section 27.7, and blocking of airborne sound with high transmission loss as detailed in Sections 27.12 to 27.17. These materials are available in the form of curtains, panels, and prefabricated partial and full enclosures, tailored to the specific characteristic of the noise source (see Fig. 27.47). These enclosures are not normally the responsibility of the building designer. It is, however, important to know that they will be used, as well as their characteristics, so that appropriate isolation can be designed into the building for the residual sound that is radiated from the enclosures.

The fundamental criteria of airborne sound insulation between dwelling units are, for grade II: (see Section 27.34 for grades)

Wall partitions	STC $\geq$ 52
Floor–ceiling assemblies	IIC $\geq$ 52

These apply where similar function spaces are contiguous, as bedroom to bedroom and living room to living room. Where this is not the case, the isolation must be revised up or down to meet the maximum sensitivity requirement.

RECOMMENDATIONS AND CRITERIA

Recommendations for background noise levels (NC criteria) are given in Table 27.8. Criteria for partition isolation (STC) and impact isolation (IIC) are given in the following sections and in Table 27.12.

27.34 Multioccupancy Residential STC/IIC Criteria

The most important and binding criteria for residential work in the United States are issued by the Department of Housing and Urban Development in conjunction with the Federal Housing Administration (HUD/FHA). The reader is referred to the latest issue of *A Guide to Airborne, Impact and Structure-Borne Noise Control* and to any subsequent HUD/FHA publications. Tables 27.13 and 27.14 give the essential data presented there in the current edition.

The recommendations are divided into grades I, II, and III. Grade II is the most important category and is applicable primarily in residential urban and suburban areas considered to have the "average" noise environment. The nighttime exterior noise levels might be about 40 to 45 dbA, and the permissible interior noise environment should not exceed NC 25–30 characteristics. Grade I is suburban, with a "quiet" noise environment characterized by a nighttime exterior noise level of about 35 to 40 dbA. *Grade I STC/IIC criteria are three points higher than those of grade II.*

Grade III recommendations are minimal and can be characterized as "noisy," with an average nighttime exterior noise level of about 55 dbA or higher. *Grade III STC/IIC recommendations are four points lower than those of grade II.*

27.35 Specific Occupancies

(a) Schools. School buildings house spaces of many kinds—classrooms, auditoriums, gym-

TABLE 27.12 **Recommended STC for Partitions; Specific Occupancies**

Type of Occupancy	*Wall, Partition, or Panel Between* Room Being Considered AND	Adjacent area	*Sound Isolation Requirement: Background Level in Room Being Considered* Quiet	Normal
Normal school buildings without extraordinary or unusual activities or requirements	Classrooms	Adjacent classrooms	STC 42	STC 40
		Corridor or public areas	STC 40	STC 38
		Kitchen and dining areas	STC 50	STC 47
		Shops	STC 50	STC 47
		Recreation areas	STC 45	STC 42
		Music rooms	STC 55	STC 50
		Mechanical equipment rooms	STC 50	STC 45
		Toilet areas	STC 45	STC 42
	Music practice rooms	Adjacent practice rooms	STC 55	STC 50
		Corridor and public areas	STC 45	STC 42
Executive areas, doctors' suites; confidential privacy requirements	Office	Adjacent offices	STC 50	STC 45
		General office areas	STC 48	STC 45
		Corridor or lobby	STC 45	STC 42
		Washrooms and toilet areas	STC 50	STC 47
Normal office; normal privacy requirements	Office	Adjacent offices	STC 40	STC 38
		Corridor, lobby, exterior	STC 40	STC 38
		Washrooms, kitchen, dining	STC 42	STC 40
Any occupancy, using rooms for group meetings	Conference rooms	Other conference rooms	STC 45	STC 42
		Adjacent offices	STC 45	STC 42
		Corridor or lobby	STC 42	STC 40
		Exterior of building	STC 40	STC 38
		Kitchen and dining areas	STC 45	STC 42
Large offices, drafting areas, banking floors, etc.	Large general office areas	Corridors, lobby, exterior	STC 38	STC 35
		Data-processing area	STC 40	STC 38
		Kitchen and dining areas	STC 40	STC 38
Motels and urban hotels, hospitals and dormitories	Bedrooms	Adjacent bedrooms[a]	STC 52	STC 50
		Bathroom[a]	STC 50	STC 45
		Living rooms[a]	STC 45	STC 42
		Dining areas	STC 45	STC 42
		Corridor, lobby, or public spaces	STC 45	STC 42

[a]Separate occupancy.

Source: Courtesy of U.S. Gypsum.

nasiums, cafeterias, shop areas, and music suites—that pose acoustics problems.

1. *Auditoriums*. All auditoriums require a sound system for some of the activities accommodated (see Figs. 26.35 and 26.38). The most difficult aspect, architecturally, is integration of the loudspeaker system into the design. To provide proper sound reinforcement, loudspeakers must be located properly without obstructions. To accomplish this, the loudspeaker system should be incorporated in the earliest drawings.

TABLE 27.13 **Criteria for Airborne Sound Insulation of Wall Partitions Between Dwelling Units**

Partition Function Between Dwellings			
Apt. A		*Apt. B*	*Grade II STC*
Bedroom	to	Bedroom	52
Living room	to	Bedroom[a]	54
Kitchen[b]	to	Bedroom[a]	55
Bathroom	to	Bedroom[a]	56
Corridor	to	Bedroom[a,c]	52
Living room	to	Living room	52
Kitchen[b]	to	Living room[a]	52
Bathroom	to	Living room	54
Corridor	to	Living room[a,c,d]	52
Kitchen	to	Kitchen[e]	50
Bathroom	to	Kitchen	52
Corridor	to	Kitchen[a,c,d]	52
Bathroom	to	Bathroom	50
Corridor	to	Bathroom[a,c]	48

[a]Whenever a partition wall might serve to separate several functional spaces, the highest criterion must prevail.

[b]Or dining or family or recreation room.

[c]It is assumed that there is no entrance door leading from corridor to living unit.

[d]Criterion applies to the partition. Doors in corridor partitions must have the rating of the partition, not vice versa.

[e]Double wall construction is recommended to minimize kitchen impact noises.

Source: Reprinted from *A Guide to Airborne, Impact, and Structure-borne Noise Control in Multifamily Dwellings* (1968). For Grade I, add 3 points; for Grade III, subtract 4 points.

In general, a school auditorium must be a multipurpose facility. It should be designed to meet speech requirements and also should be suitable for the school's music activities.

Often a modified gymnasium (gymnatorium) or cafeteria (cafetorium) functions as an auditorium. Obviously, acoustics compromises occur in such facilities. The large areas of sound-absorbing treatment in either kind of space make them unsuitable as auditoriums and most events that require speech amplification.

2. *Classrooms.* Typical classrooms are approximately 30 ft square with 10-ft ceilings. Adequate speech communication is easily achieved in a room of this size. Classroom acoustic design usually involves:

a. Locating sound-absorbing treatment to reduce classroom noise levels.
b. Ensuring adequate privacy between adjacent spaces.
c. Control of air-handling system noise.

Acoustic tile ceilings yield adequate sound absorption for most classrooms. An NRC of 0.7 is recommended (see Section 27.6).

Partition systems must produce sufficient isolation to prevent disturbance from activities in other classrooms and corridors. Such partitions should run full height from floor to ceiling slab or roof construction. If return air transfer ducts are needed, their noise reduction characteristics must be as good as the walls or doors that they penetrate (for NC data, see Table 27.8). Unit ventilators commonly used for classrooms produce approximately the required level of background noise.

3. *Music Suites.* School music programs usually range from individual instruction to band and choral concerts. The teaching spaces required for such a program include practice rooms, ensemble rooms, and large rehearsal spaces. Both room acoustics design and sound isolation are important in music suites. Privacy between adjacent spaces is critical, since simultaneous use is necessary.

4. *Dining Areas.* The activity in cafeterias or lunchrooms generates noise. The kitchen and serving areas should be separated from the eating spaces. Treat ceilings and wall areas in the cafeteria with sound-absorbing material. Unless the ceiling is completely treated with a highly efficient sound absorbing material, the environment will be unsatisfactory due to its high noise level. Minimum NRC should be 0.8.

5. *Gymnasiums.* Activities in gymnasiums create so much noise that even extensive treatment will not quiet these spaces. A quiet gymnasium probably would be unsatisfactory, since spectators are conditioned to consider the

TABLE 27.14 **Criteria for Airborne and Impact Sound Insulation of Floor–Ceiling Assemblies Between Dwelling Units**

Partition Function Between Dwellings			*Grade II*	
Apt. A		*Apt. B*	*STC*	*IIC*
Bedroom	above	Bedroom	52	52
Living room	above	Bedroom[a]	54	57
Kitchen[b]	above	Bedroom[a,c]	55	62
Family room	above	Bedroom[a,d]	56	62
Corridor	above	Bedroom[a]	52	62
Bedroom	above	Living room[e]	54	52
Living room	above	Living room	52	52
Kitchen	above	Living room[a,c]	52	57
Family room	above	Living room[a,d]	54	60
Corridor	above	Living room[a]	52	57
Bedroom	above	Kitchen[c,e]	55	50
Living room	above	Kitchen[c,e]	52	52
Kitchen	above	Kitchen[c]	50	52
Bathroom	above	Kitchen[a,c]	52	52
Family room	above	Kitchen[a,c,d]	52	58
Corridor	above	Kitchen[a,c]	48	52
Bedroom	above	Family room[e]	56	48
Living room	above	Family room[e]	54	50
Kitchen	above	Family room[e]	52	52
Bathroom	above	Bathroom[c]	50	50
Corridor	above	Corridor	48	48

[a]This arrangement requires greater impact sound insulation than the converse, where a sensitive area is above a less sensitive area.

[b]Or dining or family or recreation room.

[c]It is assumed that plumbing fixtures, appliances, and piping are installed with proper vibration isolation.

[d]The airborne STC criteria in this table apply as well to vertical partitions between these two spaces.

[e]This arrangement requires equivalent airborne sound insulation and perhaps less impact sound insulation than the converse.

Source: Reprinted from *A Guide to Airborne, Impact, and Structure-borne Noise Control in Multifamily Dwellings* (1968). For Grade I, add 3 points; for Grade II, subtract 4 points.

"noise" as an enjoyable aspect of athletic events. However, to provide a proper environment for normal sports activities, the ceiling area should be sound absorbing. In addition, if a sound amplification system is to be used, sound-absorbing wall treatment may be required to eliminate echoes that would reduce intelligibility of announcements. An NRC of 0.7 is suggested with sound-absorbent material to be ceiling mounted.

If a gymnasium will also serve as an auditorium, loudspeaker system placement needs special consideration. For example, the loudspeakers should be located above the source location for speeches and plays.

6. *Swimming Pools*. The acoustic environment of swimming pools is often chaotic. Most sound-absorbing materials disintegrate in the high-humidity conditions prevalent in pool areas. Use special sound-absorbing units that have moisture-resistant properties.

7. *Shops*. Metal, woodworking, and scenery shops in schools contain many noise sources—saws, planers, drill presses, and manual tools. Each generates high airborne and structure-borne noise levels. Consolidating noisy areas and maximizing the distance between them and quiet spaces are essential. Ceiling and wall absorptive treatment with an NRC of at least 0.75 is recommended.

(b) Churches. The basic activities of most churches combine speech and music. Thus, the church environment must be acoustically hospitable to both. The architectural plan also must respond to religious requirements including the relative positioning of pulpits, lecterns, the altar, and choir.

Successful church acoustics can be achieved by designing the overall environment for music and providing special assistance for speakers. Figure 27.48 illustrates a church that is adequately reverberant for music and includes a sound-reflecting canopy over the pulpit to direct the minister's voice to the congregation. In some larger churches, a loudspeaker located above the canopy further reinforces speech from the pulpit. The choir and organ communicate with the entire volume of the church and, therefore, benefit from the reverberant environment.

(c) Offices. Although office buildings contain public spaces, auditoriums, and restaurants, prime occupancy is in office areas. Most acoustics problems in office buildings relate to privacy—either between spaces within a single firm or between adjacent firms. Speech privacy has been discussed at length in Section 27.22, including consideration of open-plan offices. Mechanical equipment noise problems are discussed in Sections 27.27 to 27.32.

(d) Apartment Buildings. Large apartment buildings sometimes house thousands of residents. Privacy and freedom from annoyance are high on the list of tenant requirements. See HUD/FHA criteria in Section 27.34.

The performance of the partitions is compromised in many designs by careless planning of convenience outlets, medicine cabinets, and mechanical services. Moreover, direct-exhaust duct connections between apartments and back-to-back placement of medicine cabinets result in loss of privacy. Back-to-back convenience outlets must be avoided.

Installation of rugs or carpeting provides the best protection against footfall noise. Some leases now require that a tenant provide such impact-reducing floor covering over most of the floor area in his apartment. Good design also dictates that similar spaces in adjacent apartments be grouped—bedrooms next to bed-

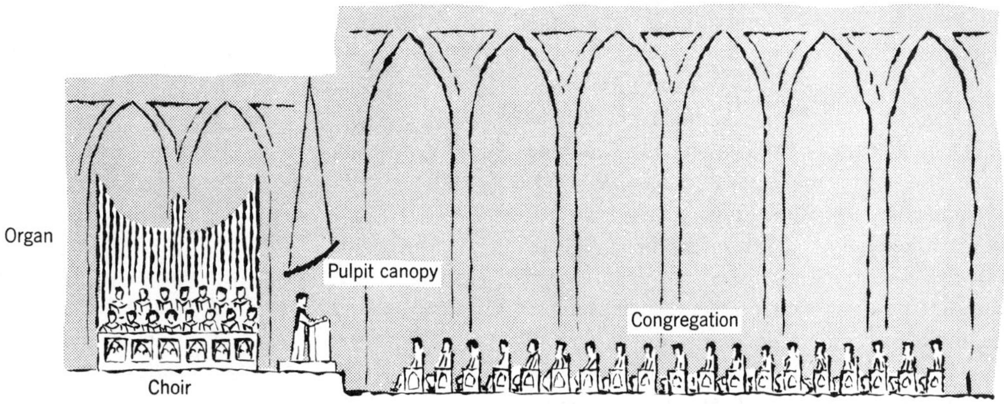

Fig. 27.48 *Church with sound-reflecting pulpit canopy.*

rooms, for example. Absorptive material in bedrooms should be ceiling mounted. An NRC of 0.6 is recommended.

Apartment house site selection seldom includes consideration of acoustics. Nevertheless, truck routes, superhighways, and airports can be annoying "neighbors." As mentioned above, cooling towers serving adjacent buildings must be considered during planning stages.

OUTDOOR ACOUSTIC CONSIDERATIONS

27.36 Sound Power and Pressure Levels in Free Space (Outdoors)

The equations of Section 26.15 are not applicable to outdoor propagation, in which the large reflective component of the indoor condition is absent. Although the propagation of sound outdoors may not appear to be of immediate importance in architectural acoustics, outdoor noise sources such as traffic, cooling tower, and aircraft are frequently loud enough to disturb activities within or immediately adjacent to a building. Conversely, the noise made by building equipment such as cooling towers, heat pumps, and even window air conditioners may be loud enough to disturb neighbors in a nearby building. For this reason it is desirable to have some basic understanding of outdoor sound propagation.

For preliminary evaluation of an outdoor noise problem, assuming a small nondirectional source on the ground, SPL can be determined from equations 26.13 and 26.14. For large sources such as cooling towers and traffic, which do not exhibit inverse square properties, sound level estimates are best made on the basis of experience and empirical data beyond the scope of this book (see Magrab, 1975; Schaudinischky, 1976). For small outdoor sources, the equipment power level can be estimated by measuring the sound power level at 5 ft and adding 15 db. Other factors, such as moisture in the air, the presence of trees, wind, and temperature gradients, will affect outdoor sound propagation to some extent, but they can be ignored except when great distances (i.e., over 1000 ft) are involved. Barriers, which are most effective outdoors, were discussed in Section 27.19.

27.37 Building Siting

As important as interior structural design is building siting vis-à-vis exterior noise sources. See also Section 3.5 and Figs. 3.18 and 3.19. Since this subject is somewhat beyond our scope, the discussion is brief. Buildings should be sited, with respect to noise sources:

1. To use natural terrain noise barriers (Fig. 27.49*a*).
2. With respect to trees as noise barriers, rely only on thick wooded areas (Fig. 27.49*b*).
3. To avoid naturally poor sites (Fig. 27.49*c*).
4. To avoid sound reflection from other buildings (Fig. 27.49*d*).

Factor 4 is also important in a multiwing building; in avoiding U shapes or other configurations where a central court becomes an echo chamber.

Where avoidance of an exterior noise source is impossible, quiet zones can be buffered from the noise by placing higher noise areas on the noisy side of the building. Thus, in a school, classrooms and offices can be buffered by a cafeteria and gym; in a residence, bedrooms by living rooms and corridors; in an office building, private offices by noisier clerical offices; and so on.

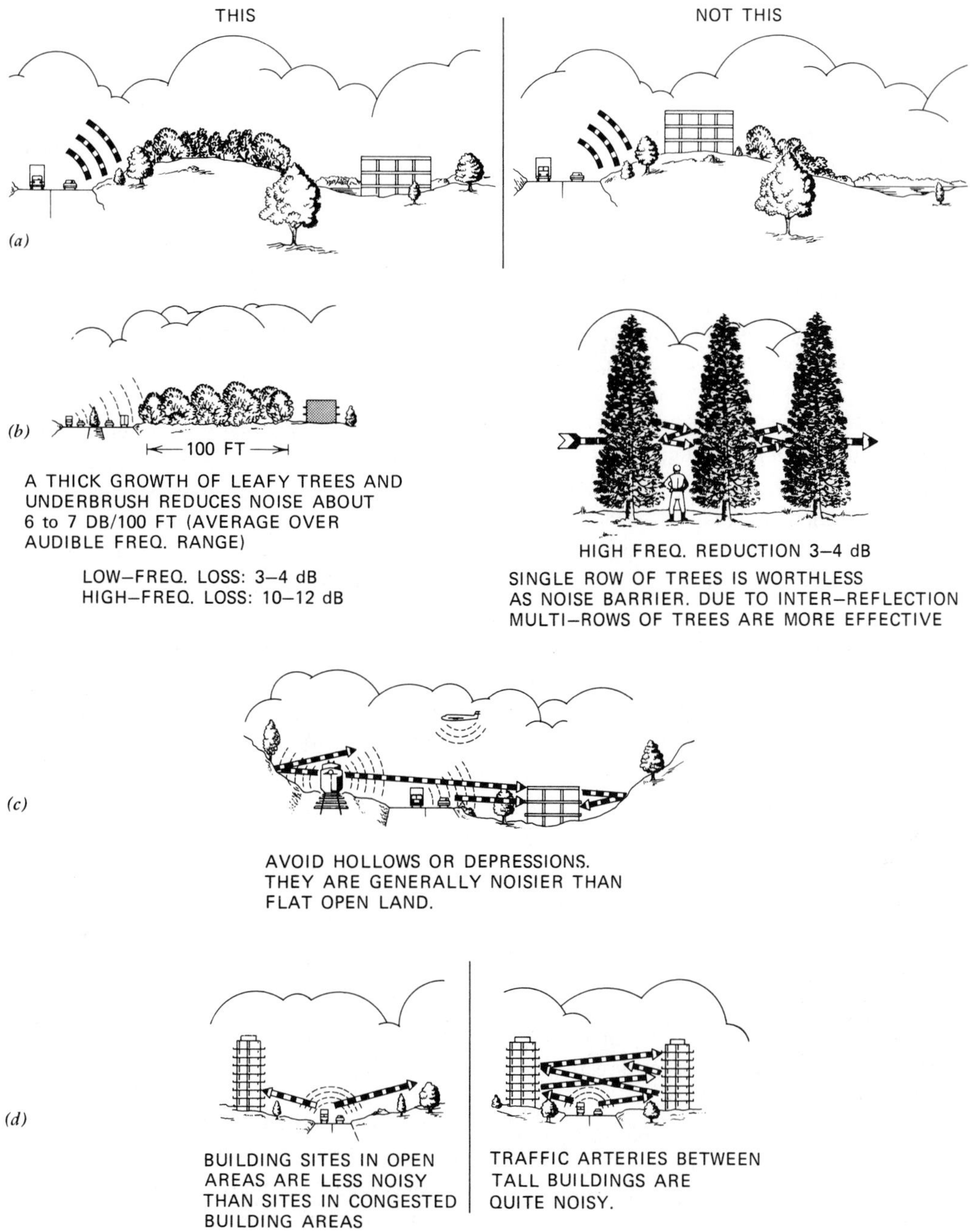

THIS
NOT THIS
(a)
(b)
100 FT
A THICK GROWTH OF LEAFY TREES AND UNDERBRUSH REDUCES NOISE ABOUT 6 to 7 DB/100 FT (AVERAGE OVER AUDIBLE FREQ. RANGE)
LOW–FREQ. LOSS: 3–4 dB
HIGH–FREQ. LOSS: 10–12 dB
HIGH FREQ. REDUCTION 3–4 dB
SINGLE ROW OF TREES IS WORTHLESS AS NOISE BARRIER. DUE TO INTER–REFLECTION MULTI–ROWS OF TREES ARE MORE EFFECTIVE
(c)
AVOID HOLLOWS OR DEPRESSIONS. THEY ARE GENERALLY NOISIER THAN FLAT OPEN LAND.
(d)
BUILDING SITES IN OPEN AREAS ARE LESS NOISY THAN SITES IN CONGESTED BUILDING AREAS
TRAFFIC ARTERIES BETWEEN TALL BUILDINGS ARE QUITE NOISY.

Fig. 27.49 (a) *Use of natural noise barriers.* (b) *Effectiveness of wooded areas as noise barriers. Noise reduction of trees.* (c) *An example of a poor building site.* (d) *Building sites near traffic arteries and other buildings. [Reprinted from* A Guide to Airborne, Impact, and Structure-Borne Noise Control in Multifamily Dwellings *(1968), p. 52.]*

REFERENCE MATERIAL

27.38 Definitions

A-Scale. Identifies the filtering system that has characteristics that roughly match the response characteristics of the human ear. Referred to as dbA.

Absorption Coefficient (α). The ratio of the sound absorbed to the sound incident on the material or device.

Anechoic Room. An anechoic room provides a free field acoustic testing environment like the out-of-doors. All the sound emanating from a source is essentially absorbed at the surfaces of the room.

Baffle or Barrier, Sound. A shielding structure or partition used to increase the effective length of a sound transmission path between two locations.

Critical Frequency. The lowest frequency at which the wavelength of a bending wave, traveling in a structure, is the same as the wavelength in air at that frequency.

Damping. Dissipation of structure-borne noise. For the most part, this is accomplished by using a material with a high internal energy-absorbing capacity (i.e., high internal damping).

Decibel (db). See "Sound Pressure Level."

Diffuse Sound Field. A sound field that at any given point is made up of sound waves of all angles of incidence.

Direct Field. The sound in a region in which all or most of the sound arrives directly from the source without reflection.

Flanking Transmission. The transmission of sound or noise from one room to another by indirect paths, rather than directly through an intervening partition.

Free Sound Field (Free Field). A field in a homogeneous, medium free from boundaries. In the free field, the sound pressure level decreases 6 db for a doubling of distance from a point source.

Impact Isolation Class (IIC). A single-figure rating that provides an estimate of the impact sound-isolating performance of a floor–ceiling assembly.

Intensity Level (IL). A measure of the acoustic power passing through a unit area expressed on a decibel scale referenced to some standard (usually 10^{-12} W/cm^2).

Loudness. The subjective human definition of the intensity of a sound.

Masking. The presence of a background noise increases the level to which a sound signal must be raised in order to be heard or distinguished. If the level of background noise is significantly higher than that of the sound signal, the signal cannot be heard. This effect is known as masking.

Mass Law. States that the transmission loss of walls (in a part of frequency range) is controlled entirely by the mass per unit area of the panel. Also that the transmission loss increases 6 db for each doubling of frequency or each doubling of the panel mass per unit area.

Noise. Any undesired sounds, usually of different frequencies, resulting in an objectionable or irritating sensation.

Noise Reduction (NR). (1) The reduction in sound pressure level caused by making some alteration to a sound source. (2) The difference in sound pressure level measured between two adjacent rooms caused by the transmission loss of the intervening wall.

Octave Band (OB). A range of frequency where the highest frequency of the band is double the lowest frequency of the band. The band is usually specified by the center frequency.

Phon. Loudness level, at a particular frequency, equal to the 1000-Hz decibel level of that equal-loudness contour.

Pink Noise. Wide frequency spectrum noise, whose amplitude drops 3 dB per octave with increasing frequency (equal energy per octave). Useful for masking.

Reverberation. The persistence or echoing of previously generated sound caused by reflection of acoustic waves from the surfaces of enclosed spaces.

Sabin. The unit of acoustic absorption. One sabin (ft^2) (m^2) is the absorption of 1 square foot (meter) of perfect sound-absorbing material.

Sound-Level Meter. An instrument for the direct measurement of sound pressure level. Sound-level meters may also incorporate octave-band filters for measuring sound directly in octave bands.

Sound Power Level (PWL). A measure of the total airborne acoustic power generated by a noise source, expressed on a decibel scale referenced to some standard (usually 10^{-12} W).

Sound Pressure Level (SPL). A measure of the air pressure change caused by a sound wave. Expressed on a decibel scale referenced to some standard (usually 0.0002 μbar).

Transmission Loss (TL). The reduction of airborne sound power that is caused by placing a wall or barrier between the reverberant sound field of a source and its receiver. Transmission loss is a property of the wall or barrier.

White Noise. Noise of wide frequency range in which the amplitude of the noise is essentially the same in all frequency bands (equal energy per frequency band).

27.39 Units and Conversions

See Table 27.15.

TABLE 27.15 **Acoustic Units and Conversions**

Units	*MKS*	*CGS*
Force	kilogram-meter/s^2 = newton	gram-cm/s^2 = dyne
Intensity	watts/meter^2	watts/cm^2
Pressure	newtons/meter^2 = pascals	dynes/cm^2 = microbars

Quantity	*In Conversion: Multiply*	*By*	*To Obtain*
Force	newtons	10^5	dynes
	dynes	10^{-5}	newtons
Intensity	watts/cm^2	10^4	watts/m^2
	watts/m^2	10^{-4}	watts/cm^2
Pressure	pascals	10	microbars
	microbar	10^{-1}	pascals

NOTE. One atmosphere = 1 bar = 10^6 μbar.

27.40 Symbols

See Table 27.16.

REFERENCES

A Guide to Airborne, Impact, and Structure-Borne Noise Control in Multifamily Dwellings (1968). U.S. Department of Housing and Urban Development, Washington, D.C.

Application of Sound Power Level Ratings (1973). Publication 303, AMCA, Arlington Heights, Ill.

Beranek, L. I. (1949). Acoustic Measurements, Wiley, New York.

Beranek, L. L. (1988). *Noise and Vibration Control*, rev. ed., INCE, Washington, D.C.

Broch, J. T. (1971). *Application of the B&K Measuring Systems to Acoustic Noise Measurements*, 2nd ed., Naerum, Denmark.

Brüel, P. V. (1951). *Sound Insulation and Room Acoustics*, Chapman & Hall, London.

Brüel, P. V., and Kjaer (1963). *Application of B&K Equipment to Architectural Acoustics*, Denmark.

Brüel, P. V., and Kjaer (1963). *Application of Sound Power*, Denmark.

Brüel, P. V., and Kjaer. *Measuring Sound*, Denmark.

Cavanaugh, W. J., Farrell, W. R., Hirtle, P. W., and Watters, B. G. (1962). "Speech Privacy in Buildings," *Journal of the Acoustical Society of America*, Vol. 34, No. 4.

Embelton, T. (1971). "Absorption Coefficients of Surfaces Calculated from Decaying Sound Fields," *Journal of the Acoustical Society of America*, Vol. 50, No. 3.

Fader, B. (1981). *Industrial Noise Control*, Wiley, New York.

Fundamentals of Noise (1971). Report NTID 300.15, U.S. Environmental Protection Agency, Washington, D.C., pp. 55–56.

Harris, C. (1979). *Handbook of Noise Control*, 2nd ed. McGraw-Hill, New York.

Information on Levels of Environmental Noise (1974). Report 550/9-74-004, U.S. Environmental Protection Agency, Washington, D.C.

Magrab, E. B. (1975). *Environmental Noise Control*, Wiley, New York.

Pearsons, K. S., and Bennet, R. L. (1974). *Handbook of Noise Rating*, National Aeronautics and Space Administration Report

ACOUSTICS

TABLE 27.16 **Symbols and Abbreviations Used Commonly in Acoustics**

A =	Total absorption, sabins; area in square inches or square feet
A_R =	Absorption in receiving room, sabins
$A_{1,2\cdots}$ =	Total absorption of each different material in a space, sabins
c =	Velocity of sound, feet per second
d =	Distance from source, meters or feet
db =	Decibel
f =	Frequency of sound, hertz (cps)
I =	Intensity, watts per square centimeter
I_a =	Absorbed energy, watts per square centimeter
I_i =	Incident energy, watts per square centimeter
I_0 =	Reference intensity, 10^{-16} watt per square centimeter
IIC =	Impact insulation class, no units
IL =	Intensity level, decibels
NC =	Noise criterion, no units
NRC =	Noise reduction coefficient, no units
NR =	Noise reduction, decibels
p =	Pressure, pascals or microbars
p_0 =	Reference base pressure 2×10^{-5} Pa
P =	Acoustic power, watts
Pa =	Pascal, unit of pressure (SI)
PWL =	Sound pressure level, decibels
R =	Room constant, square feet
r =	Distance from source, meters or feet
S =	Surface area, square feet
SPL =	Sound pressure level, decibels
STC =	Sound transmission class, no units
T_R =	Reverberation time, seconds
TL =	Transmission loss, decibels
V =	Volume, cubic feet
W,P =	Sound power, watts
W_0 =	Reference base sound power 10^{-12} watt
a, α =	Absorption coefficient (no units)
$\bar{a}$, $\bar{\alpha}$ =	Average absorption coefficient (no units)
λ =	Wavelength, feet or meters
Σ =	Sum of, or total (no units)
$\Sigma S_\alpha = \Sigma A$ =	Total absorption, sabins
Δ =	Change in a quantity or difference between two quantities

NOTE. Where definitions are expressed in feet, centimeters or meters is also understood, with proper conversion factors, and vice versa.

NASA CR-2376, Washington, D.C., pp. 55–56.

Practical Building Acoustics (1976). Sound Research Labs, Holstead, England.

Procedure for Computation of Loudness of Noise. (1968). ANSI, USAS 53.4, New York.

Quieting: A Practical Guide to Noise Control (1976). National Bureau of Standards Handbook 119, NBS, Washington, D.C.

Schaudinischky, L. H. (1976). *Sound, Man and Building,* Applied Science Publishers, London.

Schiff, M. L. (1975). "Noise Control for Draft Fans," *Specifying Engineer*, pp. 62–70.

Terminology, Laws and Application of Fan Sound Data (1972). Publication SD100-1972, Barry Blower Co.,

Ward, W. D., et al. (1968). *Proposed Damage-Risk Criterion for Impulse Noise*, Report of Working Group 57, NAS-NRC Committee on Hearing, Bioacoustics and Biomechanics (CHABA), Washington, D.C.

White, F. A. (1975). *Our Acoustic Environment*, Wiley, New York.

Young, R. W. (1965). "Revision of the Speech Privacy Calculation," *Journal of the Acoustical Society of America*, Vol. 38, No. 4.

PART X
APPENDICES

The purpose of this part is to present extensive important reference and ancillary data in a separate, easily accessed section. The appendices are entitled:

A Climate Conditions for the United States, Canada, and Mexico

B Solar Data

C Annual Solar Performance

D Sunshading

E Economic Analysis

F Sound Transmission Data for Walls

G Sound Transmission and Impact Insulation Data for Floor/Ceiling Construction

H Options for Computation Assistance

I Metrication; SI Units; Conversions

J Building Automation; Intelligent Buildings

K ASHRAE Standards for Energy Efficient Building Design

A

CLIMATIC CONDITIONS FOR THE UNITED STATES, CANADA, AND MEXICO

Prepared by ASHRAE Technical Committee 4.2, Weather Data, from data compiled from official weather stations where hourly weather observations are made by trained observers.

Station Designations

AP = airport

CO = urban areas (influenced by surroundings)

No designation = semirural, similar to airport locations

S = solar data available from National Climatic Data Center

Winter Design Conditions

For United States: percentage of the three-month period, December through February.

For Canada: percentage for January.

Prevailing winter wind lists the direction from which the wind occurs most frequently with the 97.5% winter design dry-bulb temperature.

Summer Design Conditions

For United States: percentage of the four-month period, June through September.

For Canada: percentage for July.

Prevailing summer wind lists the direction from which the wind occurs most frequently with the 2.5% summer design dry-bulb temperature.

Other stations: Many more stations in each state or province are found in the *ASHRAE 1989 Handbook of Fundamentals*, Chapter 24.

Interpretations Between Stations. As a general rule, weather data from listed stations can be adjusted for nearby locations in these ways:

*Adjustment for elevation.** For lower elevations, which tend to be warmer, *increase* the design temperatures; for higher elevations, *decrease* the design temperatures:

Dry-bulb temperature: 1F° per 200 ft of elevation

Wet-bulb temperature: 1F° per 500 ft of elevation

Adjustment for air mass (at coastlines). Along the west coast, both dry-bulb and wet-bulb design temperatures increase with distance from the ocean. Along the Gulf coast, dry-bulb temperatures increase for the first 200 to 300 miles inland, while wet-bulb temperatures decrease slightly. (Beyond this 200- to 300-mile belt, both dry-bulb and wet-bulb values decrease.)

* Elevations for many locations are given in Appendix B, Table B.15.

TABLE A.1 **Outside Design Conditions: United States and Canada**

		Winter	*Summer*			
State and Station[a]	*Prevailing Wind*	*Design Dry-Bulb (97.5%)*	*Design Dry-Bulb and Mean Coincident Wet-Bulb (2.5%)*	*Mean Daily Range*	*Design Wet-Bulb (2.5%)*	*Prevailing Wind*
ALABAMA						
Auburn		22	93/76	21	78	
Birmingham AP	NNW	21	94/75	21	77	WNW
Huntsville AP	N	16	93/74	23	77	SW
Mobile AP	N	29	93/77	18	79	N
Montgomery AP	NW	25	95/76	21	79	W
ALASKA						
Anchorage AP	SE	−18	68/58	15	59	WNW
Fairbanks AP (S)	N	−47	78/60	24	62	S
Juneau AP	N	1	70/58	15	59	W
Nome AP	N	−27	62/55	10	56	W
ARIZONA						
Flagstaff AP	NE	4	82/55	31	60	SW
Phoenix AP (S)	E	34	107/71	27	75	W
Prescott AP		9	94/60	30	65	
Tuscon AP (S)	SE	32	102/66	26	71	WNW
Winslow AP	SW	10	95/60	32	65	WSW
Yuma AP	NNE	39	109/72	27	78	WSW
ARKANSAS						
Fayetteville AP	NE	12	94/73	23	76	SSW
Fort Smith AP	NW	17	98/76	24	79	SW
Little Rock AP (S)	N	20	96/77	22	79	SSW
Texarkana AP	WNW	23	96/77	21	79	SSW
CALIFORNIA						
Bakersfield AP	ENE	32	101/69	32	71	WNW
Barstow AP	WNW	29	104/68	37	71	W
Burbank AP	NW	39	91/68	25	70	S
Eureka/Arcata AP	E	33	65/59	11	60	NW
Fresno AP (S)	E	30	100/69	34	71	WNW
Long Beach AP	NW	43	80/68	22	69	WNW
Los Angeles AP (S)	E	43	80/68	15	69	WSW
Los Angeles CO (S)	NW	40	89/70	20	71	NW
Needles AP		33	110/71	27	75	
Oakland AP	E	36	80/63	19	64	WNW
Pomona CO	E	30	99/69	36	72	W
Redding AP		31	102/67	32	69	
Sacramento AP	NNW	32	98/70	36	71	SW
San Diego AP	NE	44	80/69	12	70	WNW
San Francisco AP	S	38	77/63	20	64	NW
San Francisco CO	W	40	71/62	14	62	W
San Luis Obispo	E	35	88/70	26	71	W
Santa Barbara AP	NE	36	77/66	24	67	SW
Santa Maria AP (S)	E	33	76/63	23	64	WNW
Santa Monica CO		43	80/68	16	69	
COLORADO						
Alamosa AP		−16	82/57	35	61	

APPENDICES

TABLE A.1 **Outside Design Conditions: United States and Canada (Continued)**

		Winter	Summer			
State and Station[a]	*Prevailing Wind*	*Design Dry-Bulb (97.5%)*	*Design Dry-Bulb and Mean Coincident Wet-Bulb (2.5%)*	*Mean Daily Range*	*Design Wet-Bulb (2.5%)*	*Prevailing Wind*
Boulder		8	91/59	27	63	
Colorado Springs AP	N	2	88/57	30	62	S
Denver AP	S	1	91/59	28	63	SE
Fort Collins		−4	91/59	28	63	
Grand Junction AP (S)	ESE	7	94/59	29	63	WNW
Leadville		−4	81/51	30	55	
Pueblo AP	W	0	95/61	31	66	SE
CONNECTICUT						
Bridgeport AP	NNW	9	84/71	18	74	WSW
Hartford	N	7	88/73	22	75	SSW
New Haven AP	NNE	7	84/73	17	75	SW
DELAWARE						
Wilmington AP	WNW	14	89/74	20	76	WSW
DISTRICT OF COLUMBIA						
Washington National AP	WNW	17	91/74	18	77	S
FLORIDA						
Daytona Beach AP	NW	35	90/77	15	79	(—)
Fort Myers AP	NNE	44	92/78	18	79	W
Gainesville AP (S)	W	31	93/77	18	79	W
Jacksonville AP	NW	32	94/77	19	79	SW
Key West AP	NNE	57	90/78	9	79	SE
Miami AP (S)	NNW	47	90/77	15	79	SE
Miami Beach CO		48	89/77	10	79	
Orlando AP	NNW	38	93/76	17	78	SSW
Panama City, Tyndall AFB	N	33	90/77	14	80	WSW
Pensacola CO	NNE	29	93/77	14	79	SW
Tallahassee AP (S)	NW	30	92/76	19	78	NW
Tampa AP (S)	N	40	91/77	17	79	W
West Palm Beach AP	NW	45	91/78	16	79	ESE
GEORGIA						
Athens	NW	22	92/74	21	77	WNW
Atlanta AP (S)	NW	22	92/74	19	76	NW
Augusta AP	W	23	95/76	19	79	WSW
Columbus, Lawson AFB	NW	24	93/76	21	78	W
Macon AP	NW	25	93/76	22	78	WNW
Rome AP		22	93/76	23	78	
Savannah-Travis AP	WNW	27	93/77	20	79	SW
Valdosta-Moody AFB	WNW	31	94/77	20	79	W
HAWAII						
Hilo AP (S)	SW	62	83/72	15	74	NE
Honolulu AP	ENE	63	86/73	12	75	ENE
Wahiawa	WNW	59	85/72	14	74	E
IDAHO						
Boise AP (S)	SE	10	94/64	31	66	NW

TABLE A.1 **Outside Design Conditions: United States and Canada (Continued)**

	Winter		Summer			
State and Station	*Prevailing Wind*	*Design Dry-Bulb (97.5%)*	*Design Dry-Bulb and Mean Coincident Wet-Bulb (2.5%)*	*Mean Daily Range*	*Design Wet-Bulb (2.5%)*	*Prevailing Wind*
Lewiston AP	W	6	93/64	32	66	WNW
Moscow		0	87/62	32	64	
Pocatello AP	NE	−1	91/60	35	63	W
ILLINOIS						
Carbondale		7	93/77	21	79	
Champaign/Urbana	W	2	92/74	21	77	SSW
Chicago, O'Hare AP	WNW	−4	89/74	20	76	SW
Chicago CO		2	91/74	15	77	
Moline AP	WNW	−4	91/75	23	77	SW
Peoria AP	WNW	−4	89/74	22	76	SW
Rockford		−4	89/73	24	76	
Springfield AP	NW	2	92/74	21	77	SW
INDIANA						
Evansville AP	NW	9	93/75	22	78	SW
Fort Wayne AP	WSW	1	89/72	24	75	SW
Indianapolis AP (S)	WNW	2	90/74	22	76	SW
Lafayette		3	91/73	22	76	
Muncie		2	90/73	22	76	
South Bend AP	SW	1	89/73	22	75	SSW
Terre Haute AP	NNW	4	92/74	22	77	SSW
IOWA						
Ames (S)	NW	−6	90/74	23	76	
Burlington AP	NW	−3	91/75	22	77	SSW
Des Moines AP	N	−5	91/74	23	77	S
Dubuque	NW	−7	88/73	22	75	SSW
Iowa City	NW	−6	89/76	22	78	SSW
Mason City AP	NNW	−11	88/74	24	75	S
Sioux City AP	NW	−7	92/74	24	77	S
Waterloo		−10	89/75	23	77	S
KANSAS	N					
Dodge City AP (S)	WSW	5	97/69	25	73	SSW
Goodland AP	NNE	0	96/65	31	70	S
Manhattan, Fort Riley (S)	NNW	3	95/75	24	77	S
Topeka AP	NNW	4	96/75	24	78	S
Wichita AP		7	98/73	23	76	SSW
KENTUCKY						
Bowling Green AP	WNW	10	92/75	21	77	
Lexington AP (S)	NW	8	91/73	22	76	SW
Louisville AP		10	93/74	23	77	SW
LOUISIANA	N					
Alexandria AP	ENE	27	94/77	20	79	S
Baton Rouge AP	N	29	93/77	19	80	W
Lafayette AP	N	30	94/78	18	80	SW
Lake Charles AP (S)		31	93/77	17	79	SSW

TABLE A.1 **Outside Design Conditions: United States and Canada (Continued)**

State and Station	Prevailing Wind	Winter: Design Dry-Bulb (97.5%)	Summer: Design Dry-Bulb and Mean Coincident Wet-Bulb (2.5%)	Summer: Mean Daily Range	Summer: Design Wet-Bulb (2.5%)	Summer: Prevailing Wind
Monroe AP	N	25	96/76	20	79	S
New Orleans AP	NNE	33	92/78	16	80	SSW
Shreveport AP (S)	N	25	96/76	20	79	S
MAINE						
Bangor, Dow AFB	WNW	−6	83/68	22	71	S
Caribou AP (S)	WSW	−13	81/67	21	69	SW
Portland (S)	W	−1	84/71	22	72	S
MARYLAND						
Baltimore AP	W	13	91/75	21	77	WSW
Baltimore CO	WNW	17	89/76	17	78	S
Cumberland	WNW	10	89/74	22	76	W
Frederick AP	N	12	91/75	22	77	WNW
Salisbury (S)		16	91/75	18	77	
MASSACHUSETTS						
Boston AP (S)	WNW	9	88/71	16	74	SW
Pittsfield AP	NW	−3	84/70	23	72	SW
Worcester AP	W	4	84/70	18	72	W
MICHIGAN						
Alpena AP	W	−6	85/70	27	72	SW
Detroit	W	6	88/72	20	74	SW
Escanaba		−7	83/69	17	71	
Flint AP	SW	1	87/72	25	74	SW
Grand Rapids AP	WNW	5	88/72	24	74	WSW
Muskegon AP	E	6	84/70	21	73	SW
Sault Ste. Marie AP (S)	E	−8	81/69	23	70	SW
Traverse City AP	SSW	1	86/71	22	73	SW
MINNESOTA						
Bemidji AP	N	−26	85/69	24	71	S
Duluth AP	WNW	−16	82/68	22	70	WSW
International Falls AP	N	−25	83/68	26	70	S
Minneapolis/St. Paul AP	NW	−12	89/73	22	75	S
Rochester AP	NW	−12	87/72	24	75	SSW
MISSISSIPPI						
Columbus AFB	N	20	93/77	22	79	W
Jackson AP	NNW	25	95/76	21	78	NW
Meridian AP	N	23	95/76	22	79	WSW
Vicksburg CO		26	95/78	21	80	
MISSOURI						
Columbia AP (S)	WNW	4	94/74	22	77	WSW
Kansas City AP	NW	6	96/74	20	77	S
St Louis AP	NW	6	94/75	21	77	WSW
St Louis CO	NW	8	94/75	18	77	S
Springfield AP	NNW	9	93/74	23	77	S
MONTANA						
Billings AP	NE	−10	91/64	31	66	SW
Bozeman		−14	87/60	32	62	

TABLE A.1 **Outside Design Conditions: United States and Canada (Continued)**

		Winter	Summer			
State and Station	*Prevailing Wind*	*Design Dry-Bulb (97.5%)*	*Design Dry-Bulb and Mean Coincident Wet-Bulb (2.5%)*	*Mean Daily Range*	*Design Wet-Bulb (2.5%)*	*Prevailing Wind*
Cut Bank AP		−20	85/61	35	62	
Glasgow AP (S)	E	−18	89/63	29	66	S
Great Falls AP (S)	SW	−15	88/60	28	62	WSW
Helena AP	N	−16	88/60	32	62	WNW
Kalispell AP		−7	87/61	34	63	
Lewistown AP	NW	−16	87/61	30	63	NW
Miles City AP	NW	−15	95/66	30	68	SE
Missoula AP	ESE	−6	88/61	36	63	NW
NEBRASKA						
Grand Island AP	NNW	−3	94/71	28	74	S
Lincoln CO (S)	N	−2	95/74	24	77	S
North Platte AP (S)	NW	−4	94/69	28	72	SSE
Omaha AP	NW	−3	91/75	22	77	S
Scottsbluff AP	NW	−3	92/65	31	68	SE
NEVADA						
Elko AP	E	−2	92/59	42	62	SW
Ely AP (S)	S	−4	87/56	39	59	SSW
Las Vegas AP (S)	ENE	28	106/65	30	70	SW
Lovelock AP		12	96/63	42	65	
Reno AP (S)	SSW	10	92/60	45	62	WNW
Tonopah AP	N	10	92/59	40	62	S
Winnemucca AP	SE	3	94/60	42	62	W
NEW HAMPSHIRE						
Berlin		−9	84/69	22	71	
Concord AP		−3	87/70	26	73	SW
Portsmouth, Pease AFB	NW	2	85/71	22	74	W
NEW JERSEY	W					
Atlantic City CO		13	89/74	18	77	WSW
Newark AP	NW	14	91/73	20	76	WSW
Trenton CO	WNW	14	88/74	19	76	SW
NEW MEXICO	W					
Albuquerque AP (S)		16	94/61	27	65	W
Farmington AP	N	6	93/62	30	65	SW
Las Cruces	ENE	20	96/64	30	68	SE
Los Alamos	SE	9	87/60	32	61	
Raton AP		1	89/60	34	64	
Roswell, Walker AFB		18	98/66	33	70	SSE
Silver City AP	N	10	94/60	30	64	
Tucumcari AP		13	97/66	28	69	SW
NEW YORK	NE					
Albany AP (S)		−1	88/72	23	74	S
Binghamton AP	WNW	1	83/69	20	72	WSW
Buffalo AP	WSW	6	85/70	21	73	SW
Ithaca (S)	W	0	85/71	24	73	SW
Massena AP	W	−8	83/69	20	72	
NYC—Central Park (S)		15	89/73	17	75	

TABLE A.1 **Outside Design Conditions: United States and Canada (Continued)**

		Winter	Summer			
State and Station	Prevailing Wind	Design Dry-Bulb (97.5%)	Design Dry-Bulb and Mean Coincident Wet-Bulb (2.5%)	Mean Daily Range	Design Wet-Bulb (2.5%)	Prevailing Wind
NYC—Kennedy AP	WNW	15	87/72	16	75	SSW
Rochester AP	WSW	5	88/71	22	73	WSW
Syracuse AP	N	2	87/71	20	73	WNW
NORTH CAROLINA						
Asheville AP	NNW	14	87/72	21	74	NNW
Charlotte AP	NNW	22	93/74	20	76	SW
Greensboro AP (S)	NE	18	91/73	21	76	SW
Raleigh/Durham AP (S)	N	20	92/75	20	77	SW
Wilmington AP	N	26	91/78	18	80	SW
NORTH DAKOTA						
Bismarck AP (S)	WNW	−19	91/68	27	71	S
Fargo AP	SSE	−18	89/71	25	74	S
Minot AP	WSW	−20	89/67	25	70	S
Williston		−21	88/67	25	70	
OHIO						
Akron-Canton AP	SW	6	86/71	21	73	SW
Cincinnati CO	W	6	90/72	21	75	SW
Cleveland AP (S)	SW	5	88/72	22	74	N
Columbus AP (S)	W	5	90/73	24	75	SSW
Dayton AP	WNW	4	89/72	20	75	SW
Mansfield AP	W	5	87/72	22	74	SW
Toledo AP	WSW	1	88/73	25	75	SW
Youngstown AP	SW	4	86/71	23	73	SW
OKLAHOMA						
Norman	N	13	96/74	24	76	S
Oklahoma City AP (S)	N	13	97/74	23	77	SSW
Stillwater (S)	N	13	96/74	24	76	SSW
Tulsa AP	N	13	98/75	22	78	SSW
OREGON						
Astoria AP (S)	ESE	29	71/62	16	63	NNW
Bend		4	87/60	33	62	
Eugene AP	N	22	89/66	31	67	N
Medford AP (S)	S	23	94/67	35	68	WNW
Pendleton AP	NNW	5	93/64	29	65	WNW
Portland AP	ESE	23	85/67	23	67	NW
Portland CO		24	86/67	21	67	
Salem AP	N	23	88/66	31	68	N
PENNSYLVANIA						
Allentown AP	W	9	88/72	22	75	SW
Erie AP	SSW	9	85/72	18	74	WSW
Harrisburg AP	NW	11	91/74	21	76	WSW
Philadelphia AP	WNW	14	90/74	21	76	WSW
Pittsburgh AP	WSW	5	86/71	22	73	WSW
Pittsburgh CO		7	88/71	19	73	
Scranton/Wilkes-Barre	SW	5	87/71	19	73	WSW
State College (S)	NNW	7	87/71	23	73	WSW

TABLE A.1 **Outside Design Conditions: United States and Canada (Continued)**

	Winter		Summer			
State and Station	Prevailing Wind	Design Dry-Bulb (97.5%)	Design Dry-Bulb and Mean Coincident Wet-Bulb (2.5%)	Mean Daily Range	Design Wet-Bulb (2.5%)	Prevailing Wind
RHODE ISLAND						
Newport (S)	WNW	9	85/72	16	75	SW
Providence AP	WNW	9	86/72	19	74	SW
SOUTH CAROLINA						
Anderson		23	92/74	21	76	
Charleston AFB (S)	NNE	27	91/78	18	80	SW
Charleston CO		28	92/78	13	80	
Columbia AP	W	24	95/75	22	78	SW
Spartanburg AP		22	91/74	20	76	
SOUTH DAKOTA						
Huron AP	NNW	−14	93/72	28	75	S
Pierre AP	NW	−10	95/71	29	74	SSE
Rapid City AP (S)	NNW	−7	92/65	28	69	SSE
Sioux Falls AP	NW	−11	91/72	24	75	S
TENNESSEE						
Bristol–Tri City AP	WNW	14	89/72	22	75	SW
Chattanooga AP	NNW	18	93/74	22	77	WSW
Knoxville AP	NE	19	92/73	21	76	W
Memphis AP	N	18	95/76	21	79	SW
Nashville AP (S)	NW	14	94/74	21	77	WSW
TEXAS						
Abilene AP	N	20	99/71	22	74	SSE
Amarillo AP	N	11	95/67	26	70	S
Austin AP	N	28	98/74	22	77	S
Brownsville AP (S)	NNW	39	93/77	18	79	SE
Bryan AP		29	96/76	20	78	
Corpus Christi AP	N	35	94/78	19	80	SSE
Dallas AP	N	22	100/75	20	78	S
Del Rio, Laughlin AFB		31	98/73	24	77	
El Paso AP (S)	N	24	98/64	27	68	S
Forth Worth AP (S)	NW	22	99/74	22	77	S
Galveston AP	N	36	89/79	10	80	S
Houston AP	NNW	32	94/77	18	79	S
Houston CO		33	95/77	18	79	
Laredo AFB	N	36	101/73	23	78	SE
Lubbock AP	NNE	15	96/69	26	72	SSE
Lufkin AP	NNW	29	97/76	20	79	S
Midland AP (S)	NE	21	98/69	26	72	SSE
Port Arthur AP	N	31	93/78	19	80	S
San Angelo, Goodfellow AFB	NNE	22	99/71	24	74	SSE
San Antonio AP (S)	N	30	97/73	19	76	SSE
Sherman Perrin AFB	N	20	98/75	22	77	S
Waco AP		26	99/75	22	78	
Wichita Falls AP	NNW	18	101/73	24	76	S
UTAH						
Cedar City AP	SE	5	91/60	32	63	SW

TABLE A.1 **Outside Design Conditions: United States and Canada (Continued)**

		Winter	*Summer*			
State and Station	*Prevailing Wind*	*Design Dry-Bulb (97.5%)*	*Design Dry-Bulb and Mean Coincident Wet-Bulb (2.5%)*	*Mean Daily Range*	*Design Wet-Bulb (2.5%)*	*Prevailing Wind*
Salt Lake City AP (S)	SSE	8	95/62	32	65	N
Vernal AP		0	89/60	32	63	
VERMONT						
Barre		−11	81/69	23	71	
Burlington AP (S)	E	−7	85/70	23	72	SSW
VIRGINIA						
Charlottesville	NE	18	91/74	23	76	SW
Norfolk AP	NW	22	91/76	18	78	SW
Richmond AP	N	17	92/76	21	78	SW
Roanoke AP	NW	16	91/72	23	74	SW
WASHINGTON						
Bellingham AP	NNE	15	77/65	19	65	WSW
Olympia AP	NE	22	83/65	32	66	NE
Seattle CO (S)	N	27	82/66	19	67	N
Seattle-Tacoma AP (S)	E	26	80/64	22	64	N
Spokane AP (S)	NE	2	90/63	28	64	SW
Walla Walla AP	W	7	94/66	27	67	W
Yakima AP	W	5	93/65	36	66	NW
WEST VIRGINIA						
Beckley	WNW	4	81/69	22	71	WNW
Charleston AP	SW	11	90/73	20	75	SW
Huntington CO	W	10	91/74	22	77	SW
Wheeling	WSW	5	86/71	21	73	WSW
WISCONSIN						
Ashland		−16	82/68	23	70	
Eau Claire AP		−11	89/73	23	75	
Green Bay AP	W	−9	85/72	23	74	SW
La Cross AP	NW	−9	88/73	22	75	S
Madison AP (S)	NW	−7	88/73	22	75	SW
Milwaukee AP	WNW	−4	87/73	21	74	SSW
WYOMING						
Casper AP	NE	−5	90/57	31	61	SW
Cheyenne AP	N	−1	86/58	30	62	WNW
Lander AP (S)	E	−11	88/61	32	63	NW
Laramie AP (S)		−6	81/56	28	60	
Rock Springs AP	WSW	−3	84/55	32	58	W
Sheridan AP	NW	−8	91/62	32	65	N
Province and Station						
ALBERTA						
Calgary AP	NNW	−23	81/61	25	63	SE
Edmonton AP	E	−25	82/65	23	66	SE
Medicine Hat AP		−24	90/65	28	68	
BRITISH COLUMBIA						
Dawson Creek		−33	79/63	26	64	

TABLE A.1 **Outside Design Conditions: United States and Canada (Continued)**

		Winter	Summer			
State and Station	*Prevailing Wind*	*Design Dry-Bulb (97.5%)*	*Design Dry-Bulb and Mean Coincident Wet-Bulb (2.5%)*	*Mean Daily Range*	*Design Wet-Bulb (2.5%)*	*Prevailing Wind*
Nanaimo (S)		20	80/65	21	66	
Prince George AP (S)	N	−28	80/62	26	64	N
Prince Rupert CO		2	63/57	12	58	
Vancouver AP (S)	E	19	77/66	17	67	WNW
Victoria CO		23	73/62	16	62	
MANITOBA						
Flin Flon		−37	81/66	19	68	
Winnipeg AP (S)	W	−27	86/71	22	73	S
NEW BRUNSWICK						
Fredericton AP (S)		−11	85/69	23	71	
Saint John AP		−8	77/65	19	68	
NEWFOUNDLAND						
Corner Brook		0	73/63	17	66	
St. John's AP (S)	N	7	75/65	18	67	WSW
NORTHWEST TERR.						
Fort Smith AP (S)	NW	−45	81/64	24	66	S
Inuvik (S)		−53	77/60	21	62	
Yellowknife AP	SSE	−46	77/61	16	63	S
NOVA SCOTIA						
Halifax AP (S)		5	76/65	16	67	
Sydney AP		3	80/68	19	70	
Yarmouth AP	NW	9	72/64	15	66	S
ONTARIO						
Kitchener		−2	85/72	23	74	
Ottawa AP (S)		−13	87/71	21	73	
Sudbury AP		−19	83/67	22	70	
Toronto AP (S)	N	−1	87/72	20	74	SW
PRINCE EDWARD ISLAND						
Charlottetown AP (S)		−4	78/68	16	70	
QUEBEC						
Chicoutimi		−22	83/68	20	70	
Montréal AP (S)		−10	85/72	17	74	
Québec AP		−14	84/70	20	72	
Sept Iles AP (S)		−21	73/61	17	65	
Val d'Or AP		−27	83/68	22	70	
SASKATCHEWAN						
Prince Albert AP		−35	84/66	25	68	
Regina AP		−29	88/68	26	70	
Saskatoon AP (S)		−31	86/66	26	68	
YUKON TERRITORY						
Whitehorse AP (S)	NW	−43	77/58	22	59	SE

[a]S = solar data available. U.S. solar data available from the National Climatic Data Center, Federal Building, Ashville, NC 28801. Canadian solar data available from The Canadian Climate Center, Atmospheric Environment Service, 4905 Dufferin St., Downsview, Ontario M3H 5T 4.

TABLE A.2 **Outdoor Design Conditions: Mexico and Puerto Rico**

	Winter[a]	*Summer*[b]		
Country and Station	*Design Dry-Bulb 97.5%*	*Design Dry-Bulb 2.5%*	*Mean Daily Range*	*Design Wet-Bulb 2.5%*
MEXICO				
Guadalajara	42	91	29	67
Mérida	61	95	21	79
Mexico City	39	81	25	60
Monterrey	41	95	20	78
Vera Cruz	62	89	20	83
PUERTO RICO				
San Juan	68	88	11	80

[a]For the three-month winter period

[b]For the four-month summer period

B

SOLAR DATA

This appendix presents both *clear day* and *average day* data.

Solar Intensity and Solar Heat Gain Factors (Tables B.1–B.10). These are listed for five latitudes; solar heat gain factors occur through one layer of double-strength sheet (DSA) glass. The data are based on an "average cloudless" day at each latitude. Tables B.1 through B.10 are copyrighted by the American Society of Heating, Refrigerating, and Air Conditioning Engineers, Inc., Atlanta, Ga., and are reprinted by permission from *ASHRAE 1989 Handbook of Fundamentals.*

Solar Position and Clear Day Global Irradiance (Insolation) (Tables B.11–B.14). These are listed for four latitudes. The data are presented in conventional units for the following tilt angles (degrees above horizontal): latitude −10°, latitude, latitude +10°, latitude +20°, and vertical. To convert to SI units: Btu/h-ft^2 × 3.152 = W/m^2.

Elevation, Latitude, Average Horizontal Insolation, Average Vertical Insolation, Average Temperature, and Heating Degree Days (Table B.15). These are listed in conventional units for January and July, as well as yearly totals, for a number of locations (Fig. B.1). To convert horizontal surface (HS) and vertical surface (VS) to SI units: Btu/h ft^2 × 3.152 = W/m^2. To convert average temperature (TA) to Celsius: (°F − 32)/1.8 = °C. To convert DD(F) to DD(C): DD(F) × 9/5 = DD(C) or DD(K).

TABLE B.1 **Solar Intensity and Solar Heat Gain Factors[a] for 16° N Latitude (Conventional Units)[b]**

Date	Solar Time am	Direct Normal Btuh ft²	N	NNE	NE	ENE	E	ESE	SE	SSE	S	SSW	SW	WSW	W	WNW	NW	NNW	HOR	Solar Time pm
			Solar Heat Gain Factors, Btuh ft²																	
Jan 21	7	141	5	6	44	92	124	134	126	96	49	6	5	5	5	5	5	5	14	5
	8	262	14	15	55	147	210	240	233	189	114	25	14	14	14	14	14	14	79	4
	9	300	21	21	32	122	200	244	251	219	152	58	22	21	21	21	21	21	150	3
	10	317	26	26	27	66	150	209	233	223	178	102	31	26	26	26	26	26	203	2
	11	325	29	29	29	31	77	148	195	210	194	146	75	31	29	29	29	29	236	1
	12	327	30	30	30	30	32	73	139	184	199	184	138	72	32	30	30	30	248	12
HALF DAY TOTALS			110	112	196	461	760	1000	1096	1020	781	426	211	127	111	110	110	110	805	
Feb 21	7	182	8	17	84	138	169	172	150	103	36	8	8	8	8	8	8	8	25	5
	8	273	17	19	96	180	231	247	224	166	77	18	17	17	17	17	17	17	101	4
	9	305	23	24	64	153	214	242	233	188	110	30	23	23	23	23	23	23	174	3
	10	319	28	29	33	92	161	202	211	188	134	61	30	28	28	28	28	28	229	2
	11	326	32	32	32	37	83	136	167	172	149	102	49	33	32	32	32	32	263	1
	12	328	33	33	33	33	34	60	107	142	154	142	106	60	34	33	33	33	275	12
HALF DAY TOTALS			124	137	321	609	865	1023	1034	885	582	287	174	132	124	124	124	124	930	
Mar 21	7	201	11	53	124	172	192	183	145	82	15	10	10	10	10	10	10	10	40	5
	8	272	20	50	140	205	239	235	195	123	35	19	19	19	19	19	19	19	120	4
	9	299	26	35	109	179	218	225	197	138	57	27	26	26	26	26	26	26	192	3
	10	312	31	33	61	120	165	182	172	134	76	34	32	31	31	31	31	31	247	2
	11	318	34	35	36	53	87	114	125	116	89	55	36	35	34	34	34	34	280	1
	12	320	35	35	36	36	37	47	69	87	93	86	68	47	37	36	36	35	291	12
HALF DAY TOTALS			141	226	494	755	928	975	879	643	319	187	153	142	139	139	139	139	1025	
Apr 21	6	14	2	8	12	14	14	12	8	2	1	1	1	1	1	1	1	1	1	6
	7	197	24	94	153	187	191	167	117	45	14	13	13	13	13	13	13	13	53	5
	8	256	27	99	172	216	227	204	150	69	24	22	22	22	22	22	22	22	131	4
	9	280	31	79	149	193	208	193	147	77	31	29	29	29	29	29	29	29	197	3
	10	293	35	54	102	141	158	151	120	73	37	34	33	33	33	33	33	33	249	2
	11	299	38	40	54	72	86	88	78	60	43	38	38	36	36	36	36	37	279	1
	12	301	39	39	39	40	40	41	43	45	45	45	43	41	40	39	39	39	289	12
HALF DAY TOTALS			179	403	674	859	922	851	653	352	174	159	157	156	155	155	155	156	1057	
May 21	6	44	14	30	41	45	43	34	19	4	3	3	3	3	3	3	3	3	5	6
	7	193	50	120	168	191	185	150	92	24	16	16	16	16	16	16	16	17	62	5
	8	244	52	132	189	218	215	179	115	38	25	24	24	24	24	24	24	25	135	4
	9	268	49	116	171	198	197	167	109	45	32	30	30	30	30	30	30	32	197	3
	10	280	47	89	130	151	150	126	84	44	37	35	35	35	35	35	35	37	245	2
	11	286	47	63	79	87	83	70	52	41	40	39	38	38	38	39	39	41	273	1
	12	288	46	46	44	43	42	41	41	41	41	41	41	41	42	43	44	46	282	12
HALF DAY TOTALS			283	575	804	916	897	748	493	217	172	167	167	167	167	168	169	176	1058	
Jun 21	6	53	20	39	52	55	51	39	20	4	4	4	4	4	4	4	4	4	7	6
	7	188	62	128	172	190	179	141	80	20	16	16	16	16	16	16	16	18	64	5
	8	238	66	142	194	217	207	167	99	31	25	25	25	25	25	25	25	27	135	4
	9	261	63	130	178	198	190	154	93	37	31	31	31	31	31	31	31	33	194	3
	10	273	59	104	140	154	145	115	70	39	37	36	36	36	36	36	36	38	241	2
	11	279	57	76	90	92	82	63	46	41	40	39	39	39	39	40	41	43	268	1
	12	281	57	55	50	45	43	42	41	41	41	41	41	42	42	45	50	55	277	12
HALF DAY TOTALS			356	648	850	929	876	700	430	194	174	171	171	171	172	173	176	190	1049	
Jul 21	6	41	14	29	39	42	40	31	18	4	3	3	3	3	3	3	3	3	6	6
	7	184	51	118	164	185	179	145	88	23	16	16	16	16	16	16	16	17	62	5
	8	236	55	132	187	214	210	174	111	37	25	25	25	25	25	25	25	26	133	4
	9	259	52	117	170	196	193	163	106	44	32	31	31	31	31	31	31	33	194	3
	10	272	50	92	131	151	148	123	81	44	38	36	36	36	36	36	36	38	241	2
	11	278	49	66	81	88	83	69	52	42	41	40	39	39	39	40	40	42	269	1
	12	279	49	48	46	44	43	42	42	42	42	42	42	42	43	44	46	48	277	12
HALF DAY TOTALS			296	580	799	903	878	729	478	215	176	172	171	171	171	172	173	182	1043	
Aug 21	6	11	2	7	10	12	12	10	6	2	1	1	1	1	1	1	1	1	1	6
	7	180	26	92	145	176	180	156	109	42	15	14	14	14	14	14	14	14	53	5
	8	240	30	100	168	209	219	196	143	65	25	23	23	23	23	23	23	23	128	4
	9	266	33	82	148	190	203	187	142	74	33	30	30	30	30	30	30	30	193	3
	10	279	37	58	104	140	155	147	117	71	39	36	35	35	35	35	35	35	243	2
	11	285	40	43	57	75	86	87	76	59	44	40	39	38	38	38	38	39	273	1
	12	287	41	41	41	42	42	43	44	45	46	45	44	43	42	41	41	41	282	12
HALF DAY TOTALS			191	410	666	837	891	817	624	339	180	167	165	164	163	163	163	164	1033	
Sep 21	7	179	12	50	114	158	176	168	133	76	15	11	11	11	11	11	11	11	39	5
	8	253	21	49	134	196	227	224	186	119	36	20	20	20	20	20	20	20	116	4
	9	281	28	36	106	173	211	217	191	134	57	28	27	27	27	27	27	27	185	3
	10	295	32	34	61	118	161	178	168	132	76	35	33	32	32	32	32	32	238	2
	11	302	35	36	37	54	86	113	123	114	88	56	38	36	35	35	35	35	271	1
	12	304	36	36	37	38	39	49	69	86	93	86	69	48	39	38	37	36	282	12
HALF DAY TOTALS			146	226	475	722	885	931	842	622	319	192	159	148	145	144	144	144	991	
Oct 21	7	166	8	18	79	128	156	159	139	95	33	9	8	8	8	8	8	8	25	5
	8	259	17	20	95	174	223	237	215	159	74	19	17	17	17	17	17	17	99	4
	9	292	24	25	65	150	209	235	225	182	106	31	24	24	24	24	24	24	170	3
	10	307	29	30	34	92	158	197	205	183	130	60	31	29	29	29	29	29	224	2
	11	314	32	32	33	39	83	133	163	167	145	100	49	34	32	32	32	32	258	1
	12	316	33	33	33	34	35	60	105	139	150	138	104	60	35	34	33	33	270	12
HALF DAY TOTALS			127	141	318	592	836	986	996	852	563	283	175	136	128	127	127	127	911	
Nov 21	7	134	5	6	43	89	119	129	120	92	47	6	5	5	5	5	5	5	14	5
	8	255	15	15	55	145	206	235	228	185	111	25	15	15	15	15	15	15	78	4
	9	295	21	21	33	121	197	241	247	215	150	57	22	21	21	21	21	21	149	3
	10	312	26	26	28	67	147	206	230	220	176	100	31	26	26	26	26	26	201	2
	11	320	29	29	29	31	77	146	192	207	191	144	74	31	29	29	29	29	234	1
	12	322	30	30	30	30	32	72	137	181	196	181	137	72	32	30	30	30	246	12
HALF DAY TOTALS			112	113	197	456	749	983	1077	1001	767	420	210	128	112	112	112	112	799	
Dec 21	7	118	4	5	30	72	101	112	107	85	48	7	4	4	4	4	4	4	10	5
	8	255	13	14	41	132	198	233	231	193	124	33	13	13	13	13	13	13	69	4
	9	297	20	20	25	108	191	241	254	227	165	72	21	20	20	20	20	20	138	3
	10	315	25	25	26	56	144	208	239	233	192	117	35	25	25	25	25	25	191	2
	11	323	28	28	28	29	73	150	202	221	207	161	86	30	28	28	28	28	223	1
	12	325	29	29	29	29	30	77	149	197	212	196	149	76	30	29	29	29	234	12
HALF DAY TOTALS			104	105	159	402	710	975	1099	1050	836	484	228	125	105	104	104	104	748	
			N	NNW	NW	WNW	W	WSW	SW	SSW	S	SSE	SE	ESE	E	ENE	NE	NNE	HOR	PM

[a]Total solar heat gains for DSA glass (based on a ground reflectance of 0.20).

[b]Half-day totals computed by Simpson's rule, time interval = 10 min.

TABLE B.2 **Solar Intensity and Solar Heat Gain Factors[a] for 32° N Latitude (Conventional Units)[b]**

Date	Solar Time am	Direct Normal Btuh/ft²	N	NNE	NE	ENE	E	ESE	SE	SSE	S	SSW	SW	WSW	W	WNW	NW	NNW	HOR	Solar Time pm
			Solar Heat Gain Factors, Btuh · ft²																	
Jan 21	7	1	0	0	0	1	1	1	1	1	0	0	0	0	0	0	0	0	0	5
	8	203	9	9	29	105	160	189	189	159	103	28	9	9	9	9	9	9	32	4
	9	269	15	15	17	91	175	229	246	225	169	82	17	15	15	15	15	15	88	3
	10	295	20	20	20	41	135	209	249	250	212	141	46	20	20	20	20	20	136	2
	11	306	23	23	23	24	68	159	221	249	238	191	110	29	23	23	23	23	166	1
	12	310	24	24	24	24	25	88	174	228	246	228	174	88	25	24	24	24	176	12
HALF DAY TOTALS			79	79	107	284	570	856	1015	1014	853	553	264	112	80	79	79	79	512	
Feb 21	7	112	4	7	47	82	102	106	95	67	26	4	4	4	4	4	4	4	9	5
	8	245	13	14	65	149	205	228	216	170	95	17	13	13	13	13	13	13	64	4
	9	287	19	19	32	122	199	242	248	216	149	55	20	19	19	19	19	19	127	3
	10	305	24	24	25	62	151	213	241	232	189	112	31	24	24	24	24	24	176	2
	11	314	26	26	26	28	76	156	208	227	212	165	87	28	26	26	26	26	207	1
	12	316	27	27	27	27	29	79	155	204	221	204	155	79	29	27	27	27	217	12
HALF DAY TOTALS			100	103	201	445	735	978	1080	1010	780	452	228	122	100	100	100	100	691	
Mar 21	7	185	10	37	105	153	176	173	142	88	20	9	9	9	9	9	9	9	32	5
	8	260	17	25	107	183	227	237	209	150	62	18	17	17	17	17	17	17	100	4
	9	290	23	25	64	151	210	237	227	183	107	30	23	23	23	23	23	23	164	3
	10	304	28	28	30	87	158	202	215	195	144	70	29	28	28	28	28	28	211	2
	11	311	31	31	31	34	82	142	179	188	168	120	59	32	31	31	31	31	242	1
	12	313	32	32	32	32	33	66	122	162	176	162	122	66	33	32	32	32	252	12
HALF DAY TOTALS			124	162	359	629	875	1033	1041	888	589	326	193	136	125	124	124	124	874	
Apr 21	6	66	9	35	54	65	66	56	38	12	4	3	3	3	3	3	3	3	7	6
	7	206	17	80	146	188	200	182	136	65	16	14	14	14	14	14	14	14	61	5
	8	255	23	61	144	200	227	219	177	107	30	22	22	22	22	22	22	22	129	4
	9	278	28	36	103	168	206	212	187	133	58	29	28	28	28	28	28	28	188	3
	10	290	32	34	52	108	155	177	172	141	87	39	33	32	32	32	32	32	233	2
	11	295	35	35	36	47	83	118	135	132	108	70	40	36	35	35	35	35	262	1
	12	297	36	36	36	37	38	53	82	106	115	106	82	53	38	37	36	36	271	12
HALF DAY TOTALS			161	296	550	792	952	992	889	645	360	228	177	157	153	152	152	152	1015	
May 21	6	119	33	77	108	121	116	94	56	13	8	8	8	8	8	8	8	9	21	6
	7	211	36	111	170	202	204	174	118	42	19	18	18	18	18	18	18	19	81	5
	8	250	29	94	165	208	220	199	149	73	27	25	25	25	25	25	25	25	146	4
	9	269	33	61	128	177	198	190	155	93	37	32	31	31	31	31	31	31	201	3
	10	280	36	40	76	121	150	156	138	99	54	37	35	35	35	35	35	35	243	2
	11	285	38	39	42	59	83	99	102	90	68	47	40	39	37	37	37	37	269	1
	12	286	38	39	40	40	41	47	59	70	74	70	59	47	41	40	40	39	277	12
HALF DAY TOTALS			222	438	702	900	985	933	747	447	250	199	183	177	175	174	174	175	1098	
Jun 21	6	131	44	92	123	135	127	99	55	12	10	10	10	10	10	10	10	11	28	6
	7	210	47	122	176	204	201	168	108	35	20	20	20	20	20	20	20	21	88	5
	8	245	36	106	171	208	214	189	135	60	28	27	27	27	27	27	27	27	151	4
	9	264	35	74	137	178	193	180	139	77	35	32	32	32	32	32	32	32	204	3
	10	274	38	47	86	125	146	145	123	83	45	38	36	36	36	36	36	36	244	2
	11	279	40	41	47	64	82	91	89	75	56	43	41	40	39	39	39	39	269	1
	12	280	41	41	41	42	42	46	52	58	60	58	52	46	42	42	41	41	276	12
HALF DAY TOTALS			261	504	762	935	985	897	678	372	225	197	189	185	184	184	183	186	1122	
Jul 21	6	113	34	76	105	117	113	90	53	12	9	9	9	9	9	9	9	9	22	6
	7	203	38	111	167	198	198	169	114	41	20	19	19	19	19	19	19	19	81	5
	8	241	31	95	163	204	215	194	145	70	28	26	26	26	26	26	26	26	145	4
	9	261	34	64	129	175	195	186	150	90	37	32	32	32	32	32	32	32	198	3
	10	271	37	42	78	121	148	153	134	96	53	38	36	36	36	36	36	36	240	2
	11	277	39	40	43	60	83	98	99	88	66	47	41	40	38	38	38	38	265	1
	12	279	40	40	41	41	42	48	58	68	72	68	58	48	42	41	41	40	273	12
HALF DAY TOTALS			231	444	701	890	967	912	726	433	248	202	187	182	180	179	179	180	1088	
Aug 21	6	59	10	33	50	60	60	51	34	11	4	4	4	4	4	4	4	4	8	6
	7	190	19	79	141	179	190	172	128	61	17	15	15	15	15	15	15	15	61	5
	8	240	25	63	141	195	219	210	170	102	31	23	23	23	23	23	23	23	128	4
	9	263	30	39	104	166	200	206	181	127	57	31	29	29	29	29	29	29	185	3
	10	276	34	36	55	109	153	173	167	136	84	40	35	34	34	34	34	34	229	2
	11	282	36	37	39	50	84	116	131	127	104	69	41	38	36	36	36	36	256	1
	12	284	37	37	37	39	40	54	81	103	111	103	81	54	40	39	37	37	265	12
HALF DAY TOTALS			171	303	546	774	922	955	854	618	352	231	184	166	162	161	160	160	999	
Sep 21	7	163	10	35	96	139	159	156	128	80	20	10	10	10	10	10	10	10	31	5
	8	240	18	26	103	173	215	224	198	143	60	19	18	18	18	18	18	18	96	4
	9	272	24	26	64	146	202	227	218	177	105	31	24	24	24	24	24	24	158	3
	10	287	29	29	32	86	154	196	208	189	141	70	31	29	29	29	29	29	204	2
	11	294	32	32	32	36	81	139	174	182	163	118	59	34	32	32	32	32	234	1
	12	296	33	33	33	33	35	66	120	158	171	158	120	66	35	33	33	33	244	12
HALF DAY TOTALS			130	164	345	598	831	982	993	852	574	325	197	142	130	129	129	129	845	
Oct 21	7	99	4	7	43	74	92	96	85	60	24	5	4	4	4	4	4	4	10	5
	8	229	13	15	63	143	195	217	206	162	90	17	13	13	13	13	13	13	63	4
	9	273	20	20	33	120	193	234	239	208	144	54	21	20	20	20	20	20	125	3
	10	293	24	24	26	62	147	207	234	225	183	109	32	24	24	24	24	24	173	2
	11	302	27	27	27	29	76	152	203	221	207	160	85	29	27	27	27	27	203	1
	12	304	28	28	28	28	30	78	151	199	215	199	151	78	30	28	28	28	213	12
HALF DAY TOTALS			103	106	200	433	708	941	1038	972	753	441	226	125	104	103	103	103	679	
Nov21	7	2	0	0	0	1	1	1	1	1	1	0	0	0	0	0	0	0	0	5
	8	196	9	9	29	103	156	184	184	155	100	27	9	9	9	9	9	9	32	4
	9	263	16	16	17	90	173	225	241	221	166	80	17	16	16	16	16	16	88	3
	10	289	20	20	21	41	134	206	245	246	209	138	45	21	20	20	20	20	136	2
	11	301	23	23	23	24	67	157	218	245	234	188	109	29	23	23	23	23	165	1
	12	304	24	24	24	24	25	87	171	224	243	224	171	87	25	24	24	24	175	12
HALF DAY TOTALS			80	81	108	282	561	841	996	995	838	544	261	113	81	80	80	80	509	
Dec 21	8	176	7	7	19	84	135	163	166	143	97	31	7	7	7	7	7	7	22	4
	9	257	14	14	15	77	162	218	238	222	171	89	15	14	14	14	14	14	72	3
	10	288	18	18	18	34	127	204	246	251	216	148	52	19	18	18	18	18	119	2
	11	301	21	21	21	22	63	157	222	252	243	197	116	29	21	21	21	21	148	1
	12	304	22	22	22	22	23	89	177	232	252	232	177	89	23	22	22	22	158	12
HALF DAY TOTALS			71	71	84	227	500	792	965	986	852	578	275	107	71	71	71	71	440	
			N	NNW	NW	WNW	W	WSW	SW	SSW	S	SSE	SE	ESE	E	ENE	NE	NNE	HOR	PM

[a]Total solar heat gains for DSA glass (based on a ground reflectance of 0.20).

[b]Half day totals computed by Simpson's rule, time interval = 10 min.

APPENDICES

TABLE B.3 **Solar Intensity and Solar Heat Gain Factors[a] for 40° N Latitude (Conventional Units)[b]**

Date	Solar Time am	Direct Normal Btuh/ft²	N	NNE	NE	ENE	E	ESE	SE	SSE	S	SSW	SW	WSW	W	WNW	NW	NNW	HOR	Solar Time pm
			Solar Heat Gain Factors, Btuh/ft²																	
Jan 21	8	142	5	5	17	71	111	132	133	114	75	22	6	5	5	5	5	5	14	4
	9	239	12	12	13	74	154	205	224	209	160	82	13	12	12	12	12	12	55	3
	10	274	16	16	16	31	124	199	241	246	213	146	51	17	16	16	16	16	96	2
	11	289	19	19	19	20	61	156	222	252	244	198	118	28	19	19	19	19	124	1
	12	294	20	20	20	20	21	90	179	234	254	234	179	90	21	20	20	20	133	12
HALF DAY TOTALS			61	61	73	199	452	734	904	932	813	561	273	101	62	61	61	61	354	
Feb 21	7	55	2	3	23	40	51	53	47	34	14	2	2	2	2	2	2	2	4	5
	8	219	10	11	50	129	183	206	199	160	94	18	10	10	10	10	10	10	43	4
	9	271	16	16	22	107	186	234	245	218	157	66	17	16	16	16	16	16	98	3
	10	294	21	21	21	49	143	211	246	243	203	129	38	21	21	21	21	21	143	2
	11	304	23	23	23	24	71	160	219	244	231	184	103	27	23	23	23	23	171	1
	12	307	24	24	24	24	25	86	170	222	241	222	170	86	25	24	24	24	180	12
HALF DAY TOTALS			84	86	152	361	648	916	1049	1015	821	508	250	114	85	84	84	84	548	
Mar 21	7	171	9	29	93	140	163	161	135	86	22	8	8	8	8	8	8	8	26	5
	8	250	16	18	91	169	218	232	211	157	74	17	16	16	16	16	16	16	85	4
	9	282	21	22	47	136	203	238	236	198	128	40	22	21	21	21	21	21	143	3
	10	297	25	25	27	72	153	207	229	216	171	95	29	25	25	25	25	25	186	2
	11	305	28	28	28	30	78	151	198	213	197	150	77	30	28	28	28	28	213	1
	12	307	29	29	29	29	31	75	145	191	206	191	145	75	31	29	29	29	223	12
HALF DAY TOTALS			114	139	302	563	832	1035	1087	968	694	403	220	132	114	113	113	113	764	
Apr 21	6	89	11	46	72	87	88	76	52	18	5	5	5	5	5	5	5	5	11	6
	7	206	16	71	140	185	201	186	143	75	16	14	14	14	14	14	14	14	61	5
	8	252	22	44	128	190	224	223	188	124	41	22	21	21	21	21	21	21	123	4
	9	274	27	29	80	155	202	219	203	156	83	29	27	27	27	27	27	27	177	3
	10	286	31	31	37	92	152	187	193	170	121	56	32	31	31	31	31	41	217	2
	11	292	33	33	34	39	81	130	160	166	146	102	52	35	33	33	33	33	243	1
	12	293	34	34	34	34	36	62	108	142	154	142	108	62	36	34	34	34	252	12
HALF DAY TOTALS			154	265	501	758	957	1051	994	782	488	296	199	157	148	147	147	147	957	
May 21	5	1	0	1	1	1	1	1	0	0	0	0	0	0	0	0	0	0	0	7
	6	144	36	90	128	145	141	115	71	18	10	10	10	10	10	10	10	11	31	6
	7	216	28	102	165	202	209	184	131	54	20	19	19	19	19	19	19	19	87	5
	8	250	27	73	149	199	220	208	164	93	29	25	25	25	25	25	25	25	146	4
	9	267	31	42	105	164	197	200	175	121	53	32	30	30	30	30	30	30	195	3
	10	277	34	36	54	105	148	168	163	133	83	40	35	34	34	34	34	34	234	2
	11	283	36	36	38	48	81	113	130	127	105	70	42	38	36	36	36	36	257	1
	12	284	37	37	37	38	40	54	82	104	113	104	82	54	40	38	37	37	265	12
HALF DAY TOTALS			215	404	666	893	1024	1025	881	601	358	247	200	180	176	175	174	175	1083	
Jun 21	5	22	10	17	21	22	20	14	6	2	1	1	1	1	1	1	1	2	3	7
	6	155	48	104	143	159	151	121	70	17	13	13	13	13	13	13	13	14	40	6
	7	216	37	113	172	205	207	178	122	46	22	21	21	21	21	21	21	21	97	5
	8	246	30	85	156	201	216	199	152	80	29	27	27	27	27	27	27	27	153	4
	9	263	33	51	114	166	192	190	161	105	45	33	32	32	32	32	32	32	201	3
	10	272	35	38	63	109	145	158	148	116	69	39	36	35	35	35	35	35	238	2
	11	277	38	39	40	52	81	105	116	110	88	60	41	39	38	38	38	38	260	1
	12	279	38	38	38	40	41	52	72	89	95	89	72	52	41	40	38	38	267	12
HALF DAY TOTALS			253	470	734	941	1038	999	818	523	315	236	204	191	188	187	186	188	1126	
Jul 21	5	2	1	2	2	2	2	1	1	0	0	0	0	0	0	0	0	0	0	7
	6	138	37	89	125	142	137	112	68	18	11	11	11	11	11	11	11	12	32	6
	7	208	30	102	163	198	204	179	127	53	21	20	20	20	20	20	20	20	88	5
	8	241	28	75	148	196	216	203	160	90	30	26	26	26	26	26	26	26	145	4
	9	259	32	44	106	163	193	196	170	118	52	33	31	31	31	31	31	31	194	3
	10	269	35	37	56	106	146	165	159	129	81	41	36	35	35	35	35	35	231	2
	11	275	37	38	40	50	81	111	127	123	102	69	43	39	37	37	37	37	254	1
	12	276	38	38	38	40	41	55	80	101	109	101	80	55	41	40	38	38	262	12
HALF DAY TOTALS			223	411	666	885	1008	1003	858	584	352	248	204	186	181	180	180	181	1076	
Aug 21	6	81	12	44	68	81	82	71	48	17	6	5	5	5	5	5	5	5	12	6
	7	191	17	71	135	177	191	177	135	70	17	16	16	16	16	16	16	16	62	5
	8	237	24	47	126	185	216	214	180	118	41	23	23	23	23	23	23	23	122	4
	9	260	28	31	82	153	197	212	196	151	80	31	28	28	28	28	28	28	174	3
	10	272	32	33	40	93	150	182	187	165	116	56	34	32	32	32	32	32	214	2
	11	278	35	35	36	41	81	128	156	160	141	99	52	37	35	35	35	35	239	1
	12	280	35	35	35	36	38	63	106	138	149	138	106	63	38	36	35	35	247	12
HALF DAY TOTALS			164	273	498	741	928	1013	956	751	474	296	205	166	157	156	156	156	946	
Sep 21	7	149	9	27	84	125	146	144	121	77	21	9	9	9	9	9	9	9	25	5
	8	230	17	19	87	160	205	218	199	148	71	18	17	17	17	17	17	17	82	4
	9	263	22	23	47	131	194	227	226	190	124	41	23	22	22	22	22	22	138	3
	10	280	27	27	28	71	148	200	221	209	165	93	30	27	27	27	27	27	180	2
	11	287	29	29	29	31	78	147	192	207	191	146	77	31	29	29	29	29	206	1
	12	290	30	30	30	30	32	75	142	185	200	185	142	75	32	30	30	30	215	12
HALF DAY TOTALS			119	142	291	534	787	980	1033	925	672	396	222	137	119	118	118	118	738	
Oct 21	7	48	2	3	20	36	45	47	42	30	12	2	2	2	2	2	2	2	4	5
	8	204	11	12	49	123	173	195	188	151	89	18	11	11	11	11	11	11	43	4
	9	257	17	17	23	104	180	225	235	209	151	64	18	17	17	17	17	17	97	3
	10	280	21	21	22	50	139	205	238	235	196	125	38	22	21	21	21	21	140	2
	11	291	24	24	24	25	71	156	212	236	224	178	101	28	24	24	24	24	168	1
	12	294	25	25	25	25	27	85	165	216	234	216	165	85	27	25	25	25	177	12
HALF DAY TOTALS			88	89	152	351	623	878	1006	974	791	493	247	117	89	88	88	88	540	
Nov 21	8	136	5	5	18	69	108	128	129	110	72	21	6	5	5	5	5	5	14	4
	9	232	12	12	13	73	151	201	219	204	156	80	13	12	12	12	12	12	55	3
	10	268	16	16	16	31	122	196	237	242	209	143	50	17	16	16	16	16	96	2
	11	283	19	19	19	20	61	154	218	248	240	194	116	28	19	19	19	19	123	1
	12	288	20	20	20	20	21	89	176	231	250	231	176	89	21	20	20	20	132	12
HALF DAY TOTALS			63	63	75	198	445	721	887	914	798	551	269	101	63	63	63	63	354	
Dec 21	8	89	3	3	8	41	67	82	84	73	50	17	3	3	3	3	3	3	6	4
	9	217	10	10	11	60	135	185	205	194	151	83	13	10	10	10	10	10	39	3
	10	261	14	14	14	25	113	188	232	239	210	146	55	15	14	14	14	14	77	2
	11	280	17	17	17	17	56	151	217	249	242	198	120	28	17	17	17	17	104	1
	12	285	18	18	18	18	19	89	178	233	253	233	178	89	19	18	18	18	113	12
HALF DAY TOTALS			52	52	56	146	374	649	822	867	775	557	276	94	53	52	52	52	282	
			N	NNW	NW	WNW	W	WSW	SW	SSW	S	SSE	SE	ESE	E	ENE	NE	NNE	HOR	PM

[a]Total solar heat gains for DSA glass (based on a ground reflectance of 0.20).

[b]Half day totals computed by Simpson's rule, time interval = 10 min.

APPENDICES

TABLE B.4 **Solar Intensity and Solar Heat Gain Factors[a] for 48° N Latitude (Conventional Units)[b]**

Date	Solar Time am	Direct Normal Btuh/ · ft²	Solar Heat Gain Factors, Btuh · ft² N	NNE	NE	ENE	E	ESE	SE	SSE	S	SSW	SW	WSW	W	WNW	NW	NNW	HOR	Solar Time pm
Jan 21	8	37	1	1	4	18	29	34	35	30	20	6	1	1	1	1	1	1	2	4
	9	185	8	8	8	53	118	160	176	166	129	69	10	8	8	8	8	8	25	3
	10	239	12	12	12	22	106	175	216	223	195	136	50	12	12	12	12	12	55	2
	11	261	14	14	14	15	53	144	208	239	233	190	116	26	14	14	14	14	77	1
	12	267	15	15	15	15	16	86	171	226	245	226	171	86	16	15	15	15	85	12
	HALF DAY TOTALS		43	43	46	117	316	567	729	776	701	512	259	85	43	43	43	43	203	
Feb 21	7	4	0	0	1	3	3	3	3	2	1	0	0	0	0	0	0	0	0	5
	8	180	8	8	36	103	149	170	166	136	82	17	8	8	8	8	8	8	25	4
	9	247	13	13	16	90	168	216	230	209	155	71	14	13	13	13	13	13	66	3
	10	275	17	17	17	38	131	203	242	244	207	138	44	18	17	17	17	17	105	2
	11	288	19	19	19	20	65	158	221	249	239	192	113	27	19	19	19	19	130	1
	12	292	20	20	20	20	22	89	176	231	250	231	176	89	22	20	20	20	138	12
	HALF DAY TOTALS		68	68	107	274	541	816	968	967	813	531	261	104	68	68	68	68	395	
Mar 21	7	153	7	22	80	123	145	145	123	80	23	7	7	7	7	7	7	7	20	5
	8	236	14	15	76	154	204	222	206	158	82	15	14	14	14	14	14	14	68	4
	9	270	19	19	3	121	193	234	239	207	142	52	20	19	19	19	19	19	118	3
	10	287	23	23	24	58	146	208	237	231	189	115	33	23	23	23	23	23	156	2
	11	295	25	25	25	26	74	156	210	232	218	172	94	28	25	25	25	25	180	1
	12	298	26	26	26	26	27	83	161	211	228	211	161	83	27	26	26	26	188	12
	HALF DAY TOTALS		100	118	250	494	775	1012	1100	1014	767	465	244	126	101	100	100	100	636	
Apr 21	6	108	12	53	86	105	107	93	64	23	6	6	6	6	6	6	6	6	15	6
	7	205	15	61	132	180	199	189	148	84	18	14	14	14	14	14	14	14	60	5
	8	247	20	32	111	179	219	225	196	138	55	21	20	20	20	20	20	20	114	4
	9	268	25	26	60	141	197	223	215	176	106	33	25	25	25	25	25	25	161	3
	10	280	28	28	31	77	148	193	209	194	150	80	31	28	28	28	28	28	196	2
	11	286	31	31	31	33	78	140	181	193	177	133	69	33	31	31	31	31	218	1
	12	288	31	31	31	31	34	71	131	172	186	172	131	71	34	31	31	31	226	12
	HALF DAY TOTALS		147	242	461	724	957	1098	1081	895	605	370	226	156	141	140	140	140	875	
May 21	5	41	17	31	40	42	39	29	14	3	3	3	3	3	3	3	3	3	5	7
	6	162	35	97	141	162	160	133	85	24	12	12	12	12	12	12	12	13	40	6
	7	219	23	90	158	200	212	191	142	68	21	19	19	19	19	19	19	19	91	5
	8	248	26	54	132	190	218	214	178	113	38	25	25	25	25	25	25	25	142	4
	9	264	29	32	82	151	194	208	192	147	77	32	29	29	29	29	29	29	185	3
	10	274	33	34	39	90	145	178	184	163	116	57	35	33	33	33	33	33	219	2
	11	279	35	35	36	40	79	126	155	160	142	101	54	37	35	35	35	35	240	1
	12	280	35	35	35	36	38	63	107	139	150	139	107	63	38	36	35	35	247	12
	HALF DAY TOTALS		215	388	645	893	1065	1114	1007	749	483	316	225	184	174	173	173	174	1045	
Jun 21	5	77	35	61	76	80	72	53	24	6	5	5	5	5	5	5	5	8	12	7
	6	172	46	110	155	175	169	138	84	22	14	14	14	14	14	14	14	16	51	6
	7	220	29	101	165	204	211	187	135	60	23	21	21	21	21	21	21	21	103	5
	8	246	29	64	139	191	215	206	168	101	34	27	27	27	27	27	27	27	152	4
	9	261	31	36	91	153	190	199	180	133	66	33	31	31	31	31	31	31	193	3
	10	269	34	36	45	94	143	169	171	148	101	50	36	34	34	34	34	34	225	2
	11	274	36	36	38	44	79	118	142	145	126	88	49	38	36	36	36	36	246	1
	12	275	37	37	37	38	40	60	96	124	134	124	96	60	40	38	37	37	252	12
	HALF DAY TOTALS		257	459	722	955	1095	1102	955	678	436	299	228	197	189	188	188	191	1108	
Jul 21	5	43	18	33	42	45	41	30	15	3	3	3	3	3	3	3	3	4	6	7
	6	156	37	96	138	159	156	129	82	24	13	13	13	13	13	13	13	14	41	6
	7	211	25	90	156	196	207	186	138	66	22	20	20	20	20	20	20	20	92	5
	8	240	27	56	132	187	214	209	174	110	38	26	26	26	26	26	26	26	142	4
	9	256	30	34	83	149	191	204	187	143	75	33	30	30	30	30	30	30	184	3
	10	266	34	35	41	90	143	174	180	158	113	56	36	34	34	34	34	34	217	2
	11	271	36	36	37	42	79	124	151	156	138	99	54	38	36	36	36	36	237	1
	12	272	36	36	36	37	39	63	104	136	146	136	104	63	39	37	36	36	244	12
	HALF DAY TOTALS		223	395	646	886	1050	1092	983	730	474	315	229	190	181	179	179	180	1042	
Aug 21	6	99	13	51	81	98	100	87	60	22	7	7	7	7	7	7	7	7	16	6
	7	190	17	61	128	172	190	179	141	79	19	15	15	15	15	15	15	15	61	5
	8	232	22	34	110	174	211	216	188	132	53	23	22	22	22	22	22	22	114	4
	9	154	27	28	63	139	192	216	108	169	102	34	27	27	27	27	27	27	159	3
	10	266	30	30	33	78	145	188	203	188	144	78	33	30	30	30	30	30	193	2
	11	272	32	32	32	36	78	137	175	187	171	129	68	35	32	32	32	32	215	1
	12	274	33	33	33	33	36	71	128	167	189	167	128	71	36	33	33	33	223	12
	HALF DAY TOTALS		157	251	459	709	929	1060	1040	862	587	366	231	165	151	149	149	149	869	
Sep 21	7	131	8	21	71	108	128	128	108	71	21	8	7	7	7	7	7	7	20	5
	8	215	15	16	72	144	191	207	193	148	77	16	15	15	15	15	15	15	65	4
	9	251	20	20	34	116	184	223	227	197	136	52	21	20	20	20	20	20	114	3
	10	269	24	24	25	58	141	200	228	221	182	112	34	24	24	24	24	24	151	2
	11	278	26	26	26	28	73	151	203	223	210	166	92	29	26	26	26	26	174	1
	12	280	27	27	27	27	29	82	156	204	220	204	156	82	29	27	27	27	182	12
	HALF DAY TOTALS		105	121	240	465	729	953	1040	963	737	453	243	131	106	105	105	105	614	
Oct 21	7	4	0	0	2	3	4	4	3	2	1	0	0	0	0	0	0	0	0	5
	8	165	8	9	35	96	139	159	155	126	77	16	8	8	8	8	8	8	25	4
	9	233	14	14	16	88	161	207	220	199	148	68	15	14	14	14	14	14	66	3
	10	262	18	18	18	39	128	196	233	234	199	133	43	18	18	18	18	18	104	2
	11	274	20	20	20	21	64	153	213	241	231	186	109	27	20	20	20	20	128	1
	12	278	21	21	21	21	23	87	171	223	242	223	171	87	23	21	21	21	136	12
	HALF DAY TOTALS		71	71	108	266	519	780	925	925	779	513	256	106	72	71	71	71	391	
Nov 21	8	36	1	1	4	18	29	34	35	30	20	6	1	1	1	1	1	1	2	4
	9	179	8	8	9	52	115	156	171	161	125	67	10	8	8	8	8	8	26	3
	10	233	12	12	12	22	104	172	212	218	191	133	49	13	12	12	12	12	55	2
	11	255	15	15	15	15	52	142	204	234	228	186	114	26	15	15	15	15	77	1
	12	261	15	15	15	15	17	85	168	222	240	222	168	85	17	15	15	15	85	12
	HALF DAY TOTALS		44	44	47	117	310	555	713	760	686	502	255	85	44	44	44	44	204	
Dec 21	9	140	5	5	6	36	86	120	133	127	100	56	8	5	5	5	5	5	13	3
	10	214	10	10	10	16	91	156	194	201	179	126	49	10	10	10	10	10	38	2
	11	242	12	12	12	13	46	134	195	225	220	180	111	25	12	12	12	12	57	1
	12	250	13	13	13	13	14	81	163	215	233	215	168	81	14	13	13	13	65	12
	HALF DAY TOTALS		33	33	34	73	233	458	610	665	616	468	247	76	34	33	33	33	141	
			N	NNW	NW	WNW	W	WSW	SW	SSW	S	SSE	SE	ESE	E	ENE	NE	NNE	HOR	PM

[a]Total solar heat gains for DSA glass (based on a ground reflectance of 0.20).

[b]Half day totals computed by Simpson's rule, time interval = 10 min.

APPENDICES

TABLE B.5 **Solar Intensity and Solar Heat Gain Factors[a] for 64° N Latitude (Conventional Units)[b]**

Date	Solar Time am	Direct Normal Btuh ft^2	Solar Heat Gain Factors, Btuh · ft^2 N	NNE	NE	ENE	E	ESE	SE	SSE	S	SSW	SW	WSW	W	WNW	NW	NNW	HOR	Solar Time pm
Jan 21	10	22	1	1	1	1	9	16	20	21	19	13	5	1	1	1	1	1	1	2
	11	81	3	3	3	3	15	45	67	77	75	62	38	8	3	3	3	3	6	1
	12	100	3	3	3	3	4	33	67	89	96	89	67	33	4	3	3	3	8	12
HALF DAY TOTALS			5	5	5	6	25	79	121	142	141	119	75	23	5	5	5	5	11	
Feb 21	8	18	1	1	3	10	15	17	17	14	9	2	1	1	1	1	1	1	1	4
	9	134	5	5	6	43	89	118	128	119	90	45	6	5	5	5	5	5	13	3
	10	190	8	8	8	18	87	144	176	180	157	108	38	9	8	8	8	8	28	2
	11	215	10	10	10	11	44	122	177	202	197	160	97	20	10	10	10	10	41	1
	12	222	11	11	11	11	12	73	147	194	210	194	147	73	12	11	11	11	45	12
HALF DAY TOTALS			29	30	33	89	244	446	578	617	560	411	212	66	30	29	29	29	106	
Mar 21	7	95	4	11	47	74	90	91	79	53	17	4	4	4	4	4	4	4	9	5
	8	185	9	10	46	113	158	177	170	135	78	14	9	9	9	9	9	9	32	4
	9	227	13	13	16	88	159	203	215	194	143	64	14	13	13	13	13	13	59	3
	10	249	16	16	16	35	122	190	226	228	194	130	42	16	16	16	16	16	84	2
	11	260	17	17	17	18	60	148	209	236	228	184	109	25	17	17	17	17	99	1
	12	263	18	18	18	18	19	85	168	221	239	221	168	85	19	18	18	18	105	12
HALF DAY TOTALS			68	74	150	334	596	854	984	958	779	504	257	104	68	68	68	68	335	
Apr 21	5	27	8	18	24	27	26	20	12	2	1	1	1	1	1	1	1	1	2	7
	6	133	12	59	102	127	132	118	84	35	8	8	8	8	8	8	8	8	21	6
	7	194	14	41	113	163	189	185	153	96	25	13	13	13	13	13	13	13	51	5
	8	228	17	19	79	153	201	217	201	153	79	19	17	17	17	17	17	17	85	4
	9	248	21	21	32	111	180	219	225	197	138	55	22	21	21	21	21	21	116	3
	10	260	23	23	24	51	134	194	225	221	185	118	38	24	23	23	23	23	140	2
	11	266	24	24	24	26	68	148	202	225	214	171	99	29	24	24	24	24	155	1
	12	268	25	25	25	25	27	83	159	208	224	208	159	83	27	25	25	25	160	12
HALF DAY TOTALS			131	218	410	671	943	1150	1186	1036	763	487	273	149	121	120	120	120	651	
May 21	4	51	30	44	51	51	43	28	8	3	3	3	3	3	3	3	3	10	6	8
	5	132	48	95	125	135	125	96	50	11	9	9	9	9	9	9	9	11	26	7
	6	185	28	97	150	181	183	158	109	40	15	15	15	15	15	15	15	15	55	6
	7	218	21	63	138	189	211	201	161	94	24	19	19	19	19	19	19	19	90	5
	8	239	23	28	97	167	209	220	198	146	68	25	23	23	23	23	23	23	124	4
	9	252	26	27	45	122	183	215	215	184	123	46	27	26	26	26	26	26	152	3
	10	261	28	28	30	61	135	188	212	205	167	102	36	28	28	28	28	28	174	2
	11	265	30	30	30	32	72	141	188	207	195	154	87	33	30	30	30	30	188	1
	12	267	30	30	30	30	33	78	146	189	204	189	146	78	33	30	30	30	192	12
HALF DAY TOTALS			247	425	680	950	1177	1291	1218	985	708	465	288	191	169	168	168	176	911	
Jun 21	4	93	53	83	96	94	78	50	14	7	7	7	7	7	7	7	7	21	16	8
	5	154	62	114	148	158	145	110	55	14	12	12	12	12	12	12	12	14	39	7
	6	194	36	107	162	191	192	163	110	39	18	17	17	17	17	17	17	18	71	6
	7	221	24	71	145	193	213	200	158	89	25	22	22	22	22	22	22	22	105	5
	8	239	25	33	104	170	208	216	192	139	62	27	25	25	25	25	25	25	137	4
	9	251	28	29	51	124	181	210	208	175	115	43	29	28	28	28	28	28	165	3
	10	258	30	30	32	65	134	183	204	195	157	94	36	30	30	30	30	30	186	2
	11	262	32	32	32	34	72	137	180	196	184	144	82	35	32	32	32	32	199	1
	12	263	32	32	32	32	35	76	138	179	193	179	138	76	35	32	32	32	203	12
HALF DAY TOTALS			322	533	801	1061	1253	1317	1195	946	679	455	296	211	192	190	191	213	1021	
Jul 21	4	53	32	47	55	54	46	29	9	4	4	4	4	4	4	4	4	11	8	8
	5	128	49	94	123	133	124	95	50	11	10	10	10	10	10	10	10	11	28	7
	6	179	30	96	148	177	180	155	106	39	16	15	15	15	15	15	15	15	57	6
	7	211	22	64	137	186	207	197	157	92	25	20	20	20	20	20	20	20	92	5
	8	231	24	30	97	165	205	215	193	142	67	26	24	24	24	24	24	24	124	4
	9	245	27	28	47	121	180	211	211	179	120	46	28	27	27	27	27	27	152	3
	10	253	29	29	31	62	134	185	208	200	164	100	37	29	29	29	29	29	174	2
	11	257	31	31	31	33	72	139	185	202	191	151	86	34	31	31	31	31	187	1
	12	259	31	31	31	31	34	78	143	185	200	185	143	78	34	31	31	31	192	12
HALF DAY TOTALS			258	434	684	946	1163	1269	1193	965	697	462	292	198	177	175	175	185	918	
Aug 21	5	29	9	20	27	30	28	22	13	2	2	2	2	2	2	2	2	2	3	7
	6	123	13	58	97	121	125	111	80	34	9	9	9	9	9	9	9	9	23	6
	7	181	15	42	109	157	180	176	145	92	26	14	14	14	14	14	14	14	53	5
	8	214	19	21	78	148	193	208	192	147	76	21	19	19	19	19	19	19	87	4
	9	234	22	22	34	109	174	211	217	189	133	55	23	22	22	22	22	22	117	3
	10	246	25	25	26	52	131	188	217	214	178	114	39	25	25	25	25	25	140	2
	11	252	26	26	26	28	69	144	196	217	207	166	97	31	26	26	26	26	154	1
	12	254	27	27	27	27	29	82	155	201	217	201	155	82	29	27	27	27	159	12
HALF DAY TOTALS			142	226	410	657	914	1109	1141	997	740	478	275	158	131	130	130	130	656	
Sep 21	7	77	4	10	39	62	74	75	65	44	15	4	4	4	4	4	4	4	8	5
	8	163	10	10	43	103	143	160	154	123	71	14	10	10	10	10	10	10	31	4
	9	206	14	14	17	83	148	189	200	181	133	61	15	14	14	14	14	14	57	3
	10	229	16	16	17	35	116	179	213	214	183	123	41	17	16	16	16	16	81	2
	11	240	18	18	18	19	59	141	198	224	216	174	104	26	18	18	18	18	96	1
	12	244	19	19	19	19	21	82	160	209	227	209	160	82	21	19	19	19	101	12
HALF DAY TOTALS			71	77	142	307	547	787	910	891	731	480	249	106	72	71	71	71	324	
Oct 21	8	17	1	1	3	10	14	16	16	13	8	2	1	1	1	1	1	1	1	4
	9	122	5	5	6	40	82	109	118	110	83	42	6	5	5	5	5	5	13	3
	10	176	9	9	9	18	83	135	165	169	147	102	36	9	9	9	9	9	29	2
	11	201	11	11	11	11	43	116	167	191	186	152	92	20	11	11	11	11	41	1
	12	208	11	11	11	11	13	70	140	184	199	184	140	70	13	11	11	11	46	12
HALF DAY TOTALS			31	31	34	86	231	420	542	580	527	388	202	66	32	31	31	31	108	
Nov 21	10	23	1	1	1	1	10	17	21	22	20	14	5	1	1	1	1	1	1	2
	11	79	3	3	3	3	15	44	65	75	74	61	37	8	3	3	3	3	6	1
	12	97	4	4	4	4	4	32	66	87	93	87	66	32	4	4	4	4	8	12
HALF DAY TOTALS			5	5	5	6	26	79	120	141	140	117	74	23	6	5	5	5	11	
Dec 21	11	4	0	0	0	0	1	2	3	4	4	3	2	0	0	0	0	0	0	1
	12	16	0	0	0	0	1	5	11	14	15	14	11	5	1	0	0	0	1	12
HALF DAY TOTALS			0	0	0	0	1	5	9	11	11	10	7	3	0	0	0	0	1	
			N	NNW	NW	WNW	W	WSW	SW	SSW	S	SSE	SE	ESE	E	ENE	NE	NNE	HOR	PM

[a]Total solar heat gains for DSA glass (based on a ground reflectance of 0.20).

[b]Half day totals computed by Simpson's rule, time interval = 10 min.

TABLE B.6 **Solar Intensity and Solar Heat Gain Factors[a] for 16° N Latitude (SI Units)[b]**

			Solar Heat Gain Factors, W/m²																	
Date	Solar Time am	Direct Normal W/m²	N	NNE	NE	ENE	E	ESE	SE	SSE	S	SSW	SW	WSW	W	WNW	NW	NNW	HOR	Solar Time pm
Jan 21	7	445	17	19	138	291	390	424	397	303	155	19	17	17	17	17	17	17	43	5
	8	827	45	48	174	463	662	757	734	596	359	79	45	45	45	45	45	45	249	4
	9	948	67	67	102	384	630	770	791	690	481	183	69	67	67	67	67	67	472	3
	10	1001	82	82	86	209	474	658	737	704	563	321	97	82	82	82	82	82	640	2
	11	1025	92	92	92	96	242	467	614	663	612	462	236	96	92	92	92	92	745	1
	12	1032	95	95	95	95	100	228	438	580	628	580	438	228	100	95	95	95	782	12
HALF DAY TOTALS			348	352	618	1453	2398	3153	3458	3217	2465	1344	666	401	350	348	348	348	2539	
Feb 21	7	575	24	55	265	435	532	544	474	326	113	26	24	24	24	24	24	24	80	5
	8	862	53	60	304	567	729	778	706	525	244	56	53	53	53	53	53	53	319	4
	9	961	74	77	202	482	676	763	733	592	347	96	74	74	74	74	74	74	549	3
	10	1006	90	91	104	292	508	636	665	593	423	193	94	90	90	90	90	90	722	2
	11	1027	99	99	102	118	262	428	527	542	471	323	154	103	99	99	99	99	831	1
	12	1034	103	103	103	105	108	189	336	448	487	448	336	189	108	105	103	103	868	12
HALF DAY TOTALS			390	431	1013	1922	2730	3228	3263	2792	1836	906	547	417	393	390	390	390	2933	
Mar 21	7	634	36	167	393	544	606	578	458	260	47	33	33	33	33	33	33	33	126	5
	8	857	63	157	442	648	752	741	615	390	111	61	61	61	61	61	61	61	380	4
	9	943	84	110	343	565	689	709	622	435	180	86	82	82	82	82	82	82	606	3
	10	983	98	103	191	379	519	575	543	424	240	107	100	98	98	98	98	98	778	2
	11	1003	108	110	113	166	273	361	395	366	281	173	114	110	108	108	108	108	885	1
	12	1008	111	111	112	114	117	149	216	273	295	273	216	149	117	114	112	111	919	12
HALF DAY TOTALS			443	712	1558	2383	2926	3076	2773	2028	1006	588	483	448	440	483	437	437	3233	
Apr 21	6	44	7	24	37	43	43	37	24	7	2	2	2	2	2	2	12	2	4	6
	7	622	75	298	482	589	604	528	369	141	45	42	42	42	42	42	42	42	169	5
	8	807	85	312	543	682	718	644	473	217	74	70	70	70	70	70	70	70	413	4
	9	885	97	248	469	610	657	608	465	244	97	90	90	90	90	90	90	90	623	3
	10	924	112	171	321	444	499	476	380	231	118	109	166	106	106	106	106	106	784	2
	11	942	120	127	169	228	270	276	245	189	136	121	118	115	115	115	115	118	882	1
	12	948	123	123	124	125	126	129	135	141	143	141	135	129	126	125	124	123	911	12
HALF DAY TOTALS			565	1272	2127	2711	2909	2684	2059	1111	547	503	494	491	490	489	490	492	3333	
May 21	6	138	43	94	128	141	134	106	59	11	9	9	9	9	9	9	9	10	195	6
	7	608	157	378	531	603	583	474	290	76	49	49	49	49	49	49	49	55	17	5
	8	771	165	415	598	689	677	564	362	121	78	76	76	76	76	76	76	80	425	4
	9	845	156	366	539	626	622	526	344	141	100	96	96	96	96	96	96	100	621	3
	10	883	149	281	410	478	474	398	264	139	116	111	111	111	111	111	111	116	772	2
	11	902	147	198	248	273	262	220	166	130	126	123	120	120	120	122	124	128	862	1
	12	907	146	144	140	134	132	131	130	129	129	129	130	131	132	134	140	144	890	12
HALF DAY TOTALS			893	1813	2537	2891	2829	2360	1555	685	541	528	527	526	527	529	533	557	3338	
Jun 21	6	168	64	124	163	175	162	123	64	13	12	12	12	12	12	12	17	13	24	6
	7	593	195	404	543	598	565	445	252	63	52	52	52	52	52	52	52	57	203	5
	8	750	209	449	612	684	653	526	313	98	79	79	79	79	79	79	79	84	425	4
	9	823	199	409	560	626	601	487	294	118	98	98	98	98	98	98	98	105	613	3
	10	861	187	329	441	486	459	363	222	125	116	113	113	113	113	113	113	121	759	2
	11	879	181	241	283	290	258	200	146	130	126	122	122	122	122	125	128	136	847	1
	12	885	179	173	158	142	134	132	130	129	129	129	130	132	134	142	158	173	873	12
HALF DAY TOTALS			1122	2043	2683	2930	2763	2207	1357	612	548	540	540	540	542	545	555	599	3308	
Jul 21	6	128	43	90	122	134	127	99	55	11	9	9	9	9	9	9	9	10	17	6
	7	579	161	373	518	585	564	456	277	74	51	51	51	51	51	51	51	55	194	5
	8	743	173	415	590	676	661	549	349	110	78	78	78	78	78	78	78	83	419	4
	9	818	165	371	537	618	610	513	334	138	102	99	99	99	99	99	99	104	611	3
	10	857	157	289	413	475	467	389	257	138	118	113	113	113	113	113	113	120	759	2
	11	876	154	207	255	277	262	218	164	133	128	125	122	122	122	125	128	131	848	1
	12	882	154	151	145	138	135	134	133	132	132	132	133	134	135	138	145	151	875	12
HALF DAY TOTALS			933	1829	2521	2848	2770	2298	1507	680	554	541	539	539	540	542	547	575	3289	
Aug 21	6	136	7	21	31	37	36	31	20	6	2	2	2	2	2	2	2	2	4	6
	7	569	81	289	458	555	567	493	343	131	48	45	45	45	45	45	45	45	167	5
	8	757	94	315	531	660	691	617	451	206	79	74	74	74	74	74	74	74	404	4
	9	838	104	258	467	598	640	589	448	235	103	95	95	95	95	95	95	95	608	3
	10	879	118	183	328	443	490	464	368	224	122	114	110	110	110	110	110	110	766	2
	11	899	127	136	180	235	271	273	240	185	138	127	124	120	120	120	120	124	860	1
	12	905	129	130	130	131	132	134	139	143	145	143	139	134	132	131	130	130	889	12
HALF DAY TOTALS			603	1294	2099	2640	2810	2578	1969	1069	568	528	519	516	515	514	515	518	3258	
Sep 21	7	565	38	157	360	497	554	529	419	240	48	35	35	35	35	35	35	35	122	5
	8	797	67	156	424	618	716	705	587	374	113	64	64	64	64	64	64	64	367	4
	9	887	87	114	335	547	665	684	602	424	181	90	86	86	86	86	86	86	585	3
	10	931	101	107	193	372	507	560	529	415	240	111	103	101	101	101	101	101	752	2
	11	952	111	114	118	169	272	356	389	361	279	176	118	114	111	111	111	111	856	1
	12	958	114	114	116	118	121	153	217	272	293	272	217	153	121	118	116	114	889	12
HALF DAY TOTALS			461	712	1500	2276	2791	2937	2658	1963	1007	605	501	466	457	455	454	454	3126	
Oct 21	7	524	25	56	249	404	492	502	437	300	105	27	25	25	25	25	25	25	79	5
	8	816	55	64	299	548	702	747	677	502	234	59	55	55	55	55	55	55	313	4
	9	920	76	80	204	473	659	741	711	573	336	97	76	76	76	76	76	76	537	3
	10	968	92	94	108	291	499	621	647	577	412	189	97	92	92	92	92	92	707	2
	11	991	102	105	122	261	420	515	528	459	315	154	106	102	102	102	102	102	814	1
	12	997	105	105	105	107	111	189	330	437	474	437	330	189	111	107	105	105	850	12
HALF DAY TOTALS			402	444	1002	1869	2637	3110	3141	2689	1776	891	553	428	404	402	402	402	2872	
Nov 21	7	423	17	19	134	280	375	406	379	289	147	19	17	17	17	17	17	17	43	5
	8	806	46	49	174	456	651	742	719	583	350	78	46	46	46	46	46	46	247	4
	9	929	67	67	105	382	623	779	779	679	472	180	70	67	67	67	67	67	468	3
	10	984	83	83	87	210	470	651	727	694	554	316	98	83	83	83	83	83	635	2
	11	1009	92	92	92	98	242	462	607	654	603	455	234	98	92	92	92	92	740	1
	12	1016	96	96	96	96	101	227	432	572	619	572	432	227	101	96	96	96	775	12
HALF DAY TOTALS			352	357	620	1438	2363	3101	3396	3159	2421	1324	664	404	355	352	352	352	2520	
Dec 21	7	372	13	14	93	228	318	354	339	268	150	21	13	13	13	13	13	13	31	5
	8	803	42	43	129	416	625	734	728	609	390	105	42	42	42	42	42	42	219	4
	9	936	63	63	78	341	604	761	800	716	520	226	66	63	63	63	63	63	436	3
	10	993	78	78	81	178	445	657	753	735	605	369	112	79	78	78	78	78	602	2
	11	1019	87	87	87	92	231	474	638	698	654	506	270	94	87	87	87	87	704	1
	12	1026	91	91	91	91	95	241	469	620	670	620	469	241	95	91	91	91	739	12
HALF DAY TOTALS			328	330	501	1269	2241	3076	3468	3312	2638	1527	721	393	331	328	328	328	2361	
			N	NNW	NW	WNW	W	WSW	SW	SSW	S	SSE	SE	ESE	E	ENE	NE	NNE	HOR	PM

[a]Total solar heat gains for DSA glass (based on a ground reflectance of 0.20).

[b]Half day totals computed by Simpson's rule, time interval = 10 min.

APPENDICES

TABLE B.7 **Solar Intensity and Solar Heat Gain Factors[a] for 32° N Latitude (SI Units)[b]**

Date	Solar Time am	Direct Normal W/m^2	Solar Heat Gain Factors, W/m^2																	Solar Time pm
			N	NNE	NE	ENE	E	ESE	SE	SSE	S	SSW	SW	WSW	W	WNW	NW	NNW	HOR	
Jan 21	7	4	0	0	1	3	4	4	4	3	2	0	0	0	0	0	0	0	0	5
	8	640	28	29	93	330	505	597	596	502	326	88	29	28	28	28	28	28	102	4
	9	849	48	48	53	286	553	721	775	711	534	258	52	48	48	48	48	48	278	3
	10	931	63	63	64	129	427	659	784	788	670	444	144	64	63	63	63	63	430	2
	11	967	71	71	71	75	213	502	698	784	750	602	347	90	71	71	71	71	524	1
	12	977	74	74	74	74	79	277	548	718	777	718	548	277	79	74	74	74	556	12
HALF DAY TOTALS			249	250	338	897	1797	2700	3201	3198	2692	1746	831	353	252	249	249	249	1614	
Feb 21	7	352	13	23	148	258	323	336	299	212	83	14	13	13	13	13	13	13	30	5
	8	771	41	43	204	469	646	719	683	538	300	53	41	41	41	41	41	41	200	4
	9	905	60	60	101	386	627	764	783	680	471	174	63	60	60	60	60	60	402	3
	10	964	75	75	78	195	475	670	760	733	595	352	99	75	75	75	75	75	556	2
	11	990	83	83	83	88	240	491	656	717	670	519	275	89	83	83	83	83	652	1
	12	998	86	86	86	86	91	250	489	644	696	643	489	250	91	86	86	86	684	12
HALF DAY TOTALS			314	324	635	1405	2318	3086	3408	3188	2461	1426	718	384	316	314	314	314	2181	
Mar 21	7	583	30	116	332	483	556	545	447	276	63	29	29	29	29	29	29	29	100	5
	8	821	54	78	339	576	717	746	661	473	195	57	54	54	54	54	54	54	315	4
	9	914	74	77	203	475	662	746	716	578	339	95	74	74	74	74	74	74	516	3
	10	959	88	88	96	273	499	637	677	615	456	221	93	88	88	88	88	88	667	2
	11	980	96	96	98	108	258	447	564	592	529	380	185	101	96	96	96	96	762	1
	12	987	99	99	99	99	105	209	386	511	554	511	386	209	105	99	99	99	795	12
HALF DAY TOTALS			393	512	1131	1985	2761	3258	3284	2801	1858	1030	609	429	394	391	391	391	2757	
Apr 21	6	210	29	110	172	205	207	177	119	38	11	11	11	11	11	11	11	11	23	6
	7	649	53	253	462	593	631	575	428	204	49	45	45	45	45	45	45	45	191	5
	8	804	73	192	453	632	715	689	559	337	95	69	69	69	69	69	69	69	408	4
	9	876	89	113	324	532	649	669	591	418	183	92	87	87	87	87	87	87	593	3
	10	913	101	106	165	342	490	599	543	445	275	123	104	101	101	101	101	101	736	2
	11	932	109	109	115	149	263	371	426	415	340	222	126	113	109	109	109	109	825	1
	12	937	112	112	112	116	120	167	260	335	363	335	260	167	120	116	112	112	854	12
HALF DAY TOTALS			508	932	1734	2500	3003	3129	2806	2033	1135	721	559	496	482	479	479	479	3203	
May 21	6	374	104	244	340	381	367	295	175	39	26	26	26	26	26	26	26	28	67	6
	7	666	112	350	535	638	643	550	374	134	60	57	57	57	57	57	57	59	256	5
	8	787	93	295	519	655	694	629	470	229	86	8C	80	80	80	80	80	80	461	4
	9	849	104	194	404	558	625	601	488	294	117	100	97	97	97	97	97	97	633	3
	10	882	114	127	240	383	473	491	435	313	170	117	110	110	110	110	110	110	766	2
	11	898	121	124	132	186	260	312	322	285	215	147	126	122	118	118	118	118	847	1
	12	904	121	123	125	127	130	148	186	220	223	220	186	148	130	127	125	123	873	12
HALF DAY TOTALS			700	1382	2214	2840	3106	2943	2356	1409	788	627	577	558	551	550	548	551	3464	
Jun 21	6	412	140	289	388	425	401	314	174	39	32	32	32	32	32	32	32	35	89	6
	7	662	148	384	556	644	634	529	342	109	62	62	62	62	62	62	62	65	279	5
	8	773	115	335	540	656	677	597	427	188	89	84	84	84	84	84	84	84	476	4
	9	831	110	234	431	563	609	567	440	244	111	101	101	101	101	101	101	101	642	3
	10	863	120	150	272	395	461	458	387	261	143	120	114	114	114	114	114	114	770	2
	11	880	126	130	148	202	258	288	280	237	177	135	128	125	122	122	122	122	847	1
	12	885	128	129	130	132	134	144	164	183	191	183	164	144	134	132	130	129	871	12
HALF DAY TOTALS			824	1589	2403	2950	3108	2829	2137	1174	710	621	595	585	582	579	579	586	3538	
Jul 21	6	358	106	240	332	370	355	285	168	39	27	27	27	27	27	27	27	30	70	6
	7	640	118	349	526	623	626	534	361	129	62	59	59	59	59	59	59	61	257	5
	8	761	98	300	515	645	680	613	456	221	89	82	82	82	82	82	82	82	457	4
	9	823	107	202	405	553	615	588	475	284	117	100	100	100	100	100	100	100	626	3
	10	856	118	133	247	383	467	481	424	303	167	120	113	113	113	113	113	113	757	2
	11	873	124	128	137	191	261	308	314	277	210	147	129	125	121	121	121	121	836	1
	12	879	126	127	128	130	133	150	184	214	226	214	184	150	133	130	128	127	861	12
HALF DAY TOTALS			728	1402	2210	2809	3052	2877	2290	1365	783	637	591	574	568	566	565	568	3431	
Aug 21	6	187	30	103	159	188	189	162	108	35	12	12	12	12	12	12	12	12	25	6
	7	599	59	250	444	565	598	542	402	192	53	49	49	49	49	49	49	49	192	5
	8	756	79	200	446	614	690	662	536	321	96	74	74	74	74	74	74	74	403	4
	9	830	95	124	328	523	632	648	570	402	179	98	93	93	93	93	93	93	583	3
	10	869	106	113	175	344	482	544	526	429	266	126	110	106	106	106	106	106	722	2
	11	889	115	117	122	157	264	365	414	401	328	217	131	119	115	115	115	115	808	1
	12	894	117	117	117	122	126	170	255	324	350	324	255	170	126	122	117	117	837	12
HALF DAY TOTALS			539	957	1721	2443	2910	3014	2694	1950	1109	729	581	524	510	507	506	506	3152	
Sep 21	7	514	32	109	302	437	502	492	405	252	62	30	30	30	30	30	30	30	97	5
	8	758	57	81	324	546	679	706	626	450	190	61	57	57	57	57	57	57	304	4
	9	857	77	82	201	459	637	717	688	557	330	99	77	77	77	77	77	77	498	3
	10	905	91	91	101	270	485	617	655	596	444	220	97	91	91	91	91	91	645	2
	11	928	100	100	102	113	257	437	549	576	516	373	187	106	100	100	100	100	737	1
	12	935	103	103	103	103	110	210	379	499	540	499	379	210	110	103	103	103	769	12
HALF DAY TOTALS			409	518	1089	1888	2621	3097	3132	2689	1812	1025	620	446	411	408	408	408	2664	
Oct 21	7	312	13	24	136	233	291	301	268	190	75	14	13	13	13	13	13	13	30	5
	8	724	42	46	200	450	616	684	649	511	285	54	42	42	42	42	42	42	198	4
	9	862	63	63	104	378	608	738	755	655	454	170	65	63	63	63	63	63	395	3
	10	923	77	77	81	197	465	652	737	711	577	342	101	77	77	77	77	77	546	2
	11	952	86	86	86	91	239	481	639	697	651	505	269	93	86	86	86	86	639	1
	12	960	89	89	89	89	95	247	478	626	677	626	278	247	95	89	89	89	671	12
HALF DAY TOTALS			325	335	632	1365	2234	2968	3275	3067	2376	1390	714	394	328	325	325	325	2143	
Nov 21	7	5	0	0	1	3	4	5	4	3	2	0	0	0	0	0	0	0	0	5
	8	619	29	29	93	323	493	581	579	488	316	86	29	29	29	29	29	29	102	4
	9	829	49	49	55	284	545	709	761	697	523	253	49	49	49	49	49	49	277	3
	10	912	64	64	65	131	423	650	772	775	659	437	142	65	64	64	64	64	428	2
	11	949	72	72	72	76	213	496	688	773	739	593	342	91	72	72	72	72	521	1
	12	959	75	75	75	75	80	274	541	708	766	708	541	274	80	75	75	75	553	12
HALF DAY TOTALS			253	254	342	890	1769	2653	3143	3138	2643	1717	824	355	256	253	253	253	1607	
Dec 21	8	556	22	22	59	265	426	515	523	452	305	98	23	22	22	22	22	22	69	4
	9	812	43	43	46	244	512	686	751	701	539	282	48	43	43	43	43	43	228	3
	10	908	57	57	57	106	402	642	777	792	683	466	163	59	57	57	57	57	375	2
	11	949	66	66	66	69	200	497	700	795	766	620	367	92	66	66	66	66	468	1
	12	960	69	69	69	69	73	281	559	733	794	733	559	281	73	69	69	69	500	12
HALF DAY TOTALS			223	223	265	717	1577	2499	3043	3111	2687	1823	868	339	225	223	223	223	1389	
			N	NNW	NW	WNW	W	WSW	SW	SSW	S	SSE	SE	ESE	E	ENE	NE	NNE	HOR	PM

[a]Total solar heat gains for DSA glass (based on a ground reflectance of 0.20).

[b]Half day totals computed by Simpson's rule, time interval = 10 min.

TABLE B.8 **Solar Intensity and Solar Heat Gain Factors[a] for 40° N Latitude (SI Units)[b]**

Date	Solar Time am	Direct Normal W/m²	Solar Heat Gain Factors, W/m² N	NNE	NE	ENE	E	ESE	SE	SSE	S	SSW	SW	WSW	W	WNW	NW	NNW	HOR	Solar Time pm
Jan 21	8	446	17	17	55	223	350	417	420	358	236	60	17	17	17	17	17	17	44	4
	9	753	37	37	41	233	485	648	706	658	504	260	42	37	37	37	37	37	173	3
	10	865	51	51	51	97	390	627	761	776	671	460	161	53	51	51	51	51	303	2
	11	913	59	59	59	62	193	493	699	796	769	623	372	89	59	59	59	59	390	1
	12	926	62	62	62	62	66	293	563	740	802	740	563	283	62	62	62	62	419	12
HALF DAY TOTALS			194	194	231	628	1426	2316	2852	2941	2566	1770	860	318	196	194	194	194	1117	
Feb 21	7	175	6	10	71	127	160	167	150	107	43	6	6	6	6	6	6	6	11	5
	8	692	33	35	158	407	576	651	628	505	296	56	33	33	33	33	33	33	136	4
	9	854	52	52	70	337	587	738	773	689	496	209	55	52	52	52	52	52	309	3
	10	926	65	65	67	155	450	666	777	768	641	408	120	66	65	65	65	65	451	2
	11	958	73	73	73	77	224	504	690	769	730	579	325	86	73	73	73	73	538	1
	12	967	76	76	76	76	80	271	536	702	760	702	536	271	80	76	76	76	568	12
HALF DAY TOTALS			267	271	478	1140	2043	2888	3308	3202	2591	1602	790	361	269	267	267	267	1730	
Mar 21	7	540	27	92	295	441	514	509	425	271	69	26	26	26	26	26	26	26	83	5
	8	789	50	57	288	534	686	732	665	494	232	54	50	50	50	50	50	50	268	4
	9	889	67	70	147	429	640	749	744	625	404	127	69	67	67	67	67	67	450	3
	10	938	80	80	85	226	482	653	722	682	539	299	91	80	80	80	80	80	587	2
	11	961	88	88	88	94	247	476	623	673	622	473	244	94	88	88	88	88	673	1
	12	968	91	91	91	91	97	238	458	602	650	602	458	238	97	91	91	91	702	12
HALF DAY TOTALS			358	440	954	1777	2626	3265	3429	3055	2191	1270	694	417	360	357	357	357	2411	
Apr 21	6	282	36	144	228	275	279	241	164	56	16	15	15	15	15	15	15	15	34	6
	7	651	50	223	442	584	633	588	451	255	51	45	45	45	45	45	45	45	193	5
	8	795	69	140	402	601	706	703	594	391	130	69	67	67	67	67	67	67	389	4
	9	865	84	91	254	488	638	691	640	494	260	91	84	84	84	84	84	84	557	3
	10	901	96	99	117	291	480	589	608	538	381	177	101	96	96	96	96	96	685	2
	11	920	104	107	107	122	255	411	506	522	459	323	164	109	104	104	104	104	766	1
	12	926	106	106	106	109	114	196	341	448	486	448	341	196	114	109	106	106	794	12
HALF DAY TOTALS			487	835	1580	2390	3020	3314	3135	2466	1539	935	628	495	467	464	464	464	3020	
May 21	5	3	1	2	3	3	3	2	1	0	0	0	0	0	0	0	0	0	0	7
	6	453	113	284	403	458	446	364	223	56	33	33	33	33	33	33	33	35	96	6
	7	681	90	320	520	638	659	580	412	172	63	59	59	59	59	59	59	59	276	5
	8	787	86	230	471	629	694	655	519	295	92	80	80	80	80	80	80	80	461	4
	9	844	99	131	330	518	620	632	551	382	168	101	96	96	96	96	96	96	616	3
	10	875	107	114	171	332	467	529	513	419	262	127	111	107	107	107	107	107	737	2
	11	891	115	115	121	152	256	357	409	400	331	222	133	120	115	115	115	115	812	1
	12	896	117	117	117	121	126	171	258	329	355	329	258	171	126	121	117	117	836	12
HALF DAY TOTALS			679	1275	2102	2818	3231	3232	2778	1897	1129	780	630	568	554	550	550	552	3418	
Jun 21	5	68	32	54	68	70	63	46	20	5	4	4	4	4	4	4	4	7	8	7
	6	488	150	329	450	501	478	380	222	53	39	39	39	39	39	39	39	43	126	6
	7	681	118	355	543	648	654	562	385	145	68	65	65	65	65	65	65	67	306	5
	8	776	94	268	492	633	680	626	480	252	93	85	85	85	85	85	85	85	484	4
	9	829	105	160	359	524	607	601	507	332	142	104	100	100	100	100	100	100	633	3
	10	859	112	121	197	345	457	500	468	366	218	122	115	112	112	112	112	112	750	2
	11	874	119	123	128	166	254	332	367	347	279	188	130	124	119	119	119	119	821	1
	12	879	121	121	121	127	131	163	227	281	301	281	227	163	131	127	121	121	844	12
HALF DAY TOTALS			799	1483	2314	2968	3275	3151	2580	1649	995	743	642	602	592	588	587	594	3551	
Jul 21	5	7	3	5	7	7	6	5	2	0	0	0	0	0	0	0	0	1	1	7
	6	435	116	281	395	447	433	352	216	55	34	34	34	34	34	34	34	37	100	6
	7	656	95	321	513	625	643	564	400	166	66	62	62	62	62	62	62	62	278	5
	8	762	90	236	468	620	680	639	505	285	94	83	83	83	83	83	83	83	459	4
	9	818	102	138	333	513	610	618	537	371	165	104	99	99	99	99	99	99	611	3
	10	850	110	118	177	333	462	519	501	407	255	129	114	110	110	110	110	110	729	2
	11	866	117	120	125	157	256	352	400	389	321	217	135	123	117	117	117	117	802	1
	12	871	120	120	120	125	130	172	253	320	345	320	253	172	130	125	120	120	826	12
HALF DAY TOTALS			705	1296	2102	2792	3180	3164	2707	1842	1110	783	643	586	572	568	567	570	3395	
Aug 21	6	255	38	137	214	256	259	223	151	52	18	17	17	17	17	17	17	17	38	6
	7	603	55	223	426	557	602	557	426	222	55	49	49	49	49	49	49	49	196	5
	8	747	75	149	397	584	681	676	569	374	128	74	72	72	72	72	72	72	386	4
	9	819	89	97	259	481	621	669	618	475	251	97	89	89	89	89	89	89	549	3
	10	857	102	105	126	294	472	574	590	519	567	173	107	102	102	102	102	102	674	2
	11	876	109	109	113	130	257	403	492	505	443	313	165	116	109	109	109	109	753	1
	12	882	112	112	112	115	120	197	333	434	470	434	333	197	120	115	112	112	780	12
HALF DAY TOTALS			518	861	1571	2338	2929	3196	3015	2370	1496	932	647	524	496	492	492	492	2983	
Sep 21	7	472	28	87	265	395	460	456	381	244	66	27	27	27	27	27	27	27	80	5
	8	725	52	61	275	504	646	689	626	467	224	57	52	52	52	52	52	52	258	4
	9	830	71	73	148	413	613	717	712	599	391	129	73	71	71	71	71	71	434	3
	10	882	84	84	89	224	468	631	697	659	522	293	96	84	84	84	84	84	567	2
	11	906	92	92	92	99	245	463	604	652	603	461	242	99	92	92	92	92	651	1
	12	914	95	95	95	95	101	237	446	584	631	584	446	237	101	95	95	95	679	12
HALF DAY TOTALS			374	447	917	1683	2484	3092	3257	2918	2119	1250	699	432	376	373	373	373	2329	
Oct 21	7	153	6	10	64	113	142	148	132	94	38	7	6	6	6	6	6	38	12	5
	8	644	35	37	155	387	545	615	592	476	280	56	35	35	35	35	35	35	136	4
	9	811	54	54	73	329	567	710	743	661	476	202	57	54	54	54	54	54	305	3
	10	884	67	67	70	157	439	646	752	742	619	395	120	69	67	67	67	67	443	2
	11	917	76	76	76	80	223	491	670	745	707	562	317	89	76	76	76	76	529	1
	12	927	78	78	78	78	84	267	521	681	737	681	521	267	84	78	78	78	558	12
HALF DAY TOTALS			277	282	479	1106	1965	2771	3173	3074	2494	1555	780	369	280	277	277	277	1704	
Nov 21	8	430	17	17	55	217	339	404	406	346	228	66	18	17	17	17	17	17	44	4
	9	733	38	38	42	231	476	634	691	643	492	254	43	38	38	38	38	38	173	3
	10	846	52	52	52	99	385	617	748	763	659	145	159	54	52	52	52	52	303	2
	11	894	60	60	60	63	192	486	688	783	757	613	367	89	60	60	60	60	388	1
	12	908	63	63	63	63	67	280	555	728	789	728	555	280	67	63	63	63	418	12
HALF DAY TOTALS			197	198	235	623	1403	2273	2797	2884	2516	1738	850	319	199	197	197	197	1115	
Dec 21	8	279	9	9	25	129	212	259	264	230	157	52	10	9	9	9	9	9	20	4
	9	685	31	31	33	188	427	584	646	611	477	260	40	31	31	31	31	31	124	3
	10	825	45	45	45	78	358	594	732	755	661	462	173	47	45	45	45	45	244	2
	11	882	53	53	53	55	177	477	685	786	765	624	379	90	53	53	53	53	327	1
	12	898	56	56	56	56	60	281	560	736	798	736	560	281	60	56	56	56	356	12
HALF DAY TOTALS			165	165	178	461	1180	2046	2594	2736	2446	1757	870	298	167	165	165	165	891	
			N	NNW	NW	WNW	W	WSW	SW	SSW	S	SSE	SE	ESE	E	ENE	NE	NNE	HOR	PM

[a]Total solar heat gains for DSA glass (based on a ground reflectance of 0.20).

[b]Half day totals computed by Simpson's rule, time interval = 10 min.

APPENDICES

TABLE B.9 **Solar Intensity and Solar Heat Gain Factors[a] for 48° N Latitude (SI Units)[b]**

Date	Solar Time am	Direct Normal W/m²	Solar Heat Gain Factors, W/m²: N	NNE	NE	ENE	E	ESE	SE	SSE	S	SSW	SW	WSW	W	WNW	NW	NNW	HOR	Solar Time pm
Jan 21	8	116	4	4	13	57	90	109	110	94	63	18	4	4	4	4	4	4	7	4
	9	584	24	24	26	168	371	505	555	523	406	217	30	24	24	24	24	24	79	3
	10	754	37	37	37	69	333	554	682	702	615	429	159	39	37	37	37	37	174	2
	11	823	45	45	45	47	166	455	656	753	734	598	365	83	45	45	45	45	244	1
	12	842	48	48	48	48	51	271	541	713	772	713	541	271	51	48	48	48	269	12
HALF DAY TOTALS			134	134	144	368	996	1788	2298	2448	2212	1616	817	267	136	134	134	134	639	
Feb 21	7	11	0	1	5	8	10	11	10	7	0	0	0	0	0	0	0	0	1	5
	8	568	24	25	114	324	470	537	524	428	259	52	24	24	24	24	24	24	78	4
	9	780	42	42	49	284	530	683	727	660	488	225	45	42	42	42	42	42	210	3
	10	869	54	54	55	120	415	641	764	768	653	434	137	55	54	54	54	54	331	2
	11	908	61	61	61	64	205	498	696	787	755	607	355	84	61	61	61	61	409	1
	12	920	63	63	63	63	68	280	555	728	790	728	555	280	68	63	63	63	435	12
HALF DAY TOTALS			214	216	337	864	1708	2575	3054	3051	2565	1677	825	328	216	214	214	214	1246	
Mar 21	7	482	23	71	253	387	458	458	388	253	71	23	22	22	22	22	22	22	64	5
	8	744	44	48	239	486	644	701	651	498	257	49	44	44	44	44	44	44	214	4
	9	853	60	60	104	381	609	738	753	652	449	164	62	60	60	60	60	60	371	3
	10	906	72	72	75	183	460	656	749	728	596	363	104	72	72	72	72	72	493	2
	11	932	79	79	79	83	232	492	663	731	689	541	297	88	79	79	79	79	568	1
	12	939	81	81	81	81	86	261	509	666	720	666	509	261	86	81	81	81	594	12
HALF DAY TOTALS			317	372	790	1558	2446	3193	3471	3199	2420	1466	769	399	319	316	316	316	2006	
Apr 21	6	340	39	167	271	330	337	294	203	74	20	19	19	19	19	19	19	19	46	6
	7	646	49	191	417	567	628	595	468	264	56	45	45	45	45	45	45	45	189	5
	8	779	64	99	350	566	690	709	619	435	173	67	64	64	64	64	64	64	359	4
	9	847	79	83	191	444	621	703	679	554	335	105	79	79	79	79	79	79	507	3
	10	884	90	90	97	244	466	610	660	613	472	251	97	90	90	90	90	90	618	2
	11	902	97	97	97	105	245	443	570	609	558	419	217	103	97	97	97	97	689	1
	12	908	99	99	99	99	106	224	414	543	587	543	414	224	106	99	99	99	713	12
HALF DAY TOTALS			463	765	1454	2285	3019	3465	3411	2825	1908	1169	713	492	445	442	442	442	2761	
May 21	5	129	52	97	125	133	122	91	44	9	8	8	8	8	8	8	8	10	16	7
	6	511	112	305	443	512	504	418	267	75	38	38	38	38	38	38	38	40	125	6
	7	690	73	283	498	631	668	604	448	214	66	61	61	61	61	61	61	61	287	5
	8	782	83	170	418	599	689	675	562	358	120	79	79	79	79	79	79	79	449	4
	9	834	93	101	259	475	611	656	605	463	243	100	93	93	93	93	93	93	585	3
	10	864	103	106	124	283	458	561	579	513	366	179	109	103	103	103	103	103	690	2
	11	879	109	109	113	127	249	396	488	505	447	320	170	116	109	109	109	109	756	1
	12	884	111	111	111	114	120	198	336	439	474	439	336	198	120	114	111	111	778	12
HALF DAY TOTALS			679	1224	2035	2817	3360	3515	3176	2363	1525	997	709	579	550	546	546	549	3297	
Jun 21	5	243	111	192	241	252	228	166	75	19	17	17	17	17	17	17	17	24	38	7
	6	544	146	348	488	552	534	434	266	70	46	46	46	46	46	46	46	49	162	6
	7	693	93	317	521	642	467	591	427	188	72	67	67	67	67	67	67	67	324	5
	8	775	90	203	440	604	678	651	529	320	107	84	84	84	84	84	84	84	479	4
	9	822	98	114	286	482	600	629	567	418	208	105	98	98	98	98	98	98	610	3
	10	849	108	113	142	296	450	534	540	465	319	157	114	108	108	108	108	108	711	2
	11	864	114	114	120	138	248	373	449	456	396	279	156	121	114	114	114	114	775	1
	12	869	116	116	116	120	126	189	303	391	423	391	303	189	126	120	116	116	796	12
HALF DAY TOTALS			811	1446	2279	3013	3454	3477	3013	2140	1376	942	718	620	597	593	592	602	3495	
Jul 21	5	135	57	104	133	141	129	96	46	11	9	9	9	9	9	9	9	12	18	7
	6	492	116	302	436	501	492	407	259	75	40	40	40	40	40	40	40	43	130	6
	7	666	78	285	492	619	653	588	436	207	69	63	63	63	63	63	63	63	290	5
	8	757	86	176	416	590	675	660	547	348	119	81	81	81	81	81	81	81	448	4
	9	809	96	106	263	471	601	643	591	450	237	104	96	96	96	96	96	96	582	3
	10	839	106	109	130	285	453	550	566	500	356	177	112	106	106	106	106	106	684	2
	11	855	112	112	117	132	249	390	477	492	435	312	169	119	112	112	112	112	749	1
	12	859	114	114	114	117	124	198	329	428	462	428	329	198	124	117	114	114	771	12
HALF DAY TOTALS			705	1247	2039	2795	3311	3446	3101	2303	1496	993	721	598	570	565	565	569	3287	
Aug 21	6	311	42	161	256	310	316	273	190	70	22	21	21	21	21	21	21	21	51	6
	7	599	53	193	403	543	598	565	444	250	59	49	49	49	49	49	49	49	193	5
	8	732	69	108	347	549	665	681	593	417	168	73	69	69	69	69	69	69	358	4
	9	801	84	90	198	437	605	681	655	534	323	108	84	84	84	84	84	84	502	3
	10	839	95	95	104	247	453	593	639	592	456	245	104	95	95	95	95	95	610	2
	11	858	102	102	102	112	247	433	553	589	539	406	215	110	102	102	102	102	679	1
	12	864	104	104	104	104	113	224	404	526	568	526	404	224	113	104	104	104	702	12
HALF DAY TOTALS			496	790	1449	2237	2929	3343	3282	2720	1852	1156	728	521	475	471	471	471	2741	
Sep 21	7	414	24	66	224	342	403	403	342	224	66	24	23	23	23	23	23	23	62	5
	8	678	46	51	228	455	602	654	608	467	244	52	46	46	46	46	46	46	206	4
	9	792	63	63	107	366	581	702	716	621	430	163	66	63	63	63	63	63	358	3
	10	848	75	75	79	182	444	630	719	699	573	353	107	75	75	75	75	75	476	2
	11	876	82	82	82	88	230	476	639	704	664	524	292	93	82	82	82	82	549	1
	12	884	84	84	84	84	91	257	494	643	695	643	494	257	91	84	84	84	574	12
HALF DAY TOTALS			332	381	758	1467	2300	3007	3280	3039	2324	1429	766	412	334	331	331	331	1937	
Oct 21	7	12	0	1	5	9	11	12	11	8	3	0	0	0	0	0	0	0	1	5
	8	522	25	27	111	304	439	501	489	398	242	51	25	25	25	25	25	25	79	4
	9	734	44	44	52	276	508	652	693	629	466	216	47	44	44	44	44	44	208	3
	10	825	56	56	58	122	403	618	736	739	628	418	136	58	56	56	56	56	327	2
	11	866	64	64	64	67	203	483	673	760	729	587	345	86	64	64	64	64	403	1
	12	878	66	66	66	66	71	274	538	704	763	704	538	274	71	66	66	66	429	12
HALF DAY TOTALS			223	225	340	838	1637	2461	2918	2918	2459	1619	808	334	226	223	223	223	1233	
Nov 21	8	115	4	4	13	57	90	108	109	93	62	18	4	4	4	4	4	4	7	4
	9	565	25	25	27	165	363	492	540	509	394	211	31	25	25	25	25	25	81	3
	10	735	38	38	38	70	328	543	668	688	602	420	156	40	38	38	38	38	175	2
	11	804	46	46	46	48	165	448	645	739	720	587	358	83	46	46	46	46	244	1
	12	823	48	48	48	48	53	267	531	700	758	700	531	267	53	48	48	48	269	12
HALF DAY TOTALS			138	138	147	368	979	1752	2250	2396	2165	1583	804	268	140	138	138	138	642	
Dec 21	9	442	16	16	18	113	272	378	420	401	316	176	25	16	16	16	16	16	42	3
	10	676	30	30	30	51	286	491	613	636	563	398	156	32	30	30	30	30	119	2
	11	765	38	38	38	40	146	421	615	708	694	569	351	80	38	38	38	38	181	1
	12	789	41	41	41	41	44	256	514	679	734	679	514	256	44	41	41	41	204	12
HALF DAY TOTALS			105	105	107	231	735	1444	1924	2098	1944	1477	778	239	107	105	105	105	444	
			N	NNW	NW	WNW	W	WSW	SW	SSW	S	SSE	SE	ESE	E	ENE	NE	NNE	HOR	PM

[a]Total solar heat gains for DSA glass (based on a ground reflectance of 0.20).

[b]Half day totals computed by Simpson's rule, time interval = 10 min.

APPENDICES

TABLE B.10 **Solar Intensity and Solar Heat Gain Factors[a] for 64° N Latitude (SI Units)[b]**

Date	Solar Time am	Direct Normal W/m²	Solar Heat Gain Factors, W/m²																	Solar Time pm
			N	NNE	NE	ENE	E	ESE	SE	SSE	S	SSW	SW	WSW	W	WNW	NW	NNW	HOR	
Jan 21	10	69	2	2	2	4	29	51	63	66	58	41	16	2	2	2	2	2	4	2
	11	256	9	9	9	9	46	143	210	242	238	196	120	24	9	9	9	9	18	1
	12	316	11	11	11	11	12	104	211	280	302	280	211	104	12	11	11	11	24	12
HALF DAY TOTALS			16	16	16	18	80	250	382	449	446	374	237	72	17	16	16	16	33	
Feb 21	8	56	2	2	9	31	46	53	53	44	27	6	2	2	2	2	2	2	3	4
	9	422	16	16	18	136	280	373	403	375	285	143	18	16	16	16	16	16	41	3
	10	601	26	26	26	57	276	453	554	567	495	341	119	28	26	26	26	26	90	2
	11	679	32	32	32	34	139	386	558	638	622	506	307	64	32	32	32	32	128	1
	12	701	34	34	34	34	37	231	464	613	662	613	464	231	37	34	34	34	143	12
HALF DAY TOTALS			93	93	103	279	770	1408	1822	1947	1767	1298	668	210	[illegible]	93	93	93	334	
Mar 21	7	300	12	34	147	235	284	288	249	168	54	13	12	12	12	12	12	12	27	5
	8	582	29	31	147	357	499	559	536	427	246	44	29	29	29	29	29	29	100	4
	9	717	41	41	51	277	502	642	679	613	450	203	44	41	41	41	41	41	187	3
	10	786	49	49	50	111	386	598	714	719	613	410	132	51	49	49	49	49	264	2
	11	821	54	54	54	57	190	468	658	746	718	579	344	80	54	54	54	54	314	1
	12	831	56	56	56	56	61	269	530	696	754	696	530	269	61	56	56	56	331	12
HALF DAY TOTALS			213	234	473	1052	1879	2695	3103	3021	2457	1591	811	330	215	213	213	213	1057	
Apr 21	5	85	24	56	77	85	81	64	36	6	4	4	4	4	4	4	4	5	8	7
	6	419	38	187	320	400	416	371	266	111	26	24	24	24	24	24	24	24	67	6
	7	613	43	129	355	516	595	584	482	303	80	41	41	41	41	41	41	41	162	5
	8	720	54	60	249	483	633	685	634	484	250	60	54	54	54	54	54	54	269	4
	9	783	65	65	100	351	567	691	710	620	436	174	68	65	65	65	65	65	367	3
	10	820	72	72	76	161	421	613	709	698	582	372	119	74	72	72	72	72	441	2
	11	839	77	77	77	82	215	468	637	710	676	540	312	93	77	77	77	77	448	1
	12	846	79	79	79	79	85	262	503	655	708	655	503	262	85	79	79	79	504	12
HALF DAY TOTALS			413	687	1294	2116	2974	3628	3741	3268	2408	1537	861	471	381	378	378	378	2053	
May 21	4	160	94	139	162	161	135	87	25	10	10	10	10	10	10	10	11	31	20	8
	5	416	152	298	393	425	395	302	158	34	29	29	29	29	29	29	29	33	81	7
	6	584	90	304	474	570	579	499	343	126	49	46	46	46	46	46	46	46	175	6
	7	688	65	199	436	596	665	634	507	297	78	60	60	60	60	60	60	60	284	5
	8	754	72	89	307	527	660	694	623	459	215	78	72	72	72	72	72	72	390	4
	9	796	82	85	143	384	577	678	679	579	389	147	85	82	82	82	82	82	481	3
	10	822	89	89	94	193	427	593	668	645	528	322	114	89	89	89	89	89	549	2
	11	836	93	93	93	100	226	446	594	652	614	485	275	105	93	93	93	93	592	1
	12	841	95	95	95	95	103	248	460	597	644	597	460	248	103	95	95	95	606	12
HALF DAY TOTALS			780	1341	2145	2998	3712	4073	3841	3108	2234	1467	909	603	533	529	530	554	2875	
Jun 21	4	294	181	263	302	297	247	156	43	21	21	21	21	21	21	21	23	67	50	8
	5	485	195	360	466	498	458	346	175	44	39	39	39	39	39	39	39	45	124	7
	6	614	113	338	510	603	605	516	347	122	57	55	55	55	55	55	55	57	223	6
	7	698	74	225	457	610	672	632	498	282	80	68	68	68	68	68	68	68	331	5
	8	754	79	105	327	535	657	682	605	438	197	85	79	79	79	79	79	79	433	4
	9	791	89	93	161	393	572	662	655	552	362	137	92	89	89	89	89	89	520	3
	10	814	96	96	102	205	423	577	642	615	497	297	114	96	96	96	96	96	585	2
	11	827	100	100	100	108	228	431	568	619	581	456	258	110	100	100	100	100	627	1
	12	831	101	101	101	101	110	240	436	566	610	566	436	240	110	101	101	101	641	12
HALF DAY TOTALS			1016	1680	2526	3347	3953	4154	3769	2985	2142	1436	935	667	605	601	603	673	3219	
Jul 21	4	168	101	149	173	172	144	93	27	12	12	12	12	12	12	12	12	35	24	8
	5	405	154	297	389	420	390	298	156	36	32	32	32	32	32	32	32	36	88	7
	6	564	94	303	467	560	567	488	335	124	51	49	49	49	49	49	49	49	181	6
	7	665	69	201	431	585	652	620	495	290	80	63	63	63	63	63	63	63	289	5
	8	730	75	94	307	520	648	680	609	449	211	81	75	75	75	75	75	75	393	4
	9	771	85	88	148	382	567	665	664	566	380	146	88	85	85	85	85	85	481	3
	10	797	92	92	98	196	422	583	655	631	516	315	116	92	92	92	92	92	548	2
	11	812	96	96	96	104	226	439	582	638	601	475	271	108	96	96	96	96	590	1
	12	816	98	98	98	98	107	246	452	585	630	585	452	246	107	98	98	98	604	12
HALF DAY TOTALS			814	1370	2157	2985	3669	4004	3763	3046	2198	1457	920	626	557	553	554	582	2985	
Aug 21	5	92	28	62	85	94	89	71	40	8	6	6	6	6	6	6	6	6	11	7
	6	388	42	182	306	380	395	352	252	107	30	27	27	27	27	27	27	27	73	6
	7	570	48	132	344	494	567	555	458	289	81	45	45	45	45	45	45	45	168	5
	8	675	59	66	247	468	609	657	607	464	241	66	59	59	59	59	59	59	273	4
	9	737	70	70	107	345	549	666	683	597	420	172	74	70	70	70	70	70	368	3
	10	775	78	78	82	165	412	594	685	674	562	361	122	80	78	78	78	78	440	2
	11	794	82	82	82	89	216	456	617	686	653	523	305	99	82	82	82	82	485	1
	12	801	84	84	84	84	92	260	489	635	684	633	489	260	92	84	84	84	501	12
HALF DAY TOTALS			447	714	1293	2073	2884	3498	3600	3147	2333	1509	869	500	413	409	409	410	2069	
Sep 21	7	242	12	30	122	194	234	238	206	139	47	13	12	12	12	12	12	12	26	5
	8	513	30	33	136	324	451	505	484	387	225	45	30	30	30	30	30	30	97	4
	9	651	43	43	54	261	468	596	631	570	420	193	47	43	43	43	43	43	181	3
	10	722	52	52	53	111	366	563	672	677	577	388	131	54	52	52	52	52	255	2
	11	758	57	57	57	61	185	445	623	706	680	549	329	83	57	57	57	57	303	1
	12	769	59	59	59	59	65	260	504	660	715	660	504	260	65	59	59	59	320	12
HALF DAY TOTALS			224	241	448	968	1727	2484	2871	2810	2305	1513	787	336	227	224	224	224	1021	
Oct 21	8	54	2	2	10	30	45	52	51	42	26	6	2	2	2	2	2	2	4	4
	9	383	17	17	19	127	259	345	372	346	263	133	19	17	17	17	17	17	42	3
	10	556	28	28	28	58	261	426	520	532	465	321	115	29	28	28	28	28	91	2
	11	633	34	34	34	36	135	367	528	603	588	479	292	64	34	34	34	34	130	1
	12	655	36	36	36	36	40	222	441	581	628	581	441	222	40	36	36	36	144	12
HALF DAY TOTALS			98	98	109	273	728	1324	1711	1829	1661	1225	638	209	100	98	98	98	339	
Nov 21	10	72	2	2	2	5	31	53	67	69	62	43	17	3	2	2	2	2	4	2
	11	250	9	9	9	10	46	140	207	238	234	192	118	24	9	9	9	9	19	1
	12	307	11	11	11	11	13	102	207	274	295	274	207	102	13	11	11	11	25	12
HALF DAY TOTALS			17	17	17	19	81	248	378	444	440	369	234	71	18	17	17	17	35	
Dec 21	11	12	0	0	0	0	2	7	10	12	11	9	6	1	0	0	0	0	1	1
	12	51	1	1	1	1	2	16	34	45	48	45	34	16	2	1	1	1	3	12
HALF DAY TOTALS			1	1	1	1	3	16	28	35	36	32	22	8	1	1	1	1	2	
			N	NNW	NW	WNW	W	WSW	SW	SSW	S	SSE	SE	ESE	E	ENE	NE	NNE	HOR	PM

[a]Total solar heat gains for DSA glass (based on a ground reflectance of 0.20).

[b]Half day totals computed by Simpson's rule, time interval = 10 min.

TABLE B.11 **Solar Position and Clear Day Insolation, 32° N Latitude**

Date	Solar Time A.M.	Solar Time P.M.	Solar Position Alt	Solar Position Azm	Direct Normal	Global Irradiance (Btu/h ft²) Horiz	South-Facing Elevation Angle 22	32	42	52	90
Jan 21	7	5	1.4	65.2	1	0	0	0	0	1	1
	8	4	12.5	56.5	203	56	93	106	116	123	115
	9	3	22.5	46.0	269	118	175	193	206	212	181
	10	2	30.6	33.1	295	167	235	256	269	274	221
	11	1	36.1	17.5	306	198	273	295	308	312	245
	12		38.0	0.0	310	209	285	308	321	324	253
	Surface Daily Totals				2458	1288	1839	2008	2118	2166	1779
Feb 21	7	5	7.1	73.5	121	22	34	37	40	42	38
	8	4	19.0	64.4	247	95	127	136	140	141	108
	9	3	29.9	53.4	288	161	206	217	222	220	158
	10	2	39.1	39.4	306	212	266	278	283	279	193
	11	1	45.6	21.4	315	244	304	317	321	315	214
	12		48.0	0.0	317	255	316	330	334	328	222
	Surface Daily Totals				2872	1724	2188	2300	2345	2322	1644
Mar 21	7	5	12.7	81.9	185	54	60	60	59	56	32
	8	4	25.1	73.0	260	129	146	147	144	137	78
	9	3	36.8	62.1	290	194	222	224	220	209	119
	10	2	47.3	47.5	304	245	280	283	278	265	150
	11	1	55.0	26.8	311	277	317	321	315	300	170
	12		58.0	0.0	313	287	329	333	327	312	177
	Surface Daily Totals				3012	2084	2378	2403	2358	2246	1276
Apr 21	6	6	6.1	99.9	66	14	9	6	6	5	3
	7	5	18.8	92.2	206	86	78	71	62	51	10
	8	4	31.5	84.0	255	158	156	148	136	120	35
	9	3	43.9	74.2	278	220	225	217	203	183	68
	10	2	55.7	60.3	290	267	279	272	256	234	95
	11	1	65.4	37.5	295	297	313	306	290	265	112
	12		69.6	0.0	297	307	325	318	301	276	118
	Surface Daily Totals				3076	2390	2444	2356	2206	1994	764
May 21	6	6	10.4	107.2	119	36	21	13	13	12	7
	7	5	22.8	100.1	211	107	88	75	60	44	13
	8	4	35.4	92.9	250	175	159	145	127	105	15
	9	3	48.1	84.7	269	233	223	209	188	163	33
	10	2	60.6	73.3	280	277	273	259	237	208	56
	11	1	72.0	51.9	285	305	305	290	268	237	72
	12		78.0	0.0	286	315	315	301	278	247	77
	Surface Daily Totals				3112	2582	2454	2284	2064	1788	469
Jun 21	6	6	12.2	110.2	131	45	26	16	15	14	9
	7	5	24.3	103.4	210	115	91	76	59	41	14
	8	4	36.9	96.8	245	180	159	143	122	99	16
	9	3	49.6	89.4	264	236	221	204	181	153	19
	10	2	62.2	79.7	274	279	268	251	227	197	41
	11	1	74.2	60.9	279	306	299	282	257	224	56
	12		81.5	0.0	280	315	309	292	267	234	60
	Surface Daily Totals				3084	2634	2436	2234	1990	1690	370

TABLE B.11 **Solar Position and Clear Day Insolation, 32° N Latitude (Continued)**

Date	Solar Time A.M.	Solar Time P.M.	Solar Position Alt	Solar Position Azm	Direct Normal	Global Irradiance (Btu/h ft²) Horiz	South-Facing Elevation Angle 22	32	42	52	90
Jul 21	6	6	10.7	107.7	113	37	22	14	13	12	8
	7	5	23.1	100.6	203	107	87	75	60	44	14
	8	4	35.7	93.6	241	174	158	143	125	104	16
	9	3	48.4	85.5	261	230	220	205	185	159	31
	10	2	60.9	74.3	271	274	269	254	232	204	54
	11	1	72.4	53.3	277	302	300	285	262	232	69
	12		78.6	0.0	279	311	310	296	273	242	74
	Surface Daily Totals				3012	2558	2422	2250	2030	1754	458
Aug 21	6	6	6.5	100.5	59	14	9	7	6	6	4
	7	5	19.1	92.8	190	85	77	69	60	50	12
	8	4	31.8	84.7	240	156	152	144	132	116	33
	9	3	44.3	75.0	263	216	220	212	197	178	65
	10	2	56.1	61.3	276	262	272	264	249	226	91
	11	1	66.0	38.4	282	292	305	298	281	257	107
	12		70.3	0.0	284	302	317	309	292	268	113
	Surface Daily Totals				2902	2352	2388	2296	2144	1934	736
Sep 21	7	5	12.7	81.9	163	51	56	56	55	52	30
	8	4	25.1	73.0	240	124	140	141	138	131	75
	9	3	36.8	62.1	272	188	213	215	211	201	114
	10	2	47.3	47.5	287	237	270	273	268	255	145
	11	1	55.0	26.8	294	268	306	309	303	289	164
	12		58.0	0.0	296	278	318	321	315	300	171
	Surface Daily Totals				2808	2014	2288	2308	2264	2154	1226
Oct 21	7	5	6.8	73.1	99	19	29	32	34	36	32
	8	4	18.7	64.0	229	90	120	128	133	134	104
	9	3	29.5	53.0	273	155	198	208	213	212	153
	10	2	38.7	39.1	293	204	257	269	273	270	188
	11	1	45.1	21.1	302	236	294	307	311	306	209
	12		47.5	0.0	304	247	306	320	324	318	217
	Surface Daily Totals				2696	1654	2100	2208	2252	2232	1588
Nov 21	7	5	1.5	65.4	2	0	0	0	1	1	1
	8	4	12.7	56.6	196	55	91	104	113	119	111
	9	3	22.6	46.1	263	118	173	190	202	208	176
	10	2	30.8	33.2	289	166	233	252	265	270	217
	11	1	36.2	17.6	301	197	270	291	303	307	241
	12		38.2	0.0	304	207	282	304	316	320	249
	Surface Daily Totals				2406	1280	1816	1980	2084	2130	1742
Dec 21	8	4	10.3	53.8	176	41	77	90	101	108	107
	9	3	19.8	43.6	257	102	161	180	195	204	183
	10	2	27.6	31.2	288	150	221	244	259	267	226
	11	1	32.7	16.4	301	180	258	282	298	305	251
	12		34.6	0.0	304	190	271	295	311	318	259
	Surface Daily Totals				2348	1136	1704	1888	2016	2086	1794

NOTE: 1 Btu/h ft^2 = 3.152 W/m^2.

Source: Reprinted by permission from Peter J. Lunde, *Solar Thermal Engineering,* © 1980, John Wiley & Sons Inc., New York.

TABLE B.12 **Solar Position and Clear Day Insolation, 40° N Latitude**

	Solar Time		*Solar Position*			*Global Irradiance (Btu/h ft²)*					
							South-Facing Elevation Angle				
Date	A.M.	P.M.	*Alt*	*Azm*	*Direct Normal*	*Horiz*	*30*	*40*	*50*	*60*	*90*
Jan 21	8	4	8.1	55.3	142	28	65	74	81	85	84
	9	3	16.8	44.0	239	83	155	171	182	187	171
	10	2	23.8	30.9	274	127	218	237	249	254	223
	11	1	28.4	16.0	289	154	257	277	290	293	253
	12		30.0	0.0	294	164	270	291	303	306	263
	Surface Daily Totals				2182	948	1660	1810	1906	1944	1726
Feb 21	7	5	4.8	72.7	69	10	19	21	23	24	22
	8	4	15.4	62.2	224	73	114	122	126	127	107
	9	3	25.0	50.2	274	132	195	205	209	208	167
	10	2	32.8	35.9	295	178	256	267	271	267	210
	11	1	38.1	18.9	305	206	293	306	310	304	236
	12		40.0	0.0	308	216	306	319	323	317	245
	Surface Daily Totals				2640	1414	2060	2162	2202	2176	1730
Mar 21	7	5	11.4	80.2	171	46	55	55	54	51	35
	8	4	22.5	69.6	250	114	140	141	138	131	89
	9	3	32.8	57.3	282	173	215	217	213	202	138
	10	2	41.6	41.9	297	218	273	276	271	258	176
	11	1	47.7	22.6	305	247	310	313	307	293	200
	12		50.0	0.0	307	257	322	326	320	305	208
	Surface Daily Totals				2916	1852	2308	2330	2284	2174	1484
Apr 21	6	6	7.4	98.9	89	20	11	8	7	7	4
	7	5	18.9	89.5	206	87	77	70	61	50	12
	8	4	30.3	79.3	252	152	153	145	133	117	53
	9	3	41.3	67.2	274	207	221	213	199	179	93
	10	2	51.2	51.4	286	250	275	267	252	229	126
	11	1	58.7	29.2	292	277	308	301	285	260	147
	12		61.6	0.0	293	287	320	313	296	271	154
	Surface Daily Totals				3092	2274	2412	2320	2168	1956	1022
May 21	5	7	1.9	114.7	1	0	0	0	0	0	0
	6	6	12.7	105.6	144	49	25	15	14	13	9
	7	5	24.0	96.6	216	114	89	76	60	44	13
	8	4	35.4	87.2	250	175	158	144	125	104	25
	9	3	46.8	76.0	267	227	221	206	186	160	60
	10	2	57.5	60.9	277	267	270	255	233	205	89
	11	1	66.2	37.1	283	293	301	287	264	234	108
	12		70.0	0.0	284	301	312	297	274	243	114
	Surface Daily Totals				3160	2552	2442	2264	2040	1760	724
Jun 21	5	7	4.2	117.3	22	4	3	3	2	2	1
	6	6	14.8	108.4	155	60	30	18	17	16	10
	7	5	26.0	99.7	216	123	92	77	59	40	14
	8	4	37.4	90.7	246	182	159	142	121	97	16
	9	3	48.8	80.2	263	233	219	202	179	151	47
	10	2	59.8	65.8	272	272	266	248	224	193	74
	11	1	69.2	41.9	277	296	296	278	253	221	92
	12		73.5	0.0	279	304	306	289	263	230	98
	Surface Daily Totals				3180	2648	2434	2224	1974	1670	610

TABLE B.12 **Solar Position and Clear Day Insolation, 40° N Latitude (Continued)**

	Solar Time		Solar Position			Global Irradiance (Btu/h ft²)					
							South-Facing Elevation Angle				
Date	A.M.	P.M.	Alt	Azm	Direct Normal	Horiz	30	40	50	60	90
Jul 21	5	7	2.3	115.2	2	0	0	0	0	0	0
	6	6	13.1	106.1	138	50	26	17	15	14	9
	7	5	24.3	97.2	208	114	89	75	60	44	14
	8	4	35.8	87.8	241	174	157	142	124	102	24
	9	3	47.2	76.7	259	225	218	203	182	157	58
	10	2	57.9	61.7	269	265	266	251	229	200	86
	11	1	66.7	37.9	275	290	296	281	258	228	104
	12		70.6	0.0	276	298	307	292	269	238	111
	Surface Daily Totals				3062	2534	2409	2230	2006	1728	702
Aug 21	6	6	7.9	99.5	81	21	12	9	8	7	5
	7	5	19.3	90.0	191	87	76	69	60	49	12
	8	4	30.7	79.9	237	150	150	141	129	113	50
	9	3	41.8	67.9	260	205	216	207	193	173	89
	10	2	51.7	52.1	272	246	267	259	244	221	120
	11	1	59.3	29.7	278	273	300	292	276	252	140
	12		62.3	0.0	280	282	311	303	287	262	147
	Surface Daily Totals				2916	2244	2354	2258	2104	1894	978
Sep 21	7	5	11.4	80.2	149	43	51	51	49	47	32
	8	4	22.5	69.6	230	109	133	134	131	124	84
	9	3	32.8	57.3	263	167	206	208	203	193	132
	10	2	41.6	41.9	280	211	262	265	260	247	168
	11	1	47.7	22.6	287	239	298	301	295	281	192
	12		50.0	0.0	290	249	310	313	307	292	200
	Surface Daily Totals				2708	1788	2210	2228	2182	2074	1416
Oct 21	7	5	4.5	72.3	48	7	14	15	17	17	16
	8	4	15.0	61.9	204	68	106	113	117	118	100
	9	3	24.5	49.8	257	126	185	195	200	198	160
	10	2	32.4	35.6	280	170	245	257	261	257	203
	11	1	37.6	18.7	291	199	283	295	299	294	229
	12		39.5	0.0	294	208	295	308	312	306	238
	Surface Daily Totals				2454	1348	1962	2060	2098	2074	1654
Nov 21	8	4	8.2	55.4	136	28	63	72	78	82	81
	9	3	17.0	44.1	232	82	152	167	178	183	167
	10	2	24.0	31.0	268	126	215	233	245	249	219
	11	1	28.6	16.1	283	153	254	273	285	288	248
	12		30.2	0.0	288	163	267	287	298	301	258
	Surface Daily Totals				2128	942	1636	1778	1870	1908	1686
Dec 21	8	4	5.5	53.0	89	14	39	45	50	54	56
	9	3	14.0	41.9	217	65	135	152	164	171	163
	10	2	20.7	29.4	261	107	200	221	235	242	221
	11	1	25.0	15.2	280	134	239	262	276	283	252
	12		26.6	0.0	285	143	253	275	290	296	263
	Surface Daily Totals				1978	782	1480	1634	1740	1796	1646

NOTE: 1 Btu/h ft² = 3.152 W/m².

Source: Reprinted by permission from Peter J. Lunde, *Solar Thermal Engineering,* © 1980, John Wiley & Sons Inc., New York.

TABLE B.13 **Solar Position and Clear Day Insolation, 48° N Latitude**

Date	Solar Time A.M.	Solar Time P.M.	Solar Position Alt	Solar Position Azm	Direct Normal	Global Irradiance (Btu/h ft²) Horiz	South-Facing Elevation Angle 38	48	58	68	90
Jan 21	8	4	3.5	54.6	37	4	17	19	21	22	22
	9	3	11.0	42.6	185	46	120	132	140	145	139
	10	2	16.9	29.4	239	83	190	206	216	220	206
	11	1	20.7	15.1	261	107	231	249	260	263	243
	12		22.0	0.0	267	115	245	264	275	278	255
	Surface Daily Totals				1710	596	1360	1478	1550	1578	1478
Feb 21	7	5	2.4	72.2	12	1	3	4	4	4	4
	8	4	11.6	60.5	188	49	95	102	105	106	96
	9	3	19.7	47.7	251	100	178	187	191	190	167
	10	2	26.2	33.3	278	139	240	251	255	251	217
	11	1	30.5	17.2	290	165	278	290	294	288	247
	12		32.0	0.0	293	173	291	304	307	301	258
	Surface Daily Totals				2330	1080	1880	1972	2024	1978	1720
Mar 21	7	5	10.0	78.7	153	37	49	49	47	45	35
	8	4	19.5	66.8	236	96	131	132	129	122	96
	9	3	28.2	53.4	270	147	205	207	203	193	152
	10	2	35.4	37.8	287	187	263	266	261	248	195
	11	1	40.3	19.8	295	212	300	303	297	283	223
	12		42.0	0.0	298	220	312	315	309	294	232
	Surface Daily Totals				2780	1578	2208	2228	2182	2074	1632
Apr 21	6	6	8.6	97.8	108	27	13	9	8	7	5
	7	5	18.6	86.7	205	85	76	68	59	48	21
	8	4	28.5	74.9	247	142	149	141	129	113	69
	9	3	37.8	61.2	268	191	216	208	194	174	115
	10	2	45.8	44.6	280	228	268	260	245	223	152
	11	1	51.5	24.0	286	252	301	294	278	254	177
	12		53.6	0.0	288	260	313	305	289	264	185
	Surface Daily Totals				3076	2106	2358	2266	2114	1902	1262
May 21	5	7	5.2	114.3	41	9	4	4	4	3	2
	6	6	14.7	103.7	162	61	27	16	15	13	10
	7	5	24.6	93.0	219	118	89	75	60	43	13
	8	4	34.7	81.6	248	171	156	142	123	101	45
	9	3	44.3	68.3	264	217	217	202	182	156	86
	10	2	53.0	51.3	274	252	265	251	229	200	120
	11	1	59.5	28.6	279	274	296	281	258	228	141
	12		62.0	0.0	280	281	306	292	269	238	149
	Surface Daily Totals				3254	2482	2418	2234	2010	1728	982
Jun 21	5	7	7.9	116.5	77	21	9	9	8	7	5
	6	6	17.2	106.2	172	74	33	19	18	16	12
	7	5	27.0	95.8	220	129	93	77	59	39	15
	8	4	37.1	84.6	246	181	157	140	119	95	35
	9	3	46.9	71.6	261	225	216	198	175	147	74
	10	2	55.8	54.8	269	259	262	244	220	189	105
	11	1	62.7	31.2	274	280	291	273	248	216	126
	12		65.5	0.0	275	287	301	283	258	225	133
	Surface Daily Totals				3312	2626	2420	2204	1950	1644	874

TABLE B.13 **Solar Position and Clear Day Insolation, 48° N Latitude (Continued)**

	Solar Time		Solar Position			Global Irradiance (Btu/h ft²)					
							South-Facing Elevation Angle				
Date	A.M.	P.M.	Alt	Azm	Direct Normal	Horiz	38	48	58	68	90
Jul 21	5	7	5.7	114.7	43	10	5	5	4	4	3
	6	6	15.2	104.1	156	62	28	18	16	15	11
	7	5	25.1	93.5	211	118	89	75	59	42	14
	8	4	35.1	82.1	240	171	154	140	121	99	43
	9	3	44.8	68.8	256	215	214	199	178	153	83
	10	2	53.5	51.9	266	250	261	246	224	195	116
	11	1	60.1	29.0	271	272	291	276	253	223	137
	12		62.6	0.0	272	279	301	286	263	232	144
	Surface Daily Totals				3158	2474	2386	2200	1974	1694	956
Aug 21	6	6	9.1	98.3	99	28	14	10	9	8	6
	7	5	19.1	87.2	190	85	75	67	58	47	20
	8	4	29.0	75.4	232	141	145	137	125	109	65
	9	3	38.4	61.8	254	189	210	201	187	168	110
	10	2	46.4	45.1	266	225	260	252	237	214	146
	11	1	52.2	24.3	272	248	293	285	268	244	169
	12		54.3	0.0	274	256	304	296	279	255	177
	Surface Daily Totals				2898	2086	2300	2200	2046	1836	1208
Sep 21	7	5	10.0	78.7	131	35	44	44	43	40	31
	8	4	19.5	66.8	215	92	124	124	121	115	90
	9	3	28.2	53.4	251	142	196	197	193	183	143
	10	2	35.4	37.8	269	181	251	254	248	236	185
	11	1	40.3	19.8	278	205	287	289	284	269	212
	12		42.0	0.0	280	213	299	302	296	281	221
	Surface Daily Totals				2568	1522	2102	2118	2070	1966	1546
Oct 21	7	5	2.0	71.9	4	0	1	1	1	1	1
	8	4	11.2	60.2	165	44	86	91	95	95	87
	9	3	19.3	47.4	233	94	167	176	180	178	157
	10	2	25.7	33.1	262	133	228	239	242	239	207
	11	1	30.0	17.1	274	157	266	277	281	276	237
	12		31.5	0.0	278	166	279	291	294	288	247
	Surface Daily Totals				2154	1022	1774	1860	1890	1866	1626
Nov 21	8	4	3.6	54.7	36	5	17	19	21	22	22
	9	3	11.2	42.7	179	46	117	129	137	141	135
	10	2	17.1	29.5	233	83	186	202	212	215	201
	11	1	20.9	15.1	255	107	227	245	255	258	238
	12		22.2	0.0	261	115	241	259	270	272	250
	Surface Daily Totals				1668	596	1336	1448	1518	1544	1442
Dec 21	9	3	8.0	40.9	140	27	87	98	105	110	109
	10	2	13.6	28.2	214	63	164	180	192	197	190
	11	1	17.3	14.4	242	86	207	226	239	244	231
	12		18.6	0.0	250	94	222	241	254	260	244
	Surface Daily Totals				1444	446	1136	1250	1326	1364	1304

NOTE: 1 Btu/h ft² = 3.152 W/m².

Source: Reprinted by permission from Peter J. Lunde, *Solar Thermal Engineering*, © 1980, John Wiley & Sons Inc., New York.

TABLE B.14 **Solar Position and Clear Day Insolation, 64° N Latitude**

Date	Solar Time A.M.	Solar Time P.M.	Solar Position Alt	Solar Position Azm	Direct Normal	Global Irradiance (Btu/h ft²) Horiz	South-Facing Elevation Angle 54	64	74	84	90
Jan 21	10	2	2.8	28.1	22	2	17	19	20	20	20
	11	1	5.2	14.1	81	12	72	77	80	81	81
	12		6.0	0.0	100	16	91	98	102	103	103
	Surface Daily Totals				306	45	268	290	302	306	304
Feb 21	8	4	3.4	58.7	35	4	17	19	19	19	19
	9	3	8.6	44.8	147	31	103	108	111	110	107
	10	2	12.6	30.3	199	55	170	178	181	178	173
	11	1	15.1	15.3	222	71	212	220	223	219	213
	12		16.0	0.0	228	77	225	235	237	232	226
	Surface Daily Totals				1432	400	1230	1286	1302	1282	1252
Mar 21	7	5	6.5	76.5	95	18	30	29	29	27	25
	8	4	12.7	62.6	185	54	101	102	99	94	89
	9	3	18.1	48.1	227	87	171	172	169	160	153
	10	2	22.3	32.7	249	112	227	229	224	213	203
	11	1	25.1	16.6	260	129	262	265	259	246	235
	12		26.0	0.0	263	134	274	277	271	258	246
	Surface Daily Totals				2296	932	1856	1870	1830	1736	1656
Apr 21	5	7	4.0	108.5	27	5	2	2	2	1	1
	6	6	10.4	95.1	133	37	15	9	8	7	6
	7	5	17.0	81.6	194	76	70	63	54	43	37
	8	4	23.3	67.5	228	112	136	128	116	102	91
	9	3	29.0	52.3	248	144	197	189	176	158	145
	10	2	33.5	36.0	260	169	246	239	224	203	188
	11	1	36.5	18.4	266	184	278	270	255	233	216
	12		37.6	0.0	268	190	289	281	266	243	225
	Surface Daily Totals				2982	1644	2176	2082	1936	1736	1594
May 21	4	8	5.8	125.1	51	11	5	4	4	3	3
	5	7	11.6	112.1	132	42	13	11	10	9	8
	6	6	17.9	99.1	185	79	29	16	14	12	11
	7	5	24.5	85.7	218	117	86	72	56	39	28
	8	4	30.9	71.5	239	152	148	133	115	94	80
	9	3	36.8	56.1	252	182	204	190	170	145	128
	10	2	41.6	38.9	261	205	249	235	213	186	167
	11	1	44.9	20.1	265	219	278	264	242	213	193
	12		46.0	0.0	267	224	288	274	251	222	201
	Surface Daily Totals				3470	2236	2312	2124	1898	1624	1436
Jun 21	3	9	4.2	139.4	21	4	2	2	2	2	1
	4	8	9.0	126.4	93	27	10	9	8	7	6
	5	7	14.7	113.6	154	60	16	15	13	11	10
	6	6	21.0	100.8	194	96	34	19	17	14	13
	7	5	27.5	87.5	221	132	91	74	55	36	23
	8	4	34.0	73.3	239	166	150	133	112	88	73
	9	3	39.9	57.8	251	195	204	187	164	137	119
	10	2	44.9	40.4	258	217	247	230	206	177	157
	11	1	48.3	20.9	262	231	275	258	233	202	181
	12		49.5	0.0	263	235	284	267	242	211	189
	Surface Daily Totals				3650	2488	2342	2118	1862	1558	1356

TABLE B.14 **Solar Position and Clear Day Insolation, 64° N Latitude (Continued)**

Date	Solar Time A.M.	Solar Time P.M.	Solar Position Alt	Solar Position Azm	Direct Normal	Global Irradiance (Btu/h ft²) Horiz	South-Facing Elevation Angle 54	64	74	84	90
Jul 21	4	8	6.4	125.3	53	13	6	5	5	4	4
	5	7	12.1	112.4	128	44	14	13	11	10	9
	6	6	18.4	99.4	179	81	30	17	16	13	12
	7	5	25.0	86.0	211	118	86	72	56	38	28
	8	4	31.4	71.8	231	152	146	131	113	91	77
	9	3	37.3	56.3	245	182	201	186	166	141	124
	10	2	42.2	39.2	253	204	245	230	208	181	162
	11	1	45.4	20.2	257	218	273	258	236	207	187
	12		46.6	0.0	259	223	282	267	245	216	195
	Surface Daily Totals				3372	2248	2280	2090	1864	1588	1400
Aug 21	5	7	4.6	108.8	29	6	3	3	2	2	2
	6	6	11.0	95.5	123	39	16	11	10	8	7
	7	5	17.6	81.9	181	77	69	61	52	42	35
	8	4	23.9	67.8	214	113	131	123	112	97	87
	9	3	29.6	52.6	234	144	190	182	169	150	138
	10	2	34.2	36.2	246	168	237	229	215	194	179
	11	1	37.2	18.5	252	183	268	260	244	222	205
	12		38.3	0.0	254	188	278	270	255	232	215
	Surface Daily Totals				2808	1646	2108	2008	1860	1662	1522
Sep 21	7	5	6.5	76.5	77	16	25	24	24	23	21
	8	4	12.7	62.6	163	51	92	92	90	85	81
	9	3	18.1	48.1	206	83	159	159	156	147	141
	10	2	22.3	32.7	229	108	212	213	209	198	189
	11	1	25.1	16.6	240	124	246	248	243	230	220
	12		26.0	0.0	244	129	258	260	254	241	230
	Surface Daily Totals				2074	892	1726	1736	1696	1608	1532
Oct 21	8	4	3.0	58.5	17	2	9	9	10	10	10
	9	3	8.1	44.6	122	26	86	91	93	92	90
	10	2	12.1	30.2	176	50	152	159	161	159	155
	11	1	14.6	15.2	201	65	193	201	203	200	195
	12		15.5	0.0	208	71	207	215	217	213	208
	Surface Daily Totals				1238	358	1088	1136	1152	1134	1106
Nov 21	10	2	3.0	28.1	23	3	18	20	21	21	21
	11	1	5.4	14.2	79	12	70	76	79	80	79
	12		6.2	0.0	97	17	89	96	100	101	100
	Surface Daily Totals				302	46	266	286	298	302	300
Dec 21	11	1	1.8	13.7	4	0	3	4	4	4	4
	12		2.6	0.0	16	2	14	15	16	17	17
	Surface Daily Totals				24	2	20	22	24	24	24

NOTE: 1 Btu/h ft² = 3.152 W/m².

Source: Reprinted by permission from Peter J. Lunde, *Solar Thermal Engineering,* © 1980, John Wiley & Sons Inc., New York.

APPENDICES

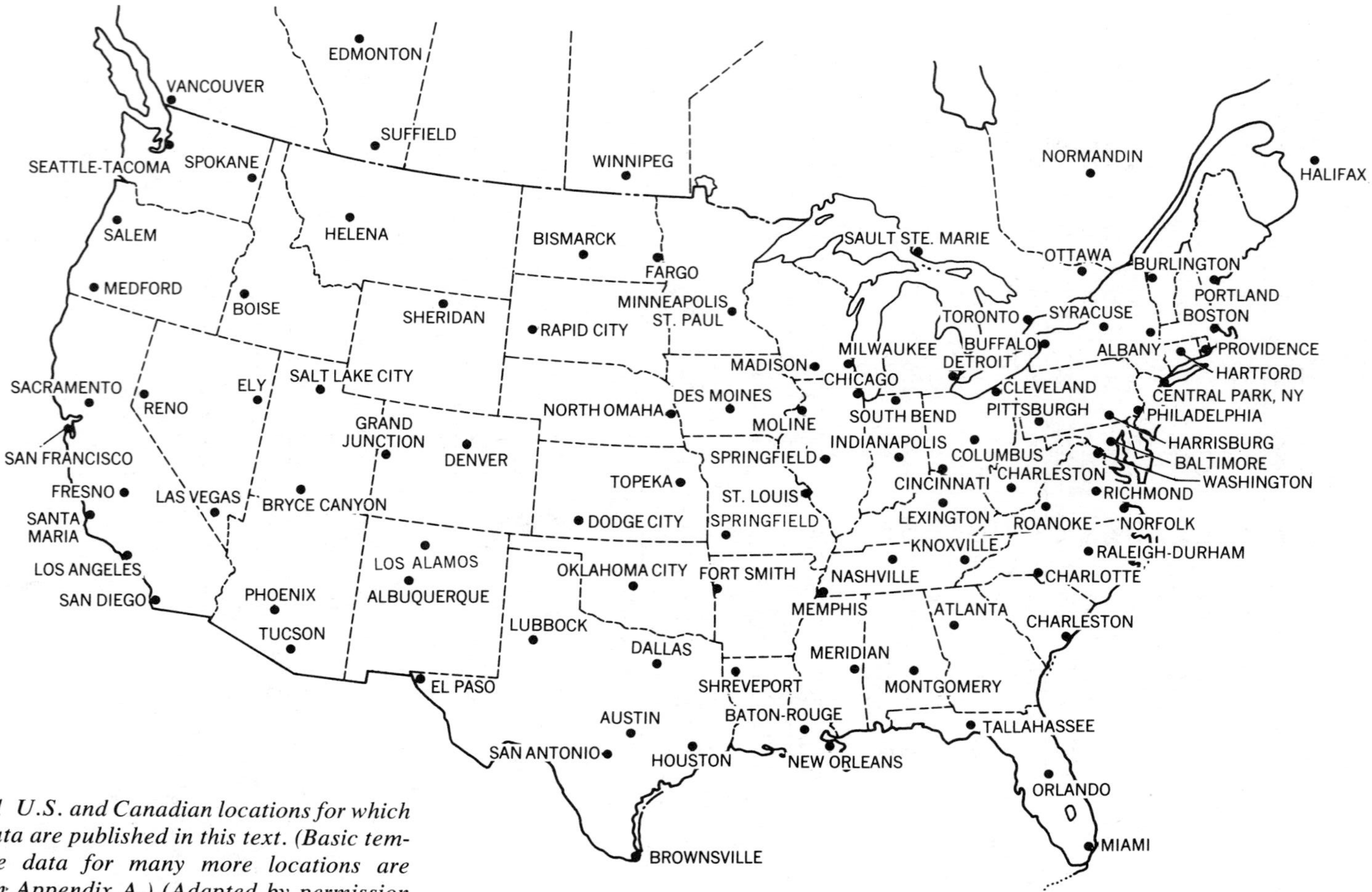

Fig. B.1 *U.S. and Canadian locations for which solar data are published in this text. (Basic temperature data for many more locations are found in Appendix A.) (Adapted by permission from J. D. Balcomb et al.,* Passive Solar Design Handbook, *Vol 3, © 1983, American Solar Energy Society, Inc., Boulder, Colo.)*

TABLE B.15 **Average Insolation, Temperature, and DD Data**

Elevation in feet, latitude in degrees north latitude, HS (horizontal surface) and VS (vertical surface) in Btu/day ft^2, TA in degrees F, D50 through D65 in degree days F.

United States

MONTGOMERY, ALABAMA Elev 203 Lat 32.3

	HS	VS	TA	D50	D55	D60	D65
Jan	752	896	48	148	256	394	556
Jul	1841	820	81	0	0	0	0
Yr	1390	946	65	445	866	1474	2269

PHOENIX, ARIZONA Elev 1112 Lat 33.4

	HS	VS	TA	D50	D55	D60	D65
Jan	1021	1462	51	78	162	285	428
Jul	2486	964	91	0	0	0	0
Yr	1371	1326	70	187	459	919	1552

TUCSON, ARIZONA Elev 2556 Lat 32.1

	HS	VS	TA	D50	D55	D60	D65
Jan	1099	1539	51	80	166	292	442
Jul	2341	922	86	0	0	0	0
Yr	1874	1307	68	214	525	1036	1752

FORT SMITH, ARKANSAS Elev 463 Lat 35.3

	HS	VS	TA	D50	D55	D60	D65
Jan	744	996	39	346	497	651	806
Jul	2065	908	82	0	0	0	0
Yr	1406	1013	61	996	1622	2405	3336

FRESNO, CALIFORNIA Elev 328 Lat 36.8

	HS	VS	TA	D50	D55	D60	D65
Jan	657	886	45	176	308	457	611
Jul	2685	1076	81	0	0	0	0
Yr	1714	1210	62	507	1021	1741	2650

LOS ANGELES, CALIFORNIA Elev 105 Lat 33.9

	HS	VS	TA	D50	D55	D60	D65
Jan	926	1293	55	21	83	186	331
Jul	2307	942	69	0	1	5	19
Yr	1596	1157	62	64	299	849	1819

SACRAMENTO, CALIFORNIA Elev 26 Lat 38.5

	HS	VS	TA	D50	D55	D60	D65
Jan	597	829	45	183	315	464	617
Jul	2688	1131	75	0	0	0	0
Yr	1646	1198	60	554	1097	1871	2843

SAN DIEGO, CALIFORNIA Elev 30 Lat 32.7

	HS	VS	TA	D50	D55	D60	D65
Jan	976	1325	55	9	58	160	314
Jul	2186	902	70	0	0	1	6
Yr	1600	1151	63	23	170	623	1507

SAN FRANCISCO, CALIFORNIA Elev 16 Lat 37.6

	HS	VS	TA	D50	D55	D60	D65
Jan	708	1023	48	82	210	363	518
Jul	2392	1034	63	0	2	21	93
Yr	1556	1156	57	202	705	1643	3042

SANTA MARIA, CALIFORNIA Elev 236 Lat 34.9

	HS	VS	TA	D50	D55	D60	D65
Jan	854	1198	51	51	150	296	450
Jul	2341	965	62	0	3	30	112
Yr	1610	1172	57	155	624	1604	3053

DENVER, COLORADO Elev 5331 Lat 39.7

	HS	VS	TA	D50	D55	D60	D65
Jan	840	1465	30	623	778	933	1088
Jul	2273	1053	73	0	0	0	0
Yr	1570	1334	50	2592	3588	4733	6016

GRAND JUNCTION, COLORADO Elev 4839 Lat 39.1

	HS	VS	TA	D50	D55	D60	D65
Jan	791	1296	27	726	880	1035	1190
Jul	2465	1094	79	0	0	0	0
Yr	1661	1346	53	2514	3412	4434	5605

HARTFORD, CONNECTICUT Elev 180 Lat 41.9

	HS	VS	TA	D50	D55	D60	D65
Jan	477	694	25	781	936	1091	1246
Jul	1649	861	73	0	0	0	0
Yr	1060	835	49	2971	3948	5075	6350

WASHINGTON, D.C. Elev 289 Lat 38.9

	HS	VS	TA	D50	D55	D60	D65
Jan	572	793	32	555	710	865	1020
Jul	1817	883	75	0	0	0	0
Yr	1210	912	54	2004	2869	3864	5010

MIAMI, FLORIDA Elev 7 Lat 25.8

	HS	VS	TA	D50	D55	D60	D65
Jan	1057	1121	67	1	4	18	53
Jul	1763	787	82	0	0	0	0
Yr	1474	941	76	3	14	55	206

ORLANDO, FLORIDA Elev 118 Lat 28.5

	HS	VS	TA	D50	D55	D60	D65
Jan	999	1151	60	13	42	105	197
Jul	1801	795	81	0	0	0	0
Yr	1488	984	72	39	126	348	733

TALLAHASSEE, FLORIDA Elev 69 Lat 30.4

	HS	VS	TA	D50	D55	D60	D65
Jan	877	1033	53	73	150	256	408
Jul	1748	786	81	0	0	0	0
Yr	1434	969	68	215	501	951	1563

ATLANTA, GEORGIA Elev 1033 Lat 33.6

	HS	VS	TA	D50	D55	D60	D65
Jan	718	884	42	246	393	546	701
Jul	1812	821	78	0	0	0	0
Yr	1347	941	61	758	1362	2150	3095

BOISE, IDAHO Elev 2867 Lat 43.6

	HS	VS	TA	D50	D55	D60	D65
Jan	485	770	29	651	806	961	1116
Jul	2613	1309	75	0	0	0	0
Yr	1499	1255	51	2420	3395	4536	5833

CHICAGO, ILLINOIS Elev 623 Lat 41.8

	HS	VS	TA	D50	D55	D60	D65
Jan	507	756	24	797	952	1107	1262
Jul	1944	984	75	0	0	0	0
Yr	1217	960	51	2954	3881	4940	6127

MOLINE, ILLINOIS Elev 594 Lat 41.4

	HS	VS	TA	D50	D55	D60	D65
Jan	535	803	22	884	1039	1194	1349
Jul	1939	974	75	0	0	0	0
Yr	1226	973	50	3191	4117	5178	6395

SPRINGFIELD, ILLINOIS Elev 614 Lat 39.8

	HS	VS	TA	D50	D55	D60	D65
Jan	585	852	27	723	877	1032	1187
Jul	2058	984	76	0	0	0	0
Yr	1304	1003	53	2558	3434	4425	5558

TABLE B.15 **Average Insolation, Temperature, and DD Data (Continued)**

INDIANAPOLIS, INDIANA Elev 807 Lat 39.7

	HS	VS	TA	D50	D55	D60	D65
Jan	496	668	28	685	840	995	1150
Jul	1806	891	75	0	0	0	0
Yr	1167	873	52	2511	3403	4421	5577

SOUTH BEND, INDIANA Elev 774 Lat 41.7

	HS	VS	TA	D50	D55	D60	D65
Jan	416	566	24	806	961	1116	1271
Jul	1852	944	72	0	1	4	6
Yr	1140	864	49	3112	4084	5199	6462

DES MOINES, IOWA Elev 965 Lat 41.5

	HS	VS	TA	D50	D55	D60	D65
Jan	581	912	19	949	1104	1259	1414
Jul	2097	1037	75	0	0	0	0
Yr	1314	1065	49	3491	4435	5510	6710

DODGE CITY, KANSAS Elev 2582 Lat 37.8

	HS	VS	TA	D50	D55	D60	D65
Jan	827	1303	31	596	750	905	1060
Jul	2295	1013	79	0	0	0	0
Yr	1562	1232	55	2132	2980	3945	5046

TOPEKA, KANSAS Elev 886 Lat 39.1

	HS	VS	TA	D50	D55	D60	D65
Jan	681	1033	28	683	837	992	1147
Jul	2128	993	78	0	0	0	0
Yr	1387	1036	54	2325	3175	4137	5243

LEXINGTON, KENTUCKY Elev 988 Lat 38.0

	HS	VS	TA	D50	D55	D60	D65
Jan	546	714	33	531	685	840	995
Jul	1850	881	76	0	0	0	0
Yr	1221	892	55	1865	2686	3632	4729

BATON ROUGE, LOUISIANA Elev 75 Lat 30.5

	HS	VS	TA	D50	D55	D60	D65
Jan	785	889	51	90	174	294	451
Jul	1746	786	82	0	0	0	0
Yr	1380	913	67	232	530	1006	1670

NEW ORLEANS, LOUISIANA Elev 10 Lat 30.0

	HS	VS	TA	D50	D55	D60	D65
Jan	835	950	53	73	150	252	403
Jul	1813	801	82	0	0	0	0
Yr	1438	943	68	197	465	887	1465

SHREVEPORT, LOUISIANA Elev 259 Lat 32.5

	HS	VS	TA	D50	D55	D60	D65
Jan	762	920	47	154	264	403	552
Jul	2014	864	83	0	0	0	0
Yr	1428	973	66	428	832	1415	2167

PORTLAND, MAINE Elev 62 Lat 43.6

	HS	VS	TA	D50	D55	D60	D65
Jan	450	689	22	884	1039	1194	1349
Jul	1659	894	68	0	1	5	27
Yr	1052	857	45	3648	4758	6039	7498

DETROIT, MICHIGAN Elev 627 Lat 42.4

	HS	VS	TA	D50	D55	D60	D65
Jan	417	585	26	760	915	1070	1225
Jul	1835	951	73	0	0	0	0
Yr	1122	869	50	2931	3890	4986	6228

SAULT STE. MARIE, MICHIGAN Elev 725 Lat 46.5

	HS	VS	TA	D50	D55	D60	D65
Jan	325	492	14	1110	1265	1420	1575
Jul	1835	1045	64	1	7	33	96
Yr	1044	861	40	4969	6198	7607	9193

MINNEAPOLIS, MINNESOTA Elev 837 Lat 44.9

	HS	VS	TA	D50	D55	D60	D65
Jan	464	768	12	1172	1327	1482	1637
Jul	1970	1071	72	0	1	5	11
Yr	1172	996	44	4584	5631	6824	8159

MERIDIAN, MISSISSIPPI Elev 308 Lat 32.3

	HS	VS	TA	D50	D55	D60	D65
Jan	744	883	47	163	274	413	575
Jul	1823	815	81	0	0	0	0
Yr	1371	933	65	510	955	1582	2388

SAINT LOUIS, MISSOURI Elev 564 Lat 38.7

	HS	VS	TA	D50	D55	D60	D65
Jan	627	898	31	581	735	890	1045
Jul	2049	959	79	0	0	0	0
Yr	1329	1006	56	1961	2762	3686	4750

SPRINGFIELD, MISSOURI Elev 1270 Lat 37.2

	HS	VS	TA	D50	D55	D60	D65
Jan	684	956	33	534	686	840	995
Jul	2063	936	78	0	0	0	0
Yr	1364	1011	56	1848	2618	3522	4571

HELENA, MONTANA Elev 3898 Lat 46.6

	HS	VS	TA	D50	D55	D60	D65
Jan	419	719	18	989	1144	1299	1454
Jul	2334	1312	68	1	3	12	33
Yr	1266	1134	43	4151	5342	6689	8190

NORTH OMAHA, NEBRASKA Elev 1325 Lat 41.4

	HS	VS	TA	D50	D55	D60	D65
Jan	634	1034	20	924	1079	1234	1389
Jul	2106	1038	75	0	1	3	7
Yr	1323	1078	49	3369	4309	5381	6601

ELY, NEVADA Elev 6253 Lat 39.3

	HS	VS	TA	D50	D55	D60	D65
Jan	819	1380	24	818	973	1128	1283
Jul	2447	1094	67	0	2	11	23
Yr	1675	1391	44	3716	4922	6291	7814

LAS VEGAS, NEVADA Elev 2178 Lat 36.1

	HS	VS	TA	D50	D55	D60	D65
Jan	978	1553	44	216	346	493	645
Jul	2588	1039	90	0	0	0	0
Yr	1866	1431	66	631	1129	1788	2601

RENO, NEVADA Elev 4400 Lat 39.5

	HS	VS	TA	D50	D55	D60	D65
Jan	800	1345	32	561	716	871	1026
Jul	2692	1167	69	0	1	5	17
Yr	1764	1439	49	2292	3345	4590	6022

ALBUQUERQUE, NEW MEXICO Elev 5312 Lat 35.0

	HS	VS	TA	D50	D55	D60	D65
Jan	1016	1562	35	459	614	769	924
Jul	2489	995	79	0	0	0	0
Yr	1830	1379	57	1497	2292	3216	4292

LOS ALAMOS, NEW MEXICO Elev 7380 Lat 35.9

	HS	VS	TA	D50	D55	D60	D65
Jan	893	1340	29	650	804	958	1113
Jul	2051	913	68	0	0	0	20
Yr	1579	1244	48	2822	3851	5043	6437

ALBANY, NEW YORK Elev 292 Lat 42.7

	HS	VS	TA	D50	D55	D60	D65
Jan	456	674	22	884	1039	1194	1349
Jul	1725	908	72	0	1	3	9
Yr	1068	843	48	3424	4428	5586	6888

TABLE B.15 **Average Insolation, Temperature, and DD Data (Continued)**

BUFFALO, NEW YORK Elev 705 Lat 42.9

	HS	VS	TA	D50	D55	D60	D65
Jan	349	465	24	815	970	1125	1280
Jul	1776	935	70	0	0	3	12
Yr	1037	780	47	3322	4363	5551	6927

NEW YORK, NEW YORK Elev 187 Lat 40.8

	HS	VS	TA	D50	D55	D60	D65
Jan	500	708	32	552	707	862	1017
Jul	1688	861	77	0	0	0	0
Yr	1101	849	55	1931	2759	3737	4848

SYRACUSE, NEW YORK Elev 407 Lat 43.1

	HS	VS	TA	D50	D55	D60	D65
Jan	385	538	24	818	973	1128	1283
Jul	1758	931	72	0	1	3	11
Yr	1037	791	48	3215	4218	5366	6678

CHARLOTTE, NORTH CAROLINA Elev 768 Lat 35.2

	HS	VS	TA	D50	D55	D60	D65
Jan	719	944	42	255	402	555	710
Jul	1831	841	79	0	0	0	0
Yr	1346	981	61	828	1451	2257	3218

RALEIGH-DURHAM, NORTH CAROLINA Elev 440 Lat 35.9

	HS	VS	TA	D50	D55	D60	D65
Jan	694	924	41	300	451	605	760
Jul	1776	832	78	0	0	0	0
Yr	1297	955	59	990	1659	2509	3514

BISMARCK, NORTH DAKOTA Elev 1647 Lat 46.8

	HS	VS	TA	D50	D55	D60	D65
Jan	467	847	8	1296	1451	1606	1761
Jul	2184	1241	71	0	2	8	18
Yr	1251	1145	41	5235	6364	7627	9044

FARGO, NORTH DAKOTA Elev 899 Lat 46.9

	HS	VS	TA	D50	D55	D60	D65
Jan	415	720	6	1367	1522	1677	1832
Jul	2120	1210	71	0	1	5	13
Yr	1206	1075	41	5485	6607	7858	9271

CINCINNATI, OHIO Elev 889 Lat 39.1

	HS	VS	TA	D50	D55	D60	D65
Jan	500	659	31	587	741	896	1051
Jul	1771	869	76	0	0	0	0
Yr	1160	858	54	2117	2973	3951	5070

CLEVELAND, OHIO Elev 804 Lat 41.4

	HS	VS	TA	D50	D55	D60	D65
Jan	388	507	27	716	871	1026	1181
Jul	1828	929	71	0	1	4	9
Yr	1093	808	50	2804	3768	4879	6154

COLUMBUS, OHIO Elev 833 Lat 40.0

	HS	VS	TA	D50	D55	D60	D65
Jan	459	606	28	670	825	980	1135
Jul	1755	876	74	0	0	0	0
Yr	1128	834	52	2524	3438	4491	5702

OKLAHOMA CITY, OKLAHOMA Elev 1302 Lat 35.4

	HS	VS	TA	D50	D55	D60	D65
Jan	801	1114	37	413	565	719	874
Jul	2128	925	82	0	0	0	0
Yr	1463	1073	60	1232	1903	2734	3695

MEDFORD, OREGON Elev 1299 Lat 42.4

	HS	VS	TA	D50	D55	D60	D65
Jan	407	565	37	417	571	725	880
Jul	2475	1207	72	0	1	3	11
Yr	1356	1033	53	1576	2505	3621	4930

SALEM, OREGON Elev 200 Lat 44.9

	HS	VS	TA	D50	D55	D60	D65
Jan	332	471	39	348	502	657	812
Jul	2142	1154	67	0	1	7	43
Yr	1130	897	52	1265	2220	3411	4852

HARRISBURG, PENNSYLVANIA Elev 348 Lat 40.2

	HS	VS	TA	D50	D55	D60	D65
Jan	536	763	30	617	772	927	1082
Jul	1764	883	76	0	0	0	0
Yr	1152	887	53	2221	3093	4086	5224

PHILADELPHIA, PENNSYLVANIA Elev 30 Lat 39.9

	HS	VS	TA	D50	D55	D60	D65
Jan	555	792	32	549	704	859	1014
Jul	1758	876	77	0	0	0	0
Yr	1170	905	55	1935	2775	3749	4865

PITTSBURGH, PENNSYLVANIA Elev 1224 Lat 40.5

	HS	VS	TA	D50	D55	D60	D65
Jan	424	553	28	679	834	989	1144
Jul	1689	857	72	0	1	3	7
Yr	1071	793	50	2635	3574	4669	5930

PROVIDENCE, RHODE ISLAND Elev 62 Lat 41.7

	HS	VS	TA	D50	D55	D60	D65
Jan	506	750	28	670	825	980	1135
Jul	1695	878	72	0	0	0	0
Yr	1114	884	50	2566	3543	4669	5972

CHARLESTON, SOUTH CAROLINA Elev 39 Lat 32.9

	HS	VS	TA	D50	D55	D60	D65
Jan	744	904	49	120	222	360	521
Jul	1799	813	80	0	0	0	0
Yr	1346	940	65	360	756	1355	2146

RAPID CITY, SOUTH DAKOTA Elev 3169 Lat 44.0

	HS	VS	TA	D50	D55	D60	D65
Jan	542	928	22	871	1026	1181	1336
Jul	2223	1161	73	1	2	8	13
Yr	1344	1177	47	3681	4749	5965	7324

KNOXVILLE, TENNESSEE Elev 981 Lat 35.8

	HS	VS	TA	D50	D55	D60	D65
Jan	621	785	41	302	449	602	756
Jul	1804	839	78	0	0	0	0
Yr	1275	909	60	1018	1671	2504	3478

MEMPHIS, TENNESSEE Elev 285 Lat 35.0

	HS	VS	TA	D50	D55	D60	D65
Jan	683	870	41	312	455	606	760
Jul	1972	879	82	0	0	0	0
Yr	1368	965	62	988	1588	2357	3227

NASHVILLE, TENNESSEE Elev 591 Lat 36.1

	HS	VS	TA	D50	D55	D60	D65
Jan	580	721	38	369	519	673	828
Jul	1891	869	80	0	0	0	0
Yr	1272	891	59	1195	1874	2720	3696

TABLE B.15 **Average Insolation, Temperature, and DD Data (Continued)**

AUSTIN, TEXAS Elev 620 Lat 30.3

	HS	VS	TA	D50	D55	D60	D65
Jan	864	1008	50	116	207	333	483
Jul	2105	865	85	0	0	0	0
Yr	1478	974	68	289	602	1088	1737

BROWNSVILLE, TEXAS Elev 20 Lat 25.9

	HS	VS	TA	D50	D55	D60	D65
Jan	913	923	60	18	51	116	225
Jul	2212	867	84	0	0	0	0
Yr	1550	917	74	44	127	325	650

DALLAS, TEXAS Elev 489 Lat 32.8

	HS	VS	TA	D50	D55	D60	D65
Jan	821	1035	45	189	312	457	608
Jul	2122	890	86	0	0	0	0
Yr	1470	1014	66	505	943	1543	2290

EL PASO, TEXAS Elev 3917 Lat 31.8

	HS	VS	TA	D50	D55	D60	D65
Jan	1125	1572	44	210	355	509	663
Jul	2450	934	82	0	0	0	0
Yr	1901	1327	63	561	1102	1810	2678

HOUSTON, TEXAS Elev 108 Lat 30.0

	HS	VS	TA	D50	D55	D60	D65
Jan	772	852	52	71	150	263	416
Jul	1828	805	83	0	0	0	0
Yr	1353	884	69	161	409	825	1434

LUBBOCK, TEXAS Elev 3241 Lat 33.6

	HS	VS	TA	D50	D55	D60	D65
Jan	1031	1497	39	343	494	648	803
Jul	2412	956	80	0	0	0	0
Yr	1768	1279	60	1069	1739	2582	3545

SAN ANTONIO, TEXAS Elev 794 Lat 29.5

	HS	VS	TA	D50	D55	D60	D65
Jan	895	1026	51	93	179	302	451
Jul	2121	863	85	0	0	0	0
Yr	1501	973	69	213	490	941	1570

BRYCE CANYON, UTAH Elev 7588 Lat 37.7

	HS	VS	TA	D50	D55	D60	D65
Jan	914	1511	20	936	1091	1246	1401
Jul	2424	1044	62	1	8	47	128
Yr	1742	1404	40	4693	6106	7675	9133

SALT LAKE CITY, UTAH Elev 4226 Lat 40.8

	HS	VS	TA	D50	D55	D60	D65
Jan	639	1017	28	683	837	992	1147
Jul	2590	1186	77	0	0	0	0
Yr	1606	1301	51	2648	3612	4725	5983

BURLINGTON, VERMONT Elev 341 Lat 44.5

	HS	VS	TA	D50	D55	D60	D65
Jan	385	572	17	1029	1184	1339	1494
Jul	1721	941	70	0	1	4	20
Yr	1023	815	44	4142	5230	6464	7876

NORFOLK, VIRGINIA Elev 30 Lat 36.9

	HS	VS	TA	D50	D55	D60	D65
Jan	678	932	41	300	450	605	760
Jul	1853	868	78	0	0	0	0
Yr	1327	990	59	974	1646	2489	3488

RICHMOND, VIRGINIA Elev 164 Lat 37.5

	HS	VS	TA	D50	D55	D60	D65
Jan	632	863	38	390	543	698	853
Jul	1774	849	78	0	0	0	0
Yr	1250	936	58	1296	2021	2909	3939

ROANOKE, VIRGINIA Elev 1175 Lat 37.3

	HS	VS	TA	D50	D55	D60	D65
Jan	660	911	36	423	577	732	887
Jul	1796	854	75	0	0	0	0
Yr	1271	958	56	1486	2277	3211	4307

SEATTLE-TACOMA, WASHINGTON Elev 400 Lat 47.4

	HS	VS	TA	D50	D55	D60	D65
Jan	262	378	38	367	521	676	831
Jul	2248	1299	65	0	2	16	80
Yr	1056	857	51	1386	2393	3662	5185

SPOKANE, WASHINGTON Elev 2365 Lat 47.6

	HS	VS	TA	D50	D55	D60	D65
Jan	315	496	25	763	918	1073	1228
Jul	2357	1368	70	0	2	7	21
Yr	1227	1068	47	3061	4150	5411	6835

CHARLESTON, WEST VIRGINIA Elev 951 Lat 38.4

	HS	VS	TA	D50	D55	D60	D65
Jan	498	638	35	483	636	791	946
Jul	1682	827	75	0	0	0	0
Yr	1125	822	55	1726	2540	3488	4590

MADISON, WISCONSIN Elev 860 Lat 43.1

	HS	VS	TA	D50	D55	D60	D65
Jan	515	822	17	1029	1184	1339	1494
Jul	1934	1009	70	0	1	5	14
Yr	1193	978	45	4086	5143	6352	7730

MILWAUKEE, WISCONSIN Elev 692 Lat 42.9

	HS	VS	TA	D50	D55	D60	D65
Jan	479	731	19	949	1104	1259	1414
Jul	1962	1017	70	0	1	4	15
Yr	1194	957	46	3774	4833	6045	7444

SHERIDAN, WYOMING Elev 3966 Lat 44.8

	HS	VS	TA	D50	D55	D60	D65
Jan	517	900	21	899	1054	1209	1364
Jul	2329	1237	70	1	2	9	28
Yr	1333	1170	45	3860	4991	6279	7708

Canada

EDMONTON, ALBERTA Elev 2220 Lat 53.6

	HS	VS	TA	D50	D55	D60	D65
Jan	324	746	4	1421	1574	1728	1883
Jul	1977	1378	62	1	7	38	117
Yr	1114	1205	36	6317	7563	9016	10650

SUFFIELD, ALBERTA Elev 2549 Lat 50.3

	HS	VS	TA	D50	D55	D60	D65
Jan	433	937	7	1333	1486	1640	1794
Jul	2173	1377	67	0	2	11	49
Yr	1239	1269	40	5500	6637	7923	9393

VANCOUVER, BRITISH COLUMBIA Elev 310 Lat 49.3

	HS	VS	TA	D50	D55	D60	D65
Jan	254	395	37	425	572	724	878
Jul	2021	1239	63	0	1	13	82
Yr	1060	916	50	1791	2781	4041	5588

WINNIPEG, MANITOBA Elev 820 Lat 49.9

	HS	VS	TA	D50	D55	D60	D65
Jan	461	1011	0	1588	1740	1893	2047
Jul	2025	1264	67	0	2	9	45
Yr	1190	1199	36	6925	8062	9338	10790

TABLE B.15 **Average Insolation, Temperature, and DD Data (Continued)**

HALIFAX, NOVA SCOTIA				Elev 136		Lat 44.6
	HS	VS	TA	D50	D55	D60	D65
Jan	456	737	26	752	900	1051	1204
Jul	1694	929	65	0	1	8	57
Yr	1076	907	46	3457	4500	5746	7211

TORONTO, ONTARIO				Elev 443		Lat 43.7
	HS	VS	TA	D50	D55	D60	D65
Jan	487	777	22	891	1041	1194	1348
Jul	1958	1035	70	0	0	3	18
Yr	1171	948	46	3842	4853	6013	7343

OTTAWA, ONTARIO				Elev 377		Lat 45.5
	HS	VS	TA	D50	D55	D60	D65
Jan	510	914	13	1169	1320	1473	1627
Jul	1875	1040	69	0	1	4	23
Yr	1158	1015	43	4912	5965	7158	8529

NORMANDIN, QUEBEC				Elev 450		Lat 48.8
	HS	VS	TA	D50	D55	D60	D65
Jan	454	921	0	1564	1719	1873	2028
Jul	1707	1031	62	1	7	40	118
Yr	1092	1053	34	7037	8308	9762	11376

Source: Reprinted by permission from J. D. Balcomb et al., *Passive Solar Design Handbook,* Vol. 3, © 1983, American Solar Energy Society, Inc., Boulder.

APPENDICES

C
ANNUAL SOLAR PERFORMANCE

This material is abridged from J. D. Balcomb et al., *Passive Solar Heating Analysis,* © 1984, American Society of Heating, Refrigerating, and Air-Conditioning Engineers, Inc., Atlanta, Ga.

This appendix lists 30 passive solar heating systems (3 water wall, 9 Trombé wall, 9 direct gain, and 9 sunspace) for the U.S. and Canadian locations shown in Fig. B.1. The *Passive Solar Heating Analysis* lists 94 systems, for about twice as many locations. The specifications for all 94 systems are listed here, so that the original source may be consulted if your passive system is not represented by one of those in this appendix.

TABLE C.1 **Characteristics of Selected Passive Solar Heating Systems**

*Part A. **Water Wall Systems***

Designation	*Thermal Storage Capacity*[a] *(Btu/ft² F)*	*Wall Thickness (in.)*	*No. of Glazings*	*Wall Surface*	*Night Insulation*
WW-A1	15.6	3	2	Normal	No
WW-A2	31.2	6	2	Normal	No
WW-A3[b]	46.8	9	2	Normal	No
WW-A4	62.4	12	2	Normal	No
WW-A5	93.6	18	2	Normal	No
WW-A6	124.8	24	2	Normal	No
WW-B1	46.8	9	1	Normal	No
WW-B2	46.8	9	3	Normal	No
WW-B3	46.8	9	1	Normal	Yes
WW-B4[b]	46.8	9	2	Normal	Yes
WW-B5	46.8	9	3	Normal	Yes
WW-C1	46.8	9	1	Selective	No
WW-C2[b]	46.8	9	2	Selective	No
WW-C3	46.8	9	1	Selective	Yes
WW-C4	46.8	9	2	Selective	Yes

*Part B. **Trombe Wall Systems: Vented***

Designation	*Thermal Storage Capacity*[c] *(Btu/ft² F)*	*Wall Thickness*[c] *(in.)*	ρck[d] *(Btu²/h-ft⁴-F²)*	*No. of Glazings*	*Wall Surface*	*Night Insulation*
TW-A1[b]	15	6	30	2	Normal	No
TW-A2[b]	22.5	9	30	2	Normal	No
TW-A3[b]	30	12	30	2	Normal	No
TW-A4[b]	45	18	30	2	Normal	No
TW-B1	15	6	15	2	Normal	No
TW-B2	22.5	9	15	2	Normal	No
TW-B3[b]	30	12	15	2	Normal	No
TW-B4	45	18	15	2	Normal	No
TW-C1	15	6	7.5	2	Normal	No
TW-C2	22.5	9	7.5	2	Normal	No
TW-C3	30	12	7.5	2	Normal	No
TW-C4	45	18	7.5	2	Normal	No
TW-D1	30	12	30	1	Normal	No
TW-D2	30	12	30	3	Normal	No
TW-D3	30	12	30	1	Normal	Yes
TW-D4[b]	30	12	30	2	Normal	Yes
TW-D5	30	12	30	3	Normal	Yes
TW-E1	30	12	30	1	Selective	No
TW-E2[b]	30	12	30	2	Selective	No
TW-E3	30	12	30	1	Selective	Yes
TW-E4	30	12	30	2	Selective	Yes

Note: Vented systems may be made to perform as unvented systems by sealing both top and bottom vent openings.

TABLE C.1 **Characteristics of Selected Passive Solar Heating Systems (Continued)**

Part C. ***Trombe Wall Systems: Unvented***

Designation	Thermal Storage Capacity[c] (Btu/ft² F)	Wall Thickness[c] (in.)	ρck[d] ($Btu^2/h\text{-}ft^4\text{-}F^2$)	No. of Glazings	Wall Surface	Night Insulation
TW-F1	15	6	30	2	Normal	No
TW-F2	22.5	9	30	2	Normal	No
TW-F3[b]	30	12	30	2	Normal	No
TW-F4	45	18	30	2	Normal	No
TW-G1	15	6	15	2	Normal	No
TW-G2	22.5	9	15	2	Normal	No
TW-G3	30	12	15	2	Normal	No
TW-G4	45	18	15	2	Normal	No
TW-H1	15	6	7.5	2	Normal	No
TW-H2	22.5	9	7.5	2	Normal	No
TW-H3	30	12	7.5	2	Normal	No
TW-H4	45	18	7.5	2	Normal	No
TW-I1	30	12	30	1	Normal	No
TW-I2	30	12	30	3	Normal	No
TW-I3	30	12	30	1	Normal	Yes
TW-I4	30	12	30	2	Normal	Yes
TW-I5	30	12	30	3	Normal	Yes
TW-J1	30	12	30	1	Selective	No
TW-J2[b]	30	12	30	2	Selective	No
TW-J3	30	12	30	1	Selective	Yes
TW-J4	30	12	30	2	Selective	Yes

Part D. ***Direct-Gain Systems***

Designation	Thermal Storage Capacity[c] (Btu/ft² F)	Mass Thickness[c] (in.)	Ratio of Mass to Glazing Area	No. of Glazings	Night Insulation
DG-B1[b]	30	2	6	2	No
DG-B2[b]	30	2	6	3	No
DG-B3[b]	30	2	6	2	Yes
DG-B1[b]	45	6	3	2	No
DG-B2[b]	45	6	3	3	No
DG-B3[b]	45	6	3	2	Yes
DG-C1[b]	60	4	6	2	No
DG-C2[b]	60	4	6	3	No
DG-C3[b]	60	4	6	2	Yes

TABLE C.1 **Characteristics of Selected Passive Solar Heating Systems (Continued)**

*Part E. **Sunspace Systems***

Designation	*Type[d]*	*Tilt (Degrees)*	*Common Wall*	*End Walls*	*Night Insulation*
SS-A1[b]	Attached	50	Masonry	Opaque	No
SS-A2	Attached	50	Masonry	Opaque	Yes
SS-A3	Attached	50	Masonry	Glazed	No
SS-A4	Attached	50	Masonry	Glazed	Yes
SS-A5	Attached	50	Insulated	Opaque	No
SS-A6	Attached	50	Insulated	Opaque	Yes
SS-A7	Attached	50	Insulated	Glazed	No
SS-A8	Attached	50	Insulated	Glazed	Yes
SS-B1[b]	Attached	90/30	Masonry	Opaque	No
SS-B2[b]	Attached	90/30	Masonry	Opaqúe	Yes
SS-B3[b]	Attached	90/30	Masonry	Glazed	No
SS-B4	Attached	90/30	Masonry	Glazed	Yes
SS-B5	Attached	90/30	Insulated	Opaque	No
SS-B6	Attached	90/30	Insulated	Opaque	Yes
SS-B7	Attached	90/30	Insulated	Glazed	No
SS-B8	Attached	90/30	Insulated	Glazed	Yes
SS-C1[b]	Semi-enclosed	90	Masonry	Common	No
SS-C2[b]	Semi-enclosed	90	Masonry	Common	Yes
SS-C3	Semi-enclosed	90	Insulated	Common	No
SS-C4	Semi-enclosed	90	Insulated	Common	Yes
SS-D1	Semi-enclosed	50	Masonry	Common	No
SS-D2	Semi-enclosed	50	Masonry	Common	Yes
SS-D3	Semi-enclosed	50	Insulated	Common	No
SS-D4	Semi-enclosed	50	Insulated	Common	Yes
SS-E1[b]	Semi-enclosed	90/30	Masonry	Common	No
SS-E2[b]	Semi-enclosed	90/30	Masonry	Common	Yes
SS-E3[b]	Semi-enclosed	90/30	Insulated	Common	No
SS-E4	Semi-enclosed	90/30	Insulated	Common	Yes

[a]Per unit of projected area.

[b]Listed in this text.

[c]The thermal storage capacity is per unit of projected area, or, equivalently, the quantity ρct. The wall thickness (t) is listed only as an appropriate guide by assuming that ρc = 30 Btu/ft^3-°F (ρ of 150 lb/ft^3 and c of 0.2 Btu/lb-°F), typical of ordinary concrete.

[d]ρck is the product of the density (lb/ft^3), specific heat (Btu/lb-°F) and thermal conductivity (Btu/h-ft-°F): see Table C.2.

[e]See Fig. C.1 for additional description.

TABLE C.2 **Thermal Properties of Various Materials**

Material	*Source*[a]	*Density,* ρ *(lb/ft*^{3}*)*	*Specific Heat,* c *(Btu/lb-°F)*	*Heat Capacity,* ρc *(Btu/ft*^{3}*-°F)*	*Conductivity,* k *(Btu/h-°F-ft)*	*Diffusivity,* $k/\rho c$ *(ft*^{2}*/h)*	ρck *(Btu*^{2}*/h-ft*^{4}*-°F*^{2}*)*
Water	1	62.3	1.0	62.3			
Concrete	2, 3	144	0.21	30.3	0.89	0.029	27.0
Concrete block	2, 3						
Heavy weight		135	0.21	28.4	0.74	0.026	21.0
Medium weight		105	0.22	23.1	0.41	0.018	9.5
Light weight		85	0.23	19.6	0.27	0.014	5.3
Brick	1, 4						
Paving		135	0.19	25.7	0.75	0.029	19.3
Face		130	0.19	24.7	0.75	0.030	18.5
Building		120	0.19	22.8	0.42	0.018	9.6
Mortar or grout	1	116	0.20	23.2	0.42	0.018	9.7
Adobe	5	80	0.20	16.0	0.38	0.024	6.1
		100		20.0	0.75	0.038	15.0
Gypsum or plasterboard	1	50	0.26	13.0	0.097	0.058	9.8
Douglas fir plywood	1	34	0.29	9.9	0.067	0.007	0.7
Hardwood	1	45	0.30	13.5	0.092	0.007	1.2

Source: Reprinted by permission from *Passive Solar Heating Analysis,* © 1984, American Society of Heating, Refrigerating, and Air-Conditioning Engineers, Inc., Atlanta, Ga.

Note: Concrete and masonry products can be manufactured in a variety of densities. The values listed are for the most commonly available products. The thermal properties of concrete and concrete block vary with density. Empirical correlations between these properties are[2] $k = 0.05e^{0.02\rho}$ and [3]$c = 1/(3.934 + 0.006\rho)$.

[a]1. *ASHRAE Handbook: 1981 Fundamentals* (American Society of Heating, Refrigerating, and Air-Conditioning Engineers, Inc., Atlanta, Ga., 1981).

2. "Calculating the Steady State *U*-Value of Concrete Masonry Walls," Expanded Shale, Clay, and Slate Institute, 4905 Del Ray Avenue, Bethesda, Md.

3. David Whiting, Albert Litvin, and Stanley E. Goodwin, "Specific Heat of Selected Concretes," *Research and Development Bulletin RD058.01B,* Portland Cement Association, 5420 Old Orchard Road, Skokie, Ill.

4. "Brick Passive Solar Heating Systems. Material Properties—Part IV," *Technical Notes on Brick Construction 43D,* Brick Institute of America, 1750 Old Meadow Road, McLean, Va. September/October 1980.

5. Benjamin T. Rogers, private communication; and W. L. Sibbitt, measurements at Los Alamos National Laboratory. Note that the thermal conductivity of adobe varies greatly with moisture content, which is part of the reason for the wide range quoted.

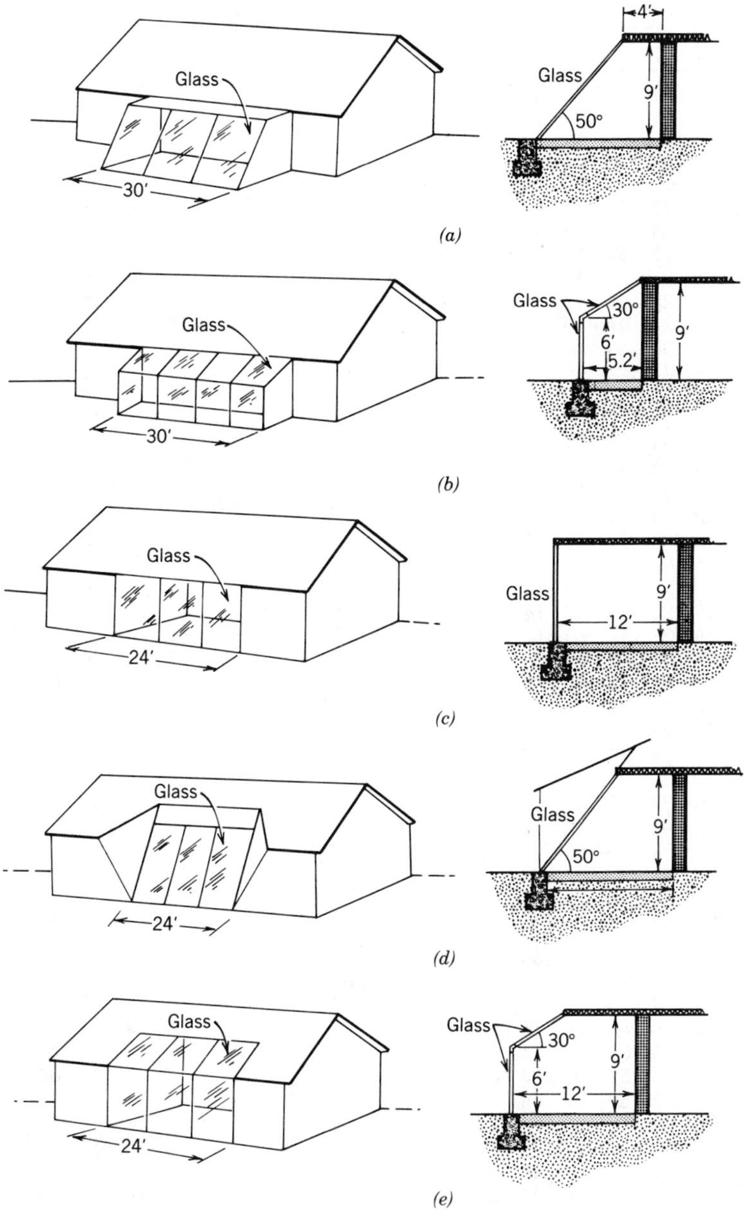

Fig. C.1 *Types of sunspaces, described in Table C.1, Part E.* (a) *and* (b) *Considered* attached *to the building;* (c), (d), *and* (e) *considered* semienclosed *by the building. The architectural detail at the sides of type* (d) *is insignificant; no shading of the sunspace by the building was accounted for in the performance estimates of Table C.2. (Reprinted by permission from J. D. Balcomb et al.,* Passive Solar Heating Analysis, *© 1989, American Society of Heating, Refrigerating, and Air-Conditioning Engineers, Inc., Atlanta, Ga.)*

TABLE C.3 **Annual Passive Heating Performance: SSF**

For each location, DD listed are DD65. For a description of SSF (Solar Savings Fraction) see Fig. 5.6, page 205; LCR (load–collector ratio) is described in Section 5.6. LCR units are Btu/DD-ft².

United States

MONTGOMERY, ALABAMA ***SSF (%)*** *2272 DD65*

Type	LCR = 200	100	70	50	40	30	25	20
WWA3	17	29	38	48	55	64	70	77
WWB4	18	36	48	60	68	78	83	89
WWC2	18	36	47	59	67	76	82	88
TWA1	17	24	29	35	40	46	51	57
TWA2	16	26	33	41	46	54	60	67
TWA3	15	26	34	43	49	58	63	70
TWA4	14	26	34	43	49	58	64	72
TWB3	14	24	31	39	45	54	60	67
TWD4	17	33	43	54	62	72	78	84
TWE2	18	34	44	55	63	72	78	84
TWF3	12	23	30	39	45	54	60	67
TWJ2	15	31	41	52	60	70	76	82
DGA1	12	21	28	34	39	45	49	53
DGA2	13	23	31	39	45	53	58	64
DGA3	15	27	37	47	55	64	70	77
DGB1	12	22	29	37	43	51	56	62
DGB2	13	24	32	41	49	58	65	71
DGB3	15	28	38	49	57	68	75	82
DGC1	14	25	34	43	50	60	66	72
DGC2	15	27	36	47	55	66	72	79
DGC3	18	32	42	55	64	75	81	87
SSA1	21	32	40	48	55	63	68	75
SSB1	17	28	35	42	48	56	62	68
SSB2	21	35	44	54	61	70	76	82
SSB3	17	26	33	40	46	54	59	66
SSC1	15	26	34	43	49	58	64	71
SSC2	17	31	41	51	58	68	74	80
SSE1	20	33	42	51	58	67	72	78
SSE2	23	40	51	62	69	78	83	88
SSE3	18	28	35	43	48	56	62	68

PHOENIX, ARIZONA ***SSF (%)*** *1556 DD65*

Type	LCR = 200	100	70	50	40	30	25	20
WWA3	33	54	66	77	83	90	93	96
WWB4	39	64	76	86	91	96	98	99
WWC2	39	63	75	85	90	95	97	99
TWA1	26	40	49	58	65	73	79	84
TWA2	29	46	56	67	74	82	87	91
TWA3	29	48	59	70	77	85	89	93
TWA4	29	48	60	71	78	86	90	94
TWB3	26	44	55	66	73	82	86	91
TWD4	35	58	70	81	87	93	96	98
TWE2	36	59	71	82	88	93	96	98
TWF3	26	44	55	67	74	82	87	92
TWJ2	33	56	68	79	86	92	95	97
DGA1	26	43	54	64	70	77	81	85
DGA2	26	45	56	67	74	82	86	90
DGA3	29	51	64	75	81	87	91	93
DGB1	27	47	60	71	77	84	87	91
DGB2	27	48	61	73	80	88	91	94
DGB3	29	53	67	79	85	91	94	96
DGC1	31	54	58	79	85	90	93	95
DGC2	30	54	68	80	86	92	95	97
DGC3	33	59	74	85	90	95	96	98
SSA1	34	52	62	73	79	86	90	94
SSB1	29	46	56	67	73	81	86	90
SSB2	36	56	67	78	84	90	93	96
SSB3	28	44	54	64	71	79	84	89
SSC1	30	49	61	72	78	86	90	94
SSC2	34	56	68	79	85	91	94	97
SSE1	36	56	67	77	83	90	93	96
SSE2	42	64	76	85	90	95	97	99
SSE3	30	47	57	67	74	82	86	91

TUCSON, ARIZONA ***SSF (%)*** *1758 DD65*

Type	LCR = 200	100	70	50	40	30	25	20
WWA3	31	52	64	75	82	89	93	96
WWB4	37	62	75	85	91	96	98	99
WWC2	37	61	74	84	90	95	97	99
TWA1	25	38	47	56	63	72	78	84
TWA2	27	44	55	65	73	81	86	91
TWA3	28	46	58	69	76	84	89	93
TWA4	27	46	58	69	77	85	89	94
TWB3	25	42	53	64	72	81	86	91
TWD4	33	56	69	80	86	93	96	98
TWE2	35	57	70	81	87	93	96	98
TWF3	24	42	53	65	73	81	86	91
TWJ2	31	54	67	78	85	92	95	97
DGA1	24	41	52	62	68	76	80	84
DGA2	24	43	54	65	72	81	85	89
DGA3	27	49	62	74	80	87	90	93
DGB1	24	44	58	69	76	83	87	91
DGB2	25	45	59	72	79	87	91	94
DGB3	27	50	65	78	85	91	94	96
DGC1	28	51	66	78	84	90	93	95
DGC2	28	51	66	79	85	92	95	97
DGC3	31	56	72	84	90	95	96	98
SSA1	33	51	62	72	79	86	90	94
SSB1	29	45	55	66	73	81	86	91
SSB2	35	55	67	77	84	90	94	97
SSB3	27	43	53	64	71	79	84	89
SSC1	28	47	59	70	77	86	90	94
SSC2	32	54	66	78	84	91	94	97
SSE1	35	55	67	77	83	90	93	96
SSE2	41	64	76	85	91	95	97	99
SSE3	30	46	56	67	73	82	86	91

FORT SMITH, ARKANSAS ***SSF (%)*** *3338 DD65*

Type	LCR = 200	100	70	50	40	30	25	20
WWA3	11	21	29	37	43	51	57	64
WWB4	12	27	37	49	57	67	73	80
WWC2	12	27	37	48	55	65	72	79
TWA1	14	19	23	27	31	36	40	45
TWA2	12	20	25	31	36	43	48	54
TWA3	11	19	26	33	38	46	51	58
TWA4	10	19	25	33	38	46	52	59
TWB3	10	18	23	30	35	43	48	54
TWD4	12	25	34	44	51	61	67	75
TWE2	13	26	35	44	52	61	67	75
TWF3	8	16	22	29	34	42	47	54
TWJ2	10	23	32	41	48	58	65	72
DGA1	8	14	19	24	28	33	36	39
DGA2	9	17	23	29	35	42	46	52
DGA3	11	21	28	37	44	53	59	66
DGB1	8	15	19	25	30	36	40	46
DGB2	9	17	23	31	36	45	51	58
DGB3	11	21	29	38	45	56	63	71
DGC1	10	18	24	30	36	44	50	56
DGC2	11	20	27	35	42	52	59	66
DGC3	13	24	32	43	51	63	70	78
SSA1	16	25	31	38	43	50	55	62
SSB1	13	21	26	33	37	44	49	55
SSB2	15	27	35	43	50	59	64	71
SSB3	12	19	25	30	35	41	46	52
SSC1	10	19	25	32	37	45	51	57
SSC2	11	23	31	40	47	56	62	69
SSE1	14	24	31	39	45	53	58	65
SSE2	16	30	39	50	57	66	72	79
SSE3	14	21	27	33	37	44	49	55

TABLE C.3 **Annual Passive Heating Performance: SSF (Continued)**

LOS ANGELES, CALIFORNIA ***SSF (%)*** *1793 DD65*

Type	LCR = 200	100	70	50	40	30	25	20
WWA3	37	58	70	81	87	93	95	98
WWB4	43	68	80	89	93	97	99	99
WWC2	42	67	79	88	93	97	98	99
TWA1	29	43	52	62	69	77	82	88
TWA2	32	49	60	71	78	86	90	94
TWA3	33	52	63	74	81	88	92	96
TWA4	32	52	64	75	82	89	93	96
TWB3	29	48	59	70	77	85	90	94
TWD4	39	62	74	85	90	95	97	99
TWE2	40	63	75	85	91	95	97	99
TWF3	29	48	59	71	78	86	90	94
TWJ2	37	60	72	83	89	94	97	98
DGA1	29	47	58	68	74	81	85	89
DGA2	29	49	60	71	78	85	89	93
DGA3	32	55	68	79	84	90	93	95
DGB1	30	52	65	76	82	88	91	94
DGB2	30	52	66	78	84	91	94	96
DGB3	33	57	72	83	89	94	96	97
DGC1	34	59	73	84	89	93	95	96
DGC2	34	58	73	84	90	95	97	98
DGC3	37	63	78	89	93	96	97	98
SSA1	41	60	71	80	86	92	95	97
SSB1	35	53	64	74	81	88	91	95
SSB2	42	64	75	84	89	94	97	98
SSB3	34	52	63	73	79	87	90	94
SSC1	33	53	65	76	83	90	93	96
SSC2	38	60	72	83	88	94	96	98
SSE1	43	64	75	84	90	94	97	98
SSE2	50	73	83	91	95	98	99	100
SSE3	36	54	65	75	81	88	92	95

SAN DIEGO, CALIFORNIA ***SSF (%)*** *1512 DD65*

Type	LCR = 200	100	70	50	40	30	25	20
WWA3	39	61	73	83	89	94	97	98
WWB4	46	71	83	91	95	98	99	100
WWC2	45	70	82	91	95	98	99	100
TWA1	30	45	55	65	72	80	85	90
TWA2	34	52	64	74	81	88	92	95
TWA3	35	55	67	78	84	91	94	97
TWA4	34	55	67	78	85	91	94	97
TWB3	31	51	62	73	80	88	92	95
TWD4	41	65	77	87	92	96	98	99
TWE2	43	66	78	88	92	97	98	99
TWF3	31	51	63	74	81	89	92	96
TWJ2	39	63	76	86	91	96	98	99
DGA1	31	51	62	72	78	84	88	91
DGA2	31	52	64	74	81	88	91	95
DGA3	34	58	71	81	87	92	94	95
DGB1	32	56	69	79	85	90	93	95
DGB2	32	56	70	81	87	93	95	97
DGB3	35	61	75	86	91	95	96	97
DGC1	37	63	77	87	91	95	96	97
DGC2	36	63	77	87	92	96	98	99
DGC3	39	68	82	91	94	97	98	98
SSA1	43	62	73	82	88	93	95	98
SSB1	37	56	67	77	83	89	93	96
SSB2	45	66	77	86	91	95	97	99
SSB3	36	54	65	75	81	88	92	95
SSC1	36	56	68	79	85	92	95	97
SSC2	40	63	75	85	91	95	97	99
SSE1	45	67	78	86	91	95	97	99
SSE2	52	75	85	92	96	98	99	100
SSE3	38	57	67	77	83	90	93	96

SACRAMENTO, CALIFORNIA ***SSF (%)*** *2845 DD65*

Type	LCR = 200	100	70	50	40	30	25	20
WWA3	18	31	39	48	55	63	68	74
WWB4	19	37	49	60	67	76	81	87
WWC2	20	37	48	59	66	75	80	85
TWA1	18	25	30	36	41	47	51	57
TWA2	17	27	34	42	47	55	60	66
TWA3	16	28	35	44	50	58	63	69
TWA4	15	27	35	44	50	58	64	70
TWB3	15	25	32	40	46	54	60	66
TWD4	18	34	44	55	62	71	76	82
TWE2	20	35	45	56	63	71	77	82
TWF3	13	24	32	40	46	54	59	66
TWJ2	17	32	42	53	60	69	74	80
DGA1	13	23	29	36	40	46	49	53
DGA2	14	25	32	40	46	54	58	64
DGA3	16	29	38	48	55	64	69	75
DGB1	13	23	31	39	44	51	55	60
DGB2	14	25	34	43	50	58	64	70
DGB3	16	30	39	50	58	67	73	79
DGC1	16	27	36	45	51	59	63	69
DGC2	16	29	39	49	56	65	70	76
DGC3	19	33	44	56	63	73	79	85
SSA1	22	33	41	48	54	61	65	71
SSB1	19	29	36	43	48	55	60	66
SSB2	22	36	45	54	60	68	73	79
SSB3	18	28	34	41	46	53	57	63
SSC1	16	27	35	44	50	58	63	69
SSC2	18	32	42	51	58	67	72	78
SSE1	21	34	42	50	56	64	68	74
SSE2	24	41	50	60	67	75	79	84
SSE3	20	30	36	43	48	55	60	65

FRESNO, CALIFORNIA ***SSF (%)*** *2657 DD65*

Type	LCR = 200	100	70	50	40	30	25	20
WWA3	18	31	40	49	56	64	69	75
WWB4	20	38	49	61	68	77	82	87
WWC2	20	38	49	60	67	76	81	86
TWA1	18	26	31	37	41	48	52	58
TWA2	18	28	35	42	48	55	60	67
TWA3	17	28	36	44	50	58	64	70
TWA4	16	28	36	44	51	59	64	71
TWB3	15	26	33	41	47	55	60	67
TWD4	19	35	45	56	63	72	77	83
TWE2	20	36	46	56	63	72	77	83
TWF3	14	25	32	40	47	55	60	66
TWJ2	17	33	43	53	60	69	75	81
DGA1	14	23	30	36	41	47	50	54
DGA2	14	25	33	41	47	54	59	65
DGA3	17	30	39	49	56	65	70	76
DGB1	14	24	31	39	45	52	56	61
DGB2	15	26	35	44	50	59	64	70
DGB3	17	30	40	51	58	68	74	80
DGC1	16	28	37	46	52	60	64	70
DGC2	17	30	39	49	56	65	71	77
DGC3	19	34	45	56	64	74	79	85
SSA1	22	34	41	49	54	61	66	72
SSB1	19	29	36	43	49	56	61	66
SSB2	22	36	45	54	60	69	74	79
SSB3	18	28	34	41	46	53	58	64
SSC1	16	28	36	44	50	59	64	70
SSC2	18	33	42	52	59	68	73	79
SSE1	22	34	42	51	57	64	69	75
SSE2	25	41	51	61	67	75	80	85
SSE3	20	30	36	44	49	56	60	66

TABLE C.3 **Annual Passive Heating Performance: SSF (Continued)**

SAN FRANCISCO, CALIFORNIA **SSF (%)** *3050 DD65*

Type	LCR = 200	100	70	50	40	30	25	20
WWA3	24	41	51	62	69	77	82	88
WWB4	27	49	61	73	80	87	91	95
WWC2	27	48	61	72	79	87	91	94
TWA1	21	31	38	46	51	59	64	71
TWA2	22	35	44	53	59	68	73	80
TWA3	21	36	46	56	63	71	77	83
TWA4	20	36	46	56	63	72	78	84
TWB3	19	33	42	52	59	68	73	79
TWD4	25	44	56	67	75	83	87	92
TWE2	26	46	57	68	75	83	88	92
TWF3	18	32	42	52	59	68	73	80
TWJ2	23	42	54	65	73	81	86	91
DGA1	18	31	40	48	54	61	65	69
DGA2	18	33	42	52	59	67	72	78
DGA3	21	38	49	60	68	76	81	86
DGB1	18	33	43	53	60	68	72	77
DGB2	19	34	45	56	64	73	79	84
DGB3	21	39	51	63	72	81	85	90
DGC1	21	38	49	61	68	76	81	85
DGC2	22	39	51	63	71	80	85	89
DGC3	24	43	57	70	78	86	90	94
SSA1	30	45	54	63	68	76	80	85
SSB1	25	39	48	57	63	70	75	81
SSB2	30	47	58	67	74	81	86	90
SSB3	24	38	46	55	61	68	73	79
SSC1	21	37	47	57	64	72	78	84
SSC2	24	42	54	64	72	80	85	90
SSE1	30	47	57	66	72	79	84	88
SSE2	35	54	65	75	81	87	91	94
SSE3	26	40	48	57	63	71	75	81

DENVER, COLORADO **SSF (%)** *6022 DD65*

Type	LCR = 100	70	50	40	30	25	20	15
WWA3	21	29	37	43	52	58	66	75
WWB4	27	38	50	58	68	75	82	89
WWC2	27	37	48	56	67	73	80	88
TWA1	19	23	27	31	37	41	46	53
TWA2	19	25	31	36	44	49	55	64
TWA3	19	26	33	39	47	52	59	69
TWA4	18	25	33	39	47	53	61	70
TWB3	17	23	30	36	43	49	56	65
TWD4	25	34	44	52	62	68	76	85
TWE2	26	35	45	52	62	69	76	85
TWF3	16	22	29	35	43	48	55	65
TWJ2	23	32	42	49	59	66	74	83
DGA1	14	19	25	28	33	37	40	45
DGA2	16	22	30	35	43	47	53	61
DGA3	20	28	37	44	54	61	68	76
DGB1	14	19	25	30	37	42	48	54
DGB2	17	23	31	37	46	52	60	68
DGB3	21	29	38	46	57	65	73	82
DGC1	17	24	31	37	46	52	59	66
DGC2	20	27	36	43	53	60	68	77
DGC3	24	33	43	52	64	72	80	88
SSA1	25	31	38	44	51	56	63	71
SSB1	21	26	33	38	45	50	56	65
SSB2	27	35	44	50	59	65	72	81
SSB3	19	25	31	35	42	46	52	61
SSC1	19	25	33	38	46	52	59	68
SSC2	23	31	41	48	57	63	71	80
SSE1	24	31	39	45	54	59	66	74
SSE2	30	40	50	57	67	73	80	87
SSE3	21	27	33	38	45	49	56	64

SANTA MARIA, CALIFORNIA **SSF (%)** *3061 DD65*

Type	LCR = 200	100	70	50	40	30	25	20
WWA3	25	44	55	67	74	82	87	92
WWB4	29	53	66	78	84	91	95	97
WWC2	29	52	65	77	84	91	94	97
TWA1	21	32	40	49	55	64	69	76
TWA2	22	37	47	57	64	73	79	85
TWA3	22	39	49	60	67	77	82	88
TWA4	22	38	49	61	68	77	83	88
TWB3	20	35	45	56	63	73	78	84
TWD4	27	48	60	72	79	87	91	95
TWE2	28	49	61	73	80	88	92	95
TWF3	19	35	45	56	64	73	79	85
TWJ2	25	46	58	70	77	86	90	94
DGA1	19	34	43	52	59	66	71	75
DGA2	19	35	46	56	64	72	77	83
DGA3	22	40	53	65	73	81	85	89
DGB1	19	35	47	59	66	74	79	83
DGB2	20	37	49	62	70	79	84	89
DGB3	22	41	55	69	77	86	90	93
DGC1	22	41	54	67	75	83	87	90
DGC2	23	42	55	69	77	85	90	93
DGC3	25	47	62	76	83	91	94	96
SSA1	32	49	60	70	76	83	87	92
SSB1	27	43	53	63	70	78	83	88
SSB2	33	52	64	74	80	88	91	95
SSB3	26	41	51	61	68	76	81	86
SSC1	23	40	50	61	69	78	83	89
SSC2	26	46	58	69	77	85	89	94
SSE1	33	52	63	73	80	87	91	94
SSE2	38	61	72	82	88	93	96	98
SSE3	28	43	53	63	70	76	83	88

GRAND JUNCTION, COLORADO **SSF (%)** *5607 DD65*

Type	LCR = 100	70	50	40	30	25	20	15
WWA3	20	27	35	41	49	55	62	71
WWB4	26	36	47	55	65	71	79	87
WWC2	26	35	46	54	64	70	77	85
TWA1	19	22	26	30	35	38	43	50
TWA2	19	24	30	35	41	46	52	61
TWA3	19	25	31	37	44	49	56	65
TWA4	18	24	31	37	45	50	57	66
TWB3	17	22	29	34	41	46	52	61
TWD4	24	32	42	49	59	65	73	81
TWE2	25	33	43	50	59	66	73	82
TWF3	15	21	28	33	40	45	52	61
TWJ2	22	30	40	47	57	63	70	79
DGA1	14	18	23	26	31	34	37	41
DGA2	16	22	28	33	40	45	50	57
DGA3	20	27	36	42	52	58	65	73
DGB1	14	18	24	28	34	38	43	49
DGB2	16	22	29	35	43	49	56	64
DGB3	20	28	37	44	54	61	69	78
DGC1	17	22	29	34	42	47	53	61
DGC2	19	26	34	41	50	56	64	73
DGC3	23	31	41	49	61	68	76	85
SSA1	24	29	36	41	48	52	58	66
SSB1	20	25	31	35	42	46	52	60
SSB2	25	33	41	47	56	62	68	77
SSB3	18	23	29	33	39	43	48	56
SSC1	18	24	31	36	43	49	55	64
SSC2	22	30	38	45	54	60	67	76
SSE1	23	29	37	42	49	54	61	69
SSE2	28	37	47	54	63	69	75	83
SSE3	20	25	31	35	41	46	51	59

TABLE C.3 **Annual Passive Heating Performance: SSF (Continued)**

HARTFORD, CONNECTICUT ***SSF (%)*** *6354 DD65*

Type	LCR = 100	70	50	40	30	25	20	15
WWA3	7	10	13	16	19	22	25	30
WWB4	9	16	23	28	36	42	49	58
WWC2	10	15	22	27	35	40	46	55
TWA1	12	13	14	15	16	17	18	20
TWA2	9	11	13	15	18	20	22	25
TWA3	8	10	13	15	18	21	24	28
TWA4	6	9	12	15	18	21	24	29
TWB3	7	10	12	15	18	20	23	28
TWD4	10	15	21	25	32	37	43	52
TWE2	10	15	21	26	32	37	43	51
TWF3	5	7	10	12	15	17	20	24
TWJ2	8	13	19	23	30	34	40	49
DGA1	4	5	6	6	7	7	6	5
DGA2	7	9	12	14	17	19	22	26
DGA3	10	14	18	22	28	32	37	45
DGB1	4	5	6	6	7	7	7	7
DGB2	7	9	12	15	18	20	23	28
DGB3	10	14	19	23	29	33	39	48
DGC1	5	7	9	10	12	13	14	15
DGC2	9	12	15	18	23	26	30	36
DGC3	12	17	22	27	33	38	45	54
SSA1	13	16	19	21	25	27	30	34
SSB1	11	13	15	17	20	22	25	28
SSB2	14	18	24	27	33	37	43	50
SSB3	10	11	14	15	17	19	21	23
SSC1	6	8	11	13	15	17	20	23
SSC2	8	12	17	21	26	30	35	42
SSE1	10	13	16	18	21	23	26	29
SSE2	13	18	24	29	35	40	45	53
SSE3	11	13	15	17	19	21	23	26

MIAMI, FLORIDA ***SSF (%)*** *206 DD65*

Type	LCR = 200	100	70	50	40	30	25	20
WWA3	82	96	99	100	100	100	100	100
WWB4	89	98	100	100	100	100	100	100
WWC2	88	98	100	100	100	100	100	100
TWA1	64	84	92	97	98	100	100	100
TWA2	73	91	96	99	100	100	100	100
TWA3	76	93	97	99	100	100	100	100
TWA4	76	93	98	99	100	100	100	100
TWB3	71	90	96	99	100	100	100	100
TWD4	84	97	99	100	100	100	100	100
TWE2	85	97	99	100	100	100	100	100
TWF3	72	91	96	99	100	100	100	100
TWJ2	83	96	99	100	100	100	100	100
DGA1	75	92	97	99	100	100	100	100
DGA2	74	92	97	99	100	100	100	100
DGA3	79	93	96	97	97	97	97	97
DGB1	81	95	98	99	99	99	99	99
DGB2	80	96	99	100	100	100	100	100
DGB3	83	96	98	98	98	98	98	98
DGC1	88	97	98	99	99	99	98	98
DGC2	86	98	99	100	100	100	100	100
DGC3	88	98	98	99	99	98	98	98
SSA1	81	95	99	100	100	100	100	100
SSB1	75	92	97	99	100	100	100	100
SSB2	84	97	99	100	100	100	100	100
SSB3	74	92	97	99	100	100	100	100
SSC1	78	94	98	99	100	100	100	100
SSC2	83	96	99	100	100	100	100	100
SSE1	85	97	99	100	100	100	100	100
SSE2	91	99	100	100	100	100	100	100
SSE3	76	93	97	99	100	100	100	100

WASHINGTON, DC ***SSF (%)*** *5010 DD65*

Type	LCR = 100	70	50	40	30	25	20	15
WWA3	12	16	22	25	31	35	40	48
WWB4	15	23	32	39	48	54	62	72
WWC2	15	23	31	38	46	52	60	69
TWA1	14	16	18	20	23	25	28	32
TWA2	13	16	20	22	27	30	34	40
TWA3	12	15	20	23	28	32	37	43
TWA4	10	14	19	23	29	33	38	45
TWB3	11	14	18	22	27	30	35	42
TWD4	15	21	29	35	43	49	56	65
TWE2	15	22	29	35	43	49	56	65
TWF3	9	12	16	20	25	28	33	40
TWJ2	13	19	27	32	40	46	53	62
DGA1	7	9	12	14	16	17	18	19
DGA2	10	13	18	21	26	29	33	39
DGA3	13	18	25	29	37	42	48	57
DGB1	7	9	12	14	17	19	21	23
DGB2	10	14	18	22	27	31	36	42
DGB3	14	19	25	31	38	44	51	61
DGC1	9	12	16	19	23	25	29	33
DGC2	12	17	22	26	33	37	43	51
DGC3	16	22	29	35	43	50	58	68
SSA1	17	21	26	29	34	38	42	49
SSB1	14	17	21	25	29	32	36	42
SSB2	18	24	31	36	43	48	54	63
SSB3	13	16	20	22	26	29	33	38
SSC1	10	14	18	21	26	30	34	40
SSC2	13	19	25	30	37	42	48	57
SSE1	15	19	24	28	33	37	41	48
SSE2	19	26	34	39	47	53	59	68
SSE3	15	18	22	24	29	31	35	41

ORLANDO, FLORIDA ***SSF (%)*** *733 DD65*

Type	LCR = 200	100	70	50	40	30	25	20
WWA3	45	68	80	89	93	97	98	99
WWB4	52	78	88	95	97	99	100	100
WWC2	52	77	87	94	97	99	100	100
TWA1	34	51	61	72	78	86	90	94
TWA2	38	59	70	80	86	92	95	98
TWA3	40	62	73	83	89	94	97	98
TWA4	39	62	74	84	90	95	97	99
TWB3	36	57	69	80	86	92	95	97
TWD4	47	72	83	91	95	98	99	100
TWE2	49	73	84	92	96	98	99	100
TWF3	36	58	70	80	87	93	95	98
TWJ2	45	70	82	90	94	98	99	100
DGA1	37	58	69	78	84	90	93	95
DGA2	36	59	71	81	86	92	95	97
DGA3	40	66	77	86	90	94	95	96
DGB1	38	64	76	85	90	94	96	97
DGB2	38	64	77	87	92	96	98	99
DGB3	41	69	82	90	94	97	97	98
DGC1	44	72	84	91	94	97	97	98
DGC2	42	71	83	92	95	98	99	100
DGC3	46	76	88	94	96	98	98	98
SSA1	46	67	78	87	92	96	98	99
SSB1	40	61	72	82	88	93	96	98
SSB2	48	71	82	90	94	98	99	100
SSB3	39	59	71	81	86	92	95	98
SSC1	41	63	75	85	90	95	97	99
SSC2	46	70	81	90	94	98	99	100
SSE1	50	72	83	91	95	98	90	100
SSE2	57	80	90	95	98	99	100	100
SSE3	41	62	73	82	88	93	96	98

TABLE C.3 **Annual Passive Heating Performance: SSF (Continued)**

TALLAHASSEE, FLORIDA ***SSF (%)*** *1563 DD65*

Type	*LCR = 200*	*100*	*70*	*50*	*40*	*30*	*25*	*20*
WWA3	24	41	52	63	70	79	84	90
WWB4	27	49	62	74	82	89	93	96
WWC2	27	49	61	73	81	88	92	96
TWA1	21	31	38	46	52	60	65	72
TWA2	22	35	44	53	60	70	75	82
TWA3	22	36	46	56	64	73	79	85
TWA4	21	36	46	57	64	74	79	86
TWB3	19	33	42	52	59	69	75	81
TWD4	25	45	57	69	76	85	89	93
TWE2	26	46	58	69	77	85	90	94
TWF3	18	32	42	52	60	69	75	82
TWJ2	23	42	54	66	74	83	88	92
DGA1	18	31	40	48	54	62	66	71
DGA2	18	33	42	53	60	68	74	79
DGA3	21	37	49	62	69	78	83	87
DGB1	18	32	43	54	61	70	75	80
DGB2	19	34	45	58	66	75	81	86
DGB3	21	38	51	65	73	83	87	92
DGC1	21	38	49	62	71	79	83	88
DGC2	21	38	51	65	73	82	87	91
DGC3	24	43	57	72	80	88	92	95
SSA1	27	42	52	62	69	77	82	88
SSB1	23	36	46	55	62	71	77	83
SSB2	28	45	56	67	74	83	87	92
SSB3	22	35	44	53	60	69	74	81
SSC1	21	37	47	57	65	74	80	86
SSC2	24	42	54	66	73	82	87	92
SSE1	28	45	55	66	73	81	86	91
SSE2	32	53	65	76	83	90	93	96
SSE3	24	37	46	56	63	71	77	83

ATLANTA, GEORGIA ***SSF (%)*** *3105 DD65*

Type	*LCR = 200*	*100*	*70*	*50*	*40*	*30*	*25*	*20*
WWA3	12	22	29	37	43	52	57	65
WWB4	12	28	38	49	57	67	74	81
WWC2	13	28	38	48	56	66	72	79
TWA1	15	20	24	28	32	37	41	46
TWA2	13	20	26	32	37	44	49	55
TWA3	12	20	26	33	39	47	52	59
TWA4	10	19	26	33	39	47	53	60
TWB3	10	18	24	31	36	43	49	55
TWD4	12	25	35	45	52	61	68	75
TWE2	13	26	35	45	52	62	68	75
TWF3	9	17	23	30	35	43	48	55
TWJ2	11	23	32	42	49	59	65	73
DGA1	9	15	20	25	29	34	37	40
DGA2	10	17	23	30	35	43	47	53
DGA3	12	21	29	38	44	54	60	67
DGB1	9	15	20	26	31	37	42	47
DGB2	10	18	24	32	37	46	52	59
DGB3	12	22	30	39	46	57	64	72
DGC1	10	19	25	32	37	45	50	57
DGC2	11	21	28	36	43	53	59	67
DGC3	13	25	33	44	52	63	71	79
SSA1	17	26	32	40	45	52	57	64
SSB1	14	22	28	34	39	46	51	57
SSB2	16	28	36	45	52	60	66	73
SSB3	13	21	26	32	37	43	48	54
SSC1	11	20	26	33	38	46	51	58
SSC2	12	24	32	41	47	57	63	70
SSE1	15	26	33	41	47	55	60	67
SSE2	17	32	41	51	58	68	74	80
SSE3	15	23	28	34	39	46	51	57

BOISE, IDAHO ***SSF (%)*** *5837 DD65*

Type	*LCR = 100*	*70*	*50*	*40*	*30*	*25*	*20*	*15*
WWA3	18	23	29	34	40	45	50	57
WWB4	22	31	40	47	56	62	69	77
WWC2	22	30	39	46	54	60	67	75
TWA1	17	20	24	26	30	33	36	41
TWA2	17	21	26	30	35	38	43	49
TWA3	16	21	27	31	37	41	46	53
TWA4	15	21	26	31	37	41	47	54
TWB3	15	20	25	29	35	39	44	50
TWD4	20	28	36	43	51	56	63	71
TWE2	21	29	37	43	51	56	63	71
TWF3	13	18	23	27	33	37	42	49
TWJ2	19	26	34	40	48	54	61	69
DGA1	12	15	19	21	25	26	28	30
DGA2	14	19	24	28	34	38	42	48
DGA3	18	24	32	37	45	50	56	64
DGB1	12	16	20	23	27	29	32	35
DGB2	15	20	26	30	36	40	46	52
DGB3	19	25	33	39	47	52	59	68
DGC1	15	19	25	28	33	37	41	45
DGC2	17	23	30	35	42	47	53	60
DGC3	21	28	37	43	52	58	65	74
SSA1	22	26	32	35	41	44	48	54
SSB1	18	22	27	31	36	39	43	49
SSB2	23	30	37	42	49	54	59	67
SSB3	17	21	25	28	33	36	39	44
SSC1	15	20	26	30	35	39	44	50
SSC2	18	25	33	38	46	51	57	65
SSE1	20	25	31	35	40	44	48	54
SSE2	25	32	40	46	53	58	64	71
SSE3	19	23	27	31	35	38	42	47

CHICAGO, ILLINOIS ***SSF (%)*** *6125 DD65*

Type	*LCR = 100*	*70*	*50*	*40*	*30*	*25*	*20*	*15*
WWA3	9	13	16	19	23	26	30	35
WWB4	12	19	26	32	40	46	53	62
WWC2	12	18	25	31	38	44	51	60
TWA1	13	14	16	17	19	20	22	24
TWA2	11	13	16	18	21	23	26	30
TWA3	9	12	16	18	22	25	28	33
TWA4	8	11	15	18	22	25	29	34
TWB3	9	11	15	17	21	24	27	32
TWD4	12	17	24	29	36	41	47	56
TWE2	12	18	24	29	36	40	47	55
TWF3	7	9	13	15	19	21	25	29
TWJ2	10	15	22	26	33	38	44	53
DGA1	5	7	8	9	10	10	11	10
DGA2	8	11	14	17	20	23	26	30
DGA3	12	16	21	25	31	35	41	48
DGB1	5	7	8	9	11	11	12	12
DGB2	8	11	15	17	21	24	28	33
DGB3	12	16	22	26	32	37	43	52
DGC1	7	9	12	13	16	17	19	21
DGC2	10	14	18	21	26	30	34	41
DGC3	14	19	25	30	37	42	49	58
SSA1	15	18	21	24	27	30	33	38
SSB1	12	15	18	20	23	25	28	32
SSB2	16	20	26	30	36	40	46	53
SSB3	11	13	16	17	20	22	24	27
SSC1	8	11	14	16	19	21	24	28
SSC2	10	15	20	24	30	34	39	46
SSE1	12	15	18	21	25	27	30	34
SSE2	15	21	27	32	39	43	49	57
SSE3	12	15	17	19	22	24	27	30

TABLE C.3 **Annual Passive Heating Performance: SSF (Continued)**

MOLINE, ILLINOIS **SSF (%)** *6399 DD65*

Type	LCR = 100	70	50	40	30	25	20	15
WWA3	9	13	16	19	24	26	30	36
WWB4	12	19	26	32	40	46	53	63
WWC2	12	18	25	31	39	44	51	60
TWA1	13	14	16	17	19	20	22	24
TWA2	11	13	16	18	21	23	26	30
TWA3	9	12	16	18	22	25	28	33
TWA4	8	11	15	18	22	25	29	34
TWB3	9	11	15	17	21	24	27	32
TWD4	12	17	24	29	36	41	47	56
TWE2	12	18	24	29	36	41	47	55
TWF3	7	9	13	15	19	21	25	29
TWJ2	10	15	22	26	33	38	44	53
DGA1	5	7	8	9	10	11	11	10
DGA2	8	11	14	17	20	23	26	30
DGA3	12	16	21	25	31	35	41	49
DGB1	5	7	8	9	11	11	12	12
DGB2	8	11	15	17	21	24	28	33
DGB3	12	16	22	26	32	37	43	52
DGC1	7	9	12	13	16	17	19	21
DGC2	10	14	18	21	26	29	34	41
DGC3	14	19	25	30	37	42	49	59
SSA1	15	18	21	24	27	30	33	38
SSB1	12	14	17	20	23	25	28	32
SSB2	15	20	26	30	36	40	46	53
SSB3	11	13	15	17	20	22	24	27
SSC1	8	11	14	16	19	21	24	29
SSC2	10	15	20	24	30	34	39	47
SSE1	12	15	18	21	24	27	30	34
SSE2	15	21	27	32	38	43	49	57
SSE3	12	15	17	19	22	24	27	30

INDIANAPOLIS, INDIANA **SSF (%)** *5585 DD65*

Type	LCR = 100	70	50	40	30	25	20	15
WWA3	9	12	16	19	22	25	29	34
WWB4	11	18	25	31	39	45	52	61
WWC2	11	18	25	30	37	43	49	58
TWA1	12	14	15	17	18	19	21	23
TWA2	10	13	15	17	20	22	25	29
TWA3	9	12	15	18	21	24	27	31
TWA4	8	11	14	17	21	24	28	33
TWB3	8	11	14	17	20	23	26	31
TWD4	11	17	23	28	35	40	46	55
TWE2	12	17	23	28	35	39	46	54
TWF3	6	9	12	14	18	20	23	28
TWJ2	10	15	21	26	32	37	43	52
DGA1	5	6	8	8	9	10	10	9
DGA2	8	11	14	16	20	22	25	29
DGA3	11	15	20	24	30	34	40	47
DGB1	5	6	8	9	10	10	11	11
DGB2	8	11	14	17	20	23	26	31
DGB3	12	16	21	25	31	36	42	51
DGC1	7	9	11	13	15	16	17	19
DGC2	10	13	17	21	25	28	33	39
DGC3	13	18	24	29	36	41	48	57
SSA1	15	18	21	24	27	30	33	38
SSB1	12	15	17	20	23	25	28	32
SSB2	15	20	26	30	36	40	46	53
SSB3	11	13	16	17	20	22	24	27
SSC1	8	10	13	15	18	20	23	27
SSC2	10	14	19	23	29	33	38	45
SSE1	12	15	18	21	24	27	30	34
SSE2	15	21	27	32	38	43	49	56
SSE3	12	15	17	19	22	24	26	30

SPRINGFIELD, ILLINOIS **SSF (%)** *5563 DD65*

Type	LCR = 100	70	50	40	30	25	20	15
WWA3	12	16	21	24	30	33	38	45
WWB4	15	23	31	38	46	53	60	70
WWC2	15	22	30	36	45	50	58	67
TWA1	14	16	18	20	22	24	27	30
TWA2	13	16	19	22	26	29	32	38
TWA3	12	15	19	22	27	30	35	41
TWA4	10	14	19	22	27	31	36	43
TWB3	11	14	18	21	26	29	33	39
TWD4	14	21	28	34	42	47	54	63
TWE2	15	21	28	34	42	47	54	63
TWF3	9	12	16	19	24	27	31	37
TWJ2	13	19	26	31	39	44	51	60
DGA1	7	9	11	13	15	16	17	17
DGA2	10	13	17	20	25	28	32	37
DGA3	14	18	24	28	35	40	47	55
DGB1	7	9	12	13	16	17	19	21
DGB2	10	14	18	21	26	29	34	40
DGB3	14	19	25	30	37	42	49	59
DGC1	9	12	15	18	21	23	26	30
DGC2	12	16	21	25	31	35	41	49
DGC3	16	22	28	34	42	48	56	66
SSA1	17	20	25	28	32	36	40	46
SSB1	14	17	21	23	28	30	34	40
SSB2	18	23	29	34	41	46	52	60
SSB3	13	15	19	21	25	27	30	35
SSC1	10	14	17	20	25	28	32	38
SSC2	13	18	24	29	36	40	46	55
SSE1	14	18	23	26	31	34	38	44
SSE2	18	25	32	37	45	50	57	65
SSE3	14	17	21	23	27	29	33	38

SOUTH BEND, INDIANA **SSF (%)** *6465 DD65*

Type	LCR = 100	70	50	40	30	25	20	15
WWA3	7	9	12	14	16	18	20	23
WWB4	9	14	21	26	33	38	44	52
WWC2	9	14	20	25	31	36	41	49
TWA1	12	12	13	14	15	16	16	17
TWA2	9	11	13	14	16	17	19	21
TWA3	8	10	12	14	16	18	20	23
TWA4	6	9	11	13	16	18	21	24
TWB3	7	9	12	13	16	18	20	23
TWD4	9	14	19	23	29	34	39	47
TWE2	10	14	19	23	29	33	38	45
TWF3	5	7	9	11	13	15	17	20
TWJ2	8	12	17	21	27	31	36	43
DGA1	4	4	5	5	5	5	4	1
DGA2	7	9	11	13	16	17	20	22
DGA3	10	14	18	21	26	29	34	41
DGB1	4	4	5	5	5	5	5	3
DGB2	7	9	12	14	16	18	21	24
DGB3	10	14	19	22	27	31	36	44
DGC1	5	7	8	9	10	10	10	10
DGC2	9	11	15	17	21	23	26	31
DGC3	12	16	21	25	31	35	41	49
SSA1	13	15	18	20	23	25	27	29
SSB1	11	13	15	17	19	20	22	25
SSB2	14	18	23	26	31	35	39	46
SSB3	10	11	13	14	16	17	18	19
SSC1	6	8	10	11	13	14	15	17
SSC2	8	11	15	18	23	26	30	36
SSE1	10	12	15	16	18	20	21	23
SSE2	13	17	23	26	32	36	40	47
SSE3	11	13	15	16	18	19	20	22

TABLE C.3 **Annual Passive Heating Performance: SSF (Continued)**

DES MOINES, IOWA **SSF (%)**						*6718 DD65*		
Type	*LCR = 100*	*70*	*50*	*40*	*30*	*25*	*20*	*15*
WWA3	11	14	19	22	27	30	34	41
WWB4	13	21	29	35	43	49	57	66
WWC2	14	20	28	34	42	47	54	64
TWA1	13	15	17	18	21	22	24	27
TWA2	12	14	17	20	23	26	29	34
TWA3	11	14	17	20	24	27	31	37
TWA4	9	13	17	20	25	28	32	39
TWB3	10	13	16	19	23	26	30	36
TWD4	13	19	26	31	39	44	51	60
TWE2	14	20	26	31	39	44	50	59
TWF3	8	11	14	17	21	24	28	33
TWJ2	11	17	24	29	36	41	48	57
DGA1	6	8	10	11	12	13	14	14
DGA2	9	12	16	18	23	25	29	34
DGA3	13	17	22	27	33	38	44	52
DGB1	6	8	10	11	13	14	15	17
DGB2	9	13	16	19	23	27	31	36
DGB3	13	18	23	28	34	39	46	56
DGC1	8	11	14	15	18	20	22	26
DGC2	11	15	20	23	29	32	38	45
DGC3	15	20	26	31	39	45	52	62
SSA1	16	19	23	25	29	32	36	41
SSB1	13	16	19	21	25	27	31	35
SSB2	16	22	27	32	38	43	48	56
SSB3	12	14	17	19	22	24	27	30
SSC1	9	12	16	18	22	25	28	33
SSC2	12	16	22	26	33	37	43	51
SSE1	13	16	20	23	27	30	34	39
SSE2	16	22	29	34	41	46	52	61
SSE3	13	16	19	21	24	26	29	33

TOPEKA, KANSAS **SSF (%)**						*5243 DD65*		
Type	*LCR = 100*	*70*	*50*	*40*	*30*	*25*	*20*	*15*
WWA3	16	21	27	32	38	43	49	57
WWB4	20	28	38	45	55	61	69	78
WWC2	20	28	37	44	53	60	67	76
TWA1	16	19	22	24	28	31	34	39
TWA2	15	19	24	27	33	36	41	48
TWA3	15	19	24	29	35	39	44	52
TWA4	14	18	24	29	35	40	45	54
TWB3	13	18	23	27	32	36	42	50
TWD4	18	26	34	41	49	55	63	72
TWE2	19	27	35	41	50	55	63	72
TWF3	12	16	21	25	31	35	40	48
TWJ2	17	24	32	38	47	53	60	70
DGA1	10	13	16	19	22	24	26	28
DGA2	13	17	22	26	32	35	40	46
DGA3	16	22	29	34	42	48	55	64
DGB1	10	13	17	19	23	26	29	34
DGB2	13	17	23	27	33	38	43	51
DGB3	17	23	30	36	44	50	58	69
DGC1	13	17	21	25	30	34	38	44
DGC2	15	21	27	32	39	44	51	60
DGC3	19	26	34	40	50	57	65	76
SSA1	20	24	30	33	39	43	48	55
SSB1	16	20	25	29	34	38	42	49
SSB2	21	28	35	40	48	53	60	68
SSB3	15	19	23	26	31	34	38	44
SSC1	14	18	23	27	33	37	43	50
SSC2	17	23	30	36	44	49	56	65
SSE1	18	23	29	33	39	43	49	56
SSE2	23	30	39	45	53	59	66	74
SSE3	17	21	25	29	33	37	41	48

DODGE CITY, KANSAS **SSF (%)**						*5053 DD65*		
Type	*LCR = 100*	*70*	*50*	*40*	*30*	*25*	*20*	*15*
WWA3	21	27	35	41	50	55	63	72
WWB4	26	36	47	55	66	72	80	88
WWC2	26	36	46	54	64	71	78	86
TWA1	19	22	27	30	35	39	44	51
TWA2	19	24	30	35	42	47	53	61
TWA3	19	25	32	37	44	50	57	66
TWA4	18	24	31	37	45	51	58	67
TWB3	17	23	29	34	41	46	53	62
TWD4	24	33	43	50	60	66	74	82
TWE2	25	33	43	50	60	66	74	83
TWF3	16	21	28	33	40	46	52	62
TWJ2	22	30	40	47	57	63	71	80
DGA1	14	18	23	27	31	34	38	42
DGA2	16	22	28	33	41	45	51	58
DGA3	20	27	36	42	52	58	65	74
DGB1	14	18	24	28	35	39	44	50
DGB2	17	22	29	35	44	49	57	65
DGB3	20	28	37	44	55	62	70	79
DGC1	17	22	29	34	42	48	55	62
DGC2	19	26	34	41	51	57	65	74
DGC3	23	31	41	50	61	69	77	86
SSA1	24	30	37	42	49	54	60	68
SSB1	20	25	31	36	43	47	53	62
SSB2	26	33	42	48	57	63	70	78
SSB3	19	24	29	33	40	44	50	57
SSC1	18	24	31	36	44	49	56	65
SSC2	22	30	39	45	55	61	68	77
SSE1	23	30	37	43	51	56	63	71
SSE2	29	38	48	55	64	70	77	85
SSE3	21	26	32	36	42	47	53	61

LEXINGTON, KENTUCKY **SSF (%)**						*4732 DD65*		
Type	*LCR = 100*	*70*	*50*	*40*	*30*	*25*	*20*	*15*
WWA3	12	16	21	25	30	34	39	45
WWB4	15	23	31	38	47	53	60	70
WWC2	15	22	30	37	45	51	58	67
TWA1	14	16	18	20	23	25	27	31
TWA2	13	16	19	22	26	29	33	38
TWA3	12	15	19	23	27	31	35	41
TWA4	11	14	19	23	28	31	36	43
TWB3	11	14	18	21	26	29	34	40
TWD4	15	21	28	34	42	47	54	64
TWE2	15	22	29	34	42	47	54	63
TWF3	9	12	16	19	24	27	32	38
TWJ2	13	19	26	32	39	45	51	61
DGA1	7	9	12	13	15	16	17	18
DGA2	10	14	17	21	25	28	32	37
DGA3	14	19	24	29	36	41	47	55
DGB1	7	10	12	14	16	17	19	21
DGB2	10	14	18	21	26	30	34	40
DGB3	14	19	25	30	37	43	50	59
DGC1	10	13	16	18	21	24	27	31
DGC2	12	17	22	26	31	36	41	49
DGC3	16	22	29	34	42	48	56	66
SSA1	17	21	25	29	34	37	41	47
SSB1	14	18	21	24	29	32	35	41
SSB2	18	24	30	35	42	47	53	61
SSB3	13	16	20	22	26	28	32	36
SSC1	10	14	18	21	25	28	32	38
SSC2	13	18	25	29	36	41	47	55
SSE1	15	19	24	27	32	35	40	46
SSE2	19	26	33	38	46	51	58	66
SSE3	15	18	21	24	28	31	34	39

TABLE C.3 **Annual Passive Heating Performance: SSF (Continued)**

BATON ROUGE, LOUISIANA ***SSF (%)*** *1670 DD65*

Type LCR =	200	100	70	50	40	30	25	20
WWA3	21	35	45	56	63	72	78	84
WWB4	23	43	56	68	75	84	89	93
WWC2	23	43	55	67	74	83	88	92
TWA1	19	28	34	41	46	54	59	65
TWA2	19	31	39	47	54	63	68	75
TWA3	19	32	40	50	57	66	72	79
TWA4	18	31	40	50	57	67	73	79
TWB3	17	29	37	46	53	62	68	75
TWD4	22	39	51	62	70	79	84	90
TWE2	23	40	52	63	70	80	85	90
TWF3	16	28	37	46	53	62	68	75
TWJ2	20	37	48	60	68	77	82	88
DGA1	15	27	34	42	47	54	58	63
DGA2	16	28	37	46	53	61	67	73
DGA3	18	33	43	55	63	72	77	83
DGB1	15	27	36	46	53	62	67	72
DGB2	16	29	39	50	58	68	74	80
DGB3	19	33	45	57	66	76	82	87
DGC1	18	32	42	53	61	71	76	81
DGC2	19	33	44	56	65	75	80	86
DGC3	21	38	50	64	73	83	88	92
SSA1	24	38	47	56	63	71	76	82
SSB1	21	33	41	50	56	65	70	77
SSB2	25	41	51	62	69	77	83	88
SSB3	20	31	39	48	54	63	68	75
SSC1	19	32	41	51	58	67	73	79
SSC2	21	37	48	59	66	76	81	87
SSE1	24	40	49	60	67	75	80	86
SSE2	28	47	59	70	77	85	89	93
SSE3	22	33	41	50	56	65	70	77

SHREVEPORT, LOUISIANA ***SSF (%)*** *2175 DD65*

Type LCR =	200	100	70	50	40	30	25	20
WWA3	17	30	39	49	56	65	71	78
WWB4	19	38	49	61	69	79	84	90
WWC2	19	37	49	60	68	78	83	89
TWA1	17	25	30	36	41	48	52	58
TWA2	17	27	34	42	48	56	61	68
TWA3	16	27	35	44	50	59	65	72
TWA4	15	27	35	44	51	60	66	73
TWB3	14	25	32	40	47	55	61	68
TWD4	18	34	45	56	63	73	79	85
TWE2	19	35	46	57	64	74	79	86
TWF3	13	24	31	40	46	55	61	68
TWJ2	16	32	42	53	61	71	77	84
DGA1	13	22	29	36	41	47	51	55
DGA2	13	24	32	40	47	55	60	66
DGA3	16	28	38	48	56	66	72	78
DGB1	13	22	30	38	45	53	58	64
DGB2	14	25	33	43	50	60	66	73
DGB3	16	29	39	50	59	70	76	83
DGC1	15	26	35	45	52	62	68	74
DGC2	16	28	38	49	57	68	74	80
DGC3	18	32	44	56	66	77	83	88
SSA1	21	33	41	49	56	64	70	76
SSB1	18	28	35	43	49	58	63	70
SSB2	21	36	45	55	62	71	77	83
SSB3	17	27	34	41	47	55	60	67
SSC1	15	27	35	44	51	60	66	73
SSC2	17	32	42	52	60	69	75	82
SSE1	21	34	43	52	59	68	73	80
SSE2	24	41	52	63	70	79	84	89
SSE3	19	29	36	44	50	58	63	70

NEW ORLEANS, LOUISIANA ***SSF (%)*** *1465 DD65*

Type LCR =	200	100	70	50	40	30	25	20
WWA3	24	41	52	63	70	79	84	89
WWB4	28	50	63	74	81	89	93	96
WWC2	28	49	62	73	80	88	92	95
TWA1	21	31	38	46	52	60	65	72
TWA2	22	35	44	54	60	70	75	81
TWA3	22	36	46	57	64	73	79	85
TWA4	21	36	46	57	64	74	79	85
TWB3	20	33	42	52	60	69	75	81
TWD4	25	45	57	69	76	85	89	93
TWE2	27	46	58	69	77	85	89	94
TWF3	19	33	42	53	60	69	75	82
TWJ2	24	43	55	66	74	83	88	92
DGA1	18	32	40	49	55	62	66	71
DGA2	19	33	43	53	60	68	73	79
DGA3	21	38	50	62	69	78	83	87
DGB1	18	33	43	54	61	70	75	80
DGB2	19	34	45	58	66	75	80	86
DGB3	22	39	51	65	73	83	87	91
DGC1	21	38	50	62	70	79	83	87
DGC2	22	39	51	65	73	82	87	91
DGC3	24	44	57	72	80	88	92	95
SSA1	28	43	52	62	69	78	82	88
SSB1	24	37	46	56	63	72	77	83
SSB2	28	46	57	68	75	83	87	92
SSB3	23	36	44	54	61	69	75	81
SSC1	22	37	47	58	65	74	79	85
SSC2	25	43	54	66	73	82	87	91
SSE1	28	45	55	67	74	82	86	91
SSE2	33	53	66	76	83	90	93	96
SSE3	25	38	47	56	63	72	77	83

PORTLAND, MAINE ***SSF (%)*** *7499 DD65*

Type LCR =	100	70	50	40	30	25	20	15
WWA3	7	10	13	15	18	21	24	28
WWB4	9	15	22	28	35	41	47	57
WWC2	9	15	21	26	33	38	45	53
TWA1	12	13	14	15	16	17	18	19
TWA2	9	11	13	15	17	19	21	24
TWA3	8	10	13	15	18	20	22	26
TWA4	6	9	12	14	18	20	23	28
TWB3	7	10	12	14	17	19	22	26
TWD4	10	15	20	25	31	36	42	50
TWE2	10	15	21	25	31	35	41	49
TWF3	5	7	10	12	15	17	19	23
TWJ2	8	13	18	22	29	33	39	47
DGA1	4	5	5	6	6	6	6	4
DGA2	7	9	12	14	17	19	21	25
DGA3	10	14	18	22	27	31	36	43
DGB1	4	5	5	6	6	7	6	5
DGB2	7	9	12	14	17	20	22	26
DGB3	11	14	19	23	28	33	38	46
DGC1	5	7	8	10	11	12	13	13
DGC2	9	12	15	18	22	25	29	34
DGC3	12	16	22	26	32	37	43	53
SSA1	13	16	19	21	24	26	29	32
SSB1	11	13	15	17	20	22	24	27
SSB2	14	18	23	27	33	37	42	49
SSB3	10	11	13	15	17	18	20	22
SSC1	6	8	10	12	14	16	18	21
SSC2	8	12	16	20	25	28	33	40
SSE1	10	13	15	17	20	22	24	27
SSE2	13	18	24	28	34	38	44	51
SSE3	11	13	15	17	19	20	22	25

TABLE C.3 **Annual Passive Heating Performance: SSF (Continued)**

BALTIMORE, MARYLAND ***SSF (%)*** *4731 DD65*

Type	LCR = 100	70	50	40	30	25	20	15
WWA3	13	18	24	28	35	39	45	53
WWB4	17	26	35	42	52	58	66	75
WWC2	17	25	34	41	50	56	64	73
TWA1	15	17	20	22	25	28	31	36
TWA2	14	17	21	25	30	33	38	44
TWA3	13	17	22	26	31	35	41	48
TWA4	12	16	22	26	32	36	42	50
TWB3	12	16	20	24	29	33	38	46
TWD4	16	23	31	38	46	52	59	69
TWE2	17	24	32	38	46	52	59	69
TWF3	10	14	19	22	28	32	37	44
TWJ2	14	21	29	35	44	49	57	66
DGA1	8	11	14	16	19	20	22	24
DGA2	11	15	19	23	28	32	37	42
DGA3	15	20	26	32	39	45	52	60
DGB1	8	11	14	17	20	22	25	29
DGB2	11	15	20	24	30	34	40	47
DGB3	15	21	27	33	41	47	55	65
DGC1	11	14	18	22	26	29	34	39
DGC2	13	18	24	29	36	41	47	56
DGC3	17	24	31	37	47	53	62	72
SSA1	18	22	27	31	37	41	46	52
SSB1	15	19	23	27	31	35	40	46
SSB2	19	26	33	38	45	51	57	66
SSB3	14	17	21	24	29	32	36	41
SSC1	12	16	20	24	30	33	38	46
SSC2	15	21	28	33	41	46	52	61
SSE1	16	21	26	31	36	40	46	53
SSE2	21	28	36	42	51	56	63	72
SSE3	16	19	23	26	31	34	38	44

BOSTON, MASSACHUSETTS ***SSF (%)*** *5622 DD65*

Type	LCR = 100	70	50	40	30	25	20	15
WWA3	10	14	18	21	26	29	33	40
WWB4	13	20	28	34	43	49	56	66
WWC2	13	20	27	33	41	47	54	63
TWA1	13	15	16	18	20	21	23	26
TWA2	11	14	17	19	23	25	28	33
TWA3	10	13	17	20	24	27	31	36
TWA4	9	12	16	19	24	27	32	38
TWB3	9	12	16	18	23	26	29	35
TWD4	13	18	25	31	38	43	50	59
TWE2	13	19	26	31	38	43	50	58
TWF3	7	10	14	16	20	23	27	33
TWJ2	11	17	23	28	35	41	47	56
DGA1	6	7	9	10	12	12	13	13
DGA2	9	12	15	18	22	25	28	33
DGA3	12	17	22	26	32	37	43	51
DGB1	6	7	9	11	12	13	14	16
DGB2	9	12	16	18	23	26	30	36
DGB3	13	17	23	27	34	39	46	55
DGC1	8	10	13	15	17	19	22	25
DGC2	11	15	19	23	28	32	37	44
DGC3	14	20	26	31	38	44	51	62
SSA1	15	19	22	25	29	32	36	41
SSB1	13	15	19	21	25	27	31	36
SSB2	16	21	27	32	38	43	48	56
SSB3	11	14	17	19	22	24	27	31
SSC1	9	12	15	17	21	24	27	32
SSC2	11	16	21	26	32	36	42	50
SSE1	13	16	20	23	27	30	34	39
SSE2	16	22	29	34	41	46	52	61
SSE3	13	16	19	21	24	26	29	33

DETROIT, MICHIGAN ***SSF (%)*** *6234 DD65*

Type	LCR = 100	70	50	40	30	25	20	15
WWA3	7	10	13	15	18	20	23	26
WWB4	9	15	22	27	35	40	46	55
WWC2	9	15	21	26	33	38	44	52
TWA1	12	13	14	15	16	16	17	18
TWA2	9	11	13	15	17	19	21	23
TWA3	8	10	13	15	17	19	22	25
TWA4	7	9	12	14	17	20	23	27
TWB3	7	10	12	14	17	19	22	26
TWD4	10	15	20	25	31	35	41	49
TWE2	10	15	20	25	31	35	40	48
TWF3	5	7	10	12	14	16	19	22
TWJ2	8	13	18	22	28	33	38	46
DGA1	4	5	6	6	6	6	5	4
DGA2	7	9	12	14	17	19	21	24
DGA3	10	14	18	22	27	31	36	43
DGB1	4	5	6	6	7	7	6	5
DGB2	7	9	12	14	17	20	22	26
DGB3	11	14	19	23	28	32	38	46
DGC1	5	7	9	10	11	12	12	13
DGC2	9	12	15	18	22	25	28	33
DGC3	12	17	22	26	32	37	43	52
SSA1	13	16	19	21	24	26	28	32
SSB1	11	13	15	17	20	21	24	26
SSB2	14	18	23	27	32	36	41	48
SSB3	10	11	13	15	17	18	20	21
SSC1	6	8	10	12	14	16	18	20
SSC2	8	12	16	20	25	28	32	39
SSE1	10	13	15	17	20	21	23	25
SSE2	13	18	24	28	33	37	43	49
SSE3	11	13	15	17	19	20	22	24

SAULT STE. MARIE, MICHIGAN ***SSF (%)*** *9201 DD65*

Type	LCR = 100	70	50	40	30	25	20	15
WWA3	5	7	8	10	11	12	13	13
WWB4	6	11	17	21	27	31	37	44
WWC2	7	11	16	20	25	29	34	40
TWA1	10	11	11	12	12	12	12	11
TWA2	8	9	10	11	12	13	13	14
TWA3	6	7	9	10	12	13	14	15
TWA4	4	6	8	10	12	13	14	16
TWB3	6	7	9	10	12	13	15	16
TWD4	7	11	16	19	25	28	33	39
TWE2	8	12	16	19	24	27	31	37
TWF3	3	5	6	8	9	10	11	12
TWJ2	6	9	14	17	22	25	30	36
DGA1	2	3	3	2	2	1	0	−4
DGA2	5	7	9	11	13	14	15	17
DGA3	9	12	15	18	23	26	30	35
DGB1	2	2	3	2	2	1	0	−2
DGB2	5	7	9	11	13	14	16	18
DGB3	9	12	16	19	24	27	32	38
DGC1	4	5	5	6	6	6	5	3
DGC2	7	9	12	14	17	19	22	25
DGC3	10	14	19	22	28	31	36	43
SSA1	12	13	15	17	19	20	21	22
SSB1	9	11	12	14	15	16	17	18
SSB2	12	16	20	23	27	30	34	39
SSB3	8	9	11	11	12	13	13	12
SSC1	4	5	7	7	8	8	8	7
SSC2	6	9	12	14	18	20	23	27
SSE1	8	9	11	12	13	13	13	11
SSE2	10	14	18	22	26	29	32	36
SSE3	10	11	12	13	14	15	15	14

TABLE C.3 **Annual Passive Heating Performance: SSF (Continued)**

MINNEAPOLIS, MINNESOTA **SSF (%)** *8165 DD65*								
Type LCR =	*100*	*70*	*50*	*40*	*30*	*25*	*20*	*15*
WWA3	7	9	12	14	17	19	21	24
WWB4	9	14	21	26	33	38	45	54
WWC2	9	14	20	25	32	36	42	50
TWA1	11	12	13	14	15	16	16	17
TWA2	9	11	13	14	16	18	19	21
TWA3	7	10	12	14	16	18	20	23
TWA4	6	8	11	13	16	19	21	25
TWB3	7	9	12	13	16	18	21	24
TWD4	9	14	19	24	30	34	40	48
TWE2	10	14	20	24	29	34	39	46
TWF3	5	7	9	11	13	15	17	20
TWJ2	7	12	17	21	27	31	37	44
DGA1	4	4	5	5	5	5	4	2
DGA2	7	9	11	13	16	18	20	23
DGA3	10	13	18	21	26	30	35	41
DGB1	3	4	5	5	6	5	5	3
DGB2	7	9	12	14	17	19	21	25
DGB3	10	14	19	22	28	31	37	44
DGC1	5	6	8	9	10	11	11	11
DGC2	8	11	15	17	21	24	27	32
DGC3	12	16	21	25	31	36	42	50
SSA1	12	15	17	19	21	23	25	28
SSB1	10	12	14	16	18	19	21	23
SSB2	13	17	22	25	30	34	38	45
SSB3	9	10	12	13	15	16	17	18
SSC1	6	8	10	11	13	14	16	18
SSC2	8	11	15	19	23	27	31	37
SSE1	9	11	13	15	17	18	20	21
SSE2	12	16	21	25	31	34	39	46
SSE3	10	12	14	15	17	18	19	20

MERIDIAN, MISSISSIPPI **SSF (%)** *2393 DD65*								
Type LCR =	*200*	*100*	*70*	*50*	*40*	*30*	*25*	*20*
WWA3	15	28	36	45	52	61	67	74
WWB4	16	34	46	58	66	75	81	87
WWC2	17	34	45	56	64	74	80	86
TWA1	16	23	28	33	38	44	49	54
TWA2	15	24	31	38	44	52	57	64
TWA3	14	25	32	41	47	55	61	68
TWA4	13	24	32	41	47	56	62	69
TWB3	13	22	29	37	43	52	57	64
TWD4	16	31	41	52	60	70	76	82
TWE2	17	32	42	53	60	70	76	83
TWF3	11	21	29	37	43	51	57	64
TWJ2	14	29	39	50	57	67	73	80
DGA1	11	20	26	32	37	43	46	50
DGA2	12	22	29	37	43	51	56	62
DGA3	14	26	35	45	52	62	68	75
DGB1	11	20	27	34	40	48	53	59
DGB2	12	22	30	39	46	56	62	69
DGB3	14	26	36	46	55	66	72	80
DGC1	13	24	31	41	47	57	63	69
DGC2	14	26	34	45	52	63	69	76
DGC3	16	30	40	52	61	73	79	86
SSA1	20	31	38	46	52	61	66	72
SSB1	16	26	33	41	46	54	59	66
SSB2	20	33	42	52	59	68	74	80
SSB3	16	25	31	39	44	52	57	63
SSC1	14	24	32	41	47	55	61	68
SSC2	15	29	38	49	56	65	71	78
SSE1	19	31	40	49	55	64	70	76
SSE2	22	38	49	60	67	76	81	87
SSE3	17	27	33	41	46	54	59	66

SAINT LOUIS, MISSOURI **SSF (%)** *4754 DD65*								
Type LCR =	*100*	*70*	*50*	*40*	*30*	*25*	*20*	*15*
WWA3	15	20	26	30	36	41	47	55
WWB4	19	27	37	44	53	59	67	76
WWC2	19	27	36	42	51	57	65	74
TWA1	16	18	21	23	27	29	32	37
TWA2	15	19	23	26	31	35	39	46
TWA3	14	18	23	27	33	37	42	50
TWA4	13	18	23	27	33	38	43	51
TWB3	13	17	22	25	31	35	40	47
TWD4	18	25	33	39	48	53	61	70
TWE2	18	26	33	39	48	53	61	70
TWF3	11	15	20	24	29	33	38	46
TWJ2	16	23	31	37	45	51	58	67
DGA1	9	12	15	18	20	22	24	26
DGA2	12	16	21	25	30	34	38	44
DGA3	16	21	28	33	41	46	53	62
DGB1	10	13	16	18	22	24	27	31
DGB2	12	17	22	26	32	36	41	49
DGB3	16	22	29	34	43	48	56	66
DGC1	12	16	20	23	28	31	35	41
DGC2	15	20	26	30	37	42	49	57
DGC3	19	25	33	39	48	54	63	73
SSA1	19	24	29	32	38	42	47	53
SSB1	16	20	24	28	33	36	41	47
SSB2	21	27	34	39	46	52	58	67
SSB3	15	18	22	25	30	33	37	43
SSC1	13	17	22	26	31	35	40	47
SSC2	16	22	29	34	42	47	54	63
SSE1	18	22	28	32	38	42	47	54
SSE2	22	29	37	43	52	57	64	73
SSE3	17	20	24	28	32	35	40	46

SPRINGFIELD, MISSOURI **SSF (%)** *4571 DD65*								
Type LCR =	*100*	*70*	*50*	*40*	*30*	*25*	*20*	*15*
WWA3	16	22	28	33	40	45	52	60
WWB4	21	30	40	47	57	63	71	80
WWC2	21	29	39	46	55	62	69	78
TWA1	17	19	23	25	29	32	36	41
TWA2	16	20	25	29	34	38	43	51
TWA3	15	20	26	30	36	41	46	55
TWA4	14	19	25	30	37	41	48	56
TWB3	14	18	24	28	34	38	44	52
TWD4	19	27	36	42	51	57	65	74
TWE2	20	28	36	43	51	57	65	74
TWF3	12	17	22	26	33	37	43	51
TWJ2	18	25	33	40	49	55	62	72
DGA1	11	14	18	20	24	26	28	30
DGA2	13	18	23	27	33	37	42	48
DGA3	17	23	30	36	44	50	57	66
DGB1	11	14	18	21	25	28	32	37
DGB2	14	18	24	28	35	39	46	54
DGB3	17	24	31	37	46	52	60	71
DGC1	13	18	22	26	32	36	41	47
DGC2	10	21	28	33	41	46	54	63
DGC3	20	27	35	42	52	59	67	78
SSA1	21	26	31	36	42	46	51	59
SSB1	17	22	27	30	36	40	45	52
SSB2	22	29	37	42	50	55	62	71
SSB3	16	20	25	28	33	37	41	48
SSC1	14	19	25	29	35	39	45	53
SSC2	18	24	32	38	46	51	58	68
SSE1	19	25	31	36	42	47	52	60
SSE2	24	32	41	47	56	62	69	77
SSE3	18	22	27	30	36	39	44	51

TABLE C.3 **Annual Passive Heating Performance: SSF (Continued)**

HELENA, MONTANA **SSF (%)**							*8197 DD65*
Type LCR = 100	*70*	*50*	*40*	*30*	*25*	*20*	*15*
WWA3	11	14	19	22	27	30	34	39
WWB4	14	21	29	35	43	49	55	64
WWC2	14	20	28	34	41	47	53	62
TWA1	13	15	17	19	21	22	24	27
TWA2	12	14	18	20	24	26	29	33
TWA3	11	14	18	20	25	28	31	36
TWA4	9	13	17	20	25	28	32	38
TWB3	10	13	16	19	23	26	30	35
TWD4	13	19	26	31	39	44	50	58
TWE2	14	20	26	31	39	43	49	57
TWF3	8	11	14	17	21	24	28	33
TWJ2	12	17	24	29	36	41	47	55
DGA1	6	8	10	11	13	13	14	14
DGA2	9	12	16	19	23	26	29	33
DGA3	13	17	23	27	34	38	44	51
DGB1	6	8	10	12	14	15	16	17
DGB2	9	13	17	20	24	27	31	37
DGB3	13	18	24	28	35	40	46	55
DGC1	8	11	14	16	19	21	23	26
DGC2	11	15	20	24	30	33	38	44
DGC3	15	20	27	32	40	45	52	61
SSA1	16	19	23	25	29	32	35	39
SSB1	13	16	19	21	25	27	30	34
SSB2	17	22	27	32	38	42	47	54
SSB3	12	14	17	19	22	24	26	29
SSC1	9	12	16	18	22	25	28	32
SSC2	12	17	22	27	33	37	42	49
SSE1	13	17	20	23	27	29	32	35
SSE2	17	23	29	34	40	45	50	57
SSE3	13	16	19	21	24	26	29	32

ELY, NEVADA **SSF (%)**							*7818 DD65*
Type LCR = 100	*70*	*50*	*40*	*30*	*25*	*20*	*15*
WWA3	18	24	32	37	46	51	58	67
WWB4	23	33	44	52	62	69	76	85
WWC2	23	32	43	50	60	67	75	83
TWA1	17	20	24	27	32	35	40	47
TWA2	17	21	27	31	38	43	49	57
TWA3	16	22	28	33	41	46	52	61
TWA4	15	21	28	33	41	46	53	63
TWB3	15	20	26	31	38	43	49	58
TWD4	21	29	39	46	56	62	70	79
TWE2	22	30	40	47	56	63	70	79
TWF3	13	18	25	29	37	42	48	57
TWJ2	19	27	37	44	53	60	68	77
DGA1	11	15	20	23	27	30	33	37
DGA2	14	19	25	30	37	41	47	54
DGA3	18	24	32	39	48	55	62	71
DGB1	11	15	20	24	30	34	39	45
DGB2	14	19	26	31	39	45	52	61
DGB3	18	25	33	40	51	58	66	76
DGC1	14	19	25	30	38	43	49	57
DGC2	17	23	31	37	46	53	60	70
DGC3	21	29	38	46	57	65	74	83
SSA1	23	29	36	40	48	52	58	66
SSB1	19	24	30	34	41	46	51	59
SSB2	25	32	41	47	55	61	68	76
SSB3	18	22	28	32	38	42	48	55
SSC1	16	21	27	32	40	45	51	60
SSC2	19	27	35	42	51	57	64	74
SSE1	22	28	36	41	49	54	60	68
SSE2	27	37	46	53	62	68	75	83
SSE3	20	24	30	34	41	45	51	58

OMAHA, NEBRASKA **SSF (%)**							*6606 DD65*
Type LCR = 100	*70*	*50*	*40*	*30*	*25*	*20*	*15*
WWA3	12	16	21	24	30	33	38	45
WWB4	15	22	31	37	47	53	60	70
WWC2	15	22	30	36	45	51	58	68
TWA1	14	16	18	20	22	24	27	30
TWA2	13	15	19	22	26	29	32	38
TWA3	11	15	19	22	27	30	35	41
TWA4	10	14	19	22	27	31	36	43
TWB3	11	14	18	21	25	29	33	40
TWD4	14	21	28	34	42	47	54	64
TWE2	15	21	28	34	42	47	54	63
TWF3	9	12	16	19	24	27	31	37
TWJ2	13	19	26	31	39	49	51	61
DGA1	7	9	11	13	15	16	17	17
DGA2	10	13	17	20	25	28	32	37
DGA3	13	18	24	28	35	40	47	55
DGB1	7	9	11	13	15	17	19	21
DGB2	10	14	18	21	26	29	34	40
DGB3	14	19	25	29	37	42	49	59
DGC1	9	12	15	18	21	23	26	30
DGC2	12	16	21	25	31	35	41	49
DGC3	16	21	28	33	42	48	56	66
SSA1	16	20	24	27	32	35	39	45
SSB1	14	17	20	23	27	30	34	39
SSB2	18	23	29	34	40	45	51	59
SSB3	12	15	18	21	24	26	29	34
SSC1	10	13	17	20	25	28	32	38
SSC2	13	18	29	29	36	40	47	55
SSE1	14	18	22	25	30	33	37	40
SSE2	18	24	31	37	44	49	56	65
SSE3	14	17	20	23	26	29	32	37

LAS VEGAS, NEVADA **SSF (%)**							*2602 DD65*
Type LCR = 200	*100*	*70*	*50*	*40*	*30*	*25*	*20*
WWA3	24	41	52	63	71	80	85	90
WWB4	28	50	63	75	82	89	93	96
WWC2	28	49	62	74	81	88	92	96
TWA1	21	31	38	46	52	61	66	73
TWA2	22	35	44	54	61	70	76	82
TWA3	22	37	47	57	64	74	79	85
TWA4	21	36	47	58	64	74	80	86
TWB3	19	33	43	53	60	69	75	82
TWD4	25	45	57	69	77	85	89	93
TWE2	27	46	58	70	77	85	90	94
TWF3	18	33	43	53	60	70	76	82
TWJ2	24	43	55	67	75	83	88	92
DGA1	18	32	40	49	55	63	67	72
DGA2	18	33	43	53	60	69	74	80
DGA3	21	38	50	62	70	78	83	88
DGB1	18	33	44	55	62	71	75	80
DGB2	19	35	46	58	67	76	81	86
DGB3	21	39	52	65	74	83	87	92
DGC1	21	38	51	63	71	79	84	88
DGC2	21	39	52	65	74	83	87	91
DGC3	24	44	58	72	80	88	92	95
SSA1	26	40	50	59	66	74	79	85
SSB1	22	35	44	53	60	69	74	80
SSB2	27	44	54	65	72	81	85	90
SSB3	21	34	42	51	57	66	71	78
SSC1	22	37	47	58	65	75	80	86
SSC2	24	43	55	66	73	82	87	92
SSE1	27	43	53	64	71	79	84	89
SSE2	31	51	63	74	80	88	91	95
SSE3	23	36	45	54	60	69	74	80

TABLE C.3 **Annual Passive Heating Performance: SSF (Continued)**

RENO, NEVADA ***SSF (%)*** *6027 DD65*

Type	*LCR = 100*	*70*	*50*	*40*	*30*	*25*	*20*	*15*
WWA3	25	33	41	48	57	63	70	78
WWB4	31	42	54	62	72	78	84	91
WWC2	31	41	53	60	70	76	83	90
TWA1	21	25	30	35	41	45	50	58
TWA2	22	28	35	40	48	53	60	69
TWA3	22	29	37	43	51	57	64	73
TWA4	21	29	37	43	52	58	65	74
TWB3	20	26	34	40	48	53	60	69
TWD4	28	38	48	56	66	72	79	87
TWE2	29	39	49	57	66	72	79	87
TWF3	19	25	33	39	47	53	60	69
TWJ2	26	36	46	54	64	70	77	85
DGA1	17	23	29	33	38	42	46	50
DGA2	19	26	34	39	47	52	58	65
DGA3	23	32	41	49	58	64	71	79
DGB1	17	23	30	36	43	48	53	60
DGB2	20	27	35	42	51	57	64	72
DGB3	24	32	43	51	62	69	76	84
DGC1	21	28	36	43	52	57	64	71
DGC2	23	31	41	48	59	65	72	80
DGC3	27	37	48	57	69	75	83	89
SSA1	29	36	43	49	56	61	67	75
SSB1	24	30	37	43	50	55	61	69
SSB2	31	39	48	55	64	69	76	83
SSB3	23	29	35	40	47	52	58	65
SSC1	22	29	37	43	51	57	64	73
SSC2	26	35	45	52	61	67	75	83
SSE1	29	36	45	51	59	64	70	78
SSE2	35	45	55	62	71	76	82	89
SSE3	25	31	38	43	50	55	61	68

LOS ALAMOS, NEW MEXICO *6437 DD65*

Type	*LCR = 100*	*70*	*50*	*40*	*30*	*25*	*20*	*15*
WWA3	20	27	35	41	49	55	62	72
WWB4	25	36	47	55	66	72	80	88
WWC2	25	35	46	54	64	71	78	86
TWA1	18	22	26	29	34	38	43	50
TWA2	18	23	29	34	41	46	52	61
TWA3	18	24	31	36	44	49	56	66
TWA4	17	23	31	36	45	50	57	67
TWB3	16	22	28	33	41	46	53	62
TWD4	23	32	42	49	59	66	74	83
TWE2	24	33	43	50	60	66	74	83
TWF3	15	20	27	32	40	45	52	62
TWJ2	21	30	40	47	57	63	71	80
DGA1	13	17	22	26	31	34	37	41
DGA2	15	21	28	33	40	45	50	58
DGA3	19	26	35	42	52	58	65	74
DGB1	13	18	23	27	34	38	44	50
DGB2	16	21	28	34	43	49	56	65
DGB3	20	27	36	43	54	62	70	80
DGC1	16	22	28	34	42	48	55	63
DGC2	18	25	33	40	50	57	65	74
DGC3	22	31	41	49	61	69	77	86
SSA1	25	31	38	44	51	56	63	71
SSB1	21	26	32	37	44	49	56	64
SSB2	27	34	43	50	59	65	72	80
SSB3	19	24	30	35	42	46	52	60
SSC1	17	23	30	36	43	49	56	65
SSC2	21	29	38	45	54	61	68	78
SSE1	24	31	39	45	53	59	65	74
SSE2	30	39	50	57	67	73	79	87
SSE3	21	27	33	37	44	49	55	63

ALBUQUERQUE, NEW MEXICO ***SSF (%)*** *4293 DD65*

Type	*LCR = 100*	*70*	*50*	*40*	*30*	*25*	*20*	*15*
WWA3	28	37	46	54	63	69	76	85
WWB4	35	47	59	67	77	83	89	94
WWC2	34	46	58	66	76	82	88	94
TWA1	22	27	33	38	45	50	56	64
TWA2	24	31	39	45	53	59	66	75
TWA3	25	32	41	48	57	63	70	79
TWA4	24	32	41	48	58	64	71	80
TWB3	22	29	38	44	53	59	66	75
TWD4	31	42	53	61	71	77	84	91
TWE2	32	43	54	62	72	78	84	91
TWF3	21	29	37	44	53	59	66	76
TWJ2	29	40	51	59	69	75	82	90
DGA1	19	26	33	37	44	48	52	58
DGA2	21	29	38	44	52	57	64	71
DGA3	25	35	46	54	64	70	76	84
DGB1	20	27	35	41	50	55	61	68
DGB2	22	30	40	47	58	64	71	79
DGB3	26	36	47	56	68	75	82	89
DGC1	23	32	42	49	60	66	72	79
DGC2	25	34	46	54	65	72	79	86
DGC3	30	40	53	63	75	82	88	93
SSA1	30	38	46	52	61	66	73	81
SSB1	25	32	40	46	54	60	67	75
SSB2	32	42	52	59	68	74	81	88
SSB3	24	30	38	43	51	57	63	72
SSC1	24	32	41	48	57	63	71	79
SSC2	29	39	50	57	67	73	80	88
SSE1	31	39	49	55	64	70	77	85
SSE2	38	49	60	67	76	82	88	93
SSE3	26	33	40	46	54	60	66	75

ALBANY, NEW YORK ***SSF (%)*** *6891 DD65*

Type	*LCR = 100*	*70*	*50*	*40*	*30*	*25*	*20*	*15*
WWA3	7	9	12	14	17	18	21	24
WWB4	8	14	21	26	33	38	45	54
WWC2	9	14	20	25	31	36	42	51
TWA1	11	12	13	14	15	15	16	17
TWA2	9	10	12	14	16	17	19	21
TWA3	7	9	12	13	16	18	20	23
TWA4	6	8	11	13	16	18	21	25
TWB3	7	9	11	13	16	18	20	24
TWD4	9	13	19	23	30	34	40	48
TWE2	9	14	19	23	29	33	39	46
TWF3	5	7	9	11	13	15	17	20
TWJ2	7	12	17	21	27	31	37	45
DGA1	3	4	5	5	5	4	4	2
DGA2	6	8	11	13	15	17	20	23
DGA3	10	13	17	21	26	29	34	41
DGB1	3	4	4	5	5	5	4	3
DGB2	6	9	11	13	16	18	21	24
DGB3	10	14	18	22	27	31	36	44
DGC1	5	6	7	8	10	10	10	10
DGC2	8	11	14	17	20	23	27	32
DGC3	12	16	21	25	31	35	41	50
SSA1	12	15	17	19	22	24	27	30
SSB1	10	12	14	16	18	20	22	25
SSB2	13	17	22	26	31	35	39	46
SSB3	9	11	12	14	15	16	18	19
SSC1	6	7	9	11	13	14	16	18
SSC2	7	11	15	18	23	26	31	37
SSE1	9	11	14	16	18	20	21	23
SSE2	12	17	22	26	32	36	41	48
SSE3	11	12	14	16	17	19	20	22

TABLE C.3 **Annual Passive Heating Performance: SSF (Continued)**

BUFFALO, NEW YORK *6931 DD65*

Type	*LCR = 100*	*70*	*50*	*40*	*30*	*25*	*20*	*15*
WWA3	5	6	8	9	11	11	12	13
WWB4	6	11	16	21	27	31	37	44
WWC2	6	11	16	19	25	29	34	41
TWA1	10	11	11	12	12	12	11	11
TWA2	8	9	10	11	12	12	13	13
TWA3	6	7	9	10	11	12	13	14
TWA4	4	6	8	9	11	12	14	16
TWB3	6	7	9	10	12	13	14	16
TWD4	7	11	15	19	24	28	32	39
TWE2	8	11	15	19	23	27	31	37
TWF3	3	5	6	7	9	10	11	12
TWJ2	5	9	13	17	22	25	29	36
DGA1	2	2	2	2	1	0	−1	−4
DGA2	5	7	9	10	12	13	14	16
DGA3	9	12	15	18	22	25	29	34
DGB1	2	2	2	2	1	0	0	−4
DGB2	5	7	9	10	12	14	15	17
DGB3	9	12	16	19	23	27	31	37
DGC1	4	4	5	5	5	5	4	2
DGC2	7	9	12	14	16	18	21	24
DGC3	10	14	18	22	27	31	35	42
SSA1	11	13	15	17	19	20	21	22
SSB1	9	11	12	13	15	16	17	18
SSB2	12	15	19	23	27	30	34	40
SSB3	8	9	11	11	12	13	13	13
SSC1	4	5	6	7	8	8	8	7
SSC2	6	8	12	14	17	20	23	27
SSE1	8	9	11	12	13	13	14	13
SSE2	10	14	18	22	26	29	33	38
SSE3	10	11	12	13	14	15	15	15

NEW YORK, NEW YORK *4851 DD65*

Type	*LCR = 100*	*70*	*50*	*40*	*30*	*25*	*20*	*15*
WWA3	10	14	19	22	27	30	35	41
WWB4	13	21	29	35	44	50	57	67
WWC2	13	20	28	34	42	48	55	64
TWA1	13	15	17	18	21	22	24	27
TWA2	11	14	17	20	23	26	30	35
TWA3	10	14	17	20	25	28	32	38
TWA4	9	13	17	20	25	28	33	39
TWB3	9	13	16	19	23	26	30	36
TWD4	13	19	26	31	39	44	51	60
TWE2	14	20	26	32	39	44	51	60
TWF3	7	11	14	17	21	24	28	34
TWJ2	11	17	24	29	36	42	48	57
DGA1	6	8	10	11	12	13	14	14
DGA2	9	12	16	18	23	25	29	34
DGA3	12	17	22	27	33	38	44	52
DGB1	6	8	10	11	13	14	16	17
DGB2	9	12	16	19	23	27	31	37
DGB3	13	17	23	28	35	40	47	56
DGC1	8	11	13	15	18	20	23	26
DGC2	11	15	20	23	29	33	38	45
DGC3	15	20	26	31	39	45	53	63
SSA1	16	19	23	26	30	33	37	43
SSB1	13	16	19	22	26	28	32	37
SSB2	17	22	28	33	39	44	50	58
SSB3	12	14	17	20	23	25	28	32
SSC1	9	12	15	18	22	25	29	34
SSC2	11	16	22	27	33	37	43	52
SSE1	13	17	21	24	28	31	35	41
SSE2	17	23	30	35	43	48	54	62
SSE3	13	16	19	22	25	27	31	35

SYRACUSE, NEW YORK ***SSF (%)*** *6681 DD65*

Type	*LCR = 100*	*70*	*50*	*40*	*30*	*25*	*20*	*15*
WWA3	5	7	9	10	12	13	14	15
WWB4	6	12	17	22	28	33	38	46
WWC2	7	11	17	21	26	30	36	43
TWA1	11	11	12	12	12	13	12	12
TWA2	8	9	10	11	13	13	14	15
TWA3	6	8	9	11	12	14	15	16
TWA4	5	7	9	10	12	14	15	18
TWB3	6	7	9	11	13	14	16	18
TWD4	7	11	16	20	25	29	34	41
TWE2	8	12	16	20	25	28	33	39
TWF3	4	5	7	8	10	11	12	13
TWJ2	6	10	14	18	23	26	31	38
DGA1	2	3	3	3	2	1	0	−3
DGA2	6	7	9	11	13	14	16	17
DGA3	9	12	16	18	23	26	30	36
DGB1	2	3	3	3	2	1	0	−2
DGB2	6	7	9	11	13	15	16	19
DGB3	9	12	16	19	24	28	32	38
DGC1	4	5	5	6	6	6	5	4
DGC2	7	9	12	14	17	19	22	25
DGC3	11	14	19	22	28	31	36	44
SSA1	12	14	16	17	19	20	22	24
SSB1	9	11	13	14	16	17	18	19
SSB2	12	16	20	23	28	31	35	41
SSB3	8	10	11	12	13	13	14	14
SSC1	4	6	7	8	9	9	9	9
SSC2	6	9	12	15	19	21	24	29
SSE1	8	10	11	13	14	14	15	15
SSE2	10	14	19	22	27	30	34	40
SSE3	10	11	12	13	15	15	16	16

CHARLOTTE, NORTH CAROLINA ***SSF (%)*** *3226 DD65*

Type	*LCR = 200*	*100*	*70*	*50*	*40*	*30*	*25*	*20*
WWA3	13	23	31	39	45	54	59	66
WWB4	13	29	40	51	59	69	75	82
WWC2	13	29	39	50	58	67	74	81
TWA1	15	20	24	29	33	38	42	47
TWA2	13	21	27	33	38	45	50	57
TWA3	12	21	27	35	40	48	54	60
TWA4	11	20	27	35	41	49	54	61
TWB3	11	19	25	32	37	45	50	57
TWD4	13	26	36	46	53	63	69	76
TWE2	14	27	37	47	54	63	69	77
TWF3	9	18	24	31	36	44	50	56
TWJ2	11	24	34	44	51	61	67	74
DGA1	9	16	21	26	30	35	39	42
DGA2	10	18	24	31	37	44	49	55
DGA3	12	22	30	39	46	55	61	68
DGB1	9	16	21	28	32	39	44	49
DGB2	10	19	25	33	39	48	54	61
DGB3	12	23	31	40	48	58	65	73
DGC1	11	20	26	33	39	47	53	59
DGC2	12	22	29	38	45	55	61	69
DGC3	14	26	35	45	53	65	72	80
SSA1	17	26	33	40	45	53	58	64
SSB1	14	22	28	35	40	47	52	58
SSB2	17	29	37	46	52	61	67	74
SSB3	14	21	26	33	37	44	49	55
SSC1	11	20	27	34	40	48	53	60
SSC2	12	25	33	42	49	58	64	72
SSE1	16	26	34	42	47	55	61	68
SSE2	18	32	42	52	59	69	74	81
SSE3	15	23	29	35	40	47	51	57

TABLE C.3 **Annual Passive Heating Performance: SSF (Continued)**

RALEIGH-DURHAM, NORTH CAROLINA ***SSF (%)*** *3520 DD65*

Type	LCR = 200	100	70	50	40	30	25	20
WWA3	11	21	28	36	41	50	55	62
WWB4	11	26	37	48	55	66	72	79
WWC2	12	26	36	46	54	64	70	77
TWA1	14	19	22	27	30	35	39	44
TWA2	12	19	24	30	35	42	47	53
TWA3	11	19	25	32	37	45	50	56
TWA4	9	18	24	32	37	45	51	57
TWB3	10	17	23	29	34	41	46	53
TWD4	11	24	33	43	50	60	66	73
TWE2	12	25	34	43	50	60	66	73
TWF3	8	16	21	28	33	41	46	52
TWJ2	10	22	31	40	47	57	63	71
DGA1	8	14	19	23	27	32	35	38
DGA2	9	16	22	29	34	41	45	51
DGA3	11	20	27	36	43	52	58	65
DGB1	8	14	19	24	29	35	39	44
DGB2	9	17	23	30	35	44	49	56
DGB3	11	21	28	37	44	55	61	70
DGC1	9	17	23	30	35	42	48	54
DGC2	10	20	27	35	41	50	57	64
DGC3	13	24	32	42	50	61	68	77
SSA1	16	24	30	37	42	50	54	61
SSB1	13	21	26	32	37	44	48	54
SSB2	15	26	34	43	49	58	63	70
SSB3	12	19	24	30	34	41	45	51
SSC1	10	18	24	31	36	44	49	56
SSC2	11	22	30	39	46	55	61	68
SSE1	14	24	31	38	44	52	57	63
SSE2	16	30	39	49	56	65	71	77
SSE3	14	21	26	32	37	43	48	54

BISMARCK, NORTH DAKOTA ***SSF (%)*** *9057 DD65*

Type	LCR = 100	70	50	40	30	25	20	15
WWA3	8	11	15	17	21	23	26	30
WWB4	11	17	24	30	37	43	49	58
WWC2	11	17	23	28	36	40	47	55
TWA1	12	13	15	16	17	18	19	21
TWA2	10	12	15	16	19	21	23	26
TWA3	9	11	14	16	20	22	25	29
TWA4	7	10	14	16	20	22	26	30
TWB3	8	10	13	16	19	21	24	29
TWD4	11	16	22	27	33	38	44	52
TWE2	11	16	22	27	33	37	43	51
TWF3	6	8	11	13	17	19	21	25
TWJ2	9	14	20	24	31	35	41	49
DGA1	5	6	7	8	8	8	8	7
DGA2	8	10	13	15	19	21	24	27
DGA3	11	15	20	23	29	33	38	45
DGB1	4	6	7	8	9	9	9	9
DGB2	8	10	14	16	20	22	25	29
DGB3	11	15	21	25	30	35	40	48
DGC1	6	8	10	12	14	15	16	17
DGC2	10	13	17	20	24	27	31	37
DGC3	13	18	24	28	35	39	46	55
SSA1	13	16	19	21	24	26	28	31
SSB1	11	13	15	17	20	22	24	27
SSB2	14	18	23	27	33	36	41	48
SSB3	10	11	13	15	17	18	20	21
SSC1	7	9	12	14	17	18	21	24
SSC2	9	13	18	22	27	31	35	42
SSE1	10	13	16	18	20	22	24	26
SSE2	13	18	24	28	34	38	43	50
SSE3	11	13	15	17	19	20	22	24

FARGO, NORTH DAKOTA ***SSF (%)*** *9278 DD65*

Type	LCR = 100	70	50	40	30	25	20	15
WWA3	6	8	11	12	15	16	18	20
WWB4	8	13	20	24	31	36	42	50
WWC2	8	13	19	23	29	33	39	46
TWA1	11	12	13	13	14	14	15	15
TWA2	9	10	12	13	15	16	17	18
TWA3	7	9	11	13	15	16	18	20
TWA4	6	8	10	12	15	16	19	21
TWB3	7	8	11	12	15	16	18	21
TWD4	8	13	18	22	28	32	37	44
TWE2	9	13	18	22	27	31	36	43
TWF3	4	6	8	10	12	13	15	17
TWJ2	7	11	16	20	25	29	34	41
DGA1	3	4	4	4	4	3	2	0
DGA2	6	8	11	12	15	16	18	20
DGA3	10	13	17	20	25	28	33	39
DGB1	3	4	4	4	4	4	3	1
DGB2	6	8	11	13	15	17	19	22
DGB3	10	13	18	21	26	30	35	41
DGC1	5	6	7	8	9	9	9	8
DGC2	8	11	14	16	20	22	25	29
DGC3	11	15	20	24	30	34	39	47
SSA1	12	14	16	17	19	21	22	24
SSB1	9	11	13	14	16	17	18	20
SSB2	12	16	20	23	28	31	36	41
SSB3	8	10	11	12	13	14	14	14
SSC1	5	7	8	10	11	12	13	14
SSC2	7	10	14	17	21	24	28	33
SSE1	8	10	12	13	14	15	16	15
SSE2	11	15	20	23	28	31	35	41
SSE3	10	11	13	14	15	16	16	17

CINCINNATI, OHIO ***SSF (%)*** *5079 DD65*

Type	LCR = 100	70	50	40	30	25	20	15
WWA3	10	13	17	21	25	28	32	38
WWB4	13	20	27	33	42	47	55	64
WWC2	13	19	27	32	40	45	52	62
TWA1	13	14	16	18	20	21	23	25
TWA2	11	14	17	19	22	25	28	32
TWA3	10	13	16	19	23	26	30	35
TWA4	9	12	16	19	23	26	31	36
TWB3	9	12	15	18	22	25	29	34
TWD4	12	18	25	30	37	42	49	58
TWE2	13	19	25	30	37	42	48	57
TWF3	7	10	13	16	20	23	26	31
TWJ2	11	16	23	28	35	40	46	55
DGA1	6	7	9	10	11	12	12	12
DGA2	9	12	15	18	21	24	27	32
DGA3	12	16	22	26	32	36	42	50
DGB1	6	7	9	10	12	13	14	14
DGB2	9	12	15	18	22	25	29	34
DGB3	12	17	23	27	33	38	44	53
DGC1	8	10	13	15	17	18	20	23
DGC2	11	15	19	22	27	31	36	42
DGC3	14	20	26	31	38	43	50	60
SSA1	15	19	23	25	29	32	36	41
SSB1	13	16	19	21	25	27	31	35
SSB2	16	22	27	32	38	42	48	56
SSB3	12	14	17	19	22	24	27	30
SSC1	8	11	15	17	21	23	26	31
SSC2	11	16	21	25	31	35	41	49
SSE1	13	16	20	23	27	30	33	38
SSE2	16	22	29	34	41	46	52	60
SSE3	13	16	19	21	24	26	29	33

TABLE C.3 **Annual Passive Heating Performance: SSF (Continued)**

CLEVELAND, OHIO **SSF (%)** *6160 DD65*

Type	*LCR = 100*	*70*	*50*	*40*	*30*	*25*	*20*	*15*
WWA3	6	9	11	13	15	16	18	20
WWB4	8	13	20	24	31	36	42	50
WWC2	8	13	19	23	29	34	39	47
TWA1	11	12	13	14	14	15	15	15
TWA2	9	10	12	13	15	16	17	19
TWA3	7	9	11	13	15	16	18	20
TWA4	6	8	10	12	15	17	19	22
TWB3	7	9	11	13	15	16	19	21
TWD4	9	13	18	22	28	32	37	45
TWE2	9	14	18	22	28	31	36	43
TWF3	5	6	8	10	12	13	15	17
TWJ2	7	11	16	20	25	29	34	41
DGA1	3	4	4	4	4	4	2	0
DGA2	7	8	11	12	15	16	18	21
DGA3	10	13	17	20	25	28	33	39
DGB1	3	4	4	5	4	4	3	1
DGB2	7	9	11	13	15	17	19	22
DGB3	10	14	18	21	26	30	35	42
DGC1	5	6	7	8	9	9	9	8
DGC2	8	11	14	16	20	22	25	29
DGC3	12	16	21	24	30	34	39	47
SSA1	13	15	18	19	22	24	26	28
SSB1	10	12	14	16	18	19	21	23
SSB2	13	17	22	25	30	34	38	44
SSB3	9	11	13	14	15	16	17	18
SSC1	5	7	9	10	11	12	13	14
SSC2	7	11	14	17	21	24	28	33
SSE1	9	12	14	15	17	18	19	20
SSE2	12	17	22	25	31	34	39	45
SSE3	11	13	14	16	17	18	19	20

COLUMBUS, OHIO **SSF (%)** *5705 DD65*

Type	*LCR = 100*	*70*	*50*	*40*	*30*	*25*	*20*	*15*
WWA3	8	11	14	16	20	22	25	29
WWB4	10	16	23	29	36	41	48	57
WWC2	10	16	22	27	34	39	46	54
TWA1	12	13	14	15	17	17	18	20
TWA2	10	12	14	16	18	20	22	25
TWA3	8	11	13	16	19	21	24	27
TWA4	7	10	13	15	19	21	24	29
TWB3	8	10	13	15	18	20	23	27
TWD4	10	15	21	26	32	37	43	51
TWE2	11	16	21	26	32	36	42	50
TWF3	6	8	11	13	16	18	20	24
TWJ2	9	13	19	23	30	34	40	48
DGA1	4	5	6	7	7	7	7	5
DGA2	7	10	12	15	18	20	22	26
DGA3	11	14	19	23	28	32	37	44
DGB1	4	5	6	7	8	8	8	7
DGB2	7	10	13	15	18	21	24	28
DGB3	11	15	20	24	29	33	39	47
DGC1	6	8	10	11	12	13	14	15
DGC2	9	12	16	19	23	26	30	35
DGC3	13	17	23	27	33	38	44	53
SSA1	14	17	20	22	25	27	30	34
SSB1	11	13	16	18	21	23	25	29
SSB2	15	19	24	28	34	38	43	50
SSB3	10	12	14	16	18	20	22	24
SSC1	7	9	11	13	16	17	19	22
SSC2	9	13	17	21	26	30	34	41
SSE1	11	14	17	19	22	24	26	29
SSE2	14	19	25	29	36	40	45	53
SSE3	12	14	16	18	20	22	24	26

OKLAHOMA CITY, OKLAHOMA **SSF (%)** *3699 DD65*

Type	*LCR = 200*	*100*	*70*	*50*	*40*	*30*	*25*	*20*
WWA3	12	22	29	37	44	52	58	65
WWB4	12	28	38	50	58	68	74	81
WWC2	13	27	38	49	56	66	73	80
TWA1	14	20	23	28	32	37	41	46
TWA2	13	20	26	32	37	44	49	55
TWA3	12	20	26	33	39	47	52	59
TWA4	10	19	26	33	39	47	53	60
TWB3	10	18	24	31	36	44	49	56
TWD4	12	25	34	45	52	62	68	76
TWE2	13	26	35	45	53	62	68	76
TWF3	9	17	23	30	35	43	48	55
TWJ2	11	23	32	42	49	59	66	73
DGA1	8	15	20	25	29	34	37	41
DGA2	9	17	23	30	35	43	47	53
DGA3	11	21	29	38	45	54	60	67
DGB1	8	15	20	26	31	37	42	47
DGB2	10	18	24	31	37	46	52	59
DGB3	12	22	29	39	46	57	64	72
DGC1	10	18	24	32	37	46	51	58
DGC2	11	21	28	36	43	53	60	68
DGC3	13	25	33	44	52	64	71	79
SSA1	16	25	32	39	44	51	57	63
SSB1	14	21	27	33	38	45	50	56
SSB2	16	27	35	44	51	60	65	72
SSB3	13	20	25	31	36	42	47	53
SSC1	11	19	26	33	38	46	52	59
SSC2	12	23	32	41	48	57	63	71
SSE1	15	25	32	40	46	54	59	66
SSE2	17	31	40	51	58	67	73	80
SSE3	15	22	27	34	38	45	50	56

MEDFORD, OREGON **SSF (%)** *4935 DD65*

Type	*LCR = 100*	*70*	*50*	*40*	*30*	*25*	*20*	*15*
WWA3	16	21	26	31	36	40	45	51
WWB4	20	28	37	43	52	58	64	72
WWC2	20	28	36	42	50	56	62	70
TWA1	17	19	22	24	28	30	33	37
TWA2	16	20	24	27	32	35	39	44
TWA3	15	19	24	28	33	37	41	47
TWA4	14	19	24	28	34	37	42	49
TWB3	14	18	23	26	31	35	40	46
TWD4	19	26	34	39	47	52	59	67
TWE2	20	26	34	40	47	52	59	66
TWF3	12	16	21	25	30	33	38	44
TWJ2	17	24	31	37	45	50	56	64
DGA1	11	14	17	19	21	23	24	25
DGA2	13	17	22	26	31	34	38	43
DGA3	17	23	29	34	41	46	52	59
DGB1	11	14	18	20	23	25	28	30
DGB2	14	18	23	27	33	37	41	47
DGB3	17	23	30	36	44	49	55	63
DGC1	13	17	22	25	30	32	36	39
DGC2	16	21	27	32	39	43	48	55
DGC3	20	26	34	40	49	54	61	69
SSA1	21	26	30	34	39	42	46	51
SSB1	18	22	26	29	34	37	41	46
SSB2	22	28	35	40	47	51	57	64
SSB3	17	20	24	27	31	34	37	42
SSC1	14	18	23	27	32	35	39	44
SSC2	17	23	30	35	42	46	52	59
SSE1	19	24	29	33	38	41	45	50
SSE2	24	31	38	44	51	55	61	68
SSE3	18	22	26	29	33	36	40	44

TABLE C.3 **Annual Passive Heating Performance: SSF (Continued)**

SALEM, OREGON ***SSF (%)*** *4854 DD65*

Type	*LCR = 100*	*70*	*50*	*40*	*30*	*25*	*20*	*15*
WWA3	15	20	25	29	35	38	43	48
WWB4	19	27	36	42	50	56	63	71
WWC2	19	27	35	41	49	54	60	68
TWA1	16	19	22	24	27	29	31	35
TWA2	16	19	23	26	30	33	37	42
TWA3	15	19	23	27	32	35	40	45
TWA4	14	18	23	27	32	36	40	47
TWB3	14	18	22	25	30	34	38	44
TWD4	18	25	32	38	46	51	57	65
TWE2	19	26	33	38	46	51	57	65
TWF3	12	16	20	24	28	32	36	42
TWJ2	16	23	30	36	43	48	55	63
DGA1	10	13	16	18	20	21	23	23
DGA2	13	17	22	25	30	33	37	41
DGA3	17	22	28	33	40	45	51	58
DGB1	11	13	17	19	22	24	26	28
DGB2	13	18	22	26	32	35	40	45
DGB3	17	23	29	35	42	47	54	62
DGC1	13	16	21	24	28	30	33	37
DGC2	16	21	26	31	37	41	46	53
DGC3	19	26	33	39	47	53	59	68
SSA1	21	25	29	33	37	40	44	49
SSB1	17	21	25	28	33	36	39	44
SSB2	22	28	34	39	46	50	55	62
SSB3	16	20	23	26	30	33	36	40
SSC1	14	18	22	25	30	33	37	42
SSC2	17	22	29	33	40	45	50	57
SSE1	19	23	28	32	36	39	43	47
SSE2	23	30	37	42	49	54	59	66
SSE3	18	22	25	28	32	35	38	42

HARRISBURG, PENNSYLVANIA ***SSF (%)*** *5226 DD65*

Type	*LCR = 100*	*70*	*50*	*40*	*30*	*25*	*20*	*15*
WWA3	11	14	19	22	28	31	36	42
WWB4	14	21	29	36	45	51	58	68
WWC2	14	20	28	34	43	48	56	65
TWA1	13	15	17	19	21	23	25	28
TWA2	12	14	18	20	24	27	30	35
TWA3	10	14	18	21	25	28	33	39
TWA4	9	13	17	20	25	29	34	40
TWB3	10	13	16	19	24	27	31	37
TWD4	13	19	26	32	40	45	52	61
TWE2	14	20	27	32	40	45	52	61
TWF3	8	11	14	17	22	25	29	35
TWJ2	11	17	24	29	37	42	49	58
DGA1	6	8	10	11	13	14	15	15
DGA2	9	12	16	19	23	26	30	35
DGA3	12	17	23	27	34	38	45	53
DGB1	6	8	10	12	14	15	16	18
DGB2	9	12	16	19	24	27	32	38
DGB3	13	18	23	28	35	40	47	57
DGC1	8	11	14	16	19	21	24	27
DGC2	11	15	20	24	29	33	39	46
DGC3	15	20	27	32	40	46	53	64
SSA1	16	19	23	26	31	34	38	44
SSB1	13	16	19	22	26	29	32	38
SSB2	17	22	28	33	40	44	50	58
SSB3	12	14	17	20	23	25	28	33
SSC1	9	12	16	19	23	26	30	35
SSC2	12	17	22	27	34	38	44	52
SSE1	13	17	21	24	29	32	36	42
SSE2	17	23	30	36	43	48	55	63
SSE3	13	16	19	22	25	28	31	36

PHILADELPHIA, PENNSYLVANIA ***SSF (%)*** *4875 DD65*

Type	*LCR = 100*	*70*	*50*	*40*	*30*	*25*	*20*	*15*
WWA3	12	17	22	26	32	36	41	49
WWB4	16	24	33	40	49	55	63	72
WWC2	16	23	32	38	47	53	61	70
TWA1	14	16	19	21	24	26	29	33
TWA2	13	16	20	23	27	31	35	41
TWA3	12	16	20	24	29	33	38	44
TWA4	11	15	20	24	29	33	39	46
TWB3	11	15	19	22	27	31	36	42
TWD4	15	22	30	35	44	49	57	66
TWE2	16	22	30	36	44	49	56	66
TWF3	9	13	17	20	25	29	34	40
TWJ2	13	20	27	33	41	47	54	63
DGA1	8	10	12	14	17	18	19	20
DGA2	10	14	18	22	26	30	34	39
DGA3	14	19	25	30	37	42	49	58
DGB1	7	10	13	15	17	19	22	24
DGB2	11	14	19	22	28	31	37	43
DGB3	14	20	26	31	39	44	52	62
DGC1	10	13	17	19	23	26	30	34
DGC2	13	17	23	27	33	38	44	52
DGC3	16	22	29	35	44	50	58	69
SSA1	17	21	26	29	34	38	43	49
SSB1	14	18	22	25	29	33	37	43
SSB2	19	24	31	36	43	48	54	63
SSB3	13	16	20	22	26	29	33	38
SSC1	11	14	19	22	27	30	35	42
SSC2	14	19	26	31	38	43	49	58
SSE1	15	20	25	28	34	37	42	48
SSE2	19	26	34	40	48	53	60	68
SSE3	15	18	22	25	29	32	36	41

PITTSBURGH, PENNSYLVANIA ***SSF (%)*** *5937 DD65*

Type	*LCR = 100*	*70*	*50*	*40*	*30*	*25*	*20*	*15*
WWA3	7	9	12	14	17	19	21	24
WWB4	9	15	21	26	33	38	45	53
WWC2	9	14	20	25	32	36	42	50
TWA1	12	12	13	14	15	16	16	17
TWA2	9	11	13	14	16	18	19	22
TWA3	8	10	12	14	17	18	21	23
TWA4	6	9	11	14	17	19	21	25
TWB3	7	9	12	14	16	18	21	24
TWD4	9	14	20	24	30	34	40	48
TWE2	10	14	20	24	30	34	39	46
TWF3	5	7	9	11	14	15	17	20
TWJ2	8	12	17	21	27	31	37	44
DGA1	4	5	5	5	5	5	4	2
DGA2	7	9	11	13	16	18	20	23
DGA3	10	14	18	21	26	30	35	41
DGB1	4	4	5	6	6	6	5	3
DGB2	7	9	12	14	17	19	21	25
DGB3	10	14	19	22	28	32	37	44
DGC1	5	7	8	9	10	11	11	11
DGC2	9	11	15	17	21	24	27	32
DGC3	12	16	21	26	32	36	42	50
SSA1	13	16	19	21	23	25	28	31
SSB1	11	13	15	17	19	21	23	26
SSB2	14	18	23	27	32	36	40	47
SSB3	10	11	13	15	17	18	19	21
SSC1	6	8	10	11	13	15	16	18
SSC2	8	12	16	19	23	27	31	37
SSE1	10	12	15	17	19	21	23	24
SSE2	13	18	23	27	33	37	42	48
SSE3	11	13	15	17	19	20	21	23

TABLE C.3 **Annual Passive Heating Performance: SSF (Continued)**

PROVIDENCE, RHODE ISLAND ***SSF (%)*** *5974 DD65*

Type	*LCR = 100*	*70*	*50*	*40*	*30*	*25*	*20*	*15*
WWA3	10	13	18	21	26	29	33	40
WWB4	13	20	28	34	43	49	56	66
WWC2	13	19	27	33	41	47	54	63
TWA1	13	14	16	18	20	21	23	26
TWA2	11	14	17	19	22	25	28	33
TWA3	10	13	16	19	24	27	31	36
TWA4	9	12	16	19	24	27	31	38
TWB3	9	12	15	18	22	25	29	35
TWD4	12	18	25	30	38	43	50	59
TWE2	13	19	25	30	38	43	50	59
TWF3	7	10	13	16	20	23	27	32
TWJ2	11	16	23	28	35	40	47	56
DGA1	6	7	9	10	11	12	13	13
DGA2	8	11	15	18	22	24	28	33
DGA3	12	16	21	26	32	37	43	51
DGB1	5	7	9	10	12	13	14	15
DGB2	9	12	15	18	22	26	30	36
DGB3	12	17	22	27	34	39	45	55
DGC1	7	10	12	14	17	19	21	25
DGC2	10	14	19	22	28	31	37	44
DGC3	14	19	26	31	38	44	52	62
SSA1	15	19	22	25	30	33	36	42
SSB1	12	15	19	21	25	28	31	36
SSB2	16	21	27	32	38	43	49	57
SSB3	11	14	17	19	22	24	27	31
SSC1	8	11	15	17	21	24	27	32
SSC2	11	16	21	25	32	36	42	50
SSE1	13	16	20	23	27	30	34	39
SSE2	16	22	29	34	42	47	53	61
SSE3	13	16	19	21	24	27	30	34

CHARLESTON, SOUTH CAROLINA ***SSF (%)*** *2150 DD65*

Type	*LCR = 200*	*100*	*70*	*50*	*40*	*30*	*25*	*20*
WWA3	18	31	40	50	57	66	72	79
WWB4	19	38	50	62	70	79	85	90
WWC2	20	38	49	61	69	78	84	89
TWA1	17	25	30	37	41	48	53	59
TWA2	17	27	34	42	48	57	62	69
TWA3	16	28	36	45	51	60	66	73
TWA4	15	27	36	45	52	61	67	74
TWB3	14	25	33	41	47	56	62	69
TWD4	18	35	45	56	64	74	79	86
TWE2	19	36	46	57	65	74	80	86
TWF3	13	24	32	41	47	56	62	69
TWJ2	17	32	43	54	62	72	77	84
DGA1	13	23	29	36	41	48	51	56
DGA2	13	24	32	41	47	55	60	67
DGA3	16	29	38	49	57	66	72	78
DGB1	13	23	31	39	45	54	59	65
DGB2	14	25	34	43	51	61	67	74
DGB3	16	29	39	51	60	71	77	83
DGC1	15	27	36	46	53	63	69	75
DGC2	16	29	38	49	58	68	75	81
DGC3	18	33	44	57	66	77	83	89
SSA1	22	33	41	50	56	65	70	76
SSB1	18	29	36	44	50	58	64	70
SSB2	22	36	46	56	63	72	77	83
SSB3	17	27	34	42	48	56	61	68
SSC1	16	28	36	45	51	60	66	73
SSC2	18	32	42	53	60	70	76	82
SSE1	21	34	43	53	60	69	74	80
SSE2	24	42	53	64	71	80	85	90
SSE3	19	29	36	44	50	58	64	70

RAPID CITY, SOUTH DAKOTA ***SSF (%)*** *7332 DD65*

Type	*LCR = 100*	*70*	*50*	*40*	*30*	*25*	*20*	*15*
WWA3	13	18	23	27	33	37	42	50
WWB4	17	25	34	40	50	56	63	73
WWC2	17	24	33	39	48	54	61	70
TWA1	15	17	19	21	24	27	30	34
TWA2	13	17	21	24	28	32	36	42
TWA3	13	17	21	25	30	34	38	45
TWA4	11	16	21	25	30	34	39	47
TWB3	11	15	20	23	28	32	37	43
TWD4	16	23	30	36	45	50	57	66
TWE2	17	23	31	37	45	50	57	66
TWF3	10	13	18	21	26	30	35	41
TWJ2	14	21	28	34	42	47	54	64
DGA1	8	11	13	15	18	19	20	21
DGA2	11	15	19	22	27	31	35	40
DGA3	14	20	26	31	38	43	50	58
DGB1	8	11	14	16	19	21	23	26
DGB2	11	15	20	23	29	33	38	44
DGB3	15	20	27	32	40	45	53	62
DGC1	10	14	18	21	25	27	31	35
DGC2	13	18	24	28	34	39	45	53
DGC3	17	23	31	36	45	51	59	69
SSA1	18	22	26	30	34	38	42	48
SSB1	15	19	22	25	30	33	37	42
SSB2	19	25	31	36	43	48	54	62
SSB3	13	16	20	23	26	29	33	37
SSC1	11	15	20	23	28	31	36	42
SSC2	14	20	27	32	39	44	50	59
SSE1	16	20	25	28	33	37	41	47
SSE2	20	27	34	40	47	52	59	67
SSE3	15	18	22	25	29	32	35	40

KNOXVILLE, TENNESSEE ***SSF (%)*** *3486 DD65*

Type	*LCR = 200*	*100*	*70*	*50*	*40*	*30*	*25*	*20*
WWA3	10	18	24	31	36	44	49	55
WWB4	9	23	32	43	50	60	66	74
WWC2	10	23	32	42	49	58	65	72
TWA1	13	17	21	24	27	32	35	39
TWA2	11	17	22	27	31	37	41	47
TWA3	10	17	22	28	33	39	44	50
TWA4	8	16	21	28	33	40	45	51
TWB3	9	15	20	26	30	37	41	47
TWD4	10	21	29	38	45	54	60	68
TWE2	11	22	30	39	46	55	60	68
TWF3	7	14	19	25	29	36	40	46
TWJ2	8	19	27	36	43	52	58	65
DGA1	7	12	16	20	23	27	29	32
DGA2	8	15	19	25	30	36	40	45
DGA3	10	18	25	32	38	47	53	60
DGB1	7	12	16	21	24	29	32	36
DGB2	8	15	20	26	31	38	43	49
DGB3	10	19	26	33	40	49	55	64
DGC1	8	15	20	25	29	36	40	45
DGC2	9	18	24	31	36	44	50	57
DGC3	12	22	29	38	45	55	62	71
SSA1	15	22	28	34	38	45	49	55
SSB1	12	19	24	29	33	39	43	49
SSB2	14	24	31	39	45	53	59	65
SSB3	12	18	22	27	31	36	40	45
SSC1	9	16	21	27	32	38	43	49
SSC2	9	19	27	35	41	49	55	62
SSE1	13	21	27	34	39	48	51	57
SSE2	14	27	35	44	51	60	65	72
SSE3	13	19	24	29	33	39	43	48

TABLE C.3 **Annual Passive Heating Performance: SSF (Continued)**

MEMPHIS, TENNESSEE **SSF (%)**						*3237 DD65*		
Type	*LCR = 200*	*100*	*70*	*50*	*40*	*30*	*25*	*20*
WWA3	11	21	27	35	41	49	54	61
WWB4	11	26	36	47	54	65	71	78
WWC2	12	26	35	46	53	63	69	77
TWA1	14	19	22	27	30	35	38	43
TWA2	12	19	24	30	35	41	46	52
TWA3	11	19	25	31	36	44	49	55
TWA4	9	18	24	31	37	44	50	56
TWB3	10	17	23	29	34	41	46	52
TWD4	11	24	32	42	49	59	65	72
TWE2	12	25	33	43	50	59	65	72
TWF3	8	16	21	28	33	40	45	51
TWJ2	10	22	30	40	47	56	62	70
DGA1	8	14	18	23	26	31	34	37
DGA2	9	16	22	28	33	40	44	50
DGA3	11	20	27	36	42	51	57	64
DGB1	8	14	19	24	28	34	38	43
DGB2	9	17	22	29	35	43	48	55
DGB3	11	21	28	37	43	54	60	69
DGC1	10	17	23	29	34	41	46	52
DGC2	11	20	26	34	40	49	56	63
DGC3	13	24	31	41	49	60	67	76
SSA1	16	24	30	37	42	49	54	60
SSB1	13	21	26	32	36	43	47	53
SSB2	15	26	34	42	49	57	63	70
SSB3	12	19	24	30	34	40	44	50
SSC1	10	18	24	31	36	43	48	54
SSC2	11	22	30	38	45	54	59	67
SSE1	14	24	30	38	43	51	56	62
SSE2	16	29	38	48	55	64	70	77
SSE3	14	21	26	32	36	43	47	53

AUSTIN, TEXAS **SSF (%)**						*1737 DD65*		
Type	*LCR = 200*	*100*	*70*	*50*	*40*	*30*	*25*	*20*
WWA3	21	37	47	58	65	75	80	86
WWB4	24	45	58	70	77	86	90	94
WWC2	24	44	57	69	76	85	89	94
TWA1	20	29	35	42	48	56	61	67
TWA2	20	32	40	49	56	65	70	77
TWA3	19	33	42	52	59	68	74	81
TWA4	18	32	42	52	60	69	75	82
TWB3	17	30	38	48	55	64	70	77
TWD4	22	41	52	64	72	81	86	91
TWE2	24	42	53	65	73	81	86	91
TWF3	16	29	38	48	55	64	70	77
TWJ2	21	39	50	62	70	79	84	90
DGA1	16	28	36	44	49	57	61	66
DGA2	16	29	38	48	55	64	69	75
DGA3	19	34	45	57	65	74	79	84
DGB1	16	29	38	48	56	64	69	75
DGB2	17	30	41	52	60	70	76	82
DGB3	19	35	46	60	68	79	84	89
DGC1	19	33	44	56	64	74	79	84
DGC2	19	34	46	59	68	77	83	88
DGC3	22	39	52	66	75	85	89	93
SSA1	25	39	48	58	65	73	78	84
SSB1	21	34	42	51	58	67	72	79
SSB2	26	42	53	63	71	79	84	89
SSB3	20	32	40	49	56	65	70	77
SSC1	19	33	42	53	60	69	75	81
SSC2	22	39	50	61	68	78	83	89
SSE1	25	41	51	62	69	77	82	88
SSE2	29	49	61	72	79	87	91	94
SSE3	22	34	43	52	58	67	72	79

NASHVILLE, TENNESSEE **SSF (%)**						*3704 DD65*		
Type	*LCR = 200*	*100*	*70*	*50*	*40*	*30*	*25*	*20*
WWA3	8	15	20	26	31	37	42	47
WWB4	7	19	28	37	44	54	60	68
WWC2	8	19	27	36	43	52	58	66
TWA1	12	16	18	21	24	27	30	33
TWA2	10	15	19	23	27	32	35	40
TWA3	8	14	19	24	28	34	38	43
TWA4	7	13	18	24	28	34	38	44
TWB3	7	13	17	22	26	32	35	41
TWD4	8	18	25	34	40	48	54	61
TWE2	9	19	26	34	40	48	54	61
TWF3	6	11	16	21	24	30	34	39
TWJ2	6	16	23	31	37	46	51	59
DGA1	6	10	13	16	18	21	23	25
DGA2	7	12	17	21	25	31	34	39
DGA3	9	16	22	28	34	41	47	54
DGB1	6	10	13	16	19	22	25	28
DGB2	7	13	17	22	26	32	36	42
DGB3	9	17	22	29	35	43	49	57
DGC1	7	12	16	20	24	29	32	36
DGC2	8	15	20	26	31	38	43	49
DGC3	10	19	25	33	39	49	55	64
SSA1	13	20	25	30	34	39	43	49
SSB1	11	17	21	25	29	34	38	42
SSB2	12	21	28	35	40	48	53	60
SSB3	10	15	19	23	27	31	35	39
SSC1	7	13	18	23	26	32	36	41
SSC2	8	16	23	30	35	43	48	55
SSE1	11	18	23	29	33	39	44	49
SSE2	12	23	31	39	45	53	59	66
SSE3	12	17	21	25	29	34	37	41

BROWNSVILLE, TEXAS **SSF (%)**						*650 DD65*		
Type	*LCR = 200*	*100*	*70*	*50*	*40*	*30*	*25*	*20*
WWA3	39	60	72	82	88	94	96	98
WWB4	45	70	82	90	94	98	99	100
WWC2	45	69	81	90	94	97	99	99
TWA1	30	45	54	64	71	79	84	89
TWA2	33	51	63	73	80	87	91	95
TWA3	34	54	66	76	83	90	93	96
TWA4	34	54	66	77	84	90	94	97
TWB3	31	50	61	72	79	87	91	95
TWD4	41	64	76	86	91	96	98	99
TWE2	42	65	77	87	92	96	98	99
TWF3	30	50	62	73	80	88	91	95
TWJ2	39	62	75	85	90	95	97	99
DGA1	31	50	61	70	76	83	87	90
DGA2	31	51	63	73	80	87	90	94
DGA3	34	57	70	80	86	91	93	95
DGB1	32	54	67	78	84	89	92	94
DGB2	32	55	69	80	86	92	95	97
DGB3	34	60	74	85	90	94	96	97
DGC1	36	62	76	85	90	94	95	97
DGC2	35	61	75	86	91	95	97	99
DGC3	39	66	80	90	94	97	98	98
SSA1	42	62	73	82	87	93	96	98
SSB1	36	55	66	76	83	89	93	96
SSB2	44	65	77	86	91	95	97	99
SSB3	35	54	65	75	81	88	92	95
SSC1	35	55	67	78	84	91	94	97
SSC2	39	62	74	84	90	95	97	99
SSE1	45	66	77	86	91	95	97	99
SSE2	51	74	85	92	96	98	99	100
SSE3	37	56	67	77	83	90	93	96

TABLE C.3 **Annual Passive Heating Performance: SSF (Continued)**

DALLAS, TEXAS ***SSF (%)*** *2297 DD65*

Type	*LCR = 200*	*100*	*70*	*50*	*40*	*30*	*25*	*20*
WWA3	17	30	39	49	56	66	72	79
WWB4	19	37	49	62	70	79	85	90
WWC2	19	37	49	61	69	78	84	89
TWA1	17	24	30	36	41	48	52	59
TWA2	16	26	33	42	48	56	62	69
TWA3	16	27	35	44	50	60	65	73
TWA4	14	26	35	44	51	60	66	74
TWB3	14	24	32	40	47	56	62	69
TWD4	18	34	45	56	64	74	79	86
TWE2	19	35	46	57	65	74	80	86
TWF3	13	23	31	40	46	56	62	69
TWJ2	16	32	42	54	61	71	77	84
DGA1	12	22	29	36	41	47	51	56
DGA2	13	24	31	40	47	55	60	66
DGA3	15	28	38	49	56	66	72	78
DGB1	12	22	30	38	45	54	59	64
DGB2	13	24	33	43	51	61	67	74
DGB3	16	28	38	50	59	71	77	83
DGC1	15	26	35	45	53	63	69	75
DGC2	15	28	38	49	57	68	74	81
DGC3	18	32	43	57	66	77	83	89
SSA1	21	32	40	49	56	64	70	76
SSB1	17	28	35	43	49	58	63	70
SSB2	21	35	45	55	62	71	77	83
SSB3	17	26	33	41	47	55	60	67
SSC1	15	27	35	44	51	60	66	73
SSC2	17	32	42	52	60	70	76	82
SSE1	20	33	42	52	59	68	74	80
SSE2	23	41	52	63	70	79	84	90
SSE3	18	28	35	43	49	58	63	70

HOUSTON, TEXAS ***SSF (%)*** *1434 DD65*

Type	*LCR = 200*	*100*	*70*	*50*	*40*	*30*	*25*	*20*
WWA3	21	37	47	58	65	74	80	86
WWB4	24	45	58	70	77	86	90	94
WWC2	24	44	57	69	76	85	89	93
TWA1	20	29	35	42	48	55	61	67
TWA2	20	32	40	49	56	65	70	77
TWA3	20	33	42	52	59	68	74	80
TWA4	18	32	42	52	59	69	75	81
TWB3	18	30	38	48	55	64	70	77
TWD4	22	41	52	64	72	81	86	91
TWE2	24	42	53	65	72	81	86	91
TWF3	16	29	38	48	55	64	70	77
TWJ2	21	38	50	62	69	79	84	89
DGA1	16	28	36	44	49	56	61	65
DGA2	16	29	38	48	55	63	69	74
DGA3	19	34	45	57	64	74	79	84
DGB1	16	28	38	48	55	64	69	74
DGB2	17	30	40	52	60	70	75	81
DGB3	19	35	46	59	68	78	83	89
DGC1	18	33	44	56	64	73	78	83
DGC2	19	34	46	59	67	77	82	88
DGC3	21	39	52	66	75	84	89	93
SSA1	25	39	48	58	65	73	78	84
SSB1	21	34	42	52	58	67	72	79
SSB2	26	42	53	64	71	79	84	89
SSB3	20	32	41	50	56	65	70	77
SSC1	19	33	42	52	60	69	75	81
SSC2	22	38	49	61	68	77	83	88
SSE1	25	41	51	62	69	77	82	88
SSE2	29	49	61	72	79	87	90	94
SSE3	22	34	43	52	59	67	73	79

EL PASO, TEXAS ***SSF (%)*** *2685 DD65*

Type	*LCR = 200*	*100*	*70*	*50*	*40*	*30*	*25*	*20*
WWA3	22	38	49	60	68	77	83	88
WWB4	25	47	60	72	80	88	92	95
WWC2	25	46	59	71	79	87	91	95
TWA1	20	29	36	44	49	58	63	70
TWA2	20	33	41	51	58	67	73	80
TWA3	20	34	44	54	61	71	77	83
TWA4	19	34	44	54	62	72	77	84
TWB3	18	31	40	50	57	67	73	79
TWD4	23	42	54	66	74	83	88	93
TWE2	24	43	55	67	75	84	88	93
TWF3	16	30	40	50	57	67	73	80
TWJ2	21	40	52	64	72	81	86	91
DGA1	16	29	37	46	52	59	64	69
DGA2	17	30	40	50	57	66	71	77
DGA3	19	35	47	59	67	76	81	86
DGB1	16	30	40	51	58	67	72	78
DGB2	17	31	42	55	63	73	79	84
DGB3	19	36	48	62	71	81	86	91
DGC1	19	35	47	59	68	77	82	86
DGC2	20	36	48	62	70	80	85	90
DGC3	22	40	55	69	78	87	91	94
SSA1	25	39	49	58	65	74	79	85
SSB1	21	34	43	52	59	68	73	80
SSB2	26	42	53	64	71	80	85	90
SSB3	20	32	41	50	56	65	71	77
SSC1	20	34	44	55	62	72	77	84
SSC2	22	40	51	63	71	80	85	90
SSE1	25	41	52	62	69	78	83	89
SSE2	29	49	61	73	80	87	91	95
SSE3	22	35	43	52	59	68	73	80

LUBBOCK, TEXAS ***SSF (%)*** *3551 DD65*

Type	*LCR = 200*	*100*	*70*	*50*	*40*	*30*	*25*	*20*
WWA3	17	30	40	50	57	67	73	80
WWB4	18	38	50	63	71	80	86	91
WWC2	19	37	49	61	69	79	85	90
TWA1	17	24	30	36	41	48	53	60
TWA2	16	26	34	42	48	57	63	70
TWA3	15	27	35	44	51	60	67	74
TWA4	14	26	35	45	52	61	67	75
TWB3	14	24	32	41	47	56	62	70
TWD4	17	34	45	57	65	75	80	87
TWE2	19	35	46	58	65	75	81	87
TWF3	12	23	31	40	47	56	63	70
TWJ2	16	32	43	54	62	72	78	85
DGA1	12	22	29	36	41	48	52	57
DGA2	13	23	32	41	47	56	61	67
DGA3	15	28	38	49	57	67	73	79
DGB1	12	22	30	39	46	55	60	66
DGB2	13	24	33	43	51	62	68	75
DGB3	15	28	39	51	60	72	78	84
DGC1	14	26	35	46	54	65	70	76
DGC2	15	28	38	50	59	70	76	82
DGC3	17	32	44	57	67	79	84	90
SSA1	21	33	41	50	56	65	70	77
SSB1	17	28	35	43	50	58	64	71
SSB2	21	35	45	55	63	72	77	84
SSB3	16	26	33	41	47	55	61	68
SSC1	15	27	35	45	52	61	67	74
SSC2	17	32	42	53	61	71	77	83
SSE1	20	34	43	53	60	69	74	81
SSE2	23	41	52	63	71	80	85	90
SSE3	18	28	36	44	50	58	64	70

TABLE C.3 **Annual Passive Heating Performance: SSF (Continued)**

SAN ANTONIO, TEXAS **SSF (%)**							*1570 DD65*	
Type	*LCR = 200*	*100*	*70*	*50*	*40*	*30*	*25*	*20*
WWA3	23	40	50	61	69	78	83	88
WWB4	26	48	61	73	80	88	92	95
WWC2	26	47	60	72	79	87	91	95
TWA1	21	30	37	45	50	59	64	71
TWA2	21	34	42	52	59	68	74	80
TWA3	21	35	45	55	62	71	77	83
TWA4	20	35	45	55	63	72	78	84
TWB3	19	32	41	51	58	67	73	80
TWD4	24	43	55	67	75	83	88	93
TWE2	26	44	56	68	75	84	88	93
TWF3	18	31	41	51	58	68	74	80
TWJ2	22	41	53	65	73	82	87	91
DGA1	17	30	38	47	53	60	64	69
DGA2	18	32	41	51	58	67	72	78
DGA3	20	36	48	60	68	77	81	86
DGB1	17	31	41	52	59	68	73	78
DGB2	18	33	44	56	64	74	79	85
DGB3	20	37	50	63	72	81	86	91
DGC1	20	36	48	60	68	77	82	86
DGC2	21	37	49	63	71	80	85	90
DGC3	23	42	56	70	78	87	91	94
SSA1	27	41	51	61	68	76	81	87
SSB1	23	36	45	54	61	70	75	82
SSB2	27	45	55	66	73	82	86	91
SSB3	22	34	43	52	59	68	73	80
SSC1	21	35	45	56	63	72	78	84
SSC2	23	41	53	64	71	80	85	91
SSE1	27	44	54	65	72	80	85	90
SSE2	31	52	64	75	82	89	92	96
SSE3	24	37	45	55	62	70	76	82

SALT LAKE CITY, UTAH **SSF (%)**							*5988 DD65*	
Type	*LCR = 100*	*70*	*50*	*40*	*30*	*25*	*20*	*15*
WWA3	19	25	32	37	44	49	55	64
WWB4	23	33	43	50	60	66	73	82
WWC2	23	32	42	49	59	65	72	80
TWA1	18	21	25	28	32	35	39	45
TWA2	18	22	28	32	38	42	47	54
TWA3	17	22	29	33	40	44	50	58
TWA4	16	22	28	33	40	45	51	60
TWB3	16	21	26	31	37	42	48	55
TWD4	22	30	39	45	55	60	68	76
TWE2	23	31	39	46	55	61	68	76
TWF3	14	19	25	30	36	41	46	54
TWJ2	20	28	37	43	52	58	65	74
DGA1	12	16	20	23	27	30	32	35
DGA2	15	20	26	30	36	41	46	52
DGA3	19	25	33	39	47	53	60	68
DGB1	13	17	21	25	30	33	37	41
DGB2	15	21	27	32	39	44	50	58
DGB3	19	26	34	40	50	56	64	73
DGC1	15	20	26	31	37	41	46	52
DGC2	18	24	31	37	45	51	58	66
DGC3	22	29	38	46	56	62	70	79
SSA1	23	28	34	38	44	48	54	60
SSB1	19	24	29	33	39	43	48	55
SSB2	24	31	39	45	52	58	64	72
SSB3	18	22	27	31	36	39	44	50
SSC1	16	22	28	32	39	43	49	57
SSC2	20	27	35	41	49	55	62	70
SSE1	22	27	34	38	45	49	55	62
SSE2	27	35	44	50	58	63	70	78
SSE3	20	24	29	33	38	42	47	53

BRYCE CANYON, UTAH **SSF (%)**							*9136 DD65*	
Type	*LCR = 100*	*70*	*50*	*40*	*30*	*25*	*20*	*15*
WWA3	16	22	30	35	43	49	56	65
WWB4	21	31	42	49	60	67	75	84
WWC2	21	30	40	48	58	65	73	82
TWA1	16	19	22	25	30	33	38	44
TWA2	15	20	25	29	36	40	46	55
TWA3	15	20	26	31	38	43	50	59
TWA4	14	19	26	31	39	44	51	60
TWB3	13	18	24	29	35	40	47	55
TWD4	19	28	37	44	54	60	68	78
TWE2	20	28	38	45	54	60	68	78
TWF3	12	17	23	27	34	39	46	55
TWJ2	17	25	35	42	51	58	65	75
DGA1	10	14	18	21	25	28	30	34
DGA2	12	17	23	28	35	39	44	51
DGA3	16	23	31	37	46	52	60	69
DGB1	10	14	18	22	27	31	36	41
DGB2	13	18	24	29	37	42	49	58
DGB3	17	23	32	38	48	55	64	74
DGC1	13	17	23	28	35	39	46	54
DGC2	15	21	29	35	43	50	58	67
DGC3	19	27	36	43	55	62	72	81
SSA1	23	29	35	40	47	52	58	66
SSB1	18	23	29	34	40	45	51	59
SSB2	24	32	40	46	55	61	68	76
SSB3	17	22	27	32	38	42	47	55
SSC1	14	19	25	30	37	42	49	58
SSC2	18	25	33	40	49	55	62	72
SSE1	21	28	35	41	48	53	60	68
SSE2	27	36	46	53	62	68	75	83
SSE3	19	24	30	34	40	45	50	58

BURLINGTON, VERMONT **SSF (%)**							*7878 DD65*	
Type	*LCR = 100*	*70*	*50*	*40*	*30*	*25*	*20*	*15*
WWA3	4	6	8	9	10	11	12	13
WWB4	6	10	16	20	26	31	36	44
WWC2	6	10	15	19	25	29	34	41
TWA1	10	11	11	11	11	11	11	10
TWA2	7	8	9	10	11	12	12	13
TWA3	6	7	8	10	11	12	13	14
TWA4	4	6	8	9	11	12	14	15
TWB3	5	7	8	10	11	12	14	16
TWD4	7	10	15	19	24	27	32	39
TWE2	7	11	15	18	23	26	31	37
TWF3	3	4	6	7	8	9	10	11
TWJ2	5	9	13	16	21	25	29	35
DGA1	2	2	2	1	1	0	−1	−5
DGA2	5	7	8	10	11	13	14	16
DGA3	8	11	15	17	22	25	29	34
DGB1	2	2	2	1	1	0	−1	−4
DGB2	5	7	9	10	12	13	15	17
DGB3	9	12	15	18	23	26	31	37
DGC1	3	4	4	4	5	4	3	2
DGC2	7	9	11	13	16	18	20	23
DGC3	10	13	18	21	26	30	35	42
SSA1	11	13	14	16	17	18	20	21
SSB1	9	10	11	13	14	15	16	17
SSB2	11	15	19	22	26	29	33	38
SSB3	8	9	10	10	11	11	12	11
SSC1	4	5	6	6	7	7	7	7
SSC2	5	8	11	14	17	19	22	27
SSE1	7	8	10	11	12	12	12	11
SSE2	9	13	17	21	25	28	32	37
SSE3	9	10	11	12	13	13	14	14

TABLE C.3 **Annual Passive Heating Performance: SSF (Continued)**

NORFOLK, VIRGINIA ***SSF (%)*** *3493 DD65*

Type	*LCR = 200*	*100*	*70*	*50*	*40*	*30*	*25*	*20*
WWA3	12	22	29	37	42	51	56	63
WWB4	12	27	37	49	56	67	73	80
WWC2	12	27	37	47	55	65	71	78
TWA1	14	19	23	27	31	36	40	45
TWA2	12	20	25	31	36	43	48	54
TWA3	11	20	26	33	38	46	51	57
TWA4	10	19	25	33	38	46	52	59
TWB3	10	18	23	30	35	42	48	54
TWD4	12	25	34	44	51	61	67	74
TWE2	13	26	35	44	51	61	67	74
TWF3	8	16	22	29	34	42	47	53
TWJ2	10	23	32	41	48	58	64	72
DGA1	8	15	19	24	28	33	36	39
DGA2	9	17	23	29	35	42	46	52
DGA3	11	21	28	37	44	53	59	66
DGB1	8	15	20	25	30	36	40	45
DGB2	9	17	23	31	36	45	50	58
DGB3	11	21	29	38	45	56	63	71
DGC1	10	18	24	31	36	44	49	56
DGC2	11	20	27	36	42	52	58	66
DGC3	13	24	33	43	51	62	70	78
SSA1	16	25	31	38	43	50	55	62
SSB1	13	21	26	33	38	44	49	55
SSB2	16	27	35	44	50	59	64	71
SSB3	13	20	25	31	35	42	46	52
SSC1	10	19	25	32	37	45	50	57
SSC2	11	23	31	40	46	56	62	69
SSE1	15	25	31	39	45	53	58	64
SSE2	16	30	40	50	57	66	72	78
SSE3	14	22	27	33	38	44	49	55

RICHMOND, VIRGINIA ***SSF (%)*** *3941 DD65*

Type	*LCR = 200*	*100*	*70*	*50*	*40*	*30*	*25*	*20*
WWA3	9	17	23	30	35	43	48	54
WWB4	8	22	31	42	49	59	65	73
WWC2	9	22	31	41	48	57	64	71
TWA1	13	17	20	24	26	31	34	38
TWA2	11	17	21	26	30	36	40	46
TWA3	9	16	21	27	32	38	43	49
TWA4	8	15	21	27	32	39	44	50
TWB3	8	15	19	25	29	36	40	46
TWD4	9	20	28	37	44	53	59	67
TWE2	10	21	29	38	45	54	60	67
TWF3	6	13	18	24	28	35	39	45
TWJ2	8	19	26	35	42	51	57	64
DGA1	6	11	15	19	22	26	28	30
DGA2	7	14	19	24	29	35	39	44
DGA3	9	18	24	31	37	46	52	59
DGB1	6	11	15	20	23	28	31	35
DGB2	8	14	19	25	30	37	42	48
DGB3	10	18	25	33	39	48	55	63
DGC1	8	14	19	24	29	35	39	44
DGC2	9	17	23	30	35	43	49	56
DGC3	11	21	28	37	44	54	61	70
SSA1	14	21	27	33	37	43	48	53
SSB1	12	18	23	28	32	38	42	47
SSB2	13	23	30	38	44	52	57	64
SSB3	11	17	21	26	29	35	39	44
SSC1	8	15	20	26	31	37	42	48
SSC2	9	19	26	34	40	48	54	61
SSE1	12	20	26	33	37	44	49	55
SSE2	13	25	34	43	49	58	64	71
SSE3	13	19	23	28	32	37	41	46

ROANOKE, VIRGINIA ***SSF (%)*** *4314 DD65*

Type	*LCR = 100*	*70*	*50*	*40*	*30*	*25*	*20*	*15*
WWA3	17	23	30	35	42	47	54	63
WWB4	22	31	41	49	59	65	73	82
WWC2	21	30	40	47	57	64	71	80
TWA1	17	20	23	26	30	33	37	43
TWA2	16	21	26	30	36	40	45	53
TWA3	16	21	27	31	38	43	49	57
TWA4	15	20	26	31	38	43	50	59
TWB3	14	19	25	29	35	40	46	54
TWD4	20	28	37	44	53	59	67	76
TWE2	21	29	38	44	53	59	67	76
TWF3	13	17	23	28	34	39	45	53
TWJ2	18	26	35	41	50	57	64	74
DGA1	11	14	18	21	25	27	30	33
DGA2	13	18	24	28	35	39	44	50
DGA3	17	23	31	37	46	52	59	67
DGB1	11	15	19	22	27	30	35	40
DGB2	14	19	25	30	37	42	48	56
DGB3	18	24	32	38	48	54	63	73
DGC1	14	18	24	28	34	38	44	51
DGC2	16	22	29	35	43	49	56	65
DGC3	20	28	36	43	54	61	70	79
SSA1	21	27	32	37	43	48	54	61
SSB1	18	22	28	32	38	42	47	55
SSB2	23	30	38	44	52	57	64	73
SSB3	17	21	26	29	35	39	44	50
SSC1	15	20	26	30	37	41	47	56
SSC2	18	25	33	39	48	53	61	70
SSE1	20	26	33	37	44	49	55	63
SSE2	25	34	43	49	58	64	71	79
SSE3	18	23	28	32	37	41	46	53

SEATTLE-TACOMA, WASHINGTON ***SSF (%)*** *5188 DD65*

Type	*LCR = 100*	*70*	*50*	*40*	*30*	*25*	*20*	*15*
WWA3	14	18	23	26	30	33	36	41
WWB4	17	25	32	38	46	51	57	64
WWC2	17	24	32	37	44	49	55	62
TWA1	16	18	20	22	25	26	28	31
TWA2	14	18	21	24	27	30	33	36
TWA3	14	17	21	24	29	31	35	39
TWA4	12	17	21	24	29	32	35	40
TWB3	13	16	20	23	27	30	34	38
TWD4	17	23	30	35	42	46	52	59
TWE2	17	24	30	35	42	46	51	58
TWF3	11	14	18	21	25	28	31	36
TWJ2	15	21	28	33	39	44	49	57
DGA1	9	12	14	16	17	18	18	18
DGA2	12	16	20	23	27	30	33	37
DGA3	16	21	27	31	37	41	46	53
DGB1	10	12	15	17	19	20	21	22
DGB2	13	16	21	24	29	32	35	40
DGB3	16	22	28	33	39	44	49	57
DGC1	12	15	19	21	24	26	28	30
DGC2	15	19	25	28	34	37	42	47
DGC3	19	24	31	36	44	49	54	62
SSA1	19	23	27	30	34	36	39	43
SSB1	16	20	23	26	30	32	35	38
SSB2	21	26	32	36	42	46	51	57
SSB3	15	18	22	24	27	29	31	34
SSC1	12	16	20	22	26	28	31	35
SSC2	15	20	26	30	36	40	45	51
SSE1	17	21	25	28	32	34	38	39
SSE2	21	27	34	38	44	48	53	58
SSE3	17	20	23	26	29	31	33	36

TABLE C.3 **Annual Passive Heating Performance: SSF (Continued)**

SPOKANE, WASHINGTON ***SSF (%)*** *6839 DD65*

Type	*LCR = 100*	*70*	*50*	*40*	*30*	*25*	*20*	*15*
WWA3	11	14	18	21	24	27	29	32
WWB4	14	20	28	33	40	45	51	58
WWC2	14	20	27	32	38	43	48	55
TWA1	14	15	17	19	21	22	23	25
TWA2	12	15	18	20	23	24	27	29
TWA3	11	14	17	20	23	25	28	31
TWA4	10	13	17	20	23	26	29	33
TWB3	10	13	17	19	22	25	28	32
TWD4	13	19	25	30	36	41	46	53
TWE2	14	20	26	30	36	40	45	52
TWF3	8	11	14	17	20	22	25	28
TWJ2	12	17	23	28	34	38	43	50
DGA1	7	9	10	11	12	12	12	11
DGA2	10	13	16	19	22	25	27	30
DGA3	13	18	23	27	33	36	41	47
DGB1	7	9	11	12	14	14	14	14
DGB2	10	13	17	20	24	26	29	33
DGB3	14	18	24	28	34	39	44	51
DGC1	9	12	14	16	19	20	21	21
DGC2	12	16	20	24	29	32	35	40
DGC3	16	21	27	32	39	43	49	56
SSA1	16	19	22	25	28	30	32	35
SSB1	13	16	19	21	24	26	28	30
SSB2	17	22	27	31	36	40	44	50
SSB3	12	15	17	19	21	22	24	25
SSC1	10	12	15	18	20	22	24	26
SSC2	12	17	22	25	30	34	38	43
SSE1	13	17	20	22	24	26	27	29
SSE2	17	22	28	32	37	41	45	50
SSE3	14	16	19	21	23	25	26	28

CHARLESTON, WEST VIRGINIA ***SSF (%)*** *4594 DD65*

Type	*LCR = 100*	*70*	*50*	*40*	*30*	*25*	*20*	*15*
WWA3	11	15	20	23	28	31	36	42
WWB4	14	21	30	36	45	50	58	67
WWC2	14	21	29	35	43	48	55	65
TWA1	14	15	18	19	21	23	25	28
TWA2	12	15	18	21	24	27	30	35
TWA3	11	14	18	21	26	29	33	38
TWA4	10	13	18	21	26	29	34	40
TWB3	10	13	17	20	24	27	31	37
TWD4	14	20	27	32	40	45	52	61
TWE2	15	20	27	32	40	45	51	60
TWF3	8	11	15	18	22	25	29	35
TWJ2	12	18	25	30	37	42	49	58
DGA1	7	9	11	12	14	14	15	15
DGA2	10	13	16	19	24	26	30	35
DGA3	13	18	23	28	34	39	45	53
DGB1	7	9	11	12	14	15	17	18
DGB2	10	13	17	20	25	28	32	38
DGB3	13	18	24	29	36	41	47	56
DGC1	9	12	15	17	20	22	24	27
DGC2	12	16	21	24	30	34	39	46
DGC3	15	21	28	33	40	46	53	63
SSA1	17	20	24	27	32	35	39	44
SSB1	14	17	20	23	27	30	33	38
SSB2	18	23	29	34	40	45	51	59
SSB3	13	15	19	21	24	27	30	34
SSC1	10	13	16	19	23	26	30	35
SSC2	12	17	23	27	34	38	44	52
SSE1	14	18	22	26	30	33	37	42
SSE2	18	24	31	37	44	49	55	63
SSE3	14	17	20	23	26	29	32	36

MADISON, WISCONSIN ***SSF (%)*** *7731 DD65*

Type	*LCR = 100*	*70*	*50*	*40*	*30*	*25*	*20*	*15*
WWA3	8	11	14	17	20	23	26	30
WWB4	10	16	24	29	37	42	49	58
WWC2	10	16	23	28	35	40	47	55
TWA1	12	13	14	15	17	18	19	20
TWA2	10	12	14	16	19	20	23	26
TWA3	8	11	14	16	19	21	24	28
TWA4	7	10	13	16	19	22	25	30
TWB3	8	10	13	15	18	21	24	28
TWD4	10	16	22	26	33	38	44	52
TWE2	11	16	22	26	33	37	43	51
TWF3	6	8	11	13	16	18	21	25
TWJ2	9	14	19	24	30	35	41	49
DGA1	4	5	6	7	8	8	7	6
DGA2	7	10	13	15	18	20	23	26
DGA3	11	15	19	23	28	32	38	45
DGB1	4	5	6	7	8	8	8	8
DGB2	8	10	13	15	19	21	24	29
DGB3	11	15	20	24	30	34	40	48
DGC1	6	8	10	11	13	14	15	16
DGC2	9	12	16	19	23	26	30	36
DGC3	13	17	23	27	34	39	45	55
SSA1	14	16	19	22	25	27	30	33
SSB1	11	13	16	18	21	22	25	28
SSB2	14	19	24	28	33	37	42	50
SSB3	10	12	14	16	18	19	21	23
SSC1	7	9	12	13	16	18	20	23
SSC2	9	13	18	21	27	30	35	42
SSE1	11	13	16	18	21	23	25	28
SSE2	14	19	25	29	35	39	45	52
SSE3	12	14	16	17	20	21	23	26

MILWAUKEE, WISCONSIN ***SSF (%)*** *7445 DD65*

Type	*LCR = 100*	*70*	*50*	*40*	*30*	*25*	*20*	*15*
WWA3	8	11	14	16	20	22	25	30
WWB4	10	16	23	29	37	42	49	58
WWC2	10	16	23	28	35	40	46	55
TWA1	12	13	15	15	17	18	19	20
TWA2	10	12	14	16	18	20	22	26
TWA3	9	11	14	16	19	21	24	28
TWA4	7	10	13	15	19	22	25	29
TWB3	8	10	13	15	18	21	24	28
TWD4	10	16	21	26	33	37	43	52
TWE2	11	16	22	26	32	37	43	51
TWF3	6	8	11	13	16	18	21	25
TWJ2	9	14	19	24	30	35	40	49
DGA1	4	6	6	7	8	8	7	6
DGA2	7	10	13	15	18	20	23	26
DGA3	11	15	19	23	28	32	37	45
DGB1	4	5	6	7	8	8	8	8
DGB2	8	10	13	15	19	21	24	28
DGB3	11	15	20	24	30	34	40	48
DGC1	6	8	10	11	13	14	15	16
DGC2	9	12	16	19	23	26	30	36
DGC3	13	17	23	27	34	39	45	54
SSA1	14	17	20	22	25	27	30	34
SSB1	11	14	16	18	21	23	25	29
SSB2	15	19	24	28	34	38	43	50
SSB3	10	12	14	16	18	19	21	23
SSC1	7	9	12	13	16	18	20	23
SSC2	9	13	18	21	26	30	35	41
SSE1	11	14	16	19	21	23	26	28
SSE2	14	19	25	29	35	39	45	52
SSE3	12	14	16	18	20	22	24	26

TABLE C.3 **Annual Passive Heating Performance: SSF (Continued)**

SHERIDAN, WYOMING **SSF (%)** *7717 DD65*

Type	*LCR = 100*	*70*	*50*	*40*	*30*	*25*	*20*	*15*
WWA3	12	17	22	26	31	35	40	47
WWB4	16	24	32	39	48	54	61	71
WWC2	16	23	31	38	46	52	59	68
TWA1	14	16	19	21	23	25	28	32
TWA2	13	16	20	23	27	30	34	40
TWA3	12	16	20	23	28	32	36	43
TWA4	11	15	20	23	29	33	37	44
TWB3	11	14	19	22	27	30	35	41
TWD4	15	22	29	35	43	48	55	65
TWE2	16	22	30	35	43	48	55	64
TWF3	9	13	17	20	25	28	33	39
TWJ2	13	20	27	32	40	46	53	62
DGA1	7	10	12	14	16	17	19	19
DGA2	10	14	18	21	26	29	33	38
DGA3	14	19	25	30	37	42	48	56
DGB1	7	10	13	15	17	19	21	23
DGB2	10	14	19	22	27	31	36	42
DGB3	14	19	26	31	38	44	51	60
DGC1	10	13	17	19	23	26	29	33
DGC2	13	17	23	27	33	37	43	51
DGC3	16	22	30	35	43	49	57	67
SSA1	17	21	25	28	33	36	40	46
SSB1	14	17	21	24	28	31	35	40
SSB2	18	24	30	35	42	46	52	60
SSB3	13	16	19	22	25	28	31	35
SSC1	11	14	18	22	26	30	34	40
SSC2	13	19	25	30	37	42	48	56
SSE1	15	19	24	27	32	35	39	44
SSE2	19	25	33	38	46	50	57	65
SSE3	15	18	21	24	28	30	34	38

SUFFIELD, ALBERTA *9423 DD65*

Type	*LCR = 100*	*70*	*50*	*40*	*30*	*25*	*20*	*15*
WWA3	11	15	19	23	27	30	34	39
WWB4	14	21	29	35	44	49	56	65
WWC2	14	21	28	34	42	47	53	62
TWA1	14	15	17	19	21	23	25	28
TWA2	12	15	18	21	24	27	30	34
TWA3	11	14	18	21	25	28	32	37
TWA4	10	13	18	21	25	29	33	38
TWB3	10	13	17	20	24	27	31	36
TWD4	14	20	27	32	39	44	50	59
TWE2	14	20	27	32	39	44	50	58
TWF3	8	11	15	18	22	25	28	33
TWJ2	12	18	24	29	37	41	48	56
DGA1	7	9	11	12	13	14	15	14
DGA2	10	13	17	19	24	26	30	34
DGA3	13	18	23	28	34	38	44	51
DGB1	7	9	11	13	15	16	17	18
DGB2	10	13	17	20	25	28	32	37
DGB3	13	18	24	29	36	41	47	55
DGC1	9	12	15	17	20	22	24	26
DGC2	12	16	21	25	30	34	39	45
DGC3	15	21	28	33	40	46	52	61
SSA1	15	18	22	24	28	30	33	37
SSB1	12	15	18	21	24	26	29	32
SSB2	16	21	27	31	37	41	46	53
SSB3	11	14	16	18	21	22	24	27
SSC1	10	13	16	19	23	25	28	33
SSC2	12	17	23	27	33	37	43	50
SSE1	13	16	19	22	25	27	30	33
SSE2	16	22	28	32	39	43	48	55
SSE3	13	16	18	20	23	25	27	30

Canada

EDMONTON, ALBERTA **SSF (%)** *10671 DD65*

Type	*LCR = 100*	*70*	*50*	*40*	*30*	*25*	*20*	*15*
WWA3	9	12	16	18	21	23	26	28
WWB4	11	18	25	30	37	42	48	56
WWC2	12	17	24	29	36	40	46	53
TWA1	12	14	15	17	18	19	20	22
TWA2	10	13	15	17	20	22	24	26
TWA3	9	12	15	17	20	23	25	28
TWA4	8	11	14	17	21	23	26	29
TWB3	8	11	14	17	20	22	25	28
TWD4	11	17	23	27	34	38	43	50
TWE2	12	17	23	27	33	37	42	49
TWF3	6	9	12	14	17	19	22	25
TWJ2	10	15	21	25	31	35	41	47
DGA1	5	7	8	9	9	9	9	7
DGA2	8	11	14	17	20	22	25	28
DGA3	12	16	21	25	30	34	39	45
DGB1	5	7	8	9	10	11	11	10
DGB2	8	11	15	17	21	24	27	30
DGB3	12	16	22	26	32	36	41	48
DGC1	7	9	12	13	15	17	18	18
DGC2	10	14	18	21	26	29	33	37
DGC3	14	19	25	29	36	41	46	53
SSA1	13	16	19	21	23	25	27	28
SSB1	11	13	16	17	20	21	23	25
SSB2	14	19	23	27	32	35	40	45
SSB3	10	12	14	15	17	18	18	19
SSC1	8	10	13	15	17	19	21	22
SSC2	10	14	19	23	28	31	35	40
SSE1	10	13	15	17	19	20	21	21
SSE2	13	18	23	27	32	36	40	45
SSE3	11	13	16	17	19	20	21	22

VANCOUVER, BRITISH COLUMBIA **SSF (%)** *5607 DD65*

Type	*LCR = 100*	*70*	*50*	*40*	*30*	*25*	*20*	*15*
WWA3	15	19	24	27	31	34	37	41
WWB4	18	25	33	39	46	51	57	65
WWC2	18	25	32	38	45	49	55	62
TWA1	16	18	21	23	25	27	29	32
TWA2	15	18	22	25	28	31	34	37
TWA3	14	18	22	25	29	32	35	40
TWA4	13	17	22	25	30	33	36	41
TWB3	13	17	21	24	28	31	35	39
TWD4	17	24	31	36	43	47	53	60
TWE2	18	24	31	36	42	47	52	59
TWF3	11	15	19	22	26	29	32	37
TWJ2	16	22	29	33	40	45	50	57
DGA1	10	12	15	17	18	19	20	19
DGA2	12	16	21	24	28	31	34	37
DGA3	16	22	28	32	38	42	47	53
DGB1	10	13	16	18	20	21	22	23
DGB2	13	17	22	25	30	33	36	41
DGB3	17	22	29	33	40	45	50	57
DGC1	13	16	20	22	26	27	29	31
DGC2	15	20	25	29	35	38	42	48
DGC3	19	25	32	37	45	49	55	62
SSA1	20	23	27	30	34	36	39	42
SSB1	16	20	24	26	30	32	35	38
SSB2	21	26	32	36	42	46	51	57
SSB3	15	19	22	24	27	29	31	34
SSC1	13	17	21	23	27	29	32	35
SSC2	16	21	27	31	37	41	45	51
SSE1	18	22	26	28	32	34	36	38
SSE2	21	28	34	38	44	48	52	58
SSE3	17	20	24	26	29	31	33	36

TABLE C.3 **Annual Passive Heating Performance: SSF (Continued)**

WINNIPEG, MANITOBA **SSF (%)** *10776 DD65*

Type	LCR = 100	70	50	40	30	25	20	15
WWA3	8	10	13	16	19	21	24	28
WWB4	10	16	23	28	36	41	48	57
WWC2	10	16	22	27	34	39	45	54
TWA1	12	13	14	15	16	17	18	19
TWA2	10	11	14	15	18	20	22	25
TWA3	8	10	13	15	18	20	23	27
TWA4	7	9	13	15	18	21	24	28
TWB3	8	10	13	15	18	20	23	27
TWD4	10	15	21	25	32	37	43	51
TWE2	11	16	21	25	32	36	42	49
TWF3	5	8	10	12	15	17	20	24
TWJ2	8	13	19	23	29	34	40	47
DGA1	4	5	6	7	7	7	6	5
DGA2	7	9	12	14	18	20	22	26
DGA3	10	14	19	22	28	32	37	44
DGB1	4	5	6	7	7	8	8	7
DGB2	7	10	13	15	18	21	24	28
DGB3	11	15	19	23	29	34	39	47
DGC1	6	7	9	10	12	13	14	15
DGC2	9	12	16	19	23	26	30	35
DGC3	12	17	22	27	34	38	45	53
SSA1	12	15	17	19	22	23	26	28
SSB1	10	12	14	16	18	20	22	24
SSB2	13	17	22	25	31	34	39	46
SSB3	9	10	12	13	15	16	17	18
SSC1	6	9	11	13	15	17	19	21
SSC2	9	12	17	20	26	29	34	40
SSE1	9	11	14	15	18	19	21	22
SSE2	12	17	22	26	31	35	40	46
SSE3	11	12	14	15	17	18	20	21

OTTAWA, ONTARIO **SSF (%)** *8461 DD65*

Type	LCR = 100	70	50	40	30	25	20	15
WWA3	8	11	14	17	20	23	26	31
WWB4	10	16	24	29	37	43	50	59
WWC2	10	16	23	28	36	41	47	56
TWA1	12	13	14	15	17	18	19	21
TWA2	10	12	14	16	19	20	23	26
TWA3	8	11	14	16	19	22	25	29
TWA4	7	10	13	16	19	22	26	31
TWB3	8	10	13	15	19	21	24	29
TWD4	10	15	22	26	33	38	44	53
TWE2	11	16	22	26	33	38	44	52
TWF3	6	8	11	13	16	18	21	26
TWJ2	9	14	19	24	31	35	41	50
DGA1	4	5	6	7	8	8	7	6
DGA2	7	10	13	15	18	20	23	27
DGA3	11	14	19	23	28	33	38	46
DGB1	4	5	6	7	8	8	9	8
DGB2	7	10	13	15	19	21	25	29
DGB3	11	15	20	24	30	34	40	49
DGC1	6	7	9	11	13	14	15	17
DGC2	9	12	16	19	24	27	31	37
DGC3	13	17	23	27	34	39	46	55
SSA1	13	16	19	21	24	26	29	33
SSB1	11	13	15	17	20	22	24	28
SSB2	14	18	23	27	33	37	42	49
SSB3	10	11	13	15	17	18	20	22
SSC1	7	9	11	13	16	18	21	24
SSC2	9	13	18	21	27	31	36	43
SSE1	10	13	16	18	21	23	25	28
SSE2	13	18	24	28	35	39	44	52
SSE3	11	13	15	17	19	21	23	25

HALIFAX, NOVA SCOTIA **SSF (%)** *7167 DD65*

Type	LCR = 100	70	50	40	30	25	20	15
WWA3	10	13	17	20	25	28	32	38
WWB4	12	19	27	33	42	47	55	64
WWC2	12	19	26	32	40	45	52	61
TWA1	13	14	16	17	19	21	23	25
TWA2	11	13	16	19	22	24	27	32
TWA3	10	13	16	19	23	26	29	35
TWA4	8	12	16	19	23	26	30	36
TWB3	9	12	15	18	22	24	28	34
TWD4	12	18	24	30	37	42	49	58
TWE2	13	18	25	30	37	42	48	57
TWF3	7	10	13	16	19	22	26	31
TWJ2	10	16	22	27	34	39	46	55
DGA1	6	7	9	10	11	11	12	11
DGA2	8	11	15	17	21	24	27	31
DGA3	12	16	21	25	32	36	42	50
DGB1	5	7	9	10	11	12	13	14
DGB2	9	12	15	18	22	25	29	34
DGB3	12	17	22	26	33	38	45	54
DGC1	7	10	12	14	16	18	20	23
DGC2	10	14	18	22	27	31	36	43
DGC3	14	19	25	30	38	43	51	60
SSA1	15	19	22	25	29	32	36	41
SSB1	13	15	19	21	25	27	30	35
SSB2	16	21	27	31	38	42	48	56
SSB3	11	14	17	19	22	24	26	30
SSC1	8	11	14	17	20	23	26	30
SSC2	11	15	21	25	31	35	41	48
SSE1	13	16	20	23	27	29	33	37
SSE2	16	22	29	34	41	46	52	59
SSE3	13	16	19	21	24	26	29	33

TORONTO, ONTARIO **SSF (%)** *7280 DD65*

Type	LCR = 100	70	50	40	30	25	20	15
WWA3	8	11	14	17	21	23	27	31
WWB4	10	17	24	30	38	43	50	59
WWC2	11	16	23	28	36	41	47	56
TWA1	12	13	15	16	17	18	19	21
TWA2	10	12	14	16	19	21	23	27
TWA3	9	11	14	16	20	22	25	29
TWA4	7	10	13	16	20	22	26	31
TWB3	8	10	13	16	19	21	24	29
TWD4	11	16	22	27	33	38	44	53
TWE2	11	16	22	27	33	38	44	52
TWF3	6	8	11	13	16	19	22	26
TWJ2	9	14	20	24	31	35	41	50
DGA1	5	6	7	7	8	8	8	7
DGA2	8	10	13	15	18	21	23	27
DGA3	11	15	19	23	29	33	38	46
DGB1	4	5	7	7	8	9	9	9
DGB2	8	10	13	16	19	22	25	29
DGB3	11	15	20	24	30	35	41	49
DGC1	6	8	10	11	13	14	16	17
DGC2	9	12	16	19	24	27	31	37
DGC3	13	17	23	28	34	39	46	55
SSA1	14	17	20	22	25	28	31	35
SSB1	11	14	16	18	21	23	26	29
SSB2	15	19	24	28	34	38	43	51
SSB3	10	12	14	16	18	20	22	24
SSC1	7	9	12	14	17	18	21	24
SSC2	9	13	18	22	27	31	36	43
SSE1	11	14	17	19	22	24	27	30
SSE2	14	19	25	30	36	40	46	53
SSE3	12	14	16	18	20	22	24	27

TABLE C.3 **Annual Passive Heating Performance: SSF (Continued)**

NORMANDIN, QUEBEC **SSF (%)**					*11364 DD65*			
Type LCR =	*100*	*70*	*50*	*40*	*30*	*25*	*20*	*15*
WWA3	6	9	11	14	17	19	21	25
WWB4	8	14	21	26	33	39	45	54
WWC2	8	14	20	25	32	36	43	51
TWA1	11	12	13	13	14	15	16	17
TWA2	8	10	12	13	16	17	19	22
TWA3	7	9	11	13	16	18	20	24
TWA4	5	8	11	13	16	18	21	25
TWB3	6	8	11	13	16	18	20	24
TWD4	8	13	19	23	30	34	40	48
TWE2	9	14	19	23	29	34	39	47
TWF3	4	6	8	10	13	15	17	20
TWJ2	7	11	17	21	27	31	37	45
DGA1	3	3	4	5	5	4	4	2
DGA2	6	8	10	13	15	17	20	23
DGA3	9	13	17	20	26	30	35	42
DGB1	3	3	4	4	5	5	5	4
DGB2	6	8	11	13	16	18	21	25
DGB3	10	13	18	21	27	31	37	45
DGC1	4	6	7	8	9	10	11	12
DGC2	8	10	14	16	20	23	27	33
DGC3	11	15	20	25	31	36	42	51
SSA1	12	14	16	18	21	23	25	27
SSB1	9	11	13	15	17	19	20	23
SSB2	12	16	21	24	30	33	38	44
SSB3	8	10	11	12	14	15	16	17
SSC1	5	7	9	11	13	14	16	18
SSC2	7	11	15	18	23	27	31	37
SSE1	8	10	13	14	16	18	19	20
SSE2	11	16	21	25	30	34	39	45
SSE3	10	11	13	14	16	17	19	20

D
SUNSHADING

D.1 Introduction

This appendix presents a graphic-analytic method for the design of fixed sunshading devices. Although it is assumed that the reader is to an extent familiar with the principles of sunshading and shading mask construction, a brief review of these topics is presented, to develop the necessary background for the design methods that will be demonstrated. It is further assumed that the reader recognizes the importance of sunshading and the importance of the effect of insolation on thermal gain (and loss), daylighting, glare, and so on. A full discussion of these topics is found in Chapters 1 and 6 of this volume.

The shadow cast by a projection from a wall depends on:

1. The shape and dimensions of the projection.
2. The direction in which the wall faces.
3. The position of the sun.

The first two factors are fixed; the third varies with the hour of the day and the date, that is, with the apparent motion of the sun.

D.2 Solar Position: Altitude and Azimuth

Refer to Fig. 19.4*a* and Tables B.11 to B.14. The position of the sun at any instant with respect to an observer on the ground is defined by its *altitude angle A* and its *azimuth angle Az*. Note that in our discussion *azimuth is measured from south*. (In other texts it is measured from north; there is no consensus on this point.) The altitude of the sun varies during the day, starting and ending at 0° at sunrise and sunset, and reaching a daily maximum at *solar* noon. (See Section D.3 for explanation of solar and clock time.) The altitude angle at noon varies from day to day, reaching a yearly maximum at the summer solstice on June 21, a minimum at the winter solstice on December 21 and a point halfway between on the vernal and autumnal equinoxes of March 21 and September 21. These seasonal changes are shown graphically on Fig. 19.4*b* and Fig. D.1. Note that the altitude change is ap-

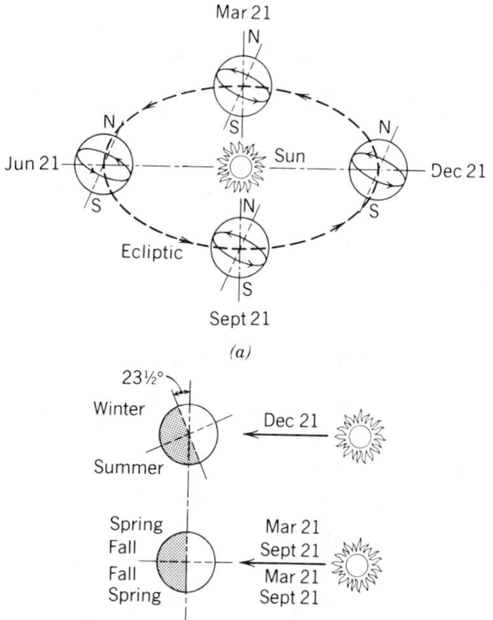

***Fig. D.1** The ecliptic is the annual path of the earth around the sun. The tilt of the earth's axis to the plane of the ecliptic results in the seasonal variations. Note in* (a) *and* (b) *that the northern hemisphere tilts away from the sun in December, resulting in winter's low sun altitude and cold weather. In June the effect is reversed.*

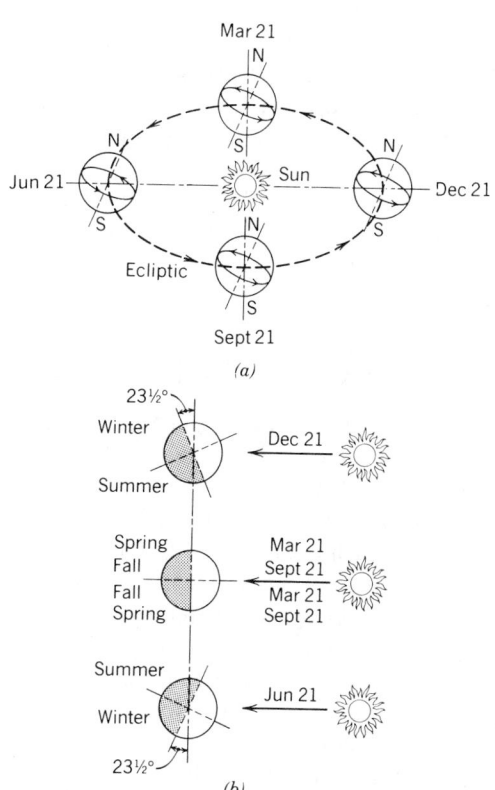

proximately 23½° every three months, making the daily change approximately a quarter-degree and the maximum difference in altitude angle about 47°, between the summer and winter solstices. See Tables 19.2 and B.11 to B.14 for accurate altitude/azimuth figures.

Referring to the ecliptic (Fig. D.1*a*), we can see that the height of the sun (its altitude angle), depends on the observer's position on the earth, specifically, his latitude. This can be expressed as follows:

$$\text{maximum solar altitude} = 113.5° - \text{latitude}$$

$$\text{equinox solar altitude} = 90° - \text{latitude}$$

$$\text{minimum solar altitude} = 66.5° - \text{latitude}$$

These relations are shown graphically in Fig. 19.4*c*.

On the equinoxes (March 21 and September 21), the sun rises due east and sets due west. In the northern hemisphere, the sun rises north of east between March 22 and September 20 and sets north of west. This is shown graphically on Fig. 19.4*b* and is projected, in Fig. D.2, for one specific latitude. Note the positions of sunrise and sunset with respect to the E–W line, to understand the explanation above.

D.3 Sun Time and Clock Time

Although for the purpose of sunshading it is not normally important to differentiate between sun time and clock time, it is important that the reader know that such a difference exists, and can be as much as an hour. The following is a brief discussion of timekeeping. The interested reader can refer to the References for a more detailed explanation. (In our discussion we will refer to the sun's motion rather than apparent motion, since relative to the observer, the sun does indeed move.)

The basis of all timekeeping is the length of the solar day, that is, one full rotation of the earth on its axis. More rigidly defined, it is the interval between two sun passages across the observer's meridian. However, because the sun moves about 1° per day on the ecliptic (360° in 365 days), it must move approximately 361° in a solar day. Furthermore, the sun's motion on the ecliptic is not uniform—it moves more rapidly when further from the earth. As a result, the actual length of the solar day varies. Since it is impractical for us to use timepieces that need daily adjustment, we use an average, or *mean solar day*, as the basis for timekeeping. The difference between mean (clock) and apparent (actual) solar time is called the equation of time and can be determined from a curve called the *analemma* or from tabulations in various sources. Furthermore, the 360° earth circumference is divided into 24 one-hour time zones, each therefore separated from the adjacent zone by 15° of longitude, which corresponds to approximately 1000 miles. Therefore an observer at any point other than directly on a time meridian (which is normally in the center of the time zone) must make a time correction of up to 30 min to correspond to actual solar time. (In some countries a reference meridian is more than 30 min from the time zone extremes. In such places, the longitude correction plus the equation of time correction can total well over an hour, and one is no

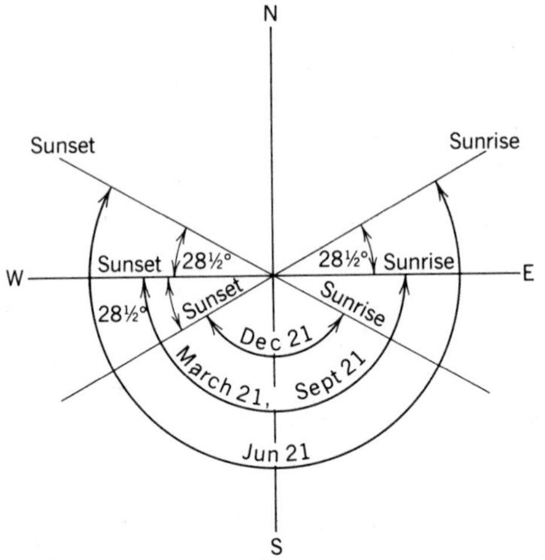

Fig. D.2 *Projected sunpaths for solstices and equinoxes at latitude 32° N (Jerusalem, Savannah, Tucson). The sun's azimuth angle, like its altitude, varies with time of day and date. The sun's azimuth path from sunrise to sunset, for this latitude is shown in horizontal projection. Sunrise on June 21 is 28½° North of East, and sunset is correspondingly North of West. On December 21 sunrise and sunset are 28½° South of East and West.*

longer able to neglect it. This situation does *not* exist in the continental United States.) Although not vital, because the length of shaded time is the same, it is good practice to make a table of mean time versus apparent solar time for any project where shading and insolation are important. In any case, the designer must remember that unless otherwise specifically noted (and this is rare), all sun charts show sun time, that is, actual solar time. This includes the well-known Libbey-Owens-Ford (LOF) *Sun Angle Calculator* (1975).

D.4 Sunpath Projection

The design of sunshading devices necessitates projecting the three-dimensional solar path onto a two-dimensional surface. There are basically two types of projection: vertical and horizontal. A version of the former is familiar to users of Waldram diagrams in daylight design. Horizontal projections are more common. The basic idea behind the horizontal projection is shown in Fig. D.3. The sky vault is represented as a hemisphere that projects as a circle, the center of

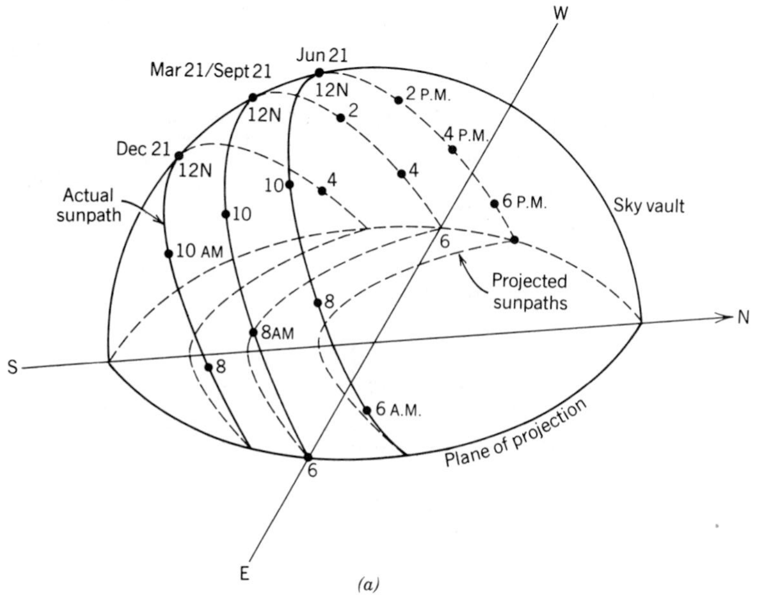

(a)

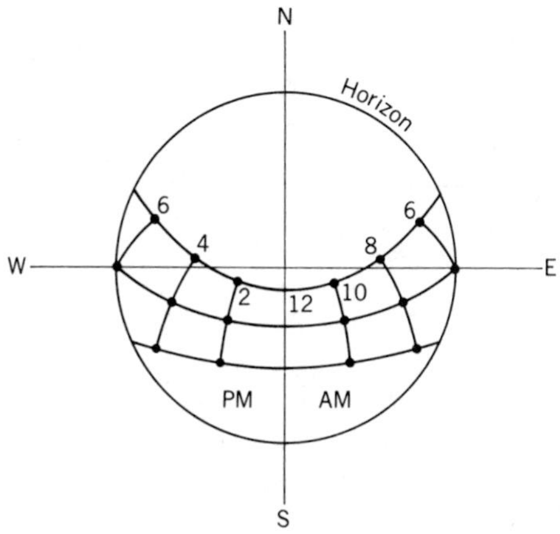

(b)

Fig. D.3 *Basis of horizontal projection of the sunpaths. The sky vault* (a) *is represented as a hemisphere, which projects as a circle* (b) *on the projection plane. The sunpaths in* (a) *project as curved lines on the projection plane* (b), *the exact shape of which depends on the projection method (Fig. D.4). The numbers on the sunpaths in* (a) *are hours: 6 A.M., 8 A.M., 10 A.M., 12 noon, 2 P.M., 4 P.M., 6 P.M. When hour points are connected between sunpaths* (b), *they form hour lines, which are useful for interpolation in completed diagrams.*

APPENDICES

which is the zenith. The observer looks down onto the hemispherical sky vault and projects the sunpaths onto the two-dimensional, projected circle. The four methods of projection are:

Gnomonic.

Orthographic.

Stereographic.

Equidistant.

The position of the observer's eye for each of these projections is shown in Fig. D.4. Spacing between circles of equal altitude is the basic difference between the four methods. In all of them, lines of equal azimuth project as straight lines emanating from the center of the circle.

1. The *gnomonic projection* is derived from the sundial. The observer is at the center of the projection; therefore low sun angles (sunrise, sunset) extend to infinity. For this reason, this method is rarely used for solar charts but is used frequently for buildings shadow studies. The method is illustrated in the sun peg shadow charts of Fig. 3.12.

2. The *orthographic projection* views the sky vault from infinity and projects as in a standard architectural plan, that is, with all projection lines parallel.

3. The *stereographic projection* is used widely throughout the world, except in the United States. The observer is fixed at the top of the hemisphere, resulting in wide-spaced, low-angle altitudes. This makes it particularly useful for northern latitudes, that is, above 40°N, since the sun at these latitudes is usually low in the sky. A special protractor is required (see Sec-

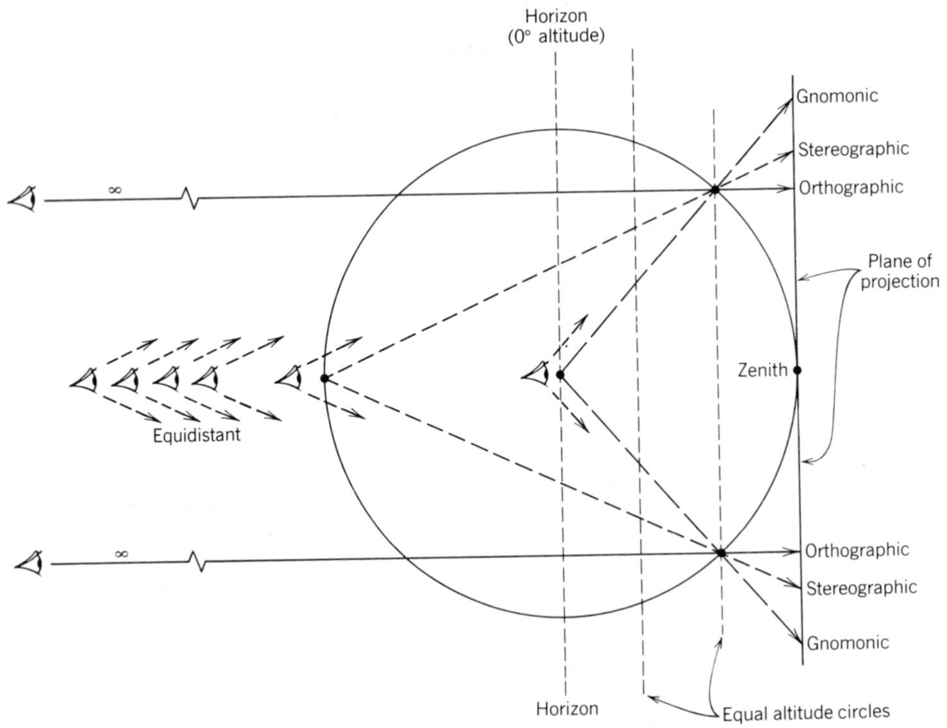

Fig. D.4 *The four standard horizontal projection methods, with observer position and projected point. Note that in the equidistant projection the observer position varies and therefore the projected point also varies. The projected point is therefore not shown. Note also that as the altitude drops and approaches the horizon, the gnomonic projection extends out to infinity.*

tion D.5). The principal advantage of this projection is that it is very simple to draw sunpaths at any scale and for any latitude without having to rely on published sunpaths. Thus for any serious sunshading design, the designer can prepare a large-scale sunpath chart at the exact latitude of the project and do accurate graphical analysis. For a detailed description of the method of preparing the charts and protractor, see Szokolay (1980, pp. 314–317).

4. The *equidistant projection* is used almost exclusively in the United States because of the ready availability and wide acceptance of the LOF sunpath charts (*Sun Angle Calculator*, 1975). These charts, one of which is illustrated in Fig. D.5, show the altitude circles as spaced equidistantly, a projection that requires moving the observer during projection, as shown in Fig. D.4. This method is equally applicable to all latitudes. Its disadvantage is the difficulty of preparing sunpaths and the required shadow angle protractor. Thus the user is essentially limited to published material, with its limitations of size and latitude. The LOF diagrams are available only in steps of 4° of latitude.

D.5 Shadow Angles and Shading Masks

The sun's altitude and azimuth angles are very useful in developing the sunpath diagrams but are much less so in defining the shadow angles cast by a projection, on a wall exposed to the sun. Since there is no universal agreement on angle nomenclature, we will very carefully define the angles involved. Refer to Fig. D.6, which shows the shadow cast by a horizontal overhang on a wall exposed to sunlight. Note that the shadow is defined by two angles: *the vertical shadow angle* (*VSA*), which indicates the position of the leading edge of the shadow as defined from the leading edge of the overhang, and the *horizontal shadow angle* (*HSA*), which defines the leading edge of a shadow cast by a vertical element (absent here, and indicated by a

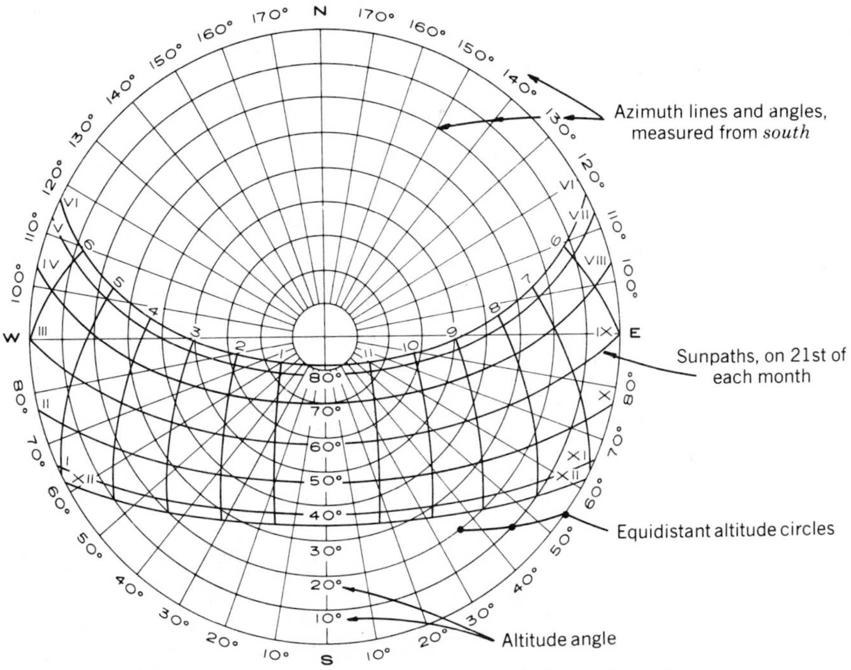

Fig. D.5 *Typical equidistant sunpath projection for 32°N latitude. This latitude is approximately that of Savannah, Georgia, Tucson, Arizona, San Diego, California, and Jerusalem, Israel. In all of these locations sun shading is desirable at least from the vernal to the autumnal equinox. [Diagram after Libbey-Owens-Ford,* Sun Angle Calculator, *(1975).]*

APPENDICES

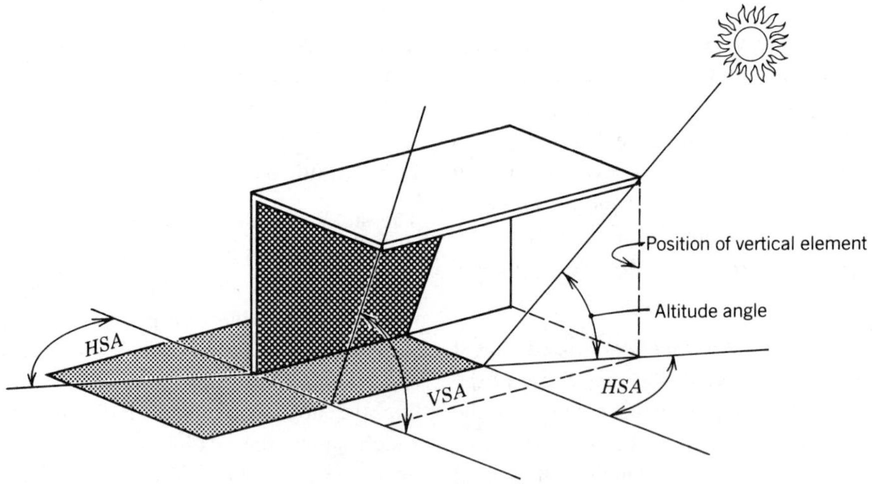

Fig. D.6 The shadow cast by a horizontal overhang is best defined by angles VSA *and* HSA, *the vertical shading angle and horizontal shading angles, respectively. VSA is also known in the United States as the profile angle.*

dotted line), as defined with respect to that element's leading edge. The terms VSA and HSA are in use throughout the world, although in the United States, the vertical shadow angle is also known widely as the *profile angle* because it is so designated on the LOF sun angle calculator. Figure D.7 shows the same information in slightly different form; line *DA* is the shadow

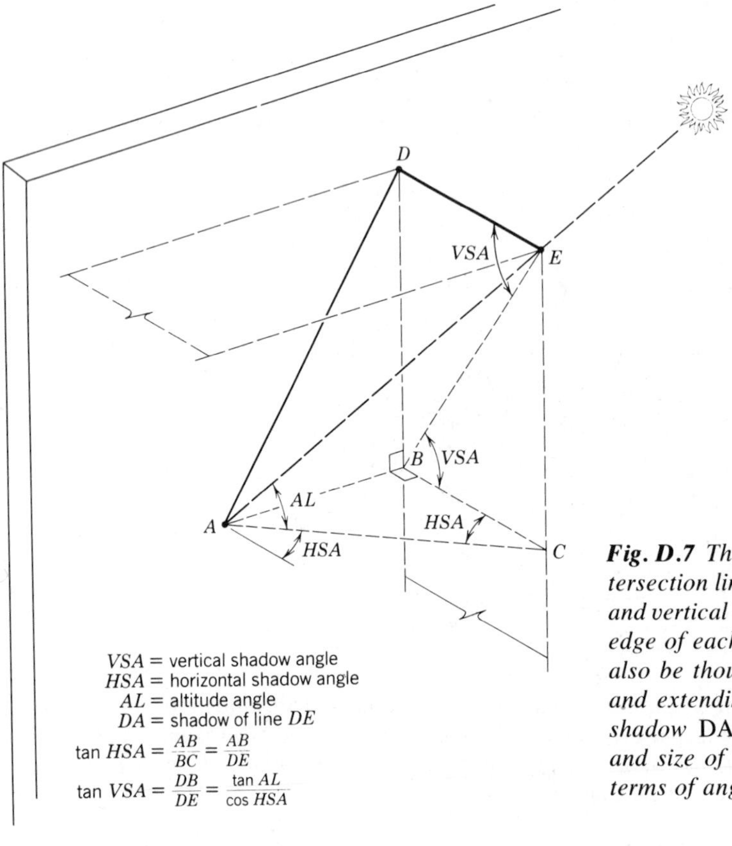

Fig. D.7 The shadow DA *cast by the intersection line* DE, *between a horizontal and vertical shading element, defines the edge of each of these shadows.* DE *can also be thought of as a pin, normal to, and extending from the wall, casting a shadow* DA *on the wall. The location and size of line* DA *are best defined in terms of angles* VSA *and* HSA.

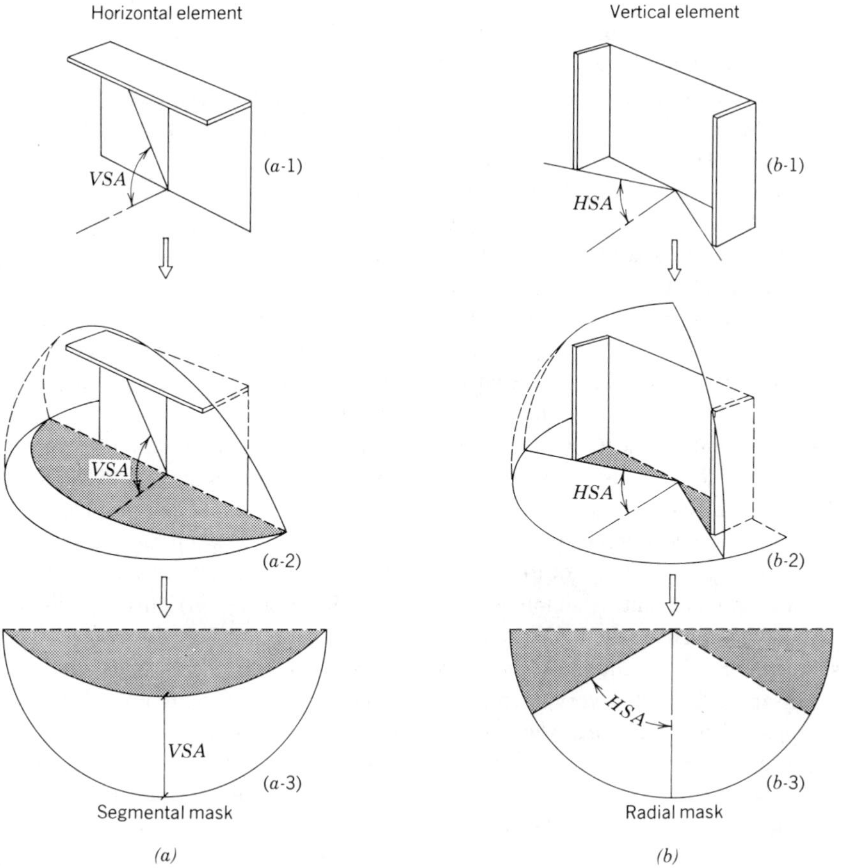

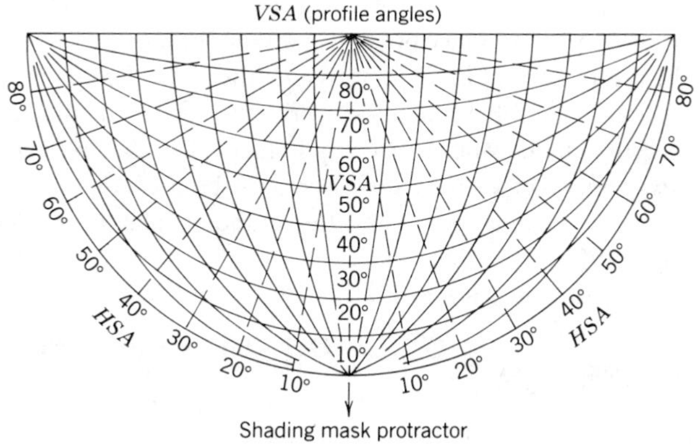

Fig. D.8 *A shading mask is the horizontal projection of the shadow cast by elements with the same* VSA *or* HSA. *Thus any* infinitely long *horizontal element with the* VSA *shown in* (a-1) *will have a shading mask as shown in* (a-3). *Similarly, any* infinitely high *vertical elements with the* HSA *shown in* (b-1) *sill have the shadow mask shown in* (b-3). *A protractor of the type shown in* (c) *is required to draw the segmental horizontal element mask properly. This device must be drawn to the same scale and with the same projection method (equidistant, stereographic) as the sunpath diagram on which it will be used. The protractor of* (c) *matches the sunpath diagram of Fig. D.5 and is drawn with equidistant projection.*

cast by line *DE*, which is the intersection between a horizontal projection and a vertical projection. This line is particularly important in determining the required extent of a shading element, as will be shown below.

Refer now to Fig. D.8. The locus of the leading edge of the shadow cast by all horizontal elements with the same VSA (profile angle) (Fig. D.8*a*-1) projects as a segmental line (Fig. D.8*a*-2), and when plotted on the projection plane (similar to the sunpath projection in Fig. D.3), shows as a segmental mask (Fig. D.8*a*-3). To draw this segment, a protractor is required, as shown in Fig. D.8*c*. Note that the angular notation on the segmental section of the protractor corresponds to VSA as shown on Fig. D.8*a*-3. Users of the LOF sun angle calculator can use the profile angle overlay to draw the required segment. It is important not to confuse the VSA with its complement. A simple way to remember this is to consider the limits of VSA. With no shading, VSA is 90° and the segment on Fig. D.8*a*-3 disappears. With a very deep overhang, VSA approaches 0° and the *unshaded* area in Fig. D.8*a*-3 shrinks. This action shows clearly on the protractor of Fig. D.8*c*.

Refer now to Fig. D.8*b*. Vertical shading elements Fig. D.8*b*-1 with the same HSA show as radii (*b*-2) and project as radial lines from the center, on the radial mask, at angle HSA from normal to the wall (*b*-3). These radial lines can be drawn with the assistance of an ordinary protractor, or by use of the protractor in Fig. D.8*c*. The reader must bear in mind that full segments and full radial lines are the shading masks of *infinitely long elements*, which, of course, do not actually exist. Shading masks that appear in the literature (Olgyay, 1975; Ramsey and Sleeper, 1981) are frequently drawn as if for infinite elements, and the reader must know how to truncate them if accurate shading design is desired.

D.6 Use of Shading Masks

Once a shading mask has been drawn, it is applicable to any latitude and exposure direction. The use of the mask is straightforward. It is drawn on some transparent medium (e.g., paper or plastic sheet), laid on top of a sunpath diagram of the proper latitude, *and drawn to the same scale*. Its centerpoint is placed on the centerpoint of the sunpath diagram, and it is rotated until its facing direction (the direction of a normal to the wall) is aligned with the appropriate azimuth line on the sunpath diagram. Assuming that the shading mask has been correctly drawn to provide 100% shading for the *entire window*, the shaded hours for the window in question can then be read directly from the sunpath diagram underneath. (This is why the shading mask must be drawn on a transparent medium.) See Fig. D.9.

It is extremely important to keep in mind that a horizontal overhang shading element the same width as the window can provide full shade for the window only when the sun is exactly opposite the window, that is, when the bearing angle is 0°. (See Fig. 19.9*a*.) This situation obtains for only an instant. See Fig. D.10. At all other times, some part of the window will be exposed to the sun. Therefore, to provide full shading for more than an instant with only a horizontal element and no vertical (side) elements, *the overhand must extend beyond the sides of the window*. The amount of such extension can be determined both graphically and analytically. Since graphic solutions are amply treated in the literature (Harkness and Mehta, 1978; Lim et al., 1979; Cowan, 1980) we will restrict ourselves to the analytic solution, leaving to the reader the parallel graphic solution. In doing this work, the reader should keep in mind the requirement that shading masks for 100% coverage must be prepared in order that the extremities of the opening be shaded.

D.7 Finite Horizontal Elements

The first design step is to establish the required *depth* of the unit, that is, the amount it projects from the wall. Although any percent coverage desired is possible, most elements are designed to give either 50 or 100% coverage. The reason for this is explained in the illustrative design example below. Figure D.11 shows how the required depth and the corresponding segmental mask are determined. The next step is to deter-

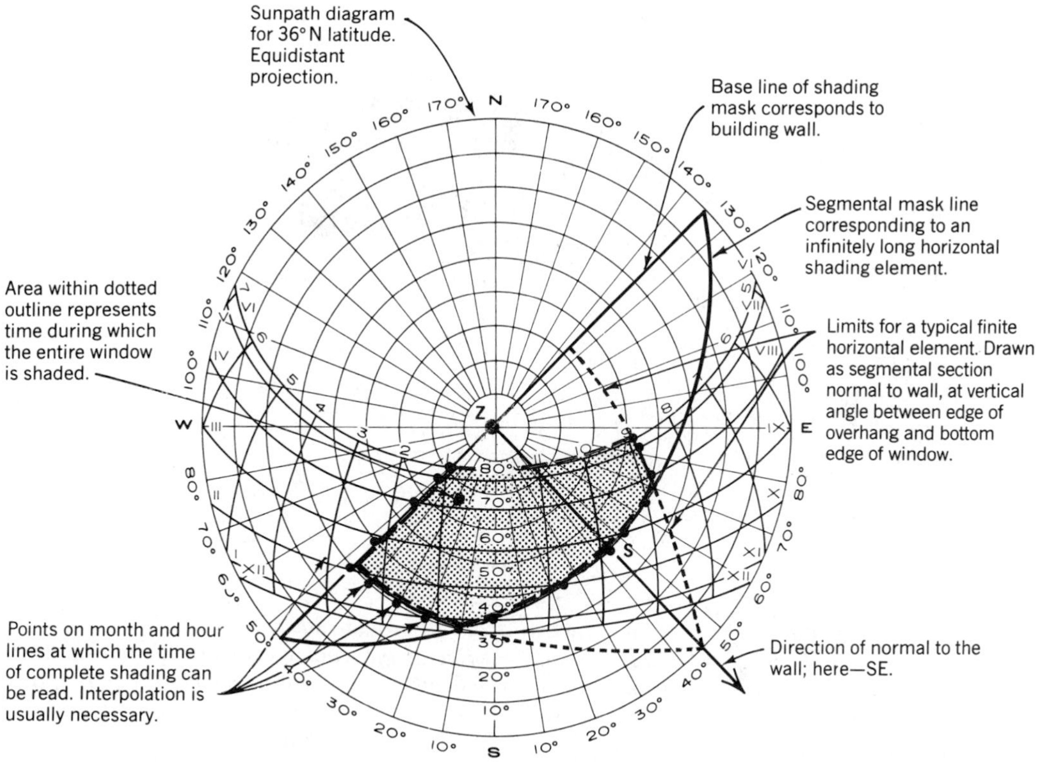
Sunpath diagram
for 36° N latitude.
Equidistant
projection.
Base line of shading
mask corresponds to
building wall.
Segmental mask line
corresponding to an
infinitely long horizontal
shading element.
Limits for a typical finite
horizontal element. Drawn
as segmental section
normal to wall, at vertical
angle between edge of
overhang and bottom
edge of window.
Area within dotted
outline represents
time during which
the entire window
is shaded.
Points on month and hour
lines at which the time
of complete shading can
be read. Interpolation is
usually necessary.
Direction of normal to the
wall; here—SE.
N
S
E
W
Z

36° N Latitude

Fig. D.9 *A typical shading mask for a horizontal overhang on a window is laid over a sunpath diagram, to permit determination of the shaded hours. The mask is drawn to provide full shading for the entire window. The hours during which the entire window is shaded are determined by the intersection of the mask perimeter with the date/hour lines of the sunpath diagram, and are tabulated on the figure. However, only at the hours falling along the normal-to-the-wall line* ZS *(where the bearing angle is 0°) and within the dotted area is a window fully shaded by an overhang element only as wide of as the window. Indeed, line* ZS *is the shading mask of such an element with respect to 100% shading of the entire window (see Fig. D.10). Note that the element shown here is not symmetrical about the center, indicating an extension to the left to provide a larger shade period after noon. This is typical with east-facing windows, and is reversed for western exposures.*

Time of Day When Window is Fully Shaded

With Required Overhang Extensions			*With Element Same Width as Window*
Date	*From*	*To*	*Time*
21 June	0900	1240	1115
21 July/May	0850	1250	1100
21 Aug./Apr.	0845	1330	1035
21 Sept./Mar.	0940	1400	0955
21 Oct./Feb.	1040	1410	—
21 Nov./Jan.	12N	1300	—
21 Dec.	1230		—

mine the required side extensions beyond the window's sides. This can best be illustrated by a practical example, using the analytic technique. First however, a few remarks about the example presented are in order.

Due to the symmetry of solar motion about the ecliptic, the position of the sun is symmetrical on both sides of its maximum/minimum positions, that is, the solstices. Thus the sun's position on May 21 is the same as on July 21, since both dates are one month from the summer solstice. As a result, any *fixed* shading device will

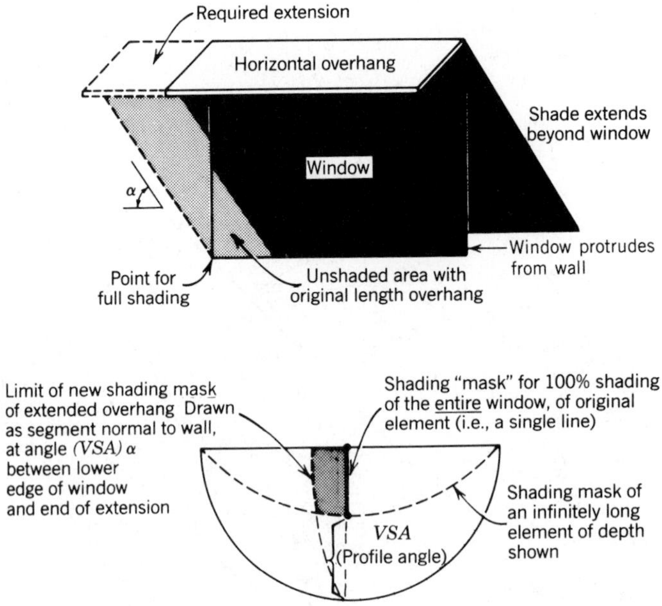

Fig. D.10 A horizontal overhang with sufficient depth to provide 100% shading (i.e., to cast a shadow reaching to the window sill line, but only as wide as the window) can provide full shading only when the sun is exactly opposite the window. At any other time (bearing angle ≠ 0), part of the window will be exposed to the sun. To compensate for this, if full shading is desired, an extension is required, as shown. For south-facing windows the extension is symmetrical on both sides. For east-facing windows an extension to the left is usually used to provide full shading for hot afternoon hours; for west-facing windows, the reverse is customary. Full shading from the low early morning or late afternoon sun is not practical with horizontal elements.

Fig. D.11 To find the required depth of a horizontal overhang, having established the required segmental shading mask (a) *it is only necessary to read* VSA *(profile angle) off the protractor in Fig. D.8*c *that corresponds to the segment in* (a) *and plot this angle as shown in* (b) *(wall section). Angle* VSA *can be plotted for 50% coverage, 100% coverage, or any other figure. The depth of the overhang can then be measured directly from the section, which is drawn to scale. Alternatively, if the overhang depth is known, the corresponding segment can be drawn.*

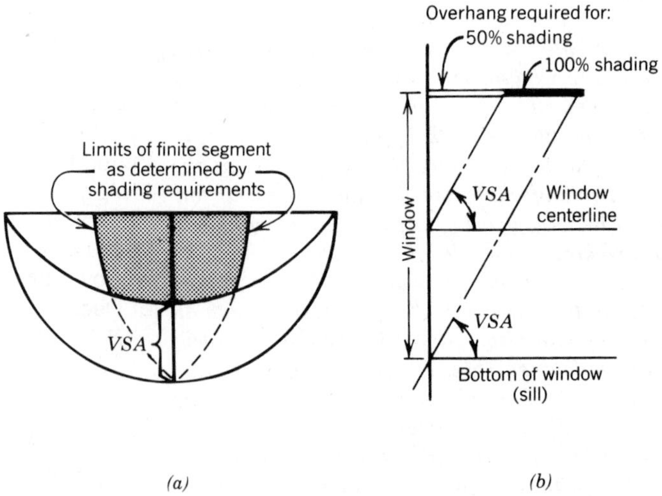

give the same shade in the spring (before 21 Jun) as in the summer (after 21 Jun). Since in many locations spring is cool, the desired summer shading will produce spring shading that *may* *not* be desirable. The solutions to this apparent dilemma are either to use a movable or variable size fixed shading device, or to compromise with the amount of shading, that is, to design for 50%

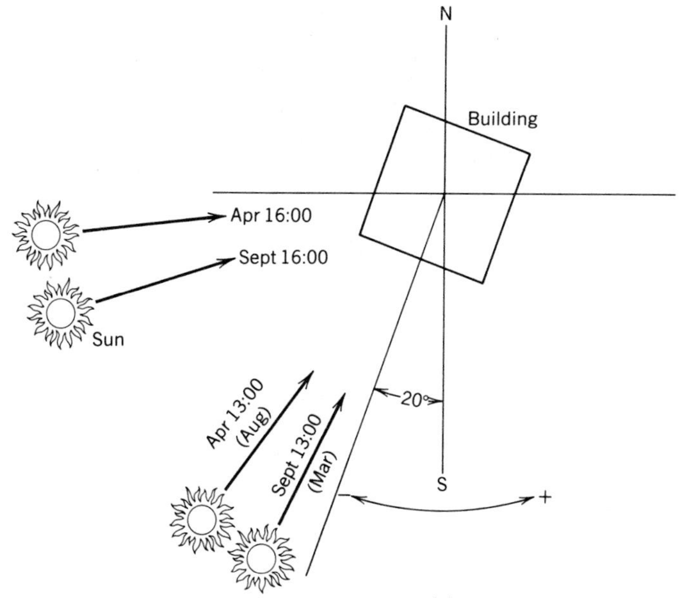

(a)

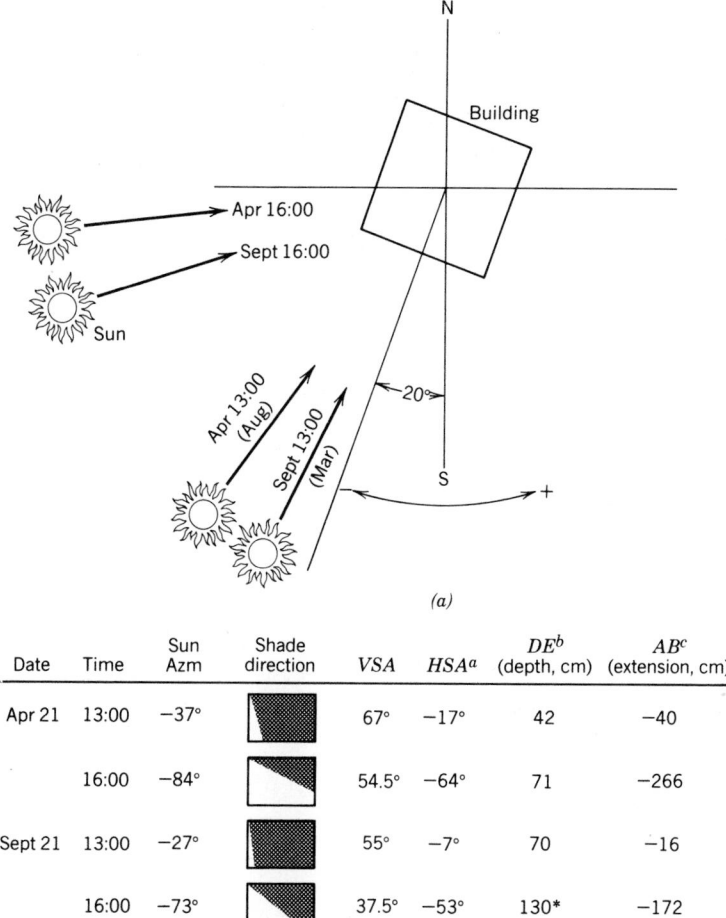

Date	Time	Sun Azm	Shade direction	*VSA*	*HSA*[a]	*DE*[b] (depth, cm)	*AB*[c] (extension, cm)
Apr 21	13:00	−37°		67°	−17°	42	−40
	16:00	−84°	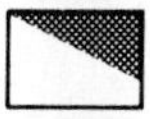	54.5°	−64°	71	−266
Sept 21	13:00	−27°	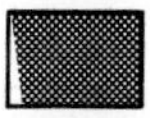	55°	−7°	70	−16
	16:00	−73°	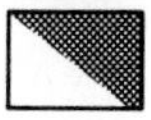	37.5°	−53°	130*	−172

[a]*HSA* = |Sun azm| - |Bldg. bearing|, i.e.,|Azm| - |20°|
[b]*DE* = Window height/tan *VSA* = 100/*VSA*
[c]*AB* = *DE* tan *HSA* = overhang depth × tan *HSA* = 130 tan *HSA*; the negative sign indicates an extension to the left.

(b)

Fig. D.12 *Sketches for solution of Example D.1 (see text). [See continuation on next page]*

shading for late summer, giving 50% insolation in early spring and 100% shading in early summer (and late spring).

The example that follows requires 100% shading for most of the summer (spring) for two reasons:

1. At the selected latitude of 32°N, sunshading is frequently desirable in the spring.
2. We wish to show that 100% shading using only a horizontal overhand can result in a very large element and that the shading characteristic of such an element can be duplicated with a much smaller combined horizontal–vertical element.

EXAMPLE D.1. Design a simple horizontal over-the-widow shading element that will provide 100% window shading from 1 P.M. to 4 P.M., from April 21 to September 21 for a window 100 cm (39 in.) high and 150 cm (59 in.) wide, in a wall facing 20° west of south. The building is at latitude 32°N.

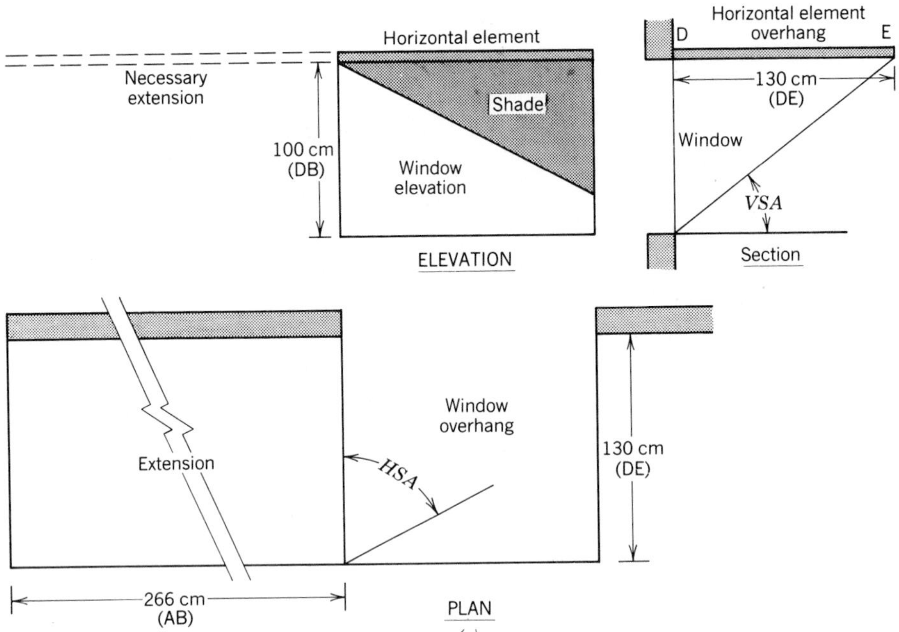

(c)

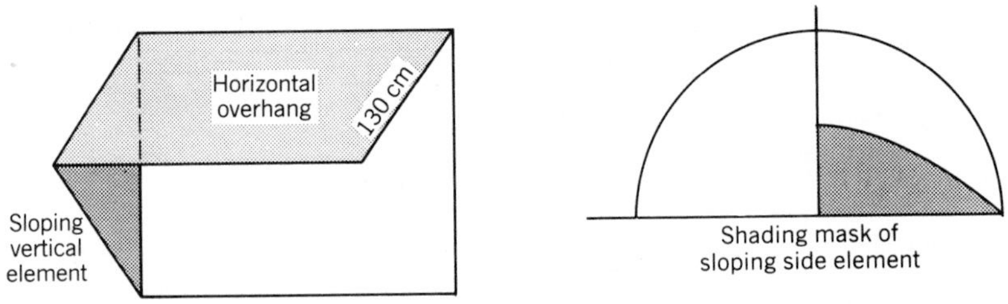

(d)

Fig. D.12 *(continued)*

ANALYTIC SOLUTION

STEP 1. The first step is to draw a sketch (Fig. D.12*a*) showing the orientation of the building. All the sketches connected with this example are shown in Fig. D.12.

STEP 2. Using either the LOF sun angle calculator, or Fig. D.5, tabulate the sun positions for the four times that bracket the shading period. Tabulate them as shown in Fig. D.12*b* and show sun directions as in Fig. D.12*a*. The purpose of establishing sun direction is to be able to immediately determine the required shading element end treatment, by sketching the direction in which the shade falls. This is shown in the fourth column of the tabulation of Fig. D.12*b*.

STEP 3. By inspection of the solar azimuth and the bearing angle (the angle between perpendicular to the wall and the vertical plane through the sun), it is apparent that all shadows fall to the right, and therefore the horizontal element must be lengthened to the left. Maximum extension is required for shading at 1600 on April 21. The next step is to establish the shadow angles *VSA* and *HSA* for each condition. These can be read directly from the LOF sun chart (profile angle = *VSA*; bearing from normal line = *HSA*) or by overlaying the protractor of Fig. D.8*c* on Fig. D.5. *HSA* need not be read off: It is simply the difference between sun azimuth and building bearing (20°) and so can be calculated directly. Fill in the required angles in the *VSA* and *HSA* columns of the table in Fig. D.12*b*.

STEP 4. At this point it is advisable to draw sketches of the window in elevation, section, and plan to understand the calculations. See Fig. D.12*c*. It is necessary for us first to establish the depth of the overhang, and with that known, to then calculate the required extension of the left. Refer now to Fig. D.7. Note carefully that the first item that we wish to calculate is the depth of the overhand, *DE*. Since

$$\tan VSA = \frac{DB}{DE}$$

and we know the window height, *DB*, to be 100 cm [39 in.], we have

$$DE = \frac{100 \text{ cm}}{\tan VSA}$$

We perform this simple calculation for every VSA, and we will select the deepest overhang as the one required. The figures are listed under column *DE* (depth) of Fig. D.12*b*.

STEP 5. Having found the overhand depth required, we can calculate the end treatment needed, that is, the amount that the overhang must be extended to cover the window. Study Fig. D.7. The extension needed is obviously *AB*, since the area *ABD* is unshaded. Therefore, if we move *D* a distance of *AB* (on the figure—to the right; in our example—to the left), point *A* will move to *B*, and shading will be complete. Since

$$\tan HSA = \frac{AB}{DE}$$

and we have already established that *DE* is 130 cm [51 in.], we have

$$AB = 130 \tan HSA$$

These four values are calculated and tabulated in Fig. D.12*b*. We find that the required extension to the left, beyond the window, is 266 cm (105 in.). At this point we can complete the sketch in Fig. D.12*c* with dimensions.

STEP 6. This step is the evaluation stage that follows every design. Our evaluation would be that if, indeed, the stated shading hours must be maintained, an overhang 130 cm (51 in.) deep and 416 cm (164 in.) wide (266 + 150) is excessively large, and we would recommend a vertical element on the left side. The type of vertical element recommended is shown in Fig. D.12*d*; it is a sloping element that reaches from the front edge of the overhang to the window. To understand how such a unit shades, refer again to Fig. D.7. Note very carefully that because *DA* is the shadow of *DE*, triangle *DAB* is the shadow area corresponding to a vertical side element *DEB*. That is, if *DEB* were solid, *DAB* would be its shade. We thus see the extremely important and useful fact that *a sloping side element like DEB is equivalent to an infinitely long extension on that side*. Thus the full segmental areas shown in shading masks can be established not by an impossible infinitely long horizontal element, but by a horizontal element the width of the window and two sloping elements. In our case, such an element would eliminate the necessity for a massive 266-cm (105 in.) side extension.

REFERENCES

Cowan, H. J., (ed.) (1980). *Solar Energy Applications in the Design of Buildings*, Applied Science Publishers, London.

Harkness, E., and Mehta, M. (1978). *Solar Radiation Control in Buildings*, Applied Science Publishers, London.

Lim, B. P., Rao, K. R., Tharmaratnam, K., and Mattar, A. M. (1979). *Environmental Factors in the Design of Building Fenestration*, Applied Science Publishers, London.

Olgyay, A., and Olgyay, V. (1951). *Solar Control and Shading Devices*, Princeton University Press, Princeton, N.J.

Ramsey, C. G., and Sleeper, H. R. (1981). *Architectural Graphic Standards*, 7th ed., AIA/Wiley, New York.

Sun Angle Calculator (1975). Libbey-Owens-Ford. Available from LOF, 811 Madison Ave., Toledo, OH 43695.

Szokolay, S. V. (1980). *Environmental Science Handbook*, Halsted/Wiley, New York.

E
ECONOMIC ANALYSIS

E.1 Economic Decisions

As has been pointed out repeatedly in this text, economic factors are a major consideration in almost every decision in the building design process. The discussion below is necessarily brief but covers the essential principles. For in-depth study of economic analyses in the design and construction process, the reader is referred to the references at the end of the appendix. Publications of the National Bureau of Standards (Marshall and Ruegg, 1980a, 1980b; Ruegg et al., 1978) are the sources of some of the principles explicated below, and are so referenced.

The most common economic questions are:

1. How can the cost of different systems *that produce the same end result* be compared? This may apply to the purchase of a single item such as a motor, to the lighting of a room, or to the choice of the type of HVAC installation for an entire system. The principle involved is the same.

2. Assuming that improving a system, existing or proposed, will reduce operating costs and that many different types of improvement are possible, what preliminary economic guidelines can be established to determine whether any proposed investment appears cost effective? Stated otherwise, how can the initial, simple rate of return of a proposed investment be determined so that it can be compared to the *minimum expected rate of return* on investment?

3. Having determined in answer to question 2 that a number of different proposals, each with its own investment and payback (savings) figures, are apparently cost effective, how can these be compared to determine which is *most* cost effective? Put another way, which proposal has the maximum *net benefit or savings*? This is the type of decision that must be made when comparing energy conservation projects, such as the lighting control systems described in Section 21.4.

4. Having established which of the proposals is most cost effective, how can the *actual* rate of return of the investment be determined? This figure, often referred to as *internal rate of return* (*IRR*), can then be compared to the required rate of return, to determine whether the investment is economically desirable.

5. What is the *payback period* of the investment?

These questions are considered individually in the discussion below.

E.2 Life-Cycle Cost (LCC)

Cost comparisons made on the basis of first cost are legitimate only when operating costs do not bear consideration, either because there are none or because the builder is not the owner. Proposals for pouring a concrete slab might represent the former situation, and speculative construction for immediate sale the latter. In such cases the technique of cost comparison is simple. However, when operating costs are involved that vary from system to system, the only reasonable way to determine and compare system costs is by *life-cycle costing*. The basic theory and calculation techniques are well known (see AIA, 1977; Marshall and Ruegg, 1980a, 1980b; Ruegg et al., 1978), so they will be reviewed only briefly here.

Stated simply, a life-cycle cost represents the total cost of an item or system over its entire life cycle, that is, the sum of first cost and all future costs, less salvage. When the life cycle of the system being studied does not correspond to that of the building in which it is installed (as is usually the case), the designer must decide which life cycle to use. For instance, a lighting system has an estimated life of 15 to 20 years, whereas the building life is usually at least double that. It is usually wiser to make the study over the shorter period, since almost certainly, changes in technology will present an entirely different picture when the time comes to replace the system. This is true of most mechanical/electrical systems being designed today.

Life-cycle cost, in equation form is

$$LCC = IC + MC + AC + OC \qquad (E.1)$$

where

LCC = life-cycle cost
IC = investment cost, (i.e., first cost minus salvage)
MC = maintenance and repair cost
AC = amortization (replacement) cost (not always included in life-cycle cost calculation)
OC = operating costs, including labor and energy

All costs are expressed in terms of present value, which is explained below. Costs can also be annualized. We use the present value format because we feel it gives a clearer answer to the economic decision questions posed above. When calculating costs, the designer must include related ancillary costs. For instance, in comparing lighting systems, the cost figures must reflect the impact on the HVAC system and the wiring system. [For lighting system cost comparisons, see IESNA (1981).]

When the life cycle is short—say two years maximum, the last three components of Equation E.1 can be assumed to be constant without introducing an intolerable error. However, for longer periods this is obviously not a valid assumption because of escalating labor and energy costs. Since these two factors weigh heavily in all economic comparisons, and since they generally exist in different proportions in the systems being considered, escalation in their costs will obviously change the economic balance when costs are projected over the equipment life. It is therefore necessary to change the usual present value formula to reflect this escalation.

The present value of future payments of 1.00, made at the *end* of a series of n periods, is

$$PV = \frac{(1+i)^n - 1}{i(1+i)^n} \qquad (E.2)$$

where

PV = present value
i = interest (discount) rate expressed as a decimal (e.g., 8% = 0.08)
n = number of periods

The factor i is more accurately the discount rate rather than the simple interest rate, the difference being that the latter reflects inflation, either actual or anticipated. If, however, the future payments escalate, the expression becomes:

$$PV = \frac{1 - \left(\frac{K}{1+i}\right)^n}{\frac{1+i}{K} - 1} \qquad (E.3)$$

where K is the escalation rate per period, and the remaining items are as in equation E.2. Thus for a 5% annual escalation, $K = 1.05$, and so on. The values of PV are tabulated in Table E.1 for interest (discount) rates of 8, 10, and 12%, and escalation rates of 3, 5, 8, and 10%. Other values can be calculated from the equations. Tables 15.5 and 16.4 were calculated assuming a 3% annual energy cost escalation.

It is apparent that the result of a life-cycle-cost analysis depends heavily on accurate forecasting when using escalation and that comparative results can readily shift not only with escalation estimates of the various components but also with the length of the life cycle.

E.3 Initial (Simple) Rate of Return

Addressing the second question in Section E.1, we wish to determine whether to examine a proposed cost-saving investment in detail. That is, is the initial (simple) rate of return sufficient? The difference between initial and *actual* rates of return is that the former is simply

$$\text{initial rate of return} = \frac{\text{annual savings}}{\text{investment}}$$

whereas the actual (internal) rate of return depends on the life of the project and the *total* savings accrued during this life cycle. The calcu-

TABLE E.1*a* **Present Value of n Future Payments Beginning at the End of the First Period**

	Interest (Discount) Rate per Period						
n	6%	8%	10%	12%	15%	20%	25%
1	0.94	0.93	0.91	0.89	0.87	0.83	0.80
2	1.83	1.78	1.74	1.69	1.63	1.53	1.44
3	2.67	2.58	2.49	2.40	2.28	2.11	1.95
4	3.47	3.31	3.17	3.04	2.86	2.59	2.36
5	4.21	3.99	3.79	3.61	3.35	2.99	2.69
6	4.92	4.62	4.36	4.11	3.78	3.33	2.95
7	5.58	5.21	4.87	4.56	4.16	3.61	3.16
8	6.21	5.75	5.34	4.97	4.49	3.84	3.33
9	6.80	6.25	5.76	5.33	4.77	4.03	3.46
10	7.36	6.71	6.15	5.65	5.02	4.19	3.57
11	7.89	7.14	6.50	5.94	5.23	4.33	3.66
12	8.83	7.54	6.81	6.19	5.42	4.44	3.73
13	8.85	7.90	7.10	6.42	5.58	4.53	3.78
14	9.30	8.24	7.37	6.63	5.72	4.61	3.82
15	9.71	8.56	7.61	6.81	5.85	4.68	3.86
16	10.11	8.85	7.82	6.97	5.95	4.73	3.89
17	10.48	9.12	8.02	7.12	6.05	4.78	3.91
18	10.83	9.37	8.20	7.25	6.13	4.81	3.93
19	11.16	9.60	8.37	7.37	6.20	4.84	3.94
20	11.47	9.82	8.51	7.47	6.26	4.87	3.95
21	11.76	10.02	8.65	7.56	6.31	4.89	3.96
22	12.04	10.20	8.77	7.65	6.36	4.91	3.97
23	12.30	10.37	8.88	7.72	6.40	4.93	3.98
24	12.55	10.53	8.99	7.78	6.43	4.94	3.98
25	12.78	10.68	9.08	7.84	6.46	4.95	3.99

lation method for the latter is given in Section E.5.

Initial rate of return is a useful criterion in that it can be very easily calculated and then compared to the *minimum* desired rate of return on investment. We emphasize *minimum*, since actual rate of return is always lower than the simple, initial rate of return, except where savings escalate during the project life cycle. If this preliminary comparison indicates that the simple rate of return exceeds the minimum by a margin of a few percent, it is likely that the project is economically feasible, and a detailed analysis can be undertaken.

E.4 Cost-Effectiveness Comparisons

This analysis is undertaken in answer to question 3, that is, several projects are being considered, all of which meet the stated criterion. Which is most desirable from a cost-effectiveness viewpoint? (If only one project is under consideration, the analysis proceeds directly to the IRR calculation described in the next section.)

The difference between life-cycle project cost and life-cycle project savings, that is, net lifetime savings, is a measure of cost effectiveness. Therefore, to compare the cost effective-

TABLE E.1*b* Present Value of *n* Future Payments Beginning at the End of the First Period and Escalating at *K* per Period

	Interest (Discount) Rate per Period 8%			
	Annual (Periodic) Escalation Rate K			
n	1.03	1.05	1.08	1.10
1	0.95	0.97	1.00	1.02
2	1.86	1.92	2.00	2.06
3	2.73	2.84	3.00	3.11
4	3.56	3.73	4.00	4.19
5	4.35	4.60	5.00	5.28
6	5.10	5.44	6.00	6.40
7	5.82	6.26	7.00	7.54
8	6.50	7.06	8.00	8.70
9	7.15	7.84	9.00	9.88
10	7.78	8.59	10.00	11.08
11	8.37	9.33	11.00	12.30
12	8.94	10.04	12.00	13.55
13	9.48	10.73	13.00	14.82
14	9.99	11.41	14.00	16.11
15	10.48	12.06	15.00	17.43
16	10.95	12.70	16.00	18.77
17	11.40	13.32	17.00	20.13
18	11.82	13.92	18.00	21.53
19	12.23	14.51	19.00	22.94
20	12.62	15.08	20.00	24.39
21	12.99	15.63	21.00	25.86
22	13.34	16.17	22.00	27.35
23	13.68	16.69	23.00	28.88
24	14.00	17.20	24.00	30.43
25	14.30	17.69	25.00	32.01

TABLE E.1*c* Present Value of *n* Future Payments Beginning at the End of the First Period and Escalating at *K* per Period

	Interest (Discount) Rate per Period 10%			
	Annual (Periodic) Escalation Rate K			
n	1.03	1.05	1.08	1.10
1	0.94	0.96	0.98	1.00
2	1.81	1.87	1.95	2.00
3	2.63	2.74	2.89	3.00
4	3.40	3.57	3.82	4.00
5	4.12	4.36	4.73	5.00
6	4.80	5.12	5.63	6.00
7	5.43	5.84	6.51	7.00
8	6.02	6.53	7.37	8.00
9	6.57	7.18	8.22	9.00
10	7.09	7.81	9.05	10.00
11	7.58	8.41	9.87	11.00
12	8.03	8.98	10.67	12.00
13	8.46	9.53	11.46	13.00
14	8.85	10.05	12.23	14.00
15	9.23	10.55	12.99	15.00
16	9.58	11.27	13.74	16.00
17	9.90	11.48	14.47	17.00
18	10.21	11.91	15.19	18.00
19	10.50	12.32	15.90	19.00
20	10.76	12.72	16.59	20.00
21	11.02	13.09	17.27	21.00
22	11.25	13.45	17.94	22.00
23	11.47	13.80	18.59	23.00
24	11.68	14.12	19.24	24.00
25	11.87	14.44	19.87	25.00

TABLE E.1*d* Present Value of *n* Future Payments Beginning at the End of First Period and Escalating at *K* per Period

	Interest (Discount) Rate per Period 12%			
	Annual (Periodic) Escalation Rate K			
n	1.03	1.05	1.08	1.10
1	.92	.94	.96	.98
2	1.77	1.82	1.89	1.95
3	2.54	2.64	2.79	2.89
4	3.26	3.41	3.66	3.82
5	3.92	4.14	4.49	4.74
6	4.52	4.82	5.29	5.64
7	5.08	5.45	6.07	6.52
8	5.59	6.05	6.82	7.38
9	6.06	6.61	7.54	8.23
10	6.49	7.13	8.23	9.07
11	6.89	7.63	8.90	9.89
12	7.26	8.09	9.55	10.69
13	7.59	8.52	10.17	11.49
14	7.90	8.92	10.77	12.26
15	8.19	9.30	11.35	13.03
16	8.45	9.66	11.91	13.78
17	8.69	9.99	12.45	14.51
18	8.91	10.31	12.97	15.23
19	9.11	10.60	13.47	15.95
20	9.30	10.87	13.95	16.64
21	9.47	11.13	14.42	17.34
22	9.63	11.37	14.87	18.00
23	9.78	11.60	15.30	18.66
24	9.91	11.81	15.72	19.31
25	10.04	12.01	16.12	19.95

ness of two projects, it is necessary simply to calculate this differential. To express this as an equation, we write:

net savings = life-cycle savings − life-cycle cost

Since this is true for all projects, a comparative differential can be calculated by using the differential in each term of the life-cycle-cost equation, that is,

$$\Delta_{\text{net}} = \Delta_{\text{savings}} - \Delta_{\text{costs}}$$

Taking a very simple example, assume the following conditions.

	Project A	*Project B*
Investment	\$10,000	\$15,000
Life cycle	10 years	8 years
Annual savings	\$2,000	\$3,200
Initial rate of return	20%	21.3%
Discount rate	12%	12%

Since project B has an apparently higher rate of return, we would formulate the differential as B minus A. Using Table E.1*a*, the present worth of lifetime savings for project B (without escalation) is \$3200 × (4.97) or \$15,904, and that of project A is \$2000 × (5.65) or \$11,120. Therefore, the differential cost benefit (CB) of B minus A is $\Delta_{\text{life savings}}$ minus $\Delta_{\text{project costs}}$, or

$$\Delta_{\text{CB}} = (\$15{,}904 - \$11{,}120) - (\$15{,}000 - \$10{,}000) = \$4784 - 5000 = -\$216$$

indicating that project A is more cost effective despite the apparently higher rate of return of project B. If we include escalation in costs, obviously the life savings of the longer life project (A) will shrink more than the savings of shorter life project B. Using Table E.1*d*, we find that the projects are about equal, with 3% annual escalation. Above that, project B is more cost effective.

E.5 Internal Rate of Return (IRR)

This figure represents the rate of return at which lifetime savings are exactly equal to lifetime costs. It cannot be calculated directly, but only by trial and error. We must assume a rate of return, calculate present value of savings, and compare PV to lifetime costs. Because of this difficulty, IRR is frequently neglected, and initial rate of return used instead, although as we saw above, this can readily give misleading results. To illustrate calculation of IRR, we will return to our very simple example and calculate IRR for projects A and B.

For project A, we need a present worth factor of \$10,000/\$2000 = 5, in 10 years. Referring to Table E.1*a*, we see that at a discount (interest) rate of 15%, the present worth factor is 5.019, indicating that IRR is slightly above this. (It is actually 15.1%.) For project B we need an 8-year present worth factor of \$15,000/\$3200 or 4.69. Referring again to Table E.1*a*, we see that this factor falls between 12 and 15%. By a series of trials with equation E.2, we arrive at 13.7%, which is 1.4% *less* than the same factor for project A, confirming the conclusion reached in the preceding step.

E.6 Payback Period

The payback period referred to in most proposals is the reciprocal of the initial rate of return and is usually referred to as the *simple* payback period. However, just as we have seen that the actual rate of return differs from the simple rate (and is usually less than it), so the actual or *discounted payback period* is different, and usually longer than the simple one. This period is, logically, the period of time required for the accumulated net savings to equal the initial investment, with all figures expressed in present value dollars. To illustrate, let us return to the example given above.

1. *Proposal A*. As in the IRR calculation, we need a present worth factor of \$10,000/\$2000, or 5, at a 12% discount rate. From Table E.1*a* we find this to be about 8.1 years. This compares to a simple payback period of 5 years. Note that this is *not* the reciprocal of internal rate of return.
2. *Proposal B*. Here we need a present worth factor of \$15,000/\$3200 or 4.7. At 12% discount rate, from Table E.1*a*, this

is about 7.3 years. This compares to a simple payback period of 4.7 years.

Note from these results the very important fact that a shorter payback period does not necessarily indicate a more cost-effective investment. Here, proposal A has a better return precisely because it yields good savings for a longer period. Summing up the results of our simple study, we have:

	Project A	*Project B*
Initial rate of return	20%	21.3%
Internal rate of return	15.1%	13.7%
Simple payback period	5 years	4.7 years
Discounted payback period	8.1 years	7.3 years

This shows very clearly that although initial rate of return and simple payback period are indications of a proposal's value, they are useless in comparative studies.

REFERENCES

American Institute of Architects (1977). *Life Cycle Cost Analysis—A Guide for Architects*, AIA, Washington, D.C.

Edison Electric Institute. *Interest Tables, 0–25%*, EEI, New York.

Illuminating Engineering Society of North America (1981). *Lighting Handbook, Application* Volume, 6th ed., Section 3, IESNA, New York.

Marshall, H. E., and Ruegg, R. T. (1980a). *Energy Conservation in Buildings: An Economics Guidebook for Investment Decisions*, National Bureau of Standards Handbook 132, NBS, Washington, D.C.

Marshall, H. E., and Ruegg, R. T. (1980b). *Simplified Energy Design Economics*, National Bureau of Standards Special Publication 544, NBS, Washington, D.C.

Ruegg, R. T., et al. (1978). *Life Cycle Costing*, National Bureau of Standards Publication 113, NBS, Washington, D.C.

APPENDICES

F

SOUND TRANSMISSION DATA FOR WALLS

To use Appendix F:

1. Find desired construction Type Code in Index F.1.
2. Find desired STC corresponding to the selected construction Type Code.
3. Refer to Table F.1 for details of wall construction.
4. Refer to Index F.2 for wall thickness weight, STC, fire rating, and the transmission losses at standard octave midpoints.

EXAMPLE. An interior masonry partition with an STC between 50 and 55 is desired. Of particular interest is the TL at 1000 Hz.

1. From Index F.1, we find Type Code "c."
2. From Index F.1, we note that constructions W6, W12b, W15b, and W22b have appropriate STC ratings.
3. From Table F.1 we decide that construction W15b is most suitable for the proposed use.
4. From Index F.2 we find that construction W15b has a transmission loss of 55 db at 1000 Hz.

INDEX F.1 **Sound Transmission Class: Walls**

Type Code:

a. *Wooden stud*
b. *Metal stud*
c. *Masonry*
d. *Concrete*
e. *Staggered stud*
f. *Plaster*
g. *Gypsum wallboard*
h. *With resilient element*
i. *Absorbent blankets or fill*
j. *Fiber board*
k. *Lead*
l. *Gypsum core board*
m. *Double wall*

STC	Type	Item No.	STC	Type	Item No.
63	d,f	W4	46	a,f	W30a
62[a]	c,f,j,m	W23	46	a,f	W31a
56	c	W7	46	a,f,k	W31c
56[a]	c,f	W8	46	a,e,g,i	W37
55[a]	b,g,i	W63	45	c,d	W10b
54[a]	c,f,m	W22b	45	a,f,i	W38
54	b,f,h	W52	45	b,f,h	W50a
53[a]	d,f	W2	45	g,l,m	W85a
53	c,f,h	W15b	44	c	W12a
52[a]	c,f	W6	44	a,e,g	W34a
51	b,f	W44a	44	a,e,g	W35a
51	b,g,h,i	W67	44	a,f,h	W40a
50	c,k	W12b	43	c,d	W10a
50	b,g,i	W60	43[a]	d,j,m	W21
50	g,i,l,m	W85b	43	a,e,f	W35a
49[a]	c,f,m	W22a	43	a,e,f	W36a
48	c,d	W9	43	b,f,k	W43b
48	a,e,f,i	W36b	42[a]	c,f	W5
48	b,f,k	W43c	42	a,f	W14a
47	d	W1	41	b,f	W43a
47	a,g,k	W28b	41	b,g	W55a
47	a,f,k	W31b	40	c,f	W13
47	a,g,h	W39a	40	a,g	W32a
46	c,f	W11a	39	a,g	W28a
46	c,f,h	W15a	36	g,l	W80

[a]Field measurement.

INDEX F.2 **Sound Transmission Loss: Walls**

Designation	*Thickness (in.)*	*Weight (lb/ft²)*	*Transmission Loss (db)*						*STC*	*Fire Rating hr*
			125	*250*	*500*	*1K*	*2K*	*4K*		
W1	3	39	35	40	44	52	58	64	47	½
W2	7	80	39	42	50	58	64	—	53	3
W4	Approx. 16	184	50	54	59	65	71	68	63	4+
W5	5½	55	34	34	41	50	66	—	42	2.5
W6	10	100	41	43	49	55	57	—	52	4+
W7	12	121	45	45	53	58	60	61	56	4+
W8	25	280	50	53	52	58	61	—	56	4+
W9	12	79	46.5	44	46	52	54	56	48	4
W10a	6	34	32	33	40	47	51	48	43	1
W10b	6	34	37	36	42	49	55	58	45	1
W11a	5¼	35.8	36	37	44	51	55	62	46	2
W12a	3¾	26.1	40	40	40	48	55	56	44	1.5
W12b	5	31	41	46	46	56	63	67	50	1.5
W13	4	21.5	39	34	38	43	48	46	40	3
W14a	5	23.4	37	42	39	44	49	49	42	4
W15a	5	27	38	37	44	51	56	59	46	3
W15b	6	31	45	44	50	55	56	59	53	4
W20	7	22.9	32	46	49	53	58	66	52	3
W21	10¼	37	41	42	46	51	52	—	43	Not available
W22a	12	100	37	41	48	60	60.5	—	49	4+
W22b	12	100	40	44	55	67.5	70	—	54	4+
W23	18	120	48	54	58	64	69	—	62	4+
W28a	5	6	21	28	35	42	45	41	39	0.5
W28b	Approx. 5⅛	12	27	37	43	52	56	—	47	0.5
W30a	5¾	13.4 to 15.7	32	37	42	47	47	63	46	0.75
W31a	5¾	13.4 to 15.7	32	37	42	48	48	63	46	0.75
W31b	Approx. 5⅞		33	41	45	52	55	65	47	0.75
W31c	Approx. 5⅞	17–19	32.5	40	43	47	50	62	46	0.75
W32a	5½	8.2	27	31	39	45	52.5	48	40	1
W34a	5	6.2	36	36	40	47	52	45	44	0.5
W35a	6½	13.4	41	41	46	49	41	54	44	1.5
W35b	5¾	15.6	48	46	48	48	48	59	43	1
W36a	6¼	11.1	36	33	42	42	41	51	43	0.75
W36b	6¼	12.8	37	37	49	50	52	66	48	1
W37	5¾	13.8	39	40	42	47.5	55	51.5	46	0.5
W38	5⅜	14.2	39	45	48	50	44	54	45	1
W39a	6¼	6.7	30	40	46	50	49	49	47	1
W40a	Approx. 6½	14.4	43	41	48	50	42	56	44	1
W43a	3⅜	12.3	27	37	43	46	39	47	41	0.75
W43b	3½	15.2	35	43	45	47	48	58	43	0.75
W43c	Approx. 3½	18.2	36	45	47	50	53	61	48	0.75
W44a	5	15.7	34	38	47	50	52	58	51	1
W50a	5¼	13	35	46	48	51	48	43	45	0.75
W52	5	19	50	52	55	56	52	60	54	1
W55a	4⅞	6	29	36	40	46	40	46	41	1
W60	3½	5.4	34	40	47	50	53	54	50	1
W63	6⅛	11.5	36	47	51	57	57	62	55	2
W67	6½	11.3	41	46	49	51	50	60	51	2
W80	2¼	10.2	34	34	37	38	39	45	36	1
W85a	5⅛	14.6	36	35	45	51	53	57	45	3
W85b	6	12.8	37	37	54	56	56	62	50	3

TABLE F.1 **Acoustic Characteristics of Walls: For Other Data See Index F.2**

Designation	*Description*	*Section Sketch*

Solid Concrete

W1 3-in.-thick solid concrete wall poured in situ in test opening. All surface cavities were sealed with thin mortar mix.

W2 6-in.-thick concrete wall with ½-in.-thick layer of plaster on both sides.

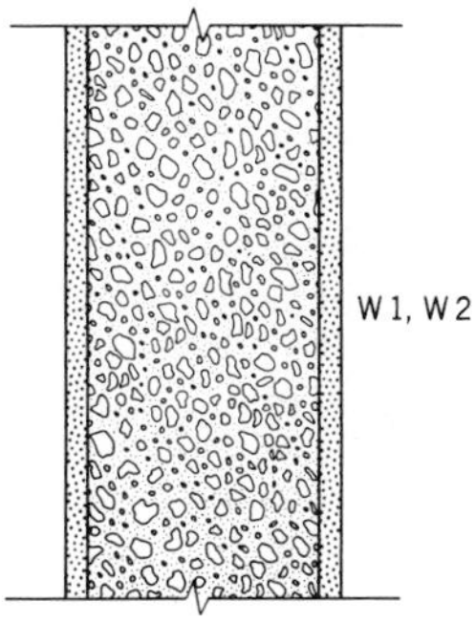

W4 Wall of 4, 6, and 8 × 8 × 16 in. sand and gravel aggregate solid concrete blocks; on each side, ¼- to ½-in.-thick layer of cement gypsum plaster and sand.

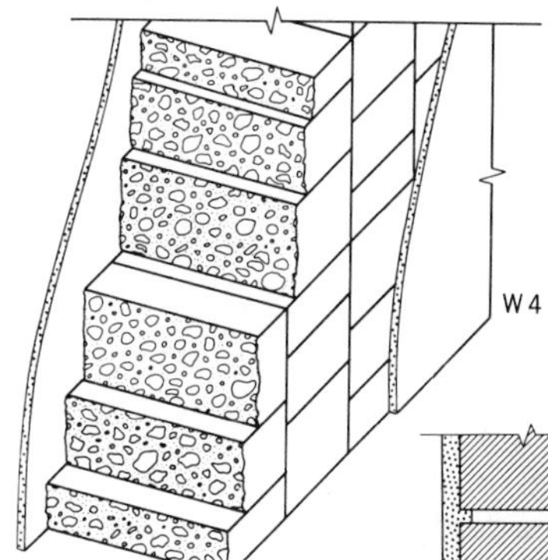

W5 4½-in.-thick brick wall with ½-in.-thick layer of plaster on each side.

W6 9-in.-thick brick wall with ½-in.-thick layer of plaster on each side

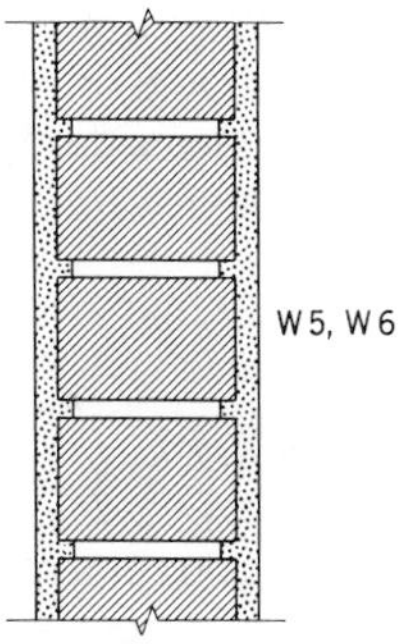

W7 12-in.-thick brick wall.

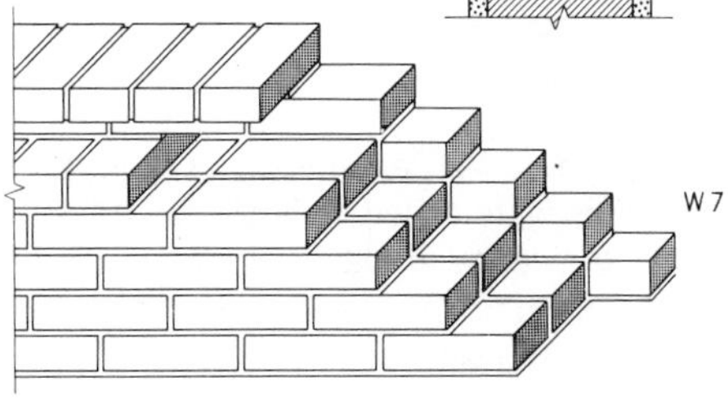

W8 24-in.-thick stone wall with ½-in.-thick layer of plaster on both sides.

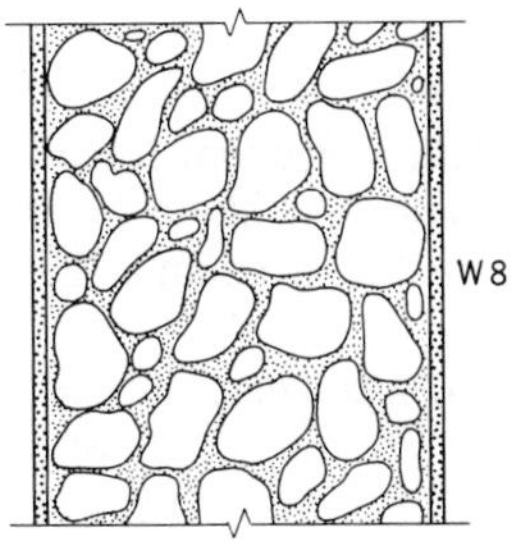

APPENDICES

TABLE F.1 **Acoustic Characteristics of Walls: For Other Data See Index F.2 (Continued)**

Designation	*Description*	*Section Sketch*

Hollow Concrete Block

W9 12-in. wall made of hollow 8 × 8 × 12 in. and 8 × 4 × 16 in. concrete blocks.

W10a 6-in. hollow concrete blocks constructed with vertical mortar joints staggered.

W10b Similar to W10a except wall painted.

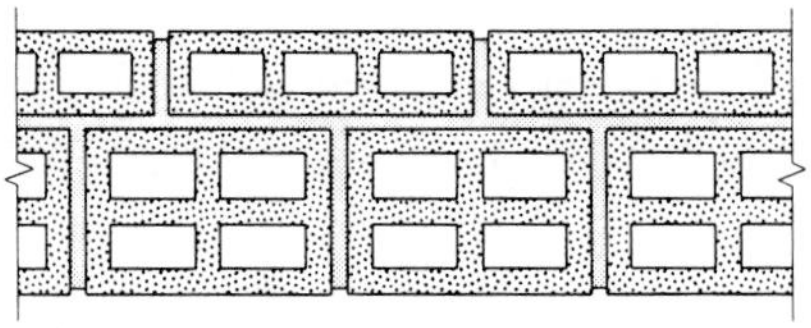

W 9

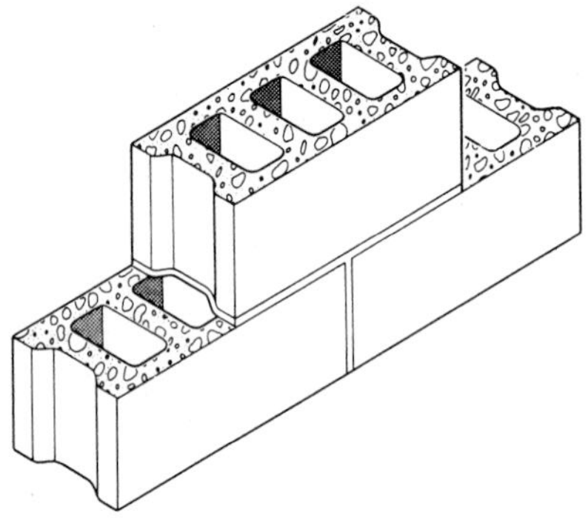

W 10

Cinder Block

W11a 4 × 8 × 16 in. hollow cinder blocks; on each side, ⅝ in. of sanded gypsum plaster.

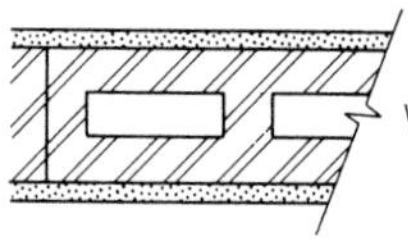

W 11a

Cement Block

W12a 3⅝ × 7¾ × 13½ in. lightweight-aggregate cement blocks with ½-in. mortar joints; three coats of masonry paint applied to each side of partition.

W12b Same as W12a except 1 × 2-in. furring strips were nailed vertically to partition on one side; $\frac{1}{16}$-in. layer of lead, 3.94 lb/ft^2, nailed to furring strips, ¼-in. plywood-covered lead with joints caulked.

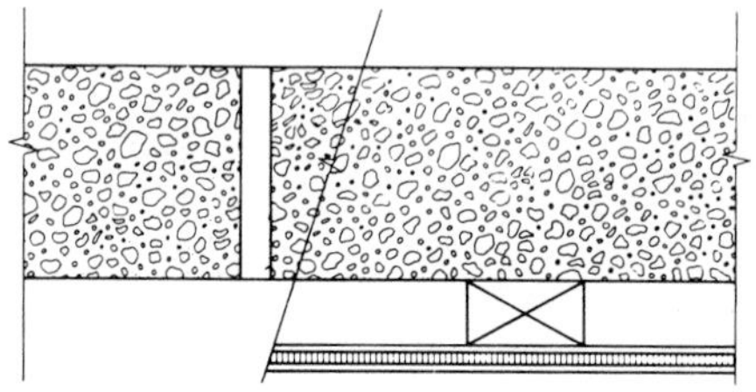

W 12a W 12b

Hollow Gypsum Block

W13 3-in. hollow gypsum blocks cemented together with ⅜-in. mortar joints; on each side, ½-in. sanded gypsum plaster.

W14a 4-in. hollow gypsum blocks cemented together with ⅜-in. mortar joints; on each side, ½-in. sanded gypsum plaster.

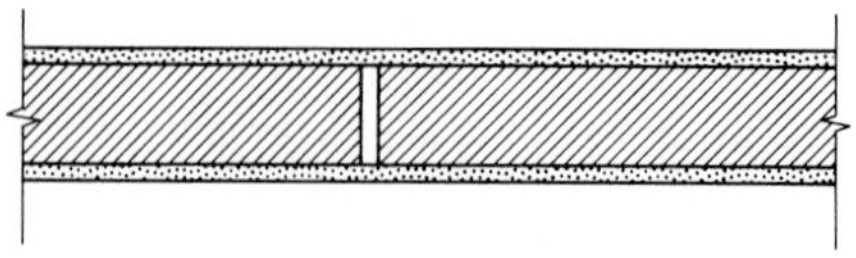

W 13
W 14a

TABLE F.1 **Acoustic Characteristics of Walls: For Other Data See Index F.2 (Continued)**

Designation *Description* *Section Sketch*

Hollow Gypsum Block, Resilient One Side, Plaster Both Sides

W15a 3 × 12 × 30 in. hollow gypsum blocks with ½-in. mortar joints. On one side 7/16-in. sanded gypsum plaster; on the other side resilient clips, spaced 18 in. on centers vertically and 16 in. on centers horizontally, held to ¾-in. metal channels 16 in. on centers, to which expanded metal lath was wire-tied; 11/16-in. sanded gypsum plaster. 1/16-in. white-coat finish applied to both sides.

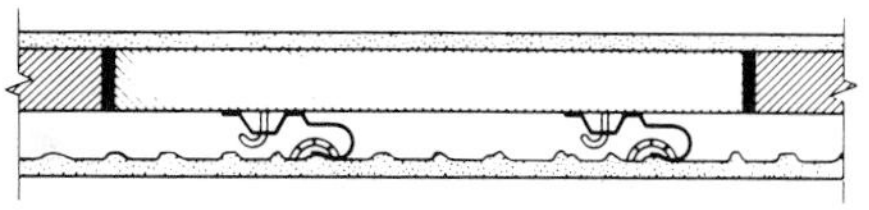

W 15,

W15b Similar to W15a except 4 × 12 × 30 in. gypsum blocks were used.

Hollow Concrete

W21 Precast concrete hollow wall panels with in situ concrete posts and beams. Panels have 1½-in.-thick concrete shells with 6¼-in. air space between them. Layer of fiberboard ½-in. thick is adhered to the exposed surfaces of the panel.

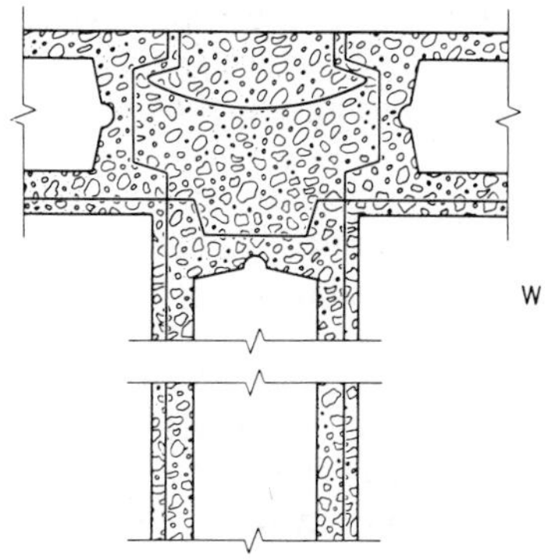

W 21

Double Walls

W22a Double wall with 4½-in.-thick brick leaves separated by a 2-in. cavity (wire ties between leaves); ½-in. plaster on exposed sides.

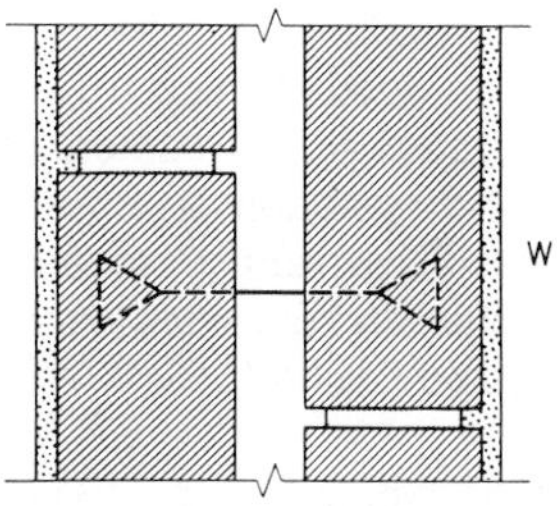

W 22

W22b Similar to W22a but no wire ties between the leaves.

W23 Double wall with 4½-in.-thick brick leaves, 6-in. cavity (no ties); on exposed sides, ½-in. plaster on 1-in.-thick wood-wool slabs mortared to the brick walls.

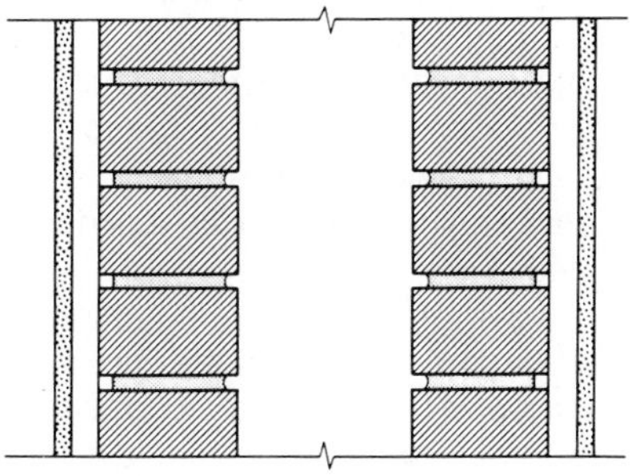

W 23

Wood Stud Walls

W28a 2 × 4 in. wooden studs, 16 in. on centers, ½-in. gypsum wallboard nailed to each side. All joints taped and finished.

W28b Similar to W28a except a layer of lead, 2.95 lb/ft^2, was laminated to each side of panel.

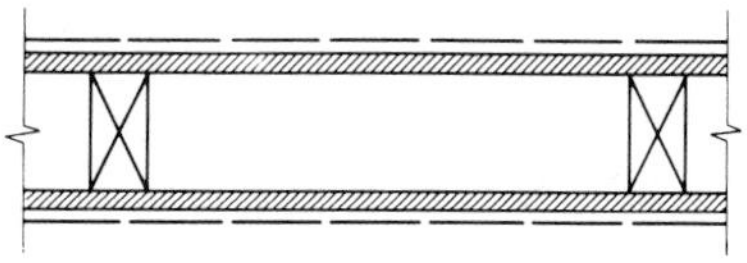

W 28

TABLE F.1 **Acoustic Characteristics of Walls: For Other Data See Index F.2 (Continued)**

Designation	*Description*	*Section Sketch*
W30a	2 × 4 in. wooden studs, 16 in. on centers, attached to 2 × 4 in. wooden floor and ceiling plates, 3/8-in. gypsum lath nailed to studs on both sides, 1/2-in. sanded plaster with white-coat finish.	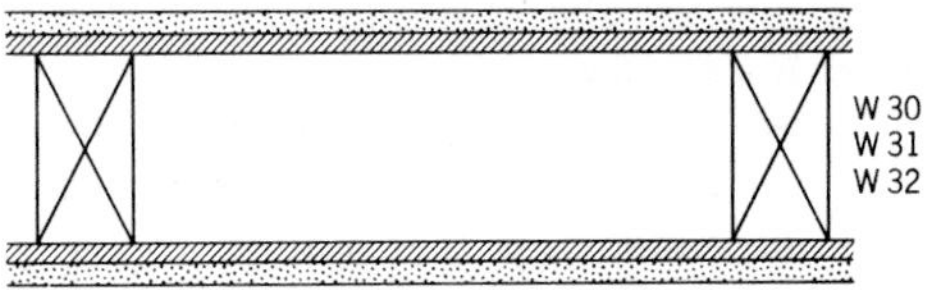
W31a	2 × 4 in. wooden studs, 16 in. on centers, 3/8-in. gypsum lath nailed to studs on both sides, 1/2-in. sanded plaster with white-coat finish.	
W31b	Similar to W31a except a 0.065-in.-thick layer of lead weighing 3.85 lb/ft^2 was laminated to each side of panel.	
W31c	Similar to W31a except a 0.13-in.-thick layer of lead weighing 7.9 lb/ft^2 was laminated to one side of panel.	
W32a	2 × 4 in. wooden studs, 16 in. on centers; on each side two layers of 3/8-in. gypsum wallboard cemented together; joints in exposed surfaces taped and finished.	
Staggered Wood Stud Walls		
W34a	2 × 3 in. wooden studs, 16 in. on centers, staggered 8 in. on centers, attached to 2 × 4 in. wooden plates at ceiling and floor; 1/2-in. gypsum wallboard nailed 7 in. on centers on both sides to studs. All joints taped and finished.	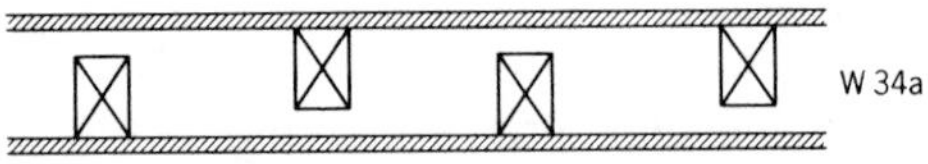
W35a	2 × 3 in. wooden studs, 16 in. on centers, staggered 8 in. on centers (attached to 2 × 4 in. wooden plates at floor and ceiling) two layers of 5/8-in. tapered-edge gypsum wallboard, first layer nailed 7 in. on centers, second layer nailed 16 in. on centers. All exposed joints taped and finished.	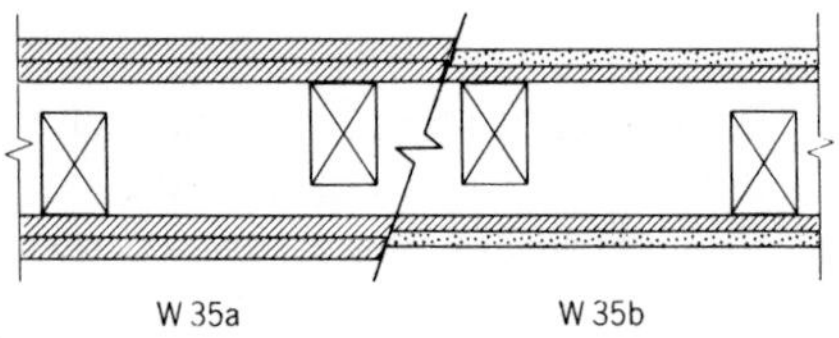
W35b	Similar to W35a except the wall was constructed with 3/8-in. perforated gypsum lath and 1/2-in. sanded gypsum plaster with white-coat finish.	

TABLE F.1 **Acoustic Characteristics of Walls: For Other Data See Index F.2 (Continued)**

Designation	*Description*	*Section Sketch*
W36a	2 × 4 in. wooden studs, 16 in. on centers, staggered 8 in. on centers, and offset ½ in. On each side ⅜-in. gypsum lath nailed to studs, ½-in. gypsum vermiculite plaster, machine applied, and a hand-applied white-coat finish.	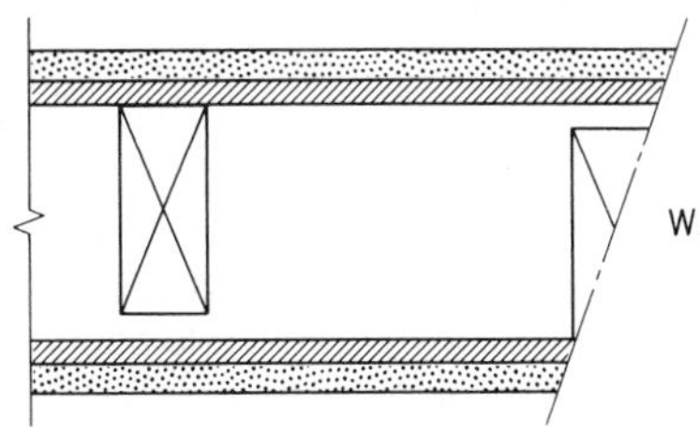 W 36a
W36b	Same as W36a except the space between the studs contained vermiculite fill with a density of 6.3 lb/ft³.	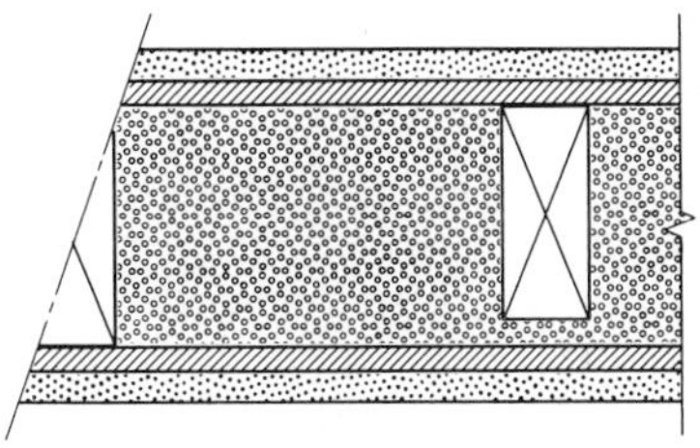 W 36b
W37	2 × 4 in. wooden studs, 16 in. on centers, staggered 8 in. on centers, attached to 2 × 4¾ in. wooden floor and ceiling plates; ½-in. gypsum wallboard nailed on both sides to studs, 0.9-in. wood-fiber wool blanket stapled on the inside of one side of the wall. All joints taped and finished.	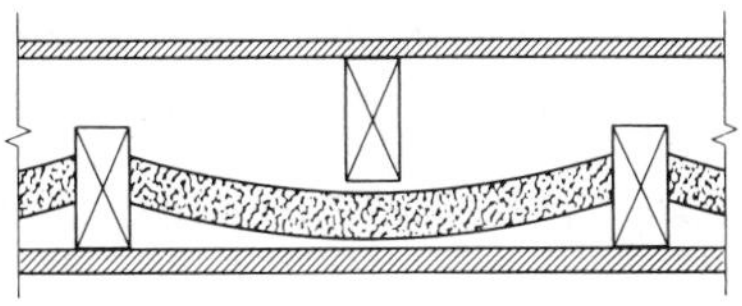 W 37
Slotted Wood Studs		
W38	2 × 4 in. slotted wooden studs, 16 in. on centers, attached to 2 × 4 in. wooden floor and ceiling plates; ⅜-in. gypsum lath nailed 7 in. on centers to studs, ½-in. gypsum plaster with white-coat finish applied to both sides. 3-in. mineral fiber batts stapled between studs.	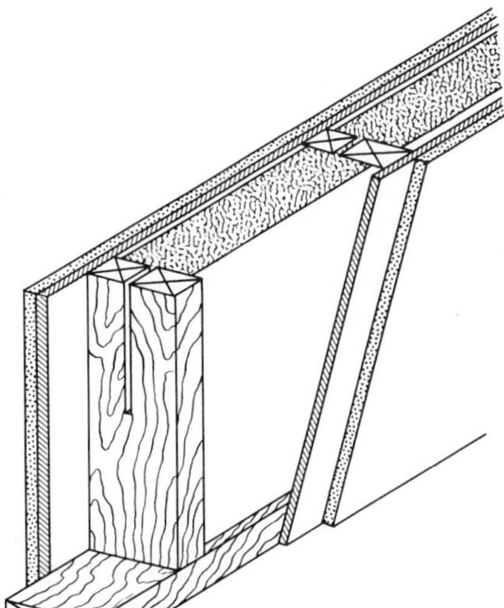 W 38

TABLE F.1 **Acoustic Characteristics of Walls: For Other Data See Index F.2 (Continued)**

Designation	*Description*	*Section Sketch*
Wood Studs; Resilient Mounting		
W39a	2 × 4 in. wooden studs, 16 in. on centers, attached to 2 × 4 in. wooden floor and ceiling plates; resilient channels nailed horizontally to both sides of studs 24 in. on centers, ⅝-in. gypsum wallboard screwed 12 in. on centers to channels. All joints taped and finished.	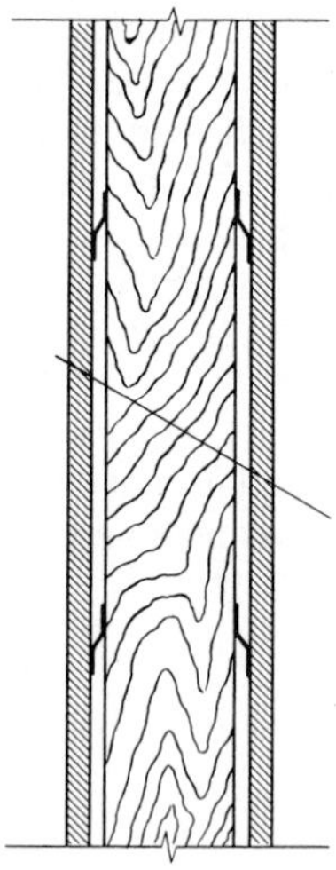 W 39a
W40a	2 × 4 in. wooden studs, 16 in. on centers; resilient clips, nailed to studs on both sides, held ⅜-in. gypsum lath, ½-in. sanded gypsum plaster with white-coat finish.	 W 40a
Steel Truss Stud Wall		
W43a	1⅝-in. steel truss studs; ⅜-in. gypsum lath, ½-in. plaster on both sides.	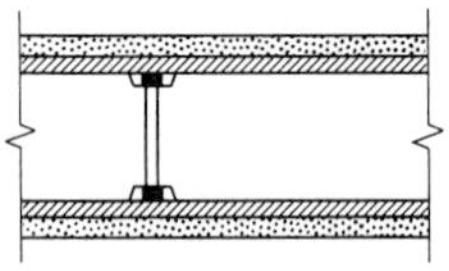 W 43 W 44
W43b	Similar to W43a except a layer of lead, 2.95 lb/ft^2, was laminated to one side of partition.	
W43c	Similar to W43a except a layer of lead, 2.95 lb/ft^2, was laminated to each side of partition.	
W44a	3¼-in. steel truss studs, 24 in. on centers, attached to metal floor and ceiling tracks; on both sides ⅜-in. perforated gypsum lath attached with wire clips wire-tied to studs, ½-in. sanded gypsum plaster.	
Steel Truss Studs; Resilient Mountings		
W50a	2½-in. steel truss studs 16 in. on centers, ⅜-in. gypsum lath attached with resilient clips to studs, ½-in. plaster applied to both sides.	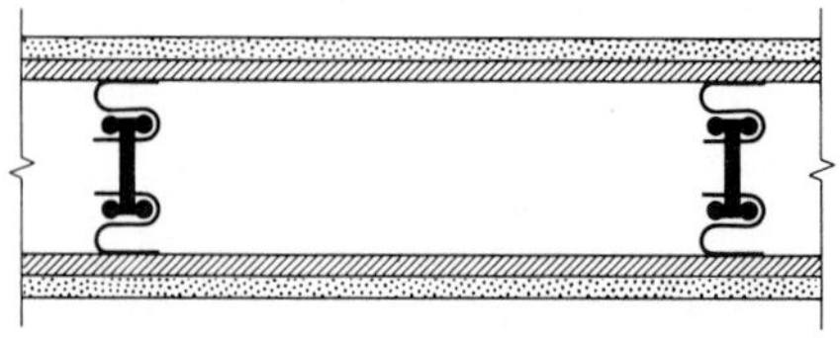 W 50a

TABLE F.1 **Acoustic Characteristics of Walls: For Other Data See Index F.2 (Continued)**

Designation	*Description*	*Section Sketch*
W52	3¼-in. steel truss studs, 16 in. on center; on each side resilient clips fastened 16 in. on centers to studs, ¼-in. metal rod wire-tied to clips, diamond mesh metal lath wire-tied to metal rods, ¾-in. sanded gypsum plaster.	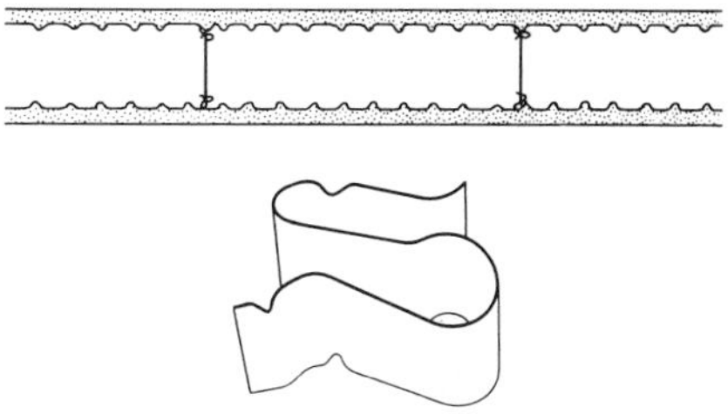 W 52
	Metal Channel Stud Wall	
W55a	3⅝-in. metal channel studs, 24 in. on centers, set into 3⅝-in. metal floor and ceiling runners; ⅝-in. gypsum wallboard screwed to studs on both sides. All joints taped and finished.	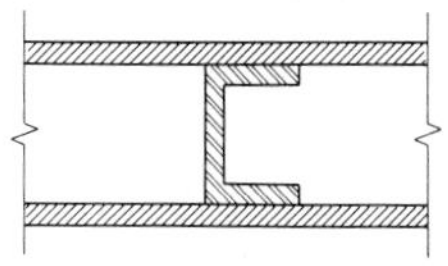 W 55a
W60	2½-in. metal channel studs, 24 in. on centers, set in 2½-in. metal floor and ceiling runners; ½-in. vinyl-coated gypsum wallboard adhesively attached and screwed to studs on both sides. All joints sealed with caulking compound. Aluminum batten strips screwed 12 in. on centers to gypsum board at joints; top and bottom finished with aluminum ceiling and base trim. 2-in. mineral fiber blankets hung between studs.	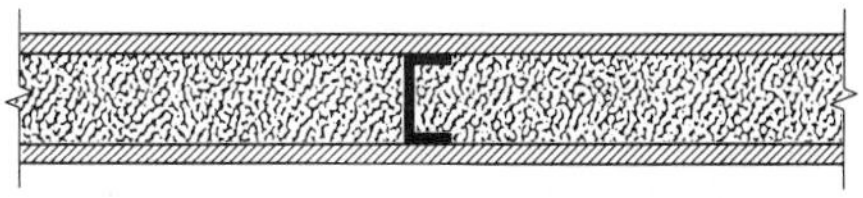 W 60
W63	3⅝-in. metal channel studs, 24 in. on centers, set into 3⅝-in. metal runners, which were attached through continuous beads of nonsetting resilient caulking compound to floor and ceiling, respectively. Two layers of ⅝-in. gypsum wallboard attached to both sides of studs. First layer screwed 8 in. on centers at joints and 12 in. on centers in field; second layer laminated and screwed 24 in. on centers to first layer, with joints staggered 24 in. 1½-in.-thick mineral fiber felt, 3 lb/ft^3, stapled between studs. All exposed joints taped and finished. The ¼-in. clearance around the perimeter closed with a nonsetting resilient caulking compound.	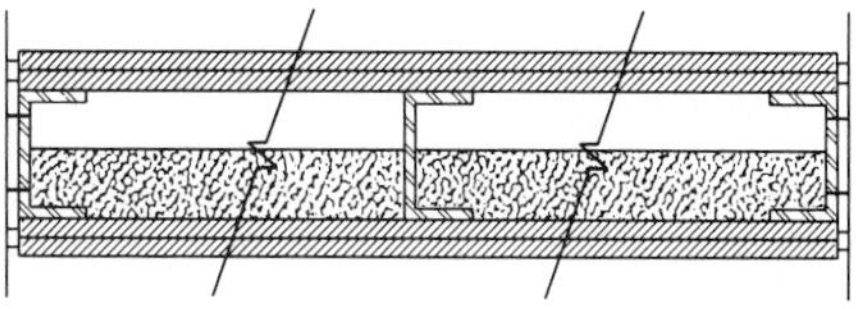 W63
W67	3⅝-in. metal channel studs, 24 in. on centers, set in 3⅝-in. metal floor and ceiling runners; ⅝-in. gypsum wallboard screwed to studs on both sides. On one side, resilient channels screwed horizontally, 24 in. on centers to inner layer; ⅝-in. gypsum wallboard screwed to channels. On the other side, ⅝-in. gypsum wallboard laminated directly to inner layer. 3-in. mineral fiber blankets hung between studs. All exposed joints taped and finished.	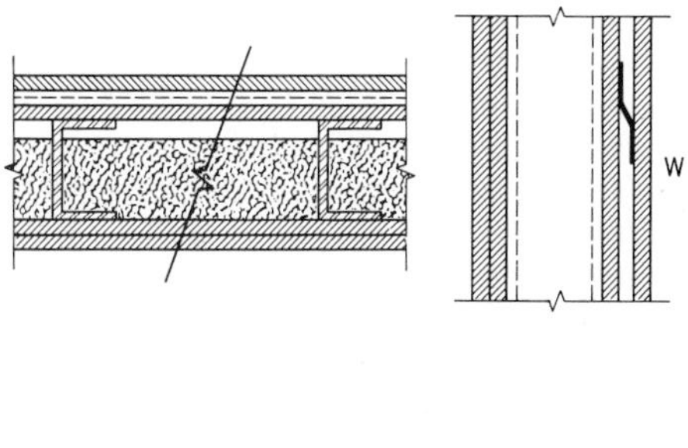 W 67

APPENDICES

TABLE F.1 **Acoustic Characteristics of Walls: For Other Data See Index F.2 (Continued)**

Designation	*Description*	*Section Sketch*
Gypsum Partitions		
W80	24-in.-wide panels constructed of 1 × 24 in. gypsum core board offset 1½ in. at edges to form tongue-and-groove edge; ⅝-in. vinyl-faced gypsum wallboard laminated to both sides of core board. Panels inserted into two-piece metal floor and ceiling tracks. Gypsum to gypsum screws at ¼ and ½ points along vertical edges of face boards.	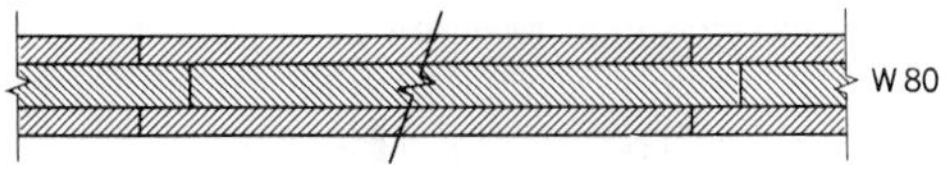
W85a	Double wall with 1⅜-in. air space. Each leaf consisted of 24-in.-wide panels of ⅝-in. gypsum core board strips, 7½ and 4⅜ in. wide, offset 1½ in. at edges to form tongue and groove; ⅝-in., vinyl-faced, gypsum wallboard laminated to both sides of core board strips. Panels screwed 12 in. on centers to 1¼ × 1 in. angle floor and ceiling runners.	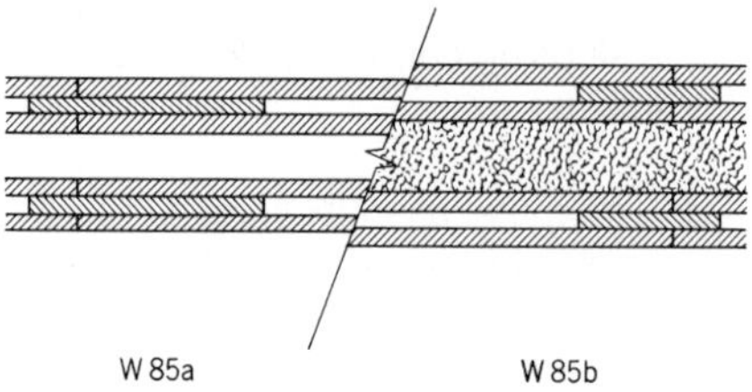
W85b	Similar to W85a except space between leaves was 2⅛ in. and contained 2-in. mineral fiber blankets stapled to one leaf. ¼-in. perimeter clearance closed with a nonsetting resilient caulking compound. Vertical face layer joints sealed with joint compound.	

G

SOUND TRANSMISSION AND IMPACT INSULATION DATA FOR FLOOR/CEILING CONSTRUCTION

To use Appendix G:

1. Find desired construction Type Code in Index G.1 or G.2. Index G.1 lists the constructions by STC. Index G.2 lists the constructions by IIC.
2. Find desired STC/IIC ratings corresponding to selected Type Code.
3. Refer to Table G.1 for details of construction.
4. Refer to Index G.3 for thickness weight, STC, IIC, fire rating, and transmission loss at standard octave midpoints.

INDEX G.1 **Sound Transmission Class: Floor/Ceiling Constructions**

Type Code:

a. *Wooden joist*
b. *Metal joist*
c. *Concrete or masonry*
d. *Plaster ceiling*
e. *Gypsum board ceiling*
f. *With resilient elements*
h. *With carpeting*
i. *With absorbent blankets*
j. *With separate ceiling joists*

STC	*IIC*	*Type*	*Item No.*	*STC*	*IIC*	*Type*	*Item No.*
55[a]	57[a]	c,d,f	F14	46[a]	42[a]	c,d	F23
54[a]	64[a]	c,d,f	F17b	45	44	a,e,f	F44
52	80	a,e,i,j	F48	44	42	c,f	F3-2d
51[a]	53[a]	c,d,f	F10	44	41	c	F3-1a
51[a]	48[a]	c,d	F7a	44	80	c,h	F2-1a
50[a]	53[a]	c,e,f	F25	44	29	c	F1-c
50[a]	51[a]	c,e	F27	44	25	c	F1a
50[a]	48[a]	c,d	F16	43[a]	43[a]	a,d	F32b
49[a]	48[a]	c,d	F9	42[a]	32[a]	c	F22
48[a]	47[a]	c,d	F12	40	32	a,e,i	F40a
48	33	b,c,d	F60a	39[a]	37[a]	a,e	F34
47	62	b,c,e	F58	37	33	a,e	F38a
47[a]	42[a]	c,d	F24	37	32	a,e	F39a
47	59	b,c,d,h	F57b	34[a]	32[a]	a,e	F30
47	37	b,c,d	F57a	29[a]	32[a]	a,e	F35a
46	74	b,c,d	F60c	29	56	a,e,h	F35b

[a]Field measurement.

EXAMPLE. A standard (simple) wooden joist floor–ceiling construction is required with a minimum IIC of 55. Any STC above 35 is acceptable.

1. Since IIC is the determining factor, refer to Index G.2. Basic Type Code is "a."
2. From index G.2 we note that none of the listed "a" constructions gives an IIC of 55. However, from Section 27.25 we note that the addition of carpeting will add 10 to 27 points to the IIC, making items F34, F38a, F39a, and F30 all suitable. We have discounted special and resilient constructions, because of the requirement for standard (simple) construction.
3. From Table G.1 we select construction F30 as being most appropriate for application of carpeting.
4. From Index G.3 we have (without carpet): thickness 9½ in., STC 34, IIC 32 (+ carpet), fire rating ¼ h. With high pile carpeting on foam rubber pad, this construction will have IIC > 55, STC > 35.

INDEX G.2 **Impact Insulation Class: Floor/Ceiling Constructions**

Type Code:

a. *Wooden joist*
b. *Metal joist*
c. *Concrete or masonry*
d. *Plaster ceiling*
e. *Gypsum board ceiling*
f. *With resilient ceiling element*
g. *With resilient floor element*
h. *With carpeting*
i. *With absorbent blankets*
j. *With separate ceiling joists.*

IIC	*STC*	*Type*	*Item No.*	*IIC*	*STC*	*Type*	*Item No.*
80	52	a,e,h,i,j	F48	44	45	a,e,f	F44
80	44[b]	c,h	F2-1a	43[a]	43[a]	a,d,g	F32b
74	46	b,c,d,h	F60c	42[a]	47[a]	c,d	F24
64[a]	54[a]	c,d,g	F17b	42[a]	46[a]	c,d	F23
62	47	b,c,e,h	F58	42	44[b]	c,g	F3-2(d)
59	47[b]	b,c,d,h	F57b	41	44[b]	c	F3-1(a)
57[a]	55[a]	c,d,g	F14	37	47	b,c,d	F57a
56[a]	29[b]	a,e,h	F35b	37[a]	39[a]	a,e	F34
53[a]	51[a]	c,d,g	F10	33	48	b,c,d	F60a
53[a]	50[a]	c,e,g	F25	33	37	a,e	F38a
51[a]	50[a]	c,e	F27	32[a]	42[a]	c	F22
48[a]	51[a]	c,d	F7a	32	40	a,e,i	F40a
48[a]	50[a]	c,d	F16	32	37	a,e	F39a
48[a]	49[a]	c,d,g	F9	32[a]	34[a]	a,e	F30
47[a]	48[a]	c,d	F12	32[a]	29[a]	a,e	F35a
				29	44[b]	c	F1c
				25	44	c	F1a

[a]Field measurement.
[b]Estimated on the basis of similar structures.

INDEX G.3 **Floor/Ceiling Sound Transmission and Construction Data**

Designation[a]	Thickness (in.)	Weight (lb/ft^2)	Transmission Loss (db)						STC	IIC	Fire Rating (h)
			125	250	500	1K	2K	4K			
F1a	4	53	59	67	75	78	77	75	44	25	1
F1c	4⅛	54	59	66	73	76	74	—	44	29	1
F2-1(a)	4⅜	54	38	36	32	23	—	—	44	80	1
F3-1(a)	4½	55	59	64	58	42	26	—	44	41	1
F3-11(d)	4⅛	53	65	67	72	67	48	40	44	42	1
F7a	5½	61	66	70	68	66	59	37	51	48	2
F9	6⅝	65	63	69	71	68	55	38	49	48	2
F10	8¼	90	60	61	59	54	45	—	51	53	2½
F12	10	62	68	65	67	66	61	50	48	47	3
F14	9½	83	59	58	55	48	42	—	55	57	3
F16	8⅛	65	56	60	60	62	54	—	50	48	2
F17b	9¼	57	56	50	44	40	30	—	54	64	2
F22	6¼	28	76	69	80	79	75	—	42	32	¾
F23	9½	45	71	73.5	72	71	66	54	46	42	¾
F24	10¼	65	67	66	67	66	62	—	47	42	¾
F25	10	45	65	61	57	51	39	—	50	53	¾
F27	7⅝	50	62	66	61	58	46	—	50	51	¾
F30	9½	7	79	79	80	76	69	—	34	32	¼
F32b	11	12	69	72	68	64	55	—	43	43	¾
F34	10¼	9.9	83	80	71	57	52	45	39	37	1
F35a	10	9.2	83	83	72	54	52	43	29	32	—
F35b	10	9.2	58	44	38	24	23	—	29	56	—
F38a	11¾	9	86.5	85.5	85	80	68.5	—	37	33	½
F39a	11⅞	9.5	80	80	81	71	66	58	37	32	1
F40a	11⅞	10	78	84	80	70	65	58	40	32	1
F44	10½	10.1	71	74	70	63.5	61	—	45	44	¾
F48	12⅜	10.7	34	28	20	17	18	—	52	80	¾
F57a	18 9/16	23.2	79	72	72	70	63.5	—	47	37	3
F57b	18 9/16	23.2	59	43	32	18	—	—	47	59	3
F58	21½	20.4	56	41	33	23	17	—	47	62	1
F60a	11	38.2	66	68.5	74	76	75	—	48	33	1½
F60c	11⅝	39	46	38	33.5	25	20	9	46	74	1½

[a]Material extracted from *A Guide to Airborne, Impact and Structure-Borne Noise Control in Multi-Family Dwellings,* HUD/FHA/NBS, 1971.

APPENDICES

TABLE G.1 **Acoustic Characteristics of Floors: For Acoustic Data see Index G.3**

Code[a]	*Description*	*Section Sketch*

Reinforced Concrete Slab

F1a 4-in.-thick reinforced concrete slab, isolated from support structure. Concrete was reinforced with 6 × 6 in. No. 6 AWG reinforcing mesh placed at the centerline horizontal plane of the slab. All surface cavities were sealed with a thin mortar mix.

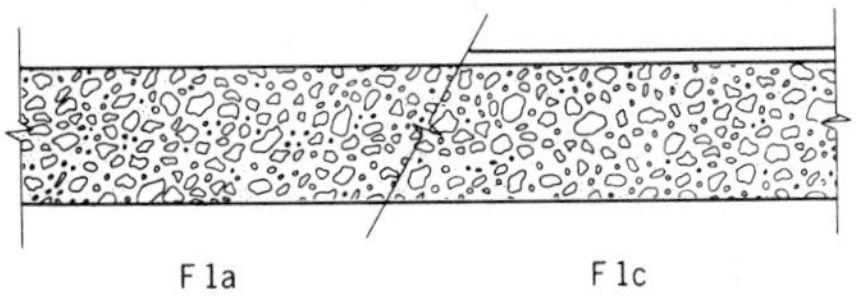

F1c Same as F1a except ⅛-in.-thick vinyl tile was adhered to concrete.

Reinforced Concrete with Floor Coverings

See also F1c above.

F2-1(a) 4-in.-thick reinforced concrete slab with carpeting and pad. The carpeting was of ¼-in. wool loop pile with ⅛-in. woven jute backing, 0.49 lb/ft^2; the foam rubber pad was ¼ in. thick and weighed 0.53 lb/ft^2.

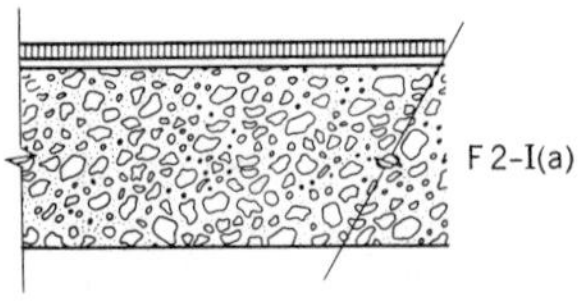

F3-1(a) 4-in. reinforced concrete slab with ½ × 9 × 9 in. oak blocks, 1.8 lb/ft^2, set in mastic.

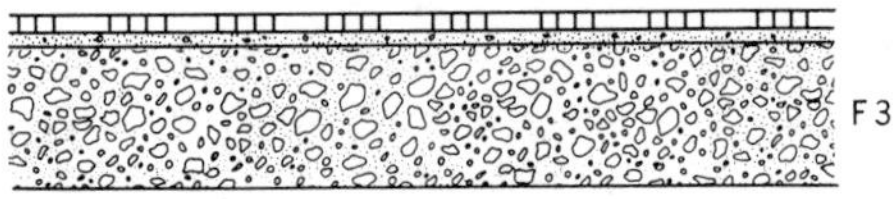

F3-2(d) 4 in. concrete slab with ⅛-in. cork.

F7a 4⅜-in.-thick reinforced concrete slab. On the floor side, ¾-in.-thick, sand-cement screed with ⅛-in. linoleum floor covering. On the ceiling side, ⅜-in. layer of plaster.

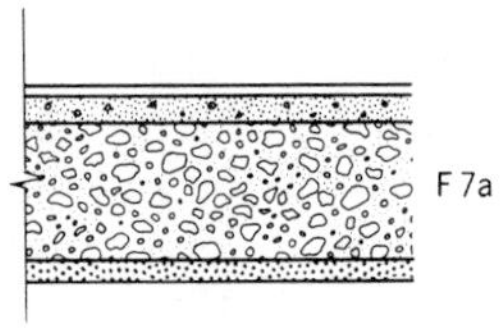

F9 4⅜-in.-thick reinforced concrete slab. On the floor side, ½-in.-thick layer of bitumen with ½-in.-thick soft wood fiberboard, which was covered with a thin layer of bitumen with sand and a ¾-in.-thick sand-cement screed. On the ceiling side ⅜-in. layer of plaster.

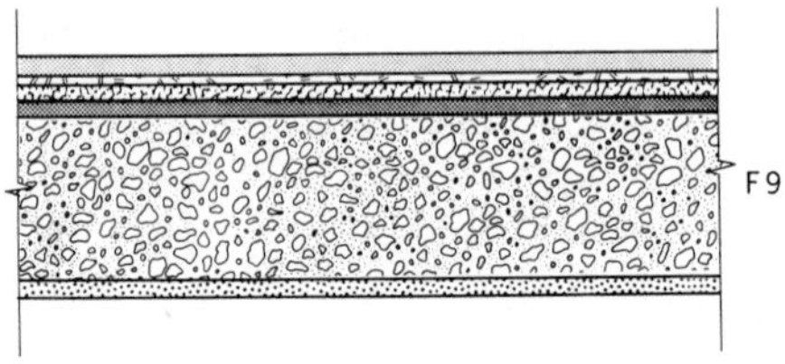

Reinforced Concrete Slab, Floating Floor

F10 5-in.-thick reinforced concrete. On the floor side, 1½-in.-thick wire mesh reinforced sand-cement screed floating on ½-in.-thick bitumen-bonded, glass-wool quilt covered with building paper. On the screed, ½-in.-thick pitch-mastic with a linoleum floor covering. On the ceiling side, ½-in. layer of plaster.

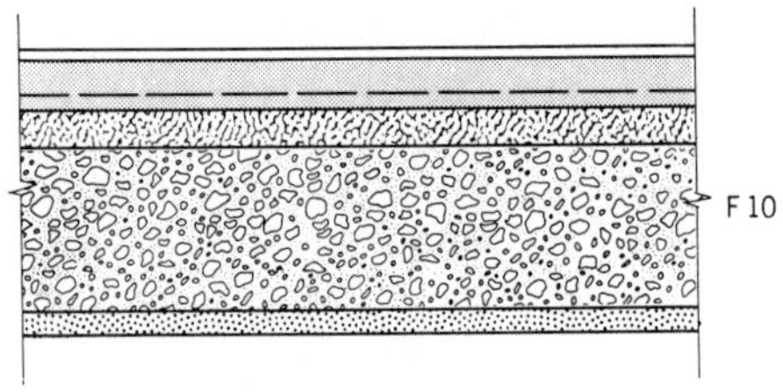

F12 4⅜-in.-thick reinforced concrete slab. On the floor side, ¾-in.-thick sand-cement screed. On the ceiling side, brick wire mesh, suspended 4 in. with wire hangers, held ⅞-in. gypsum plaster.

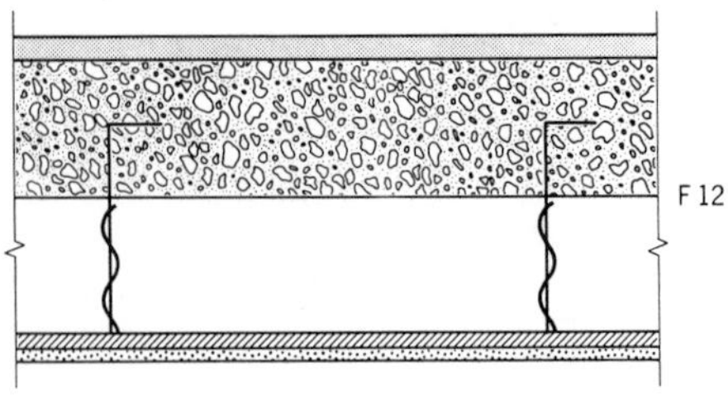

TABLE G.1 **Acoustic Characteristics of Floors: For Acoustic Data see Index G.3 (Continued)**

Code[a]	*Description*	*Section Sketch*
F14	6-in.-thick reinforced concrete slab. On the floor side, ¾-in.-thick tongue-and-groove wood flooring nailed to 1½ × 2 in. wooden battens, 16 in. on centers, floating on 1-in.-thick, glass-wool quilt. On the ceiling side, ½-in. layer of plaster.	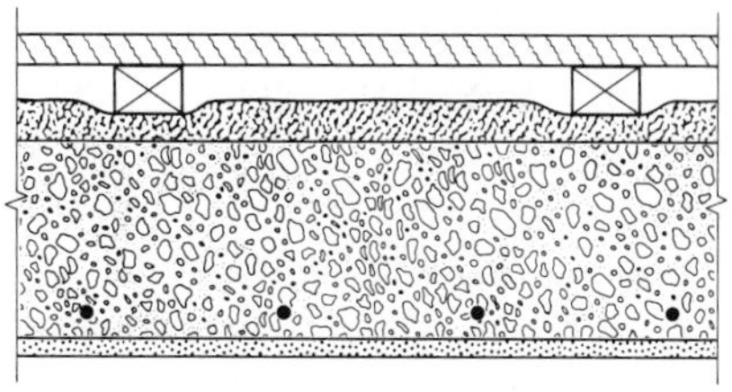 F 14
	Concrete with Hollow Blocks	
F16	5 × 10 in. hollow masonry blocks, 14 in. on centers, with spaces between blocks filled with 5-in.-thick reinforced concrete. On the floor side, ⅞-in.-thick wood blocks adhered to 1½-in.-thick, sand-cement screed. On the ceiling side, ¾-in. layer of plaster.	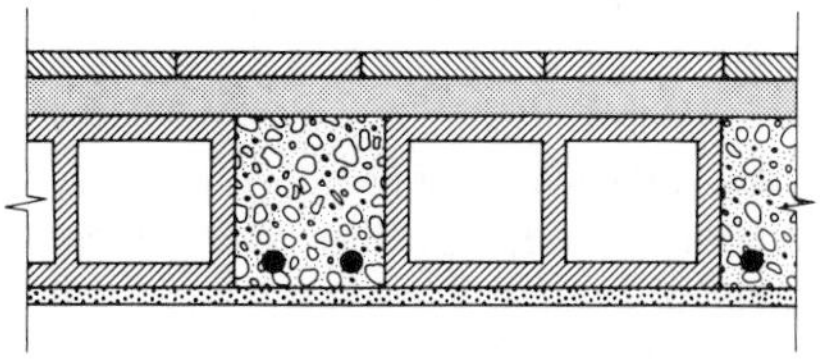 F 16
F17b	4 × 12½ in. hollow masonry blocks, 15½ in. on centers, with spaces between blocks filled with 4-in.-thick reinforced concrete. On the floor side, 2-in.-thick, sand-cement screed; linoleum on 1-in.-thick wood flooring nailed to 1 × 2 in. wooden battens, spaced 15½ in. on centers, floating on glass-wool quilt, approximately 1 in. thick. On the ceiling side, ¾-in. layer of plaster.	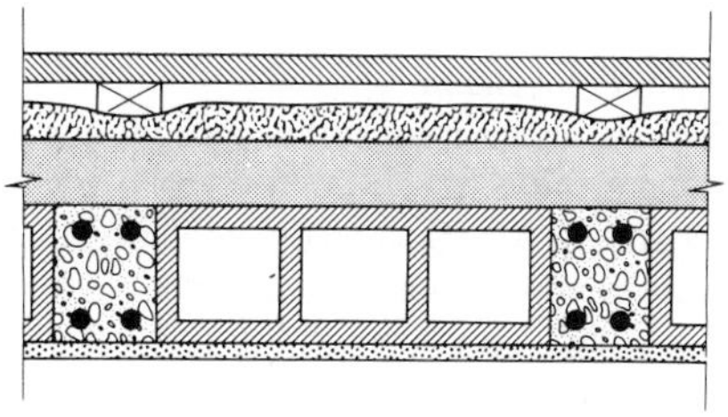 F 17b
	Concrete Channel Slab	
F22	Prefabricated concrete channel slabs mortared together 20 in. on centers. Each slab had a 3-in.-deep trapezoidal channel with bases of 11 and 14¾ in. On the floor side, ¾-in.-thick, sand-cement finish.	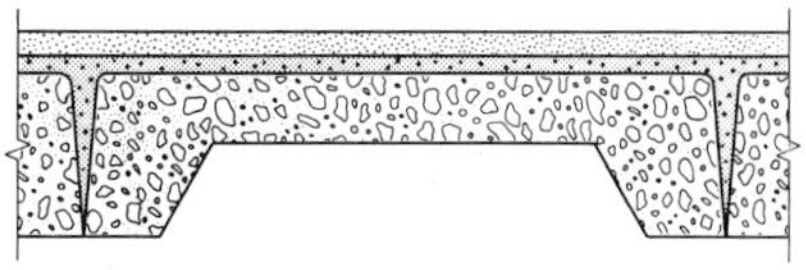 F 22
	Ribbed Concrete	
F23	7¼-in. ribbed concrete floor. Ribs were 5¼ × 3¾ in., spaced 21 in. on centers, with 1 × 2 in. wooden nailing strips cast into ends. On the floor side, the slab was 2 in. thick with a ¾-in.-thick, sand-cement screed. On the ceiling side, ⅝-in.-thick wooden laths nailed to nailing strips, held ⅝-in.-thick plaster.	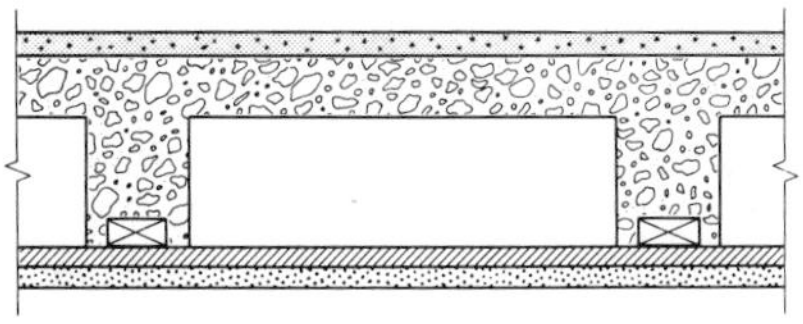 F 23
	Concrete Channel Beam	
F24	7-in. precast trapezoidal concrete channel beams, 14 in. on centers, with spaces between beams filled with sand-cement mix. On the floor side, 1½-in.-thick, sand-cement screed with 1-in.-thick, wood-block floor	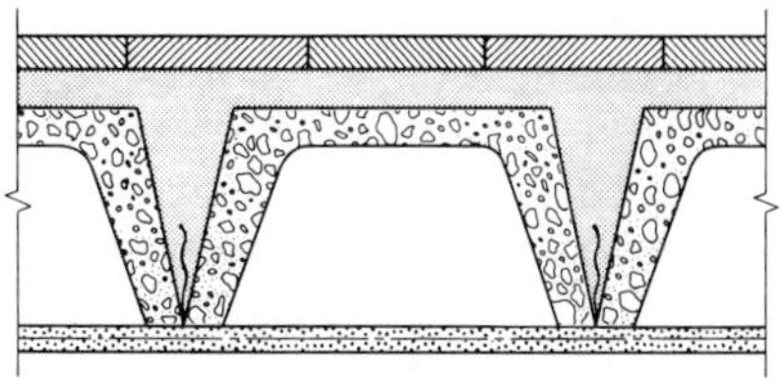 F 24

TABLE G.1 **Acoustic Characteristics of Floors: For Acoustic Data see Index G.3 (Continued)**

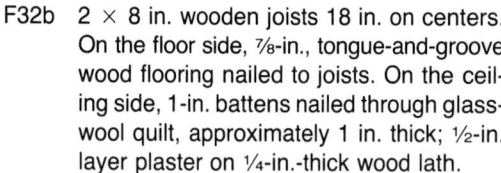

Code[a]	*Description*	*Section Sketch*
	covering. On the ceiling side, approximately 3/4-in.-thick layer of plaster on expanded metal lath.	
	Precast Concrete Beam, Floating Floor	
F25	5-in. precast concrete channel beams, 14 1/2 in. on centers, with spaces between beams filled with a sand-cement mix. On the floor side, 7/8-in.-thick, tongue-and-groove wood flooring nailed to 1 × 2 in. wooden battens, 20 in. on centers, on approximately 1-in.-thick, glass-wool quilt on 3/4-in.-thick, sand-cement screed. On the ceiling side, 1/8-in. layer of plaster on 3/8-in. gypsum wallboard nailed to 1 × 2 in. wooden battens spaced 14 1/2 in. on centers.	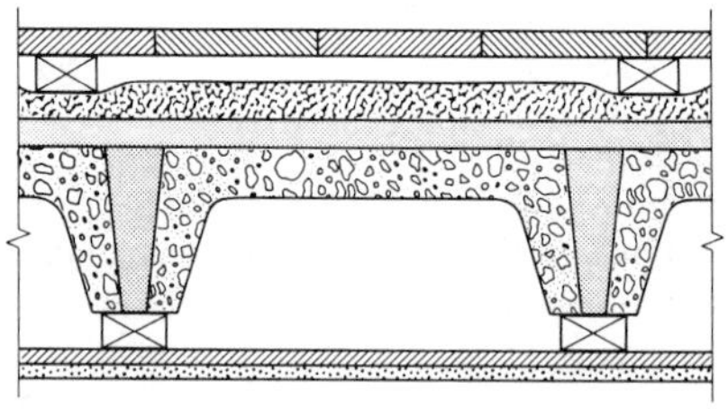 F 25
	Hollow Concrete Beam	
F27	5-in. precast trapezoidal hollow concrete beams, 14 1/2 in. on centers, with bases of 14 and 12 1/2 in. Spaces between beams filled with sand-cement mix. On the floor side, 1-in.-thick, sand-cement screed with 3/16-in. cork tile floor covering. On the ceiling side, 3/8-in.-thick gypsum wallboard attached to 1 × 2 in. wooden battens held by metal clips.	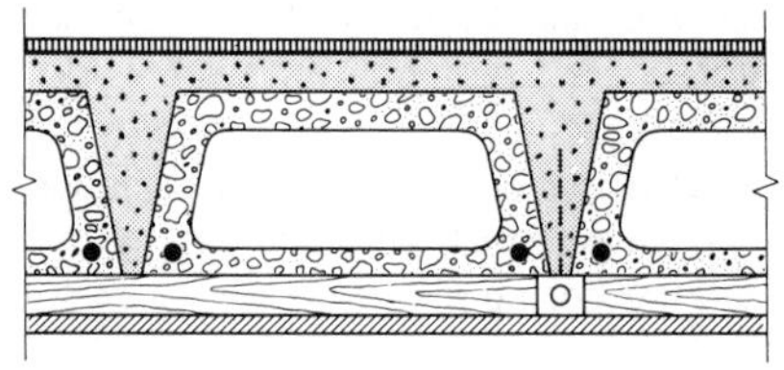 F 27
	Wooden Joist	
F30	2 × 8 in. wooden joists 16 in. on centers. On the floor side, 7/8-in., tongue-and-groove flooring nailed to joists; on the ceiling side, 3/8-in. gypsum wallboard nailed to joists with joints sealed.	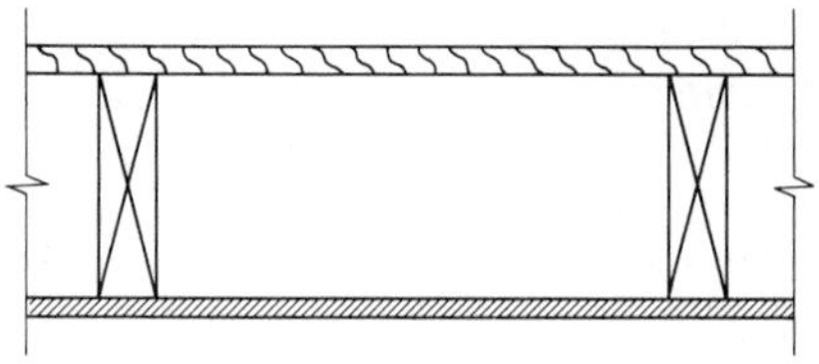 F 30
F32b	2 × 8 in. wooden joists 18 in. on centers. On the floor side, 7/8-in., tongue-and-groove wood flooring nailed to joists. On the ceiling side, 1-in. battens nailed through glass-wool quilt, approximately 1 in. thick; 1/2-in. layer plaster on 1/4-in.-thick wood lath.	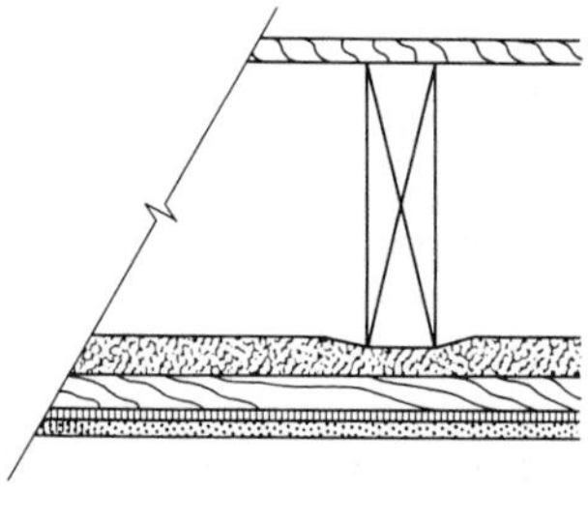 F 32b
F34	2 × 8 in. wooden joists, 16 in. on centers. On the floor side, 1/2-in.-thick C-D plywood nailed 8 in. on centers to joists, 25/32-in.-thick hard wood flooring on plywood. On the ceiling side, 1/2-in.-thick gypsum wallboard nailed 6 in. on centers to joists. All joints taped and finished; ceiling tile adhered to gypsum board.	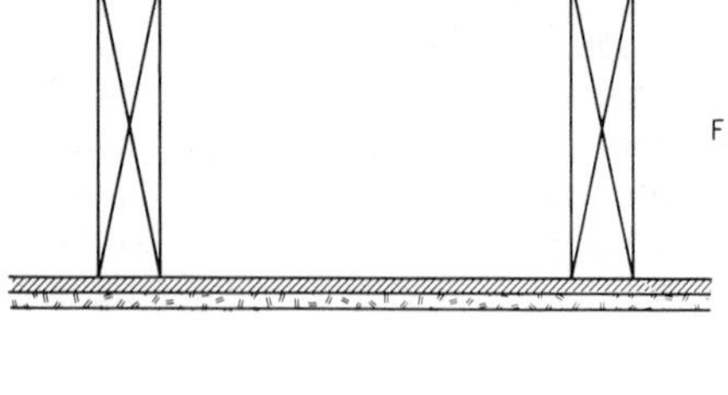 F 34

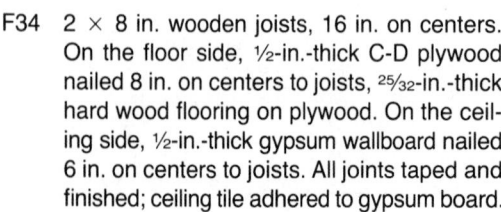

TABLE G.1 **Acoustic Characteristics of Floors: For Acoustic Data see Index G.3 (Continued)**

Code[a]	*Description*	*Section Sketch*

F35a 2 × 8 in. wooden joists, 16 in. on centers. On the floor side, 1½-in.-thick, tongue-and-groove wood fiberboard nailed to joists, vinyl tile floor covering. On the ceiling side, ½-in.-thick gypsum wallboard nailed 6 in. on centers to joists. All joints taped and finished.

F35b Similar to F35a except fiberboard covered with carpet and pad.

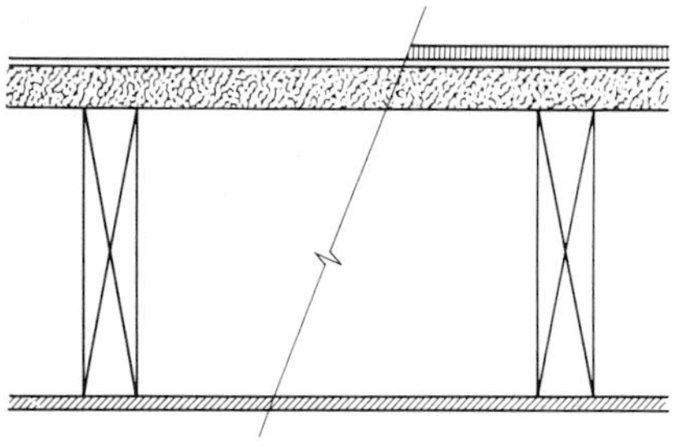

F 35a F 35b

F38a 2 × 10 in. wooden floor joists spaced 16 in. on centers. ⅝-in. fir plywood subfloor nailed to joists 8 in. on centers; ½-in. plywood underlayment nailed to subfloor with joints staggered to miss joints of the subfloor; ⅛ × 9 × 9 in. vinyl asbestos tile glued to underlayment. On the ceiling side, ½-in. gypsum wallboard nailed 12 in. on centers with all joints and nailheads taped and finished.

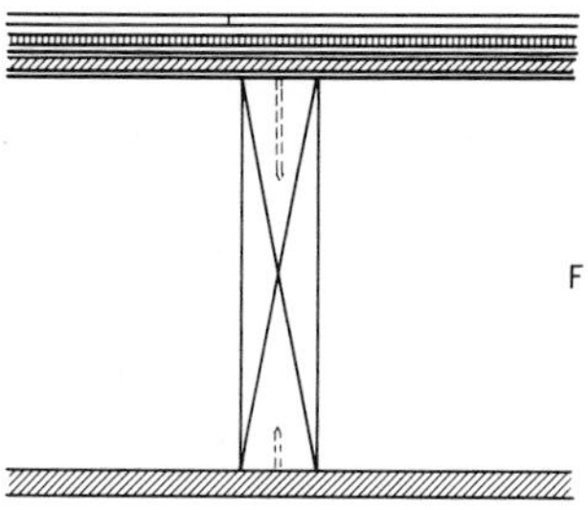

F 38a

F39a 2 × 10 in. wooden joists 16 in. on centers. On the floor side, ½-in.-thick plywood subfloor nailed 6 in. on centers along edges and 10 in. on centers in field, building paper underlayment, 25⁄32 × 2¼ in. oak wood flooring nailed at each joist intersection and midway between joists. On the ceiling side, ⅝-in.-thick gypsum wallboard, nailed 6 in. on centers to joists. All joints taped and finished.

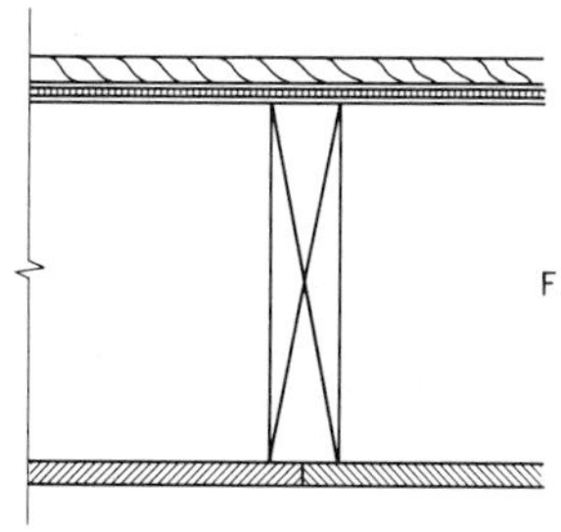

F 39a

Wooden Joist, Resilient Ceiling

F44 2 × 8 in. wooden joists 16 in. on centers. On the floor side, ¾-in.-thick wood subfloor, layer of building paper, and ¾-in.-thick tongue-and-groove fir finish flooring. On the ceiling side, resilient runners bridged across joists and nailed 12 in. on centers to joists; ⅝-in.-thick gypsum wallboard screwed to resilient runners. All joints taped and finished.

F 44

Wooden Joist with Insulation

F40a 2 × 10 in. wooden joists 16 in. on centers with 3-in.-thick mineral fiber batts stapled between joists. On the floor side, ½-in.-thick plywood subfloor nailed 6 in. on centers along edges and 10 in. on centers in field, building paper underlayment, 25⁄32 ×

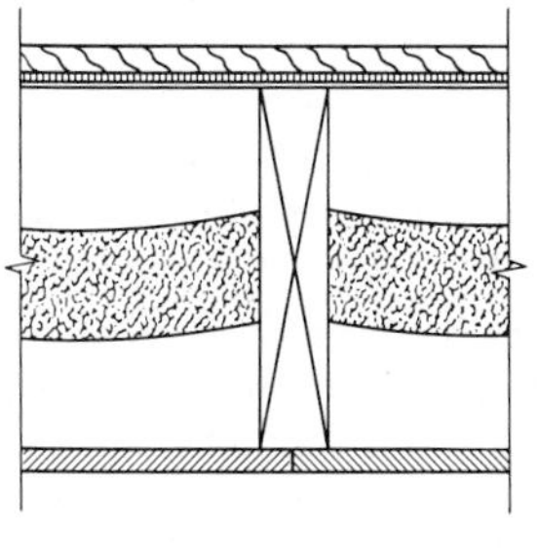

F 40a

TABLE G.1 **Acoustic Characteristics of Floors: For Acoustic Data see Index G.3 (Continued)**

Code[a]	*Description*	*Section Sketch*
	$2\frac{1}{4}$ in. oak wood flooring nailed at each joist intersection and midway between joists. On the ceiling side, $\frac{5}{8}$-in.-thick gypsum wallboard nailed 6 in. on centers to joists. All joints taped and finished.	
F48	2 × 8 in. wooden joists, 16 in. on centers. On the floor side, $1\frac{1}{8}$-in.-thick regular C-D rough plywood nailed 6 in. on centers along periphery and 16 in. on centers at other bearings, plywood covered with an all-hair pad (40 oz/yd^2) and all-wool pile (44 oz/yd^2) carpet. The total weight of the carpet was 4.14 lb/yd^2 and the total thickness was $\frac{3}{8}$ in. On the ceiling side, 2 × 4 in. wooden joists, 16 in. on centers, staggered 8 in. on centers relative to the floor joists, 3-in.-thick fibered glass blankets stapled between ceiling joists, $\frac{5}{8}$-in.-thick gypsum wallboard nailed to ceiling joists. All joints taped and finished; entire periphery of panel caulked and sealed. The ceiling was supported independently from the floor structure.	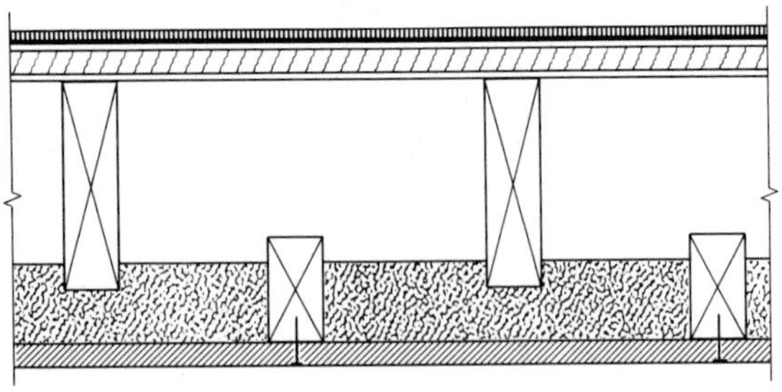 F 48
	Steel Joist with Concrete Floor	
F57a	$2\frac{1}{2}$-in.-thick perlite concrete, 72 lb/ft^3 on 28-gauge corrugated steel units supported by 14-in. steel bar joists; $\frac{1}{8}$-in.-thick asphalt tile cemented to concrete. On the ceiling side, $\frac{3}{4}$-in. furring channels, $13\frac{1}{2}$ in. on centers, wire-tied to joists, 3.4 lb/yd^2 diamond mesh metal lath wire-tied to furring channels, $\frac{9}{16}$-in. coat of plaster with $\frac{1}{16}$-in., white-coat finish.	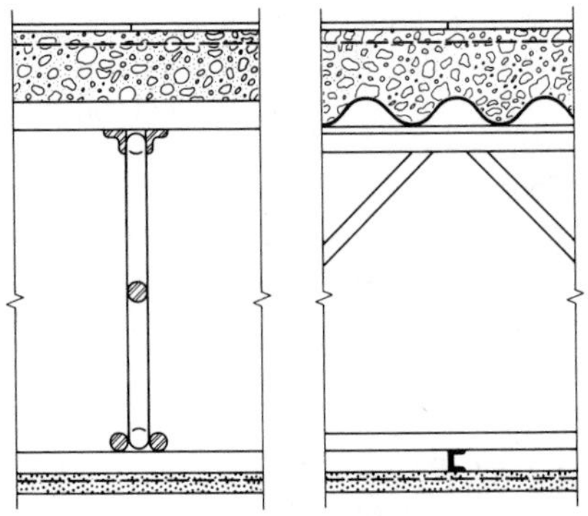 F 57
F57b	Same as F57a except carpet and pad in lieu of asphalt tile.	
F58	18-in. steel joists 16 in. on centers. On the floor side, $\frac{5}{8}$-in.-thick C-D rough plywood nailed to joists, $1\frac{5}{8}$-in.-thick foamed concrete, 100 lb/ft^3, slab constructed on the plywood; concrete covered with an all-hair pad (40 oz/yd^2) and an all-wool pile (44 oz/yd^2) carpet. Total weight of the carpet, 4.14 lb/yd^2, total thickness, $\frac{3}{8}$ in. On the ceiling side, $\frac{5}{8}$-in.-thick gypsum wallboard nailed to joists. All joints taped and finished; entire periphery of panel caulked and sealed.	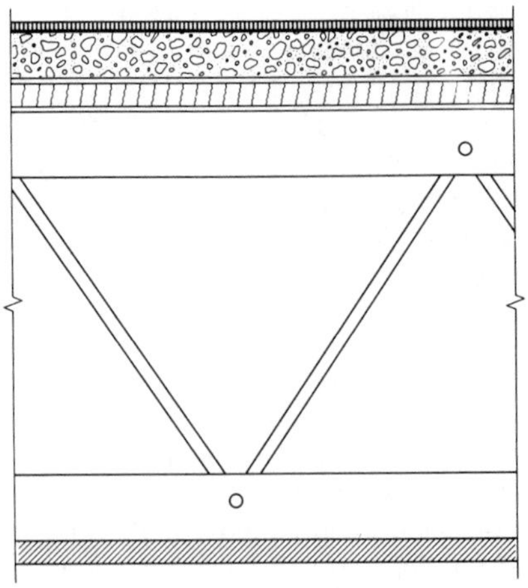 F 58

TABLE G.1 **Acoustic Characteristics of Floors: For Acoustic Data see Index G.3 (Continued)**

Code[a]	*Description*	*Section Sketch*
F60a	7-in. steel bar joists spaced 27 in. on centers. On the floor side, 3/8-in. metal rib lath attached to top of joists, and 2-in.-thick poured concrete floor. On the ceiling side, 3/4-in. metal furring channels wire-tied to joists 16 in. on centers; 3/8 × 16 × 48 in. plain gypsum lath held with wire clips and sheet metal end joint clips; 7/16-in. sanded gypsum plaster and 1/16-in., white-coat finish.	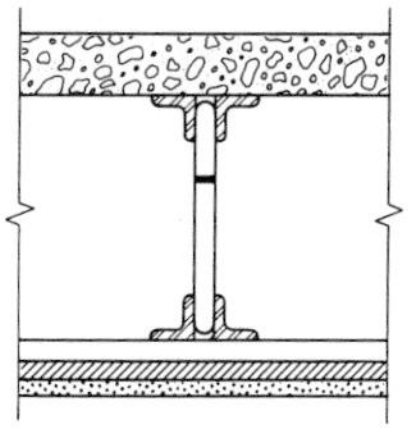 F 60
F60c	Structure F60a with nylon carpeting and foam rubber pad placed on the floor. The carpet pad had an uncompressed thickness of 1/4 in., backed with a woven jute fiber cloth. The carpet had 1/8-in. woven backing and 1/4-in. looped pile spaced 7 loops per inch with a total thickness of 3/8 in.	

[a]Material extracted from *A Guide to Airborne, Impact and Structure-Borne Noise Control in Multi-Family Dwellings,* HUD/FHA/NBS, 1971.

H

OPTIONS FOR COMPUTATION ASSISTANCE

This book presents "manual" calculation procedures—simple enough to require only a hand-held calculator and an informed and patient user. It is by now obvious, however, that some very time-consuming procedures are required for the detailed environmental analysis of buildings. If a machine can be made to perform repetitious calculations automatically, its human user can be freed to do higher-grade tasks. Three degrees of such automatic computation assistance are available. In order of increasing complexity, they are the handheld *programmable calculator*, the *micro-* (or "personal") *computer*, and the *mainframe computer*. It is important to realize that with each step toward greater complexity, the user gets faster and more thorough analysis. But each step also requires a greater purchase cost for the hardware (machine) and greater training time for the user—whether to learn how to run the software (prepared programs, which are themselves more expensive with each step) or to learn how to program the machine. There is also decreasing mobility with each step. Hand-held calculators may be small enough to carry in a wallet, and programmable calculators (which should be accompanied by a printer unit) can be stuffed into a briefcase. Some personal computers are portable, but again, the question of a printer unit arises. Mainframe computers may be accessible by telephone line, but printed analysis requires access to fixed-in-place equipment. Finally, with each step, a typical user of prepared software becomes more remote from the calculation process—for better or for worse.

The point of this introduction, then, is to help you select the appropriate tool for the task. Quick approximations require no more than a hand-held calculator; thorough analysis of many alternative designs for a given problem will be more efficiently done with automatic computation.

There is a rapidly growing array of software for the automatic calculations associated with building energy performance. Table H.1 presents a guide to the selection of software for building thermal analysis. If your architectural idea cannot be reduced to numbers, it cannot be assisted by computer analysis.

H.1 Hand-Held Calculators

This book presents manual procedures that require a common device, the hand-held calculator. It should have trigonometric functions and one or two memories. For acoustics, logarithms are essential as well. Powered by batteries or by photovoltaic cells, hand-held calculators are truly portable, and inexpensive enough for every professional to own at least one.

H.2 Microcomputers

These machines are commonplace in architectural offices, since their services include word processing, graphics, billing, and other project management functions in addition to their role in a building's technical analysis. There are three broad categories of microcomputer software available for building design assistance:

1. *Specific issue design tools*
   This software is aimed at particular issues, such as thermal analysis, daylighting, acoustics, fire protection, etc., that arise

Table H.1 **Considerations in Choosing Computer Software for Heating or Cooling Analysis**

Part A. ***Facts and Figures***
- Package name and version
- Price
- Systems available for (applications)
- Required supporting software
- Memory requirements
- Diskette capacity required
- Utility programs included
- Internal format, records, types
- Portability
- User skill level required
- System upgrade policy and cost

Part B. ***Qualitative Factors***
- Documentation
  - Organization for learning—tutorials, examples
  - Organization for reference
  - Readability
  - Includes all needed information
  - Table of contents and index
- Menus (optional)
  - Ease of use
  - Includes all needed information
- Ease of use
  - Initial startup and installation
  - Conversion of external data
  - Operator use
  - Application implementation (speed, accuracy, etc.)
- Error recovery
  - From input error
  - Restart from interruption
  - From data media damage (bad disk)
- Support
  - For initial startup
  - For system improvement

Part C. ***Technical Factors***
- General technical user concerns
  - Speed and convenience
  - Validity and accuracy
  - Ability to modify analysis
  - Source code availability
- Building types
  - Residential buildings
  - Commercial and nonresidential buildings
  - Underground or below-grade considerations
  - Special conditions
- Weather and climate
  - Weather data source
  - Availability of weather data files
  - Ease of generating weather data
- Building or system loads analysis
  - Method of analysis
  - Orientation and shading
  - Number of building zones
  - Building schedules and internal gains
  - Temperature setpoints and deadbands
- Solar features or systems analysis
  - Method of analysis
  - Orientation and shading
  - System types modeled
  - Component properties and ranges
  - Multiple solar features
  - System limitations
- Report generation
  - Summary and detail report formats
  - Flexibility of report formats
  - Ability to generate custom reports
  - Ability to print reports to disk files
- Special features
  - Supports graphics output
  - Terminals supported
  - Printers or plotters supported
  - Other special features

Source: M. Steven Baker, *Computer-Aided Solar Design* presentation at Oregon Solar Conference, September 1984.

during the design process. Some programs overlap between issues; relating daylighting to thermal performance is particularly common.

2. *Compliance tools*
   Some state governments certify software that demonstrates how a building design complies with the state energy code; this can include thermal, lighting and electrical criteria.
3. *Whole-building design tools*
   These are performance simulations, usually concerned with energy consumption. Preliminary design tools can work with rules-of-thumb that yield approximate results, quickly guiding the designer to more efficient solutions. The tools for evaluating a finished building design include both monthly and yearly energy consumption and life-cycle costs.

The following list of software sources and programs in each of the above categories is presented in the approximate order in which these issues are encountered within this text—that is, design strategies first, acoustics last. We recognize the danger of immediate obsolescence of such lists.

(a) Design Strategies

CLIMATE CONSULTANT. Prof. Murray Milne, Graduate School of Architecture, UCLA. 405 Hilgard Avenue, Los Angeles, California, 90024-1467. A detailed data base which can be queried by users, this is a sophisticated elaboration on the climate design strategy method given in Figs. 2.9 and 2.10.

(b) Solar Position and Performance

CALPAS. Berkeley Solar Group, 454 Santa Clara Ave., Oakland, California, 94610. A series of programs for daylighting, passive and other thermal analysis. Available for the IBM PC. The buyer receives the program, user's manual, one year of user support services, a full-day seminar, five weather files, and a newsletter. (A mainframe computer subscription service is also available.)

F-CHART. F-Chart Software (4406 Fox Bluff Rd., Middleton, WI 53562). The programs allow analysis of four collector types and eight system types (including passive direct gain and storage wall systems). They run on either Apple or IBM machines.

UISUN. Prof. Bruce Haglund, Department of Architecture, University of Idaho. Moscow, Idaho, 83843. Rapid feedback to designers of passive solar heated buildings, using rule-of-thumb assumptions early on, more detailed data later in the design process. Performs the detailed Solar Load Ratio analysis (for which Appendix C is an approximation) to obtain Solar Savings Fraction. Runs on IBM compatible machines.

MICROPAS. Enercomp, Inc., 123 C Street, Davis, California, 95616. This is an hourly simulation of energy performance for residential and small commercial buildings. It runs on IBM PCs.

SOLAR 5. Prof. Murray Milne, Graduate School of Architecture, UCLA. 405 Hilgard Avenue, Los Angeles, California, 90024-1467. Using information about the climate, orientation, geometry and shading configuration, users get a three-dimensional plot of building energy performance.

SUNDAY. Ecotope, Inc., 2812 E. Madison, Seattle WA 98112. This daily building load simulation gives surprisingly accurate yearly results for a relatively quick and simple program. It is particularly applicable to superinsulated, direct gain solar or conventional buildings designs.

SUNPATCH. P.O. Box 14523, San Luis Obispo, California. Cal Poly Prof. David Lord developed this program that both calculates sun position and illustrates sun penetration, for as many as 32 windows simultaneously.

AESOP. Designers Software Exchange, Graduate School of Design, Harvard University, Cambridge MA 02138. Used for designing an exterior shading device, the program also calculates sunrise, sunset, and solar position.

(c) HVAC.

HVAC&R Software Directory

ASHRAE, 1791 Tullie Circle, Atlanta, Georgia 30329. An annual bibliography of computer programs in heating, ventilating, air-conditioning and refrigerating.

(d) Fire Protection

HAZARD 1.0. One-Stop Data Shop, National Fire Protection Association. Batterymarch Park, Quincy, Mass., 02269. The user specifies the fire and its growth, the building and its characteristics, occupants' locations and characteristics. The program then calculates the development of fire effects over time, occupant behaviors, activation of detectors, and predicted fatal injury or survival of every occupant.

(e) Daylighting.

MICROLITE 1.0.

Designers Software Exchange (DSE), Graduate School of Design, Harvard University,

Cambridge, MA 02138. Calculates illuminance or daylight factor at any point within a room, utilizing sky component (SC) and reflected components (IRC and ERC).

SUPERLITE 1.01.
Distributed by J. J. Kim, School of Architecture, Arizona State University, Tempe, AZ 85287. Versions for PC and mainframe. Accepts nonrectangular spaces, side lighting and skylights, clear and overcast sky, light shelves and overhangs, and exterior obstructions. PC version yields only numerical results; mainframe version also produces contour plots.

(f) Lighting. There are many good lighting programs available today that will perform all of the required calculations rapidly and accurately. Some will also plot output in a variety of forms and a few will produce three-dimensional graphics. Indeed, because of the very wide choice available the lighting designer seeking to acquire a program should have guidelines by which to make a selection. One interesting fact is that some lighting manufacturers make excellent programs available free of charge. These programs usually come with a complete data base containing the manufacturer's fixtures. However, a few of these programs permit users to enter their own data, and by so doing the program becomes nonproprietary. A few important characteristics to look for when selecting a program are:

1. Clearly written adequate documentation.
2. Menu driven.
3. Data entry echo and printout.
4. Capability to stop at any point and continue at a later date, without data loss.
5. Clear error messages in plain English.
6. Large-capacity data base, with capability of machine reading of IES format files.
7. Rapid calculation of both zonal cavity and point illuminance, surface exitances, ESI, VCP.
8. Capability of considering several fixture types per area and of subdividing large areas.
9. Capability of calculating coefficients from candela data, and printing CU table.
10. Perform annual-owning and life-cycle cost analysis.
11. Produce graphic output in the form of isolux maps, shaded graphics, fixture layout.
12. Automatic periodic saves.

A few readily available programs:

1. LUMEN-MICRO, LUMEN$, LUMEN DATA, LUMEN POINT. Lighting Technologies, 2540 Frontier, Suite 107 Boulder, CO 80301-2458. A very comprehensive group of sophisticated lighting analysis programs for interior and exterior lighting. Zonal cavity and point-by-point interior lighting calculations, and point exterior calculations. Built-in plus user-created data base, economic analyses, VCP, RVP, ESI, exitance calculations, contouring, and shaded perspective graphics of interior illuminances.

2. ICON/ECON. Cooper Lighting, POB 1207, Americus, GA 31709. Zonal cavity, indoor point-by-point illuminance, ESI, LEF, VCP. Printout of proportionally spaced fixture rows. Exterior area, street, and floodlighting calculations.

3. LIGHT. Elite Software, POB 1194, Bryan, TX 77806. Zonal cavity calculations, economic analyses with modified life-cycle calculations. Accepts user photometrics.

4. CALA (*C*omputer *A*ided *L*ighting *A*nalysis). Holophane Division, Manville Sales Corporation, 214 Oakwood Avenue, Newark, OH 43055. Comprehensive interior calculations including zonal cavity and point-by-point, plus drawing plan and perspective views. Exterior point calculations; area and floodlighting. Built-in plus user photometric data base.

5. Data*Light 1. General Electric, Lamp Technical Services, Nela Park, OH 44112. Specialized program for calculation of lighting levels using MR-16 halogen sources.

6. POINT; ISOPOINT. Lighting Analysts Inc., 10572 Mountain Road, Littleton, CO 80127. Interior and exterior point calculations;

interior without interflectance. Isolux (isofootcandle) charts for any fixture in data base. User photometrics.

7. MICRO-SITE-LITE, MICRO-EYE-LITE, MICRO-COST-LITE. Lighting Sciences, Inc., 7830 East Evans Road, Scottsdale, AZ 85260. A family of computer programs for indoor and outdoor lighting analyses. Point calculations, user photometrics, cost comparisons, and tabular/graphic output of illuminance, exitance, ESI, VCP, RVP.

8. CALFS. Rosser Fabrap International, 524 Peachtree Street, Atlanta, GA 30308. Specialty lighting program whose purpose is to produce lighting fixture schedules. Extensive data base provides alternative luminaires for the specified unit.

9. LUM-H. Globe Lighting, Gardena, CA. Calculates zonal cavity illuminances, utilization, and exitance coefficients when zonal flux is input. Accepts user photometrics. Also does basic cost analyses.

10. PHOTO TOOLS. Chas. H. Loch, 2726 Winding Trail Place, Boulder, CO 80302. A series of photometric utilities that produce a complete photometric luminaire report, including polar candela plots, cd table, CU table, and roadway luminaire data.

(g) Acoustic.

From TPM Software, POB 77, Cabin John, MD 20818.

1. TL/STC: for computing and modeling sound transmission loss and STC.
2. OPTRT 60: for computing the absorption and reverberation times of an enclosed space.
3. DNLAIR, DNLROAD: community noise calculation and assessment.
4. AISI: evaluation of Articulation Index (AI) and Speech Intelligibility (SI) for open and confined spaces.

(h) Graphic Plots

MATLAB. The Math Works, Inc., 21 Eliot Street, South Natick MA 01760. Contains graphics utility for performing contour and three-dimensional mesh plots with user-defined parameters.

(i) Compliance Tools

WATTSUN. Washington State Energy Office, 809 Legion Way S.E., Olympia, Washington 98504-1211. This calculates heat loss and provides annual space heating energy budget, to demonstrate compliance with several Pacific northwest jurisdictions.

CALRES. California Energy Commission, Publications Unit 1516 9th Street, Sacramento, California 95814. Calculates the annual heating and cooling energy use for residence designs, including the effects of solar gain and thermal mass, to demonstrate compliance with California's energy code.

(j) Whole-building Tools

ADM-DOE. ADM Associates, Inc., 3299 Ramos Circle, Sacramento, California 95827. A microcomputer translation of the mainframe DOE-2 series, which is described in Section H.3.

MICRO-DOE2. Acrosoft International Inc., 9745 East Hampden Avenue, Suite 230, Denver, Colorado 80231. Another microcomputer translation of the mainframe DOE-2 series.

H.3 Mainframe Computers

Most universities, many municipal governments, and some larger architectural firms own these machines. Many other professionals buy access to big computers through time sharing. The more complicated thermal analysis programs require extensive training for operators, and as much as several days to generate the input data. When a detailed performance comparison is required for several alternative designs on larger buildings, these programs are appropriate.

Most mainframe computer programs for thermal analysis require hourly data for input; this usually requires the purchase of a tape of data for a typical meteorological year (TMY), available from the U.S. Department of Commerce, National Climatic Center (Asheville, NC 28801) for a variety of locations. The resulting energy analysis is thus based on a very detailed, sequential analysis of the building's use of energy for all purposes. As can be imagined, considerable detail about the building's configuration and equipment must be included as input for this analysis.

The most widely used of the mainframe computer programs include:

DOE series, often considered the standard evaluation technique; contact National Energy Software Center, Argonne National Laboratory, 9700 South Cass Avenue, Argonne, Illinois 60439.

BLAST series (Building Loads Analysis and System Thermodynamics); contact BLAST Support Office, 1206 West Green Street, Urbana, IL 61801. (These programs have been compared in an analysis done at the Solar Energy Research Institute, Golden, Colorado.)

CALPAS series, developed by the Berkeley Solar Group, at the address listed in Section H.2.

TRNSYS, a detailed simulation program for active solar systems, developed at the Solar Energy Laboratory, University of Wisconsin, Madison.

SUPERLITE, which includes daylighting in overall energy analysis; developed at the Lawrence Berkeley Laboratory, Berkeley, California. The microcomputer version is described in Section H.2.

Again, these large, detailed, complex programs are not for the casual user. It is more likely that a consulting engineer rather than an architect will utilize them. They represent a sophisticated opportunity for analyzing energy performance and choosing between alternative designs.

REFERENCES

ASHRAE (1990). *ASHRAE Journal's HVAC & R Software Directory,* American Society of Heating, Refrigerating, and Air-Conditioning Engineers, Inc., Atlanta, Ga.

F-Chart Software (1980). *Solar Energy Programs,* F-Chart Software, 4406 Fox Bluff Road, Middleton, WI 53562.

Novitski, B. J. (1991). "Energy Conservation Software," in *Architecture,* May 1991, Affiliated Publications, New York.

The DOE-2 User News, The Simulation Research Group, Applied Science Division. Lawrence Laboratory, Berkeley California.

I

METRICATION; SI UNITS; CONVERSIONS

I.1 General Comments on Metrication

The building profession has been slower than most in adopting the metric (more accurately the Système International, i.e., SI) system of units for many reasons, a discussion of which is not relevant here. The change will come. Many of the major professional societies use a mixture of both systems because certain units are so entrenched that changing them is an almost insurmountable task. However, the advantages of the SI system are well known to all and need no justification here. In this book we have followed current industry practice—a mixture of the two systems. For this reason tables of conversions and approximations are presented below to enable the reader to live with both systems. Also given below are some useful facts that should make use of the SI (metric) system a bit easier.

I.2 SI Nomenclature, Symbols

For a full discussion of the SI system the reader is referred to *AIA Metric Building and Construction Guide*, edited by S. Braybrooke (Wiley, New York, 1980). The units in common use include the basic SI units; meter, kilogram, and second (MKS), plus a host of derived, supplementary, and non-SI units such as pascal (pressure), radian (solid angle), and kilowatt-hour (energy), respectively. Also, multiple and submultiple units such as liter, metric ton, and millibar are so common that they stand as separate units instead of being expressed as $10^{-3}m^3$, 10^3kg and $10^{-2}Pa$. For this reason we omit a detailed analysis of the subject and simply supply data that by experience we have found useful.

Table I.1 lists prefixes with their accepted symbols. Symbols do not change in plural. That

TABLE I.1 **These Prefixes May Be Applied to All SI Units**

Multiples and Submultiples	*Prefixes*	*Symbols*
1 000 000 000 = 10^9	giga	G
1 000 000 = 10^6	mega	M[a]
1 000 = 10^3	kilo	k[a]
100 = 10^2	hecto[b]	h
10 = 10	deka	da
0.1 = 10^{-1}	deci	d
0.01 = 10^{-2}	centi	c[a]
0.001 = 10^{-3}	milli	m[a]
0.000 001 = 10^{-6}	micro	μ[a]
0.000 000 001 = 10^{-9}	nano	n
0.000 000 000 001 = 10^{-12}	pico	p

[a]Most commonly used.
[b]A hectare is a square hectometer (i.e., 10^4 m^2).

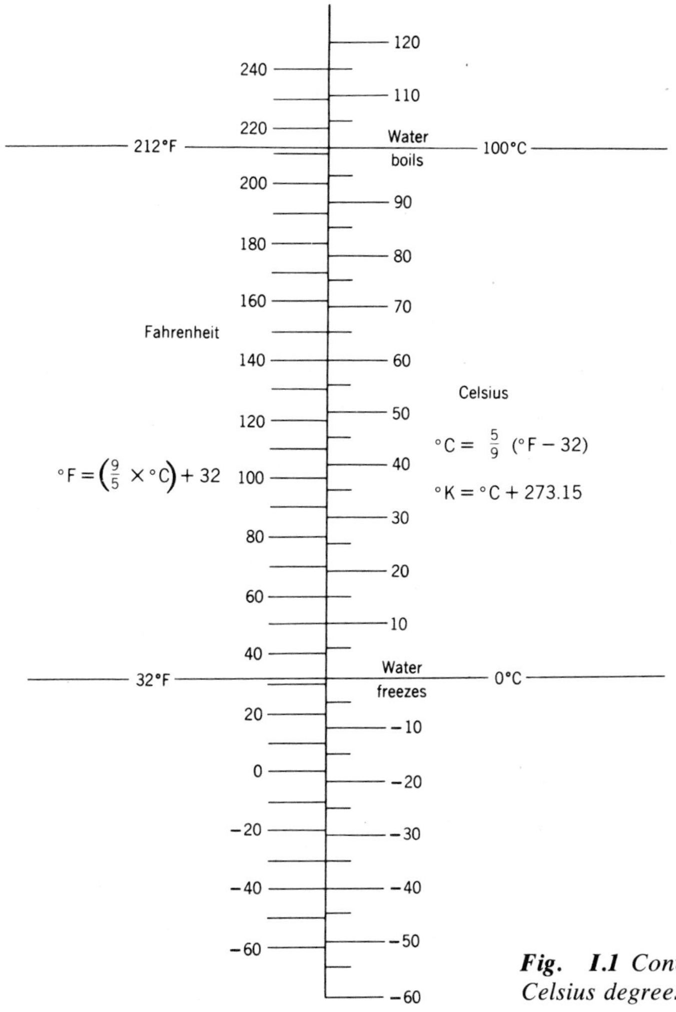

***Fig. I.1** Conversion: Fahrenheit degrees–Celsius degrees.*

is, 6 millimeters is written 6 mm and 20 kilograms is written 20 kg. All units and prefixes except Fahrenheit and Celsius are uncapitalized when written out, as in megaton or meter.

I.3 Common Usage

1. *Length:* meter (m), kilometer (km), millimeter (mm), micrometer (μm), nanometer (nm).
2. *Area:* square meter (m^2), hectare (ha).
3. *Volume:* cubic meter (m^3), liter (L).
4. *Flow:* cubic meters per second (m^3/s).
5. *Velocity, Airflow:* meters per second (m/s).
6. *Weight:* kilogram (kg), gram (g).

The SI system clearly differentiates between mass (kg) and force ($kg{\cdot}m/s^2$), the latter being given a separate name and symbol, newton (N). Weight is not used, because it is a force that depends on acceleration and is therefore variable. However, the construction industry in large measure continues to use the terms mass, weight, and force interchangeably.

7. *Force:* newton (N), kilonewton (kN). A newton is the force required to accelerate 1 kg at 1.0 m/s^2.
8. *Pressure:* pascal (Pa), kilopascal (kPa). A pascal is a newton per square meter (N/m^2).

9. *Energy, Work, Quantity of Heat:* joule (J), kilojoule (kJ), megajoule (MJ). A joule is a watt-second (W·s).
10. *Temperature:* degree Celsius (°C), degree kelvin (K).

The SI unit (kelvins) is not used as commonly as °C. (Celsius is the accepted term; "centigrade" is obsolete). The Celsius and Kelvin scales are subdivided equally but start at different points, that is, 0 K is −273.15°C. Therefore, to determine Kelvin from Celsius, simply add 273.15. Increments are equal because of equal subdivision; that is, a change of 10 K is the same as a 10°C change. Because of its special importance, a separate Fahrenheit/Celsius conversion chart is given in Fig. I.1.

11. *Illumination:* see Table 18.1.
12. *Acoustics:* see Table 27.15.
13. *CGS/MKS Conversions:* see Table 27.15.
14. *Abbreviations:* see Table I.2.
15. *Approximations:* see Table I.3.

I.4 Conversion Factors

Table I.4 is an alphabetized list of useful conversion factors. Because normal use involves a hand calculator, we have avoided scientific notation and used decimal notation instead; that is, we write 0.00378, not 3.78×10^{-3} or 3.78E-3.

TABLE I.2 **Typical Abbreviations: All Systems of Units**

atmospheres	atm	kilopascals	kPa
British thermal units	Btu	kilowatts	kW
British thermal units per hour	Btu/h	kilowatt-hours	kWh
calorie	cal	liters	L
cubic feet	cf, ft^3	liters per second	L/s
cubic feet per minute	cfm, ft^3/min	megajoules	MJ
cubic feet per second	cfs, ft^3/s	meganewtons	MN
cubic meters	m^3	megapascals	MPa
feet	ft	meters	m
feet per second	fps, ft/s	meters per second	m/s
gallons	gal	miles per hour	mph
gallons per hour	gph, gal/h	millimeters	mm
gallons per minute	gpm, gal/min	millimeters of mercury	mm Hg
grams	g	newtons	N
hectares	ha	ounces	oz
horsepower	hp	pounds	lb
inches	in.	pounds of force	lbf
inches of mercury	in. Hg	pounds per cubic foot	lb/ft^3
joules	J	second	sec, s
kilocalories	kcal	square feet	ft^2
kilograms	kg	square inches	in.^{2}
kilograms per second	kg/s	square meters	m^2
kilojoules	kJ	watts	W
kilometers	km	watts per square meter	W/m^2
kilometers per hour	kph, km/h	yards	yd
kilonewtons	kN		

TABLE I.3 **Common Approximations**

Approximate Common Equivalents			
1 inch	= 25 millimeters	1 millimeter	= 0.04 inch
1 foot	= 0.3 meter	1 liter	= 61 cubic inches
1 yard	= 0.9 meter	1 meter	= 3.3 feet
1 mile	= 1.6 kilometers	1 meter	= 1.1 yards
1 square inch	= 6.5 square centimeters	1 kilometer	= 0.6 mile
1 square foot	= 0.09 square meter	1 square centimeter	= 0.16 square inch
1 square yard	= 0.8 square meter	1 square meter	= 11 square feet
1 acre	= 0.4 hectare	1 square meter	= 1.2 square yards
1 cubic inch	= 16 cubic centimeters	1 hectare	= 2.5 acres
1 cubic foot	= 0.03 cubic meter	1 cubic centimeter	= 0.06 cubic inch
1 cubic yard	= 0.8 cubic meter	1 cubic meter	= 35 cubic feet
1 quart	= 1 liter	1 cubic meter	= 1.3 cubic yards
1 gallon	= 0.004 cubic meter	1 liter	= 1 quart
1 ounce	= 28 grams	1 cubic meter	= 250 gallons
1 pound	= 0.45 kilogram	1 kilogram	= 2.2 pounds
1 horsepower	= 0.75 kilowatt	1 kilowatt	= 1.3 horsepower

TABLE I.4 **Useful Conversion Factors: Alphabetized**

Multiply	*By*	*To Get*
acres	4047	square meters
atmospheres	33.93	feet of water
atmospheres	29.92	inches of mercury
atmospheres	760.0	millimeters of mercury
Btu (energy)	0.252	kilocalories
	1.055	kilojoules
Btu/h (power)	0.2928	watts
Btu/h/ft^2 (energy transfer)	3.152	watts per square meter
BtuF (heat capacity)	1.897	kilojoules per kelvin[a]
Btu/lb°F (specific heat)	4.182	kilojoules per kilogram per kelvin[a]
Btu/h/Fft (thermal conductivity[b])	1.729	watts per kelvin[a] per meter
Btu/hFft^2 (conductance[c])	5.673	watts per kelvin[a] per square meter
Btu/Fday (building load coefficient, BLC)	0.022	watts per kelvin[a]
Btu/Fdayft^2 (load–collector ratio, LCR)	0.236	watts per kelvin[a] per square meter
cubic feet	0.028	cubic meters
cubic feet	7.481	gallons
cubic feet	28.32	liters
cubic feet per minute	0.472	liters per second
cubic feet per second	2.832	liters per second
cubic inches	16.39	cubic centimeters
cubic meters	35.32	cubic feet
cubic meters	1.308	cubic yards
cubic meters	264.2	gallons
cubic yards	0.765	cubic meters
feet	0.305	meters
feet	304.8	millimeters
feet per second	0.3048	meters per second
foot-pounds of force per second	1.356	watts

TABLE I.4 **Useful Conversion Factors: Alphabetized (Continued)**

Multiply	*By*	*To Get*
gallons	3.785	liters
gallons her hour	0.00152	liters per second
gallons per minute	0.0022	cubic feet per second
gallons per minute	0.06308	liters per second
grams	0.035	ounces (avoirdupois)
hectares	2.471	acres
horsepower	0.746	kilowatts
horsepower	746	watts
inches	25.4	millimeters
inches of mercury	0.033	atmospheres
inches of mercury	1.133	feet of water
inches of mercury (60°F)	3377	newtons per square meter
inches of mercury	0.491	pounds per square inch
inches of water	0.002458	atmospheres
inches of water	0.036	pounds per square inch
inches of water (60°F)	248.8	newtons per square meter
kilocalories	3.968	British thermal units
kilocalories	4190	joules
kilograms	2.205	pounds
kilograms per cubic meter	1.686	pounds per cubic yard
kilograms per square meter	0.0033	feet of water
kilograms per square meter	0.0029	inches of mercury
kilograms per square meter	0.205	pounds per square foot
kilograms per square meter	0.001422	pounds per square inch
kilojoules	0.948	British thermal units
kilojoules per kilogram	0.430	British thermal units per pound
kilometers	0.621	miles
kilometers per hour	0.621	miles per hour
kilonewtons	0.1004	tons of force
kilonewtons	224.8	pounds of force
kilopascals	20.89	pounds of force per square foot
kilowatts	1.341	horsepower
kilowatt-hours	3.6	megajoules
liters	0.03532	cubic feet
liters	61.02	cubic inches
liters	0.2642	gallons
liters	1.057	quarts
liters per second	2.119	cubic feet per minute
liters per second	951.0	gallons per hour
liters per second	15.85	gallons per minute
megajoules	0.278	kilowatt-hours
meganewtons	100.36	tons of force
megapascals	145.04	pounds of force per square inch
megapascals	9.324	tons of force per square foot
meters	3.281	feet
meters	1094	yards
meters per second	196.86	feet per minute
meters per second	2.237	miles per hour
miles	1.609	kilometers
miles per hour	1.609	kilometers per hour
miles per hour	0.447	meters per second

TABLE I.4 **Useful Conversion Factors: Alphabetized (Continued)**

Multiply	*By*	*To Get*
milliliters	0.061	cubic inches
milliliters	0.035	fluid ounces
millimeters	0.039	inches
millimeters of mercury	133.3	newtons per square meter
million gallons per day	18.94	cubic meters per hour
newtons	0.225	pounds of force
ounces (avoirdupois)	28.35	grams
ounces (fluid)	28.41	milliliters
pounds	0.454	kilograms
pounds of force	4.448	newtons
pounds of force per square foot	47.88	pascals
pounds of force per square inch	6.895	kilopascals
pounds per cubic foot	16.02	kilograms per cubic meter
pounds per cubic yard	0.593	kilograms per cubic meter
pounds per square foot	4.882	kilograms per square meter
square feet	0.0929	square meters
square inches	645.2	square millimeters
square kilometers	0.386	square miles
square meters	10.76	square feet
square meters	1.196	square yards
square miles	2.590	square kilometers
square yards	0.836	square meters
tons of force	9.964	kilonewtons
tons of force per square foot	107.25	kilopascals
tons of force per square inch	15.44	megapascals
torr (millimeters of mercury at 0°C)	133.3	newtons per square meter
watts	3.412	British thermal units per hour
watts	0.738	foot-pounds of force per second
watts per square meter	0.317	British thermal units per square foot
yards	0.914	meters

Add your own conversion factors here.

Multiply	*By*	*To Get*
Multiply	*By*	*To Get*
Multiply	*By*	*To Get*

[a]°K or °C.
[b]Thermal conductivity (K).
[c]Thermal conductance (C) or transmittance (U).

J
BUILDING AUTOMATION; INTELLIGENT BUILDINGS

J.1 Building Automation System (BAS)

(a) General. The trend in modern construction, except for small or simple structures, is clearly to use integrated system design plus centralized monitoring and control of building systems. The subsystems almost always included in a building automation system (BAS) are HVAC, energy management, and lighting control. Inclusion of security, life safety (fire alarm, fire control and surpression, plus emergency aspects of vertical transportation), material handling, and some aspects of communications depends on the specific needs of the building. This trend toward building automation, which previously had been economically justifiable only in large owner-user facilities, is today not only economically feasible but very nearly an economic necessity because of high labor costs and the relatively low cost of computer and microprocessor controls. What the rapid advances in microelectronics and computer technology have done is to make possible and practical detailed, multipoint monitoring and control in real time, the result of which (assuming proper programming and BAS operation) is a highly cost-efficient and environmentally appropriate facility. Indeed, the advantages of such an arrangement are so great that retrofitting of existing buildings, which is obviously more cost intensive than new construction, has become a major industry.

Because the technique of building automation has grown so rapidly in the past few years, the terminology surrounding it and its equipment suffers from a lack of standardization and uniformity. Thus terms such as *building automation, intelligent buildings, computerized building control, integrated computer control, supervisory data systems, integrated building control, facilities management system,* and so on, are used almost interchangeably. The expression *building automation system* (BAS) will be used in the discussion that follows. Since every BAS is by definition programmed, any discussion necessarily involves basic computer terminology in addition to that which specifically applies to BAS. A short glossary should prove helpful in this regard.

(b) Glossary

Algorithm. Detailed description, often in flow-diagram form, of the method for solving a problem. The outline used for computer or microprocessor programming.

Alphanumeric. Characters, which are either alphabet letters, symbols, or numbers.

Analog Data. Numerical information about physical variables given in representative, continuously variable form (e.g., dial meters, temperature gauges, etc.).

Artificial Intelligence. Branch of computer science concerned with computer programs that "learn" on the basis of experience and modify their own behavior accordingly.

Baud. Rate at which information is transmitted in terms of discrete events per second.

Building Automation Systems (BAS). Building-wide computer-based system that monitors and controls selected aspects of building operations. The BAS is understood to include all of the equipment involved and their interconnections.

Central Processing Unit (CPU). That portion of a computer, microprocessor, or programmable controller that processes data and executes programs. The "brains" of the computer.

APPENDICES

Converter. General term for a device that converts data or energy from one form to another. Data converters are analog to digital (ADC) or digital to analog (DAC). Energy-form converters (e.g., temperature or illuminance to voltage) are properly called transducers.

Digital Data. Data in digital form, usually binary.

Direct Digital Control (DDC). Control arrangement that processes input data according to a programmed algorithm and yields a control function (e.g., the action of a local microprocessor controller or a computer control function).

Duplex, Duplex Transmission. Simultaneous bidirectional data transmission.

Energy Management System (EMS). Computer-based system for monitoring and controlling facility energy use. Frequently part of a BAS.

Hardware. All physical equipment associated with a computer system, as opposed to software.

I/O. Input/output.

Load Control, Load Management. Arrangement whereby energy-using devices are monitored and controlled. Part of EMS.

Local Area Network (LAN). Signal network providing intercomputer communications.

Logic. Factors involved in the design of a CPU.

Microprocessor. A CPU, usually in the form of a single integrated circuit (IC), preprogrammed to perform specific functions in response to specific input.

Modem. Device that connects a data-originating device to a communication line; from "*mo*dulator–*dem*odulator."

Multiplex(er) (MX, MUX). Use of a single communications line for simultaneous transmission of multiple signals.

Operating System. Software, which controls input to, output from, and operation of, the CPU.

Point. General term to describe either a sensor/monitor location or a specific control signal in a BAS.

Programmable Controller. Electronic device containing a microprocessor, a programming means or device, and I/O interfaces. See Fig. 16.14 and Section 16.16.

Protocol. Conventions governing format and timing of signaling between communication devices (e.g., between computers in area networks).

RAM, ROM, PROM, EPROM. *r*andom *a*ccess *m*emory, *r*ead-*o*nly *m*emory, *p*rogrammable *ROM*, *e*rasable *PROM*.

Real time. Arrangement in which a computer receives and processes data without introduction of any deliberate time delay (i.e., effectively, immediate response.)

RS-232. Standard for interconnecting digital communications devices, such as computers, printers, and monitors.

Software. Computer program; as opposed to hardware.

Stand-Alone. Descriptive term for a control device that can perform independently and usually can also interconnect as a subsystem controller in a larger system.

UPS. uninterrupted power supply.

(c) System Arrangement. A block diagram of a general BAS is given in Fig. J.1. Any specific BAS may contain more or fewer monitors, sensors, dependent and stand-alone local controllers, and interconnected systems but the overall arrangement remains the same. Actually, depending on the equipment manufacturer and the system complexity, some of the functions and levels can be telescoped or grouped. We will however discuss the general case. To concretize the discussion we will use fire control and evacuation system terminology as a relatively simple straightforward example of overall system design, function and interconnections.

Level 5 is the lowest system level. It contains automatic operation passive sensors such flame, smoke, ionization, temperature, and water-flow detectors; physical condition indicators such as smoke and fire door position sensors; and man-

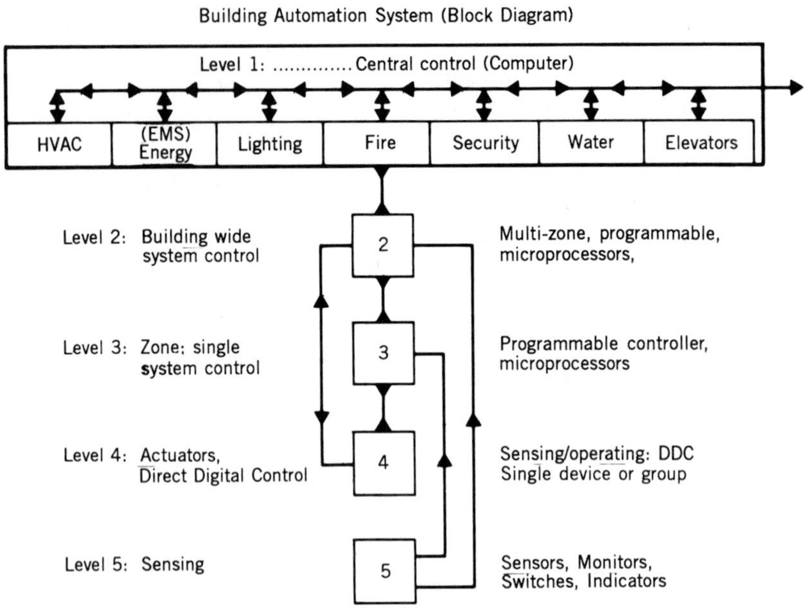

Fig. J.1 *Overall block diagram of a BAS (building automation system). In an actual system the various levels and devices can be combined and telescoped by use of equipment that combines several level functions in a single housing.*

ual fire alarm pull stations. All yield a digital signal (on–off) which is transmitted to level 3. As level 5 is purely a sensor/monitor level, it does not receive any control or operating signals.

Level 4 contains sensor/operators or actuators (i.e., devices that both transmit a condition signal and receive an operation signal). In our example, this level could contain fire and smoke door actuators, smoke vent controls, sprinkler system valves, and fan control (override) switches. This last item can either reside at this level or can be controlled through the HVAC system via level 1, 2, or 3. Control at this level is of a single device or block.

Level 3 could represent a floor, zone, or area controller, with either relay circuitry or microprocessor control (programmable controller), depending on system complexity. This level receives information from levels 4 and 5, processes it, returns control information to level 4, operates specific local audible and visible alarms directly, and sends alarm and condition information to the level 2 building controller and to the master controller (computer) at level 1. It also receives and processes signals from levels 1 and 2, containing such information as alarm reset, alarm device operation for building emergency evacuation, and any reprogram or reset function.

Level 2 contains the building's central fire and evacuation system (life safety system) controls. This level can contain some of the level 1 programming logic relating to firemen's control of elevator operation and HVAC fan functioning, or it can simply act as an input/output (I/O) device to level 1 for these functions, again depending on system design and complexity. In some facilities these life safety system functions may be controlled from a separate console that is interconnected with the BAS. In a multibuilding facility, this level contains the individual building fire system central control, interconnections to other building fire-system controllers, as well as external connections for city alarm, and external supervisory services.

Level 1 is the central computer console. This level receives status reports on individual devices from level 4 or 5; on areas, zones, and subsystems from level 3; and on the entire fire/

evacuation system from level 2. Hard-copy printout of all alarms and periodic status reports are generally made at this level, although they can also be accomplished at level 2. Video display terminals (VDT) at level 1 can be used to view the status of any area and all the devices in the fire/evacuation system (Fig. J.2) or any other system (Fig. J.3) graphically or tabularly (Fig. J.4). Pictorial graphics for nontechnical end users are also possible (Fig. J.5). At this level interconnection is made to HVAC, elevator, security, and lighting systems in preprogrammed alarm modes, with manual override possible. A fire emergency scenario (Fig. J.2), depending on level 1 programming, might interconnect with the HVAC, elevator, security, water, electrical, and lighting systems to activate exhaust and pressurizing fans and deactivate supply fans, place elevators in fire-evacuation mode (see Section 24.5), override access barriers to permit unhindered facility evacuation, connect fire pumps to emergency power feeders, disconnect high-voltage feeders, and activate emergency lighting, all via preprogrammed intersystem functions. The system operator could then view any part of any system and reactivate or override preprogrammed functions as desired.

Fig. J.2 The central computer's monitor screen shows a selected building area and its zoning (Johnson Plaza East, third floor, zones 1 to 6), all of the area's fire/evacuation system devices and their status, and the control system's present functioning. The illustrated screen shows an active alarm condition with messages: elevators being recalled to lobby; floor is now pressurized; audio evacuation messages active; CALL FIRE DEPARTMENT. *(Courtesy of Johnson Controls.)*

Returning to Fig. J.1, we can consider a much more complex, nonemergency system; energy. Here level 5 senses and continuously monitors inside and outside temperature and humidity, duct air temperature, space pressurization, airflow, and the operating and load status of all fans, compressors, and refrigeration devices. Level 4 contains monitors and DDC controls for valves, boilers, chillers, fans, humidifiers, dampers, cooling towers, and compressors (see Fig. J.6). Levels 2 and 3 contain electric load managers for duty cycling and demand limiting (see Section 14.13), HVAC controls for economizers, run-time optimizers, supply air and supply water reset, chilled and condenser water reset, enthalpy switching, load sequencing, and the like. These functions are performed either by a stand-alone EMS controller (see Fig. 16.14*b*) or by individual demand, HVAC, and chiller load managers controlled by level 1 logic. Finally, level 1 contains the programming for all of the foregoing functions, any of which can be displayed at the VDT terminal at any time. The system operator can then reprogram any function after analyzing system performance (Fig. J.2) at the level 1 computer and obtain immediate visual and hard-copy readout of the changes. A discussion of building automation controls as applied to HVAC systems is given in Section 7.3*h*.

J.2 Intelligent Buildings

Although the term "intelligent building" seems to attribute a human characteristic to an inanimate object, the term's actual meaning, at least as defined by the Intelligent Buildings Institute (*Guideline*, 1987), is "a building which provides

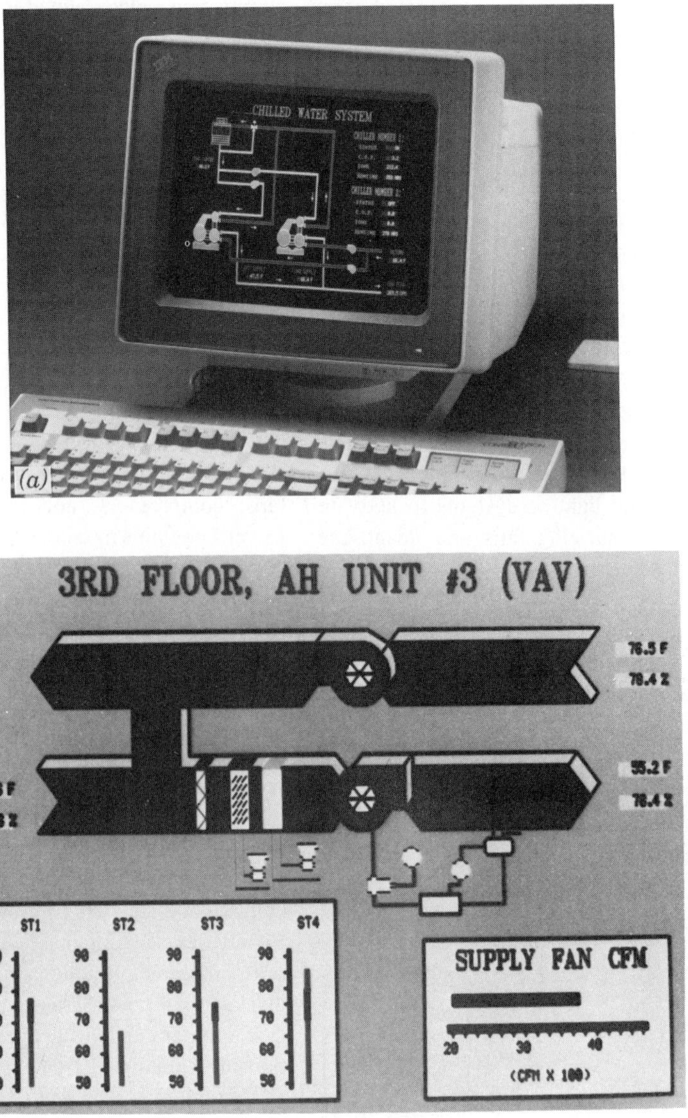

Fig. J.3 (a) *Computer screen shows any section of any building system and the status of all components. Here we see chiller status (on–off, COP, tons, run time), water temperature and flow in supply and return lines, and condenser water temperature.* (b) *Portion of the system showing status of air-handling unit 3. (Courtesy of Johnson Controls.)*

a productive and cost-effective environment through optimization of its four basic elements—structure, systems, services and management—and the interrelationships between them. . . . Optimal building intelligence is the matching of solutions to occupant need.'' In the framework of such a definition, the intelligence of a building resides essentially in its designers' and operators' abilities. An intelligent building is, therefore, not necessarily packed with electronic systems, nor even one with an extensive BAS, unless it is necessitated by the building program and can demonstrate its cost-effectiveness. Rather, it is a building designed with fore-

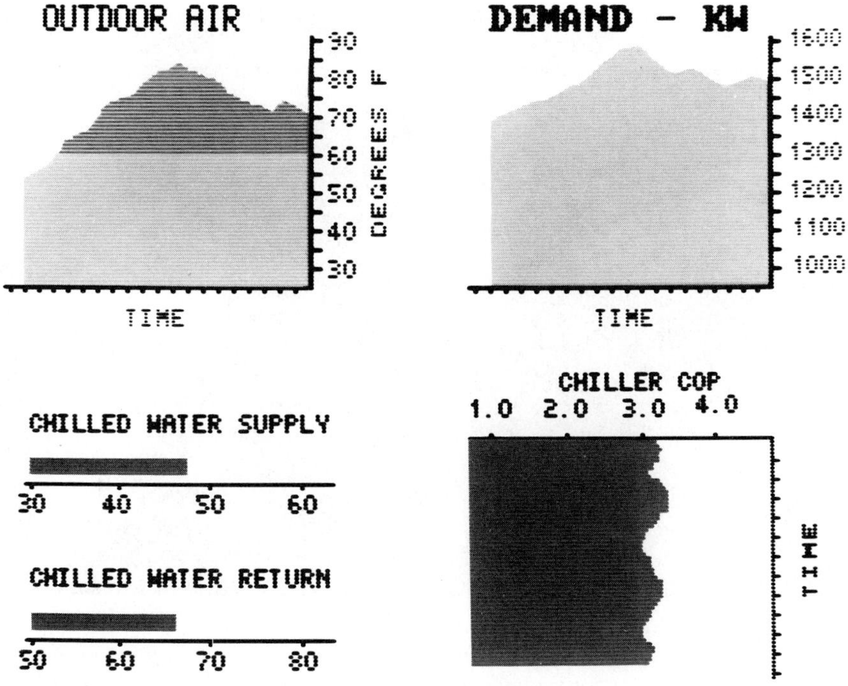

(a)

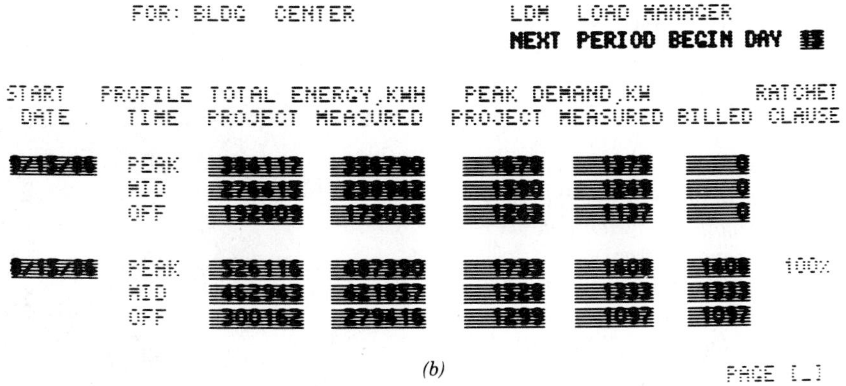

(b)

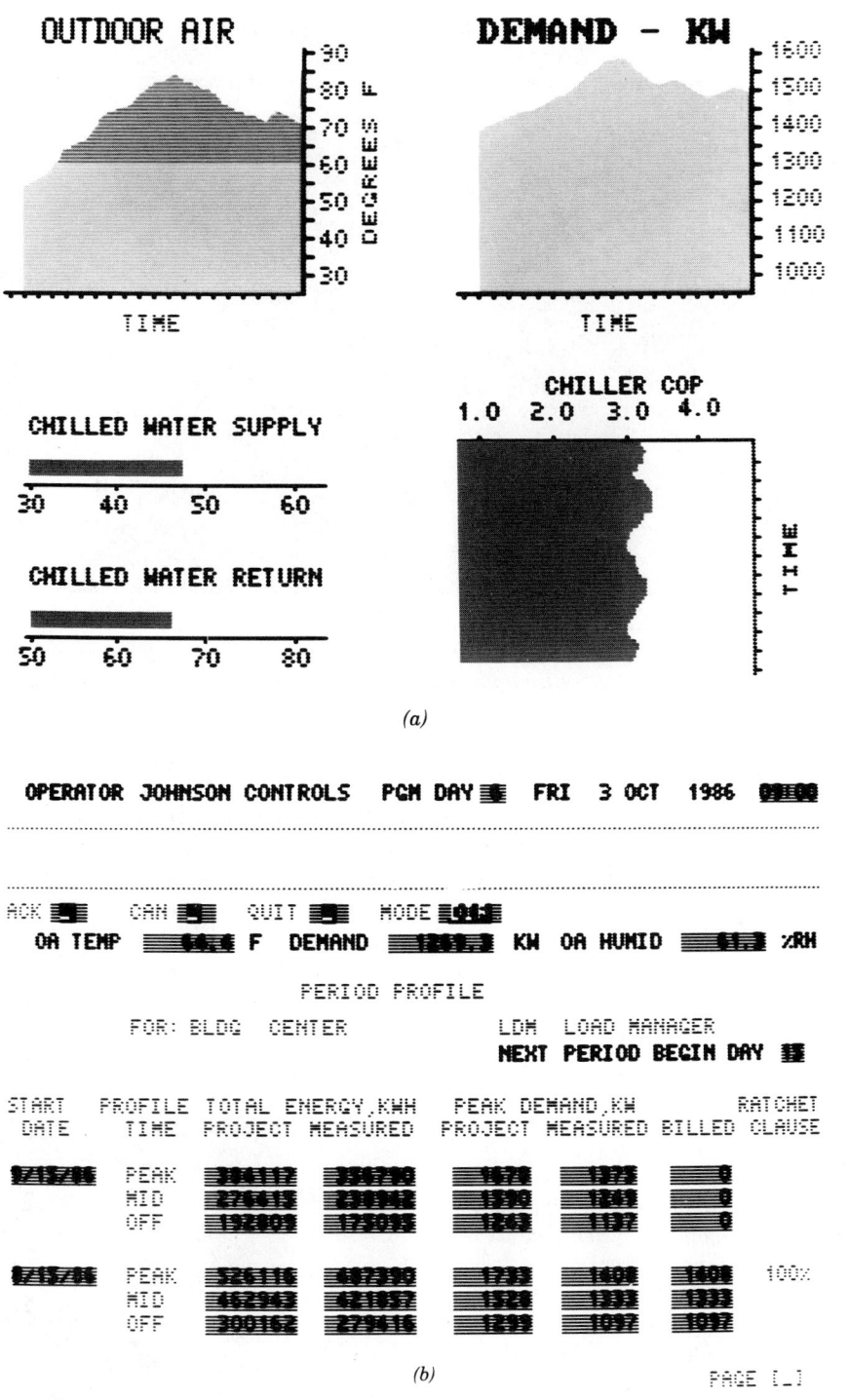

Fig. J.4 *The status of various aspects of the building's M-E systems can be displayed individually or grouped* (a). *Similarly, detailed data on any aspect of operation can be displayed* (b) *and printed, as desired. (Courtesy of Johnson Controls.)*

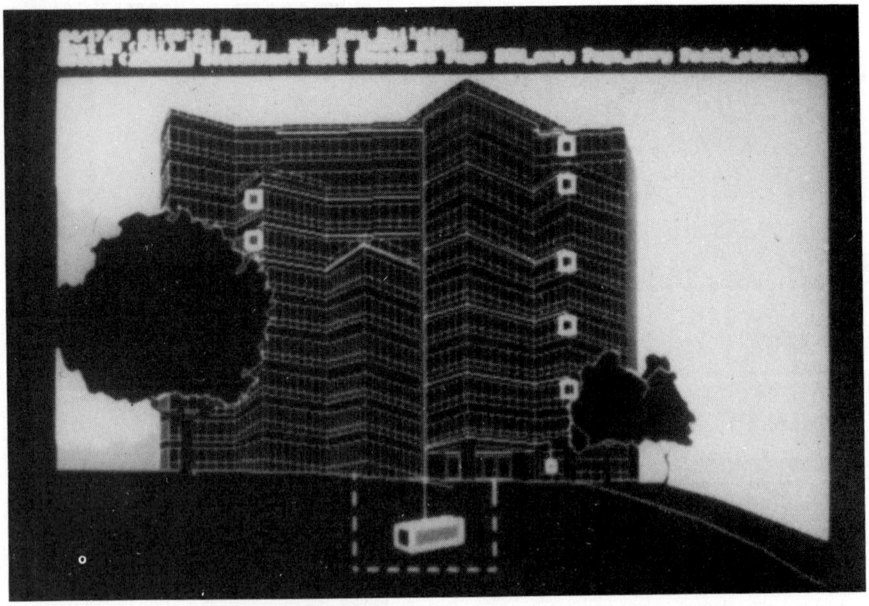

Fig. J.5 It is frequently desirable for both demonstration and training purposes to utilize pictorial graphics in a nontechnical mode. In this illustration a building system is superimposed on a visual building graphic in order to concretize and simplify a technical concept. (Courtesy of CSI Control Systems International.)

thought and perhaps some inspiration, both of which are necessary to satisfy immediate and predictable need, and yet also to anticipate the occupants' future requirements, since modern building usage is rarely static. Referring to the IBI-defined characteristics of an intelligent building and addressing each of the four components individually, but in the framework of our study, we can comment as follows.

(a) Building Structure. These are essentially architectural and marketing considerations, since as has been pointed out repeatedly, there is a vast difference between speculative construction for immediate return, construction for rental, and owner–user buildings. Since every additional inch of slab-to-slab height and every square foot of utility closet space in-

Fig. J.6 Illustrated units are a stand-alone direct digital control actuator (level 4) for use in a VAV (variable air volume) system and its room temperature and setpoint transmitter (level 5). This unit provides adjustable setpoint, local indication, night setback, morning warmings, and adjustable flow setting for a VAV terminal. It communicates back to level 2 (or 1) via standard data links and accepts and executes level 2 or 1 control settings for its functions when not operated in stand-alone configuration. It can also monitor conditioned air use by a terminal for cost apportionment. (Courtesy of Barber-Coleman Co., Environmental Control Division.)

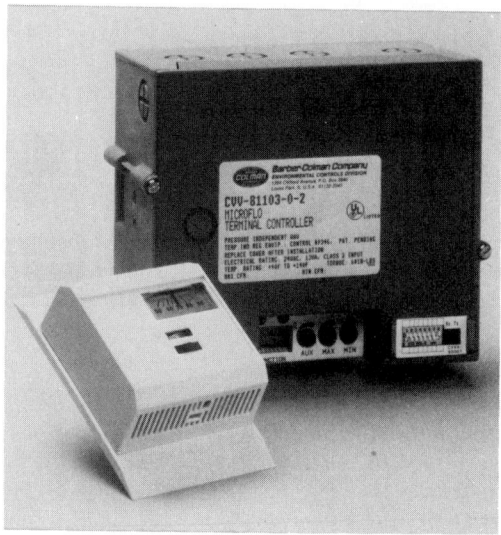

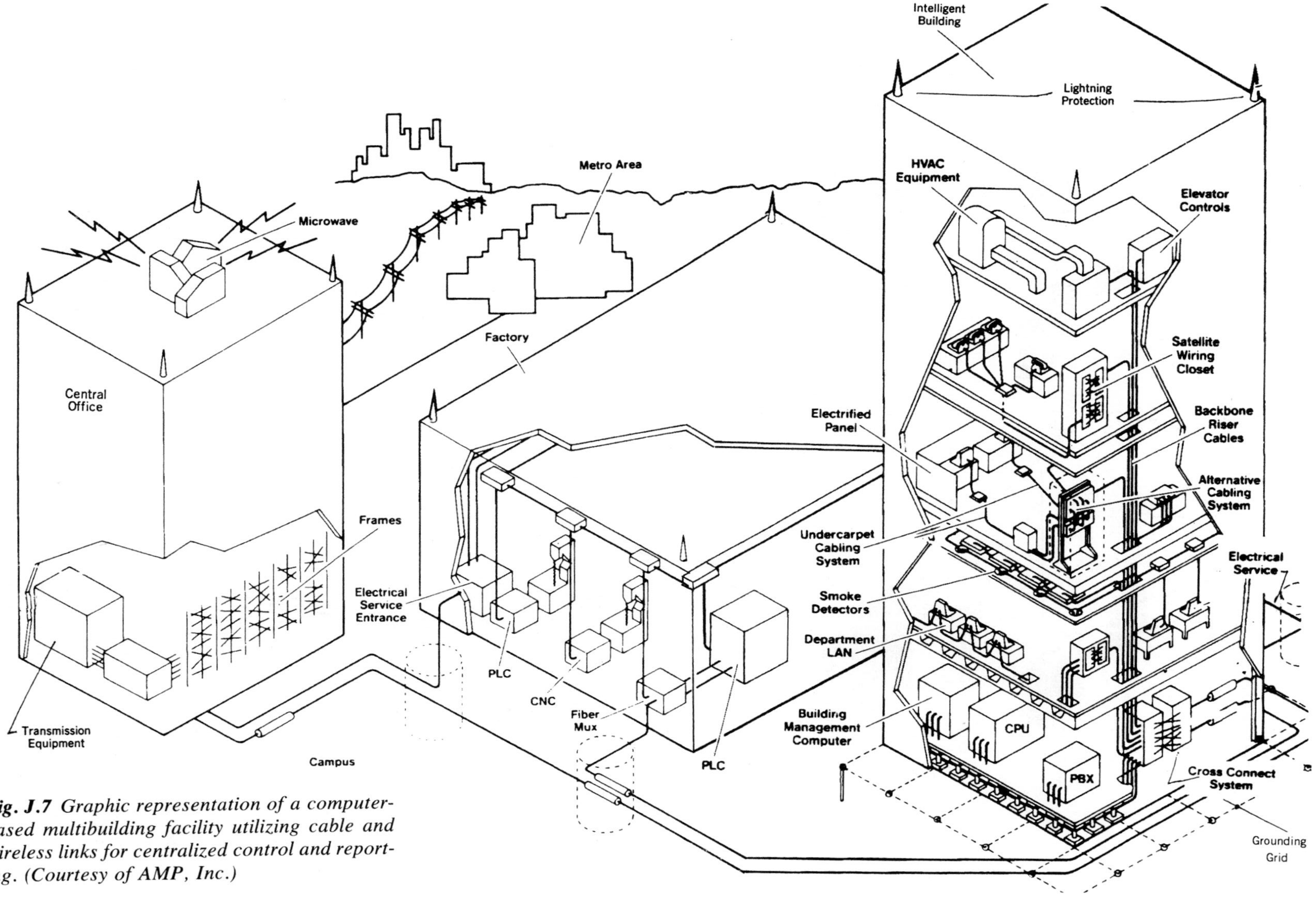

Fig. J.7 *Graphic representation of a computer-based multibuilding facility utilizing cable and wireless links for centralized control and reporting. (Courtesy of AMP, Inc.)*

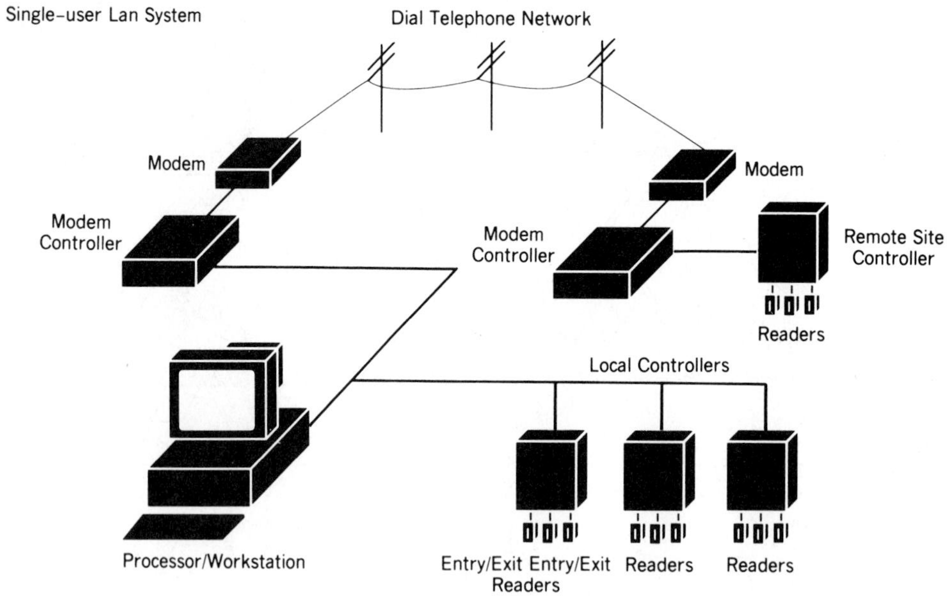

Fig. J.8 Computer control of a single system (security) at multiple locations via telephone link, in a single-user local area network (LAN). (Courtesy of CSI Control Systems International.)

creases construction costs or reduces occupant space, the speculative builder will be interested only if an immediate return is likely; the rental builder will consider it in the light of his or her immediate rental prospects; and the informed owner–user builder will analyze the rate of return for initial investment and decide accordingly. These comments apply all the more when referring to additional costs for unseen items such as spare ducts and conduits and "frills" such as fixtures, finishes, and furnishings.

(b) Building Systems. This is the area of high-tech equipment such as BAS, plus all the individual subsidiary systems. One difficulty lies in the highly proprietary nature of all of this equipment, making integration between systems and future alterations difficult if not problematic. Second, the resistance of some major manufacturers to open protocol (i.e., interequipment communication compatibility) places another difficult problem into the design process. Experience has shown clearly, however, that in the very fast-moving building systems market, today's high-tech can become tomorrow's albatross, and therefore planning for future requirements means intelligent preparation as much as immediate provision.

(c) Building Services and Management. With the exception of communication facilities, these items are not within our scope, and with respect to communications, which are the lifeline of all modern commercial and industrial facilities, the foregoing remarks are equally applicable.

In summary, then, the intelligent building is essentially the building that is designed with foresight. Hindsight will determine the degree of intelligence.

J.3 Multibuilding Facilities

The advantages in the use of a single building-wide BAS are redoubled when applied to a multibuilding facility as compared to precomputer technology. A graphic representation of such a multibuilding system is shown in Fig. J.7. The intelligent building BAS controls its building systems and interconnects via telephone cables (and microwave link) to other buildings in the network and to the central office. It matters little whether the various buildings in the network are geographically concentrated in campus fashion as would be the case in an industrial park, a university campus, or a multibuilding industrial facility, or spread out; the control, monitoring, and alarm functions remain the same. The same principle of building interconnection can be applied to a particular system rather than the BAS. Thus Fig. J.8 shows a single processor–workstation (PC-type computer) controlling the security aspect of a number of facilities via controllers, modems, and telephone cable connections in a single-user local area network (LAN) system. This workstation can then be incorporated into a multisystem, multiuser computer network.

BIBLIOGRAPHY

Bernaden, J., and Neubauer, R. (eds.) (1988). *The Intelligent Building Sourcebook*, Fairmont Press, Lilburn, Ga.

Guideline (1987). Intelligent Buildings Institute, *Technical Data Bulletin: Intelligent Buildings* (1988). ASHRAE, Atlanta, Ga.

K

ASHRAE STANDARDS FOR ENERGY EFFICIENT BUILDING DESIGN

K.1 Criteria for Components of the Envelope: Nonresidential Buildings

The design guidelines in this section are taken from ASHRAE/IES Standard 90.1-1989, *Energy Efficient Design of New Buildings Except Low-Rise Residential Buildings,* published by the American Society of Heating, Refrigerating, and Air-Conditioning Engineers, Inc., Atlanta, Ga. This standard presents two types of design procedures: the system/component method and the building energy cost budget method. Both methods rely heavily on the use of microcomputers, although there is a "prescriptive path" within the system/component method which allows the designer to quickly check the window areas and the U values of the major components of a building envelope. This quick prescriptive path is presented here. First, the features of the computer-based methods should be understood. The two procedures are diagrammed in Fig. K.1.

The building energy cost budget method is based on the quasi-free market cost of energy to the consumer. Although we may question the wisdom of basing today's energy design decisions on today's massive subsidies of nonrenewable energy sources, the method is especially recommended by ASHRAE for buildings that have high internal heat gains, unusual operating schedules, or innovative energy designs. With this method, credit is more readily given for daylighting, passive solar heating, heat recovery, better zonal temperature control, thermal storage, and off-peak electrical energy use. Further, this method provides a way to judge the relative effectiveness of several design alternatives in reducing energy costs. First, an *energy cost budget* is determined; for a given building size and type, this budget will vary only with climate, the number of stories, and the choice of energy performance simulation tools. This budget is based on monthly energy usage simulations of either a "prototype" or a "reference" building, with detailed conventional assumptions of its internal loads, orientation, shape, envelope, and HVAC system characteristics. Then, monthly energy simulations of the actual design are made. The yearly total energy use simulations, and resulting costs, of the budget and the actual building are compared; if the actual building energy costs are lower, the design complies with this Standard. Software to assist the energy cost budget analysis is provided when the standard is purchased from ASHRAE.

The system performance method presents requirements for maximum overall thermal transmittance of each major component of the envelope: roofs, opaque walls, windows, and floors. The requirements vary with several climate variables, such as heating and cooling degree days, and with average annual insolation on vertical surfaces at each orientation. Therefore, the calculations are so complex as to require the use of a computer, and software to assist the system performance analysis is provided when the standard is purchased from ASHRAE. By-hand cal-

Fig. K.1. *Alternative methods of achieving compliance.*

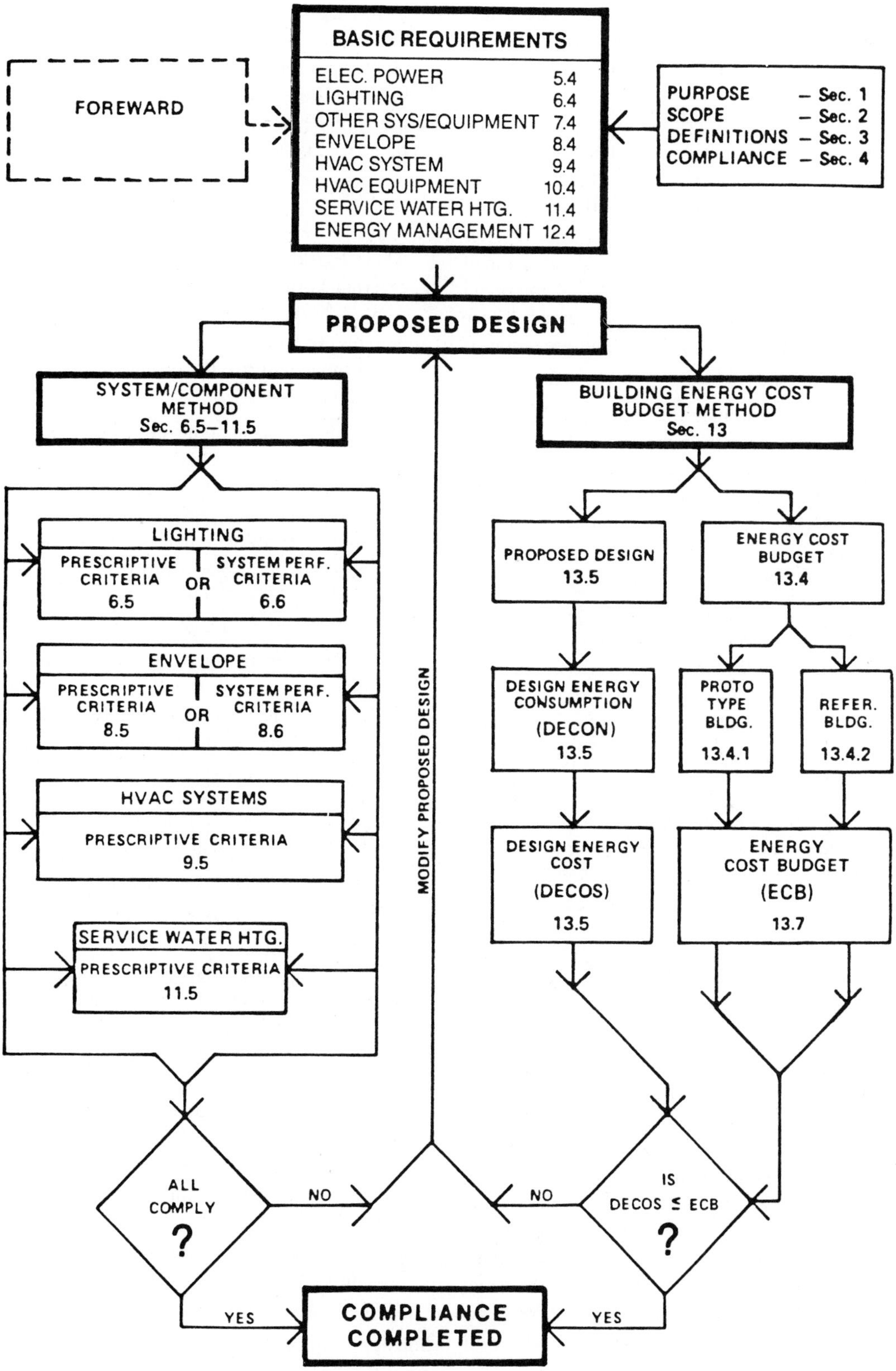
BASIC REQUIREMENTS
ELEC. POWER 5.4
LIGHTING 6.4
OTHER SYS/EQUIPMENT 7.4
ENVELOPE 8.4
HVAC SYSTEM 9.4
HVAC EQUIPMENT 10.4
SERVICE WATER HTG. 11.4
ENERGY MANAGEMENT 12.4
FOREWARD
PURPOSE – Sec. 1
SCOPE – Sec. 2
DEFINITIONS – Sec. 3
COMPLIANCE – Sec. 4
PROPOSED DESIGN
SYSTEM/COMPONENT METHOD Sec. 6.5–11.5
BUILDING ENERGY COST BUDGET METHOD Sec. 13
LIGHTING
PRESCRIPTIVE CRITERIA 6.5
OR
SYSTEM PERF. CRITERIA 6.6
ENVELOPE
PRESCRIPTIVE CRITERIA 8.5
OR
SYSTEM PERF. CRITERIA 8.6
HVAC SYSTEMS
PRESCRIPTIVE CRITERIA 9.5
SERVICE WATER HTG.
PRESCRIPTIVE CRITERIA 11.5
MODIFY PROPOSED DESIGN
PROPOSED DESIGN 13.5
ENERGY COST BUDGET 13.4
DESIGN ENERGY CONSUMPTION (DECON) 13.5
PROTO TYPE BLDG. 13.4.1
REFER. BLDG. 13.4.2
DESIGN ENERGY COST (DECOS) 13.5
ENERGY COST BUDGET (ECB) 13.7
ALL COMPLY ?
NO
NO
IS DECOS ≤ ECB ?
YES
YES
COMPLIANCE COMPLETED

APPENDICES

culations for electric lighting density are somewhat less complex, and are explained in Section 20.5c. The necessary data are given in Table K.2 Parts C to F.

The prescriptive criteria method is a relatively fast and simple way to check for building envelope compliance with the standard. However, it is somewhat more restrictive than either the energy budget cost or the system performance methods. For each major climate zone of the United States, an Alternate Component Package (ACP) table has been prepared. These appear in Tables K.4 to K.35. The procedure is as follows.

STEP 1. Determine the appropriate ACP Tables K.4 to K.35 for your climate. If your location is not listed, compare your location's climate data with those of nearby locations (see Table K.1) to choose the listed location best matching yours.

STEP 2. Determine the maximum allowable percent fenestration. Note that this compares total window area to total gross *wall* area (not floor area). In the appropriate ACP table, you must enter the applicable horizontal line, and applicable vertical column, to find maximum percent fenestration. These substeps are therefore necessary:

a. Find the internal load density (ILD), which is the sum of the lighting power density (LPD), the equipment power density (EPD), and the occupant load adjustment (OLA):

$$\text{ILD} = \text{LPD} + \text{EPD} + \text{OLA}$$

and enter your ACP table in the appropriate one of the three ranges of ILD shown. The LPD, lighting power density, is determined either by estimating from Table K.2, Part A, or by using actual electric lighting W/ft^2, if known. The EPD, equipment power density, may be estimated from Table K.2, Part B. The OLA, occupant load adjustment, will ordinarily be 0.0 W/ft^2, since a value of 0.6 W/ft^2 sensible heat gain is built into these ACP tables. If your occupant heat gain rate is markedly different, adjust from 0.6 W/ft^2 accordingly; see Table 5.8, page 214, where 0.6 W/ft^2 is equivalent to 2 Btu/h-ft^2.

b. Select the external shading projection factor (PF) and continue into the ACP table on the appropriate one of the three PF ranges listed. (If no external shading projections are used in the proposed design, enter the PF range 0 to 0.259.)

$$\text{PF} = \frac{P_d}{H}$$

where P_d is the depth (in feet or inches) of the external horizontal projection and H is the sum of the height (same units as P_d) of the window and the vertical distance from the top of the window to the bottom of the external shading device.

c. Select the shading coefficient of the window (SC_x), including the internal, integral, and external shading devices EXCEPT the device whose PF has already been determined. SC values are found in Tables 4.18 to 4.20, pages 164 to 168. Then continue on into the ACP table in the appropriate one of the five SC_x ranges listed.

d. To enter the vertical "fenestration" columns of your ACP table, select either "base case" (no thermal credit for daylighting) or "perimeter daylighting," in which case automatic daylighting controls for electric lighting must be used. In many cooler locations, the perimeter daylighting thermal credits are not available, although the daylighting contribution to energy used by electric lights is still recognized in lighting power allowances.

e. Select appropriate fenestration type, based on the thermal transmittance value (U_{of}) of the entire fenestration assembly (including frames, sash, edge effects, and spacers). The ranges shown in the ACP tables generally correspond to single glazing, double glazing, and "more": triple glazing, or high-performance glazing with low-emissivity films, and so on. In some fenestration options, the visible light transmittance (VLT) of the fenestration must not be less than the shading coefficient of the glazing, not considering any shading devices.

Substeps a through e end in the selection of the maximum percent fenestration area (of gross wall area) allowable in this climate by this standard.

STEP 3. Determine the maximum U_{ow} for the opaque wall assembly. Here, the mass of the wall is influential. For ordinary lightweight construction [where the heat capacity (HC) is less than 5 Btu/ft^2-°F], simply use the value indicated as the maximum U_{ow}. For more massive walls, the maximum U_{ow} will depend on the same range of internal load density (ILD) as was used to select the percent fenestration. (Note that HC, Btu/ft^2-°F = density, lb/ft^3 × thickness, ft × specific heat, Btu/lb-°F; these quantities for individual materials are found in Table 4.2, page 136.) When HC = 5 or more, the U_{ow} depends on whether the wall insulation is inside or integral with the wall mass (INT INS) or outside the wall mass (EXT INS, in which case the mass can better reduce the impact of internal heat gains). Then, using the percent fenestration (of wall area) of your design (or the maximum percent allowable, whichever is less) find the maximum U_{ow}, interpolating as necessary. If your percent fenestration is less than the minimum listed, select the U_{ow} for the smallest percent fenestration listed. Similarly, if your percent fenestration is greater than the maximum percent fenestration listed, select the U_{ow} for the largest percent fenestration listed.

STEP 4. Determine the minimum R value for insulation on walls below grade, or for a slab on grade. There is but one value shown for the minimum R value of insulation for walls down to one story below grade. (No insulation is required either for walls more than one story below grade, or for any below-grade walls where there are fewer than 3000 HDD65.) For an unheated slab on grade (building's heat source is above the slab, not integral with or below it), the minimum R depends on the *minimum distance listed,* either (1) first vertically downward to the bottom of the slab, then horizontally either beneath the slab or outward from the building, or (2) vertically downward from the top of the slab. For a heated slab, an R value of 2 must be added to these minimum R values.

STEP 5. Note the maximum overall thermal transmittance for the roof assembly, U_{or}. This includes the gross area of a roof assembly exposed to outside air (or unconditioned space) and includes skylight surfaces but excludes service openings such as for plumbing vents or fans. The roof assembly includes all roof or ceiling components through which heat may flow between indoor and outdoor (or unconditioned) environments. Skylights used in conjunction with automatic daylight controls on all-electric lighting in the area served by the skylights can be *excluded* from the U_{or} calculation, up to the maximum percent roof areas listed in Table K.3. However, these minimum areas can be increased by 50% if a shading device is used that blocks over 50% of the solar gain through the skylight during the peak cooling design condition (usually, July or August). The information on HDD65 and CDH80 for your location are found in Table K.1. The visible light transmittance (VLT) for skylights should only be interpolated between the 0.5 and 0.75 values shown in Table K.3. Note that *vertical glazed surfaces* such as monitors or clerestories are not considered skylights, but instead are to be included in the wall fenestration calculations.

STEP 6. Note the maximum overall thermal transmittance value for walls adjacent to unconditioned space. This includes both the opaque and glazed elements of such a wall.

STEP 7. Note the maximum overall thermal transmittance value for the floor assembly, over outside or unconditioned space, U_{off}. The floor assembly includes all floor components through which heat may flow between indoor and outdoor (or unconditioned) environments.

TABLE K.1 **Degree Days and Annual Average Insolation on Vertical Surfaces**[a]

City	HDD50	HDD65	(CDH74)[b]	VSN	VSEW	VSS	CDD50	CDD65	CDH80	DR	NO HRS 8AM–4PM $T < 55$	NO HRS 8AM–4PM $55 \leq T \leq 69$	ACP Table
ALABAMA													
Birmingham	765	2882	(21533)	464	789	908	5182	1825	6272	17.5	720	760	K.9
Mobile	164	1580	(28912)	486	816	919	6478	2419	7479	16.6	408	774	K.12
Montgomery	491	2261	(24564)	462	823	981	5821	2116	8473	19.5	609	734	K.12
ALASKA													
Adak	3562	8913	(0)	280	434	652	124	0	0	9.9	2754	156	K.35
Anchorage	5301	10540	(35)	272	538	926	236	0	0	13.8	2398	521	K.35
Annette Island	2545	7277	(308)	285	482	739	756	12	0	10.1	2169	719	K.35
Bethel	8285	13449	(238)	252	453	789	312	0	0	14.3	2555	347	—[c]
Big Delta	9355	14069	(582)	249	527	989	777	16	25	19.0	2141	606	—[c]
Fairbanks	9841	14414	(752)	241	492	919	922	19	8	18.2	2083	682	—[c]
Gulkana	8865	13846	(106)	257	522	943	498	4	6	18.5	2225	615	—[c]
Juneau	4223	9350	(34)	254	410	642	348	0	0	12.7	2367	540	K.35
King Salmon	6843	11992	(64)	270	499	860	330	4	6	15.4	2395	502	—[c]
Kodiak	3775	8896	(335)	276	509	852	360	6	0	10.6	2519	384	K.35
McGrath	9967	14868	(405)	246	467	841	578	3	0	15.9	2265	596	—[c]
Nome	9061	14418	(272)	242	478	871	119	0	0	9.0	2710	210	—[c]
Summit	9210	14530	—	247	488	893	155	0	0	13.6	2616	298	—[c]
Yakutat	4486	9714	(1)	247	402	650	248	0	0	9.3	2471	439	K.35
ARIZONA													
Phoenix	90	1382	(54404)	488	1116	1310	7830	3647	34521	21.2	373	746	K.17
Prescott	1477	4462	(3828)	473	1090	1334	3385	895	3973	24.0	1021	725	—[c]
Tucson	178	1601	(35946)	500	1112	1280	6822	2769	19657	21.9	399	716	K.14
Winslow	1695	4603	(10636)	471	1092	1338	3708	1141	7347	27.7	1130	634	—[c]
Yuma	43	782	(62507)	493	1151	1330	8921	4186	37892	23.5	247	697	—[c]
ARKANSAS													
Fort Smith	1149	3394	(23485)	462	842	1005	5307	2077	10413	22.4	925	547	K.22
Little Rock	912	3091	(23839)	465	831	981	5351	2055	8450	17.5	865	626	—[c]
CALIFORNIA													
Arcata	582	5020	(19)	407	724	926	1038	1	0	8.9	1396	1509	—[c]
Bakersfield	305	2194	(29954)	474	1053	1211	5879	2294	15447	28.9	645	848	K.13
China Lake	409	2444	(26531)	468	1091	1312	6222	2782	26739	27.4	582	772	K.14
Daggett	237	1916	(40146)	475	1102	1309	6516	2720	22302	27.0	472	841	K.14
El Toro	32	1577	(6921)	486	977	1163	4764	834	2391	22.3	215	1474	K.8
Fresno	492	2700	(19366)	459	1029	1199	5070	1803	13085	31.8	780	785	K.11
Long Beach	54	1483	(7755)	482	956	1144	4947	900	1616	16.1	263	1502	K.8

Los Angeles	3	1494	(4306)	482	962	1146	4456	472	136	14.1	145	1849	K.8
Mount Shasta	1947	5583	(2097)	419	909	1153	2395	556	2073	16.2	1544	756	K.26
Oakland	157	2922	(435)	453	909	1102	2792	82	23	16.4	770	1905	K.6
Point Mugu	8	2193	(1232)	477	936	1131	3435	145	70	12.3	209	2146	K.6
Red Bluff	589	2884	(22618)	428	951	1177	5110	1930	14404	29.5	860	810	K.11
Sacramento	381	2753	(12556)	444	987	1185	4274	1171	7315	34.6	834	990	K.11
San Diego	2	1275	(4643)	490	950	1121	4865	662	383	11.5	102	1911	K.8
San Francisco	186	3238	(265)	454	941	1146	2496	73	204	20.2	782	1796	K.7
Santa Maria	138	3041	(131)	476	950	1128	2663	92	513	20.9	414	2016	K.7
Sunnyvale	142	2708	(1430)	456	947	1145	3112	204	421	16.9	610	1794	K.6
COLORADO													
Colorado Springs	2587	5996	(3658)	435	976	1321	2557	491	2075	24.0	1357	725	K.26
Denver	2652	6083	(5908)	428	971	1321	2611	567	2934	25.5	1329	739	K.26
Eagle	4232	8317	(248)	432	976	1296	1480	90	1008	35.4	1650	660	K.33
Grand Junction	2616	5701	(12097)	438	1003	1303	3611	1221	6147	27.4	1383	518	K.28
Pueblo	2223	5285	(10983)	442	992	1309	3384	971	5899	27.5	1077	720	K.26
CONNECTICUT													
Hartford	2953	6277	(4840)	384	646	834	2857	706	2197	23.7	1459	598	K.24
DELAWARE													
Wilmington	2133	5084	(8167)	414	726	921	3602	1078	2188	17.2	1289	617	K.23
DISTRICT OF COLUMBIA													
Washington	2004	4828	(10000)	419	724	905	3734	1083	3592	18.6	1205	657	K.23
FLORIDA													
Apalachicola	163	1366	(26051)	508	887	971	6967	2695	8289	14.3	322	778	K.13
Daytona	81	787	(27107)	503	860	953	7404	2635	5252	14.8	177	641	K.13
Jacksonville	206	1357	(24148)	495	849	943	7045	2721	7488	16.4	354	674	K.13
Miami	3	185	(39401)	527	874	936	9338	4045	9166	12.4	55	259	K.15
Orlando	33	532	(33985)	511	881	974	8288	3312	9757	17.1	131	571	K.16
Tallahassee	307	1721	(25185)	495	845	944	6462	2401	7323	16.1	421	747	K.13
Tampa	37	575	(33677)	518	890	974	7985	3047	8905	14.9	147	592	K.13
West Palm Beach	10	177	(35175)	519	846	906	9203	3904	10324	13.1	22	308	K.15
GEORGIA													
Atlanta	866	3070	(16803)	467	807	930	4837	1566	3799	17.6	915	749	K.10
Augusta	664	2584	(19510)	468	803	933	5458	1904	6904	21.3	690	774	K.9
Macon	514	2330	(24443)	476	822	939	5769	2111	8097	18.7	667	787	K.12
Savannah	410	1967	(22773)	474	805	926	6112	2194	6308	16.6	529	725	K.12
HAWAII													
Barbers Point	0	3	—	592	978	965	9314	3842	3617	11.2	1	97	K.4
Hilo	0	0	(14770)	557	817	805	8494	3019	1112	11.0	0	153	K.4
Honolulu	0	0	(36213)	588	953	932	9625	4150	4537	9.8	0	69	K.4
Lihue	0	2	(25602)	567	895	893	9219	3746	1912	9.6	0	140	K.4

TABLE K.1 **Degree Days and Annual Average Insolation on Vertical Surfaces**[a] ***(continued)***

City	HDD50	HDD65	(CDH74)[b]	VSN	VSEW	VSS	CDD50	CDD65	CDH80	DR	NO HRS 8AM–4PM		ACP Table
											$T < 55$	$55 \le T \le 69$	
IDAHO													
Boise	2276	5667	(7979)	399	916	1228	2828	744	4512	28.9	1542	647	K.26
Lewiston	2015	5426	(7923)	370	729	988	2709	645	4121	29.7	1467	748	K.23
Pocatello	3404	7075	(3286)	405	935	1262	2330	526	3293	32.8	1681	546	K.32
ILLINOIS													
Chicago	3000	6151	(8196)	402	729	936	3339	1015	3190	16.6	1426	613	K.24
Moline	3085	6250	(8616)	405	736	959	3204	894	2808	19.5	1357	640	K.24
Springfield	2490	5448	(12438)	422	768	962	3675	1158	4038	20.2	1260	600	K.27
INDIANA													
Evansville	1948	4625	(14797)	426	736	890	4063	1265	4288	18.4	1141	611	K.27
Fort Wayne	3023	6145	(6816)	395	664	826	3096	743	1629	17.7	1400	601	K.24
Indianapolis	2624	5620	(9082)	407	692	851	3430	951	2263	18.0	1375	602	K.24
South Bend	3038	6280	(6606)	396	690	857	2917	684	1840	21.1	1415	635	K.24
IOWA													
Burlington	3009	6094	(9712)	419	802	1030	3393	1002	2598	17.1	1354	649	K.24
Des Moines	3275	6447	(10484)	413	788	1027	3116	812	2383	17.5	1423	667	K.29
Mason City	4311	7735	(6687)	400	783	1053	2708	658	1882	20.8	1548	610	K.31
Sioux City	3608	6750	(10117)	406	794	1064	3326	993	3488	18.6	1438	602	K.29
KANSAS													
Dodge City	2280	5131	(18470)	450	942	1196	4008	1384	7186	26.0	1252	637	K.28
Goodland	2757	6090	(10079)	434	935	1228	3047	905	5147	26.3	1358	625	K.26
Topeka	2458	5201	(16640)	434	837	1068	4120	1388	5212	22.3	1192	608	K.27
KENTUCKY													
Covington	2154	5030	(9281)	408	687	843	3656	1057	2638	18.3	1316	661	K.23
Lexington	1921	4649	(11208)	425	729	872	3904	1157	2853	15.6	1211	618	K.27
Louisville	1851	4539	(13268)	424	727	883	4144	1357	4716	17.6	1192	636	K.27
LOUISIANA													
Baton Rouge	237	1573	(26938)	488	806	889	6682	2543	8814	17.2	440	677	K.12
Lake Charles	214	1455	(28573)	489	795	864	6849	2615	7883	14.8	396	668	K.12
New Orleans	179	1392	(32758)	497	838	923	6840	2578	7380	15.1	324	789	K.12
Shreveport	447	2265	(28295)	484	843	954	6022	2365	10039	18.1	687	697	K.12
MAINE													
Bangor	4132	7998	(1232)	378	693	950	1853	243	454	21.5	1721	669	K.31
Caribou	5297	9483	(920)	357	649	922	1410	121	203	18.1	1862	692	K.34
Portland	3531	7305	(1104)	376	643	856	1946	245	399	19.6	1604	665	K.30

MARYLAND													
Baltimore	2020	4946	(15162)	419	739	932	3683	1134	3825	18.6	1268	593	K.23
Patuxent	1418	4002	(11022)	429	758	943	4180	1289	2966	12.9	1118	729	K.20
MASSACHUSETTS													
Boston	2416	5775	(5358)	387	659	849	2810	695	1601	16.5	1495	713	K.23
MICHIGAN													
Alpena	4282	8164	(837)	371	661	862	1928	335	894	17.3	1707	695	K.31
Detroit	2799	5997	(4889)	390	676	858	3199	922	2238	18.8	1404	632	K.24
Flint	3471	6917	(2929)	379	641	811	2502	473	921	18.1	1563	634	K.30
Grand Rapids	3392	6777	(4555)	390	688	872	2680	590	1461	22.2	1562	622	K.29
Sault Sainte Marie	5087	9282	(844)	359	640	858	1399	119	246	21.0	1838	733	K.34
Traverse City	3934	7654	(2968)	369	642	818	2193	438	1124	21.0	1651	679	K.30
MINNESOTA													
Duluth	5797	9918	(849)	355	633	886	1511	157	258	20.0	1882	650	K.34
International Falls	6414	10535	(1614)	351	669	962	1473	119	167	22.0	1870	656	K.34
Minneapolis	4563	8060	(6806)	380	709	972	2751	773	2509	20.7	1620	566	K.31
Rochester	4544	8100	(3901)	383	691	927	2360	442	590	18.8	1584	652	K.31
MISSISSIPPI													
Jackson	546	2424	(25152)	481	833	942	5927	2330	8789	17.2	646	640	K.12
Meridian	546	2446	(23827)	480	811	905	5723	2148	9508	20.2	613	719	K.12
MISSOURI													
Columbia	2225	4994	(14475)	431	790	972	3940	1234	4242	21.5	1189	633	K.27
Saint Louis	2111	4860	(17843)	432	797	983	4193	1467	5379	18.7	1124	614	K.27
Springfield	1839	4509	(16262)	446	812	982	4115	1311	4170	20.4	1215	544	K.27
MONTANA													
Billings	3627	7156	(6167)	380	814	1160	2544	598	2695	25.6	1650	617	K.29
Cut Bank	4718	8941	(1378)	357	768	1150	1368	117	702	27.6	1834	672	K.31
Dillon	4140	8210	(1690)	386	838	1187	1564	159	784	28.6	1814	639	K.31
Glasgow	5082	8828	(4762)	361	752	1115	2272	543	2642	26.0	1688	570	K.34
Great Falls	3728	7454	(3574)	366	776	1133	2199	450	1886	26.7	1684	641	K.30
Helena	3926	7817	(2547)	372	771	1098	1911	328	1771	28.3	1784	651	K.30
Lewistown	4027	8089	(2127)	368	753	1084	1629	216	1270	29.8	1740	673	K.31
Miles City	4435	7989	(10094)	374	800	1156	2694	773	4364	26.9	1588	565	K.31
Missoula	3492	7560	(1068)	363	704	957	1629	221	1513	30.8	1843	658	K.30
NEBRASKA													
Grand Island	3315	6477	(11957)	420	843	1115	3309	996	4580	24.5	1431	611	K.29
North Platte	3447	6905	(8494)	419	880	1183	2731	715	3468	26.2	1514	592	K.32
Omaha	2981	5968	(10509)	414	806	1066	3618	1130	3883	19.6	1355	586	K.24
Scottsbluff	3335	6900	(7339)	413	861	1168	2603	693	3745	28.3	1457	620	K.32

TABLE K.1 **Degree Days and Annual Average Insolation on Vertical Surfaces**[a] ***(continued)***

City	HDD50	HDD65	(CDH74)[b]	VSN	VSEW	VSS	CDD50	CDD65	CDH80	DR	NO HRS 8AM–4PM $T < 55$	NO HRS 8AM–4PM $55 \le T \le 69$	ACP Table
NEVADA													
Elko	3345	7178	(3840)	420	1000	1332	1997	355	4065	37.8	1540	569	K.32
Ely	3683	7666	(659)	432	1014	1350	1650	157	1317	30.1	1529	683	K.32
Las Vegas	449	2399	(43153)	456	1136	1417	6567	3043	26408	25.5	604	719	K.14
Lovelock	2438	5845	(7845)	418	1094	1452	2813	745	6659	34.7	1358	606	K.26
Reno	2181	5841	(2192)	428	1068	1401	2180	365	4059	39.3	1306	752	K.26
Tonopah	2308	5652	(5888)	427	1130	1502	2742	611	3777	28.4	1257	660	K.26
Winnemucca	2774	6471	(4284)	418	1014	1350	2264	486	6366	41.0	1383	608	K.26
Yucca Flats	1664	4802	—	450	1112	1399	3378	1041	11568	35.9	1004	670	—[c]
NEW HAMPSHIRE													
Concord	3742	7425	(1993)	375	630	824	2254	463	1865	22.6	1533	683	K.30
NEW JERSEY													
Lakehurst	2174	5265	(5594)	407	712	917	3299	915	3019	20.5	1312	645	K.23
Newark	2027	4956	(9107)	406	710	912	3556	1009	2487	17.7	1325	644	K.23
NEW MEXICO													
Albuquerque	1633	4423	(11012)	469	1105	1361	3942	1257	5705	25.3	1148	703	K.21
Clayton	2138	5176	(6793)	457	1019	1310	3122	685	2093	20.0	1150	770	K.26
Roswell	1008	3486	(20216)	490	1081	1280	4536	1539	11135	26.1	825	677	K.21
Truth or Consequences	1074	3592	(14599)	488	1113	1326	4457	1500	6882	23.4	889	744	K.21
Tucumcari	1344	3922	(14754)	470	1046	1300	4451	1554	8424	26.9	914	710	K.21
NEW YORK													
Albany	3488	6770	(2998)	395	719	942	2812	619	1308	19.7	1487	605	K.29
Binghampton	3885	7397	(1646)	370	592	733	2373	410	672	18.5	1657	662	K.30
Buffalo	3213	6721	(3044)	371	609	746	2476	509	779	19.2	1571	697	K.29
Massena	4583	8397	(2012)	380	708	942	2026	365	913	20.9	1674	627	K.31
New York (Central Pk)	1986	5022	(9537)	392	650	817	3273	834	911	12.5	1335	790	K.23
New York (LAG)	1986	5022	(9189)	392	650	817	3273	834	911	12.5	1335	790	K.23
Rochester	3482	6995	(3764)	374	622	771	2557	595	1642	20.1	1612	608	K.29
Syracuse	3448	6856	(3516)	371	611	764	2579	513	926	20.2	1521	730	K.29
NORTH CAROLINA													
Asheville	1407	4203	(6292)	449	782	946	3442	763	1298	21.1	1083	915	K.19
Cape Hatteras	635	2745	(12440)	460	819	972	4978	1613	2039	10.0	765	820	K.9
Charlotte	1086	3412	(15159)	456	809	968	4698	1549	4299	19.6	892	777	K.20
Cherry Point	569	2556	(13152)	461	826	996	5277	1788	3614	15.2	690	757	K.9
Greensboro	1261	3760	(11020)	449	810	994	4274	1298	3642	17.4	1018	718	K.20
Raleigh	1131	3509	(11845)	445	774	935	4485	1389	3697	16.5	918	740	K.20
NORTH DAKOTA													
Bismarck	5196	8992	(4554)	371	766	1114	2175	496	2067	27.8	1724	556	K.34

Fargo	5582	9242	(4284)	371	751	1077	2388	573	2288	22.2	1730	546	K.34
Minot	5336	9178	(4157)	358	724	1059	2064	431	1570	24.5	1800	581	K.34
OHIO													
Akron	2881	6172	(4808)	396	664	812	2845	661	1100	17.3	1460	680	K.24
Columbus	2424	5493	(7490)	401	671	819	3195	789	2268	22.6	1375	708	K.23
Dayton	2573	5549	(8280)	408	696	855	3367	868	1346	17.1	1388	611	K.23
Toledo	3132	6514	(5081)	393	676	853	2791	698	1794	17.8	1500	652	K.24
Youngstown	3129	6557	(2975)	383	624	760	2593	546	1128	21.4	1523	679	K.24
OKLAHOMA													
Oklahoma City	1417	3825	(22978)	465	875	1053	4901	1834	8878	20.8	980	733	K.21
Tulsa	1429	3732	(26468)	453	820	991	5244	2072	10065	19.7	983	591	K.22
OREGON													
Astoria	1080	5226	(30)	350	588	782	1357	29	145	12.3	1571	1236	K.18
Medford	1531	4893	(6151)	405	814	1005	2681	568	4081	32.9	1442	749	K.19
North Bend	629	4678	(58)	392	740	977	1429	2	0	11.8	1351	1553	—[c]
Portland	1151	4577	(1851)	364	647	841	2321	272	1086	22.8	1421	1060	K.18
Redmond	2535	6665	(1972)	395	835	1127	1573	228	2390	34.4	1631	695	K.25
Salem	1128	4926	(1081)	373	680	874	1849	172	1224	29.3	1499	916	K.18
PENNSYLVANIA													
Allentown	2692	5760	(5804)	401	682	864	3105	698	1146	17.0	1335	710	K.24
Avoca (Scranton)	2931	6236	(3774)	389	646	811	2823	652	1547	19.7	1505	705	K.24
Erie	3006	6426	(2222)	384	646	792	2527	472	378	14.8	1532	716	K.25
Harrisburg	2302	5251	(9071)	404	687	864	3518	992	2860	20.1	1342	648	K.23
Philadelphia	2044	4923	(8896)	408	701	889	3661	1065	3172	17.1	1286	646	K.23
Pittsburgh	2773	5907	(5009)	392	642	780	2989	648	1040	19.0	1426	700	K.24
RHODE ISLAND													
Providence	2610	6022	(3631)	393	677	874	2756	693	1284	16.8	1429	684	K.24
SOUTH CAROLINA													
Charleston	435	2194	(23283)	467	796	925	5722	2005	5249	16.4	570	767	K.12
Columbia	694	2666	(21805)	467	816	953	5613	2110	8541	19.5	741	705	K.12
Greenville	907	3220	(14069)	459	814	971	4563	1400	3494	17.7	866	851	K.10
SOUTH DAKOTA													
Huron	4820	8351	(8677)	390	769	1044	2718	774	3739	24.5	1630	545	K.31
Pierre	4028	7358	(10444)	392	822	1147	3079	934	5262	24.2	1564	557	K.31
Rapid City	3672	7229	(8286)	394	819	1142	2581	663	3477	28.2	1530	572	K.29
Sioux Falls	4240	7683	(8638)	394	778	1078	2811	779	3029	20.2	1553	599	K.31
TENNESSEE													
Chattanooga	1232	3595	(17017)	444	738	869	4652	1541	5079	17.6	1050	684	K.20
Knoxville	1283	3818	(14214)	446	762	898	4455	1514	3840	17.8	1076	703	K.20
Memphis	1034	3259	(24504)	460	806	935	5319	2069	7807	19.2	865	851	K.22
Nashville	1165	3609	(18543)	443	749	863	4583	1552	5078	18.2	897	749	K.20

TABLE K.1 **Degree Days and Annual Average Insolation on Vertical Surfaces**[a] ***(continued)***

City	HDD50	HDD65	(CDH74)[b]	VSN	VSEW	VSS	CDD50	CDD65	CDH80	DR	NO HRS 8AM–4PM		ACP Table
											$T < 55$	$55 \le T \le 69$	
TEXAS													
Abilene	792	2714	(31904)	494	924	1066	5968	2416	13206	21.5	760	648	K.13
Amarillo	1592	4331	(15718)	471	1013	1253	4113	1377	6763	23.9	1109	680	K.21
Austin	271	1735	(35180)	503	877	972	6873	2862	14093	19.3	564	664	K.13
Brownsville	35	642	(42529)	547	908	908	8531	3664	12218	14.8	191	422	K.16
Corpus Christi	106	889	(42515)	529	906	946	8200	3508	13109	17.2	249	543	K.16
Del Rio	186	1397	(42032)	511	903	1008	7376	3112	14870	19.8	474	732	K.13
El Paso	522	2605	(22966)	503	1133	1306	5617	2225	13224	21.3	660	735	K.13
Fort Worth	605	2354	(30148)	485	875	994	6174	2448	13682	20.5	673	772	K.13
Houston	195	1346	(30474)	490	805	883	7215	2891	10569	18.2	352	703	K.12
Kingsville	49	874	(42877)	527	881	922	8302	3652	15512	19.2	260	523	K.16
Laredo	65	842	(52560)	532	900	936	8827	4130	25225	21.4	286	598	—[c]
Lubbock	1173	3643	(18218)	488	1070	1267	4754	1749	9827	25.1	917	743	K.21
Lufkin	370	1846	(30370)	492	848	942	6667	2668	11737	21.5	478	681	K.12
Midland	634	2573	(24201)	504	1079	1247	5695	2159	11177	25.9	698	729	K.13
Port Arthur	167	1416	(31667)	497	824	900	6888	2662	8837	17.4	384	697	K.12
San Angelo	538	2110	(32735)	503	944	1076	6522	2619	14621	20.6	641	619	K.13
San Antonio	261	1579	(36179)	510	878	955	7170	3013	13841	20.1	462	690	K.13
Sherman	699	2708	(29686)	476	862	996	5844	2378	12065	20.2	785	721	K.13
Waco	488	2166	(36671)	495	874	972	6676	2879	15658	21.1	651	622	K.13
Wichita Falls	984	3049	(34510)	480	911	1077	5708	2299	14487	18.8	802	723	—[c]
UTAH													
Bryce Canyon	4709	9288	(31)	445	1063	1386	899	4	69	30.0	1660	841	K.33
Cedar City	2592	5888	(5036)	447	1054	1342	2802	624	3119	27.1	1392	629	K.26
Salt Lake City	2570	5975	(9898)	422	975	1266	3011	941	7030	29.1	1426	586	K.26
VERMONT													
Burlington	4211	7932	(2562)	382	698	925	2118	365	490	18.3	1697	637	K.31
VIRGINIA													
Norfolk	1185	3609	(13677)	443	792	964	4636	1586	4554	15.0	1014	685	K.20
Richmond	1322	3895	(12260)	430	745	923	4225	1323	4021	17.6	996	716	K.20
Roanoke	1520	4192	(9275)	433	763	946	3986	1183	3306	19.0	1148	713	K.20
WASHINGTON													
Olympia	1546	5550	(341)	351	619	819	1550	79	466	26.4	1577	985	K.18
Seattle/Tacoma	1382	5281	(1050)	350	621	828	1683	106	256	16.5	1700	982	K.18
Spokane	2983	6727	(3453)	363	758	1064	2094	363	1595	25.3	1669	640	K.25
Whidbey Island	1179	5274	(59)	344	630	878	1403	22	7	14.8	1671	1169	K.18
Yakima	2323	5877	(4118)	373	790	1091	2370	449	3285	31.2	1413	703	K.25

TABLE K.2 **ASHRAE/IES Standard 90.1-1989: Electric Lighting and Power Estimating (*continued*)**

Area/Activity	*UPD W/ft²*	*Note*	*Area/Activity*	*UPD W/ft²*	*Note*
Hotel/conference center			Post office		
Banquet room/multipurpose	2.4	c	Lobby	1.1	
Bathroom/powder room	1.2		Sorting & mailing	2.1	
Guest room	1.4		Service station/auto repair	1.0	
Public area	1.2		Theater		
Exhibition hall	2.6		Performance arts	1.5	
Conference/meeting	1.8	c	Motion picture	1.0	
Lobby	1.9		Lobby	1.5	
Reception desk	2.4		Retail establishments (Merchandising & circulation area)		
Laundry			Applicable to all lighting, including accent and display lighting, installed in merchandising and circulation areas		
Washing	0.9		Type A	5.6	h
Ironing & sorting	1.3		Type B	3.2	h
Museum & gallery			Type C	3.3	h
General exhibition	1.9		Type D	3.1	h
Inspection/restoration	3.9		Type E	2.8	h
Storage (artifacts)			Type F	2.7	h
Inactive	0.6		Mall concourse	1.4	
Active	0.7		Retail support areas		
			Tailoring	2.1	
			Dressing/fitting rooms	1.4	

Part E. System Performance Unit Lighting Power Allowance, Indoor Athletic Areas[i]

Area/Activity	*UPD W/ft²*	*Note*	*Area/Activity*	*UPD W/ft²*	*Note*
Seating area, all sports	0.4		Handball/racquetball/squash		
Badminton			Club	1.3	
Club	0.5		Tournament	2.6	
Tournament	0.8		Hockey, ice		
Basketball/volleyball			Amateur	1.3	
Intramural	0.8		College or professional	2.6	
College	1.3		Skating rink		
Professional	1.9		Recreational	0.6	
Bowling			Exhibition/professional	2.6	
Approach area	0.5		Swimming		
Lanes	1.1		Recreational	0.9	
Boxing or wrestling (platform)			Exhibition	1.5	
Amateur	2.4		Under water	1.0	
Professional	4.8		Tennis		
Gymnasium			Recreational (class III)	1.3	
General exercising & recreation only	1.0		Club/college (class II)	1.9	
			Professional (class I)	2.6	
			Tennis, table		
			Club	1.0	
			Tournament	1.6	

WEST VIRGINIA													
Charleston	1816	4587	(8790)	409	667	798	3712	1008	3054	20.8	1215	704	K.23
WISCONSIN													
Eau Claire	4751	8285	(3886)	376	683	923	2545	603	1898	18.2	1565	661	K.31
Green Bay	4310	8039	(2453)	380	696	947	2172	426	957	22.1	1604	651	K.31
La Crosse	3838	7243	(6795)	386	701	937	2786	716	2121	18.9	1568	644	K.29
Madison	4009	7466	(3343)	391	717	955	2559	542	1329	19.1	1511	658	K.31
Milwaukee	3586	7121	(3313)	396	724	941	2427	487	1013	17.1	1587	618	K.29
WYOMING													
Casper	3824	7617	(4370)	403	961	1343	2177	495	2699	29.8	1670	535	K.32
Cheyenne	3435	7218	(2087)	416	906	1267	1963	271	1040	26.4	1618	608	K.32
Rock Springs	4407	8391	(1043)	411	1012	1395	1698	207	702	29.1	1828	552	K.33
Sheridan	3605	7366	(4115)	387	806	1133	2074	360	2105	30.8	1650	574	K.30
OTHER LOCATIONS OUTSIDE U.S.A.													
Guantanamo Bay, CU	0	0	—	612	1045	1018	11071	5596	18452	15.5	0	17	K.5
Koror Island, PN	0	0	(70454)	662	890	827	11435	5960	14548	9.5	0	0	K.5
Kwajalein Island, PN	0	0	(72338)	678	961	888	11635	6160	16217	8.2	0	0	K.5
San Juan, PR	0	0	(55224)	608	963	931	10648	5173	11563	12.7	0	14	K.5
Wake Island, PN	0	0	(57637)	609	1002	977	10869	5394	10167	9.7	0	0	K.5

Source: Reprinted by permission from *ASHRAE/IES Standard 90.1-1989, Energy Efficient Design of New Buildings Except Low Rise Residential Buildings*, © American Society of Heating, Refrigerating, and Air-Conditioning Engineers, Inc., Atlanta, Ga.

[a]HDD50 Average annual heating degree days, base 50 F
HDD65 Average annual heating degree days, base 65 F
VSN Average annual insolation, vertical north facing surface, Btu/ft²-day
VSEW Average annual insolation, vertical east (west) facing surface, Btu/ft²-day
VSS Average annual insolation, vertical south facing surface, Btu/ft²-day
CDD50 Average annual cooling degree days, base 50 F
CDH80 Average annual cooling degree hours, base 80 F
DR Daily range in warmest month between day high and night low temperatures, F°
NO HRS 8 AM–4 PM Average annual hours at these temperatures during this time period

[b](CDH74) Average annual cooling degree hours, base 74 F, to nearest location listed; From Proposed BSR/ASHRAE Standard 90.2 P, *Energy Efficient Design of New Low-Rise Residential Buildings*.

[c]See ASHRAE/IES Standard 90.1-1989 for ACP table at this location.

APPENDICES

TABLE K.2 **ASHRAE/IES Standard 90.1-1989: Electric Lighting and Power Estimating**

Part A. ***Prescriptive Unit Lighting Power Allowance (ULPA), W/ft² Gross Lighted Area of Total Building[a]***

Building Type or Space Activity	*0 to 2000 ft²*	*2000 to 10,000 ft²*	*10,001 to 25,000 ft²*	*25,001 to 50,000 ft²*	*50,001 to 250,000 ft²*	*> 250,000 ft²*
Food service						
Fast food cafeteria	1.50	1.38	1.34	1.32	1.31	1.30
Leisure dining/bar	2.20	1.91	1.71	1.56	1.46	1.40
Offices	1.90	1.81	1.72	1.65	1.57	1.50
Retail[b]	3.30	3.08	2.83	2.50	2.28	2.10
Mall concourse multistore service	1.60	1.58	1.52	1.46	1.43	1.40
Service establishment	2.70	2.37	2.08	1.92	1.80	1.70
Garages	0.30	0.28	0.24	0.22	0.21	0.20
Schools						
Preschool/elementary	1.80	1.80	1.72	1.65	1.57	1.50
Jr. high/high school	1.90	1.90	1.88	1.83	1.76	1.70
Technical/vocational	2.40	2.33	2.17	2.01	1.84	1.70
Warehouse storage	0.80	0.66	0.56	0.48	0.43	0.40

Part B. ***Average Receptacle Power Densities***

Building Type	*W/ft²*
1. Assembly	0.25
2. Office	0.75
3. Retail	0.25
4. Warehouse	0.10
5. School	0.50
6. Hotel or motel	0.25
7. Restaurant	0.10
8. Health	1.00
9. Multi-family	0.75

Part C. ***System Performance Unit Lighting Power Allowance, Common Activity Areas***

Area/Activity	*UPD W/ft²*	*Note*	*Area/Activity*	*UPD W/ft²*	*Note*
Auditorium	1.6	c	Garage		
Corridor	0.8	d	Auto & pedestrian circulation area	0.3	
Classroom/lecture hall	2.0		Parking area	0.2	
Electrical/mechanical equipment room			Laboratory	2.3	
General	0.7	d	Library		
Control rooms	1.5	d	Audio visual	1.1	
Food service			Stack area	1.5	
Fast food/cafeteria	1.3		Card file & cataloging	1.6	
Leisure dining	2.5	e	Reading area	1.9	
Bar/lounge	2.5	e	Lobby (general)		
Kitchen	1.4		Reception & waiting	1.0	
Recreation/lounge	0.7		Elevator lobbies	0.8	
Stair			Atrium (multi-story)		
Active traffic	0.6		First 3 floors	0.7	
Emergency exit	0.4		Each additional floor	0.2	
Toilet & washroom	0.8		Locker room & shower	0.8	

TABLE K.2 **ASHRAE/IES Standard 90.1-1989: Electric Lighting and Power** (*continued*)

Area/Activity	*UPD W/ft²*	*Note*
Office category 1		
Enclosed offices, all open plan offices without partitions or with partitions[f] lower than 4.5 ft below the ceiling.		
Reading, typing and filing	1.8	g
Drafting	2.6	g
Accounting	2.1	g
Office category 2		
Open plan offices 900 ft² or larger with partitions[f] 3.5 to 4.5 ft below the ceiling.		
Offices less than 900 ft² shall use category 1.		
Reading, typing and filing	1.9	d
Drafting	2.9	d
Accounting	2.4	d

Area/Activity	*UPD W/ft²*	*Note*
Office category 3		
Open plan offices 900 ft² or larger with partitions[f] higher than 3.5 ft below the ceiling. Offices less than 900 ft² shall use category 1.		
Reading, typing and filing	2.2	
Drafting	3.4	d
Accounting	2.7	d
Common activity areas		
Conference/meeting room	1.8	c
Computer/office equipment	2.1	
Filing, inactive	1.0	
Mail room	1.8	
Shop (non-industrial)		
Machinery	2.5	
Electrical/electronic	2.5	
Painting	1.6	
Carpentry	2.3	
Welding	1.2	
Storage & warehouse		
Inactive storage	0.3	
Active storage, bulky	0.3	
Active storage, fine	1.0	
Material handling	1.0	
Unlisted space	0.2	

Part D. ***System Performance Unit Lighting Power Allowance, Specific Buildings***

Area/Activity	*UPD W/ft²*	*Note*	*Area/Activity*	*UPD W/ft²*	*Note*
Airport, bus and rail station			Hospital/nursing home		
Baggage area	1.0		Corridor	1.3	d
Concourse/main thruway	0.9		Dental suite/examination/treatment	1.6	
Ticket counter	2.5		Emergency	2.3	
Waiting & lounge area	1.2		Laboratory	1.9	
Bank			Lounge/waiting room	0.9	
Customer area	1.1		Medical supplies	2.4	
Banking activity area	2.8		Nursery	2.0	
Barber & beauty parlor	2.0		Nurse station	2.1	
Church, synagogue, chapel			Occupational therapy/physical therapy		
Worship/congregational	2.5		Patient room		
Preaching & sermon/choir	2.7		Pharmacy		
Dormitory			Radiology		
Bedroom	1.1		Surgical & O.B. suites		
Bedroom with study	1.4		General area		
Study hall	1.8		Operating room		
Fire & police department			Recovery		
Fire engine room	0.7				
Jail cell	0.8				

TABLE K.2 **ASHRAE/IES Standard 90.1-1989: Electric Lighting and Power Estimating (*continued*)**

Part F. ***Area Factors***

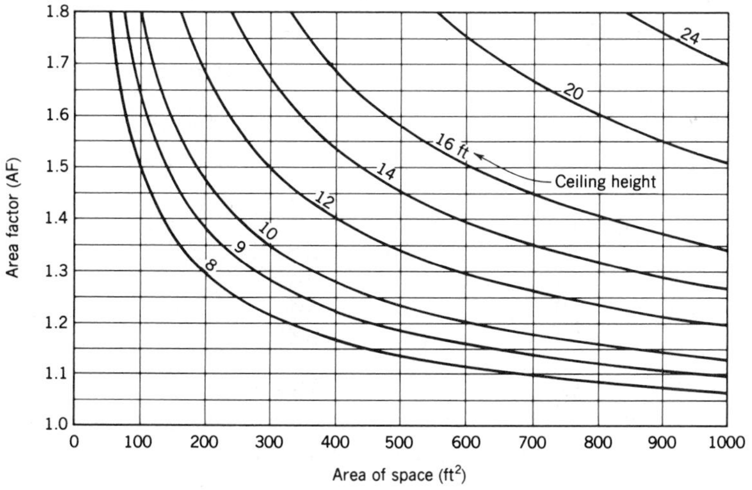

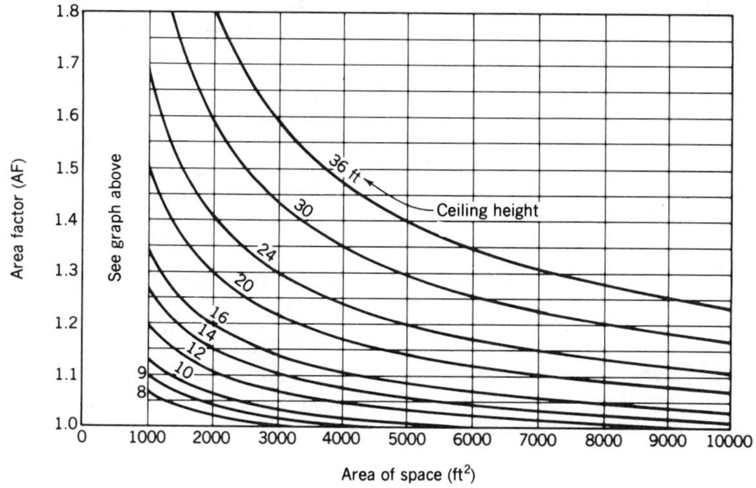

Source: Reprinted by permission from ASHRAE/IES Standard 90.1-1989, *Energy Efficient Design of New Buildings Except Low-Rise Residential Buildings,* © 1989 by the American Society of Heating, Refrigerating, and Air-Conditioning Engineers, Inc., Atlanta, Ga.

[a]This prescriptive table is intended primarily for core and shell (i.e., speculative) buildings or during the preliminary design phase (i.e., when the space uses are less than 80% defined). The values in this table are not intended to represent the needs of all buildings within the types listed.

[b]Includes general, merchandising, and display lighting.

[c]A 1.5 power adjustment factor is applicable for multi-functional spaces.

[d]Area factor of 1.0 shall be used for these spaces.

[e]UPD includes lighting power required for clean-up purposes.

[f]Not less than 90% of all work stations shall be individually enclosed with partitions of at least the height described.

[g]Area factor shall not exceed 1.55.

[h]Retail establishment classifications:
Type A: jewelry merchandising
Type B: fine apparel and accessories, china, crystal and silver, art galleries
Type C: "warehouse type" display of general apparel, varieties, stationery, books, sporting goods, hobby, cameras, gifts, luggage
Type D: "department store type" display of general apparel, varieties, stationery, books, sporting goods, hobby, cameras, gifts, luggage
Type E: Food and miscellaneous: bakeries, hardware and housewares, grocery, appliances, furniture
Type F: Service establishments

Area factor of 1.0 shall be used for all indoor athletic spaces.

TABLE K.3 **ASHRAE/IES Standard 90.1-1989: Maximum % Skylight Area of Roof Area**

Part A. ***Skylight VLT (Visible Light Transmission) = 0.75***

Building Location			Range of Lighting Power Density (W/ft^2)				
HDD65	*CDH80*	*Light Level (fc)*	*<1.00*	*1.01–1.50*	*1.51–2.00*	*2.01–2.50*	*>2.50*
0–3000	0–10,000	30	2.3	3.1	3.9	4.7	4.7
		50	3.1	4.3	5.5	6.7	6.7
		70	4.3	5.5	6.7	7.9	7.9
0–3000	>10,000	30	2.2	2.8	3.4	4.0	4.0
		50	2.3	3.1	3.9	4.7	4.7
		70	2.9	4.1	5.3	6.5	6.5
>3000	All	30	2.3	3.4	4.5	5.6	5.6
		50	2.5	4.0	5.5	7.0	7.0
		70	2.8	4.6	6.4	8.2	8.2

Part B. ***Skylight VLT (Visible Light Transmission) = 0.50***

Building Location			Range of Lighting Power Density (W/ft^2)				
HDD65	*CDH80*	*Light Level (fc)*	*<1.00*	*1.01–1.50*	*1.51–2.00*	*2.01–2.50*	*>2.50*
0–3000	0–10,000	30	3.6	4.8	6.0	7.2	7.2
		50	4.8	6.6	8.4	10.2	10.2
		70	6.6	8.4	10.2	12.0	12.0
0–3000	>10,000	30	3.3	4.2	5.1	6.0	6.0
		50	3.6	4.8	6.0	7.2	7.2
		70	4.2	6.0	7.8	9.6	9.6
>3000	All	30	3.6	5.1	6.6	8.1	8.1
		50	3.9	6.0	8.1	10.2	10.2
		70	4.2	6.9	9.6	12.3	12.3

Source: Reprinted by permission from ASHRAE/IES Standard 90.1-1989, *Energy Efficient Design of New Buildings Except Low-Rise Residential Buildings,* © 1989 by the American Society of Heating, Refrigerating, and Air-Conditioning Engineers, Inc., Atlanta, Ga.

TABLE K.4 **ACP for Hawaii (Barbers Point, Hilo, Honolulu, Lihue)**

Alternate Component Package, for ASHRAE/IES Standard 90.1-1989

HDD50 = 0
CDD65 = 3001-4500
VSEW = >800

FENESTRATION (Max. Pct. Fenestr.)

INTERNAL LOAD DENSITY (ILD) RANGE	PROJECTION FACTOR (PF) RANGE	SHADING COEFF RANGE (SCx)	BASE CASE Uof 0.81 to 0.46	BASE CASE Uof 0.45 to 0.39	BASE CASE Uof 0.38 to 0	PERIMETER DAYLIGHTING N/A	PERIMETER DAYLIGHTING VLT>=SC
0 - 1.50	0.000 - 0.249	1.000 - 0.71	24	31	33		
		0.709 - 0.60	26	37	39		
		0.599 - 0.50	27	41	44		
		0.499 - 0.38	29	45	50		
		0.379 - 0.25	30	52	59		
		0.249 - 0.00	32	59	70		
	0.250 - 0.499	1.000 - 0.71	28	42	45		
		0.709 - 0.60	30	47	52		
		0.599 - 0.50	31	52	58		
		0.499 - 0.38	32	56	64		
		0.379 - 0.25	33	61	72		
	0.500 +	1.000 - 0.71	31	50	55		
		0.709 - 0.60	32	56	63		
		0.599 - 0.50	33	59	68		
		0.499 - 0.38	33	62	73		
1.51 - 3.00	0.000 - 0.249	1.000 - 0.71	22	27	28		
		0.709 - 0.60	25	32	34		
		0.599 - 0.50	27	37	39		
		0.499 - 0.38	29	42	45		
		0.379 - 0.25	32	49	54		
		0.249 - 0.00	35	60	69		
	0.250 - 0.499	1.000 - 0.71	27	37	39		
		0.709 - 0.60	30	43	46		
		0.599 - 0.50	32	48	52		
		0.499 - 0.38	33	53	59		
		0.379 - 0.25	35	60	69		
	0.500 +	1.000 - 0.71	31	46	49		
		0.709 - 0.60	33	52	57		
		0.599 - 0.50	35	57	63		
		0.499 - 0.38	36	61	70		
3.01 - 3.50	0.000 - 0.249	1.000 - 0.71	21	25	26		
		0.709 - 0.60	24	30	31		
		0.599 - 0.50	26	34	36		
		0.499 - 0.38	28	39	42		
		0.379 - 0.25	31	47	51		
		0.249 - 0.00	35	58	66		
	0.250 - 0.499	1.000 - 0.71	26	34	36		
		0.709 - 0.60	29	40	43		
		0.599 - 0.50	31	45	49		
		0.499 - 0.38	33	51	56		
		0.379 - 0.25	35	58	66		
	0.500 +	1.000 - 0.71	30	43	46		
		0.709 - 0.60	33	49	54		
		0.599 - 0.50	34	54	60		
		0.499 - 0.38	36	59	67		

Daylight Sensing Controls

OPAQUE WALL (Max. Uow)

LIGHT WEIGHT WALL | MASS WALL

ILD 0 - 1.50:

Uow (HC<5)	HC RANGE	PCT FEN	INT INS	EXT INS
0.092	HC >= 5	24	0.096	0.12
	HC >= 10	24	0.10	0.14
	HC >= 15	24	0.11	0.14
	HC >= 5	73	0.096	0.12
	HC >= 10	73	0.10	0.14
	HC >= 15	73	0.11	0.14

ILD 1.51 - 3.00:

Uow (HC<5)	HC RANGE	PCT FEN	INT INS	EXT INS
0.092	HC >= 5	22	0.098	0.13
	HC >= 10	22	0.11	0.16
	HC >= 15	22	0.12	0.17
	HC >= 5	70	0.097	0.13
	HC >= 10	70	0.11	0.16
	HC >= 15	70	0.12	0.17

ILD 3.01 - 3.50:

Uow (HC<5)	HC RANGE	PCT FEN	INT INS	EXT INS
0.092	HC >= 5	21	0.098	0.14
	HC >= 10	21	0.11	0.17
	HC >= 15	21	0.13	0.18
	HC >= 5	67	0.098	0.13
	HC >= 10	67	0.11	0.17
	HC >= 15	67	0.12	0.18

OTHER CRITERIA

Minimum R-Value					Max Uo
WALL BELOW GRADE:		9		ROOF:	0.064
UNHEATED SLAB ON GRADE:	24"	36"	48"	WALL ADJACENT TO UNCOND SPACE:	0.14
Horizontal	16	13	11	FLOOR OVER	
Vertical	8	6	4	UNCOND. SPACE:	0.056

Source: Adapted with permission from ASHRAE/IES Standard 90.1-1989, *Energy Efficient Design of New Buildings Except Low-Rise Residential Buildings,* © 1989 by the American Society of Heating, Refrigerating, and Air-Conditioning Engineers, Inc., Atlanta, Ga.

TABLE K.5 **ACP for Puerto Rico (also Guantanamo Bay, CU; Kwajalein, Koror, Wake Islands)**

Alternate Component Package, for ASHRAE/IES Standard 90.1-1989

HDD50 = 0
CDD65 = >4500
VSEW = >845

FENESTRATION (Max. Pct. Fenestr.)

INTERNAL LOAD DENSITY (ILD) RANGE	PROJECTION FACTOR (PF) RANGE	SHADING COEFF RANGE (SCx)	BASE CASE Uof 1.15 to 0.82	BASE CASE Uof 0.81 to 0	BASE CASE N/A	PERIMETER DAYLIGHTING Uof 1.15 to 0.82	PERIMETER DAYLIGHTING Uof 0.81 to 0	PERIMETER DAYLIGHTING VLT>=SC Uof 0.81 to 0
0 - 1.50	0.000 - 0.249	1.000 - 0.71	14	14		17	16	17
		0.709 - 0.60	19	18		21	21	22
		0.599 - 0.50	23	22		26	26	26
		0.499 - 0.38	29	28		33	32	33
		0.379 - 0.25	42	39		49	46	47
		0.249 - 0.00	81	72		96	82	85
	0.250 - 0.499	1.000 - 0.71	21	21		25	24	25
		0.709 - 0.60	28	27		33	31	32
		0.599 - 0.50	35	33		41	39	40
		0.499 - 0.38	45	42		53	49	51
		0.379 - 0.25	69	62		81	72	74
	0.500 +	1.000 - 0.71	30	29		35	33	34
		0.709 - 0.60	40	38		47	44	45
		0.599 - 0.50	50	47		60	55	56
		0.499 - 0.38	67	61		80	71	73
1.51 - 3.00	0.000 - 0.249	1.000 - 0.71	12	12		19	19	21
		0.709 - 0.60	16	15		25	24	26
		0.599 - 0.50	19	19		31	30	32
		0.499 - 0.38	24	23		40	38	40
		0.379 - 0.25	35	33		60	54	58
		0.249 - 0.00	66	60		100	93	100
	0.250 - 0.499	1.000 - 0.71	18	17		29	28	30
		0.709 - 0.60	23	23		39	37	39
		0.599 - 0.50	29	28		49	46	49
		0.499 - 0.38	37	35		65	59	62
		0.379 - 0.25	56	52		96	83	90
	0.500 +	1.000 - 0.71	25	24		42	39	42
		0.709 - 0.60	33	31		57	52	55
		0.599 - 0.50	41	39		73	66	69
		0.499 - 0.38	55	50		95	82	88
3.01 - 3.50	0.000 - 0.249	1.000 - 0.71	9	9		17	17	19
		0.709 - 0.60	12	12		23	22	25
		0.599 - 0.50	15	14		28	27	31
		0.499 - 0.38	18	18		36	34	38
		0.379 - 0.25	26	25		55	50	55
		0.249 - 0.00	49	46		100	86	96
	0.250 - 0.499	1.000 - 0.71	14	13		26	25	29
		0.709 - 0.60	18	17		35	33	37
		0.599 - 0.50	22	21		45	42	46
		0.499 - 0.38	28	27		60	54	59
		0.379 - 0.25	42	39		88	77	85
	0.500 +	1.000 - 0.71	19	18		38	36	40
		0.709 - 0.60	25	24		52	48	52
		0.599 - 0.50	31	30		68	61	66
		0.499 - 0.38	41	38		87	76	84

Daylight Sensing Controls

OPAQUE WALL (Max. Uow)

ILD 0 - 1.50:

LIGHT WEIGHT WALL Uow (HC<5)	MASS WALL HC RANGE	PCT FEN	INT INS	EXT INS
1.00	HC >= 5	14	1.00	1.00
	HC >= 10	14	1.00	1.00
	HC >= 15	14	1.00	1.00
	HC >= 5	96	1.00	1.00
	HC >= 10	96	1.00	1.00
	HC >= 15	96	1.00	1.00

ILD 1.51 - 3.00:

LIGHT WEIGHT WALL Uow (HC<5)	MASS WALL HC RANGE	PCT FEN	INT INS	EXT INS
1.00	HC >= 5	12	1.00	1.00
	HC >= 10	12	1.00	1.00
	HC >= 15	12	1.00	1.00
	HC >= 5	96	1.00	1.00
	HC >= 10	96	1.00	1.00
	HC >= 15	96	1.00	1.00

ILD 3.01 - 3.50:

LIGHT WEIGHT WALL Uow (HC<5)	MASS WALL HC RANGE	PCT FEN	INT INS	EXT INS
1.00	HC >= 5	9	1.00	1.00
	HC >= 10	9	1.00	1.00
	HC >= 15	9	1.00	1.00
	HC >= 5	96	1.00	1.00
	HC >= 10	96	1.00	1.00
	HC >= 15	96	1.00	1.00

OTHER CRITERIA

	Minimum R-Value		
WALL BELOW GRADE:	0		
UNHEATED SLAB ON GRADE:	24"	36"	48"
Horizontal	0	0	0
Vertical	0	0	0

	Max Uo
ROOF:	0.057
WALL ADJACENT TO UNCOND SPACE:	1.0
FLOOR OVER UNCOND. SPACE:	0.40

Source: Adapted with permission from ASHRAE/IES Standard 90.1-1989, *Energy Efficient Design of New Buildings Except Low-Rise Residential Buildings,* © 1989 by the American Society of Heating, Refrigerating, and Air-Conditioning Engineers, Inc., Atlanta, Ga.

APPENDICES

TABLE K.6 **ACP for Oakland, Point Mugu, and Sunnyvale, CA**

Alternate Component Package, for ASHRAE/IES Standard 90.1-1989

HDD50 = 1 - 1000
CDD65 = 0 - 300
VSEW = > 845
HDD65 = 1-3000

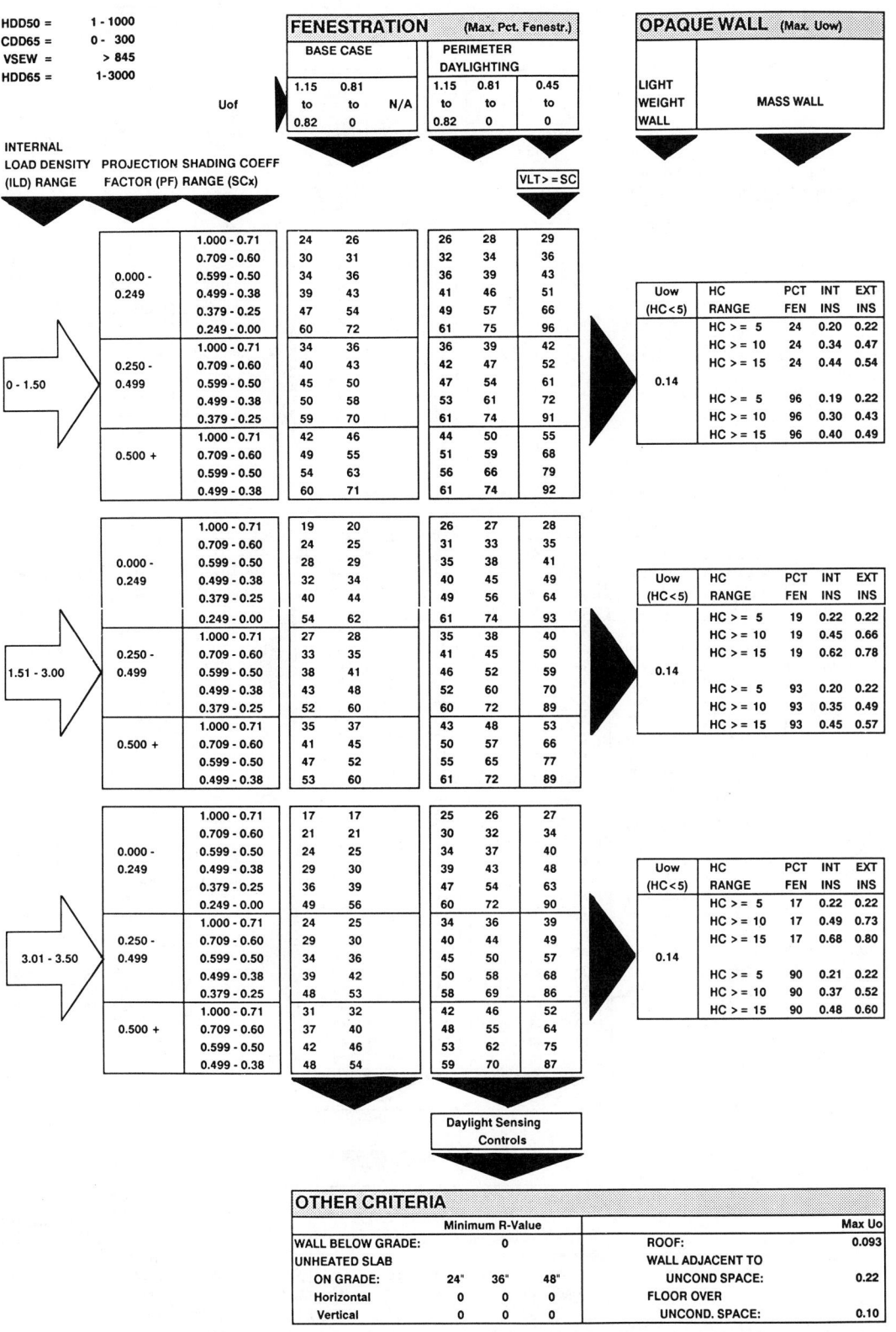

FENESTRATION (Max. Pct. Fenestr.)

INTERNAL LOAD DENSITY (ILD) RANGE	PROJECTION FACTOR (PF)	SHADING COEFF RANGE (SCx)	BASE CASE Uof 1.15 to 0.82	BASE CASE Uof 0.81 to 0	BASE CASE N/A	PERIMETER DAYLIGHTING Uof 1.15 to 0.82	PERIMETER DAYLIGHTING Uof 0.81 to 0	PERIMETER DAYLIGHTING Uof 0.45 to 0 (VLT>=SC)
0 - 1.50	0.000 - 0.249	1.000 - 0.71	24	26		26	28	29
		0.709 - 0.60	30	31		32	34	36
		0.599 - 0.50	34	36		36	39	43
		0.499 - 0.38	39	43		41	46	51
		0.379 - 0.25	47	54		49	57	66
		0.249 - 0.00	60	72		61	75	96
	0.250 - 0.499	1.000 - 0.71	34	36		36	39	42
		0.709 - 0.60	40	43		42	47	52
		0.599 - 0.50	45	50		47	54	61
		0.499 - 0.38	50	58		53	61	72
		0.379 - 0.25	59	70		61	74	91
	0.500 +	1.000 - 0.71	42	46		44	50	55
		0.709 - 0.60	49	55		51	59	68
		0.599 - 0.50	54	63		56	66	79
		0.499 - 0.38	60	71		61	74	92
1.51 - 3.00	0.000 - 0.249	1.000 - 0.71	19	20		26	27	28
		0.709 - 0.60	24	25		31	33	35
		0.599 - 0.50	28	29		35	38	41
		0.499 - 0.38	32	34		40	45	49
		0.379 - 0.25	40	44		49	56	64
		0.249 - 0.00	54	62		61	74	93
	0.250 - 0.499	1.000 - 0.71	27	28		35	38	40
		0.709 - 0.60	33	35		41	45	50
		0.599 - 0.50	38	41		46	52	59
		0.499 - 0.38	43	48		52	60	70
		0.379 - 0.25	52	60		60	72	89
	0.500 +	1.000 - 0.71	35	37		43	48	53
		0.709 - 0.60	41	45		50	57	66
		0.599 - 0.50	47	52		55	65	77
		0.499 - 0.38	53	60		61	72	89
3.01 - 3.50	0.000 - 0.249	1.000 - 0.71	17	17		25	26	27
		0.709 - 0.60	21	21		30	32	34
		0.599 - 0.50	24	25		34	37	40
		0.499 - 0.38	29	30		39	43	48
		0.379 - 0.25	36	39		47	54	63
		0.249 - 0.00	49	56		60	72	90
	0.250 - 0.499	1.000 - 0.71	24	25		34	36	39
		0.709 - 0.60	29	30		40	44	49
		0.599 - 0.50	34	36		45	50	57
		0.499 - 0.38	39	42		50	58	68
		0.379 - 0.25	48	53		58	69	86
	0.500 +	1.000 - 0.71	31	32		42	46	52
		0.709 - 0.60	37	40		48	55	64
		0.599 - 0.50	42	46		53	62	75
		0.499 - 0.38	48	54		59	70	87

Perimeter daylighting: Daylight Sensing Controls

OPAQUE WALL (Max. Uow)

ILD 0 - 1.50:

LIGHT WEIGHT WALL Uow (HC<5)	MASS WALL HC RANGE	PCT FEN	INT INS	EXT INS
0.14	HC >= 5	24	0.20	0.22
	HC >= 10	24	0.34	0.47
	HC >= 15	24	0.44	0.54
	HC >= 5	96	0.19	0.22
	HC >= 10	96	0.30	0.43
	HC >= 15	96	0.40	0.49

ILD 1.51 - 3.00:

LIGHT WEIGHT WALL Uow (HC<5)	MASS WALL HC RANGE	PCT FEN	INT INS	EXT INS
0.14	HC >= 5	19	0.22	0.22
	HC >= 10	19	0.45	0.66
	HC >= 15	19	0.62	0.78
	HC >= 5	93	0.20	0.22
	HC >= 10	93	0.35	0.49
	HC >= 15	93	0.45	0.57

ILD 3.01 - 3.50:

LIGHT WEIGHT WALL Uow (HC<5)	MASS WALL HC RANGE	PCT FEN	INT INS	EXT INS
0.14	HC >= 5	17	0.22	0.22
	HC >= 10	17	0.49	0.73
	HC >= 15	17	0.68	0.80
	HC >= 5	90	0.21	0.22
	HC >= 10	90	0.37	0.52
	HC >= 15	90	0.48	0.60

OTHER CRITERIA

	Minimum R-Value				Max Uo
WALL BELOW GRADE:	0			ROOF:	0.093
UNHEATED SLAB ON GRADE:	24"	36"	48"	WALL ADJACENT TO UNCOND SPACE:	0.22
Horizontal	0	0	0	FLOOR OVER UNCOND. SPACE:	0.10
Vertical	0	0	0		

Source: Adapted with permission from ASHRAE/IES Standard 90.1-1989, *Energy Efficient Design of New Buildings Except Low-Rise Residential Buildings,* © 1989 by the American Society of Heating, Refrigerating, and Air-Conditioning Engineers, Inc., Atlanta, Ga.

TABLE K.7 **ACP for San Francisco and Santa Maria, CA**

Alternate Component Package, for ASHRAE/IES Standard 90.1-1989

HDD50 =	1-1000
CDD65 =	0- 300
VSEW =	>845
HDD65 =	1-3000

FENESTRATION (Max. Pct. Fenestr.)

INTERNAL LOAD DENSITY (ILD) RANGE	PROJECTION FACTOR (PF) RANGE	SHADING COEFF RANGE (SCx)	BASE CASE Uof 0.81 to 0.46	BASE CASE Uof 0.45 to 0.39	BASE CASE Uof 0.38 to 0	PERIMETER DAYLIGHTING Uof 0.81 to 0.46	PERIMETER DAYLIGHTING Uof 0.45 to 0	PERIMETER DAYLIGHTING Uof 0.38 to 0 (VLT>=SC)
0 - 1.50	0.000 - 0.249	1.000 - 0.71	21	22	23	23	25	25
		0.709 - 0.60	26	28	28	28	31	31
		0.599 - 0.50	30	33	33	32	36	37
		0.499 - 0.38	35	39	40	37	43	44
		0.379 - 0.25	42	50	52	45	55	57
		0.249 - 0.00	55	72	76	57	77	82
	0.250 - 0.499	1.000 - 0.71	30	32	33	32	36	37
		0.709 - 0.60	35	40	40	38	44	45
		0.599 - 0.50	40	46	47	43	51	53
		0.499 - 0.38	46	55	56	48	60	63
		0.379 - 0.25	54	69	72	56	74	79
	0.500 +	1.000 - 0.71	38	43	43	40	47	48
		0.709 - 0.60	44	52	53	47	57	59
		0.599 - 0.50	49	60	62	52	65	69
		0.499 - 0.38	55	70	73	57	75	80
1.51 - 3.00	0.000 - 0.249	1.000 - 0.71	17	18	18	23	25	26
		0.709 - 0.60	21	22	22	28	31	32
		0.599 - 0.50	25	26	26	33	36	38
		0.499 - 0.38	29	31	32	38	43	45
		0.379 - 0.25	37	41	42	46	55	59
		0.249 - 0.00	51	61	63	60	77	84
	0.250 - 0.499	1.000 - 0.71	25	26	26	32	35	37
		0.709 - 0.60	30	32	32	39	43	46
		0.599 - 0.50	34	37	38	44	51	54
		0.499 - 0.38	40	45	45	49	60	64
		0.379 - 0.25	49	58	59	58	74	81
	0.500 +	1.000 - 0.71	32	34	34	41	47	49
		0.709 - 0.60	38	42	42	48	57	60
		0.599 - 0.50	43	49	50	53	65	70
		0.499 - 0.38	49	58	60	59	75	82
3.01 - 3.50	0.000 - 0.249	1.000 - 0.71	15	15	15	22	24	25
		0.709 - 0.60	18	19	19	27	29	31
		0.599 - 0.50	22	22	22	31	34	36
		0.499 - 0.38	26	27	27	36	41	44
		0.379 - 0.25	32	36	36	44	53	57
		0.249 - 0.00	45	53	54	57	73	82
	0.250 - 0.499	1.000 - 0.71	21	22	22	31	34	36
		0.709 - 0.60	26	27	27	37	42	44
		0.599 - 0.50	30	32	32	42	49	52
		0.499 - 0.38	35	38	39	47	57	62
		0.379 - 0.25	43	50	51	56	71	78
	0.500 +	1.000 - 0.71	27	29	29	39	44	48
		0.709 - 0.60	33	36	36	46	54	58
		0.599 - 0.50	38	42	43	51	63	68
		0.499 - 0.38	44	50	51	57	72	79

Perimeter Daylighting: Daylight Sensing Controls

OPAQUE WALL (Max. Uow)

LIGHT WEIGHT WALL / MASS WALL

ILD 0 - 1.50:

LIGHT WEIGHT WALL Uow (HC<5)	MASS WALL HC RANGE	PCT FEN	INT INS	EXT INS
0.13	HC >= 5	21	0.17	0.21
	HC >= 10	21	0.27	0.36
	HC >= 15	21	0.36	0.38
	HC >= 5	82	0.17	0.21
	HC >= 10	82	0.24	0.34
	HC >= 15	82	0.33	0.36

ILD 1.51 - 3.00:

LIGHT WEIGHT WALL Uow (HC<5)	MASS WALL HC RANGE	PCT FEN	INT INS	EXT INS
0.13	HC >= 5	17	0.21	0.21
	HC >= 10	17	0.36	0.46
	HC >= 15	17	0.48	0.51
	HC >= 5	84	0.17	0.21
	HC >= 10	84	0.28	0.37
	HC >= 15	84	0.38	0.40

ILD 3.01 - 3.50:

LIGHT WEIGHT WALL Uow (HC<5)	MASS WALL HC RANGE	PCT FEN	INT INS	EXT INS
0.13	HC >= 5	15	0.21	0.21
	HC >= 10	15	0.37	0.49
	HC >= 15	15	0.52	0.56
	HC >= 5	82	0.17	0.21
	HC >= 10	82	0.29	0.37
	HC >= 15	82	0.39	0.41

OTHER CRITERIA

	Minimum R-Value				Max Uo
WALL BELOW GRADE:	7			ROOF:	0.088
UNHEATED SLAB ON GRADE:	24"	36"	48"	WALL ADJACENT TO UNCOND SPACE:	0.21
Horizontal	11	9	8	FLOOR OVER UNCOND. SPACE:	0.094
Vertical	6	5	4		

Source: Adapted with permission from ASHRAE/IES Standard 90.1-1989, *Energy Efficient Design of New Buildings Except Low-Rise Residential Buildings,* © 1989 by the American Society of Heating, Refrigerating, and Air-Conditioning Engineers, Inc., Atlanta, Ga.

TABLE K.8 **ACP for El Toro, Long Beach, Los Angeles, and San Diego, CA**

Alternate Component Package, for ASHRAE/IES Standard 90.1-1989

HDD50 = 1 - 1000
CDD65 = 301 - 1150
VSEW = > 845

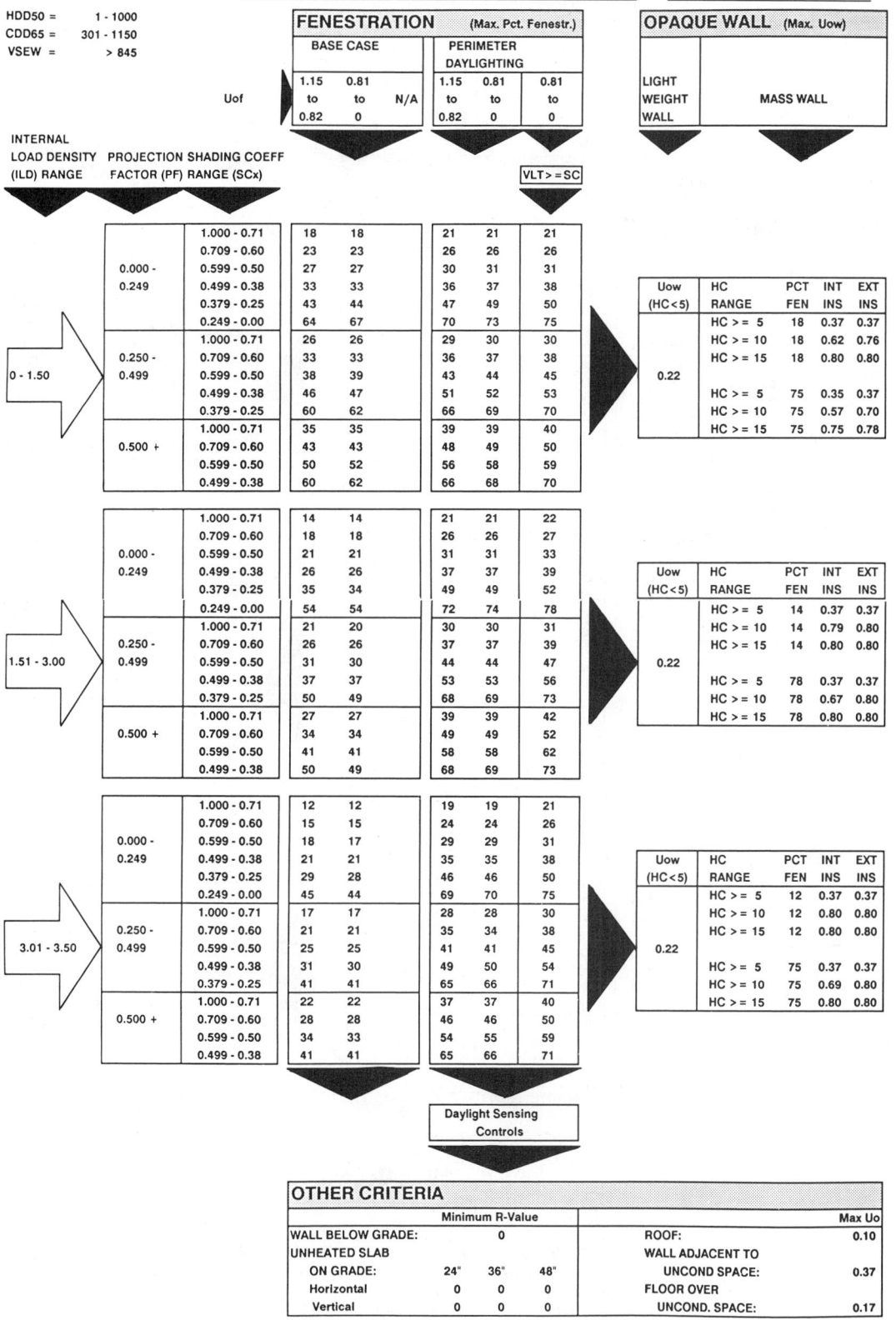

FENESTRATION (Max. Pct. Fenestr.)

Internal Load Density (ILD) Range	Projection Factor (PF) Range	Shading Coeff Range (SCx)	Base Case Uof 1.15 to 0.82	Base Case Uof 0.81 to 0	Base Case Uof N/A	Perimeter Daylighting Uof 1.15 to 0.82	Perimeter Daylighting Uof 0.81 to 0	Perimeter Daylighting Uof 0.81 to 0 (VLT>=SC)
0 - 1.50	0.000 - 0.249	1.000 - 0.71	18	18		21	21	21
		0.709 - 0.60	23	23		26	26	26
		0.599 - 0.50	27	27		30	31	31
		0.499 - 0.38	33	33		36	37	38
		0.379 - 0.25	43	44		47	49	50
		0.249 - 0.00	64	67		70	73	75
	0.250 - 0.499	1.000 - 0.71	26	26		29	30	30
		0.709 - 0.60	33	33		36	37	38
		0.599 - 0.50	38	39		43	44	45
		0.499 - 0.38	46	47		51	52	53
		0.379 - 0.25	60	62		66	69	70
	0.500 +	1.000 - 0.71	35	35		39	39	40
		0.709 - 0.60	43	43		48	49	50
		0.599 - 0.50	50	52		56	58	59
		0.499 - 0.38	60	62		66	68	70
1.51 - 3.00	0.000 - 0.249	1.000 - 0.71	14	14		21	21	22
		0.709 - 0.60	18	18		26	26	27
		0.599 - 0.50	21	21		31	31	33
		0.499 - 0.38	26	26		37	37	39
		0.379 - 0.25	35	34		49	49	52
		0.249 - 0.00	54	54		72	74	78
	0.250 - 0.499	1.000 - 0.71	21	20		30	30	31
		0.709 - 0.60	26	26		37	37	39
		0.599 - 0.50	31	30		44	44	47
		0.499 - 0.38	37	37		53	53	56
		0.379 - 0.25	50	49		68	69	73
	0.500 +	1.000 - 0.71	27	27		39	39	42
		0.709 - 0.60	34	34		49	49	52
		0.599 - 0.50	41	41		58	58	62
		0.499 - 0.38	50	49		68	69	73
3.01 - 3.50	0.000 - 0.249	1.000 - 0.71	12	12		19	19	21
		0.709 - 0.60	15	15		24	24	26
		0.599 - 0.50	18	17		29	29	31
		0.499 - 0.38	21	21		35	35	38
		0.379 - 0.25	29	28		46	46	50
		0.249 - 0.00	45	44		69	70	75
	0.250 - 0.499	1.000 - 0.71	17	17		28	28	30
		0.709 - 0.60	21	21		35	34	38
		0.599 - 0.50	25	25		41	41	45
		0.499 - 0.38	31	30		49	50	54
		0.379 - 0.25	41	41		65	66	71
	0.500 +	1.000 - 0.71	22	22		37	37	40
		0.709 - 0.60	28	28		46	46	50
		0.599 - 0.50	34	33		54	55	59
		0.499 - 0.38	41	41		65	66	71

Perimeter Daylighting: Daylight Sensing Controls

OPAQUE WALL (Max. Uow)

ILD 0 - 1.50:

Light Weight Wall Uow (HC<5)	Mass Wall HC Range	PCT FEN	INT INS	EXT INS
0.22	HC >= 5	18	0.37	0.37
	HC >= 10	18	0.62	0.76
	HC >= 15	18	0.80	0.80
	HC >= 5	75	0.35	0.37
	HC >= 10	75	0.57	0.70
	HC >= 15	75	0.75	0.78

ILD 1.51 - 3.00:

Light Weight Wall Uow (HC<5)	Mass Wall HC Range	PCT FEN	INT INS	EXT INS
0.22	HC >= 5	14	0.37	0.37
	HC >= 10	14	0.79	0.80
	HC >= 15	14	0.80	0.80
	HC >= 5	78	0.37	0.37
	HC >= 10	78	0.67	0.80
	HC >= 15	78	0.80	0.80

ILD 3.01 - 3.50:

Light Weight Wall Uow (HC<5)	Mass Wall HC Range	PCT FEN	INT INS	EXT INS
0.22	HC >= 5	12	0.37	0.37
	HC >= 10	12	0.80	0.80
	HC >= 15	12	0.80	0.80
	HC >= 5	75	0.37	0.37
	HC >= 10	75	0.69	0.80
	HC >= 15	75	0.80	0.80

OTHER CRITERIA

Minimum R-Value					Max Uo
WALL BELOW GRADE:	0			ROOF:	0.10
UNHEATED SLAB ON GRADE:	24"	36"	48"	WALL ADJACENT TO UNCOND SPACE:	0.37
Horizontal	0	0	0	FLOOR OVER UNCOND. SPACE:	0.17
Vertical	0	0	0		

Source: Adapted with permission from ASHRAE/IES Standard 90.1-1989, *Energy Efficient Design of New Buildings Except Low-Rise Residential Buildings*, © 1989 by the American Society of Heating, Refrigerating, and Air-Conditioning Engineers, Inc., Atlanta, Ga.

TABLE K.9 **ACP for Augusta, GA, Birmingham, AL, Cape Hatteras and Cherry Point, NC**

Alternate Component Package, for ASHRAE/IES Standard 90.1-1989

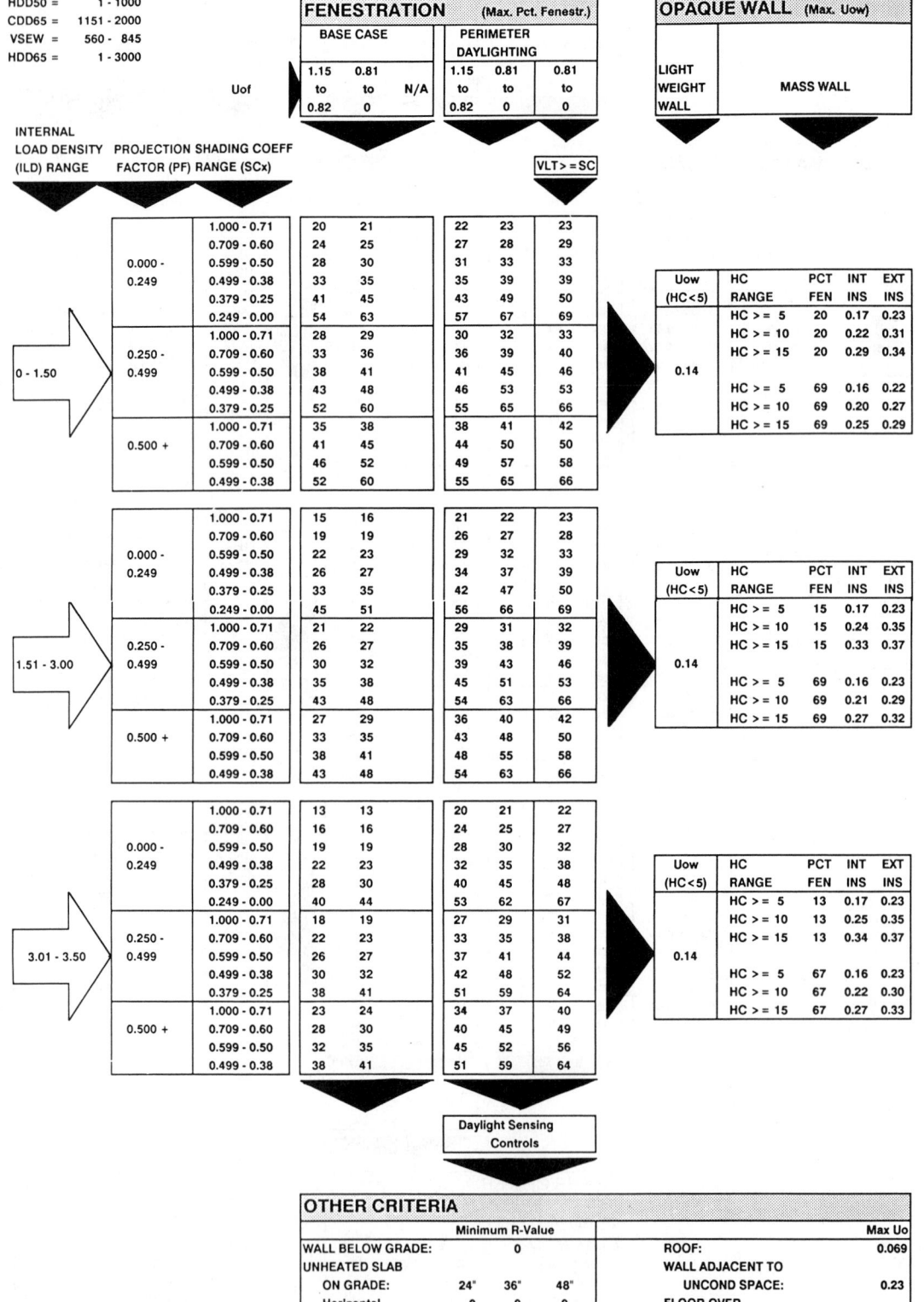

HDD50 = 1 - 1000
CDD65 = 1151 - 2000
VSEW = 560 - 845
HDD65 = 1 - 3000

FENESTRATION (Max. Pct. Fenestr.)

INTERNAL LOAD DENSITY (ILD) RANGE	PROJECTION FACTOR (PF) RANGE	SHADING COEFF RANGE (SCx)	BASE CASE Uof 1.15 to 0.82	BASE CASE Uof 0.81 to 0	BASE CASE N/A	PERIMETER DAYLIGHTING Uof 1.15 to 0.82	PERIMETER DAYLIGHTING Uof 0.81 to 0	PERIMETER DAYLIGHTING Uof 0.81 to 0 (VLT> =SC)
0 - 1.50	0.000 - 0.249	1.000 - 0.71	20	21		22	23	23
		0.709 - 0.60	24	25		27	28	29
		0.599 - 0.50	28	30		31	33	33
		0.499 - 0.38	33	35		35	39	39
		0.379 - 0.25	41	45		43	49	50
		0.249 - 0.00	54	63		57	67	69
	0.250 - 0.499	1.000 - 0.71	28	29		30	32	33
		0.709 - 0.60	33	36		36	39	40
		0.599 - 0.50	38	41		41	45	46
		0.499 - 0.38	43	48		46	53	53
		0.379 - 0.25	52	60		55	65	66
	0.500 +	1.000 - 0.71	35	38		38	41	42
		0.709 - 0.60	41	45		44	50	50
		0.599 - 0.50	46	52		49	57	58
		0.499 - 0.38	52	60		55	65	66
1.51 - 3.00	0.000 - 0.249	1.000 - 0.71	15	16		21	22	23
		0.709 - 0.60	19	19		26	27	28
		0.599 - 0.50	22	23		29	32	33
		0.499 - 0.38	26	27		34	37	39
		0.379 - 0.25	33	35		42	47	50
		0.249 - 0.00	45	51		56	66	69
	0.250 - 0.499	1.000 - 0.71	21	22		29	31	32
		0.709 - 0.60	26	27		35	38	39
		0.599 - 0.50	30	32		39	43	46
		0.499 - 0.38	35	38		45	51	53
		0.379 - 0.25	43	48		54	63	66
	0.500 +	1.000 - 0.71	27	29		36	40	42
		0.709 - 0.60	33	35		43	48	50
		0.599 - 0.50	38	41		48	55	58
		0.499 - 0.38	43	48		54	63	66
3.01 - 3.50	0.000 - 0.249	1.000 - 0.71	13	13		20	21	22
		0.709 - 0.60	16	16		24	25	27
		0.599 - 0.50	19	19		28	30	32
		0.499 - 0.38	22	23		32	35	38
		0.379 - 0.25	28	30		40	45	48
		0.249 - 0.00	40	44		53	62	67
	0.250 - 0.499	1.000 - 0.71	18	19		27	29	31
		0.709 - 0.60	22	23		33	35	38
		0.599 - 0.50	26	27		37	41	44
		0.499 - 0.38	30	32		42	48	52
		0.379 - 0.25	38	41		51	59	64
	0.500 +	1.000 - 0.71	23	24		34	37	40
		0.709 - 0.60	28	30		40	45	49
		0.599 - 0.50	32	35		45	52	56
		0.499 - 0.38	38	41		51	59	64

Perimeter daylighting: Daylight Sensing Controls

OPAQUE WALL (Max. Uow)

LIGHT WEIGHT WALL / MASS WALL

ILD 0 - 1.50:

LIGHT WEIGHT WALL Uow (HC<5)	HC RANGE	PCT FEN	INT INS	EXT INS
0.14	HC >= 5	20	0.17	0.23
	HC >= 10	20	0.22	0.31
	HC >= 15	20	0.29	0.34
	HC >= 5	69	0.16	0.22
	HC >= 10	69	0.20	0.27
	HC >= 15	69	0.25	0.29

ILD 1.51 - 3.00:

LIGHT WEIGHT WALL Uow (HC<5)	HC RANGE	PCT FEN	INT INS	EXT INS
0.14	HC >= 5	15	0.17	0.23
	HC >= 10	15	0.24	0.35
	HC >= 15	15	0.33	0.37
	HC >= 5	69	0.16	0.23
	HC >= 10	69	0.21	0.29
	HC >= 15	69	0.27	0.32

ILD 3.01 - 3.50:

LIGHT WEIGHT WALL Uow (HC<5)	HC RANGE	PCT FEN	INT INS	EXT INS
0.14	HC >= 5	13	0.17	0.23
	HC >= 10	13	0.25	0.35
	HC >= 15	13	0.34	0.37
	HC >= 5	67	0.16	0.23
	HC >= 10	67	0.22	0.30
	HC >= 15	67	0.27	0.33

OTHER CRITERIA

	Minimum R-Value				Max Uo
WALL BELOW GRADE:	0			ROOF:	0.069
UNHEATED SLAB ON GRADE:	24"	36"	48"	WALL ADJACENT TO UNCOND SPACE:	0.23
Horizontal	0	0	0	FLOOR OVER UNCOND. SPACE:	0.10
Vertical	0	0	0		

Source: Adapted with permission from ASHRAE/IES Standard 90.1-1989, *Energy Efficient Design of New Buildings Except Low-Rise Residential Buildings,* © 1989 by the American Society of Heating, Refrigerating, and Air-Conditioning Engineers, Inc., Atlanta, Ga.

Alternate Component Package, for ASHRAE/IES Standard 90.1-1989

HDD50 = 1 - 1000
CDD65 = 1151 - 2000
VSEW = 560 - 845
HDD65 = > 3000

FENESTRATION (Max. Pct. Fenestr.)

INTERNAL LOAD DENSITY (ILD) RANGE	PROJECTION FACTOR (PF) RANGE	SHADING COEFF RANGE (SCx)	BASE CASE Uof 0.81 to 0.46	BASE CASE Uof 0.45 to 0.39	BASE CASE Uof 0.38 to 0	PERIMETER DAYLIGHTING Uof 0.81 to 0.46	PERIMETER DAYLIGHTING Uof 0.45 to 0	PERIMETER DAYLIGHTING Uof 0.38 to 0 (VLT>=SC)
0 - 1.50	0.000 - 0.249	1.000 - 0.71	16	16	17	17	18	19
		0.709 - 0.60	19	20	21	21	23	24
		0.599 - 0.50	22	24	25	24	27	28
		0.499 - 0.38	26	29	29	28	32	34
		0.379 - 0.25	32	38	39	35	42	44
		0.249 - 0.00	43	55	58	46	61	66
	0.250 - 0.499	1.000 - 0.71	22	24	24	24	26	28
		0.709 - 0.60	26	29	30	29	33	34
		0.599 - 0.50	30	34	35	33	38	40
		0.499 - 0.38	35	41	42	37	45	48
		0.379 - 0.25	42	52	55	45	58	62
	0.500 +	1.000 - 0.71	28	31	32	30	35	37
		0.709 - 0.60	33	38	39	36	43	45
		0.599 - 0.50	37	45	46	40	50	53
		0.499 - 0.38	42	53	55	45	58	63
1.51 - 3.00	0.000 - 0.249	1.000 - 0.71	12	12	13	17	18	20
		0.709 - 0.60	15	16	16	21	22	25
		0.599 - 0.50	18	19	19	24	27	29
		0.499 - 0.38	21	22	23	28	32	35
		0.379 - 0.25	26	29	30	35	41	46
		0.249 - 0.00	37	44	46	47	61	68
	0.250 - 0.499	1.000 - 0.71	17	18	18	24	26	29
		0.709 - 0.60	21	22	23	28	32	35
		0.599 - 0.50	24	26	27	32	38	42
		0.499 - 0.38	28	32	32	37	45	50
		0.379 - 0.25	35	41	43	45	57	65
	0.500 +	1.000 - 0.71	22	24	24	30	34	38
		0.709 - 0.60	27	29	30	35	42	47
		0.599 - 0.50	31	35	35	40	49	55
		0.499 - 0.38	35	41	43	45	58	65
3.01 - 3.50	0.000 - 0.249	1.000 - 0.71	10	10	10	16	16	19
		0.709 - 0.60	12	13	13	19	20	23
		0.599 - 0.50	14	15	15	22	24	28
		0.499 - 0.38	17	18	18	26	29	33
		0.379 - 0.25	22	24	25	32	38	44
		0.249 - 0.00	31	37	38	43	56	66
	0.250 - 0.499	1.000 - 0.71	14	15	15	22	24	27
		0.709 - 0.60	17	18	18	26	29	34
		0.599 - 0.50	20	22	22	30	35	40
		0.499 - 0.38	24	26	26	34	41	48
		0.379 - 0.25	30	34	35	41	52	62
	0.500 +	1.000 - 0.71	18	19	19	27	31	36
		0.709 - 0.60	22	24	24	32	38	44
		0.599 - 0.50	25	28	29	36	45	52
		0.499 - 0.38	30	34	35	41	53	62

Daylight Sensing Controls (required for Perimeter Daylighting)

OPAQUE WALL (Max. Uow)

LIGHT WEIGHT WALL / MASS WALL

ILD 0 - 1.50:

Uow (HC<5)	HC RANGE	PCT FEN	INT INS	EXT INS
0.13	HC >= 5	16	0.15	0.21
	HC >= 10	16	0.20	0.26
	HC >= 15	16	0.24	0.29
	HC >= 5	66	0.15	0.20
	HC >= 10	66	0.19	0.25
	HC >= 15	66	0.23	0.27

ILD 1.51 - 3.00:

Uow (HC<5)	HC RANGE	PCT FEN	INT INS	EXT INS
0.13	HC >= 5	12	0.16	0.21
	HC >= 10	12	0.22	0.31
	HC >= 15	12	0.29	0.34
	HC >= 5	68	0.16	0.21
	HC >= 10	68	0.20	0.28
	HC >= 15	68	0.26	0.31

ILD 3.01 - 3.50:

Uow (HC<5)	HC RANGE	PCT FEN	INT INS	EXT INS
0.13	HC >= 5	10	0.16	0.21
	HC >= 10	10	0.22	0.31
	HC >= 15	10	0.30	0.35
	HC >= 5	66	0.16	0.21
	HC >= 10	66	0.21	0.28
	HC >= 15	66	0.26	0.31

OTHER CRITERIA

	Minimum R-Value				Max Uo
WALL BELOW GRADE:	7			ROOF:	0.072
UNHEATED SLAB ON GRADE:	24"	36"	48"	WALL ADJACENT TO UNCOND SPACE:	0.21
Horizontal	11	9	8	FLOOR OVER UNCOND. SPACE:	0.094
Vertical	6	5	4		

Source: Adapted with permission from ASHRAE/IES Standard 90.1-1989, *Energy Efficient Design of New Buildings Except Low-Rise Residential Buildings,* © 1989 by the American Society of Heating, Refrigerating, and Air-Conditioning Engineers, Inc., Atlanta, Ga.

TABLE K.11 **ACP for Red Bluff, Fresno, and Sacramento, CA**

Alternate Component Package, for ASHRAE/IES Standard 90.1-1989

HDD50 = 1 - 1000
CDD65 = 1151 - 2000
VSEW = > 845

FENESTRATION (Max. Pct. Fenestr.)

INTERNAL LOAD DENSITY (ILD) RANGE	PROJECTION FACTOR (PF) RANGE	SHADING COEFF (SCx)	BASE CASE Uof 1.15 to 0.82	BASE CASE Uof 0.81 to 0	BASE CASE Uof N/A	PERIMETER DAYLIGHTING Uof 1.15 to 0.82	PERIMETER DAYLIGHTING Uof 0.81 to 0	PERIMETER DAYLIGHTING Uof 0.81 to 0 (VLT>=SC)
0 - 1.50	0.000 - 0.249	1.000 - 0.71	19	20		21	22	22
		0.709 - 0.60	24	25		26	27	28
		0.599 - 0.50	28	29		30	32	33
		0.499 - 0.38	33	35		35	38	39
		0.379 - 0.25	41	45		44	49	50
		0.249 - 0.00	56	65		59	69	70
	0.250 - 0.499	1.000 - 0.71	27	29		30	32	32
		0.709 - 0.60	33	35		36	39	39
		0.599 - 0.50	38	41		41	45	46
		0.499 - 0.38	44	49		47	53	54
		0.379 - 0.25	54	62		57	66	68
	0.500 +	1.000 - 0.71	35	38		38	41	42
		0.709 - 0.60	42	46		45	50	51
		0.599 - 0.50	48	54		51	58	59
		0.499 - 0.38	55	63		57	67	68
1.51 - 3.00	0.000 - 0.249	1.000 - 0.71	15	15		20	21	22
		0.709 - 0.60	18	19		25	26	27
		0.599 - 0.50	22	23		29	30	32
		0.499 - 0.38	26	27		34	36	38
		0.379 - 0.25	33	35		42	47	49
		0.249 - 0.00	47	53		58	67	70
	0.250 - 0.499	1.000 - 0.71	21	22		28	30	31
		0.709 - 0.60	26	27		34	37	39
		0.599 - 0.50	30	32		39	43	45
		0.499 - 0.38	36	38		46	51	53
		0.379 - 0.25	45	50		55	64	67
	0.500 +	1.000 - 0.71	28	29		36	39	41
		0.709 - 0.60	34	36		43	48	50
		0.599 - 0.50	39	42		49	56	59
		0.499 - 0.38	45	50		56	65	68
3.01 - 3.50	0.000 - 0.249	1.000 - 0.71	13	13		19	20	21
		0.709 - 0.60	16	16		23	24	26
		0.599 - 0.50	19	19		27	29	31
		0.499 - 0.38	22	23		32	34	37
		0.379 - 0.25	29	31		40	44	48
		0.249 - 0.00	42	46		55	64	68
	0.250 - 0.499	1.000 - 0.71	18	19		27	28	30
		0.709 - 0.60	22	23		32	35	37
		0.599 - 0.50	26	27		37	41	44
		0.499 - 0.38	31	33		43	48	52
		0.379 - 0.25	39	43		53	61	66
	0.500 +	1.000 - 0.71	24	25		34	37	40
		0.709 - 0.60	29	31		41	45	49
		0.599 - 0.50	34	36		47	53	57
		0.499 - 0.38	40	43		53	61	66

Daylight Sensing Controls

OPAQUE WALL (Max. Uow)

ILD 0 - 1.50:

LIGHT WEIGHT WALL: Uow (HC<5)	MASS WALL: HC RANGE	PCT FEN	INT INS	EXT INS
0.14	HC >= 5	19	0.17	0.22
	HC >= 10	19	0.23	0.31
	HC >= 15	19	0.30	0.35
	HC >= 5	70	0.17	0.22
	HC >= 10	70	0.22	0.29
	HC >= 15	70	0.27	0.31

ILD 1.51 - 3.00:

LIGHT WEIGHT WALL: Uow (HC<5)	MASS WALL: HC RANGE	PCT FEN	INT INS	EXT INS
0.14	HC >= 5	15	0.18	0.22
	HC >= 10	15	0.24	0.33
	HC >= 15	15	0.32	0.37
	HC >= 5	70	0.17	0.22
	HC >= 10	70	0.22	0.30
	HC >= 15	70	0.28	0.33

ILD 3.01 - 3.50:

LIGHT WEIGHT WALL: Uow (HC<5)	MASS WALL: HC RANGE	PCT FEN	INT INS	EXT INS
0.14	HC >= 5	13	0.18	0.22
	HC >= 10	13	0.24	0.34
	HC >= 15	13	0.32	0.37
	HC >= 5	68	0.17	0.22
	HC >= 10	68	0.23	0.31
	HC >= 15	68	0.29	0.34

OTHER CRITERIA

	Minimum R-Value				Max Uo
WALL BELOW GRADE:	0			ROOF:	0.059
UNHEATED SLAB ON GRADE:	24"	36"	48"	WALL ADJACENT TO UNCOND SPACE:	0.22
Horizontal	0	0	0	FLOOR OVER UNCOND. SPACE:	0.10
Vertical	0	0	0		

Source: Adapted with permission from ASHRAE/IES Standard 90.1-1989, *Energy Efficient Design of New Buildings Except Low-Rise Residential Buildings*, © 1989 by the American Society of Heating, Refrigerating, and Air-Conditioning Engineers, Inc., Atlanta, Ga.

TABLE K.12 **ACP for Louisiana, Mississippi, Charleston and Columbia, SC, Macon and Savannah, GA, Mobile and Montgomery, AL, Houston, Lufkin, and Port Arthur TX**

Alternate Component Package, for ASHRAE/IES Standard 90.1-1989

HDD50 = 1 - 1000
CDD65 = 2001 - 3250
VSEW = 560 - 845
HDD65 = 1 - 3000

FENESTRATION (Max. Pct. Fenestr.)

INTERNAL LOAD DENSITY (ILD) RANGE	PROJECTION FACTOR (PF)	SHADING COEFF RANGE (SCx)	BASE CASE Uof 1.15 to 0.82	BASE CASE Uof 0.81 to 0	BASE CASE N/A	PERIMETER DAYLIGHTING Uof 1.15 to 0.82	PERIMETER DAYLIGHTING Uof 0.81 to 0	PERIMETER DAYLIGHTING Uof 0.81 to 0 (VLT>=SC)
0 - 1.50	0.000 - 0.249	1.000 - 0.71	19	19		21	22	22
		0.709 - 0.60	23	24		26	27	27
		0.599 - 0.50	27	28		30	32	32
		0.499 - 0.38	32	34		36	38	38
		0.379 - 0.25	41	44		45	49	50
		0.249 - 0.00	59	66		63	72	73
	0.250 - 0.499	1.000 - 0.71	26	27		29	31	31
		0.709 - 0.60	32	34		36	38	38
		0.599 - 0.50	38	40		41	44	45
		0.499 - 0.38	44	47		48	53	54
		0.379 - 0.25	55	61		60	67	69
	0.500 +	1.000 - 0.71	34	36		38	40	41
		0.709 - 0.60	41	44		46	49	50
		0.599 - 0.50	48	52		52	58	59
		0.499 - 0.38	56	61		60	68	69
1.51 - 3.00	0.000 - 0.249	1.000 - 0.71	15	15		21	21	23
		0.709 - 0.60	18	18		26	27	28
		0.599 - 0.50	22	22		30	31	33
		0.499 - 0.38	26	27		36	38	40
		0.379 - 0.25	34	35		46	49	52
		0.249 - 0.00	49	53		65	72	76
	0.250 - 0.499	1.000 - 0.71	21	21		29	30	32
		0.709 - 0.60	26	26		36	38	40
		0.599 - 0.50	30	31		42	45	47
		0.499 - 0.38	36	37		49	53	56
		0.379 - 0.25	46	49		61	68	72
	0.500 +	1.000 - 0.71	27	28		38	40	42
		0.709 - 0.60	33	35		46	49	52
		0.599 - 0.50	39	41		53	58	61
		0.499 - 0.38	46	49		61	68	72
3.01 - 3.50	0.000 - 0.249	1.000 - 0.71	12	12		19	20	22
		0.709 - 0.60	15	15		24	25	27
		0.599 - 0.50	18	18		28	29	32
		0.499 - 0.38	22	22		33	35	38
		0.379 - 0.25	28	29		43	46	50
		0.249 - 0.00	42	45		61	68	73
	0.250 - 0.499	1.000 - 0.71	17	17		27	28	31
		0.709 - 0.60	21	22		33	35	38
		0.599 - 0.50	25	26		39	42	45
		0.499 - 0.38	30	31		46	50	54
		0.379 - 0.25	39	41		57	64	69
	0.500 +	1.000 - 0.71	23	23		36	37	41
		0.709 - 0.60	28	29		43	46	50
		0.599 - 0.50	33	34		50	54	59
		0.499 - 0.38	39	41		57	64	69

Perimeter daylighting: Daylight Sensing Controls

OPAQUE WALL (Max. Uow)

ILD 0 - 1.50:

LIGHT WEIGHT WALL Uow (HC<5)	MASS WALL HC RANGE	PCT FEN	INT INS	EXT INS
0.15	HC >= 5	19	0.17	0.24
	HC >= 10	19	0.23	0.33
	HC >= 15	19	0.30	0.36
	HC >= 5	73	0.17	0.24
	HC >= 10	73	0.22	0.30
	HC >= 15	73	0.28	0.33

ILD 1.51 - 3.00:

LIGHT WEIGHT WALL Uow (HC<5)	MASS WALL HC RANGE	PCT FEN	INT INS	EXT INS
0.15	HC >= 5	15	0.17	0.24
	HC >= 10	15	0.24	0.36
	HC >= 15	15	0.34	0.39
	HC >= 5	76	0.17	0.24
	HC >= 10	76	0.23	0.33
	HC >= 15	76	0.30	0.36

ILD 3.01 - 3.50:

LIGHT WEIGHT WALL Uow (HC<5)	MASS WALL HC RANGE	PCT FEN	INT INS	EXT INS
0.15	HC >= 5	12	0.17	0.24
	HC >= 10	12	0.25	0.36
	HC >= 15	12	0.34	0.39
	HC >= 5	73	0.17	0.24
	HC >= 10	73	0.23	0.33
	HC >= 15	73	0.31	0.37

OTHER CRITERIA

Minimum R-Value				Max Uo	
WALL BELOW GRADE:	0			ROOF:	0.066
UNHEATED SLAB ON GRADE:	24"	36"	48"	WALL ADJACENT TO UNCOND SPACE:	0.24
Horizontal	0	0	0	FLOOR OVER UNCOND. SPACE:	0.11
Vertical	0	0	0		

Source: Adapted with permission from ASHRAE/IES Standard 90.1-1989, *Energy Efficient Design of New Buildings Except Low-Rise Residential Buildings,* © 1989 by the American Society of Heating, Refrigerating, and Air-Conditioning Engineers, Inc., Atlanta, Ga.

TABLE K.13 **ACP for Apalachicola, Daytona Beach, Jacksonville, Tallahassee, and Tampa, FL, Bakersfield, CA, Abilene, Austin, Del Rio, El Paso, Fort Worth, Midland, San Angelo, San Antonio, Sherman, and Waco, TX**

Alternate Component Package, for ASHRAE/IES Standard 90.1-1989

HDD50 = 1 - 1000
CDD65 = 2001 - 3250
VSEW = > 845
CDH80 = 0 - 18000
HDD65 = 1 - 3000

FENESTRATION (Max. Pct. Fenestr.)

INTERNAL LOAD DENSITY (ILD) RANGE	PROJECTION FACTOR (PF)	SHADING COEFF RANGE (SCx)	BASE CASE Uof 1.15 to 0.82	BASE CASE Uof 0.81 to 0	BASE CASE N/A	PERIMETER DAYLIGHTING Uof 1.15 to 0.82	PERIMETER DAYLIGHTING Uof 0.81 to 0	PERIMETER DAYLIGHTING Uof 0.81 to 0 (VLT>=SC)
0 - 1.50	0.000 - 0.249	1.000 - 0.71	17	18		19	19	20
		0.709 - 0.60	22	22		23	24	25
		0.599 - 0.50	25	26		27	29	29
		0.499 - 0.38	30	32		33	34	35
		0.379 - 0.25	39	42		42	45	46
		0.249 - 0.00	56	63		59	67	68
	0.250 - 0.499	1.000 - 0.71	25	26		27	28	29
		0.709 - 0.60	31	32		33	35	35
		0.599 - 0.50	36	38		38	41	42
		0.499 - 0.38	42	45		45	49	50
		0.379 - 0.25	53	59		57	64	65
	0.500 +	1.000 - 0.71	33	34		35	37	38
		0.709 - 0.60	40	42		43	46	47
		0.599 - 0.50	46	50		49	54	55
		0.499 - 0.38	54	60		57	64	65
1.51 - 3.00	0.000 - 0.249	1.000 - 0.71	14	14		19	19	20
		0.709 - 0.60	17	18		23	24	25
		0.599 - 0.50	21	21		27	28	30
		0.499 - 0.38	25	26		32	34	36
		0.379 - 0.25	32	34		42	45	47
		0.249 - 0.00	48	52		60	67	70
	0.250 - 0.499	1.000 - 0.71	20	20		27	28	29
		0.709 - 0.60	25	26		33	34	36
		0.599 - 0.50	29	30		38	40	43
		0.499 - 0.38	35	37		45	49	51
		0.379 - 0.25	45	49		57	63	67
	0.500 +	1.000 - 0.71	26	27		35	37	39
		0.709 - 0.60	33	34		42	45	48
		0.599 - 0.50	38	40		49	53	56
		0.499 - 0.38	45	49		57	64	67
3.01 - 3.50	0.000 - 0.249	1.000 - 0.71	12	12		17	18	19
		0.709 - 0.60	15	15		22	22	24
		0.599 - 0.50	18	18		25	26	29
		0.499 - 0.38	21	22		30	32	35
		0.379 - 0.25	28	29		39	42	45
		0.249 - 0.00	42	45		57	64	68
	0.250 - 0.499	1.000 - 0.71	17	18		25	26	28
		0.709 - 0.60	21	22		31	32	35
		0.599 - 0.50	25	26		36	38	41
		0.499 - 0.38	30	32		43	46	49
		0.379 - 0.25	39	42		54	60	64
	0.500 +	1.000 - 0.71	23	23		33	35	37
		0.709 - 0.60	28	29		40	43	46
		0.599 - 0.50	33	35		47	51	54
		0.499 - 0.38	39	42		54	60	65

Daylight Sensing Controls

OPAQUE WALL (Max. Uow)

ILD 0 - 1.50:

LIGHT WEIGHT WALL Uow (HC<5)	MASS WALL HC RANGE	PCT FEN	INT INS	EXT INS
0.15	HC >= 5	17	0.17	0.23
	HC >= 10	17	0.21	0.30
	HC >= 15	17	0.27	0.32
	HC >= 5	68	0.17	0.22
	HC >= 10	68	0.21	0.28
	HC >= 15	68	0.25	0.30

ILD 1.51 - 3.00:

LIGHT WEIGHT WALL Uow (HC<5)	MASS WALL HC RANGE	PCT FEN	INT INS	EXT INS
0.15	HC >= 5	14	0.17	0.24
	HC >= 10	14	0.22	0.32
	HC >= 15	14	0.30	0.35
	HC >= 5	70	0.17	0.23
	HC >= 10	70	0.21	0.30
	HC >= 15	70	0.27	0.32

ILD 3.01 - 3.50:

LIGHT WEIGHT WALL Uow (HC<5)	MASS WALL HC RANGE	PCT FEN	INT INS	EXT INS
0.15	HC >= 5	12	0.17	0.24
	HC >= 10	12	0.23	0.33
	HC >= 15	12	0.30	0.36
	HC >= 5	68	0.17	0.23
	HC >= 10	68	0.21	0.30
	HC >= 15	68	0.27	0.33

OTHER CRITERIA

	Minimum R-Value				Max Uo
WALL BELOW GRADE:	0			ROOF:	0.058
UNHEATED SLAB ON GRADE:	24"	36"	48"	WALL ADJACENT TO UNCOND SPACE:	0.24
Horizontal	0	0	0	FLOOR OVER UNCOND. SPACE:	0.11
Vertical	0	0	0		

Source: Adapted with permission from ASHRAE/IES Standard 90.1-1989, *Energy Efficient Design of New Buildings Except Low-Rise Residential Buildings,* © 1989 by the American Society of Heating, Refrigerating, and Air-Conditioning Engineers, Inc., Atlanta, Ga.

TABLE K.14 **ACP for Tucson, AZ, China Lake and Daggett, CA, and Las Vegas, NV**

Alternate Component Package, for ASHRAE/IES Standard 90.1-1989

HDD50 = 1 - 1000
CDD65 = 2001 - 3250
VSEW = > 845
CDH80 = > 18000

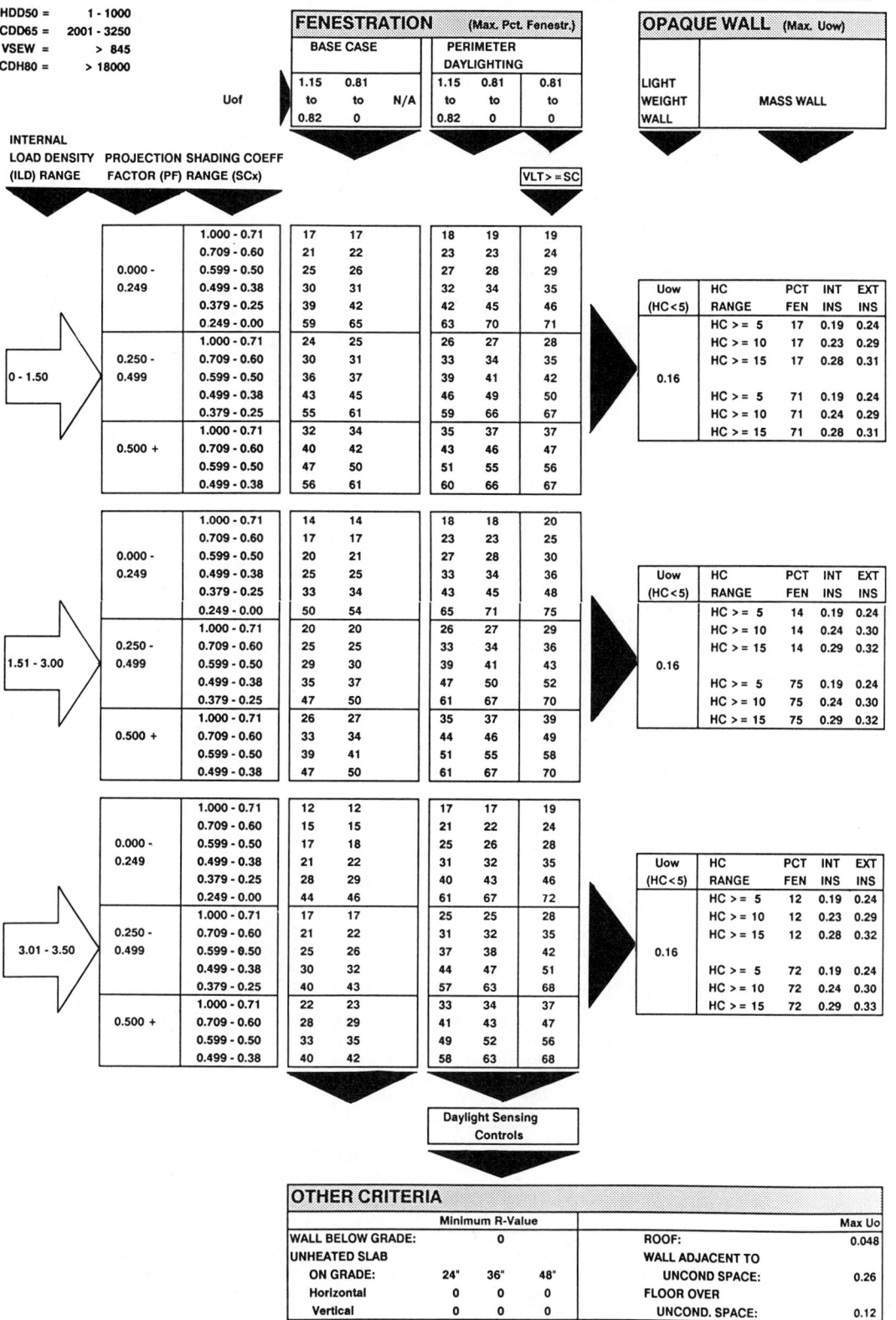

FENESTRATION (Max. Pct. Fenestr.)

Internal Load Density (ILD) Range	Projection Factor (PF) Range	Shading Coeff (SCx)	Base Case Uof 1.15 to 0.82	Base Case Uof 0.81 to 0	Base Case Uof N/A	Perimeter Daylighting Uof 1.15 to 0.82	Perimeter Daylighting Uof 0.81 to 0	Perimeter Daylighting Uof 0.81 to 0 (VLT>=SC)
0 - 1.50	0.000 - 0.249	1.000 - 0.71	17	17		18	19	19
		0.709 - 0.60	21	22		23	23	24
		0.599 - 0.50	25	26		27	28	29
		0.499 - 0.38	30	31		32	34	35
		0.379 - 0.25	39	42		42	45	46
		0.249 - 0.00	59	65		63	70	71
	0.250 - 0.499	1.000 - 0.71	24	25		26	27	28
		0.709 - 0.60	30	31		33	34	35
		0.599 - 0.50	36	37		39	41	42
		0.499 - 0.38	43	45		46	49	50
		0.379 - 0.25	55	61		59	66	67
	0.500 +	1.000 - 0.71	32	34		35	37	37
		0.709 - 0.60	40	42		43	46	47
		0.599 - 0.50	47	50		51	55	56
		0.499 - 0.38	56	61		60	66	67
1.51 - 3.00	0.000 - 0.249	1.000 - 0.71	14	14		18	18	20
		0.709 - 0.60	17	17		23	23	25
		0.599 - 0.50	20	21		27	28	30
		0.499 - 0.38	25	25		33	34	36
		0.379 - 0.25	33	34		43	45	48
		0.249 - 0.00	50	54		65	71	75
	0.250 - 0.499	1.000 - 0.71	20	20		26	27	29
		0.709 - 0.60	25	25		33	34	36
		0.599 - 0.50	29	30		39	41	43
		0.499 - 0.38	35	37		47	50	52
		0.379 - 0.25	47	50		61	67	70
	0.500 +	1.000 - 0.71	26	27		35	37	39
		0.709 - 0.60	33	34		44	46	49
		0.599 - 0.50	39	41		51	55	58
		0.499 - 0.38	47	50		61	67	70
3.01 - 3.50	0.000 - 0.249	1.000 - 0.71	12	12		17	17	19
		0.709 - 0.60	15	15		21	22	24
		0.599 - 0.50	17	18		25	26	28
		0.499 - 0.38	21	22		31	32	35
		0.379 - 0.25	28	29		40	43	46
		0.249 - 0.00	44	46		61	67	72
	0.250 - 0.499	1.000 - 0.71	17	17		25	25	28
		0.709 - 0.60	21	22		31	32	35
		0.599 - 0.50	25	26		37	38	42
		0.499 - 0.38	30	32		44	47	51
		0.379 - 0.25	40	43		57	63	68
	0.500 +	1.000 - 0.71	22	23		33	34	37
		0.709 - 0.60	28	29		41	43	47
		0.599 - 0.50	33	35		49	52	56
		0.499 - 0.38	40	42		58	63	68

Daylight Sensing Controls

OPAQUE WALL (Max. Uow)

ILD 0 - 1.50:

Light Weight Wall: Uow (HC<5)	Mass Wall: HC Range	PCT FEN	INT INS	EXT INS
0.16	HC >= 5	17	0.19	0.24
	HC >= 10	17	0.23	0.29
	HC >= 15	17	0.28	0.31
	HC >= 5	71	0.19	0.24
	HC >= 10	71	0.24	0.29
	HC >= 15	71	0.28	0.31

ILD 1.51 - 3.00:

Light Weight Wall: Uow (HC<5)	Mass Wall: HC Range	PCT FEN	INT INS	EXT INS
0.16	HC >= 5	14	0.19	0.24
	HC >= 10	14	0.24	0.30
	HC >= 15	14	0.29	0.32
	HC >= 5	75	0.19	0.24
	HC >= 10	75	0.24	0.30
	HC >= 15	75	0.29	0.32

ILD 3.01 - 3.50:

Light Weight Wall: Uow (HC<5)	Mass Wall: HC Range	PCT FEN	INT INS	EXT INS
0.16	HC >= 5	12	0.19	0.24
	HC >= 10	12	0.23	0.29
	HC >= 15	12	0.28	0.32
	HC >= 5	72	0.19	0.24
	HC >= 10	72	0.24	0.30
	HC >= 15	72	0.29	0.33

OTHER CRITERIA

	Minimum R-Value				Max Uo
WALL BELOW GRADE:	0			ROOF:	0.048
UNHEATED SLAB ON GRADE:	24"	36"	48"	WALL ADJACENT TO UNCOND SPACE:	0.26
Horizontal	0	0	0	FLOOR OVER UNCOND. SPACE:	0.12
Vertical	0	0	0		

Source: Adapted with permission from ASHRAE/IES Standard 90.1-1989, *Energy Efficient Design of New Buildings Except Low-Rise Residential Buildings*, © 1989 by the American Society of Heating, Refrigerating, and Air-Conditioning Engineers, Inc., Atlanta, Ga.

TABLE K.15 **ACP for Miami and West Palm Beach, FL**

Alternate Component Package, for ASHRAE/IES Standard 90.1-1989

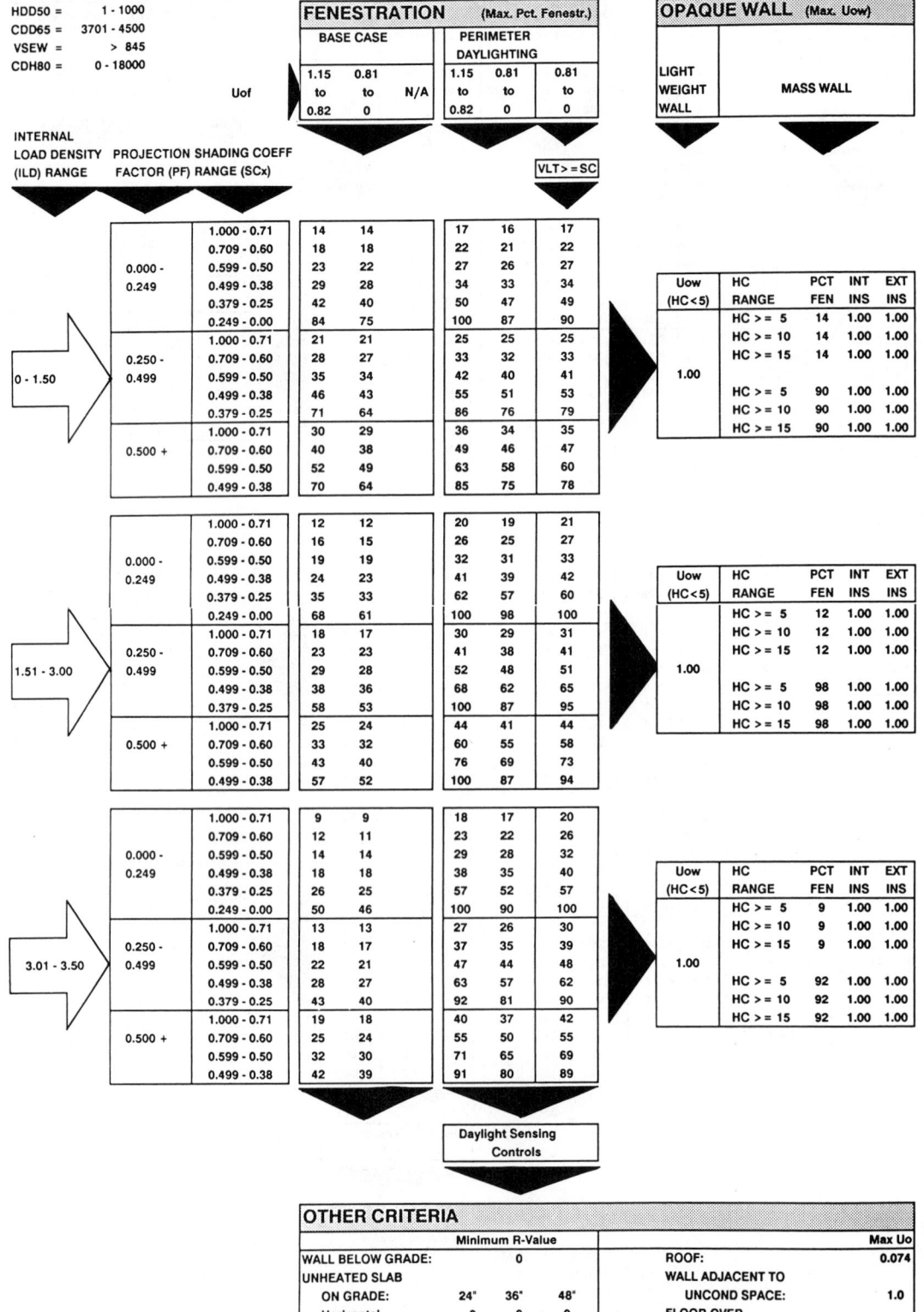

HDD50 =	1 - 1000
CDD65 =	3701 - 4500
VSEW =	> 845
CDH80 =	0 - 18000

FENESTRATION (Max. Pct. Fenestr.)

INTERNAL LOAD DENSITY (ILD) RANGE	PROJECTION FACTOR (PF) RANGE	SHADING COEFF RANGE (SCx)	BASE CASE Uof 1.15 to 0.82	BASE CASE Uof 0.81 to 0	BASE CASE Uof N/A	PERIMETER DAYLIGHTING Uof 1.15 to 0.82	PERIMETER DAYLIGHTING Uof 0.81 to 0	PERIMETER DAYLIGHTING Uof 0.81 to 0 (VLT>=SC)
0 - 1.50	0.000 - 0.249	1.000 - 0.71	14	14		17	16	17
		0.709 - 0.60	18	18		22	21	22
		0.599 - 0.50	23	22		27	26	27
		0.499 - 0.38	29	28		34	33	34
		0.379 - 0.25	42	40		50	47	49
		0.249 - 0.00	84	75		100	87	90
	0.250 - 0.499	1.000 - 0.71	21	21		25	25	25
		0.709 - 0.60	28	27		33	32	33
		0.599 - 0.50	35	34		42	40	41
		0.499 - 0.38	46	43		55	51	53
		0.379 - 0.25	71	64		86	76	79
	0.500 +	1.000 - 0.71	30	29		36	34	35
		0.709 - 0.60	40	38		49	46	47
		0.599 - 0.50	52	49		63	58	60
		0.499 - 0.38	70	64		85	75	78
1.51 - 3.00	0.000 - 0.249	1.000 - 0.71	12	12		20	19	21
		0.709 - 0.60	16	15		26	25	27
		0.599 - 0.50	19	19		32	31	33
		0.499 - 0.38	24	23		41	39	42
		0.379 - 0.25	35	33		62	57	60
		0.249 - 0.00	68	61		100	98	100
	0.250 - 0.499	1.000 - 0.71	18	17		30	29	31
		0.709 - 0.60	23	23		41	38	41
		0.599 - 0.50	29	28		52	48	51
		0.499 - 0.38	38	36		68	62	65
		0.379 - 0.25	58	53		100	87	95
	0.500 +	1.000 - 0.71	25	24		44	41	44
		0.709 - 0.60	33	32		60	55	58
		0.599 - 0.50	43	40		76	69	73
		0.499 - 0.38	57	52		100	87	94
3.01 - 3.50	0.000 - 0.249	1.000 - 0.71	9	9		18	17	20
		0.709 - 0.60	12	11		23	22	26
		0.599 - 0.50	14	14		29	28	32
		0.499 - 0.38	18	18		38	35	40
		0.379 - 0.25	26	25		57	52	57
		0.249 - 0.00	50	46		100	90	100
	0.250 - 0.499	1.000 - 0.71	13	13		27	26	30
		0.709 - 0.60	18	17		37	35	39
		0.599 - 0.50	22	21		47	44	48
		0.499 - 0.38	28	27		63	57	62
		0.379 - 0.25	43	40		92	81	90
	0.500 +	1.000 - 0.71	19	18		40	37	42
		0.709 - 0.60	25	24		55	50	55
		0.599 - 0.50	32	30		71	65	69
		0.499 - 0.38	42	39		91	80	89

Perimeter Daylighting: Daylight Sensing Controls

OPAQUE WALL (Max. Uow)

ILD 0 - 1.50:

LIGHT WEIGHT WALL Uow (HC<5)	MASS WALL HC RANGE	PCT FEN	INT INS	EXT INS
1.00	HC >= 5	14	1.00	1.00
	HC >= 10	14	1.00	1.00
	HC >= 15	14	1.00	1.00
	HC >= 5	90	1.00	1.00
	HC >= 10	90	1.00	1.00
	HC >= 15	90	1.00	1.00

ILD 1.51 - 3.00:

LIGHT WEIGHT WALL Uow (HC<5)	MASS WALL HC RANGE	PCT FEN	INT INS	EXT INS
1.00	HC >= 5	12	1.00	1.00
	HC >= 10	12	1.00	1.00
	HC >= 15	12	1.00	1.00
	HC >= 5	98	1.00	1.00
	HC >= 10	98	1.00	1.00
	HC >= 15	98	1.00	1.00

ILD 3.01 - 3.50:

LIGHT WEIGHT WALL Uow (HC<5)	MASS WALL HC RANGE	PCT FEN	INT INS	EXT INS
1.00	HC >= 5	9	1.00	1.00
	HC >= 10	9	1.00	1.00
	HC >= 15	9	1.00	1.00
	HC >= 5	92	1.00	1.00
	HC >= 10	92	1.00	1.00
	HC >= 15	92	1.00	1.00

OTHER CRITERIA

	Minimum R-Value				Max Uo
WALL BELOW GRADE:	0			ROOF:	0.074
UNHEATED SLAB ON GRADE:	24"	36"	48"	WALL ADJACENT TO UNCOND SPACE:	1.0
Horizontal	0	0	0	FLOOR OVER UNCOND. SPACE:	0.40
Vertical	0	0	0		

Source: Adapted with permission from ASHRAE/IES Standard 90.1-1989, *Energy Efficient Design of New Buildings Except Low-Rise Residential Buildings*, © 1989 by the American Society of Heating, Refrigerating, and Air-Conditioning Engineers, Inc., Atlanta, Ga.

TABLE K.16 ACP for Orlando, FL, Brownsville, Corpus Christi, and Kingsville, TX

Alternate Component Package, for ASHRAE/IES Standard 90.1-1989

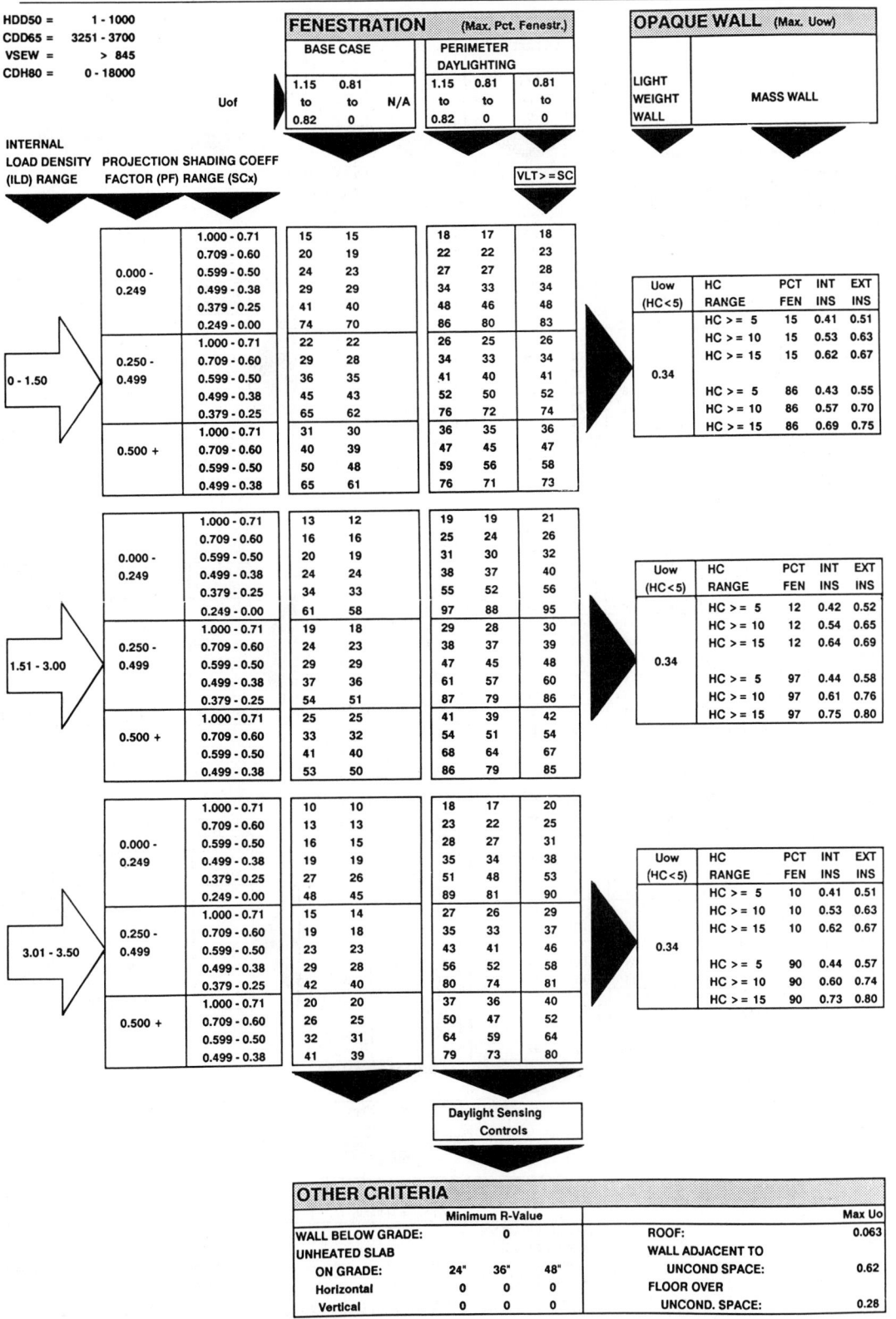

HDD50 =	1 - 1000
CDD65 =	3251 - 3700
VSEW =	> 845
CDH80 =	0 - 18000

FENESTRATION (Max. Pct. Fenestr.) — **OPAQUE WALL** (Max. Uow): LIGHT WEIGHT WALL / MASS WALL

ILD 0 - 1.50

INTERNAL LOAD DENSITY (ILD) RANGE	PROJECTION FACTOR (PF) RANGE	SHADING COEFF (SCx)	Base Case Uof 1.15 to 0.82	Base Case Uof 0.81 to 0	Base Case N/A	Perimeter Daylighting Uof 1.15 to 0.82	Perimeter Daylighting Uof 0.81 to 0	Perimeter Daylighting Uof 0.81 to 0, VLT>=SC
0 - 1.50	0.000 - 0.249	1.000 - 0.71	15	15		18	17	18
		0.709 - 0.60	20	19		22	22	23
		0.599 - 0.50	24	23		27	27	28
		0.499 - 0.38	29	29		34	33	34
		0.379 - 0.25	41	40		48	46	48
		0.249 - 0.00	74	70		86	80	83
	0.250 - 0.499	1.000 - 0.71	22	22		26	25	26
		0.709 - 0.60	29	28		34	33	34
		0.599 - 0.50	36	35		41	40	41
		0.499 - 0.38	45	43		52	50	52
		0.379 - 0.25	65	62		76	72	74
	0.500 +	1.000 - 0.71	31	30		36	35	36
		0.709 - 0.60	40	39		47	45	47
		0.599 - 0.50	50	48		59	56	58
		0.499 - 0.38	65	61		76	71	73

Uow (HC<5)	HC RANGE	PCT FEN	INT INS	EXT INS
0.34	HC >= 5	15	0.41	0.51
	HC >= 10	15	0.53	0.63
	HC >= 15	15	0.62	0.67
	HC >= 5	86	0.43	0.55
	HC >= 10	86	0.57	0.70
	HC >= 15	86	0.69	0.75

ILD 1.51 - 3.00

ILD RANGE	PF RANGE	SCx	Base Case 1.15 to 0.82	Base Case 0.81 to 0	Base Case N/A	Perimeter Daylighting 1.15 to 0.82	Perimeter Daylighting 0.81 to 0	Perimeter Daylighting 0.81 to 0, VLT>=SC
1.51 - 3.00	0.000 - 0.249	1.000 - 0.71	13	12		19	19	21
		0.709 - 0.60	16	16		25	24	26
		0.599 - 0.50	20	19		31	30	32
		0.499 - 0.38	24	24		38	37	40
		0.379 - 0.25	34	33		55	52	56
		0.249 - 0.00	61	58		97	88	95
	0.250 - 0.499	1.000 - 0.71	19	18		29	28	30
		0.709 - 0.60	24	23		38	37	39
		0.599 - 0.50	29	29		47	45	48
		0.499 - 0.38	37	36		61	57	60
		0.379 - 0.25	54	51		87	79	86
	0.500 +	1.000 - 0.71	25	25		41	39	42
		0.709 - 0.60	33	32		54	51	54
		0.599 - 0.50	41	40		68	64	67
		0.499 - 0.38	53	50		86	79	85

Uow (HC<5)	HC RANGE	PCT FEN	INT INS	EXT INS
0.34	HC >= 5	12	0.42	0.52
	HC >= 10	12	0.54	0.65
	HC >= 15	12	0.64	0.69
	HC >= 5	97	0.44	0.58
	HC >= 10	97	0.61	0.76
	HC >= 15	97	0.75	0.80

ILD 3.01 - 3.50

ILD RANGE	PF RANGE	SCx	Base Case 1.15 to 0.82	Base Case 0.81 to 0	Base Case N/A	Perimeter Daylighting 1.15 to 0.82	Perimeter Daylighting 0.81 to 0	Perimeter Daylighting 0.81 to 0, VLT>=SC
3.01 - 3.50	0.000 - 0.249	1.000 - 0.71	10	10		18	17	20
		0.709 - 0.60	13	13		23	22	25
		0.599 - 0.50	16	15		28	27	31
		0.499 - 0.38	19	19		35	34	38
		0.379 - 0.25	27	26		51	48	53
		0.249 - 0.00	48	45		89	81	90
	0.250 - 0.499	1.000 - 0.71	15	14		27	26	29
		0.709 - 0.60	19	18		35	33	37
		0.599 - 0.50	23	23		43	41	46
		0.499 - 0.38	29	28		56	52	58
		0.379 - 0.25	42	40		80	74	81
	0.500 +	1.000 - 0.71	20	20		37	36	40
		0.709 - 0.60	26	25		50	47	52
		0.599 - 0.50	32	31		64	59	64
		0.499 - 0.38	41	39		79	73	80

Uow (HC<5)	HC RANGE	PCT FEN	INT INS	EXT INS
0.34	HC >= 5	10	0.41	0.51
	HC >= 10	10	0.53	0.63
	HC >= 15	10	0.62	0.67
	HC >= 5	90	0.44	0.57
	HC >= 10	90	0.60	0.74
	HC >= 15	90	0.73	0.80

Daylight Sensing Controls (Perimeter Daylighting)

OTHER CRITERIA

Minimum R-Value					Max Uo
WALL BELOW GRADE:	0			ROOF:	0.063
UNHEATED SLAB ON GRADE:	24"	36"	48"	WALL ADJACENT TO UNCOND SPACE:	0.62
Horizontal	0	0	0	FLOOR OVER UNCOND. SPACE:	0.28
Vertical	0	0	0		

Source: Adapted with permission from ASHRAE/IES Standard 90.1-1989, *Energy Efficient Design of New Buildings Except Low-Rise Residential Buildings,* © 1989 by the American Society of Heating, Refrigerating, and Air-Conditioning Engineers, Inc., Atlanta, Ga.

TABLE K.17 **ACP for Phoenix, AZ**

Alternate Component Package, for ASHRAE/IES Standard 90.1-1989

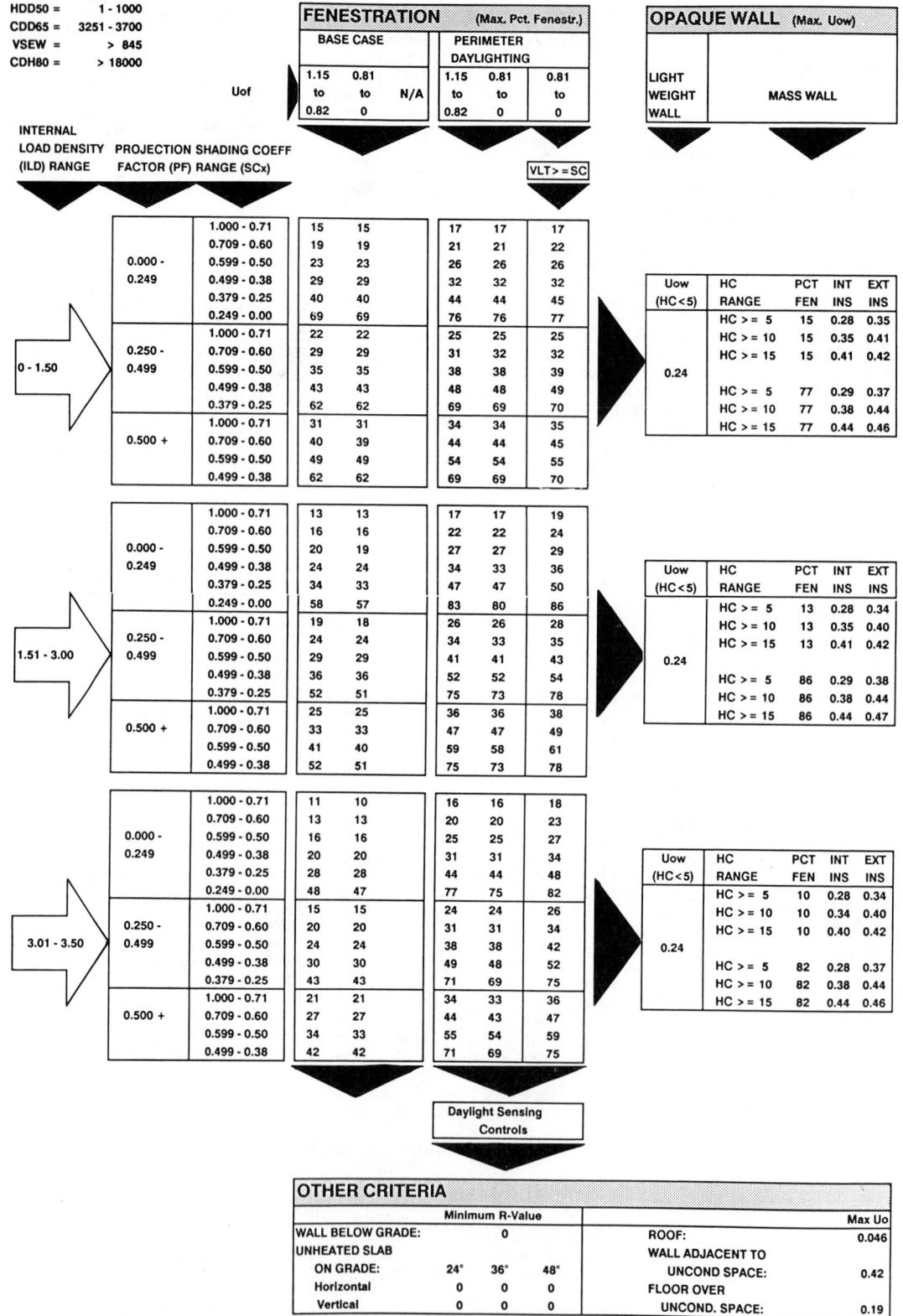

HDD50 =	1 - 1000
CDD65 =	3251 - 3700
VSEW =	> 845
CDH80 =	> 18000

FENESTRATION (Max. Pct. Fenestr.)

INTERNAL LOAD DENSITY (ILD) RANGE	PROJECTION FACTOR (PF)	SHADING COEFF RANGE (SCx)	BASE CASE Uof 1.15 to 0.82	BASE CASE Uof 0.81 to 0	BASE CASE N/A	PERIMETER DAYLIGHTING Uof 1.15 to 0.82	PERIMETER DAYLIGHTING Uof 0.81 to 0	PERIMETER DAYLIGHTING Uof 0.81 to 0 (VLT>=SC)
0 - 1.50	0.000 - 0.249	1.000 - 0.71	15	15		17	17	17
		0.709 - 0.60	19	19		21	21	22
		0.599 - 0.50	23	23		26	26	26
		0.499 - 0.38	29	29		32	32	32
		0.379 - 0.25	40	40		44	44	45
		0.249 - 0.00	69	69		76	76	77
	0.250 - 0.499	1.000 - 0.71	22	22		25	25	25
		0.709 - 0.60	29	29		31	32	32
		0.599 - 0.50	35	35		38	38	39
		0.499 - 0.38	43	43		48	48	49
		0.379 - 0.25	62	62		69	69	70
	0.500 +	1.000 - 0.71	31	31		34	34	35
		0.709 - 0.60	40	39		44	44	45
		0.599 - 0.50	49	49		54	54	55
		0.499 - 0.38	62	62		69	69	70
1.51 - 3.00	0.000 - 0.249	1.000 - 0.71	13	13		17	17	19
		0.709 - 0.60	16	16		22	22	24
		0.599 - 0.50	20	19		27	27	29
		0.499 - 0.38	24	24		34	33	36
		0.379 - 0.25	34	33		47	47	50
		0.249 - 0.00	58	57		83	80	86
	0.250 - 0.499	1.000 - 0.71	19	18		26	26	28
		0.709 - 0.60	24	24		34	33	35
		0.599 - 0.50	29	29		41	41	43
		0.499 - 0.38	36	36		52	52	54
		0.379 - 0.25	52	51		75	73	78
	0.500 +	1.000 - 0.71	25	25		36	36	38
		0.709 - 0.60	33	33		47	47	49
		0.599 - 0.50	41	40		59	58	61
		0.499 - 0.38	52	51		75	73	78
3.01 - 3.50	0.000 - 0.249	1.000 - 0.71	11	10		16	16	18
		0.709 - 0.60	13	13		20	20	23
		0.599 - 0.50	16	16		25	25	27
		0.499 - 0.38	20	20		31	31	34
		0.379 - 0.25	28	28		44	44	48
		0.249 - 0.00	48	47		77	75	82
	0.250 - 0.499	1.000 - 0.71	15	15		24	24	26
		0.709 - 0.60	20	20		31	31	34
		0.599 - 0.50	24	24		38	38	42
		0.499 - 0.38	30	30		49	48	52
		0.379 - 0.25	43	43		71	69	75
	0.500 +	1.000 - 0.71	21	21		34	33	36
		0.709 - 0.60	27	27		44	43	47
		0.599 - 0.50	34	33		55	54	59
		0.499 - 0.38	42	42		71	69	75

Daylight Sensing Controls

OPAQUE WALL (Max. Uow)

ILD 0 - 1.50:

LIGHT WEIGHT WALL Uow (HC<5)	MASS WALL HC RANGE	PCT FEN	INT INS	EXT INS
0.24	HC >= 5	15	0.28	0.35
	HC >= 10	15	0.35	0.41
	HC >= 15	15	0.41	0.42
	HC >= 5	77	0.29	0.37
	HC >= 10	77	0.38	0.44
	HC >= 15	77	0.44	0.46

ILD 1.51 - 3.00:

LIGHT WEIGHT WALL Uow (HC<5)	MASS WALL HC RANGE	PCT FEN	INT INS	EXT INS
0.24	HC >= 5	13	0.28	0.34
	HC >= 10	13	0.35	0.40
	HC >= 15	13	0.41	0.42
	HC >= 5	86	0.29	0.38
	HC >= 10	86	0.38	0.44
	HC >= 15	86	0.44	0.47

ILD 3.01 - 3.50:

LIGHT WEIGHT WALL Uow (HC<5)	MASS WALL HC RANGE	PCT FEN	INT INS	EXT INS
0.24	HC >= 5	10	0.28	0.34
	HC >= 10	10	0.34	0.40
	HC >= 15	10	0.40	0.42
	HC >= 5	82	0.28	0.37
	HC >= 10	82	0.38	0.44
	HC >= 15	82	0.44	0.46

OTHER CRITERIA

	Minimum R-Value				Max Uo
WALL BELOW GRADE:	0			ROOF:	0.046
UNHEATED SLAB ON GRADE:	24"	36"	48"	WALL ADJACENT TO UNCOND SPACE:	0.42
Horizontal	0	0	0	FLOOR OVER UNCOND. SPACE:	0.19
Vertical	0	0	0		

Source: Adapted with permission from ASHRAE/IES Standard 90.1-1989, *Energy Efficient Design of New Buildings Except Low-Rise Residential Buildings*, © 1989 by the American Society of Heating, Refrigerating, and Air-Conditioning Engineers, Inc., Atlanta, Ga.

TABLE K.18 **ACP for Astoria, Portland, and Salem, OR, Olympia, Seattle, and Whidbey Island, WA**

Alternate Component Package, for ASHRAE/IES Standard 90.1-1989

HDD50 = 1001 - 1750
CDD65 = 0 - 500
VSEW = 560 - 845

FENESTRATION (Max. Pct. Fenestr.) — Base Case by Uof; Perimeter Daylighting: N/A (VLT> = SC)

INTERNAL LOAD DENSITY (ILD) RANGE	PROJECTION FACTOR (PF) RANGE	SHADING COEFF RANGE (SCx)	Uof 0.81 to 0.46	Uof 0.45 to 0.39	Uof 0.38 to 0	Perimeter Daylighting N/A	VLT> = SC
0 - 1.50	0.000 - 0.249	1.000 - 0.71	24	31	33		
		0.709 - 0.60	26	37	39		
		0.599 - 0.50	27	41	44		
		0.499 - 0.38	29	45	50		
		0.379 - 0.25	30	52	59		
		0.249 - 0.00	32	59	70		
	0.250 - 0.499	1.000 - 0.71	28	42	45		
		0.709 - 0.60	30	47	52		
		0.599 - 0.50	31	52	58		
		0.499 - 0.38	32	56	64		
		0.379 - 0.25	33	61	72		
	0.500 +	1.000 - 0.71	31	50	55		
		0.709 - 0.60	32	56	63		
		0.599 - 0.50	33	59	68		
		0.499 - 0.38	33	62	73		
1.51 - 3.00	0.000 - 0.249	1.000 - 0.71	22	27	28		
		0.709 - 0.60	25	32	34		
		0.599 - 0.50	27	37	39		
		0.499 - 0.38	29	42	45		
		0.379 - 0.25	32	49	54		
		0.249 - 0.00	35	60	69		
	0.250 - 0.499	1.000 - 0.71	27	37	39		
		0.709 - 0.60	30	43	46		
		0.599 - 0.50	32	48	52		
		0.499 - 0.38	33	53	59		
		0.379 - 0.25	35	60	69		
	0.500 +	1.000 - 0.71	31	46	49		
		0.709 - 0.60	33	52	57		
		0.599 - 0.50	35	57	63		
		0.499 - 0.38	36	61	70		
3.01 - 3.50	0.000 - 0.249	1.000 - 0.71	21	25	26		
		0.709 - 0.60	24	30	31		
		0.599 - 0.50	26	34	36		
		0.499 - 0.38	28	39	42		
		0.379 - 0.25	31	47	51		
		0.249 - 0.00	35	58	66		
	0.250 - 0.499	1.000 - 0.71	26	34	36		
		0.709 - 0.60	29	40	43		
		0.599 - 0.50	31	45	49		
		0.499 - 0.38	33	51	56		
		0.379 - 0.25	35	58	66		
	0.500 +	1.000 - 0.71	30	43	46		
		0.709 - 0.60	33	49	54		
		0.599 - 0.50	34	54	60		
		0.499 - 0.38	36	59	67		

Daylight Sensing Controls

OPAQUE WALL (Max. Uow) — LIGHT WEIGHT WALL / MASS WALL

ILD 0 - 1.50:

Uow (HC<5)	HC RANGE	PCT FEN	INT INS	EXT INS
0.092	HC >= 5	24	0.096	0.12
	HC >= 10	24	0.10	0.14
	HC >= 15	24	0.11	0.14
	HC >= 5	73	0.096	0.12
	HC >= 10	73	0.10	0.14
	HC >= 15	73	0.11	0.14

ILD 1.51 - 3.00:

Uow (HC<5)	HC RANGE	PCT FEN	INT INS	EXT INS
0.092	HC >= 5	22	0.098	0.13
	HC >= 10	22	0.11	0.16
	HC >= 15	22	0.12	0.17
	HC >= 5	70	0.097	0.13
	HC >= 10	70	0.11	0.16
	HC >= 15	70	0.12	0.17

ILD 3.01 - 3.50:

Uow (HC<5)	HC RANGE	PCT FEN	INT INS	EXT INS
0.092	HC >= 5	21	0.098	0.14
	HC >= 10	21	0.11	0.17
	HC >= 15	21	0.13	0.18
	HC >= 5	67	0.098	0.13
	HC >= 10	67	0.11	0.17
	HC >= 15	67	0.12	0.18

OTHER CRITERIA

Minimum R-Value					Max Uo
WALL BELOW GRADE:		9		ROOF:	0.064
UNHEATED SLAB				WALL ADJACENT TO	
ON GRADE:	24"	36"	48"	UNCOND SPACE:	0.14
Horizontal	16	13	11	FLOOR OVER	
Vertical	8	6	4	UNCOND. SPACE:	0.056

Source: Adapted with permission from ASHRAE/IES Standard 90.1-1989, *Energy Efficient Design of New Buildings Except Low-Rise Residential Buildings*, © 1989 by the American Society of Heating, Refrigerating, and Air-Conditioning Engineers, Inc., Atlanta, Ga.

TABLE K.19 **ACP for Asheville, NC, and Medford, OR**

Alternate Component Package, for ASHRAE/IES Standard 90.1-1989

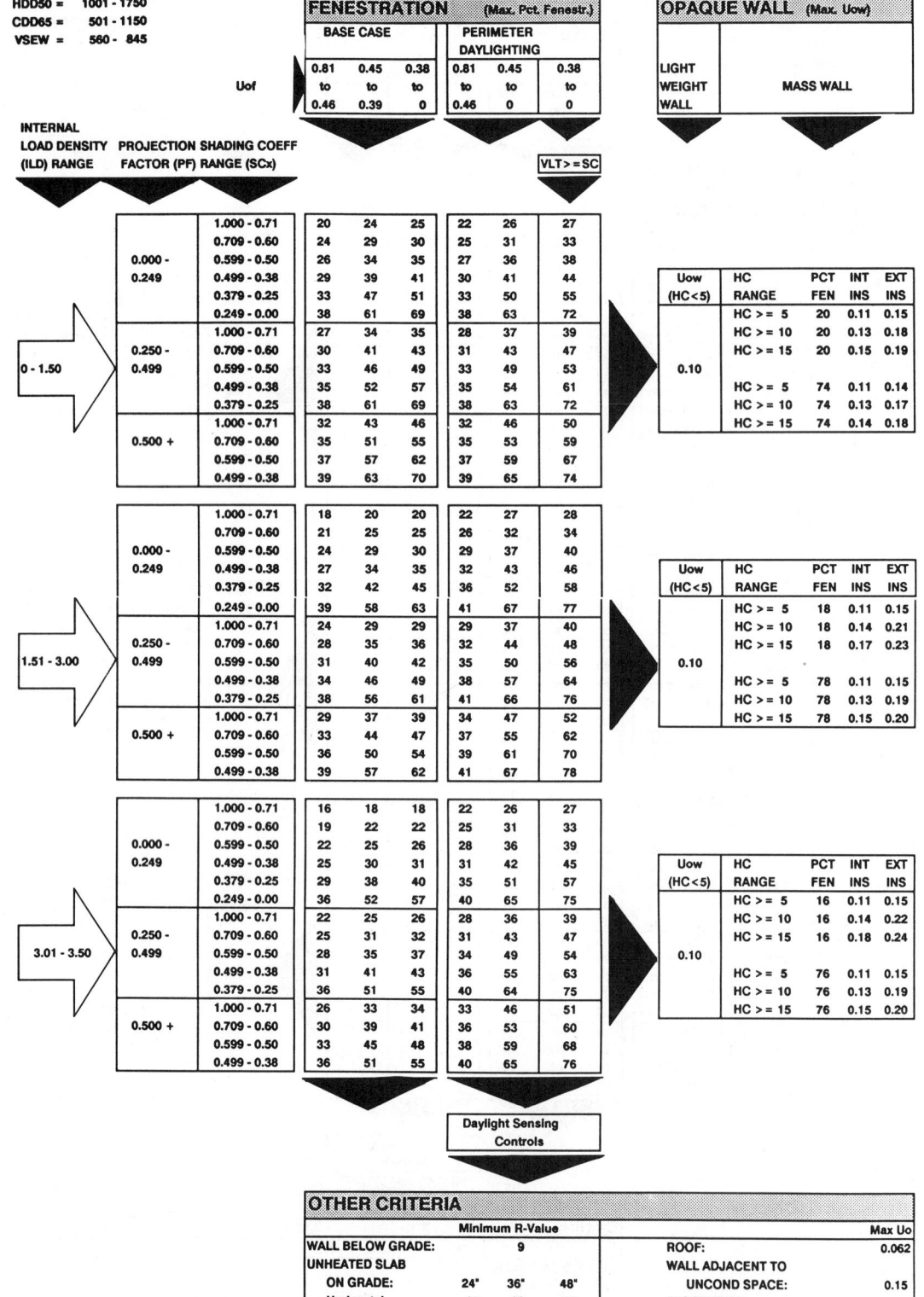

HDD50 = 1001 - 1750
CDD65 = 501 - 1150
VSEW = 560 - 845

FENESTRATION (Max. Pct. Fenestr.)

INTERNAL LOAD DENSITY (ILD) RANGE	PROJECTION FACTOR (PF)	SHADING COEFF RANGE (SCx)	BASE CASE Uof 0.81 to 0.46	BASE CASE Uof 0.45 to 0.39	BASE CASE Uof 0.38 to 0	PERIMETER DAYLIGHTING Uof 0.81 to 0.46	PERIMETER DAYLIGHTING Uof 0.45 to 0	PERIMETER DAYLIGHTING Uof 0.38 to 0 (VLT>=SC)
0 - 1.50	0.000 - 0.249	1.000 - 0.71	20	24	25	22	26	27
		0.709 - 0.60	24	29	30	25	31	33
		0.599 - 0.50	26	34	35	27	36	38
		0.499 - 0.38	29	39	41	30	41	44
		0.379 - 0.25	33	47	51	33	50	55
		0.249 - 0.00	38	61	69	38	63	72
	0.250 - 0.499	1.000 - 0.71	27	34	35	28	37	39
		0.709 - 0.60	30	41	43	31	43	47
		0.599 - 0.50	33	46	49	33	49	53
		0.499 - 0.38	35	52	57	35	54	61
		0.379 - 0.25	38	61	69	38	63	72
	0.500 +	1.000 - 0.71	32	43	46	32	46	50
		0.709 - 0.60	35	51	55	35	53	59
		0.599 - 0.50	37	57	62	37	59	67
		0.499 - 0.38	39	63	70	39	65	74
1.51 - 3.00	0.000 - 0.249	1.000 - 0.71	18	20	20	22	27	28
		0.709 - 0.60	21	25	25	26	32	34
		0.599 - 0.50	24	29	30	29	37	40
		0.499 - 0.38	27	34	35	32	43	46
		0.379 - 0.25	32	42	45	36	52	58
		0.249 - 0.00	39	58	63	41	67	77
	0.250 - 0.499	1.000 - 0.71	24	29	29	29	37	40
		0.709 - 0.60	28	35	36	32	44	48
		0.599 - 0.50	31	40	42	35	50	56
		0.499 - 0.38	34	46	49	38	57	64
		0.379 - 0.25	38	56	61	41	66	76
	0.500 +	1.000 - 0.71	29	37	39	34	47	52
		0.709 - 0.60	33	44	47	37	55	62
		0.599 - 0.50	36	50	54	39	61	70
		0.499 - 0.38	39	57	62	41	67	78
3.01 - 3.50	0.000 - 0.249	1.000 - 0.71	16	18	18	22	26	27
		0.709 - 0.60	19	22	22	25	31	33
		0.599 - 0.50	22	25	26	28	36	39
		0.499 - 0.38	25	30	31	31	42	45
		0.379 - 0.25	29	38	40	35	51	57
		0.249 - 0.00	36	52	57	40	65	75
	0.250 - 0.499	1.000 - 0.71	22	25	26	28	36	39
		0.709 - 0.60	25	31	32	31	43	47
		0.599 - 0.50	28	35	37	34	49	54
		0.499 - 0.38	31	41	43	36	55	63
		0.379 - 0.25	36	51	55	40	64	75
	0.500 +	1.000 - 0.71	26	33	34	33	46	51
		0.709 - 0.60	30	39	41	36	53	60
		0.599 - 0.50	33	45	48	38	59	68
		0.499 - 0.38	36	51	55	40	65	76

Perimeter daylighting: Daylight Sensing Controls

OPAQUE WALL (Max. Uow)

ILD 0 - 1.50:

LIGHT WEIGHT WALL Uow (HC<5)	MASS WALL HC RANGE	PCT FEN	INT INS	EXT INS
0.10	HC >= 5	20	0.11	0.15
	HC >= 10	20	0.13	0.18
	HC >= 15	20	0.15	0.19
	HC >= 5	74	0.11	0.14
	HC >= 10	74	0.13	0.17
	HC >= 15	74	0.14	0.18

ILD 1.51 - 3.00:

LIGHT WEIGHT WALL Uow (HC<5)	MASS WALL HC RANGE	PCT FEN	INT INS	EXT INS
0.10	HC >= 5	18	0.11	0.15
	HC >= 10	18	0.14	0.21
	HC >= 15	18	0.17	0.23
	HC >= 5	78	0.11	0.15
	HC >= 10	78	0.13	0.19
	HC >= 15	78	0.15	0.20

ILD 3.01 - 3.50:

LIGHT WEIGHT WALL Uow (HC<5)	MASS WALL HC RANGE	PCT FEN	INT INS	EXT INS
0.10	HC >= 5	16	0.11	0.15
	HC >= 10	16	0.14	0.22
	HC >= 15	16	0.18	0.24
	HC >= 5	76	0.11	0.15
	HC >= 10	76	0.13	0.19
	HC >= 15	76	0.15	0.20

OTHER CRITERIA

Minimum R-Value					Max Uo
WALL BELOW GRADE:		9		ROOF:	0.062
UNHEATED SLAB ON GRADE:	24"	36"	48"	WALL ADJACENT TO UNCOND SPACE:	0.15
Horizontal	15	13	10	FLOOR OVER UNCOND. SPACE:	0.064
Vertical	7	6	4		

Source: Adapted with permission from ASHRAE/IES Standard 90.1-1989, *Energy Efficient Design of New Buildings Except Low-Rise Residential Buildings,* © 1989 by the American Society of Heating, Refrigerating, and Air-Conditioning Engineers, Inc., Atlanta, Ga.

TABLE K.20 **ACP for Charlotte, Greensboro, and Raleigh, NC, Chattanooga, Knoxville, and Nashville, TN, Norfolk, Richmond, and Roanoke, VA, and Patuxent, MD**

Alternate Component Package, for ASHRAE/IES Standard 90.1-1989

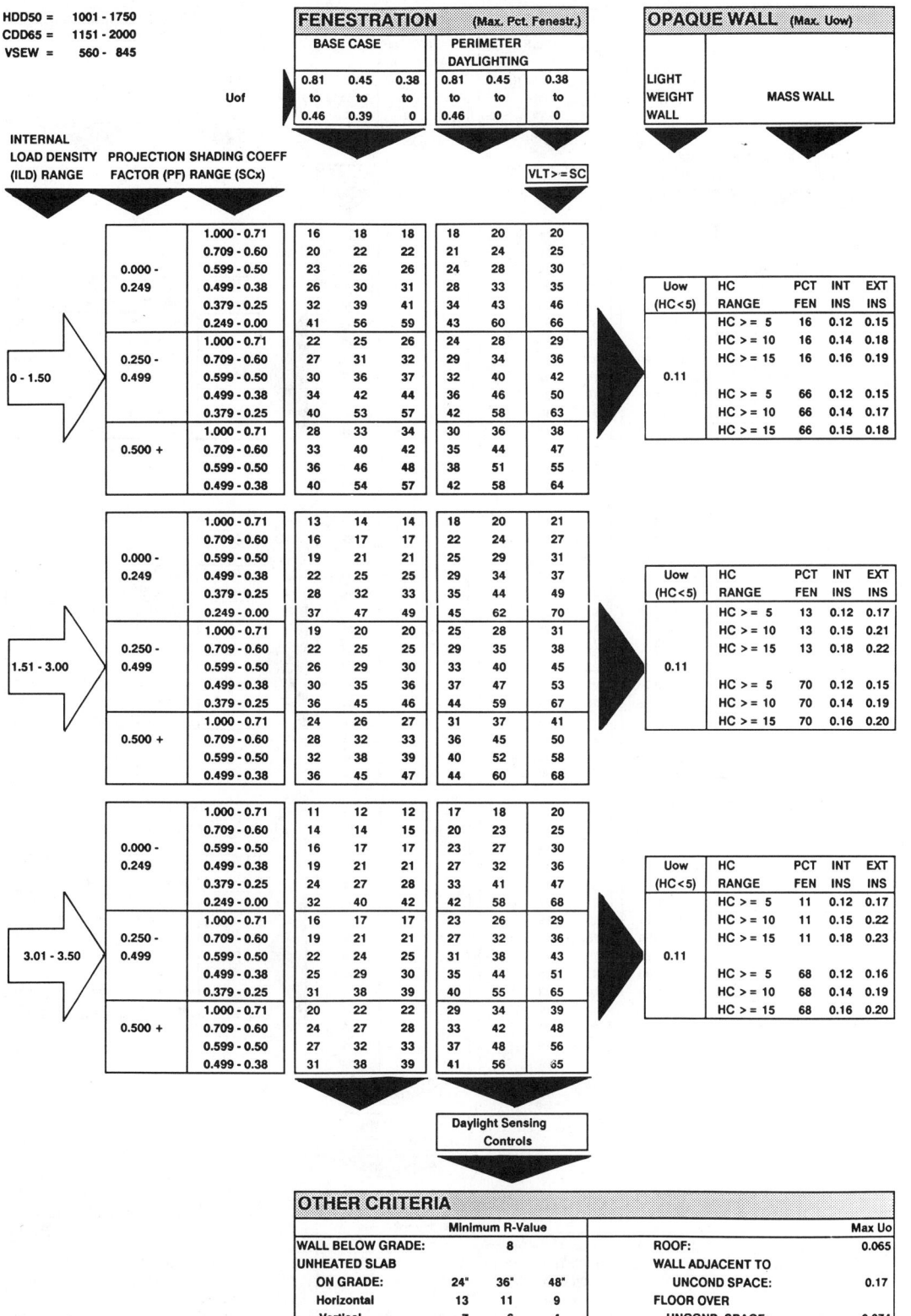

HDD50 = 1001 - 1750
CDD65 = 1151 - 2000
VSEW = 560 - 845

FENESTRATION (Max. Pct. Fenestr.)

INTERNAL LOAD DENSITY (ILD) RANGE	PROJECTION FACTOR (PF)	SHADING COEFF RANGE (SCx)	BASE CASE Uof 0.81 to 0.46	BASE CASE Uof 0.45 to 0.39	BASE CASE Uof 0.38 to 0	PERIMETER DAYLIGHTING Uof 0.81 to 0.46	PERIMETER DAYLIGHTING Uof 0.45 to 0	PERIMETER DAYLIGHTING Uof 0.38 to 0 (VLT>=SC)
0 - 1.50	0.000 - 0.249	1.000 - 0.71	16	18	18	18	20	20
		0.709 - 0.60	20	22	22	21	24	25
		0.599 - 0.50	23	26	26	24	28	30
		0.499 - 0.38	26	30	31	28	33	35
		0.379 - 0.25	32	39	41	34	43	46
		0.249 - 0.00	41	56	59	43	60	66
	0.250 - 0.499	1.000 - 0.71	22	25	26	24	28	29
		0.709 - 0.60	27	31	32	29	34	36
		0.599 - 0.50	30	36	37	32	40	42
		0.499 - 0.38	34	42	44	36	46	50
		0.379 - 0.25	40	53	57	42	58	63
	0.500 +	1.000 - 0.71	28	33	34	30	36	38
		0.709 - 0.60	33	40	42	35	44	47
		0.599 - 0.50	36	46	48	38	51	55
		0.499 - 0.38	40	54	57	42	58	64
1.51 - 3.00	0.000 - 0.249	1.000 - 0.71	13	14	14	18	20	21
		0.709 - 0.60	16	17	17	22	24	27
		0.599 - 0.50	19	21	21	25	29	31
		0.499 - 0.38	22	25	25	29	34	37
		0.379 - 0.25	28	32	33	35	44	49
		0.249 - 0.00	37	47	49	45	62	70
	0.250 - 0.499	1.000 - 0.71	19	20	20	25	28	31
		0.709 - 0.60	22	25	25	29	35	38
		0.599 - 0.50	26	29	30	33	40	45
		0.499 - 0.38	30	35	36	37	47	53
		0.379 - 0.25	36	45	46	44	59	67
	0.500 +	1.000 - 0.71	24	26	27	31	37	41
		0.709 - 0.60	28	32	33	36	45	50
		0.599 - 0.50	32	38	39	40	52	58
		0.499 - 0.38	36	45	47	44	60	68
3.01 - 3.50	0.000 - 0.249	1.000 - 0.71	11	12	12	17	18	20
		0.709 - 0.60	14	14	15	20	23	25
		0.599 - 0.50	16	17	17	23	27	30
		0.499 - 0.38	19	21	21	27	32	36
		0.379 - 0.25	24	27	28	33	41	47
		0.249 - 0.00	32	40	42	42	58	68
	0.250 - 0.499	1.000 - 0.71	16	17	17	23	26	29
		0.709 - 0.60	19	21	21	27	32	36
		0.599 - 0.50	22	24	25	31	38	43
		0.499 - 0.38	25	29	30	35	44	51
		0.379 - 0.25	31	38	39	40	55	65
	0.500 +	1.000 - 0.71	20	22	22	29	34	39
		0.709 - 0.60	24	27	28	33	42	48
		0.599 - 0.50	27	32	33	37	48	56
		0.499 - 0.38	31	38	39	41	56	65

Perimeter daylighting: Daylight Sensing Controls

OPAQUE WALL (Max. Uow)

ILD RANGE	LIGHT WEIGHT WALL Uow (HC<5)	MASS WALL HC RANGE	PCT FEN	INT INS	EXT INS
0 - 1.50	0.11	HC >= 5	16	0.12	0.15
		HC >= 10	16	0.14	0.18
		HC >= 15	16	0.16	0.19
		HC >= 5	66	0.12	0.15
		HC >= 10	66	0.14	0.17
		HC >= 15	66	0.15	0.18
1.51 - 3.00	0.11	HC >= 5	13	0.12	0.17
		HC >= 10	13	0.15	0.21
		HC >= 15	13	0.18	0.22
		HC >= 5	70	0.12	0.15
		HC >= 10	70	0.14	0.19
		HC >= 15	70	0.16	0.20
3.01 - 3.50	0.11	HC >= 5	11	0.12	0.17
		HC >= 10	11	0.15	0.22
		HC >= 15	11	0.18	0.23
		HC >= 5	68	0.12	0.16
		HC >= 10	68	0.14	0.19
		HC >= 15	68	0.16	0.20

OTHER CRITERIA

	Minimum R-Value				Max Uo
WALL BELOW GRADE:	8			ROOF:	0.065
UNHEATED SLAB ON GRADE:	24"	36"	48"	WALL ADJACENT TO UNCOND SPACE:	0.17
Horizontal	13	11	9	FLOOR OVER UNCOND. SPACE:	0.074
Vertical	7	6	4		

Source: Adapted with permission from ASHRAE/IES Standard 90.1-1989, *Energy Efficient Design of New Buildings Except Low-Rise Residential Buildings,* © 1989 by the American Society of Heating, Refrigerating, and Air-Conditioning Engineers, Inc., Atlanta, Ga.

APPENDICES

TABLE K.21 **ACP for Albuquerque, Roswell, Truth or Consequences, and Tucumcari, NM, Oklahoma City, OK, Amarillo and Lubbock, TX**

Alternate Component Package, for ASHRAE/IES Standard 90.1-1989

HDD50 = 1001 - 1750
CDD65 = 1151 - 2000
VSEW = > 845

FENESTRATION (Max. Pct. Fenestr.)

INTERNAL LOAD DENSITY (ILD) RANGE	PROJECTION FACTOR (PF)	SHADING COEFF RANGE (SCx)	BASE CASE Uof 0.81 to 0.46	BASE CASE Uof 0.45 to 0.39	BASE CASE Uof 0.38 to 0	PERIMETER DAYLIGHTING Uof 0.81 to 0.46	PERIMETER DAYLIGHTING Uof 0.45 to 0	PERIMETER DAYLIGHTING Uof 0.38 to 0 (VLT>=SC)
0 - 1.50	0.000 - 0.249	1.000 - 0.71	14	15	16	15	17	17
		0.709 - 0.60	17	19	19	18	21	22
		0.599 - 0.50	19	22	23	20	24	25
		0.499 - 0.38	22	26	27	23	28	30
		0.379 - 0.25	26	33	35	28	36	38
		0.249 - 0.00	33	47	50	35	49	55
	0.250 - 0.499	1.000 - 0.71	19	22	23	21	24	26
		0.709 - 0.60	23	27	28	24	30	31
		0.599 - 0.50	26	32	33	27	34	37
		0.499 - 0.38	29	37	39	30	40	43
		0.379 - 0.25	33	46	49	34	49	54
	0.500 +	1.000 - 0.71	24	29	30	26	32	34
		0.709 - 0.60	28	35	37	29	38	41
		0.599 - 0.50	31	41	43	32	44	47
		0.499 - 0.38	34	47	50	35	50	55
1.51 - 3.00	0.000 - 0.249	1.000 - 0.71	11	12	12	15	16	18
		0.709 - 0.60	14	15	15	18	20	22
		0.599 - 0.50	16	18	18	20	23	26
		0.499 - 0.38	19	21	22	23	28	31
		0.379 - 0.25	23	27	28	28	35	40
		0.249 - 0.00	31	39	42	35	49	57
	0.250 - 0.499	1.000 - 0.71	16	17	18	20	24	26
		0.709 - 0.60	19	22	22	24	29	32
		0.599 - 0.50	22	25	26	27	33	37
		0.499 - 0.38	25	30	31	30	39	44
		0.379 - 0.25	30	38	40	35	48	55
	0.500 +	1.000 - 0.71	20	23	24	25	31	35
		0.709 - 0.60	24	28	29	29	37	42
		0.599 - 0.50	27	33	34	32	43	49
		0.499 - 0.38	30	38	40	35	49	57
3.01 - 3.50	0.000 - 0.249	1.000 - 0.71	10	10	10	14	15	17
		0.709 - 0.60	12	13	13	16	18	21
		0.599 - 0.50	14	15	15	19	22	25
		0.499 - 0.38	16	18	18	21	26	29
		0.379 - 0.25	20	23	24	26	33	38
		0.249 - 0.00	27	34	35	33	46	54
	0.250 - 0.499	1.000 - 0.71	13	15	15	19	22	25
		0.709 - 0.60	16	18	18	22	26	30
		0.599 - 0.50	19	21	22	25	31	36
		0.499 - 0.38	21	25	26	28	36	42
		0.379 - 0.25	26	32	34	32	44	53
	0.500 +	1.000 - 0.71	17	19	20	23	29	33
		0.709 - 0.60	20	24	24	27	34	40
		0.599 - 0.50	23	28	29	30	39	46
		0.499 - 0.38	26	33	34	32	45	54

Perimeter daylighting: Daylight Sensing Controls

OPAQUE WALL (Max. Uow)

LIGHT WEIGHT WALL / MASS WALL

ILD 0 - 1.50:

Uow (HC<5)	HC RANGE	PCT FEN	INT INS	EXT INS
0.10	HC >= 5	14	0.11	0.16
	HC >= 10	14	0.14	0.20
	HC >= 15	14	0.16	0.21
	HC >= 5	55	0.11	0.16
	HC >= 10	55	0.14	0.19
	HC >= 15	55	0.16	0.21

ILD 1.51 - 3.00:

Uow (HC<5)	HC RANGE	PCT FEN	INT INS	EXT INS
0.10	HC >= 5	11	0.12	0.16
	HC >= 10	11	0.15	0.22
	HC >= 15	11	0.18	0.24
	HC >= 5	57	0.12	0.16
	HC >= 10	57	0.14	0.20
	HC >= 15	57	0.17	0.22

ILD 3.01 - 3.50:

Uow (HC<5)	HC RANGE	PCT FEN	INT INS	EXT INS
0.10	HC >= 5	10	0.12	0.16
	HC >= 10	10	0.15	0.23
	HC >= 15	10	0.19	0.25
	HC >= 5	54	0.12	0.16
	HC >= 10	54	0.14	0.21
	HC >= 15	54	0.17	0.22

OTHER CRITERIA

	Minimum R-Value				Max Uo
WALL BELOW GRADE:		8		ROOF:	0.059
UNHEATED SLAB				WALL ADJACENT TO	
ON GRADE:	24"	36"	48"	UNCOND SPACE:	0.16
Horizontal	14	12	10	FLOOR OVER	
Vertical	7	6	4	UNCOND. SPACE:	0.070

Source: Adapted with permission from ASHRAE/IES Standard 90.1-1989, *Energy Efficient Design of New Buildings Except Low-Rise Residential Buildings,* © 1989 by the American Society of Heating, Refrigerating, and Air-Conditioning Engineers, Inc., Atlanta, Ga.

APPENDICES

TABLE K.22 **ACP for Fort Smith, AR, Tulsa, OK, and Memphis, TN**

Alternate Component Package, for ASHRAE/IES Standard 90.1-1989

HDD50 = 1001 - 1750
CDD65 = 2001 - 3250
VSEW = 560 - 845

FENESTRATION (Max. Pct. Fenestr.)

INTERNAL LOAD DENSITY (ILD) RANGE	PROJECTION FACTOR (PF) RANGE	SHADING COEFF RANGE (SCx)	BASE CASE Uof 0.81 to 0.46	BASE CASE Uof 0.45 to 0.39	BASE CASE Uof 0.38 to 0	PERIMETER DAYLIGHTING Uof 0.81 to 0.46	PERIMETER DAYLIGHTING Uof 0.45 to 0	PERIMETER DAYLIGHTING Uof 0.38 to 0 (VLT>=SC)
0 - 1.50	0.000 - 0.249	1.000 - 0.71	14	15	15	16	17	17
		0.709 - 0.60	17	19	19	19	21	22
		0.599 - 0.50	20	22	22	22	25	26
		0.499 - 0.38	24	26	27	26	29	31
		0.379 - 0.25	30	35	36	32	38	41
		0.249 - 0.00	41	52	54	43	57	62
	0.250 - 0.499	1.000 - 0.71	20	21	22	22	24	25
		0.709 - 0.60	24	27	27	26	30	31
		0.599 - 0.50	28	31	32	30	35	37
		0.499 - 0.38	32	38	39	35	42	44
		0.379 - 0.25	39	49	51	42	54	58
	0.500 +	1.000 - 0.71	25	28	29	28	32	33
		0.709 - 0.60	30	35	36	33	39	41
		0.599 - 0.50	34	41	43	37	46	49
		0.499 - 0.38	39	49	51	42	54	58
1.51 - 3.00	0.000 - 0.249	1.000 - 0.71	11	11	11	15	16	18
		0.709 - 0.60	13	14	14	19	20	22
		0.599 - 0.50	16	17	17	22	24	27
		0.499 - 0.38	19	20	20	25	29	32
		0.379 - 0.25	24	27	27	32	38	43
		0.249 - 0.00	34	41	42	43	56	65
	0.250 - 0.499	1.000 - 0.71	15	16	16	21	23	26
		0.709 - 0.60	19	20	20	26	29	32
		0.599 - 0.50	22	24	24	29	34	38
		0.499 - 0.38	26	29	29	34	41	46
		0.379 - 0.25	32	38	39	41	53	61
	0.500 +	1.000 - 0.71	20	21	22	27	31	35
		0.709 - 0.60	24	27	27	32	38	43
		0.599 - 0.50	28	32	32	37	45	51
		0.499 - 0.38	32	38	39	42	53	61
3.01 - 3.50	0.000 - 0.249	1.000 - 0.71	9	9	9	14	14	17
		0.709 - 0.60	11	11	11	17	18	21
		0.599 - 0.50	13	14	14	20	21	25
		0.499 - 0.38	15	16	17	23	26	30
		0.379 - 0.25	20	22	22	29	34	40
		0.249 - 0.00	28	33	34	39	51	61
	0.250 - 0.499	1.000 - 0.71	12	13	13	19	21	24
		0.709 - 0.60	15	16	16	23	26	30
		0.599 - 0.50	18	19	20	27	31	36
		0.499 - 0.38	21	23	24	31	37	43
		0.379 - 0.25	27	31	32	37	48	57
	0.500 +	1.000 - 0.71	16	17	17	24	28	32
		0.709 - 0.60	20	22	22	29	34	40
		0.599 - 0.50	23	26	26	33	40	48
		0.499 - 0.38	27	31	31	37	48	57

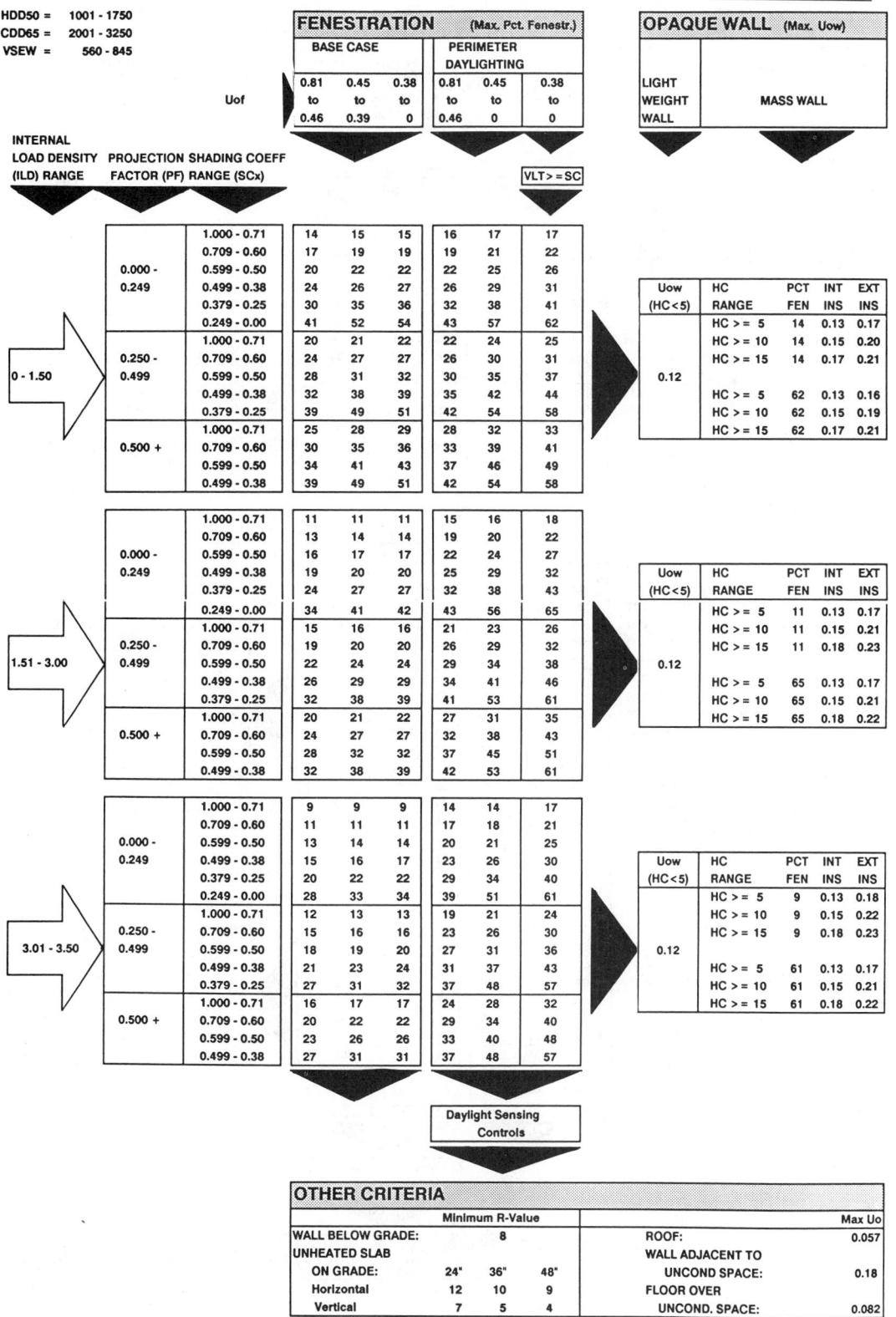

Perimeter Daylighting → Daylight Sensing Controls

OPAQUE WALL (Max. Uow)

ILD 0 - 1.50:

LIGHT WEIGHT WALL Uow (HC<5)	MASS WALL HC RANGE	PCT FEN	INT INS	EXT INS
0.12	HC >= 5	14	0.13	0.17
	HC >= 10	14	0.15	0.20
	HC >= 15	14	0.17	0.21
	HC >= 5	62	0.13	0.16
	HC >= 10	62	0.15	0.19
	HC >= 15	62	0.17	0.21

ILD 1.51 - 3.00:

LIGHT WEIGHT WALL Uow (HC<5)	MASS WALL HC RANGE	PCT FEN	INT INS	EXT INS
0.12	HC >= 5	11	0.13	0.17
	HC >= 10	11	0.15	0.21
	HC >= 15	11	0.18	0.23
	HC >= 5	65	0.13	0.17
	HC >= 10	65	0.15	0.21
	HC >= 15	65	0.18	0.22

ILD 3.01 - 3.50:

LIGHT WEIGHT WALL Uow (HC<5)	MASS WALL HC RANGE	PCT FEN	INT INS	EXT INS
0.12	HC >= 5	9	0.13	0.18
	HC >= 10	9	0.15	0.22
	HC >= 15	9	0.18	0.23
	HC >= 5	61	0.13	0.17
	HC >= 10	61	0.15	0.21
	HC >= 15	61	0.18	0.22

OTHER CRITERIA

	Minimum R-Value				Max Uo
WALL BELOW GRADE:		8		ROOF:	0.057
UNHEATED SLAB ON GRADE:	24"	36"	48"	WALL ADJACENT TO UNCOND SPACE:	0.18
Horizontal	12	10	9	FLOOR OVER UNCOND. SPACE:	0.082
Vertical	7	5	4		

Source: Adapted with permission from ASHRAE/IES Standard 90.1-1989, *Energy Efficient Design of New Buildings Except Low-Rise Residential Buildings,* © 1989 by the American Society of Heating, Refrigerating, and Air-Conditioning Engineers, Inc., Atlanta, Ga.

TABLE K.23 **ACP for Delaware, New Jersey, Washington, DC, Baltimore, MD, Boston, MA, New York, NY, Columbus and Dayton, OH, Covington, KY, Charleston, WV, Harrisburg and Philadelphia, PA, and Lewiston, ID**

Alternate Component Package, for ASHRAE/IES Standard 90.1-1989

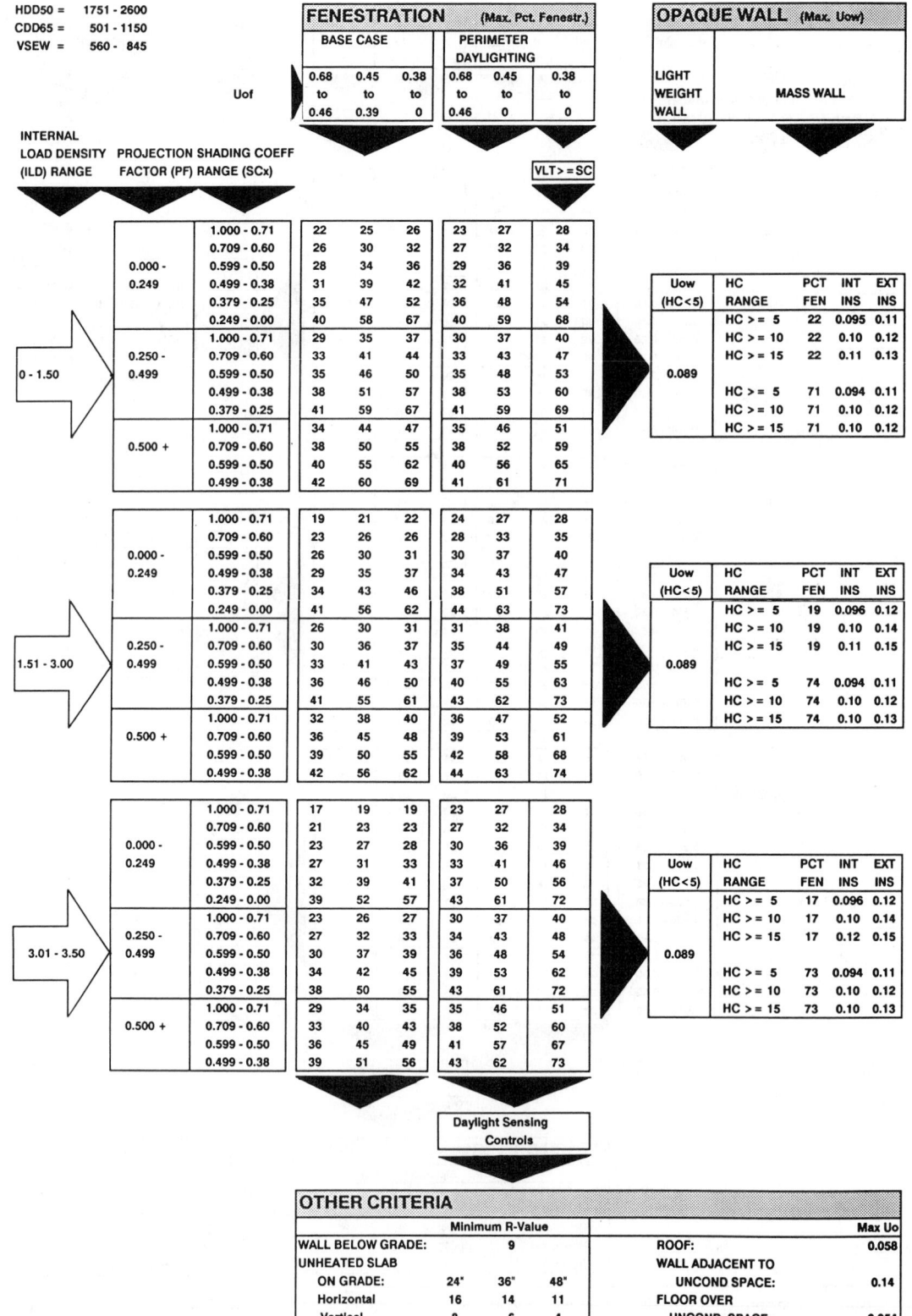

HDD50 = 1751 - 2600
CDD65 = 501 - 1150
VSEW = 560 - 845

FENESTRATION (Max. Pct. Fenestr.)

INTERNAL LOAD DENSITY (ILD) RANGE	PROJECTION FACTOR (PF)	SHADING COEFF RANGE (SCx)	Base Case Uof 0.68 to 0.46	Base Case Uof 0.45 to 0.39	Base Case Uof 0.38 to 0	Perimeter Daylighting Uof 0.68 to 0.46	Perimeter Daylighting Uof 0.45 to 0	Perimeter Daylighting Uof 0.38 to 0 (VLT>=SC)
0 - 1.50	0.000 - 0.249	1.000 - 0.71	22	25	26	23	27	28
		0.709 - 0.60	26	30	32	27	32	34
		0.599 - 0.50	28	34	36	29	36	39
		0.499 - 0.38	31	39	42	32	41	45
		0.379 - 0.25	35	47	52	36	48	54
		0.249 - 0.00	40	58	67	40	59	68
	0.250 - 0.499	1.000 - 0.71	29	35	37	30	37	40
		0.709 - 0.60	33	41	44	33	43	47
		0.599 - 0.50	35	46	50	35	48	53
		0.499 - 0.38	38	51	57	38	53	60
		0.379 - 0.25	41	59	67	41	59	69
	0.500 +	1.000 - 0.71	34	44	47	35	46	51
		0.709 - 0.60	38	50	55	38	52	59
		0.599 - 0.50	40	55	62	40	56	65
		0.499 - 0.38	42	60	69	41	61	71
1.51 - 3.00	0.000 - 0.249	1.000 - 0.71	19	21	22	24	27	28
		0.709 - 0.60	23	26	26	28	33	35
		0.599 - 0.50	26	30	31	30	37	40
		0.499 - 0.38	29	35	37	34	43	47
		0.379 - 0.25	34	43	46	38	51	57
		0.249 - 0.00	41	56	62	44	63	73
	0.250 - 0.499	1.000 - 0.71	26	30	31	31	38	41
		0.709 - 0.60	30	36	37	35	44	49
		0.599 - 0.50	33	41	43	37	49	55
		0.499 - 0.38	36	46	50	40	55	63
		0.379 - 0.25	41	55	61	43	62	73
	0.500 +	1.000 - 0.71	32	38	40	36	47	52
		0.709 - 0.60	36	45	48	39	53	61
		0.599 - 0.50	39	50	55	42	58	68
		0.499 - 0.38	42	56	62	44	63	74
3.01 - 3.50	0.000 - 0.249	1.000 - 0.71	17	19	19	23	27	28
		0.709 - 0.60	21	23	23	27	32	34
		0.599 - 0.50	23	27	28	30	36	39
		0.499 - 0.38	27	31	33	33	41	46
		0.379 - 0.25	32	39	41	37	50	56
		0.249 - 0.00	39	52	57	43	61	72
	0.250 - 0.499	1.000 - 0.71	23	26	27	30	37	40
		0.709 - 0.60	27	32	33	34	43	48
		0.599 - 0.50	30	37	39	36	48	54
		0.499 - 0.38	34	42	45	39	53	62
		0.379 - 0.25	38	50	55	43	61	72
	0.500 +	1.000 - 0.71	29	34	35	35	46	51
		0.709 - 0.60	33	40	43	38	52	60
		0.599 - 0.50	36	45	49	41	57	67
		0.499 - 0.38	39	51	56	43	62	73

Perimeter Daylighting → Daylight Sensing Controls

OPAQUE WALL (Max. Uow)

LIGHT WEIGHT WALL / MASS WALL

ILD 0 - 1.50:

Uow (HC<5)	HC RANGE	PCT FEN	INT INS	EXT INS
0.089	HC >= 5	22	0.095	0.11
	HC >= 10	22	0.10	0.12
	HC >= 15	22	0.11	0.13
	HC >= 5	71	0.094	0.11
	HC >= 10	71	0.10	0.12
	HC >= 15	71	0.10	0.12

ILD 1.51 - 3.00:

Uow (HC<5)	HC RANGE	PCT FEN	INT INS	EXT INS
0.089	HC >= 5	19	0.096	0.12
	HC >= 10	19	0.10	0.14
	HC >= 15	19	0.11	0.15
	HC >= 5	74	0.094	0.11
	HC >= 10	74	0.10	0.12
	HC >= 15	74	0.10	0.13

ILD 3.01 - 3.50:

Uow (HC<5)	HC RANGE	PCT FEN	INT INS	EXT INS
0.089	HC >= 5	17	0.096	0.12
	HC >= 10	17	0.10	0.14
	HC >= 15	17	0.12	0.15
	HC >= 5	73	0.094	0.11
	HC >= 10	73	0.10	0.12
	HC >= 15	73	0.10	0.13

OTHER CRITERIA

Minimum R-Value					Max Uo
WALL BELOW GRADE:		9		ROOF:	0.058
UNHEATED SLAB ON GRADE:	24"	36"	48"	WALL ADJACENT TO UNCOND SPACE:	0.14
Horizontal	16	14	11	FLOOR OVER UNCOND. SPACE:	0.054
Vertical	8	6	4		

Source: Adapted with permission from ASHRAE/IES Standard 90.1-1989, *Energy Efficient Design of New Buildings Except Low-Rise Residential Buildings*, © 1989 by the American Society of Heating, Refrigerating, and Air-Conditioning Engineers, Inc., Atlanta, Ga.

TABLE K.24 **ACP for Connecticut, Rhode Island, Allentown, Pittsburgh, and Scranton, PA, Akron, Toledo, and Youngstown, OH, Fort Wayne, Indianapolis, and South Bend, IN, Detroit, MI, Chicago and Moline IL, Burlington, IA, and Omaha, NE**

Alternate Component Package, for ASHRAE/IES Standard 90.1-1989

HDD50 = 2601 - 3200
CDD65 = 501 - 1150
VSEW = 560 - 845

FENESTRATION (Max. Pct. Fenestr.)

INTERNAL LOAD DENSITY (ILD) RANGE	PROJECTION FACTOR (PF)	SHADING COEFF RANGE (SCx)	BASE CASE Uof 0.68 to 0.46	BASE CASE Uof 0.45 to 0.39	BASE CASE Uof 0.38 to 0	PERIMETER DAYLIGHTING Uof 0.68 to 0.46	PERIMETER DAYLIGHTING Uof 0.45 to 0	PERIMETER DAYLIGHTING Uof 0.38 to 0 (VLT>=SC)
0 - 1.50	0.000 - 0.249	1.000 - 0.71	22	26	27	23	27	28
		0.709 - 0.60	25	30	32	25	32	34
		0.599 - 0.50	27	34	37	27	35	38
		0.499 - 0.38	29	38	42	30	39	44
		0.379 - 0.25	32	45	50	32	45	51
		0.249 - 0.00	36	53	62	36	53	62
	0.250 - 0.499	1.000 - 0.71	28	35	37	28	36	40
		0.709 - 0.60	31	40	44	31	41	46
		0.599 - 0.50	33	44	49	32	45	51
		0.499 - 0.38	34	48	54	34	49	56
		0.379 - 0.25	37	54	62	36	54	63
	0.500 +	1.000 - 0.71	32	43	47	32	44	49
		0.709 - 0.60	34	48	53	34	48	56
		0.599 - 0.50	36	52	59	35	52	60
		0.499 - 0.38	37	55	64	37	55	64
1.51 - 3.00	0.000 - 0.249	1.000 - 0.71	19	22	22	23	27	29
		0.709 - 0.60	23	26	27	26	32	35
		0.599 - 0.50	25	30	31	29	36	40
		0.499 - 0.38	28	34	37	31	41	45
		0.379 - 0.25	32	41	45	34	47	54
		0.249 - 0.00	37	52	59	39	56	66
	0.250 - 0.499	1.000 - 0.71	25	30	31	29	37	40
		0.709 - 0.60	29	35	38	32	42	47
		0.599 - 0.50	31	40	43	34	46	53
		0.499 - 0.38	34	44	49	36	50	59
		0.379 - 0.25	37	52	58	39	56	66
	0.500 +	1.000 - 0.71	30	38	40	33	44	50
		0.709 - 0.60	33	43	47	36	50	58
		0.599 - 0.50	36	48	53	37	53	63
		0.499 - 0.38	38	52	59	39	57	67
3.01 - 3.50	0.000 - 0.249	1.000 - 0.71	18	19	20	23	27	28
		0.709 - 0.60	21	23	24	26	32	34
		0.599 - 0.50	23	27	28	28	36	39
		0.499 - 0.38	26	31	33	31	40	45
		0.379 - 0.25	30	38	41	34	46	53
		0.249 - 0.00	36	48	54	38	55	66
	0.250 - 0.499	1.000 - 0.71	23	27	28	28	36	39
		0.709 - 0.60	26	32	34	31	41	47
		0.599 - 0.50	29	36	39	33	45	52
		0.499 - 0.38	32	41	44	35	49	58
		0.379 - 0.25	35	48	53	38	55	65
	0.500 +	1.000 - 0.71	28	34	36	32	43	50
		0.709 - 0.60	31	39	42	35	48	57
		0.599 - 0.50	33	44	48	36	52	62
		0.499 - 0.38	36	48	54	38	56	67

Perimeter daylighting: Daylight Sensing Controls

OPAQUE WALL (Max. Uow)

LIGHT WEIGHT WALL / MASS WALL

ILD 0 - 1.50:

Uow (HC<5)	HC RANGE	PCT FEN	INT INS	EXT INS
0.082	HC >= 5	22	0.085	0.10
	HC >= 10	22	0.091	0.11
	HC >= 15	22	0.095	0.11
	HC >= 5	64	0.085	0.098
	HC >= 10	64	0.089	0.10
	HC >= 15	64	0.092	0.10

ILD 1.51 - 3.00:

Uow (HC<5)	HC RANGE	PCT FEN	INT INS	EXT INS
0.082	HC >= 5	19	0.086	0.10
	HC >= 10	19	0.093	0.12
	HC >= 15	19	0.10	0.12
	HC >= 5	67	0.085	0.10
	HC >= 10	67	0.090	0.11
	HC >= 15	67	0.094	0.11

ILD 3.01 - 3.50:

Uow (HC<5)	HC RANGE	PCT FEN	INT INS	EXT INS
0.082	HC >= 5	18	0.086	0.11
	HC >= 10	18	0.094	0.12
	HC >= 15	18	0.10	0.13
	HC >= 5	67	0.085	0.10
	HC >= 10	67	0.090	0.11
	HC >= 15	67	0.095	0.11

OTHER CRITERIA

	Minimum R-Value				Max Uo
WALL BELOW GRADE:		10		ROOF:	0.053
UNHEATED SLAB ON GRADE:	24"	36"	48"	WALL ADJACENT TO UNCOND SPACE:	0.13
Horizontal	17	14	11	FLOOR OVER UNCOND. SPACE:	0.048
Vertical	8	6	4		

Source: Adapted with permission from ASHRAE/IES Standard 90.1-1989, *Energy Efficient Design of New Buildings Except Low-Rise Residential Buildings,* © 1989 by the American Society of Heating, Refrigerating, and Air-Conditioning Engineers, Inc., Atlanta, Ga.

TABLE K.25 **ACP for Erie, PA, Redmond, OR, Spokane and Yakima, WA**

Alternate Component Package, for ASHRAE/IES Standard 90.1-1989

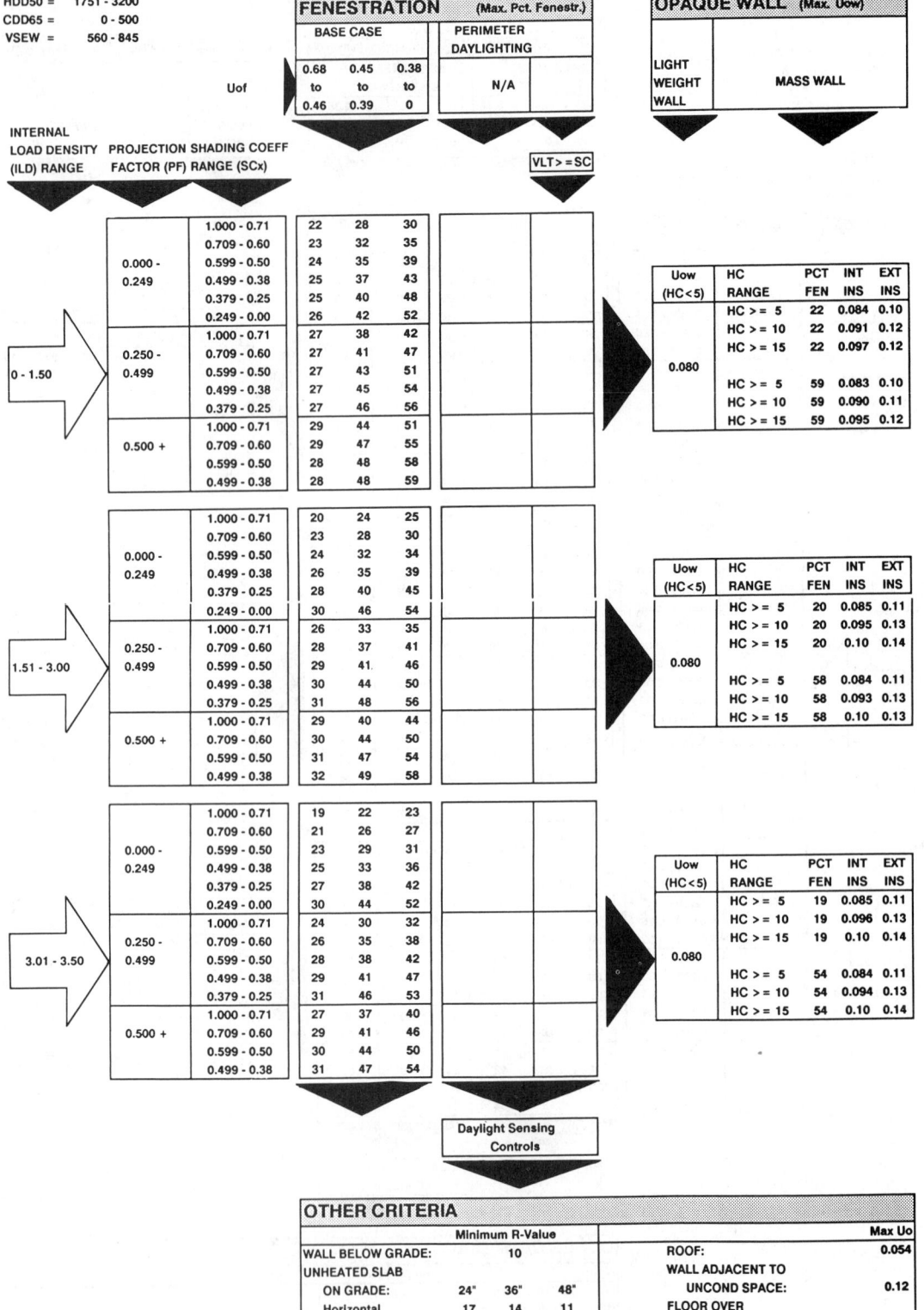

HDD50 = 1751 - 3200
CDD65 = 0 - 500
VSEW = 560 - 845

FENESTRATION (Max. Pct. Fenestr.)

INTERNAL LOAD DENSITY (ILD) RANGE	PROJECTION FACTOR (PF) RANGE	SHADING COEFF RANGE (SCx)	BASE CASE Uof 0.68 to 0.46	BASE CASE Uof 0.45 to 0.39	BASE CASE Uof 0.38 to 0	PERIMETER DAYLIGHTING N/A	VLT>=SC
0 - 1.50	0.000 - 0.249	1.000 - 0.71	22	28	30		
		0.709 - 0.60	23	32	35		
		0.599 - 0.50	24	35	39		
		0.499 - 0.38	25	37	43		
		0.379 - 0.25	25	40	48		
		0.249 - 0.00	26	42	52		
	0.250 - 0.499	1.000 - 0.71	27	38	42		
		0.709 - 0.60	27	41	47		
		0.599 - 0.50	27	43	51		
		0.499 - 0.38	27	45	54		
		0.379 - 0.25	27	46	56		
	0.500 +	1.000 - 0.71	29	44	51		
		0.709 - 0.60	29	47	55		
		0.599 - 0.50	28	48	58		
		0.499 - 0.38	28	48	59		
1.51 - 3.00	0.000 - 0.249	1.000 - 0.71	20	24	25		
		0.709 - 0.60	23	28	30		
		0.599 - 0.50	24	32	34		
		0.499 - 0.38	26	35	39		
		0.379 - 0.25	28	40	45		
		0.249 - 0.00	30	46	54		
	0.250 - 0.499	1.000 - 0.71	26	33	35		
		0.709 - 0.60	28	37	41		
		0.599 - 0.50	29	41	46		
		0.499 - 0.38	30	44	50		
		0.379 - 0.25	31	48	56		
	0.500 +	1.000 - 0.71	29	40	44		
		0.709 - 0.60	30	44	50		
		0.599 - 0.50	31	47	54		
		0.499 - 0.38	32	49	58		
3.01 - 3.50	0.000 - 0.249	1.000 - 0.71	19	22	23		
		0.709 - 0.60	21	26	27		
		0.599 - 0.50	23	29	31		
		0.499 - 0.38	25	33	36		
		0.379 - 0.25	27	38	42		
		0.249 - 0.00	30	44	52		
	0.250 - 0.499	1.000 - 0.71	24	30	32		
		0.709 - 0.60	26	35	38		
		0.599 - 0.50	28	38	42		
		0.499 - 0.38	29	41	47		
		0.379 - 0.25	31	46	53		
	0.500 +	1.000 - 0.71	27	37	40		
		0.709 - 0.60	29	41	46		
		0.599 - 0.50	30	44	50		
		0.499 - 0.38	31	47	54		

Daylight Sensing Controls

OPAQUE WALL (Max. Uow)

LIGHT WEIGHT WALL | MASS WALL

ILD 0 - 1.50:

Uow (HC<5)	HC RANGE	PCT FEN	INT INS	EXT INS
0.080	HC >= 5	22	0.084	0.10
	HC >= 10	22	0.091	0.12
	HC >= 15	22	0.097	0.12
	HC >= 5	59	0.083	0.10
	HC >= 10	59	0.090	0.11
	HC >= 15	59	0.095	0.12

ILD 1.51 - 3.00:

Uow (HC<5)	HC RANGE	PCT FEN	INT INS	EXT INS
0.080	HC >= 5	20	0.085	0.11
	HC >= 10	20	0.095	0.13
	HC >= 15	20	0.10	0.14
	HC >= 5	58	0.084	0.11
	HC >= 10	58	0.093	0.13
	HC >= 15	58	0.10	0.13

ILD 3.01 - 3.50:

Uow (HC<5)	HC RANGE	PCT FEN	INT INS	EXT INS
0.080	HC >= 5	19	0.085	0.11
	HC >= 10	19	0.096	0.13
	HC >= 15	19	0.10	0.14
	HC >= 5	54	0.084	0.11
	HC >= 10	54	0.094	0.13
	HC >= 15	54	0.10	0.14

OTHER CRITERIA

	Minimum R-Value				Max Uo
WALL BELOW GRADE:	10			ROOF:	0.054
UNHEATED SLAB				WALL ADJACENT TO	
ON GRADE:	24"	36"	48"	UNCOND SPACE:	0.12
Horizontal	17	14	11	FLOOR OVER	
Vertical	8	6	4	UNCOND. SPACE:	0.047

Source: Adapted with permission from ASHRAE/IES Standard 90.1-1989, *Energy Efficient Design of New Buildings Except Low-Rise Residential Buildings,* © 1989 by the American Society of Heating, Refrigerating, and Air-Conditioning Engineers, Inc., Atlanta, Ga.

TABLE K.26 **ACP for Colorado Springs, Denver, and Pueblo, CO, Boise, ID, Goodland, KS, Clayton, NM, Lovelock, Reno, Tonopah, and Winnemucca, NV, Cedar City and Salt Lake City, UT, and Mount Shasta, CA**

Alternate Component Package, for ASHRAE/IES Standard 90.1-1989

HDD50 = 1751 - 3200
CDD65 = 0 - 1150
VSEW = > 845

FENESTRATION (Max. Pct. Fenestr.)

INTERNAL LOAD DENSITY (ILD) RANGE	PROJECTION FACTOR (PF)	SHADING COEFF RANGE (SCx)	BASE CASE Uof 0.68 to 0.46	BASE CASE Uof 0.45 to 0.39	BASE CASE Uof 0.38 to 0	PERIMETER DAYLIGHTING Uof 0.68 to 0.46	PERIMETER DAYLIGHTING Uof 0.45 to 0	PERIMETER DAYLIGHTING Uof 0.38 to 0 (VLT>=SC)
0 - 1.50	0.000 - 0.249	1.000 - 0.71	22	24	25	23	26	27
		0.709 - 0.60	25	29	31	26	31	33
		0.599 - 0.50	27	33	35	28	35	38
		0.499 - 0.38	30	38	41	30	39	43
		0.379 - 0.25	33	45	49	33	46	52
		0.249 - 0.00	37	54	62	37	55	63
	0.250 - 0.499	1.000 - 0.71	29	35	36	29	37	39
		0.709 - 0.60	32	40	43	32	42	46
		0.599 - 0.50	34	45	49	34	46	52
		0.499 - 0.38	36	50	55	36	51	58
		0.379 - 0.25	38	56	64	38	56	66
	0.500 +	1.000 - 0.71	34	43	47	34	45	50
		0.709 - 0.60	36	49	54	36	51	58
		0.599 - 0.50	38	54	60	38	55	63
		0.499 - 0.38	39	58	66	39	58	68
1.51 - 3.00	0.000 - 0.249	1.000 - 0.71	19	20	21	23	26	28
		0.709 - 0.60	22	25	25	26	31	33
		0.599 - 0.50	25	29	30	29	36	39
		0.499 - 0.38	28	33	35	32	41	45
		0.379 - 0.25	32	41	44	36	48	54
		0.249 - 0.00	39	52	58	40	58	68
	0.250 - 0.499	1.000 - 0.71	25	29	30	30	37	40
		0.709 - 0.60	29	35	36	33	43	47
		0.599 - 0.50	32	39	42	36	47	54
		0.499 - 0.38	35	44	48	38	52	60
		0.379 - 0.25	39	52	58	41	59	69
	0.500 +	1.000 - 0.71	31	37	39	35	46	51
		0.709 - 0.60	34	43	46	38	52	59
		0.599 - 0.50	37	48	52	40	56	66
		0.499 - 0.38	40	54	59	41	60	71
3.01 - 3.50	0.000 - 0.249	1.000 - 0.71	17	18	18	22	26	27
		0.709 - 0.60	20	22	23	26	31	33
		0.599 - 0.50	22	26	26	28	35	38
		0.499 - 0.38	25	30	31	31	39	44
		0.379 - 0.25	30	37	39	35	47	53
		0.249 - 0.00	36	48	53	40	57	67
	0.250 - 0.499	1.000 - 0.71	23	26	26	29	36	39
		0.709 - 0.60	26	31	32	32	42	46
		0.599 - 0.50	29	35	37	35	46	52
		0.499 - 0.38	32	40	43	37	51	59
		0.379 - 0.25	36	48	52	40	57	68
	0.500 +	1.000 - 0.71	28	33	34	34	45	50
		0.709 - 0.60	32	39	41	37	50	58
		0.599 - 0.50	34	44	47	39	54	65
		0.499 - 0.38	37	49	53	41	59	70

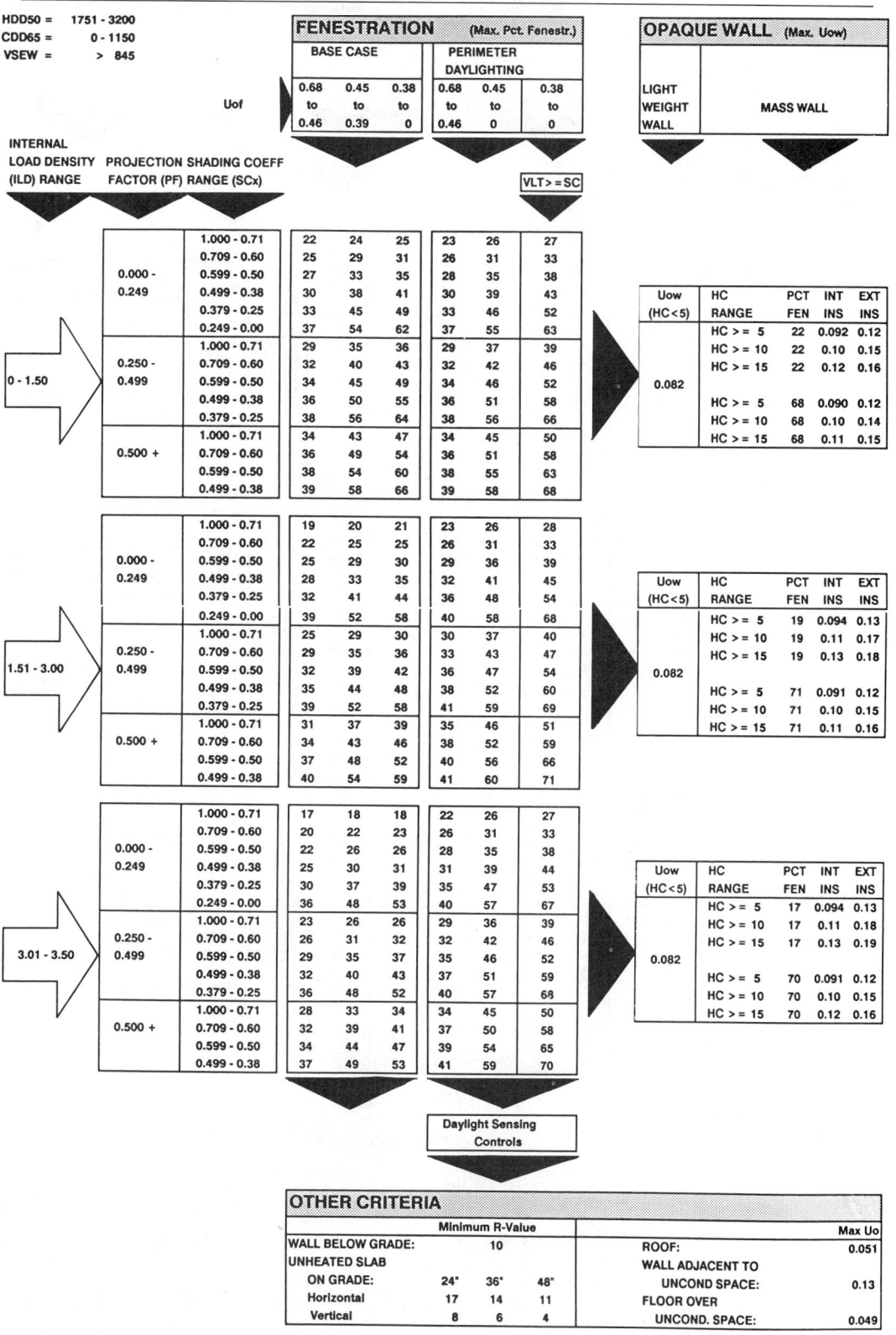

Perimeter daylighting: Daylight Sensing Controls

OPAQUE WALL (Max. Uow)

LIGHT WEIGHT WALL / MASS WALL

ILD 0 - 1.50:

Uow (HC<5)	HC RANGE	PCT FEN	INT INS	EXT INS
0.082	HC >= 5	22	0.092	0.12
	HC >= 10	22	0.10	0.15
	HC >= 15	22	0.12	0.16
	HC >= 5	68	0.090	0.12
	HC >= 10	68	0.10	0.14
	HC >= 15	68	0.11	0.15

ILD 1.51 - 3.00:

Uow (HC<5)	HC RANGE	PCT FEN	INT INS	EXT INS
0.082	HC >= 5	19	0.094	0.13
	HC >= 10	19	0.11	0.17
	HC >= 15	19	0.13	0.18
	HC >= 5	71	0.091	0.12
	HC >= 10	71	0.10	0.15
	HC >= 15	71	0.11	0.16

ILD 3.01 - 3.50:

Uow (HC<5)	HC RANGE	PCT FEN	INT INS	EXT INS
0.082	HC >= 5	17	0.094	0.13
	HC >= 10	17	0.11	0.18
	HC >= 15	17	0.13	0.19
	HC >= 5	70	0.091	0.12
	HC >= 10	70	0.10	0.15
	HC >= 15	70	0.12	0.16

OTHER CRITERIA

Minimum R-Value					Max Uo
WALL BELOW GRADE:	10			ROOF:	0.051
UNHEATED SLAB				WALL ADJACENT TO UNCOND SPACE:	0.13
ON GRADE:	24"	36"	48"	FLOOR OVER UNCOND. SPACE:	0.049
Horizontal	17	14	11		
Vertical	8	6	4		

Source: Adapted with permission from ASHRAE/IES Standard 90.1-1989, *Energy Efficient Design of New Buildings Except Low-Rise Residential Buildings,* © 1989 by the American Society of Heating, Refrigerating, and Air-Conditioning Engineers, Inc., Atlanta, Ga.

TABLE K.27 **ACP for Springfield, IL, Evansville, IN, Lexington and Louisville, KY, Columbia, St. Louis, and Springfield, MO, and Topeka, KS**

Alternate Component Package, for ASHRAE/IES Standard 90.1-1989

HDD50 = 1751 - 3200
CDD65 = 1151 - 2000
VSEW = 560 - 845

FENESTRATION (Max. Pct. Fenestr.)

Internal Load Density (ILD) Range	Projection Factor (PF)	Shading Coeff Range (SCx)	Base Case Uof 0.81 to 0.46	Base Case Uof 0.45 to 0.39	Base Case Uof 0.38 to 0	Perimeter Daylighting Uof 0.81 to 0.46	Perimeter Daylighting Uof 0.45 to 0	Perimeter Daylighting Uof 0.38 to 0 (VLT>=SC)
0 - 1.50	0.000 - 0.249	1.000 - 0.71	17	19	19	18	20	21
		0.709 - 0.60	20	23	23	21	25	26
		0.599 - 0.50	22	27	27	24	29	31
		0.499 - 0.38	25	31	32	27	34	36
		0.379 - 0.25	30	39	41	31	42	46
		0.249 - 0.00	37	54	58	38	57	63
	0.250 - 0.499	1.000 - 0.71	22	26	27	24	29	30
		0.709 - 0.60	26	32	33	27	35	37
		0.599 - 0.50	29	37	39	30	40	43
		0.499 - 0.38	32	43	45	33	46	50
		0.379 - 0.25	36	52	57	37	56	62
	0.500 +	1.000 - 0.71	27	34	36	29	37	40
		0.709 - 0.60	31	41	43	32	44	48
		0.599 - 0.50	34	47	50	35	50	55
		0.499 - 0.38	37	53	57	38	56	63
1.51 - 3.00	0.000 - 0.249	1.000 - 0.71	14	15	15	18	21	22
		0.709 - 0.60	17	19	19	22	25	27
		0.599 - 0.50	19	22	22	25	30	32
		0.499 - 0.38	22	26	27	28	35	38
		0.379 - 0.25	27	33	35	33	44	49
		0.249 - 0.00	35	47	50	40	59	68
	0.250 - 0.499	1.000 - 0.71	19	21	22	25	29	32
		0.709 - 0.60	23	26	27	28	36	39
		0.599 - 0.50	26	31	32	31	41	46
		0.499 - 0.38	29	36	38	34	47	53
		0.379 - 0.25	34	45	48	39	57	66
	0.500 +	1.000 - 0.71	24	28	29	30	38	42
		0.709 - 0.60	28	34	35	33	45	51
		0.599 - 0.50	31	39	41	36	51	58
		0.499 - 0.38	34	46	48	39	58	67
3.01 - 3.50	0.000 - 0.249	1.000 - 0.71	12	13	13	17	19	21
		0.709 - 0.60	14	16	16	20	24	26
		0.599 - 0.50	17	19	19	23	28	31
		0.499 - 0.38	19	22	23	26	33	37
		0.379 - 0.25	24	29	30	31	41	47
		0.249 - 0.00	31	41	43	38	56	66
	0.250 - 0.499	1.000 - 0.71	16	18	18	23	28	31
		0.709 - 0.60	20	22	23	27	33	38
		0.599 - 0.50	22	26	27	29	38	44
		0.499 - 0.38	25	31	32	32	44	51
		0.379 - 0.25	30	39	41	37	54	64
	0.500 +	1.000 - 0.71	21	24	24	28	36	40
		0.709 - 0.60	24	29	30	31	42	49
		0.599 - 0.50	27	34	35	34	48	56
		0.499 - 0.38	30	39	41	37	55	65

Perimeter daylighting: Daylight Sensing Controls

OPAQUE WALL (Max. Uow) — Light Weight Wall / Mass Wall

ILD 0 - 1.50:

Uow (HC<5)	HC RANGE	PCT FEN	INT INS	EXT INS
0.093	HC >= 5	17	0.099	0.12
	HC >= 10	17	0.11	0.14
	HC >= 15	17	0.12	0.14
	HC >= 5	63	0.099	0.11
	HC >= 10	63	0.10	0.13
	HC >= 15	63	0.11	0.13

ILD 1.51 - 3.00:

Uow (HC<5)	HC RANGE	PCT FEN	INT INS	EXT INS
0.093	HC >= 5	14	0.10	0.13
	HC >= 10	14	0.11	0.15
	HC >= 15	14	0.12	0.16
	HC >= 5	68	0.099	0.12
	HC >= 10	68	0.11	0.14
	HC >= 15	68	0.11	0.14

ILD 3.01 - 3.50:

Uow (HC<5)	HC RANGE	PCT FEN	INT INS	EXT INS
0.093	HC >= 5	12	0.10	0.13
	HC >= 10	12	0.11	0.16
	HC >= 15	12	0.13	0.17
	HC >= 5	66	0.099	0.12
	HC >= 10	66	0.11	0.14
	HC >= 15	66	0.11	0.14

OTHER CRITERIA

	Minimum R-Value		
WALL BELOW GRADE:	9		
UNHEATED SLAB ON GRADE:	24"	36"	48"
Horizontal	16	13	11
Vertical	8	6	4

	Max Uo
ROOF:	0.056
WALL ADJACENT TO UNCOND SPACE:	0.14
FLOOR OVER UNCOND. SPACE:	0.057

Source: Adapted with permission from ASHRAE/IES Standard 90.1-1989, *Energy Efficient Design of New Buildings Except Low-Rise Residential Buildings,* © 1989 by the American Society of Heating, Refrigerating, and Air-Conditioning Engineers, Inc., Atlanta, Ga.

TABLE K.28 **ACP for Dodge City, KS, and Grand Junction, CO**

Alternate Component Package, for ASHRAE/IES Standard 90.1-1989

HDD50 = 1751 - 3200
CDD65 = 1151 - 2000
VSEW = > 845

FENESTRATION (Max. Pct. Fenestr.)

Internal Load Density (ILD) Range	Projection Factor (PF) Range	Shading Coeff Range (SCx)	Base Case, Uof 0.81 to 0.46	Base Case, Uof 0.45 to 0.39	Base Case, Uof 0.38 to 0	Perimeter Daylighting, Uof 0.81 to 0.46	Perimeter Daylighting, Uof 0.45 to 0	Perimeter Daylighting, Uof 0.38 to 0 (VLT>=SC)
0 - 1.50	0.000 - 0.249	1.000 - 0.71	15	17	17	16	19	19
		0.709 - 0.60	18	21	21	19	23	24
		0.599 - 0.50	20	24	25	21	26	28
		0.499 - 0.38	22	28	30	23	30	32
		0.379 - 0.25	26	35	38	27	37	41
		0.249 - 0.00	31	47	52	32	49	56
	0.250 - 0.499	1.000 - 0.71	20	24	25	21	26	28
		0.709 - 0.60	23	29	31	24	32	34
		0.599 - 0.50	25	34	36	26	36	39
		0.499 - 0.38	28	39	42	29	41	45
		0.379 - 0.25	31	47	51	32	49	55
	0.500 +	1.000 - 0.71	24	32	33	25	34	37
		0.709 - 0.60	27	37	40	28	40	44
		0.599 - 0.50	30	42	46	30	45	50
		0.499 - 0.38	32	48	52	32	50	56
1.51 - 3.00	0.000 - 0.249	1.000 - 0.71	12	13	14	16	18	20
		0.709 - 0.60	15	17	17	19	22	24
		0.599 - 0.50	17	20	20	21	26	28
		0.499 - 0.38	19	23	24	23	30	34
		0.379 - 0.25	23	29	31	27	37	42
		0.249 - 0.00	30	41	44	33	50	58
	0.250 - 0.499	1.000 - 0.71	17	19	20	21	26	29
		0.709 - 0.60	20	24	24	24	31	35
		0.599 - 0.50	22	27	29	26	35	40
		0.499 - 0.38	25	32	34	29	41	47
		0.379 - 0.25	29	40	43	33	49	57
	0.500 +	1.000 - 0.71	21	25	26	25	33	37
		0.709 - 0.60	24	30	32	28	39	45
		0.599 - 0.50	27	35	37	30	44	51
		0.499 - 0.38	29	40	43	33	50	58
3.01 - 3.50	0.000 - 0.249	1.000 - 0.71	11	11	12	15	17	19
		0.709 - 0.60	13	14	14	17	21	23
		0.599 - 0.50	15	17	17	20	24	27
		0.499 - 0.38	17	20	20	22	28	32
		0.379 - 0.25	21	25	27	26	35	41
		0.249 - 0.00	27	36	38	31	47	56
	0.250 - 0.499	1.000 - 0.71	15	16	17	20	24	27
		0.709 - 0.60	17	20	21	23	29	33
		0.599 - 0.50	19	24	24	25	33	39
		0.499 - 0.38	22	28	29	27	38	45
		0.379 - 0.25	26	35	37	31	46	55
	0.500 +	1.000 - 0.71	18	22	22	24	31	36
		0.709 - 0.60	21	26	27	26	37	43
		0.599 - 0.50	24	30	32	29	41	49
		0.499 - 0.38	26	35	37	31	47	56

Daylight Sensing Controls (applies to Perimeter Daylighting)

OPAQUE WALL (Max. Uow)

Light Weight Wall / Mass Wall

ILD 0 - 1.50:

Uow (HC<5)	HC RANGE	PCT FEN	INT INS	EXT INS
0.090	HC >= 5	15	0.097	0.12
	HC >= 10	15	0.11	0.14
	HC >= 15	15	0.12	0.15
	HC >= 5	56	0.097	0.12
	HC >= 10	56	0.10	0.14
	HC >= 15	56	0.11	0.15

ILD 1.51 - 3.00:

Uow (HC<5)	HC RANGE	PCT FEN	INT INS	EXT INS
0.090	HC >= 5	12	0.098	0.13
	HC >= 10	12	0.11	0.16
	HC >= 15	12	0.13	0.17
	HC >= 5	58	0.097	0.12
	HC >= 10	58	0.11	0.15
	HC >= 15	58	0.12	0.16

ILD 3.01 - 3.50:

Uow (HC<5)	HC RANGE	PCT FEN	INT INS	EXT INS
0.090	HC >= 5	11	0.099	0.13
	HC >= 10	11	0.11	0.16
	HC >= 15	11	0.13	0.17
	HC >= 5	56	0.097	0.12
	HC >= 10	56	0.11	0.14
	HC >= 15	56	0.12	0.15

OTHER CRITERIA

Minimum R-Value				Max Uo	
WALL BELOW GRADE:	9			ROOF:	0.053
UNHEATED SLAB ON GRADE:	24"	36"	48"	WALL ADJACENT TO UNCOND SPACE:	0.14
Horizontal	16	14	11	FLOOR OVER UNCOND. SPACE:	0.055
Vertical	8	6	4		

TABLE K.29 **ACP for Albany, Buffalo, Rochester, and Syracuse, NY, Grand Rapids, MI, La Crosse and Milwaukee, WI, Des Moines and Sioux City, IA, Grand Island, NE, Rapid City, SD, and Billings, MT**

Alternate Component Package, for ASHRAE/IES Standard 90.1-1989

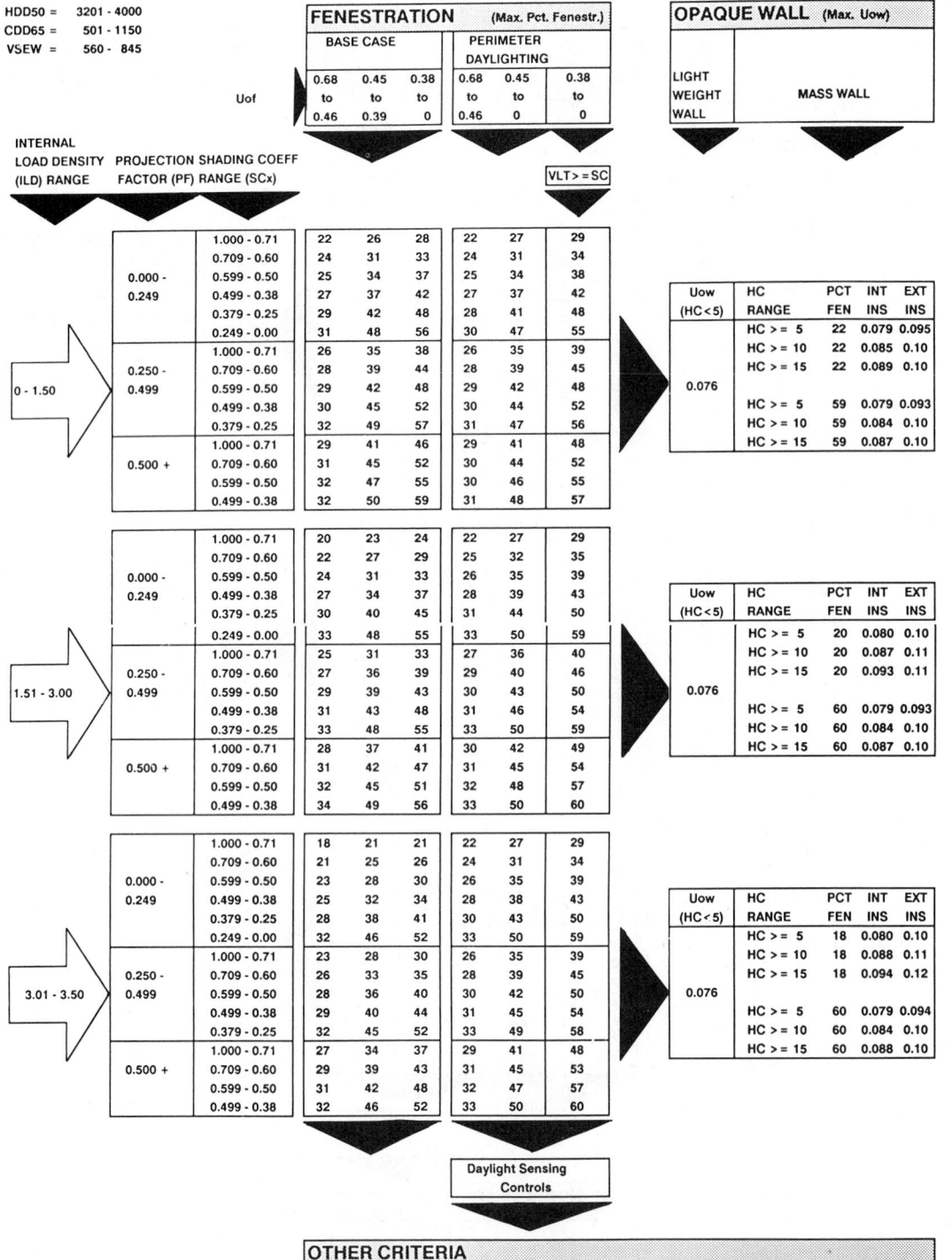

HDD50 = 3201 - 4000
CDD65 = 501 - 1150
VSEW = 560 - 845

FENESTRATION (Max. Pct. Fenestr.)

Internal Load Density (ILD) Range	Projection Factor (PF) Range	Shading Coeff Range (SCx)	Base Case Uof 0.68 to 0.46	Base Case Uof 0.45 to 0.39	Base Case Uof 0.38 to 0	Perimeter Daylighting Uof 0.68 to 0.46	Perimeter Daylighting Uof 0.45 to 0	Perimeter Daylighting Uof 0.38 to 0 (VLT>=SC)
0 - 1.50	0.000 - 0.249	1.000 - 0.71	22	26	28	22	27	29
		0.709 - 0.60	24	31	33	24	31	34
		0.599 - 0.50	25	34	37	25	34	38
		0.499 - 0.38	27	37	42	27	37	42
		0.379 - 0.25	29	42	48	28	41	48
		0.249 - 0.00	31	48	56	30	47	55
	0.250 - 0.499	1.000 - 0.71	26	35	38	26	35	39
		0.709 - 0.60	28	39	44	28	39	45
		0.599 - 0.50	29	42	48	29	42	48
		0.499 - 0.38	30	45	52	30	44	52
		0.379 - 0.25	32	49	57	31	47	56
	0.500 +	1.000 - 0.71	29	41	46	29	41	48
		0.709 - 0.60	31	45	52	30	44	52
		0.599 - 0.50	32	47	55	30	46	55
		0.499 - 0.38	32	50	59	31	48	57
1.51 - 3.00	0.000 - 0.249	1.000 - 0.71	20	23	24	22	27	29
		0.709 - 0.60	22	27	29	25	32	35
		0.599 - 0.50	24	31	33	26	35	39
		0.499 - 0.38	27	34	37	28	39	43
		0.379 - 0.25	30	40	45	31	44	50
		0.249 - 0.00	33	48	55	33	50	59
	0.250 - 0.499	1.000 - 0.71	25	31	33	27	36	40
		0.709 - 0.60	27	36	39	29	40	46
		0.599 - 0.50	29	39	43	30	43	50
		0.499 - 0.38	31	43	48	31	46	54
		0.379 - 0.25	33	48	55	33	50	59
	0.500 +	1.000 - 0.71	28	37	41	30	42	49
		0.709 - 0.60	31	42	47	31	45	54
		0.599 - 0.50	32	45	51	32	48	57
		0.499 - 0.38	34	49	56	33	50	60
3.01 - 3.50	0.000 - 0.249	1.000 - 0.71	18	21	21	22	27	29
		0.709 - 0.60	21	25	26	24	31	34
		0.599 - 0.50	23	28	30	26	35	39
		0.499 - 0.38	25	32	34	28	38	43
		0.379 - 0.25	28	38	41	30	43	50
		0.249 - 0.00	32	46	52	33	50	59
	0.250 - 0.499	1.000 - 0.71	23	28	30	26	35	39
		0.709 - 0.60	26	33	35	28	39	45
		0.599 - 0.50	28	36	40	30	42	50
		0.499 - 0.38	29	40	44	31	45	54
		0.379 - 0.25	32	45	52	33	49	58
	0.500 +	1.000 - 0.71	27	34	37	29	41	48
		0.709 - 0.60	29	39	43	31	45	53
		0.599 - 0.50	31	42	48	32	47	57
		0.499 - 0.38	32	46	52	33	50	60

Daylight Sensing Controls

OPAQUE WALL (Max. Uow)

ILD Range	Light Weight Wall Uow (HC<5)	Mass Wall HC Range	PCT FEN	INT INS	EXT INS
0 - 1.50	0.076	HC >= 5	22	0.079	0.095
		HC >= 10	22	0.085	0.10
		HC >= 15	22	0.089	0.10
		HC >= 5	59	0.079	0.093
		HC >= 10	59	0.084	0.10
		HC >= 15	59	0.087	0.10
1.51 - 3.00	0.076	HC >= 5	20	0.080	0.10
		HC >= 10	20	0.087	0.11
		HC >= 15	20	0.093	0.11
		HC >= 5	60	0.079	0.093
		HC >= 10	60	0.084	0.10
		HC >= 15	60	0.087	0.10
3.01 - 3.50	0.076	HC >= 5	18	0.080	0.10
		HC >= 10	18	0.088	0.11
		HC >= 15	18	0.094	0.12
		HC >= 5	60	0.079	0.094
		HC >= 10	60	0.084	0.10
		HC >= 15	60	0.088	0.10

OTHER CRITERIA

	Minimum R-Value		
WALL BELOW GRADE:	10		
UNHEATED SLAB ON GRADE:	24"	36"	48"
Horizontal	18	15	11
Vertical	8	6	4

	Max Uo
ROOF:	0.049
WALL ADJACENT TO UNCOND SPACE:	0.12
FLOOR OVER UNCOND. SPACE:	0.044

Source: Adapted with permission from ASHRAE/IES Standard 90.1-1989, *Energy Efficient Design of New Buildings Except Low-Rise Residential Buildings,* © 1989 by the American Society of Heating, Refrigerating, and Air-Conditioning Engineers, Inc., Atlanta, Ga.

TABLE K.30 **ACP for Portland, ME, Concord, NH, Binghampton, NY, Flint and Traverse City, MI, Great Falls, Helena, and Missoula, MT, and Sheridan, WY**

Alternate Component Package, for ASHRAE/IES Standard 90.1-1989

HDD50 = 3201 - 4000
CDD65 = 0 - 500
VSEW = 560 - 845

FENESTRATION (Max. Pct. Fenestr.)

INTERNAL LOAD DENSITY (ILD) RANGE	PROJECTION FACTOR (PF)	SHADING COEFF RANGE (SCx)	BASE CASE Uof 0.68 to 0.46	BASE CASE Uof 0.45 to 0.29	BASE CASE Uof 0.28 to 0	PERIMETER DAYLIGHTING N/A	PERIMETER DAYLIGHTING VLT>=SC
0 - 1.50	0.000 - 0.249	1.000 - 0.71	20	27	32		
		0.709 - 0.60	21	30	39		
		0.599 - 0.50	21	31	44		
		0.499 - 0.38	21	33	49		
		0.379 - 0.25	22	35	57		
		0.249 - 0.00	22	36	64		
	0.250 - 0.499	1.000 - 0.71	23	34	46		
		0.709 - 0.60	23	36	53		
		0.599 - 0.50	23	37	59		
		0.499 - 0.38	23	38	64		
		0.379 - 0.25	23	38	69		
	0.500 +	1.000 - 0.71	24	38	57		
		0.709 - 0.60	24	39	64		
		0.599 - 0.50	24	40	69		
		0.499 - 0.38	23	39	72		
1.51 - 3.00	0.000 - 0.249	1.000 - 0.71	19	24	27		
		0.709 - 0.60	21	27	33		
		0.599 - 0.50	22	30	38		
		0.499 - 0.38	23	32	43		
		0.379 - 0.25	24	36	52		
		0.249 - 0.00	26	40	64		
	0.250 - 0.499	1.000 - 0.71	23	31	38		
		0.709 - 0.60	24	34	45		
		0.599 - 0.50	25	37	51		
		0.499 - 0.38	26	39	57		
		0.379 - 0.25	26	41	66		
	0.500 +	1.000 - 0.71	25	36	49		
		0.709 - 0.60	26	39	56		
		0.599 - 0.50	26	41	62		
		0.499 - 0.38	27	42	68		
3.01 - 3.50	0.000 - 0.249	1.000 - 0.71	18	22	24		
		0.709 - 0.60	20	25	30		
		0.599 - 0.50	21	28	34		
		0.499 - 0.38	22	31	40		
		0.379 - 0.25	24	35	49		
		0.249 - 0.00	26	39	61		
	0.250 - 0.499	1.000 - 0.71	22	29	35		
		0.709 - 0.60	23	32	42		
		0.599 - 0.50	24	35	47		
		0.499 - 0.38	25	37	53		
		0.379 - 0.25	26	40	62		
	0.500 +	1.000 - 0.71	24	34	45		
		0.709 - 0.60	25	37	52		
		0.599 - 0.50	26	39	58		
		0.499 - 0.38	27	41	64		

Daylight Sensing Controls

OPAQUE WALL (Max. Uow)

LIGHT WEIGHT WALL / MASS WALL

ILD 0 - 1.50:

Uow (HC<5)	HC RANGE	PCT FEN	INT INS	EXT INS
0.072	HC >= 5	20	0.075	0.092
	HC >= 10	20	0.080	0.10
	HC >= 15	20	0.084	0.10
	HC >= 5	72	0.075	0.094
	HC >= 10	72	0.080	0.10
	HC >= 15	72	0.085	0.11

ILD 1.51 - 3.00:

Uow (HC<5)	HC RANGE	PCT FEN	INT INS	EXT INS
0.072	HC >= 5	19	0.075	0.098
	HC >= 10	19	0.082	0.11
	HC >= 15	19	0.088	0.11
	HC >= 5	68	0.075	0.099
	HC >= 10	68	0.082	0.11
	HC >= 15	68	0.089	0.12

ILD 3.01 - 3.50:

Uow (HC<5)	HC RANGE	PCT FEN	INT INS	EXT INS
0.072	HC >= 5	18	0.075	0.10
	HC >= 10	18	0.083	0.11
	HC >= 15	18	0.090	0.12
	HC >= 5	64	0.075	0.10
	HC >= 10	64	0.083	0.11
	HC >= 15	64	0.090	0.12

OTHER CRITERIA

	Minimum R-Value		
WALL BELOW GRADE:	11		
UNHEATED SLAB ON GRADE:	24"	36"	48"
Horizontal	18	15	11
Vertical	8	6	4

	Max Uo
ROOF:	0.049
WALL ADJACENT TO UNCOND SPACE:	0.11
FLOOR OVER UNCOND. SPACE:	0.040

Source: Adapted with permission from ASHRAE/IES Standard 90.1-1989, *Energy Efficient Design of New Buildings Except Low-Rise Residential Buildings,* © 1989 by the American Society of Heating, Refrigerating, and Air-Conditioning Engineers, Inc., Atlanta, Ga.

TABLE K.31 ACP for Bangor, ME, Burlington, VT, Massena, NY, Alpena, MI, Eau Claire, Green Bay, and Madison, WI, Mason City, IA, Minneapolis and Rochester, MN, Huron, Pierre, and Sioux Falls, SD, Cut Bank, Dillon, Lewistown, and Miles, MT

Alternate Component Package, for ASHRAE/IES Standard 90.1-1989

HDD50 = 4001 - 5000
CDD65 = 0 - 1150
VSEW = 560 - 845

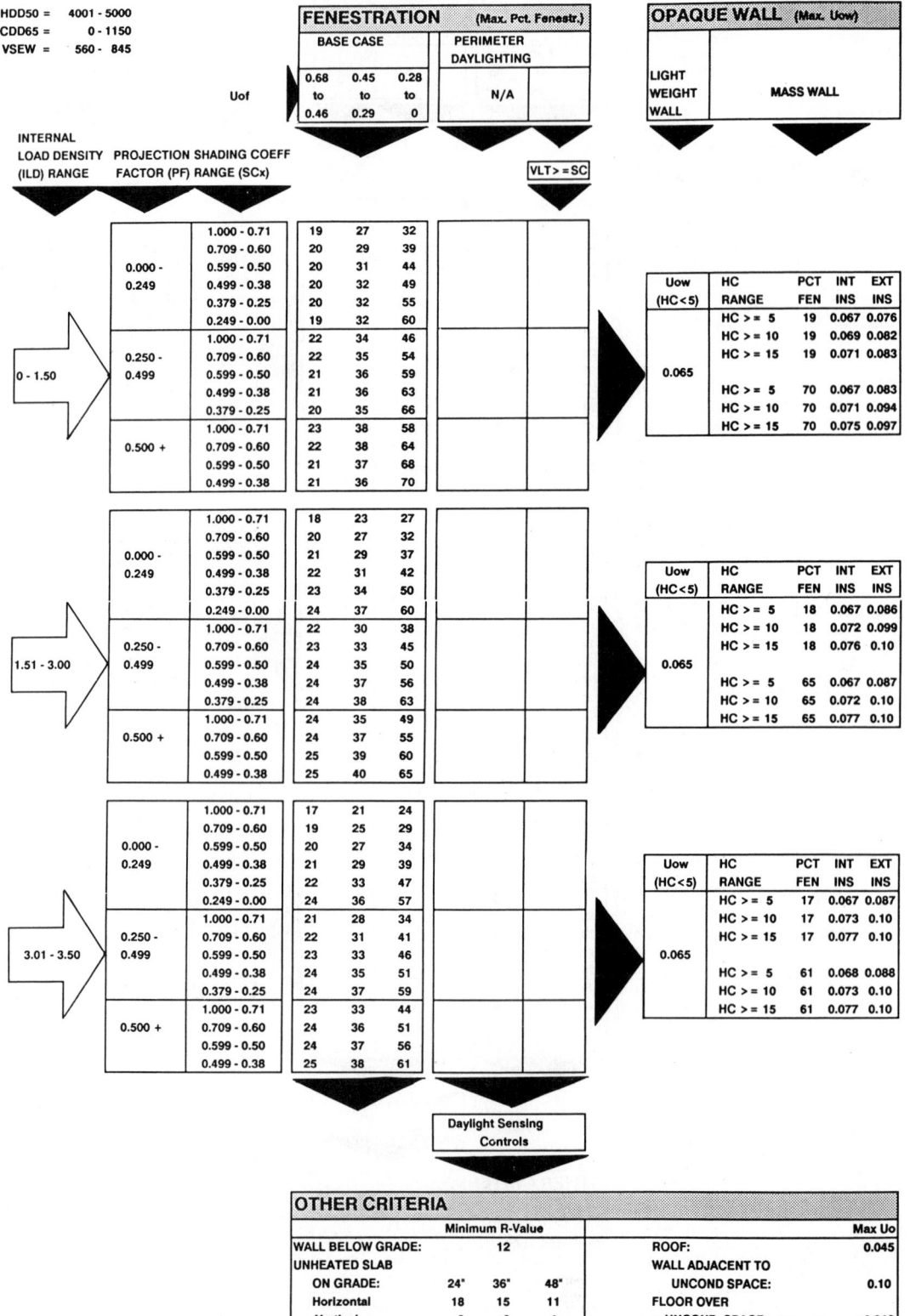

FENESTRATION (Max. Pct. Fenestr.) — Base Case by Uof; Perimeter Daylighting: N/A (VLT> =SC; Daylight Sensing Controls)

OPAQUE WALL (Max. Uow) — Light Weight Wall; Mass Wall

INTERNAL LOAD DENSITY (ILD) RANGE	PROJECTION FACTOR (PF)	SHADING COEFF RANGE (SCx)	Uof 0.68 to 0.46	Uof 0.45 to 0.29	Uof 0.28 to 0	PERIMETER DAYLIGHTING N/A	
0 - 1.50	0.000 - 0.249	1.000 - 0.71	19	27	32		
		0.709 - 0.60	20	29	39		
		0.599 - 0.50	20	31	44		
		0.499 - 0.38	20	32	49		
		0.379 - 0.25	20	32	55		
		0.249 - 0.00	19	32	60		
	0.250 - 0.499	1.000 - 0.71	22	34	46		
		0.709 - 0.60	22	35	54		
		0.599 - 0.50	21	36	59		
		0.499 - 0.38	21	36	63		
		0.379 - 0.25	20	35	66		
	0.500 +	1.000 - 0.71	23	38	58		
		0.709 - 0.60	22	38	64		
		0.599 - 0.50	21	37	68		
		0.499 - 0.38	21	36	70		

Uow (HC<5)	HC RANGE	PCT FEN	INT INS	EXT INS
0.065	HC >= 5	19	0.067	0.076
	HC >= 10	19	0.069	0.082
	HC >= 15	19	0.071	0.083
	HC >= 5	70	0.067	0.083
	HC >= 10	70	0.071	0.094
	HC >= 15	70	0.075	0.097

INTERNAL LOAD DENSITY (ILD) RANGE	PROJECTION FACTOR (PF)	SHADING COEFF RANGE (SCx)	Uof 0.68 to 0.46	Uof 0.45 to 0.29	Uof 0.28 to 0	PERIMETER DAYLIGHTING N/A	
1.51 - 3.00	0.000 - 0.249	1.000 - 0.71	18	23	27		
		0.709 - 0.60	20	27	32		
		0.599 - 0.50	21	29	37		
		0.499 - 0.38	22	31	42		
		0.379 - 0.25	23	34	50		
		0.249 - 0.00	24	37	60		
	0.250 - 0.499	1.000 - 0.71	22	30	38		
		0.709 - 0.60	23	33	45		
		0.599 - 0.50	24	35	50		
		0.499 - 0.38	24	37	56		
		0.379 - 0.25	24	38	63		
	0.500 +	1.000 - 0.71	24	35	49		
		0.709 - 0.60	24	37	55		
		0.599 - 0.50	25	39	60		
		0.499 - 0.38	25	40	65		

Uow (HC<5)	HC RANGE	PCT FEN	INT INS	EXT INS
0.065	HC >= 5	18	0.067	0.086
	HC >= 10	18	0.072	0.099
	HC >= 15	18	0.076	0.10
	HC >= 5	65	0.067	0.087
	HC >= 10	65	0.072	0.10
	HC >= 15	65	0.077	0.10

INTERNAL LOAD DENSITY (ILD) RANGE	PROJECTION FACTOR (PF)	SHADING COEFF RANGE (SCx)	Uof 0.68 to 0.46	Uof 0.45 to 0.29	Uof 0.28 to 0	PERIMETER DAYLIGHTING N/A	
3.01 - 3.50	0.000 - 0.249	1.000 - 0.71	17	21	24		
		0.709 - 0.60	19	25	29		
		0.599 - 0.50	20	27	34		
		0.499 - 0.38	21	29	39		
		0.379 - 0.25	22	33	47		
		0.249 - 0.00	24	36	57		
	0.250 - 0.499	1.000 - 0.71	21	28	34		
		0.709 - 0.60	22	31	41		
		0.599 - 0.50	23	33	46		
		0.499 - 0.38	24	35	51		
		0.379 - 0.25	24	37	59		
	0.500 +	1.000 - 0.71	23	33	44		
		0.709 - 0.60	24	36	51		
		0.599 - 0.50	24	37	56		
		0.499 - 0.38	25	38	61		

Uow (HC<5)	HC RANGE	PCT FEN	INT INS	EXT INS
0.065	HC >= 5	17	0.067	0.087
	HC >= 10	17	0.073	0.10
	HC >= 15	17	0.077	0.10
	HC >= 5	61	0.068	0.088
	HC >= 10	61	0.073	0.10
	HC >= 15	61	0.077	0.10

Daylight Sensing Controls

OTHER CRITERIA

Minimum R-Value					Max Uo
WALL BELOW GRADE:	12			ROOF:	0.045
UNHEATED SLAB ON GRADE:	24"	36"	48"	WALL ADJACENT TO UNCOND SPACE:	0.10
Horizontal	18	15	11	FLOOR OVER UNCOND. SPACE:	0.040
Vertical	8	6	4		

Source: Adapted with permission from ASHRAE/IES Standard 90.1-1989, *Energy Efficient Design of New Buildings Except Low-Rise Residential Buildings,* © 1989 by the American Society of Heating, Refrigerating, and Air-Conditioning Engineers, Inc., Atlanta, Ga.

TABLE K.32 **ACP for North Platte and Scottsbluff, NE, Casper and Cheyenne, WY, Pocatello, ID, Elko and Ely, NV**

Alternate Component Package, for ASHRAE/IES Standard 90.1-1989

HDD50 = 3201 - 4000
CDD65 = 0 - 1150
VSEW = > 845

FENESTRATION (Max. Pct. Fenestr.)

INTERNAL LOAD DENSITY (ILD) RANGE	PROJECTION FACTOR (PF)	SHADING COEFF RANGE (SCx)	BASE CASE Uof 0.68 to 0.46	BASE CASE Uof 0.45 to 0.29	BASE CASE Uof 0.28 to 0	PERIMETER DAYLIGHTING Uof 0.68 to 0.46	PERIMETER DAYLIGHTING Uof 0.45 to 0	PERIMETER DAYLIGHTING Uof 0.28 to 0 (VLT>=SC)
0 - 1.50	0.000 - 0.249	1.000 - 0.71	19	26	30	19	26	32
		0.709 - 0.60	20	28	36	19	29	38
		0.599 - 0.50	20	30	41	19	30	43
		0.499 - 0.38	20	32	46	19	31	48
		0.379 - 0.25	20	33	53	19	32	54
		0.249 - 0.00	20	33	60	19	32	57
	0.250 - 0.499	1.000 - 0.71	23	34	44	22	34	47
		0.709 - 0.60	23	36	52	21	35	55
		0.599 - 0.50	22	37	57	21	36	60
		0.499 - 0.38	22	37	62	21	35	64
		0.379 - 0.25	21	37	67	20	34	66
	0.500 +	1.000 - 0.71	24	39	56	22	38	60
		0.709 - 0.60	23	39	64	22	38	67
		0.599 - 0.50	23	39	68	21	37	71
		0.499 - 0.38	22	39	71	21	36	71
1.51 - 3.00	0.000 - 0.249	1.000 - 0.71	18	22	24	20	27	32
		0.709 - 0.60	20	25	30	21	30	38
		0.599 - 0.50	21	28	34	22	32	44
		0.499 - 0.38	22	31	40	23	34	50
		0.379 - 0.25	23	34	48	23	36	57
		0.249 - 0.00	25	38	59	24	38	64
	0.250 - 0.499	1.000 - 0.71	22	30	35	24	35	47
		0.709 - 0.60	24	33	42	24	37	55
		0.599 - 0.50	24	35	48	24	38	61
		0.499 - 0.38	25	37	54	25	39	66
		0.379 - 0.25	26	40	62	25	40	70
	0.500 +	1.000 - 0.71	25	35	46	25	40	60
		0.709 - 0.60	26	38	53	25	41	68
		0.599 - 0.50	26	40	59	25	41	73
		0.499 - 0.38	26	41	65	25	41	75
3.01 - 3.50	0.000 - 0.249	1.000 - 0.71	16	19	21	20	26	31
		0.709 - 0.60	18	23	26	21	30	38
		0.599 - 0.50	20	25	30	22	32	43
		0.499 - 0.38	21	28	35	23	34	49
		0.379 - 0.25	23	32	43	24	36	57
		0.249 - 0.00	25	36	55	25	38	64
	0.250 - 0.499	1.000 - 0.71	21	27	31	23	35	45
		0.709 - 0.60	22	30	37	24	37	53
		0.599 - 0.50	23	32	43	24	38	60
		0.499 - 0.38	24	35	48	25	39	66
		0.379 - 0.25	25	38	56	25	40	70
	0.500 +	1.000 - 0.71	23	32	41	25	39	58
		0.709 - 0.60	24	35	48	25	40	67
		0.599 - 0.50	25	37	53	25	40	72
		0.499 - 0.38	26	39	58	26	41	74

Daylight Sensing Controls

OPAQUE WALL (Max. Uow)

ILD 0 - 1.50:

LIGHT WEIGHT WALL Uow (HC<5)	MASS WALL HC RANGE	PCT FEN	INT INS	EXT INS
0.073	HC >= 5	19	0.076	0.098
	HC >= 10	19	0.083	0.11
	HC >= 15	19	0.088	0.11
	HC >= 5	71	0.076	0.10
	HC >= 10	71	0.084	0.11
	HC >= 15	71	0.091	0.12

ILD 1.51 - 3.00:

LIGHT WEIGHT WALL Uow (HC<5)	MASS WALL HC RANGE	PCT FEN	INT INS	EXT INS
0.073	HC >= 5	18	0.077	0.10
	HC >= 10	18	0.087	0.13
	HC >= 15	18	0.096	0.13
	HC >= 5	75	0.078	0.10
	HC >= 10	75	0.086	0.12
	HC >= 15	75	0.094	0.13

ILD 3.01 - 3.50:

LIGHT WEIGHT WALL Uow (HC<5)	MASS WALL HC RANGE	PCT FEN	INT INS	EXT INS
0.073	HC >= 5	16	0.077	0.11
	HC >= 10	16	0.088	0.13
	HC >= 15	16	0.097	0.14
	HC >= 5	74	0.078	0.10
	HC >= 10	74	0.086	0.12
	HC >= 15	74	0.095	0.13

OTHER CRITERIA

	Minimum R-Value				Max Uo
WALL BELOW GRADE:	11			ROOF:	0.049
UNHEATED SLAB				WALL ADJACENT TO	
ON GRADE:	24"	36"	48"	UNCOND SPACE:	0.11
Horizontal	18	15	11	FLOOR OVER	
Vertical	8	6	4	UNCOND. SPACE:	0.041

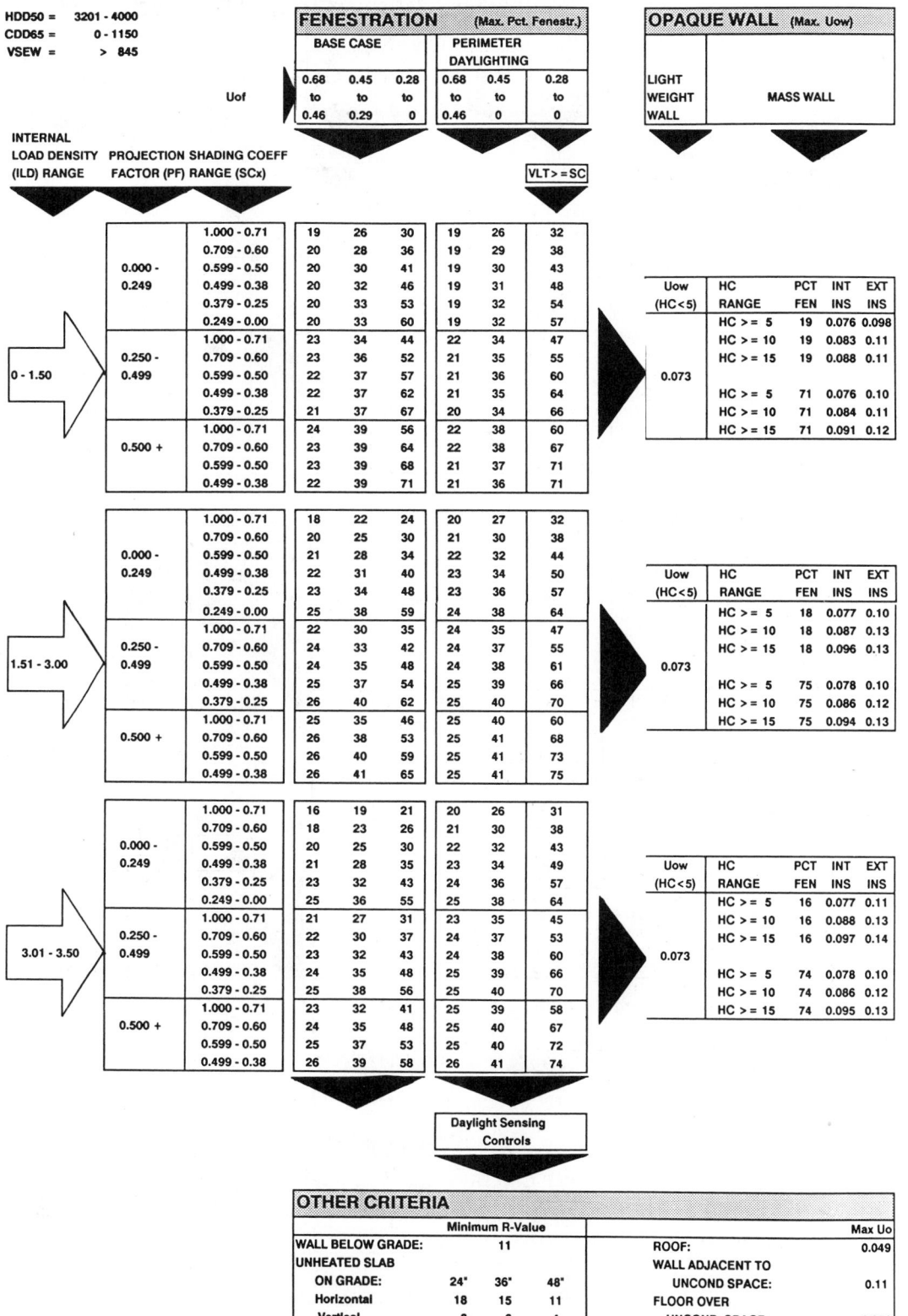

Source: Adapted with permission from ASHRAE/IES Standard 90.1-1989, *Energy Efficient Design of New Buildings Except Low-Rise Residential Buildings*, © 1989 by the American Society of Heating, Refrigerating, and Air-Conditioning Engineers, Inc., Atlanta, Ga.

TABLE K.33 **ACP for Eagle, CO, Rock Springs, WY, and Bryce Canyon, UT**

Alternate Component Package, for ASHRAE/IES Standard 90.1-1989

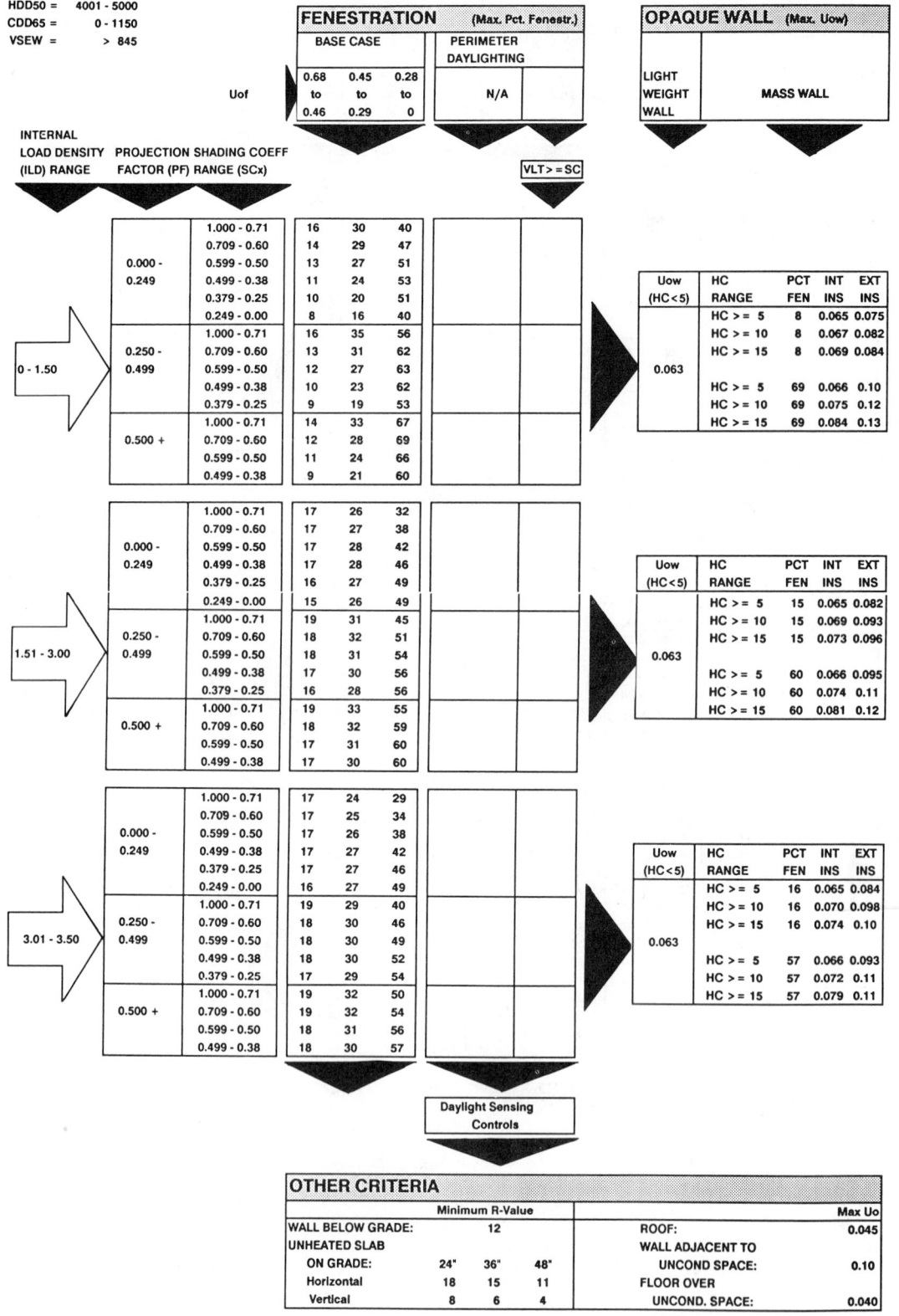

HDD50 = 4001 - 5000
CDD65 = 0 - 1150
VSEW = > 845

FENESTRATION (Max. Pct. Fenestr.)

INTERNAL LOAD DENSITY (ILD) RANGE	PROJECTION FACTOR (PF) RANGE	SHADING COEFF RANGE (SCx)	BASE CASE Uof 0.68 to 0.46	BASE CASE Uof 0.45 to 0.29	BASE CASE Uof 0.28 to 0	PERIMETER DAYLIGHTING N/A	PERIMETER DAYLIGHTING VLT>=SC
0 - 1.50	0.000 - 0.249	1.000 - 0.71	16	30	40		
		0.709 - 0.60	14	29	47		
		0.599 - 0.50	13	27	51		
		0.499 - 0.38	11	24	53		
		0.379 - 0.25	10	20	51		
		0.249 - 0.00	8	16	40		
	0.250 - 0.499	1.000 - 0.71	16	35	56		
		0.709 - 0.60	13	31	62		
		0.599 - 0.50	12	27	63		
		0.499 - 0.38	10	23	62		
		0.379 - 0.25	9	19	53		
	0.500 +	1.000 - 0.71	14	33	67		
		0.709 - 0.60	12	28	69		
		0.599 - 0.50	11	24	66		
		0.499 - 0.38	9	21	60		
1.51 - 3.00	0.000 - 0.249	1.000 - 0.71	17	26	32		
		0.709 - 0.60	17	27	38		
		0.599 - 0.50	17	28	42		
		0.499 - 0.38	17	28	46		
		0.379 - 0.25	16	27	49		
		0.249 - 0.00	15	26	49		
	0.250 - 0.499	1.000 - 0.71	19	31	45		
		0.709 - 0.60	18	32	51		
		0.599 - 0.50	18	31	54		
		0.499 - 0.38	17	30	56		
		0.379 - 0.25	16	28	56		
	0.500 +	1.000 - 0.71	19	33	55		
		0.709 - 0.60	18	32	59		
		0.599 - 0.50	17	31	60		
		0.499 - 0.38	17	30	60		
3.01 - 3.50	0.000 - 0.249	1.000 - 0.71	17	24	29		
		0.709 - 0.60	17	25	34		
		0.599 - 0.50	17	26	38		
		0.499 - 0.38	17	27	42		
		0.379 - 0.25	17	27	46		
		0.249 - 0.00	16	27	49		
	0.250 - 0.499	1.000 - 0.71	19	29	40		
		0.709 - 0.60	18	30	46		
		0.599 - 0.50	18	30	49		
		0.499 - 0.38	18	30	52		
		0.379 - 0.25	17	29	54		
	0.500 +	1.000 - 0.71	19	32	50		
		0.709 - 0.60	19	32	54		
		0.599 - 0.50	18	31	56		
		0.499 - 0.38	18	30	57		

Daylight Sensing Controls

OPAQUE WALL (Max. Uow)

ILD 0 - 1.50:

LIGHT WEIGHT WALL Uow (HC<5)	MASS WALL HC RANGE	PCT FEN	INT INS	EXT INS
0.063	HC >= 5	8	0.065	0.075
	HC >= 10	8	0.067	0.082
	HC >= 15	8	0.069	0.084
	HC >= 5	69	0.066	0.10
	HC >= 10	69	0.075	0.12
	HC >= 15	69	0.084	0.13

ILD 1.51 - 3.00:

LIGHT WEIGHT WALL Uow (HC<5)	MASS WALL HC RANGE	PCT FEN	INT INS	EXT INS
0.063	HC >= 5	15	0.065	0.082
	HC >= 10	15	0.069	0.093
	HC >= 15	15	0.073	0.096
	HC >= 5	60	0.066	0.095
	HC >= 10	60	0.074	0.11
	HC >= 15	60	0.081	0.12

ILD 3.01 - 3.50:

LIGHT WEIGHT WALL Uow (HC<5)	MASS WALL HC RANGE	PCT FEN	INT INS	EXT INS
0.063	HC >= 5	16	0.065	0.084
	HC >= 10	16	0.070	0.098
	HC >= 15	16	0.074	0.10
	HC >= 5	57	0.066	0.093
	HC >= 10	57	0.072	0.11
	HC >= 15	57	0.079	0.11

OTHER CRITERIA

	Minimum R-Value				Max Uo
WALL BELOW GRADE:		12		ROOF:	0.045
UNHEATED SLAB ON GRADE:	24"	36"	48"	WALL ADJACENT TO UNCOND SPACE:	0.10
Horizontal	18	15	11	FLOOR OVER UNCOND. SPACE:	0.040
Vertical	8	6	4		

Source: Adapted with permission from ASHRAE/IES Standard 90.1-1989, *Energy Efficient Design of New Buildings Except Low-Rise Residential Buildings,* © 1989 by the American Society of Heating, Refrigerating, and Air-Conditioning Engineers, Inc., Atlanta, Ga.

TABLE K.34 **ACP for Caribou, ME, Sault St. Marie, MI, Duluth and International Falls, MN, Bismarck, Fargo, and Minot, ND, and Glasgow, MT**

Alternate Component Package, for ASHRAE/IES Standard 90.1-1989

HDD50 = 5001 - 6500
CDD65 = 0 - 1150
VSEW = 560 - 845

FENESTRATION (Max. Pct. Fenestr.)

INTERNAL LOAD DENSITY (ILD) RANGE	PROJECTION FACTOR (PF) RANGE	SHADING COEFF RANGE (SCx)	BASE CASE Uof 0.68 to 0.46	BASE CASE Uof 0.45 to 0.29	BASE CASE Uof 0.28 to 0	PERIMETER DAYLIGHTING N/A	PERIMETER DAYLIGHTING VLT>=SC
0 - 1.50	0.000 - 0.249	1.000 - 0.71	19	28	37		
		0.709 - 0.60	19	30	43		
		0.599 - 0.50	19	31	47		
		0.499 - 0.38	19	31	51		
		0.379 - 0.25	18	31	55		
		0.249 - 0.00	18	30	57		
	0.250 - 0.499	1.000 - 0.71	20	33	50		
		0.709 - 0.60	20	34	56		
		0.599 - 0.50	20	34	59		
		0.499 - 0.38	19	33	61		
		0.379 - 0.25	19	32	62		
	0.500 +	1.000 - 0.71	21	35	59		
		0.709 - 0.60	20	35	63		
		0.599 - 0.50	20	34	65		
		0.499 - 0.38	19	33	65		
1.51 - 3.00	0.000 - 0.249	1.000 - 0.71	18	25	31		
		0.709 - 0.60	19	28	36		
		0.599 - 0.50	20	29	41		
		0.499 - 0.38	20	31	46		
		0.379 - 0.25	21	33	52		
		0.249 - 0.00	21	34	58		
	0.250 - 0.499	1.000 - 0.71	21	31	42		
		0.709 - 0.60	21	33	48		
		0.599 - 0.50	22	34	52		
		0.499 - 0.38	22	34	56		
		0.379 - 0.25	22	35	61		
	0.500 +	1.000 - 0.71	22	34	51		
		0.709 - 0.60	22	35	57		
		0.599 - 0.50	22	36	60		
		0.499 - 0.38	22	36	63		
3.01 - 3.50	0.000 - 0.249	1.000 - 0.71	18	24	29		
		0.709 - 0.60	19	27	34		
		0.599 - 0.50	20	29	39		
		0.499 - 0.38	21	30	44		
		0.379 - 0.25	21	33	50		
		0.249 - 0.00	22	35	58		
	0.250 - 0.499	1.000 - 0.71	21	30	39		
		0.709 - 0.60	22	32	46		
		0.599 - 0.50	22	33	50		
		0.499 - 0.38	22	35	55		
		0.379 - 0.25	23	36	60		
	0.500 +	1.000 - 0.71	22	33	48		
		0.709 - 0.60	23	35	54		
		0.599 - 0.50	23	36	58		
		0.499 - 0.38	23	36	61		

Daylight Sensing Controls

OPAQUE WALL (Max. Uow)

LIGHT WEIGHT WALL / MASS WALL

ILD 0 - 1.50:

LIGHT WEIGHT WALL Uow (HC<5)	HC RANGE	PCT FEN	INT INS	EXT INS
0.058	HC >= 5	18	0.059	0.065
	HC >= 10	18	0.060	0.068
	HC >= 15	18	0.061	0.069
	HC >= 5	65	0.059	0.070
	HC >= 10	65	0.061	0.075
	HC >= 15	65	0.063	0.077

ILD 1.51 - 3.00:

LIGHT WEIGHT WALL Uow (HC<5)	HC RANGE	PCT FEN	INT INS	EXT INS
0.058	HC >= 5	18	0.059	0.072
	HC >= 10	18	0.062	0.079
	HC >= 15	18	0.064	0.081
	HC >= 5	63	0.059	0.072
	HC >= 10	63	0.062	0.079
	HC >= 15	63	0.065	0.081

ILD 3.01 - 3.50:

LIGHT WEIGHT WALL Uow (HC<5)	HC RANGE	PCT FEN	INT INS	EXT INS
0.058	HC >= 5	18	0.059	0.073
	HC >= 10	18	0.062	0.080
	HC >= 15	18	0.065	0.082
	HC >= 5	61	0.059	0.072
	HC >= 10	61	0.062	0.080
	HC >= 15	61	0.065	0.082

OTHER CRITERIA

Minimum R-Value					Max Uo
WALL BELOW GRADE:	13			ROOF:	0.040
UNHEATED SLAB				WALL ADJACENT TO	
ON GRADE:	24"	36"	48"	UNCOND SPACE:	0.10
Horizontal	18	15	11	FLOOR OVER	
Vertical	8	6	4	UNCOND. SPACE:	0.040

Source: Adapted with permission from ASHRAE/IES Standard 90.1-1989, *Energy Efficient Design of New Buildings Except Low-Rise Residential Buildings,* © 1989 by the American Society of Heating, Refrigerating, and Air-Conditioning Engineers, Inc., Atlanta, Ga.

TABLE K.35 **ACP for Adak, Anchorage, Annette Island, Juneau, Kodiak, and Yakutat, AK**

Alternate Component Package, for ASHRAE/IES Standard 90.1-1989

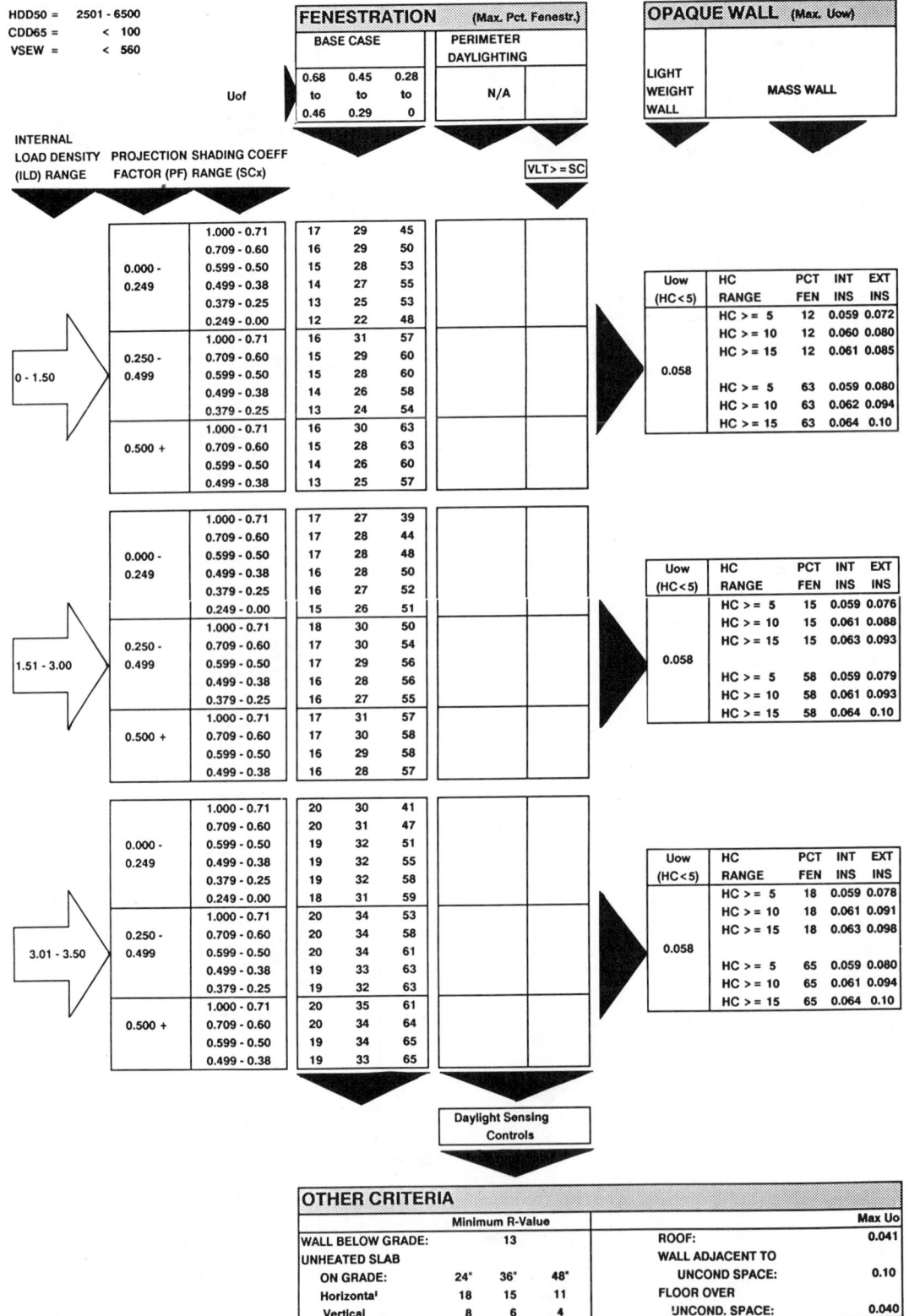

HDD50 = 2501 - 6500
CDD65 = < 100
VSEW = < 560

FENESTRATION (Max. Pct. Fenestr.)

INTERNAL LOAD DENSITY (ILD) RANGE	PROJECTION FACTOR (PF)	SHADING COEFF RANGE (SCx)	BASE CASE Uof 0.68 to 0.46	BASE CASE Uof 0.45 to 0.29	BASE CASE Uof 0.28 to 0	PERIMETER DAYLIGHTING N/A	VLT>=SC
0 - 1.50	0.000 - 0.249	1.000 - 0.71	17	29	45		
		0.709 - 0.60	16	29	50		
		0.599 - 0.50	15	28	53		
		0.499 - 0.38	14	27	55		
		0.379 - 0.25	13	25	53		
		0.249 - 0.00	12	22	48		
	0.250 - 0.499	1.000 - 0.71	16	31	57		
		0.709 - 0.60	15	29	60		
		0.599 - 0.50	15	28	60		
		0.499 - 0.38	14	26	58		
		0.379 - 0.25	13	24	54		
	0.500 +	1.000 - 0.71	16	30	63		
		0.709 - 0.60	15	28	63		
		0.599 - 0.50	14	26	60		
		0.499 - 0.38	13	25	57		
1.51 - 3.00	0.000 - 0.249	1.000 - 0.71	17	27	39		
		0.709 - 0.60	17	28	44		
		0.599 - 0.50	17	28	48		
		0.499 - 0.38	16	28	50		
		0.379 - 0.25	16	27	52		
		0.249 - 0.00	15	26	51		
	0.250 - 0.499	1.000 - 0.71	18	30	50		
		0.709 - 0.60	17	30	54		
		0.599 - 0.50	17	29	56		
		0.499 - 0.38	16	28	56		
		0.379 - 0.25	16	27	55		
	0.500 +	1.000 - 0.71	17	31	57		
		0.709 - 0.60	17	30	58		
		0.599 - 0.50	16	29	58		
		0.499 - 0.38	16	28	57		
3.01 - 3.50	0.000 - 0.249	1.000 - 0.71	20	30	41		
		0.709 - 0.60	20	31	47		
		0.599 - 0.50	19	32	51		
		0.499 - 0.38	19	32	55		
		0.379 - 0.25	19	32	58		
		0.249 - 0.00	18	31	59		
	0.250 - 0.499	1.000 - 0.71	20	34	53		
		0.709 - 0.60	20	34	58		
		0.599 - 0.50	20	34	61		
		0.499 - 0.38	19	33	63		
		0.379 - 0.25	19	32	63		
	0.500 +	1.000 - 0.71	20	35	61		
		0.709 - 0.60	20	34	64		
		0.599 - 0.50	19	34	65		
		0.499 - 0.38	19	33	65		

Daylight Sensing Controls

OPAQUE WALL (Max. Uow)

LIGHT WEIGHT WALL | MASS WALL

ILD 0 - 1.50:

Uow (HC<5)	HC RANGE	PCT FEN	INT INS	EXT INS
0.058	HC >= 5	12	0.059	0.072
	HC >= 10	12	0.060	0.080
	HC >= 15	12	0.061	0.085
	HC >= 5	63	0.059	0.080
	HC >= 10	63	0.062	0.094
	HC >= 15	63	0.064	0.10

ILD 1.51 - 3.00:

Uow (HC<5)	HC RANGE	PCT FEN	INT INS	EXT INS
0.058	HC >= 5	15	0.059	0.076
	HC >= 10	15	0.061	0.088
	HC >= 15	15	0.063	0.093
	HC >= 5	58	0.059	0.079
	HC >= 10	58	0.061	0.093
	HC >= 15	58	0.064	0.10

ILD 3.01 - 3.50:

Uow (HC<5)	HC RANGE	PCT FEN	INT INS	EXT INS
0.058	HC >= 5	18	0.059	0.078
	HC >= 10	18	0.061	0.091
	HC >= 15	18	0.063	0.098
	HC >= 5	65	0.059	0.080
	HC >= 10	65	0.061	0.094
	HC >= 15	65	0.064	0.10

OTHER CRITERIA

	Minimum R-Value		
WALL BELOW GRADE:	13		
UNHEATED SLAB ON GRADE:	24"	36"	48"
Horizontal	18	15	11
Vertical	8	6	4

	Max Uo
ROOF:	0.041
WALL ADJACENT TO UNCOND SPACE:	0.10
FLOOR OVER UNCOND. SPACE:	0.040

Source: Adapted with permission from ASHRAE/IES Standard 90.1-1989, *Energy Efficient Design of New Buildings Except Low-Rise Residential Buildings,* © 1989 by the American Society of Heating, Refrigerating, and Air-Conditioning Engineers, Inc., Atlanta, Ga.

K.2 Low-Rise Residential Design Criteria for Energy Conservation

The following criteria are taken from ASHRAE Standard 90A–1980, *Energy Conservation in New Building Design*. This standard is frequently updated; series 90.1 deals with all buildings except low-rise residential; series 90.2 deals with low-rise residential buildings.

An important basis of these criteria is the concept of an average U value for all portions of vertical walls, called U_{overall} walls or U_o walls; similarly, an average roof U value is called U_o roofs. This value is calculated in the same way as any other weighted average:

$$U_o \text{ walls} = \frac{(U_{\substack{\text{opaque}\\\text{wall}}} \times A_{\substack{\text{opaque}\\\text{wall}}}) + (U_{\text{window}} \times A_{\text{window}}) + \cdots}{A_o \text{ walls}}$$

where A_o includes *all* the areas of all vertical exposed surfaces (walls, windows, doors, etc.)—in other words, the sum of all the areas listed in the numerator of the equation. Note that *each* wall component that is thermally different from other wall components is treated separately in the equation; thus $(U_{\text{wall}} \times A_{\text{wall}})$ might become $(U_{\text{masonry wall}} \times A_{\text{masonry wall}})$ plus $(U_{\text{frame wall}} \times A_{\text{frame wall}})$, and so on.

Similarly,

$$U_o \text{ roofs} = \frac{(U_{\substack{\text{opaque}\\\text{roof}}} \times A_{\substack{\text{opaque}\\\text{roof}}}) + (U_{\text{skylight}} \times A_{\text{skylight}}) + \cdots}{A_o \text{ roofs}}$$

where A_o roofs is, again, the sum of all horizontal or sloped surface areas, from the numerator.

Residential (type A) buildings. This includes detached one- and two-family dwellings (A_1) and multifamily dwellings that are *three stories or less* (A_2). (For four or more floors, see ASHRAE Standard 90.1-1989.)

Roof/ceiling: Figure K.2 shows the *maximum* value of U_o roofs for buildings of type A_1 and A_2 that are heated and/or mechanically cooled. For heating degree days, see Appendix B. There is an exception for "cathedral" ceilings, where the finished interior surface is essentially the exposed underside of the structural roof deck. In this case, for any geographic location, U_o roofs maximum = 0.08 Btu/ft²-h-°F (0.45 W/m²-°C; units are interchangeable with W/m²-K).

Walls: Figure K.3 shows the *maximum* value of U_o walls for buildings of types A_1 and A_2

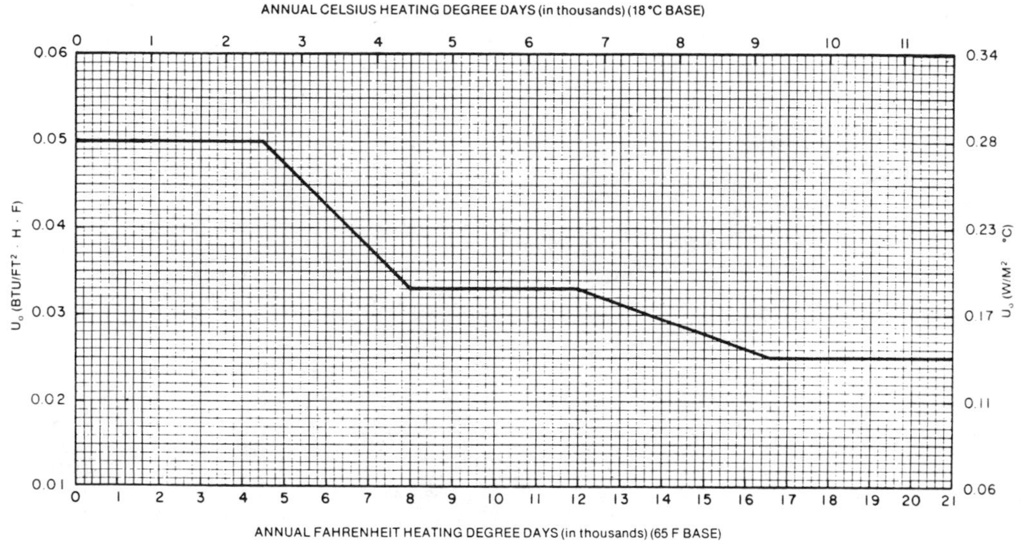

Fig. K.2 *U_o values for roofs/ceilings, type A_1 and A_2 (small residential) buildings. Copyright © by the American Society of Heating, Refrigerating, and Air-Conditioning Engineers, Inc., Atlanta, Ga. Reprinted by permission from* ASHRAE Standard 90A-1980.

that are heated and/or mechanically cooled. In the special case of warmer climates below 500 heating degree days Fahrenheit (278 DD Celsius), buildings that are mechanically cooled have a maximum U_o wall as follows: type A_1, 0.30 Btu/ft²-h-°F (1.70 W/m²-°C); type A_2, 0.38 Btu/ft²-h-°F (2.15 W/m²-°C).

Slab floors on grade: Figure K.4 shows the minimum required thermal resistance (*R* value) for *perimeter insulation,* which extends downward from the top of the slab for a minimum distance of 24 in. (0.6 m). Alternatively, such insulation should extend from the top of the slab to the bottom, then horizontally below the slab for a minimum distance of 24 in. (0.6 m).

Exposed floors. Two conditions are covered here. Heated and/or mechanically cooled spaces with floors that are exposed to *exterior* conditions should have, at maximum, a floor U_o as shown in Fig. K.2 (same as for roofs). However, for floors over protected, but unheated spaces, the maximum U_o is as shown in Fig. K.5.

Infiltration. Air leakage rates are based on a pressure differential equivalent to the effect of a 25-mph (11.1-m/s) wind. Technically, this is a pressure differential of 1.57 lb/ft² (75 Pa).

Windows should have a maximum air infiltration rate of 0.5 cfm per foot (7.74×10^{-4} m³/s per meter) of sash crack.

Doors for residential living units should have a maximum air infiltration rate of 0.5 cfm per square foot (2.54×10^{-3} m³/s per square meter) of door area.

Doors for other usages should have a maximum air infiltration rate of 11 cfm per linear foot (1.7×10^{-2} m³/s per linear meter) of door crack.

Other joints in the exterior envelope that can be sources of air leakage (around window and door frames, between walls and foundation, walls and roof, at utility penetrations, etc.) should be caulked, gasketed, weather stripped, or otherwise sealed.

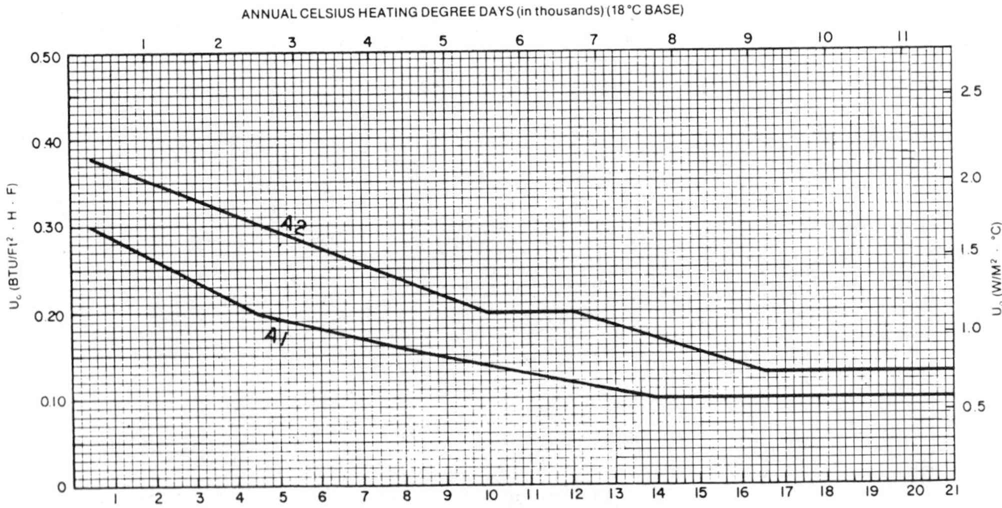

Fig. K.3 *U_o values for walls, type A_1 and A_2 (small residential) buildings. Copyright © by the American Society of Heating, Refrigerating, and Air-Conditioning Engineers, Inc., Atlanta, Ga. Reprinted by permission from* ASHRAE Standard 90A-1980.

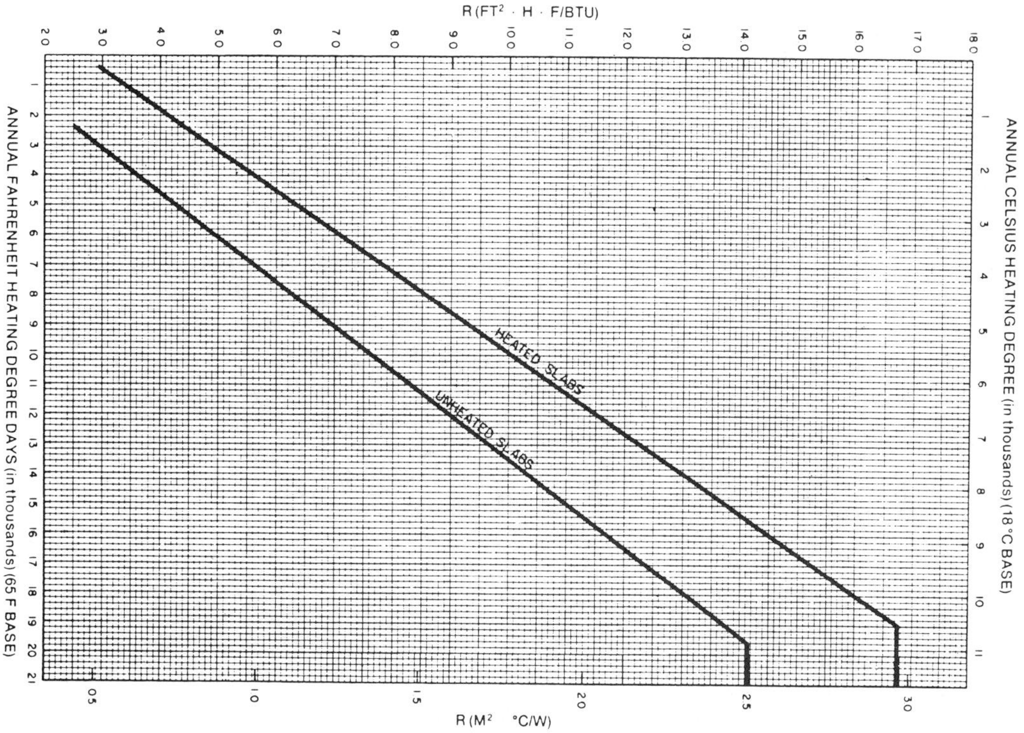

Fig. K.4 R values for slab floors on grade. Copyright © by the American Society of Heating, Refrigerating, and Air-Conditioning Engineers, Inc., Atlanta, Ga. Reprinted by permission from ASHRAE Standard 90A-1980.

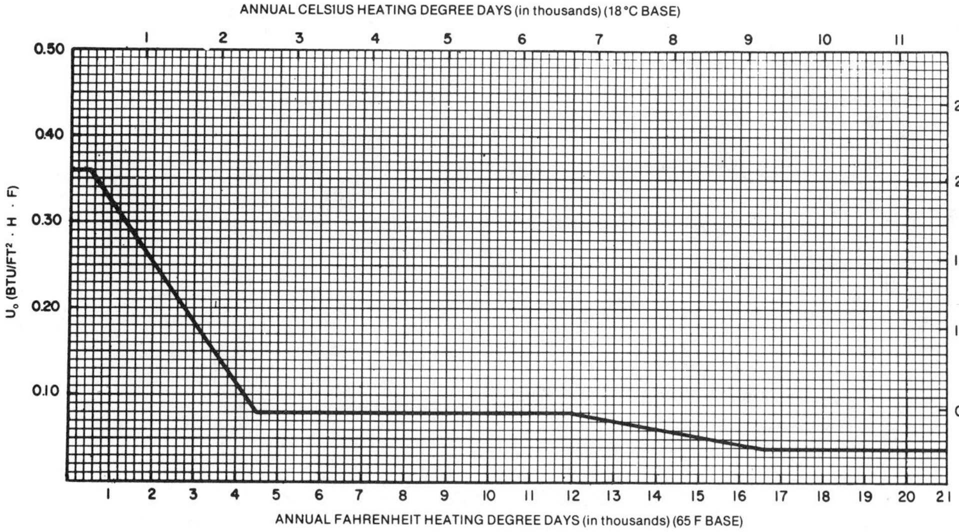

Fig. K.5 U_o values for floors over unheated spaces. Copyright © by the American Society of Heating, Refrigerating, and Air-Conditioning Engineers, Inc., Atlanta, Ga. Reprinted by permission from ASHRAE Standard 90A-1980.

INDEX